Conversion Factors

Mass	1 kg = 1000 g
	1 g = 10^{-3} kg
	1 u = 1.66×10^{-27} kg $\leftrightarrow$ 931.5 MeV/c^2
	1 slug = 14.6 kg
Length	1 m = 100 cm = 1000 mm
	1 foot = 1 ft = 0.305 m
	1 inch = 1 in. = 2.54 cm
	1 mile = 1 mi = 5280 ft = 1609 m
	1 light-year (ly) = 9.46×10^{15} m
Time	1 min = 60 s
	1 h = 60 min = 3600 s
	1 day = 24 h = 8.64×10^4 s
	1 yr = 365.24 days = 3.16×10^7 s
Force	1 N = 0.225 lb
	1 dyne = 10^{-5} N
	1 lb = 4.45 N = 16 oz
Energy	1 J = 10^7 erg = 0.239 cal
	1 cal = 4.186 J
	1 kW · h = 3.60×10^6 J
	1 eV = 1.60×10^{-19} J
Power	1 [illegible]
	1 [illegible]
Pressure	1 Pa = 1 N/m^2
	1 atmosphere = 1 atm = 1.01×10^5 Pa = 14.7 lb/in.^{2}
	1 millimeter of Hg = 1 mm Hg = 133 Pa
	1 bar = 10^5 Pa
	1 lb/in.^{2} = 6.89×10^3 Pa
Speed/velocity	1 m/s = 3.60 km/h
	1 mi/h = 0.447 m/s
	100 mi/h $\approx$ 45 m/s
Temperature	$T(\text{K}) = T(°\text{C}) + 273.15$
	$T(°\text{F}) = \frac{9}{5}T(°\text{C}) + 32$
	$T(°\text{C}) = \frac{5}{9}[T(°\text{F}) - 32]$
Angle	$180° = \pi$ rad

Greek Alphabet

Letter	Symbol (uppercase)	Symbol (lowercase)	Letter	Symbol (uppercase)	Symbol (lowercase)
alpha	A	α	nu	N	ν
beta	B	β	xi	Ξ	ξ
gamma	Γ	γ	omicron	O	o
delta	Δ	δ	pi	Π	π
epsilon	E	ε	rho	P	ρ
zeta	Z	ζ	sigma	Σ	σ
eta	E	η	tau	T	τ
theta	Θ	θ	upsilon	Υ	υ
iota	I	ι	phi	Φ	ϕ
kappa	K	κ	chi	X	χ
lambda	Λ	λ	psi	Ψ	ψ
mu	M	μ	omega	Ω	ω

Trigonometric Relations

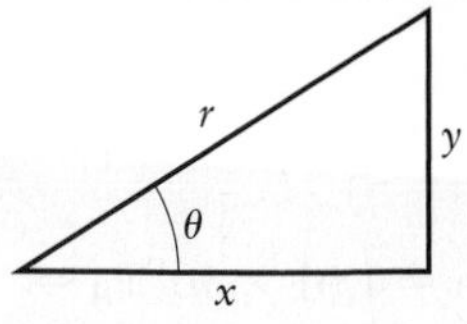

$\sin\theta = \frac{y}{r}$ $\cos\theta = \frac{x}{r}$ $\tan\theta = \frac{y}{x}$

Pythagorean Theorem: $x^2 + y^2 = r^2$

$\sin^2\theta + \cos^2\theta = 1$

$\sin(\theta + 90°) = \cos\theta$

$\cos(\theta + 90°) = -\sin\theta$

$\sin(0) = 0$	$\sin(\pi/4) = 1/\sqrt{2}$	$\sin(\pi/2) = 1$	$\sin(\pi) = 0$
$\cos(0) = 1$	$\cos(\pi/4) = 1/\sqrt{2}$	$\cos(\pi/2) = 0$	$\cos(\pi) = -1$

Degrees and Radians

$2\pi \text{ rad} = 360°$

$1 \text{ revolution} = 1 \text{ rev} = 360° = 2\pi \text{ rad}$

$1° = 60 \text{ arcmin} = 60'$

$1 \text{ arcmin} = 60 \text{ arcsec} = 60''$

$45° = \pi/4 \text{ rad}$

$90° = \pi/2 \text{ rad}$

Powers of 10

Prefix	Factor
atto (a)	10^{-18}
femto (f)	10^{-15}
pico (p)	10^{-12}
nano (n)	10^{-9}
micro (μ)	10^{-6}
milli (m)	10^{-3}
centi (c)	10^{-2}
deci (d)	10^{-1}
kilo (k)	10^{3}
mega (M)	10^{6}
giga (G)	10^{9}
tera (T)	10^{12}
peta (P)	10^{15}
exa (E)	10^{18}

Useful Mathematical Formulas

Circumference of a circle $= 2\pi r$

Area of a circle $= \pi r^2$

Area of a triangle $= \frac{1}{2}(\text{base} \times \text{height})$

Surface area of a sphere $= 4\pi r^2$

Volume of a sphere $= \frac{4}{3}\pi r^3$

Volume of a cylinder $= \pi r^2 L$

$$ax^2 + bx + c = 0 \Rightarrow x = \frac{-b \pm \sqrt{b^2 - 4ac}}{2a}$$

Exponential Functions and Logarithms

$x = e^y \Leftrightarrow \ln x = y$

$e^{xy} = e^x e^y$

$\ln(xy) = \ln x + \ln y$

$\ln\left(\frac{x}{y}\right) = \ln x - \ln y$

$x = 10^y \Leftrightarrow \log x = y$

$e^0 = 1$

$\ln(1) = 0$

$10^0 = 1$

$\log(1) = 0$

$\log(10) = 1$

COLLEGE PHYSICS

Reasoning and Relationships

SECOND EDITION

Hybrid Edition

Nicholas J. Giordano

Purdue University

BROOKS/COLE
CENGAGE Learning

Australia • Brazil • Japan • Korea • Mexico • Singapore • Spain • United Kingdom • United States

College Physics: Reasoning and Relationships, **Second Edition**
Hybrid Edition

Nicholas J. Giordano

Publisher, Physical Sciences: Mary Finch

Publisher, Physics and Astronomy: Charles Hartford

Senior Development Editor: Susan Dust Pashos

Development Editor, Market Specialist: Nicole Mollica

Associate Development Editor: Brandi Kirksey

Editorial Assistant: Brendan Killion

Senior Media Editor: Rebecca Berardy Schwartz

Marketing Manager: Jack Cooney

Marketing Coordinator: Julie Stefani

Marketing Communications Manager: Darlene Macanan

Senior Content Project Manager: Cathy Brooks

Senior Art Director: Cate Rickard Barr

Print Buyer: Diane Gibbons

Manufacturing Planner: Douglas Bertke

Rights Acquisition Specialist: Shalice Shah-Caldwell

Production Service: Lachina Publishing Services

Text Designer: Roy Neuhaus

Cover Designer: Roy Neuhaus

Cover Image: Feature—Lights © Stephan Jansen/epa/Corbis

Compositor: Lachina Publishing Services

Library of Congress Control Number: 2011933489

Hybrid Student Edition:
ISBN-13: 978-1-111-57137-5
ISBN-10: 1-111-57137-6

Hybrid Instructor's Edition:
ISBN-13: 978-1-133-11357-7
ISBN-10: 1-133-11357-5

Brooks/Cole
20 Channel Center Street
Boston, MA 02210
USA

Cengage Learning is a leading provider of customized learning solutions with office locations around the globe, including Singapore, the United Kingdom, Australia, Mexico, Brazil, and Japan. Locate your local office at **international.cengage.com/region**

Cengage Learning products are represented in Canada by Nelson Education, Ltd.

For your course and learning solutions, visit **www.cengage.com**

Purchase any of our products at your local college store or at our preferred online store **www.cengagebrain.com**

Printed in the United States of America
1 2 3 4 5 6 7 15 14 13 12

Brief Table of Contents

VOLUME 1

Mechanics, Waves, & Thermal Physics

VOLUME 2

Electromagnetism, Optics, & Modern Physics

Contents

© Rubberball/Mike Kemp/Getty Images

© Joseph Van Os, Getty Images

CHAPTER 4

Forces and Motion in Two and Three Dimensions 81

CHAPTER 5

Circular Motion and Gravitation 114

CHAPTER 6

Work and Energy 146

CHAPTER 7

Momentum, Impulse, and Collisions 181

CHAPTER 8

Rotational Motion 213

CHAPTER 9

Energy and Momentum of Rotational Motion 249

CHAPTER 10

Fluids 275

The Africa Image Library/Alamy

CHAPTER 11

Harmonic Motion and Elasticity 311

CHAPTER 12

Waves 337

© John Elk III/Alamy

CHAPTER 13

Sound 362

CHAPTER 14

Temperature and Heat 388

CHAPTER 15

Gases and Kinetic Theory 423

CHAPTER 16

Thermodynamics 448

CHAPTER 17

Electric Forces and Fields 483

CHAPTER 18

Electric Potential 515

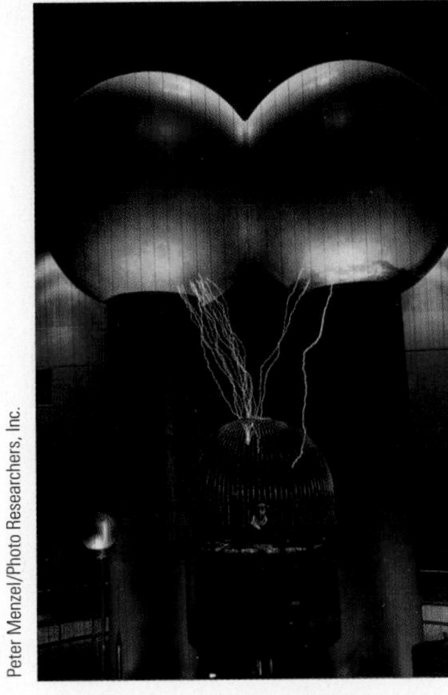

Peter Menzel/Photo Researchers, Inc.

CHAPTER 19

Electric Currents and Circuits 549

© Amazing Images/Alamy

CHAPTER 20

Magnetic Fields and Forces 588

© Lynn Goldsmith/Corbis

CHAPTER 21

Magnetic Induction 626

CHAPTER **22**

Alternating-Current Circuits and Machines 659

CHAPTER **23**

Electromagnetic Waves 694

© Martin Barraud/Alamy

CHAPTER **24**

Geometrical Optics 725

© Oleksiy Maksymenko/Alamy

CHAPTER 25

Wave Optics 764

NASA

CHAPTER 26

Applications of Optics 800

NASA/CXC/CfA/R. Kraft et al.

CHAPTER 27

Relativity 836

CHAPTER 28

Quantum Theory 870

CHAPTER 29

Atomic Theory 899

European Southern Observatory/Photo Researchers, Inc.

CHAPTER 30

Nuclear Physics 932

CMS Experiment © 2009 CERN

CHAPTER 31

Physics in the 21st Century 967

Preface

Changing the Way Students View Physics

The *relationships* between physics and other areas of science are rapidly becoming stronger and are transforming the way all fields of science are understood and practiced. Examples of this transformation abound, particularly in the life sciences. Many students of college physics are engaged in majors relating to the life sciences, and the manner in which they need and will use physics differs from only a few years ago. For their benefit, and for the benefit of students in virtually all technical and even nontechnical disciplines, textbooks must place a greater emphasis on how to apply the *reasoning* of physics to real-world examples. Such examples come quite naturally from the life sciences, but many everyday objects are filled with good applications of fundamental physics principles as well.

Goals of This Book

Reasoning and Relationships

Students often view physics as merely a collection of loosely related equations. We who teach physics work hard to overcome this perception and help students understand how our subject is part of a broader science context. But what does "understanding" in this context really mean? Many physics textbooks assume understanding will result if a solid problem-solving methodology is introduced early and followed strictly. Students in this model can be viewed as successful if they deal with a representative collection of quantitative problems. However, physics education research has shown that students can succeed in such narrow problem-solving tasks and at the same time have fundamentally flawed notions of the basic principles of physics. For these students, physics *is* simply a collection of equations and facts without a firm connection to the way the world works. Although students do need a solid problem-solving framework, I believe such a framework is only one component to learning physics. For real learning to occur, students must know how to reason and must see the relationships between the ideas of physics and their direct experiences. Until the reasoning is sound and the relationships are clear, fundamental learning will remain elusive.

The central theme of this book is to weave reasoning and relationships into the way we teach introductory physics. Three important results of this approach are the following:

1. Early in the text, common student misconceptions about physics and how to study are addressed.
2. Reasoning and relationships examples and problems coach students on how to reason correctly.
3. A systematic approach to problem solving provides the model for how students should approach quantitative problems.

1. Addressing Student Misconceptions. Many students come to their college physics course with a common set of pre-Newtonian misconceptions about physics. I believe the best way to help students overcome these misconceptions is to address them directly and help students see where and how their pre-Newtonian ideas fail. Accordingly, *College Physics: Reasoning and Relationships* devotes Chapter 2 to a qualitative and conceptual discussion of Newton's laws of motion and what they tell us about the relationship between forces and motion. The goal is to arm students with an understanding of this relationship in order to address many of their pre-Newtonian misconceptions and prepare for the discussion of the application of those laws in Chapter 3 and beyond. Armed with an understanding of the proper relationship between kinematics and forces, students can then reason about a variety of problems in mechanics, providing instructors with the flexibility to introduce a wider variety of problems much sooner in the course.

This approach also presents students with a model for how to study for a course in the sciences. Too often, students believe that a focus on "formulas" and their application to a narrow set of drill problems results in understanding of the material. The qualitative overview in Chapter 2 models for students the keys to the course: Begin with an understanding of key physics principles, identify the *relationships* between those principles, and apply *reason* to the selection of equations needed to solve quantitative problems, as well as analyze the solution itself.

2.5 Why Did It Take Newton to Discover Newton's Laws?

Newton's second law tells us that the acceleration of an object is given by $\vec{a} = (\sum \vec{F})/m$, where $\sum \vec{F}$ is the total force acting on the object. In the simplest situations, there may only be one or two forces acting on an object, but in some cases there may be a very large number of forces acting on an object. Multiple forces can make things appear to be very complicated, which is perhaps why the correct laws of motion—Newton's laws—were not discovered sooner.

Forces on a Swimming Bacterium

Figure 2.34 shows a photo of the single-celled bacterium *Escherichia coli*, usually referred to as *E. coli*. An individual *E. coli* propels itself by moving thin strands of protein that extend away from its body (rather like a tail) called flagella. Most *E. coli* possess several flagella as in the photo in Figure 2.34A, but to understand their function, we consider a single flagellum as sketched in Figure 2.34B. A flagellum is fairly rigid, and because it has a spiral shape, one can think of it as a small propeller. An *E. coli* bacterium moves about by rotating this propeller, thereby exerting a force $\vec{F}_w$ on the nearby water. According to Newton's third law, the water exerts a force $\vec{F}_E$ of equal magnitude and opposite direction on the *E. coli* as sketched in Figure 2.34B. One might be tempted to apply Newton's second law with the force $\vec{F}_E$ and conclude that the *E. coli* will move with an acceleration that is proportional to this force. However, this conclusion is incorrect because we have not included the forces from the water on the body of the *E. coli*. These forces are also indicated in Figure 2.34B; to properly describe the total force from the water, we must draw in many force vectors, pushing the *E. coli* in virtually all directions. At the molecular level, we can understand these forces as follows. Water is composed of molecules that are in constant motion and bombard the *E. coli* from all sides. Each time a water molecule collides with the *E. coli*, the molecule exerts a force on the bacterium, much like the collision of the baseball and bat in Figure 2.32. As we saw in that case, the two colliding objects both experience a recoil force, another example of action–reaction forces; hence, the *E. coli* and the water molecule exert forces on each other. An individual *E. coli* is not very large, but a water molecule is much smaller than the bacterium, and the force from one such collision will have only a small effect on the

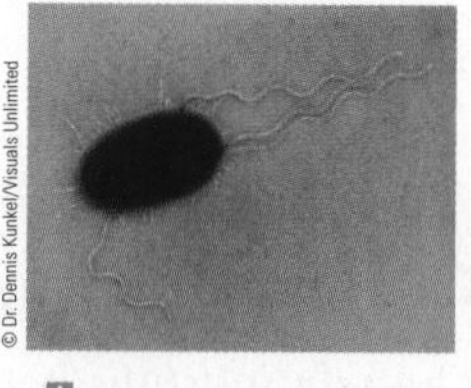

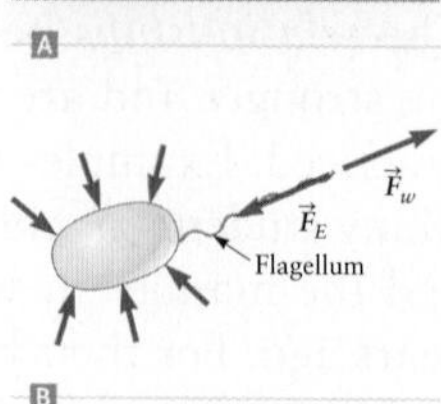

▲ **Figure 2.34** A *E. coli* use the action–reaction principle to propel themselves. An individual *E. coli* bacterium is a few micrometers in diameter (1 μm = 0.000001 m). B The flagellum exerts a force $\vec{F}_w$ on the water, and the water exerts a force $\vec{F}_E = -\vec{F}_w$ on the *E. coli*. There are also forces from the water molecules (indicated by the other red arrows) that act on the cell body.

From Chapter 2, page 43

2. Reasoning and Relationships Problems. Many interesting real-world physics problems cannot be solved exactly with the mathematics appropriate for a college physics course, but they can often be handled in an approximate way using the simple methods (based on algebra and trigonometry) developed in such a course. Professional physicists are familiar with these back of-the-envelope calculations. For instance, we may want to know the approximate force on

PROBLEM SOLVING Dealing with Reasoning and Relationships Problems

1. **RECOGNIZE THE PRINCIPLE.** Determine the key physics ideas that are central to the problem and that connect the quantity you want to calculate with the quantities you know. In the examples found in this section, this physics involves motion with constant acceleration.
2. **SKETCH THE PROBLEM.** Make a drawing that shows all the given information and everything else you know about the problem. For problems in mechanics, your drawing should include all the forces, velocities, and so forth.
3. **IDENTIFY THE RELATIONSHIPS.** Identify the important physics relationships; for problems concerning motion with constant acceleration, they are the relationships between position, velocity, and acceleration in Table 3.1. For many reasoning and relationships problems, values for some of the essential unknown quantities may not be given. You must then use common sense to make reasonable estimates for these quantities. Don't worry or spend time trying to obtain precise values of every quantity (such as the amount that the child's knees flex in Fig. 3.24). Accuracy to within a factor of 3 or even 10 is usually fine because the goal is to calculate the quantity of interest to within an order of magnitude (a factor of 10). Use the Internet, the library, and (especially) your own intuition and experiences.
4. **SOLVE.** Since an exact mathematical solution is not required, cast the problem into one that is easy to solve mathematically. In the examples in this section, we were able to use the results for motion with constant acceleration.
5. Always *consider what your answer means* and check that it makes sense.

As is often the case, practice is a very useful teacher.

From Chapter 3, page 72

From Chapter 3, page 72

EXAMPLE 3.7 ® Cars and Bumpers and Walls

Consider a car of mass 1000 kg colliding with a rigid concrete wall at a speed of 2.5 m/s (about 5 mi/h). This impact is a fairly low-speed collision, and the bumpers on a modern car should be able to handle it without significant damage to the car. Estimate the force exerted by the wall on the car's bumper.

RECOGNIZE THE PRINCIPLE

The motion we want to analyze starts when the car's bumper first touches the wall and ends when the car is stopped. To treat the problem approximately, we *assume* the force on the bumper is constant during the collision period, so the acceleration is also constant. We can then use our expressions from Table 3.1 to analyze the motion. Our strategy is to first find the car's acceleration and then use it to calculate the associated force exerted by the wall on the car from Newton's second law.

SKETCH THE PROBLEM

Figure 3.25 shows a sketch of the car along with the force exerted by the wall on the car. There are also two vertical forces—the force of gravity on the car and the normal force exerted by the road on the car—but we have not shown them because we are concerned here with the car's horizontal (x) motion, which we can treat using $\sum F = ma$ for the components of force and acceleration along x.

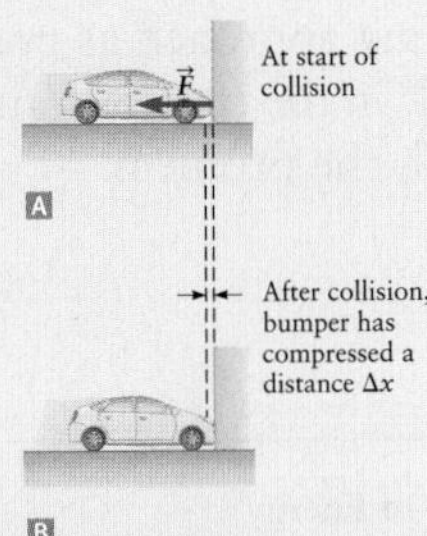

At start of collision
A
After collision, bumper has compressed a distance Δx
B

▲ **Figure 3.25** Example 3.7. A When a car collides with a wall, the wall exerts a force F on the bumper. This force provides the acceleration that stops the car. B During the collision and before the car comes to rest, the bumper deforms by an amount Δx. The car travels this distance while it comes to a complete stop.

IDENTIFY THE RELATIONSHIPS

To find the car's acceleration, we need to estimate either the stopping time or the distance Δx traveled during this time. Let's take the latter approach. We are given the initial velocity ($v_0 = 2.5$ m/s) and the final velocity ($v = 0$). Both of these quantities are in Equation 3.4:

$$v^2 = v_0^2 + 2a(x - x_0) \quad (1)$$

a skydiver's knees when she hits the ground. The precise value of the force depends on the details of the landing, but we are often interested in only an approximate (usually order-of-magnitude) answer. We call such problems ***reasoning and relationships*** problems because solving them requires us to identify the key physics relationships and quantities needed for the problem and that we also estimate values of some important quantities (such as the mass of the skydiver and how she flexes her knees) based on experience and common sense.

Reasoning and relationships problems provide physical insight and a chance to practice critical thinking (reasoning), and they can help students see more clearly the fundamental principles associated with a problem. A truly unique feature of this book is the inclusion of these problems in both the worked examples in the text and the problems in Enhanced WebAssign. The reasoning and relationships problems in Enhanced WebAssign are assignable, offer

From ENHANCED WebAssign

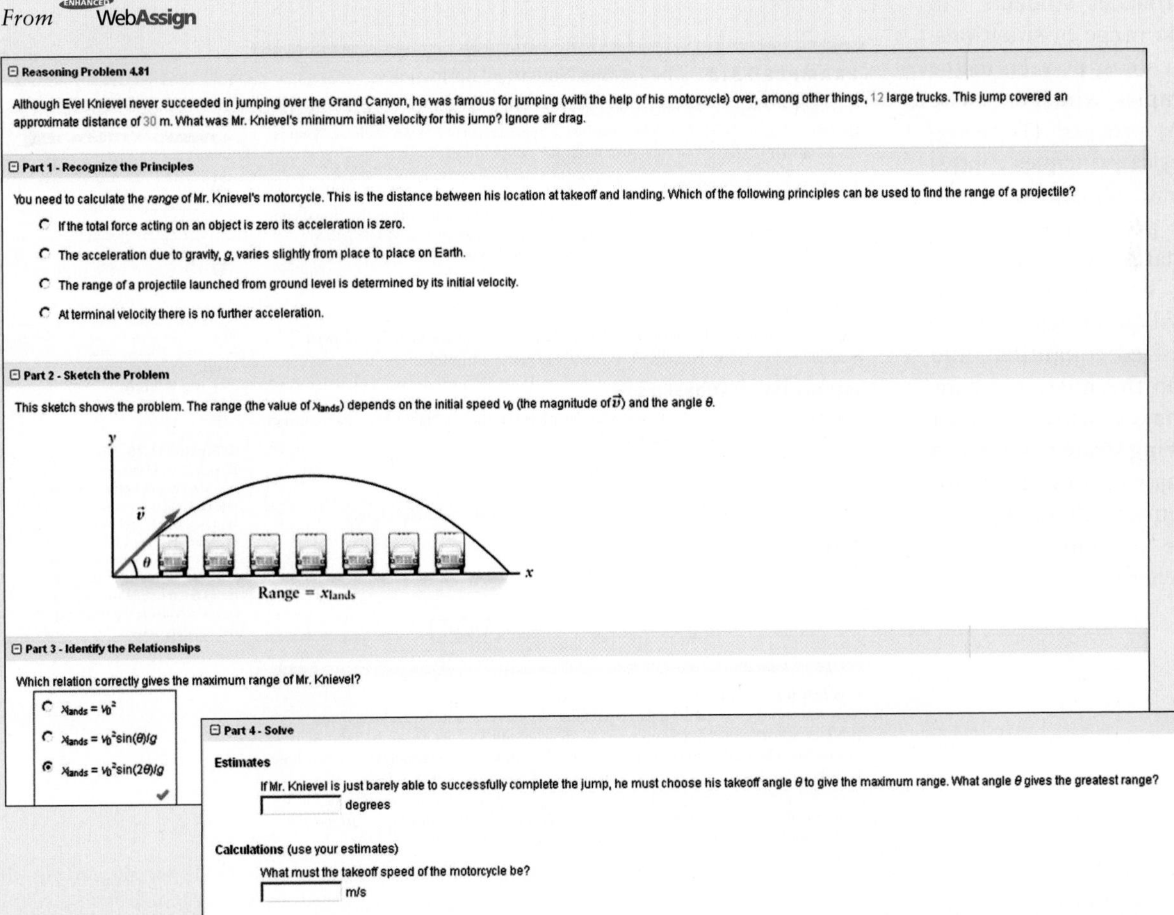

⊟ Reasoning Problem 4.81
Although Evel Knievel never succeeded in jumping over the Grand Canyon, he was famous for jumping (with the help of his motorcycle) over, among other things, 12 large trucks. This jump covered an approximate distance of 30 m. What was Mr. Knievel's minimum initial velocity for this jump? Ignore air drag.
⊟ Part 1 - Recognize the Principles
You need to calculate the range of Mr. Knievel's motorcycle. This is the distance between his location at takeoff and landing. Which of the following principles can be used to find the range of a projectile?
If the total force acting on an object is zero its acceleration is zero.
The acceleration due to gravity, g, varies slightly from place to place on Earth.
The range of a projectile launched from ground level is determined by its initial velocity.
At terminal velocity there is no further acceleration.
⊟ Part 2 - Sketch the Problem
This sketch shows the problem. The range (the value of x_lands) depends on the initial speed v0 (the magnitude of v) and the angle θ.
y
v
θ
x
Range = x_lands
⊟ Part 3 - Identify the Relationships
Which relation correctly gives the maximum range of Mr. Knievel?
x_lands = v0²
x_lands = v0²sin(θ)/g
x_lands = v0²sin(2θ)/g
⊟ Part 4 - Solve
Estimates
If Mr. Knievel is just barely able to successfully complete the jump, he must choose his takeoff angle θ to give the maximum range. What angle θ gives the greatest range?
degrees
Calculations (use your estimates)
What must the takeoff speed of the motorcycle be?
m/s
⊟ Part 5 - What Does It Mean?
Evel Knievel's actual jump covered about 135 feet (about 41 m). A minimum takeoff speed of 20 m/s would have allowed him to successfully cover the required distance. 20 m/s is equivalent to about 45 mph, which is certainly a realistic value for a motorcycle to achieve. In fact, his motorcycle probably had a much greater initial speed. Our calculation has ignored the effect of air drag - in reality, air drag would have been very important in this case and we would need to include it to get an accurate value of Mr. Knievel's initial speed. One more point—our calculation did not account for the fact that to be successful Mr. Knievel must clear the front corner of the first truck and the back corner of the last truck. This means that his range would have to be a bit larger than simply the distance covered by the trucks.

detailed conditional feedback, and provide students with the opportunity to truly internalize the problem-solving process. These detailed tutorials take students step-by-step through the solution. The steps include useful hints to guide student thinking and extensive feedback to assist when student responses show a lack of understanding.

In order to ensure that the structure of the problems and the feedback offered is as specific to my approach as possible, I personally adapted those problems for Enhanced WebAssign. I authored all of the feedback in the tutorials and designed them to have the same "feel" as if I were in the room helping the student reason through the problem.

3. A Systematic Approach to Problem Solving. Although reasoning and relationships problems are used to help students develop a broad understanding of physics, this book also contains a strong component of traditional quantitative problem solving. Quantitative problems are a component of virtually all college physics courses, and students can benefit by developing a systematic approach to such problems. *College Physics: Reasoning and Relationships* therefore places extra emphasis on step-by-step approaches students can use in a wide range of situations. This approach can be seen in the worked examples, which use a five step solution process: (1) recognize the physics principles central to the problem, (2) draw a sketch showing the problem and all the given information, (3) identify the relationships between the known and unknown quantities, (4) solve for the desired quantity, and (5) ask what the answer means and if it makes sense. Explicit problem-solving strategies are also given for major classes of quantitative problems, such as applying the principle of conservation of mechanical energy.

PROBLEM SOLVING Applying the Principle of Conservation of Energy to Problems of Rotational Motion

1. **RECOGNIZE THE PRINCIPLE.** The mechanical energy of an object is conserved only if all the forces that do work on the object are conservative forces. Only then can we apply the conservation of mechanical energy condition in Equation 9.10.
2. **SKETCH THE PROBLEM.** Always use a sketch to collect your information concerning the initial and final states of the system.
3. **IDENTIFY THE RELATIONSHIPS.** Find the initial and final kinetic and potential energies of the object. They will usually involve the initial and final linear and angular velocities, and the initial and final heights, as for the rolling ball in Figure 9.7. You will usually also need to find the moment of inertia.
4. **SOLVE** for the unknown quantities using the conservation of energy condition,
$$KE_i + PE_i = KE_f + PE_f$$
The kinetic energy terms must account for both the translational and the rotational kinetic energies (Eq. 9.5):
$$KE_{\text{total}} = \underbrace{\tfrac{1}{2}mv_{CM}^2}_{\text{translational } KE} + \underbrace{\tfrac{1}{2}I\omega^2}_{\text{rotational } KE}$$
5. Always *consider what your answer means* and check that it makes sense.

From Chapter 9, page 255

EXAMPLE 13.3 The Lowest Note of a Clarinet

A clarinet is a musical instrument we can model as a pipe open at one end and closed at the other (Fig. 13.10A). A standard clarinet is approximately $L = 66$ cm long. What is the frequency of the lowest tone such a clarinet can produce?

RECOGNIZE THE PRINCIPLE

According to our work with pipes, there must be a pressure antinode at the closed end of the clarinet and a pressure node at the open end. Frequency and wavelength are related by $v_{\text{sound}} = f\lambda$, so the standing wave with the lowest frequency will be the one with the longest wavelength.

SKETCH THE PROBLEM

Figure 13.10B shows the clarinet modeled as a pipe open at one end. The allowed standing waves will be just like those found in Figures 13.7 and 13.8.

IDENTIFY THE RELATIONSHIPS

The standing wave with the longest wavelength is sketched in Figure 13.7. Its frequency is given by Equation 13.14, with $n = 1$:

$$f_1 = \frac{v_{\text{sound}}}{4L}$$

This is the fundamental frequency for the clarinet with all tone holes closed.

SOLVE

The speed of sound in air (at room temperature) is $v_{\text{sound}} = 343$ m/s and the length of the tube is $L = 0.66$ m, so we have

$$f_1 = \frac{v_{\text{sound}}}{4L} = \frac{343 \text{ m/s}}{4(0.66 \text{ m})} = \boxed{130 \text{ Hz}}$$

► ***What does it mean?***
This result is the lowest frequency—that is, the lowest "note"—a clarinet can produce according to our simple model. To play other notes, the musician uncovers one of the many different tone holes along the body of the clarinet. If an open hole is sufficiently large, the pressure node moves from the end of the clarinet to a spot very near the tone hole (Fig. 13.10C), thereby shortening the wavelength of the fundamental standing wave and increasing f_1. In this way, a musician can use the many tone holes to produce many different notes.

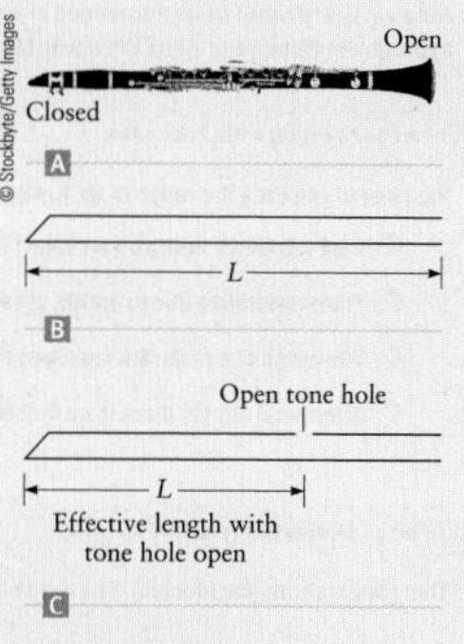

▲ **Figure 13.10** Example 13.3. A A clarinet. B Physicist's model of a clarinet, as a pipe closed at one end and open at the other. The standing waves will be the same as shown in Figures 13.7 and 13.8. C If a tone hole is opened, it will act as the open end of the pipe as far as the standing wave is concerned, changing the frequency of the standing waves.

From Chapter 13, page 373

Changes in the Second Edition

The entire text was reviewed and many of the derivations rewritten to improve clarity and strengthen the connection to basic principles. As an example, the discussion of electric field lines (Chapter 17) was rewritten to better emphasize how they represent the electric field. Many electric circuit problems were added (Chapter 19) to give students more practice in the *qualitative* analysis of circuits and a better intuitive understanding of Kirchhoff's rules. The section on cameras (Chapter 26) was completely reworked to emphasize digital cameras, how they work, and how they can be so tiny. Good graphical representations are important, especially in problems that are inherently three-dimensional, and we improved the way vectors are represented in three-dimensional drawings. This is particularly important in discussions of magnetism (Chapter 20), the propagation of electromagnetic waves (Chapter 23), and rotational motion (Chapter 9).

A new worked example and a new Concept Check were added to nearly every chapter, giving students additional problem-solving exemplars. All of the end-of-chapter questions and problems were systematically reviewed and edited for clarity, and new reasoning and relationships problems (typically three per chapter) were added.

Chapter	New Worked Examples	New Concept Checks
1	—	1.3 Getting the Units Right
2	2.2 Average and Instantaneous Velocity of a Car	2.5 Finding the Velocity
3	3.2 Car Chase	3.5 Normal Forces
4	4.10 Sliding up a Hill	4.2 Falling Balls
5	5.11 A Double Star System	5.3 Net Force and Circular Motion
6	6.2 Kinetic Energy of an Asteroid	6.5 A 1000-Calorie Hill
7	7.3 Impulse and Playing Golf	7.2 Momentum and Kinetic Energy of Two Particles
		7.9 Explosions and the Center of Mass
8	8.9 Which Stick Falls Faster?	8.5 Carrying Your Share of the Weight
9	9.4 Rolling up a Hill	9.5 Rolling, Rolling, Rolling
10	10.9 How Can a Submarine Float?	10.4 Floating in a Lake
11	11.5 Energy of a Simple Harmonic Oscillator	11.2 Analyzing a Simple Harmonic Oscillator
		11.5 Amplitude and Energy of a Simple Harmonic Oscillator
12	12.7 Bending a Note on a Guitar	12.1 Waves on a String
13	13.5 Two Organ Pipes	13.6 Using Beats to Tune a Guitar
14	14.4 Heating a Bucket of Water	14.4 Heating Your Piano
15	15.5 Physics of a Weather Balloon	15.1 Mass of a Water Molecule
16	16.9 A Magnetic Heat Engine	16.8 Heat Engine or Refrigerator?
17	17.8 Electric Field Between Two Sheets of Charge	17.4 Electric Forces and Neutral Objects
18	18.5 Measuring the Charge on an Electron	18.7 Forces on a Capacitor
19	19.9 An *RC* Circuit in Your Car	19.6 A Resistor Puzzle

Chapter	New Worked Examples	New Concept Checks
20	20.2 Magnetic Force on an Electron 20.5 Magnetic Force on a Wire	20.9 Forces on a Current Loop
21	21.4 Moving Magnets and Induced Currents	21.7 Energy in an *LR* Circuit
22	22.5 Analyzing an *LR* Circuit	22.2 AC Power in the U.S.
23	23.5 Noise on Your Radio	23.7 Polarization of a Wave
24	24.9 Designing the Lenses for a Pair of Glasses	24.2 Using Snell's Law 24.10 Ray Tracing With a Diverging Lens
25	25.7 Rayleigh Criterion for a Marksman	25.4 Color of a *Very* Thin Soap Film
26	26.5 Galileo's Telescope	26.4 Making a Better Microscope
27	27.7 A Remarkable Proton	27.6 The Fastest Proton Ever Known
28	28.4 Photons From Your Cell Phone	28.5 Wavelengths of Electrons and Protons
29	29.6 Filling the 4*f* Subshell	29.4 Electron Configuration of an Ion
30	30.10 Application of a Radioactive Tracer	30.5 Making a New Element
31	31.5 Annihilation of a Positron	31.3 Conserving Lepton Number

Additional Features of this Book

Diagrams with Additional Explanatory Labeling

Every college physics textbook contains line art with labeling. This book adds another layer of labeling that explains the phenomenon being illustrated, much as an instructor would explain a process or relationship in class. This additional labeling is set off in a different style.

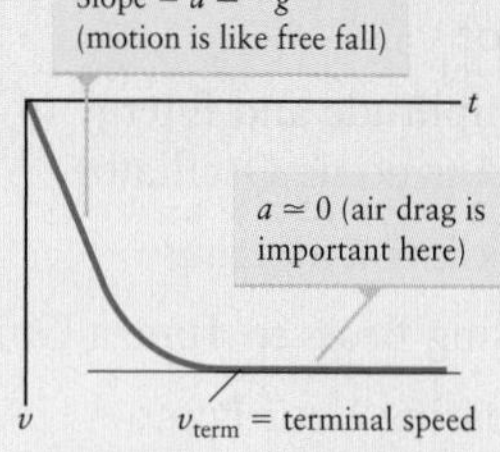

▲ **Figure 3.27** The skydiver's motion is initially like free fall; compare with Figure 3.16A. Eventually, however, air drag becomes as large as the force of gravity and the skydiver reaches her terminal speed v_{term}.

From Chapter 3, page 74

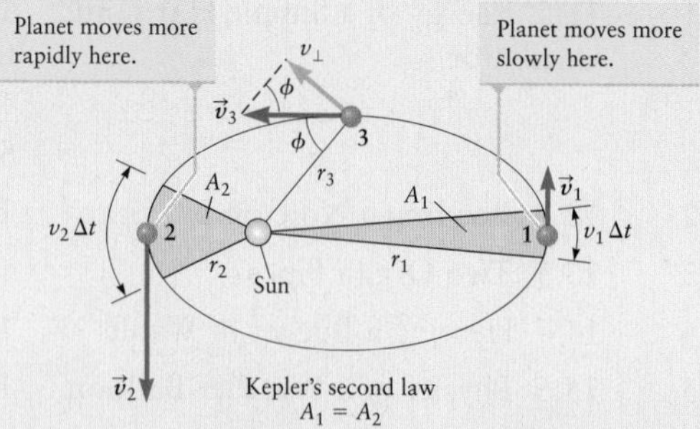

▲ **Figure 9.15** When a planet follows an elliptical orbit, the speed is greatest when the planet is closest to the Sun. According to Kepler's second law, a planet sweeps out equal areas in equal time intervals as it moves about the Sun. Hence, $A_1 = A_2$.

From Chapter 9, page 263

Concept Checks

At various points in the chapter are conceptual questions, called "Concept Checks." These questions are designed to make the student reflect on a fundamental issue. They may involve interpreting the content of a graph or drawing a new graph to predict a relationship between quantities. Many Concept Check questions are in multiple-choice format to facilitate their use in audience response systems. Answers to Concept Checks are given at the end of the book. Full explanations of each answer are given in the *Instructor's Solutions Manual*.

CONCEPT CHECK 6.6 Spring Forces and Newton's Third Law

Figure 6.27 shows two identical springs. In both cases, a person exerts a force of magnitude F on the right end of the spring. The left end of the spring in Figure 6.27A is attached to a wall, while the left end of the spring in Figure 6.27B is held by another person, who exerts a force of magnitude F to the left. Which statement is true?

(1) The spring in Figure 6.27A is stretched half as much as the spring in Figure 6.27B.
(2) The spring in Figure 6.27A is stretched twice as much as the spring in Figure 6.27B.
(3) The two springs are stretched the same amount.

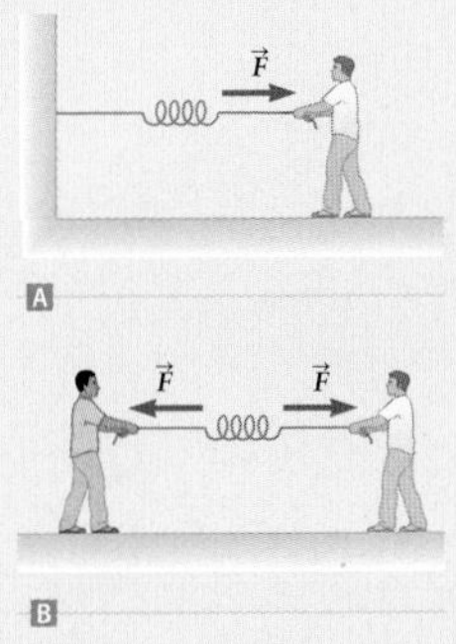

▲ **Figure 6.27** Concept Check 6.6.

From Chapter 6, page 169

Insights

Each chapter contains several special marginal comments called "Insights" that add greater depth to a key idea or reinforce an important message. For instance, Insight 3.3 emphasizes the distinction between weight and mass, and Insight 16.1 explains why diesel engines are inherently more efficient than conventional gasoline internal combustion engines.

Insight 16.1
EFFICIENCY OF A DIESEL ENGINE
A diesel engine is similar to a gasoline internal combustion engine (Example 16.8). One difference is that a gasoline engine ignites the fuel mixture with a spark from a spark plug, whereas a diesel engine ignites the fuel mixture purely "by compression" (without a spark). The compression of the fuel mixture is therefore much greater in a diesel engine, which leads to a higher temperature in the hot reservoir. According to Equation 16.21 and Example 16.8, this higher temperature gives a higher theoretical limit on the efficiency of a diesel engine.

From Chapter 16, page 468

Chapter Summaries

Each concept is described in its own panel, often with an explanatory diagram. This format helps students organize information for review and further study. Concepts are classified into two major groups: (1) *Key Concepts and Principles* and (2) *Applications*.

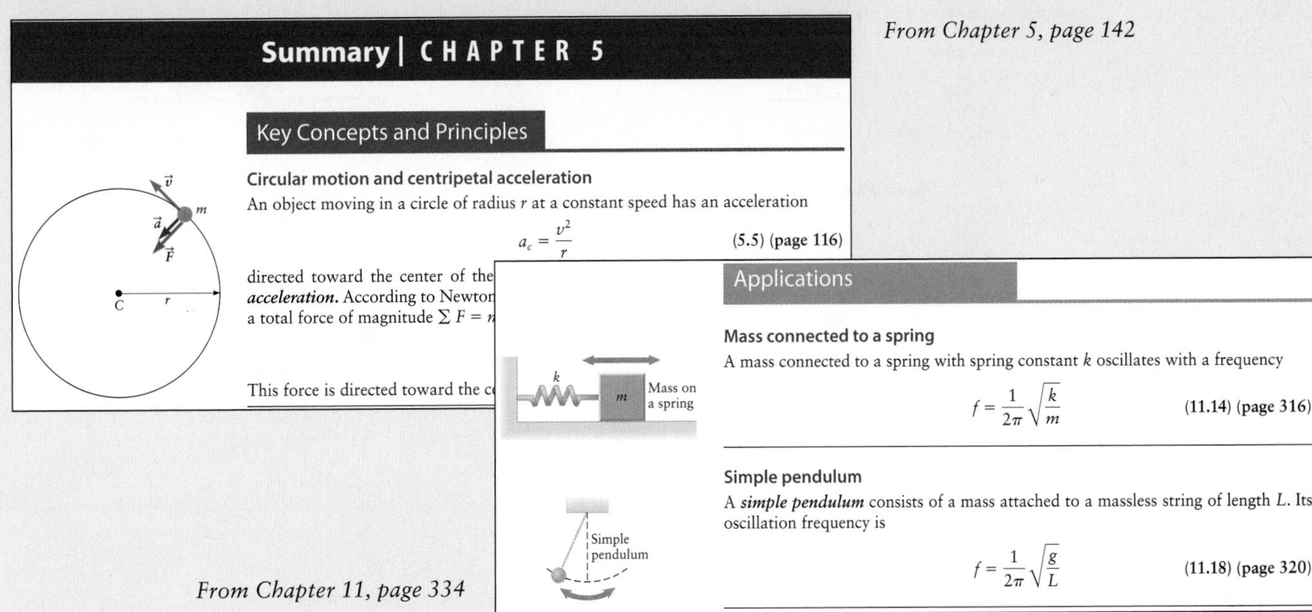

Summary | CHAPTER 5

Key Concepts and Principles

Circular motion and centripetal acceleration
An object moving in a circle of radius r at a constant speed has an acceleration

$$a_c = \frac{v^2}{r} \quad (5.5)\ (\text{page } 116)$$

directed toward the center of the ... ***acceleration***. According to Newto... a total force of magnitude $\sum F = m$...

This force is directed toward the c...

Applications

Mass connected to a spring
A mass connected to a spring with spring constant k oscillates with a frequency

$$f = \frac{1}{2\pi}\sqrt{\frac{k}{m}} \quad (11.14)\ (\text{page } 316)$$

Simple pendulum
A ***simple pendulum*** consists of a mass attached to a massless string of length L. Its oscillation frequency is

$$f = \frac{1}{2\pi}\sqrt{\frac{g}{L}} \quad (11.18)\ (\text{page } 320)$$

From Chapter 5, page 142

From Chapter 11, page 334

End-of-Chapter Questions

Approximately 20 questions at the end of each chapter ask students to reflect on and strengthen their understanding of conceptual issues. These questions are suitable for use in recitation sessions or other group work. Questions of special interest to life science students are marked ⓧ. Answers to questions designated SSM are provided in the *Student Companion & Problem-Solving Guide*, and all questions are answered in the *Instructor's Solutions Manual*.

22. The author once had a small car with a "hatchback," a rear door that opens upward. One day, he forgot to latch it before driving on the highway, and he found that at high speeds the rear door lifted up as in Figure Q10.22B. Explain why.

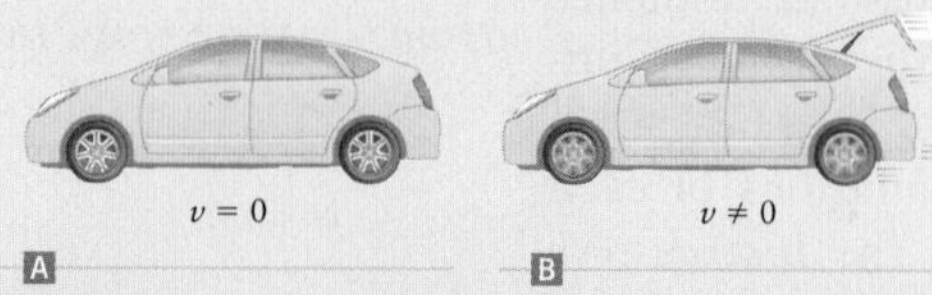

Figure Q10.22

From Chapter 10, page 309

Problems in ENHANCED WebAssign

Homework problems are designed to match the examples that are worked throughout each chapter. Most of these problems are grouped according to the matching chapter section. A final set of "Additional Problems" contains problems that bring together ideas from across the chapter or from multiple chapters. A chart at the end of each chapter in this text lists the reasoning and relationships problems, problems supported by reasoning tutorials, and problems whose solutions appear in the *Student Companion & Problem-Solving Guide*.

63. You work for a moving company and are given the job of pulling two large boxes of mass m_1 = 120 kg and m_2 = 290 kg using ropes as shown in Figure P3.63. You pull very hard, and the boxes are accelerating with a = 0.22 m/s². What is the tension in each rope? Assume there is no friction between the boxes and the floor.

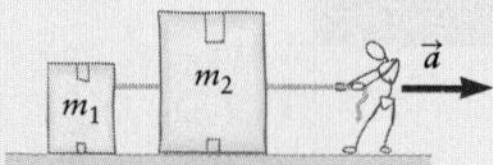

Figure P3.63

74. A man (m_1 = 90 kg) is standing on a railroad flat car that is 9.0 m long and of unknown mass m_2. The man and the car are initially at rest on a level track (Fig. P7.74A), and the wheels of the car are frictionless. The man then runs to the opposite end of the car and the car moves a distance 1.5 m to the left (Fig. P7.74B). Estimate the mass of the car.

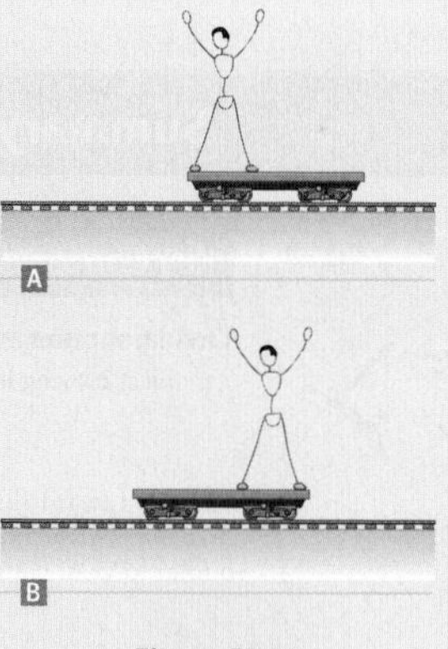

Figure P7.74

Ancillaries

Using Technology to Enhance Learning

ENHANCED WebAssign **Enhanced WebAssign for Giordano's *College Physics: Reasoning and Relationships*, Second Edition**

Exclusively from Cengage Learning, Enhanced WebAssign® offers an extensive online program for physics to encourage the practice that's so critical for concept mastery. The meticulously crafted pedagogy and exercises in our proven texts become even more effective in Enhanced WebAssign. Enhanced WebAssign includes the Cengage YouBook, a highly customizable, interactive eBook.

WebAssign includes:

- All of the quantitative end-of-chapter problems for the second edition
- Reasoning and Relationships tutorials with feedback to address common misconceptions
- Quick Prep: 67 algebra and trigonometry math remediation tutorials
- Concept Checks
- Active Figure animations with assignable questions
- PhET simulations
- The Cengage YouBook

WebAssign's customizable and interactive eBook, the Cengage YouBook, lets you tailor the textbook to fit your course and connect with your students. You can remove and rearrange chapters in the table of contents and tailor assigned readings that match your syllabus exactly. Powerful editing tools let you change as much as you'd like—or leave it just like it is. You can highlight key passages or add sticky notes to pages to comment on a concept in the reading, and then share any of these individual notes and highlights with your students, or keep them personal. You can also edit narrative content in the textbook by adding a text box or striking out text. With a handy link tool, you can drop in an icon at any point in the eBook that lets you link to your own lecture notes, audio summaries, video lectures, or other files on a personal website or anywhere on the Web. A simple YouTube widget lets you easily find and embed videos from YouTube directly into eBook pages. There is a light discussion board that lets students and instructors find others in their class and start a chat session. The Cengage YouBook helps students go beyond just reading the textbook. Students can also highlight the text, add their own notes, and bookmark the text. Animations play right on the page at the point of learning so that they're not speed bumps to reading but true enhancements.

Please visit **www.webassign.net/brookscole** to view an interactive demonstration of Enhanced WebAssign.

PowerLecture with JoinIn and ExamView for Giordano's *College Physics: Reasoning and Relationships*, Second Edition

PowerLecture™

This DVD provides the instructor with dynamic media tools for teaching. Create, deliver, and customize tests (both print and online) in minutes with ExamView® Computerized Testing. The *Instructor's Solutions Manual*, which includes all solutions to questions and problems, is available as Microsoft® Word documents on the PowerLecture DVD. JoinIn™ "clicker" content helps you turn your lectures into an interactive learning environment that promotes conceptual understanding. Available exclusively for higher education from our partnership with Turning Technologies, JoinIn is the easiest way to turn your lecture into a personal, fully interactive experience for your students! By using JoinIn, you can pose text-specific questions to a large group, gather results, and display students' answers seamlessly. To further facilitate an engaging lecture environment,

Microsoft® PowerPoint® lecture slides and figures from the book are also included on this DVD, as well as animations and movies.

Additional Instructor Resources

The *Instructor's Solutions Manual* was authored by Michael Meyer (Michigan Technological University) and James Heath (Austin Community College). For the second edition, Meyer and Heath have reviewed all of the solutions and provided additional support for many of the solutions.

A *Test Bank* created by Ed Oberhofer (University of North Carolina at Charlotte and Lake-Sumter Community College) is available on the *PowerLecture*™ CD-ROM as editable electronic files or via the ExamView® test software. The file contains questions in multiple-choice format for all chapters of the text. Instructors may print and duplicate pages for distribution to students.

Student Resources

Student Companion & Problem-Solving Guide by Richard Grant (Roanoke College) will prove to be an essential study resource. For each chapter, it contains a summary of problem-solving techniques (following the text's methodology), a list of frequently-asked questions students often have when attempting homework assignments, selected solutions to problems, solved Capstone Problems representing typical exam questions, and a set of MCAT review questions with explanations of strategies behind the answers.

Physics Laboratory Manual, Third Edition by David Loyd (Angelo State University) supplements the learning of basic physical principles while introducing laboratory procedures and equipment. Each chapter includes a prelaboratory assignment, objectives, an equipment list, the theory behind the experiment, experimental procedures, graphing exercises, and questions. A laboratory report form is included with each experiment so that the student can record data, calculations, and experimental results. Students are encouraged to apply statistical analysis to their data. A complete *Instructor's Manual* is also available to facilitate use of this lab manual.

Physics Laboratory Experiments, Seventh Edition by Jerry D. Wilson (Lander College) and Cecilia A. Hernández (American River College). This market-leading manual for the first-year physics laboratory course offers a wide range of class-tested experiments designed specifically for use in small to midsize lab programs. A series of integrated experiments emphasizes the use of computerized instrumentation and includes a set of "computer-assisted experiments" to allow students and instructors to gain experience with modern equipment. This option also enables instructors to determine the appropriate balance between traditional and computer-based experiments for their courses. By analyzing data through two different methods, students gain a greater understanding of the concepts behind the experiments. The Seventh Edition is updated with the latest information and techniques involving state-of-the-art equipment and a new Guided Learning feature addresses the growing interest in guided-inquiry pedagogy. Fourteen additional experiments are also available through custom printing.

Acknowledgments

Creating a textbook is an enormous job requiring the assistance of many people. To all these people, I extend my sincere thanks.

Contributors

Raymond Hall of California State University, Fresno, and Richard Grant of Roanoke College contributed many interesting and creative questions and problems.

Accuracy Reviewers

David Aaron, *South Dakota State University*
Brad Abbott, *University of Oklahoma*
Phil Adams, *Louisiana State University*
David Cole, *Northern Arizona University*
Andrew Cornelius, *University of Nevada—Las Vegas*
Shelly Lesher, *University of Wisconsin—La Crosse*
Mark Lucas, *Ohio University*
Michel Pleimling, *Virginia Tech*
Michael Pravica, *University of Nevada—Las Vegas*
Mahdi Sanati, *Texas Tech University*
Michael Strauss, *University of Oklahoma*
Wayne Trail, *Southwest Oklahoma State University*
Richard Vallery, *Grand Valley State University*
David Young, *Louisiana State University*

Pre-Revision Reviewers

The following individuals provided helpful comments on the first edition and valuable advice in preparing the revision.

David Aaron, *South Dakota State University*
Gerald Cleaver, *Baylor University*
Christopher Coffin, *Oregon State University*
Alejandro Garcia, *University of Washington*
Eitan Gross, *University of Arkansas*
James Heath, *Austin Community College*
Leo Piilonen, *Virginia Tech*
David Young, *Louisiana State University*

The following individuals provided useful and detailed feedback toward the revision of questions and problems.

Kevin Haglin, *St. Cloud State University*
Marko Horbatsch, *York University*
Sylvio May, *North Dakota State University*
Michael Meyer, *Michigan Technological University*
David Sokoloff, *University of Oregon*

First Edition Manuscript Reviewers

Jeffrey Adams, *Montana State University*
Anthony Aguirre, *University of California—Santa Cruz*
David Balogh, *Fresno City College*
David Bannon, *Oregon State University*
Phil Baringer, *University of Kansas*
Natalie Batalha, *San Jose State University*
Mark Blachly, *Arsenal Technical High School*
Gary Blanpied, *University of South Carolina*
Ken Bolland, *The Ohio State University*
Scott Bonham, *Western Kentucky University*
Marc Caffee, *Purdue University*
Lee Chow, *University of Central Florida*
Song Chung, *William Patterson University*
Alice Churukian, *Concordia College*
Thomas Colbert, *Augusta State University*
David Cole, *Northern Arizona University*
Sergio Conetti, *University of Virginia*
Gary Copeland, *Old Dominion University*
Doug Copely, *Sacramento City College*
Robert Corey, *South Dakota School of Mines & Technology*
Andrew Cornelius, *University of Nevada—Las Vegas*
Nimbus Couzin, *Indiana University, Southeast*
Carl Covatto, *Arizona State University*
Thomas Cravens, *University of Kansas*
Sridhara Dasu, *University of Wisconsin—Madison*
Timir Datta, *University of South Carolina*
Susan DiFranzo, *Hudson Valley Community College*
David Donnelly, *Texas State University*
Sandra Doty, *The Ohio State University*
Steve Ellis, *University of Kentucky*
Len Finegold, *Drexel University*
Carl Fredrickson, *University of Central Arkansas*
Joe Gallant, *Kent State University, Warren Campus*
Kent Gee, *Brigham Young University*
Bernard Gerstman, *Florida International University*
James Goff, *Pima Community College*
Richard Grant, *Roanoke College*
William Gregg, *Louisiana State University*
James Guinn, *Georgia Perimeter College, Clarkston*
Richard Heinz, *Indiana University—Bloomington*
John Hopkins, *Pennsylvania State University*
Karim Hossain, *Edinboro University of Pennsylvania*
Linda Jones, *College of Charleston*
Alex Kamenev, *University of Minnesota*
Daniel Kennefick, *University of Arkansas*
Aslam Khalil, *Portland State University*
Jeremy King, *Clemson University*
Randy Kobes, *University of Winnipeg*
Raman Kolluri, *Camden County College*
Ilkka Koskelo, *San Francisco State University*
Fred Kuttner, *University of California—Santa Cruz*
Richard Ledet, *University of Louisiana—Lafayette*
Alexander Lisyansky, *Queens College, City University of New York*
Carl Lundstedt, *University of Nebraska—Lincoln*
Donald Luttermoser, *East Tennessee State University*
Steven Matsik, *Georgia State University*
Sylvio May, *North Dakota State University*
Bill Mayes, *University of Houston*
Arthur McGum, *Western Michigan University*
Roger McNeil, *Louisiana State University*
Rahul Mehta, *University of Central Arkansas*
Charles Meitzler, *Sam Houston State University*
Michael Meyer, *Michigan Technological University*
Vesna Milosevic-Zdjelar, *University of Winnipeg*
John Milsom, *University of Arizona*
Wouter Montfrooij, *University of Missouri*
Ted Morishige, *University of Central Oklahoma*
Halina Opyrchal, *New Jersey Institute of Technology*
Michelle Ouellette, *California Polytechnic State University—San Louis Obispo*
Kenneth Park, *Baylor University*
Galen Pickett, *California State University—Long Beach*
Dinko Pocanic, *University of Virginia*
Amy Pope, *Clemson University*
Michael Pravica, *University of Nevada—Las Vegas*
Laura Pyrak-Nolte, *Purdue University*
Mark Riley, *Florida State University*
Mahdi Sanati, *Texas Tech University*
Cheryl Schaefer, *Missouri State University*

Alicia Serfaty de Markus, *Miami Dade College—Kendall Campus*
Marc Sher, *College of William & Mary*
Douglas Sherman, *San Jose State University*
Marllin Simon, *Auburn University*
Chandralekha Singh, *University of Pittsburgh*
David Sokoloff, *University of Oregon*
Noel Stanton, *Kansas State University*
Donna Stokes, *University of Houston*
Carey Stronach, *Virginia State University*
Chun Fu Su, *Mississippi State University*
Daniel Suson, *Texas A&M University—Kingsville*
Doug Tussey, *Pennsylvania State University*
John Allen Underwood, *Austin Community College—Rio Grande Campus*
James Wetzel, *Indiana University-Purdue University—Fort Wayne*
Lisa Will, *Arizona State University*
Gerald T. Woods, *University of South Florida*
Guoliang Yang, *Drexel University*
David Young, *Louisiana State University*
Michael Yurko, *Indiana University-Purdue University—Indianapolis*
Hao Zeng, *State University of New York at Buffalo*
Nouredine Zettili, *Jacksonville State University*

Extra Credit

Special thanks go to Lachina Publishing Services, especially Jennifer Bonnar, Jeanne Lewandowski, and Aaron Kantor, for keeping up with innumerable changes during the production stages of this book. Kathleen M. Lafferty once again edited the copy with diligence. Roy Neuhaus brightened the interior with a new design. Steve McEntee provided valuable updates to the art program. Greg Gambino continued creating the "Stick Dude" art for worked examples and end-of-chapter questions that is so popular with reviewers and students. Dena Digilio Betz once again went the extra mile when needed to find photos from unusual sources. Katie Huha and Martha Hall of PreMediaGlobal assisted in securing permissions. Margaret Pinette assisted in the preparation of this Hybrid Edition.

I would also like to thank all the team at Cengage Learning, including Michelle Julet, Mary Finch, Charles Hartford, Cate Barr, Cathy Brooks, Nicole Mollica, Jack Cooney, Brandi Kirksey, Brendan Killion, Shalice Shah-Caldwell, Diane Gibbons, and Doug Bertke, for their fine work in the development, production, and promotion of this new edition. Special recognition goes to Rebecca Berardy Schwartz for her skillful management of the new reasoning tutorials, and again to Susan Pashos for her unwavering and tireless support. I also thank my wife Pat for her continued patience and encouragement.

No textbook is perfect for every student or for every instructor. It is my hope that both students and instructors will find some useful, stimulating, and even exciting material in this book and that you will all enjoy physics as much as I do.

Nicholas J. Giordano

About the Author

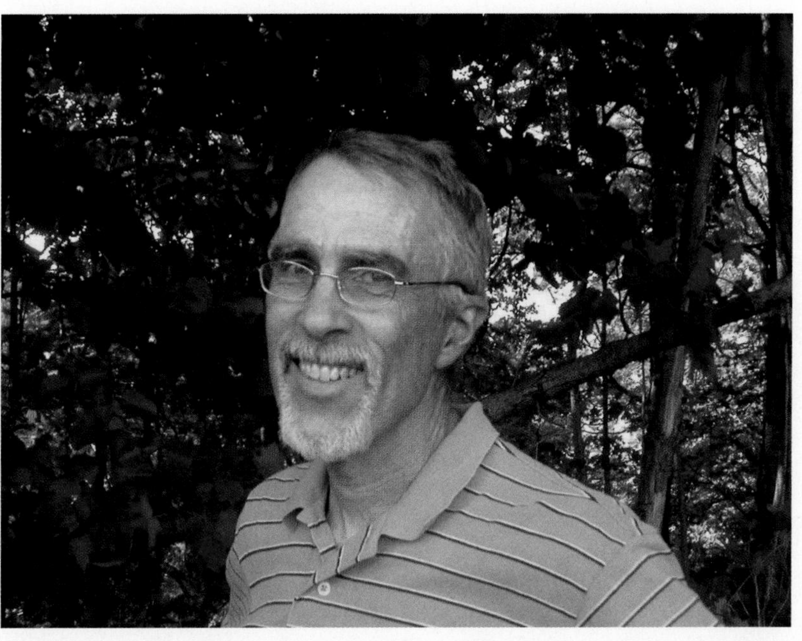

Nicholas J. Giordano obtained his B.S. at Purdue University and his Ph.D. at Yale University. He has been on the faculty at Purdue since 1979, served as an Assistant Dean of Science from 2000 to 2003, and in 2004 was named the Hubert James Distinguished Professor of Physics. His research interests include the properties of nanoscale metal structures, nanofluidics, science education, and biophysics, along with musical acoustics and the physics of the piano. Dr. Giordano earned a Computational Science Education Award from the Department of Energy in 1997, and he was named Indiana Professor of the Year by the Carnegie Foundation for the Advancement of Teaching and the Council for the Advancement and Support of Education in 2004. His hobbies include distance running and restoring antique pianos, and he is an avid baseball fan. He also just finished a book on the physics of the piano.

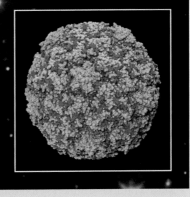

CHAPTER 1

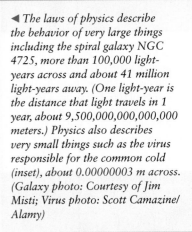

◀ *The laws of physics describe the behavior of very large things including the spiral galaxy NGC 4725, more than 100,000 light-years across and about 41 million light-years away. (One light-year is the distance that light travels in 1 year, about 9,500,000,000,000,000 meters.) Physics also describes very small things such as the virus responsible for the common cold (inset), about 0.00000003 m across. (Galaxy photo: Courtesy of Jim Misti; Virus photo: Scott Camazine/Alamy)*

Introduction

OUTLINE

1.1 The Purpose of Physics

1.2 Problem Solving in Physics: Reasoning and Relationships

1.3 Dealing with Numbers

1.4 Physical Quantities and Units of Measure

1.5 Dimensions and Units

1.6 Algebra and Simultaneous Equations

1.7 Trigonometry

1.8 Vectors

▲ **Figure 1.1** Physics is the science of matter and energy, and the interactions between them. At the center of this galaxy lies a massive black hole, containing a very large amount of matter and energy.

▲ **Figure 1.2** Isaac Newton (1642–1727) developed the laws of mechanics we study in the first part of this book. Newton was also a great mathematician and invented much of calculus.

1.1 The Purpose of Physics

This book is about the field of science known as physics. Let's therefore begin by considering what the word *physics* means. One popular definition is

***physics*: the science of matter and energy, and the interactions between them.**

Matter and energy are fundamental to all areas of science; thus, physics is truly a foundational subject. The principles of physics form the basis for understanding chemistry, biology, and essentially all other areas of science. These principles enable us to understand phenomena ranging from the very small (atoms, molecules, and cells) to the very large (planets and galaxies; Fig. 1.1). Indeed, the word *physics* has its origin in the Greek word for "nature." For this reason, an alternative and much broader definition is

***physics*: the study of the natural or material world and phenomena; natural philosophy.**

An important part of this definition is the term *natural philosophy*. In a sense, physics is the oldest of the sciences, and at one time *all* scientists were physicists. Many famous thinkers, including Aristotle, Galileo, and Isaac Newton (Fig. 1.2), were such natural philosophers/physicists.

So, if physics is the science of matter and energy, how does one actually study and learn physics? Our primary objective is to learn how to predict and understand the way matter and energy behave; that is, we want to predict and understand how the universe works. Physics is organized around a collection of ***physical laws***. In the first part of this book, we learn about Newton's laws, which are concerned with the motion of mechanical objects such as cars, baseballs, and planets. Later we'll encounter physical laws associated with a variety of other phenomena including heat, electricity, magnetism, and light. Our job is to learn about these physical laws (sometimes called ***laws of nature***) and how to use them to predict the workings of the universe: how objects move, how electricity flows, how light travels, and more. These physical laws are usually expressed mathematically, so much of our work will involve mathematics. However, good physics is more than just good mathematics; an appreciation of the basic concepts and how they fit together is essential.

In addition to *predicting how* the world works, we would also like to *understand why* it works the way it does. Making predictions requires us to apply the physical laws to a particular situation and work out the associated mathematics to arrive at specific predictions. Understanding the world is more difficult because it involves understanding where the physical laws "come from." This "come from" question is so difficult because physical law comes into being in the following way. Initially, someone formulates a hypothesis that describes all that is known about a particular phenomenon. For example, Newton probably formulated such an initial hypothesis for the laws of motion. One must then show that this hypothesis correctly describes *all* known phenomena. For Newton, it meant that his proposed laws of motion had to account correctly for the motion of apples, rocks, arrows, and all other terrestrial objects. In addition, a successful hypothesis is often able to explain things that were previously not understood. In Newton's case, his hypothesis was able to explain celestial motion (the motion of the planets and the Moon, as sketched in Fig. 1.3), a problem that was unsolved before Newton's time. Only after a hypothesis passes such tests does it qualify as a law or principle of physics.

This process through which a hypothesis becomes a law of physics means that there is no way to prove that such a physical law is correct. It is always possible that a future experiment or observation will reveal a flaw or limitation in a particular "law." This step is part of the scientific process because the discovery of such flaws leads to the discovery of new laws and new insights into nature.

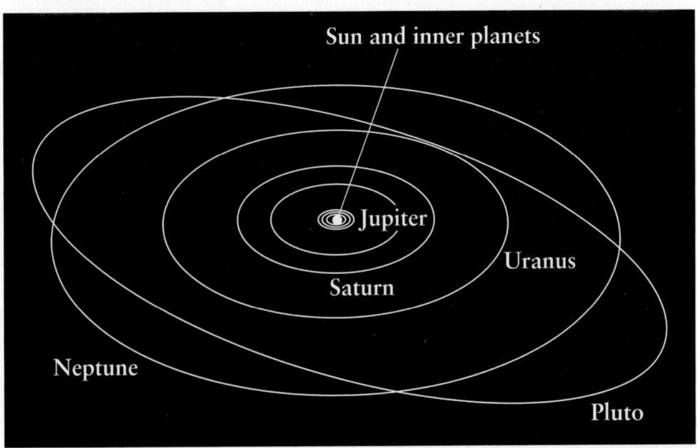

◀ **Figure 1.3** Newton showed that the motion of the planets around our Sun can be explained by the same laws of motion that describe the movement of terrestrial objects, such as baseballs and automobiles. This discovery unified human understanding of the motion of terrestrial and celestial objects.

Although this process of constructing and testing hypotheses can lead us to the laws of physics, it does not necessarily give us an understanding of why these laws are correct. That is, why has nature chosen a particular set of physical laws instead of a different set? Insight into this question can often be gained by examining the form of a physical law and the predictions it leads to. We'll do that as we proceed through this book and in this way get important glimpses into how nature works. The important point is that the problem of *predicting how* the world works is different from *understanding why* it works the way it does. Our goal is to do both.

1.2 Problem Solving in Physics: Reasoning and Relationships

According to our definitions in Section 1.1, physics involves predicting how matter and energy behave in different situations. Making such predictions usually requires that we apply a general physical law to a particular case, a process known as ***problem solving***. Problem solving is an essential part of physics, and we'll spend a great deal of time learning how to do it in many different situations. In some ways, learning the art of problem solving is like learning how to play the piano or hit a golf ball: it takes practice. Just as in playing a musical instrument or learning a sport, certain practices are the keys to good problem solving. Following these practices consistently will help lead to a thorough understanding of the underlying concepts.

Problem solving often involves some mathematical calculations. Many problems we encounter are ***quantitative problems***, which give quantitative information about a situation and require a precise mathematical calculation using that information. An example is to calculate the time it takes an apple to fall from a tree to the ground below, given the initial height of the apple. A problem may also involve the application of a key concept in a nonmathematical way. We use such ***concept checks*** to test your general understanding of a particular physical law or concept and how it is applied. We will also encounter ***reasoning and relationship problems***, which will often require you to identify important information that might be "missing" from an initial description of the problem. For example, you might be asked to calculate the forces acting on two cars when they collide, given only the speed of the cars just before the collision. From an understanding of the relationship between force and motion, you would need to recognize that additional information is required to solve this problem. (In this problem, that additional information is the mass of the cars and the way the car bumpers deform in the collision.) It is your job to use common sense and experience to estimate realistic values of these "missing" quantities for typical cars and then use these values to compute an answer. Because such estimated values vary from case to case (e.g., not all cars have the same mass), an approximate mathematical solution and an approximate numerical answer are usually sufficient for such reasoning and relationship problems.

An ability to deal with these three different types of problems is essential to gaining a thorough understanding of physics and for success in any science- or technology-related field.

PROBLEM SOLVING Problem-Solving Strategies

We'll encounter problems involving different laws of physics, including Newton's laws of mechanics, the laws of electricity and magnetism, and quantum theory. Although these problems involve many different situations, they can all be attacked using the same basic problem-solving strategy.

1. **RECOGNIZE THE** key physics **PRINCIPLES** that are central to the problem. For example, one problem might involve the principle of conservation of energy, whereas another might require Newton's action–reaction principle. The ability to recognize the central principles requires a conceptual understanding of the laws of physics, how they are applied, and how they are interrelated. Such knowledge and skill are obtained from experience, practice, and careful study.

2. **SKETCH THE PROBLEM.** A diagram showing all the given information, the directions of any forces, and so forth is valuable for organizing your thoughts. A good diagram will usually contain a coordinate system to be used in measuring the position of an object and other important quantities.

3. **IDENTIFY THE** important **RELATIONSHIPS** among the known and unknown quantities. For example, Newton's second law gives a relationship between force and motion, and is thus the key to analyzing the motion of an object. This step in the problem-solving process may involve several parts (substeps), depending on the nature of the problem. For example, problems involving collisions may involve steps that aren't necessary for a problem in magnetism. When dealing with a reasoning and relationship problem, one of these substeps may involve identifying the "missing" information or quantities and then estimating their values.

4. **SOLVE** for the unknown quantities using the relationships in step 3.

5. **WHAT DOES IT MEAN?** Does your answer make sense? Take a moment to think about your answer and reflect on the general lessons to be learned from the problem.

1.3 Dealing with Numbers

Scientific Notation

During the course of problem solving, we will often encounter numbers that are very large or very small. ***Scientific notation*** was invented as a convenient way to abbreviate extremely large or extremely small numbers. We can understand how scientific notation works by using it to express some of the numbers found in Table 1.1, which lists some important lengths and distances. As you probably know, lengths and distances can be measured in units of ***meters***. We'll say more about meters and other units of measure in the next section. For now, we can rely on your intuitive notion of length and that 1 meter (abbreviated "m") is approximately the distance from the floor to the doorknob for a typical door. The height of a typical adult male is about 1.8 m, and the diameter of a compact disc (CD) or digital video disc (DVD) is about 0.12 m (Fig. 1.4).

The numerical values in Table 1.1 span a tremendous range. The largest is the distance from Earth to the Sun, which is 150,000,000,000 m. This number is so large and contains so many zeros to the left of the decimal point[1] that in Table 1.1 we have written it in scientific notation. When written in this way, the distance from Earth to the Sun is 1.5×10^{11} m. The exponent has the value 11 because the deci-

[1] In numbers like this one, the decimal point is usually not written explicitly. If we were to include the decimal point, however, this number would be written as 150,000,000,000. m.

TABLE 1.1 Some Common Lengths and Distances

Quantity	Length or Distance in Meters (m)	Quantity	Length or Distance in Meters (m)
Diameter of a proton	1×10^{-15}	Height of the author	1.80
Diameter of a red blood cell*	8×10^{-6}	Height of the Empire State Building	443.2
Diameter of a human hair (removed from the author's head)	5.5×10^{-5}	Distance from New York City to Chicago	1,268,000
Thickness of a sheet of paper	6.4×10^{-5}	Circumference of the Earth	4.00×10^{7}
Diameter of a CD or DVD	0.12	Distance from the Earth to the Sun	1.5×10^{11}

*Red blood cells are not spherical, but are shaped more like flat plates. The value given here is the approximate diameter of the plate.

mal point in the number 150,000,000,000 is 11 places to the right of its location in the number 1.5. Likewise, the distance from New York to Chicago is 1,268,000 m, which is 1.268×10^6 m. Here the decimal point in the number 1,268,000 is six places to the right of its location in the number 1.268. Scientific notation is also a useful way to write very small numbers. For example, the diameter of a hair taken from the author's head is 0.000055 m, which is written as 5.5×10^{-5} m in scientific notation (Table 1.1). Here the exponent has the value -5 because the decimal point in 0.000055 is five places to the left of its location in the number 5.5.

Writing numbers in scientific notation

These rules for expressing a number in scientific notation can be summarized as follows. Move the decimal point in the original number to obtain a new number between 1 and 10. Count the number of places the decimal point has been moved; this number will become the exponent of 10 in scientific notation. If you started with a number greater than 10 (such as 150,000,000,000), the exponent of 10 is positive (1.5×10^{11}). If you started with a number less than 1 (such as 0.000055), the exponent is negative (5.5×10^{-5}).

EXAMPLE 1.1 Population of the Earth

The number of people living on the Earth in 2007 is estimated to have been approximately 6,600,000,000. Express this number in scientific notation.

RECOGNIZE THE PRINCIPLE

To write this number using scientific notation, we must compare the location of the decimal point in the number 6,600,000,000 with its location in the number 6.6.

SKETCH THE PROBLEM

We must move the decimal point nine places to get a number between 1 and 10:

$$6{,}\underbrace{600{,}000{,}000}_{\text{9 places}}$$

IDENTIFY THE RELATIONSHIPS AND SOLVE

We started with a number greater than 10. Using our rules for expressing a number in scientific notion, the population of the Earth in 2007 was

$$6{,}600{,}000{,}000 = \boxed{6.6 \times 10^9}$$

► ***What does it mean?***

The exponent in this answer is 9 because the decimal point in the number 6,600,000,000 is nine places to the right of its location in the number 6.6.

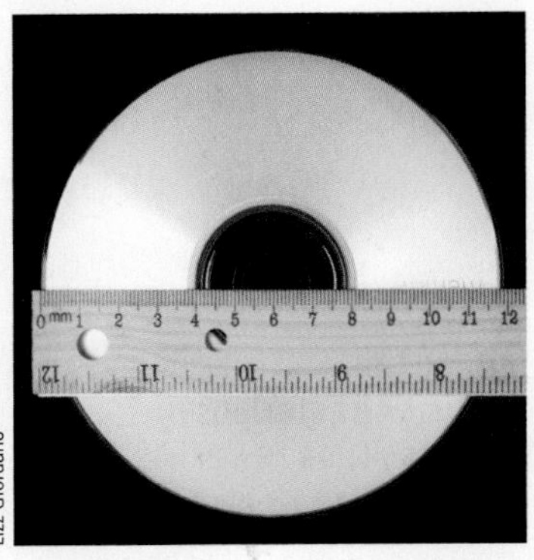

Lizz Giordano

▲ **Figure 1.4** A compact disc has a diameter of approximately 0.12 m.

Significant Figures

Suppose you are asked to measure the width of a standard piece of paper such as a page in this book. The answer is approximately 0.216 meter, as you can confirm for yourself using a ruler. However, you should realize that this value is not exact. Different pieces of paper will be slightly different in size, and your measurement will also have some uncertainty (some experimental error). There is an uncertainty associated with all such measured values, including all the values in Table 1.1, and these uncertainties affect the way we write a numerical value. For our piece of paper, we have written the width using three ***significant figures***. The term *significant* refers to the number of digits that are meaningful with regard to the accuracy of the value. In this particular case, writing the value as 0.216 m implies that the true value lies between 0.210 m and 0.220 m, and that it is likely to be close to 0.216 m. We should not be surprised to find a value of 0.215 m or 0.217 m, though.

When writing a number in scientific notation, the number of digits that are written depends on the number of significant figures. For our piece of paper, we would write 2.16×10^{-1} m and would say that all three digits are significant. There can be some ambiguity when dealing with digits that are zero. For example, the length of a particular race in the sport of track and field is 100 meters (the distance between the starting line and the finish). In terms of significant figures, you might think that this number has only one significant figure, which would imply that the length of the race falls within the rather large range of 0 to 200 m. In this example, however, the distance between the start and finish lines will be measured quite precisely, so we should suspect the value to have at least three significant figures, with the true value lying between 99 and 101 meters. If this accuracy is in fact correct, the value "100 meters" actually contains three significant figures; the zeros here are significant because they are indicative of the accuracy. In such cases, we must determine the number of significant figures from the context of the problem or from other information. This ambiguity does not arise with scientific notation; the length of our race would be written as 1.00×10^2 m, which indicates explicitly that the zeros are significant.

Significant figures are also important in calculations. As an example, Table 1.1 lists the thickness of a typical sheet of paper as 6.4×10^{-5} m. This value—which has two significant figures—was measured by the author for a piece of paper of the kind used in this book. Suppose a book were to contain 976 such sheets. We could then compute the thickness of the book (without the covers) as

$$\text{thickness of book} = \text{number of sheets of paper} \times \text{thickness of one sheet}$$
$$= 976 \times 0.000064 \text{ m} = 0.062464 \text{ m} \quad (1.1)$$

Insight 1.1
SIGNIFICANT FIGURES IN NUMBERS LESS THAN ONE
The number of significant figures in the value 0.062464 is *five*. For a number smaller than 1, the zeros immediately to the right of the decimal point are *not* significant figures. Hence, the number 0.000064 has just two significant figures.

The final result in Equation 1.1 is written with five significant figures, as you might read from a calculator used to do this multiplication. Because the number of significant figures indicates the expected accuracy of a value, writing this answer with five significant figures implies that we know the thickness of the book to very high accuracy. However, since the thickness of one page is known with an accuracy of only two significant figures, the final value of this calculation—the thickness of the entire book—will actually have an accuracy of only two significant figures. So, we should round the result in Equation 1.1 to two significant figures:

$$0.062464 \text{ m rounded to two significant figures} = 0.062 \text{ m} \quad (1.2)$$

Determining the number of significant figures in a calculation involving multiplication or division

Rules for significant figures in calculations involving *multiplication and division.*

1. Use the full accuracy of all known quantities when doing the computation. In Equation 1.1, we used the number of sheets as given to three

significant figures and the thickness per sheet, which is known to two significant figures.

2. At the *end* of the calculation, round the answer to the number of significant figures present in the *least* accurate starting quantity. In Equation 1.1, the least accurate starting value was the thickness of a single sheet, which was known to two significant figures, so we rounded our final answer to two significant figures in Equation 1.2.

Notice that the rounding in Equation 1.2 took place at the *end* of the calculation. Some calculations involve a sequence of several separate computations, and rounding can sometimes cause trouble if it is applied at an intermediate stage of the computation. For example, suppose we want to use the result from Equation 1.2 to find the height of a stack of 12 such books. We could multiply the answer from Equation 1.2 by 12 to get the height of the stack. We thus have

$$\begin{aligned}\text{height of stack} &= 12 \times (0.062\text{ m}) = 0.744\text{ m}\\ &= 0.74\text{ m (rounding to two significant figures)}\end{aligned} \tag{1.3}$$

Here we have rounded our final answer to two significant figures because the starting value—the thickness of one book—is known to two significant figures. On the other hand, if we use the unrounded value of the thickness of one book from Equation 1.1, we get

$$\begin{aligned}\text{height} &= 12 \times (0.062464\text{ m}) = 0.749568\text{ m}\\ &= 0.75\text{ m (rounding to two significant figures)}\end{aligned} \tag{1.4}$$

In the last step in Equation 1.4, we again rounded to two significant figures. Comparing the results in Equations 1.3 and 1.4, we see that the final answers differ by 0.01 m. This difference is due to ***round-off error***, which can happen when we round an answer too soon in the course of a multistep calculation. Such errors can be avoided by carrying along an extra significant figure through intermediate steps in a computation and then performing the final rounding at the very end. In this example, we should keep three significant figures for the answer from Equation 1.1 (carrying an extra digit); using this three-significant-figure value in Equation 1.3 would then have given us the correct answer of 0.75 m.

The procedures we have just described for dealing with significant figures apply to calculations that involve multiplication and division. Computations involving addition or subtraction require a different approach to determine the final accuracy.

Determining the number of significant figures in a calculation involving addition or subtraction

Rule for significant figures in calculations involving *addition or subtraction*.

The location of the least significant digit in the answer is determined by the location of the least significant digit in the starting quantity that is known with the *least accuracy*.

Consider the addition of the numbers 4.52 and 1.2. The least accurate of these numbers is 1.2, so the least significant digit here and in the final answer is one place to the right of the decimal point. Pictorially, we have

Least significant digit in the number 1.2, and in the final sum, is one place to the right of the decimal point.

$$\begin{array}{r} 4.52 \\ +1.2 \\ \hline 5.7 \end{array}$$

The value of this digit is unknown, so the answer is not known to this accuracy.

Hence, in this example, the final answer has two significant figures.

An example involving subtraction is shown below, where we consider the calculation 4.52 − 1.2. The starting numbers contain three and two significant figures, respectively, and the final answer contains two. Pictorially,

Least significant digit in the number 1.2, and in the final answer, is one place to the right of the decimal point.	$\begin{array}{r} 4.52 \\ -1.2 \\ \hline 3.3 \end{array}$	The value of this digit is unknown, so the answer is not known to this accuracy.

As another example, consider the subtraction of two numbers whose difference is very small, such as 4.52 − 4.1. Now we have

Least significant digit in the final answer.	$\begin{array}{r} 4.52 \\ -4.1 \\ \hline 0.4 \end{array}$	The value of this digit is unknown, so the answer is not known to this accuracy.

In this case, the number of significant figures in the final answer is *smaller* than the number of significant figures in either starting number.

Some numbers that appear in a calculation are known exactly. For example, there are 60 seconds in 1 minute. This value is *exact*, so it really should be thought of as 60.000000 . . . in a calculation. The number of significant figures in a calculation with such numbers is then determined by the accuracy (i.e., the number of significant figures) with which other quantities in the calculation are known.

In most of the calculations and problems in this book, we'll round answers to two significant figures.

EXAMPLE 1.2 Volume of a Blood Cell

Consider a red blood cell, and for simplicity assume it is spherical with radius $r_1 = 5.0 \times 10^{-6}$ m.

(a) What is the volume of this cell?

(b) Suppose a second red blood cell has a radius $r_2 = 5.1 \times 10^{-6}$ m. What is the *difference* in the volume of the two cells?

RECOGNIZE THE PRINCIPLE

The volume of a sphere of radius r is $V = \frac{4}{3}\pi r^3$. We must evaluate V for both values of the radius, r_1 and r_2. Since r_1 and r_2 are both given with two significant figures, the results for V must also have two significant figures.

SKETCH THE PROBLEM

The radii r_1 and r_2 are almost the same.

IDENTIFY THE RELATIONSHIPS AND SOLVE

(a) Using the value of r_1 given above, the volume of the first cell V_1 is[2]

$$V_1 = \tfrac{4}{3}\pi r_1^3 = \tfrac{4}{3}\pi(5.0 \times 10^{-6}\text{ m})^3 = 5.236 \times 10^{-16}\text{ m}^3$$

Insight 1.2

A MISUSE OF SIGNIFICANT FIGURES

It is often quoted that normal human body temperature is 98.6°F. This value is measured with the Fahrenheit temperature scale. This temperature was originally measured in experiments that employed the Celsius scale. When measured in Celsius, it was found that healthy people exhibit temperatures in the range of 36°C to 38°C, with 37°C being typical. So, the value for body temperature on the Celsius scale is 37°C and thus has *two* significant figures. However, when this value was converted to the Fahrenheit scale it was specified as 98.6°F, with *three* significant figures. To be consistent with significant figures (and with the accuracy of the original experimental value and the variation from person to person), this value should actually be rounded to two significant figures, which would give 99°F. Using the value of 98.6°F with the *incorrect* number of significant figures implies that a difference of 0.1°F is significant, but that is not the case at all!

[2]Notice that this answer is given in units of m³ = cubic meters. We'll say more about this and other such units below.

The value of r_1 is given to two significant figures, so we must round our final answer to two significant figures:

$$V_1 = \boxed{5.2 \times 10^{-16}\ \text{m}^3}$$

(b) A similar calculation gives the volume of the second cell:

$$V_2 = \tfrac{4}{3}\pi r_2^3 = \tfrac{4}{3}\pi(5.1 \times 10^{-6}\ \text{m})^3 = 5.6 \times 10^{-16}\ \text{m}^3$$

where we have again rounded to two significant figures. The difference in the volumes of the two cells is

$$\Delta V = V_2 - V_1 = (5.6 \times 10^{-16} - 5.2 \times 10^{-16})\ \text{m}^3$$

$$\Delta V = 0.4 \times 10^{-16}\ \text{m}^3 = \boxed{4 \times 10^{-17}\ \text{m}^3}$$

▸ ***What does it mean?***

The given values of r_1 and r_2 contained two significant figures, so the results for V_1 and V_2 also contained two significant figures. However, according to the rule for significant figures in subtraction, the value for ΔV has only one significant figure.

CONCEPT CHECK 1.1 Significant Figures

Give the number of significant figures for all the lengths and distances in Table 1.1.

1.4 Physical Quantities and Units of Measure

To conduct experiments and test physical theories, we must be able to *measure* various physical quantities. Typical quantities of importance in physics include the ***distance*** traveled by a baseball, the ***time*** it takes an apple to fall from a tree, and the ***mass*** of the Earth. These three types of quantities—distance (or equivalently, length), time, and mass—play central roles in physics. In order to measure a particular physical quantity, we must have a ***unit of measure*** for that quantity. Consider, for example, the quantity length. Table 1.1 lists a variety of lengths and distances, with values given in ***meters***, a unit of measure widely used in scientific work. The meter was initially defined in terms of the circumference of the Earth and was subsequently redefined and maintained using a carefully constructed metal bar composed of platinum and iridium (Fig. 1.5). In 1960, the meter was redefined again, this time in terms of the wavelength of light emitted by krypton atoms. Other units, such as inches and feet, can be used to measure length, and scientists involved in metrology (the science of measurement) have determined how different units of length, including meters, inches, and feet, are related.

Courtesy NIST

▲ **Figure 1.5** Prior to 1960, the length of the standard meter was maintained using a platinum–iridium bar like those shown here from the National Institute of Standards and Technology.

It is often necessary to convert the value of a particular length from one unit to another. For example, the width of a page of this book is approximately 0.216 m. We can express this width in inches using the appropriate ***conversion factor***, which relates the two units of interest. For instance, a length of 1 inch (abbreviated "in.") is equal to 2.54×10^{-2} m. The conversion between inches and meters can be expressed in equation form as 1 in. = 2.54×10^{-2} m or in fractional form as

$$\frac{1\ \text{in.}}{2.54 \times 10^{-2}\ \text{m}}$$

If a page is 0.216 m wide, its width in inches is

$$0.216\ \text{m} = 0.216\ \text{m} \times \overbrace{\frac{1\ \text{in.}}{2.54 \times 10^{-2}\ \text{m}}}^{\text{conversion factor}} = 8.50\ \text{in.} \tag{1.5}$$

In terms of only the units, the conversion in Equation 1.5 has the form

$$\text{meters} \times \frac{\text{inches}}{\text{meters}} = \cancel{\text{meters}} \times \frac{\text{inches}}{\cancel{\text{meters}}} = \text{inches} \tag{1.6}$$

Converting from one unit to another

so the "unwanted" units—in this case meters—cancel. The conversion of units always has this form, with the unwanted units canceling to leave the desired unit.

This example of units conversion involves units of length, but the same approach applies to other types of units (e.g., converting from hours to seconds). A set of commonly used conversion factors is given inside the front cover of this book. These conversion factors either are exact numbers or are known with very high precision, so they can generally be treated as exact numbers when determining the number of significant figures in a calculation.

EXAMPLE 1.3 Converting Units

Find the number of inches in 1 mile.

RECOGNIZE THE PRINCIPLE

Inches and miles are two units used to measure length. They are not used often in scientific work, but are often encountered in everyday activities. To convert from miles to inches, we use our knowledge that 1 mile is equal to exactly 5280 feet and that there are exactly 12 inches in 1 foot.

SKETCH THE PROBLEM

No sketch is needed.

IDENTIFY THE RELATIONSHIPS AND SOLVE

The conversion from miles to inches looks like

$$1 \text{ mile} \times \frac{5280 \text{ feet}}{1 \text{ mile}} \times \frac{12 \text{ inches}}{1 \text{ foot}} = \boxed{63{,}360 \text{ inches}} \tag{1}$$

In terms of the units, this conversion has the form

$$1 \text{ mile} = \cancel{\text{mile}} \times \frac{\cancel{\text{feet}}}{\cancel{\text{mile}}} \times \frac{\text{inches}}{\cancel{\text{foot}}} = \text{inches}$$

► ***What does it mean?***

The answer (Eq. 1) is *exact* because the number of feet in 1 mile and the number of inches in 1 foot are both exact values, determined by the definitions of these units and not by measurement.

▲ **Figure 1.6** Some of the most accurate clocks are based on the properties of cesium atoms and the light they emit. These clocks gain or lose less than 10^{-8} s each day.

Units of Time and Mass

A particularly important physical quantity is ***time***. As you probably know, time is measured in units of seconds, minutes, hours, and so forth. Table 1.2 lists the values of a number of important time intervals. In most scientific work, time is measured in units of seconds (denoted by "s"). The value of the second is based on the frequency of light emitted by cesium atoms. Figure 1.6 shows a cesium atomic clock from the National Institute of Standards and Technology.

A third important physical quantity is ***mass***. Intuitively, mass is related to the amount of material contained in an object; a slightly better definition will be given when we discuss Newton's laws of motion in Chapter 2. Mass can be measured in units called kilograms and in several other units, including grams and slugs. Table 1.3 lists the masses of several common objects in units of kilograms (abbreviated "kg"). By international agreement, the current standard of mass is a piece of a spe-

TABLE 1.2 Some Common Time Intervals

Quantity	Time in Seconds (s)
Time for light to travel 1 meter	3.34×10^{-9}
Time between heartbeats (approximate)	1
Time for light to travel from the Sun to Earth	500
1 day	86,400
1 month (30 days)	2,592,000
Human life span (approximate)	2×10^{9}
Age of the universe	5×10^{17}

TABLE 1.3 Some Common Masses

Object	Mass in Kilograms (kg)
Electron	9.1×10^{-31}
Proton	1.7×10^{-27}
Red blood cell	1×10^{-13}
Mosquito	1×10^{-5}
Typical person	70
Automobile	1200
Earth	6.0×10^{24}
Sun	2.0×10^{30}

cially configured metal alloy composed of platinum and iridium stored in Paris under extremely well controlled conditions (Fig. 1.7).

The SI System of Units

To be useful, units of measure must be standardized. Everyone should agree on the length of 1 meter, the length of 1 second, and the amount of mass in 1 kilogram. Without such an agreement, there would be no way to conduct science or commerce since we could not accurately compare quantities measured by different individuals. This standardization has been established by international agreements. One such agreement established the ***Système International d'Unités***, usually referred to as the SI system of units. It employs meters, seconds, and kilograms as the primary units of length, time, and mass. We generally use the SI system of units throughout this book and usually abbreviate meters as m, seconds as s, and kilograms as kg.

Two other systems of units are often encountered. The CGS system uses the units of centimeters (cm) for length, grams (g) for mass, and seconds (s) for time. Another common system, the U.S. customary system, measures length in feet, mass in slugs, and time in seconds. These systems are summarized in Table 1.4.

Courtesy of BIPM

▲ **Figure 1.7** The standard kilogram is a cylinder of metal composed of an alloy of platinum and iridium, and is stored in France. This photo shows a copy that is kept in a vacuum inside two separate glass jars. (It is treated very carefully!)

Powers of 10 and Prefixes

We have already seen how to use scientific notation to work with numbers that are very large or very small. Another way to deal with such numbers is to add a ***prefix*** to the unit of measure. Table 1.5 lists the common prefixes; when attached to a unit, a prefix multiplies the unit by the amount indicated in the table. These prefixes can be applied to any unit of measure. For example, the prefix ***milli*** is equivalent to a factor of 10^{-3}, so 1 millimeter is 10^{-3} m $= 0.001$ m. Another common prefix is ***kilo***, which is equivalent to a factor of $10^3 = 1000$. One kilometer is thus 1000 m, whereas 1 kilosecond is 1000 s. Indeed, you can see that the SI unit of mass contains this prefix since 1 kilogram = 1000 grams.

TABLE 1.4 Units of Length, Mass, and Time

Dimension	SI	CGS	U.S. Customary Units
length	meter (m)	centimeter (cm)	foot (ft)
mass	kilogram (kg)	gram (g)	slug
time	second (s)	second (s)	second (s)

TABLE 1.5 Prefixes That Can Be Attached to Units of Measure

Prefix	Power of 10	Symbol
atto	10^{-18}	a
femto	10^{-15}	f
pico	10^{-12}	p
nano	10^{-9}	n
micro	10^{-6}	μ
milli	10^{-3}	m
centi	10^{-2}	c
deci	10^{-1}	d
deka	10^{1}	da
hecto	10^{2}	h
kilo	10^{3}	k
mega	10^{6}	M
giga	10^{9}	G
tera	10^{12}	T
peta	10^{15}	P
exa	10^{18}	E

EXAMPLE 1.4 Units Conversion and Scientific Notation

How many seconds are in 1 year? Express your answer in scientific notation.

RECOGNIZE THE PRINCIPLE

To compute the number of seconds in 1 year, we use our knowledge that there are 365 days in a year (except for leap years), 24 hours in 1 day, 60 minutes in 1 hour, and 60 seconds in 1 minute.

SKETCH THE PROBLEM

No sketch is needed.

IDENTIFY THE RELATIONSHIPS AND SOLVE

The conversion from years to seconds is

$$1 \text{ year} = 1 \text{ year} \times \frac{365 \text{ days}}{1 \text{ year}} \times \frac{24 \text{ hours}}{1 \text{ day}} \times \frac{60 \text{ minutes}}{1 \text{ hour}} \times \frac{60 \text{ seconds}}{1 \text{ minute}}$$

$$1 \text{ year} = 31{,}536{,}000 \text{ s}$$

Expressed in scientific notation,

$$1 \text{ year} = \boxed{3.1536 \times 10^{7} \text{ s}}$$

► ***What does it mean?***

Note that this value is exact (since there are exactly 60 seconds in 1 minute, etc.).

CONCEPT CHECK 1.2 Using Prefixes and Powers of 10

Write the values of all the lengths in Table 1.1 using a prefix from Table 1.5.

1.5 Dimensions and Units

We have discussed the physical quantities length, time, and mass and their units in some detail, but they are only three of the many different types of physical quantities we encounter in physics. In many cases, the units for these other quantities are *derived* from the units of length, time, and mass. In the SI system, *all* the units needed for the study of mechanics can be derived from the ***primary units*** meters, seconds, and kilograms. For example, the volume of an object is measured in units of cubic meters, which is written as m^3. Hence, the unit of volume is a derived unit since it can be expressed in terms of the primary unit (meters). Likewise, the density of an object is equal to its mass divided by its volume, so density is measured in units of

$$\text{density} = \frac{\text{mass}}{\text{volume}} \Rightarrow \frac{\text{kg}}{\text{m}^3} \tag{1.7}$$

The units of density are thus derived from the SI units kilogram and meter.

This distinction between derived units and the primary units of meters, seconds, and kilograms also carries over to our fundamental view of physical quantities. For example, what is length? We all have a clear intuitive notion of length, but it is very difficult to give a formal definition of the term *length* that does not rely on some other physical quantity or idea.[3] Likewise, it is difficult to give a satisfactory formal definition of the quantity we call time. The best we can do is take a set of primary

[3]Defining length is like trying to give a definition of every word in a dictionary. The first definitions have to use words that are not yet defined!

physical quantities as "given" and then use them to build our theories of physics. For our studies of mechanics, we need only three primary physical quantities—length, time, and mass—and we will build all other necessary quantities—such as force, velocity, energy, pressure, and work—from these primary quantities.

In later chapters, we will learn about four more primary units associated with electricity, magnetism, and heat. There are thus seven primary units in the SI system. All other units can be derived from the seven primary units.

Dimensions and Dimensional Analysis

In our calculation of the volume of a blood cell in Example 1.2, we expressed the answer in units of cubic meters. Hence, we calculated both a numerical value and the units of the answer. This pattern is followed for most calculations because most answers are meaningless unless a unit of measure is specified. Calculating and then checking the units of an answer are important parts of the problem-solving process. This checking process, called ***dimensional analysis***, can be carried out in a very general way using the primary physical quantities. For example, the distance from New York City to Chicago has the ***dimensions*** of length and can be measured in units of meters. If we denote these primary dimensional quantities as **L** (length), **T** (time), and **M** (mass), we can express the dimensions of any combination of these quantities in terms of **L**, **T**, and **M**. Suppose the answer for a particular problem can be written as a length divided by a time. The SI units of the answer would then be the units for length (meters) divided by the units for time (seconds), or m/s. We would express the dimensions of this answer as **L/T**. The dimensions of an answer are independent of the particular units used to measure length, time, and so forth.

EXAMPLE 1.5 Using Dimensional Analysis to Check an Answer

Suppose we have performed a calculation of the position of an object (x) as a function of time (t) and arrived at the result

$$x = vt \qquad (1)$$

Use dimensional analysis to find the dimensions of the quantity v.

RECOGNIZE THE PRINCIPLE

The dimensions must be the same on both sides of the equal sign in Equation (1), so the dimensions of x must equal the dimensions of the product vt. If we rearrange the equation to solve for v, we get $v = x/t$. Hence, the dimensions of v must equal the dimensions of x divided by t.

SKETCH THE PROBLEM

No sketch is necessary.

IDENTIFY THE RELATIONSHIPS AND SOLVE

The dimensions of position x (i.e., distance) is **L**, and the dimensions of t is **T**. The dimensions of v must therefore be

$$\text{dimensions of } v = \boxed{\frac{\mathbf{L}}{\mathbf{T}}}$$

► ***What does it mean?***

In the SI system, v could be measured in units of meters per second, or m/s. It is always a good idea to check the dimensions of an answer; this check can sometimes reveal errors in a calculation.

CONCEPT CHECK 1.3 Getting the Units Right

The area of a surface can be measured in units of square meters (m^2). Which of the following combinations of units could also be used to measure area? (More than one answer may be correct.)

(a) m · cm (b) ft^3/m^2 (c) $in.^2$ (d) m · ft

1.6 Algebra and Simultaneous Equations

The problem solving we encounter in this book involves three basic types of mathematics: algebra, trigonometry, and vectors. The next few sections are intended as quick refreshers on these topics. A mathematical review is also given in Appendix B.

When solving a problem, we often need to deal with one or two equations containing one or two unknown quantities. Our job is usually to solve for the unknown quantities. Suppose we are given a single equation containing two variables a and T and are asked to find a in terms of T. The equation might be

$$5a = T - 10$$

The goal is to isolate a on one side of the equal sign (usually the left). In this example, we can divide both sides by the constant factor 5 to get the solution

$$a = \frac{T}{5} - 2$$

The numerical value of T might be already known so that the value of a can now be found, or a result expressed in terms of T may be sufficient, depending on the problem.

A slightly more complicated case is with two equations,

$$5a = T - 10 \tag{1.8}$$

$$7a = T + 24 \tag{1.9}$$

A set of equations like this one might arise in a problem in which there are two unknown quantities, a and T. The goal now is to solve for both unknowns. According to the rules of algebra, to solve for two unknown quantities we must have (at least) two equations, so Equations 1.8 and 1.9 should contain all the mathematical information we need to solve for these unknowns. To work out a solution, we can first rearrange Equation 1.8 to obtain a in terms of T:

$$a = \frac{T}{5} - 2 \tag{1.10}$$

Substituting into Equation 1.9 gives

$$7a = T + 24$$

$$7\left(\frac{T}{5} - 2\right) = T + 24$$

$$\frac{7T}{5} - 14 = T + 24 \tag{1.11}$$

We now have a single equation containing only the unknown T. We can find T by moving all the terms involving T to one side to get

$$\frac{7T}{5} - T = 24 + 14$$

$$\frac{7T - 5T}{5} = 38$$

$$\frac{2T}{5} = 38$$

$$T = \frac{5 \times 38}{2} = 95 \qquad (1.12)$$

This value of T can then be inserted back into Equation 1.10 to find the value of a:

$$a = \frac{T}{5} - 2 = \frac{95}{5} - 2 = 17$$

Checking the Units of an Answer

Algebra can be used to solve an equation or a set of equations to find the values of variables such as a and T in the examples above. In physics, these variables represent physical quantities, such as mass, or force, or velocity. Physical quantities possess units, and the algebra we use to compute a and T will also give the value of the associated unit. For example, suppose we have computed the density of an object (represented by the variable ρ) and arrived at the answer

$$\rho = \frac{M}{V} \qquad (1.13)$$

Here, M and V both have units that combine to determine the units of ρ. If we focus only on the units in Equation 1.13, we have the relation

$$\text{units of } \rho = \frac{\text{units of } M}{\text{units of } V}$$

M is the mass of an object so its units are kilograms, and V is a volume with units of cubic meters. The units of ρ are then

$$\text{units of } \rho = \frac{\text{units of } M}{\text{units of } V} = \frac{\text{kg}}{\text{m}^3} \qquad (1.14)$$

Insight 1.3
DIMENSIONS OF DENSITY
In terms of the dimensions, Equation 1.14 reads M/L^3 = mass divided by $(\text{length})^3$.

This answer is indeed the correct unit for density (compare with Equation 1.7).

In Example 1.5, we saw how dimensional analysis can be used to check an answer, and exactly the same approach can be used when calculating and checking units. For example, in our calculation of the density, we might have made an error and obtained the result

$$\rho = \frac{M^2}{V} \quad \textbf{(wrong!!)} \qquad (1.15)$$

The units in this case would work out as

$$\text{units of } \rho = \frac{\text{units of } M^2}{\text{units of } V} = \frac{\text{kg}^2}{\text{m}^3} \quad \textbf{(wrong!!)}$$

Since we know that kg^2/m^3 is not the correct unit for density, this result tells us immediately that there must be an error in the calculation that led to Equation 1.15. It is *always* useful to check the units (and dimensions) of your answers.

1.7 Trigonometry

We often need to deal with angles and the location of an object within a coordinate system. This is most easily done with trigonometry, right triangles, and the

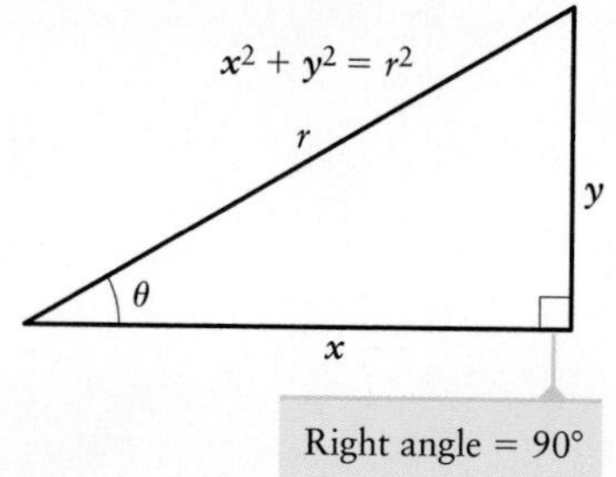

▲ **Figure 1.8** The Pythagorean theorem is a relation between the lengths of the sides of a right triangle.

Pythagorean theorem. Figure 1.8 shows a right triangle with sides x, y, and r, where r is the hypotenuse. According to the Pythagorean theorem,

$$x^2 + y^2 = r^2 \tag{1.16}$$

The trigonometric functions sine, cosine, and tangent are defined as

$$\sin\theta = y/r \tag{1.17}$$

$$\cos\theta = x/r \tag{1.18}$$

$$\tan\theta = y/x \tag{1.19}$$

The Pythagorean theorem (Equation 1.16) implies that

$$\sin^2\theta + \cos^2\theta = 1 \tag{1.20}$$

for any value of the angle θ. Equation 1.20 also leads to a number of other trigonometric relations (called ***identities***), some of which are listed inside the back cover of this book and in Appendix B.

Equations 1.17 through 1.19 tell us how to calculate the sine, cosine, and tangent of an angle from the lengths of the sides of the associated right triangle. We can also use this approach to find the value of the angle. For example, if we know the values of y and r, we can solve for the angle:

$$\sin\theta = y/r$$

$$\theta = \sin^{-1}(y/r) \tag{1.21}$$

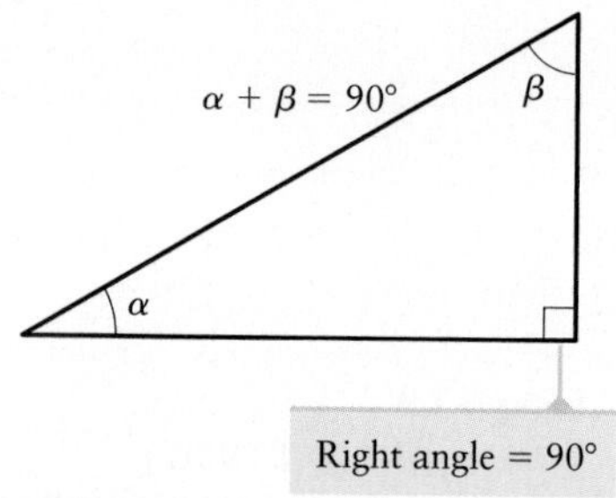

▲ **Figure 1.9** The sum of the three interior angles of a triangle is 180°. Hence, $\alpha + \beta = 90°$.

The function $\sin^{-1}$ is the ***inverse sine*** function, also known as the ***arcsine*** (= arcsin = $\sin^{-1}$). In words, Equation 1.21 says that θ equals the angle whose sine is y/r. Scientific calculators all contain this function (as a single "button"). If the values of y and r are known, you only take the ratio y/r and then compute the arcsine of this value to find θ in Equation 1.21. There are also functions for the inverse cosine ($\cos^{-1}$ = arccos) and inverse tangent ($\tan^{-1}$ = arctan). From Equations 1.18 and 1.19, we have

$$\theta = \cos^{-1}(x/r) \tag{1.22}$$

and

$$\theta = \tan^{-1}(y/x) \tag{1.23}$$

We often need to deal with triangles like the one in Figure 1.9, where we have a right triangle with interior angles α and β. The sum of a triangle's three interior angles is always 180°. For the triangle in Figure 1.9, one of the angles is 90° (because it is a right triangle), so

$$\alpha + \beta = 90° \tag{1.24}$$

Two angles whose sum is 90° are called complementary angles. From the definitions of the sine and cosine functions in Equations 1.17 and 1.18, we have the relations

$$\sin\alpha = \cos\beta$$

$$\cos\alpha = \sin\beta$$

for any pair of complementary angles α and β.

An angle of approximately 57° corresponds to 1 radian.

The angle θ here is approximately 1 radian.

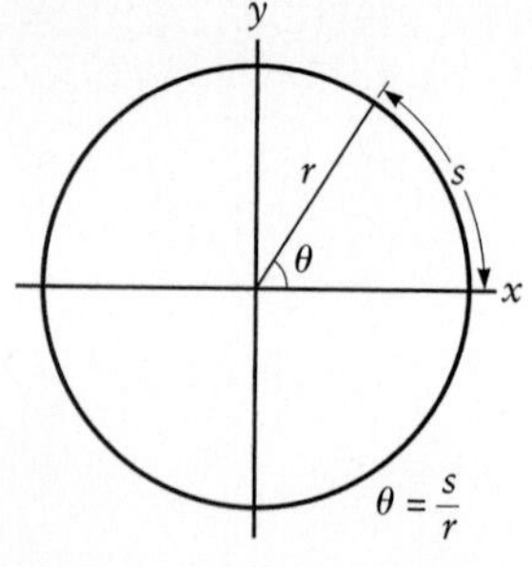

▲ **Figure 1.10** The angle θ equals the ratio of the length s measured along the circular arc, divided by the radius r of the circle. This ratio gives the value of θ in units of radians.

Measuring Angles

The value of an angle is often measured in units of ***degrees***; indeed, we have already used this unit when discussing trigonometry. Another common unit for measuring angles is the ***radian*** (abbreviated "rad"). The conversion between radians and degrees is accomplished just like any other units conversion problem, using the relation

$$360° = 2\pi \text{ rad}$$

or

$$1° = \frac{2\pi}{360} \text{ rad} \tag{1.25}$$

At this point, you might ask why we need two different units—degrees and radians—to measure angles. To explain the mathematical origin of the radian unit, Figure 1.10 shows a circle of radius r along with a "radius line" that makes an angle θ with the x axis. The arc length s is measured *along the circle* from the x axis to the point where the radius line meets the circle. The angle θ when measured in radians is equal to the ratio of the arc length to the radius:

$$\theta = \frac{s}{r} \quad (\theta \text{ measured in radians}) \tag{1.26}$$

This connection between the angle and the length measured along a circular arc is very useful in work on circular motion (in which a particle moves along a circular path) and rotational motion (in which an object spins about an axis). The relation between s, r, and θ in Equation 1.26 only holds when s and r are measured in the same units (e.g., both measured in meters) and the angle is measured in radians. Most calculators can be set to work with angles in either degrees or radians, but you should always be clear about which way your calculator is set!

EXAMPLE 1.6 The 3-4-5 Right Triangle and Designing a Roof

The roof of a house has the form of a right triangle with sides of length 3.0 m, 4.0 m, and 5.0 m (Fig. 1.11). Find the interior angles θ and ϕ of the triangle. Express your answers in degrees and radians.

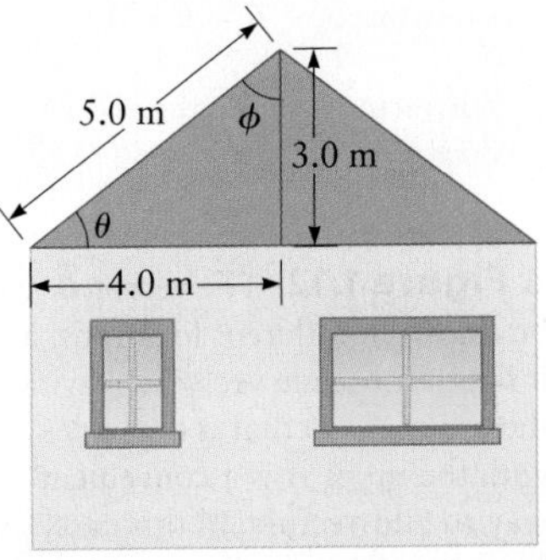

▲ **Figure 1.11** Example 1.6.

RECOGNIZE THE PRINCIPLE

Given the lengths of the sides of a right triangle, we can find the values of the interior angles using the $\sin^{-1}$, $\cos^{-1}$, and $\tan^{-1}$ functions, Equations 1.21 through 1.23. The interior angles θ and ϕ are also complementary (Eq. 1.24).

SKETCH THE PROBLEM

Figure 1.11 shows the problem.

IDENTIFY THE RELATIONSHIPS AND SOLVE

The right triangle in Figure 1.11 has sides $x = 4.0$ m, $y = 3.0$ m, and $r = 5.0$ m. We can compute the value of θ using the inverse sine function ($\sin^{-1}$) in Equation 1.21:

$$\theta = \sin^{-1}(y/r) = \sin^{-1}(3.0/5.0) = \boxed{37°} \tag{1}$$

where we have rounded to two significant figures. To express this value in radians, we convert units:

$$\theta = 37° \times \frac{2\pi \text{ rad}}{360°} = \boxed{0.64 \text{ rad}}$$

The interior angle ϕ is complementary to θ (see Eq. 1.24) and has a value

$$\phi = 90 - \theta = \boxed{53°} = 53° \times \frac{2\pi \text{ rad}}{360°} = \boxed{0.93 \text{ rad}}$$

► ***What does it mean?***

In computing the $\sin^{-1}$ function in Equation (1), you must be sure that you know how your calculator is set and whether it is giving an answer in degrees or radians.

Insight 1.4
INVERSE TRIGONOMETRIC FUNCTIONS

The inverse sine function used in Example 1.6 can be understood as follows. If $\theta = \sin^{-1}(w)$, then w is equal to the sine of the angle θ; that is, $w = \sin\theta$. When computing an inverse trigonometric function, be sure to take note of the units; that is, you should know if your calculator is set for degrees or for radians.

1.8 Vectors

In this book, we deal with two different types of mathematical quantities. One type includes quantities such as time and mass, which are described by a simple number (with a unit, of course) and are called *scalars*. Quantities described by a simple

magnitude, such as the mass of a car or the number of seconds in 1 year, are scalars. Some physical quantities, however, cannot be described completely in this way. For example, the velocity with which the wind is blowing is a ***vector***. A vector quantity has both a magnitude and a ***direction***. In diagrams and figures, it is often useful to represent a vector quantity by an arrow. The arrow's length indicates the magnitude of the vector (e.g., how fast the wind is blowing), while the arrow's direction indicates the direction of the vector relative to a chosen coordinate system (e.g., the direction in which the wind is blowing). Many quantities in physics are vectors, and we often need to add, subtract, and perform multiplication with vectors.

Vectors have a magnitude and a direction.

Adding Vectors

Representing vectors as arrows leads to a convenient way to think about the addition of two vectors. Figure 1.12 shows two vectors $\vec{A}$ and $\vec{B}$; here we use a notation in which vector variables are indicated by placing arrows over a boldface symbol. A vector has a magnitude and a direction, so both the lengths and the directions of the arrows in Figure 1.12 are important. However, when we draw a vector in this manner, the location of the arrow is somewhat arbitrary. We can place the arrow in different locations, and as long as it has the same length and the same direction, it is still the same vector. In Figure 1.12, we have drawn $\vec{B}$ several times; in one case, the "tail" of $\vec{B}$ is located at the "tip" of $\vec{A}$. The sum of $\vec{A}$ and $\vec{B}$ is then the vector $\vec{C}$, which connects the tail of $\vec{A}$ with the tip of $\vec{B}$. Mathematically,

$$\vec{C} = \vec{A} + \vec{B} \tag{1.27}$$

These are the same vector $\vec{B}$ even though they are drawn at different locations.

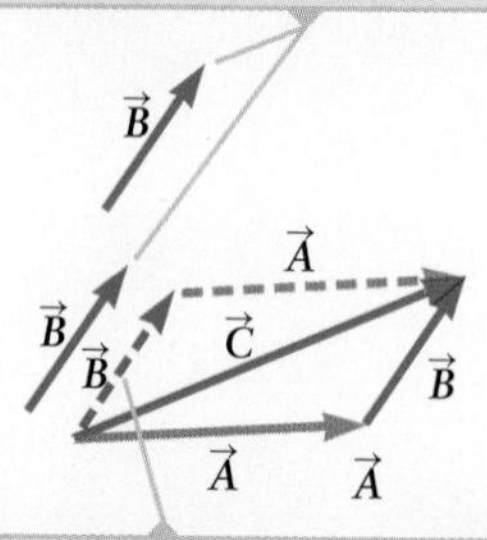

The dashed versions of $\vec{B}$ and $\vec{A}$ show that $\vec{A} + \vec{B} = \vec{B} + \vec{A}$.

For vector addition the order does not matter.

▲ **Figure 1.12** The vector $\vec{B}$ can be drawn in different locations, and it is still the same vector. Drawing the tail of $\vec{B}$ so that it coincides with the tip of $\vec{A}$ is a convenient way to add vectors. In this case, $\vec{C} = \vec{A} + \vec{B}$.

A useful way to understand Equation 1.27 is to think of these vectors as displacements or movements. The vector $\vec{A}$ then represents movement from the tail of $\vec{A}$ to the tip of $\vec{A}$, while the vector $\vec{B}$ represents movement from the tail of $\vec{B}$ to the tip of $\vec{B}$. When these two movements are added together, we get the total displacement represented by $\vec{C}$.

Multiplying a Vector by a Scalar and Subtracting Vectors

Before dealing with vector subtraction, it is useful to first consider the multiplication of a vector by a scalar. Scalars are simply numbers like 2 or 5.7. Multiplication of a vector by a scalar changes the vector's length. Figure 1.13 shows a vector $\vec{A}$ multiplied by a scalar K:

$$\vec{B} = K\vec{A} \tag{1.28}$$

If the scalar K is greater than 1 ($K > 1$), the vector $\vec{B}$ is longer than $\vec{A}$, while if K is positive and less than 1, $\vec{B}$ is shorter than $\vec{A}$. If the scalar K is negative, then $\vec{A}$ and $\vec{B}$ are in *opposite directions*.

We can now see how to do vector subtraction. Subtracting the vector $\vec{B}$ from the vector $\vec{A}$ is equivalent to adding the vectors $\vec{A}$ and $-\vec{B}$ (see Fig. 1.14):

$$\vec{A} - \vec{B} = \vec{A} + (-\vec{B}) \tag{1.29}$$

$\vec{B} = 0.5 \times \vec{A}$

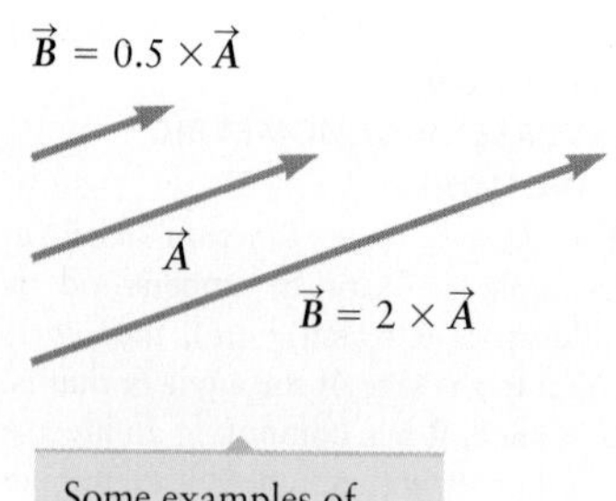

Some examples of $\vec{B} = K\vec{A}$ with different values of K.

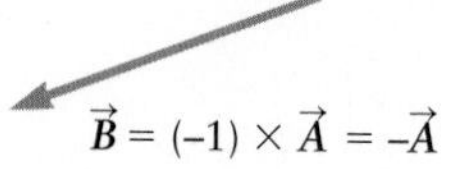

$\vec{B} = (-1) \times \vec{A} = -\vec{A}$

▲ **Figure 1.13** When a vector (such as $\vec{A}$) is multiplied by a scalar, the result is a vector that is parallel to the original vector but with a different length.

Vectors and Components

When performing calculations involving vector quantities, it is often useful to deal with vectors in their component form. Figure 1.15 shows a vector $\vec{A}$ drawn with its tail at the origin of an $x-y$ coordinate system. This vector has ***components*** along the x and y directions, as shown. The components A_x and A_y are the projections of $\vec{A}$ along x and y.

The vector $\vec{A}$ in Figure 1.15 can be described mathematically in two different ways. The first involves the angle θ this vector makes with the x axis along with the length of $\vec{A}$, which we can denote either by A or by $|\vec{A}|$. The second is to give its components along x and y. According to Figure 1.15, the components of $\vec{A}$ are

$$A_x = A\cos\theta = |\vec{A}|\cos\theta \tag{1.30}$$

$$A_y = A\sin\theta = |\vec{A}|\sin\theta \tag{1.31}$$

With Equations 1.30 and 1.31, we can calculate the components if we know the magnitude (i.e., length) and direction (θ) of a vector. We can also work backward from the components to find the magnitude and direction. Applying some trigonometry to Figure 1.15 gives

$$A = |\vec{A}| = \sqrt{A_x^2 + A_y^2} \tag{1.32}$$

$$\tan\theta = A_y/A_x$$

or

$$\theta = \tan^{-1}(A_y/A_x) \tag{1.33}$$

Hence, if we know the components A_x and A_y, we can find A (the length of the vector) and θ (the direction).

Let's again consider the addition of two vectors, but now work in terms of the components. Figure 1.16 shows two vectors $\vec{A}$ and $\vec{B}$ added graphically, and we have also indicated the values of each vector's components. To add two vectors, we simply add their components. That is, to compute the vector sum $\vec{C} = \vec{A} + \vec{B}$, we add components so that

$$C_x = A_x + B_x \quad \text{and} \quad C_y = A_y + B_y \tag{1.34}$$

Subtraction of two vectors is handled in a similar way. If $\vec{C} = \vec{A} - \vec{B}$, then, in terms of components,

$$C_x = A_x - B_x \quad \text{and} \quad C_y = A_y - B_y \tag{1.35}$$

Multiplication of a vector by a scalar K can also be done using components. If $\vec{B} = K\vec{A}$, we then have (Fig. 1.17)

$$B_x = KA_x \quad \text{and} \quad B_y = KA_y \tag{1.36}$$

In our discussion of vector components, we have so far dealt with only the x and y components. Our results can be generalized to include a third coordinate direction, z, for cases in which we must deal with three-dimensional space. For example, when adding two vectors, Equation 1.34 becomes

$$C_x = A_x + B_x \quad C_y = A_y + B_y \quad C_z = A_z + B_z \tag{1.37}$$

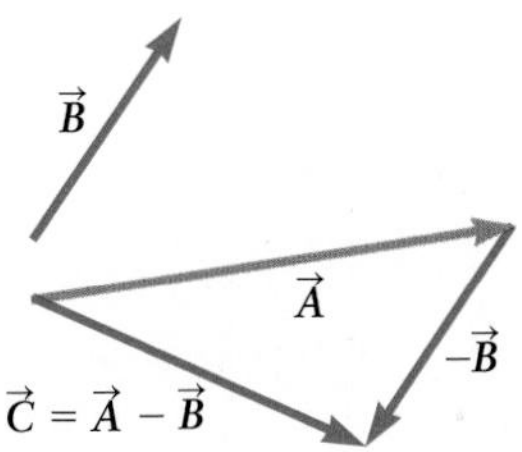

▲ **Figure 1.14** Vector subtraction. Calculating $\vec{A} - \vec{B}$ is equivalent to computing the sum of $\vec{A}$ and $-\vec{B}$.

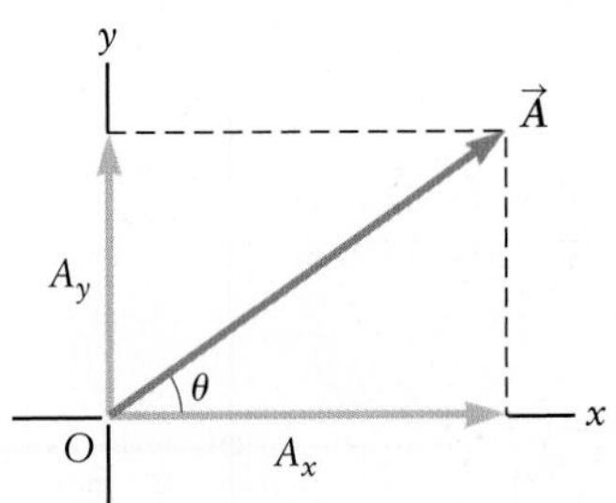

▲ **Figure 1.15** The components of a vector are the projections of the vector onto the coordinate axes. Here we show a vector that lies in the x–y plane, so it has components along the x and y axes. Some vectors have components along the z axis (not shown here).

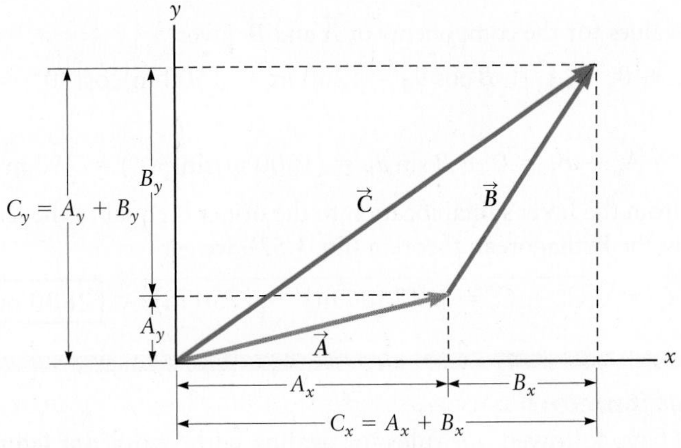

▲ **Figure 1.16** To compute the sum of two vectors, we add their components. Here we add the x components of $\vec{A}$ and $\vec{B}$ to get the x component of $\vec{C}$. We follow the same procedure to find the y component of $\vec{C}$.

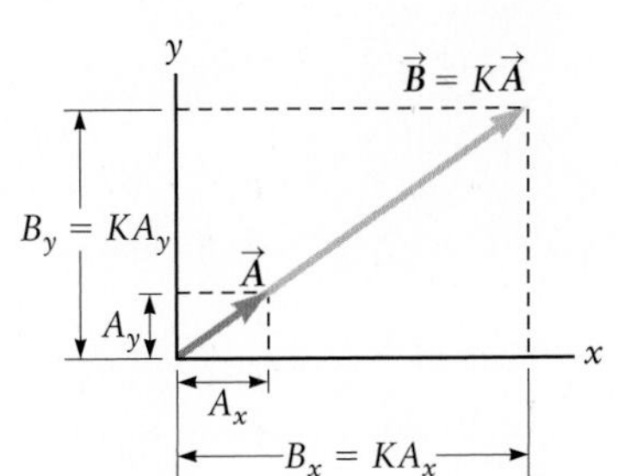

▲ **Figure 1.17** To multiply a vector $\vec{A}$ by a scalar K, we multiply each component of the vector by K.

EXAMPLE 1.7 Hiking and Vectors

A hiker is walking through the woods. He starts from his campsite, which is at the origin in Figure 1.18. He initially travels along the x axis in Figure 1.18A, and his initial displacement of 1200 m takes him to the tip of vector $\vec{A}$; hence, $\vec{A}$ has a length of 1200 m and is parallel to the x direction. He then turns and moves along the path described by vector $\vec{B}$ in Figure 1.18A. This vector has a length of 1500 m and makes an angle $\theta_B = 30°$ with the x axis. When the hiker stops, how far is he from the campsite?

RECOGNIZE THE PRINCIPLE

To find the final location of the hiker, we must add the vectors $\vec{A}$ and $\vec{B}$. We can do so by adding the components of these vectors: $\vec{C} = \vec{A} + \vec{B}$ (Eq. 1.34).

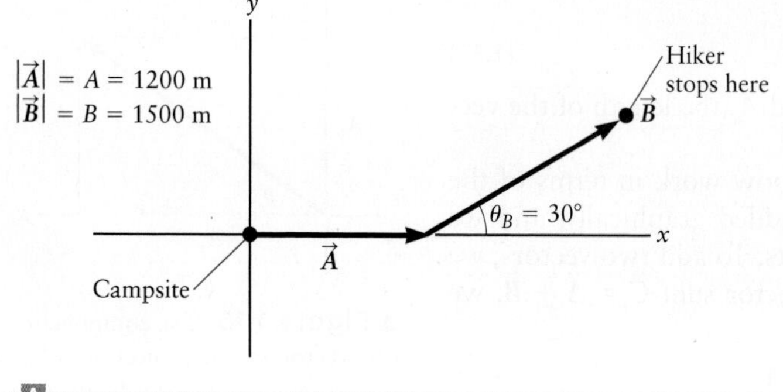

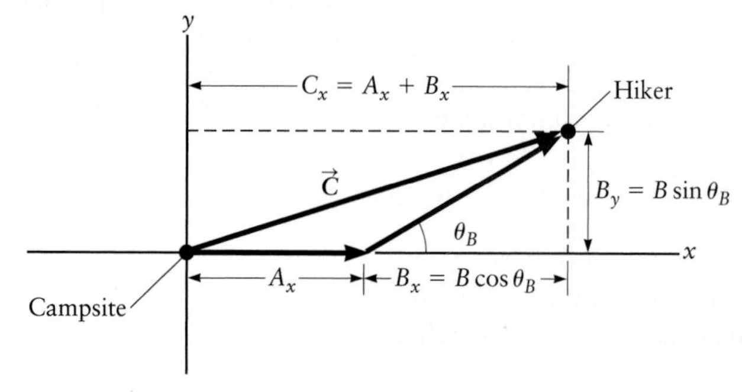

▲ **Figure 1.18** Example 1.7.

SKETCH THE PROBLEM

Figure 1.18 illustrates the problem. Figure 1.18B also shows the components of the various vectors, which will be needed in the solution.

IDENTIFY THE RELATIONSHIPS

According to Figure 1.18B, we can find the distance of the hiker to the campsite (i.e., the origin) if we know the components of the vectors $\vec{A}$ and $\vec{B}$ along the x and y axes. The vector $\vec{A}$ is parallel to x, so

$$A_x = A = 1200 \text{ m} \quad A_y = 0$$

where we again denote the length of the vector $\vec{A}$ by A. For $\vec{B}$, the trigonometry sketched in Figure 1.18B gives

$$B_x = B\cos\theta_B \quad B_y = B\sin\theta_B$$

The hiker's final location is described by the vector $\vec{C}$, which is equal to the sum of the vectors $\vec{A}$ and $\vec{B}$:

$$\vec{C} = \vec{A} + \vec{B}$$

In terms of components,

$$C_x = A_x + B_x \qquad C_y = A_y + B_y$$

SOLVE

Inserting our values for the components of $\vec{A}$ and $\vec{B}$ gives

$$C_x = A_x + B_x = A_x + B\cos\theta_B = 1200 \text{ m} + (1500 \text{ m})\cos(30°) = 2500 \text{ m}$$

and

$$C_y = A_y + B_y = 0 + B\sin\theta_B = (1500 \text{ m})\sin(30°) = 750 \text{ m}$$

The distance from the hiker's final location to the origin is equal to the length of the vector $\vec{C}$. From the Pythagorean theorem (Eq. 1.32), we get

$$C = \sqrt{C_x^2 + C_y^2} = \sqrt{(2500 \text{ m})^2 + (750 \text{ m})^2} = \boxed{2600 \text{ m}}$$

▶ *What does it mean?*

Note that we have followed our rules for dealing with significant figures; the lengths of the vectors $\vec{A}$ and $\vec{B}$ and of the angle θ were given to two significant figures, so the answer C was also given with two significant figures. We'll work with vectors and their components in many problems. At this time, it might be valuable to review your trigonometry (see Appendix B).

Summary | CHAPTER 1

What is physics?

Physics is the science of matter and energy, and the interactions between them. The laws of physics enable us to predict and understand the way the universe works.

Problem solving

Problem solving is an essential part of physics. Problems may be quantitative with a precise answer, or they may be conceptual. For some problems, you may need to use common-sense reasoning to estimate the values of important quantities.

Scientific notation and significant figures

Scientific notation is used to express very large and very small numbers. We can also use prefixes to write very large or very small numbers, such as 1 micrometer = 1 μm = 1×10^{-6} meter and 1 megasecond = 1 Ms = 1×10^{6} seconds.

The accuracy of a quantity is reflected in the number of ***significant figures*** used to express its value. The result of a calculation should be expressed with an appropriate number of significant figures.

Units of measure

The ***primary units*** of mechanics involve length, time, and mass. Most scientific work employs the SI system of units, in which length is measured in ***meters***, time is measured in ***seconds***, and mass is measured in ***kilograms***. The values of the standard meter, the standard second, and the standard kilogram are established by international agreement and are a foundation for all scientific work.

Physical quantities and dimensions

We cannot give a definition of the concepts of length, time, and mass, but must instead treat them as "givens." The definitions of all other physical quantities encountered in mechanics can be derived from these three primary quantities. A total of seven primary physical quantities are needed to describe all of physics; the other four are connected with electricity, magnetism, and heat.

The ***dimensions*** of all quantities in mechanics can be expressed in terms of length L, mass M, and time T. Dimensional analysis involves checking that the dimensions of an answer correctly match the dimensions of the quantity being calculated.

The mathematics of physics

Several types of mathematics are used in this book: algebra, trigonometry, and vectors. Algebra is essential for solving systems of equations. When dealing with motion or other problems in physics, we often need to express position or movement in terms of a ***coordinate system*** (usually the x–y–z set of coordinate axes). Trigonometry and vectors are extremely useful in such calculations. A ***vector*** quantity has both a ***magnitude*** and a ***direction***. Vectors can be added graphically or in terms of their ***components***. Appendix B contains a quick review of algebra, trigonometry, and vectors.

Questions

SSM = answer in *Student Companion & Problem-Solving Guide* = life science application

1. Suppose a friend told you that the density of a sphere is $\frac{4}{3}\pi r^4$, where r is the radius of the sphere. Compute the dimensions of this expression and show that it cannot be correct.

2. Discuss the difficulties of giving a definition of the concept of "time" without using any other scientific terms.

3. SSM Which of the following are units of volume?

cubic meters	acres/m^2
mm $\times$ mi^2	hours $\times$ mm^3/s
kiloseconds $\times$ ft^2	mm $\times$ cm $\times$ mm^2/ft
kg^2 $\times$ cm	

4. What physical property, process, or situation is described by the following combinations of primary units? (For example, the combination of units kilograms per square meter can be used to measure the mass per unit area, sometimes called the areal density, of a thin sheet of metal.)

m/s	m/s^2
m^3/s	m^2/s
kg/m	

5. The standard meter used to be a metal bar composed of platinum and iridium. The current standard is based on the measured speed of light and an optical device called an interferometer. What advantages does the new standard have over the old?

6. SSM Which of the following quantities have the properties of a vector and which have the properties of a scalar?

mass	density
velocity	temperature
displacement (change in position)	

7. Give and discuss one advantage and one disadvantage of the SI system compared with the U.S. customary system of units.

8. **Astronomical distances.** Spiral galaxies like our own Milky Way measure approximately 2×10^5 light-years in diameter, where 1 light-year (ly) equals the distance traveled by light in 1 year. Our nearest-neighbor galaxy is the great spiral galaxy Andromeda, which has been determined to be approximately 2.5 million ly away. Construct a scale model to show the sizes of the Milky Way and Andromeda galaxies using two pie tins each 20 cm in diameter. To keep the model to scale, how far apart would these pie tins need to be separated?

9. A football field in the United States is 120 yards long (including the end zones). With how many significant figures do you think this value should be written?

Problems available in 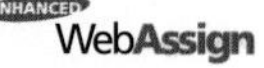

Section 1.3 Dealing with Numbers
Problems 1–15

Section 1.4 Physical Quantities and Units of Measure

Section 1.5 Dimensions and Units
Problems 16–31

Section 1.6 Algebra and Simultaneous Equations
Problems 32–35

Section 1.7 Trigonometry
Problems 36–60

Additional Problems
Problems 61–75

List of Enhanced Problems

Problem number	Solution in *Student Companion & Problem-Solving Guide*	Reasoning & Relationships Problem	Reasoning Tutorial in Enhanced WebAssign
1.1		✓	
1.3		✓	✓
1.4		✓	
1.12	✓		
1.13		✓	✓
1.14		✓	
1.17		✓	
1.18		✓	✓
1.19	✓		
1.34	✓		
1.36	✓		
1.60	✓		
1.61		✓	✓
1.62	✓	✓	
1.64		✓	
1.67		✓	
1.68		✓	
1.69		✓	
1.70	✓	✓	
1.75		✓	

CHAPTER

2

▶ *The laws of mechanics describe the motion of objects such as this skateboarder. (© Rubberball/Mike Kemp/Getty Images)*

Motion, Forces, and Newton's Laws

OUTLINE

We begin our study of physics with the field known as ***mechanics***. This area of physics is concerned with the motion of objects such as rocks, balloons, cars, water, planets, and skateboarders. To understand mechanics, we must be able to answer two questions. First, what *causes* motion? Second, given a particular situation, *how* will an object move?

The laws of physics that deal with the motion of terrestrial objects were developed over the course of many centuries, culminating with the work of Isaac Newton and the formulation of what are known as Newton's laws of motion. Newton's laws are a cornerstone of physics and are the basis for nearly everything we do in the first part of this book. Even so, it is useful to begin with a brief discussion of some of the ideas that dominated physics before Newton's time. Those ideas originated many centuries ago with the Greek philosopher–scientist Aristotle. Although Aristotle was a leading scientist of his era and was certainly a very deep thinker, we now know that many of his ideas about how objects move were not correct. Why, then, should we now be concerned with such historical (and incorrect) notions about motion? We examine Aristotle's ideas because many students begin this course with some of the *same* incorrect notions. A good way to reach a thorough understanding of the physics of motion is to consider the origins of Aristotle's ideas and identify where they go wrong.

2.1 Aristotle's Mechanics

Aristotle began by identifying two general types of motion: (1) celestial motion, the motion of things like the planets, the Moon, and the stars; and (2) terrestrial motion, the motion of "everyday" objects such as rocks and arrows. We all know that while rocks and other terrestrial objects can move in a variety of ways, they usually come to rest eventually. On the other hand, it appears that celestial objects never come to a stop. According to Aristotle, terrestrial objects move only when something acts on them directly. A rock moves because some other object, such as a person's hand, acts on it to make it move. In contrast, there does not seem (at least to Aristotle) to be anything acting on the Moon to cause its motion. Aristotle thus raised a key point: the motions of celestial and terrestrial objects look very different. A theory of mechanics must explain why.

Aristotle asserted that the "natural" state for a terrestrial object is a state of rest and that this is why terrestrial objects move only when acted on by another object. In modern terminology, Aristotle claimed that an object only moves when acted on by a ***force***. A force is simply a push or a pull, and Aristotle believed that a push or a pull on one object is always produced by a second object. Furthermore, he claimed that a force could only exist if the two objects are in direct contact.

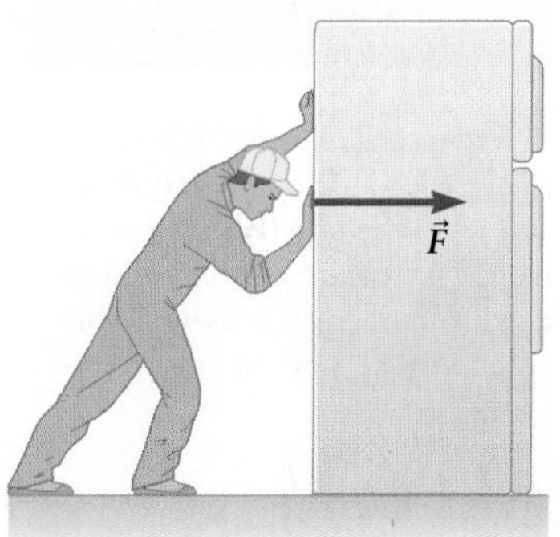

▲ **Figure 2.1** This person is exerting a force, denoted by $\vec{F}$, on the refrigerator.

Hence, Aristotle believed that (1) motion is caused by forces and (2) forces are produced by contact with other objects. These ideas *seem* quite reasonable when we consider something like a refrigerator being pushed along a level floor as in Figure 2.1. Our everyday experience is that the refrigerator only moves while someone is pushing on it. If the person stops pushing, the refrigerator quickly comes to a stop. The person pushing on the refrigerator exerts a force directly on the refrigerator, and the person is in contact with the refrigerator. So far, so good.

A key aspect of mechanics is the notion of force. We have already noted that a force is simply a push or a pull on an object. A force has both a magnitude and a direction, so *force is a vector quantity*, often denoted by $\vec{F}$. The magnitude of a force is the strength of the push or the pull, while the direction of this vector is the direction of the push or the pull.

Force is a *vector* quantity. It has a magnitude and a direction.

We next need to consider how to describe motion. One quantity we normally associate with motion is ***velocity***. Velocity is also a vector quantity and is usually represented by the symbol $\vec{v}$. A careful mathematical definition of velocity is given in the next section. For our discussion of Aristotle's ideas, it is enough to know that the magnitude of $\vec{v}$ is the distance that an object travels (as might be measured in meters) per second. The direction of the vector $\vec{v}$ is the direction of motion (Fig. 2.2).

Aristotle proposed that force and velocity are directly connected. In the mathematical language of today, he would have written

$$\vec{v} = \frac{\vec{F}}{R} \quad \text{(INCORRECT!)} \tag{2.1}$$

In words, this relationship says that the velocity $\vec{v}$ of an object is proportional to the force $\vec{F}$ that acts on it, with the constant of proportionality being related to a quantity R, which is the resistance to motion. We'll refer to this relationship as "Aristotle's law of motion." We also caution that this relationship is *not* a correct law of physics. It does, however, seem to explain the motion of the refrigerator in Figure 2.1. If no force is exerted on the refrigerator, then $\vec{F} = 0$. According to Equation 2.1, the velocity $\vec{v}$ is then also zero and hence the refrigerator does not move. When a person

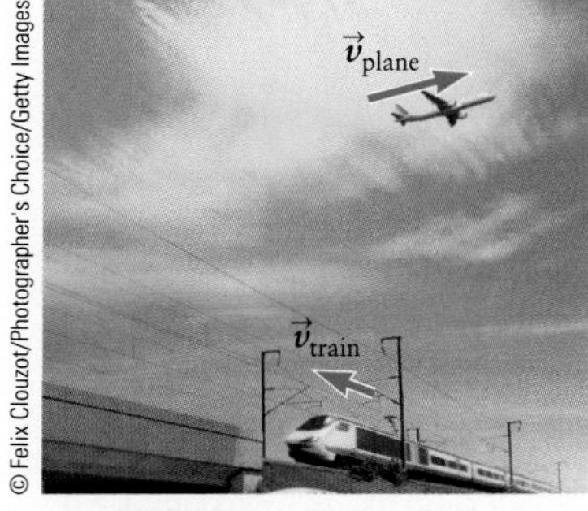

▲ **Figure 2.2** The velocities of an airplane and a train are vectors, having both magnitude and direction.

▲ **Figure 2.3** After it leaves the pitcher's hand, the only forces acting on the baseball are a force from gravity and a force from the air through which it moves. This second force is called air drag.

▲ **Figure 2.4** An archer shoots an arrow straight upward. While the arrow is traveling on the upward part of its trajectory, the forces acting on it—from gravity and from air drag—are both directed downward. Hence, these forces are not parallel to the velocity of the arrow while it is traveling upward.

pushes on the refrigerator, $\vec{F}$ is nonzero and Aristotle's theory (Eq. 2.1) predicts that $\vec{v}$ is therefore also nonzero. Since $\vec{v}$ and $\vec{F}$ are both vectors, Equation 2.1 asserts that the velocity and the force are always in the same direction.

The Failures of Aristotle's Ideas about Mechanics

Aristotle's law of motion—Equation 2.1—*seems* to explain the motion of the refrigerator in Figure 2.1, but it does *not* work as well in other situations. Consider the motion of a thrown baseball (Fig. 2.3). After a baseball leaves the thrower's hand, it continues to move until it hits the ground, is struck by a bat, or is caught by someone. This motion is *not* consistent with Equation 2.1. Recall Aristotle's assertion that forces are always caused by direct contact with another object. While the ball is traveling through the air, there does not appear to be another object in contact with it, so according to this reasoning the force on the baseball is zero after it leaves the thrower. If the force is zero, Equation 2.1 says that the ball's velocity should also be zero. However, we know from experience that a thrown baseball continues to move! One way to attempt an escape from this dilemma is to recognize that the ball also experiences a force due to gravity. Unfortunately (at least for Aristotle), the gravitational force does not fit into Aristotle's ideas about forces because there is nothing in "direct" contact with the baseball to produce this force. Even if we overlook this difficulty, we still have to consider that the gravitational force on the ball is directed downward (toward the surface of the Earth) and hence the gravitational force and the velocity are not in the same direction for the ball in Figure 2.3. Figures 2.3 and 2.4 both illustrate situations that contradict Equation 2.1, which predicts that force and velocity are always parallel. It is particularly puzzling (at least for Aristotle) that objects such as baseballs and arrows can be traveling *upward* when the forces acting on them are directed *downward*.

More problems with Aristotle's law of motion can be seen when we consider falling objects. If someone simply drops a baseball, it will fall to the ground. There is nothing in "direct contact" with the ball to cause it to fall. What, then, makes it move from the person's hand? Because the case of falling objects is quite common, Aristotle gave it special attention. He proposed that there is a special force that does not require contact with anything. This force is the ***weight*** of the object, which Aristotle believed was equivalent to what we now call mass.[1] Aristotle also thought that the resistance factor R in Equation 2.1 is a property of the substance through which the object moves. For an object dropped near the Earth's surface, this substance is air. This reasoning leads one to expect that if Equation 2.1 is correct, a heavy object will fall faster than a light one when both are dropped in the same medium. Thanks to experiments performed by Galileo (Fig. 2.5), we know that is *not* the case. Galileo showed that light objects fall at the same rate as heavy objects.

Despite all these difficulties, Aristotle's ideas have a certain appeal. They are based on the notion that an object's velocity is directly proportional to the force acting on the object, which at first glance seems plausible. However, we have seen a number of very simple situations in which this notion fails badly. In the next few sections, we discuss the connection between force and motion in more detail and arrive at the laws of motion discovered by Isaac Newton. We'll then see how Newton's laws overcome the difficulties that frustrated Aristotle.

Although we must reject Aristotle's theory of motion, it is still very instructive to understand where his theory goes wrong and why. Some of the everyday ideas that *you* have brought to this course may be similar to those of Aristotle. Understanding precisely when and why those ideas fail will help you understand and appreciate the correct way to describe motion.

[1]We'll see in the next chapter that mass and weight are *not* the same. Recall that mass is a fundamental physical quantity, as explained in Chapter 1. Weight is defined carefully in Chapter 3.

CONCEPT CHECK 2.1 Force and Motion

Do the following examples of motion appear to be *consistent* or *inconsistent* with Aristotle's law of motion?

(a) A hockey puck sliding along an icy horizontal surface
(b) A car coasting along a level road
(c) A piano that is pushed across a room

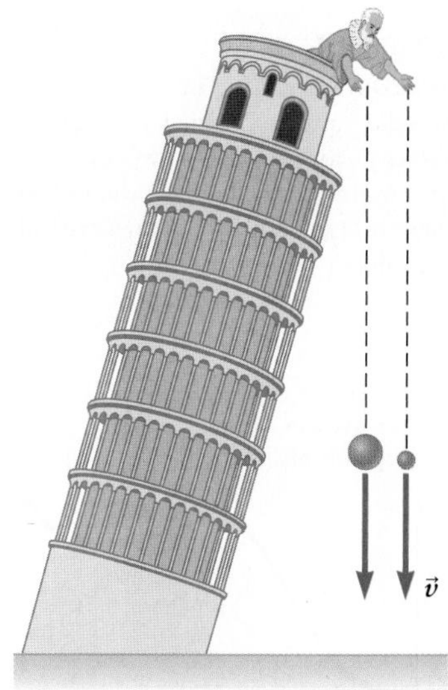

▲ **Figure 2.5** Galileo is reported to have used the Leaning Tower of Pisa in Italy in his studies of falling objects. Although the story may not be strictly accurate, Galileo certainly did conduct experiments showing that light and heavy objects fall at the same rate.

2.2 What Is Motion?

It is now time to consider how motion is described and measured in a precise, mathematical sense. The concept of motion is more complicated than you might first guess, and we'll need several quantities—***position***, ***velocity*** (which we have already encountered), and ***acceleration***—to describe it fully.

Let's first consider motion along a straight line, called ***one-dimensional motion***. A hockey puck sliding on a horizontal icy surface is a good example of this type of motion, and the top portion of Figure 2.6A shows what would happen if we took a multiple-exposure photo of such a hockey puck. Figure 2.6A is called a ***motion diagram***, a multiple-exposure photograph or similar sketch that shows the location of an object at regularly spaced instants in time. These exposures are captured at evenly spaced time intervals, and we can use them to construct the graph of the puck's position as a function of time shown in Figure 2.6B. Here we measure position as the distance from the origin on the x axis in Figure 2.6A to the center of the hockey puck. For one-dimensional motion, this distance, which we denote by x, completely specifies the position of the object. Notice in Figure 2.6B that the x axis is now vertical as we plot the position (x) as a function of time (t), which is plotted along the horizontal axis.

Velocity and Speed

The distance between adjacent dots on the x axis in Figure 2.6A shows how far the puck has moved during each time interval. We have already mentioned that velocity $\vec{v}$ is a vector quantity. The magnitude of $\vec{v}$ is called the ***speed***, the distance traveled per unit of time, while the direction of $\vec{v}$ gives the direction of the motion (in this example, $\vec{v}$ is directed to the right). Speed is a *scalar* quantity; it does not have a direction. In SI units, position is measured in meters and time is measured in seconds, so velocity and speed are both measured in meters per second, or simply m/s.

Speed and velocity are related quantities, but they are *not* the same. Speed tells how fast an object is moving, and it is always a positive quantity (or perhaps zero). The velocity contains this information and in addition tells the *direction* of motion. For the hockey puck in Figure 2.6A, the direction may be positive (motion to the right, toward larger or more positive values of x) or negative (motion to the left, toward smaller or more negative values of x). For one-dimensional motion, the direction of the velocity vector must lie parallel to the x axis. Thus, in cases involving one-dimensional motion, we need only deal with the *component* of the velocity parallel to x. This component can be positive, negative, or zero.

Vector quantities are usually written with arrows overhead, so the velocity vector is generally written as $\vec{v}$. For one-dimensional motion, the velocity vectors have only one component, and we can refer to this component as simply v, without an arrow. The *sign* of v (either positive or negative) then gives the direction of the velocity. Thus, for motion in one dimension,

$$\text{speed} = |v| \quad \text{(one dimension)} \tag{2.2}$$

In words, this expression says that speed is equal to the magnitude of the velocity. For two- or three-dimensional motion,

$$\text{speed} = |\vec{v}| \quad \text{(two or three dimensions)} \tag{2.3}$$

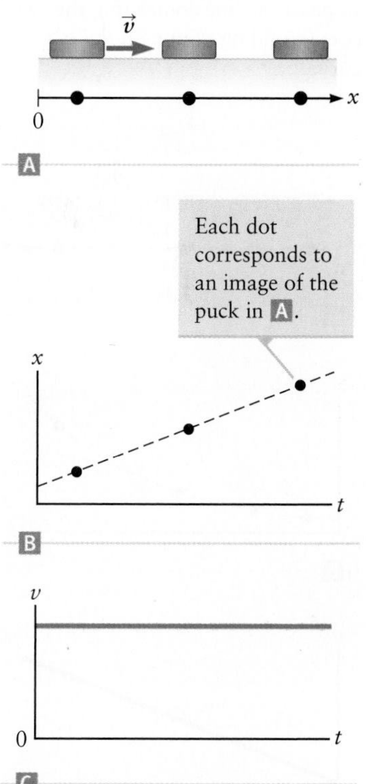

▲ **Figure 2.6** A Multiple images of a hockey puck traveling across an icy surface. B Plot of the position x of the puck as a function of time. The dots correspond to the images of the puck in part A. C Velocity v of the puck as a function of time.

Insight 2.1
SPEED AND VELOCITY
Speed is equal to the *magnitude* of the velocity. For motion in two or three dimensions, velocity is a vector. In these cases, speed is the *magnitude* of the velocity vector $\vec{v}$.

where the vertical bars again indicate that the speed is equal to the magnitude of the velocity vector. In either case (Eq. 2.2 or 2.3), speed is always a positive quantity or zero, but never negative.

The hockey puck in Figure 2.6A is sliding at a constant speed, so its velocity has a constant value as shown in the velocity–time (v–t) graph in Figure 2.6C. In this case, the velocity is positive, which means that the direction of motion is toward increasing values of x (i.e., to the right).

How Is an Object's Velocity Related to Its Position?

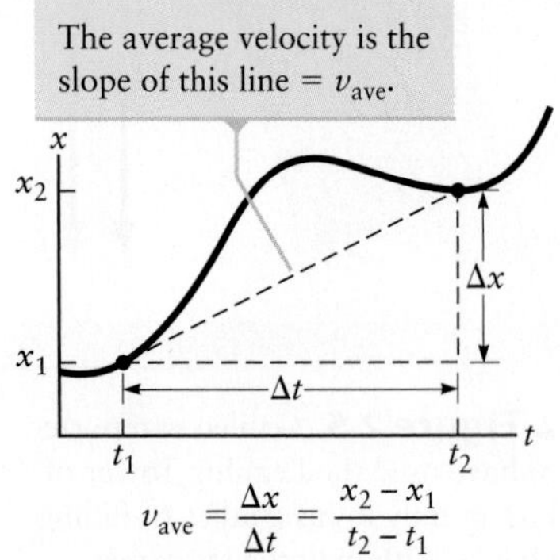

▲ **Figure 2.7** Hypothetical plot of an object's position as a function of time. The average velocity during the time interval from t_1 to t_2 is the slope of the line connecting the two corresponding points on the x–t curve.

Velocity is the change in position per unit time. Since time is measured in seconds, it is natural to think about the position at 1-s time intervals as suggested by the dots in Figure 2.6B. Alternatively, we could consider a particular time interval that begins at time t_1 and ends at time t_2 so that the size of the time interval is $\Delta t = t_2 - t_1$. The change in position during a particular time interval is called the ***displacement***. For one-dimensional motion, displacement is denoted by $\Delta x = x_2 - x_1$, where x_1 is the ***initial position***, the position at the beginning of the time interval (t_1), and x_2 is the ***final position***, the position at the end of the interval (t_2). The ***average velocity*** during this time interval is

$$v_{ave} = \frac{x_{final} - x_{initial}}{t_{final} - t_{initial}} = \frac{x_2 - x_1}{t_2 - t_1}$$

$$v_{ave} = \frac{\Delta x}{\Delta t} \tag{2.4}$$

According to Equation 2.4, the average velocity is the slope of the line segment that connects the positions at the beginning and end of the time interval. This is illustrated in the hypothetical x–t graph in Figure 2.7.

Another example of motion along a line is the case of a rocket-powered car traveling on a flat road. Let's assume the car is at rest when our clock reads zero. At $t = 0$, the driver turns on the rocket engine and the car begins to move in a straight-line path, along a horizontal axis we denote as x. Figure 2.8A is a motion diagram showing the position of the car at evenly spaced instants in time. The x axis is along the road, and the position of the car at a particular instant is the distance from the origin to the center of the car. The corresponding position–time graph for the car is shown in Figure 2.8B, where the dots mark the car's position at evenly spaced time intervals. In this case, the spacing between dots increases as the car travels. So, the car moves a greater distance during each successive (and equal) time interval and hence the speed of the car increases with time. The car moves toward increasing values of x, so the velocity is again positive and v increases smoothly with time as shown in Figure 2.8C. The precise shape of the v–t curve will depend on the way the engine fires, a problem we will consider in Chapter 3.

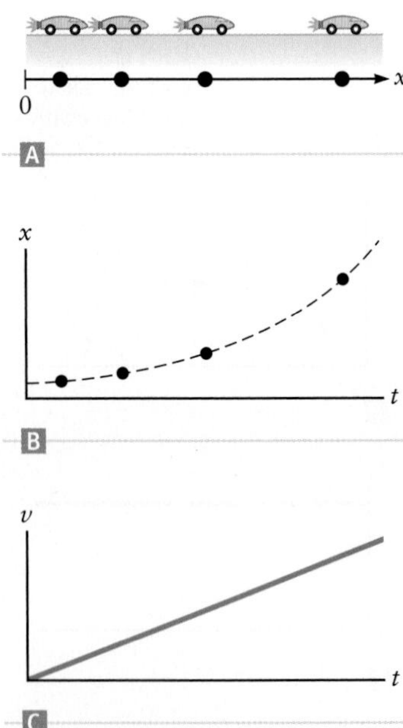

▲ **Figure 2.8** A Motion diagram for a rocket-propelled car traveling along a horizontal road. B Position as a function of time for the rocket-powered car. C Velocity of the car as a function of time.

Average Velocity and Instantaneous Velocity

Figure 2.9A shows the position as a function of time for a hypothetical object moving in a straight line, using dots to mark the position at the beginning and end of a time interval that starts at $t = 1.0$ s and ends at $t = 2.0$ s. According to Equation 2.4, the average velocity during this time interval is just the displacement during the interval divided by the length of the interval. Figure 2.9A shows that this average velocity is the *average slope* of the position–time curve (i.e., the slope of the line connecting the start to the end of the entire interval).

With this approach, though, we lose all details about what happens in the middle of the interval. In Figure 2.9A, the slope of the x–t curve varies considerably as we move through the interval from $t = 1.0$ s to $t = 2.0$ s. If we want to get a more accurate description of the object's motion at a particular instant within this time interval, say at $t = 1.5$ s, it is better to use a smaller interval. How small an interval should we use? Intuitively, we expect that using a smaller interval will give a better measure of

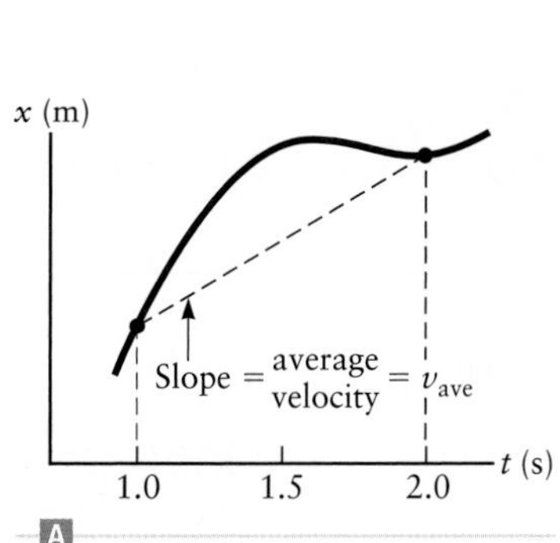

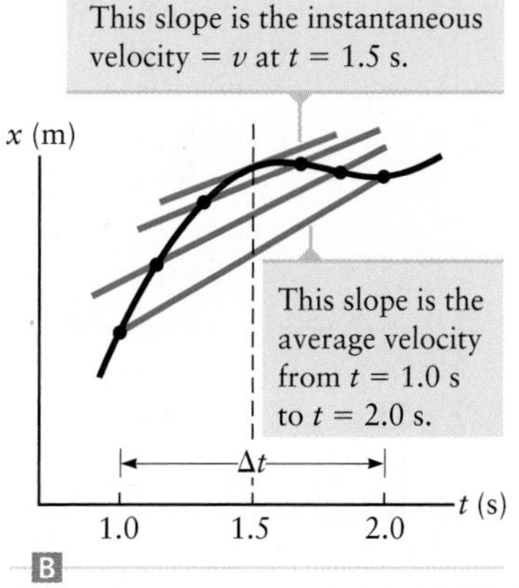

◀ **Figure 2.9** A The average velocity during a particular time interval is the slope of the line connecting the start of the interval to the end of the interval. B The instantaneous velocity at a particular time is the slope of the x–t curve at that time. The instantaneous velocity in the middle of a time interval is not necessarily equal to the average velocity during the interval. The instantaneous velocity is defined as the limit of the slope over a time interval Δt as $\Delta t \rightarrow 0$.

the motion at a particular point in time. From Figure 2.9B, we see that as we take ever smaller time intervals we are actually calculating the slope of the position–time curve at the point of interest (here at $t = 1.5$ s). The slope of the position–time curve at the point of interest is called the ***instantaneous velocity***. For one-dimensional motion, the instantaneous velocity v is the *slope of the position–time (x–t) curve* and is given by[2]

$$v = \lim_{\Delta t \rightarrow 0} \frac{\Delta x}{\Delta t} \tag{2.5}$$

Definition of instantaneous velocity

For the example shown in Figure 2.9, v is not constant. Rather, it varies with time over the interval from $t = 1.0$ s to $t = 2.0$ s.

The difference between the average and instantaneous values can be understood in analogy with a car's speedometer. The speedometer reading gives your instantaneous speed, the magnitude of your instantaneous velocity at a particular moment in time. If you are taking a long drive, your average speed will generally be different because the average value will include periods at which you are stopped in traffic, passing other cars, and so forth. In many cases, such as in discussions with a police officer, the instantaneous value will be of greatest interest.

The instantaneous velocity gives a mathematically precise measure of how the position is changing at a particular moment, making it much more useful than the average velocity. For this reason, from now on in this book we refer to the instantaneous velocity as simply the "velocity," and we denote it by v as in Equation 2.5.

CONCEPT CHECK 2.2 Estimating the Instantaneous Velocity

Is there an instant in time in Figure 2.9 at which the instantaneous velocity is zero? If so, what is the approximate value of t at which $v = 0$?

EXAMPLE 2.1 Average Velocity of a Bicycle

Consider the multiple images in Figure 2.10, showing a bicyclist moving along a level road. Find the average velocity of the bicyclist during the interval from $t = 2.0$ s to $t = 3.0$ s.

RECOGNIZE THE PRINCIPLE

The average velocity during a particular time interval is the slope of the position–time graph during that interval, $v_{ave} = \Delta x/\Delta t$ (Eq. 2.4). The time interval is given, so we need to deduce the displacement Δx from Figure 2.10.

(continued) ▶

[2]Here the term "lim" (limit) means to take the ratio $\Delta x/\Delta t$ as the quantity Δt approaches zero.

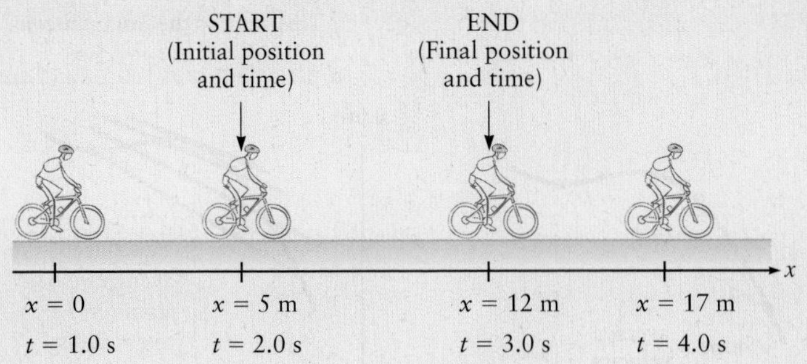

▲ **Figure 2.10** Example 2.1.

Insight 2.2

ALWAYS DRAW A PICTURE

The solution to most problems in physics starts with a picture. Begin by drawing a picture showing all the given information. The picture should show coordinate axes and any other information that seems relevant to the problem.

SKETCH THE PROBLEM

Our first step is to draw a picture that contains all the relevant information; this step is essential for organizing our thoughts and seeing connections. The images in Figure 2.10 form the heart of the picture, but because we want to extract some quantitative information, we have added the x axis, with its origin at the bicyclist's position at $t = 1.0$ s. Using this coordinate axis, we can read off the value of the bicycle's position at the times of interest.

IDENTIFY THE RELATIONSHIPS

The average velocity is the bicyclist's displacement during a particular time interval divided by the length of the interval (Eq. 2.4). We have therefore added arrows to our picture that mark the positions at the start ($t = 2.0$ s) and end ($t = 3.0$ s) of the interval of interest. We have

$$v_{\text{ave}} = \frac{\Delta x}{\Delta t} = \frac{x_{\text{final}} - x_{\text{initial}}}{t_{\text{final}} - t_{\text{initial}}} \tag{1}$$

SOLVE

Inserting values from Figure 2.10 into Equation (1), we get

$$v_{\text{ave}} = \frac{x(t = 3.0 \text{ s}) - x(t = 2.0 \text{ s})}{t_{\text{final}} - t_{\text{initial}}} = \frac{(12 \text{ m} - 5 \text{ m})}{(3.0 \text{ s} - 2.0 \text{ s})} = \boxed{7 \text{ m/s}}$$

▶ ***What does it mean?***

Each bicycle in Figure 2.10 corresponds to a point on the x–t graph, with coordinates given by reading off the values of x and t. Once we had the values of x and t, we then found the average velocity through the relation $v_{\text{ave}} = \Delta x/\Delta t$.

EXAMPLE 2.2 Average and Instantaneous Velocity of a Car

A car is traveling on a long, straight road. It starts at the origin at $t = 0$ and moves with a velocity of $+9.0$ m/s for 30 s. After stopping for 10 s at a crosswalk to allow a pedestrian to cross the road, it then moves at $+6.0$ m/s for another 20 s. (**a**) What is the final position of the car? (**b**) What is the car's average velocity?

RECOGNIZE THE PRINCIPLE

The car's motion can be broken into three separate intervals, with values for the velocity of $+9.0$ m/s, zero, and $+6.0$ m/s, respectively. During each of these intervals the velocity is constant, so the displacement during each interval will be related to the velocity by $\Delta x = v\Delta t$. The final position of the car will be the sum of the displacements during the three intervals, and the average velocity will be the total displacement divided by the total time.

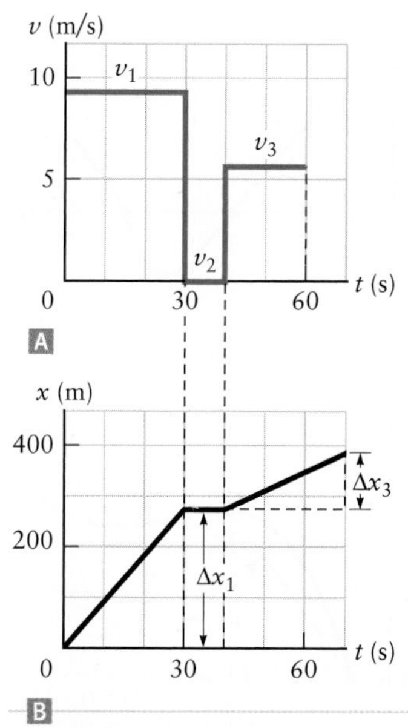

▲ **Figure 2.11** Example 2.3. A Velocity versus time for the car. B Position versus time.

SKETCH THE PROBLEM

Figure 2.11A shows the velocity as a function of time, and Figure 2.11B shows how the displacement varies with time.

IDENTIFY THE RELATIONSHIPS AND SOLVE

(**a**) The displacement during the first interval (between $t = 0$ and 30 s) is $\Delta x_1 = v_1 \Delta t_1$, where $v_1 = +9.0$ m/s and $\Delta t_1 = 30$ s. We thus find

$$\Delta x_1 = v_1 \Delta t_1 = (9.0 \text{ m/s})(30 \text{ s}) = 270 \text{ m}$$

In a similar way, we get $\Delta x_2 = v_2 \Delta t_2 = (0)(10 \text{ s}) = 0$ and $\Delta x_3 = v_3 \Delta t_3 = (6.0 \text{ m/s})(20 \text{ s}) = 120$ m. The total displacement is thus

$$\Delta x_{\text{total}} = \Delta x_1 + \Delta x_2 + \Delta x_3 = (270 + 0 + 120) \text{ m} = \boxed{390 \text{ m}}$$

(**b**) The average velocity during the entire interval is (using Eq. 2.4)

$$v_{\text{ave}} = \frac{\Delta x_{\text{total}}}{\Delta t_{\text{total}}} = \frac{390 \text{ m}}{(30 + 10 + 20) \text{ s}} = \boxed{6.5 \text{ m/s}}$$

► ***What does it mean?***

The average velocity during the entire interval is not equal to the instantaneous velocity during any portion of the motion. The value of the v_{ave} pertains to the entire interval, while the instantaneous velocity describes the motion at a particular instant in time.

In Example 2.1, we saw how to use observations of the position as a function of time to compute an object's average velocity. It is also very useful to be able to deduce the instantaneous velocity from the position–time behavior. We can do so by estimating the slope of the position–time curve at different values of t and using these estimates to make a qualitative plot of the velocity–time relation. This approach is illustrated in Example 2.3.

EXAMPLE 2.3 Computing Velocity Using a Graphical Method

A hypothetical object moves according to the x–t graph in Figure 2.12A. This object is initially (when t is near t_1) moving to the right, in the "positive" x direction. The object reverses direction near t_2, and it is again moving to the right at the end when t is near t_4. (**a**) Sketch the qualitative behavior of the velocity of the object as a function of time using a graphical approach. (**b**) Estimate the average velocity during the interval between $t_1 = 1.0$ s and $t_2 = 2.5$ s.

RECOGNIZE THE PRINCIPLE

For part (**a**), we want to find the velocity—which means the *instantaneous* velocity—so we need to estimate the slope of the x–t curve as a function of time. For part (**b**), the average velocity over the interval $t = 1.0$ s to $t = 2.5$ s is the slope of the x–t curve during this interval.

SKETCH THE PROBLEM

Figure 2.12B shows the x–t graph with lines drawn tangent to the x–t curve at various instants. The slopes of these tangent lines are the velocities at times $t_1, t_2, \ldots$ in Figure 2.12A.

(continued) ►

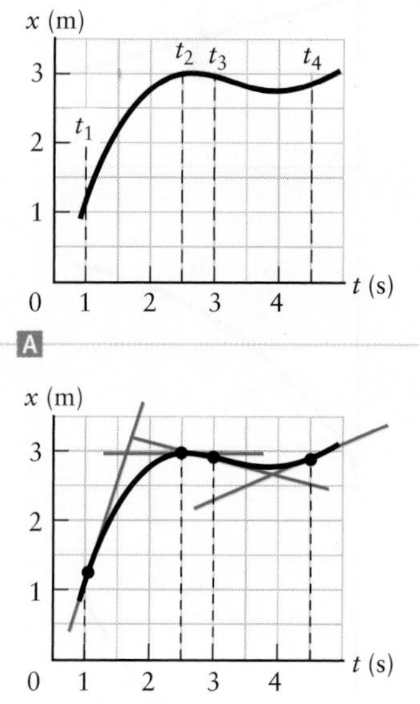

▲ **Figure 2.12** Example 2.3. A Hypothetical position–time graph. The slopes of the tangent lines in B are equal to the velocity at various instants in time.

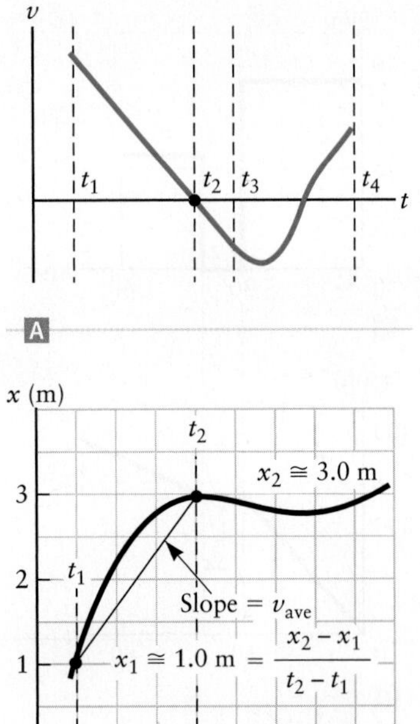

▲ **Figure 2.13** Example 2.3. A Qualitative plot of the velocities obtained from the slopes in Figure 2.12B. B Calculation of the average velocity during the interval between $t_1 = 1.0$ s and $t_2 = 2.5$ s.

IDENTIFY THE RELATIONSHIPS AND SOLVE

(a) At $t = t_1$, the $x-t$ slope is large and positive, so our result for v in Figure 2.13A is large and positive at $t = t_1$. At $t = t_2$, the $x-t$ slope is approximately zero and hence v is near zero. At $t = t_3$, the object is moving toward smaller values of position x, so the slope of the x–t curve and hence also the velocity are negative. Finally, at $t = t_4$, the object is again moving to the right as x is increasing with time, so v is again positive. After estimating the $x-t$ slope at these places, we can construct the smooth v–t curve shown in Figure 2.13A, which shows the *qualitative* behavior of the velocity as a function of time.

(b) To estimate the average velocity between t_1 and t_2, we refer to Figure 2.13B, which shows a line segment that connects these two points on the position–time graph. The slope of this segment is the average velocity:

$$v_{ave} = \frac{\Delta x}{\Delta t} = \frac{x_2 - x_1}{t_2 - t_1}$$

Reading the values of x_1, x_2, t_1, and t_2 from that graph, we find

$$v_{ave} = \frac{x_2 - x_1}{t_2 - t_1} = \frac{(3.0\text{ m}) - (1.0\text{ m})}{(2.5\text{ s}) - (1.0\text{ s})} = \boxed{1.3\text{ m/s}}$$

► ***What does it mean?***

The (instantaneous) velocity is the slope of the position–time graph. To find the qualitative behavior of the velocity, we found approximate values by drawing lines tangent to the x–t curve at several places and estimating their slopes. The value of v at a particular value of t is always equal to the slope of the x–t curve at that time.

CONCEPT CHECK 2.3 The Relation between Velocity and Position

For which of the position–time graphs in Figure 2.14 are the following statements true?

(a) The velocity increases with time.
(b) The velocity decreases with time.
(c) The velocity is constant (does not change with time).

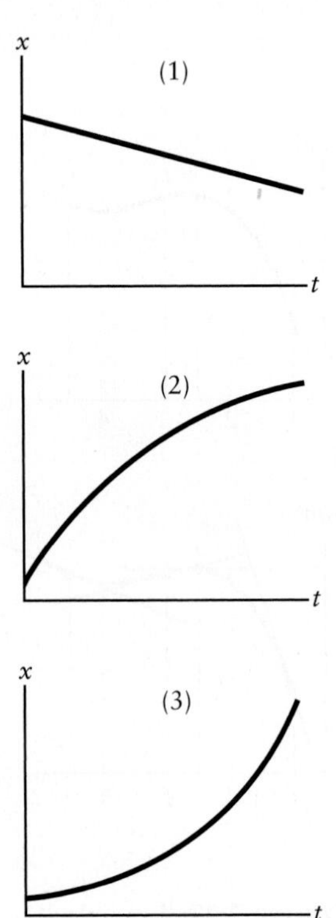

▲ **Figure 2.14** Concept Check 2.3.

Acceleration

We have seen that two quantities—position and velocity—are important for describing the motion of an object. One additional quantity, ***acceleration***, will play a central role in our theory of motion. Acceleration is related to how the velocity changes with time. Consider again our rocket-powered car from Figure 2.8. In Figure 2.15, we resketch the v–t plot that we derived in Figure 2.8C and notice again that the car's velocity increases as time proceeds. Acceleration is defined as the rate at which the velocity is changing. If the velocity changes by an amount Δv over the time interval Δt, the ***average acceleration*** during this interval is

$$a_{ave} = \frac{\Delta v}{\Delta t} \tag{2.6}$$

As with the velocity, we are usually concerned with the acceleration at a particular instant in time, which leads us to consider the acceleration in the limit of very small time intervals. We thus define the ***instantaneous acceleration*** as

$$a = \lim_{\Delta t \to 0} \frac{\Delta v}{\Delta t} \tag{2.7}$$

The instantaneous acceleration a equals the *slope* of the v–t curve at a particular instant in time. The SI unit of acceleration is m/s^2 (meters per second squared).

We have now introduced several quantities associated with motion, including position, displacement, velocity, and acceleration. These quantities are connected in a mathematical sense through Equations 2.5 and 2.7. We also showed important graphical relationships: velocity is the slope of the position–time graph, while acceleration is the slope of the velocity–time graph. You might now be wondering if we'll continue this progression and consider the slope of the acceleration and so forth. The answer is that acceleration is as far as we need to go; x, v, and a are all we need in our formulation of a complete theory of motion. Newton's laws of motion (Section 2.4) will show us why.

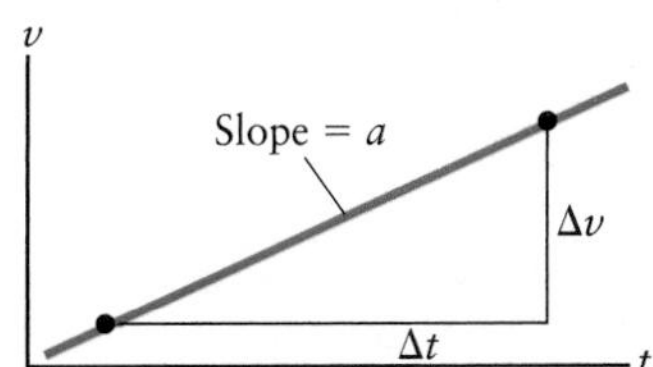

▲ **Figure 2.15** Acceleration is the slope of the velocity–time curve. For the case shown here, in which v varies linearly with time, the average acceleration is equal to the instantaneous acceleration.

EXAMPLE 2.4 Acceleration of a Sprinter

Consider a sprinter (Fig. 2.16A) running a 100-m dash. Figure 2.16B shows the velocity–time graph for the sprinter. Use a graphical approach to calculate the corresponding acceleration–time graph. What is the approximate value of the sprinter's maximum acceleration, and when does it occur?

RECOGNIZE THE PRINCIPLE

Acceleration is the slope of the velocity–time graph, so we must estimate this slope at enough different values of t to be able to make a qualitative plot of the sprinter's acceleration as a function of time.

SKETCH THE PROBLEM

In Figure 2.16B, we have drawn in several lines tangent to the v–t curve at various times. The slopes of these tangent lines give the acceleration.

IDENTIFY THE RELATIONSHIPS AND SOLVE

We have measured approximate values of the slopes of the tangent lines in Figure 2.16B, and the results are plotted in Figure 2.17. This acceleration–time graph is only qualitative (approximate). More accurate results would be possible had we started with a more detailed graph of the velocity.

▶ *What does it mean?*

The shape of the acceleration–time curve in Figure 2.17 is consistent with expectations. The largest value of the acceleration (about 10 m/s^2) occurs at the start of the race when the sprinter is "getting up to speed"; that is where the v–t slope is largest. At the end of the race as the runner crosses the finish line, she slows down and eventually comes to a stop with $v = 0$ at the far right in Figure 2.16B. During that time, her velocity is decreasing with time, so her acceleration is negative.

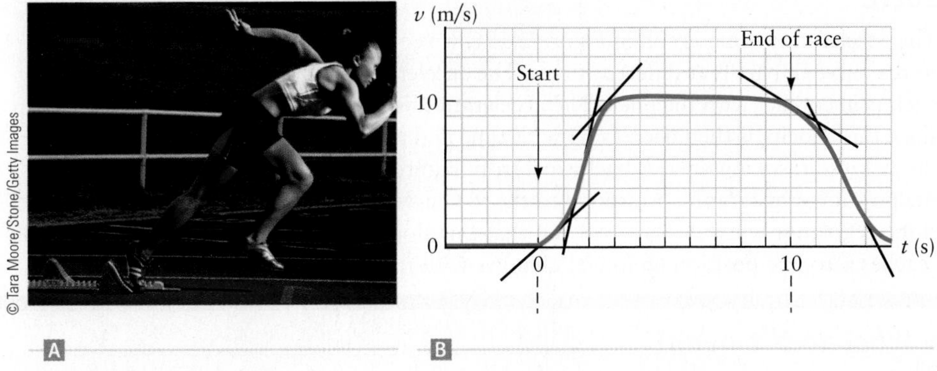

▲ **Figure 2.16** Example 2.4. A A sprinter and B her velocity–time graph.

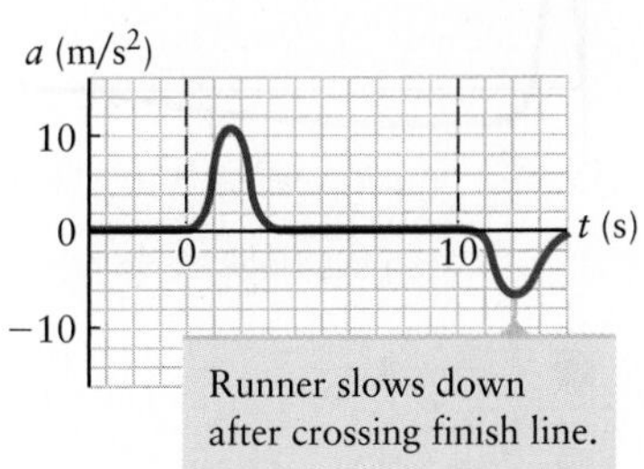

▲ **Figure 2.17** Example 2.4. The corresponding acceleration–time graph is given qualitatively.

The Relation between Velocity and Acceleration

An interesting result in Example 2.4 is that the maximum velocity and the maximum acceleration do *not* occur at the same time. It is tempting to think that if the "motion" is large, both v and a will be large, but this notion is *incorrect*. Acceleration is the slope—the rate of change—of the velocity with respect to time. The time at which the rate of change of velocity is greatest may not be the time at which the velocity itself is greatest.

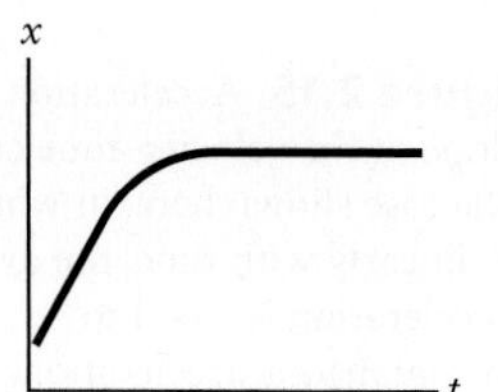

▲ **Figure 2.18** Concept Check 2.4. What type of motion is described by this position–time graph?

CONCEPT CHECK 2.4 Analyzing a Position–Time Graph

The position–time curve of a hypothetical object is shown in Figure 2.18. Which of the following scenarios might be described by this x–t graph?

(a) A car starting from rest when a traffic light turns green.
(b) A car slowing to a stop when a traffic light turns red.
(c) A bowling ball rolling toward the pins.
(d) A runner slowing from top speed to a stop at the end of a race.

EXAMPLE 2.5 Sliding to a Stop

Consider a hockey puck that starts from rest. At $t = 0$, the puck is struck by a hockey stick, which starts the puck into motion with a high velocity. The puck then slides a very long distance before coming to rest. Draw qualitative plots of the puck's position, velocity, and acceleration as functions of time.

RECOGNIZE THE PRINCIPLES

We first use our experience with sliding objects to deduce the qualitative shape of the velocity–time curve. The puck starts from rest, so initially $v = 0$. The velocity then increases quickly to a high value when the stick is in contact with the puck. After the puck leaves the stick, the velocity gradually decreases as the puck slides to a stop. From the v–t curve, we can get the position and acceleration as functions of time.

SKETCH THE PROBLEM

In Figure 2.19, imagine drawing part B first (qualitative behavior of velocity). The puck reaches a high velocity very quickly and then falls to zero at long times.

IDENTIFY THE RELATIONSHIPS

To find the acceleration, we must find the slope of the v–t curve as a function of time. To get the position, we must find an x–t curve whose slope gives this velocity–time graph.

SOLVE

The acceleration–time graph in Figure 2.19C was obtained from estimates of the slope of the curve in part B at different times. The acceleration is *positive* and large when the stick is in contact with the puck. The acceleration is *negative* as the puck slides to a stop since the velocity is then decreasing with time and the v–t slope is negative. To deduce the x–t result, we must work backward from Figure 2.19B so as to make the slope of the position–time graph correspond to the v–t curve. We do so by noting that the x–t slope is greatest when v is greatest and that this slope is small when v is small. Once v reaches zero, the position no longer changes with time.

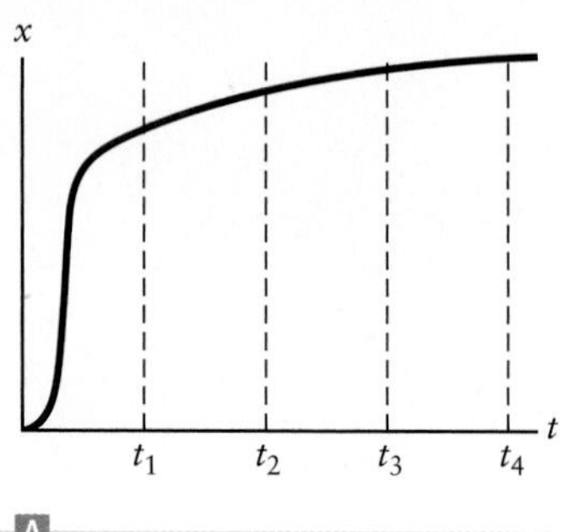

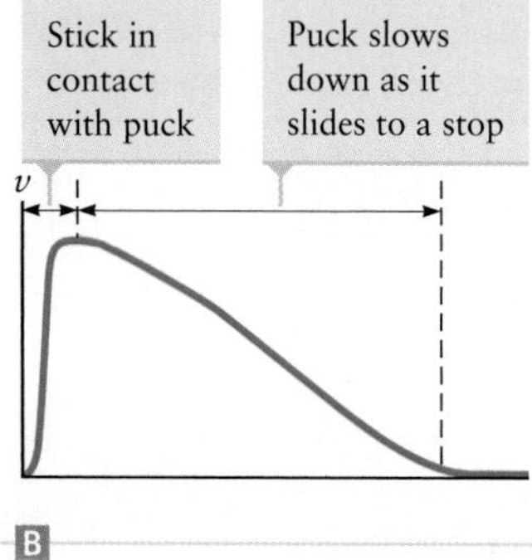

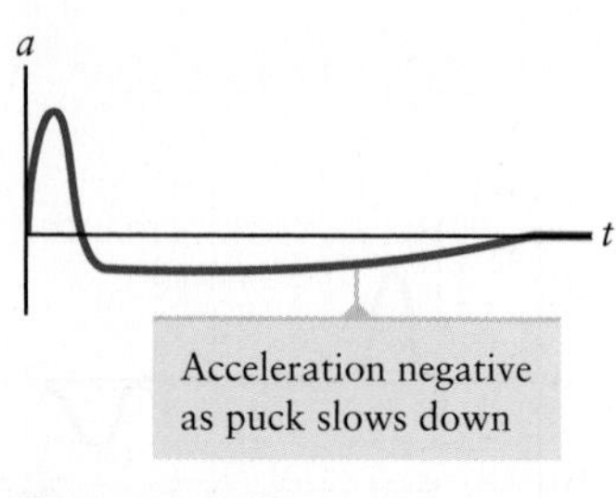

▲ **Figure 2.19** Example 2.5. Motion of a hockey puck as described by graphs of its position, velocity, and acceleration versus time.

▶ ***What does it mean?***

These graphs of the position and velocity can also be appreciated using the "time-lapsed" sketches in Figure 2.20. These sketches show the location of the puck at

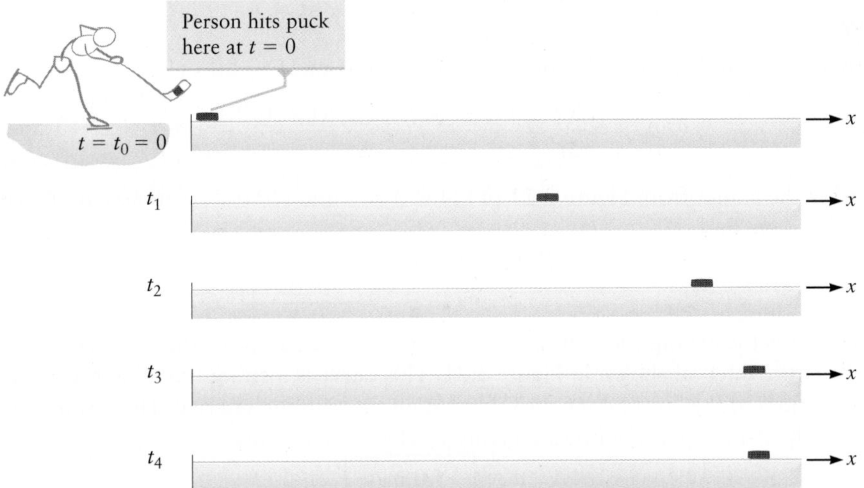

▲ **Figure 2.20** Example 2.5. Time-lapse sketches of the hockey puck's motion.

evenly spaced instants in time, beginning at $t_0 = 0$ when it is struck by the hockey stick and then at $t_1, t_2, \ldots$ until the puck comes to rest at t_4. These times are equally spaced along the time axis (Fig. 2.19A), so the time intervals $\Delta t = t_1 - t_0 = t_2 - t_1$, ... are all equal. The displacements during these intervals are *not* equal. For example, the displacement during the first interval (from t_0 to t_1) is much larger than the displacement during the second interval. In words, the puck moves farther—because it has a higher velocity—during the first interval. Likewise, at the end of the time period, the puck's velocity is small, so the distance traveled between t_3 and t_4 is much smaller than the distance traveled between t_0 and t_1.

EXAMPLE 2.6 Stopping in a Hurry

A driver is in a hurry, and her car is traveling along a straight road with velocity $v_0 = 20$ m/s (about 40 mi/h) when she spots a problem in the road ahead and applies the brakes. The car then slows to a stop according to the velocity–time curve in Figure 2.21. Find the average acceleration during the interval from $t = 0.0$ s to $t = 4.0$ s and the instantaneous acceleration at $t = 2.0$ s.

RECOGNIZE THE PRINCIPLE

The average acceleration is the average slope of the velocity–time curve *during the time interval* of interest. The instantaneous acceleration is the slope of the v–t curve *at the point* of interest.

SKETCH THE PROBLEM

Figure 2.21 describes the problem and contains all the information we need to solve it.

IDENTIFY THE RELATIONSHIPS

From the definition of average acceleration in Equation 2.6, we have

$$a_{ave} = \frac{\Delta v}{\Delta t} \tag{1}$$

We can read the values of Δv and Δt from Figure 2.21.

(continued) ▶

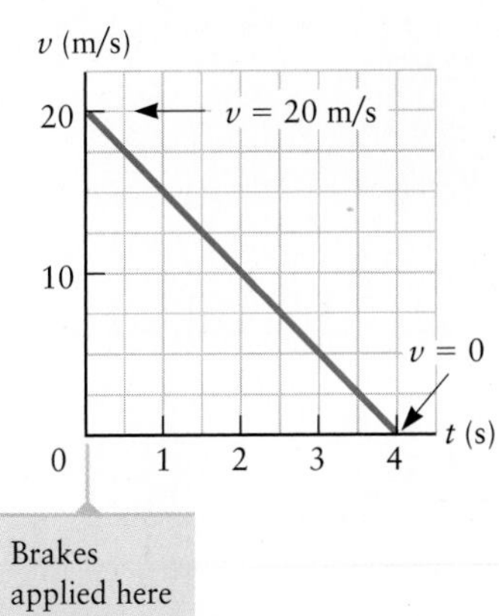

▲ **Figure 2.21** Example 2.6.

SOLVE

Applying Equation (1) over the interval from $t = 0.0$ s to $t = 4.0$ s gives

$$a_{\text{ave}} = \frac{\Delta v}{\Delta t} = \frac{v(t = 4.0 \text{ s}) - v(t = 0.0 \text{ s})}{\Delta t}$$

Inserting the values from Figure 2.21 of $v(4.0 \text{ s}) = 0$ and $v(0.0 \text{ s}) = 20$ m/s along with $\Delta t = 4.0$ s, we find

$$a_{\text{ave}} = \frac{(0 - 20) \text{ m/s}}{4.0 \text{ s}} = \boxed{-5.0 \text{ m/s}^2}$$

This result is the average acceleration, so it is the average slope of the v–t curve during the entire interval shown in Figure 2.21. This curve is a straight line, so the average slope is equal to the "instantaneous" slope at all times in the interval. The instantaneous slope of the v–t curve is the instantaneous acceleration a; hence

$$a = \boxed{-5.0 \text{ m/s}^2}$$

► ***What does it mean?***

The average and instantaneous acceleration in this example are both negative because the velocity *decreases* during the time interval of interest. The velocity is positive (Fig. 2.21) during the entire interval, so the velocity and acceleration are in opposite directions: the velocity is positive (along the $+x$ direction), while the acceleration is negative (along the $-x$ direction).

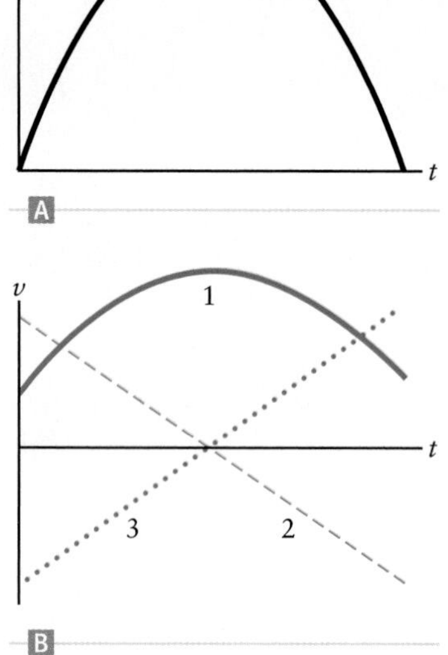

▲ **Figure 2.22** Concept Check 2.5.

CONCEPT CHECK 2.5 Finding the Velocity

Figure 2.22A shows the position–time graph for an object. Which of the curves in Figure 2.22B describes the qualitative behavior of the object's velocity as a function of time?

(a) Curve 1 (b) Curve 2 (c) Curve 3

2.3 The Principle of Inertia

We spent much of Section 2.1 discussing Aristotle's ideas about motion, and you should now be convinced that his theory of motion has some serious flaws. Nevertheless, it is worthwhile to consider how Aristotle might have explained the motion of the rocket-powered car in Figure 2.8. He probably would have claimed that the car moves by virtue of the force exerted on it by the engine, with a velocity $v = F/R$ as predicted from Equation 2.1. If the rocket engine is then turned off, however, the force would vanish ($F = 0$) and, according to the same argument, the car would stop immediately. Your intuition should tell you that this prediction is not correct; the car would instead continue along, for at least a short period, after the engine is turned off. That is, the car will coast for a while before coming to rest.

The difficulties with Aristotle's ideas about motion all come down to his belief that force and velocity are directly linked. This linkage is expressed in Equation 2.1, which states that if an object has a nonzero velocity, then according to Aristotle there must be a nonzero force acting on the object at that time. However, our example with a rocket-powered car shows that force and velocity are *not* linked in this way because it is possible for the car to have a nonzero velocity along the x direction even when the engine is turned off and the force is zero. The correct connection between force and motion is instead based on a direct linkage between force and *acceleration*. This connection to acceleration is at the heart of Newton's laws of motion.

Before we can really appreciate Newton, however, we must first consider the work of Galileo. Although Galileo did not arrive at the correct laws of motion, his experiments on the motion of terrestrial objects showed that an object can move even if

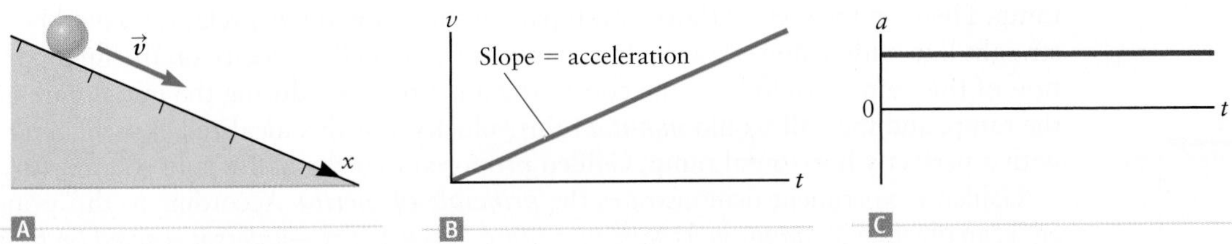

▲ **Figure 2.23** A Ball rolling down a simple incline. B Galileo found that the velocity of the ball increases linearly with time. The slope of this line is the acceleration in C.

there is *zero* total force acting on it. This result led to the discovery of the ***principle of inertia***, which is a cornerstone of Newton's laws.

Galileo's Experiments on Motion

Aristotle was not able to explain how an object could move when there did not appear to be any force acting on it. A good example is the somewhat idealized case of a hockey puck sliding on a very smooth, horizontal, icy surface. Our intuition tells us that the puck will slide a very long way before coming to rest; in fact, if we could somehow make the surface "perfectly" icy so that there is absolutely no friction whatsoever, our intuition suggests that the puck would slide forever.

Galileo did not carry out experiments with a hockey puck on an icy surface. Instead, his experiments involved a ball rolling on an incline. If the ball is very hard, like a billiard ball, and the surface of the incline is also very hard and smooth, the effect of friction on the ball's motion is very small. Hence, this situation is actually quite similar to that of our idealized hockey puck.

Galileo experimented with the motion of a ball on an incline as sketched in Figure 2.23. He found that when a ball is released from rest, its velocity varies with time as shown in Figure 2.23B. This is another example of one-dimensional motion, with the direction of motion denoted by an x axis lying along the incline as shown. The velocity of the ball increases as the ball rolls down the incline, and Galileo found that the magnitude of the velocity increases *linearly* with time. Since acceleration is the slope of the v–t curve, the acceleration is *constant* and *positive*.

Galileo then repeated this experiment, but this time with the ball rolling *up* an incline with the same tilt angle (Fig. 2.24A). When he gave the ball some initial velocity, he found that it rolled up the incline with the v–t curve shown in Figure 2.24B. Again he observed that the velocity varies *linearly* with time, but now the slope is *negative*. In fact, the slope in this case is equal in magnitude to the slope found for the down-tilted incline. Hence, the acceleration was opposite in sign but equal in magnitude to that found when the ball rolled down the incline. Galileo performed this experiment with many inclines and with many different tilt angles, and he always observed that the velocity varied linearly with time, with the slope of the $v-t$ graph being determined by the tilt angle. Moreover, the acceleration (the slope of the v–t relation) when a ball rolled up a particular incline was always equal in magnitude, but opposite in sign, when compared with the acceleration when the ball rolled down the same incline. He then reasoned that if the tilt of the incline were precisely zero (a perfectly level surface), the slope of the v–t line and hence the acceleration would be zero. Galileo therefore asserted that on a level surface the ball would roll with a *constant velocity*.

This result may seem obvious to you, since it means that a ball placed at rest on a horizontal surface will remain motionless, with zero velocity. Such a ball has a constant velocity because $v = 0$ is a constant! It was, however, the genius of Galileo to realize that his experiment implied the result sketched in Figure 2.25. Here, a ball is rolling along a different sort of ramp. The initial portion of the ramp is sloped like the incline in Figure 2.23, but the final portion is perfectly horizontal and extremely long, with only part of it shown here. Galileo realized that for this type of ramp the ball would have a positive velocity when it completes the first portion of the

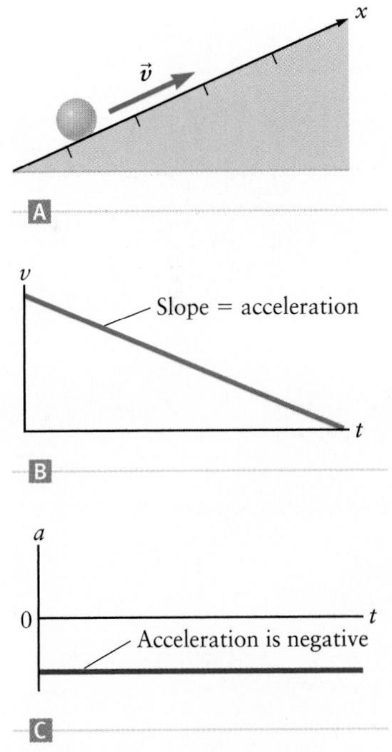

▲ **Figure 2.24** A Ball rolling up an incline with the same angle as in Figure 2.23. B Galileo found that the velocity of the ball decreases linearly with time. C The slope of this line—the acceleration—has the same magnitude, but the opposite sign, as the acceleration when the ball rolls down the incline (Fig. 2.23C).

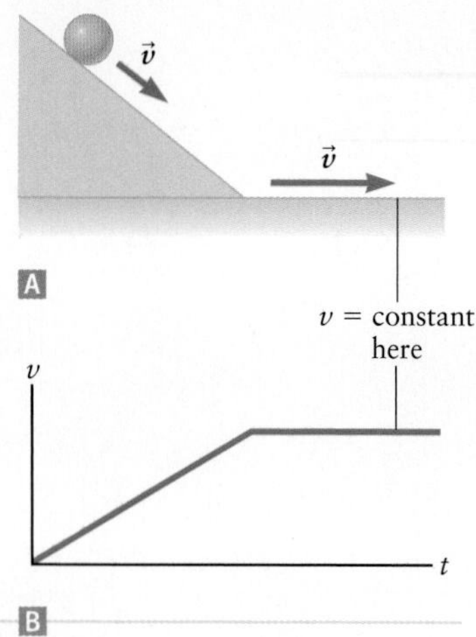

▲ **Figure 2.25** When a ball rolls along this two-part incline, its velocity increases linearly with time while on the initial (sloped) portion, and v is then constant on the final (flat) portion.

ramp. Then, on the second (horizontal) part of the ramp, the v–t relation would be a straight line with a slope of zero. In other words, the ball's velocity on the final portion of this ramp would have the constant value produced during the initial part of the ramp, and the ball would *maintain* this velocity. For this idealized case that ends with a perfectly horizontal ramp, Galileo proposed that the ball would roll *forever*.

Galileo's experiment demonstrates the ***principle of inertia***. According to this principle, an object will *maintain its state of motion*—its velocity—*unless it is acted on by a force*. On a horizontal surface, there is no force in the direction of motion, so the velocity is constant and the ball rolls forever. With his discovery of the principle of inertia, Galileo broke Aristotle's link between velocity and force. Galileo showed that one can have motion (a nonzero velocity) *without* a force. This discovery does not answer the question of exactly how force is linked to motion, however. The answer to that question was provided by Newton. (It is interesting that Newton was born in 1643, the year after Galileo died. In a very direct sense, Newton thus built on the work of Galileo.)

2.4 Newton's Laws of Motion

Newton's laws of motion are three separate statements about how things move.

Newton's First Law

Newton's first law is a careful statement about the principle of inertia that we described in the previous section in connection with Galileo's experiments.

> ***Newton's first law of motion:*** **If the total force acting on an object is zero, the object will maintain its velocity forever.**

If the total force acting on an object is *zero*, the object will move with a constant velocity. In other words, such an object will move with a constant speed along a particular direction and will continue this motion—with the same speed in the same direction—forever, or as long as the total force acting on it is zero.[3] This is another way of stating the principle of inertia, and you can see that it is essentially what Galileo found in his experiments with rolling balls. The total force acting on one of Galileo's balls rolling on a horizontal surface is zero, so in this ideal case the ball will roll forever.

Inertia and Mass

Now that we have introduced the principle of inertia, you should be curious about the term *inertia* and precisely what it means. The inertia of an object is a measure of its *resistance to changes in motion*. This resistance to change depends on the object's ***mass***. It is not possible to give a "first principles" definition of the term *mass*, but your intuitive notion is very useful. The mass of an object is a measure of the amount of matter it contains. Objects that contain a large amount of matter have a larger mass and a greater inertia than objects containing a small amount of matter. The SI unit of mass is the kilogram (kg). Mass is an intrinsic property of an object. For example, the mass m of an object is independent of its location; m is the same on the Earth's surface as on the Moon and in distant space. In addition, an object's mass does not depend on its velocity or acceleration. Although an object's mass does not appear explicitly in Newton's first law, it has an essential role in Newton's second law.

Newton's first law may seem surprising to you. After all, terrestrial objects always come to rest eventually. The difference here, and a key to appreciating Newton's first law, is that an object will move with a constant velocity only in the ideal case that the

[3]This result may be surprising because it may *appear* to contradict your intuition. This result only holds if the total force on an object is *precisely zero*, however, and that can be hard to achieve in practice.

total force is precisely zero. For terrestrial objects, it is very difficult to find such ideal cases because it is difficult to completely eliminate all forces, especially frictional forces. Hence, terrestrial objects come to rest because of the forces that act on them.

Two other hypothetical cases may help make Newton's first law fit with your intuition. Case 1: Imagine a frozen lake with an extremely smooth surface. A hockey puck sliding on such a perfectly flat and icy surface experiences a very tiny frictional force and will therefore slide for a very long distance before coming to rest. If this frictional force could be made to vanish, the puck would slide forever (if the lake were large enough). Of course, actual icy surfaces are not perfect; there will always be a small amount of friction, which would make the puck eventually come to rest. Case 2: Imagine a spaceship that is coasting (i.e., with its engines turned off) someplace in the universe very far from any stars or planets. Such a spaceship would experience only a very small gravitational force (from the nearest stars and planets). If the nearest stars and planets were very far away, this force would be negligible and the spaceship would move with a constant velocity, in accord with Newton's first law.

Newton's Second Law

Newton's second law of motion: **In many situations, several different forces act on an object simultaneously. The total force on the object is the sum of these individual forces, $\vec{F}_{\text{total}} = \sum \vec{F}$. The acceleration of an object with mass m is then given by**

$$\vec{a} = \frac{\sum \vec{F}}{m} \tag{2.8}$$

Newton's second law tells us how an object will move when acted on by a force or by a collection of forces. *This law is our link between force and motion.* The acceleration of an object is directly proportional to the total force that acts on it. Newton's second law, Equation 2.8, is often written in the equivalent form

$$\sum \vec{F} = m\vec{a}$$

Keep in mind that the term $\sum \vec{F}$ in Newton's second law is the *total* force on the object. As you might imagine, in most cases there are several forces acting on an object. So, we have to add them all up, and that resulting vector sum is the force $\sum \vec{F}$ in Newton's second law (see Fig. 2.26). You should also recall that vectors must be added according to the vector arithmetic procedures described in Chapter 1. The direction of the acceleration $\vec{a}$ is then parallel to the direction of the total force $\sum \vec{F}$.

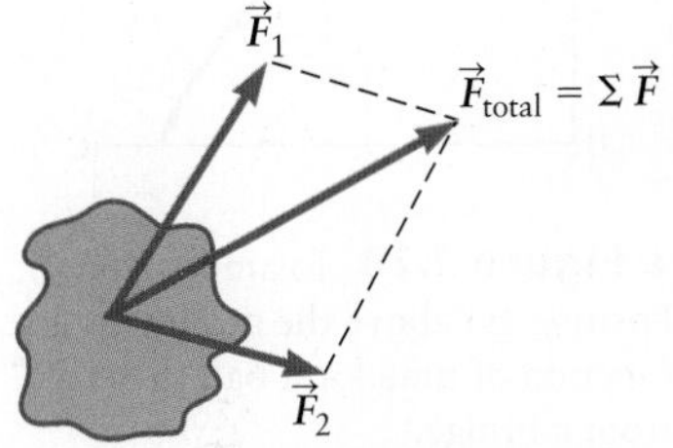

▲ **Figure 2.26** When several forces act on an object, the vector sum of these forces $\sum \vec{F}$ determines the acceleration according to Newton's second law.

In the SI system of units, force is measured in units called newtons (N). We can use Newton's second law to express this unit in terms of the primary SI units. Mass is measured in units of kilograms, while acceleration has the units meters per second squared. In terms of only the units, Equation 2.8 can be rearranged to read

$$\text{force} = \text{mass} \times \text{acceleration}$$

$$\text{newtons} = \text{kg} \times \frac{\text{m}}{\text{s}^2}$$

The value of the newton as a unit of force is therefore

$$1\,\text{N} = 1\,\text{kg} \cdot \text{m/s}^2 \tag{2.9}$$

We'll spend the next dozen chapters or so exploring applications of Newton's second law. Many applications start by determining the total force acting on an object from all sources. The object's acceleration can then be calculated using Equation 2.8. Acceleration is the change in velocity per unit time, so it is possible to use the acceleration to deduce the velocity. In a similar manner, since velocity is the change in position per unit time, one can use the velocity to find the object's position as a function of time. In this way, an object's acceleration, velocity, and position can all be found.

EXAMPLE 2.7 Using Newton's Second Law

A single force of magnitude 6.0 N acts on a stone of mass 1.1 kg. Find the acceleration of the stone.

RECOGNIZE THE PRINCIPLE

The force on the stone and its acceleration are related through Newton's second law,

$$\vec{a} = \frac{\sum \vec{F}}{m} \tag{1}$$

where $\sum \vec{F}$ is the total force on the stone, of magnitude 6.0 N as given.

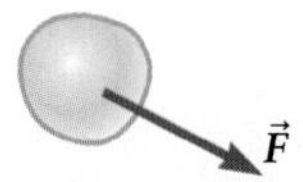

▲ **Figure 2.27** Example 2.7.

SKETCH THE PROBLEM

We begin by drawing a picture, showing all the forces acting on the stone (Fig. 2.27). Since there is only a single force in this example, it is also the total force.

IDENTIFY THE RELATIONSHIPS AND SOLVE

Using Newton's second law, Equation (1), the magnitudes of the acceleration and force are related by

$$a = \frac{\sum F}{m} = \frac{6.0 \text{ N}}{1.1 \text{ kg}} = \frac{6.0 \text{ kg} \cdot \text{m/s}^2}{1.1 \text{ kg}} = \boxed{5.5 \text{ m/s}^2} \tag{2}$$

The direction of the stone's acceleration is parallel to that of the total force, so the direction of $\vec{F}$ in Figure 2.27 gives the direction of $\vec{a}$.

▶ *What does it mean?*

Notice how the units in Equation (2) combine, as the factors of kilograms on the top and bottom cancel to give an answer with units of meters per second squared, as expected for acceleration. You should always check the units of an answer in this way. An incorrect result for the units usually indicates an error in the calculation.

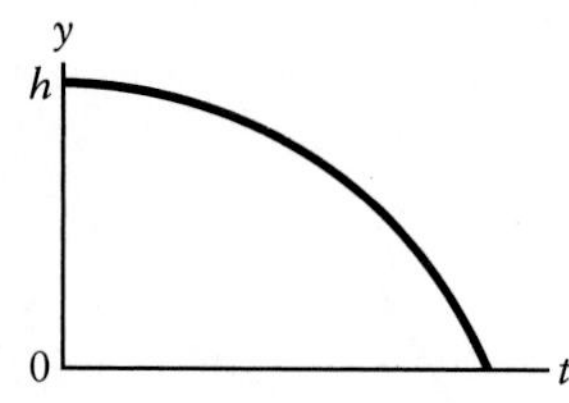

▲ **Figure 2.28** Example 2.8. Position (y) above the ground as a function of time for a ball that falls from a bridge.

EXAMPLE 2.8 Motion of a Falling Object

A ball is dropped from a bridge onto the ground below. The height of the ball above the ground as a function of time is shown in Figure 2.28. Use a graphical approach to find, as functions of time, the qualitative behavior of (**a**) the velocity of the ball, (**b**) the acceleration of the ball, and (**c**) the total force on the ball.

RECOGNIZE THE PRINCIPLE

Velocity is the slope of the position–time curve, so we can find v from the slope of the y–t curve in Figure 2.28. Notice that here we use y to measure the position, taking the place of x in previous examples. Acceleration is the slope of the velocity–time curve, so we can find the behavior of a once we have the behavior of the velocity. We can then find the total force on the ball through Newton's second law, $\sum \vec{F} = m\vec{a}$.

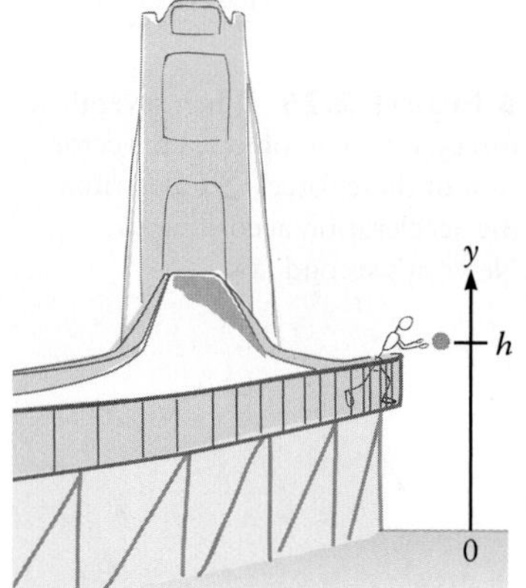

▲ **Figure 2.29** Example 2.8. A ball is dropped from a bridge, starting from rest ($v = 0$) at a height h above the ground. The motion of this ball is described by the y–t graph in Figure 2.28.

SKETCH THE PROBLEM

Figure 2.28 shows how the position of the ball varies with time, and Figure 2.29 shows a picture of the ball as it falls from the bridge. The motion of the ball is one-dimensional, falling directly downward from the bridge to the ground, and its position can be measured by its height y above the ground. Figure 2.29 also shows the y coordinate axis, with its origin at ground level.

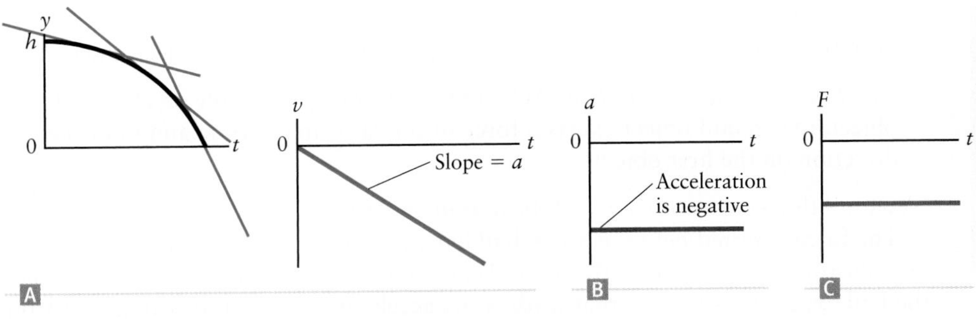

◀ **Figure 2.30** Example 2.8. **A** Position as a function of time for the falling ball from Figure 2.28. The slopes of the tangent lines give the ball's velocity at three instants in time. The velocity of the ball as a function of time is obtained from plotting these slopes. **B** The ball's acceleration is constant. **C** The force on the ball is proportional to the ball's acceleration.

IDENTIFY THE RELATIONSHIPS AND SOLVE

(**a**) Velocity is the change in position per unit time, so the value of v at any particular time is the slope of the y–t graph at that instant. We can obtain this slope graphically by following the approach in Example 2.3 and drawing lines tangent to the y–t curve at various times in Figure 2.30A. Using estimates for the slopes of these lines leads to the qualitative velocity–time graph shown alongside the y–t curve. Note that the velocity is always negative since y decreases monotonically with time. Also, the magnitude of the velocity, which is the speed, becomes larger and larger as the ball falls. (The plot here shows v as a function of time up until just before the ball reaches the ground.)

(**b**) Acceleration is the slope of the v–t curve; that slope is constant in Figure 2.30A, leading to the qualitative acceleration–time graph in Figure 2.30B. The acceleration is negative because the value of v is decreasing (the value of v becomes more negative with time). The magnitude of a is approximately constant during this period.

(**c**) To compute the force on the ball, we use Newton's second law. According to Equation 2.8, the total force is proportional to the acceleration. Even if we do not know the sources of all the forces on the ball, we can still compute the total force from Newton's second law. We can rearrange Newton's second law (Eq. 2.8) as $\sum \vec{F} = m\vec{a}$ and thus arrive at the qualitative force–time graph in Figure 2.30C.

▶ ***What does it mean?***

Given the behavior of the position as a function of time, we can deduce (by estimating slopes) the qualitative behavior of both the velocity and the acceleration as functions of time. The behavior of the force can then be found by using Newton's second law. For this falling ball, the total force is negative (Fig. 2.30C), meaning that the force is directed downward, along the $-y$ direction. This force is just the gravitational force acting on the ball.

After leaving the bow, $\vec{v}$ is upward while the total force on the arrow and the acceleration of the arrow are both downward.

▲ **Figure 2.31** As the arrow travels upward, its velocity is upward but the total force on the arrow is downward, so $\vec{v}$ and $\sum \vec{F}$ are in opposite directions.

Newton's Second Law and the Directions of $\vec{v}$ and $\vec{a}$

Newton's second law can be written as $\sum \vec{F} = m\vec{a}$, so the acceleration of an object $\vec{a}$ is always parallel to the total force $\sum \vec{F}$ acting on the object. However, *the velocity and total force need not be in the same direction*. For example, when an object such as the arrow in Figure 2.31 is fired upward, its velocity is "upward," along the $+y$ direction in the figure. The total force on this object is due almost entirely to gravity (which we'll discuss more in later chapters) and is downward, along $-y$. Hence, in this case, $\vec{v}$ and $\sum \vec{F}$ are in *opposite* directions. While the total force and acceleration are *always* parallel, the direction of the velocity can be different.

Newton's Third Law

The term $\sum \vec{F}$ appearing in Newton's second law is the total force acting on the object whose motion we are studying or calculating. This force must come from somewhere; in fact, it is *always* produced by other objects. For example, when a

A

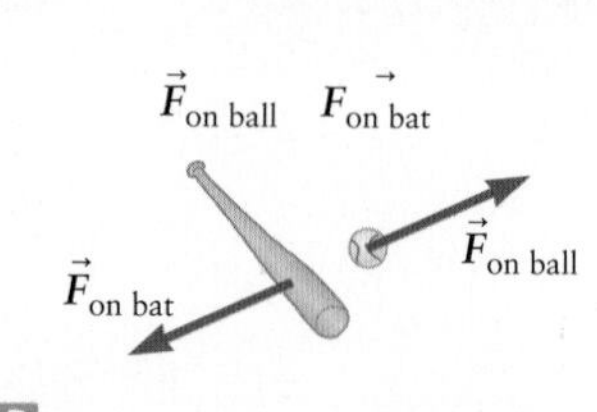

B

▲ **Figure 2.32** A When a bat strikes a baseball, the bat exerts a large force on the ball. B According to Newton's third law, the ball exerts a reaction force—a force of equal magnitude and opposite direction—on the bat.

baseball is struck by a bat, the force on the ball is due to the action of the bat. Newton's third law is a statement about what happens to the "other" object, the bat.

> ***Newton's third law of motion:*** **When one object exerts a force on a second object, the second object exerts a force of the same magnitude and opposite direction on the first object.**

Newton's third law is often called the ***action–reaction principle.***

The forces exerted between a baseball bat and a ball (Fig. 2.32) provide an excellent illustration of Newton's third law. When the bat is in contact with the baseball, the ball experiences a force that leads to its acceleration, as can be calculated from Newton's second law. At the same time, the bat experiences a force acting on it that comes from the ball. You may have already encountered this force; it is most noticeable when you hit the ball a little away from the "sweet spot" of the bat. The point of Newton's third law is that forces *always* come in such pairs. The force exerted by the bat on the ball and the force exerted by the ball acting back on the bat are known as an action–reaction pair of forces. According to Newton's third law, these two forces are always *equal in magnitude* and *opposite in direction*, and they must act on *different* objects.

Which Law Do We Use?

Let's now look ahead to how we will actually use Newton's laws to calculate the motion of various objects. Newton's second law tells us how to calculate an object's acceleration. For example, if we want to study the motion of a baseball, we need to calculate the total force on the ball and then insert this $\sum \vec{F}$ into Newton's second law (Eq. 2.8) to find the ball's acceleration. This total force will often have contributions from several sources; for a baseball, there may be a force from the impact with a bat along with other forces. The main points are that the total force $\sum \vec{F}$ in Equation 2.8 is the total force on *just the ball* and that Newton's second law enables us to calculate the acceleration of *just the ball.* On the other hand, Newton's third law tells us that forces always come in pairs. Hence, if we were somehow able to calculate or measure the force exerted by a baseball bat on a ball, Newton's third law tells us that there must be a corresponding reaction force that acts *on the bat*. It is extremely useful to know about this reaction force since it will contribute to the total force *on the bat*.

Newton's three laws of motion are the foundation for nearly everything that we do in the first part of this book. In this section, we have given some background for each of the three laws so that you can appreciate what they tell us about the nature of motion. We'll show in subsequent chapters that Newton's laws contain many other ideas and concepts, such as energy and momentum, that are essential to our thinking about the physical world.

The forces in an action–reaction pair act on different objects.

EXAMPLE 2.9 Action–Reaction

A person pushes a refrigerator across the floor of a room. The person exerts a force $\vec{F}_1$ on the refrigerator. From Newton's third law, we know that $\vec{F}_1$ is part of an action–reaction pair of forces. What is the reaction force to $\vec{F}_1$?

RECOGNIZE THE PRINCIPLE

The force on the refrigerator $\vec{F}_1$ is caused by the action of the person *on the refrigerator*. According to Newton's third law, the reaction force must be equal in magnitude to $\vec{F}_1$ but in the opposite direction. The reaction force must also act on a different object (it cannot act on the refrigerator).

SKETCH THE PROBLEM

Figure 2.33 shows the problem.

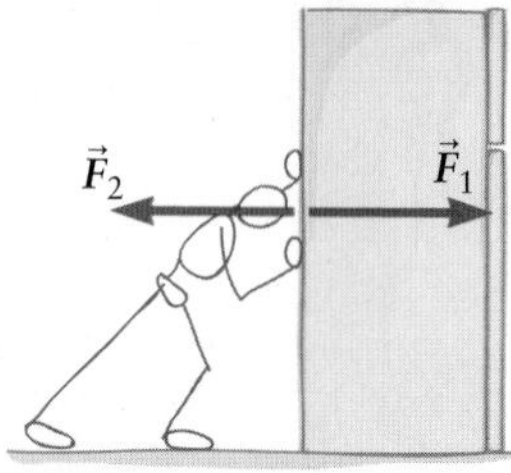

▲ **Figure 2.33** Example 2.9. Newton's third law, the action–reaction principle, applies to people and refrigerators, too. Here the person exerts a force $\vec{F}_1$ on the refrigerator while the refrigerator exerts a force $\vec{F}_2$ on the person.

IDENTIFY THE RELATIONSHIPS AND SOLVE

The reaction force $\vec{F}_2$ is the action of the refrigerator *on the person* as shown in Figure 2.33. According to Newton's third law, the forces in an action–reaction pair have equal magnitudes and are in opposite directions, so $\vec{F}_2 = -\vec{F}_1$.

▶ ***What does it mean?***

Forces *always* come in action–reaction pairs. The two forces in an action–reaction pair always act on *different* objects.

CONCEPT CHECK 2.6 Action–Reaction Force Pairs

Which of the following is *not* an action–reaction pair of forces? (More than one answer may be correct.)

(a) The force exerted by a pitcher on a baseball and the force exerted by the ball when it hits the bat
(b) When you lean against a wall, the force exerted by your hands on the wall and the force exerted by the wall on your hands
(c) In Figure 2.32, the force exerted by the ball on the bat and the force exerted by the bat on the player's hands

2.5 Why Did It Take Newton to Discover Newton's Laws?

Newton's second law tells us that the acceleration of an object is given by $\vec{a} = (\sum \vec{F})/m$, where $\sum \vec{F}$ is the total force acting on the object. In the simplest situations, there may only be one or two forces acting on an object, but in some cases there may be a very large number of forces acting on an object. Multiple forces can make things appear to be very complicated, which is perhaps why the correct laws of motion—Newton's laws—were not discovered sooner.

Forces on a Swimming Bacterium

Figure 2.34 shows a photo of the single-celled bacterium *Escherichia coli*, usually referred to as *E. coli*. An individual *E. coli* propels itself by moving thin strands of protein that extend away from its body (rather like a tail) called flagella. Most *E. coli* possess several flagella as in the photo in Figure 2.34A, but to understand their function, we consider a single flagellum as sketched in Figure 2.34B. A flagellum is fairly rigid, and because it has a spiral shape, one can think of it as a small propeller. An *E. coli* bacterium moves about by rotating this propeller, thereby exerting a force $\vec{F}_w$ on the nearby water. According to Newton's third law, the water exerts a force $\vec{F}_E$ of equal magnitude and opposite direction on the *E. coli* as sketched in Figure 2.34B. One might be tempted to apply Newton's second law with the force $\vec{F}_E$ and conclude that the *E. coli* will move with an acceleration that is proportional to this force. However, this conclusion is incorrect because we have not included the forces from the water on the body of the *E. coli*. These forces are also indicated in Figure 2.34B; to properly describe the total force from the water, we must draw in many force vectors, pushing the *E. coli* in virtually all directions. At the molecular level, we can understand these forces as follows. Water is composed of molecules that are in constant motion and bombard the *E. coli* from all sides. Each time a water molecule collides with the *E. coli*, the molecule exerts a force on the bacterium, much like the collision of the baseball and bat in Figure 2.32. As we saw in that case, the two colliding objects both experience a recoil force, another example of action–reaction forces; hence, the *E. coli* and the water molecule exert forces on each other. An individual *E. coli* is not very large, but a water molecule is much smaller than the bacterium, and the force from one such collision will have only a small effect on the

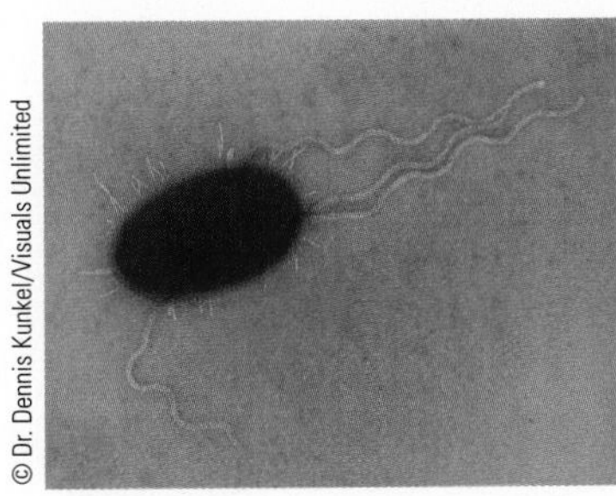

A

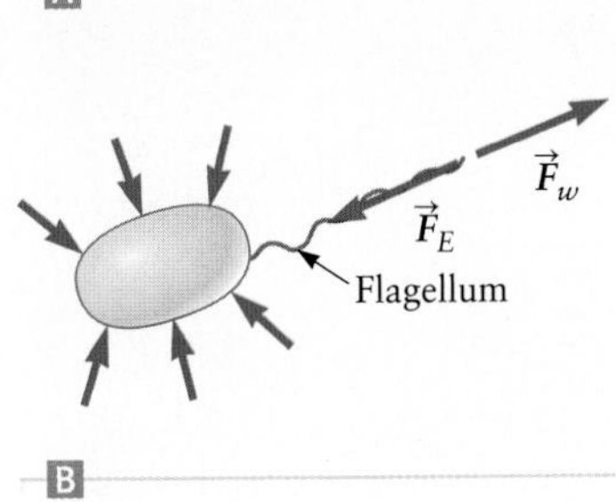

B

▲ **Figure 2.34** A *E. coli* use the action–reaction principle to propel themselves. An individual *E. coli* bacterium is a few micrometers in diameter (1 μm = 0.000001 m). B The flagellum exerts a force $\vec{F}_w$ on the water, and the water exerts a force $\vec{F}_E = -\vec{F}_w$ on the *E. coli*. There are also forces from the water molecules (indicated by the other red arrows) that act on the cell body.

E. coli. However, because there are many water molecules and many such collisions, the sum of the forces from all the water molecule collisions is quite substantial.

Suppose the bacterium is moving toward the left in Figure 2.34B. There will then be more collisions with the water molecules ahead of it (on the left) than with those behind it, resulting in a larger collision force on the bacterium from the left than from the right. This extra force on the front edge of the bacterium is the same type of force you feel when you try to move your body while swimming through water. This resistive force, due to collisions with the water molecules, brings the *E. coli* to a stop when the flagellum stops rotating.

This example with *E. coli* emphasizes again that the force in Newton's second law is the *total* force on the object. It is essential that we account for *all* the forces on an object when applying Newton's second law. Accounting for and calculating all the collision forces on an *E. coli* bacterium is a very complex problem. In fact, we'll learn about other ways to deal with the motion of an *E. coli* and similar problems when we consider the process of *diffusion* in Chapter 15.

2.6 Thinking about the Laws of Nature

In Chapter 1, we described in general terms how ideas and theories can eventually lead to the discovery of a "law" of physics. Let's now discuss this process in a little more detail and consider how Newton's laws came to be.

Discovery of a New Law of Physics

During his lifetime and for many years after, Aristotle's ideas were, in a sense, laws of physics. As we have seen, however, these ideas were not able to explain some important observations, which ultimately led to Galileo's work on the principle of inertia and to Newton. Although we'll never know for sure how Newton arrived at his discoveries, it seems likely that he developed his ideas by first hypothesizing his laws of motion. He then tested his ideas by comparing their predictions with the behavior observed in the experiments of Galileo and others. When Newton found discrepancies, he would rework his theories until eventually they were able to describe correctly the motion of everything that had been studied up to that time. Newton also found that his theories were able to explain the motion of the Moon and the planets. This finding was a tremendous result because prior to Newton there was really no satisfactory theory or explanation of celestial motion. After thus showing that his theory could successfully predict the motion of a wide variety of objects under a wide range of conditions, Newton proposed that they are true "laws" of nature. Other scientists then used Newton's ideas to predict motion in situations that had not yet been studied and tested these predictions with new experiments. When Newton's theories passed these tests, they were accepted as "the" laws of physics, replacing what had come before.

After Newton, What Next?

After Newton's laws were accepted as the "true" laws of motion, what then? Newton's laws are limited to describing how things move. Although this is a broad area of physics, many problems concerning matter and energy are not covered by Newton's laws. Such problems include the behavior of light and electricity, and we require additional theories and laws to describe these phenomena. Also, while we may accept Newton's three laws of motion as the laws of physics that describe mechanics, it is important to continue testing these laws in new situations. In the vast majority of cases, Newton's laws have passed these tests, but around the beginning of the 20th century it was found that Newton's laws do not correctly predict the behavior of electrons and protons, and of atoms and molecules. This monumental discovery eventually led to the development of new laws of physics known as quantum mechanics. What, then, happened to Newton's laws? They were not discarded; instead, physicists realized that Newton's ideas work extremely well for describing motion in what is now known as the ***classical*** regime. This regime includes terrestrial-scale objects, such as rocks and cars, and it even extends to *E. coli* and to planets. Newton's laws break down in the ***quantum*** regime of electrons, protons, and atoms, however, and in that regime a different law of nature—quantum mechanics—must be used.

If Newton's laws break down when applied in certain regimes or to certain types of objects, are we still justified in referring to them as laws of nature? One could take the point of view that a law of nature must always hold true and apply to all situations, with no exceptions. Unfortunately, *no* current law of physics passes this test! All the presently known laws of physics are known to fail or to be inadequate in some regime or another. A more widely accepted viewpoint is that a law of physics must correctly describe all behavior in a particular regime of nature. It is believed that Newton's laws provide an accurate description of all motion in the regime of classical physics, which includes terrestrial objects and extends to things like planets, the Sun, and individual cells.

Although the process of testing and retesting the laws of physics may lead to the discovery of new laws, it is usually found that the law being tested works fine. Such testing, however, can still lead to a better understanding of physics and can reveal important new insights. For example, Newton's three laws do not mention anything about the notion of ***energy***. Nevertheless, we'll see that the concept of energy and many other useful ideas are contained within Newton's laws. Applying Newton's laws to new situations can help us discover such unanticipated results.

Summary | CHAPTER 2

Key Concepts and Principles

Motion

A complete description of an object's motion involves its ***position***, ***velocity***, and ***acceleration***. For motion in one dimension (along a line), position can be specified by a single quantity, x. The ***instantaneous velocity*** is related to changes in position by

$$v = \lim_{\Delta t \to 0} \frac{\Delta x}{\Delta t} \qquad \text{(2.5) (page 29)}$$

The instantaneous velocity is usually referred to as simply the "velocity."

The ***instantaneous acceleration*** is given by

$$a = \lim_{\Delta t \to 0} \frac{\Delta v}{\Delta t} \qquad \text{(2.7) (page 32)}$$

and is usually referred to simply as the "acceleration."

Position, velocity, and acceleration are vector quantities. For motion in two or three dimensions, we must consider how the directions of these quantities relate to our chosen coordinate axes. This topic is explored further in Chapter 4.

Newton's laws of motion

The connection between ***force*** and motion is at the heart of ***mechanics***. A force is a push or a pull. Force is a vector. Given the forces acting on an object, we can calculate how it will move using Newton's three laws of motion.

- ***Newton's first law***: If the total force acting on an object is zero, the object will move with a constant velocity.
- ***Newton's second law***:

$$\vec{a} = \frac{\sum \vec{F}}{m} \qquad \text{(2.8) (page 39)}$$

Forces thus cause acceleration.

(Continued)

- *Newton's third law*: For every action (force), there is a reaction (force) of equal magnitude and opposite direction.

All forces come in ***action–reaction pairs.*** The two forces in an action–reaction pair act on different objects.

Applications

The relationships between position, velocity, and acceleration

The ***instantaneous velocity*** at a particular time t is the slope of the position–time curve at that time (Fig. 2.35).

The ***average velocity*** during a particular time interval is equal to the slope of the line connecting the start and end of that interval on the position–time curve (Fig. 2.35).

The ***instantaneous acceleration*** equals the slope of a plot of the instantaneous velocity v as a function of time (Fig. 2.36).

The ***average acceleration*** during a time interval is equal to the slope of the line connecting the start and end of that interval on the velocity–time curve (Fig. 2.36).

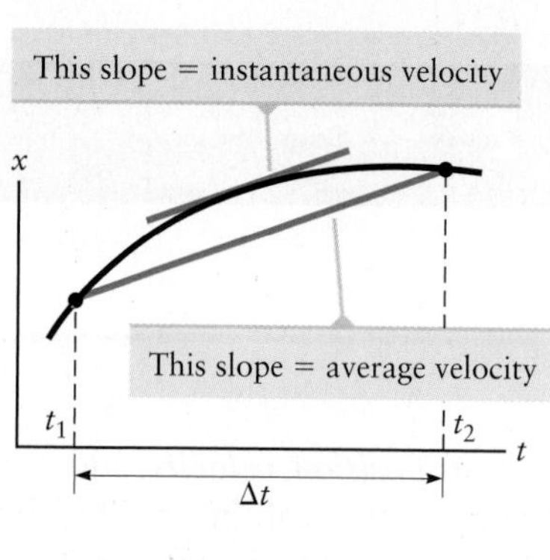

▲ **Figure 2.35**

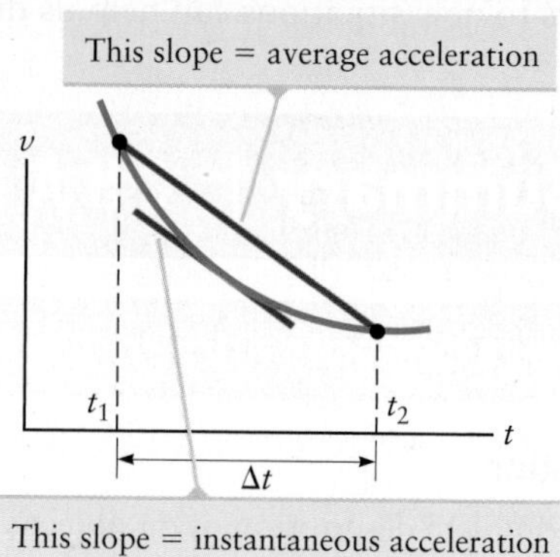

▲ **Figure 2.36**

Questions

SSM = answer in *Student Companion & Problem-Solving Guide*   ✕ = life science application

1. Make a hypothetical sketch of a velocity–time graph in which the velocity is always positive, but the acceleration is always negative. Give a physical example of such motion.
2. A ball is thrown into the air with an upward velocity of 10 m/s. A short time later, it is caught on its way down, also with a speed of 10 m/s.
   (a) Draw the velocity–time graph for this situation.
   (b) Draw the acceleration–time graph for this situation.
   (c) Many people would describe the ball's motion as "slowing down" until it reaches the top of its flight and then "speeding up" on the downward trip. However, a correct graph in part (b) should show a consistently negative (downward) acceleration. Explain how a negative acceleration can mean both "speeding up" and "slowing down."
3. When a car collides with a wall, a force on the car causes it to stop. Identify an action–reaction pair of forces involving the car.
4. You push on a refrigerator, as in Figure 2.1, but the refrigerator does not move. Hence, even though you are applying a nonzero force, the acceleration is still zero. Explain why this does not contradict Newton's second law.
5. A car is traveling on an icy road that is extremely slippery. The driver finds that she is not able to stop or turn the car. Explain this situation in terms of the principle of inertia (Newton's first law).
6. A car is initially at rest on an icy road that is extremely slippery. The driver finds that he is unable to get the car to drive away because the wheels simply spin when he tries to accelerate. Explain this situation in terms of the principle of inertia (Newton's first law).
7. Give an example of motion for which the average velocity is zero, but the speed is never zero.

8. Give an example of motion for which the average velocity is equal to the instantaneous velocity. *Hint*: Sketch the velocity–time graph.

9. SSM **Abracadabra!** A magician pulls a tablecloth off of a set table with one swift, graceful motion. Amazingly, the fine china, glassware, and silverware are practically undisturbed. Although amazing, this feat is not an illusion. Describe the behavior of the plates and glasses in terms of the principle of inertia (Newton's first law).

10. Give examples of motion matching the following descriptions. (a) The velocity is positive and the acceleration is positive. (b) The velocity is negative and the acceleration is negative. (c) The velocity is positive and the acceleration is negative.

11. A car starts from the origin at $t = 0$. At some later time, is it possible for the car's velocity to be positive but its displacement from the origin to be zero? Explain and give an example.

12. The acceleration of an object that falls freely under the action of gravity near the Earth's surface is negative and constant. (a) Does the object's instantaneous acceleration equal its average acceleration? (b) Draw the corresponding velocity–time graph. (c) Does the instantaneous velocity equal the average velocity? Explain why or why not.

13. Consider again Example 2.8. Draw plots of the position, velocity, and acceleration as functions of time, starting from when the ball is released and ending *after* the ball hits the ground. Indicate the time at which the ball hits the ground.

14. Make a qualitative sketch of the position y as a function of time for the center of a yo-yo (the point at the middle of the axle). Also make sketches of the velocity and acceleration as functions of time. Is the total force on the yo-yo zero or nonzero? Explain how you can tell from your graphs.

15. SSM Figure Q2.15 shows a motion diagram for a rocket-powered car. The photos are taken at 1.0-s intervals. Make qualitative plots of the position, velocity, acceleration, and force on the car as functions of time.

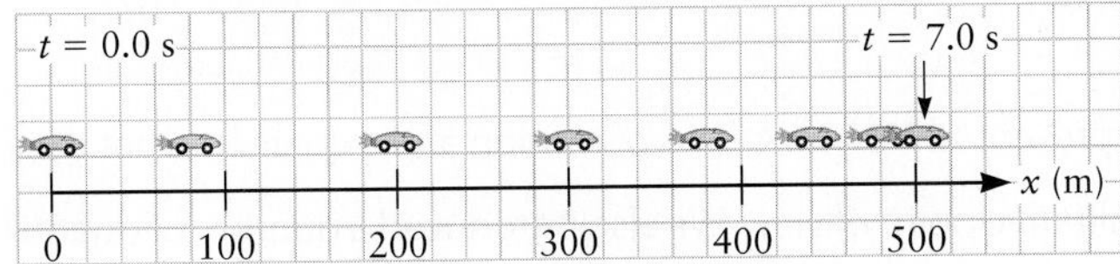

Figure Q2.15

16. A person stands on level ground and throws a baseball straight upward, into the air. (a) Does the person exert a force on the ball while it is in his hand? After it leaves his hand? (b) Does the ball ever exert a force on the person? If so, what is the direction of this force? (c) If your answer to part (b) is yes, why does the person not accelerate?

17. Consider the motion of the Moon as it orbits the Earth. (a) Is the Moon's acceleration zero or nonzero? Explain. (b) If the Moon has a nonzero acceleration, what force is responsible?

18. Consider the motion of a marble as it falls to the bottom of a jar of honey. Experiments show (see also Chapter 3) that the marble moves with a constant velocity. Applying Newton's first law, does that mean that no forces are acting on the marble?

19. According to Newton's first law (the principle of inertia), if there is no force exerted on an object, the object will move with constant velocity. Consider the following examples of motion. Is the velocity of the object constant? What are the forces acting on each object? In cases in which the velocity is constant, explain why Newton's first law applies.
(a) A hockey puck sliding on an ice-covered surface that is horizontal and frictionless
(b) A mug of root beer sliding down a bar, whose surface is slippery but not frictionless
(c) A car skidding on a flat, desert highway

20. Three blocks rest on a table as shown in Figure Q2.20. Identify three action–reaction pairs of forces.

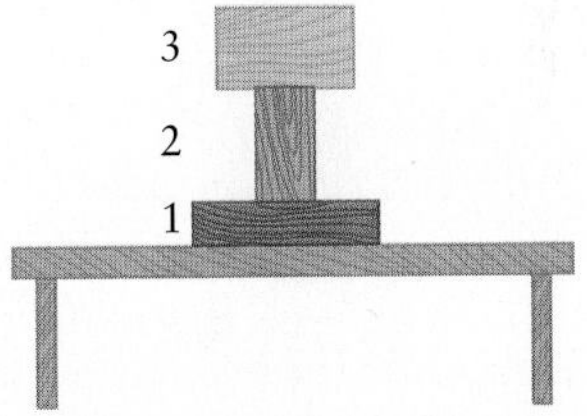

Figure Q2.20

21. Two football players start running at opposite ends of a football field (opposite goal lines), run toward each other, and then collide at the center of the field. They start from rest and are running at top speed when they collide. (a) Draw a graph showing the position as a function of time for both players. (b) Draw a graph showing their velocities as functions of time.

22. A person is riding in a car traveling on a straight level road. What is the direction of the *net* force on the person if the car is (a) speeding up, (b) slowing down, (c) has a constant speed?

Problems available in ENHANCED WebAssign

Section 2.2 What Is Motion?
Problems 1–37

Section 2.3 The Principle of Inertia
Problems 38–41

Section 2.4 Newton's Laws of Motion
Problems 42–50

Additional Problems
Problems 51–60

List of Enhanced Problems

Problem number	Solution in *Student Companion & Problem-Solving Guide*	Reasoning & Relationships Problem	Reasoning Tutorial in ENHANCED WebAssign
2.11		✓	
2.13	✓		
2.33	✓		
2.38	✓		
2.49	✓		
2.52	✓		
2.54	✓	✓	✓

CHAPTER

3

▸ *These railroad tracks carry trains across the desert in southern California. They are very straight and very long, so the motion of the trains they carry is indeed one-dimensional. These tracks can be viewed as a coordinate axis complete with "tick marks"! (© Richard T. Norwitz/National Geographic/ Getty Images)*

Forces and Motion in One Dimension

OUTLINE

In Chapter 2, we presented Newton's laws of motion and explained what they mean from a qualitative point of view. We also introduced a number of quantities, including displacement, velocity, and acceleration, that are essential for describing motion. Our next job is to apply Newton's laws to calculate how things move in various situations. In this chapter, we consider motion in one dimension, that is, motion along a straight line. You might wonder why we devote an entire chapter to such an idealized case; after all, the world is not one dimensional, and in most situations, objects move along two- or three-dimensional trajectories. Even so, it is very useful to start with one-dimensional examples. The mathematics is simpler in this case, and the approaches we develop for dealing with motion in one dimension can be applied quite directly to motion in higher dimensions.

3.1 Motion of a Spacecraft in Interstellar Space

Figure 3.1 shows a hypothetical spacecraft traveling somewhere in distant space, traveling along a straight-line path from one galaxy to another. Since the spacecraft is moving along a line, this is an example of one-dimensional motion. To describe this motion, we must first choose a coordinate system with which to measure the displacement, velocity, and acceleration of the spacecraft. The spacecraft's motion is along the line that connects the two galaxies, so we choose that line to be our coordinate axis. We have drawn this axis in Figure 3.1 and labeled it the x axis. We have also placed the origin at the starting galaxy and taken the "positive" direction along the x axis as headed toward the destination.

If our spacecraft is very far from any planets or stars, the force of gravity on it will be very small. To make this example as simple as possible, let's assume this gravitational force is zero. We now want to calculate how the spacecraft will move in two different cases.

In the first case, the engine is turned off so that the spacecraft is "coasting." Since the engine is off and the gravitational force is zero, there are no forces acting on the spacecraft. Newton's first law tells us that when the total force acting on an object is zero, the object has a constant velocity. Hence, our spacecraft moves with a constant velocity; that is, it moves in a straight-line path with a constant speed. Examples of such constant-velocity motion are given in Figures 3.2 and 3.3, which show graphs of how the velocity and position might look as functions of time. We learned in Chapter 2 that displacement and velocity are vectors, having both magnitude and direction. For the cases involving one-dimensional motion that we consider here and throughout this chapter, these vectors all lie along the chosen coordinate axis. In general, we could write these quantities using our usual vector notation with vector arrows, but with one-dimensional motion, we can simplify the notation and drop the vector arrows because we specify direction by the sign of each quantity (+ or −). This notation corresponds to specifying the *components* of the displacement and velocity along the coordinate axis.

Figure 3.2 shows the behavior of the velocity and position of our spacecraft when the velocity is zero, while Figure 3.3 shows the behavior for a constant nonzero velocity. Recall that v is the slope of the position–time curve; in both Figures 3.2 and 3.3, the velocity is constant, so the x–t relations are both linear. Although Newton's first law tells us that the velocity is constant, it does not tell us the value of v. This value is determined by forces applied to the spacecraft at earlier times. In this problem, a force might have been applied to the spacecraft before the engine was turned off, giving a nonzero acceleration prior to $t = 0$ and causing the velocity in Figure 3.3 to be nonzero.

Insight 3.1

ALWAYS START WITH A PICTURE

The first step in attacking any problem involving motion is to draw a picture. This picture should always show the coordinate axes you have chosen to use for the problem.

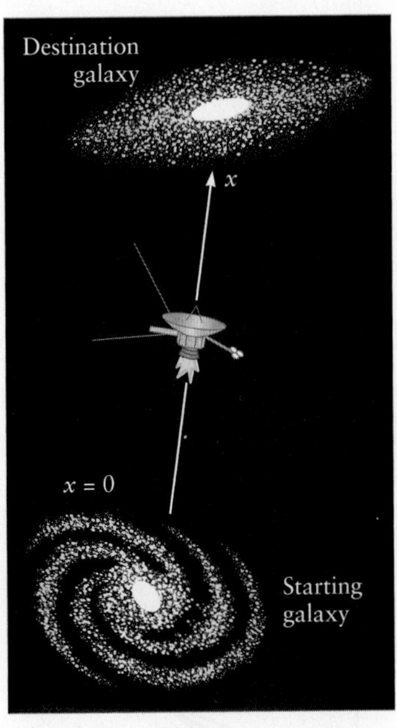

▲ **Figure 3.1** This hypothetical spacecraft is following a one-dimensional trajectory along the x axis.

Motion with a Constant Nonzero Acceleration

Let's next consider what happens when the spacecraft's engine is turned on. There is now a force exerted on the spacecraft, so we need to use Newton's second law to

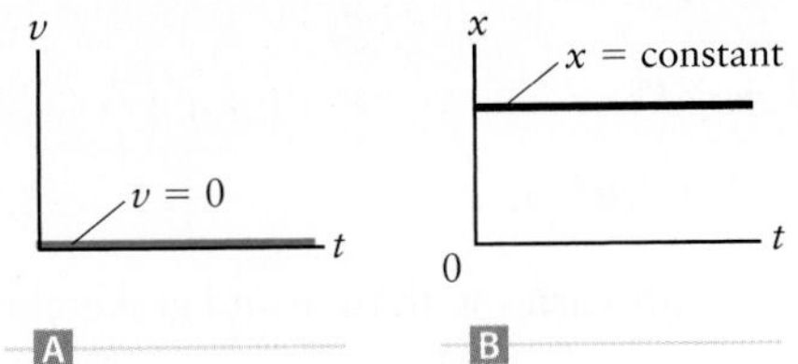

▲ **Figure 3.2** If the spacecraft's velocity is zero, its position x is a constant. This is one example of motion with a constant velocity.

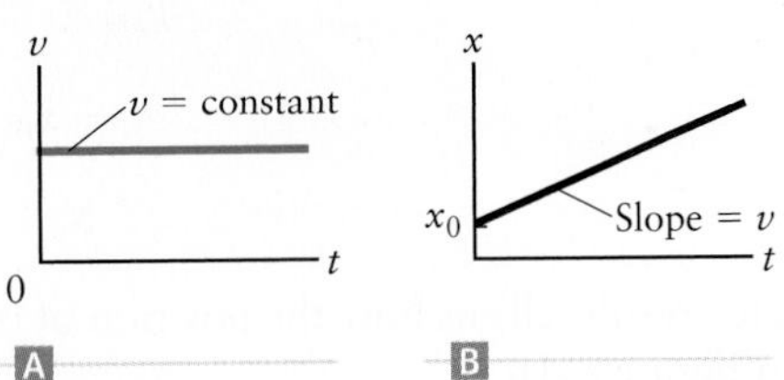

▲ **Figure 3.3** Motion with a constant and nonzero velocity. Since v is equal to the slope of the position–time curve, the graph of x as a function of t is a straight line.

determine the resulting motion. Recall that according to Newton's second law the acceleration is given by $a = \sum F/m$. For our case of one-dimensional motion, both the total force $\sum F$ and the acceleration a are parallel to the x axis in Figure 3.1 (so they are the components of the force and acceleration along x). In this example, there is only a single force (due to the engine); in the simplest case, the engine will be designed to produce a constant force on the spacecraft, so the force term $\sum F$ in Newton's second law is a constant. For simplicity, we also assume the mass m of the spacecraft is constant.[1] We can then use Newton's second law to calculate the acceleration and find that a is also a constant. Since acceleration is the slope of the velocity–time curve, the v–t relation must be a *straight line*. This relationship is shown in Figure 3.4A; we have

Velocity versus time for motion with a constant acceleration

$$v = v_0 + at \tag{3.1}$$

In Equation 3.1, v_0 is the "initial" velocity of the spacecraft, that is, the velocity at $t = 0$; it is useful to think of v_0 as the velocity at the moment the stopwatch used to describe this problem is started. The value of v_0 depends on what happened prior to $t = 0$; for example, one might imagine that the engine was turned on for a while to launch the spacecraft and then turned off before $t = 0$. In most situations (and problems!), we must rely on someone else (such as the person who writes the problem or the pilot of the spacecraft) to provide the value of v_0.

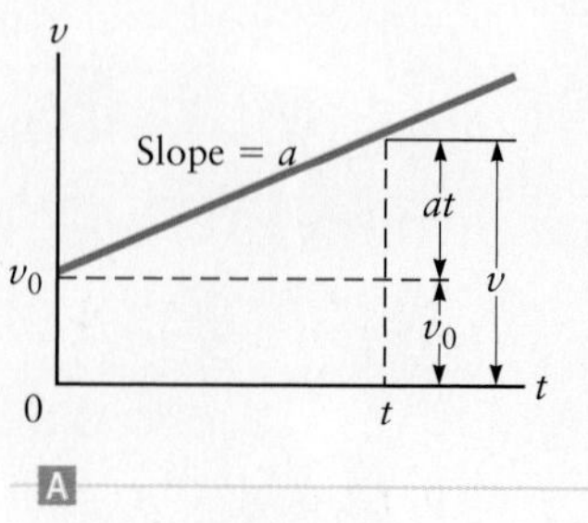

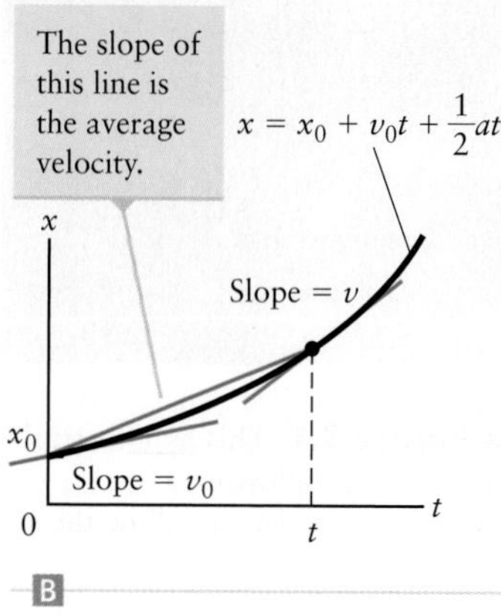

▲ Figure 3.4 A The slope of the velocity–time graph is equal to the acceleration, so when the acceleration a is constant, the v–t plot is a straight line. B The corresponding graph of position versus time. The slope of the x–t plot at a certain point on the curve is equal to the instantaneous velocity v.

To give a complete description of the spacecraft's motion, we must also derive its position as a function of time. We already have the velocity as a function of time from Equation 3.1, and we know that v is the slope of the x–t curve. To derive the x–t relationship we use the approach shown in Figure 3.4B. In Chapter 2, we learned that the instantaneous velocity v is the slope of the x–t curve at a particular point on the curve, whereas the average velocity over a particular time interval is found by taking the slope of the line that spans the interval. One such time interval is shown in Figure 3.4B. The slope of the line connecting $t = 0$ with some general time t is

$$v_{ave} = \frac{x - x_0}{t}$$

where x_0 is the value of x at $t = 0$ (the "initial" position). We can also calculate the average velocity from Equation 3.1; since v varies linearly with time, the average velocity during a particular time interval is the average of the instantaneous velocities at the start and end of the interval. Hence,

$$v_{ave} = \frac{v(t = 0) + v(t)}{2}$$

We can now insert our result for $v(t)$ from Equation 3.1:

$$v_{ave} = \frac{v(t = 0) + v(t)}{2} = \frac{v_0 + v_0 + at}{2}$$

$$v_{ave} = v_0 + \tfrac{1}{2}at \tag{3.2}$$

Equating this expression to our previous result $v_{ave} = (x - x_0)/t$ and doing a little rearranging leads to

$$v_{ave} = \frac{x - x_0}{t} = v_0 + \tfrac{1}{2}at$$

$$x - x_0 = v_0t + \tfrac{1}{2}at^2$$

Position versus time for motion with a constant acceleration

$$x = x_0 + v_0t + \tfrac{1}{2}at^2 \tag{3.3}$$

This result tells us how the position of our spacecraft varies with time and is sketched in Figure 3.4B.

[1]Most rocket engines work by expelling gas, so the total mass of the spacecraft, including the engine and fuel, will generally decrease with time. Here, however, we ignore that effect.

Relations for Motion with Constant Acceleration

It is useful to recap how we used Newton's second law to analyze the motion of our spacecraft. The first step was to determine the total force on the object. For our spacecraft problem, this force was simply a given; for example, the engine designer might have told us the value. We then used Newton's second law $a = \sum F/m$ to calculate the acceleration a. Because the total force and m were both constants, a is also constant. Acceleration is the slope of the v–t relation; hence, the velocity is a linear function of t. The final step was to deduce the x–t relation; that analysis was sketched in Figure 3.4B and led to Equation 3.3. Keep in mind that these results for the position and velocity of an object with constant acceleration (Eqs. 3.3 and 3.1, respectively) are not "new" laws of physics; both are a direct result of Newton's second law.

It is rare to find situations in real life in which the acceleration is *precisely* constant, so Equations 3.1 and 3.3 will usually not give an exact description of the motion of an object. However, in many interesting cases the acceleration is very nearly constant, and these cases are well described by the constant-acceleration relations in Equations 3.1 and 3.3.

For future reference, Table 3.1 collects these results and also lists one more very useful relationship. Equations 3.1 and 3.3 give the velocity and position as functions of time. We can eliminate t from these two relations to get a single equation that relates x and v without direct reference to t. We first use Equation 3.1 to express t in terms of v and a:

$$v = v_0 + at$$

$$t = \frac{v - v_0}{a}$$

We next insert this result for t into Equation 3.3:

$$x = x_0 + v_0 t + \frac{1}{2}at^2 = x_0 + v_0\left(\frac{v - v_0}{a}\right) + \frac{1}{2}a\left(\frac{v - v_0}{a}\right)^2$$

Collecting and combining terms gives

$$x = x_0 + \frac{v_0 v}{a} - \frac{v_0^2}{a} + \frac{1}{2}\frac{v^2}{a} - \frac{v_0 v}{a} + \frac{1}{2}\frac{v_0^2}{a} = x_0 + \frac{1}{2a}(v^2 - v_0^2)$$

We can now rearrange to reach our desired result:

$$v^2 = v_0^2 + 2a(x - x_0) \tag{3.4}$$

which is also listed in Table 3.1. This relation is interesting because it does not contain time (t); it only involves position, velocity, and acceleration. We'll encounter many cases in which this result is very handy.

TABLE 3.1 Equations for Motion with Constant Acceleration

Equation Number	Mathematical Relation	Variables
3.3	$x = x_0 + v_0 t + \frac{1}{2}at^2$	position, acceleration, and time
3.1	$v = v_0 + at$	velocity, acceleration, and time
3.4	$v^2 = v_0^2 + 2a(x - x_0)$	position, velocity, and acceleration

Note: In cases in which the position variable in a problem is y, the same equations apply with y inserted in place of x.

Relations between x, v, a, and t for motion with a constant acceleration

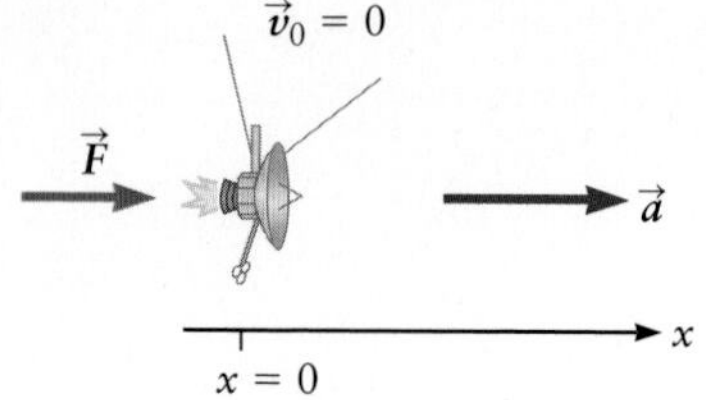

▲ **Figure 3.5** Example 3.1. Sketch of a spacecraft moving in response to a constant force F along the $+x$ direction.

Insight 3.2

WORKING WITH UNITS

In our calculations in Example 3.1, we canceled and combined units as reviewed in Chapter 1. When dealing with the units of force, it is simplest to express the newton in terms of the fundamental units of meters, kilograms, and seconds before canceling and combining units. (Recall that $1\text{ N} = 1\text{ kg}\cdot\text{m/s}^2$.) You should always include the units in your calculations and check that they correctly match the quantity being calculated.

EXAMPLE 3.1 Accelerating into Space

Let's see how the basic relations between x, v, and t in Table 3.1 can be used to calculate the motion of our spacecraft in a particular case. We assume the spacecraft has mass $m = 200$ kg and is initially at rest. The engine is turned on at $t = 0$, and from this time forward there is a constant force $F = 2000$ N on the spacecraft. (**a**) What is the velocity of the spacecraft at $t = 40$ s? (**b**) How far does the spacecraft travel during this time? (**c**) What is the velocity of the spacecraft when it reaches a distance 1000 m from where it started?

RECOGNIZE THE PRINCIPLE

The only force on the spacecraft is the force F from the engine, so this force is equal to $\sum F$ in Newton's second law. The force is constant; hence, the acceleration is also constant and we can find its value using Newton's second law. We can then use our relations for motion with constant acceleration from Table 3.1.

SKETCH THE PROBLEM

Figure 3.5 shows the spacecraft moving along the x direction, with a force F from the engine. The force and therefore also the acceleration are both along the $+x$ direction. We have also noted that the spacecraft is initially at rest (hence $v_0 = 0$) and have chosen the origin for the x axis to be the spacecraft's location when $t = 0$, so $x_0 = 0$.

IDENTIFY THE RELATIONSHIPS

Using Newton's second law along with the given values of the mass and the component of the force along x gives

$$a = \frac{\sum F}{m} = \frac{2000\text{ N}}{200\text{ kg}}$$

The unit of force is newtons (N), with $1\text{ N} = 1\text{ kg}\cdot\text{m/s}^2$. We thus have

$$a = 10\,\frac{\text{kg}\cdot\text{m/s}^2}{\text{kg}}$$

Canceling the unit of kilograms that appears in both the numerator and denominator we get

$$a = 10\text{ m/s}^2$$

We can now use our relations for motion with a constant acceleration from Table 3.1. For part (**a**) of the problem, we are given t and asked to find the velocity. Equation 3.1 involves v, t, and other variables whose values we already know, namely a and v_0, so we can use it to solve for v. Likewise, for parts (**b**) and (**c**) of this problem, we use the relations in Table 3.1 that contain the quantity we want to find, along with the quantities whose values we know.

SOLVE

(**a**) We compute the velocity at $t = 40$ s using Equation 3.1:

$$v = v_0 + at = 0 + (10\text{ m/s}^2)(40\text{ s}) = \boxed{400\text{ m/s}}$$

(**b**) We are given the value of t (= 40 s) and asked to find x. We use Equation 3.3 because it contains x, t, and other quantities we know, namely a and v_0. So,

$$x = x_0 + v_0 t + \tfrac{1}{2}at^2 = 0 + 0 + \tfrac{1}{2}(10\text{ m/s}^2)(40\text{ s})^2 = \boxed{8000\text{ m}}$$

(**c**) To deal with this part of the problem, we note that we are given the value of the position and asked to find the velocity; there is no mention of time. So, we can use Equation 3.4 in Table 3.1 to get

$$v^2 = v_0^2 + 2a(x - x_0) = 0 + 2(10\text{ m/s}^2)(1000\text{ m}) = 2.0 \times 10^4\text{ m}^2/\text{s}^2$$

$$v = \boxed{140\text{ m/s}}$$

▶ ***What does it mean?***
Problem solving with the relations for motion with constant acceleration involves one or more of Equations 3.1, 3.3, and 3.4. Choosing the appropriate relation depends on what quantities are known and what you wish to calculate. It is usually best to choose the equation that contains only one unknown quantity.

EXAMPLE 3.2 Car Chase

A car is traveling at a constant velocity $v = 20$ m/s on a long, straight road. Unfortunately, it is near a school where the speed limit is 10 m/s, and when the car passes a police officer, the officer immediately gives chase. If the police car starts from rest and has a constant acceleration $a_2 = 2.5$ m/s^2, how long does it take the officer to catch the speeder?

RECOGNIZE THE PRINCIPLE

Both cars are moving with a constant acceleration, so we can use the result for the displacement from Table 3.1, $x = x_0 + v_0t + \frac{1}{2}at^2$, for both. We need to find the value of t at which the displacements of the two cars are equal, indicating that the police car has caught the speeder.

SKETCH THE PROBLEM

Figure 3.6 shows the displacement of both cars, where the speeder is car 1 and the police car is car 2.

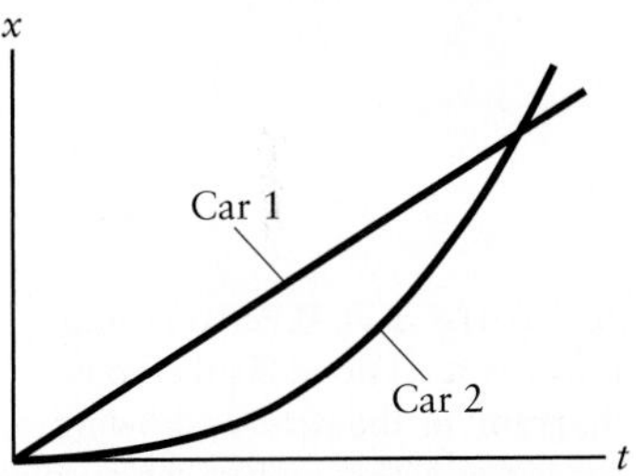

▲ **Figure 3.6** Example 3.2.

IDENTIFY THE RELATIONSHIPS

The acceleration of the speeder is zero ($a_1 = 0$), so it moves with a constant velocity equal to $v_{1,0} = 20$ m/s, while the police car (car 2) has zero initial velocity ($v_{2,0} = 0$) and a nonzero acceleration $a_2 = 2.5$ m/s^2. Both cars start at the origin ($x_{1,0} = x_{2,0} = 0$), so their displacements are

$$x_1 = x_{1,0} + v_{1,0}t + \tfrac{1}{2}a_1t^2 = v_{1,0}t \quad (1)$$

$$x_2 = x_{2,0} + v_{2,0}t + \tfrac{1}{2}a_2t^2 = \tfrac{1}{2}a_2t^2 \quad (2)$$

SOLVE

Setting the displacements in Equations (1) and (2) equal ($x_1 = x_2$) and solving for t, we get

$$v_{1,0}t = \tfrac{1}{2}a_2t^2 \quad (3)$$

$$t = \frac{2v_{1,0}}{a_2} = \frac{2(20 \text{ m/s})}{2.5 \text{ m/s}^2} = \boxed{16 \text{ s}}$$

▶ ***What does it mean?***
Equation (3) has another solution for t, namely $t = 0$, because car 1 initially passes the police car (car 2) at $t = 0$.

3.2 Normal Forces and Weight

We next consider some examples in which the force of gravity plays an important role. Figure 3.7A shows a person standing on the floor of a room. The person is standing still, so her velocity and acceleration are both zero. Since the acceleration is zero and, according to Newton's second law, $\sum \vec{F} = m\vec{a}$, the total force on the person is zero. This statement is true, but it is not the whole story.

In this case there are two forces acting on the person. One is the gravitational force exerted by the Earth; this force is called the ***weight*** of the person and is denoted in Figure 3.7 by the vector $\vec{F}_{grav}$. When an object of mass m is located near the surface of the Earth, the gravitational force on the object is directed downward (toward the Earth) and has a magnitude

Weight is the force of gravity exerted by the Earth on an object.

$$|\vec{F}_{grav}| = mg \tag{3.5}$$

We'll see in Chapter 5 how this result for the force of gravity near the Earth's surface is a consequence of Newton's law of universal gravitation. For now, we note that (1) the value of g is approximately the same for all locations near the surface of the Earth, with $g \approx 9.8 \text{ m/s}^2$; (2) because weight is due to the gravitational attraction of the Earth (or whichever planet or star on which the object is located), the weight of an object will be different if it is taken to another planet or to the Moon; (3) the value of g is independent of the mass of the object, so the weight of an object (Eq. 3.5) is proportional to its mass; and (4) for reasons we'll see shortly, g is commonly referred to as the ***acceleration due to gravity***.

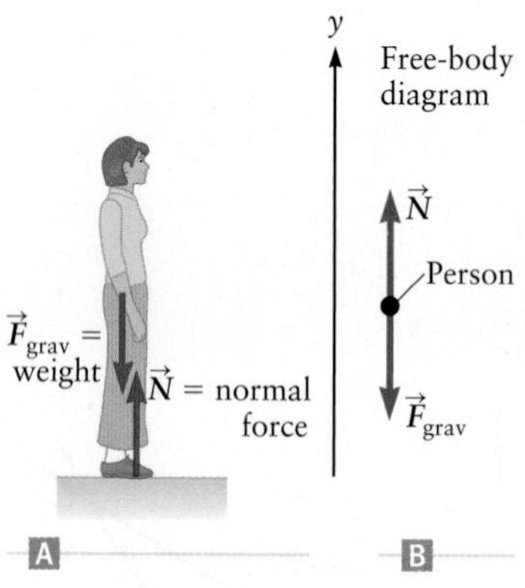

▲ **Figure 3.7** A Person standing on a level floor. B Free-body diagram for the person, showing all the forces acting on her: the normal force $\vec{N}$ and the force of gravity $\vec{F}_{grav}$. In a free-body diagram, the object is often denoted by a "dot."

The weight $\vec{F}_{grav}$ of an object is a force and can thus be measured in units of newtons. It is often convenient to use the coordinate system shown in Figure 3.7, where we follow common convention and label the vertical axis as y. The gravitational force lies along the vertical direction with a component along y given by

$$F_{grav} = -mg \tag{3.6}$$

The negative sign here indicates that the force is in the "negative" y direction, or *down*, toward the center of the Earth.

Besides the gravitational force, the other force acting on the person in Figure 3.7 is exerted by the floor on the bottoms of her feet. This force is called a ***normal force*** because it acts in a direction perpendicular ("normal") to the plane of contact with the floor. Normal forces occur whenever the surfaces of two objects come into contact. Here the normal force is directed upward, as indicated by the upward arrow labeled $\vec{N}$. The person in Figure 3.7 is at rest with an acceleration $a = 0$. Using Newton's second law for components of the force and acceleration along y, we get

$$\sum F = -mg + N = ma = 0$$

Hence, in this case $N = mg$. In this case, the normal force is equal in magnitude and opposite in direction to the person's weight.

CONCEPT CHECK 3.1 Where Is the Reaction?

There are two forces acting on the person in Figure 3.7A: the force of gravity $\vec{F}_{grav}$ and $\vec{N}$, the normal force from the floor. These forces are equal in magnitude and opposite in direction. Are they an action–reaction pair? Explain why or why not. If not, identify the reaction force to $\vec{N}$.

Free-Body Diagrams

Drawing a picture is an important part of the problem-solving process. This picture should include coordinate axes along with other information such as the values of forces and the initial velocity. In preparation for an analysis using Newton's second law, you should also construct a simplified diagram showing all the forces acting on each object involved in the problem. In Figure 3.7, these forces are the normal force and the force of gravity acting on the person. The diagram that shows them, Figure 3.7B, is called a ***free-body diagram***. The recommended procedures for constructing a free-body diagram are as follows.

Always draw a free-body diagram.

PROBLEM SOLVING **Constructing Free-Body Diagrams and Applying Newton's Laws**

1. **RECOGNIZE** the objects of interest. List all the forces acting on each one.

2. **SKETCH THE PROBLEM.**
   - Start with a drawing that shows all the objects of interest in the problem along with all the forces acting on each one, as in Figure 3.7A.
   - Make a separate sketch showing the forces acting on each object, which is the free-body diagram for that object. For clarity, you can represent each object by a simple dot.
   - Forces in a free-body diagram should be represented by arrows that show the direction of each force. In each free-body diagram, show only the forces acting on that particular object.

3. **IDENTIFY THE RELATIONSHIPS.** A force in a free-body diagram may be known (i.e., its direction and magnitude may be given), or it may be unknown. Represent unknown quantities by a symbol, such as $\vec{N}$ in Figure 3.7. Typically, equations derived from Newton's second law are used to solve for these unknowns.

4. **SOLVE.** The information contained in a free-body diagram can be used in writing Newton's second law for an object. A few algebraic steps will then lead to values for the unknown quantities.

5. Always *consider what your answer means* and check that it makes sense.

Acceleration and Apparent Weight

The normal force acting on the person in Figure 3.7 is equal in magnitude to her weight (mg). This result is often found for objects at rest on a level surface. However, in many situations the normal force on an object is not equal to its weight. Such a situation is sketched in Figure 3.8A, which shows a person standing in a moving elevator. There are again two forces acting on the person (Fig. 3.8B), the force of gravity (the person's weight) and the normal force exerted by the floor of the elevator on the bottoms of his feet. However, this situation is not the same as in our previous example because the elevator in Figure 3.7A has an acceleration $\vec{a}$, which is also the acceleration of the person inside.

Applying Newton's second law to the motion of the person gives

$$m\vec{a} = \sum \vec{F} = \vec{N} + \vec{F}_{grav}$$

where m is the person's mass. These vectors are along y, so we can write this expression in terms of the y components of the acceleration and force vectors:

$$ma = +N - mg \tag{3.7}$$

The signs here indicate the directions of the normal force and the force of gravity. As usual, these directions are defined by our choice of the coordinate axis (y) in Figure 3.8, which shows that the "positive" direction is upward. This choice for the positive direction also applies to the acceleration. Solving Equation 3.8 for the normal force, we find

$$N = mg + ma$$

Hence, if the acceleration is positive (directed upward in Fig. 3.8), the normal force is greater than the weight mg, while if the acceleration is negative (downward), the normal force is less than mg. In words, when you are in an elevator that is just starting to move to a higher floor, the acceleration is positive (directed upward) and you "feel" heavier. The force exerted by the floor of the elevator on the bottoms of your feet is greater than your true weight mg. Likewise, if you are in an elevator that is just beginning to move to a lower floor, your acceleration is negative (downward), so N is smaller than mg and you feel lighter.

The normal force acting on the bottoms of a person's feet as in Figure 3.8 is called the ***apparent weight***. Loosely speaking, the only way that you are sensitive to the value of your weight is through the normal force that supports you. Consider (hypothetically) what would happen if the elevator in Figure 3.8 were to malfunction

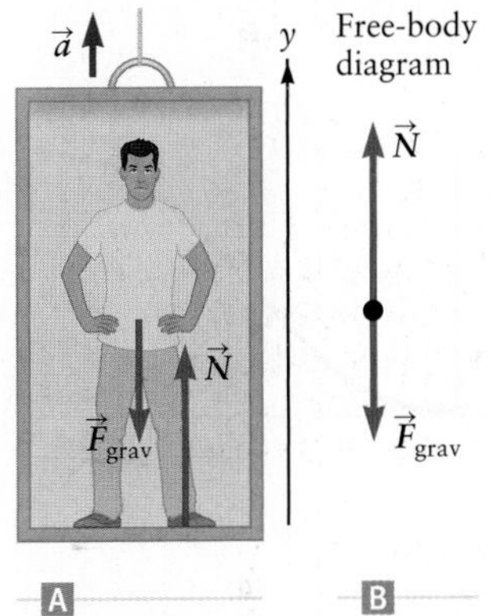

▲ **Figure 3.8** A When a person stands in an elevator, there are only two forces acting on him: $\vec{N}$ and $\vec{F}_{grav}$. B Free-body diagram for the person.

Insight 3.3
WEIGHT AND MASS
Weight and mass are closely related but they are not the same. For an object on or near the Earth's surface, the weight has a magnitude of $F_{grav} = mg$ (Eq. 3.5), so F_{grav} and m are proportional. However, weight and mass are fundamentally different quantities. The mass of an object is an intrinsic property of the object, whereas the weight of an object is a force whose magnitude varies depending on the location of the object. For example, the magnitude of your weight on Earth is different from its value on the Moon, but your mass is the same wherever you are located.

▲ **Figure 3.9** These young people are experiencing apparent "weightlessness" in a special NASA plane. Notice the padding.

and fall freely down the elevator shaft. This kind of arrangement is similar to the special airplane used by NASA to train astronauts (Fig. 3.9). That airplane flies in a long dive, during which time the plane and the astronauts fall together, so the normal force on the astronauts is zero. Their apparent weight is then zero, and the astronauts feel "weightless." During such a dive, their apparent weight is zero even though their true weight is not zero (and the force of gravity is also *not* zero).

All Forces Come from Interactions

Newton's third law implies that every force on an object arises from an interaction with a second object. This notion that forces come from interactions leads to the concept of an action–reaction pair of forces. Newton also understood that two objects could exert forces on each other even when they are not in direct contact. The gravitational force is a good example. The Earth exerts a gravitational force on objects such as the person in the elevator in Figure 3.8 or the young people in a free-fall dive in Figure 3.9, even though these people are not in contact with the planet. The exertion of forces between objects that are not in direct contact is known as ***action-at-a-distance***. You probably know that the gravitational force from the Sun is responsible for the Earth's orbital motion. The space between the Sun and the Earth is largely a vacuum, and it is certainly not obvious how this force can make itself felt through such an empty region. Action-at-a-distance is a fundamental aspect of nature that bothered many scientists after it was proposed by Newton, and it took the genius of Einstein to explain how this works for the gravitational force. The notion of action-at-a-distance applies to other forces, including electric and magnetic forces, as we'll discuss later.

Another important idea is contained in the notion of "direct contact" between the floor and the person's feet in Figure 3.7, where it certainly seems like the floor and the person are in contact. But what does this mean at a microscopic scale? We know that the floor and the person's feet are composed of atoms. Can two atoms actually "touch"? Contact forces are a result of electric forces between atoms that are in very close proximity. This atom–atom interaction cannot be described by Newton's laws, but must be treated using quantum mechanics (Chapters 28 and 29).

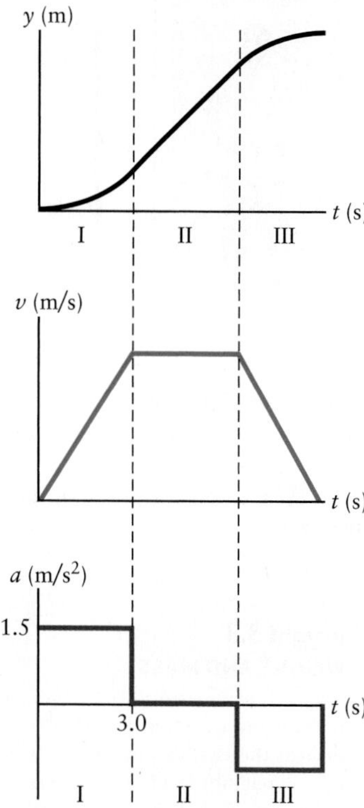

▲ **Figure 3.10** Example 3.3. Position, velocity, and acceleration versus time for an elevator as it travels from the first to the fifth floor of a building. In period I, the elevator is accelerating from rest. In period II, the acceleration is zero and the elevator moves with a constant velocity. During period III, the elevator slows to a stop.

EXAMPLE 3.3 Traveling Up

An elevator begins from rest at the first floor of a building. It starts a journey to the fifth floor by moving with an acceleration of 1.5 m/s^2 for 3.0 s; thereafter, it moves with a constant velocity until it gets near to its stopping point. Find the apparent weight of a person of mass m who is traveling in the elevator at different times during the motion.

RECOGNIZE THE PRINCIPLE

The person's apparent weight will depend on the acceleration. During its journey, the elevator must go through three stages of motion: (I) a period of positive acceleration (with $a = 1.5$ m/s^2), (II) a period of constant velocity (with $a = 0$), and (III) a period of negative acceleration while it slows to a stop.

SKETCH THE PROBLEM

Figure 3.8 shows a person in an elevator and defines our coordinate system; the motion is again along y (the vertical direction). Figure 3.8B is our free-body diagram.

IDENTIFY THE RELATIONSHIPS

Figure 3.10 contains qualitative plots of the position, velocity, and acceleration of the elevator versus time during the three phases of the motion. In each phase, we apply the relations for motion with constant acceleration from Table 3.1.

The apparent weight of a person in the elevator depends on the acceleration, so it is different in periods I, II, and III in Figure 3.10. Using Newton's second law and the components of the forces along y from the free-body diagram in Figure 3.8B gives

$$ma = \sum F = +N + F_{grav} = +N - mg \qquad (1)$$

The minus sign in front of the last term indicates (again) that the force of gravity is directed downward.

SOLVE

Rearranging Equation (1) gives the person's apparent weight as $N = ma + mg$. The person's true weight is mg, while his apparent weight during period I is

$$N = m(a + g) = m(1.5\ \text{m/s}^2 + 9.8\ \text{m/s}^2) = m(11.3\ \text{m/s}^2)$$

Comparing N to the true weight, we find

$$\frac{N}{mg} = \frac{m(11.3\ \text{m/s}^2)}{mg} = \frac{11.3\ \text{m/s}^2}{g} = \frac{11.3\ \text{m/s}^2}{9.8\ \text{m/s}^2} \approx 1.2$$

or

$$N = \boxed{1.2(mg)}$$

Hence, during the acceleration period (I) the person's apparent weight N is approximately 20% larger than his true weight.

During period II, the elevator's acceleration is zero, so $N = mg$ and the person's apparent weight is equal to his true weight. During period III, the acceleration is negative. The normal force is again $N = ma + mg$, and since a is now negative, the normal force and thus also the person's apparent weight are smaller than mg. The precise value of N depends on the value of a during period III (Fig. 3.10).

▸ ***What does it mean?***

The value of the apparent weight depends on an object's acceleration. The apparent weight can be greater than, equal to, or less than the true weight.

What Is "Mass"?

So far, we have only considered the force of gravity on terrestrial objects, and we have seen that this force has a very simple form

$$F_{grav} = -mg \qquad (3.8)$$

where (as usual) we take the "positive" direction to be upward. However, this simplicity tends to hide a very mysterious fact. According to Equation 3.8, the force of gravity depends on the mass of the object. We originally encountered mass in connection with the concept of inertia and Newton's laws of motion. According to Newton's second law, the mass of an object determines how it will move in response to forces. Here is the puzzle: if mass is the property of an object that determines how it moves (as calculated using Newton's second law), why does the *same* quantity determine the magnitude of the force of gravity? Even Newton was not able to answer this profound question, which we'll discuss more in Chapters 5 and 27. The quantity called "mass" that enters into Newton's second law is called the ***inertial mass*** of an object, while the "mass" that enters into the gravitational force is referred to as the object's ***gravitational mass***. Physicists believe that the inertial mass is precisely equal to the gravitational mass, which suggests a fundamental connection between gravitation and Newton's laws of motion. We'll explain this connection when we study relativity in Chapter 27.

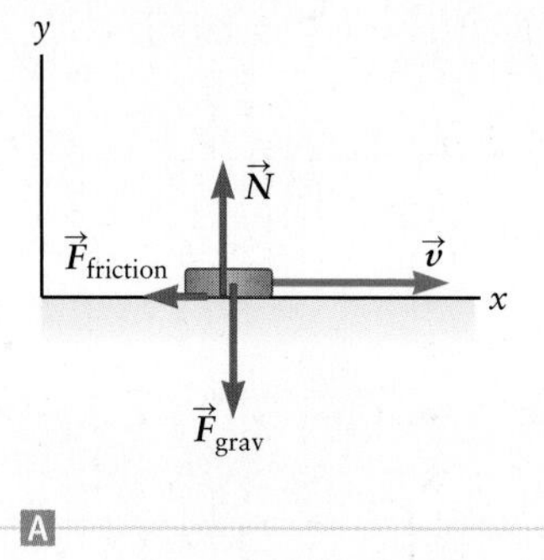

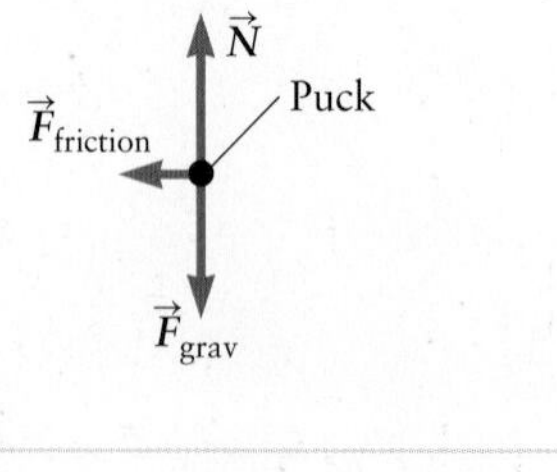

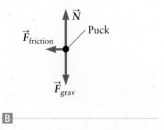

▲ **Figure 3.11** A Three forces act on this hockey puck: the force of gravity (its weight), the normal force from the ice, and the force of friction. B Free-body diagram for the puck.

3.3 Adding Friction to the Mix

One of Aristotle's central beliefs was that terrestrial objects all tend toward a natural state of rest. Indeed, that is what we commonly observe in the world around us because objects generally do spend most of their time at rest relative to us as observers. However, Newton's explanation of this behavior was quite different from the one provided by Aristotle. According to Newton, when an object comes to a state of rest, it does so because a force acts on it, and this force is often due to a phenomenon called friction.

Kinetic Friction

Figure 3.11 shows a hockey puck sliding on an icy surface. While this surface is very slippery, there is a small amount of friction, and we now consider how the force of friction eventually causes the puck to stop.

Figure 3.11B shows a free-body diagram containing all the forces acting on our hockey puck.[2] The puck's motion is along the horizontal direction x, and we assume the puck is sliding toward the right in the $+x$ direction. There is only one force along the horizontal, the force of friction. Because the puck's velocity is along $+x$ (to the right in Fig. 3.11), the force of friction will oppose this motion and hence be directed along $-x$ (to the left). There are also two forces in the vertical direction, y: the weight of the puck (the force of gravity) and the normal force exerted by the icy surface acting on the puck. This example is our first one involving forces acting in more than one direction, and we have not yet discussed how this affects the use of Newton's laws. For now, we appeal to your intuition and argue that the two forces directed along y, the normal force and the weight of the puck, will cancel each other. This is very similar to the cancellation we found in connection with Figure 3.7. We thus have $N - mg = 0$, or $N = mg$ for the hockey puck.

If the motion of our puck is along x, why did we need to mention forces along y? The reason is that the normal force is intimately connected with the force of friction. Experiments show that for a sliding object, the magnitudes of these two forces are related by

Force of kinetic friction

$$F_{friction} = \mu_K N \tag{3.9}$$

The quantity μ_K is called the ***coefficient of kinetic friction***. The frictional force in Equation 3.9 describes cases in which two surfaces are in motion (slipping) with respect to each other. As a "coefficient," μ_K is a pure number without dimensions or units, and its value depends on the surface properties. For a hockey puck on an icy surface, μ_K is relatively small and the frictional force is very small; the value of μ_K might typically be $\mu_K = 0.05$. The coefficient of kinetic friction for two rough surfaces would be much larger, with $\mu_K = 0.3$ being a common value. Table 3.2 lists coefficients of kinetic and static friction (discussed below) for some common materials. These measured values of μ_K show that the frictional force depends on the nature of the surfaces that are in contact.

Note that the magnitude of the frictional force in Equation 3.9 depends on the normal force. If the normal force is increased—say, by piling a large additional mass on top of our hockey puck—the frictional force is larger than if the extra mass is not present. We must emphasize that Equation 3.9 is not a "law" of nature; it is merely an approximate relation that has been found to work well in a wide variety of cases. To derive a fundamental law or theory of friction would require us to consider in detail the atomic interactions that occur when two surfaces are in contact. This problem is very complicated and is currently a topic of much research.

[2]Actually, we have omitted the force from air drag (sometimes called air resistance), but that will usually be very small for a hockey puck.

TABLE 3.2 Typical Values for the Coefficients of Kinetic Friction and Static Friction for Some Common Materials

Surfaces	μ_K	μ_S	Surfaces	μ_K	μ_S
rubber on dry concrete	0.8	0.9	ski on snow	0.05	0.10
rubber on wet concrete	0.5	0.6	glass on glass	0.4	0.9
rubber on glass	0.7	0.9	Teflon on Teflon	0.04	0.04
hockey puck on ice	0.05	0.10	metal on metal	0.5	0.6
wood on wood	0.3	0.5	bone on bone (with joint fluid)	0.015	0.02

Notes: The values for metal on metal are only approximate. The coefficients in all cases depend on the smoothness of the surfaces.

Analyzing Motion in the Presence of Friction

Suppose we know the value of μ_K and that the hockey puck in Figure 3.11 is given some initial velocity v_0. How long will the puck slide before coming to rest? We can apply Newton's second law along with the free-body diagram in Figure 3.11B to calculate the puck's acceleration. In this problem, the motion in which we are interested is along x, so we need to focus on the forces along the x direction. Writing Newton's second law for motion along x, we get

$$ma = \sum F_x = F_{\text{friction}} = -\mu_K N$$

All these terms involve components of acceleration and force along the horizontal (x) direction. The negative sign in the last term indicates that the frictional force is directed along $-x$, because friction acts to oppose the motion of the puck. In words, the velocity in Figure 3.11 is to the right, so the frictional force acts to the left. The normal force in this case is equal to mg, and we have

$$ma = -\mu_K N = -\mu_K mg$$
$$a = -\mu_K g$$

Since μ_K and g are both constants, the puck's acceleration is constant, so we can apply our general relations for motion with constant acceleration in Table 3.1. However, we must decide which one of these relations to use. In this problem, we are asked to find when the puck stops, and we'll know that the puck has come to a stop when $v = 0$. We therefore need a relation that involves v and t. Examining Table 3.1, we see that Equation 3.1 involves these quantities. Using this relation, we have

$$v = v_0 + at = v_0 - \mu_K gt = 0$$

and solving for t gives

$$t = \frac{v_0}{\mu_K g} \tag{3.10}$$

It is always a good idea to work out the units of an answer, even if no numbers are given. The coefficient of friction is dimensionless, so the units of the term on the right in Equation 3.10 are (m/s)/(m/s^2) = s, as expected because the result is a time. Notice also that we did not need to know the mass of the puck; the value of m canceled when calculating the acceleration.

Another question that we might have asked is, "How far will the puck slide before coming to a stop?" We could find the answer in two different ways: (1) by using the value of t found in Equation 3.10 and inserting it into Equation 3.3 in Table 3.1 or (2) by using Equation 3.4. Let's use the second approach. When the puck stops, its velocity is $v = 0$, and inserting this into Equation 3.4 gives

$$v^2 = v_0^2 + 2a(x - x_0) = 0$$

We can then solve for the distance traveled:

$$2a(x - x_0) = -v_0^2$$

$$(x - x_0) = -\frac{v_0^2}{2a} = \frac{v_0^2}{2\mu_K g}$$

Static Friction

The case of the sliding hockey puck in Figure 3.11 involves surfaces that are in motion (slipping) relative to each other, so the frictional force in such cases is referred to as kinetic friction. Situations in which the relevant surfaces are not slipping involve what is known as ***static friction***. An example involving static friction is given in Figure 3.12, which shows a refrigerator at rest on a level floor. This situation is similar to the hockey puck in Figure 3.11, but with an additional force exerted by a person pushing on the refrigerator. We know intuitively that when the force exerted by the person is small, the refrigerator will not move. Since the acceleration is zero, applying Newton's second law along the horizontal direction leads to

$$ma = \sum F_x = F_{push} + F_{friction} = 0$$

In this case, the force exerted by the person and the force of friction cancel. The force of static friction is sufficiently strong that no relative motion (no slipping) occurs. In terms of magnitudes, we have

$$|F_{push}| = |F_{friction}|$$

However, the value of F_{push} can vary and the refrigerator will still remain at rest. That is, we can push a lot or a little or not at all, without the refrigerator starting into motion. In all these cases, the frictional force exactly cancels F_{push}. The only way this can happen is if the magnitude of the frictional force varies depending on the value of F_{push}, as described by

Force of static friction

$$|F_{friction}| \leq \mu_S N \qquad (3.11)$$

where μ_S is the ***coefficient of static friction.***

The term *static* again means that the two surfaces (the floor and the bottom of the refrigerator) are not moving relative to each other. According to Equation 3.11, the magnitude of the force of static friction can take any value up to a maximum of $\mu_S N$. If F_{push} in Figure 3.12 is small, the force of static friction is small and will precisely cancel F_{push} so that the total horizontal force is zero. If F_{push} is increased, the force of static friction increases and again precisely cancels F_{push}. However, the magnitude of the static friction force has an upper limit of $\mu_S N$.

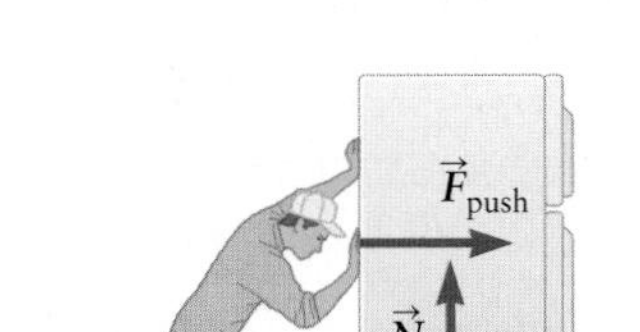

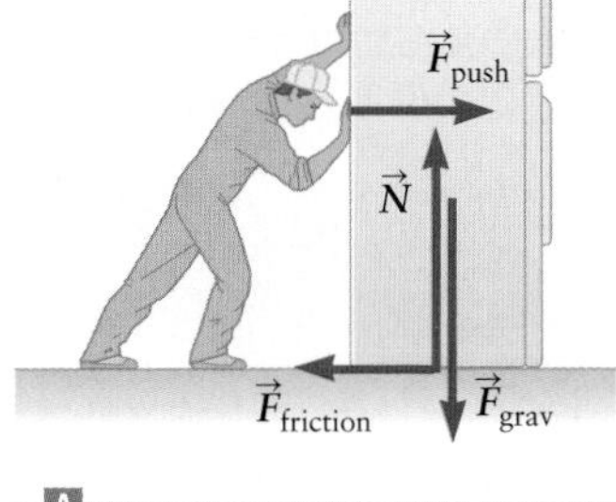

Free-body diagram

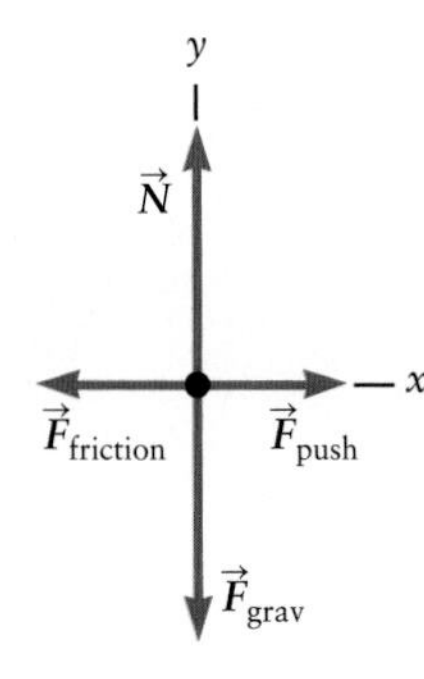

▲ **Figure 3.12** A Four forces act on this refrigerator. The force of static friction acts along the horizontal and opposes the movement of the refrigerator relative to the floor. B Free-body diagram for the refrigerator. Here we assume the force F_{push} is small enough that the refrigerator does not tip.

Comparing Kinetic Friction and Static Friction

To describe friction fully, we must consider both the kinetic and static cases, with two different coefficients of friction. In the kinetic case, the two surfaces are slipping relative to each other and the frictional force has the simple value $F_{friction} = \mu_K N$ (Eq. 3.9). The static case is a bit more complicated; here, the frictional force varies so as to cancel the other force(s) on the object, up to a maximum value of $\mu_S N$ (Eq. 3.11). For a given combination of surfaces, it usually happens that $\mu_S > \mu_K$ (see Table 3.2), so the maximum force of static friction is greater than the force of kinetic friction. This confirms an experiment you have probably done yourself. The force required to start something into motion—that is, the minimum value of F_{push} required to just barely move the refrigerator in Figure 3.12—is larger than the force required to keep it moving subsequently at a constant velocity.

EXAMPLE 3.4 Sliding Away

A wooden crate sits on the floor of a flatbed truck that is initially at rest, and the coefficient of static friction between the crate and the floor of the truck is $\mu_S = 0.25$. The driver of the truck wants to accelerate to a velocity of 20 m/s within 4.0 s. If she does so, will the crate slide off the back of the truck? Assume the truck moves with a constant acceleration.

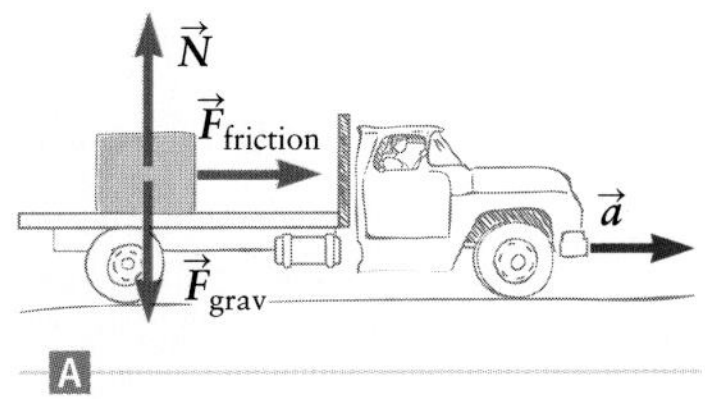

Free-body diagram

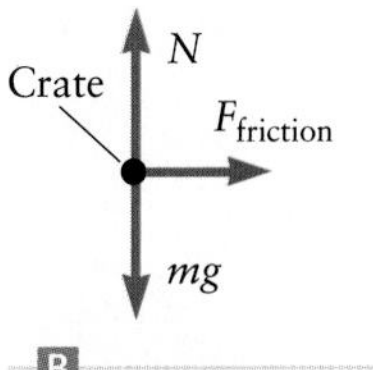

▲ **Figure 3.13** Example 3.4. A The force of friction on the crate is directed to the right, in the direction of motion of the truck and the crate. The frictional force acts to prevent the crate from slipping relative to the truck. B Free-body diagram for the crate.

RECOGNIZE THE PRINCIPLE

Figure 3.13A shows that the only horizontal force on the crate is the force of static friction. The static frictional force is not opposing any other force. Instead, static friction is opposing the tendency of the crate to slide along the floor of the truck. That is, static friction opposes any *relative motion* of the two surfaces. If there is no relative motion and the crate does not slip, it will be moving along with the truck. Hence, in this example the force of friction is what causes the crate to accelerate! Since there is an upper limit to the magnitude of the static frictional force (Eq. 3.11), there is an upper limit to this horizontal force on the crate, giving an upper limit to the acceleration that static friction can provide to the crate. If this acceleration limit is smaller than the truck's actual acceleration, the crate will not be able to keep up with the truck and the crate will slide off the back.

SKETCH THE PROBLEM

Figure 3.13B shows a free-body diagram for the crate. The vertical forces on the crate are its weight (directed downward) and the normal force from the floor of the truck (directed upward). As the truck accelerates, there is also a horizontal force on the crate caused by static friction from the floor of the truck.

IDENTIFY THE RELATIONSHIPS

Using Newton's second law, the force of static friction leads to a maximum acceleration of the crate in the x direction of

$$a_{\text{crate, max}} = \frac{F_{\text{friction, max}}}{m} = \frac{\mu_S N}{m} \tag{1}$$

Calculating the normal force between the crate and floor of the truck is done in the same way as with the hockey puck in Figure 3.11. Hence, the magnitude of the normal force acting on the crate is $N = mg$, where m is the mass of the crate. Substituting into Equation (1) leads to

$$a_{\text{crate, max}} = \frac{\mu_S N}{m} = \frac{\mu_S mg}{m} = \mu_S g \tag{2}$$

To calculate the acceleration of the truck as determined by the driver, we note that the truck is undergoing motion with constant acceleration and choose Equation 3.1 from Table 3.1 because this relation involves velocity, acceleration, and time. The truck starts from rest, so $v_0 = 0$. Inserting this into Equation 3.1 gives

$$v = v_0 + a_{\text{truck}} t = a_{\text{truck}} t$$

$$a_{\text{truck}} = \frac{v}{t} \tag{3}$$

SOLVE

We now need to calculate the value of the truck's acceleration in Equation (3) and compare it with the maximum possible acceleration of the crate in Equation (2). Solving for a_{truck}, we find

$$a_{\text{truck}} = \frac{v}{t} = \frac{20 \text{ m/s}}{4.0 \text{ s}} = 5.0 \text{ m/s}^2$$

Evaluating Equation (2) for the maximum possible acceleration of the crate gives

$$a_{\text{crate, max}} = \mu_S g = (0.25)(9.8 \text{ m/s}^2) = 2.5 \text{ m/s}^2$$

The acceleration of the truck is thus greater than the maximum possible frictional acceleration of the crate, so the crate will *not* be able to keep up with the truck and it will slide off the back.

(continued) ▶

▸ ***What does it mean?***
Normally, friction either prevents an object from moving or brings it to a stop. However, in this case friction is responsible for the motion of the crate because friction always opposes the relative motion of the surfaces in contact.

CONCEPT CHECK 3.2 **Trucks and Crates**

The driver of the truck in Example 3.4 is in a hurry and wants to drive at the highest speed possible to her destination without losing the crate. Is there a maximum speed at which she can drive without having the crate slide off the truck?

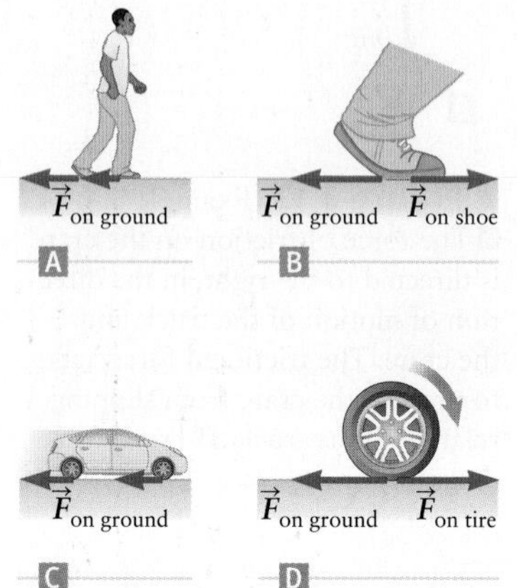

▲ **Figure 3.14** Frictional forces are essential for walking and rolling.

The Role of Friction in Walking and Rolling

It is interesting that the force of static friction on the crate in Figure 3.13 actually *causes* the crate's motion. This happens in other cases too. Figure 3.14A shows a person walking on level ground. As the person "pushes" off during each step, the bottoms of his shoes exert a force $\vec{F}_{\text{on ground}}$ on the ground. If the shoes do not slip, this force is due to static friction since the shoes do not move relative to the ground. According to Newton's third law, there is a reaction force $\vec{F}_{\text{on shoe}}$ on the person's shoes (Fig. 3.14B), and this force "propels" a person as he walks. If the surface were extremely slippery and there were no frictional force, then $\vec{F}_{\text{on ground}}$ and $\vec{F}_{\text{on shoe}}$ would both be zero and the person would slip when he attempted to move. Hence, the force of friction makes walking and running possible.

A similar set of forces is found when a wheel rolls along level ground. If a car's tire does not slip, friction between the tire and the ground leads to a force $\vec{F}_{\text{on ground}}$ on the ground (Fig. 3.14C). There is a reaction force on the tire $\vec{F}_{\text{on tire}}$ (Fig. 3.14D), and this force propels the car forward. Friction thus also plays a key role in rolling motion.

3.4 Free Fall

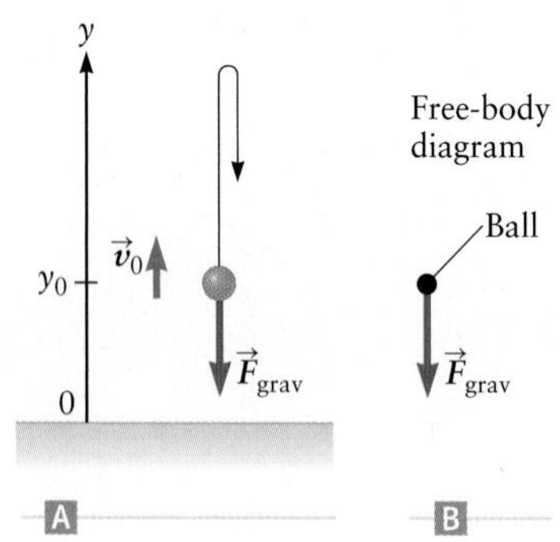

▲ **Figure 3.15** A ball is thrown upward beginning at an initial height y_0 with initial velocity $\vec{v}_0$. After leaving the thrower's hand, the ball is in free fall and the only force acting on the ball is the force of gravity. Even though this force is directed downward, the ball still moves up initially due to the initial velocity imparted by the thrower.

In Chapter 2, we discussed how several quantities, including velocity and acceleration, are needed when describing the motion of an object. Indeed, the relationship and differences between velocity and acceleration are absolutely crucial for understanding motion, as we now illustrate for a type of motion known as ***free fall***. An example of free fall is the motion of a ball that is thrown upward, reaches some maximum height, and then falls back down to Earth. After the ball leaves the thrower's hand, it follows the familiar trajectory sketched in Figure 3.15. Let's analyze this trajectory and consider how the acceleration, velocity, and position of the ball all vary with time.

We begin with the acceleration. Before the ball leaves the thrower's hand, there are forces from gravity and from her hand acting on the ball, and these forces lead to an upward acceleration as she gives the ball an upward velocity. In Figure 3.15, we take $t = 0$ to be the instant just *after* the ball leaves her hand. From that instant until the ball hits the ground, the only forces acting on the ball are the force of gravity and a force due to air drag. In many cases, the force from air drag is very small, so we'll ignore it here.

We choose a coordinate system that measures the position as the height y of the ball above the ground, with the origin chosen to be at ground level. Using Newton's second law, we have

$$a = \frac{F_{\text{grav}}}{m} = \frac{-mg}{m} = -g \tag{3.12}$$

In words, the negative sign here means that the force of gravity is directed downward, in the $-y$ direction. We can again use our relations for motion with constant acceleration (Eqs. 3.1 and 3.3):

$$v = v_0 + at = v_0 - gt \tag{3.13}$$

$$y = y_0 + v_0 t + \tfrac{1}{2}at^2 = y_0 + v_0 t - \tfrac{1}{2}gt^2 \tag{3.14}$$

Sketches of how v and y vary with time are shown in Figure 3.16. The velocity varies linearly with time with a slope of $-g$ and an intercept equal to v_0, the velocity of the ball at $t = 0$ (just after it leaves the thrower's hand). The position varies quadratically with t, with a positive intercept y_0 that is the height of the ball when it leaves the thrower's hand.

The sketches in Figure 3.16 reveal some very important features of freefall. Initially, the ball's velocity is *positive*, whereas its acceleration is *negative*. That is, the velocity vector is directed *upward*, and the acceleration vector is directed *downward*. In fact, the ball's acceleration is downward during the *entire* time (Fig. 3.16C and Eq. 3.12). The force on the ball and therefore its acceleration are both negative; they are both directed along $-y$ throughout the motion. Since the acceleration is the slope of the v–t curve, the v–t relation must have a negative and constant slope throughout as shown in Figure 3.16A. Although this slope is always negative, the velocity itself can have a value that is positive, negative, or zero. The velocity is positive while the ball is on the upward (initial) part of its trajectory and negative while it is on the downward (final) part. The velocity is zero, shown by the v–t curve crossing through zero in Figure 3.16A, at the instant the ball reaches its highest point along y (the highest point on its trajectory).

To understand the behavior of the displacement, recall again that v is the slope of the y–t curve, so while the ball is on the way up and v is positive, the y–t curve has a positive slope. The ball reaches its highest point when the velocity is zero, and the y–t slope at that instant is zero. While the ball is on the way down, the velocity is negative, so the slope of the y–t curve is negative.

An *extremely* important lesson from this problem is that the acceleration and velocity can be in *different* directions; that is, they can have different signs. Depending on the situation, they can have either the same signs (both positive or both negative) or one can be positive while the other is negative.

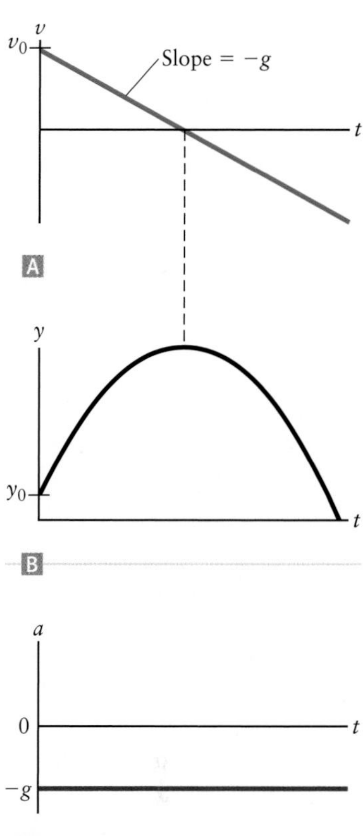

▲ **Figure 3.16** A For a ball in free fall, the acceleration is constant, so the velocity varies linearly with time. The slope of the v–t line is $-g$. B Height of the ball y versus time. The ball reaches its maximum height when $v = 0$. C The acceleration of the ball is constant and nonzero ($a = -g$) during the entire time, even at the highest point on the trajectory.

CONCEPT CHECK 3.3 The Direction of Velocity and Acceleration

For which of the following situations is the velocity parallel to the acceleration, and for which are $\vec{v}$ and $\vec{a}$ in opposite directions?

(a) The hockey puck sliding to a stop in Figure 3.11.
(b) The crate on the flatbed truck in Figure 3.13.
(c) An apple falling from a tree.

EXAMPLE 3.5 Maximum Height of a Projectile

Consider the motion of the ball in Figure 3.15 and assume the ball has an initial speed of 12 m/s (about 25 mi/h). Calculate (**a**) the time it takes for such a ball to reach its maximum height, (**b**) the maximum height reached by the ball, (**c**) the time the ball spends in the air, and (**d**) the ball's velocity just prior to hitting the ground. For simplicity, assume the initial position of the ball is at ground level.

RECOGNIZE THE PRINCIPLE

Since the acceleration is constant for free fall, we can use the relations from Table 3.1 or, equivalently, Equations 3.13 and 3.14. For each part of the problem, we select the

(continued) ▶

relation that contains the quantity we want to find along with quantities that we know. For part (**a**), we know that the velocity is zero when the ball reaches the maximum height, so we select Equation 3.13. We attack the other parts of the problem in a similar way.

SKETCH THE PROBLEM

Figure 3.15 contains all the information we need, including a free-body diagram. This free-body diagram applies at *all* points on the trajectory. While the ball is traveling through the air, the only force exerted on it is the force of gravity. (We again assume the force exerted by air drag is very small and can be neglected.)

IDENTIFY THE RELATIONSHIPS

The ball starts at ground level, so $y_0 = 0$. The initial velocity of the ball is $v_0 = 12$ m/s. (**a**) At the top of the trajectory, the velocity is $v = 0$. From Equation 3.13, we get

$$v_{top} = v_0 - gt_{top} = 0$$

$$t_{top} = \frac{v_0}{g}$$

(**b**) Once we know the time needed to get to the top of the trajectory, we can find the ball's position at the top of its path using Equation 3.14:

$$y_{max} = y_0 + v_0 t_{top} - \tfrac{1}{2} g t_{top}^2$$

SOLVE

(**a**) Inserting the given value of v_0 into our result for t_{top} yields

$$t_{top} = \frac{12 \text{ m/s}}{9.8 \text{ m/s}^2} = \boxed{1.2 \text{ s}}$$

(**b**) Substituting the value we just found for t_{top} into the equation for y_{max} gives

$$y_{max} = 0 + (12 \text{ m/s})(1.2 \text{ s}) - \tfrac{1}{2}(9.8 \text{ m/s}^2)(1.2 \text{ s})^2 = \boxed{7.3 \text{ m}}$$

IDENTIFY THE RELATIONSHIPS

(**c**) When the ball reaches the ground, its position will be $y = 0$. Using Equation 3.14, we have

$$y = y_0 + v_0 t - \tfrac{1}{2} g t^2 = 0 \quad (1)$$

We set $t = t_{ground}$ and take $y_0 = 0$:

$$y = 0 + v_0 t_{ground} - \tfrac{1}{2} g t_{ground}^2 = 0$$

$$t_{ground} = \frac{2v_0}{g}$$

(**d**) To find the velocity of the ball just before it lands, we use Equation 3.13 because it contains v and t. We want the velocity at time $t = t_{ground}$.

SOLVE

(**c**) Inserting the given value for v_0 into our result for t_{ground} yields

$$t_{ground} = \frac{2(12 \text{ m/s})}{9.8 \text{ m/s}^2} = \boxed{2.4 \text{ s}}$$

(**d**) Substituting the value we just found for t_{ground} into Equation 3.13 for the velocity gives

$$v = v_0 - gt_{ground} = 12 \text{ m/s} - (9.8 \text{ m/s}^2)(2.4 \text{ s}) = \boxed{-12 \text{ m/s}}$$

▸ What does it mean?

In this example, we assumed the initial position of the ball was $y_0 = 0$. A nonzero value would be more realistic, but would complicate the mathematics in part (c) since solving Equation (1) with a nonzero value of y_0 would require us to solve a quadratic equation.

Motion of a Dropped Ball

Example 3.5 gave us an interesting result. The ball's speed just before it hit the ground is equal to the ball's initial speed (although the ball's *velocity* is positive at the start and negative at the end). This result is no accident; it is true because the position y is a symmetric function with respect to the time at which the ball reaches its highest point (see Fig. 3.16B). The time spent on the way up is equal to the time spent falling back to the ground. We can see why by comparing the results for t_{top} and t_{ground} in Example 3.5, where we found

$$t_{top} = \frac{v_0}{g} \quad \text{and} \quad t_{ground} = \frac{2v_0}{g}$$

Hence, t_{ground} is precisely twice t_{top}. This symmetry is an important property of motion with constant acceleration, although it is not found if the acceleration varies with time. For example, when air drag is important, the acceleration is not constant and the position–time relation is no longer symmetric with respect to t_{top}.

EXAMPLE 3.6 Symmetry of Free Fall

Consider an apple that drops from a tree, and assume the apple falls from a height that is the same as the maximum height reached in the previous example. Calculate (**a**) how long it takes the apple to reach the ground and (**b**) its velocity the instant before it hits the ground.

RECOGNIZE THE PRINCIPLE

Since the apple is in free fall, the only force on it is due to gravity and the acceleration is constant. We can thus (yet again) use the expressions for motion with a constant acceleration in Table 3.1.

SKETCH THE PROBLEM

Figure 3.17 shows the problem. The apple starts at $y_0 = 7.3$ m (the maximum height reached by the ball in Example 3.5), with an initial velocity of zero. It undergoes free fall until it reaches the ground.

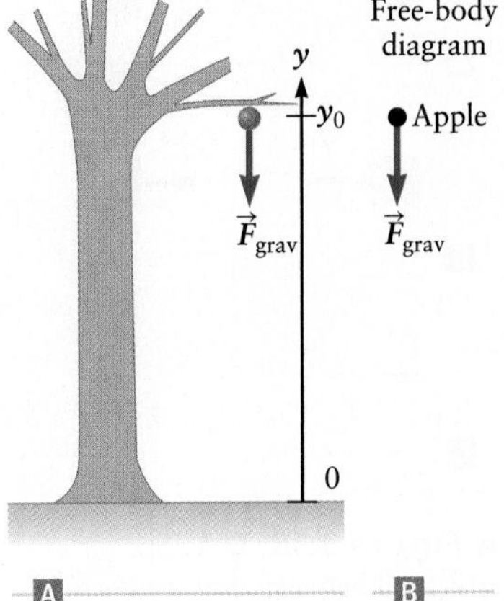

▲ **Figure 3.17** Example 3.6. The acceleration of an apple during free fall is the same as for the thrown ball in Figure 3.15, but the initial velocity ($v = 0$) is different.

IDENTIFY THE RELATIONSHIPS

(**a**) The apple reaches the ground when $y = 0$, so to calculate when the apple reaches this point we choose the relation that contains both position and time (Eq. 3.14 or its equivalent, Eq. 3.3). Applying Equation 3.14, we have

$$y = y_0 + v_0 t - \tfrac{1}{2}gt^2 \qquad (1)$$

The initial velocity is $v_0 = 0$ because the apple starts from rest.

SOLVE

(**a**) Substituting $v_0 = 0$ and $y = 0$ into Equation (1) and solving for t gives

$$y = y_0 - \tfrac{1}{2}gt_{ground}^2 = 0$$

$$t_{ground} = \sqrt{\frac{2y_0}{g}} = \sqrt{\frac{2(7.3 \text{ m})}{9.8 \text{ m/s}^2}} = \boxed{1.2 \text{ s}}$$

(continued) ▸

IDENTIFY THE RELATIONSHIPS

(b) We now need to evaluate the velocity at the value of t found in part (a). For the apple's velocity the instant before it hits the ground, we have (Eq. 3.13)

$$v = v_0 - gt_{\text{ground}} = -gt_{\text{ground}}$$

where again $v_0 = 0$.

SOLVE

(b) Inserting the value of t from part (a) gives

$$v = -(9.8 \text{ m/s}^2)(1.2 \text{ s}) = \boxed{-12 \text{ m/s}}$$

► ***What does it mean?***

This result for the velocity is the same as we found for the thrown ball in Example 3.5, part (d). That is no accident: the motion *on the way down* is precisely the same in the two cases.

CONCEPT CHECK 3.4 Free Fall

Consider an acorn as it falls from the top of a tall tree. Take the y axis to be in the vertical direction, with "up" being the $+y$ direction. Which of the following statements is true?

(a) The velocity is negative, the acceleration is negative, and the speed is increasing.

(b) The velocity is negative, the acceleration is positive, and the speed is decreasing.

(c) The velocity is negative and the speed is negative.

3.5 Cables, Strings, and Pulleys: Transmitting Forces from Here to There

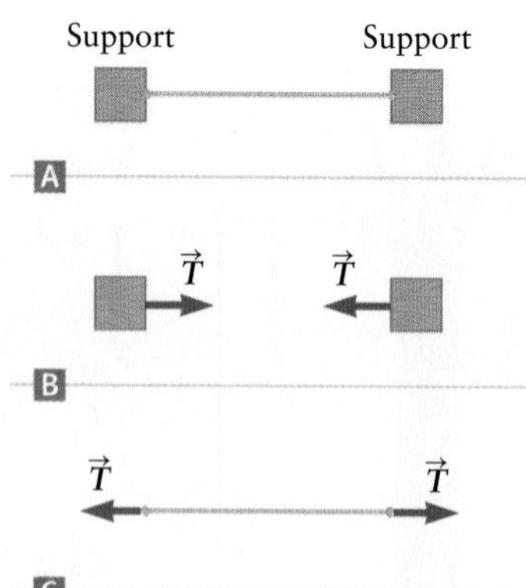

▲ **Figure 3.18** A A rope stretched between two supports. B The rope exerts a force of magnitude T on both supports, the one on the left and the one on the right. C According to Newton's third law, each support exerts a force of magnitude T on the rope.

Strings, cables, and ropes exert a force on the objects they are connected to due to their ***tension***. To understand these tension forces, we begin by considering an elevator supported by a cable. To choose a cable that is strong enough to support the elevator, you need to understand the forces acting on the cable. When a cable (or rope or string, etc.) is pulled tight, it exerts a force on whatever it is connected to. This force is illustrated in Figure 3.18; the ends of the rope both exert a force of magnitude T on the supports to which they are connected (Fig. 3.18B). At the same time, Newton's third law (the action–reaction principle) tells us that the supports both exert a reaction force of magnitude T on the ends of the rope (Fig. 3.18C). In words, T is equal to the tension in the rope.

The reasoning in Figure 3.18 also applies to the elevator in Figure 3.19; the cable exerts an upward force T on the elevator compartment and a separate (downward) force of magnitude T on whatever it is connected to at the top of the elevator shaft.

Tension Forces

To deal with the motion of the elevator compartment in Figure 3.19, we start with the free-body diagram in Figure 3.19B. There are two forces acting on the compartment, the force of gravity and a force from the cable that has a magnitude T (recall that T is the tension in the cable). If the elevator has an acceleration a along the vertical direction and m is the mass of the compartment plus passengers, we can apply Newton's second law for motion along the vertical direction to write

$$ma = \sum F = F_{grav} + T = -mg + T$$

The tension in the cable is thus

$$T = mg + ma$$

This answer is quite reasonable; if the elevator is at rest and the acceleration is zero, the force exerted by the cable must only support the weight of the elevator compartment and $T = mg$. If a is positive, a larger tension is required.

We next consider the forces acting on the cable. We have just seen that the cable exerts an upward force T on the elevator. According to Newton's third law—the action–reaction principle—the elevator must exert a force of equal magnitude and opposite direction on the cable. This is the origin of the downward force T acting on the cable in the free-body diagram in Figure 3.19C. There is a second force T_C acting on the cable, which comes from its connection to the top of the elevator shaft (to a motor, etc., located there). Usually the mass of a cable is very small when compared with the mass of other parts of the system, so for simplicity we will assume the cable is massless. If the mass of the cable is zero, the force of gravity on the cable will also be zero. Applying Newton's second law to our elevator cable in Figure 3.19C and assuming its mass is zero, we find

$$m_{cable}a = (0)a = +T_C - T = 0$$

$$T_C = T$$

Hence, for this cable, and for all such massless cables, the forces acting on the two ends are equal and both equal T, the tension in the cable. Put another way, for a massless cable the tension is the same at all points along the cable.

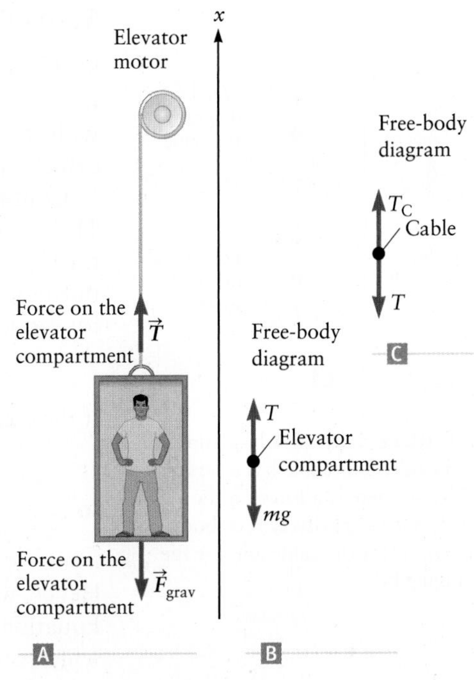

▲ **Figure 3.19** A When an elevator is lifted by a cable, there are two forces on the elevator compartment, the gravitational force and the tension from the cable. This tension force is part of an action–reaction pair; the other force in the pair acts downward on the cable. B Free-body diagram for the elevator compartment. C Free-body diagram for the cable.

Let's now take on the more complicated situation in Figure 3.20, which shows a hanging spider holding onto its dinner (a bug). Here we have two "cables," both composed of spider silk, and their tensions (denoted T_1 and T_2) will be different. Figure 3.20 also shows free-body diagrams for the spider, the bug, and the lower piece of silk. The tension in the lower piece of silk is T_2. This silk is attached to the spider, so there is a force of magnitude T_2 directed downward on the spider, and there is also a corresponding reaction force T_2 directed upward on the silk. The silk exerts a separate force of magnitude T_2 directed upward on the bug, and there is a reaction force T_2 acting downward on the silk. There are similar forces due to the tension T_1 in the upper piece of silk.

The spider and bug have masses m_{spider} and m_{bug} and a common acceleration a. Let's now find the tension in each piece of silk, assuming the silk is massless. We begin by applying Newton's second law to the bug in Figure 3.20. The free-body diagram for the bug shows that the forces in this case are just the force of gravity and the tension T_2, so from Newton's second law for motion along y we have

$$m_{bug}a = \sum F = F_{grav,\ bug} + T_2 = -m_{bug}g + T_2$$

$$T_2 = m_{bug}(g + a)$$

We next write Newton's second law for the spider:

$$m_{spider}a = \sum F = F_{grav,\ spider} + T_1 - T_2 = -m_{spider}g + T_1 - T_2$$

$$T_1 = m_{spider}(g + a) + T_2$$

Inserting our result for T_2 leads to

$$T_1 = m_{spider}(g + a) + m_{bug}(g + a) = (m_{spider} + m_{bug})(g + a)$$

This result for T_1 could also have been obtained by realizing that because the spider and the bug are connected and move together, we can think them as just a single object of mass $(m_{spider} + m_{bug})$. Applying Newton's second law to this composite object leads to the same result.

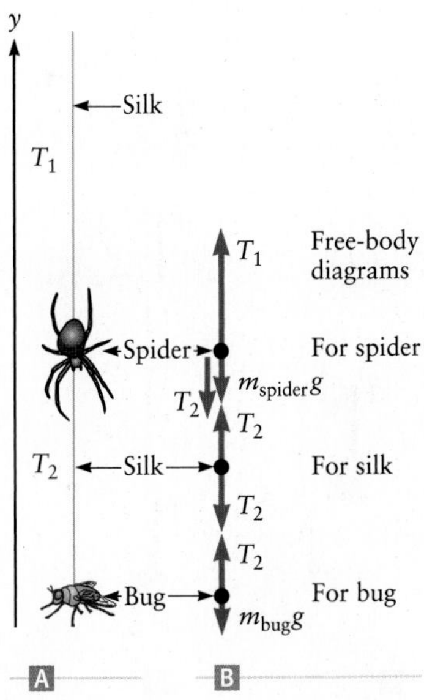

▲ **Figure 3.20** A A spider hanging and holding onto a bug using a strand of silk. B Free-body diagrams showing the forces acting on the spider, the bug, and the lower pieces of silk. There are several action–reaction force pairs in this problem.

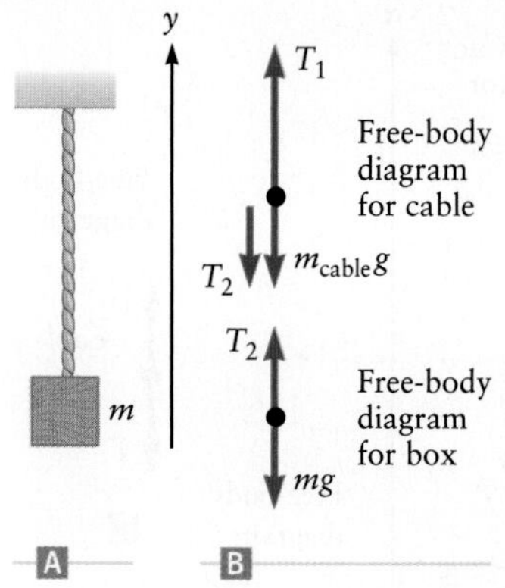

▲ **Figure 3.21** A A box attached to a cable. This cable has a nonzero mass, so there is a force on the cable due to gravity. B Free-body diagrams for the cable and for the hanging box.

Some Cables Are Not Massless

So far, we have assumed cables and ropes are massless. This assumption is often justified since the mass of a realistic cable or rope is usually very small compared with that of other parts of the system. Let's now see how to deal with cases when a cable's mass is not zero.

Consider the cable of mass m_{cable} in Figure 3.21, holding up a box of mass m. There are now three forces acting on the cable: one tension force from the support at the top (T_1), a second tension force from the box tied to the bottom end (T_2), and a third force due to the weight of the cable. Writing Newton's second law for the cable for motion along the y direction gives

$$m_{cable}a = \sum F = +T_1 - T_2 - m_{cable}g$$

If the cable is not accelerating, then $a = 0$ and

$$T_1 = T_2 + m_{cable}g \tag{3.15}$$

so

$$T_1 > T_2$$

Hence, we now have two different values for the tension in the cable. According to Equation 3.15, the tension T_1 at the top is larger than the tension at the bottom, which makes sense because the tension at the top must support the weight of the cable. We can also see what is required for a cable to be approximated as "massless." From Equation 3.15, the difference in the tensions at the top and bottom is equal to the weight of the cable. A cable can thus be approximated as massless if its weight is small compared to either tension.

Using Pulleys to Redirect a Force

Cables and ropes are an efficient way to transmit force from one place to another and are often used together with a device called a pulley. A simple pulley is shown in Figure 3.22A; it is just a wheel free to spin on an axle through its center, and it is arranged so that a rope or cable runs along its edge without slipping. For simplicity, we assume both the rope and the pulley are massless. Typically, a person pulls on one end of the rope so as to lift an object connected to the other end. Here the pulley changes the direction of the force associated with the tension in the rope as illustrated in Figure 3.22B, which shows the rope "straightened out" (i.e., with the pulley removed). In either case—with or without the pulley in place—the person exerts a force F on one end of the rope, and this force is equal to the tension. The tension is the same everywhere along this massless rope, so the other end of the rope exerts a force of magnitude T on the object. A comparison of the two arrangements in Figure 3.22—one with the pulley and one without—suggests the tension in the rope is the same in the two cases, and in both cases the rope transmits a force of magnitude $T = F$ from the person to the object.

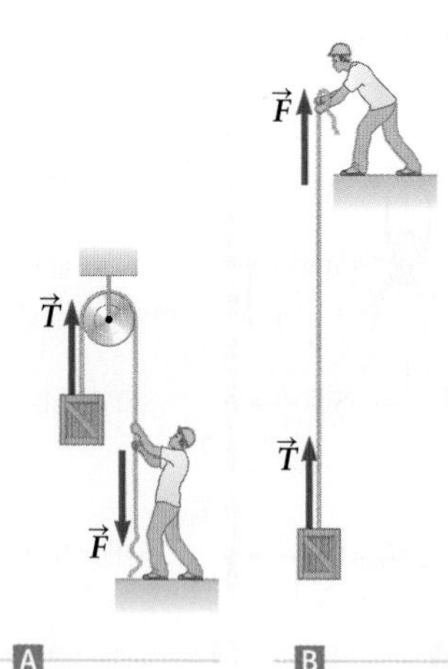

▲ **Figure 3.22** A A simple pulley. If the person exerts a force F on a massless string, there is a tension T in the string, and this tension force can be used to lift an object. B The string "straightened out." The pulley simply redirects the force.

CONCEPT CHECK 3.5 Normal Forces

Consider the man pulling downward on the rope in Figure 3.22A. Is the normal force of the floor on the man (a) greater than, (b) less than, or (c) equal to the man's weight?

Amplifying Forces

A pulley can do much more than simply redirect a force. Figure 3.23 shows a pulley configured as a block and tackle, a device used to lift heavy objects. To analyze this case, we have to think carefully about the force between the string and the surface of the pulley. We have already mentioned that the rope does not slip along this surface. There is a frictional force between the rope and the surface of the pulley wherever the two are in contact, which is all along the bottom half of the pulley in Figure 3.23. Since the rope does not slip relative to the pulley, the rope exerts a force on the pulley as if the rope and pulley were actually attached to each other. First consider the part of the rope held by the person: it exerts an upward force on the pulley, equal in magnitude to the tension T (Fig. 3.23B). The same thing happens with the section of the rope on the left that is tied to the ceiling; this part of the rope also exerts a force on the pulley, and this force is again equal to the tension. Hence, there is a force $2T$ acting upward on the pulley. The interesting point is that the person exerts a force of only T on the rope, but the block and tackle converts it into a force of magnitude $2T$ *on the pulley*. This force on the pulley can then be used to lift an object attached to the pulley, such as the wooden crate that hangs from the pulley's axle in Figure 3.23A. A block and tackle can thus *amplify forces*. This particular arrangement enables the person to lift twice as much weight as would be possible if he used the simpler pulley arrangement in Figure 3.22.

The block and tackle in Figure 3.23 amplifies forces by a factor of two, and more complex block-and-tackle devices can amplify an applied force by even greater factors. This result might disturb you a little because it appears to violate an intuitive notion that "you can't get something for nothing." Although this statement is not a formal law of physics, you might still wonder what Newton would have thought. A key feature of the block and tackle in Figure 3.23 is that while it amplifies the force by a factor of two, the displacement of the object is *reduced* by a factor of two. If the person pulls the end of the rope upward through a distance L in Figure 3.23A, the body of the pulley and hence also the mass attached at the bottom move a distance of only $L/2$. In fact, this reduction of the displacement is a feature of all block and tackles. While it is possible to amplify a force by a factor that can be much larger than unity, the corresponding displacement is always reduced by the same factor. You do *not* get something for nothing.

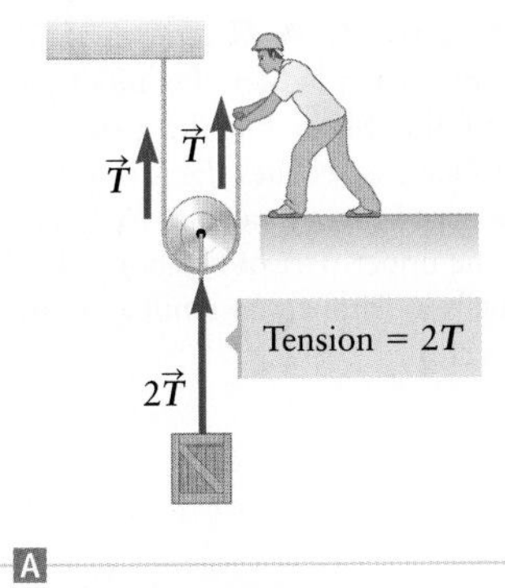

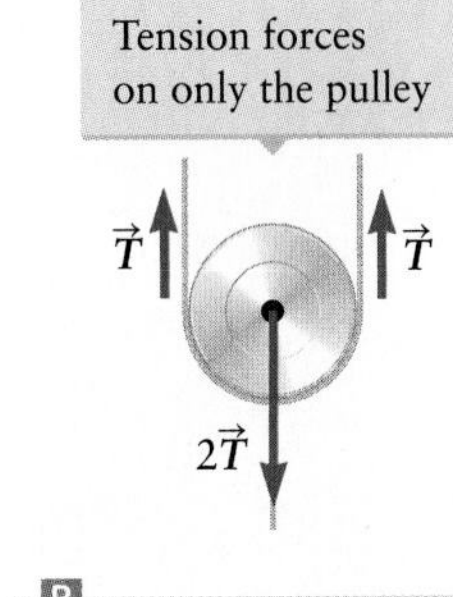

▲ **Figure 3.23** A A block-and-tackle arrangement. The upward force on the pulley, from both sides of the rope, is $2T$. This force can be used to lift an object attached to the pulley. B Close-up of rope in contact with the pulley.

3.6 Reasoning and Relationships: Finding the Missing Piece

You may have the impression that physics always involves a lot of precise mathematics. However, it is sometimes very useful to solve a problem approximately, perhaps because some important quantities (such as the initial velocity of an object) are not known precisely or because an approximate answer is sufficient. The reasoning involved in an approximate analysis can often give clear insight into a problem, without the "distraction" of a lot of mathematics. In this section, we practice such ***reasoning and relationships problems.***

Dealing with these problems involves two challenges that have not been present in the problems we have faced so far. First, we may need to identify important information that is "missing" from an initial description of the problem. For example, you might be asked to calculate the forces acting on two cars when they collide, given only the speed of the cars just before the collision. From an understanding of the relationship between force and motion, you would have to *recognize that additional information is needed* to solve this problem. In this problem, that information is the mass of the cars and the way the car bumpers deform in the collision. (We'll actually consider this problem later in this section.) It would then be your job to use common sense to *make reasonable estimates* of these "missing" quantities for typical cars and use your values to compute an answer. Because such estimated values vary from case to case (not all cars have the same mass or the same type of bumper), an approximate mathematical solution and an approximate numerical answer is usually sufficient. Let's now consider a few problems of this kind.

Jumping off a Ladder

Consider the case of a child who jumps off a ladder or playground structure for fun. We want to find the approximate force between the child and the ground when she

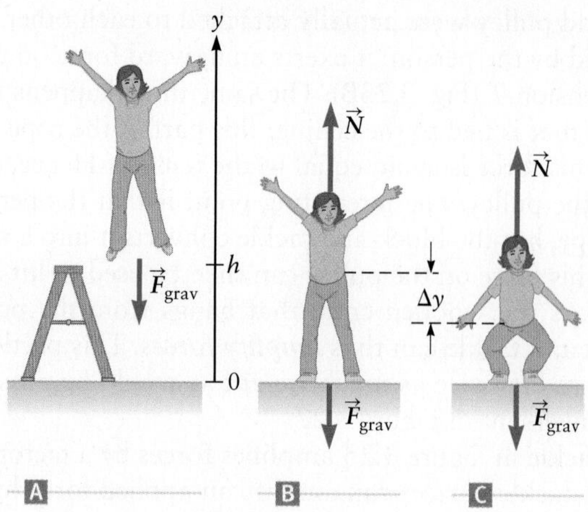

▶ **Figure 3.24** A When a child jumps from a ladder, she undergoes free fall until she hits the ground. B When she reaches the ground, she experiences a normal force N. C The upper part of her body then travels a distance Δy while coming to a stop.

hits the ground and what strategy she should follow to minimize this force so as to make the landing as "soft" as possible. This type of problem would be of interest to someone who designs playgrounds or to a doctor who deals with children's injuries. The same sort of problem arises in various sports and activities, such as basketball and skydiving. For example, a skydiver might want to know how much force her legs will have to withstand on landing.

As usual, we begin by making a drawing for the problem, Figure 3.24. The initial part of the motion is just free fall as the child falls from the ladder to the ground. When she strikes the ground, the ground exerts a force on the child, which causes her to stop. Let's first deal with the initial motion, the free-fall phase. For simplicity, we assume the child steps off the ladder with an initial speed of zero ($v_0 = 0$). While she is falling, her motion is described by our expressions for constant acceleration in Table 3.1, and we use Equation 3.4 with an acceleration $a = -g$ to calculate her velocity just before she hits the ground. We choose the origin ($y = 0$) to be at ground level as in Figure 3.24A and the child's initial position to be $y_0 = h$, where h is the height of the ladder. We can thus calculate v_{land}, the speed of the child just before she hits the ground at $y = 0$:

$$v_{land}^2 = v_0^2 + 2a(y - y_0) = 0 + 2(-g)(0 - h) = 2gh$$

$$v_{land} = \sqrt{2gh} \tag{3.16}$$

This result is her *speed* at the end of the free-fall period; her velocity will, of course, be directed downward.

When the child reaches the ground, she begins the second phase of her motion. There are now two forces acting on her (Fig. 3.24B): gravity and the normal force from the ground N, a quantity that we want to calculate. The precise value of N and how it varies with time depends on how the child lands. To get an approximate value, we *assume* N is constant during this period. This is only an approximation to the true behavior, but it should lead to a good approximate value for the average normal force. Our goal is obtain a value for N that is accurate to within an order of magnitude (i.e., to within a factor of 10).

Identifying the "Missing" Information and Making an Estimate. We now need to consider in a little more detail the child's motion just after hitting the ground. During this time, her legs will flex at the knees, and the upper portion of her body travels a distance Δy while coming to a stop. We know intuitively that if she lands with stiff knees (very little bending), the normal force will be much higher and will act over a shorter time than if her knees flex a lot. The maximum flex possible is roughly the distance from her feet to her knees, which is about $\Delta y \approx 0.3$ m for an average child. We can use Equation 3.4 again to describe her motion while she comes to a stop.

Since the distance traveled during the deceleration phase is $y - y_0 = -\Delta y$ (see Figs. 3.24B and C), we have

$$v^2 = v_0^2 + 2a(y - y_0) = v_0^2 + 2a(-\Delta y) \tag{3.17}$$

Here v is the speed at the end of the deceleration phase, which is zero because the child comes to a stop. In Equation 3.17, v_0 is the speed at the *start* of the deceleration phase and is thus equal to her speed at the *end* of the free-fall phase, v_{land} from Equation 3.16. With these values, we can use Equation 3.17 to calculate the acceleration during this part of the motion. Once we have the acceleration, we can use Newton's second law to find the normal force N exerted by the ground.

Rearranging Equation 3.17 to solve for the acceleration, we find

$$v^2 = 0 = v_0^2 + 2a(y - y_0) = v_{\text{land}}^2 + 2a(-\Delta y)$$

$$a = \frac{v_{\text{land}}^2}{2\,\Delta y} \tag{3.18}$$

This is the acceleration of the child just after she hits the ground and before she comes to a complete stop. Using Newton's second law during this period gives

$$ma = \sum F = F_{\text{grav}} + N = -mg + N$$

$$N = mg + ma$$

Although it seems natural to take m to be the mass of the child, it may actually be more accurate to let m be the mass of only her upper body because that is what is decelerating. However, because we are aiming for an approximate answer we'll let m be her total mass. Inserting the result for the acceleration from Equation 3.18, we can solve for the normal force acting on the child:

$$N = mg + ma = mg + \frac{mv_{\text{land}}^2}{2\,\Delta y}$$

According to Equation 3.16, $v_{\text{land}}^2 = 2gh$. Therefore,

$$N = mg + \frac{mgh}{\Delta y} = mg\left(1 + \frac{h}{\Delta y}\right) \tag{3.19}$$

For a typical "experiment" using a ladder of height $h = 2$ m and a child with $\Delta y = 0.3$ m, Equation 3.19 predicts

$$N = mg\left(1 + \frac{h}{\Delta y}\right) = mg\left(1 + \frac{2}{0.3}\right) \approx mg \times 8$$

The normal force is thus about eight times the weight (mg) of the child.

We emphasize that this is only an approximate solution. In our calculation, we assumed the stopping acceleration is constant, and in practice that may not be exactly the case. Another key to the analysis was our estimate of the stopping distance Δy. Our estimate was $\Delta y = 0.3$ m; a value of 0.03 m (about 1 inch) seems much too small, whereas a value of 3 m is unrealistically large. Our value of Δy should be acccurate to better than a factor of 10, which should lead to similar accuracy in the final answer. Here and in most reasoning and relationships problems, our goal is to get the correct order of magnitude for the answer.

Besides giving a numerical estimate of the magnitude of the impact force, Equation 3.19 also provides insight on how to minimize that force. We see that a larger stopping distance Δy leads to a smaller value of N. This is why playgrounds and playground equipment are made of soft materials and why the steering column of a car is designed to collapse on impact.

The example in Figure 3.24 is intended to give some insight into how to deal with reasoning and relationships problems. A general strategy is presented on page 72.

PROBLEM SOLVING Dealing with Reasoning and Relationships Problems

1. **RECOGNIZE THE PRINCIPLE.** Determine the key physics ideas that are central to the problem and that connect the quantity you want to calculate with the quantities you know. In the examples found in this section, this physics involves motion with constant acceleration.

2. **SKETCH THE PROBLEM.** Make a drawing that shows all the given information and everything else you know about the problem. For problems in mechanics, your drawing should include all the forces, velocities, and so forth.

3. **IDENTIFY THE RELATIONSHIPS.** Identify the important physics relationships; for problems concerning motion with constant acceleration, they are the relationships between position, velocity, and acceleration in Table 3.1. For many reasoning and relationships problems, values for some of the essential unknown quantities may not be given. You must then use common sense to make reasonable estimates for these quantities. Don't worry or spend time trying to obtain precise values of every quantity (such as the amount that the child's knees flex in Fig. 3.24). Accuracy to within a factor of 3 or even 10 is usually fine because the goal is to calculate the quantity of interest to within an order of magnitude (a factor of 10). Use the Internet, the library, and (especially) your own intuition and experiences.

4. **SOLVE.** Since an exact mathematical solution is not required, cast the problem into one that is easy to solve mathematically. In the examples in this section, we were able to use the results for motion with constant acceleration.

5. Always *consider what your answer means* and check that it makes sense.

As is often the case, practice is a very useful teacher.

EXAMPLE 3.7 Ⓡ Cars and Bumpers and Walls

Consider a car of mass 1000 kg colliding with a rigid concrete wall at a speed of 2.5 m/s (about 5 mi/h). This impact is a fairly low-speed collision, and the bumpers on a modern car should be able to handle it without significant damage to the car. Estimate the force exerted by the wall on the car's bumper.

RECOGNIZE THE PRINCIPLE

The motion we want to analyze starts when the car's bumper first touches the wall and ends when the car is stopped. To treat the problem approximately, we *assume* the force on the bumper is constant during the collision period, so the acceleration is also constant. We can then use our expressions from Table 3.1 to analyze the motion. Our strategy is to first find the car's acceleration and then use it to calculate the associated force exerted by the wall on the car from Newton's second law.

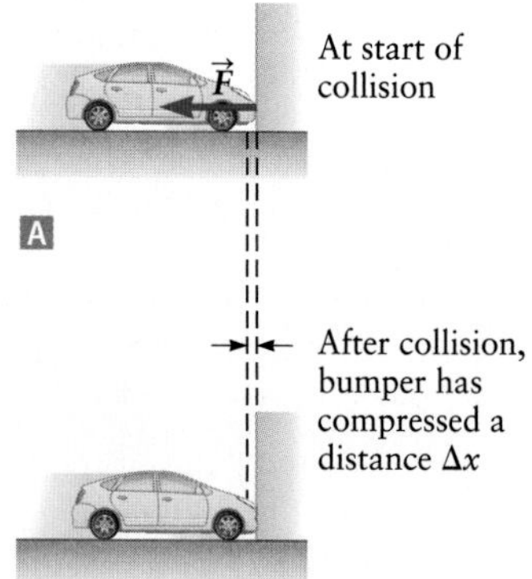

▲ **Figure 3.25** Example 3.7. **A** When a car collides with a wall, the wall exerts a force F on the bumper. This force provides the acceleration that stops the car. **B** During the collision and before the car comes to rest, the bumper deforms by an amount Δx. The car travels this distance while it comes to a complete stop.

SKETCH THE PROBLEM

Figure 3.25 shows a sketch of the car along with the force exerted by the wall on the car. There are also two vertical forces—the force of gravity on the car and the normal force exerted by the road on the car—but we have not shown them because we are concerned here with the car's horizontal (x) motion, which we can treat using $\sum F = ma$ for the components of force and acceleration along x.

IDENTIFY THE RELATIONSHIPS

To find the car's acceleration, we need to estimate either the stopping time or the distance Δx traveled during this time. Let's take the latter approach. We are given the initial velocity ($v_0 = 2.5$ m/s) and the final velocity ($v = 0$). Both of these quantities are in Equation 3.4:

$$v^2 = v_0^2 + 2a(x - x_0) \quad (1)$$

We can use this to find the acceleration, provided we can make an estimate of $\Delta x = x - x_0$. Intuition tells us that for this low-speed collision, the bumper will deform only a little. We'll guess that a deformation of $\Delta x \approx 0.1$ m is a reasonable estimate and expect it to be correct to within an order of magnitude. We do not suggest that you try this experiment at home, but you have likely witnessed it! A value $\Delta x = 0.01$ m $= 1$ cm seems too small, whereas $\Delta x = 1$ m is much too large; our estimate of 0.1 m is somewhere in the middle.

SOLVE

The final velocity is $v = 0$ because the car comes to rest at the end of the collision. Inserting this into Equation (1) leads to

$$v^2 = v_0^2 + 2a(\Delta x) = 0$$

$$a = -\frac{v_0^2}{2\,\Delta x}$$

The acceleration is negative because it is directed opposite to the initial velocity. Using $v_0 = 2.5$ m/s and our estimate $\Delta x = 0.1$ m, the magnitude of the acceleration is

$$|a| = \frac{v_0^2}{2\,\Delta x} = \frac{(2.5\text{ m/s})^2}{2(0.1\text{ m})} = 30\text{ m/s}^2$$

Applying Newton's second law for the mass $m = 1000$ kg, this acceleration requires a force of magnitude

$$|F| = m|a| = \boxed{3 \times 10^4\text{ N}}$$

► ***What does it mean?***

A key to this problem was our estimate of the stopping distance Δx. Using some intuition and common sense, we estimated Δx with an accuracy of about an order of magnitude; this level of accuracy is usually sufficient for a reasoning and relationships problem.

CONCEPT CHECK 3.6 Force on a Car Bumper

How large is the force exerted by the wall on the car bumper in Example 3.7 compared with the weight of the car?

3.7 Parachutes, Air Drag, and Terminal Speed

We have spent considerable time discussing problems in which the acceleration is constant. Such cases are very useful because the mathematics of how the position and velocity vary with time is fairly simple and also because these cases are a good approximation to many real-life examples. In this section, we consider some interesting and important cases in which the acceleration is *not* constant.

In our discussions of the motion of a ball through the air, we claimed that the force of air drag can often be neglected, with the excuse that it is generally small compared with the force of gravity for objects moving slowly through our atmosphere. However, as the speed of an object increases, the magnitude of the air drag force increases rapidly. You have probably felt the force of air drag by holding your hand outside the window of a moving car. This force is barely noticeable at low speeds, but can be large at highway speeds. The drag force also depends on how you hold your hand; that is, the force depends on the exposed surface area of the moving

Insight 3.4

DRAG FORCES DEPEND ON THE SHAPE OF THE OBJECT

According to Equation 3.20, the force due to air drag is given by $F_{\text{drag}} = \frac{1}{2}\rho A v^2$. The drag force thus depends on the size of the object through its frontal area A. This expression for the drag force is only approximate. In a more accurate treatment, the drag force also depends on the aerodynamics of the object. For example, a smoothly shaped object like a race car will have a smaller drag force than a truck with a boxy shape having the same area and speed. This difference can be accounted for through the improved relation for F_{drag}:

$$F_{\text{drag}} = \tfrac{1}{2}C_D\rho A v^2$$

where C_D is called the "drag coefficient." The value of C_D depends on the aerodynamic shape. For a "normal" shape, such as a boxy truck, $C_D \approx 1$, but for more streamlined shapes, the drag coefficient can be much smaller than 1.

object. Hence, the drag force is generally most significant for objects moving at high speeds, or with large surface areas perpendicular to the direction of motion, or both.

To get a rough idea of the importance of air drag, let's consider again the motion of a ball. To be specific, consider a baseball hit directly upward with an initial speed of 45 m/s, which is typical of what a professional player can do (about 100 mi/h). If we ignore air drag, we can use results from Section 3.4 for the motion of a thrown ball (Fig. 3.15 and Eq. 3.14) and predict that the ball will spend approximately 9.2 s in the air. For a real ball, the correct answer is about half as long. (We encourage you to observe this situation for yourself at a future baseball game.) The error is due to neglect of the force from air drag.

A high-accuracy calculation of the force due to air drag is extremely complicated, and there is no simple general "law" or formula that can be used. For common cases such as a baseball moving with speed v through the air, a good estimate of the magnitude of the drag force is

Force due to air drag

$$F_{\text{drag}} = \tfrac{1}{2}\rho A v^2 \tag{3.20}$$

where ρ is the density of the air and A is the exposed area of the moving object; that is, A is the cross-sectional area perpendicular to the direction of motion. The direction of the drag force is always opposite to the direction of the velocity. Equation 3.20 is only approximate, but it is generally useful for speeds between about 1 m/s and 100 m/s.

The form of Equation 3.20 makes good intuitive sense. The drag force is proportional to the density of the fluid through which the object is moving. Here this fluid is air, but if we were considering a fluid such as water, the density and hence the drag force would increase. The drag force also increases as the exposed cross-sectional area A of the object is made larger (larger parachutes are better than small ones) and increases rapidly with the speed of the object.

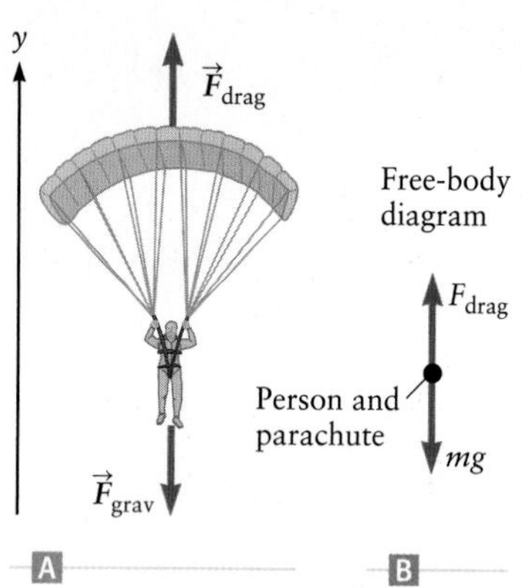

▲ **Figure 3.26** A There are two forces acting on this skydiver: the force of gravity and an upward force due to air drag. B Free-body diagram for the skydiver.

Skydiving and Air Drag

Consider a skydiver who uses a parachute (Fig. 3.26). The free-body diagram shows two forces acting on the skydiver, the force of gravity (her weight) and the force of air drag. Using Newton's second law for motion along y, we find

$$ma = \sum F = F_{\text{grav}} + F_{\text{drag}}$$

Inserting the expression from Equation 3.20 for F_{drag} gives

$$ma = -mg + \tfrac{1}{2}\rho A v^2 \tag{3.21}$$

There is no simple way to solve Equation 3.21 to get the velocity as a function of time, but we can understand the general behavior as sketched in Figure 3.27. We assume the skydiver opens her parachute immediately after exiting the plane at $t = 0$, so the drag force is acting on her throughout the jump. When she leaves the plane, her velocity in the vertical direction is initially very small, so we have $v \approx 0$ in Equation 3.21, which leads to

$$ma = -mg + \tfrac{1}{2}\rho A v^2 \approx -mg$$
$$a \approx -g$$

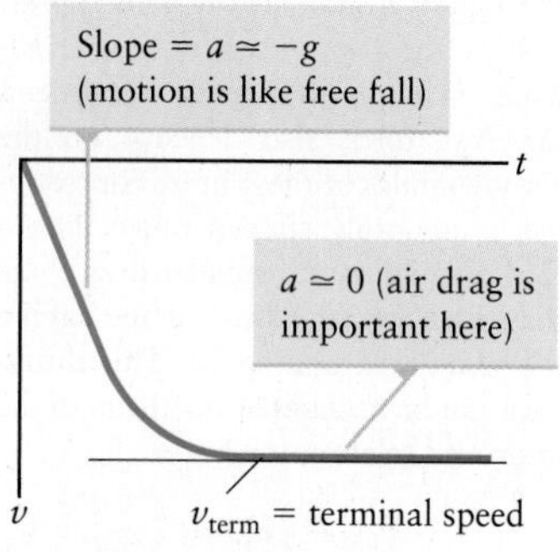

▲ **Figure 3.27** The skydiver's motion is initially like free fall; compare with Figure 3.16A. Eventually, however, air drag becomes as large as the force of gravity and the skydiver reaches her terminal speed v_{term}.

Hence, the initial motion is approximately the same as if no air drag were present. We can then use our standard result for motion with a constant acceleration (Eq. 3.1), with $v_0 = 0$ and $a = -g$, to get $v \approx -gt$. The skydiver's speed thus increases with time. However, as this speed increases she will eventually reach a time when the drag force is not small, and we must then include the effect of air drag. In fact, the speed will eventually become large enough that F_{drag} is nearly equal in magnitude, and opposite in direction, to the force of gravity so that the total force on the skydiver is very close to zero. From Newton's second law, we know that her acceleration is then also approximately zero and her speed is constant as sketched in Figure 3.27. This constant speed is known as the ***terminal speed***, and the skydiver maintains this speed for the remainder of her jump. (The word *terminal* refers to the end of the skydiver's jump and not to the end of something else.) When the skydiver reaches her terminal speed, the total force is zero and she is moving at a constant speed equal to v_{term}, and

$$\sum F = 0 = F_{\text{grav}} + F_{\text{drag}} = -mg + \tfrac{1}{2}\rho A v_{\text{term}}^2$$

Solving for the magnitude of v_{term}, we find

$$v_{\text{term}} = \sqrt{\frac{2mg}{\rho A}} \tag{3.22}$$

For a skydiver of mass 60 kg with a parachute of area 100 m^2, and using the density of air (ρ = 1.3 kg/m^3), we find a terminal speed v_{term} = 3.0 m/s. This is approximately 7 mi/h, which should be an acceptably gentle landing.

The message here is that air drag *cannot* always be neglected. The drag force increases rapidly with velocity. Air drag is generally small for slowly moving objects, but it is very important in some cases.

CONCEPT CHECK 3.7 Acceleration of a Skydiver

Figure 3.27 shows a skydiver's velocity as a function of time. Which of the plots in Figure 3.28 shows the qualitative behavior of the skydiver's acceleration as a function of time?

(a) Plot (A) (b) Plot (B) (c) Plot (C) (d) Plot (D)

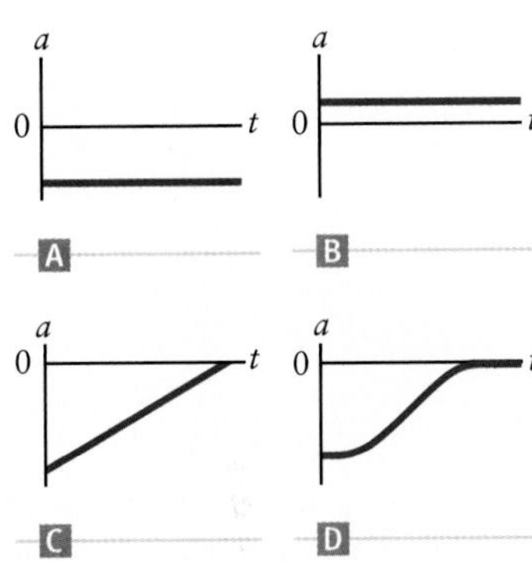

▲ **Figure 3.28** Concept Check 3.7.

EXAMPLE 3.8 Ⓡ Air Drag on a Car

Estimate the drag force on a car moving at 50 mi/h and at 150 mi/h and compare it to the weight of a typical car with m = 1000 kg. The density of air is ρ = 1.3 kg/m^3. *Hint*: Treat this as a reasoning and relationships problem and consider what essential information is missing from the description of the problem.

RECOGNIZE THE PRINCIPLE

The drag force depends on the car's speed v, the density of the air ρ, and the frontal area A (the cross-sectional area perpendicular to the direction of motion, Fig. 3.29). We are given v and ρ, but not A. Hence, A is the "missing" information, and we must estimate a typical car's frontal area.

SKETCH THE PROBLEM

The problem is shown in Figure 3.29.

IDENTIFY THE RELATIONSHIPS

The frontal area A depends on the car's size and shape. When viewed from the front, a typical car is about 2 m wide and 1.5 m tall, so we estimate A to be approximately 3 m^2.

SOLVE

The magnitude of the drag force is given by Equation 3.20,

$$F_{\text{drag}} = \tfrac{1}{2}\rho A v^2$$

To evaluate F_{drag}, it is convenient to express the speed in units of meters per second. For 50 mi/h, we find

$$50\ \frac{\text{mi}}{\text{h}} = 50\ \frac{\text{mi}}{\text{h}} \times \frac{1600\ \text{m}}{1\ \text{mi}} \times \frac{1\ \text{h}}{3600\ \text{s}} = 22\ \text{m/s}$$

(continued) ▶

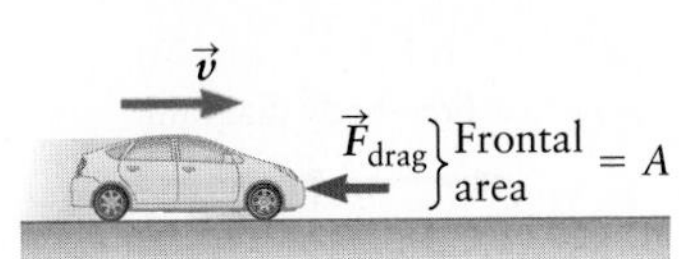

▲ **Figure 3.29** Example 3.8. The drag force on a car depends on the frontal area A of a cross section perpendicular to the car's motion. The drag force always opposes the velocity.

Inserting this value into our expression for F_{drag} along with our values of ρ and A leads to

$$F_{drag} = \tfrac{1}{2}\rho A v^2 = \tfrac{1}{2}(1.3\ \text{kg/m}^3)(3\ \text{m}^2)(22\ \text{m/s})^2 = \boxed{1000\ \text{N}}$$

The mass of the car is 1000 kg, so its weight is $mg = 9800$ N, which is about 10 times larger than this drag force. Repeating the calculation for a speed of 150 mi/h (= 67 m/s), we find $F_{drag} = \boxed{10{,}000\ \text{N}}$. This is about a factor of 10 larger than the result at 50 mi/h; the difference can be traced to the strong dependence of the drag force on the speed. At 150 mi/h, the air drag force on a car is approximately equal to the car's weight.

► *What does it mean?*
These estimates for the drag force ignore the aerodynamics of the car because our simple expression for the drag force (Eq. 3.20) does not account for the effect of an object's aerodynamic shape. (See Insight 3.4.) For this reason, Equation 3.20 provides only an approximate value of F_{drag}. Even so, our results show that the air drag force on a car is quite substantial at high speeds. That is why a car's fuel efficiency decreases significantly at speeds greater than about 50 mi/h.

3.8 Life as a Bacterium

In this section, we consider the motion of a small object in a liquid, such as the *E. coli* bacterium discussed in Chapter 2. Bacteria do not move as fast as baseballs or skydivers, and the expression for the drag force in Equation 3.20 is not very accurate for a swimming *E. coli*. For this case—and for other situations involving small things moving slowly (less than about 1 m/s or so)—the drag force is described much more accurately by an expression worked out in the 19th century by George Stokes. He applied Newton's laws to this problem and showed that the drag force on a spherical object moving slowly through a fluid is given by

Drag force in a liquid

$$\vec{F}_{drag} = -Cr\vec{v} \tag{3.23}$$

where C is a constant that depends on the properties of the fluid and r is the radius of the object. For water, $C \approx 0.02\ \text{N} \cdot \text{s/m}^2$. The negative sign in Equation 3.23 indicates that the drag force is always directed opposite to the velocity. Strictly speaking, Equation 3.23 applies only to spherical objects. Of course, most bacteria are not spherical (!), but we can usually approximate their shape at least roughly in this way.

Let's now consider how *E. coli* move about in a fluid. Recall from Section 2.5 that an *E. coli* possesses one or more flagella that it rotates to propel itself (Fig. 3.30). Writing Newton's second law for motion along the horizontal direction in Figure 3.30, we have

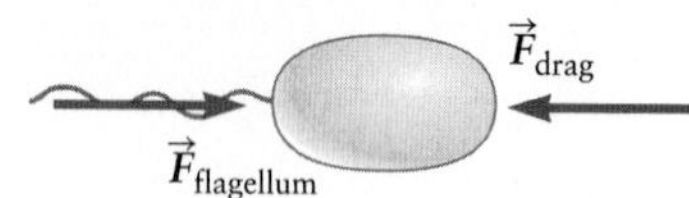

Free-body diagram

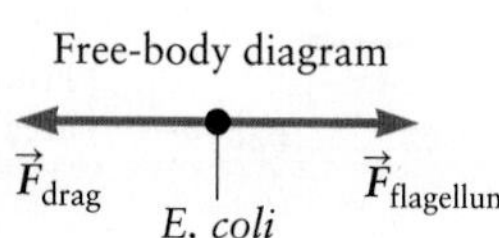

▲ Figure 3.30 The two dominant forces on an *E. coli* are from its flagellum (propeller) and from the drag force. These forces are in opposite directions.

$$ma = \sum F = F_{flagellum} + F_{drag}$$
$$ma = F_{flagellum} - Crv \tag{3.24}$$

The drag force on a bacterium is very large compared to its weight, so these microbes cannot move very rapidly. In fact, their acceleration is always so small that the two terms on the right-hand side of Equation 3.24 are both much larger in magnitude than ma in any realistic situation. In other words, the forces on the bacterium are both large, but they are in opposite directions and they almost exactly cancel. To a very good approximation, we can therefore simply set the right-hand side equal to zero and solve for the velocity, with the result

$$v = \frac{F_{flagellum}}{Cr} \tag{3.25}$$

Hence, in this case when the flagellum stops turning and the force stops, the bacterium stops moving immediately; it does not coast!

It is common for thrown balls and sliding hockey pucks to coast since the frictional and drag forces they encounter are usually small. However, for a bacterium in water the drag force is substantial and leads to very different behavior. In either case, though, the object's motion is described correctly and accurately by Newton's laws.

Summary | CHAPTER 3

Key Concepts and Principles

Forces and motion

A force is a push or a pull exerted by one object on another. Force is a ***vector*** quantity.

All calculations of motion start with Newton's second law. For one-dimensional motion, the forces and acceleration are directed along a line, and Newton's second law reads

$$\sum F = ma$$

A ***free-body diagram*** is a very useful first step in the application of Newton's second law.

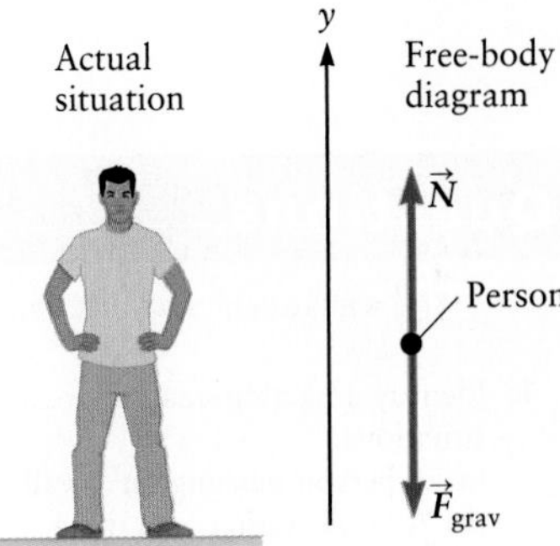

Applications

Motion with constant acceleration

A common and extremely important case is motion with constant acceleration. When a is constant, the position, velocity, and acceleration are related by

$$x = x_0 + v_0 t + \tfrac{1}{2}at^2 \qquad \textbf{(3.3) (page 50)}$$

$$v = v_0 + at \qquad \textbf{(3.1) (page 50)}$$

$$v^2 = v_0^2 + 2a(x - x_0) \qquad \textbf{(3.4) (page 51)}$$

Common types of forces

Gravity

Near the surface of the Earth, the gravitational force on an object of mass m is directed along the vertical direction and is given by

$$F_{grav} = -mg \qquad \textbf{(3.8) (page 57)}$$

The negative sign here indicates that F_{grav} is downward. This force is also called the ***weight*** of the object. When an object is moving freely under the action of gravity, the motion is called ***free fall*** and the acceleration is $a = -g$.

Normal force

When two surfaces are in contact, there is a normal force between them. This force is perpendicular to the contact surface.

Friction

When two surfaces are in contact, a frictional force opposes motion of the surfaces relative to each other. If the surfaces are slipping relative to each other, the opposing force is ***kinetic friction*** given by

$$F_{friction} = \mu_K N \qquad \textbf{(3.9) (page 58)}$$

where N is the normal force. If the surfaces are not slipping, the resistive force is ***static friction*** given by

$$|F_{friction}| \leq \mu_S N \qquad \textbf{(3.11) (page 60)}$$

(Continued)

Drag forces

When an object moves through a fluid such as air or water, there is a drag force arising from contact between the fluid molecules and the moving object. We consider drag forces in two regimes:

- For motion through the air, the magnitude of the drag force is described approximately by

$$F_{\text{drag}} = \tfrac{1}{2}\rho A v^2 \qquad (3.20)\ (\text{page } 74)$$

This regime is appropriate for skydivers, cars, and baseballs.

- For motion of cells and bacteria through water, the drag force is given by

$$\vec{F}_{\text{drag}} = -Cr\vec{v} \qquad (3.23)\ (\text{page } 76)$$

Questions

SSM = answer in *Student Companion & Problem-Solving Guide*

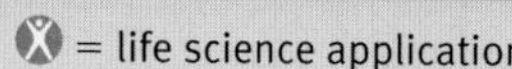

= life science application

1. Identify an action–reaction pair of forces in each of the following situations.
(a) A person pushing on a wall
(b) A book resting on a table
(c) A hockey puck sliding across an icy surface
(d) A car accelerating from rest (*Hint*: Consider the tires.)
(e) An object undergoing free fall in a vacuum
(f) A basketball player jumping to dunk a basketball
(g) A person throwing a baseball

2. Give an example of motion in which the acceleration and the velocity are in opposite directions.

3. Give an example of motion in one dimension for an object that starts at the origin at $t = 0$. At some time later, the displacement from the origin is zero but the velocity is not zero. Draw the corresponding position–time and velocity–time graphs.

4. Two objects are released simultaneously from the same height. The objects are both spherical with the same radius, but their masses differ by a factor of two. (a) Ignoring air drag, which object strikes the ground first? Or do they strike at the same time? Explain. (b) Including air drag, which object strikes the ground first? Or do they strike at the same time? Explain. (c) Find two household objects that might be used to demonstrate part (b).

5. Two mountain climbers are suspended by a rope as shown in Figure Q3.5. Which rope is most in danger of breaking? That is, which rope has the greatest tension?

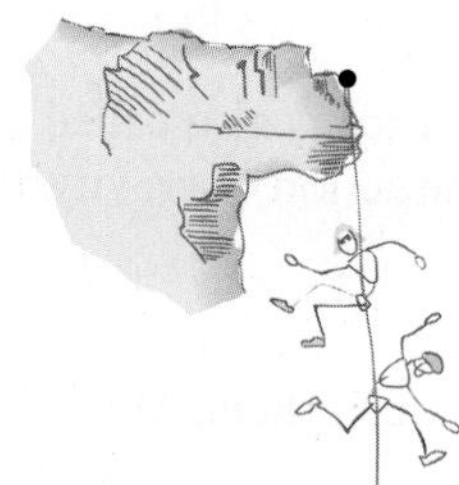

Figure Q3.5

6. You are riding in a car that starts from rest, accelerates for a short distance, and then moves with constant velocity. Explain why you feel a force from the back of your seat only while the car is accelerating and not while you are moving with a constant velocity.

7. Two balls are thrown from a tall bridge. One is thrown upward with an initial velocity $+v_0$, while the other is thrown downward with an initial velocity $-v_0$. Which one has the greater speed just before it hits the ground?

8. SSM (a) Suppose a tire rolls without slipping on a horizontal road. Explain the role friction plays in this motion. What two surfaces are involved in this frictional force? Is it static friction or kinetic friction?
(b) Suppose a racecar driver wants to get his car started very quickly and "burns rubber" as he leaves the starting line. Is he exploiting static friction or kinetic friction?
(c) Normally, the coefficient of kinetic friction between two surfaces is smaller than the coefficient of static friction for the same two surfaces. Assuming that is true for the case of friction between a car tire and a road, explain why the car will accelerate faster if the driver does not "burn rubber."

9. The lower piece of silk in Figure 3.20 is acted on by two forces, $+T_2$ at the upper end and $-T_2$ at the lower end. These two forces are equal in magnitude and opposite in direction. Are they an action–reaction force pair? Explain why or why not.

10. Devise a block-and-tackle arrangement that amplifies the applied force by a factor of three.

11. Imagine that you are a passenger on the International Space Station. While on the station, you are "weightless" and so is everything else, including wrenches and furniture. Even though everything appears weightless, explain why mass still matters. Why does it still take longer for a given force to move a massive object a certain distance than to move an object with a very small mass the same distance?

12. Imagine a skydiver who waits a long time before opening her parachute. For simplicity, assume she moves along a straight line. (a) In what direction is the skydiver moving (what is the direction of her velocity) just before she opens her parachute? (b) Suppose that when she finally opens her parachute, she is traveling faster than her terminal speed with the parachute open. What is the direction of the skydiver's acceleration just after she opens her parachute?

13. Explain why the air bags in a car reduce the forces on a passenger in the event of an accident.

14. For the books in Figure Q3.14, there is a force F from the table supporting the book on the bottom. Identify two action–reaction force pairs for each of the following objects: (a) the table,

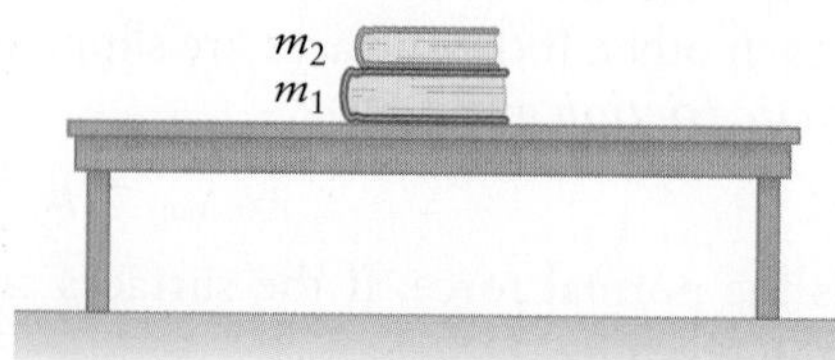

Figure Q3.14

(b) the book on top, (c) the book on the bottom, and (d) the Earth.

15. The bacterium in Figure 3.30 experiences a force due to drag from the surrounding fluid. Does the reaction force to F_{drag} act on (a) the flagellum, (b) the body of the bacterium, or (c) the fluid?

16. SSM Two balls of the same diameter are dropped simultaneously from a very tall bridge. One ball is solid lead, and the other is hollow plastic and has a much smaller mass than the solid lead ball. Use a free-body diagram to explain why the solid lead ball reaches the ground first. *Hint*: Include the air drag force in your analysis.

17. Consider a string with one end tied to a tall ceiling and the other end hanging freely. Explain why the tension at the bottom of the string is smaller than the tension at the top.

18. In our discussion of the block and tackle in Figure 3.23, we claimed that when the right end of the string is lifted through a distance L, the body of the pulley is lifted a distance $L/2$. Give a geometric argument to prove this claim.

19. An astronaut measures her mass and her weight on Earth and again when she reaches the Moon. Which one of these quantities changes as a result of this trip and which one does not? Explain.

20. One end of a string is tied to the ceiling of an elevator, and the other end is tied to a rock (Fig. Q3.20). The elevator is moving in such a way that the tension in the string is zero. (a) What is the acceleration of the rock? (b) Explain why you cannot find the velocity of the rock, even though you can find its acceleration.

Figure Q3.20

Problems available in ENHANCED WebAssign

Section 3.1 Motion of a Spacecraft in Interstellar Space
Problems 1–26

Section 3.2 Normal Forces and Weight
Problems 27–36

Section 3.3 Adding Friction to the Mix
Problems 37–47

Section 3.4 Free Fall
Problems 48–59

Section 3.5 Cables, Strings, and Pulleys: Transmitting Forces from Here to There
Problems 60–65

Section 3.6 Reasoning and Relationships: Finding the Missing Piece
Problems 66–70

Section 3.7 Parachutes, Air Drag, and Terminal Speed
Problems 71–78

Section 3.8 Life as a Bacterium
Problems 79–80

Additional Problems
Problems 81–99

List of Enhanced Problems

Problem number	Solution in *Student Companion & Problem-Solving Guide*	Reasoning & Relationships Problem	Reasoning Tutorial in ENHANCED WebAssign
3.11	✓		
3.23	✓		
3.34	✓		
3.41	✓		
3.53	✓		
3.63	✓		
3.66		✓	✓
3.67		✓	✓
3.68		✓	✓
3.70	✓	✓	
3.71		✓	✓
3.73		✓	
3.75		✓	
3.77	✓		
3.79	✓	✓	✓
3.80		✓	
3.81		✓	✓
3.87		✓	✓
3.90		✓	
3.94		✓	
3.96		✓	
3.97	✓		
3.98		✓	
3.99		✓	

CHAPTER

4

◀ *These Adelie penguins diving from an iceberg demonstrate motion in two dimensions. In this chapter, we learn how to describe and predict their motion.* (© *Joseph Van Os, Getty Images*)

Forces and Motion in Two and Three Dimensions

In Chapter 3, we considered how to apply Newton's laws to deal with motion in one dimension (along a line). We now build on our understanding of motion in one dimension and deal with forces and motion in two and three dimensions. Newton's laws are again our foundation, and we continue to describe motion in terms of displacement, velocity, and acceleration. Now, though, we must allow for the vector nature of force, displacement, velocity, and acceleration. Recall that vector quantities have both a magnitude and a direction. For one-dimensional motion, we chose a positive direction at the start of a problem and used an algebraic sign (+ or −) to denote the directions of the displacement, velocity, and acceleration. For motion in two or three dimensions, we must express the directions of these vectors according to our chosen coordinate axes. Once that is done, we can use the same principles and problem-solving techniques we developed in Chapter 3.

OUTLINE

4.1 Statics

We begin with Newton's second law in vector form,

Newton's second law

$$\sum \vec{F} = m\vec{a} \tag{4.1}$$

where $\sum \vec{F}$ is the total force acting on the object. When applying Newton's second law, we generally start by determining all the individual forces acting on the object and construct a free-body diagram, showing these forces in graphical form. These individual forces must be added *as vectors* to determine the total force $\sum \vec{F}$ acting on the object. Once the total force is known, Equation 4.1 can be used to compute the object's acceleration, which then leads to its velocity and displacement.

Conditions for Translational Equilibrium

Let's first consider problems in which the velocity and acceleration are both zero. This area of mechanics is known as ***statics***, and in such situations we say that an object is in ***translational equilibrium***. The problem of when an object is or is not in translational equilibrium is very important for many engineering applications. (Imagine being on a bridge or in a building that is not in translational equilibrium, but is accelerating instead!) To simplify our terminology, we often drop the word *translational* and refer to such objects as being in "static equilibrium" or just "equilibrium."

From Equation 4.1, we see that if the acceleration is zero, the total force on the object $\sum \vec{F}$ must also be zero. Since there may be several separate forces $\vec{F}_1, \vec{F}_2, \vec{F}_3, \ldots$ acting on an object, the sum of all these forces must obey

$$\sum \vec{F} = \vec{F}_1 + \vec{F}_2 + \vec{F}_3 + \cdots = 0 \tag{4.2}$$

which is known as the ***condition for translational equilibrium***.[1]

Let's apply the condition for translational equilibrium to a situation we encountered in Chapter 3. Figure 4.1 shows a person attempting to push a refrigerator across a level floor.[2] The person is applying only a small force, so the refrigerator is at rest with zero acceleration. There are four forces acting on the refrigerator. Two of these forces are in the vertical direction (y): the force of gravity and the normal force acting between the floor and the refrigerator. Two other forces are acting in the horizontal direction (x): the force exerted by the person and the force of static friction.

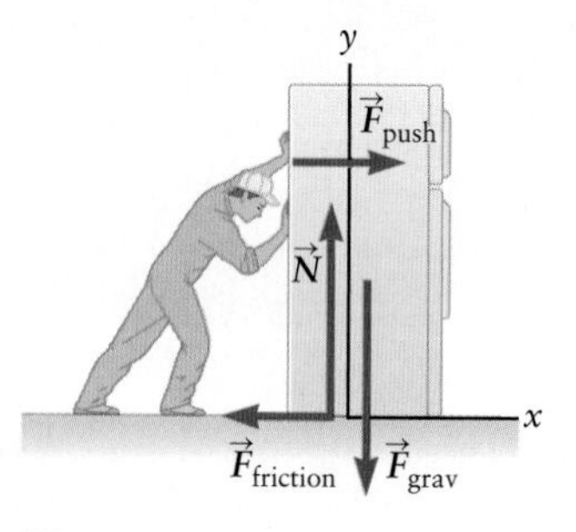

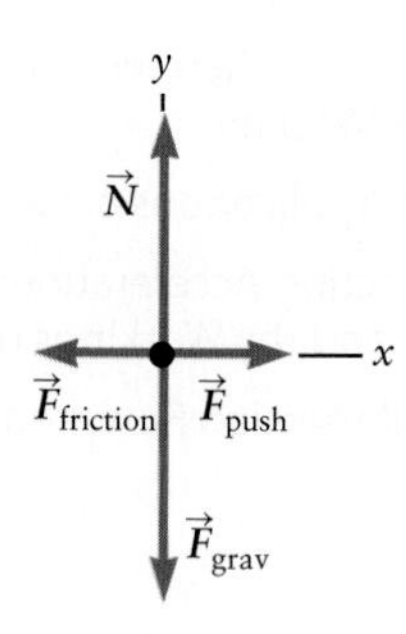

▲ **Figure 4.1** A Four forces are acting on this refrigerator. Two of them, the normal force $\vec{N}$ exerted by the floor and the force of gravity $\vec{F}_{grav}$, lie along the vertical direction (y). The other two forces, $\vec{F}_{push}$ exerted by the person and the force of friction $\vec{F}_{friction}$, lie along the horizontal direction (x). B Free-body diagram for the refrigerator.

To apply Equation 4.2 to this case, we first draw a free-body diagram (Fig. 4.1B). This diagram also shows our coordinate system, with the usual horizontal and vertical axes denoted x and y. Our next step is to express the forces in terms of their components along x and y. In this example, that work is already done for us since each of the four forces is directed along either x or y. We can then proceed to apply Equation 4.2. In this case, Equation 4.2 actually contains two relations, one for the x components of the forces and another for the y components:

$$\sum F_x = 0 \quad \text{and} \quad \sum F_y = 0 \tag{4.3}$$

Hence, for the refrigerator to be in equilibrium, the sum of all the forces along x must be zero and the sum of all forces along y must be zero. From Figure 4.1, we have $N - mg = 0$ for the y components and $F_{push} - F_{friction} = 0$ for the x components. These results are identical to what we derived in Section 3.3. This analysis could then be the first step in finding the minimum value of F_{push} needed to just break the equilibrium condition and accelerate the refrigerator.

Figure 4.2 shows a case in which the forces do not all align with the x or y directions. Here a sled is stuck in the snow, and a child is attempting to extract it by pull-

[1]When we discuss objects in translational equilibrium, we will also assume their velocity is zero.

[2]For simplicity, we assume the refrigerator does not tip.

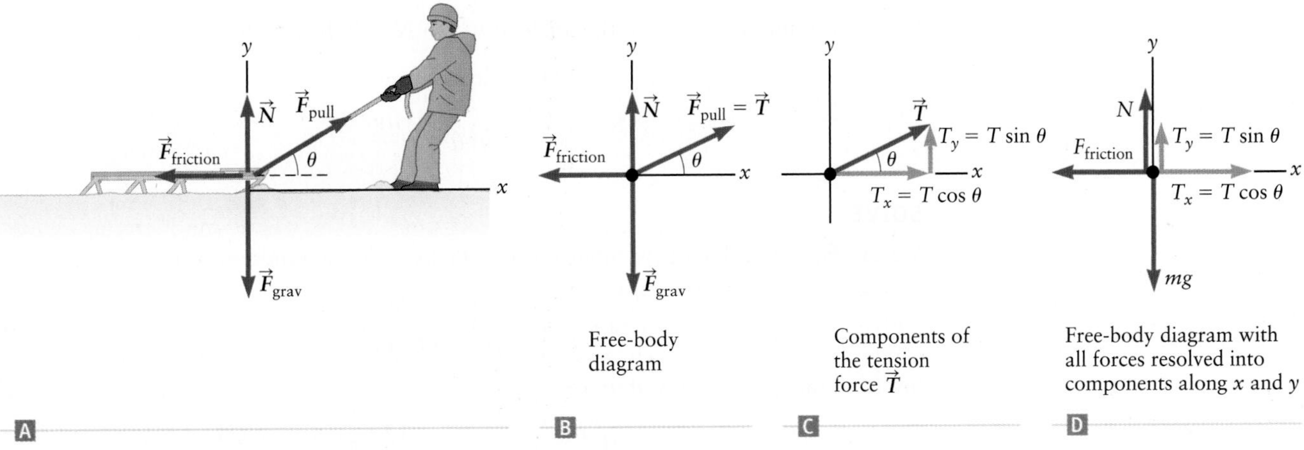

▲ **Figure 4.2** A The force exerted by the rope on the sled has components along the horizontal and the vertical directions. B Free-body diagram for the sled. C Resolving the tension force into x and y components. D Modified free-body diagram that can be used to write equations for the force components in both the x and y directions.

ing with a rope that makes an angle θ with respect to the x axis. If the child pulls with a force $\vec{F}_{pull}$, the force on the sled due to the tension in the rope is $\vec{T} = \vec{F}_{pull}$. The free-body diagram in Figure 4.2B shows that several of the forces are parallel to our chosen coordinate axes. The normal force is parallel to y, the force of gravity is along $-y$, and the force of friction is parallel to the x axis. However, the force exerted by the rope on the sled has components along both x and y. These components are shown in Figure 4.2C, where we also show how the vector $\vec{T}$ and its components along x and y form a right triangle with components $T_x = T\cos\theta$ along x and $T_y = T\sin\theta$ along y.

In Figure 4.2D, we have redrawn the free-body diagram with $\vec{T}$ replaced by its components, so all the forces in the free-body diagram are now parallel to the coordinate axes. We next apply the conditions for equilibrium in two dimensions (Eq. 4.3). For the forces along x,

$$\sum F_x = T_x - F_{friction} = T\cos\theta - F_{friction} = 0$$

whereas for the forces along y,

$$\sum F_y = N - mg + T_y = N - mg + T\sin\theta = 0 \tag{4.4}$$

In Example 4.1, we show how this line of analysis can be used.

EXAMPLE 4.1 Trying to Move a Sled

Suppose the sled in Figure 4.2 has a mass of 12 kg and the child is exerting a force of magnitude $F_{pull} = T = 200$ N. Find the angle θ at which the normal force N becomes zero. What is happening physically when $N = 0$?

RECOGNIZE THE PRINCIPLE

We need to apply the conditions for static equilibrium to the sled. Given the mass of the sled and the magnitude of F_{pull}, these conditions cannot be satisfied if the angle θ is too large.

SKETCH THE PROBLEM

Figure 4.2 shows all the forces on the sled and their components along x and y.

IDENTIFY THE RELATIONSHIPS

From Equation 4.4, for the forces along the vertical (y),

$$\sum F_y = +N - mg + T\sin\theta = N - mg + F_{pull}\sin\theta = 0$$

(continued) ▶

We want to find the value of the angle at which $N = 0$. Inserting this condition leads to

$$F_{\text{pull}} \sin \theta = mg$$

$$\sin \theta = \frac{mg}{F_{\text{pull}}}$$

SOLVE

We can find the value of the angle θ using the inverse sine function ($\sin^{-1}$):

$$\theta = \sin^{-1}\left(\frac{mg}{F_{\text{pull}}}\right)$$

Inserting the given values of m and F_{pull} gives us

$$\theta = \sin^{-1}\left(\frac{mg}{F_{\text{pull}}}\right) = \sin^{-1}\left[\frac{(12 \text{ kg})(9.8 \text{ m/s}^2)}{200 \text{ N}}\right]$$

$$\theta = \sin^{-1}(0.59) = \boxed{36°}$$

▸ ***What does it mean?***

The normal force is due to contact between the sled and the ground. When the child pulls at an angle of 36°, the normal force goes to zero, indicating that contact between the sled and the ground is being lost.

CONCEPT CHECK 4.1 Pulling on the Sled

What will happen in Example 4.1 if the angle θ is made even larger than 36°?

(a) The sled will not move.
(b) The sled will be lifted off the ground.
(c) The rope will break.

© Bettmann/Corbis

A

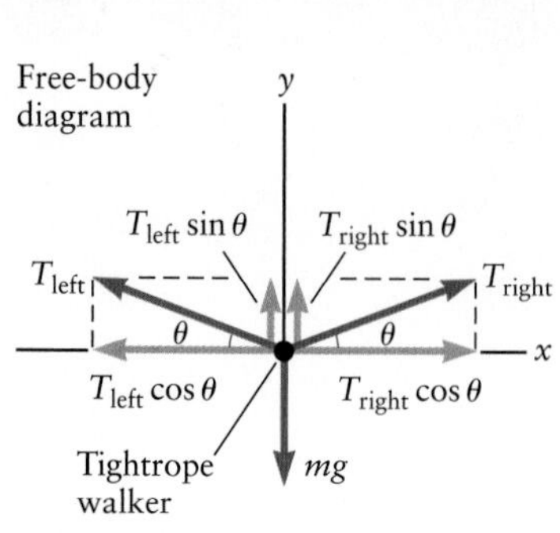

B

▲ **Figure 4.3** A Both sections of the rope exert a tension force on the portion at the center where the tightrope walker is standing. B Free-body diagram for a tightrope walker located at the rope's center point.

A Tightrope Walker in Equilibrium

Figure 4.3 shows an interesting equilibrium situation in which several forces are involved. The photo in part A shows a person standing at the middle of a tightrope. Suppose the walker has a mass $m = 60$ kg and the tension in the rope is $T = 800$ N. What angle θ does the rope make with the x axis?

We construct, as usual, a free-body diagram (Fig. 4.3B). Since the tightrope walker and the rope are assumed to be at rest, we can apply our conditions for translational equilibrium to the small piece of rope on which the tightrope walker stands. Imagine the rope is composed of two separate pieces, left and right, tied together with a small knot, with the tightrope walker standing on the knot. These two pieces of rope have tensions T_{right} and T_{left}. The walker is located at the midpoint of the rope, so you should suspect that these tensions are equal. (In the end, we'll set both of them equal to T.) However, when analyzing the equilibrium conditions, it is useful to distinguish between the tensions on both sides of the knot.

The free-body diagram in Figure 4.3B shows the forces acting on the knot. There are tension forces from the left and right sections of the rope as well as a downward force $\vec{F}_{\text{grav}}$ whose magnitude is equal to the weight of the tightrope walker (mg). Our next step is to pick a coordinate system; we choose the usual x–y coordinate axes along the horizontal and vertical directions, respectively. We then express all the forces in terms of their components along x and y as shown in Figure 4.3B. With these components, we can apply the conditions for translational equilibrium in Equation 4.3 to get

$$\sum F_x = +T_{\text{right}} \cos \theta - T_{\text{left}} \cos \theta = 0 \tag{4.5}$$

$$\sum F_y = +T_{\text{right}} \sin \theta + T_{\text{left}} \sin \theta - mg = 0 \tag{4.6}$$

From Equation 4.5,

$$T_{\text{right}} \cos\theta = T_{\text{left}} \cos\theta$$

and hence

$$T_{\text{right}} = T_{\text{left}}$$

The tensions in the two sides of the rope are thus equal (as we had already expected), and we can write $T_{\text{right}} = T_{\text{left}} = T$, where T is "the" tension in the rope. From Equation 4.6, we see that there are two vertical components of the tension forces, one from the left side of the rope and one from the right. These forces act to support the tightrope walker. Inserting $T_{\text{right}} = T_{\text{left}} = T$ into Equation 4.6 gives

$$+T_{\text{right}} \sin\theta + T_{\text{left}} \sin\theta = 2T\sin\theta = mg$$

We can now solve for θ:

$$\sin\theta = \frac{mg}{2T}$$

$$\theta = \sin^{-1}\left(\frac{mg}{2T}\right)$$

Inserting the given values of m and T leads to

$$\theta = \sin^{-1}\left(\frac{mg}{2T}\right) = \sin^{-1}\left[\frac{(60\text{ kg})(9.8\text{ m/s}^2)}{2(800\text{ N})}\right] = 22°$$

PROBLEM SOLVING Plan of Attack for Problems in Statics

1. **RECOGNIZE THE PRINCIPLE.** For an object to be in static equilibrium, the sum of all the forces on the object must be zero. This principle leads to Equation 4.2, which can be applied to calculate any unknown forces in the problem.
2. **SKETCH THE PROBLEM.** It is usually a good idea to show the given information in a picture, which should include a coordinate system. Figures 4.1 through 4.3 and the following examples provide guidance and advice on choosing coordinate axes.
3. **IDENTIFY THE RELATIONSHIPS.**
   - Find all the forces acting on the object that is (or should be) in equilibrium and construct a free-body diagram showing all the forces on the object.
   - Express all the forces on the object in terms of their components along x and y.
   - Apply the conditions $\sum F_x$ and $\sum F_y = 0$.
   - Forces along the z direction, if any, should also be included in your free-body diagram, and the additional condition $\sum F_z = 0$ must be applied.
4. **SOLVE.** Solve the equations resulting from step 3 for the unknown quantities. The number of equations must equal the number of unknown quantities.
5. Always *consider what your answer means* and check that it makes sense.

EXAMPLE 4.2 Stuck in the Mud

Your car is stuck in the mud, and you ask a friend to help you pull it free using a cable. You tie one end of the cable to your car and then pull on the other end with a force of 1000 N. Unfortunately, the car does not move. Your friend then suggests you tie the other end of the cable to a tree as shown from above in Figure 4.4. Although you are

(continued) ▶

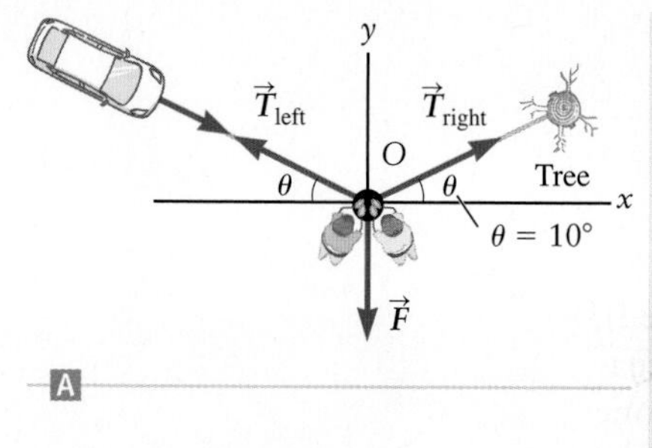

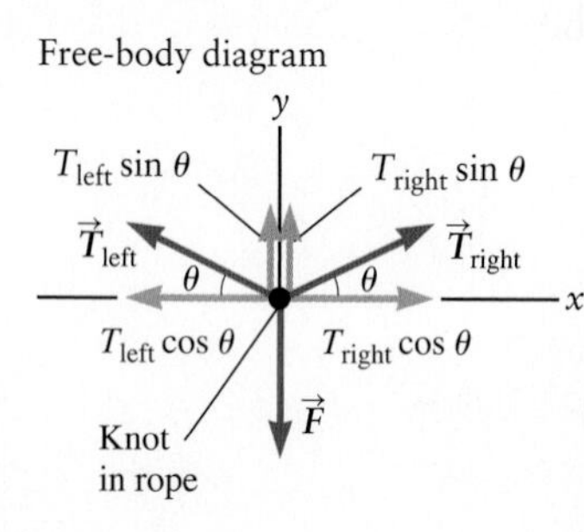

▲ **Figure 4.4** Example 4.2. A Pulling on a car. B Free-body diagram for the "imaginary knot" at the center of the cable. These drawings show top views of the predicament. (*Note*: The angle θ is not drawn to scale.)

skeptical that your friend's idea will help, you try it anyway and find that when a force $\vec{F}$ with the same magnitude (1000 N) is applied to the middle of the cable in the direction shown in Figure 4.4, you are able to pull the car free. Why does this work? Assume an angle $\theta = 10°$ as indicated in Figure 4.4.

RECOGNIZE THE PRINCIPLE

Our objective is to calculate the tension in the cable given the force $\vec{F}$ exerted by you and your friend as shown in Figure 4.4. The value we find for the tension T will *not* necessarily be the same as the magnitude of $\vec{F}$. We imagine the car is just on the verge of moving, so we can apply our conditions for equilibrium to this situation. We need to think carefully in selecting the object to which we should apply these conditions. We could apply them to the car, but that would not help calculate T. However, if we consider the forces acting on the cable at point O (the point at which you and your friend exert your force), we can then find the tension in terms of the applied force $\vec{F}$. This approach is similar to the problem of the tightrope walker in Figure 4.3, where we considered the forces applied to an imaginary knot in the rope.

SKETCH THE PROBLEM

We follow the steps listed in the Plan of Attack for Problems in Statics and apply the conditions for static equilibrium to the cable at point O in Figure 4.4A. Part B of the figure shows a free-body diagram with the forces acting on the knot at point O.

IDENTIFY THE RELATIONSHIPS

Three forces act on the cable at point O: the force $\vec{F}$ and the tension forces from each side of the cable. Choosing the x–y coordinate system shown in the figure, we calculate the components of these three forces along x and y, and then apply Equation 4.3. Along y,

$$\sum F_y = +T_{\text{right}} \sin \theta + T_{\text{left}} \sin \theta - F = 0 \tag{1}$$

Since we have a single continuous cable and the angles on the two sides are equal, the tensions in the left and right portions of the cable are the same, $T_{\text{right}} = T_{\text{left}} = T$. Inserting this result into Equation (1) leads to

$$2T \sin \theta - F = 0 \tag{2}$$

SOLVE

Rearranging Equation (2) to solve for T, we find

$$T = \frac{F}{2 \sin \theta}$$

Inserting the value of θ (Fig. 4.4) gives

$$T = \frac{F}{2 \sin \theta} = \frac{F}{2 \sin(10°)} \approx \boxed{2.9 \times F}$$

▶ ***What does it mean?***

The tension in the cable is thus *larger* than F! By tying the cable to a tree, T is larger than it would have been had you simply pulled on one end of the cable with the same force F. This arrangement essentially *amplifies* the force applied to the car. Recall that we encountered another example of amplifying forces in Chapter 3 when we discussed the block-and-tackle device. We'll say more about amplifying forces when we discuss work and energy in Chapter 6.

Static Equilibrium and Frictional Forces

Consider the situation in Figure 4.5 in which a car is parked on a steep hill. Imagine it is winter, when the surface of the hill might be slippery, and we want to determine if we can park the car safely or if it will instead slide down the hill. That is, we want to know if the car will be in equilibrium. Following our general strategy in the Plan of Attack for Problems in Statics, Figure 4.5B shows the free-body diagram for the car. This diagram contains three forces: the force from gravity $\vec{F}_{grav}$, the normal force $\vec{N}$ exerted by the surface, and the force of friction $\vec{F}_{friction}$, which is directed up the hill. Our next step is to choose a coordinate system. It might seem natural to choose coordinate axes that are along the horizontal and the vertical. Although this choice is a reasonable one, let's instead choose axes that are parallel and perpendicular to the plane defined by the hill as shown in Figure 4.5. With this choice of coordinate axes, the normal force is purely along y, whereas the frictional force is along x; we only have to worry about finding the components of the gravitational force.[3] Another reason for choosing x to be parallel to the hill is that were the car to move (as might happen in future problems), its velocity and acceleration would be purely along x. The trigonometry needed to find the components of the gravitational force $\vec{F}_{grav}$ along x and y is shown in Figure 4.5B.

Using the forces and components from Figure 4.5B, we apply the conditions for translational equilibrium. Along x,

$$\sum F_x = F_{grav} \sin\theta - F_{friction} = 0$$

where we have taken the $+x$ direction to be downward along the hill. The magnitude of the force of gravity is $F_{grav} = mg$; inserting this we get

$$mg \sin\theta - F_{friction} = 0 \tag{4.7}$$

Along y,

$$\sum F_y = N - F_{grav} \cos\theta = N - mg \cos\theta = 0$$

which leads to

$$N = mg \cos\theta \tag{4.8}$$

Unlike many of our previous problems, the normal force here is *not* equal to mg. The value of N depends on the angle of the hill and will be much smaller than the weight if θ is large.

If we know the coefficient of friction for the car's tires with the hill and also the angle of the hill, we can determine if the car will slip. From Equation 4.7, we find that the minimum frictional force required to keep the car from slipping is

$$F_{friction}(\text{required}) = mg \sin\theta \tag{4.9}$$

This result is static friction, and we know from Chapter 3 that the force of static friction can be as large as $\mu_S N$ but no larger. If the required force of friction in Equation 4.9 is greater than $\mu_S N$, the car will slip. On the other hand, if the required frictional force in Equation 4.9 is equal to $\mu_S N$, the car will just barely be in equilibrium. In this case,

$$\mu_S N = \mu_S mg \cos\theta = F_{friction}(\text{required}) = mg \sin\theta$$

where we have used the result for N from Equation 4.8. Solving for θ gives

$$\mu_S mg \cos\theta = mg \sin\theta$$

$$\frac{\sin\theta}{\cos\theta} = \frac{\mu_S mg}{mg}$$

$$\tan\theta = \mu_S \tag{4.10}$$

[3]Using coordinate axes oriented along the vertical and horizontal directions would lead to precisely the *same* answer as found with the axes chosen in Figure 4.5, although the algebra corresponding to Equations 4.7 and 4.8 would be a little different.

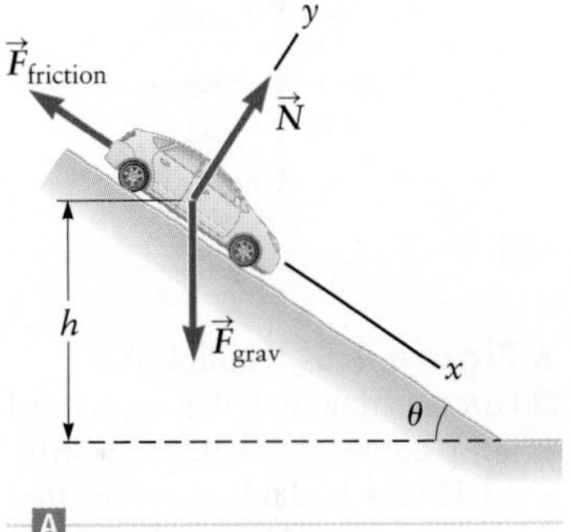

Free-body diagram

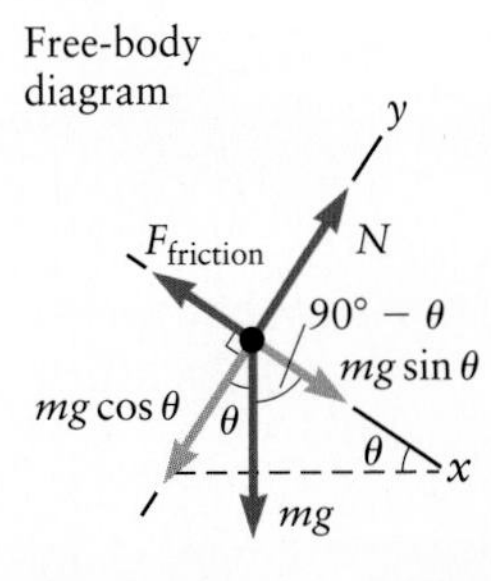

▲ **Figure 4.5** A Three forces are acting on the car. For the car to be in translational equilibrium, the three forces must add to zero. We choose the x and y axes to be parallel and perpendicular (respectively) to the plane defined by the hill. B We then express all the forces in terms of their components along these two coordinate directions. Notice that the magnitude of the force of gravity is mg.

Let's estimate the value of θ for a car just barely in equilibrium on a snowy street. The coefficient of static friction for rubber tires with snow varies with the way the snow is compacted, but is typically in the neighborhood of $\mu_S \approx 0.2$. Inserting this value into Equation 4.10 leads to $\theta \approx 10°$. This angle corresponds to a steep but realistic street, so our result suggests that a car parked on such a street might not always be in equilibrium under snowy conditions.

EXAMPLE 4.3 Supporting a Flag

A flag of mass $m = 3.5$ kg is suspended by a single cable attached at the top corner as shown in Figure 4.6A. Suppose a stiff breeze exerts a horizontal force of 75 N on the flag. Find the angle θ at which the cable hangs from the flagpole. For simplicity, assume the flag stays flat as sketched in Figure 4.6A.

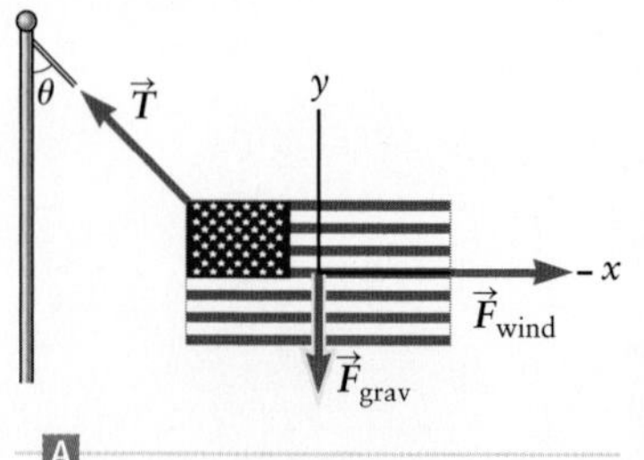

Free-body diagram

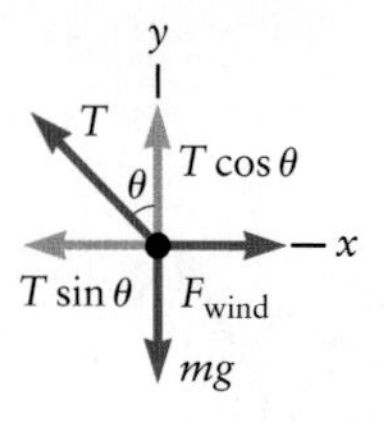

▲ **Figure 4.6** Example 4.3. **A** Forces acting on a flag suspended at a top corner by a cable in a stiff wind. **B** Free-body diagram for the flag.

RECOGNIZE THE PRINCIPLE

We want the flag to be in static equilibrium. The unknowns in this problem are the magnitude of the tension T and the angle θ. Since we have two unknowns, we must use the conditions for equilibrium in two dimensions (Eq. 4.3) to obtain two equations to solve the problem.

SKETCH THE PROBLEM

Figure 4.6 shows the forces acting on the flag: the tension force $\vec{T}$ from the cable, the force $\vec{F}_{wind}$ from the wind, and the force of gravity $\vec{F}_{grav}$. The picture shows our choice of coordinate axes with x and y along the horizontal and vertical, respectively. The free-body diagram shows the components of all the forces along x and y.

IDENTIFY THE RELATIONSHIPS

Writing the condition for equilibrium for the force components along x gives us

$$\sum F_x = F_{wind} - T\sin\theta = 0$$

$$T\sin\theta = F_{wind} \tag{1}$$

Along y,

$$\sum F_y = T\cos\theta - F_{grav} = 0$$

$$T\cos\theta = F_{grav} = mg \tag{2}$$

SOLVE

We can solve for θ by taking the ratio of Equations (1) and (2):

$$\frac{T\sin\theta}{T\cos\theta} = \frac{F_{wind}}{mg}$$

$$\tan\theta = \frac{F_{wind}}{mg}$$

Inserting the given values of the force exerted by the wind and of m leads to

$$\theta = \tan^{-1}\left(\frac{F_{wind}}{mg}\right) = \tan^{-1}\left[\frac{75\text{ N}}{(3.5\text{ kg})(9.8\text{ m/s}^2)}\right] = \boxed{65°}$$

► ***What does it mean?***

Our work on this problem began with a general picture together with a free-body diagram. The same approach will be useful when we apply Newton's second law to calculate how objects move, beginning with projectiles in Section 4.2.

Insight 4.1

STATICS IN THREE DIMENSIONS

All the examples in this section have been two dimensional, with vector forces in an x–y plane. Many situations can be treated in this way by choosing the x–y plane to match the geometry of the problem. Problems in which the force vectors lie in three-dimensional space can be treated using the same basic approach. We use the two conditions for equilibrium in Equation 4.3 and add a corresponding relation for the forces along the z direction, $\sum F_z = 0$.

4.2 Projectile Motion

We are now ready to consider objects that are in motion and learn how to calculate their displacement, velocity, and acceleration. In this section, we use ***projectile motion*** to illustrate the basic ideas. Projectile motion is the motion of a baseball, arrow, rock, or similar object that is thrown (i.e., *projected*), typically through the air. For the simplest type of projectile motion, the only force acting on the projectile is the force of gravity, and we ignore (at least for now) the force from air drag.

It is convenient to use the coordinate system sketched in Figure 4.7, with x along the horizontal and y along the vertical direction. The force of gravity has components

$$F_{\text{grav},\,x} = 0 \qquad F_{\text{grav},\,y} = -mg \tag{4.11}$$

Writing Newton's second law (Eq. 4.1) in component form, we have

$$\sum F_x = ma_x \qquad \sum F_y = ma_y \tag{4.12}$$

For the case of simple projectile motion, the only force is due to gravity, so Equation 4.11 gives the total force. Inserting this into Newton's second law (Eq. 4.12) gives

$$\sum F_x = 0 = ma_x$$
$$a_x = 0 \tag{4.13}$$

and

$$\sum F_y = -mg = ma_y$$
$$a_y = -g \tag{4.14}$$

Hence, the acceleration is constant along both x and y. In fact, the component of the acceleration along x is zero.

Now comes a crucial point: the motions along x and y are *independent* of each other. According to Equation 4.12, the acceleration along x is determined by the force along x, while the acceleration along y is determined by the force along y. For the gravitational force, the components of the force along x and y are independent of each other (Eq. 4.11), so the accelerations along x and y are also independent. What's more, according to Equation 4.13 the motion of a projectile along x is completely equivalent to one-dimensional motion along the x axis with an acceleration $a_x = 0$. Hence, the results found in Chapter 3 for motion in one dimension can be applied to give the x component of the position and the x component of the velocity for a projectile. Likewise, according to Equation 4.14, the motion of our projectile along the y direction is equivalent to one-dimensional motion along y with an acceleration $a_y = -g$. Hence, we can again use our results for one-dimensional motion to calculate the y component of the position and the y component of the velocity of a projectile. In fact, the motion along y is the same as simple free fall, which we studied in Chapter 3. As a result, *projectile motion is simply two cases of motion with constant acceleration*: one along x and one along y. The relationships among displacement, velocity, acceleration, and time given in Table 3.1 for constant acceleration thus apply directly to projectile motion.

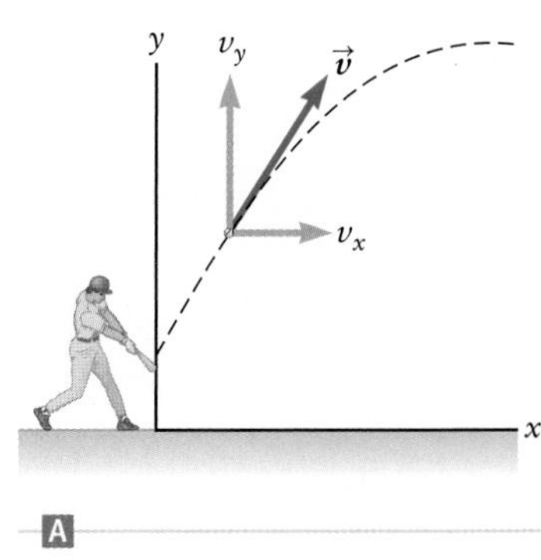

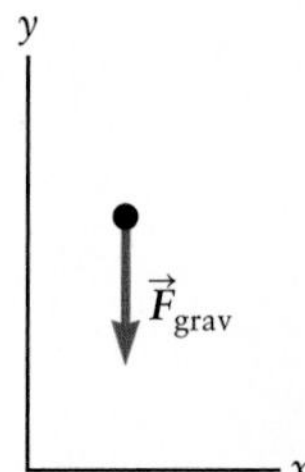

▲ **Figure 4.7** A The dashed line shows the trajectory of a thrown or batted baseball. B Free-body diagram for the ball. Throughout the entire course of the motion, the force on the ball $\vec{F}_{\text{grav}}$ is directed downward with a magnitude equal to mg.

Rolling Off a Cliff

An example of projectile motion is shown in Figure 4.8, which shows a car rolling off a cliff. Suppose the car leaves the cliff with a velocity of 10 m/s directed along the horizontal. If the cliff has a height h = 20 m, where and when will the car land? To attack this problem, we begin, as usual, with a picture that establishes our coordinate system and contains the starting information. In Figure 4.8, we choose x

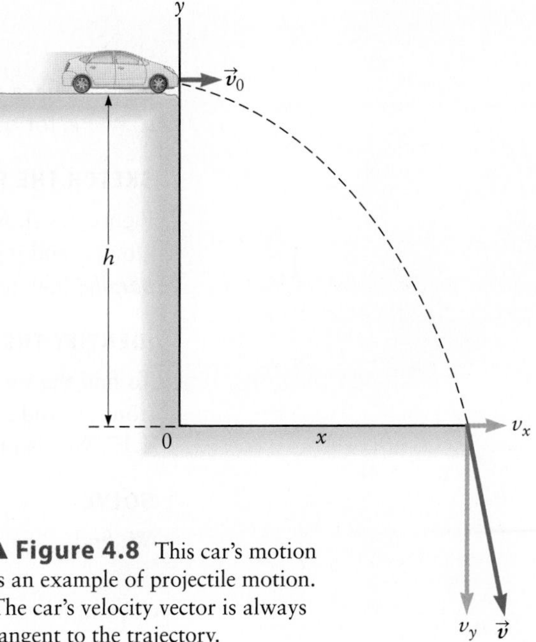

▲ **Figure 4.8** This car's motion is an example of projectile motion. The car's velocity vector is always tangent to the trajectory.

and y to be horizontal and vertical, with the origin at the bottom of the cliff. We then apply our relations for motion with constant acceleration from Table 3.1. Here we are given information on where the car starts and are asked to find where and when it lands, so we choose the relation that involves position and time. Writing this equation twice, once for the horizontal motion and once for the vertical motion, we have

$$x = x_0 + v_{0x}t + \tfrac{1}{2}a_x t^2 = 0 + v_{0x}t + 0 = v_{0x}t \tag{4.15}$$

$$y = y_0 + v_{0y}t + \tfrac{1}{2}a_y t^2 = h + 0 - \tfrac{1}{2}gt^2 = h - \tfrac{1}{2}gt^2 \tag{4.16}$$

where the initial position of the car is $x_0 = 0$ and $y_0 = h$ (the height of the cliff). The initial velocity components are $v_{0x} = 10$ m/s and $v_{0y} = 0$ since the car is moving horizontally at the moment it leaves the cliff. Notice again that the acceleration along x is zero (so $a_x = 0$ and v_x is constant) and that $a_y = -g$. Equations 4.15 and 4.16 give the position of the car as a function of time. This position is a vector quantity and has components x and y that vary with time. We can calculate when the car lands by computing the value of t when the car reaches $y = 0$ because this point is ground level (the bottom of the cliff). From Equation 4.16, we have

$$y = h - \tfrac{1}{2}gt^2 = 0$$

$$t = \sqrt{\frac{2h}{g}} = \sqrt{\frac{2(20\text{ m})}{9.8\text{ m/s}^2}} = 2.0\text{ s} \tag{4.17}$$

To find where the car lands, we use this value of t in Equation 4.15:

$$x = v_{0x}t = (10\text{ m/s})(2.0\text{ s}) = 20\text{ m}$$

EXAMPLE 4.4 Driving Off a Cliff

For the car in Figure 4.8, find the x and y components of the velocity and also the speed of the car just before it hits the ground.

RECOGNIZE THE PRINCIPLE

We use the relations for motion with constant acceleration from Table 3.1. We are interested in the velocity and have to deal with two equations, one for the horizontal component of the velocity v_x and another for the vertical component v_y. So,

$$v_x = v_{0x} + a_x t = v_{0x} \tag{1}$$

$$v_y = v_{0y} + a_y t = v_{0y} - gt \tag{2}$$

Here we have also inserted values for the components of the acceleration, $a_x = 0$ and $a_y = -g$, for our projectile.

SKETCH THE PROBLEM

Figure 4.8 shows the trajectory of the car and also the velocity and its components along x and y just before the car hits the ground. Notice that the direction of $\vec{v}$ is *parallel* (i.e., tangent) to the trajectory curve.

IDENTIFY THE RELATIONSHIPS

To find the velocity when the car reaches the ground, we evaluate v_x and v_y in Equations (1) and (2) at the value of t at which the car hits the ground as found in Equation 4.17. We also use the given values of the initial velocity, $v_{0x} = 10$ m/s and $v_{0y} = 0$.

SOLVE

We find

$$v_x = v_{0x} = \boxed{10\text{ m/s}}$$

and

$$v_y = v_{0y} - gt = 0 - (9.8 \text{ m/s}^2)(2.0 \text{ s}) = \boxed{-20 \text{ m/s}}$$

The speed v of the car is the magnitude of the velocity vector:

$$v = \sqrt{v_x^2 + v_y^2} = \sqrt{(10 \text{ m/s})^2 + (-20 \text{ m/s})^2} = \boxed{22 \text{ m/s}}$$

► ***What does it mean?***
Notice that we have calculated the velocity and speed the instant *before* the car strikes the ground. When it contacts the ground, there will be a normal force exerted by the ground and also a frictional force, and the car's behavior will no longer be described by simple projectile motion.

Independence of the Vertical and Horizontal Motion of Projectiles

The independence of a projectile's vertical and horizontal motions leads to some interesting results. In the example with the car in Figure 4.8, this independence means that the time it takes the car to reach the ground (Eq. 4.17) is independent of the motion along x. This time is the same if the car is initially traveling very fast (large v_{0x}) or very slow, or even if the car is just nudged from the cliff ($v_{0x} \approx 0$).

This result is demonstrated in the time-lapse photos shown in Figure 4.9 showing the motion of two falling balls that are released from the same height at the same time. The ball on the left is simply dropped, while the ball on the right is given an initial velocity along the horizontal direction x. The ball on the left falls directly downward, whereas the ball on the right follows a parabolic path characteristic of projectile motion. Although the balls have very different velocities along the horizontal direction, at each "flashpoint" in the photo the two balls are always at the same height, which shows that their displacements and velocities along y are the same. This result confirms that the motion along y does *not* depend on the velocities along the horizontal direction, in accordance with Equations 4.13 through 4.16.

CONCEPT CHECK 4.2 Falling Balls

When the balls in Figure 4.9 reach the ground, which one has the larger speed?
(a) The red ball does.
(b) The yellow ball does.
(c) They have the same speed when they reach the ground.

The independence of the motion along x and y for a projectile is central to the following hypothetical situation. A monkey has escaped from a zoo and is hiding in a tree. The zookeeper wants to capture the monkey without hurting him, so the zookeeper plans to shoot a tranquilizer dart at the monkey (Fig. 4.10). This very clever monkey is watching closely as the rifle is aimed at him. When the monkey observes a flash of light from the rifle (indicating that it has been fired), the monkey releases his grip on the tree and begins to fall. The monkey does this so that he will fall some distance vertically while the dart is traveling on its way to him. The monkey believes his falling will cause the dart to pass over his head and thereby miss him, but will the dart really miss the monkey?

Figure 4.10 shows the monkey and his predicament. Part A of the figure shows what would happen if we could (hypothetically) turn off gravity for the duration of the problem. That is, we assume $g = 0$. With this assumption, the acceleration of both the dart and the monkey are zero along both x and y, and both the dart and the monkey move with constant velocities. The dart thus moves in a straight line that takes it from the rifle to the monkey (as sketched in Fig. 4.10A), while the monkey

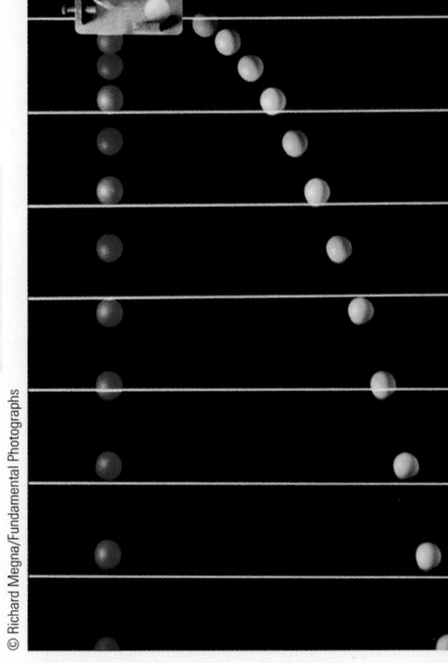

▲ **Figure 4.9** Time-lapse photos of two falling balls. The ball on the left falls straight downward, whereas the ball on the right has a nonzero component of velocity along the horizontal direction. The two balls strike the ground at the same time.

does not move at all since his initial velocity is zero. Hence, with gravity turned off, the dart would hit the monkey, as you should have expected.

Now consider the case with gravity turned on. The trajectories of the dart and the monkey in this case are shown in Figure 4.10B. They have the same acceleration: $a_x = 0$, $a_y = -g$, so the vertical motions of the dart and the monkey are both described by

$$y = y_0 + v_{0y}t - \tfrac{1}{2}gt^2$$

In words, we can write this result for y as

$$y = \overbrace{y_0 + v_{0y}t}^{\text{displacement without gravity}} \quad \overbrace{-\tfrac{1}{2}gt^2}^{\text{displacement due to gravity}} \tag{4.18}$$

Equation 4.18 emphasizes that the effect of gravity on y is the *same* for the dart and the monkey. The trajectories of both fall a distance $\Delta y = -\frac{1}{2}gt^2$ below the trajectories in the absence of gravity in Figure 4.10A. Hence, gravity causes the dart and the monkey to fall the same distance, so their trajectories still intersect. The dart *will* hit the monkey. The point of this story is that in projectile motion, the motion along the vertical (y) is *independent* of the horizontal motion. The dart and the monkey fall the same amount due to the acceleration caused by gravity, even though their horizontal velocities are very different.

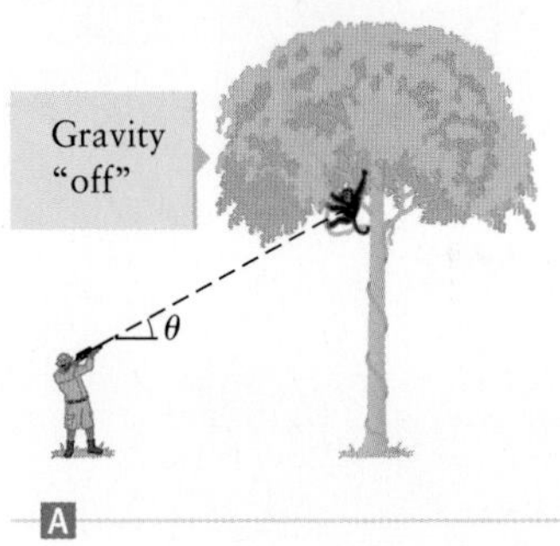

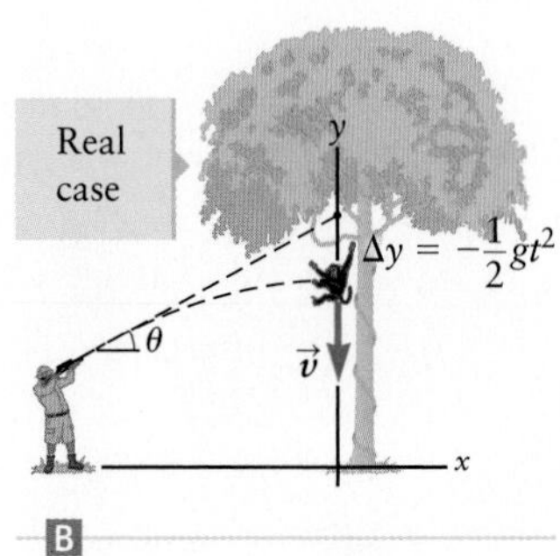

▲ **Figure 4.10** A If gravity were "turned off," this dart would follow a straight-line path to the monkey. B When gravity is turned on (which is usually the case!), the dart and the monkey are both accelerated downward and they fall the same distance Δy. The dart thus hits the monkey, but at a location below the monkey's initial position.

Projectile Motion and Target Practice

Figure 4.11 shows a sharpshooter firing a bullet horizontally at a target a distance L away. From our previous arguments, we know that because of the acceleration caused by the gravitational force, the bullet falls a distance $\Delta y = -\frac{1}{2}gt^2$ below the straight-line trajectory it would have followed were gravity (hypothetically) turned off (Eq. 4.18).

It is interesting to work out the value of Δy for a typical case. On a shooting range, the distance to the target might be 100 m or more, so let's take $L = 100$ m. The speed of the bullet when it leaves the rifle is typically in the range of 400 m/s to 800 m/s; we'll assume a value in the middle of this range and take $v_{0x} = 600$ m/s (about 1200 mi/h) for our calculation. To compute Δy, we must first calculate the time it takes the bullet to reach the target. Assuming simple projectile motion (so that there is no force due to air drag), we can use Equation 4.15 to find

$$x = v_{0x}t = L$$

which leads to

$$t = \frac{L}{v_{0x}} = \frac{100 \text{ m}}{600 \text{ m/s}} = 0.17 \text{ s}$$

For Δy, we then get

$$\Delta y = -\tfrac{1}{2}gt^2 = -\tfrac{1}{2}(9.8 \text{ m/s}^2)(0.17 \text{ s})^2 = -0.14 \text{ m}$$

You may now wonder how a sharpshooter can ever hit the bull's-eye. If he aims directly at the center of the target, the bullet in this example will pass 0.14 m = 14 cm below the bull's-eye. The resolution of this "paradox" is that a sharpshooter does not actually aim directly at the target. The sighting telescope on a rifle is calibrated so that when the rifle is "sighted" on the bull's-eye, the barrel is in reality

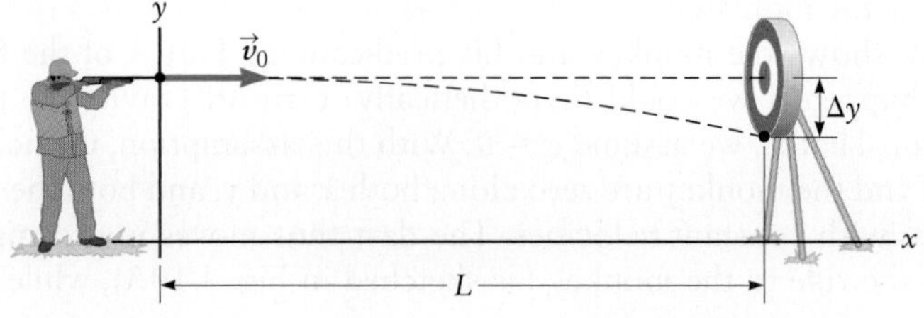

▶ **Figure 4.11** This bullet's initial velocity is horizontal. Gravity, however, causes the bullet to "fall" a distance Δy while it travels to the target.

aimed a distance Δy *above* the target. Since the value of Δy depends on the distance to the target and the speed of the bullet, the sighting telescope must be adjusted for the particular target distance and bullet in use. Hence, the rifle "knows" physics (or, rather, the rifle maker and sharpshooter do).

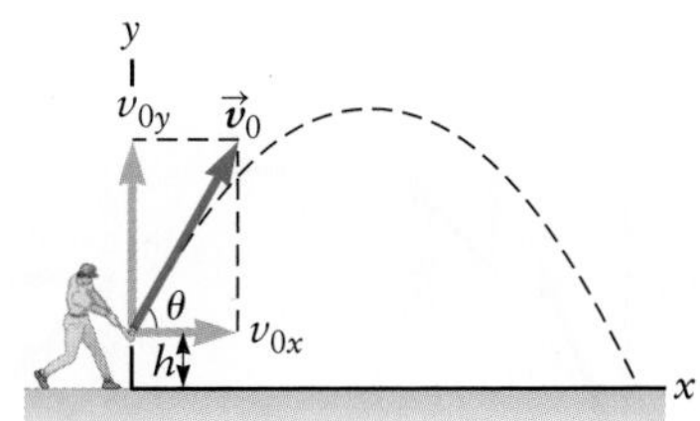

▲ **Figure 4.12** Trajectory of a batted baseball.

Motion of a Baseball: Calculating the Trajectory and the Velocity

Let's now consider the motion of a batted baseball. Our projectile—the baseball—starts from some initial position with a specified initial velocity, and we want to calculate its trajectory. The problem is sketched in Figure 4.12. The ball has an initial position with $x_0 = 0$ and $y_0 = h$, where h is the height of the ball when it leaves the bat. The ball is hit with an initial speed v_0 at an angle θ with respect to the horizontal, so the initial components of its velocity are $v_{0x} = v_0 \cos\theta$ and $v_{0y} = v_0 \sin\theta$. Our relations for motion with constant acceleration then have the form

$$x = x_0 + v_{0x}t + \tfrac{1}{2}a_x t^2 = 0 + v_{0x}t + 0 = v_0(\cos\theta)t \tag{4.19}$$

and

$$y = y_0 + v_{0y}t + \tfrac{1}{2}a_y t^2 = h + v_{0y}t - \tfrac{1}{2}gt^2 = h + v_0(\sin\theta)t - \tfrac{1}{2}gt^2 \tag{4.20}$$

To completely describe the motion, we also need the velocity:

$$v_x = v_{0x} + a_x t = v_0 \cos\theta \tag{4.21}$$

$$v_y = v_{0y} + a_y t = v_0 \sin\theta - gt \tag{4.22}$$

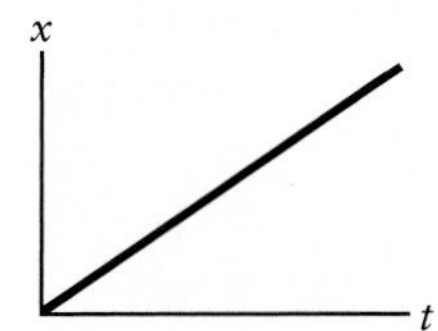

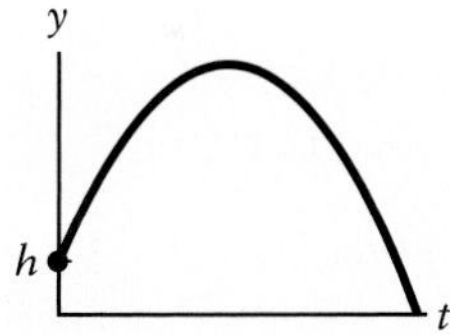

▲ **Figure 4.13** Plots of the x and y coordinates of a baseball as it travels along its trajectory.

Figures 4.13 and 4.14 show the components of the ball's position and velocity as functions of time. The results for the x direction are similar to what we would find for the dart and bullet in Figures 4.10 and 4.11. Since there is no acceleration along the horizontal direction, v_x is constant and x varies linearly with t. For the motion along y, we have an acceleration of $-g$; hence, y varies quadratically with t, and the path followed by the baseball is a parabola in the x–y plane as sketched in Figure 4.12. This path is symmetric in the following sense: the trajectory followed on the way toward the highest point is a mirror image of the trajectory on the way down. Moreover, the time the ball spends traveling from some initial height (such as h) to the top is the same as the time it spends traveling back down to that same height.

It is instructive to show the velocity components at different points along the trajectory as plotted in Figure 4.15. This is just another way of displaying the results contained in Equations 4.21 and 4.22, and in Figure 4.14. The velocity also displays a symmetry like that found for the trajectory. For example, the value of v_y at some point on the way up (i.e., at some particular height) is equal in magnitude but opposite in sign compared with v_y when the ball is at the same height on the way down. In other words, if a baseball begins its motion with speed v_0 at an initial height $y = h$, it will have the same speed when it is at this height on the way back down.

CONCEPT CHECK 4.3 Trajectory of a Baseball

For the baseball in Figure 4.12, at what point(s) on its trajectory is

(a) the acceleration zero?
(b) the vertical component of the velocity zero?
(c) the force on the ball zero?
(d) the speed zero?

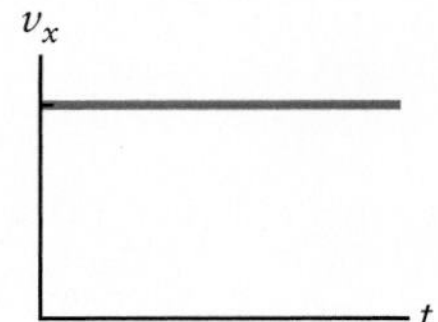

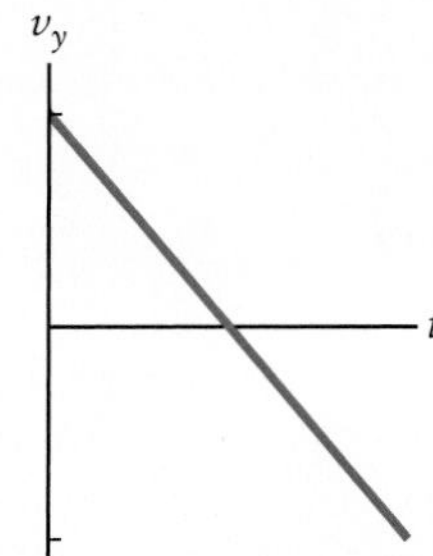

▲ **Figure 4.14** Plots of the velocity components v_x and v_y as a baseball travels along its trajectory.

Motion of a Baseball: Analyzing the Results

Let's now consider a few quantitative examples involving the baseball's trajectory in Figures 4.12 through 4.15. To be somewhat realistic, we assume the ball has an initial speed of 45 m/s (about 100 mi/h) with $\theta = 30°$ and it starts at an initial height of $h = 1.0$ m.

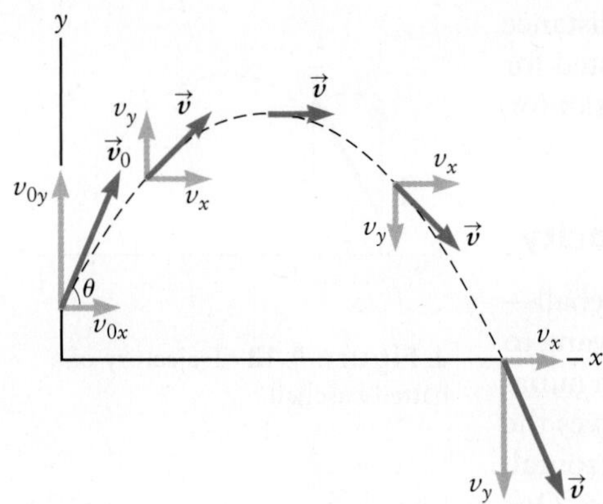

▲ **Figure 4.15** Components of the velocity at various points along the trajectory of a baseball. Compare with m Figure 4.14.

Top of the Trajectory (Maximum Height). Suppose we want to find both the height and the speed of the ball when it reaches the uppermost point on its trajectory. We can find the time at which it reaches this point by knowing that $v_y = 0$ at that instant. Inserting this into Equation 4.22 gives

$$v_y = 0 = v_0 \sin\theta - gt_{top}$$

$$t_{top} = \frac{v_0 \sin\theta}{g} \tag{4.23}$$

Using Equation 4.20, the ball's height at this moment is

$$y_{top} = h + v_0(\sin\theta)t_{top} - \tfrac{1}{2}gt_{top}^2 \tag{4.24}$$

Inserting our initial conditions into these results, we get

$$t_{top} = \frac{v_0 \sin\theta}{g} = \frac{(45\text{ m/s})\sin(30°)}{9.8\text{ m/s}^2} = 2.3\text{ s}$$

and

$$y_{top} = h + v_0(\sin\theta)t_{top} - \tfrac{1}{2}gt_{top}^2$$

$$y_{top} = (1.0\text{ m}) + (45\text{ m/s})\sin(30°)(2.3\text{ s}) - \tfrac{1}{2}(9.8\text{ m/s}^2)(2.3\text{ s})^2 = 27\text{ m}$$

The ball's speed at this moment will be

$$v_{top} = \sqrt{v_x^2 + v_y^2} = \sqrt{v_x^2 + 0} = v_x = v_0 \cos\theta \tag{4.25}$$

since $v_y = 0$ at the top. Indeed, we could have obtained this result for the speed directly from Figure 4.14. Evaluating Equation 4.25, we find

$$v_{top} = v_0 \cos\theta = (45\text{ m/s})\cos(30°) = 39\text{ m/s}$$

Landing (End of Trajectory). To find the velocity of the baseball just before it hits the ground, we can use the fact that ground level is at $y = 0$. Using Equation 4.20, we get

$$y_{lands} = 0 = h + (v_0 \sin\theta)t_{lands} - \tfrac{1}{2}gt_{lands}^2 \tag{4.26}$$

which gives us a quadratic equation to solve for t_{lands}. The solution[4] is $t_{lands} = 4.64$ s. (Here we are keeping three significant figures to minimize rounding errors in the next calculation.) To find the speed at this moment, we first need to find the component of the velocity along y, which we can obtain from Equation 4.22:

$$v_{lands,\,y} = v_0 \sin\theta - gt_{lands}$$

$$v_{lands,\,y} = (45\text{ m/s})\sin(30°) - (9.8\text{ m/s}^2)(4.64\text{ s}) = -23\text{ m/s}$$

We can now compute the speed using $v_{lands} = \sqrt{v_{lands,\,x}^2 + v_{lands,\,y}^2}$. The component v_x is constant (see Eq. 4.21 and Fig. 4.14), so $v_{lands,\,x} = v_x = v_0 \cos\theta = 39$ m/s as calculated above. Inserting these results we find

$$v_{lands} = \sqrt{v_{lands,\,x}^2 + v_{lands,\,y}^2} = \sqrt{(39\text{ m/s})^2 + (-23\text{ m/s})^2} = 45\text{ m/s}$$

We can use these results to check for the symmetry of the trajectory mentioned earlier. For a symmetric trajectory, t_{lands} should be twice the time it takes to reach the top of the trajectory found from Equation 4.23, and the speed just before the ball reaches the ground should be equal to the initial speed. To be strictly correct, we should notice that in this example the trajectory of the ball does not end at its initial

[4]Question 9 at the end of this chapter asks you to use the quadratic formula to solve for t_{lands}.

height because it starts a small distance above the ground at $y_0 = h = 1.0$ m but lands at $y = 0$. Here, however, h is sufficiently small that to within the two significant figures used in most of our numerical evaluations, t_{lands} is equal to twice the time it takes to reach the peak of the trajectory (t_{top}) and the speed at landing is equal to the initial speed. In Example 4.5, we show that the time of flight is exactly symmetric for objects projected from ground level.

Range of a Projectile. In our example with the baseball, we assumed the ball was hit with a particular value of the initial angle θ. It is interesting to consider how the trajectory varies as a function of θ. If you were a baseball coach, you might want to instruct your players on the value of θ they should use to get the ball to travel as far as possible. (This would be applied physics at its best.) This horizontal distance traveled by the ball is called the ***range***. To solve this problem, we must calculate the distance traveled by a projectile as a function of angle θ and then find the value of θ that maximizes this distance. For simplicity, let's assume the baseball starts from ground level so that $y_0 = h = 0$. The time the baseball spends in the air can be found from Equation 4.26:

$$y_{\text{lands}} = 0 = h + (v_0 \sin\theta)t_{\text{lands}} - \tfrac{1}{2}gt_{\text{lands}}^2 = (v_0 \sin\theta)t_{\text{lands}} - \tfrac{1}{2}gt_{\text{lands}}^2$$

$$t_{\text{lands}} = \frac{2v_0 \sin\theta}{g} \tag{4.27}$$

The ball then lands at the location

$$x_{\text{lands}} = v_{0x}t_{\text{lands}} = v_0(\cos\theta)\frac{2v_0 \sin\theta}{g} = \frac{2v_0^2 \sin\theta \cos\theta}{g}$$

and x_{lands} is also equal to the range. Using the trigonometric identity $\sin\theta \cos\theta = \frac{1}{2}\sin(2\theta)$ from Appendix B, we find

$$x_{\text{lands}} = \text{range} = \frac{v_0^2 \sin(2\theta)}{g} \tag{4.28}$$

For a given value of the initial speed v_0, the largest value of the range will be found when $\sin(2\theta) = 1$ since that is the largest possible value of the sine function. So, $2\theta = 90°$, or $\theta = 45°$ gives the maximum range. This result applies to all projectiles that start and end at the same height, provided that the only force acting on the projectile is the force of gravity. When the force from air drag is significant, the range is no longer given by Equation 4.28.

EXAMPLE 4.5 Symmetry of Projectile Motion

Consider a projectile that starts and ends at ground level. Show that the time spent traveling to the point of maximum height y_{top} is equal to the time spent moving from y_{top} back to the ground.

RECOGNIZE THE PRINCIPLE

We want to prove this result in the general case, so we return to Equations 4.23 and 4.27. We need to compare t_{top} and t_{lands} and show that t_{lands} is exactly twice t_{top}.

SKETCH THE PROBLEM

Figures 4.13, 4.14, and 4.16 describe the problem. For this example, the initial height is $y_0 = 0$ because the projectile starts at ground level. We have already mentioned that the trajectory of a projectile is a symmetric function of time, which is why the y–t graph in Figure 4.16 is drawn as a symmetric parabola.

(continued) ▶

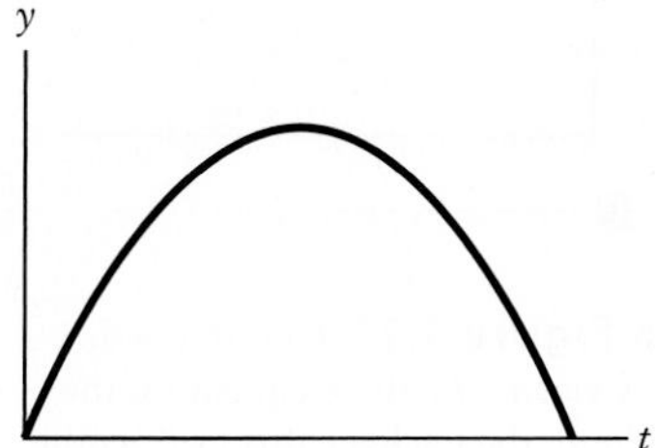

▲ **Figure 4.16** Example 4.5. Vertical position y versus t for a projectile that starts at ground level.

IDENTIFY THE RELATIONSHIPS

From Equation 4.23, we have

$$t_{top} = \frac{v_0 \sin\theta}{g} \tag{1}$$

Applying Equation 4.27 to find when the projectile lands gives

$$t_{lands} = \frac{2v_0 \sin\theta}{g} \tag{2}$$

SOLVE

By comparing Equations (1) and (2), we see that $t_{lands} = 2t_{top}$. Therefore, the time spent traveling down is *precisely* equal to the time spent traveling to the top.

▸ What does it mean?

This result confirms our earlier claim that the trajectory for a projectile that begins and ends at the same height is symmetric.

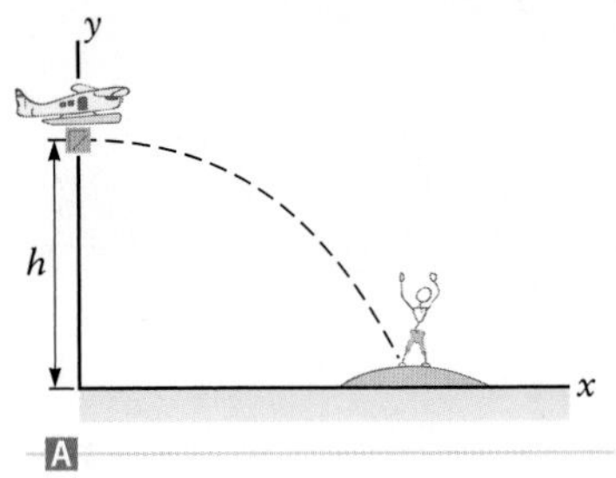

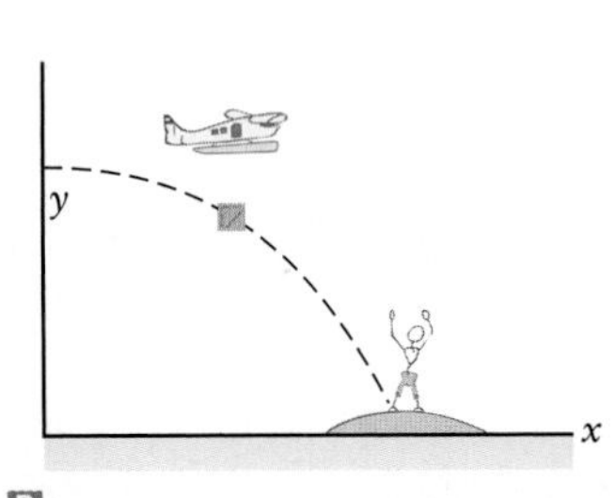

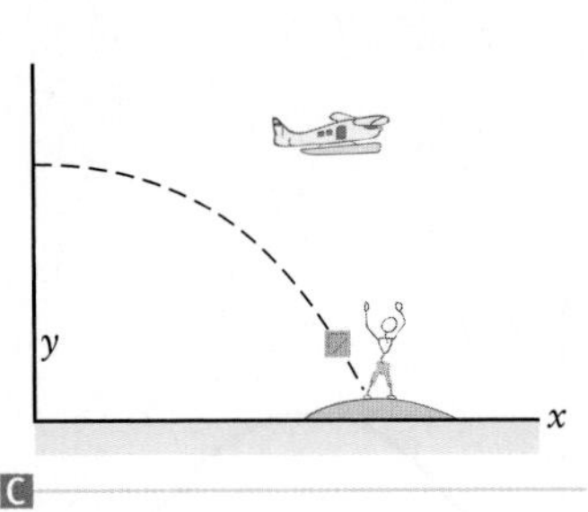

▲ **Figure 4.17** Example 4.6. As viewed by the recipient on the island, the package dropped by the airplane is a projectile that has an initial velocity equal to the velocity of the airplane. The package then follows the parabolic trajectory as sketched.

EXAMPLE 4.6 Dropping a Payload

An airplane is carrying relief supplies to a person stranded on an island. The island is too small to land on, so the pilot decides to drop the package of supplies as she flies horizontally over the island. (**a**) Find the time the package spends in the air. (**b**) If the airplane is flying horizontally at an altitude h at speed v_{plane}, where should the pilot release the package?

RECOGNIZE THE PRINCIPLE

For this projectile to hit the target, the pilot must release the package *before* she is directly over the island. We first compute how long it takes the package to fall a distance h and land on the island. This answer will give us the time the package spends in the air. We can then calculate how far along x the package travels during this time, and that will tell us how far in advance of the island the pilot should release the package.

SKETCH THE PROBLEM

The trajectory of the package is sketched in Figure 4.17. Just after its release (Fig. 4.17A), the package has a velocity along the horizontal equal to the speed of the airplane. Since the package is a projectile, the horizontal (x) component of its velocity is constant while the package falls, so the velocity of the package along the x direction is always the same as the airplane's velocity. The package is therefore always directly under the airplane as it falls to the island as shown in Figure 4.17. This example again shows the independence of the horizontal and vertical motions for a projectile.

IDENTIFY THE RELATIONSHIPS AND SOLVE

(**a**) The time it takes the package to reach the ground can be calculated from the displacement along y:

$$y = y_0 + v_{0y}t + \tfrac{1}{2}a_y t^2$$

The initial height of the package is $y_0 = h$, and at the instant the package is released, $v_{0y} = 0$. The package's acceleration is $a_y = -g$. We want to find when the package reaches $y = 0$, so we have

$$y = h - \tfrac{1}{2}gt_{lands}^2 = 0$$

Solving for the time gives

$$t_{\text{lands}} = \sqrt{\frac{2h}{g}}$$

(b) During the time the package is in the air, the airplane and the package both travel a horizontal distance

$$x = x_0 + v_{0x}t = v_{\text{plane}}t_{\text{lands}} = v_{\text{plane}}\sqrt{\frac{2h}{g}}$$

▶ ***What does it mean?***

The pilot should release the package when she is this distance (horizontally) from the island.

4.3 A First Look at Reference Frames and Relative Velocity

In Example 4.6, we considered the motion of the airplane and the falling package from the point of view of an observer on the ground. According to such an observer, the airplane is moving along the $+x$ direction with a velocity $\vec{v}_{\text{plane}}$ and the package moves along the parabolic path shown in Figure 4.17. Another way to view the problem is from the point of view of the pilot. According to her, the package simply falls straight down to Earth below (Fig. 4.18), while the island is moving along $-x$; that is, the island moves (!) relative to the airplane. Although the trajectories viewed by these two observers are different, the time the package takes to fall is the *same* from either point of view.

This is an example of using two different reference frames. A ***reference frame*** is an observer's choice of coordinate system for making measurements. In this example, one reference frame is at rest relative to the ground and the other reference frame (containing the airplane and its pilot) is moving with a constant velocity relative to the ground. Newton's laws give a correct description of the motion in any reference frame that moves with a constant velocity. We discuss what happens in other types of reference frames in Section 4.5.

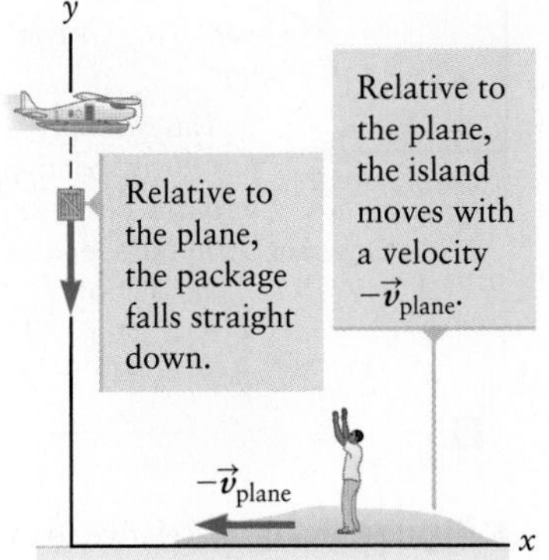

▲ **Figure 4.18** As viewed by the pilot of the airplane, the package falls directly downward because the package is always directly beneath the airplane. When viewed in the pilot's reference frame, the island is moving with a velocity $-\vec{v}_0 = -\vec{v}_{\text{plane}}$.

Relative Velocity

When we discuss or calculate the velocity of an object such as the package in Figures 4.17 and 4.18, we are always considering the velocity *relative* to a particular coordinate system or observer; that is, we consider the velocity in a particular reference frame. So, given $\vec{v}$ in one reference frame, how do we find the velocity in a different reference frame? Let's first consider this problem for motion in one dimension.

Figure 4.19A shows two cars traveling at constant velocities along the x direction. According to an observer (and reference frame) at rest on the sidewalk, the cars have velocities $\vec{v}_1$ and $\vec{v}_2$. Now consider a reference frame at rest with respect to car 1; this reference frame is defined by the coordinate axes x' and y' in Figure 4.19B that "travel along" with car 1. What is the velocity of car 2 in this new reference frame with axes x' and y'? The velocity of car 2 *relative to car 1* is just $\vec{v}_2' = \vec{v}_2 - \vec{v}_1$. Furthermore,

velocity of car 2 relative to observer = velocity of car 2 relative to car 1
+ velocity of car 1 relative to observer

This is the general way velocities in different reference frames—that is, the velocities seen by different observers—are related.

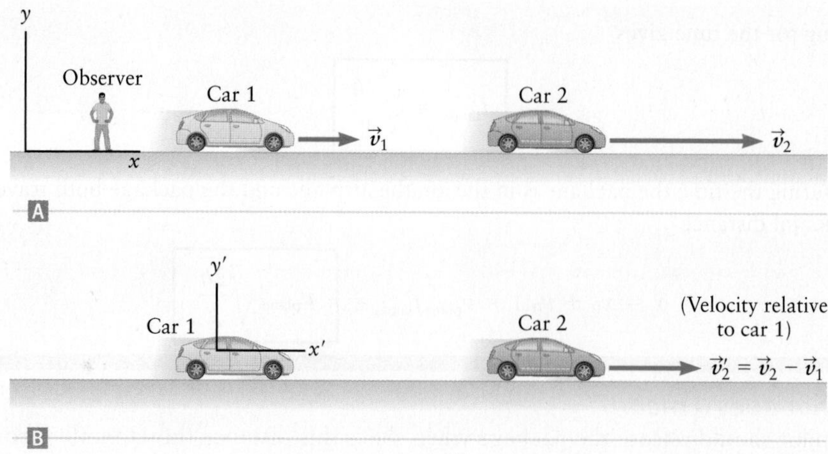

▶ **Figure 4.19** A Two cars moving along a straight road (i.e., in one dimension). B The velocity of car 2 relative to car 1 is $\vec{v}_2 - \vec{v}_1$.

The same ideas concerning relative velocities apply for motion in two dimensions. Figure 4.20 shows two people playing in a river: one is floating along with the current, while the other is swimming from one riverbank to the other. Relative to an observer on the shore, the water in this river moves at a velocity $\vec{v}_{current}$ along the x direction, and this current affects the motion of the two people in the river. The person floating along with the current (Fig. 4.20A) moves with a velocity equal to the velocity of the water. In terms of the velocity components in Figure 4.20A,

$$v_{floater,\,x} = v_{current}$$

$$v_{floater,\,y} = 0$$

The person in Figure 4.20B is swimming so as to travel across the river. Let's assume this swimmer is moving relative to the water with a velocity of magnitude v_0 in the y direction. Since the swimmer is carried along with the current, she will also have a nonzero velocity along x, and this velocity component will be equal to $v_{current}$. In words,

$$\begin{pmatrix}\text{velocity of swimmer relative}\\ \text{to observer on shore}\end{pmatrix} = \begin{matrix}\text{velocity of swimmer relative to water}\\ +\text{ velocity of water relative to observer}\end{matrix}$$

In terms of components, we have

$$v_{swimmer,\,x} = v_{current} \tag{4.29}$$

$$v_{swimmer,\,y} = v_0$$

According to an observer on the shore in Figure 4.20B, the swimmer's velocity is the sum of the two components in Equation 4.29, so the speed of the swimmer is

$$v_{swimmer} = \sqrt{v_{swimmer,\,x}^2 + v_{swimmer,\,y}^2} = \sqrt{v_{current}^2 + v_0^2} \tag{4.30}$$

Moreover, *relative to this observer*, the swimmer does not travel directly across the river, but instead moves at an angle θ. From Figure 4.20B, this angle is given through the relation

$$\tan\theta = \frac{v_{swimmer,\,y}}{v_{swimmer,\,x}} \tag{4.31}$$

The swimmer's velocity is thus different for an observer on the shore (Eqs. 4.30 and 4.31) than for the observer who floats along with the water current in Figure 4.20A.

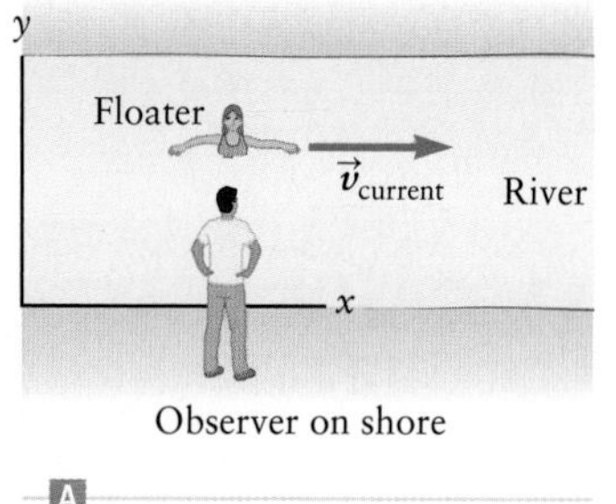

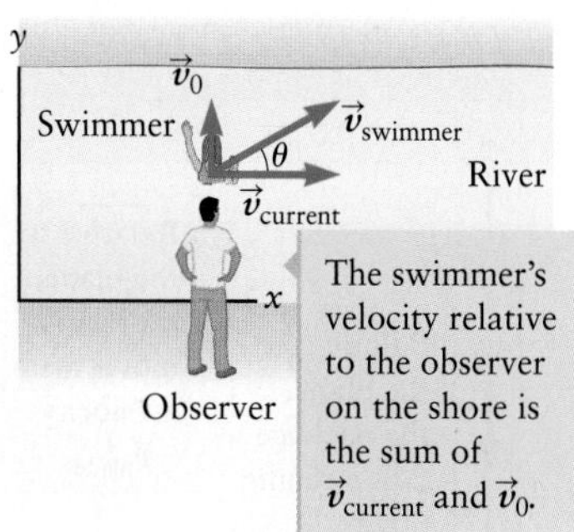

▲ **Figure 4.20** A Relative to an observer on the shore, a person floating along in a river has a velocity equal to the velocity of the river's current, $\vec{v}_{current}$. B As viewed by the observer on the shore, this swimmer's velocity is equal to the sum of the swimmer's velocity relative to the water plus the velocity of the river's current.

EXAMPLE 4.7 Swimming Across a River

Suppose the river's current in Figure 4.21 has a speed $v_{\text{current}} = 2.0$ m/s. To travel directly across the river *as viewed by an observer on the shore*, at what velocity (along the direction of the river) should the person swim?

RECOGNIZE THE PRINCIPLE

Let the components of the swimmer's velocity *relative to the water* be v_x and v_y (see Fig. 4.21). The velocity of the swimmer relative to an observer on the shore is equal to the velocity of the swimmer relative to the water *plus* the velocity of the water relative to the observer. These velocity vectors are shown in Figure 4.21, which also shows their vector sum. Relative to the observer on the shore, we thus have

$$v_{\text{swimmer},\,x} = v_{\text{current}} + v_x \qquad (1)$$

$$v_{\text{swimmer},\,y} = v_y$$

To swim directly across the river, we must have $v_{\text{swimmer},x} = 0$.

SKETCH THE PROBLEM

Figure 4.21 shows the problem, including the components of the swimmer's velocity.

IDENTIFY THE RELATIONSHIPS AND SOLVE

Inserting $v_{\text{swimmer},x} = 0$ into Equation (1), we find

$$v_{\text{current}} + v_x = 0$$

$$v_x = -v_{\text{current}}$$

Using the given value of v_{current}, we get

$$v_x = \boxed{-2.0 \text{ m/s}} \qquad (2)$$

▶ ***What does it mean?***

The negative sign in Equation (2) means that the swimmer must swim against the current. Notice that the value of v_y is not important; as long as it is not zero, the swimmer will be able to swim across the river.

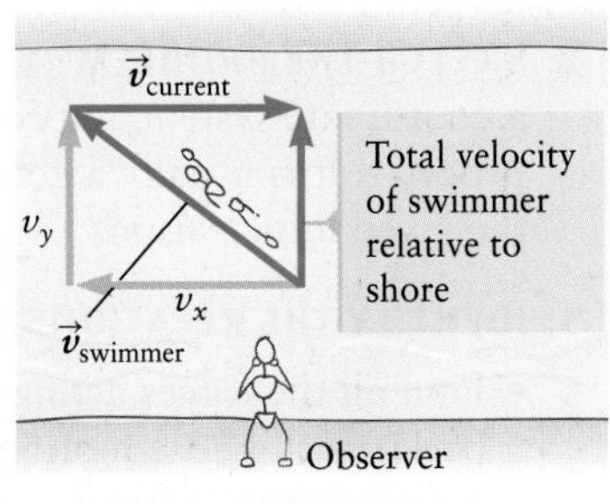

▲ **Figure 4.21** Example 4.7.

4.4 Further Applications of Newton's Laws

We have now worked through a number of applications of Newton's second law, both in one dimension (in Chapter 3) and in our work on projectile motion in Section 4.2. With that experience, we can formulate a general strategy for attacking such problems. We will apply this problem-solving strategy as we explore a number of additional applications of Newton's laws in the rest of this chapter.

PROBLEM SOLVING with Newton's Second Law

1. **RECOGNIZE THE PRINCIPLE.** To use Newton's second law to find the acceleration, velocity, and position of an object, we need to consider all the forces acting on an object and compute the total force. For an object in static equilibrium, the total force must be zero; now, for an object that is not in equilibrium, the total force is equal to $m\vec{a}$.

(continued) ▶

2 **SKETCH THE PROBLEM.** Your picture should define a coordinate system and contain all the forces in the problem. It is usually a good idea to also show all the given information.

3. **IDENTIFY THE RELATIONSHIPS.**
   - Find all the forces acting on the object of interest (the object whose motion you wish to describe) and construct a free-body diagram.
   - Express all the forces in terms of their components along x and y.
   - Apply Newton's second law (Eq. 4.1) in component form:

$$\sum F_x = ma_x \quad \text{and} \quad \sum F_y = ma_y$$

   - If the acceleration is constant along x or y (or both), you can apply the kinematic equations from Table 3.1.

4. **SOLVE.** Solve the equations resulting from step 3 for the unknown quantities in terms of the known quantities. The number of equations must equal the number of unknown quantities.

5. Always *consider what your answer means* and check that it makes sense.

Traveling Down a Hill

A typical problem involving Newton's second law is sketched in Figure 4.22, which shows a sled traveling down a snowy hillside. If the sled starts from rest at the top of the hill, how long does it take to reach the bottom? For simplicity, we ignore friction between the sled and the snow and assume the hill makes a constant angle θ with the horizontal. There are then only two forces acting on the sled, the force of gravity and the normal force exerted by the hill's surface.

Following our problem-solving strategy, we next add a coordinate system to our diagram. We know that the sled's motion is along the hill, so we choose x to be directed parallel to the hill, with x increasing downward. The y axis is then perpendicular to the hill's surface. We next draw a free-body diagram (Fig. 4.22B), which shows the components of the forces along x and y. The normal force is already along y, but the force of gravity on the sled has components along both x and y. Using some trigonometry, we can get those components (Fig. 4.22B). We now apply Newton's second law for motion along x:

$$\sum F_x = mg \sin\theta = ma_x \tag{4.32}$$

Applying Newton's second law along y, we have

$$\sum F_y = N - mg\cos\theta = ma_y \tag{4.33}$$

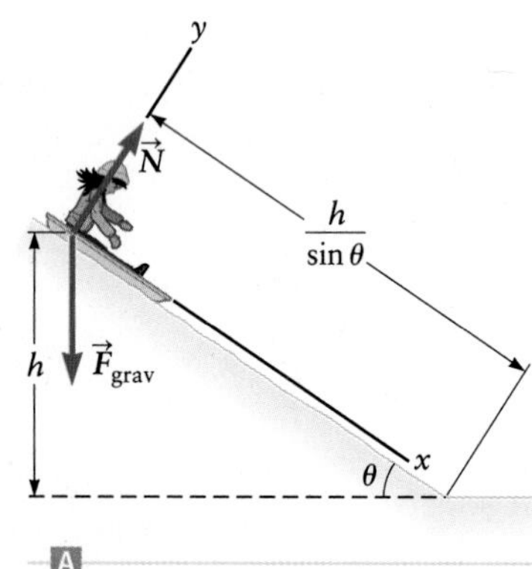

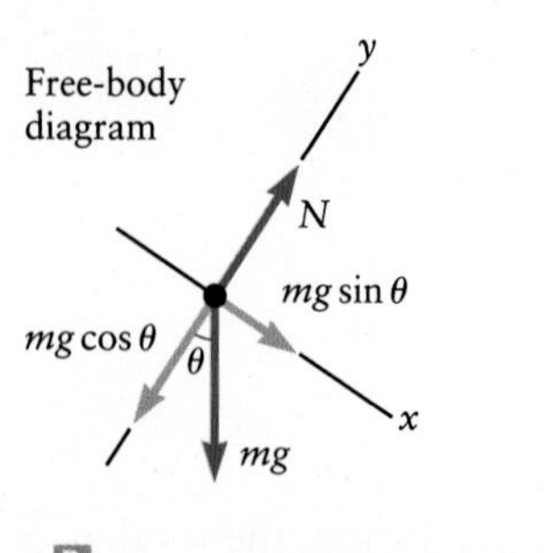

▲ **Figure 4.22** A If this hill is frictionless, only two forces are acting on the sled: a normal force $\vec{N}$ exerted by the hill and the force of gravity $\vec{F}_{grav}$. B Free-body diagram for the sled.

Since the sled is moving purely along the x direction, the acceleration along y must be zero. Hence, $a_y = 0$, and we could use Equation 4.33 to calculate the magnitude of the normal force, N. However, in this particular problem we are interested in the motion along the hillside (along x), so all we need is the acceleration a_x. We can get this acceleration by solving Equation 4.32:

$$a_x = g\sin\theta \tag{4.34}$$

This acceleration is constant, and we can use our relation for motion with constant acceleration:

$$x = x_0 + v_{0x}t + \tfrac{1}{2}a_x t^2$$

The sled starts at $x_0 = 0$ and begins from rest, so $v_{0x} = 0$. Inserting the acceleration from Equation 4.34 gives

$$x = \tfrac{1}{2}a_x t^2 = \tfrac{1}{2}g(\sin\theta)t^2 \tag{4.35}$$

If the hill has a vertical height h, the distance to the bottom of the hill as measured along the slope is $h/\sin\theta$ (see Fig. 4.22A). Inserting this into Equation 4.35, we find

$$x = \tfrac{1}{2}g(\sin\theta)t^2 = \frac{h}{\sin\theta}$$

when the sled is at the bottom of the hill. The time it takes to reach the bottom is then

$$t_{\text{bottom}} = \frac{\sqrt{2h/g}}{\sin\theta} \tag{4.36}$$

EXAMPLE 4.8 Sledding Down the Hill

For the sled in Figure 4.22, find the velocity when it reaches the bottom of the hill.

RECOGNIZE THE PRINCIPLE

We wish to find the components of the sled's velocity along the x and y directions as defined by the coordinate axes in Figure 4.22. Here, the x direction is parallel to the hill and y is perpendicular. From Newton's second law, we can find the acceleration along x (Eq. 4.34) and then apply the relations for motion with a constant acceleration (Table 3.1) to get the velocity.

SKETCH THE PROBLEM

Figure 4.22 describes the problem.

IDENTIFY THE RELATIONSHIPS

The sled is moving along the hill (along x), so the component of the velocity along y (i.e., perpendicular to the hill) is zero; that is,

$$\boxed{v_y = 0}$$

We already calculated the acceleration along x in Equation 4.34, where we found $a_x = g\sin\theta$. The acceleration is constant, so the velocity along x is again given by one of our relations for motion with constant acceleration,

$$v_x = v_{0x} + a_x t$$

Inserting the result for a_x from Equation 4.34 gives

$$v_x = v_{0x} + a_x t = g(\sin\theta)t \tag{1}$$

SOLVE

The sled reaches the bottom at t_{bottom}, which we found in Equation 4.36. Inserting t_{bottom} into Equation (1) leads to

$$v_{x,\text{ bottom}} = g(\sin\theta)t_{\text{bottom}} = g\sin\theta\,\frac{\sqrt{2h/g}}{\sin\theta} = \boxed{\sqrt{2gh}}$$

► ***What does it mean?***

While the speed at the bottom depends on the height h of the hill, it is *independent* of the angle θ of the hill. We show in Chapter 6 that this is a general property of the gravitational force and is connected with the notion of potential energy.

EXAMPLE 4.9 Towing a Car

Your car breaks down, and you ask a friend to help you move it to the nearest repair shop. You need to move your car up a hill, so your friend is going to use his car to pull your vehicle using a rope as shown in Figure 4.23A. If the angle of the hill is $\theta = 5.0°$ and you pull your car ($m = 1200$ kg) with a constant velocity, what is the tension in the rope? Assume all friction can be neglected.

(continued) ►

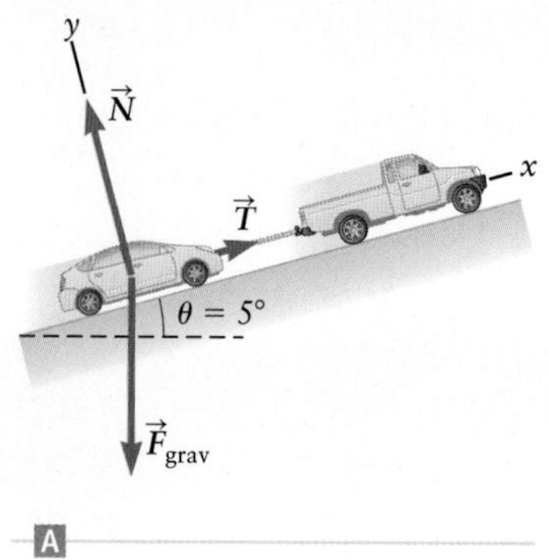

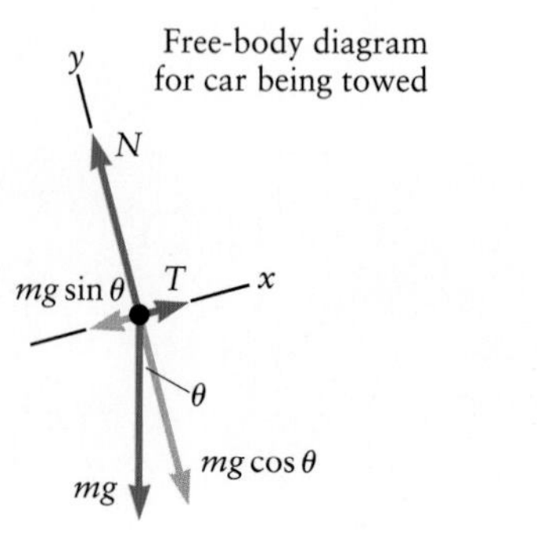

▲ **Figure 4.23** Example 4.9. **A** Three forces are acting on this car: a normal force $\vec{N}$ exerted by the road, the force of gravity $\vec{F}_{grav}$, and a force due $\vec{T}$ to the cable. **B** Free-body diagram for the towed car.

RECOGNIZE THE PRINCIPLE

We can use Newton's second law to calculate the motion of the cars (Fig. 4.23A). Since the cars are moving with constant velocity, the acceleration along the hill (the x direction) is $a_x = 0$; hence, the sum of all the forces along the hill must be zero. From this we can solve for the tension T in the rope.

SKETCH THE PROBLEM

Figure 4.23B shows a free-body diagram for the car being towed. The forces acting on it are the tension from the rope, the force of gravity, and the normal force exerted by the hill's surface. The figure also shows the components of these forces along the hill and perpendicular to the hill. The $+x$ direction is along the hill's surface and up the incline, while y is perpendicular to x.

IDENTIFY THE RELATIONSHIPS

Writing Newton's second law for motion along the x direction (along the hill), we have

$$\sum F_x = -mg \sin \theta + T = ma_x$$

Since the velocity is constant, a_x is zero, which leads to

$$-mg \sin \theta + T = 0$$

$$T = mg \sin \theta$$

SOLVE

Inserting the given values of m and θ gives

$$T = (1200 \text{ kg})(9.8 \text{ m/s}^2)\sin(5.0°) = \boxed{1000 \text{ N}}$$

▶ ***What does it mean?***

This tow rope only needs to withstand 1000 N (about 250 lb) to pull the car at a 5.0° angle without breaking, which is much less than the weight of the car.

EXAMPLE 4.10 Sliding Up a Hill

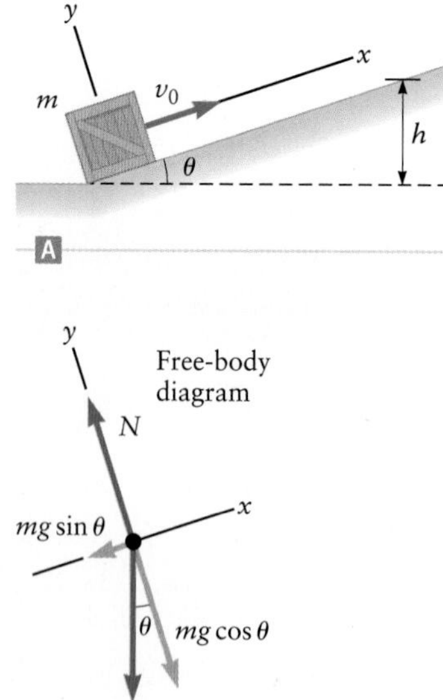

▲ **Figure 4.24** Example 4.10.

Consider a box of mass m that slides *up* a hill as sketched in Figure 4.24A. The hill is frictionless and makes an angle θ with the horizontal, and the box is given an initial speed $v_0 = 5.0$ m/s. What is the maximum height reached by the box?

RECOGNIZE THE PRINCIPLE

Following our problem-solving strategy for Newton's second law, we begin by noting that since the hill is frictionless, there are only two forces on the box, the force of gravity (magnitude mg) and the normal force (N) from the hill. These are the forces we will use in Newton's second law to find the acceleration, velocity, and position of the box. When the box reaches its maximum height, its velocity will be zero.

SKETCH THE PROBLEM

Figure 4.24 shows the problem, including a free-body diagram. We choose a coordinate system with y perpendicular to the hill and x along it, with the $+x$ direction being uphill.

IDENTIFY THE RELATIONSHIPS

We use the free-body diagram and the components of the forces along x and y in Figure 4.24B to write Newton's second law (Eq. 4.1) for both the x and y direction:

$$\sum F_x = -mg \sin \theta = ma_x \quad (1)$$

$$\sum F_y = N - mg \cos \theta = ma_y$$

We are only concerned with motion along the x direction and can get the acceleration along x from Equation (1):

$$a_x = -g \sin\theta \tag{2}$$

This acceleration is constant, so we can apply our usual relations from Table 3.1. We want to find the displacement when the velocity v_x of the box is zero, and the relation containing both of these quantities is

$$v_x^2 = v_0^2 + 2a_x(x - x_0) \tag{3}$$

with a_x given in Equation (2).

SOLVE

The block starts at the origin, so $x_0 = 0$. The velocity of the box will be zero when it reaches its highest point; setting $v_x = 0$ in Equation (3) and solving for the displacement, we find

$$0 = v_0^2 + 2a_x(x - 0)$$

$$x = -\frac{v_0^2}{2a_x} = \frac{v_0^2}{2g\sin\theta}$$

If the maximum height of the block is h (Fig. 4.24A), we can use some trigonometry to write

$$\frac{h}{x} = \sin\theta$$

and solving for h gives

$$h = x\sin\theta = \frac{v_0^2}{2g\sin\theta}\sin\theta = \frac{v_0^2}{2g}$$

$$h = \frac{v_0^2}{2g} = \frac{(5.0 \text{ m/s})^2}{2(9.8 \text{ m/s}^2)} = \boxed{1.3 \text{ m}}$$

► *What does it mean?*

The maximum height of the block depends only on its initial speed and is independent of the angle θ. In fact, the maximum height here is the same as for a projectile thrown straight up! We will explain why that happens when we consider mechanical energy in Chapter 6.

Adding the Frictional Force

In the previous examples involving motion along a slope, we have assumed friction is negligible. Now let's see how to include it. Returning to our problem of a sled sliding down a snowy hill (Fig. 4.22), we add friction between the sled and the hill. The revised free-body diagram is shown in Figure 4.25, which contains a frictional force of magnitude $F_{\text{friction}} = \mu_K N$ directed up the incline (along $-x$). This force is kinetic friction because the sled is slipping relative to the hill. To calculate the frictional force, we must first find the normal force N, which we can get from Equation 4.33:

$$\sum F_y = N - mg\cos\theta = ma_y$$

Since the acceleration along y (perpendicular to the hill) is zero,

$$\sum F_y = N - mg\cos\theta = 0$$

$$N = mg\cos\theta \tag{4.37}$$

We next apply Newton's second law for motion along the hill:

$$\sum F_x = mg\sin\theta - F_{\text{friction}} = mg\sin\theta - \mu_K N = ma_x$$

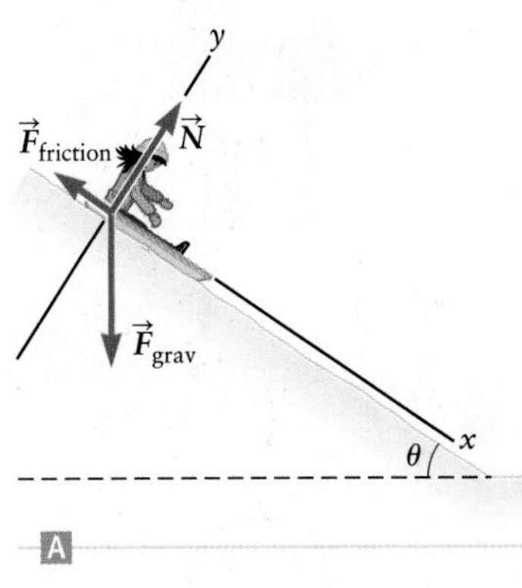

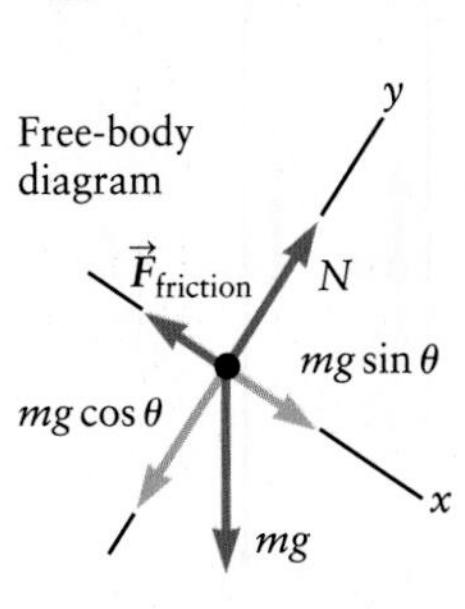

▲ **Figure 4.25** For this sled, there is a frictional force $\vec{F}_{\text{friction}}$ from the sled slipping on the snow. Compare with Figure 4.22.

The acceleration down the incline thus depends on the normal force. Using the result for the normal force from Equation 4.37 leads to

$$a_x = g \sin \theta - \frac{\mu_K N}{m} = g \sin \theta - \frac{\mu_K mg \cos \theta}{m} = g(\sin \theta - \mu_K \cos \theta) \quad (4.38)$$

Let's now estimate quantitatively the effect of friction. Suppose we have a hill with $\theta = 10°$, which is a typical value for a reasonably steep sledding hill. Without friction, $\mu_K = 0$ in Equation 4.38 and we get $a_x = g \sin \theta = 1.7 \text{ m/s}^2$. A typical value for the coefficient of kinetic friction between a wooden ski and snow is $\mu_K \approx 0.05$, and using this value in Equation 4.38 we find $a_x \approx 1.2 \text{ m/s}^2$. Hence, friction reduces the acceleration (and therefore also the speed at any moment) by about 30%.

Pulleys and Cables

In Chapter 3, we found that pulleys could be used to change the direction of a force and also to amplify forces. Let's now consider the associated motion in more detail. Figure 4.26 shows a common situation of two crates connected by a rope that passes over a pulley. For simplicity, we assume the rope is very light and the pulley is frictionless and massless. We want to find how the crates move; that is, we want to calculate their acceleration.

As usual, we need to apply Newton's second law. The forces on each crate are shown in Figure 4.26. Recall that for an ideal (massless) string, the tension is the same throughout the string, so the magnitudes of the tension forces acting on the two crates are the same. We next need to establish a coordinate system, and for this problem the choice requires some thought. Because the two crates are connected by the string, they must move together; otherwise, the string would break. Therefore, the magnitudes of their displacements, velocities, and accelerations must be the same. With that in mind, we choose the x coordinate direction to be *parallel to the string* as sketched in Figure 4.26B. This coordinate axis *follows the string* as it loops around the pulley. It may be useful to think of this coordinate "axis" as a sort of "railroad track" along which the two crates move. If we take the "positive" direction for this track to be upward for m_1, it will then be downward for m_2 as shown in the figure. Although the "positive" directions are thus opposite for the two crates, the motion is still essentially one dimensional, so we can use our familiar results for problems involving one-dimensional motion. This choice of a coordinate axis whose direction is different for the two crates will make our mathematical solution of Newton's second law a bit simpler because now the acceleration of the two crates will be the same from the start.

To find the acceleration, we first write Newton's second law for m_1. Since the forces are all along the string direction x, the acceleration will also be along this direction. We denote the total force on m_1 as $\sum F_1$, so Newton's second law for this crate reads

$$\sum F_1 = m_1 a = +T - m_1 g \quad (4.39)$$

where we have written the acceleration as simply a (because the acceleration is the same for m_1 and m_2). The forces on the right side of Equation 4.41 are $+T$ (since the tension force is directed along the $+x$ direction for m_1) and $-m_1 g$ (since the force of gravity is along $-x$). For m_2,

$$\sum F_2 = m_2 a = -T + m_2 g \quad (4.40)$$

Here the tension term is $-T$ because the tension force on m_2 is directed along $-x$ for this crate. The sign is reversed with respect to Equation 4.39 because the x direction is reversed for m_2. We now have two equations and two unknowns—the acceleration and the tension—so we can solve for both a and T. To solve for the acceleration, we add Equations 4.39 and 4.40, which eliminates T and gives

$$m_1 a + m_2 a = +T - m_1 g - T + m_2 g$$

$$(m_1 + m_2)a = -m_1 g + m_2 g$$

$$a = \frac{(m_2 - m_1)g}{m_1 + m_2} \quad (4.41)$$

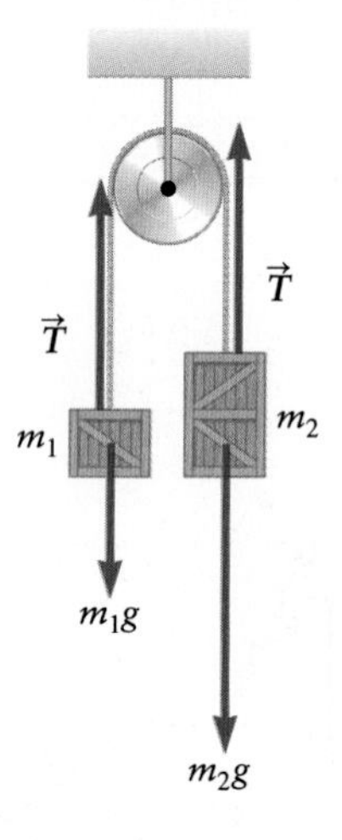

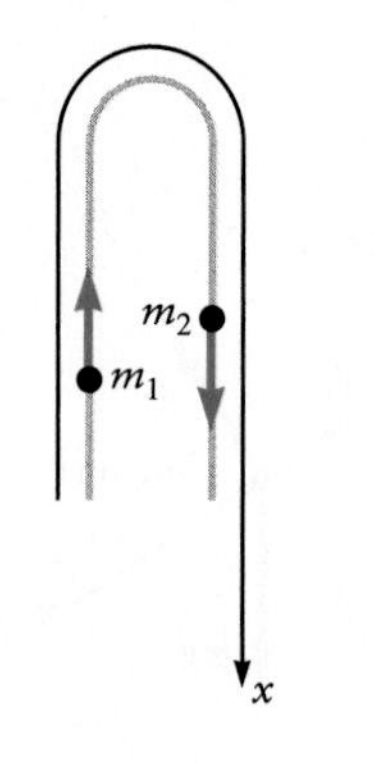

▲ **Figure 4.26** A Forces on two crates connected by a rope-and-pulley system. B The coordinate axis (the x axis) follows the string as it bends around the pulley.

To solve for the tension, we insert this result for a back into Equation 4.39 to find

$$T = m_1 a + m_1 g = m_1(a + g)$$

$$T = m_1\left[\frac{(m_2 - m_1)g}{m_1 + m_2} + g\right] = m_1 g\left[\frac{m_2 - m_1 + m_2 + m_1}{m_1 + m_2}\right]$$

$$T = \frac{2m_1 m_2 g}{m_1 + m_2}$$

If you are uneasy about our choice of the x direction in this problem, look again at Equation 4.41. There we see that if $m_2 > m_1$, the acceleration is positive and the block with the larger mass (m_2) moves downward as expected. On the other hand, if $m_2 < m_1$, Equation 4.41 tells us that the acceleration is negative, which simply means that m_2 now accelerates upward, again as we should expect. The *physical answer* does not depend on how we choose the "positive" direction.

EXAMPLE 4.11 More Crates and a Pulley

A crate sits on a frictionless table and is connected to a second crate by a string that passes over a pulley as shown in Figure 4.27. If the pulley is frictionless and massless and the string is also massless, find the acceleration of the crates and the tension in the string.

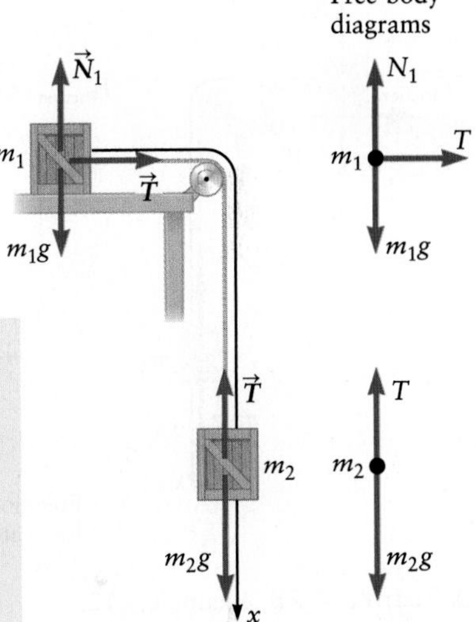

▲ **Figure 4.27** Example 4.11. Another arrangement of crates and a pulley. We assume there is no frictional force between the table and crate 1. Notice that we have chosen the x direction to follow the string.

RECOGNIZE THE PRINCIPLE

We follow our problem-solving strategy for applying Newton's laws. To find the acceleration of the crates, we must consider all the forces acting on them. Our sketch (Fig. 4.27) shows all these forces. We write Newton's second law for each crate and solve for the unknown quantities.

SKETCH THE PROBLEM

Figure 4.27 shows the forces acting on both crates and indicates our choice of coordinate system. As in Figure 4.26, we take the x direction to follow the string around the pulley. That is, we choose the $+x$ direction to be horizontal and to the right for the crate on the table, and down along the string for the crate that is suspended in midair.

IDENTIFY THE RELATIONSHIPS

Using the forces shown in Figure 4.27, Newton's second law for the crate on the table reads

$$\sum F_x = m_1 a = +T \quad (1)$$

For the crate that hangs from the string, Newton's second law gives us

$$\sum F_x = m_2 a = m_2 g - T \quad (2)$$

SOLVE

To find the acceleration, we eliminate T by adding Equations (1) and (2):

$$m_1 a + m_2 a = +T + m_2 g - T = m_2 g$$

$$a = \boxed{\frac{m_2 g}{m_1 + m_2}}$$

Solving for T then gives, from Equation (1),

$$m_1 a = +T$$

$$T = m_1 a = \boxed{\frac{m_1 m_2 g}{m_1 + m_2}}$$

(continued) ▶

▸ What does it mean?

Our approach to this problem follows our usual pattern. We start with a picture and construct free-body diagrams for the objects of interest (the two crates). We then write Newton's second law for each crate and solve for the unknowns (a and T).

EXAMPLE 4.12 Crates with Friction

Suppose there is a frictional force between the crate m_1 and the table in Example 4.11 (Fig. 4.27). Write the new Newton's second law equation for this crate's acceleration along the horizontal direction.

RECOGNIZE THE PRINCIPLE

We follow the same approach as taken in Example 4.11. Now our sketch (Fig. 4.28) contains an additional force due to friction on crate 1. This frictional force has a magnitude of $\mu_K N_1$, where N_1 is the normal force between crate 1 and the table.

SKETCH THE PROBLEM

Figure 4.28 shows the new situation along with free-body diagrams for both crates. This frictional force acts in the $-x$ direction since it opposes the motion of crate 1.

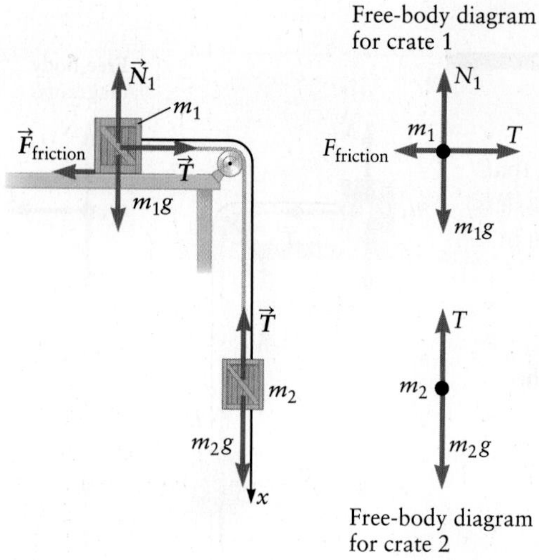

▲ **Figure 4.28** Example 4.12. We add friction to the system in m Figure 4.27, so there is now a frictional force between crate 1 and the table.

IDENTIFY THE RELATIONSHIPS AND SOLVE

When we add the frictional force, Newton's second law for the crate on the table becomes

$$\sum F_x = m_1 a = +T - \mu_K N_1$$

▸ What does it mean?

In this problem we have an additional unknown, the normal force N_1. This normal force can be found by considering the forces on m_1 along the y direction, leading to $N_1 = m_1 g$.

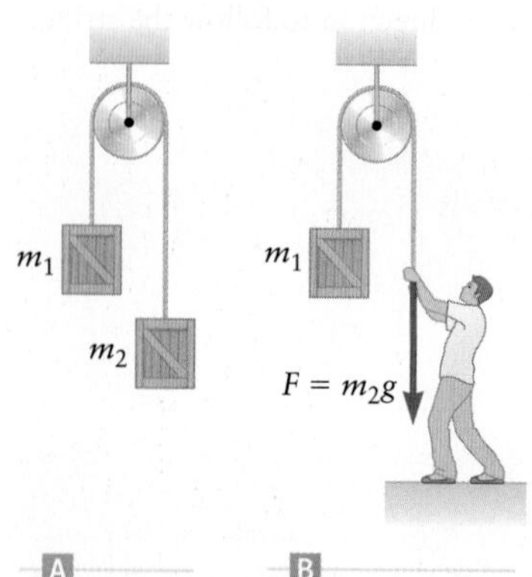

▲ **Figure 4.29** Concept Check 4.4. Do these two systems have the same acceleration?

CONCEPT CHECK 4.4 Tension in a String

Consider the two pulley arrangements in Figure 4.29. In both cases, a crate of mass m_1 is attached to one end of a string. In Figure 4.29A, the other end of the string is tied to a crate of mass m_2, so there is a force from gravity of magnitude $F_{2,\text{grav}} = m_2 g$ acting on this crate. In Figure 4.29B, crate m_2 is replaced by a person who pulls on the string, and we suppose he pulls with a force *equal* to $F_{2,\text{grav}} = m_2 g$. Will the acceleration of m_1 be the same in these two cases?

4.5 Detecting Acceleration: Reference Frames and the Workings of the Ear

Suppose you are traveling in an airplane initially flying at a constant horizontal velocity and the pilot decides to accelerate slightly. You immediately sense this acceleration. How did you do it? The human ear contains structures that act as sensitive acceleration sensors, and we can understand how they work using Newton's second law.

Before we consider the workings of an actual ear, let's first consider a simple mechanical device you could build to detect and measure an airplane's acceleration. Our device, shown in Figure 4.30, consists of a rock tied to one end of a string, with the other end of the string fastened to the airplane's ceiling. When the acceleration is zero, the string hangs vertically, which is consistent with a Newton's law analysis. Two forces act on the rock—the force of gravity and the tension force from the string—and in Figure 4.30A these forces are both along the vertical (y). The acceleration along y is zero, so these forces must add to zero with the result $T = mg$. There are no forces along the horizontal; the acceleration along x is zero, and the rock moves with a constant velocity.

When the airplane is accelerating along the horizontal (x), the rock will have the same acceleration as the airplane and the string hangs at an angle θ as sketched in Figure 4.30B. The value of θ depends on the airplane's acceleration. Writing Newton's laws for motion along x and y leads to (Fig. 4.30C)

$$\sum F_x = ma_x = T \sin \theta$$

$$\sum F_y = ma_y = T \cos \theta - mg = 0 \tag{4.42}$$

since the acceleration along y is again zero. We can use this relation for a_y to find the tension in the string,

$$T \cos \theta = mg$$

$$T = \frac{mg}{\cos \theta}$$

and then get the acceleration along x:

$$a_x = \frac{T \sin \theta}{m} = \frac{mg}{\cos \theta} \frac{\sin \theta}{m} = g \tan \theta \tag{4.43}$$

Whenever the airplane has a nonzero acceleration, the angle θ is nonzero and the string hangs at an angle from the vertical. By measuring this angle, we can use Equation 4.43 to find the value of the acceleration a_x.

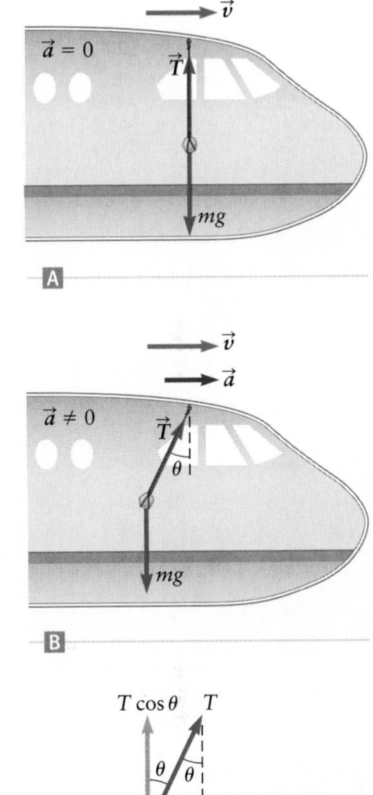

▲ **Figure 4.30** A rock hanging by a string acts as an accelerometer, a device that measures acceleration. A When the system (here an airplane cabin) moves with a constant velocity, the string hangs vertically. B When the system is accelerating, the string hangs at a nonzero angle θ, and the value of θ depends on the acceleration. C Vertical and horizontal components of the forces on the rock.

CONCEPT CHECK 4.5 Effect of Mass on an Accelerometer

How will the angle of the accelerometer string in Figure 4.30 depend on the *mass* of the rock?

(a) The angle is larger for a rock with a larger mass.
(b) The angle is independent of the mass of the rock.
(c) The angle is smaller for a rock with a larger mass.

The Accelerometer in Your Ear

Your ears contain a structure that acts very much like the simple accelerometer in Figure 4.30. A region within the ear called the utricle contains hair cells that project up into a gelatinous layer filled with a very thick fluid (Fig. 4.31). This gelatinous layer also contains some small masses called otoliths (also known as "ear stones"). When your head—that is, the utricle—accelerates along the horizontal, the gelatinous layer lags behind a small amount in the same way the string in Figure 4.30B hangs behind when the airplane accelerates. This displacement of the gelatinous layer causes the hair cells to deflect (just like the string in Fig. 4.30B), and the hair cells send signals to the brain when they are bent in this way. Hence, the brain is "told" that the ear is being accelerated. Although the ear is certainly more complex than a rock on a string, the accelerometers in Figures 4.30 and 4.31 are remarkably similar.

Inertial Reference Frames

Let's analyze the accelerometer in Figure 4.30 once more, this time from a slightly different point of view. In Equation 4.42, we applied Newton's second law from the point of view of an observer outside the airplane, a person who might have viewed the rock and string through a window as the airplane flew past. Now imagine how things look to an observer sitting inside the airplane. When the acceleration is zero and the string hangs vertically as in Figure 4.30A, an observer inside the airplane would see two forces acting vertically on the rock: gravity and the tension in the string. This observer would say that the acceleration along y is zero and hence that $T = mg$. The observer inside the airplane would also say that there is no force along x and there is also no acceleration along x, so Newton's second law works just fine. Hence, when the airplane moves with a constant velocity, an observer inside the airplane would be in complete agreement with the observer outside the airplane. Both would find that Newton's second law describes things perfectly.

The situation is different when the airplane has a nonzero acceleration relative to an observer outside the airplane. Free-body diagrams for this case are shown in Figure 4.32. Our two observers—one outside the airplane and one inside—would draw the same free-body diagrams. However, they would disagree when they use these free-body diagrams to apply Newton's second law:

$$\sum F_x = T \sin\theta = ma_x \tag{4.44}$$

The outside observer would say that the acceleration a_x is nonzero and that Newton's second law works fine since θ is nonzero (Fig. 4.32A). The observer inside the airplane would also see that the string hangs at an angle and hence would conclude that there is a force $T \sin\theta$ along the x direction. According to this observer, though, there is *no acceleration along* x. That is, *as viewed by an observer riding inside the airplane*, the rock is simply at rest; there is no velocity or acceleration along the x direction *relative to this observer*. The observer inside the airplane would therefore say that there is a force along x but no acceleration along x, so the left-hand side of Equation 4.44 is not zero, whereas the right-hand side is zero. Our inside observer would thus be tempted to conclude that Newton's second law does not work inside airplanes!

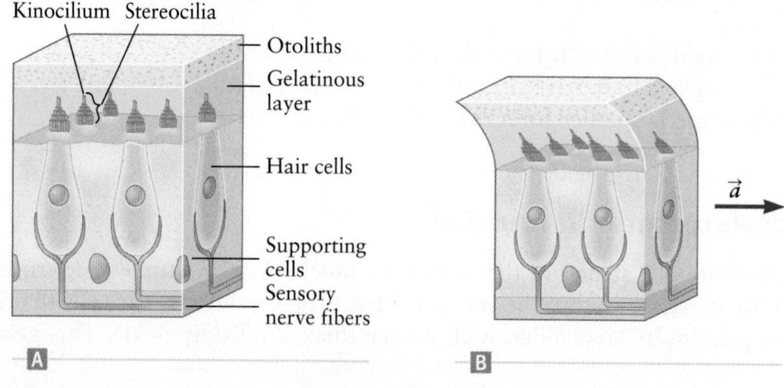

▶ **Figure 4.31** **A** Hair cells within the utricle in a human ear. These cells act like the accelerometer in Figure 4.30. **B** When your head is accelerated, the hair cells are deflected from the vertical, producing a signal that is sent to your brain.

The correct interpretation of this situation is that the fault lies not with Newton, but with the observer inside the airplane who is viewing things in an *accelerating reference frame*. Newton's second law can only be applied in nonaccelerated reference frames, also called ***inertial reference frames***. An inertial reference frame can be moving, but it must be moving with a constant velocity relative to other inertial frames. Suppose an observer in one inertial reference frame (frame 1) conducts a test of Newton's second law by measuring the forces on an object and the acceleration of the object. He would then (in agreement with Newton's second law) find that

$$\vec{F}_{\text{measured}} = m\vec{a}_{\text{observed}} \tag{4.45}$$

Now consider a second observer who is moving at a constant velocity relative to frame 1, so his reference frame (frame 2) is an inertial frame. Observer 2 would measure the same horizontal force as observer 1 because he will observe the tension in the string in Figure 4.32 to be the same. Observer 2 will also measure the *same* acceleration as observer 1; since acceleration is the rate of change of velocity, adding the constant velocity of frame 2 does not change the acceleration. Hence, Equation 4.45 would still be satisfied. However, if an observer in a *noninertial* reference frame (frame 3) were to conduct the same test, he would measure the same force but a *different* acceleration. As in the case of the airplane in Figure 4.32, the acceleration in frame 3 would differ from that found in frame 1 by an amount equal to the acceleration of the noninertial frame. Hence, to the observer in frame 3, it would appear that Newton's laws fail. Again, Newton's laws can only be used in inertial reference frames.

The problem of inertial and noninertial reference frames has a deep connection to Newton's laws and to other theories of motion. We'll explore this connection further when we discuss gravitation and relativity in Chapter 27.

Observer outside airplane: inertial frame

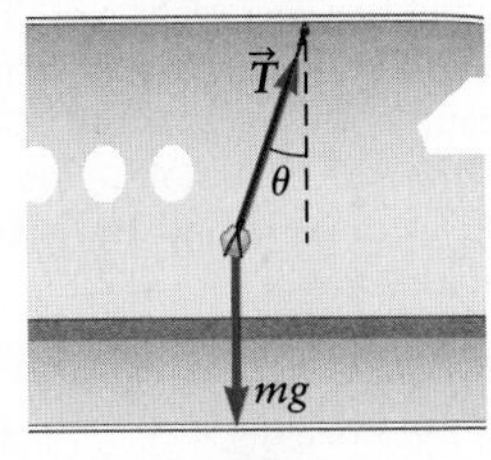

A

Observer inside airplane: noninertial frame

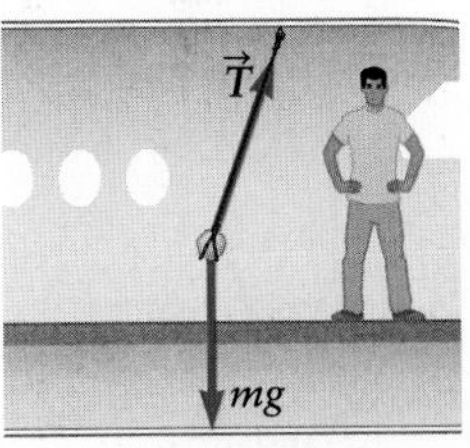

B

▲ **Figure 4.32** A An observer in an inertial reference frame (i.e., an observer at rest, outside the airplane) finds that there is a horizontal component to the total force (because the string hangs at an angle) and also measures that the rock is accelerating, in agreement with Newton's second law. B An observer inside the airplane accelerates along with the rock. According to this noninertial observer, the rock's acceleration is zero since the rock is at rest relative to this observer. However, this observer still finds that there is a horizontal component to the total force on the rock. According to this observer, there is a horizontal force on the rock but no horizontal acceleration, so this observer claims that Newton's second law does not work!

CONCEPT CHECK 4.6 Newton's Laws and Reference Frames

Give an example showing that Newton's first law—the principle of inertia—does not hold in a noninertial reference frame.

4.6 Projectile Motion Revisited: The Effect of Air Drag

In Section 4.2, we discussed "ideal" projectile motion in which the only force acting on a projectile is the force of gravity. Real projectiles generally also experience a force due to air drag, and in some cases this drag force can have a very significant effect. Let's consider the drag force in a few cases and estimate its effects.

We begin with the motion of a batted baseball and calculate how far it would travel in the absence of air drag. We worked through this problem in Section 4.2, where we saw that the range of the ball is (Eq. 4.28)

$$x_{\text{range}} = \text{range} = \frac{v_0^2 \sin(2\theta)}{g}$$

We also showed that for a given value of the initial speed v_0, the maximum range occurs when $\theta = 45°$. A skilled baseball player can set a ball into motion with an initial speed of $v_0 = 45$ m/s (about 100 mi/h). Inserting this value into our relation for x_{range} gives a range of 210 m, or about 680 ft. This distance is far longer than a batted baseball has ever traveled. (For a typical baseball field, the distance from home plate to the centerfield fence is about 120 m, and only the longest home runs travel this far.) Our calculation must have omitted something important: the force of air drag.

Insight 4.2
APPLYING NEWTON'S SECOND LAW IN THREE DIMENSIONS
In our applications of Newton's second law, we have dealt with many examples, such as projectiles, that move in two dimensions. For such cases, Newton's second law can be written as

$$\sum F_x = ma_x \text{ and } \sum F_y = ma_y$$

and we can deal with forces and acceleration along x and y. To deal with a problem that involves forces or motion in three dimensions, we must also use

$$\sum F_z = ma_z$$

The basic approach to such problems is the same as in two dimensions.

In Chapter 3, we saw that for an object like a baseball, a bullet, a cannon shell, or an airplane, the magnitude of the air drag force is given approximately by

$$F_{\text{drag}} = \tfrac{1}{2}\rho A v^2 \tag{4.46}$$

where ρ is the density of air (1.3 kg/m^3), A is the cross-sectional area of the object, and v is its speed. (See Insight 3.4 on page 73 for more on the drag force.) For a baseball[5] moving at 45 m/s, Equation 4.46 gives $F_{\text{drag}} = 5.4$ N. The baseball's weight is approximately $mg = 1.4$ N, so the drag force here is much *greater* than the force of gravity on the ball. We certainly cannot ignore the drag force in this case.

Although we have a mathematical expression for the drag force in Equation 4.46, it is difficult to work out a simple expression for the trajectory of a projectile (such as a baseball) when this force is included. However, such trajectories can be readily calculated with computer simulation methods, so we can describe how drag affects the shape of the trajectory and the range. We have already seen that for an ideal projectile (i.e., one without air drag), the maximum range occurs when the initial velocity makes an angle of 45° with the horizontal. Drag can substantially alter this result, and in either direction. For a baseball, the maximum range is found with a much smaller angle. The "best" angle depends on the initial speed of the ball and is typically around 35°. On the other hand, for a powerful artillery gun, the artillery shell can travel very high into the atmosphere where the density of the air is much less than at ground level, which greatly reduces the drag force on the shell. To achieve maximum range, these guns are usually aimed well above 45°, and the artillery shell spends a large amount of time at high altitudes where the air density is lower than it is near sea level.

Effect of Air Drag on a Bicycle

Air drag is very important for vehicles such as cars and bicycles. Indeed, most of us have felt the force from air drag when riding a bicycle even at rather low speeds. Consider the bicyclist in Figure 4.33 riding downhill on a frictionless bicycle so that the only forces along the incline are due to gravity and air drag. If the slope of the hill is $\theta = 5.0°$, how fast will the bicycle coast?

Figure 4.33 shows the forces acting on the bicycle. We take the x direction to be along the incline. Writing Newton's second law for motion along x, we get

$$\sum F_x = F_{\text{grav},x} + F_{\text{drag}}$$

The component of the gravitational force along x is

$$F_{\text{grav},x} = mg \sin\theta \tag{4.47}$$

Using Equation 4.46, the drag force is

$$F_{\text{drag}} = -\tfrac{1}{2}\rho A v_x^2 \tag{4.48}$$

where the negative sign indicates that F_{drag} is directed up the incline (along $-x$). When the bicyclist is coasting at her terminal velocity, v is constant, so the total force along x must be zero. Using Equations 4.47 and 4.48, we thus find

$$\sum F_x = mg\sin\theta - \tfrac{1}{2}\rho A v_x^2 = 0$$

$$v_x = \sqrt{\frac{2mg\sin\theta}{\rho A}}$$

To get a feeling for the value of this terminal (coasting) velocity, we consider a bicycle plus rider of mass 100 kg, with an approximate frontal area of 1 m^2. For a hill with slope angle $\theta = 5.0°$, a moderately steep hill, we have a speed

$$v_x = \sqrt{\frac{2mg\sin\theta}{\rho A}} = \sqrt{\frac{2(100\text{ kg})(9.8\text{ m/s}^2)\sin(5.0°)}{(1.3\text{ kg/m}^3)(1\text{ m}^2)}} = 10\text{ m/s}$$

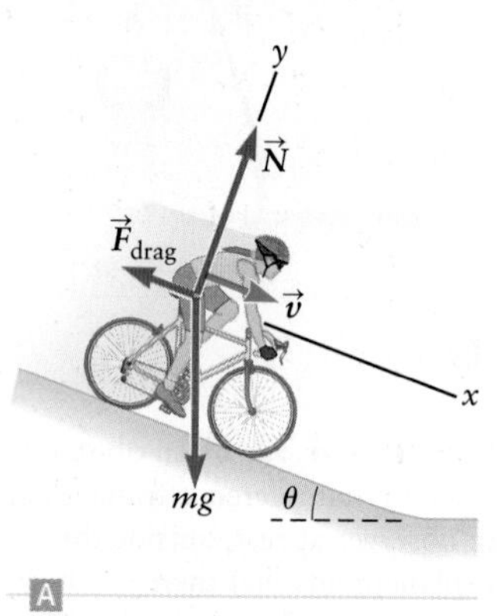

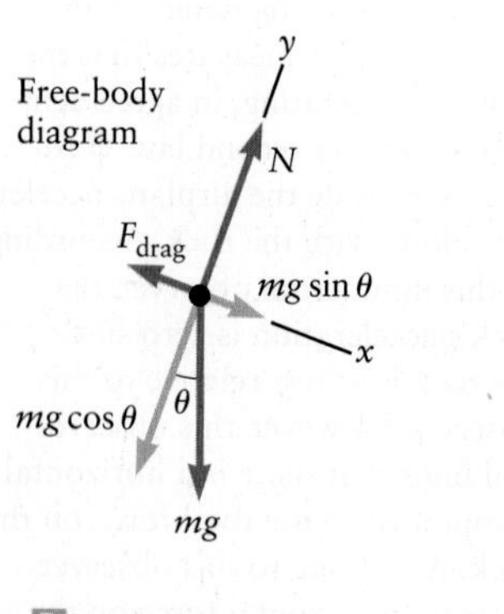

▲ **Figure 4.33** A When a bicycle coasts down a long hill, the bicycle eventually reaches its terminal velocity. The forces acting on the bicycle are the normal force $\vec{N}$ exerted by the hill's surface, the force of gravity $\vec{F}_{\text{grav}}$, and the drag force $\vec{F}_{\text{drag}}$. B Free-body diagram with forces resolved along x and y.

[5]The frontal area of the baseball is $A = \pi r^2$, where the radius of a baseball is $r = 3.6$ cm.

(Here we keep just one significant figure, since the area is only estimated with this accuracy.) This is about 25 mi/h. How does this answer compare with your experience?

Summary | CHAPTER 4

Key Concepts and Principles

Translational equilibrium

When the total force on an object is zero, Newton's second law tells us that the acceleration is also zero. In other words, all the vector components of $\vec{F}_{total} = \sum \vec{F}$ and $\vec{a}$ are zero. Such an object is in ***translational equilibrium***.

Analyzing motion in two and three dimensions

We use Newton's second law to calculate the acceleration of an object:

$$\sum \vec{F} = m\vec{a} \qquad \text{(4.1) (page 82)}$$

The acceleration can then be used to find the velocity and displacement.

Inertial and noninertial reference frames

A ***reference frame*** is an observer's choice of coordinate system, including an origin, for making measurements. An ***inertial reference frame*** is one that moves with a constant velocity, whereas a ***noninertial*** reference frame is one that is accelerating. Newton's laws are only obeyed in inertial reference frames. They are not obeyed in noninertial reference frames.

Applications

Projectile motion

In many cases, the force of air drag is small. The force on a projectile near the Earth's surface, such as a baseball, is then just the force of gravity, and the acceleration has a constant magnitude g and is directed downward. The displacement and velocity for such a projectile are described by the relations for motion with constant acceleration.

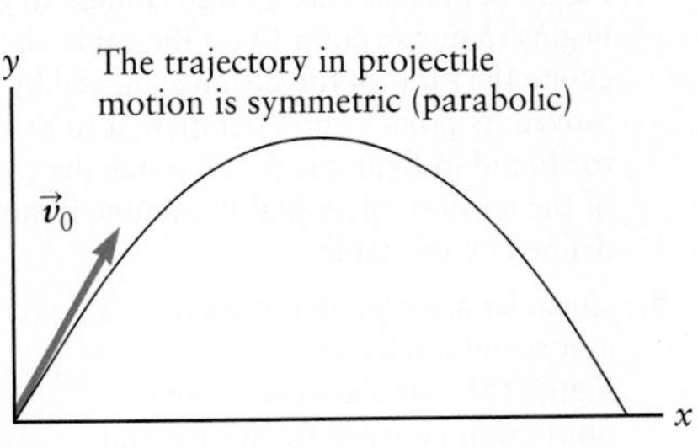

y
The trajectory in projectile motion is symmetric (parabolic)
$\vec{v}_0$
x

For projectile motion, the vertical and horizontal motions are independent. Two objects can have very different horizontal velocities, but they still fall at the same rate.

The trajectory for simple projectile motion starting from the origin is symmetric in two ways:

- It is symmetric in time. The time spent traveling to the point of maximum height is equal to the time spent falling back down to the initial height.
- It is symmetric in space. The trajectory has a parabolic shape.

How the body detects acceleration

The ear contains a sensitive acceleration detector that enables the brain to know when the head is accelerating.

Questions

SSM = answer in *Student Companion & Problem-Solving Guide* ⓧ = life science application

1. At what angle should a ball be thrown so that it has the maximum range? Ignore air drag.

2. Two children are having a snowball fight, and one is trying to land a snowball on a friend who is hiding behind a wall (Fig. Q4.2). The child on the left can only throw snowballs with one value of the initial speed and finds that for one throwing angle α that is less than 45° he is able to have his snowball land on his friend. Show that there is a corresponding throwing angle β greater than 45° that yields the same value of the range and find a general relation between α and β. *Hint:* Consider the range of a projectile as calculated in Equation 4.28 (i.e., ignoring the effect of air drag), and assume that the snowballs in Figure Q4.2 start and land at the same height above the ground.

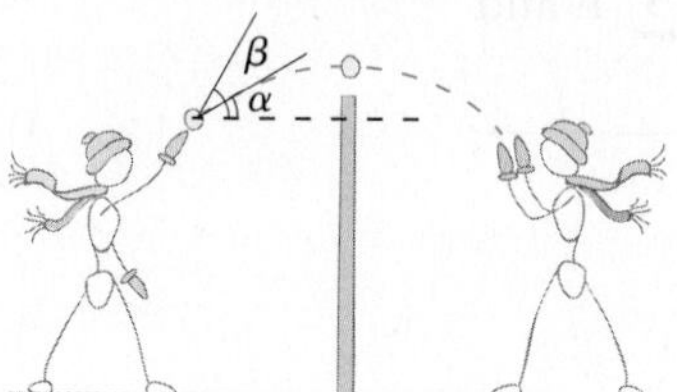

Figure Q4.2

3. An object has an initial velocity $\vec{v}_i$ and a final velocity $\vec{v}_f$ as sketched in Figure Q4.3. Sketch the approximate direction of the acceleration $\vec{a}$ of the object during this time interval.

Figure Q4.3

4. During a particular time interval, an object has an acceleration $\vec{a}$ in the direction shown in Figure Q4.4. If $\vec{v}_i$ is the initial velocity of the object, draw a vector $\vec{v}_f$ that might be the velocity of the object at the end of the interval.

Figure Q4.4

5. The arrangement in Example 4.2 (p. 85) makes it possible to amplify a force. (Compare this example to the block and tackle in Fig. 3.23.) Amplifying forces in this way comes at a price. To appreciate this price, imagine that the force on the car in Figure 4.4 is just barely large enough to move the car. If the car begins to move, point O on the cable also begins to move. Calculate the ratio of the distance moved by the car to the distance moved by point O and compare it to the force amplification factor found in Example 4.2. Assume the car moves in the direction of the tension force; that is, assume it moves along the direction defined by the cable.

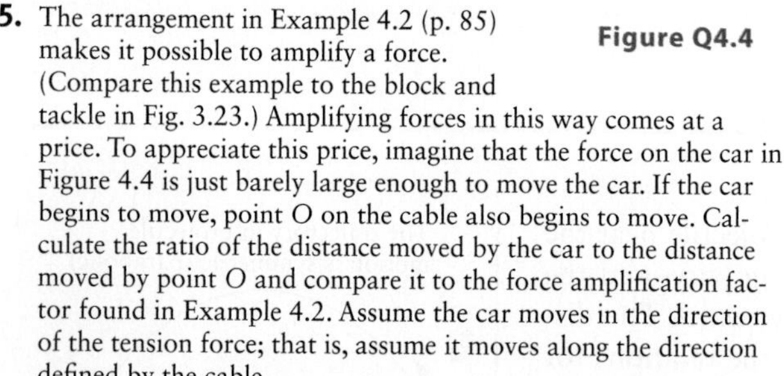

6. Consider a wedge that rests on a horizontal table as shown in Figure Q4.6 and assume there is no friction between the wedge and the table. A force is now applied vertically as shown in the figure. You might expect this force will cause the wedge to move to the left. Explain why the wedge does not move, even though there is no friction between it and the table.

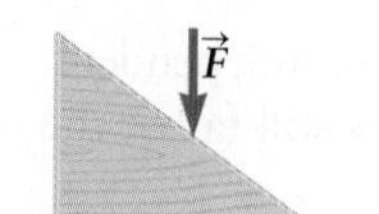

Frictionless table

Figure Q4.6

7. SSM Consider again the wedge in Question 6, but now assume a block is placed onto it as shown in Figure Q4.7. There is again no friction between the wedge and the table, and there is also no friction between the block and the wedge. Explain why in this case the wedge *will* be accelerated to the left. How does this situation differ from that in Question 6? *Hint:* Compare the free-body diagrams for the wedge in the two cases.

Frictionless table

Figure Q4.7

8. The rock-on-a-string accelerometer in Figure 4.30 can be used to measure acceleration along a horizontal direction. Discuss how you could use it to also measure the vertical component of the acceleration.

9. **Time of flight.** Equation 4.26 gives a relation for the time at which a batted baseball lands. Use the quadratic formula to find the solutions for t_{lands}. (See Appendix B if you don't remember the quadratic formula.) Since you are solving a quadratic equation, there are *two* solutions for t_{lands}. Find numerical values for both solutions and give a physical interpretation of the results. *Hint:* Consider how the problem would be changed if the ball was initially directed "backward" along the trajectory in Figure 4.12. Assume $h = 1.0$ m, $v_0 = 45$ m/s, and $\theta = 30°$.

10. Explain why the string in Figure 4.30 hangs vertically if the velocity is constant. Consider both an airplane moving horizontally and one moving with a nonzero velocity component along y.

11. ⓧ Explain why you cannot run unless there is a frictional force between your feet and the ground.

12. **Peel out!** The author has a small pickup truck. He finds that it is much easier to "burn rubber" (i.e., spin the back wheels so that they slip relative to the road surface) when the truck is empty than when it is carrying a heavy load. Explain why.

13. Give an example in which the magnitude of the instantaneous velocity is always larger than the average velocity.

14. Give an example in which the acceleration is perpendicular to the velocity.

15. Give an example of motion in which the instantaneous velocity is zero but the acceleration is not zero.

16. A rock is thrown up in the air in such a way that its speed is zero at the top of its trajectory. Where does the rock land?

17. Give three examples in which the force of friction on an object has the same direction as the velocity of the object.

18. Ball 1 is thrown with an initial speed v_0 at an angle θ relative to the horizontal (x) axis. Ball 2 is thrown at the same angle, but with an initial speed of $2v_0$. (a) If ball 1 is in the air for a time t_1, how long is ball 2 in the air? (b) If ball 1 reaches a maximum height h_1, what is the maximum height reached by ball 2 in terms of h_1?

19. Ball 1 rolls off the edge of a table and falls to the floor below, while ball 2 is dropped from the same height (Fig. Q4.19). Which ball reaches the ground first? Explain why your answer does, or does not, depend on the velocity of ball 1 just before it leaves the table.

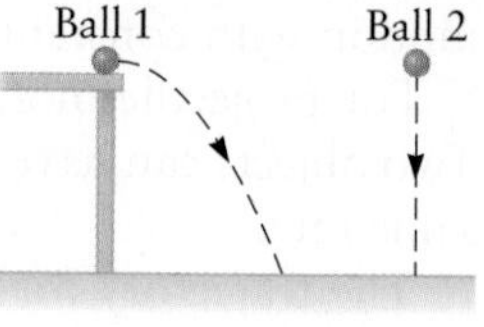

Figure Q4.19

20. SSM Two balls are thrown into the air with the same initial speed, directed at the same initial angle with respect to the horizontal. Ball 1 has a mass five times the mass of ball 2, and the force of air drag is negligible.
(a) Which ball has the larger acceleration as it moves through the air?
(b) Which ball lands first?
(c) Which ball reaches the greatest height?
(d) For which ball is the force of gravity larger at the top of the trajectory?

Problems available in ENHANCED WebAssign

Section 4.1 Statics
Problems 1–14

Section 4.2 Projectile Motion
Problems 15–34

Section 4.3 A First Look at Reference Frames and Relative Velocity
Problems 35–42

Section 4.4 Further Applications of Newton's Laws
Problems 43–55

Section 4.5 Detecting Acceleration: Reference Frames and the Workings of the Ear
Problems 56–58

Section 4.6 Projectile Motion Revisited: The Effect of Air Drag
Problems 59–63

Additional Problems
Problems 64–92

List of Enhanced Problems

Problem number	Solution in *Student Companion & Problem-Solving Guide*	Reasoning & Relationships Problem	Reasoning Tutorial in ENHANCED WebAssign
4.3	✓		
4.22	✓		
4.29		✓	
4.40	✓		
4.50		✓	✓
4.53	✓		
4.55		✓	
4.58	✓	✓	✓
4.59	✓	✓	
4.61		✓	
4.62		✓	
4.63		✓	✓
4.64	✓		
4.70		✓	
4.79	✓		
4.81		✓	✓
4.84	✓		
4.87		✓	
4.89		✓	✓
4.91		✓	
4.92		✓	

CHAPTER 5

► *The motion of this satellite as it orbits the Earth is nearly circular and also involves the gravitational force. We study both of these topics in this chapter. (Erik Simonsen/ Photographer's Choice/Getty Images)*

Circular Motion and Gravitation

OUTLINE

In the past few chapters, we studied the connection between forces and motion, and we learned how to apply Newton's laws in a variety of situations. In many cases, those situations involved motion in one dimension. While we have also studied two- and three-dimensional motion, most of those cases could be treated as a combination of one-dimensional problems. For example, in our work on projectile motion, we were able to treat the horizontal and vertical motions as essentially separate one-dimensional problems. In this chapter, we consider a different type of motion, called **circular motion**, in which the acceleration is far from constant and that *cannot* be reduced to a one-dimensional problem. Circular motion is found in many situations, such as a car traveling around a turn, a roller coaster traveling near the top or bottom of its track, a centrifuge, and the Earth orbiting the Sun. Our studies of orbital motion and the forces that make it possible will also lead us to explore the gravitational force in more detail. We'll then encounter Kepler's laws of orbital motion and see how they describe the motion of planets, moons, and satellites. These studies of gravitation will give us new insights into the quantity g, the acceleration due to gravity near the Earth's surface.

5.1 Uniform Circular Motion

Figure 5.1 shows a top-down view of a person running around a circular track. For simplicity, let's assume our runner is moving with a constant speed so that the magnitude of her velocity is constant. The direction of her velocity $\vec{v}$, however, changes as she moves around the track since the velocity vector is always directed tangent to the circle. Her velocity is thus not constant, even though her speed is constant. Circular motion at constant speed is known as ***uniform circular motion***.

The runner in Figure 5.1 is moving at a constant speed v, so the time it takes her to travel once around the track (to run one complete lap) is equal to the circumference of the track divided by v. If the track has a radius r, the circumference is $2\pi r$ and the time to complete one lap is

$$T = \frac{2\pi r}{v} \tag{5.1}$$

This time T is called the ***period*** of the motion.

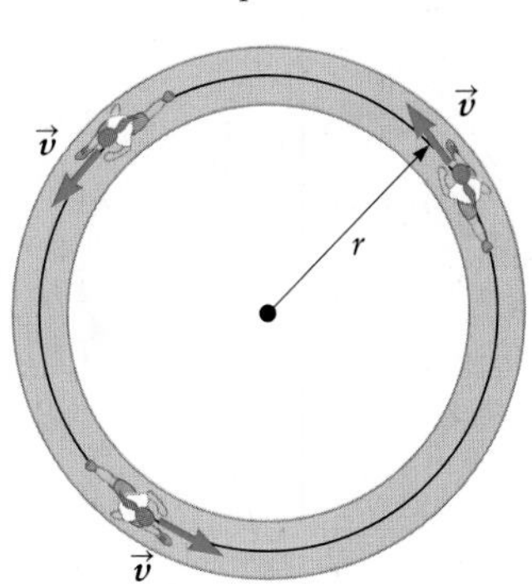

▲ **Figure 5.1** This runner is moving with a constant speed around a circular track. While her speed is constant, her velocity is not constant. The direction of $\vec{v}$ changes as she moves around the track.

EXAMPLE 5.1 A Bug on a Compact Disc

A bug is sitting on the edge of a compact disc of radius $r = 6.0$ cm (Fig. 5.2). The bug undergoes uniform circular motion as the CD spins. **(a)** If the bug traverses this circle exactly six times in precisely 1 s, what is the period of the motion? **(b)** What is the bug's speed?

RECOGNIZE THE PRINCIPLE

The period T is the time it takes to travel once around the circle, so we can find T from the given information. We can then use the relation between period and v in Equation 5.1 to find the bug's speed.

SKETCH THE PROBLEM

Figure 5.2 shows the problem. The bug moves in a circle of radius r.

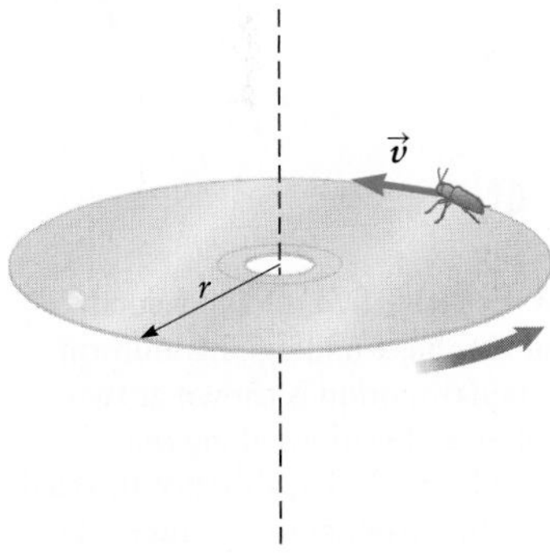

▲ **Figure 5.2** Example 5.1. The bug sitting on this CD is undergoing uniform circular motion.

IDENTIFY THE RELATIONSHIPS AND SOLVE

(a) Since the bug completes six trips each 1 s, we have

$$T = \frac{1.0 \text{ s}}{6 \text{ trips around circle}} = \frac{0.17 \text{ s}}{1 \text{ trip}}$$

The period is the time to complete one trip around the circle, so the bug's period is

$$T = \boxed{0.17 \text{ s}}$$

(b) The bug is undergoing uniform circular motion, so its speed v is constant and equal to the total distance traveled divided by the time. A distance d equal to six times the circumference is traveled in 1 s, so $d = 6 \times (2\pi r)$ while $t = 1.0$ s. Hence,

$$v = \frac{d}{t} = \frac{6(2\pi)(0.060 \text{ m})}{1.0 \text{ s}} = \boxed{2.3 \text{ m/s}}$$

► ***What does it mean?***

The period of circular motion is equal to the time it takes to move around the circle once. Hence, T does not depend on the bug's location on the CD; the period is the same for a bug at the edge of the CD or near the center. The speed of the bug, however, does depend on its location (the value of r). The speed is largest at the edge (a large value of r) and decreases as the bug moves toward the center.

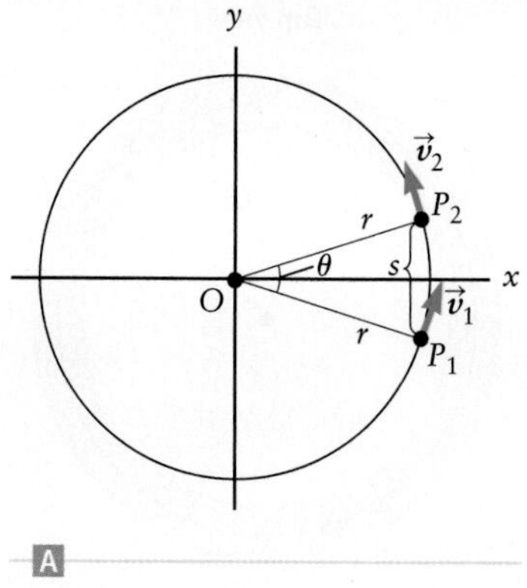

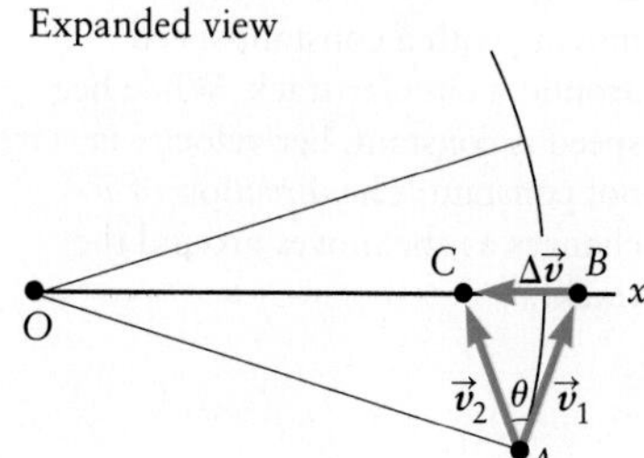

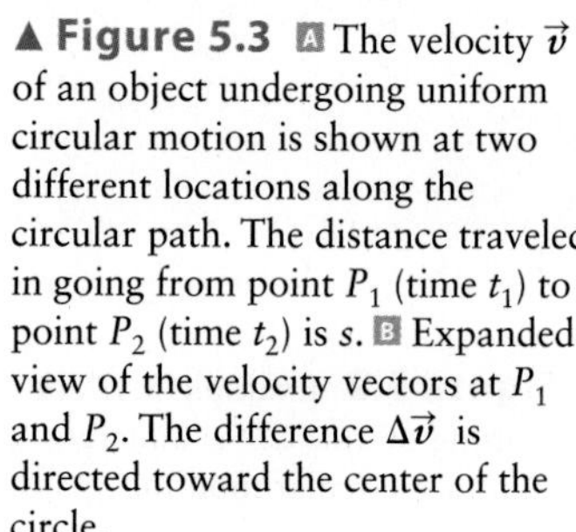

▲ **Figure 5.3** A The velocity $\vec{v}$ of an object undergoing uniform circular motion is shown at two different locations along the circular path. The distance traveled in going from point P_1 (time t_1) to point P_2 (time t_2) is s. B Expanded view of the velocity vectors at P_1 and P_2. The difference $\Delta\vec{v}$ is directed toward the center of the circle.

Centripetal Acceleration

Recall from Chapter 2 that the acceleration $\vec{a}$ of an object is proportional to the change in $\vec{v}$ over the course of a time interval Δt:

$$\vec{a} = \lim_{\Delta t \to 0} \frac{\Delta\vec{v}}{\Delta t} \tag{5.2}$$

For an object undergoing uniform circular motion, the velocity is changing with time, so $\Delta\vec{v}$ is definitely not zero. Uniform circular motion therefore involves a *nonzero acceleration.*

To calculate this acceleration, let's consider how the velocity of the runner in Figure 5.1 changes as she moves around the track. Figure 5.3A shows her velocity vectors at two nearby locations P_1 and P_2, and Figure 5.3B shows the corresponding change in velocity $\Delta\vec{v}$ over a short time interval Δt. We can use the geometry in Figure 5.3B to estimate the average acceleration $\vec{a}_{ave} = \Delta\vec{v}/\Delta t$ during this interval. If Δt is small, the instantaneous acceleration (Eq. 5.2) is approximately equal to $\vec{a}_{ave}$. We can therefore write

$$\vec{a} \approx \vec{a}_{ave} = \frac{\Delta\vec{v}}{\Delta t} \tag{5.3}$$

Let's first consider the direction of $\vec{a}$. During the time interval $\Delta t = t_2 - t_1$, our runner travels a distance $s = v\,\Delta t$ along the circumference of the circle. (Recall that the speed v is the magnitude of both $\vec{v}_1$ and $\vec{v}_2$.) The velocity vectors $\vec{v}_1$ and $\vec{v}_2$ are both tangent to the circle and hence are perpendicular to the radius lines drawn at P_1 and P_2 in Figure 5.3A. In Figure 5.3B, we have redrawn the velocity vectors so that they have a common "tail point." When drawn in this way, the velocity vectors $\vec{v}_1$ and $\vec{v}_2$ form two sides of the triangle labeled ABC. The third side of this triangle is $\Delta\vec{v} = \vec{v}_2 - \vec{v}_1$ and is directed toward the center of the circle. This result is true at all points along our runner's circular path. Since the acceleration vector is parallel to $\Delta\vec{v}$ (see Eq. 5.3), the acceleration of an object undergoing uniform circular motion is always directed *toward the center of the circle.*

To compute the magnitude of this acceleration, we again make use of triangle ABC in Figure 5.3B. Two sides of this triangle are formed by $\vec{v}_1$ and $\vec{v}_2$, so these sides are both of length v, while the other side of this triangle has a length Δv. Triangle ABC has the same interior angles as the triangle defined by O, P_1, and P_2 in Figure 5.3A, so these triangles are similar. Two sides of triangle OP_1P_2 are along the radius of the circle and are thus of length r, while the other side (between points P_1 and P_2) has a length $s = v\Delta t$, and when Δt is small, this arc approaches a straight line. Triangles ABC and OP_1P_2 are similar, so the ratios of their corresponding sides are equal; hence,

$$\frac{\Delta v}{v} = \frac{s}{r} = \frac{v(\Delta t)}{r}$$

$$\Delta v = \frac{v^2(\Delta t)}{r} \tag{5.4}$$

According to Equation 5.3, the magnitude of the acceleration is equal to the ratio $\Delta v/\Delta t$, so we can rearrange the result in Equation 5.4 to get

$$a_c = \frac{\Delta v}{\Delta t} = \frac{v^2(\Delta t)/r}{\Delta t}$$

Magnitude of the centripetal acceleration

$$a_c = \frac{v^2}{r} \tag{5.5}$$

This acceleration, called the ***centripetal acceleration***,[1] is usually denoted by the symbol a_c. Although this derivation started with the average acceleration in Equation 5.3, the result becomes exact for a very small time interval $\Delta t \to 0$; hence, Equation 5.5 gives the instantaneous centripetal acceleration.

[1] The adjective *centripetal* means "center seeking."

Circular Motion and Forces

The result for the acceleration in Equation 5.5 applies to any object undergoing circular motion. Such an object has an acceleration of magnitude $a_c = v^2/r$, and this acceleration is directed toward the center of the circle. From Newton's second law, we know that accelerations are caused by forces, so for an object undergoing uniform circular motion, we have

$$\sum \vec{F} = m\vec{a}$$

In terms of magnitudes,

$$\sum F = ma_c = \frac{mv^2}{r} \quad (5.6)$$

Force required for uniform circular motion

which is the force required to make an object of mass m travel with speed v in a circle of radius r. This force must be directed toward the center of the circle.

As an example, consider a rock tied to a string and suppose a person is twirling the string so that the rock moves in a circle. For simplicity, let's assume this experiment is being done by an astronaut in distant space (Fig. 5.4), so the only force on the rock comes from the string (i.e., all gravitational forces are negligible). This rock is undergoing circular motion; hence, its motion is described by Equation 5.6. The only force comes from the tension T in the string, so

$$\sum F = T = \frac{mv^2}{r}$$

For the rock to travel in a circle, the tension must have this value. (Note that T here is the tension in the string, not the period of the motion.)

When an object undergoes circular motion, there must be a force of magnitude mv^2/r acting on it. For the rock in Figure 5.4A, this force is the tension in the string, but in other situations, the force might be due to gravity, friction, or some other source. Without such a force, the object *cannot* undergo circular motion.

What happens if the string in Figure 5.4 suddenly breaks? The force on the rock would then be zero, and according to Newton's first law, the rock would then move away in a straight-line path—that is, with a constant velocity (Fig. 5.4B). The rock does not move radially outward, nor does it "remember" its circular trajectory. The only way the rock can move in a circle is if there is a force that makes it do so, and that force is provided by the string.

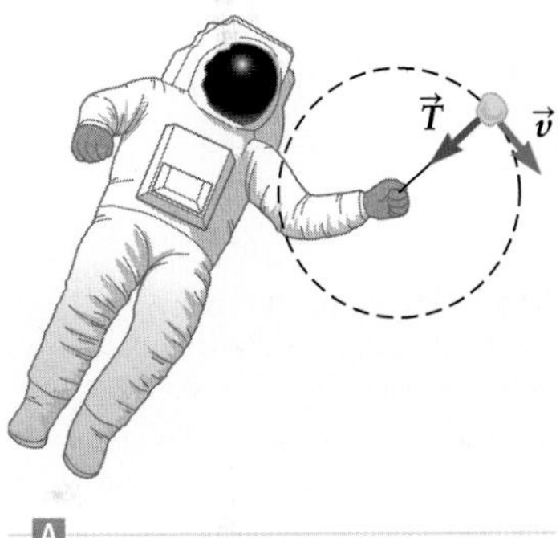
$\vec{T}$
$\vec{v}$
A

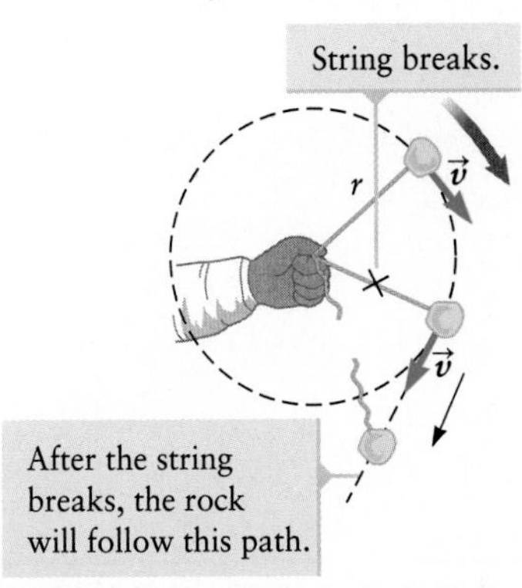
String breaks.
r
$\vec{v}$
$\vec{v}$
After the string breaks, the rock will follow this path.
B

▲ **Figure 5.4** A An astronaut in distant space twirls a rock on a string. The only force on the rock is due to the tension in the string (all gravitational forces are negligible). B If $T = mv^2/r$, the rock will undergo uniform circular motion. If the string breaks, the rock will move along a straight line (obeying Newton's first law).

CONCEPT CHECK 5.1 Velocity and Acceleration in Circular Motion

Consider the rock in Figure 5.4A as it undergoes uniform circular motion (before the string breaks). Which of the following statements is correct? (More than one statement may be correct.)

(a) The direction of $\vec{a}$ changes as the rock moves around the circle.
(b) The direction of $\vec{v}$ changes as the rock moves around the circle.
(c) $\vec{a}$ and $\vec{v}$ are always perpendicular.
(d) $\vec{a}$ and $\vec{v}$ are always parallel.

EXAMPLE 5.2 Centripetal Acceleration of a Compact Disc

Suppose the bug in Example 5.1 has a mass $m = 5.0$ g and sits on the edge of a compact disc of radius 6.0 cm. The CD is spinning such that the bug travels around its circular path three times per second. Find (**a**) the centripetal acceleration of the bug and (**b**) the total force on the bug. Also identify the *origin* of the force that enables the bug to move in a circle.

(continued) ▶

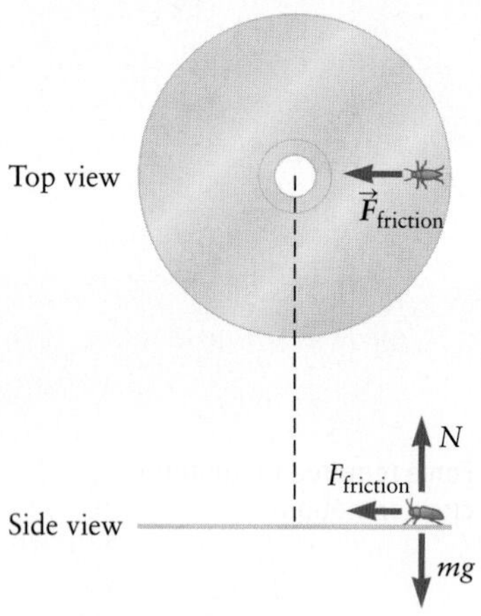

▲ **Figure 5.5** Example 5.2. There are three forces acting on the bug on this CD. The force responsible for the centripetal acceleration is provided by the force of static friction and is directed toward the center of the circle.

RECOGNIZE THE PRINCIPLE

Because the bug's acceleration along the vertical direction is zero, the normal force N and the force of gravity mg must cancel; hence, $N = mg$. The bug is undergoing circular motion, so we know a third force must provide the force required to produce the centripetal acceleration a_c. This third force keeps the bug from slipping relative to the CD and is just the frictional force. To make the bug undergo uniform circular motion, this force must be directed toward the center of the circle.

SKETCH THE PROBLEM

The forces acting on our bug are shown at the bottom of Figure 5.5.

IDENTIFY THE RELATIONSHIPS

To find the centripetal acceleration, we need to determine the speed of the bug, which we can find from its period and the radius of its circular path. We can then compute the magnitude of a_c and the force needed to produce it.

SOLVE

(a) The bug traverses the circle three times in 1 s, so it travels a distance equal to three times the circumference each second. Its speed is thus

$$v = \frac{3 \times (2\pi r)}{t} = \frac{3(2\pi)(6.0 \text{ cm})}{1.0 \text{ s}} = 110 \text{ cm/s} = 1.1 \text{ m/s}$$

From Equation 5.5, the centripetal acceleration is

$$a_c = \frac{v^2}{r} = \frac{(1.1 \text{ m/s})^2}{0.060 \text{ m}} = \boxed{20 \text{ m/s}^2}$$

(b) The required force is (Eq. 5.6)

$$F = ma_c = (5.0 \text{ g})(20 \text{ m/s}^2) = (5.0 \times 10^{-3} \text{ kg})(20 \text{ m/s}^2)$$

$$F = 0.10 \text{ kg} \cdot \text{m/s}^2 = \boxed{0.10 \text{ N}}$$

This force is produced by friction between the bug's feet and the surface of the CD.

► ***What does it mean?***

Even though the bug is moving, the force in part **(b)** is static friction because the bug is not slipping relative to the CD's surface.

Now that we have analyzed some examples of circular motion, it is time to outline a general way to approach such problems.

PROBLEM SOLVING Analyzing Circular Motion

1. **RECOGNIZE THE PRINCIPLE.** An object can move in a circle only if there is a force $F = mv^2/r$ directed toward the center of the circle.
2. **SKETCH THE PROBLEM.** Make a drawing that shows the path followed by the object of interest. This drawing should identify the circular part of the path, the radius of this circle, and the center of the circle.
3. **IDENTIFY THE RELATIONSHIPS.**
   - Find all the forces acting on the object; as in our applications of Newton's laws in Chapters 3 and 4, a free-body diagram is often very useful.
   - Using your drawing and free-body diagram, find the components of the forces that are directed *toward the center* of the circle and find the components perpendicular to this direction.

- Apply Newton's second law $\sum F = ma$ for motion toward the center of the circle and (if necessary) in the perpendicular direction. The total force directed toward the center of the circle is (Eq. 5.6)

$$\sum F_{\text{center}} = ma_c = \frac{mv^2}{r}$$

4. **SOLVE FOR THE QUANTITIES OF INTEREST.** For example, the centripetal acceleration is (Eq. 5.5)

$$a_c = \frac{v^2}{r}$$

5. Always *consider what your answer means* and check that it makes sense.

When discussing the centripetal acceleration and the associated forces, it is often convenient to use a coordinate system in which the positive direction is toward the center of the circle. The "positive" direction changes as the object moves around the circle (as for the bug in Fig. 5.5). An advantage of this approach is that the centripetal acceleration is always positive, which can eliminate some extra minus signs in your equations.

Centripetal Acceleration of a Turning Car: What Are the Forces?

The difficulties faced by the bug in Example 5.2 are similar to those encountered by the driver of a car. While cars do not usually travel in perfect circles, a car making a turn or rounding a bend in a road is traveling in an approximate circle. Over a short distance, any curved path can be approximated by a circle, and the radius of this circle is called the ***radius of curvature***. Let's analyze the motion of a car that travels along a short section of such a circle as sketched in Figure 5.6. As it travels around a bend in this road, the car travels in a circular path, so there must be a force on the car directed toward the center of the circle (point C in Figure 5.6A). Continuing with our problem-solving strategy for circular motion (step 2), the sketch in Figure 5.6A shows the circular path followed by the car as the dashed curve. The center of this circle is at point C, and its radius is r. For step 3, the sketch in Figure 5.6B shows all the forces on the car along with a free-body diagram. There are two forces along the vertical direction, the force of gravity and the normal force exerted by the road on the car, while the only horizontal force is the force of friction F_{friction} from the road on the car. This force is static friction because we assume the tires are not slipping relative to the road. Applying Newton's second law along y, we get

$$\sum F_y = +N - mg = ma_y$$

Since the acceleration along y is zero, this result leads to $N = mg$. The only force directed toward the center of the circle is F_{friction}, and inserting it into Equation 5.6 gives

$$F_{\text{friction}} = \frac{mv^2}{r}$$

Given the values of m, v, and r, we can now calculate the force required to make the turn successfully, without leaving the road.

Notice that the required force increases as the car's speed increases. There is an upper limit on the force of static friction; hence, there is a maximum speed at which the car can make the turn without slipping. The maximum frictional force is $F_{\text{friction,max}} = \mu_S N = \mu_S mg$, so the condition for this maximum speed is

$$\frac{mv^2}{r} = F_{\text{friction, max}} = \mu_S mg$$

Solving for the maximum speed gives

$$v = \sqrt{\mu_S g r} \quad (5.7)$$

Let's put in some realistic numbers to see how the result in Equation 5.7 compares with our everyday experience. For a typical intersection in a residential

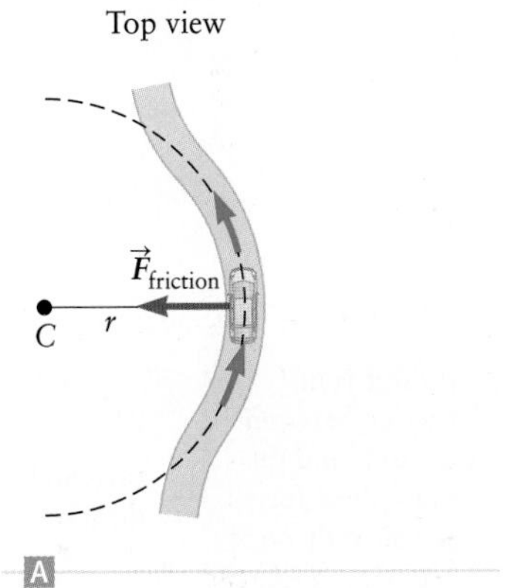

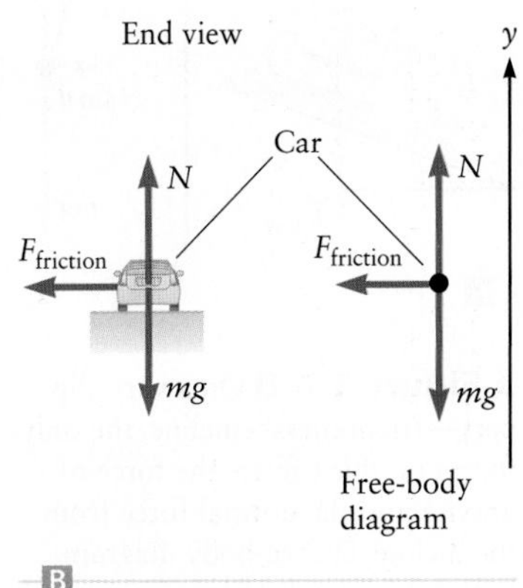

▲ **Figure 5.6** This car moves in an approximately circular path as it travels along a curve in the road. Friction between the car's tires and the road provides the force that causes the centripetal acceleration. **A** Top view. **B** End view showing forces on the car and a free-body diagram.

neighborhood, the effective radius of a turn might be $r = 10$ m, and the coefficient of friction between rubber and a road surface is typically $\mu_S = 0.6$. Inserting these values into Equation 5.7 gives

$$v = \sqrt{\mu_S g r} = \sqrt{0.6(9.8 \text{ m/s}^2)(10 \text{ m})} = 8 \text{ m/s}$$

which is approximately 15 mi/h. Hence, our physics does indeed give a good description of "real-life" driving.

CONCEPT CHECK 5.2 High-Speed Driving

Consider two vehicles, a compact car and a large truck, that are both traveling around the turn in the road in Figure 5.6 and assume the coefficient of friction between the tires and the road is the same for both. Which vehicle can drive through the turn at the largest speed without slipping?

(a) The car
(b) The truck
(c) The nonslipping speed is the same.

A Car on a Banked Turn: Analyzing the Forces

The maximum speed for making a turn successfully can be increased by banking the turn as shown in Figure 5.7. Let's analyze this situation, but for simplicity, we assume there is no friction between the tires and the road. (We'll add friction to this problem in Example 5.3.) We want to calculate the speed at which a car can successfully travel through such a turn without slipping off the road.

We again follow our problem-solving strategy for circular motion. The car travels in a circle, so there must be a force $F = ma_c$ directed toward the center of its circular path. For step 2, the car is shown in Figure 5.7A. The overall circular path is the same as for the car in Figure 5.6, and we again take r to be the radius of curvature. For step 3, the forces on the car are shown in Figure 5.7, which also gives a free-body diagram. We have assumed there is no friction, so the only forces on the car are due to gravity and the normal force exerted by the road on the car. We determine the components of the force along the vertical and along the direction toward the center of the circle (horizontal). Due to the banking, there is now a component of the normal force directed toward the center of the circle. From Figure 5.7B, the component of the force directed toward the center of the circle is $N \sin\theta$, and this must be equal to the mass of the car times the centripetal acceleration. For a car of mass m traveling at a speed v, we have

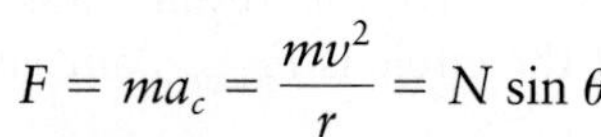

$$F = ma_c = \frac{mv^2}{r} = N \sin\theta$$

which can be solved for the speed of the car:

$$v = \sqrt{\frac{Nr \sin\theta}{m}} \tag{5.8}$$

At this speed, the car will just be able to negotiate the turn without sliding up or down the banked road. To complete our calculation of the speed v, we must know the value of the normal force, which we can determine by considering the forces along the vertical direction in Figure 5.7B. Since the acceleration along the vertical direction is zero, the total force along y must be zero:

$$\sum F_y = N \cos\theta - mg = 0$$

Hence,

$$N = \frac{mg}{\cos\theta}$$

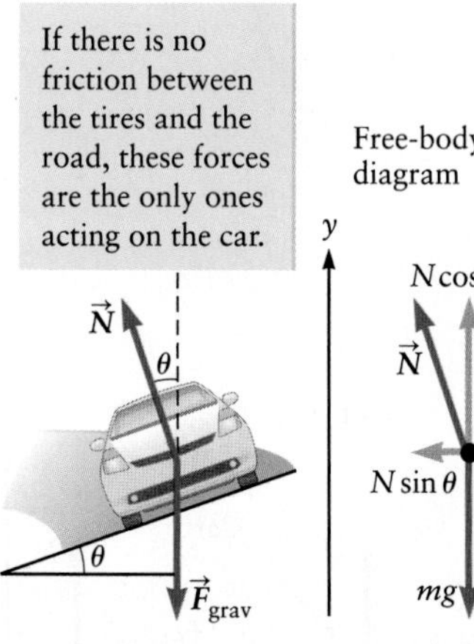
If there is no friction between the tires and the road, these forces are the only ones acting on the car.
Free-body diagram
y
N cos θ
N
θ
N sin θ
θ
F grav
mg
A
B

▲ **Figure 5.7** A On a very slippery—frictionless—incline, the only forces on this car are the force of gravity and the normal force from the incline. B Free-body diagram with forces resolved into horizontal and vertical components. The horizontal component of the normal force is $N \sin\theta$. This is the only horizontal force on the car, and it provides the centripetal acceleration that enables the car to move in a circular path as it travels around a banked curve.

Inserting this result for N into Equation 5.8, we find

$$v = \sqrt{\frac{Nr\sin\theta}{m}} = \sqrt{\frac{mg}{\cos\theta}\frac{r\sin\theta}{m}} = \sqrt{gr\tan\theta}$$

When $\theta = 0$, this road is flat, and the result for the maximum speed is then zero because $\tan(0) = 0$. Therefore, you cannot turn on a very icy (frictionless) unbanked road without slipping, as you probably already knew.

EXAMPLE 5.3 Traveling through a Banked Turn with Friction

Consider again the problem of a car making a turn on a banked road, but now let's add friction to the situation. If the coefficient of static friction between the tires and the road is μ_S, what is the maximum speed at which the car can safely negotiate a turn of radius r with a banking angle θ?

RECOGNIZE THE PRINCIPLE

Because the car is traveling in a circle, the total force on the car must be $\sum F = mv^2/r$ directed toward the center of the circle. We must now include the force of friction when computing $\sum F$.

SKETCH THE PROBLEM

Figure 5.8A shows our car along with all the forces—the force of gravity, the normal force from the road, and the force of static friction—acting on it. Since we want to calculate the maximum safe speed, the force of friction will be directed *down* the incline so as to oppose any slipping of the car that would otherwise be directed up the plane.

IDENTIFY THE RELATIONSHIPS

The components of the forces are shown in parts B and C of Figure 5.8. We have two unknown quantities, the speed of the car and the normal force. We can get two relations by applying Newton's second law along the vertical and horizontal directions, just as we did for the case without friction. We can then solve for N and v.

SOLVE

The car is not slipping up or down the incline, so the acceleration along y is zero. Hence, the total force along y must be zero, and from Figure 5.8, we have

$$\sum F_y = 0 = +N\cos\theta - F_{\text{friction}}\sin\theta - mg$$

If the car is on the verge of slipping, the frictional force will have a magnitude of $F_{\text{friction}} = \mu_S N$. Hence,

$$N\cos\theta - \mu_S N\sin\theta - mg = 0$$

and we can rearrange to find the normal force

$$N = \frac{mg}{\cos\theta - \mu_S\sin\theta}$$

The total force along the horizontal direction provides the centripetal acceleration; in the coordinate system in Figure 5.8, we find

$$\sum F_x = -N\sin\theta - F_{\text{friction}}\cos\theta = -N\sin\theta - \mu_S N\cos\theta$$

(continued) ▶

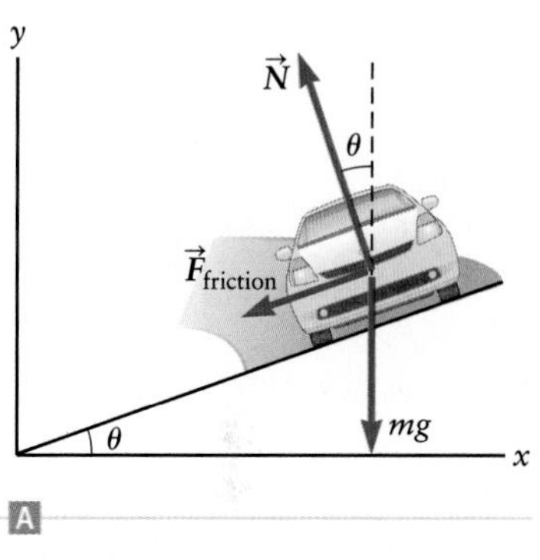

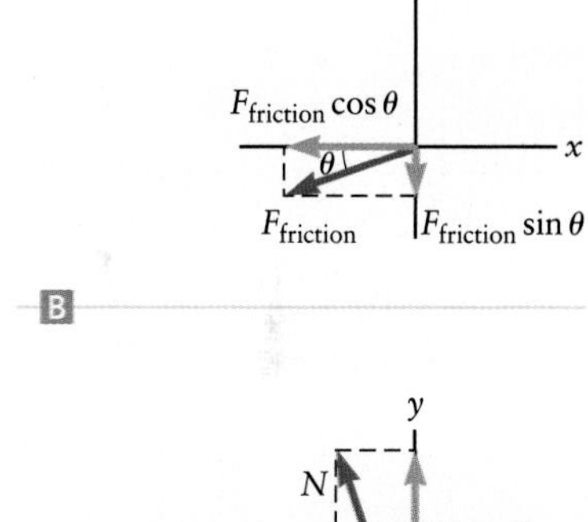

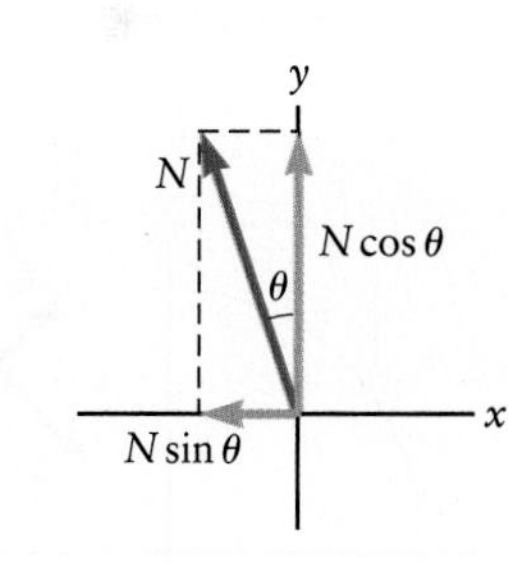

▲ **Figure 5.8** Example 5.3. A When friction is added to the road surface in Figure 5.7, there is an additional force (static friction) parallel to the incline. The horizontal and vertical components of B the frictional force and C the normal force are shown. The total horizontal force on the car is the sum of the horizontal components of the frictional force and the normal force.

The minus signs indicate that these forces are along the $-x$ direction. However, the center of the circle is also in this direction, so we have

$$N \sin \theta + \mu_S N \cos \theta = ma_c = \frac{mv^2}{r}$$

Solving for v then gives

$$v = \sqrt{\frac{Nr(\sin \theta + \mu_S \cos \theta)}{m}}$$

Inserting our result for the normal force leads to

$$v = \sqrt{\frac{mg}{\cos \theta - \mu_S \sin \theta} \frac{r(\sin \theta + \mu_S \cos \theta)}{m}} = \boxed{\sqrt{gr \frac{\mu_S \cos \theta + \sin \theta}{\cos \theta - \mu_S \sin \theta}}}$$

▸ *What does it mean?*

For an automobile racetrack, a banking of 20° is common and the tracks are quite large; the radius of the turn could be 200 m. Inserting these values along with a coefficient of friction of $\mu_S = 0.6$, we find $v = 50$ m/s, which is approximately 110 mi/h; this is a typical speed in auto racing, so our model seems reasonable. Obtaining an even higher speed for this value of the track radius would require a larger coefficient of friction, a larger banking angle, or both.

5.2 Examples of Circular Motion

In all our examples so far, we have assumed the speed of the object undergoing circular motion is constant. In this case, the centripetal acceleration is the total acceleration. In many situations, however, the speed is not constant, resulting in *nonuniform* circular motion. In such cases, the total acceleration has two components as indicated in Figure 5.9. One component is directed along the circumference, that is, tangent to the circle. The magnitude of the tangential acceleration is nonzero only when v is changing with time. The other component of the acceleration is directed toward the center of the circle and is just the centripetal acceleration we derived in Equation 5.5. Even when v is not constant, our results from Section 5.1 for the centripetal acceleration and the associated force are still valid. Whenever an object is moving in a circle, there must be a component of the acceleration equal to a_c directed toward the center of the circle, so there must also be an associated force equal to mv^2/r. Hence, we can use the results from Section 5.1 to analyze a great many situations involving circular motion; we are *not* limited to situations in which the speed is constant.

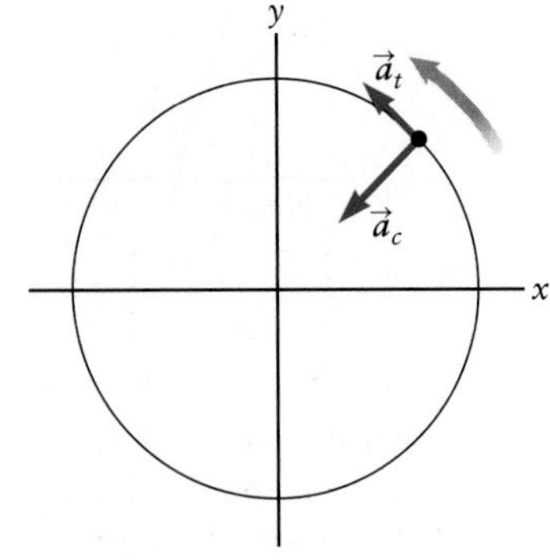

▲ **Figure 5.9** In general, an object traveling in a circle might not move with a constant speed. If the speed is not constant, there is a nonzero tangential acceleration $\vec{a}_t$ directed tangent to the circle. There is still a centripetal acceleration $\vec{a}_c$ directed toward the center of the circle, with magnitude $a_c = mv^2/r$.

Twirling a Rock on a String: What Is the Tension in the String?

Figure 5.10A shows a simple experiment in circular motion you have probably performed yourself: a rock tied to string is twirled in a vertical circle. The force of gravity acting on the rock makes it very difficult to keep the speed constant. Intuitively, we know that the speed will tend to be highest when the rock is at the bottom of the circle and lowest when the rock is at the top. Because the speed varies with time, the tangential acceleration is nonzero. Even so, the centripetal acceleration, the component of the acceleration directed toward the center of the circle, is still equal to $a_c = v^2/r$, where v is the speed at the point of interest.

Let's assume we know the speed v of the rock and consider how to calculate the tension T in the string. The answer will depend on the rock's location, so let's first deal with the case when the rock is at the bottom of the circle (Fig. 5.10B). Here the two forces on the rock, gravity and tension, are in opposite directions. These two forces together must add up to produce the centripetal acceleration

$$\sum F_{center} = +T_{bottom} - mg = \frac{mv^2}{r}$$

Solving for the tension gives

$$T_{bottom} = m\left(\frac{v^2}{r} + g\right) \tag{5.9}$$

Hence, the tension when the rock is at the bottom of the circle is large enough to overcome the gravitational force (mg) and also provide the required centripetal acceleration (mv^2/r).

When the rock is at the top of the circle as shown in Figure 5.10C, the gravitational force is parallel to the tension. The total force directed toward the center of the circle is now (recall that we take the positive direction to be toward the center of the circle)

$$\sum F_{center} = T_{top} + mg$$

The centripetal acceleration is also directed toward the center, so

$$\sum F_{center} = T_{top} + mg = \frac{mv^2}{r}$$

which leads to

$$T_{top} = m\left(\frac{v^2}{r} - g\right) \tag{5.10}$$

Comparing the results for T_{bottom} and T_{top} we see that if the speed v is the same at the top and bottom, the tension is smaller at the top. From Equation 5.10, we can also see that there is a minimum value of the speed that will keep the string taut. The tension in a string cannot be negative because a string cannot "push" on an object. As v is made smaller and smaller, however, the tension T_{top} in Equation 5.10 will eventually become zero. That will happen when

$$T_{top} = 0 = m\left(\frac{v^2}{r} - g\right)$$

$$\frac{v^2}{r} = g$$

$$v = \sqrt{gr} \tag{5.11}$$

If the speed is made smaller than this value, the string will become slack and circular motion is no longer possible.

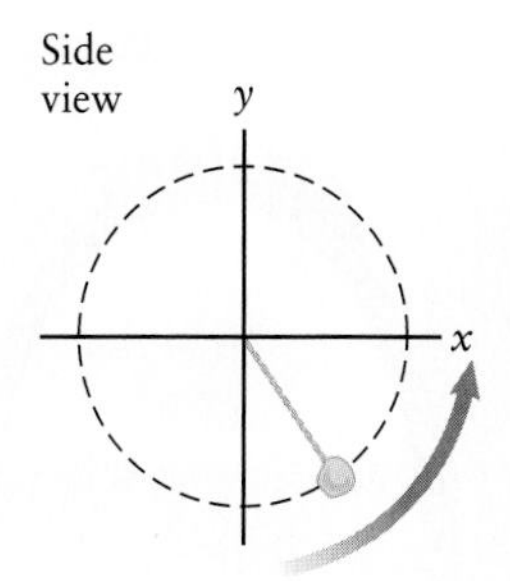

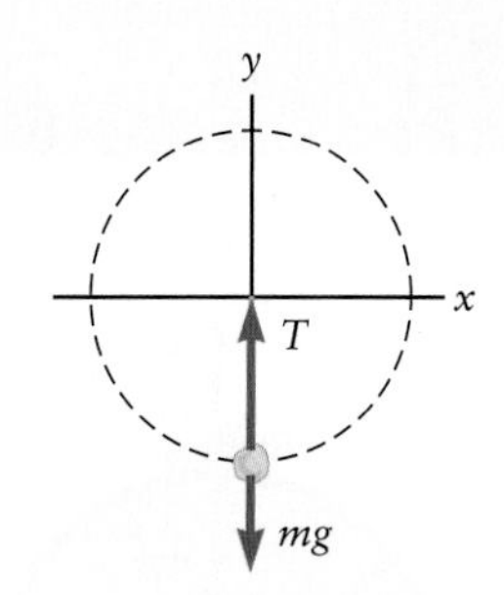

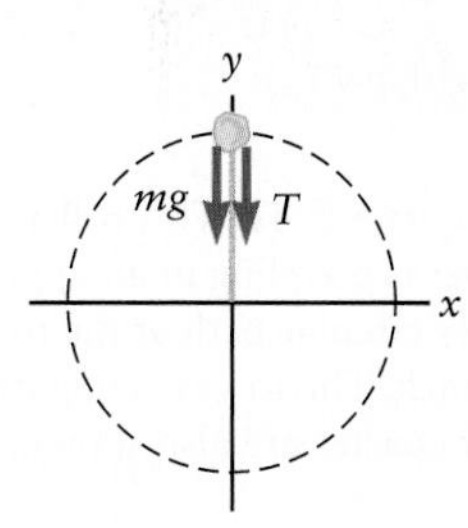

▲ **Figure 5.10** A This rock is moving in a vertical circle. There are two forces acting on it: the force of gravity and the tension from the string. B Forces on the rock when it is at the bottom of the circle. C Forces on the rock when it is at the top of its circular path.

Circular Motion and Amusement Park Activities: Maximum Speed of a Roller Coaster

The analysis of the rock's motion in Figure 5.10 can be applied to various amusement park activities. For example, when a roller coaster is near the minimum or maximum points on its track, its path is approximately circular. At these points, there must be an acceleration of magnitude v^2/r directed toward the center of the circle, where r is the track's radius of curvature. The roller coaster in Figure 5.11 is at the highest point on the track; let's calculate the maximum speed it can have without leaving the track. (This calculation would presumably be of interest to the riders.) Figure 5.11 shows the circular path followed by the roller coaster as well as the forces in the problem, gravity and the normal force exerted by the track on the roller coaster. The gravitational force is toward the center of the circle while the normal force is directed upward and is thus away from the center of the circle. The total force directed toward the center of the circle is

$$\sum F_{center} = mg - N$$

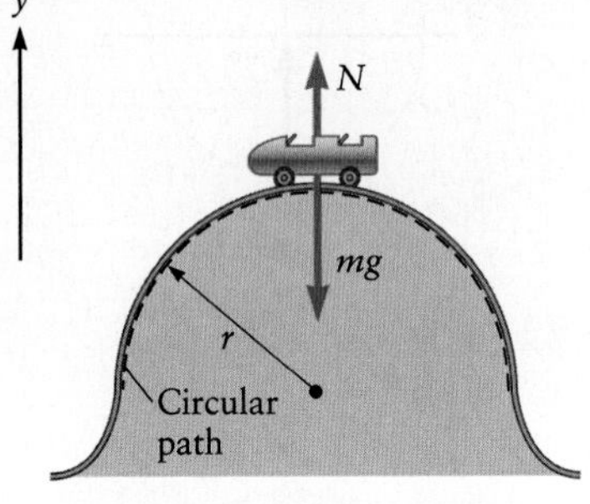

▲ **Figure 5.11** This roller coaster is traveling in an approximately circular path at the top of the track. The forces acting on the roller coaster are also shown.

which is related to the centripetal acceleration by (Eq. 5.6)

$$\sum F_{\text{center}} = mg - N = ma_c = \frac{mv^2}{r} \tag{5.12}$$

To find the maximum safe speed, we first rearrange Equation 5.12 to get the normal force:

$$N = m\left(g - \frac{v^2}{r}\right)$$

From this expression, we can see that as the speed v increases, the normal force N decreases. Eventually, we will reach a value of v at which N is zero:

$$N = m\left(g - \frac{v^2}{r}\right) = 0$$

which occurs when the speed satisfies

$$g - \frac{v^2}{r} = 0$$

$$v = \sqrt{gr} \tag{5.13}$$

If the speed is increased beyond this value, the value of N at the top of the track would have to be *negative*. Since a normal force can only push (but not pull), this scenario is impossible. Such a negative solution for N tells us that the roller coaster will leave the track if its speed exceeds $\sqrt{gr}$. Notice also that the normal force is the apparent weight of the roller coaster, and it will be your apparent weight if you are a passenger. Hence, when the speed is $v = \sqrt{gr}$, you will feel "weightless."

You might notice that our result (Eq. 5.13) for the speed at which the roller coaster leaves the track is the same as that found for the rock-on-a-string example in Figure 5.10C. There we saw (Eq. 5.11) that the string will go slack when the speed at the top is less than $v = \sqrt{gr}$. It is no accident that these two results are the same; in both cases, we are dealing with circular motion (so the acceleration is v^2/r for both), and in both cases, we have special situations in which the only force is gravity (because the tension in the string is zero and the normal force from the track is zero). In that sense, these problems are the "same."

EXAMPLE 5.4 Apparent Weight on a Roller Coaster

Consider the roller coaster in Figure 5.12 at the lowest point on its track, where the radius of curvature is $r = 20$ m. If the apparent weight of a passenger on the roller coaster is 3.0 times her true weight, what is the speed of the roller coaster?

RECOGNIZE THE PRINCIPLE

The apparent weight is the normal force on the object (or person) undergoing circular motion. So, we need to find the value of v at which $N = 3.0 \times mg$.

SKETCH THE PROBLEM

Figure 5.12 shows the circle followed by the roller coaster; also shown are the forces on the roller coaster, gravity and the normal force. Since the roller coaster is at the low point of the track, the normal force is directed toward the center of the circular arc defined by the track, and gravity is downward, away from the center.

IDENTIFY THE RELATIONSHIPS

We apply Newton's second law to relate the normal force to the speed of the roller coaster. Applying our relation for circular motion (Eq. 5.6), we have

$$\sum F_{\text{center}} = +N - mg = ma_c = \frac{mv^2}{r}$$

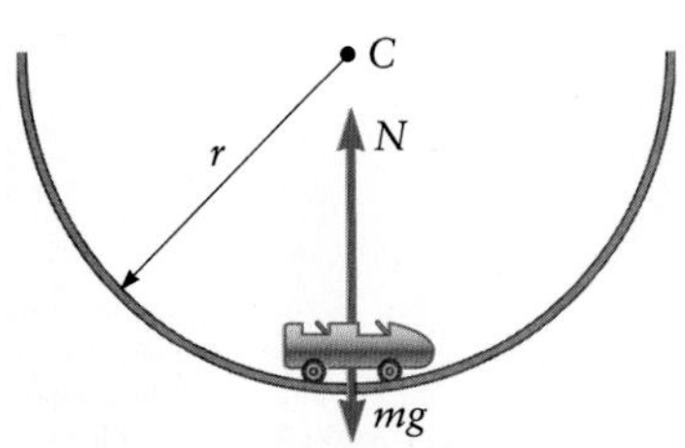

▲ **Figure 5.12** Example 5.4. Forces on a roller coaster at the bottom of its track.

The apparent weight is three times the true weight, so $N = 3.0 \times mg$ and

$$N - mg = 3.0(mg) - mg = 2.0(mg) = \frac{mv^2}{r}$$

SOLVE

Solving for v, we find

$$v = \sqrt{2.0(gr)}$$

For a track of radius of $r = 20$ m, this result gives

$$v = \sqrt{2.0(gr)} = \sqrt{2.0(9.8 \text{ m/s}^2)(20 \text{ m})} = \boxed{20 \text{ m/s}}$$

▶ *What does it mean?*

Our result for v is about 45 mi/h, which is a typical roller-coaster speed. Notice the similarity of this problem to the rock-on-a-string example in Figure 5.10B. The normal force on the roller coaster plays the role of tension in the string.

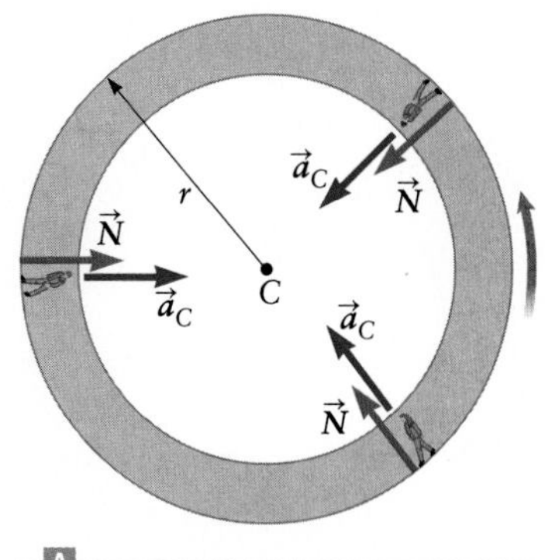

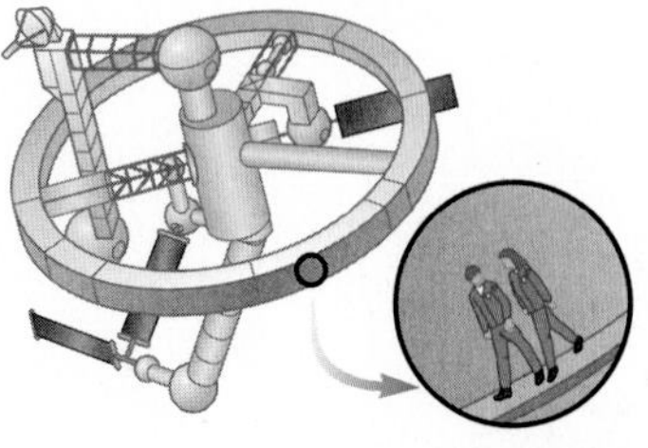

▲ **Figure 5.13** A circular space station that rotates about its central axis (C) generates "artificial gravity" for its passengers.

"Artificial Gravity" and a Rotating Space Station

A practical use of circular motion is sketched in Figure 5.13, which shows a hypothetical space station that uses circular motion to produce "artificial gravity" for its inhabitants. The station spins about a central axis so that the edge of the station, where the passengers spend most of their time, undergoes circular motion. The passengers in Figure 5.13 all experience a centripetal acceleration directed toward the center of the station (i.e., toward the center of the circle). This acceleration is produced by the normal force from the station floor on each passenger. For each passenger, $N = mv^2/r$, where v is the speed of the station edge and r is the radius of the station.

The passengers would have a slightly different interpretation of the situation. They would still be aware of the normal force acting on the bottoms of their shoes, but to them this force would be like the gravitational force they feel when on the surface of the Earth. In fact, if $N = mv^2/r = mg$, this force of "artificial gravity" would be the same as the real gravitational force on the Earth and the passengers would feel right at "home"!

EXAMPLE 5.5 Designing a Space Station

Consider a rotating space station similar to the one in Figure 5.13. If the radius of the station is $r = 40$ m, how many times per minute must the station rotate to produce a force due to "artificial gravity" equal to 30% of that found on the Earth?

RECOGNIZE THE PRINCIPLE

Figure 5.13A shows the circular path that the passengers follow. The only force is the normal force N, directed toward the center of the circle. We can therefore apply Equation 5.6 to get

$$\sum F = N = \frac{mv^2}{r} \quad (1)$$

We want the normal force to be 30% of the gravitational force on the Earth, which means $N = 0.30 \times mg$. We can use this expression together with Equation (1) to find the required speed v.

SKETCH THE PROBLEM

Figure 5.13 describes the problem and shows the forces on a passenger.

(continued) ▶

IDENTIFY THE RELATIONSHIPS AND SOLVE

Inserting the desired value of N into Equation (1) gives

$$N = 0.30(mg) = \frac{mv^2}{r}$$

and solving for v we find

$$v = \sqrt{0.30(gr)} \tag{2}$$

The time T for one revolution is equal to the distance traveled divided by the speed. The distance traveled is just the circumference of the station, so

$$T = \frac{2\pi r}{v} = \frac{2\pi r}{\sqrt{0.30(gr)}} = 2\pi\sqrt{\frac{r}{0.30g}} = 2\pi\sqrt{\frac{40 \text{ m}}{0.30(9.8 \text{ m/s}^2)}} = 23 \text{ s}$$

The number of revolutions per minute (rpm) is then

$$\frac{1 \text{ revolution}}{23 \text{ s}} \times \frac{60 \text{ s}}{\text{min}} = \boxed{2.6 \text{ rpm}}$$

► ***What does it mean?***

The speed in Equation (2) is independent of the person's mass. Hence, all passengers (regardless of their mass) experience an "artificial gravity" that is 30% of their weight on the Earth.

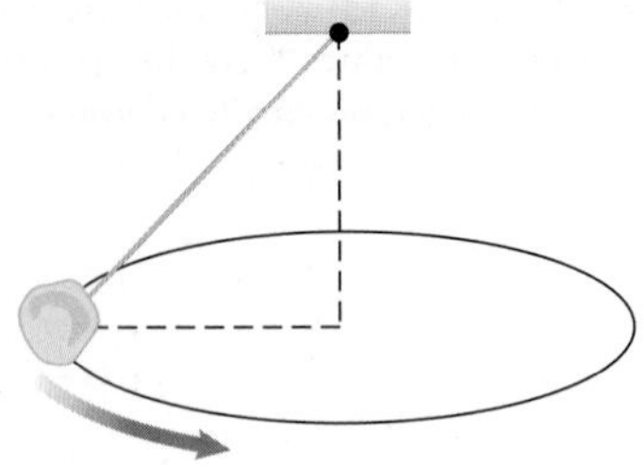

▲ **Figure 5.14** Concept Check 5.3.

CONCEPT CHECK 5.3 Net Force and Circular Motion

The rock in Figure 5.14 is suspended by a string tied to the ceiling and is undergoing uniform circular motion in a horizontal plane. What is the direction of the *net force* on the rock?

(a) It is along the string.
(b) It is downward (from gravity).
(c) It is toward the center of the circle.
(d) The net force is zero.

Physics of a Centrifuge

While a space station with "artificial gravity" has not yet been built (except in movies), a similar device called a ***centrifuge*** is used in many laboratories. A centrifuge can be used to remove or separate particles that are in suspension in a liquid. A simple centrifuge design is shown in Figure 5.15. A test tube containing the liquid of interest is rotated so as to undergo circular motion; when compared to the space station in Figure 5.13, the test tube plays the part of a spoke on the station.

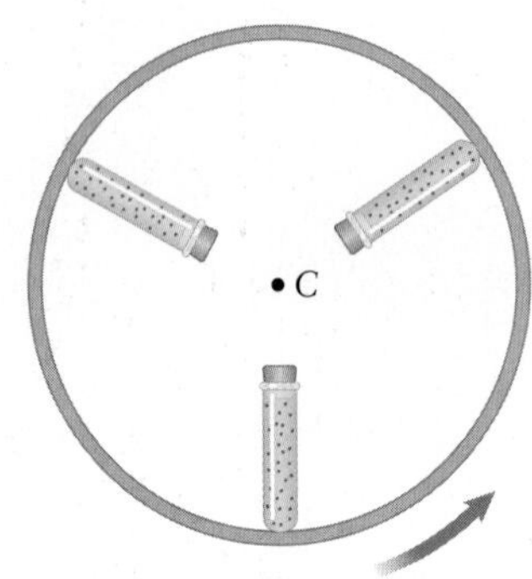

▲ **Figure 5.15** A centrifuge rotates at an extremely high rate about its axis at C, producing a large centripetal acceleration for the contents of the centrifuge. Here, the contents are a liquid that might contain cells in suspension. The result is that the cells move to the outer end of the tube.

The liquid in Figure 5.15 might be blood containing a particular type of cell. How long will it take to separate the cells from the blood? That is, how long will it take a cell to move from an initial location near the center of the centrifuge tube to the outer end of the tube?

Consider a cell that is already located at the end of the test tube so that it is in contact with the bottom of the tube. The circular motion of this cell is just like the motion of a passenger in the space station in Figure 5.13; hence, there is a normal force $N = mv^2/r$ exerted by the end of the tube on the cell. The cell thus experiences a force of "artificial gravity" $F_{AG} = mv^2/r$ much like the artificial gravity felt by a passenger on the space station. What's more, this "effective force" due to artificial gravity acts on the cell even when it is not at the end of the test tube, and this force will cause a cell to move toward the bottom (the outer end). When viewed by an observer moving along with the test tube, it seems as if the cell is simply falling "vertically" to the bottom of the tube in response to the force of artificial gravity (Fig. 5.16A).

We have called F_{AG} an "effective force" because things appear different when viewed by an observer who is not rotating with the centrifuge. From the point of view of a stationary observer, the motion of the cell appears as shown in Figure 5.16B. A cell that is initially located away from the end of the test tube moves in an expanding arclike trajectory since there is nothing to provide the force required to make it move in a circle (until it reaches the end of the tube). The cell would move in a straight line (compare to the rock in Fig. 5.4) if not for the walls of the test tube and the drag force from the liquid.

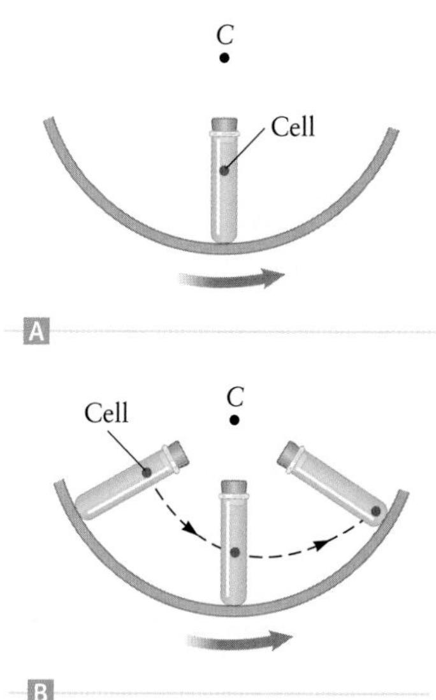

▲ **Figure 5.16** A A cell in the centrifuge moves as if there is a force due to "artificial gravity" acting on the cell. B When viewed by an observer who is not spinning with the centrifuge, the cell moves in a path that spirals outward as indicated by the dashed line.

Inertial and Noninertial Reference Frames Applied to a Centrifuge

While the motion of the cell in a centrifuge may be interpreted differently by different observers (as in Fig. 5.16), all observers will agree that the cell moves outward along the test tube, eventually coming to rest at the end of the tube. We have seen, however, that one observer (the cell) will deduce the presence of a force F_{AG}, whereas the other (stationary) observer will not. Which interpretation is correct? Is there actually a force F_{AG} or not? We mentioned in Chapter 4 that Newton's laws of motion should be used only by observers in *inertial reference frames*. You will recall that an inertial reference frame is one that moves with a constant velocity. In this example, the stationary observer is in an inertial reference frame, while Figure 5.16A corresponds to a noninertial reference frame. Only the stationary observer can use Newton's second law, and that observer's conclusion is correct: there is *no force* pushing the cell outward along the test tube. For this reason, F_{AG} as observed by the cell—that is, by the noninertial observer—is called a ***fictitious force***. Observers in rotating, and hence noninertial, reference frames often describe motion in terms of such fictitious forces.

Why should we ever want to use a noninertial reference frame? Why not stick to inertial frames, where we know that Newton's laws can be used? The answer is that sometimes motion appears simpler when viewed in a noninertial frame. For example, with our cell in a centrifuge, the path seen by a noninertial observer (Fig. 5.16A) is very simple. The cell just undergoes free fall to the bottom of the test tube, and it moves *as if* there were a force F_{AG} directed along the "vertical" (i.e., along the tube). This motion is certainly simpler than the curved path seen by an inertial observer as sketched in Figure 5.16B. It is thus sometimes useful to analyze motion from a noninertial reference frame. However, in such an analysis one must always account for the fictitious forces, such as F_{AG}, because these forces are necessary to make the inertial and noninertial results agree.

To calculate how long it will take for the cell in Figure 5.16 to be removed from the solution, let's take the point of view of the cell as it falls to the bottom of the test tube due to the force of artificial gravity, $F_{AG} = mv^2/r$. We must recognize that another force on the cell opposes its fall: the drag force we encountered in Chapter 3. There we saw that for a spherical object moving slowly through a liquid, there is a drag force described by Stokes's law (Eq. 3.23),

$$\vec{F}_{\text{drag}} = -Cr_{\text{cell}}\vec{v}_{\text{cell}} \tag{5.14}$$

where r_{cell} is the cell's radius and v_{cell} is the cell's speed along the test tube. Notice that v_{cell} is *not* the same as the speed v of the end of the test tube (v is the speed of the outer rim of the spinning centrifuge). Here, C is a constant that depends on the viscosity (the resistance to flow) of the fluid. For blood, C has a value of approximately 0.075 kg/(m · s). The drag force becomes larger as the speed v_{cell} of the cell increases, and the situation is very similar to the problem of a falling skydiver we encountered in Chapter 3. There we found that the skydiver reaches a terminal speed at which the drag force balances the force of gravity. For the cell in a centrifuge, the drag force in Equation 5.14 will balance F_{AG} and the cell will move at a constant speed along the test tube. This balance condition is

$$F_{AG} + F_{\text{drag}} = F_{AG} - Cr_{\text{cell}}v_{\text{cell}} = 0$$

Insight 5.1

USING DIFFERENT REFERENCE FRAMES

It is often useful to analyze a situation from different viewpoints. In Chapter 4, we took this approach when we considered the behavior of the mass-on-a-string accelerometer in a moving airplane. These different viewpoints are called ***reference frames***. A reference frame is simply a coordinate system, but it is necessary to specify the motion of the coordinate axes. For the cell in a centrifuge (Fig. 5.15), one reference frame moves with the cell as it undergoes circular motion (Fig. 5.16A). Hence, these coordinate axes rotate along with the arms of the centrifuge. This problem can also be analyzed from the point of view of an observer who is watching the centrifuge from the "outside" (Fig. 5.16B).

Since $F_{AG} = mv^2/r$, we have

$$\frac{mv^2}{r} = Cr_{\text{cell}}v_{\text{cell}}$$

Solving for the speed along the test tube gives

$$v_{\text{cell}} = \frac{mv^2}{Cr_{\text{cell}}r} \tag{5.15}$$

It is important to distinguish the tangential speed of the centrifuge (v) from the settling speed of the cell (v_{cell}), and the radius of the cell (r_{cell}) from the radius of the centrifuge (r).

CONCEPT CHECK 5.4 Inertial and Noninertial Reference Frames

We have seen that an observer in an inertial reference frame can use Newton's laws, while observers in a noninertial frame must include fictitious forces. Identify the following reference frames as either inertial or noninertial and explain your answers.

(a) A car traveling on a level road at a constant velocity
(b) A car traveling up a steep mountain road at a constant velocity
(c) A child sitting on a rotating merry-go-round
(d) An apple falling from a tree

EXAMPLE 5.6 Operation of a Centrifuge

Calculate the settling speed v_{cell} of a cell of radius 10 μm and mass 10^{-5} mg ($= 10^{-11}$ kg) in blood using Equation 5.15. Assume the centrifuge has a radius of 10 cm and a rotation rate of 3000 revolutions per minute. How long must the centrifuge run to make the cell settle out? Assume $C = 0.075$ kg/(m · s) for blood.

RECOGNIZE THE PRINCIPLE

The radius and mass of a cell are never known precisely and vary from cell to cell. Moreover, Stokes's relation for the drag force, Equation 5.14, applies exactly only for a spherical object, and cells are usually not precisely spherical! Even so, we can use Equation 5.15 to get an approximate value for the speed at which a cell moves. The only unknown quantity in Equation 5.15 is v, the speed of the centrifuge. We can find v from the period of the centrifuge, along with the relation between period and v in Equation 5.1.

SKETCH THE PROBLEM

Figure 5.17 shows the problem from the point of view of the cell. According to this noninertial observer, the force of artificial gravity ($\vec{F}_{AG}$) causes an acceleration a_c "downward" along the centrifuge tube.

IDENTIFY THE RELATIONSHIPS AND SOLVE

Our centrifuge makes 3000 revolutions in 1 min, so the time to make 1 revolution is $T = 1$ min/3000 $= 0.020$ s, and the circumferential speed is

$$v = \frac{2\pi r}{T} = \frac{2\pi(0.10\text{ m})}{0.020\text{ s}} = 31\text{ m/s}$$

Inserting this in Equation 5.15 gives

$$v_{\text{cell}} = \frac{mv_c^2}{Cr_{\text{cell}}r} = \frac{(1 \times 10^{-11}\text{ kg})(31\text{ m/s})^2}{[0.075\text{ kg/(m}\cdot\text{s)}](1.0 \times 10^{-5}\text{ m})(0.10\text{ m})} = \boxed{0.1\text{ m/s}}$$

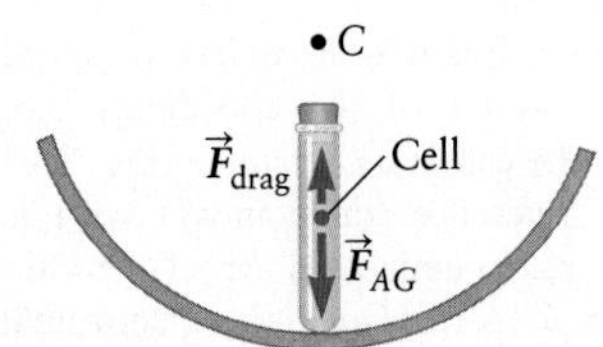

▲ **Figure 5.17** Example 5.6. A cell in a centrifuge moves as if there is a force of artificial gravity acting "downward" on the cell.

The length of the centrifuge tube will be limited by the radius of the centrifuge (Fig. 5.15). As a rough estimate, we can assume the tube length L is about half the radius of the centrifuge, so the cell in this case must travel a distance of at most $L = 5$ cm $= 0.05$ m. Using the value found for v_{cell}, the time required for the cell to move this distance is

$$t = \frac{L}{v_{\text{cell}}} = \frac{0.05 \text{ m}}{0.1 \text{ m/s}} = \boxed{0.5 \text{ s}}$$

▸ ***What does it mean?***

A centrifuge with a similar speed is used to separate blood cells from plasma. Note that our analysis used the value of v as measured at the end of the centrifuge tube, while most cells will begin their motion closer to the center of the centrifuge. Our calculation will thus overestimate F_{AG} and v_{cell}. Even allowing for this, and for variations in cell size, our analysis shows that it only takes a few seconds of spinning to completely separate the cells from the plasma.

5.3 Newton's Law of Gravitation

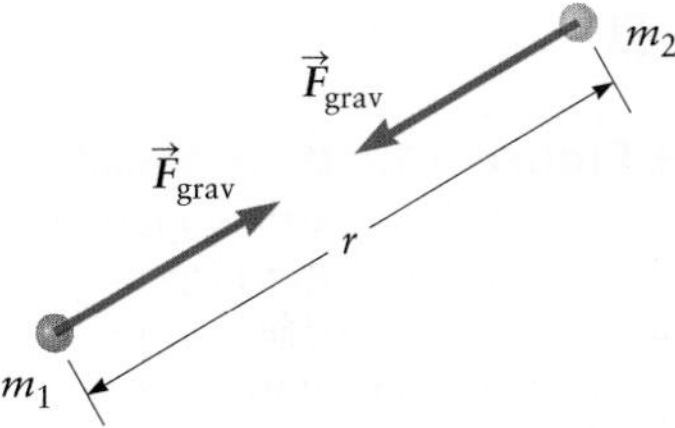

▲ **Figure 5.18** The gravitational force between two point masses m_1 and m_2 that are separated by a distance r is given by Equation 5.16. Notice that here the distance r is not the radius of a circle.

The orbital motions of planets and moons are circular (to a good approximation) in many cases, and the force responsible for this circular motion is gravity. We have already encountered the force of gravity as it acts on objects near the Earth's surface. Now we need to consider the gravitational force in more general situations.

Newton's three laws of motion are the foundation of our study of mechanics. However, Newton discovered another law of nature, the law of gravitation, that is perhaps just as important as his laws of motion. Newton's law of gravitation plays a key role in physics for two reasons. First, it allows us to calculate and understand the motions of a wide variety of objects, including baseballs, apples, and planets. Second, Newton's application of his law of gravitation to the motion of planets and moons was the first time that physics was applied (successfully) to describe the motion of the solar system. His work showed for the first time that the laws of physics apply to *all* objects and had a profound effect on how people viewed the universe.

Newton's law of gravitation: **There is a gravitational attraction between *any* two objects. If the objects are point masses m_1 and m_2, separated by a distance r (Fig. 5.18), the magnitude of the gravitational force is**

$$F_{\text{grav}} = \frac{Gm_1m_2}{r^2} \tag{5.16}$$

Newton's law of gravitation

where $G = 6.67 \times 10^{-11}$ N · m²/kg² is a constant of nature known as the universal gravitational constant. The gravitational force is *always attractive*. Every mass attracts every other mass.

The gravitational force law, Equation 5.16, is "symmetric." That is, the magnitude of the gravitational force exerted by mass 1 on mass 2 is equal to the magnitude of the gravitational force exerted by mass 2 on mass 1. Since the forces are both attractive, this result is precisely what we would expect from Newton's third law; the two gravitational forces are an *action–reaction pair* because they are equal in magnitude and opposite in direction and they act on different members of the pair of objects.

Gravitation and the Orbital Motion of the Moon

The gravitational force is responsible for the motion of planets, moons, asteroids, and comets. For example, the Moon follows an approximately circular path as it orbits the Earth (Fig. 5.19). This circular motion requires a force, which is provided by gravity. As a check, we can use Equation 5.6 to calculate the force required to make the Moon move in its circular orbit and then compare it with the gravitational

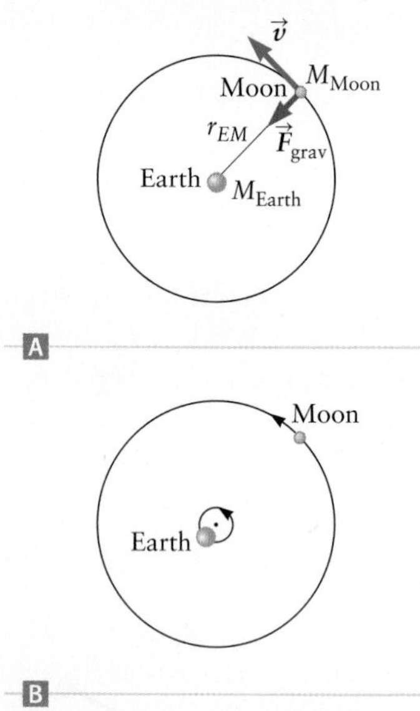

▲ **Figure 5.19** A The Moon's orbit around the Earth is approximately circular. The centripetal acceleration is provided by the Earth's gravitational force. To a good approximation, the Earth in this picture can be assumed to be fixed in space. B In a more accurate description, the Earth also moves in a circular "orbit" due to the gravitational force of the Moon on the Earth. This sketch is not to scale; the center of this orbit is inside the Earth.

force calculated from Newton's law of gravitation, Equation 5.16. The force required to make the Moon move in a circle is (from Eq. 5.6)

$$F = \frac{M_{Moon}v^2}{r_{EM}}$$

where M_{Moon} is the mass of the Moon and r_{EM} is the distance from the Earth to the Moon (the radius of the Moon's orbit). To find the speed v of the Moon, we notice that it completes one orbit and travels a distance of $2\pi r_{EM}$ in approximately T = 27.3 days;[2] hence,

$$v = \frac{2\pi r_{EM}}{27.3 \text{ days}} = \frac{2\pi r_{EM}}{T}$$

The required force is thus

$$F = \frac{M_{Moon}v^2}{r_{EM}} = \frac{M_{Moon}(2\pi r_{EM}/T)^2}{r_{EM}} = \frac{4\pi^2 M_{Moon} r_{EM}}{T^2} \tag{5.17}$$

Inserting the known values (see Table 5.1) $M_{Moon} = 7.35 \times 10^{22}$ kg and $r_{EM} = 3.85 \times 10^8$ m along with $T = 27.3$ days $= 2.36 \times 10^6$ s, we find

$$F = 2.0 \times 10^{20} \text{ N} \tag{5.18}$$

This is the magnitude of the force needed to make the Moon follow its observed circular orbit. To confirm that this force is actually provided by gravity, let's calculate the gravitational force exerted by the Earth on the Moon from Equation 5.16 and show that it is indeed equal to the result for F in Equation 5.18. The mass of the Earth is (from Table 5.1) $M_{Earth} = 5.98 \times 10^{24}$ kg, and inserting the other values given above into Equation 5.16 gives

$$F_{grav} = \frac{GM_{Earth}M_{Moon}}{r_{EM}^2} = 2.0 \times 10^{20} \text{ N} \tag{5.19}$$

which does agree with the expected value of F in Equation 5.18.

When he was developing his theories of motion and gravitation, Newton almost certainly carried out these same calculations and obtained the results in Equations 5.18 and 5.19. That the force of gravity on the Moon is precisely equal to the force required to make the Moon move in its circular orbit must have been very strong evidence for Newton that his theories were indeed correct.

Insight 5.2

ORBITAL MOTION OF THE MOON AND THE EARTH

In our analysis of the Moon's orbital motion, we have assumed the Moon orbits around a "fixed" Earth. That is, we have assumed the Earth does not move at all as the Moon moves in a circle of radius r_{EM}. Although that is a good approximation, it is only an approximation because it ignores two aspects of the Earth's motion. (1) The Earth orbits the Sun. That orbital motion is much slower than the Moon's orbital speed, so it still makes sense (it is mathematically extremely accurate) in Equation 5.17 to treat the Earth as fixed. (2) The Earth also "orbits" the Moon. That is, as the Moon moves in its circular orbit, the Earth moves in a corresponding circular orbit as shown in Figure 5.19B. The radius of the Earth's circular orbit is much smaller than that of the Moon. In fact, the center of the Earth's circle is inside the Earth(!) and is located at a spot called the *center of mass* that we'll discuss in Chapter 7.

Applying Newton's Law of Gravitation: Calculating the Value of g

In Chapter 3, we introduced the "constant" g and learned that the gravitational force on an object near the Earth's surface has the magnitude $F_{grav} = mg$. Let's now see how this result is contained in the universal law of gravitation. Strictly speaking, Equation 5.16 applies only to the case of two "point" masses, two objects that are very small compared to the distance between them. This assumption works for our calculation of the gravitational force between the Earth and the Moon, but we certainly cannot consider the Earth to be a point mass in relation to an object on its surface. To appreciate the problem, consider the force exerted by the Earth's gravity on a person standing on the surface as sketched in Figure 5.20A. We would like to use Equation 5.16 to calculate this force, but what value should we use for the distance r? Some parts of the Earth are very close to the person (just beneath his feet), whereas other parts are quite far away (as much as twice the radius of the Earth). The answer is that when dealing with a spherical object, one can usually calculate the gravitational force it exerts on another object *as if* all the sphere's mass were located at its center. (See Insight 5.3.)

[2]In our examples involving planetary and satellite motion, we will carry three significant figures through the calculation (one more than usual in this book) to avoid problems due to rounding errors.

TABLE 5.1 Solar System Data: Properties of Several Objects in the Solar System, Including the Planets and Two of the Largest Dwarf Planets (Pluto and Eris)

Name	Mean Orbital Radius ($\times 10^{11}$ m)	Orbital Period (years)	Radius of Object ($\times 10^6$ m)	Mass ($\times 10^{24}$ kg)	Orbital Eccentricity
Mercury	0.579	0.240	2.44	0.330	0.21
Venus	1.08	0.615	6.05	4.87	0.007
Earth	1.50	1.00	6.37	5.98	0.017
Mars	2.28	1.88	3.39	0.644	0.093
Jupiter	7.78	11.9	71.5	1900	0.048
Saturn	14.3	29.4	60.3	568	0.054
Uranus	28.7	83.8	25.6	86.6	0.047
Neptune	45.0	164	24.8	102	0.009
Pluto	59.1	248	1.14	0.0131	0.25
Eris	100	560	2.4	—	0.44
Earth's Moon	3.85×10^8 m	27.3 days	1.74	0.0735	0.055
Sun			6.96×10^8 m	1.99×10^{30} kg	

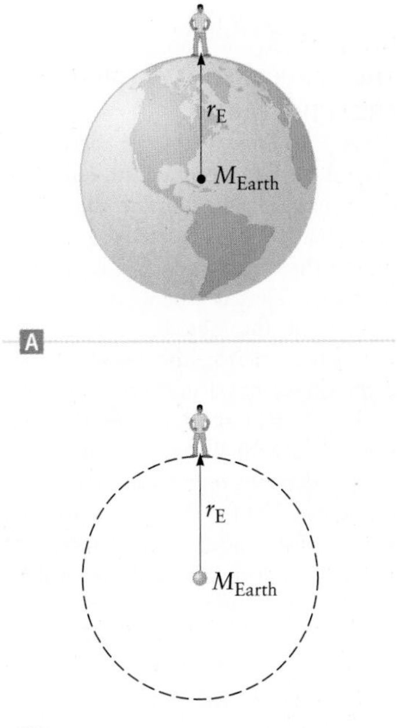

▲ **Figure 5.20** Assuming a sphericaly symmetric Earth (which is a good approximation), the gravitational force exerted by the Earth on an object (i.e., a person) on the surface as shown in A is equal to the force one would find if all the mass of the Earth were located at its center as shown in B.

The Earth's shape is very close to spherical, and the gravitational force exerted by the Earth on the person in Figure 5.20 can be calculated as if all the Earth's mass is at its center as sketched in Figure 5.20B. We can therefore use Equation 5.16 to calculate this force, with the separation r equal to the distance from the person to the center of the Earth; this distance is just the Earth's radius, r_E. We thus have

$$F_{grav} = \frac{GM_{Earth}M_{person}}{r_E^2} \tag{5.20}$$

This force is the weight of the person, which we have denoted by $M_{person}g$, where g is the "acceleration due to the Earth's gravity." Hence, we can use Equation 5.20 to calculate the value of g. Examining Equation 5.20, we see that it has the form $M_{person}g$ provided that g is given by

$$g = \frac{GM_{Earth}}{r_E^2} \tag{5.21}$$

The value of g is a function of only the Earth's mass and radius, and the universal gravitational constant G. The value of g is thus the same for all terrestrial objects near the Earth's surface. Inserting the known values of M_{Earth} and r_E (Table 5.1) along with G, we find

g and Newton's law of gravitation

$$g = \frac{GM_{Earth}}{r_E^2} = \frac{(6.67 \times 10^{-11}\ \text{N} \cdot \text{m}^2/\text{kg}^2)(5.98 \times 10^{24}\ \text{kg})}{(6.37 \times 10^6\ \text{m})^2}$$

$$g = 9.8\ \text{m/s}^2$$

which is the value we have already been using for g.

This calculation shows us where the value of g "comes from," and it also shows that g is not really a constant. The Earth happens to be approximately spherical, so all objects on the surface of the planet (i.e., all terrestrial objects) are approximately the same distance from the center and hence all have approximately the same value of g. However, we see from Equation 5.21 that g—and hence the weight of an object—will change if the object is moved farther from the Earth (e.g., by climbing a mountain). It also shows that the weight of an object will be different on another planet or on the Moon than it is on the Earth.

Insight 5.3
THE GRAVITATIONAL FORCE FROM THE EARTH
When dealing with a spherically shaped object, the gravitational force on a mass located outside the object can usually be calculated by assuming *all* the object's mass is located at its center. That is true as long as the density of the object is spherically symmetric. Hence, the result applies if the object has a uniform (constant) density. It also applies when the density varies with depth, as long as the density depends only on distance from the center. The result thus applies to a more realistic model of the Earth, consisting of a core with a different density than the crust.

EXAMPLE 5.7 What Would You Weigh on the Moon?

What is your weight on the Moon? Compare it with your weight on the Earth.

RECOGNIZE THE PRINCIPLE

Your weight on the Moon is just the Moon's gravitational force when you are located on its surface. Hence, we need to evaluate Equation 5.20 using the mass of the Moon and the radius of the Moon in place of those quantities for the Earth. (These data are listed in Table 5.1.)

SKETCH THE PROBLEM

This problem is described by Figure 5.20, but with the Earth replaced by the Moon.

IDENTIFY THE RELATIONSHIPS AND SOLVE

The author has a mass of approximately $M_{\text{author}} = 70$ kg; inserting this value into Equation 5.20 gives

$$F_{\text{grav}} = \frac{GM_{\text{moon}}M_{\text{author}}}{r_M^2} = \frac{(6.67 \times 10^{-11}\ \text{N} \cdot \text{m}^2/\text{kg}^2)(7.35 \times 10^{22}\ \text{kg})(70\ \text{kg})}{(1.74 \times 10^6\ \text{m})^2}$$

$$F_{\text{grav}} = 110\ \text{N}$$

The author's weight on the Earth is $M_{\text{author}}g = (70\ \text{kg})g = 690$ N, so his weight on the Moon is approximately one-sixth of his weight on the Earth.

▸ What does it mean?
The *ratio* of the weight on the Moon to the weight on the Earth is *independent* of the mass of the object because both are proportional to the mass of the object. All objects weigh less on the Moon by a factor of approximately one-sixth.

EXAMPLE 5.8 The Force of Gravity in a Very Tall Building

When this book was written, the tallest building in the world was the Burj Khalifa building in Dubai (Fig. 5.21A), with the top being a distance $h = 830$ m above the bottom. Find the ratio of your weight at the top of the building to your weight at ground level.

RECOGNIZE THE PRINCIPLE

At the top of the Burj Khalifa building, your distance from the center of the Earth is $r_E + h$, where r_E is the Earth's radius (Fig. 5.21B). Hence, the force of gravity (your weight) is slightly smaller than when you are on the ground. Note that because h is much smaller than r_E, we must carry extra significant figures in this calculation.

SKETCH THE PROBLEM

We want to compare the force of gravity at two locations as sketched in Figure 5.21B. The force of gravity depends on the distance from the center of the Earth.

IDENTIFY THE RELATIONSHIPS

At ground level, you are a distance r_E from the center of the Earth, so your weight is

$$F_{\text{grav}}\ (\text{ground level}) = \frac{GM_{\text{Earth}}M_{\text{person}}}{r_E^2}$$

At the top of the building, you are a distance $r_E + h$ from the center of the Earth, so

$$F_{\text{grav}}\ (\text{top}) = \frac{GM_{\text{Earth}}M_{\text{person}}}{(r_E + h)^2}$$

The ratio of these two forces is

$$\frac{F_{\text{grav}}\text{ (top)}}{F_{\text{grav}}\text{ (ground level)}} = \frac{GM_{\text{Earth}}M_{\text{person}}/(r_E + h)^2}{GM_{\text{Earth}}M_{\text{person}}/r_E^2} = \left(\frac{r_E}{r_E + h}\right)^2$$

SOLVE

Inserting values for the Earth's radius and the height of the Burj Khalifa building, we find

$$\frac{F_{\text{grav}}\text{(top)}}{F_{\text{grav}}\text{(ground level)}} = \left(\frac{r_E}{r_E + h}\right)^2 = \left[\frac{6.37 \times 10^6 \text{ m}}{(6.37 \times 10^6 + 830) \text{ m}}\right]^2 = \boxed{0.99974}$$

► ***What does it mean?***

This result is independent of the mass of the person; all objects weigh less at the top of the Burj Khalifa building by approximately 0.026%. Sensitive instruments called gravimeters can measure changes as small as 1 part in 10^9 in the gravitational force, so this difference can be easily measured.

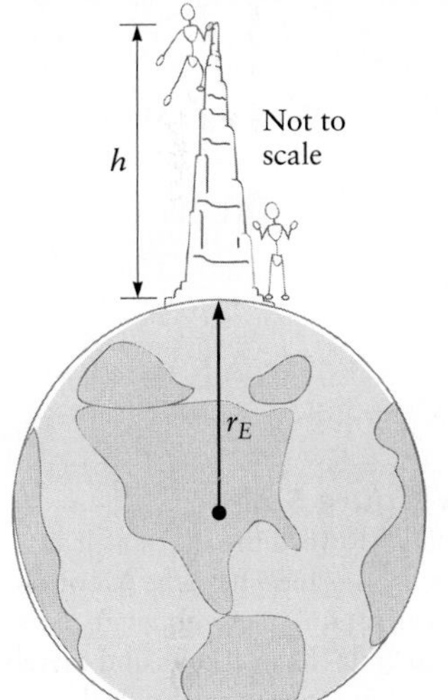

▲ **Figure 5.21** Example 5.8. A The Burj Khalifa building in Dubai is the tallest building in the world. B A person at the top of the Burj Khalifa Building is a distance $r_E + h$ from the center of the Earth, where h is the height of the building.

CONCEPT CHECK 5.5 Gravity on Another Planet

You travel to another planet in our solar system and find that your weight is twice as large as it is on Earth. If the radius of this planet is twice the Earth's radius, is the mass of the planet (a) two times the Earth's mass, (b) four times the Earth's mass, (c) half of the Earth's mass, or (d) eight times the Earth's mass?

Measuring *G*: The Cavendish Experiment

The first precision measurement of the force of gravity between two terrestrial objects was carried out in a famous experiment by Henry Cavendish, who measured the gravitational force between two lead spheres. This experiment is very difficult for the following reason. The Cavendish spheres each had a diameter of 1 m and a mass of approximately 5900 kg. We have already seen that the gravitational force exerted by a spherical mass can be calculated as if all the mass is at the center. With that in mind, let's calculate the gravitational force between two point particles of mass 5900 kg, separated by a distance $r = 1.0$ m. Using Equation 5.16 with $m_1 = m_2 = 5900$ kg gives

$$F_{\text{grav}} = \frac{Gm_1m_2}{r^2} = \frac{(6.67 \times 10^{-11}\ \text{N} \cdot \text{m}^2/\text{kg}^2)(5900\ \text{kg})(5900\ \text{kg})}{(1.0\ \text{m})^2}$$

$$F_{\text{grav}} = 2.3 \times 10^{-3}\ \text{N} \tag{5.22}$$

This force is a quite small, only a little larger than the weight of a mosquito! That's why a high-precision measurement of the gravitational force between two terrestrial objects is a very challenging experiment.

Cavendish did his work more than 200 years ago, but the same basic experimental design is still used today to study gravitational forces. The Cavendish apparatus (Fig. 5.22) uses a "dumbbell" with two spheres each of mass m_1 at the ends, suspended at the middle by a very thin wire. Another pair of spheres, each of mass m_2, is then brought very close to the dumbbell masses, and the force of gravity causes the dumbbell rod to rotate. The rotation angle θ in Figure 5.22 depends on the force, so by measuring θ, Cavendish was able to determine the gravitational force. Since the values of m_1 and m_2 and their separation can also be measured, Cavendish was able to deduce the value of G from Equation 5.16.

The quantity G is our first encounter with a *constant of nature*. Laws of physics, such as Newton's law of gravitation, often contain such constants, and the only way to know their values is through experimentation. It is interesting that Cavendish's method for measuring G is still the basis for most experimental studies of gravitation.

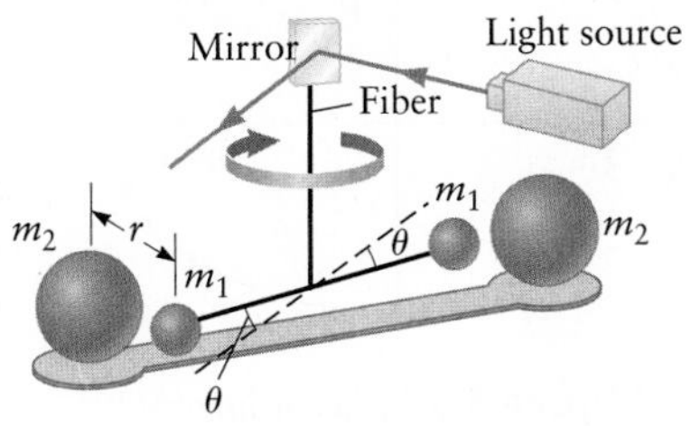

▲ **Figure 5.22** Cavendish used an apparatus like this one to measure the force of gravity between terrestrial objects.

Insight 5.4

GRAVITY IS A WEAK FORCE

In our calculations of the gravitational force between two terrestrial-scale objects, such as the two spheres in the Cavendish experiment, we found that the force of gravity is much weaker than typical normal and tension forces. In later chapters, we'll see that in most cases gravity is also much weaker than electric, magnetic, and nuclear forces. For this reason, gravitation is considered the weakest of the fundamental forces in nature.

Newton's Apple

Newton's discovery of the law of gravitation had a profound impact on our view of the universe. To appreciate this impact, it is useful to recall the allegory of Newton's apple. The story has several versions, but the general idea is that Newton was sitting under an apple tree, thinking about the motion of the Moon, when an apple fell from the tree and struck him on the head. According to the story, this jolt led to the "discovery" of gravity. The point of this story is not simply that falling apples undergo the motion we have called "free fall." Rather, it is that the acceleration of an apple falling from a tree and the acceleration of the Moon as it moves in a circular orbit around the Earth are *basically the same* (Fig. 5.23). They are caused by the same force (gravity), which has the same direction (toward the center of the Earth), and if the Moon and the apple were at the same distance from the Earth, their accelerations would be equal. This connection is not obvious because for circular motion the acceleration is perpendicular to the velocity, whereas the acceleration of the apple is parallel to $\vec{v}$. It took the genius of Newton to make this connection.

Prior to Newton, it was widely believed that the motion of celestial objects, such as the Moon, is fundamentally different from the motion of terrestrial objects, such as apples. Newton showed that the motion of falling apples and the motion of the Moon are caused by the same force and governed by the same laws of motion. So, one can use Newton's laws to understand the motion of the solar system and beyond, a regime that had previously been "off-limits" to such scientific study.

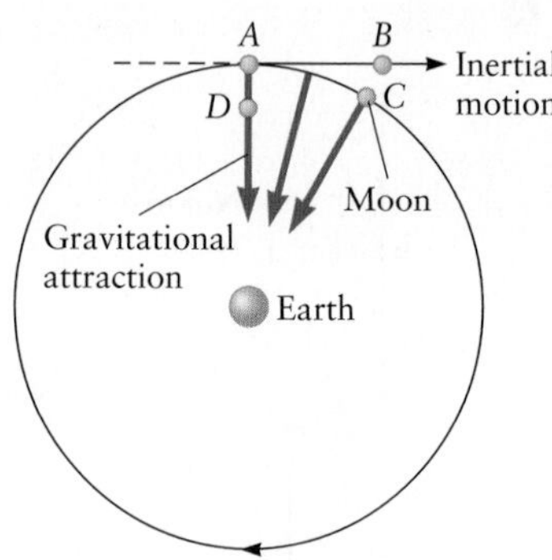

▲ **Figure 5.23** The Moon "falls" toward the Earth as it travels in its orbit. Here, the Moon's acceleration due to gravity causes it to "fall" from *B* to *C* and thereby travel in a circular orbit. If the Moon's initial velocity were zero, it would simply fall from *A* to *D*, in the same way that Newton's apple fell from its tree.

5.4 Planetary Motion and Kepler's Laws

Of the many important astronomers prior to the time of Newton, one of the most famous is Johannes Kepler (1571–1630). During and prior to his lifetime, a number of astronomers recorded the positions of the Moon, planets, comets, and so forth as functions of time and showed that bodies in the solar system move in an extremely precise fashion. Kepler studied these results very carefully and found that the motion of the Moon and planets could be described by what are now known as ***Kepler's laws*** of planetary motion. Kepler's laws are not "laws of nature" in the sense of Newton's laws of motion or Newton's law of gravitation. Rather, Kepler's laws are mathematical rules that describe motion in the solar system. These rules were inferred by Kepler from the available astronomical observations, but he was not able to give a scientific explanation or derivation. It remained for Newton to show how his three fundamental laws of motion together with his law of gravitation explain Kepler's laws.

Kepler's First Law

Kepler's first law is a statement about the shapes of orbits. Aristotle and many other early scientists believed that orbits are circular. The main reason for this belief was that circles are the most "perfect" shape for such a curve, and early thinkers believed nature must be perfect. By Kepler's time, however, it was well established[3] that while planetary orbits are approximately circular, they are definitely not perfect circles. Kepler showed that all the planetary orbits are *elliptical*. Furthermore, for the motion of planets around the Sun, the Sun is not at the center of the ellipse; instead, it is at one of the foci of the ellipse as illustrated in Figure 5.24. This discovery must have been quite a shock because the off-center placement of the Sun would seem to violate nature's tendency for symmetry.

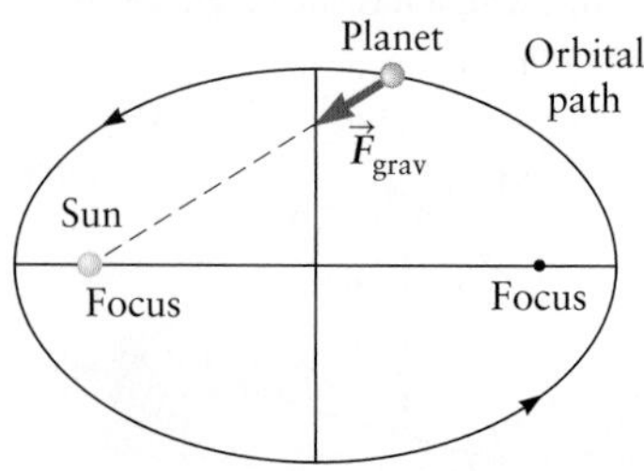

▲ **Figure 5.24** According to Kepler's first law, the planetary orbits are elliptical. An ellipse has two foci, which are offset from the center of the ellipse. The Sun is located at one of the foci.

For the planets known to Kepler, as well for the Moon, the orbits deviate only a small amount from perfect circles (by much less than the orbit in Fig. 5.24). For example, the distance between the Earth and the Sun varies by only ±3% during the

[3]At that time, only the six innermost planets in our solar system had been discovered: Mercury, Venus, Earth, Mars, Jupiter, and Saturn.

course of an orbit. Nevertheless, the deviation from a perfect circle is certainly real, and it had been accurately measured by Kepler's time.

Kepler's first law is the statement that planetary orbits are elliptical. This statement also applies to the Moon because the Moon is really just a "planet" belonging to the Earth, and Kepler's first law also applies to artificial satellites that have been launched into orbit around the Earth. Because a circle is a special case of an ellipse, Kepler's first law also allows circles, but real orbits generally deviate at least a small amount from a perfect circular shape. Soon after he developed his law of gravitation, Newton was able to use it to derive Kepler's first law and thus explain the shapes of planetary orbits.[4]

> ***Kepler's first law of planetary motion:*** **Planets move in elliptical orbits.**

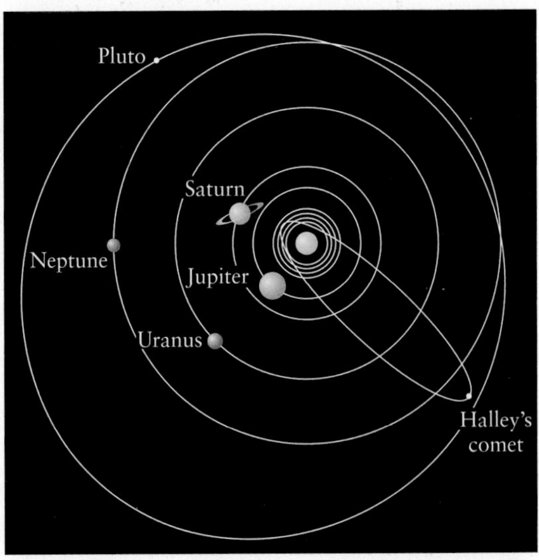

▲ **Figure 5.25** Scale drawing of the planetary orbits in our solar system. The orbits of most planets are nearly circular. The orbit of the dwarf planet Pluto deviates substantially from a circle. The orbit of Halley's comet, which is highly elliptical, is also shown.

Most comets have highly elliptical orbits. Perhaps the best known case is Halley's comet. A scale drawing of the orbit of Halley's comet is shown in Figure 5.25. This figure includes the orbit of the dwarf planet Pluto, which is also noticeably elliptical. In fact, Pluto's orbit is such that it actually spends a substantial amount of time inside Neptune's orbit.

> **CONCEPT CHECK 5.6** Acceleration of a Comet
>
> Sketch the direction of the acceleration vector for Halley's comet at various points in its orbit in Figure 5.25.

Kepler's Second Law

Kepler's second law concerns the speed of a planet as it moves around its orbit. This law applies to a planet in our solar system moving about our Sun as well as to a planet in another solar system moving about its sun. For a perfectly circular orbit, this speed is constant. However, for an elliptical orbit, the speed is smallest when the planet is farthest from its sun, whereas v is largest when the planet is nearest its sun (Fig. 5.26). To understand Kepler's second law, it is useful to consider a line drawn from the planet to its sun. This line moves along with the planet and sweeps out area as the planet moves. According to Kepler's second law, this line sweeps out area at a constant rate. Let A_1 be the area swept out in time Δt when the planet is near its sun and A_2 be the amount of area swept out in the same amount of time Δt when the planet is somewhere else in its orbit. Kepler's second law then states that $A_1 = A_2$. For this statement to be true, the planet must (as we have already noted) speed up when it is nearest its sun. We'll discuss Kepler's second law again in Chapter 9, where we'll see that it is closely connected with the ***angular momentum*** of the planet.

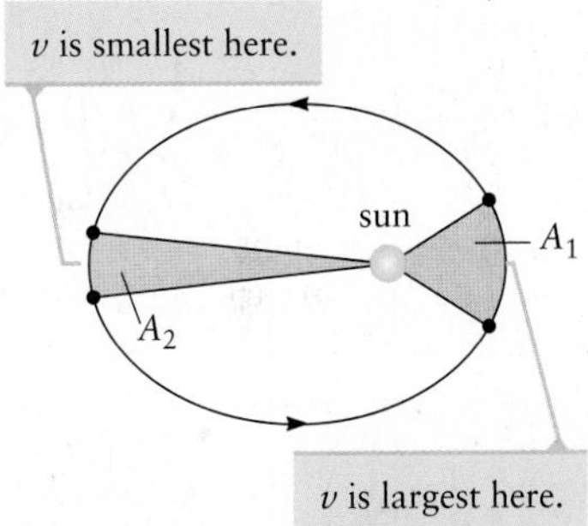

▲ **Figure 5.26** Kepler's second law is called the "equal areas" law. A line that extends from the sun to a planet sweeps out equal areas during equal time intervals as a planet moves around its orbit. If the time required for the planet to sweep out area A_1 is equal to the time associated with area A_2, the areas will be equal.

> ***Kepler's second law of planetary motion:*** **A line connecting a planet to its sun sweeps out equal areas in equal times as the planet moves around its orbit.**

Kepler's Third Law

Kepler's third law relates the timing of an orbit to the size of the orbit. It is simplest to derive Kepler's third law for the special case of a perfectly circular orbit, although it also applies to elliptical orbits. For a circular orbit, the speed of the planet is

---

[4]Nonclosed trajectories are also possible. For example, objects that pass near the Sun once and never return move along either a parabolic or a hyperbolic path. These trajectories, including the elliptical orbits of the planets, are ***conic sections***. Kepler's first law is sometimes worded to reference these nonclosed trajectories. The only closed trajectories—that is, the only true orbits—are elliptical, however.

constant. If the orbit has a radius r, the time it takes to complete one orbit equals the distance traveled (the circumference) divided by the speed:

$$T = \frac{2\pi r}{v} \tag{5.23}$$

Since the gravitational force of the sun on the planet is equal to ma_c, where a_c is the centripetal acceleration, we get (compare with Eqs. 5.16 and 5.17)

$$F_{grav} = \frac{GM_{sun}M_{planet}}{r^2} = ma_c = \frac{M_{planet}v^2}{r} \tag{5.24}$$

From Equation 5.23, we have $v = 2\pi r/T$, and substituting for v gives

$$\frac{M_{planet}v^2}{r} = \frac{M_{planet}(2\pi r/T)^2}{r} = \frac{4\pi^2 M_{planet}r}{T^2} \tag{5.25}$$

We can now combine Equations 5.24 and 5.25 to get a relation between the orbital time (the period T) and the orbital radius r. We find

$$\frac{GM_{sun}M_{planet}}{r^2} = \frac{4\pi^2 M_{planet}r}{T^2}$$

and solving for T gives

$$T^2 = \left(\frac{4\pi^2}{GM_{sun}}\right)r^3 \tag{5.26}$$

which is Kepler's third law. In words, Equation 5.26 states that the square of the orbital period is proportional to the cube of the orbital radius. This result does not depend on the mass of the planet. It applies to all orbiting bodies, including planets, comets, and spacecraft. Jupiter and Saturn, for example, possess many moons, and the orbital periods and radii of these moons must obey Equation 5.26, with M_{sun} replaced by the mass of the appropriate planet.

Kepler's third law: **The square of the period of an orbit is proportional to the cube of the orbital radius.**

EXAMPLE 5.9 Neptune's Orbit

The Earth completes one orbit about the Sun in 1 year and has an orbital radius of 1.50×10^{11} m (see Table 5.1). If the orbital radius of Neptune is 4.50×10^{12} m, what is the period of Neptune's orbit?

RECOGNIZE THE PRINCIPLE

We could solve this problem by simply evaluating Equation 5.26 using the given radius of Neptune's orbit. Here, though, we take a different approach by applying Equation 5.26 first to Neptune and next to the Earth, and then computing the ratio of their orbital periods. Notice that the orbits of the Earth and Neptune are both very close to circular.

SKETCH THE PROBLEM

No sketch is needed.

IDENTIFY THE RELATIONSHIPS

Applying Equation 5.26 to Neptune, we have

$$T^2_{Neptune} = \left(\frac{4\pi^2}{GM_{Sun}}\right)r^3_{Neptune}$$

with a similar result for the Earth. Taking the ratio gives

$$\frac{T^2_{\text{Neptune}}}{T^2_{\text{Earth}}} = \frac{\left(\dfrac{4\pi^2}{GM_{\text{Sun}}}\right) r^3_{\text{Neptune}}}{\left(\dfrac{4\pi^2}{GM_{\text{Sun}}}\right) r^3_{\text{Earth}}} = \frac{r^3_{\text{Neptune}}}{r^3_{\text{Earth}}} \tag{1}$$

SOLVE

Inserting the orbital radius of Neptune and of the Earth from Table 5.1 gives

$$\frac{T^2_{\text{Neptune}}}{T^2_{\text{Earth}}} = \left(\frac{4.50 \times 10^{12}\text{ m}}{1.50 \times 10^{11}\text{ m}}\right)^3 = 2.70 \times 10^4$$

$$T^2_{\text{Neptune}} = 2.70 \times 10^4 \times T^2_{\text{Earth}}$$

$$T_{\text{Neptune}} = \boxed{160 \times T_{\text{Earth}}}$$

Since $T_{\text{Earth}} = 1$ year, it takes Neptune approximately $\boxed{160 \text{ years}}$ to complete one orbit.

▶ ***What does it mean?***

Neptune thus takes more than a century to complete one orbit! When doing a calculation, it is sometimes mathematically simpler to use the ratio of two similar quantities. In this example, the factors of G and M_{Sun} canceled when the ratio was taken in Equation (1), simplifying the calculation.

Satellite Orbits around the Earth

Many satellites, including most that carry astronauts, are in what is often called a "low Earth" orbit. You may wonder why any orbit would be called "low," but the reason for this expression becomes clear when we calculate the radius of a typical orbit. We can do so using Kepler's third law (Eq. 5.26) if we know the period of the orbit. The period for a satellite in low Earth orbit, such as the International Space Station, is approximately 90 min, so $T = (90 \text{ min})(60 \text{ s/min}) = 5400$ s. Kepler's third law, with the mass of the Earth as that of the central body, gives

$$T^2 = \left(\frac{4\pi^2}{GM_{\text{Earth}}}\right) r^3$$

Solving for the radius of the orbit, we find

$$r = \left(\frac{GM_{\text{Earth}} T^2}{4\pi^2}\right)^{1/3} \tag{5.27}$$

Inserting values for the mass of the Earth and T, we get

$$r = \left(\frac{GM_{\text{Earth}} T^2}{4\pi^2}\right)^{1/3} = \left[\frac{(6.67 \times 10^{-11}\text{ N}\cdot\text{m}^2/\text{kg}^2)(5.98 \times 10^{24}\text{ kg})(5400\text{ s})^2}{4\pi^2}\right]^{1/3}$$

$$r = 6.66 \times 10^6 \text{ m} \tag{5.28}$$

The radius of the Earth is $r_E = 6.37 \times 10^6$ m, so these satellites orbit at a height above the Earth's surface that is only 5% of the Earth's radius. This orbit is illustrated in the scale drawing in Figure 5.27. Even though it may seem low when compared to the Earth's radius, the tallest mountain on the Earth (Mount Everest) is only about 0.1% of r_E above sea level, so the satellites are in no danger of colliding with any mountains!

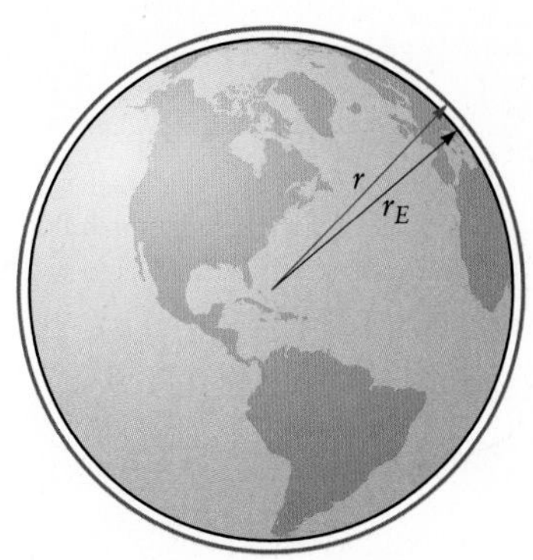

▲ **Figure 5.27** Scale drawing of a "low Earth" orbit. The orbit is shown in red. On this scale, it is barely distinguishable from the surface of the Earth.

EXAMPLE 5.10 Geosynchronous Orbits

Earth-orbiting satellites used for transmitting telephone and television signals travel in *geosynchronous* orbits. These orbits have a period of 1 day, so these satellites move in synchrony with the Earth's rotation and are thus always at the same position in the sky relative to a person on the ground. Sending signals to and from these satellites is thus greatly simplified because an antenna such as the author's satellite TV dish can be aligned only once and then needs no further adjustment. Calculate the orbital radius of a satellite in a geosynchronous orbit.

RECOGNIZE THE PRINCIPLE

We have already done a very similar problem in our discussion of a satellite in low Earth orbit. We can thus apply Equation 5.27 to the geosynchronous case using a period $T = 1$ day.

SKETCH THE PROBLEM

No sketch is needed.

IDENTIFY THE RELATIONSHIPS AND SOLVE

The period of a geosynchronous orbit is 1 day. Hence, the period $T = 1$ day = (1 day)(24 h/day)(3600 s/h) = 86,000 s. Inserting this result into Equation 5.27 gives

$$r = \left(\frac{GM_{\text{Earth}}T^2}{4\pi^2}\right)^{1/3}$$

$$= \left[\frac{(6.67 \times 10^{-11}\ \text{N}\cdot\text{m}^2/\text{kg}^2)(5.98 \times 10^{24}\ \text{kg})(8.6 \times 10^4\ \text{s})^2}{4\pi^2}\right]^{1/3}$$

$$r = \boxed{4.2 \times 10^7\ \text{m}}$$

► *What does it mean?*

The radius of the Earth is 6.37×10^6 m, so the radius of a geosynchronous orbit is about seven times larger than the radius of the Earth. It is thus much larger than the low Earth orbit in Figure 5.27, which is why much more fuel is required to launch a satellite into geosynchronous orbit than into low Earth orbit.

Kepler's Laws, Putting a Satellite into Orbit, and the Origin of the Solar System

Kepler's three laws of planetary motion apply to all types of gravitationally produced orbital motion, including the motion of planets and comets about the Sun and the motion of moons and satellites about a planet. Indeed, one can also think of a freely falling apple as an example of gravitationally produced motion. It is fascinating that such a variety of motion can be produced by a single force. These different types of motion result from the different ways in which these objects are initially set into motion. The dropped apple is released from rest, so its initial velocity is zero. For the Moon to follow a nearly circular orbit, it must have been given the proper "initial" velocity at some point in the distant past. For planets such as the Earth and Jupiter to be moving in orbits about the Sun that are now approximately circular, it was necessary that the planets be set into motion with the proper velocity. This problem brings us to some very interesting questions concerning the origin and evolution of the solar system. It is now believed that the solar system was originally a rotating mass of gas and that this gas gradually condensed to form the planets. The rotational motion of the original gas cloud then led to the approximately circular orbits we now observe.

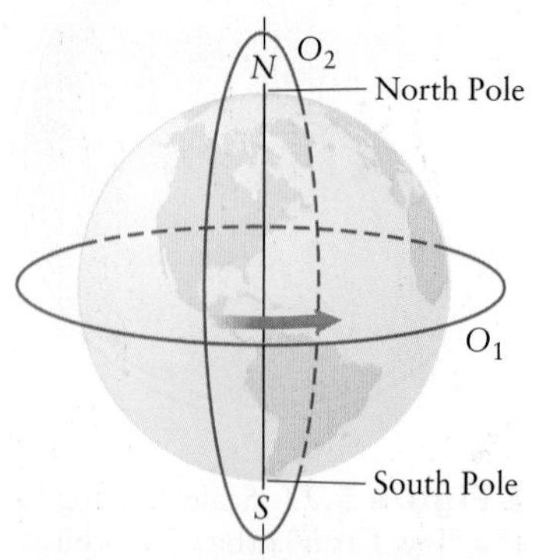

▲ **Figure 5.28** It is possible to place a satellite into an equatorial orbit (O_1) or a polar orbit (O_2). A polar orbit takes a satellite over the Earth's North and South Poles.

The problem of setting up an orbit of the desired shape is also encountered when a satellite is launched from the Earth. NASA usually launches satellites into equatorial orbits that are approximately parallel to the Earth's equator as sketched in Figure 5.28. We have always considered or computed orbital speeds as measured with respect to a stationary observer. The Earth's surface is not stationary, however; rather, it moves with a substantial speed v_R due to the Earth's rotation. When a satellite is launched toward the east in an equatorial orbit, it starts with a speed v_R and the rocket engines then add to this speed as the satellite is put into orbit. If the satellite were launched into a different orbit—say, into an equatorial orbit to the west—it would not be able to take advantage of v_R and the launch would require more fuel (and therefore more money).

CONCEPT CHECK 5.7 Apparent Weight and Earth's Rotational Motion

A person at the equator moves in a circle due to the Earth's rotational motion. How does this circular motion affect the person's apparent weight?

(a) The circular motion causes the apparent weight to be *larger* than it would be if the Earth were not rotating.
(b) The circular motion causes the apparent weight to be *smaller* than it would be if the Earth were not rotating.
(c) There is no effect on the apparent weight.

EXAMPLE 5.11 A Double Star System

A double star system is one in which two stars orbit around each other due to their mutual gravitational attraction. In the simplest case, the stars have the same mass m and move in circular orbits with the same radius r and speed v (Fig. 5.29), with the center of each orbit at the point midway between the two stars. Find the orbital period T of such a double star system. Express your answer in terms of r, m, and G.

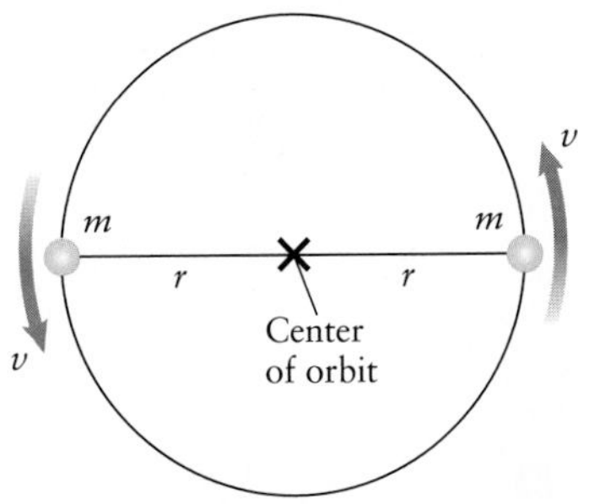

▲ **Figure 5.29** Example 5.11. A double star system.

RECOGNIZE THE PRINCIPLE

The gravitational force on one of the stars can be calculated from Newton's law of universal gravitation. This force provides the centripetal acceleration of that star. We can use Newton's second law (Eq. 5.6) to find the orbital period, following the approach we used in deriving Kepler's third law (Eq. 5.26).

SKETCH THE PROBLEM

Figure 5.29 shows the problem.

IDENTIFY THE RELATIONSHIPS

Both stars have mass m and are separated by a distance $2r$, so the gravitational force on one of the stars is

$$F_{\text{grav}} = \frac{Gmm}{(2r)^2} = \frac{Gm^2}{4r^2} \tag{1}$$

The force required to make a star move in a circular orbit is mv^2/r, where the speed is equal to the circumference of an orbit divided by the period, $v = 2\pi r/T$. Setting this force equal to the gravitational force in Equation (1) leads to

$$\frac{Gm^2}{4r^2} = \frac{mv^2}{r} = \frac{m(2\pi r/T)^2}{r} = \frac{4\pi^2 mr}{T^2} \tag{2}$$

SOLVE

Solving Equation (2) for the period, we get

$$\frac{Gm^2}{4r^2} = \frac{4\pi^2 mr}{T^2}$$

$$T^2 = \frac{16\pi^2 r^3}{Gm}$$

$$T = \boxed{4\pi\sqrt{\frac{r^3}{Gm}}}$$

(continued) ▶

► ***What does it mean?***
The period thus depends on both the mass of the stars and the orbital radius. In their studies of double star systems, astronomers use measurements of this period to learn about both of these quantities.

Courtesy of Nova Scotia Economic & Rural Development & Tourism (both)

► **Figure 5.30** An example of tides. The ocean level can vary substantially between periods of low and high tide. There are generally two high tides every 24 h, although in some regions factors such as the shape of the ocean basin lead to only one high tide and one low tide each day.

Moon's gravitational force is larger here because the ocean is closer to the Moon.

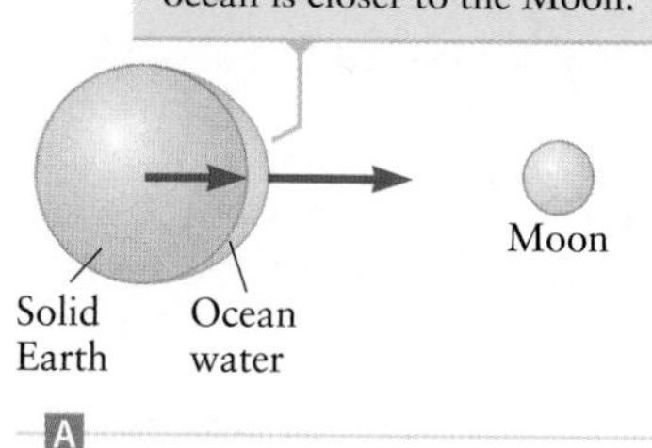

Force is larger on the solid Earth because it is closer to the Moon.

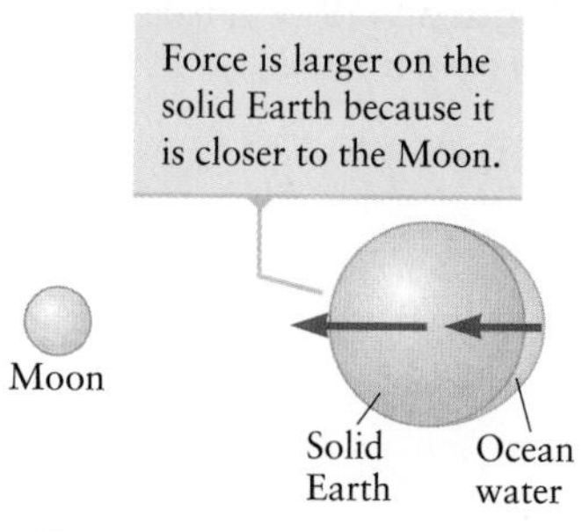

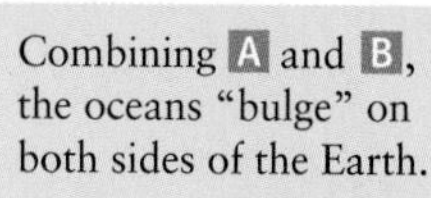

Combining A and B, the oceans "bulge" on both sides of the Earth.

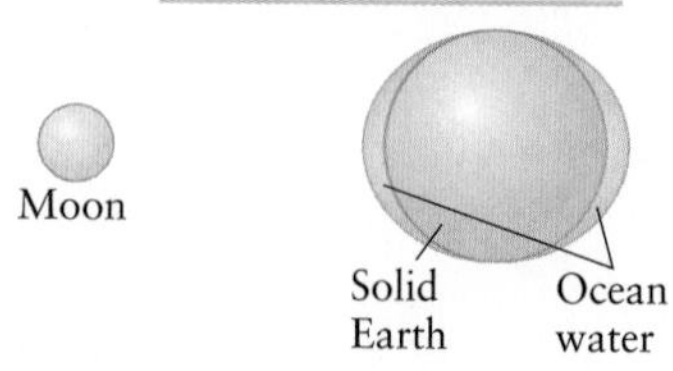

▲ **Figure 5.31** The Moon's gravitational force on the oceans is responsible for the tides. (Not to scale.)

5.5 Moons and Tides

The Origin of Tides

You have probably noticed that the level of the Earth's oceans fluctuates up and down every day; these fluctuations are called tides (Fig. 5.30). Tides are due to the Moon's gravitational force on the oceans. Although the Moon exerts a gravitational force on both the ocean water and the solid Earth, the gravitational force decreases with distance, so the resulting acceleration is slightly higher for the water nearest the Moon (Fig. 5.31A). Thus, the water nearest the Moon "falls toward" the Moon slightly faster than does the solid Earth. The result is a "bulge" in the ocean, which is just a high tide.

Figure 5.31 explains why there is a high tide when the Moon is overhead. However, there are actually *two* high tides every day in most parts of the world, one when the Moon is overhead and another 12 h later when it is on the opposite side of the Earth. The origin of this second high tide is explained in Figure 5.31B. Because the gravitational force of the Moon decreases with increasing distance, the acceleration of the solid Earth is larger than the acceleration of the ocean on the far side of the Earth. As a result, the solid Earth "falls toward" the Moon faster than the ocean water on the far side and the ocean now bulges away from the Moon, producing another high tide.

The Sun also affects the tides, although its effect is smaller than the Moon's. When the Sun and the Moon are aligned and are on the same side of the Earth, their gravitational forces add, giving a higher high tide than produced by the Moon alone.

5.6 Deep Notions Contained in Newton's Law of Gravitation

The Inverse Square Law

A key feature of the gravitational force is the manner in which it varies with distance. According to Equation 5.16, the gravitational force between two objects falls off as the square of the distance between them:

$$F_{\text{grav}} = \frac{Gm_1m_2}{r^2} \propto \frac{1}{r^2} \qquad (5.29)$$

Mathematically, Equation 5.29 is called an "inverse square" law. A number of other forces in nature, such as the force between two electric charges, fall off as the square of the distance and are thus also described by inverse square laws. Why do many natural forces follow this pattern? Imagine that an object—for example, the Sun—possesses gravitational "field lines" that emanate radially outward from it as sketched in Figure 5.32. We also imagine that the number of these lines is proportional to the mass of the object. When these lines intersect another object, such as the Earth, there is a force on that object directed parallel to the lines and hence toward the original object; for example, there is a force on the Earth directed toward the Sun.

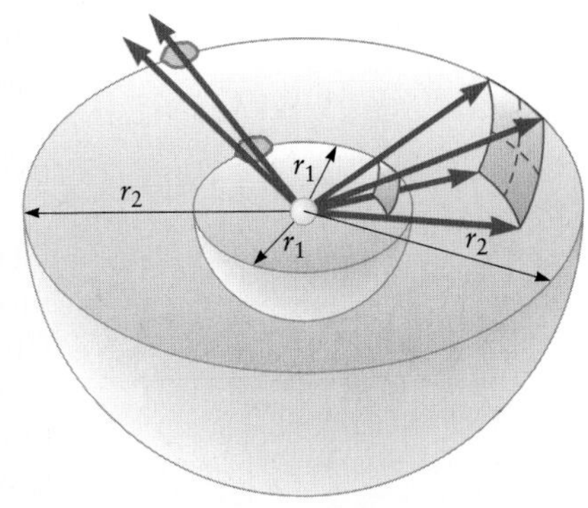

▲ **Figure 5.32** Lines of gravitational force. According to the field line picture, gravitational force acts along lines that emanate from all objects. The gravitational force exerted by one object on another is proportional to the number of lines intersecting the second object. This number falls as $1/r^2$ and thus accounts for the inverse square force law in Equation 5.29. Here, the object shown in blue (which might be the Earth) is located at two different distances from the central object in yellow (which might be the Sun). The number of lines intercepted by the blue mass is smaller when the separation increases because the field lines emanate radially and expand into a larger area (tan squares).

Because the force lines emanate in three-dimensional space, the number of lines over a given area—that is, the number of lines that intercept the Earth in Figure 5.32—falls off with distance as $1/r^2$. Hence, this picture explains the inverse square dependence in Equation 5.29. This result implies that gravity follows an inverse square law because we live in a three-dimensional space. It also means that we should expect other forces described by a field line picture to have the same inverse square dependence, which does seem to be the way nature works.

Do such field lines really exist? So far, no one has devised a way to "see" the lines; the best we can do is to observe the resulting force that the lines are presumed to cause. The field line picture explains the general form of Newton's law of gravitation, so there is strong evidence in favor of this model. However, while the field line picture is very attractive, it leads to other questions. The field lines that emanate from the Sun in Figure 5.32 and eventually reach the Earth must travel through the nearly perfect vacuum of space. Other observations indicate that the gravitational force is felt even when a perfect vacuum separates two objects. How can "something," like a gravitational field line, exist in a vacuum? Does a field line have a mass? If the Sun were to suddenly move, how fast would the corresponding change in its gravitational field be felt on the Earth?

The gravitational force is an example of what is called ***action at a distance.*** Newton's theory of gravitation tells us that action at a distance does indeed occur, but it does not tell us *how* it occurs. The field line picture was invented to answer this "how" question, but it does not provide a complete answer. For some forces, such as the electric force, there is a theoretical explanation of the nature of the associated field lines. Einstein's theories of relativity answer many of these questions in the case of the gravitational field, as we'll describe in Chapter 27.

Gravitation and Mass

A very important feature of the gravitational force law is that F_{grav} is linearly proportional to the mass m of each object. This quantity m is sometimes referred to as the ***gravitational mass*** of an object. We first encountered the concept of mass in Newton's second law of motion ($\sum \vec{F} = m\vec{a}$), and the m in Newton's second law is often called the ***inertial mass*** of the object. As far as physicists can tell, the gravitational mass of an object is *precisely equal* to the inertial mass. That $m_{\text{grav}} = m_{\text{inertial}}$ suggests a deep connection between gravitation and inertia (and hence motion). Newton did not understand why there should be a connection; it was not explained until the work of Einstein, whose theories showed why the inertial and gravitational masses are the same.

Einstein also considered the effect of a gravitational field on the motion of light. Consider how we would apply Newton's second law to the motion of a "particle" of mass m_p that experiences a gravitational force caused by another object, such as the Sun. We would write Newton's second law as

$$m_p a_p = \frac{GM_{\text{Sun}} m_p}{r^2}$$

where a_p is the acceleration of the particle and r is the distance from it to the Sun. Strictly speaking, the mass m_p on the left side is the inertial mass of the particle, whereas the mass m_p on the right side is the gravitational mass. Because they are equal, however, we can cancel these factors to get

$$a_p = \frac{GM_{Sun}}{r^2} \tag{5.30}$$

Hence, the acceleration is independent of the mass of the particle. We have seen this result many times before (e.g., in free fall), but there is a new point to make. Given that the acceleration is independent of the mass, does this result also apply to a "particle" that has no mass? As we'll see in Chapter 28, light can be described in terms of particles called ***photons*** that have no mass. Even though a photon's mass is zero, Equation 5.30 suggests that it is still accelerated by gravity. That is indeed the case, although a correct theoretical description of this acceleration requires Einstein's general theory of relativity. We'll say more about this acceleration and about the equivalence of gravitational and inertial mass in Chapter 27.

Summary | CHAPTER 5

Key Concepts and Principles

Circular motion and centripetal acceleration

An object moving in a circle of radius r at a constant speed has an acceleration

$$a_c = \frac{v^2}{r} \qquad \text{(5.5) (page 116)}$$

directed toward the center of the circle. The quantity a_c is called the ***centripetal acceleration.*** According to Newton's second law, this acceleration must be caused by a total force of magnitude $\sum F = ma_c$, so

$$\sum F = \frac{mv^2}{r} \qquad \text{(5.6) (page 117)}$$

This force is directed toward the center of the circle.

Newton's law of gravitation

There is a gravitational force

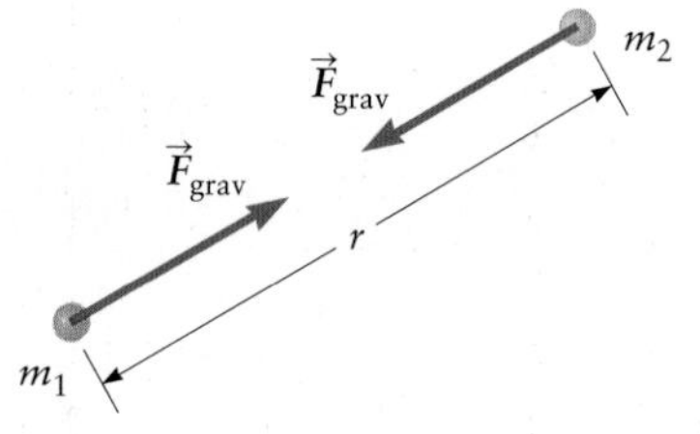

$$F_{grav} = \frac{Gm_1m_2}{r^2} \qquad \text{(5.16) (page 129)}$$

between any two objects. This force is always attractive.

Newton's law of gravitation is an example of an ***inverse square law.*** This $1/r^2$ dependence suggests a field line model of gravity and tells us something about the geometry of the universe. Gravitation is also an example of "action at a distance." Other forces, including electric forces, exhibit this property.

Applications

There are many examples of circular motion, including roller coasters, cars traveling on a curved road, and centrifuges. In all cases, the total force on an object undergoing uniform circular motion must equal $F = mv^2/r$ and be directed toward the center of the circle.

The motion of the Moon as it orbits the Earth and of the planets as they orbit the Sun are examples of (nearly) circular motion. The force in these cases is due to gravitation. Near the Earth's surface, the magnitude of the gravitational force on an object of mass m is $F_{grav} = mg$ and is directed toward the center of the Earth.

Kepler's laws

Kepler deduced three laws of planetary motion: (1) planetary orbits are elliptical, (2) a planetary orbit sweeps out equal areas in equal times, and (3) the square of the orbital period is proportional to the cube of the average orbital radius. These laws apply to planets orbiting a sun and also to satellites and moons orbiting a planet.

Kepler's laws, and hence the motions of planets, moons, comets, and other objects in the solar system, are all explained by Newton's law of gravitation, together with Newton's laws of motion. The circular motion of the Moon and the free fall of an apple look different, but they are due to the same force.

Questions

SSM = answer in *Student Companion & Problem-Solving Guide* ⓧ = life science application

1. Give an example of motion in which (a) the magnitude of the instantaneous velocity is always larger than the average velocity and (b) the instantaneous velocity is never parallel to the instantaneous acceleration.

2. SSM In Example 5.3, we considered a car traveling on a banked turn with friction. Draw free-body diagrams for the car when the speed is low and when the speed is high, and explain why they are different. *Hint*: Consider the direction of the frictional force in the two cases.

3. In a reference listing found on the Internet, it is stated that $g = 9.80665$ m/s^2. Discuss why it is not correct to think that the "exact" value of g can be given with this accuracy. Indeed, is there an "exact" value of g?

4. Consider the Cavendish experiment in Figure 5.22. When he designed this experiment, Cavendish had to decide how large to make the spheres. If they are made larger, they will have a larger mass, which, according to Equation 5.16, will make the gravitational attraction larger and the force therefore easier to measure. If the spheres are made larger, however, the distance between their centers will necessarily increase, which makes the force smaller. Assuming the spheres in Figure 5.22 all have the same radii, suppose the value of r is increased by a factor of two. Will that change increase or decrease the force, and by what factor? Assume that the spheres all have the same constant density.

5. Explain why a geosynchronous satellite cannot remain directly overhead for an observer in Boston. *Hint*: Consider first an observer at the North Pole.

6. Kepler's second law, the statement that a planet sweeps out equal areas in equal times, can be derived by a geometrical argument. To see how one might construct such a geometrical proof, consider the simpler case of a planet moving with constant velocity as shown in Figure Q5.6. The points A, B, C, . . . are spaced at equal time intervals. Show that this planet obeys Kepler's second law; that is, show that it sweeps out equal areas in equal times. *Hint*: Calculate the areas of triangles OAB, OBC, and so on.

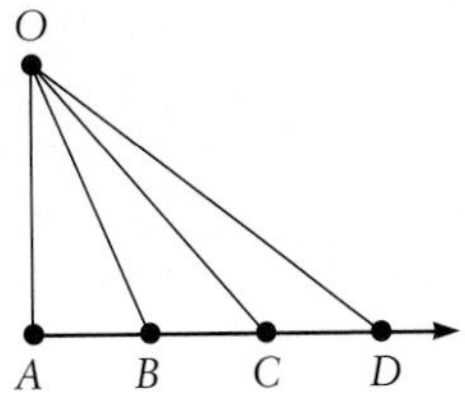

O
A B C D

Figure Q5.6

7. It is sometimes claimed by astrologers (but not by astronomers!) that because the Moon dramatically affects the seas of the world, as evidenced by the tides, the Moon must also affect individual people because more than 60% of an average adult's mass is water. Does that claim make sense? Use Newton's law of universal gravitation to make an estimate of the forces involved to justify your answer.

8. What force makes it possible for a car to move along a curved road? A straight, flat road?

9. Astronomers have discovered that some stars orbit around regions of space that contain large amounts of cosmic dust. This dust provides the gravitational force that causes the stars to move in a circular orbit. Suppose the density of dust is constant in a certain region and several stars orbit through this dust. Find a relation between the orbital period T and orbital radius R for these stars and show that in this case T is independent of R. Compare your result to Kepler's third law. Studies of these stars give astronomers a way to determine the distribution of cosmic dust in the universe.

10. Reconnaissance satellites (often called spy satellites) travel in very low orbits so that their cameras can take photos of objects on the Earth's surface with the highest possible resolution. Explain why such satellites are often launched into polar orbits; when in these orbits, satellites travel over both the north and south poles of the Earth.

11. It is believed that much of the mass in the universe is carried by what is called dark matter, matter that does not emit enough visible light or other radiation to be detected by conventional telescopes. However, dark matter does exert a gravitational force on other objects in the universe. Explain how dark matter might be detected and studied through the observation of "normal" matter (such as conventional stars).

12. When a planet orbits around a star, the star also moves in an "orbit." Since it is much more massive than the planet, the star's orbital radius r_{star} is much smaller than that of the planet r_{planet}. Work out how the ratio r_{star}/r_{planet} depends on the ratio of the masses. Discuss how this effect could be used to detect the presence of planets in distant solar systems. For simplicity, assume circular orbits.

13. NASA uses a specially equipped airplane (called the "Vomit Comet") to provide a simulated zero-gravity environment for training and experiments. This airplane flies in a long, parabolic path. Explain how a passenger can feel "weightless" near the top of the parabola.

14. You are a prospector looking for gold by taking high-precision measurements of the acceleration due to gravity, g, at different points on the Earth's surface. In one region, you find that g is slightly higher than its average value. Are you standing over what might be a deposit of gold or over an underground lake? Explain.

15. How does your weight on a ship in the middle of the ocean compare with your weight when you are standing on solid ground? Explain why they are not the same.

16. The Sun exerts an overall force on the Earth many times greater than that of the Moon. How can it then be that the ocean tides are primarily due to the influence of the Moon and, to a much lesser extent, the Sun?

17. SSM The difference in the gravitational force is only about 10% less on an object that is in a low Earth orbit than it is for the same object on the ground. Why is it that an astronaut in orbit experiences weightlessness?

18. An astronaut on the peak of a mountain on the Moon fires a rifle along the horizontal direction. Is it possible, given a sufficient initial speed for the bullet, that the bullet might hit her in the back? Explain how it could happen.

19. **Pluto's mass.** In 1978, it was discovered that Pluto had a moon of its own. The moon was given the name Charon (now known to be one of three; see Fig. Q5.19). After the discovery of this moon, the hitherto unknown mass of Pluto was calculated to a precision of less than 1%. How did the discovery of Charon allow the mass to be determined?

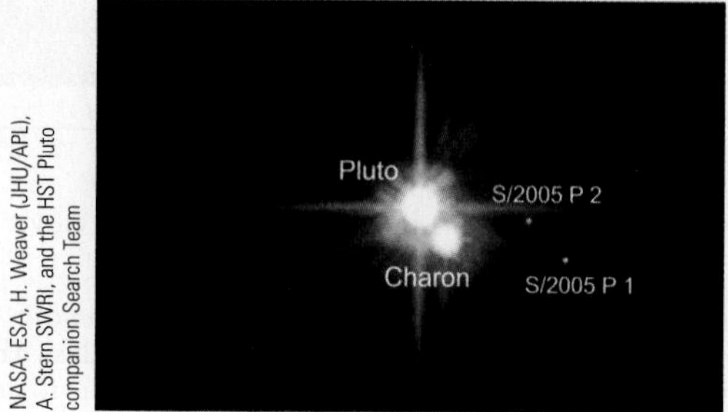

NASA, ESA, H. Weaver (JHU/APL), A. Stern SWRI, and the HST Pluto companion Search Team

Figure Q5.19 Pluto and its three moons imaged by the Hubble Space Telescope.

20. **A coffee centrifuge?** In one popular demonstration, a full cup of hot coffee is placed on a platform suspended by strings to the lecturer's hand as seen in Figure Q5.20. With some practice, the platform, coffee and all, can be made to rotate in a vertical circle. How does the coffee stay in the cup? If the rotation of the coffee and cup is sustained for some time, what would happen to any grinds that happen to be mixed in the coffee?

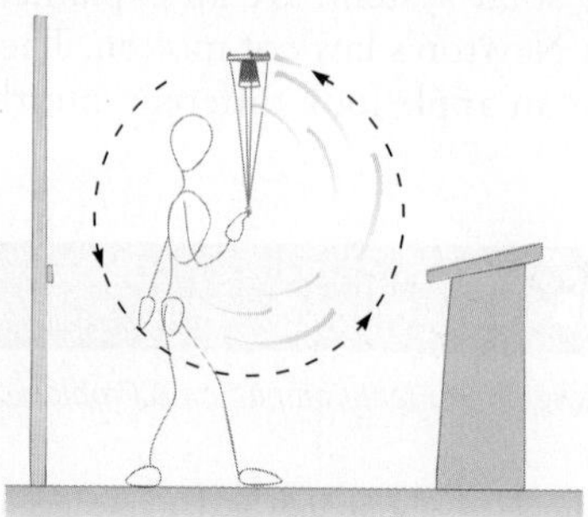

Figure Q5.20

Problems available in ENHANCED WebAssign

Section 5.1 Uniform Circular Motion
Problems 1–15

Section 5.2 Examples of Circular Motion
Problems 16–34

Section 5.3 Newton's Law of Gravitation
Problems 35–48

Section 5.4 Planetary Motion and Kepler's Laws
Problems 49–55

Section 5.5 Moons and Tides
Problems 56–59

Additional Problems
Problems 60–78

List of Enhanced Problems

Problem number	Solution in *Student Companion & Problem-Solving Guide*	Reasoning & Relationships Problem	Reasoning Tutorial in ENHANCED WebAssign
5.3		✓	✓
5.4		✓	
5.11	✓		
5.13		✓	
5.14		✓	
5.21	✓		
5.30		✓	✓
5.32		✓	
5.33		✓	
5.34	✓		
5.41		✓	✓
5.42		✓	✓
5.43	✓		
5.51	✓		
5.56		✓	
5.57		✓	
5.59	✓		
5.63		✓	✓
5.64	✓	✓	
5.66		✓	✓
5.67		✓	
5.72		✓	✓
5.73	✓		
5.74		✓	
5.77		✓	
5.78		✓	

CHAPTER

6

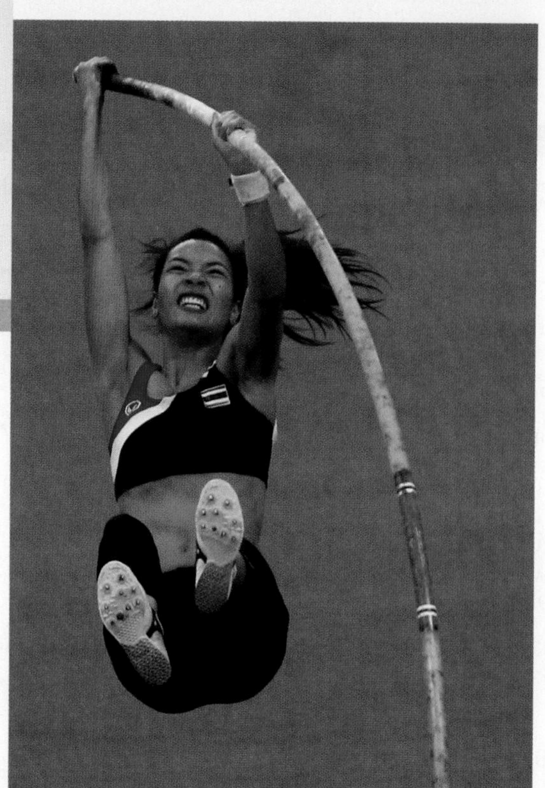

▶ *This pole vaulter uses her kinetic and potential energies in making her vault. In this chapter, we explore the connection between mechanical energy and motion. (© Mark Dadswell/Getty Images)*

Work and Energy

OUTLINE

So far in this book, our discussions of motion have been based on very direct applications of Newton's laws. For example, to predict the motion of an object, we used Newton's second law ($\sum \vec{F} = m\vec{a}$) to calculate the acceleration; from there, we worked out the velocity and position and how they vary with time. Such direct applications of Newton's laws can take us a long way, but there is a lot more to mechanics than simply forces and acceleration. In this and the next few chapters, we will explore some very important concepts and principles that are, in a sense, "hidden" just beneath the surface of Newton's second law. This chapter is based on the concepts of ***work*** and ***energy***, and we'll see how Newton's second law leads to definitions of these quantities. We will also consider the concept of energy as applied to an individual particle and to systems of particles or objects. We'll find that the total energy of an isolated system is constant with time, which will lead us to the general principle of ***conservation of energy***. This principle has an important role in many fields, including engineering, physiology, and physics.

6.1 Force, Displacement, and Work

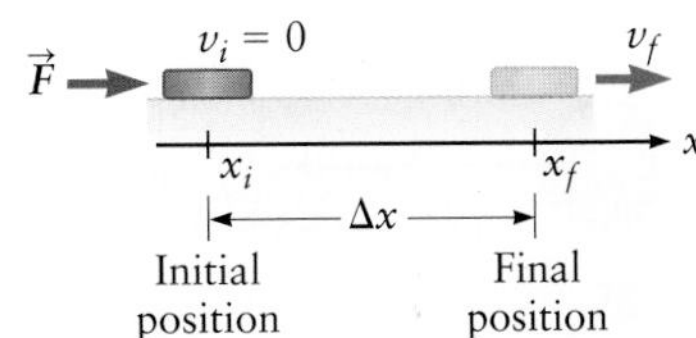

▲ **Figure 6.1** When a force $\vec{F}$ acts on this hockey puck, the puck accelerates. We also say that this force does work on the puck.

According to Newton's second law ($\sum \vec{F} = m\vec{a}$), the acceleration of a particle is proportional to the total applied force. A large force thus gives a large acceleration, while a small force gives a small acceleration. Given enough time, though, a small acceleration can still produce a large velocity and large displacement, so there is a sort of trade-off between force and time. Our intuition tells us that a small force acting for a sufficiently long time can have the same effect as a large force acting over a shorter time, but how long a time is required? The notion of "work" gives a way to answer this and other similar questions.

Consider a hockey puck of mass m placed on a frictionless, horizontal surface as sketched in Figure 6.1. The puck is initially at rest, and we want to consider how a constant force of magnitude F applied horizontally will set the puck into motion. Since the force is constant, we can use our results for motion with constant acceleration from Chapter 3. Applying Equation 3.4 and denoting the initial and final velocities by v_i and v_f, respectively, we can write

$$v_f^2 = v_i^2 + 2a\Delta x \tag{6.1}$$

where Δx is the displacement. The puck starts from rest, so $v_i = 0$, giving $v_f^2 = 2a\Delta x$. Using Newton's second law, we also have $F = ma$. Putting it all together leads to

$$v_f^2 = 2a\Delta x = \frac{2}{m}(ma)\Delta x = \frac{2}{m}(F\Delta x)$$

which we can rearrange to read

$$F\Delta x = \tfrac{1}{2}mv_f^2 \tag{6.2}$$

In words, this result means that if we want to accelerate an object that starts from rest to a particular velocity, we can exert a large force over a short distance or a small force over a large distance. As long as the product of force and displacement is the same, the object will reach the same final velocity. The product $F\,\Delta x$ in Equation 6.2 is called ***work***, and the work done by the applied force F on the hockey puck in Figure 6.1 is

$$W = F\,\Delta x \tag{6.3}$$

In SI units, we have

$$\text{work} = \text{force} \times \text{displacement} = \text{newtons} \cdot \text{meters} = \text{N} \cdot \text{m}$$

The unit N · m is called a joule (abbreviated "J").

The definition of work given in Equation 6.3 only applies for one-dimensional motion in which a constant force F is applied along the direction of motion. In two or three dimensions, we must consider that force $\vec{F}$ and displacement $\Delta\vec{r}$ are both vectors. In this case (Fig. 6.2), the work done on the particle is

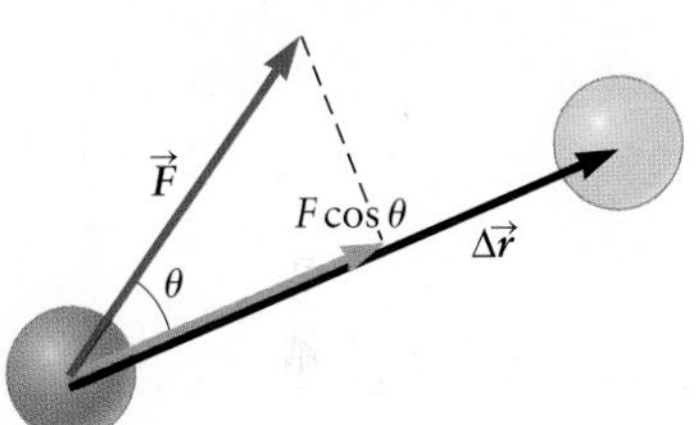

▲ **Figure 6.2** The force $\vec{F}$ acts on this particle while the particle moves through a displacement $\Delta\vec{r} = \vec{r}_f - \vec{r}_i$. The work done by the force is given by $W = F(\Delta r)\cos\theta$.

$$W = F(\Delta r)\cos\,\theta \tag{6.4}$$

Definition of work

where θ is the angle between the force and the particle's displacement. The factor F in Equation 6.4 denotes the magnitude of the force, while Δr is the magnitude of the displacement. Although force and displacement are both vectors, work is a scalar. Also, W can be positive, negative, or zero and does not have a direction.

W Depends on the Direction of the Force Relative to the Displacement

Figure 6.2 shows that the factor $F\cos\theta$ in Equation 6.4 is equal to the component of the force along the direction of the displacement. Hence, the work W equals the component of the force *along* the displacement multiplied by the magnitude of the displacement. In Figure 6.2, this component of the force is parallel to $\Delta\vec{r}$, and W is positive. A similar case in one dimension is shown in Figure 6.3A, in which the angle

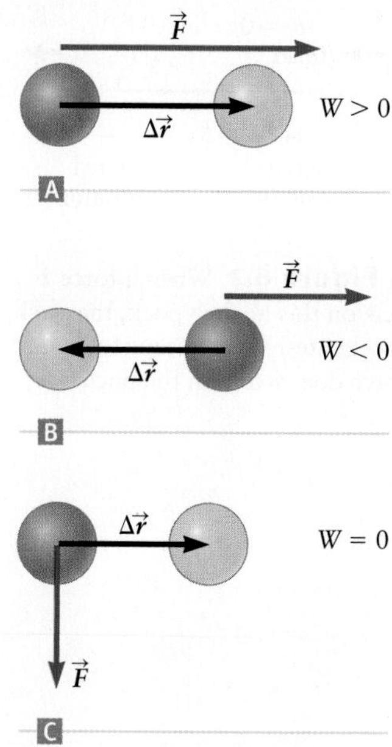

▲ **Figure 6.3** A The work done on an object is positive when the applied force and the object's displacement are parallel. B W is negative if the force and displacement are antiparallel. C When the force is perpendicular to the displacement, as in uniform circular motion, W is zero.

between the applied force and the displacement is zero (which was also the case for the hockey puck in Fig. 6.1). The cos θ factor in Equation 6.4 is then equal to unity, because $\cos(0) = 1$, and this expression is equivalent to the relation for work in one dimension (Eq. 6.3).

Figure 6.3B shows a situation in which the force on an object and the object's displacement are in opposite directions; that is, $\vec{F}$ and $\Delta\vec{r}$ are *anti*parallel. This object might be a hockey puck sliding across a surface with friction, since the force due to friction is directed opposite to the displacement. We then have $\theta = 180°$, giving $\cos\theta = -1$ in Equation 6.4, and the work done by friction on the puck is *negative*. Another important case is shown in Figure 6.3C, in which the force is perpendicular to the displacement. This situation often arises because, as we learned in Chapter 5, the force responsible for uniform circular motion is perpendicular to the direction of motion. When $\vec{F}$ and $\Delta\vec{r}$ are perpendicular, $\theta = 90°$ and $\cos\theta = 0$ in Equation 6.4; in this case, the work done on the object is $W = 0$.

The relationship between force, displacement, and work are central to this chapter and can be summarized as follows.

Key concepts: The relationship between force, displacement, and work

- Work is done *by* a force $\vec{F}$ acting *on* an object.
- The work W depends on the force acting on the object and on the object's displacement, according to Equation 6.4.
- The value of W depends on the direction of $\vec{F}$ relative to the object's displacement.
- W may be positive, negative, or zero, depending on the angle θ between the force and the displacement. W is a scalar; it is *not* a vector.
- If the displacement is zero (the object does not move), then $W = 0$, even though the force may be very large.

How Physics Uses the Term *Work*

The term *work* is used in everyday conversation. It is crucial to see how the everyday definition of this term differs from the "physics definition" in Equations 6.3 and 6.4. One important difference is that W (the "physics" work) can be negative. An example with $W < 0$ is sketched in Figure 6.3B, where the value of W is negative because the force and displacement are in opposite directions. As a result, the final speed of the object is less than its initial speed. In general, we can say that if $W > 0$, an object will "speed up"; if $W < 0$, it will "slow down."

Although we have not yet defined the concept of energy, we'll soon see how the motion of an object is related to its energy. Situations in which the force *increases* the energy of the object correspond to a positive value of W. Conversely, it is possible for a force to *reduce* the energy of an object (as with friction and the hockey puck in Fig. 6.3B); in such cases, the work done on the object is negative.

CONCEPT CHECK 6.1 Force, Displacement, and Work

Figure 6.3 shows hypothetical cases in which the force is (a) parallel, (b) antiparallel, and (c) perpendicular to the displacement. Identify which case applies to the following situations.

(1) An apple falling from a tree
(2) A satellite in geosynchronous orbit around the Earth
(3) A car skidding to a stop on a horizontal road

EXAMPLE 6.1 Towing a Car

A tow truck pulls a car of mass 1200 kg to a repair shop several miles away. As part of this journey, the tow truck and car must travel on a highway, and when entering the

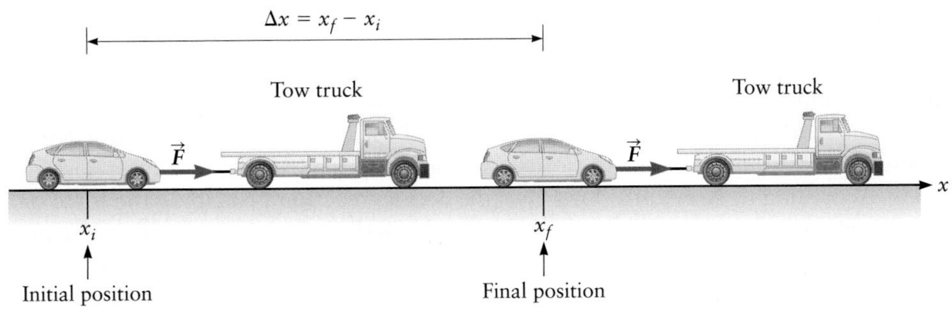

◀ **Figure 6.4** Example 6.1. The work done in towing this car along a straight road is equal to $F\,\Delta x$.

highway they accelerate uniformly as they merge into traffic. They have an initial speed of 10 m/s when they enter the merge ramp and a final speed of 30 m/s when they leave the merge lane and enter the highway. The length of the merge lane is 20 m. Find the work done on the car.

RECOGNIZE THE PRINCIPLE

To calculate W, we need to know the displacement of the car, the force on the car, and the angle between the force and the displacement. The displacement is equal to the length of the merge lane, whereas the force is directed along the lane; hence, $\theta = 0$ in Equation 6.4. We can obtain the force on the car from the acceleration together with Newton's second law.

SKETCH THE PROBLEM

Figure 6.4 shows the problem. The car and tow truck have a constant acceleration along the direction of the road. The displacement and the force are both directed along the merge lane, which we can take as the x direction.

IDENTIFY THE RELATIONSHIPS

To calculate the acceleration, we use the relations for motion with constant acceleration from Chapter 3. We denote the initial velocity as v_i and the initial position as x_i. From Equation 3.4, we have

$$v_f^2 = v_i^2 + 2a(x_f - x_i)$$

The initial speed is given as $v_i = 10$ m/s and the final speed as $v_f = 30$ m/s, and the displacement during this time is $(x_f - x_i) = 20$ m. Solving for the acceleration, we get

$$a = \frac{v_f^2 - v_i^2}{2(x_f - x_i)} = \frac{(30\text{ m/s})^2 - (10\text{ m/s})^2}{2(20\text{ m})} = 20\text{ m/s}^2$$

We now use Newton's second law to find the force needed to produce this acceleration:

$$\sum F = ma = (1200\text{ kg})(20\text{ m/s}^2) = 2.4 \times 10^4\text{ N}$$

Here the total force is equal to the total horizontal force, which we denote by F.

SOLVE

The work done on the car is equal to F times the displacement:

$$W = F\,\Delta x = F(x_f - x_i) = (2.4 \times 10^4\text{ N})(20\text{ m})$$

$$W = 4.8 \times 10^5\text{ N}\cdot\text{m} = \boxed{4.8 \times 10^5\text{ J}}$$

Notice that the units of work are joules (J).

▶ ***What does it mean?***

This result is the work done by the tow truck on the car. Notice that W is positive and that the car's final speed is greater than its initial speed.

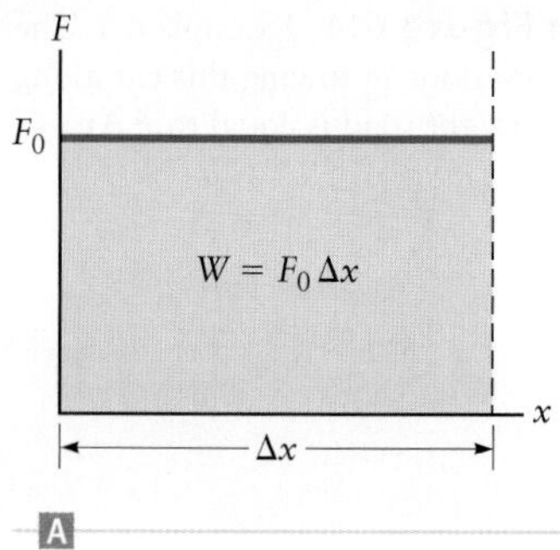
F
F0
W = F0 Δx
x
Δx
A

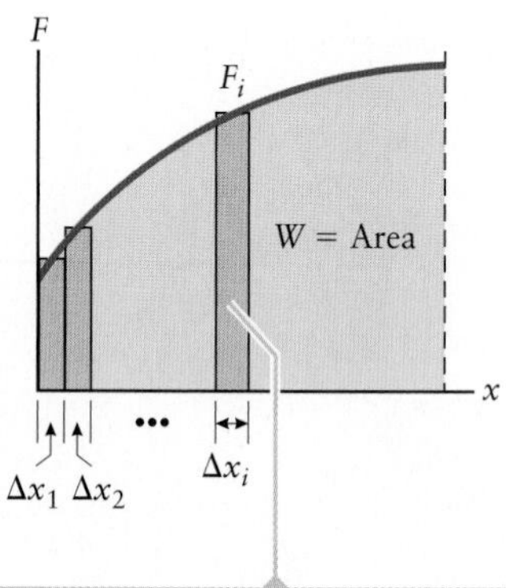
F
Fi
W = Area
x
Δx1 Δx2
Δxi
The work done during each small step Δx_i is equal to $F\Delta x_i$, which is just the area of a shaded box.
The total work done is the total area under the curve.

▲ **Figure 6.5** Work is equal to the area under a graph of force versus displacement. A The force is constant and equals the area of the shaded rectangle. B The force is not constant, but the work done by this force is still equal to the area under the curve of F versus x.

What Does the Work?

In Example 6.1, we found that a certain amount of work $W = 4.8 \times 10^5$ J was done, and we could say that the tow truck does this work *on the car*. We might also ask about the work done *by the car on the tow truck*. The work done on the tow truck is also given by Equation 6.4, where F is now the force exerted by the car on the truck. According to Newton's third law (the action–reaction principle), the force on the tow truck is equal in magnitude and opposite in direction to the force on the car. The force on the truck is thus opposite to its displacement, so the work done on the truck is *negative*. In general, when an agent applies a force to an object and does an amount of work W *on that object*, the object will do an amount of work equal to $-W$ "back" *on the agent*.

We can use the same ideas to discuss the work done in cases in which several forces, from different agents, act on an object. It is often useful to talk about the work done by each separate agent, that is, the work done separately by each force, which can be calculated using Equation 6.4, with F being the force exerted by the particular agent of interest.

Graphical Analysis and Work Done by a Variable Force

In our discussions of work, we have so far assumed the force is constant. To see how to deal with situations in which the force is not constant, it is useful to plot the force as a function of the displacement. When F is constant (Fig. 6.5A), this graph is simply a horizontal line. For simplicity, let's assume we have a case of one-dimensional motion, so the force and displacement are both along the x axis. The work done is then equal to the force times the displacement, and W is equal to the area under this force–displacement plot. Now consider a case in which the force is not constant as sketched in Figure 6.5B. We can (mathematically) think of the motion as a sequence of displacements Δx_1, Δx_2, Δx_3, and so forth, with the force approximately constant during each of these displacements. For each displacement, we can calculate the work using Equation 6.3 and then sum the results to get the total work done. We can see from Figure 6.5B that this process is equivalent to calculating the area under the entire force–displacement curve. Hence, when dealing with a nonconstant force, the work W is equal to the area under the force–displacement graph. In practice, this area can be estimated by dividing the area under the force–displacement graph into a series of rectangles as in Figure 6.5B and then computing the areas of these rectangles.

6.2 Kinetic Energy and the Work–Energy Theorem

In our derivation of Equation 6.2, we assumed the object started from rest so that its initial velocity was $v_i = 0$. Let's consider that problem again, but without making this assumption. Returning to Equation 6.1, we have

$$v_f^2 = v_i^2 + 2a\Delta x = v_i^2 + 2a(x_f - x_i) \tag{6.5}$$

where the initial velocity v_i is not zero and the displacement is $\Delta x = x_f - x_i$, with x_f being the final position. Notice again that we are using the subscript i to denote the initial position and initial velocity and subscript f to denote the final values. Rearranging Equation 6.5 gives

$$a(x_f - x_i) = \frac{v_f^2 - v_i^2}{2} \tag{6.6}$$

We now want to calculate the work done on the object as it moves from the initial position x_i to the final position x_f. Multiplying both sides of Equation 6.6 by m and using the definition of work $W = F\Delta x$ (Equation 6.3) leads to

$$W = ma(x_f - x_i) = m\left(\frac{v_f^2 - v_i^2}{2}\right)$$

$$W = \tfrac{1}{2}mv_f^2 - \tfrac{1}{2}mv_i^2 \tag{6.7}$$

The expression $\frac{1}{2}mv^2$ that appears on the right-hand side of Equation 6.7 is an *extremely* important quantity. When an object of mass m has speed v, it has energy $\frac{1}{2}mv^2$ due to its motion. We call this ***kinetic energy*** (*KE*):

$$KE = \tfrac{1}{2}mv^2 \tag{6.8}$$

Kinetic energy

"Energy" is a new concept for us in mechanics, and we can use Equation 6.7 to gain some understanding of it. Equation 6.7 tells us that the kinetic energy of an object can be changed by doing work on the object. If we write the change in kinetic energy as ΔKE, Equation 6.7 can be written

$$W = \tfrac{1}{2}mv_f^2 - \tfrac{1}{2}mv_i^2 = KE_f - KE_i$$

$$W = \Delta KE \tag{6.9}$$

Work–energy theorem

The work done on an object is thus equal to the *change* in its kinetic energy. This relation tells us how work, and hence also force and displacement, are connected to the kinetic energy of an object. Equation 6.9 is called the ***work–energy theorem.*** According to this result, the units of work and energy are the same, so energy is measured in joules. Another commonly used unit of energy is the calorie, with 1 cal equal to 4.186 J.

EXAMPLE 6.2 Kinetic Energy of an Asteroid

Asteroids are constantly bombarding the Earth and often explode as they heat up while passing through the atmosphere. Consider an asteroid with radius $r = 5$ m and a mass of 5×10^6 kg. Such asteroids have a speed of about $v = 10^4$ m/s when they enter the Earth's atmosphere. What is the kinetic energy of this asteroid? Is that a lot of energy?

RECOGNIZE THE PRINCIPLE

The mass and speed of the asteroid are given, so we simply need to evaluate its kinetic energy using Equation 6.8.

SKETCH THE PROBLEM

No figure is necessary.

IDENTIFY THE RELATIONSHIPS AND SOLVE

The kinetic energy is given by

$$KE = \tfrac{1}{2}mv^2$$

Inserting the given values of m and v, we find

$$KE = \tfrac{1}{2}mv^2 = \tfrac{1}{2}(5 \times 10^6)(10^4 \text{ m/s})^2 = \boxed{3 \times 10^{14} \text{ J}}$$

Here we have kept only one significant figure in our answer because the initial values were only given to that accuracy. This is a lot of energy; it is about five times the energy released by the atomic bomb that was dropped on Hiroshima, Japan, in 1945.

(continued) ▶

▸ ***What does it mean?***

The hypothetical asteroid we have considered here is not very large—about the size of a small bus—but it carries a *lot* of kinetic energy and can thus do a lot of damage. Data obtained from the U.S. Air Force Defense Support Program suggest that an asteroid explosion with this much energy occurs more than once a year! Fortunately, such explosions usually take place in the upper atmosphere and do little damage on the Earth.

EXAMPLE 6.3 Kinetic Energy of a Falling Object

A rock of mass m is dropped from the top of a tall building of height h. Find (**a**) the kinetic energy and (**b**) the speed of the rock just before it reaches the ground.

RECOGNIZE THE PRINCIPLE

The only force on the rock is the gravitational force, and as shown in the coordinate system of Figure 6.6, this force and the rock's displacement are both along the $-y$ direction. We have $F_{grav} = -mg$ where the negative sign indicates that the direction of this force is downward. This is a case of one-dimensional motion with the force parallel to the displacement, so we can apply Equation 6.3 to find the work done on the rock. We can then use the work–energy theorem to find the final kinetic energy and speed of the rock.

SKETCH THE PROBLEM

Figure 6.6 shows the problem. We have also drawn in a coordinate system along with a free-body diagram for the rock.

IDENTIFY THE RELATIONSHIPS

The initial position of the rock is at $y_i = h$ and the final location is at ground level, $y_f = 0$, so the displacement of the rock is

$$\Delta y = y_f - y_i = 0 - h = -h$$

The work done on the rock is equal to the force times the displacement (Eq. 6.3):

$$W = F_{grav}\,\Delta y = (-mg)(-h) = mgh$$

Using the work–energy theorem (Eq. 6.9), we can write

$$W = \Delta KE = KE_f - KE_i$$

SOLVE

(**a**) The rock is initially at rest ($v_i = 0$), so its initial kinetic energy is $KE_i = \frac{1}{2}mv_i^2 = 0$, which leads to

$$W = mgh = \Delta KE = KE_f - KE_i = KE_f - 0$$

$$KE_f = \boxed{mgh} \quad (1)$$

Since the force and displacement are parallel (both downward), W is positive.

(**b**) The final speed v_f of the rock is related to its final kinetic energy by (Eq. 6.8):

$$KE_f = \tfrac{1}{2}mv_f^2$$

Inserting the result for KE_f from Equation (1) and solving for v_f gives

$$mgh = \tfrac{1}{2}mv_f^2$$

$$v_f^2 = 2gh$$

$$v_f = \boxed{\sqrt{2gh}} \quad (2)$$

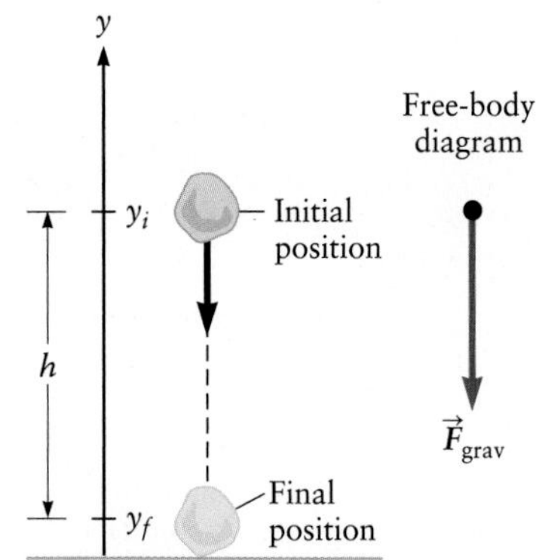

▲ **Figure 6.6** Example 6.3. When a rock falls vertically, the gravitational force is parallel to the rock's displacement. The work done by gravity on the rock is thus positive, with $W = mgh$.

▶ ***What does it mean?***

These results could also have been obtained using the methods for dealing with forces and acceleration that we developed in Chapter 3. Because the acceleration is constant, the displacement and velocity are related by (Eq. 3.4)

$$v_f^2 = v_i^2 + 2a(y_f - y_i) \tag{3}$$

The acceleration of the rock is $a = -g$, and (as we have already noted) $v_i = 0$, $y_i = h$, and $y_f = 0$. Inserting into Equation (3) gives

$$v_f = \sqrt{2gh}$$

This result is precisely the same as we found using the work–energy approach (Eq. 2). For this particular problem, both approaches are straightforward to carry out, but we'll soon encounter cases that are much easier to solve with the work–energy method.

CONCEPT CHECK 6.2 Dependence of the Kinetic Energy on Mass and Speed

Consider a rock of mass m traveling at speed v. Suppose the mass is doubled and the speed is cut in half. Which of the following statements is true?

(a) The kinetic energy is the same.
(b) The kinetic energy increases.
(c) The kinetic energy decreases.

EXAMPLE 6.4 Speed of a Skier

Consider a person of mass m skiing down an extremely icy, frictionless ski slope. This specially constructed ski slope has the shape of an inclined plane of angle α as shown in Figure 6.7. If the skier starts from an initial height h, what is the work done by gravity on the skier while she skis to the bottom of the slope? Use this result to find the speed of the skier when she reaches the bottom, assuming she starts from rest.

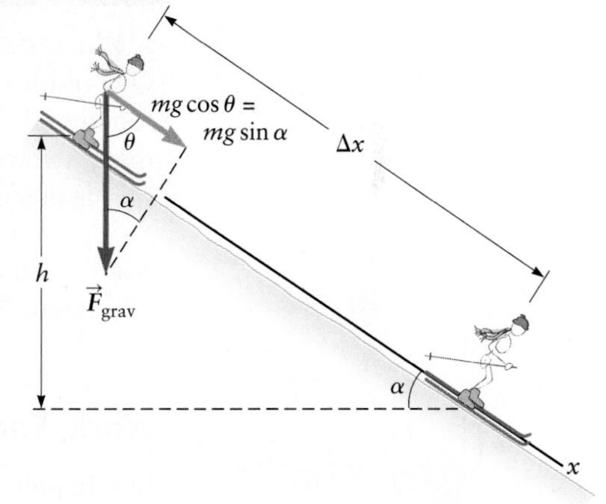

RECOGNIZE THE PRINCIPLE

From Equation 6.4 and Figure 6.2, the work done by gravity on the skier equals the component of the gravitational force along the slope (parallel to the displacement) multiplied by the displacement. The force makes an angle θ with the displacement (see Fig. 6.7). This angle is *not* the same as the angle of the incline α.

▲ **Figure 6.7** Example 6.4. The gravitational force is not parallel to this skier's displacement. The work done by gravity on the skier is equal to the component of the gravitational force along the slope multiplied by the skier's displacement.

SKETCH THE PROBLEM

Figure 6.7 shows the problem. Because the motion of the skier is along the incline, we take the coordinate axis to be parallel to the slope, with the positive x direction downward, parallel to the displacement of the skier.

IDENTIFY THE RELATIONSHIPS

The work done by the gravitational force is (Eq. 6.4)

$$W = F_{\text{grav}}\, \Delta x \cos\theta \tag{1}$$

The angle θ the force makes with the direction of the skier's displacement is related to the angle of the slope by $\theta = 90° - \alpha$ because they are two of the interior angles of a

(continued) ▶

right triangle. The displacement of the skier Δx is the distance traveled along the slope. From the trigonometry in Figure 6.7, we have

$$\frac{h}{\Delta x} = \sin \alpha$$

$$\Delta x = \frac{h}{\sin \alpha}$$

SOLVE

Inserting this result for Δx into Equation (1) and using $\theta = 90° - \alpha$, we find

$$W = F_{\text{grav}}\, \Delta x \cos \theta = mg\left(\frac{h}{\sin \alpha}\right)\cos(90° - \alpha)$$

This result can be simplified if we use the trigonometric identity $\cos(90° - \alpha) = \sin \alpha$:

$$W = mg\frac{h}{\sin \alpha}\cos(90° - \alpha) = mg\frac{h}{\sin \alpha}\sin \alpha = \boxed{mgh} \qquad (2)$$

To find the final speed of the skier, we use the work–energy theorem and that her initial speed and hence also her initial kinetic energy are zero:

$$W = KE_f - KE_i = \tfrac{1}{2}mv_f^2$$

Using our result for W (= mgh) and solving for v_f, we find

$$W = mgh = \tfrac{1}{2}mv_f^2$$

$$v_f = \boxed{\sqrt{2gh}}$$

► ***What does it mean?***

Our result for the work W done by gravity in Equation (2) has *precisely* the same functional form as the work done on the rock in Example 6.3. The work done by the gravitational force on a rock that falls vertically through a height h (Example 6.3) is precisely equal to the work done by gravity on a skier who moves along a slope through the same vertical height h. In the next section, we'll see that this result is no accident and that it has some profound consequences.

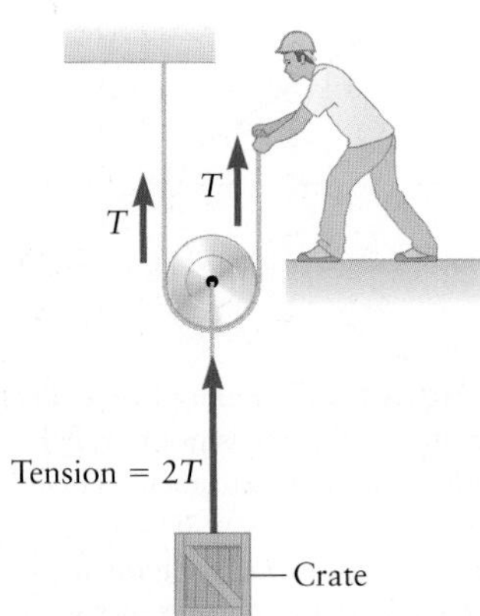

▲ **Figure 6.8** The person applies a force T to the rope. The block and tackle amplifies this force, and the total force applied to the crate by the rope is $2T$. However, when he moves the end of the rope a distance L, the crate moves a distance of only $L/2$.

Work, Energy, and Amplifying Forces

In Chapter 3, we discussed the block-and-tackle device in Figure 6.8, and showed that it amplifies forces by a factor of two. If a person applies a force F to the rope, the tension in the rope is $T = F$. Because the pulley is suspended by two portions of the rope, the upward force on the pulley is $2T$ and the pulley exerts a total force $2T$ on the object to which it is connected, that is, the crate in Figure 6.8. Let's now consider how this process of force amplification affects the work done by the person. If the person lifts his end of the rope through a distance L, the pulley will be raised by half that amount, that is, a distance of $L/2$. You can see why by noticing that when the person pulls his end of the rope up a distance L, the rope on the left side of the pulley will be shortened by an amount $L/2$, and this is also the distance that the pulley moves upward.

When the pulley moves upward by a distance $L/2$, the crate is displaced by the same amount. The work done *by the pulley on the crate* is equal to the total force of the pulley on the crate ($2T$) multiplied by the displacement of the crate, which is $L/2$:

$$W_{\text{on crate}} = 2T(L/2) = TL$$

At the same time, the work done *by the person on the rope* is equal to the force that he exerts on the rope (T) multiplied by the displacement of the rope, which is L, so

$$W_{\text{on rope}} = FL = TL$$

Thus, the work done on the rope is precisely equal to the work done on the crate.

There are several important messages from this example.

1. The work done by the person is effectively "transferred" to the crate.
2. Forces can be amplified, but work cannot be increased in this way. The force exerted by the person is increased by the block and tackle such that the force on the crate is amplified by a factor of two. However, the work is *not* amplified.
3. We'll see that this result—that work cannot be amplified—is a consequence of the principle of conservation of energy.

These points are all in accord with our observation in Chapter 3 that you "cannot get something for nothing." There, we also observed that pulleys can amplify a force, but they always do so by decreasing the associated displacement. Now we can go a bit further. According to the work–energy theorem, $W = \Delta KE$, which suggests that we can "convert" or "trade" work to kinetic energy and vice versa. The fact that work cannot be amplified suggests that this trade does not increase the amount of energy that is available. These qualitative ideas are leading us to a very important fundamental principle of physics, the principle of conservation of energy.

6.3 Potential Energy and Conservation of Energy

In Examples 6.3 and 6.4, we calculated the work done by the gravitational force in two different situations. In one case (Example 6.3), described by the path in Figure 6.9A, a rock of mass m fell vertically from a height h to ground level ($y = 0$), and we found that the work done by gravity on the rock is $W = mgh$. In the second case (Example 6.4), a skier of mass m skied down a slope, starting from an initial height h above the bottom of the slope, and we found that the work done by gravity is given by $W = mgh$. So, the work done by gravity was precisely the same, even though the rock and the skier followed very different paths. In fact, when an object of mass m follows *any* path that moves through a vertical distance h, the work done by the gravitational force is *always* equal to $W = mgh$. This amazing fact is connected with the notion of ***potential energy***.

Figure 6.9 shows several different paths that might be followed by an object as it moves between the same initial and final positions. The work done by the gravitational force as an object moves from a particular initial location (i) to a particular final location (f) is completely *independent of the path* the object takes to move from i to f. This statement is true for all conceivable paths and can be understood as follows. An object near the Earth's surface has a potential energy (PE) that depends only on the object's height h. The PE depends only on where the object is located and does not depend on how the object got there. This potential energy is a result of the gravitational force between the Earth and the object. Strictly speaking, both the force and the potential energy are properties of the "system" that is composed of the Earth and the object. Although we will often speak of the gravitational potential energy of an object, it should be understood that this PE is a property of the object and Earth together. A general feature of potential energy is that it is always a property of the *system* of particles or objects that interact through the underlying force.

The potential energy associated with a particular force is related to the work done by that force on an object as the object moves from one position to another. If this work is W, the change in the potential energy ΔPE is defined as

$$\Delta PE = PE_f - PE_i = -W \tag{6.10}$$

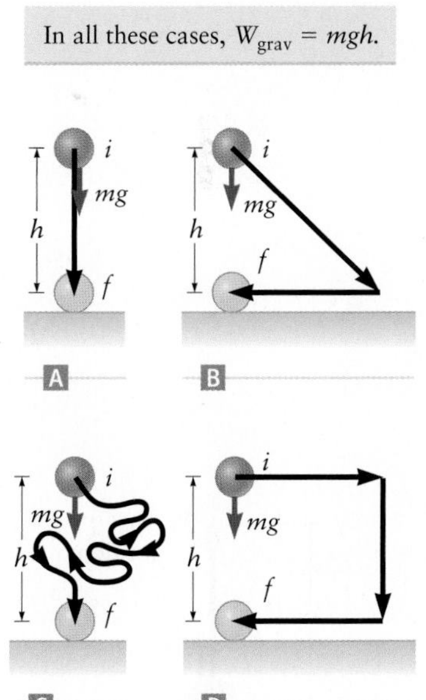

▲ **Figure 6.9** The work done by gravity is independent of the path taken. The initial and final points are the same in parts A, B, C, and D, so the work done is the same in all four cases. In addition, the work done by gravity on objects near the Earth's surface, W_{grav}, depends only on the change in height h.

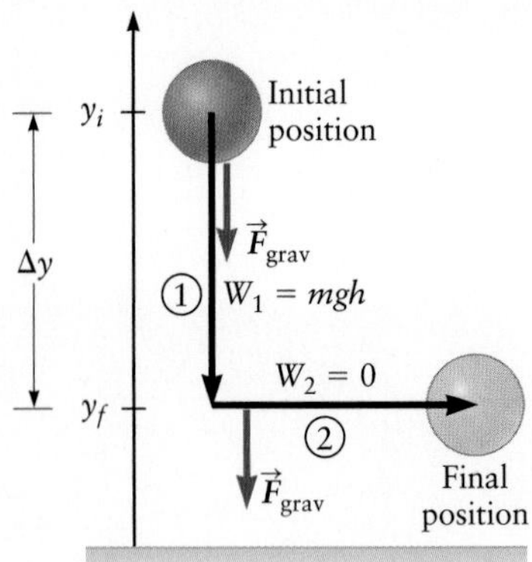

▲ **Figure 6.10** The work done by gravity along part 1 of this path is $W_1 = mgh$, whereas the work along part 2 is $W_2 = 0$ because the gravitational force and the object's displacement are perpendicular along this part of the path.

Since W is a scalar, potential energy is also a scalar. For the case of the gravitational force, we can thus find the change in the potential energy by calculating the work done in moving from an initial height y_i to a final height y_f. The work done is independent of the path, so we might as well use the simple path sketched in Figure 6.10 for the calculation of W. This path moves from our initial point to the final point in two steps. In the first step, we start at y_i and move downward (vertically) to the height y_f, while the second step moves horizontally to the final point. The work done by gravity during the second step is zero because the gravitational force and the object's displacement are perpendicular. We are thus left only with the work done during the first step, which is

$$W = F_{grav}\,\Delta y$$

The force is $F_{grav} = -mg$ (since the gravitational force is downward, along the $-y$ direction), and the displacement is $\Delta y = y_f - y_i$, so

$$W = F_{grav}\,\Delta y = (-mg)(y_f - y_i) \tag{6.11}$$

Using our relation for potential energy in Equation 6.10, the change in the potential energy is

$$\Delta PE = PE_f - PE_i = -W$$

$$\Delta PE = mg(y_f - y_i) \tag{6.12}$$

Another way to write this result is to say that the initial potential energy is $PE_i = mgy_i$ and the final potential energy is $PE_f = mgy_f$. Or, we could simply say that the gravitational potential energy of the object when it is at a height y is

Gravitational potential energy for an object near the Earth's surface

$$PE_{grav} = mgy \tag{6.13}$$

Our derivation of Equation 6.13 used the fact that $F_{grav} = -mg$, so this result for the gravitational potential energy applies only for objects near the surface of the Earth. In Section 6.4, we'll consider the gravitational potential energy of planets, moons, and other objects in the solar system.

Potential Energy Is Stored Energy

The result for the gravitational potential energy in Equation 6.13 tells us that the potential energy increases as the object is moved higher and decreases when the object is moved lower, which agrees with our intuition. If an object is moved to a greater height, it is capable of producing a greater kinetic energy if it were to fall back to a lower height, where its potential energy is smaller. This example also shows that potential energy is *stored energy*. By "allowing" an object to fall, the energy stored as gravitational potential energy can be converted to kinetic energy. We can increase the gravitational PE of an object by increasing its height as sketched in Figure 6.11. As the object moves higher, the work done *by gravity* is negative because the gravitational force is directed downward while the displacement of the object is upward. Now, the object moves higher because there is another force acting on it. We call this an "external" force F_{ext}, and it might be produced by your hand as you lift the object. This external force is directed upward, so the work done by F_{ext} on the object is positive. In fact, the work done by this external force, W_{ext}, is equal in magnitude but opposite in sign compared to the work done by gravity. Hence, we have stored an amount of energy W_{ext} as gravitational potential energy. We can recover this energy by letting the object fall back down to its initial height, thus gaining kinetic energy.

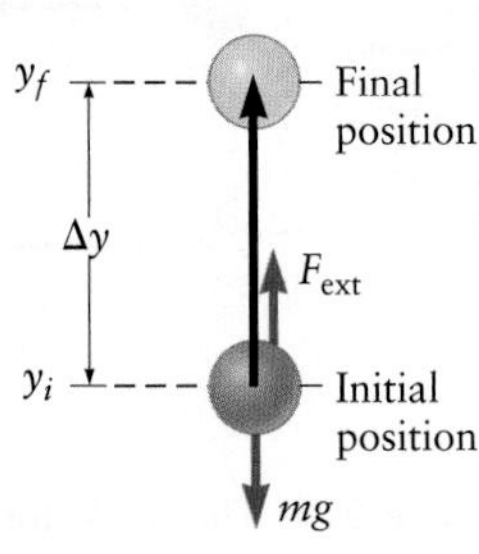

▲ **Figure 6.11** An external agent (perhaps a person's hand) lifts an object by exerting a force F_{ext} on the object. This agent does an amount of work $W_{ext} = F_{ext}\,\Delta y$ on the object. This energy is stored as potential energy of the object.

Potential Energy and Conservative Forces

We have seen that the work done by the gravitational force depends only on the initial and final locations and not on the path taken to get from one point to the other. We say that the work done by gravity is *independent of the path* taken. This property allowed us to define a potential energy function that depends only on the height of the object being studied. Many other forces besides gravity have the property that the work done is independent of the path, and they all have a corresponding potential energy function. Such forces are called ***conservative forces.*** All conservative forces can be used to store energy as potential energy. Other types of forces do not have potential energy functions and are called ***nonconservative forces***. We'll spend the rest of this section learning how to apply and understand the gravitational potential energy in Equation 6.13. We then explore how these ideas apply to other types of conservative and nonconservative forces in subsequent sections.

We have introduced several new and important concepts connected with potential energy, and it is worth summarizing them now.

Key Facts about Potential Energy and Conservative Forces

- Potential energy is a result of the force(s) that act on an object. Since forces always come from the interaction of two objects, *PE* is a property of the objects (the "system") involved in the force.
- Potential energy is energy that an object or system has by virtue of its *position.*
- Potential energy is *stored energy*; it can be converted to kinetic energy.
- Potential energy is a scalar; its value can be positive, negative, or zero, but it does not have a direction.
- For forces that are associated with a potential energy (and hence have a potential energy function), the work done is independent of the path taken. Such forces are called *conservative forces.*
- Some forces, such as friction and air drag, are *nonconservative forces*. These forces do not have potential energy functions and cannot be used to store energy. The work done by a nonconservative force depends on the path taken by the object of interest.

Potential Energy, the Work–Energy Theorem, and Conservation of Energy

Let's now reexamine the work–energy theorem and consider how potential energy fits into our ideas about work and energy. We first wrote the work–energy theorem as (compare with Eq. 6.9)

$$W = \Delta KE = KE_f - KE_i \tag{6.14}$$

Here, W is the work done by *all* the forces acting on the object of interest. In many cases, some of or all these forces are associated with a potential energy. If so, the potential energy associated with a particular force is related to work done by that force, with $\Delta PE = -W$ (Eq. 6.10). If *all* the work is done by conservative forces such as gravity, then in Equation 6.14 we can replace the work done by that force with the negative of the change in the potential energy, and we have

$$W = -\Delta PE = -(PE_f - PE_i) = KE_f - KE_i$$

which can be rearranged to read

$$KE_i + PE_i = KE_f + PE_f \tag{6.15}$$

Conservation of mechanical energy

This version of the work–energy theorem applies to situations in which all the forces are conservative forces, and it contains potential energy and kinetic energy together. In words, Equation 6.15 says that the sum of the potential and kinetic energies of an object is *conserved*. That is, the sum of the potential and kinetic energies when the object is at its initial location (i) is equal to the sum of the potential and kinetic energies at the final location (f). This sum of the potential and kinetic energies is called the ***total mechanical energy***.

Equation 6.15 is thus a statement of the ***conservation of mechanical energy***. In our derivation, we assumed only conservative forces are acting on the object of interest, so that all the forces involved possess a potential energy function. Frictional forces do not satisfy this requirement, and we'll learn how to deal with them in Section 6.5. Nevertheless, in many situations all the forces in a problem are connected with potential energy. In these cases, the principle of conservation of mechanical energy in Equation 6.15 is a very powerful tool for understanding, analyzing, and predicting motion.

Conservation of Mechanical Energy and the Speed of a Snowboarder

Let's apply the conservation of energy relation in Equation 6.15 to analyze the motion of a snowboarder who is sliding down an extremely icy (frictionless) hill (Fig. 6.12). If she starts from a height $h = 20$ m above the bottom of the hill with initial speed $v_i = 15$ m/s, how fast will she be traveling at the bottom of the hill?

The only forces acting on the snowboarder are the force of gravity and the normal force from the hill. (Recall that the hill is very slippery and the frictional force is zero.) The normal force is perpendicular to the displacement of the snowboarder, so it does no work on her. The only remaining force is gravity, which is a conservative force, so the mechanical energy of the snowboarder is conserved. As the snowboarder moves down the hill, her potential energy is converted to kinetic energy, resulting in a larger velocity at the bottom of the hill. Applying Equation 6.15, we have

$$KE_i + PE_i = KE_f + PE_f$$

$$\tfrac{1}{2}mv_i^2 + mgy_i = \tfrac{1}{2}mv_f^2 + mgy_f \qquad (6.16)$$

where v_f is her velocity at the bottom of the hill and y_f is her height at the bottom. If we place the origin of the y axis at the bottom of the hill, then $y_i = h$ and $y_f = 0$. Inserting these values into Equation 6.16 and solving for v_f, we find

$$\tfrac{1}{2}mv_i^2 + mgh = \tfrac{1}{2}mv_f^2 + mg(0)$$

$$\tfrac{1}{2}v_i^2 + gh = \tfrac{1}{2}v_f^2$$

$$v_f = \sqrt{v_i^2 + 2gh} = \sqrt{(15 \text{ m/s})^2 + 2(9.8 \text{ m/s}^2)(20 \text{ m})} = 25 \text{ m/s}$$

The final speed v_f does *not* depend on the angle of the hill; it only depends on the height of the hill. That is because the gravitational potential energy depends only on the snowboarder's height, and changes in this potential energy are independent of the path she follows. So, the final speed will be the same for the two different hills in parts A and B of Figure 6.12, provided they have the same height.

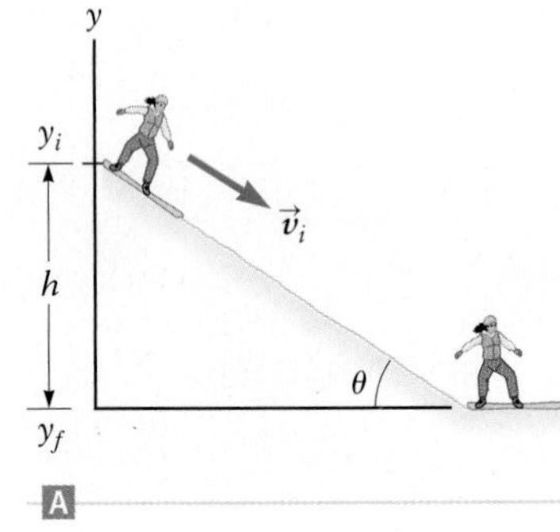

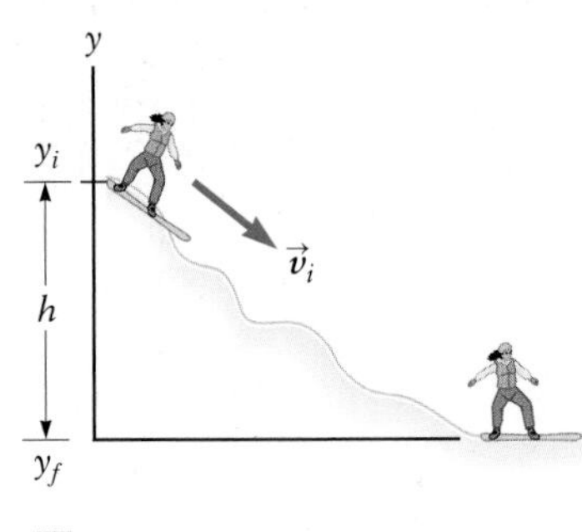

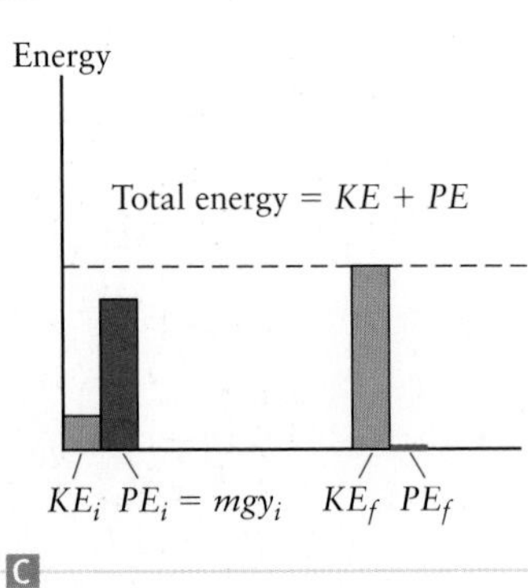

▲ **Figure 6.12** A and B These two hills have the same height h, so even though they have very different shapes, the work done by gravity in the two cases is the same: $W[\text{hill A}] = W[\text{hill B}]$. The change in the potential energy is therefore the same in the two cases ($\Delta PE_A = \Delta PE_B$). C These bar charts show the initial and final kinetic and potential energies of the snowboarder.

Charting the Energy

A useful way to illustrate the conservation of energy is with bar charts. The bar charts in Figure 6.12C show the kinetic and potential energies of the snowboarder at the start and the end of the slope. These charts show how the initial potential energy of the snowboarder is converted to kinetic energy. Notice also how the sum of the kinetic and potential energies is the same at the start and end, which is just another way to represent the conservation of energy.

Why Is the Principle of Conservation of Energy Useful?

Using conservation of energy principles made the calculation with the snowboarder in Figure 6.12 very straightforward. For the hill with the simple shape in Figure 6.12A, we could have done this problem in Chapter 4 by applying Newton's second law ($\sum \vec{F} = m\vec{a}$). Because this hill is just a ramp, the component of the gravitational force along the hill is constant; the acceleration is therefore also constant, so we could have used our relations for motion with constant acceleration to find the final velocity. We did not really need to use conservation of energy principles as expressed by Equation 6.15 to solve this problem. However, most snowboard hills do not have such a simple shape. The hill in Figure 6.12B has a much more interesting and realistic shape, as it undulates several times as the snowboarder travels to the bottom. Because the hill's angle is not constant, the acceleration is definitely not constant and the methods we employed in Chapter 4 will not work here. However, this complicated hill can easily be handled with the conservation of energy approach. To find the final velocity of the snowboarder, we only need to know the height of the hill; the shape of the hill does not have any effect on the final velocity, and our calculation thus applies to any hill, regardless of its shape. This example illustrates the power of the conservation of energy approach; this approach will allow us to tackle many problems that cannot be handled using the methods of Chapter 4.

PROBLEM SOLVING Applying the Principle of Conservation of Mechanical Energy

Most applications of conservation of energy ideas can be attacked using the following steps.

1. **RECOGNIZE THE PRINCIPLE.** Start by finding the object (or system of objects) whose mechanical energy is conserved.
2. **SKETCH THE PROBLEM,** showing the initial and final states of the object. This sketch should also contain a coordinate system, including an origin, with which to measure the potential energy.
3. **IDENTIFY THE RELATIONSHIPS.** Find expressions for the initial and final kinetic and potential energies. One or more of these energies may involve unknown quantities.
4. **SOLVE.** Equate the initial and final mechanical energies and solve for the unknown quantities.
5. Always *consider what your answer means* and check that it makes sense.

This approach can only be used when mechanical energy is conserved, that is, when all the forces that do work on an object are conservative forces. We'll consider how to extend this approach to deal with nonconservative forces such as friction in Section 6.5.

EXAMPLE 6.5 Roller-Coaster Physics

Consider the roller coaster in Figure 6.13. The car starts at position A with an initial height $h_A = 20$ m and speed $v_A = 15$ m/s. Find (**a**) the speed of the roller coaster when it reaches the top of the track at position B, where $h_B = 25$ m, and (**b**) the speed of the roller coaster when it reaches the bottom of the track, where $h_C = 0$. (**c**) Make a qualitative sketch of the roller-coaster car's kinetic energy and potential energy as a function of position along the track. Also make a corresponding energy bar chart for the system at points A, B, and C. Ignore the rotational motion of the wheels and assume all frictional drag on the roller coaster is negligible.

RECOGNIZE THE PRINCIPLE

We follow the steps outlined in the problem-solving strategy on conservation of energy. Step 1: The object of interest is the roller-coaster car; its mechanical energy is conserved

(continued) ▶

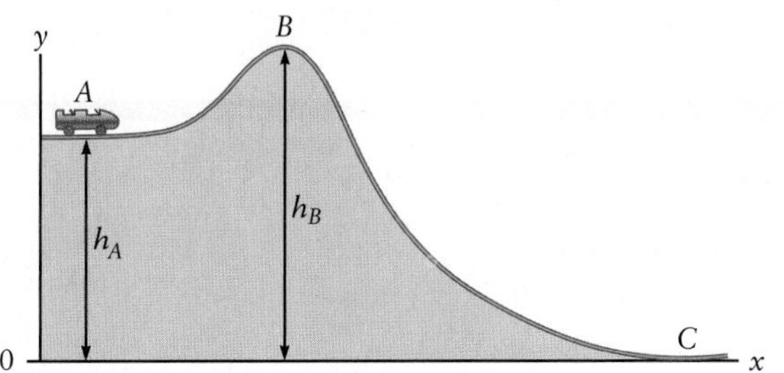

◀ **Figure 6.13** Example 6.5.

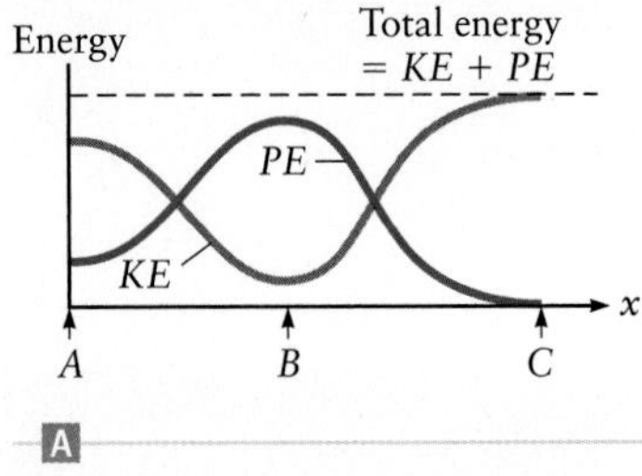

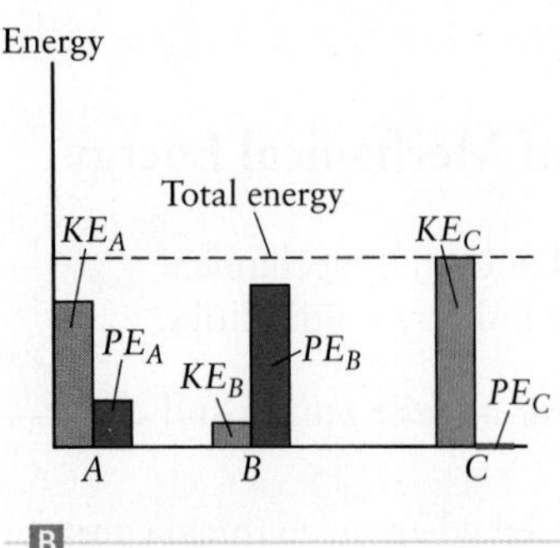

▲ **Figure 6.14** Behavior of the kinetic and potential energies of the roller-coaster car in Example 6.5.

because the only force that does work on the car is gravity, which is a conservative force.

SKETCH THE PROBLEM

Step 2: Figure 6.13 shows the initial position of the car along with the other positions of interest (points B and C). We take the origin of the y axis to be at the bottom of the track.

IDENTIFY THE RELATIONSHIPS

Step 3: We next apply the conservation of energy relation (Eqs. 6.15 and 6.16) and then (step 4) solve for the speed at B and C.

SOLVE

(a) Applying the conservation of energy relation (Eq. 6.15) at locations A and B gives

$$KE_A + PE_A = KE_B + PE_B$$
$$\tfrac{1}{2}mv_A^2 + mgh_A = \tfrac{1}{2}mv_B^2 + mgh_B$$

We now see that the mass of the roller coaster cancels. Solving for v_B, we find

$$\tfrac{1}{2}v_A^2 + gh_A = \tfrac{1}{2}v_B^2 + gh_B$$
$$v_B^2 = v_A^2 + 2gh_A - 2gh_B = v_A^2 + 2g(h_A - h_B)$$
$$v_B = \sqrt{v_A^2 + 2g(h_A - h_B)}$$

Inserting the given values of v_A, h_A, and h_B leads to

$$v_B = \sqrt{(15 \text{ m/s})^2 + 2(9.8 \text{ m/s}^2)(20 \text{ m} - 25 \text{ m})} = \boxed{11 \text{ m/s}}$$

(b) We again use Equation 6.15, but now we apply it at locations A and C, which leads to

$$v_C = \sqrt{v_A^2 + 2g(h_A - h_C)} = \sqrt{(15 \text{ m/s})^2 + 2(9.8 \text{ m/s}^2)(20 \text{ m} - 0)} = \boxed{25 \text{ m/s}}$$

(c) Figure 6.14A shows the qualitative behavior of the kinetic and potential energies as functions of x, the position along the track, while bar charts showing the kinetic and potential energies at points A, B, and C are given in Figure 6.14B. The kinetic energy is largest at the lowest point on the track (C), while the potential energy is largest when the roller-coaster car is at the highest point (B). The sum $KE + PE$ is the total mechanical energy of the car and is constant throughout.

▸ ***What does it mean?***
The value of the final velocity v_C is independent of the shape of the roller-coaster track and depends only on the initial and final heights. However, this result assumes the roller coaster will actually make it past the top of the track. If the height at location B were made much larger, the roller coaster would not make it to the top of the track and hence would not be able to travel to location C.

CONCEPT CHECK 6.3 Roller Coasters and Conservation of Energy

Consider the roller coaster in Example 6.5. How would you use conservation of energy ideas to calculate the value of h_B for which the roller coaster would barely make it over the highest point of the track?

EXAMPLE 6.6 Driving through a Loop-the-Loop Track

The author's daughter likes to relax by driving fast. One particularly relaxing case involves loop-the-loop tracks like the one shown in Figure 6.15. Although few roads for automobiles are constructed with this shape, it is a common design for roller coasters.

The middle of the track has a circular shape with radius $r = 15$ m, and the car enters and leaves along the horizontal portions at ground level. If a driver wants to coast all the way through without losing contact with the track when she reaches the top, what is the smallest velocity she can have when she enters the track at point A? As was the case with the roller coaster in Figure 6.13, ignore the rotational motion of the car's wheels and assume all friction is negligible.

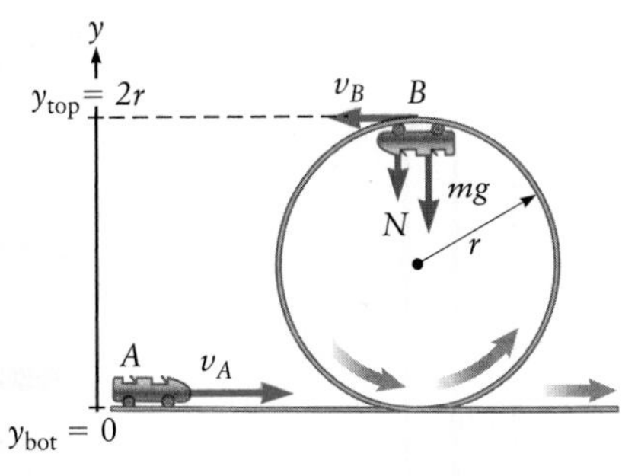

Free-body diagram for roller coaster on top of track

▲ **Figure 6.15** Example 6.6. A car enters this loop-the-loop track from the left at point A and coasts all the way through. The middle portion of the track has a circular shape with radius r.

RECOGNIZE THE PRINCIPLE

The only force that does work on the car is gravity (because the normal force exerted by the track is always perpendicular to the car's path), so the car's mechanical energy is conserved. We therefore follow the problem-solving strategy for problems in which energy is conserved as applied (step 1) to the car. Step 2: Figure 6.15 shows the car at the initial point of its trip (point A). You should suspect that the most dangerous part of the trip will be when the car is upside down at the top of the track (point B), so we take this point as the final position. We can then follow steps 3 and 4 of the problem-solving strategy (see the solution below) to find the speed v_B of the car at the top of the track. However, just knowing v_B will not tell us if the car will stay on the track. For the car to stay on the track at point B, the normal force N exerted by the track on the car must always be greater than zero. If the car has a very high speed at the top of the track, then N will be large. As this speed is reduced, N decreases. Eventually, if v_B is too small, the normal force is zero, signaling that the car has just lost contact. We can use the condition that $N = 0$ to find the minimum velocity at the top and then use conservation of energy principles to find the corresponding velocity at the bottom of the track at point A.

SKETCH THE PROBLEM

Figure 6.15 shows the car at its initial and final positions, along with a free-body diagram corresponding to the final position at B.

IDENTIFY THE RELATIONSHIPS

There are two forces acting on the car at point B—the normal force exerted by the track and the gravitational force—and both are directed downward. For an object undergoing uniform circular motion, the total force must be directed toward the center of the circle and have a magnitude mv^2/r. We thus have

$$\frac{mv_B^2}{r} = N + mg$$

Since the car is just barely staying on the track, the normal force will be very small, so we can take $N = 0$. Solving for v_B then gives

$$v_B = \sqrt{gr}$$

This is the speed at the top that will just keep the car on the track. We now use the conservation of energy condition to work back and find the corresponding speed at the bottom of the track. Using conservation of mechanical energy, we have

$$PE_A + KE_A = PE_B + KE_B$$

$$mgy_A + \tfrac{1}{2}mv_A^2 = mgy_B + \tfrac{1}{2}mv_B^2 \qquad (1)$$

If we take $y = 0$ at the bottom of the track, then $y_A = 0$ and $y_B = 2r$ (see Fig. 6.15).

SOLVE

Inserting these values of y_A and y_B and our result for v_B into Equation (1) gives

$$\tfrac{1}{2}mv_A^2 = mgy_B + \tfrac{1}{2}mv_B^2 = mg(2r) + \tfrac{1}{2}m(gr) = \tfrac{5}{2}mgr$$

(continued) ▶

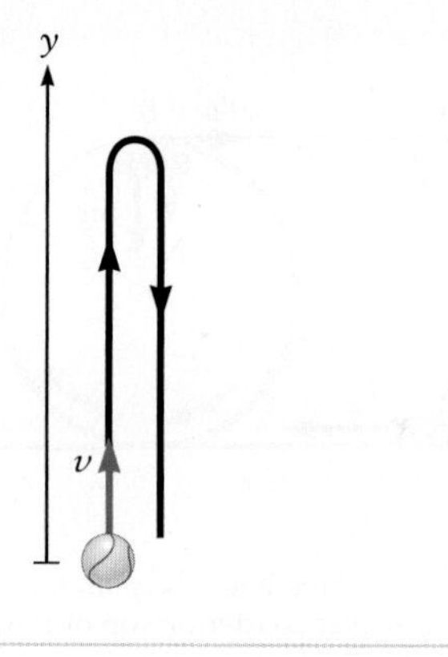

and solving for v_A gives

$$v_A^2 = 5gr$$

$$v_A = \sqrt{5gr}$$

We then insert the value of the radius of the track to get

$$v_A = \sqrt{5gr} = \sqrt{5(9.8 \text{ m/s}^2)(15 \text{ m})} = \boxed{27 \text{ m/s}}$$

▶ ***What does it mean?***

This speed is approximately 60 mi/h, which is a typical speed for a roller coaster. This also suggests that this stunt should be possible for an ordinary automobile, but the author's daughter has (supposedly) not yet attempted it.

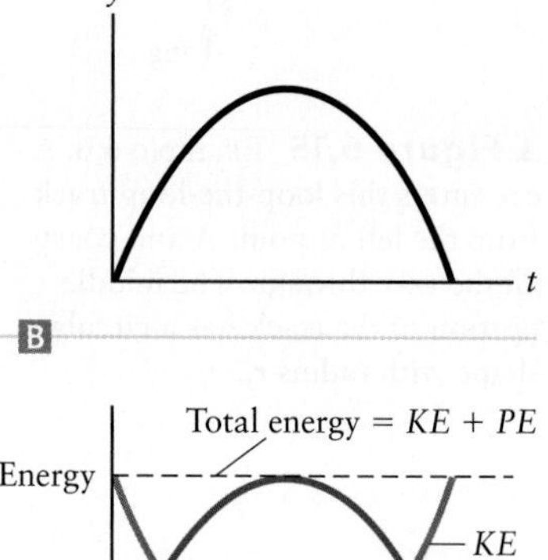

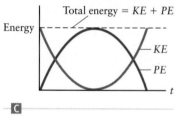

▲ **Figure 6.16** A When a ball is thrown upward (as a simple projectile), the height of the ball varies parabolically with time as sketched in B, the position versus time graph. C The *KE*, *PE*, and total mechanical energy vary with time as shown. Since the total mechanical energy is conserved, the sum $KE + PE$ is constant.

Projectile Motion and Conservation of Energy

It is instructive to apply conservation of energy principles to projectile motion and consider how the kinetic and potential energies vary as an object moves along its trajectory. Figure 6.16A shows an example in which an object, such as a baseball, is thrown upward and then returns to the Earth. When the position y of the ball is plotted as a function of time t, we find the familiar parabolic relation between y and t sketched in Figure 6.16B. The behavior of the ball's energy as a function of time is shown in Figure 6.16C. Since the gravitational potential energy is $PE = mgy$, the potential energy also varies parabolically with time. Figure 6.16C shows the kinetic energy and the total mechanical energy, which is just the sum $PE + KE$. The total mechanical energy is *conserved*, so the sum $KE + PE$ is constant and does not change during the course of the motion. The kinetic energy thus varies as shown, in an "inverted" parabolic manner.

Only Changes in Potential Energy Matter

In our discussion of the gravitational potential energy, we found that $PE_{\text{grav}} = mgh$, where h is the height of the object (Eq. 6.13). However, in our calculations involving potential energy, we have always ended up computing changes in the potential energy as an object moves from an initial position to a final position. Recall the conservation of energy condition (Eq. 6.15):

$$KE_i + PE_i = KE_f + PE_f$$

When this relation is used to compute either the final kinetic energy or the change in the kinetic energy, the answer depends only on the potential energy difference $\Delta PE = PE_f - PE_i$. Indeed, our initial definition of potential energy in Equation 6.10 was in terms of the work done in moving an object from one place to another, and this original definition only referred to *changes* in potential energy.

The idea that only changes in potential energy are physically relevant can be appreciated if we consider the variation of the potential energy as a ball is released from a height h above the ground and falls to ground level (Fig. 6.17). If we use a

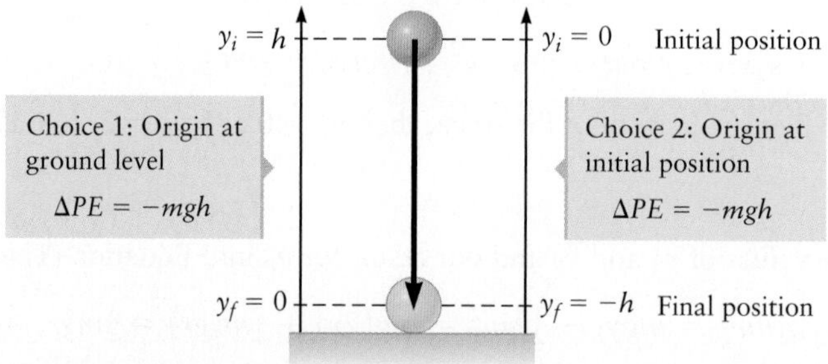

▶ **Figure 6.17** Only changes in the potential energy matter. Here, we consider the change in the gravitational potential energy using two different choices for the origin on the vertical axis. On the left, we take the origin at ground level, while on the right, the origin is at the initial position of the object. The change in the potential energy is the same; it does not depend on the choice of the origin.

coordinate axis with the origin at ground level (the coordinate axis on the left in Fig. 6.17), we have $y_i = h$ and $y_f = 0$, so the change in potential energy is

$$\Delta PE = PE_f - PE_i = mgy_f - mgy_i = mg(0 - h) = -mgh \quad (6.17)$$

On the other hand, we could just as well have chosen the origin of our coordinate system to be at the initial position of the ball (the axis shown on the right in Fig. 6.17). In this case, we would have $y_i = 0$ and $y_f = -h$, and the change in potential energy would be

$$\Delta PE = PE_f - PE_i = mgy_f - mgy_i = mg(-h - 0) = -mgh \quad (6.18)$$

The change in the potential energy in Equations 6.17 and 6.18 is *precisely the same*, and we can use either coordinate system to calculate the speed of the ball just before it reaches the ground or in any other calculation. Choosing a different origin for our coordinate system is very similar to adding or subtracting a constant from the potential energy at all locations. The results obtained from the application of the conservation of energy condition, Equation 6.15, depend only on changes in the potential energy, never on your choice of the "zero level" of the potential energy.

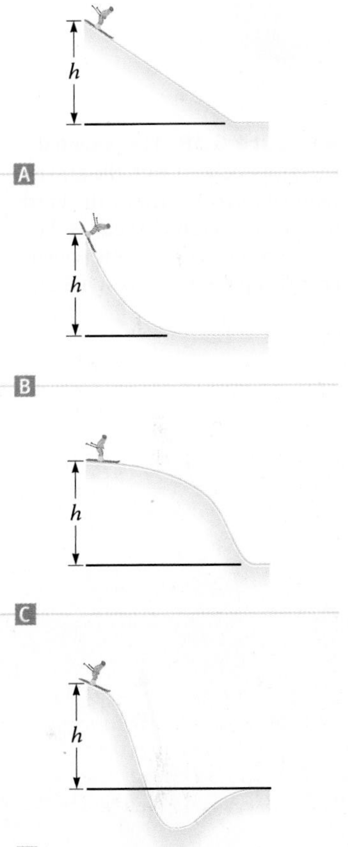

▲ **Figure 6.18** Concept Check 6.4.

CONCEPT CHECK 6.4 Skiing Down a Hill

Four expert skiers travel down the ski trails shown in Figure 6.18. All have the same initial height and the same final height at the bottom, on the far right in each sketch, and there is no friction on any of the trails. Which of the following statements are true? (More than one answer may be correct.) Assume air drag can be neglected.

(1) The skiers all arrive at the bottom at the same time.
(2) The skiers all arrive at the bottom with the same speed.
(3) The skier in Figure 6.18C probably arrives at the far right last.

CONCEPT CHECK 6.5 A 1000-Calorie Hill

The author's local bicycling club often rides along a particularly long and steep stretch of road the cyclists have named "Thousand Calorie Hill." This name suggests that the change in potential energy, and hence the amount of energy expended, in riding up this hill is 1000 Calories. The Calorie is a unit used to measure the energy in food, with 1 Calorie = 1000 calories (lowercase "c") = 4186 J. Do you think this name could be a reasonable one for any biking hill? *Hint*: Estimate the change in gravitational potential energy of a person who climbs the steepest hill in your neighborhood.

6.4 More Potential Energy Functions

In the last section, we used the potential energy associated with the gravitational force near the Earth's surface to illustrate the key ideas connected with potential energy. Let's now consider the potential energy functions associated with some other forces.

Gravitational Potential Energy in the Solar System

For objects near the surface of the Earth, the force of gravity exerted by the Earth is $F_{grav} = -mg$, and in the last section we showed that the potential energy associated with this force is $PE_{grav} = mgy$ (Eq. 6.13). In Chapter 5, we introduced Newton's law of gravitation, which describes the more general case of the gravitational force that exists between two objects separated by *any* distance. According to Newton's law of

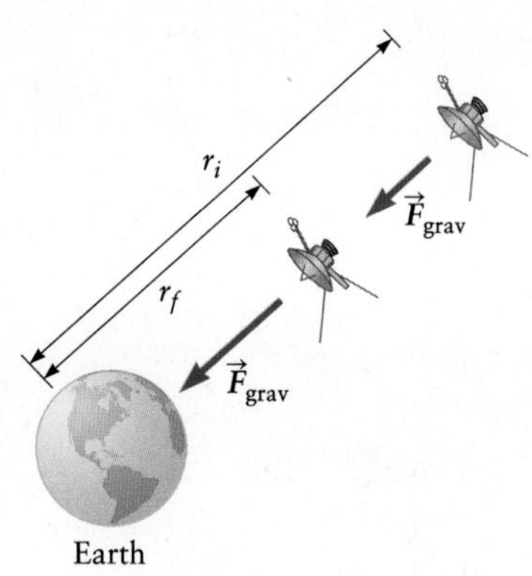

▲ **Figure 6.19** The potential energy associated with the gravitational force exerted by the Earth on this spacecraft changes as the separation r changes. This potential energy is given by Equation 6.20.

gravitation, the gravitational force between two objects of mass m_1 and m_2 separated by a distance r is (Eq. 5.16)

$$F_{grav} = -\frac{Gm_1m_2}{r^2} \tag{6.19}$$

The negative sign here indicates that the force is attractive. In a typical example, the mass m_1 might be the Earth, and m_2 could be a spacecraft that is either leaving the Earth or approaching the Earth as sketched in Figure 6.19. If this spacecraft is initially a distance r_i from the Earth and moves closer to a final distance r_f, the gravitational force does work on the spacecraft. According to the connection between work and potential energy (Eq. 6.10), there is a change in the gravitational potential energy of the spacecraft. While we will not give a detailed derivation here, the gravitational potential energy of two objects separated by a distance r is

$$PE_{grav} = -\frac{Gm_1m_2}{r} \tag{6.20}$$

The negative sign here means that the potential energy is lowered as the objects are brought closer together. The change in potential energy is then

$$\Delta PE_{grav} = -\frac{Gm_1m_2}{r_f} - \left(-\frac{Gm_1m_2}{r_i}\right) = -\frac{Gm_1m_2}{r_f} + \frac{Gm_1m_2}{r_i}$$

Gravitational Potential Energy: Launching a Satellite into Space

The result for the gravitational potential energy in Equation 6.20 can be used to analyze the motion of planets, moons, and satellites, including a calculation of the ***escape speed*** of a satellite. Suppose you want to launch a projectile of mass m_p into distant space. NASA currently does so by attaching a rocket engine to the projectile. However, several scientists have considered the possibility of simply "firing" the projectile from a very large cannon. (This possibility was described in a story by Jules Verne in 1865.) What is the minimum initial velocity required so that a projectile will completely escape from the Earth? To calculate this velocity, we ignore the drag force due to the atmosphere, so the only force acting on the projectile after it leaves the cannon is the gravitational force. This force is conservative, so we can apply the principle of conservation of energy to the projectile's motion. The instant after it leaves the cannon, the projectile is just above the Earth's surface with speed v_i at distance r_i from the Earth's center (Fig. 6.20A), and v_i is the unknown quantity we wish to calculate. The projectile eventually travels very far from the Earth; if it just barely escapes, the final speed is $v_f \approx 0$ and the final distance is very large, so $r_f \approx \infty$ (Fig. 6.20B).

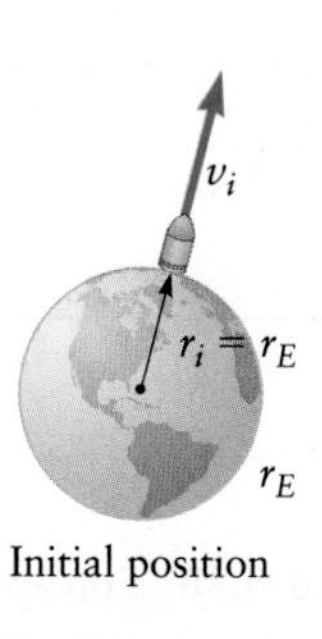

A

We now apply the conservation of energy condition with the potential energy given by Equation 6.20. Let M_{Earth} be the mass of the Earth and m_p the mass of the projectile. We then have

$$KE_i + PE_i = KE_f + PE_f$$

$$\frac{1}{2}m_pv_i^2 - \frac{GM_{Earth}m_p}{r_i} = \frac{1}{2}m_pv_f^2 - \frac{GM_{Earth}m_p}{r_f} \tag{6.21}$$

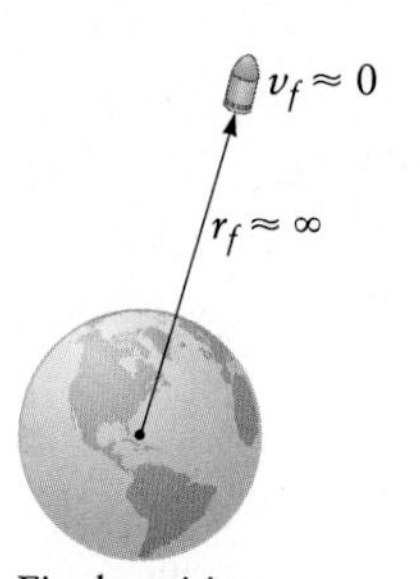

B

▲ **Figure 6.20** Calculating the escape velocity. A Initial position and speed of the projectile. B Final position and speed.

The projectile begins on the surface of the Earth, so it is initially a distance $r_i = r_E$ from the center of the Earth, where r_E is the radius of the Earth. Since $r_f \approx \infty$, the final potential energy is

$$PE_f = -\frac{GM_{Earth}m_p}{r_f} \approx 0$$

Inserting these results along with the final speed $v_f \approx 0$ into the conservation of energy relation Equation 6.21, we find

$$\frac{1}{2}m_p v_i^2 - \frac{GM_{\text{Earth}}m_p}{r_E} = \frac{1}{2}m_p v_f^2 - \frac{GM_{\text{Earth}}m_p}{r_f} = 0$$

$$\frac{1}{2}m_p v_i^2 = \frac{GM_{\text{Earth}}m_p}{r_E}$$

$$v_i = \sqrt{\frac{2GM_{\text{Earth}}}{r_E}} \tag{6.22}$$

which is called the escape speed. This is the minimum speed an object at the Earth's surface must have (ignoring air drag) to completely escape from the gravitational attraction of the Earth. Inserting values for the mass and radius of the Earth gives

$$v_i = \sqrt{\frac{2GM_{\text{Earth}}}{r_E}} = \sqrt{\frac{2(6.67 \times 10^{-11}\ \text{N}\cdot\text{m}^2/\text{kg}^2)(6.0 \times 10^{24}\ \text{kg})}{6.4 \times 10^6\ \text{m}}} = 1.1 \times 10^4\ \text{m/s}$$

which is approximately 24,000 mi/h.

Insight 6.1

GRAVITATIONAL POTENTIAL ENERGY

We have now derived two different expressions for the gravitational potential energy. The result $PE_{\text{grav}} = mgy$ (Eq. 6.13) is the potential energy associated with the Earth's gravitational force for objects near its surface. The result $PE_{\text{grav}} = -Gm_1m_2/r$ (Eq. 6.20) is the potential energy associated with the gravitational force between any two objects separated by a distance r. Equation 6.20 is required for problems involving motion in the solar system.

Escape velocity

The escape speed is different for objects that are fired from different planets or from the Moon because that speed depends on the mass and radius of the planet or the Moon. Of course, to launch a real satellite, NASA should not rely on our simple calculation because the effect of air drag on such a projectile as it moves through the atmosphere cannot be ignored.

Elastic Forces and Potential Energy: Springs

Hooke's law

One important type of potential energy is associated with springs and other elastic objects. A simple spring is a tight coil of wire as sketched in Figure 6.21. In its "relaxed" state with no force applied to its end, the spring rests as shown in Figure 6.21A. We might now pull on the spring with a force F_{pull}, causing the spring to stretch as shown in Figure 6.21B. When it is stretched, the spring itself exerts a force F_{spring} that opposes the stretching. Likewise, if you were to push on the spring with a force F_{push} so as to compress it (Fig. 6.21C), the spring exerts a force back on you in opposition. The force exerted by the spring when it is either stretched or compressed is

$$F_{\text{spring}} = -kx \tag{6.23}$$

where x is the amount the end of the spring is displaced from its equilibrium position, which is assumed to be at $x = 0$. The quantity k is called the ***spring constant***, and it has units of newtons per meter. The value of k depends on the properties of the spring (the number of coils, the stiffness of the wire, etc.) and is thus different for different springs. The negative sign in Equation 6.23 indicates that the force exerted by the spring always *opposes* the displacement of the end of the spring. If the end of the spring is pulled toward $+x$, the force exerted by the spring is directed along $-x$, whereas if the spring is pushed toward $-x$, the spring exerts a force along $+x$.

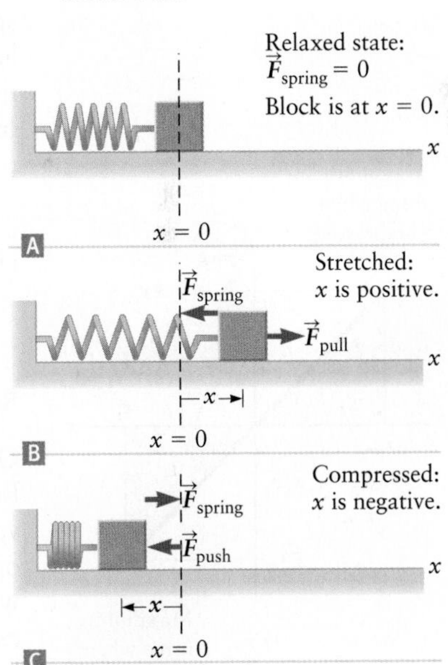

▲ **Figure 6.21** The force exerted by a spring is always opposite to the displacement of the end of the spring. A When a spring is unstretched and uncompressed, the force exerted by the spring is zero. B This spring is stretched by pulling it to the right, so the force exerted by the spring is to the left. C The spring is compressed, and the force exerted by the spring is to the right.

Equation 6.23, called ***Hooke's law***, provides an accurate description of spring forces in a wide variety of situations. Hooke's law is not a "law" of physics in the sense of Newton's laws. Instead, Hooke's law is an empirical relation that is accurate for most springs. Springs like the simple coils of wire sketched in Figure 6.21 are rather idealized; real springs can be far more complex. Hooke's law, however, applies to a variety of elastic objects. For example, a diving board is a type of spring: it exerts a force on a diver given by Equation 6.23, with x equal to the distance the diving board is deflected (Fig. 6.22A). A tennis racket also acts like a spring: the force exerted by the racket on a tennis ball is again given by an expression like Equation 6.23, with x equal to the deflection of the strings (Fig. 6.22B).

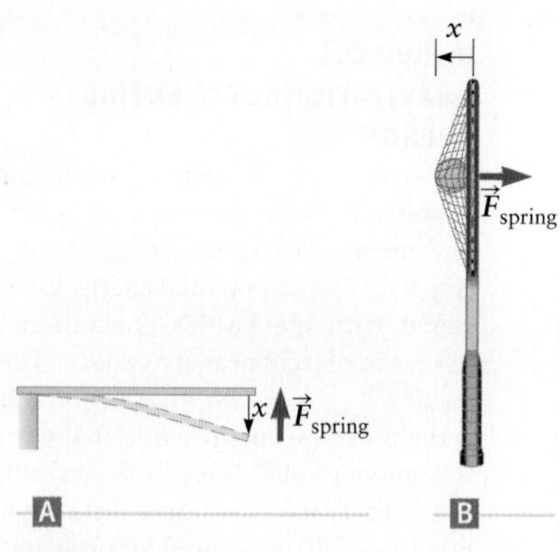

▲ **Figure 6.22** Objects such as **A** a diving board and **B** a tennis racket obey Hooke's law (Eq. 6.23), with the force proportional to the displacement x.

Potential Energy Stored in a Spring

According to Hooke's law, the force exerted by a spring is proportional to x, the amount the spring is stretched or compressed. To calculate the potential energy associated with this force, we return to the definition of potential energy (Eq. 6.10),

$$\Delta PE = -W$$

with W now the work done by the Hooke's law force (Eq. 6.23). When you pull or push on a spring to stretch it or compress it as in part B or C of Figure 6.21, W is the work done *by the spring* on your hand.

The force F_{spring} is not a constant, but varies as the displacement changes. In Figure 6.5, we showed that the work done by a variable force is equal to the area under the force–displacement curve. This curve for a spring is shown in Figure 6.23. We assume the spring is initially in its relaxed state (not stretched or compressed), so $x_i = 0$. If the spring is then stretched an amount $x = x_s$, the work done by the spring is the area under the curve $F_{spring} = -kx$. This area is just a triangle of height $F_s = kx_s$, and the area of this triangle is $\frac{1}{2}F_s x_s$ (see Fig. 6.23). Inserting the expression for F_{spring} from Hooke's law, the area of the triangle is $\frac{1}{2}F_s x_s = \frac{1}{2}kx_s^2$. We must now be careful to get the correct sign for the work done by the spring. In this example, the spring is being stretched, so according to Figure 6.21B the force exerted by the spring is directed to the left, whereas the displacement of the end of the spring is to the right. Since the force and the displacement are in opposite directions, the work done by the spring is negative, and

$$W = -\tfrac{1}{2}kx_s^2$$

Here, W is the work done by the spring as it is *stretched* an amount x_s starting from its relaxed state. From Figure 6.23, we see that the work done in *compressing* the spring is equal to the area under a similar triangular region, the triangle in the upper left in Figure 6.23. The work done by the spring when it is compressed is again negative because the force exerted by the spring is directed opposite to the displacement of the end of the spring. (See Fig. 6.21C.) Hence, if a spring is stretched or compressed an amount x, the work done by the spring is

$$W_{spring} = -\tfrac{1}{2}kx^2$$

Using the definition of potential energy, we then get

$$PE_{spring} = \tfrac{1}{2}kx^2 \qquad (6.24)$$

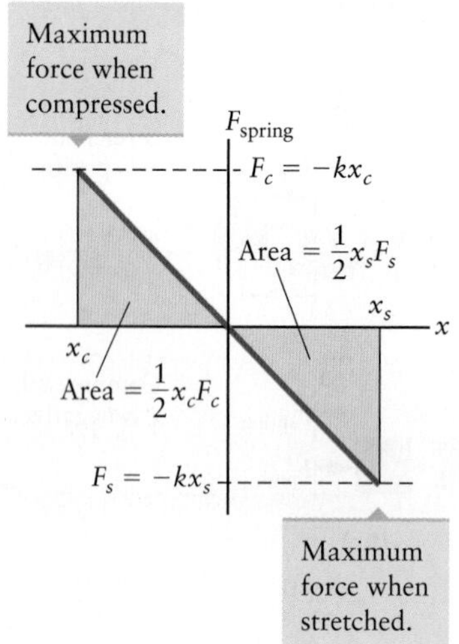

▲ **Figure 6.23** The work done by a variable force is equal to the area under the force–displacement graph. For a spring, this area is a triangle. When the spring is stretched, $x = x_s$ and W is the area of the triangle at the lower right. When the spring is compressed, $x = x_c$ and W is the area of the triangle at the upper left.

Spring Forces and Potential Energy: A Recap

Our results for the elastic (spring) force in Equation 6.23 and potential energy in Equation 6.24 are collected in Figure 6.24, which shows how F_{spring} and PE_{spring} vary as a spring is stretched and compressed. The force exerted by the spring always opposes the displacement, so F_{spring} can be either positive or negative, depending on whether the spring is stretched or compressed. The elastic potential energy is smallest ($PE_{spring} = 0$) when a spring is in its relaxed state ($x = 0$), and PE_{spring} always increases as a spring is either stretched or compressed.

EXAMPLE 6.7 Force Exerted by a Bungee Cord

A bungee cord is a kind of spring, and the force exerted by the cord on a person attached to it is given by $F_{spring} = -kh$, where h is the amount the cord is stretched (Hooke's law, Eq. 6.23). Consider a person of mass $m = 75$ kg attached to a bungee cord with spring constant $k = 50$ N/m. When the jump is completed, the person is at rest, suspended by the cord (Fig. 6.25B). How much is the cord stretched at that time?

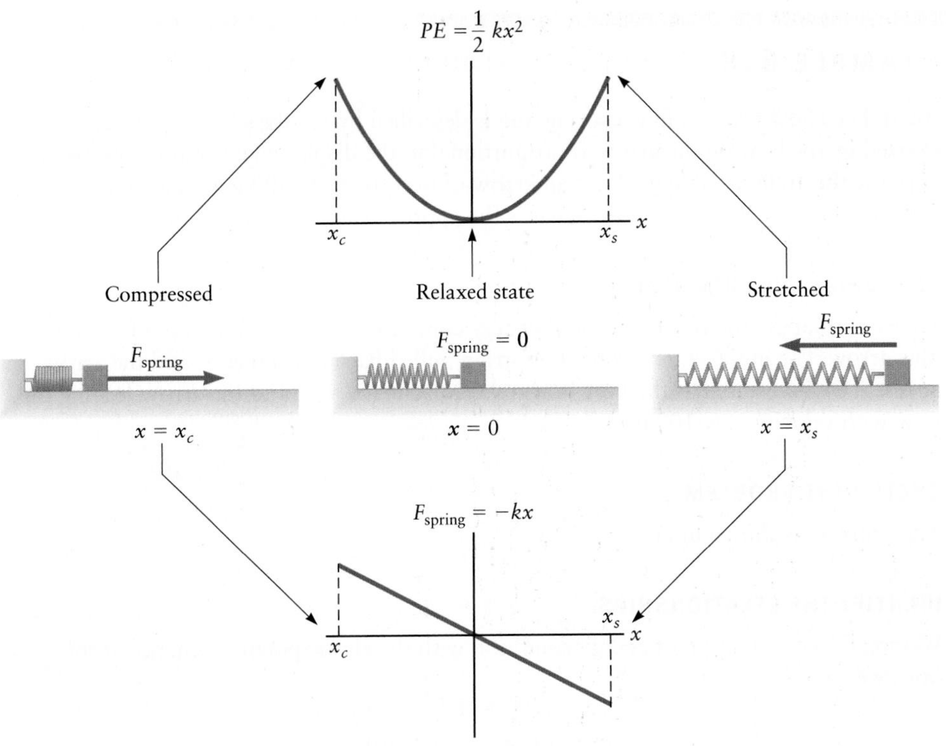

◀ **Figure 6.24** Variation of the potential energy and force as a spring is compressed and stretched.

RECOGNIZE THE PRINCIPLE

At the end of the jump and after the person has stopped oscillating up and down, the person is at rest, so his acceleration is zero. Hence, according to Newton's second law, the total force must be zero.

SKETCH THE PROBLEM

Figure 6.25B shows the forces acting on the person when the cord is stretched: the force of gravity and the spring force from the bungee cord.

IDENTIFY THE RELATIONSHIPS

Applying Newton's second law for motion in the vertical direction for the person in Figure 6.25B, we have

$$\sum F = kh + F_{grav} = kh - mg = ma = 0 \quad (1)$$

The spring force is $+kh$, where h is the amount the bungee cord is stretched; the positive sign is here because this force is directed upward.

SOLVE

Solving Equation (1) for h, we get

$$kh = mg$$

$$h = \frac{mg}{k} = \frac{(75 \text{ kg})(9.8 \text{ m/s}^2)}{50 \text{ N/m}} = \boxed{15 \text{ m}}$$

▶ ***What does it mean?***

This result should be of interest to the bungee cord company. For safety reasons, it would want to be sure that the customer is suspended well above the ground.

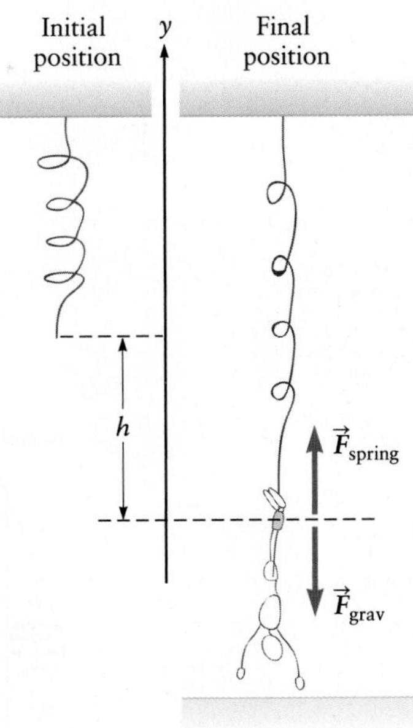

▲ **Figure 6.25** Example 6.7. **A** The bungee cord in its relaxed state (neither stretched or compressed). **B** The bungee jumper is hanging at rest, suspended by the cord.

EXAMPLE 6.8 Ⓡ Spring Constant of an Archer's Bow

An archer's bow behaves like a spring and is described by Hooke's law. The force exerted by the bow on the arrow is proportional to the displacement of the bowstring. Suppose the archer is able to shoot an arrow of mass $m = 0.050$ kg with a velocity of 100 m/s just as it leaves the bow. What is the approximate spring constant of her bow?

RECOGNIZE THE PRINCIPLE

We want to apply the conservation of energy condition to the combination of the bow plus arrow. The initial state has the bowstring pulled back a distance x_i and the arrow at rest. The final state has the bow in its relaxed state ($x_f = 0$) and the arrow leaving the bow with velocity $v_f = 100$ m/s.

SKETCH THE PROBLEM

The problem is shown in Figure 6.26A.

IDENTIFY THE RELATIONSHIPS

Writing the conservation of energy condition with the elastic potential energy of the bow, we have

$$KE_i + PE_i = KE_f + PE_f$$

$$\tfrac{1}{2}mv_i^2 + \tfrac{1}{2}kx_i^2 = \tfrac{1}{2}mv_f^2 + \tfrac{1}{2}kx_f^2 \qquad (1)$$

SOLVE

Inserting our values for the initial arrow velocity ($v_i = 0$) and final bow stretch ($x_f = 0$) into Equation (1) leaves us with an equation involving k and the initial bow stretch x_i. Solving for the spring constant k, we find

$$0 + \tfrac{1}{2}kx_i^2 = \tfrac{1}{2}mv_f^2 + 0$$

$$k = \frac{mv_f^2}{x_i^2} \qquad (2)$$

We are given the mass of the arrow and its final speed v_f. To find the spring constant k, we also need to know x_i, the amount the bow stretches, but this information was not given. Relying on intuition and experience, we estimate $x_i \approx 0.1$ m. A value of

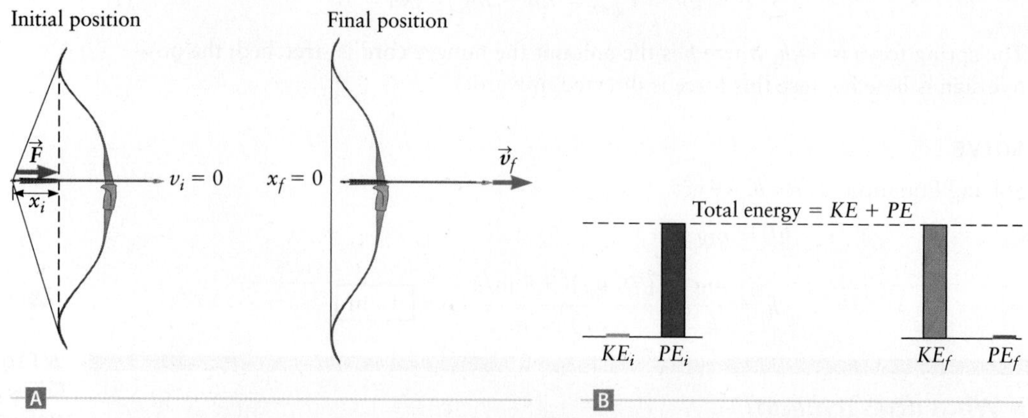

▲ **Figure 6.26** Example 6.8. A A bow acts like a spring with potential energy given by $PE = \tfrac{1}{2}kx^2$, where x is the distance the bowstring is displaced. B Bar charts showing the initial and final kinetic and potential energies of the arrow and bow.

$x = 0.01$ m $= 1$ cm seems much too small, whereas $x = 1$ m is too large. Our estimate is in the middle. Evaluating Equation (2) for k then gives

$$k = \frac{mv_f^2}{x_i^2} = \frac{(0.050 \text{ kg})(100 \text{ m/s})^2}{(0.1 \text{ m})^2} = \boxed{5 \times 10^4 \text{ N/m}}$$

▶ *What does it mean?*

Notice how we followed the general steps in our strategy for dealing with conservation of energy problems. We began with a picture and identified the initial and final states of the bow plus arrow. We then wrote the conservation of energy relation, Equation (1), and solved for the quantity of interest (k).

Figure 6.26B shows bar charts for the initial and final kinetic and potential energies. The initial potential energy is converted into the final kinetic energy, and the total energy $KE + PE$ is conserved.

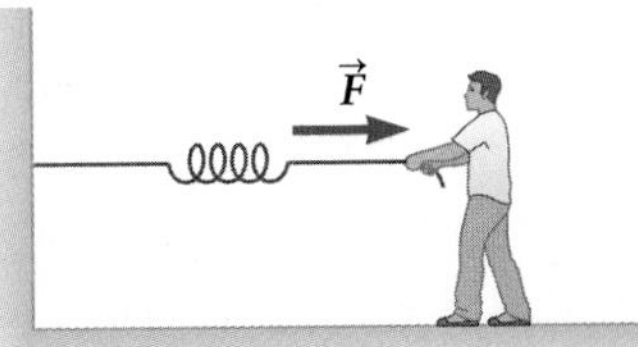

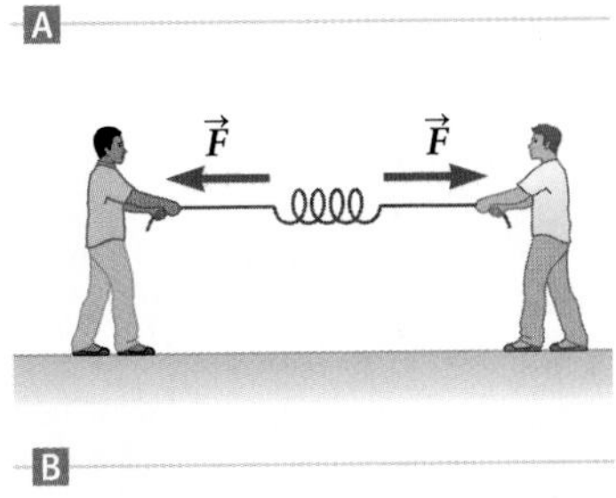

▲ **Figure 6.27** Concept Check 6.6.

CONCEPT CHECK 6.6 Spring Forces and Newton's Third Law

Figure 6.27 shows two identical springs. In both cases, a person exerts a force of magnitude F on the right end of the spring. The left end of the spring in Figure 6.27A is attached to a wall, while the left end of the spring in Figure 6.27B is held by another person, who exerts a force of magnitude F to the left. Which statement is true?

(1) The spring in Figure 6.27A is stretched half as much as the spring in Figure 6.27B.

(2) The spring in Figure 6.27A is stretched twice as much as the spring in Figure 6.27B.

(3) The two springs are stretched the same amount.

Total Potential Energy with Multiple Forces

In all our problems so far, there has only been one type of force and hence only one type of potential energy function to deal with. Often, however, several different forces—such as a gravitational force and a spring force—may be acting together on an object. In these cases, we must be sure to account for the total potential energy from all forces when applying the principle of conservation of energy. For example, if both gravity near the Earth's surface and a spring force are acting on an object, the total potential energy is

$$PE_{\text{total}} = PE_{\text{grav}} + PE_{\text{spring}} = mgh + \tfrac{1}{2}kx^2$$

This type of situation is illustrated in the next example.

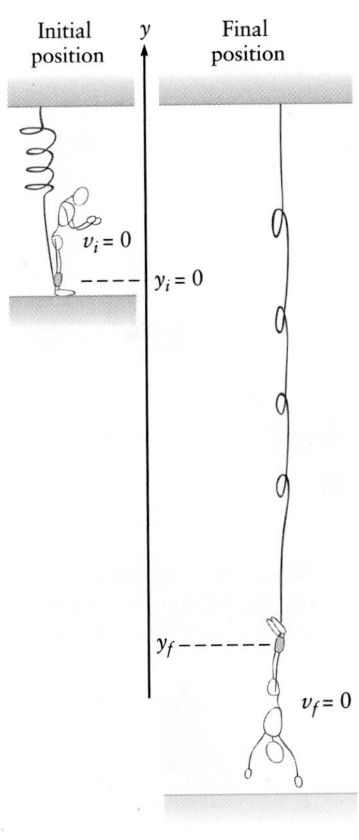

▲ **Figure 6.28** Example 6.9. **A** This bungee jumper is initially at rest with the cord unstretched. **B** At his lowest point, the bungee jumper's position is a distance y_f below the starting point.

EXAMPLE 6.9 Potential Energy from a Spring Plus Gravity

Consider again the bungee jumper from Example 6.7. He is initially standing on the bungee platform with the cord unstretched (Fig. 6.28A). He then jumps from the platform and travels downward (Fig. 6.28B). How much is the cord stretched when the jumper reaches his lowest point? The mass of the jumper is $m = 75$ kg, and the spring constant of the bungee cord is $k = 50$ N/m.

RECOGNIZE THE PRINCIPLE

We again apply the principle of conservation of energy, but now we have to deal with the potential energy associated with both gravity and the bungee cord. Initially, the cord is in its relaxed state, so all the potential energy is from gravity. In the final

(continued) ▶

configuration, the cord is stretched, so we have to include the contributions to the potential energy from both this "spring" and from gravity.

SKETCH THE PROBLEM

Figure 6.28 describes the problem. This sketch shows both the initial and final states of the system.

IDENTIFY THE RELATIONSHIPS

According to the conservation of energy condition,

$$KE_i + PE_{\text{spring}, i} + PE_{\text{grav}, i} = KE_f + PE_{\text{spring}, f} + PE_{\text{grav}, f}$$

where we have written the gravitational and cord (i.e., spring) potential energies as separate terms. The jumper begins at rest ($v_i = 0$) and also ends at rest ($v_f = 0$), so the initial and final kinetic energies are both zero. The cord is initially relaxed, so the initial potential energy of the cord is zero. We take the origin ($y = 0$) at the initial position of the jumper, so the initial gravitational potential energy is also zero. Using the coordinate system in Figure 6.28, the final position of the jumper is at $y = y_f$. Inserting these values into the conservation of energy relation leads to

$$PE_{\text{spring}, i} + PE_{\text{grav}, i} = PE_{\text{spring}, f} + PE_{\text{grav}, f}$$

$$0 + 0 = \tfrac{1}{2}ky_f^2 + mgy_f$$

SOLVE

Solving for y_f, we get

$$\tfrac{1}{2}ky_f^2 = -mgy_f$$

$$y_f = -\frac{2mg}{k} = -\frac{2(75 \text{ kg})(9.8 \text{ m/s}^2)}{50 \text{ N/m}} = \boxed{-29 \text{ m}}$$

► ***What does it mean?***

The value of y_f is negative because the final position of the jumper is below the starting point at $y = 0$. Comparing this result with Example 6.7, we see that the jumper in Figure 6.28B is at a lower point than the person in Figure 6.25B. What are the velocity and acceleration of the jumper when he reaches the lowest point in Figure 6.28B? Is this lowest point a point of static equilibrium?

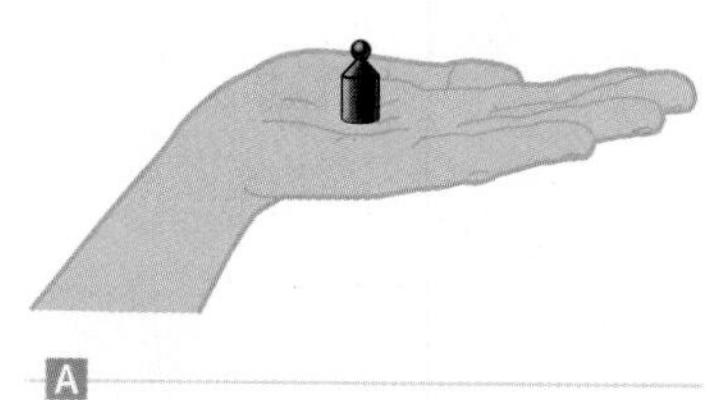

When holding a massive object, your skin and muscles are compressed like a spring.

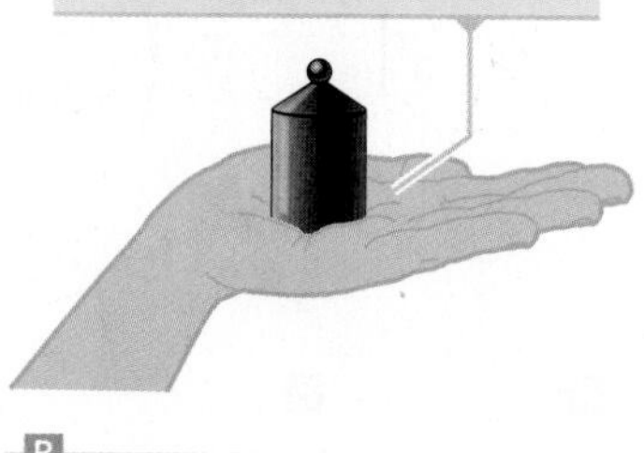

▲ **Figure 6.29** When you support an object with your hand, your muscles deform much like a spring, producing an elastic force on the object.

Elastic Forces and the "Feeling" of Holding a Heavy Object

According to the definition of mechanical work (Eq. 6.3), W is zero if the displacement is zero, even though the force on the object might be very large. This definition is quite different from our "everyday" concept of work. Consider what happens when you hold an object in your hand (Fig. 6.29A). If the object is at rest, the displacement is zero, and the work done by your hand on the object is zero. Hence, $W = 0$ if the object has a very small mass or even a very large mass. However, you also know intuitively that supporting a very massive object will eventually make your hand "tired," and in everyday language we would say that holding something requires an expenditure of "energy" by your hand.

If the "physics" work done on the object is zero, however, why do your muscles get tired? One reason is that your muscles consist of many parallel fibers, and when they apply a force, these fibers are constantly slipping relative to one another. Muscles also contain small motorlike molecules that pull and move each fiber relative to its neighbors to counteract this slipping. (We say more about these motors in Section 6.8.) Because the molecular motors exert a force on the muscle fibers as they move along a fiber, the motors do a nonzero amount of "physics" work, and the associated energy comes from chemical reactions in the muscle. You sense this expenditure of chemical energy in your muscles, and that is why you must exert "energy" to support an object.

6.5 Conservative versus Nonconservative Forces and Conservation of Energy

In our derivation of the potential energy associated with gravity near the surface of the Earth (Eq. 6.13), we mentioned that W is independent of the path taken by the object. This path independence is an essential characteristic of the work done by a conservative force. The corresponding change in the potential energy depends only on the object's initial and final positions. Changes in the potential energy, and hence in the value of W, are independent of the path taken. Many forces satisfy this requirement of path independence, including gravitation and spring (Hooke's law) forces. As already mentioned, such forces are called conservative forces.

The work done by a conservative force is independent of the path.

Although potential energy is an extremely useful concept, some forces—called ***nonconservative forces***—cannot be associated with a potential energy. Unlike conservative forces, the work done by a nonconservative force *does* depend on the path taken. In this section, we'll deal with nonconservative forces.

The work done by a nonconservative force does depend on the path taken.

The Work Done by Friction Depends on the Path

Friction is a good example of a nonconservative force. Consider the work done by friction as an object moves along the path sketched in Figure 6.30, where a crate is pushed across a table from an initial location x_i to a final location x_f. This can be done by pushing the crate directly from x_i to x_f (path A in Fig. 6.30) or by taking an indirect path (path B). Since we are dealing with kinetic friction, $F_{\text{friction}} = -\mu_K N$, where N is the normal force between the table and the crate. The work done by friction along path A is

$$W_A = F_{\text{friction}}\,\Delta x = -\mu_K N(x_f - x_i)$$

Here, W_A is negative because friction opposes the motion. We can calculate the work done in moving along path B in the same way. The work done is

$$W_B = -\mu_K N\ (\text{length of path } B)$$

The length of path B is much longer than that of path A (see Fig. 6.30), so W_A and W_B are certainly not equal. Hence, the work done by friction depends on the path taken by the object. From Equation 6.10, it is therefore impossible to have a single value for the change in a corresponding potential energy, so we conclude that such a potential energy cannot exist. It is thus impossible to have a potential energy associated with the frictional force. Other nonconservative forces have the same property: the work done by a nonconservative force depends on the path taken by the object, so it is not possible to have a corresponding potential energy function.

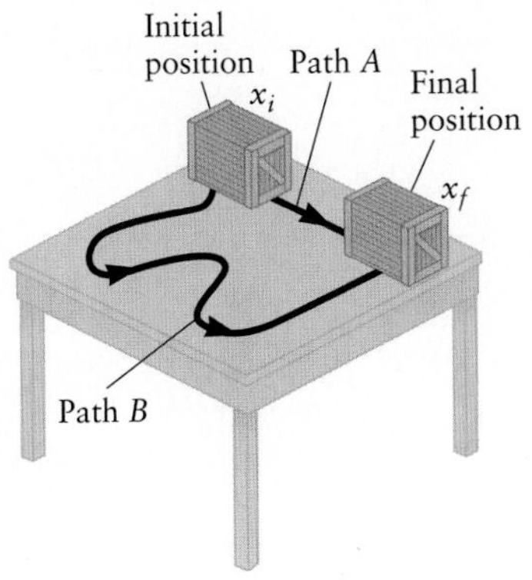

▲ **Figure 6.30** The work done by friction is larger for path B than it is for path A because path B is longer. Since the work done by friction depends on the path taken, friction is a nonconservative force.

The Work–Energy Theorem Revisited: Including Nonconservative Forces

If we cannot define potential energy for a nonconservative force, how do we think about conservation of energy in the presence of nonconservative forces? To answer this question, it is simplest to return to our original version of the work–energy theorem (Eq. 6.9),

$$W_{\text{total}} = \Delta KE = KE_f - KE_i \tag{6.25}$$

which tells us that the change in kinetic energy is equal to the *total* work done on the object. This total work can be due to several different forcess. Let's split it into two terms, one for the work done by all the conservative forces and another for the work done by the nonconservative forces:

$$W_{\text{total}} = W_{\text{con}} + W_{\text{noncon}}$$

According to the definition of potential energy (Eq. 6.10), the work done by the conservative forces is related to the change in the potential energy:

$$-W_{\text{con}} = \Delta PE = PE_f - PE_i$$

Collecting these results and inserting them into Equation 6.25 leads to

Work–energy theorem including nonconservative forces

$$KE_i + PE_i + W_{\text{noncon}} = KE_f + PE_f \tag{6.26}$$

which looks just like our previous conservation of energy result (Eq. 6.15) but with an additional term on the left for the work done by nonconservative forces. Equation 6.26 is our general conservation of energy/work–energy theorem with nonconservative forces included. The sum of the kinetic and potential energies is the total ***mechanical energy*** of an object. According to Equation 6.26, the final mechanical energy ($KE_f + PE_f$) is equal to the initial mechanical energy ($KE_i + PE_i$) plus the work done by any nonconservative forces that act on the object.

EXAMPLE 6.10 Work Done by Friction on a Skier

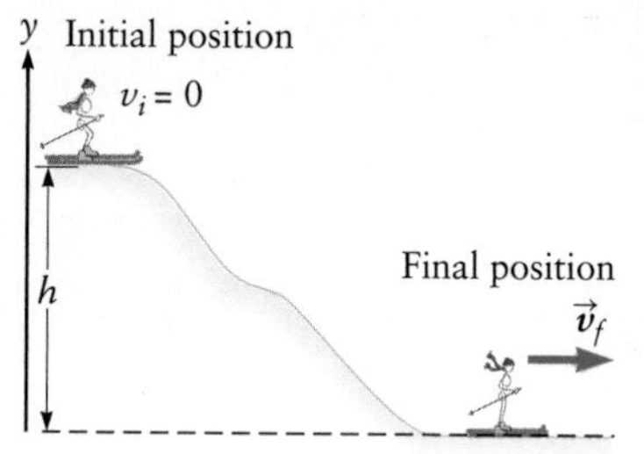

▲ **Figure 6.31** Example 6.10.

A skier of mass $m = 60$ kg starts from rest at the top of a ski hill of height $h = 50$ m (Fig. 6.31). When she reaches the bottom, she has speed $v_f = 25$ m/s. How much work is done by friction on the skier as she travels down the hill?

RECOGNIZE THE PRINCIPLE

Since the ski hill does not have a simple shape (it is not a simple inclined plane), we cannot calculate the work done by friction using the definition of work in Equation 6.3 ($W_{\text{friction}} = F_{\text{friction}} \Delta x$). Friction is the only nonconservative force in the problem, however, so $W_{\text{friction}} = W_{\text{noncon}}$, and we can find the work done by friction using the conservation of energy relation

$$KE_i + PE_i + W_{\text{friction}} = KE_f + PE_f \tag{1}$$

SKETCH THE PROBLEM

Figure 6.31 shows the initial and final states of the skier.

IDENTIFY THE RELATIONSHIPS

We first rearrange Equation (1):

$$W_{\text{friction}} = KE_f + PE_f - (KE_i + PE_i) \tag{2}$$

In Equation (2), the potential energy is due to gravity; choosing the origin of the vertical axis at the bottom of the hill, we get $PE_i = mgh$ and $PE_f = 0$. The skier starts from rest, so the initial kinetic energy is zero and the final kinetic energy is $\frac{1}{2}mv_f^2$.

SOLVE

Inserting all this information into Equation (2) leads to

$$W_{\text{friction}} = \tfrac{1}{2}mv_f^2 - mgh = \tfrac{1}{2}(60\text{ kg})(25\text{ m/s})^2 - (60\text{ kg})(9.8\text{ m/s}^2)(50\text{ m})$$

$$W_{\text{friction}} = \boxed{-1.1 \times 10^4\text{ J}}$$

► ***What does it mean?***

The work done by friction is negative because the frictional force opposes the skier's motion along the hill. Friction thus causes the final speed of the skier to be less than it would have been without friction.

Conservation of Energy of a System

We have mentioned the concept of "conservation of energy" several times, in several different contexts. We now consider the principle of conservation of energy from a very general point of view. Suppose a system of particles or objects exert forces on one another as they move about. These forces may be conservative (perhaps gravity and elastic forces) or nonconservative (friction or air drag). While there are many forces involving particles within this system, let's assume there are no forces on any of these particles from outside the system so that the total energy of the system will be conserved. As the particles move around, we would expect some potential energy to be converted to kinetic energy or vice versa. It is also possible for mechanical energy (i.e., elastic potential energy or kinetic energy) to be converted to another form of energy such as heat energy, electrical energy, or chemical energy that we describe later in this book. Nevertheless, the total energy of the entire system will be conserved.

Of course, some agent from outside this system may exert a force on one of the particles within the system. If this outside agent does an amount of work W_{ext} on the particle, the total energy of the system will change by an amount W_{ext}. This change in the system's energy might be positive or negative, depending on the sign of W_{ext}. In this way, energy can be exchanged between an external agent and the system. If a certain amount of energy is added to the system, the same amount of energy must be removed from the agent, so we again arrive at the general principle of ***conservation of total energy***.

6.6 The Nature of Nonconservative Forces: What Is Friction Anyway?

When all the forces acting on an object are conservative forces, the total mechanical energy is conserved:

$$KE_i + PE_i = KE_f + PE_f$$

Intuitively, the idea that energy is conserved should not be surprising. If energy were not conserved, it would be possible to do some very incredible things—such as generate limitless amounts of energy and construct mechanical perpetual motion machines—and physicists have found no evidence that such things are possible.

We have also seen that when a nonconservative force such as friction is present, the principle of conservation of energy can be expressed as (Eq. 6.26)

$$KE_i + PE_i + W_{noncon} = KE_f + PE_f$$

or

$$\underbrace{(KE_f + PE_f)}_{\text{final mechanical energy}} - \underbrace{(KE_i + PE_i)}_{\text{initial mechanical energy}} = W_{noncon} \quad (6.27)$$

In most situations, the work done by friction is negative because the frictional force is usually directed opposite to the displacement. So, Equation 6.27 tells us that in such cases the final mechanical energy is less than the initial mechanical energy. A simple illustration of this situation is given in Figure 6.32, which shows a block sliding along a rough surface. The block has some initial kinetic energy, so the initial mechanical energy is not zero. Friction causes the block to come to a stop, so the final kinetic energy and the final mechanical energy are both zero. Hence, the mechanical energy of the block is not conserved; by the end of this process, the mechanical energy of the block has changed by an amount W_{noncon}. Let's now consider precisely where the energy associated with W_{noncon} goes.

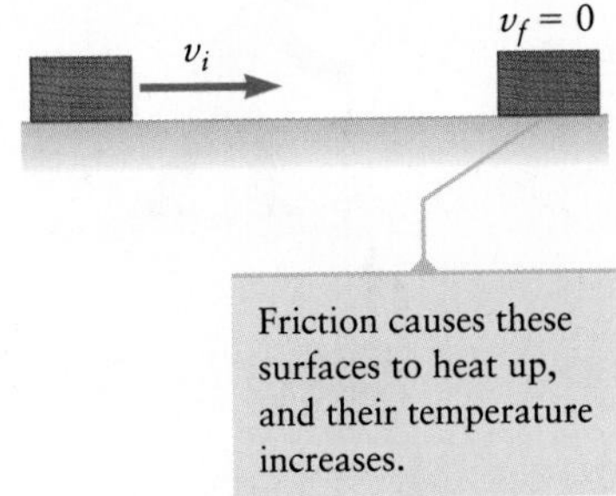
v_i
$v_f = 0$
Friction causes these surfaces to heat up, and their temperature increases.

▲ **Figure 6.32** When friction does work on a sliding block, the block's mechanical energy is converted into heat energy, resulting in a temperature increase at the surfaces of the block and the floor.

For the sliding block in Figure 6.32, the energy associated with W_{noncon} goes into heating up the surface of the block and the surface of the floor. This increase in the temperature of the two surfaces can be understood in terms of the motion of the atoms at the surfaces. All substances are composed of atoms in constant motion. The chemical bonds that hold the atoms together are analogous to springs, which are stretched and compressed as the atoms move and vibrate. When a substance is heated and its temperature is increased, these vibrations become larger in size and the atoms move faster. Since they move faster, the atoms have an increased kinetic energy, and there is also an increase in the potential energy of the chemical bonds/springs. Hence, the initial mechanical energy of the block in Figure 6.32 does not disappear, but shows up as an increase in the energy of the atoms at the surfaces. Even in the presence of friction, the *total* energy is still conserved.

This conclusion leads to another, and deeper, question. If the energy W_{noncon} has simply gone into the mechanical energy (kinetic and potential) of the atoms, is it possible to somehow return all this energy back to its original form? That is, can we convert this added mechanical energy of the heated atoms back into the mechanical energy of the block? We'll see in Chapter 16 that it is possible to transfer *some* of the energy from the heated atoms into kinetic energy of the block. However, it is *not* possible to transfer *all* this energy back to the block.

This energy that flows in connection with a temperature difference is called ***heat.*** Heat plays a central role in the branch of physics called ***thermodynamics***, which we will explore in Chapters 14 through 16. Although mechanical energy is easily converted completely into heat as with the sliding block in Figure 6.32, it is not nearly as simple to convert heat energy back into mechanical energy. After all, if the block in Figure 6.32 were initially at rest and we then heated it, we would not expect the block to begin sliding along the floor!

These observations also suggest why friction is called a "nonconservative" force. Conservative forces are associated with potential energy, and we have seen that potential energy is stored energy that can be "retrieved" and then converted freely into other forms of energy. The heat energy produced by friction cannot be retrieved in this manner, and it cannot all be transferred back to other forms of energy.

6.7 Power

In Section 6.2, we found that a block and tackle is able to amplify the force that can be applied to an object. However, this amplification comes at the cost of a reduction in the object's displacement. For the particular block and tackle in Figure 6.8, the force amplification factor is two and the displacement is reduced by the same factor. The work done on the object is therefore the *same* as the work done by the person as he pulls on the rope. We can view the person and the crate in Figure 6.8 as two "systems"; the block and tackle "transfers" energy from the person to the crate, and according to the principle of conservation of energy, the energy expended by the person—that is, the work done by the person—must equal the energy gained by the crate.

More complicated block-and-tackle arrangements can give much greater force amplification factors, so in principle a person could use such a device to lift an extremely massive object. While the force amplification factor can be very large, the displacement is always "deamplified" by the same factor. Hence, it may take a very long time to lift a massive object through an appreciable distance. Time enters into work–energy ideas through the concept of ***power***. Consider the person in Figure 6.33 who is using a rope to lift a crate very slowly. Work is done on the crate as it moves from the floor to a height h, which changes the potential energy of the crate by an amount mgh. This energy comes from the person as he pulls on the other end of the rope and does an amount of work $W = mgh$ on the rope. If this work is expended during a time t, the average power exerted by the person is defined as

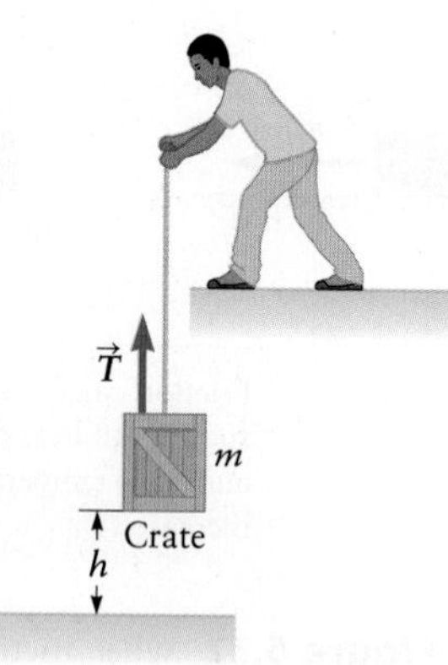

▲ **Figure 6.33** This person is expending a certain amount of energy per unit time as he lifts the crate.

$$P_{ave} = \frac{W}{t} \tag{6.28}$$

TABLE 6.1 Typical Values for the Power Output and Consumption of Some Common Appliances and Devices

Device	Power Output or Consumption
Lightbulb	20 W
Video game console	200 W
Automobile engine (small car)	100 hp = 7.5×10^4 W
Automobile engine (race car)	700 hp = 5.2×10^5 W
Smart phone (when talking)	1 W
Elite bicycle racer	400 W

The units of power are joules per second; this unit has been given the name watt (W) after James Watt (1736–1819), a developer of the steam engine. So,

$$1 \text{ W} = 1 \text{ J/s}$$

According to Equation 6.28, power is the *rate* at which work is done. Since work is a way to transfer energy from one system to another, power is also equal to the energy output per unit time of a device. For example, the specifications for a motor or engine often include the maximum power the engine is able to produce. For historical reasons, the power of an automobile is usually given in terms of an odd unit called "horsepower." Presumably, 1 horsepower (hp) is equal to the power that a typical ("standard"?) horse is able to expend. You can convert between watts and horsepower using the relation

$$1 \text{ hp} = 745.7 \text{ W} = 745.7 \text{ J/s}$$

In addition to mechanical devices, the concept of power also applies to chemical and electrical processes and devices. Indeed, there is a power output or input associated with many of the electrical devices in your house. For example, a typical lightbulb is rated at 20 W = 20 J/s. This rating means that the lightbulb consumes 20 J of electrical energy for every second it is turned on. Likewise, devices such as computers and DVD players also have a "power rating." Table 6.1 lists the power output and consumption of a number of common devices. The distinction between power "consumption" and power "output" is important. For example, a lightbulb consumes a certain amount of electrical energy (for which you pay the utility company), and it outputs a certain amount of energy in the form of visible light along with a certain amount of energy as heat.

EXAMPLE 6.11 Comparing Mechanical and Electrical Energy

Consider a lightbulb rated for a power $P_{\text{bulb}} = 20$ W. This lightbulb consumes a certain amount of electrical energy in 1 second of use. Suppose we want to generate the same amount of energy by dropping the lightbulb from height h to ground level. If the mass of the lightbulb is $m = 0.050$ kg, what is h?

RECOGNIZE THE PRINCIPLE

We apply conservation of energy principles to the lightbulb. The change in potential energy of the lightbulb is $\Delta PE = -mgh$. This energy must equal (in magnitude) the energy used by the lightbulb in 1 s.

(continued) ▶

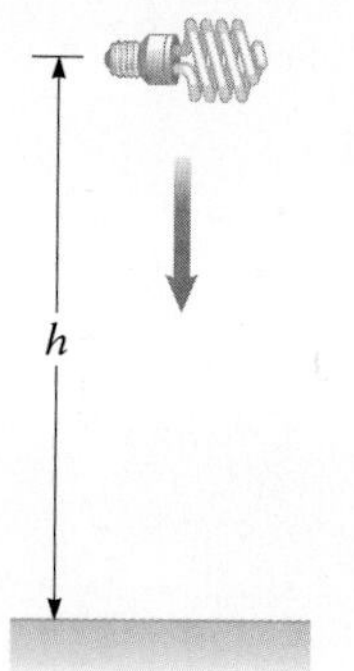

▲ **Figure 6.34** Example 6.11. A falling lightbulb.

SKETCH THE PROBLEM

Figure 6.34 describes the problem.

IDENTIFY THE RELATIONSHIPS

Using the definition of power (Eq. 6.28), the energy consumed by the lightbulb in time t is

$$W = P_{\text{bulb}}t$$

Setting this equal to the change in potential energy when the lightbulb falls gives

$$mgh = P_{\text{bulb}}t$$

SOLVE

Solving for h, we find

$$h = \frac{P_{\text{bulb}}t}{mg} = \frac{(20\ \text{W})(1.0\ \text{s})}{(0.050\ \text{kg})(9.8\ \text{m/s}^2)} = \boxed{41\ \text{m}}$$

► ***What does it mean?***

This result is a surprisingly large height: what we normally consider to be a rather small electrical power rating (of the lightbulb) corresponds to a substantial amount of mechanical energy. Imagine what h would be if the lightbulb were turned on for 1 day!

Power and Velocity

Consider a "device" such as a motor or a person that is moving an object from one place to another. This device is expending an amount of power given by Equation 6.28. For one-dimensional motion with the force parallel to the displacement Δx, we can write this relation as

$$P_{\text{ave}} = \frac{W}{t} = \frac{F\,\Delta x}{t}$$

Since the average velocity is equal to $v_{\text{ave}} = \Delta x/t$, we have

$$P_{\text{ave}} = Fv_{\text{ave}}$$

This relationship between power and velocity also holds for the instantaneous velocity v and the instantaneous power P:

$$P = Fv \tag{6.29}$$

If the device, such as a motor, is capable of expending a certain amount of power, Equation 6.29 shows that there is a trade-off between force and velocity. That is, with a fixed value of P, a motor can exert a large force (as might be required to move a heavy object) while moving an object at a small velocity, or the motor can exert a small force (as needed to move a light object) while moving the object at a large velocity. Such trade-offs involving force and velocity are quite common.

Power, Force, and Efficiency

The trade-off between force and velocity in Equation 6.29 limits the *maximum* possible force consistent with Newton's laws and conservation of mechanical energy. *Real* motors and other devices usually do not attain this theoretical limit, however. For example, an automobile engine might stop running ("stall") at a very low velocity, or there might be some frictional forces in the problem. To account for this behavior, we define the efficiency ε by

$$F = \varepsilon\frac{P}{v} \tag{6.30}$$

The efficiency cannot be larger than unity. We'll see an example involving efficiency in the next section, when we discuss the operation of molecular motors.

6.8 Work, Energy, and Molecular Motors

In Section 6.7, we discussed how our ideas about work, energy, and power can be used to understand the behavior of motors and similar devices. The same ideas apply to the molecular motors that transport materials within and between cells. Several different types of molecular motors have been discovered, one example of which is sketched in Figure 6.35. This motor is based on a molecule called myosin that moves along long filaments composed of actin molecules.

Myosin is believed to move by "walking" along a filament in steps, much like a person walks on two legs. This process takes place in muscles, where the filaments form bundles. A myosin molecule is also attached to its own muscle filament, so as it walks, the myosin drags one muscle filament relative to another. The operation of your muscles is produced by these molecular motors.

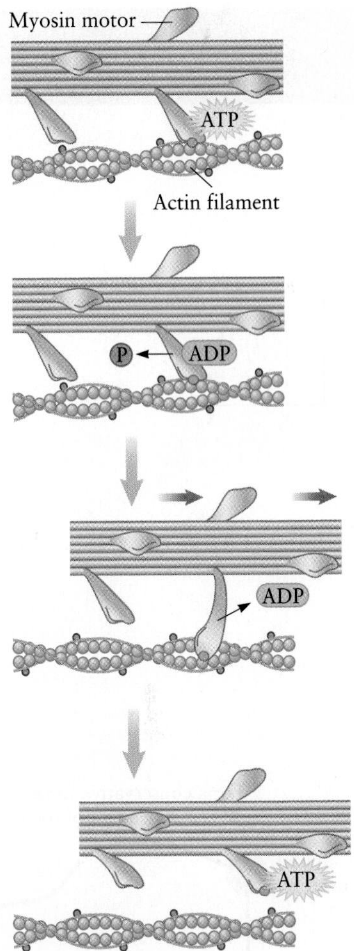

▲ **Figure 6.35** Some molecular motors move by "walking" along long strands of a protein called actin. These motors are the subject of much current research. We can use work–energy principles to understand their behavior.

Calculating the Force Exerted by a Molecular Motor

The precise biochemical reactions involved in the myosin walking motion are not completely understood. However, we do know that each step has a length of approximately 5 nm (5×10^{-9} m) and that the energy for this motor comes from a chemical reaction in which the molecule ATP reacts to form ADP. (ATP is shorthand for adenosine triphosphate, and ADP is adenosine diphosphate.) The energy released when a single ATP molecule breaks down is $E_{ATP} \approx 5 \times 10^{-20}$ J. With this information, we can use work–energy ideas to calculate the maximum force this motor can achieve.

The energy E_{ATP} is stored as chemical potential energy. One ATP molecule is consumed during each step, so the molecular motor uses an amount of potential energy E_{ATP} during each step. During the course of a single step, the potential energy of the motor decreases by E_{ATP} while it does an amount of work W_{motor}. Energy is conserved, so

$$W_{motor} = E_{ATP}$$

The work done by the motor is also equal to the motor force times the length of a single step Δx:

$$W_{motor} = F_{motor}\,\Delta x = E_{ATP} \tag{6.31}$$

$$F_{motor} = \frac{E_{ATP}}{\Delta x}$$

Inserting the values of E_{ATP} and Δx given above, we find

$$F_{motor} = \frac{E_{ATP}}{\Delta x} = \frac{5 \times 10^{-20}\text{ J}}{5 \times 10^{-9}\text{ m}} = 1 \times 10^{-11}\text{ N}$$

for the force generated by the motor. This force is 10 piconewtons, which may seem like a very small force. However, it is actually very substantial compared with other "molecular-scale" forces such as the weight of a typical molecule. Indeed, the weight of a single ATP molecule is about 1×10^{-23} N, which is *many* orders of magnitude smaller than the force exerted by this molecular motor!

Our calculation of F_{motor} assumes all the ATP potential energy is converted to work by the motor, that is, that the efficiency of the motor is $\varepsilon = 1$ (a "perfect" motor). Real motors will have a smaller efficiency as some of the chemical potential energy will go into vibrations, other molecular motions of the actin filament, and so forth. For this reason, our value for F_{motor} is an upper limit on the force that can be produced by a myosin motor. Evidence suggests that real molecular motors operate with a typical efficiency $\varepsilon > 0.5$, so our result for F_{motor} provides a fairly good description of real myosin motors.

Summary | CHAPTER 6

Key Concepts and Principles

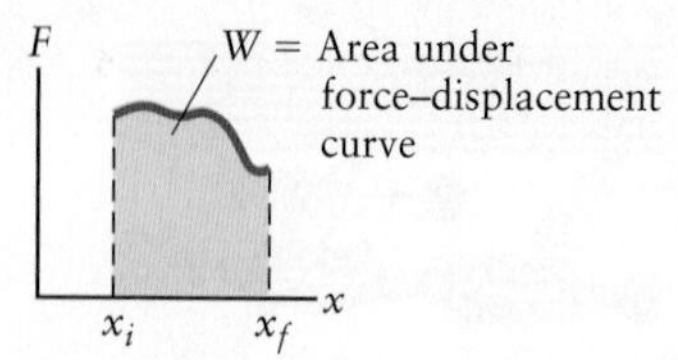

Work

When a force acts on an object, the force does ***work*** on the object:

$$W = F(\Delta r)\cos\theta \qquad \text{(6.4) (page 147)}$$

The work done by an applied force can be interpreted in a graphical manner as the area under the force–displacement curve.

Kinetic energy

Kinetic energy is energy of motion and is given by

$$KE = \tfrac{1}{2}mv^2 \qquad \text{(6.8) (page 151)}$$

Work and kinetic energy are related through the ***work–energy theorem***

$$W = \Delta KE \qquad \text{(6.9) (page 151)}$$

where W is the work done on the object.

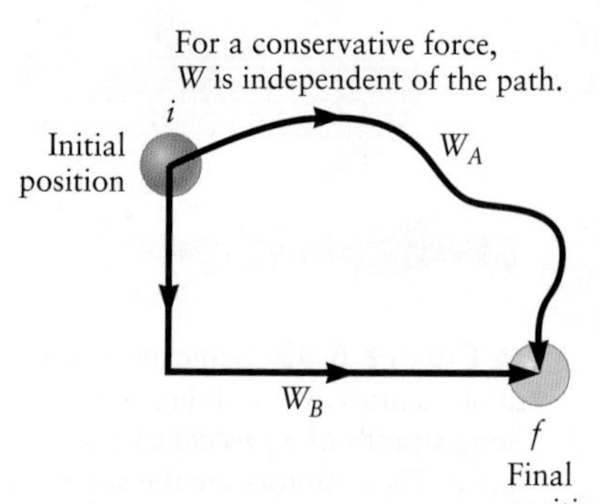

Potential energy

Potential energy is stored energy associated with a particular force or forces. The potential energy associated with a force is related to the work done by this force:

$$\Delta PE = PE_f - PE_i = -W \qquad \text{(6.10) (page 155)}$$

The work done by a ***conservative force*** is independent of the path.

Conservation of energy

The principle of conservation of energy can be written as

$$KE_i + PE_i + W_{\text{noncon}} = KE_f + PE_f \qquad \text{(6.26) (page 172)}$$

where W_{noncon} is the work done on the system by nonconservative forces.

Applications

Hooke's law

The force exerted by a spring or similar elastic object is

$$F = -kx \qquad \text{(6.23) (page 165)}$$

where k is the ***spring constant*** and x is the amount that the spring is stretched or compressed.

Potential energy functions

Gravitational potential energy for an object near the Earth's surface:

$$PE_{\text{grav}} = mgy \qquad \text{(6.13) (page 156)}$$

Gravitational potential energy of two masses (general case):

$$PE_{\text{grav}} = -\frac{Gm_1m_2}{r} \qquad (6.20)\ (\text{page } 164)$$

Elastic potential energy:

$$PE_{\text{spring}} = \tfrac{1}{2}kx^2 \qquad (6.24)\ (\text{page } 166)$$

Questions

SSM = answer in *Student Companion & Problem-Solving Guide* ⊗ = life science application

1. Consider an object undergoing uniform circular motion. We know from Chapter 5 that circular motion is caused by a force that is directed toward the center of the circle. What is the work done by this force during circular motion?
2. Explain why kinetic energy can never be negative, but the potential energy can be positive, negative, or zero.
3. A heavy crate is pushed across a level floor at a constant velocity. What is the total work done on the crate by all forces? How does your answer change if the crate is being pushed up a hill?
4. Consider a satellite in an elliptical orbit around the Earth. What is the sign (positive or negative) of the power associated with the gravitational force of the Earth on the satellite? *Hint*: The answer depends on the position of the satellite along its orbit. Where in the orbit is the power zero?
5. Consider an apple that falls from a branch to the ground below. At what moment is the kinetic energy of the apple largest? At what moment is the gravitational potential energy largest?
6. SSM For the bungee jumper in Example 6.9 (page 169), plot the following quantities as a function of time: (a) the gravitational potential energy of the jumper, (b) the kinetic energy of the jumper, (c) the potential energy associated with the bungee cord, and (d) the total mechanical energy.
7. The work done by friction is usually negative, as in the case of an object that slides to a stop. The work done by friction does not have to be negative, however. Give an example in which the work done by friction is positive.
8. Motor 1 does 20 J of work in 10 s, while motor 2 does 5 J of work in 1 s. Which motor produces the greatest power?
9. According to Equation 6.29, power is related to force and velocity by $P = Fv$. Is it possible for P to be negative? Give an example and explain what it means for P to be negative.
10. Give two examples in which the work done on an object by friction is negative.
11. Give two examples in which a nonzero force acts on an object, but the work done by that force is zero.
12. Explain how a mosquito can have a greater kinetic energy than a baseball.
13. A hockey puck slides along a level surface, eventually coming to rest. Is the energy of the hockey puck conserved? Is the total energy of the universe conserved? Discuss where the initial kinetic energy of the puck "goes."
14. A spring is initially compressed. It is then released, sending an object flying off the spring. Discuss this process in terms of conservation of energy. What and "where" is the initial energy, and where is the energy found at the end?
15. Tarzan starts at rest near the top of a tall tree. He then swings on a vine and reaches the base of the tree, where he jumps off the vine and comes to rest on the ground. Discuss and compare Tarzan's total energy at (a) the top of the tree, (b) just before he jumps off the vine at the bottom, and (c) when he is standing on the ground. Is Tarzan's total energy the same in all three places? Explain how total energy is conserved in going from (a) to (b) to (c).
16. During the course of each day, you move from place to place and do work on various objects. Where does the energy for these everyday activities come from? Explain.
17. SSM When a rubber ball is dropped onto a concrete floor, it bounces to a height that is slightly lower than its initial height. Compare the ball's (a) initial mechanical energy to (b) its energy just before hitting the ground, to (c) the total energy just after bouncing off the ground, and to (d) the total energy when it reaches it final height. Is the ball's mechanical energy conserved? Explain and discuss using the principle of conservation of total energy.
18. Construct bar charts showing the kinetic energy and potential energy of the projectile in Figure 6.16. Show these energies when the projectile has just left the ground, when it is at the highest point of its trajectory, and just before it returns to the ground. Explain how these charts illustrate the principle of conservation of mechanical energy.
19. Construct bar charts showing the kinetic energy and potential energy for the bungee jumper in Example 6.9 at various points during his jump. Explain how these charts illustrate the principle of conservation of mechanical energy.

Problems available in ENHANCED WebAssign

Section 6.1 Force, Displacement, and Work
Problems 1–14

Section 6.2 Kinetic Energy and the Work–Energy Theorem
Problems 15–30

Section 6.3 Potential Energy and Conservation of Energy
Problems 31–51

Section 6.4 More Potential Energy Functions
Problems 52–66

Section 6.5 Conservative Versus Nonconservative Forces and Conservation of Energy
Problems 67–71

Section 6.7 Power
Problems 72–77

Section 6.8 Work, Energy, and Molecular Motors
Problems 78–83

Additional Problems
Problems 84–103

List of Enhanced Problems

Problem number	Solution in *Student Companion & Problem-Solving Guide*	Reasoning & Relationships Problem	Reasoning Tutorial in ENHANCED WebAssign
6.4	✓		
6.22		✓	
6.23		✓	✓
6.26		✓	
6.28	✓		
6.37	✓		
6.40		✓	
6.42	✓		
6.44		✓	✓
6.54	✓		
6.63		✓	✓
6.64		✓	
6.65		✓	✓
6.66		✓	✓
6.67	✓		
6.75		✓	
6.77		✓	✓
6.80		✓	
6.83	✓	✓	
6.84	✓		
6.85		✓	✓
6.86		✓	
6.87	✓	✓	
6.94		✓	
6.101	✓		
6.102		✓	
6.103		✓	

CHAPTER 7

◀ *Collisions are important in many everyday situations. In this chapter, we will learn how to analyze different types of collisions. (TRL Ltd./Photo Researchers, Inc.)*

Momentum, Impulse, and Collisions

OUTLINE

Until now, we have treated all objects as "point particles," objects whose mass is located at a single point in space. The point particle assumption is very useful and is the correct way to deal with many situations. However, some important kinds of motion cannot be dealt with in this way. For example, objects such as baseballs and boomerangs can spin or rotate as they move along their trajectory. To deal with such motion, we must consider the ball or boomerang as an extended object, taking size and shape into account. In this and the next two chapters, we show how Newton's laws can be used to treat the motion of such extended objects. Conceptually, we will imagine that our baseball is composed of many very small "pieces." If the pieces are small enough, we can treat each as a point particle using the methods developed in previous chapters. The different pieces of an object exert forces on one another—when you push on one side of a baseball, the rest of the ball moves, too—so we must also deal with the forces that inevitably exist in such a *system* of interacting particles. We first encountered the notion of a system in Chapter 6, where we found that the total energy of a closed system (i.e., a system that is not acted on by any outside forces) is conserved. In this chapter, we introduce another property called ***momentum*** and discover that the total momentum of a closed system of particles is also conserved. This principle of ***conservation of momentum*** provides a very powerful tool in the analysis of motion in a wide variety of situations. It will lead us to the concept of ***center of mass*** and provide a general way to deal with collisions between particles.

7.1 Momentum

The variables velocity and acceleration are needed to describe the motion of a single particle. When dealing with a collection of particles, it is useful to define an additional quantity, ***momentum***. The momentum $\vec{p}$ of a single particle of mass m moving with a velocity $\vec{v}$ is defined as

The momentum of a particle depends on its mass and velocity.

$$\vec{p} = m\vec{v} \tag{7.1}$$

The momentum of a particle thus is along the same direction as the velocity. Notice also that $\vec{p}$ is proportional to the mass of the particle. We'll soon see how this relation for the momentum arises from Newton's second law.

EXAMPLE 7.1 The Momentum of a Bullet

Consider a bullet that has a mass of 10 g and a speed of 600 m/s. (**a**) Find the momentum of the bullet. (**b**) Find the speed of a baseball (m = 0.14 kg) that has the same momentum as the bullet.

RECOGNIZE THE PRINCIPLE

The magnitude of the momentum can be found directly from Equation 7.1, given the mass and speed of the objects of interest. The goal of this example is to get a sense of the momentum of two "everyday" objects.

SKETCH THE PROBLEM

No figure is necessary.

IDENTIFY THE RELATIONSHIPS

The speed and mass of the bullet are both given, so we can apply Equation 7.1,

$$\vec{p} = m\vec{v}$$

SOLVE

(**a**) In terms of magnitudes, the momentum of the bullet is

$$p = (0.010 \text{ kg})(600 \text{ m/s}) = \boxed{6.0 \text{ kg} \cdot \text{m/s}}$$

The direction of $\vec{p}$ is parallel to the velocity of the bullet.

(**b**) If a baseball has the same momentum as this bullet, we have

$$\vec{p}_{\text{ball}} = \vec{p}_{\text{bullet}}$$

In terms of the magnitudes, we then get

$$m_{\text{ball}}v_{\text{ball}} = m_{\text{bullet}}v_{\text{bullet}}$$

Hence, the speed of the ball is

$$v_{\text{ball}} = \frac{m_{\text{bullet}}v_{\text{bullet}}}{m_{\text{ball}}}$$

Inserting the numerical values given above, we find

$$v_{ball} = \frac{m_{bullet}v_{bullet}}{m_{ball}} = \frac{(0.010\text{ kg})(600\text{ m/s})}{0.14\text{ kg}} = \boxed{43\text{ m/s}}$$

▸ *What does it mean?*

This speed is approximately 100 mi/h and is about the speed that can be reached by a professional baseball pitcher. The momentum of a particle is the *product* of its mass and velocity, so a particular value of $\vec{p}$ can be obtained with a small mass moving at a high velocity or with a large mass moving with a smaller velocity. The momentum carried by a baseball thrown by a major-league pitcher is about the same as the momentum of a bullet!

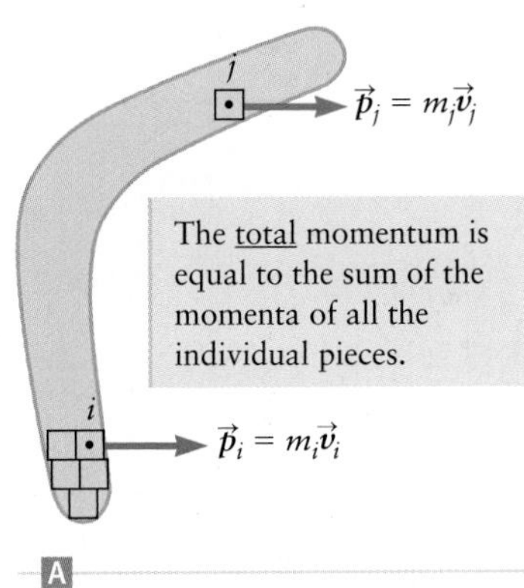

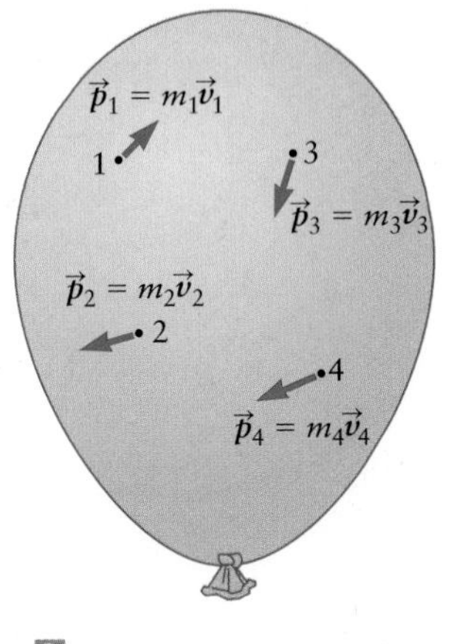

▲ **Figure 7.1** A A solid object such as a boomerang can be thought of as a system composed of very small pieces, that is, a collection of point particles. B The gas in a balloon is a system of molecules. The total momentum is the sum of the individual momenta of all the molecules.

Momentum of a System of Particles

To find the total momentum of a system of particles, we take the sum of the momenta of all the individual pieces of the system. For example, if we want to find the total momentum of a boomerang, we can imagine it as a collection of many small pieces as illustrated in Figure 7.1A. Each piece can be treated as a point particle, and the momentum of each piece is given by Equation 7.1. The total momentum of a collection of such pieces is then

$$\vec{p}_{total} = \Sigma_i \vec{p}_i = \Sigma_i m_i\vec{v}_i \tag{7.2}$$

where the sums are over all the individual particles (pieces) in the system. Such a system of particles may be the pieces of a solid object, such as a boomerang, or pieces of a nonsolid object, such as the molecules in a gas (Fig. 7.1B). In either case, the total momentum of the system is given by Equation 7.2.

EXAMPLE 7.2 Momentum of Two Cars

Two cars are on course for a collision in one dimension (Fig. 7.2). The small car has a mass of 1200 kg and a speed of 20 m/s, and the large car has a mass of 2000 kg and a speed of 15 m/s in the opposite direction. What is the total momentum of the two cars?

RECOGNIZE THE PRINCIPLE

We can find the momentum of each car using Equation 7.1 and then add the results to get the total momentum. We must include the sign of the velocity (+ or −) of each car when calculating their momenta.

SKETCH THE PROBLEM

Figure 7.2 describes the problem. The cars are both moving along the x axis.

IDENTIFY THE RELATIONSHIPS

To calculate the total momentum, we can treat each car as a point particle. We then have (Eq. 7.2)

$$\vec{p}_{total} = \Sigma_i \vec{p}_i = \Sigma_i m_i\vec{v}_i = m_1\vec{v}_1 + m_2\vec{v}_2$$

where m_1 and m_2 are the masses of the cars and $\vec{v}_1$ and $\vec{v}_2$ are the cars' velocities. In this example, the motion is one dimensional, so we can work in terms of the x components of $\vec{p}$ and $\vec{v}$. From Figure 7.2, we have $v_1 = +20$ m/s for the smaller car and $v_2 = -15$ m/s for the larger car. Here, the signs *do* matter.

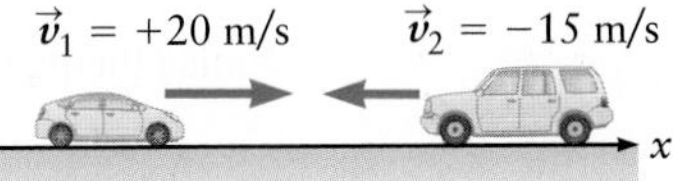

▲ **Figure 7.2** Example 7.2.

(continued) ▸

SOLVE

The total momentum of our system of two cars is thus

$$p_{\text{total}} = m_1 v_1 + m_2 v_2$$

$$p_{\text{total}} = (1200 \text{ kg})(20 \text{ m/s}) + (2000 \text{ kg})(-15 \text{ m/s}) = \boxed{-6000 \text{ kg} \cdot \text{m/s}}$$

▸ ***What does it mean?***

Even though the smaller car has a higher speed, the magnitude of the momentum of the larger car is greater because of its greater mass. For this reason, the total momentum is in the $-x$ direction, that is, toward the left in Figure 7.2.

CONCEPT CHECK 7.1 Momentum and Kinetic Energy

Consider a rock of mass m moving with a speed v. Suppose the mass of the rock is doubled, while the speed is cut in half. Which of the following statements is true?

(a) The momentum of the rock is the same, but its kinetic energy is smaller.
(b) The momentum and kinetic energy of the rock are both unchanged.
(c) The momentum of the rock is greater, and its kinetic energy is smaller.

CONCEPT CHECK 7.2 Momentum and Kinetic Energy of Two Particles

Two particles of different mass have the same kinetic energy. Which one has the larger momentum?

(a) The particle with the smaller mass
(b) The particle with the larger mass
(c) They have the same momentum.

7.2 Force and Impulse

Suppose a force acts on a particle of mass m. Let's consider how this force affects the momentum of the particle. If $\vec{F}$ is the total force acting on the particle, from Newton's second law we then have $\vec{F} = m\vec{a}$. For the moment, we assume the force and hence also the acceleration are constant. During the course of a very short time interval Δt, the acceleration is related to changes in the velocity according to $\vec{a} = \Delta\vec{v}/\Delta t$. Combining these two facts gives

$$\vec{F} = m\vec{a} = m\frac{\Delta\vec{v}}{\Delta t} = m\frac{\vec{v}_f - \vec{v}_i}{\Delta t} \tag{7.3}$$

Here, we imagine that the particle has a velocity $\vec{v}_i$ just before the force is applied and that after a time Δt the velocity is $\vec{v}_f$. From the definition of momentum in Equation 7.1, the initial momentum of the particle is $\vec{p}_i = m\vec{v}_i$ and the final momentum is $\vec{p}_f = m\vec{v}_f$. Inserting this all into Equation 7.3 and rearranging gives

$$\vec{F}\,\Delta t = m(\vec{v}_f - \vec{v}_i) = \vec{p}_f - \vec{p}_i = \Delta\vec{p} \tag{7.4}$$

where $\Delta\vec{p}$ is the change in the momentum. The product $\vec{F}\,\Delta t$ is called the ***impulse.*** According to Equation 7.4, when a force acts on a particle, the change in momentum of the particle is equal to the impulse:

Impulse theorem

$$\text{impulse} = \vec{F}\,\Delta t = \Delta\vec{p} \tag{7.5}$$

Equation 7.5 is called the ***impulse theorem***, and it follows directly from Newton's laws. Impulse is a vector, so its direction is parallel to the total force on the particle. Because impulse is the product of force and time, a particular value of the impulse can be obtained with a small force acting for a long time or with a large force acting for a short time.

Impulse Associated with a Variable Force

In our derivation of the relation between impulse and momentum in Equation 7.5, we assumed the force acts for a very short time Δt and the force is constant during this time interval as in Figure 7.3A. In many cases involving impulse and momentum, however, the force is not constant. Such a case is illustrated in Figure 7.4, which shows a tennis ball being struck by a racket. The force exerted by the racket on the ball has the form sketched in Figure 7.3B: the magnitude of the force grows from zero to some maximum value and then decreases to zero as the ball leaves the racket. For the simpler case of a constant force, the magnitude of the impulse is simply $F\,\Delta t$ (Eq. 7.5), which is the area under the corresponding force–time curve in Figure 7.3A. When the force is not constant, we can calculate the impulse by dividing the entire time period into many small intervals and using Equation 7.5 for each interval as illustrated in Figure 7.3B. The total impulse is then the sum of the impulses during each interval and is equal to the area under the complete force–time curve:

$$\text{impulse} = \text{area under the force–time curve} = \Delta\vec{p} \tag{7.6}$$

In both cases—with a constant force as in Figure 7.3A or a force that varies with time as in Figure 7.3B—the impulse is equal to the change in the momentum of the object. Hence, the impulse theorem we initially derived in Equation 7.5 for the case of a constant force also applies for the impulse produced by a force that varies with time.

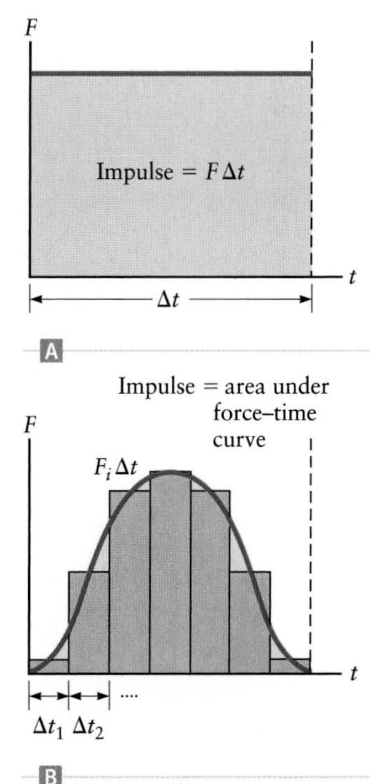

▲ **Figure 7.3** Impulse is equal to the area under the force−time graph. **A** Impulse for a constant force. **B** When the force varies with time, the impulse during a time interval Δt_i is approximately equal to the area of a rectangular bar of height F_i.

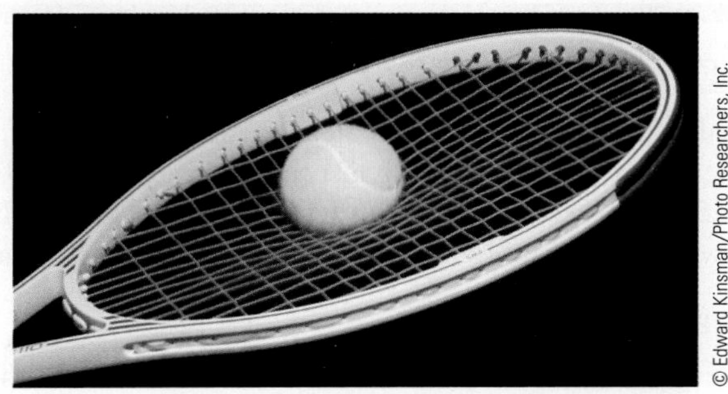

▲ **Figure 7.4** The force exerted by a tennis racket on a ball is usually not constant. In such cases, the impulse is still equal to the area under the force−time graph as illustrated in Figure 7.3B.

EXAMPLE 7.3 Impulse and Playing Golf

A golf ball (m = 0.046 kg) is hit from a tee (Fig. 7.5). If the ball has speed v_f = 40 m/s just after it is hit, find (**a**) the magnitude of the impulse imparted by the golf club on the ball and (**b**) the work done by the club on the ball.

RECOGNIZE THE PRINCIPLE

From the impulse theorem, the impulse imparted to the ball is equal to the change in momentum. We can find the change in momentum since the initial and final velocities are known. The work done on the ball is equal to the change in its kinetic energy.

SKETCH THE PROBLEM

Figure 7.5 shows the ball just after it leaves the tee, when its speed is v_f.

(continued) ▶

▲ **Figure 7.5** Example 7.3.

IDENTIFY THE RELATIONSHIPS AND SOLVE

(a) Using the impulse theorem (Eq. 7.5), we have

$$\text{impulse} = \Delta\vec{p} = \vec{p}_f - \vec{p}_i = m\vec{v}_f - m\vec{v}_i$$

The ball is initially at rest, so $\vec{v}_i = 0$. The magnitude of the impulse is thus

$$\text{impulse} = mv_f$$

Inserting the given mass and final speed, we get

$$\text{impulse} = mv_f = (0.046 \text{ kg})(40 \text{ m/s}) = \boxed{1.8 \text{ kg} \cdot \text{m/s}}$$

(b) According to the work–energy theorem, the work done on the ball equals the change in kinetic energy. The initial kinetic energy is zero, so

$$W = \Delta KE = \tfrac{1}{2}mv_f^2 - \tfrac{1}{2}mv_i^2 = \tfrac{1}{2}mv_f^2$$

Inserting the given values of m and v_f gives

$$W = \tfrac{1}{2}mv_f^2 = \tfrac{1}{2}(0.046 \text{ kg})(40 \text{ m/s})^2 = \boxed{37 \text{ J}}$$

▶ ***What does it mean?***

The initial momentum is zero, so the direction of the impulse is parallel to the final velocity $\vec{v}_f$ in Figure 7.5.

EXAMPLE 7.4 Impulse and a Home Run

Consider a baseball player who hits a baseball as sketched in Figure 7.6. The ball travels from the pitcher toward the batter with a speed of 45 m/s (about 100 mi/h). The batter then hits it directly back at the pitcher with a speed of 50 m/s. If the ball has a mass of 0.14 kg, what is the impulse imparted by the bat to the baseball?

RECOGNIZE THE PRINCIPLE

We are given the initial and final velocities of the baseball, so we can find the impulse using Equation 7.5. We must be careful to use the correct signs (+ and −) for the velocities.

SKETCH THE PROBLEM

Figure 7.6 shows the problem. The initial velocity is along $-x$ because the ball is moving to the left, and the final velocity is positive (to the right along $+x$).

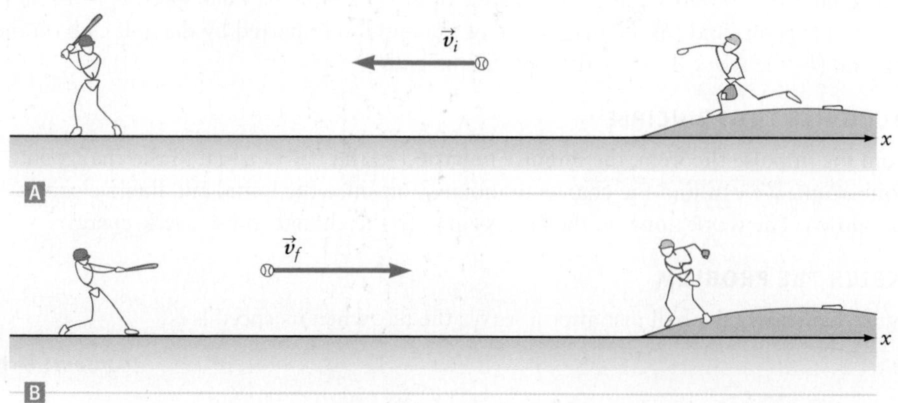

▶ **Figure 7.6** Example 7.4. A The ball has some initial momentum as it travels to the batter and B some final value of momentum after being hit by the bat. The impulse imparted to the baseball equals the change in its momentum. Here, $\Delta\vec{p}$ is equal to the final momentum minus the initial momentum.

IDENTIFY THE RELATIONSHIPS

We begin with the impulse theorem:

$$\text{impulse} = \Delta\vec{p} = m\vec{v}_f - m\vec{v}_i \quad (1)$$

This problem is one dimensional since the initial and final velocities are both along the line that extends from the batter to the pitcher, the x direction in Figure 7.6. The force exerted by the bat on the ball is also parallel to this direction. We can thus work simply in terms of the components of the velocity and momentum vectors along x. Using the coordinate system in Figure 7.6, we have $v_i = -45$ m/s and $v_f = +50$ m/s.

SOLVE

Inserting our values of m, v_i, and v_f into Equation (1), the impulse is

$$\Delta p = mv_f - mv_i = (0.14 \text{ kg})(50 \text{ m/s}) - (0.14 \text{ kg})(-45 \text{ m/s}) = \boxed{13 \text{ kg} \cdot \text{m/s}}$$

► *What does it mean?*

The initial velocity and the initial momentum are both in the $-x$ direction. The bat then imparts a large positive impulse to the ball, so the final momentum of the ball is positive, directed toward the right in Figure 7.6B.

Impulse and the Average Force

For cases such as the force exerted by a tennis racket on a ball in Figure 7.4, it is usually very difficult to calculate the precise form of the force–time curve, so it is also difficult to apply Equation 7.6 to find the impulse. However, analyzing the impulse in a graphical manner is very useful in a different way. Often, the interaction time—the length of time during which the force is nonzero—is very short. For example, when a baseball is hit by a bat, high-speed photography (Fig. 7.7) shows that the bat and ball are in contact for approximately $\Delta t = 0.001$ s. We can approximate the force–time curve by assuming F is constant during this short interaction time and write

$$\text{impulse} = \vec{F}_{\text{ave}}\,\Delta t = \Delta\vec{p} \quad (7.7)$$

where $\vec{F}_{\text{ave}}$ is the *average* interaction force as indicated graphically in Figure 7.8. The value of F_{ave} is chosen so that the area under the rectangular region bounded by F_{ave} in Figure 7.8 is equal to the area under the true force–time curve. In many cases, this average force is all we are interested in. For example, if we assume the baseball in Example 7.4 is in contact with the bat for a time $\Delta t = 0.001$ s, the average force is related to the change in momentum by

$$\vec{F}_{\text{ave}}\,\Delta t = \Delta\vec{p}$$

We can then use the result for the impulse found in Example 7.4 to calculate the size of this force. In terms of the magnitudes, we get

$$F_{\text{ave}} = \frac{\Delta p}{\Delta t} = \frac{13 \text{ kg} \cdot \text{m/s}}{0.001 \text{ s}} \approx 1 \times 10^4 \text{ N} \quad (7.8)$$

The force on the baseball is thus *extremely* large. The weight of the ball is $mg = (0.14 \text{ kg})g = 1.4$ N, so the force exerted by the bat in Equation 7.8 is about 10,000 times the ball's weight! Such collision forces are often very large because the corresponding interaction times are usually very short.

▲ **Figure 7.7** A baseball is in contact with a bat for approximately 1 ms (= 0.001 s). During this short time, the force exerted by the bat on the ball is very large.

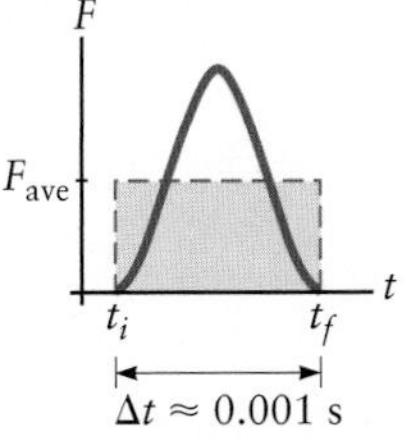

▲ **Figure 7.8** We can find the average force associated with an impulse by approximating the force–time relation by a constant (F_{ave}) during the time interval over which the force acts. The dashed horizontal line shows the (constant) average force over this time interval.

Minimizing Collision Forces

In Section 3.6, we considered a child jumping from a tall ladder to the ground below and analyzed the forces on the child's legs during the landing. We can also view that example as an impulse problem. The child has a certain initial momentum just before she hits the ground, and her final momentum is zero because she comes to rest. The

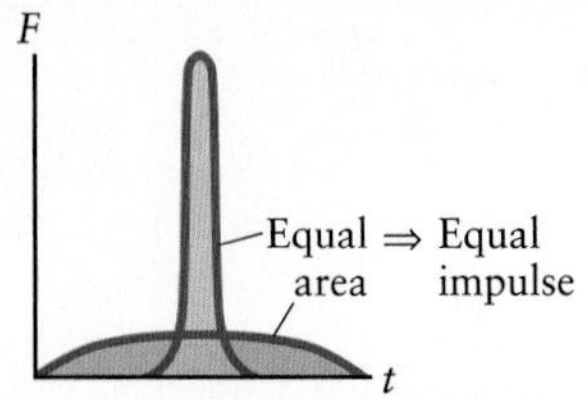

▲ **Figure 7.9** Impulse is equal to the area under the force–time curve. A particular value of the impulse can be produced by a small force acting for a long time or by a large force acting for a short time.

impulse exerted by the ground on the child can be calculated in the same way we found the impulse on the batted baseball in Example 7.4. In Chapter 3, we discussed strategies for minimizing the force on the child, and we found that it is safest for her to allow her legs to bend and flex as much as possible. By doing so, she is making the interaction time Δt as long as possible. Our understanding of impulse and force (Eq. 7.7) also explains why increasing Δt will reduce the average force F_{ave}. The effect of Δt on the force is illustrated graphically in Figure 7.9. A particular value of the impulse (the area under the force–time curve) can be obtained with a large force acting for a short time or with a small force acting for a long time. There are many other examples of this relationship between force and collision time. For example, the air bags in your car are designed to take advantage of this trade-off. When your car stops suddenly, as in an accident, the air bags inflate and fill the region between you (the driver) and the steering wheel and front windshield. Your collision with an air bag then involves a much longer interaction time than if you were to collide with the steering column or windshield. This longer collision time leads to a smaller force, and a safer accident.

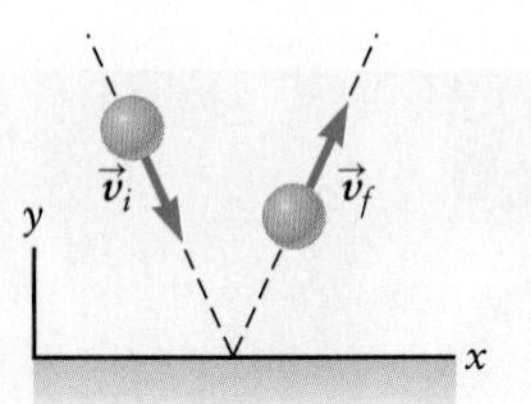

▲ **Figure 7.10** Concept Check 7.3.

CONCEPT CHECK 7.3 Impulse on a Bouncing Ball

Consider the collision of a bouncing ball with a floor as sketched in Figure 7.10. What is the direction of the impulse imparted by the floor to the ball?

(a) Horizontal and to the right in Figure 7.10, $\Delta\vec{p} = \rightarrow$
(b) Vertical and upward, $\Delta\vec{p} = \uparrow$
(c) At 45° from the vertical direction, $\Delta\vec{p} = \nearrow$

7.3 Conservation of Momentum

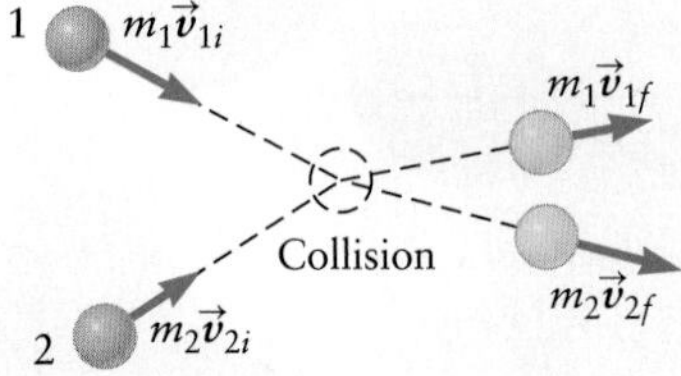

▲ **Figure 7.11** When two objects collide, the total momentum just after the collision is equal to the total momentum just before the collision. Momentum is conserved.

It is useful to apply our ideas about momentum and the impulse theorem to the motion of a system of particles. Consider the system comprised of just two colliding point particles sketched in Figure 7.11. For the moment, we ignore any rotational or spinning motion. With this assumption, let's use the notion of impulse to analyze how a collision affects the subsequent motion of the two particles.

When two objects collide, there is an associated collision force. Let $\vec{F}_1$ be the force exerted by object 2 on object 1 and $\vec{F}_2$ be the force exerted by object 1 on object 2. These forces are an action–reaction pair, so according to Newton's third law, they must be equal in magnitude and opposite in direction. Hence,

$$\vec{F}_1 = -\vec{F}_2 \tag{7.9}$$

If this collision takes place over a time interval Δt, the impulse imparted to object 1 is $\vec{F}_1\,\Delta t$. According to the impulse theorem, this impulse equals the change in momentum of object 1, so we have

$$\Delta\vec{p}_1 = \vec{p}_{1f} - \vec{p}_{1i} = \vec{F}_1\,\Delta t \tag{7.10}$$

Here, $\vec{p}_{1i}$ is the (initial) momentum of object 1 just before the collision and $\vec{p}_{1f}$ is the (final) momentum just after the collision. Likewise, the impulse imparted to object 2 is $\vec{F}_2\,\Delta t$, which equals the change in the momentum of object 2:

$$\Delta\vec{p}_2 = \vec{p}_{2f} - \vec{p}_{2i} = \vec{F}_2\,\Delta t \tag{7.11}$$

Since $\vec{F}_1 = -\vec{F}_2$ and the interaction times Δt are the same in Equations 7.10 and 7.11, the impulse imparted to object 1 is equal in magnitude but opposite in sign to the impulse imparted to object 2. We thus have

$$\Delta\vec{p}_1 = \vec{F}_1\,\Delta t = -\vec{F}_2\,\Delta t = -\Delta\vec{p}_2$$

$$\Delta\vec{p}_1 = -\Delta\vec{p}_2 \tag{7.12}$$

We can also write Equation 7.12 as

Conservation of momentum in a collision

$$\Delta\vec{p}_1 + \Delta\vec{p}_2 = \Delta\vec{p}_{total} = 0 \tag{7.13}$$

Hence, the collision does *not* change the *total* momentum of the two particles. Whatever momentum is lost or gained by one particle in the collision is gained or lost by the other. The *total momentum of the system is conserved.*

Conservation of Momentum for a System of Many Particles

Our derivation of the conservation of momentum in Equation 7.13 was based on a simple system of just two particles undergoing a single collision. We can apply the same argument to a system containing a very large number of particles, in which the particles undergo many collisions with one another. Further, let's assume this system is *closed*, meaning that there are no external forces exerted on any of the particles; that is, the only forces on a particle are exerted by other particles within the system. These internal forces are due to collisions between particles in the system; each collision can be viewed as in Figure 7.11, and the sum of the momenta of the two particles is the same before and after the collision (Eq. 7.13). The same argument applies to all the collisions between particles in the system, so we conclude that the total momentum of the entire system is conserved.

This result is very similar to what we found when we considered the total energy of a closed system in Chapter 6. Recall that the particles in a closed system can exchange energy among one another in various ways, but the particles do not gain energy from or lose energy to objects outside the system. The total energy of the system is thus conserved. The same notion applies to the total momentum. The particles in a closed system do not experience any forces from objects external to the system. The momentum gained or lost by one particle in a collision is lost or gained by other particles in the system, and the total momentum of the system is conserved.

This analysis has so far assumed the system is composed of a collection of particles that are traveling more or less independently, except for occasional collisions. It also applies to a solid object such as a boomerang, a football, or a planet. We can think of a solid object as being composed of many point particles; these individual particles are subject to strong forces from other particles in the system that hold the object together. These forces will always come in action–reaction pairs, so for any two interacting particles, we always have $\vec{F}_1 = -\vec{F}_2$, just as in Equation 7.9. The same reasoning then tells us that any momentum lost by one particle will be gained by the other. Hence, we again conclude that the total momentum of a system, in this case a solid object, is conserved, provided there are no external forces acting on the object.

Momentum Conservation and External Forces

We have just shown that for a closed system of particles, the interaction forces between particles do not change the total momentum of the system. These interactions cause some particles to lose momentum while others gain momentum, but the total momentum of the system stays the same. However, these arguments apply only to the internal forces, that is, the forces that act between particles within the system. There may also be forces acting from outside the system, called external forces. As an example, consider again the two-particle system in Figure 7.11 and assume these two particles are asteroids in outer space. If these two asteroids are near a star, the gravitational force from the star is an external force on this two-asteroid system. This external force causes the asteroids to accelerate toward the star and hence changes the total momentum of the two asteroids.

In practice, it is difficult to find situations in which the external forces on a system are exactly zero. For example, when dealing with any system of terrestrial objects, the Earth's gravitational force is always present as an external force. However, our analysis of the collision between a baseball and a bat (Eq. 7.8) showed that collision forces are often much larger than this gravitational force. For the baseball–bat collision, the collision force was about 10,000 times larger than the weight of the ball. Thus, while the total momentum of the bat–baseball system is not conserved precisely during this collision, it will change by only about 0.01% (1 part in 10,000).

For this reason, the notion of momentum conservation is still a very useful way to analyze most collisions.

7.4 Collisions

In this section, we apply the principle of momentum conservation to analyze several different types of collisions. In general, a collision changes the particle velocities; the velocities just after the collision are different from the velocities just before the collision. The kinetic energy of a particle is proportional to the square of its speed, so the kinetic energy of any particular particle is also changed by the collision. Collisions fall into two general types, depending on what happens to the total kinetic energy of the two particles involved. In an ***elastic collision***, the system's kinetic energy is conserved: the total kinetic energy of the two particles after the collision is equal to their total kinetic energy before the collision. In contrast, ***inelastic collisions*** are collisions in which some kinetic energy is lost. In both types of collisions, *momentum is conserved*.

The term *elastic* can help us understand how and why a collision affects the kinetic energy. For example, an extremely elastic ball (such as a Super Ball) is compressed in a collision, and this compression stores energy in the ball just as energy is stored in a compressed spring. In an ideal Super Ball, all this potential energy is turned back into kinetic energy when the ball uncompresses ("springs back") at the end of the collision. Collisions involving such objects will therefore tend to be elastic. On the other hand, consider a collision involving a ball composed of soft putty or clay. Such a ball does not "spring back" at the end of the collision, causing the kinetic energy after the collision to be smaller than the initial kinetic energy. The collision is thus inelastic.

We next analyze a few examples of these different types of collisions and show how conservation principles regarding momentum (for all collisions) and kinetic energy (for elastic collisions) can be used to predict the motions of two colliding particles. Our analyses will be guided by the following general approach.

PROBLEM SOLVING Analyzing a Collision

1. **RECOGNIZE THE PRINCIPLE.** The momentum of a system is conserved only when the external forces are zero. For a collision between two particles, the system is just the two particles. When external forces are zero or negligible, one can use the principle of conservation of momentum to analyze a collision.

2. **SKETCH THE PROBLEM.** Make a sketch of the system, showing the coordinate axes and (where possible) the initial and final velocities of the particles in the system.

3. **IDENTIFY THE RELATIONSHIPS.**
   - Express the conservation of momentum condition (Eq. 7.13) for the system.
   - Is kinetic energy conserved?

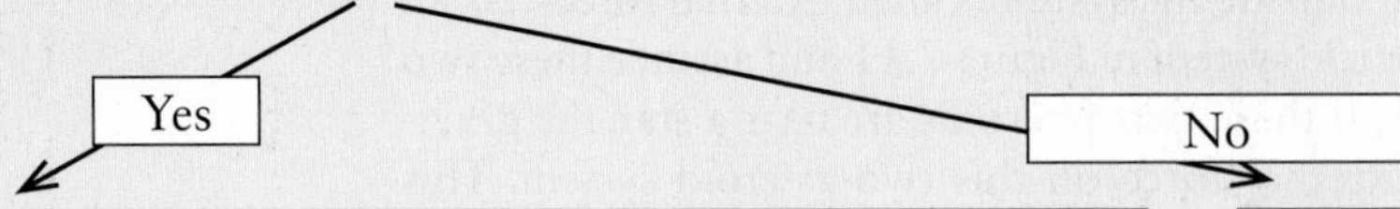

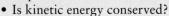

Elastic collision

4. **SOLVE.**
   - Express the conservation of kinetic energy for both particles.
   - Use the equations describing conservation of momentum and kinetic energy to solve for unknown quantities, such as the final velocities.

Inelastic collision

4. **SOLVE** for unknown quantities using the equations derived from the conservation of momentum.

5. Always *consider what your answer means* and check that it makes sense.

Elastic Collisions in One Dimension

Let's apply the above strategy for dealing with collisions to an elastic collision in one dimension involving particles with masses m_1 and m_2 that are subject to no external forces. Step 1: Because the external forces are zero, the momentum of the system (the two colliding particles) will be conserved. Step 2 is the sketch in Figure 7.12, which shows the particles and their velocities before and after the collision. The particles are moving along the x axis before the collision with velocities v_{1i} and v_{2i}; since this problem is one dimensional, we only need to be concerned with the x components of the velocities. After the collision, the velocities are v_{1f} and v_{2f}. The total momentum of the two particles is conserved, so we have

$$\overbrace{m_1v_{1i} + m_2v_{2i}}^{\text{initial momentum}} = \overbrace{m_1v_{1f} + m_2v_{2f}}^{\text{final momentum}} \tag{7.14}$$

This is an elastic collision, so the total kinetic energy of the system is also conserved, which gives

$$\overbrace{\tfrac{1}{2}m_1v_{1i}^2 + \tfrac{1}{2}m_2v_{2i}^2}^{\text{initial kinetic energy}} = \overbrace{\tfrac{1}{2}m_1v_{1f}^2 + \tfrac{1}{2}m_2v_{2f}^2}^{\text{final kinetic energy}} \tag{7.15}$$

In a typical collision, the initial velocities are known, and we have two equations—Equations 7.14 and 7.15—to solve for the two unknowns, the final velocities v_{1f} and v_{2f}.

Insight 7.1

ALWAYS BEGIN BY IDENTIFYING THE SYSTEM

When applying the principle of conservation of momentum, it is important to first identify the system. It is usually best to choose the system so that all the important forces act between different parts of the system and the external forces are zero or at least small. If the external forces are zero, the total momentum of the system will be conserved exactly.

Elastic collision in one dimension

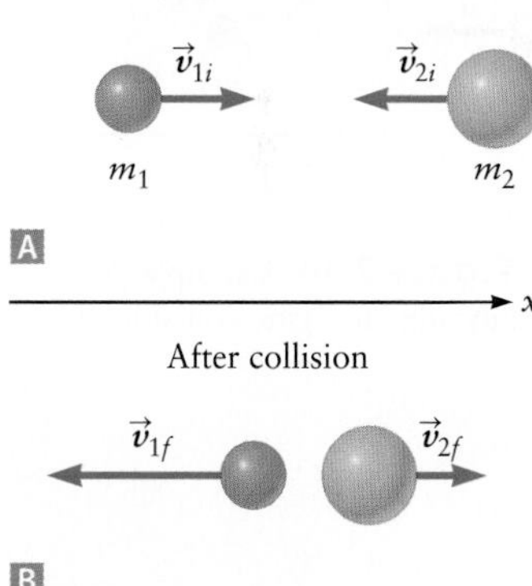

▲ **Figure 7.12** In an elastic collision, both momentum and kinetic energy are conserved. **A** Velocities just before a one-dimensional collision. **B** Velocities just after the collision.

A Collision between Two Billiard Balls

Let's now use Equations 7.14 and 7.15 to analyze a one-dimensional elastic collision between two billiard balls, assuming one of the balls is initially at rest and the other has an initial speed v_0 as sketched in Figure 7.13A. Because our problem is one dimensional, the collision is "head-on" and the initial and final velocities are all along the same direction. For simplicity we'll ignore all effects of spin in this problem. In the game of billiards, all balls have the same mass m. Inserting $v_{1i} = v_0$ and $v_{2i} = 0$ into our relations for the momentum (Eq. 7.14) and the kinetic energy (Eq. 7.15), we find

$$mv_0 = mv_{1f} + mv_{2f} \tag{7.16}$$

$$\tfrac{1}{2}mv_0^2 = \tfrac{1}{2}mv_{1f}^2 + \tfrac{1}{2}mv_{2f}^2 \tag{7.17}$$

Solving for v_{2f} in Equation 7.16 gives

$$v_{2f} = v_0 - v_{1f} \tag{7.18}$$

and inserting this into Equation 7.17, we get

$$\tfrac{1}{2}mv_0^2 = \tfrac{1}{2}mv_{1f}^2 + \tfrac{1}{2}m(v_0 - v_{1f})^2$$

Canceling the factors of $\frac{1}{2}m$ and solving for v_{1f} leads to

$$v_0^2 = v_{1f}^2 + (v_0 - v_{1f})^2 = v_{1f}^2 + v_0^2 - 2v_0v_{1f} + v_{1f}^2$$

Subtracting v_0^2 from each side and collecting terms gives

$$0 = v_{1f}^2 - 2v_0v_{1f} + v_{1f}^2 = 2v_{1f}^2 - 2v_0v_{1f}$$

$$2v_{1f}^2 = 2v_0v_{1f} \tag{7.19}$$

This equation has two solutions. One solution is $v_{1f} = 0$, and using Equation 7.18 then gives $v_{2f} = v_0$. Physically, ball 1 comes to a complete stop after the collision and ball 2 moves away with a velocity of v_0 afterward (Fig. 7.13B). This solution is familiar to many pool players.

The second solution of Equation 7.19 is $v_{1f} = v_0$, and using Equation 7.18 leads to $v_{2f} = 0$. Hence, with this solution, ball 2 stays at rest and ball 1 simply passes through it. You can think of this as a collision between two transparent balls. While you may not have expected this solution for a collision problem, it certainly does

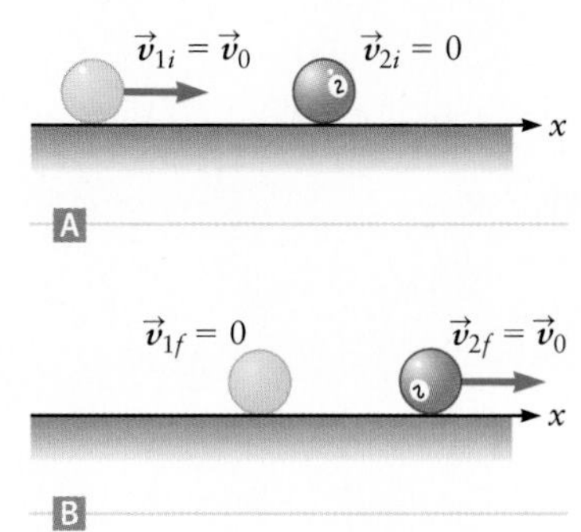

▲ **Figure 7.13** Elastic collision in one dimension between two billiard balls. **A** One of the balls is initially at rest. **B** After the collision, the other ball is at rest.

conserve both kinetic energy and momentum. This solution would only apply if the collision force between the two balls is zero, and that does not occur in any real collision between two objects.

EXAMPLE 7.5 Elastic Collision between Balls of Different Mass

Consider again an elastic collision between two balls with one initially at rest (Fig. 7.13A), but this time assume the balls have different masses. Repeat the analysis of this collision and find the final velocities of both balls.

RECOGNIZE THE PRINCIPLE

We can again apply conservation of momentum (Eq. 7.14) and conservation of kinetic energy (Eq. 7.15). Our goal is to understand how the relative masses of the particles affect the final velocities. We follow the steps outlined in the problem-solving box on analyzing a collision. Step 1: The external forces are zero, so the momentum of the colliding particles will be conserved.

SKETCH THE PROBLEM

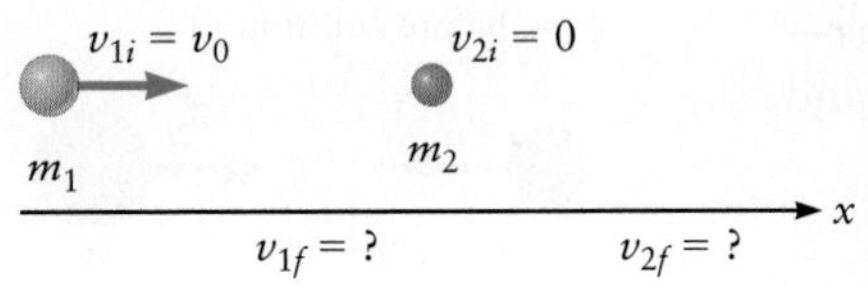

▲ **Figure 7.14** Example 7.5. Analyzing an elastic collision.

Step 2: Figure 7.14 shows the problem. We denote the masses as m_1 and m_2. One of them is initially at rest, so the velocities before the collision are $v_{1i} = v_0$ and $v_{2i} = 0$. The final velocities (denoted v_{1f} and v_{2f}) are unknowns that we wish to find. This information is all collected in Figure 7.14.

IDENTIFY THE RELATIONSHIPS

We continue with our problem-solving strategy. Step 3: The conservation of momentum condition gives

$$\overbrace{m_1 v_{1i} + m_2 v_{2i}}^{\text{initial momentum}} = m_1 v_0 = \overbrace{m_1 v_{1f} + m_2 v_{2f}}^{\text{final momentum}} \tag{1}$$

We are given that kinetic energy is conserved, so we have an elastic collision.

SOLVE

Step 4: Setting the initial and final kinetic energies equal, we find

$$\overbrace{\tfrac{1}{2}m_1 v_{1i}^2 + \tfrac{1}{2}m_2 v_{2i}^2}^{\text{initial } KE} = \tfrac{1}{2}m_1 v_0^2 = \overbrace{\tfrac{1}{2}m_1 v_{1f}^2 + \tfrac{1}{2}m_2 v_{2f}^2}^{\text{final } KE} \tag{2}$$

We next rearrange the momentum conservation relation (Eq. 1) to solve for the final velocity of mass 2:

$$v_{2f} = \frac{m_1(v_0 - v_{1f})}{m_2} \tag{3}$$

Substituting this result into the conservation of kinetic energy expression, Equation (2):

$$\frac{1}{2}m_1 v_0^2 = \frac{1}{2}m_1 v_{1f}^2 + \frac{1}{2}m_2\left[\frac{m_1(v_0 - v_{1f})}{m_2}\right]^2$$

We can now solve for the final velocity of mass 1 and then use Equation (3) to find the final velocity of mass 2. After some algebra, we find two solutions. Solution 1 is

$$v_{1f} = v_0 \quad \text{and} \quad v_{2f} = 0$$

In this solution, there is no collision at all because mass 1 simply passes through mass 2; that is, the final velocity of mass 1 is equal to its initial velocity, while mass 2 is at rest before and after the collision. This particular solution is only possible if the collision force between the two balls is zero, which does not occur in a real collision. There is a second solution for the final velocities that does describe a real collision. It is

$$v_{1f} = \left(\frac{m_1 - m_2}{m_1 + m_2}\right)v_0 \quad \text{and} \quad v_{2f} = \left(\frac{2m_1}{m_1 + m_2}\right)v_0$$

▸ *What does it mean?*

These results for the final velocities of the two balls correspond to an actual collision. One important case is when m_2 is very small compared with m_1, which leads to

$$v_{1f} = \left(\frac{m_1 - m_2}{m_1 + m_2}\right)v_0 \approx \left(\frac{m_1 - 0}{m_1 + 0}\right)v_0 = v_0$$

Hence, the final velocity of a very massive particle (m_1) is approximately unchanged by the collision with a very light particle (m_2). When a car collides with a mosquito, the car's velocity barely changes.

The Power of Conservation Principles

In dealing with the elastic collisions in Figures 7.12 through 7.14, we started with the conditions for conservation of momentum and conservation of kinetic energy. For a one-dimensional collision, this process leads to two equations (Eqs. 7.14 and 7.15). In a typical problem, we might know the initial velocities of the two objects and be asked to find their final velocities. Hence, the two equations are all we need to solve for these two unknowns. This result is remarkable: it means that we can work out the outcome of the collision without knowing *anything* about the force that acts between the two objects during the collision. For an elastic collision in one dimension, the conditions for conservation of kinetic energy and momentum *completely* determine the results. This example illustrates that conservation laws are an extremely powerful way to get from general principles to specific results.

Inelastic Collisions in One Dimension

We have just analyzed some collisions in which kinetic energy was conserved. However, in many collisions the total kinetic energy after a collision is smaller than the kinetic energy just prior to the collision. These collisions are called *inelastic*. Although some portion of the kinetic energy of the two colliding objects is "lost" in such a collision, the *total* energy of the universe is still conserved. This "lost" kinetic energy goes into other forms of energy, such as potential energy and heat energy. Although kinetic energy is not conserved in an inelastic collision, the total momentum is still conserved. Hence, for a one-dimensional inelastic collision, we can still use the conservation of momentum relation

$$m_1 v_{1i} + m_2 v_{2i} = m_1 v_{1f} + m_2 v_{2f} \tag{7.20}$$

However, this equation does not give enough information to allow us to solve for the final velocities of both objects. We have only one relation, but two unknown quantities, v_{1f} and v_{2f}. So, to analyze an inelastic collision fully, we need additional information.

Completely Inelastic Collisions

The simplest type of inelastic collision is one in which the two objects stick together after the collision so that the objects have the same final velocity. This is called a ***completely inelastic collision***. A good example of a completely inelastic collision is one in which two cars lock bumpers. Figure 7.15A shows two cars of mass m_1 and m_2 coasting on a straight, one-dimensional road with velocities v_{1i} and v_{2i}. The cars collide and lock bumpers, and they have the same velocity v_f after the collision (Fig. 7.15B). The cars are moving in a horizontal direction (x), and if they are coasting, there are no external forces on the cars in this direction. Hence, the total momentum along x is conserved and we can apply our problem-solving strategy for dealing with collisions. Writing the condition for conservation of momentum along x gives

$$m_1 v_{1i} + m_2 v_{2i} = m_1 v_f + m_2 v_f \tag{7.21}$$

In this case, we have only one unknown, the final velocity v_f, so the conservation of momentum relation in Equation 7.21 is all we need. Solving for v_f, we have

$$m_1 v_{1i} + m_2 v_{2i} = (m_1 + m_2) v_f$$

Completely inelastic collision in one dimension

$$v_f = \frac{m_1 v_{1i} + m_2 v_{2i}}{m_1 + m_2} \tag{7.22}$$

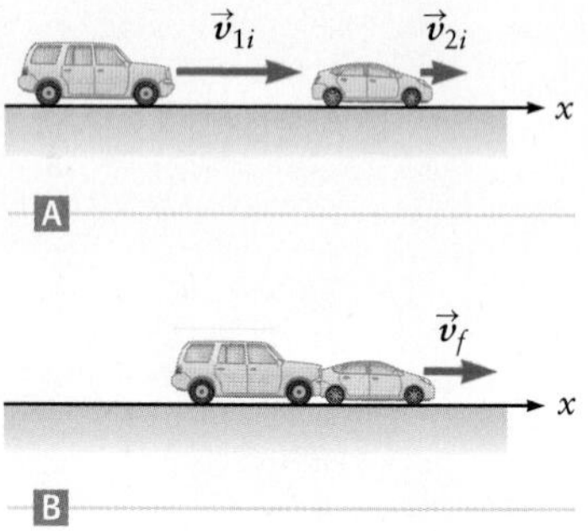

▲ **Figure 7.15** A Two cars on a collision course in one dimension. B In a completely inelastic collision, the two objects stick together after the collision.

EXAMPLE 7.6 A Completely Inelastic Collision between Two Cars

Two cars, an SUV of mass 2500 kg and a compact model of mass 1200 kg, are coasting along a straight road. Their initial velocities along this coordinate direction are 40 m/s and 10 m/s, respectively, so the SUV is overtaking the compact car. When they collide, the cars lock bumpers. Find the cars' velocity just after the collision.

RECOGNIZE THE PRINCIPLE

The cars "stick together" after the collision, so this collision is completely inelastic, and we can use our problem-solving strategy for collisions. Step 1: We first note that the two cars are the system. Since they are both coasting, there are no external forces along the horizontal direction (x), and the total momentum along x is conserved.

SKETCH THE PROBLEM

Step 2: The collision is described in Figure 7.15. It is a one-dimensional collision, so the given initial velocities are the velocity components along our coordinate direction (x), which we choose as parallel to the road.

IDENTIFY THE RELATIONSHIPS

Step 3: The cars have masses $m_1 = 2500$ kg (the SUV) and $m_2 = 1200$ kg (the compact car), with initial velocities $v_{1i} = 40$ m/s and $v_{2i} = 10$ m/s (Fig. 7.15A). The final velocity of the two cars together is unknown, and we denote it by v_f. Continuing with our problem-solving strategy, conservation of momentum leads to the results in Equations 7.21 and 7.22. This collision is inelastic, so the kinetic energy of the two cars is not conserved.

SOLVE

Inserting the given values of the masses and initial velocities into Equation 7.22 leads to

$$v_f = \frac{m_1 v_{1i} + m_2 v_{2i}}{m_1 + m_2} = \frac{(2500 \text{ kg})(40 \text{ m/s}) + (1200 \text{ kg})(10 \text{ m/s})}{2500 \text{ kg} + 1200 \text{ kg}} = \boxed{30 \text{ m/s}}$$

▶ *What does it mean?*

The final velocity lies between the two initial velocities, but v_f is closer to the initial velocity of the larger car because it had a larger fraction of the initial momentum.

CONCEPT CHECK 7.4 Hitting a Parked Car

A careless driver is coasting at 20 m/s (about 40 mi/h) along a straight, level road in a car of mass 1400 kg when she collides with a parked car. The parked car is in neutral, and its brakes are off. The two cars lock bumpers and move off together. If the final velocity is 10 m/s, what is the mass of the parked car?

(a) 700 kg (b) 1400 kg (c) 2800 kg

Inelastic Collisions: What Happens to the Kinetic Energy?

We have already mentioned that kinetic energy is not conserved in an inelastic collision. Let's compare the initial and final kinetic energies for the specific case in Example 7.6. The kinetic energy before the collision is

$$KE_i = \tfrac{1}{2}m_1v_{1i}^2 + \tfrac{1}{2}m_2v_{2i}^2$$

Inserting the values of the masses and velocities from Example 7.6 leads to

$$KE_i = \tfrac{1}{2}m_1v_{1i}^2 + \tfrac{1}{2}m_2v_{2i}^2 = \tfrac{1}{2}(2500 \text{ kg})(40 \text{ m/s})^2 + \tfrac{1}{2}(1200 \text{ kg})(10 \text{ m/s})^2$$

$$KE_i = 2.1 \times 10^6 \text{ J}$$

The cars stick together after the collision, so they have the same final velocity v_f. The final kinetic energy is therefore

$$KE_f = \tfrac{1}{2}m_1v_{1f}^2 + \tfrac{1}{2}m_2v_{2f}^2 = \tfrac{1}{2}(m_1 + m_2)v_f^2$$

Inserting the value of the final velocity from Example 7.6 gives

$$KE_f = \tfrac{1}{2}(m_1 + m_2)v_f^2 = \tfrac{1}{2}(2500 \text{ kg} + 1200 \text{ kg})(30 \text{ m/s})^2 = 1.7 \times 10^6 \text{ J}$$

The final kinetic energy is thus about 20% smaller than the initial kinetic energy. This "missing" energy would be converted to several other forms, including heat and sound.

Now consider a completely inelastic "head-on" collision involving two objects. For example, suppose two cars of equal mass are moving with equal speeds, but in opposite directions. In this case, $v_{1i} = -v_{2i}$ and $m_1 = m_2 = m$. Inserting these values into the result for v_f for a completely inelastic collision (Eq. 7.22) gives

$$v_f = \frac{m_1v_{1i} + m_2v_{2i}}{m_1 + m_2} = \frac{m(v_{1i} - v_{1i})}{2m} = 0$$

Hence, the two objects come to a complete stop after the collision, and the final kinetic energy is zero. *All* the initial kinetic energy is converted into other forms of energy.

Collisions in Two Dimensions

Dealing with collisions in two dimensions involves the same basic ideas as in one dimension, except that the final velocity of each object involves two unknowns: the two components of the velocity vector. Hence, for two colliding objects, there are usually four unknowns (two velocity components for each object). The momentum conservation relation then becomes

$$\vec{p}_{1i} + \vec{p}_{2i} = \vec{p}_{1f} + \vec{p}_{2f}$$

where $\vec{p}_{1i}$ is the initial momentum of object 1, $\vec{p}_{1f}$ is its final momentum, and so forth. It is useful to write this relation in terms of components:

$$p_{1ix} + p_{2ix} = p_{1fx} + p_{2fx} \quad \text{and} \quad p_{1iy} + p_{2iy} = p_{1fy} + p_{2fy} \tag{7.23}$$

To analyze a two-dimensional collision, we can again follow the basic approach outlined in our problem-solving strategy for analyzing a collision, using the momentum conservation relations in Equation 7.23. We must then include whatever additional information is available. This additional information might be that the

► **Figure 7.16** This crater (known as Meteor Crater) was produced when an asteroid struck the Earth in eastern Arizona about 49,000 years ago. This crater is about 1.2 km across.

Jim Wark/Visuals Unlimited/ Getty Images

collision is elastic and hence that total kinetic energy is conserved, or we might have some information regarding the final velocity of one of the objects.

A Collision in Two Dimensions: A Rocket, an Asteroid, and Saving the Earth

Our planet is constantly being bombarded with asteroids. These asteroids are usually very small and disintegrate completely as they enter the Earth's atmosphere. On occasion, however, a very large asteroid reaches the Earth's surface and causes a great deal of damage. It is believed that such a collision formed Meteor Crater, an impact crater 1.2 km in diameter located in eastern Arizona (Fig. 7.16). It has been proposed that a similar collision involving a much larger asteroid led to the extinction of the dinosaurs, and NASA is concerned about what might happen to life on the Earth when another large asteroid strikes the Earth at some future time. Scientists have therefore considered how to use rockets to deflect incoming asteroids before they reach the Earth. The collision between such a rocket and an asteroid is sketched in Figure 7.17. For simplicity, we assume the rocket hits the asteroid and then becomes embedded in it, so this collision is completely inelastic. Given the masses of the rocket and asteroid and their initial velocities, we want to calculate the common final velocity of the two. From that result, we can determine if the asteroid will subsequently strike the Earth.

We can analyze this collision using our problem-solving strategy for dealing with collisions. Step 1: The system consists of the two colliding "particles," the rocket and the asteroid. The external forces on this system are due to gravity from the Sun and the Earth, which will be much smaller than the collision force when the rocket hits the asteroid. Hence, to very good accuracy, the momentum of this system will be conserved. Step 2: We choose the coordinate system in Figure 7.17 with the initial velocity of the asteroid in the $+y$ direction, aimed directly toward the Earth. The asteroid's initial velocity is then $v_{ax} = 0$, $v_{ay} = v_{a0}$. The rocket approaches the asteroid with an initial velocity in the $+x$ direction, so the rocket's initial velocity is $v_{rx} = v_{r0}$, $v_{ry} = 0$. The rocket "sticks to" the asteroid after the collision, so the rocket and asteroid have the same final velocity, and we denote the components of this final velocity by v_{fx} and v_{fy}. Step 3: We can now write the momentum conservation relations for our system of the rocket plus the asteroid. For the momentum along the x direction, we have

$$\overbrace{m_a v_{ax} + m_r v_{rx}}^{\text{initial momentum along } x} = m_r v_{r0} = \overbrace{(m_a + m_r) v_{fx}}^{\text{final momentum along } x}$$

The corresponding result for the momentum along y is

$$\overbrace{m_a v_{ay} + m_r v_{ry}}^{\text{initial momentum along } y} = m_a v_{a0} = \overbrace{(m_a + m_r) v_{fy}}^{\text{final momentum along } y}$$

Step 4: Solving for the components of the final velocity, we find

$$v_{fx} = \frac{m_r v_{r0}}{m_a + m_r} \quad (7.24)$$

$$v_{fy} = \frac{m_a v_{a0}}{m_a + m_r} \quad (7.25)$$

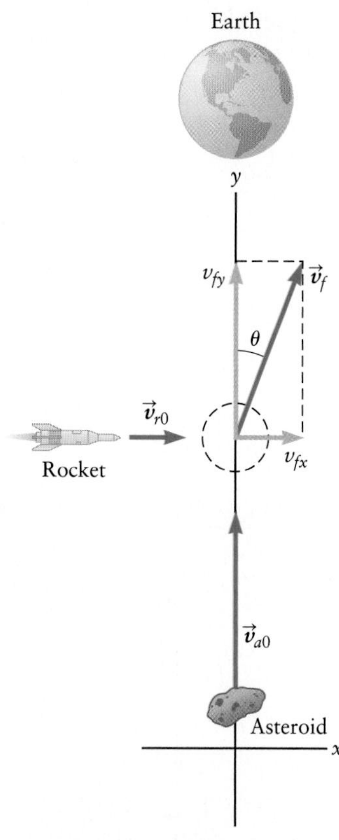

▲ **Figure 7.17** The collision between a rocket and an asteroid could be used to deflect an asteroid that is heading toward the Earth. If the rocket embeds (sticks) in the asteroid, the collision would be completely inelastic.

We can now calculate the final speed using $v_f = \sqrt{v_{fx}^2 + v_{fy}^2}$ and express the final direction in terms of the angle $\theta = \tan^{-1}(v_{fx}/v_{fy})$ shown in Figure 7.17.

Now let's put in some realistic values to see if this asteroid deflection plan can actually work. A rocket of reasonable size has an approximate mass $m_r = 2 \times 10^6$ kg, and a large asteroid has mass $m_a = 1 \times 10^{18}$ kg. If we assume the approach velocity of the rocket is similar to the velocity attained when NASA sent the Apollo spacecrafts to the Moon, then $v_{r0} = 1000$ m/s. We also assume the asteroid's initial velocity is equal to the Earth's orbital velocity relative to the Sun, which is approximately 3×10^4 m/s. Inserting all these values into Equations 7.24 and 7.25 leads to

$$v_{fx} = \frac{m_r v_{r0}}{m_a + m_r} = \frac{(2 \times 10^6 \text{ kg})(1000 \text{ m/s})}{(1 \times 10^{18} + 2 \times 10^6) \text{ kg}} = 2 \times 10^{-9} \text{ m/s} \tag{7.26}$$

$$v_{fy} = \frac{m_a v_{a0}}{m_a + m_r} = \frac{(1 \times 10^{18} \text{ kg})(3 \times 10^4 \text{ m/s})}{(1 \times 10^{18} + 2 \times 10^6) \text{ kg}} = 3 \times 10^4 \text{ m/s}$$

These values for the final velocity of the asteroid plus rocket show that asteroid deflection is a very tough problem. The value of v_{fx} is *extremely* small because the mass of the asteroid is much, much larger than the mass of the rocket. In addition, v_{fy} is essentially equal to the initial velocity of the asteroid, again because the rocket is much smaller than the asteroid. The real test of success (or failure) can be determined by calculating the angle through which the asteroid is deflected. That angle is

$$\theta = \tan^{-1}\left(\frac{v_{fx}}{v_{fy}}\right) = \tan^{-1}\left(\frac{2 \times 10^{-9}}{3 \times 10^4}\right)$$

$$\theta = 7 \times 10^{-14} \text{ rad} = 4 \times 10^{-12} \text{ degrees} \tag{7.27}$$

This angle is very small. It is not easy to deflect an asteroid.

CONCEPT CHECK 7.5 Saving the Earth

The deflection angle in the rocket–asteroid collision in Figure 7.17 (and Eq. 7.27) is very small. Which of the following changes would increase the deflection angle? More than one choice may apply.

(a) Increase the mass of the rocket.
(b) Decrease the mass of the rocket.
(c) Increase the speed of the rocket.
(d) Decrease the speed of the rocket.
(e) Pick a slower asteroid.
(f) Options (a), (c), and (e) would all work.
(g) Options (b) and (d) would both work.

EXAMPLE 7.7 Ⓡ Forces in an Automobile Collision

Two cars approach an intersection as shown schematically in Figure 7.18. One of the drivers ignores a stop sign, resulting in a right-angle collision with the cars locked together after the collision. What is the approximate average force between the two cars during the collision? Assume they each have mass $m = 1000$ kg and they both are traveling with an initial speed of 20 m/s (about 45 mi/h).

RECOGNIZE THE PRINCIPLE

This problem requires us to combine what we know about force and impulse together with the principle of conservation of momentum. The geometry is similar to the rocket–asteroid problem in Figure 7.17, and we can take the same approach to calculating the final velocity $\vec{v}_f$ of the two cars. From $\vec{v}_f$, we can use the impulse theorem to get the impulse exerted by one car on the other. If we then estimate the collision time, we can get the average force exerted by one car on the other during the collision.

(continued) ▶

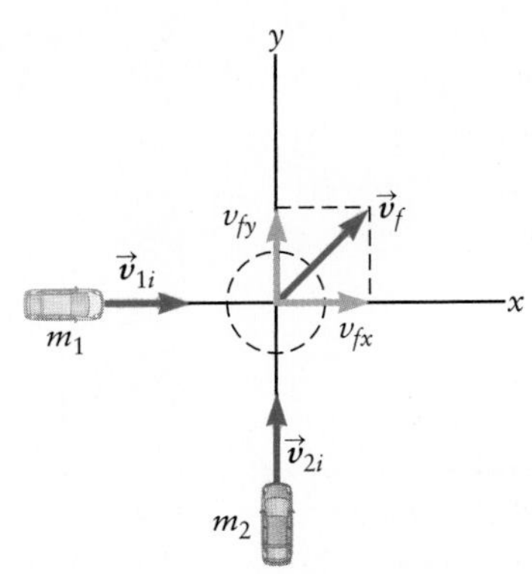

▲ **Figure 7.18** Example 7.7.

SKETCH THE PROBLEM

Figure 7.18 shows the problem. The initial velocities are along both x (car 1) and y (car 2), so we will have to account for momentum conservation in two dimensions.

IDENTIFY THE RELATIONSHIPS

Noticing the directions of $\vec{v}_{1i}$ and $\vec{v}_{2i}$ in Figure 7.18, we can write the components of the initial velocities of the two cars as $v_{1x} = 20$ m/s, $v_{1y} = 0$, and $v_{2x} = 0$, $v_{2y} = 20$ m/s. The collision is a completely inelastic right-angle collision, just as with the rocket and asteroid in Figure 7.17, so we can modify Equations 7.24 and 7.25 to get

$$v_{fx} = \frac{mv_{1x}}{m + m} = \frac{v_{1x}}{2}$$

$$v_{fy} = \frac{mv_{2y}}{m + m} = \frac{v_{2y}}{2}$$

Inserting the given values for the initial velocities, we find

$$v_{fx} = v_{fy} = 10 \text{ m/s} \tag{1}$$

We can now use the impulse theorem to relate the collision force exerted on one of the cars to the change in the momentum of that car.

SOLVE

Writing the impulse theorem (Eq. 7.5) for the x component of the momentum of car 2, we have

$$\text{impulse on car 2 along } x = \Delta p_{2x}$$

Expressing the impulse along x in terms of the average force along x (Eq. 7.7) gives

$$F_{\text{ave, }x}\,\Delta t = \Delta p_{2x}$$

The initial momentum of car 2 along x is zero, which leads to

$$F_{\text{ave, }x}\,\Delta t = \Delta p_{2x} = mv_{fx} - mv_{ix} = mv_{fx}$$

$$F_{\text{ave, }x} = \frac{mv_{fx}}{\Delta t}$$

For our final step, we need to know the collision time Δt. We estimate a value of approximately $\Delta t = 0.1$ s. As usual, we aim here for accuracy to within a factor of 10. A value of $\Delta t = 1$ s seems too long, whereas 0.01 s seems too short. We found v_{fx} in Equation (1). Using that result, we get

$$F_{\text{ave, }x} = \frac{mv_{fx}}{\Delta t} = \frac{(1000 \text{ kg})(10 \text{ m/s})}{(0.1 \text{ s})} = \boxed{1 \times 10^5 \text{ N}}$$

▶ ***What does it mean?***

This force is quite large, as we should expect; cars are massive objects. Notice also that this is just the x component of the collision force. From the symmetry of this problem in terms of x and y, however, you should suspect that the y component of the collision force equals $F_{\text{ave, }x}$.

CONCEPT CHECK 7.6 How Big Is the Collision Force?

In Example 7.7, we estimated the collision force in an automobile accident. Which of the following statements is correct?

(a) The magnitude of the collision force is much larger than a car's weight.
(b) The magnitude of the collision force is much smaller than a car's weight.
(c) The magnitude of the collision force is approximately equal to a car's weight.

7.5 Using Momentum Conservation to Analyze Inelastic Events

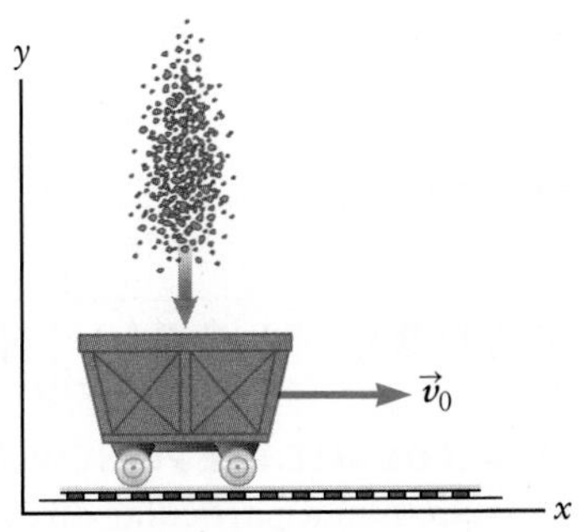

▲ **Figure 7.19** This system consists of the railroad car and the load of gravel dropped into it. The momentum of this system along the x direction is conserved.

Momentum is the product of mass and velocity. In all our examples so far, the mass has been constant (that is, unchanging). When the momentum of an object or system with constant mass is conserved, the velocity is constant. In some situations, though, the mass can change with time, and it is interesting to consider how the principle of conservation of momentum applies in such cases.

Figure 7.19 shows an empty railroad car coasting at speed v_0 along a level track. The track is straight and lies along the x direction. A load of gravel is then dropped into the car as it moves along. Given the masses of the empty car m_c and the load of gravel m_g, what is the speed of the railroad car after the gravel is dropped into it?

We can treat the problem as a "collision" involving the gravel and the railroad car and apply the principle of momentum conservation. The railroad car moves on a straight track, so its motion is one dimensional, along the x direction in Figure 7.19. If we assume there is no frictional force on the car (from bearings in the wheels, etc.), there are no external forces along x, and the component of the momentum along this direction is conserved. The initial momentum of the car along x is $m_c v_0$. Prior to landing in the car, the gravel has no momentum along x, so the total initial momentum of the car plus the gravel along x is also $m_c v_0$. After the gravel is added to the car, the car plus gravel has a velocity v_f, so the momentum conservation relation has the form

$$\overbrace{m_c v_0}^{\text{initial momentum along } x} = \overbrace{(m_c + m_g) v_f}^{\text{final momentum along } x} \tag{7.28}$$

Solving for the final velocity gives

$$v_f = \frac{m_c v_0}{m_c + m_g}$$

The value of v_f is always smaller than v_0, so the railroad car will always slow down after the gravel is dropped into it.

Our analysis so far has only considered momentum along x; we should also consider what happens to the momentum along y during this collision. Your first expectation may be that the momentum along y should also be conserved. Before the collision, the gravel has momentum along the $-y$ direction, whereas the car has no momentum along y. After the collision, the motion is solely along x, so the final momentum along y is zero. Hence, the momentum of the gravel plus the car along y is *not* conserved in this example. Why not? When the gravel hits the car, a vertical force is exerted by the car on the gravel, bringing the gravel to a stop along the y direction in the car. A corresponding normal force exerted by the railroad track on the car prevents the car from falling through the track and moving along y. The force exerted by the track on our system (the car plus gravel) is an external force; it imparts an impulse to the system and hence produces a change in the system's momentum along y. For this reason, the momentum of this system along y is not conserved. The momentum of a system is conserved *only* if the forces from outside the system are negligible. In this example, there is an external force along the y direction, so the momentum of the system along this direction is not conserved. However, the momentum of the system along the x direction is still conserved since there are no external forces along x.

Applying the Principle of Conservation of Momentum to Inelastic Events

The principle of momentum conservation can be applied to inelastic events, such as the railroad car in Figure 7.19, using a problem-solving strategy similar to the one we have used for dealing with collisions. That strategy was introduced with collisions in mind, so it is useful to state a new version written specifically for inelastic events. The basic ideas are the same: whenever the external forces on a system are zero or negligible, the total momentum of the system is conserved.

PROBLEM SOLVING Analyzing an Inelastic Event

1. **RECOGNIZE THE PRINCIPLE.** The momentum of a system in a particular direction is conserved only when the net external force in that direction is zero or negligible.
2. **SKETCH THE PROBLEM.** Make a sketch of the system, including a coordinate system, and use the given information to determine (where possible) the initial and final velocities of the system's components.
3. **IDENTIFY THE RELATIONSHIPS.**
   - Express the conservation of momentum condition for the direction(s) identified in step 1.
   - Use any given information to determine the increase or decrease of the kinetic energy.
4. **SOLVE** for the unknown quantities, such as the final velocities or the force exerted on one of the particles in the system.
5. Always *consider what your answer means* and check that it makes sense.

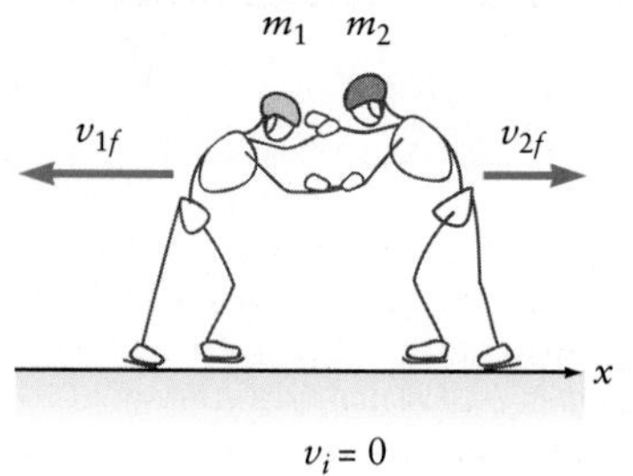

▲ **Figure 7.20** Example 7.8.

EXAMPLE 7.8 Momentum on Ice

Two angry children are fighting while standing on a very slippery icy surface (Fig. 7.20). Their masses are $m_1 = 25$ kg and $m_2 = 40$ kg, and they are initially at rest. The larger child pushes the smaller child, who then slides away to the left with a speed of 1.5 m/s. What is the final velocity of the larger child?

RECOGNIZE THE PRINCIPLE

We follow the strategy outlined in the problem-solving box on inelastic events. Step 1: The two children are the "system." The ice is very slippery, so the external forces in the horizontal (x) direction are negligible and the total momentum along x is conserved.

SKETCH THE PROBLEM

Continuing with our problem-solving strategy (step 2), the problem is sketched in Figure 7.20, which shows the initial and final velocities. Initially, the children are at rest, so $v_{1i} = v_{2i} = 0$, and we denote their final velocities v_{1f} and v_{2f}.

IDENTIFY THE RELATIONSHIPS

Step 3: The total initial momentum of this system along x is zero, and the final momentum along x is $m_1v_{1f} + m_2v_{2f}$. The final momentum must equal the initial momentum:

$$\overbrace{0}^{\text{initial momentum}} = \overbrace{m_1v_{1f} + m_2v_{2f}}^{\text{final momentum}}$$

SOLVE

Solving for the final velocity of child 2 gives

$$v_{2f} = -\frac{m_1v_{1f}}{m_2} = -\frac{(25\text{ kg})(-1.5\text{ m/s})}{40\text{ kg}} = \boxed{0.94\text{ m/s}}$$

► ***What does it mean?***

The final velocity of child 1 is to the left, so his velocity is negative, which leads to a positive velocity (to the right) for child 2.

Inelastic Processes Are Similar to Collisions

Most inelastic processes are very similar to collisions. Consider again the two children in Example 7.8. In Figure 7.21A, the children are initially at rest, whereas after pushing on each other they are traveling in opposite directions (Fig. 7.21B). A closely related situation is sketched in parts C and D of Figure 7.21. Here, the children are initially moving toward each other; then they collide and stick together, which is just a completely inelastic collision. Comparing these two scenarios, we see that the inelastic event in which the children push each other apart is just like a completely inelastic collision in "reverse." In both scenarios, the total momentum is conserved.

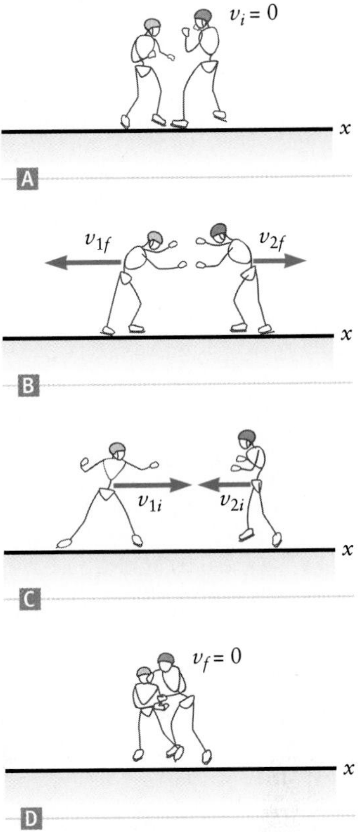

▲ **Figure 7.21** A and B: An inelastic event in which two children on ice push on each other and move apart. C and D: An inelastic collision in which the two children come together and "collide."

CONCEPT CHECK 7.7 Children and Forces in an Inelastic Event

Consider again the two children in Example 7.8, who push on each other as they fight. If the push lasts for 0.50 s, what is the magnitude of the average force exerted by the larger child on the smaller one?

(a) 25 N (b) 38 N (c) 75 N (d) Zero

Splitting Asteroids

Let's take another look at the daunting task of deflecting an asteroid that threatens the Earth. In our previous analysis, we tried to deflect the asteroid by arranging for an inelastic collision with a rocket (Fig. 7.17) and found that it is difficult to deflect an asteroid very much with a rocket of reasonable size. We will therefore try a different approach in which a large bomb splits the asteroid into two pieces as illustrated in Figure 7.22. This is another example of an inelastic event, and we can attack it using our problem-solving strategy for inelastic events. The two pieces of the asteroid are the system, and because the external forces are very small (we assume we are far from the Sun, so its gravitational force is small), the momentum of this system is conserved. For simplicity, we assume the two pieces are of equal mass m and choose the coordinate axis x to pass through the center of the asteroid (Fig. 7.22A). Before the explosion, the x component of the velocity is zero, so the momentum along x is also zero. Since there are no external forces on the system (the asteroid plus the bomb), the final momentum along x is also zero. Hence, the two pieces will separate after the explosion with velocities that are equal in magnitude and opposite in direction. We therefore take $\pm v_f$ to be the x components of the velocity of the two pieces of the asteroid after the explosion (Fig. 7.22B).

To calculate v_f, we need to know something about the final kinetic energy. Suppose an energy E_{exp} is released in the explosion. If we are optimistic, we might hope all this energy goes into the final kinetic energy.

$$\tfrac{1}{2}mv_f^2 + \tfrac{1}{2}mv_f^2 = mv_f^2 = E_{exp}$$

Here, there are two terms equal to $\frac{1}{2}mv_f^2$ because the asteroid is split into two pieces, each of mass m. We can now solve for the final velocity and find

$$v_f = \sqrt{\frac{E_{exp}}{m}} \quad (7.29)$$

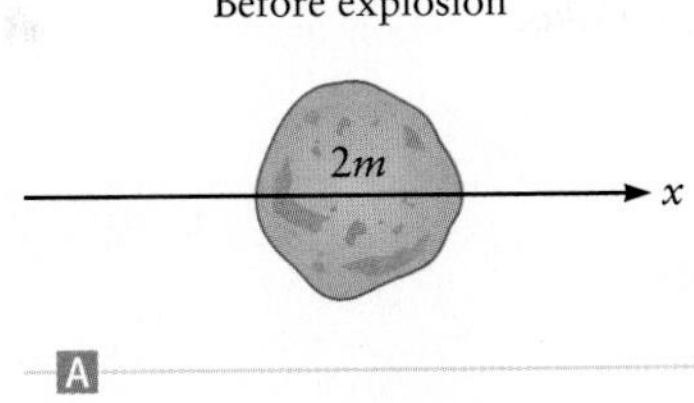

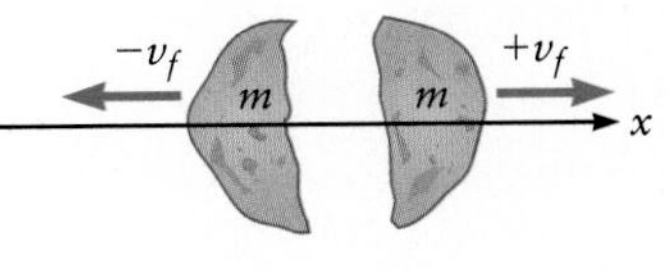

▲ **Figure 7.22** When this asteroid is split into two pieces by an explosion, the explosion releases energy, so the kinetic energy of the system—the pieces of the asteroid—is larger after the explosion. However, the forces from outside the system are zero, so the total momentum is still conserved.

To decide if this approach will succeed, we need to estimate the value of v_f with some realistic numbers. The mass of a typical large asteroid is 1×10^{18} kg, so if the asteroid is split into two equal-sized pieces, we would have $m = 5 \times 10^{17}$ kg. We must next consider the energy we can obtain from the bomb because it will be providing the final kinetic energy of the two asteroid pieces. The energy released by one of the atomic bombs that was dropped in World War II was approximately 1×10^{17} J. Let's assume we can increase that value by a factor of 10 with a contemporary thermonuclear explosive, so $E_{exp} = 1 \times 10^{18}$ J. Inserting these values into Equation 7.29 gives

$$v_f = \sqrt{\frac{E_{exp}}{m}} = \sqrt{\frac{1 \times 10^{18}\text{ J}}{5 \times 10^{17}\text{ kg}}} \approx 1\text{ m/s}$$

which is *much* larger than the lateral velocity we found for a collision with a rocket (Eq. 7.26). Hence, splitting the asteroid with a bomb is the better approach for achieving a significant deflection. Even in this case, though, v_f is not very large, so the task is still not easy. Notice that the momentum of our asteroid system is conserved since the *external* forces on it are zero. Although the bomb's explosion certainly does exert forces on the asteroid, the bomb is part of the system, so these are *internal* forces. The kinetic energy is *not* conserved because the energy of the bomb goes into the final kinetic energy of the system, the two pieces of the asteroid.

7.6 Center of Mass

In the previous sections, we distinguished between *internal* and *external* forces, that is, between forces that act between the particles in aa system and forces that come from outside the system. This distinction is important for understanding the motion of a composite object, that is, an object composed of many point particles such as the boomerang in Figure 7.1A. Before we attempt to deal with an object as complicated as a boomerang, let's consider the simpler case of two point masses connected by a light rod (Fig. 7.23A). We can write Newton's second law for the two combined masses as

$$\Sigma \vec{F} = \vec{F}_1 + \vec{F}_2 = m_1\vec{a}_1 + m_2\vec{a}_2 \tag{7.30}$$

where $\vec{F}_1$ is the total force acting on m_1 and $\vec{F}_2$ is the total force acting on m_2. A portion of the force acting on m_1 will come from its interaction with m_2 and is an internal force. The rest of the force on m_1 will come from outside the system and is thus an external force. The same argument applies to the force on m_2. Let's therefore write the total force on the left-hand side of Equation 7.30 as a sum of the external and internal forces:

$$\overbrace{\Sigma \vec{F}}^{\text{total force}} = \overbrace{\Sigma \vec{F}_{ext}}^{\text{external forces}} + \overbrace{\Sigma \vec{F}_{int}}^{\text{internal forces}}$$

The internal forces always come in action–reaction pairs; for the example in Figure 7.23A, the force exerted by m_1 on m_2 is equal in magnitude and opposite in sign to the force exerted by m_2 on m_1. Hence, when we consider the entire system, the internal forces will cancel so that

$$\Sigma \vec{F}_{int} = 0 \tag{7.31}$$

While we have derived this result for a system of just two particles, Equation 7.31 is true for a system with any number of particles. The internal forces always come in action–reaction pairs, so the sum of all the internal forces is always zero. Thus, Equation 7.30 becomes

$$\Sigma \vec{F}_{ext} = m_1\vec{a}_1 + m_2\vec{a}_2$$

If we denote the total mass of the system as $m_1 + m_2 = M_{tot}$, we have

$$\Sigma \vec{F}_{ext} = m_1\vec{a}_1 + m_2\vec{a}_2 = M_{tot}\vec{a}_{CM} \tag{7.32}$$

where $\vec{a}_{CM}$ is the acceleration of what is called the ***center of mass*** of the system. Equation 7.32 has precisely the same form as Newton's second law for a point particle. In words, it says that the total external force ($\sum \vec{F}_{ext}$) on a system of particles is equal to the mass of the system (M_{tot}) multiplied by a new quantity called the acceleration of the center of mass.

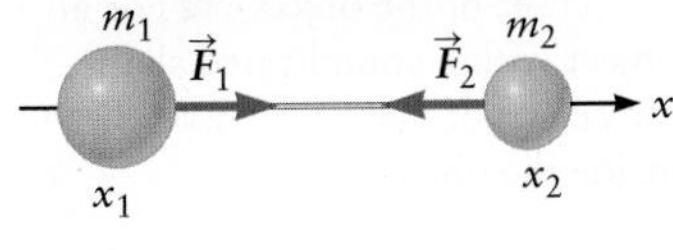

What Is the Center of Mass and How Is It Useful?

So what is the "center of mass"? According to Equation 7.32, the acceleration of the center of mass is related to the accelerations of all the pieces of the system. We can rearrange Equation 7.32 to find $\vec{a}_{CM}$ for a system of just two particles moving in one dimension. Solving for $\vec{a}_{CM}$ gives

$$\vec{a}_{CM} = \frac{m_1\vec{a}_1 + m_2\vec{a}_2}{M_{tot}}$$

This relation suggests that we can express the position of the center of mass along the x coordinate direction in terms of the positions of all the pieces of the system:

$$x_{CM} = \frac{m_1x_1 + m_2x_2}{M_{tot}} \tag{7.33}$$

Here, x_1 and x_2 are the positions of the two particles measured along the x axis in Figure 7.23A, and x_{CM} is the position of the center of mass along x. If the two particles have equal masses m, according to Equation 7.33 the center of mass is at

$$x_{CM} = \frac{m_1x_1 + m_2x_2}{M_{tot}} = \frac{mx_1 + mx_2}{2m} = \frac{x_1 + x_2}{2}$$

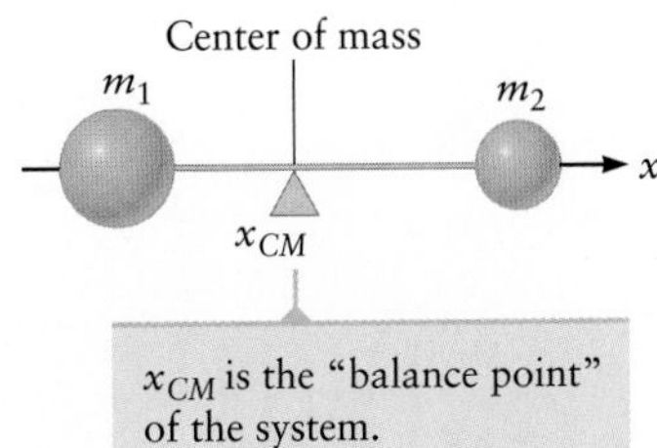

▲ **Figure 7.23** A A system consisting of two particles connected by a massless rod. The forces exerted by each particle on the other (through the rod) are internal forces. B The center of mass of the two point particles is the "balance" point along a line that connects them.

The center of mass of this system is thus at the point midway between the two particles. On the other hand, if one particle is more massive than the other, the center of mass is closer to the one with greater mass. Intuitively, you can think of the center of mass as the "balance point"; that is, the center of mass is the point on the connecting rod in Figure 7.23B where the system would be "in balance."

This result for the center of mass for two particles can be generalized to a system of many particles in two or three dimensions. The result is

$$x_{CM} = \frac{\sum_i m_ix_i}{\sum_i m_i} = \frac{\sum_i m_ix_i}{M_{tot}}$$
$$y_{CM} = \frac{\sum_i m_iy_i}{\sum_i m_i} = \frac{\sum_i m_iy_i}{M_{tot}} \tag{7.34}$$

Center of mass

with a similar expression for z_{CM}. Here, the sums are over all the particles (i) in the system. The total mass of the system is just

$$M_{tot} = \sum_i m_i$$

When using Equation 7.34 to find the location of the center of mass, you must first choose a coordinate system (with an origin). The values of x_{CM} and y_{CM} then refer to the same coordinate system.

For a collection of several masses, computing the center of mass requires a sum involving the positions and masses of the objects. For example, if there are three particles of equal mass m arranged in plane as shown in Figure 7.24, the coordinates of the center of mass are given by

$$x_{CM} = \frac{\sum_i m_ix_i}{\sum_i m_i} = \frac{m_1x_1 + m_2x_2 + m_3x_3}{m_1 + m_2 + m_3} = \frac{m(0) + m(0) + m(L)}{3m} = \frac{L}{3}$$

and

$$y_{CM} = \frac{\sum_i m_iy_i}{\sum_i m_i} = \frac{m_1y_1 + m_2y_2 + m_3y_3}{m_1 + m_2 + m_3} = \frac{m(L) + m(0) + m(0)}{3m} = \frac{L}{3}$$

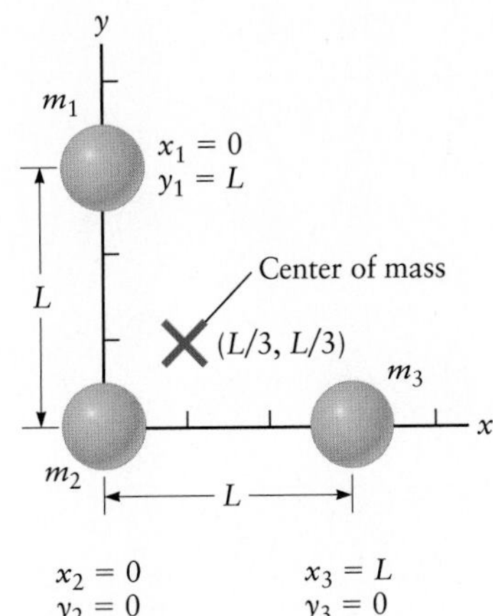

▲ **Figure 7.24** The center of mass of a collection of particles is given by Equation 7.34.

The calculation of the center of mass of a solid object, such as a baseball bat or a boomerang, requires a sum over all the pieces of the object; that is, a sum over all the individual "point particles" that make up the object. Such a calculation can be

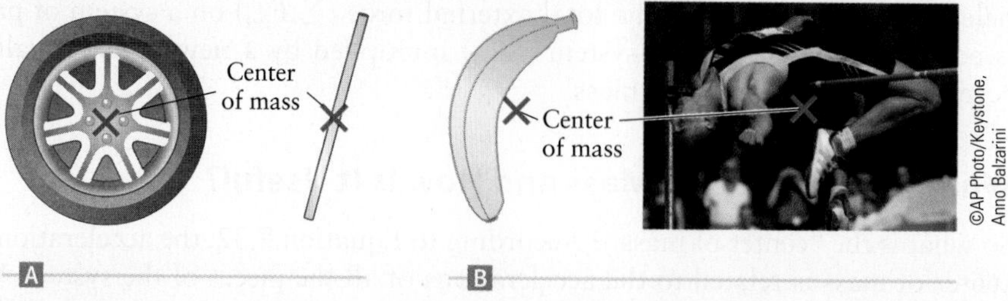

▶ **Figure 7.25** Approximate location of the center of mass of various objects. A For a symmetric object, the center of mass is located at the center of symmetry (i.e., the center of the object). B For an object with a complicated shape, the center of mass may not lie inside the object.

complicated, especially if the mass is not distributed uniformly or if the shape is complex. Even so, the intuitive notion of "balance point" can be very helpful in locating the center of mass. Thus, for a symmetric object such as a simple rod or wheel, the center of mass is at the center of symmetry of the object. Notice also that the center of mass need not be located within an object. The center of mass of a banana may not be inside the banana (Fig. 7.25B).

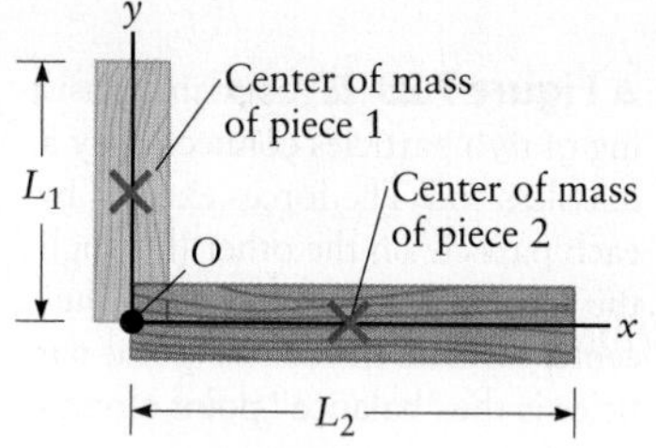

▲ **Figure 7.26** Example 7.9. This bracket is composed of two separate pieces. Each piece is a simple board, one lies along x and one along y, and they overlap near the origin.

EXAMPLE 7.9 Center of Mass of a Bracket

You work for a carpenter and are building a bracket composed of two straight sections of wood as sketched in Figure 7.26. You want to show your boss that physics can be useful, so you decide to calculate the position of the center of mass of the bracket. If the two sections of wood have lengths $L_1 = 1.2$ m and $L_2 = 2.0$ m and masses $m_1 = 0.50$ kg and $m_2 = 1.0$ kg, what is the location of the center of mass of the entire bracket?

RECOGNIZE THE PRINCIPLE

We could find the center of mass of this system of particles by applying the definition in Equation 7.34 and summing over all the pieces that make up the bracket. However, a simpler approach uses the notion of the center of mass as a balance point. With that in mind, we first find the center of mass of each straight section of the bracket. We can treat these two pieces as particles of mass m_1 and m_2 located at their respective centers of mass and then compute the location of the overall center of mass of this two-particle system.

SKETCH THE PROBLEM

Figure 7.26 shows the problem.

IDENTIFY THE RELATIONSHIPS

For each piece of wood, we use the balance-point notion of center of mass to tell us that the center of mass of each piece will be at each respective center as indicated in Figure 7.26. Thus, the center of mass of piece 1 is at

$$x_{1,\,CM} = 0 \quad \text{and} \quad y_{1,\,CM} = \frac{L_1}{2} = 0.60 \text{ m}$$

and for piece 2, we have

$$x_{2,\,CM} = \frac{L_2}{2} = 1.0 \text{ m} \quad \text{and} \quad y_{2,\,CM} = 0$$

SOLVE

We can now treat this bracket as a system of two particles of mass m_1 and m_2 at these two locations and use the relations for the center of mass coordinates in Equation 7.34 to find the center of mass of the entire bracket:

$$x_{CM} = \frac{m_1 x_{1,\,CM} + m_2 x_{2,\,CM}}{m_1 + m_2} = \frac{(0.50 \text{ kg})(0) + (1.0 \text{ kg})(1.0 \text{ m})}{(0.50 \text{ kg}) + (1.0 \text{ kg})} = \boxed{0.67 \text{ m}}$$

$$y_{CM} = \frac{m_1 y_{1,\,CM} + m_2 y_{2,\,CM}}{m_1 + m_2} = \frac{(0.50 \text{ kg})(0.60 \text{ m}) + (1.0 \text{ kg})(0)}{(0.50 \text{ kg}) + (1.0 \text{ kg})} = \boxed{0.20 \text{ m}}$$

▶ ***What does it mean?***
Notice that the center of mass of the bracket is outside the bracket. When an object has a simple shape (e.g., a wheel, a ball, or a straight piece of wood), the center of mass is at the center of symmetry of the object. For more complicated shapes, Equation 7.34 must be used to find the center of mass.

CONCEPT CHECK 7.8 Finding the Center of Mass

A sledgehammer has a massive metal head attached to a much lighter wooden handle (Fig. 7.27). Which location is most likely to be the sledgehammer's center of mass?

(a) As shown in the left drawing, midway along the handle
(b) As shown in the center drawing, on the metal head
(c) As shown in the right drawing, slightly below the head

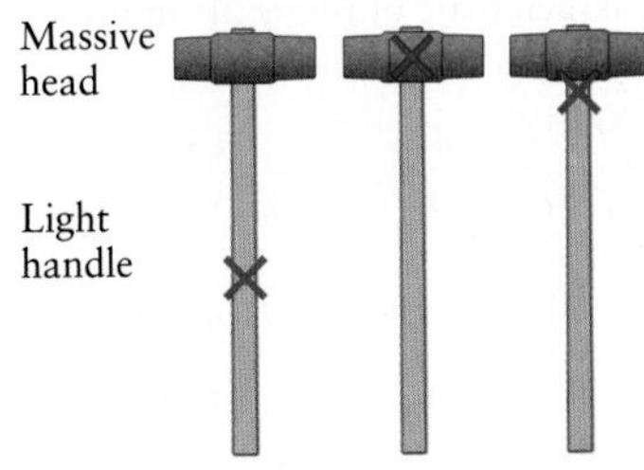

▲ **Figure 7.27** Concept Check 7.8.

Motion of the Center of Mass

Now that we have seen how to calculate the location of the center of mass, let's consider how it *moves*. Figure 7.28 shows two ice skaters; they are initially very close together, so they form a system with the center of mass initially located as shown in part A. We also assume the skaters begin at rest, so their center of mass is initially at rest. If the skaters then push very forcefully on each other, they will move apart as sketched in Figure 7.28B.

If the ice is very slippery (i.e., frictionless), the skaters will move with constant velocities after they separate since there are no forces on either skater along the horizontal (x) direction. Moreover, the *external* force on this system along x is zero the entire time, so the x component of the total momentum of the two skaters is conserved. The initial velocities of the two skaters are zero, so the initial momentum of the system (before the push) is zero and the final momentum must also be zero. The final momentum of the system can be written in terms of the final velocities of the two skaters (see Fig. 7.28), which gives

$$p_{\text{final}} = m_1 v_{1f} + m_2 v_{2f} = 0 \tag{7.35}$$

Since the skaters move with constant velocities and they both start at the origin (being initially very close together), their positions after the push are given by $x_1 = v_{1f}t$ and $x_2 = v_{2f}t$. We have

$$m_1 v_{1f} + m_2 v_{2f} = m_1 \frac{x_1}{t} + m_2 \frac{x_2}{t}$$

Combining this with Equation 7.35, we find

$$m_1 \frac{x_1}{t} + m_2 \frac{x_2}{t} = 0$$

If we now cancel the factor of t, we obtain

$$m_1 x_1 + m_2 x_2 = 0 \tag{7.36}$$

According to Equation 7.34, the center of mass of the two skaters is given by

$$x_{CM} = \frac{m_1 x_1 + m_2 x_2}{m_1 + m_2}$$

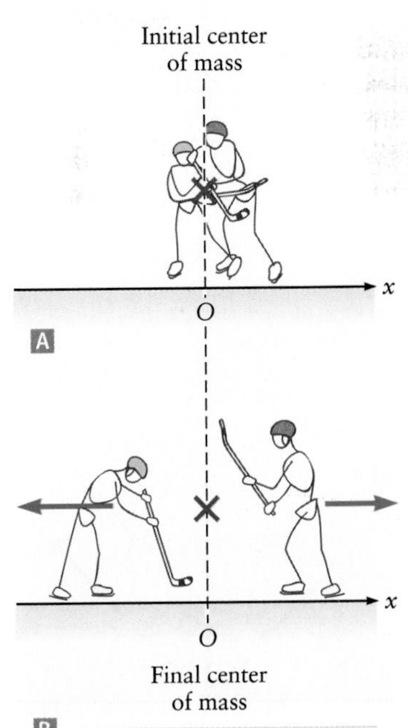

▲ **Figure 7.28** A These two skaters are initially at rest and there are no external forces along the horizontal direction (x). B After they push on each other, the skaters move apart but the center of mass does not move.

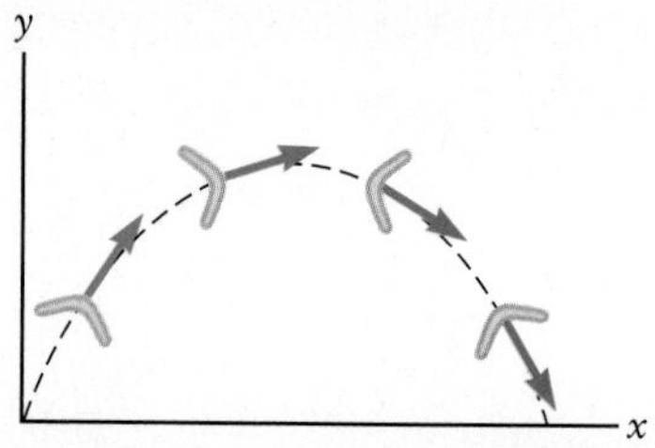

▲ **Figure 7.29** An object such as a boomerang may rotate or tumble when it is thrown, but the center of mass follows a parabolic trajectory characteristic of projectile motion.

Using Equation 7.36, the numerator in this expression for x_{CM} is equal to zero. So, the center of mass of the two skaters is at $x_{CM} = 0$. Moreover, this result applies at all values of t. Hence, even though the skaters move after the push, their center of mass does not move. It stays in the same location, before and after the push.

We could have anticipated this result from our original derivation of the center of mass with Equation 7.32, where we found

$$\sum \vec{F}_{ext} = M_{tot}\vec{a}_{CM} \tag{7.37}$$

which is simply Newton's second law written for the center of mass. In words, Equation 7.37 says that a system of particles moves as if all the mass (M_{tot}) is located at the center of mass and that the center of mass motion is caused *only by the external forces* acting on the system.

Translational Motion of a System

Consider the motion of a complicated object such as a boomerang (Fig. 7.29). A boomerang has an asymmetrical shape, which causes it to spin in a complicated manner when thrown. This complicated motion can be viewed as a combination of ***translational motion*** of the center of mass together with **rotational motion** as the boomerang spins about an axis that extends through its center of mass. Translational motion is often referred to as ***linear motion***. Here the term *linear* does not mean motion along a straight line; rather, it means motion in which the center of mass moves along a path such as the parabolic trajectory we found in projectile motion. Translational motion is different from rotational motion, in which an object spins or rotates about a particular axis. According to Equation 7.37, the translational motion of any system of particles is described by Newton's second law as applied to an equivalent particle of mass M_{tot}. This equivalent particle is located at the center of mass, and for the purposes of the translational motion, we can treat the motion as if all the mass were located at the center of mass. We can then use everything we know about the motion of a point particle to analyze and predict the motion of more complicated objects such as boomerangs. For example, if the force due to air drag is small, the center of mass of the boomerang will move in a simple parabolic trajectory as calculated in Chapter 4. Even though the boomerang may appear to spin in a very complicated manner as it moves through the air, the center of mass motion will be *precisely the same* as that of a simple point particle.

The concept of center of mass allows us to deal with the translational motion of any complicated object. According to Equation 7.37, the center of mass always moves as if all the mass were located at the center of mass of the object. The rotational motion of the object can then be treated separately as we'll do in Chapters 8 and 9.

CONCEPT CHECK 7.9 Explosions and the Center of Mass

A bomb is initially located at the origin ($x = 0$). The bomb then explodes, breaking into two pieces with masses m and $3m$. The pieces fly away in opposite directions along the x axis. Some time later, the smaller piece is at $x = -6.0$ m. If all friction is negligible, where is the larger piece at that moment?

(a) $x = 2.0$ m (b) $x = 3.0$ m (c) $x = 12$ m (d) $x = 18$ m

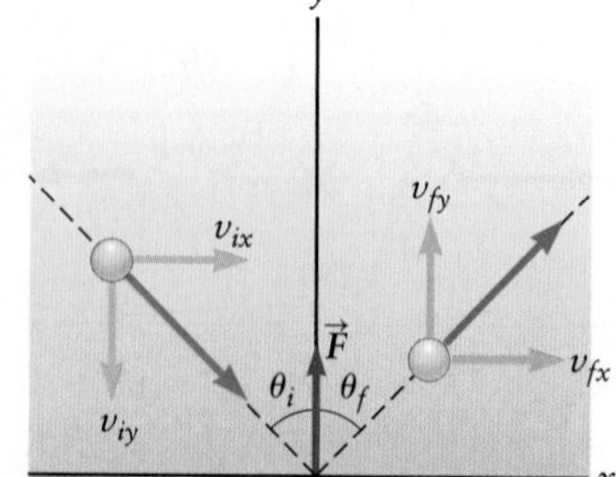

Top view of a cue ball bouncing from the edge of a pool table.

▲ **Figure 7.30** When this cue ball bounces from a wall, the force $\vec{F}$ of the wall on the ball is along y, so the momentum of the ball along x is conserved. Here, $\vec{F}$ is a normal force directed perpendicular to the wall.

7.7 A Bouncing Ball and Momentum Conservation

Consider a pool player who wants to bounce the cue ball off a "rail" at the edge of the table as sketched in Figure 7.30. A good player must be able to predict how a ball will bounce off the rail. The ball approaches the rail at a certain angle θ_i (Fig. 7.30), and we want to calculate the outgoing angle θ_f. We can attack this problem using the principle of momentum conservation.

When the ball collides with the rail, the force exerted by the rail changes the momentum of the ball. For a typical ball and table, the cue ball has nearly the same speed after the collision as it had before hitting the rail, so for simplicity let's assume the initial and final speeds are equal. The final kinetic energy of the ball will therefore equal the initial kinetic energy, so we could also think of this collision as an elastic process. The force of the rail on the cue ball is just a normal force and is directed perpendicular to the rail.[1] Using the coordinate system in Figure 7.30, this normal force is along the y direction and will therefore impart an impulse to the ball in the y direction. There is no force on the ball along the x direction, so the momentum of the ball along x is conserved, and we can write

$$mv_{ix} = mv_{fx}$$

Hence, the initial and final components of the velocity along x are equal:

$$v_{ix} = v_{fx} \tag{7.38}$$

Since the initial and final kinetic energies of the ball are equal, we also have

$$\tfrac{1}{2}mv_i^2 = \tfrac{1}{2}mv_f^2$$

$$\tfrac{1}{2}m\sqrt{v_{ix}^2 + v_{iy}^2} = \tfrac{1}{2}m\sqrt{v_{fx}^2 + v_{fy}^2} \tag{7.39}$$

which leads to

$$\sqrt{v_{ix}^2 + v_{iy}^2} = \sqrt{v_{fx}^2 + v_{fy}^2}$$

$$v_{ix}^2 + v_{iy}^2 = v_{fx}^2 + v_{fy}^2 \tag{7.40}$$

From Equation 7.38, we have $v_{ix} = v_{fx}$, so the terms involving these quantities in Equation 7.40 cancel, leaving

$$v_{iy}^2 = v_{fy}^2$$

which gives

$$v_{fy} = \pm v_{iy}$$

We thus have two solutions for the final y component of the velocity, and we have to decide which one describes the reflection in Figure 7.30. The cue ball bounces from the rail,[2] so the y component of the final velocity is directed opposite to that component of the initial velocity. The solution that describes the reflection is therefore $v_{fy} = -v_{iy}$. The direction of the cue ball's final velocity in Figure 7.30 is then

$$\theta_f = \tan^{-1}\left|\frac{v_{fx}}{v_{fy}}\right| = \tan^{-1}\left|\frac{v_{ix}}{v_{iy}}\right| = \theta_i$$

The outgoing angle is thus equal to the incoming angle. This result, called a ***specular reflection***, is no surprise to an experienced pool player. This kind of reflection will come up again when we study the reflection of light from a mirror in Chapter 24.

7.8 The Importance of Conservation Principles in Physics

Conservation of momentum is the second example of a ***conservation principle*** we have encountered in this book; conservation of energy was a key part of Chapter 6. Conservation principles are important for several reasons. First, they allow us to analyze problems in a very general and powerful way. For example, we saw that for a one-dimensional elastic collision, the combination of conservation of momentum and conservation of kinetic energy completely determines the outcome of the

[1]Things can be more complicated if the ball has sidespin, which can lead to a force on the ball along x.

[2]The other solution corresponds to a ball that passes through the rail, much like some of the solutions we saw in Section 7.4 in which two particles "pass through" each other instead of colliding.

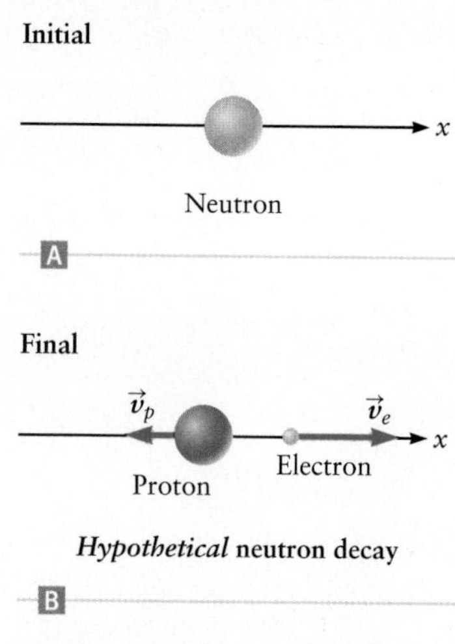

▲ **Figure 7.31** Diagram of a hypothetical decay of a neutron into a proton and an electron. A The neutron is initially at rest and then B decays into a proton and an electron. If there are only two outgoing particles, they must be emitted in opposite directions; otherwise, momentum cannot be conserved. This leads to a single value for the speed of the outgoing electron. Neutrons do not decay in this way.

collision. Analyzing the collision in this way shows that the outcome is a result of these general conservation principles rather than the details of the interaction forces.

Second, conservation principles are extremely general statements about the physical world. While we derived conservation of momentum starting from Newton's laws, this principle is much more general than that. In later chapters, we will study situations in which Newton's laws fail, such as the quantum regime of atoms and subatomic particles, and the astrophysical regime of massive stars and black holes. Although Newton's laws fail in those situations, momentum is still conserved. Indeed, the principle of conservation of momentum is believed to be an exact law of physics, valid in all regimes.

Third, careful tests of conservation principles can sometimes lead to new discoveries. The discovery of an elementary particle called the ***neutrino*** was based quite closely on ideas from this chapter. A subatomic particle called the ***neutron*** was discovered in the early 1930s and was soon recognized as an important component of the atomic nucleus. It was also found that an isolated neutron—that is, a neutron removed from a nucleus—is an unstable particle. When the decay of an isolated neutron was first studied, the details of the decay process were very puzzling. It was initially believed that a neutron decays to leave behind just a proton and an electron:

$$\text{neutron} \rightarrow \text{proton} + \text{electron} \qquad \text{(incomplete!)} \tag{7.41}$$

We have labeled this reaction as "incomplete" for the following reason. This decay process, in which one particle (the neutron) decays to form two other particles (a proton and an electron), is very similar to the exploding asteroid in Figure 7.22. The decaying neutron releases a certain amount of energy E_{decay} (analogous to the bomb that splits the asteroid), which goes into the kinetic energy of the reaction products. There are no external forces acting on the system, so we also know that momentum must be conserved. With conservation of momentum together with knowledge of the reaction energy, we can calculate the final velocities of the proton and electron, just as we calculated the final speeds of the asteroid's pieces in Figure 7.22B.

Let's now analyze the neutron decay reaction in a little more detail.[3] To conserve momentum, the proton and electron in Figure 7.31 must travel away in opposite directions, along a common axis we call x. The problem is thus one dimensional, and we only need to find the components of the proton and electron velocities along x. For simplicity, we assume the neutron was at rest prior to its decay, so the total momentum of the proton plus electron after the collision must be zero. If the electron's velocity along x is v_e and the proton's velocity along x is v_p, conservation of momentum along x gives

$$m_e v_e + m_p v_p = 0 \tag{7.42}$$

The conservation of energy condition can be written as

$$\tfrac{1}{2}m_e v_e^2 + \tfrac{1}{2}m_p v_p^2 = E_{\text{decay}} \tag{7.43}$$

We thus have two conservation conditions (two equations) and two unknowns, the final velocities of the electron and the proton. Hence, these final velocities are fully determined: their precise values are just the solution of these two equations. If Equation 7.41 is really the correct decay reaction, all the electrons must have the same outgoing speed.

When the neutron decay reaction was studied, a very different result was observed. The speed of the outgoing electron was *not* always a single, precise value as expected from the analysis in Equations 7.42 and 7.43. Instead, a range of speeds was observed. This result was very puzzling, and some physicists suggested that either conservation of momentum or conservation of energy might not apply to subatomic particles.

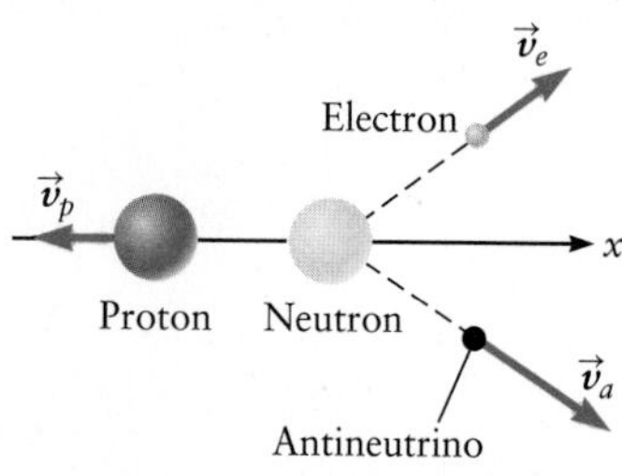

▲ **Figure 7.32** Diagram of the actual decay of a neutron. There are three outgoing particles: a proton, an electron, and an antineutrino. The speed of the outgoing electron is not the same for every decay.

[3]We will learn in Chapter 27 that our Newton's law expressions for momentum and kinetic energy begin to break down when applied to the motion of neutrons, protons, electrons, and neutrinos moving at high speeds. However, these expressions are still a useful approximation and will take us to the correct conclusions regarding the decay of the neutron.

The correct explanation was not this drastic, however. It was eventually found that a neutron actually decays into *three* particles (Fig. 7.32):

$$\text{neutron} \rightarrow \text{proton} + \text{electron} + \text{antineutrino} \qquad \text{(correct!)}$$

The third particle was an entirely new type of particle called an ***antineutrino***. Adding the contributions of the antineutrino to the final momentum and kinetic energy in Equations 7.42 and 7.43 leads to a range of possible values for the final velocity of the electron, as found in the experiments. In fact, this experiment, together with its analysis in terms of conservation laws, marked the discovery of the antineutrino and led to the observation of a closely related particle called the neutrino. The point of this story is that a very careful test of conservation principles led to an entirely unexpected discovery.

Summary | CHAPTER 7

Key Concepts and Principles

Momentum

The ***momentum*** of a particle is given by

$$\vec{p} = m\vec{v} \qquad (7.1)\ \text{(page 182)}$$

The total momentum of a system of particles is equal to the sum of the momenta of the individual particles.

Impulse theorem

When a constant force acts on an object during a time interval Δt, it imparts an ***impulse*** to the object equal to $\vec{F}\,\Delta t$. The impulse is equal to the area under the force–time curve. The impulse imparted to an object is equal to the change in the momentum of the object:

$$\text{impulse} = \vec{F}\,\Delta t = \Delta\vec{p} \qquad (7.5)\ \text{(page 184)}$$

Conservation of momentum

If there is no external force on a system of particles, the total momentum of the system is constant. In such a case, momentum is ***conserved.***

Applications

Collisions

The two objects involved in a collision can be thought of as a "system." During a collision, forces external to the system are usually very small and the total momentum of the system after the collision is equal to the total momentum of the system before the collision. The momentum is conserved.

- In an ***elastic collision***, both momentum and kinetic energy are conserved.
- In an ***inelastic collision***, momentum is conserved but kinetic energy is not conserved.
- In a ***completely inelastic collision***, the two objects stick together after the collision. Momentum is still conserved.

(*Continued*)

Center of mass

Most real objects have a size and shape and can be thought of as being composed of many separate pieces. Such extended objects possess a ***center of mass***, and the motion of this point is very simple. For any object, no matter how complicated, the center of mass moves in response to the total external force acting on the object. This motion follows Newton's laws as if all the mass were located at the center of mass. For a system of two particles, the x coordinate of the center of mass is given by

$$x_{CM} = \frac{m_1 x_1 + m_2 x_2}{m_1 + m_2} \quad \text{(7.33) (page 203)}$$

with a similar expression for y_{CM} and z_{CM}. For a system of many particles, the result is

$$x_{CM} = \frac{\sum_i m_i x_i}{\sum_i m_i} = \frac{\sum_i m_i x_i}{M_{tot}} \quad \text{(7.34) (page 203)}$$

where M_{tot} is the total mass of the system.

Questions

SSM = answer in *Student Companion & Problem-Solving Guide*

= life science application

1. SSM A bomb that is initially at rest breaks into several pieces of approximately equal mass, two of which are shown with their velocity vectors in Figure Q7.1. Use conservation of momentum to determine if there might be other pieces of the bomb not shown in the figure. Assume there is only one missing piece and estimate the direction it is traveling after the explosion.

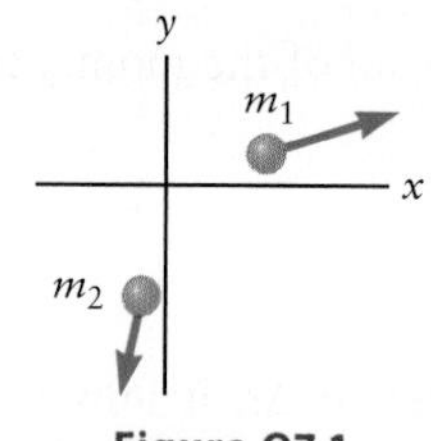

Figure Q7.1

2. The boxes in Figure Q7.2 all have the same mass and size. What is the approximate location of the center of mass of the two boxes in case 1, the three boxes in case 2, and the four boxes in case 3?

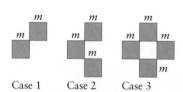

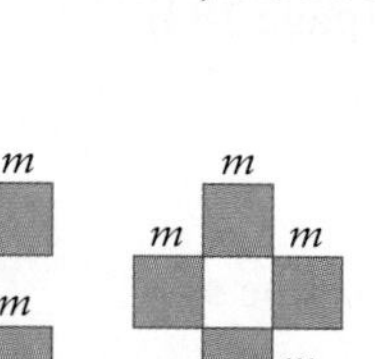

Figure Q7.2

3. What is the approximate location of the center of mass of the boomerang in Figure Q7.3?

4. Explain why the center of mass of an object (such as the boomerang in Fig. Q7.3) may not be "inside" the object. We discussed how the center of mass can be thought of as the "balance point" of an object. How might you use the idea of a balance point to locate the center of mass in cases in which the center of mass is outside the object?

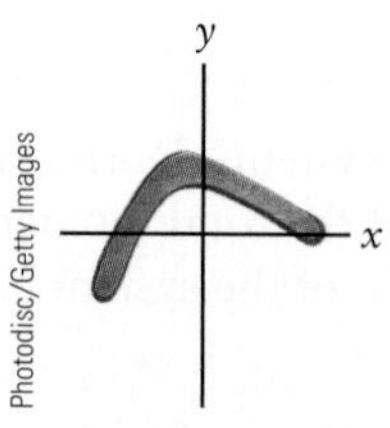

Figure Q7.3
Questions 3 and 4.

5. If the magnitude of the momentum of an object is increased by a factor of three, by what factor is the kinetic energy changed? Assume the mass of the object is unchanged.

6. Two objects of different mass have the same kinetic energy. Which one has the larger momentum?

7. A baseball has a certain momentum p_{ball} and a certain kinetic energy KE_{ball}. One golf ball (ball 1) is given a momentum equal to p_{ball}, and a second golf ball (ball 2) is given a kinetic energy equal to KE_{ball}. Which golf ball has the higher speed?

8. Consider the momentum of an object that is undergoing projectile motion. Is its momentum conserved? If the answer is no, is any component of the momentum conserved? Explain.

9. In which case is the momentum change of a baseball largest, (a) a pitch caught by a catcher, (b) a baseball thrown by a pitcher, or (c) a baseball hit by a batter? Explain your answer.

10. **Deep impact?** Consider the collision between an asteroid and a rocket as discussed in connection with Figure 7.17. We have already calculated the deflection angle that results from the collision with the rocket (Eq. 7.27), but we need to also determine if this deflection will actually save the Earth. If we assume the collision takes place near the Moon, will the asteroid miss the Earth? Ignore the effect of the Earth's gravitational force on the asteroid.

11. Finish the analysis of the force in the automobile collision in Example 7.7. Find the average collision force along y and the magnitude of the total collision force.

12. A cue ball bounces off the rail of a pool table as shown in Figure Q7.12. The momentum of the ball changes. (a) What is the source of the impulse? (b) That is, what is the source of the force? (c) Explain why the answers to (a) and (b) are closely(!) related.

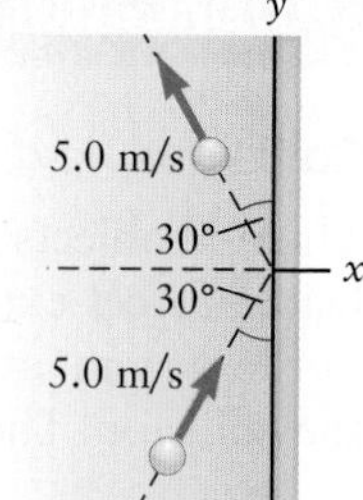

Figure Q7.12

13. Two crates are on an icy, frictionless, horizontal surface. One of the crates is then given a push and collides with the other. If the crates stick together after the collision, which of the following quantities is conserved in the collision, (a) horizontal momentum, (b) kinetic energy, (c) both, or (d) neither?

14. SSM A tennis ball and a ball of soft clay are dropped from the same height onto the floor below. The force from the floor

produces an impulse on each ball. If the balls have the same mass, which impulse is larger? Explain.

15. When you fire a gun, you experience a recoil force due to the impulse produced by the bullet on the barrel of the gun. Is it possible to design a gun in which the recoil force is zero? If so, explain how you could do it. Why do you think guns are not made in this way?

16. Two masses m_1 and m_2 have the same momentum. If the masses are not equal, which statement(s) are correct?
(a) The velocities are the same.
(b) The velocities are in the same direction.
(c) The kinetic energies are the same.

17. In the event of a two-car, head-on collision, would you rather be riding in a heavy car (large mass) or a small car? Justify your answer in terms of the impulse and force on a car during a collision.

18. Explain why a padded dashboard makes a car safer in the event of an accident.

19. An astronaut is working in distant space (outside his spaceship and far from any planets or stars) and floating freely when he accidently throws a wrench. (a) Is the astronaut's momentum conserved (i.e., is his momentum the same before and after he throws the wrench)? (b) Is the wrench's momentum conserved? (c) Is there a system in this situation whose momentum is conserved?

20. As an apple falls from a tree, its speed increases, so the magnitude of its momentum also increases. Explain why this is consistent with the principle of conservation of momentum.

Problems available in ENHANCED WebAssign

Section 7.1	Momentum Problems 1–4
Section 7.2	Force and Impulse Problems 5–19
Section 7.3	Conservation of Momentum
Section 7.4	Collisions Problems 20–31
Section 7.5	Using Momentum Conservation to Analyze Inelastic Events Problems 32–40
Section 7.6	Center of Mass Problems 41–49
Section 7.7	A Bouncing Ball and Momentum Conservation Problems 50–56
Additional Problems	Problems 57–74

List of Enhanced Problems

Problem number	Solution in *Student Companion & Problem-Solving Guide*	Reasoning & Relationships Problem	Reasoning Tutorial in ENHANCED WebAssign
7.4	✓		
7.8		✓	
7.9		✓	
7.11	✓		
7.14		✓	✓
7.15		✓	✓
7.16		✓	✓
7.17	✓	✓	
7.18		✓	
7.19		✓	✓
7.25	✓		
7.30	✓		
7.40	✓		
7.44	✓		
7.45		✓	✓
7.46		✓	
7.49		✓	
7.51	✓		
7.53		✓	✓
7.55		✓	
7.63		✓	✓
7.70		✓	
7.71	✓		
7.73		✓	
7.74		✓	✓

CHAPTER

◀ *Rotating objects have played a major role in science and technology for many years. Each of these windmills can provide the electrical power for about 500 homes. (Jon Boyes/ Getty Images)*

Rotational Motion

In previous chapters, we learned how to apply Newton's laws to deal with translational motion in many different situations, and in this work, we generally assumed our objects were "point particles." A point particle is one in which all the mass is located at an infinitesimal point in space. Real objects like soccer balls, cars, and windmills are definitely not point particles. For these objects, mass is distributed over a certain region of space; that is, they have a certain size and shape. Although the point particle treatment is the correct way to describe the *translational* motion of all objects, this approach does not allow us to deal with the ***rotational motion*** of an object. Our job in this chapter and Chapter 9 is to learn how to use Newton's laws to describe and analyze rotational motion. To do so, we must treat objects more realistically by accounting for their size and shape.

OUTLINE

Top view of a rod rotating in the x–y plane. The rotation axis z is perpendicular to the plane of this figure.

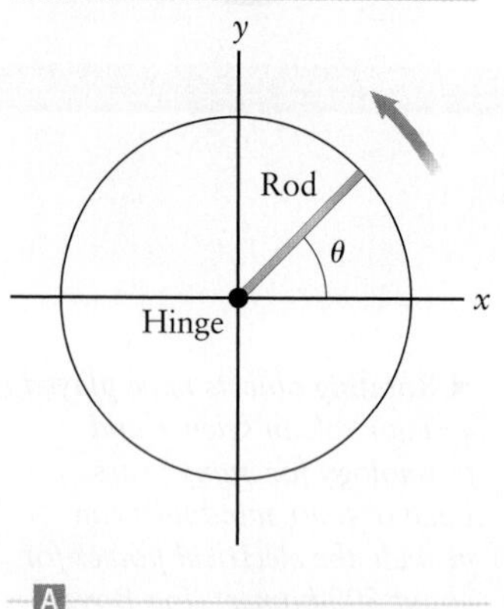

Perspective view of a CD in the x–y plane. The rotation axis is along z.

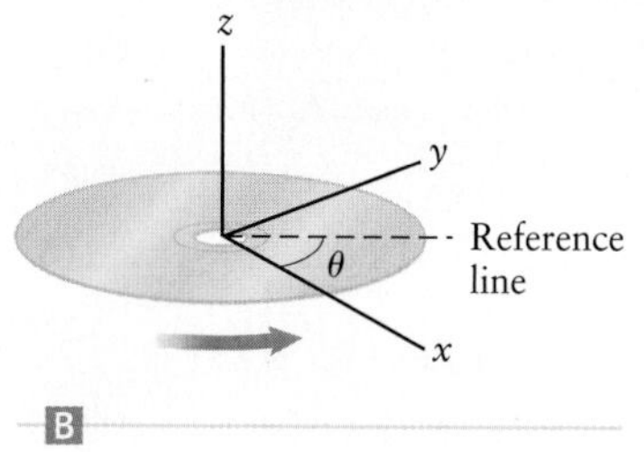

▲ **Figure 8.1** For both A a rotating rod and B a spinning compact disc, the angular position is described by the angle θ. In both of these examples, the rotation axis is the z coordinate axis.

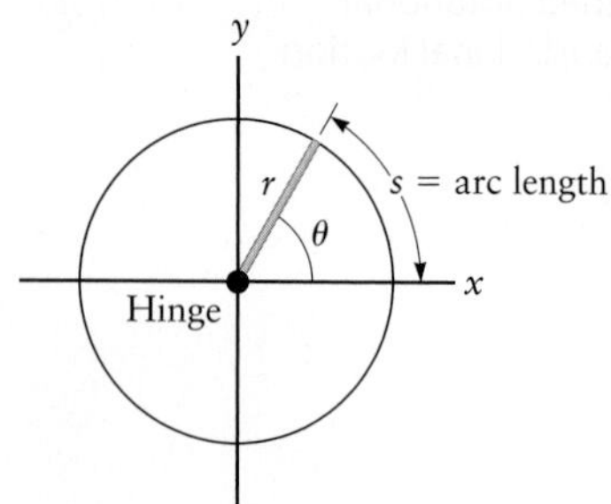

▲ **Figure 8.2** The arc length s measured along the circle is proportional to the angle $\theta = s/r$, measured in radians.

8.1 Describing Rotational Motion

In our studies of translational motion in Chapters 2 and 3, we introduced the concepts of position, velocity, and acceleration. Because translational motion usually involves motion along a trajectory or line in space, it is commonly referred to as ***linear motion***. Hence, the translational velocity and acceleration are often referred to as the linear velocity and linear acceleration. To deal with rotational motion, we now need to define the analogous rotational quantities: ***angular position***, ***angular velocity***, and ***angular acceleration***. We encountered angular position in Chapter 5 when we studied circular motion. Circular motion is related to rotational motion: in both cases, an object or a point on an object moves in a circular path. A simple particle undergoing circular motion as in Chapter 5 thus closely resembles a point on a rotating wheel or cylinder.

For problems involving rotational motion, the first step is to identify the ***rotation axis***. Figure 8.1 shows two different objects undergoing rotational motion. The rod in Figure 8.1A is hinged at one end so that it can rotate about the z axis. The angle θ gives the angular position of the rod, that is, the angle the rod makes with the x axis. Figure 8.1B shows a compact disc spinning about an axis passing through its center, and we again take this rotation axis to be along the z direction. The angular position of the CD can be specified by the angle θ that a reference line on the CD makes with the x axis. For both the hinged rod and the CD, the angle θ tells us all we need to know about the object's angular position.

Figure 8.2 shows the motion of our hinged rod in a little more detail. Suppose the moving end of the rod travels a distance s along its circular path. At the same time, the rod sweeps out an angle θ as sketched in Figure 8.2. If the rod has length r, the distance s and the angle θ are related by

$$\theta = \frac{s}{r} \tag{8.1}$$

where the angle θ is measured in ***radians*** (rad). Of course, angles can also be measured in degrees, and we use both units throughout this book. An angle of 360° is equivalent to 2π radians (one complete circle). (The measurement of angles using units of degrees and radians was discussed in Chapter 1. You might want to review that material now.)

CONCEPT CHECK 8.1 Measuring Angles in Degrees and Radians

What is the approximate value of the angle θ between the rod and the x axis in Figure 8.2?

(a) 90° (b) 60° (c) 0°
(d) π (e) $\pi/2$ (f) $\pi/3$
(g) both values (a) and (e) (h) both values (b) and (f)

Angular Velocity and Acceleration

To describe the rotational motion of a hinged rod or a CD, we need a variable that describes how the angular position θ is changing with time. That variable is the ***angular velocity***, denoted by ω (lowercase Greek omega). The angular velocity ω is a measure of how rapidly the angle θ is changing with time. If we are interested in the rotational motion over a time interval Δt, we can define the average angular velocity as

$$\omega_{\text{ave}} = \frac{\Delta\theta}{\Delta t} \tag{8.2}$$

where $\Delta\theta$ is the change in the angle θ during this time interval. We can also define an instantaneous angular velocity given by

$$\omega = \lim_{\Delta t \to 0} \frac{\Delta \theta}{\Delta t} \tag{8.3}$$

In words, ω is the slope of the θ–t curve, which can be calculated by considering the change $\Delta\theta$ during a time interval Δt as this interval is allowed to shrink to zero ($\Delta t \to 0$). As with linear motion, the instantaneous angular velocity equals the average angular velocity when ω is constant. Angular velocity ω is usually measured in units of radians per second (rad/s), but can also be expressed in terms of revolutions per minute (rpm), where one revolution about the rotation axis corresponds to an angular change of 360°.

Because ω is called the angular *velocity*, you should expect it to be a vector quantity, just as the linear velocity $\vec{v}$ is a vector. All vectors have both a magnitude and a direction. The magnitude of the angular velocity is given by Equation 8.3; how do we determine its direction? The direction of ω is determined by the rotation axis. Figure 8.3 shows a rotating circular disc lying in the x–y plane with a rotation axis along $+z$. If the angle θ increases with time, the angular velocity is positive according to the definition in Equation 8.3. Hence, a counterclockwise rotation corresponds to a positive value of ω. A clockwise rotation would then give a negative value for ω.

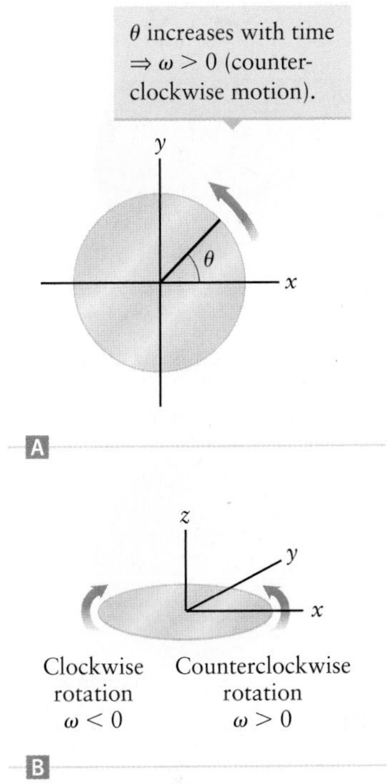

θ increases with time $\Rightarrow \omega > 0$ (counterclockwise motion).

y

θ

x

A

z

y

x

Clockwise rotation $\omega < 0$

Counterclockwise rotation $\omega > 0$

B

▲ **Figure 8.3** An increasing value of θ corresponds to counterclockwise motion. Hence, the angular velocity is positive if the object is rotating counterclockwise.

Acceleration is a very important quantity when dealing with translational motion, and the same is true with rotational motion. We now define the ***angular acceleration***, denoted by α (lowercase Greek alpha). The average angular acceleration during a time interval Δt is related to changes in ω by

$$\alpha_{\text{ave}} = \frac{\Delta \omega}{\Delta t} \tag{8.4}$$

while the instantaneous angular acceleration is defined by

Definition of angular acceleration

$$\alpha = \lim_{\Delta t \to 0} \frac{\Delta \omega}{\Delta t} \tag{8.5}$$

Thus, α is the slope of the ω–t curve and has units of rad/s^2. As with linear motion, the instantaneous angular acceleration equals the average angular acceleration when α is constant.

A good example of rotational motion is a ceiling fan (Fig. 8.4). Suppose the fan is spinning (i.e., rotating) at a constant angular velocity of 85 rpm. Let's calculate the angular velocity in rad/s and also find the total rotation angle after exactly 5 minutes.

To convert from units of rpm to rad/s, we know that one complete revolution corresponds to 2π radians, so calculating ω gives (to two significant figures)

$$\omega = 85 \frac{\text{rev}}{\text{min}} \times \frac{2\pi \text{ rad}}{1 \text{ rev}} \times \frac{1 \text{ min}}{60 \text{ s}} = 8.9 \text{ rad/s}$$

In this example, ω is constant, so the instantaneous angular velocity equals the average angular velocity. We can therefore rearrange Equation 8.3 to get the angular displacement $\Delta\theta$:

$$\Delta\theta = \omega \, \Delta t$$

Inserting our value for ω and using $\Delta t = 5 \text{ min} = 300 \text{ s}$, we have

$$\Delta\theta = \omega \, \Delta t = \left(\frac{8.9 \text{ rad}}{\text{s}}\right)(300 \text{ s}) = 2700 \text{ rad}$$

Note that $\Delta\theta = 2700$ rad corresponds to $(2700 \text{ rad}) \times (1 \text{ rev}/2\pi \text{ rad}) = 430$ revolutions.

▲ **Figure 8.4** When this ceiling fan is turned on, the blades undergo rotational motion.

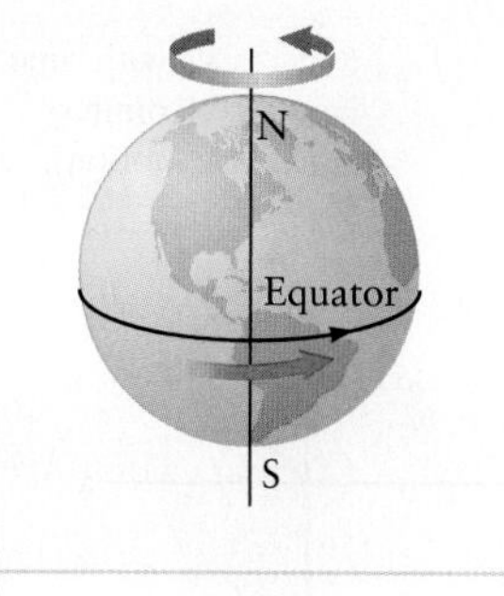

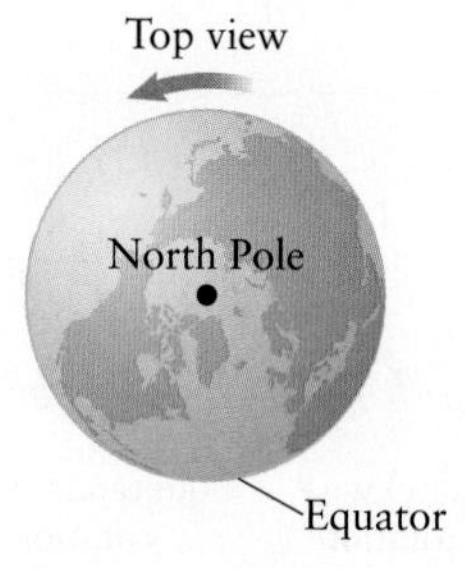

▲ **Figure 8.5** Example 8.1. The Earth undergoes rotational motion as it spins about its axis.

EXAMPLE 8.1 Angular Velocity of the Earth

The Earth rotates about the axis that connects the North and South Poles as sketched in Figure 8.5A. Find the Earth's angular velocity in rad/s.

RECOGNIZE THE PRINCIPLE

The angular velocity of the Earth is constant, so $\omega = \omega_{ave}$, which can be calculated from the angle $\Delta\theta$ through which the Earth rotates in a given amount of time Δt using Equation 8.2 or 8.3.

SKETCH THE PROBLEM

Figure 8.5B shows a top view of how the Earth's equator rotates (compare with Figs. 8.1 through 8.3).

IDENTIFY THE RELATIONSHIPS

The Earth completes one rotation ($\Delta\theta = 1$ rev $= 2\pi$ rad) every day, so its angular velocity is simply 1.0 rev/day.

SOLVE

Expressing 1.0 rev/day in rad/s, we find

$$\omega = \frac{\Delta\theta}{\Delta t} = \frac{1.0 \text{ rev}}{1 \text{ day}} \times \frac{1 \text{ day}}{24 \text{ h}} \times \frac{1 \text{ h}}{60 \text{ min}} \times \frac{1 \text{ min}}{60 \text{ s}} \times \frac{2\pi \text{ rad}}{\text{rev}} = \boxed{7.3 \times 10^{-5} \text{ rad/s}}$$

▶ ***What does it mean?***

The Earth's angular velocity is constant, so its average angular velocity equals the instantaneous angular velocity. Also, a point on the Earth's equator moves in a circle with constant speed and thus undergoes uniform circular motion as discussed in Chapter 5.

EXAMPLE 8.2 Angular Acceleration of a DVD

A DVD used to view a movie does not spin with a constant angular velocity. Instead, it spins most rapidly when images are being read from regions nearest the rotation axis (the inner "tracks" near the center of the disc) and slowest when the images are played from regions near the edge of the disc (the outer "tracks"). If the angular velocity decreases uniformly from $\omega = 1500$ rpm to 630 rpm as the DVD is scanned from the innermost to the outermost track, what is the angular acceleration? Assume the DVD contains a movie that lasts 110 min.

RECOGNIZE THE PRINCIPLE

We are given that the angular velocity decreases uniformly, which means that the angular acceleration is constant; hence, $\alpha = \alpha_{ave}$. We can compute α from the change in ω during the 110 min it takes to play the DVD.

SKETCH THE PROBLEM

Figure 8.6 shows the tracks on a DVD. These tracks form one very long "spiral" that begins near the center of the disc and winds out to the edge.

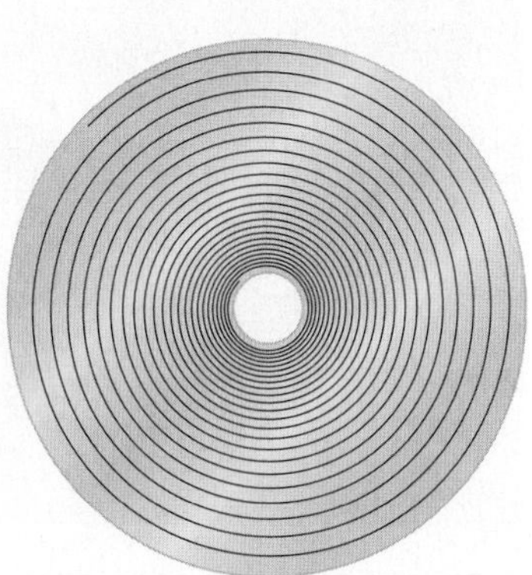

▲ **Figure 8.6** Example 8.2. The information on a DVD is stored in a very long spiral "track."

IDENTIFY THE RELATIONSHIPS

The angular velocity ω decreases from 1500 rpm to 630 rpm as the DVD player moves from the inside to the outside tracks. Because α is constant, we can use Equation 8.4

with a time interval that is not infinitesimally small. The angular acceleration is then given by

$$\alpha = \alpha_{ave} = \frac{\Delta\omega}{\Delta t} = \frac{\omega_f - \omega_i}{\Delta t}$$

SOLVE

The initial angular velocity is $\omega_i = 1500$ rpm and the final angular velocity is $\omega_f = 630$ rpm, which leads to

$$\alpha = \frac{\omega_f - \omega_i}{\Delta t} = \frac{(630 - 1500)\text{ rpm}}{110\text{ min}} = -7.9\text{ rev/min}^2$$

We now need to convert the units and express the answer in rad/s^2:

$$\alpha = \frac{-7.9\text{ rev}}{\text{min}^2} \times \frac{2\pi\text{ rad}}{\text{rev}} \times \left(\frac{1\text{ min}}{60\text{ s}}\right)^2 = \boxed{-1.4 \times 10^{-2}\text{ rad/s}^2}$$

▶ ***What does it mean?***

The information on a DVD is stored in a very "tight" spiral (Fig. 8.6), so at any particular point on the disc, the path containing the video information is extremely close to being a perfect circle. During one revolution, the length of each of these approximately circular paths is just the circle's circumference, a distance smaller near the center of the DVD than at regions near the edge. The amount of information stored in a circular track is proportional to the length of a track, so a circle near the center contains less information than a circle near the edge. To compensate for this difference, the DVD player spins faster when reading images and sound from near the center since the information must be processed and displayed at a constant rate.

Angular and Centripetal Acceleration Are Different

The angular acceleration α defined in Equation 8.5 is *different* from the centripetal acceleration associated with circular motion discussed in Chapter 5. This difference can be appreciated from Figure 8.2. The free end of the rod moves like a particle moving around the circle with a constant linear speed v. This particle is undergoing uniform circular motion, and its angular position θ increases at a constant rate. Hence, the particle's angular velocity is constant, and a constant value for ω means the angular acceleration $\alpha = 0$ (according to Eq. 8.5). At the same time, the particle is undergoing uniform circular motion, so from Chapter 5 we know that it also has centripetal acceleration $a_c = v^2/r$ (Eq. 5.5). This centripetal acceleration refers to the linear motion (and the linear acceleration) of the particle, whereas the angular acceleration α is concerned with the associated angular motion. Hence, the centripetal acceleration a_c is *not* zero, even though the angular acceleration α does equal zero.

The Period of Rotational Motion

When an object such as a compact disc, a DVD, or the Earth is rotating with a particular value of ω, the angular velocity is the *same* for all points on the object. This is one property that makes ω such a useful quantity for describing rotational motion. While the angular velocity is the same for all points, the linear velocity can be different for different points on the rotating object. To illustrate, we consider again the motion of a compact disc as it rotates about its axis in Figure 8.7. If we focus on the motion of a general point P on the CD, we can relate the magnitude of the linear velocity v of this point to the angular velocity ω. If the angular velocity is constant, the CD moves through ω radians every second. Since one revolution corresponds to 2π radians, the CD moves through $\omega/(2\pi)$ complete revolutions every second. Equivalently, we can

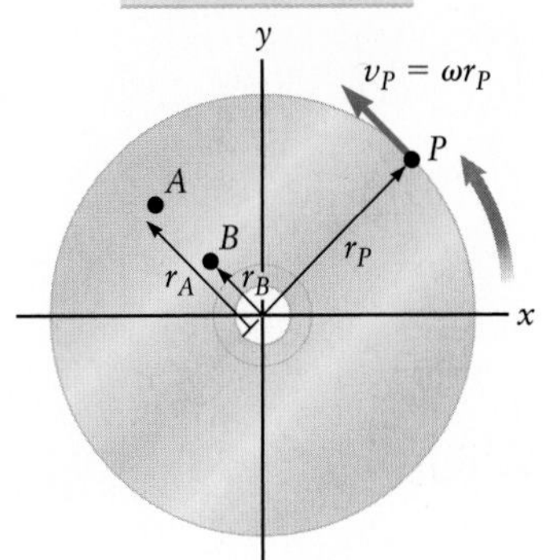

▲ **Figure 8.7** When an object such as this compact disc undergoes rotational motion, the linear velocity of a point on the object depends on its distance from the rotation axis, but ω is the same for all points on the object.

say that each complete revolution takes $2\pi/\omega$ seconds. The time required to complete one revolution is called the ***period*** of the motion, denoted by T:

Period and angular velocity

$$\text{period} = T = \frac{2\pi}{\omega} \tag{8.6}$$

We encountered the notion of period in Chapter 5 in connection with circular motion. The relation between period and angular velocity in Equation 8.6 assumes ω is constant over the time it takes to complete one revolution (one period).

The Connection between Linear and Rotational Motion

Point P on our CD in Figure 8.7 lies on a circle of radius r_P. As the CD rotates, this point completes one revolution of the circle during each period of the motion, so it travels a distance $2\pi r_P$ every T seconds. Its velocity thus has a magnitude of $2\pi r_P/T$. Using Equation 8.6, we have

$$v_P = \frac{2\pi r_P}{T} = \frac{2\pi r_P}{2\pi/\omega} = \omega r_P$$

This result applies just as well to any point on the CD if r_P is replaced by the distance r between that point and the rotation axis. The *linear* velocity of a point a distance r from the rotation axis is related to the *angular* velocity by

Relation between linear speed and angular speed

$$v = \omega r \tag{8.7}$$

Equation 8.7 gives a very general and useful relation between the linear velocity and the angular velocity. To be precise, this relation really only involves the linear and angular *speeds* because here we have not accounted for the directions of these two types of velocities.

Let's compare the linear speeds of points A and B on the CD sketched in Figure 8.7. These points are located distances r_A and r_B, respectively, from the axis of the CD. Using Equation 8.7, we find $v_A = \omega r_A$ and $v_B = \omega r_B$. Point A is farther from the axis (so $r_A > r_B$), so the linear speed of point A will be *greater* than that of point B, even though they are both attached to the same CD. Physically, the reason for this difference is the different values of the radius. Point A is farther from the rotation axis, so it travels on a larger circle than point B. The motions of the two points have the same period, however, so point A must travel faster to complete its circle in the same amount of time (T). While the linear speeds at points A and B are thus *not* equal, the angular velocities at these two points are the *same*.

The relationship between the angular and translational speeds in Equation 8.7 can be extended to the accelerations. For linear motion, we showed in Chapter 2 that when the acceleration is constant, the instantaneous acceleration is related to changes in velocity by (Eqs. 2.6 and 2.7):

$$a = \frac{\Delta v}{\Delta t}$$

If a is constant, the angular acceleration is also constant, so we can write (from Eq. 8.5)

$$\alpha = \frac{\Delta \omega}{\Delta t}$$

The ratio a/α is therefore

$$\frac{a}{\alpha} = \frac{\Delta v/\Delta t}{\Delta \omega/\Delta t} = \frac{\Delta v}{\Delta \omega}$$

From Equation 8.7, we can write $\omega = v/r$, which leads to $\Delta\omega = \Delta v/r$. We thus get

$$\frac{a}{\alpha} = \frac{\Delta v}{\Delta \omega} = \frac{\Delta v}{\Delta v/r} = r$$

The *angular* acceleration α and the *linear* acceleration a of a point a distance r from the rotation axis are therefore related by

$$a = \alpha r \tag{8.8}$$

Relation between linear acceleration and angular acceleration

CONCEPT CHECK 8.2 Angular Velocity and Period

Consider two merry-go-rounds (Fig. 8.8) whose radii differ by a factor of two. If they have the same angular velocities, how do their periods compare?

(a) The larger merry-go-round has a shorter period.
(b) The larger merry-go-round has a longer period.
(c) The periods are the same.

▲ **Figure 8.8** Concept Check 8.2.

EXAMPLE 8.3 Ⓡ Speed of a Merry-Go-Round

Estimate the angular velocity of a typical merry-go-round and find the corresponding linear speed of a point on its outer edge.

RECOGNIZE THE PRINCIPLE

The angular velocity is related to the period of the motion (Eq. 8.6), and it is also related to the linear speed v of a point on a rotating object (the merry-go-round) through Equation 8.7.

SKETCH THE PROBLEM

Figure 8.9 shows a top view of a rotating merry-go-round. A point on the edge moves in a circle with r equal to the radius of the merry-go-round.

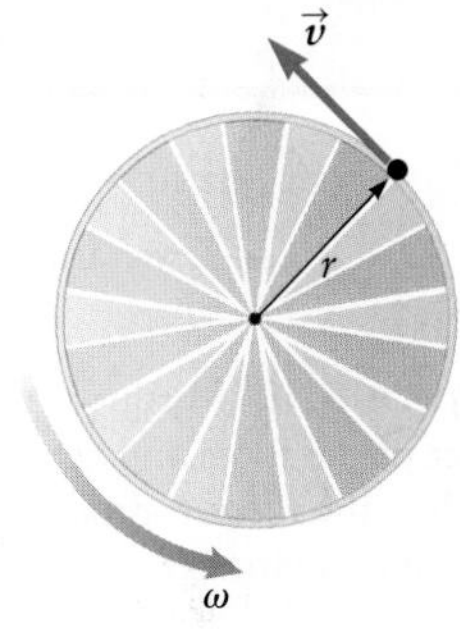

▲ **Figure 8.9** Example 8.3.

IDENTIFY THE RELATIONSHIPS

According to Equation 8.6, the angular velocity and period are related by

$$\omega = \frac{2\pi}{T}$$

To find ω, we must therefore estimate the value of T for a typical merry-go-round. The linear speed is related to ω by

$$v = \omega r$$

Hence, to find v, we must know the radius r of the merry-go-round. The value of r is not given, so we must also estimate r for a typical case.

SOLVE

A typical merry-go-round (Fig. 8.8) might have a radius of 3 m and move at a rate such that it completes one revolution in about 5 s. Hence, the period of the motion is $T = 5$ s. Working to one significant figure (to match our estimates of r and T), these values lead to

$$\omega = \frac{2\pi}{T} = \frac{2\pi}{5\text{ s}} = \boxed{1\text{ rad/s}}$$

The linear speed of a point on the edge of the merry-go-round is then

$$v = \omega r = (1\text{ rad/s})(3\text{ m}) = \boxed{3\text{ m/s}}$$

(continued) ▶

► *What does it mean?*
The angular velocity is the same for all points on the merry-go-round, but the linear speed depends on r, the distance from the rotation axis. Notice also how the units for v here work out to be m/s. As we discussed in Chapter 1 (see also Appendix B), radians are a dimensionless unit, so they cancel automatically.

8.2 Torque and Newton's Laws for Rotational Motion

Our next job is to explore the connection between force and rotational motion. In particular, we want to know how forces give rise to angular acceleration; this approach is very similar to the one we took in Chapter 2 when we dealt with the linear (i.e., translational) motion of a particle.

Torque and Lever Arm

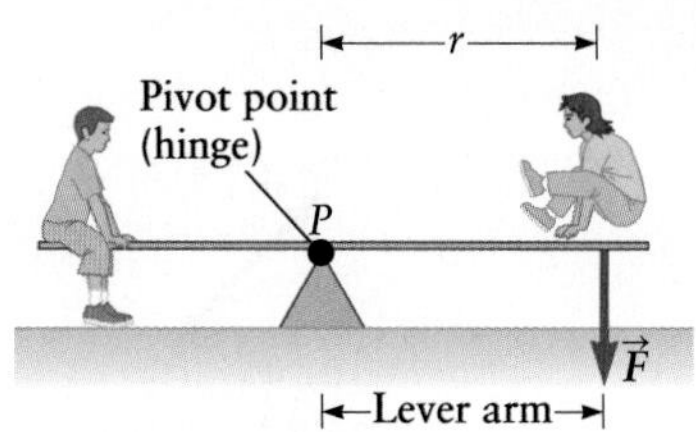

▲ **Figure 8.10** Torque is the product of the force and the lever arm. Here the force $\vec{F}$ produces a torque of magnitude Fr.

Figure 8.10 shows a child sitting on one end of a seesaw while his friend prepares to jump onto the opposite side. The seesaw is supported near the center by a hinge that allows it to rotate about its point of support P. Hence, both ends of the seesaw, and indeed all points along it, move on circular arcs as the seesaw rotates. When the child jumps on board, her weight gives a force on the seesaw, and when the seesaw is horizontal (as drawn here), this force is directed perpendicular to the seesaw. This force makes the seesaw rotate, causing an angular acceleration of the seesaw. To predict the angular acceleration we must consider both the magnitude of this force and *where* it is applied to the seesaw.

From our playground experiences, we know that applying a force far from the support point P is more effective at producing rotation than applying the same force close to P. We therefore define a quantity called ***torque***, which equals the product of the applied ***force*** and the ***distance*** it is applied from the support point. Since the seesaw is able to rotate about this point, point P is often called the ***pivot point***. In the simplest cases, such as this seesaw, the force is directed *perpendicular* to a line connecting its point of application to the pivot point. We then define the ***lever arm*** as the distance between the pivot point and the place where the force acts. Later, we'll describe how the lever arm is defined in more general situations. The torque in this simple case is given by

$$\tau = Fr \tag{8.9}$$

where r is the length of the lever arm. Torque is the fundamental "force-type" quantity with regard to rotational motion; it is analogous to force in translational motion. The units of τ are N · m (force times distance).

We can use the seesaw in Figure 8.10 to understand why torque depends on both the force and the lever arm. For a force of a given magnitude, it is easier to rotate the seesaw if r is large than if r is small. That is, a given force produces a larger angular acceleration when it is applied farther from the pivot point.

Relating Torque and Angular Acceleration

Our next job is to derive the relation between torque and the angular motion it produces. Consider Figure 8.11, which shows a hinged rod to which a ball of mass m is attached at one end. The other end of the rod at point P is held in place, but the hinge allows the rod to rotate about P. If a force $\vec{F}_{\text{applied}}$ is applied to the ball, the rod and ball will rotate. Let's now apply Newton's second law to analyze this motion.

For simplicity, we assume the rod is extremely light (massless), so all the rotating mass is concentrated in the ball. Applying Newton's second law to the ball, we have

$$\sum \vec{F} = m\vec{a}$$

where $\sum \vec{F}$ is the total force. There are two forces acting on the ball: our applied force $\vec{F}_{\text{applied}}$ and another force $\vec{F}_{\text{rod}}$ from the rod. We thus have

$$\sum \vec{F} = \vec{F}_{\text{applied}} + \vec{F}_{\text{rod}} = m\vec{a} \tag{8.10}$$

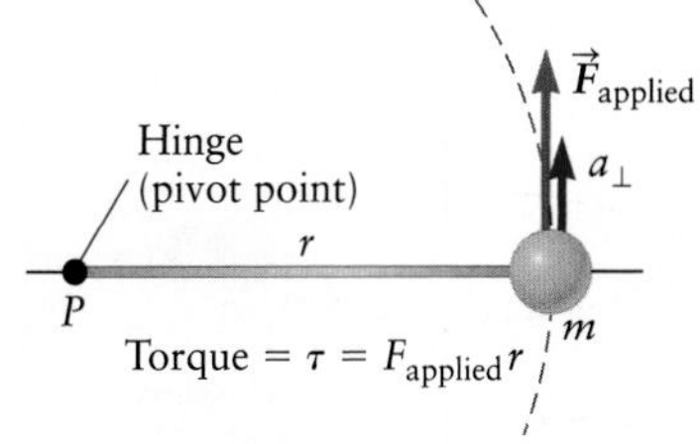

▲ **Figure 8.11** Only a perpendicular force can make an object rotate. The force $\vec{F}_{\text{applied}}$ accelerates this ball in the perpendicular direction.

Let's begin with the case in which $\vec{F}_{\text{applied}}$ is directed perpendicular to the rod as shown in Figure 8.11. The rod is rigid, so the ball will move in a circular arc along the direction of $\vec{F}_{\text{applied}}$. The force $\vec{F}_{\text{rod}}$ is along the direction of the rod, perpendicular to the ball's circular path. As a result, $\vec{F}_{\text{rod}}$ does not contribute to the ball's angular acceleration. *Only forces with a component perpendicular to the rod can contribute to the angular acceleration.* With that in mind, consider only the perpendicular components in Equation 8.10 of the force on the ball and its acceleration along this circular path. We call this acceleration $a_\perp$ because it denotes the acceleration perpendicular to the rod as in Figure 8.11. Likewise, we denote the perpendicular component of the applied force as $F_\perp$. The components of Equation 8.10 along this perpendicular direction then read

$$F_\perp = ma_\perp$$

The component $a_\perp$ is a linear (translational) acceleration, and we saw in Section 8.1 that this linear acceleration is proportional to the angular acceleration of the ball as it moves along its circular path. Using that result (Eq. 8.8) here leads to

$$F_\perp = ma_\perp = m\alpha r$$

where r is the distance from the hinge to the ball. Multiplying both sides by r and doing a little rearranging gives

$$F_\perp r = (m\alpha r)r = (mr^2)\alpha \tag{8.11}$$

Comparing with Equation 8.9, you should recognize that the left-hand side of Equation 8.11 is the torque on the ball; it is the product of the force and the lever arm r. The right-hand side contains mass and (angular) acceleration, so Equation 8.11 looks a lot like Newton's second law. Indeed, it *is* Newton's second law for rotational motion, and it is usually written as

Newton's second law for rotational motion

$$\sum \tau = I\alpha \tag{8.12}$$

where $\sum \tau$ is the total torque, that is, the sum of all the different torques acting on the object. Here we have introduced a new quantity I called the ***moment of inertia***. In rotational motion, I plays the role that mass does in translational motion. For the system considered here, consisting of a ball attached to a massless hinged rod, we can compare Equations 8.11 and 8.12 to find that the moment of inertia is

$$I = mr^2 \quad \text{(massless hinged rod/massive ball)} \tag{8.13}$$

Newton's Second Law for Rotational Motion and the Analogy with Translational Motion

Newton's second law for rotational motion, $\sum \tau = I\alpha$, and the associated definitions of torque and moment of inertia have many applications. Before exploring them, let's review and recap three major points.

First, *torque* plays the role of force in rotational motion. Torque depends on the magnitude of the force and on *where* the force is applied relative to the pivot point. There may be multiple forces acting on an object; hence, there may be multiple lever arms and multiple torques, all involving a single pivot point.

Second, the *moment of inertia* enters into rotational motion in the same way that mass enters into translational motion. We found (Eq. 8.13) that the moment of inertia of a ball at the end of a massless rod of length r is $I = mr^2$. For an object

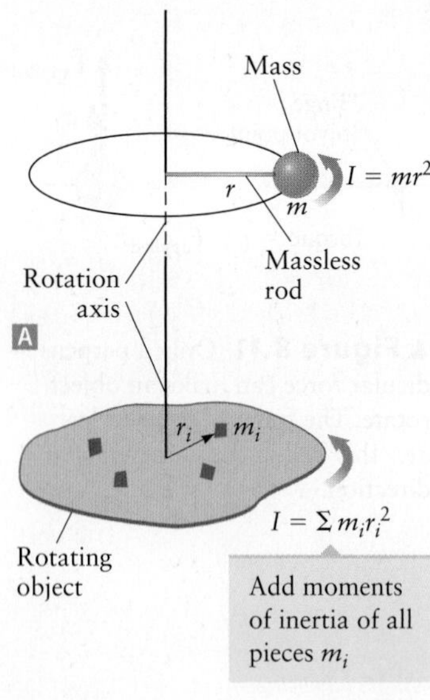

▲ **Figure 8.12** A Moment of inertia of a single point mass rotating about an axis: $I = mr^2$. B The moment of inertia of an extended object is the sum of the moments of inertia of all the individual pieces of the object. The moment of inertia of each piece can be calculated as in part A.

TABLE 8.1 Quantities Used to Describe Linear Motion and Rotational Motion

Linear Motion	Rotational Motion
Linear position $= x$	Angular position $= \theta$
Linear velocity $= v$	Angular velocity $= \omega$
Linear acceleration $= a$	Angular acceleration $= \alpha$
Force $= F$	Torque $= \tau$
Mass $= m$	Moment of inertia $= I$

composed of many pieces of mass m_i located at distances r_i from the pivot point, the total moment of inertia *of the object* is just the sum of the moments of inertia of the individual pieces (Fig. 8.12); hence,

$$I = \sum_i m_i r_i^2 \tag{8.14}$$

The moment of inertia of an object thus depends on the object's mass and on how that mass is distributed relative to the rotation axis. We'll apply Equation 8.14 and consider the moment of inertia for various types of objects in Section 8.4.

Third, Newton's second law for translational motion leads to *Newton's second law for rotational motion*:

$$\sum \tau = I\alpha$$

We'll spend the rest of this chapter exploring how to apply this relation.

The similarities between translational and rotational motion are reinforced in Table 8.1, where we list some of the quantities used to describe each.

Torques and Lever Arms Revisited: A More General Definition

We have defined torque for cases in which the applied force is purely in the perpendicular direction (Figs. 8.10 and 8.11), but, in general, the force may be applied in a nonperpendicular direction. One such example is shown in Figure 8.13, again for our hinged rod with a ball attached at the end. Here the applied force $\vec{F}_{\text{applied}}$ acts at an angle ϕ with respect to the rod. Since the ball is attached to the rod, the ball is again restricted to move along the circular arc sketched in Figure 8.13. Only forces directed along this arc (perpendicular to the rod) contribute to the torque. For this reason, we can ignore the force of the rod on the ball when calculating the torque; this force does not contribute to the angular acceleration. The perpendicular force is $F_\perp = F_{\text{applied}} \sin \phi$, and the torque in this case is

$$\tau = F_\perp r = (F_{\text{applied}} \sin \phi) r$$

Definition of torque

$$\tau = F_{\text{applied}} r \sin \phi \tag{8.15}$$

Here $F_\perp$ is the component of the force that is perpendicular to the rod and r is the distance between the pivot point and the point where the force acts. Equation 8.15 is our general definition of torque, and it applies even when the force is not directed perpendicular to the lever arm. When the force is directed perpendicular to the lever

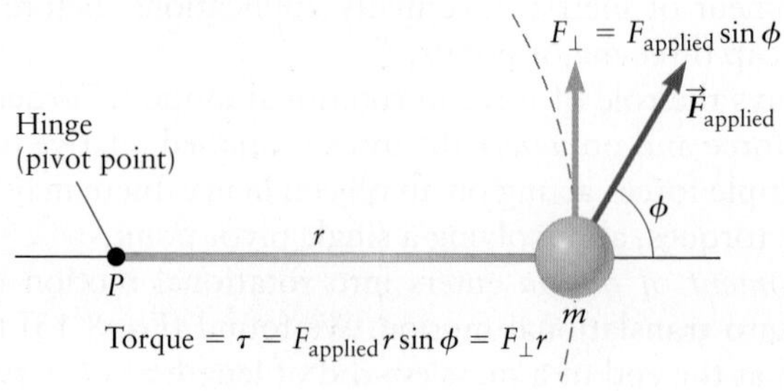

► **Figure 8.13** Only the perpendicular component of $\vec{F}_{\text{applied}}$ contributes to the torque. Any force components parallel to the rod do not contribute to the torque.

arm, $\phi = 90°$; hence, $\sin\phi = 1$ and $F_\perp = F_{\text{applied}}$. When the force is parallel to the lever arm, $\phi = 0$ and the torque is zero. This conclusion makes sense because a force directed parallel to the lever arm ($\phi = 0$) cannot cause an object to rotate.

Two Ways to Think about Torque

The general relation for the torque in Equation 8.15 can be illustrated graphically in two different ways. The perpendicular component of the applied force is $F_\perp = F_{\text{applied}} \sin\phi$, so we can group the terms in Equation 8.15 as

$$\tau = \overbrace{(F_{\text{applied}} \sin\phi)}^{F_\perp} r = F_\perp r \tag{8.16}$$

which is illustrated in Figure 8.14A; in words, this says that torque equals the product of the perpendicular component of the force ($F_\perp$) and the distance from the pivot point to the point where the force acts on the object (r). This product of perpendicular force and distance is equivalent to our definition of torque in the simpler cases in Figures 8.10 and 8.11. The dashed line in Figure 8.14A also shows that you can find the angle ϕ by extending the "radius line" beyond the point where the force acts; ϕ is the angle between this extension and the force vector.

On the other hand, we can group the terms in Equation 8.15 as

$$\tau = F_{\text{applied}} \overbrace{(r \sin\phi)}^{r_\perp} = F_{\text{applied}} r_\perp \tag{8.17}$$

which is illustrated in parts B and C of Figure 8.14; in words, this equation says that torque equals the product of the applied force (F_{applied}) and the perpendicular distance $r_\perp$. This perpendicular distance is the distance from the line defined by the force vector (parts B and C of Fig. 8.14) to the rotation axis as measured on a line that is perpendicular to both. The perpendicular distance $r_\perp = r \sin\phi$ is the lever arm in the general case of a force applied at an angle ϕ to the rotation axis. These two ways of writing the torque (Eqs. 8.16 and 8.17) are completely equivalent. The general expression for the lever arm is

$$r_\perp = r \sin\phi \tag{8.18}$$

When the force is purely perpendicular and $\phi = 90°$, this expression reduces to just r as in the cases we considered in Figures 8.10 and 8.11. On the other hand, if $\phi = 0$, the lever arm is zero, corresponding to a force applied parallel to the "radius line" (Fig. 8.14D). Such a parallel force cannot cause an object to rotate and thus cannot produce a torque.

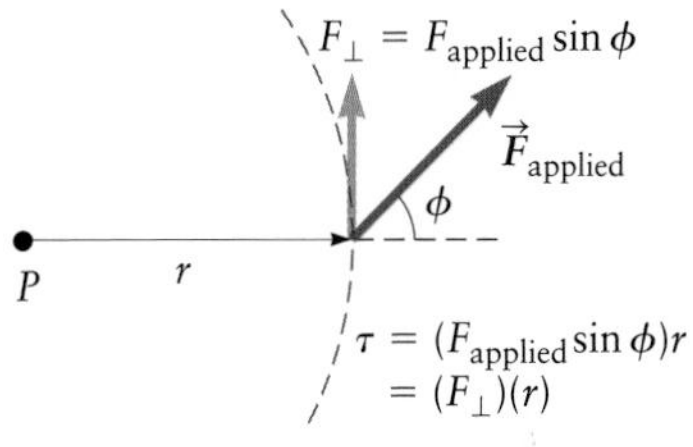

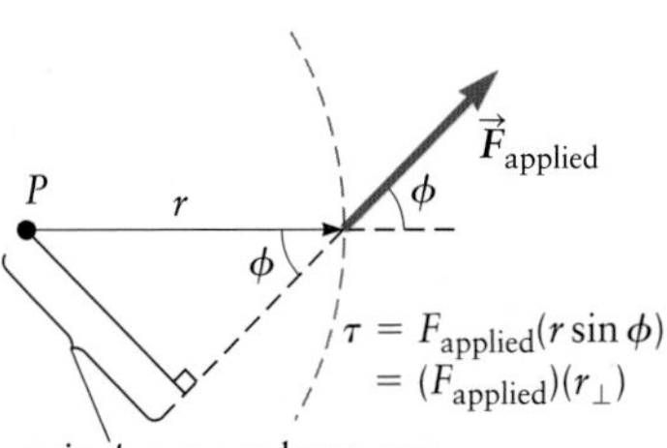

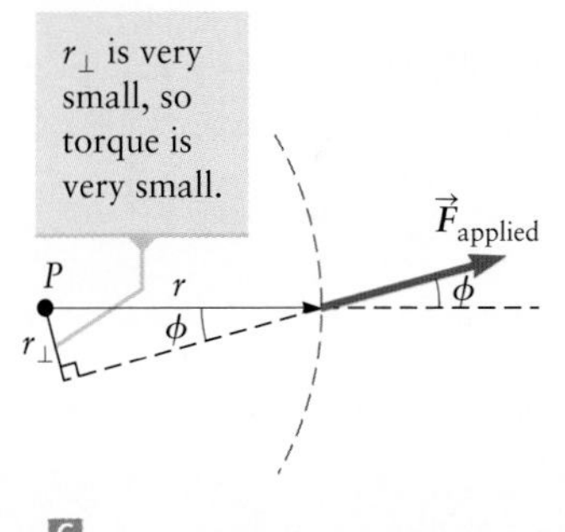

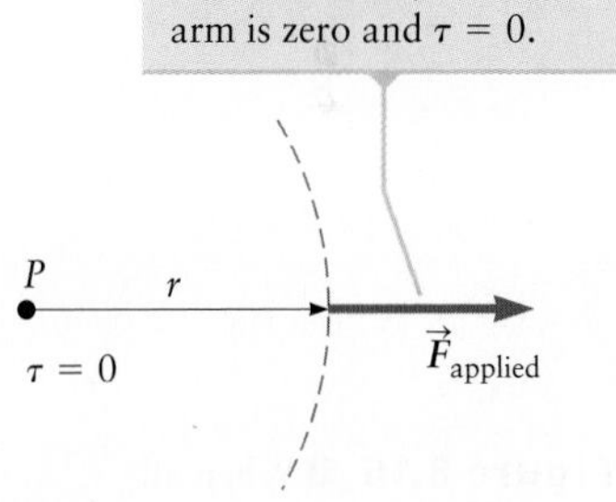

▲ **Figure 8.14** Torque is given by $\tau = F_{\text{applied}}(r \sin\phi) = F_\perp r = F_{\text{applied}} r_\perp$.

> **CONCEPT CHECK 8.3 Forces on a Torque Wrench**
>
> A torque wrench is a special type of wrench that limits the magnitude of the torque it is applying to a bolt, helping the user avoid breaking a bolt or stripping a thread. The torque wrench in Figure 8.15 has a long handle, and we suppose one user applies a force $F_1 = 50$ N at a distance of 10 cm from the head of the wrench. If a second user applies a force F_2 at a point 25 cm from the head, what value of F_2 will produce the same torque? Assume forces F_1 and F_2 are both applied perpendicular to the wrench handle.
>
> (a) 10 N (b) 20 N (c) 50 N (d) 100 N (e) 125 N

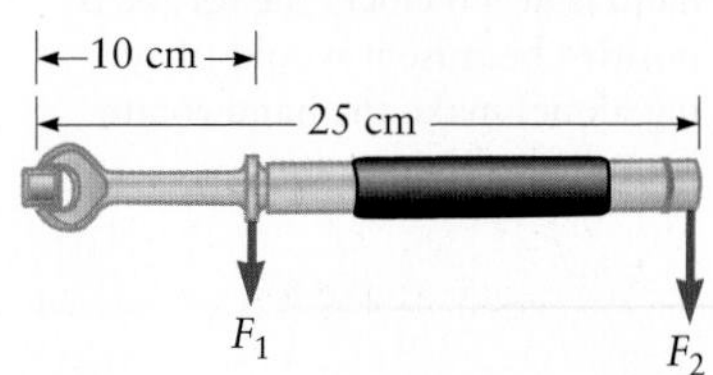

▲ **Figure 8.15** Concept Check 8.3. Forces on a torque wrench.

Center of Gravity, Center of Mass, and the Direction of Torque

Consider the motion of the hands of a large outdoor clock such as one might find near the top of a clock tower as represented very schematically in Figure 8.16. Let's calculate the torque on one clock hand due to the force of gravity. For simplicity, we first assume (unrealistically) that all the mass m is located at the end of the hand, which makes the clock hand just like the ball attached to a massless rod in Figures 8.11 and 8.13. Using Equation 8.15, the magnitude of the torque is

$$|\tau| = F_{\text{grav}} r \sin \phi$$

Figure 8.16 shows the clock hand at two different positions. When the hand is at "3 o'clock" (i.e., horizontal), the gravitational force is perpendicular to the hand and $\phi = 90°$. If L is the length of the clock hand, we have

$$|\tau| = F_{\text{grav}} r \sin \phi = mgL \sin(90°)$$

$$|\tau| = mgL \tag{8.19}$$

which is the *magnitude* of the torque. We must also pay attention to its *direction.* When we have a single rotation axis, as is the case here, the direction of the torque is specified by its sign. Recall our convention for the positive and negative directions for rotations in Figure 8.3. Those conventions also give the positive and negative directions for τ. A *positive* torque is one that would produce a counterclockwise rotation, whereas a *negative* torque would produce a clockwise rotation. In this example, with the clock hand at 3 o'clock, the torque from gravity is negative since this torque would, if it acted alone, cause the clock hand to move downward and hence rotate clockwise. If the clock hand were instead at the 9 o'clock position (Fig. 8.16B), the torque would have the same magnitude as found in Equation 8.19 (because the angle ϕ is again a right angle), but τ would now be positive because it would cause a counterclockwise rotation of the hand.

Having all the mass located at the end of the clock hand is rather unrealistic, even for a physics problem, so let's now make the more reasonable assumption that the mass is distributed uniformly along the clock hand. This situation is shown in Figure 8.17, with the clock hand located at 3 o'clock. To calculate the torque, we want to apply our relation for τ in Equation 8.15, but we immediately run into a problem. Torque involves the product of force and lever arm, and in this case, with the force in a purely perpendicular direction, the lever arm lies along the clock hand. The length of the lever arm depends on *where* the force acts. The gravitational force acts wherever there is mass, so it acts at all points along the hand. We could imagine the hand to be broken up into many infinitesimally small pieces and then add up the torques on each piece to get the total torque, but a much more convenient approach is to use the concept of ***center of gravity***.

When calculating the torque due to the gravitational force, one can assume all the force acts at a single location called the object's center of gravity. For our clock hand and for all other examples we encounter in this book, the center of gravity of an object is located at its center of mass,[1] defined in Chapter 7. For our uniform

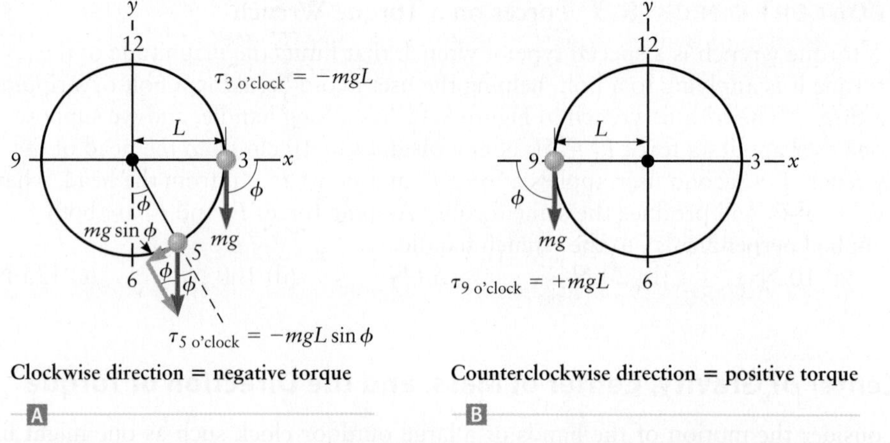

► **Figure 8.16** A When all mass of this clock hand is at the end, the torque due to gravity is $\tau = -mgL \sin \phi$. When the clock hand is at 3 o'clock, $\phi = 90°$, whereas when the hand is at 5 o'clock, this angle is smaller, so the magnitude of the torque is smaller. B When the clock hand is at 9 o'clock, the torque is positive because it would (if acting alone) make the hand rotate counterclockwise.

[1]The center of gravity and the center of mass of an object are usually at the same point, but that is not the case when the force of gravity varies with position, as in planetary motion. We do not treat such cases in this text.

clock hand, the center of mass is at the center of the hand. The length of the lever arm is thus $L/2$ as indicated in Figure 8.17, and the torque with the clock hand at 3 o'clock is

$$\tau = -mgr \sin \phi = -mg\left(\frac{L}{2}\right)\sin(90°) = -\frac{mgL}{2}$$

because again the angle $\phi = 90°$. The negative signs here account for gravity's tendency to make the clock hand rotate clockwise, which, according to our convention, is the negative direction.

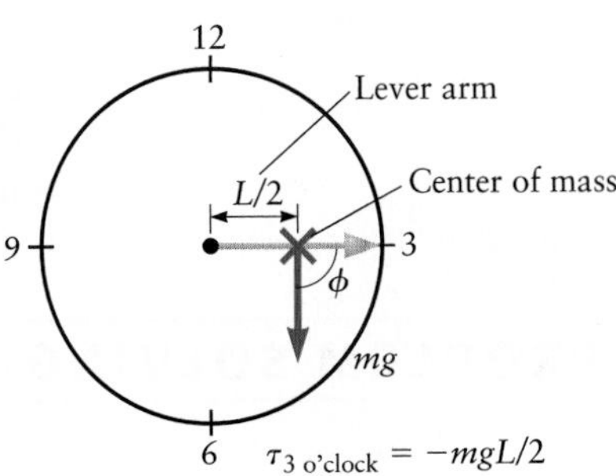

▲ **Figure 8.17** When the mass of the clock hand is distributed uniformly, the torque is calculated by letting the force of gravity act at the center of mass.

CONCEPT CHECK 8.4 Determining the Sign of the Torque

Figure 8.18 shows several rotating objects and the forces applied to them. According to our convention for finding the sign of τ, which of these torques are positive and which are negative?

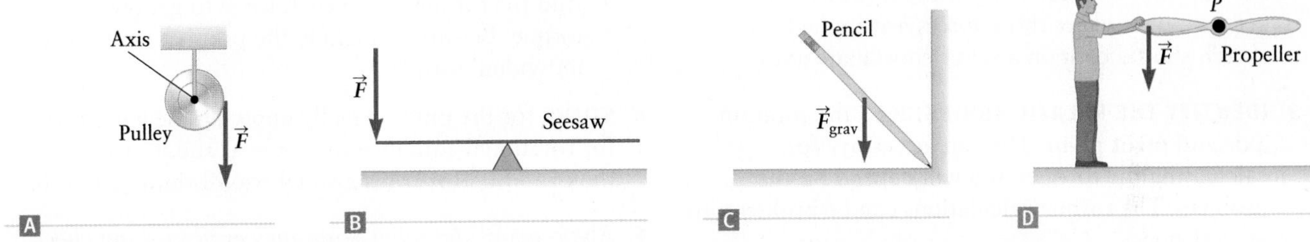

▲ **Figure 8.18** Concept Check 8.4.

8.3 Rotational Equilibrium

We showed in Chapter 4 that for an object to be in equilibrium with respect to its translational motion, the linear acceleration $\vec{a}$ must be zero. According to Newton's second law, $\sum \vec{F} = m\vec{a}$, so if an object is in translational equilibrium, the total force must be zero, $\sum \vec{F} = 0$. Also, since the acceleration is zero, an object in translational equilibrium will have a constant velocity $\vec{v}$. If that velocity is zero, we say that the object is in static equilibrium with respect to its translational motion.

We now want to extend these ideas to the notion of ***rotational equilibrium***. An object that is in equilibrium with regard to both its translational and its rotational motion must have a linear acceleration of zero *and* an angular acceleration that is zero. The condition that the total force is zero, $\sum \vec{F} = 0$, is *not* sufficient to guarantee that both these requirements are satisfied. See, for instance, Figure 8.19, which shows two forces applied to an airplane propeller. The forces $\vec{F}_1$ and $\vec{F}_2$ are applied horizontally, and we suppose they are equal in magnitude but opposite in direction, so their vector sum is zero. Although the condition for translational equilibrium along x is thus satisfied, this propeller will not be in rotational equilibrium. Even though the sum of these two forces is zero, they still produce a net torque on the propeller and lead to an angular acceleration in the clockwise direction.

For an object to be in complete equilibrium, the angular acceleration must be zero. From Newton's second law for rotational motion (Eq. 8.12), the total torque must be zero:

$$\sum \tau = \tau_1 + \tau_2 + \tau_3 + \cdots = 0 \tag{8.20}$$

For the problems we consider in this book, all the torques in this equation refer to a particular rotation axis. The same ideas can be applied to the rotation about two or three different axes simultaneously, but we leave that for a more advanced course.

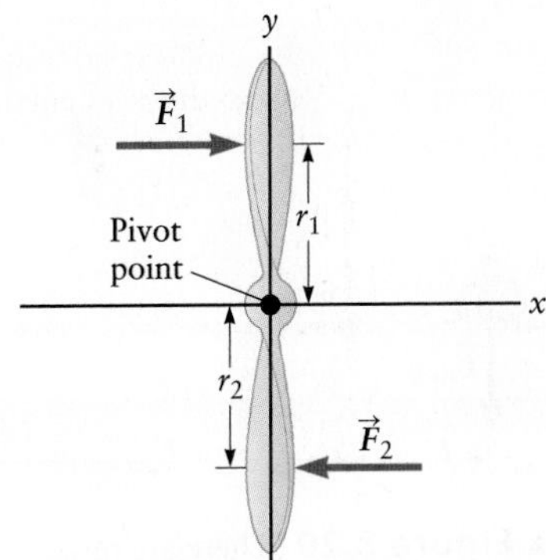

▲ **Figure 8.19** Although the horizontal forces on this propeller may add up to zero, they will still produce a nonzero torque. Hence, the propeller will not be in rotational equilibrium.

To apply the condition for rotational equilibrium in Equation 8.20, we must calculate the total torque on an object from all the applied forces. The following procedure describes how to calculate the torque and apply the conditions for equilibrium in any situation.

PROBLEM SOLVING Applying the Conditions for Translational and Rotational Equilibrium

1. **RECOGNIZE THE PRINCIPLE.** An object is in translational static equilibrium if its linear acceleration and linear velocity are both zero. An object is in rotational static equilibrium if its angular acceleration and angular velocity are both zero.

2. **SKETCH THE PROBLEM.** Make a drawing showing the object of interest along with all the forces that act on it and where those forces are applied. This sketch should contain a set of coordinate axes.

3. **IDENTIFY THE RELATIONSHIPS.** Find the rotation axis and pivot point. They are necessary for calculating the torques and will depend on the problem. The torque calculations can be broken into several steps:
   - For each force, first determine the lever arm, the perpendicular distance from the pivot point to the point at which the force acts.
   - Calculate the magnitude of the torque using $\tau = Fr \sin \phi$ (Eq. 8.15). Here, F is the magnitude of the force.
   - Determine the sign of τ. If this force *acting alone* would produce a *counterclockwise* rotation, the torque is *positive*. If the force would cause a *clockwise* rotation, the torque is *negative*.
   - Add the torques from each force to get the total torque. Be sure to include the proper sign for each individual torque.

4. **SOLVE** for the unknowns by applying the condition for rotational equilibrium $\sum \tau = 0$ and, if necessary, the condition for translational equilibrium $\sum \vec{F} = 0$.

5. Always *consider what your answer means* and check that it makes sense.

When we applied Newton's laws to the translational motion of a point particle (in Chapters 3 and 4), we began with a free-body diagram showing all the forces acting on the particle. When dealing with rotational motion, we also start with a force diagram, but we cannot simply show the object as a point. We must consider all the forces on the object and the *places on the object where these forces act*. This information is needed to determine the torque associated with each force.

Choosing a pivot point is also an essential part of dealing with torques and rotational motion, especially in equilibrium problems. In many cases, such as the seesaw in Figure 8.10, there is a "natural" choice for the pivot point, but sometimes there may be more than one plausible choice. Fortunately, for an object in rotational equilibrium, you may choose *any* spot to be the pivot point without affecting the final answer. However, some choices will lead to a simpler torque equation. Since torque is equal to the product of force and lever arm, forces whose lever arms are zero will not contribute to the torque. These forces will then not appear in the torque equation. This is illustrated in Example 8.4.

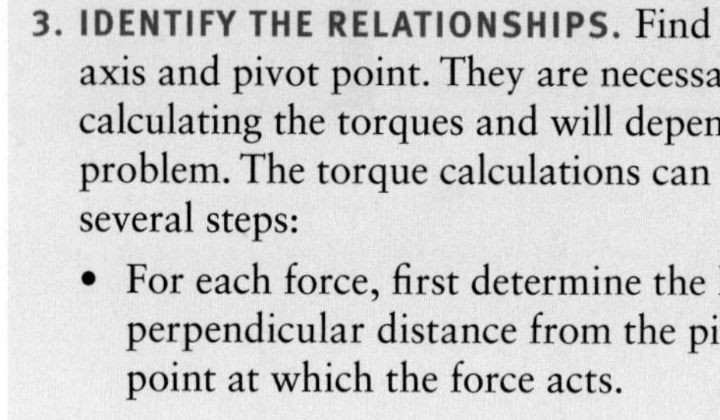

▲ **Figure 8.20** There are three forces acting on the lever (assuming it is very light so that its weight can be ignored). The normal force F_N exerted by the support acts at the pivot point P, so this force does not produce a torque about P.

Rotational Equilibrium of a Lever: Amplifying Forces

Figure 8.20 shows a lever being used to lift a very heavy rock of mass m. Let's calculate the minimum force needed to just begin lifting the rock and then find how it compares to the rock's weight.

Since the rock is just barely being lifted from the ground, both its acceleration and its angular acceleration are essentially zero and we can apply our procedure for dealing with objects in rotational equilibrium to the lever. Figure 8.20 serves as our drawing (step 2 in our problem-solving procedure). It shows all the forces on the lever and the places at which they act. For simplicity, we assume the lever is much lighter than the rock, so we can ignore the force of gravity on the lever. That still leaves three forces acting on the lever: the force $\vec{F}_{\text{rock}}$ from the rock on the lever (the weight of the rock), the force $\vec{F}_{\text{person}}$ applied by the person on the other end of the lever, and the force $\vec{F}_N$ of the support on the lever (at point P). We next (step 3) identify the pivot point: we choose the pivot point to be the spot at which the lever is supported (point P in Fig. 8.20), with the rotation axis perpendicular to the plane of the picture. The lever arm associated with $\vec{F}_{\text{rock}}$ is L_{rock}, while the lever arm for the torque associated with $\vec{F}_{\text{person}}$ is L_{person}. The lever arm for the force from the support is zero because this force acts at the pivot point.

Using the results from Figure 8.20 for the lever arms and forces, the condition for rotational equilibrium is

$$\sum \tau = \tau_{\text{rock}} + \tau_{\text{person}} + \tau_N = 0$$

$$+|F_{\text{rock}}|L_{\text{rock}} - |F_{\text{person}}|L_{\text{person}} + 0 = 0 \tag{8.21}$$

The signs in front of these terms come from our convention for positive and negative torques. The force from the rock has a magnitude $|F_{\text{rock}}|$ and is directed downward, so it gives a counterclockwise (positive) torque (hence the positive sign in front of this term in Eq. 8.21). To balance this torque, the force exerted by the person on the lever's other end must be directed downward, with a magnitude $|F_{\text{person}}|$. This force produces a clockwise (negative) torque (hence the negative sign in front of that term in Eq. 8.21). Since F_{rock} is due to the weight of the rock, $|F_{\text{rock}}| = mg$. Inserting into Equation 8.21 leads to

$$mgL_{\text{rock}} - |F_{\text{person}}|L_{\text{person}} = 0 \tag{8.22}$$

We want to find the force F_{person} that is just sufficient to balance the rock on the lever, and we can find its value by solving Equation 8.22 (step 4):

$$|F_{\text{person}}| = mg\frac{L_{\text{rock}}}{L_{\text{person}}}$$

This expression gives the magnitude of F_{person}, but we also need to know its sign (i.e., direction). From Figure 8.20, we can see that the person exerts a downward force on the lever (as we also assumed in our calculation of the torque), so F_{person} must be negative. We get

$$F_{\text{person}} = -mg\frac{L_{\text{rock}}}{L_{\text{person}}} \tag{8.23}$$

What does this result mean? Equation 8.23 shows that the magnitude of the force exerted by the person (F_{person}) can be smaller than the weight of the rock. If the lever arm extending to the person is longer than the lever arm to the rock (i.e., if $L_{\text{person}} > L_{\text{rock}}$), this simple lever will *amplify* the force exerted by the person. Of course, this result is probably what you expected if you have ever used a screwdriver to pry something open.

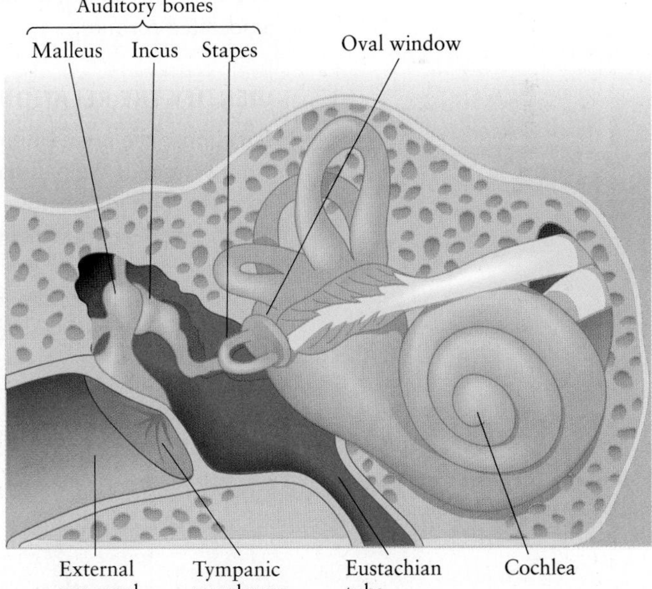

▲ **Figure 8.21** The incus bone in the ear acts as a lever as it transmits forces to the inner ear. The pivot point for the incus lever is near where the incus meets the malleus.

Amplification of Forces in the Ear

Many mechanical devices, including the human ear, use a lever to amplify forces. As you may know, three tiny bones help carry sound through the middle ear as shown in Figure 8.21. One of them, the incus, is supported by a sort of hinge so that it is able to transmit forces from one end to the other, just like the lever in Figure 8.20. The analysis given above applies also to the forces at the ends of the incus. The lever arm from this hinge to the end in contact with the stapes bone is about a factor of three shorter than the end connected to the malleus bone. By Equation 8.23, the lever associated with the incus bone amplifies the forces associated with a sound vibration by the same factor of three.

EXAMPLE 8.4 Painters and Physics

A housepainter is using a board of length $L = 3.0$ m that sits on two supports as a scaffold (Fig. 8.22). The painter wants to reach very far to one side of the board so that he can paint as much area as possible without moving the scaffolding. **(a)** How far can the painter in Figure 8.22 walk to the right-hand side of the board before it tips? The mass of the painter is 80 kg, and the mass of the board is 30 kg. **(b)** What is the force of the right-hand support on the board when the board just begins to tip? Assume the painter is always standing straight up, so his center of mass is always above his feet.

RECOGNIZE THE PRINCIPLE

The board does not quite tip, so it is always in rotational equilibrium. The total torque must therefore equal zero. The board is also in translational equilibrium, so the total force on the board must also equal zero.

SKETCH THE PROBLEM

Figure 8.22 shows all the forces on the board: the forces exerted by both supports F_R and F_L, the force of gravity on the board F_B, and the force of gravity on the painter F_P. The figure also shows where each force acts, information that is needed to calculate the associated torques.

IDENTIFY THE RELATIONSHIPS

We choose the pivot point to be at P because that is the point around which the board would rotate if the painter moves too far to the right. When the painter is standing as far to the right as possible at a distance x from the pivot point, the board will just barely begin to tip, which means that F_L, the force exerted by the left-hand support, is *zero*.

We next compute the torques using the information in Figure 8.22 and apply the condition for rotational equilibrium $\sum \tau = 0$. Using the lever arms from Figure 8.22

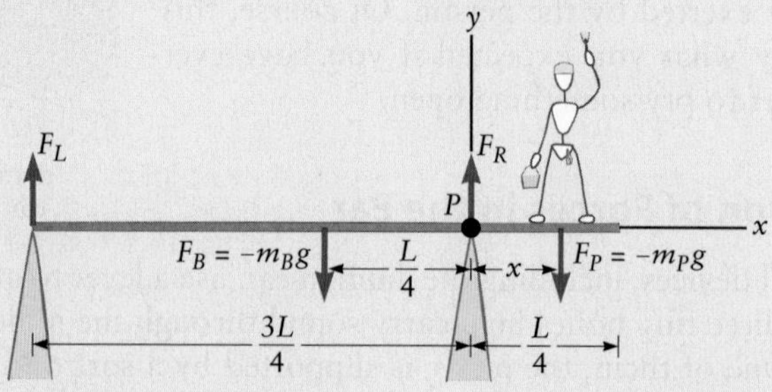

▶ **Figure 8.22** Example 8.4. How far to the right from point P can this painter go without tipping the board?

and $F_B = -m_B g$ (the weight of the board) and $F_P = -m_P g$ (the weight of the painter), we have

$$\Sigma\tau = 0 = -F_L\left(\frac{3}{4}L\right) + F_R(0) + m_B g\left(\frac{L}{4}\right) - m_P g(x) \quad (1)$$

The signs of these torque terms are determined using the convention that a counterclockwise rotation is in the positive direction. For example, the last term on the right in Equation (1) is the torque from the weight of the painter; this torque would produce a clockwise rotation and thus is negative.

The next step is to write the condition for translational equilibrium $\sum \vec{F} = 0$. The forces on the board are all along the y direction. The force condition is then

$$\sum F = F_L + F_R + F_B + F_P = 0 \quad (2)$$

Notice that some of these terms, such as $F_B = -m_B g$ and $F_P = -m_P g$, are themselves negative. Inserting $F_L = 0$ and our values of F_B and F_P into Equation (2) gives

$$F_R - m_B g - m_P g = 0 \quad (3)$$

SOLVE

(a) We can now solve Equation (1) to find x, the location of the painter. Inserting $F_L = 0$ because the board is on the verge of tipping, we find

$$m_B g\left(\frac{L}{4}\right) - m_P g(x) = 0$$

$$x = \frac{m_B g(L/4)}{m_P g} = \frac{m_B L}{4m_P}$$

Inserting the given values of the masses of the board and painter along with L gives

$$x = \frac{m_B L}{4m_P} = \frac{(30\text{ kg})(3.0\text{ m})}{4(80\text{ kg})} = \boxed{0.28\text{ m}}$$

The painter should thus remain a distance less than 0.28 m from the pivot point.

(b) To find the force from the right-hand support, we rearrange Equation (3) and get

$$F_R = m_B g + m_P g = (m_B + m_P)g = (80\text{ kg} + 30\text{ kg})(9.8\text{ m/s}^2)$$

$$F_R = \boxed{1100\text{ N}}$$

► ***What does it mean?***

Since F_L is zero, the right-hand support must support the entire weight of the system. Hence, the force from the right-hand support must equal the weight of the board plus the weight of the painter, as found in the answer to part (**b**).

CONCEPT CHECK 8.5 Carrying Your Share of the Weight

The two workers in Figure 8.23 are carrying a long, heavy steel beam. Which one is exerting a larger force on the object? How can you tell?

(a) The worker on the left (b) The worker on the right

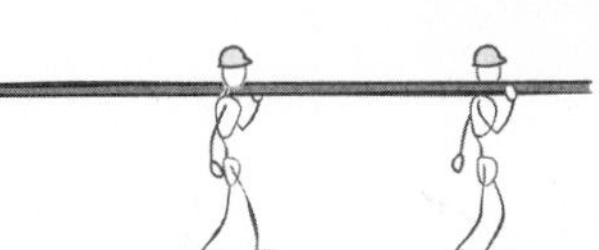

▲ **Figure 8.23** Concept Check 8.5.

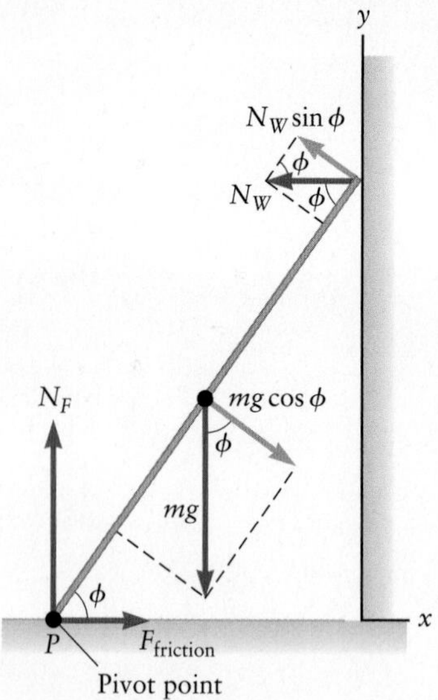

▲ **Figure 8.24** Example 8.5. Forces acting on a leaning ladder. We assume there is no friction between the wall and the top end of the ladder.

EXAMPLE 8.5 Leaning on a Wall

Consider the stability of a ladder of mass m and length L leaning against a wall as shown in Figure 8.24. A painter wants to be sure the ladder will not slip across the floor. He knows that the frictional force exerted by the floor on the bottom end of the ladder is needed to keep the ladder in equilibrium, and he also knows that the coefficient of static friction between the ladder and the floor is $\mu_S = 0.50$. What is the angle ϕ at which the ladder will just become unstable and slide across the floor? For simplicity, assume there is no frictional force between the ladder and the wall.

RECOGNIZE THE PRINCIPLE

The ladder is in equilibrium, so we can apply the general conditions for static equilibrium $\sum \tau = 0$ and $\sum \vec{F} = 0$. The angle ϕ plays a key role: if we take the bottom of the ladder (point P) as our pivot point, the torque on the ladder due to gravity increases as ϕ becomes smaller because the lever arm increases. At some critical value of the angle, this torque will be large enough that the ladder will start to move.

SKETCH THE PROBLEM

Figure 8.24 shows all the forces acting on the ladder along with their components along x and y. There are normal forces exerted by the floor (N_F) and the wall (N_W), the force of friction exerted by the floor ($F_{friction}$), and the ladder's weight (mg). The force from the floor acts on the end of the ladder at point P, while the force of gravity acts at the ladder's center of mass.

IDENTIFY THE RELATIONSHIPS

Let's choose the base of the ladder at P as the pivot point. By doing so, we make the torques due to the forces from the floor (N_F and $F_{friction}$) zero because these forces act at P and hence their lever arms are zero. Using the lever arms and angles from Figure 8.24, the torque condition is

$$\sum \tau = 0 = +N_W L \sin\phi - mg\left(\frac{L}{2}\right)\sin(90^\circ - \phi)$$

$$0 = +N_W L \sin\phi - mg\left(\frac{L}{2}\right)\cos(\phi) \quad (1)$$

where we have used the general definition of torque (Eq. 8.15) along with the trigonometric identity $\sin(90° - \phi) = \cos(\phi)$. The appearance of the cosine term in Equation (1) can be understood from the expression for torque in Equation 8.16. The torque due to the ladder's weight is the product of $L/2$ (the distance from the center of mass to the pivot point) and the component of the force of gravity *perpendicular* to the ladder. Figure 8.24 shows that this component involves $\cos\phi$.

Since the ladder is in static equilibrium, the sum of all forces along both x and y must be zero, which leads to

$$\sum F_x = 0 = -N_W + F_{friction} \quad (2)$$

$$\sum F_y = 0 = N_F - mg = 0 \quad (3)$$

SOLVE

From Equation (3), we can solve for the normal force from the floor and get $N_F = mg$. The maximum frictional force is $F_{friction} = \mu_S N_F$, so we have

$$F_{friction} = \mu_S N_F = \mu_S mg$$

and, according to Equation (2), this result is also equal to N_W. Inserting into Equation (1) leads to

$$+N_W L \sin\phi - mg\left(\frac{L}{2}\right)\cos\phi = 0$$

$$\mu_S mgL \sin\phi = mg\left(\frac{L}{2}\right)\cos\phi$$

$$\tan\phi = \frac{1}{2\mu_S}$$

Using the given value $\mu_S = 0.50$, we find

$$\tan\phi = \frac{1}{2\mu_S} = \frac{1}{2(0.50)} = 1.0 \quad (4)$$

$$\phi = \boxed{45°}$$

► ***What does it mean?***

For a smaller angle, the frictional force would not be able to keep the ladder in equilibrium and the ladder would slip. A sensible painter would choose a steeper angle than 45°.

CONCEPT CHECK 8.6 Adding Friction between the Ladder and the Wall

Suppose we add friction between the wall and the upper end of the ladder in Example 8.5. How would that change the result for ϕ?

(a) The minimum safe value of ϕ would increase.
(b) The minimum safe value of ϕ would decrease.
(c) The minimum safe value of ϕ would stay the same.

Pushing on a Crate: When Will It Tip?

Another example of an object in static equilibrium is shown in Figure 8.25A. A cubical crate of length L sits at rest on a level floor as a person pushes on one side of the crate. If the person applies a very large force of magnitude F_{person} horizontally at the top edge and the frictional force $F_{friction}$ at the bottom edge is large, the crate will tip as sketched in Figure 8.25B. Let's consider how to calculate the value of F_{person} at which the crate just begins to tip.

When the crate is on the verge of tipping, it is still in static equilibrium, so the total torque and the total force are both zero. There are four forces acting on the crate: (1) the force of gravity, which acts at the center of mass; (2) the force of static friction, which acts along the bottom surface of the crate; (3) the normal force exerted by the floor, which acts at the surface in contact with the floor; and (4) the force applied by the person, F_{person}. Any tipping will involve rotation about the point P located at the corner of the crate, so let's choose that as our pivot point. The corresponding lever arms are indicated in Figure 8.25A; the lever arm associated with the frictional force is zero, so it will not contribute to the torque. If the crate is on the verge of rotating about P, the normal force exerted by the floor will then be acting along a line passing through P and its lever arm is also zero. The total torque on the crate is then

$$\sum \tau = 0 = \underbrace{mg(L/2)}_{\text{gravity}} - \underbrace{F_{person}(L)}_{\text{person}} + \underbrace{N(0)}_{\text{normal force}} + \underbrace{F_{friction}(0)}_{\text{friction}} = 0 \quad (8.24)$$

where the corresponding lever arm factors are all given in parentheses. The lever arm for the gravitational force is $L/2$ because this force acts at the center of mass, which is in the middle of the crate, a distance $L/2$ from the crate's side. The lever

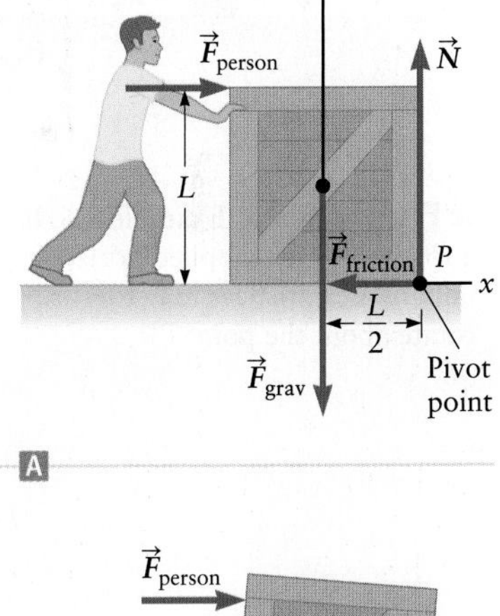

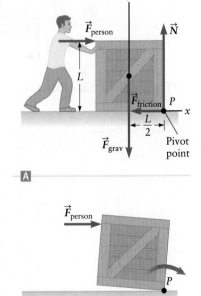

▲ **Figure 8.25** A Forces on a crate. As the person pushes harder and harder, the crate will eventually tip and B rotate about a pivot point P at the corner of the crate.

arm for the torque from F_{person} is L because the person is pushing at the top edge of the crate. The signs in Equation 8.24 can again be understood using our convention for positive and negative torque. For example, a positive value of the applied force (F_{person} directed to the right, along $+x$) gives a clockwise rotation and hence a *negative* torque. We can solve Equation 8.24 for F_{person} and find

$$mg(L/2) = F_{person}(L)$$

$$F_{person} = \frac{mg}{2}$$

which is half the weight of the crate. If F_{person} exceeds this value, the crate will tip.

This problem of a tipping crate is very similar to the problem of a car rolling over in a traffic accident. This subject is of great interest to automobile designers and is explored in the next example.

EXAMPLE 8.6 Ⓡ Cars and Static Equilibrium

Figure 8.26 shows the forces acting on a car in a collision. The force $\vec{F}_{collision}$ comes from another car, and for simplicity we assume this force is directed horizontally and acts at the center of the body of the car. Estimate the magnitude of $\vec{F}_{collision}$ for which the car in Figure 8.26 will just roll over. Express the answer as a ratio $|\vec{F}_{collision}|/mg$, where m is the mass of the car, and use the dimensions of a typical car in your calculation.

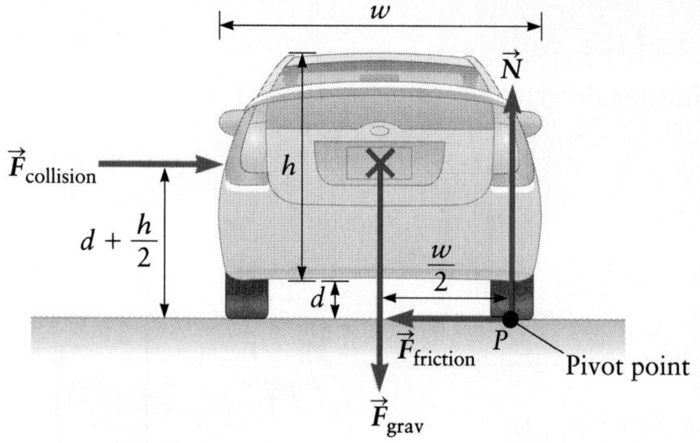

▲ **Figure 8.26** Example 8.6. If a force $\vec{F}_{collision}$ is applied horizontally to a car, the car may tip (i.e., rotate about the point P).

RECOGNIZE THE PRINCIPLE

We wish to calculate the minimum force $\vec{F}_{collision}$ that will just barely cause the car to tip. This problem is very similar to the tipping crate in Figure 8.25. When the car tips, it will rotate about point P in Figure 8.26, so we take that as the pivot point.

SKETCH THE PROBLEM

Figure 8.26 shows all the forces on the car, including the force of friction between the tires and road ($F_{friction}$) and the force of gravity ($F_{grav} = mg$). It also shows the points at which these forces act.

IDENTIFY THE RELATIONSHIPS

Although this problem is similar to the tipping crate in Figure 8.25, there are some important differences. Because the car sits on four wheels, we must account for the height of the wheels when finding the car's center of mass. Indeed, we are not given any information about the dimensions of the car's body (w and h in Fig. 8.26) or the wheels (d), so we'll have to estimate their values. First, though, let's express the conditions for equilibrium in terms of w, h, and d.

We treat the car as a simple "box" of width w and height h that is a distance d off the ground and ignore the mass in the wheels. There are four forces on the car as indicated in Figure 8.26. The condition for rotational equilibrium is

$$\sum \tau = 0 = \underbrace{-F_{collision}\left(d + \frac{h}{2}\right)}_{\text{collision force}} + \underbrace{mg\left(\frac{w}{2}\right)}_{\text{gravity}} + \underbrace{N(0)}_{\text{normal force}} + \underbrace{F_{friction}(0)}_{\text{friction}} \quad (1)$$

where the lever arm factors are given in parentheses. The term involving $F_{collision}$ is the torque produced by the collision force. The lever arm in this case is the distance from the center of the car to the ground ($d + h/2$) since this force acts at the middle of the car, at its center of mass. This force is directed to the right (along $+x$) and would, if it acted alone, produce a clockwise rotation; hence, this torque is *negative*.

SOLVE

We can rearrange Equation (1) to get

$$F_{\text{collision}}\left(d + \frac{h}{2}\right) = mg\left(\frac{w}{2}\right)$$

Dividing $F_{\text{collision}}$ by the weight of the car leads to

$$\boxed{\frac{F_{\text{collision}}}{mg} = \frac{w}{2d + h}} \quad (1)$$

▸ What does it mean?

If $F_{\text{collision}}/mg$ is small, a relatively small force is enough to cause a rollover accident, whereas a large ratio means that a larger impact force is needed to cause a rollover. Equation (1) shows that a large value of d—that is, a large ground clearance—will make a rollover more likely. For a typical compact car that sits relatively low to the ground, we estimate[2] $w = 170$ cm, $d = 20$ cm, and $h = 120$ cm, which gives the ratio

$$\frac{F_{\text{collision}}}{mg} = \frac{170 \text{ cm}}{2(20 \text{ cm}) + 120 \text{ cm}} = 1.1 \quad \text{(compact car)}$$

For an SUV, we estimate $w = 170$ cm, $d = 40$ cm, and $h = 140$ cm, which leads to

$$\frac{F_{\text{collision}}}{mg} = \frac{170 \text{ cm}}{2(40 \text{ cm}) + 140 \text{ cm}} = 0.77 \quad \text{(SUV)}$$

Hence, according to this measure, the SUV is more likely to roll over. Of course, real accidents are more complicated than this simple model, but current government rollover ratings are based in part on a similar calculation.

8.4 Moment of Inertia

In Section 8.2, we introduced the rotational form of Newton's second law:

$$\sum \tau = I\alpha \quad (8.25)$$

In all our work on rotational equilibrium, $\alpha = 0$, so the right-hand side of Equation 8.25 was zero and we did not have to worry about the moment of inertia I. To tackle cases in which the angular acceleration is not zero, we must consider the moment of inertia in more detail.

Equation 8.14 defines the moment of inertia for a general object composed of many pieces of mass m_i as

$$I = \textstyle\sum_i m_i r_i^2 \quad (8.26)$$

where each piece i is a distance r_i from the axis of rotation. The moment of inertia of an object depends on its mass *and* on how this mass is distributed with respect to the rotation axis. The value of I for a particular object thus depends on the choice of rotation axis, as illustrated using the simple "object" in Figure 8.27. Two point particles, both with mass m, are connected by a very light (massless) rod of length L. Figure 8.27A shows a rotation about the center of the rod. Applying Equation 8.26, the moment of inertia in this case is

$$I_a = \textstyle\sum_i m_i r_i^2 = m_1 r_1^2 + m_2 r_2^2 \quad (8.27)$$

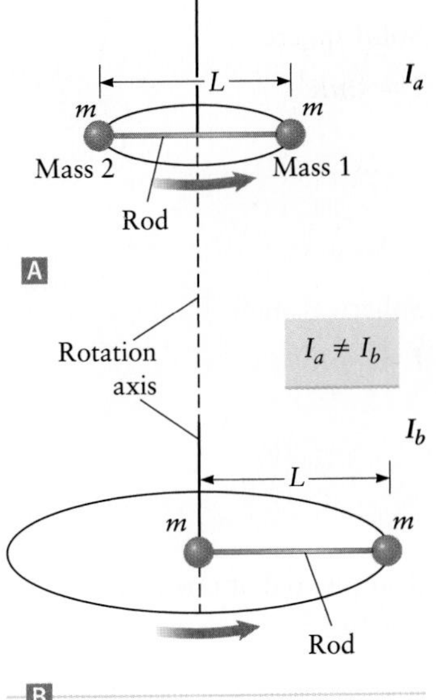

▲ **Figure 8.27** The moment of inertia depends on the choice of rotation axis.

[2]These estimates of w, d, and h were obtained from the specifications of actual vehicles.

Both particles have mass m; hence, $m_1 = m_2 = m$, and they are both a distance $L/2$ from the rotation axis, so $r_1 = r_2 = L/2$. Inserting this information into Equation 8.27 leads to

$$I_a = m(L/2)^2 + m(L/2)^2 = mL^2/2 \tag{8.28}$$

This object can also rotate about an axis that passes through one end of the rod (Fig. 8.27B). The moment of inertia in this case is again given by Equation 8.27, but now $r_2 = 0$ (because the axis runs through mass 2) while $r_1 = L$. The moment of inertia is thus

$$I_b = m_1 r_1^2 + m_2 r_2^2 = m(L)^2 + m(0)^2 = mL^2 \tag{8.29}$$

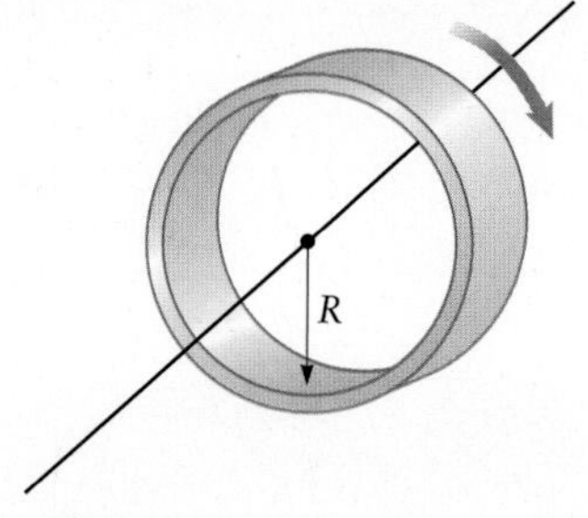

▲ **Figure 8.28** For a simple hoop rotating about its geometric center, all the mass is a distance R from the rotation axis.

The moments of inertia I_a and I_b (Eqs. 8.28 and 8.29) are different, showing that the moment of inertia of an object does indeed depend on the choice of axis.

Using Equation 8.26 to find the moment of inertia of a complicated object can be difficult. Fortunately, though, the results for a variety of common shapes are known, and some are listed in Table 8.2. For example, the moment of inertia of a solid sphere of mass m and radius R is $I_{sphere} = 2mR^2/5$ for an axis that passes through the center of the sphere. This result applies to *all* solid spheres. Likewise, the moment of inertia of a disc or solid cylinder rotating about its central axis is $I_{disc} = mR^2/2$, where m is the total mass and R is the radius. The result is different from the moment of inertia of a hoop, which is $I_{hoop} = mR^2$ for an axis perpendicular to the plane of the hoop and passing through its center as indicated in Table 8.2. This case with a hoop is particularly simple because all the mass is a distance R from the rotation axis (Fig. 8.28). In this case, the sum in Equation 8.26 can be evaluated as

$$I_{hoop} = \Sigma_i m_i r_i^2 = (\Sigma_i m_i)R^2 \tag{8.30}$$

TABLE 8.2 Moment of Inertia for Some Common Objects

Object	Shape	Object	Shape
Hoop $I = mR^2$	R	Rod pivoted at center $I = \frac{1}{12}mL^2$	L
Solid sphere $I = \frac{2}{5}mR^2$	R	Pulley/cylinder/disc $I = \frac{1}{2}mR^2$	R
Spherical shell $I = \frac{2}{3}mR^2$	R	Wheel or hollow cylinder $I = \frac{1}{2}m(R_{max}^2 + R_{min}^2)$	R_{min}, R_{max}
Rod pivoted at one end $I = \frac{1}{3}mL^2$	L	Solid square plate with axis perpendicular to plate $I = \frac{1}{6}mL^2$	L, L

Note: In each case, m is the total mass of the object.

All the mass of the hoop is at the same distance R from the rotation axis at the center, so we can pull this factor out of the summation. The sum $\Sigma_i m_i$ is the total mass m of the hoop, so Equation 8.30 becomes $I_{\text{hoop}} = mR^2$.

Table 8.2 also lists the moment of inertia for a uniform rod of length L and shows how the moment of inertia depends on the axis of rotation, just as we found for the simple object in Figure 8.27. If a rod rotates about a perpendicular axis that runs through the center of the rod, then $I_{\text{rod center}} = mL^2/12$, whereas if the rotation axis runs through one end of the rod, the moment of inertia is $I_{\text{rod end}} = mL^2/3$.

As an example, let's calculate the moment of inertia of a compact disc. A typical CD has a mass of about 9.0 g and a diameter of 12 cm. If we ignore the relatively small area of the hole in its center, a CD is just a disc of mass 0.0090 kg and radius 0.060 m. Inserting these values into the expression for a disc in Table 8.2 gives

$$I_{\text{disc}} = \frac{mR^2}{2} = \frac{(0.0090\text{ kg})(0.060\text{ m})^2}{2} \tag{8.31}$$

$$I_{\text{disc}} = 1.6 \times 10^{-5}\text{ kg}\cdot\text{m}^2 \tag{8.32}$$

This is the moment of inertia for rotations about an axis that runs through the center and is perpendicular to the plane of the disc. This is the axis appropriate for a CD player.

EXAMPLE 8.7 Moment of Inertia of a Dirty CD

A rather large bug of mass 5.0 g manages to land on the edge of your favorite video-game CD (mass 9.0 g and radius 6.0 cm). How does that affect the CD's moment of inertia? (Ignore the hole in the center of the CD.)

RECOGNIZE THE PRINCIPLE

To calculate the total moment of inertia of the bug plus the CD, we start with the general definition of the moment of inertia in Equation 8.26:

$$I_{\text{total}} = \Sigma_i m_i r_i^2$$

We now split this expression into two terms, one for the clean CD and one for the contribution of the bug:

$$I_{\text{total}} = \Sigma_i m_i r_i^2 = (\Sigma_i m_i r_i^2)_{\text{CD}} + (\Sigma_i m_i r_i^2)_{\text{bug}}$$

We calculated the first term (for the clean CD) in Equation 8.32. The bug (the second term) acts just as a point mass, as treated in connection with the object in Figure 8.11.

SKETCH THE PROBLEM

The CD, bug included, is shown in Figure 8.29.

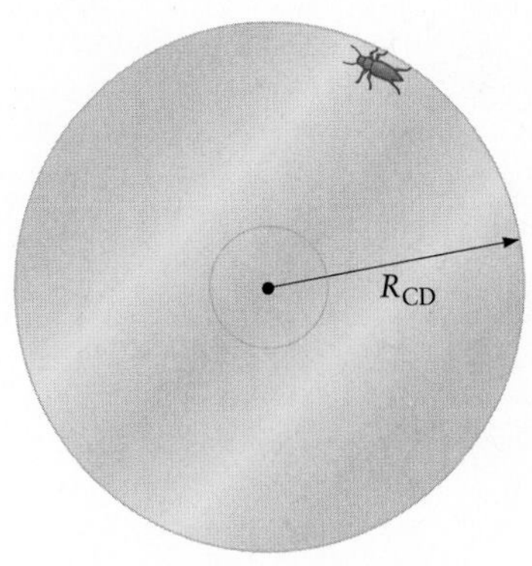

▲ **Figure 8.29** Example 8.7. The total moment of inertia of this "dirty" compact disc is the sum of the moment of inertia of the clean CD plus the moment of inertia of the bug.

IDENTIFY THE RELATIONSHIPS

A clean CD is simply a disc; its moment of inertia is $I_{\text{CD}} = m_{\text{CD}}R_{\text{CD}}^2/2$ and was calculated in Equations 8.31 and 8.32. For the contribution of the bug, we notice that all the bug's mass is located at the edge of the CD, so according to Equation 8.26 $I_{\text{bug}} = m_{\text{bug}}r_{\text{bug}}^2 = m_{\text{bug}}R_{\text{CD}}^2$. Combining these results gives

$$I_{\text{total}} = \frac{m_{\text{CD}}R_{\text{CD}}^2}{2} + m_{\text{bug}}R_{\text{CD}}^2$$

SOLVE

Inserting all the given values, we find

$$I_{\text{total}} = \frac{(0.0090\text{ kg})(0.060\text{ m})^2}{2} + (0.0050\text{ kg})(0.060\text{ m})^2 = \boxed{3.4 \times 10^{-5}\text{ kg}\cdot\text{m}^2}$$

(continued) ▶

► ***What does it mean?***
Comparing this result with the moment of inertia of a clean CD (Eq. 8.32), we see that the bug increases the moment of inertia by about a factor of two. This increase could well affect the way your videogame player is able to spin a CD. It pays to keep your CDs clean.

8.5 Rotational Dynamics

We are now ready to tackle some problems involving rotational dynamics. The basis for all our calculations will be

$$\sum \tau = I\alpha \tag{8.33}$$

We can predict the rotational motion of an object by calculating both the total torque and the moment of inertia and using Equation 8.33 to find the angular acceleration α. We can then find the angular velocity and angular displacement of the object. Often, the angular acceleration is constant, as in cases in which the applied torque is constant. The results for ω and θ then have particularly simple forms. In fact, these mathematical forms are already quite familiar to us. You should recall that α is the slope of the $\omega - t$ curve (see Eq. 8.5). If the angular acceleration has the constant value α, then ω must vary linearly with time:

$$\omega = \omega_0 + \alpha t \tag{8.34}$$

Equation 8.34 should remind you of the behavior of the linear velocity v as a function of time in situations in which the linear acceleration a is constant, as we studied in Chapter 3. Likewise, for this case of constant angular acceleration, the angular displacement has the familiar form

Rotational motion with constant angular acceleration

$$\theta = \theta_0 + \omega_0 t + \tfrac{1}{2}\alpha t^2 \tag{8.35}$$

We can also combine Equations 8.34 and 8.35 to get a relation that does not involve time; the result is

$$\omega^2 = \omega_0^2 + 2\alpha(\theta - \theta_0) \tag{8.36}$$

These relations, Equations 8.34 through 8.36, are very similar to the results for linear velocity and displacement we encountered for translational motion when the linear acceleration is constant. They are extremely useful, so we have collected them together in Table 8.3.

Angular Motion of a Compact Disc

Let's now use the relations in Table 8.3 to tackle a few problems involving rotational motion. Consider yet again the motion of a compact disc. Suppose a CD is inserted into the optical drive of a computer at $t = 0$ and accelerates uniformly to reach its operational angular velocity of 500 rpm in 5.0 s. Let's calculate (a) the angular acceleration of the CD, (b) the total angular displacement of the CD during this period, and (c) the torque the optical drive must exert on the CD.

TABLE 8.3 Kinematic Relations for Constant Angular Acceleration α and Corresponding Relations for Linear Motion with Constant Acceleration a

Rotational Motion	Equation Number	Translational Motion
$\theta = \theta_0 + \omega_0 t + \frac{1}{2}\alpha t^2$	8.35	$x = x_0 + v_0 t + \frac{1}{2}at^2$
$\omega = \omega_0 + \alpha t$	8.34	$v = v_0 + at$
$\omega^2 = \omega_0^2 + 2\alpha(\theta - \theta_0)$	8.36	$v^2 = v_0^2 + 2a(x - x_0)$

To find the angular acceleration, we note that the initial angular velocity is zero ($\omega_0 = 0$) and that the final angular velocity is

$$\omega = 500 \text{ rpm} = 500 \frac{\text{rev}}{\text{min}} \times \frac{2\pi \text{ rad}}{1 \text{ rev}} \times \frac{1 \text{ min}}{60 \text{ s}} = 52 \text{ rad/s}$$

Inserting this result into Equation 8.34 and solving for α gives

$$\alpha = \frac{\omega - \omega_0}{t} = \frac{(52 \text{ rad/s}) - 0}{5.0 \text{ s}} = 10 \text{ rad/s}^2 \tag{8.37}$$

The angular displacement during this interval can be obtained from Equation 8.35,

$$\theta = \theta_0 + \omega_0 t + \tfrac{1}{2}\alpha t^2 = 0 + 0 + \tfrac{1}{2}(10 \text{ rad/s}^2)(5.0 \text{ s})^2 = 130 \text{ rad}$$

The torque that must be supplied by the CD player can be found from Equation 8.33,

$$\sum \tau = I\alpha$$

We have already calculated the moment of inertia of a CD (Eq. 8.32). Using that result together with the angular acceleration from Equation 8.37, we find

$$\sum \tau = I\alpha = (1.6 \times 10^{-5} \text{ kg} \cdot \text{m}^2)(10 \text{ rad/s}^2) = 1.6 \times 10^{-4} \text{ N} \cdot \text{m}$$

EXAMPLE 8.8 ® Motion of a Frisbee

You and a friend are playing catch with a Frisbee (Fig. 8.30). Estimate the total angular displacement of the Frisbee as it travels from you to your friend. Assume the Frisbee's angular velocity is constant.

RECOGNIZE THE PRINCIPLE

We are given that the Frisbee moves with a constant angular velocity as it travels through the air; this makes sense since we know from experience that when a Frisbee reaches its destination, it is usually still spinning quite rapidly. If ω is constant, the angular acceleration $\alpha = 0$. We can then use the relations in Table 8.3 to find the total angular displacement.

SKETCH THE PROBLEM

The problem is shown in Figure 8.30. The top of the figure shows the rotation axis of the Frisbee and the angular displacement θ.

IDENTIFY THE RELATIONSHIPS

We can find the total angular displacement from Equation 8.35 provided we can estimate the initial angular velocity ω_0 and the time the Frisbee spends in the air. (We already know that $\alpha = 0$.) Based on experience, we estimate that the Frisbee spends about 5 s in the air. It is a bit more difficult to estimate the Frisbee's angular velocity, but a reasonable guess is 10 rev/s. This value should be correct to within an order of magnitude; a value of 1 rev/s seems much too slow, and 100 rev/s would be quite fast. (The maximum rotation rate for most car engines is about 100 rev/s, and we don't expect a Frisbee to rotate that rapidly.)

SOLVE

Our estimate for the angular velocity is thus

$$\omega_0 \approx 10 \text{ rev/s} = 10 \frac{\text{rev}}{\text{s}} \times \frac{2\pi \text{ rad}}{\text{rev}} = 60 \text{ rad/s}$$

(continued) ▶

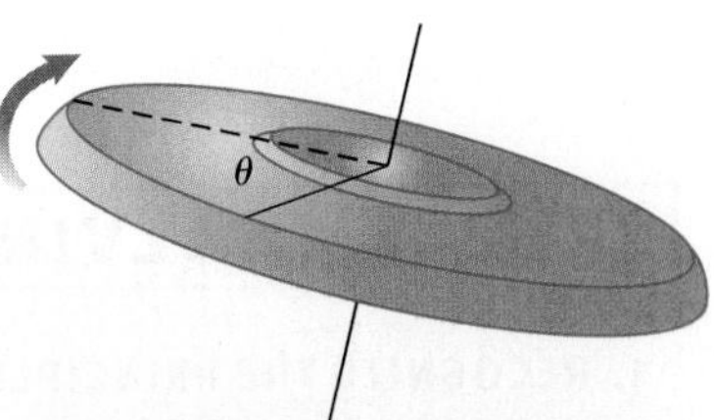

▲ **Figure 8.30** Example 8.8. A Frisbee spins about its axis with an angular velocity that is approximately constant during the course of a "throw." This rotational motion is independent of the Frisbee's translational motion.

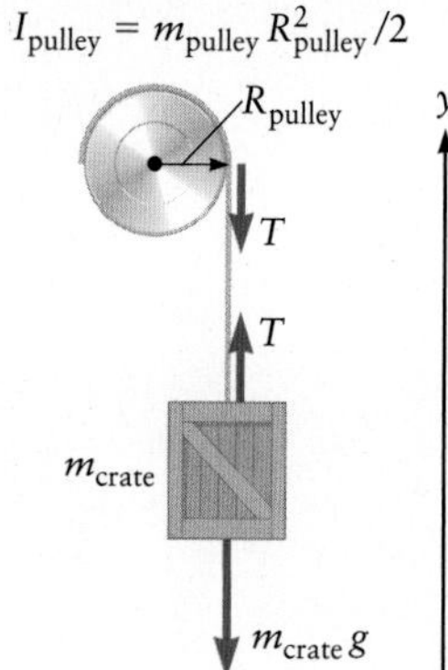

▲ **Figure 8.31** The torque on this pulley is due to the tension in the rope. The rotation axis is the axle of the pulley, so the torque equals the tension multiplied by the radius of the pulley.

(Notice that our value for ω_0 contains only one significant figure because it is only an estimate.) Using Equation 8.35, in a time $t = 5$ s the Frisbee will undergo an angular displacement of

$$\Delta\theta = \theta - \theta_0 = \omega_0 t = (60 \text{ rad/s})(5 \text{ s}) = \boxed{300 \text{ rad}}$$

▶ ***What does it mean?***
One revolution corresponds to $2\pi \approx 6.3$ rad, so 300 rad is about 50 revolutions.

Pulling on a Pulley: Real Pulleys with Mass

In previous chapters, we discussed pulleys and explained how they can be used to amplify forces. However, our previous treatments have assumed extremely light (massless) pulleys. Now we are ready to deal with the more realistic case of a pulley with mass.

Figure 8.31 shows a crate suspended by a very light (massless) rope. The rope is wound around a pulley, so as the crate is accelerated by the force of gravity, the pulley rotates. We now want to calculate the motion of the crate and pulley.

To calculate the linear acceleration of the crate and the angular acceleration of the pulley, we apply Newton's second law for linear motion ($\sum \vec{F} = m\vec{a}$) and Newton's second law for rotational motion ($\sum \tau = I\alpha$, Eq. 8.33). We take an approach similar to the one we used to deal with problems involving rotational equilibrium.

PROBLEM SOLVING Applying Newton's Second Law for Rotational Motion

1. **RECOGNIZE THE PRINCIPLE.** According to Newton's second law for rotational motion ($\sum \tau = I\alpha$), torque causes an angular acceleration. To find α, we must therefore find the total torque ($\sum \tau$) and the moment of inertia I.

2. **SKETCH THE PROBLEM.** Make a drawing showing all the objects of interest along with the forces that act on them and where those forces are applied. This sketch should contain a set of coordinate axes (x and y) for translational motion.

3. **IDENTIFY THE RELATIONSHIPS.** Determine the rotation axis and pivot point for calculating the torques for objects that may rotate. Then
   - Find the total torque on the objects that are undergoing rotational motion. Calculate the torque using the method employed in our work on rotational statics. This torque will then be used in Newton's second law for rotational motion, $\sum \tau = I\alpha$.
   - Calculate the sum of the forces on the objects that are undergoing linear motion for use in Newton's second law, $\sum \vec{F} = m\vec{a}$.
   - Check for a relation between linear acceleration and angular acceleration. There will usually be a connection between a and α through a relation such as Equation 8.8.

4. **SOLVE** for the quantities of interest (which might be a, α, or both) using Newton's second law for linear motion ($\sum \vec{F} = m\vec{a}$) and rotational motion ($\sum \tau = I\alpha$).

5. Always *consider what your answer means* and check that it makes sense.

Motion of a Pulley and Crate: Example of Combined Translational and Rotational Motion

We demonstrate the above problem-solving approach by applying it to calculate the crate's translational motion and the pulley's rotational motion in Figure 8.31. This figure serves as our sketch; it shows all the forces on the crate (gravity and the force from tension in the rope). Since the same rope is attached to the pulley, the torque on the pulley is also due to this tension. (We omit the force of gravity on the pulley because it is balanced by the force from the pulley's support, and neither of these forces produces a torque on the pulley.) The pulley rotates about its axle, so we pick that as the rotation axis.

To find the torque on the pulley, we first note that the tension in the rope will make the pulley rotate clockwise. According to our convention, the torque is negative. The force acts at the edge of the pulley, so the lever arm equals R_{pulley}. We thus have

$$\sum \tau = -TR_{\text{pulley}} = I_{\text{pulley}}\alpha \tag{8.38}$$

The pulley is a disc and from Table 8.2 its moment of inertia is $I_{\text{pulley}} = m_{\text{pulley}}R^2_{\text{pulley}}/2$.

We next write Newton's second law for the translational motion of the crate. Taking the $+y$ direction as the "positive" direction for forces, we have

$$\sum F_y = +T - m_{\text{crate}}g = m_{\text{crate}}a \tag{8.39}$$

Here, a is the acceleration of the crate along the y direction.

There are three unknowns—the tension T in the rope, the acceleration a of the crate, and the angular acceleration α of the pulley—but only two equations (Eqs. 8.38 and 8.39). We therefore need one more relation among these unknowns. We now recognize that the linear acceleration of the crate equals the linear acceleration of a point on the edge of the pulley. We have already discussed this situation in connection with Eq. 8.8 and found that a and α are related by

$$a = \alpha R_{\text{pulley}} \tag{8.40}$$

Equation 8.40 is actually a relation between the magnitudes of the linear and angular accelerations. In this problem, we must also be careful about the signs of a and α. Equation 8.40 treats these signs correctly for the crate and the pulley in Figure 8.31: a negative acceleration of the crate (i.e., a negative value of a) corresponds to a clockwise angular acceleration of the pulley and hence a negative value of α.

We can now solve these three relations (Eqs. 8.38 through 8.40) to find the accelerations of the crate and pulley. We first use Equation 8.40 to eliminate a from Equation 8.39:

$$T - m_{\text{crate}}g = m_{\text{crate}}a = m_{\text{crate}}\alpha R_{\text{pulley}} \tag{8.41}$$

Equation 8.38 can be simplified by inserting the expression for I_{pulley}

$$-TR_{\text{pulley}} = I_{\text{pulley}}\alpha = \tfrac{1}{2}m_{\text{pulley}}R^2_{\text{pulley}}\alpha$$

and canceling one factor of R_{pulley}:

$$T = -\tfrac{1}{2}m_{\text{pulley}}R_{\text{pulley}}\alpha$$

Using this result, we can eliminate T from Equation 8.41:

$$\overbrace{-\tfrac{1}{2}m_{\text{pulley}}R_{\text{pulley}}\alpha}^{T} - m_{\text{crate}}g = m_{\text{crate}}\alpha R_{\text{pulley}}$$

Now we can solve for α:

$$\alpha(m_{\text{crate}}R_{\text{pulley}} + \tfrac{1}{2}m_{\text{pulley}}R_{\text{pulley}}) = -m_{\text{crate}}g$$

$$\alpha = \frac{-m_{\text{crate}}g}{m_{\text{crate}}R_{\text{pulley}} + \tfrac{1}{2}m_{\text{pulley}}R_{\text{pulley}}} = \frac{-g/R_{\text{pulley}}}{1 + [m_{\text{pulley}}/(2m_{\text{crate}})]}$$

This result for the angular acceleration of the pulley can be used in Equations 8.40 and 8.39 to find the linear acceleration of the crate and the tension. Since α is constant, the angular velocity and displacement of the pulley are given by the relations for motion with a constant angular acceleration in Table 8.3.

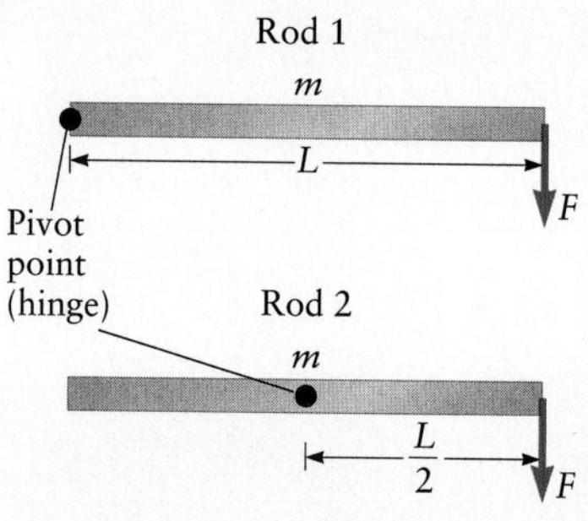

▲ **Figure 8.32** Concept Check 8.7.

CONCEPT CHECK 8.7 Angular Acceleration of a Rotating Rod

A rod of mass m and length L is hinged at one end, and a force F is applied at one end as shown in Figure 8.32 for rod 1. An identical rod is mounted on an axle so that it rotates about its center, and a force F is applied to its end as shown for rod 2 in Figure 8.32. If the forces are the same, which rod has the greater angular acceleration? What is the ratio of the angular accelerations?

(a) The angular accelerations are the same, $\alpha_1/\alpha_2 = 1$.
(b) Rod 1 has the larger angular acceleration, with $\alpha_1/\alpha_2 = 2$.
(c) Rod 1 has the larger angular acceleration, with $\alpha_1/\alpha_2 = 4$.
(d) Rod 2 has the larger angular acceleration, with $\alpha_2/\alpha_1 = 2$.
(e) Rod 2 has the larger angular acceleration, with $\alpha_2/\alpha_1 = 4$.

EXAMPLE 8.9 Which Stick Falls Faster?

Two uniform flat sticks are held at the same angle and with one end of each stick resting in the corner of a wall as shown in Figure 8.33A. The sticks have the same mass m, but lengths L and $2L$. If the sticks are released simultaneously, which one hits the floor first?

RECOGNIZE THE PRINCIPLE

We can use Newton's law for rotational motion to calculate the angular acceleration of each stick. The stick for which the magnitude of α is larger will have the larger angular speed at all times and thus will reach the floor first.

SKETCH THE PROBLEM

Figure 8.33B shows the gravitational force acting on the shorter stick and has the information we need to calculate the associated torque.

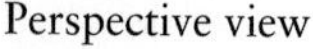

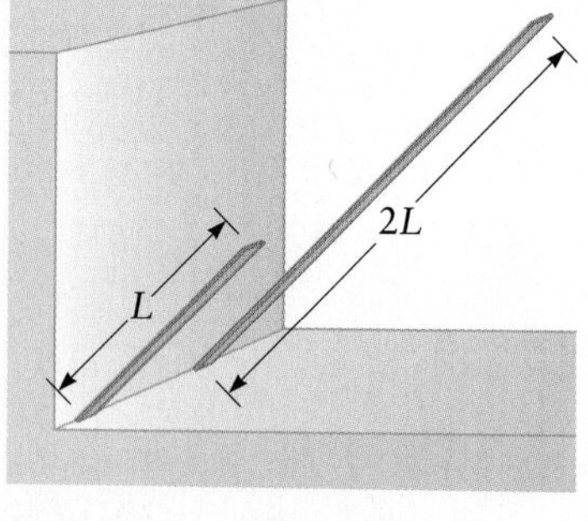

A

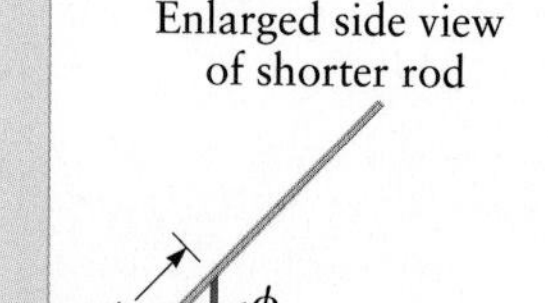

B

▲ **Figure 8.33** Example 8.9.

IDENTIFY THE RELATIONSHIPS

Let's calculate α for the stick of length L. The only torque τ is from gravity. Using Newton's second law for rotational motion, we have $\tau = I\alpha$, where the moment of inertia of the stick is $I = \frac{1}{3}mL^2$ (from Table 8.2, modeling the stick as a rod). The stick will rotate clockwise in Figure 8.33B, so both τ and α will be negative. The torque can be calculated by assuming the force of gravity acts at the center of mass of the stick, a distance $L/2$ from the end. The torque is thus (Fig. 8.33B and Eq. 8.15)

$$\tau = Fr \sin \phi = -mg\left(\frac{L}{2}\right)\cos\theta$$

Inserting this result into Newton's second law for rotational motion gives

$$\tau = -mg\left(\frac{L}{2}\right)\cos\theta = I\alpha = \frac{1}{3}mL^2\alpha$$

and we can solve for the angular acceleration:

$$\alpha = -\frac{3g\cos\theta}{2L} \tag{1}$$

SOLVE

The angular acceleration for the longer stick can be found by inserting $2L$ in place of L in Equation (1). The magnitude of the angular acceleration will thus be *larger* for the *shorter* stick, and the shorter stick will reach the floor first.

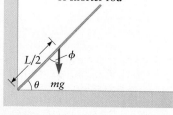

> ▶ ***What does it mean?***
> You may recall from Chapter 3 that Galileo showed that balls of different mass fall at the same rate, so you might have guessed that the two sticks in Figure 8.33A would reach the floor at the same time. For rotational motion, however, the free end of the longer stick has farther to travel, so it reaches the floor last.

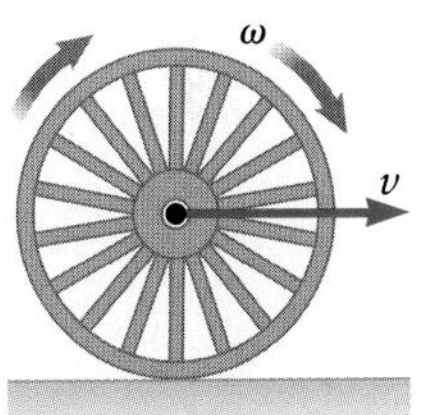

▲ **Figure 8.34** A rolling object undergoes translational motion with a center of mass velocity v and simultaneous rotational motion with angular velocity ω.

8.6 Combined Rotational and Translational Motion

The combination of rotational and translational motion arises in many common situations, including a rolling wheel and the motion of a baseball bat.

Rolling Motion

An extremely common situation that combines translation and rotation is ***rolling motion.*** Figure 8.34 shows a simple wheel as it rolls along the ground. Here the center of a wheel is moving horizontally with a constant velocity of magnitude v, but a point on the edge of the wheel is certainly not moving with a constant velocity. In fact, if the wheel is rolling without slipping, the point of the wheel in contact with the ground is *at rest* during the instant it is in contact.

As a wheel rolls, it rotates about an axis that runs along its axle, and this rotational motion is described by an angular velocity ω. To understand how ω is related to the linear velocity v of the center of the wheel, consider the series of "snapshots" of the wheel in Figure 8.35. Points on the wheel's rim are labeled, and corresponding labels where these points touch the ground as the wheel rolls are also shown. The wheel begins (Fig. 8.35A) with point 1 in contact with the ground, and when it completes one full revolution, this point is again touching the ground (Fig. 8.35F). Since the wheel is always in contact with the ground and does not slip, the distance traveled *as measured along the ground* must equal the *circumference* of the wheel, $2\pi R$. This is also the distance traveled by the center of the wheel, so if the rotation period is T, the speed of the center is

$$v = \frac{2\pi R}{T} \tag{8.42}$$

At the same time, the wheel has completed one full revolution (2π rad), so its angular velocity is

$$\omega = \frac{2\pi}{T}$$

Combing this with Equation 8.42 gives

$$v = \omega R \quad \text{(rolling motion)} \tag{8.43}$$

This relation connects the linear velocity of the center of the wheel with its angular velocity and is true for any object that rolls without slipping. It is interesting that this relation (Eq. 8.43) has the same form as Equation 8.7, but even though these equations are related, the ideas are slightly different. The result in Equation 8.7 involves the angular velocity of a rotating object and the linear speed of a *point on the edge* of the object. The result for rolling motion in Equation 8.43 connects the angular velocity with the velocity of the *center* (i.e., the axis) of a rolling object. The arguments contained in Figure 8.35 can also be used to relate the acceleration of the center of a rolling object to its angular acceleration:

$$a = \alpha R \quad \text{(rolling motion)} \tag{8.44}$$

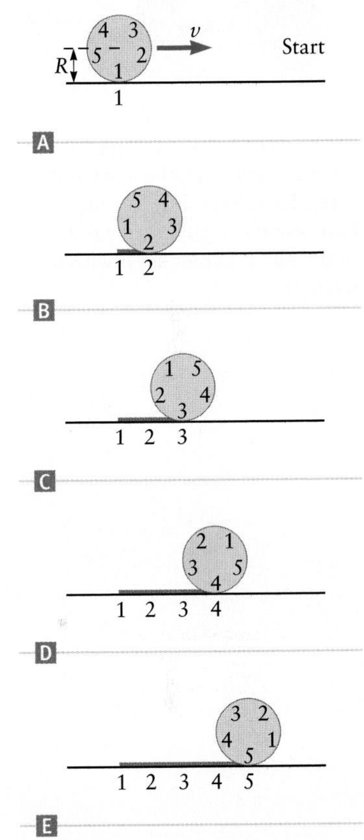

▲ **Figure 8.35** Successive "snapshots" as a wheel rolls to the right. A wheel that rolls without slipping travels a linear distance of $2\pi R$ during each complete revolution of the wheel. Each complete revolution occurs in a time equal to the period, T. This leads to the relation $v = \omega R$ between the translational velocity of the center of the wheel and the wheel's angular velocity.

It is interesting that friction plays an essential role in rolling motion. Indeed, an object cannot roll without it! It may seem a bit strange to claim that an object cannot move *without* friction, but consider a car that is initially at rest on an icy, frictionless surface. If you try to drive away, the wheels will spin freely because there is no frictional force to prevent them from slipping relative to the ice. On normal pavement (*with* friction), the force of friction between the road and the surface of the wheels enables the car to accelerate when the wheels turn. Usually, the wheels do not slip relative to the pavement, so this is static friction; if the driver tries to accelerate very quickly, however, the wheels may slip relative to the pavement bringing kinetic friction into play. Hence, for rolling motion to occur, friction is *required*.

▲ **Figure 8.36** Concept Check 8.8. On some cars, the front and rear wheels have different radii. How are the angular velocities of the two different-sized wheels related?

CONCEPT CHECK 8.8 Angular Velocity of Rolling Wheels

Consider a car whose rear wheels are much larger than the front wheels as shown in Figure 8.36. If the car is traveling up a hill, which wheels will have a larger angular velocity?

(a) The front wheels will.
(b) The rear wheels will.
(c) The front and the rear wheels will have the same angular velocity.

EXAMPLE 8.10 Ⓡ Angular Velocity of an Automobile Wheel

Consider a car traveling on a highway at a speed of 50 mi/h (about 23 m/s). Estimate the angular velocity of the car's wheels.

RECOGNIZE THE PRINCIPLE

A car's wheels are rolling without slipping (we hope!), so we can apply Equation 8.43 to relate the linear velocity v and the angular velocity ω.

SKETCH THE PROBLEM

This problem is illustrated in Figures 8.34 and 8.35.

IDENTIFY THE RELATIONSHIPS

The velocity of the center of a wheel equals the overall speed of the car, $v = 23$ m/s. The angular velocity of a wheel can then be obtained using the relation $v = \omega R$ (Eq. 8.43). This relation involves the radius R of the wheel, so we need to estimate the value of R. For a typical car, we estimate $R \approx 40$ cm $= 0.40$ m.

SOLVE

Using these values gives

$$\omega = \frac{v}{R} = \frac{23 \text{ m/s}}{0.4 \text{ m}} = \boxed{60 \text{ rad/s}}$$

► ***What does it mean?***

This angular velocity corresponds to

$$\omega = (60 \text{ rad/s}) \times (1 \text{ rev}/2\pi \text{ rad}) = 10 \text{ rev/s}$$

so the wheels undergo approximately 10 revolutions each second.

Sweet Spot of a Baseball Bat

The general motion of any object can be described as a combination of linear and rotational motion. The center of mass plays an important role in problems involving such combined motions. The linear motion of an object can be calculated from Newton's second law,

$$\sum \vec{F} = m\vec{a}$$

with $\vec{a}$ being the acceleration of the center of mass of the object. As far as the linear (i.e., translational) motion is concerned, the object moves as if it were a point particle with all its mass located at the center of mass. The rotational motion then follows from

$$\sum \tau = I\alpha$$

An interesting example of this principle involves a baseball bat and the problem of how best to hit a baseball or softball. You may know from experience that if you hit a ball off the very end of a bat or very near your hands, you feel a significant and often uncomfortable vibration at your hands. There is, however, a particular spot on the bat called the "sweet spot," and if the ball strikes the bat at that location, your hands feel essentially no vibration. A ball hit from the sweet spot seems to rebound effortlessly from the bat and, all else being equal, will travel farther than a ball that hits the bat away from this location. Suppose a baseball-playing friend wants to use physics to calculate the location of the sweet spot on his favorite bat. How would you help him?

Figure 8.37 shows the problem. As the player swings, the bat undergoes rotational motion about the batter's hands at point P. In practice, it is not quite "pure" rotational motion because a batter will often move the bat handle forward a small amount while he swings. Since this point P usually moves very little compared with the other end of the bat, it is a good approximation to assume P is fixed while the bat simply rotates about P. When the ball strikes the bat, the ball exerts a force F_b on the bat due to the ball–bat collision, producing a torque on the bat. In Figure 8.37A, this force F_b is shown acting at the sweet spot, a distance d from the center of mass of the bat.

The special property of the sweet spot is that the force F_b does not cause the handle of the bat to move. If the batter holds the bat *very lightly*, the force exerted by his hands on the bat is approximately zero and the only significant horizontal force on the bat is the force F_b exerted by the ball. This force will cause the bat's center of mass to move to the left in Figure 8.37B and at the same time will make the bat rotate. If the bat's translational and rotational motions are related in just the right way (by Eqs. 8.43 and 8.44), the bat will undergo a sort of rolling motion with point P just like the point of a wheel in contact with the ground. This behavior is found only if the force is applied at the sweet spot. In this case, no force is needed by the player's hands to keep the end of the bat at P from moving.

To find the location of the sweet spot (the value of d in Fig. 8.37A), we need to calculate the linear acceleration of the center of mass a_{CM} and the angular acceleration α of the bat. Newton's second law for the bat's translational motion reads

$$\sum F = F_b = ma_{CM} \tag{8.45}$$

We consider only motion along the direction of F_b, so Equation 8.45 involves just the components of the force and acceleration along this direction. If we assume a bat with a uniform cross section (a simple rod, another physics simplification), the center of mass is at the center of the bat. The force exerted by the ball on the bat also produces a torque on the bat. To analyze the rotational motion, let's *not* use point P in Figure 8.37 as the pivot point. Instead, let's take the pivot point to be at the bat's center of mass so that the lever arm is d in Figure 8.37A. Newton's second law for the rotational motion is then

$$\sum \tau = F_b d = I\alpha \tag{8.46}$$

For point P on the bat handle in Figure 8.37 to stay at rest, the bat must rotate in the same way as the rolling wheel in Figure 8.35, which happens only if the linear speed of the center of mass "matches" the angular velocity according to $\omega = v_{CM}/R$.

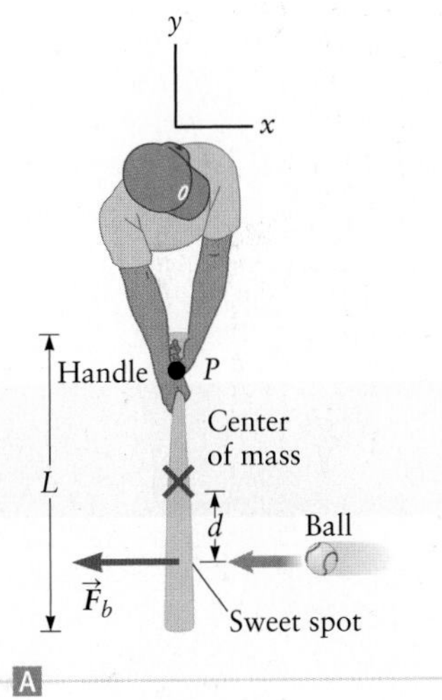

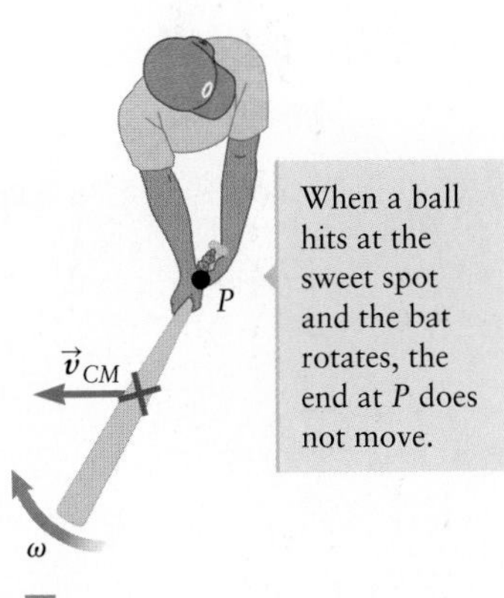

▲ **Figure 8.37** A When a baseball hits at the sweet spot of the bat, the velocity at the handle (point P) is zero. B The handle's velocity will be zero if the velocity v_{CM} of the center of mass and the angular velocity ω are related by $v_{CM} = \omega R = \omega(L/2)$, which will only be true if the linear acceleration and the angular acceleration are related by $a_{CM} = \alpha(L/2)$.

Here, R is the distance from the center of mass to point P, so $R = L/2$, where L is the length of the bat, and

$$\omega = \frac{v_{CM}}{R} = \frac{v_{CM}}{L/2}$$

If the velocities are related in this way, the angular and linear accelerations will also be related in the same way (see Eq. 8.44):

$$\alpha = \frac{a_{CM}}{L/2} \tag{8.47}$$

If we model the bat as a rod pivoted about its center, its moment of inertia from Table 8.2 is $I = mL^2/12$. Inserting our results for I and α into Equation 8.46 leads to

$$\sum \tau = F_b d = I\alpha = \frac{mL^2}{12}\frac{a_{CM}}{(L/2)} = \frac{mLa_{CM}}{6}$$

Combining this result with Equation 8.45 gives

$$F_b d = (ma_{CM})d = \frac{mLa_{CM}}{6}$$

which we can solve to find the location of the sweet spot:

$$d = \frac{L}{6}$$

Hence, the sweet spot for a uniform bat is located a distance $L/6$ from the center of the bat. A real bat is not uniform, but its sweet spot can be found in a similar way.

Summary | CHAPTER 8

Key Concepts and Principles

Angular displacement, velocity, and acceleration

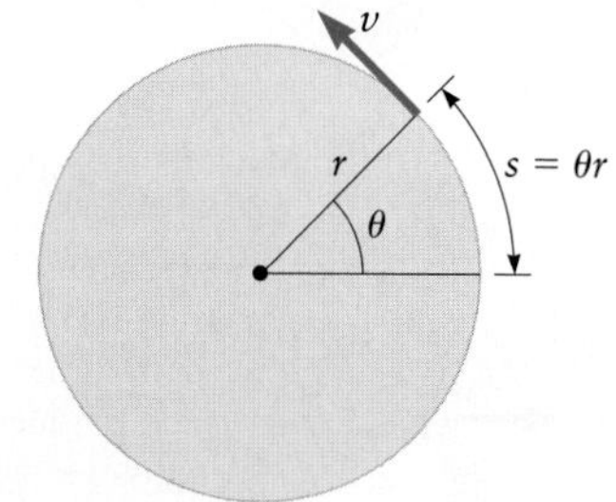

Rotational motion is described by an ***angular displacement*** θ (measured in radians), an ***angular velocity*** ω (in rad/s), and an ***angular acceleration*** α (in rad/s^2). They are related to the linear motion of a point on the object by

$$\theta = \frac{s}{r} \quad \textbf{(8.1) (page 214)}$$

$$v = \omega r \quad \textbf{(8.7) (page 218)}$$

$$a = \alpha r \quad \textbf{(8.8) (page 219)}$$

Torque

Torque plays the role of force in cases of rotational motion. The torque produced by an applied force is

$$\tau = Fr \sin \phi \quad \textbf{(8.15) (page 222)}$$

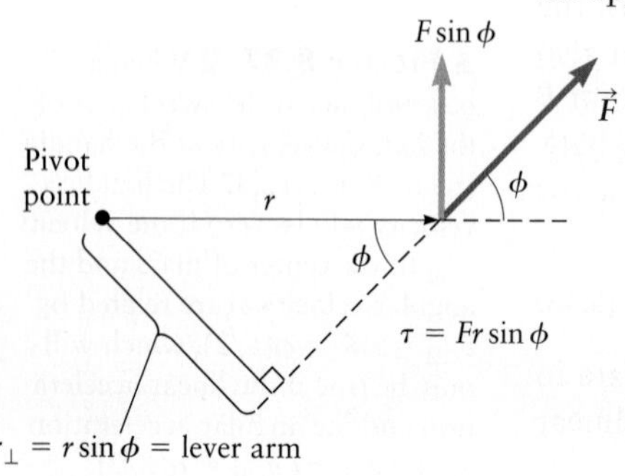

This relation for the torque can be written in two equivalent ways. One way is

$$\tau = F_\perp r \quad \textbf{(8.16) (page 223)}$$

where $F_\perp$ is the component of the force perpendicular to the line defined by r, the distance from the rotation axis to the point on the object where the force acts. The torque can also be written as

$$\tau = Fr_\perp \quad \textbf{(8.17) (page 223)}$$

where F is the applied force and

$$r_{\perp} = r \sin \phi \qquad (8.18)\ (\text{page } 223)$$

The distance $r_{\perp}$ is called the ***lever arm***; it is the distance from the rotation axis to the line of action of the force, measured along a line that is perpendicular to both.

Newton's second law for rotational motion

$$\sum \tau = I\alpha \qquad (8.12)\ (\text{page } 221)$$

The ***moment of inertia*** I of an object depends on its mass and shape. For a collection of particles of mass m_i, the moment of inertia is

$$I = \sum_i m_i r_i^2 \qquad (8.14)\ (\text{page } 222)$$

where r_i is the distance from the rotation axis to m_i. The moment of inertia is proportional to the total mass and depends on its distribution relative to the rotation axis. Values of I for some common shapes are listed in Table 8.2.

Applications

Rotational equilibrium

An object is in translational equilibrium and rotational equilibrium if its linear acceleration and angular acceleration are both zero. For that to be true, both the total force and the total torque on the object must be zero.

Rotational motion

Calculations of the rotational dynamics start with Newton's second law for rotational motion,

$$\sum \tau = I\alpha \qquad (8.12)\ (\text{page } 221)$$

When the angular acceleration is constant, the solutions for the angular displacement and velocity as functions of time are (Table 8.3)

$$\theta = \theta_0 + \omega_0 t + \tfrac{1}{2}\alpha t^2 \qquad (8.35)\ (\text{page } 236)$$

$$\omega = \omega_0 + \alpha t \qquad (8.34)\ (\text{page } 236)$$

$$\omega^2 = \omega_0^2 + 2\alpha(\theta - \theta_0) \qquad (8.36)\ (\text{page } 236)$$

These relations are very similar to the results we found for linear motion with constant linear acceleration in Chapter 3.

Rolling motion

Rolling motion is an example of combined translational and rotational motion. If an object rolls without slipping, the angular velocity and angular acceleration are related to the linear velocity and acceleration of the center of mass by

$$v = \omega R \qquad (8.43)\ (\text{page } 241)$$

$$a = \alpha R \qquad (8.44)\ (\text{page } 241)$$

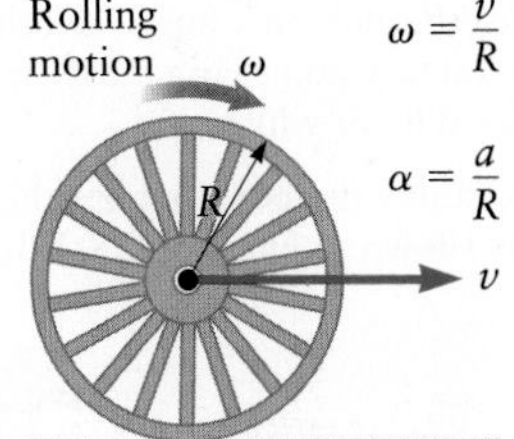

Questions

SSM = answer in *Student Companion & Problem-Solving Guide* = life science application

1. Explain why placing your hands behind your head rather than placing them on your stomach makes sit-ups more difficult.

2. Two disks, each of mass m, have radii R and $2R$ and are mounted on axles that pass through their centers. (a) Which disk has the larger moment of inertia? (b) If a force of magnitude F is applied at the edge of each disk as shown in Figure Q8.2, which disk will experience the larger torque? (c) Which disk will have the larger angular acceleration?

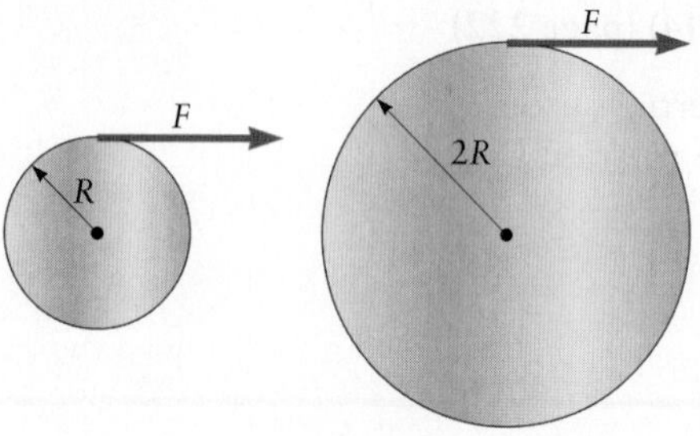

Figure Q8.2

3. A hoop and a disk have the same radius and the same moment of inertia. What is the ratio of their masses?

4. According to Table 8.2, the moment of inertia for a rod that rotates about an axis perpendicular to the rod and passing through one end is $I = ML^2/3$; if the axis passes through the center of the rod, then $I = ML^2/12$. Give a physical explanation for this difference in terms of the way the mass of the rod is distributed with respect to the axis in the two cases.

5. The tree in Figure Q8.5 experiences a large torque due to the force of gravity, yet it is in static equilibrium. What additional torque balances the torque due to gravity?

Figure Q8.5

6. In our analysis of a rollover car accident in Example 8.6, we treated the car in a very simplified way. Develop a more realistic model of a car with special attention to the height of its center of mass and use it to compute the force required to tip over the car. (See Fig. 8.26.)

7. SSM Two golfers team up to win a golf tournament and are awarded a solid gold golf club. To divide their winnings, they balance the club on a finger and then cut the club into two pieces at the balance point with a hacksaw. Was the winning gold split evenly? Why or why not?

8. A mechanic's toolbox contains the two wrenches and two screwdrivers shown in Figure Q8.8. While working on her vintage car, the mechanic comes across a stubborn and rusted bolt. Which wrench would be the most useful in breaking the bolt free? Why is one better than the other? She also encounters a screw that is similarly stuck tight. Which screwdriver should she choose? Is one better than the other? Why?

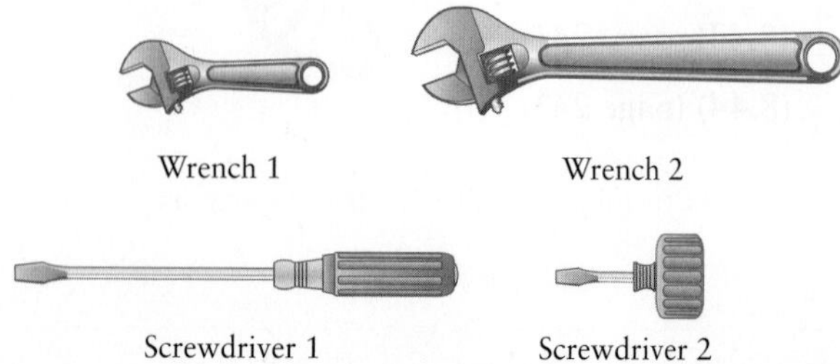

Figure Q8.8

9. A snap of the wrist will usually detach a paper towel from a roll. Why does that almost always work for a full roll, yet an almost empty roll often gives much more than a single sheet?

10. Two metersticks, one with a weight attached to its end, are held with one of their ends in the corner formed by the floor and a wall as depicted in Figure Q8.10. If the ends of the metersticks are let go at the same time and start with the same angle with respect to the floor, which one will hit the floor first? Why?

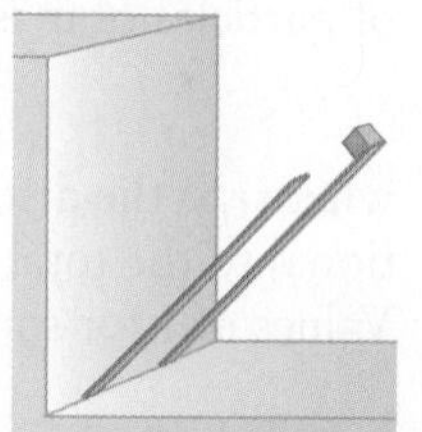

Figure Q8.10

11. **Three types of levers.** Levers can be classified as first, second, or third class, depending on how the load force ($\vec{F}_L$) and effort force ($\vec{F}_E$) are configured with respect to the pivot point or fulcrum as shown in Figure Q8.11. Determine the lever classification of the following items: (a) a crowbar, (b) a pair of pliers, (c) a clawed hammer pulling a nail, (d) a wheelbarrow (Fig. P8.37), (e) a seesaw (Fig. 8.10), (f) a fishing pole, (g) a light switch, (h) a gate with closing spring attached to hinge, and (i) biceps raising the forearm (Fig. P8.79).

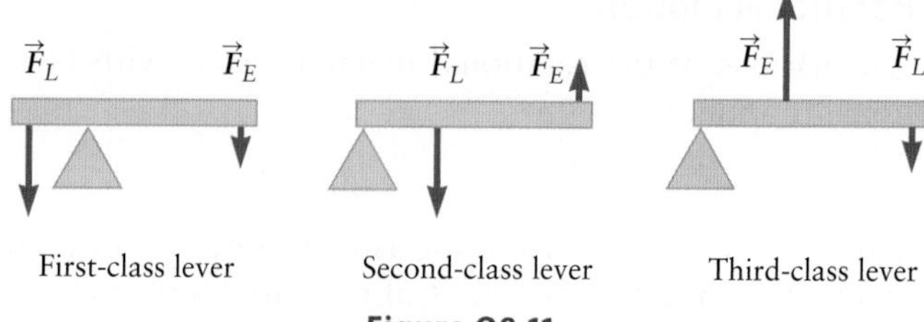

Figure Q8.11

12. In about 200 BCE, Archimedes made the statement, "Give me a lever long enough and a place to stand, and I will move the Earth." Assume he can exert a force equal to his weight on the Earth (about 900 N). Design a lever that would enable him to give the Earth an acceleration of 1 m/s^2.

13. A tennis racket possesses a sweet spot in close analogy to that of a baseball bat. One might expect that rackets are designed so that the sweet spot is in the center of the strings. Somewhat surprisingly, that is not the case. The approximate location of a racket's sweet spot can be found from the simple experiment sketched in Figure Q8.13. The racket is suspended by a string tied to the handle and is then struck (gently) by a ball or a light blow. When this force is exerted at the sweet spot, the handle will not rebound. Perform this experiment yourself and show that the sweet spot is rather far from the center of the strings. Explain how you could add or subtract mass from different spots on the racket to move the sweet spot to this desired location.

Figure Q8.13

14. A tree has a large branch that grows horizontally out from the trunk. If the branch doubles in length and doubles in diameter, by what factor does the torque on the branch due to gravity increase?

15. SSM A tree has two large branches that grow out horizontally from the trunk in opposite directions. One branch has length L and diameter d, whereas the other has length $L/2$. If the magnitude of the torque due to gravity is the same on the two branches, what is the diameter of the shorter branch?

16. A force of magnitude F is applied to a rod at various spots and at various angles (Fig. Q8.16). In which case, (1), (2), or (3), is the magnitude of the torque largest? Smallest?

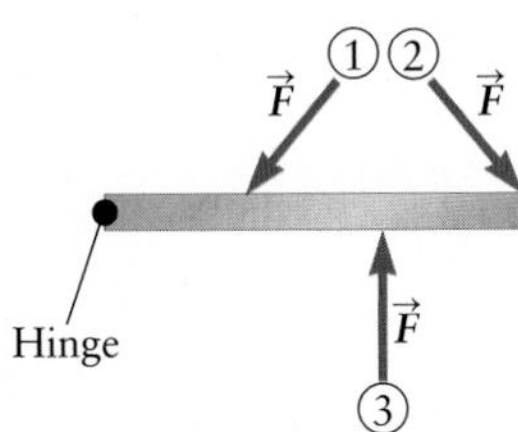

Figure Q8.16

17. Bodybuilders do an exercise called "curls" in which they lift a weight by bending at the elbow. Explain why the torque on the elbow is largest when the arm is horizontal.

18. Explain why it is easier (it requires a smaller force) to open a large jar than a small one.

19. Consider a wheel that rolls without slipping such as the wheel in Figure 8.34. We have seen that slipping is prevented by the force of static friction between the wheel and the pavement. Can this force do work on the wheel?

20. In Section 8.6, we found that for a uniform baseball bat, the sweet spot is a distance $L/6$ from the center of the bat. A real bat is not uniform; it is much thinner in the handle than in the barrel (the end of the bat farthest from point P in Fig. 8.37). Estimate how that affects the location of the sweet spot.

Problems available in ENHANCED WebAssign

Section 8.1 Describing Rotational Motion
Problems 1–15

Section 8.2 Torque and Newton's Laws for Rotational Motion
Problems 16–24

Section 8.3 Rotational Equilibrium
Problems 25–40

Section 8.4 Moment of Inertia
Problems 41–50

Section 8.5 Rotational Dynamics
Problems 51–64

Section 8.6 Combined Rotational and Translational Motion
Problems 65–73

Additional Problems
Problems 74–84

List of Enhanced Problems

Problem number	Solution in *Student Companion & Problem-Solving Guide*	Reasoning & Relationships Problem	Reasoning Tutorial in ENHANCED WebAssign
8.13	✓		
8.18	✓		
8.22		✓	✓
8.23	✓	✓	
8.24		✓	✓
8.25		✓	✓
8.32		✓	
8.37		✓	✓
8.38		✓	
8.39		✓	
8.40	✓		
8.42	✓		
8.45		✓	✓
8.48		✓	
8.50		✓	
8.52	✓		
8.54		✓	
8.58	✓		
8.60		✓	✓
8.61		✓	
8.67	✓		
8.71		✓	
8.76	✓		
8.80		✓	✓
8.84		✓	

CHAPTER

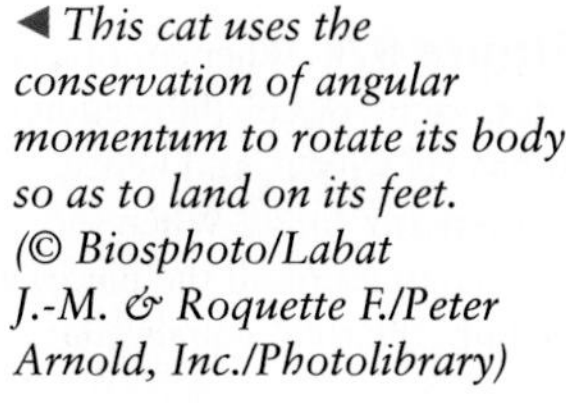

Energy and Momentum of Rotational Motion

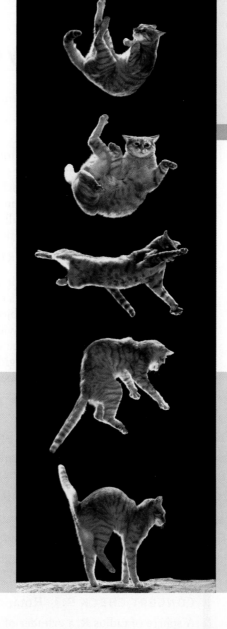

◀ *This cat uses the conservation of angular momentum to rotate its body so as to land on its feet. (© Biosphoto/Labat J.-M. & Roquette F./Peter Arnold, Inc./Photolibrary)*

In Chapter 8, we introduced the notion of torque and saw that torques give rise to rotational motion in much the same way that forces give rise to linear motion. In this chapter, we consider the kinetic energy associated with rotation and derive a work–energy theorem that relates torque and rotational kinetic energy. We then explore situations involving the conservation of mechanical energy, this time with the rotational kinetic energy included. We also introduce the idea of angular momentum, the rotational analog of linear momentum. In Chapter 7, we saw that the general principle of momentum conservation (for linear momentum) has a fundamental place in physics. Angular momentum has a similar role: the conservation of angular momentum is important in many situations, ranging from planetary orbits to the structure of atoms (as we'll see in Chapter 29) and the periodic table, and it also explains the mystery of how cats are almost always able to land on their feet.

OUTLINE

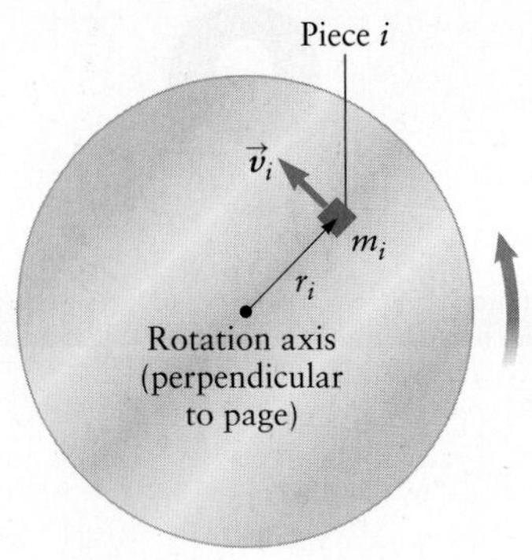

▲ **Figure 9.1** When an object rotates, each "piece" (labeled here as i) has velocity $\vec{v}_i$ and hence kinetic energy $\frac{1}{2}m_i v_i^2$. When the kinetic energies of all the pieces are added up, the total equals $\frac{1}{2}I\omega^2$.

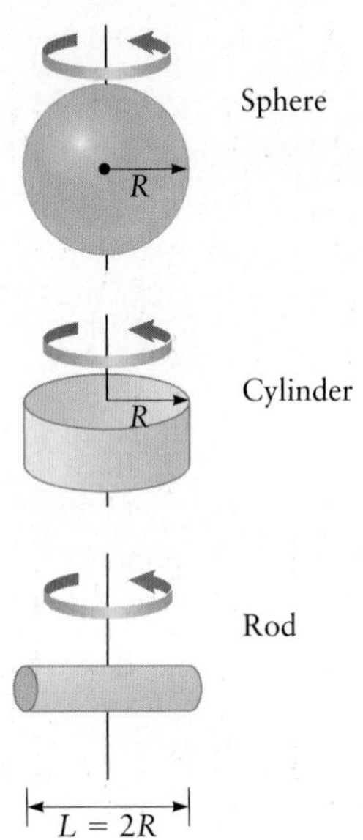

▲ **Figure 9.2** Concept Check 9.1.

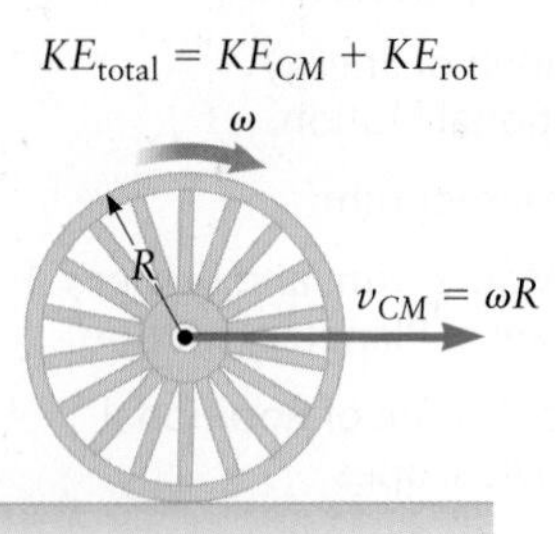

▲ **Figure 9.3** For an object such as a rolling wheel, the total KE is the sum of the translational KE associated with the center of mass velocity ($\frac{1}{2}mv_{CM}^2$) and the KE of rotation about the center of mass ($\frac{1}{2}I\omega^2$). For a wheel that rolls without slipping, $\omega = v_{CM}/R$.

9.1 Kinetic Energy of Rotation

In Chapter 6, we learned that the kinetic energy of a point particle of mass m moving with a linear speed v is

$$KE = \tfrac{1}{2}mv^2 \tag{9.1}$$

We now wish to generalize Equation 9.1 to include the contribution of rotation to the kinetic energy of an extended object. We can view such objects as being composed of many small pieces, and each piece can be thought of as a point particle with a kinetic energy given by Equation 9.1. The total kinetic energy can be found by adding up the kinetic energies of all the pieces. If each piece has mass m_i and moves with linear speed v_i, the total kinetic energy is

$$KE = \textstyle\sum_i \tfrac{1}{2} m_i v_i^2 \tag{9.2}$$

This is the total kinetic energy of any moving object, including objects that are undergoing rotational motion. This result can be simplified if we assume the pieces in Equation 9.2 are part of a rigid object undergoing simple rotational motion. For an object rotating about a fixed axis with an angular velocity ω, we have, from Chapter 8 (Eq. 8.7), $v_i = \omega r_i$ where r_i is the distance of the ith piece from the rotation axis (Fig. 9.1). Because this object is rigid, all pieces have the same angular velocity, and we can rearrange Equation 9.2 to get

$$KE_{rot} = \textstyle\sum_i \tfrac{1}{2} m_i v_i^2 = \sum_i \tfrac{1}{2} m_i(\omega r_i)^2 = \sum_i \tfrac{1}{2} m_i r_i^2 \omega^2$$

$$KE_{rot} = \tfrac{1}{2}(\textstyle\sum_i m_i r_i^2)\omega^2 \tag{9.3}$$

In Equation 9.3, we pulled the factor of ω^2 out of the summation since ω is the same for all pieces of the object. The term in parentheses in Equation 9.3 is the moment of inertia I (recall that $I = \sum_i m_i r_i^2$, Eq. 8.14). We thus have

$$KE_{rot} = \tfrac{1}{2}I\omega^2 \tag{9.4}$$

This kinetic energy is due to the rotational motion of an extended object. You should see, however, that KE_{rot} has as its origin the translational kinetic energies of all the individual pieces of the object, as given by Equation 9.1.

CONCEPT CHECK 9.1 Rotational Kinetic Energy

A sphere of radius R, a cylinder of radius R, and a rod of length $L = 2R$ all have the same mass, and all three are rotating about an axis through their centers (Fig. 9.2). If the objects all have the same angular velocity, which one has the greatest rotational kinetic energy, (a) the sphere, (b) the cylinder, or (c) the rod?

The Total Kinetic Energy of an Object Is the Sum of the Rotational and Translational Kinetic Energies

The rotational kinetic energy of an object undergoing "pure" rotation about a fixed axis is $KE_{rot} = \frac{1}{2}I\omega^2$ (Eq. 9.4). Rolling objects such as the wheel in Figure 9.3 undergo both translational and rotational motions. In this case, the *total* kinetic energy has contributions from both the rotational motion and the overall translational motion. In practice, it is usually desirable to use an axis that passes through the center of mass. For an object of mass m whose center of mass moves with a linear speed v_{CM}, the total kinetic energy is then given by

$$KE_{total} = \overbrace{\tfrac{1}{2}mv_{CM}^2}^{\text{translational } KE} + \overbrace{\tfrac{1}{2}I\omega^2}^{\text{rotational } KE} \tag{9.5}$$

which is the total kinetic energy of *any* moving object, including rolling wheels and spinning Frisbees. The first term on the right of the equal sign in Equation 9.5 is the kinetic energy of translation, whereas the second term is the rotational kinetic energy. In Equation 9.5, ω is the angular velocity about a rotation axis that passes through the center of mass.

EXAMPLE 9.1 Energy of a Wind Farm

A promising and environmentally friendly method of generating electricity is a wind turbine (Fig. 9.4A). Wind turns the propellers, which in turn drive an electrical generator that produces electricity. A typical propeller unit has three blades (usually composed of a lightweight carbon fiber composite material) that are about 30 m long and rotate at 30 rpm. Calculate the rotational kinetic energy of this propeller unit. Approximate each blade as a uniform rod of mass 300 kg.

RECOGNIZE THE PRINCIPLE

The total kinetic energy of the propeller unit equals the sum of the rotational kinetic energies of the three blades. The kinetic energy of each blade can be found from Equation 9.4, using the moment of inertia for a uniform rod rotating about an axis that passes through one end, $I_{rod} = \frac{1}{3}mL^2$, where m is the mass of a blade and L is its length (Table 8.2).

SKETCH THE PROBLEM

The problem is sketched in Figure 9.4B. The three blades have the same angular velocity.

IDENTIFY THE RELATIONSHIPS

The total rotational kinetic energy of the propeller unit is

$$KE_{rot} = \tfrac{1}{2}I_{tot}\omega^2 \quad (1)$$

where I_{tot} is the total moment of inertia of all three blades. We treat each blade as a uniform rod that rotates about an axis through one end. The total moment of inertia of three propeller blades is thus

$$I_{tot} = 3(mL^2/3) = mL^2 = (300\text{ kg})(30\text{ m})^2 = 2.7 \times 10^5\text{ kg}\cdot\text{m}^2$$

where we have inserted the given values for m and L. The angular velocity is

$$\omega = 30\,\frac{\text{rev}}{\text{min}} \times \frac{2\pi\text{ rad}}{\text{rev}} \times \frac{1\text{ min}}{60\text{ s}} = 3.1\text{ rad/s}$$

SOLVE

Inserting these values into our expression for the rotational kinetic energy (Eq. 9.4) gives

$$KE_{rot} = \tfrac{1}{2}I_{tot}\omega^2 = \tfrac{1}{2}(2.7 \times 10^5\text{ kg}\cdot\text{m}^2)(3.1\text{ rad/s})^2 = \boxed{1.3 \times 10^6\text{ J}}$$

► What does it mean?

If this much energy can be extracted each second, the power generated by the wind turbine will be

$$P = \frac{KE_{rot}}{\Delta t} = \frac{1.3 \times 10^6\text{ J}}{1\text{ s}} = 1.3 \times 10^6\text{ W} = 1.3\text{ MW}$$

which is enough energy to power about 400 homes.

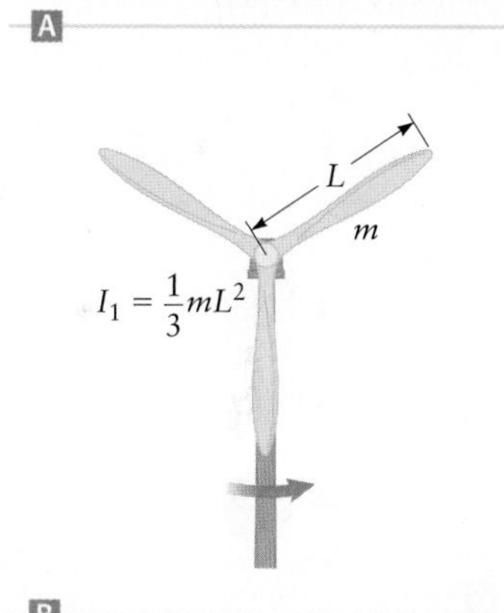

▲ **Figure 9.4** Example 9.1.

Rolling Motion and the Distribution of Kinetic Energy

Rolling objects have both rotational and translational kinetic energy. Let's consider how that energy is distributed. For example, if you are designing the wheels for a skateboard and want the skateboard to go as fast as possible, you will want to minimize the fraction of kinetic energy that goes into the rotational motion of each wheel, since this will leave more energy for the translational motion of the skateboard.

A rolling wheel has translational kinetic energy $\frac{1}{2}mv_{CM}^2$ and rotational kinetic energy $\frac{1}{2}I\omega^2$. If the wheel has the shape of a disk, the moment of inertia is $I = \frac{1}{2}mR^2$, where m is the wheel's mass and R is its radius (Table 8.2). Combining all these results, we find

$$KE_{\text{tot}} = \frac{1}{2}mv_{CM}^2 + \frac{1}{2}I\omega^2 = \frac{1}{2}mv_{CM}^2 + \frac{1}{2}\left(\frac{mR^2}{2}\right)\omega^2 = \frac{1}{2}mv_{CM}^2 + \frac{1}{4}mR^2\omega^2$$

In Chapter 8, we found that a wheel's angular velocity and center of mass velocity are related by $\omega = v_{CM}/R$ (Eq. 8.43). Inserting this information leads to

$$KE_{\text{tot}} = \frac{1}{2}mv_{CM}^2 + \frac{1}{4}mR^2\omega^2 = \frac{1}{2}mv_{CM}^2 + \frac{1}{4}mR^2\left(\frac{v_{CM}}{R}\right)^2$$

$$KE_{\text{tot}} = \overbrace{\tfrac{1}{2}mv_{CM}^2}^{\text{translational } KE} + \overbrace{\tfrac{1}{4}mv_{CM}^2}^{\text{rotational } KE} = \overbrace{\tfrac{3}{4}mv_{CM}^2}^{\text{total}} \qquad (9.6)$$

The wheel's rotational kinetic energy is thus $\frac{1}{4}mv_{CM}^2$; this is half as large as the translational kinetic energy and one third of the total kinetic energy. These fractions are *independent* of the wheel's radius since all factors of R have canceled in deriving Equation 9.6. The only assumption we made is that the moment of inertia is given by $I = mR^2/2$, so our results are true for any object that has the shape of a wheel or disk. One might have expected that for a smaller wheel (a smaller radius R), less of the total kinetic energy would go into rotation, but that is not the case. The *fraction* of the total kinetic energy associated with rotation is *independent* of R. The only way to reduce this fraction is to change the moment of inertia, that is, to change the wheel's shape so as to reduce its moment of inertia.

Notice that the distribution of the kinetic energy in Equation 9.6 applies only to the wheels. The rest of a skateboard does not rotate, and the kinetic energy of that part of the skateboard comes just from translational motion ($KE = \frac{1}{2}mv_{CM}^2$).

Torque and Rotational Kinetic Energy: Rotational Version of the Work–Energy Theorem

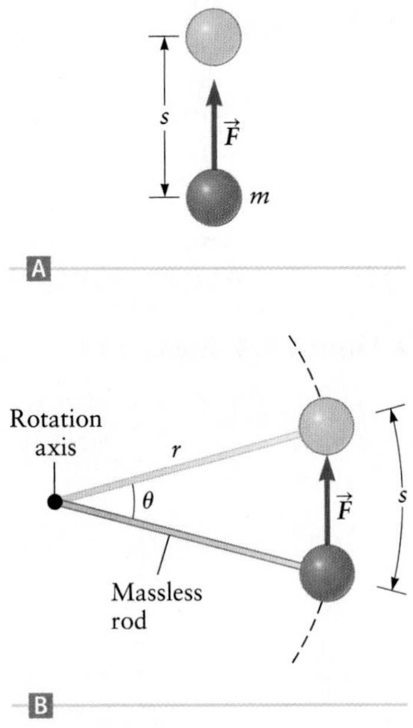

▲ **Figure 9.5** A When this mass moves a distance s while acted on by the force F, the work done is $W = Fs$. B If the mass is attached to a light (massless) rod so that the mass and the rod now rotate, the work done is also equal to $\tau\theta$.

In Chapter 6, we introduced the notion of ***work*** and showed that the work done on an object equals the change in the object's translational kinetic energy. Let's now derive a similar result involving torque and rotational kinetic energy.

Consider a one-dimensional situation in which a force of magnitude F pushes an object of mass m through a distance s as sketched in Figure 9.5A. Since the force here is parallel to the displacement, the work equals the product of the force and the displacement; hence, $W = Fs$. Applying the work–energy theorem (Eq. 6.9), W is equal to the change in kinetic energy of the object:

$$W = Fs = \Delta KE = \tfrac{1}{2}mv_f^2 - \tfrac{1}{2}mv_i^2 \qquad (9.7)$$

where v_i and v_f are the object's initial and final speeds. Now suppose this object is held by a very light (massless) rod so that the object rotates as sketched in Figure 9.5B, with a rotational radius of r. If the angle θ is very small, the force F produces a torque $\tau = Fr$ about this axis. If the object moves a linear distance s, there is a corresponding angular displacement $\theta = s/r$ according to the definitions of angles and radians (Eq. 1.26). Inserting a factor of r in the denominator and numerator of Equation 9.7, we get

$$W = Fs = \overbrace{(Fr)}^{\tau}\overbrace{\left(\frac{s}{r}\right)}^{\theta}$$

$$W = \Delta KE = \tau\theta \qquad (9.8)$$

Because it is moving along a circle, the object is undergoing rotational motion with an angular velocity related to its linear velocity by $v_i = \omega_i r$ and $v_f = \omega_f r$ for the initial and final velocities, respectively. Inserting into Equation 9.7 gives

$$\Delta KE = \tfrac{1}{2}mv_f^2 - \tfrac{1}{2}mv_i^2 = \tfrac{1}{2}m(\omega_f r)^2 - \tfrac{1}{2}m(\omega_i r)^2$$

$$\Delta KE = \tfrac{1}{2}(mr^2)\omega_f^2 - \tfrac{1}{2}(mr^2)\omega_i^2$$

The term mr^2 is just the moment of inertia of the object (from Eq. 8.13), so the change in kinetic energy is

$$\Delta KE = \tfrac{1}{2}I\omega_f^2 - \tfrac{1}{2}I\omega_i^2$$

Combining this expression with Equation 9.8 gives

$$W = \tau\theta = \Delta KE = \tfrac{1}{2}I\omega_f^2 - \tfrac{1}{2}I\omega_i^2 \qquad (9.9)$$

Work–energy theorem for rotational motion

This result is the ***work–energy theorem*** for rotational motion. It says that the product of the torque τ and the angular displacement θ equals the change in the rotational kinetic energy.

EXAMPLE 9.2 Physics on the Playground

A child is given the job of getting a merry-go-round up to an enjoyable speed by pushing on its edge as she runs along the outer circumference as sketched in Figure 9.6. If she pushes with a force of 30 N and the merry-go-round starts from rest, how far will she have to run to get the merry-go-round to rotate at a rate of 0.20 rev/s (which corresponds to an angular velocity of 1 rev every 5.0 s = ω_f = 1.3 rad/s)? Assume the merry-go-round is a disk of mass m = 100 kg and radius R = 2.0 m. Express your answer in terms of the child's angular displacement during her run.

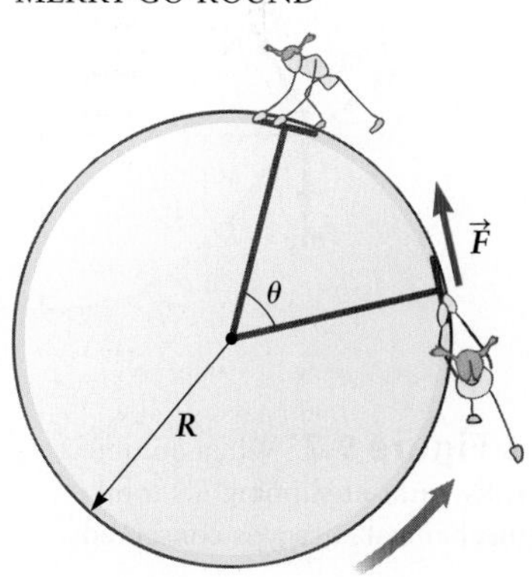

▲ **Figure 9.6** Example 9.2. This child pushes on the merry-go-round with a force F while it moves through an angular displacement θ.

RECOGNIZE THE PRINCIPLE

Since the force and radius of the merry-go-round are given, we can find the applied torque. The initial and final angular velocities are also given, so we can calculate the change in the rotational kinetic energy. The rotational work–energy theorem (Eq. 9.9) then relates the change in kinetic energy to the angular displacement, which is what we want to find.

SKETCH THE PROBLEM

Figure 9.6 shows the problem.

IDENTIFY THE RELATIONSHIPS

The torque exerted by the child is $\tau = FR$ since the associated lever arm is the radius R of the merry-go-round. (See Fig. 9.6.) If the child travels an angular displacement θ as she pushes the merry-go-round, the work done is $W = \tau\theta$. The initial kinetic energy is zero (because the merry-go-round starts from rest), whereas the final kinetic energy is $\tfrac{1}{2}I\omega_f^2$. Inserting all these results into the work–energy theorem (Eq. 9.9) leads to

$$W = \tau\theta = FR\theta = \Delta KE = \tfrac{1}{2}I\omega_f^2$$

$$\theta = \frac{I\omega_f^2}{2FR}$$

SOLVE

The merry-go-round is a disk, so its moment of inertia is $I = mR^2/2$. (See Table 8.2.) Inserting this equation and the given values of F, R, and ω_f gives

$$\theta = \frac{I\omega_f^2}{2FR} = \frac{(mR^2/2)\omega_f^2}{2FR} = \frac{mR\omega_f^2}{4F} = \frac{(100\text{ kg})(2.0\text{ m})(1.3\text{ rad/s})^2}{4(30\text{ N})} = \boxed{2.8\text{ rad}}$$

(continued) ▶

▶ ***What does it mean?***
This answer is less than half of a revolution (since 1 rev corresponds to $2\pi \approx$ 6.2 rad), so the child will not have to run very far.

9.2 Conservation of Energy and Rotational Motion

In Chapter 6, we showed that if all the forces that do work on an object are conservative forces, the total mechanical energy of the object is conserved. In such cases

Conservation of total mechanical energy

$$\overbrace{KE_i + PE_i}^{\text{initial mechanical energy}} = \overbrace{KE_f + PE_f}^{\text{final mechanical energy}} \tag{9.10}$$

Here *KE* denotes the *total* kinetic energy (including translation and rotation) given by Equation 9.5 and *PE* is the object's potential energy.

Consider a solid ball (a sphere) that starts from rest and rolls down a hill as sketched in Figure 9.7. Let's apply the principle of conservation of energy to calculate its speed when it reaches the bottom. In this example, there are three forces acting on the ball: the normal force exerted by the hill, the force of friction between the ball and the hill, and the force of gravity. The normal force does not do any work on the ball because this force is always directed perpendicular to the ball's displacement. The force of friction is necessary for the ball to roll without slipping. However, friction does not do any work on the ball because the contact point where the ball meets the surface does not slip. Hence, friction does not play a role in the conservation of energy condition (Eq. 9.10) when an object rolls without slipping (see also Insight 9.1). The only force that does work on the ball in Figure 9.7 is the force of gravity.

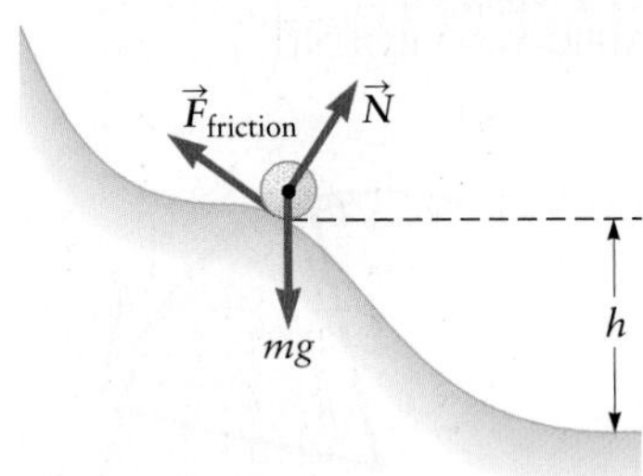

▲ **Figure 9.7** When an object rolls without slipping, its total mechanical energy is conserved.

We saw in Chapter 6 that the work done by gravity equals the change in gravitational potential energy $PE_{\text{grav}} = mgh$. Hence, the conservation of mechanical energy condition for the ball becomes

$$KE_i + PE_i = KE_f + PE_f$$

$$\tfrac{1}{2}mv_i^2 + \tfrac{1}{2}I\omega_i^2 + mgh_i = \tfrac{1}{2}mv_f^2 + \tfrac{1}{2}I\omega_f^2 + mgh_f \tag{9.11}$$

where the subscript i denotes the ball's initial state (at the top of the hill) and the subscript f denotes the ball's final state (at the bottom). The speeds v_i and v_f in Equation 9.11 are the center of mass speeds of the ball, which (until this point) we have denoted by v_{CM}. (To avoid the excessive use of subscripts, in the rest of this chapter we often drop the subscript "*CM*" in places where it should be obvious.)

The ball started from rest, so $v_i = \omega_i = 0$. The ball is also rolling without slipping, so $\omega_f = v_f/R$, where R is the ball's radius. Inserting this all into Equation 9.11 and taking $h_f = 0$ and $h_i = h$ (the height of the hill), we find

$$\overbrace{\tfrac{1}{2}mv_i^2}^{0} + \overbrace{\tfrac{1}{2}I\omega_i^2}^{0} + \overbrace{mgh_i}^{mgh} = \tfrac{1}{2}mv_f^2 + \tfrac{1}{2}I\omega_f^2 + \overbrace{mgh_f}^{0}$$

$$mgh = \tfrac{1}{2}mv_f^2 + \tfrac{1}{2}I\omega_f^2 = \tfrac{1}{2}mv_f^2 + \tfrac{1}{2}I\left(\frac{v_f}{R}\right)^2 \tag{9.12}$$

Inserting the moment of inertia of a solid sphere ($I = 2mR^2/5$) from Table 8.2 leads to

$$mgh = \frac{1}{2}mv_f^2 + \frac{1}{2}I\left(\frac{v_f}{R}\right)^2 = \frac{1}{2}mv_f^2 + \frac{1}{2}\frac{2mR^2}{5}\left(\frac{v_f}{R}\right)^2$$

$$mgh = \frac{1}{2}mv_f^2 + \frac{1}{5}mv_f^2 = \frac{7}{10}mv_f^2$$

Insight 9.1

FRICTION DOES NO WORK ON A ROLLING OBJECT

Consider an object that rolls without slipping, such as the ball in Figure 9.7. Slipping is prevented by the frictional force between the surface of the hill and the ball. This frictional force acts at the point of contact, and because the ball is not slipping, this point is always at rest. Work is the product of force and displacement. Since the contact point is at rest, its displacement is zero, and the work done by the frictional force is zero in such cases.

Solving for the speed v_f at the bottom, we have

$$v_f = \sqrt{\frac{10gh}{7}} \tag{9.13}$$

In getting this result, we have relied only on the fact that the mechanical energy of the ball is conserved. We did not need to know anything about the hill other than its height h.

Our application of the principle of conservation of energy to the rolling ball in Figure 9.7 is similar to the way we used conservation of energy principles in Chapter 6. Our general strategy for dealing with these problems is as follows.

PROBLEM SOLVING Applying the Principle of Conservation of Energy to Problems of Rotational Motion

1. **RECOGNIZE THE PRINCIPLE.** The mechanical energy of an object is conserved only if all the forces that do work on the object are conservative forces. Only then can we apply the conservation of mechanical energy condition in Equation 9.10.
2. **SKETCH THE PROBLEM.** Always use a sketch to collect your information concerning the initial and final states of the system.
3. **IDENTIFY THE RELATIONSHIPS.** Find the initial and final kinetic and potential energies of the object. They will usually involve the initial and final linear and angular velocities, and the initial and final heights, as for the rolling ball in Figure 9.7. You will usually also need to find the moment of inertia.
4. **SOLVE** for the unknown quantities using the conservation of energy condition,
$$KE_i + PE_i = KE_f + PE_f$$
The kinetic energy terms must account for both the translational and the rotational kinetic energies (Eq. 9.5):
$$KE_{\text{total}} = \underbrace{\tfrac{1}{2}mv_{CM}^2}_{\text{translational } KE} + \underbrace{\tfrac{1}{2}I\omega^2}_{\text{rotational } KE}$$
5. Always *consider what your answer means* and check that it makes sense.

CONCEPT CHECK 9.2 Rolling versus Sliding

Consider a ball that rolls down a hill as compared to an object (such as a skier) that slides down a frictionless hill of the same height. Which of the following statements is correct regarding their speeds at the bottom of the hill?

(a) The ball has a higher linear speed because it does not slip.
(b) The sliding object has a higher speed because the hill is frictionless.
(c) The ball and skier have the same speed at the bottom of the hill.
(d) The ball has a smaller linear speed because some of its kinetic energy goes into rotational motion.

EXAMPLE 9.3 Motion of a Yo-Yo

Consider the yo-yo in Figure 9.8. The yo-yo starts from rest and then moves downward in the manner you have probably practiced at least a few times. Find the speed of the yo-yo after it has moved 50 cm below its starting location. Assume the yo-yo is a solid disk with the string wound around the outer edge.

RECOGNIZE THE PRINCIPLE

There are two forces acting on the yo-yo: the force exerted by the string and the force of gravity. The force exerted by the string plays a role very similar to that of friction for the rolling ball in Figure 9.7, and the string does no work on the yo-yo. That is, the

(continued) ▶

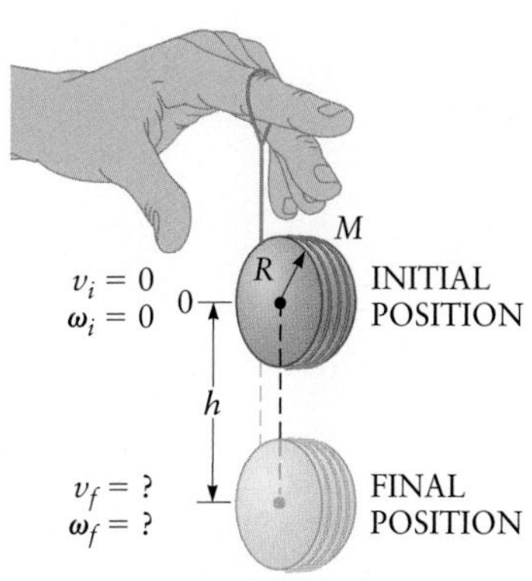

▲ **Figure 9.8** Example 9.3. A yo-yo traveling down a string.

yo-yo "rolls down" the string in the same way that the ball rolls without slipping down the hill. The only force that does work on the yo-yo is the force of gravity. Since gravity is a conservative force, we can apply the principle of conservation of energy.

SKETCH THE PROBLEM

Figure 9.8 shows the initial and final positions of the yo-yo.

IDENTIFY THE RELATIONSHIPS

The conservation of energy relation for the yo-yo is

$$KE_i + PE_i = KE_f + PE_f$$

The kinetic energy has contributions from both translational and rotational motion:

$$\tfrac{1}{2}mv_i^2 + \tfrac{1}{2}I\omega_i^2 + PE_i = \tfrac{1}{2}mv_f^2 + \tfrac{1}{2}I\omega_f^2 + PE_f$$

The potential energy is due to gravity; taking the starting height to be $h_i = 0$, the final height is $h_f = -h$ ($= -0.50$ m as given above). This yo-yo is a disk of mass m and radius R, so its moment of inertia is $I = mR^2/2$ (from Table 8.2). Inserting all this into the conservation of energy relation leads to

$$\frac{1}{2}mv_i^2 + \frac{1}{2}\left(\frac{mR^2}{2}\right)\omega_i^2 = \frac{1}{2}mv_f^2 + \frac{1}{2}\left(\frac{mR^2}{2}\right)\omega_f^2 + mg(-h) \quad (1)$$

The yo-yo starts from rest, so $v_i = \omega_i = 0$. The yo-yo is also (as already noted) "rolling down" the string, so the final linear and angular velocities are related by $\omega_f = v_f/R$.

SOLVE

Putting all this information into Equation (1), we find

$$0 = \frac{1}{2}mv_f^2 + \frac{1}{2}\left(\frac{mR^2}{2}\right)\left(\frac{v_f}{R}\right)^2 - mgh = \frac{1}{2}mv_f^2 + \frac{1}{4}mv_f^2 - mgh$$

$$0 = \frac{3}{4}mv_f^2 - mgh$$

$$\frac{3}{4}mv_f^2 = mgh$$

$$v_f = \sqrt{\frac{4gh}{3}}$$

Inserting the given value of h, we find

$$v_f = \sqrt{\frac{4gh}{3}} = \sqrt{\frac{4(9.8\text{ m/s}^2)(0.50\text{ m})}{3}} = \boxed{2.6\text{ m/s}}$$

► ***What does it mean?***

Notice that v_f does not depend on either the mass or the radius of the yo-yo. A large yo-yo and a small yo-yo will move at the same speed; that is, they will "fall" at the same rate. Hence, Galileo could have used yo-yos in his experiments at the Leaning Tower of Pisa (Chapter 2).

CONCEPT CHECK 9.3 Designing a Faster Yo-Yo

Suppose you want to design and market a faster yo-yo. That is, you want to make a yo-yo that will have a higher speed than that found in Example 9.3. How could you do it?

(a) Make the yo-yo radius larger.
(b) Make the yo-yo radius smaller.
(c) Add mass near the axis so that the yo-yo is shaped more like a sphere.
(d) Hollow out the center so that the yo-yo is shaped like a cylindrical shell.

EXAMPLE 9.4 Rolling up a Hill

A bowling ball rolls without slipping up an incline as shown in Figure 9.9. If the ball has speed v_i = 12 m/s at the base of the incline, what height h_f on the incline will the bottom of the ball reach?

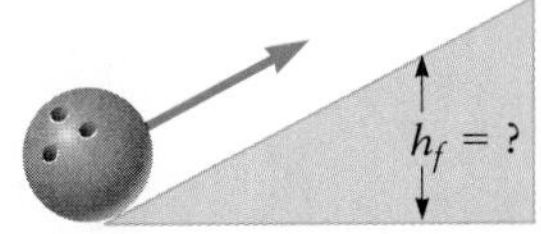

▲ **Figure 9.9** Example 9.4.

RECOGNIZE THE PRINCIPLE

The only force that does work on the ball is gravity, which is a conservative force. The mechanical energy of the ball is therefore conserved, and the initial energy is equal to the final energy.

SKETCH THE PROBLEM

Figure 9.9 shows the problem.

IDENTIFY THE RELATIONSHIPS

We can write

$$KE_i + PE_i = KE_f + PE_f \tag{1}$$

where the subscripts i and f denote the initial state (with the ball at the bottom of the incline) and the final state (with the ball at its highest point), respectively. If we measure height from the bottom of the ramp, $h_i = 0$, and h_f is the quantity we wish to find. The initial kinetic energy is

$$KE_i = \frac{1}{2}mv_i^2 + \frac{1}{2}I\omega_i^2$$

where m is the mass of the ball and its moment of inertia is $I = \frac{2}{5}mR^2$. The radius of the ball is R, and its initial angular velocity is $\omega_i = v_i/R$. When the ball reaches its highest point, its speed is zero, so the final kinetic energy is zero. Inserting all this into Equation (1) leads to

$$KE_i + PE_i = \frac{1}{2}mv_i^2 + \frac{1}{2}\left(\frac{2}{5}mR^2\right)\left(\frac{v_i}{R}\right)^2 + mg(0) = KE_f + PE_f = 0 + mgh_f \tag{2}$$

SOLVE

Solving Equation (2) for the final height gives

$$gh_f = \frac{1}{2}v_i^2 + \frac{1}{5}v_i^2 = \frac{7}{10}v_i^2$$

$$h_f = \frac{7v_i^2}{10g} = \frac{7(12 \text{ m/s})^2}{10(9.8 \text{ m/s}^2)} = 10 \text{ m}$$

► ***What does it mean?***

The maximum height of a projectile that does not have any rotational kinetic energy would be $h_f = v_i^2/(2g)$, so a rolling ball will reach higher than a projectile with the same initial speed because the ball has some initial rotational kinetic energy that is converted to potential energy at the highest point.

9.3 Angular Momentum

In Chapter 7, we defined the *linear momentum* for a point particle of mass m moving with velocity $\vec{v}$ as

$$\vec{p} = m\vec{v} \tag{9.14}$$

We showed that the momentum of a system of particles is conserved provided there is no net external force on the system. It should not surprise you to find that a rotating object has a property called angular momentum. In nearly all the rotational motion examples we have discussed so far, the angular velocity has been defined with respect to a single rotation axis that does not change direction during the motion. In such cases, the ***angular momentum*** is given by

Angular momentum for rotation about a fixed axis

$$L = I\omega \tag{9.15}$$

This equation looks very much like Equation 9.14: the moment of inertia I is analogous to mass, and angular velocity is analogous to linear velocity. One difference is that the linear momentum $\vec{p}$ is a vector, whereas L in Equation 9.15 is not. This difference is due to our simplified treatment of rotations. In Section 9.4, we'll generalize Equation 9.15 to a vector form with a true angular velocity vector when we discuss gyroscopes. In this section, we assume the rotation axis keeps a fixed direction so that we can use the scalar (nonvector) form in Equation 9.15.

So far, we have always associated an angular velocity with objects that rotate about a fixed axis, but the notion of angular momentum also applies to more general cases. Figure 9.10 shows a particle moving freely through space. Even though this particle is not constrained to move in a rotational fashion, it can still have angular momentum. For example, if we choose a particular pivot point P, the particle will at any given instant be moving tangent to the circular arc of radius r sketched in Figure 9.10. At this instant, the motion of the particle is the same as that of an object that rotates along this circular arc. We can therefore define an angular velocity $\omega = v_\perp/r$, where $v_\perp$ is the component of the particle's velocity perpendicular to the radius of the arc and r is the distance from the particle to the point P. The moment of inertia is given by $I = mr^2$ (Eq. 8.13), so using the definition of angular momentum (Eq. 9.15), we have

$$L = I\omega = mr^2\left(\frac{v_\perp}{r}\right)$$

$$L = mrv_\perp \tag{9.16}$$

This result for L applies to any moving particle. We can always choose a hypothetical pivot point and rotation axis, even if the particle is not moving in a complete closed path about the axis. The angular momentum with respect to that axis is then proportional to $v_\perp$, the component of the velocity perpendicular to the line that connects the particle to the pivot point. The result for angular momentum in Equation 9.16 will be important when we wish to calculate the total angular momentum of a system of particles.

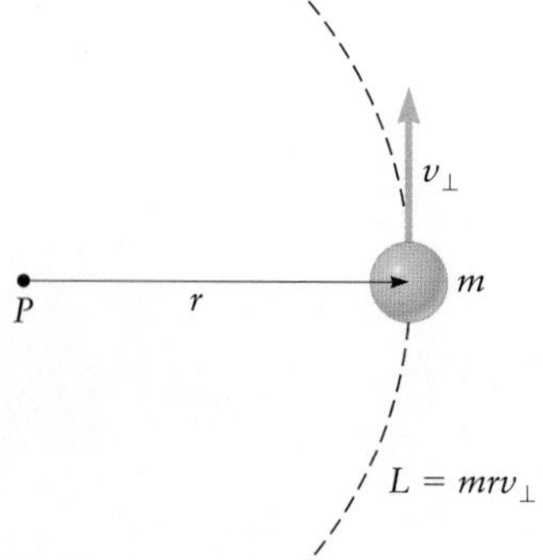

▲ **Figure 9.10** The angular momentum of a moving particle about a rotation axis that passes through point P is $L = mrv_\perp$, where $v_\perp$ is the component of the velocity perpendicular to the line that connects the particle to the pivot point P. Here the rotation axis is perpendicular to the plane of the drawing.

Conservation of Angular Momentum and a Spinning Skater

A rotating object with an angular momentum $L = I\omega$ will maintain its angular momentum provided no external torques act on the object. We say that such an object's angular momentum is *conserved*. This notion of angular momentum conservation also applies to a system of objects. If there are no *external* torques on a system, the total angular momentum of the system will be conserved.

A familiar example in which angular momentum conservation plays a key role is the motion of the figure skater in Figure 9.11. The skater is initially spinning about a vertical axis with an angular velocity ω_i (Fig. 9.11A). If she then pulls her arms and legs in very close to the midline of her body, we know that her spin rate will increase so that her final angular velocity ω_f is much larger than ω_i. This behavior can be understood in terms of the skater's angular momentum.

During the time she is spinning, no torques are acting on the skater. The only forces acting on her are gravity and the normal force from the ice, and they do not exert a torque on the skater. (We assume the ice is very slippery, so the frictional force is negligible.) The skater's angular momentum is therefore conserved:

$$L_i = L_f$$

Her initial angular momentum is $L_i = I_i\omega_i$, where the subscript i indicates the skater's initial state. When she pulls her arms and legs close to her midline, she is redistributing her mass so that it is closer to the axis of rotation, changing her moment of inertia to a new value I_f. Recall that the moment of inertia is equal to a sum over all the individual pieces of an object:

$$I = \sum mr^2$$

If an object changes shape so that some of the individual pieces are moved closer to the axis of rotation, the values of r for these pieces are reduced and the overall moment of inertia decreases. That is precisely what happens with the skater in Figure 9.11B. When she pulls her arms and legs in close to her rotation axis, these "pieces" have smaller values of r and her moment of inertia becomes smaller. Her total angular momentum is conserved; thus,

$$L_i = I_i\omega_i = L_f = I_f\omega_f \tag{9.17}$$

Pulling her arms in close to her body reduces her moment of inertia, so $I_f < I_i$ and hence $\omega_f > \omega_i$. She spins *faster*. This increase in her angular velocity depends on how much she can reduce her moment of inertia; that is, it depends on how closely she can pull her arms and legs in line with the rest of her body.

▲ **Figure 9.11** There are no external torques acting on the skater, so her angular momentum is conserved. Her angular velocity increases when she pulls her arms and legs close to her body, but her angular momentum stays the same.

Problem Solving with Angular Momentum

We saw in Chapter 7 that linear momentum and momentum conservation are extremely useful when analyzing collisions and other situations that involve a system of particles. The same is true of angular momentum. Typically, a system such as the skater in Figure 9.11 is moving in some manner with an initial angular momentum. There is then some sort of "change" in the system (such as the skater pulling in her arms and legs) that leaves the system in a final state of motion. If the external torque on the system is zero, the total angular momentum must be conserved. Hence, $L(\text{final}) = L(\text{initial})$, which we can use to find the final angular velocity or some other quantity. The general approach to such problems is as follows.

PROBLEM SOLVING Applying the Principle of Conservation of Angular Momentum

1. **RECOGNIZE THE PRINCIPLE.** If the external torque on a system is zero, the angular momentum is conserved.

2. **SKETCH THE PROBLEM.** Always use a sketch to collect your information concerning the initial and final states of the system.

3. **IDENTIFY THE RELATIONSHIPS.**
   - The system of interest and its initial and final states depend on the problem. The system might be a single object (such as the skater in Figure 9.11) or a collection of objects.
   - Express the initial and final angular velocities and moments of inertia.
   - Apply any information concerning the initial and final mechanical energies. (Such information will only be available in some cases.)

4. **SOLVE** for the quantities of interest using the principle of conservation of angular momentum,

$$L_i = I_i\omega_i = L_f = I_f\omega_f$$

5. Always *consider what your answer means* and check that it makes sense.

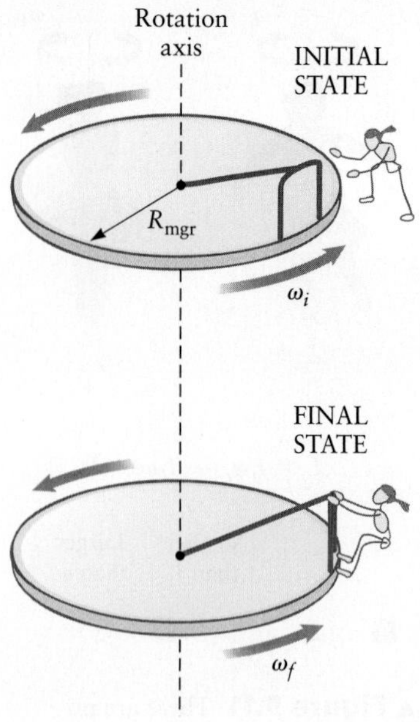

▲ **Figure 9.12** Example 9.5. A child steps onto a rotating merry-go-round.

EXAMPLE 9.5 Angular Momentum on the Playground

A merry-go-round of mass m_{mgr} and radius R_{mgr} is rotating freely as sketched in Figure 9.12. A child then steps onto the edge of the merry-go-round and notices that the angular velocity of the merry-go-round has decreased slightly. If the child's mass is m_c, what is the ratio of the final angular velocity to the initial angular velocity?

RECOGNIZE THE PRINCIPLE

The merry-go-round and the child are a "system," and although there are external forces acting on them (such as gravity), for a well-oiled merry-go-round the torque on this system will be negligible and the angular momentum will be conserved.

SKETCH THE PROBLEM

Figure 9.12 shows the system just before the child steps onto the merry-go-round (the initial state) and just after (the final state).

IDENTIFY THE RELATIONSHIPS

Let I_{mgr} be the moment of inertia of the merry-go-round and ω_i be its angular velocity before the child gets on. Also let I_c be the moment of inertia of the child, and her initial angular velocity is zero. The initial angular momentum is then

$$L_i = I_{mgr}\omega_i + I_c(0) = I_{mgr}\omega_i$$

We can approximate the merry-go-round as a large disk, with moment of inertia $I = \frac{1}{2}mR^2$, so we have $I_{mgr} = \frac{1}{2}m_{mgr}R_{mgr}^2$. The child has a mass m_c and is located a distance R_{mgr} from the axis of rotation, so her moment of inertia is $I_c = \sum mr^2 = m_c R_{mgr}^2$. After the child steps onto the merry-go-round, the angular momentum of the system is

$$L_f = I_{mgr}\omega_f + I_c\omega_f$$

where ω_f is the (common) final angular velocity. Both the child and the merry-go-round have the same final angular velocity.

SOLVE

Angular momentum is conserved, so $L_i = L_f$, which gives

$$L_i = I_{mgr}\omega_i = L_f = I_{mgr}\omega_f + I_c\omega_f = (I_{mgr} + I_c)\omega_f$$

Solving for the final angular velocity, we find

$$\omega_f = \frac{I_{mgr}\omega_i}{I_{mgr} + I_c}$$

The ratio of the initial and final angular velocities is then

$$\frac{\omega_f}{\omega_i} = \frac{I_{mgr}}{I_{mgr} + I_c} \qquad (1)$$

Inserting the results for the moments of inertia into Equation (1) gives

$$\frac{\omega_f}{\omega_i} = \frac{I_{mgr}}{I_{mgr} + I_c} = \frac{m_{mgr}R_{mgr}^2/2}{(m_{mgr}R_{mgr}^2/2) + m_c R_{mgr}^2} = \boxed{\frac{m_{mgr}}{m_{mgr} + 2m_c}}$$

▶ ***What does it mean?***

This example is similar to an inelastic collision between two objects as we encountered in Chapter 7. The two "objects"—the merry-go-round and the child—stick together and move with the same angular final velocity, just as the two objects in a completely inelastic collision stick together after colliding.

EXAMPLE 9.6 Jumping off a Merry-Go-Round

Suppose the child in Example 9.5 is initially standing on the edge of the merry-go-round and both are initially at rest. The child then jumps off the merry-go-round, which begins to rotate. If she jumps in a direction tangent to the edge of the merry-go-round as in Figure 9.13A and has a speed of 3.0 m/s while she is in the air, what is the final angular velocity of the merry-go-round? Assume the mass of the child is $m_c = 30$ kg, the merry-go-round has mass $m_{\text{mgr}} = 100$ kg, and the merry-go-round's radius is $R_{\text{mgr}} = 2.0$ m. Also assume the merry-go-round is frictionless.

RECOGNIZE THE PRINCIPLE

Just as in Example 9.5, the external torques on the merry-go-round and child in Figure 9.13 are negligible. Hence, the angular momentum of this system is conserved.

SKETCH THE PROBLEM

Figure 9.13 shows the initial and final states of the system. In Figure 9.13B, the child has just jumped off the merry-go-round and is still moving through the air.

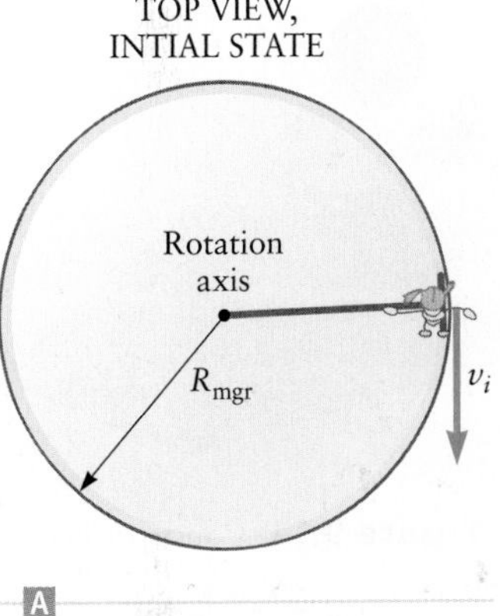

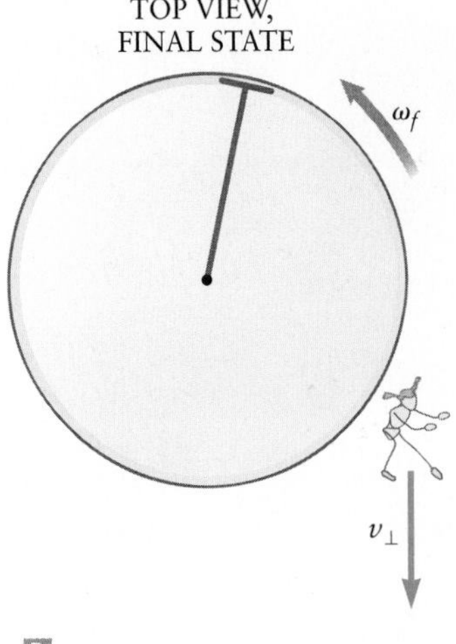

▲ **Figure 9.13** Example 9.6. A child jumps off a stationary merry-go-round. The final angular momentum of the child is $m_c R_{\text{mgr}} v_\perp$.

IDENTIFY THE RELATIONSHIPS

The initial state of the system is just prior to when the child jumps; everything is at rest at that moment, so the initial angular momentum is zero. The final state is the instant just after the child has jumped and before she reaches the ground. *When the child is moving through the air as a projectile, she still has angular momentum* as described in Figure 9.10 and Equation 9.16, which we must account for when we compute the final angular momentum of the system.

If the final angular velocity of the merry-go-round is ω_{mgr}, its final angular momentum is $I_{\text{mgr}}\omega_{\text{mgr}}$, where the moment of inertia of the merry-go-round is as described in Example 9.5. The final angular momentum of the child can be found from Equation 9.16. The distance from the child to the rotation axis is R_{mgr}, and her velocity perpendicular to the radius of the merry-go-round is $v_\perp$ (= 3.0 m/s as given in the problem statement). The child's final angular momentum is thus

$$L_c = m_c R_{\text{mgr}} v_\perp$$

Combining all these results, the final angular momentum of the system is the sum of the child's angular momentum plus that of the merry-go-round:

$$L_f = L_{\text{mgr}} + L_c = I_{\text{mgr}}\omega_{\text{mgr}} + m_c R_{\text{mgr}} v_\perp$$

Inserting the moment of inertia of the merry-go-round ($I_{\text{mgr}} = \frac{1}{2}m_{\text{mgr}}R_{\text{mgr}}^2$) leads to

$$L_f = \tfrac{1}{2}m_{\text{mgr}}R_{\text{mgr}}^2\omega_{\text{mgr}} + m_c R_{\text{mgr}} v_\perp$$

SOLVE

The final angular momentum equals the initial angular momentum, which is zero, so

$$\tfrac{1}{2}m_{\text{mgr}}R_{\text{mgr}}^2\omega_{\text{mgr}} + m_c R_{\text{mgr}} v_\perp = 0$$

Solving for the final angular velocity of the merry-go-round leads to

$$\omega_{\text{mgr}} = -\frac{m_c R_{\text{mgr}} v_\perp}{m_{\text{mgr}}R_{\text{mgr}}^2/2} = -\frac{2m_c v_\perp}{m_{\text{mgr}}R_{\text{mgr}}}$$

(continued) ▶

The negative sign here indicates that the final angular velocities of the child and merry-go-round are in opposite directions. Inserting the given values of the various quantities, we find

$$\omega_{\text{mgr}} = -\frac{2m_c v_\perp}{m_{\text{mgr}} R_{\text{mgr}}} = -\frac{2(30 \text{ kg})(3.0 \text{ m/s})}{(100 \text{ kg})(2.0 \text{ m})} = \boxed{-0.90 \text{ rad/s}}$$

▸ ***What does it mean?***

An important point of this example is that an object moving in a straight line (here, the child after she jumps from the merry-go-round) can still have angular momentum with respect to a particular rotation axis.

CONCEPT CHECK 9.4 Where Does the Torque Come From?

In Example 9.6, we treated the merry-go-round plus child as a system (Fig. 9.13) whose angular momentum is conserved. The merry-go-round is initially at rest, so its initial angular momentum is zero, but at the end it is rotating, so its final angular momentum is nonzero. The angular momentum of the merry-go-round alone thus changes, which can only happen if there is a torque on the merry-go-round. What force provides this torque?

(a) The force of gravity on the merry-go-round
(b) The force of air drag on the merry-go-round
(c) The force produced by the child when she jumps off
(d) The force of friction at the axle of the merry-go-round

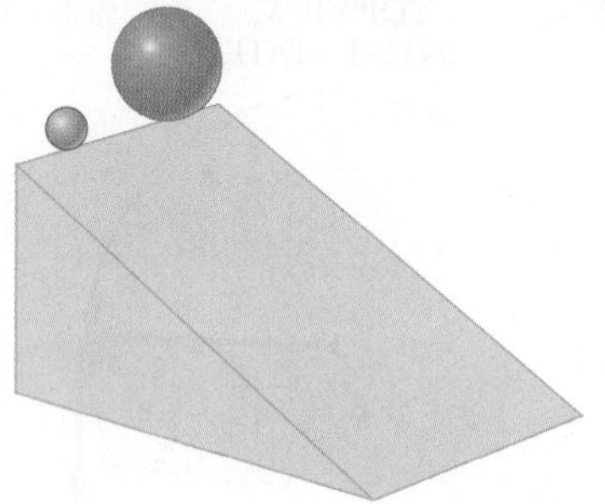

▲ **Figure 9.14** Concept Check 9.5.

CONCEPT CHECK 9.5 Rolling, Rolling, Rolling

Two balls with the same mass but different radii roll down a ramp (Fig. 9.14). If the balls both start from rest, which one has the larger angular momentum when they reach the bottom of the ramp?

Angular Momentum and Kinetic Energy

Let's now return to the problem of the spinning ice skater in Figure 9.11 and consider how her rotational kinetic energy varies during the spin. Our skater began with her arms and legs extended away from her body's rotation axis and then pulled them in close to her body, increasing her angular velocity. In terms of her angular momentum, we wrote (Eq. 9.17)

$$L_i = I_i\omega_i = L_f = I_f\omega_f$$

which can be rearranged to give

$$\omega_f = \frac{I_i}{I_f}\omega_i \tag{9.18}$$

Since her final moment of inertia (with her arms and legs in) is smaller than her initial moment of inertia (arms and legs out), $\omega_f > \omega_i$ and she spins faster at the end.

Rotational kinetic energy is $KE = \frac{1}{2}I\omega^2$, so the skater's final rotational kinetic energy is $KE_f = \frac{1}{2}I_f\omega_f^2$. Inserting ω_f from Equation 9.18 and using the skater's initial kinetic energy of $KE_i = \frac{1}{2}I_i\omega_i^2$ leads to

$$KE_f = \frac{1}{2}I_f\omega_f^2 = \frac{1}{2}I_f\left(\frac{I_i}{I_f}\omega_i\right)^2 = \frac{I_i}{I_f}\left(\frac{1}{2}I_i\omega_i^2\right)$$

$$KE_f = \frac{I_i}{I_f}KE_i \tag{9.19}$$

Because the skater's initial moment of inertia is larger than her final moment of inertia ($I_i > I_f$), Equation 9.19 shows that her final kinetic energy is *greater* than her initial kinetic energy. Even though her angular momentum is conserved, her kinetic energy is *not conserved*.

Where does the additional kinetic energy of the skater come from? That is, what force does work on the skater? This force comes from the skater herself as she pulls on her arms. Her arms undergo a displacement as they are pulled inward, and the work done (W) equals the product of this displacement and the component of the force parallel to the displacement. From the work–energy theorem, W equals the increase in the skater's kinetic energy.

9.4 Angular Momentum and Kepler's Second Law of Planetary Motion

In Chapter 5, we learned about Kepler's three laws of planetary motion. According to Kepler's first law, planets follow elliptical orbits about the Sun, and his second law states that a planet moving about its orbit sweeps out equal areas in equal times. If we observe the motion of the planet during two separate time intervals of equal duration Δt, we can consider the areas A_1 and A_2 swept out during these intervals as sketched in Figure 9.15. During an interval Δt, the planet travels a distance $v\,\Delta t$. When the planet is far from the Sun (at point 1 in Fig. 9.15), its speed is less than when the planet is close to the Sun (at point 2). However, Kepler's second law states that the areas that are swept out are the same; the areas A_1 and A_2 in Figure 9.15 are thus equal. This equality holds for all points along the orbit; the area swept out by the planet in a time Δt is the same for all positions along the orbit. We'll now show that Kepler's second law is intimately connected with the angular momentum of the planet.

Angular Momentum of an Orbiting Planet

The planet orbiting the Sun in Figure 9.15 has angular momentum, just like the moving particle in Figure 9.10. Applying Equation 9.16, the planet's angular momentum is given by

$$L_{\text{planet}} = m_{\text{planet}} r v_{\perp} \tag{9.20}$$

where r is the distance from the planet to the Sun and $v_{\perp}$ is the component of the velocity perpendicular to the line connecting the planet and the Sun. As the planet moves around its orbit, both r and $v_{\perp}$ change but the planet's angular momentum stays constant for the reason shown in Figure 9.16. The planet is now our system, and the only force on this system is the force of gravity from the Sun. The gravitational

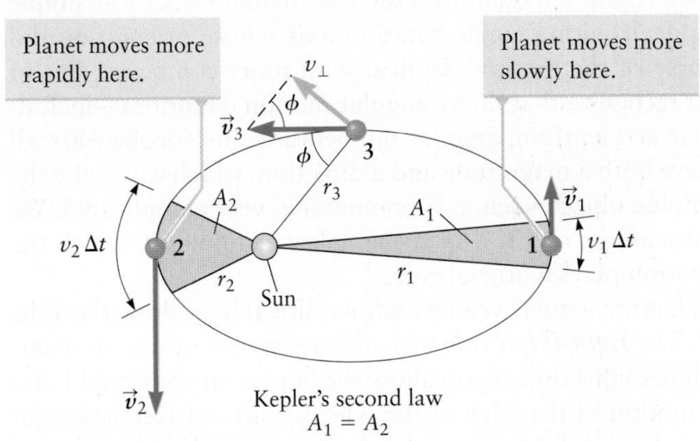

▲ **Figure 9.15** When a planet follows an elliptical orbit, the speed is greatest when the planet is closest to the Sun. According to Kepler's second law, a planet sweeps out equal areas in equal time intervals as it moves about the Sun. Hence, $A_1 = A_2$.

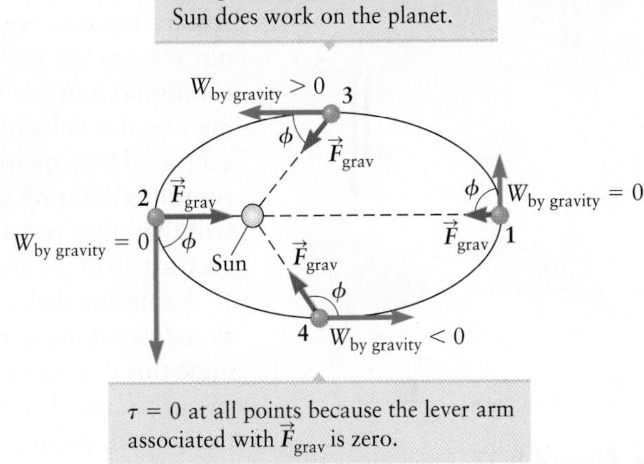

▲ **Figure 9.16** The force of gravity on the planet is directed along the orbital radius. This force cannot exert a torque on the planet, but it can do work on the planet.

force is always directed along the radial line that runs from the planet to the Sun. The axis for the planet's rotational motion passes through the Sun, so the torque due to the gravitational force is zero. The angular momentum of the planet is conserved.

To see how conservation of angular momentum of the planet leads to Kepler's second law, consider the locations along the orbit where the planet is farthest and closest to the Sun, at points 1 and 2 in Figure 9.15. At these positions, the planet's velocity is perpendicular to the line between the Sun and the planet, so $v_\perp = v_1$ at point 1 and $v_\perp = v_2$ at point 2. Using Equation 9.20, we get

$$L_1 = m_{\text{planet}} r_1 v_1 = m_{\text{planet}} r_2 v_2 = L_2 \tag{9.21}$$

$$r_1 v_1 = r_2 v_2 \tag{9.22}$$

The regions A_1 and A_2 swept out by the planet in Figure 9.15 are approximately triangular,[1] so the area A_1 is given by

$$A_1 = \tfrac{1}{2}(\text{base})(\text{height}) = \tfrac{1}{2}\,(v_1 \Delta t)\,(r_1)$$

Using Equation 9.22, we now have

$$A_1 = \tfrac{1}{2}(\Delta t) r_1 v_1 = \tfrac{1}{2}(\Delta t) r_2 v_2 = A_2$$

The areas A_1 and A_2 are thus equal; this relation is Kepler's second law.

We have already noted that because the gravitational force from the Sun acts along the planet's orbital radius, this force cannot exert a torque on the planet. This result is true for all points along the orbit. Gravity, however, *can* do work on the planet. The work done during a time interval Δt is equal to

$$W_{\text{by gravity}} = Fd\cos\phi$$

where d is the displacement during this interval and the angle ϕ is indicated in Figure 9.16. At positions 1 and 2, this angle is 90°, so $\cos\phi = 0$ and the work done is zero, but at other places along the orbit, such as locations 3 and 4, the work done is not zero. The work done by gravity causes the planet to speed up, thus increasing its kinetic energy as it approaches the Sun. As the planet moves farther from the Sun during the course of its orbit, the work done by gravity is negative (at point 4 in Fig. 9.16), causing the planet to slow down.

9.5 The Vector Nature of Rotational Motion: Gyroscopes

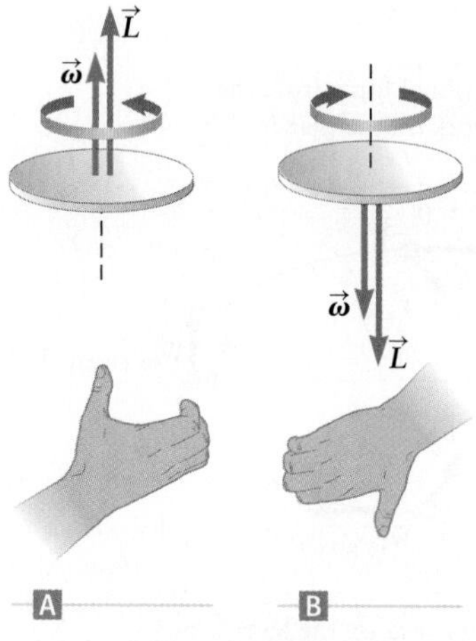

To this point, we have treated rotational quantities such as angular velocity as simple scalars because we always dealt with a single rotation axis whose orientation did not change during the course of the motion. To deal with more complex types of rotational motion, we must recognize that many angular motion quantities—including angular velocity, angular acceleration, angular momentum, and torque—are all vectors. These quantities have both a magnitude and a direction, which can make the rotational motion of a complex object, such as a boomerang, very complicated. We will therefore restrict our discussion to a few of the simplest examples in which the vector nature of angular motion plays a crucial role.

A rotating disk or wheel has an angular velocity whose direction is along the axle, as indicated in Figure 9.17. The ***right-hand rule*** provides a convenient way to determine this direction. According to this rule, if you allow the fingers on your right hand to curl in the direction of motion of the edge of the wheel, your outstretched right thumb will point in the direction of $\vec{\omega}$, the angular velocity vector. The direction of other angular quantities such as the angular momentum $\vec{L}$ is also given by the right-

$\vec{L}$ $\vec{\omega}$ A $\vec{\omega}$ $\vec{L}$ B

▲ **Figure 9.17** Application of the right-hand rule to determine the direction of the angular velocity $\vec{\omega}$ and angular momentum $\vec{L}$ of a rotating disk or wheel.

[1]This approximation for the area becomes more and more accurate as the time interval Δt is made smaller and smaller.

hand rule. For the cases we consider in this book, the angular momentum is parallel to the angular velocity, with

$$\vec{L} = I\vec{\omega} \tag{9.23}$$

When we say that angular momentum is conserved, we mean that both the *magnitude* and *direction* of $\vec{L}$ are conserved. That is the principle behind the operation of a ***gyroscope***. There are many ways to build a gyroscope; one design (Fig. 9.18) employs a spinning wheel mounted on a special frame that allows the axle to rotate freely. Because of the way the wheel is mounted in this frame, the torque on the wheel is zero, even when the frame is moved or rotated. Hence, even if the gyroscope frame is moved or rotated, the wheel's orientation does not change because the angular momentum vector is conserved: $\vec{L}$ of the wheel maintains this constant direction. The orientation of a gyroscope thus provides a "direction finder," a tool extremely useful to pilots and navigators. All airplanes have gyroscopes so that pilots can be certain of their orientation and also the direction in which their plane is moving. Of course, modern aeronautical gyroscopes are more complicated than the one in Figure 9.18, but the principle is the same.

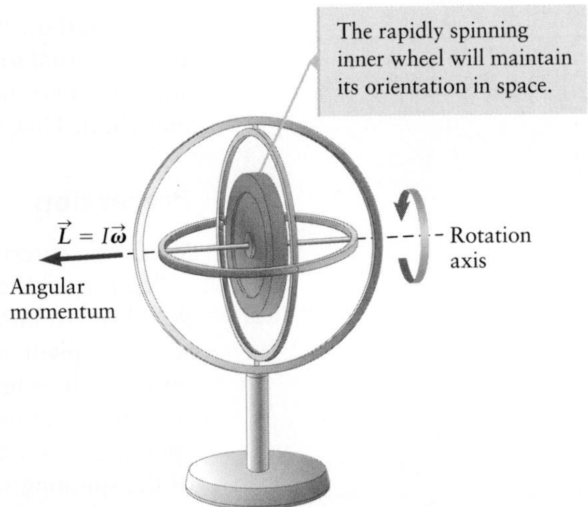

▲ **Figure 9.18** The rotating wheel in this gyroscope is mounted on a system of axles that are free to rotate independently, even when the base of the gyroscope is moved or rotated. Conservation of angular momentum keeps the direction of the inner wheel's axis fixed in space.

The Earth as a Gyroscope

The Earth itself acts as a gyroscope, rotating once each day about an axis that runs between the geographic North and South Poles. This rotational motion is often referred to as "spin," and it produces the "spin angular momentum" of the Earth, which should be distinguished from the orbital angular momentum we discussed in connection with Kepler's second law (Fig. 9.15). The Earth moves in a nearly circular orbit around the Sun, and the Earth's rotational axis is not perpendicular to the orbital plane. Instead, the Earth's spin rotation axis is tilted approximately 23.5° away from the normal to this plane as shown in Figure 9.19. Applying the right-hand rule to the Earth's rotational motion in Figure 9.19 shows that the Earth's spin angular momentum is directed along its rotation axis and points in the direction of the North Pole. There are no external torques on the Earth; its spin angular momentum is therefore conserved, and the rotation axis remains tilted at a fixed angle with respect to the orbital plane. So, for part of each year, the northern half of the Earth tilts toward the Sun, and during the other part of the year, the southern half tilts toward the Sun. The part of the Earth that tilts toward the Sun receives slightly more energy from sunlight, leading to the seasons we observe.

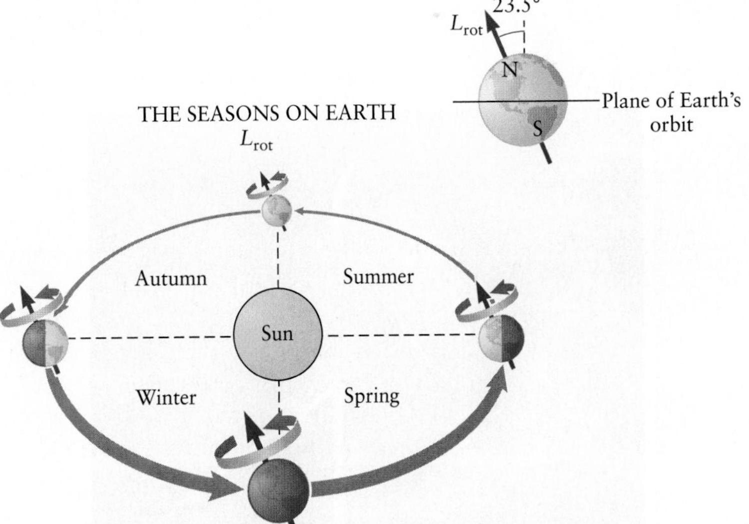

▲ **Figure 9.19** The Earth's spin angular momentum $\vec{L}_{rot}$ lies along a direction parallel to its north–south axis. The vector $\vec{L}_{rot}$ is tilted at an angle of approximately 23.5° away from the normal to the Earth's orbital plane. The seasons indicated in the figure are those in the northern hemisphere.

Angular Momentum and the Stability of a Spinning Wheel

If you have a bicycle wheel that is not mounted on a bicycle frame, you can perform an experiment that demonstrates gyroscopic motion. A nonrotating wheel is very hard to "balance" on its edge and will quickly fall over if left alone. A rolling wheel, however, is much more stable and can travel for quite a long distance before it eventually tips over and falls to one side. The extra stability of a rolling wheel is caused by its angular momentum. According to the right-hand rule, the angular momentum $\vec{L}$ of the wheel is directed along the rotation axis (i.e., it is horizontal). If the external torque on the rolling wheel were exactly zero, then $\vec{L}$ would remain in this direction forever and the wheel would never fall over. In reality, however, there is always some small external torque, mainly from friction, causing the wheel to fall eventually. Even so, its angular momentum makes a rolling wheel much more stable than a nonrotating wheel. This experiment is one you *should* try at home!

Precession

When the external torque on a rotating object is nonzero, it can lead to an effect called ***precession***. A popular device that demonstrates this effect is shown in Figure 9.20A; it is a bicycle wheel mounted on an axle, with one end of the axle resting on a rotatable pivot attached to a vertical pole. As you might expect, this "system" (composed of the wheel plus axle) can be extremely unstable; if the wheel is not rotating, the wheel and its axle will immediately fall over. On the other hand, if the wheel is spinning about the axle, its stability is greatly enhanced. The angular momentum of the spinning wheel prevents it from falling over, even though the force of gravity exerts a large external torque on the system.

We can calculate the torque on the system of the wheel plus axle by assuming the entire gravitational force acts at the center of mass (Chapter 8). Most of the mass of the system in Figure 9.20A is in the wheel, so the system's center of mass is close to the center of the wheel as shown in the figure. To compute the direction of the gravitational torque $\vec{\tau}$ we use another version of the right-hand rule. The procedure is sketched in Figure 9.20B. (1) Start from the pivot point, which in this case is point P on the rotatable pivot at the base of the axle. With P as an origin, place the fingers of your right hand along the rotation axis of the wheel, that is, along the vector $\vec{r}$ in Figure 9.20B. Point your fingers toward the center of the wheel, which is the place where the force $\vec{F}_{grav}$ acts. (2) Curl your fingers in the direction of the force $\vec{F}_{grav}$. Your outstretched right thumb then points in the direction of $\vec{\tau}$ as indicated in Figure 9.20B.

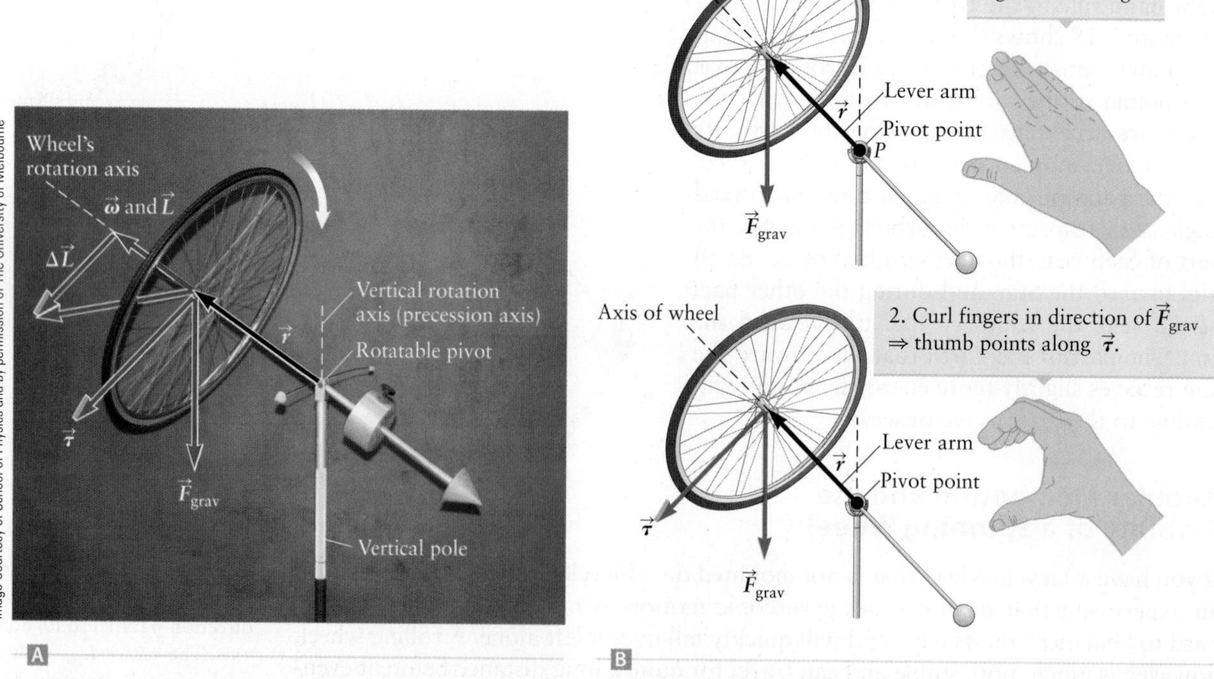

▲ **Figure 9.20** A There is a torque on this bicycle wheel due to the force of gravity. The direction of the torque is given by applying the right-hand rule. This torque makes the axis of the wheel precess. Note that the change in angular momentum $\Delta\vec{L}$ is parallel to $\vec{\tau}$. B Application of the right-hand rule to find the direction of the torque due to gravity.

We must next consider how this torque affects the angular momentum of this system (wheel plus axle). In Chapter 7, we showed that an applied force $\vec{F}$ changes the linear momentum $\vec{p}$ of an object. This leads to the *impulse theorem*, involving force and impulse (Eq. 7.5):

$$\vec{F}\,\Delta t = \Delta\vec{p}$$

An analogous result is true for rotational motion: an applied torque produces a change in the angular momentum. We have

$$\vec{\tau}\,\Delta t = \Delta\vec{L} \tag{9.24}$$

so if a constant torque $\vec{\tau}$ acts for a time Δt, there is a change $\Delta\vec{L}$ in the angular momentum. Applying this result to the wheel plus axle in Figure 9.20, we find that the torque due to the gravitational force is perpendicular to the rotation axis of the wheel, which is also the direction of the angular momentum vector $\vec{L}$. This torque causes the wheel's angular momentum to turn in the direction of $\vec{\tau}$ as indicated in Figure 9.20A. However, as the rotation axis of the wheel turns, the direction of the torque also changes. In fact, $\vec{\tau}$ and hence also $\Delta\vec{L}$ from Equation 9.24 are *always perpendicular* to $\vec{L}$. As a result, the wheel plus axle rotate continuously about a vertical rotation axis (the pole in Fig. 9.20). This movement is called *precession* and is an example of the type of motion that can occur when the angular momentum and the applied torque are not parallel.

9.6 Cats and Other Rotating Objects

In this section, we discuss two interesting examples of rotational motion based on the principle of conservation of angular momentum.

Rotating Cats

A widely known piece of folklore is the claim that a falling cat always lands on its feet. Although we certainly do not want to encourage the unethical treatment of animals, you have probably observed that a falling cat is usually able to rotate its body so as to land feet first, even if it begins its fall with its feet pointing upward. This example of rotational motion is shown in Figure 9.21 and in the opening photo for this chapter. The cat in Figure 9.21A is not rotating when it begins its fall, so its initial angular momentum is zero. Since there are no external torques on the cat while it is falling, its angular momentum must stay constant and hence is always zero during the course of its fall. How can the cat manage to rotate even though it has no angular momentum?

Cats accomplish this feat by changing their shape while they are falling. To understand this process, consider separately the motion of the cat's front section (near its head) and its hindquarters (near its tail). In Figure 9.21B, the cat has pulled its front legs close to its body; just as for the figure skater in Figure 9.11, this reduces the moment of inertia of the cat's front section (I_{front} is small). At the same time, the cat's tail and rear legs are extended away from its body, so the moment of inertia of the rear part of its body is large (I_{rear} is large). The cat then "swivels" at its middle as shown in Figure 9.21C, with the front half of the body rotating in one direction and the hindquarters in the *opposite* direction. The total angular momentum is still zero, but because the front and rear moments of inertia are different, the angular velocities of the front and rear parts of the cat's body are different. Since $I_{\text{rear}} > I_{\text{front}}$, we know from the example with the skater that $\omega_{\text{rear}} < \omega_{\text{front}}$. As a result, the rotation angle for the front half is much larger than for the hindquarters. Figure 9.21D shows the cat's upper body rotating by approximately 180°, whereas the lower body has rotated much less.

The cat then repeats the same process in parts D and E of Figure 9.21, pulling its hind legs and tail close to its body while extending its front legs. Then $I_{\text{rear}} < I_{\text{front}}$, and the cat can now swivel at the middle with $\omega_{\text{rear}} > \omega_{\text{front}}$. This swivel allows the

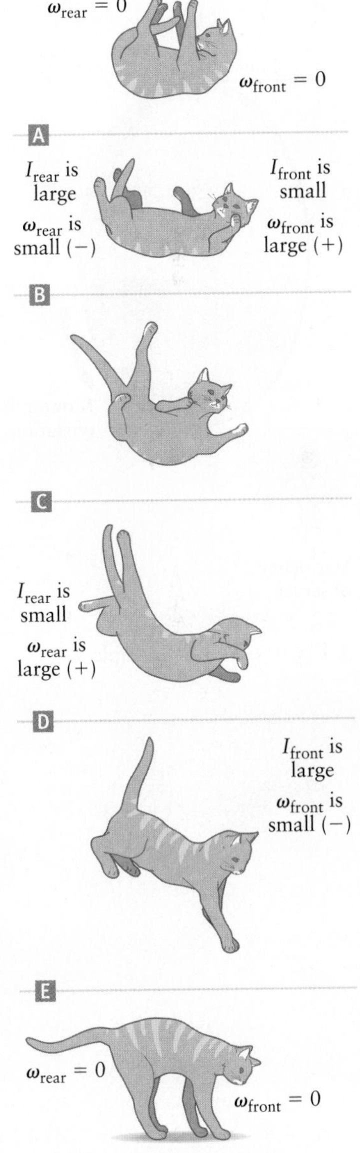

▲ **Figure 9.21** The angular momentum of this cat is conserved as it falls to the ground. The cat is still able to rotate its body so as to land safely, however.

cat to rotate its hindquarters through approximately 180° as in Figure 9.21D while rotating the front part of its body very little. The cat is then able to extend its legs and land safely in Figure 9.21F.

The point of this example is not that cats have a keen understanding of angular momentum, but rather that a system whose total angular momentum is zero can still rotate in some interesting ways. All we needed to explain this cat's motion was the concept that angular momentum is the product of the moment of inertia and the angular velocity.

EXAMPLE 9.7 An Astronaut's Treadmill

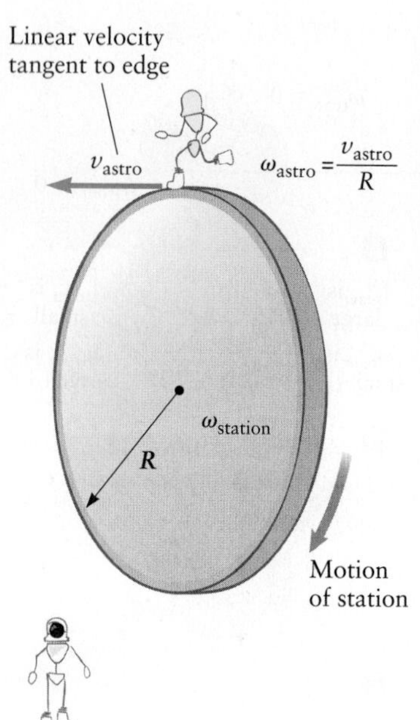

▲ **Figure 9.22** Example 9.7.

A circular space station (Figure 9.22) is initially at rest (i.e., not rotating), when an astronaut begins his daily exercise in which he runs laps around the station with linear speed $v_{astro} = 3.0$ m/s. This is the speed that would be measured by a stationary observer outside the station and is directed *tangentially*, along the edge of the station. Find the resulting angular velocity of the station. The mass of the astronaut is 80 kg, the mass of the station is 1.0×10^5 kg, and (for simplicity) assume the station has the shape of a simple disk with radius $R = 30$ m.

RECOGNIZE THE PRINCIPLE

The astronaut and the space station form a system. There are no external torques, so its angular momentum must be conserved. If part of the system starts to rotate in one direction (i.e., the astronaut runs in one direction), another part of the system must rotate in the opposite direction. We can use the principle of conservation of angular momentum to find how the angular velocities of the astronaut and the station are related.

SKETCH THE PROBLEM

Figure 9.22 shows the problem. The astronaut and station rotate in opposite directions.

IDENTIFY THE RELATIONSHIPS

The station and astronaut are initially at rest relative to the observer in Figure 9.22, so the system's total angular momentum is zero throughout the astronaut's workout. Let $\omega_{station}$ be the angular velocity of the station when the astronaut is running. As he runs, the astronaut's motion is similar to that of the child on (or near) the edge of the merry-go-round in Figure 9.13 (compare also with Fig. 9.10). We can therefore find the astronaut's angular momentum L_{astro} from Equation 9.16, which gives

$$L_{astro} = m_{astro} R v_{astro}$$

The total angular momentum of the astronaut plus the station is then

$$L_{total} = L_{astro} + L_{station} = m_{astro} R v_{astro} + I_{station}\omega_{station} = 0 \quad (1)$$

We approximate the station as a disk, so its moment of inertia (from Table 8.2) is

$$I_{station} = I_{disk} = \tfrac{1}{2} m_{station} R^2$$

SOLVE

Using this result in Equation (1) leads to

$$0 = m_{astro} R v_{astro} + \tfrac{1}{2} m_{station} R^2 \omega_{station}$$

$$\omega_{station} = -\frac{m_{astro} R v_{astro}}{m_{station} R^2/2} = -\frac{2 m_{astro} v_{astro}}{m_{station} R}$$

Inserting the values of the various quantities gives

$$\omega_{station} = -\frac{2(80 \text{ kg})(3.0 \text{ m/s})}{(1.0 \times 10^5 \text{ kg})(30 \text{ m})} = \boxed{-1.6 \times 10^{-4} \text{ rad/s}}$$

▶ ***What does it mean?***

The angular velocity of the astronaut is

$$\omega_{astro} = \frac{v_{astro}}{R} = \frac{3.0 \text{ m/s}}{30 \text{ m}} = +0.10 \text{ rad/s}$$

The angular velocity of the station is much smaller because the station is much more massive than the astronaut and has a larger moment of inertia. Notice that $\omega_{station}$ is negative because the station rotates in the direction opposite that of the astronaut.

CONCEPT CHECK 9.6 Rotational Motion When the Total Angular Momentum Is Zero

Explain what aspects of the motion in Example 9.7 are similar to the rotating cat in Figure 9.21.

Angular Momentum and Motorcycles

As a final example of angular momentum conservation, we consider the motion of a motorcycle stunt rider who jumps through the air as shown in Figure 9.23. Let's examine the combined translational and rotational motions of the system consisting of the motorcycle plus the rider. In both jumps shown in Figure 9.23, the system's center of mass moves as a projectile, following a parabolic trajectory as we established in Chapter 4. The rotational motions are different in the two cases, however. In Figure 9.23A, the motorcycle frame maintains a fixed angular orientation throughout the jump, and the motorcycle lands on its rear wheel. Throughout the jump, the angular velocity of the system (the motorcycle and rider) is zero, so the angular displacement of the motorcycle frame is also zero. This example shows conservation of angular momentum: the angular momentum of the entire system, and also of the motorcycle frame, is constant and zero throughout the jump.

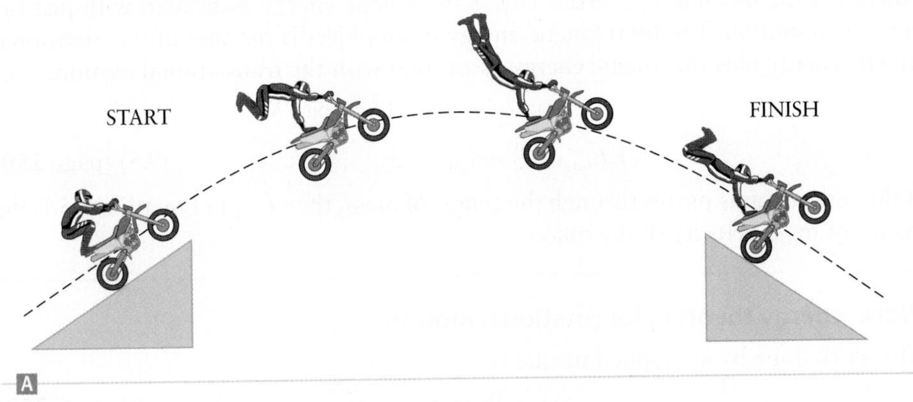

◀ **Figure 9.23** A In this jump, the orientation of the motorcycle frame changes very little during the course of the jump. B Here the rider reduced the angular velocity of the rear wheel after leaving the ground (by adjusting the throttle). To conserve angular momentum about the axis that runs along the rear axle, the frame then began to rotate counterclockwise.

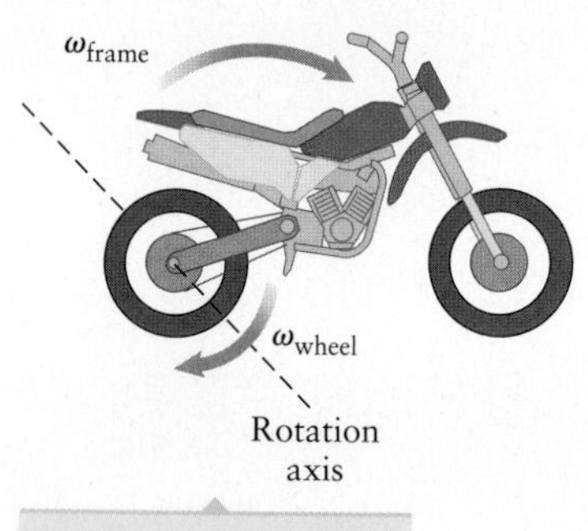

▲ **Figure 9.24** The total angular momentum of a motorcycle is conserved during a jump. The total angular momentum has contributions from the wheels and the frame.

In the jump shown in Figure 9.23B, the motorcycle again begins with zero angular velocity, but then *during the jump* the motorcycle frame rotates. This rotation allows the motorcycle to land simultaneously on both wheels, with the motorcycle frame "sloping down," leading to a gentler (and presumably safer) landing since the landing is usually onto a downward-sloping ramp. How did the rider cause the angular velocity of the motorcycle frame to change from its value of zero at the start of the jump to a nonzero value during the jump?

The answer to this question is well known to stunt riders: during the jump, they adjust the throttle so as to change the angular velocity of the rear wheel as it rotates about its axle, thus changing the wheel's angular momentum. There are no external torques on the system, so the total angular momentum of the system must remain constant. The total angular momentum is equal to the sum of the angular momentum of the wheels plus the angular momentum of the frame plus the angular momentum of the rider. In this case, we can take the rotation axis to be the rear axle, which is perpendicular to the plane of Figure 9.24. If the angular momentum of the rear wheel is decreased (by adjusting the throttle), the remaining angular momentum must increase. This causes the frame to rotate clockwise about the rear axle as in Figures 9.23B and 9.24, and is another nice application of physics!

Summary | CHAPTER 9

Key Concepts and Principles

Rotational kinetic energy

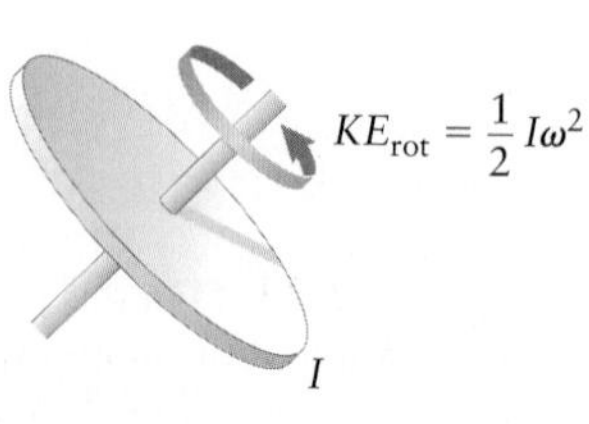

The ***rotational kinetic energy*** of an object is

$$KE_{rot} = \tfrac{1}{2}I\omega^2 \qquad (9.4)\ (\text{page } 250)$$

where I is the moment of inertia. This is the kinetic energy associated with just the rotational motion. The total kinetic energy of an object is the sum of this rotational kinetic energy plus the kinetic energy associated with the translational motion:

$$KE_{total} = \underbrace{\tfrac{1}{2}mv_{CM}^2}_{\text{translational } KE} + \underbrace{\tfrac{1}{2}I\omega^2}_{\text{rotational } KE} \qquad (9.5)\ (\text{page } 250)$$

If the rotation axis passes through the center of mass, then v_{CM} in Equation 9.5 is the center of mass velocity of the object.

Work–energy theorem for rotational motion

The work done by an applied torque is

$$W = \tau\theta \qquad (9.8)\ (\text{page } 252)$$

where θ is the angular displacement. The work–energy theorem applied to rotational motion is

$$W = \tau\theta = \Delta KE = \tfrac{1}{2}I\omega_f^2 - \tfrac{1}{2}I\omega_i^2 \qquad (9.9)\ (\text{page } 253)$$

Angular momentum

A rotating object has an ***angular momentum*** L. If the direction of the rotation axis is fixed, then L is given by

$$L = I\omega \qquad (9.15)\ (\text{page } 258)$$

If the external torque on a system is zero, the total angular momentum of the system is conserved.

Vector nature of angular momentum

The direction of the angular velocity $\vec{\omega}$ is given by the ***right-hand rule***, and the angular momentum vector $\vec{L}$ is

$$\vec{L} = I\vec{\omega} \qquad \text{(9.23) (page 265)}$$

When the fingers of your right hand curl in the direction of motion, your outstretched right thumb points along $\vec{\omega}$ and $\vec{L}$.

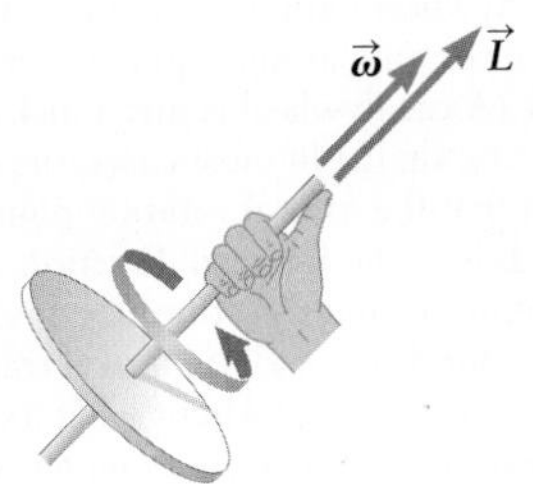

Applications

Conservation of energy

Conservation of energy principles can be applied to rotational motion. When all the forces that do work on an object are conservative forces, the total mechanical energy is conserved. When the potential energy is due to gravity, as for an object that rolls up or down an incline, conservation of mechanical energy leads to

$$KE_i + PE_i = KE_f + PE_f$$

$$\tfrac{1}{2}mv_i^2 + \tfrac{1}{2}I\omega_i^2 + mgh_i = \tfrac{1}{2}mv_f^2 + \tfrac{1}{2}I\omega_f^2 + mgh_f \qquad \text{(9.11) (page 254)}$$

The kinetic energy given here is the total kinetic energy, including the rotational contribution.

Conservation of angular momentum

Angular momentum and its conservation play important roles in the motion of many objects, including figure skaters, cats, motorcycles, and gyroscopes.

Questions

SSM = answer in *Student Companion & Problem-Solving Guide*

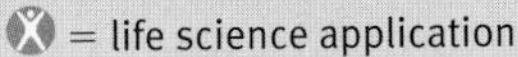

1. A solid wood ball is rotating about an axis that passes through its center. If its angular speed is doubled, (a) by what factor does its rotational kinetic energy change? (b) By what factor does its angular momentum change?

2. A pitcher throws a baseball (mass $m = 0.14$ kg, $r = 7.4$ cm) with linear speed v and angular velocity ω. Which of the following statements is correct? Explain your answer. (a) The total kinetic energy of the baseball can be calculated knowing only m and v. (b) The total kinetic energy of the baseball can be calculated knowing only m, r, and ω. (c) To calculate the total kinetic energy of the baseball, we need to know only v and ω.

3. Two identical spheres start from rest at the top of a long ramp. One ball rolls down the ramp without slipping, while the other slides down without rolling. Which one reaches the bottom first? Explain your answer.

4. Two objects having different shapes start from rest from the top of a long ramp. Both then roll without slipping down the ramp, and one reaches the bottom before the other. Which object has the smaller moment of inertia? Explain your answer.

5. SSM Consider the translational and rotational kinetic energies of a disc that rolls without slipping. Show that the ratio of these energies is independent of the size (the radius) of the wheel.

6. You are given two objects and a ramp. Explain how you could measure the ratio of the moments of inertia of the two objects. Assume the two objects are both round (so that both roll), have the same mass and radius, and are solid (no holes in the center, etc.). Consider only objects that meet these criteria and with the shapes in Table 8.2. *Hint*: Both objects can roll without slipping down the ramp, and you have the equipment needed to measure the speeds of the objects as they roll.

7. Tightrope walkers often carry a long pole (Fig. Q9.7). Explain why.

8. Is it possible for the total force on a system to be nonzero but the torque to be zero? Give an example.

9. You are given the job of designing a new high-tech skateboard. For it to travel as fast as possible, you want to minimize the kinetic energy in its wheels. All else being equal (i.e., the mass and radius of the wheels), is it better to make the wheels in the shape of a disk, a sphere, or a conventional wheel with spokes?

Figure Q9.7

10. Explain why the rim brakes on a bicycle wheel are located at the outer edge of the wheel instead of near the hub. *Hint*: Consider the work–energy theorem (Eq. 9.9) and how the magnitude of the

torque produced by a given force varies when the force is applied at different distances from the axis of rotation.

11. The flywheel within a car engine is a form of gyroscope, and when rotating at high rpm, it carries significant angular momentum. (A car flywheel is just a disk attached to the engine's main rotating shaft.) In most cases, the engine of a race car is mounted such that the axis of rotation points in the forward direction and $\vec{L}$ is in the forward direction. Because angular momentum (magnitude and direction) is conserved, what happens to the car as it rounds the end of an oval track in a counterclockwise direction (turning left)? About what axis will the car tend to twist? If instead the engine were mounted such that the axis of rotation was parallel to the axles of the car, along which axis would the car then tend to twist? *Hint*: Consider the gyroscope and the rotation of precession. In auto racing, the cars typically go around the oval track in a counterclockwise direction as viewed from above. Would race cars be engineered differently if races were conducted in a clockwise direction?

12. Discuss how and if the moment of inertia of the Earth associated with its orbital motion around the Sun changes as the Earth moves through its orbit. How are these changes connected to changes in the Earth's speed?

13. Many kilograms of asteroids land on the Earth each year. Discuss how they affect the length of a day.

14. Consider a child who is jumping on a trampoline as shown in Figure Q9.14. Explain how the child can have zero angular momentum, but still rotate to either the right or the left while she is in the air.

15. The cat in Figure 9.21 undergoes free fall due to the force of gravity. Explain why we did not consider gravity when we discussed how the cat in Figure 9.21 managed to land on its feet.

Photodisc/Getty

Figure Q9.14

16. The barrel of a rifle contains spiral grooves that impart some spin to a bullet as it leaves the barrel. Explain why this spin improves the rifle's accuracy.

17. SSM (a) Why do most small helicopters have a rotor on their tail (Fig. Q9.17)? (b) Many large helicopters have two large rotors. Why do these helicopters not need a tail rotor? Do you think the two large propellers rotate in the same direction? Explain.

U.S. Navy/Mass Communication Specialist Seaman Jon Dasbach

Figure Q9.17

18. Why is it possible for a gymnast to perform more rotations during an airborne somersault when she is in a tuck position than when her body is straight (called the layout position)?

19. Some early spacecrafts used tape recorders to store data. These recorders have spools of magnetic tape that are rotated in spindles to move the tape past "heads" that record or read data. Explain why a tape recorder can affect a spacecraft's motion.

20. When a spinning figure skater pulls her arms close to her body (Fig. 9.11), she spins faster. If she then pushes her arms away from her body, her angular velocity decreases. Her rotational kinetic energy also decreases. Explain why her kinetic energy decreases. What force or torque does the corresponding work?

Problems available in ENHANCED WebAssign

List of Enhanced Problems

Problem number	Solution in *Student Companion & Problem-Solving Guide*	Reasoning & Relationships Problem	Reasoning Tutorial in ENHANCED WebAssign
9.4		✓	
9.6		✓	✓
9.7		✓	✓
9.8		✓	
9.9		✓	
9.11		✓	✓
9.12	✓		
9.30		✓	
9.31	✓		
9.32		✓	
9.34		✓	✓
9.37		✓	
9.40		✓	
9.43	✓	✓	
9.44		✓	
9.45		✓	✓
9.48	✓		
9.50	✓		
9.51		✓	✓
9.56	✓		
9.59	✓		
9.61		✓	✓
9.64	✓		
9.72		✓	
9.73		✓	
9.76		✓	
9.81		✓	✓

◀ *In this chapter, we use Newton's laws to describe and predict the motion of fluids, such as the water in Victoria Falls or the air above it. (The Africa Image Library/Alamy)*

Fluids

In previous chapters, we learned how to apply Newton's laws to the motion of solid objects, first dealing with point particles and then with extended objects that may rotate. In this chapter, we turn our attention to the behavior of ***fluids***. A fluid may be either a liquid or a gas. We say that a fluid *flows* from one place to another, and the shape of a fluid varies according to its container. To describe the mechanics of fluids, we need several new quantities, but our analysis is still based on Newton's laws of motion. Understanding the mechanics of fluids is certainly an important topic, since most of the Earth's surface is covered with a liquid (water), and the atmosphere is a gas (air). Moreover, the motion of fluids, including the flow of blood in the body and the movement of water within plants, is essential to most living things.

OUTLINE

10.1 Pressure and Density

We want to apply Newton's second law to analyze and predict the behavior of a fluid. Recall that Newton's second law is

$$\sum \vec{F} = m\vec{a}$$

When we apply this relation to describe the motion of a particle, m is the mass of the particle, $\vec{a}$ is its acceleration, and $\sum \vec{F}$ is the total force acting on the particle. To treat the motion of a fluid, what should we use for the mass m? If we want to describe the water in a river, should m be the mass of the entire river? The idea of using a single acceleration variable $\vec{a}$ to describe the entire river does not make sense either since different parts of the river might be moving in different directions.

We must therefore rewrite Newton's second law in a form designed with fluids in mind. We need to define some new quantities to use in place of our familiar variables force and mass. The first of these new quantities is ***pressure***. The pressure in a fluid is connected with the force that the fluid exerts on a particular surface. This concept is illustrated in Figure 10.1A, which shows a submarine with a window. Seawater exerts a force on the window. This force is related to the pressure P in the seawater by

Definition of pressure

$$P = \frac{F}{A} \tag{10.1}$$

where F is the magnitude of the force exerted by the seawater on the window and A is the area of the window. This force is *perpendicular* to the window's surface. The units of pressure correspond to force divided by area, which in SI units is newtons/square meter (N/m^2). This unit is called the ***pascal*** (abbreviated Pa). In the U.S. customary system of units, pressure is measured in pounds per square inch ($lb/in.^2$). You may have encountered that unit when you last checked the pressure in an automobile or bicycle tire.

Although force is a vector quantity, pressure is not. When an object is submerged in a static fluid, the fluid exerts a force on the object that is always directed perpendicular to the object's surface. As a result, there is never any doubt about the direction of the force in Equation 10.1. This point is illustrated in Figure 10.1B, which shows a small, sealed box filled with a liquid at pressure P. The fluid within the box produces an outward force on *all* the sides of the box. If the box is small, the pressure will be approximately constant throughout the box.[1] For any particular wall of the box (area A), the total force on the wall is equal to $F = PA$.

The pressure in the Earth's atmosphere near sea level has the approximate value

Value of atmospheric pressure

$$P_{atm} = 1.01 \times 10^5 \text{ Pa} \tag{10.2}$$

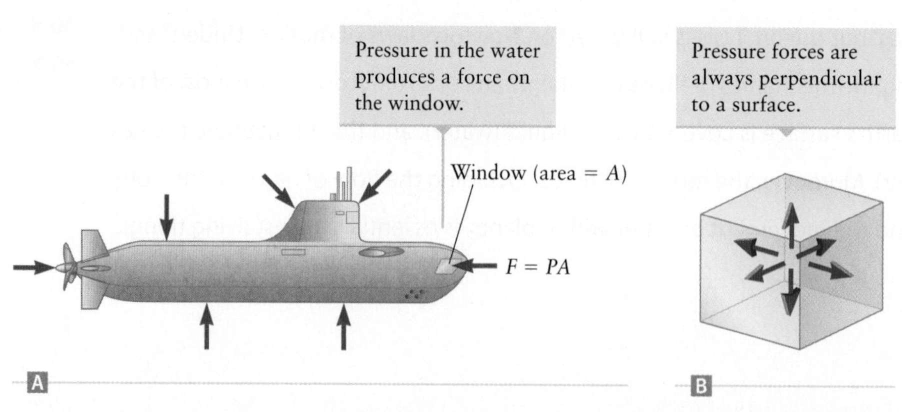
Pressure in the water produces a force on the window.
Window (area = A)
$F = PA$
Pressure forces are always perpendicular to a surface.
A
B

▶ **Figure 10.1** A The pressure P in the surrounding water produces a force on a submarine's shell. This force is everywhere directed perpendicular to the surface of the submarine. The force on a submarine window is $F = PA$, where A is the window's area. B When a container is filled with a fluid at a pressure P, the fluid produces an outward force that is perpendicular to the walls of the container.

[1]The pressure in a fluid actually varies from place to place within the container according to depth as we'll see in Section 10.2. For a small box, however, this variation will be small and P will be nearly constant.

This value of P_{atm} is approximate because the pressure in the atmosphere varies in response to changes in the weather and other factors (such as altitude).

Atmospheric Pressure

Atmospheric pressure (Eq. 10.2) corresponds to surprisingly large forces for areas you encounter every day. For example, a sheet of paper (like a page of this book) has an area of about 500 cm² = 0.05 m². Rearranging Equation 10.1, the force on one side of this sheet of paper is

$$F = PA = P_{atm}A = (1.01 \times 10^5 \text{ Pa})(0.05 \text{ m}^2)$$

$$F = 5 \times 10^3 \left(\frac{\text{N}}{\text{m}^2}\right)\text{m}^2 = 5 \times 10^3 \text{ N}$$

which is a very substantial force; it is equal to the combined weight of several large people! You might at first be skeptical of this result and ask why a sheet of paper does not fold or bend under such a large force. The answer is that the air exerts a force on *both sides* of the sheet of paper as sketched in Figure 10.2. The pressure in the air on both sides is equal to P_{atm}. These forces are in opposite directions, so the total force on the paper is zero. This cancellation occurs in many situations, including the force due to air pressure on your body.

$F = P_{atm}A$

$F = P_{atm}A$

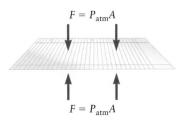

▲ **Figure 10.2** The atmosphere produces a downward force of magnitude $P_{atm}A$ on this sheet of paper, where A is the area of the sheet. The air underneath the sheet produces a force of the same magnitude directed upward.

Vacuum and the Magdeburg Experiment

The balance of forces associated with air pressure can be altered by removing air from inside an enclosed volume with a device called a vacuum pump. This was first demonstrated 400 years ago by Otto von Guericke. He constructed two metal hemispheres as shown in Figure 10.3A; when fitted together, they form a Magdeburg sphere, so named in honor of the town where von Guericke lived. One hemisphere contained a small tube connected to a valve through which air was removed from inside the two hemispheres by a pump that von Guericke invented. The region inside where the pressure was zero, or very small, is known as a ***vacuum***.

To understand von Guericke's experiment, let's analyze the somewhat simpler case of two Magdeburg plates sketched in Figure 10.3B. The air outside produces an inward force equal to $P_{atm}A$, where A is the plate's area. Because there is no air on the inside, $P_{inside} \approx 0$ and there is no outward force from air pressure. There will be a similar inward force $P_{atm}A$ on the other plate, and the net result is a substantial force holding the plates together. The force required to pull the plates apart is calculated in Example 10.1.

ICONISMUS XI

A

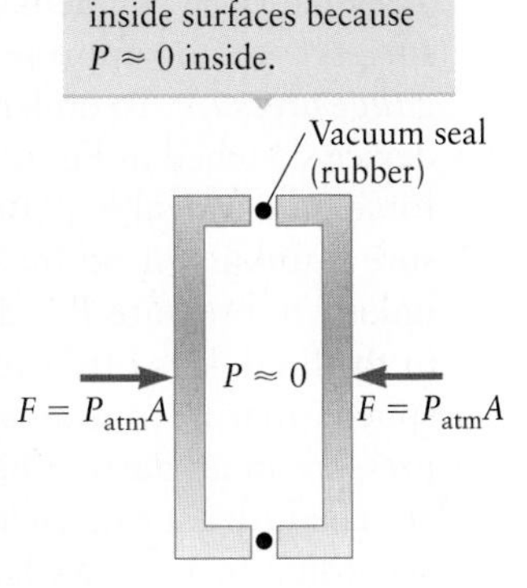

There is no force on the inside surfaces because $P \approx 0$ inside.
Vacuum seal (rubber)
$P \approx 0$

B

◀ **Figure 10.3** A Otto von Guericke demonstrated the force associated with atmospheric pressure using this apparatus. He made two tight-fitting hemispheres and then removed the air from the inside using a vacuum pump, which he invented. Von Guericke employed hemispheres about 1 m in diameter and used horses to (attempt to) pull them apart! B A similar von Guericke apparatus, using two flat plates in place of the hemispheres.

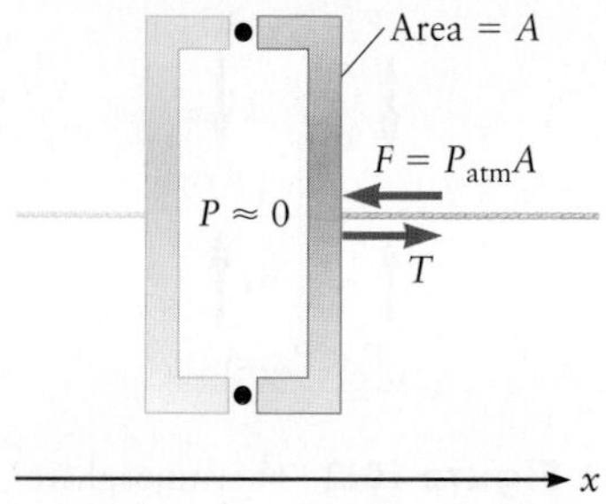

▲ **Figure 10.4** Example 10.1.

EXAMPLE 10.1 Force in the Magdeburg Demonstration

Suppose the Magdeburg plates in Figure 10.3B have an area of $A = 1.0\ \text{m}^2$. What force is required to pull the plates apart?

RECOGNIZE THE PRINCIPLE

This problem involves an application of the relation between pressure, area, and force. The force due to air pressure on one plate is $F = P_{atm}A$ (Eq. 10.1).

SKETCH THE PROBLEM

Figure 10.4 shows the horizontal forces acting on one of the plates.

IDENTIFY THE RELATIONSHIPS

A cable with tension T is attached to the Magdeburg plate on the right in Figure 10.4. There are two forces: a force $P_{atm}A$ to the left due to atmospheric pressure and a force T to the right. (We can ignore the weight of the plate because it will be much smaller than the other two forces.) If the plates are just barely being pulled apart, we can take the acceleration of the plate on the right to be zero.

SOLVE

Applying Newton's second law gives

$$\sum F = +T - P_{atm}A = ma = 0$$

where these forces are along the horizontal. This leads to

$$T = P_{atm}A = (1.01 \times 10^5\ \text{Pa})(1.0\ \text{m}^2) = \boxed{1.0 \times 10^5\ \text{N}}$$

▶ ***What does it mean?***

This force is the combined weight of about 10 cars! It's no wonder the painting in Figure 10.3A shows a sizable team of horses pulling on the Magdeburg plates.

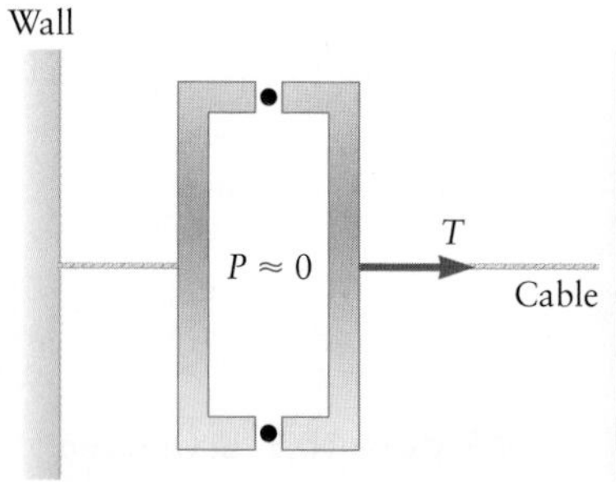

▲ **Figure 10.5** Concept Check 10.1.

CONCEPT CHECK 10.1 Magdeburg Plates and Newton's Third Law

When we calculated the force required to pull apart the Magdeburg plates in Example 10.1, we actually calculated the force exerted by the atmosphere on *one* of the plates and found that $F = 1.0 \times 10^5$ N. Suppose the other plate is attached to a rigid wall as shown in Figure 10.5. Now what tension T is required in the cable on the right to pull apart the plates?

(a) 1.0×10^5 N (b) 2.0×10^5 N (c) 0

Gauge Pressure versus Absolute Pressure

The quantity P we defined in Equation 10.1 is often referred to as the ***absolute*** pressure. A common approach to measuring pressure leads to a related quantity called the ***gauge pressure***. To understand the difference, consider the simple pressure-measuring device sketched in Figure 10.6. This device determines the pressure by measuring the force on a movable plate. One side of this plate is exposed to the fluid whose pressure is unknown, so the force on this side of the plate is equal to the product of the unknown pressure P_{abs} and the plate's area. Notice that P_{abs} is the *absolute* pressure in the fluid. It is often convenient to expose the other side of the plate to the atmosphere, that is, to atmospheric pressure. For example, a gauge used to measure the pressure in an automobile tire is constructed in this way. The movable plate is also attached to a spring (with spring constant k) exerting a force $-kx_{spring}$ on the plate, according to Hooke's law (Chapter 6). Here x_{spring} is the displacement of the plate from a reference position where the spring is unstretched and uncompressed. The

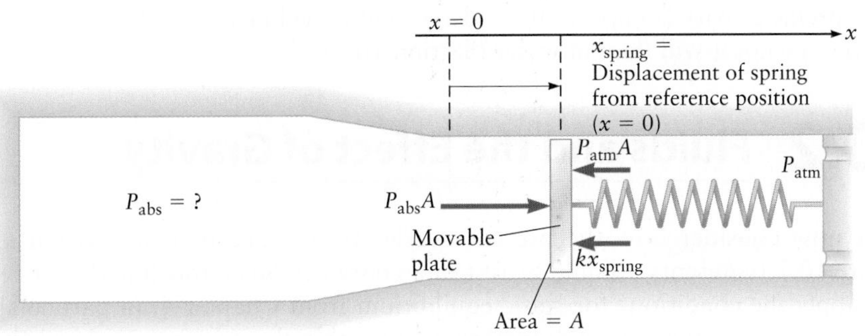

◀ **Figure 10.6** Simple design for a pressure gauge. The unknown pressure P_{abs} on the left leads to a force $F = P_{abs}A$ on a movable plate inside the gauge. The other side of the plate is attached to a spring and is exposed to the atmosphere. The deflection of the spring (x) is proportional to the difference in the pressures $P_{abs} - P_{atm}$. This difference is the gauge pressure.

total force on the plate inside the gauge is due to the pressures on both sides plus the force exerted by the spring. If we take the $+x$ direction to be to the right in the figure, we have

$$\sum F = +P_{abs}A - P_{atm}A - kx_{spring}$$

If the gauge plate is in equilibrium, the total force must be zero and we get

$$P_{abs}A - P_{atm}A - kx_{spring} = 0$$

$$x_{spring} = (P_{abs} - P_{atm})\left(\frac{A}{k}\right)$$

The deflection of the gauge x_{spring} is thus proportional to the *difference* between the pressure one wants to measure (P_{abs}) and atmospheric pressure. This difference is called the gauge pressure, $P_{gauge} = P_{abs} - P_{atm}$. Hence, the absolute pressure is related to the gauge pressure by

Gauge pressure

$$P_{abs} = P_{gauge} + P_{atm}$$

Many common pressure gauges work this way and thus measure the gauge pressure. For the remainder of our work in this chapter, the term *pressure* means the absolute pressure unless we specifically note otherwise.

CONCEPT CHECK 10.2 Pressure in an Automobile Tire

The pressure in an automobile tire is usually quoted (in the United States) in units of pounds per square inch (lb/in.^{2}), and in a typical tire the gauge pressure is P_{gauge} = 25 lb/in.^{2}. What is the value of the gauge pressure in pascals?

(a) 2.7×10^3 Pa (b) 1.0×10^5 Pa (c) 1.7×10^5 Pa (d) 110 Pa

Density

Another quantity we need to describe fluids is the ***density***, denoted by ρ (lowercase Greek rho). The density of a substance is equal to its mass per unit volume,

$$\rho = \frac{M}{V} \tag{10.3}$$

where M is the mass of a sample of the substance and V is the sample volume. This definition of ρ assumes the density is constant throughout the volume V.

The density of a substance can vary according to temperature and may also depend on pressure. However, for most liquids—including water and oil—density is approximately independent of pressure. Such fluids are called ***incompressible***. The densities of some common substances are listed in Table 10.1.

Although most liquids are incompressible, that is not the case for gases. The density of a gas depends strongly on pressure; that is, gases are ***compressible***, and we'll show in Section 10.3 how that affects their behavior. The densities of some common gases at atmospheric pressure are listed in Table 10.1.

The SI unit of density is the kilogram per cubic meter, or kg/m^3 (= mass/volume). Another way to express the density of a substance is in terms of its ***specific gravity***, which is the ratio of the density of the substance to the density of water. We'll see

TABLE 10.1 Densities of Some Common Solids, Liquids, and Gases[a]

Substance	Density (kg/m^3)
SOLIDS	
Ice (at 0°C)	917
Aluminum	2,700
Lead	11,300
Platinum	21,500
Gold	19,300
Tungsten	19,300
Steel	7,800
Concrete	2,000
Bone	1,500–2,000
Polystyrene	100
Glass	2,500
Wood (spruce)	400
Wood (balsa wood)	120
Wood (oak)	750
LIQUIDS	
Water (0°C)	999.8
Water (4°C)	1,000.0
Water (20°C)	999.9
Seawater	1,025
Ethanol	790
Mercury	13,600
Oil	700–900
GASES	
Air (0°C)	1.29
Air (20°C)	1.20
Air (100°C)	0.84
Helium (0°C)	0.18
Hydrogen (0°C)	0.090
Carbon dioxide (25°C)	1.80

[a]Unless otherwise noted, all values are for room temperature and pressure.

that specific gravity is important for determining the buoyancy of an object, that is, whether or not it will float in water (Section 10.4).

10.2 Fluids and the Effect of Gravity

Let's now consider how the pressure varies from place to place within a fluid. Figure 10.7 represents a liquid at rest in a container. Since this liquid is at rest, we can apply the conditions for static equilibrium from Chapter 4. In particular, consider the forces acting on a region of the fluid within the imaginary cubical "box" in Figure 10.7A. The system is at rest, so we can treat the fluid cube as a "simple" mass. For this cubical box, each wall (or face) has edge length h and area $A = h^2$.

We now apply our condition for translational equilibrium to the mass of fluid within the box. Because this mass is at rest, the total force on it must be zero. There are contributions to the total force from the pressure outside the box. These forces are all directed inward on the walls of the box as indicated in Figure 10.7. The forces along the horizontal directions are equal in magnitude and opposite in direction, so they cancel. However, the forces on the top and bottom of the box do not cancel. Although these forces are in opposite directions, they cannot be equal in magnitude because the pressure force at the bottom must support both the pressure force at the top and the weight of the fluid in the box. Hence, the pressure at the bottom must be greater than the pressure at the top. In general, *pressure always increases as one goes deeper into a fluid.*

To calculate how pressure depends on depth, note that the force on the bottom of the box is upward with $F_{bot} = P_{bot}A$, where P_{bot} is the pressure at the bottom face of the box. The force on the top of the box is $F_{top} = -P_{top}A$, where the negative sign indicates that this force is downward. There is also a gravitational force on the fluid mass due to its weight. This force is along the vertical with $F_{grav} = -mg$, where m is the fluid mass inside the box. The total force along y is then

$$\sum F_y = P_{bot}A - P_{top}A - mg = 0 \tag{10.4}$$

This total force is zero because the system is in translational equilibrium. From the definition of density (Eq. 10.3), the mass of the fluid in the box is $m = \rho V$, and the volume of the box is equal to the area A times the box height h, so $V = Ah = h^3$ and

$$m = \rho V = \rho A h = \rho h^3$$

Inserting this into Equation 10.4 leads to

$$P_{bot}A - P_{top}A - mg = P_{bot}h^2 - P_{top}h^2 - \rho h^3 g = 0$$

$$P_{bot} - P_{top} = \rho g h$$

We can also write this result as

$$P = P_0 + \rho g h \tag{10.5}$$

Here P_0 is the pressure at a "reference point" in the fluid and P is the pressure at a depth h relative to the reference point as sketched in Figure 10.8. We have derived Equation 10.5 with the use of a cubical box, but this result is not limited to fluid "boxes." In words, pressure increases as one goes to greater depths because a particular section of fluid must support the weight of the fluid above it.

When a fluid has a surface open to the atmosphere, it is usually convenient to take the reference point at this surface. In this case, $P_0 = P_{atm}$ as shown in Figure 10.8B. No matter where the reference point is chosen (Fig. 10.8A or B), a positive value of h in Equation 10.5 corresponds to a location that is *deeper* than the reference location.

Equation 10.5 is a general relation that tells how the pressure varies with depth in an *incompressible* fluid since our derivation assumed the density of the fluid is constant; that is, ρ does not vary with depth in the fluid. Equation 10.5 can therefore be applied only to liquids, not to gases. We'll see how to deal with compressible fluids (gases) after we consider a few examples involving liquids.

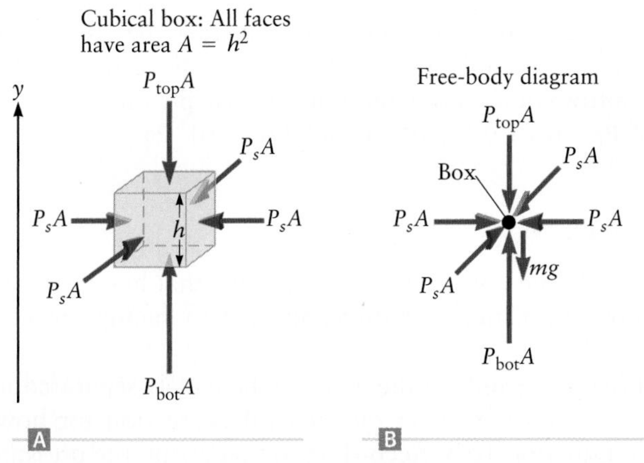

◀ Figure 10.7 A An imaginary cubical box enclosing a region of fluid. Pressure from the fluid pushes on each face of the box. The pressure forces on the sides of the box (magnitude P_sA) cancel. B Free-body diagram for the box.

EXAMPLE 10.2 Pressure in the Marianas Trench

The Marianas Trench, a region at the bottom of the Pacific Ocean, is believed to be the deepest spot on the ocean floor. In 1960, a U.S. Navy submersible vehicle went to a spot in this trench that is 10,700 m below sea level. Suppose this submarine had a tiny window of area 1.0 cm² so that sailors could enjoy the view. Find the force on this window due to the water pressure.

RECOGNIZE THE PRINCIPLE

The force on the window depends on the pressure through $F = PA$. We thus need to find the pressure in the Marianas Trench, which we can do using the dependence of pressure on depth (Eq. 10.5).

SKETCH THE PROBLEM

Figure 10.8B describes the problem, with the submarine at point B.

IDENTIFY THE RELATIONSHIPS

We take the ocean surface as the reference point for calculating the pressure using Equation 10.5. Therefore, $P_0 = P_{atm}$, the depth is $h = +10{,}700$ m, and the density of seawater is given in Table 10.1.

SOLVE

From Equation 10.5 and using $P_0 = P_{atm}$, we find the pressure to be

$$P = P_{atm} + \rho g h = (1.01 \times 10^5 \text{ Pa}) + (1025 \text{ kg/m}^3)(9.8 \text{ m/s}^2)(10{,}700 \text{ m})$$
$$P = 1.1 \times 10^8 \text{ Pa}$$

which is about 1000 times atmospheric pressure. The force on the submarine's window is

$$F = PA = (1.1 \times 10^8 \text{ Pa})(1.0 \text{ cm}^2) = (1.1 \times 10^8 \text{ N/m}^2)(1.0 \times 10^{-4} \text{ m}^2)$$
$$F = \boxed{1.1 \times 10^4 \text{ N}}$$

▶ *What does it mean?*
This force is approximately the weight of a small automobile. The result explains why submarines do not have large windows.

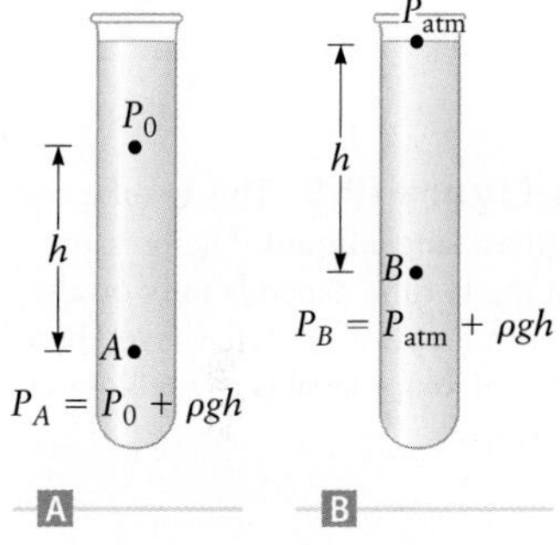

▲ Figure 10.8 Illustration of the relation $P = P_0 + \rho g h$. In A, the reference point is at a particular point inside the liquid, whereas in B, the reference point is at the surface in contact with the atmosphere, so $P_0 = P_{atm}$. In both cases, h is the depth relative to the reference point.

CONCEPT CHECK 10.3 Pressure at the Bottom of a Swimming Pool

The pressure increases as one goes deeper within a liquid, and in Example 10.2 we found that the pressure increase can be quite large in deep locations in the oceans. This pressure change can also be quite noticeable in a swimming pool.

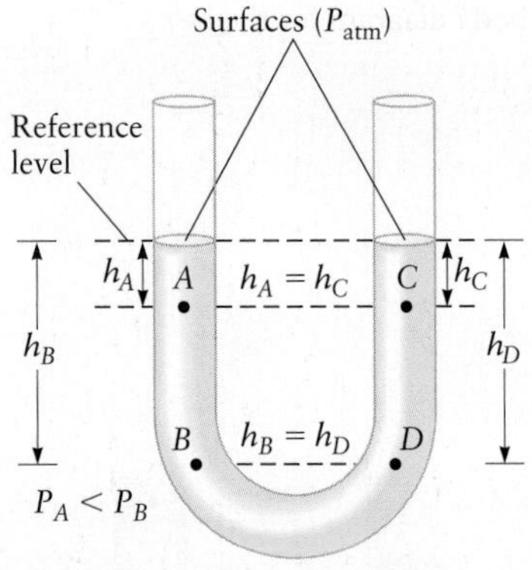

▲ **Figure 10.9** This U-tube contains a single liquid. The pressure in the U-tube depends only on the depth below a reference level; here the reference level is at the liquid's surface.

When you dive to the bottom of a pool, you can sense the pressure increase with your ears. For a pool with a depth of 3.0 m, what is the difference between the pressure at the bottom of the pool and atmospheric pressure?
(a) 2.9×10^4 Pa (b) 3.0×10^5 Pa (c) 3.0×10^3 Pa

Pressure in a U-Tube

Figure 10.9 shows a liquid contained in a glass tube that has a "U" shape; this container is called, not surprisingly, a U-tube, and is very handy for a variety of fluid experiments.

In a U-tube, different regions of the fluid can be widely separated in space. However, this separation does not affect our general expression for how the pressure varies with depth, Equation 10.5. According to that result, the pressure in the liquid in Figure 10.9 depends only on the depth h below a reference level. The pressures are the *same* at two spots on different sides of the U-tube that are at the *same* depth below the surface; hence, $P_A = P_C$ and $P_B = P_D$ in Figure 10.9. This result is true as long as the U-tube is filled with only one type of liquid. Example 10.3 shows how to analyze a U-tube containing two different liquids.

EXAMPLE 10.3 Two Liquids in a U-Tube

The U-tube in Figure 10.10 has both ends open to the atmosphere. The tube is filled with two liquids of different densities. These liquids are immiscible; that is, they do not mix. If liquid 1 has a higher density than liquid 2, the level of liquid 2 is higher than the level of liquid 1 by some distance d. Suppose liquid 1 is water ($\rho_1 = 1000$ kg/m^3); liquid 2 is a type of oil with $\rho_2 = 700$ kg/m^3 and is filled to a height $y_R = 0.10$ m above the oil-water boundary (Fig. 10.10). Find d.

RECOGNIZE THE PRINCIPLE

For the U-tube containing a single type of fluid in Figure 10.9, pressure depends only on depth, so at any particular depth below the surface, the pressure on the two sides of the tube is the same. This result is true only when the U-tube is filled with a single liquid so that the density is constant throughout, a key assumption in the derivation of Equation 10.5. In Figure 10.10, the density changes as we pass from one liquid to the other, so we *cannot* use Equation 10.5 to compare the pressures in the different liquids. We can, however, use Equation 10.5 to compare the pressures at different points within the same connected piece of liquid. Consider the pressure at point B in Figure 10.10. This point is inside liquid 1 and is just below the boundary that separates the two liquids. Because we consider point B to be at the same depth as point A (also located in liquid 1, but on the other side of the U-tube), the pressures at points A and B are the same.

We next consider point C; this point is in liquid 2, just above the boundary between the liquids. Because points B and C are extremely close together, the pressures at these two locations must be the same. You can see how that is true by imagining a thin (massless) plastic sheet stretched along the boundary between liquids 1 and 2. Because the system is at rest, this sheet is in translational equilibrium; hence, the total force on the sheet must be zero. That can only be true if the pressure is the same on the top and bottom surfaces of the sheet. We have thus concluded that the pressure is the same at points A, B, and C; that is, $P_A = P_B = P_C$. This condition will allow us to find the value of d.

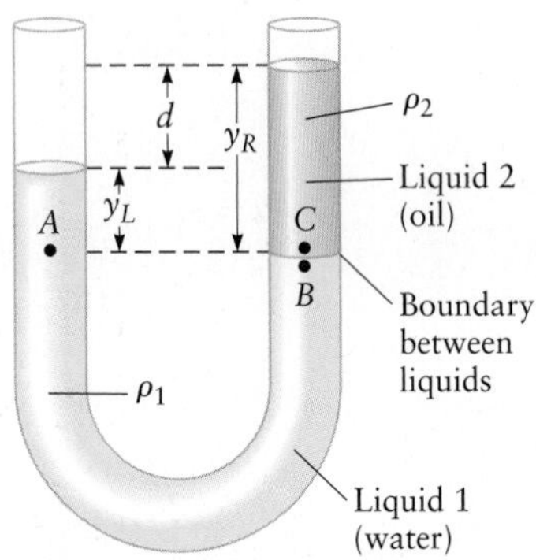

▲ **Figure 10.10** Example 10.3. When a U-tube contains more than one type of liquid, one must apply Equation 10.5 separately within each liquid region. Points B and C are on opposite sides of the boundary between the fluids. These points are extremely close together and are at the same height as point A.

SKETCH THE PROBLEM

Figure 10.10 describes the problem.

IDENTIFY THE RELATIONSHIPS

In liquid 1, the pressure at the surface is atmospheric pressure since this surface is in contact with the atmosphere; let's take this surface as our reference point in applying Equation 10.5. If y_L is the distance from the point A to the surface, we get

$$P_A = P_{atm} + \rho_1 g y_L$$

Likewise, in liquid 2,

$$P_C = P_{atm} + \rho_2 g y_R$$

where y_R is the distance from point C to the surface of liquid 2. Combining these results and using $P_A = P_C$ gives

$$P_A = P_{atm} + \rho_1 g y_L = P_C = P_{atm} + \rho_2 g y_R$$

$$\rho_1 g y_L = \rho_2 g y_R$$

$$y_L = y_R \frac{\rho_2}{\rho_1}$$

The difference in heights of the two surfaces is then

$$d = y_R - y_L = y_R\left(1 - \frac{\rho_2}{\rho_1}\right) = y_R \frac{\rho_1 - \rho_2}{\rho_1}$$

SOLVE

Inserting the values of the densities of water (ρ_1) and oil (ρ_2) given above and using the given value of y_R, we find

$$d = y_R \frac{\rho_1 - \rho_2}{\rho_1} = (0.10 \text{ m})\frac{(1000 - 700)(\text{kg/m}^3)}{1000 \text{ kg/m}^3} = \boxed{0.030 \text{ m}}$$

► ***What does it mean?***

When two fluids are in contact with each other and are at rest, the pressure must be the same on the two sides of the interface.

Barometers, Vacuums, and Measuring Pressure

U-tubes and similar types of containers can be used in many different ways. Example 10.3 shows that when a U-tube is filled with two different (and immiscible) liquids, the open surfaces of the two liquids will sit at different heights. This result can be used to measure the density of a liquid. If the density ρ_1 of liquid 1 in Figure 10.10 is known, a measurement of d as calculated in Example 10.3 then allows us to find the value of ρ_2.

A U-tube can also be used to measure pressure differences. For example, consider the situation in Figure 10.11, which shows a U-tube whose ends are connected to two different containers of gas at pressures P_L and P_R. When partially filled with liquid, the U-tube can be used to measure the pressure difference $P_L - P_R$ by applying Equation 10.5. Taking our reference point to be the top surface of the liquid on the right side of the U-tube, we find

$$P_L = P_R + \rho g h$$

where h is the difference in the heights of the two surfaces of the liquid. Hence, the difference in pressure is

$$\Delta P = P_L - P_R = \rho g h \tag{10.6}$$

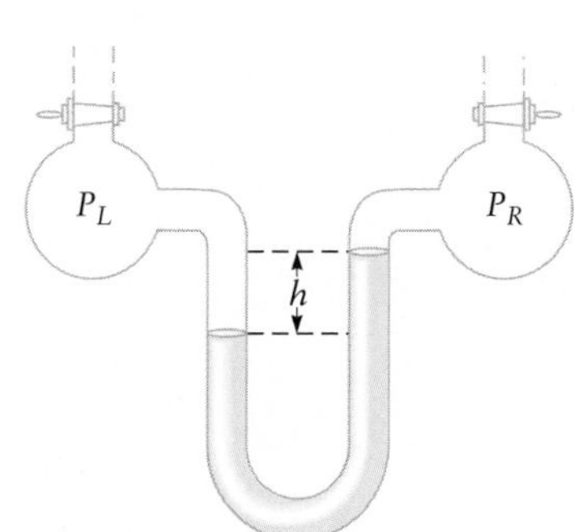

▲ **Figure 10.11** This U-tube can be used to measure the difference in pressure, $\Delta P = P_L - P_R$.

Figure 10.12 shows another way to measure pressure. Here we have a straight tube, partially filled with liquid and closed at the top end, with the open end at the bottom inserted into a dish containing the same liquid. The pressure at the top of the tube is zero (this can be done with a vacuum pump), while the open surface of the

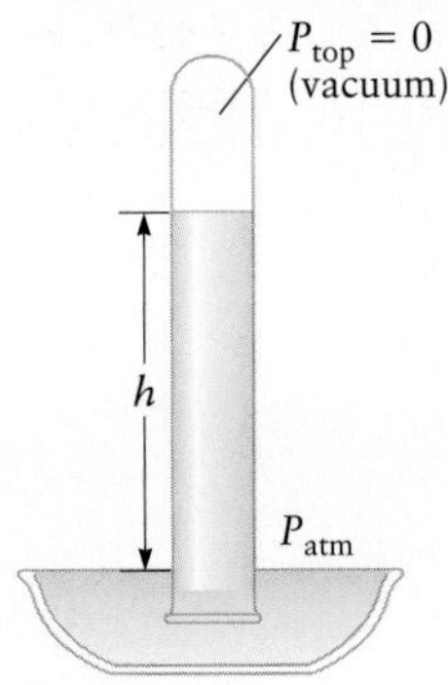

▲ **Figure 10.12** Design of a simple barometer.

liquid in the dish is exposed to the atmosphere. Applying Equation 10.5 and using the top surface of the liquid in the tube as our reference point, we have

$$P_{atm} = P_{top} + \rho g h$$

where h is the height of the liquid column in the tube. Since $P_{top} = 0$, we find

$$P_{atm} = \rho g h \tag{10.7}$$

Hence, if we measure h, we can use the device in Figure 10.12 as a ***barometer*** and measure the absolute pressure of the atmosphere.

Let's now design such a barometer and assume we are going to use water as the liquid. We want to calculate how tall the tube must be to measure the pressure in the atmosphere. We can rearrange Equation 10.7 to solve for the height of the water column h; inserting the values of atmospheric pressure and the density of water from Table 10.1 gives

$$h = \frac{P_{atm}}{\rho g} \tag{10.8}$$

$$h = \frac{1.01 \times 10^5 \text{ Pa}}{(1000 \text{ kg/m}^3)(9.8 \text{ m/s}^2)} = 10 \text{ m} \tag{10.9}$$

This barometer would require quite a long tube to measure atmospheric pressure; it would certainly not fit into a normal-sized room. That is why most barometers do not employ water. Instead, they use mercury, which has a much higher density than water, giving a smaller value of h in Figure 10.12 and Equation 10.8.

EXAMPLE 10.4 Designing a Mercury Barometer

Consider again the barometer in Figure 10.12, but now assume the liquid in the tube and dish is mercury. If the pressure to be measured is atmospheric pressure, what is h?

RECOGNIZE THE PRINCIPLE

We can use the results for pressure in an incompressible fluid and the reasoning that led to Equation 10.8, but now we replace water with mercury, so we have a different value for the density.

SKETCH THE PROBLEM

Figure 10.12 describes the problem.

IDENTIFY THE RELATIONSHIPS AND SOLVE

Using the density of mercury (which is a liquid at room temperature) from Table 10.1 along with Equation 10.8 leads to

$$h = \frac{P_{atm}}{\rho g} = \frac{1.01 \times 10^5 \text{ Pa}}{(13{,}600 \text{ kg/m}^3)(9.8 \text{ m/s}^2)} = \boxed{0.76 \text{ m}}$$

► ***What does it mean?***

This size is much more manageable than the water barometer we considered above (Eq. 10.9). Two old-fashioned units of pressure are "inches" and "millimeters of mercury." Both units refer to the value of h measured with a mercury barometer.

Units for Measuring Pressure

For most scientific work, pressure is measured in pascals, and we have seen that 1 Pa = 1 N/m^2. In the United States, we often encounter the unit pounds per square inch (lb/in.^{2}), especially when dealing with automobile tires. The pressure unit "inches" is often heard on a TV weather report. This unit originates with a barometer like the one in Example 10.4, where the pressure in the atmosphere can be determined by giving the value of h (measured in inches) for a mercury-filled barometer. Likewise, the barometer height h can be measured in millimeters, which leads to the unit "millimeters of mercury," (sometimes referred to as simply mm). Another common unit is the atmosphere (atm), which is the pressure in the Earth's atmosphere at sea level under normal weather conditions. These units are summarized in Table 10.2, which also gives the value of normal atmospheric pressure in the different units.

TABLE 10.2 Units Used for Measuring Pressure

Pressure Unit	Value of Atmospheric Pressure
Pascal	1.01×10^5 Pa
Pounds per square inch	14.7 lb/in.^{2}
Inches	29.9 in.
Millimeters of mercury	760 mm
Atmosphere	1 atm

Pumping a Liquid

Our discussion of barometers led us to Equation 10.9, showing that a water-filled barometer is not very practical. This result also has an important implication for pumping water from place to place. Figure 10.13 shows a well that brings water to the surface from deep underground. If we are going to use a pump to move the water, where is the best place to put the pump? Two possibilities are shown in Figure 10.13; in one case, the pump is located at the top of the well, and in the other, it is located underground at the bottom of the well. Placing it at the top of the well has several advantages; for example, it is simpler to fix the pump if it breaks down. There is a limitation with this location, however. If the pump is placed at the top, it must draw water up from the ground by establishing a very low pressure in the top of the pipe that carries water to the surface. This arrangement is just like the barometer in Figure 10.12, where a very low pressure (a vacuum) in the top of the tube and the atmospheric pressure at the dish act together to push water up the column. Suppose the water pump in Figure 10.13A generates a perfect vacuum at the top ($P_{top} = 0$). Since soil is porous to air, the pressure at the underground surface of the water is approximately atmospheric pressure. Following the approach we used in deriving Equation 10.8, we have, from Figure 10.13A,

$$P_{bot} = P_{top} + \rho g h$$
$$P_{atm} = 0 + \rho g h$$

which leads to

$$h = \frac{P_{atm}}{\rho g}$$

For water, then, $h = 10$ m as we found in Equation 10.9. This is the *maximum* height to which a vacuum pump can "draw" water. The design in Figure 10.13A can thus only be used if the well depth is less than 10 m.

Most water wells, however, are much deeper than 10 m. For example, the well at the author's home is approximately 50 m deep. These cases are handled by placing the pump at the bottom of the well as in Figure 10.13B. Now the pump is not a vacuum pump, but instead generates a large pressure, much greater than atmospheric pressure. We can again apply the relation for pressure in an incompressible fluid (Eq. 10.5) and get

$$P_{bot} = P_{top} + \rho g h$$

where h is again the depth, and the pressure at the top, P_{top}, is now equal to atmospheric pressure. To satisfy this relation, the pressure at the bottom of the well pipe must be larger than P_{atm}. There is no theoretical limit on the magnitude of P_{bot}; a realistic pump can generate pressures of many times atmospheric pressure, so the design in Figure 10.13B can easily pump water from typical well depths. Similar considerations arise when water must be moved or pumped to different vertical levels, such as in the water system of a tall building. (Many buildings are much taller than 10 m!)

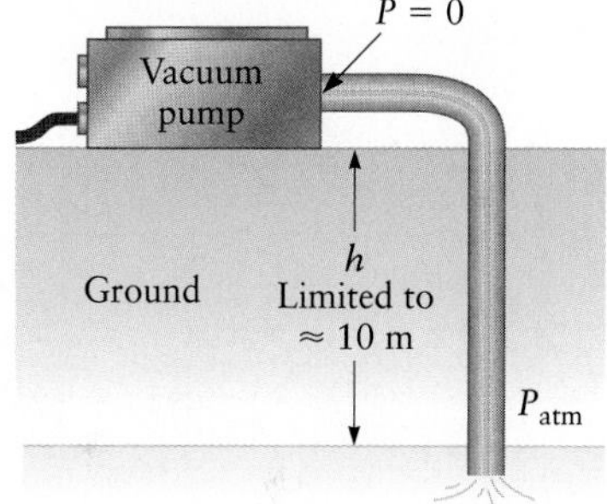

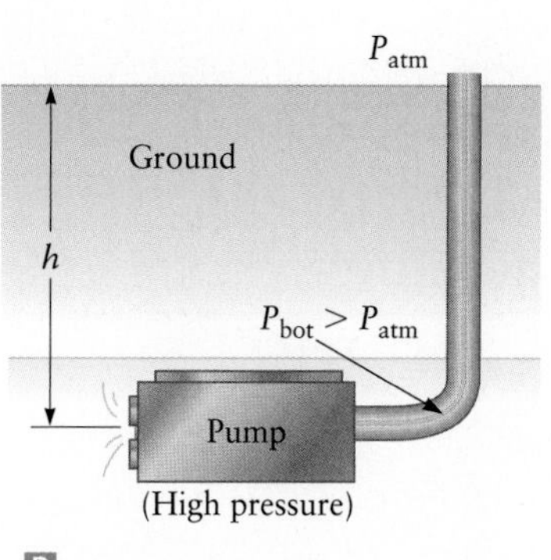

▲ **Figure 10.13** A A vacuum pump is used to pump water from a well (underground) to the surface. Since soil is porous, the pressure at the underground surface of the water is approximately P_{atm}. B Most wells use a pump placed at the bottom of the well. This pump produces a high pressure that pushes the water to the surface.

Pressure in a Compressible Fluid

Equation 10.5 tells how pressure varies with depth, but it applies only to an incompressible fluid, that is, a liquid such as water or mercury. Gases (such as air in the atmosphere) are compressible fluids and follow a different relation between pressure and depth because the density of a gas depends greatly on pressure, whereas the density of a liquid is approximately constant. The pressure in a compressible fluid still

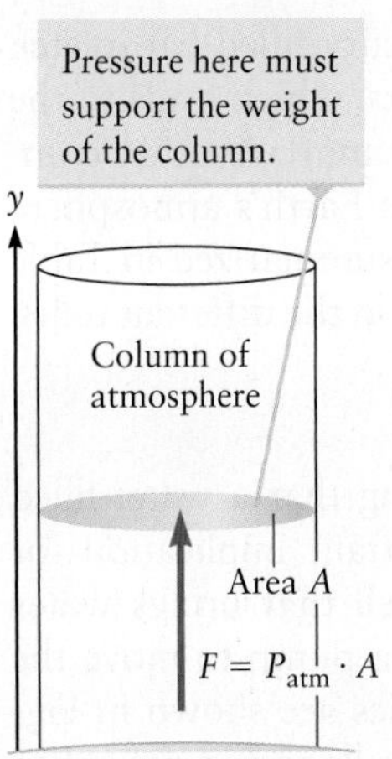

▶ **Figure 10.14** For a fluid at rest in a gravitational field, the pressure at a particular point in the fluid must support the weight of the fluid above that point. For this reason, the pressure in a fluid always increases with depth within the fluid. This result applies to both liquids and gases.

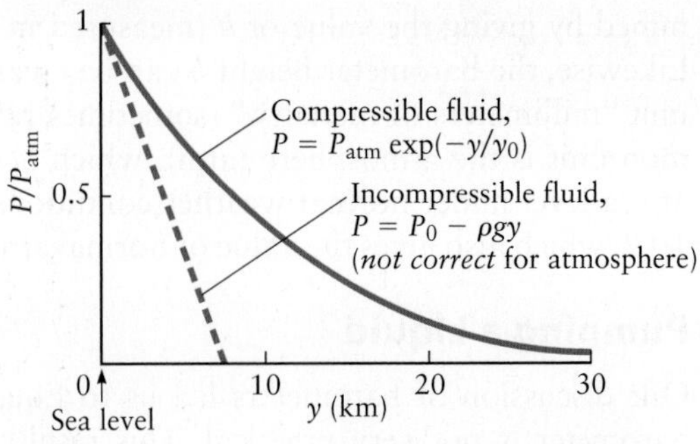

▲ **Figure 10.15** Variation of pressure in the Earth's atmosphere with height y above sea level.

increases with depth as explained in Figure 10.14. As an example, the pressure at any point in the Earth's atmosphere must support the weight of all the fluid (air) above it.

If we consider a column of fluid of area A, the upward force supporting this column is $F = P_{atm}A$. For a liquid with constant density, this analysis leads to Equation 10.5, which gives $P = P_{atm} + \rho gh$, where h is the *depth* in the fluid measured from the surface. In Figure 10.14, we denote the *height* above the surface as y, which is related to the depth by $y = -h$; we thus have for a liquid

$$P = P_0 - \rho gy \tag{10.10}$$

which is plotted as the straight dashed line in Figure 10.15.

This result does *not* apply to a gas because as y increases and the pressure drops, the density decreases. The weight of the fluid column is therefore smaller for a compressible fluid, and the pressure decreases more slowly as y increases than for an incompressible fluid. The pressure as a function of height y for a compressible fluid is also plotted in Figure 10.15 as the solid curve. This result can be derived using Newton's laws and is

$$P = P_0 \exp\left(\frac{-y}{y_0}\right) \tag{10.11}$$

where P_0 is the pressure at a reference point, y is the height as measured from the reference point, and y_0 is a constant that depends on the gas. For air with a reference point at sea level, $P_0 = P_{atm}$ and $y_0 \approx 1.0 \times 10^4$ m. The height y_0 is the altitude at which the pressure falls to $e^{-1} = 0.37 = 37\%$ of its value at the surface. This result applies for low regions of the atmosphere, where the temperature is approximately constant. Dealing with pressure variations over a wider altitude range is more complicated since the air temperature can vary substantially and the air density also depends on temperature.

Whether we are dealing with a liquid or a gas, the pressure in a fluid must vary with depth and P must increase as one goes deeper into the fluid. In both cases, the pressure difference between the top and bottom surfaces of an arbitrary volume of fluid (Fig. 10.7) must support the weight of the fluid within the volume, so the pressure increases as depth increases.

EXAMPLE 10.5 Pressure at the Top of Mount Everest

The peak of Mount Everest is 8850 m above sea level. Estimate the pressure in the atmosphere at this location.

RECOGNIZE THE PRINCIPLE

This example is a direct application of the result in Equation 10.11 for the pressure as a function of altitude in a compressible fluid (air).

SKETCH THE PROBLEM

Figure 10.15 shows the variation of pressure with altitude y in the atmosphere as described mathematically by Equation 10.11.

IDENTIFY THE RELATIONSHIPS AND SOLVE

The altitude at the top of Mount Everest is $y = 8850$ m. Using this value in Equation 10.11 with $y_0 = 1.0 \times 10^4$ m as given above leads to

$$P = P_{atm} \exp\left(\frac{-y}{y_0}\right) = (1.01 \times 10^5 \text{ Pa})\exp\left(\frac{-8850 \text{ m}}{1.0 \times 10^4 \text{ m}}\right) = \boxed{4.2 \times 10^4 \text{ Pa}}$$

► ***What does it mean?***

This result is about 40% of atmospheric pressure. According to the kinetic theory of gases (Chapter 15), the density of a gas is proportional to the pressure, assuming its temperature is constant. The temperature at the top of Mount Everest is usually a bit below (!) the temperature at sea level, which reduces the pressure on Mount Everest by about another 15%. Hence, the air density at the top of Mount Everest is less than 40% of the density at sea level. That is why mountain climbers usually bring their own oxygen when scaling this mountain.

10.3 Hydraulics and Pascal's Principle

For a liquid at rest, the pressure varies with depth according to

$$P = P_0 + \rho g h \tag{10.12}$$

Now imagine that something is done to change the pressure at a particular point in the liquid; to be specific, suppose the pressure P_0 is changed by an amount ΔP. According to Equation 10.12, the pressure everywhere in the liquid must change by an equal amount. This result is known as ***Pascal's principle.***

Hydraulics: Pascal's principle

> ***Pascal's principle:*** **If the pressure at one location in a closed container of fluid changes, this change is transmitted equally to *all* locations in the fluid.**

While Equation 10.12 (which is the same as Eq. 10.5) was derived for a liquid (an incompressible fluid), Pascal's principle is also true for a gas (a compressible fluid) as long as the gas is held in a closed container and is in static equilibrium. Pascal discovered this principle before the work of Newton. However, looking back at our derivation of Equation 10.5, we see that this principle is a direct consequence of Newton's laws.

Designing a Hydraulic Lift: Amplifying Forces

Pascal's principle is the basis of ***hydraulics***. A simple version of a hydraulic lift is shown in Figure 10.16A. It is shaped like a U-tube and is typically filled with a liquid such as oil, with pistons at both ends of the U-tube. These pistons are movable, but they fit tightly to keep the oil from leaking out. The key point is that the two ends of the tube have different areas. Here the area A_L on the left is much smaller than the area A_R on the right.

Now imagine that a very large object such as a car sits on the piston on the right. To keep the car and its piston from moving, or to lift the car very slowly off the ground, we apply a force F_L to the left as sketched in Figure 10.16A. What force F_L is required to keep the system in static equilibrium? If the car has mass m, the car's weight produces a downward force on the right-hand piston of magnitude $F_R = mg$.

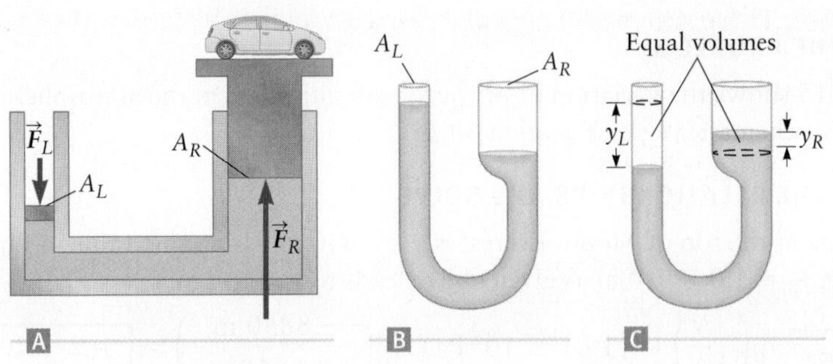

► **Figure 10.16** A A hydraulic jack uses a liquid (usually oil) to amplify an applied force. Here the force on the right-hand piston is greater than the force applied to the piston on the left. B and C show how the liquid moves as the left-hand piston is depressed. The volume of liquid that leaves the left-hand tube must equal the volume that appears in the right-hand tube.

From the general relation between force and pressure (Eq. 10.1), this force will lead to an increase in the pressure on the right, with

$$\Delta P = \frac{F_R}{A_R}$$

According to Pascal's principle, this pressure change is transmitted throughout the fluid, giving an increased force of magnitude $(\Delta P)A_L$ on the left-hand piston. To keep this piston from moving, there must be an external force of equal magnitude F_L applied downward on the piston. Hence,

$$F_L = (\Delta P)A_L$$

or

$$F_L = F_R \frac{A_L}{A_R} \tag{10.13}$$

The piston on the right is much larger than the one on the left ($A_R > A_L$), so

$$F_L < F_R$$

The force on the left-hand piston is thus smaller than the force on the right-hand piston. A relatively small force applied on the left can therefore lift a very massive object (the car) on the right. In other words, the hydraulic device in Figure 10.16 *amplifies forces*. This principle is used by auto mechanics every day.

The simple system in Figure 10.16 illustrates the principles of hydraulics, but practical systems are usually made as sketched in Figure 10.17, where a pump is used to produce a high pressure in the liquid in a tube connected to an elevator. The tube leading from the pump plays the role of the left-hand tube in Figure 10.16 and is smaller than the elevator tube. Here the large object to be lifted is the elevator compartment.

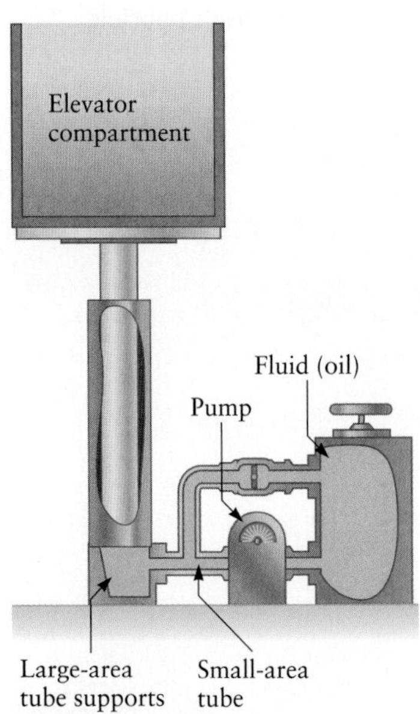

▲ **Figure 10.17** Schematic of a hydraulic system used for an elevator.

Work–Energy Analysis of a Hydraulic System

Now let's analyze the hydraulic lift in terms of work and energy. We start with the configuration in Figure 10.16B and suppose the piston on the right moves upward through a small distance y_R. The work done by the right-hand piston on the automobile is equal to the product of the force and the distance traveled:

$$W_R = F_R y_R \tag{10.14}$$

We can also consider the work done by the force F_L on the left-hand piston. That piston moves downward a distance y_L, so this work is

$$W_L = F_L y_L \tag{10.15}$$

Since we expect that the hydraulic lift will conserve mechanical energy, you should suspect that the work done *on* the left-hand piston should equal the work done *by* the right-hand piston; that is, $W_L = W_R$. Let's now show that this is indeed true.

When the two pistons move, liquid (oil) is displaced from one side of the U-tube to the other as shown in parts B and C of Figure 10.16. Because the liquid is incompressible, the volume of oil that travels to the right must equal the volume that leaves from the left. These volumes are equal to the area of the tube times the height of the oil, so

$$A_L y_L = A_R y_R \tag{10.16}$$

and

$$y_L = y_R \frac{A_R}{A_L} \tag{10.17}$$

The piston with the *smaller* area (on the left) thus moves through a *larger* distance. Combining this result with our result for the forces from Equation 10.13 leads to

$$F_R y_R = \left(F_L \frac{A_R}{A_L}\right)\left(y_L \frac{A_L}{A_R}\right) = F_L y_L$$

The term on the far left, $F_R y_R$, equals W_R (Eq. 10.14), while the term on the far right, $F_L y_L$, is W_L (Eq. 10.15). We thus have

$$W_R = W_L$$

Hence, the work done *on the piston on the left* equals the work done *by the piston on the right*, and mechanical energy is indeed conserved. A hydraulic device can amplify forces, but in doing so, it must "de-amplify" the displacement (Eq. 10.17).

EXAMPLE 10.6 Designing a Hydraulic Lift

You are given the job of designing a hydraulic device for lifting large objects as in Figure 10.16. You want this device to be able to lift an SUV of mass $m_{SUV} = 2000$ kg, when operated with an applied force of just 750 N (the approximate weight of the author). The two pistons of the hydraulic lift are cylindrical, and suppose the piston lifting the SUV has a radius of 0.50 m. Find the radius of the other piston.

RECOGNIZE THE PRINCIPLE

From Pascal's principle, the pressure increases on the two sides of a hydraulic tube are equal. This leads to Equation 10.13, which (in words) states that the ratio of the forces equals the ratio of the piston areas. Given the ratio of the weight of the SUV and the applied force, we can thus find the ratio of the piston areas.

SKETCH THE PROBLEM

Figure 10.16 shows the problem. The piston lifting the SUV must be the one with the larger area.

IDENTIFY THE RELATIONSHIPS

To lift the SUV, we must amplify the applied force of 750 N by a factor of $m_{SUV}g/(750\text{ N}) = (2000\text{ kg})(9.8\text{ m/s}^2)/(750\text{ N}) = 26$; this is the ratio of the forces F_R/F_L in Figure 10.16A. Rearranging Equation 10.13 leads to

$$\frac{F_R}{F_L} = \frac{A_R}{A_L} = 26$$

The sides of the lift are cylindrical, so for each piston $A = \pi r^2$, and

$$\frac{A_R}{A_L} = \frac{\pi r_R^2}{\pi r_L^2} = \frac{r_R^2}{r_L^2} = 26$$

SOLVE

Solving for the radius of the left-hand piston gives

$$r_L = \frac{r_R}{\sqrt{26}} = \frac{0.50\text{ m}}{\sqrt{26}} = \boxed{0.098\text{ m}}$$

(continued) ▶

► *What does it mean?*

This hydraulic system will amplify force by a factor of 26 and will de-amplify displacement by the same factor. The operator will have to displace 26 m of oil through the left-hand piston to lift the car through 1 m.

10.4 Buoyancy and Archimedes's Principle

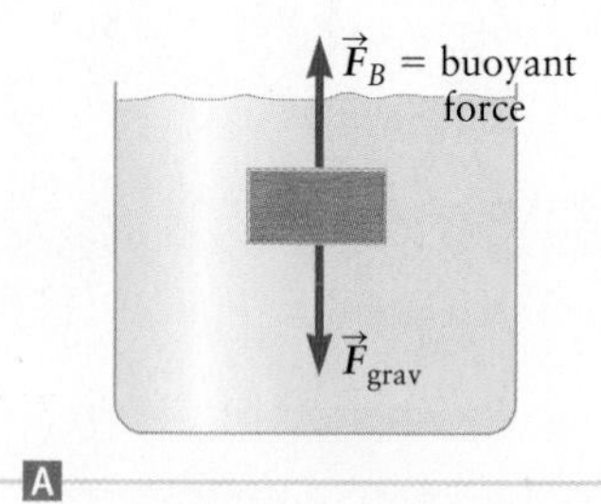

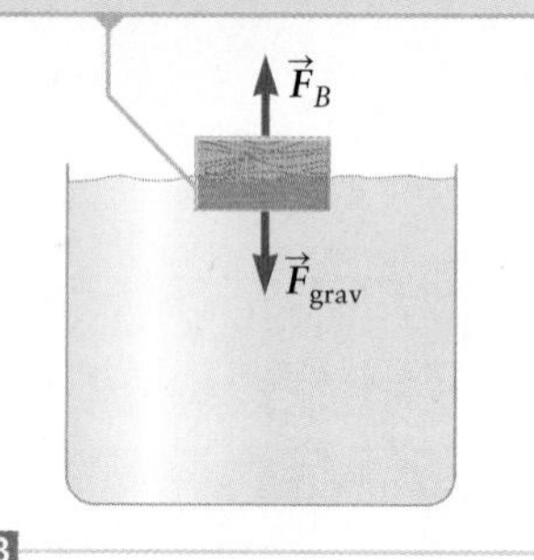

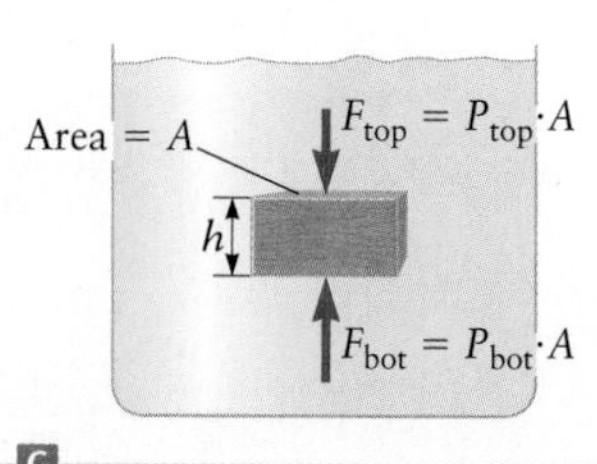

▲ **Figure 10.18** According to Archimedes's principle, an object that is immersed in a fluid will experience a buoyant force that is directed upward, whether A the object is fully immersed in the fluid or B the object floats at the surface and is partially exposed. In both cases, the buoyant force on the object equals the weight of the fluid displaced by the object. C The buoyant force is due to the increase in pressure with depth.

Archimedes lived in the third century BC and is known for several contributions to science and mathematics. While his work on fluids was done long before the time of Newton, Archimedes made some very important discoveries involving forces in fluids that anticipated Newton's laws. Consider an object that is either fully immersed or is floating with a portion exposed above the surface of the fluid as sketched in parts A and B of Figure 10.18. In either case, ***Archimedes's principle*** applies.

> ***Archimedes's principle:*** **When an object is immersed in a fluid, the fluid exerts an upward force on the object that is equal to the weight of the fluid displaced by the object.**

This upward force on the object is called the ***buoyant force***, and it is due to the increase of pressure with depth in a fluid. The pressure at the bottom surface of an object is always greater than the pressure at the top, leading to a larger (upward) force on the bottom surface. The difference between these two forces is the buoyant force.

We can calculate the buoyant force (and thus prove Archimedes's principle) by applying our relation for the variation of pressure with depth (Eq. 10.5) to the object in Figure 10.18C. Here, a solid box is completely submerged in a liquid of density ρ_L. For simplicity, we assume this box is rectangular with height h and top or bottom area A. There are pressure forces on all faces of the box, the sides as well as the top and bottom. The horizontal forces on opposite sides of the box are equal in magnitude and opposite in direction and thus cancel. The force on the top is downward with magnitude $P_{top}A$, whereas the force on the bottom is upward and of magnitude $P_{bot}A$. The total force *exerted by the liquid* on the box is along the vertical direction, with

$$F = +P_{bot}A - P_{top}A$$

If we assume the fluid is a liquid of density ρ_L, according to Equation 10.5 the pressures at the bottom and top are related by

$$P_{bot} = P_{top} + \rho_L gh$$

where h is the height of the box. Combining these results gives

$$F = +P_{bot}A - P_{top}A = (P_{top} + \rho_L gh)A - P_{top}A = \rho_L hAg \tag{10.18}$$

Hence, the force exerted by the liquid is upward; this is the buoyant force. Since the factor (hA) in Equation 10.18 equals the volume V of the box, this result is usually written as

$$F_B = \rho_L Vg \tag{10.19}$$

The mass of the liquid displaced by the box is $\rho_L V$, so the term $\rho_L Vg$ in Equation 10.19 equals the weight of the displaced liquid. In words, Equation 10.19 says that the buoyant force F_B equals the weight of the *liquid* displaced by the box. This is another statement of Archimedes's principle.

Examples and Applications of Archimedes's Principle

Figure 10.18 shows Archimedes's principle at work in two different cases. In Figure 10.18A, a box of volume V is fully immersed in the fluid. The buoyant force equals the weight of the displaced fluid, and because the box is completely immersed, the volume of displaced fluid equals the volume V of the box. The weight of the displaced fluid is thus $F_B = \rho_L g V$, where ρ_L is the density of the fluid. If the weight of this box is exactly equal to F_B, the box will remain where it is, fully submerged in the liquid. If the box is heavier than F_B, it will sink to the bottom. It might also happen that the weight of the box is smaller than the weight of the liquid it displaces, which could happen with an empty box or with a boat. In such a case, the buoyant force for a fully submerged box would be *greater* than the weight of the box, and the box would rise to the surface and float (Fig. 10.18B). A portion of the box would then be exposed above the surface of the liquid so that the weight of the *displaced* liquid (the amount below the "waterline") equals the weight of the box. For that to occur, the average density of the box in Figure 10.18B must be less than the density of the liquid (just as the average density of a boat must be less than the density of water).

As an application of Archimedes's principle, consider the situation in Figure 10.19. This box might be very heavy and submerged in a lake, and you might have the job of lifting it to the surface. The minimum force required to lift the box is called the ***apparent weight***.

Figure 10.19 shows the forces acting on the box. They are the force of gravity (the weight of the box), the force F_{lift} you are exerting as you lift the box, and the buoyant force F_B from the water. According to Archimedes's principle, F_B is the total force exerted by the fluid on the box. Since F_{lift} is the minimum force necessary to barely lift the box, we can take the acceleration of the box to be zero. Newton's second law for the box's motion along the vertical direction is then

$$\sum F = F_{\text{grav}} + F_B + F_{\text{lift}} = -mg + \rho_L V g + F_{\text{lift}} = 0$$

where the buoyant force F_B is given by Archimedes's principle, Equation 10.19. The apparent weight of the box is equal to F_{lift}. Solving for F_{lift}, we find

$$F_{\text{lift}} = mg - \rho_L V g \tag{10.20}$$

The apparent weight is thus smaller than the true weight (mg) by an amount equal to the buoyant force. You have probably already done this type of experiment in a swimming pool. To estimate the magnitude of the buoyant force, suppose the box in Figure 10.19 has a volume of 0.10 m^3 (about the size of a small suitcase) and is filled with lead. Its weight is

$$\text{true weight} = mg = \rho_{\text{lead}} V g$$

since the mass of the box is equal to the density of lead (ρ_{lead}) times the volume (V). Using the density of lead from Table 10.1, we find that the box has a true weight of

$$\text{true weight} = \rho_{\text{lead}} V g = (11{,}300 \text{ kg/m}^3)(0.10 \text{ m}^3)(9.8 \text{ m/s}^2) = 1.1 \times 10^4 \text{ N}$$

From Equation 10.20, the apparent weight of the lead-filled box equals

$$\text{apparent weight} = \rho_{\text{lead}} V g - \rho_L V g = (\rho_{\text{lead}} - \rho_L) V g$$

where ρ_L is the density of the liquid. If we assume the liquid is water (which has a density of 1000 kg/m^3), we find

$$\text{apparent weight} = (\rho_{\text{lead}} - \rho_L) V g$$
$$= (11{,}300 - 1000)(\text{kg/m}^3)(0.10 \text{ m}^3)(9.8 \text{ m/s}^2)$$
$$\text{apparent weight} = 1.0 \times 10^4 \text{ N}$$

Hence, in this case, the buoyant force makes the apparent weight about 10% smaller than the true weight.

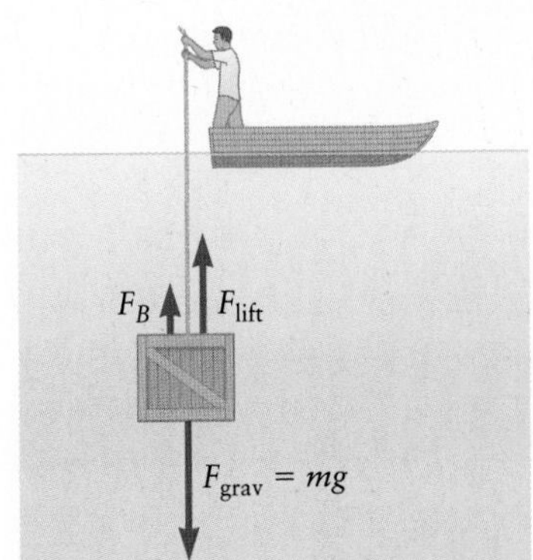

▲ **Figure 10.19** The buoyant force causes the apparent weight (F_{lift}) to be smaller than the true weight (mg).

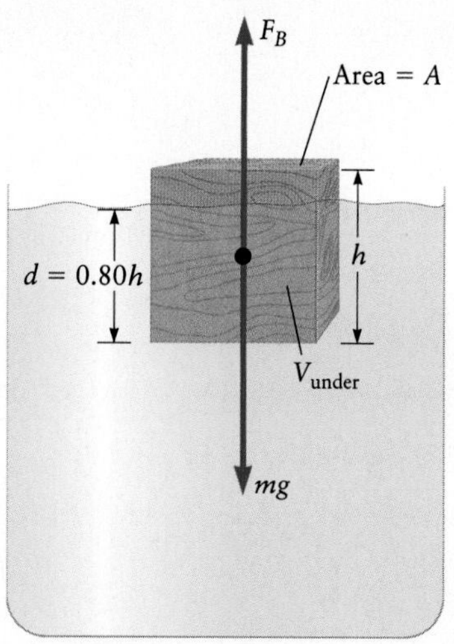

▲ **Figure 10.20** Example 10.7.

EXAMPLE 10.7 Floating in Water

A rectangular block of wood is floating at the surface of a container filled with water as shown in Figure 10.18B. You observe that 80% of the block is under water, whereas the rest is above the surface. Find the density of the wood.

RECOGNIZE THE PRINCIPLE

According to Archimedes's principle, the buoyant force equals the weight of the water displaced by the block. This upward buoyant force must equal the weight of the block in order for the block to remain in translational equilibrium.

SKETCH THE PROBLEM

Figure 10.20 shows the problem. The block of wood has a total height h. Since 80% of the block is submerged, the height of the portion under water is $d = 0.80h$.

IDENTIFY THE RELATIONSHIPS

The top and bottom of the block each have an area A, and the volume of the block that is under water is $V_{under} = dA$. The block thus displaces a volume of water equal to V_{under}, and the buoyant force equals the weight of this displaced water:

$$F_B = \rho_{water} V_{under} g = \rho_{water} dAg$$

The block is floating, so this buoyant force must equal (in magnitude) the gravitational force on the block:

$$F_{grav} = m_{wood} g = \rho_{wood} V_{total} g$$

where $V_{total} = hA$ is the total volume of the block.

SOLVE

Setting F_{grav} equal to the buoyant force F_B leads to

$$F_B = \rho_{water} dAg = \rho_{wood} V_{total} g = \rho_{wood} hAg$$

$$\rho_{water} d = \rho_{wood} h$$

$$\rho_{wood} = \rho_{water} \frac{d}{h} \quad (1)$$

Using $d = 0.80h$ together with the density of water from Table 10.1, we find

$$\rho_{wood} = \rho_{water} \frac{d}{h} = (1000 \text{ kg/m}^3)(0.80) = \boxed{800 \text{ kg/m}^3}$$

▸ What does it mean?

We see from Equation (1) that if an object is floating in a liquid, the object must be less dense than the liquid (as we claimed in our discussion of Fig. 10.18B). Recall that the specific gravity of a substance equals the ratio of its density to the density of water. Hence, a substance will float in water if its specific gravity is less than 1.

CONCEPT CHECK 10.4 Floating in a Lake

A block of wood is initially floating in a lake at 4°C with 20% of the block exposed above the surface of the water (Fig. 10.20). The temperature of the water then increases to 20°C. Will the block move (a) up or (b) down relative to the surface of the water?

CONCEPT CHECK 10.5 Buoyant Force and a Floating Ship

A ship floats because of the buoyant force on it. A typical ship is composed of steel, and we see from Table 10.1 that the density of steel is much *greater* than the density of water. How can the buoyant force support a steel ship?

(a) Real ships are made from special low-density metals.
(b) A ship will float if its average density (including both the steel and the air inside) is less than the density of water.
(c) A ship floats because it is watertight and does not leak.

Archimedes's Principle Holds for Objects of Any Shape, in Both Incompressible and Compressible Fluids

When we derived Archimedes's result for the buoyant force, Equation 10.19, we assumed the submerged (or floating) object has the shape of a simple rectangular block (as in Figs. 10.18 and 10.20). We also used our relation for the variation of the pressure with depth in a liquid ($P = P_0 + \rho gh$). However, Archimedes's principle applies to objects of *all* shapes that are immersed in gases as well as in liquids. Archimedes's principle thus applies to a very wide variety of situations.

EXAMPLE 10.8 Operation of a Helium-Filled Balloon

NASA is experimenting with balloons for use on ultralong-duration (100-day) flights around the Earth. Consider a balloon that has the shape of a cylindrical chamber (Fig. 10.21) that is h = 35 m tall, has radius r = 30 m, and is made of extremely light plastic. (The balloon in Figure 10.22 has a more realistic shape.) If our model cylindrical balloon is filled with helium gas, what is the mass of the maximum payload it could lift from the Earth's surface?

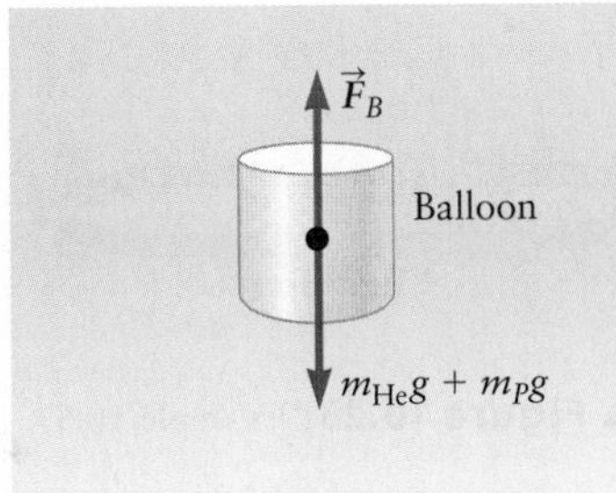

▲ **Figure 10.21** Example 10.8.

RECOGNIZE THE PRINCIPLE

The weight of the maximum payload equals the buoyant force on the balloon minus the weight of the balloon itself. We can find the buoyant force using Archimedes's principle from the volume of the balloon and the density of the air it displaces.

SKETCH THE PROBLEM

This problem is described by Figure 10.21. The balloon is fully immersed in the atmosphere.

IDENTIFY THE RELATIONSHIPS

The buoyant force on the balloon equals the weight of the air the balloon displaces. If the balloon has a volume V, this weight is

$$\text{weight of air displaced} = \rho_{air}Vg = F_B$$

The "skin" of the balloon is composed of a very light plastic, so we'll assume its mass is very small compared with both the mass of the payload m_P and the mass of the helium gas inside the balloon. This helium gas has mass m_{He}, which is given by $m_{He} = \rho_{He}V$, where ρ_{He} is the density of helium gas. The buoyant force must support the weight of the helium and the weight of the payload, so

$$F_B = \rho_{air}Vg = m_{He}g + m_Pg = \rho_{He}Vg + m_Pg$$
$$m_Pg = \rho_{air}Vg - \rho_{He}Vg$$
$$m_P = (\rho_{air} - \rho_{He})V$$

The volume of a cylindrical balloon is equal to the area of the base multiplied by the height, $V = \pi r^2 h$.

(continued) ▶

▲ **Figure 10.22** Example 10.8. This large NASA balloon is filled with helium gas and used to lift scientific payloads to high altitudes.

SOLVE

Using the densities of air and helium from Table 10.1 along with the dimensions of the balloon, we find

$$m_P = (\rho_{air} - \rho_{He})V = (\rho_{air} - \rho_{He})(\pi r^2 h)$$

$$= (1.29 - 0.18)(\text{kg/m}^3)\pi(30\text{ m})^2(35\text{ m})$$

$$m_P = \boxed{1.1 \times 10^5\text{ kg}}$$

► ***What does it mean?***

This payload is quite large; most scientific payloads would have a much smaller mass. Why would NASA make the balloon so large for a relatively small payload? In this example, we have calculated the mass of the largest possible payload for the balloon at sea level. Air is a compressible fluid, so its density decreases as the balloon moves up in the atmosphere and the lower density at high altitudes decreases the lifting capability of the balloon. NASA designs the balloon so that it can lift the payload at the desired "cruising" altitude, so the balloon will need to have a much greater lift force at ground level.

EXAMPLE 10.9 How Can a Submarine Float?

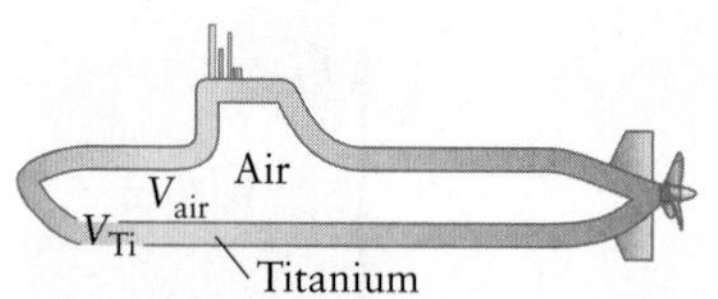

▲ **Figure 10.23** Example 10.9.

Submarines are made of metal, and all metals have a density greater than that of water. So how can a submarine float? Concept Check 10.5 asks for the basic answer, but here we analyze this problem further. Figure 10.23 shows a hypothetical submarine, with an outer metal shell and an inner cavity of air. Assume the metal is titanium (a low-density metal used to make submarines) and the air is at atmospheric pressure (so that the people inside can breathe normally!). What fraction of the submarine must be occupied by the air cavity if the submarine is able to float just below the surface of the water?

RECOGNIZE THE PRINCIPLE

According to Archimedes's principle, the buoyant force on the submarine is equal to the weight of the water it displaces. That force must be equal in magnitude to the total weight of the submarine, which includes the titanium shell and the air in the cavity. The density of titanium is 4500 kg/m³.

SKETCH THE PROBLEM

Figure 10.23 shows the problem. We denote the volume of titanium by V_{Ti} and the volume of air by V_{air}.

IDENTIFY THE RELATIONSHIPS

We denote the density of titanium and air by ρ_{Ti} and ρ_{air}, respectively. The volume of water displaced is $V_{Ti} + V_{air}$, so the weight of the displaced water is

$$\text{weight of water} = \rho_{water}(V_{Ti} + V_{air})g = F_B$$

which is equal to the buoyant force F_B. The total weight of the submarine is the weight of the titanium ($\rho_{Ti}V_{Ti}g$) plus the weight of the air ($\rho_{air}V_{air}g$). The submarine is assumed to be floating under the surface of the water, so its total weight equals the buoyant force:

$$F_B = \rho_{water}(V_{Ti} + V_{air})g = \rho_{Ti}V_{Ti}g + \rho_{air}V_{air}g \quad (1)$$

SOLVE

We can rearrange Equation (1) to solve for V_{air}. Collecting terms in V_{air} and V_{Ti}, we find

$$V_{air}(\rho_{water} - \rho_{air}) = V_{Ti}(\rho_{Ti} - \rho_{water})$$

$$\frac{V_{air}}{V_{Ti}} = \frac{\rho_{Ti} - \rho_{water}}{\rho_{water} - \rho_{air}}$$

Inserting known values of the densities (for seawater from Table 10.1 and titanium from the problem statement), we get

$$\frac{V_{air}}{V_{Ti}} = \frac{\rho_{Ti} - \rho_{water}}{\rho_{water} - \rho_{air}} = \frac{(4500 - 1025)\text{ kg/m}^3}{(1025 - 1.3)\text{ kg/m}^3} = \boxed{3.4}$$

keeping two significant figures in the answer.

► ***What does it mean?***

The volume of the air cavity in the hypothetical submarine in Figure 10.23 must therefore be almost 3.5 times larger than the volume of the titanium shell. Real submarines have a more complicated design, with a ballast tank that can be filled with either water or air to adjust the total weight of the submarine and thus determine if it will float at or underneath the surface.

10.5 Fluids in Motion: Continuity and Bernoulli's Equation

So far in this chapter, we have only considered fluids that are at rest. We are now ready to tackle some problems in which a fluid is in motion. To keep our discussion of fluid dynamics as simple as possible, we make the following assumptions. (1) The fluid density is constant; that is, the fluid is incompressible. (2) The flow is steady; that is, the fluid velocity is independent of time (although it can still vary from point to point in space). (3) There is no friction, that is, no *viscosity*. (We'll discuss viscosity later in this chapter.) (4) We assume there are no "smoke rings" or other kinds of complex flow patterns such as turbulence. Fluids that satisfy these assumptions are called ***ideal fluids***. Although we thus consider only a few of the simplest situations, they will still illustrate some key aspects of fluid dynamics.

Figure 10.24 shows a fluid flowing through a pipe. Because of the simple geometry of this pipe, we need only consider the velocity of the fluid along the pipe, in the x direction in Figure 10.24. We denote the fluid's speed at the pipe inlet (on the left) as v_L and the speed at the pipe outlet (on the right) as v_R. We now want to consider how v_L and v_R are related.

The amount of fluid that flows through the pipe must be *conserved*; this means that the mass of fluid flowing in from the left during a particular time interval Δt must equal the mass flowing out at the right during this interval. In other words, the rate at which the fluid flows into the pipe must equal the rate at which the fluid flows out of the pipe. This is called the ***principle of continuity***.

The basic idea behind the principle of continuity is that "what goes in must come out." In other words, the amount of fluid that leaves the pipe each second must equal the amount that flows in. So, since the outlet in Figure 10.24 is smaller than the inlet, the outlet speed (v_R) must be larger than the inlet speed. A small-diameter flow stream with a large speed can carry the same amount of fluid per unit time as a large-diameter flow stream with a small speed. That is why the speed of water coming from a garden hose increases when you put your thumb over part of the end of the hose (thus decreasing the outlet area).

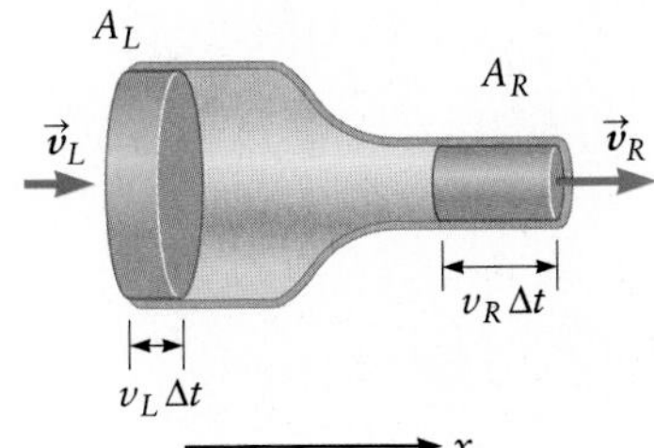

▲ **Figure 10.24** According to the principle of continuity, the rate at which liquid flows into a pipe must equal the rate at which it flows out.

Let's now give a more careful analysis of the principle of continuity. For an incompressible fluid, the *volume* of fluid flowing into a pipe must equal the volume of fluid that flows out (because mass is equal to density times volume, and the density is constant). This inflowing volume is shown in Figure 10.24; it equals the volume of fluid that flows into the pipe from the left during a time interval Δt, filling a cylinder of length $v_L \Delta t$. Since the inlet area is A_L, the volume of fluid flowing into the pipe is

$$V_{in} = (v_L \Delta t)A_L = v_L A_L \Delta t \tag{10.21}$$

Notice that here the capital V is a *volume*; the lowercase v is a *speed*. Using the same argument, the fluid flowing out of the pipe is

$$V_{out} = (v_R \Delta t)A_R = v_R A_R \Delta t \tag{10.22}$$

According to the principle of continuity, the amount that flows into the pipe must equal the amount that flows out, so

$$V_{in} = V_{out}$$

$$v_L A_L \Delta t = v_R A_R \Delta t$$

which leads to

Equation of continuity

$$v_L A_L = v_R A_R \tag{10.23}$$

This result is called the ***equation of continuity***.

The product of fluid speed and area in Equation 10.23 is also called the ***flow rate***, which we denote as Q:

$$\text{flow rate} = Q = vA \tag{10.24}$$

Hence, the equation of continuity (Eq. 10.23) says that the incoming flow rate equals the outgoing flow rate. The SI units of flow rate are cubic meters per second (m^3/s), but liters per second are also widely used. In U.S. customary units, one might use gallons per minute (gal/min). The flow rate from a garden hose is typically 2 to 5 gal/min.

CONCEPT CHECK 10.6 Flow Rate in the Aorta

The flow rate of blood through the aorta is typically 0.080 liter/s = 80 cm^3/s. If it takes 1 minute for all your blood to pass through the aorta, is the volume of blood in your body (a) 4.8 liters, (b) 10 liters, (c) 8.0 liters, or (d) 1.0 liter?

Bernoulli's Equation

Let's again consider an ideal fluid moving through a pipe as sketched in Figure 10.25 and examine the flow from the point of view of work and energy. The fluid in Figure 10.25 is flowing from left to right. We denote the pressure and speed of the fluid at the far left as P_1 and v_1, and we suppose the left end of the pipe is at a height h_1 above the reference level for gravitational potential energy. Likewise, the pressure, fluid speed, and height at the far right are P_2, v_2, and h_2. Figure 10.25 shows two snapshots of the fluid as it moves through the pipe. The first snapshot (Fig. 10.25A) is at $t = 0$, and the second snapshot (Fig. 10.25B) is a short time Δt later. During this time, the "boundaries" of this portion of the fluid have moved a certain distance; the boundary on the left moved a distance Δx_1 and the boundary on the right moved a distance Δx_2.

We now apply the work–energy theorem; our goal is to get a relationship between the pressure, speed, and height of the fluid on the two sides in Figure 10.25. There are external forces on the fluid associated with the pressure on the far left end and with the pressure on the far right end. The work done on the fluid by the pressure on the left equals the force on the boundary at the left multiplied by the distance this boundary travels. This force F_1 is related to the pressure on the left by $F_1 = P_1 A_1$, so the work done on the fluid is

$$W_1 = F_1 \Delta x_1 = P_1 A_1 \Delta x_1$$

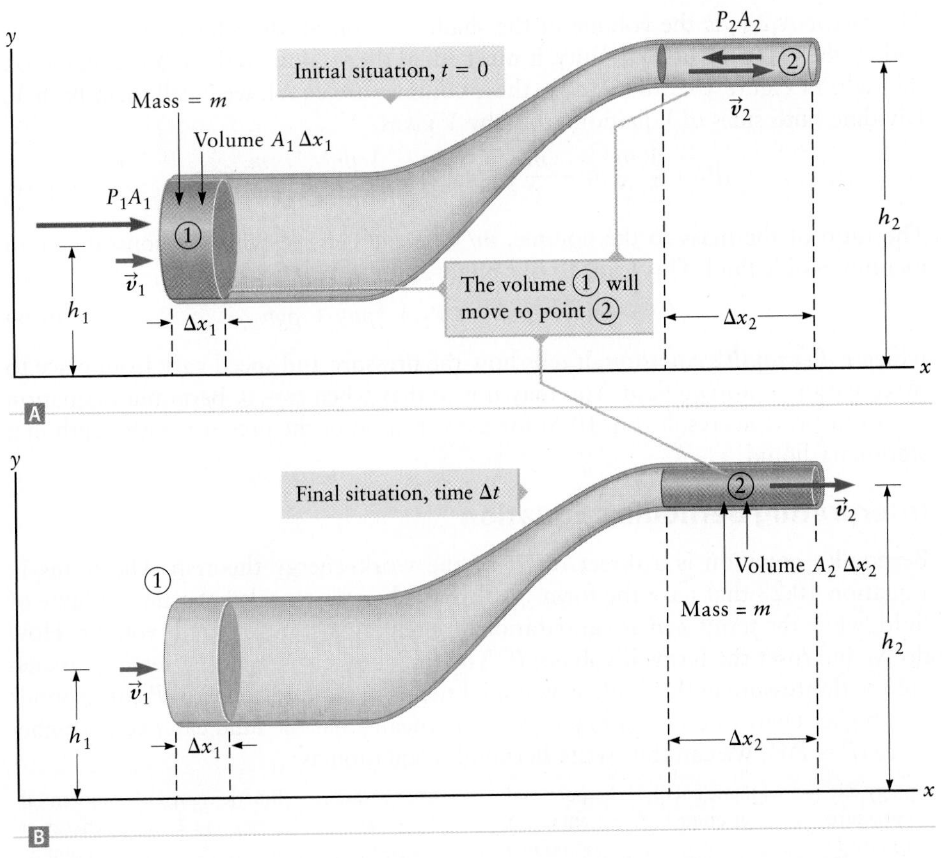

Figure 10.25 Bernoulli's equation tells how the pressure, speed, and height at one location in an ideal fluid are related to P, v, and h at any other location. As the fluid flows through this pipe, the net effect is to transport the shaded volume from the left end in A to the right end in B. The change in the energy of the entire system is then just the difference between the energies of these two shaded volumes.

To find the work done on the fluid by the pressure on the far right end, we note that because the force is a pressure force, it is directed to the left (in the $-x$ direction). The right-hand boundary of the fluid moves to the right, so the force and displacement are in opposite directions and thus the work done on the fluid is negative. We have

$$W_2 = -F_2\,\Delta x_2 = -P_2A_2\,\Delta x_2$$

The total work done on the fluid is the sum of W_1 and W_2, and according to the work–energy theorem, this sum equals the change in the mechanical energy of the fluid:

$$W_{\text{total}} = W_1 + W_2 = P_1A_1\,\Delta x_1 - P_2A_2\,\Delta x_2 = \Delta KE + \Delta PE \qquad (10.25)$$

Figure 10.25 shows snapshots of the initial and final states of the system. The *only* difference between these two snapshots is that the volume of fluid shown as shaded has been moved from the left end of the pipe to the right end. So, the change in energy of the fluid in going from the snapshot in Figure 10.25A to that in Figure 10.25B is equal to the change in the energy of just this shaded region. If the shaded volume of fluid has mass m, its initial mechanical energy (kinetic plus potential) is

$$KE_i + PE_i = \tfrac{1}{2}mv_1^2 + mgh_1$$

whereas the final mechanical energy is

$$KE_f + PE_f = \tfrac{1}{2}mv_2^2 + mgh_2$$

The change in the mechanical energy is therefore

$$\Delta KE + \Delta PE = \tfrac{1}{2}mv_2^2 + mgh_2 - \left(\tfrac{1}{2}mv_1^2 + mgh_1\right) \qquad (10.26)$$

Inserting this expression into Equation 10.25 leads to

$$W_{\text{total}} = P_1A_1\,\Delta x_1 - P_2A_2\,\Delta x_2 = \Delta KE + \Delta PE$$
$$P_1A_1\,\Delta x_1 - P_2A_2\,\Delta x_2 = \tfrac{1}{2}mv_2^2 + mgh_2 - \left(\tfrac{1}{2}mv_1^2 + mgh_1\right)$$
$$P_1A_1\,\Delta x_1 + \tfrac{1}{2}mv_1^2 + mgh_1 = P_2A_2\,\Delta x_2 + \tfrac{1}{2}mv_2^2 + mgh_2 \qquad (10.27)$$

The factor $A_1\Delta x_1$ is the volume of the shaded region on the left in Figure 10.25A, and by the principle of continuity, it must equal the volume of the shaded region on the right in Figure 10.25B. Because these volumes are equal, we'll call them both V. Dividing both sides of Equation 10.27 by V gives

$$P_1 + \frac{1}{2}\frac{m}{V}v_1^2 + \frac{m}{V}gh_1 = P_2 + \frac{1}{2}\frac{m}{V}v_2^2 + \frac{m}{V}gh_2$$

The ratio of the mass to the volume, m/V, is the density, which is constant in an incompressible fluid. This leads to our final result

Bernoulli's equation

$$P_1 + \tfrac{1}{2}\rho v_1^2 + \rho g h_1 = P_2 + \tfrac{1}{2}\rho v_2^2 + \rho g h_2 \qquad (10.28)$$

which is ***Bernoulli's equation.*** It tells how the pressure and speed vary from place to place within a *moving* fluid. You may notice that when $v = 0$, Bernoulli's equation gives our previous result (Eq. 10.5) for the variation of the pressure with depth in a stationary liquid.

Interpreting Bernoulli's Equation

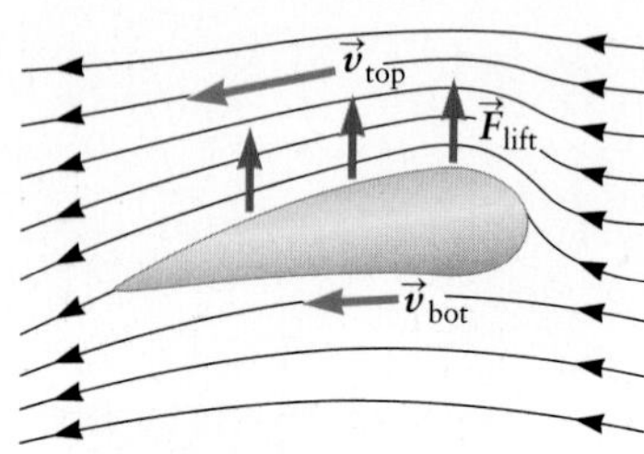

▲ **Figure 10.26** Schematic of the airflow pattern around an airplane wing. Because of the shape of the wing, the air speed is lower under the bottom of the wing than over the top. According to Bernoulli's equation, this difference leads to a higher pressure on the bottom and hence to an upward force called "lift."

Bernoulli's equation is a direct result of the work–energy theorem. The terms in Equation 10.28 that have the form $\frac{1}{2}\rho v^2$ are the kinetic energies of a unit volume of fluid, while the terms $\rho g h$ are gravitational potential energies per unit volume. How do we interpret the terms involving P? The pressure terms represent energy associated with pressure in the fluid; as you might expect, a higher value of P corresponds to a higher energy because a higher pressure means that the fluid can exert a higher force ($F = PA$). We can thus write Bernoulli's equation as

energy due to pressure at point 1		kinetic energy at point 1		potential energy at point 1		energy due to pressure at point 2		kinetic energy at point 2		potential energy at point 2
P_1	$+$	$\frac{1}{2}\rho v_1^2$	$+$	$\rho g h_1$	$=$	P_2	$+$	$\frac{1}{2}\rho v_2^2$	$+$	$\rho g h_2$

In words, Bernoulli's equation says that the total mechanical energy of the fluid is conserved as it travels from place to place (point 1 to point 2), but some of this energy can be "converted" from kinetic energy to potential energy and vice versa as the fluid moves.

Applications of Bernoulli's Equation

One of the most important applications of Bernoulli's equation involves the operation of an airplane wing. Figure 10.26 shows the pattern of airflow near a wing. Most airplane wings have an asymmetric shape, with a "bulge" on the top surface. This bulge causes the air speed to be slightly greater at the top of the wing compared with the air speed at the bottom of the wing. To see why the speed of the air on the top is greater, notice that the air flowing over the top must travel a greater distance to get around the wing (so it must travel faster) than the air that flows around the bottom. According to Bernoulli's equation (Eq. 10.28), an *increase* in the fluid speed leads to a *decrease* in pressure. Hence, the pressure on the bottom of the wing is greater than the pressure on the top. The total force due to air pressure on the wing is

$$F_{total} = P_{bot}A - P_{top}A = (P_{bot} - P_{top})A \qquad (10.29)$$

Since $P_{bot} > P_{top}$, this force is *upward*; it is the "lift" force on a wing that enables an airplane to fly.

Let's now consider the lift on an airplane wing in a little more detail. Suppose the area of the wing is 60 m^2, the air speed over the top surface of the wing is $v_{top} = 250$ m/s (about 500 mph), and the air speed under the bottom of the wing is $v_{bot} = 200$ m/s, all typical values for a commercial jet. To calculate the lift using Equation 10.29, we need to find the difference in pressure between the bottom and top of the wing. Applying Bernoulli's equation to a point at the bottom of the wing (point 1 in Eq. 10.28) and a point at the top (point 2), we have

$$P_{bot} + \tfrac{1}{2}\rho v_{bot}^2 + \rho g h_{bot} = P_{top} + \tfrac{1}{2}\rho v_{top}^2 + \rho g h_{top}$$

Insight 10.1

AIRPLANE WINGS IN AIR

In our derivation of Bernoulli's equation (Eq. 10.28), we assumed the fluid is incompressible. We then applied Bernoulli's equation to calculate the lift on an airplane wing (Eq. 10.29). The fluid in that case is air, which is compressible, but Bernoulli's equation can still be used to get an approximate value of the lift for the following reason. The pressure difference between the top and bottom of the wing (Eq. 10.30) is about 15% of atmospheric pressure, so there is a corresponding 15% variation in the air density between the bottom and top. Hence, to within this accuracy of about 15%, air under these conditions can be treated as having a constant density and our estimate for the lift force should thus be reliable to about this accuracy.

which can be rearranged to give the pressure difference across the wing:

$$P_{bot} - P_{top} = \tfrac{1}{2}\rho v_{top}^2 - \tfrac{1}{2}\rho v_{bot}^2 + \rho g h_{top} - \rho g h_{bot} = \tfrac{1}{2}\rho(v_{top}^2 - v_{bot}^2) + \rho g(h_{top} - h_{bot})$$

Here ρ is the density of air, which we know from Table 10.1. (For simplicity, we use the value of ρ at sea level and 0°C; to be more accurate, we would need to know the precise altitude and temperature.) We also know the speeds; our last step is to deal with the difference in height from top to bottom. For a typical wing of a large commercial jet, this difference is about 1 m, so we take $h_{top} - h_{bot} = 1$ m. Inserting all these values gives

$$P_{bot} - P_{top} = \tfrac{1}{2}\rho(v_{top}^2 - v_{bot}^2) + \rho g(h_{top} - h_{bot})$$
$$= \tfrac{1}{2}(1.29 \text{ kg/m}^3)[(250 \text{ m/s})^2 - (200 \text{ m/s})^2] + (1.29 \text{ kg/m}^3)(9.8 \text{ m/s}^2)(1 \text{ m})$$
$$P_{bot} - P_{top} = 1.5 \times 10^4 \text{ Pa} + 13 \text{ Pa} \tag{10.30}$$

In the last step, we have written the pressure difference as a sum of two separate terms; the first is due to the difference in air speeds, and the second term is due to the difference in heights between the top and bottom of the wing. For this rapidly moving airplane wing, where the air speeds are large and the height difference is small, the potential energy terms (ρgh) make only a very small contribution in Bernoulli's equation. Hence, the pressure difference in Equation 10.30 and thus also the lift are due almost entirely to the speed terms in Bernoulli's equation.

To complete our calculation of the lift on the wing, we insert the pressure difference from Equation 10.30 into our expression for the lift (Eq. 10.29). We find

$$F_{lift} = (P_{bot} - P_{top})A = (1.5 \times 10^4 \text{ Pa})(60 \text{ m}^2) = 9.0 \times 10^5 \text{ N}$$

The mass of a medium-sized commercial jet is about 5×10^4 kg (weight $= mg \approx 5 \times 10^5$ N), so this lift force is indeed sufficient for the jet to fly.

The problem of lift on a real airplane wing is complicated. The Bernoulli effect (Eq. 10.28) is responsible for much of the lift, but other effects are important and must be taken into account in the design of a real airplane. To measure the lift on prototype designs, many airplane makers employ wind tunnels (Fig. 10.27).

National Museum of USAF/U.S. Air Force Photo

▲ **Figure 10.27** The aerodynamic forces on airplanes and other objects are often studied in wind tunnels. The Wright brothers used this wind tunnel to study the forces on wings they designed.

EXAMPLE 10.10 Ⓡ Speed of Water from a Faucet

Figure 10.28 shows the water system for a typical house. The source is a water tower located at some height above the houses it serves. In this particular water tower, the surface of the water in the tower is open to the atmosphere and is 30 m above the home shown in the figure. If a faucet in the home is then opened, what is the speed of the water as it comes out of the faucet?

RECOGNIZE THE PRINCIPLE

We use Bernoulli's equation (Eq. 10.28) to find the speed of the water coming out of the faucet. We take point 1 to be at the surface of the water in the tower and point 2 to be at the faucet.

SKETCH THE PROBLEM

Figure 10.28 describes the problem.

(continued) ▶

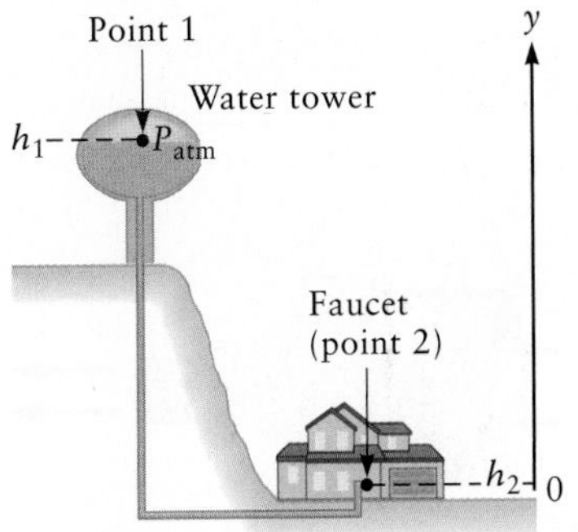

▲ **Figure 10.28** Example 10.10.

▲ **Figure 10.29** The narrowing of this stream of water as it flows from a faucet can be understood using Bernoulli's equation and the principle of continuity.

IDENTIFY THE RELATIONSHIPS

Bernoulli's equation reads

$$P_1 + \tfrac{1}{2}\rho v_1^2 + \rho g h_1 = P_2 + \tfrac{1}{2}\rho v_2^2 + \rho g h_2$$

and we take locations 1 and 2 as shown in Figure 10.28. We wish to solve for the speed v_2 at the faucet, so we must find values for the pressures at the tower and faucet (P_1 and P_2, respectively), the heights of the tower and faucet (h_1 and h_2, respectively), and the speed at the tower (v_1). The water tower is open to the atmosphere, so the pressure there is atmospheric pressure and $P_1 = P_{atm}$. Likewise, the outlet of the faucet is open to the atmosphere, so it is also at atmospheric pressure and $P_2 = P_{atm}$. We can take the origin of our y coordinate axis to be at the faucet, leading to $h_2 = 0$ and $h_1 = 30$ m. We must now deal with v_1, the speed of the top surface of the water in the tower. The area of this top surface is much larger than the cross-sectional area of the faucet. As a result, this surface will fall very slowly as water flows from the faucet, and it is a good approximation to take $v_1 = 0$. This argument is equivalent to using the equation of continuity (Eq. 10.23) with the area of the water surface in the tower being much greater than the cross-sectional area of the faucet.

SOLVE

Putting all these quantities into Bernoulli's equation, we find

$$P_{atm} + 0 + \rho g h_1 = P_{atm} + \tfrac{1}{2}\rho v_2^2 + 0$$

$$\rho g h_1 = \tfrac{1}{2}\rho v_2^2$$

$$v_2 = \sqrt{2gh_1} \tag{1}$$

Inserting our value for the height of the water tower gives

$$v_2 = \sqrt{2gh_1} = \sqrt{2(9.8\ \text{m/s}^2)(30\ \text{m})} = \boxed{24\ \text{m/s}}$$

▸ ***What does it mean?***

Our result for v_2 in Equation (1) is precisely the same as the speed of an object that falls freely from height h_1, as found in Chapters 3 and 6. That is no accident. The water that flows from the water tower to the faucet in Figure 10.28 is also undergoing free fall, and we should expect it to have the same final speed we would find for a rock or a baseball falling through the same distance.

Figure 10.29 shows another example of water undergoing free fall as it flows from a faucet. Notice that the water stream becomes narrower (has a smaller cross-sectional area) the farther one gets from the faucet. This narrowing is a result of the principle of continuity together with Bernoulli's equation. According to the equation of continuity (Eq. 10.23), we have

$$v_1 A_1 = v_2 A_2$$

for any two points in a flowing stream of liquid. The speed is greater at the bottom of the stream in Figure 10.29 than at the top, so the cross-sectional area must be smaller.

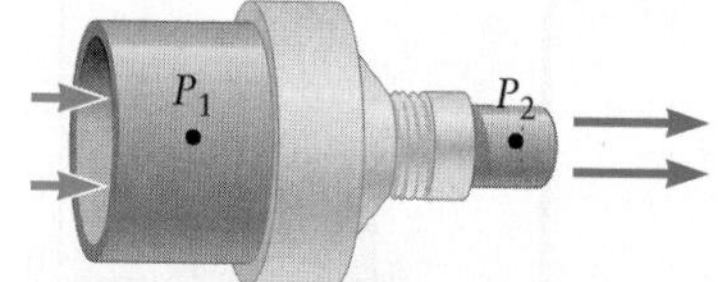

▲ **Figure 10.30** Concept Check 10.7.

CONCEPT CHECK 10.7 Pressure in a Garden Hose

Water flows through two garden hoses, one large and one small, connected as shown in Figure 10.30. How are the pressures in the two hoses related?

(a) $P_1 > P_2$ (b) $P_1 = P_2$ (c) $P_1 < P_2$

10.6 Real Fluids: A Molecular View

In all our work on moving fluids in Section 10.5, we ignored frictional forces internal to the fluid and between the fluid and the walls of the container (i.e., a pipe). These ***viscous*** forces depend on the fluid velocity and are often negligible when this velocity is small. Bernoulli's equation therefore works well when the fluid viscosity is small (e.g., when the fluid is air), when the fluid is flowing slowly, or both. However, in many situations viscosity is important. To understand the origin of viscosity and several other effects in fluids, we must consider fluids at a molecular scale.

Viscosity and Poiseuille's Law

Consider a fluid flowing through a tube as sketched in Figure 10.31. Fluid molecules near the wall of the tube experience strong forces from the molecules in the wall. In fact, these forces are so strong that in most cases the fluid molecules next to the wall are almost bound to it. You may think of the fluid in the tube as a series of cylindrical layers, a few of which are sketched in Figure 10.31A. Fluid molecules in layers near the wall move relatively slowly due to their attraction to the wall, whereas those in layers far from the wall move with a higher speed.

The motion of each fluid layer is strongly influenced by adjacent layers. Whenever the layers move at different speeds, there is a force between them very similar to the force of kinetic friction between two solid surfaces sliding past each other (e.g., a hockey puck sliding across a horizontal surface). The fluid layer denoted as layer 2 in Figure 10.31A is moving faster than the adjacent layer 1. As a result, layer 2 exerts a force on layer 1 that acts to "drag" layer 1 along with layer 2. This force is due to a property of fluids called ***viscosity***. In a "thick" fluid such as honey, drag forces are large and the fluid is said to have a large viscosity. "Thin" fluids such as water or methanol have a much smaller viscosity.

Viscous forces between adjacent fluid layers lead to the pattern of flow speeds shown in Figure 10.31B, where the fluid speed is plotted as a function of position within a pipe. Layers near the center of the tube have the highest speeds, and the flow speed approaches zero at the walls.

For a viscous fluid to move through a tube, the force due to the pressure difference across the ends of the tube must overcome the viscous force. If the pipe has a circular cross section with radius r and length L, and the pressures at the ends are P_1 and P_2, a careful application of Newton's second law shows that the *average* fluid speed in the tube is given by ***Poiseuille's law***,

$$v_{ave} = \frac{P_1 - P_2}{8\eta L} r^2 \tag{10.31}$$

Here η (the Greek letter eta) is the viscosity of the fluid. Values of the viscosity for some common fluids are given in Table 10.3. According to Poiseuille's law, increasing the pressure difference $(P_1 - P_2)$ increases v_{ave} and hence increases the flow rate. Most importantly, decreasing the tube radius r decreases the flow speed; in a smaller

TABLE 10.3 Viscosities of Some Common Fluids

Fluid	Viscosity (N · s/m²)
Water (20°C)	1.0×10^{-3}
Blood (at body temperature, 37°C)	2.7×10^{-3}
Air (20°C)	1.8×10^{-5}
Honey (20°C)	1000
Methanol (30°C)	5.1×10^{-3}
Ethanol (30°C)	1.0×10^{-3}

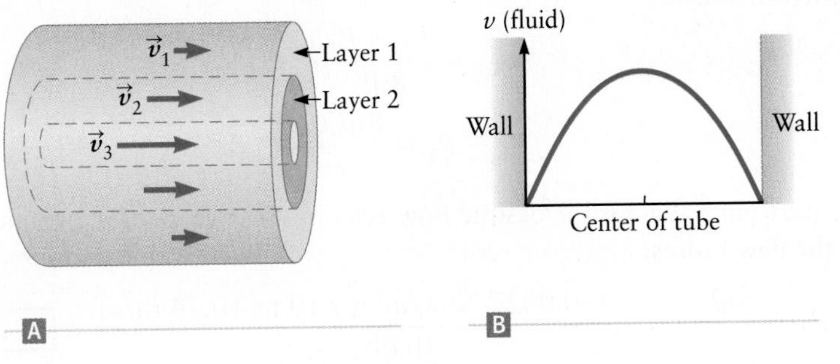

◀ **Figure 10.31** A When a fluid flows through a tube or pipe, viscosity causes fluid near the center to have a higher speed than fluid near the walls, as plotted schematically in B.

tube, a greater fraction of the fluid is close to the walls, so the viscous drag from the walls has a greater effect.

Poiseuille's law is useful for understanding fluid flow in various situations, including the flow of cooling water in an engine, the flow of crude oil through a long pipeline, and the flow of blood through the body. In many cases, one is interested in the flow rate Q (which we defined in Eq. 10.24); here Q is the rate at which fluid passes into or out of a tube or pipe. In our discussion of the principle of continuity and the volume flow into and out of a pipe, we assumed the flow velocity was constant across the radius of the pipe and found that the volume flow rate is $Q = vA$. For the case in Figure 10.31, in which the flow speed is not constant, we must use the average flow speed v_{ave}. Hence, we have

$$Q = v_{ave}A \tag{10.32}$$

where A is the area of the tube or pipe.

EXAMPLE 10.11 Blood Flow through an Artery

Consider the flow of blood through the large artery that extends from the heart to the lungs. The radius of this artery is typically 2.0 mm, and it is about 10 cm long. **(a)** If the pressure difference across the ends of the artery is 400 Pa (about 0.4% of atmospheric pressure), what is the average speed of the blood? **(b)** Now suppose this artery becomes somewhat narrower, as often happens with age. If the radius is reduced to 1.5 mm, what pressure difference is required to maintain the same average speed as in part (a)?

RECOGNIZE THE PRINCIPLE

This problem is a direct application of Poiseuille's law (Eq. 10.31), the relation between pressure difference and average speed for the flow of a viscous fluid (blood).

SKETCH THE PROBLEM

Figure 10.32 describes the problem.

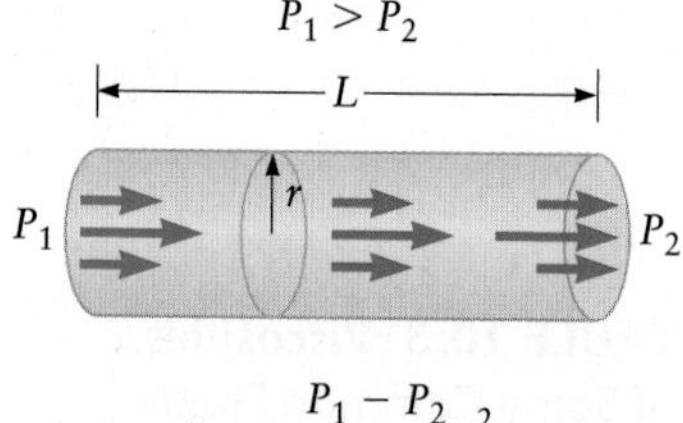

▲ Figure 10.32 Example 10.11. Poiseuille's law tells how the average flow speed depends on the dimensions of the tube, the viscosity, and the pressure difference $P_1 - P_2$.

IDENTIFY THE RELATIONSHIPS

According to Poiseuille's law, the average flow speed is related to the pressure across the artery ($P_1 - P_2$) by

$$v_{ave} = \frac{P_1 - P_2}{8\eta L} r^2$$

where r is the radius of the artery and η is the viscosity.

SOLVE

(a) For blood (viscosity $\eta = 0.0027\ \text{N}\cdot\text{s/m}^2$ as listed in Table 10.3) flowing in an artery of radius $r = 2.0$ mm with a pressure difference $P_1 - P_2 = 400$ Pa, we find

$$v_{ave} = \frac{P_1 - P_2}{8\eta L} r^2 = \frac{(400\ \text{Pa})}{8(0.0027\ \text{N}\cdot\text{s/m}^2)(0.10\ \text{m})}(0.0020\ \text{m})^2 = \boxed{0.74\ \text{m/s}}$$

(b) To find the effect of changing the radius, we rearrange Poiseuille's law to solve for the pressure difference:

$$v_{ave} = \frac{P_1 - P_2}{8\eta L} r^2$$

$$P_1 - P_2 = \frac{8\eta L v_{ave}}{r^2}$$

Now we insert our value for the desired flow velocity—the value of v_{ave} found in part (a)—and the new radius:

$$P_1 - P_2 = \frac{8\eta L v_{ave}}{r^2} = \frac{8(0.0027\ \text{N}\cdot\text{s/m}^2)(0.10\ \text{m})(0.74\ \text{m/s})}{(0.0015\ \text{m})^2} = \boxed{710\ \text{Pa}}$$

▸ ***What does it mean?***

When the radius is decreased, the pressure must increase from 400 Pa to 710 Pa—quite a substantial change—to maintain the same flow speed. This example shows how what might seem to be a small change in an artery's size can lead to high blood pressure and the associated health risks.

Viscosity and Stokes's Law

Viscosity leads to a frictional force between adjacent "layers" of a moving fluid and also to a frictional force on the wall in Figure 10.31. The wall attracts molecules of the fluid, and the fluid molecules in turn exert a reaction force on the wall (Newton's third law at work again). This force on the wall is also the source of the drag force we considered in Chapter 3. According to Stokes's law (Eq. 3.23), the drag force on a spherical object of radius r that moves slowly through a fluid with a viscosity η is

$$\vec{F}_{\text{drag}} = -6\pi\eta r\vec{v} \tag{10.33}$$

where $\vec{v}$ is now the velocity of the object in the fluid. In Chapter 3, we wrote this relation as $\vec{F}_{\text{drag}} = -Cr\vec{v}$, where C is a constant; now we see that the drag coefficient C is proportional to the viscosity η. This is the *same* viscosity that enters into Poiseuille's law. The negative sign in Equation 10.33 indicates that the drag force is in the opposite direction as the object's velocity.

The Stokes's law force in Equation 10.33 is the key to understanding an important method for separating molecules according to their size. This method, based on a process called ***electrophoresis***, is shown schematically in Figure 10.33. Here we imagine a fluid containing a collection of molecules of different sizes; in a typical application, they are biological molecules such as DNA, RNA, or proteins. An electric force is then applied that pushes the molecules through the fluid such that the magnitude of the electric force is the same on each molecule.[2] Each molecule also experiences viscous drag. The Stokes drag force in Equation 10.33 applies to spherical objects, so it does not apply precisely to a real biomolecule. The basic physics of viscous drag certainly *does* apply, however, and each molecule in Figure 10.33 experiences a drag force whose magnitude depends on the molecule's size and shape. For these biomolecules, the drag force is given by the relation

$$\vec{F}_{\text{drag}} = -B\vec{v} \tag{10.34}$$

where B is a factor that depends on the size of the molecule. The value of B is a constant for any given kind of molecule—that is, for a particular type of DNA or RNA or for a particular virus—but it is different for different molecules.

If the electric force on a molecule is F_E, Newton's second law becomes

$$\sum F = F_E - Bv = ma$$

where we assume one-dimensional motion so that the forces and v are all along a single direction. This net force accelerates a molecule until it reaches a terminal speed (as discussed in Chapter 3), at which time the molecule's speed is constant and its acceleration is zero, leading to

$$\sum F = F_E - Bv_{\text{term}} = 0$$

$$v_{\text{term}} = \frac{F_E}{B}$$

Since B is different for different molecules, the terminal speed v_{term} also varies. Hence, different molecules travel through the fluid with different speeds, providing a way

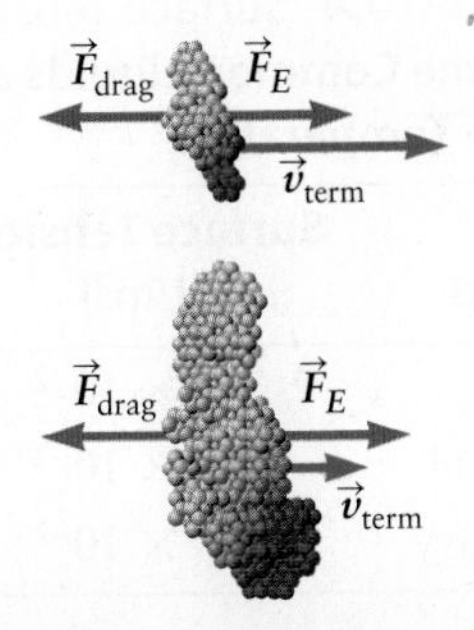

The electric forces are the same, but the drag coefficients are different, so the terminal speed is *larger* for the *smaller* molecule.

▲ **Figure 10.33** When an electric field is applied to a typical biomolecule in solution, the electric force F_E on the molecule is usually the same for all molecules because they usually have the same electric charge. The molecules also experience a drag force due to the viscosity of the fluid. This drag force depends on the size of the molecules, so different molecules move with different terminal speeds. This analysis is very similar to our calculation of the terminal speed of a skydiver in Chapter 3.

[2]The origin of this force is explained in Chapter 17, where we discuss how an electric field exerts a force on an ion.

▲ **Figure 10.34** A and B Possible results for a small amount of liquid sitting on a solid surface. C Surface tension allows certain creatures to "walk" on the surface of a liquid.

to separate molecules by type. This difference is the basis for one method of DNA "fingerprinting."

Surface Tension

Consider a small amount of a liquid placed on the surface of a solid such as water sitting on a sheet of glass. Parts A and B of Figure 10.34 show two possible results; in part A, the water forms a spherical droplet, whereas in part B, the water spreads out as a flat layer. Which of these cases is found in real life? *Both* are possible, depending on the intermolecular forces. In most liquids, the liquid molecules bind more tightly to one another than to air. That is, molecules in the liquid "prefer" to be in contact with one another rather than in contact with molecules in the air. For this reason, droplets of such liquids form a shape giving the smallest possible surface area with the air, and that shape is a spherical drop.

This result can be expressed mathematically by saying that the fluid has a ***surface tension*** γ, where γ is a measure of the energy required to increase the surface area. The surface tensions of some typical fluids are given in Table 10.4. The SI unit of surface tension is joules per square meter (J/m^2).

TABLE 10.4 Surface Tensions of Some Common Liquids at Room Temperature

Liquid	Surface Tension (J/m^2)
Water	7.3×10^{-2}
Ethanol	2.3×10^{-2}
Mercury	48×10^{-2}

Surface tension is responsible for effects such as the one shown in Figure 10.34C, where a bug is "walking" on water. As with the case of a spherical drop, the surface of the water takes on a shape that keeps the surface area as small as possible, given that it also must support the bug's weight. Here, that shape is a nearly flat surface, which supports the bug's feet rather than letting them penetrate deep under the original surface. The weight of the bug is very small, so the forces that bind the water molecules together at the surface are able to overcome the added force due to the bug's weight.

Capillary Pressure

We have just stated that the molecules of a liquid such as water are attracted more strongly to other water molecules than to air molecules. When a liquid is placed in a very narrow tube, the shape of the surface also depends on the forces between the molecules in the liquid and the walls of the tube. Figure 10.35 shows water in a narrow glass tube, called a capillary tube. The surface of the liquid in a capillary tube is *not* flat; rather, the water's surface (called the meniscus) in a glass capillary tube is concave upward. Figure 10.35A also shows a molecular-scale view. Water molecules are attracted to the glass, and this attraction pulls the surface of the water up a small amount at the walls of the tube, producing the concave shape of the meniscus.

The molecules in some liquids do not bind well to the surfaces of a capillary tube. The case of mercury is shown in Figure 10.35B; here the mercury atoms are attracted more strongly to one another than to the walls, leading to a convex meniscus. Hence, the shape of the meniscus in a capillary tube depends on the molecular forces between the liquid and the walls of the tube.

The attraction of water molecules in Figure 10.35A to the walls of a glass capillary tube leads to a force that "pulls" the water upward in the tube. This force, which is associated with the ***capillary pressure***, causes the curvature of the water's surface (the meniscus), and the magnitude of this effect is proportional to the surface tension. In the simplest case, like the one shown in Figure 10.35A, the capillary pressure for a liquid in a tube of radius r is given by

$$P_{cap} = \frac{2\gamma}{r} \tag{10.35}$$

where γ is again the surface tension (Table 10.4). This result applies to water in a glass tube, where the fluid molecules are strongly attracted to the walls of the tube. For other cases, such as mercury in a glass tube (Fig. 10.35B), the capillary pressure is given by a slightly more complicated relation than Equation 10.35 and is more than we want to discuss in this book.

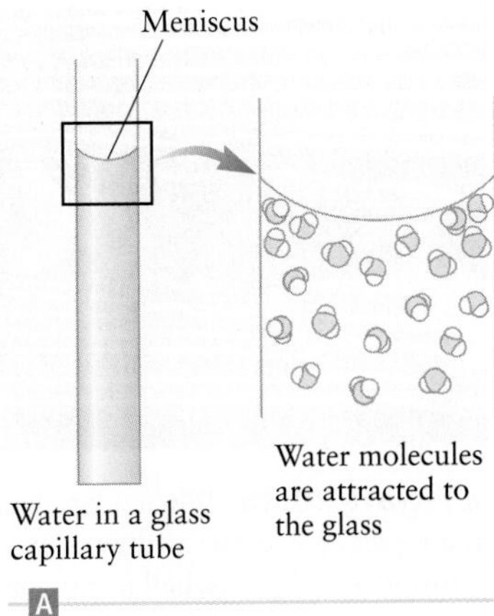

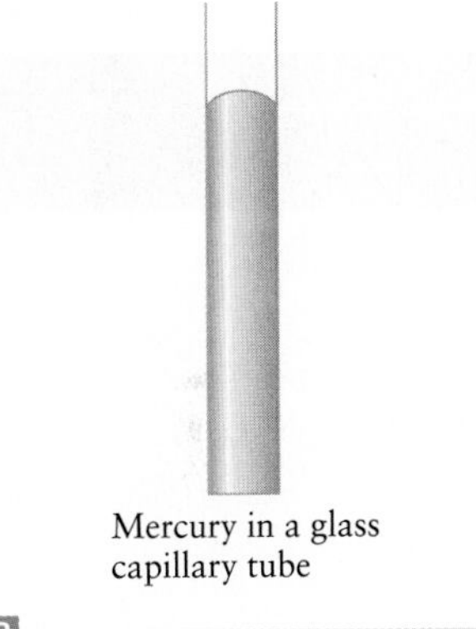

▲ **Figure 10.35** A When a thin glass capillary tube contains water, the shape of the water surface (the meniscus) is concave upward. This shape is caused by the attraction of the water molecules to the surface of the tube. B For mercury in a glass tube, the meniscus is convex.

How Plants Use Capillary Pressure

As an example of capillary pressure in action, let's consider how a tree moves water from its roots to its top branches. This problem is similar to the one we considered in Figure 10.13, where we discussed how to pump water from an underground well to the surface. There we saw that with a pressure difference of P_{atm}, one can only pump water up to a maximum height $h = (P_{atm}/\rho g) = 10$ m.

A tree is faced with a similar problem as it "pumps" water from its roots to its leaves. A tree also has a maximum pressure difference of P_{atm} to work with; hence, one might suspect a tree to have the same limitation on the maximum pumping height. So, with 10 m equal to about 30 ft, a tree cannot be taller than 30 ft and still move water from its roots to its top branches in this way. We know, though, that trees can be taller than 30 ft. How do they do it? The answer is that they use capillary pressure. The tubes that take water from tree's roots to its branches vary in size, but a radius of 1 μm (= 10^{-6} m) is quite common. According to Equation 10.35, and consulting Table 10.4 for the surface tension of water, we find a capillary pressure of

$$P_{cap} = \frac{2\gamma}{r} = \frac{2(7.3 \times 10^{-2}\ \text{J/m}^2)}{1 \times 10^{-6}\ \text{m}} = 1 \times 10^5\ \text{Pa} \tag{10.36}$$

The case of a tree pumping water to its branches is then similar to the pump in Figure 10.13A, but now the capillary pressure adds to the atmospheric pressure, which enables the tree to pump water to a greater height. Recall that $P_{atm} = 1.01 \times 10^5$ Pa, so the capillary pressure in Equation 10.36 is just as great as atmospheric pressure. Narrower tubes can produce larger pressures. That is how trees do it.[3]

10.7 Turbulence

In our discussion of moving fluids and Bernoulli's equation, we explained how the pattern of airflow around an airplane wing produces lift on the wing, and we drew a hypothetical flow pattern in Figure 10.26. The flow pattern shown there is particularly simple. In many situations, fluid flow patterns can be much more complex and very difficult to predict. Indeed, in the development of the airplane, designers such as the Wright brothers used experiments in wind tunnels (Fig. 10.27) to study lift forces on wings of different shapes. Similar studies can now be done with large-scale

[3]The precise details of this process in trees are still a matter of much research. Although our discussion is certainly a simplified version, capillary pressure is the basic key to the process.

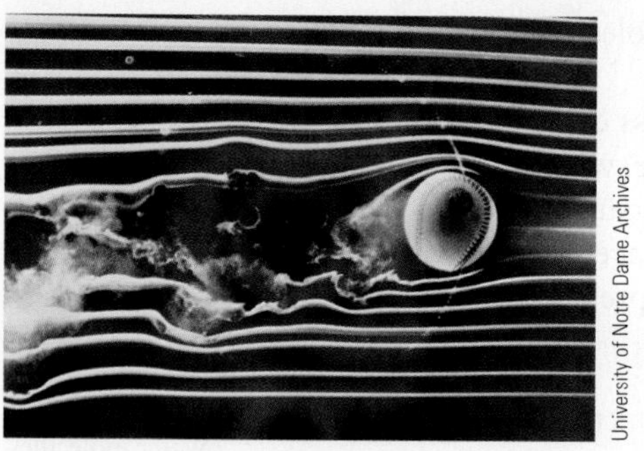

▲ **Figure 10.36** When a baseball travels at a typical speed, the airflow around the ball is turbulent. This baseball is moving from left to right.

computer simulations, but wind tunnels are still widely used to explore the aerodynamic forces on cars, bicycles, airplanes, and many other objects.

Figure 10.36 shows an example of the types of complex flow patterns that occur near even simple objects. This photo shows the flow of air around a baseball. When the speed of the incoming air is greater than about 25 m/s (about 50 mi/h), ***vortices*** are created in the baseball's wake. These vortices are circulating patterns of air, much like small tornados.

When fluid speeds are large, the flow patterns can be very complex and fluctuate erratically with time, a behavior known as ***turbulence***. Such complex flows play a role in a wide range of problems, including combustion, the weather (tornados and hurricanes), and the noise produced by rapidly moving cars and airplanes.

Summary | CHAPTER 10

Key Concepts and Principles

Pressure and density

The mechanics of fluids is described using the quantities ***pressure*** and ***density***. Pressure is equal to the force per unit area,

$$P = \frac{F}{A} \qquad \text{(10.1) (page 276)}$$

Pressure can be associated with the force on the surface of a container, or it can involve a surface contained within the fluid. The pressure in the atmosphere at the Earth's surface is approximately

$$P_{\text{atm}} = 1.01 \times 10^5 \text{ Pa} \qquad \text{(10.2) (page 276)}$$

The density of a substance is given by

$$\rho = \frac{M}{V} \qquad \text{(10.3) (page 279)}$$

An ***incompressible*** fluid is one for which the density is independent of pressure; most liquids are incompressible. The density of a gas varies with pressure; hence, gases are ***compressible*** fluids.

Dependence of pressure on depth

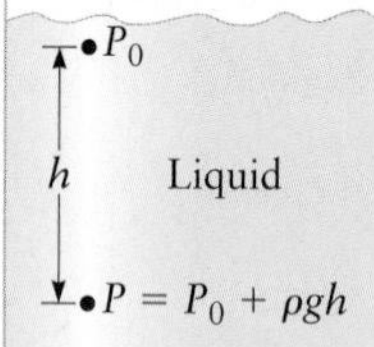

The pressure in a liquid varies with depth as

$$P = P_0 + \rho g h \qquad \text{(10.5) (page 280)}$$

The pressure increases as one moves deeper in a fluid because the pressure below must support the weight of the fluid above.

Real fluids

Fluids are composed of molecules, which helps explain the origin of ***viscosity*** (i.e., friction), ***surface tension***, and ***capillary pressure***. ***Poiseuille's law*** relates the average flow speed to the viscosity of the fluid and the pressure difference across the ends of a pipe:

$$v_{\text{ave}} = \frac{P_1 - P_2}{8\eta L} r^2 \qquad \text{(10.31) (page 301)}$$

where $P_1 - P_2$ is the pressure difference and η is the viscosity. The corresponding flow rate is

$$Q = v_{ave}A \qquad (10.32)\ (\text{page } 302)$$

Surface tension γ is the energy needed to increase the surface area of a fluid. The surface tension leads to an extra pressure when a fluid is confined to a narrow capillary. This capillary pressure is

$$P_{cap} = \frac{2\gamma}{r} \qquad (10.35)\ (\text{page } 305)$$

Applications

Pascal's principle

Pressure changes at one location in a closed container of fluid are transmitted equally to all locations in the fluid. This principle is the basis of ***hydraulics***. Hydraulic devices, such as a lift, are able to amplify forces. However, they de-amplify the corresponding displacement so that the work done is conserved.

Archimedes's principle

When an object is immersed in a fluid, the fluid exerts an upward force on the object that is equal to the weight of the fluid that is displaced by the object. This upward force is called the ***buoyant force***.

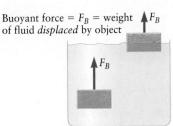

Buoyant force = F_B = weight of fluid *displaced* by object

F_B

F_B

Archimedes's principle and buoyancy

Principle of continuity

When a fluid flows through a pipe (or similar geometry), the ***equation of continuity*** tells how the speed v of a fluid is related to the cross-sectional area A of the pipe:

$$v_1A_1 = v_2A_2 \qquad (10.23)\ (\text{page } 296)$$

where the subscripts 1 and 2 denote two different sections of the pipe.

The work–energy theorem leads to ***Bernoulli's equation***, which relates the speed, pressure, and height at two different locations in an ideal fluid:

$$P_1 + \tfrac{1}{2}\rho v_1^2 + \rho g h_1 = P_2 + \tfrac{1}{2}\rho v_2^2 + \rho g h_2 \qquad (10.28)\ (\text{page } 298)$$

According to Bernoulli's equation, when the flow speed increases, the pressure must decrease, and vice versa. This leads to the lift force on an airplane wing.

Questions

SSM = answer in *Student Companion & Problem-Solving Guide* (X) = life science application

1. Design a hydraulic lift that could be used to move a car. Assume the input piston the person would push on has an area of 0.10 m^2. If you want a force of 300 N applied by the person to lift a car, estimate the area of the piston supporting the car.

2. (X) Explain how Archimedes's principle can be used to measure a person's percentage body fat. *Hint*: Fat has a different density than other body tissue.

3. The specific gravity of a substance is the ratio of its density to the density of water. Among other applications, measurements of the specific gravity are used by mineralogists to determine the composition of a gemstone. What are the units of specific gravity?

4. SSM Figure Q10.4 is a photograph of a graduated cylinder filled with four fluids. Starting with mercury on the bottom and going up, we have salt water, water, and vegetable oil. In addition, a solid object rests at the interface between each liquid. At the bottom is a steel ball bearing, next an egg, followed by a block of wood, and a table-tennis ball on top. (a) Rank these eight substances in terms of their densities from the highest to the lowest. (b) Which items have a specific gravity greater than 1? Less than 1?

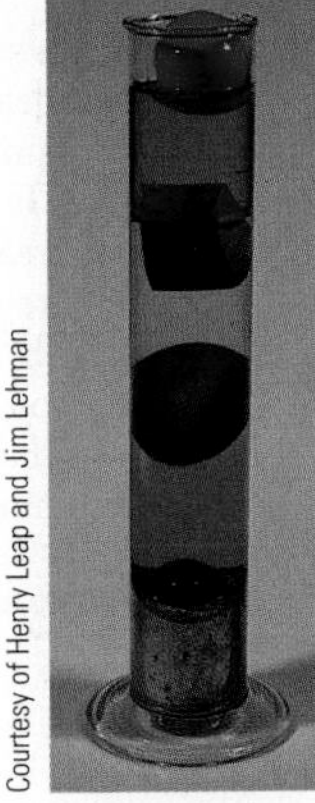

Courtesy of Henry Leap and Jim Lehman

Figure Q10.4

5. Explain why an ice cube floats in water. For ordinary water (i.e., not salt water) at room temperature, what fraction of an ice cube is above the water?

6. When water flows slowly from a faucet or pipette, the stream narrows and then eventually breaks into droplets (Fig. Q10.6). Explain why. *Hint*: Consider the effect of surface tension.

7. Explain why some people are able to float in ocean water but not in a freshwater lake (see Fig. Q10.7). (A freshwater lake contains rainwater.)

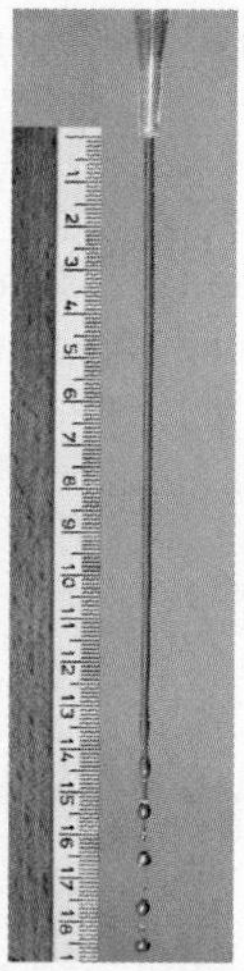

Figure Q10.7

Figure Q10.6

8. A cargo ship travels from the Mississippi River into the Gulf of Mexico. Will the ship sink or rise with respect to the waterline as it moves from the river to ocean water? Why?

9. Three containers are filled with water to the same height (Fig. Q10.9). For which container is the pressure at the bottom the greatest? Or, are the pressures the same? Explain.

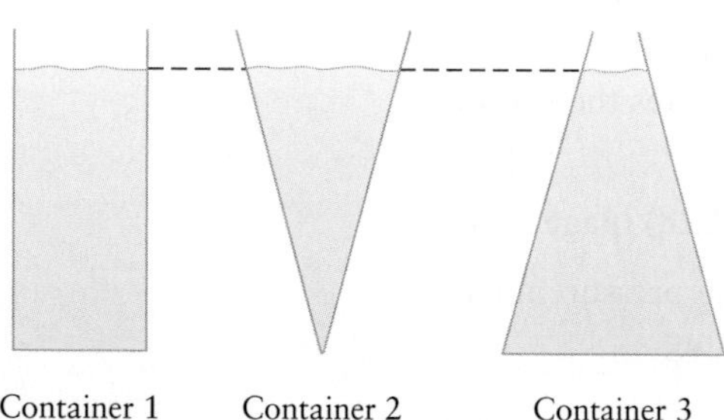

Figure Q10.9

10. Why does the "lava" in a Lavalamp (Fig. Q10.10) rise and then fall?

11. Consider a swimming pool and a tall graduated cylinder (like the one in Fig. Q10.4), both containing water and filled to the same height. (a) Is the pressure at the bottom of the pool greater than, less than, or equal to the pressure at the bottom of the graduated cylinder? (b) Is the total force exerted by the water on the bottom of the pool greater than, less than, or equal to the total force on the bottom of the graduated cylinder?

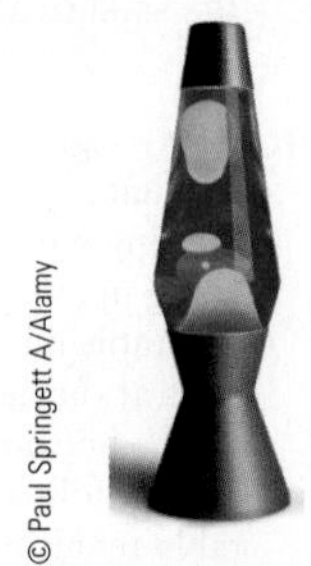

Figure Q10.10

12. Why are dams thicker at the bottom than at the top?

13. Consider a glass tube that has the shape shown in Figure Q10.13. It is filled with a fluid and is closed at the end on the left. If P_A is the pressure at point A, P_B the pressure at point B, and so forth, rank the pressures P_A, P_B, . . . from lowest to highest.

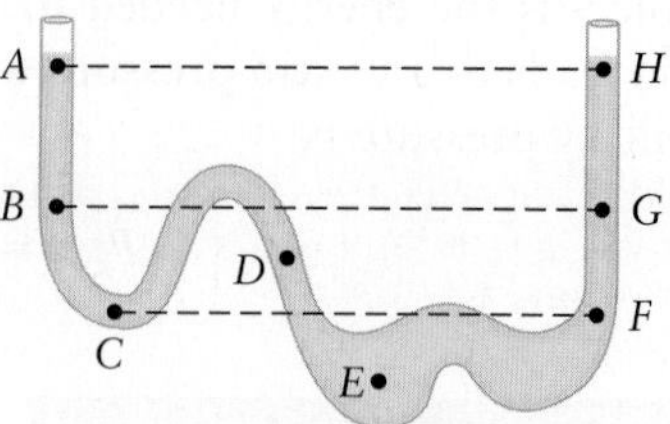

Figure Q10.13

14. Two blocks with the same volume but different densities are in a lake as shown in Figure Q10.14. Block m_1 floats at the surface with a portion above the water level, and block m_2 floats underwater. (a) Which block has the greatest mass? (b) For which block is the buoyant force greatest?

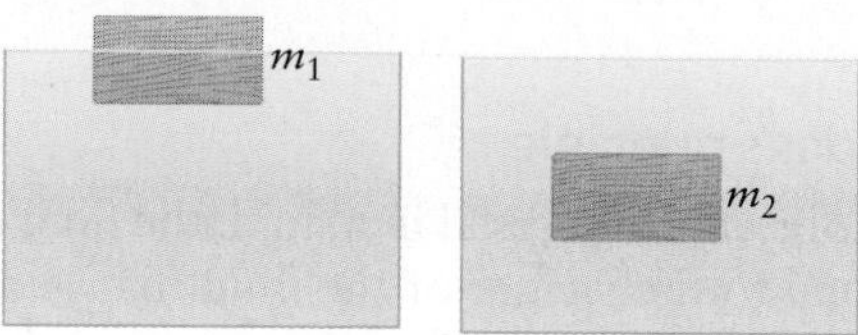

Figure Q10.14

15. When traveling in a commercial airplane, you are sometimes given snacks, such as peanuts, in a sealed bag. The bag is often "bulging" much more than when you purchase a bag of peanuts in a store. Explain why.

16. SSM Figure Q10.16 shows a popular demonstration involving a moving fluid. Here an air "jet" is aimed upward and levitates a small object such as a table-tennis ball. The ball is drawn *to the center of the jet*, where the speed is greatest. This behavior is found when the air jet is directed vertically (Fig. Q10.16 left) and also when the jet is directed at an angle (Fig. Q10.16 right). Use Bernoulli's equation to explain this behavior.

Figure Q10.16

17. Explain how there can be dust on a moving fan blade.

18. An altimeter is a device used to measure altitude. Most altimeters are based on measurements of air pressure. Explain how they work and estimate how much the pressure changes for a 100-m change in altitude. Discuss how changes in the weather can affect an altimeter reading.

19. In preparation for a bike tour of several countries, you ship your bicycle by air freight to Europe. When you arrive, you find that the tires of your bicycle have both blown out. Explain why that happened.

20. Explain how the specific gravity of a substance is connected with the buoyant force when an object is immersed in water.

21. The region near the South Pole in Antarctica has an altitude of about 3000 m, making the air pressure lower than at sea level. Explain why it is more difficult for an airplane to take off from the South Pole than from an airfield at sea level.

22. The author once had a small car with a "hatchback," a rear door that opens upward. One day, he forgot to latch it before driving on the highway, and he found that at high speeds the rear door lifted up as in Figure Q10.22B. Explain why.

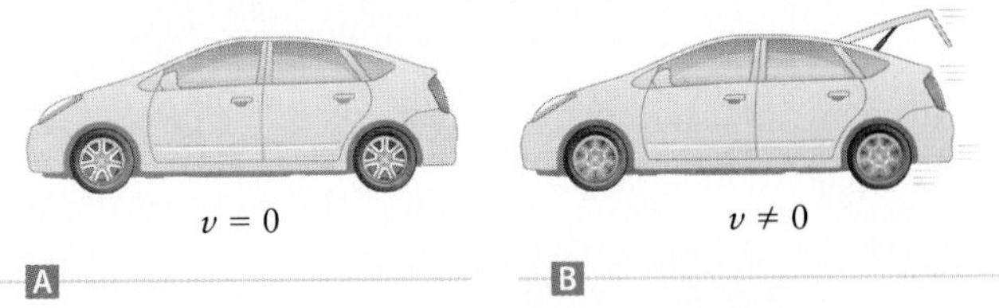

Figure Q10.22

Problems available in ENHANCED WebAssign

Section 10.1 Pressure and Density
Problems 1–20

Section 10.2 Fluids and the Effect of Gravity
Problems 21–35

Section 10.3 Hydraulics and Pascal's Principle
Problems 36–39

Section 10.4 Buoyancy and Archimedes's Principle
Problems 40–56

Section 10.5 Fluids in Motion: Continuity and Bernoulli's Equation
Problems 57–67

Section 10.6 Real Fluids: A Molecular View
Problems 68–77

Additional Problems
Problems 78–99

List of Enhanced Problems

Problem number	Solution in *Student Companion & Problem-Solving Guide*	Reasoning & Relationships Problem	Reasoning Tutorial in ENHANCED WebAssign
10.5		✓	
10.7		✓	✓
10.12	✓		
10.14		✓	✓
10.15		✓	
10.19		✓	✓
10.25		✓	
10.26		✓	
10.27	✓		
10.31		✓	✓
10.33		✓	
10.38	✓		
10.44		✓	✓
10.48	✓	✓	✓
10.59	✓		
10.60		✓	✓
10.62		✓	
10.65	✓	✓	
10.71		✓	
10.72	✓		
10.75		✓	
10.76		✓	
10.87		✓	✓
10.91	✓		
10.98		✓	
10.99		✓	

◀ *The physics of oscillations and elasticity are keys to the operation of this bungee cord (and the safety of the jumper!). (David Wall/Alamy)*

Harmonic Motion and Elasticity

In this chapter, we consider ***harmonic motion***, a type of motion in which an object moves along a repeating path over and over and over again. The universe is filled with examples of harmonic motion, such as the pendulum motion of a child on a playground swing and the motion of the Earth as it orbits the Sun. While there are many different examples of harmonic motion, certain features involving force and energy are common to all. We'll also find a close connection to many features of circular and rotational motion we have studied in previous chapters.

OUTLINE

11.1 General Features of Harmonic Motion

Figure 11.1A shows a person hanging from a spring. This spring is a little bit like the bungee cord in the photo at the start of this chapter, except that the spring in Fig. 11.1 is tightly coiled so that the force from the spring is the dominant force acting on the person (e.g., this force dominates over the force of gravity). Figure 11.1A shows several "snapshots" of the person as he "vibrates" up and down due to the force from the bungee spring.

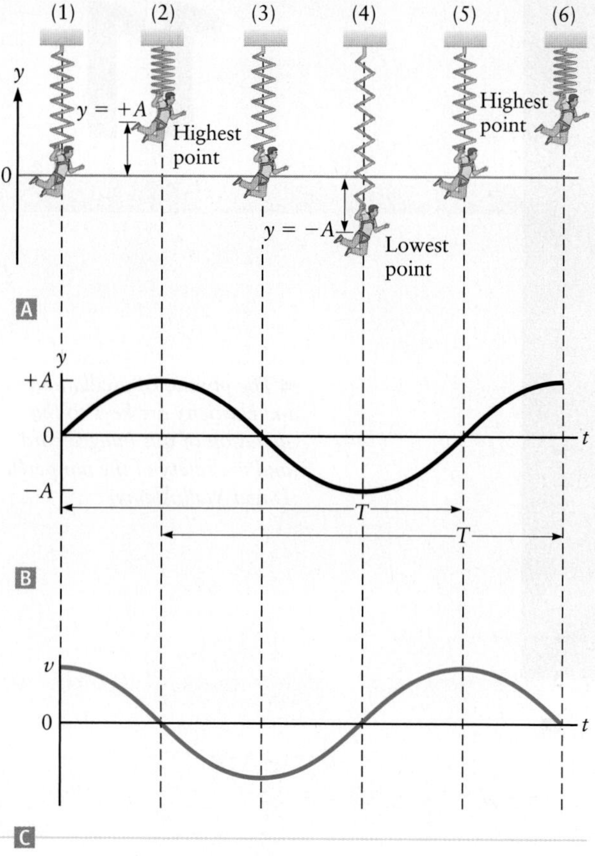

▲ **Figure 11.1** A This bungee jumper undergoes simple harmonic motion as he oscillates up and down. B Plot of the jumper's position y versus time. The period T of the motion is the time it takes to make one complete up-and-down oscillation. C Plot of the jumper's velocity versus time.

We can describe the position of the person on the bungee spring by measuring his location y on a vertical axis, and Figure 11.1B shows a plot of y as a function of time. Notice how each point along the y–t plot matches the corresponding physical location of the person in Figure 11.1A. This is an example of ***oscillatory motion***; the jumper's motion varies in a repeating fashion as he moves up and down.

To describe oscillatory motion mathematically, we use the notions of position, velocity, and acceleration, but we also need some additional quantities that relate to the repeating nature of the motion. One such variable is the ***period*** T, which is the *repeat time* of the motion. We have already encountered this quantity in our discussion of circular motion in Chapter 5. The period of our bungee jumper is the time it takes him to travel through one complete "cycle" of the motion, so it is the time interval between adjacent maxima of the position y when plotted as a function of time (Fig. 11.1B). The period T is also equal to the time interval between adjacent minima of y.

Another way to characterize harmonic motion is in terms of its ***frequency***. The frequency f is the number of oscillations that occur in one unit of time; hence, f is the number of cycles completed in 1 second. Since the period is the number of seconds required to perform one cycle, frequency and period are related by

$$f = \frac{1}{T} \tag{11.1}$$

The units of frequency are cycles per second, which are called ***hertz*** (Hz).

EXAMPLE 11.1 Measuring the Period

Consider a hypothetical system for which the position y oscillates as sketched in Figure 11.2. Estimate the period and frequency of the motion.

RECOGNIZE THE PRINCIPLE

The period T is the time required for one *complete* oscillation, and we can estimate T from Figure 11.2. The frequency is then given by $f = 1/T$.

SKETCH THE PROBLEM

Figure 11.2 describes the problem.

IDENTIFY THE RELATIONSHIPS

We can estimate the period from the spacing along the time axis between two corresponding (i.e., "equivalent") points of the plot in Figure 11.2. For example, we could choose two adjacent maxima, such as points A and D, or we could choose two adjacent minima (points B and E) or any other pair of adjacent "repeated" points (such as C and F).

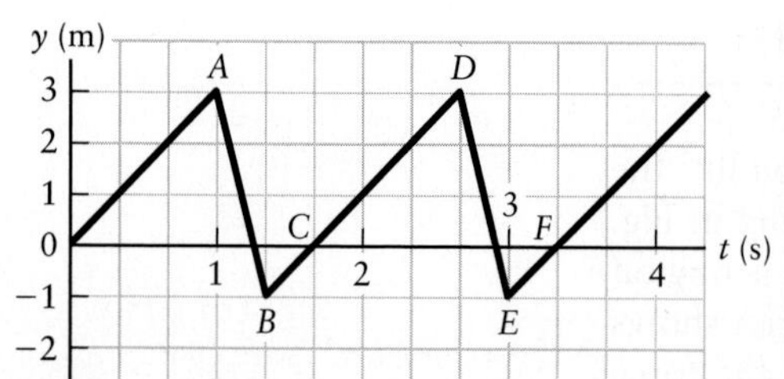

▲ **Figure 11.2** Example 11.1.

SOLVE

From Figure 11.2, the time between points A and D is approximately

$$T \approx \boxed{1.7 \text{ s}}$$

The frequency is then

$$f = \frac{1}{T} = \frac{1}{1.7\ \text{s}} = \boxed{0.59\ \text{Hz}}$$

► ***What does it mean?***
Notice that the units of f are hertz.

Simple Harmonic Motion

In many cases, the motion of an oscillator is described by a simple sinusoidal variation with time,

$$y = A\sin(2\pi ft) \tag{11.2}$$

For our bungee jumper (Fig. 11.1A), y is the location along the vertical axis, and Equation 11.2 describes how y varies with time in Figure 11.1B. Systems that oscillate in a sinusoidal manner are called ***simple harmonic oscillators***, and they exhibit ***simple harmonic motion***. You might think that the behavior described by Equation 11.2 is too simple to describe any real system, but simple harmonic motion is, in fact, extremely common.

The quantity A in Equation 11.2 is called the ***amplitude*** of the motion. The bungee jumper in Figure 11.1A moves back and forth between the values $y = \pm A$ as indicated in the figure.

Figure 11.1C shows how the velocity of the bungee jumper varies with time. The velocity also "oscillates" between positive and negative values as the bungee jumper moves, and the frequency of these oscillations is the same as the frequency associated with the position y. However, the maximum values of the velocity do *not* occur when y has its maximum values. Instead, the largest values of v occur when $y = 0$. Likewise, the jumper's velocity is zero when he is at his maximum and minimum heights, at $y = \pm A$ (snapshots 2, 4, and 6 in Fig. 11.1A).

CONCEPT CHECK 11.1 Acceleration of an Object Undergoing Simple Harmonic Motion

Consider the acceleration of an object that is undergoing simple harmonic motion. Is this acceleration zero or nonzero? If it is nonzero, there must be a nonzero force acting on the object. What forces act on the bungee jumper in Figure 11.1?

The Connection between Simple Harmonic Motion and Circular Motion

A DVD spinning at a constant rate is another example of periodic, repeating motion. There is a close relationship between uniform circular motion as followed by a point on the edge of the disc and simple harmonic motion.

A particle at the edge of the DVD moves at a constant speed we call v_c around a circle of radius A (Fig. 11.3). We can specify the particle's position using a reference line drawn on the disc. If θ is the angle this reference line makes with the x axis, then θ increases with time according to

$$\theta = \omega t \tag{11.3}$$

where ω is the angular velocity of the corresponding rotational motion (Chapter 8). As the particle completes one full trip around the circle, θ varies from zero to 2π (measured in radians), corresponding to one full period of this harmonic motion. From Equation 11.3, we thus have after one full period ($t = T$):

$$\theta = 2\pi = \omega t = \omega T$$

Solving for T gives

$$T = \frac{2\pi}{\omega}$$

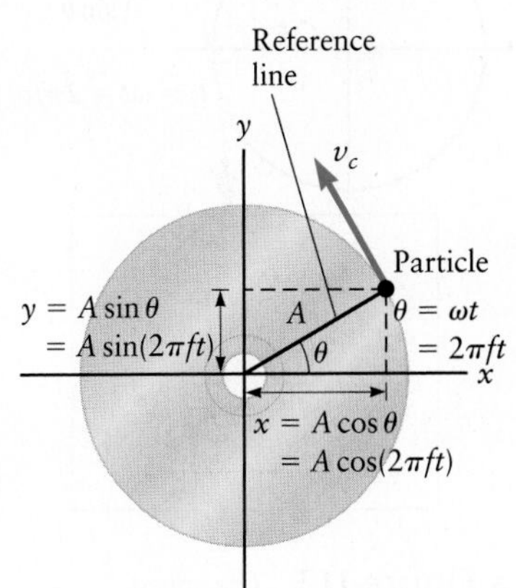

▲ **Figure 11.3** When a particle moves in a circle with constant speed v_c, both the x and y components of the particle's position undergo simple harmonic motion.

Both curves describe simple harmonic motion.

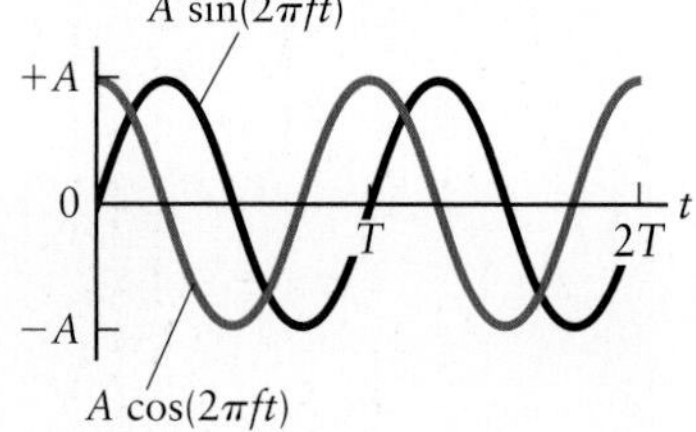

▲ **Figure 11.4** A sine-function time dependence (Eq. 11.5) and a cosine-function time dependence (Eq. 11.6) both describe simple harmonic motion. They differ only by a shift along the horizontal (time) axis.

Comparing this result with our definition of frequency in Equation 11.1, we see that harmonic motion can be described in terms of either an angular velocity ω or an equivalent frequency f, with

$$f = \frac{\omega}{2\pi} \tag{11.4}$$

The frequency f is the number of cycles a system completes in 1 second. The disc in Figure 11.3 moves through an angular displacement of 2π during each cycle, so Equation 11.4 says (in words) that the system completes an angular displacement of ω each second. For this reason, ω is also called the ***angular frequency***.

Simple Harmonic Motion: Position as a Function of Time. The connection between circular motion and simple harmonic motion becomes even clearer when we describe the motion of the particle in Figure 11.3 in terms of its coordinates in the x–y plane. For the moment, let's follow the y coordinate. According to the trigonometry in Figure 11.3, the y component of the particle's position is $y = A \sin\theta$. Since $\theta = \omega t$ (Eq. 11.3), we have

$$y = A \sin\theta = A \sin(\omega t)$$

Using $\omega = 2\pi f$ (Eq. 11.4), we also get

$$y = A \sin(2\pi f t) \tag{11.5}$$

which is precisely the same as the motion of a simple harmonic oscillator in Equation 11.2 with amplitude A.

There is nothing special about the y direction. We could just as easily make the connection between circular motion and simple harmonic motion using the x component of the particle's position. The x coordinate of the particle in Figure 11.3 is $x = A \cos\theta = A \cos(\omega t)$, which leads to

$$x = A \cos(2\pi f t) \tag{11.6}$$

Although this equation looks a bit different from our original expression for simple harmonic motion (Eqs. 11.2 and 11.5), it describes the same basic motion. This can be seen from Figure 11.4, which shows both a sine function (Eqs. 11.2 and 11.5) and a cosine function (Eq. 11.6). Both functions describe simple harmonic motion with period $T = 1/f$. The only difference is a "shift" along the time axis. One expression (the sine function) begins at $y = 0$ when $t = 0$, while the other (the cosine function) starts at a maximum when $t = 0$. In terms of the actual motion of our bungee jumper in Figure 11.1, the sine function describes a jumper who starts at the "midpoint" ($y = 0$) at $t = 0$, whereas the cosine function describes a jumper who starts at the highest point ($y = +A$). The only difference is in the way in which the system is initially set into motion.

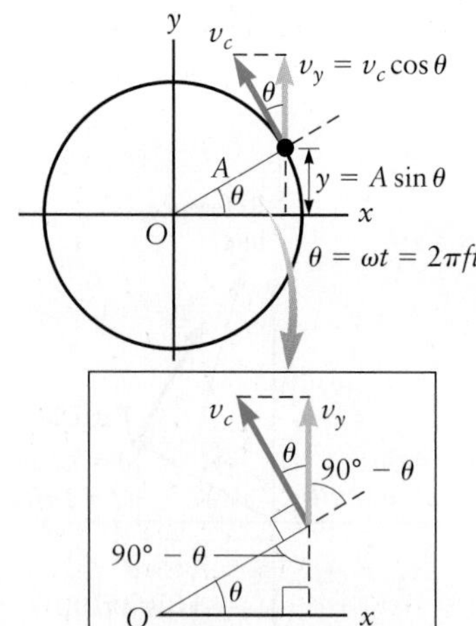

▲ **Figure 11.5** The speed v_c of the particle as it moves along the circle is constant, and the component of the velocity along y is $v_y = v_c \cos\theta$. Hence, v_y varies with time according to $v_y = v_c \cos(\omega t) = v_c \cos(2\pi f t)$.

Simple Harmonic Motion: Velocity as a Function of Time. We just showed that we can derive the position of a simple harmonic oscillator by taking either the x or y component of the position of a particle moving in a circle. We can also use the connection with circular motion to derive the velocity of a simple harmonic oscillator. Let's again follow the y coordinate of a particle moving in a circle (Fig. 11.5), so the velocity of our oscillator is v_y, the y component of the particle's velocity. Although the speed of the particle along the circular path in Figure 11.5 is constant, the y component v_y of its velocity is *not constant*. From Figure 11.5, we get

$$v_y = v_c \cos\theta = v_c \cos(\omega t)$$

$$v_y = v_c \cos(2\pi f t) \tag{11.7}$$

The particle travels once around the circle (a distance of $2\pi A$) in a time equal to one period, so its speed is $v_c = 2\pi A/T$. Using Equation 11.1 relating the period and frequency gives

$$v_c = \frac{2\pi A}{T} = 2\pi f A \tag{11.8}$$

Inserting this into Equation 11.7, we obtain

$$v_y = 2\pi fA\cos(2\pi ft)$$

Here v_y is the velocity of our simple harmonic oscillator, which we can simply call v.

We thus find that the position and velocity of a simple harmonic oscillator can be described by

$$y = A\sin(2\pi ft) \qquad v = 2\pi fA\cos(2\pi ft) \tag{11.9}$$

Position and velocity of a simple harmonic oscillator

(Compare Eq. 11.9 with the plots in Fig. 11.1.) This result for the velocity applies when we use the y component of circular motion to describe simple harmonic motion. We could also have used the x component, which leads to (see Eq. 11.6)

$$x = A\cos(2\pi ft) \qquad v = -2\pi fA\sin(2\pi ft) \tag{11.10}$$

Position and velocity of a simple harmonic oscillator

The relations in Equation 11.9 and 11.10 both describe simple harmonic motion.

CONCEPT CHECK 11.2 Analyzing a Simple Harmonic Oscillator

The position of a simple harmonic oscillator is given by $y = 25\sin(0.22t)$, where we use SI units for position and time. What is the period of this oscillator?

(a) 4.5 s (b) 0.22 s (c) 25 s (d) 29 s

11.2 Examples of Simple Harmonic Motion

Mass on a Spring

One of the simplest examples of simple harmonic motion is sketched in Figure 11.6, which shows a block of mass m attached to a spring, with the opposite end of the spring attached to a wall. This is very similar to the person in Figure 11.1 except the block in Figure 11.6 moves on a frictionless, horizontal surface.

Let's apply Newton's second law to determine the horizontal motion of the block in Figure 11.6. Recall from Chapter 6 that the force exerted by a spring is given by Hooke's law,

$$F_{spring} = -kx \tag{11.11}$$

where x is the amount the spring is stretched or compressed away from its "natural" relaxed length. The spring constant k is a measure of the spring's strength and is different for different springs. However, the general relation in Equation 11.11 is the same for all springs, provided only that they are not stretched or compressed an excessive amount.

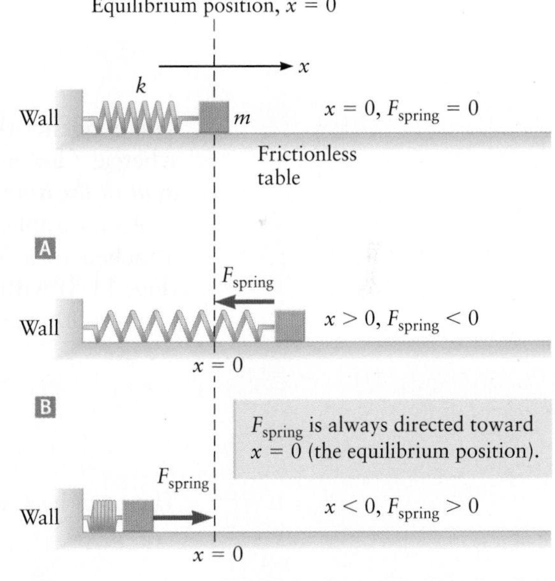

▲ **Figure 11.6** This mass on a spring is a simple harmonic oscillator. A When the block is at the equilibrium position ($x = 0$), the force exerted by the spring on the block is zero. B When the block is displaced to the right ($x > 0$), the force on the block is directed to the left. C When the block is displaced to the left ($x < 0$), the force is directed to the right. The force exerted by the spring on the block is thus always directed toward $x = 0$, the equilibrium point of the system.

The equilibrium position of the block in Figure 11.6 is at $x = 0$; when the block is at this position, the spring is in its relaxed state and the force exerted by the spring on the block is zero. If the block is displaced to the right so that $x > 0$, the force exerted by the spring on the block, according to Hooke's law (Eq. 11.11), is in the negative direction (i.e., to the left). Likewise, if the block is displaced to the left so that $x < 0$, the spring force is to the right. This force is called a ***restoring force*** because it always opposes the displacement away from the equilibrium position. Whenever there is a restoring force described by Equation 11.11, the system will exhibit simple harmonic motion.

To calculate the frequency of the oscillator, we use Newton's second law, $F = ma$, where F is the total force in the horizontal direction (parallel to x) in Figure 11.6. The force is given by Hooke's law (Eq. 11.11), so we have

$$F = -kx = ma \tag{11.12}$$

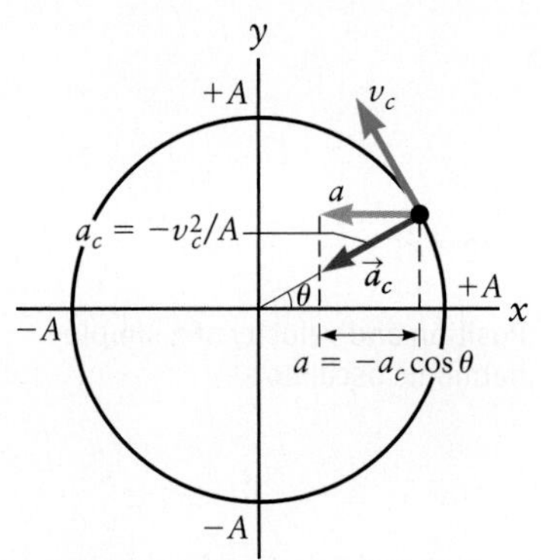

▲ **Figure 11.7** This particle undergoes uniform circular motion; its acceleration has magnitude $a_c = v_c^2/A$ and is directed toward the center of the circle. The acceleration of the corresponding simple harmonic motion is the component of $\vec{a}_c$ along one of the axes; here we use the x direction because we are modeling the oscillator in Figure 11.6.

To apply Equation 11.12, we need to find the particle's position and acceleration. We can do so by using the connection with circular motion sketched in Figure 11.7, which shows a particle moving at constant speed v_c in a circle of radius A. The x component of the particle's position is $x = A\cos\theta = A\cos(2\pi f t)$ as in Equation 11.10. The particle is moving in a circle, so it has a centripetal acceleration $a_c = v_c^2/A$ directed toward the center of the circle (since A is the radius of the circle). Here the simple harmonic motion is described by the particle's position along x, so we need to find the x component of the acceleration; from Figure 11.7, we get

$$a = -a_c \cos\theta = -\frac{v_c^2}{A}\cos\theta$$

Inserting this expression and our result for x into Equation 11.12 gives

$$-kx = -k(A\cos\theta) = ma = m\left(-\frac{v_c^2}{A}\cos\theta\right)$$

$$kA = \frac{mv_c^2}{A} \tag{11.13}$$

From Equation 11.8, we have $v_c = 2\pi f A$; using this in Equation 11.13 leads to

$$kA = \frac{mv_c^2}{A} = \frac{m(2\pi f A)^2}{A} = 4\pi^2 m f^2 A$$

We can now solve for the frequency f of this simple harmonic oscillator:

$$f = \frac{1}{2\pi}\sqrt{\frac{k}{m}} \tag{11.14}$$

The frequency thus increases as the spring constant k is increased (a "stiffer" spring), whereas f becomes smaller as the mass m is increased. Notice also that f is *independent of the amplitude.*

As an application of Equation 11.14, consider the motion of a small block attached to a Slinky, a common child's toy. A Slinky is a loosely coiled spring (Fig. 11.8) with a spring constant of about $k = 1$ N/m. A small toy might have a mass of 300 g, so we take $m = 0.3$ kg. Inserting these values into Equation 11.14, we get

$$f = \frac{1}{2\pi}\sqrt{\frac{k}{m}} = \frac{1}{2\pi}\sqrt{\frac{1\ \text{N/m}}{0.3\ \text{kg}}}$$

The units here are

$$\sqrt{\frac{\text{N}}{\text{m}\cdot\text{kg}}} = \sqrt{\frac{\text{kg}\cdot\text{m/s}^2}{\text{m}\cdot\text{kg}}} = \sqrt{\frac{1}{\text{s}^2}} = \frac{1}{\text{s}}$$

which is just the unit of frequency, hertz. Our result for the frequency is thus

$$f = \frac{1}{2\pi}\sqrt{\frac{1}{0.3}}\ \text{Hz} = 0.3\ \text{Hz}$$

with a corresponding period of

$$T = \frac{1}{f} = \frac{1}{0.3\ \text{Hz}} = 3\ \text{s}$$

How does this value compare with what you found the last time you played with a Slinky?

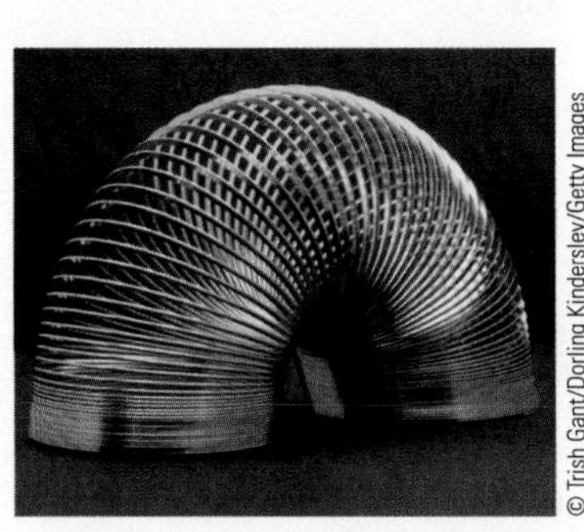

▲ **Figure 11.8** A Slinky has a spring constant of about $k = 1$ N/m.

EXAMPLE 11.2 Frequency of a Simple Harmonic Oscillator

The mass in Figure 11.9A is attached to a single spring with spring constant k, and according to Equation 11.14 it will undergo simple harmonic motion with a frequency

$$f_{\text{single}} = \frac{1}{2\pi}\sqrt{\frac{k}{m}}$$

If a second identical spring is then added as in Figure 11.9B, what is the new oscillation frequency?

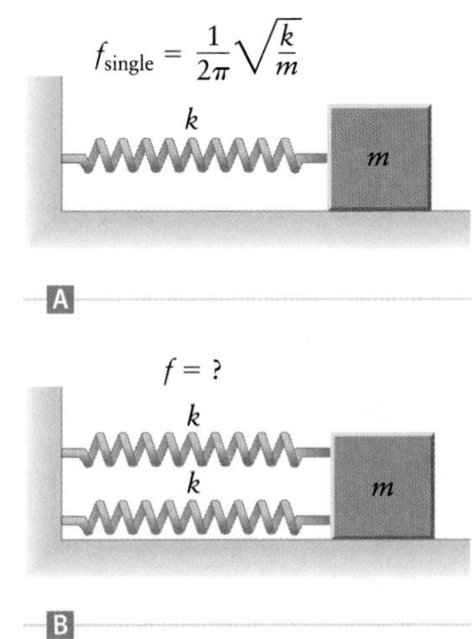

▲ **Figure 11.9** Example 11.2.

RECOGNIZE THE PRINCIPLE

We need to consider how the restoring force on the mass depends on its displacement. If we can show that this restoring force has the same form as Hooke's law, we can then adapt Equation 11.14, which gives the frequency for the case of a single spring.

SKETCH THE PROBLEM

Figure 11.9 shows the problem.

IDENTIFY THE RELATIONSHIPS

Let the position of the mass be $x = 0$ at the location where the two springs are both in their relaxed state; that is the equilibrium point. If the block in Figure 11.9B is then displaced to one side by an amount x, *each* spring will exert a force on the block according to Hooke's law. The *total* horizontal force on the block is then

$$F_{\text{both springs}} = \sum F = -2kx$$

This expression has precisely the same form as for the case of a single spring (compare with Eq. 11.11), so we can think of the two-spring combination as just a single "effective" spring with spring constant $k_{\text{new}} = 2k$.

SOLVE

For a single spring with spring constant k_{new}, we can calculate the oscillation frequency using Equation 11.14 and find

$$f_{\text{both springs}} = \frac{1}{2\pi}\sqrt{\frac{k_{\text{new}}}{m}} = \boxed{\frac{1}{2\pi}\sqrt{\frac{2k}{m}}}$$

► ***What does it mean?***

The frequency for the case of a single spring is

$$f_{\text{single}} = \frac{1}{2\pi}\sqrt{\frac{k}{m}}$$

so we can write our result for the two-spring combination as

$$f_{\text{both springs}} = \sqrt{2}\left(\frac{1}{2\pi}\sqrt{\frac{k}{m}}\right) = \boxed{\sqrt{2}\,f_{\text{single}}}$$

The frequency with two springs is thus $\sqrt{2}$ times higher than with a single spring.

Mass on a Vertical Spring: Bungee Jumping Revisited

The result for the frequency in Equation 11.14 was derived for a mass–spring system in which the mass moves along a frictionless, *horizontal* surface. The bungee jumper in Figure 11.1 is a lot like a mass on a spring, but in this case the mass hangs *vertically* and there are now two forces acting on the mass, the force from the spring and the force of gravity. Interestingly, the oscillation frequency for a mass on a vertical spring is precisely the *same* as the frequency for a mass on a horizontal spring. A mass hanging on a vertical spring is another example of simple harmonic motion, with an oscillation frequency given by Equation 11.14.

EXAMPLE 11.3 Ⓡ Analyzing a Bungee Oscillator

Consider the person attached to a bungee cord as shown in the photo at the start of this chapter. During the first part of the jump, the bungee cord is slack and the person undergoes (enjoys) free fall, but after a while, the person oscillates up and down in response to the forces of the bungee spring and gravity. Assume the bungee plus person can be modeled as a mass on a vertical, massless spring. For a person of average mass (e.g., your own mass), estimate (**a**) the frequency of the motion and (**b**) the spring constant of the bungee cord.

RECOGNIZE THE PRINCIPLE

The frequency of the motion is related to the period (the repeat time) by $f = 1/T$. To estimate the period, we appeal to your intuition (or experience) with bungee jumping. Once we have an approximate value for the frequency, we can use the relation for the frequency in Equation 11.14 along with the typical mass of a person to find k. This problem thus requires you to estimate two key quantities. The ability to make such estimates gives insight into how physics applies to real-world situations.

SKETCH THE PROBLEM

Figure 11.10 shows the person oscillating up and down and indicates the amplitude of the oscillation.

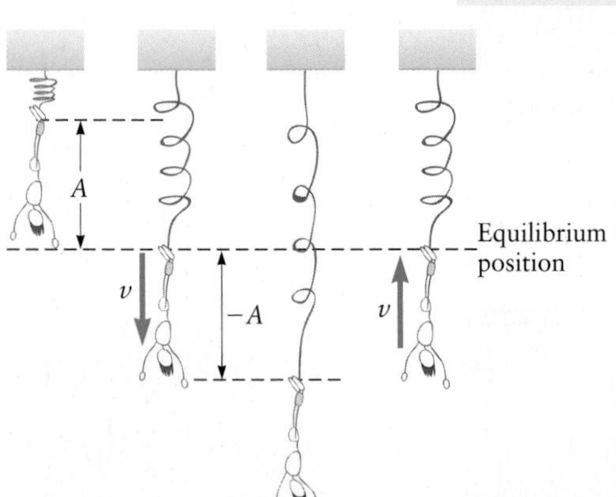

▲ **Figure 11.10** Example 11.3.

IDENTIFY THE RELATIONSHIPS

(**a**) The period of a bungee system depends on the length of the cord and the height of the jump. In a "typical" jump at a carnival, the person travels a distance $A = 10$ m below the equilibrium point (Fig. 11.10) and the period is about 5 s (based on the author's experience). These values are only approximate; other bungee jump experiments could be different.

(**b**) The mass of a typical person is about $m = 80$ kg.

SOLVE

(**a**) If the period of the oscillation is $T = 5$ s, the frequency is (Eq. 11.1)

$$f = \frac{1}{T} = \frac{1}{5\text{ s}} = \boxed{0.2\text{ Hz}}$$

(**b**) A bungee oscillator is just a mass on a spring, and its frequency is (Eq. 11.14)

$$f = \frac{1}{2\pi}\sqrt{\frac{k}{m}}$$

Rearranging to solve for k and inserting our values for f and m, we find

$$\sqrt{\frac{k}{m}} = 2\pi f$$

$$k = 4\pi^2 f^2 m = 4\pi^2(0.2\text{ Hz})^2(80\text{ kg}) = \boxed{100\text{ N/m}}$$

We include only one significant figure in our answer because the period and frequency were only known to one significant figure.

▶ ***What does it mean?***
Other bungee jumps will have different heights, and their bungee cords will have different values of k. However, we can always use our experience (and common sense) to estimate quantities such as the period and frequency, and thus apply our results for simple harmonic motion.

The Simple Pendulum

A pendulum is another example of a simple harmonic oscillator. Pendulums can be constructed in several ways; Figure 11.11 shows one example. Perhaps the simplest pendulum is made by tying a rock of mass m to one end of a string, with the other end of the string fastened to a support as sketched in Figure 11.12. If the diameter of the rock is much smaller than the length L of the string and the string is massless, this system is a simple pendulum.

Figure 11.12A shows sketches of the rock (also called the pendulum "bob") as it moves along its path, a circular arc of radius L. It is convenient to measure the bob's displacement in terms of y, the displacement along this circular arc. To examine the pendulum motion in terms of Newton's second law, we have drawn the two forces acting on the bob: gravity and tension from the string. (We ignore the air drag force.) The force along (parallel to) the circular arc is responsible for the simple harmonic motion. Figure 11.12A shows $F_{parallel}$ at three instants in time. The bottom point on the trajectory ($y = 0$) is the equilibrium point of the pendulum, the point where $F_{parallel} = 0$. If the bob were placed at that location and given no initial velocity, it would remain there, in translational (and rotational) equilibrium (Chapters 4 and 8). As the bob moves back and forth past this equilibrium point, the magnitude and direction of $F_{parallel}$ vary with time. When the bob is on the right in Figure 11.12A, where $y > 0$, the force is directed to the left (the negative direction). Likewise, if the bob is on the left, where $y < 0$, then $F_{parallel}$ is to the right, the positive direction. Because this force always opposes the bob's displacement away from the equilibrium point, this is another example of a restoring force. (Compare with the restoring force in Fig. 11.6.)

▲ **Figure 11.11** A child's tire swing is a good approximation to a simple pendulum.

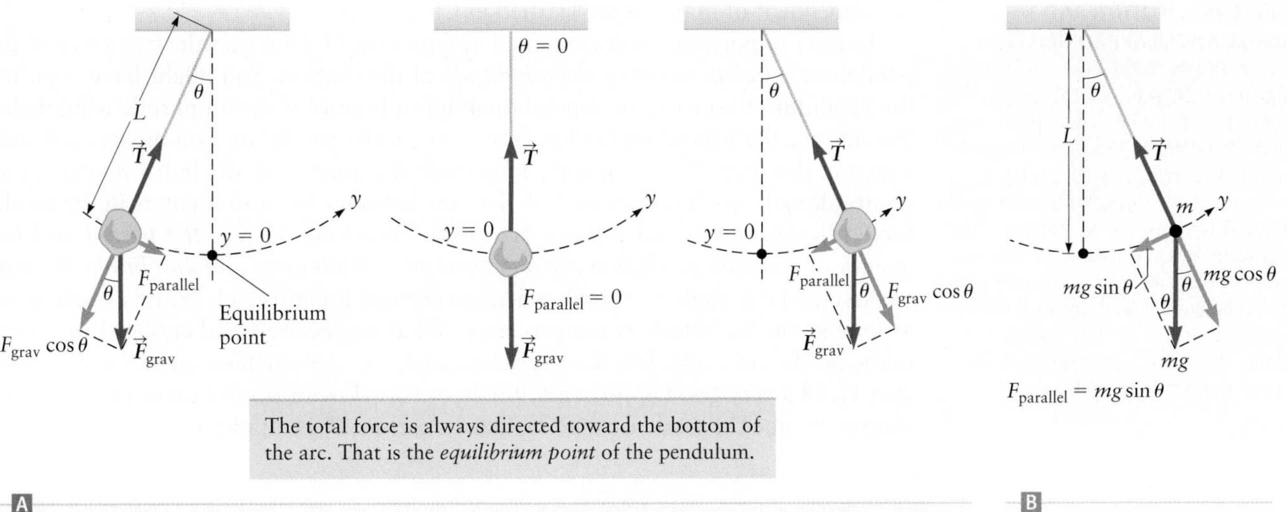

▲ **Figure 11.12** A This simple pendulum consists of a pendulum bob of mass m tied to the end of a very light string. There are two forces on the bob, from gravity and from the tension in the string. B Their vector sum is directed tangent to the circular arc along which the bob swings. This total (net) force is always directed back toward the bottom of the arc, the equilibrium point of the pendulum.

For a simple pendulum, the restoring force is due to the component of the gravitational force directed along the bob's path. Figure 11.12B shows that for a pendulum of length L with a bob of mass m, this restoring force is

$$F_{\text{restore}} = F_{\text{parallel}} = -mg \sin\theta \tag{11.15}$$

If the angle θ is small, the function $\sin\theta$ is approximately equal to θ when the angle is measured in radians. That is, $\sin\theta \approx \theta$. In addition, this angle is related to the displacement by $\theta = y/L$ according to the general definition of angle as measured in radians. Inserting all this into Equation 11.15 leads to

$$F_{\text{parallel}} = -mg\theta = -\frac{mgy}{L} \tag{11.16}$$

Newton's second law for motion along the circular arc ($F_{\text{parallel}} = ma$) leads to:

$$F_{\text{parallel}} = -\frac{mgy}{L} = ma \tag{11.17}$$

This relation has exactly the same form we found in Equation 11.12 when we applied Newton's second law to the mass on a spring. In both cases, the restoring force is proportional to the displacement from equilibrium. Comparing Equations 11.12 and 11.17, the only difference is that the spring constant k for the mass on a spring is replaced by the factor mg/L. We can therefore use the result for the oscillation frequency of a mass on a spring (Eq. 11.14) if we replace k by mg/L, which leads to

$$f = \frac{1}{2\pi}\sqrt{\frac{k}{m}} = \frac{1}{2\pi}\sqrt{\frac{mg/L}{m}}$$

Frequency of a simple pendulum

$$f = \frac{1}{2\pi}\sqrt{\frac{g}{L}} \tag{11.18}$$

Hence, given the length of the pendulum, its frequency can be calculated. This frequency is *independent* of the mass of the bob, which may at first seem surprising; you might have expected that a heavier bob will move more slowly and have a lower frequency than a lighter bob. However, the restoring force is due to gravity and is thus proportional to the bob's mass (since its weight is equal to mg). The pendulum's motion is independent of m for the same reason the velocity of a freely falling object is independent of mass (as we learned in Chapter 3).

Another important result contained in Equation 11.18 is that the frequency of the pendulum is *independent of the amplitude* of the motion. You might have expected the pendulum frequency to depend on amplitude since if the amplitude is increased (by starting the pendulum farther from $y = 0$), the pendulum bob must travel over a greater distance. However, if the amplitude A is increased, the bob's velocity at all points along its path is increased by the same amount because A enters in the results for y (displacement) and v (velocity) in the same way. Hence, *the period and frequency of a simple pendulum are independent of both mass and amplitude*. Because we assumed the angle θ is small when we derived Equation 11.18, this result holds as long as the amplitude is not too large. What angles are small enough? In typical cases, pendulum angles less than approximately 30° are small enough to make Equation 11.18 a very good approximation. In our work with pendulum oscillators, we'll always assume the angle is small so that Equation 11.18 applies.

Insight 11.1

HARMONIC MOTION AND THE SMALL ANGLE APPROXIMATION

For a simple pendulum, the restoring force is (Eq. 11.15) $F_{\text{restore}} = -mg \sin\theta$. When the pendulum angle θ is small, this force is given approximately by (Eq. 11.16) $F_{\text{restore}} = -mg \sin\theta \approx -mg\theta$, with $\sin\theta \approx \theta$ when θ is measured in radians. You can verify this approximation yourself by using a calculator to compare θ and $\sin\theta$, and you'll find that they differ by less than 10% when θ is smaller than about 45°. The approximation gets better and better as θ gets smaller.

EXAMPLE 11.4 The Frequency of a Playground Swing

If a pendulum swing on a playground has length $L = 3.0$ m, what are the frequency and period of the swing? Assume the pendulum angle is small.

RECOGNIZE THE PRINCIPLE

A pendulum swing is approximately a simple pendulum as in Figures 11.11 and 11.12, and its frequency is given by (Eq. 11.18)

$$f = \frac{1}{2\pi}\sqrt{\frac{g}{L}} \quad (1)$$

SKETCH THE PROBLEM

Figure 11.12 describes the problem.

IDENTIFY THE RELATIONSHIPS

We are given the pendulum length, so we can use Equation (1) to find the frequency. The period is related to the frequency by $T = 1/f$ (Eq. 11.1).

SOLVE

Substituting the given value of L into Equation (1) gives

$$f = \frac{1}{2\pi}\sqrt{\frac{g}{L}} = \frac{1}{2\pi}\sqrt{\frac{9.8 \text{ m/s}^2}{3.0 \text{ m}}} = 0.29\left(\frac{1}{\text{s}}\right) = \boxed{0.29 \text{ Hz}}$$

The period can be found from our general relation between frequency and period,

$$T = \frac{1}{f} = \frac{1}{0.29 \text{ Hz}} = \boxed{3.4 \text{ s}}$$

► ***What does it mean?***

These results are, as we have already noted, independent of mass, so children of different size will swing at the same frequency.

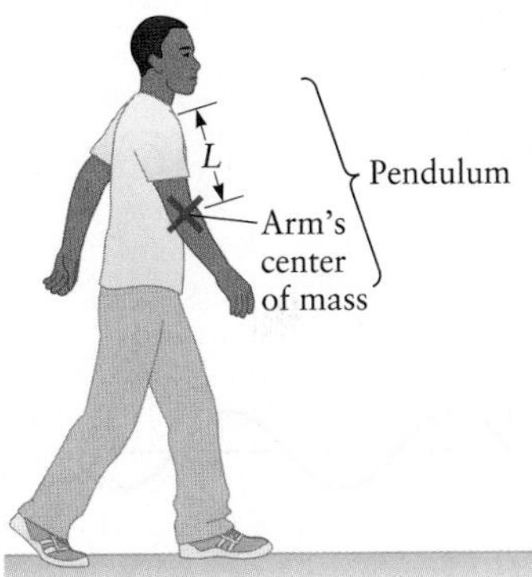

▲ **Figure 11.13** When you walk, each of your arms swings like a pendulum.

The Human Arm as a Pendulum

In Example 11.4, we calculated the period of a pendulum. The same ideas can be used to understand the way your arms swing when you walk. If your arms swing freely, they act as pendulums. Let's now estimate their period of oscillation. The arm is not quite a simple pendulum because all the mass is not at the end. We can, however, get an approximate value of the period by assuming all the mass is at the center of mass, roughly halfway along your arm (Fig. 11.13). An adult's arm is about 60 cm long, so the center of mass is about $L = 30$ cm $= 0.30$ m from the shoulder. Taking 0.30 m as the length of a pendulum, the frequency is (Eq. 11.18)

$$f = \frac{1}{2\pi}\sqrt{\frac{g}{L}} = \frac{1}{2\pi}\sqrt{\frac{9.8 \text{ m/s}^2}{0.30 \text{ m}}} = 0.91 \text{ Hz}$$

which corresponds to a period

$$T = \frac{1}{f} = 1.1 \text{ s}$$

The next time you go for a walk, you should compare this result with the way you swing your arms.

The Torsional Oscillator

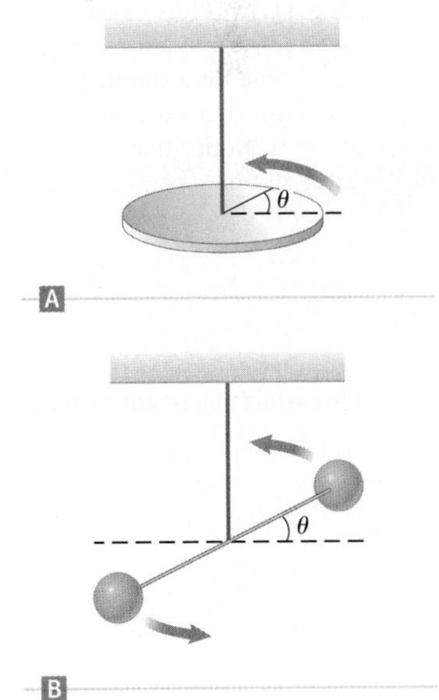

▲ **Figure 11.14** Two examples of torsional oscillators. These objects rotate about a vertical (y) axis, and the torque exerted by the fibers tends to restore the systems to their equilibrium positions.

Another system displaying simple harmonic motion is the ***torsional oscillator***. Figure 11.14 shows two examples: in part A of the figure, a circular disk is suspended by a thin wire attached to the center of the disk, and in part B, a rod is attached to masses at each end, with the rod suspended by a wire attached to its center. The wire used in both of these examples is called a ***torsion fiber***, and when it is twisted, it exerts a torque τ on whatever is connected to it, the disk in Figure 11.14A or the rod in Figure 11.14B. The torque is proportional to the twist angle θ, with

$$\tau = -\kappa\theta \tag{11.19}$$

Here, κ (the Greek letter kappa) is called the ***torsion constant*** and is a property of the particular torsion fiber. The negative sign in Equation 11.19 indicates that it is a *restoring* torque. When the disk is rotated to positive angles, the torque tends to make the disk turn back toward the "negative" angular direction, whereas if the disk is rotated to negative angles, the torque acts toward the direction of positive angles. This is just like the restoring force for the mass on a spring in Figure 11.6 and for the pendulum in Figure 11.12. Since Equation 11.19 has the same mathematical form as Hooke's law for a spring (Eq. 11.11), the torsional pendulum is also a simple harmonic oscillator, with a frequency given by

$$f = \frac{1}{2\pi}\sqrt{\frac{\kappa}{I}} \tag{11.20}$$

where I is the moment of inertia of the object attached to the fiber (the disk in Fig. 11.14A or the two masses and rod in Fig. 11.14B). This result is similar to what we found for a mass on a spring (Eq. 11.14), but the frequency of the torsional oscillator depends on the moment of inertia I of the system instead of on its mass.

Features Common to All Simple Harmonic Oscillators

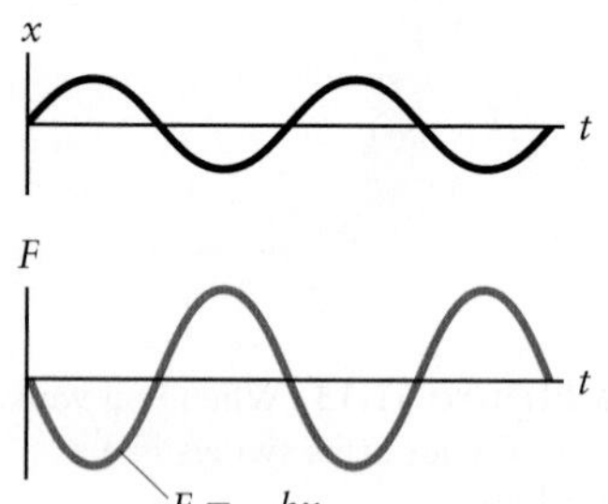

▲ **Figure 11.15** Plots of the position x and restoring force F as functions of time for a simple harmonic oscillator that starts at $x = 0$ at $t = 0$. Notice that $F = -kx$ at all times.

All simple harmonic oscillators have a number of features in common. First, their motion exhibits a sinusoidal time dependence as in Equation 11.9 or, equivalently, Equation 11.10. The variables in these equations are called "position" (x or y) and "velocity" (v). For a simple pendulum and a mass on a spring, x and v are the ordinary position and velocity, whereas for the torsional oscillator, they are the angular position (the angle θ in Fig. 11.14) and the corresponding angular velocity. The position and velocity of *all* simple harmonic oscillators have these mathematical forms.

A second feature common to all simple harmonic oscillators is that they all involve a restoring force. Figure 11.15 shows the time dependence of both the position (which we denote by x) and the restoring force F. For the horizontal mass on a spring, F is the force exerted by the spring on the mass, whereas for a simple pendulum, F is the component of the gravitational force along the path of the pendulum bob. For the torsional oscillator, we would instead plot the torque exerted by the torsion fiber. Both the displacement and the restoring force or torque vary sinusoidally with time, and F always has a sign *opposite* that of the displacement. This is the same negative sign we encountered in Hooke's law (Eq. 11.11) and for the pendulum (Eq. 11.16). This is a key feature of a restoring force and is essential for producing harmonic motion. Whenever an object experiences a *restoring* force with a magnitude proportional to the displacement from the equilibrium position, the object will undergo simple harmonic motion.

Properties of a restoring force

The Frequency of a Simple Harmonic Oscillator Is Independent of the Amplitude

A very important feature of all simple harmonic oscillators is that the frequency is *independent* of the amplitude A. The frequency depends on quantities such as the length of the pendulum string (Fig. 11.12), the mass and spring constant in Figure 11.6, or the moment of inertia and torsion constant in Figure 11.14. Hence, this frequency does depend on the properties of the oscillator, but f does *not* depend on the amplitude of the motion. The amplitude is determined entirely by how the oscillator is initially set into motion.

CONCEPT CHECK 11.3 Connection between Velocity and Force for a Simple Harmonic Oscillator

Figure 11.16 shows the displacement of a simple harmonic oscillator as a function of time. (a) At which point(s) is the velocity zero? (b) At which point(s) is the magnitude of the force a maximum?

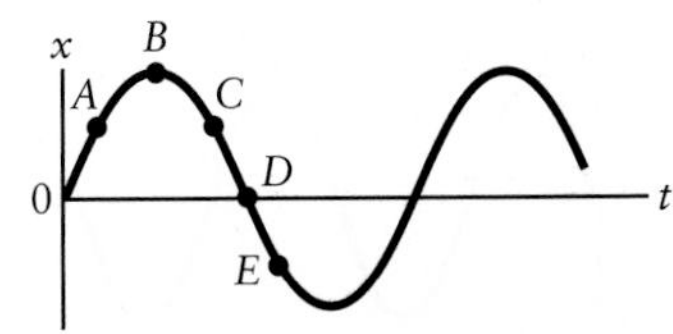

▲ **Figure 11.16** Concept Check 11.3.

11.3 Harmonic Motion and Energy

Let's next consider how the kinetic and potential energies of the simple harmonic oscillator in Figure 11.17 vary with time. The mechanical energy is the sum of the kinetic energy ($\frac{1}{2}mv^2$) and potential energy ($\frac{1}{2}kx^2$, the spring potential energy). When the mass is at $x = 0$ (parts A and C of Fig. 11.17), the potential energy is zero because the spring is relaxed, but the speed is a maximum; hence, the kinetic energy is large. When the displacement is largest, $x = \pm A$ (parts B and D of Fig. 11.17), and the potential energy is large, the speed is zero, so the kinetic energy is zero. The energy thus "oscillates" back and forth between the kinetic and potential energies as shown in Figure 11.17E.

For an ideal oscillator, there is no friction, and the total mechanical energy is conserved. Hence, the sum of the kinetic and potential energies in Figure 11.17E is constant. This also means that the maximum kinetic energy (Fig. 11.17A) must equal the maximum potential energy (Fig. 11.17B). If the maximum speed is v_{max}, the maximum kinetic energy is $\frac{1}{2}mv_{max}^2$ and the maximum potential energy is $\frac{1}{2}kA^2$ (because the maximum displacement is $x_{max} = A$). We thus have

$$\frac{1}{2}mv_{max}^2 = KE_{max} = PE_{max} = \frac{1}{2}kA^2 \tag{11.21}$$

which gives a useful relation between the maximum speed and displacement.

These results for the time dependence of the kinetic energies and potential energies have been derived for a mass on a horizontal spring, but they apply to all simple harmonic oscillators. The *KE* and *PE* of any simple harmonic oscillator vary with

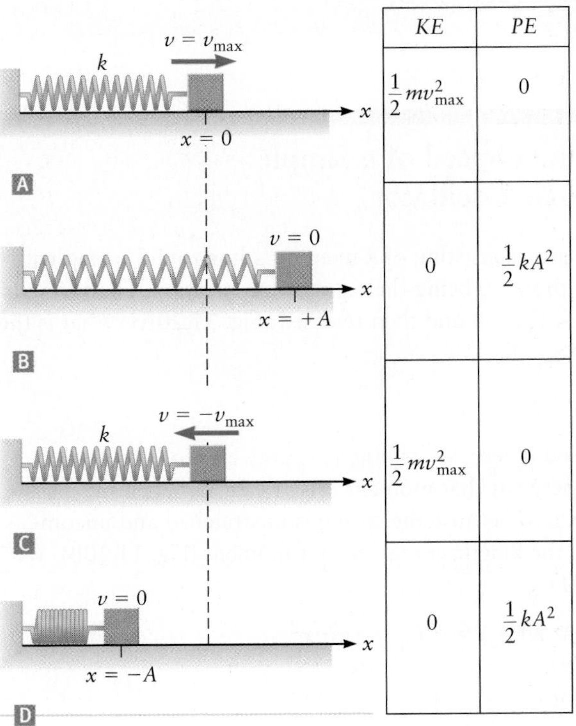

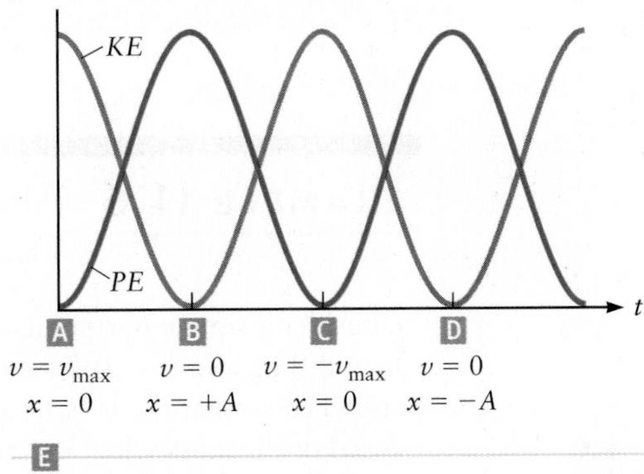

▲ **Figure 11.17** The kinetic and potential energies of a mass on a spring oscillate with time. The sum $KE + PE$ is constant, as the mechanical energy of a simple harmonic oscillator is conserved.

time in the manner sketched in Figure 11.17, and the total mechanical energy is conserved.

CONCEPT CHECK 11.4 Energy of a Simple Harmonic Oscillator

Figure 11.18 shows the displacement of a simple harmonic oscillator as a function of time. The graph in Figure 11.18B shows either the kinetic or the potential energy. Which one?

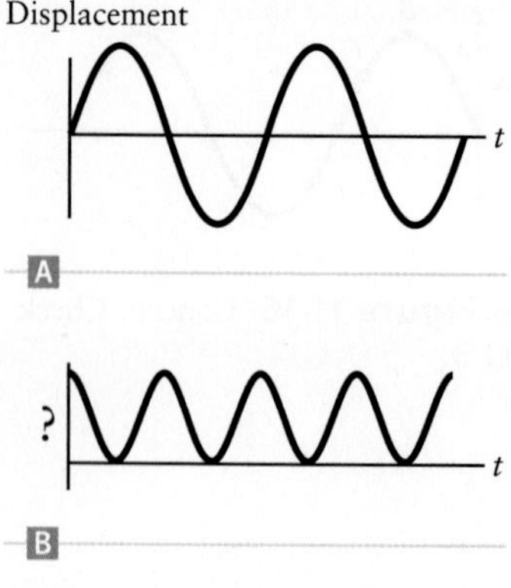

▲ **Figure 11.18** Concept Check 11.4.

EXAMPLE 11.5 Energy of a Simple Harmonic Oscillator

Figure 11.19 shows the velocity of a simple harmonic oscillator (like the one in Figure 11.17) as a function of time. At which of the labeled points *A*, *B*, *C*, *D*, and *E* on the curve is the kinetic energy greatest? At what point is it smallest?

RECOGNIZE THE PRINCIPLE

The kinetic energy of the oscillator is just the kinetic energy of the mass, which is $KE = \frac{1}{2}mv^2$.

SKETCH THE PROBLEM

Figure 11.19 shows the problem.

IDENTIFY THE RELATIONSHIPS AND SOLVE

From Figure 11.19, we see that the magnitude of the velocity is greatest at point *D* (even though v is negative). The kinetic energy is thus greatest at point *D*. Likewise, v is zero at point *B*, so the kinetic energy is smallest at this point.

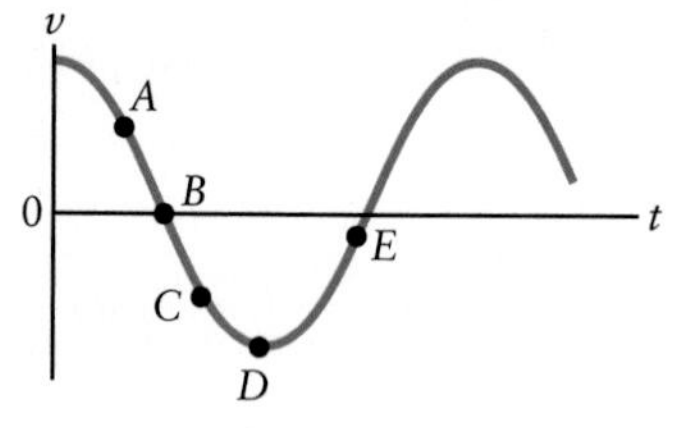

▲ **Figure 11.19** Example 11.5.

► ***What does it mean?***

The total energy of the oscillator must be conserved (i.e., constant), so the potential energy will be smallest when the speed is greatest. For the points marked in Figure 11.19, the speed is greatest at point *D*. Likewise, the potential energy is largest when the speed is zero, which occurs at point *B*.

EXAMPLE 11.6 Maximum Speed of a Simple Harmonic Oscillator

Consider a simple harmonic oscillator consisting of a mass on a horizontal spring with $m = 3.0$ kg and $k = 500$ N/m, with $x = 0$ being the equilibrium position. The spring is stretched so that the mass is at $x_i = 0.10$ m and then released (Fig. 11.20A). What is the speed of the mass when it reaches $x = 0$?

RECOGNIZE THE PRINCIPLE

Just as the mass is released, its speed is zero, so its kinetic energy is zero. The total energy thus equals the potential energy at that moment (Fig. 11.20A). When the mass is at $x = 0$, the potential energy is zero because the spring is unstretched and uncompressed, so the total energy equals the kinetic energy at that moment (Fig. 11.20B). We thus have (compare with Eq. 11.21)

$$\frac{1}{2}mv^2 = KE_{\text{max}} = PE_{\text{max}} = \frac{1}{2}kx_i^2 \tag{1}$$

We can use this relation to solve for v.

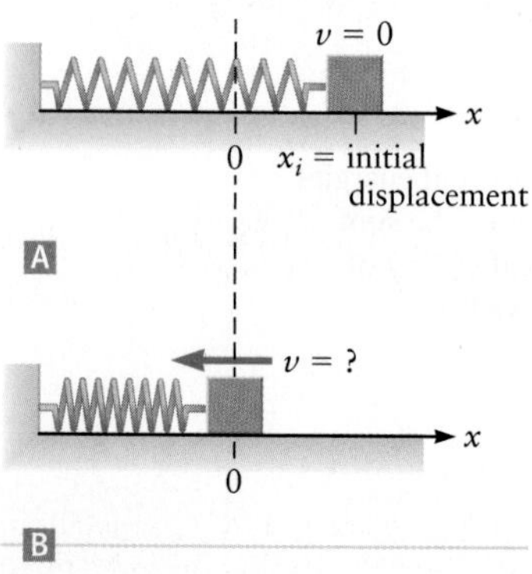

▲ **Figure 11.20** Example 11.6.

SKETCH THE PROBLEM

Figure 11.20 shows the system when the block is released and when it passes through $x = 0$.

IDENTIFY THE RELATIONSHIPS

We can rearrange Equation (1) to get

$$mv^2 = kx_i^2$$

$$v^2 = \frac{k}{m}x_i^2$$

$$v = \sqrt{\frac{k}{m}}x_i$$

SOLVE

The initial displacement is $x_i = 0.10$ m, and the values of m and k are given; we find

$$v = \sqrt{\frac{k}{m}}x_i = \sqrt{\frac{500\ \text{N/m}}{3.0\ \text{kg}}}(0.10\ \text{m}) = \boxed{1.3\ \text{m/s}}$$

▸ ***What does it mean?***

The maximum potential energy occurs when the displacement is greatest, whereas the maximum kinetic energy is found when the displacement is zero. This result applies to all simple harmonic oscillators, including pendulums and torsional oscillators.

CONCEPT CHECK 11.5 Amplitude and Energy of a Simple Harmonic Oscillator

If the amplitude of a simple harmonic oscillator is doubled, by what factor does the *total* energy increase?

(a) no change
(b) by a factor of two
(c) by a factor of four
(d) by a factor of eight

11.4 Stress, Strain, and Hooke's Law

According to Hooke's law, the force exerted by a spring is directly proportional to the amount that the spring is stretched or compressed, with

$$F = -kx \tag{11.22}$$

Hooke's law applies not only to the "ideal" springs found in physics books, but to many other situations as well. Consider a solid bar of metal as sketched in Figure 11.21. If you "squeeze" the bar an amount x, it will "push back." The magnitude of this "push-back" force depends on how much you squeeze and is given by Hooke's law (Eq. 11.22). The value of the constant k depends on the size of the bar and the material it is made of. This push-back force also means that virtually all objects act as springs when squeezed, leading to many applications of our results for simple harmonic motion.

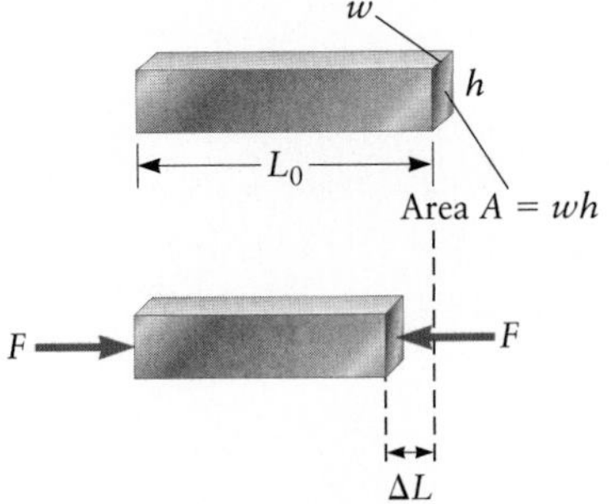

▲ **Figure 11.21** The forces F applied to the ends of this metal bar compress the bar by an amount ΔL. Tensile forces would stretch the bar instead.

In Figure 11.21, a force of magnitude F is applied to each end of the metal bar, and the bar changes length by an amount ΔL. The force in Figure 11.21 is a ***compressive*** force, making ΔL negative. If the force directions are reversed so that the bar is stretched (ΔL positive), we call F a ***tensile*** force. The value of ΔL depends on the material; a soft material compresses or stretches more than a stiff material. "Stiffness" is characterized by ***Young's modulus***, Y. In Figure 11.21, the area of the end of

Insight 11.2
ELASTIC FORCES AND HOOKE'S LAW
Hooke's law for a spring (Eq. 11.22) contains a negative sign, whereas Equation 11.23 does not, but we have still called them both "Hooke's law." The difference is due to how F is defined in the two cases. For the spring described by Equation 11.22, F is the force exerted *by* the spring, whereas in the definition of Young's modulus, Equation 11.23, F is the force *on* the bar. According to Newton's third law, the force exerted *by* the bar equals $-F$, so Equation 11.23 is indeed equivalent to Hooke's law for a spring.

TABLE 11.1 Values of Young's Modulus, Shear Modulus, and Bulk Modulus for Some Common Materials

Material	Young's Modulus, Y (Pa)	Shear Modulus, S (Pa)	Bulk Modulus, B (Pa)
Aluminum	7.0×10^{10}	2.4×10^{10}	7.1×10^{10}
Steel	2.0×10^{11}	8.1×10^{10}	1.4×10^{11}
Copper	1.3×10^{11}	4.8×10^{10}	1.4×10^{11}
Glass (Pyrex)	6.2×10^{10}	2.7×10^{9}	2.6×10^{10}
Human bone	9×10^{9} (compression) 2×10^{10} (when stretched)	8×10^{10}	
Diamond	1.2×10^{12}	4.8×10^{10}	6.2×10^{11}
Quartz	1.0×10^{11}	3.1×10^{10}	3.6×10^{10}
Teflon	4×10^{8}		6×10^{9}
Ice			9×10^{9}
Air (at atmospheric pressure)			1.0×10^{5}
Water			2.2×10^{9}
Mercury (liquid)			2.5×10^{10}

Note: These values generally vary slowly with temperature. The numbers given here are for room temperature.

the bar is $A = wh$, and L_0 is the bar's original length. The magnitude of the force F and the change in length ΔL are related by

Stress, strain, and Hooke's law

$$\frac{F}{A} = Y\frac{\Delta L}{L_0} \tag{11.23}$$

In words, Equation 11.23 says that the change in length is directly proportional to the applied force. What's more, if the material is pulled so that it is elongated by an amount ΔL (i.e., if the directions of the forces in Fig. 11.21 are reversed), Equation 11.23 still holds. Because the amount the bar is stretched or compressed is directly proportional to the magnitude of the applied force, this relation is equivalent to Hooke's law (Eq. 11.22).

The ratio F/A that appears in Equation 11.23 is called the ***stress***. By considering the force applied per unit of cross-sectional area, we can more easily compare the properties of bars of different size. For example, if we consider a second bar of the same material and length whose area is larger by a factor of two, a force twice as large is required to produce the same change in length ΔL. The stress F/A would be the same in the two cases, though. Likewise, the ratio $\Delta L/L_0$ is called the ***strain***. This ratio is the fractional change in the length. The value of Young's modulus depends on the material; values for some common materials are listed in Table 11.1. Since the ratio $\Delta L/L$ on the right-hand side of Equation 11.23 is unitless, Y has units of force/area = N/m^2 = Pa.

Elastic versus Plastic Deformations

Young's modulus describes how a material stretches or compresses in response to an applied force. We call Y an ***elastic constant*** because, according to Equation 11.23, the material (such as the bar in Fig. 11.21) will return to its original length when the force F is removed. That is what is meant by "elastic" behavior. Such behavior is found for most materials provided the strain is not too large. If the strain exceeds a certain value, however, the simple linear relation between stress and strain in Equation 11.23 will no longer be found. The value of the strain at which that happens is called the ***elastic limit***, and depends on the material. When the strain is pushed beyond this point, the material will no longer return to its original length even when F is completely removed. A typical plot of stress versus strain is shown in Figure 11.22. When the elastic limit is exceeded, the material will be permanently deformed.

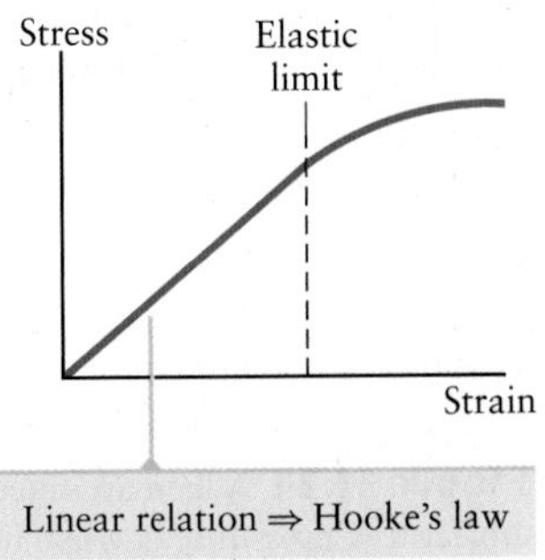
When the elastic limit is exceeded, the stress–strain curve is no longer linear.
Stress
Elastic limit
Strain
Linear relation ⇒ Hooke's law

▲ **Figure 11.22** When a material is strained (i.e., compressed, stretched, or sheared), it will initially follow a linear stress–strain relation; the magnitude of the applied force (or stress) is proportional to the amount the material is strained. When the strain exceeds a certain value, called the elastic limit, this linear relation is no longer followed. Such large stresses usually cause the object to be permanently deformed.

EXAMPLE 11.7 Ⓡ Stretching a Guitar String

When expressed in SI units, the numerical values of elastic constants are typically very large (see Table 11.1), making it difficult to get an intuitive feeling for how much an object will stretch or compress in response to typical forces. So, consider the following "everyday" problem. A guitar string is composed of steel and is under a tension of 200 N when mounted on the instrument. How much does the string stretch in response to this tension?

RECOGNIZE THE PRINCIPLE

When a force F is applied to the string, it becomes longer by an amount ΔL with (Eq. 11.23)

$$\frac{F}{A} = Y\frac{\Delta L}{L_0} \tag{1}$$

To use Equation (1) to find ΔL, we must have values for the length and area of the string. The value of Young's modulus is given in Table 11.1, and F is just the tension, which is given.

SKETCH THE PROBLEM

We can use Figure 11.21, with the bar replaced by a guitar string.

IDENTIFY THE RELATIONSHIPS

Based on experience, the length of a typical guitar string is about $L_0 = 0.6$ m and the radius is about $r = 0.5$ mm.

SOLVE

Rearranging Equation (1) to find ΔL, we get

$$\Delta L = \frac{FL_0}{YA} = \frac{FL_0}{Y\pi r^2}$$

where $A = \pi r^2$ is the cross-sectional area of the string and F is the tensile force, given as 200 N. Inserting the value of Young's modulus for steel from Table 11.1 along with the other values estimated above, we get

$$\Delta L = \frac{FL_0}{Y\pi r^2} = \frac{(200 \text{ N})(0.6 \text{ m})}{(2.0 \times 10^{11} \text{ N/m}^2)\pi(0.0005 \text{ m})^2} = 8 \times 10^{-4} \text{ m} = \boxed{0.8 \text{ mm}}$$

▶ ***What does it mean?***

Although 0.8 mm is not a large value, it is certainly noticeable to a guitarist turning the tuning peg of a guitar.

The Shear Modulus

An applied force may deform a material in ways besides compressing or stretching, leading to other elastic constants besides Young's modulus. Consider a shearing force applied to a metal bar as shown in Figure 11.23; this force tends to make two parallel faces of the bar shift laterally by an amount Δx. The shear stress is defined as F/A, where A is the cross-sectional area of the end of the bar; and the shear strain is

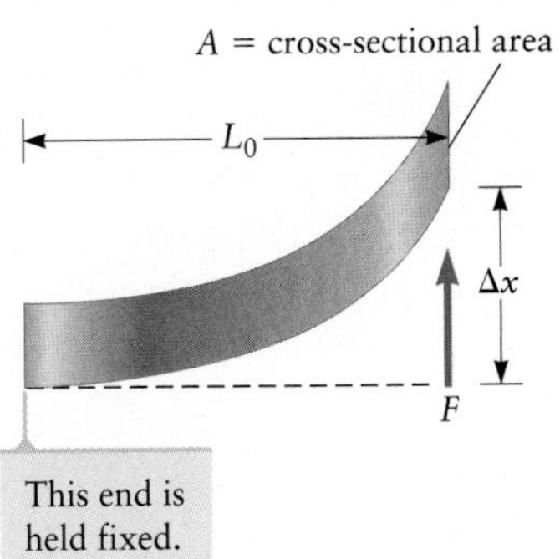
A = cross-sectional area
L_0
Δx
F
This end is held fixed.

▲ **Figure 11.23** A shear force (related to the shear stress) acts to bend this bar.

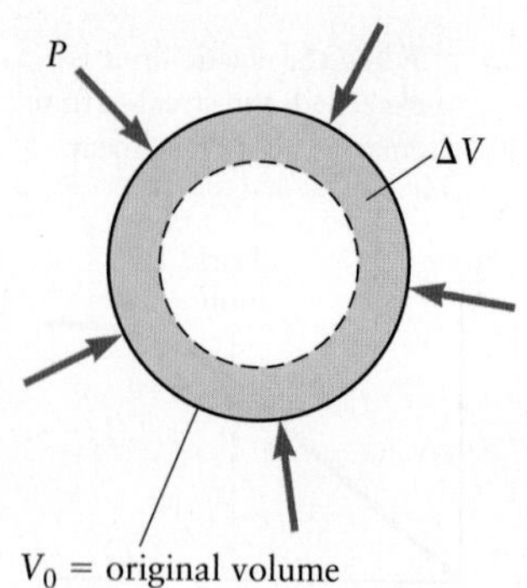

▲ **Figure 11.24** When an object is subject to a pressure (as when it is immersed in a fluid), the pressure force will compress the object, making the volume smaller by an amount ΔV.

defined as $\Delta x/L_0$, where L_0 is the length of the bar (Fig. 11.22). The shear stress and shear strain are related by

$$\frac{F}{A} = S\frac{\Delta x}{L_0} \tag{11.24}$$

where S is the ***shear modulus***. Table 11.1 lists the shear modulus for several materials.

This relation is very similar to the expression that involves a compressive stress and Young's modulus (Eq. 11.23). In both cases, the deformation of the bar is directly proportional to the stress, so both relations are equivalent to Hooke's law. These relations between elastic stress and strain describe many materials, so many material objects will act as simple Hooke's law springs when acted on by a compressive or tensile force or by a shear force. We can therefore apply our results for simple springs to a variety of situations.

The Bulk Modulus

Figure 11.24 shows an object immersed in a fluid. The pressure P in the fluid leads to a force on the surface of the object. This force is directed normal to the object's surface and causes the object's volume to change by an amount ΔV. If V_0 is the original volume of the object, the pressure and volume change are related by

$$P = -B\frac{\Delta V}{V_0} \tag{11.25}$$

where B is an elastic constant called the ***bulk modulus***. Note that the pressure and volume change are directly proportional. The presence of the negative sign in Equation 11.25 means that a positive pressure will produce a negative value of ΔV, as we should expect.

In our discussions of Young's and shear moduli, we have assumed the object in question is a solid, not a liquid or a gas. Compressive, tensile, and shear stresses (Figs. 11.21 and 11.23) can only be applied to an object that maintains its shape without the help of a container. In contrast, the concept of a bulk modulus applies to solids, liquids, and gases because all change volume in response to changes in pressure.

Elastic Properties and Simple Harmonic Motion

We learned in Section 11.2 that when an object is subject to a restoring force proportional to the displacement, the object can undergo simple harmonic motion. The same type of motion is found with the metal bars in Figures 11.21 and 11.23. If the end of the bar is displaced—that is, deformed an amount ΔL by compressing, stretching, or shearing the bar—it experiences a restoring force proportional to ΔL. If the displacement is caused by a compressive or tensile stress, the restoring force involves Young's modulus Y; for a shear displacement, the restoring force involves the shear modulus S. Some distortions produce both compression and shear, so the total restoring force may involve a combination of Y and S. In all these cases, the restoring force is proportional to the displacement and is thus described by Hooke's law. For this reason, a vibrating metal bar is closely analogous to a mass on a spring and generally behaves as a simple harmonic oscillator.

EXAMPLE 11.8 Ⓡ Oscillation Frequency of a Diving Board

When a diving board bends, the restoring force due to the board's elastic properties can be written as

$$F_{\text{board}} = -kx \tag{1}$$

where x is the displacement of the end of the board (Fig. 11.25) and k is the "spring" constant of the diving board. Estimate k for a typical diving board and use your result to find the oscillation frequency when a diver is standing at the end of the board.

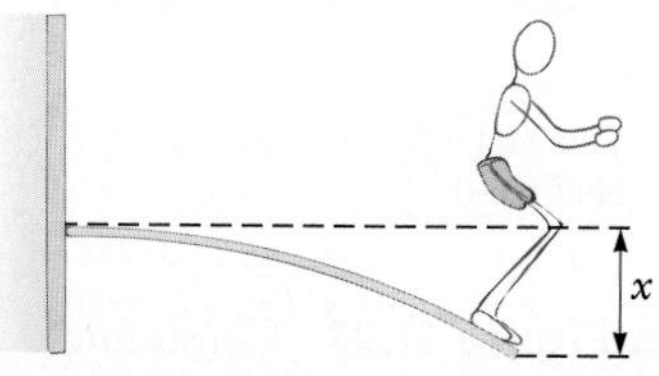

▲ **Figure 11.25** Example 11.8.

RECOGNIZE THE PRINCIPLE

We can use Equation (1) to find k if we know how much a diving board will bend in response to a known force. Assuming the mass of the board is small, when a diver of mass m stands at the end of the board, the force on the board is about mg.

SKETCH THE PROBLEM

Figure 11.25 shows the diving board bent (displaced) an amount x as the diver stands at the end of the board.

IDENTIFY THE RELATIONSHIPS

A typical person has a mass of about $m = 80$ kg. When he or she stands at rest at the end of a diving board, the displacement of the board is typically $x = 10$ cm (Fig. 11.25). Of course, this value will vary from board to board, but $x = 1$ m seems too large and $x = 1$ cm is much smaller than common experience indicates. Applying Newton's second law to the diver in Figure 11.25 leads to

$$\sum F = F_{\text{board}} - mg = -kx - mg = 0$$

$$k = -\frac{mg}{x}$$

SOLVE

Using $m = 80$ kg and taking $x = -0.1$ m (a negative value because the board is displaced downward), we find

$$k = -\frac{(80\text{ kg})(9.8\text{ m/s}^2)}{(-0.1\text{ m})} = \boxed{8000\text{ N/m}}$$

If we assume the mass of the diving board is much smaller than the mass of the person (which is only a very rough approximation), the oscillation frequency is the same as that of a simple mass-on-a-spring system (Eq. 11.14),

$$f = \frac{1}{2\pi}\sqrt{\frac{k}{m}}$$

Inserting our values for k and m gives

$$f = \boxed{2\text{ Hz}}$$

► ***What does it mean?***

The period for the diver's oscillation is thus about $T = 1/f = 0.5$ s. You should compare this value with the value you expected based on your intuition or the last time you visited a swimming pool.

11.5 Damping and Resonance

So far, we have ignored the effect of friction on the motion of a harmonic oscillator. If there is no friction, an oscillator set into motion will oscillate forever. Of course, most oscillators eventually come to rest as their vibrations decay with time. The friction in an oscillating system is referred to as ***damping***, and we now consider the motion of a ***damped harmonic oscillator***.

Newton's second law can be used to derive the displacement y of a damped oscillator as a function of time. However, the mathematics is more complicated than we can tackle here, so let's focus on the qualitative behavior. Consider a pendulum swing moving through air. Curve 3 in Figure 11.26 shows how the displacement varies with time when the damping is weak, that is, when there is only a small amount of friction. This plot applies to any weakly damped harmonic oscillator, so

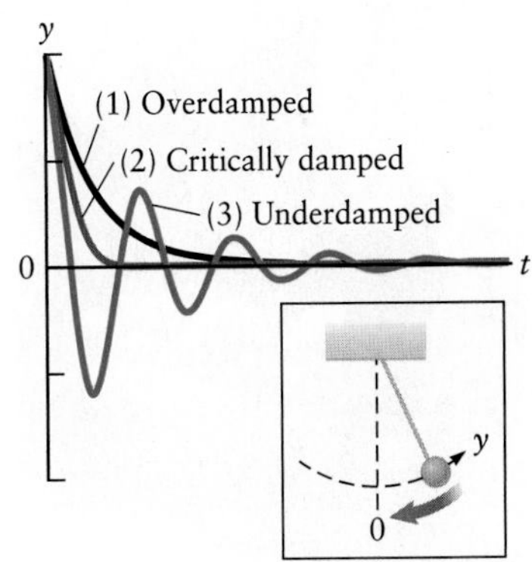

▲ **Figure 11.26** When a damped oscillator is given a nonzero displacement and then released, it can exhibit three different types of behavior: (1) overdamped, (2) critically damped, and (3) underdamped.

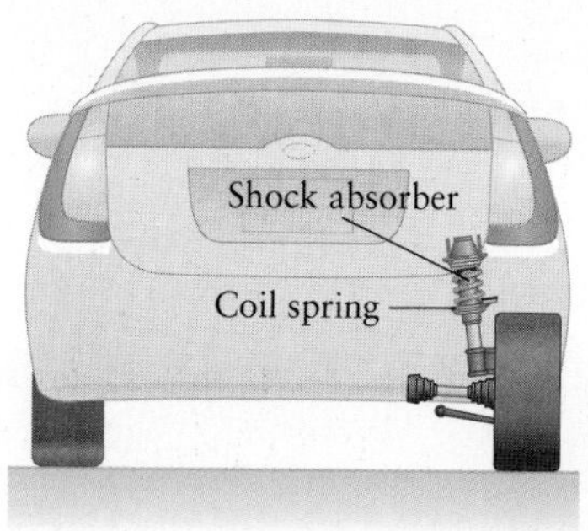

▲ **Figure 11.27** The struts on a car consist of a coil spring and a shock absorber, and act as springs with damping. The goal is usually to have a car respond to bumps in the road as a critically damped oscillator.

it describes the back-and-forth swinging of a pendulum or the motion of a mass on a horizontal surface (Fig. 11.6) when the surface is very slippery. The system still oscillates as the displacement alternates between positive and negative values, but the amplitude of the oscillation gradually decreases with time. The system undergoes many oscillations before damping eventually brings it to rest. This type of motion is called an ***underdamped*** oscillation.

CONCEPT CHECK 11.6 Sources of Damping in a Real Oscillator

Consider a real pendulum or a real mass-on-a-spring system as in Figure 11.6. What are some possible sources of damping in these systems?

When friction is very large, the oscillator is said to be ***overdamped***. The resulting displacement as a function of time in this case is shown as curve 1 in Figure 11.26. This type of motion would be found if the pendulum mass were moving through a very thick fluid (such as honey) rather than air. When the bob of an overdamped pendulum is pulled to one side and then released, the bob swings *very* slowly back to the bottom ($y = 0$) as plotted in Figure 11.26 (curve 1).

The underdamped and overdamped regimes are distinguished in the following way. In the underdamped case, the displacement always passes through zero at least once, and usually many times, before the system comes to rest. In contrast, an overdamped system released from rest does not swing past the bottom, but moves to the equilibrium point without going past it. At the boundary between these two cases the system is ***critically damped***. There, the displacement falls to zero as rapidly as possible *without* moving past the equilibrium position ($y = 0$) as in curve 2 in Figure 11.26.

These different categories of damping have important implications for applications. For example, the struts in a car are essentially springs that support the car's body as shown in Figure 11.27. Struts enable the tires to move up and down according to bumps in the road without directly passing these vibrations to the car's body (and the passengers). When the car hits a bump, the strut's springs are compressed, corresponding to the initial value of the displacement in Figure 11.26. To make the ride as comfortable as possible, the struts are designed to be critically damped, which allows the body of the car to return to its original height as quickly as possible. With underdamped (worn-out!) struts, a car oscillates up and down after hitting a bump, resulting in an uncomfortable ride.

The Driven Oscillator

In many applications, an oscillator is subjected to a force at regular intervals, or even continuously. A familiar example—a child being pushed on a swing—is shown in Figure 11.28. This particular pendulum is "driven" by the force that the parent exerts on the swing. We have already seen (Eq. 11.18) that a pendulum–swing has an oscillation frequency of

$$f_{\text{nat}} = \frac{1}{2\pi}\sqrt{\frac{g}{L}}$$

which is often referred to as the ***natural frequency*** of the swing; it is not necessarily the same as the frequency associated with the driving force. Suppose the parent applies a small push to the swing at regular time intervals. If the period of this pushing is T_{drive}, we can define a ***driving frequency*** $f_{\text{drive}} = 1/T_{\text{drive}}$. The amplitude of the oscillations of the pendulum–swing depend on the driving frequency as sketched in Figure 11.29.

If the frequency of the force from the parent does not match the pendulum's natural frequency—that is, if f_{drive} is not close to f_{nat}—the pendulum amplitude is small, even if the parent applies a large force. The amplitude is largest when the driving

▲ **Figure 11.28** When the parent pushes this child, the swing acts as a driven oscillator.

frequency is close to the natural frequency of a system. That happens when the parent in Figure 11.28 times her pushing to match the natural swinging frequency of the pendulum. This phenomenon is called ***resonance***. Whenever the frequency of the driving force matches the natural frequency of an oscillator, the amplitude of oscillation will be large. Figure 11.29 shows the behavior for cases of weak, medium, and strong damping. The resonance curve is narrowest and tallest when the damping is weakest, and it becomes lower and wider as the damping is increased. The peak of the resonance curve occurs at the ***resonant frequency***; for weak damping, the resonant frequency is very close to the natural frequency f_{nat} of the oscillator.

A harmonic oscillator for which damping is important is the vibrating element in a loudspeaker, the so-called speaker cone. Different musical notes have different frequencies, and when a loudspeaker is used to play music, the speaker cone is subject to a force whose frequency is determined by the frequencies present in the music. According to Figure 11.29, if the speaker cone oscillator is weakly damped, it will respond strongly to music near its natural frequency and relatively weakly to frequencies that are far from f_{nat}. A weakly damped speaker will thus produce loud sounds at some frequencies and weak sounds at others, which is *not* how one would like a loudspeaker to work. Instead, a speaker should have a frequency response that is as constant (uniform) as possible. One way to accomplish that is to make the speaker cone heavily damped. The cone of a good loudspeaker, however, should also move fairly easily so that it produces sound efficiently, hence the damping cannot be too large. This trade-off is one of the challenges faced by the speaker designer. Similar design challenges are faced in other applications of harmonic oscillators, such as the struts for a car.

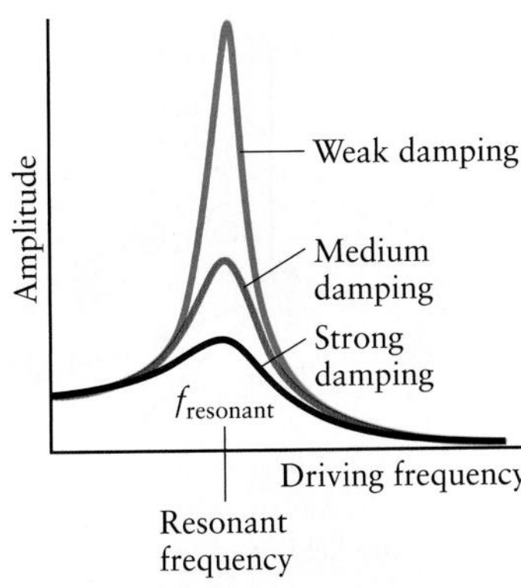

▲ **Figure 11.29** The amplitude of a driven oscillator depends on the damping. For weak damping, the oscillator will have a large amplitude when the driving frequency is near the resonant frequency, $f_{resonant}$. For a weakly damped oscillator, the resonant frequency is very close to the natural frequency.

CONCEPT CHECK 11.7 Damping of a Bridge

A structure such as a bridge or a building can oscillate in response to forces from the wind or an earthquake. Do you think these structures should be designed to be (a) underdamped, (b) critically damped, or (c) overdamped?

11.6 Detecting Small Forces

The phenomena of oscillatory motion and elasticity are closely connected and come together in the problem of detecting small forces.

The Cavendish Experiment

Let's consider an experiment to measure the force of gravity between two terrestrial objects. In Chapter 5, we found that the force of gravity between two massive and closely spaced lead spheres is about $F_{grav} = 2 \times 10^{-3}$ N (Eq. 5.22). This is a small force, and we need to be clever if we want to measure it experimentally. We might try to measure this force by connecting m_1 to a spring as sketched in Figure 11.30. The gravitational force exerted by m_2 on m_1 will cause the spring to stretch an amount x, and by measuring x, we can determine the magnitude of the force. Using Hooke's law for the spring, we have (considering magnitudes only) $F_{grav} = kx$. A soft spring might have a spring constant of about $k = 1$ N/m. Using this value and $F_{grav} = 2 \times 10^{-3}$ N, we can calculate the expected value of x:

$$x = \frac{F_{grav}}{k} = \frac{2 \times 10^{-3}\ \text{N}}{1\ \text{N/m}} = 0.002\ \text{m} = 2\ \text{mm}$$

If we want to measure F_{grav} with an accuracy of 1%, we would need to measure x with a precision of 0.02 mm. This is about the diameter of a human hair, so this

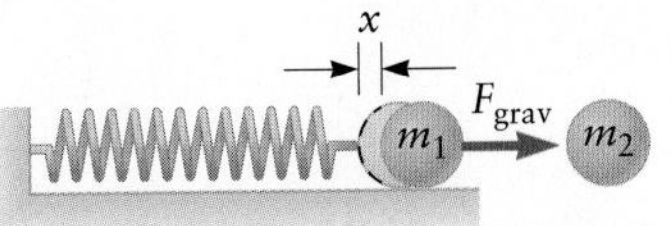

The gravitational force exerted by m_2 on m_1 will stretch the spring a distance x.

▲ **Figure 11.30** In theory, one could measure the gravitational force between two masses m_1 and m_2 by attaching m_1 to a spring and then observing how much the spring is stretched (x) when m_2 is brought nearby. This experimental design is not recommended because the displacement of the spring will be extremely small.

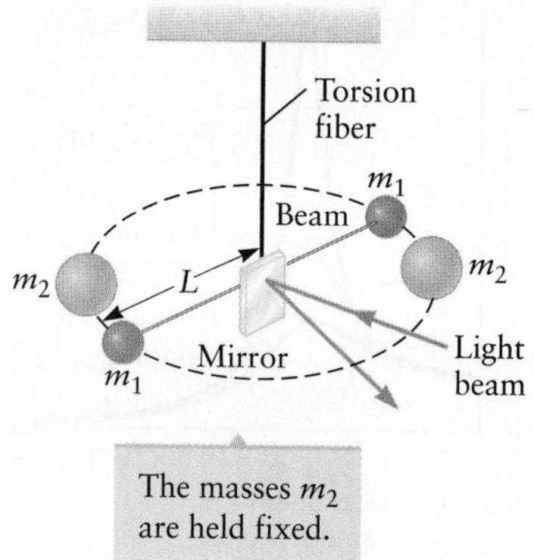
Torsion fiber
m_1
Beam
m_2
m_2
L
Mirror
m_1
Light beam
The masses m_2 are held fixed.

▲ **Figure 11.31** The Cavendish apparatus. The force of gravity exerted by the large masses m_2 on the small masses m_1 causes the torsion fiber to twist. The twist angle can be measured by attaching a small mirror and observing the reflection of a light beam.

method is not a very promising approach for a high precision measurement of F_{grav}. We need a different strategy.

Figure 11.31 shows how Henry Cavendish tackled this problem in an experiment he conducted about 200 years ago. He mounted two metal spheres, each of mass m_1, on a light rod and suspended the rod by a thin wire. This setup is very similar to the torsional oscillator in Figure 11.14B. Cavendish then brought two other objects, each of mass m_2, near the original spheres as shown in Figure 11.31, and the force of gravity exerted by the objects m_2 on the spheres m_1 caused the torsion fiber to twist. We saw in Section 11.2 that when a torsion fiber is twisted by an angle θ, it exerts a torque $\tau = -\kappa\theta$, where κ is the torsion constant of the fiber. Torsion fibers can have very small diameters; they can be as thin as a human hair (about 20 μm in diameter), making them *extremely* easy to twist. In fact, a value of $\kappa = 1 \times 10^{-9}$ N·m is quite common for a wire of this diameter.

With such small values of the torsion constant, Cavendish was able to measure F_{grav} accurately. However, Cavendish also had to deal with the fact that the system in Figure 11.31 acts as a torsional *oscillator*. When the two masses m_2 are brought close to the spheres on the torsion fiber, the gravitational force will set this pendulum into oscillation with a frequency (Eq. 11.20)

$$f = \frac{1}{2\pi}\sqrt{\frac{\kappa}{I}} \tag{11.26}$$

where I is the moment of inertia of the rotating part of the apparatus. Recall that the torsion constant κ is very small—that is what makes the Cavendish experiment so sensitive to small forces—but a very small value of κ in Equation 11.26 gives a very low frequency. In fact, frequencies of 0.0001 Hz are typical! That is *quite* a low frequency. The period of such an oscillator is $T = 1/f \approx 1 \times 10^4$ s, which is about 3 hours. Such a torsion oscillator is also a weakly damped system (because the friction in a torsion fiber is very small), so its oscillations will decay very slowly with time (Fig. 11.26). An experimenter will thus have to wait many periods before the oscillations stop and the twist angle can be measured. These very slow and weakly damped oscillations make the Cavendish experiment quite challenging. Legend has it that Cavendish performed his experiments very late on Sunday evenings, when the interfering vibrations from traffic on neighboring roads were smallest.

The Atomic Force Microscope

We next consider a modern invention that uses a type of spring to measure very small forces. The ***atomic force microscope*** (AFM) is illustrated in Figure 11.32A. A small bar, typically a fraction of a millimeter in length (about the diameter of the period at the end of this sentence), is fashioned so that it has an even smaller sharp tip at one

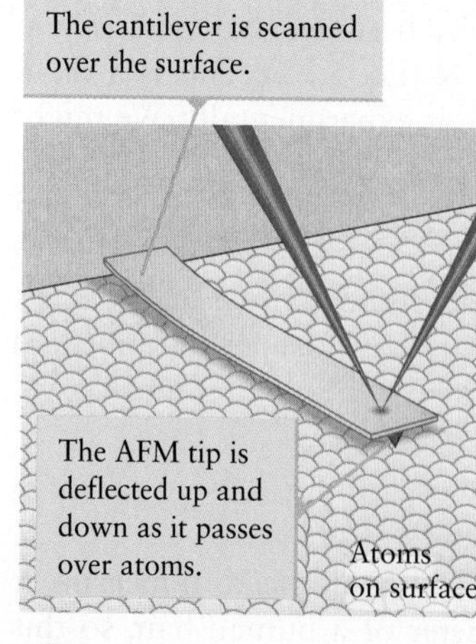
The cantilever is scanned over the surface.
The AFM tip is deflected up and down as it passes over atoms.
Atoms on surface
A

B
Image originally created by IBM Corporation

► **Figure 11.32** A Schematic design of an atomic force microscope (AFM). A very sharp tip is mounted on the end of a flexible bar called a "cantilever." The cantilever moves up and down as the atoms on the surface exert forces on the tip. The vertical deflection of the end of the cantilever is then used to construct an image of the surface profile. The red beams denote laser light reflected from the cantilever to measure its deflection. B An AFM image showing the atoms on a nickel surface. Each bump is an individual nickel atom.

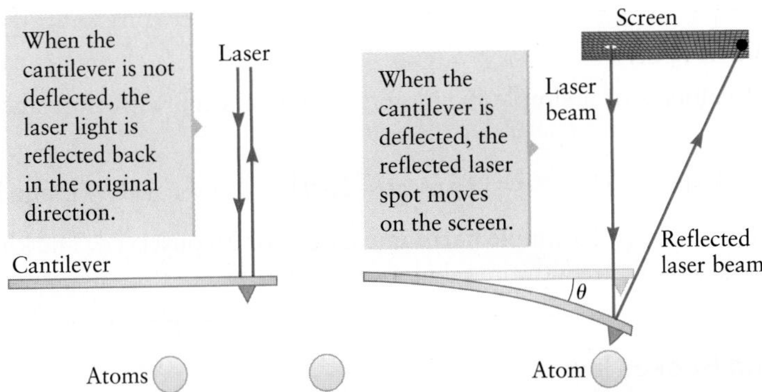

◀ **Figure 11.33** The deflection of an atomic force microscope cantilever is often measured by using the deflection of a laser beam. The location of the laser spot on a distant screen depends on the angle θ of the cantilever tip from the horizontal. By measuring the location of the laser spot, you can thus measure how much the cantilever is bent when it approaches atoms on the surface being studied.

end. This bar, called a cantilever, bends in response to any forces that act on it. These forces usually act at the tip, so the cantilever bends up and down as shown in the figure. This microscope works by scanning the tip of the cantilever over the surface of a material that is to be studied. The tip is *very* sharp, with a width at the end of only a few atoms. Although the tip usually does not directly touch the surface atoms, it is attracted to them, and the magnitude of the attractive force depends on the distance between the tip and the nearest surface atom. If an atom is "high" (i.e., sticks out and is close to the tip), the force on the tip is large, causing the cantilever to be deflected. By measuring this deflection as a function of the cantilever's position, one can make an image of the surface as shown in Figure 11.32B. An atomic force microscope is extremely sensitive to small forces. This sensitivity, along with the sharpness of the tip, makes it possible to resolve features as small as single atoms.

The cantilever behaves just like the diving board in Example 11.8 and can be treated as a simple spring obeying Hooke's law. A force of magnitude $F = kx$ is thus required to deflect the tip by a distance x. The value of k can be estimated from the shear modulus of whatever material is used to make the cantilever (using Eq. 11.24). For a typical cantilever, $k = 0.1$ N/m. The force between the tip and the surface[1] is typically 1×10^{-9} N, so the deflection of the cantilever is

$$x = \frac{F}{k} = \frac{1 \times 10^{-9}\text{ N}}{0.1\text{ N/m}} = 1 \times 10^{-8}\text{ m}$$

This very small deflection can be measured by reflecting a laser beam from the cantilever as shown in Figure 11.33. An AFM thus uses what is basically a simple spring to measure very small forces and displacements, thereby obtaining images of the atoms on the surface of a solid.

Summary | CHAPTER 11

Key Concepts and Principles

Frequency and period

Any motion that repeats with time is ***harmonic***. The repeat time is called the ***period***. The ***frequency*** (measured in hertz) is equal to the number of cycles that are completed each second. Frequency and period are related by

$$f = \frac{1}{T} \qquad \text{(11.1) (page 312)}$$

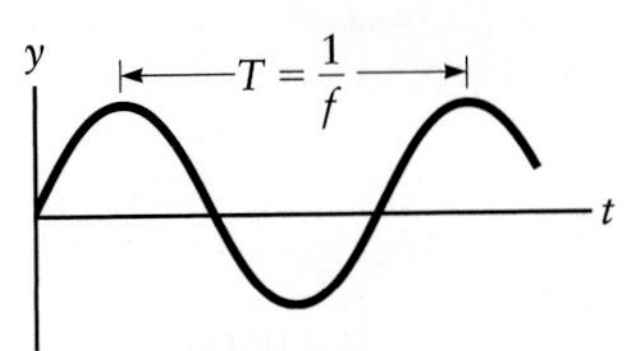

(Continued)

[1]We'll discuss this force in Chapter 29.

Simple harmonic motion

The position and velocity of a simple harmonic oscillator vary with time according to

$$y = A\sin(2\pi ft) \qquad v = 2\pi fA\cos(2\pi ft) \tag{11.9 (page 315)}$$

The total mechanical energy of a simple harmonic oscillator is conserved. The kinetic and potential energies oscillate with time, but their sum is constant.

Stress, strain, and Hooke's law

Elastic objects stretch or compress in response to applied forces. For a compressive or tensile force, the deformation is

$$\frac{F}{A} = Y\frac{\Delta L}{L_0} \tag{11.23 (page 326)}$$

where Y is an elastic constant called ***Young's modulus***. The ratio F/A is called the ***stress***, and $\Delta L/L_0$ is called the ***strain***. Other types of deformations (such as shear strains) are described by other elastic constants.

Damping and resonance

Most real harmonic oscillators are affected by friction, causing the oscillations to be damped. There are three classes of damped oscillators: ***underdamped*** (weak damping), ***overdamped*** (large damping), and ***critically damped***. When a damped harmonic oscillator is subject to a time-dependent driving force, the amplitude will be largest when the driving frequency is close to the natural frequency of the oscillator. This is called ***resonance***.

Applications

Mass connected to a spring

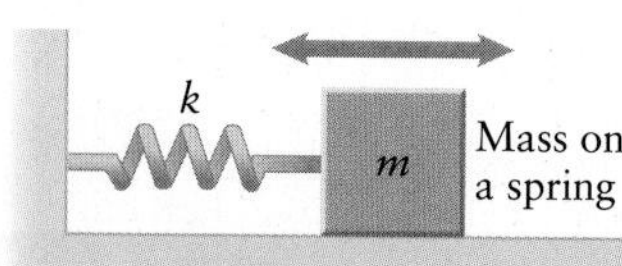

A mass connected to a spring with spring constant k oscillates with a frequency

$$f = \frac{1}{2\pi}\sqrt{\frac{k}{m}} \tag{11.14 (page 316)}$$

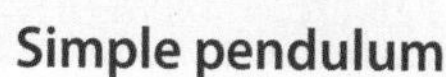

Simple pendulum

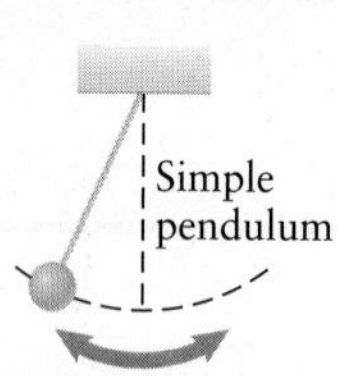

A ***simple pendulum*** consists of a mass attached to a massless string of length L. Its oscillation frequency is

$$f = \frac{1}{2\pi}\sqrt{\frac{g}{L}} \tag{11.18 (page 320)}$$

Torsional oscillator

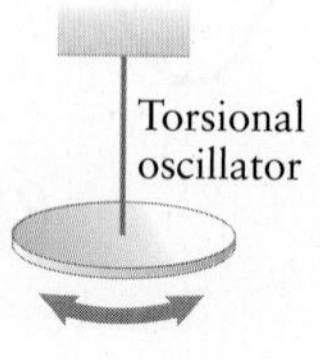

An object with a moment of inertia I suspended by a torsion fiber is a ***torsional oscillator***. Its frequency is

$$f = \frac{1}{2\pi}\sqrt{\frac{\kappa}{I}} \tag{11.20 (page 322)}$$

where κ is the ***torsion constant*** of the fiber.

Properties of simple harmonic oscillators

All simple harmonic oscillators have the following properties:

- The frequency is independent of amplitude.
- The force or torque that causes the oscillatory motion is a restoring force whose magnitude is proportional to the displacement of the oscillator from its equilibrium position. Such a restoring force is described by Hooke's law.

Questions

SSM = answer in *Student Companion & Problem-Solving Guide*

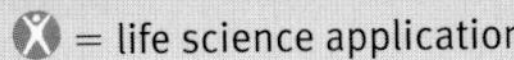 = life science application

1. A friend of yours is asked to design a swing using two ropes attached to the branch of a tree, and he shows you the design as sketched in Figure Q11.1. Explain why this design will not work very well. That is, why it is important that both of the ropes be the same length?

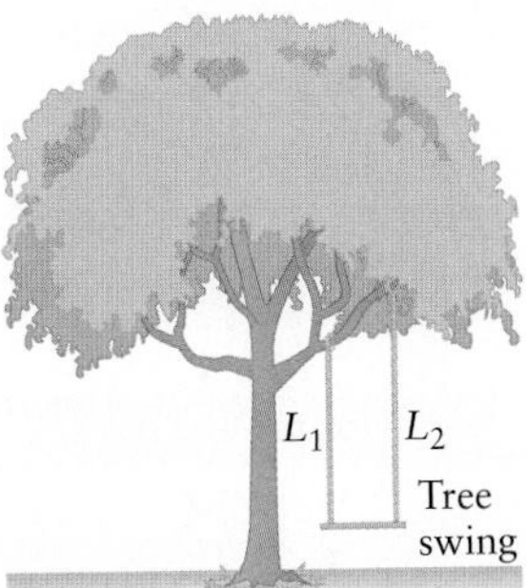

Figure Q11.1 Why is it desirable to have $L_1 = L_2$?

2. A mass hangs from a vertical spring and is initially at rest. A person then pulls down on the mass, stretching the spring. Does the total mechanical energy of this system (the mass plus the spring) increase, decrease, or stay the same? Explain your answer.

3. Design an experiment to measure the torsion constant κ of a torsion fiber. The value of κ is needed in the analysis of the Cavendish experiment (Sec. 11.6).

4. SSM Consider a simple pendulum that is used as a clock. (a) What should the length of the pendulum be to make one oscillation (one "tick" of the clock) every second when it is at sea level? (b) Will this clock "speed up" or "slow down" when it is taken to the top of Mount Everest? (c) How long will it take this clock to make 60 ticks on Mount Everest?

5. If a spring with spring constant k_0 is cut in half, what is the spring constant of one of the pieces?

6. A car mounted on struts is like a mass on a spring. If you ignore damping, how will the frequency of the oscillations change if passengers (or a heavy load) are added to the car? Will the frequency increase, decrease, or stay the same?

7. A basketball player dribbles the ball as she moves along the court. What is a typical value for the frequency of this oscillatory motion?

8. When a group of marching soldiers reach a bridge, they often "break stride" and do not walk "in step" across the bridge. Explain why.

9. Two metal rods with the same length and cross-sectional area are both subjected to a compressive force F. If the length of rod 1 changes by more than the length of rod 2, which one has the larger Young's modulus?

10. Two solid rods are made of the same material and have the same cross-sectional areas, and their lengths differ by a factor of two. If a compressive force F is applied to both, what is the ratio of their changes in length?
(a) The longer rod changes length by more, by a factor of two.
(b) The shorter rod changes length by more, by a factor of two.
(c) The lengths change by the same amount.

11. Make a sketch of how the position and kinetic energy of a harmonic oscillator vary with time. The period of the oscillations of KE can be determined from the separation in time of adjacent maxima in your KE sketch. Show that the frequency of the KE oscillations is equal to twice the frequency of the displacement oscillations (Eq. 11.14). Explain why.

12. In Section 11.3, we discussed the total mechanical energy of a mass-on-a-spring oscillator. The result in Equation 11.21 shows that the total energy is proportional to the square of the amplitude. Derive the corresponding results for the total mechanical energy of a simple pendulum and for a torsional oscillator and show that they are also proportional to the square of the amplitude.

13. SSM Use energy considerations to derive the oscillation frequency for a mass-on-a-spring oscillator. *Hint*: The maximum potential energy stored in the spring must be equal to the maximum kinetic energy of the mass.

14. Derive the frequency of oscillation of a torsional oscillator, Equation 11.20. *Hint*: Consider how the restoring force depends on the twist angle and use Newton's second law for rotational motion ($\tau = I\alpha$).

15. Pendulum clocks use a pendulum to keep time (Fig. Q11.15). This type of clock can be adjusted by varying the length of the pendulum. How should the length be adjusted in the following cases?
(a) The clock is running slow.
(b) The clock is running fast.
(c) The clock is running fine at sea level, but needs to be moved to Denver.

Figure Q11.15

16. Many musicians use a metronome to help keep time when playing music. Some metronomes use a simple pendulum. If a metronome clicks 100 times per minute, what is the length of the pendulum?

17. In Chapter 3, we learned how the apparent weight of a person in an elevator depends on the acceleration of the elevator. Consider a simple pendulum that is placed in an elevator. Does the elevator's acceleration affect the pendulum's period? If the pendulum is accelerating downward with magnitude g, what is the period of the pendulum?

18. The windshield wipers on your car are an example of periodic motion. What is the approximate period of the motion?

19. Arrange these materials according to their Young's modulus from smallest to largest: (a) Jell-O, (b) steel, (c) wood, (d) diamond.

20. Discuss the difference between the Young's modulus of a material and the strength of a material. Does a large value of Y guarantee a strong material?

Problems available in ENHANCED WebAssign

Section 11.1 General Features of Harmonic Motion
Problems 1–11

Section 11.2 Examples of Simple Harmonic Motion
Problems 12–33

Section 11.3 Harmonic Motion and Energy
Problems 34–45

Section 11.4 Stress, Strain, and Hooke's Law
Problems 46–59

Section 11.5 Damping and Resonance
Problems 60–62

Section 11.6 Detecting Small Forces
Problem 63

Additional Problems
Problems 64–73

List of Enhanced Problems

Problem number	Solution in *Student Companion & Problem-Solving Guide*	Reasoning & Relationships Problem	Reasoning Tutorial in ENHANCED WebAssign
11.10	✓		
11.16	✓		
11.18	✓	✓	✓
11.21		✓	✓
11.28		✓	✓
11.37	✓		
11.45		✓	✓
11.57	✓		
11.60	✓		
11.61		✓	✓
11.64	✓		
11.68		✓	✓
11.73	✓		

◀ *There are many types of waves, including waves on the surface of the ocean. (©Royalty-Free/Corbis/Jupiterimages)*

Waves

In this chapter, we build on our work from Chapter 11 as we study ***wave motion***. A wave is *a moving disturbance that transports energy* from one place to another *without transporting matter*. This definition of a wave leaves two basic questions unanswered: (1) *what* is it that is "disturbed" by the wave, and (2) *how* is it disturbed? These questions will be answered as we proceed through this chapter. The motion associated with a wave disturbance often has a repeating form, so wave motion has much in common with simple harmonic motion.

In this chapter, we concentrate on aspects of wave motion that are common to all types of waves. The next chapter will then focus on a particularly common type of wave, *sound*. Chapters 23 through 26 are devoted to electromagnetic waves (including light), so many of the ideas that we discuss now will be revisited and developed further in those chapters.

OUTLINE

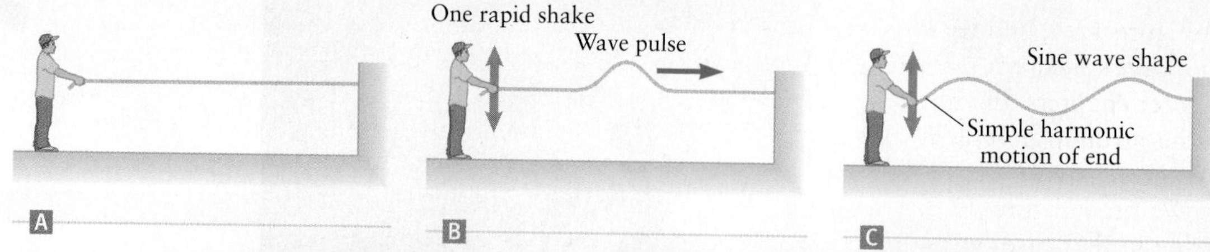

▲ **Figure 12.1** A A long string. B When the end of a string is shaken once up and down, a wave pulse that travels horizontally along the string is generated. C Repeated shaking of the string generates a periodic wave.

12.1 What Is a Wave?

Figure 12.1A shows a person holding one end of a flexible string, with the other end tied to a wall. When the person shakes his end of the string up and down, he creates a "disturbance" that moves horizontally along the string. The shape of this disturbance depends on how the person shakes the string. In Figure 12.1B, he gives the end of the string a single shake, producing a single ***wave pulse***. In Figure 12.1C, the end of the string is shaken up and down in a simple harmonic manner, creating a ***periodic wave*** (described by a sine or cosine function). These disturbances are examples of ***waves***. Since portions of the string are moving, these portions have a nonzero velocity; hence, kinetic energy is associated with the wave. There is also elastic potential energy in the string as it stretches while the wave pulse travels by. The wave carries this energy as it travels.

Waves carry energy from one location to another. In Figure 12.1, a wave on the string gains energy from the person's hand as the force exerted by the hand does work on the string. The wave then carries this energy to the right as the wave moves along the string. Although a wave thus transmits energy from one place to another, it does *not* transport matter along with the energy. Pieces of the string do not move horizontally along with the wave disturbance. In particular, the left end of the string in Figure 12.1 does not move to the right along with the wave.

Figure 12.2 shows a wave pulse as it travels along a string. As with the wave pulse in Figure 12.1A, we suppose a person is holding the end of the string on the far left (out of the picture). We then imagine that the person gives one large up-and-down shake to the left end of the string, creating a single pulse that moves to the right as sketched in Figure 12.2. We say that this pulse ***propagates*** to the right. Parts A through C of Figure 12.2 show the string's shape at three successive times. Several points along the string are labeled to show how they move upward and then back down to the original, undisturbed configuration as the wave pulse travels by.

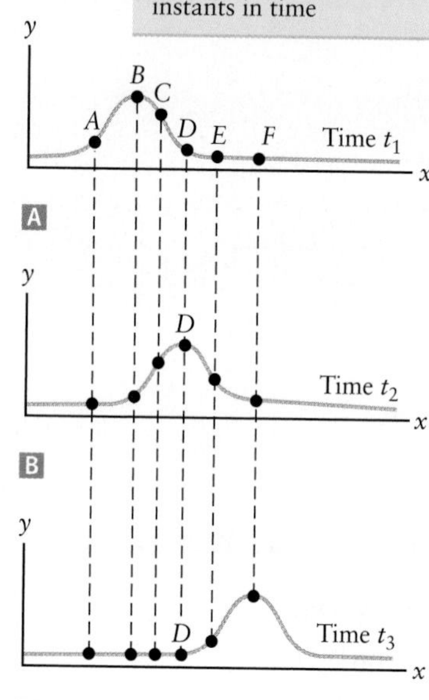

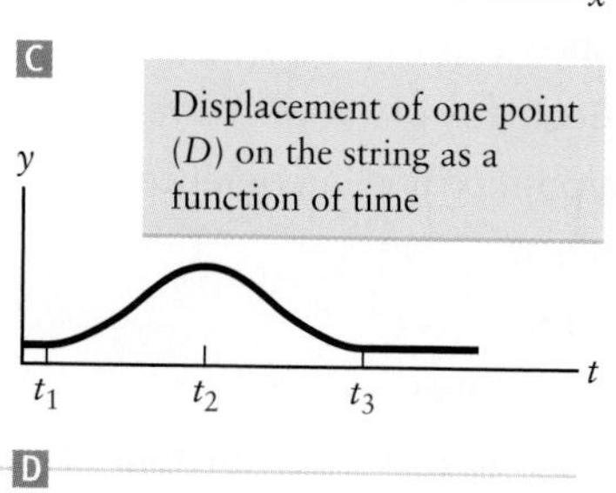

▲ **Figure 12.2** A, B, and C: Successive snapshots of a wave pulse traveling along a string with $t_1 < t_2 < t_3$. The wave pulse maintains its shape as it moves. This wave on a string is a transverse wave. Any particular point on the string (such as point D) moves in a direction perpendicular to the velocity of the wave. D Displacement of point D on the string as a function of time.

Let's now focus on the motion of point D on the string as the pulse moves by. The displacement of point D is plotted as a function of time in Figure 12.2D. This displacement is along the y direction and is thus *perpendicular* to the direction of propagation. At early times, when $t < t_1$, point D is at rest as the wave pulse has not yet arrived. When the pulse arrives, point D on the string first moves upward (Fig. 12.2B) and then moves back down to its original, undisturbed position (Fig. 12.2C). Other points on the string move in a similar manner. While points on the string are set into motion when the wave travels by, these points do not travel horizontally along with the wave. Hence, we again see that this wave *transports energy without transporting matter*. This is a key feature of *all* waves.

Before the wave pulse in Figure 12.2 was generated, the string was in a "relaxed" state, lying parallel to the x (horizontal) axis. When the pulse arrives, it displaces the string from this state, so we can say that it disturbs the string. All waves are a "disturbance" of some type, and the "thing" that is disturbed by the wave is called its ***medium***. In many cases, this medium is a material substance such as the string in Figure 12.1, and these waves are called ***mechanical waves***. It is also possible for the wave's medium to be a field, such as the electromagnetic field we'll learn about in later chapters. In either case, a wave must disturb something, and that something is the medium in which the wave travels. For a mechanical wave, the medium is a substance we can describe using Newton's laws of motion.

We used Figure 12.2 to compare and contrast the motion of the medium (the string in that case) with the motion of the wave. In Figure 12.2, pieces of the string move along the vertical direction (y) while the wave travels along a horizontal direction (x). Waves like this one, in which the motion of the medium is perpendicular to the direction of wave propagation, are called ***transverse waves***.

If the motion of the medium is parallel to the wave's propagation direction, the waves are called ***longitudinal***. For example, it is possible to generate longitudinal waves on a spring such as a Slinky as illustrated in Figure 12.3. Here the person holding the Slinky shakes the end back and forth along the x direction, generating a disturbance in which the loops of the Slinky alternate between being tighter (more compressed) and looser (more stretched) than normally found when the Slinky is at rest.

In nearly all cases, waves can be classified as either transverse or longitudinal, according to the motion of the medium.

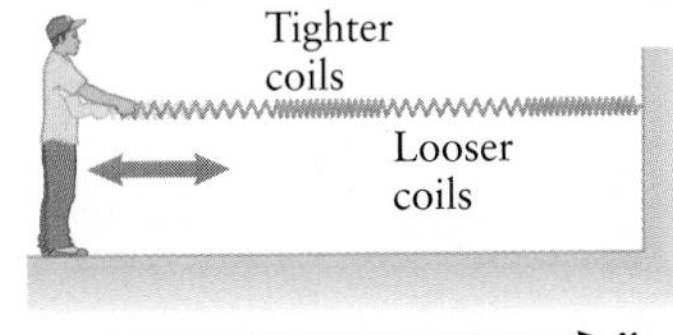

▲ **Figure 12.3** A Slinky (a loosely coiled spring) is stretched and held horizontally. Shaking one end of the Slinky back and forth in the horizontal direction generates a longitudinal wave. At some places, the Slinky coils are compressed an extra amount relative to those in an undisturbed Slinky, but at other places the coils are looser. These tight and loose places move along the Slinky, forming the wave.

EXAMPLE 12.1 Determining the Velocity of a Wave

Figure 12.4 shows three snapshots of a wave pulse traveling along a string. This pulse was generated by shaking the left end of the string, generating a disturbance that propagated to the right. (**a**) Use Figure 12.4 to estimate the speed of the wave. (**b**) Estimate the speed of point A on the string at $t = 1.5$ s as the wave pulse passed by that point.

RECOGNIZE THE PRINCIPLE

(**a**) The wave in Figure 12.4 is a single pulse. We can determine the velocity of this pulse by noting the position of its peak along the x direction as a function of time. (**b**) To estimate the velocity of a point *on the string*, we examine the motion of that point along the vertical (y) direction.

SKETCH THE PROBLEM

Figure 12.4 shows the wave pulse at three different times as well as the position y of point A.

IDENTIFY THE RELATIONSHIPS AND SOLVE

(**a**) From Figure 12.4, the peak of the wave pulse is at $x \approx 1.8$ m when $t = 1.0$ s and has moved to $x \approx 3.8$ m when $t = 3.0$ s. The average velocity during this time interval is thus along the $+x$ direction with

$$v = \frac{\Delta x}{\Delta t} = \frac{(3.8 - 1.8)\text{ m}}{(3.0 - 1.0)\text{ s}} = \boxed{1.0\text{ m/s}}$$

This is the speed *of the wave*.

(**b**) Comparing the snapshots in Figure 12.4 at $t = 1.0$ s and 2.0 s, we see that point A on the string is near its undisturbed position with $y \approx 0.03$ m at $t = 1.0$ s and that this point is at the peak of the wave when $t = 2.0$ s, where we find $y \approx 0.20$ m. The velocity of point A at $t = 1.5$ s is thus approximately

$$v_A = \frac{\Delta y}{\Delta t} = \frac{(0.20 - 0.03)\text{ m}}{(2.0 - 1.0)\text{ s}} = \boxed{0.17\text{ m/s}}$$

(continued) ▶

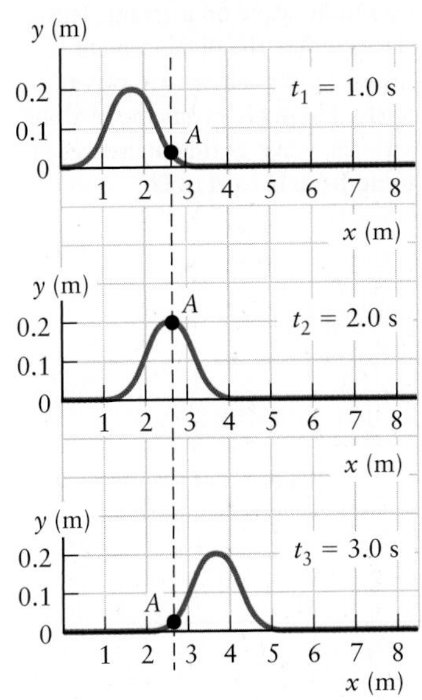

▲ **Figure 12.4** Example 12.1.

▶ What does it mean?
This is a transverse wave because the velocity of point A on the string is along y, which is perpendicular to the velocity of the wave. In general, the velocity of a point on the string will *not* have the same magnitude as the velocity of the wave. The velocity of a point on the string also will not be constant; here we have made an estimate of v_A during the time the wave is passing the point of interest. Before and after the wave pulse arrives, the velocity of a point on the string will be zero.

12.2 Describing Waves

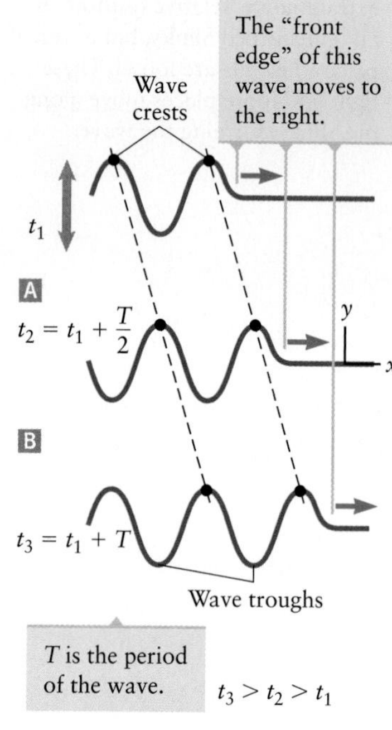

▲ **Figure 12.5** Snapshots of a periodic wave on a string. The places where the displacement is largest (and positive) are wave crests. The wave crests move along with the wave as time increases in going from A to B to C.

Wave pulses like those in Figures 12.2 and 12.4 are often found in nature, but it is also common to find periodic waves. A periodic wave can be generated by shaking the end of a string up and down repeatedly, in a periodic fashion. In the case shown in Figure 12.5, we imagine that a person (not shown in the figure) is shaking the far left end of the string in such a way that this end is undergoing simple harmonic motion along the y (vertical) direction. Figure 12.5 then shows snapshots of the resulting periodic wave as it travels along. These successive snapshots show the "front edge" of the wave, making the overall motion of the wave clear. However, if we wait until this front edge moves out of the picture to the right, all we will see is a repeating wave disturbance with a "sine-wave" form. If we examine just a single snapshot, we find places where the string displacement y is largest and positive, called wave ***crests***. Likewise, at some places called wave ***troughs***, y is equally large in magnitude but negative in sign.

Two wave crests in Figure 12.5 are marked with large dots to emphasize the motion of the wave. The wave's speed can be measured by following the position of a particular wave crest as a function of time, similar to what we did in Example 12.1.

The wave pulses in Figure 12.2 and 12.4 are examples of nonperiodic waves. In these cases, the wave disturbance is limited to a small region of space. In contrast, a periodic wave as in Figure 12.5 extends over a very wide region of space, and the displacement of the medium varies in a repeating and often sinusoidal manner as a function of both position and time. For example, if we examine a single snapshot of a periodic wave—say, in Figure 12.5C—we see that the string displacement y varies in a periodic manner as a function of *position* along the string x. We can also monitor the string displacement at a particular point on the string and find that the displacement also varies in a periodic manner as a function of *time*. Hence, *a periodic wave involves repeating motion as a function of both space and time.*

The "Equation" of a Wave

The wave in Figure 12.5 was generated by shaking the left end of the string up and down. For simplicity, let's assume the displacement at the left end vibrates as a simple harmonic oscillator with $y_{\text{end}} = A\sin(2\pi ft)$, where f is the frequency of this harmonic motion. This periodic shaking generates a periodic wave that travels along the string as sketched in Figure 12.5. The string's displacement is given by the relation

Equation of a periodic wave

$$y = A\sin\left(2\pi ft - \frac{2\pi x}{\lambda}\right) \tag{12.1}$$

where λ (the Greek letter lambda) is called the ***wavelength*** as explained below.

Equation 12.1 is a mathematical description of a periodic wave. For a wave on a string, it tells how the transverse displacement y of a point on the string varies as a function of both time t and location x along the string. It applies to many different types of waves and describes how a wave travels (propagates) away from its source. For the wave on a string in Figure 12.5, the source is the person's hand at the left end of the string. Equation 12.1 does not include what happens when the wave reaches the right end of the string. We'll learn about that in Sections 12.6 and 12.7, but first let's explore the meaning of Equation 12.1 in a little more detail.

Equation 12.1 has much in common with the displacement of a simple harmonic oscillator (Eq. 11.2). Like a simple harmonic oscillator, a periodic wave has a frequency f and an amplitude A. The frequency is related to the "repeat time" for the wave and can be understood using the snapshots in Figure 12.5. A point on the string that is at a crest in Figure 12.5A moves down to a trough in snapshot B and then returns to a crest in snapshot C, constituting one period T of the motion. The frequency of the wave is then given by $f = 1/T$. Since the sine function in Equation 12.1 repeats itself as the value of t increases, this behavior is all contained in Equation 12.1.

The meaning of the wavelength λ can be understood from Figure 12.6. Here again we show a snapshot of a wave on a string. This plot corresponds to Equation 12.1, showing the transverse displacement y as a function of position x along the string at a particular value of t. Figure 12.6 shows that y is a repeating function of position. That is, if we advance x by a distance equal to λ, the value of y is repeated. One can thus think of λ as the *repeat distance* of the wave. Since the sine function in Equation 12.1 repeats itself as x varies to either increasing or decreasing values, this behavior is contained in Equation 12.1. Periodic waves thus have both a repeat *time* and a repeat *distance*. A periodic wave is a combination of two simple harmonic motions: one as a function of time and the other as a function of space.

Figure 12.6 also illustrates the meaning of the amplitude A of a wave. The wave displacement y varies between $\pm A$, so the wave crests have a height $y = +A$. At the troughs, $y = -A$.

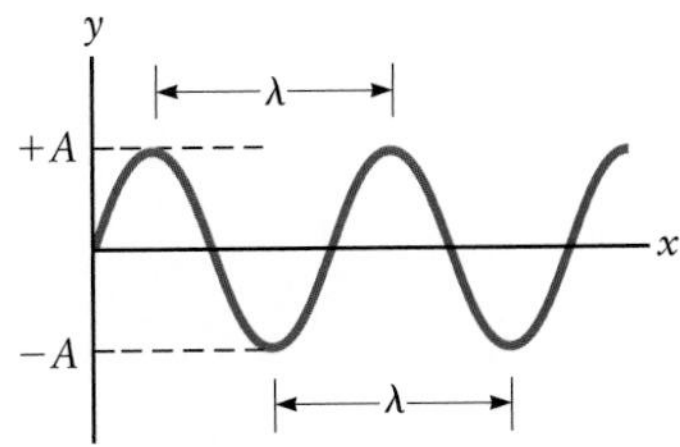

▲ **Figure 12.6** Snapshot of a periodic wave on a string. This is just a plot of Equation 12.1 showing the displacement associated with a wave as a function of x at a particular value of t. The wavelength λ is the "repeat distance." The wavelength can be measured from any point on the wave to the next "equivalent" point. The displacement oscillates between $+A$ and $-A$, where A is the amplitude.

Speed of a Wave

The mathematical description of a wave in Equation 12.1 contains three important wave properties: frequency, wavelength, and amplitude. Let's now consider how these are connected with the velocity of the wave. The snapshots in Figure 12.5 show that the wave moves a distance $\Delta x = \lambda$ during one period of the motion ($\Delta t = T$), which also follows from the definitions of wavelength and period. The speed of the wave is therefore given by

$$v = \frac{\Delta x}{\Delta t} = \frac{\lambda}{T}$$

Because frequency is related to the period by $f = 1/T$, we have

$$v = \frac{\lambda}{1/f}$$

or

$$v = f\lambda \tag{12.2}$$

The speed of a periodic wave is related to its frequency and wavelength.

Information regarding the *direction* of the wave is also contained in Equation 12.1. To see how, we focus on the motion of a wave crest. A crest occurs whenever the sine function in Equation 12.1 has its maximum value, and, because $\sin(\pi/2) = 1$, one of these maxima will occur when the argument of the sine function is equal to $\pi/2$. So,

$$2\pi ft - \frac{2\pi x}{\lambda} = \frac{\pi}{2} \tag{12.3}$$

Hence, at $t = 0$, a crest will be found at

$$2\pi f(0) - \frac{2\pi x_{\text{crest}}}{\lambda} = \frac{\pi}{2}$$

$$x_{\text{crest}} = -\frac{\lambda}{4} \tag{12.4}$$

After one period, $t = T$, and using Equation 12.3 we get

$$2\pi fT - \frac{2\pi x_{\text{crest}}}{\lambda} = \frac{\pi}{2}$$

$$x_{\text{crest}} = -\frac{\lambda}{4} + f\lambda T = -\frac{\lambda}{4} + vT$$

where we have also used the relation $v = f\lambda$. Comparing this expression with Equation 12.4, we see that x_{crest} has become larger, indicating that the wave has moved to the right. Hence, the velocity of the wave described by Equation 12.1 is positive. (The wave has moved in the $+x$ direction.)

A similar analysis shows that a wave that moves to the left is described by

$$y = A \sin\left(2\pi ft + \frac{2\pi x}{\lambda}\right) \tag{12.5}$$

Equations 12.1 and 12.5 describe periodic waves that move in opposite directions. Other similar functions can also be used to describe such traveling waves. For example,

$$y = A \cos\left(2\pi ft - \frac{2\pi x}{\lambda}\right) \tag{12.6}$$

describes a wave that moves in the $+x$ direction, whereas

$$y = A \cos\left(2\pi ft + \frac{2\pi x}{\lambda}\right) \tag{12.7}$$

is a wave that travels in the $-x$ direction. All four cases (Eqs. 12.1, 12.5, 12.6, and 12.7) describe waves with the same frequency, wavelength, speed, and amplitude. The only differences involve the direction of wave propagation (the sign of v) and the shape of the string at $t = 0$ (either a sine function or a cosine).

Interpreting the equation of a periodic wave

Many properties of a periodic wave can be read directly from the equation of the wave:

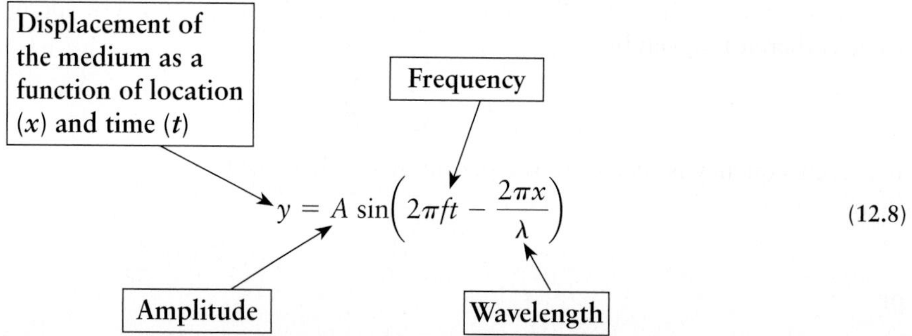

(12.8)

CONCEPT CHECK 12.1 Waves on a String

A certain wave on a string is found to have a frequency of 500 Hz and a wavelength of 0.60 m. If a new wave of frequency 250 Hz is then established on this string and the wave speed does not change, is the new wavelength (a) 0.60 m, (b) 0.30 m, or (c) 1.2 m? *Hint*: You do not need a calculator!

EXAMPLE 12.2 Finding the Properties of a Wave

Figure 12.7 shows snapshots of a wave on a string at two very closely spaced times, separated by $\Delta t = 0.050$ s. Assuming the wave is moving to the right, find the wavelength, amplitude, frequency, and velocity of the wave.

RECOGNIZE THE PRINCIPLE

The wavelength λ is the distance between wave crests, which we can read from Figure 12.7. We can also read the amplitude from the figure. By comparing the two snapshots, we can deduce the wave speed v and then find the frequency from the relation $f\lambda = v$.

SKETCH THE PROBLEM

Figure 12.7 describes the problem. We have also labeled two wave crests A and B, which can be used to follow how the wave moves.

IDENTIFY THE RELATIONSHIPS AND SOLVE

From Figure 12.7, the distance between the wave crests at A and B is approximately

$$\lambda = \boxed{0.65 \text{ m}}$$

We can also estimate the amplitude from the height of a crest (or trough) and find

$$A = \boxed{0.40 \text{ m}}$$

To find the velocity, we must compare snapshots at different times. These snapshots were taken at very closely spaced times, so we can assume the wave has only had time to move a small amount. Wave crest A in the first snapshot (at t_1) then corresponds to crest A' in the second snapshot. From Figure 12.7, this crest has moved a distance of approximately 0.25 m in a time interval $t_2 - t_1 = \Delta t = 0.050$ s, so the wave velocity is

$$v = \frac{\Delta x}{\Delta t} = \frac{0.25 \text{ m}}{0.050 \text{ s}} = \boxed{5.0 \text{ m/s}}$$

The velocity is positive because the wave is moving along the $+x$ direction. We can find the frequency using Equation 12.2. We get

$$f = \frac{v}{\lambda} = \frac{5.0 \text{ m/s}}{0.65 \text{ m}} = \boxed{7.7 \text{ Hz}}$$

► ***What does it mean?***

All the properties of a wave can be read off from snapshots like those in Figure 12.7 if you know what to look for!

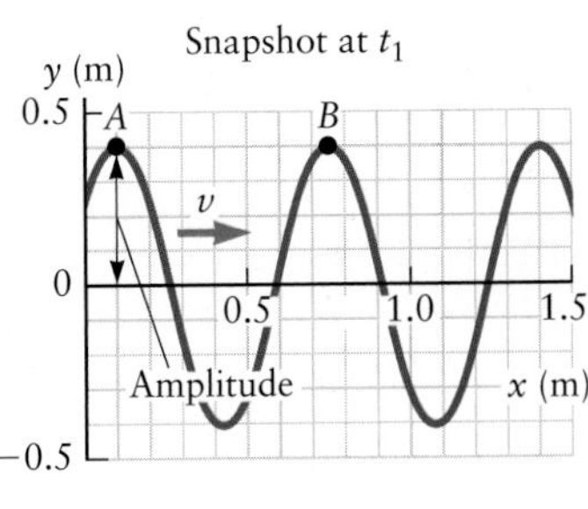

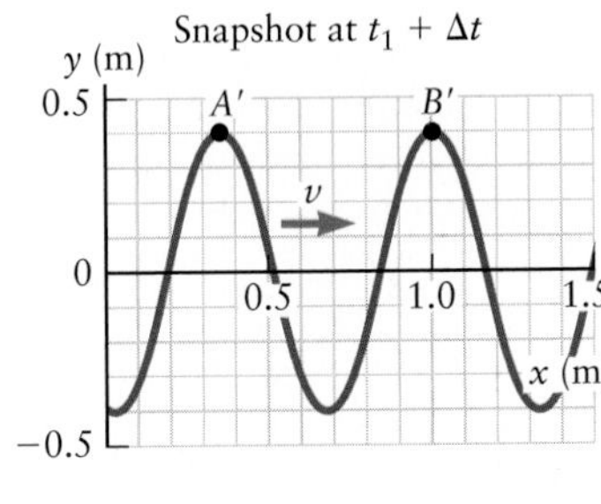

▲ **Figure 12.7** Example 12.2. Snapshots of a wave on a string at time t_1 and at a short time later, $t_1 + \Delta t$.

EXAMPLE 12.3 Analyzing the Equation of a Wave

Suppose a wave on a string is described by the equation

$$y = (0.040 \text{ m})\cos(25t - 30x) \quad (1)$$

where the units of t are seconds and x is given in meters. Find the frequency, wavelength, and velocity of this wave.

RECOGNIZE THE PRINCIPLE

We can read all the properties of the wave in Equation (1) by following the outline in Equation 12.8.

SKETCH THE PROBLEM

No figure is needed.

IDENTIFY THE RELATIONSHIPS AND SOLVE

We compare Equation (1) with Equation 12.8 and read off values for the corresponding quantities. The factor that multiplies t in Equation (1) equals $2\pi f$, so we have

$$2\pi f = 25 \text{ Hz}$$

(continued) ►

The units here are hertz because that is the unit for frequency f in the SI system. Solving for f gives

$$f = \frac{25 \text{ Hz}}{2\pi} = \boxed{4.0 \text{ Hz}}$$

Likewise, the factor multiplying x in both Equation 12.8 and Equation (1) equals $2\pi/\lambda$. Setting $2\pi/\lambda = 30 \text{ m}^{-1}$ and solving for the wavelength, we find

$$\lambda = \frac{2\pi}{30 \text{ m}^{-1}} = \boxed{0.21 \text{ m}}$$

According to Equation 12.2 the speed of the wave is

$$v = f\lambda = (4.0 \text{ Hz})(0.21 \text{ m}) = \boxed{0.84 \text{ m/s}}$$

▸ ***What does it mean?***

Comparing the given equation again with Equation 12.1, we see that the velocity of this wave is positive; it moves in the $+x$ direction.

CONCEPT CHECK 12.2 Properties of a Wave

Figure 12.8 shows snapshots of several different waves on a string. The four strings are identical (the same tensions, etc.). Which wave has (a) the longest wavelength, (b) the highest frequency, and (c) the smallest amplitude?

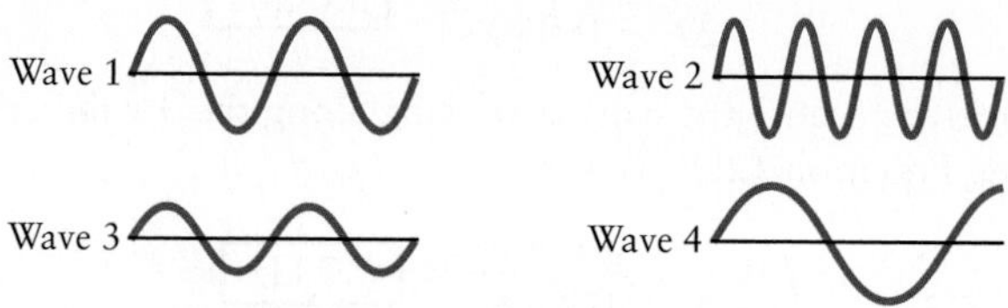

▲ **Figure 12.8** Concept Check 12.2.

12.3 Examples of Waves

We have discussed some general properties of waves and how they behave in time and space. Let's next consider some specific types of waves.

Waves on a String

Waves on a string are mechanical waves, and the medium that is disturbed is the string. For a transverse wave on a string, the velocity depends on the tension in the string and the string's mass per unit length, denoted by μ. This is not surprising; we expect that a wave will travel faster in a lighter string (one with smaller μ) since a lighter string will have a greater acceleration (and velocity) for a given value of the tension. Likewise, the wave speed is increased when the tension is increased (when the string is more "taut").

Let's denote the tension in the string by F_T, deviating from our usual symbol for tension so as not to confuse it with the period of the wave. The speed of a wave on a string can be calculated using Newton's laws of motion and is

$$v = \sqrt{\frac{F_T}{\mu}} \tag{12.9}$$

The speed of this wave does *not* depend on its frequency. The frequency is determined by how rapidly the end of the string is shaken, but the wave speed in Equation 12.9 is completely independent of f. In fact, the speed of transverse waves on a string is the *same* for periodic and nonperiodic waves (wave pulses).

EXAMPLE 12.4 Speed of a Wave on a Piano String

A typical string in the author's grand piano is composed of steel ($\rho = 7800$ kg/m^3) with radius $r = 0.50$ mm and has a tension of about 650 N. What is the speed of a wave on such a string?

RECOGNIZE THE PRINCIPLE

We can find the wave speed from $v = \sqrt{F_T/\mu}$ (Eq. 12.9). The tension F_T is given, and we can find the mass per unit length μ from the known density of steel.

SKETCH THE PROBLEM

No figure is needed.

IDENTIFY THE RELATIONSHIPS

A string is just a long cylinder; if it has length L and radius r, its volume is $\pi r^2 L$. Density ρ is mass per unit volume, so the string's mass is $m = \pi r^2 L \rho$. The density of steel is given as 7800 kg/m^3. The mass per unit length equals m/L:

$$\mu = \frac{m}{L} = \pi r^2 \rho = \pi(5.0 \times 10^{-4}\text{ m})^2(7800\text{ kg/m}^3) = 6.1 \times 10^{-3}\text{ kg/m}$$

SOLVE

Inserting this value of μ into Equation 12.9 along with the given value of the tension, we find

$$v = \sqrt{\frac{F_T}{\mu}} = \sqrt{\frac{650\text{ N}}{6.1 \times 10^{-3}\text{ kg/m}}} = \boxed{330\text{ m/s}}$$

▶ *What does it mean?*

This speed is more than 700 miles per hour! The speed is high because of the high tension. (You should compare the tension of 650 N to your own weight.)

CONCEPT CHECK 12.3 Waves on a Violin String

The tension in a violin string is 60 N, and the speed of a wave on this string is 300 m/s. If the tension is reduced to 30 N, will the new speed of a wave on this string be (a) 150 m/s, (b) 210 m/s, or (c) 600 m/s?

Sound

Sound is a mechanical wave that can travel through virtually any material substance, including gases, liquids, and solids. By far, the most familiar case is sound in a gas such as air. Sound waves can be excited in a gas by a loudspeaker as sketched in Figure 12.9. Here the surface of the speaker moves back and forth along the horizontal direction (x). As it moves, the speaker collides with nearby air molecules, and because the speaker is moving parallel to x, these collisions affect the x component of the velocity of the air molecules. The displacement of the air molecules associated with the sound wave is thus also along x. This displacement alternates between the positive and negative x directions due to the back-and-forth motion of the speaker. The displacement of the medium (the air molecules) is parallel to the wave's propagation direction, so sound

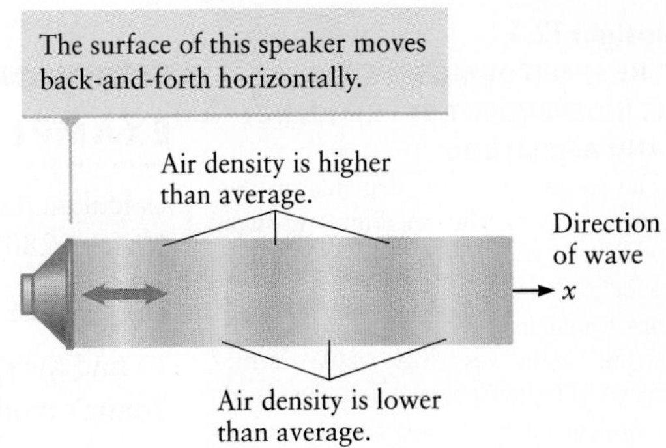
The surface of this speaker moves back-and-forth horizontally.
Air density is higher than average.
Direction of wave
x
Air density is lower than average.

▲ Figure 12.9 Sound is a longitudinal wave. Within this wave are regions where the density of air molecules is high, separated by regions where it is lower. Compare with the compressed and loose regions of the Slinky in Figure 12.3.

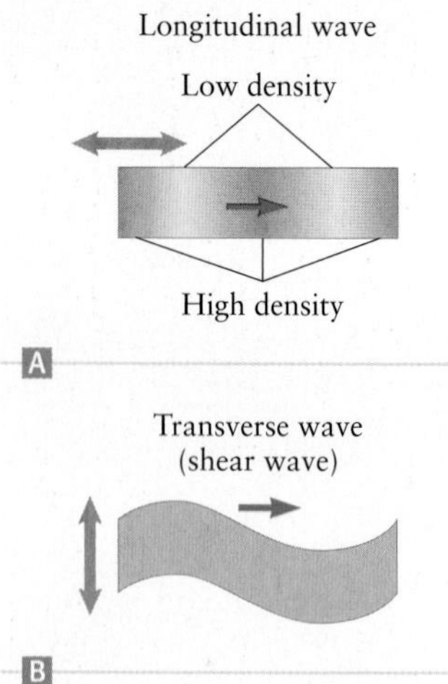

▲ **Figure 12.10** A solid, such as a metal bar, can carry A longitudinal waves or B transverse waves. In this sketch of a transverse wave, the amplitude is greatly exaggerated.

is a longitudinal wave. The speed of sound depends on the properties of the medium, such as the gas density. For air at room temperature, the speed of sound is approximately 343 m/s. This speed is independent of the frequency of the sound wave and applies to both periodic waves (steady tones) and nonperiodic waves (short "clicks" or "beeps").

Sound in a liquid or a solid also travels as longitudinal waves, with the displacements of the molecules in the liquid or solid being parallel to the propagation direction. The speed of sound is generally smallest for gases and highest for solids.

Wave Propagation in a Solid

We have just mentioned that sound (longitudinal waves) can propagate in a solid. It is also possible to have transverse waves in a solid. The motions associated with both longitudinal and transverse waves are sketched in Figure 12.10. Longitudinal waves involve motion of atoms parallel to the direction of the wave (compare Fig. 12.10A with Fig. 12.9), and such waves are therefore also called "sound." In fact, when a sound wave from air impinges on a solid, a corresponding sound wave propagates into the solid. The speed of sound waves in a solid depends on the solid's elastic properties, and sound travels faster in "stiffer" materials. For a thin bar of material, the speed of sound is given by

$$v = \sqrt{\frac{Y}{\rho}} \qquad (12.10)$$

where Y is Young's modulus (Chapter 11) and ρ is the density of the material.

Transverse waves in a solid involve motion of atoms in directions perpendicular to the wave velocity (Fig. 12.10B); the velocity of these transverse waves is not given by Equation 12.10, but has a different (and slightly more complicated) dependence on the shear modulus and other elastic constants (Chapter 11). In general, the speed of these transverse waves is lower than the speed of longitudinal waves. The key points here are that *both* transverse *and* longitudinal waves can propagate in a solid and that the longitudinal waves are called *sound* because they are just like the sound waves that propagate in a gas or a liquid.

While transverse waves can travel through a solid, they *cannot* travel through liquids or gases. The displacements associated with a transverse wave involve a shearing motion (a shear stress). When a solid is subject to a shear stress, it is bent as in Figure 12.10B and then "springs back." Liquids and gases *flow* and thus do not spring back, so there is no restoring force to produce the oscillations necessary for a transverse wave. Hence, only longitudinal waves (called sound) can propagate in a liquid or a gas. This fact has some important consequences that we'll explore in Section 12.9.

Insight 12.1
THE SPEED OF MOST WAVES IS INDEPENDENT OF FREQUENCY AND AMPLITUDE

The speed of a wave depends on the properties of the medium through which it travels. This speed varies widely, from the relatively low values found for a wave on a Slinky or string to the very high value of the speed of light. In most cases, the speed is independent of both the frequency and amplitude. This speed is also the same for periodic and nonperiodic waves (wave pulses). Although the wave speed is usually independent of frequency, when we study light and optics we will see some important cases in which the speed does depend on f.

EXAMPLE 12.5 Speed of Sound in a Solid

Calculate the speed of longitudinal waves (sound) in a steel bar. Use data from Table 10.1 (p. 280) and Table 11.1 (p. 326).

RECOGNIZE THE PRINCIPLE

To find the speed of sound in steel, we apply Equation 12.10, $v = \sqrt{Y/\rho}$, where Y is Young's modulus.

SKETCH THE PROBLEM

No figure is necessary.

IDENTIFY THE RELATIONSHIPS

From Tables 10.1 and 11.1, the density of steel is $\rho = 7800\ \text{kg/m}^3$, and Young's modulus is $Y = 2.0 \times 10^{11}$ Pa.

SOLVE

Using these values in Equation 12.10 gives

$$v = \sqrt{\frac{Y}{\rho}} = \sqrt{\frac{2.0 \times 10^{11}\ \text{Pa}}{7800\ \text{kg/m}^3}} = \boxed{5100\ \text{m/s}}$$

▶ *What does it mean?*

This speed is *much* faster than that of sound in air, which is about 340 m/s.

Visible Light and Other Electromagnetic Waves

Visible light is a type of ***electromagnetic wave***. Other types include radio waves, X-rays, gamma rays, and microwaves. These are not mechanical waves, but are electrical and magnetic disturbances that can propagate even in a vacuum, requiring no mechanical medium. Because the electric and magnetic disturbances (the electric and magnetic fields) are always perpendicular to the propagation direction, electromagnetic waves are transverse.

Electromagnetic waves are classified according to their frequency, so visible light waves fall into a different range of frequencies than radio waves, X-rays, and so forth. The speed of an electromagnetic wave in a vacuum c is independent of the frequency of the wave and has the value $c = 3.00 \times 10^8$ m/s. We'll discuss the properties of electromagnetic waves in detail in Chapter 23.

Water Waves

A wave can be generated on the surface of water by simply dropping a rock into a still body of water. After the rock strikes the water (Fig. 12.11A), waves propagate outward as shown in Figure 12.11B. The up-and-down motion of the water's surface is probably familiar to you. If you were to carefully watch a bug floating on the surface, though, you would find that the motion of the water's surface is actually *both* transverse *and* longitudinal as illustrated in Figure 12.12. A water surface wave is one the few types of waves that cannot be classified as purely transverse or longitudinal. However, most waves fall into one of these two categories, making it a useful classification scheme.

CONCEPT CHECK 12.4 What Determines the Properties of a Wave?

What determines the (1) frequency, (2) amplitude, and (3) wavelength of a wave, (a) the properties of the wave's medium or (b) the way the wave is generated (i.e., the wave's source)?

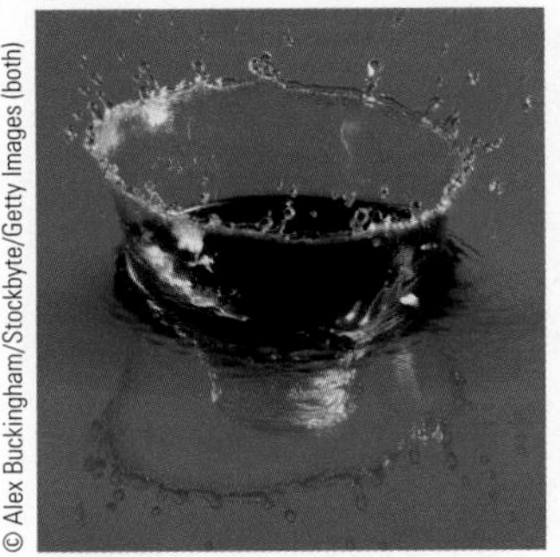

▲ **Figure 12.11** A Dropping a rock into a pool of water generates a water wave. B Waves generated by a disturbance on the surface of a pool of water. The wave crests and troughs move outward as expanding circles.

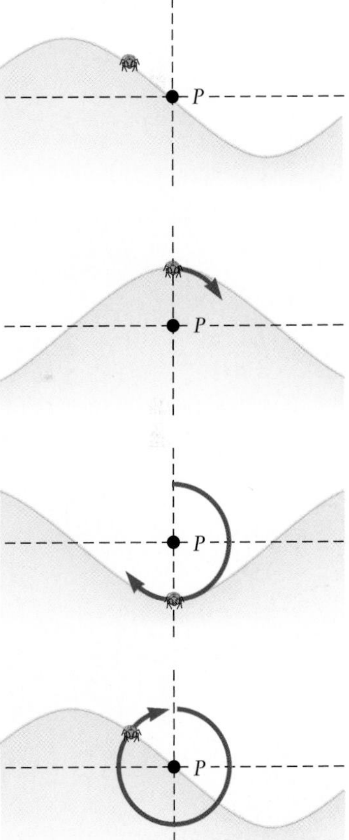

▲ **Figure 12.12** The motion associated with a wave on the surface of water has both transverse and longitudinal components. A bug on the surface of the water moves up and down—and also forward and backward—as the wave passes by, which can be seen by comparing the bug's position with that of point P.

12.4 The Geometry of a Wave: Wave Fronts

Spherical Waves

Consider a sound wave generated at the center of a large room by a small explosion. The resulting sound wave travels away from the source in a three-dimensional fashion as sketched in Figure 12.13A. The wave crests form concentric spheres centered at the source, so this is called a ***spherical wave***. These crests are also called ***wave fronts***. The direction of wave propagation is always perpendicular to the surface of a wave front. This direction is indicated by arrows called ***rays*** in Figure 12.13.

Each wave crest carries a certain amount of energy as it travels away from the source. Several quantities provide a measure of the amount of this energy, one of which is the ***power***. The total power carried by a wave equals the energy emitted by the source (and carried away by the wave) per unit time. The units of power are thus the units of energy divided by time, or joules/second = watts, just as we found for mechanical power in Chapter 6. A closely related quantity is the ***intensity*** of the wave, which is defined as the power carried by the wave over a unit area of wave front. The units of intensity are the units of power/area = watts/m^2.

Let's now follow one spherical wave front as it travels away from the source in Figure 12.13A. The amount of energy carried by a particular wave front is determined when it is created at the source, so once the wave front is emitted, the total energy it contains is fixed. As the wave front travels away from the origin and its spherical surface expands, this fixed amount of energy (and also power) is spread over a larger and larger surface. The power divided by the area thus becomes smaller, and the intensity of the wave falls as one moves away from the source. At a distance r from the source, the area of a spherical wave front is equal to $4\pi r^2$ (the surface area of a sphere of radius r), so we have

$$I = \frac{\text{power}}{\text{area}} = \frac{P}{4\pi r^2}$$

The intensity of a spherical wave thus falls with distance as

$$I \propto \frac{1}{r^2} \quad \text{(spherical wave)} \tag{12.11}$$

This dependence of the wave intensity on r is found whenever the wave front expands spherically in three-dimensional space, and it arises in many cases involving sound (as in this example) and light (such as the light waves that emanate from the Sun).

Plane Waves

Wave fronts are not always spherical. Another important case is a ***plane wave***, in which the wave fronts are (flat) planes as sketched in Figure 12.13B. For a "perfect" plane wave, each wave crest extends over an infinite plane in space, and these planes

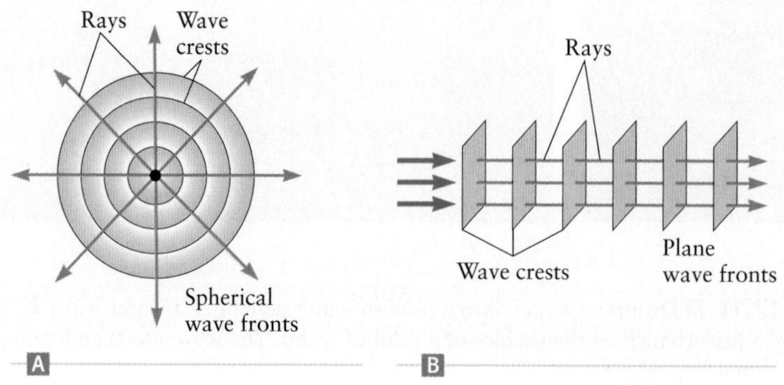

► **Figure 12.13** The geometry of a wave can be described by the shape of its wave fronts. A A spherical wave has spherical wave fronts that are concentric with the source of the wave. B The wave fronts of an ideal plane wave are parallel, flat, and infinite in extent. This idealization is not realized in practice, but is approximated by the light from a laser.

are perpendicular to the direction of propagation. Of course, such a perfect plane wave is an idealization; a real wave will not be spread over an infinite region of space. The light wave produced by a laser, however, is a very good approximation to an ideal plane wave. (We'll see other ways to generate approximate plane waves when we study geometrical optics in Chapter 24.) The wave fronts produced by a laser are very nearly flat and extend over the width of the laser beam. These wave fronts spread very little as they travel, so the wave intensity is approximately constant over long propagation distances. In contrast to the case of a spherical wave, where the intensity falls off as the square of the distance from the source (Eq. 12.11), the intensity of an *ideal* plane wave is constant, independent of distance from the source.

Intensity and Amplitude of a Wave

For all types of waves (and all types of wave fronts), the intensity of a wave is related to its amplitude A by

$$I \propto A^2$$

The reason I is proportional to A^2 can be seen by considering waves on a Slinky. The Slinky is just a spring, and the potential energy associated with a spring equals $\frac{1}{2}kx^2$, where x is the amount the spring is stretched or compressed. For a wave on a Slinky, the magnitude of this stretching and compressing is equal to the displacement of the Slinky from its undisturbed state and is thus proportional to the amplitude A of the wave. Hence, the energy carried by a wave and its intensity are both proportional to A^2.

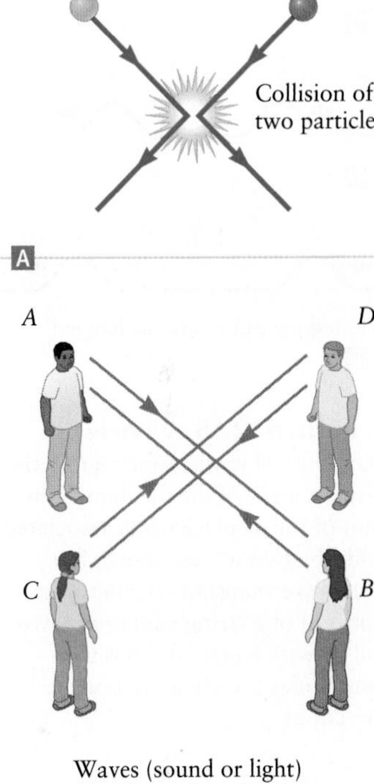
Collision of two particles
A
A
D
C
B
Waves (sound or light) pass through each other.
B

▲ **Figure 12.14** A When two particles collide, their momenta change. B If we replace the particles by waves (such as sound or light), the two waves will pass through each other. The sound waves involved in two conversations (between persons A and B and between persons C and D) do not affect each other; they propagate independently of each other.

12.5 Superposition and Interference

All our discussions so far have involved the case of a single wave propagating by itself. We now want to consider what happens when two or more waves travel at the same time through the same medium. That is, what happens when two waves collide?

When two particles collide, the momentum of each particle changes (although the total momentum of the two particles is conserved; see Chapter 7). Hence, a collision has a big effect on the motion of the particles as illustrated in Figure 12.14A, but the behavior of two colliding waves is quite different. Waves almost always propagate *independently* of one another. A wave can travel through a particular region of space without affecting the motion of another wave traveling through the same region. This behavior is illustrated in Figure 12.14B, which shows sound waves crossing in space (as in a typical classroom). Persons A and B are having a conversation, and their sound waves travel independently of the sound waves of persons C and D. The same sort of behavior is found if persons A and C shine flashlights toward their partners. The light from the flashlight at A travels independently of the light from the flashlight at C. This independence is a result of the ***principle of superposition.*** When applied to wave motion, this principle can be stated as follows.

Principle of superposition

Principle of superposition

When two (or more) waves are present, the displacement of the medium is equal to the sum of the displacements of the individual waves.

For example, if we have two periodic waves with displacements y_1 and y_2 that are each described by a relation such as Equation 12.1, the total displacement of the medium is the sum $y_1 + y_2$. These two displacements each describe traveling waves, and the presence of one of the waves does not affect the frequency, amplitude, or velocity of the other. Hence, the two waves travel independently.

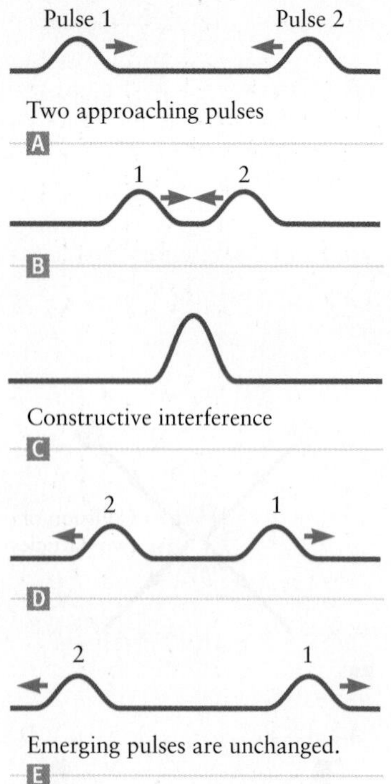

▲ **Figure 12.15** When two waves travel in the same region, the displacement of the medium is the sum of the displacements associated with the two waves. Shown are successive snapshots (from top to bottom) of a string containing two pulses with equal (and positive) amplitudes traveling in opposite directions.

Constructive and Destructive Interference

Figure 12.15 shows two wave pulses traveling in opposite directions. These pulses have equal and positive amplitudes, and the successive snapshots of the string show how the two pulses pass through each other. By the time of the final snapshot, the pulses are again isolated. Notice that the shapes of the final pulses in snapshot E are the same as the initial pulse shapes in snapshot A. According to the superposition principle, the displacement of the medium is always equal to the sum of the displacements produced by the two individual pulses, so when the two pulses overlap completely as in Figure 12.15C, the maximum displacement of the string is twice the amplitude of a single pulse. This is an example of ***constructive interference***. The two waves add together to produce a displacement that is larger than the displacement of either of the individual waves.

Figure 12.16 shows snapshots of another experiment with two wave pulses on a string, but this time the displacement of one string is downward and that of the other is upward. Here again the two waves pass through each other, and the pulse shapes at the end in snapshot E are the same as the initial pulse shapes in snapshot A. The displacement of the string is again (according to the superposition principle) equal to the sum of the displacements of the individual waves. However, since the waves have opposite amplitudes when the pulses overlap in snapshot C, the total displacement of the string is *zero*! This is an example of ***destructive interference***. The total displacement is smaller than the individual displacements of the two waves.

It is interesting to consider the effect of interference on the energy carried by the two waves in Figure 12.16. If you were to examine the string at the moment of perfect destructive interference in Figure 12.16C, you might think that there are no waves present and hence no propagation of energy, but even though there is a moment when the displacement of the string is zero, its velocity is *not* zero. At the instant of snapshot C, all the energy of each wave is contained in the kinetic energy of the medium (the string). In this way, two waves can interfere destructively and still carry energy independently. The term *interference* is thus a bit misleading because the presence of one wave does not "interfere" with the propagation of the other in any way.

Interference of Periodic Waves

Figure 12.17 shows an example of superposition with two periodic waves. Here two water waves are produced by dropping two rocks a short distance apart onto the surface of a still pool. Each rock produces a pattern of circular wave crests and troughs that travel away from the initial disturbance. Figure 12.17A shows the wave

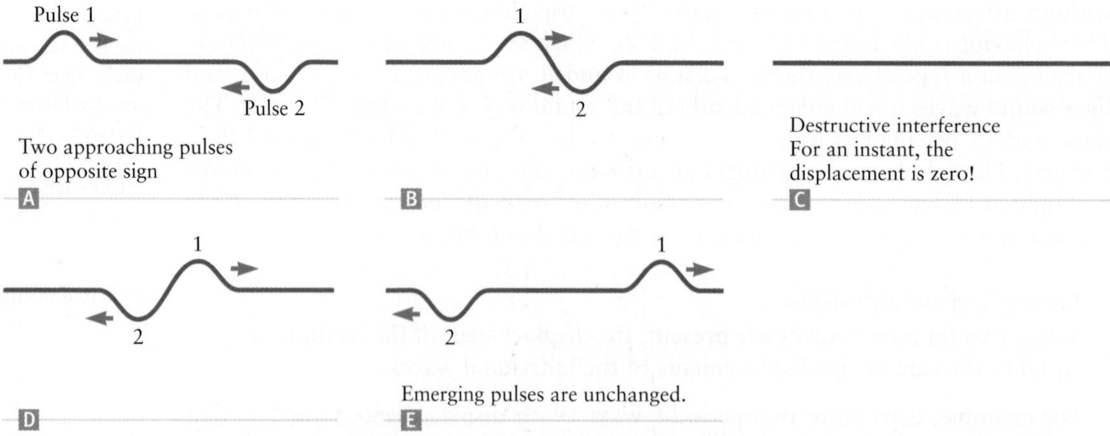

▲ **Figure 12.16** When the amplitudes of two interfering wave pulses are equal in magnitude but opposite in sign, the total displacement will be zero for an instant (snapshot C). This illustration is an example of destructive interference.

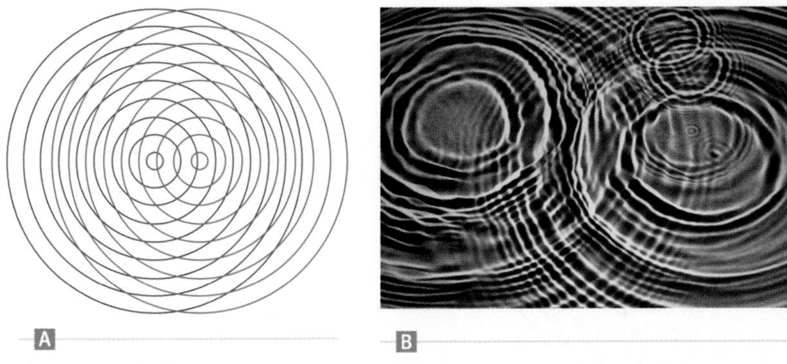
© Brand-X Pictures/Jupiterimages
A
B

◀ **Figure 12.17** A Interference of two overlapping water surface waves. Each source emits a circular pattern of wave crests and troughs. B Photo of interfering water waves on the surface of still water.

crests of these two waves separately. According to the principle of superposition, the total displacement of the water surface is the sum of the displacements of the individual waves. There is constructive interference at all points where two wave crests (or two wave troughs) in Figure 12.17A overlap, and the displacement of the surface is largest at these points. Likewise, there is destructive interference at points where a crest of one wave overlaps with a trough of the other, and at these points the displacement of the surface is small or zero. The resulting surface pattern is shown in Figure 12.17B. This ***interference pattern*** has an interesting geometry.

Insight 12.2

TWO WAVES WILL TRAVEL INDEPENDENTLY THROUGH THE SAME REGION OF SPACE

The terms *constructive interference* and *destructive interference* suggest the presence of one wave can affect the propagation of another, but quite the opposite is true. According to the principle of superposition, two waves propagate through the same region of space independently of each other. One wave does not destroy the other!

12.6 Reflection

Consider a wave pulse traveling on a string as sketched in Figure 12.18. Initially, this pulse is traveling to the right with a positive amplitude. It then ***reflects*** from the right-hand end of the string, and in the final snapshots it is traveling to the left. Another example of reflection is given in Figure 12.19, which shows a light wave from a laser reflecting from a mirror. The incoming wave approaches from the lower left; reflection from the mirror causes the wave to travel away toward the upper left edge of the photo.

In both examples, the wave on a string in Figure 12.18 and the light wave in Figure 12.19, reflection changes the propagation direction of the wave. It is useful to show this change using rays to indicate the direction of energy flow (compare with Fig. 12.13). Rays for the light wave in Figure 12.19 are shown in more detail in Figure 12.20. The rays change direction when a wave reflects from the boundary of the medium.

The rays in Figure 12.20 make an incoming (initial) angle θ_i and an outgoing (reflected) angle θ_r with a line drawn perpendicular to the surface. To understand how a wave changes direction when it is reflected, consider the component of a wave's velocity perpendicular to the surface, along with the component parallel to the surface. The perpendicular component of the wave velocity reverses direction; that is, the component of the wave that was initially traveling to the right in Figure 12.20 travels to the left after being reflected. This is just like the case with the string in Figure 12.18, where the wave is reflected back in the opposite direction. On the other hand, the parallel component of the wave in Figure 12.20 is not affected by the reflection. The final result is then as sketched in Figure 12.20, with the angle of incidence equal to the angle of reflection:

$$\theta_i = \theta_r \tag{12.12}$$

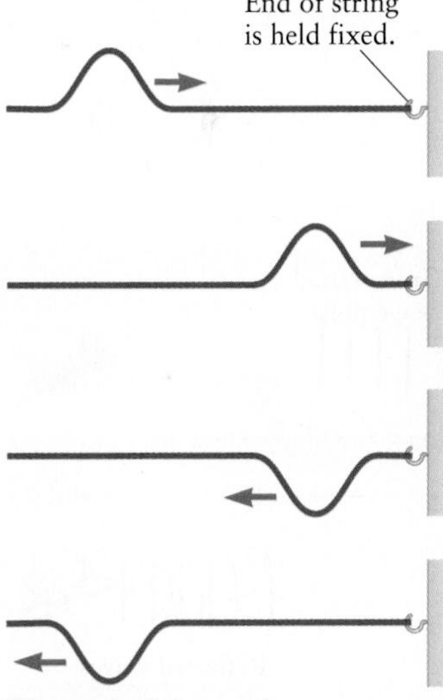
End of string is held fixed.
Wave pulse is inverted after reflection.

▲ **Figure 12.18** When a wave reaches a boundary (such as the end of a string), it can be reflected. When the end of the string is held fixed by tying it to a wall, the displacement of the reflected wave is reversed; here the displacement is changed from positive to negative.

Waves are usually reflected when they reach the boundaries of the medium, and this reflection process can sometimes (but not always) *invert* the wave's displacement. For example, in Figure 12.18, the wave reflected from the end of a string that is fastened to the wall is inverted; the reflected wave pulse has a negative amplitude, whereas the incoming wave had a positive amplitude.

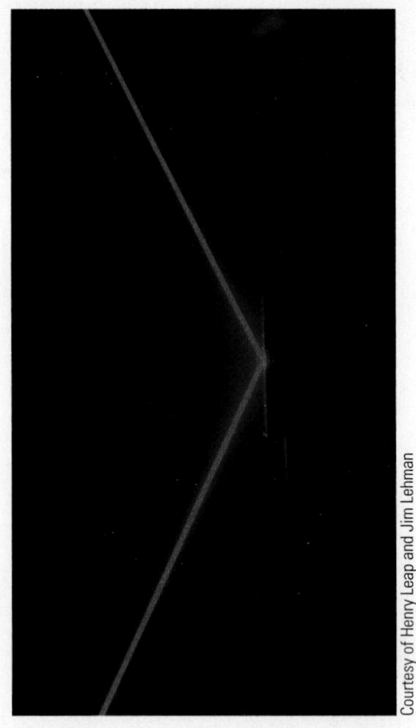

▲ **Figure 12.19** A light wave can be reflected when it strikes a boundary such as a mirror.

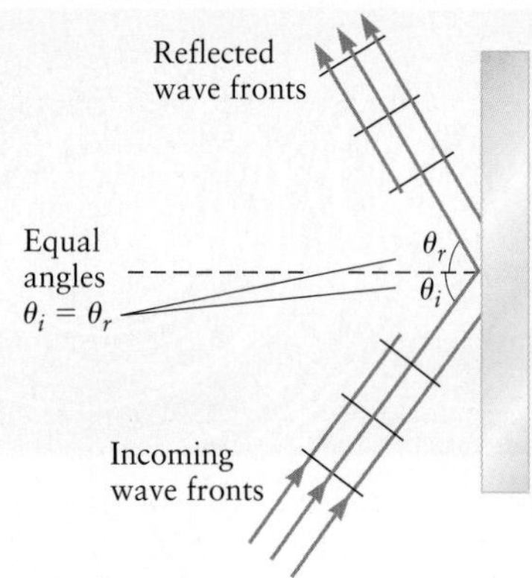

▲ **Figure 12.20** When a plane wave such as light, a water wave, or a sound wave reflects from a flat boundary, the angle of incidence θ_i is equal to the angle of reflection θ_r.

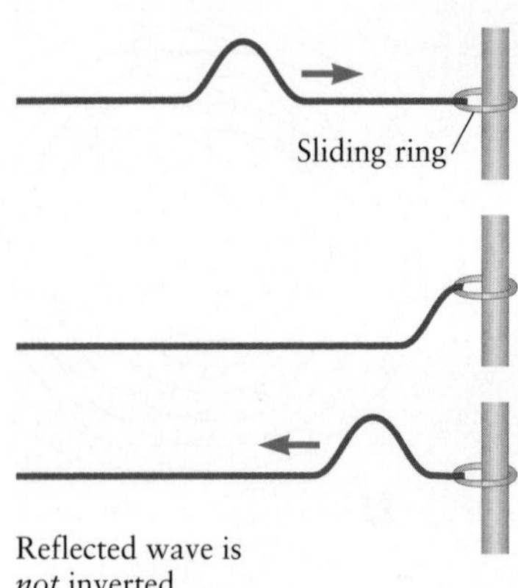

▲ **Figure 12.21** When the end of a string is attached to a ring that can slide up and down (e.g., along a pole), the reflected wave is not inverted.

While waves are often inverted on reflection, that is not always the case. Figure 12.21 shows a wave on a string where the end of the string is not fastened rigidly to the wall, but is able to slide up and down. Now the reflected wave is not inverted. In this example and for other cases of wave reflection, the properties of the medium at the boundary determine if the reflected wave will be inverted or not.

Radar

One interesting application of wave reflection is ***radar***, a technique that uses radio waves to detect objects remotely and determine their distance (Fig. 12.22). A radio wave pulse is sent from a transmitting antenna and reflects from some distant object such as an airplane. If the reflecting surface of the airplane is large and flat, the wave will be reflected at a single angle as in Figure 12.20. For radar applications, however, the surfaces on the object are usually smaller than (or comparable to) the wavelength of the radio wave, and in this case, reflected waves travel outward in all directions. A portion of the reflected wave will thus arrive back at the original transmitter, where it is detected. The distance to an object is determined by measuring the time delay Δt between the original outgoing wave pulse and the returning one. If the speed of the wave is v, the total round-trip distance traveled by the wave is $L_{tot} = v\,\Delta t$, which is twice the distance from the transmitter to the object. So, by measuring Δt, a radar system uses wave reflection to measure the distance to a remote object. By using a rotating antenna, the direction of the object can also be determined. The amplitude of the reflected wave gives information about the object's size; a large object reflects more of the wave energy than a small object and thus gives a larger signal at the detecting antenna.

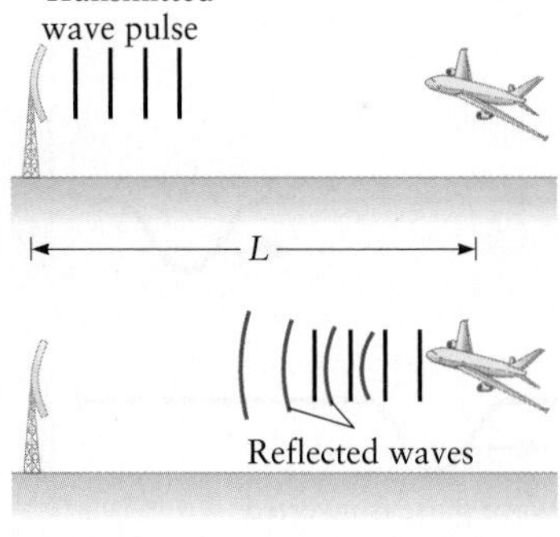

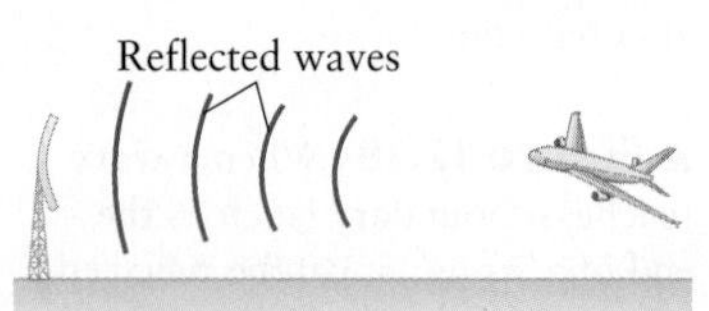

▲ **Figure 12.22** Radar uses reflected waves to measure the location of a distant object.

12.7 Refraction

In our discussions of wave fronts (Section 12.4), we assumed the speed of the wave is the same at all places in the medium. Figure 12.23A shows such an example for a plane sound wave traveling through a solid. Here the wave speed is the same everywhere, and this plane wave propagates along a straight line from left to right. In some cases, however, the wave speed may be different at different locations. A hypothetical

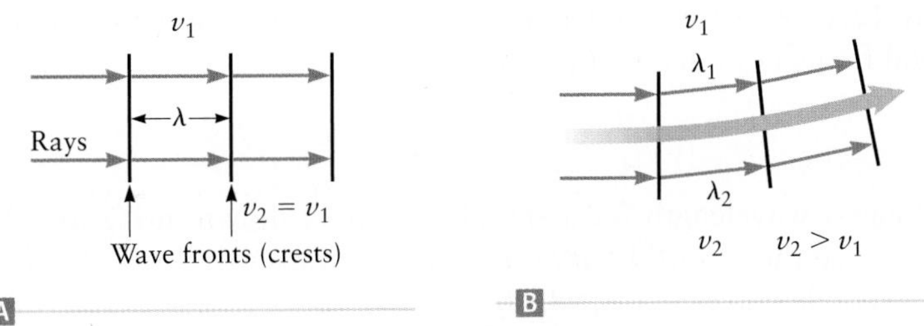

◀ **Figure 12.23** A When the wave speed is the same throughout the medium, the rays follow straight lines. B When the wave speed is different in different regions (here $v_2 > v_1$), the rays are "bent" and the waves are refracted.

case is sketched in Figure 12.23B, which shows a wave traveling through a material in which the wave speed is slightly greater at the bottom than at the top region of the figure ($v_2 > v_1$). The solid black lines show wave crests, and by definition they are separated by a wavelength. The wave frequency is the same everywhere (its value is determined by the frequency of the source), and because $\lambda = v/f$, the wavelength—and hence the distance between crests—is slightly longer in the bottom region. As a result, the wave crests tilt progressively as the wave propagates to the right, and the wave direction (the direction of energy flow) curves upward, as indicated by the blue rays, an effect called ***refraction.*** Refraction is the *change in direction of a wave due to a change in its speed.* We'll see examples of refraction when we study seismic waves in the Earth (Section 12.9) and again when we study optics.

12.8 Standing Waves

Consider waves traveling back and forth along a string of length L with both ends held fixed (i.e., tied to walls as sketched in Figure 12.24A). What will the string displacement look like as a function of both time t and position x along the string? For a very long string carrying a single periodic wave traveling to the right, the string displacement y is given by a relation such as Equation 12.1. Because the ends of the string in Figure 12.24A are fastened to rigid supports, however, the displacement must be zero at both ends. That is, the string displacement y must be zero at both $x = 0$ and $x = L$. These conditions can be satisfied by a periodic wave such as Equation 12.1 only for certain wavelengths. For example, suppose the wavelength is equal to precisely twice the length of the string so that $\lambda = 2L$. The wave will then "fit" onto the string as shown in Figure 12.24B, which shows several superimposed snapshots of the vibrating string. While the two ends are at rest with $y = 0$, the middle portion of the string oscillates up and down with a large amplitude.

The vibration in Figure 12.24B is called a ***standing wave*** because the outline of the wave appears to stand still. Such a wave pattern can be obtained by the interference of two waves, both with wavelength $\lambda = 2L$, traveling in *opposite* directions. These waves travel back and forth along the string and are reflected when they reach the ends of the string. According to the principle of superposition, the total string displacement is the sum of the displacements of these two waves, and that sum gives the pattern in Figure 12.24B. Points where the string displacement is always zero are called ***nodes*** of the standing wave. The points where the displacement is largest are called ***antinodes.*** In Figure 12.24B, there is an antinode at the center of the string, while there are nodes at both ends.

Other possible standing waves that fit on the string are shown in parts C and D of Figure 12.24. For the wave in Figure 12.24C, we have $\lambda = L$, whereas in Figure 12.24D, we find $\lambda = 2L/3$. We can generate many other standing waves by employing smaller and smaller wavelengths. The longest possible wavelength for a standing wave is the one shown in Figure 12.24B, corresponding to the *smallest* possible standing wave frequency. This is called the ***fundamental frequency*** of the string, denoted by f_1. The frequency of a standing wave is related to its wavelength by

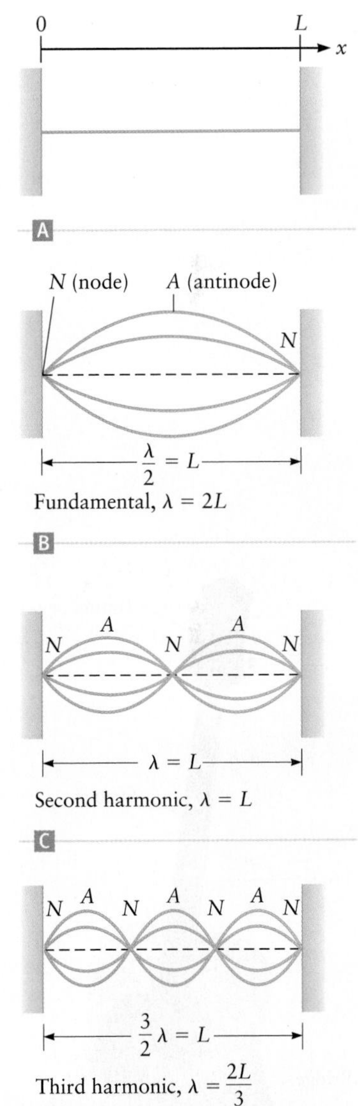

▲ **Figure 12.24** Standing waves on a string, such as on a guitar or a piano. A The string at rest. B through D Standing waves with different wavelengths. Here, N and A denote nodes and antinodes of the standing waves.

$f = v/\lambda$ (Eq. 12.2). According to Figure 12.24B, the wavelength corresponding to the fundamental frequency equals $2L$, so we have

$$f_1 = \frac{v}{2L} \tag{12.13}$$

The next longest wavelength for a standing wave is shown in Figure 12.24C. This vibration is called the ***second harmonic***, and its frequency is $f_2 = v/L$. There is a pattern to the possible wavelengths, which is

$$\lambda_n = \frac{2L}{n} \quad \text{with } n = 1, 2, 3, \ldots \quad \text{(standing waves on a string with fixed ends)} \tag{12.14}$$

In Equation 12.14, $n = 1$ gives the wavelength of the fundamental, $n = 2$ gives the second harmonic, and so forth. The corresponding pattern for the frequencies is

$$f_n = \frac{v}{\lambda_n} \tag{12.15}$$

which leads to $f_n = nv/2L$ and

$$f_n = nf_1 \quad n = 1, 2, 3, \ldots \quad \text{(standing waves on a string with fixed ends)} \tag{12.16}$$

The allowed standing wave frequencies are thus *integer multiples* of the fundamental frequency f_1.

> **CONCEPT CHECK 12.5 Nodes and Antinodes**
>
> Consider a standing wave on a string with a frequency equal to the fourth harmonic (Eq. 12.16 with $n = 4$). How many nodes and antinodes does this standing wave possess?

Musical Tones

Standing waves on strings are closely connected to musical tones. Many musical instruments, including guitars, violins, and pianos, employ strings as a vibrating element. A guitar string is stretched under tension from the tuning peg at the top of the neck to a point where it is fastened to the bridge and top plate of the instrument as shown in Figure 12.25. The player uses his or her finger to press the string down against a metal strip called a "fret" so that a length L of the string extending from the bridge to the fret is free to vibrate. The section of the guitar string between the fret and the bridge is thus similar to the string in Figure 12.24.

A plucked guitar string vibrates according to all the possible standing wave patterns with frequencies $f_1, f_2, f_3, \ldots$ (Eq. 12.16) all at the same time. The total string motion is, according to the superposition principle, a sum of the vibrations at these individual frequencies, although the vibrations at the lowest frequencies are usually largest. These string vibrations cause the body of the guitar to vibrate, producing the sound we hear. The frequencies of these sound waves are the standing wave frequencies of the guitar string. The sound of a single note is thus composed of many individual frequencies, which are related according to the harmonic sequence in Equation 12.16.

The ***pitch*** of a note is determined by its fundamental frequency. A guitar player can change the pitch by changing the length L of the vibrating section of the string, which can be done by holding the string against a different fret. According to our relation between the fundamental frequency and L, Equation 12.13, the pitch is inversely proportional to the length of the string. Two notes whose fundamental frequencies differ by a factor of two are said to be separated by an ***octave***. According to Equation 12.13, reducing L by a factor of two increases the frequency by one octave. This relation between musical pitch, the notes of a musical scale, and the length of a vibrating string was first discovered by Pythagoras, the mathematician famous for his work with right triangles.

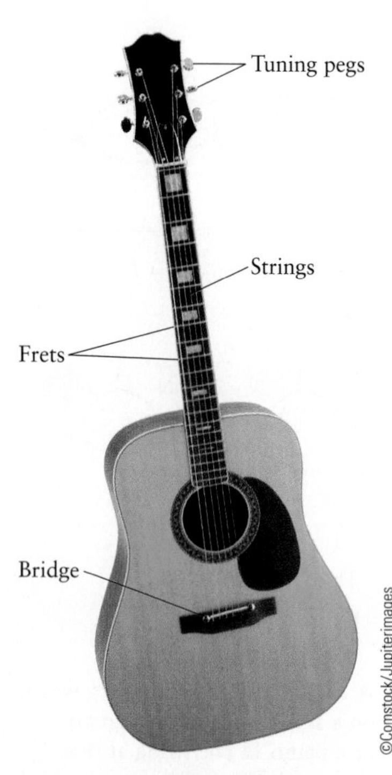

▲ **Figure 12.25** A guitar uses standing waves on a string to generate particular frequencies.

EXAMPLE 12.6 Playing a Guitar

Guitarists often play several notes simultaneously by plucking or strumming two or more strings at the same time. Two notes whose fundamental frequencies are in the ratio 1.5/1 = 3/2 are known to sound pleasing when played together. This musical interval is known as a "perfect fifth." Suppose you wish to play these notes on two similar guitar strings, that is, strings that have the same value for the wave speed, v. What is the ratio of the vibrating lengths of the two strings?

RECOGNIZE THE PRINCIPLE

The fundamental frequency of a guitar string is $f_1 = v/2L$. We can apply this relation to both strings and then calculate the ratio of the lengths needed to get a frequency ratio of 3/2.

SKETCH THE PROBLEM

The problem is described by Figures 12.24 and 12.25.

IDENTIFY THE RELATIONSHIPS

Let L and L' be the vibrating lengths of the two strings. According to the relation between fundamental frequency and string length (Eq. 12.13), the fundamental frequencies are $f_1 = v/2L$ and $f_1' = v/2L'$. The ratio of the fundamental frequencies is thus

$$\frac{f_1'}{f_1} = \frac{v/2L'}{v/2L} = \frac{L}{L'}$$

SOLVE

Inserting the desired value for the ratio of the frequencies, we get

$$\frac{f_1'}{f_1} = \frac{3}{2} = \frac{L}{L'} = \boxed{1.5}$$

▶ ***What does it mean?***
This type of analysis can also be used to calculate how to position the frets on a guitar (see Question 8 at the end of this chapter). Changing the vibrating length of a guitar string by "one fret" changes the frequency so as to take you to the next note in the musical scale (e.g., from B to C or from C to C-sharp).

EXAMPLE 12.7 Bending a Note on a Guitar

The pitch of a guitar note is related to the length L of the vibrating portion of the string and its tension. The length is determined by the fret spacing (Fig. 12.25), whereas the tension is adjusted with the tuning pins. A guitar player can also change the pitch by "bending" a note. Bending is illustrated in Figure 12.26; the guitarist pushes the string to one side, increasing the tension and thus increasing the pitch. By doing this as a note is being played, the pitch changes continuously during the note, an effect used by many guitar players. Suppose a note is "bent" from E to F, a change in frequency of 5.9%. By what factor has the tension been increased?

RECOGNIZE THE PRINCIPLE

The pitch of a note corresponds to its fundamental frequency, with (Eq. 12.13)

$$f_1 = \frac{v}{2L} \tag{1}$$

(continued) ▶

C. Betz

▲ **Figure 12.26** Example 12.7. "Bending" a note on a guitar.

To give a 5.9% change in the pitch, we therefore require a 5.9% change in the wave speed v. The wave speed depends on the tension, so we need to find the change in tension required to give this change in v.

SKETCH THE PROBLEM

Figure 12.26 shows the problem.

IDENTIFY THE RELATIONSHIPS

The wave speed varies with the tension F_T as $v \propto \sqrt{F_T}$ (Eq. 12.9). Combining Equation 12.9 with Equation (1), the frequency, wave speed, and tension for the note E are related by $f_1(\text{E}) \propto v(\text{E}) \propto \sqrt{F_T(\text{E})}$, with a similar relation for the note F. We can thus write

$$\frac{f_1(\text{F})}{f_1(\text{E})} = \sqrt{\frac{F_T(\text{F})}{F_T(\text{E})}} \tag{2}$$

SOLVE

The frequencies differ by 5.9%, so their ratio $f_1\,(\text{F})/f_1(\text{E})$ is 1.059. Inserting this information into Equation (2) and squaring both sides leads to

$$1.059^2 = \frac{F_T(\text{F})}{F_T(\text{E})} = \boxed{1.12}$$

► ***What does it mean?***

Here we considered a case in which the "bending" moves from one note to the next on a musical scale, but bending is often used to change the pitch by only a "fraction" of a note interval. Bending a guitar note requires strong fingers.

12.9 Seismic Waves and the Structure of the Earth

Waves can be used to probe the structure of the Earth. Any large mechanical disturbance such as an earthquake or nuclear explosion acts as a source of ***seismic waves***, another term for waves that propagate through the Earth. There are three main types of seismic waves as sketched in Figure 12.27A. One is called an S wave

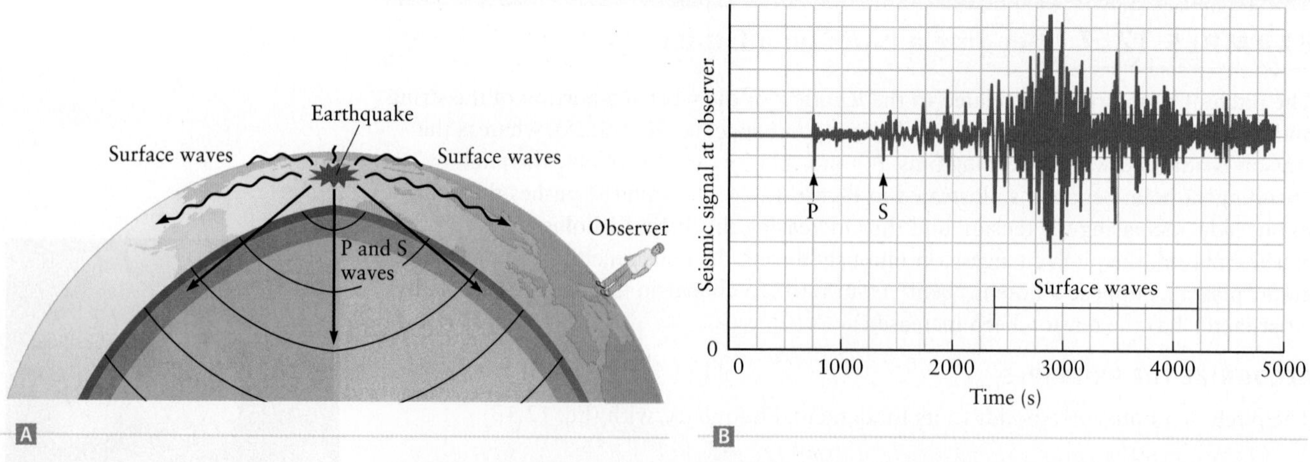

▲ **Figure 12.27** A When an earthquake occurs, several different types of seismic waves are generated, including P waves and S waves (both of which travel through the Earth) and surface waves. B These waves have different speeds, so they reach an observer at different times.

(S for "shear"). Like shear waves in a metal bar, S waves are transverse waves (Fig. 12.10B), with the displacement of the solid Earth perpendicular to the direction of propagation. Another type of seismic wave is called a P wave (P for "pressure"). P waves are longitudinal sound waves (see Fig. 12.10A and Section 12.3). Although S and P waves have different speeds, they both propagate into the body of the Earth. A third type of wave travels on the surface. These surface waves travel more slowly than S and P waves and are similar to water waves on the surface of a lake, which we described in Section 12.3.

Seismic waves can be detected by an instrument called a ***seismograph***, which measures the displacement of the Earth as a function of time. A typical seismograph signal is shown in Figure 12.27B. The P wave pulses arrive first because they travel fastest, followed by the S waves and then the surface waves. The seismograph can also measure the direction of the displacements associated with each of these wave pulses, which can be used to distinguish the different types of waves. By knowing the speeds of the three types of waves, the distance from the observer to the earthquake in Figure 12.27A can be determined.

This picture correctly describes things when the earthquake is not too far from the observer. When the earthquake and the observer are on opposite sides of the Earth, however, the observer's seismograph does not detect any S wave component. The reason is shown in Figure 12.28A. The center of the Earth contains a region called the *core*, and the area of the Earth where the S waves do not reach is in the "shadow" of the core. S waves do not propagate through the core, whereas P waves can. S waves are shear (transverse) waves, and we saw in Section 12.3 that they cannot travel in a liquid. The existence of the S wave shadow provided the first evidence that the Earth's core contains a liquid. It is now known that the core contains liquid metals, such as iron, at a very high temperature. Seismographic recordings provide the only way to measure the diameter of the core and study its properties.

Figure 12.28B shows in greater detail, in a more realistic sketch, how S and P waves propagate through the Earth. We again see the S wave shadow made by the core and also notice that the S and P waves do *not* travel in straight lines. Instead, they follow curved paths that bend toward the Earth's surface because of refraction (Section 12.7). The speeds of S and P waves depend on the composition and density of the Earth, and these speeds increase as one goes deeper into the Earth. Refraction causes the seismic waves to be deflected toward the surface, in the direction of smaller wave speed as shown in Figure 12.23B.

The amount of bending of the P and S waves depends on how the wave velocities vary from place to place within the Earth. Because it is impossible to measure the composition deep in the interior of the Earth directly, it is not possible to predict ahead of time precisely how the waves will travel. However, data taken with seismographs at the Earth's surface can be used to infer the paths actually taken by S and P waves. These inferred paths give information on how the composition of the Earth varies with depth. In this way, it was discovered that the Earth is composed of several different solid layers, two of which are called the ***crust*** and the ***mantle***, with different compositions. It has also been found that the core is not simply a spherical fluid region. Rather, the core has a solid central region (the inner core) surrounded by a fluid outer core (Fig. 12.28B). This surrounding fluid produces the S wave shadow described above.

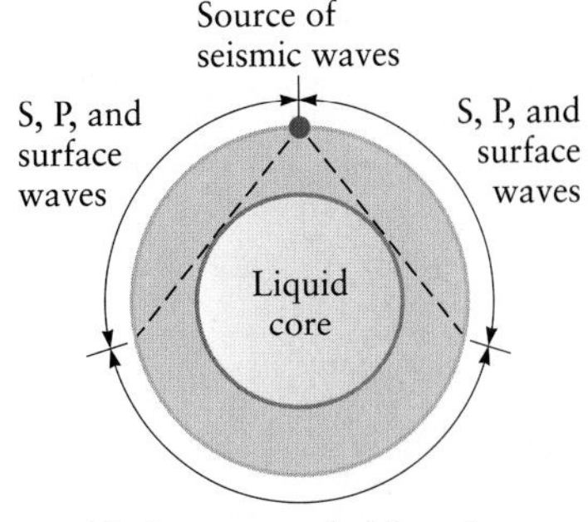

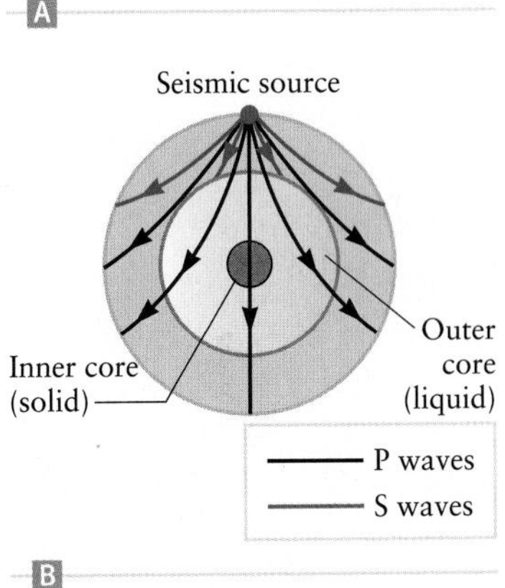

▲ **Figure 12.28** A The core of the Earth contains liquid, and transverse waves do not propagate through a liquid. Therefore, S waves do not travel through the core, and observers who are in the core's "shadow" do not record S waves from a source on the opposite side of the Earth. B Due to refraction, P and S waves do not travel in straight lines.

EXAMPLE 12.8 Ⓡ Measuring the Size of the Earth's Core

From extensive seismograph measurements, it is known that the angle θ defining the edge of the S wave shadow in Figure 12.29 is approximately 39°. Use this angle to estimate the radius of the Earth's liquid core. For simplicity, ignore the refraction (bending) of S and P waves and assume they travel in straight lines.

(continued) ▶

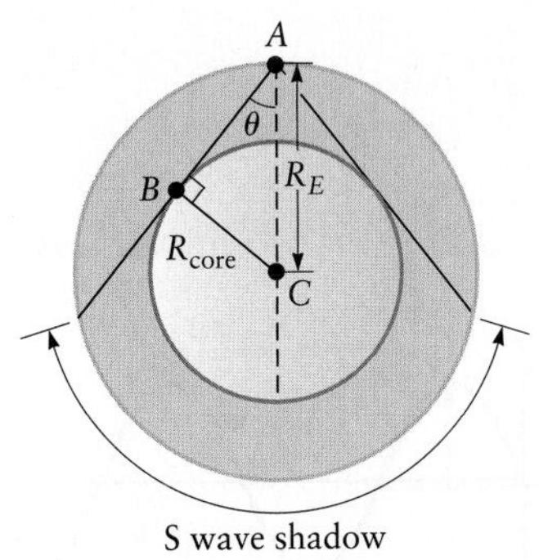

▲ **Figure 12.29** Example 12.8.

RECOGNIZE THE PRINCIPLE

We are assuming straight-line propagation, so we can apply the geometry in Figure 12.29. S waves that just graze the outer edge of the core follow a ray from A to B. The triangle ABC is a right triangle, and the side BC is the radius of the core, R_{core}, which is what we want to find.

SKETCH THE PROBLEM

Figure 12.29 shows the geometry of the problem.

IDENTIFY THE RELATIONSHIPS

The hypotenuse of triangle ABC is the radius of the Earth R_E, so we have

$$R_{core} = R_E \sin \theta$$

SOLVE

Inserting the known radius of the Earth ($R_E = 6.4 \times 10^6$ m, Table A.3) along with the given value of θ, we find

$$R_{core} = R_E \sin \theta = (6.4 \times 10^6 \text{ m})\sin(39°) = \boxed{4.0 \times 10^6 \text{ m}}$$

► ***What does it mean?***

Our result for R_{core} is about 60% of the Earth's radius. The core thus makes up a relatively large fraction of the Earth's volume.

Summary | CHAPTER 12

Key Concepts and Principles

Waves

A ***wave*** is a disturbance that transports energy from one place to another without transporting matter. Waves travel through a medium; for mechanical waves, the medium is a material substance. For ***transverse waves***, the displacement of the medium is perpendicular to the velocity of the wave. With ***longitudinal waves***, the displacement is parallel to the wave velocity. Fluids (liquids and gases) cannot support transverse waves.

Frequency and period

A wave can consist of a "single" disturbance—a wave pulse—or the disturbance can have a repetitive form. A periodic wave has a characteristic repeat time called the ***period*** T. The ***frequency*** of a periodic wave is related to the period by $f = 1/T$.

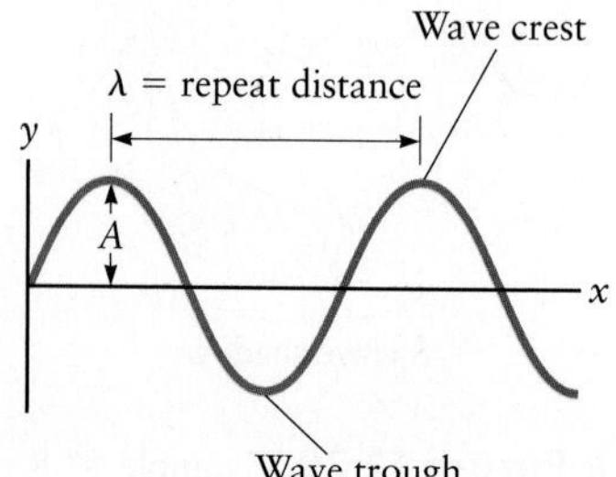

Displacement of a wave

The displacement associated with a periodic wave is often described by

$$y = A \sin\left(2\pi ft - \frac{2\pi x}{\lambda}\right) \quad \text{(12.1) (page 340)}$$

Here A is the amplitude of the wave, f is the frequency, and λ is the wavelength.

The frequency, wavelength, and velocity of a wave are related by

$$v = f\lambda \quad (12.2)\ (\text{page } 341)$$

Superposition

According to the ***principle of superposition***, when two waves are present simultaneously, the total displacement of the medium is the sum of the displacements of the individual waves. One consequence of this principle is that two waves can "pass through" each other.

Wave fronts and rays

A ***wave front*** is a surface that describes how the energy carried by a wave travels through space. ***Rays*** indicate the direction of energy flow and are perpendicular to the wave fronts.

Reflection and refraction

Waves can be reflected when they reach the boundary of the medium in which they travel. ***Refraction*** is the change in direction of a wave due to a change in velocity. The wave velocity can change at a boundary or within the medium.

Applications

Examples of waves

There are many examples of waves. The velocity of a wave depends on the properties of the medium.

- ***Waves on a string*** are transverse waves and are employed in many musical instruments.
- ***Sound*** is a longitudinal wave that can exist in solids, liquids, and gases. (See also Chapter 13.)
- ***Light*** is an electromagnetic wave. Because they are not mechanical waves, light and other electromagnetic waves can travel through a vacuum as well as through a material medium such as air.

Interference and standing waves

When two waves overlap in space, they are said to ***interfere***. One example of interference is the phenomenon of ***standing waves***. A point along a standing wave where the displacement of the medium is always zero is called a ***node***, and the points where the displacement is largest are ***antinodes***. Standing waves are used to produce musical tones in stringed instruments such as guitars and pianos. The lowest possible frequency of the standing wave (corresponding to the longest possible wavelength) is called the ***fundamental frequency***. For a string held rigidly at both ends, the standing waves form a series of ***harmonics*** with frequencies

$$f_n = nf_1 \quad \text{with } n = 1, 2, 3, \ldots \quad \text{(standing waves on a string with fixed ends)}$$

(12.16) (page 354)

where $f_1 = v/(2L)$ is the fundamental frequency.

Standing waves on a string

$\lambda = 2L$

Fundamental

$\lambda = L$

Second harmonic

Node

Antinodes

Seismic waves

Seismic waves provide information about the Earth's inner structure.

Questions

SSM = answer in *Student Companion & Problem-Solving Guide* = life science application

1. If the frequency of a wave is doubled, how does the wavelength change?

2. Identify the medium for each of the following waves:
   (a) Sound waves in a room
   (b) Waves on the surface of a pond
   (c) Radio waves
   (d) Light waves
   (e) Waves created when a guitar string is plucked
   (f) Seismic waves in the Earth

3. Identify the following waves as either longitudinal or transverse or both. Use a drawing to illustrate how the motion of the wave's medium is related to the direction of propagation of the wave.
   (a) Sound waves in a room
   (b) Waves on a guitar string
   (c) Seismic waves in the Earth
   (d) Waves on the surface of a pond

4. In a sound wave in a solid, the atoms in the solid move parallel to the propagation direction. Does that means the atoms travel along with the wave? Explain why or why not.

5. A key property of a wave is that it transports energy without transporting mass. Explain how this is possible, using work–energy ideas (i.e., $W = F\,\Delta x$). *Hint*: Consider the "leading edge" of a wave.

6. Devise or describe an experiment that uses standing waves to measure the velocity of a wave.

7. A bass guitar has longer strings than a regular guitar. Explain why these longer strings allow a bass guitarist to play "lower" notes.

8. SSM The neck of a guitar is designed with ***frets*** as shown in Figure Q12.8. A player can hold a string against one of the frets and thus shorten the vibrating length L of the string. In this way, a particular string can be used to play notes with different frequencies. The frequency of a note is determined by the fundamental standing wave mode (Fig. 12.24), and this frequency depends on L. In a musical scale, the frequencies of adjacent notes (i.e., between C and C-sharp) are in the ratio of $2^{1/12}/1 \approx 1.059$; this ratio produces the commonly used 12-tone musical scale. The artisans who make guitars position the frets according to the "rule of 18ths." According to this rule, each fret is positioned so that depressing adjacent frets shorten the vibrating length of the string by $L/18$. (See Fig. Q12.8.) Show that the rule of 18ths produces note frequencies that fit approximately on the desired musical scale.

Figure Q12.8

9. Important properties of a wave on a string include the (a) wavelength, (b) frequency, (c) amplitude, and (d) speed. Which of these properties is independent of the others? Explain.

10. Figure Q12.10 shows a child holding one end of a long Slinky. If the child wants to generate a longitudinal wave in the Slinky, in what direction should he shake the end of the Slinky, (a) in a horizontal direction or (b) in a vertical direction?

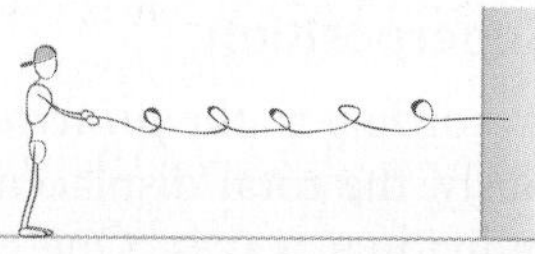

Figure Q12.10

11. A student sends a transverse wave pulse down a stretched piece of string attached to a wall. A moment later, she sends a second wave pulse. Is there any way she could alter the conditions to make the second wave overtake the first wave (i.e., before it is reflected back)? What would happen if she pulls harder on the string?

12. A rope of linear density of about 1 kg per meter is hung from the ceiling. The lower free end is then oscillated, producing waves that travel up the rope. Describe the velocity of the wave as it travels up the rope. Does it change? Why or why not? *Hint*: Is the tension the same at all points on the string?

13. SSM A woman wakes up just after 2 AM. Wondering what woke her, she goes to the window and looks around. Noting nothing out of the ordinary, she sits back down on her bed and all of a sudden a loud explosion rattles her window. The next day, she reads that an underground munitions dump exploded about 10 km from her home. She thinks that perhaps she woke up because she experienced a psychic precognition of the explosion. Can you think of a physical explanation for what might have awoken her the few moments before the sound from the blast hit her window?

14. When a sound wave passes from air into a solid object, which of the following wave properties stays the same and which change? Explain.
   (a) The amplitude
   (b) The frequency
   (c) The wave speed
   (d) The wavelength

15. Equation 12.1 describes a wave that is traveling to the right, and Equation 12.5 corresponds to a wave of the same frequency and wavelength that is moving to the left. Show that the sum of these two waves gives a standing wave. That is, show that this sum has nodes and antinodes as sketched in the standing waves in Figure 12.24.

16. Consider a water wave as illustrated in Figure 12.11. This wave front is circular. How does the intensity of this wave vary as it moves away from its source? *Hint*: You should find that $I \propto r^p$. What is the value of p?

17. Which property of a wave is independent of its amplitude, (a) frequency, (b) wavelength, or (c) speed?

18. Most modern guitars use steel strings, whereas many historical instruments use strings made of animal gut. Suppose the strings on a guitar that was originally strung with gut are replaced with steel strings of the same diameter. Will the tension in the steel strings be greater or less than in the gut strings?

19. When you tune a guitar, what causes the frequency to change?

20. For guitars and violins, some strings are "simple" strings, and others contain a winding around a string core. Which type of string, simple or wound, is used for the low-frequency notes? Explain why.

21. **Seismic waves in the Earth.** Surface seismic waves generated by an earthquake arrive at a distant observer after the arrival of P waves and S waves (Fig. 12.27B), partly because surface waves travel more slowly than both P and S waves. Even if all these waves had the same velocities, however, the surface waves would still arrive last. Explain why.

Problems available in ENHANCED WebAssign

Section 12.1 What Is a Wave?
Problems 1–5

Section 12.2 Describing Waves
Problems 6–20

Section 12.3 Examples of Waves
Problems 21–34

Section 12.4 The Geometry of a Wave: Wave Fronts
Problems 35–38

Section 12.5 Superposition and Interference
Problems 39–41

Section 12.6 Reflection
Problems 42–44

Section 12.7 Refraction
Problem 45

Section 12.8 Standing Waves
Problems 46–61

Section 12.9 Seismic Waves and the Structure of the Earth
Problems 62–63

Additional Problems
Problems 64–75

List of Enhanced Problems

Problem number	Solution in *Student Companion & Problem-Solving Guide*	Reasoning & Relationships Problem	Reasoning Tutorial in ENHANCED WebAssign
12.1	✓		
12.3		✓	
12.4		✓	✓
12.10	✓		
12.11		✓	✓
12.25	✓	✓	✓
12.34		✓	✓
12.35	✓		
12.39	✓		
12.41		✓	✓
12.44	✓		
12.45	✓		
12.49	✓		
12.62	✓		
12.72	✓		
12.74		✓	
12.75		✓	✓

CHAPTER

13

▶ *Sound is an important part of everyday life. (© John Elk III/Alamy)*

Sound

OUTLINE

In Chapter 12, we discussed aspects of wave motion that are common to all types of waves. In this chapter, we focus on one particular type of wave, ***sound***. Sound is an especially important example of waves; it is central to hearing, speech, and music. We'll consider how sound waves are generated, how they propagate, and how they are detected by the ear. We also describe some medical applications of sound.

13.1 Sound Is a Longitudinal Wave

Sound waves can travel through gases, liquids, and solids. Our everyday experience is mainly with sound that travels through air, so let's begin with a molecular-scale view of sound in a gas. Air is composed mainly of oxygen and nitrogen molecules that are in constant random motion, even in "still" air. A typical air molecule at room temperature has a speed of about 500 m/s. We refer to this motion as the "background velocity" and denote this velocity by $\vec{v}_0$; the direction of this velocity is different for every molecule.

Figure 13.1 shows a tube filled with air, with a loudspeaker at one end of the tube. When the loudspeaker is turned off, the density of air molecules is the same throughout the tube (Fig. 13.1A). When the loudspeaker is turned on, a surface of

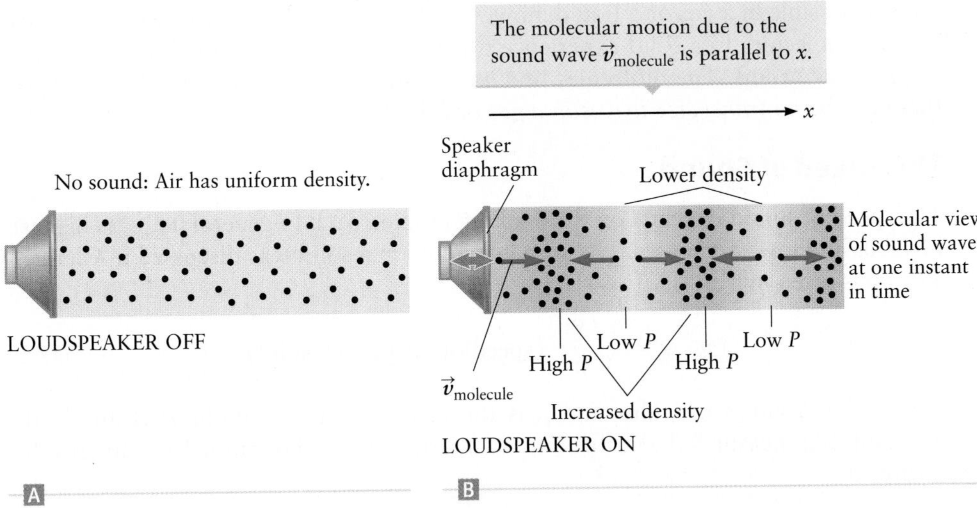

◀ **Figure 13.1** Molecular-scale view of sound in a gas. **A** An air column with no sound present. The molecules are in motion, but their velocities are in random directions and the air molecules are distributed uniformly throughout the container. **B** A sound wave generated by a loudspeaker. Each molecule now has an additional contribution to its total velocity. This contribution is denoted by $\vec{v}_{molecule}$ and is parallel to x, the direction in which the sound wave propagates.

the loudspeaker (called the diaphragm) moves back and forth along the x direction, generating a sound wave (Fig. 13.1B). As the diaphragm moves, it collides with air molecules, imparting a small extra velocity $\vec{v}_{molecule}$ to the molecules. Because the diaphragm moves back and forth along the x direction, $\vec{v}_{molecule}$ alternates along $\pm x$. At some instants, $\vec{v}_{molecule}$ at a particular location is directed to the right; at other instants, $\vec{v}_{molecule}$ at this same location is toward the left. The total velocity of a molecule is the sum of its background velocity $\vec{v}_0$ and the extra velocity $\vec{v}_{molecule}$ associated with the sound wave. This extra velocity is *not* the velocity of the sound wave; $\vec{v}_{molecule}$ is the *velocity of the medium* (the air) associated with the sound wave.

The green arrows in Figure 13.1B show $\vec{v}_{molecule}$ at several points along the wave. This snapshot only applies at a particular instant in time. Since $\vec{v}_{molecule}$ is parallel to x, the "extra" motion of the molecules—that is, the displacement of the medium—is parallel to the direction of propagation of the wave. For this reason, sound is a *longitudinal wave*.

Because the displacement alternates between the $+x$ and $-x$ directions as we move along the sound wave (along the x axis in Fig. 13.1B), the density of molecules is increased in some regions and decreased in others. Oscillations in the density cause oscillations in the pressure, so we can also view sound as a *pressure wave*.

A plot of the pressure associated with a sound wave at a particular instant in time is shown in Figure 13.2. The regions of high density and high pressure are called regions of ***condensation***. Places of low density and low pressure are called regions of ***rarefaction***. Adjacent regions of high pressure are separated by a distance equal to the wavelength λ, and adjacent regions of low pressure are separated by the same distance. The speed of the sound wave is related to its frequency and wavelength by

$$v_{sound} = f\lambda \tag{13.1}$$

It is crucial to distinguish the velocity of the wave (v_{sound} in Eq. 13.1) from the velocity $\vec{v}_{molecule}$ of particles within the medium. They are *not* the same.

Figures 13.1 and 13.2 describe a sound wave in air, and the picture is similar for sound traveling through a solid or a liquid. The sound wave causes a displacement of molecules in the solid or liquid parallel to the propagation direction, so sound is a longitudinal wave in solids and liquids as well as in gases.

In a solid, there can also be transverse waves that propagate through the solid, in which the molecular motion is perpendicular to the direction of wave propagation (such as some of the seismic waves described in Chapter 12). Such transverse waves

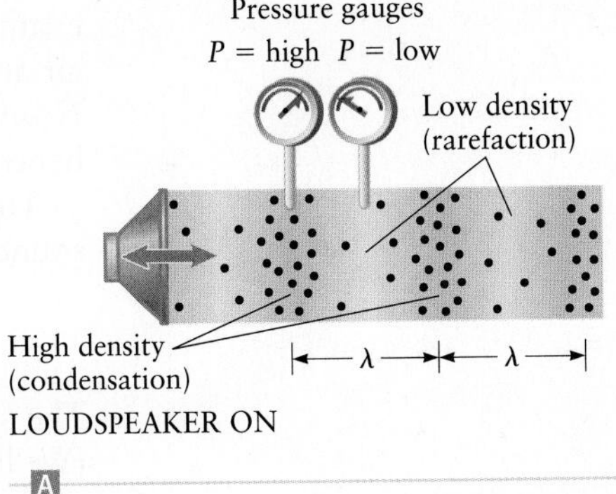

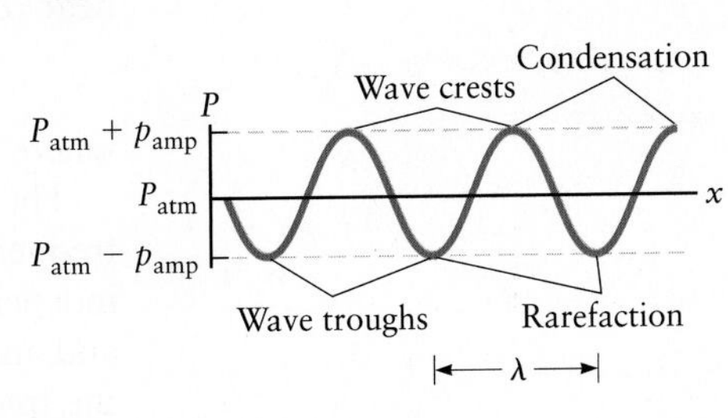

▲ **Figure 13.2** **A** The pressure in a gas is highest in the regions with the highest density. **B** For a sound wave in air, the total pressure is the sum of atmospheric pressure P_{atm} and the oscillating pressure associated with the sound wave.

TABLE 13.1 Speed of Sound in Some Common Materials

Material	Speed of Sound (m/s)
Air	343
Helium gas	1000
Hydrogen gas	1330
Oxygen gas	330
Carbon dioxide gas	269
Water	1480
Seawater	1520
Ethyl alcohol	1160
Aluminum	5100
Steel	5790
Lead	1960
Concrete	3100
Glass	5600

Note: All values are for 20°C and a pressure of 1 atm.

are *not* possible in a gas or a liquid, however. Transverse waves are possible in solids because molecules in a solid experience transverse forces due to the bonds between neighboring atoms. The molecules in a gas or liquid are not bound in this way, so there can be *no transverse waves in a gas or liquid.*

The Speed of Sound

The speed of sound depends on the medium (Table 13.1). In general, v_{sound} is highest in solids and lowest in gases. The speed of sound in a solid was discussed in Chapter 12 and is given by (Eq. 12.10)

$$v_{\text{sound}} = \sqrt{\frac{Y}{\rho}} \quad \text{(speed of sound in a solid)} \tag{13.2}$$

where Y is Young's modulus and ρ is the density. Young's modulus is an elastic constant that measures the stiffness of a material. Hence, Equation 13.2 can also be written as

$$v_{\text{sound}} = \sqrt{\frac{\text{stiffness}}{\text{density}}} \tag{13.3}$$

Equation 13.3 explains qualitatively how the speed of sound varies when we compare solids, liquids, and gases. A stiffer material (larger Young's modulus) generates a larger restoring force for a given displacement, leading to a larger particle acceleration and ultimately a greater speed of sound. Solids are stiffer than liquids, and liquids are stiffer than gases, which explains why the speed of sound is highest in solids and lowest in gases.

The density factor in the denominator in Equation 13.3 means that the speed of sound increases as the medium becomes "lighter." For example, lead has a very high density, and the speed of sound in lead is much less than that found for aluminum, which has a low density (for a metal). A similar effect is found with gases; for example, the speed of sound in helium gas is much greater than the speed of sound in air at the same pressure and temperature. This result can again be understood from Newton's second law: for a given force, a smaller mass has a larger acceleration and hence a higher speed.

The stiffness of a liquid is measured by its bulk modulus (Chapter 11), and for sound in a liquid one finds

$$v_{\text{sound}} = \sqrt{\frac{B}{\rho}} \quad \text{(speed of sound in a liquid)} \tag{13.4}$$

The speed of sound in a gas depends on its composition and also on temperature. (We'll discuss the properties of gases in Chapter 15.) The most important case is air near room temperature and pressure, for which

$$v_{\text{sound}} \approx 343 + 0.6(T - 20°\text{C})\ \text{m/s} \quad \text{(speed of sound in air)} \tag{13.5}$$

where T is the temperature in Celsius.

The general relation $v_{\text{sound}} = f\lambda$ (Eq. 13.1) connects the speed of sound with the frequency and wavelength. The speed of sound is a property of the medium and is independent of both the frequency and the amplitude of the wave. The frequency and amplitude of a sound wave depend on how the sound is generated. For example, the frequency of the sound wave in Figure 13.1 equals the oscillation frequency of the loudspeaker diaphragm. This frequency can vary over an extremely wide range, so there is no absolute lowest or highest frequency for a sound wave. The range of normal human hearing extends from approximately 20 Hz to about 20,000 Hz. These limits are only approximate and vary from person to person. The upper limit also tends to decrease with age due to a gradual stiffening of the eardrum and the other flexible connections within the ear. Sound waves can have both higher and lower frequencies than this 20–20,000 Hz range. Frequencies below 20 Hz are called ***infrasonic*** and can be generated and detected by large animals such as elephants. People can sometimes sense such very low frequencies as vibrations rather than as sound detected by the ears. Frequencies above 20,000 Hz are called ***ultrasonic*** and can be heard by many species, such as dogs and bats.

Musical Tones and Pitch

A sound wave described by a single frequency is called a ***pure tone***. Although it is possible to generate a pure tone, most sounds are a combination of many pure tones. For example, the sounds associated with speech or with the banging of a drum are complex combinations of pure tones.

In Section 12.8, we discussed standing waves on a guitar string and saw that they have frequencies given by $f_1, 2f_1, 3f_1, \ldots$. This pattern is called a ***harmonic sequence***, and f_1 is called the ***fundamental frequency***. Many sounds are a combination of frequencies that are harmonically related. For example, with an acoustic guitar, the standing wave vibrations of a string are transferred to the body of the guitar, and the vibrating surfaces of the guitar act like the loudspeaker diaphragm in Figure 13.1. The result is a combination tone containing frequencies $f_1, 2f_1, 3f_1, \ldots$.

One way to characterize such a combination tone is by a property called ***pitch***. Pitch is a subjective quality that depends on how the human brain processes sound. This processing is complicated and is still not completely understood. Under most conditions, the pitch of a pure tone is equal to its frequency, whereas the pitch of a combination of harmonically related tones (as would be produced by a plucked guitar string) is associated with the fundamental frequency. More complex sounds, as might be produced by an orchestra, are usually interpreted by the brain as a collection of tones with different pitches.

EXAMPLE 13.1 The Wavelength of Sound in Air

The frequency range that can normally be detected by a human ear is 20–20,000 Hz. What is the corresponding range of wavelengths for sound in air?

RECOGNIZE THE PRINCIPLE

The wavelength is related to the speed of sound and the frequency through $v_{\text{sound}} = f\lambda$ (Eq. 13.1). Given v_{sound} and f, we can thus find λ.

SKETCH THE PROBLEM

No figure is necessary.

IDENTIFY THE RELATIONSHIPS

Solving for the wavelength in terms of the speed of sound and the frequency gives

$$\lambda = \frac{v_{\text{sound}}}{f}$$

Here v_{sound} is the speed of sound in air at room temperature; according to Table 13.1 and Equation 13.5, $v_{\text{sound}} = 343$ m/s.

SOLVE

For the lower end of the given frequency range, we have

$$\lambda = \frac{v_{\text{sound}}}{f} = \frac{343 \text{ m/s}}{20 \text{ Hz}}$$

$$\lambda = 17 \frac{\text{m/s}}{\text{Hz}} = 17 \frac{\text{m/s}}{1/\text{s}} = \boxed{17 \text{ m}}$$

(continued) ▶

For the high-frequency end of the range of human hearing,

$$\lambda = \frac{v_{\text{sound}}}{f} = \frac{343 \text{ m/s}}{20{,}000 \text{ Hz}} = 0.017 \text{ m} = \boxed{1.7 \text{ cm}}$$

► *What does it mean?*

The ear canal is a few centimeters in length, which is comparable to the wavelength of sound at the high-frequency end of the range that humans can hear.

13.2 Amplitude and Intensity of a Sound Wave

The oscillating pressure associated with a sound wave is sketched in Figure 13.2B, where the amplitude of the pressure oscillation is denoted p_{amp}. When a sound wave reaches your ear, the pressure on the outside of your eardrum is the sum of atmospheric pressure plus the oscillating sound pressure, $P_{\text{atm}} + p$, whereas the pressure inside your eardrum is P_{atm} (Fig. 13.3). The total force on your eardrum equals the difference in these two pressures multiplied by the area of the eardrum A. Hence, the total force equals pA, and your ear detects the sound pressure p while being insensitive to P_{atm}.

The ***intensity*** I of the sound wave equals the power carried by the wave per unit area of wave front. We showed in Chapter 12 that the intensity of a wave is proportional to the square of the amplitude, so for sound we have

Intensity is proportional to the square of the amplitude.

$$I \propto p_{\text{amp}}^2 \tag{13.6}$$

In many cases, sound waves propagate outward in all directions away from their source. In Section 12.4, we discussed the spherical wave fronts of such waves and found that the intensity at a distance r from the source varies as

$$I \propto \frac{1}{r^2} \tag{13.7}$$

Combining this expression with Equation 13.6, we see that the pressure amplitude of a spherical sound wave falls with distance as $p_{\text{amp}}^2 \propto 1/r^2$, which is equivalent to

$$p_{\text{amp}} \propto \frac{1}{r} \tag{13.8}$$

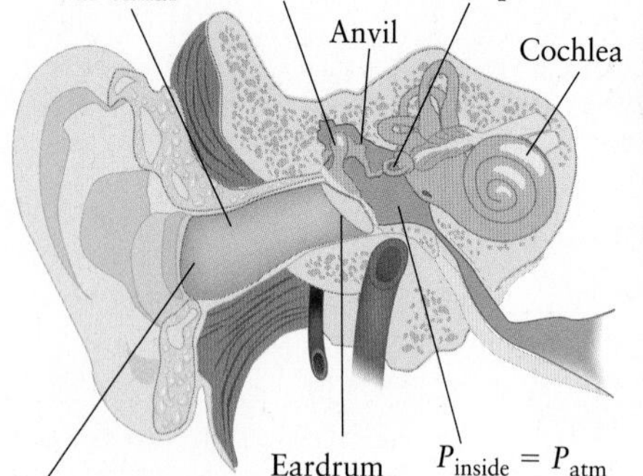

Ear canal
Hammer
Stirrup
Anvil
Cochlea
Eardrum
$P_{\text{inside}} = P_{\text{atm}}$
$P_{\text{outside}} = P_{\text{atm}} + p$

▲ **Figure 13.3** Your ear detects sound through the pressure on the eardrum. The oscillating pressure of the sound wave produces an oscillating force on the eardrum.

CONCEPT CHECK 13.1 The Intensity of a Sound Wave Falls Off with Increasing Distance from the Source

A chirping bird sits at the top of a tree, and the chirping sound has an intensity I_0 at a certain distance from the bird. If the distance to the bird is increased by a factor of three, is the new intensity (a) $I_0/3$, (b) $I_0/9$, or (c) I_0?

Decibels

The sensitivity of the ear depends on frequency. In the most favorable range of frequencies, the ear can detect sound intensities as small as about 10^{-12} W/m². Because values this small are rather cumbersome to write, another unit, the ***decibel*** (abbreviated dB), is widely used to measure sound intensity. This unit is computed in the following way. Suppose the intensity of a sound wave as measured in watts per square meter is I. We now introduce a constant $I_0 \equiv 1.00 \times 10^{-12}$ W/m², which is approximately the lowest intensity detectable by the human ear. We then define a quantity called the ***sound intensity level*** β (the Greek letter beta) by

Threshold for human hearing and the decibel scale

$$\beta = 10 \log\left(\frac{I}{I_0}\right) \quad \text{(sound intensity level in decibels)} \tag{13.9}$$

The intensity level β is given in decibels, but you can see from Equation 13.9 that since β is defined through a ratio of intensities, it is, in a strict mathematical sense, unitless! When working with β and Equation 13.9, it is helpful to recall the properties of common (base 10) logarithms as reviewed in Appendix B. In particular, if $a = 10^b$, then $b = \log(a)$. We also have $\log(1) = 0$, $\log(10) = 1$, $\log(100) = 2$, and so forth.

To see how the definition of sound intensity level in Equation 13.9 is applied, consider a sound that is just at the threshold of human hearing, that is, a sound with intensity $I = 1.0 \times 10^{-12}\ \text{W/m}^2$. The intensity level in decibels is then

$$\beta = 10 \log\left(\frac{I}{I_0}\right) = 10 \log\left(\frac{1.0 \times 10^{-12}\ \text{W/m}^2}{1.00 \times 10^{-12}\ \text{W/m}^2}\right) = 10 \log(1)$$

Because $\log(1) = 0$, we find

$$\beta = 0\ \text{dB}$$

The intensity of this sound is thus 0 dB = "zero decibels." An intensity level of $\beta = 0$ dB does *not* mean that there is an absence of sound. It merely indicates that the intensity I is equal to the reference intensity I_0, which corresponds to a very faint sound, a sound at the limit of human hearing.

Now suppose we have another sound of intensity $I = 1.0 \times 10^{-11}\ \text{W/m}^2 = 10 \times I_0$, which is 10 times the nominal threshold for human hearing. The intensity level of this sound as measured in decibels is

$$\beta = 10 \log\left(\frac{I}{I_0}\right) = 10 \log\left(\frac{10 \times I_0}{I_0}\right) = 10 \log(10)$$

Because $\log(10) = \log(10^1) = 1$, we get

$$\beta = 10\ \text{dB}$$

which is the result for a sound that has an intensity 10 times the threshold for hearing. If a sound has an intensity 100 times this threshold ($I = 1.0 \times 10^{-10}\ \text{W/m}^2$), we have

$$\beta = 10 \log\left(\frac{I}{I_0}\right) = 10 \log\left(\frac{100 \times I_0}{I_0}\right) = 10 \log(100)$$

Because $\log(100) = 2$, we get

$$\beta = 20\ \text{dB}$$

Thus, for every increase in intensity by a factor of 10, the intensity level as measured in decibels increases by 10 dB. The intensity levels of some familiar sounds are listed in Table 13.2.

CONCEPT CHECK 13.2 Sound Intensity Level and Decibels

What is the intensity in watts per square meter of a sound with a sound intensity level of 50 dB?

(a) $1.0 \times 10^{-12}\ \text{W/m}^2$
(b) $1.0 \times 10^{-10}\ \text{W/m}^2$
(c) $1.0 \times 10^{-7}\ \text{W/m}^2$
(d) $5.0 \times 10^{-12}\ \text{W/m}^2$

TABLE 13.2 Intensity Level of Some Common Sounds

Sound	Intensity Level (dB)	Sound	Intensity Level (dB)
Threshold of pain for people	130	City traffic	70
Jet takeoff at a distance of 100 m	90–120	Normal conversation	60–70
Rock concert	110–120	Whisper	20
Thunder (nearby)	110	Pin dropping	10
Factory noise	80	Threshold of human hearing	0

CONCEPT CHECK 13.3 How Loud Is It?

The sound intensity level at a rock concert can be as high as 120 dB, whereas normal speech corresponds to about 70 dB. What is the ratio of the intensities of these two sounds?

(a) 190 (b) 50 (c) 5 (d) 10^5

Human Perception of Sound

Decibels are based on a ratio involving I_0 and logarithms, and you might now wonder why this way of measuring sound intensity was invented. The motivation comes from a feature of human hearing illustrated in Figure 13.4, which shows several contour lines obtained from extensive studies of human hearing. The lowest line plots the intensity of the minimum detectable sound as a function of the sound frequency. In the range of 500 Hz to 5000 Hz, this curve is roughly constant (horizontal) with a value close to $I_0 = 1.00 \times 10^{-12}$ W/m^2. (See the right-hand scale in Fig. 13.4.) When expressed in terms of intensity level in decibels (the left-hand scale), this curve corresponds to approximately 0 dB. The other contour curves show how the intensity varies for a fixed value of the *perceived* loudness, with each successive contour curve corresponding to an increase by a factor of two in the loudness. These results were obtained by asking many people to judge sounds of different loudness, and it was found that doubling the perceived loudness corresponds to moving from one contour to the next. Figure 13.4 shows averages obtained from many different individuals, and these contour curves represent normal human hearing. Loudness is a subjective quality that depends on how the ear functions at different frequencies, so it is fascinating that different people can agree on a loudness scale!

Over the range of 250 Hz to 8000 Hz, the contours in Figure 13.4 have intensity levels β that differ by about 10 dB. Hence, a doubling of the perceived loudness corresponds to about a 10-dB increase in the intensity level, regardless of the initial loudness. That is why the sound intensity level measured in decibels with its logarithmic definition is a very useful way to describe the intensity of a sound.

The highest contour curve in Figure 13.4 has a loudness at which the ear is damaged. Such damage occurs at an intensity level $\beta \approx 120$ dB, which is approximately 120 dB larger than the threshold for hearing. Each change in the intensity level by 10 dB corresponds to a factor of 10 change in the intensity, so an increase of β by

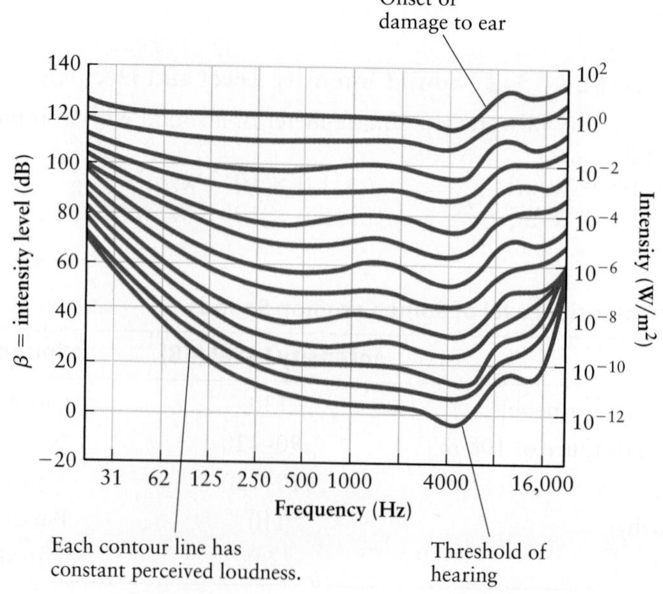

▶ **Figure 13.4** Contours of constant perceived loudness as a function of frequency for normal human hearing.

120 dB corresponds to an increase in the intensity by a factor of 10^{12}! Hence, as a mechanical device, the ear is able to function over a very wide range of forces.

Another important feature of Figure 13.4 is that the ear's sensitivity to sound pressure decreases at low frequencies. For example, to have the same perceived loudness (and thus fall on the same contour curve), a sound at 60 Hz must have a larger intensity than a sound at 1000 Hz.

The Ear as a Pressure Detector

Quantitative measurements of sound often involve values of the intensity, but it is also interesting to consider the corresponding values of the sound pressure. Intensity is proportional to the square of the pressure amplitude (Eq. 13.6), so there is a direct relation between the threshold intensity I_0 for human hearing and the smallest pressure oscillation that can be detected by the ear. A sound intensity $I_0 = 1.00 \times 10^{-12}$ W/m^2 corresponds to a pressure amplitude $p_0 = 2 \times 10^{-5}$ Pa. (This result can be obtained by applying Newton's laws to calculate the molecular motion of a sound wave, but the derivation is more than we can include here.) Hence, the smallest pressure amplitude that can be detected by the human ear is about $p_{amp} = p_0 = 2 \times 10^{-5}$ Pa. Since atmospheric pressure is $P_{atm} = 1.01 \times 10^5$ Pa, the ear is therefore able to detect variations in pressure that are $p_0/P_{atm} = (2 \times 10^{-5} \text{ Pa})/(1.01 \times 10^5 \text{ Pa}) = 2 \times 10^{-10}$ times smaller than atmospheric pressure! The ear is thus an extremely impressive mechanical device.

EXAMPLE 13.2 The Pressure Amplitude at a Rock Concert

According to Table 13.2, the sound intensity level at a rock concert can reach values as high as 120 dB. What is the approximate value of the corresponding pressure amplitude? The pressure amplitude at $\beta = 0$ dB is $p_0 = 2 \times 10^{-5}$ Pa.

RECOGNIZE THE PRINCIPLE

We first use the definition of decibels (Eq. 13.9) to convert the given sound intensity level β to an intensity value I. We can then use the relationship between I and p_{amp} to find the corresponding pressure amplitude.

SKETCH THE PROBLEM

No figure is necessary.

IDENTIFY THE RELATIONSHIPS

Using Equation 13.9, we can convert the intensity level at a rock concert ($\beta = 120$ dB) into intensity I in units of W/m^2. We have

$$\beta = 120 = 10 \log\left(\frac{I_{rc}}{I_0}\right)$$

where I_{rc} is the sound intensity at the rock concert (in watts per square meter) and I_0 is known to be 1.00×10^{-12} W/m^2. This leads to

$$\log\left(\frac{I_{rc}}{I_0}\right) = 12 \quad (1)$$

According to the properties of logarithms, if $\log(a) = b$, then $a = 10^b$. Applying this fact to Equation (1), we get

$$\frac{I_{rc}}{I_0} = 10^{12}$$

$$I_{rc} = 10^{12} I_0 \quad (2)$$

(continued) ▶

Now use the relation between the intensity I and pressure amplitude p_{amp}. At an intensity I_0 (corresponding to 0 dB), the pressure amplitude is $p_0 = 2 \times 10^{-5}$ Pa. We also know that intensity is proportional to the *square* of the pressure amplitude (Eq. 13.6). We can therefore replace the intensities in Equation (2) by the squares of the corresponding pressure amplitudes and write

$$p_{rc}^2 = 10^{12} p_0^2$$

where p_{rc} is the pressure amplitude at the rock concert.

SOLVE

Inserting the value of p_0 leads to

$$p_{rc} = \sqrt{10^{12}}\, p_0 = (10^6)(2 \times 10^{-5}\text{ Pa}) = \boxed{20\text{ Pa}}$$

▸ *What does it mean?*

Although a rock concert produces quite a loud sound, the resulting pressure amplitude is still only a very tiny fraction of atmospheric pressure ($P_{atm} = 1.01 \times 10^5$ Pa).

13.3 Standing Sound Waves

In Figures 13.1 and 13.2, we considered sound waves that travel in one direction inside a tube. We next want to consider what happens when these waves reach the end of the tube. This is the key to understanding how many musical instruments work, including flutes, trombones, and organs, and it is very similar to the case of standing waves on a guitar string studied in Chapter 12.

Standing Waves in a Pipe Closed at Both Ends

A standing wave is created when two waves travel in opposite directions in the same medium. As a first example, consider sound waves propagating back and forth in a pipe that is closed at both ends. A standing wave is possible only for certain wavelengths; some number of sound waves must "fit into" or match the length of the pipe. To understand this matching process, we must consider how a sound wave behaves at the ends of the pipe. Because sound is a longitudinal wave, the motion of the molecules would normally be directed along the pipe (the x direction in Fig. 13.5). The wall prevents this motion, however; for example, the right-hand wall in Figure 13.5 prevents the adjacent molecules from moving to the right. These molecules will also not move to the left (away from the wall) because that would produce a low-pressure region that pulls them back, so the velocity of the air molecules ($\vec{v}_{molecule}$ in Fig. 13.1) is zero at a wall, that is, at the closed end of a pipe. Because $\vec{v}_{molecule} = 0$ at a wall, the displacement of the molecules $d_{molecule}$ is also zero at the ends of the pipe.

A standing sound wave must therefore have zero displacement at both ends of the pipe, leading to the results in Figure 13.5. Figure 13.5A shows the displacement pattern ($d_{molecule}$) for a possible standing wave. These curves are the standing wave envelopes; that is, they show how the amplitude of the oscillation varies with position along the pipe. Places where the displacement is zero are called displacement ***nodes***, while the displacement is largest at the displacement ***antinodes.*** The wavelength of this standing wave is $2L$, where L is the length of the pipe. The standing wave pattern in Figure 13.5A is identical to the pattern found for a standing wave on a string (Fig. 13.5D; compare with Fig. 12.24B). Both have nodes at the ends of a pipe or string and an antinode in the center. The molecular velocity forms a standing wave described by the same pattern (Fig. 13.5B). The amplitude of the molecular velocity is zero at the ends of the pipe and large at the center.

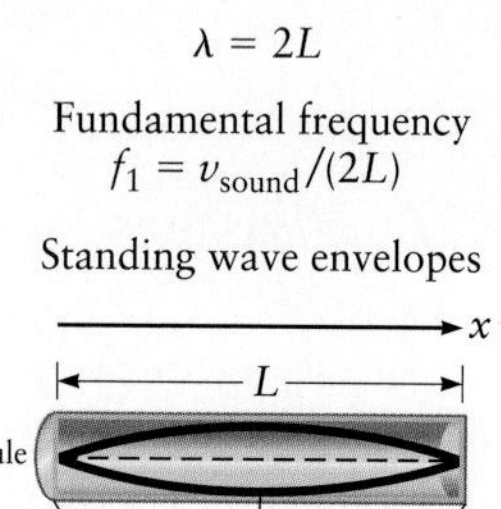

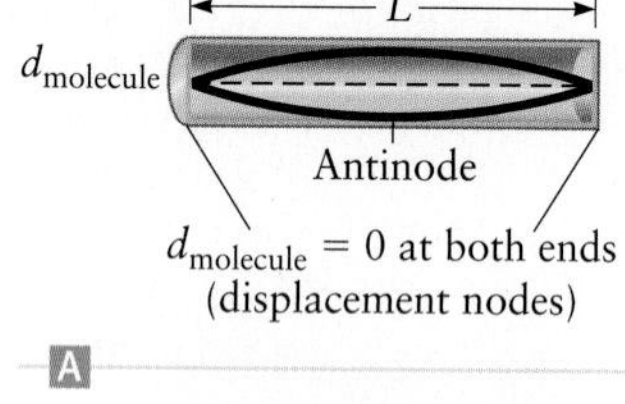

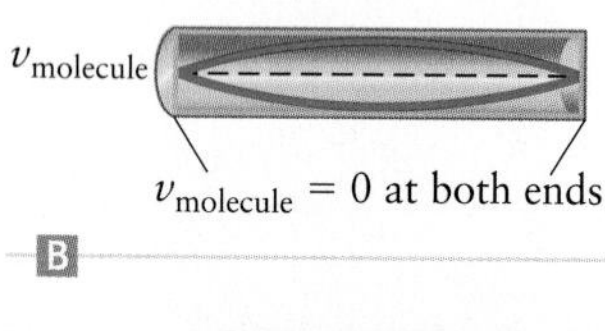

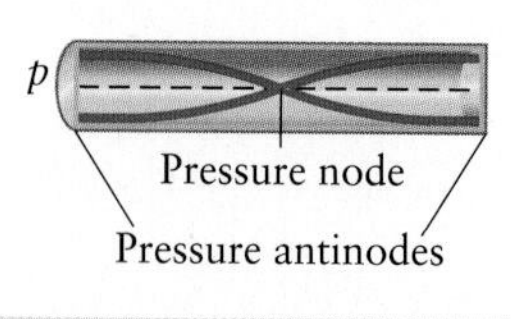

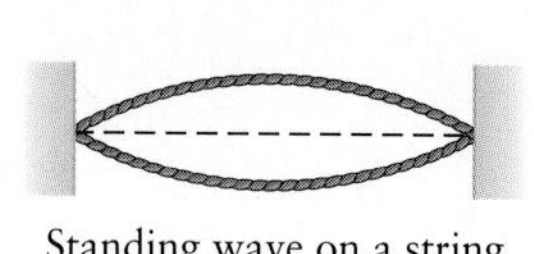

▲ **Figure 13.5** Standing sound waves in a pipe that is closed at both ends. A Molecular displacement, B velocity, and C sound pressure for the lowest-frequency standing wave. This is the fundamental frequency of the pipe. D Displacement for a standing wave on a string; it is identical to the molecular displacement in part A.

A sketch of the sound pressure p is shown in Figure 13.5C. The behavior of the pressure at the ends is different from that of the molecular velocity and displacement. We have already mentioned that the air molecules cannot move past the ends of the pipe (because of the walls). During a portion of the wave's cycle, air molecules that are initially in the interior of the pipe move toward a wall, which increases the pressure at the wall to a large value because the molecules are "trapped" there. During the opposite portion of the wave's cycle, these air molecules move away from the wall, producing a low pressure at the wall. As a result, pressure oscillations are large at a closed end, forming a *pressure antinode* in the standing wave (Fig. 13.5C). Hence, both the displacement and pressure form standing waves, with the nodes of displacement located at the antinodes of p and vice versa.

Other possible standing wave patterns for the displacement and pressure are shown in Figure 13.6 for standing waves with $\lambda = L$ and $\lambda = 2L/3$. These standing wave patterns are identical to those found for a wave on a string (Fig. 12.24).

There is a mathematical pattern to the standing waves. Figures 13.5 and 13.6 show the three standing waves with the longest possible wavelengths: $\lambda = 2L$, L, and $2L/3$. There are many more possible values of the wavelength (an infinite number), and they follow the pattern

$$\lambda_n = 2L,\ L,\ \frac{2L}{3},\ \frac{2L}{4},\ \frac{2L}{5},\ldots$$

or

$$\lambda_n = \frac{2L}{n} \quad \text{with } n = 1, 2, 3, \ldots \quad \text{(standing waves in a pipe closed at both ends)} \tag{13.10}$$

Since frequency and wavelength are connected by $v_{\text{sound}} = f\lambda$, the frequencies of these standing waves are

$$f_n = n\frac{v_{\text{sound}}}{2L} \tag{13.11}$$

$$f_n = nf_1 \quad \text{with } n = 1, 2, 3, \ldots \quad \text{(standing waves in a pipe closed at both ends)} \tag{13.12}$$

where $f_1 = v_{\text{sound}}/(2L)$ is called the ***fundamental frequency*** (also known as the first harmonic). The standing wave with a frequency $f_2 = 2f_1$ is called the second harmonic, the standing wave with frequency f_3 is called the third harmonic, and so on.

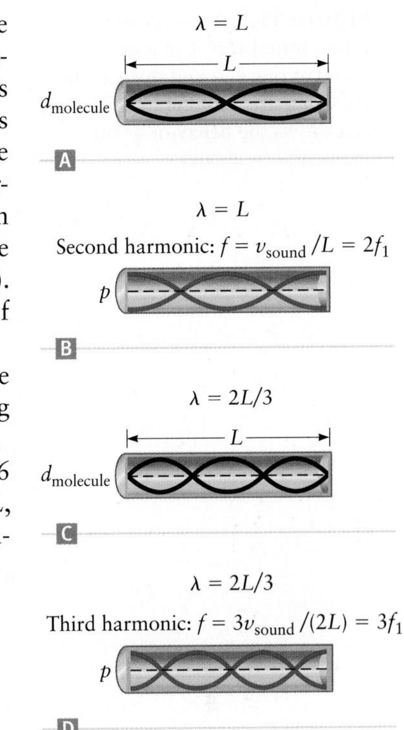

▲ **Figure 13.6** A and B: Displacement and pressure for a standing wave in a pipe closed at both ends, at the frequency of the second harmonic. C and D: Displacement and pressure at the frequency of the third harmonic.

Standing Waves in a Pipe Open at One End and Closed at the Other

Many musical instruments, including trumpets and clarinets, have an opening at one end. Let's therefore consider standing waves in a pipe that is *open* at one end and closed at the other. To analyze this situation, we must consider what happens to a sound wave at the open end of the pipe. We have already seen that there is a displacement node and a pressure antinode at the closed end of a pipe (Fig. 13.5). The behavior at an open end is quite different. At an open end, a pipe connects to the rest of the room, where the pressure is equal to atmospheric pressure P_{atm}. This outside volume is much larger than the volume of air in the pipe, so the flow of air molecules out of or into the pipe due to a sound wave has a negligible effect on the outside pressure. The pressure at the open end of the pipe is therefore held fixed at P_{atm} by the large body of air in the room, and the pressure amplitude of a standing sound wave at an open end must be $p_{\text{amp}} = 0$. Hence, there is a pressure *node* at the open end of the pipe.

Figures 13.7 and 13.8 show some of the possible standing waves for a pipe open at one end and closed at the other. There is a pressure node at the open end and (as we have seen in Fig. 13.5C) a pressure antinode at the closed end. The behaviors of

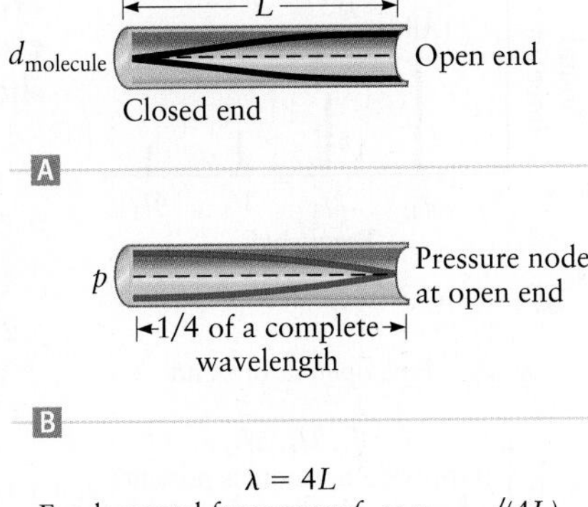

▲ **Figure 13.7** Standing sound waves in a pipe that is closed at one end and open at the other. A Molecular displacement and B pressure at the fundamental frequency of the pipe. There is a pressure node at the open end, which makes the standing wave frequencies different from those found in a pipe that is closed at both ends (Figs. 13.5 and 13.6).

▶ **Figure 13.8** Some possible standing sound waves in a pipe that is closed at one end and open at the other. Compare with Figure 13.7, which shows the behavior at the fundamental frequency.

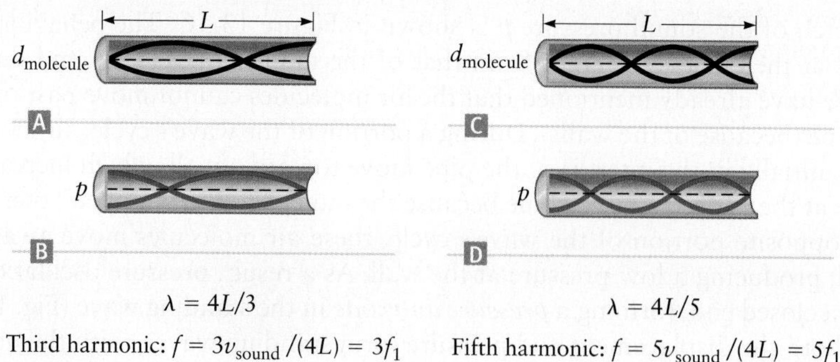

the sound pressure and displacement are linked, so there is a displacement antinode at the open end and a displacement node at the closed end. Standing sound waves must always satisfy these conditions at the ends of such a pipe. The standing wave with the longest possible wavelength is shown in Figure 13.7; it has $\lambda = 4L$. The next standing waves (in order of decreasing wavelength) are shown in Figure 13.8; they have wavelengths $\lambda = 4L/3$ and $\lambda = 4L/5$. These and the other possible standing waves for this tube follow the pattern

$$\lambda_n = 4L, \frac{4L}{3}, \frac{4L}{5}, \frac{4L}{7}, \ldots$$

$$\lambda_n = \frac{4L}{n} \quad \text{with } n = 1, 3, 5, \ldots \quad \text{(standing waves in a pipe open at one end)} \tag{13.13}$$

The frequencies of these standing waves are $f_n = v_{sound}/\lambda_n$, so

$$f_n = n\frac{v_{sound}}{4L} \tag{13.14}$$

$$f_n = nf_1 \quad \text{with } n = 1, 3, 5, \ldots$$

$$\text{(standing waves in a pipe open at one end)} \tag{13.15}$$

and the fundamental frequency is $f_1 = v_{sound}/(4L)$. Comparing with Equations 13.11 and 13.12, the fundamental frequency is lower for the pipe with one end open (if we compare two pipes of the same length L), and the pattern of harmonics is also different. For the pipe with both ends closed, the allowed frequencies form the sequence f_1, $2f_1$, $3f_1$, . . . , whereas with one end open, the sequence is f_1, $3f_1$, $5f_1$, ... Hence, the even-numbered harmonics are missing for a pipe that is open at one end. Figure 13.9 shows the pattern of standing wave frequencies in the two cases.

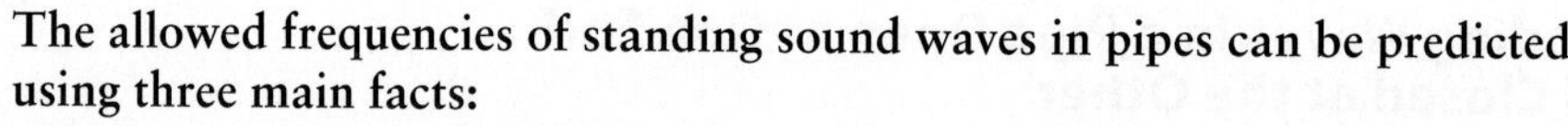

The allowed frequencies of standing sound waves in pipes can be predicted using three main facts:

1. The *closed* end of a pipe is a pressure *antinode* for the standing wave.
2. The *open* end of a pipe is a pressure *node* for the standing wave.
3. A pressure node is always a displacement antinode, and a pressure antinode is always a displacement node.

Pipe closed at both ends

$f = f_1, 2f_1, 3f_1, \ldots$
(All harmonics present)

Sound intensity

f_1 $2f_1$ $3f_1$ $4f_1$
Frequency

A

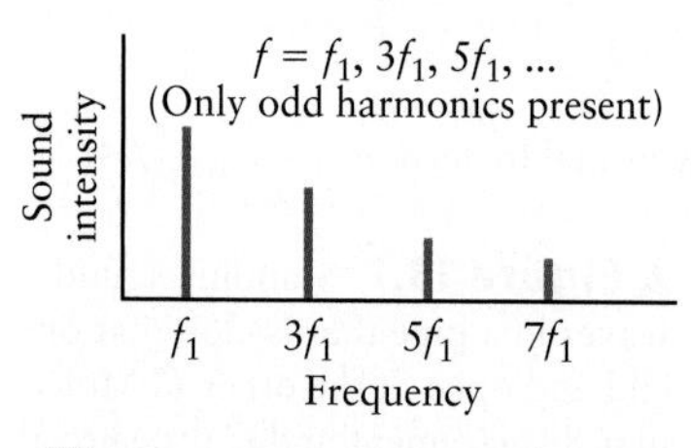

▲ **Figure 13.9** The pattern of standing wave frequencies in a pipe depends on whether the ends of the pipe are open or closed.

Composition of a Real Musical Tone

Pipes like those shown schematically in Figures 13.5 through 13.8 are the basis of many musical instruments, including flutes, trumpets, and organs. The sound we hear originates as a standing sound wave inside the instrument. This standing wave does not have a single frequency, but is a combination (a *superposition*) of standing waves with different frequencies. The allowed standing wave frequencies follow a pattern such as Equation 13.11 or 13.14, depending on whether the ends of the instrument are open or closed. Hence, these instruments produce combination tones as described above in our discussion of pitch. These combination tones are often said to have a property called ***timbre***, or ***tone color***. The musical timbre of a tone depends on the mix of frequencies (as in Fig. 13.9) and on the relative amplitudes of these different frequency components. The timbre is different for different instruments, and it also depends on how the instrument is played. (A "wailing" saxophone will have a different tone color than one played less forcefully.) These variations in timbre help make musical tones interesting to a listener.

EXAMPLE 13.3 The Lowest Note of a Clarinet

A clarinet is a musical instrument we can model as a pipe open at one end and closed at the other (Fig. 13.10A). A standard clarinet is approximately $L = 66$ cm long. What is the frequency of the lowest tone such a clarinet can produce?

RECOGNIZE THE PRINCIPLE

According to our work with pipes, there must be a pressure antinode at the closed end of the clarinet and a pressure node at the open end. Frequency and wavelength are related by $v_{sound} = f\lambda$, so the standing wave with the lowest frequency will be the one with the longest wavelength.

SKETCH THE PROBLEM

Figure 13.10B shows the clarinet modeled as a pipe open at one end. The allowed standing waves will be just like those found in Figures 13.7 and 13.8.

IDENTIFY THE RELATIONSHIPS

The standing wave with the longest wavelength is sketched in Figure 13.7. Its frequency is given by Equation 13.14, with $n = 1$:

$$f_1 = \frac{v_{sound}}{4L}$$

This is the fundamental frequency for the clarinet with all tone holes closed.

SOLVE

The speed of sound in air (at room temperature) is $v_{sound} = 343$ m/s and the length of the tube is $L = 0.66$ m, so we have

$$f_1 = \frac{v_{sound}}{4L} = \frac{343 \text{ m/s}}{4(0.66 \text{ m})} = \boxed{130 \text{ Hz}}$$

► *What does it mean?*

This result is the lowest frequency—that is, the lowest "note"—a clarinet can produce according to our simple model. To play other notes, the musician uncovers one of the many different tone holes along the body of the clarinet. If an open hole is sufficiently large, the pressure node moves from the end of the clarinet to a spot very near the tone hole (Fig. 13.10C), thereby shortening the wavelength of the fundamental standing wave and increasing f_1. In this way, a musician can use the many tone holes to produce many different notes.

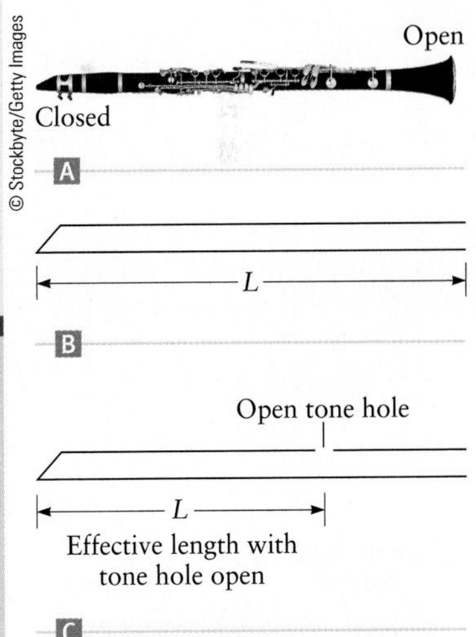

▲ **Figure 13.10** Example 13.3. A A clarinet. B Physicist's model of a clarinet, as a pipe closed at one end and open at the other. The standing waves will be the same as shown in Figures 13.7 and 13.8. C If a tone hole is opened, it will act as the open end of the pipe as far as the standing wave is concerned, changing the frequency of the standing waves.

CONCEPT CHECK 13.4 Designing a Clarinet

Suppose you design a new type of clarinet that is half the length of the ordinary clarinet in Example 13.3. How will this change affect the frequency of the lowest note on this new instrument?

(a) The frequency of the lowest note will decrease by a factor of two.
(b) The frequency of the lowest note will increase by a factor of two.
(c) The frequency of the lowest note will not change.

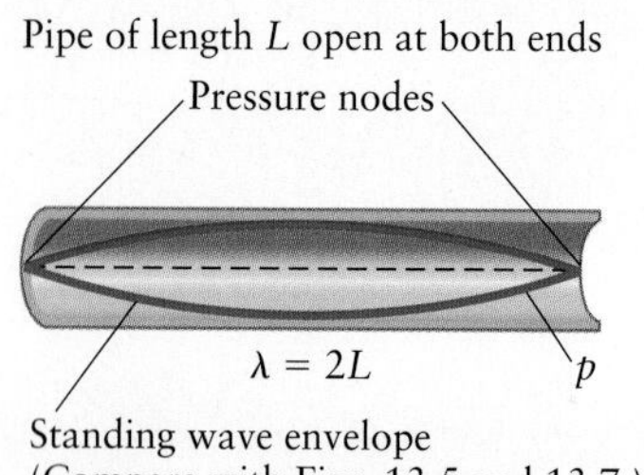

▲ **Figure 13.11** Example 13.4.

EXAMPLE 13.4 Standing Waves in a Pipe Open at Both Ends

Consider a hypothetical musical instrument that uses a pipe open at both ends. If the pipe is $L = 1.0$ m long, what is the longest wavelength that can produce a standing wave? Sketch how the pressure of this standing wave varies along the tube.

RECOGNIZE THE PRINCIPLE

A pipe open at both ends must have pressure nodes at both ends as explained on page 371.

SKETCH THE PROBLEM

Figure 13.11 shows the problem.

IDENTIFY THE RELATIONSHIPS AND SOLVE

The standing wave with the longest wavelength is sketched in Figure 13.11. For a pipe of length L, the wavelength is $\lambda = 2L$. We thus have

$$\lambda = 2L = 2(1.0 \text{ m}) = \boxed{2.0 \text{ m}}$$

► ***What does it mean?***

Comparing Figure 13.11 with the standing wave pattern for a pipe with both ends closed (Fig. 13.5), we find that the fundamental frequency is the same for these two cases.

CONCEPT CHECK 13.5 Playing a Funky Organ

An organ uses standing sound waves in pipes of different lengths to produce different notes. Your friend is an organist, and you play a trick on him by filling his organ pipes with helium gas. The speed of sound in helium is higher than the speed of sound in air by a factor of $v_{\text{helium}}/v_{\text{air}} = 2.9$. (See Table 13.1.) How will your prank affect the tones produced by the organ?

(a) The notes will all go up in frequency by this factor.
(b) The notes will all go down in frequency by this factor.
(c) There will be no change.

An experiment similar to this one is often performed at carnivals, when a performer fills his vocal tract with helium gas and then demonstrates his "new" voice. The pitch of the human voice is determined by the frequencies of standing waves in the vocal tract.

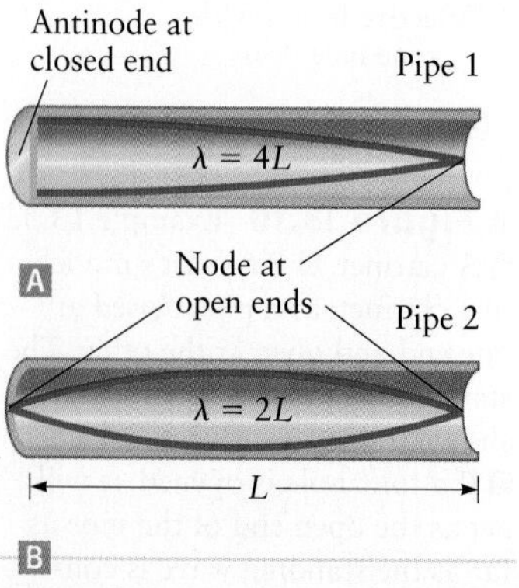

EXAMPLE 13.5 Two Organ Pipes

Two organ pipes have the same length and are both filled with air. Pipe 1 is open at one end and closed at the other, while pipe 2 is open at both ends. Which has the lowest fundamental frequency?

RECOGNIZE THE PRINCIPLE

For standing sound waves in a pipe, there must be a pressure antinode at a closed end and a pressure node at an open end. Using these facts, we can determine the fundamental frequencies for each pipe.

SKETCH THE PROBLEM

Figure 13.12 shows the standing wave envelopes for the pressure for each pipe.

Antinode at closed end — Pipe 1 — $\lambda = 4L$ — A — Node at open ends — Pipe 2 — $\lambda = 2L$ — L — B

▲ **Figure 13.12** Example 13.5.

IDENTIFY THE RELATIONSHIPS AND SOLVE

Figure 13.12A shows the standing wave envelope with the longest wavelength for pipe 1, and Figure 13.12B shows the envelope for pipe 2. Notice the pressure antinode at the closed end and pressure nodes at the open ends. The wavelength is longer for pipe 1, so pipe 1 has the lowest fundamental frequency.

► ***What does it mean?***

We could determine the pipe having the standing wave with the longest wavelength, and hence the lowest frequency, without any mathematics. Plots of the standing wave envelopes in Figure 13.12 were all we needed.

13.4 Beats

When two tones are played at the same time, the principle of superposition tells us that the total sound pressure is the sum of the sound pressures of the individual tones. Figure 13.13A plots the variation of pressure as a function of time for two hypothetical pure tones, one at 500 Hz and the other at 600 Hz. Figure 13.13B shows the sum of these two pressure waves, which would be the total sound pressure when the two tones are played simultaneously. This figure also shows how the amplitude of this combination pressure wave varies back and forth (oscillates) from a large value when the two waves interfere constructively to near zero when there is nearly destructive interference. These amplitude oscillations are called ***beats***. The frequency of these oscillations in this case is 100 Hz, which is called the ***beat frequency***. In general, whenever two tones of frequencies f_1 and f_2 are played simultaneously, they "beat against" each other with a beat frequency of

$$f_{\text{beat}} = |f_1 - f_2| \tag{13.16}$$

The human ear is very sensitive to these beats, which are heard as variations in loudness with a frequency of f_{beat}. Most people are able to detect beat frequencies as low as about 0.2 Hz, even when the two simultaneous frequencies are near 1000 Hz or even higher. Hence, beats can be used to detect very small frequency differences, provided the tones are played simultaneously.

Beats are familiar to most musicians and can be used to tune an instrument. When the different instruments in an orchestra are "in tune," they each produce precisely the same fundamental frequency f_1 when the musicians are playing notes with the same pitch. Musicians adjust their instruments (prior to the performance!) by listening to the beat frequency between their own tone and a reference tone, and making that beat frequency as small as possible.

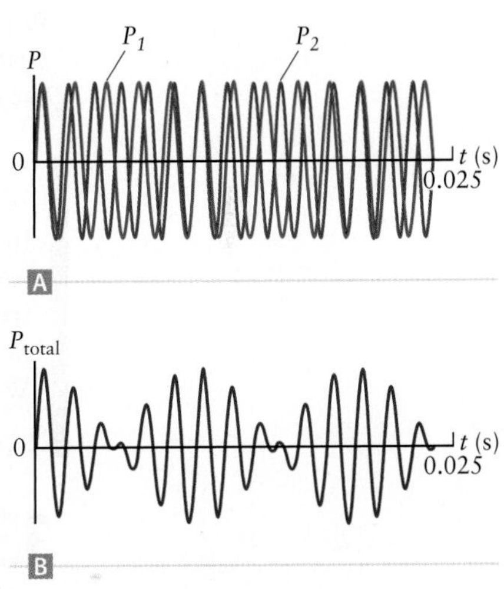

▲ **Figure 13.13** Beating of two sound waves. **A** Pressure as a function of time for two separate sound waves of frequency 500 Hz and 600 Hz. **B** The sum of these two pressure waves produces a pattern of "beats" in which the amplitude of the resulting wave oscillates with time.

EXAMPLE 13.6 Beat Frequency of Two Guitar Tones

A standard guitar has six strings, so it is possible to play six tones at any particular moment. The fundamental frequency of a particular string is determined by the standing wave with the lowest possible frequency. This lowest standing wave frequency depends on the vibrating length of the string as discussed in Chapter 12 and illustrated in Figure 13.5D. By holding a string against a particular fret (Fig. 12.25), it is possible to vary the vibrating length of the string so that two different strings have the same fundamental frequency and thus play the same note. Now suppose two strings that *should* produce the same note are actually found to produce 4 beats per second. If one note has a frequency of 247 Hz, what is the frequency of the other note?

(continued) ►

RECOGNIZE THE PRINCIPLE

If the two fundamental frequencies are f_1 and f_2, the beat frequency is $f_{beat} = |f_1 - f_2|$ (Eq. 13.16). Given the value of f_{beat} and that one of the frequencies is $f_1 = 247$ Hz, we can solve for f_2.

SKETCH THE PROBLEM

No figure is necessary.

IDENTIFY THE RELATIONSHIPS AND SOLVE

If there are 4 beats per second, the beat frequency is $f_{beat} = 4$ Hz. We know that $f_1 = 247$ Hz and $f_{beat} = |f_1 - f_2|$, so the other tone must have a frequency of either $247 + 4 = \boxed{251 \text{ Hz}}$ or $247 - 4 = \boxed{243 \text{ Hz}}$. If we are only allowed to listen to the beats, there is no way to tell which of these answers applies to a particular guitar.

What does it mean?

In practice, a guitarist will listen to the beats and adjust the tension of one of the strings to make the beat frequency as low as possible. (Recall from Chapter 12 that the fundamental frequency of waves on a string depends on the string tension.)

CONCEPT CHECK 13.6 Using Beats to Tune a Guitar

In Example 13.6, we considered the beats produced by two guitar strings and found that given only the beat frequency, there are two possible solutions for the frequency produced by the second string: $f_2 = 243$ Hz and $f_2 = 251$ Hz. Suppose the beat frequency increases when the tension in the first string is increased. Which of these solutions for f_2 is now the correct one?

(a) 243 Hz (b) 251 Hz (c) We still can't tell.

13.5 Reflection and Scattering of Sound

A standing wave can be viewed as a combination of two waves traveling in opposite directions. For the standing sound wave in a pipe, these waves travel back and forth as they are reflected from the ends of the pipe. Let's now examine the phenomenon of reflection for other geometries and also consider a related process called ***scattering***.

Figure 13.14 shows a sound wave with a plane wave front reflecting from a flat surface. It is very similar to the reflecting wave fronts we sketched in Figure 12.20. Such a ***plane wave*** reflects in a mirrorlike manner from a flat surface. The incoming angle θ_i, called the ***angle of incidence*** (see Fig. 13.14), equals the outgoing angle θ_r, called the ***angle of reflection***. We call this reflection mirrorlike because this situation is also encountered when light reflects from a mirror.

A mirrorlike reflection will occur whenever a plane wave encounters a flat surface, as long as the size of the reflecting surface is large compared to the wavelength of sound.[1] A typical sound might have a frequency of 200 Hz, in which case the wavelength is $\lambda = v_{sound}/f = (343 \text{ m/s})/(200 \text{ Hz}) = 1.7$ m. For sound in a typical room, most of the room surfaces are not large compared to this wavelength, so most of the reflections are not like the one shown in Figure 13.14. Instead, the reflected sound waves behave as in Figure 13.15, which shows an incoming plane wave reflecting from a small object. Here, there is at least a small amount of reflected sound in all directions. It is often called scattered sound because, loosely speaking, the original wave is "scattered" in many different directions. This result is found whenever a

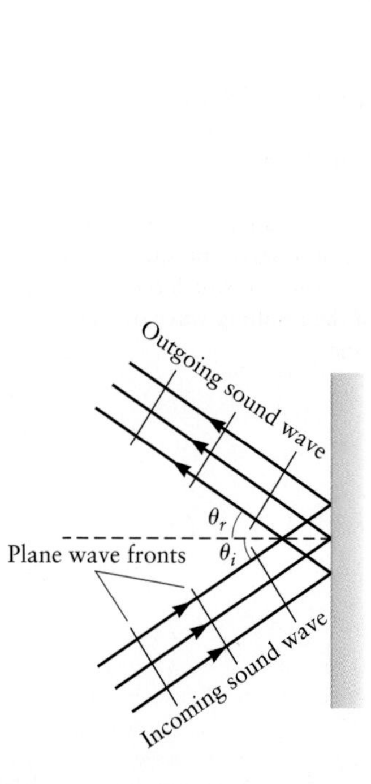

Outgoing sound wave

θ_r

θ_i

Plane wave fronts

Incoming sound wave

▲ **Figure 13.14** Reflection of a sound wave with plane wave fronts from a flat wall. The outgoing (reflected) angle θ_r is equal to the incoming (incident) angle θ_i.

[1]This requirement does *not* apply to a sound wave traveling in a very narrow pipe (as in Fig. 13.5 or with the clarinet in Fig. 13.10). When the diameter of the pipe is small compared to wavelength, the sound wave can only be reflected backward because it literally has nowhere else to go.

sound wave encounters an object whose size is approximately equal to, or smaller than, the wavelength of the sound. In such cases, reflected sound waves propagate out in virtually all directions from the scattering source.

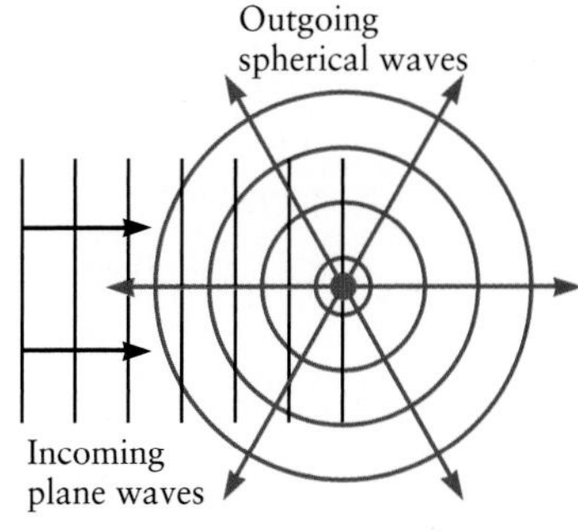

▲ **Figure 13.15** When a wave reflects from a small object (denoted by the red dot), the reflected wave travels outward in all directions. When the object is very small compared to the wavelength, the outgoing wave fronts are approximately spherical.

13.6 The Doppler Effect

Suppose the siren of an ambulance emits a pure tone of frequency f_{source}. If the ambulance is parked so that it is not moving relative to a listener standing nearby, the sound waves emitted by the siren will be as sketched in Figure 13.16A. Here we show two wave crests as they move from the source to the observer (listener) at speed v_{sound}. The source emits one wave crest during each period of the wave (one period $= 1/f_{source}$), so consecutive crests are separated by a distance equal to the wavelength $\lambda_{source} = v_{sound}/f_{source}$. The crests arrive at the observer spaced in time by $\lambda_{source}/v_{sound} = 1/f_{source}$, so one wave crest arrives at the observer every $1/f_{source}$ seconds. In other words, the frequency *as heard by the observer* is equal to f_{source}. The frequency the observer measures thus equals the frequency emitted by the siren, as you should have expected.

Let's now use the same approach to analyze the case of a siren moving toward the observer. The siren emits sound waves whose crests again travel toward the observer. These wave crests move at a speed v_{sound} *relative to the medium*, which in this case is the air. Once a sound wave leaves the source, the wave's motion is determined by the

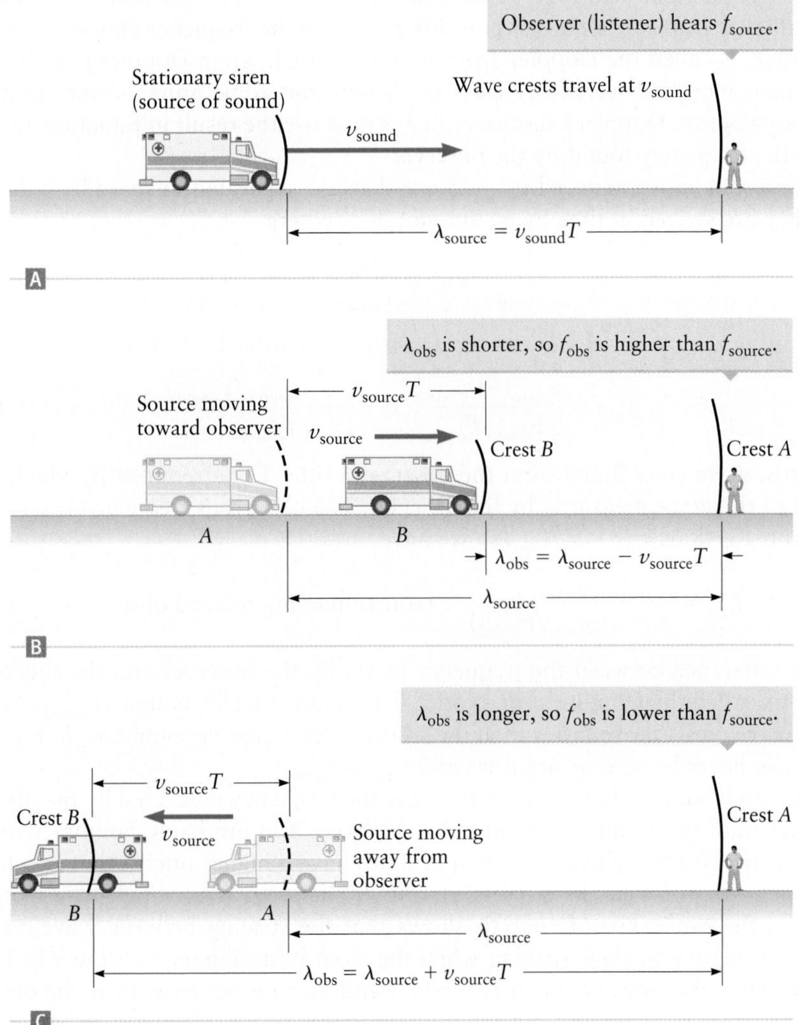

◀ **Figure 13.16** The Doppler effect. A When a source of sound is at rest relative to a stationary observer, the frequency heard by the observer is equal to the frequency emitted by the source. B When the source is moving toward the observer, the observed frequency is higher than the emitted frequency. C When the source is moving away from the observer, the observed frequency is lower than the emitted frequency.

medium; the observer in Figure 13.16B is stationary relative to the air, and he will find that the wave crests move at v_{sound} no matter what speed the source may have.

We must now account for the motion of the siren between the emission of each wave crest. Figure 13.16B shows two successive wave crests along with the position of the siren when it emitted these crests. This snapshot shows the siren at point B, having just emitted the wave crest also labeled B. The previous wave crest labeled A was emitted when the siren was at location A, one period earlier. The frequency of the siren, as would be measured by someone sitting in the ambulance, is still f_{source}, so the siren emits wave crests with the same period $T = 1/f_{\text{source}}$ as when the ambulance is not moving. Wave crest A has traveled for this amount of time at speed v_{sound}, so it has traveled a distance $v_{\text{sound}}T = v_{\text{sound}}/f_{\text{source}} = \lambda_{\text{source}}$ from the point where it was emitted.

Let's call the speed of the ambulance (and attached siren) v_{source}. By the time the siren emits wave crest B, it has traveled a distance $v_{\text{source}}T$ since emitting crest A. The spacing between crest A and crest B is thus smaller than was the case with a stationary siren. From Figure 13.16B, we see that the new distance between crests is

$$\lambda_{\text{obs}} = v_{\text{sound}}T - v_{\text{source}}T$$

The period T is equal to $1/f_{\text{source}}$, so

$$\lambda_{\text{obs}} = \frac{v_{\text{sound}} - v_{\text{source}}}{f_{\text{source}}} \tag{13.17}$$

Because the distance between crests is smaller when the ambulance moves toward the observer than when it is stationary, the crests arrive more frequently at the observer and the frequency as *measured by the observer* is *higher*. This phenomenon, in which the frequency heard by an observer is different from the frequency emitted by a moving source, is called the ***Doppler effect*** in honor of Christian Doppler (1803–1853), the person who discovered it. (We'll see below that some animals were using the effect long before Doppler's discovery.) Let's now use the result in Equation 13.17 to derive the frequency found by the observer.

Doppler effect

Wave crests move at speed v_{sound}. Since they are separated in space by a distance λ_{obs}, they are spaced in time by an amount T_{obs}, where

$$T_{\text{obs}} = \frac{\lambda_{\text{obs}}}{v_{\text{sound}}}$$

Substituting the observed wavelength λ_{obs} from Equation 13.17 leads to

$$T_{\text{obs}} = \frac{v_{\text{sound}} - v_{\text{source}}}{v_{\text{sound}}\, f_{\text{source}}} = \frac{1 - (v_{\text{source}}/v_{\text{sound}})}{f_{\text{source}}} \tag{13.18}$$

In words, wave crest B arrives at the observer a time T_{obs} after crest A, which is the period of the wave measured by the observer. The observer's frequency is related to Equation 13.18 by $f_{\text{obs}} = 1/T_{\text{obs}}$, so we have

$$f_{\text{obs}} = \frac{f_{\text{source}}}{1 - (v_{\text{source}}/v_{\text{sound}})} \quad \text{(source moving toward observer)} \tag{13.19}$$

The difference between the frequency heard by the observer and the source frequency is called the ***Doppler shift***. Notice that Figure 13.16B assumes $v_{\text{sound}} > v_{\text{source}}$, so the wave crests move faster than the source. Hence, the denominator in Equation 13.19 can never be zero or negative, and $f_{\text{obs}} > f_{\text{source}}$.

This result for f_{obs} shows that in this case the frequency measured by the observer is higher than the frequency emitted by the siren. You are likely familiar with this result; a siren traveling toward you appears to have a higher pitch—that is, a higher frequency—than when the siren is at rest. If the siren is moving away from the observer, the same type of analysis shows that the spacing between wave crests is increased relative to their spacing when the siren is stationary, as shown in Figure 13.16C. Here the source emits wave crest A and then moves away from the observer by a distance $v_{\text{source}}T$ before emitting the next wave crest B. The distance between the adjacent wave crests is now increased, causing the frequency heard by the observer to be lower. Taking the same approach that led to Equation 13.19, we find that the result for f_{obs} in this case is

$$f_{obs} = \frac{f_{source}}{1 + (v_{source}/v_{sound})} \quad \text{(source moving away from observer)} \tag{13.20}$$

Let's estimate the size of the Doppler shift for a typical case. We assume the source is a siren whose frequency when at rest is $f_{source} = 500$ Hz. Suppose the siren is on a fire truck racing to an emergency at a speed of 27 m/s (about 60 mi/h). For a listener standing on a nearby sidewalk, what frequency does the siren have? If the truck is moving toward the observer, the Doppler-shifted frequency is given by Equation 13.19, where $v_{source} = 27$ m/s. Inserting these values along with $v_{sound} = 343$ m/s into Equation 13.19 gives

$$f_{obs} = \frac{f_{source}}{1 - (v_{source}/v_{sound})} = \frac{500 \text{ Hz}}{1 - [(27 \text{ m/s})/(343 \text{ m/s})]} \tag{13.21}$$

$$f_{obs} = 540 \text{ Hz}$$

Hence, the Doppler shift is $f_{obs} - f_{source} = 40$ Hz, which is nearly 10% of the source frequency. In terms of notes on a musical scale, this value is approximately the separation between two consecutive whole notes (i.e., "do" and "re").

So far, we have considered the Doppler effect for a stationary observer. Motion of the observer also gives a Doppler shift. In the most general case, in which the observer has a speed v_{obs} and the source a speed v_{source}, the observed frequency is

$$f_{obs} = f_{source}\left[\frac{1 \pm (v_{obs}/v_{sound})}{1 \pm (v_{source}/v_{sound})}\right] \tag{13.22}$$

Here again the speeds v_{obs} and v_{source} are measured relative to the medium, and both these speeds are assumed to be less than v_{sound}. The choices of plus and minus signs in Equation 13.22 depend on whether the source and observer are moving toward or away from each other. The signs can always be remembered by noting that the frequency measured by the observer f_{obs} is *increased when the observer and source are moving toward each other*, whereas f_{obs} is *decreased when they move away from each other*.

EXAMPLE 13.7 Doppler Shift, Coming and Going

A fire truck approaches you from one direction, passes by, and then continues on away from you, with the siren emitting sound the entire time. You have a friend (a music student) with a very keen sense of pitch who tells you that the frequency of the siren drops by precisely 50 Hz when the truck passes your location on the street. If the truck has a speed of 20 m/s, what is the frequency of the siren at rest? Take the speed of sound to be $v_{sound} = 343$ m/s.

RECOGNIZE THE PRINCIPLE

This problem is similar to our analysis of the Doppler shift for a moving truck (Eq. 13.21), but now we have to take into account *two* Doppler shifts, one as the truck moves toward you and another as it moves away from you. The observed frequency will be shifted upward to a higher frequency when the truck is moving toward you and downward to a lower frequency after the truck passes. Your friend has used his sense of pitch to estimate the difference between the "coming" and "going" frequencies.

SKETCH THE PROBLEM

The problem is described by parts B and C of Figure 13.16.

(continued) ▸

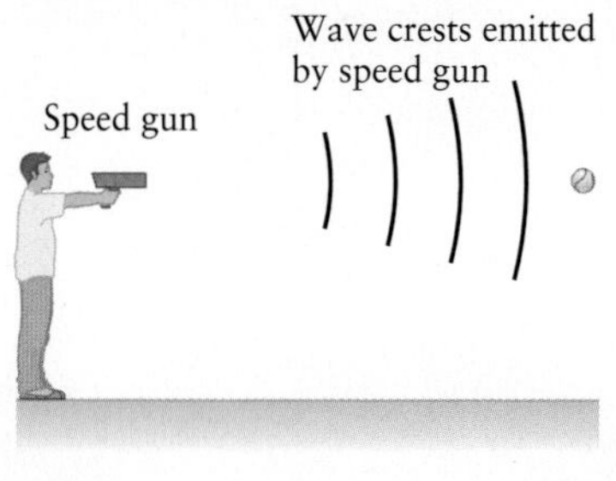

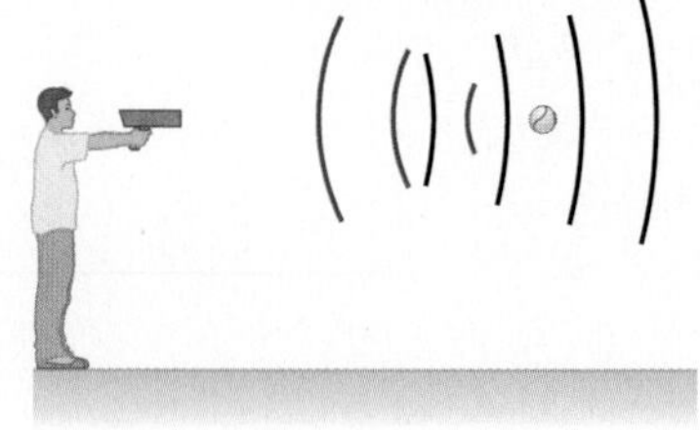

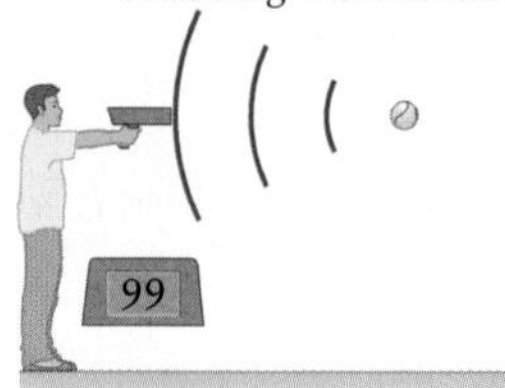

▲ **Figure 13.17** A "speed gun" (sometimes also called a "radar gun") uses the Doppler effect to measure the speed of a moving object. A Waves are emitted by the speed gun and strike the object. B These waves scatter from the object, and some of the scattered waves return to the speed gun. C The speed gun measures the difference between the frequencies of the emitted and scattered waves.

IDENTIFY THE RELATIONSHIPS

When the truck is approaching, the observed frequency is (from Eq. 13.19)

$$f_{\text{coming}} = \frac{f_{\text{source}}}{1 - (v_{\text{source}}/v_{\text{sound}})}$$

and as it moves away the frequency is (from Eq. 13.20)

$$f_{\text{going}} = \frac{f_{\text{source}}}{1 + (v_{\text{source}}/v_{\text{sound}})}$$

Your friend senses the difference $\Delta f = f_{\text{coming}} - f_{\text{going}}$, which is given by

$$\Delta f = \frac{f_{\text{source}}}{1 - (v_{\text{source}}/v_{\text{sound}})} - \frac{f_{\text{source}}}{1 + (v_{\text{source}}/v_{\text{sound}})} \tag{1}$$

SOLVE

We now want to solve for f_{source}. Rearranging Equation (1) leads to

$$\Delta f = f_{\text{source}}\left[\frac{1}{(1 - v_{\text{source}}/v_{\text{sound}})} - \frac{1}{1 + (v_{\text{source}}/v_{\text{sound}})}\right]$$

$$f_{\text{source}} = \frac{\Delta f}{\left[\frac{1}{(1 - v_{\text{source}}/v_{\text{sound}})} - \frac{1}{1 + (v_{\text{source}}/v_{\text{sound}})}\right]}$$

Inserting the given values of Δf, v_{source}, and v_{sound}, we get

$$f_{\text{source}} = \frac{50 \text{ Hz}}{\left[\frac{1}{1 - [(20 \text{ m/s})/(343 \text{ m/s})]} - \frac{1}{1 + [(20 \text{ m/s})/(343 \text{ m/s})]}\right]} = \boxed{430 \text{ Hz}}$$

▸ ***What does it mean?***

The Doppler shift can also be used to measure the speed of the source, as we explore in the next subsection.

Speed Guns, Bats, and the Doppler Effect

So far, we have discussed the Doppler effect using the example of sound waves from a moving siren, but the Doppler effect applies to other types of waves as well. For example, a "speed gun" used in sporting events such as baseball (Fig. 13.17) uses the Doppler effect with electromagnetic waves to measure the speed of a moving object. Electromagnetic waves from the speed gun are scattered by the ball, and some of the scattered waves then return to the speed gun. (Compare with the scattered waves in Fig. 13.15.) As the waves strike it, the ball acts like a moving "observer," whereas for the scattered waves, the ball acts like a moving source. For the case shown in Figure 13.17, with the ball moving toward the observer, the ball's motion causes the scattered wave crests to be more closely spaced than if the ball were at rest, just as we found with the moving siren in Figure 13.16B, and there is again a Doppler shift. The speed gun uses the Doppler-shifted frequency of the scattered waves to deduce v_{source}, the speed of the ball.

▲ **Figure 13.18** Some bats use reflected sound to judge the location and motion of nearby objects, including their prey.

Bats employ an approach similar to that used by the speed gun to help them navigate as they fly (Fig. 13.18). Bats generate pulses of sound waves at frequencies well above the range of human hearing (in the ultrasonic range). These ultrasonic sound pulses are reflected or scattered by objects in the bat's environment, and the bat uses the reflected sound to judge the location and motion of nearby objects. Bats have a very keen sense of timing and are able to determine accurately the time delay between an emitted pulse and its reflection. A long reflection time means the object is far away, and a small reflection time means the object is nearby (similar to the way radar works as discussed in Chapter 12). In addition to the reflection time, some bats also use the Doppler effect.[2] Suppose the bat's ultrasonic sound wave is reflected by a moth moving through the air.[3] The reflected sound is Doppler-shifted, and the bat uses this frequency shift to sense the moth's speed and obtain a meal.

Moving Sources and Shock Waves

In all our examples involving moving sources of sound, the speed of the source has (so far) been much less than the speed of sound, but it is also possible for the speed of a wave's source to be greater than the speed of the wave. For example, a jet can fly faster than the speed of sound. Figure 13.19 shows the wave crests of sound emitted by a jet in three cases. Figure 13.19A shows what happens when the jet's speed is less than the speed of sound; this arrangement of wave crests is the one we considered in our analysis of the Doppler effect in Figure 13.16. Figure 13.19B shows what happens when the jet's speed is equal to the speed of sound. The jet now travels at the same speed as the wave crests, so the crests in front are not able to move away from it. The wave crests at the front of the jet thus "pile up" at this point. According to the principle of superposition, the total sound pressure is the sum of the pressures due to the individual sound waves, so there is a very large sound intensity at the front of the jet. Figure 13.19C shows the situation when the jet moves faster than the speed of sound. Sound waves emitted at past times pile up along a conical envelope (which appears as two sides of a triangle in the projection in Fig. 13.19C). At any point on this envelope, many different sound waves, emitted at many different previous times, arrive simultaneously, producing a very large sound intensity and forming a ***shock wave***. Shock waves are the source of the ***sonic booms*** that can be heard when a supersonic jet passes overhead.

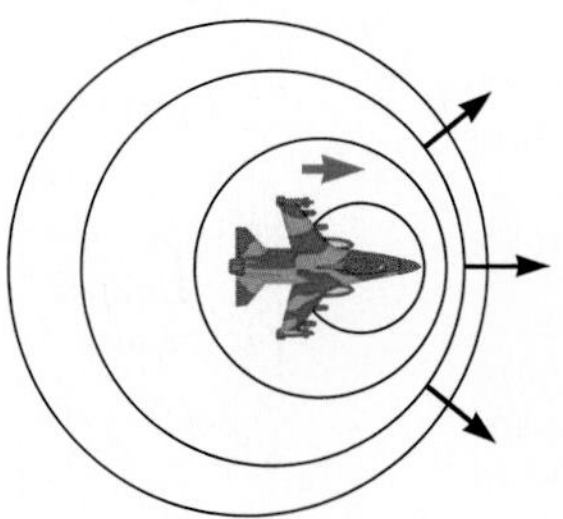

$v_{jet} < v_{sound}$
The jet does not keep up with the wave crests.
A

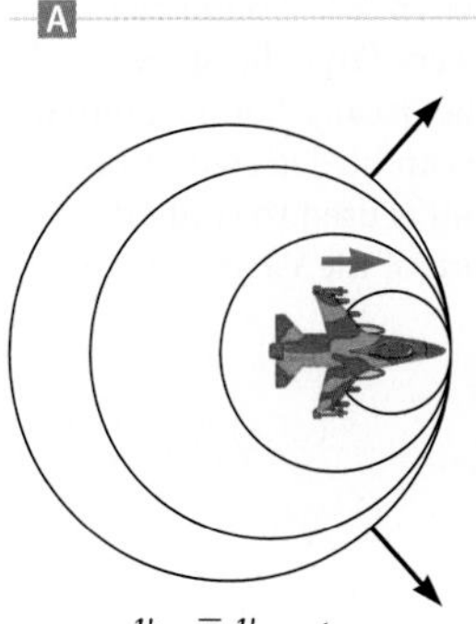

$v_{jet} = v_{sound}$
The jet keeps up with the wave crests, so the wave crests "pile" up.
B

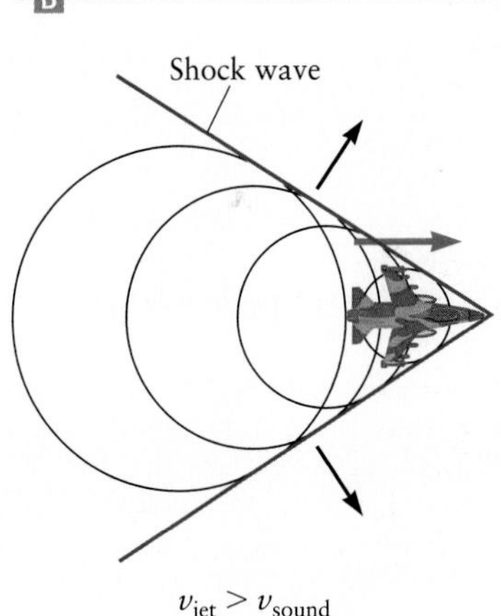

▲ **Figure 13.19** Wave fronts emitted by a moving source. A The speed of the source is less than the speed of sound, leading to the Doppler effect (Fig. 13.16). B When the speed of the source is equal to the speed of sound, the leading part of the wave front never "escapes" from the source. C When the speed of the source is greater than the speed of sound, shock waves are created. The direction of the shock wave depends on the ratio v_{source}/v_{sound}.

13.7 Applications

Applications involving sound are found in many places, including the home (such as the earphones of your MP3 player), in scientific research, and in medicine.

Using Sound to Study Global Warming

The speed of sound depends on a material's temperature. Table 13.1 lists v_{sound} in some typical materials at room temperature. For example, sound travels through salt water with a typical speed $v_{sound} \approx 1500$ m/s, and this value increases by approximately 4.0 m/s when the temperature is increased by 1°C. This effect is now being used in studies of the oceans. Many scientists believe that the Earth's average surface temperature is increasing, an effect known as global warming. Because changes in the Earth's average temperature as small as 1°C can greatly affect life on this planet, it is very important to confirm that such climate change is actually occurring and, if so, to measure the rate of warming. Studies of global warming are very difficult due to natural fluctuations in the Earth's temperature from place to place and from year to year. For example, the fluctuations of the average temperature over the surface of the entire Earth are typically 0.2°C from one year to the next.

A clever approach called "acoustic thermometry" has been developed to study global warming. This approach uses the temperature dependence of the speed of sound to measure changes in the temperature of the oceans. Special sound generators and detectors have been developed to send sound waves through the oceans across very large distances. For example, sound generated in the Pacific Ocean near Hawaii

[2]You can learn more about bats and the Doppler effect in C. F. Moss and S. R. Sinha, "Neurobiology of echolocation in bats," *Current Opinion in Neurobiology* 13:751–758 (2003).

[3]Because the bat is also moving, there is also a Doppler effect due to the bat's motion. The bat is able to compensate for Doppler shifts caused by its motion and isolate the shift due solely to the moth's motion.

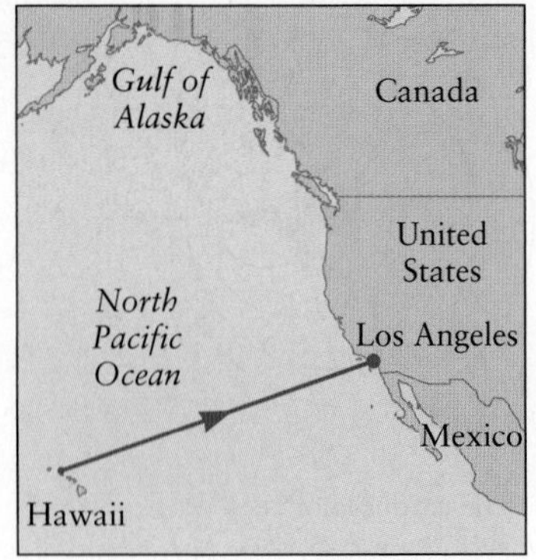

▲ **Figure 13.20** Sound can travel for very large distances through the oceans. Sound emitted in Hawaii can be detected in Los Angeles and is used to study the temperature of the Pacific Ocean.

can be detected when it reaches the coast of California (Fig. 13.20). The speed of sound in salt water is approximately 1500 m/s and the distance from Honolulu to Los Angeles is about 4100 km, so it takes about

$$t = \frac{L}{v_{\text{sound}}} = \frac{4.1 \times 10^6 \text{ m}}{1500 \text{ m/s}} = 2700 \text{ s} \approx 45 \text{ min}$$

for sound to travel this distance. The actual travel time will depend on the average temperature of seawater over this distance. Global warming would produce changes in this average temperature, so acoustic thermometry is ideal for studies of global warming. The travel time for sound can be measured very accurately; current precision approaches 20 ms (0.02 s), making it possible to measure very accurately how the average ocean temperature varies over the course of days and years. These experiments are ongoing and should yield important clues for understanding climate change.

EXAMPLE 13.8 Temperature Changes Measured by Acoustic Thermometry

What is the smallest temperature change that can be measured using acoustic thermometry? Assume the sound generator and detector are located near Honolulu and Los Angeles and are thus separated by $L = 4100$ km, and assume the sound travel time can be measured with an accuracy of 0.02 s. The speed of sound in seawater varies by 4.0 m/s for each 1.0°C change in temperature.

RECOGNIZE THE PRINCIPLE

The travel time t is related to the travel distance and the speed of sound by $t = L/v_{\text{sound}}$. To change the travel time t by 0.02 s, there must be a corresponding change in v_{sound}. This change in v_{sound} can then be used to find the change in temperature.

SKETCH THE PROBLEM

No figure is needed.

IDENTIFY THE RELATIONSHIPS

On an "average" day, the travel time for this example is approximately

$$t = \frac{L}{v_{\text{sound}}} = \frac{4.1 \times 10^6 \text{ m}}{1500 \text{ m/s}} = 2700 \text{ s}$$

Now suppose the experiment is performed on two consecutive days and the travel times are measured to be $t_1 = 2700.00$ s and $t_2 = 2700.02$ s. We can rearrange the relation between t and v_{sound} to solve for the value of v_{sound} on each day. The results are

$$v_1 = \frac{L}{t_1} = \frac{4.1 \times 10^6 \text{ m}}{2700.00 \text{ s}} = 1518.519 \text{ m/s}$$

and

$$v_2 = \frac{L}{t_2} = \frac{4.1 \times 10^6 \text{ m}}{2700.02 \text{ s}} = 1518.507 \text{ m/s}$$

Here we have kept extra significant figures because we now need to compare v_1 and v_2. The difference in these velocities is

$$v_1 - v_2 = 0.012 \text{ m/s}$$

SOLVE

This velocity difference corresponds to a temperature change ΔT of

$$\Delta T = (0.012 \text{ m/s}) \times \frac{1.0°\text{C}}{4.0 \text{ m/s}} = \boxed{0.003°\text{C}}$$

▸ ***What does it mean?***
Acoustic thermometry can thus be used to measure *extremely* small changes in the average temperature of the ocean. By placing the transmitters and receivers in different locations, the temperature in different ocean regions can be measured.

Imaging with Ultrasound

Another important application of sound is the technique of ***ultrasonic imaging***. This technique uses sound waves to obtain images inside a material. The material could be virtually anything (such as an airplane wing), but the most familiar use of ultrasonic imaging produces views inside the human body. An instrument that generates sound is placed on the skin (Fig. 13.21A), and sound waves propagate into the body. A portion of the sound energy is reflected by bones and other tissue, and the reflected waves are used to construct an image (Fig. 13.21B). The depth of the bone (or whatever causes the reflection) is determined by the time it takes the sound to travel from the generator to the bone and then return to the detector.

Some medical ultrasonic imaging devices also use the Doppler effect. When ultrasound is used to construct an image of a blood vessel, sound waves are reflected by cells moving in the blood. The frequency of these reflected sound waves is Doppler-shifted as we found with the speed gun in Figure 13.17. By measuring this Doppler shift, physicians can measure the blood velocity. This information can be used to determine if blood is flowing properly and can reveal the presence of blockages or constrictions.

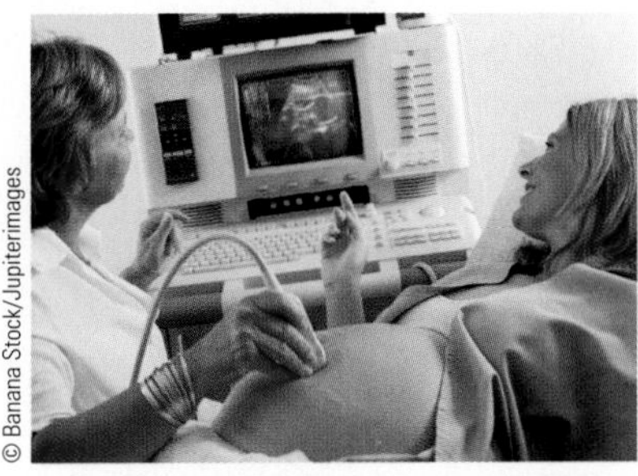
© Banana Stock/Jupiterimages

A

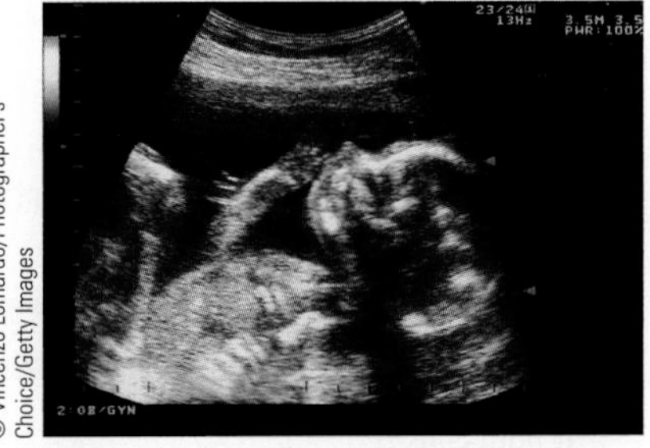
© Vincenzo Lomardo/Photographer's Choice/Getty Images

B

▲ **Figure 13.21** Ultrasonic imaging (ultrasound) uses reflected sound waves to reveal structures inside living tissue. A A transducer (a device that can both emit and detect sound) is placed on the abdomen of a pregnant woman. B Reflected waves are detected by the transducer and used to construct an image of the fetus.

CONCEPT CHECK 13.7 Ultrasonic Imaging

A typical ultrasonic imaging device employs sound with a frequency of 10 MHz. If the speed of sound in tissue is approximately 1500 m/s, is the wavelength of these sound waves (a) 1.5 mm, (b) 0.15 mm, or (c) 1.5 cm?

Summary | CHAPTER 13

Key Concepts and Principles

Sound is a longitudinal wave

The molecular motion associated with a sound wave is parallel to the wave velocity, so sound is a longitudinal wave. Sound can also be viewed as a ***pressure wave***. Regions of high pressure are regions of condensation, whereas low-pressure regions are regions of rarefaction. The frequency and wavelength of a periodic sound wave are related by

$$v_{\text{sound}} = f\lambda \qquad \text{(13.1) (page 363)}$$

The speed of sound depends on the properties of the medium and is generally high in solids and low in gases. The speed of sound in air at atmospheric pressure and room temperature is approximately 343 m/s.

Human hearing

The normal range of human hearing is approximately 20–20,000 Hz. Sounds with lower frequencies are called ***infrasonic***, and sounds with higher frequencies are ***ultrasonic***.

(*Continued*)

Sound intensity

The ***intensity*** of a sound wave is proportional to the square of the amplitude. Typically, the human ear can hear sounds with intensities as small as 1×10^{-12} W/m^2, which corresponds to a pressure amplitude of about 2×10^{-5} Pa. Sound intensity is often measured in ***decibels*** (dB). The intensity level of a sound in decibels is given by

$$\beta = 10 \log\left(\frac{I}{I_0}\right) \qquad \textbf{(13.9) (page 367)}$$

where $I_0 = 1.00 \times 10^{-12}$ W/m^2. The human ear can function (without damage) at intensity levels up to about 120 dB.

Applications

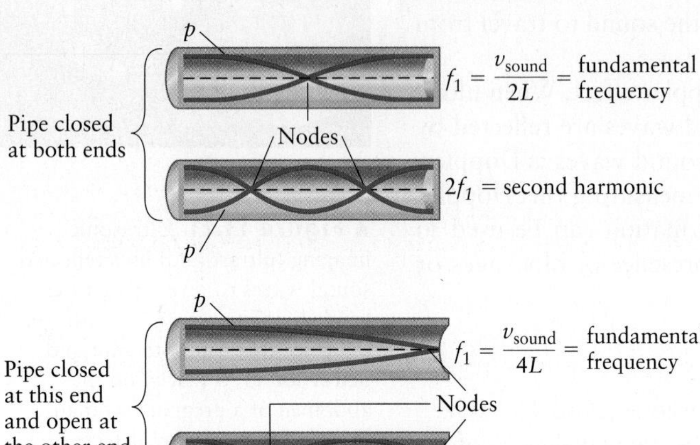

Sound in pipes

Sound propagating back and forth within a pipe forms ***standing waves*** when the wavelength of the sound (or a multiple of the wavelength) matches the length of the pipe. Standing waves have pressure ***nodes*** where the pressure amplitude is zero and pressure ***antinodes*** where the pressure amplitude is largest. The open end of a pipe is always a pressure node, while the closed end is always a pressure antinode.

For a pipe closed at both ends, the possible standing waves have frequencies

$$f_n = n\frac{v_{sound}}{2L} = nf_1 \quad \text{with } n = 1, 2, 3, \ldots \qquad \textbf{(13.11 and 13.12) (page 371)}$$

where $f_1 = v_{sound}/(2L)$ is called the ***fundamental frequency***, the lowest possible standing wave frequency. For a pipe that is closed at one end and open at the other (as in many musical instruments), the standing wave frequencies are

$$f_n = n\frac{v_{sound}}{4L} = nf_1, \quad \text{with } n = 1, 3, 5, \ldots \qquad \textbf{(13.14 and 13.15) (page 372)}$$

with $f_1 = v_{sound}/(4L)$.

Beats

When two sound waves that have nearly the same frequency are combined, a listener will hear ***beats***. Beats are an oscillation of the overall sound intensity at a beat frequency equal to $|f_1 - f_2|$, where f_1 and f_2 are the frequencies of the two sound waves.

Doppler shift

When the source of a sound is moving, the frequency observed by a listener is ***Doppler-shifted***. If the frequency of the source when stationary is f_{source}, then when the source moves at speed v_{source} and the listener moves at speed v_{obs}, the frequency heard by the listener is

$$f_{obs} = f_{source}\left[\frac{1 \pm (v_{obs}/v_{sound})}{1 \pm (v_{source}/v_{sound})}\right] \qquad \textbf{(13.22) (page 379)}$$

The choices of plus and minus signs in Equation 13.22 depend on whether the source and observer are moving toward or away from each other. When using Equation 13.22, choose the signs so that the frequency measured by the observer f_{obs} is increased when the observer and source are moving toward each other, and f_{obs} is decreased when they move away from each other.

Questions

SSM = answer in *Student Companion & Problem-Solving Guide* ⓧ = life science application

1. Rank each substance according to how fast sound travels in that medium, from that with the largest velocity to that with the slowest.
(a) Air
(b) Water
(c) Steel
(d) Aluminum

2. ⓧ A typical dog can hear sounds with frequencies as high as 45 kHz. Assuming they can hear sounds as low as 40 Hz, make a qualitative sketch of the sensitivity curve for a dog (i.e., the canine version of the dashed curve in Fig. 13.4).

3. Musicians who play the guitar and other stringed instruments use beats to tune their instruments. The value of the beat frequency tells them the difference in the frequencies of two strings, but it does not tell them which one is higher (or lower). Describe an experiment musicians could do to resolve this question.

4. Two sound waves of equal frequency interfere constructively at a listener. Which of the following statements is true?
(a) The frequency is increased by the interference.
(b) The sound amplitude is increased by the interference.
(c) The listener will hear beats.

5. How is it possible for a singer to shatter a wine glass by singing a note? Which is more important in this process, the frequency or intensity of the singing, or are they both important?

6. In this chapter, we discussed the Doppler effect that occurs when a source of sound moves directly toward or away from a listener. Do you think there will be a Doppler shift if the velocity of the source is perpendicular to a line that runs to the listener (Fig. Q13.6)? Explain why or why not.

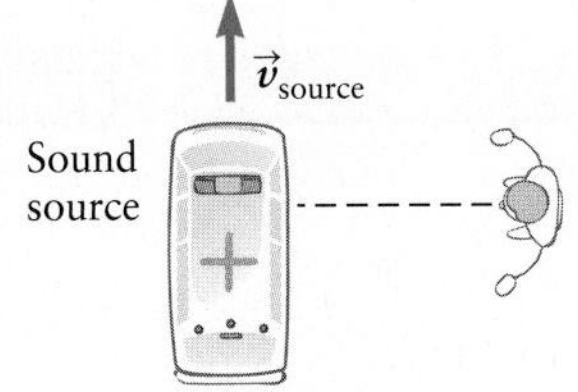

Figure Q13.6

7. ⓧ Explain why the sounds emitted by large animals usually have a lower frequency than those from small animals. *Hint*: Consider the standing waves in the vocal tract.

8. ⓧ **Ocean in a shell.** What is it that you hear when you hold a seashell (or any closed chamber like a cup or jar) to your ear? We are almost always surrounded by sound that spans every frequency (white noise). What happens if you cup your hands around an ear and vary the size of the chamber? The pitch should change as you vary the chamber. What are you hearing and why?

9. Consider the method of acoustic thermometry using the ocean. In Example 13.8, we saw that very small changes in the travel time of sound can be detected and can be used to observe very small changes in the ocean temperature. There are, however, complications that we did not mention in that example. It turns out that the velocity of sound in water also depends on the salinity of the water. Ocean water contains about 3.5% salt by volume, and the speed of sound changes by about 1.4 m/s for each 0.1% change in the salt concentration. Investigate (using the Internet or some other reference) how much the salinity of ocean water fluctuates over time and discuss how much this fluctuation will affect the measurement of temperature changes with this technique.

10. Use the general relation for the speed of sound in Equation 13.3 together with the ideal gas law (see Chapter 15 if you need a refresher on the ideal gas law) to show that the speed of sound in a gas can be written as $v_{sound} = \sqrt{P/\rho}$ where P is the pressure. Can you give an intuitive explanation of why v_{sound} increases as the pressure is increased?

11. The first television remote controllers (invented in the 1950s) used ultrasonic waves. Some people claimed they could change channels by simply jiggling their keys. Is that possible? Explain why or why not.

12. SSM ⓧ A buzzer generates sound by vibrating with a frequency of 440 Hz in air. If this buzzer is placed underwater, what frequency would a fish hear? How does placing the buzzer underwater affect the wavelength of the sound?

13. One day, while hiking in a particular spot in your favorite cave, you notice that it takes a certain time t for an echo to return to you. The next day, you find that the echo time at the same spot is shorter. How might that happen?

14. When a police car is at rest, its siren has a frequency of 750 Hz. You are a thief, and just after finishing a night's work, you hear a siren with a frequency of 700 Hz. Your fellow thief tells you to relax and assures you that there is nothing to worry about. Is she right?

15. A radio produces sound waves in air, but they can also travel from air into water so that they can be heard by a swimmer. Is the frequency of the sound in air the same or different from the frequency in water? Is the wavelength the same or different? Explain.

16. ⓧ The manufacturer of a set of earplugs advertises that they will reduce the sound intensity level by 40 dB. By what factor do they reduce the intensity?

17. SSM When you blow across the top of a cola bottle, you can generate sound (Figure Q13.17). As you drink more and more of the cola and lower the liquid level, how does the frequency of the sound change?

Figure Q13.17

18. When you tune a flute or trombone, what causes the frequency to change?

19. Noise-canceling headphones make use of destructive interference of sound waves to eliminate "noise," while still playing music. Explain how they work.

20. If the note middle C is played on trumpet and on a piano, you can identify them as the same note but can certainly tell the difference. What feature of these sounds tells you that they are the same note? What feature of these sounds tells you that they come from different instruments?

Problems available in ENHANCED WebAssign

Section 13.1 Sound Is a Longitudinal Wave
Problems 1–13

Section 13.2 Amplitude and Intensity of a Sound Wave
Problems 14–30

Section 13.3 Standing Sound Waves
Problems 31–44

Section 13.4 Beats
Problems 45–50

Section 13.5 Reflection and Scattering of Sound
Problems 51–52

Section 13.6 The Doppler Effect
Problems 53–61

Section 13.7 Applications
Problems 62–63

Additional Problems
Problems 64–76

List of Enhanced Problems

Problem number	Solution in *Student Companion & Problem-Solving Guide*	Reasoning & Relationships Problem	Reasoning Tutorial in ENHANCED WebAssign
13.8	✓		
13.17	✓	✓	
13.22		✓	✓
13.30	✓		
13.36	✓		
13.40		✓	✓
13.42		✓	
13.43		✓	✓
13.44		✓	✓
13.47	✓		
13.50		✓	✓
13.51		✓	✓
13.52	✓		
13.54	✓		
13.62		✓	
13.63	✓		
13.66	✓		
13.73		✓	✓
13.76		✓	✓

CHAPTER

14

► *Although she may not be thinking about it, temperature and heat are important for this camper! (© Extreme Sports Photo/Alamy)*

Temperature and Heat

OUTLINE

All our work on mechanics in Chapters 2 through 13 has been based on Newton's laws of motion. The central thread of that work was *force*. We discussed various types of forces (such as gravity) and how these forces cause a particle to move in various situations. A careful consideration of Newton's laws led us to other important concepts such as work and energy, but the main theme was the concept of force. In this chapter and Chapters 15 and 16, we move to a different area of physics known as ***thermodynamics***, and our focus now shifts to *energy*. The key pillar of thermodynamics is the principle of conservation of energy. Although conservation of energy certainly has an important place in mechanics, it will now play an even more central role. In a nutshell, thermodynamics is about the transfer of energy between *systems* of particles and about the way changes in the energy of a system affect its properties.

For thermodynamics, we need several new quantities to describe the properties of systems and their interactions. Two of these quantities—temperature and heat—are the main subjects of this chapter.

14.1 Thermodynamics: Applying Physics to a "System"

A thermodynamic *system* contains multiple particles, usually a very large number of particles. Examples are a balloon full of gas molecules, a drop of water, and a sugar cube, each of which contains many atoms and molecules. The particles in a system are able to exchange energy with one another (via collisions), and systems are able to exchange energy with other systems.

Figure 14.1A shows a balloon full of oxygen molecules. What quantities are useful for describing the physics of this system? We could describe its properties by specifying the position and velocity of every oxygen molecule. We would then want to determine the forces exerted by each molecule in the balloon on all the other molecules, along with all the external forces (e.g., the forces from outside the system such as gravity). With all this information, we could use Newton's laws to calculate the motion of each individual oxygen molecule. Although we could take this approach in principle, it is not feasible in practice because of the extremely large number of atoms or molecules in a typical sample of gas. Moreover, even if it were feasible, we are really not interested in this level of detail. We can describe all the important properties of this system in a much simpler way.

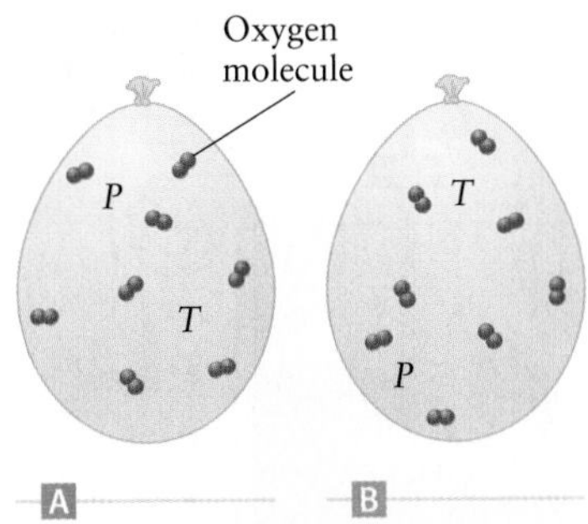

▲ **Figure 14.1** These two hypothetical balloons have the same volume and gas density. Two such systems having equal density, pressure, and temperature are said to be in the same state, even though the precise positions and velocities of the gas molecules are different.

Figure 14.1B shows a second balloon that also contains oxygen gas. The two balloons in Figure 14.1 are "identical" in that they are the same size and shape and contain the same number of molecules. The positions and velocities of the molecules in the two balloons will certainly not be the same, however, so these two systems will not be identical on the molecular scale. Fortunately, we are usually not interested in knowing precisely where all the molecules are located. It is much more useful to know the values of various properties of the system *as a whole*.

To describe the physics of a system fully, only a small number of systemwide properties such as pressure and temperature are needed. Also called ***macroscopic*** properties, they describe the behavior on a scale that is much larger than the scale of an individual particle. Macroscopic properties contrast with ***microscopic*** variables such as the position and velocity of a particular molecule within the system. Newton's laws provide a description at the level of these microscopic variables. Our task in this chapter is to describe a system of many particles, such as a gas-filled balloon (Fig. 14.1) or a piece of steel, at the macroscopic level. We also explore the connection between microscopic and macroscopic descriptions and use Newton's laws to gain insight into the meanings of the macroscopic properties.

14.2 Temperature and Heat

Temperature is not contained in or derivable from Newton's laws, which is one reason it is not easy to give a "simple" definition of temperature in the same way we gave definitions of velocity and acceleration in our work on mechanics. We can approach the notion of temperature in several different ways, each yielding useful insights.

Temperature is connected with the "hotness" or "coldness" of a system, so let's first consider a macroscopic definition. Figure 14.2A shows two hypothetical systems, which might be balloons filled with gas. If each system is initially isolated from its surroundings for a long period of time, each will have its own temperature. In general, these two initial temperatures will not be the same, and we denote them by T_1 and T_2. Suppose $T_1 > T_2$; in ordinary language, we would say that system 1 is initially hotter than system 2. If we then bring the systems into contact (Fig. 14.2B), energy is transferred spontaneously from system 1 (the hotter system) to system 2 (the colder system), and their temperatures will change. If we wait for a long time (Fig. 14.2C), the two systems will reach ***thermal equilibrium***. They then have the

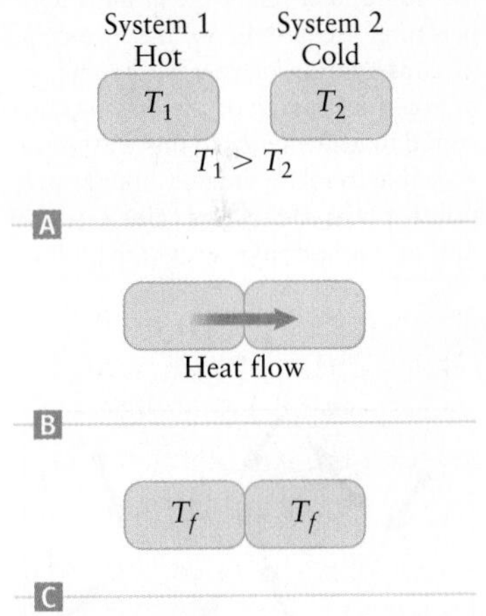

▲ **Figure 14.2** Heat (also referred to as heat energy) is energy transferred from one system to another due to a temperature difference. A Two systems at different temperatures. B Energy flows from the hotter system at temperature T_1 to the cooler system at temperature T_2 when they are brought into contact. C If the systems are left in contact for a long time, they will eventually come into thermal equilibrium and have the same final temperature.

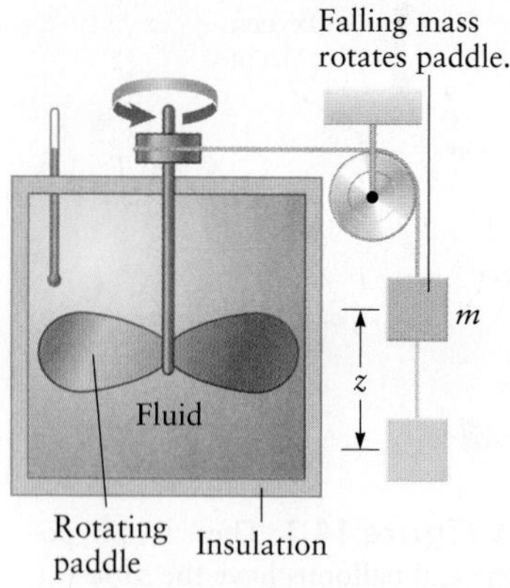

▲ **Figure 14.3** Apparatus used to measure the relation between heat energy and mechanical energy. See Insight 14.1.

Insight 14.1
THE EQUIVALENCE OF MECHANICAL ENERGY AND HEAT ENERGY

In the 1840s, James Joule measured the "mechanical equivalent" of heat energy with an apparatus like the one shown in Figure 14.3. As a mass falls through a height z, its potential energy is used to rotate a propeller in a liquid. The propeller's motion heats the liquid, increasing the liquid's temperature. According to the principle of conservation of energy, the change in potential energy of the mass is converted to heat energy. In this way, Joule was able to relate the unit used at that time for heat energy (the calorie) to the unit of mechanical energy (the joule).

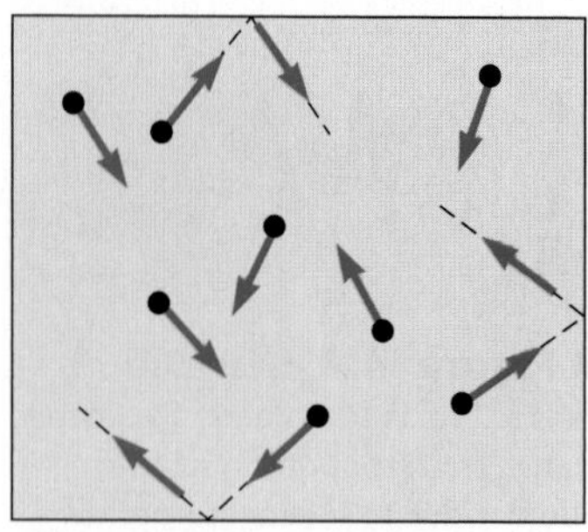

▲ **Figure 14.4** The pressure in a gas comes from collisions of gas molecules with the surface of the container. Each collision produces a force on the container walls. If the temperature is increased, the molecular speed increases, causing the collision forces and hence the pressure to increase.

same final temperature T_f, which lies somewhere between T_1 and T_2. The energy that flows between the two systems is called ***heat*** or ***heat energy***.

> ***Heat*** **is energy that passes from one system to another by virtue of a temperature difference.**

The terms *heat* and *heat energy* are often used interchangeably, especially in everyday conversation. When used in physics, these terms always refer to the transfer of energy between systems. Hence, you may also hear the term *heat flow* used to describe the process by which heat energy is transferred from one object to another. According to the principle of conservation of energy, the amount of heat energy that "leaves" system 1 in Figure 14.2B must equal the amount of heat energy that "enters" system 2. This transfer of energy can take place in different ways, but the direction of transfer—the direction of heat flow—depends only on the temperature difference.

Units of Heat

Heat is measured with the SI unit of energy, the joule (J). A unit called the ***calorie*** is also widely used in measurements involving heat. The term *calorie* comes from the 1700s when some scientists thought that heat was a sort of fluid (called a "caloric fluid") that flowed between objects. Although it is not immediately obvious that the energy that flows into your finger when you touch a hot object is the same type of physical quantity as the kinetic energy of a falling brick, experiments in the mid-1800s using an apparatus like the one in Figure 14.3 showed that heat is, in fact, a form of energy.

While most physicists measure heat in units of joules, the calorie is widely used, especially in chemistry. These units are related by

$$1 \text{ cal} = 4.186 \text{ J} \tag{14.1}$$

A unit named the "Calorie" (uppercase "C") is used to measure the energy content of foods. This "food calorie" is actually equal to 1000 cal (lowercase "c").

Temperature: A Microscopic Picture

Our definition of temperature tells how to *compare* the temperatures of two different systems. We simply place the two systems in contact and then observe if energy is transferred from one to the other and in which direction (Fig. 14.2). This procedure, however, does not give us a way to measure the *value* of the temperature of a particular system. One way to measure the temperature is to place the system of interest in contact with a gas-filled container held at constant volume and density. The pressure in the gas is caused by collisions of gas atoms and molecules with the walls of the container (Fig. 14.4). If the temperature is increased, the average speed of the gas atoms increases, causing the pressure to increase also. We'll see in Chapter 15 that for a dilute gas (one with a low density), the temperature is linearly proportional to the gas pressure, so measurement of the pressure gives a direct way to find the temperature (Fig. 14.5A). Such a device is called a ***gas thermometer***.

The picture in Figure 14.4 gives insight into the microscopic meaning of temperature. The temperature of a system of particles (such as a gas) is related to the average particle speed. A high speed corresponds to a high temperature, and a low speed corresponds to a low temperature. Because high speeds also mean high kinetic energies, this relation suggests that the temperature of a system can be increased by adding energy to it. This is another example of the connection between temperature and energy. We'll derive the relation between temperature and energy for a gas in Chapter 15.

Temperature Scales

Three different ***temperature scales*** are in common use: the ***Fahrenheit*** scale, the ***Celsius*** scale, and the ***Kelvin*** scale. The Kelvin scale is the one employed in the SI system of units, but all three scales are widely used, so you should be able to con-

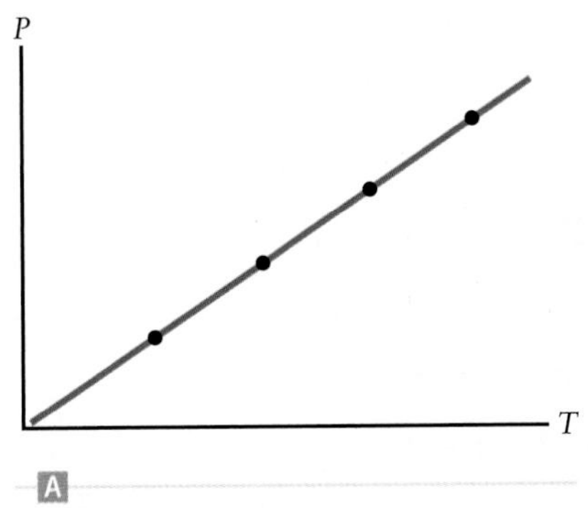

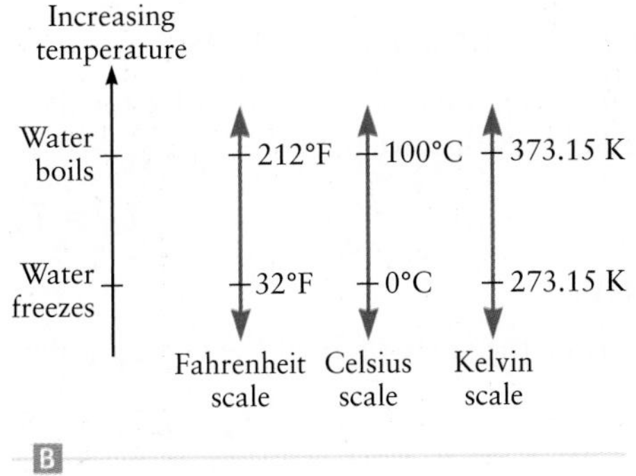

◀ **Figure 14.5** A The pressure in a dilute gas (such as air) is proportional to the temperature. B Temperature can be measured with the Fahrenheit, Celsius, or Kelvin scale.

vert temperature values from one scale to any other. This conversion process can be understood from Figure 14.5B.

The Celsius scale is defined so that at atmospheric pressure, water freezes at exactly 0°C and boils at precisely 100°C. In words, these temperatures are "zero degrees Celsius" and "100 degrees Celsius." On the Fahrenheit scale, these processes occur at 32°F and 212°F, respectively, referred to as "32 degrees Fahrenheit" and so on. On the Kelvin scale, these temperatures are 273.15 K and 373.15 K, respectively. These values are referred to as "273.15 kelvin" and so on; by convention, the term *degrees* is not used in connection with the Kelvin scale.

The relations between the different temperature scales are illustrated in Figure 14.5B. Because all three scales are linear, the values of the temperatures at which water freezes and boils on each scale define how to convert a temperature value from one scale to another. For example, to convert from a value on the Celsius scale T_C to a value on the Kelvin scale T_K, we have (according to Fig. 14.5B)

$$T_K = T_C + 273.15 \quad \text{(conversion from Celsius to Kelvin)} \tag{14.2}$$

Notice that the sizes of a temperature unit are equal in the Celsius and Kelvin scales, so a change of 1 "degree" on the Celsius scale is equal to a change of 1 kelvin on the Kelvin scale.

Conversions involving the Fahrenheit scale are not quite as simple because a 1° change on the Fahrenheit scale is not equal to a 1° change on the Celsius scale. From Figure 14.5B, we see that a change of 100°C (going from freezing water to boiling water) corresponds to a change of 212 − 32 = 180°F. A temperature change of 1°C is thus equivalent to a change of 180/100 = 9/5 = 1.8°F. We also know that 0°C is equivalent to 32°F (the freezing point of water), which leads to the relation

$$T_F = \tfrac{9}{5}T_C + 32 \quad \text{(conversion from Celsius to Fahrenheit)} \tag{14.3}$$

EXAMPLE 14.1 The Boiling Temperature of Liquid Nitrogen

At atmospheric pressure, liquid nitrogen boils at a temperature of 77 K (Fig. 14.6). Convert this value to the Celsius and Fahrenheit scales.

RECOGNIZE THE PRINCIPLE

Given the temperature of liquid nitrogen, we can use Figure 14.5B and Equations 14.2 and 14.3 to find the corresponding temperatures on the Celsius and Fahrenheit scales.

SKETCH THE PROBLEM

Figure 14.5B shows the relationship between the Kelvin, Celsius, and Fahrenheit scales.

(continued) ▶

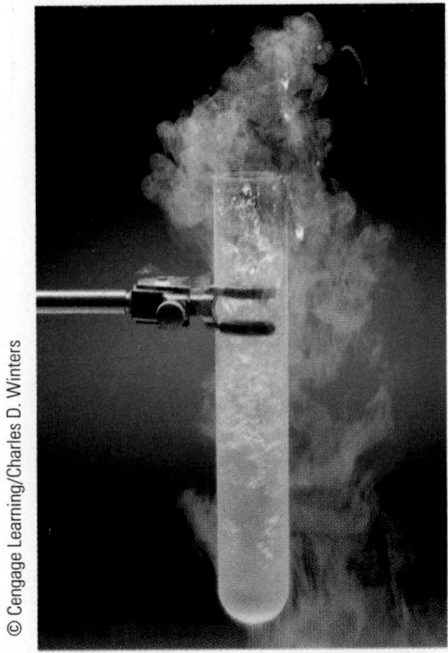

▲ **Figure 14.6** Example 14.1. At atmospheric pressure, nitrogen boils at 77 K.

IDENTIFY THE RELATIONSHIPS

Rearranging Equation 14.2 to solve for the value on the Celsius scale (T_C) in terms of the Kelvin value, we have

$$T_C = T_K - 273$$

SOLVE

Inserting the value of T_K at the boiling point of nitrogen gives

$$T_C = T_K - 273 = 77 - 273 = \boxed{-196°\text{C}}$$

To get the value on the Fahrenheit scale, we use Equation 14.3:

$$T_F = \tfrac{9}{5}T_C + 32 = \tfrac{9}{5}(-196) + 32 = \boxed{-321°\text{F}}$$

► ***What does it mean?***

Equations 14.2 and 14.3 can be used to convert between any two temperature scales. If you forget the equations, you can always work out the conversions if you remember the temperatures at which water freezes and boils.

Very High and Very Low Temperatures: What Are the Limits?

The temperature axes in Figure 14.5B have arrows at both ends to indicate that they extend to both higher and lower temperatures. Many important temperature values lie well above the boiling point of water or well below the freezing point of water (Table 14.1). Several of these temperatures have negative values on either the Fahrenheit or Celsius scales (or both). Indeed, the boiling temperature of nitrogen, which was considered in Example 14.1, falls at such negative temperatures. However, there is a limit at the bottom ends of these scales; there are no negative values of temperature on the Kelvin scale listed in Table 14.1. The value $T_K = 0$ K is called ***absolute zero***; in Chapter 16, we'll see why 0 K is a very special temperature.

TABLE 14.1 Values of Some Important Temperatures on the Kelvin, Celsius, and Fahrenheit Scales

	Temperature Scale		
	K	°C	°F
Interior of the Sun	$\approx 1 \times 10^7$	$\approx 1 \times 10^7$	$\approx 2 \times 10^7$
Center of the Earth	≈7000	≈7000	≈13,000
Surface of the Sun	≈6000	≈6000	≈10,000
Lead melts	600	327	621
Water boils	373.15	100	212
Human body temperature	310	37	98.6
Water freezes	273.15	0	32
Nitrogen boils	77	−196	−321
Intergalactic space	3	−270	−454
Absolute zero	0	−273.15	−459.67

14.3 Thermal Equilibrium and the Zeroth Law of Thermodynamics

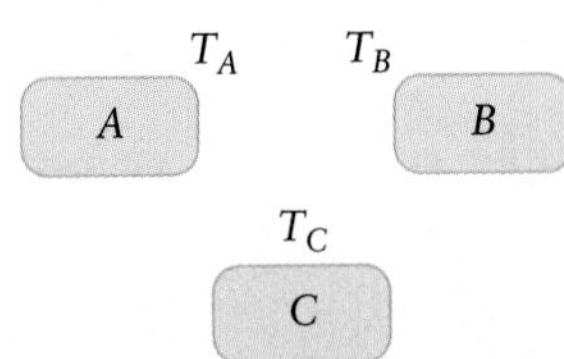

ZEROTH LAW: If A and B are in thermal equilibrium and if B and C are in thermal equilibrium, then A and C are also in thermal equilibrium and $T_A = T_B = T_C$.

▲ **Figure 14.7** The zeroth law of thermodynamics involves the relationships between thermal equilibrium and temperature for three different systems.

Let's consider the transfer of energy between the three systems sketched in Figure 14.7. We assume these three systems are initially isolated, with temperatures T_A, T_B, and T_C. Suppose we bring systems A and B into contact and find that there is no heat flow; that is, there is no transfer of energy between them. Using our definitions of temperature and heat, we would conclude that $T_A = T_B$. We then repeat this exercise with systems B and C, and again we find that there is no heat flow between these two systems; hence, $T_B = T_C$. What, then, would we say about the temperatures of A and C? The obvious conclusion is that $T_A = T_C$, which is called the ***zeroth law of thermodynamics***.[1]

> ***Zeroth law of thermodynamics:*** **If two systems A and B are in thermal equilibrium (so that $T_A = T_B$) and systems B and C are in thermal equilibrium ($T_B = T_C$), systems A and C are also in thermal equilibrium ($T_A = T_C$).**

Although this statement may seem trivial, it has some very important implications. In words, the zeroth law of thermodynamics tells us that the basic concept of temperature is meaningful and that temperature is a unique property of a system that is allowed to come into thermal equilibrium. Indeed, we took these facts for granted in our discussion of temperature and heat in Section 14.2. It is actually an amazing fact that *only one property* of a system, its temperature, determines when heat energy will flow between it and another system.

We can illustrate one implication of the zeroth law pictorially. Figure 14.8A shows an endless waterfall in which water circulates around a closed loop, flowing downhill during each part of the loop. Of course, such a waterfall is impossible because it would violate what we know about gravitational potential energy and the principle of conservation of energy. Figure 14.8B shows an "equivalent" (and still impossible) situation, this time involving heat flow. Here heat flows from system 1 to system 2, then flows from system 2 to system 3, and finally completes the loop by

NOT POSSIBLE!

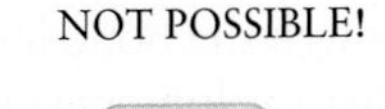
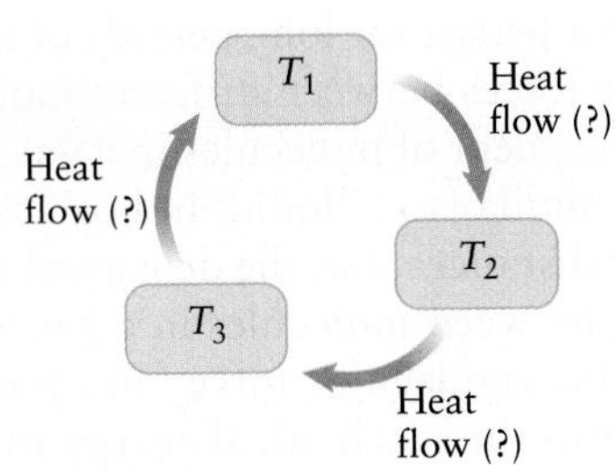

◄ **Figure 14.8** A In this famous lithograph *Waterfall* by M. C. Escher, water flows downhill endlessly around a closed loop. That is not possible in the real world because it is not consistent with the principle of conservation of energy. (It would violate the properties of gravitational potential energy.) B The zeroth law of thermodynamics states that this "circular" flow of heat energy between three different systems is *not possible*.

[1]You might wonder why physicists would start counting such laws at the number zero. The logical need for the zeroth law of thermodynamics was realized only after the first, second, and third laws of thermodynamics were already discovered and named.

flowing from system 3 to system 1. The zeroth law and our understanding of heat flow tell us that such a circular flow of heat is impossible because heat must always flow from high temperature to low temperature.

14.4 Phases of Matter and Phase Changes

The three phases of matter—***solid***, ***liquid***, and ***gas***—are illustrated in Figure 14.9. The atoms in many solids are arranged in an orderly and repeating pattern called a crystalline lattice (Fig. 14.9A), in which each atom is held in place by the forces exerted by neighboring atoms. These forces are a result of chemical bonds within the solid. For example, ionic bonds hold Na^+ and Cl^- within a lattice to form sodium chloride (NaCl), and covalent bonds hold silicon (Si) atoms in a crystalline lattice. The bonds between neighboring atoms in a solid behave essentially as "springs," giving a force similar to the Hooke's law force discussed in Chapter 11. If a particular bond is stretched or compressed, the atom experiences a restoring force proportional to its displacement. This force is directed back toward the "ideal" location of the atom within the lattice.

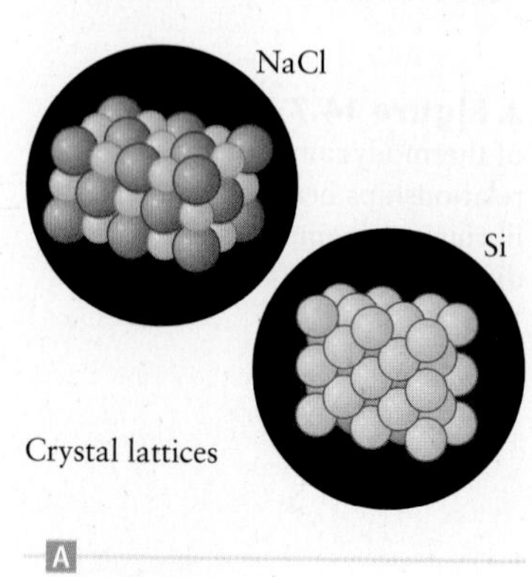

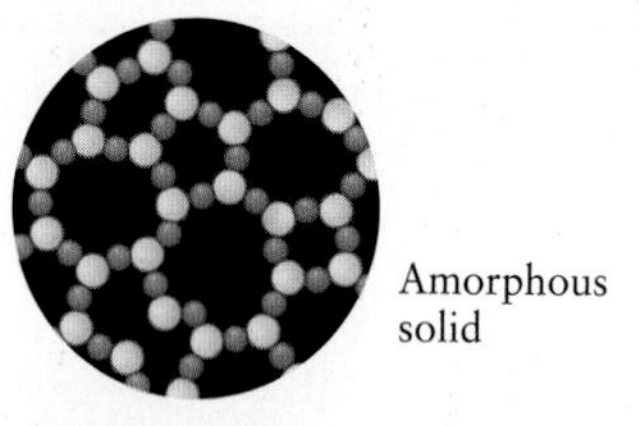

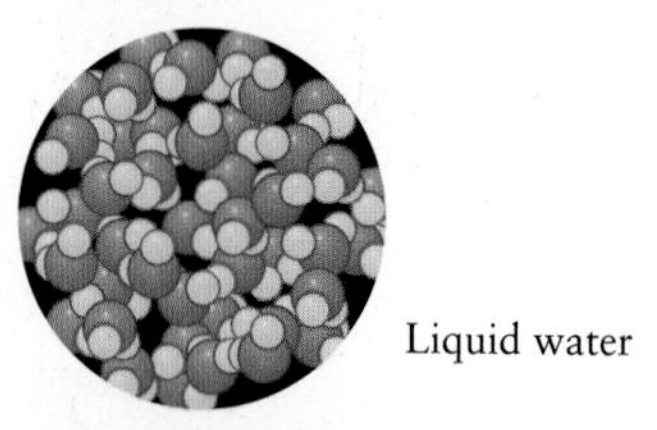

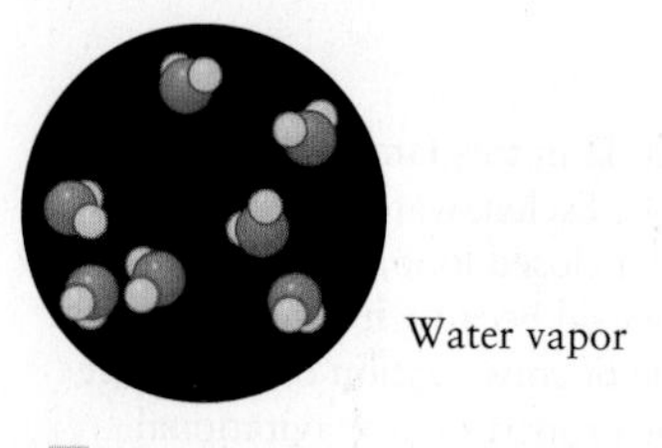

▲ **Figure 14.9** There are three phases of matter: solid, liquid, and gas. The atoms in a solid can be arranged in A a crystal lattice or B a somewhat random spatial arrangement called an amorphous solid. C In a liquid and D in a gas, the atoms do not have a regular arrangement and are able to move about.

Figure 14.9A is a simplified picture, showing all the atoms in their ideal lattice positions. More realistically, the atoms vibrate about these positions as simple harmonic oscillators. The kinetic and potential energies associated with these oscillations can be understood using the ideas we encountered in Chapter 11.

Figure 14.9B shows another example of a solid phase called an amorphous solid. A familiar example is common window glass, which is composed mainly of silicon and oxygen. The atomic arrangement in an amorphous solid does not have the regular repeating structure found in a crystal. Even so, the atoms are still held in place by chemical bonds that behave as springs obeying Hooke's law.

Figure 14.9C shows the atomic arrangement found in a liquid. The difference between liquids and amorphous solids is evident when the atomic arrangement is followed for a period of time. The atoms in a liquid are not held in fixed locations by the forces from neighboring atoms. Instead, atoms in a liquid are able to move about, and particular atoms that happen to be close neighbors at one moment are usually not neighbors a short time later. This motion at the atomic scale enables a liquid to flow. Although the bonds between different molecules in water (or in other liquids) do not persist for long periods of time, some potential energy is still associated with the forces between different molecules and atoms in a liquid.

The arrangement of molecules in a gas is shown in Figure 14.9D. In some ways, a gas is very similar to a liquid. In both cases, the molecules are able to move easily over long distances, but the density of a gas is much lower than that of a liquid. The spacing between molecules in a gas is therefore larger, and the magnitude of the average intermolecular force (and potential energy) is much smaller. For this reason, most of the mechanical energy in a gas is found in the kinetic energies of the molecules.

Internal Energy

Consider a collection of H_2O molecules. What determines if these molecules form a solid, a liquid, or a gas? Figure 14.9 describes the difference between these phases in terms of the positions and motions of the molecules. Let's instead consider the energies of molecules when the system is in these different phases. We have already mentioned that the kinetic energy of a typical molecule is largest in the gas and smallest in the solid, so we can expect the mechanical energy of the molecules to be different in the different phases. This energy is called the ***internal energy*** of the system and is denoted by U. The internal energy of a system of molecules is the sum of the potential energies associated with all the intermolecular bonds plus the kinetic energies of all the molecules. The value of U increases as we go from the solid to the liquid to the gas phase. In general, the internal energy of all systems increases as the temperature is increased.

Phase Changes

The transformation of a solid to a liquid, and of a liquid to a gas, are examples of ***phase changes***. Phase changes can be produced by changing the temperature or by changing the system's pressure as shown in the ***phase diagram*** of a substance. The phase diagram of H_2O (Fig. 14.10A) shows the phases found at different pressures and temperatures. For example, H_2O is a gas at low pressures and moderate to high temperatures, whereas it is a solid at low temperatures and moderate to high pressures. Phase changes occur at each of the solid brown lines in Figure 14.10A. For example, if we start with H_2O in the solid phase (ice) above a pressure of about 600 Pa and then increase the temperature (following, for example, path A in Fig. 14.10A), we eventually cross into the liquid phase. This phase change is called ***melting***. If instead we were to start in the liquid phase and then reduce the temperature, we would observe the phase change called ***freezing***. The phase diagram of H_2O includes other phase changes between liquid and gas (evaporation) and between solid and gas (sublimation). The line that separates liquid and gas ends at the ***critical point***. There is also a ***triple point***, where the solid, liquid, and gas phase regions all meet. The temperature of the triple point for water is 273.16 K.

Other substances have similar phase diagrams. For example, Figure 14.10B shows the phase diagram for CO_2. The overall arrangement of the phases is similar to the case of H_2O, although the values of temperature and pressure at which the phase changes occur vary from one substance to another. Table 14.2 lists the melting and evaporation temperatures of some common substances at atmospheric pressure.

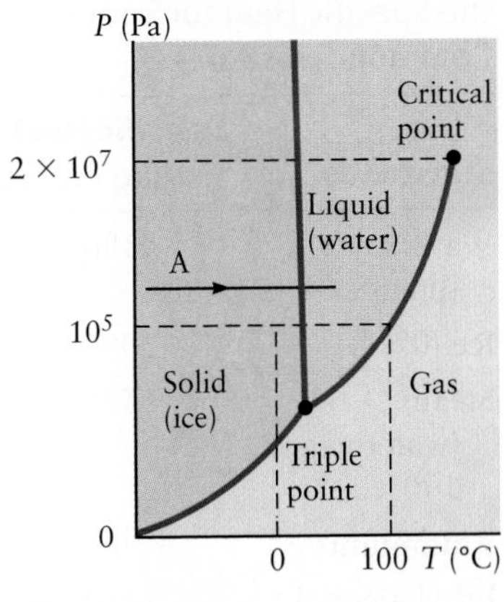

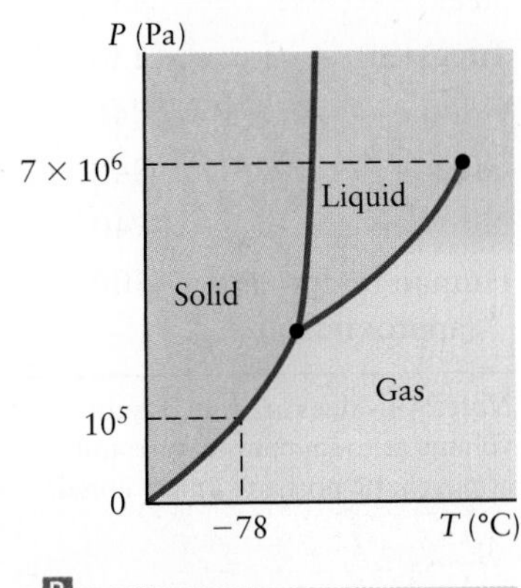

▲ **Figure 14.10** A phase diagram shows the temperatures and pressures at which phase changes occur (solid brown lines). Phase diagram for A H_2O and B CO_2. Note that 10^5 Pa is approximately atmospheric pressure.

Specific Heat and Heat Capacity

The phase diagram in Figure 14.10A shows a path labeled A along which the system can be taken from a low to a high temperature. In practice, this change could be accomplished by adding heat energy to the system by placing the system in contact with a second system at a higher temperature so that heat energy can flow between them. How much energy is required? Let's first consider what happens to a system when heat is added, but the system does not undergo a phase change. For example, our system might be a liquid (water) whose temperature increases but does not reach the evaporation (boiling) temperature. Let Q denote the heat energy added to the system and ΔT be the change in temperature. The ratio of these two quantities is called the ***heat capacity***:

$$\text{heat capacity} = \frac{Q}{\Delta T} \tag{14.4}$$

TABLE 14.2 Melting and Evaporation Temperatures for Some Common Substances at Atmospheric Pressure

Substance	Melting Temperature (K)	Evaporation Temperature (K)
Water	273	373
Aluminum	933	2720
Gold	1340	2930
Ethyl alcohol	159	351
Carbon dioxide	195	217
Nitrogen	63	77

TABLE 14.3 Values of the Specific Heat for Some Common Substances

Substance	Specific Heat J/(kg · K)
Water (liquid, 15°C)	4186
Ice (0°C)	2090
Steam (water vapor, 100°C)	1520
Aluminum	900
Ethyl alcohol	2450
Gold	129
Steel	450
Tungsten	130
Granite	840
Glass	840
Nitrogen gas	740
Human body (approximate)	3500

Note: All values are for constant volume at room temperature and atmospheric pressure unless noted.

This ratio does not take account of the mass of the system; a large system has a large heat capacity because a large amount of energy Q is required to change the temperature by a given amount. (Likewise, a small system has a small heat capacity.) It is therefore useful to define a closely related quantity called the ***specific heat*** c by

$$\text{specific heat} = c = \frac{Q}{m\,\Delta T} \tag{14.5}$$

The specific heat takes account of the size of the system, so c is the same for any sample of a certain substance (at a particular temperature and pressure), regardless of its mass.

We can rearrange Equation 14.5 as

$$Q = cm\,\Delta T \tag{14.6}$$

Knowing the value of c thus allows us to calculate how much heat energy is needed to increase the temperature of a system of mass m by an amount ΔT. By convention, a positive value of Q corresponds to heat energy *added to* the system, and a negative value of Q indicates that energy flows *out of* the system. In our mechanical picture involving internal energy, adding or subtracting energy changes the internal energy U of the system (the "internal" kinetic and potential energies of the atoms and molecules).

Values of the specific heat for a few common materials are given in Table 14.3. The values in Table 14.3 were (unless noted) determined at room temperature and pressure, with the volume of the system kept fixed; such values are called the specific heat at *constant volume*. The specific heat can vary with temperature, but these changes are usually small. Notice that c is always positive. Adding energy to a system (a positive value of Q in Eq. 14.6) always increases its temperature (a positive ΔT).

Values of the specific heat vary widely among different substances and tend to be larger for liquids than for solids or gases. In Table 14.3, the values of c are listed in SI units, J/(kg · K). You will often find values of specific heat quoted in terms of calories (instead of joules), especially in chemistry books, so you need to be able to convert between these two units. From Table 14.3, we have $c = 4186$ J/(kg · K) for water. Converting the joules to calories (according to Eq. 14.1), we have

$$c = 4186\,\frac{\text{J}}{\text{kg}\cdot\text{K}} \times \frac{1\text{ cal}}{4.186\text{ J}} \times \frac{1\text{ kg}}{1000\text{ g}} = 1.000\,\frac{\text{cal}}{\text{g}\cdot\text{K}} \quad \text{(specific heat of water)}$$

The value of the specific heat of water thus has a particularly "simple" (and easy to remember) value when expressed using calories. Indeed, the calorie was originally defined through the specific heat of water.

CONCEPT CHECK 14.1 Specific Heat of the Human Body

Even though the human body is not a pure substance, we can still define its specific heat through Equation 14.5. According to Table 14.3, the human body's specific heat is only a little smaller than the specific heat of water and is much larger than the specific heat of the other substances listed. Why?

Why Is Specific Heat Important?

With knowledge of the specific heat, we can calculate how the temperature of an object will change when a certain amount of heat energy is added or removed. One practical application is in understanding how oceans and lakes affect the weather. The specific heat of water is much greater than the specific heat of most other substances. (Compare the values in Table 14.3.) A large body of water thus has a very large heat capacity relative to the surrounding air and soil. As a result, the temperature of an ocean or lake changes relatively little even when relatively large amounts of heat energy are added in the summer or taken away in winter, and thus it moderates ("smooths out") temperature fluctuations.

The value of the specific heat also gives insight into the internal energy U of a substance. According to the principle of conservation of energy, if an amount of heat Q is added to a system, then U must increase by that amount. The specific heat tells how the added heat changes the temperature; hence, c tells how changes in internal energy and temperature are related. This concept is important for understanding how the microscopic atomic motions in a substance change with temperature; we'll discuss this subject more in Chapter 15.

Calorimetry

We can find the specific heat of a system by adding a known amount of heat energy Q and measuring the resulting temperature change ΔT (Eq. 14.5). A measurement of this kind is called ***calorimetry*** and is shown schematically in Figure 14.11. The system of interest is placed in contact with a reference system; the reference system is often very large, in which case it is called a ***thermal reservoir***. For example, the system of interest might be a hot piece of metal, and the thermal reservoir could be a large bucket of water into which the metal is dropped (to cool it). When defining the specific heat, Q is the total heat energy added to the system (the metal); in this example, Q would come entirely from the thermal reservoir (the water). Most calorimetry experiments are designed to control the sources of heat flow into the system. In words, the system of interest is "thermally isolated" from all but the thermal reservoir.

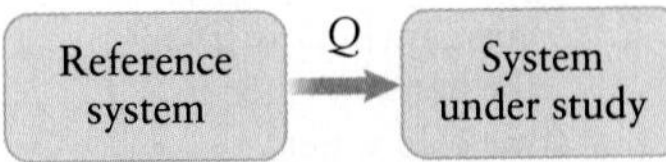

▲ **Figure 14.11** In a typical calorimetry experiment, the system of interest is arranged so that the only heat flow into or out of the system is from a reference system.

The general principles of calorimetry can also be used to analyze processes that do not involve a large thermal reservoir. For example, when a cold ice cube is dropped into a glass of warm water, heat flows between the water and the ice, changing the temperature of both. These temperature changes depend on the initial temperatures of both systems and their specific heats. In terms of conservation of energy, all the energy that leaves one system enters the other system, and that is the basis of a general way to attack all problems in calorimetry.

PROBLEM SOLVING Dealing with Specific Heat in Calorimetry

1. **RECOGNIZE THE PRINCIPLE.** All calorimetry problems are based on the principle of conservation of energy.
2. **SKETCH THE PROBLEM.** Your sketch should show the system or systems of interest and indicate how heat flows between them.
3. **IDENTIFY THE RELATIONSHIPS.** Determine (if possible) the initial and final temperatures of the system(s) and the heat energy Q added to each system. If energy flows into a system, Q is positive; if energy flows out of a system, Q is negative.
4. **SOLVE.** Apply $Q = cm\,\Delta T$ (Eq. 14.6) to relate Q and ΔT for each system and solve for the quantities of interest.

   Important note: Step 4 (and Eq. 14.6) assumes there are no phase changes; for example, the systems do not melt or evaporate. We'll see how to analyze such cases when we discuss the latent heat associated with phase changes later in this section.
5. Always *consider what your answer means* and check that it makes sense.

EXAMPLE 14.2 Specific Heat in the Blacksmith's Shop: Part 1

A blacksmith uses the heat energy from an oven to heat steel and other metals. The blacksmith also uses cold water to cool a hot horseshoe to room temperature. A steel horseshoe of mass $m_{\text{shoe}} = 0.50$ kg at a temperature of 800 K ("red hot") is dropped into a bucket of water ($m_{\text{water}} = 1.0$ kg) at 300 K (about room temperature). What is the final temperature of the horseshoe and the water? For simplicity, assume none of the water evaporates, so none of the water is converted to steam.

(continued) ▶

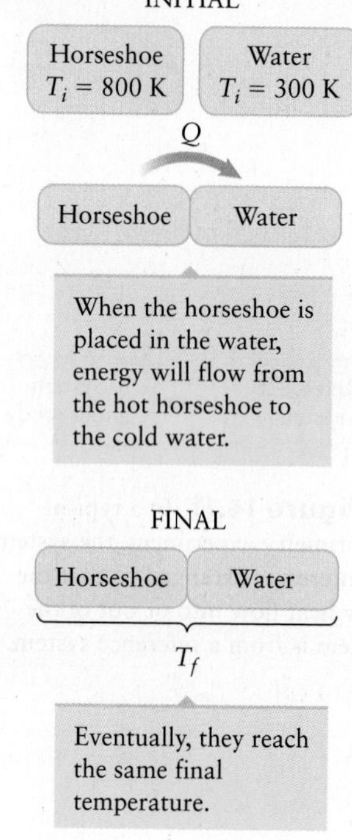
INITIAL
Horseshoe $T_i = 800$ K
Water $T_i = 300$ K
Q
Horseshoe
Water
When the horseshoe is placed in the water, energy will flow from the hot horseshoe to the cold water.
FINAL
Horseshoe
Water
T_f
Eventually, they reach the same final temperature.

▲ **Figure 14.12** Example 14.2.

RECOGNIZE THE PRINCIPLE

In this example, our two systems are the horseshoe and the water. When the horseshoe is dropped into the bucket, heat energy flows from the horseshoe to the water (Fig. 14.12). Because the horseshoe is completely immersed in the water, we can assume there is no other inflow or outflow of energy involving either system. We denote the energy that flows from the horseshoe into the water as Q. From conservation of energy principles, the energy "added" to the horseshoe is $-Q$; the negative sign here indicates that the final energy of the horseshoe is less than its initial energy. The initial temperatures of the horseshoe and the water are given, and the common final temperature T_f is the quantity we wish to find. We can calculate T_f by applying the specific heat relation between Q and ΔT for both systems.

SKETCH THE PROBLEM

Figure 14.12 shows the two systems along with their initial temperatures and the direction of heat flow.

IDENTIFY THE RELATIONSHIPS

We denote the specific heats of the horseshoe and the water by c_{steel} and c_{water}, respectively, and obtain their values from Table 14.3. Applying Equation 14.6 to the horseshoe, we get

$$Q_{shoe} = c_{steel} m_{shoe} \Delta T_{shoe} = c_{steel} m_{shoe}(T_f - T_{shoe,\,i}) \tag{1}$$

Because the final temperature is less than the initial temperature ($T_{shoe,\,i}$) of the steel horseshoe, Q_{shoe} will be negative as anticipated above. Applying the same relation to the water, we get

$$Q_{water} = c_{water} m_{water}(T_f - T_{water,\,i}) \tag{2}$$

Because the water temperature increases ($T_f > T_{water,\,i}$), Q_{water} will be positive. Energy is conserved, so

$$Q_{shoe} + Q_{water} = 0$$

and

$$Q_{shoe} = -Q_{water}$$

SOLVE

Using our results from Equations (1) and (2) and solving for T_f leads to

$$c_{steel} m_{shoe}(T_f - T_{shoe,\,i}) = -c_{water} m_{water}(T_f - T_{water,\,i})$$

$$T_f(c_{steel} m_{shoe} + c_{water} m_{water}) = c_{steel} m_{shoe} T_{shoe,\,i} + c_{water} m_{water} T_{water,\,i}$$

$$T_f = \frac{c_{steel} m_{shoe} T_{shoe,\,i} + c_{water} m_{water} T_{water,\,i}}{c_{steel} m_{shoe} + c_{water} m_{water}}$$

Inserting the given values of the various quantities, including the specific heats from Table 14.3, gives

$$T_f = \frac{[450\ \text{J/(kg}\cdot\text{K)}](0.50\ \text{kg})(800\ \text{K}) + [4186\ \text{J/(kg}\cdot\text{K)}](1.0\ \text{kg})(300\ \text{K})}{[450\ \text{J/(kg}\cdot\text{K)}](0.50\ \text{kg}) + [4186\ \text{J/(kg}\cdot\text{K)}](1.0\ \text{kg})}$$

$$T_f = \boxed{330\ \text{K}}$$

▸ *What does it mean?*

The final temperature is quite close to the initial temperature of the water and far from the initial temperature of the horseshoe. That is because the specific heat of water is much greater than the specific heat of steel, so the water acts as a kind of thermal reservoir in this case.

EXAMPLE 14.3 Maintaining a Safe Body Temperature

A bicycle rider of mass $m = 60$ kg generates a power of approximately 400 W when riding at a moderate pace. Of this energy, about 25% is used to propel the bicycle, and the remaining 75% goes into heat generated within his body. If none of this heat were lost to the surroundings, how much would the cyclist's body temperature increase after 1 hour of riding?

RECOGNIZE THE PRINCIPLE

In this calorimetry problem, the system of interest is the cyclist along with an "internal" source (the cyclist's metabolism) that adds energy Q to this system. We must first find Q and then use it to compute the temperature rise of the system with Equation 14.6.

SKETCH THE PROBLEM

Figure 14.13 shows the system (the cyclist). An amount of heat Q is generated within, increasing his temperature.

IDENTIFY THE RELATIONSHIPS

Power (P) is energy per unit time, so the energy Q added to the cyclist in 1 h is $Q = P\,\Delta t$. Here P is the fraction of the power that goes into heat, which is $P = 0.75(400 \text{ W}) = 300$ W, with $\Delta t = 1 \text{ h} = 3600$ s. We thus get

$$Q = P\,\Delta t = (300 \text{ W})(3600 \text{ s}) = 1.08 \times 10^6 \text{ J} \quad (1)$$

where we have kept three significant figures to avoid roundoff errors below. Equation (1) is related to the change in temperature of the cyclist by (using Eq. 14.6)

$$\Delta T = \frac{Q}{cm}$$

SOLVE

Inserting Q from Equation (1) along with the given value of m and the specific heat of the human body (Table 14.3) gives

$$\Delta T = \frac{Q}{cm} = \frac{1.08 \times 10^6 \text{ J}}{[3500 \text{ J/(kg}\cdot\text{K)}](60 \text{ kg})} = 5.1 \text{ K} = \boxed{5.1°\text{C}}$$

► ***What does it mean?***

When a person's body temperature increases more than about 4°C (corresponding to a change ΔT of about $4°\text{C} \times \frac{9}{5} \approx 7°\text{F}$), thus giving a body temperature above about 106°F, there is danger of heatstroke. It is therefore important that our cyclist, or anyone else engaged in strenuous exercise, have some way to pass this generated heat to his surroundings. We'll consider that problem in further examples below.

© Image Source/Jupiterimages

▲ **Figure 14.13** Example 14.3. **A** When exercising (or even when just sitting on a couch), a person generates heat. **B** This heat will raise the body's temperature unless it escapes through the skin.

Latent Heat

The specific heat tells how much the temperature will increase if a certain amount of heat is added to a system, but the definition of specific heat assumes the system does not change from one phase to another as the heat is added. When a system undergoes a phase change such as melting or evaporation, we must also account for the ***latent heat***. For example, for a block of ice at 0°C to undergo a phase change to liquid water at 0°C, an amount of heat equal to the ***latent heat of fusion*** must be added. Likewise, when water at a temperature of 0°C freezes to become ice at 0°C, this latent heat must be removed.

The latent heat associated with a phase change is connected with the energies of the atoms within the substance. For a solid, substantial energy is stored in the atomic bonds holding the atoms in place. These bonds store potential energy in

TABLE 14.4 Latent Heats for Various Phase Changes

Substance	Latent Heat of Fusion (kJ/kg)	Latent Heat of Vaporization (kJ/kg)
Water	334	2,200
Aluminum	397	11,000
Gold	67	870
Ethyl alcohol	104	850
Copper	205	5,070
Lead	23	870

a manner analogous to springs that obey Hooke's law (Chapter 6). For a solid to melt, energy is required to overcome this potential energy and break these bonds, enabling the atoms and molecules to move about freely as in a liquid. That energy is provided by the latent heat.

By convention, the latent heat of fusion L_{fusion} is a positive quantity; for a system of mass m to melt, an amount of heat

$$Q_{melt} = +mL_{fusion} \tag{14.7}$$

must be added. Likewise, when a system freezes (also called fusion), we have

$$Q_{freeze} = -mL_{fusion}$$

The negative sign indicates that energy must be removed from a system to make it freeze.

The ***latent heat of vaporization*** is associated with the phase change from a liquid to a gas (called vaporization, evaporation, or boiling) and from a gas to a liquid (condensation). To vaporize an amount of liquid of mass m, we must add an energy

$$Q_{vaporization} = mL_{vaporization} \tag{14.8}$$

and for condensation, we have

$$Q_{condensation} = -mL_{vaporization}$$

Here the negative sign indicates that energy must be removed from a system to make it condense. On an atomic scale, the latent heat of vaporization can be understood as follows. When a liquid evaporates to become a gas, the weak atom–atom interactions in the liquid become even weaker in the gas as the atoms are moved farther apart. This process requires the addition of energy, now through the latent heat of vaporization.

Table 14.4 lists the latent heats of fusion and vaporization for some common substances. Notice that for a given substance, the latent heat of fusion is *not* equal to the latent heat of vaporization.

Calorimetry: Including the Latent Heat

Figure 14.14A shows the phase diagram for a hypothetical substance. Let's consider how much heat we must add to move along the path indicated by the horizontal dashed line. Starting from a point in the solid region of the phase diagram (point A), we add heat. This increases the temperature as shown in region I in Figure 14.14B, where we plot the temperature as a function of the amount of added heat Q_{added}. The temperature increase ΔT in the solid phase is described by the specific heat through $\Delta T = Q/(cm)$ (Eq. 14.6), with c equal to the specific heat of the solid.

Point B in Figure 14.14A is at the melting transition. If we continue to add heat, the system undergoes a phase change from solid to liquid and the amount of heat needed is given by the latent heat of fusion (Eq. 14.7). As this latent heat is added and the system is melting, the temperature remains constant (region II in Fig.

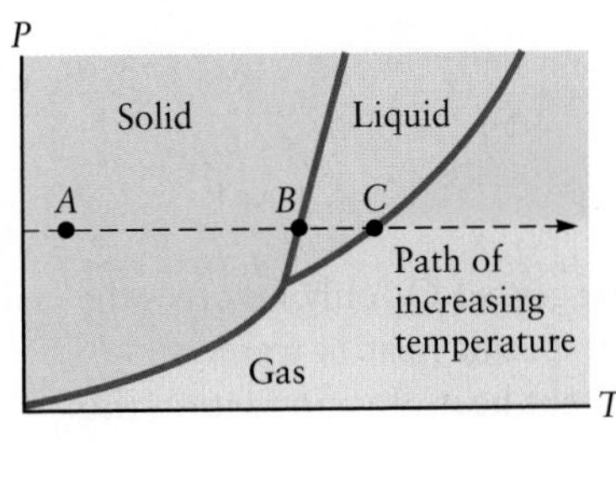

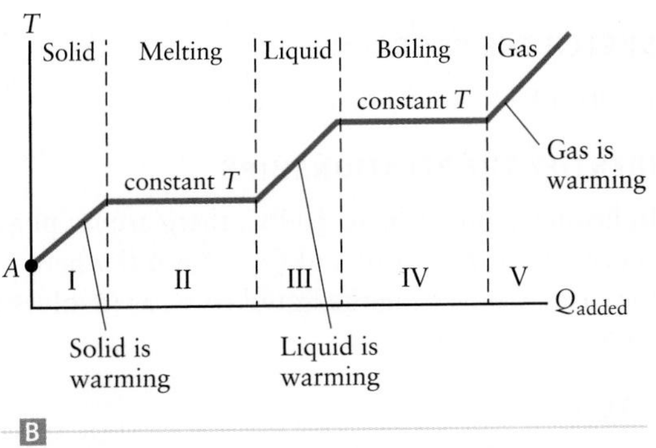

◀ **Figure 14.14** A Phase diagram of a typical system. B Results of a calorimetry experiment in which T is plotted as a function of the energy Q_{added} added to the system. The slopes in regions I, III, and V depend on the specific heat of the solid, liquid, and gas respectively. The slope is zero in regions II and IV.

14.14B). If this substance is water, we would be converting ice at 0°C into water at 0°C. When the system is completely melted, adding further heat increases the temperature according the specific heat of the liquid (region III). Eventually, the phase change to a gas is reached at point C. Now an amount of energy corresponding to the latent heat of vaporization must be added, and during this time, the temperature is again constant (region IV). If this substance is water, this latent heat converts water at 100°C to gas (water vapor) also at 100°C. Adding more heat would then increase the temperature of the gas.

Calorimetry calculations involving phase changes are similar to the problems we considered in Examples 14.2 and 14.3. The only difference is that we must now allow for the energy associated with the latent heat of any phase changes that occur.

PROBLEM SOLVING Dealing with Both Specific Heat and Latent Heat in Calorimetry

1. **RECOGNIZE THE PRINCIPLE.** All calorimetry problems are based on the principle of conservation of energy.
2. **SKETCH THE PROBLEM.** Your sketch should show the system or systems of interest and indicate how heat flows between them.
3. **IDENTIFY THE RELATIONSHIPS.** Determine (if possible) the initial and final temperatures of the system(s) and the heat energy Q added to each system. If energy flows into a system, Q is positive; if energy flows out of a system, Q is negative.
4. **SOLVE.**
   - Use $Q = cm\,\Delta T$ (Eq. 14.6) to relate Q and ΔT for temperature changes that do not involve a phase change.
   - If there is a phase change, the latent heat ($Q = mL$, Eqs. 14.7 and 14.8) must be added to or subtracted from the system, depending on the phase change.
5. Always *consider what your answer means* and check that it makes sense.

EXAMPLE 14.4 Heating a Bucket of Water

A bucket contains 2.0 kg of water (about 2 L) at 30°C (about room temperature). How much energy must be added to convert all the water to vapor at 100°C?

RECOGNIZE THE PRINCIPLE

Our system is the water, initially a liquid at 30°C (Fig. 14.15). An amount of heat energy Q_1 is added to take the liquid to 100°C, and then an additional heat energy Q_2 is needed to evaporate the liquid.

(continued) ▶

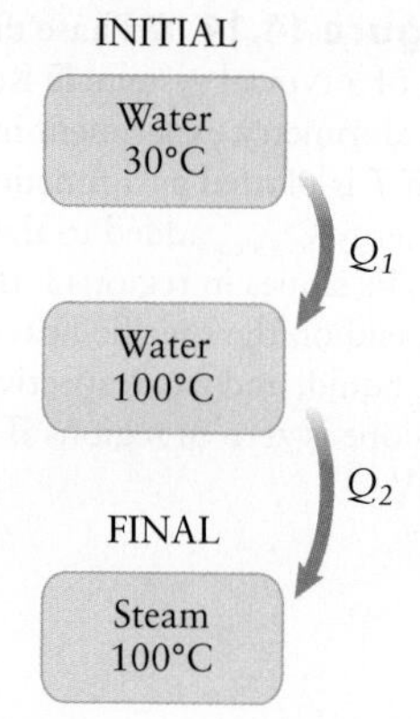

▲ **Figure 14.15** Example 14.4.

SKETCH THE PROBLEM

Figure 14.15 shows the problem.

IDENTIFY THE RELATIONSHIPS

In heating from 30°C to 100°C, there are no phase changes, so Q_1 only involves the specific heat (Eq. 14.6) and $Q_1 = cm\,\Delta T$, where c is the specific heat of water and m is the mass. Converting the liquid to a gas involves the latent heat of vaporization (Eq. 14.8), so $Q_2 = mL_{\text{vaporization}}$.

SOLVE

Using the value of c for water from Table 14.3, the given value of m, and $\Delta T = 100°\text{C} - 30°\text{C} = 70\text{ K}$, we find

$$Q_1 = cm\Delta T = [4186\text{ J/(kg}\cdot\text{K)}](2.0\text{ kg})(70\text{ K}) = 5.9 \times 10^5\text{ J}$$

Using the latent heat of water from Table 14.4 leads to

$$Q_2 = mL_{\text{vaporization}} = (2.0\text{ kg})(2200\text{ kJ/kg}) = 4.4 \times 10^6\text{ J}$$

The total heat energy required to convert all the water to steam is thus

$$Q_{\text{total}} = Q_1 + Q_2 = [(5.9 \times 10^5) + (4.4 \times 10^6)]\text{ J} = \boxed{5.0 \times 10^6\text{ J}}$$

▶ ***What does it mean?***

Whenever there is a phase change, we must account for the latent heat. In this case, the latent heat is the largest contribution to Q_{total}.

EXAMPLE 14.5 Specific Heat in the Blacksmith's Shop: Part 2

Consider again the blacksmith in Example 14.2, who uses a bucket of water ($m_{\text{water}} = 1.0$ kg, $T_{\text{water},\,i} = 300$ K) to cool a hot steel horseshoe. This time, he has a much larger horseshoe ($m_{\text{shoe}} = 2.0$ kg, $T_{\text{shoe},\,i} = 800$ K) and notices that when he drops the horseshoe in the water, some of the water evaporates as steam. Assuming the final temperature of the horseshoe plus water plus steam is 100°C (= 373 K), how much of the water is converted into steam?

RECOGNIZE THE PRINCIPLE

When the horseshoe is dropped into the water, an amount of heat energy Q flows from the horseshoe to the water (Fig. 14.16). From conservation of energy principles, the energy "added" to the horseshoe is $-Q$; the negative sign indicates that the final energy of the horseshoe is less than its initial energy. The initial temperatures of the horseshoe and the water are given. In the end, an unknown amount of the water undergoes a phase change to steam, and the horseshoe, water, and steam have a common final temperature $T_f = 100°\text{C} = 373$ K. We can find how much water is converted to steam using the principle of conservation of energy.

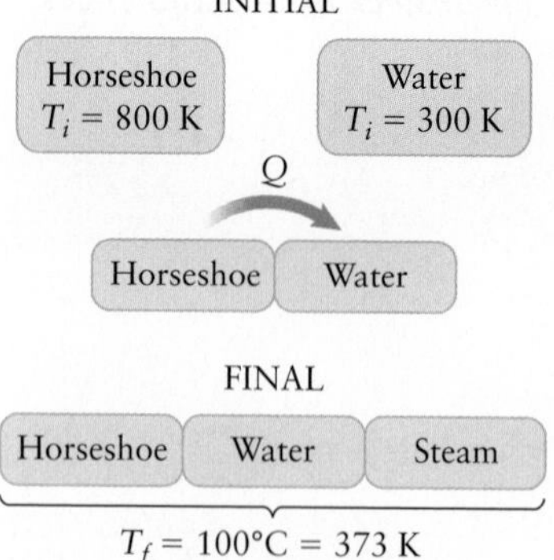

▲ **Figure 14.16** Example 14.5.

SKETCH THE PROBLEM

Figure 14.16 shows the horseshoe and the water along with their initial temperatures and the direction of heat flow.

IDENTIFY THE RELATIONSHIPS

The only unknown is the mass of water converted to steam. The heat that flows out of the horseshoe can be found from the specific heat of the horseshoe and its temperature change. Applying Equation 14.6, we get (compare with Example 14.2)

$$Q_{\text{shoe}} = c_{\text{steel}} m_{\text{shoe}}(T_f - T_{\text{shoe},\,i})$$

Consulting Table 14.3 for c_{steel} and inserting the given values, we find

$$Q_{shoe} = [450\ \text{J}/(\text{kg}\cdot\text{K})](2.0\ \text{kg})(373\ \text{K} - 800\ \text{K}) = -3.84 \times 10^5\ \text{J} \quad (1)$$

where we have kept three significant figures to avoid roundoff errors below. The water begins at $T_{water,\,i}$, and it all reaches a final temperature of 373 K. The heat required to bring all the liquid to this temperature is

$$Q_{water} = c_{water} m_{water}(T_{water,\,f} - T_{water,\,i})$$

Using c_{water} from Table 14.3 and inserting the given values, we find

$$Q_{water} = [4186\ \text{J}/(\text{kg}\cdot\text{K})](1.0\ \text{kg})(373\ \text{K} - 300\ \text{K}) = 3.06 \times 10^5\ \text{J} \quad (2)$$

Next we write the conservation of energy condition. The sum of the energy Q_{shoe} plus the total energy that flows into the water must equal zero. The energy that flows into the water has two contributions, one from the energy needed to heat all the water to 373 K as done in Equation (2) and another from the latent heat of the water converted to steam. We thus have (compare again with Example 14.2)

$$aQ_{shoe} + Q_{water} + Q_{steam} = 0 \quad (3)$$

We can use the results from Equations (1) and (2) to find Q_{steam}:

$$Q_{steam} = -Q_{shoe} - Q_{water} = (3.84 \times 10^5\ \text{J}) - (3.06 \times 10^5\ \text{J}) = 8 \times 10^4\ \text{J} \quad (4)$$

We want to find the mass of water m_{steam} converted to steam; it is related to Q_{steam} by (Eq. 14.8)

$$Q_{steam} = m_{steam} L_{vaporization}$$

SOLVE

Solving for m_{steam} and inserting the value of Q_{steam} from Equation (4) and the latent heat from Table 14.4, we find

$$m_{steam} = \frac{Q_{steam}}{L_{vaporization}} = \frac{8 \times 10^4\ \text{J}}{2.2 \times 10^6\ \text{J/kg}} = \boxed{0.04\ \text{kg}}$$

▶ ***What does it mean?***

The key to this problem, and to all calorimetry, is conservation of energy. In this example, it is expressed in Equation (3).

CONCEPT CHECK 14.2 Comparing Heat Energy with a Typical Kinetic Energy

The total heat energy required to cool the horseshoe in Example 14.5 is $|Q_{shoe}| = 3.8 \times 10^5$ J. Suppose this energy could be converted to the kinetic energy of a baseball ($m \approx 0.2$ kg). Would the speed of the baseball be (a) 50 m/s (about 100 mi/h), (b) 5 m/s, or (c) 2000 m/s?

14.5 Thermal Expansion

Changing the temperature of a system can affect many of its properties. In particular, the size of a system will usually change with temperature, an effect called ***thermal expansion***. Consider an object such as a piece of metal as in Figure 14.17A, with length L_0. If the temperature of the object increases by ΔT, the length changes by an amount ΔL. This change in length is proportional to the change in temperature ΔT and to the ***coefficient of linear thermal expansion*** α:

$$\frac{\Delta L}{L_0} = \alpha\,\Delta T \quad (14.9)$$

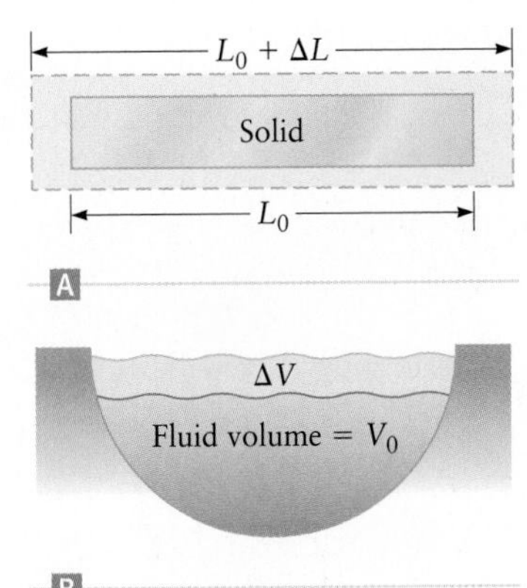

▲ **Figure 14.17** A If a solid object is heated, its length increases by an amount ΔL proportional to the coefficient of thermal expansion α. B Any substance (solid, liquid, or gas) may change volume when it is heated. This change in volume ΔV is proportional to the coefficient of volume expansion β.

TABLE 14.5 Coefficients of Linear and Volume Expansion

Substance	α (10^{-6} K^{-1})	β (10^{-6} K^{-1})
Ethyl alcohol	—	1120
Mercury	—	182
Water	—	210
Ice (0°C)	51	153
Pyrex	3.3	9.8
Glass (ordinary)	9	27
Aluminum	22.5	69
Gold	14	42
Cement	12	36
Iron	12	36
Steel	12	36
Platinum	8.8	26
Invar[a]	0.9	2.7
Wood (typical)	5.4	16
Brass	18.7	56
Copper	17	51

Note: Values are for room temperature and pressure unless noted.

[a]Invar is an alloy of iron and nickel that has very low values of the coefficients of thermal expansion, which makes it useful in applications.

As indicated in Figure 14.17A, the object's width and height also change. These changes are given by expressions similar to Equation 14.9, so both change by amounts proportional to α and ΔT, which leads to a ***volume expansion*** ΔV given by

$$\frac{\Delta V}{V_0} = \beta\,\Delta T \tag{14.10}$$

where V_0 is the initial volume and β is the ***coefficient of volume expansion.*** Values of α and β for a variety of substances are given in Table 14.5. Values of α are only given for solids because liquids and gases cannot (on their own) maintain a fixed shape as in Figure 14.17A. We can still define a coefficient of volume expansion (see Fig. 14.17B) for liquids and gases, however, and a few values are listed in Table 14.5.

CONCEPT CHECK 14.3 Linear Expansion (α) and Volume Expansion (β) Are Related

You can see from Table 14.5 that a substance with a large value of α also has a large value of β. Why?

EXAMPLE 14.6 Thermal Expansion of a Metal Bar

Consider a meterstick composed of platinum. It might be a copy of the standard meterstick as formerly employed to maintain the SI unit of length (Chapter 1). By what amount does the length of this meterstick change if the temperature increases by 1.0 K?

RECOGNIZE THE PRINCIPLE

This problem is a direct application of the definition of the thermal expansion coefficient. Given the change in temperature and the initial length, the change in length is $\Delta L = L_0 \alpha\, \Delta T$ (Eq. 14.9).

SKETCH THE PROBLEM

Figure 14.17A describes the problem.

IDENTIFY THE RELATIONSHIPS AND SOLVE

From Table 14.5, we have $\alpha = 8.8 \times 10^{-6}\ \text{K}^{-1}$ for platinum. Inserting into Equation 14.9 gives

$$\Delta L = L_0\alpha\,\Delta T = (1.0\ \text{m})(8.8 \times 10^{-6}\ \text{K}^{-1})(1.0\ \text{K}) = \boxed{8.8 \times 10^{-6}\ \text{m}}$$

► ***What does it mean?***

This increase in the bar's length is rather small (a bit smaller than the diameter of a typical human hair). For some applications, however, changes of this magnitude cannot be tolerated, especially when high precision is required. That is why standards of length that use the wavelength of light (instead of the length of a metal bar) have been developed. We'll explain in Chapter 25 how the wavelength of light can be measured very accurately.

EXAMPLE 14.7 Area Expansion

Consider a square hole in an aluminum plate (Fig. 14.18). If the aluminum is heated, will this hole become smaller or larger?

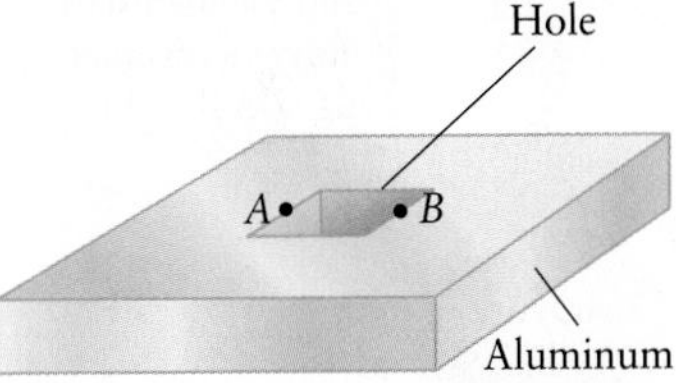

▲ **Figure 14.18** Example 14.7. What happens to the area of a hole when this plate is heated?

RECOGNIZE THE PRINCIPLE

A metal bar becomes longer when it is heated; hence, the ends of the bar are farther apart. Therefore, any two points on the bar will also be farther apart after heating.

SKETCH THE PROBLEM

Figure 14.19 shows how two points on a metal bar move farther apart when the bar is heated.

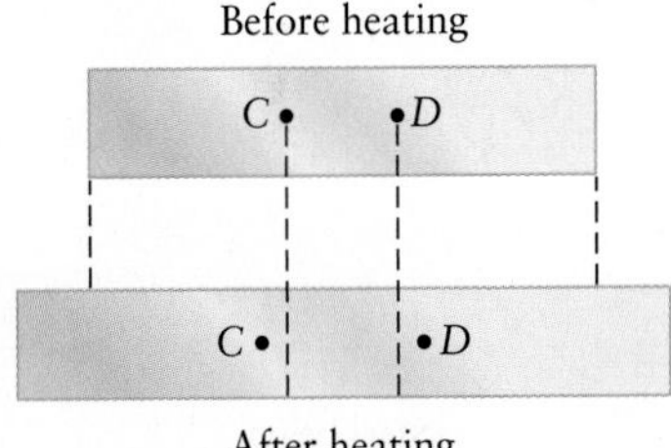

▲ **Figure 14.19** Example 14.7. When a bar expands, points *C* and *D* move farther apart.

IDENTIFY THE RELATIONSHIPS AND SOLVE

Figure 14.18 shows two points *A* and *B* at the edges of a square hole. These points will be farther apart after the plate is heated, so the hole will get $\boxed{\textit{larger}}$.

► ***What does it mean?***

The same reasoning applies to a circular hole and explains why heating the metal lid on a glass jar makes it easier to remove a "stuck" lid. The metal lid expands when it is heated. The glass jar also expands, but because the coefficient of linear thermal expansion of most metals is greater than the coefficient for glass (Table 14.5), the opening in the metal lid expands more than the glass jar, making it easier to remove the lid.

Effects of Thermal Expansion

The thermal expansion of the metal bar in Example 14.6 is not large, but the effect of thermal expansion can be quite significant in certain situations. For example, consider the problem of designing a large bridge. One of the first bridges built to span the Mississippi River near St. Louis, the Eads Bridge, was built with steel in the 1870s. This bridge was one of the first large-scale applications of steel. The bridge used several spans, each with a length of about 500 ft (about 150 m). Let's calculate

▲ **Figure 14.20** Many bridges contain expansion joints so that they can expand and contract without breaking away from their supports or forming unsafe gaps.

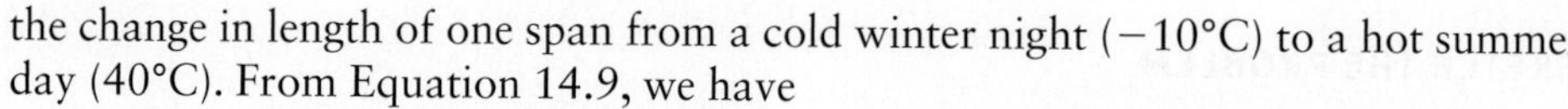

the change in length of one span from a cold winter night ($-10°C$) to a hot summer day ($40°C$). From Equation 14.9, we have

$$\Delta L = L_0 \alpha \, \Delta T = L_0 \alpha (T_f - T_i)$$

The temperature change is $40°C - (-10°C) = 50°C$, which corresponds to a temperature change of 50 K. Using the coefficient of linear thermal expansion for steel from Table 14.5 gives

$$\Delta L = (150 \text{ m})(12 \times 10^{-6} \text{ K}^{-1})(50 \text{ K}) = 0.090 \text{ m} \quad (14.11)$$

This change in length is about 3.5 in. and is thus quite significant; it would certainly not be desirable for a bridge that carries trains or cars to have a gap of this size. For this reason, bridges, buildings, and many other structures are designed with special expansion joints that allow for expansion and contraction without forming unsafe gaps or breaking (Fig. 14.20).

Another example of thermal expansion is sketched in Figure 14.21. A mercury thermometer uses the thermal expansion of mercury to measure changes in temperature. When T changes, the mercury (which is a liquid) expands or contracts, whereas the glass holding the mercury expands or contracts relatively little because the coefficients of thermal expansion of glass are small. The result is a change ΔL in the level of the meniscus in the thermometer tube (Fig. 14.21). Changes in the meniscus position are proportional to changes in temperature, just as are changes in the length of the bar in Example 14.6. The magnitudes of the changes, though, are much larger for the thermometer because there is a (relatively) large pool of mercury in the bulb of the thermometer, and all this mercury expands into a very narrow tube when the temperature increases.

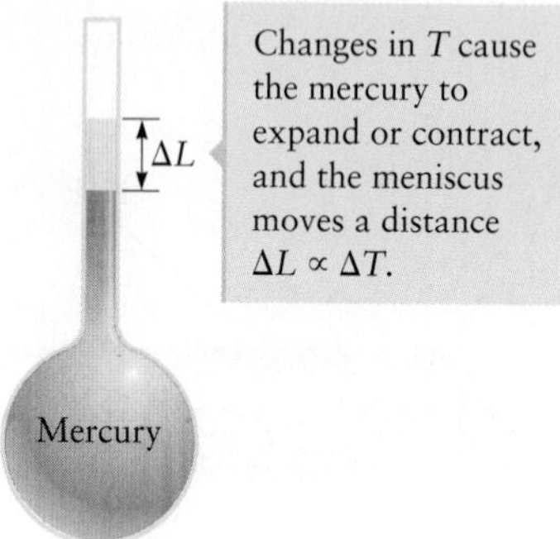

Changes in T cause the mercury to expand or contract, and the meniscus moves a distance $\Delta L \propto \Delta T$.
ΔL
Mercury

▲ **Figure 14.21** A mercury thermometer uses the thermal expansion of liquid mercury to measure temperature.

CONCEPT CHECK 14.4 Heating Your Piano

A piano string is held at a particular tension T so that its standing wave frequencies (Chapter 13) correspond to the note middle C (fundamental frequency approximately 262 Hz). Suppose the string is heated while the temperature of the rest of the piano does not change. Will the fundamental frequency of the string go up or down?

EXAMPLE 14.8 Thermal Expansion and Forces on a Bridge

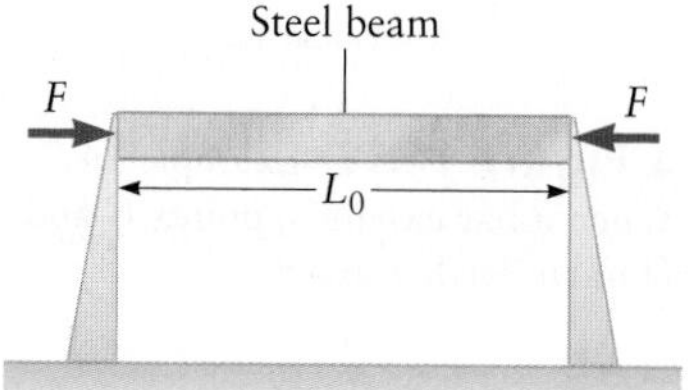

▲ **Figure 14.22** Example 14.8.

Figure 14.22 shows a steel bridge, similar to the one considered in connection with Equation 14.11. In that calculation, we found that a steel beam with initial length $L_0 = 150$ m will expand an amount $\Delta L = 0.090$ m when heated by 50 K. Suppose this steel beam has cross-sectional area $A = 1.0 \text{ m} \times 0.10 \text{ m} = 0.10 \text{ m}^2$ and is part of a bridge. Also suppose the supports of the bridge hold the ends of the steel beam fixed in place so that the ends cannot move. What force must the supports exert on the ends of the beam to keep it from expanding?

RECOGNIZE THE PRINCIPLE

If there are no forces on the ends of the beam, it will expand by an amount ΔL when it is heated. To prevent this expansion, the supports must exert a force that compresses the beam by an amount ΔL. This compression is described by the elastic properties of the beam and is proportional to Young's modulus (Chapter 11).

SKETCH THE PROBLEM

Figure 14.22 shows the problem along with the forces exerted by the supports on the ends of the beam.

IDENTIFY THE RELATIONSHIPS

If a force F is applied to each end of the bar, it will compress an amount ΔL. From Equation 11.23,

$$\frac{F}{A} = Y\frac{|\Delta L|}{L_0} \quad (1)$$

where Y is Young's modulus and A is the cross-sectional area of the bar. We have used the absolute value of ΔL because the applied forces compress the bar, reducing its length (making ΔL negative). The bridge supports hold the ends fixed, so this change in length must compensate for the change in length due to thermal expansion in Equation 14.11. Hence, $|\Delta L| = 0.090$ m in Equation (1).

SOLVE

Young's modulus for steel is 2.0×10^{11} Pa (Table 11.1). Solving for the force, we get

$$F = AY\frac{|\Delta L|}{L_0} = (0.10 \text{ m}^2)(2.0 \times 10^{11} \text{ Pa})\frac{(0.090 \text{ m})}{(150 \text{ m})} = \boxed{1.2 \times 10^7 \text{ N}}$$

▸ ***What does it mean?***

This force is quite substantial and is why metal beams and bridges sometimes "buckle." Architects and mechanical engineers are paid to prevent such buckling.

Thermal Expansion of Water

Table 14.5 lists the coefficients of thermal expansion for a number of different substances near room temperature. Strictly speaking, α and β both vary with temperature, but this variation is often small and can usually be ignored. Therefore, the length and volume of an object both change linearly with temperature, and we can usually use Equations 14.9 and 14.10 to calculate changes in length and volume even when ΔT is large (i.e., many tens or even hundreds of kelvins). There is, however, one important substance whose expansion is not a simple linear function. Figure 14.23 shows how the density of H_2O varies over a wide temperature range, and Figure 14.23B shows an expanded view near the freezing temperature ($T = 0°$C). At temperatures well above freezing, the density decreases as T increases. That is just the usual thermal expansion; an expanding sample has a lower density. At temperatures near and above room temperature (around 20°C), this expansion can be described by Equation 14.10 with $\beta \approx 2.1 \times 10^{-4}$ K^{-1}. Just above freezing, however, the density of water exhibits a maximum and cannot be described by Equation 14.10 with a constant value of β. What's more, the density of the solid (ice) near freezing is less than the density of the liquid, and H_2O thus contracts when it melts. That is why ice floats in water, as described by Archimedes's principle (Chapter 10). Most substances do not behave in this way; a solid generally does not float in its own liquid. This unusual behavior of water has many important consequences. For example, in winter a lake freezes first at the surface, which is good for ice skaters. It also enables fish to live comfortably in the water underneath and still have access to food at the lake bottom.

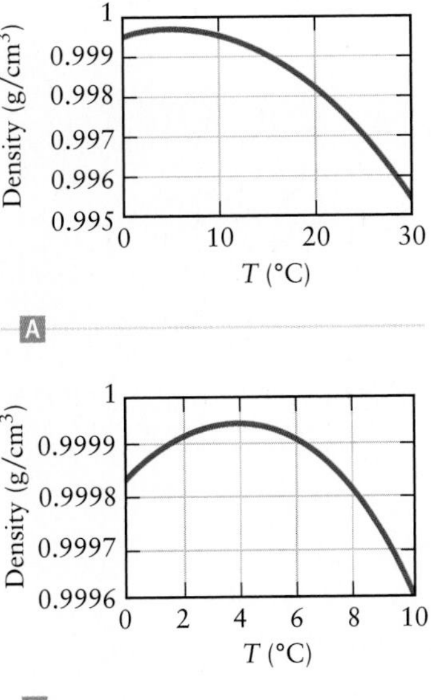

▲ **Figure 14.23** A Density of water as a function of temperature. B Expanded view of part A near the freezing temperature (0°C).

EXAMPLE 14.9 Effect of Freezing on a Cell

The fluid inside a cell is composed mainly of water, so when a cell freezes, the fluid inside expands. This expansion causes a large ("outward") pressure on the cell wall and usually causes the cell to rupture. Consider a spherical cell with initial radius $r_i = 5.0$ μm.

(continued) ▸

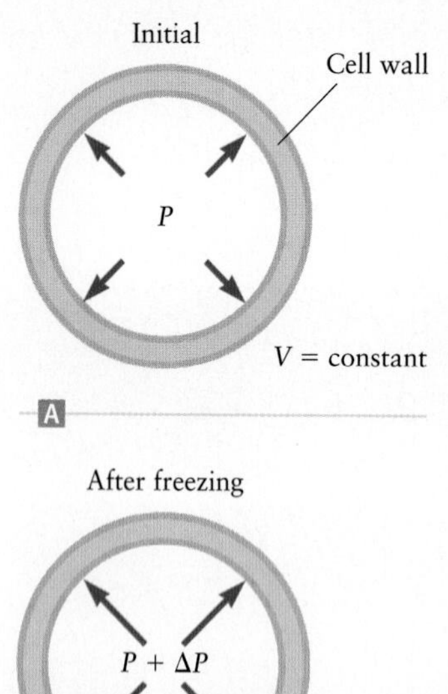

▲ **Figure 14.24** Example 14.9. If water in a cell freezes, it produces an increased pressure on the cell wall.

Calculate the increase in pressure inside the cell when it freezes. Assume the fluid inside the cell is mainly water and use the data for water from Figure 14.23; the density of water at 0°C is 999.8 kg/m^3 (= 0.9998 g/cm^3), and the density of ice is 917.0 kg/m^3.

RECOGNIZE THE PRINCIPLE

When the water in the cell freezes, its volume expands by an amount ΔV_{ice}, which we can calculate from the given values of the densities of water and ice. We then need to calculate what pressure must be applied to the cell to compress it back to its original volume; this pressure is the extra pressure that would be exerted on the cell wall (assuming the cell does not rupture). We can find it using the bulk modulus of ice (Chapter 11).

SKETCH THE PROBLEM

Figure 14.24 describes the problem. The volume expansion that would occur due to freezing must be compensated for by an increase in the pressure.

IDENTIFY THE RELATIONSHIPS

If ΔV_{ice} is the amount the water would expand when it freezes, the pressure P from the cell wall must produce a volume change $\Delta V = -\Delta V_{ice}$. In Chapter 11, we learned that the pressure P required to compress a material to give a volume change ΔV is (Eq. 11.25)

$$P = -B\frac{\Delta V}{V_i} \quad (1)$$

where B is an elastic constant called the bulk modulus. In this example, we need to compress the ice, so ΔV is negative and P in Equation (1) will be positive. From Table 11.1, the bulk modulus of ice is approximately $B = 9 \times 10^9$ Pa.

SOLVE

Mass is equal to the product of density and volume, so $m = \rho V$. The mass of the cell is the same at the start (i) and end (f), giving

$$\rho_i V_i = \rho_f V_f \quad (2)$$

The cell is (for simplicity) spherical; hence,

$$V_i = \tfrac{4}{3}\pi r_i^3 = \tfrac{4}{3}\pi(5.0 \times 10^{-6}\text{ m})^3 = 5.2 \times 10^{-16}\text{ m}^3$$

Inserting this result into Equation (2) along with the given initial and final densities, we find

$$V_f = \frac{\rho_i}{\rho_f}V_i = \frac{999.8\text{ kg/m}^3}{917.0\text{ kg/m}^3}(5.2 \times 10^{-16}\text{ m}^3) = 5.7 \times 10^{-16}\text{ m}^3$$

The change of volume of the cell is thus

$$\Delta V_{ice} = V_f - V_i = 5 \times 10^{-17}\text{ m}^3$$

Using the relation in Equation (1), we then get

$$P = -B\frac{(-\Delta V_{ice})}{V} = (9 \times 10^9\text{ Pa})\frac{5 \times 10^{-17}\text{ m}^3}{5.2 \times 10^{-16}\text{ m}^3} = \boxed{9 \times 10^8\text{ Pa}}$$

► ***What does it mean?***

Atmospheric pressure is only about 1×10^5 Pa, so this pressure is nearly 10,000 times atmospheric pressure! That is why cells usually rupture when they freeze.

14.6 Heat Conduction

Heat is the energy that flows between systems at different temperatures. This energy transfer can take place in three ways: ***conduction***, ***convection***, and ***radiation***. We now consider heat conduction; we'll discuss the other two methods of heat transfer in the following sections.

Figure 14.25A shows a solid bar connecting two separate systems. These systems might be large pieces of metal, and we assume they are at different temperatures T_1 and T_2 so that heat energy flows through the bar from one end to the other. If $T_1 > T_2$, energy flows from the system on the left (the hotter one) to the one on the right, and we say that the bar ***conducts*** heat. The bar itself cannot be described by a single temperature value. Instead, the bar's temperature varies smoothly from T_1 on the left to T_2 on the right (Fig. 14.25B).

We can understand this heat flow in terms of the vibrations of atoms in the bar. (If the bar is made of metal, electrons outside the ion cores in the bar will also contribute to heat flow.) In any particular region within the bar, atoms closer to the hot end (at the left) are at a slightly higher temperature than the neighboring atoms that are nearer the cold end (the right). The atoms at a higher temperature have a larger vibration amplitude, corresponding to larger potential and kinetic energies. As the atoms vibrate, some of the extra vibrational energy of "hotter" atoms is transferred to nearby "colder" atoms, similar to the way a wave transfers energy from place to place without transferring matter; energy thus "flows" from the hot end of the bar to the cold end.

The amount of energy that flows through the bar in Figure 14.25 depends on the bar's area A, length L, and the temperature difference $\Delta T = T_1 - T_2$. Heat flow also depends on a property of the bar called the ***thermal conductivity***, κ. Heat flow is measured by the energy Q that arrives per unit time t at the cold end. This heat flow is the rate at which energy leaves the hot end of the bar and arrives at the cold end and is given by

$$\frac{Q}{t} = \kappa A \frac{\Delta T}{L} \tag{14.12}$$

Values of the thermal conductivity for a variety of materials are listed in Table 14.6. Values of κ vary widely, with small values for gases (such as air) and high values for metals.

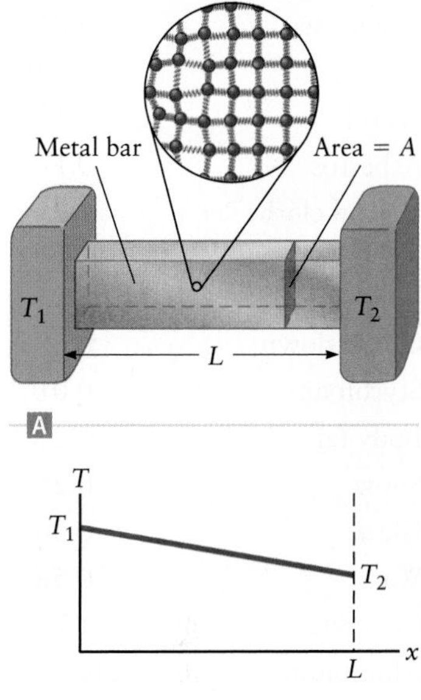

▲ **Figure 14.25** A When the ends of a solid bar are placed in contact with systems at different temperatures, the bar conducts energy from the hot end to the cold end. B Plot of the temperature as a function of position along the bar.

CONCEPT CHECK 14.5 Thermal Conductivity of Gases

The values in Table 14.6 show that the thermal conductivities of gases are generally much smaller than the thermal conductivities of solids. Use the microscopic picture of heat conduction involving atomic vibrations (Fig. 14.25) to explain why.

CONCEPT CHECK 14.6 Designing a Better Building

The walls of buildings are made from many different materials, including concrete, wood, glass, and steel. If solid walls with the same area and thickness are made from these materials, which would allow the most heat conduction through the wall for a given temperature difference, (a) concrete, (b) wood, (c) glass, or (d) steel?

TABLE 14.6 Thermal Conductivities of Some Common Substances

Material	κ W/(m · K)
Argon gas	0.016
Air	0.024
Wood	0.15
Asbestos	0.17
Cotton cloth (approximate)	0.2
Wool	2.2
Goose down	0.025
Styrofoam	0.010
Body fat	0.20
Snow	0.25
Glass	0.80
Water	0.56
Concrete	1.7
Aluminum	240
Iron	79
Steel	50
Copper	400
Silver	430
Ice (0°C)	2.2

Note: All substances are at atmospheric pressure and room temperature unless noted.

EXAMPLE 14.10 Conduction of Heat through a Windowpane

Consider a glass windowpane (Fig. 14.26) with area $A = 0.50$ m^2, thickness $L = 4.0$ mm, and temperature difference $\Delta T = 30$°C between the two sides (i.e., between the inside of the house and the outside). What is the total amount of heat energy that flows through the windowpane in one day?

RECOGNIZE THE PRINCIPLE

We are given the dimensions of the windowpane, and we know the thermal conductivity of glass from Table 14.6. We can therefore use Equation 14.12 to calculate the rate at which heat flows through the windowpane, and from that we can calculate the total energy that flows through in one day.

SKETCH THE PROBLEM

Figure 14.26 describes the problem.

IDENTIFY THE RELATIONSHIPS AND SOLVE

The value of a temperature difference on the Celsius and Kelvin scales is the same, so $\Delta T = 30$ K. Inserting these values into Equation 14.12 with the value of the thermal conductivity κ for glass from Table 14.6, we find

$$\frac{Q}{t} = \kappa A \frac{\Delta T}{L} = \left(0.80 \frac{\text{W}}{\text{m} \cdot \text{K}}\right)(0.50 \text{ m}^2)\frac{30 \text{ K}}{0.0040 \text{ m}} = 3000 \text{ W} \quad (1)$$

The total energy that flows out through the window in one day can be calculated from Equation (1) using $t = 1$ day $= 86{,}400$ s. We get

$$Q = (3000 \text{ W})t = (3000 \text{ J/s})(86{,}400 \text{ s}) = \boxed{2.6 \times 10^8 \text{ J}} \quad (2)$$

► ***What does it mean?***

This energy comes (ultimately) from the house's furnace. In the case of an electric furnace, the cost of electrical energy is typically about \$0.02 for 10^6 J. The total cost of the energy Q in Equation (2) is thus approximately

$$\text{cost} = (2.6 \times 10^8 \text{ J})(\$0.02/10^6 \text{ J}) = \$5$$

A cost of \$5 per window per day is quite substantial because most houses have many windows; that is why energy-efficient windows are popular.

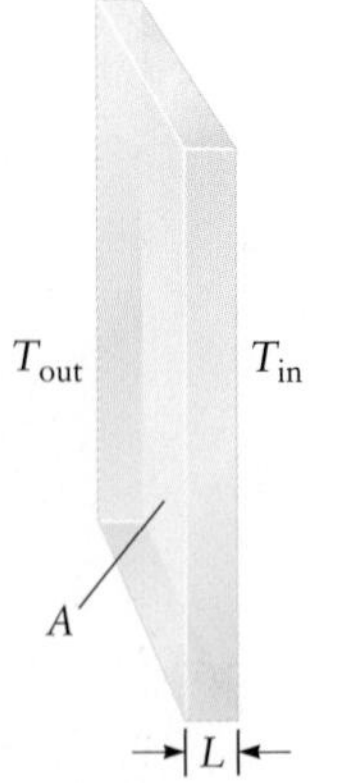

▲ **Figure 14.26** Conduction of heat through a windowpane.

CONCEPT CHECK 14.7 Dependence of Heat Conduction on Size

Suppose the height and width of the windowpane in Figure 14.26 are both increased by a factor of two. By what factor does the heat flow through the window change?

(a) There is no change.
(b) It increases by a factor of two.
(c) It increases by a factor of four.

EXAMPLE 14.11 Heat Flow and Cooling Your Coffee

The author likes to bring a thermos bottle of coffee to his office each morning. The walls and lid of a thermos bottle have low thermal conductivities, so the rate of heat flow out of the bottle is small (Fig. 14.27A). Suppose the coffee inside has mass $m = 0.30$ kg and an initial temperature of 65°C. Coffee is mostly water, so its specific heat is about the same as that of water. If the average rate of heat flow through the walls of the thermos bottle is $Q/t = 3.0$ W, approximately how long will it take the coffee to cool to a final temperature of 55°C? Take room temperature to be 25°C.

RECOGNIZE THE PRINCIPLE

We use the general approach outlined in the problem-solving strategies on calorimetry. We identify two systems, the coffee and the room around it. These systems are initially at temperatures $T_{\text{coffee}, i} = 65°\text{C}$ and $T_{\text{room}} = 25°\text{C}$, and the final temperature of the coffee is $T_{\text{coffee}, f} = 55°\text{C}$. Because the room is much larger than the coffee, the temperature of the room does not change significantly. (The room acts as a thermal reservoir.) The heat energy that must flow from the coffee into the room to produce this change in temperature can be found from the specific heat of the coffee and Equation 14.6. This energy equals the rate of heat flow (given to be $Q/t = 3.0$ W) multiplied by the total time.

SKETCH THE PROBLEM

Figure 14.27 describes the problem and shows the direction of heat flow.

IDENTIFY THE RELATIONSHIPS

Using the definition of specific heat (Eq. 14.6) gives

$$Q = cm\,\Delta T = cm(T_{\text{coffee}, f} - T_{\text{coffee}, i})$$

Inserting the given values for the temperatures, mass, and specific heat of water (Table 14.3), we find

$$Q = [4186\ \text{J/(kg}\cdot\text{K)}](0.30\ \text{kg})(55 - 65)\ \text{K} = -1.26 \times 10^4\ \text{J} \quad (1)$$

where we have again kept three significant figures to avoid roundoff errors below. The negative sign indicates that energy flows out of the coffee. Here, Q equals the rate at which heat energy travels through the walls of the thermos bottle multiplied by the total time. This rate is related to Q by $rate = |Q|/t = 3.0$ W, so we have

$$|Q| = (rate)(t_{\text{total}})$$

SOLVE

Solving for the time and using the result for Q from Equation (1) leads to

$$t_{\text{total}} = \frac{|Q|}{rate} = \frac{1.26 \times 10^4\ \text{J}}{3.0\ \text{W}} = \boxed{4200\ \text{s}}$$

► ***What does it mean?***

This result is more than 1 hour, which gives plenty of time to enjoy the coffee. A better thermos bottle would have a smaller heat flow rate and would thus keep the coffee warm longer.

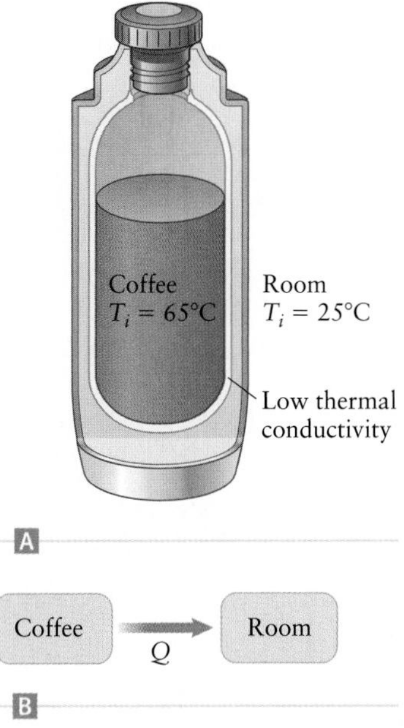

▲ **Figure 14.27** Example 14.11. A The coffee in this thermos bottle is initially much warmer than room temperature. B Energy flows via conduction from the coffee to the room, thus cooling the coffee.

Why Do Metals "Feel" Cold?

The concepts of thermal conductivity and thermal resistance can clarify a well-known aspect of heat conduction. Experience tells us that on a wintry day, an outdoor metal flagpole feels colder to the touch than a piece of Styrofoam. If the flagpole and the Styrofoam are at the same outdoor temperature, why do they feel different?

The temperature-sensitive nerves in your skin sense the difference between your inside body temperature and your skin temperature. We thus have a heat-flow problem involving a series of materials (Fig. 14.28). The thermal conductivity κ of a metal is high (Table 14.6), whereas κ for Styrofoam is low. So, according to the definition of thermal conductivity (Eq. 14.12), when your finger is in contact with a metal, the rate of heat flow Q/t from your finger to the metal is much larger than

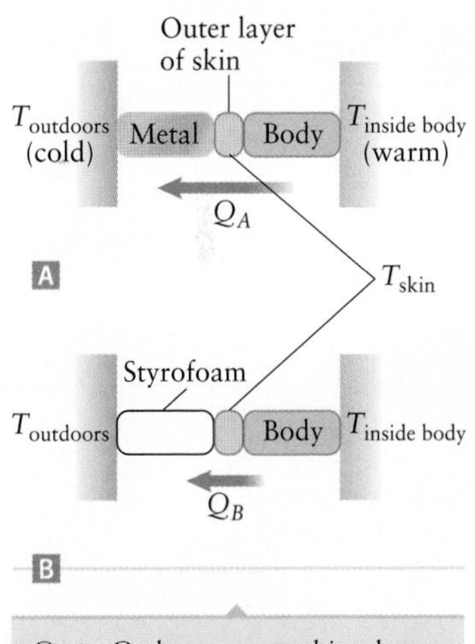

▲ **Figure 14.28** Heat flowst from inside your body into your surroundings. A If the outer layer of skin is in contact with a metal (a good thermal conductor), the rate of heat flow is greater than in B, where the skin is in contact with Stryofoam (a poor thermal conductor). The higher heat-flow rate in part A makes the outer layer of skin colder than in part B.

if you were touching Styrofoam instead. This increased heat flow into the metal causes your skin to have a lower temperature than with Styrofoam. That is why a metal feels colder to the touch than Styrofoam.

14.7 Convection

Convective heat flow is based on the phenomenon of thermal expansion. Consider a pot of water being heated on a stove (Fig. 14.29A). If the pot is heated gently so that no gas bubbles are formed, heat energy is carried through the water in two ways. One way is by conduction as discussed in Section 14.6, but much more energy can transported via the motion of water sketched in the figure. Water near the bottom becomes hotter due to the conduction of heat from the burner. Thermal expansion then causes the density of this warmer water to be less than the density of the colder water that is above it in the pot. This warm low-density water moves upward due to the buoyant force associated with Archimedes's principle. As the warm water moves upward, it cools through conduction of heat to the cooler part of the pot and the air above; eventually, this water returns to the bottom of the pot, following the circular flow pattern in Figure 14.29A. A similar convective heat-flow pattern involving the air in a house causes the upper floors to be warmer than the lower floors. Convective flow also plays an important role in transporting heat energy within the oceans and the atmosphere.

Wind Chill

A process similar to convection leads to the "wind chill" effect familiar on a cold winter day. Your skin is moist, and water from it evaporates slowly, placing a small amount of water vapor into the layer of air next to your skin. Evaporation is a phase change, requiring an amount of energy equal to the latent heat. This energy comes from your skin and flows to the nearby air, causing your skin to cool (Fig. 14.30A). This process is called ***evaporative cooling***. In our previous discussion of the latent heat of vaporization of water, we assumed water was at the boiling temperature (100°C), but there is a latent heat of vaporization at all points along the liquid–gas phase change line in Figure 14.10. Evaporative cooling can thus occur over a wide temperature range.

On a windy day, the layer of water-vapor-filled air near your skin is quickly carried away, leaving a "dry" layer in its place. The rate of evaporation is faster into this dry layer than it would be into a moist layer of air. As a result, more latent heat is removed from your skin than if the air layer were filled with water vapor (Fig. 14.30B). Your skin thus loses energy faster on a windy day than on a calm day.

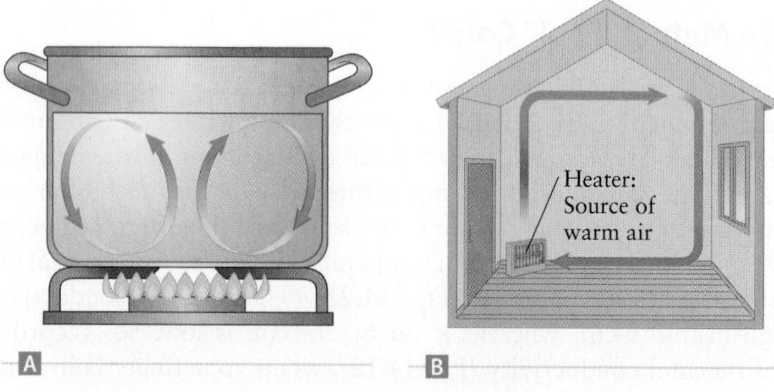

▶ **Figure 14.29** Convection transports energy through the movement of matter. This movement is caused by thermal expansion.

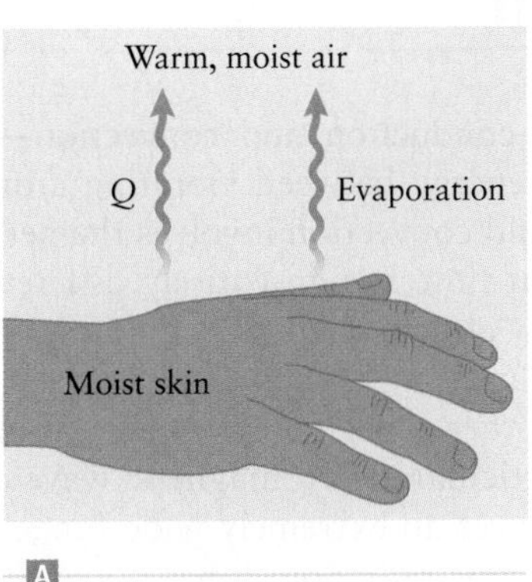

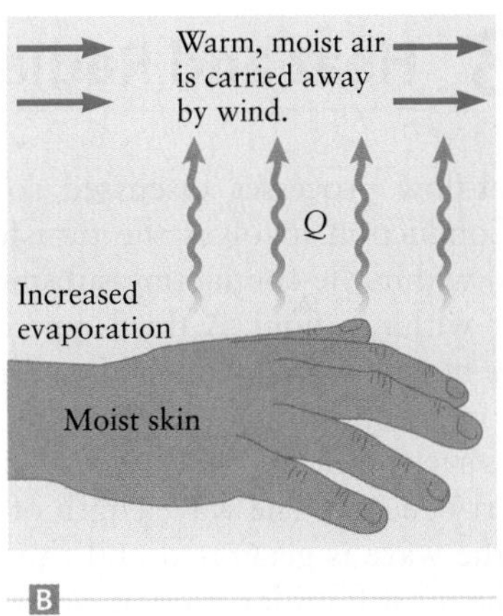

◀ **Figure 14.30** A When moisture evaporates from the body, the latent heat of evaporation leaves your skin, cooling your body and forming a layer of moist and relatively warm air near your skin. B On a windy day, this moist layer of air is rapidly blown away, increasing the evaporation rate. This evaporation increases the amount of latent heat removed from your body and lowers the skin temperature relative to part A.

EXAMPLE 14.12 Cooling Your Body by Sweating

The bicycle rider in Example 14.3 generated 300 J of heat each second. If all this energy is removed from his body by sweating, how much water would he lose each hour by sweating?

RECOGNIZE THE PRINCIPLE

When sweat evaporates from a cyclist's skin, it carries away an amount of energy equal to the latent heat of vaporization. The motion of the bicycle provides the "wind" that carries away the water vapor from a cyclist's skin.

SKETCH THE PROBLEM

Heat energy flows from the cyclist into the surrounding air as sketched in Figure 14.31.

IDENTIFY THE RELATIONSHIPS

Let m be the mass of water evaporated as sweat in 1 h. The corresponding latent heat is

$$Q = mL_{\text{vaporization}} \tag{1}$$

which must equal the amount of heat generated by the cyclist. The cyclist generates 300 J each second, so the total heat generated in 1 h (= 3600 s) is

$$\text{total heat generated} = (300\text{ J})(3600\text{ s}) = 1.1 \times 10^6\text{ J}$$

SOLVE

Setting the heat generated equal to the latent heat in Equation (1) gives

$$Q = mL_{\text{vaporization}} = 1.1 \times 10^6\text{ J}$$

Solving for m and using the latent heat of vaporization of water from Table 14.4, we get

$$m = \frac{1.1 \times 10^6\text{ J}}{2.2 \times 10^6\text{ J/kg}} = \boxed{0.50\text{ kg}}$$

► ***What does it mean?***

This mass corresponds to about half a liter of water, which is a typical amount of water a person should consume during 1 h of strenuous exercise.

▲ **Figure 14.31** Example 14.12. Convection or just wind carries moist air away from this cyclist, increasing the rate at which sweat evaporates.

14.8 Heat and Radiation

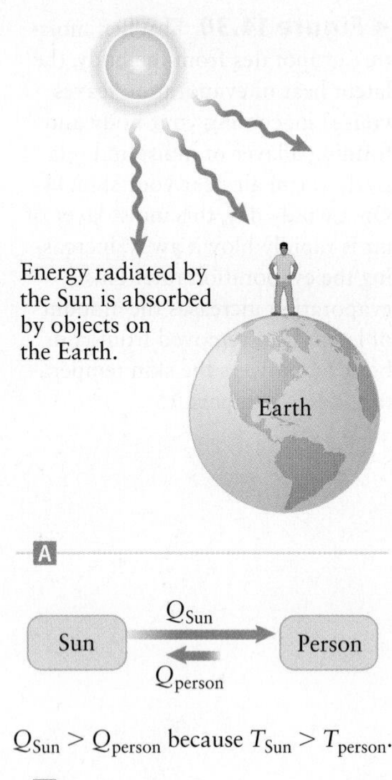
Energy radiated by the Sun is absorbed by objects on the Earth.
Earth
A
Sun
Q_{Sun}
Person
Q_{person}
$Q_{Sun} > Q_{person}$ because $T_{Sun} > T_{person}$.
B

▲ **Figure 14.32** A Electromagnetic radiation from the Sun is absorbed by a person on the Earth. B The person also emits radiation, so there is a flow of energy in both directions, from the Sun to the person and from the person to the Sun.

Both heat-flow processes discussed so far—conduction and convection—involve matter. Conduction involves the transfer of energy between vibrating atoms (and electrons) within the conducting substance, and convection involves the net motion of matter within a fluid. A third type of heat flow has an entirely different basis. ***Radiative*** heat flow involves energy carried by electromagnetic radiation.

Electromagnetic radiation is a type of wave and is thus characterized by frequency, wavelength, and wave speed. Like other types of waves, electromagnetic waves carry energy. The wavelength of a particular electromagnetic wave depends on how the wave is generated and can vary over an extremely wide range. Visible light is the most familiar type of electromagnetic radiation; it has a wavelength in the range of about 400 nm to 700 nm. Other types of electromagnetic radiation correspond to different wavelengths, including ultraviolet and infrared radiation which are also important in heat flow.

Radiative heat flow involves two objects, which might be the Sun and a person on the Earth as sketched in Figure 14.32A. If these two objects are at different temperatures, there is a net transfer of energy—heat energy—from the hotter object (in this case the Sun) to the cooler one (the person).

To understand radiative heat transfer, we must consider several questions. First, why does an object, such as the Sun in Figure 14.32, emit electromagnetic radiation, and how is this radiation absorbed by another object? Second, what are the frequencies of the radiated electromagnetic waves, and how do they depend on the temperature of the emitting object? Finally, how does the amount of radiation absorbed by the cooler object depend on the temperature of that object?

Electromagnetic radiation is generated whenever a particle with an electric charge, such as an electron, proton, or ion, vibrates or undergoes an acceleration in some way. Hence, the atomic vibrations sketched in Figure 14.25A generate electromagnetic radiation that carries away energy. The vibration amplitude of the atoms in an object depends on temperature, so the radiated energy depends on temperature. This radiation—that is, the energy in these electromagnetic waves—is absorbed by another object when that radiation produces a force on the electric charges in the object.

Radiation and the Notion of a "Blackbody"

When electromagnetic radiation, including visible light, bombards an object, some of the radiation energy is absorbed and some is reflected. When you look at an object such as a wall or this book, your eye detects light reflected by the object, which determines the color of the object. An object that completely absorbs all visible light appears black. Physicists have extended this notion and defined a ***blackbody*** as an object that absorbs all electromagnetic radiation at all frequencies. A perfect blackbody does not exist, but the concept of an ideal blackbody is very useful.

The total amount of energy an object radiates depends on its temperature. An object with surface area A and temperature T radiates energy at a rate Q/t given by the ***Stefan–Boltzmann law,***

Stefan–Boltzmann law

$$\frac{Q}{t} = \sigma e A T^4 \tag{14.13}$$

where $\sigma = 5.67 \times 10^{-8}\ \text{W/(m}^2 \cdot \text{K}^4)$ is called Stefan's constant. The factor e in Equation 14.13 is the ***emissivity*** of the object, a measure of how efficiently it radiates energy. The value of the emissivity must be less than or equal to 1. Objects for which $e = 1$ are perfect blackbodies. Hence, a blackbody is not only a perfect absorber of radiation, it is also the best possible radiator of energy.

The Stefan–Boltzmann law gives the rate at which energy is emitted as electromagnetic radiation—that is, the radiated power—at all wavelengths. This power is

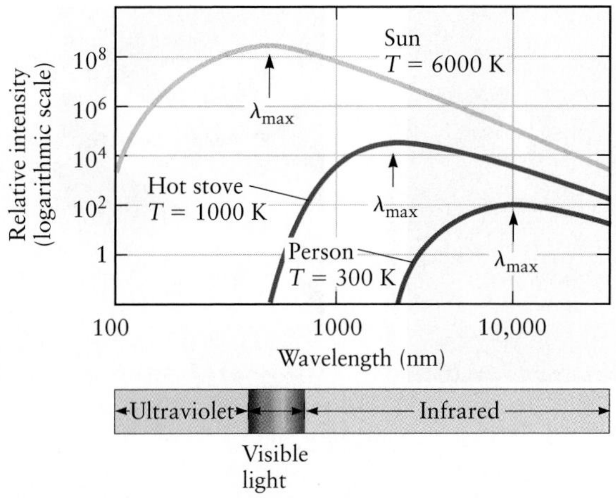

▲ **Figure 14.33** The wavelength at which an object emits radiation most strongly (λ_{max}) depends on its temperature, and that determines the "color" at which it glows. Notice how the total amount of radiation emitted increases very rapidly as T increases.

distributed as a function of wavelength as shown in Figure 14.33. For an object at temperature T, the radiated power is greatest at the wavelength λ_{max} given by ***Wien's law***:

$$\lambda_{max} = (2.90 \times 10^{-3}\ \text{m} \cdot \text{K})/T \tag{14.14}$$

Wien's law

Figure 14.33 shows a plot of the radiated power as a function of wavelength for blackbody objects at different temperatures. According to Wien's law, the wavelength λ_{max} at which the radiated power is greatest depends on the temperature of the object. The Sun has a surface temperature of approximately 6000 K, and λ_{max} falls in the range of visible light. Some stars are much hotter than our Sun, with temperatures as high as 20,000 K, and for these stars λ_{max} is shorter than the visible range of wavelengths. These stars appear blue to the human eye because they emit more radiation at the blue end of the visible range than at the red end. For a somewhat cooler object, such as the burner of an electric stove at a typical temperature of 1000 K, λ_{max} lies in the infrared. The stove burner appears red because it emits some light at the red end of the visible spectrum in addition to its infrared radiation.

EXAMPLE 14.13 Radiation Emitted by the Human Body

The human body emits electromagnetic radiation according to the body's temperature. Estimate λ_{max} for the human body.

RECOGNIZE THE PRINCIPLE

The peak intensity of the blackbody radiation from an object occurs at a wavelength that depends on its temperature according to Wien's law. Knowing the temperature of the body, we can find λ_{max}.

SKETCH THE PROBLEM

No sketch is needed.

IDENTIFY THE RELATIONSHIPS

The temperature of the human body is approximately 37°C, which must be converted to the Kelvin scale before insertion into Wien's law. We find $T = 37 + 273.15 = 310$ K.

(continued) ▶

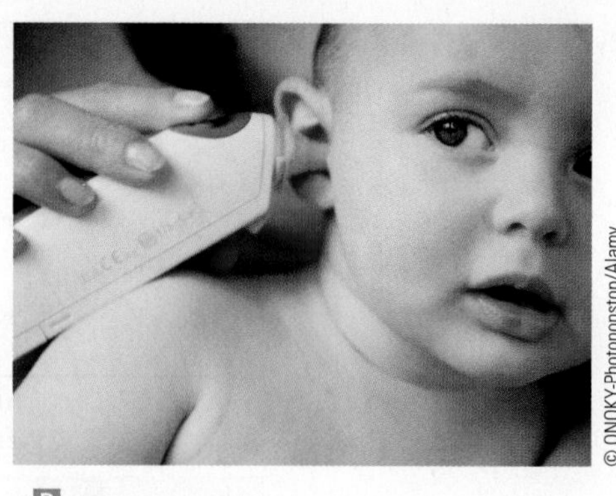

▶ **Figure 14.34** Example 14.13. A Night-vision goggles convert the infrared radiation emitted by an object to visible light, making it possible to see objects in the dark. B An "ear thermometer" uses the blackbody radiation from the eardrum to find the body's temperature.

SOLVE

Using this value of the temperature in Wien's law (Eq. 14.14), we find

$$\lambda_{max} = (2.90 \times 10^{-3}\ \text{m} \cdot \text{K})/T = (2.90 \times 10^{-3}\ \text{m} \cdot \text{K})/(310\ \text{K}) = \boxed{9.4 \times 10^{-6}\ \text{m}}$$

▶ ***What does it mean?***

The wavelength λ_{max} lies far into the infrared region of the spectrum (9400 nm; see Fig. 14.33). It is well outside the visible range, so your eye cannot detect this emitted radiation. Night-vision equipment is able to detect this radiation and convert it into wavelengths that can be detected by the human eye (Fig. 14.34A). Also, the "ear thermometers" used by medical professionals (Fig. 14.34B) measure your body temperature by detecting the infrared radiation emitted by your eardrum.

Insight 14.2

THE RELATION BETWEEN RADIATION AND ABSORPTION

The same relation gives the amount of radiation emitted by an object at a particular temperature (Eq. 14.13) and the amount of radiation absorbed by an object from an environment at that same temperature (Eq. 14.15). This follows from the notion of thermal equilibrium. If an object is in thermal equilibrium with its environment, both are at the same temperature. The total energy of the object is then constant, and the rate at which the object emits energy (through radiation) must equal the rate at which it absorbs energy (also through radiation) from its environment. Hence,

$$\left(\frac{Q}{t}\right)_{\text{absorbed}} = \left(\frac{Q}{t}\right)_{\text{emitted}}$$

The Stefan–Boltzmann Law and Heat Flow

The Stefan–Boltzmann law (Eq. 14.13), which gives the total power radiated by an object at temperature T, has several noteworthy features. First, the total power varies as the fourth power of the temperature, so the radiated energy increases very rapidly as T is increased. Second, the power is proportional to the emissivity e; this factor is equal to unity for an ideal blackbody, but is smaller than 1 for any real object. In general, the value of e depends on the detailed properties of the material and will even be a function of frequency. Many objects are not far from the blackbody limit, however, so the Stefan–Boltzmann law with an emissivity factor of $e = 1$ provides an approximate description of most radiating objects. Third, the Stefan–Boltzmann law applies to all objects, at all temperatures. The Sun radiates energy that reaches a person on the Earth, but the person also radiates energy that reaches the Sun (Fig. 14.32B). Because the hotter object has a larger emitted power than the cooler one, the net transfer of energy is from the hotter to the cooler object, as we expect for heat flow. Fourth, the Stefan–Boltzmann law also describes how radiation is *absorbed* by an object. When an object of surface area A is located in an environment that has a temperature T, the heat absorbed by the object is

$$\left(\frac{Q}{t}\right)_{\text{absorbed}} = \sigma e A T^4 \tag{14.15}$$

Radiation from the Sun and the Temperature of the Earth

Let's apply the Stefan–Boltzmann law to the energy radiated by the Sun. The Sun has a surface temperature of approximately 6000 K, and its emissivity $e \approx 1$. The radius of the Sun is $r = 7.0 \times 10^8$ m, and the surface area through which the radiation is emitted is the area of a sphere with this radius, $A = 4\pi r^2 = 6.2 \times 10^{18}\ \text{m}^2$. Inserting these values into the Stefan–Boltzmann law (Eq. 14.13), we have

$$\frac{Q}{t} = \sigma e A T^4 = [5.67 \times 10^{-8}\ \text{W}/(\text{m}^2 \cdot \text{K}^4)](1)(6.2 \times 10^{18}\ \text{m}^2)(6000\ \text{K})^4$$

$$\frac{Q}{t} = 4.6 \times 10^{26}\ \text{W} \tag{14.16}$$

which (not surprisingly) is an enormous amount of power.

We can use our result for the heat radiated by the Sun to compute the Earth's temperature. This calculation is based on the principle of conservation of energy. According to Figure 14.35, energy arrives at the Earth via radiation from the Sun, and the Earth itself then radiates energy back into space.[2] The amount of energy radiated by the Earth depends on its temperature T_E, and the value of T_E must be such that the total energy *radiated by* the Earth is equal to the energy it *absorbs* from the Sun. We'll leave the details of this calculation to the next example.

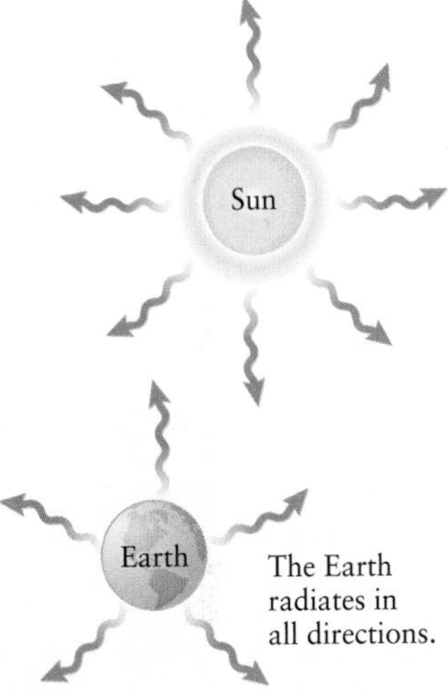

▲ **Figure 14.35** The Earth absorbs energy through radiation from the Sun and emits radiation according to its own temperature. The balance of these energies determines the temperature of the Earth.

EXAMPLE 14.14 Ⓡ Temperature of the Earth

Estimate the temperature of the Earth according to the picture of heat flow in Figure 14.35. Assume the Earth is a perfect blackbody so that $e = 1$.

RECOGNIZE THE PRINCIPLE

We need to compute the amount of energy radiated by the Sun that is absorbed by the Earth, Q_{Sun}, and the energy radiated by the Earth, Q_{Earth}. We have already calculated the total power radiated by the Sun in Equation 14.16, but only a fraction of this power reaches the Earth. That fraction is given by the area of the Earth's *disk* (not the spherical surface area of the Earth) divided by the surface area of a sphere that contains the Earth's orbit (Fig. 14.36).

SKETCH THE PROBLEM

Figure 14.36 shows the problem.

IDENTIFY THE RELATIONSHIPS

Denoting the radius of the Earth by r_E, the area of the Earth's disk is $A_E = \pi r_E^2$. If r_S is the Earth–Sun distance, the surface area of the sphere that contains the Earth's orbit (Fig. 14.36) is $A_S = 4\pi r_S^2$, and the fraction of the Sun's radiated power that strikes the Earth is the ratio A_E/A_S. The radiated power from the Sun that actually intercepts the Earth is thus

$$\frac{Q_{\text{Sun}}}{t} = \frac{A_E}{A_S}\frac{Q_{\text{total}}}{t} = \frac{\pi r_E^2}{4\pi r_S^2}\frac{Q_{\text{total}}}{t}$$

where Q_{total}/t is given in Equation 14.16. Inserting values for r_E and r_S from Appendix A, Table A.3, and our previous result for Q_{total}/t, we obtain

$$\frac{Q_{\text{Sun}}}{t} = \frac{\pi r_E^2}{4\pi r_S^2}\frac{Q_{\text{total}}}{t} = \frac{\pi(6.4 \times 10^6\ \text{m})^2}{4\pi(1.5 \times 10^{11}\ \text{m})^2}(4.6 \times 10^{26}\ \text{W}) = 2.1 \times 10^{17}\ \text{W} \tag{1}$$

The energy radiated by the Earth is also given by the Stefan–Boltzmann law, Equation 14.13. The factor A in this application of the Stefan–Boltzmann law is the area of the

(continued) ▶

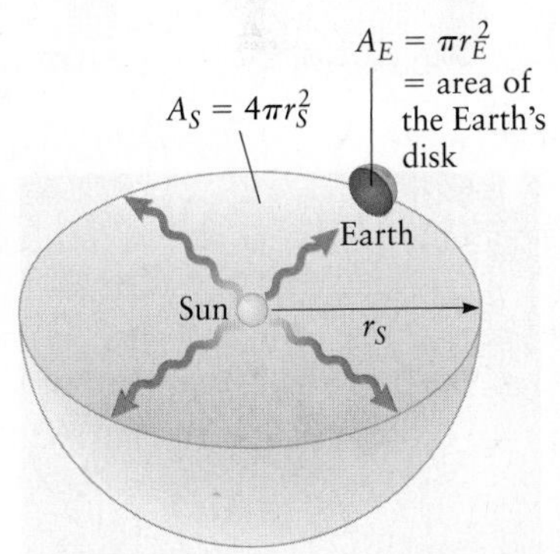

▲ **Figure 14.36** Example 14.14. The Sun emits radiation that propagates outward with circular wave fronts. The spherical wave front that intercepts the Earth has an area $A_S = 4\pi r_S^2$, where r_S is the Earth–Sun distance. The area of the Earth exposed to the Sun is $A_E = \pi r_E^2$, where r_E is the radius of the Earth. Notice that A_E is *not* the (spherical) surface area of the Earth.

[2]Energy also reaches the Earth from "deep space." In Chapter 31, we'll see that regions of space that appear dark (i.e., without stars) radiate energy as a blackbody at a temperature of 3 K. The energy that reaches the Earth in this way is negligible compared to the radiation from the Sun.

Earth's spherical surface, which is $4\pi r_E^2$. (It is *not* the same as the area of the Earth's disk in Fig. 14.36.) The temperature T_E of the Earth is the quantity that we wish to find. Inserting all these values into the Stefan–Boltzmann law gives

$$\frac{Q_{\text{Earth}}}{t} = \sigma e A T_E^4$$

SOLVE

The rate at which the Earth absorbs energy must equal the rate at which it radiates; otherwise, the Earth would be heating up or cooling down over time. Hence,

$$\frac{Q_{\text{Sun}}}{t} = \frac{Q_{\text{Earth}}}{t} = \sigma e A T_E^4$$

Solving for T_E leads to

$$T_E^4 = \frac{Q_{\text{Sun}}/t}{\sigma e A}$$

$$T_E = \left(\frac{Q_{\text{Sun}}/t}{\sigma e A}\right)^{1/4}$$

Taking the Earth's emissivity as $e = 1$, inserting Q_{Sun}/t from Equation (1), and using $r_E = 6.4 \times 10^6$ m gives

$$T_E = \left[\frac{2.1 \times 10^{17}\ \text{W}}{[5.67 \times 10^{-8}\ \text{W}/(\text{m}^2 \cdot \text{K}^4)](1)(4\pi r_E^2)}\right]^{1/4} = \boxed{290\ \text{K}}$$

► ***What does it mean?***

Our result for T_E is approximately 20°C. In obtaining this result, we assumed the Earth is a perfect blackbody, so $e = 1$. In fact, the Earth absorbs only about 70% of the radiation from the Sun and reflects the rest. In addition, our calculation actually gives the temperature of the atmosphere, which is somewhat cooler than the Earth's surface. Even with these approximations, our result is quite close to the actual value, confirming that the Sun is responsible for maintaining the Earth's temperature.

36,4°C

35

30

25

21,3°C

A

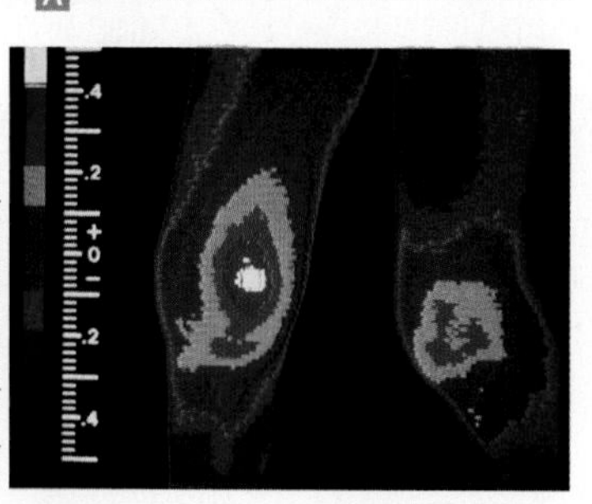

B

▲ **Figure 14.37** All objects, including people, emit radiation according to their temperature. With special detection equipment, this radiation can be used to deduce how the temperature of the skin varies from place to place on the body. **A** The skin temperature is lowest at the nose. **B** An arthritic knee (left) is warmer than a healthy one.

Medical Uses of Heat Radiation

An important application of heat radiation is medical thermography. This technique measures the heat radiation emitted by the body and forms an image based on the wavelength at which the radiated power is largest. According to Wien's law (Eq. 14.14), this wavelength depends on temperature, so these images show where a body is warmest and where it is coolest. Some examples are given in Figure 14.37; in these images, whiter regions are warmer than redder areas. Figure 14.37A shows that the nose is the coolest place on a person's face (not a surprising result!). There can be significant variations in temperature near joints; the image in Figure 14.37B shows arthritis associated with a knee joint. Other medical problems can be detected in this way; for example, cancerous tumors are often warmer than normal tissue. Other medical devices, such as ear thermometers (Example 14.13 and Fig. 14.34B), also use the human body's blackbody radiation.

The Greenhouse Effect

As a first approximation, we can treat most real objects as perfect blackbodies. Such was the approach that we took in Example 14.14, when we estimated the temperature of the Earth. A more realistic model of the Earth's temperature must go beyond this approximation, which leads us to consider the ***greenhouse effect***.

Although a complete discussion of the greenhouse effect is too complex to include here, we can understand the qualitative behavior using Figure 14.38, which shows how radiated energy is absorbed and reemitted by the Earth. This sketch is similar to Figure 14.35, but now we have included the effect of the atmosphere. We also indicate separately the infrared and visible radiation coming from the Sun; the atmosphere is not an ideal blackbody, and it absorbs these radiations differently. The Earth's atmosphere allows most of the radiation in the visible part of the spectrum through to the surface, whereas it absorbs much of the infrared radiation, allowing only a little of it to reach the Earth. Because much of the Sun's radiation is in the visible range, most of the energy from the Sun still passes through to the Earth's surface.

When we estimated the temperature of the Earth in Example 14.14, we had to account for radiation emitted by the Earth. Because its temperature is much lower than that of the Sun, most of the power radiated by the Earth is in the infrared (Fig. 14.33). The atmosphere absorbs some of this radiation, effectively trapping it and not letting this energy escape back into space. This pattern of energy flow is similar to what takes place in a greenhouse, where the glass roof plays the role of the atmosphere. (The analogy with a true greenhouse is not perfect, but the effect is still widely referred to as "the" greenhouse effect.) The infrared radiation trapped by the atmosphere makes the Earth's surface warmer than would be found if the atmosphere were not present.

The absorption of infrared radiation by the atmosphere is caused by certain molecules such as CO_2, which are called ***greenhouse gases.*** If the concentration of these gases in the atmosphere increases, the atmosphere absorbs infrared radiation more strongly and traps even more of the Earth's emitted radiation. The result is an increase in the Earth's temperature. It appears that the burning of fossil fuels is now causing an increase in the concentration of greenhouse gases, and many scientists believe that it will cause a significant increase in the Earth's temperature in the coming decades.

The details of radiative heat flow can be quite involved; even Figure 14.38 is a highly simplified picture of the heat-flow processes that determine the Earth's temperature. Our simple blackbody approximation is a very useful first step, but more accurate models are much more complex.

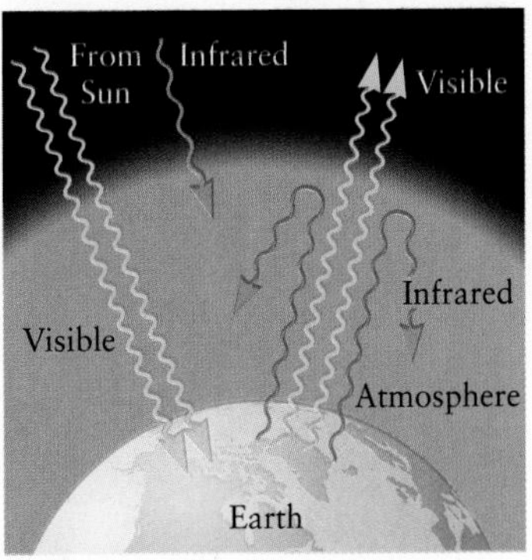

▲ **Figure 14.38** Greenhouse model of radiative heat flow between the Sun, the Earth, and the Earth's atmosphere.

Summary | CHAPTER 14

Key Concepts and Principles

Zeroth law of thermodynamics

If two systems A and B are in thermal equilibrium ($T_A = T_B$) and systems B and C are in thermal equilibrium ($T_B = T_C$), systems A and C are also in thermal equilibrium ($T_A = T_C$).

In words, the zeroth law of thermodynamics asserts that temperature is a unique property of a system that determines when two systems are in thermal equilibrium. Three different ***temperature scales*** are in common use: ***Fahrenheit***, ***Celsius***, and ***Kelvin***.

(Continued)

Heat

Heat flow is the transfer of energy from one object to another by virtue of a temperature difference. The energy transferred is also called heat energy or simply heat. Heat flow can occur in three ways: ***conduction***, ***convection***, and ***radiation***.

- ***Heat conduction*** is determined by the ***thermal conductivity*** κ, with

$$\frac{Q}{t} = \kappa A \frac{\Delta T}{L} \qquad \text{(14.12) (page 409)}$$

  where Q is the heat energy transferred in time t, L is the length of the system, and A is the cross-sectional area of the system.
- ***Convective heat flow*** involves the movement of mass and is caused by thermal expansion and contraction.
- ***Radiative heat transfer*** involves electromagnetic radiation. The total amount of energy an object radiates depends on its temperature. An object with a surface area A and temperature T radiates energy at a rate Q/t given by the ***Stefan–Boltzmann law***,

$$\frac{Q}{t} = \sigma e A T^4 \qquad \text{(14.13) (page 414)}$$

  The wavelength at which this radiation is strongest depends on temperature through ***Wien's law***,

$$\lambda_{\text{max}} = (2.90 \times 10^{-3}\ \text{m} \cdot \text{K})/T \qquad \text{(14.14) (page 415)}$$

  Wien's law explains why the color of a glowing object depends on its temperature.

Applications

Specific heat and latent heat

If an amount of heat energy Q is added to a substance of mass m, the temperature of the system increases by an amount ΔT, where

$$Q = cm\,\Delta T \qquad \text{(14.6) (page 396)}$$

and c is the ***specific heat*** of the substance. Equation 14.6 applies provided there is not a phase change. If there is a phase change such as melting or evaporation, a certain amount of energy called the ***latent heat*** must be added to or taken from the system.

Thermal expansion

When the temperature of a substance changes, its physical dimensions usually change. Changes in length are proportional to the ***coefficient of thermal expansion*** α, with

$$\frac{\Delta L}{L_0} = \alpha\,\Delta T \qquad \text{(14.9) (page 403)}$$

and changes in volume are proportional to the ***coefficient of volume expansion*** β

$$\frac{\Delta V}{V_0} = \beta\,\Delta T \qquad \text{(14.10) (page 404)}$$

Questions

SSM = answer in *Student Companion & Problem-Solving Guide*   ⓧ = life science application

1. When the metal cap on a glass jar is stuck (i.e., when the cap is too tight to open), it sometimes helps to hold the jar and cap under hot water. Explain why.

2. Consider a hole in a block of steel such as the cylinder in a steel engine block. Will the diameter of this hole get larger or smaller when the steel is heated? By how much will it change?

3. Common bulb thermometers use liquid mercury (the familiar silver liquid) or liquid methanol (with red dye for coloring) as the working substance enclosed in glass. What are the limitations in using these devices? Could a mercury or methanol thermometer measure the temperature of dry ice? (CO_2 is a solid below $-78°C$ at atmospheric pressure.) Why or why not? Could either type of thermometer measure the temperature of molten lead?

4. A bar of steel is attached to a bar of brass side by side to form one thicker bar of the same length (Fig. Q14.4). You find that when you heat this bar of two metals, it bends. Why? Will it bend toward the brass or toward the steel? Which way would it bend if the bar were cooled instead of heated?

Figure Q14.4

5. The pressure–temperature phase diagram for a typical substance is shown in Figure Q14.5. The line that separates the solid and liquid phases has a positive slope; that is, the melting temperature increases as the pressure is increased. Give a qualitative explanation of this observation, assuming that a material expands when it is heated.

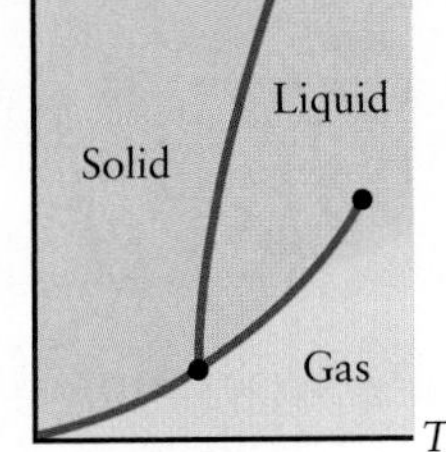

Figure Q14.5

6. ⓧ When you step out of a swimming pool on a windy day, you usually feel chilly. Explain why.

7. ⓧ On a hot summer day, an electric fan can make a difference in comfort. The fan does not cool the air, however. In fact, it slightly raises the temperature of the air that it circulates. How is it that a fan can make you cooler?

8. Most internal combustion engines in modern automobiles have a cooling system that uses a circulating fluid. In designing such a cooling system, should the liquid coolant have a small or a large heat capacity? Which would be best, a large or a small thermal conductivity?

9. ⓧ Consider a winter mountain survival scenario. Why is it a bad idea to eat snow, even if you have no way of melting it for drinking?

10. Given the value of α (e.g., for steel), can you calculate the value of β? Derive a general relation between α and β.

11. ⓧ Explain why getting burned with a certain mass of steam at 100°C is more painful and harmful than getting burned with the same mass of water (liquid) at 100°C.

12. ⓧ When you take a cookie sheet out of the oven, you use a pot holder because touching the hot cookie sheet would result in injury. The air in the oven, however, is at exactly the same temperature. Why doesn't the air burn your hand the way the cookie sheet would?

13. ⓧ You pour a cup of coffee, intending to drink it a few minutes later. If you want your coffee to be as hot as possible at this later time, should you add the cream when you pour the coffee or just before you drink it? Why?

14. SSM ⓧ Fruit trees bloom in the spring, and their blossoms eventually turn into oranges, apples, and so forth. Blossoms are susceptible to cold weather, so when there is the possibility of a frost, orchard owners often spray their trees with water. Explain how this practice can help prevent the blossoms from freezing.

15. Metal cookware made from steel or aluminum often has a copper bottom. Why?

16. A potato will bake faster if a nail is stuck through it. Why? Which would make the potato cook faster, an aluminum nail or a steel nail?

17. ⓧ Explain why a layer of fur (Fig. Q14.17) is a very poor conductor of heat.

Figure Q14.17

18. There are many different types of glass; all are composed mainly of SiO_2, but the addition of other elements can change the color and other properties substantially. One type of glass, called Pyrex, incorporates small amounts of the element boron, which gives it a very small coefficient of thermal expansion. Most oven-safe dishes are made from Pyrex because it tends not to crack when heated or cooled. Explain why.

19. Explain why it takes longer to cook a hard-boiled egg in Denver than in San Francisco. *Hint*: Consider the phase diagram of water and how the boiling temperature depends on pressure.

20. SSM A good-quality thermos bottle is double-walled and evacuated between these walls, and the internal surfaces are like mirrors with a silver coating. This configuration combats heat loss from all three transfer methods and keeps the bottle's contents—your coffee—hot. Which design feature of a thermos bottle minimizes heat transfer by radiation, by conduction, and by convection?

21. ⓧ Table 14.6 lists the thermal conductivities of several substances. Give qualitative arguments to explain the following questions. (a) Why is the thermal conductivity of goose down about the same as the thermal conductivity of air? (b) Goose down is very effective for making winter coats. Explain why goose down must be kept dry to be effective. Can you predict the value of the thermal conductivity when it is soaked with water?

22. ⓧ Heat energy is "lost" from the body through the skin, so Q_{lost} is proportional to the area of the skin. Heat energy is produced throughout the body as food is metabolized, muscles flex, and so on, so $Q_{produced}$ is proportional to the body volume. Explain why these two facts tend to make it easier for a large animal to stay warm compared with a small animal.

23. An ice cube can exist for many minutes while floating in water, even when the temperature of the water is greater than 0°C. Explain why.

24. ⓧ On a hot summer day, why is it easier to keep cool by wearing white clothing instead of black clothing?

25. When a wire is strung tightly between two supports, it is found that transverse waves on the wire have speed v. If the wire is heated, does the wave speed increase or decrease? Explain.

26. ⓧ Why does a lake freeze first at its surface?

27. Climates near an ocean tend to be milder than climates far inland. Explain why.

Problems available in ENHANCED WebAssign

Section 14.2 Temperature and Heat
Problems 1–8

Section 14.4 Phases of Matter and Phase Changes
Problems 9–18

Section 14.5 Thermal Expansion
Problems 19–35

Section 14.6 Heat Conduction
Problems 36–41

Section 14.7 Convection
Problems 42–43

Section 14.8 Heat and Radiation
Problems 44–53

Additional Problems
Problems 54–76

List of Enhanced Problems

Problem number	Solution in *Student Companion & Problem-Solving Guide*	Reasoning & Relationships Problem	Reasoning Tutorial in ENHANCED WebAssign
14.8	✓		
14.15	✓		
14.16		✓	
14.18		✓	
14.20		✓	✓
14.21		✓	
14.26	✓		
14.31		✓	
14.33		✓	✓
14.37		✓	
14.39	✓		
14.40		✓	
14.42	✓		
14.46		✓	
14.49		✓	
14.50	✓		
14.51		✓	
14.52		✓	✓
14.53		✓	
14.54	✓	✓	✓
14.55		✓	
14.59	✓		
14.71		✓	
14.72		✓	
14.73		✓	
14.74		✓	✓
14.75		✓	✓
14.76		✓	

CHAPTER

15

◀ *The properties of the air in these balloons and of other gases are described by an area of physics called kinetic theory. (© Travelwide/Alamy)*

Gases and Kinetic Theory

In this chapter, we apply Newton's laws to analyze the behavior of gases, the simplest phase of matter. In Chapter 14, we argued that systems containing many particles are so complicated that an approach using Newton's laws is prohibitively difficult and does not give a useful understanding of problems pertaining to heat flow and temperature. However, one special type of system—a dilute gas—is simple enough that Newton's laws *can* be applied. The atoms and molecules in a dilute gas are far apart, so they collide with one another only rarely. Hence, it is possible to apply Newton's laws in ways not possible with liquids or solids. Analysis of the properties of a dilute gas leads to important connections between temperature and motion on an atomic scale. Many properties of dilute gases are independent of composition, meaning that the same theory applies to a dilute gas of hydrogen, helium, oxygen, and so on. These ideas led to the notion of an ***ideal gas*** and an area of physics called ***kinetic theory***. This theory has a very rich and interesting history and has seen contributions from many physicists, including Newton and Einstein.

OUTLINE

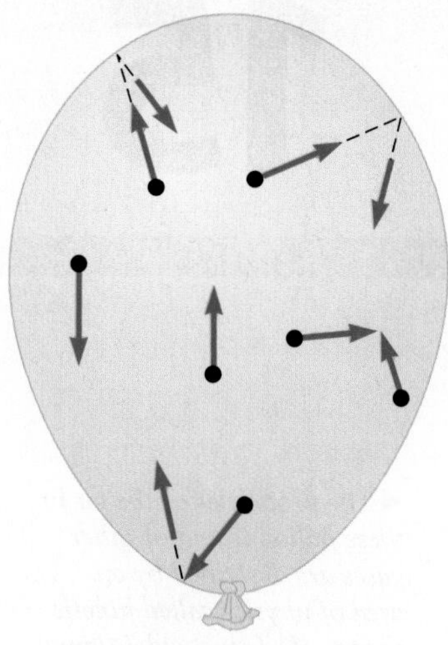

▲ **Figure 15.1** The molecules in a gas are in constant motion as they collide with one another and with the walls of their container.

15.1 Molecular Picture of a Gas

Consider a balloon containing a pure substance, so all the particles inside the balloon are identical (Fig. 15.1). These particles might be atoms, such as a balloon full of helium atoms, or molecules, such as carbon dioxide. For simplicity, we often refer to the particles as "molecules," with the understanding that all our ideas apply just as well to gases composed of single atoms, such as helium or argon.

The molecules in Figure 15.1 are all in motion, colliding with other gas molecules and with the walls of the container. The snapshot in Figure 15.1 raises a number of questions. (1) How often does a gas molecule collide with other molecules? (2) Are these collisions elastic or inelastic? (3) How are *microscopic* properties of the gas such as the molecular velocities related to *macroscopic* properties of the system such as temperature and pressure? (4) How many molecules are there in a typical sample of gas? Our job in this chapter is to answer all these questions.

A typical sample of gas contains a very large number of particles, usually on the order of 10^{23} molecules (or more). This very large number would seem to make the application of Newton's laws hopelessly complex. Actually, though, this large number of particles simplifies things by allowing us to apply statistical arguments, and because the number of particles is so large, the accuracy of a statistical analysis will be very good.

As you probably suspect, the number 10^{23} just mentioned is connected with ***Avogadro's number***, N_A. This pure number has the value

Avogadro's number

$$N_A = 6.02 \times 10^{23} \tag{15.1}$$

where N_A is the number of particles in a ***mole***.[1] Avogadro's number plays a central role in connecting the microscopic world of atoms and molecules to the macroscopic world. The ***atomic mass*** of an element is the mass (usually quoted in grams) of a natural sample of the element containing 1 mole of atoms. For example, the atomic mass of carbon is $M_C = 12.011$ g (sometimes the units are given as grams per mole). We denote atomic mass using a capital letter to distinguish it from the mass of a single carbon atom m_C; the units of m_C are grams per atom, or simply grams. From the definition of atomic mass,

$$m_C = \frac{M_C}{N_A} = \frac{12.011 \text{ g/mole}}{6.02 \times 10^{23} \text{ atoms/mole}} = 2.0 \times 10^{-23} \text{ g} = 2.0 \times 10^{-26} \text{ kg} \tag{15.2}$$

In Section 15.2, we'll show that a typical sample of gas particles contains on the order of 1 mole of particles. Thus, in discussions of kinetic theory and the properties of gases, N_A and factors like 10^{23} appear often.

EXAMPLE 15.1 Ⓡ The Number of Molecules in a Cup of Coffee

How many molecules are in a typical cup of coffee? Coffee is composed mainly of water, so for simplicity assume the sample contains only water molecules.

RECOGNIZE THE PRINCIPLE

We can calculate the number of moles of water molecules in a cup of coffee if we know the total mass of water M_{total} and the mass of 1 mole of water molecules M_{water}. We can find M_{water} from the atomic masses of hydrogen and oxygen, and although M_{total} is not given, we can use our experience to estimate a value.

[1] Avogadro's number, N_A, is defined as the number of atoms in a sample containing exactly 12 g of the carbon-12 isotope. We'll discuss isotopes in Chapter 30.

SKETCH THE PROBLEM

No figure is necessary.

IDENTIFY THE RELATIONSHIPS

A typical cup of coffee contains about 300 g (300 milliliters) of water, so we take $M_{total} = 300$ g. Water has the molecular formula H_2O (two hydrogen atoms plus one oxygen atom). The atomic mass of hydrogen is $M_H = 1.01$ g, and the atomic mass of oxygen is $M_O = 16.00$ g. (See the periodic table at the back of this book.) The mass of 1 mole of water molecules is thus

$$M_{water} = 2M_H + M_O = 2(1.01 \text{ g}) + 16.00 \text{ g} = 18.02 \text{ g}$$

which can also be written as $M_{water} = 18.02$ g/mole.

SOLVE

To find the number of moles in our sample, we divide M_{total} by the mass of 1 mole. The number of moles of water molecules in our sample of 300 g is thus

$$n_{water} = \frac{M_{total}}{M_{water}} = \frac{300 \text{ g}}{18.02 \text{ g/mole}} = 17 \text{ moles}$$

and the number of molecules N_{water} is

$$N_{water} = n_{water} \times N_A = (17 \text{ moles})(6.02 \times 10^{23} \text{ molecules/mole})$$

$$N_{water} = \boxed{1.0 \times 10^{25} \text{ molecules}}$$

► ***What does it mean?***

The number of water molecules in this sample is within two orders of magnitude of Avogadro's number. We'll find similar results for the numbers of atoms or molecules in typical macroscopic samples of gases, liquids, and solids.

CONCEPT CHECK 15.1 Mass of a Water Molecule

What is the mass of a single water molecule?

(a) 18 g (b) 3.0×10^{-23} kg (c) 3.3×10^{-27} kg (d) 3.0×10^{-26} kg

15.2 Ideal Gases: An Experimental Perspective

Experimental studies of the properties of gases began around the time of Newton and led to the discovery of a number of different "gas laws." These laws provided the clues that eventually gave rise to kinetic theory and relate various macroscopic properties of a gas such as its temperature and pressure. Most of these gas laws apply accurately only for dilute samples, which means that the spacing between molecules is much larger than the size of an individual molecule. Fortunately, that is not a serious restriction because many important cases, including air at room temperature, can be closely approximated as a dilute gas.

Avogadro's law

For a sample of gas at constant pressure and temperature, the volume is proportional to the number of molecules in the sample.

Avogadro's law

Avogadro's law tells us that the average spacing between gas particles is constant, so the density is also constant, provided the pressure and temperature are held fixed.

Boyle's law

For a sample of gas at constant temperature, the product of the pressure and volume is constant.

In mathematical terms, Boyle's law is just

Boyle's law

$$PV = \text{constant} \quad (\text{at constant } T) \tag{15.3}$$

According to Boyle's law, if we increase the pressure in a sample of gas held at constant temperature, the volume must decrease so that the product PV stays fixed. On the other hand, if we decrease the pressure, the volume must then increase, again keeping PV constant. In terms of Figure 15.1, Boyle's law says that changing the pressure changes the average spacing between particles.

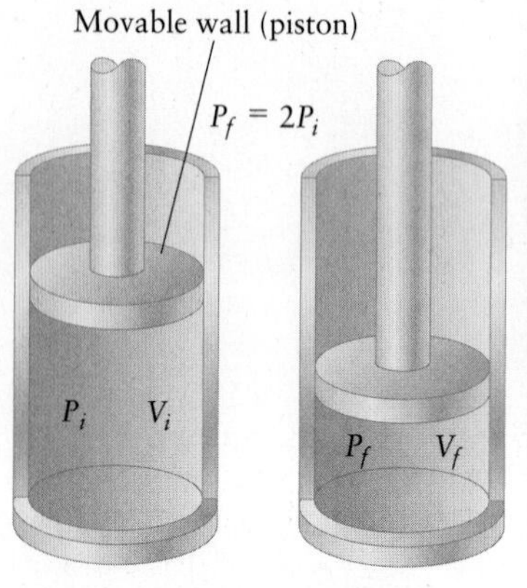

▲ **Figure 15.2** Concept Check 15.2.

CONCEPT CHECK 15.2 Density of a Dilute Gas

A cylinder (Fig. 15.2) contains a dilute gas at pressure P_i and volume V_i. One end wall of the cylinder is movable (forming a piston); moving this wall changes the volume of the gas. Suppose the gas is compressed so that the final pressure is twice the initial pressure, $P_f = 2P_i$. If the temperature is constant (so that Boyle's law applies), what is the final density?

(a) The final density is twice the initial density.
(b) The final density is half the initial density.
(c) The final density is equal to the initial density.

An important part of Boyle's law is the constant on the right-hand side of Equation 15.3. This constant must be proportional to the number of particles in the gas to be consistent with Avogadro's law. Experiments show that this constant also depends on temperature, as described by two more gas laws.

Charles's law

For a sample of gas at constant pressure, if the temperature is changed by a small amount ΔT, the volume changes by an amount ΔV, with

Charles's law

$$\Delta V \propto \Delta T \tag{15.4}$$

Gay-Lussac's law

For a sample of gas held in a container with constant volume, changes in pressure are proportional to changes in temperature.

Gay-Lussac's law

$$\Delta P \propto \Delta T \tag{15.5}$$

We have actually encountered Gay-Lussac's law in our discussion of gas thermometers in Chapter 14, where we found a linear relation between P and T in a dilute gas (Fig. 14.5A).

Absolute Temperature and the Kelvin Scale

Charles's law and Gay-Lussac's law both include temperature, and it is interesting to ask how Charles and Gay-Lussac were able to define and measure temperature in their work. Charles and Gay-Lussac used ***reference temperatures*** determined by the properties of various substances. Two convenient temperatures are those at which ice melts and at which water boils. Charles and Gay-Lussac used these reference temperatures together with a material property such as thermal expansion (Chapter 14) to interpolate between and extrapolate beyond the reference range. That is how a mercury thermometer works; the basic principle is shown in Figure 15.3.

This approach to measuring temperature was very useful, but it led to new questions. (1) What materials and properties can be used to make a thermometer (besides mercury), and can we be sure that all such thermometers will give the same results?

(2) According to Equation 15.5, changes in P for a dilute gas are proportional to changes in T. What happens to the temperature when $P = 0$, and does such a pressure (or temperature) even make sense?

Let's consider question 2 first. Careful measurements of pressure and temperature for a variety of gases give results like those sketched in Figure 15.4A; as temperature is reduced, the pressure decreases in a linear fashion (in proportion to changes in T), according to Gay-Lussac's law (Eq. 15.5). Any particular sample of gas will eventually condense and form a liquid, at which point the pressure falls below the linear P–T relation that holds at high temperature (Fig. 15.4B). This result does not conflict with Gay-Lussac's law because that relation applies only for dilute samples, and a gas is not dilute near its condensation temperature. Even though the linear behavior breaks down at low temperatures, we can still extrapolate the linear region from high temperature and pressure down to low pressures and find the temperature at which the pressure "would" be zero. If we do that with different samples of different gases, we find that they all extrapolate to zero pressure at the same temperature, $T = -273.15°C$ (Fig. 15.4A). This result applies to all gases, regardless of composition, and also holds for samples with different densities. Pressure is defined as the magnitude of the force per unit area exerted by a fluid on a surface (see Eq. 10.1), so it is not possible for P to be negative. This fact helped lead to the creation of the Kelvin temperature scale. The Kelvin scale is defined so that $T = 0$ K when the pressure of a dilute gas becomes zero, which is also called the ***absolute zero*** of temperature.

Because the linear relation between P and T is a universal property of dilute gases, it can be used as a universal way to measure temperature. One can calibrate a ***gas thermometer*** by measuring P at two reference temperatures (such as the melting point of ice and the boiling point of water) and thereby obtain an absolute calibration of the P–T relation for that particular thermometer. Most importantly, a gas thermometer can be made using any type of dilute gas.

The laws of Charles and (especially) Gay-Lussac thus lead to the notion of absolute temperature and provide an experimental method for measuring it. When we deal with gases and kinetic theory, it is most convenient to work with the Kelvin temperature scale.

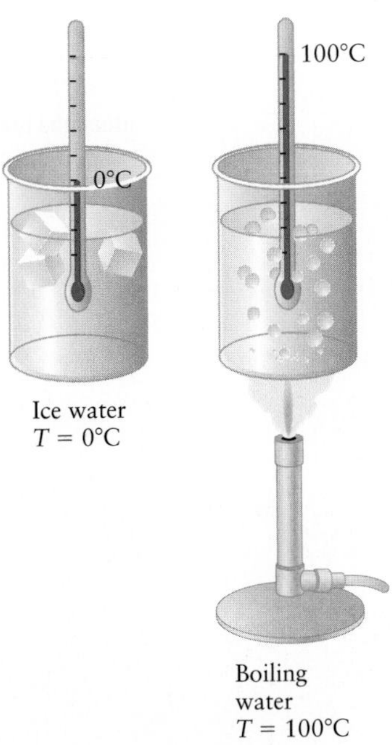

▲ **Figure 15.3** A mercury thermometer uses thermal expansion of mercury to measure changes in temperature. This thermometer is calibrated using known reference temperatures such as the melting and boiling temperatures of water.

The Ideal Gas Law

Each of the gas laws discussed above concern a particular aspect of the behavior of a dilute gas. These laws are all contained in a single relation known as the ***ideal gas law.***

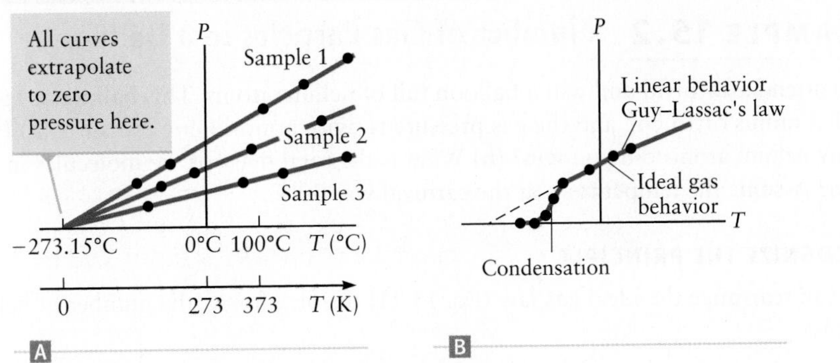

▲ **Figure 15.4** A Experiments show that the pressure of all dilute gases varies linearly with temperature when the gas is well above the condensation temperature. *All* the data at high temperatures for *all* gases extrapolate to $P = 0$ at the *same* temperature, defined to be $T = 0$ on the Kelvin scale. B At sufficiently low temperatures, all gases condense and the pressure then drops below the linear P–T dependence.

Ideal gas law

For a dilute gas composed of any substance,

Ideal gas law

$$PV = nRT \tag{15.6}$$

This relation applies to a sample of a dilute gas containing n moles of gas particles, and R is the ***universal gas constant*** with the value

$$R = 8.31 \frac{\text{J}}{\text{mole} \cdot \text{K}} \tag{15.7}$$

We can also express the ideal gas relation in terms of the number of gas particles N instead of the number of moles n. Using the definition of the mole and Avogadro's number (Eq. 15.1), the total number of particles is $N = nN_A$, so $n = N/N_A$. Substituting for n in Equation 15.6, we get

$$PV = nRT = \frac{N}{N_A} RT$$

$$PV = N\left(\frac{R}{N_A}\right)T \tag{15.8}$$

The factor R/N_A is the ratio of the gas constant to Avogadro's number; it is called ***Boltzmann's constant*** k_B and has the value

$$k_B = \frac{R}{N_A} \tag{15.9}$$

$$k_B = 1.38 \times 10^{-23}\ \text{J/K} \tag{15.10}$$

Using the number of gas particles N and Boltzmann's constant, the ideal gas law (Eq. 15.8) can be written as

$$PV = Nk_BT \tag{15.11}$$

The two versions of the ideal gas law, Equations 15.6 and 15.11, are equivalent; the only difference is that one involves the number of moles n, whereas the other involves the number of particles N.

The ideal gas law contains the laws of Avogadro, Boyle, Charles, and Gay-Lussac all as special cases. For example, when the temperature and number of particles are kept fixed, the ideal gas law predicts that the product PV is a constant, which is just Boyle's law (Eq. 15.3). Most importantly, the ideal gas law applies to all dilute gases, with the same value of the gas constant R and the same value of Boltzmann's constant k_B.

EXAMPLE 15.2 Number of Gas Particles in a Balloon

You attend a carnival and win a balloon full of helium atoms. This balloon is spherical with a radius of 15 cm, and the gas pressure is twice atmospheric pressure. (**a**) How many helium atoms did you win? (**b**) What is the total mass of the molecules in the balloon? Assume the temperature at the carnival was 20°C.

RECOGNIZE THE PRINCIPLE

We can rearrange the ideal gas law (Eq. 15.11) to solve for N, the number of helium atoms:

$$N = \frac{PV}{k_BT} \tag{1}$$

We are given the pressure, volume, and temperature, so we can find N. The total mass can then be found using the atomic mass of helium.

SKETCH THE PROBLEM

No figure is needed.

IDENTIFY THE RELATIONSHIPS

The pressure is twice atmospheric pressure, and $P_{atm} = 1.01 \times 10^5$ Pa, so for the balloon we get $P = 2P_{atm} = 2.0 \times 10^5$ Pa. The sample is a sphere of radius 0.15 m; hence, the volume is

$$V = \tfrac{4}{3}\pi r^3 = \tfrac{4}{3}\pi(0.15 \text{ m})^3 = 0.014 \text{ m}^3$$

Converting the temperature to the Kelvin scale, we have $T = 273 + 20 = 293$ K.

SOLVE

(a) Inserting these values for the pressure, volume, and temperature into our expression for N, Equation (1), gives

$$N = \frac{PV}{k_BT} = \frac{(2.0 \times 10^5 \text{ Pa})(0.014 \text{ m}^3)}{(1.38 \times 10^{-23} \text{ J/K})(293 \text{ K})} = \boxed{6.9 \times 10^{23} \text{ atoms}}$$

which corresponds to

$$n = \frac{N}{N_A} = \frac{6.9 \times 10^{23} \text{ atoms}}{6.02 \times 10^{23} \text{ atoms/mole}} = 1.1 \text{ moles}$$

(b) The total mass m_{total} of the helium atoms equals the number of moles n multiplied by the mass of 1 mole. According to the periodic table, the mass of 1 mole of helium atoms is 4.00 g. Combining this with the answer for n from part (a) gives

$$m_{total} = n \times (4.00 \text{ g/mole}) = (1.1 \text{ moles})(0.00400 \text{ kg/mole}) = \boxed{0.0044 \text{ kg}}$$

► ***What does it mean?***

The total mass is equal to the mass of about two (U.S.) pennies. When using the ideal gas law in either form (Eq. 15.6 or 15.11), T must be expressed in terms of the absolute temperature, the temperature on the Kelvin scale. The answer for part (a) illustrates again that the number of molecules in a typical macroscopic sample of gas is on the order of Avogadro's number.

CONCEPT CHECK 15.3 The Ideal Gas Law and the Effect of Composition

Suppose the balloon in Example 15.2 contains nitrogen molecules instead of helium atoms. How does that affect the number of particles N in the balloon?

(a) There is no change. The number of nitrogen molecules would be equal to the number of helium atoms found in Example 15.2.

(b) The number of nitrogen molecules is twice the number of helium atoms found in Example 15.2.

(c) The number of nitrogen molecules is half the number of helium atoms found in Example 15.2.

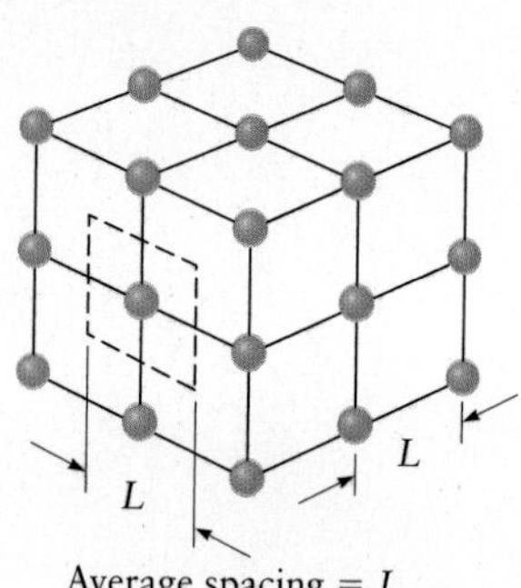

▲ **Figure 15.5** Example 15.3. To find the average molecular spacing, we can imagine the molecules are arranged in a regular array. While the molecules in any real sample of gas will be arranged in a much more random manner, each molecule will occupy, on average, a volume L^3, which is just the volume of a cube of length L.

EXAMPLE 15.3 Particle Spacing in a Gas

What is the average distance between helium atoms in Example 15.2?

RECOGNIZE THE PRINCIPLE

To find the *average* spacing, we can imagine that the atoms are arranged in a regular manner as sketched in Figure 15.5, with the distance between neighboring atoms given by L. This arrangement has atoms at the corners of cubes with edges of length L. Hence, there is a volume L^3 associated with each atom. The total gas volume is then NL^3, where the number of atoms N was found in Example 15.2.

SKETCH THE PROBLEM

Figure 15.5 shows the atoms in a regular cubic arrangement. The atoms in a real gas will not be arranged in such a simple way, but if the average distance between atoms is L, the average volume associated with each atom will still be L^3.

IDENTIFY THE RELATIONSHIPS AND SOLVE

The total gas volume from Example 15.2 is $V = 0.014\ \text{m}^3$. Setting this volume equal to NL^3 and using the result for the number of atoms N found in that example, we get $NL^3 = V$ and

$$L = \left(\frac{V}{N}\right)^{1/3} = \left(\frac{0.014\ \text{m}^3}{6.9 \times 10^{23}}\right)^{1/3} = \boxed{2.7 \times 10^{-9}\ \text{m}}$$

► What does it mean?

The radius of an atom is about 5×10^{-11} m, so the spacing between particles in a gas is *much* larger than the size of an atom. The spacing between atoms in a solid or liquid is typically 0.1 nm ($= 0.1 \times 10^{-9}$ m). The average distance between molecules in this balloon, and in a typical dilute gas at room temperature, is thus about 30 times larger than the spacing between atoms in a solid or liquid.

CONCEPT CHECK 15.4 Changing the Pressure in a Gas

The helium-filled balloon in Example 15.2 contained $n = 1.1$ moles of helium atoms at room temperature and a pressure $P = 2 \times P_{atm}$ (twice atmospheric pressure). Atoms are added to the balloon so that the pressure doubles to $4 \times P_{atm}$ while the temperature and volume are kept fixed. Is the number of moles in the balloon (a) 1.1 moles, (b) 2.2 moles, (c) 4.4 moles, or (d) 0.55 mole?

EXAMPLE 15.4 Effect of Temperature on the Pressure of an Ideal Gas

A jar containing a gas of nitrogen molecules has a pressure $P_{atm} = 1.01 \times 10^5$ Pa at a temperature of 20°C. This jar is then put in a refrigerator where the temperature is 0°C. If the lid of the jar is circular with a radius of 5.0 cm, how much force is needed to overcome the pressure that is holding the lid closed?

RECOGNIZE THE PRINCIPLE

When the gas is cooled, the pressure P_{in} inside the jar will decrease according to the ideal gas law and will then be less than the pressure outside. The force required to open the jar will equal the difference in pressures multiplied by the area of the lid.

SKETCH THE PROBLEM

Figure 15.6 describes the problem.

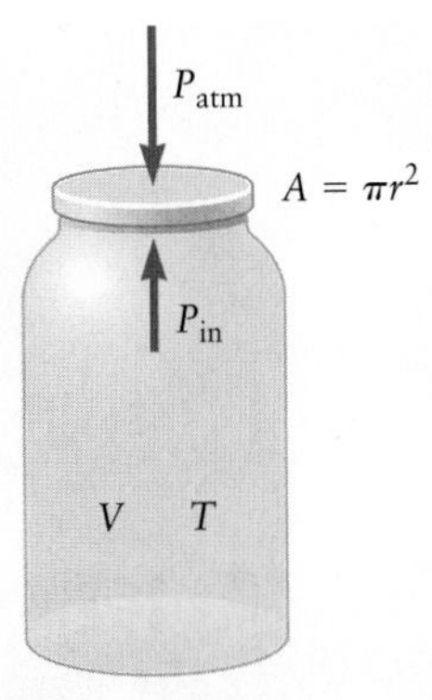

▲ **Figure 15.6** Example 15.4. The force on the lid of a jar is proportional to the difference in pressure between outside and inside the jar.

IDENTIFY THE RELATIONSHIPS

The ideal gas law reads $PV = Nk_BT$, and in this example, the volume V and the number of molecules N are held fixed. Even though we do not know the values of N and V, we can still write

$$\frac{P}{T} = \frac{Nk_B}{V} = \text{constant}$$

which is just another way of writing Gay-Lussac's law, Eq. 15.5. If we denote the initial pressure and temperature inside as $P_{in,\,i}$ and T_i and denote the final values as $P_{in,\,f}$ and T_f, we can now write

$$\frac{P_{in,\,i}}{T_i} = \frac{P_{in,\,f}}{T_f}$$

Rearranging to solve for the final pressure gives

$$P_{in,\,f} = \left(\frac{P_{in,\,i}}{T_i}\right)T_f$$

SOLVE

Inserting the given values of the initial pressure ($P_{in,\,i} = 1.01 \times 10^5$ Pa) and temperature ($T_i = 293$ K), along with $T_f = 0°\text{C} = 273$ K, we find

$$P_{in,\,f} = \left(\frac{P_{in,\,i}}{T_i}\right)T_f = \left(\frac{1.01 \times 10^5 \text{ Pa}}{293 \text{ K}}\right)(273 \text{ K}) = 9.4 \times 10^4 \text{ Pa}$$

With a pressure P_{atm} on the outside of the lid and a smaller pressure $P_{in,\,f}$ inside, the force required to open the lid is

$$F = (P_{atm} - P_{in,\,f})A$$

The area of the lid is $A = \pi r^2$, so we have

$$F = (P_{atm} - P_{in,\,f})A = (P_{atm} - P_{in,\,f})(\pi r^2)$$

$$F = (1.01 \times 10^5 \text{ Pa} - 0.94 \times 10^5 \text{ Pa})\pi(5.0 \times 10^{-2} \text{ m})^2 = \boxed{55 \text{ N}}$$

▸ ***What does it mean?***

The pressure $P_{in,\,f}$ is only slightly smaller than the initial pressure because the change in temperature is only a small fraction of the absolute temperature. Even though the resulting pressure difference across the lid is small, the force is still substantial. You could do this experiment at home.

© Mark Burnett/Alamy

▲ **Figure 15.7** Example 15.5. Launching a weather balloon.

EXAMPLE 15.5 Physics of a Weather Balloon

Consider a weather balloon that has a volume of 5.0 m^3 when released from a location where the temperature is 25°C and the pressure is $P_{atm} = 1.01 \times 10^5$ Pa. If the balloon eventually reaches an altitude of 16 km where the temperature is −30°C and the pressure is $0.10 \times P_{atm}$, what is the final volume of the balloon?

RECOGNIZE THE PRINCIPLE

When the balloon reaches its final altitude, its temperature and pressure are different from their initial values, so its final volume will also be different. We can use the ideal gas law to find the final volume.

SKETCH THE PROBLEM

Figure 15.7 shows a typical weather balloon just after its release.

IDENTIFY THE RELATIONSHIPS

The ideal gas law can be rearranged to read

$$n = \frac{PV}{RT}$$

This relation must hold at the initial (i) and final (f) locations, so we can write

$$\frac{P_iV_i}{RT_i} = \frac{P_fV_f}{RT_f} \qquad (1)$$

The temperatures are $T_i = 273 + 25 = 298$ K and $T_f = 273 - 30 = 243$ K.

SOLVE

We can solve Equation (1) for the final volume; inserting the given values of the pressure and temperature, we find

$$V_f = \frac{P_iT_fV_i}{P_fT_i} = \frac{P_{atm}(243\text{ K})(5.0\text{ m}^3)}{0.10 \times P_{atm}(298\text{ K})} = \boxed{41\text{ m}^3}$$

► ***What does it mean?***
When both the temperature and pressure of a gas change, we must use the ideal gas law to find the resulting change in volume. In this case, the reduction in temperature tends to make the volume smaller, whereas the reduction in pressure makes the volume larger. The latter effect dominates, so the balloon has a much larger volume at its final altitude.

15.3 Ideal Gases and Newton's Laws

We now return to some of the issues raised at the very beginning of the chapter and consider how to account for the gas laws by applying Newton's laws to microscopic molecular motions.

Pressure Comes from Collisions with Gas Molecules

The molecules in a gas move with different speeds and in all possible directions. To simplify things, we will (for the moment) consider just one typical molecule of mass

m and take its speed v to be the average molecular speed. Figure 15.8 shows this molecule as it is initially traveling along the $-x$ direction; it then collides with a container wall and travels back along the $+x$ direction. In a typical gas, such collisions are elastic, so the speed is the same before and after the collision. From Chapter 7, we know that the change in momentum $\Delta\vec{p}$ of the molecule is proportional to the force exerted by the wall on the molecule. According to the impulse theorem (Eq. 7.5), we have

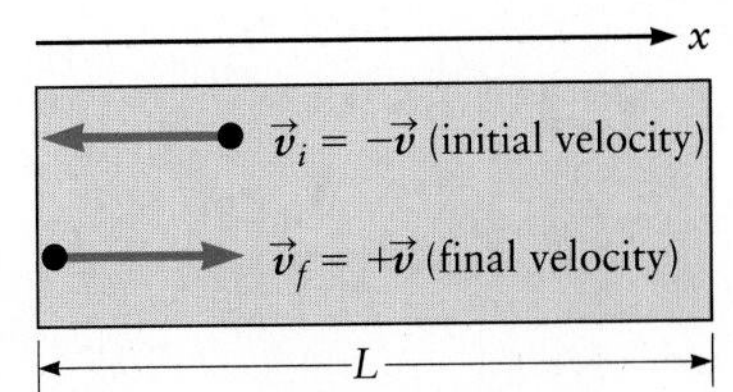

▲ **Figure 15.8** When a gas molecule collides with a wall, the force exerted by the wall leads to an impulse on the molecule, causing the molecule's momentum to change. The corresponding force exerted by the molecule on the wall contributes to the gas pressure.

$$\Delta\vec{p} = \vec{F}\,\Delta t \tag{15.12}$$

The initial and final momenta in Figure 15.8 are along the x axis, and the force from the wall is parallel to the x direction. We can therefore write Equation 15.12 in terms of the components along x. The initial momentum is $-mv$, and the final momentum is $+mv$. Inserting these terms into Equation 15.12, we find

$$\Delta p = p_f - p_i = +mv - (-mv) = F\Delta t$$

$$2mv = F\Delta t$$

Solving for the force gives

$$F = \frac{2mv}{\Delta t} \tag{15.13}$$

Equation 15.13 gives the force exerted by the wall on the molecule. According to Newton's third law (the action–reaction principle), this is also the force exerted by the molecule on the wall, and it contributes to the pressure. To find the total pressure, we must add up the collision forces from all the molecules that strike the wall. Figure 15.8 shows one hypothetical molecule initially moving in the $-x$ direction; there will also be molecules moving in the $+x$, $+y$, $-y$, $+z$, and $-z$ directions. We thus have six different possible directions, so for a system with a total of N molecules, we can expect that one-sixth of them are initially moving along $-x$ and are thus candidates for striking the left-hand wall in Figure 15.8. Of these $N/6$ molecules, only those within a distance $v\,\Delta t$ strike the wall during a time interval Δt because only those molecules reach the wall during this time. The container in Figure 15.8 has a length L, so the number of molecules that collide with the wall during this time interval is

$$\frac{N}{6} \times (\text{fraction that reaches wall}) = \frac{N}{6}\frac{v\Delta t}{L} \tag{15.14}$$

The total force exerted by all these molecules on the wall is the product of the force exerted by one molecule (Eq. 15.13) and the number of such molecules (Eq. 15.14):

$$F_{\text{total}} = \left(\frac{2mv}{\Delta t}\right)\left(\frac{N}{6}\frac{v\Delta t}{L}\right) = \frac{Nmv^2}{3L}$$

Pressure is defined as the total force on the wall divided by the area A, so

$$P = \frac{F_{\text{total}}}{A} = \frac{Nmv^2}{3LA}$$

The volume of the container is $V = LA$, so the pressure can also be written as

$$P = \frac{Nmv^2}{3V}$$

or

$$PV = N\frac{mv^2}{3} \tag{15.15}$$

The Microscopic Basis of Temperature

Equation 15.15 relates a *microscopic* property of the molecules, their speed v, to a *macroscopic* property, the pressure in the gas. We also notice that the mathematical form of this result is *very* similar to the ideal gas law. According to Equation 15.11,

$$PV = Nk_BT$$

Comparing this expression with Equation 15.15, we see that they agree provided that

$$\frac{mv^2}{3} = k_BT \tag{15.16}$$

The kinetic energy of a molecule is $KE = \frac{1}{2}mv^2$, so Equation 15.16 can be written as $\frac{2}{3}KE = k_BT$ or equivalently,

Average kinetic energy of a gas atom

$$KE = \tfrac{3}{2}k_BT \tag{15.17}$$

While we derived this result for a dilute gas, it also holds for the average kinetic energy of atoms and molecules in a liquid or solid.

In our derivation of Equation 15.15, we assumed all the molecules move with the same speed v. An exact analysis allows for different molecules moving at different speeds and shows that v in Equations 15.15 and 15.16 is nearly equal to the average molecular speed (the difference is only a few percent). We can therefore still think of v as being the speed of a typical molecule.

In words, Equation 15.17 says that the kinetic energy of a typical molecule is proportional to the temperature of the gas. This result is remarkable because it tells how temperature is related to the microscopic motion of the gas molecules. We have thus derived the ideal gas law from Newton's laws of mechanics, provided that temperature is related to the molecular kinetic energy by Equation 15.17. This result answers one of the big questions raised in Section 15.1 because we are now able to relate the macroscopic quantities in the ideal gas law to the microscopic quantities that appear in Newton's laws.

It is interesting to use Equation 15.16 to find the speed of a typical molecule in the atmosphere. Rearranging to solve for v, we find

$$v = \sqrt{\frac{3k_BT}{m}} \tag{15.18}$$

The Earth's atmosphere is composed mainly of nitrogen, so let's calculate v for N_2 in the atmosphere. The mass of a nitrogen molecule is $m = 2(14.01 \text{ g})/N_A = 4.7 \times 10^{-23} \text{ g} = 4.7 \times 10^{-26}$ kg. At room temperature, we have $T \approx 293$ K (corresponding to 20°C). Inserting these values into Equation 15.18 gives

$$v_{N_2} = \sqrt{\frac{3k_BT}{m}} = \sqrt{\frac{3(1.38 \times 10^{-23} \text{ J/K})(293 \text{ K})}{4.7 \times 10^{-26} \text{ kg}}} = 510 \text{ m/s} \tag{15.19}$$

which is approximately 1100 mi/h! Molecules having this speed are striking your body right now as you read this book.

15.4 Kinetic Theory

The calculations in Section 15.3 take us to the heart of the mechanics of a dilute gas, a topic called ***kinetic theory***. This theory is based on a small number of assumptions that were used either explicitly or implicitly in Section 15.3:

1. Gas atoms and molecules spend most of their time moving freely; that is, they move with a constant speed in a straight-line path. Collisions with other atoms and molecules in the gas are very infrequent.
2. Newton's laws can be used to describe the motion of individual gas particles.
3. The collisions between gas atoms and molecules, and with the walls of the container, are elastic. The average distance between collisions is called the ***mean free path***, ℓ. The value of ℓ depends on the density of gas particles, their size, and the temperature. The mean free path is not the same as the average spacing between particles (Fig. 15.9). In Example 15.3, we found that the average spacing between molecules in a typical dilute gas is approximately 3×10^{-9} m. The mean free path for a molecule in air is about 100×10^{-9} m, so in this case, ℓ is about 30 times the average spacing.
4. Even a dilute gas contains a very large number of atoms or molecules, so we can use statistical analyses to calculate its properties. We have already discussed the large number of particles present in a typical gas and have seen that it is on the order of Avogadro's number.

These assumptions about the nature of an ideal gas are all satisfied very accurately by a dilute gas and lead to the ideal gas law. We next consider some other properties of ideal gases.

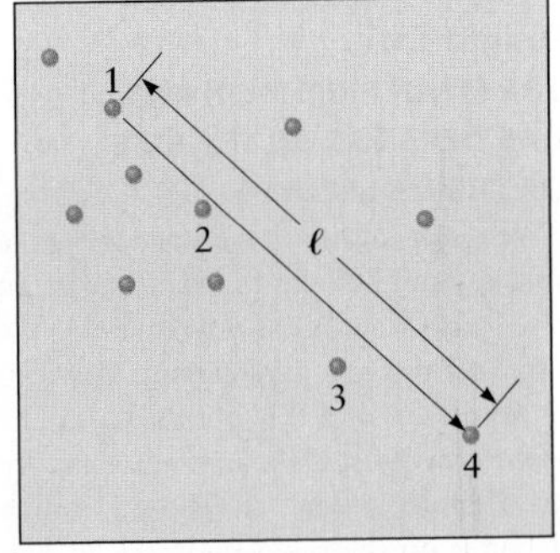

The mean free path depends on the molecular size.

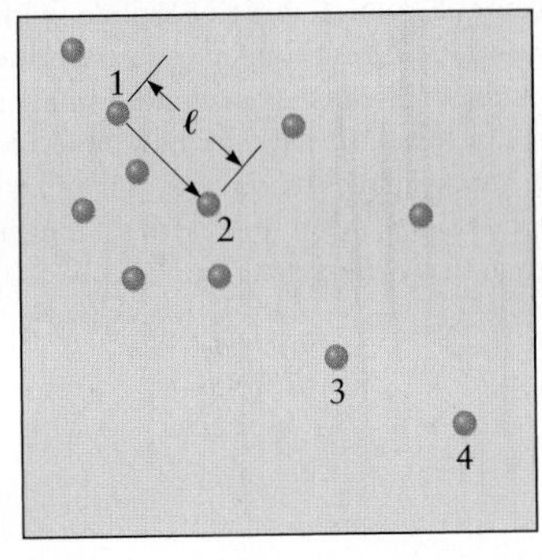

▲ **Figure 15.9** The average distance a gas molecule travels between collisions is called the mean free path. A If the molecules are very small, the mean free path is much longer than the average spacing between them. B As molecular size increases, the mean free path becomes shorter. See Insight 15.1.

Internal Energy of an Ideal Gas

The particles in an ideal gas spend most of their time moving freely, during which time the force on a particular molecule from all the other molecules is essentially zero. The mechanical energy then equals the kinetic energy, which (as we learned in Chapters 6 and 8) has contributions from both the translational motion ($\frac{1}{2}mv^2$) and the rotational motion ($\frac{1}{2}I\omega^2$). For a monatomic gas such as helium, the rotational kinetic energy doesn't contribute to the gas properties and can be ignored. (For polyatomic gases such as N_2, one must include the rotational contribution; see below.) The translational kinetic energy is $KE_{\text{trans}} = \frac{1}{2}mv^2$. We calculated this kinetic energy in Equation 15.17:

$$KE_{\text{trans}} = \tfrac{3}{2}k_BT \tag{15.20}$$

The total energy of a system of N gas atoms is thus

$$KE_{\text{total}} = N(KE_{\text{trans}}) = \tfrac{3}{2}Nk_BT$$

which is the total energy of a monatomic ideal gas. In a sense, this energy is held "internally" by gas atoms and is not evident if we are simply given a container of the gas. For that reason, it is called the ***internal energy***, denoted by U. Hence, for a monatomic ideal gas, we have

$$U = \tfrac{3}{2}Nk_BT \tag{15.21}$$

According to Equation 15.21, the internal energy of an ideal gas depends *only* on temperature and the number of gas atoms N. It is independent of the molecular mass and pressure.

We can also express the internal energy in terms of the number of moles n in the system instead of the number of atoms N. The number of atoms is $N = n \times N_A$, so Equation 15.21 becomes

$$U = \tfrac{3}{2}(nN_A)k_BT = \tfrac{3}{2}n(N_Ak_B)T$$

From the definition of Boltzmann's constant (Eq. 15.9), we have $N_Ak_B = R$, so

$$U = \tfrac{3}{2}nRT \tag{15.22}$$

Insight 15.1

THE MEAN FREE PATH IN A GAS DEPENDS ON THE SIZE OF THE MOLECULES

Given our atomic-scale picture of a gas, you might have expected the mean free path to be approximately the same as the average spacing between molecules, but that is not the case. The mean free path is the average distance traveled between collisions, which depends on the molecular size. In Figure 15.9A, a hypothetical molecule (molecule 1) travels near molecules 2 and 3 before colliding with molecule 4. If the molecules begin at the same places but are larger (Fig. 15.9B), molecule 1 would initially collide with molecule 2, resulting in a shorter value of the collision distance and a smaller mean free path. The size of a molecule is much smaller than the average spacing, so the mean free path is longer than the average spacing.

The results in Equations 15.21 and 15.22 for the internal energy apply only for a monatomic ideal gas. For other systems such as liquids, solids, nondilute gases, and polyatomic ideal gases, the contribution from the potential energy is important and makes the internal energy a more complicated function than the simple result in Equation 15.21. We'll discuss a few such cases later in this chapter and also in Chapter 16, where the internal energy plays an important role in our work on thermodynamics.

CONCEPT CHECK 15.5 Internal Energy of Two Gases

One balloon contains exactly 3 moles of helium atoms, and a second balloon contains exactly 3 moles of argon atoms. The mass of an argon atom is about 9 times the mass of a helium atom. If the internal energies of these two systems are equal, what can we conclude?

(a) They must have the same volume.
(b) They must have the same pressure.
(c) They must have the same temperature.
(d) They must have the same pressure, volume, and temperature.

EXAMPLE 15.6 Ⓡ Internal Energy of a Gas: How Big Is It?

Helium-filled balloons are often sold at carnivals. What is the approximate internal energy of the gas in such a balloon?

RECOGNIZE THE PRINCIPLE

To calculate the internal energy of a dilute gas, we must know its temperature and the number of moles n. Neither value is given, but we can use our experience to find typical values. The temperature is about room temperature, so $T \approx 290$ K. A typical balloon has a volume of about $V = 20$ L $= 0.02$ m^3 (Fig. 15.10), and the pressure is near atmospheric pressure ($P_{atm} = 1.01 \times 10^5$ Pa); using this information, we can use the ideal gas law to find n.

SKETCH THE PROBLEM

We can estimate the balloon's volume from the photo in Figure 15.10.

IDENTIFY THE RELATIONSHIPS

We first rearrange the ideal gas law (Eq. 15.6) to solve for n:

$$n = \frac{PV}{RT}$$

Inserting our estimates for P, V, and T leads to

$$n = \frac{P_{atm}V}{RT} = \frac{(1.01 \times 10^5 \text{ Pa})(0.02 \text{ m}^3)}{[8.31 \text{ J/(mole} \cdot \text{K)}](290 \text{ K})} = 0.8 \text{ mole}$$

SOLVE

Using this value for n in our relation for the internal energy (Eq. 15.22) gives

$$U = \tfrac{3}{2}nRT = \tfrac{3}{2}(0.8 \text{ mole})[8.31 \text{ J/(mole} \cdot \text{K)}](290 \text{ K}) = \boxed{3000 \text{ J}}$$

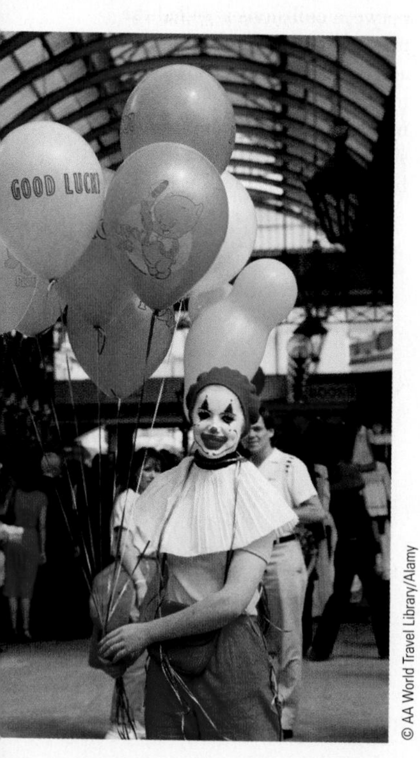

▲ **Figure 15.10** Example 15.6.

> ▸ ***What does it mean?***
> This result is much larger than the kinetic energy of a baseball traveling at 100 mi/h, so it is a very significant amount of energy! It is quite surprising that a simple balloon of helium gas contains such a large amount of energy "internally."

Specific Heat of an Ideal Gas

In Chapter 14, we learned that when heat is added to a system, the temperature changes according to the value of the system's specific heat. Let's apply kinetic theory to find the specific heat of an ideal gas. Heat energy added to an ideal gas causes the molecular kinetic energy to increase, thus increasing the internal energy. So, if an amount of heat energy ΔQ is added to an ideal gas, the internal energy must increase by that amount and $\Delta Q = \Delta U$. Using Equation 15.21, we can find the corresponding change in temperature ΔT. We have

$$\Delta Q = \Delta U = \tfrac{3}{2} N k_B \Delta T$$

The heat capacity of the system (Chapter 14) is equal to $\Delta Q / \Delta T$, so

$$\text{heat capacity} = \frac{\Delta Q}{\Delta T} = \tfrac{3}{2} N k_B$$

This is the heat capacity at *constant volume* because our derivation depended on Figure 15.8, where we assumed the walls of the container were fixed in place so that the system's volume is constant. For a system containing 1 mole of atoms, N is equal to Avogadro's number ($N = N_A$); since $N_A k_B = R$ (Eq. 15.9), we thus get

$$\text{heat capacity of 1 mole} = \frac{\Delta Q}{\Delta T} = \tfrac{3}{2} N_A k_B = \tfrac{3}{2} R$$

which is also called the ***specific heat*** per mole at constant volume and is denoted C_V:

$$C_V = \tfrac{3}{2} R \tag{15.23}$$

Specific heat per mole of a monatomic ideal gas

In Chapter 14, we discussed the specific heat *per unit mass*, denoted by c. Here we denote the specific heat *per mole* by C_V (an uppercase C).

Polyatomic Gases

Our results for the internal energy (Eqs. 15.21 and 15.22) and the specific heat (Eq. 15.23) apply only for monatomic ideal gases. The derivations of those results assumed all the internal energy is in the form of translational kinetic energy. For gases composed of polyatomic molecules such as O_2 or N_2 or NH_3, we must also include the energy associated with the rotational and vibrational motion of the molecules. The internal energy of these gases can also be found using kinetic theory, but the analysis is more than we can include here. The internal energy and the specific heat per mole of such gases are both greater than the values for monatomic gases.

Distribution of Speeds in a Gas

So far, we have dealt with the speed of a typical (average) molecule. The speeds of individual molecules in a given sample are not all the same, however. Some molecules have speeds greater than the typical speed v in Equation 15.18, whereas others have lower speeds. The molecular speed distribution in a gas is an important part of kinetic theory. We will not calculate this distribution here, but it is instructive to consider it qualitatively because it gives rise to some important effects.

Figure 15.11 shows the distribution of molecular speeds in a typical case. Called the ***Maxwell–Boltzmann*** distribution, it gives the probability that a molecule will

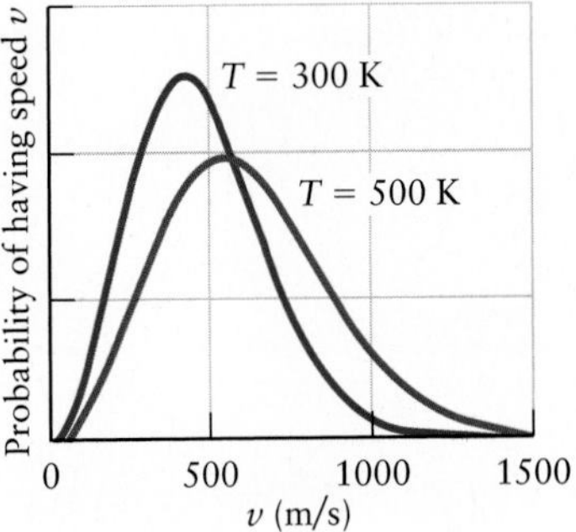

▲ **Figure 15.11** Distribution of speeds for a dilute N_2 gas. The vertical scale is the relative probability of finding a molecule moving with a particular speed v.

have a particular speed. The typical speed v found in Equation 15.18 is near the average speed, but a significant number of molecules have speeds that are greater or lower by a factor of two or even more. The distribution varies with temperature, and if the temperature is increased, the entire distribution curve shifts to higher speeds. This temperature dependence makes intuitive sense (because "hotter" molecules have greater kinetic energy) and is responsible for the temperature dependence of the average molecular speed in Equation 15.18.

EXAMPLE 15.7 (Absence of) Helium in the Atmosphere

Experiments show that a helium atom released into the atmosphere escapes to outer space, never to return. To understand how a helium atom is able to escape from the Earth's gravitational attraction, begin by calculating the typical speed v of a helium atom in a balloon held at atmospheric pressure and temperature. Comparing v with the escape speed for an object near the Earth's surface as calculated in Chapter 6 shows that v is smaller than the escape speed, so it would seem that a "typical" helium will not be able to escape. Given that fact, how can the distribution of molecular speeds in Figure 15.11 explain why all helium atoms do, eventually, escape from the atmosphere?

RECOGNIZE THE PRINCIPLE

We can find the speed v of a typical helium atom using the same approach we applied to a nitrogen molecule (which led to the result in Eq. 15.19). If v is greater than the escape speed, a typical helium atom will be able to escape the Earth's gravitation attraction, just as would any projectile fired from the surface of the Earth. We must also recognize that, according to the Maxwell–Boltzmann speed distribution (Fig. 15.11), some helium atoms will have a speed much greater than v, and we also need to consider the behavior of these atoms.

SKETCH THE PROBLEM

The Maxwell–Boltzmann distribution (Fig. 15.11) is a key to this problem.

IDENTIFY THE RELATIONSHIPS

According to Equation 15.18, the speed of an atom or molecule in a gas depends on the particle's mass. The mass of a helium atom (see the periodic table) is $m_{He} = 4.0 \text{ g}/N_A = 6.6 \times 10^{-27}$ kg.

SOLVE

Using Equation 15.18 at room temperature ($T = 293$ K), we get

$$v_{He} = \sqrt{\frac{3k_BT}{m_{He}}} = \sqrt{\frac{3(1.38 \times 10^{-23} \text{ J/K})(293 \text{ K})}{6.6 \times 10^{-27} \text{ kg}}} = \boxed{1400 \text{ m/s}}$$

► ***What does it mean?***

This speed is greater than the typical speed of N_2 molecules in the atmosphere found in Equation 15.19 because helium has a much smaller mass. In Chapter 6, we found that a projectile at the Earth's surface with a speed greater than about 11,000 m/s has enough kinetic energy to "escape" from the Earth's gravitational attraction, so the speed v of a typical helium atom is *less* than the escape speed. However, according to the distribution of molecular speeds in Figure 15.11, some helium atoms will have speeds much greater than the typical speed. A small fraction of helium atoms will be moving faster than the escape speed and hence will be able to escape from the atmosphere. All helium atoms released into the atmosphere have some probability of achieving such a high speed, and when they do, they will escape into space. That is why our atmosphere does not contain any helium gas. The helium used in carnival balloons and other applications was obtained from trapped underground deposits. What about atoms and molecules heavier than helium? A small percentage of such heavier molecules, including O_2 and N_2, will also escape from the Earth, but their rate of escape is extremely small; hence, these gases remain in the atmosphere indefinitely. (They are also replenished by natural processes.)

15.5 Diffusion

A molecule moving through a gas follows a zigzag path as it collides successively with other molecules (Fig. 15.12). This type of motion is called ***diffusion***. Similar motion is found for molecules moving through a liquid and through certain types of solids, including cell membranes.

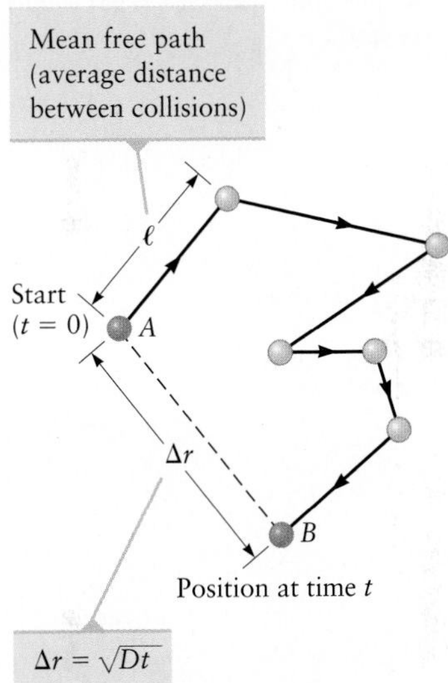

▲ **Figure 15.12** The zigzag motion of a molecule in a gas or liquid is called diffusion. (Similar motion is found for an impurity molecule moving through a solid.) The average length of each step is the mean free path. Each change of direction of a molecule is caused by a collision. The steps are in random directions, so after many steps (i.e., many collisions), the magnitude of the net displacement (Δr) is much smaller than if there had been no collisions.

Each particle in a gas follows a different random zigzag path, but these different trajectories can be described in an average way in terms of the typical molecular speed v between collisions and the average distance ℓ between collisions. The speed of a typical molecule is given by Equation 15.18, so v depends on both the temperature and the mass of the molecule. The distance ℓ is the mean free path we introduced earlier in this chapter. The value of the mean free path depends on the gas density; if the density is high, the gas molecules are close together, the probability for a collision is high, and ℓ is short. If the density is low, the collision probability is also low and ℓ is long. An analysis of the collision mechanics shows that the mean free path varies inversely with the density. Because the density is proportional to the number of molecules per unit volume N/V, the average distance between collisions varies as

$$\ell \propto \frac{V}{N} \tag{15.24}$$

for a gas in which N particles occupy a total volume V. For air at room temperature and pressure, the mean free path is approximately $\ell \approx 1 \times 10^{-7}\ \text{m} = 100\ \text{nm}$, which is about 30 times greater than the average spacing between gas molecules.

In most situations, we are interested in how the molecule travels over a relatively long distance such as when a gas molecule travels from one side of a room to the other. Figure 15.12 shows a typical diffusive path taken as a molecule travels from an initial point A to a final location B. The total distance traveled by a molecule is much longer than the length of the direct path that connects A and B. For this reason, diffusive motion between two points is much slower than motion without all the collisions.

Although we can never predict the precise path followed by a particular molecule, diffusion usually involves a large number of molecules making a large number of zigzag steps, so statistical arguments give an accurate description of an average random walk path. The magnitude of the average displacement Δr of a molecule after many steps taken over time t is

Distance traveled by a diffusing particle

$$\Delta r = \sqrt{Dt} \tag{15.25}$$

Insight 15.2

THE GRAVITATIONAL FORCE IS NOT IMPORTANT IN KINETIC THEORY

In all our discussions of the molecular motions in kinetic theory (e.g., Fig. 15.12), we have ignored the effect of the gravitational force on the molecular trajectories. This approximation is very good because the molecular speeds are so large. From mechanics (Chapter 6), when a particle falls through a distance L, the change of gravitational potential energy is mgL. If the particle is initially at rest, the final speed is $v = \sqrt{2gL}$. For $L = 1$ m, this speed is about 4 m/s, which is much less than the speed of a typical molecule in the atmosphere (we found $v \approx 500$ m/s for N_2; see Eq. 15.19). The effect of gravity on molecular trajectories in a typical gas is thus very small.

TABLE 15.1 Values of Some Typical Diffusion Constants

Diffusing Molecule or Particle	Medium	D (m²/s)
Oxygen (O_2)	Water	1×10^{-9}
Oxygen (O_2)	Air	2×10^{-5}
Nitrogen (N_2)	Air	2×10^{-5}
Hydrogen (H_2)	Air	6×10^{-5}
Oxygen (O_2)	Tissue	2×10^{-11}
Water (H_2O)	Water	3×10^{-9}
Sucrose	Water	5×10^{-10}
DNA	Water	1×10^{-12}
Hemoglobin	Water	7×10^{-11}
Pollen grain (5×10^{-7} m diameter)	Water	1×10^{-9}

Notes: All substances are at room temperature and pressure. Except for the pollen grain, the diffusing objects are all molecules.

The quantity D is called the ***diffusion constant.*** The result above can also be written as

$$t = \frac{(\Delta r)^2}{D} \tag{15.26}$$

giving the time t required for an average molecule to diffuse a distance Δr.

The value of the diffusion constant depends on both v and ℓ, so it depends on temperature and on the mass of the diffusing particle as well as on the properties of the medium. The values of some typical diffusion constants are listed in Table 15.1. Diffusive motion is found in many situations, including molecules moving through a gas or liquid, and through a porous membrane such as a cell membrane.

A key feature of the result for Δr (Eq. 15.25) is that the net displacement is proportional to $\sqrt{t}$, the *square root* of the time. That is in contrast to motion with a constant velocity, for which the displacement increases *linearly* with time. This square root dependence means that a diffusing particle moves relatively slowly through the surrounding medium, as we explore in Example 15.8.

EXAMPLE 15.8 Diffusion in a Gas

You are at work in a chemical laboratory when a dangerous gas accidentally begins to leak from a container. Approximately how long do you have to evacuate the room safely, assuming the gas travels by diffusion alone?

RECOGNIZE THE PRINCIPLE

If the gas leaks from the center of the room, you will want to be gone long before a typical molecule has diffused to the door. This diffusion time is $t = (\Delta r)^2/D$ (Eq. 15.26). The diffusion distance Δr is not given, so to get the approximate time, we must estimate the size of a typical room.

SKETCH THE PROBLEM

Figure 15.12 describes the problem, with the distance from A to B equal to the size of the room.

IDENTIFY THE RELATIONSHIPS AND SOLVE

The diffusion constant for a typical molecule in air is $D \approx 2 \times 10^{-5}\ \text{m}^2/\text{s}$ (Table 15.1), and the size of a typical room is about $\Delta r = 3$ m. These values lead to

$$t = \frac{(\Delta r)^2}{D} = \frac{(3\ \text{m})^2}{2 \times 10^{-5}\ \text{m}^2/\text{s}} = \boxed{5 \times 10^5\ \text{s}}$$

▶ ***What does it mean?***

This result is more than 100 h, so diffusion is a very slow process. It would thus seem that you have plenty of time to escape. In most cases, however, convection currents in the room will disperse the molecules much more quickly. In addition, some molecules will have speeds much greater than the typical speed (Fig. 15.11) and these will also arrive much sooner than the typical molecule considered here.

Using Diffusion in Medicine

The slow process of diffusion is employed in a medical application called ***transdermal drug delivery***. Sometimes a drug should be delivered into the body in a slow and steady manner, which can be accomplished by placing a "patch" containing the drug in direct contact with the skin (Fig. 15.13A). The drug diffuses through the patch membrane, through the skin, and into the body. By careful choice of the membrane thickness, the drug can be made to enter the body slowly, over the course of many days or weeks. Such a slow delivery is preferred in certain cases such as the use of nicotine to help a person stop smoking.

CONCEPT CHECK 15.6 Designing a Transdermal Patch

The membrane in a transdermal patch (Fig. 15.13B) is typically about 200 μm thick. Suppose the thickness is doubled. How does that affect the rate with which the drug diffuses through the membrane?

(a) The rate is doubled.
(b) The rate is cut in half.
(c) The rate is reduced by a factor of four.

Isotope Separation

Another important application of diffusion is in the separation of different ***isotopes*** of a given element. As an example, consider the nucleus of a uranium atom. The two most common types of uranium nuclei are denoted $^{235}_{92}\text{U}$ and $^{238}_{92}\text{U}$; here the lower number is the number of protons in the nucleus, and the upper number is the number of neutrons plus protons. These two different nuclei contain different numbers of neutrons and are called isotopes of uranium. (We'll say much more about the properties of different nuclei in Chapter 30.) Naturally occurring uranium contains a mixture of $^{235}_{92}\text{U}$ and $^{238}_{92}\text{U}$, but only $^{235}_{92}\text{U}$ is able to undergo the nuclear reactions required in power plants (and nuclear bombs). To make nuclear fuel, the $^{235}_{92}\text{U}$ atoms must be removed from a uranium sample without having the other isotope come along with them. This process is called ***isotope separation*** and is used to make what is called "enriched" uranium. One way to accomplish isotope separation is with diffusion.

A diffusing molecule follows a zigzag path as sketched in Figure 15.12. On each straight portion, the molecule moves with a constant speed; for diffusion through a gas, the average speed on these different zigs and zags is just v in Equation 15.18. Since v increases as the molecular mass decreases, a light atom or molecule diffuses faster than a heavy one.

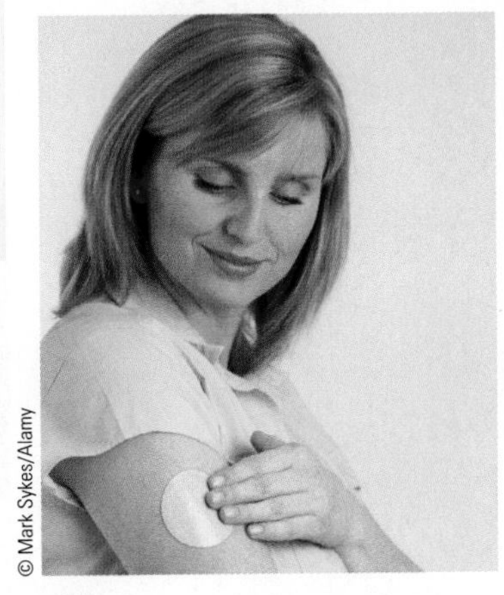

© Mark Sykes/Alamy

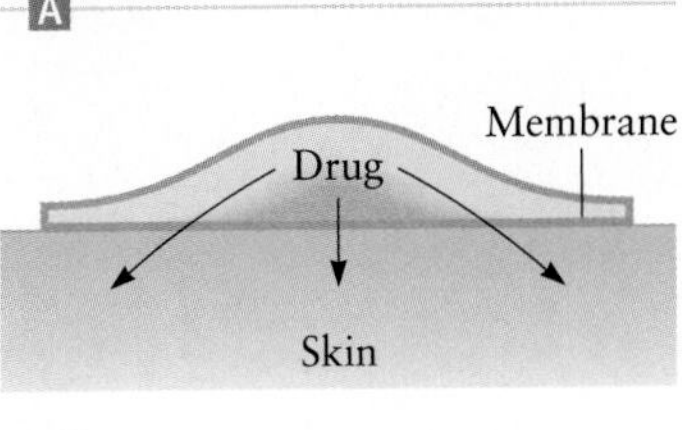

▲ **Figure 15.13** A A transdermal patch delivers drugs into the body at a slow, steady rate. B Structure of a transdermal patch.

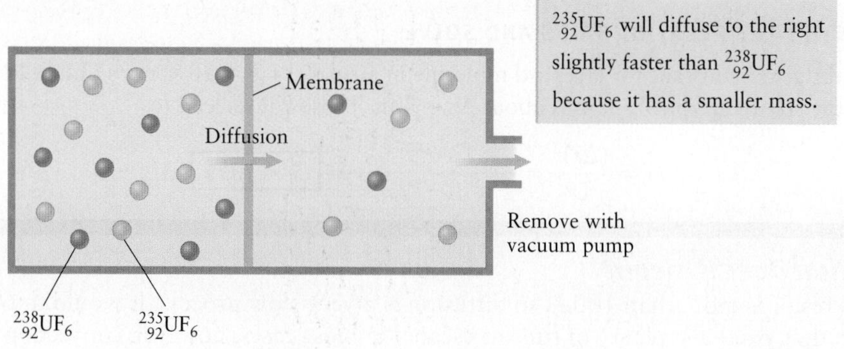

▲ **Figure 15.14** The process of diffusion can be used to separate different isotopes of uranium. For simplicity, we show roughly equal numbers of $^{235}_{92}UF_6$ and $^{238}_{92}UF_6$ molecules on the left. Usually, the initial concentration of $^{235}_{92}UF_6$ is much smaller than the concentration of $^{238}_{92}UF_6$.

Diffusion experiments with uranium usually employ molecules of uranium hexafluoride containing either ^{235}U or ^{238}U and denoted as $^{235}_{92}UF_6$ and $^{238}_{92}UF_6$ respectively. Since $^{235}_{92}UF_6$ has a slightly smaller mass, it diffuses slightly faster than a molecule of $^{238}_{92}UF_6$. This difference can be used to separate isotopes as shown in Figure 15.14. The left side of this system contains an equal mixture of $^{235}_{92}UF_6$ and $^{238}_{92}UF_6$, and in the middle there is a medium (perhaps a thin membrane like the one used in a transdermal patch; Fig. 15.13) through which the molecules can diffuse. The lighter $^{235}_{92}UF_6$ molecules (with the larger diffusion constant) move through the membrane slightly faster than the $^{238}_{92}UF_6$ molecules, so there is a slightly higher concentration of $^{235}_{92}UF_6$ on the right compared with the left. Hence, we have separated (at least partially) the different isotopes; additional diffusion steps can be used to make the separation more and more complete. A similar separation process takes place in plants when they process two isotopes of carbon, $^{12}_{6}C$ and $^{13}_{6}C$, via photosynthesis.

CONCEPT CHECK 15.7 Relation of the Diffusion Distance to Time

Suppose we double the size of the room in Example 15.8. How does that affect the time required for an average molecule to diffuse across the room?

(a) The time is doubled.
(b) The time does not change.
(c) The time increases by a factor of four.

Diffusion, Brownian Motion, and the Discovery of Atoms

Figure 15.12 shows the characteristic random-walk path followed by a diffusing particle. This figure is only schematic; there is no way to observe the motion of a molecule in a gas or liquid in such fine detail. It is possible, however, to observe the diffusive motion of larger objects. This experiment was actually carried out nearly 200 years ago by botanist Robert Brown in his studies of the motion of pollen grains in water. The grains were about 1 μm in size, much larger than a molecule but still small enough to make diffusion observable with a microscope. Brown found that pollen grains do indeed follow random zigzag paths like those sketched in Figure 15.12, a motion now called ***Brownian motion*** in his honor. Just as for the molecule in Figure 15.12, the zigzag motion of Brown's pollen grains was caused by collisions with molecules in the fluid. Brown made his observations before the discovery of atoms and molecules, and there was no real explanation of this motion during his lifetime. In hindsight, his experiments provided some of the very first evidence for the existence of atoms and molecules, and detailed studies of Brownian motion led to the first accurate measurements of Avogadro's number.

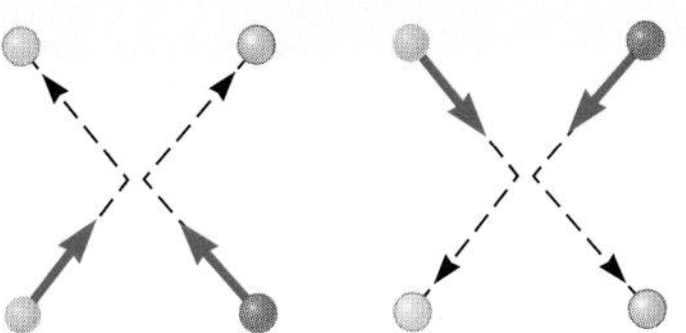

▲ **Figure 15.15** An elastic collision between two molecules in a gas is time reversible. Both forward and backward collision processes satisfy all the laws of physics (including conservation of energy and momentum).

The Arrow of Time

Consider a video recording of an elastic collision of two particles; they might be two molecules in a dilute gas. We could play this video normally or in reverse (Fig. 15.15), and both cases yield motion consistent with our intuition. In fact, both collision trajectories are consistent with Newton's laws.

Now consider the mixing of two gases A and B in Figure 15.16. Here we begin at $t = 0$ with all the molecules of type A on the left side of a container and all the molecules of type B on the right side. After a period of time, the gases mix, and there is an equal chance of finding either type of molecule anywhere in the container. If we were to videotape this mixing process, the first few frames of the video would look like Figure 15.16A and the final frames would look like Figure 15.16B. If we were to play this video backward, it would show molecules in a fully mixed gas (Fig. 15.16B) spontaneously unmix so as to end up with the two separate regions in Figure 15.16A. Your intuition should tell you that such behavior is not possible, and indeed, such unmixing of two gases does not occur in real life. In this process, the direction of time does matter.

The hypothetical unmixing process involves many collisions between molecules, and we could use Newton's laws to analyze these collisions one at a time. Each collision is time reversible, just as with the collisions in Figure 15.15, so each individual unmixing collision would appear to be perfectly reasonable. Only when we string them all together do we run into problems.

This discussion reveals a deep aspect of physics. Newton's laws are time reversible, but many processes in nature are not! Processes that are not time reversible are possible in systems that contain a very large number of particles (such as a gas). The statistical improbability of different systemwide configurations such as the mixed and unmixed gas pictures in Figure 15.16 makes certain processes such as the unmixing of a gas time irreversible. We'll see in Chapter 16 that time-irreversible behavior of systems with many particles also plays a role in the area of physics called ***thermodynamics***.

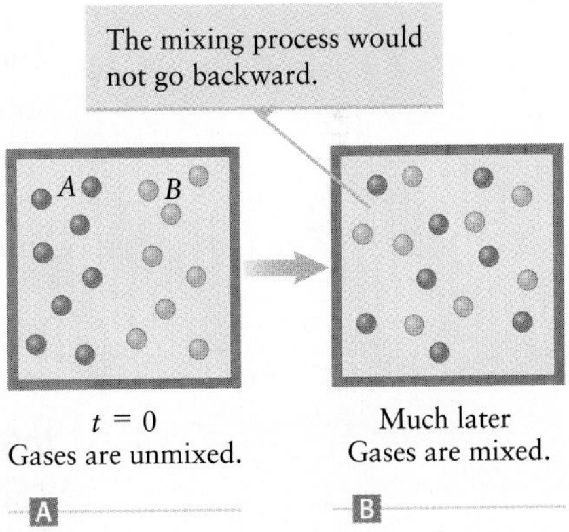

The mixing process would not go backward.

A B

$t = 0$ Gases are unmixed.

Much later Gases are mixed.

A B

▲ **Figure 15.16** The mixing of two gases is not time reversible. A Two different types of molecules are initially separated (placed on different sides of a container). B As time passes, the molecules mix, with both types of molecules spread evenly (with equal probability) throughout the container.

15.6 Deep Puzzles in Kinetic Theory

In this chapter, we focused on the successes of classical kinetic theory. That theory is based on an application of classical mechanics—Newton's laws of motion—to the properties of gases. Classical kinetic theory is able to explain the ideal gas law, the meaning of temperature, and the universal values of the specific heats of dilute gases. However, certain properties of simple gases are *not* explained by classical kinetic theory. These failures were important in the history of physics and helped lead to the development of quantum theory. A kinetic theory based on quantum rather than classical mechanics successfully explains all the properties of dilute gases.

Another puzzle concerns the notion of an absolute temperature scale. What happens in a dilute gas, or any other phase of matter, when it is cooled to absolute zero? Do the atoms in such a gas or solid stop all motion? These important conceptual issues are not addressed by classical kinetic theory. We'll explore these issues later in this book, after we learn about thermodynamics and quantum theory.

Summary | CHAPTER 15

Key Concepts and Principles

Kinetic theory

Kinetic theory is the application of Newton's laws to the mechanics of a gas of particles. This theory leads to the ***ideal gas law***; for a dilute gas composed of any substance,

$$PV = nRT \quad \text{(15.6) (page 428)}$$

where P is the pressure, V is the volume, n is the number of moles of particles, and R is the ***universal gas constant*** with the value

$$R = 8.31 \frac{\text{J}}{\text{mole} \cdot \text{K}} \quad \text{(15.7) (page 428)}$$

The ideal gas law can also be written in the form

$$PV = Nk_BT \quad \text{(15.11) (page 428)}$$

where N equals the number of particles (i.e., molecules) and k_B is ***Boltzmann's constant***,

$$k_B = 1.38 \times 10^{-23} \text{ J/K} \quad \text{(15.10) (page 428)}$$

Internal energy

The temperature of a dilute gas is related to the average kinetic energy of a molecule by

$$KE = \tfrac{3}{2}k_BT \quad \text{(15.17) (page 434)}$$

The ***internal energy*** of a monatomic ideal gas is equal to the total kinetic energy of the molecules and is given by

$$U = \tfrac{3}{2}Nk_BT \quad \text{(15.21) (page 435)}$$

For an ideal gas, U is independent of the molecular mass and pressure.

Applications

Gas laws

The ideal gas law contains several other gas laws as special cases.

Avogadro's law:
For a sample of gas at constant pressure and temperature, the volume is proportional to the number of molecules in the sample.

Boyle's law:
For a sample of gas at constant temperature, the product of the pressure and the volume is constant:

$$PV = \text{constant} \quad \text{(15.3) (page 426)}$$

Charles's law:
For a sample of gas at constant pressure, if the temperature is changed by a small amount ΔT, the volume changes by an amount ΔV, with

$$\Delta V \propto \Delta T \qquad (15.4)$$ (page 426)

Gay-Lussac's law:
For a sample of gas held in a container with constant volume, changes in pressure are proportional to changes in temperature:

$$\Delta P \propto \Delta T \qquad (15.5)$$ (page 426)

Specific heat of a gas

The specific heat per mole of a monatomic ideal gas at constant volume is given by

$$C_V = \tfrac{3}{2}R \qquad (15.23)$$ (page 437)

Diffusion

The random motion of individual gas molecules is described by ***diffusion***. The average net distance that a molecule diffuses in time t is given by

$$\Delta r = \sqrt{Dt} \qquad (15.25)$$ (page 439)

where D is the ***diffusion constant.***

Questions

SSM = answer in *Student Companion & Problem-Solving Guide* ⓧ = life science application

1. Show that Avogadro's law, Charles's law, and Gay-Lussac's law all follow from the ideal gas law (Eq. 15.6).
2. In Example 15.6, we claimed that the total kinetic energy of the gas is greater than the kinetic energy of a baseball. Show that this claim is in fact true by working out the value of $\frac{1}{2}mv^2$ with realistic values of m and v for a baseball.
3. If you were to hammer on a piece of metal, the temperature of the metal would rise. Why? Describe from a microscopic point of view.
4. Is it possible to heat a can of soup by continuously lifting and dropping it on a hard surface? (Assume the can will take the abuse and not rupture.) Would it make a difference if the can were encased in an excellent thermal insulator?
5. Consider two identical containers, one full of helium and the other full of oxygen. If the molecules of both gases have the same average speed, what is the ratio of their temperatures? (Approximate both as ideal gases.)
6. SSM In Chapter 11, we learned about the bulk modulus B of a substance. The bulk modulus is related to changes in volume by

$$\frac{\Delta V}{V} = -\frac{\Delta P}{B}$$

Use the ideal gas law to calculate the bulk modulus for an ideal gas. Assume the temperature is held constant.

7. An automobile cooling system circulates a liquid called antifreeze (which is mostly water) and typically operates at a pressure higher than atmospheric pressure. Why is that desirable?
8. A family drives from a low valley to a high-altitude mountain meadow for a picnic. When they arrive at the meadow, they find that their bag of potato chips has burst open and their bag of pretzels looks like an inflated balloon. Explain the behavior of the packaging.
9. SSM A container of gas under pressure has a pinhole leak. If the content of the container is oxygen, the pressure decreases at a certain rate as the gas escapes. If the content is hydrogen gas, however, we find that the rate of pressure drop is greater under the same conditions. Why?
10. The diffusion constant D of a particle in a gas or liquid depends on temperature. If the temperature is increased, does D increase or decrease? Explain why.
11. In Example 15.8 we saw that diffusion is a very slow process. Calculate the *total distance* (not the net displacement Δr) traveled by a diffusing molecule in Example 15.8.
12. Compare the act of making tea with a tea bag in cold water versus steeping it in hot water. Which method will make the tea stronger faster? Why?
13. Consider an ideal gas composed of helium atoms. The collisions between atoms are similar to the collisions between billiard balls.

That is, whenever the distance between two atoms is less than $2r$ (where r is the radius of a helium atom), the two atoms undergo a collision. Show that the mean free path between collisions ℓ is proportional to $1/\rho$, where ρ is the density of the gas. *Hint*: Consider the motion of just one atom and assume the others are all "frozen" in place.

14. The mass of 1 mole of carbon atoms of isotope number 12 (so-called carbon 12) is defined to be exactly 12 g. The *atomic mass unit* is defined as one-twelfth the mass of a single carbon-12 atom. Use this fact to calculate the conversion factor that relates atomic mass units to kilograms.

15. In Concept Check 15.6, we considered how to slow down the release of a drug in a transdermal patch by increasing the thickness of the membrane. Which of the following strategies will decrease the rate at which drug molecules enter the body?
(a) Lower the concentration of molecules in the patch.
(b) Increase the temperature.
(c) Use larger drug molecules.

16. The air in a room consists mainly of N_2 and O_2 molecules. Which has the greater average speed? *Hint*: You do not have to calculate the average speed of either molecule to answer this question.

17. Why does the pressure in a car tire increase when the car is driven?

Problems available in ENHANCED WebAssign

Section 15.1 Molecular Picture of a Gas
Problems 1–5

Section 15.2 Ideal Gases: An Experimental Perspective
Problems 6–24

Section 15.3 Ideal Gases and Newton's Laws

Section 15.4 Kinetic Theory
Problems 25–33

Section 15.5 Diffusion
Problems 34–45

Additional Problems
Problems 46–60

List of Enhanced Problems

Problem number	Solution in *Student Companion & Problem-Solving Guide*	Reasoning & Relationships Problem	Reasoning Tutorial in ENHANCED WebAssign
15.4		✓	✓
15.5	✓		
15.10		✓	✓
15.14	✓		
15.20		✓	
15.22		✓	✓
15.23	✓		
15.26	✓		
15.32	✓		
15.35	✓		
15.39	✓	✓	
15.42		✓	
15.43		✓	
15.47		✓	
15.49		✓	✓
15.51		✓	
15.54		✓	
15.57	✓		
15.58		✓	✓
15.59		✓	

CHAPTER 16

► *Heat engines have a long and interesting history. This is a reconstruction of a steam engine invented by Denis Papin of France, circa 1690. (CNAM, Conservatoire Nationale des Arts et Metiers, Paris/Archives Charmet/The Bridgeman Art Library)*

Thermodynamics

OUTLINE

In Chapter 14, we discussed heat energy and how the flow of heat into or out of a system can affect its properties. In Chapter 15, we developed the kinetic theory of a dilute gas and saw how the mechanics of molecular motion is connected to the temperature of a gas. The central thread of Chapters 14 and 15 was the principle of conservation of energy. In this chapter, we continue along that thread, exploring the area of physics called ***thermodynamics***. Our goal now is more ambitious: we wish to understand the rules that govern the flow of heat and the exchange of other forms of energy between all types of systems, including gases, liquids, and solids. We are especially interested in *fundamental limits* on how heat can flow from one system to another. This understanding will lead us to some very surprising and powerful results, including the impossibility of building a perpetual motion machine and the best possible efficiency of an engine.

16.1 Thermodynamics Is About the Way a System Exchanges Energy with Its Environment

Two systems are in the same state if their macroscopic properties are the same.

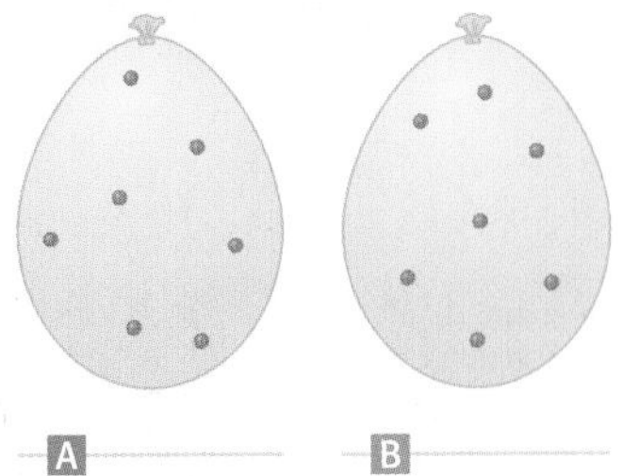

▲ **Figure 16.1** A system of many particles such as the gas in a balloon can be described by macroscopic quantities such as temperature and pressure.

Thermodynamics is concerned with the properties of systems composed of many particles such as a gas, liquid, or solid. We need variables for describing the state of a system and for describing the interactions of one system with another. In mechanics, we describe the state of motion of a particle in terms of its position, velocity, and acceleration, and the interactions between particles are described by forces. However, a typical system such as 1 mole of a dilute gas contains far too many values of position, velocity, and acceleration to keep track of, even with any computers that are likely to be available in your lifetime. In addition, even if we had such a computer, it would be virtually impossible to make sense of all that information.

Fortunately, we can use a small number of macroscopic quantities such as temperature, pressure, and volume to describe systems and their interactions. For example, we might have two balloons containing helium gas as sketched in Figure 16.1. These balloons have the same volume V and contain the same number N of helium atoms, with the same temperature T and pressure P. Although P, V, N, and T are each the same in the two balloons, the precise locations of the helium atoms and their velocities are certainly *not* the same, so these balloons are not microscopically identical. They are nevertheless macroscopically identical, and their properties as systems will be the same. For example, they will have the same values for the velocity of sound and will exhibit the same condensation temperature. Thermodynamics is based on a description involving such macroscopic quantities; the extra detail of the microscopic variables is not needed.

Our gas balloon may interact with the air around it as sketched in Figure 16.2. In such cases, when one of the systems (the air around the balloon) is very much larger than the other (the balloon), the larger system is often referred to as the ***environment***. The interactions between two systems or between a system and its environment generally involve forces and the transfer of energy. These interactions can be described in terms of the work done by one system on another or in terms of the heat that flows between them. One goal of thermodynamics is to understand how energy is exchanged between systems.

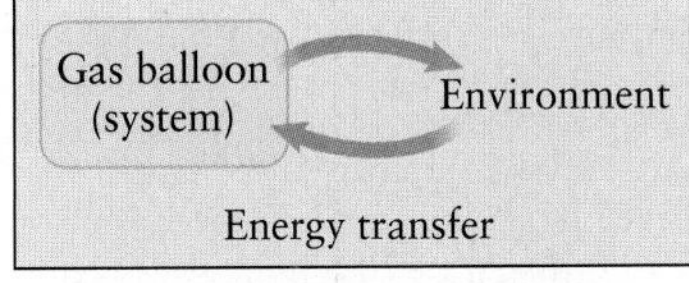

▲ **Figure 16.2** Thermodynamics is concerned with the interactions of a system with its environment. These interactions transfer energy between the two.

As with Newton's laws of mechanics, thermodynamics is based on a small set of physical laws. The laws of thermodynamics each seem quite "innocent" or obvious. One of them (the first law) is a restatement of the principle of conservation of energy, and another (the second law) is a simple statement about the relation between heat flow and temperature. When the laws of thermodynamics are put together, however, they yield remarkable results, including fundamental limits on what is and is not possible.

16.2 The Zeroth Law of Thermodynamics and the Meaning of Temperature

There are four laws of thermodynamics, beginning with the "zeroth" law, which is a statement about the concept of temperature. We encountered the zeroth law in our study of heat in Chapter 14, but it is important to repeat it here.

Zeroth law of thermodynamics

Zeroth law of thermodynamics

If two systems A and B are both in thermal equilibrium with a third system, then A and B are in thermal equilibrium with each other.

This statement of the zeroth law is based on the notion of ***thermal equilibrium*** and hence involves the properties of heat. You should recall from Chapter 14 that heat is the energy that flows from one system to another as a result of a temperature difference. We can use this fact as the basis of an experiment to tell if two systems, which we can call A and B, are in thermal equilibrium with each other. The experiment begins with the two systems separated from each other so that no heat can flow between them. We then allow the systems to interact by putting them into direct contact so that heat can flow between them via conduction (Chapter 14). If our experiment finds that heat flows between A and B, the two systems were not initially in thermal equilibrium; that is, they had different temperatures prior to the experiment. The notions of thermal equilibrium and temperature are thus intimately connected. Two systems in thermal equilibrium with each other have the same temperature. So, the zeroth law of thermodynamics can be restated as follows.

Zeroth law of thermodynamics: alternative form

Zeroth law of thermodynamics (restatement)

Suppose the temperature of system A is equal to the temperature of system C; that is, $T_A = T_C$ (so that A and C are in thermal equilibrium). In addition, the temperature of system B is equal to the temperature of system C; that is, $T_B = T_C$. Then $T_A = T_B$.

The zeroth law is illustrated schematically in Figure 16.3.

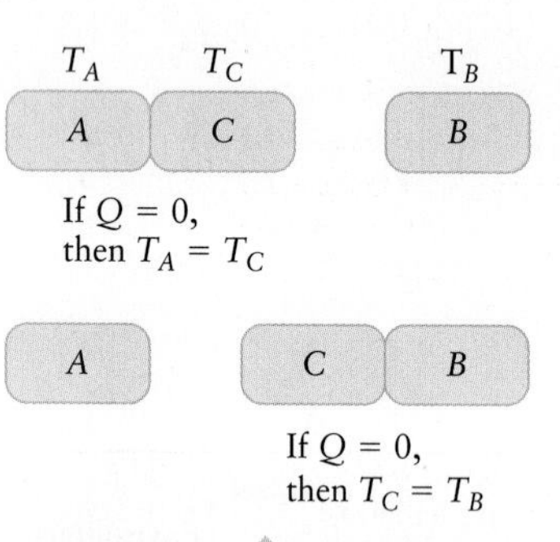

T_A
T_C
T_B
A
C
B
If $Q = 0$, then $T_A = T_C$
A
C
B
If $Q = 0$, then $T_C = T_B$
ZEROTH LAW: If $T_A = T_C$ and $T_C = T_B$, then $T_A = T_B$.

▲ **Figure 16.3** The zeroth law of thermodynamics involves the notion of thermal equilibrium as applied to three (or more) systems. See also Figure 14.7.

From the point of view of arithmetic, the statements of the zeroth law regarding the temperatures of three systems seem quite obvious. So why does it deserve to be called a law of physics? The answer is that the zeroth law declares that a property called *temperature* does indeed exist (!) and that this quantity determines the way heat flows between systems. This profound statement cannot be derived from Newton's laws or any other laws of physics.

16.3 The First Law of Thermodynamics and the Conservation of Energy

One property central to thermodynamics is ***internal energy***, denoted by U. We first introduced internal energy in Chapter 14; it is the total energy associated with all the particles in a system. For an ideal monatomic gas, all this energy resides in the kinetic energy of the particles, whereas in other systems such as liquids and solids, the chemical bonds or interactions between the particles also contribute to U. Physicists would say that these bonds and interactions give rise to potential energy because they can be modeled as springlike forces obeying Hooke's law. (See spring potential energy, Equation 6.24.) A chemist would say that the bonds store chemical energy because this energy can be released (or gained) in chemical reactions. Both views are correct. We'll adopt the physicist's point of view of the internal energy of a system as the sum of the molecular kinetic energies plus the potential energies due to the interactions between particles within the system.[1]

The internal energy of a system is a function of macroscopic variables such as temperature, pressure, and volume. For example, in Chapter 15 we showed that the internal energy of a monatomic ideal gas containing N atoms at temperature T is

$$U = \tfrac{3}{2}Nk_BT \qquad (16.1)$$

where k_B is Boltzmann's constant. It is useful to write Equation 16.1 in terms of the number of moles n in the gas instead of the number of atoms. The result is

$$U = \tfrac{3}{2}nRT \qquad (16.2)$$

[1]By definition, internal energy does not include the kinetic energy of the system "as a whole," that is, the kinetic energy associated with either the motion of the center of mass or the rotational motion of the system.

where R is the gas constant. The internal energy of an ideal gas thus depends only on temperature and the number of particles and is independent of the pressure and volume. Because this internal energy function is so mathematically simple, we'll use the monatomic ideal gas in many of our thermodynamic examples; however, the same basic ideas apply to all systems, although with more complicated expressions for U.

Consider a system in some initial state described by temperature T_i, pressure P_i, and volume V_i as shown in Figure 16.4A. We now allow the system to interact with its environment as shown schematically in Figure 16.4B. These interactions with the environment might result in some heat flow into or out of the system. Forces might also be exerted between our system and the environment, causing the system to do work on the environment or the environment to do work on the system (Fig. 16.4B). As a result of the interactions, the system ends up in some final state (Fig. 16.4C), with a temperature T_f, pressure P_f, and volume V_f. This change in our system is called a ***thermodynamic process***. The first law of thermodynamics is an application of the principle of conservation of energy to such processes.

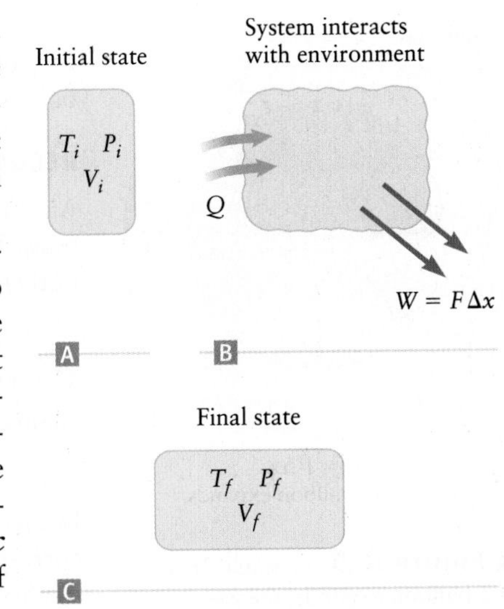

▲ **Figure 16.4** Schematic description of a thermodynamic process. A A system begins in some initial state. B The system then interacts with its environment. This interaction involves the flow of heat (Q) and mechanical work (W). C The system is then left in some final state.

First law of thermodynamics

If an amount of heat Q flows *into* a system from its environment and an amount of work W is done *by* the system on its environment, the internal energy of the system changes by an amount

$$\Delta U = U_f - U_i = Q - W \quad (16.3)$$

The First Law of Thermodynamics: The Meaning of Q and W

There are two ways to change the energy of a system. One way is to arrange for heat to flow into (or out of) the system. By convention, a positive value for Q means that heat flows into the system (and energy is added); hence, the internal energy increases (ΔU is positive) as indicated by Equation 16.3. Another way to change the internal energy is through the action of an external force, that is, a force exerted on the system by the environment. We saw in Chapter 6 that the work done *on* a system is equal to the change in the energy of the system. By convention, a positive value of W in Equation 16.3 means that a positive amount of work is done *by* the system on its environment, so if W is positive, the internal energy of a system must decrease. To apply Equation 16.3 correctly, you must remember that Q is the heat energy that flows *into the system*, whereas W is the work done *by the system*.

Definitions of Q and W in the first law of thermodynamics

Nowadays, we take it for granted that heat is a form of energy, so it is natural to include heat energy along with kinetic and potential energy whenever we discuss the principle of conservation of energy. However, when the science of thermodynamics was being developed around 200 years ago, the nature of "heat" was a matter of much debate. For example, when you place your hand over a flame, you know that "something" flows from the flame to your hand. It is not obvious that the "stuff" that flows from the flame to your hand is similar to the kinetic energy of a moving baseball or the potential energy of a compressed spring. Eventually, physicists discovered that heat is indeed a form of energy, so heat can be measured with either joules (the SI unit for energy) or calories (an alternative unit of energy defined in Chapter 14).

EXAMPLE 16.1 The First Law of Thermodynamics Applied to an Ideal Gas

Consider a flexible balloon containing $n = 3.5$ moles of an ideal monatomic gas such as helium. The balloon is placed over a flame so that $Q = 700$ J of heat flows into the

(continued) ▶

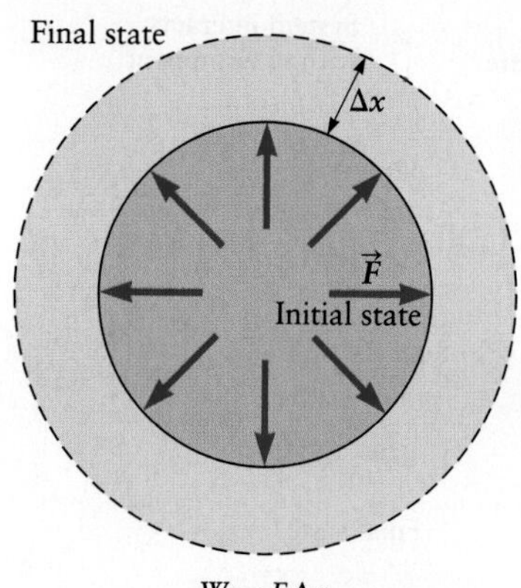

$W = F\Delta x$
If $W > 0$, balloon expands.

▲ **Figure 16.5** Example 16.1. If a balloon expands, the gas inside does work on the surface of the balloon. The thermodynamic quantity W is defined as the work done by the system (the gas) on its environment (the wall of the balloon), so for the balloon in this sketch, $W > 0$.

gas, and its temperature increases by 10 K. How much work is done by the gas in the balloon on its environment during this process?

RECOGNIZE THE PRINCIPLE

A central thread of thermodynamics is energy and how it is exchanged between systems. The principle of conservation of energy is thus at the heart of all problem solving in thermodynamics. From the first law of thermodynamics,

$$\Delta U = Q - W \tag{1}$$

The internal energy of a monatomic ideal gas is given by $U = \frac{3}{2}nRT$ (Eq. 16.2), so changes in U are related to changes in temperature by

$$\Delta U = \tfrac{3}{2}nR\,\Delta T \tag{2}$$

In this example, the change in temperature is known, so we can use it to calculate the corresponding change in the internal energy ΔU. Because Q is given, we can then use the first law of thermodynamics, Equation (1), to find the work W done by the gas.

SKETCH THE PROBLEM

Figure 16.5 shows the initial and final states of the balloon. An expanding balloon does work on its environment because the pressure force and displacement are parallel, corresponding to a positive value of W in the first law of thermodynamics.

IDENTIFY THE RELATIONSHIPS

We first find the change in the internal energy using Equation (2):

$$\Delta U = \frac{3}{2}nR\,\Delta T = \frac{3}{2}(3.5 \text{ moles})\left(8.31 \frac{\text{J}}{\text{mole}\cdot\text{K}}\right)(10 \text{ K}) = 440 \text{ J}$$

Notice that this value is positive. The internal energy of an ideal gas increases when its temperature is increased.

SOLVE

Inserting this result for ΔU with the given value of Q into Equation (1) leads to

$$W = Q - \Delta U = (700 - 440) \text{ J} = \boxed{260 \text{ J}}$$

► ***What does it mean?***

Recall again that W is the work done *by* the gas in the balloon *on* its environment. For W to be positive, the force exerted by the gas in the balloon must be parallel to the associated displacement. Here that is the displacement Δx of the surface of the balloon, whereas the force F is produced by the pressure of the gas on this surface (see Fig. 16.5). A positive value for W thus means that the balloon *expands*, as shown in the figure.

CONCEPT CHECK 16.1 A Thermodynamic Process with $W = Q$

Consider a thermodynamic process involving a monatomic ideal gas in which, by some coincidence, $W = Q$. Which statement best describes the process?

(a) The temperature increases.
(b) The temperature decreases.
(c) The temperature does not change.
(d) The balloon expands.
(e) The balloon contracts.
(f) The volume of the balloon does not change.

The Signs of Q and W in the First Law of Thermodynamics

A common pitfall or source of confusion involves the signs of Q and W in Equation 16.3. These signs are such an important part of thermodynamics that it is worth repeating and emphasizing a few crucial points.

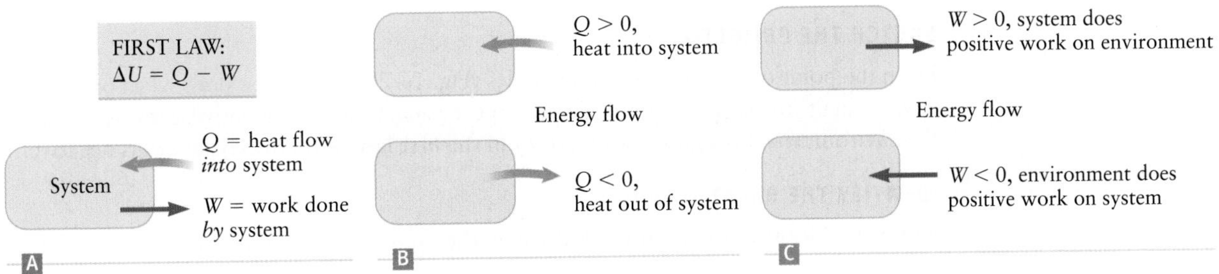

▲ **Figure 16.6** A In the first law of thermodynamics, Q is defined as the heat added to the system and W is the work done by the system. B Meaning of positive and negative Q. C Meaning of positive and negative W.

The first law of thermodynamics reads

$$\Delta U = Q - W \tag{16.4}$$

and is indicated pictorially in Figure 16.6A. By convention, a *positive* value of Q indicates that heat flows *into* the system, which means that the energy of the system *increases*; hence, the change in the internal energy is positive. In any particular case, it is possible for Q to have a *negative* value, and if so, heat energy flows *out of* the system *into* the environment and the internal energy of the system decreases. These two cases are shown schematically in Figure 16.6B.

By convention, a *positive* value of W indicates that the system does a positive amount of work *on* its environment. The internal energy of the system must therefore decrease while the energy of the environment increases (Fig. 16.6C). It is also possible for W to have a *negative* value, which leads to an increase in the internal energy of the system.

These sign conventions for Q and W are a crucial part of the first law of thermodynamics. You can always remember (or re-derive) the correct signs in Equation 16.4 by using energy conservation principles as illustrated in Figure 16.6.

EXAMPLE 16.2 Thermodynamics and Your Diet

A human being is an example of a thermodynamic system. A person riding a bicycle at a moderate speed (Fig. 16.7A) does mechanical work on the pedals at a rate of about

$$P = \frac{W}{t} \approx 100 \text{ W} \tag{1}$$

At the same time, she gives off some heat to her surroundings. Because this heat is flowing out of the cyclist and into the environment, the term Q in the first law of thermodynamics is negative, and measurements on typical cyclists give

$$\frac{Q}{t} = -400 \text{ W} \tag{2}$$

Suppose a cyclist rides for 1 h. What is the change in her internal energy?

RECOGNIZE THE PRINCIPLE

The principle of conservation of energy is at the heart of all problem solving in thermodynamics. From the first law of thermodynamics,

$$\Delta U = Q - W$$

To calculate the change in the internal energy, we need to find both Q and W. The value of Q/t is given, and W can be found from the power exerted by the cyclist.

(continued) ▶

A

Q is negative because heat flows from the cyclist to the environment.

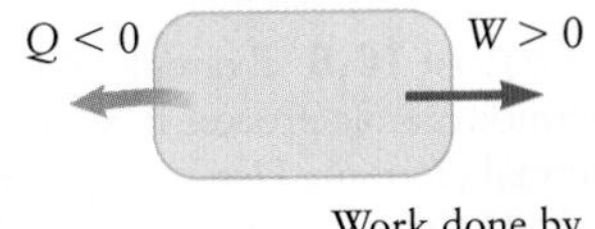

Work done by cyclist on pedals

B

▲ **Figure 16.7** Example 16.2.

SKETCH THE PROBLEM

From the point of view of thermodynamics (Fig. 16.7B), the cyclist does work on her environment (through the force exerted on the pedals), so W is positive. Heat is lost to the environment through sweating, so Q in the first law of thermodynamics is negative.

IDENTIFY THE RELATIONSHIPS

In 1 h ($t = 3600$ s), the total work done by the cyclist is, using Equation (1),

$$W = Pt = (100\ \text{W})(3600\ \text{s}) = +3.6 \times 10^5\ \text{J}$$

where we have included the positive sign explicitly to emphasize that the cyclist does a positive amount of work on the pedals. In 1 h, the total heat Q is

$$Q = (-400\ \text{W})(3600\ \text{s}) = -1.4 \times 10^6\ \text{J}$$

Because heat flows from the cyclist to her environment, Q is negative.

SOLVE

Inserting these values of W and Q into the first law gives

$$\Delta U = Q - W = (-1.4 \times 10^6\ \text{J} - 3.6 \times 10^5\ \text{J}) = \boxed{-1.8 \times 10^6\ \text{J}} \quad (3)$$

► ***What does it mean?***

The negative value of ΔU means that the internal energy of the cyclist decreases as the result of her exercise, as expected. The cyclist can replenish this energy by eating. For example, according to dietary tables, the total energy available from a typical fast-food hamburger is about 2.0×10^6 J. So, our cyclist would have to eat about one hamburger to fuel her hour of exercise.

16.4 Thermodynamic Processes

Thermal Reservoirs

Energy can flow via conduction from one system to another when the two systems are in contact. For example, suppose an ice cube is placed outside on a sidewalk on a warm day (Fig. 16.8). Because the ice cube is colder than its environment, heat flows from the air and the sidewalk into the ice cube and Q is positive. Eventually, the ice cube melts and the puddle of water reaches a final temperature equal to the temperature of its surroundings. In this example, the environment around the ice cube acts as a ***thermal reservoir.*** By definition, a reservoir is much larger than the system to which it is connected, so there is essentially no change in the temperature of the reservoir when heat enters or leaves it. The notion of a thermal reservoir comes up often in discussions of thermodynamic processes.

▲ **Figure 16.8** Example of a thermodynamic process involving a thermal reservoir. Here the system is the H_2O (the ice plus water). The air and the surroundings act as a thermal reservoir; their temperature does not change during the thermodynamic process in which the ice cube melts.

Calculating the Work Done in a Thermodynamic Process

To understand how to calculate the work associated with a thermodynamic process, consider a container filled with gas (Fig. 16.9A). One wall of this container is a movable piston. Due to its pressure, the gas exerts a force on the piston, moving the piston to the right through a distance Δx (Fig. 16.9B). If F is the force exerted by the gas on the piston, the work done by the gas on the piston is

$$W = F\,\Delta x \quad (16.5)$$

The piston is part of the environment of the gas (the piston might be connected to other components of an engine), so W is also the work done by the gas on its environment.

If the pressure in the gas is P and the area of the piston is A, the force on the piston is $F = PA$ and the work done by the gas on the piston is (using Eq. 16.5)

$$W = F\,\Delta x = PA\,\Delta x$$

The product $A\,\Delta x$ equals the change in volume of the gas ΔV, so we have

$$W = P\,\Delta V \tag{16.6}$$

We have derived this result for a gas in a piston-type container, but the result is very general. The only restriction on Equation 16.6 is that the force and hence also the pressure are constant. Whenever the volume of a system changes by a small amount ΔV, the work done by the system is equal to $P\,\Delta V$.

We can apply this result to cases in which the volume changes by a large amount by viewing the process as plotted in Figure 16.10. Here the initial (i) and final (f) states of the system are indicated by points in the P–V plane. A particular thermodynamic process such as the expansion of the gas in Figure 16.9B corresponds to a line or curve in the P–V plane that connects the initial state in which the system has pressure P_i and volume V_i to the final state with pressure P_f and volume V_f. The entire process produces a large change in the volume, but we can also view it as a series of small changes and apply Equation 16.6 to each small change (Fig. 16.10A). The total work done during the entire process thus equals *the area under the corresponding curve in the P–V plane*.

In the hypothetical process in Figure 16.10A, the final volume is greater than the initial volume, so the gas expands. The force exerted by the system is thus parallel to Δx (Fig. 16.9A and B), and the work done by the gas on the environment is positive ($W > 0$). The work done by a system can also be negative. For W to be negative, the force exerted by the gas on the piston must be opposite to the piston's displacement. In Figure 16.9C, the piston moves to the left, which corresponds to compression of the gas, giving a negative value of ΔV in Equation 16.6. This situation can occur if there is an external force on the system as might be exerted by a person pushing on the piston. Such a process is shown in the P–V plane in Figure 16.10B. The magnitude of the work done by the system during this process is still equal to the area under the P–V curve, but we are traversing this curve from right to left, so ΔV is negative and hence W is negative.

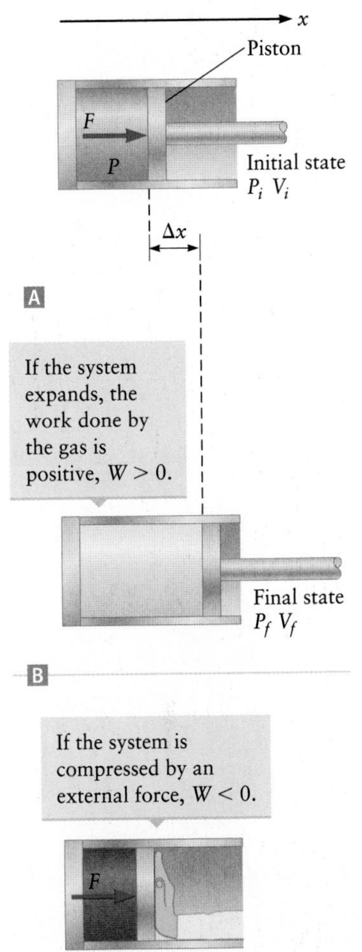

▲ **Figure 16.9** A gas expands as the piston moves to the right in going from A to B. The work done by the gas on the piston is thus positive, and $W > 0$ in Equation 16.5. C If an outside agent exerts a force on the piston so as to compress the gas, the work done *by the gas* is still $W = F\,\Delta x$ (where F is the force exerted by the gas on the piston), but now $W < 0$.

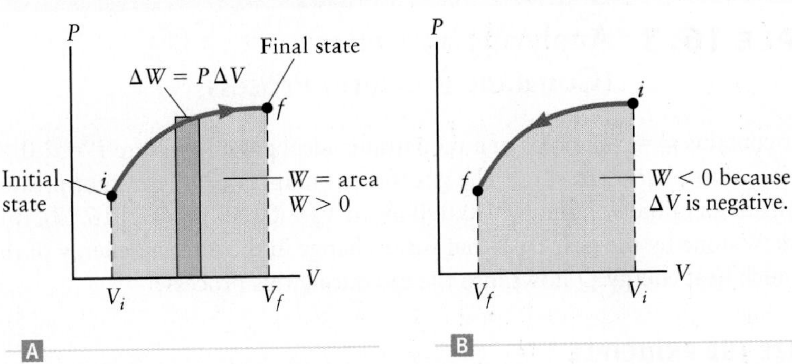

▲ **Figure 16.10** The work W is the area under the path in the P–V plane. A In this case, the gas expands ($V_f > V_i$) and does a positive amount of work on its environment ($W > 0$). B Here the gas is compressed ($V_f < V_i$), and $W < 0$.

CONCEPT CHECK 16.2 Interpreting a P–V Diagram

Consider a thermodynamic process involving an ideal gas that is described by the P–V diagram in Figure 16.11. Which of the following statements is true?

(a) The gas contracts, and W (the work done by the gas) is negative.
(b) The gas pressure drops, so W is negative.
(c) The gas expands, and W is positive.

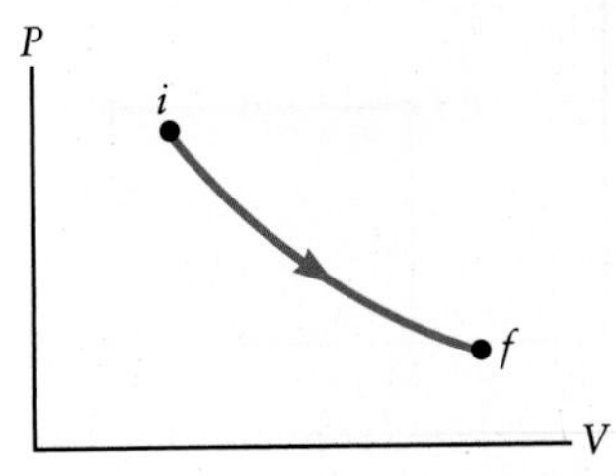

▲ **Figure 16.11** Concept Check 16.2.

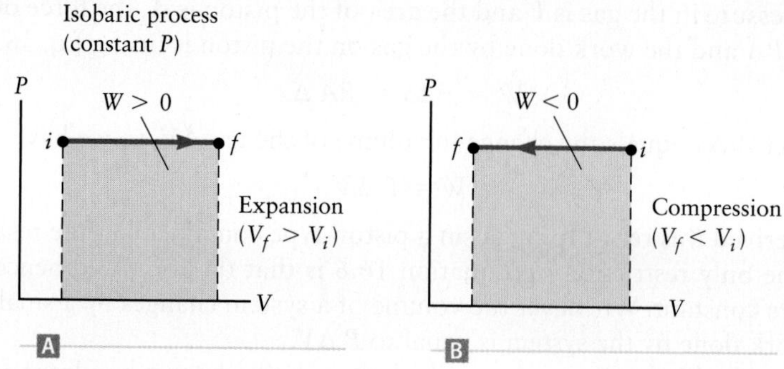

▲ **Figure 16.12** A P–V diagram for an isobaric (constant pressure) expansion. B An isobaric compression.

Examples of Thermodynamic Processes

Thermodynamic processes are classified according to how quantities such as P, V, and T change or remain constant during the course of the process.

Changes at Constant Pressure (Isobaric). If the pressure is constant during a process, the process is called ***isobaric***. When viewed on a P–V diagram, an isobaric process is a horizontal line as shown in Figure 16.12. In the hypothetical process shown in Figure 16.12A, the system expands and the work done by the gas is positive. The work W is equal to the area under the P–V curve, and because P is constant, we can write

$$W = P\,\Delta V = P(V_f - V_i) \quad \text{(isobaric process)} \tag{16.7}$$

For an expansion, $V_f > V_i$ so W is positive, whereas for a contraction, $\Delta V < 0$ and the work done is negative (Fig. 16.12B). An isobaric process may involve some heat flow, as we illustrate in Example 16.3.

EXAMPLE 16.3 Analyzing an Isobaric (Constant-Pressure) Process

A cylinder contains $n = 5.0$ moles of a monatomic ideal gas at pressure $P = 2.0 \times 10^5$ Pa (about twice atmospheric pressure). The gas is then compressed at constant pressure (isobarically) from an initial volume $V_i = 0.090\ \text{m}^3$ to $V_f = 0.040\ \text{m}^3$ (Fig. 16.13). **(a)** What is the work W done by the gas? **(b)** What is the change in the internal energy of the gas? **(c)** How much heat energy Q flows into the gas during this process?

RECOGNIZE THE PRINCIPLE

We can find the work done by the gas using $W = P\,\Delta V$ (Eq. 16.7) along with the given initial and final volumes. The change in the internal energy can be calculated using $U = \frac{3}{2}nRT$ for an ideal gas (Eq. 16.2) together with the initial and final temperatures, which can be found using the ideal gas law. When we know W and ΔU, we can use the first law of thermodynamics to get Q.

SKETCH THE PROBLEM

A P–V diagram describing the process is given in Figure 16.13. The work done by the gas is the area under the curve that connects the initial (i) and final (f) states.

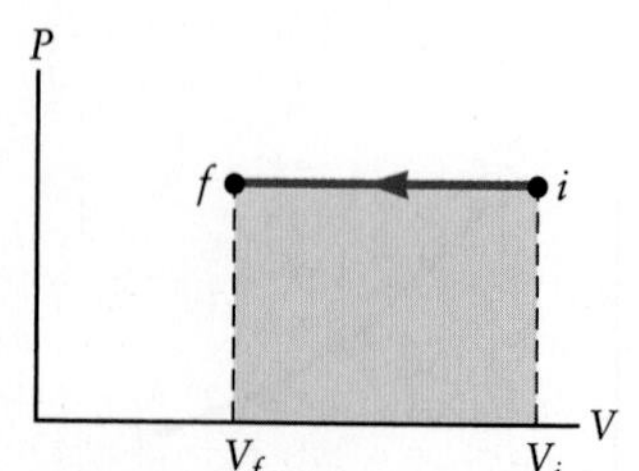

▲ **Figure 16.13** Example 16.3.

IDENTIFY THE RELATIONSHIPS AND SOLVE

(a) The magnitude of W is the area under the P–V curve in Figure 16.13. We find

$$W = P\,\Delta V = P(V_f - V_i) = (2.0 \times 10^5 \text{ Pa})(0.040 \text{ m}^3 - 0.090 \text{ m}^3)$$

$$W = \boxed{-1.0 \times 10^4 \text{ J}}$$

Because $V_f < V_i$, ΔV is negative and W is negative.

(b) The internal energy of an ideal gas is

$$U = \tfrac{3}{2}nRT \qquad (1)$$

so to find the change in U, we must first calculate the initial and final temperatures. Rearranging the ideal gas law $PV = nRT$ to solve for T gives

$$T_i = \frac{PV_i}{nR} = \frac{(2.0 \times 10^5 \text{ Pa})(0.090 \text{ m}^3)}{(5.0 \text{ moles})[8.31 \text{ J/(mole}\cdot\text{K)}]} = 430 \text{ K}$$

For the final temperature, we have

$$T_f = \frac{PV_f}{nR} = \frac{(2.0 \times 10^5 \text{ Pa})(0.040 \text{ m}^3)}{(5.0 \text{ mole})[8.31 \text{ J/(mole}\cdot\text{K)}]} = 190 \text{ K}$$

Using these values with Equation (1) leads to

$$\Delta U = \tfrac{3}{2}nR\,\Delta T = \tfrac{3}{2}nR(T_f - T_i)$$

$$\Delta U = \tfrac{3}{2}(5.0 \text{ moles})[8.31 \text{ J/(mole}\cdot\text{K)}](190 \text{ K} - 430 \text{ K}) = \boxed{-1.5 \times 10^4 \text{ J}}$$

(c) According to the first law of thermodynamics,

$$\Delta U = Q - W$$

Using the values of W and ΔU from parts (a) and (b), we find

$$Q = \Delta U + W = (-1.5 \times 10^4 \text{ J}) + (-1.0 \times 10^4 \text{ J}) = \boxed{-2.5 \times 10^4 \text{ J}}$$

▶ ***What does it mean?***

In an isobaric process, the pressure stays constant and the work can be found using $W = P\,\Delta V$. The temperature may or may not change, depending on the value of W and on how much heat Q flows into or out of the system.

Changes at Constant Temperature (Isothermal). Consider the expansion of an ideal gas in Figure 16.14A, in which the gas moves the piston to the right. Let's also assume the walls of the container are in contact with a thermal reservoir (surrounded with air at room temperature or immersed in water at a particular temperature). Contact with the thermal reservoir keeps the temperature of the gas constant during the course of the expansion. A thermodynamic process in which the temperature is constant is called an ***isothermal process***.

Even though the temperature is constant in an isothermal process, the pressure and volume can both change. Figure 16.14B shows an isothermal process for an ideal gas, drawn as a path on a P–V diagram. The ideal gas law is $PV = nRT$, so if the temperature is constant, the pressure and volume are related by

$$P = \frac{nRT}{V} \propto \frac{1}{V} \quad \text{(isothermal process)} \qquad (16.8)$$

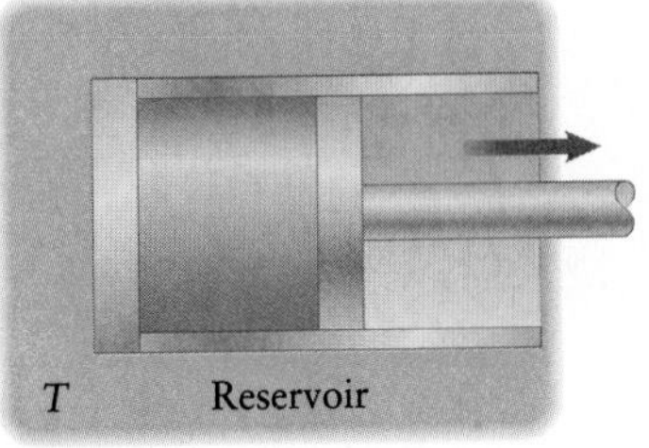

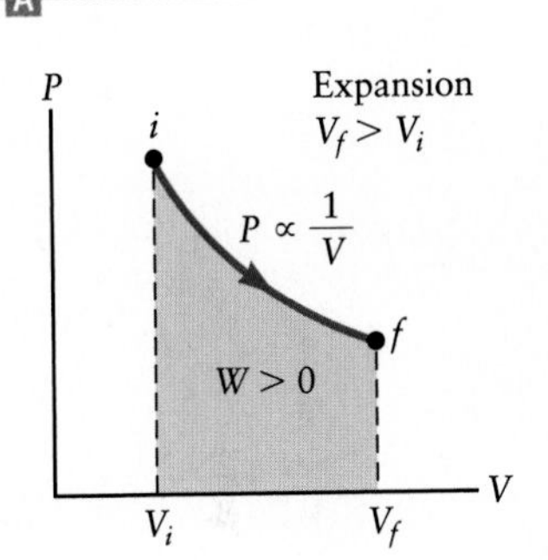

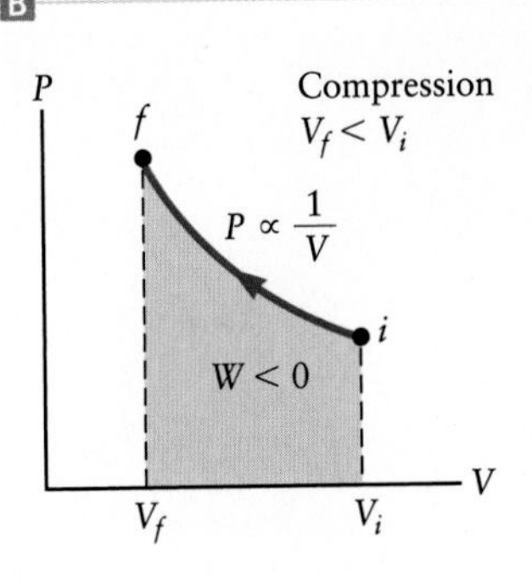

▲ **Figure 16.14** A In an isothermal process, the temperature of the system is held fixed, usually because it is in contact with a thermal reservoir. B Isothermal expansion of an ideal gas in the P–V plane. C Isothermal compression.

which is the curve plotted in Figure 16.14B. The work done by the gas is again equal to the area under the P–V curve; for an expansion described by Equation 16.8 with T held constant, the result is[2]

$$W = nRT \ln\left(\frac{V_f}{V_i}\right) \quad \text{(isothermal process)} \tag{16.9}$$

Here, "ln" is the natural logarithm function, which is reviewed in Appendix B. In an expansion the volume increases, so $V_f > V_i$ and the logarithmic term in Equation 16.9 is positive. Hence, the work done by the gas is positive in this case as we would expect from Figure 16.14B and from our previous discussions of the work done by an expanding gas. Conversely, for an isothermal compression (Fig. 16.14C), the factor $\ln(V_f/V_i)$ in Equation 16.9 is negative and the work done by the gas is negative.

EXAMPLE 16.4 Analyzing an Isothermal (Constant-Temperature) Process

Consider the isothermal expansion of an ideal gas sketched in Figure 16.14B. The system contains $n = 4.0$ moles of argon gas at room temperature. If the system then expands to double its initial volume, what are (**a**) the work done by the system and (**b**) the amount of heat that flows into the system from its surroundings?

RECOGNIZE THE PRINCIPLE

Our discussion of isothermal processes has emphasized the work done (Eq. 16.9), but there may also be heat flow either into or out of the system during the process. In this example, the system expands to fill a greater volume. For the gas to maintain a constant temperature (because we have an isothermal process), some heat Q must flow into the gas from its environment. We can calculate Q using the first law.

SKETCH THE PROBLEM

Figure 16.14B describes an isothermal expansion. The work W is the area under the P–V curve, and because the gas expands, W is positive.

(a) IDENTIFY THE RELATIONSHIPS

The work done by an ideal gas in an isothermal process is given by Equation 16.9:

$$W = nRT \ln\left(\frac{V_f}{V_i}\right)$$

The system is at room temperature ($T \approx 293$ K) and the final volume is twice the initial volume ($V_f = 2V_i$), so

$$W = (4.0 \text{ moles})[8.31 \text{ J/(mole} \cdot \text{K)}](293 \text{ K})\ln\left(\frac{2V_i}{V_i}\right)$$

SOLVE

Evaluating this expression for W gives

$$W = \boxed{6800 \text{ J}}$$

[2]This result can be derived by using calculus to compute the area under the curve specified by Equation 16.8.

(b) IDENTIFY THE RELATIONSHIPS

To find the heat that flows into the gas from its environment during this expansion, we use the first law of thermodynamics,

$$\Delta U = Q - W$$

The internal energy of a monoatomic ideal gas (such as argon) is $U = \frac{3}{2}nRT$. Because we are dealing with an isothermal process, the final temperature is equal to the initial temperature, and there is no change in the internal energy.

SOLVE

We thus have $\Delta U = 0$, and the first law of thermodynamics gives

$$\Delta U = 0 = Q - W$$

Using the result for W from part (**a**), we find

$$Q = W = \boxed{6800\ \text{J}}$$

▸ ***What does it mean?***

In this example, the internal energy of the system does not change. Hence, as expected from the principle of conservation of energy, the energy added to the system (Q) must equal the energy that leaves the system (W).

Changes with $Q = 0$ (Adiabatic). Sometimes, a system is thermally isolated from its surroundings as shown schematically in Figure 16.15A. For example, the system might be the liquid in a thermos bottle. The wall of a thermos bottle contains a vacuum, making the heat flow through the wall extremely small. If there is no heat flowing into or out of a system during a thermodynamic process, then $Q = 0$ and the process is called *adiabatic*.

To analyze an adiabatic process for a particular system, we need to know the relation between temperature, pressure, and volume. Let's again consider an example with an ideal gas (because we know its internal energy function). The path for an adiabatic process in the P–V plane can be calculated using the ideal gas law and the expression for the internal energy (Eq. 16.2). We will omit the derivation, but the result is

$$PV^{\gamma} = \text{constant}$$

where $\gamma = \frac{5}{3}$ for a dilute monatomic gas.[3] The pressure thus varies as

$$P \propto \frac{1}{V^{\gamma}} \quad \text{(adiabatic process)} \tag{16.10}$$

On a P–V diagram, an adiabatic path (Fig. 16.15B) is steeper than the path for an isothermal process that begins from the same initial state. (Recall that an isothermal path follows $P \propto 1/V$; Eq. 16.8.)

The value of W for an adiabatic process is again equal to the area under the P–V curve (Fig. 16.15B). According to the first law of thermodynamics, $\Delta U = Q - W$, and because $Q = 0$ for an adiabatic process, we have

$$\Delta U = -W \quad \text{(adiabatic process)} \tag{16.11}$$

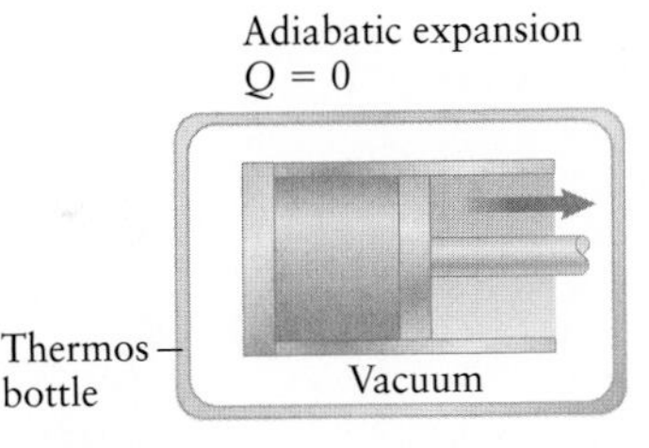

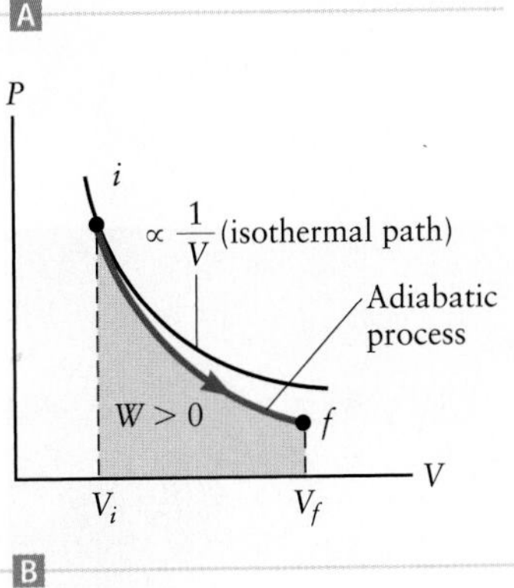

▲ **Figure 16.15** A In an adiabatic process, no heat flows into or out of the system ($Q = 0$). B Adiabatic expansion of an ideal gas in the P–V plane.

[3]The parameter γ is the ratio of the specific heats at constant pressure and constant volume. This ratio depends on the type of molecules in the gas. For a monatomic gas, $\gamma = \frac{5}{3}$, and for a diatomic gas, $\gamma = \frac{7}{5}$.

Isochoric (constant-volume) process: $V_i = V_f$

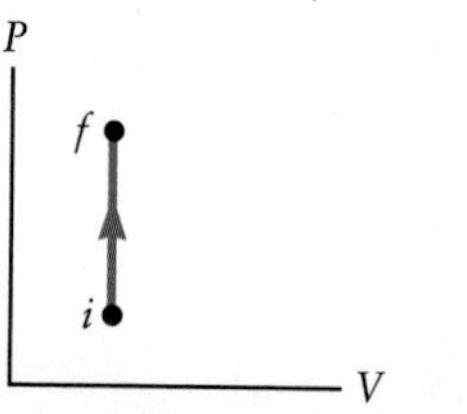

▲ **Figure 16.16** In an isochoric process, the volume does not change.

A hypothetical adiabatic process is the expansion of a gas against a piston. As the gas expands, it does a positive amount of work on the piston, and $W > 0$. According to Equation 16.11, the internal energy of the gas must decrease. For a monatomic ideal gas, $U = \frac{3}{2}Nk_BT$, meaning that the temperature decreases; that is, the gas cools. Although $Q = 0$ in an adiabatic process, other properties of the system such as the temperature can change during the process.

Changes at Constant Volume (Isochoric). An ***isochoric*** process is one for which the volume does not change; it is also called an ***isovolumetric*** process. The corresponding path on a P–V plot is shown in Figure 16.16. A key feature of all isochoric processes is discussed in Example 16.5.

CONCEPT CHECK 16.3 Heating a Gas at Constant Volume

Figure 16.17 shows P–V plots for two different isochoric (constant-volume) processes. In one case, the pressure increases, whereas in the other, the pressure decreases. In which case does the temperature increase?

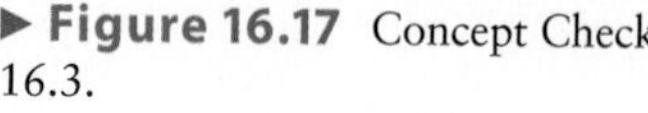

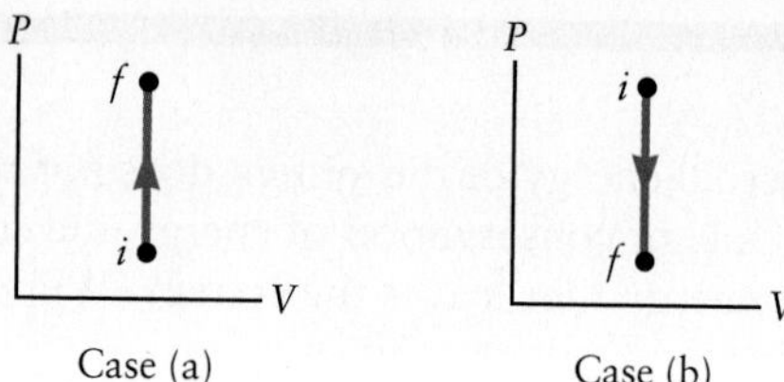

▶ **Figure 16.17** Concept Check 16.3.

EXAMPLE 16.5 A Constant-Volume (Isochoric) Process

Consider an ideal gas that is heated under isochoric conditions. What is the work done by the gas?

RECOGNIZE THE PRINCIPLE

Because the volume is kept fixed, an isochoric process is a vertical line in the P–V plane.

SKETCH THE PROBLEM

The P–V diagram for this process is shown in Figure 16.16. In this example, the gas is heated, so the final pressure is greater than the initial pressure.

IDENTIFY THE RELATIONSHIPS AND SOLVE

The work done by the gas is the area under this curve, which is zero. So, for this process (and for *all* isochoric processes), $\boxed{W = 0}$.

▶ ***What does it mean?***

When analyzing a thermodynamic process, it is always useful to consider the P–V diagram. The work done by the system is equal to the area under the P–V curve.

▲ **Figure 16.18** Concept Check 16.4.

CONCEPT CHECK 16.4 What Kind of Process Is It?

A bear has a very thick fur coat, providing very good thermal insulation from its environment (Fig. 16.18). When the bear hibernates, is it an example of an isothermal or an adiabatic process, or both? Explain.

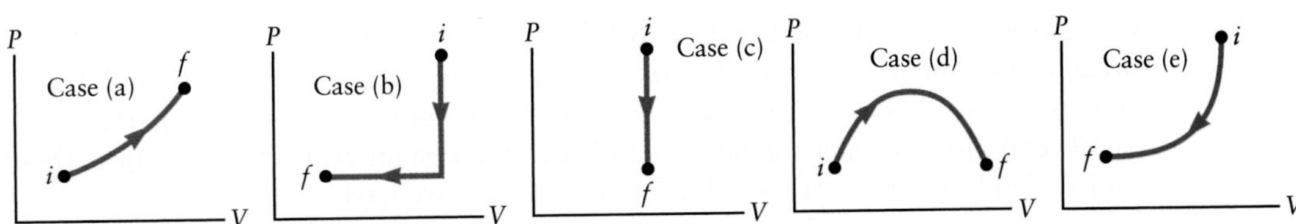

▲ **Figure 16.19** Concept Check 16.5.

CONCEPT CHECK 16.5 Work Done in a Thermodynamic Process

Figure 16.19 shows several thermodynamic processes. Without doing a calculation, determine if W is positive, negative, or zero for each process.

Some Properties of Q and W

A thermodynamic process can be described by a path on the P–V diagram that takes the system from some initial state i to a final state f. Many paths connect a particular initial state to a given final state; Figure 16.20 shows two paths that connect states i and f. One of these paths passes through point A, and we'll refer to it as "path A." The other passes through point B ("path B"). Let's consider how ΔU, W, and Q compare for these two paths, that is, for these two different thermodynamic processes.

We can write the change in internal energy as

$$\Delta U = U_f - U_i \tag{16.12}$$

In words, the change in U depends only on the internal energies of the initial and final states, so ΔU is the same for *any* path that connects states i and f. This result is reminiscent of our work in Chapter 6, where we learned that changes in the potential energy are independent of the path taken.

The work done in going from state i to state f along path A is the sum of the work done in going from i to A and the work done in going from A to f; hence,

$$W_A = W_{iA} + W_{Af}$$

Here W_{iA} denotes the work done on the path from i to A in Figure 16.20 and so forth for W_{Af}. Work is the area under the P–V curve, so

$$W_{iA} = P\,\Delta V = P_i(V_A - V_i)$$

The work along the portion of the path from A to f is zero ($W_{Af} = 0$) because there is no change in the volume. The total work done along path A is thus

$$W_A = W_{iA} + W_{Af} = P_i(V_A - V_i)$$

The volume is the same at point A and f ($V_A = V_f$); hence,

$$W_A = P_i(V_f - V_i) \tag{16.13}$$

We can find the work done along path B in a similar way:

$$W_B = W_{iB} + W_{Bf}$$

with $W_{iB} = 0$ because there is no volume change on this portion of the path, and

$$W_{Bf} = P_f(V_f - V_B)$$

The result for the work on path B is then

$$W_B = W_{iB} + W_{Bf} = P_f(V_f - V_B)$$

Using $V_B = V_i$, this expression becomes

$$W_B = P_f(V_f - V_i) \tag{16.14}$$

The change in internal energy is the same: $\Delta U_A = U_f - U_i = \Delta U_B$ **but** $W_A \neq W_B$ and $Q_A \neq Q_B$

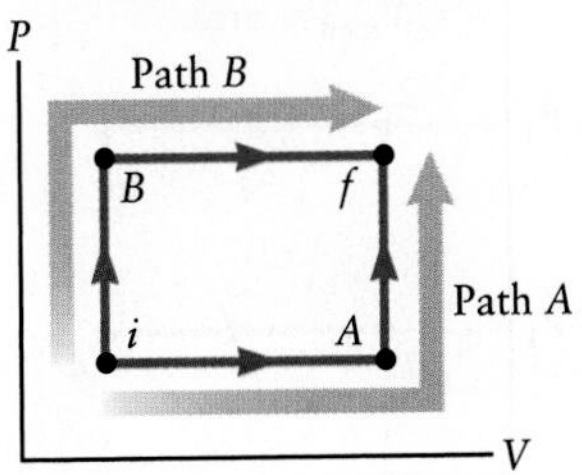

▲ **Figure 16.20** Two different thermodynamic processes (two paths in the P–V plane) can connect the same initial and final states.

Comparing the results for W_A (Eq. 16.13) and W_B (Eq. 16.14), we see that the work done by the system along path A is *not* the same as the work done along path B. Unlike changes in the internal energy, the amount of work done *depends on the path.*

We can also consider the heat Q added to the system as it travels along paths A and B. Applying the first law of thermodynamics, we have

$$\Delta U_A = Q_A - W_A \quad \text{(along path } A\text{)} \tag{16.15}$$

$$\Delta U_B = Q_B - W_B \quad \text{(along path } B\text{)} \tag{16.16}$$

We have already seen that changes in the internal energy are independent of the path, so $\Delta U_A = \Delta U_B$. We have also just shown that the work done is different along the two paths, so $W_A \neq W_B$. Using these facts with Equations 16.15 and 16.16, we see that

$$Q_A \neq Q_B$$

In words, the heat added to the system also *depends on the path* taken.

The principle of conservation of energy thus leads us to the following conclusions.

Properties of U, W, and Q for thermodynamic processes

1. **The internal energy of a system depends only on the current state of the system. Changes in the internal energy are thus independent of the path taken in a thermodynamic process. The change in internal energy ΔU depends only on the internal energies of the initial and final states.**
2. **The work done during a thermodynamic process depends on the path taken. Even with the same initial and final states, two different thermodynamic paths can have different values of W.**
3. **The heat added to a system during a thermodynamic process depends on the path taken. Even with the same initial and final states, two different thermodynamic paths can have different values of Q.**

EXAMPLE 16.6 Analyzing a Cyclic Process

Consider the thermodynamic process described in Figure 16.21. This P–V diagram is similar to the one in Figure 16.20 except that now we begin and end at the same point (the same state). The system in Figure 16.21 starts in state 1, moves as indicated to state 2, state 3, and state 4, and then moves back to state 1. This set of changes is called a ***cyclic process*** because the system returns to its initial state. Find the work done during this process.

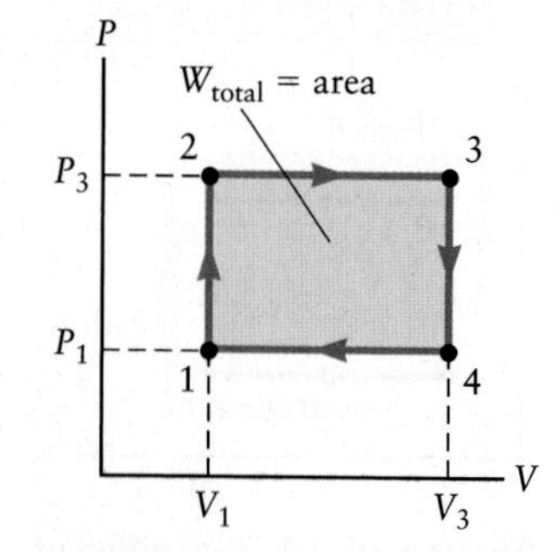

RECOGNIZE THE PRINCIPLE

We can use $W = P\,\Delta V$ (Eq. 16.6) to find the work done along each portion of the path. Adding those results gives the total work done for the entire process.

SKETCH THE PROBLEM

Figure 16.21 shows the P–V diagram.

▲ Figure 16.21 Example 16.6. A cyclic process is one that returns to its original state.

IDENTIFY THE RELATIONSHIPS

The volume does not change along the path from state 1 to state 2 or along the path from state 3 to state 4, so $\Delta V = 0$ and the work done in these parts of the process is zero. Using $W = P\,\Delta V$ for the path from state 2 to state 3 gives

$$W_{23} = P\,\Delta V = P_3(V_3 - V_1)$$

Likewise, along the path from state 4 to state 1, we get

$$W_{41} = P_1(V_1 - V_3)$$

Notice that W_{41} is negative because $V_1 < V_3$.

SOLVE

Combining these results gives

$$W_{total} = W_{23} + W_{41} = P_3(V_3 - V_1) + P_1(V_1 - V_3) = \boxed{(P_3 - P_1)(V_3 - V_1)}$$

► ***What does it mean?***

From the geometry of Figure 16.21, W_{total} is just the area enclosed by the path in the P–V diagram.

Work Done by a Cyclic Process

Example 16.6 considered the work done in a cyclic process and showed that W is equal to the area enclosed by the path on the P–V diagram. The process analyzed in that example had an especially simple path, but the general result applies to all cyclic processes. The work done in any cyclic process equals the area enclosed by the path in the corresponding P–V diagram (Fig. 16.22). We'll use this result when we analyze the performance of heat engines and other devices in Section 16.6.

▲ **Figure 16.22** In any cyclic thermodynamic process, the work done by the system equals the area enclosed by the path on the P–V diagram.

CONCEPT CHECK 16.6 Thermodynamic Processes

Figure 16.23 shows several thermodynamic processes. Which of these processes could be (a) an isothermal expansion, (b) an adiabatic compression, (c) an isobaric expansion, and (d) an isobaric compression?

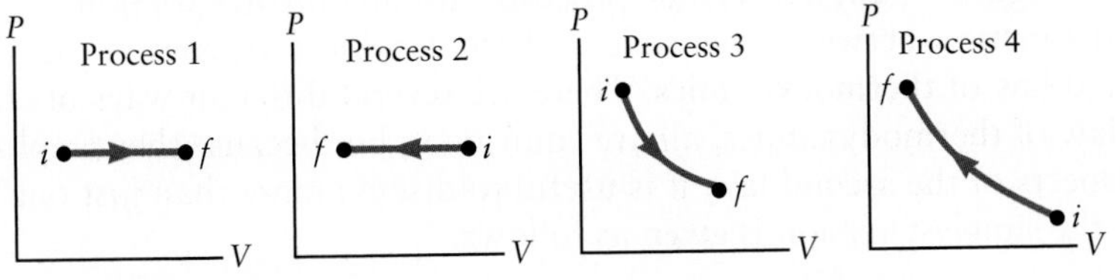

◄ **Figure 16.23** Concept Check 16.6.

16.5 The Second Law of Thermodynamics

Figure 16.24A shows an isothermal expansion of a dilute gas in which the system starts in an initial state i and moves to a final state f. The piston moves to the right as the gas does some amount of work $W > 0$, and during this time the gas absorbs some heat $Q > 0$ from the environment. This example shows a thermodynamically ***reversible*** process. We could imagine a reversed process in which the gas begins in state f and is isothermally compressed back to state i (Fig. 16.24B). During this

Reversible versus irreversible processes

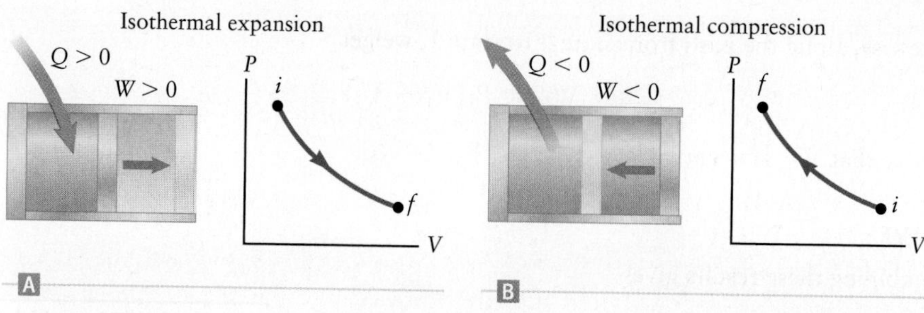

▶ **Figure 16.24** Schematic of a thermodynamically reversible process. A Isothermal expansion from state i to state f. B Isothermal compression from state f back to state i.

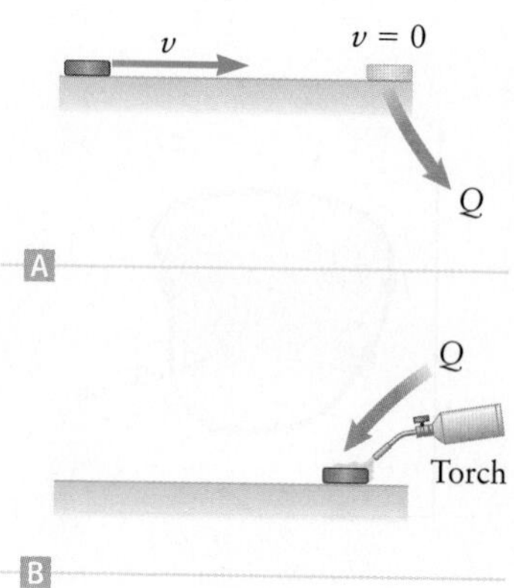

▲ **Figure 16.25** A A hockey puck sliding across a floor with friction is an example of a thermodynamically irreversible process. B Heating a hockey puck will *not* cause it to move back to its original position!

compression, the gas does a negative amount of work ($W < 0$) on the piston because an outside agent must push on the piston to compress the gas. At the same time, heat would flow out of the gas into the surroundings. The work done by the gas during the compression is equal in magnitude but opposite in sign compared with W during the expansion. Also, the amount of heat that flows out of the gas during compression is equal in magnitude to the heat that flows in during expansion. Hence, both the system and the environment are brought back to precisely their original states, which is the key feature of a thermodynamically reversible process.

An example of an ***irreversible*** process is sketched in Figure 16.25. Here a hockey puck is initially sliding to the right on a horizontal surface. There is some friction between the puck and the surface, causing the puck to come to a stop as all its initial kinetic energy is converted to heat energy. This heat energy flows from the region where the puck is in contact with the horizontal surface into its surroundings (Fig. 16.25A). Suppose we now try to reverse this process by arranging for heat to flow from the surroundings into the puck, perhaps with a torch as sketched in Figure 16.25B. Intuition tells us that this will not cause the puck to begin moving! Of course, we could start the puck in motion again by striking it with a hockey stick, but for a process to be thermodynamically reversible, it must be possible to return both the system (the puck) and its surroundings (the rest of the universe) back to their original states. There is nothing we could do to reverse the process in Figure 16.25; it is simply *not possible* to take the puck and the rest of the universe back to precisely their original states. This is the essential feature of an irreversible process. And, as in Figure 16.25, irreversible processes usually involve friction.

The distinction between reversible and irreversible processes is the subject of the second law of thermodynamics. There are several different ways of stating the second law of thermodynamics; all are equivalent, but because they emphasize different aspects of the second law, it is useful to discuss more than just one of them. Perhaps the simplest version is given as follows.

Second law of thermodynamics

Second law of thermodynamics

Heat flows spontaneously from a warm object to a colder one. It is not possible for heat to flow spontaneously from a cold object to a warmer one.

This statement certainly seems obvious. The part about heat flowing from a warm object to a colder one is implied by the zeroth law of thermodynamics. Two crucial aspects of the second law go beyond the zeroth law, however. The first is the notion of a *spontaneous* process. This is a process that takes place "on its own" without work done by any added or external forces. For example, the flow of heat from the surroundings into the ice cube in Figure 16.8 is a spontaneous process. The other crucial part of the second law of thermodynamics is the statement that heat cannot flow spontaneously from a cold object to a warmer one. This statement leads to many important results, including the impossibility of building a perpetual motion machine.

16.6 Heat Engines and Other Thermodynamic Devices

One of the discoverers of the second law of thermodynamics was an engineer named Sadi Carnot (1796–1832), who was interested in how the science of thermodynamics could be applied to practical devices such as steam engines and other marvels of the Industrial Revolution that were being developed around that time. Carnot's engineering approach leads to a rather different way of thinking about the second law of thermodynamics.

Imagine you are an engineer and you have been asked to invent the world's best heat engine. A heat engine takes heat energy, as might be generated by a coal-burning furnace, and converts this energy into work. For example, this work might rotate a shaft as would be useful in a steamboat, locomotive, or automobile. With that in mind, we can draw a heat engine in the highly schematic form shown in Figure 16.26A. An amount of heat Q_H is extracted from the furnace and fed into some sort of mechanical device, which then does an amount of work W. It is your engineering challenge to design the mechanical device. The principle of conservation of energy and the first law of thermodynamics tell us that the work done cannot be larger than the energy input Q_H, but you might hope to design and patent a device that reaches this limit. The world's best heat engine would perform an amount of work $W = Q_H$, and from the principle of conservation of energy, you would be absolutely guaranteed that no one could design an engine better than that one.

That is the problem Carnot tackled. Although he was not able to construct the heat engine in Figure 16.26A, Carnot was able to show that such an engine is *impossible*! He showed that it is simply not possible to construct a device whose sole effect is to convert heat into work, that is, to convert a given amount of heat energy completely into mechanical energy. Instead, all heat engines must be diagrammed as in Figure 16.26B. Here we again have a mechanical device that takes an amount of heat Q_H from a hot reservoir and does an amount of work W. However, Carnot showed that all real heat engines must also expel some heat energy Q_C to a colder reservoir.

Applying conservation of energy to the heat engine in Figure 16.26B, we have

$$W = Q_H - Q_C \tag{16.17}$$

Your goal as an engine designer is get the most work (the largest W) possible out of the engine, so we define the ***efficiency*** e of the engine as

$$e = \frac{W}{Q_H} \tag{16.18}$$

Combining Equation 16.18 with Equation 16.17 gives

$$e = \frac{W}{Q_H} = \frac{Q_H - Q_C}{Q_H}$$

$$e = 1 - \frac{Q_C}{Q_H} \tag{16.19}$$

When comparing heat engines, a larger value of e means that the engine converts a larger fraction of the input energy Q_H into work.

Carnot's Engine

To maximize the efficiency e, one must make Q_C as small as possible, which makes W as large as possible. Carnot considered this problem for a heat engine that makes

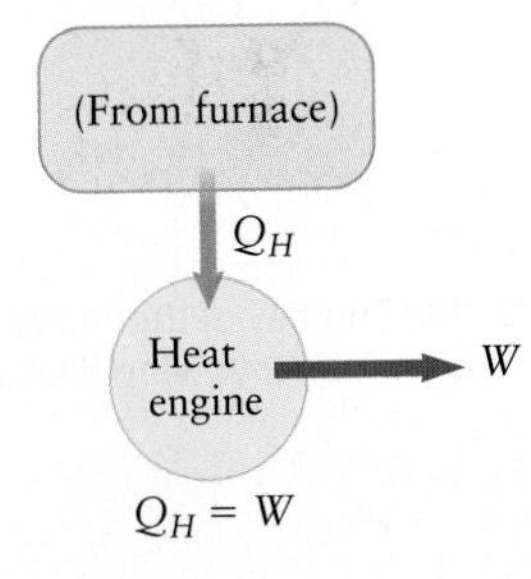

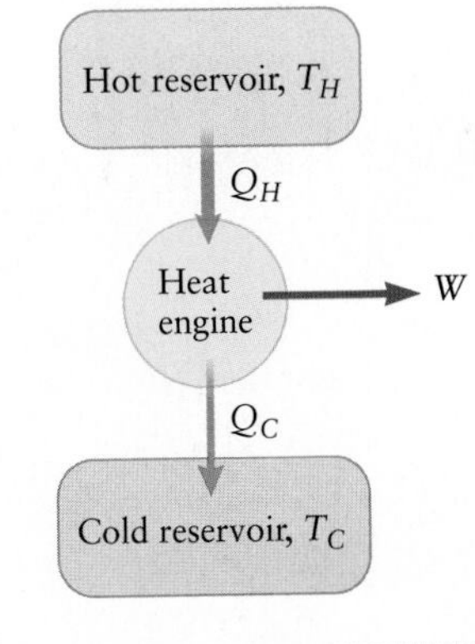

▲ **Figure 16.26** Schematic diagrams for a heat engine. A A hypothetical engine, which the laws of thermodynamics tell us is *not* possible. B All real heat engines must expel some heat to a cold reservoir.

use of the reversible compression and expansion of an ideal gas. Because Carnot's engine uses only reversible processes, it is an example of a reversible heat engine. He was able to prove that all reversible heat engines have the same efficiency as his particular design. Carnot also showed that for a hot reservoir at temperature T_H (providing the heat input Q_H) and a cold reservoir at temperature T_C, the heat expelled to the cold reservoir satisfies

$$\frac{Q_C}{Q_H} = \frac{T_C}{T_H} \tag{16.20}$$

Here the temperatures T_H and T_C must be measured in Kelvin units. Combining Equation 16.20 with the definition of efficiency (Eq. 16.19) leads to

Efficiency of a reversible heat engine

$$e = 1 - \frac{T_C}{T_H} \tag{16.21}$$

Equation 16.21 is an amazing result: it applies to *all* reversible engines, no matter how they are constructed. Carnot did not stop here, but discovered one more crucial point: no engine can have an efficiency that is better than the efficiency of a reversible engine. In practice, all heat engines will always be irreversible to some extent, so Carnot's result sets an *absolute limit* on the efficiency of all real heat engines. Carnot's results are the basis for an alternative statement of the second law of thermodynamics.

Second law of thermodynamics (alternative form)

The efficiency of a reversible heat engine is given by

Second law of thermodynamics: formulation involving heat engines

$$e = 1 - \frac{T_C}{T_H}$$

No heat engine can have a greater efficiency than this.

We have now seen two statements of the second law, this one involving the efficiency of heat engines and another (in Section 16.5) involving the direction of heat flow. Although it is not obvious, these two statements are *completely equivalent.* Given either one of them as a starting point, the other one can be derived!

The result in Equation 16.21 shows that the efficiency of an engine depends on the temperatures of the reservoirs. If the temperature of the cold reservoir is zero ($T_C = 0$ K), the efficiency $e = 1$. In this case, we also have $Q_C = 0$, and it would thus seem that we have created the hypothetical heat engine in Figure 16.26A—a heat engine that expels no heat—which we said was impossible. One way out of this apparent contradiction is to hypothesize that it is impossible to lower the temperature of a reservoir to $T = 0$ K, the temperature called absolute zero. We'll see this hypothesis again in Section 16.8 when we discuss the third law of thermodynamics.

EXAMPLE 16.7 Efficiency of a Steam Engine

Steamboats were used extensively in the 1800s to carry people and goods (Fig. 16.27). These boats used heat engines: a hot reservoir contained steam ($T_H \approx 200°C$) that was heated by burning wood or coal, and the cold reservoir was provided by the air or river water ($T_C \approx 10°C$). The engine turned a shaft that drove the boat's paddle wheels. Suppose one of these steam engines does $W = 10{,}000$ J of work each second. If it is a reversible heat engine, how much heat Q_H does it absorb from the hot reservoir in 1 s?

RECOGNIZE THE PRINCIPLE

The efficiency of any heat engine is defined as (Eq. 16.18)

$$e = \frac{W}{Q_H} \tag{1}$$

For a reversible heat engine, Carnot showed that e is also equal to

$$e = 1 - \frac{T_C}{T_H} \tag{2}$$

We can use these two relations to find the heat absorbed from the hot reservoir Q_H in terms of the given quantities W, T_H, and T_C.

▲ **Figure 16.27** Example 16.7. Steamboats in the 1800s used heat engines to drive their paddle wheels.

SKETCH THE PROBLEM

Figure 16.27 shows a picture of a real steamboat, but the diagram of a heat engine in Figure 16.26B is more useful for analyzing the operation of the engine.

IDENTIFY THE RELATIONSHIPS

Combining Equations (1) and (2) gives

$$\frac{W}{Q_H} = 1 - \frac{T_C}{T_H}$$

Rearranging to solve for Q_H, we have

$$Q_H = \frac{W}{1 - (T_C/T_H)} \tag{3}$$

SOLVE

When using Equation (3), we must be careful to use values of the temperature on the Kelvin scale; Carnot's result in Equation (2) assumes the temperatures are measured on this scale. Converting the given values of T_H and T_C gives $T_H = (200 + 273)\text{ K} = 473\text{ K}$ and $T_C = (10 + 273)\text{ K} = 283\text{ K}$. Inserting these values of the reservoir temperatures and the given value of W into Equation (3), we find

$$Q_H = \frac{W}{1 - (T_C/T_H)} = \frac{10{,}000\text{ J}}{1 - (283\text{ K}/473\text{ K})} = \boxed{25{,}000\text{ J}}$$

▶ ***What does it mean?***

Because the engine did 10,000 J of work each second, the energy expelled to the cold reservoir in that time was $Q_C = Q_H - W = 15{,}000$ J. Most of the energy extracted from the hot reservoir was expelled to the cold reservoir and was thus "wasted." When analyzing the operation of a heat engine, temperature values should always be expressed in Kelvin units.

EXAMPLE 16.8 Ⓡ Efficiency of a Modern Automobile Engine

The internal combustion engine found in most cars is a type of heat engine (Fig. 16.28). A hot reservoir created by exploding gasoline provides heat energy, causing the gas in a piston to expand and do work as it rotates the driveshaft of the engine. Some heat is then expelled to a cold reservoir (the air) via the radiator and the exhaust. What is the maximum possible efficiency of such an engine?

(continued) ▶

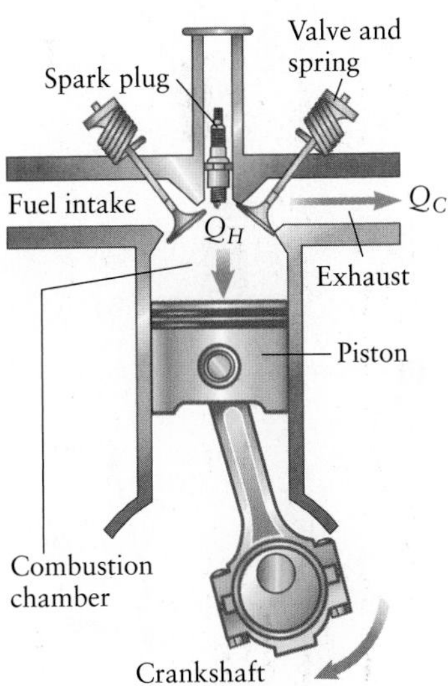

▲ **Figure 16.28** Example 16.8. A real automobile engine is a type of heat engine following the general schematic form in Figure 16.26B. The laws of thermodynamics place rigorous limits on the efficiency of all real heat engines.

Insight 16.1

EFFICIENCY OF A DIESEL ENGINE

A diesel engine is similar to a gasoline internal combustion engine (Example 16.8). One difference is that a gasoline engine ignites the fuel mixture with a spark from a spark plug, whereas a diesel engine ignites the fuel mixture purely "by compression" (without a spark). The compression of the fuel mixture is therefore much greater in a diesel engine, which leads to a higher temperature in the hot reservoir. According to Equation 16.21 and Example 16.8, this higher temperature gives a higher theoretical limit on the efficiency of a diesel engine.

RECOGNIZE THE PRINCIPLE

A real internal combustion engine is irreversible, so its efficiency will always be less than that of a reversible engine. This irreversibility is caused by friction between the moving parts inside the engine and by other factors. Nevertheless, an upper limit on the efficiency is set by the reversible case (Eq. 16.21)

$$e = 1 - \frac{T_C}{T_H} \tag{1}$$

To apply Equation (1), we need to know the temperatures of the two reservoirs T_C and T_H; these values are not given, so we need to use some common sense and knowledge about this type of engine to find approximate values.

SKETCH THE PROBLEM

The schematic heat engine in Figure 16.26B describes the problem. The corresponding hot and cold reservoirs for a real automobile engine are shown in Figure 16.28.

IDENTIFY THE RELATIONSHIPS

The cold reservoir (the radiator and exhaust) is at air temperature, so we estimate $T_C \approx 300$ K. For the hot reservoir, we can use reference data (from the Internet or elsewhere) to find that the temperature of exploding gas in a piston is about $T_H \approx 600$ K.

SOLVE

Inserting our estimates for the reservoir temperatures into Equation (1) gives

$$e = 1 - \frac{T_C}{T_H} = 1 - \frac{300 \text{ K}}{600 \text{ K}} = \boxed{0.5} \tag{2}$$

▶ ***What does it mean?***

Even in the best possible case, an automobile engine can only convert about half the chemical energy in the gasoline into useful work. This limit on the efficiency could be improved by increasing the temperature of the hot reservoir (the temperature in the piston), which can be accomplished by using a suitable blend of gasoline with a higher octane rating. Such high-performance engines and fuels are more expensive than typical engines and fuels, however.

In a real engine, a substantial amount of energy is lost to friction. Only about 25% of the chemical energy in the gasoline is typically converted into work.

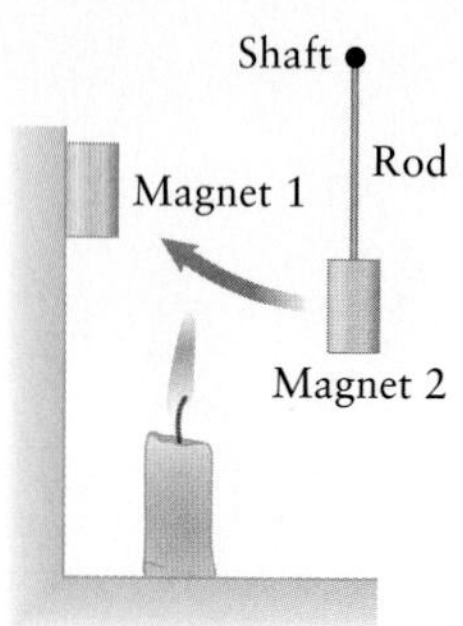

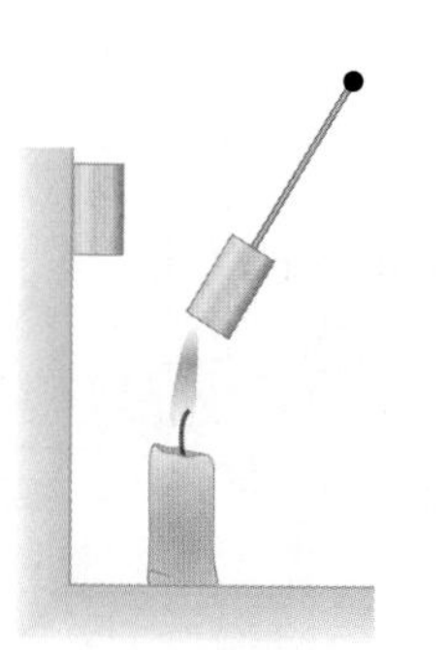

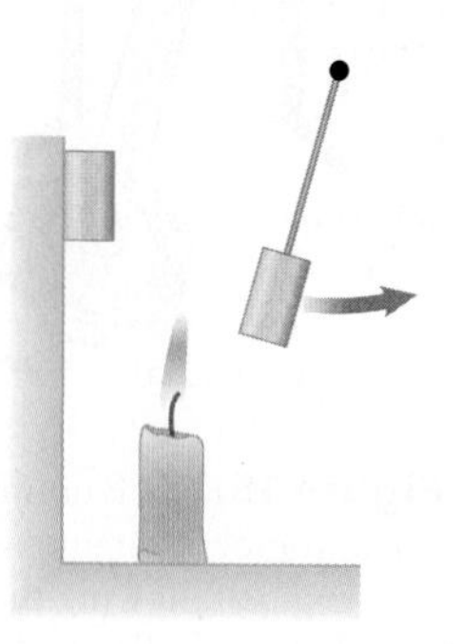

▲ **Figure 16.29** Example 16.9. A simple heat engine.

EXAMPLE 16.9 A Magnetic Heat Engine

Figure 16.29 shows a heat engine that uses two permanent magnets and a candle. Magnet 1 is attached to a wall, while magnet 2 swings on a rod that can take it over the candle's flame. Although we have not yet discussed permanent magnets (see Chapter 20), we can still analyze this engine; we only need to know that these two permanent magnets attract each other except when one of them is very hot, in which case the hot one ceases to behave as a permanent magnet. This heat engine works as follows. (1) Magnet 2 is initially far from the flame of the candle (Fig. 16.29A). Magnet 2 is attracted to magnet 1, causing the rod to rotate clockwise. (2) Magnet 2 eventually passes over the flame and is heated so that it is no longer attracted to magnet 1 (Fig. 16.29B), and the torque from gravity causes the rod to rotate back counterclockwise (Fig. 16.29C). (3) Magnet 2 then moves away from the flame, it cools to room temperature, and its permanent magnetism is restored. It is then attracted to magnet 1, and the cycle repeats. This system thus behaves as an engine that can do work via the motion of the rod. What are the hot and cold reservoirs of this heat engine?

RECOGNIZE THE PRINCIPLE

Without the heat from the candle, magnet 2 would always be attracted to magnet 1 and the rod would never swing back counterclockwise. At the same time, this engine would not work if magnet 2 were not able to cool back to room temperature, which it does through its interaction with the surrounding air.

SKETCH THE PROBLEM

Figure 16.29 shows the problem.

IDENTIFY THE RELATIONSHIPS AND SOLVE

The heat from the candle and the cool air are both necessary for this engine to work.

The candle acts as the hot reservoir and the air acts as the cold reservoir.

▶ *What does it mean?*

Many mechanical devices can be viewed as heat engines and analyzed using the principles discovered by Carnot. Room temperature is about 300 K. Although the temperature of a candle's flame depends on the type of wax and other factors, it is typically around 1500 K. The maximum efficiency of this heat engine is thus about $e = 1 - T_C/T_H = 1 - (300\text{ K})/(1500\text{ K}) = 0.8$. The actual efficiency will be less than this value because the magnet will probably not reach 1500 K when it is nearest the flame.

Perpetual Motion and the Second Law of Thermodynamics

The construction of a perpetual motion machine has been the quest of many inventors. Various laws of physics imply that such machines are impossible. One popular perpetual motion machine design sketched in Figure 16.30 makes use of the hypothetical (and impossible) heat engine in Figure 16.26A. The machine in Figure 16.30 takes some heat energy Q from a hot reservoir (the atmosphere) and converts it all to an amount of work W. This work is then used to power a car. As the car moves, it loses energy to friction due to air drag, the wheels, and so on. This loss causes all the energy W to be converted to heat energy, which returns to the atmosphere. The net effect is that the car has moved, whereas the energy in the atmosphere is back to its original value; hence, the process can be repeated, and the car can move on forever.

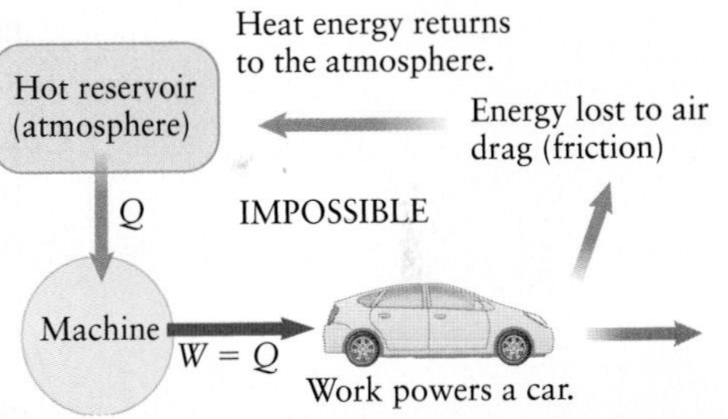

▲ **Figure 16.30** Hypothetical (and *impossible*) design of a perpetual motion machine. This machine is not possible because it violates the second law of thermodynamics.

Carnot's work and the second law of thermodynamics tell us that the perpetual motion machine in Figure 16.30 is *impossible* because a real heat engine must always expel some heat energy to a cold reservoir. Hence, all the heat energy Q extracted from the atmosphere cannot later be returned to the atmosphere. In this way, thermodynamics rules out many hypothetical perpetual motion designs. In fact, the second law along with conservation of energy principles rule out *all* perpetual motion machines.

The Carnot Cycle

We have so far represented heat engines in a highly schematic form as in the block diagram in Figure 16.26B and have ignored how one might construct a real engine. Carnot showed how to use an ideal gas and piston to make a reversible heat engine. Figure 16.31A shows the operation of Carnot's engine as a path in the P–V plane. The corresponding changes of the gas and piston are shown in Figure 16.31B, where

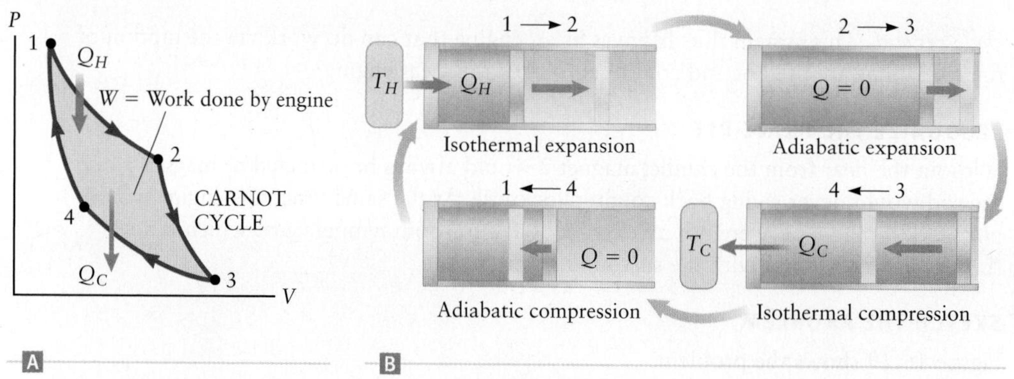

▲ **Figure 16.31** An example of a thermodynamically reversible heat engine. This engine uses a cyclic process called a Carnot cycle. A The work done during a Carnot cycle is equal to the area enclosed on the *P–V* diagram. B The four steps in the Carnot cycle.

we show only the mechanical part of the heat engine; the thermal reservoirs could be air or water held at temperatures T_H and T_C.

The system begins in state 1 on the *P–V* diagram. It then moves progressively to states 2, 3, and 4 and finally back to state 1.

Change from state 1 to state 2: The system is placed in contact with the hot reservoir, during which time the gas absorbs an amount of heat Q_H and expands to state 2. Because the system is in contact with the hot reservoir, this process is an *isothermal expansion* at temperature T_H.

Change from state 2 to state 3: The system is isolated from its surroundings (detached from the hot reservoir) and allowed to expand to state 3. Because no heat is absorbed from or expelled to the environment, this process is an *adiabatic expansion.*

Change from state 3 to state 4: We place the system in contact with the cold reservoir, and an amount of heat Q_C flows out of the gas and into that reservoir. This change causes the gas volume to decrease and is an *isothermal compression* because it occurs at the fixed temperature T_C of the cold reservoir.

Change from state 4 to state 1: The system is again isolated from its surroundings and *compressed adiabatically* back to its initial state 1.

This cyclic path on the *P–V* diagram is called a ***Carnot cycle***. Each step in the cycle involves a reversible process, and we saw in Section 16.4 how each of these processes (isothermal expansions and compressions, and adiabatic expansions and compressions) can be analyzed. The total work done by the gas in one cycle is the area enclosed by the path on the *P–V* diagram. This area is the total work *W* done by the heat engine during one complete cycle and is indicated in Figure 16.31A for the Carnot cycle.

CONCEPT CHECK 16.7 Work Done in a Carnot Cycle

Consider the steps in the Carnot cycle in Figure 16.31B. During which steps is the work *W* done by the system on the piston positive and during which steps is *W* negative?

Thermodynamic Devices: The Refrigerator

The Carnot heat engine in Figure 16.31 is reversible because each step in the cycle involves a reversible process. It is therefore possible to run this engine in reverse, that is, to run "backward" along the *P–V* path through states $1 \rightarrow 4 \rightarrow 3 \rightarrow 2 \rightarrow 1$. An amount of heat Q_C is then extracted from the cold reservoir, an amount of heat Q_H is expelled to the hot reservoir, and a certain amount of work W is done on the gas as indicated schematically in Figure 16.32A. This diagram is very similar to the heat engine in Figure 16.26B except that all the arrows showing the direction of energy flow are reversed.

Running a heat engine in reverse may seem like a crazy thing to do, but it is actually of great practical importance. Such a device is called a ***refrigerator.*** In a typical case (Fig. 16.32B), an electric motor does an amount of work W on a refrigeration "unit," which extracts an amount of heat Q_C from inside the refrigerator, keeping your food cold. The thermodynamic details of a refrigeration unit are more than we can discuss here, but they can be based on (you guessed it) cyclic processes involving the compression and expansion of gases. A refrigerator expels some heat Q_H to a hot reservoir, which in this case is the air around the refrigerator. Hence, refrigerators warm the air around them, usually underneath or behind the appliance. Applying conservation of energy to the refrigerator in Figure 16.32, we have

$$W + Q_C = Q_H \tag{16.22}$$

Carnot's results apply in this case, just as they apply to a heat engine, so we again have (from Eq. 16.20)

$$\frac{Q_C}{Q_H} = \frac{T_C}{T_H} \tag{16.23}$$

A good refrigerator extracts a great deal of heat Q_C from the cold reservoir while using as little work W as possible. The performance of a refrigerator is therefore measured in terms of the ***refrigeration efficiency*** e_{refrig} (also called the coefficient of performance), where

$$e_{\text{refrig}} = \frac{Q_C}{W} \tag{16.24}$$

The goal of a refrigerator designer is to make e_{refrig} as large as possible. Thermodynamics, however, places a limit on this efficiency. As with heat engines, the highest efficiency is found with a reversible refrigerator, and we can use Equations 16.22 and 16.23 to calculate e_{refrig}. Combining those relations, we find

$$W + Q_C = Q_H = Q_C\frac{T_H}{T_C}$$

$$W = Q_C\left(\frac{T_H}{T_C} - 1\right) \tag{16.25}$$

Inserting this result for W into the definition of refrigeration efficiency (Eq. 16.24) leads to

$$e_{\text{refrig}} = \frac{Q_C}{W} = \frac{Q_C}{Q_C[(T_H/T_C) - 1]} = \frac{1}{(T_H/T_C) - 1}$$

$$e_{\text{refrig}} = \frac{T_C/T_H}{1 - (T_C/T_H)} = \frac{T_C}{T_H - T_C} \tag{16.26}$$

A refrigerator is used to keep things cold, so T_C is usually low, but a low value of T_C leads to a small value for e_{refrig}. We can see that by inserting $T_C \approx 0$ into Equation 16.26, which gives

$$e_{\text{refrig}} = \frac{T_C}{T_H - T_C} = \frac{0}{T_H - 0} = 0$$

So, when the cold reservoir is at a very low temperature, the efficiency is also very low. That is an unavoidable consequence of the second law of thermodynamics.

A refrigerator is a heat engine run in reverse.

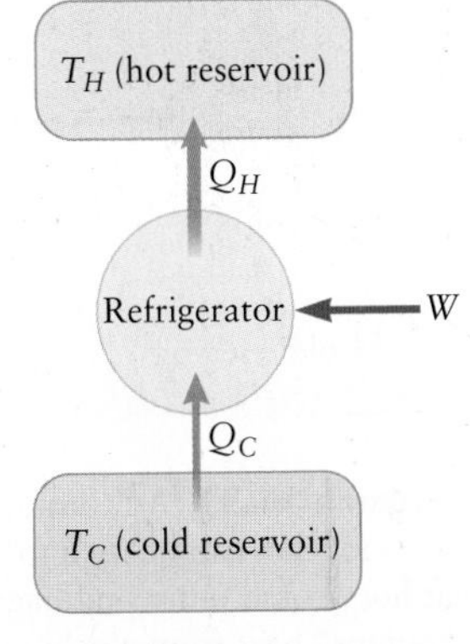

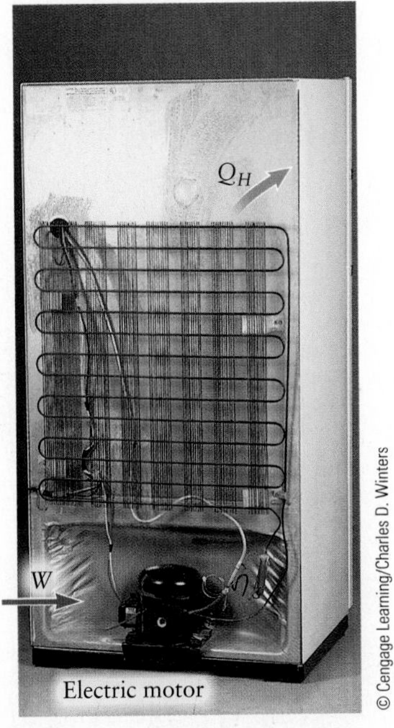

© Cengage Learning/Charles D. Winters

▲ **Figure 16.32** A When a heat engine is operated in "reverse," it functions as a refrigerator. It extracts heat energy from a cold reservoir that could then be used to store food. B In a real refrigerator, the cold reservoir is the air inside the refrigerator, and the hot reservoir is the room air.

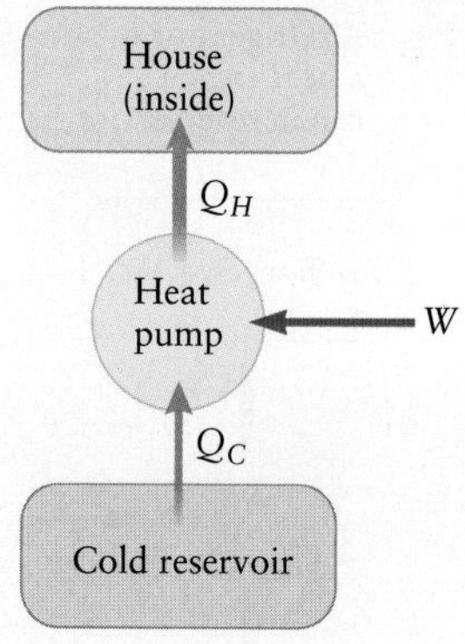

▲ **Figure 16.33** A heat pump is a device that can be used to heat homes and other buildings. It "pumps" heat from a cold reservoir into the building (the hot reservoir).

The schematic diagram of a refrigerator in Figure 16.32A shows that heat is removed from a cold reservoir and expelled to a hot reservoir. Hence, a refrigerator causes heat energy to flow from a cold object to a warmer one. Doesn't that violate the second law of thermodynamics? The answer is no; it does not violate the second law because the second law applies to the *spontaneous* flow of heat energy. The heat flow in a refrigerator is possible only because of the work W done, so the heat flow in Figure 16.32A is not spontaneous.

Heat Pumps

We have seen that a refrigerator must emit some heat; this heat is Q_H in Figure 16.32. A device called a ***heat pump*** uses this principle as part of a furnace for heating a house (Fig. 16.33). An amount of heat Q_C is extracted from a cold reservoir, which might be the outside air or perhaps water from underground. This device literally "pumps heat" from the cold reservoir into your house.

For a heat pump to be effective, you want the amount of heat that goes into your house to be as large as possible while at the same time keeping W as small as possible. So, the efficiency of a heat pump is defined as

$$e_{\text{pump}} = \frac{Q_H}{W} \tag{16.27}$$

and a good heat pump should have a large value of e_{pump}. Equations 16.22 and 16.23 apply to a heat pump as well as to a refrigerator (because a heat pump is thermodynamically identical to a refrigerator; compare Figs. 16.32A and 16.33), and we can use them to calculate e_{pump} for a reversible heat pump. They lead to

$$W = Q_H - Q_C = Q_H - Q_H \frac{T_C}{T_H}$$

$$W = Q_H[1 - (T_C/T_H)]$$

Inserting this result into the definition of e_{pump} (Eq. 16.27), we get

$$e_{\text{pump}} = \frac{Q_H}{W} = \frac{Q_H}{Q_H[1 - (T_C/T_H)]}$$

$$e_{\text{pump}} = \frac{1}{1 - (T_C/T_H)} \tag{16.28}$$

Efficiency values much greater than 1 are mathematically possible and achievable in practice, as we explore in Example 16.10.

EXAMPLE 16.10 Ⓡ Heating the Author's House

The author uses a heat pump to keep his house warm in the winter. The cold reservoir is water that comes from an underground well. What is the approximate efficiency of such a heat pump, assuming it is reversible?

RECOGNIZE THE PRINCIPLE

To find the efficiency of the heat pump (Eq. 16.28), we must know the reservoir temperatures. These temperatures are not given, so we need to estimate values from the description of the heat pump.

SKETCH THE PROBLEM

No sketch is needed.

IDENTIFY THE RELATIONSHIPS

The "hot" reservoir is the temperature inside the house, which is about 20°C = 293 K = T_H. The temperature of the cold reservoir is the temperature of the water in the well; for cold water from underground, it is about 10°C = 283 K = T_C.

SOLVE

Using these values for the reservoir temperatures, the efficiency is

$$e_{\text{pump}} = \frac{1}{1 - T_C/T_H} = \frac{1}{1 - (283\text{ K})/(293\text{ K})} \approx \boxed{29}$$

▸ ***What does it mean?***

The work W that acts as the "input" to the heat pump (Fig. 16.33) is provided by an electric motor. With the value of e_{pump} for this heat pump, each 1 J of energy used to run the electric motor leads to about 29 J of heat energy "pumped" into the house, which is quite a good deal! An alternative scheme, called "electric resistive heating," produces a maximum of 1 J of heat for each 1 J of electrical energy consumed. Hence, a heat pump can be a very efficient way to heat a house. Notice, however, that we have calculated the efficiency of a *reversible* heat pump. The efficiency of a real heat pump will be lower than this value.

CONCEPT CHECK 16.8 Heat Engine or Refrigerator?

Figure 16.34 shows the P–V diagrams for two cyclic thermodynamic processes. Which one describes a heat engine, and which one describes a refrigerator?

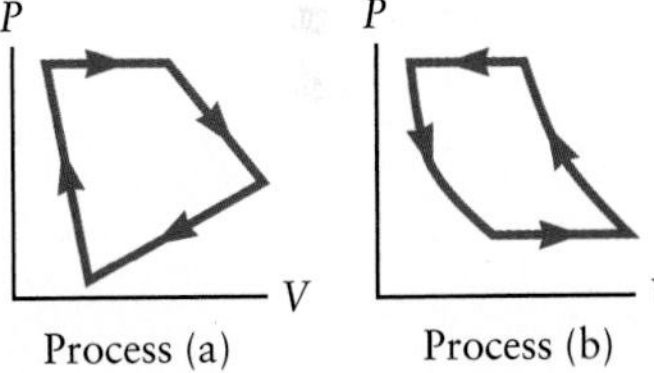

▲ **Figure 16.34** Concept Check 16.8.

16.7 Entropy

Before the work of Carnot and other scientists who developed the theory of thermodynamics, the nature of heat energy was not completely understood. Some physicists proposed (incorrectly) that heat was a sort of "fluid" that can flow from one object to another, much like water can flow from one container to another. This picture implied that an object "contains" only a certain amount of heat; if that amount of heat were somehow removed, it would then be impossible to extract any more heat from the object. We now know that this picture is not correct. However, there is a quantity called ***entropy*** that does have some of these properties.

Entropy, denoted by S, is a macroscopic property of a system much like pressure and temperature. If a small amount of heat Q flows into a system, its entropy changes by

$$\Delta S = \frac{Q}{T} \tag{16.29}$$

where T is the temperature of the system. The units of entropy are thus joules per kelvin. Let's apply Equation 16.29 to compute the total entropy change when heat flows between the two systems in Figure 16.35. The quantity Q in Equation 16.29 is the heat that flows *into* a system; in Figure 16.35, an amount of heat Q flows into system 2, so the entropy of this system changes by

$$\Delta S_2 = \frac{+Q}{T_2} \tag{16.30}$$

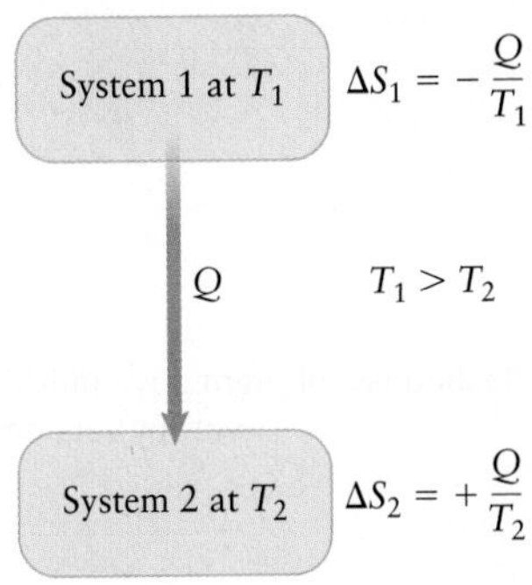

▲ **Figure 16.35** Whenever heat flows into or out of a system, the entropy of the system changes by an amount $\Delta S = Q/T$. The entropy change of a system can be positive or negative, depending on the direction of heat flow (the sign of Q).

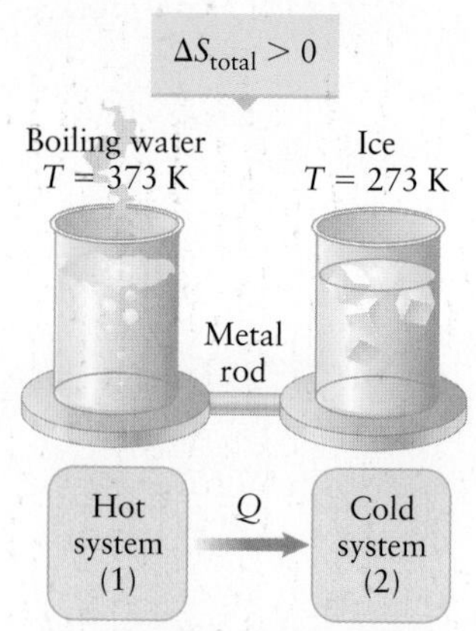

▲ **Figure 16.36** Heat flow from a hot system (1) to a cold system (2). The entropy change for the entire process, including both systems, is ΔS_{total}.

Likewise, an amount of heat Q flows *out* of system 1, so we have

$$\Delta S_1 = \frac{-Q}{T_1} \tag{16.31}$$

Hence, if heat flows from system 1 to system 2 as shown in Figure 16.35, the entropy of system 2 increases and the entropy of system 1 decreases.

A typical heat-flow experiment might involve a system at 373 K (the boiling temperature of water) in contact with a system at 273 K (the freezing temperature of water) as sketched in Figure 16.36. In this example, heat flows via conduction through a metal rod connecting the two systems. If a total amount of heat Q flows from the hot system (system 1) to the cold system (system 2), the total change in entropy is (using Eqs. 16.30 and 16.31)

$$\Delta S_{\text{total}} = \Delta S_1 + \Delta S_2 = \frac{-Q}{T_1} + \frac{Q}{T_2}$$

$$\Delta S_{\text{total}} = Q\left(-\frac{1}{T_1} + \frac{1}{T_2}\right) \tag{16.32}$$

Inserting $T_1 = 373$ K and $T_2 = 273$ K gives

$$\Delta S_{\text{total}} = \frac{-Q}{T_1} + \frac{Q}{T_2} = Q\left(-\frac{1}{373\text{ K}} + \frac{1}{273\text{ K}}\right) \tag{16.33}$$

The quantity in parentheses is positive, so ΔS_{total} is also positive.

Now suppose the temperatures of the two systems are nearly equal; in fact, let's assume we have an isothermal process so that $T_1 = T_2$. According to Equation 16.32, the total entropy change of both systems combined is zero. Such an isothermal process is reversible and gives an example of a process in which the entropy does not change. In fact, the total entropy change is zero for *all* reversible processes:

$$\Delta S_{\text{total}} = 0 \quad \text{(for a reversible process)} \tag{16.34}$$

We could also say that the total entropy change of the universe is zero for any reversible process.

The heat-flow process in Figure 16.36 involves systems at different temperatures and is irreversible because heat flows spontaneously from the hot to the cold system but not in the reverse direction (according to the second law of thermodynamics). According to Equation 16.32, ΔS_{total} in this case is always positive, which leads to the result

$$\Delta S_{\text{total}} > 0 \quad \text{(for an irreversible process)} \tag{16.35}$$

Equation 16.35 applies to *all* irreversible processes.

The results in Equations 16.34 and 16.35 form the basis for another statement of the second law of thermodynamics.

Second law of thermodynamics: entropy version

Second law of thermodynamics (entropy version)

In any thermodynamic process, $\Delta S_{\text{universe}} \geq 0$.

We now have three separate versions of the second law of thermodynamics. All are equivalent; any one version can be used to prove the other two. Each version gives different insights into the meaning and consequences of the second law.

Entropy: A Microscopic View

Let's now describe entropy in terms of a microscopic picture. Consider an ice cube placed in a warm environment. The ice cube is initially at a lower temperature than its environment, so heat will flow from the surroundings into the ice cube, causing it to melt. Because heat flows into the ice cube, its entropy will increase (according to Eq. 16.29). A molecular-scale view of this process is shown in Figure 16.37. How can we tell from this picture that the entropy has changed?

Before the ice cube melts, the water molecules are in regular positions characteristic of a solid, whereas after melting, the molecules are in the disordered arrangement characteristic of a liquid state. The system thus becomes much more *random* after the heat has been added. In microscopic terms, *entropy is a measure of the amount of disorder* or randomness in a system. Increasing the temperature of a gas or liquid increases the kinetic energy of the molecules and increases the amount of disorder. Adding heat to a system always increases its entropy.

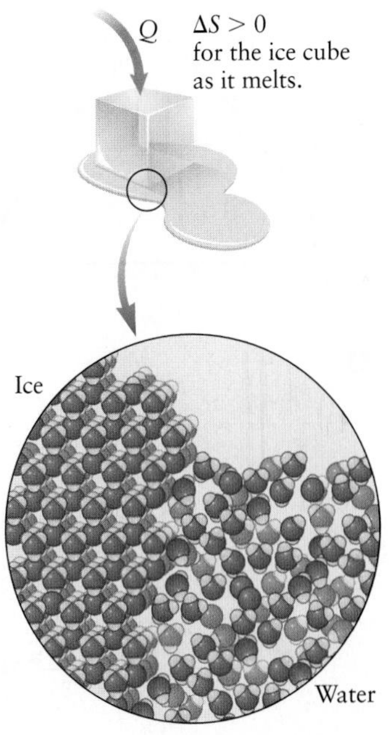

▲ **Figure 16.37** When the entropy of a system increases, the system becomes more disordered.

Entropy and Probability

Let's consider a system of gas molecules in a box and ask how the molecules are distributed. Are they all on one side of the box, are they distributed randomly throughout the box, or are they distributed in some other way? For simplicity, we can divide our box into two halves and ask how many molecules are in each half at any given "snapshot" in time. By symmetry, a particular molecule is equally likely to be on one side as on the other, so the probability of finding any given molecule on the left side is $p = \frac{1}{2}$. For a system containing only a single molecule, we thus have two possibilities—we call these two possible "states" of the system—as shown in Figure 16.38A.

We can do the same analysis with 2 molecules. Because each molecule can be on either the right or left side of the box, we now have $2 \times 2 = 4$ possible states for the system—four different possible arrangements of the molecules—as shown in Figure 16.38B. Figure 16.38C shows all the different possible states with 4 molecules. Notice how the number of possible states grows very rapidly as we add molecules to the system. With 4 molecules, we already have 16 possible states, and by the time

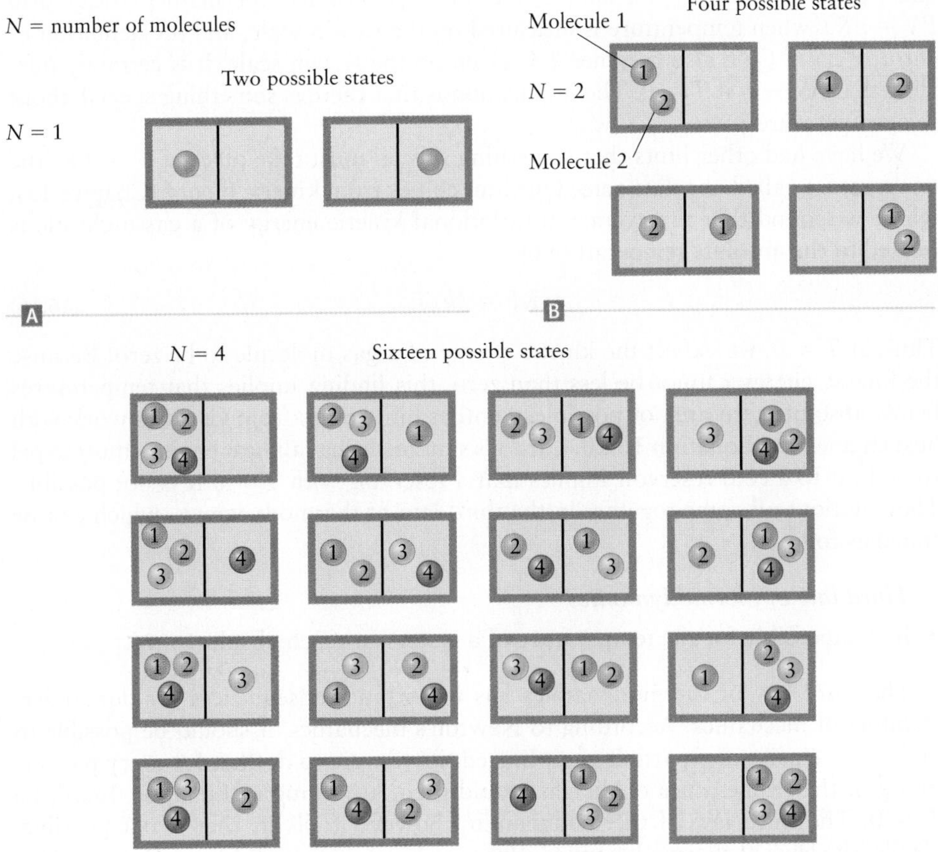

◀ **Figure 16.38** A A box containing a single gas molecule. We divide the box into two equal parts; the molecule might be found in either of these two possible "states." B If the box contains 2 molecules, they can be arranged in the two parts of the box in four different ways. C With 4 molecules, there are 16 different ways to arrange the molecules. There are thus 16 different possible states of the system, and all are equally likely to occur.

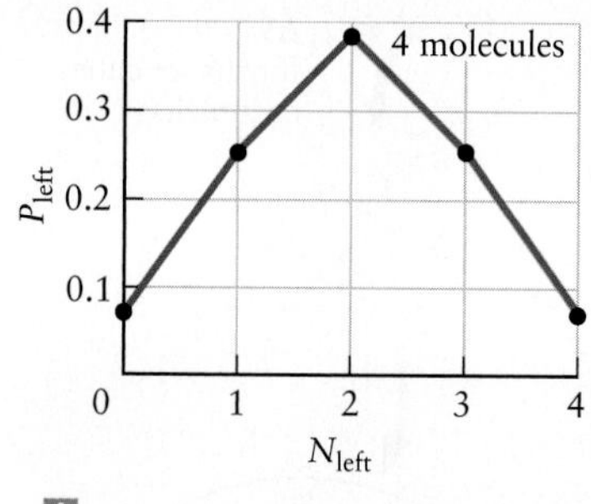

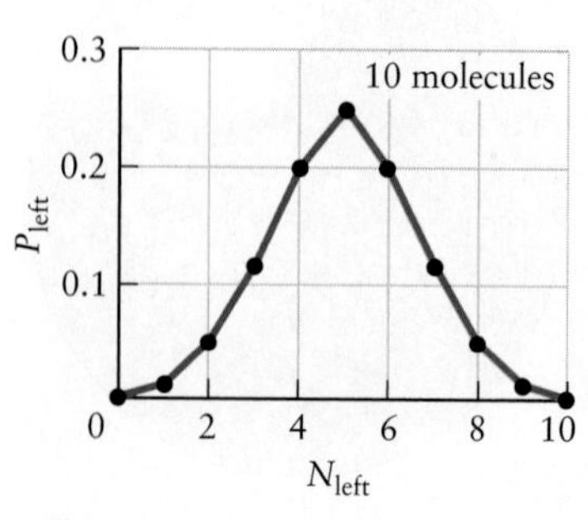

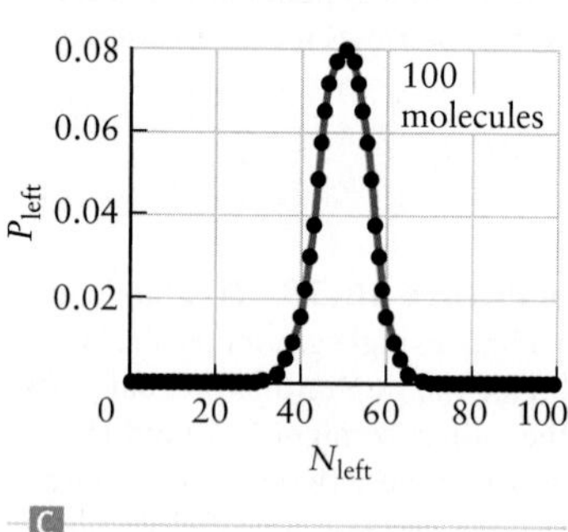

▲ **Figure 16.39** Probability for finding different numbers of molecules (N_{left}) on the left side of the box. A For 4 molecules. When $N_{left} = 2$, there are equal numbers of molecules on both sides. B For 10 molecules. C For 100 molecules. The greatest probability is found when $N_{left} = 50$, which means that the number of molecules on the left equals the number of molecules on the right.

we reach 10 molecules, the number of states is more than 1000. A key assumption of thermodynamics is that each of these states is equally likely. With this assumption, we can calculate P_{left}, the probability of finding a molecule on the left side of the box; when $P_{left} = 0.5$, the molecules are equally distributed on the two sides. Hence, the expected random distribution of molecules corresponds to values of P_{left} near 0.5. Figure 16.39 shows results for 4 molecules, 10 molecules, and 100 molecules. As the number of molecules grows, the probability peak becomes narrower, meaning that the probability of finding the molecules fairly equally distributed becomes very high.

What do the probability curves in Figure 16.39 have to do with entropy? Large values of the entropy correspond to a greater amount of randomness in a system. The results in Figures 16.38 and 16.39 show that there are many more ways of distributing the gas molecules so that there are equal numbers on the two sides of the box—that is, so that $P_{left} = 0.5$—than if all the molecules are on one side or on the other. As the number of molecules becomes very large, the most likely state of the system (the state with the highest probability) will be one in which the number of molecules on the left and on the right are approximately equal. Since there are many molecular arrangements that achieve this condition, this state has the greatest randomness and hence the largest entropy.

16.8 The Third Law of Thermodynamics and Absolute Zero

We defined the three common temperature scales (Fahrenheit, Celsius, and Kelvin) in Chapter 14. The Kelvin scale is most closely connected to kinetic theory and thermodynamics. For example, the ideal gas law has the mathematical form $PV = nRT$ when temperature is measured on the Kelvin scale, and the definition of entropy $\Delta S = Q/T$ also assumes T is given on the Kelvin scale. It is certainly puzzling that $\Delta S \to \infty$ if $T = 0$. The implication is that there is something special about this temperature.

We have had other hints that something special must take place at $T = 0$ K, the temperature called absolute zero. One hint comes from kinetic theory (Chapter 15), where we found that the average translational kinetic energy of a gas molecule is related to the absolute temperature by

$$KE = \tfrac{3}{2}k_B T \tag{16.36}$$

Thus, at $T = 0$, we expect the kinetic energy of a gas molecule to be zero! Because the kinetic energy cannot be less than zero, this finding implies that temperatures below absolute zero are not possible. Another hint comes from Carnot's work with heat engines and Equation 16.20. Carnot's statement that all heat engines must expel some heat to a cold reservoir implies that a reservoir with $T = 0$ K is not possible. These notions all come together in the third law of thermodynamics, which can be stated as follows.

Third law of thermodynamics

Third law of thermodynamics

It is impossible for the temperature of a system to reach absolute zero.

The third law of thermodynamics has important consequences for our understanding of mechanics. According to Newton's mechanics, it should be possible to make the velocity of a particle zero. Indeed, if we were to do that for every particle in a gas, the temperature of the gas would then, according to Equation 16.36, be $T = 0$. The third law of thermodynamics, however, tells us that is not possible. Hence, the laws of thermodynamics appear to be at odds with Newton's laws! This conflict was not resolved until the development of quantum mechanics in the early twentieth century.

16.9 Thermodynamics and Photosynthesis

In our work on thermodynamics, we made extensive use of the ideal gas because its properties, such as the internal energy and the ideal gas law, have simple mathematical forms. Thermodynamics, however, can be applied to many different physical systems. In this section, we show how thermodynamic arguments can be applied to photosynthesis.

Figure 16.40 represents the process of photosynthesis in a way that shows its similarity to a heat engine. An amount of heat Q_H from the Sun enters a plant, and (in the spirit of a heat engine) the plant "does an amount of work" W. At the same time, the plant expels some heat Q_C to its surroundings (the cold reservoir). In physics and engineering, the heat engine is a mechanical device that does work (e.g., by turning a shaft). In photosynthesis, a plant stores chemical energy in certain molecules, so the product of the photosynthetic heat engine in Figure 16.40 is actually potential energy stored in chemical bonds. This chemical energy can be used later to produce mechanical work when, for instance, an animal eats the plant, so the analogy with a heat engine still holds.

Sun

Q_H

Plant → W

Q_C

Atmosphere

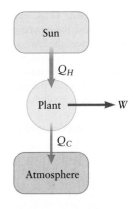

▲ **Figure 16.40** Photosynthesis is a thermodynamic process and can be viewed as a type of heat engine. Compare this diagram with Figure 16.26B.

Let's now analyze the efficiency of a photosynthetic heat engine. Assuming it is a reversible engine, this efficiency depends on the temperatures of the reservoirs and is given by $e = 1 - (T_C/T_H)$ (Eq. 16.21). The cold reservoir is the air or soil around the plant; hence, $T_C \approx 300$ K. The Sun serves as the hot reservoir, so our first estimate for the temperature of the hot reservoir is the temperature of the surface of the Sun, which is $T_H \approx 6000$ K as mentioned in Chapter 14. (The temperature of the hot reservoir is not the temperature of the plant. The leaves are *not* at 6000 K!) Inserting these values into Equation 16.21 gives

$$e_{\text{photosyn}} = 1 - \frac{T_C}{T_H} = 1 - \frac{300\text{ K}}{6000\text{ K}} = 0.95$$

A real photosynthetic engine does not approach this efficiency for several reasons, one of which can be seen from Figure 16.41. Radiation emitted by the Sun is distributed over a range of wavelengths throughout the infrared, visible, and ultraviolet regions (the blackbody distribution in Fig. 14.33). Plants absorb solar radiation mainly in the range of wavelengths $\lambda \approx 400$–700 nm and do not absorb much of the radiation outside this range. A mathematical analysis shows that plants absorb only about 50% of the Sun's total radiation, thus reducing the maximum efficiency of photosynthesis to approximately $0.95 \times 0.5 \approx 48\%$. There are other inefficiencies in photosynthesis, and the actual efficiency of the entire process is somewhat lower than 48%.[4] The important point is that thermodynamics and the physics of blackbody radiation place absolute physical limits on the efficiency of a real process.

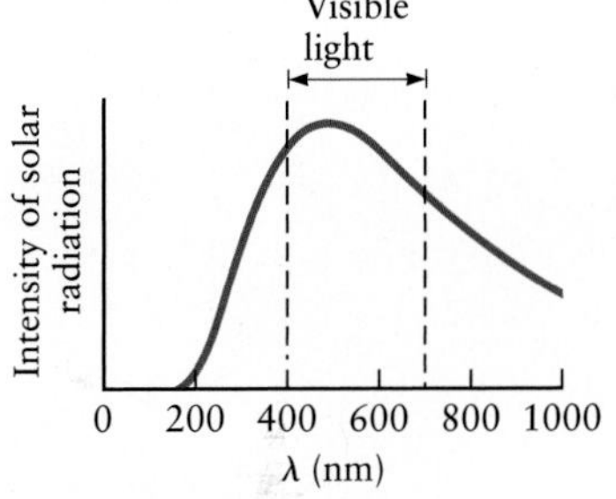

▲ **Figure 16.41** Plants absorb light only in the visible portion of the spectrum. Visible radiation corresponds to only about half the total radiated energy from the Sun. Shown is the blackbody distribution from Figure 14.33, but here it is plotted on a linear scale.

16.10 Converting Heat Energy to Mechanical Energy and the Origin of the Second Law of Thermodynamics

One key tenet of thermodynamics is that heat is a form of energy. The second law of thermodynamics, however, also tells us that heat energy differs from kinetic and potential energy in a subtle way. If you are given a certain amount of energy in the

[4]For comparison, solar cells convert sunlight into electrical energy with an efficiency of typically 15% to 20%. Photosynthesis is thus more efficient than manufactured solar cells.

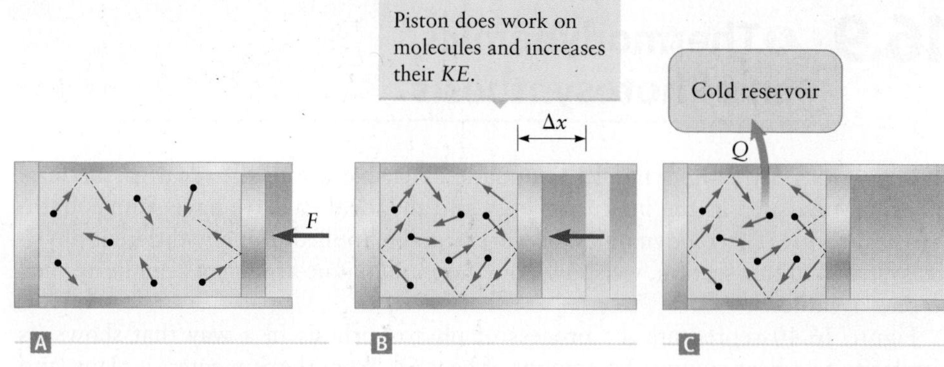

▶ **Figure 16.42** A Molecular-scale view of how the movement of a piston changes the energy of a gas. B When the gas is compressed, the piston does a net positive amount of work on all the gas molecules with which it collides. C This increased energy can later be transferred (via conduction) to a colder reservoir.

form of the kinetic energy of a baseball or the potential energy of a compressed spring, it is possible to convert all this energy to heat energy. Suppose, however, you are given the same amount of heat energy and asked to convert it to mechanical energy. The second law of thermodynamics as applied to a heat engine tells us that *only a fraction* of this energy can be converted to mechanical energy; this fraction is equal to the efficiency of a reversible heat engine (Eq. 16.21). A conversion from mechanical energy to heat can be done with 100% efficiency; why can't we convert heat energy to work with the same 100% efficiency?

To gain some insight into this difference, we first consider an example in which mechanical energy is converted to heat: the compression of an ideal gas. A molecular-scale view of this process is given in Figure 16.42, showing gas compressed by moving a piston. As the piston moves to the left (Figs. 16.42A and B), it does work on the gas molecules with which it collides. The work done by an external force acting on the piston is $F\,\Delta x$; this amount of energy is converted to additional kinetic energy of the molecules and shows up thermodynamically as an increase in the internal energy of the gas. If we now place this warm gas in contact with a colder thermal reservoir, heat flows from the gas into the reservoir (Fig. 16.42C). In this way, we could convert all the mechanical energy used to move the piston to heat energy.

Now consider the possibility (actually, the *impossibility*) of running the process in Figure 16.42 backward. To run in reverse, all the gas molecules in Figure 16.42C would have to collide with the piston in synchrony so as to exert a force on the piston to the right. The randomness of the gas makes such perfect synchrony impossible, however; some of the molecules may collide with the piston in this way, but the inherent disorder associated with heat energy prevents all the molecules from simultaneously "giving energy back" to the piston. For this reason, it is not possible to extract all the molecular kinetic energy from the gas. The disordered motion of the gas molecules makes it impossible to convert heat energy completely to mechanical energy.

Summary | CHAPTER 16

Key Concepts and Principles

Thermodynamics is about the properties of systems

Thermodynamics is concerned with the behavior of systems containing many particles such as a liquid, a gas, or a solid. Such systems interact with their environment, and the laws of thermodynamics describe these interactions.

Zeroth law of thermodynamics

If two systems A and B are both in thermal equilibrium with a third system, then A and B are in thermal equilibrium with each other.

First law of thermodynamics

If an amount of heat Q flows *into* a system from its environment and an amount of work W is done *by* the system on its environment, the internal energy of the system changes by an amount

$$\Delta U = U_f - U_i = Q - W \qquad \text{(16.3) (page 451)}$$

Second law of thermodynamics

The second law of thermodynamics can be stated in several different ways.

- ***Second law of thermodynamics (formulation involving heat flow):*** Heat flows spontaneously from a warm object to a colder one. It is not possible for heat to flow spontaneously from a cold object to a warmer one.
- ***Second law of thermodynamics (formulation involving heat engines):*** The efficiency of a reversible heat engine operating between reservoirs at temperatures T_H and T_C is

$$e = 1 - \frac{T_C}{T_H} \qquad \text{(16.21) (page 466)}$$

  No heat engine can have a greater efficiency.
- ***Second law of thermodynamics (entropy version):*** In any thermodynamic process, $\Delta S_{\text{universe}} \geq 0$.

Third law of thermodynamics

It is impossible for the temperature of a system to reach absolute zero.

Applications

Thermodynamic processes

A ***thermodynamic process*** takes a system from some initial state with an initial temperature T_i, pressure P_i, and volume V_i to a final state described by T_f, P_f, and V_f. Such a process can be described as a path on a ***P–V diagram***. The work done by the system on its environment is the area under this P–V curve. An ***isothermal process*** is one in which the temperature is constant ($T_i = T_f$). In an ***adiabatic process***, there is no heat flow into or out of the system ($Q = 0$).

Heat engines

A ***heat engine*** takes heat energy from a hot reservoir, expels heat energy to a cold reservoir, and does an amount of work on its environment. The ***efficiency of a heat engine*** is defined as

$$e = \frac{W}{Q_H} = \frac{Q_H - Q_C}{Q_H}$$

(Continued)

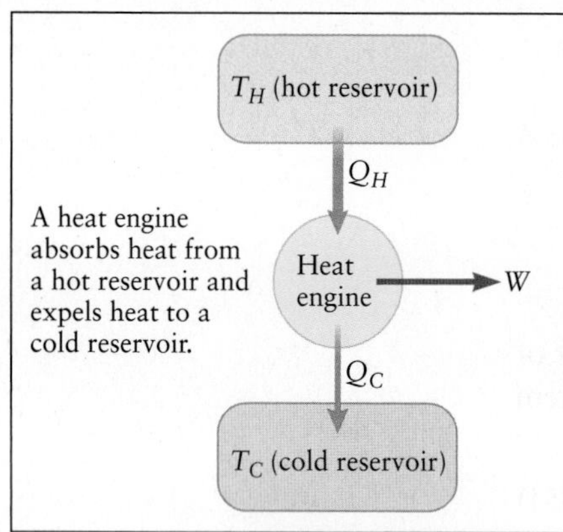

A heat engine absorbs heat from a hot reservoir and expels heat to a cold reservoir.

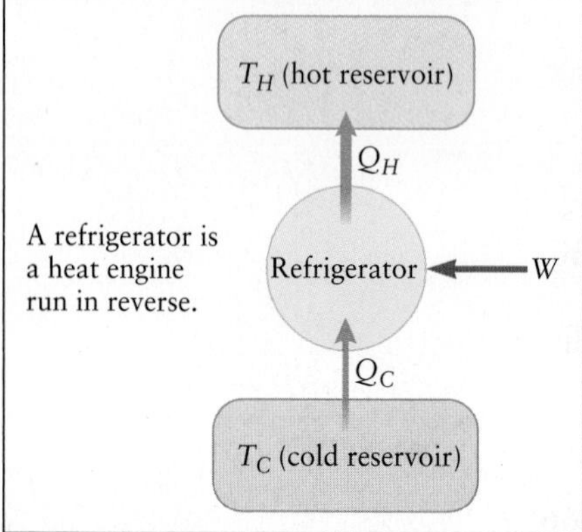

A refrigerator is a heat engine run in reverse.

The efficiency of a thermodynamically reversible heat engine is given by the laws of thermodynamics as

$$e = 1 - \frac{Q_C}{Q_H} = 1 - \frac{T_C}{T_H} \qquad \text{(16.19 and 16.21) (pages 465 and 466)}$$

No heat engine can have a greater efficiency.

Refrigerators and heat pumps

In a ***refrigerator***, work is done by the environment on the system, which then extracts heat from a cold reservoir and expels heat to a hot reservoir. The ***efficiency of a refrigerator*** is $e_{\text{refrig}} = Q_C/W$. A thermodynamically reversible refrigerator has an efficiency of

$$e_{\text{refrig}} = \frac{T_C/T_H}{1 - (T_C/T_H)} = \frac{T_C}{T_H - T_C} \qquad \text{(16.26) (page 471)}$$

No refrigerator can have a greater efficiency.

Heat pumps use the same thermodynamic arrangement as in a refrigerator to heat your house. The hot reservoir in the case of a heat pump is the interior of your house.

Questions

SSM = answer in *Student Companion & Problem-Solving Guide* ⓧ = life science application

1. ⓧ Two blocks, one of wood and the other aluminum, are at room temperature. When you touch them, the aluminum feels cold and the wood feels warm. Why? Is there a temperature at which both blocks would feel the same temperature to you? Explain.

2. It is a very hot day, and your friend decides to cool his apartment by opening the door of his refrigerator. Use thermodynamics to answer the following question: Will this method cool his apartment?

3. Can heat flow either into or out of a substance without the substance changing temperature? If not, why? If so, explain and give an example of such a process.

4. Consider an ideal gas that is expanding in a cylinder with a piston. As the gas expands, the piston moves such that a constant pressure is maintained. Does the temperature rise or fall? Does heat flow into or out of the gas in this process?

5. What quantity is represented by an area enclosed by a curve on a *P–V* diagram? What device is described by following such a curve in a clockwise direction? If the process follows the curve counterclockwise, what kind of device is being represented?

6. Consider a window air conditioner in terms of the thermodynamic block diagram for a refrigerator in Figure 16.32A. Identify the hot reservoir, the cold reservoir, and the source of mechanical work.

7. Explain how a heat pump can deliver more heat energy into a house than the pump consumes as work.

8. SSM Consider a thermodynamic process in which an ice cube melts to become a liquid (water). Is the change in internal energy of the H_2O positive or negative? Is the change in entropy of the H_2O positive or negative?

9. (a) Many heat engines need both a working fluid and a source of heat, often from a chemical process like combustion. Identify the working fluid and the source of heat for a steam engine like the one in Figure 16.27. (b) Many refrigerators require a working fluid and some device that does work. Identify the working fluid and the device that does work for a refrigerator like the one in Figure 16.32B. (c) Do *all* heat engines have a working fluid? Explain.

10. Figure Q16.10 shows a thermodynamic process in which a system first expands and is then compressed. Indicate the area on the *P–V* diagram that corresponds to the work done by the system.

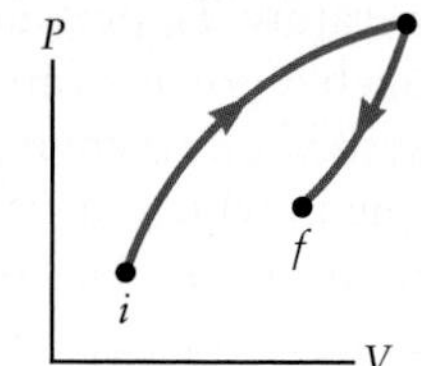

Figure Q16.10

11. In Section 16.4, we asserted that the work done in any cyclic process is equal to the area enclosed by the path on a *P–V* plot. In Example 16.6, we proved this assertion for a cyclic process described by a rectangular path. Prove this result for a path of any shape. *Hint*: Use a graphical argument involving area.

12. Give an example of a mechanical process that is reversible.

13. Give an example of a mechanical process that is irreversible. *Hint*: Include friction.

14. Suppose the second law of thermodynamics is *not* true and it is possible for heat to flow spontaneously from a cold object to a warmer one. Show that this process would allow one to construct a perpetual motion machine. *Hint*: Consider a machine that functions in two or more separate steps, with one step being the flow of heat from a cold reservoir to a hot reservoir. Then, in a separate step, energy from the hot reservoir is used in some way.

15. Consider the entropy changes that occur during the Carnot cycle sketched in Figure 16.31. For each stage of the cycle, is the entropy change of the engine positive, negative, or zero? What is the total entropy change of the engine, and of the environment, during one complete cycle? Is there anything that would prevent a real engine from having the same entropy changes for each stage of its cycle?

16. Under what conditions will the efficiency of a Carnot engine approach 100%?

17. If an ideal gas is compressed adiabatically, does the internal energy of the gas increase or decrease?

18. SSM Explain why heat pumps that use air as the cold reservoir are most efficient in mild climates. (They are called "air-to-air" heat pumps.) Why does a heat pump that uses water from deep underground as the cold reservoir have a thermodynamic advantage over air-to-air heat pumps when the weather is very cold?

19. In Example 16.10, we found that for typical temperatures of the hot and cold reservoirs, the theoretical efficiency of a home heat pump can be quite large. Investigate the efficiency obtained with real heat pumps (i.e., do some research on the Internet). How does the efficiency of a real heat pump compare with the theoretical limit in Equation 16.28? What factors make the efficiency of a real heat pump smaller than this theoretical limit?

20. Two ideal heat engines A and B operate using the same hot reservoir. It is found that engine B is more efficient than engine A. Explain how that can be.

21. Consider the common expressions (a) "There's no such thing as a free lunch" and (b) "You can't even break even." Which of the laws of thermodynamics might these expressions refer to? Explain.

22. A hydroelectric power plant converts potential energy of water stored behind a dam to electrical energy with an efficiency close to 100%. A coal-burning (or nuclear) power plant only approaches 40% for conversion of the chemical (or fission) energy to electricity. Why?

Problems available in ENHANCED WebAssign

Section 16.3 The First Law of Thermodynamics and the Conservation of Energy
Problems 1–8

Section 16.4 Thermodynamic Processes
Problems 9–25

Section 16.6 Heat Engines and Other Thermodynamic Devices
Problems 26–45

Section 16.7 Entropy
Problems 46–52

Additional Problems
Problems 53–68

List of Enhanced Problems

Problem number	Solution in *Student Companion & Problem-Solving Guide*	Reasoning & Relationships Problem	Reasoning Tutorial in ENHANCED WebAssign
16.3	✓		
16.17	✓		
16.18		✓	✓
16.23	✓		
16.31	✓		
16.36		✓	✓
16.44	✓		
16.45	✓	✓	✓
16.47	✓		
16.52		✓	
16.54	✓		
16.56		✓	✓
16.66		✓	✓
16.68		✓	

CHAPTER

◀ *This engraving shows Benjamin Franklin during one of his famous experiments with kites in thunderstorms. Those experiments helped prove that lightning involves the motion of electric charge. This motion and the associated electric and magnetic phenomena are the subject of Chapters 17 through 23. (© World History Archive/ Alamy)*

Electric Forces and Fields

In this and the next two chapters, we focus on electric phenomena before moving on to describe magnetism in Chapter 20. Electric and magnetic forces both act on electric charges and currents. Electric charges and currents also act as *sources* of electric and magnetic fields. Electricity and magnetism are really a single unified phenomenon often referred to as ***electromagnetism***, and this unified picture leads to a theory of electromagnetic radiation. Light is a type of electromagnetic radiation, so electromagnetism also forms the basis for the theory of ***optics***, which is discussed in Chapters 24 through 26. Our study of electricity and magnetism will thus take us quite a long way!

OUTLINE

A

B

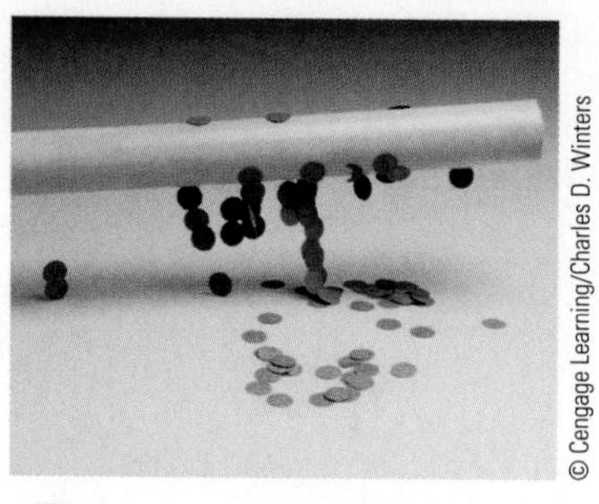

C

▲ **Figure 17.1** A Amber is the dried, hardened sap from certain trees. Small creatures or bits of plant material are often preserved inside. B When a plastic or amber rod is rubbed with fur, the rod acquires an electric charge. C A charged rod attracts small bits of paper and other objects.

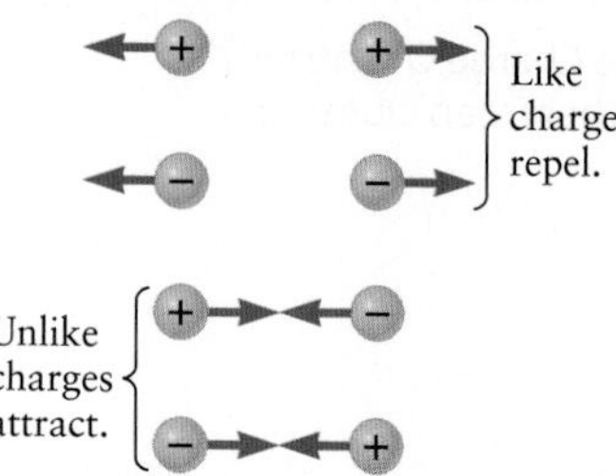

▲ **Figure 17.2** The electric force between two like charges (charges with the same sign) is repulsive, whereas the force between unlike charges (those of opposite sign) is attractive.

17.1 Evidence for Electric Forces: The Observational Facts

The discovery of electricity is generally credited to the Greeks and is thought to have occurred around 2500 years ago, which is approximately the era in which Aristotle lived. The Greeks observed electric charges and the forces between them in a variety of situations. Many of their observations made use of a material called amber, a plastic-like substance formed by allowing the sap from certain trees to dry and harden (Fig. 17.1A). The Greeks found that after amber is rubbed with a piece of animal fur, the amber can attract small pieces of dust. The same effect can be demonstrated with a piece of plastic and small bits of paper. This experiment (parts B and C of Fig. 17.1) is quite striking because neither the fur nor the plastic rod ordinarily attracts paper. Somehow, the process of rubbing the two together "creates" an attractive force. Moreover, this force occurs even when the amber and the paper are *not* in contact, and there would be an attractive force even if the plastic rod and the paper were placed in a vacuum (although the Greeks did not try that).

The Greek word for amber, "elektron," is the origin of the terms *electricity* and *electron*. The Greeks did not discover the electron; even though the notion of a small charge-carrying particle was hypothesized by a number of early philosophers and scientists, there was no clear experimental demonstration of the existence of the electron until the late 1880s. The long history of scientific work on electricity is very interesting, but we will skip most of the details and simply state a few key facts that may already be familiar to you.

- *There are two types of electric charge*, called ***positive*** and ***negative***. The subatomic particle called a ***proton*** has a positive charge, and an ***electron*** has a negative charge.
- *Charge comes in quantized units. All* protons carry the *same* amount of charge $+e$, and *all* electrons carry a charge $-e$. We will discuss how charge is measured and the unit of electric charge below.
- *Like charges repel each other; unlike charges attract.* The electric force between two objects is repulsive if the objects carry "like" charge, that is, if both are positively charged or both are negatively charged (Fig. 17.2). The electric force is attractive if the two objects carry "unlike" charge. The expression "unlike charges" means that one charge is positive and the other is negative. The terms *like* and *unlike* thus refer to the *signs* of the charges, not to their magnitudes.
- *Charge is conserved.* The *total* charge on an object is the sum of all the individual charges (protons and electrons) carried by the object. The total charge can be positive, negative, or zero. Charge can move from place to place, and from one object to another, but the total charge of the universe does not change.

What Is Electric Charge?

You are likely familiar with the term *electric charge* or simply *charge*. Charge is a fundamental property of matter: the amount of charge that is "on" or "carried by" a particle determines how the particle reacts to electric and magnetic fields. In many ways, it is similar to the quantity we call *mass*; the mass of a particle determines how it reacts when a force acts on it. The mass of a particle is a measure of the amount of matter it carries, whereas the charge of a particle is a measure of the amount of "electric-ness" it carries. Charge and mass are both primary entries in our dictionary of physics terms, but it is not possible to give definitions for these fundamental entries in our dictionary.

In the SI system of units, charge is measured in ***coulombs*** (C) in honor of French physicist Charles de Coulomb (1736–1806). The charge on a single electron is[1]

$$\text{electron charge} = -e = -1.60 \times 10^{-19}\ \text{C} \tag{17.1}$$

and the charge carried by a single proton is

$$\text{proton charge} = +e = +1.60 \times 10^{-19}\ \text{C} \tag{17.2}$$

We use the symbol e to denote the *magnitude* of the charge on an electron or a proton. We will always take e to be a *positive* quantity ($e = +1.60 \times 10^{-19}$ C). We use the symbols q and Q to denote charge in general, such as the total charge on a bit of paper. When discussing a charged particle, it is common to say that the particle "has" a charge q or that it "carries" a charge q. These two expressions simply mean that the total charge of the particle is q.

17.2 Electric Forces and Coulomb's Law

Imagine that you travel to a distant star where you discover a new type of force acting between particles. The force can be attractive or repulsive, and it is *extremely* large. In fact, the force between two objects each having a mass of only 1 g is large enough to hold an entire planet in orbit around this alien star! This force sounds very different from anything you have encountered before, but it *already exists* on the Earth. It is the *electric force* between two charged objects.

Consider two charged objects that are so tiny that they can be modeled as point particles. If the charges carried by the two objects are q_1 and q_2 and they are separated by a distance r (Fig. 17.3), the electric force between the objects can be written as

$$F = k\frac{q_1 q_2}{r^2} \tag{17.3}$$

Equation 17.3 is called ***Coulomb's law***. The constant k has the value

$$k = 8.99 \times 10^9\ \text{N} \cdot \text{m}^2/\text{C}^2 \tag{17.4}$$

Do not confuse the constant k in Equation 17.4 with the spring constant in Hooke's law. The value of F in Equation 17.3 can be positive or negative, depending on whether q_1 and q_2 are like or unlike charges. Strictly speaking, the magnitude of the electric force is $|F|$, the absolute value of the result in Equation 17.3. The direction of the electric force on each of the charges is along the line that connects the two charges and is illustrated in Figure 17.3 for the case of two like charges. As already mentioned, this force is repulsive for like charges, whereas the force is attractive for unlike charges.

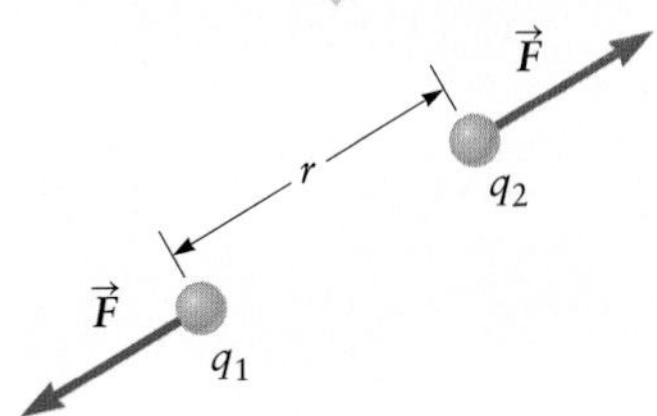

▲ **Figure 17.3** The electric force between two point charges q_1 and q_2 is given by Coulomb's law.

The value of F in Equation 17.3 applies only for two point charges, but it is a good approximation whenever the sizes of the particles are much smaller than their separation r. Later in this chapter, we'll see how to build on Equation 17.3 to find the electric force in other situations, including nonpoint particles.

The mathematical form of Equation 17.3 is very similar to that of Newton's law of gravitation, with the constant k playing a role analogous to the gravitational constant G. Another way to write Coulomb's law is

$$F = \frac{q_1 q_2}{4\pi\varepsilon_0 r^2} \tag{17.5}$$

where ε_0 is called the ***permittivity of free space***, having the value

$$\varepsilon_0 = 8.85 \times 10^{-12}\ \frac{\text{C}^2}{\text{N} \cdot \text{m}^2} \tag{17.6}$$

[1]Here we give the value of e to just three significant figures, but it is known to much greater precision than that. The latest experiments give $e = 1.60217653 \times 10^{-19}$ C.

The values of ε_0 and k are related by

$$\frac{1}{4\pi\varepsilon_0} = k$$

so these two forms of Coulomb's law, Equations 17.3 and 17.5, are completely equivalent. Coulomb's law in its slightly more complicated form in Equation 17.5 is useful because other equations of electromagnetism have simpler forms when written in terms of ε_0 instead of k. For calculating the electric force between charges, either form of Coulomb's law (Eq. 17.3 or 17.5) can be used.

Important Features of Coulomb's Law

We have already noted that Coulomb's law is very similar in form to Newton's universal law of gravitation. Both laws exhibit a $1/r^2$ dependence on the separation of the two particles. Therefore, a negative charge can move in a circular orbit around a positive charge, just like a planet orbiting the Sun, and that was an early model for the hydrogen atom. There is one very important difference between the two forces: gravity is always an attractive force, whereas the electric force in Coulomb's law can be either attractive or repulsive.

The magnitude of F in Equations 17.3 and 17.5 is the magnitude of the force exerted on *each* of the particles. That is, a force of magnitude F is exerted on charge q_1, and a force of equal magnitude and opposite direction is exerted on q_2. We should expect such a pair of forces based on Newton's third law, the action–reaction principle.

Let's now return to the claim made at the beginning of this section and show that electric forces can be extremely large. Suppose you are given a box holding 1 g (0.001 kg) of pure electrons. The mass of a single electron is $m_e = 9.11 \times 10^{-31}$ kg, so the total number of electrons N_e in the box is

$$N_e = \frac{1 \times 10^{-3}\text{ kg}}{9.11 \times 10^{-31}\text{ kg/electron}} = 1.1 \times 10^{27}$$

Each of these electrons carries a charge $-e$, so the total charge is

$$Q_{\text{total}} = N_e(-e) = (1.1 \times 10^{27})(-1.60 \times 10^{-19}\text{ C}) = -1.8 \times 10^{8}\text{ C}$$

Now suppose there are two of these boxes, each with charge Q_{total}, separated by a distance $r = 1$ m, and we assume each box is small so that it can be modeled as a point particle. The magnitude of the total electric force is (Eq. 17.3)

$$F = \frac{kQ_{\text{total}}Q_{\text{total}}}{r^2} = \frac{(8.99 \times 10^{9}\text{ N}\cdot\text{m}^2/\text{C}^2)(-1.8 \times 10^{8}\text{ C})(-1.8 \times 10^{8}\text{ C})}{(1\text{ m})^2}$$

$$F = 3 \times 10^{26}\text{ N} \tag{17.7}$$

which is an *enormous* force. It is nearly a million times larger than the gravitational force exerted between the Sun and the Earth. All that from just two small containers of electrons!

The value found for F in Equation 17.7 is so large that you may be a little skeptical: if the electric force between two small pieces of matter is this large, there must be many staggering consequences. For example, the electric forces acting within ordinary matter, which contains both electrons and protons, must be huge. Why don't these forces dominate everyday life? The resolution of this apparent paradox is that it is essentially impossible to obtain a box containing only electrons. You probably know that a neutral atom contains equal numbers of electrons and protons. In fact, the term *neutral* means that the total charge is zero. If our two boxes had contained equal numbers of electrons and protons, their *total* charges would have been $Q_{\text{total}} = 0$ and the force in Equation 17.7 would be zero.

Ordinary matter always consists of equal, or nearly equal, numbers of electrons and protons. The total charge is then zero or very close to zero. At the atomic and molecular scale, however, it is common to have the positive and negative charges (nuclei and electrons) separated by a small distance. In this case, the electric force is not zero, and these electric forces are responsible for holding matter together. That is why solids can be quite strong.

EXAMPLE 17.1 Using Coulomb's Law to Clean the Air

Coal-burning power plants produce large amounts of potential pollution in the form of small particles (soot). Modern smokestacks use devices called scrubbers to remove these particles from the smoke they emit. Scrubbers use a two-step process: electrons are first added to the soot particle, and an electric force then pulls the particle out of the smoke stream (Fig. 17.4). In this example, we analyze the force on a soot particle after electrons are added.

Consider a soot particle of mass m_{soot} = 1.0 nanogram (= 1.0×10^{-9} g = 1.0×10^{-12} kg), which corresponds to a diameter of a few microns. Some number of electrons have been added to give the particle a total charge q_{soot}. Suppose the collector has a total charge $q_{collector} = 1.0 \times 10^{-6}$ C and is small enough to be treated as a point charge at the rim of the smokestack. **(a)** If the separation between the collector and the soot particle is $r = 0.10$ m, what is the value of q_{soot} so that the electric force exerted on the particle is equal to its weight? **(b)** How many electrons must be added to the soot particle?

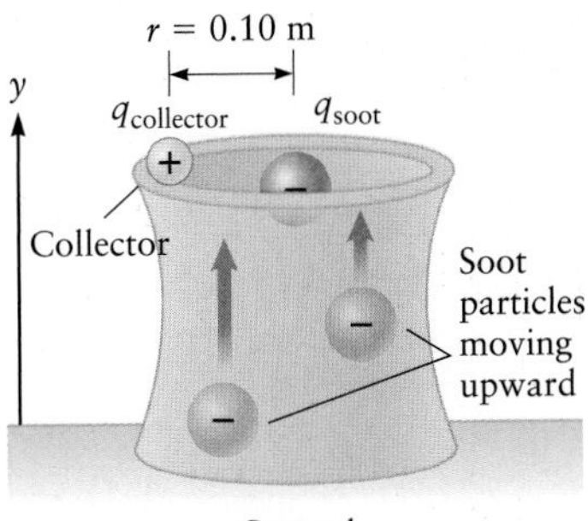

▲ **Figure 17.4** Example 17.1. A dust particle in a smokestack is attracted to the electric charge on the collector of a smokestack scrubber.

RECOGNIZE THE PRINCIPLE

The collector charge is assumed to be a point particle and the soot particle is also very small, so we can apply Coulomb's law for two point particles (Eq. 17.3). The electric force between the collector (charge $q_{collector}$) and the soot particle (charge q_{soot}) has a magnitude

$$F_{elec} = \frac{kq_{collector}q_{soot}}{r^2} \tag{1}$$

SKETCH THE PROBLEM

Figure 17.4 shows the soot particle and the collector charge. They are separated by a distance r, just like the charges in Figure 17.3.

IDENTIFY THE RELATIONSHIPS

We want the electric force on the particle in Equation (1) to be equal in magnitude to the particle's weight, so

$$F_{elec} = \frac{kq_{collector}q_{soot}}{r^2} = mg$$

SOLVE

(a) Solving for the charge q_{soot} leads to

$$q_{soot} = \frac{mgr^2}{kq_{collector}}$$

Inserting the given values of m and r, we find

$$q_{soot} = \frac{mgr^2}{kq_{collector}} = \frac{(1.0 \times 10^{-12}\text{ kg})(9.8\text{ m/s}^2)(0.10\text{ m})^2}{(8.99 \times 10^9\text{ N}\cdot\text{m}^2/\text{C}^2)(1.0 \times 10^{-6}\text{ C})}$$

$$q_{soot} = \boxed{1.1 \times 10^{-17}\text{ C}}$$

(continued) ▶

(b) The charge on a single electron has a magnitude $e = 1.60 \times 10^{-19}$ C (Eq. 17.1), so our value of q_{soot} corresponds to

$$N = \frac{q_{\text{soot}}}{e} = \frac{1.1 \times 10^{-17}\ \text{C}}{1.60 \times 10^{-19}\ \text{C/electron}} = \boxed{69 \text{ electrons}}$$

which is the number of extra electrons added to the soot particle.

► ***What does it mean?***

It is amazing that only this small number of electrons is needed for the scrubber to successfully remove a soot particle from air. The total number of atoms in the particle is about 4×10^{14} (see Question 13 at the end of this chapter), so the fraction of "unbalanced" electrons on the particle is only $69/(4 \times 10^{14}) \approx 2 \times 10^{-13}$!

Superposition of Electric Forces

So far, we have discussed the electric force between two point particles; let's now consider how to use Coulomb's law to deal with more complicated charge distributions. Suppose two particles of charge q_1 and $2q_1$ are separated by a distance L as shown in Figure 17.5. What is the force on a third charge q_2 placed midway between these two charges?

Electric forces and the principle of superposition

We can deal with this problem by first using Coulomb's law to find the force exerted by charge q_1 on q_2 and then using Coulomb's law a second time to calculate the force exerted by the charge $2q_1$ on q_2. The total force on q_2 is the sum of these two separate contributions, which is an example of the ***principle of superposition.*** In general, the total force on q_2 is just the sum of the individual forces exerted on q_2 by all the other charges in the problem. Because force is a vector, we must be careful to add these forces as vectors.

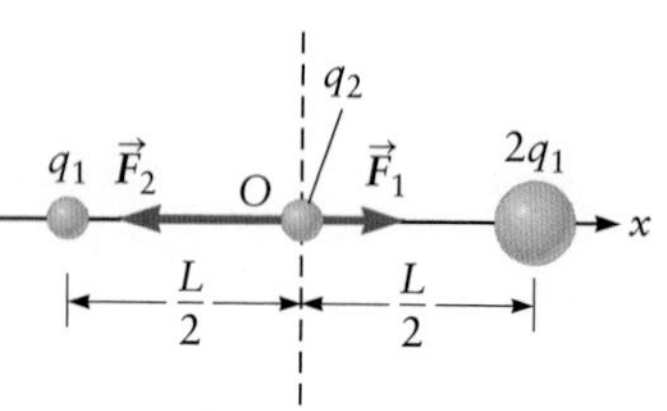

▲ **Figure 17.5** The total force on q_2 is equal to the sum of the forces from charges q_1 and $2q_1$. This is the principle of superposition at work.

From Figure 17.5, the separation between q_2 and q_1 is $L/2$. Writing Coulomb's law for this pair of charges, we have

$$F_1 = \frac{kq_1q_2}{(L/2)^2} = \frac{4kq_1q_2}{L^2} \tag{17.8}$$

Here a positive value corresponds to a repulsive force (since like charges repel), and in the coordinate system in Figure 17.5 it corresponds to a force on q_2 in the $+x$ direction. Hence, F_1 in Equation 17.8 is the component of the force along the x axis, and the component along y is zero.

We can deal with the force from the charge $2q_1$ in a similar way. The separation of the charges is again $L/2$, so, applying Coulomb's law (Eq. 17.5), we find

$$F_2 = \frac{k(2q_1)q_2}{(L/2)^2} = \frac{8kq_1q_2}{L^2} \tag{17.9}$$

From the geometry of Figure 17.5 we can see that this result is the x component of the force on q_2 because the electric force acts along the line that connects the two charges and here that line is along the x direction. For this reason, the y component of the force on q_2 is again zero. Equation 17.9 gives the magnitude of the force F_2. If they are "like" charges (i.e., if q_1 and q_2 are both positive or both negative so that the product q_1q_2 is positive), the force F_2 is in the $-x$ direction.

The total force on q_2 is the sum of Equations 17.8 and 17.9, but we must account for the direction (the sign) of each. Assuming all charges are of "like sign," the result is

$$F_{\text{total}} = +\frac{4kq_1q_2}{L^2} - \frac{8kq_1q_2}{L^2} = -\frac{4kq_1q_2}{L^2}$$

The negative sign here tells us that if the product $q_1q_2 > 0$ (consistent with our "like sign" assumption), the force exerted on q_2 is along the $-x$ direction. This example leads to the following strategy.

PROBLEM SOLVING Calculating Forces with Coulomb's Law

1. **RECOGNIZE THE PRINCIPLE.** The electric force on a charged particle can be found using Coulomb's law together with the principle of superposition.
2. **SKETCH THE PROBLEM.** Construct a drawing (including a coordinate system) and show the location and charge for each object in the problem. Your drawing should also show the directions of all the electric forces—$\vec{F}_1$, $\vec{F}_2$, and so forth—on the particle(s) of interest.
3. **IDENTIFY THE RELATIONSHIPS.** Use Coulomb's law (Eq. 17.3 or 17.5) to find the magnitudes of the forces $\vec{F}_1, \vec{F}_2, \ldots$ acting on the particle(s) of interest.
4. **SOLVE.** The total force on a particle is the sum (the superposition) of all the individual forces from steps 2 and 3. Add these forces *as vectors* to get the total force. When adding these vectors, it is usually simplest to work in terms of the components of $\vec{F}_1$, $\vec{F}_2, \ldots$ along the coordinate axes.
5. Always *consider what your answer means* and check that it makes sense.

CONCEPT CHECK 17.1 Superposition and the Direction of the Electric Force

An electron and a proton are placed on the x axis as shown in Figure 17.6. Another electron (electron 2 in Fig. 17.6) is placed to the right of the proton as shown. Is the *direction* of the electric force on electron 2 (a) in the $+x$ direction or (b) in the $-x$ direction, or (c) is the force zero?

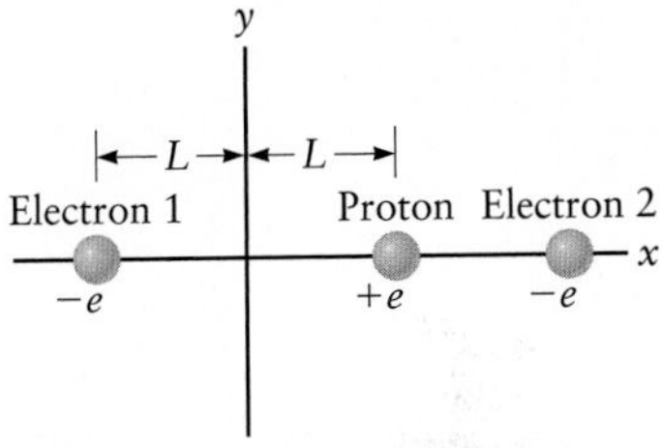

▲ **Figure 17.6** Concept Check 17.1.

EXAMPLE 17.2 An Electric Dipole

Consider two point particles of charge $+q$ and $-q$ (with q positive) arranged as shown in Figure 17.7A. This arrangement of charge, called an ***electric dipole***, is found in many molecules, so it is important in chemistry as well as physics. Suppose a third charge Q is placed on the x axis as shown in Figure 17.7A. Assuming Q is positive, what are the magnitude and the direction of the total electric force on Q?

RECOGNIZE THE PRINCIPLE

Our approach is based on Coulomb's law and the principle of superposition, as outlined in the "Calculating Forces with Coulomb's Law" problem-solving strategy.

(continued) ▶

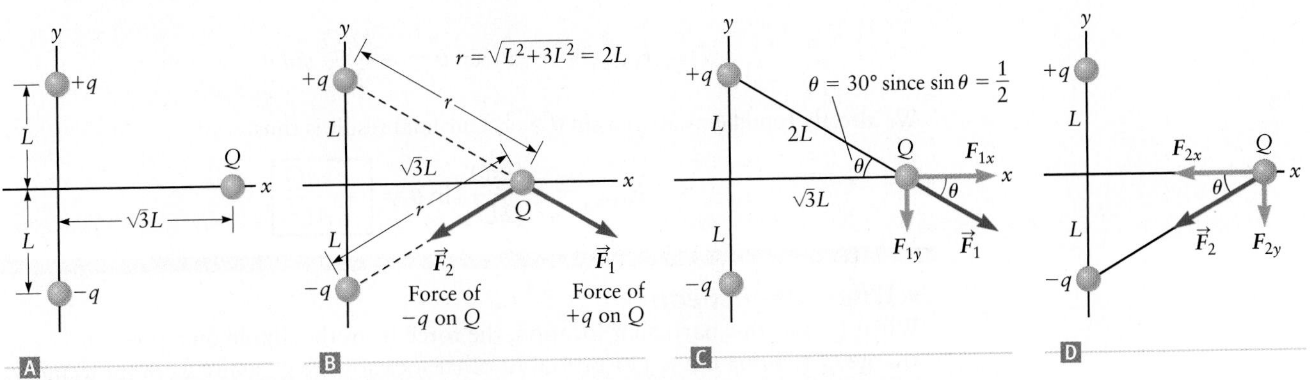

▲ **Figure 17.7** Example 17.2. Calculation of the force exerted by a dipole (charges $+q$ and $-q$) on a third charge Q.

SKETCH THE PROBLEM

The sketch in Figure 17.7A includes the coordinate axes and the positions of all the charges in the problem. We next draw in vectors $\vec{F}_1$ and $\vec{F}_2$ showing the forces on charge Q in Figure 17.7B. The force of charge $+q$ on charge Q is represented by $\vec{F}_1$. These charges are both positive, so the force is repulsive (away from $+q$) and $\vec{F}_1$ is directed to the lower right of the figure. The force exerted by $-q$ on Q is represented by $\vec{F}_2$. Because Q and $-q$ are unlike charges, this force is attractive (toward $-q$).

IDENTIFY THE RELATIONSHIPS

We apply Coulomb's law (Eq. 17.3) to find the magnitudes of $\vec{F}_1$ and $\vec{F}_2$. Using the Pythagorean theorem in Figure 17.7B, the distance from $+q$ to Q is $r = \sqrt{L^2 + 3L^2} = 2L$. Inserting into Equation 17.3 gives

$$F_1 = \frac{kqQ}{r^2} = \frac{kqQ}{(2L)^2} = \frac{kqQ}{4L^2}$$

The distance between $-q$ and Q is also r, and the magnitude of $\vec{F}_2$ is

$$F_2 = \frac{kqQ}{r^2} = \frac{kqQ}{4L^2} = F_1 \tag{1}$$

SOLVE

We now add $\vec{F}_1$ and $\vec{F}_2$ *as vectors* to get the total force on Q. The components of these vectors are shown in parts C and D of Figure 17.7. For the components of $\vec{F}_1$, we have

$$F_{1x} = F_1 \cos\theta \qquad F_{1y} = -F_1 \sin\theta \tag{2}$$

where the angle θ is defined in Figure 17.7C. This angle is contained in a right triangle with $\sin\theta = \frac{1}{2}$, so $\theta = 30°$. Note that F_{1y} is negative due to the direction of $\vec{F}_1$ in parts B and C of Figure 17.7. We next write the components of $\vec{F}_2$. From Figure 17.7D, we get

$$F_{2x} = -F_2 \cos\theta \qquad F_{2y} = -F_2 \sin\theta$$

We now add the components of $\vec{F}_1$ and $\vec{F}_2$ to get the total force. For the x components, we find

$$F_{\text{total},\,x} = F_{1x} + F_{2x} = F_1 \cos\theta - F_2 \cos\theta$$

Since $F_1 = F_2$ by Equation (1), we have

$$F_{\text{total},\,x} = \boxed{0}$$

For the y components,

$$F_{\text{total},\,y} = F_{1y} + F_{2y} = -F_1 \sin\theta - F_2 \sin\theta$$

$$F_{\text{total},\,y} = -2F_1 \sin\theta = -\frac{kqQ}{2L^2}\sin\theta$$

We already found $\theta = 30°$, so $\sin\theta = \frac{1}{2}$. Our final result is thus

$$F_{\text{total},\,y} = -\frac{kqQ}{2L^2}\sin\theta = \boxed{-\frac{kqQ}{4L^2}}$$

► ***What does it mean?***

When Q is at this particular location, the force from the dipole on a positive charge Q is along the $-y$ direction. At other locations (e.g., away from the x and y axes), the force will have nonzero components along both x and y. Had the charge

Q instead been $-Q$ (negative), the force would have the same magnitude as found here, but directed along $+y$. It is also remarkable that a *neutral object* (the dipole consisting of charges $+q$ and $-q$) can exert an electric force on the charge Q as occurs in Figure 17.1C (where the pieces of paper are neutral) and in many other situations. We will soon explain why.

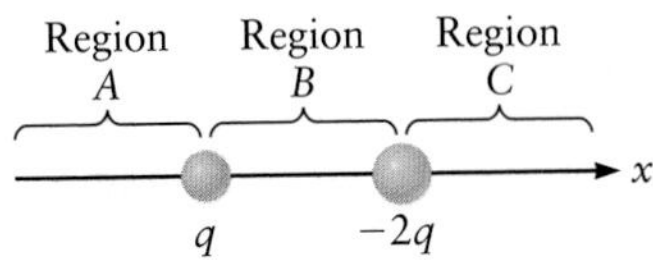

▲ **Figure 17.8** Concept Check 17.2.

CONCEPT CHECK 17.2 Force on a Point Charge

Two particles of charge q and $-2q$ are located as shown in Figure 17.8. A third charge Q is placed on the x axis and it is found that the total electric force on Q is zero. In which region in Figure 17.8 is Q located?

(a) region A (b) region B (c) region C

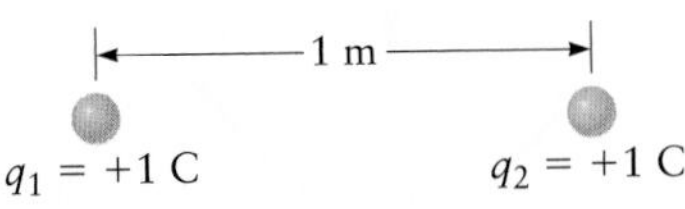

▲ **Figure 17.9** Concept Check 17.3.

CONCEPT CHECK 17.3 Holding Together Two Charges

Two particles, both with charge $q = +1$ C, are located 1 m apart as sketched in Figure 17.9. If q_1 is held fixed in place, what is the magnitude of the force that must be exerted on q_2 to keep it from moving?

(a) 9×10^9 N (b) 1 N (c) zero (d) 9×10^{18} N

CONCEPT CHECK 17.4 Electric Forces and Neutral Objects

Is it possible to have a nonzero electric force between two neutral objects? If your answer is yes, give an example.

17.3 The Electric Field

Coulomb's law gives the electric force between a pair of charges, but there is another way to describe electric forces. Imagine a single isolated point charge, that is, a point charge very far from any other charges as in Figure 17.10A. The presence of this charge produces an ***electric field.*** The electric field is a *vector*—it has a magnitude and a direction—and is denoted by $\vec{E}$. Figure 17.10 shows the magnitude and direction of the electric field near both a positive and negative point charge. For a positive point charge, the electric field is directed radially outward (Fig. 17.10A), whereas for a negative charge the electric field is directed inward, toward the charge (Fig. 17.10B).

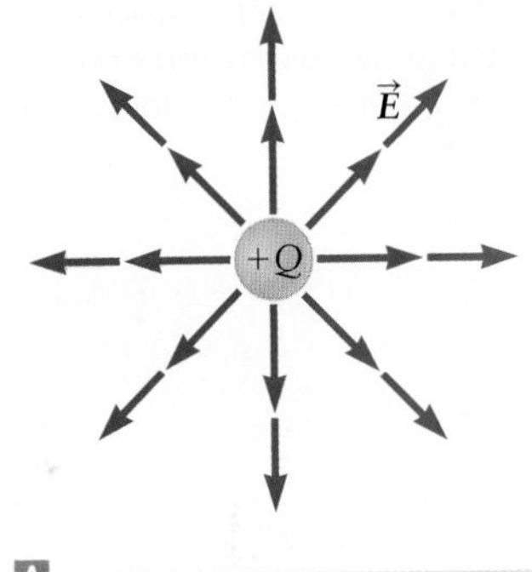

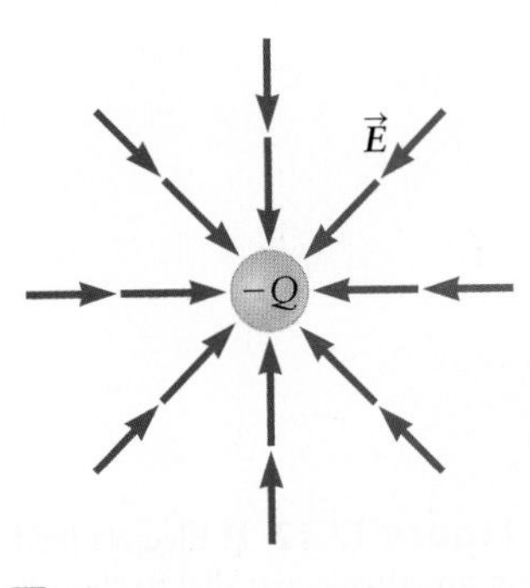

▲ **Figure 17.10** Electric field near a point charge placed at the origin. **A** If the charge is positive, the electric field is directed outward away from the charge, while **B** if the charge is negative, the field lines are directed inward toward the charge.

Using a Test Charge to Measure $\vec{E}$

Consider a particular point in space where the electric field is $\vec{E}$ (Fig. 17.11). This field might be produced by a point charge or by some other arrangement of charges. If we now place a charge q at this location, it experiences an electric force given by

$$\vec{F} = q\vec{E} \tag{17.10}$$

The electric force is thus either parallel to $\vec{E}$ (if q is positive) or antiparallel (if q is negative). The charge q in Figure 17.11 (and Eq. 17.10) is called a ***test charge.*** By measuring the force on a test charge, we can infer the magnitude and direction of the electric field at the location of the test charge. Because force is measured in units of newtons (N) and charge in coulombs (C), the electric field (according to Eq. 17.10) has units of newtons/coulombs.

The electric field $\vec{E}$ is related to electric force by Equation 17.10, so we can use a Coulomb's law calculation of the force to find the electric field in many situations. For example, let's calculate the electric field a distance r from the charge Q in Figure 17.11. Here q is the test charge, and we want to calculate the electric field produced by Q. According to Coulomb's law, the magnitude of the electric force exerted on the test charge q is

$$F = \frac{kQq}{r^2}$$

Inserting this expression into our relation for the electric field (Eq. 17.10) gives

$$F = \frac{kQq}{r^2} = qE$$

which leads to

Electric field of a point charge

$$E = \frac{kQ}{r^2} \tag{17.11}$$

This result is the magnitude of the electric field a distance r from a point charge Q. The direction of $\vec{E}$ is along the line that connects the charge producing the field to the point where the field is measured. The electric field is directed outward away from Q when Q is positive (as sketched in Fig. 17.11) and inward toward Q when Q is negative.

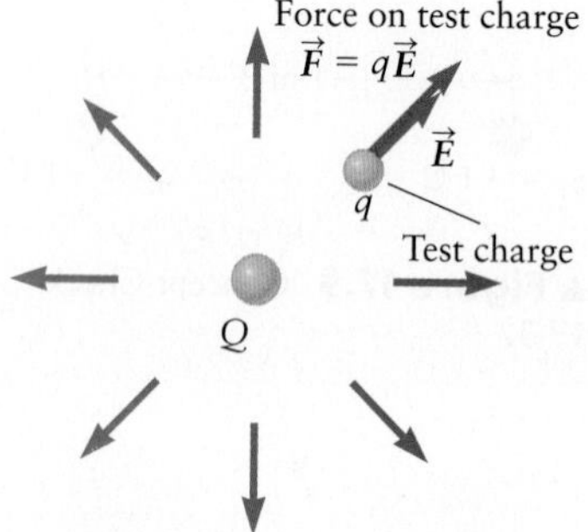

▲ **Figure 17.11** The electric field at a particular point in space is related to the electric force on a test charge q placed at that location by $\vec{F} = q\vec{E}$.

Why Do We Need the Electric Field?

You may now be wondering why we need the concept of an electric field. What does this field enable us to do or understand that we cannot do with the Coulomb's law force between two point charges in Equations 17.3 and 17.5? The answer is that the electric field of the point charge Q in Figure 17.10A is always "there" even if there is no second charge (or test charge) to experience the electric force. Any charge or collection of charges will produce an electric field $\vec{E}$. The magnitude and direction of $\vec{E}$ at a particular location tells what would happen if a charge were placed at that location, which implies that the electric field exists even if there were no test charge present to measure it.

This notion of an electric field helps explain how the Coulomb force can act between two charges that are separated by a large distance or a vacuum. The electric field due to one of the charges at the location of the other produces the electric force that acts between them. The concept of an electric field is also essential for understanding electromagnetic waves (Chapter 23).

Electric Field Lines and the Inverse Square Law

The electric field vectors in Figure 17.10 are one way to visualize the electric field. Another is with *electric field lines,* a set of continuous lines that are always parallel to the electric field (Fig. 17.12A). Two examples of electric field lines are given in parts B and C of Figure 17.12. Part B shows the electric field lines of a positive point charge (compare with Fig. 17.10A), and part C shows the electric field lines near a dipole. A plot of electric field lines does not show the magnitude of the field directly,

► **Figure 17.12** A Electric field lines are always parallel to the electric field. B and C Examples of electric field lines.

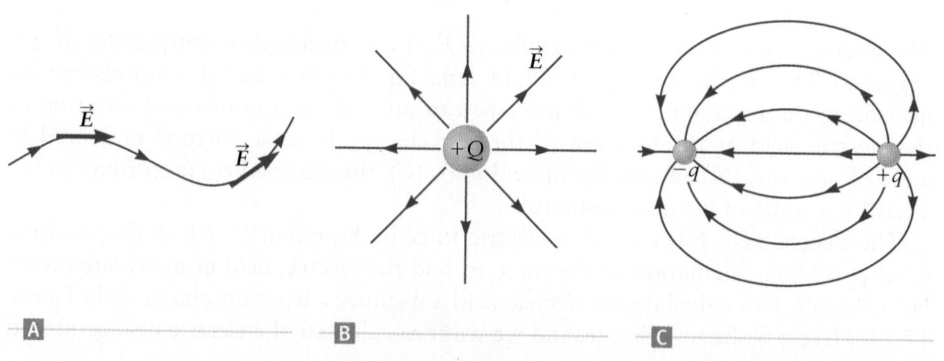

but changes in field strength can be inferred from the spacing of the field lines. For example, the field lines are most closely spaced near the charges in Figures 17.12B and C, where the magnitude of $\vec{E}$ is largest. Regions where the field line spacing is large are regions where the field is small.

Besides being a very useful way to display the electric field, the notion of field lines provides some deeper insight into Coulomb's law. According to Coulomb's law, the force between two point charges falls off as $1/r^2$, where r is the separation between the two charges. In a similar way, the electric field produced by a point charge (Eq. 17.11) also varies as $1/r^2$. The electric force thus obeys an ***inverse square law*** just as we found for the gravitational force in Chapter 5 (Fig. 5.32). Roughly speaking, a charged particle "sets up" field lines in its neighborhood, and the density of field lines is proportional to the amount of charge on the particle. These field lines are associated with an electric field that produces a force on nearby charges, and the magnitude of that force on a nearby "test" charge at a particular location is proportional to the density of field lines at that location. For a point charge Q as in Figure 17.13, the density of field lines—that is, the number of field lines per unit area of space—falls as $1/r^2$ as one goes away from the charge. An equivalent statement is that because the field lines emanating from Q spread out in a three-dimensional space, the area penetrated by any given set of field lines grows as r^2, thus reducing the density of field lines by the same factor as shown by the geometry in Figure 17.13. The notion of field lines thus explains *why* the electric field of a point charge follows an inverse square law.

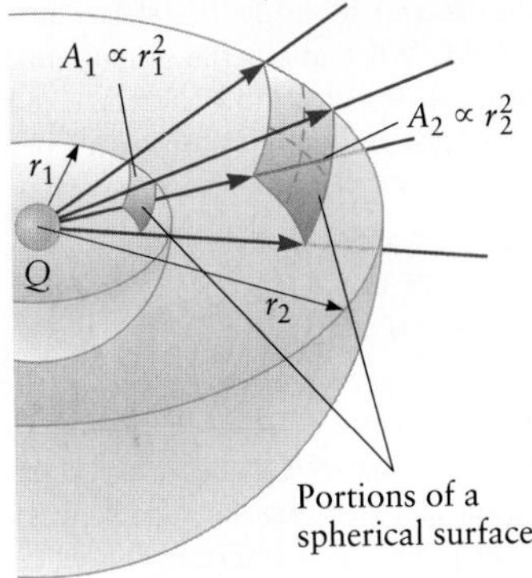

▲ **Figure 17.13** As electric field lines emanate outward from a point charge, they intercept a larger and larger surface area. These surfaces are spherical, so their areas increase with radius as $A \propto r^2$.

EXAMPLE 17.3 Electrons in a Television Tube

The first televisions employed cathode-ray tubes (CRTs), which use electrons to produce an image on a screen. (CRTs are still in use, but are being replaced by flat-panel displays.) In a CRT, a beam of electrons is produced at one end of the tube (Fig. 17.14) by a device called an "electron gun." The electrons have a very small velocity when they leave the vicinity of the gun and then move through a second region, where an electric field accelerates them to the final speed necessary for the CRT to operate properly. The electric field in a typical CRT has a value of 8.0×10^5 N/C, and assume it is constant throughout the acceleration region. If the acceleration region has length $L = 2.0$ cm, what is the speed of an electron when it exits this region?

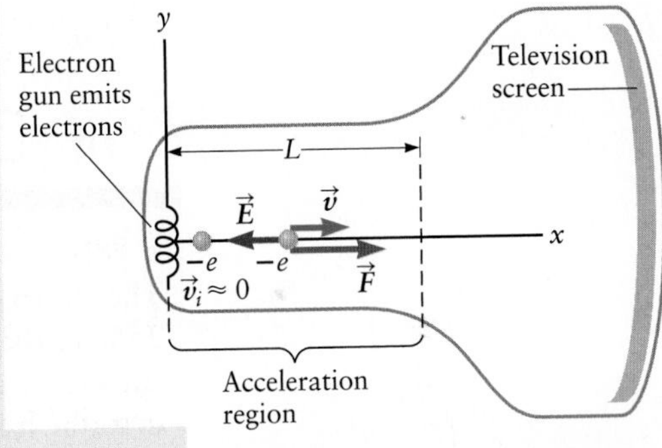

▲ **Figure 17.14** Example 17.3. An old-fashioned television picture tube. (Not to scale.)

RECOGNIZE THE PRINCIPLE

The electric field in the acceleration region produces a force on the electron according to Equation 17.10. The force F and the electron's displacement L are parallel, and since the force is constant the work done by the electric force on the electron is $W = FL$. We can then use the work–energy theorem to calculate the final kinetic energy of the electron when it leaves the accelerator.

SKETCH THE PROBLEM

Figure 17.14 shows the directions of the electric field and the force. Since the charge on an electron is negative, the electric field must be directed to the left so as to give a force in the $+x$ direction (to the right).

IDENTIFY THE RELATIONSHIPS

The horizontal axis in Figure 17.14 is the $+x$ direction, and the electric field in the acceleration region is along $-x$ (to the left). We denote the x component of the electric field as E, so the value of E is negative. The charge on an electron is $-e$, so the x component of the force exerted on an electron is $F = -eE$. (Since the value of E is negative,

(continued) ▶

Insight 17.1

WHAT ABOUT THE FORCE OF GRAVITY?

In Example 17.3, we considered only the electric force on the electrons in a CRT. What about the gravitational force on the electron? The magnitude of the gravitational force is $m_e g$, where m_e is the mass of an electron. We have

$$F_{grav} = m_e g = (9.11 \times 10^{-31}\ \text{kg})(9.8\ \text{m/s}^2) = 8.9 \times 10^{-30}\ \text{N} \quad (1)$$

We mentioned that the electric field in a typical CRT has a magnitude $E = 8 \times 10^5$ N/C. The electric force on an electron thus has a magnitude $F_E = |-eE|$, and inserting the values of e and E gives $F_E = 1 \times 10^{-13}$ N. The gravitational force found in Equation (1) is thus *much* smaller than this electric force, so gravity will not affect the operation of a CRT, thus reinforcing our earlier claim that typical electric forces are much larger than gravitational forces.

F has a positive value; see Fig. 17.14.) This is the component of the force along x; the y component of the force in the acceleration region is zero. The work done on the electron by the electric force is

$$W = FL = -eEL$$

According to the work–energy theorem, this result equals the change in the electron's kinetic energy. Because the problem specified that the electron begins with a very small initial velocity, the initial kinetic energy is approximately zero. If the electron's mass is m_e and its final speed (when it leaves the acceleration region) is v_f, we have

$$KE_f - KE_i = W = -eEL$$

$$KE_f = \tfrac{1}{2}m_e v_f^2 = -eEL \quad (1)$$

SOLVE

We can now solve for the final speed of the electron using Equation (1):

$$v_f^2 = -\frac{2eEL}{m_e}$$

$$v_f = \sqrt{-\frac{2eEL}{m_e}} \quad (2)$$

The negative sign inside the square root might be a "sign" of trouble, but here it is correct. As we have already mentioned, the electric field is in the $-x$ direction and hence the value of E is negative. Inserting values of the various factors in Equation (2), including the electron mass from Appendix A, gives

$$v_f = \sqrt{-\frac{2eEL}{m_e}} = \sqrt{-\frac{2(1.60 \times 10^{-19}\ \text{C})(-8.0 \times 10^5\ \text{N/C})(0.020\ \text{m})}{9.11 \times 10^{-31}\ \text{kg}}}$$

$$v_f = \boxed{7.5 \times 10^7\ \text{m/s}}$$

► *What does it mean?*

The electrons in a CRT have quite a high speed. In fact, our result for v_f is about 25% of the speed of light![2] This high value of v_f is needed so that the electrons can create light when they the strike the television screen. In Section 17.5, we'll describe how the electric field in the acceleration region is generated.

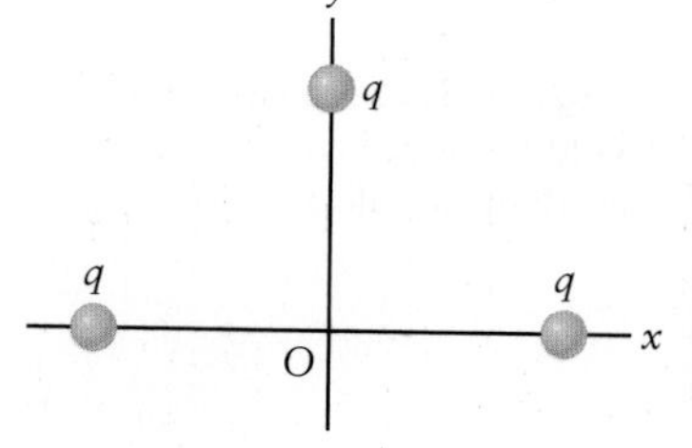

▲ **Figure 17.15** Concept Check 17.5.

CONCEPT CHECK 17.5 Direction of an Electric Field

Three equal and positive charges q are located equidistant from the origin as shown in Figure 17.15. Is the direction of the electric field at the origin (a) along $+x$, (b) along $-x$, (c) along $+y$, or (d) along $-y$?

EXAMPLE 17.4 Electric Field of a Dipole

Consider an electric dipole consisting of two charges $+q$ and $-q$ as sketched in Figure 17.16A. They might be an electron and a proton, or negative and positive ions that make up part of a molecule. Make a qualitative sketch showing the magnitude and direction of the electric field at points A and B in the vicinity of this dipole.

[2]The speed of our CRT electrons is so high that a high-accuracy calculation of their motion requires the use of special relativity (Chapter 27). Including relativity would change our result here by about 6%.

RECOGNIZE THE PRINCIPLE

Figure 17.16A shows the electric fields created by the two point charges. To find the total field $\vec{E}_{total}$ at points A and B, we need to add the electric fields from the two point charges (according to the principle of superposition).

SKETCH THE PROBLEM

Figure 17.16 shows the problem.

IDENTIFY THE RELATIONSHIPS AND SOLVE

Figure 17.16B shows the fields $\vec{E}_{+q}$ and $\vec{E}_{-q}$ at point A from the two point charges. The total field at that point, $\vec{E}_{total}$, is the vector of sum of $\vec{E}_{+q}$ and $\vec{E}_{-q}$ and is also shown in Figure 17.16B. Figure 17.16C uses the same type of analysis to find the field at point B.

▶ ***What does it mean?***

The charges $+q$ and $-q$ form an electric dipole, and Figure 17.12C shows the field lines in their vicinity. Many molecules, including H_2O, act as electric dipoles, and the forces between such molecules involve the dipole electric field. The magnetic field produced by a magnetic dipole follows a similar pattern.

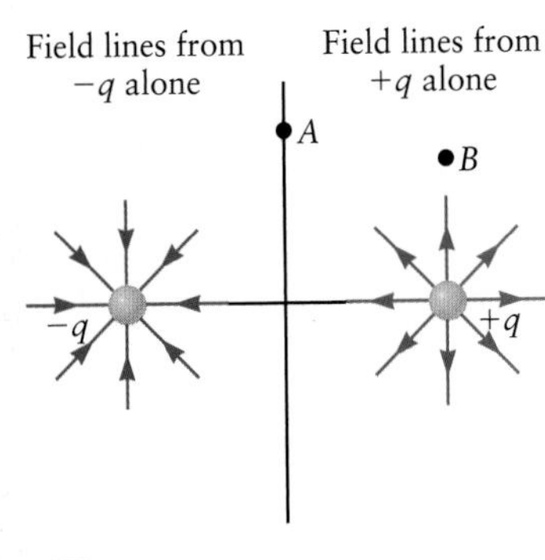

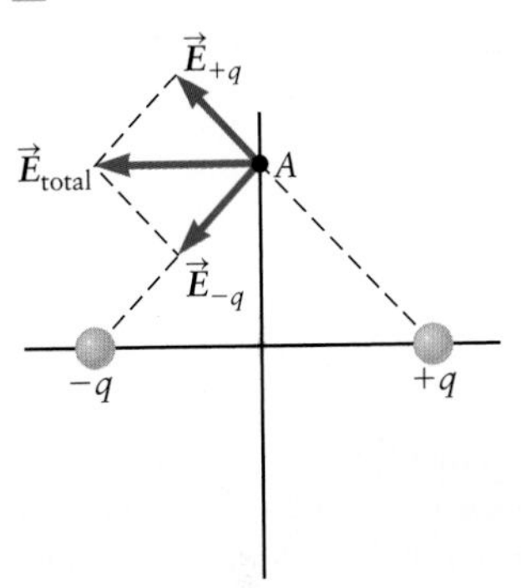

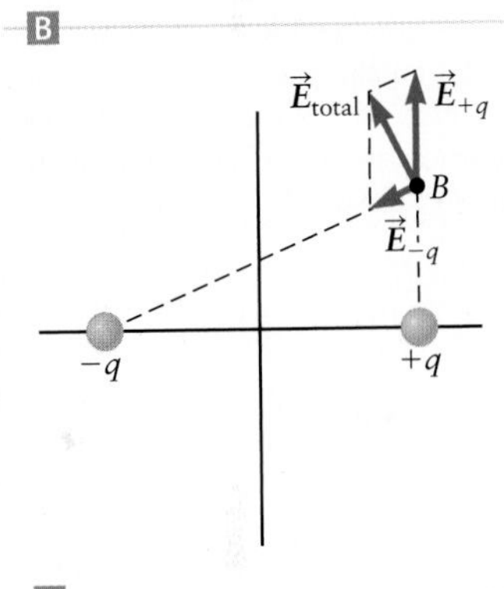

▲ **Figure 17.16** Example 17.4.

EXAMPLE 17.5 Electric Field from Two Point Charges

Two point charges with $Q_1 = 4.0\ \mu C$ and $Q_2 = -2.0\ \mu C$ are arranged as shown in Figure 17.17A. If $L = 3.0$ cm, what is the electric field at the origin?

SKETCH THE PROBLEM

Figure 17.17B shows the directions of the electric fields produced by the two charges along with the total field $\vec{E}_{total}$.

RECOGNIZE THE PRINCIPLE

The magnitude of the electric field from each charge can be calculated using Equation 17.11. The electric field at the origin is the vector sum of the field from each charge. We need to find both the magnitude and direction of $\vec{E}_{total}$.

IDENTIFY THE RELATIONSHIPS

The field from Q_1 has a magnitude (Eq. 17.11)

$$E_1 = \frac{kQ_1}{L^2} = \frac{(8.99 \times 10^9\ \text{N} \cdot \text{m}^2/\text{C}^2)(4.0 \times 10^{-6}\ \text{C})}{(0.030\ \text{m})^2} = 4.0 \times 10^7\ \text{N/C}$$

The field from Q_2 has a magnitude

$$E_2 = \frac{k|Q_2|}{L^2} = \frac{(8.99 \times 10^9\ \text{N} \cdot \text{m}^2/\text{C}^2)(2.0 \times 10^{-6}\ \text{C})}{(0.030\ \text{m})^2} = 2.0 \times 10^7\ \text{N/C}$$

(continued) ▶

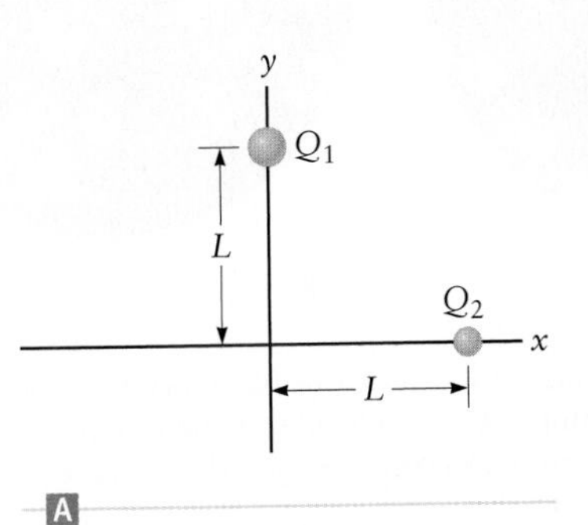

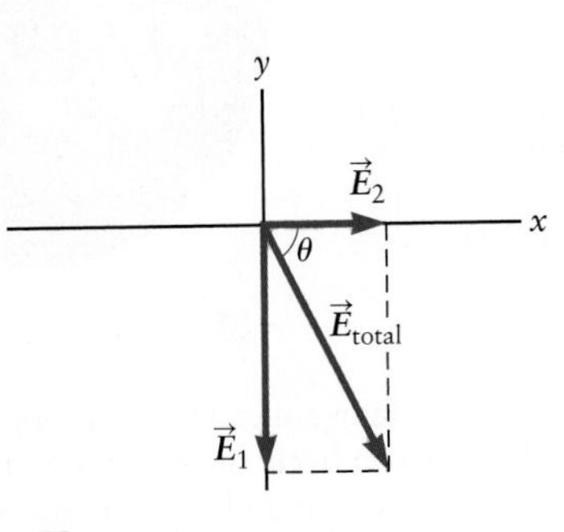

◀ **Figure 17.17** Example 17.5.

SOLVE

Adding the electric fields as vectors is done graphically in Figure 17.17B. The magnitude of the total field E_{total} is

$$E_{total} = \sqrt{E_1^2 + E_2^2} = \sqrt{(4.0 \times 10^7 \text{ N/C})^2 + (2.0 \times 10^7 \text{ N/C})^2}$$

$$E_{total} = \boxed{4.5 \times 10^7 \text{ N/C}}$$

From the trigonometry in Figure 17.17B, the angle that $\vec{E}_{total}$ makes with the x axis is

$$\theta = \tan^{-1}\left(\frac{E_1}{E_2}\right) = \tan^{-1}\left(\frac{4.0 \times 10^7 \text{ N/C}}{2.0 \times 10^7 \text{ N/C}}\right) = \boxed{63°}$$

► ***What does it mean?***

When adding the fields from different charges, the fields must always be added as vectors.

Insight 17.2

DRAWING ELECTRIC FIELD LINES

When you draw the electric field lines produced by a collection of point charges, the field lines can only begin or end on charges. As with the dipole in Figure 17.12C, electric field lines begin at a positive charge and end at a negative charge.

17.4 Conductors, Insulators, and the Motion of Electric Charge

In Section 17.1, we described how the ancient Greeks used amber and fur to observe electric forces. To understand those experiments, we need to understand how charges can be transferred from one material to another.

The Structure of Matter from an "Electrical" Viewpoint

An atomic-scale picture of a *metal* such as copper is given in Figure 17.18A. Each Cu atom by itself is electrically neutral, with equal numbers of protons and electrons. When these neutral atoms come together to form a piece of copper metal, one or more electrons from each atom are able to "escape" from its "parent" atom and move freely through the entire piece of metal. These electrons are called ***conduction electrons***. They leave behind positively charged ion cores that are bound in place and not mobile. In addition, a piece of metal can accept some additional electrons or

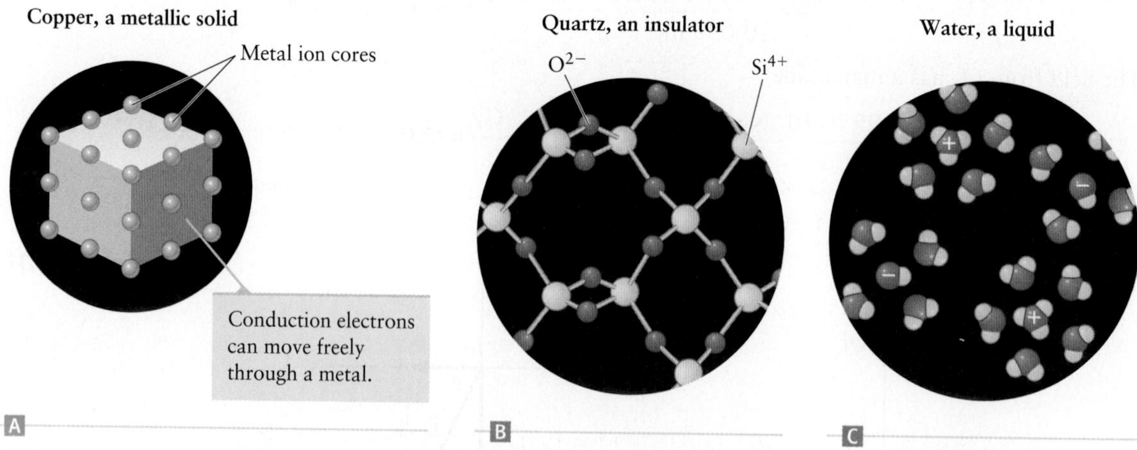

▲ **Figure 17.18** A In a metal such as copper, some of the electrons (called conduction electrons) are able to move freely through the entire sample. B In an insulator such as quartz, all the electrons are bound to ions and there are no conduction electrons. C Liquids and gases contain relatively small numbers of mobile charges (mainly ions).

let some of the conduction electrons be taken away, so the entire piece of metal can acquire a net negative or positive charge.

Amber, plastic, and quartz are ***insulators***; electrons in these materials are *not* able to move freely through the material. A typical example is quartz, which is composed of SiO_2 molecules, each of which is a combination of one Si^{+4} and two O^{-2} ions. (See Fig. 17.18B.) These ions are bonded in place within the solid and are not able to move about. Unlike in metals, electrons cannot "escape" from these ions, and there are no conduction electrons available to carry charge through the solid. Extra electrons can be placed onto a piece of quartz, but they are not able to move about freely. Instead, an added electron would tend to stay in one spot (wherever it was initially placed), although the small amount of moisture that is often present will allow excess charge to gradually move about on the surface.

An atomic-scale picture of another important type of matter is shown in Figure 17.18C. A drop or cup of water consists of a collection of H_2O molecules, each of which has a total charge of zero and is thus electrically neutral. Each water molecule consists of two H atoms and one O atom; these atoms and their electrons are usually tightly bound to each other within a particular molecule. If all the atoms and electrons were bound in this way, there would be no "free" charges available to carry charge throughout the liquid, and water would behave as a perfect insulator. However, a few water molecules in any sample are always dissociated into free H^+ and OH^- ions. These ions can carry charge from place to place in much the same way that conduction electrons carry charge in a metal. (H^+ ions may also combine with neutral water molecules to form hydronium ions, H_3O^+, which along with OH^- ions can carry charge from place to place in a sample of water.) Most samples of water also contain impurities that contribute additional ions such as Na^+ or Cl^-, which act as mobile charge carriers.

The picture in a gas is similar to that found in a liquid; the constituent atoms and molecules are mostly neutral, but a few are present as ions able to carry charge from place to place. A few free electrons are usually present in a gas as well.

Placing Charge on an Insulator

To understand the electrical behavior of an object, we must understand what happens when charge is placed on or taken off the object. Let's begin with the case of an insulator such as quartz, and, for simplicity, assume there is no moisture on it and the air around it is very dry. If a few electrons are placed onto such an insulator, they will stay where they are put (Fig. 17.19A) because there are no free ions in an insulator and the insulator does not allow the extra electrons to move about.

In real life, excess charge will not stay on an insulator indefinitely. If an insulator contains some excess electrons, they will attract stray positive ions from the surrounding air; these stray ions will move onto the insulator and cause it to become neutral, or the electrons may combine with the ions to form neutral atoms (Fig. 17.19B). The amount of time needed to neutralize the charge depends on the amount of moisture in the air. It can be many minutes on a dry (usually winter) day or only a few seconds on a humid day. Any small amount of moisture on an insulator's surface forms a thin layer of water that enables charge to move from one spot to another.

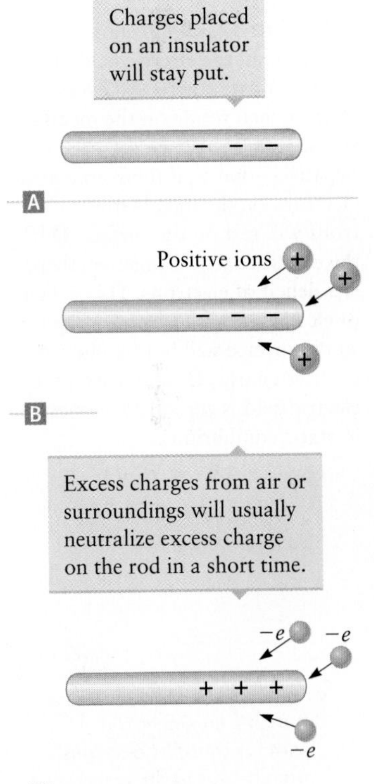
Charges placed on an insulator will stay put.
A
Positive ions
B
Excess charges from air or surroundings will usually neutralize excess charge on the rod in a short time.
−e
−e
−e
C

▲ **Figure 17.19** A When excess electrons are placed on an insulator, this charge generally stays where it is placed for a period of time (unlike in a metal, where the charge will move immediately). B and C Stray ions or excess electrons in the surrounding air are attracted to the insulator's surface, so the insulator's excess charge will eventually be neutralized.

Excess Charge Goes to the Surface of a Metal

Unlike the case of an insulator, electrons can move easily through a metal. So, if we place electrons on one spot of a piece of metal, there is no guarantee they will stay there, and we might imagine that they could go anywhere within the metal. In fact, any excess electrons on a piece of metal will all be distributed *on the surface* of the metal as sketched in Figure 17.20A. To understand this result, consider a piece of metal that is electrically isolated from the rest of the universe so that no new charge can jump onto or off the metal. If we wait long enough, all mobile charges (electrons) in the metal will come to rest and be in static equilibrium. Now imagine that some

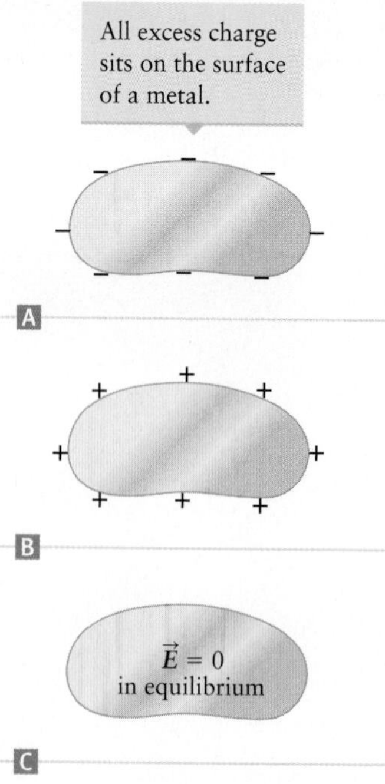
All excess charge sits on the surface of a metal.
A
B
$\vec{E} = 0$ in equilibrium
C

▲ **Figure 17.20** In static equilibrium, the electric field inside a metal must be zero. All excess charges then reside on the metal's surface. A If the excess charge is negative—that is, if there are extra electrons on the metal—these electrons will rest on the surface. B If the excess charge is positive, there is a deficit of electrons. This deficit of electrons will be at the surface, so the surface will have a net positive charge. C In all cases, the electric field is zero inside a metal in static equilibrium.

of these excess electrons are not on the surface but are in the interior instead. These electrons would each produce an electric field, causing other electrons in the metal to move. This result, however, contradicts our assumption that the system is in static equilibrium and means that the excess electrons cannot be inside the metal. Hence, for a metal in equilibrium, any excess electrons must be *at the surface* and ***the electric field is zero inside a metal in equilibrium*** (Fig. 17.20C).

This argument applies only to the *excess* charges placed on the metal. The interior of a neutral, uncharged metal contains equal numbers of positive and negative charges. These charges are present inside the metal at all times, and their electric fields cancel, leaving zero electric field inside the metal. Only the excess charge resides at the metal's surface.

In Figure 17.20A, some extra electrons have been added to a metal, so the surface has a net negative charge. In Figure 17.20B, we imagine that some electrons have been removed from the metal, so its net charge is positive. It is customary to draw this situation as if there are positive charges at the surface of the metal, even though that does *not* correspond to placing positive ions or protons onto the metal!

Charging an Object by Rubbing It

The Greeks discovered electricity when they rubbed a piece of amber with fur. The rubbing caused charge to be transferred between the amber and the fur, and we now know that electrons moved from the fur to the amber. The amber thus acquired a net negative charge (an excess of electrons), while the fur was left with a net positive charge (an excess number of positive ions) as illustrated in Figure 17.21.

There is nothing unique about amber; a similar result is found if a rubber or plastic rod is rubbed with fur. Likewise, there is nothing special about fur. For example, a glass rod can be charged by rubbing it with a silk cloth, although in this case the glass acquires a net positive charge because electrons leave the glass and move to the silk. The mechanical process of rubbing is thus a general way to move electrons from one material to another. This is the basis of the Van de Graaff generator (Fig. 17.22), which is found in many science museums.

We mentioned in connection with Figure 17.1 that a charged object, such as a piece of amber, can be used to attract and move small pieces of paper and similar objects. Let's consider this attraction in a little more detail. Suppose we have a rubber rod and a small piece of paper, with both electrically neutral at the start of the experiment (Fig. 17.23). If the rubber rod is rubbed with fur, it will acquire a nega-

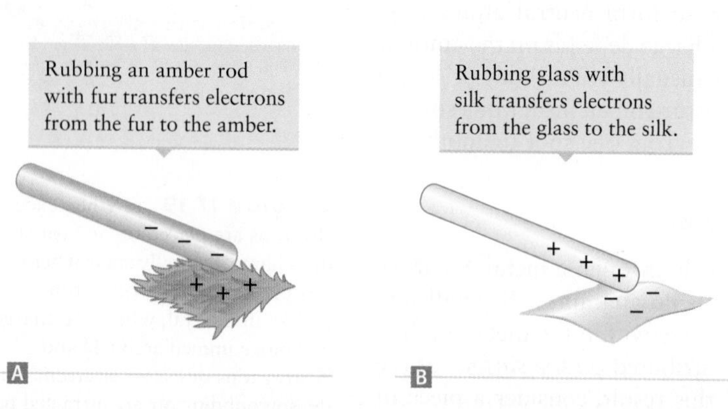
Rubbing an amber rod with fur transfers electrons from the fur to the amber.
A
Rubbing glass with silk transfers electrons from the glass to the silk.
B

▲ **Figure 17.21** A When a piece of amber or a plastic rod is rubbed with fur, the amber or plastic becomes negatively charged. B A glass rod rubbed with silk becomes positively charged.

Tony Freeman/Photo Edit

▲ **Figure 17.22** A Van de Graaff generator produces a large electric charge by rubbing an internal rubber belt. This belt transfers electric charge to the metal sphere at the top of the generator.

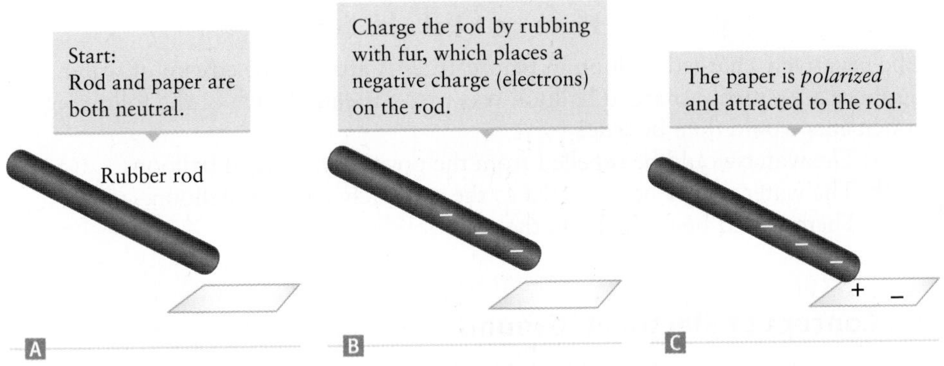
Start:
Rod and paper are both neutral.
Rubber rod
Charge the rod by rubbing with fur, which places a negative charge (electrons) on the rod.
The paper is *polarized* and attracted to the rod.
A
B
C

◀ **Figure 17.23** A A rubber rod that is initially neutral can B be charged negatively by rubbing it with fur. C This negatively charged rod can then polarize a nearby object such as a small piece of paper, resulting in electrical attraction between the rod and the paper.

tive charge. If this charged rod is then brought near the paper, the paper is attracted to the rod. This effect is caused by electric forces between charges on the rod and the paper. The rod acquired its excess charge by rubbing, but the paper has a net charge of zero, so why is an electric force exerted on the paper? Although the paper is electrically neutral, the electric field of the rod causes some movement of the charges in the paper. The rod is negatively charged, so electrons in the paper are repelled, and positive ions in the paper are attracted. Although these electrons and ions are not free to move very far, they can move a small amount while still being bound inside the paper. The electrons thus move a short distance away from the rod and the positive ions move toward the rod as sketched in Figure 17.23C, leaving a net positive charge on the portion of the paper nearest the rod and a net negative charge on the opposite side of the paper. The paper is then said to be ***polarized***.

Because unlike charges attract, the positive side of the paper is attracted to the negatively charged rod. At the same time, the negative side of the paper is repelled from the rod. The positively charged side of the paper is closer to the rod, so because the electric force decreases as the distance between the charges increases, the magnitude of the attractive force is larger than the magnitude of the repulsive force. The net result is that the paper is *attracted* to the rod. In this way, there can be an electric force on an object even when the object is electrically neutral, provided the object is polarized.

The same principle is illustrated in Figure 17.24A, which shows water flowing from a container. The stream of water would ordinarily flow vertically downward due to the force of gravity, but a charged balloon placed near the stream exerts a force on the water and deflects the stream sideways. The electric field from the charged balloon polarizes the water molecules (Fig. 17.24B), leading to an attractive force as found with the paper in Figure 17.23.

Insight 17.3

WHY DOES RUBBING TWO OBJECTS MAKE THEM CHARGED?

To understand why rubbing two objects together transfers electrons from one object to the other, we appeal to the concept of *electronegativity* that arises in chemistry. Electronegativity is a measure of the tendency of an atom to attract bonding electrons. When two objects are rubbed together, atoms from the two materials are brought into contact. Unless they are identical atoms in identical materials, their electronegativities will be different, causing the atoms in one material to "steal" electrons from the other material.

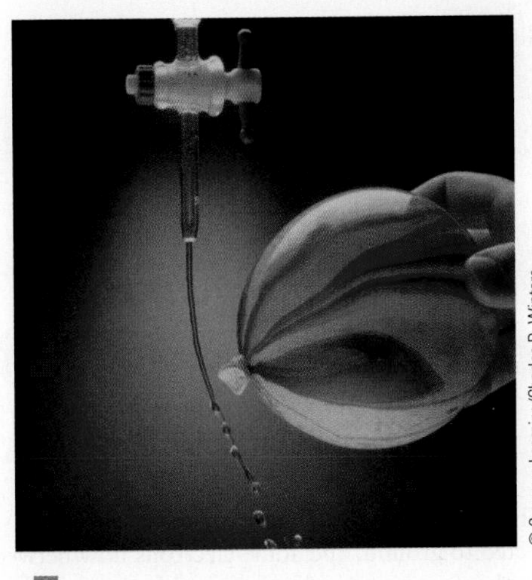
A

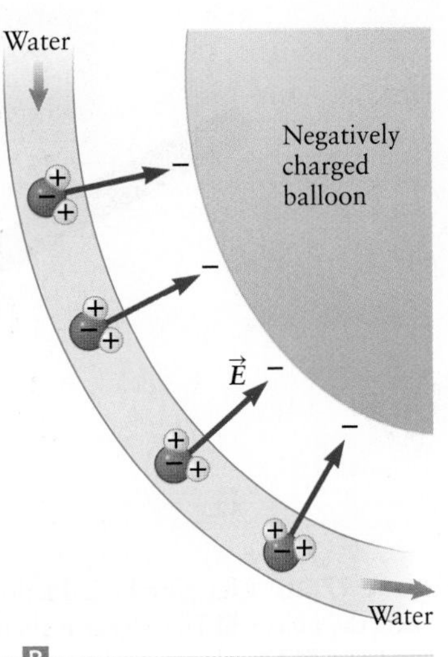
Water
Negatively charged balloon
$\vec{E}$
Water
B

◀ **Figure 17.24** A A stream of water is deflected by a nearby charged balloon. B The balloon polarizes the water, similar to the piece of paper in Figure 17.23C.

CHARGING AN OBJECT BY CONTACT
Before contact
Contact
After contact

▲ **Figure 17.25** If one charged object touches a second object, the second object will usually acquire some of the excess charge. Hence, the second object is charged "by contact."

CONCEPT CHECK 17.6 Force on a Polarized Object

The negatively charged balloon in Figure 17.24A attracts the stream of water. Suppose a positively charged balloon was used instead. Which of the following statements would then be true?

(a) The water would be repelled from the positively charged balloon.
(b) The water would be attracted to the positively charged balloon.
(c) There would be no force on the water.

The Concept of Electrical "Ground"

Suppose a rubber rod is charged by rubbing so that it acquires some extra electrons and is then placed on a table. If we were to wait for some period of time (the time required depends on the amount of moisture in the room), we would find that the rod has lost its excess charge. If we were to watch very carefully with sensitive electronic measuring equipment, we would find that the excess electrons originally on the rod move from the rod to the table and eventually into the ground below. The ground is generally quite moist, so it conducts charge well. Because the ground is literally everywhere, it provides a common path that excess charge can use to flow from one spot to another. In fact, the term *ground* is used generically to denote the path or destination of such excess charges, even if the true path does not involve any dirt. This notion of ***electrical ground*** plays an important role in many situations.

Charging by Contact and by Induction

Suppose you are given a negatively charged rubber rod and are asked to transfer some of this charge to a piece of metal. You could accomplish this task by simply bringing the rubber rod into contact with the metal (Fig. 17.25). The negative charge on the rod is caused by an excess of electrons, and some of these electrons will move to the metal when it touches the rod.

Now suppose you are given the same negatively charged rubber rod, but this time you are asked to give the metal a net *positive* charge. This task might seem to be impossible because the excess charges on the rod are negative, but an approach called ***charging by induction*** will accomplish it. This approach makes use of polarization and the properties of an electrical ground. The negatively charged rod is first brought near the metal (Fig. 17.26A), polarizing the metal. That is, some of the conduction electrons in the metal move to the opposite side—away from the rod—due to the repulsion of like charges, leaving the side of the metal near the rod with

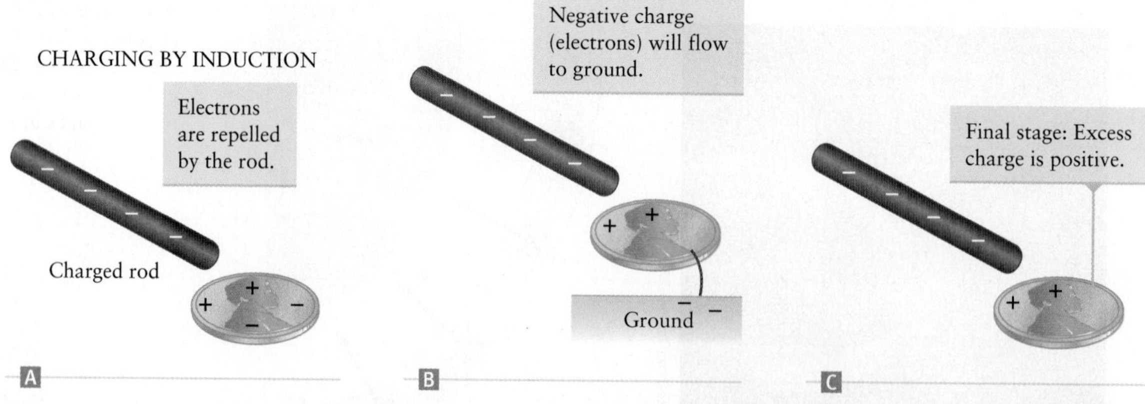
CHARGING BY INDUCTION
Electrons are repelled by the rod.
Charged rod
Negative charge (electrons) will flow to ground.
Ground
Final stage: Excess charge is positive.
A
B
C

▲ **Figure 17.26** Charging by induction. A The object is first brought near a charged rod, polarizing the object. B The object is then connected to ground, and some electrons flow between it and ground. C The object is left with an excess charge when it is disconnected from ground.

a net positive charge. A connection is then made from the piece of metal to electrical ground using a metal wire, so electrons are able to move even farther from the charged rod by moving into the electrical ground region (Fig. 17.26B). The final step in the process is to remove the grounding wire, leaving the original piece of metal with a net positive charge (Fig. 17.26C). Notice that we have not put protons or any other positive charges on the metal. Instead, we have *removed* electrons. The final result is that the metal has a net positive charge.

17.5 Electric Flux and Gauss's Law

We can calculate the electric field due to a point charge using Equation 17.11. In principle, we can use this result to deal with any conceivable distribution of charges by treating the distribution as a collection of point charges and using superposition to find the total field. Fortunately, there is a simpler way to deal with complex charge distributions, based on ***Gauss's law***. To explain Gauss's law, we must first define a quantity called the ***electric flux***. In words, we say that there is an electric flux through a surface whenever electric field lines pass through the surface. In the simplest case, with the electric field directed perpendicular to the surface, the magnitude of the flux is the product of the electric field and the area of the surface. Electric flux is denoted by the symbol Φ_E.

Some examples of electric flux calculations are given pictorially in Figure 17.27. For simplicity, these examples assume the electric field is constant in both magnitude and direction. In Figure 17.27A, $\vec{E}$ is perpendicular to a flat surface having total area A, and the flux is $\Phi_E = EA$. In this case, the flux is just the magnitude of the electric field multiplied by the area of the surface. Flux is a scalar, just an ordinary number; it is *not* a vector.

If $\vec{E}$ is parallel to the surface (Fig. 17.27B), no electric field lines cross the surface and $\Phi_E = 0$. In general, if the electric field makes an angle θ with the direction normal to the surface (Fig. 17.27C), the magnitude of the flux is proportional to the component of the field perpendicular to the surface, and

$$\Phi_E = EA\cos\theta \tag{17.12}$$

The three examples in Figure 17.27A–C all involve simple, flat surfaces. We'll usually be interested in the flux through a closed surface such as a box or a sphere. The flux due to a constant field $\vec{E}$ through these closed surfaces is shown in parts D and E of Figure 17.27. By convention, the flux through a surface is *positive* if the field is directed *out* of the region contained by the surface, whereas the flux is *negative* if $\vec{E}$ is directed *into* the region. For the cases with a constant field in parts D and E of Figure 17.27, the total flux through the entire closed surface is *zero* in each case. For the box in Figure 17.27D, there is a negative flux through the face of the box on the left and a positive flux through the face on the right. The total flux is the sum of these two contributions and is zero. Likewise, there is a negative flux through the left side of the spherical surface in Figure 17.27E and a positive flux through the right side, and the total flux is again zero.

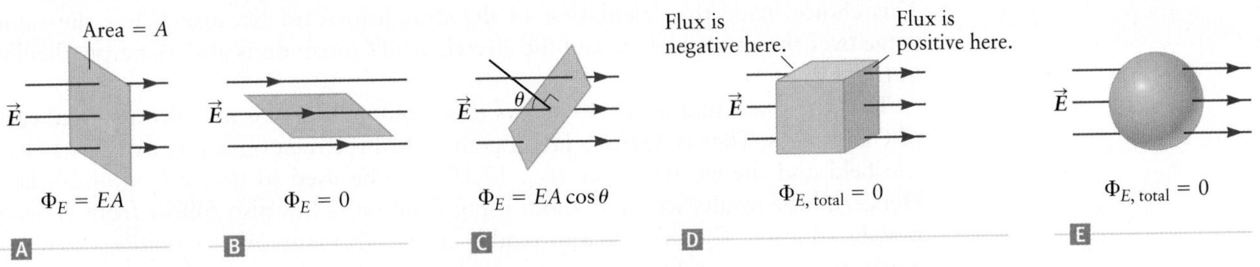

▲ **Figure 17.27** Finding the electric flux through a surface.

Gauss's Law

Gauss's law asserts that the electric flux through any closed surface is proportional to the total charge q inside the surface, with

Gauss's law

$$\Phi_E = \frac{q}{\varepsilon_0} \tag{17.13}$$

The constant of proportionality that relates flux and charge is ε_0, the same physical constant that enters Coulomb's law (Eq. 17.5).

Since electric flux depends on the magnitude and direction of the electric field, the left-hand side of Equation 17.13 depends on $\vec{E}$ while the right-hand side depends on the charge. We would like to use Gauss's law to calculate $\vec{E}$, but it is not immediately obvious how to do so. In particular, Φ_E is the total flux through a closed surface, so its value depends on the magnitude and direction of the electric field at *all* points on the surface. To see how to use Gauss's law, it is simplest to consider some examples.

Using Gauss's Law to Find $\vec{E}$ for a Point Charge

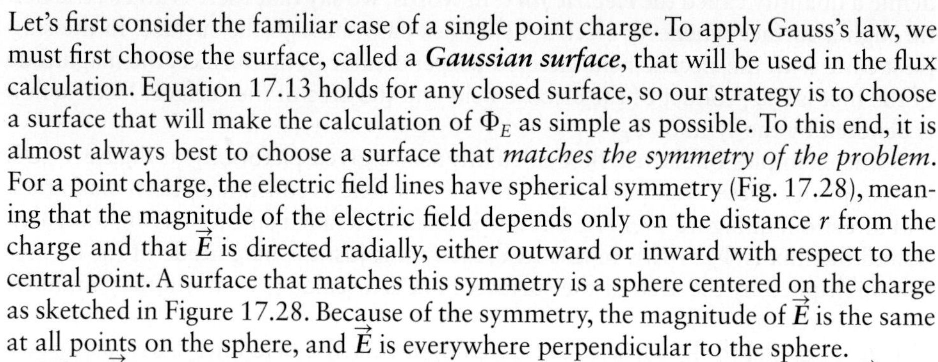

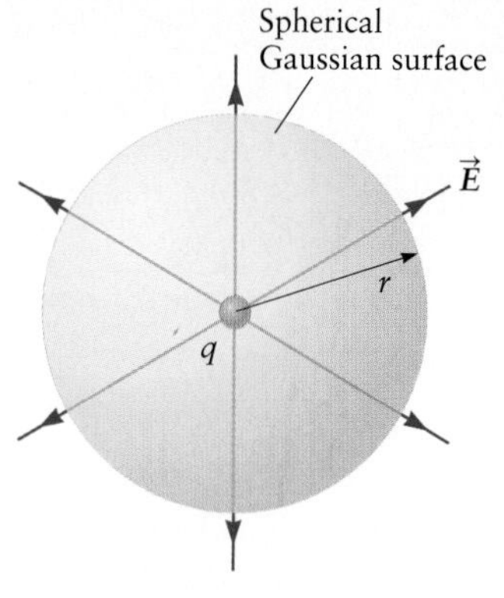

▲ **Figure 17.28** To calculate the electric field near a point charge using Gauss's law, we choose a spherical Gaussian surface centered on the point charge.

Let's first consider the familiar case of a single point charge. To apply Gauss's law, we must first choose the surface, called a ***Gaussian surface***, that will be used in the flux calculation. Equation 17.13 holds for any closed surface, so our strategy is to choose a surface that will make the calculation of Φ_E as simple as possible. To this end, it is almost always best to choose a surface that *matches the symmetry of the problem.* For a point charge, the electric field lines have spherical symmetry (Fig. 17.28), meaning that the magnitude of the electric field depends only on the distance r from the charge and that $\vec{E}$ is directed radially, either outward or inward with respect to the central point. A surface that matches this symmetry is a sphere centered on the charge as sketched in Figure 17.28. Because of the symmetry, the magnitude of $\vec{E}$ is the same at all points on the sphere, and $\vec{E}$ is everywhere perpendicular to the sphere.

When $\vec{E}$ is perpendicular to a surface, the flux is equal to the magnitude of $\vec{E}$ multiplied by the area of the surface. So, for the flux in Figure 17.28, we have

$$\Phi_E = EA_{\text{sphere}}$$

where A_{sphere} is the area of our spherical Gaussian surface. If the radius of this sphere is r, then $A_{\text{sphere}} = 4\pi r^2$ and the flux is

$$\Phi_E = 4\pi r^2 E$$

According to Gauss's law, this flux is proportional to the total charge contained within the surface. Using Equation 17.13, we have

$$\Phi_E = 4\pi r^2 E = \frac{q}{\varepsilon_0}$$

We can now solve for E and find

$$E = \frac{q}{4\pi\varepsilon_0 r^2} \tag{17.14}$$

which agrees with our previous result obtained using Coulomb's law (Eq. 17.11). The key to this application of Gauss's law was our choice of the Gaussian surface. This choice made the calculation of Φ_E straightforward because E has the same value over the entire surface and the electric field's direction is always perpendicular to the surface.

The result in Equation 17.14 shows that Coulomb's law is actually a special case of Gauss's law. That is, Gauss's law together with the relationship between the electric field and the electric force (Eq. 17.10) can be used to derive Coulomb's law. Hence, all the results we have found using Coulomb's law also follow from Gauss's law. In addition, Gauss's law can readily deal with many other situations that are mathematically very difficult or awkward to attack with Coulomb's law. A general strategy for applying Gauss's law is summarized below.

PROBLEM SOLVING Applying Gauss's Law

1. **RECOGNIZE THE PRINCIPLE.** To apply Gauss's law, you must first calculate the electric flux through a well-chosen Gaussian surface.
2. **SKETCH THE PROBLEM.** Your drawing should show the charge distribution. Using the symmetry of this distribution, sketch the electric field. The Gaussian surface should match the symmetry of the electric field.
3. **IDENTIFY THE RELATIONSHIPS.** Your Gaussian surface should satisfy one or more of the following conditions:
   - $\vec{E}$ has constant magnitude over all or much of the surface and makes a constant angle with the surface. The most convenient surface is one that is perpendicular to $\vec{E}$ at all (or most) points.
   - $\vec{E}$ may be zero over a portion of the surface. The flux through that part of the surface is zero.
   - $\vec{E}$ may be parallel to some part of the surface. The flux through that part of the surface is zero.
4. **SOLVE.** First calculate the total electric flux through the entire Gaussian surface. Then find the total electric charge inside the surface and apply Gauss's law (Eq. 17.13) to solve for $\vec{E}$.
5. Always *consider what your answer means* and check that it makes sense.

EXAMPLE 17.6 Electric Field from a Spherical Charge

Consider the uniform spherical ball of charge in Figure 17.29A, with total charge Q and radius r_b. Find the electric field at points outside the ball.

RECOGNIZE THE PRINCIPLE

We follow the procedure described in the "Applying Gauss's Law" problem-solving strategy. The key is to find a Gaussian surface that matches the symmetry of the charge distribution (step 1 in the strategy). The charge is spherically distributed in Figure 17.29A, so the electric field must also have spherical symmetry.

SKETCH THE PROBLEM

Step 2 in the strategy: The spherical symmetry means that $\vec{E}$ must be directed radially pointing either away from or toward the center of the ball of charge. If Q is positive, the field will be as sketched in Figure 17.29B.

IDENTIFY THE RELATIONSHIPS

Step 3: Choose a Gaussian surface. Because of the spherical symmetry, the magnitude of $\vec{E}$ will depend only on the distance from the center of the ball of charge. We therefore pick a spherical Gaussian surface as sketched in Figure 17.29A. The electric field will have a constant magnitude at all points on this surface, and $\vec{E}$ will be perpendicular to the surface. This geometry is similar to what we had in Figure 17.28 when we applied Gauss's law to the case of a point charge.

SOLVE

We next calculate the electric flux through our chosen Gaussian surface. The Gaussian surface is a sphere of radius r, so its surface area is $A_{\text{sphere}} = 4\pi r^2$. Denoting the

(continued) ▶

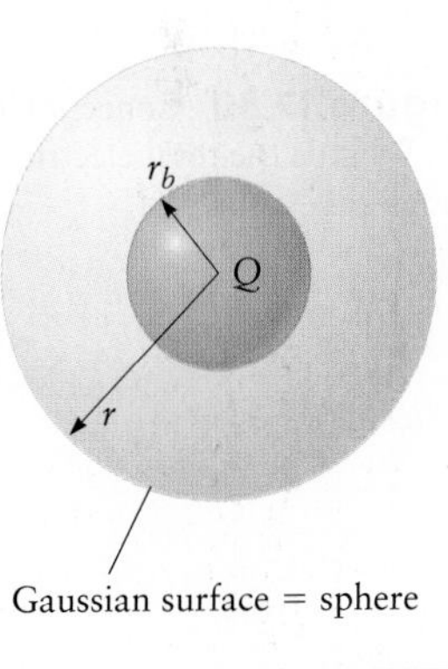

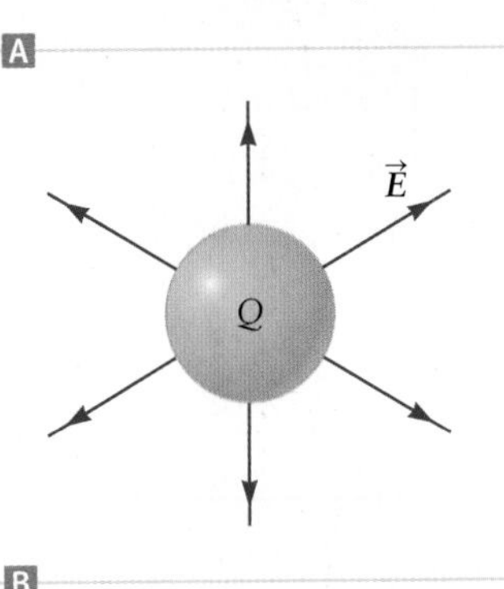

▲ **Figure 17.29** Example 17.6.

magnitude of the electric field on the sphere by E and again noting that $\vec{E}$ is everywhere perpendicular to the surface, the electric flux is

$$\Phi_E = EA_{\text{sphere}} = 4\pi r^2 E$$

Gauss's law relates this flux to the total charge contained within the spherical surface. We are interested in locations outside the ball ($r > r_b$), so the charge inside the Gaussian surface is the total charge Q of the ball. We thus have

$$\Phi_E = 4\pi r^2 E = \frac{Q}{\varepsilon_0}$$

which leads to

$$\boxed{E = \frac{Q}{4\pi\varepsilon_0 r^2}}$$

Since $1/(4\pi\varepsilon_0) = k$, we can also write this result as

$$\boxed{E = \frac{kQ}{r^2}}$$

▸ What does it mean?

Our result for E is identical to the electric field of a point charge of magnitude Q located at the origin (Eq. 17.11). In fact, the electric field from *any* spherical distribution of charge is the *same* as the field from a point charge with the same total charge. Note that this result applies *only outside* the ball of charge.[3]

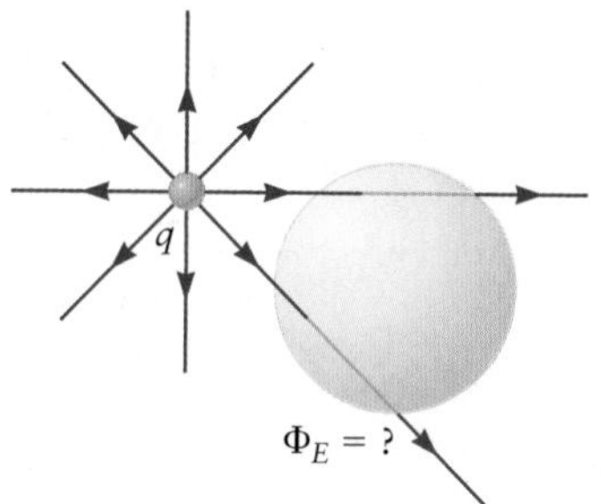

▲ Figure 17.30 Concept Check 17.7. What is the total electric flux through this surface?

CONCEPT CHECK 17.7 Finding the Electric Flux

Consider a point charge q located outside a closed surface as sketched in blue in Figure 17.30. The electric field lines from the charge penetrate the surface. What is the total electric flux through this surface?

(a) q/ε_0 (b) $-q/\varepsilon_0$ (c) 0

More Applications of Gauss's Law

Consider a very long, straight line of charge as sketched in Figure 17.31A. The line of charge has total length L (where L is very large) and total charge Q. How can we find the electric field produced by this charge distribution?

The first two steps in the "Applying Gauss's Law" problem-solving strategy involve recognizing the symmetries of the charge distribution and of the electric field it produces so that you can choose a Gaussian surface. For step 2, the sketch in Figure 17.31B shows two observers who are studying the charged line and its electric field. These two observers are looking at the charged line from opposite directions. Because it is very long, the charged line will look the same to these observers, so the electric field must also look the same. Suppose (hypothetically) observer A finds that $\vec{E}$ points away from him and makes an angle θ with the x axis as sketched in Figure 17.31C. Since the observers are symmetrically arranged, observer B must find a similar result (Fig. 17.31D), with the electric field pointing away from him and also making an angle θ with the x axis. However, the two observers *must agree* on the magnitude and direction of $\vec{E}$, and they can only agree if $\theta = 90°$. The symmetry of the line of charge means that $\vec{E}$ must be *perpendicular to the line*. The electric field could be directed away from the line or toward the line, but symmetry requires that it be perpendicular (Fig. 17.31E).

[3]We found a similar result in connection with gravitation in Chapter 5, where we noted that the gravitational force due to a spherically symmetric mass is the same as if all the mass were located at the center of the sphere.

The next step is to choose a Gaussian surface that matches this symmetry. We wish (if possible) for $\vec{E}$ to be perpendicular to the surface, so we choose a cylinder centered on the line of charge (Fig. 17.31F). This cylinder can be divided into a curved section plus two flat, circular ends. The electric field is parallel to the circular ends, so the electric flux through the ends is zero. We want the magnitude of $\vec{E}$ to be the same at all points on the curved part of the cylinder. The magnitude of E will be constant on this surface provided the charged line runs along the central axis of the cylinder.

Next, we calculate the flux through our chosen Gaussian surface. For a cylinder of radius r and length h, the area of the curved part of the cylinder is $A = 2\pi rh$. The flux through this curved surface is thus

$$\Phi_E = EA = 2\pi Erh$$

We now find the total charge inside the Gaussian surface and then apply Gauss's law. The total charge within the cylinder in Figure 17.31F is equal to the charge per unit length Q/L multiplied by the length h of the cylinder, so

$$q = \frac{Q}{L}h$$

Applying Gauss's law then gives

$$\Phi_E = 2\pi Erh = \frac{q}{\varepsilon_0} = \frac{Qh/L}{\varepsilon_0}$$

We can now solve for the magnitude of E:

$$E = \frac{Q}{2\pi\varepsilon_0 Lr} \tag{17.15}$$

CONCEPT CHECK 17.8 Electric Field from a Line of Charge

If the charge Q on the line in Figure 17.31F is negative, is the direction of the electric field (a) perpendicular to the line and directed away from it, (b) perpendicular to the line and directed toward it, or (c) parallel to the line?

EXAMPLE 17.7 Field from a Flat Sheet of Charge

Consider a large, flat sheet of charge as sketched in Figure 17.32A. If this sheet has a positive charge per unit area σ, find the electric field produced by the sheet.

RECOGNIZE THE PRINCIPLE

We again follow the steps outlined in the "Applying Gauss's Law" problem-solving strategy. We begin by considering the symmetry of the charge distribution and of the resulting electric field so that we can choose a Gaussian surface.

SKETCH THE PROBLEM

We again imagine how this charge distribution and field would appear to different observers who look at them from different directions (Fig. 17.32B). These observers must agree on the direction of $\vec{E}$, and as was the case with the charged line in Figure 17.31, the electric field must be perpendicular to the plane. If the charge on the plane is positive, the electric field points away from the plane.

IDENTIFY THE RELATIONSHIPS

We next choose a Gaussian surface. The cylinder's axis in Figure 17.32A is oriented perpendicular to the plane. The flux through the curved sidewall of the surface is zero

(continued) ▶

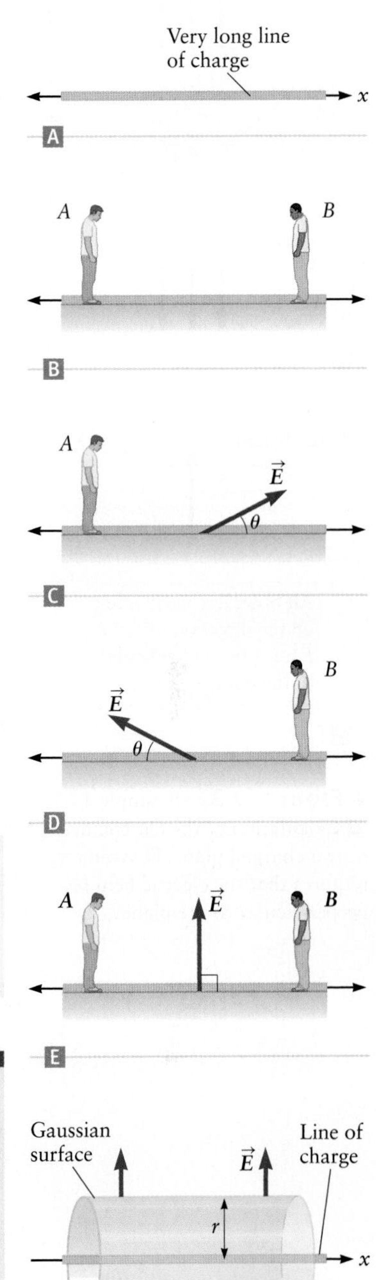

▲ **Figure 17.31** A Calculation of the electric field near a line of charge. B–E Symmetry requires that the electric field be perpendicular to the line. The magnitude of E depends only on the distance r from the line. F If we choose a cylindrical Gaussian surface, the electric field is constant in magnitude at the walls of the cylinder and $\vec{E}$ is parallel to the ends of the cylinder.

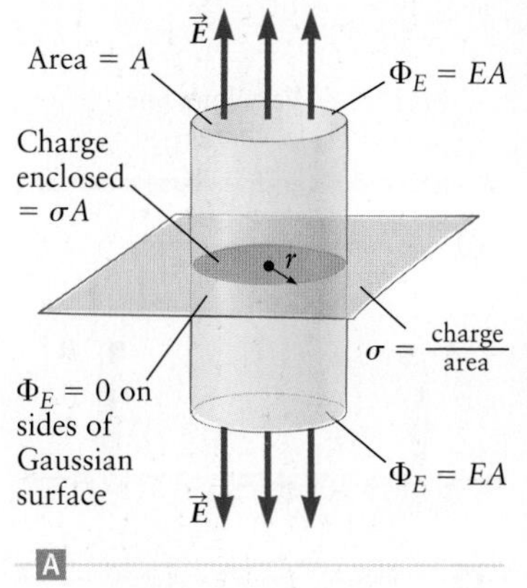

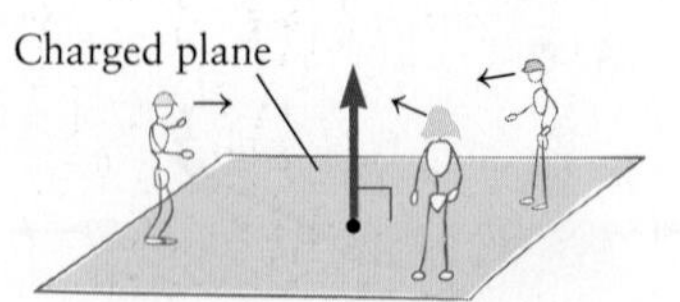

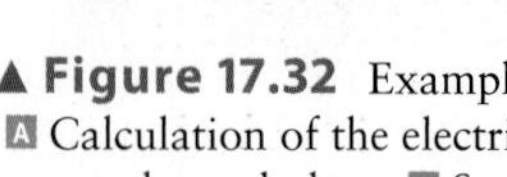

▲ **Figure 17.32** Example 17.7. A Calculation of the electric field near a charged plane. B Symmetry requires that the electric field be perpendicular to the plane.

because $\vec{E}$ is parallel to this part of the cylinder. The electric field $\vec{E}$ is perpendicular to the ends of the cylinder, so if the ends each have an area A, the flux is $\Phi_E = EA$ through each end. For a positively charged plane, the electric flux through each end of the cylinder is positive because the electric field lines pass outward through the cylinder. The total electric flux through the cylinder is thus

$$\Phi_E = 2EA$$

where the factor of two is needed to include the flux through both ends.

SOLVE

The total charge inside the Gaussian surface is equal to the charge per unit area on the sheet σ multiplied by the cross-sectional area A of the cylinder, so $q = \sigma A$. Inserting our results into Gauss's law leads to

$$\Phi_E = \frac{q}{\varepsilon_0}$$

$$2EA = \frac{\sigma A}{\varepsilon_0}$$

Solving for the electric field, we get

$$E = \boxed{\frac{\sigma}{2\varepsilon_0}}$$

► ***What does it mean?***

The magnitude of the electric field is constant, independent of distance from the plane. For this reason, a large, flat sheet of charge is a very convenient way to produce a uniform electric field. Such charged sheets are used often in practical devices such as the accelerator of the CRT in Figure 17.14.

EXAMPLE 17.8 Electric Field between Two Sheets of Charge

Two infinite planes have charge per unit area $+\sigma$ and $-\sigma$, respectively, as sketched in Figure 17.33A. Find the electric field at points B, C, and D.

RECOGNIZE THE PRINCIPLE

We used Gauss's law to find the field from a single infinite plane of charge in Example 17.7. We found that the field is perpendicular to the plane with magnitude $E = \sigma/2\varepsilon_0$ and is constant, independent of distance from the plane. Using the principle of superposition, we can apply that result to find the field due to two planes of charge.

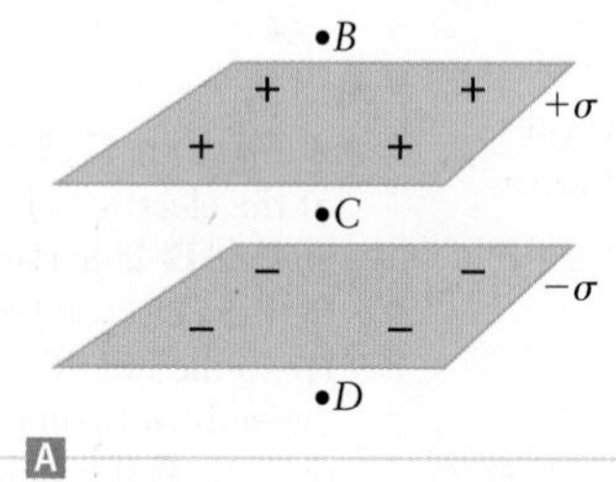

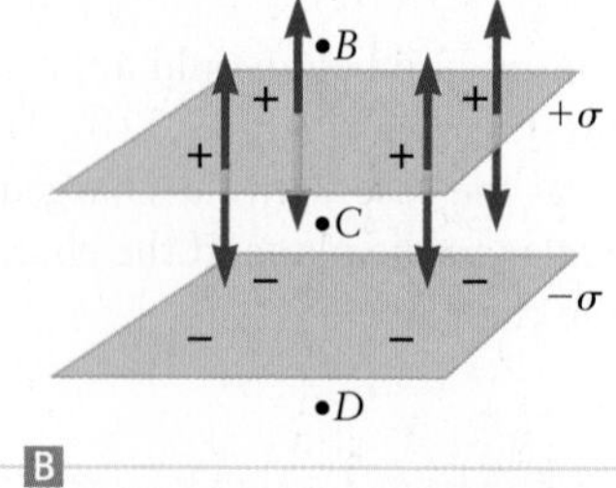

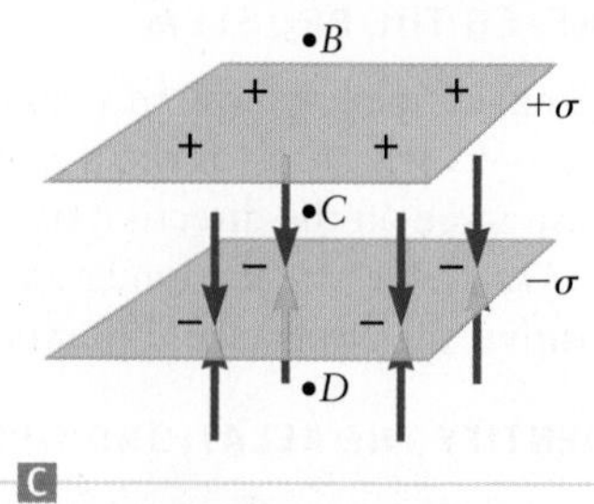

▲ **Figure 17.33** Example 17.8. A Two charged planes. For each plane, σ has the same magnitude, but the charges are of opposite sign. B Electric field due to the positively charged plane. C Electric field due to the negatively charged plane.

SKETCH THE PROBLEM

Parts B and C of Figure 17.33 show the electric fields produced by each plane. The field is directed away from the positively charged plane and toward the negatively charged plane.

IDENTIFY THE RELATIONSHIPS AND SOLVE

The total electric field at any point is the sum of the fields of the two planes. From Figure 17.33B and C we see that these two fields are in opposite directions at points B and D, so $\boxed{E_B \text{ and } E_D \text{ are both zero}}$. The fields of the two planes are in the same direction at point C, so the magnitude of E_C is twice the field from a single plane, resulting in $\boxed{E_C = \sigma/\varepsilon_0}$.

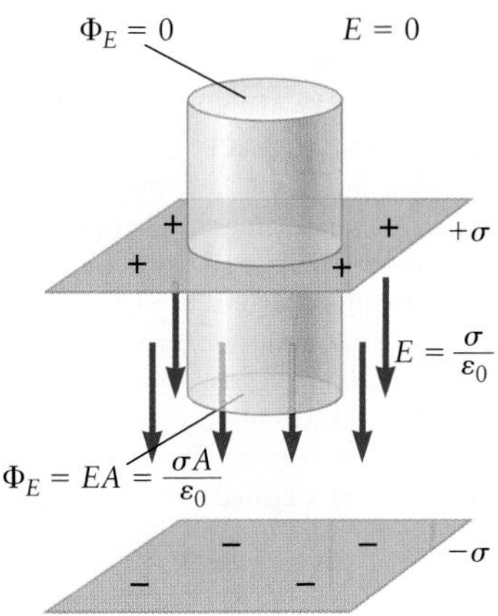

▲ **Figure 17.34** Example 17.8. Using Gauss's law to check the result for the electric field of two charged planes.

► ***What does it mean?***

We can verify our result by applying Gauss's law, $\Phi_E = q/\varepsilon_0$, to this problem using the cylindrical Gaussian surface in Figure 17.34, which also shows the field E_C at point C derived above. The fluxes through the sides of the cylinder and through the upper end are zero. The flux through the end between the planes is $\Phi_E = E_C A$, so the left-hand side of Gauss's law (Eq. 17.13) is $\Phi_E = (\sigma/\varepsilon_0)A$, where A is the area of one end of the cylinder. The charge enclosed in this Gaussian surface is $q = \sigma A$, so the right-hand side of Gauss's law is $(q/\varepsilon_0) = \sigma A/\varepsilon_0$ also. We have just shown that our result for the electric field does indeed satisfy Gauss's law.

Electric Field Near a Charged Metal Plate

A large sheet of charge, usually involving a charged metal plate, is used in many applications. We know that all the excess charge sits on the surface of a metal, so if a charge Q is placed on a thin metal plate, it will reside equally on the two flat surfaces as shown in Figure 17.35. The two surfaces thus both act as sheets of charge, and each electric field is described by the result in Example 17.7. Since these two sheets carry equal charge, their fields cancel in the region between them and the electric field inside the metal is zero (as stated in Section 17.4).

Some applications use two charged metal plates as sketched in Figure 17.36. In this case, the charges on the two plates attract each other and all the charge on each plate resides on the inner surface of the plate, with none on the outer surfaces, as shown in the figure. This arrangement of charges is very similar to the one analyzed in Example 17.8, and the results we found there for the electric field will also apply in Figure 17.36. If the plates in Figure 17.36 have charges $\pm Q$ and areas A, their charges per unit area are $\sigma = Q/A$ and the magnitude of the electric field produced by the inner surface of a single plate will be

$$E = \frac{\sigma}{2\varepsilon_0} \quad \text{(magnitude of electric field from a } \textit{single charged plane}\text{)} \tag{17.16}$$

The two inner surfaces both contribute to the total field in the region between the plates, so the electric field in that region will be twice as large, with

$$E = \frac{\sigma}{\varepsilon_0} \quad \text{(magnitude of electric field } \textit{between two charged plates}\text{)} \tag{17.17}$$

as found in Example 17.8. In terms of the charge Q on each plate, the electric field between the plates is

$$E = \frac{Q}{\varepsilon_0 A} \tag{17.18}$$

The electric field elsewhere, including inside the two plates, will be zero, again as shown in Example 17.8. We will use this result in Chapters 18 and 19 when we analyze the properties of capacitors.

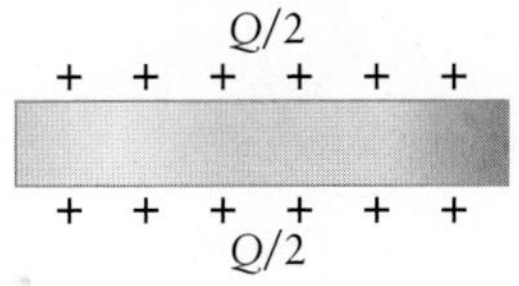

$Q/2$ + + + + + +

+ + + + + + $Q/2$

▲ **Figure 17.35** For a single thin metal plate, any excess charge is distributed evenly on the two sides.

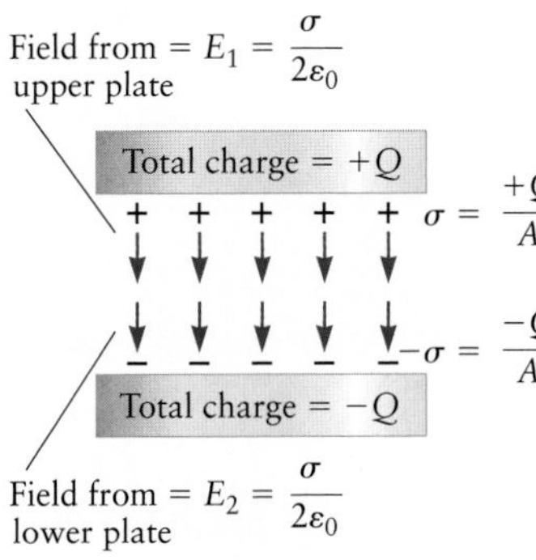

Field from upper plate $= E_1 = \frac{\sigma}{2\varepsilon_0}$

Total charge = $+Q$ $\quad \sigma = \frac{+Q}{A}$

$\sigma = \frac{-Q}{A}$ Total charge = $-Q$

Field from lower plate $= E_2 = \frac{\sigma}{2\varepsilon_0}$

▲ **Figure 17.36** Calculation of the electric field between two thin metal plates. This arrangement is called a parallel-plate capacitor.

Insight 17.4
GAUSS'S LAW APPLIES FOR *ANY* GAUSSIAN SURFACE
In applying Gauss's law, we have considered highly symmetric charge distributions. We then looked for Gaussian surfaces on which the electric field was either constant or zero. However, Gauss's law applies to *any* surface and *any* charge distribution, even ones with little or no symmetry. The total electric flux through a closed surface depends *only* on the total charge enclosed.

EXAMPLE 17.9 Electric Fields in a Smokestack Scrubber

In Example 17.1, we discussed a very simple model for a smokestack scrubber that removes undesirable particles by first adding some excess electrons and then using electric forces to pull the particles out of the air. Consider again a soot particle of mass $m_{soot} = 1.0 \times 10^{-12}$ kg that travels upward in a smokestack. The scrubber in Example 17.1 added 69 electrons to the particle, giving it a charge of $q_{soot} = 1.1 \times 10^{-17}$ C. Assume the electric field in the scrubber is produced by two parallel, square plates of width $L = 1.0$ m and separation $d = 0.010$ m, with charges $\pm Q$. (**a**) What must be the value of the electric field between the plates so that the force on the soot particle is equal to the weight of the particle? (A real scrubber would use a collection of many pairs of such plates in parallel.) (**b**) What charge Q on the scrubber's plates is required to produce the electric field in part (a)?

RECOGNIZE THE PRINCIPLE

We can find the magnitude of E by setting the electric force $q_{soot}E$ on a soot particle equal to its weight. We can then use our result for the electric field between two charged plates (Eq. 17.18) to relate the electric field to the charge $\pm Q$ on each plate.

SKETCH THE PROBLEM

The scrubber is shown in Figure 17.37. This arrangement is similar to the two metal plates in Figure 17.36, with $\vec{E}$ perpendicular to the plates.

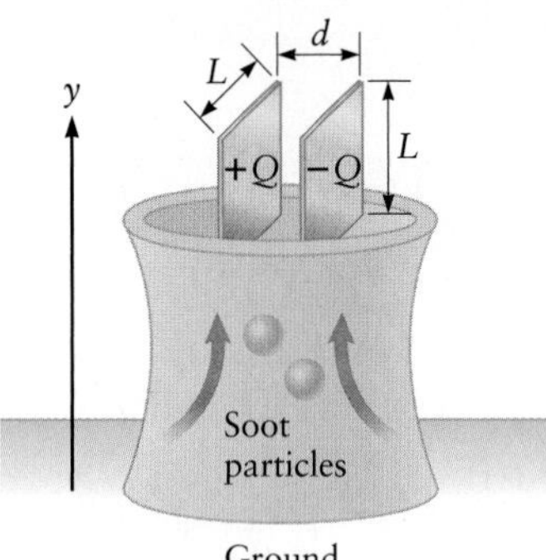

▲ **Figure 17.37** Example 17.9.

IDENTIFY THE RELATIONSHIPS

(**a**) The electric force on a soot particle is $F_{elec} = q_{soot}E$, where E is the magnitude of the field between the scrubber plates. This force is designed to equal the particle's weight, so we have

$$F_{elec} = q_{soot}E = m_{soot}g$$

SOLVE

Solving for E gives

$$E = \frac{m_{soot}g}{q_{soot}}$$

Inserting the mass and charge of the soot particle leads to

$$E = \frac{m_{soot}g}{q_{soot}} = \frac{(1.0 \times 10^{-12}\ \text{kg})(9.8\ \text{m/s}^2)}{1.1 \times 10^{-17}\ \text{C}} = \boxed{8.9 \times 10^5\ \text{N/C}}$$

► ***What does it mean?***
This electric field is quite large, but still less than the electric field at which air molecules are torn apart, an effect called dielectric breakdown (Chapter 18). Realistic scrubber designs do employ large electric fields.

IDENTIFY THE RELATIONSHIPS

(**b**) The electric field between two metal scrubber plates with charge $\pm Q$ is (Eq. 17.18)

$$E = \frac{Q}{\varepsilon_0 A}$$

For square plates with sides of length L, we have $A = L^2$ and

$$E = \frac{Q}{\varepsilon_0 L^2}$$

SOLVE

Solving for the charge Q, we find

$$Q = \varepsilon_0 EL^2$$

Inserting the given dimensions of our plates and the value of E from part (a) gives

$$Q = \varepsilon_0 EL^2 = [8.85 \times 10^{-12}\ \mathrm{C^2/(N \cdot m^2)}](8.9 \times 10^5\ \mathrm{N/C})(1.0\ \mathrm{m})^2$$

$$Q = \boxed{7.9 \times 10^{-6}\ \mathrm{C}}$$

▸ ***What does it mean?***

Our analysis of a scrubber has so far only considered the force and electric field that acts on the soot particle. We'll consider how to place the necessary charge Q on the plates when we discuss electric potential energy in Chapter 18.

17.6 Applications: DNA Fingerprinting

Electric forces and fields are used in many applications, some of which are very old. The smokestack scrubber used for pollution control in Examples 17.1 and 17.9 was invented about 100 years ago. Let's now consider a relatively new application, the technique of DNA fingerprinting, that is widely used in biology. The DNA molecules to be studied are first broken into fragments using enzymes (Fig. 17.38A; a job for biochemists). The fragments have a range of sizes and form ions in solution with typical charges between $-100e$ and $+100e$. Each specific DNA molecule forms its own unique set of fragments. The DNA fragments are next placed at one end of a long tube containing a gel (Fig. 17.38B). An electric field is then turned on in the gel tube; this field can be produced using two metal plates as we analyzed in Section 17.5. Since the DNA fragments are ions, they experience an electric force that pushes them through the gel. In Figure 17.38B, we assume the ions are negatively charged, so if the electric field is to the left, the force $\vec{F}_{\text{elec}} = q\vec{E}$ is to the right.

In addition to the electric force, there is also a drag force on the ions due to collisions with molecules in the gel. This drag force was described in Chapter 3; for a particle of radius r and velocity $\vec{v}$, the drag force is given by (Eq. 3.23)

$$\vec{F}_{\text{drag}} = -Cr\vec{v} \tag{17.19}$$

where C is a constant that depends on the properties of the gel. The DNA fragment ions move through the gel in response to the forces $\vec{F}_{\text{elec}}$ and $\vec{F}_{\text{drag}}$. At first, their velocity is very small, so the drag force is also small and the fragment ions accelerate. After a short time, their velocity increases to the point at which $\vec{F}_{\text{drag}}$ is equal in magnitude and opposite in direction relative to $\vec{F}_{\text{elec}}$. The total force is then zero, and the ions move at a constant velocity through the gel in a process very similar to a skydiver moving through the air at her terminal speed (Chapter 3). When an ion in the gel travels at its terminal speed, we thus have $\vec{F}_{\text{elec}} + \vec{F}_{\text{drag}} = 0$. Inserting our results for the components of these forces along the gel tube gives

$$qE - Crv = 0$$

Solving for the ion's speed, we find

$$Crv = qE$$

$$v = \frac{qE}{Cr} \tag{17.20}$$

The speed of the ion thus depends on its charge q and its size r, so different DNA fragments move through the gel at different speeds, with the smallest fragments moving fastest (assuming similar values of q). After some time, the fragments are

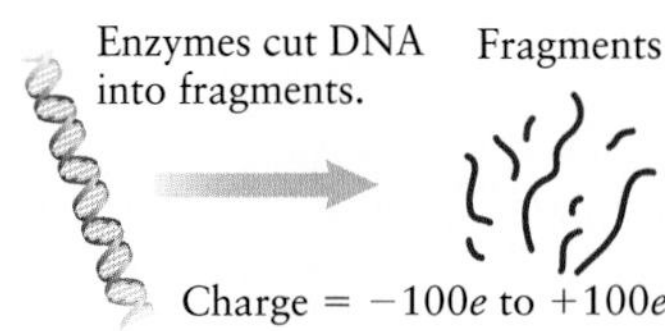

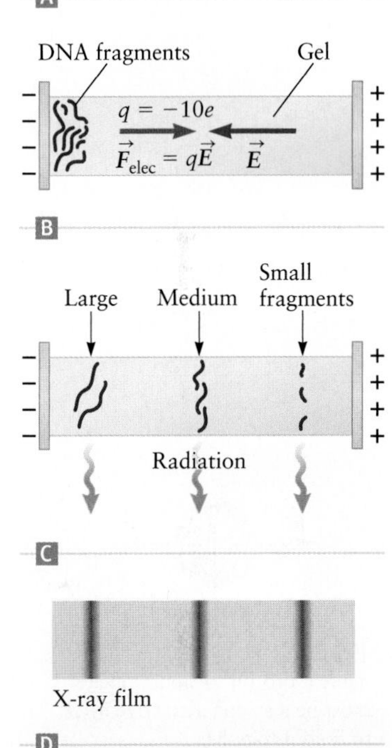

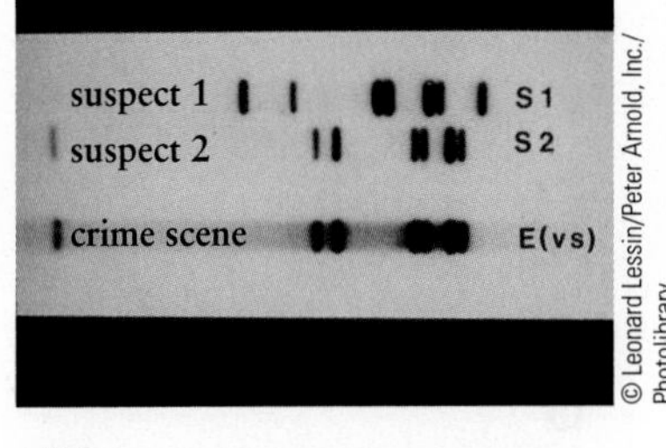

© Leonard Lessin/Peter Arnold, Inc./Photolibrary

▲ **Figure 17.38** A–D DNA fingerprinting uses the electric force on fragments of DNA to separate the fragments according to their size and charge. E In a forensic analysis, the DNA fingerprints of suspects are compared to a "known" sample (i.e., one collected at the crime scene). Here it appears that suspect 2 is guilty.

distributed along the gel tube as sketched in Figure 17.38C, and the fragments are thus separated according to their size.

The final step in this process is to record the fragment locations, which is done by attaching radioactive isotopes (see Chapter 30) to the DNA either before or after the fragments are made. These isotopes emit radiation that can be detected with a photographic detector or in other ways to record the position of the fragments within the gel tube (Fig. 17.38D).

The technique of separating ions according to their size by using electric forces to push the ions through a gel is called ***electrophoresis***. This method has become an important tool in biology and forensics (Fig. 17.38E).

17.7 "Why Is Charge Quantized?" and Other Deep Questions

In Section 17.1, we stated some "well-known" facts about electric charges. Let's now reexamine those facts in light of what we have learned about electric forces and fields.

How Are Protons Held Together?

We have mentioned many times that electric charge is carried by subatomic particles called protons and electrons. Indeed, we tend to take the existence of these particles for granted today, but what do they "look like" inside? Let's consider the proton first. We have seen that electric forces can be very large, and we know that like charges repel. If a proton is a small ball of positive charge as sketched in Figure 17.39A, the repulsive electric forces between different parts of a proton will tend to make these parts fly apart. How, then, is the proton held together?

Experiments have shown that a proton is actually an assembly of three particles called ***quarks*** as illustrated schematically in Figure 17.39B. The proton is composed of two particular types of quarks called "up" and "down." Each proton consists of two up quarks, each with charge $+\frac{2}{3}e$, and one down quark of charge $-\frac{1}{3}e$. The total charge of a proton is thus $\frac{2}{3}e + \frac{2}{3}e - \frac{1}{3}e = +e$. That a proton is composed of three separate quarks does not really answer our initial question about how protons are held together because electric forces between the quarks will still tend to push them apart, especially since the up quarks are both positively charged. The large repulsive electric forces between quarks are overcome by an even more powerful attractive force called the ***strong force***, which is a fundamental force of nature just as gravity and electromagnetism are fundamental forces. What's more, the strong force is so powerful that an individual quark is not able to escape from a proton under any circumstances. This is why we only see integral multiples of $\pm e$ in nature.

Asking simple questions about electric forces and Coulomb's law has thus led us to some deep questions about the fundamental structure of matter. The stability of matter relies on a force that prior to the early 1900s was not even known to exist! We'll say more about the strong force and quarks in Chapters 30 and 31.

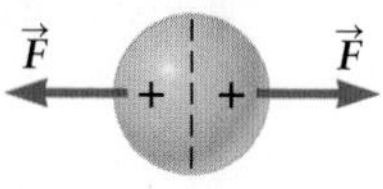

The Coulomb force tends to make a proton fly apart. There must be a strong attractive force to hold it together.

A

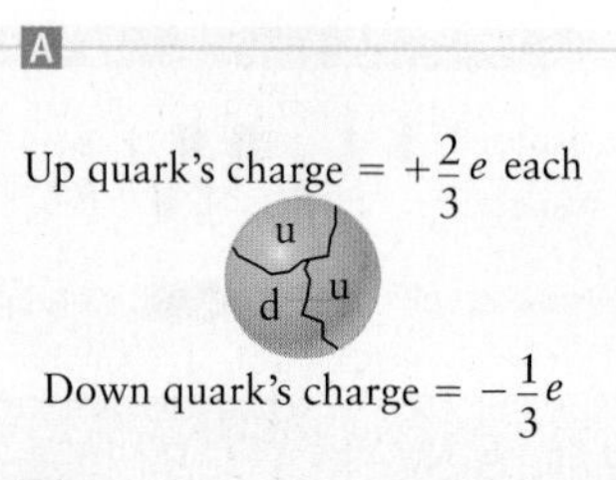

B

▲ **Figure 17.39** A A "ball" of electric charge will tend to fly apart due to the electric forces between different parts of the ball. If this ball is to be in mechanical equilibrium, there must be another force that overcomes the electric force and holds the ball together. B A proton is such a charged ball, made up of three quarks held together by the strong force.

Electrons Are Point Charges

We have just described the internal structure of the proton in terms of quarks. You might now expect that the electron has a similar internal structure, but nature has another surprise for us. As far as physicists can tell, the electron has *no* internal structure; the mass and charge of an electron are located at an infinitesimal "point" in space. Your everyday intuition will probably have difficulty accepting that a particle can be a true "point" in space. There is no way to explain it except to say that this is how nature seems to have designed the electron!

Conservation of Charge

Electric charge is conserved. That is, the total charge of the universe is constant. This does *not* mean that the total number of protons or electrons in the universe is constant. Indeed, in Section 7.8 we mentioned that in certain situations a neutron can decay to form a proton, an electron, and an antineutrino. (See also Chapter 31.) The neutron is neutral, so the charge before this decay is zero. The total charge of all the decay products is also zero since the antineutrino is neutral and the proton and electron carry charges of $+e$ and $-e$, respectively. Hence, the net charge is the same before and after the decay, even though the number of charged particles changes.

The principle of ***conservation of charge*** is believed to be an exact law of nature. No violations of this principle have ever been observed, but there is no fundamental understanding of why. Again, all we can say is that this is the way the laws of physics seem to operate.

Summary | CHAPTER 17

Key Concepts and Principles

Electric charge and Coulomb's law

There are two types of electric charge, positive and negative. Protons have a positive charge $+e$, and electrons carry a negative charge $-e$. Charge is measured in ***coulombs*** (C), with

$$e = 1.60 \times 10^{-19}\ \text{C} \qquad \text{(17.1) and (17.2) (page 485)}$$

The electric force between two point charges q_1 and q_2 is given by ***Coulomb's law.*** If the charges are separated by a distance r, this force has a magnitude

$$F = k\frac{q_1 q_2}{r^2} = \frac{q_1 q_2}{4\pi\varepsilon_0 r^2} \qquad \text{(17.3) and (17.5) (page 485)}$$

and is directed along the line joining the two charges.

The electric force is repulsive for like charges (charges with the same sign) and attractive for unlike charges (those of opposite sign). Here, k is a constant of nature with the value

$$k = 8.99 \times 10^{9}\ \text{N}\cdot\text{m}^2/\text{C}^2 \qquad \text{(17.4) (page 485)}$$

and ε_0 is a universal constant called the permittivity of free space, with the value

$$\varepsilon_0 = 8.85 \times 10^{-12}\ \frac{\text{C}^2}{\text{N}\cdot\text{m}^2} \qquad \text{(17.6) (page 485)}$$

Like charges repel

$\vec{F}$ q_2 r $\vec{F}$ q_1

$|\vec{F}| = \frac{kq_1q_2}{r^2} = \frac{q_1q_2}{4\pi\varepsilon_0 r^2}$

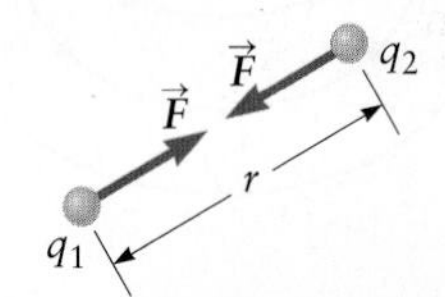

Unlike charges attract

Electric fields

The ***electric field*** $\vec{E}$ can be measured using a test charge q and is proportional to the electric force, with

$$\vec{F} = q\vec{E} \qquad \text{(17.10) (page 491)}$$

The electric field due to a point charge Q can be determined from Coulomb's law. Its magnitude is

$$E = \frac{kQ}{r^2} \qquad \text{(17.11) (page 492)}$$

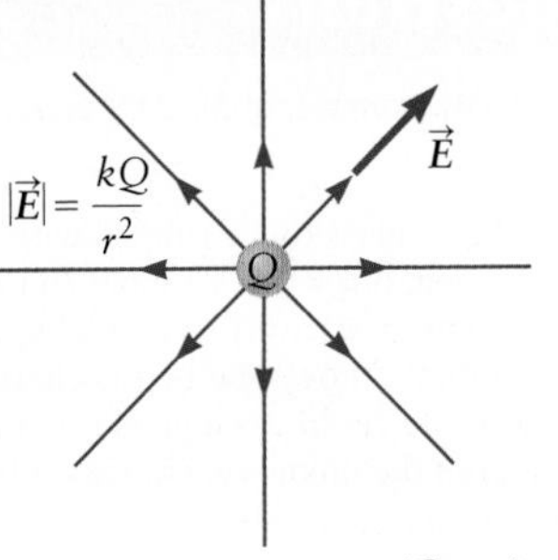

(Continued)

$\Phi_E = EA\cos\theta$

Electric flux and Gauss's law

When an electric field of magnitude E passes through a surface of area A while making an angle θ with the normal to the surface, the ***electric flux*** through the surface is

$$\Phi_E = EA\cos\theta \qquad \text{(17.12) (page 501)}$$

Gauss's law relates the electric flux Φ_E through a closed surface to the amount of electric charge inside the surface:

$$\Phi_E = \frac{q}{\varepsilon_0} \qquad \text{(17.13) (page 502)}$$

Gauss's law gives a powerful way to calculate the electric field in cases that are very symmetric. Coulomb's law can be derived from Gauss's law.

Conservation of charge

A fundamental law of nature is that electric charge is conserved. Charge is also quantized; it comes in discrete amounts, such as electrons and protons, which have charges $-e$ (electron) and $+e$ (proton).

Applications

Electric field lines near a dipole

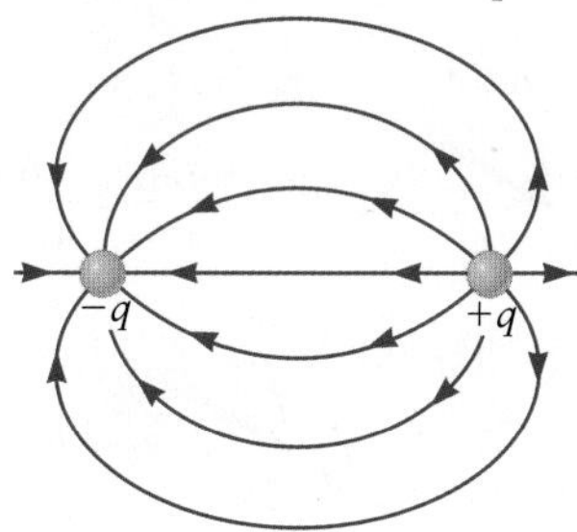

Principle of superposition

Coulomb's law gives the electric field due to a single point charge (Eq. 17.11). For a collection of point charges, the principle of superposition states that the total electric field is the vector sum of the fields produced by the individual charges.

Conductors and insulators

In a metal, some of the electrons are free to move, whereas in an insulator, there are no mobile charges. For a metal in static equilibrium, the electric field inside is zero and all excess charges reside on the metal's surface. The electric field inside an insulator need not be zero.

DNA fingerprinting

The technique of ***electrophoresis*** uses the electric force to separate ions according to their size and charge. This is the basis of DNA fingerprinting.

Questions

SSM = answer in *Student Companion & Problem-Solving Guide* ⓧ = life science application

1. When a glass rod is rubbed with silk, the rod becomes positively charged, but when a rubber rod is rubbed with fur, the rubber becomes negatively charged. Suppose you have a charged object but don't know whether its charge is positive or negative. Explain how you could use a glass rod and piece of silk to determine the sign of the unknown charge on the object.

2. Suppose (hypothetically) that two electric field lines crossed. What would that mean for a test charge placed at the crossing point? Use your result to explain why two electric field lines cannot cross.

3. Determine the sign of each charge q_a, q_b, and q_c in Figure Q17.3.

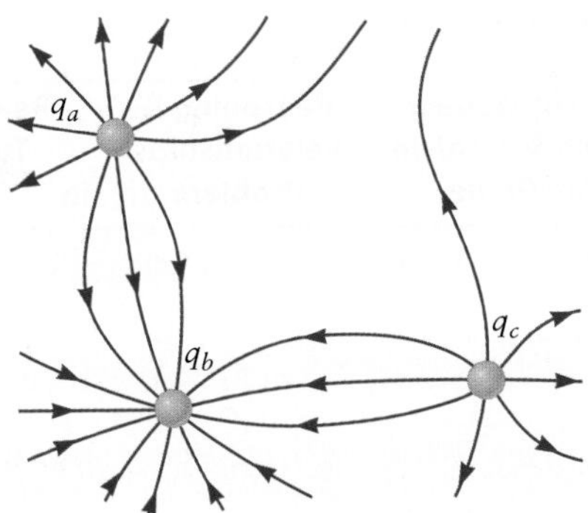

Figure Q17.3 Questions 3 and 4.

4. Consider the charges q_a, q_b, and q_c in Figure Q17.3. In terms of magnitude, which charge is the greatest? Which is the smallest? How do you know?

5. SSM Explain how two objects can be attracted due to an electric force, even when *one* object has *zero* net charge.

6. Explain how two objects can be attracted due to an electric force, even when *both* objects have *zero* net charge.

7. If a charged point particle is repelled from a glass rod that has been charged by rubbing with silk, what is the sign of the charge on the particle?

8. The children in Figure Q17.8 are rubbing balloons on their hair and then placing the balloons on the wall and ceiling. If the rubbing process puts excess electrons on the balloon, how does the balloon then stay attached to the ceiling or wall?

Figure Q17.8 Questions 8 and 14.

9. The end of a charged rubber rod will attract small pellets of Styrofoam that, having made contact with the rod, will move violently away from it. Describe why that happens.

10. Two point particles, each of charge Q_1, are located on the x axis at $x = \pm L$. Another particle having charge Q_2 is now placed at the origin. (a) Show that the total force on charge Q_2 is zero. (b) Suppose Q_2 is now moved a very small distance away from the origin, along the x axis. What is the direction of the electric force on Q_2? (c) Based on your result in part (b), explain why the origin is, or is not, a point of *stable* equilibrium for Q_2.

11. Under normal atmospheric conditions, the surface of the Earth is negatively charged. What is the direction of the electric field near the Earth's surface?

12. When two objects (such as a glass rod and a silk cloth) are rubbed together, electrons can be transferred from one to the other. Can protons also be transferred? Explain why or why not.

13. In Example 17.1, we mentioned that there are approximately 4×10^{14} atoms in a dust particle of mass 1 ng. Estimate this number for yourself.

14. Explain why the child's hair in Figure Q17.8 is "standing on end."

15. SSM Describe at least three ways in which Coulomb's law is similar to Newton's law of gravitation. Discuss at least two ways in which the two laws are different.

16. Explain why it is desirable that a test charge used to measure a field be small in comparison with the field's source charge.

17. A charge is placed inside a partially inflated balloon. If the balloon is then further inflated to a larger volume and the associated surface area increases, does the electric flux though the balloon's surface change? Why or why not?

18. The electron was not discovered until well after the laws of electricity were determined. Protons were discovered even later, so scientists such as Coulomb and Gauss did not know that electric charge is quantized. How do you think that affected their way of thinking about electric phenomena?

19. An ion is released from rest in a region in which the electric field is nonzero. If the ion moves in a direction antiparallel (opposite) to the direction of the electric field, is the ion positively charged or negatively charged?

20. Figure Q17.20 shows the electric field lines outside several Gaussian surfaces, but does not show the electric charges inside. In which cases is the net charge inside positive, negative, or perhaps zero? How do you know?

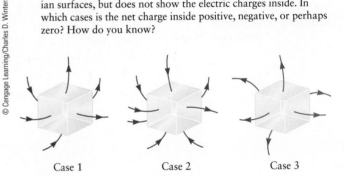

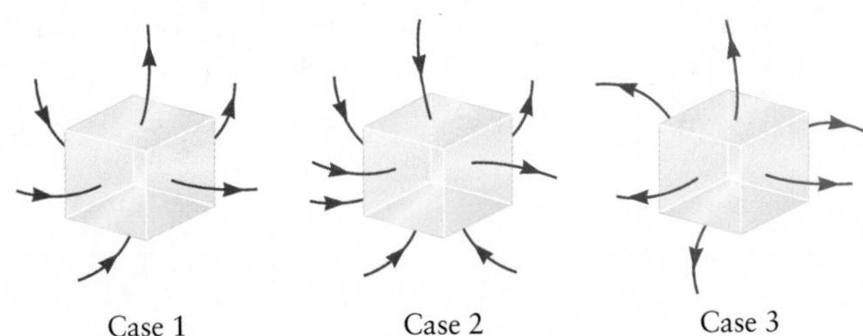

Figure Q17.20

Problems available in Enhanced WebAssign

Section 17.1 Evidence for Electric Forces: The Observational Facts
Problems 1–9

Section 17.2 Electric Forces and Coulomb's Law
Problems 10–34

Section 17.3 The Electric Field
Problems 35–51

Section 17.4 Conductors, Insulators, and the Motion of Electric Charge
Problems 52–54

Section 17.5 Electric Flux and Gauss's Law
Problems 55–73

Section 17.6 Applications: DNA Fingerprinting
Problems 74–76

Section 17.7 "Why Is Charge Quantized?" and Other Deep Questions
Problems 77–78

Additional Problems
Problems 79–98

List of Enhanced Problems

Problem number	Solution in *Student Companion & Problem-Solving Guide*	Reasoning & Relationships Problem	Reasoning Tutorial in Enhanced WebAssign
17.4	✓	✓	✓
17.6		✓	✓
17.15		✓	✓
17.18		✓	
17.22	✓		
17.23		✓	✓
17.31	✓		
17.34		✓	✓
17.49	✓		
17.50		✓	
17.53	✓		
17.60	✓		
17.76	✓	✓	
17.77	✓		
17.83		✓	✓
17.90	✓		

CHAPTER

18

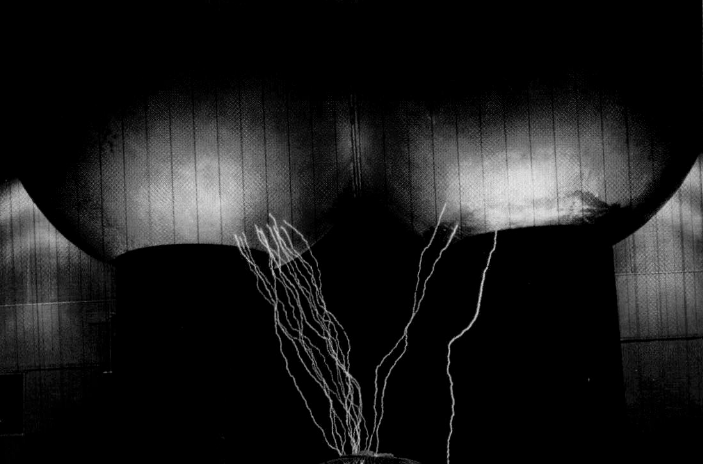

◀ *These sparks are one example of the motion of electric charges. This motion is determined by how the electric potential energy varies from place to place. (Peter Menzel/ Photo Researchers, Inc.)*

Electric Potential

In Chapter 17, we discussed electric forces and electric fields, and learned how to calculate them using Coulomb's law and Gauss's law. From Chapter 6, we know that when a force acts on an object, the force may do mechanical work on the object. It should thus be no surprise that electric forces can do work on a charged object. Electrical work is related to ***electric potential energy***, which is analogous in many ways to gravitational potential energy. Using electric potential energy, we will define a closely related quantity called the ***electric potential***. These quantities and the underlying ideas are quite similar to mechanical potential energy and will lead us again to the principle of conservation of energy that we explored in Chapter 6 with mechanical systems. Extending these ideas to electric forces and systems will give us a more complete picture of electric phenomena and provide the basis for our work on electric circuits in Chapter 19.

OUTLINE

When q is positive, the electric force $\vec{F}$ is parallel to $\vec{E}$.

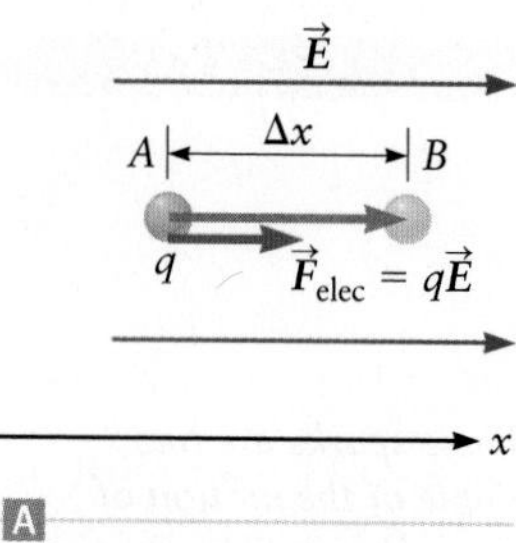

To move q from B to A, an external agent must exert a force $\vec{F}_{ext}$ to overcome the electric force.

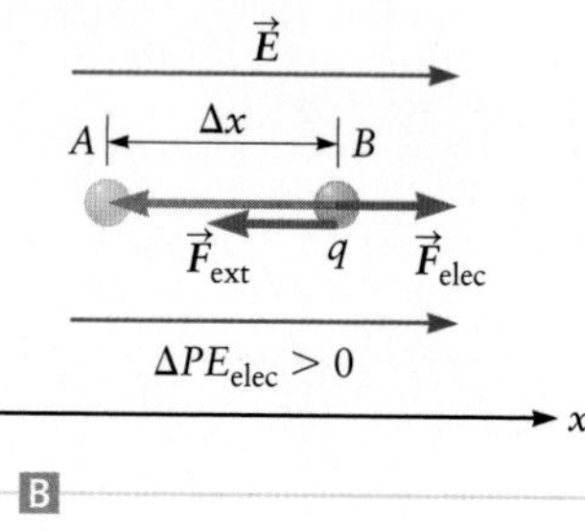

If the charge moves back from A to B, the system gives up potential energy.

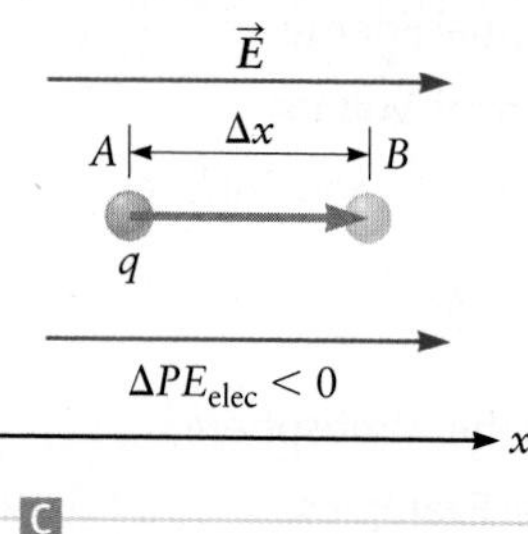

▲ **Figure 18.1** A The electric force on a charged particle can do work on the particle. B and C This work is connected with changes in the electric potential energy PE_{elec} of the particle. Here we assume a positively charged particle ($q > 0$).

18.1 Electric Potential Energy

In our discussion of force and Newton's laws of mechanics, we were led naturally to conservative forces and potential energy in Chapter 6. Many of the same ideas apply to electric forces and Coulomb's law, and our main job in this chapter is to introduce ***electric potential energy*** and the closely related concept of ***electric potential.***

Let's begin by reviewing the relationship between force and work. Figure 18.1A shows a region of space in which the electric field is constant, so $\vec{E}$ has the same magnitude and direction at all points. A point charge q in this region experiences an electric force

$$\vec{F}_{elec} = q\vec{E} \tag{18.1}$$

If the charge q is positive, this force is parallel to $\vec{E}$ as shown in Figure 18.1A.

Suppose this charge moves a distance Δx, starting at point A and ending up at point B, and for simplicity we assume this displacement is parallel to the electric force $\vec{F}_{elec}$. According to the definition of work in Chapter 6, the work done by the electric force on the charge is

$$W = F\,\Delta x \tag{18.2}$$

The electric force is conservative, so the work done on the charge is independent of the path it takes to go from A to B. We can now define the ***electric potential energy,*** which we denote by PE_{elec}. From Chapter 6, the change in potential energy associated with a particular conservative force is equal to $-W$, where W is the work done by that force (Eq. 6.10). So, if the electric force does an amount of work W on a charged particle, the change in electric potential energy is

$$\Delta PE_{elec} = -W \tag{18.3}$$

Combining this equation with Equations 18.1 and 18.2, the change in electric potential energy when the charged particle moves from A to B in Figure 18.1A is

$$\Delta PE_{elec} = -W = -F_{elec}\,\Delta x = -qE\,\Delta x \tag{18.4}$$

Equation 18.4 gives the *change* in potential energy as the charge moves through a displacement Δx, in a region where the electric field is parallel to the displacement. The change in potential energy depends on the starting and ending locations, but does *not* depend on the path taken. In Figure 18.1A, the displacement is along a line, but the charge may move from A to B along many other paths without affecting ΔPE_{elec}. All conservative forces, including the electric force, have this property: changes in the potential energy and the work done are independent of the path taken between two points. In the simple example shown in Figure 18.1A, the electric field is constant, but the same results hold when $\vec{E}$ varies with position. In this more general case, the potential energy function will have a different mathematical form than the one in Equation 18.4, but the electric force is still conservative.

Potential Energy Is Stored Energy

Electric potential energy is stored through the "potential" effect of the electric field on an electric charge. This effect is illustrated in Figure 18.1B, where the charge q is moved from point B to point A by a force $\vec{F}_{ext}$ directed to the left; this force is exerted by an external "agent," which might be your hand. The electric force on q is directed to the right (assuming q is positive), so the displacement is opposite to the electric force. The work done *by the electric field* on the particle is thus *negative,* and according to Equations 18.3 and 18.4 the change in the electric potential energy is *positive*. In words, we have stored a positive amount of energy in the system that is composed of the charge q and the electric field. That energy came from the external agent (and the external force $\vec{F}_{ext}$) that moved the charge from B to A.

The entire process can be reversed if we allow the charge to move from A back to B again (Fig. 18.1C). Now the electric field does a positive amount of work on the particle because the electric force and the particle's displacement are parallel and the change in the potential energy is negative (Eq. 18.4). Hence, stored energy is taken out of the system, and this energy might show up as an increase in the kinetic energy of the particle when it reaches B.

Potential Energy of Two Point Charges

Let's now consider the electric potential energy for the slightly more complicated case of two charges q_1 and q_2 in Figure 18.2A. From Coulomb's law, for two point charges separated by a distance r, the magnitude of the electric force exerted by q_1 on q_2 is

$$F = \frac{kq_1q_2}{r^2} \tag{18.5}$$

If we now assume they are like charges (i.e., both positive or both negative), the force is repulsive. The solid curve sketched in Figure 18.2B shows how F varies as a function of the separation r. If q_2 is brought closer to q_1, the electric force is directed opposite to the displacement (Fig. 18.2A), so the work done *by the electric field* on q_2 is negative. From our relation between work and potential energy (Eq. 18.3), we know that the electric potential energy *increases*, so $\Delta PE_{elec} > 0$. Mathematically, it can be shown that this electric potential energy is given by

$$PE_{elec} = \frac{kq_1q_2}{r} \tag{18.6}$$

We can also write PE_{elec} in terms of the permittivity of free space ε_0. From Chapter 17, $k = 1/(4\pi\varepsilon_0)$; using this expression in Equation 18.6 leads to

$$PE_{elec} = \frac{q_1q_2}{4\pi\varepsilon_0 r} \tag{18.7}$$

We can use either Equation 18.6 or 18.7 to calculate PE_{elec} for two point charges.

A plot of PE_{elec} is shown as the solid curve in Figure 18.2C. If the charges are initially separated by a distance r_i and then brought together to a final separation r_f, the change in the potential energy is

$$\Delta PE_{elec} = PE_{elec,\,f} - PE_{elec,\,i} = \frac{kq_1q_2}{r_f} - \frac{kq_1q_2}{r_i} \tag{18.8}$$

Only changes in potential energy are important, so we could add a constant value to PE_{elec} in Equation 18.6 without changing ΔPE_{elec}. Conventional practice, however, is to use Equation 18.6 without any added constant, so PE_{elec} approaches zero when the two charges are very far apart (i.e., when r becomes infinitely large). The electric force also approaches zero in this limit.

The general behavior of the electric potential energy for two like charges is shown as the solid curve in Figure 18.2C. In this case, PE_{elec} is positive and becomes larger and larger as the charges are brought together. On the other hand, if the charges have opposite sign, the electric force is attractive; that is, it is negative as plotted by the dashed curve in Figure 18.2B. The potential energy is then also negative (according to Eq. 18.6) and is shown by the dashed curve in Figure 18.2C.

If q_1 and q_2 are both positive, the electric force is repulsive and the work $W_E < 0$.

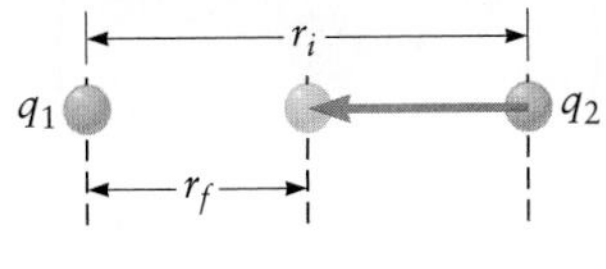

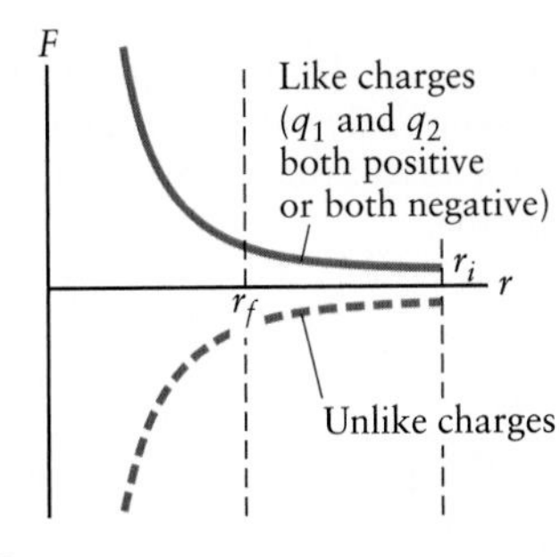

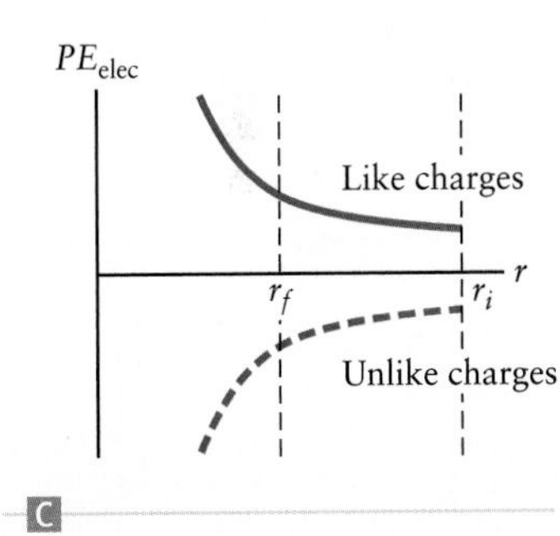

▲ **Figure 18.2** A Two point charges exert an electric force on each other. If the separation between the two charges changes, this force does work. B Plot of the force and C electric potential energy as functions of the separation between q_1 and q_2. In B, a negative value of F indicates an attractive force.

Insight 18.1

SIMILARITY BETWEEN ELECTRIC AND GRAVITATIONAL POTENTIAL ENERGY

The electric potential energy of two point charges (Eq. 18.6) falls to zero as the separation between the charges is made very large (infinite). This behavior is similar to that of the gravitational potential energy between two masses (Eq. 6.20), which also varies as $1/r$ and falls to zero when the two masses are very far apart.

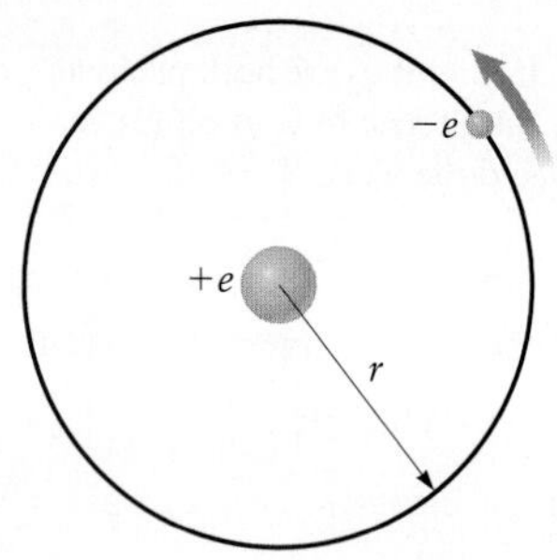

▲ **Figure 18.3** Example 18.1. A simplified model of an atom, with an electron in orbit around a proton.

EXAMPLE 18.1 Potential Energy of a Hydrogen Atom

A very simple model of a hydrogen atom is sketched in Figure 18.3, which shows an electron traveling in a circular orbit around a proton. (We'll discuss more realistic models of an atom in Chapter 29.) If the distance between the electron and the proton is $r = 5.0 \times 10^{-11}$ m, what is the electric potential energy of this atom?

RECOGNIZE THE PRINCIPLE

We can treat the electron and proton as two point charges. We know their charges and the separation r is given, so we can calculate the potential energy using Equation 18.6.

SKETCH THE PROBLEM

Figure 18.3 shows the problem.

IDENTIFY THE RELATIONSHIPS

The electric potential energy is given by Equation 18.6, with $q_1 = +e$ (the proton) and $q_2 = -e$ (the electron). We have

$$PE_{\text{elec}} = \frac{kq_1q_2}{r} = \frac{k(+e)(-e)}{r} = -\frac{ke^2}{r}$$

This potential energy is negative because the electron and proton have opposite charges. This energy holds the atom together and is related to its ionization energy, the energy required to separate an electron from the rest of the atom. In this case, it is the energy needed to move the electron an infinite distance from the proton.

SOLVE

Inserting the given value of r, we find

$$PE_{\text{elec}} = -\frac{ke^2}{r} = -\frac{(8.99 \times 10^9 \text{ N}\cdot\text{m}^2/\text{C}^2)(1.6 \times 10^{-19} \text{ C})^2}{5.0 \times 10^{-11} \text{ m}}$$

$$PE_{\text{elec}} = \boxed{-4.6 \times 10^{-18} \text{ J}}$$

► ***What does it mean?***

Our calculation has considered only the potential energy of the electron, but a better calculation of the ionization energy must also include the electron's kinetic energy (Section 29.1). Even so, let's compare our value of PE_{elec} with the measured ionization energy of hydrogen. It is conventional to quote positive values for the ionization energy, so our estimate is 4.6×10^{-18} J. The actual value for a hydrogen atom is approximately 2.2×10^{-18} J, so our simple calculation gives a value within about a factor of two of the correct answer.

Electric Potential Energy and Superposition

Our result for the electric potential energy of two point charges (Eq. 18.6) can be applied to other situations by using the principle of superposition. For example, if we have a collection of point charges, the total potential energy is the sum of the potential energies of each pair of charges. Note again that this electric potential energy arises from the interaction (the electric force) between two charges, so when we speak of potential energy we must keep in mind that it is the potential energy of a *system* of two particles.

The principle of superposition can be used to prove another important result. We know that the Coulomb's law force between two point charges is conservative, with a potential energy function given by Equation 18.6. More complicated charge distributions can always be treated as a collection of point charges arranged in some particular manner. We can therefore use superposition to conclude that electric forces between *any* conceivable collection of charges will always be conservative, no matter how complex the charge distribution or electric field may be.

Using the Principle of Conservation of Energy

In Chapter 6, we applied the principle of conservation of energy to solve various problems in mechanics. The approach we used there can be applied to problems involving electric potential energy.

PROBLEM SOLVING **Applying the Principle of Conservation of Energy to Problems with Electric Potential Energy**

1. **RECOGNIZE THE PRINCIPLE.** First determine the system whose energy is conserved. It might be two point charges, a collection of point charges, or a single charge in an electric field.
2. **SKETCH THE PROBLEM.** Draw a figure showing the initial and final states of all the charges in the system.
3. **IDENTIFY THE RELATIONSHIPS.** Identify the change in electric potential energy. For a charge in a constant electric field, you can find ΔPE_{elec} using $\Delta PE_{\text{elec}} = -qE\,\Delta x$ (Eq. 18.4), whereas for two point charges we have $PE_{\text{elec}} = kq_1q_2/r$ (Eq. 18.6).
4. **SOLVE.** The next step depends on the problem.
   - If there are no external forces on the system, the sum of the electric potential energy plus kinetic energy is conserved and $KE_i + PE_i = KE_f + PE_f$.
   - If there is a nonzero external force, the work done by that force will equal the change in potential plus kinetic energy.
   - The electric force is conservative, so the work done by the electric force is independent of the path and equals $-\Delta PE_{\text{elec}}$.
5. Always *consider what your answer means* and check that it makes sense.

We'll apply this general approach in the next two examples.

EXAMPLE 18.2 Moving Charges Around

Consider two point charges $+q$ and $-q$ separated by a distance $2r$ as shown in Figure 18.4. A third particle of charge Q is taken from rest very far away by an external force and brought to rest at the origin along the path shown in the figure. How much work is required to push or pull Q against the electric forces from the other charges?

RECOGNIZE THE PRINCIPLE

Start with step 1 in our strategy for applying the principle of conservation of energy to problems with electric potential energy. The charge Q moves in an electric field produced by the charges $\pm q$; these three charges are the system. The initial and final velocities of Q are zero, so its kinetic energy is zero at both the start and end. According to the principle of conservation of energy, the total work done on Q (the work done by the electric forces plus the work done by the external agent) must be zero. The work done by one is therefore the negative of the work done by the other. So, the work done on Q by the external agent that moves it to the origin must equal ΔPE_{elec}, the change in its electric potential energy.

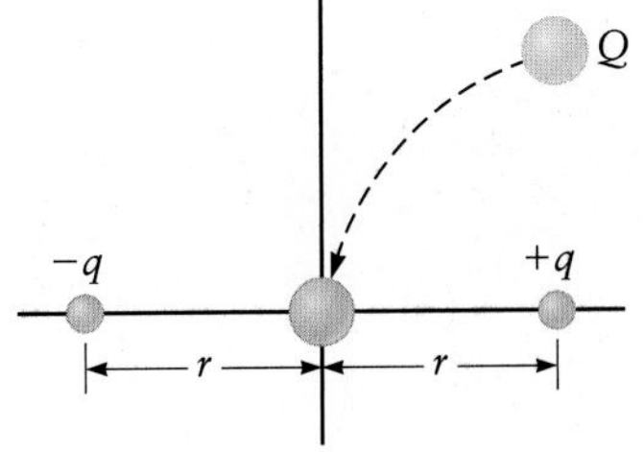

▲ **Figure 18.4** Example 18.2. A charge Q is taken from very far away and moved to the origin. Like the gravitational force, the electric force is conservative; for both forces, the work done is independent of the path taken.

SKETCH THE PROBLEM

Step 2 in our strategy is the sketch in Figure 18.4, which shows one possible path the charge Q might take to get to the origin.

(continued) ▶

IDENTIFY THE RELATIONSHIPS

Step 3: We use the result for the change in potential energy in Equation 18.8 twice, once for the pair of charges $+q$ and Q, and again for the pair $-q$ and Q. The charge Q begins very far from the others, so r_i is very large (infinite) in both cases, whereas the final location has $r_f = r$ for both pairs.

SOLVE

Step 4: Solve for the change in electric potential energy:

$$\Delta PE_{\text{elec}} = \frac{kqQ}{r} - \frac{kqQ}{r_i} + \frac{k(-q)Q}{r} - \frac{k(-q)Q}{r_i}$$

The terms involving r_i are each zero because r_i is infinite, whereas the other two terms (involving r) have opposite signs and cancel. Hence, $\Delta PE_{\text{elec}} = 0$, and the work done by the external agent that moves the charge to the origin is $\boxed{\text{zero}}$.

► *What does it mean?*

The work required to move Q to a point midway between two particles of charge $+q$ and $-q$ is zero because the two particles have opposite charge, making their contributions to the total potential energy cancel. Note also that ΔPE_{elec} is independent of the path taken by Q.

EXAMPLE 18.3 Moving Charges (Again)

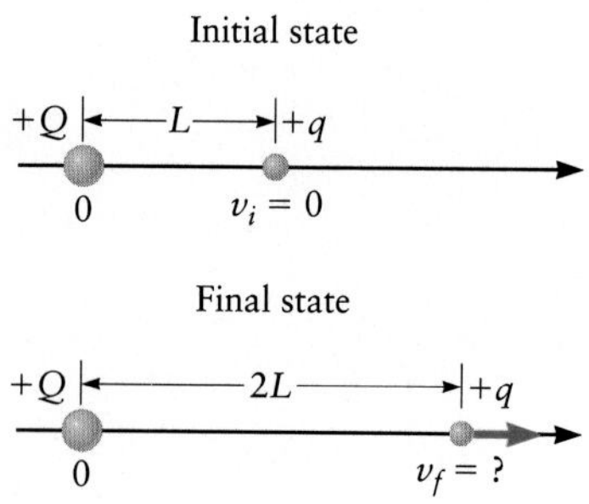

Initial state

$+Q$ —L— $+q$

0 $v_i = 0$

Final state

$+Q$ —$2L$— $+q$

0 $v_f = ?$

▲ **Figure 18.5** Example 18.3.

A point charge q with mass m is initially at rest, a distance L from a second charge Q (Fig. 18.5). Both charges are positive, so they repel each other. Charge Q is held fixed at the origin, whereas q is released and travels to the right. What is the speed of charge q when it reaches a distance $2L$ from the origin?

RECOGNIZE THE PRINCIPLE

We again follow our strategy for applying the principle of conservation of energy to problems with electric potential energy. Our system is just the two charges Q and q. Because there are no external forces on the system, its energy must be conserved. We can thus write

$$KE_i + PE_i = KE_f + PE_f \tag{1}$$

Because charge Q is held fixed, KE is the kinetic energy of charge q, and PE is the electric potential energy of the two charges.

SKETCH THE PROBLEM

Step 2: Figure 18.5 shows the problem, including both the initial and final states of the system.

IDENTIFY THE RELATIONSHIPS

Step 3: We can find the change in potential energy using $PE_{\text{elec}} = kqQ/r$ (Eq. 18.6). We can then calculate the change in kinetic energy and, from that, the final speed. Charge q is initially at rest, so the initial kinetic energy is $KE_i = 0$. The initial separation between the two charges is L, so the initial potential energy is (Eq. 18.6)

$$PE_i = PE_{\text{elec},\,i} = \frac{kqQ}{L}$$

The final separation is $2L$, so

$$PE_f = PE_{\text{elec},f} = \frac{kqQ}{2L}$$

Denoting the final speed of charge q by v_f, the final kinetic energy is

$$KE_f = \tfrac{1}{2}mv_f^2$$

SOLVE

Step 4: Inserting all this information in Equation (1) leads to

$$KE_i + PE_{\text{elec},i} = 0 + \frac{kqQ}{L} = KE_f + PE_{\text{elec},f} = \frac{1}{2}mv_f^2 + \frac{kqQ}{2L}$$

Solving for v_f gives

$$\frac{1}{2}mv_f^2 = \frac{kqQ}{L} - \frac{kqQ}{2L} = \frac{kqQ}{2L}$$

$$v_f^2 = \frac{2}{m}\left(\frac{kqQ}{2L}\right) = \frac{kqQ}{mL}$$

$$v_f = \boxed{\sqrt{\frac{kqQ}{mL}}}$$

▶ ***What does it mean?***

Electric forces are conservative, so we can use conservation of energy principles to analyze many situations involving electric forces and electric potential energy.

CONCEPT CHECK 18.1 **Moving a Charge from Place to Place**

A point charge Q is held fixed at the origin while an external force $\vec{F}_{ext}$ is used to move a second charge q from point A to point B along the three different paths shown in Figure 18.6. The charge q begins from rest and has zero velocity when it reaches B. Is the work done *by the external force* largest along (a) Path I, (b) Path II, or (c) Path III, or (d) is the work done by $\vec{F}_{ext}$ the same for all three paths?

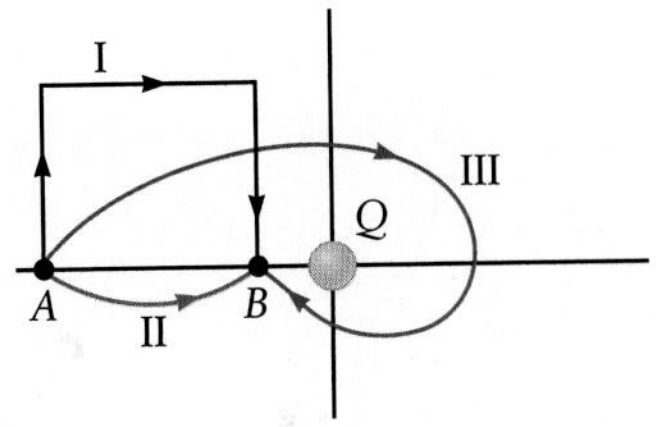

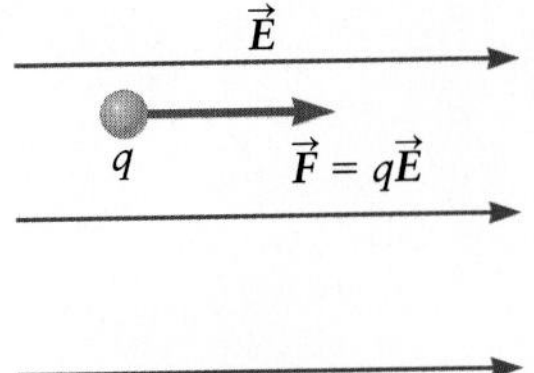

▲ **Figure 18.6** Concept Check 18.1.

18.2 Electric Potential: Voltage

In Chapter 17, we introduced the notion of a *test charge* in our definition of the electric field. We can measure the electric field at a particular location in space by placing a test charge q at the location of interest (Fig. 18.7) and then measuring the force on the test charge. The electric force is related to the electric field by $\vec{F} = q\vec{E}$, so we can deduce the field from the measured force using (Eq. 17.10)

$$\vec{E} = \frac{\vec{F}}{q} \tag{18.9}$$

▲ **Figure 18.7** The electric field at a particular point in space can be determined by measuring the force on a test charge q placed at that point. In a similar way, changes in the potential energy of a test charge as it is moved from place to place can be used to measure changes in the electric potential.

The value of this electric field does *not* depend on the test charge; the test charge merely gives us a convenient way to measure the field.

We now treat the electric potential energy in a similar manner. We place a test charge q at a particular location and measure the potential energy of this charge. We then define the ***electric potential*** V through the relation

Electric potential V

$$V = \frac{PE_{\text{elec}}}{q} \tag{18.10}$$

The electric potential is often referred to as simply "the potential." The SI unit of potential is the volt (V) in honor of Alessandro Volta, 1745–1827. According to Equation 18.10, the volt is related to other SI units by

$$\text{units of electric potential: } 1\text{ V} = 1\text{ J/C} = 1\text{ N}\cdot\text{m/C} \tag{18.11}$$

The volt is widely employed in the measurement of many electrical quantities. In particular, the electric field is usually measured in units of volts per meter. We thus have several different ways to express the units of the electric field. The ones most commonly used are

$$\text{units of electric field: } 1\text{ V/m} = 1\text{ N/C} \tag{18.12}$$

Using an Electric Potential Difference to Accelerate Charged Particles

Many of the important concepts associated with electric potential are at work in an old-fashioned television picture tube (a CRT; see Example 17.3). A CRT uses a beam of electrons to produce an image on a screen. As shown in Figure 18.8, this electron beam is produced by passing a stream of electrons through a region called the "accelerator," consisting of two parallel metal plates containing small holes that allow the electrons to enter and exit the central region. Some external electronic circuitry is attached to the metal plates, placing positive charge on one plate and negative charge on the other. We calculated the electric field between two parallel charged plates in Chapter 17 and found that for plates of a given charge and area this field is constant (Eq. 17.17).

Let's first analyze the case of a positive test charge moving through the accelerator shown in Figure 18.8A. Here the electric field is directed from left to right, which is the $+x$ direction. The electric field has magnitude E, and the plates are separated by a distance L. We imagine that the test charge q is initially placed at the entry hole on the left. The magnitude of the electric force on the test charge is $F = qE$, and the charge moves a distance L from left to right. The work done on the test charge by the electric field is thus $W = FL = qEL$. From Equation 18.4, the corresponding change in electric potential energy is

$$\Delta PE_{\text{elec}} = -qEL$$

From the definition of electric potential (Eq. 18.10), the change in the potential is

$$\Delta V = \frac{\Delta PE_{\text{elec}}}{q} = -EL \tag{18.13}$$

We say that there is a ***potential difference*** ΔV between the two plates. The electric field is directed to the right in Figure 18.8A, so the negative sign in Equation 18.13 means that the plate on the left is at a higher potential than the plate on the right.

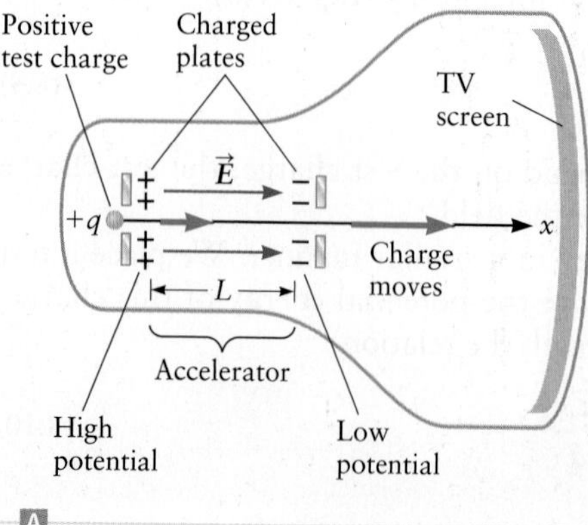

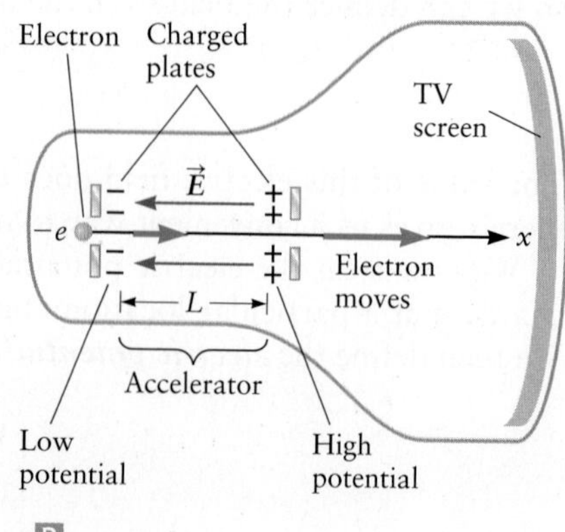

▶ **Figure 18.8** Electric field and potential inside a CRT. A Hypothetical arrangement using the electric field to accelerate a positive test charge $+q$. B Realistic CRT, with an electron moving through the accelerator from left to right.

Now we replace the test charge by an electron and consider the motion of the electron as it travels through the accelerator. Since an electron is negatively charged, we must reverse the direction of the electric field; the electric field is now negative because it is along the $-x$ direction in Figure 18.8B. This change is needed because we want the electric force on the electron to push it from left to right through the accelerator region. Because the electron is negatively charged, it will tend to move from the region of low potential (the plate on the left) to the region of high potential (the plate on the right) in Figure 18.8B.

Electrons enter the accelerator with a very small velocity, so for simplicity we'll assume their initial velocity is $v_i = 0$. We can use ΔV to find their final velocity v_f as they leave the accelerator. Conservation of energy tells us that

$$KE_i + PE_{\text{elec}, i} = KE_f + PE_{\text{elec}, f}$$

The initial kinetic energy is zero (since $v_i = 0$), so we have

$$KE_f = \tfrac{1}{2}mv_f^2 = PE_{\text{elec}, i} - PE_{\text{elec}, f} = -\Delta PE_{\text{elec}} \quad (18.14)$$

The change in potential energy is related to the change in electric potential. Using Equation 18.10 gives

$$\Delta PE_{\text{elec}} = q\,\Delta V = -e\,\Delta V$$

Inserting this result into Equation 18.14 leads to

$$\tfrac{1}{2}mv_f^2 = -\Delta PE_{\text{elec}} = e\,\Delta V \quad (18.15)$$

$$v_f = \sqrt{\frac{2e\,\Delta V}{m}}$$

The potential difference in a CRT is typically 15,000 V; inserting this value along with the values of the charge and mass of an electron, we find

$$v_f = \sqrt{\frac{2e\,\Delta V}{m}} = \sqrt{\frac{2(1.60 \times 10^{-19}\text{ C})(1.5 \times 10^4\text{ V})}{9.11 \times 10^{-31}\text{ kg}}} = 7.3 \times 10^7\text{ m/s}$$

So, these electrons are moving quite rapidly.[1] This high velocity and high kinetic energy enables the electrons to produce light when they strike the specially coated screen at the opposite face of the CRT.

Relating the Electric Potential to the Electric Field

In Figure 18.8, we assumed the electric field is constant in the accelerator region, which is a good assumption in that case. In many situations, though, the electric field varies with position, so let's now ask how the electric field and the electric potential are related in these more general cases. We start by considering a very small region of space so that the electric field is approximately constant and apply the approach sketched in Figure 18.9, where a test charge q moves a distance Δx parallel to $\vec{E}$. The force on this charge is $F = qE$, and the work done by the field on the charge is $W = F\,\Delta x = qE\,\Delta x$. The change in electric potential energy during this process is then (from Eq. 18.3)

$$\Delta PE_{\text{elec}} = -W = -qE\,\Delta x \quad (18.16)$$

Using Equation 18.10 leads to an electric potential difference

$$\Delta V = \frac{\Delta PE_{\text{elec}}}{q}$$

which gives

$$\Delta V = -E\,\Delta x \quad (18.17)$$

[1]Our result for v_f is about 25% of the speed of light. We show in Chapter 27 that Newton's mechanics begins to break down at such high speeds. The corrections can be calculated using the theory of special relativity, and they amount to a few percent for these electrons.

The electric field is proportional to the rate of change of V with position.

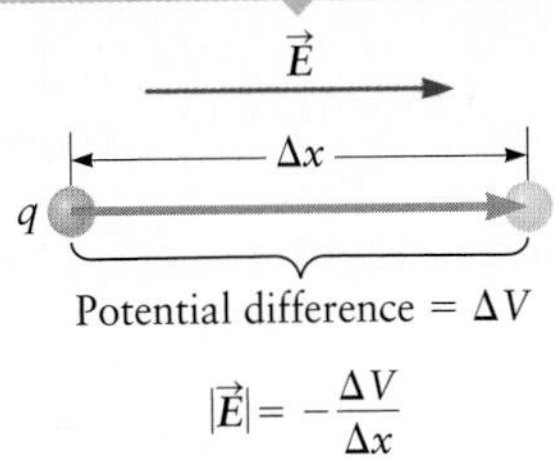

▲ **Figure 18.9** The magnitude and direction of the electric field are related to how the electric potential V changes with position.

Equation 18.17 is a general relation between the electric field E and the variation in electric potential V with position. Strictly speaking, this relation only holds for small steps Δx during which the field is constant. We can, however, combine the potential changes ΔV from many such small steps to find the change in the potential over a large distance. We can also use Equation 18.17 to deduce the electric field from knowledge of how the potential changes with position. Rearranging that equation gives

Relation between E and changes in V with position

$$E = -\frac{\Delta V}{\Delta x} \tag{18.18}$$

In words, this says that the magnitude of the electric field is largest in regions where V is changing rapidly (and ΔV is large). Conversely, the electric field is zero in regions where V is constant. Notice that because of the negative sign in Equation 18.18 the electric field is directed from regions of high potential to regions of low potential.

The relation between the electric field and changes of the potential in Equation 18.18 involves the component of the electric field in the direction parallel to the displacement Δx. The electric field is a vector, so if we want to find E in a particular direction, we must consider how the potential V changes along that direction.

We now have a relation between the electric field and the variation of potential V with position. This relation holds in simple cases when the electric field is constant as in the CRT in Figure 18.8 and in more complicated cases in which the electric field varies with position.

EXAMPLE 18.4 Electric Field in a CRT

Consider the accelerator region of the CRT in Figure 18.8. If the potential between the two plates is 15,000 V and the plate separation is $L = 2.0$ cm, what is the electric field in this region? Assume the electric field is constant in the accelerator.

RECOGNIZE THE PRINCIPLE

We can find the electric field using the relation between the electric potential and E in Equation 18.18,

$$E = -\frac{\Delta V}{\Delta x} \tag{1}$$

SKETCH THE PROBLEM

Figure 18.10 shows the problem. Because the electron has a negative charge, it moves from left to right, from low to high electric potential. The corresponding charge on the plates is also shown.

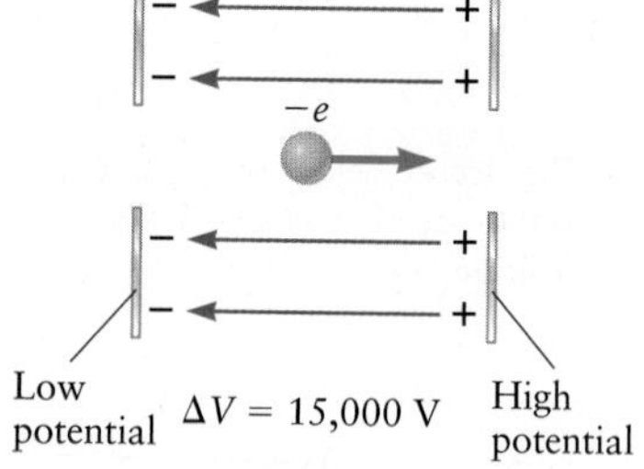

$\vec{E}$

$-e$

Low potential $\Delta V = 15{,}000$ V High potential

▲ **Figure 18.10** Example 18.4.

IDENTIFY THE RELATIONSHIPS AND SOLVE

The electric field is constant in the accelerator, so we can take ΔV to be the total potential difference, and Δx is the distance between the plates; we thus have $\Delta V = 15{,}000$ V and $\Delta x = L = 2.0$ cm. Inserting these values in Equation (1), we find

$$E = -\frac{\Delta V}{\Delta x} = -\frac{1.5 \times 10^4 \text{ V}}{0.020 \text{ m}} = \boxed{-7.5 \times 10^5 \text{ V/m}}$$

▶ ***What does it mean?***

The negative sign indicates that the field is directed to the left in Figure 18.10, from high to low potential. The electric field vector always points from regions of high V to low V because $\vec{E}$ is parallel to the direction a positive test charge would move if it were placed at that location.

EXAMPLE 18.5 Measuring the Charge on an Electron

Around 1900, Robert Millikan began an experiment to measure the charge of an individual electron. The experiment involved small charged oil drops that were "suspended" by the electric field between two metal plates (Fig. 18.11). Millikan adjusted the potential between the two plates so that an individual drop was held stationary. On occasion, an electron would leave or join the drop, changing the electric force and causing the drop to move. By studying the motion of these drops, Millikan was able to show that the drop's charge was quantized by amounts $\pm e$, and he was able to deduce the value of e. Suppose a drop of radius 10 μm had a mass of about 2.0×10^{-12} kg and was held stationary between two plates a distance 0.50 mm apart. If the potential difference between the plates was $\Delta V = 1000$ V, how many excess electrons were on the drop?

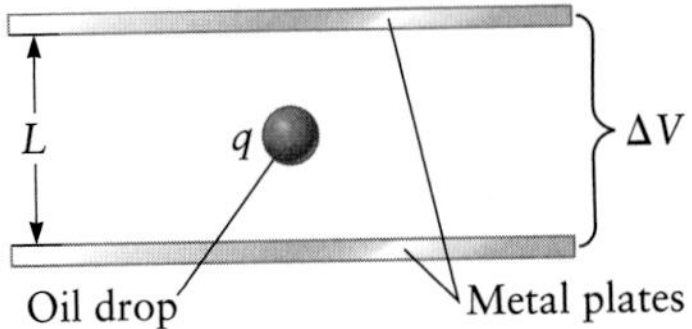

▲ **Figure 18.11** Example 18.5. Millikan's oil drop experiment.

RECOGNIZE THE PRINCIPLE

We are given the potential across the plates and their separation, so we can find the electric field E between the plates. If the charge on the drop is q, the electric force on it will have a magnitude qE. Setting this equal (in magnitude) to the gravitational force on a drop will allow us to find q and hence the number of electrons on the drop.

SKETCH THE PROBLEM

Figure 18.11 shows the problem.

IDENTIFY THE RELATIONSHIPS

The magnitude of the electric field between the plates is related to the plate separation L by $E = |\Delta V|/L$ (Eq. 18.18), so the magnitude of the electric force on the drop is

$$F_E = qE = \frac{q|\Delta V|}{L} \quad (1)$$

The magnitude of the gravitational force on the drop is

$$F_g = mg \quad (2)$$

SOLVE

Setting $F_E = F_g$ and solving for q leads to

$$\frac{q|\Delta V|}{L} = mg$$

$$q = \frac{mgL}{|\Delta V|}$$

The number of electrons on the drop will be $N = q/e$. Inserting the given values, we find

$$N = \frac{q}{e} = \frac{mgL}{e|\Delta V|} = \frac{(2.0 \times 10^{-12}\text{ kg})(9.8\text{ m/s}^2)(5.0 \times 10^{-4}\text{ m})}{(1.6 \times 10^{-19}\text{ C})(1000\text{ V})}$$

$$N = \boxed{61\text{ electrons}}$$

▶ ***What does it mean?***

It is amazing that the charge of only 61 electrons gives a strong enough force to suspend an oil drop that is nearly visible to the naked eye, again illustrating that the electric force is quite strong when compared to the force of gravity. It is also interesting that this experiment was carried out only about 100 years ago. Before that time, scientists did not even know the charge on an electron!

The Electron-volt as a Unit of Energy

We often deal with the motion of electrons or protons, both of which have a charge of magnitude e. Also, we are often concerned with the energy gained or lost as an electron or ion moves through a potential difference measured in volts as with the example in Figure 18.10. For these reasons, it is convenient to define a unit of energy called the ***electron-volt***, the energy gained or lost when an electron travels through a potential difference of 1 V. The electron-volt is abbreviated as "eV"; it is not part of the SI system of units, but we can relate the electron-volt to the joule. Using Equation 18.10, we have

$$V = \frac{PE_{\text{elec}}}{q}$$

which can be rearranged as

$$PE_{\text{elec}} = qV$$

In words, this equation reads

$$\text{energy} = (\text{charge})(\text{potential})$$

Applying this definition to that of the electron volt as a unit of energy, we find

$$1 \text{ eV} = e\,\Delta V = (1.60 \times 10^{-19} \text{ C})(1 \text{ V})$$

$$1 \text{ eV} = 1.60 \times 10^{-19} \text{ J} \tag{18.19}$$

In Example 18.4 with the CRT, the electrons gained a kinetic energy $e\,\Delta V$ = 15,000 eV as they traveled through the accelerator.

> **CONCEPT CHECK 18.2** Electron-volts and Potential Changes
>
> In Figure 18.10, an electron moved from left to right through a region in which the electric potential changed by ΔV = 15,000 V. The corresponding change in potential energy was $\Delta PE_{\text{elec}} = -e\,\Delta V = -15{,}000$ eV. Suppose the electron is replaced by an oxygen ion O^{2-}. What is the change in potential energy of the oxygen ion as measured in electron-volts, (a) −7500 eV, (b) −15,000 eV, or (c) −30,000 eV?

Electric Potential Due to a Point Charge

In Equation 18.6, we found the electric potential energy of two point charges. We can combine that with the definition of electric potential (Eq. 18.10) to obtain the electric potential a distance r away from a single point charge q:

$$V = \frac{kq}{r} \tag{18.20}$$

Because the constants k and ε_0 are related by $k = 1/(4\pi\varepsilon_0)$, we can also write the potential due to a point charge as

$$V = \frac{q}{4\pi\varepsilon_0 r} \tag{18.21}$$

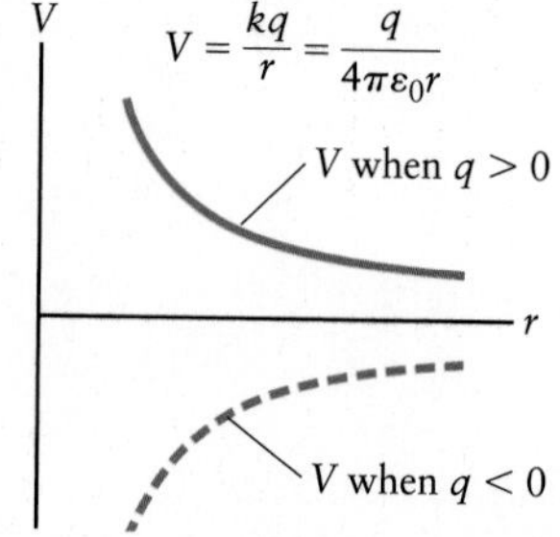

▲ **Figure 18.12** Electric potential near a point charge. The solid curve is for a positive point charge, and the dashed curve is for a negative point charge.

We show this result as the solid curve in Figure 18.12 for a positive charge ($q > 0$) and as the dashed curve for a negative charge ($q < 0$).

These results reinforce what we learned about potential energy and the behavior of like and unlike charges (Fig. 18.2C). Suppose a charge $q > 0$ is placed at the origin; we'll call it a "source charge" because it produces a potential given by Equations 18.20 and 18.21. We then place a positive test charge nearby. Because both the source charge and the test charge are positive, they repel, and the test charge experiences a force that carries it "downhill" along the solid potential curve in Figure 18.12, away from the origin. On the other hand, if the source charge is negative, the potential is described by the dashed curve in Figure 18.12. A positive test charge would be attracted to this source charge and carried "downhill" along the dashed potential curve.

Only Changes in Electric Potential Matter

Only *changes* in potential energy are important, so we must always be clear about where the "reference point" for potential energy is chosen. Because electric potential is proportional to electric potential energy, we must also pay attention to the reference point for electric potential. When dealing with a point charge and Equation 18.20, the standard convention is to choose $V = 0$ at an infinite distance from the source charge, that is, when $r = \infty$. In many other problems, the usual choice is to take the Earth as $V = 0$ because (from Chapter 17) the Earth is a good conductor of charge. This is the origin of the term ***electrical ground*** or, simply, ***ground*** and the convention that the ground is a point where $V = 0$.

Electric Potential and Field Near a Metal

Consider a solid metal sphere as in Figure 18.13A that carries an excess positive charge q, with the charge at rest. To understand the behavior of the electric potential and field, recall from Chapter 17 that the excess electric charge on a metal resides at the surface and the field inside the metal is zero. What about the electric field outside the sphere? Since the metal is spherical, the field must have spherical symmetry; hence, E outside the sphere must depend only on the distance r from the center of the sphere. We showed in Example 17.6 that the field outside any spherical ball of charge is given by

$$E = \frac{kq}{r^2} = \frac{q}{4\pi\varepsilon_0 r^2} \tag{18.22}$$

where q is the total charge and r is the distance from the center of the ball. This result is precisely the same as that for the electric field due to a point charge. Equation 18.22 holds only outside the metal sphere; the field inside is zero. The magnitude of the field, both inside and outside, is plotted in Figure 18.13B as a function of the distance r from the center of the sphere.

The field outside the metal sphere is identical to the field of a point charge, so the potential outside the sphere is the same as the potential due to a point charge. We thus have (from Eq. 18.20)

$$V = \frac{kq}{r} \quad \text{for } r > r_{\text{sphere}} \tag{18.23}$$

The field inside the sphere is zero, so a test charge placed inside the metal would experience no force. The work done in moving a test charge around inside the metal is thus also zero. From the relation between work and electric potential energy (Eq. 18.3), this means that both the electric potential energy and the electric potential are *constant inside a metal*. Neither the potential energy nor the electric potential can jump discontinuously (see Insight 18.2), so the potential at all locations in the metal sphere must equal the potential at the outer edge of the sphere, which we can obtain from Equation 18.23. Hence,

$$V = \frac{kq}{r_{\text{sphere}}} \quad \text{everywhere inside a metal sphere} \tag{18.24}$$

Figure 18.13C shows the electric potential as a function of distance r from the sphere's center.

This example reviews and illustrates several important points. (1) The excess charge on a metal in equilibrium always resides at the surface, as we established in Chapter 17. (2) Because the electric field inside a metal is zero, the potential is constant throughout a piece of metal. (3) The electric field lines approach perpendicular to the surface of the metal sphere, Figure 18.13A, as required by the spherical

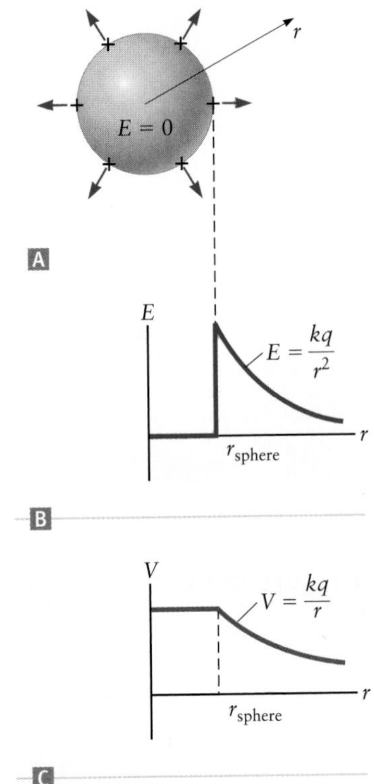

▲ **Figure 18.13** A Electric field lines near a charged metal sphere. B Magnitude of the electric field as a function of the distance r from the center of the sphere. C Electric potential as a function of r.

symmetry of this electric field. Moreover, electric field lines *always* approach perpendicular to a metal surface of any shape, spherical or not. Why? Suppose an electric field at the surface of a metal somehow had a component parallel to the surface. This parallel field component would produce a force on any charge at the surface, causing it to move (because excess charge moves very easily in a conductor). This statement, however, contradicts our assumption that all the excess charge is in static equilibrium. Thus, for a piece of metal with any shape, electric field lines near a metal in equilibrium must approach perpendicular to the metal's surface.

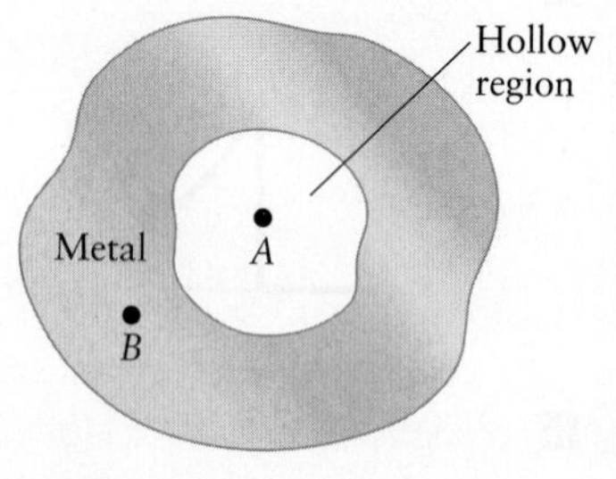

▲ **Figure 18.14** Concept Check 18.3.

CONCEPT CHECK 18.3 Electric Field and Potential Inside a Metal

Consider a metal shell with a hollow region inside (Fig. 18.14). If the shell has a total charge Q, which of the following statements is true? (More than one statement may be correct.)

(a) The electric field is zero inside the hollow part of the shell (e.g., at point A).
(b) The electric potential is zero inside the hollow part of the shell (e.g., at point A).
(c) The electric field is zero in the metal part of the shell (e.g., at point B).
(d) The electric potential is zero in the metal part of the shell (e.g., at point B).

Electric Field Near a Lightning Rod

The stories about Benjamin Franklin's experiments with kites during thunderstorms are famous (even though we don't know for sure if these stories are true!). Franklin understood the connection between lightning and electricity and also showed the value of attaching a metal rod, called a lightning rod (which he also invented), to the top of a building as in Figure 18.15A. When lightning strikes the rod, a large amount of charge (electrons) moves onto the rod; the rod is connected to the ground by a metal wire, allowing this charge to move safely to a region of low potential (the ground) through the metal wire rather than travel destructively through other portions of the building. Let's now consider why a lightning strike is "attracted" to a lightning rod.

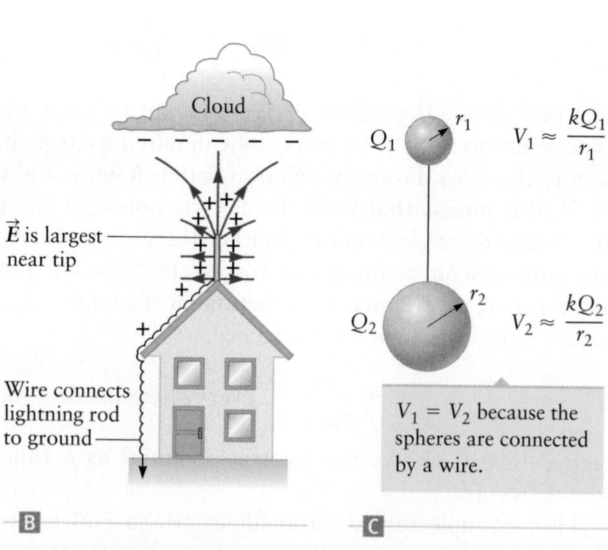

▲ **Figure 18.15** A Lightning rod on the roof of a building. B For the lightning rod to function properly, a wire must connect the rod to the ground. C A simplified model of a lightning rod. The small metal sphere represents the sharp tip of the rod.

Figure 18.15B gives a sketch of the electric field lines near a positively charged lightning rod. The field is largest near the sharp tip of the rod and smaller near the flat sides. That is another general property of the electric field near a metal: E is always largest near sharp metal corners and smaller near flat, metal surfaces. We can explain this by applying Equation 18.23 for the electric potential near a metal sphere to the simplified model of a lightning rod shown in Figure 18.15C. Here the rod is modeled as two metal spheres connected by a metal wire. The smaller sphere (radius r_1) represents the tip of the lightning rod, and the larger sphere (radius r_2) represents the flatter body of the rod. We also assume a net positive charge on the system. A portion Q_1 of this charge sits on the surface of the smaller sphere, and the rest of the charge Q_2 sits on the large sphere. The potential due to each sphere can be obtained from Equation 18.23. Strictly speaking, the result in Equation 18.23 applies only to a single sphere far from any other charges. However, if the two spherical parts of our model lightning rod are not too close together, we can use Equation 18.23 to calculate the potential of each sphere. Because they are connected by a wire, the two spheres are part of a single piece of metal and must therefore have the same electric potential. Applying Equation 18.23 gives

$$V(\text{sphere 1}) = \frac{kQ_1}{r_1} = V(\text{sphere 2}) = \frac{kQ_2}{r_2}$$

Canceling common factors, we find that the charges on the two spheres are related by

$$\frac{Q_1}{r_1} = \frac{Q_2}{r_2} \tag{18.25}$$

We can use this equation to find the relative magnitudes of the electric fields produced by the two spheres. The electric field outside each sphere is given by $E = kQ/r^2$ (Eq. 18.22). For the field just outside sphere 1,

$$E_1 = \frac{kQ_1}{r_1^2}$$

and just outside sphere 2, we get

$$E_2 = \frac{kQ_2}{r_2^2}$$

The ratio of these fields is thus

$$\frac{E_1}{E_2} = \frac{kQ_1/r_1^2}{kQ_2/r_2^2} = \frac{Q_1 r_2^2}{Q_2 r_1^2}$$

Inserting the ratio of the charges from Equation 18.25, we find

$$\frac{E_1}{E_2} = \frac{Q_1}{Q_2}\frac{r_2^2}{r_1^2} = \frac{r_1}{r_2}\frac{r_2^2}{r_1^2}$$

$$\frac{E(\text{small sphere})}{E(\text{large sphere})} = \frac{E_1}{E_2} = \frac{r_2}{r_1} \tag{18.26}$$

Hence, the field is *larger* near the surface of the *smaller* sphere (because $r_2 > r_1$), and the field is largest near the sharp edges of a charged piece of metal.

This result explains how a lightning rod works. The large electric field near the tip causes nearby air molecules to be ionized through an effect known as ***dielectric breakdown*** (discussed in more detail in Section 18.5). The resulting mobile electrons and ions in the air near the tip are then able to carry charge from the atmosphere (the lightning) to the rod. The key point is that dielectric breakdown occurs preferentially near the sharp tip of the lighting rod, so the large current from a lightning bolt is carried only to the rod and not to the rest of the building.

Insight 18.2

NEITHER THE POTENTIAL ENERGY NOR THE ELECTRIC POTENTIAL CAN JUMP DISCONTINUOUSLY

We can combine Equations 18.2 and 18.3 to get a relation between a force and the corresponding potential energy. When moving a particle between two points A and B that are a distance Δx apart, we have

$$F\,\Delta x = W = -\Delta PE$$

Solving for F, we get

$$F = -\frac{\Delta PE}{\Delta x}$$

If the two points are close together, Δx is very small. The change in potential energy ΔPE must therefore also be very small; otherwise, the force would be extremely large. This argument applies as the displacement becomes infinitesimal (in the limit $\Delta x \rightarrow 0$) and means that the electric potential energy cannot jump discontinuously when a particle moves between two points A and B an infinitesimal distance apart. Since electric potential is proportional to electric potential energy (Eq. 18.10), the electric potential also cannot jump discontinuously.

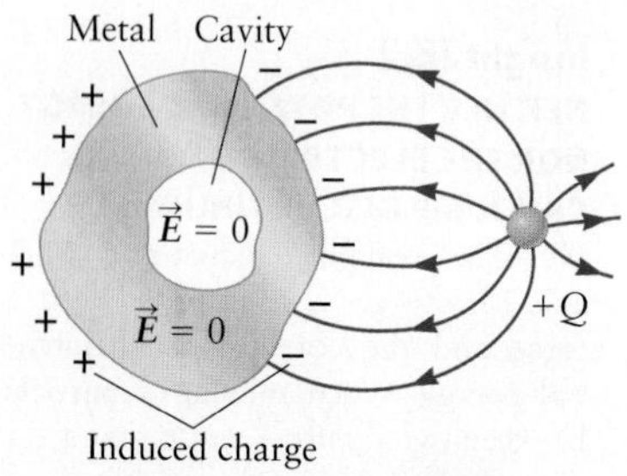

▲ **Figure 18.16** Shielding of the electric field in a cavity inside a metal.

Shielding Out the Electric Field

Our results for the electric field and potential near and inside the metal sphere in Figure 18.13 can be used to design a region in which the electric field is zero. This process is called "shielding out" the electric field. We already know how to make the electric field zero: we simply go inside a metal. This approach is sketched in Figure 18.16, where we show a piece of metal that contains a hollow cavity. A charge Q placed near the metal produces an electric field in its vicinity, but the metal prevents the field from penetrating into the cavity. The electric field in the cavity is zero no matter what charges or electric fields are present outside the metal.

A physical picture of this shielding is given in Figure 18.16. The field from a positive charge Q pulls electrons to the surface of the metal nearest Q, leaving a net positive charge on the surface farthest from Q. The fields of these induced charges resting on the outer surface of the metal precisely cancel the field from Q inside the metal and inside the cavity. This method of shielding has many applications, one of which you may have already experienced. During a thunderstorm, large and potentially deadly electric fields are associated with every "bolt" of lightning, but when you are in your car during a thunderstorm, you are protected from these large fields. The shell of your car acts as the piece of metal in Figure 18.16 while you sit in the cavity, and the shielding effect of the cavity protects you from the large electric fields of a lightning strike.

18.3 Equipotential Lines and Surfaces

Insight 18.3
EQUIPOTENTIAL SURFACES AND GRAVITY

The meaning of an equipotential surface can also be understood by analogy with the gravitational potential energy at different places on a mountain. As you climb up a mountain, your gravitational potential energy PE_{grav} increases, whereas PE_{grav} decreases when you go downhill. However, PE_{grav} stays constant if you follow a path that stays at a constant altitude. Such a path lies on a surface of constant gravitational potential energy and is thus an equipotential surface for gravitation.

We have used drawings of electric field lines extensively to show how the electric field varies in space around various charge distributions. Another very useful way to visualize the electric field is through plots of ***equipotential surfaces***. These are contours on which the electric potential is constant. Consider again the electric field in the region between two large, flat plates that carry charges $+q$ and $-q$ as sketched in Figure 18.17A. To be specific, assume they are the accelerator plates in the CRT in Figure 18.8 and the potential difference between the two plates is 15,000 V. The electric field between the plates is constant and is directed from the plate with positive charge to the plate with negative charge. According to our relation between $\vec{E}$ and the electric potential (Eq. 18.17), the electric potential varies as we move from one plate to the other. We can now draw an imaginary surface between the plates such that $V = 10{,}000$ V at all points on this surface. Likewise, we can imagine a surface on which the potential is 5000 V (or any other value of interest). Several such surfaces are indicated in Figure 18.17B, labeled by their values of the potential. In this example, these surfaces are planes parallel to the charged plates.

Equipotential surfaces (also called equipotential *lines* in two-dimensional plots) are a very useful way to visualize the electric field and how it varies in space. Equipotential surfaces and lines are always perpendicular to the direction of the electric

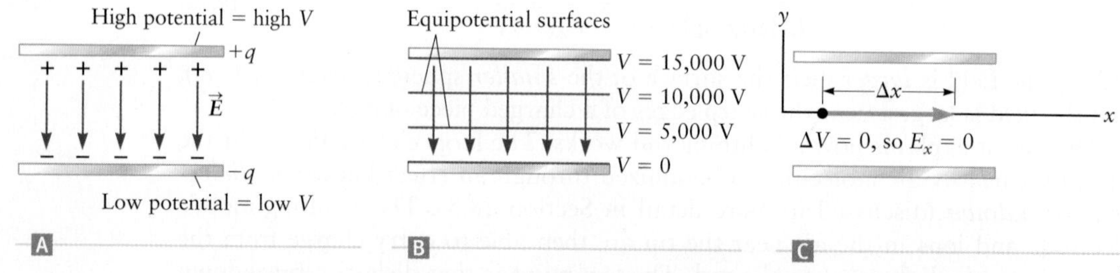

▲ **Figure 18.17** A Electric field between two parallel, charged metal plates. B In this case, the contours of equal V, called equipotential surfaces, are planes parallel to the metal plates. C The electric potential is constant along the x direction, so $E_x = 0$.

field. This property follows from the relationship between $\vec{E}$ and V. In Section 18.2, we showed that for a displacement along a particular direction, the change in V is related to the component of the electric field in that direction. If we consider a displacement along the x direction as in Figure 18.17C, the component of the field along x is

$$E_x = -\frac{\Delta V}{\Delta x} \tag{18.27}$$

By definition, V is constant and $\Delta V = 0$ for motion parallel to an equipotential surface. From Equation 18.27, the electric field component parallel to the surface is thus zero. Hence, $\vec{E}$ is always perpendicular to an equipotential surface.

Some equipotential surfaces in the neighborhood of a point charge are shown in Figure 18.18. In this case, the electric field lines emanate radially outward from the charge. The equipotential surfaces are perpendicular to $\vec{E}$ and are thus a series of concentric spheres, with different spheres corresponding to different values of V.

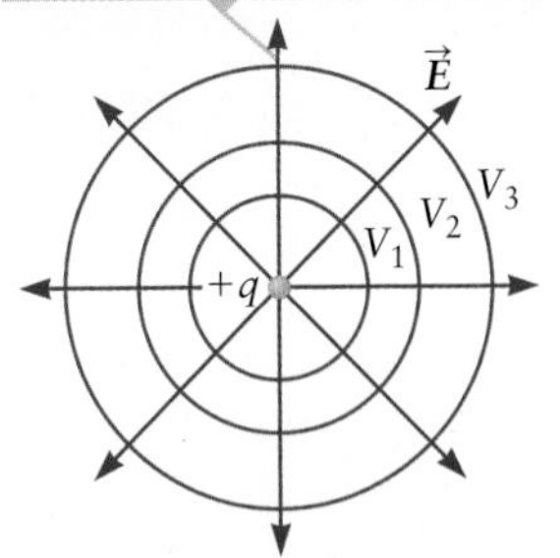

▲ **Figure 18.18** The equipotential surfaces around a point charge are spheres. These circles are projections of the spherical surfaces onto the plane of the drawing.

CONCEPT CHECK 18.4 Equipotential Surfaces and the Magnitude of E

Figure 18.19 shows equipotential surfaces near a hypothetical charge distribution. Is the magnitude of the electric field largest at (a) point A, (b) point B, or (c) point C? *Hint*: Because $E = -\Delta V/\Delta x$, the field is largest in regions and directions where the potential is changing most rapidly.

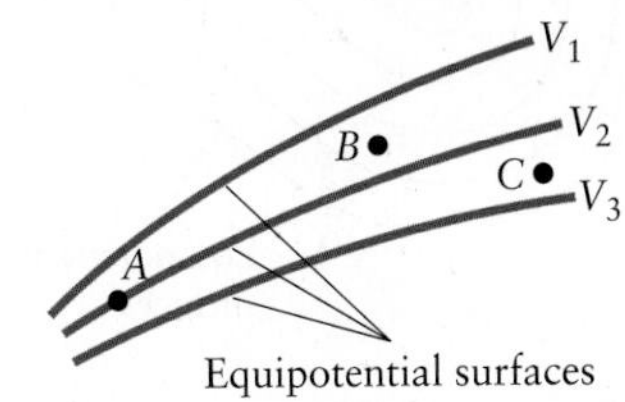

▲ **Figure 18.19** Concept Check 18.4.

EXAMPLE 18.6 Equipotential Surfaces around a Line of Charge

Consider a uniform line of positive charge as shown in Figure 18.20A. Make a plot showing some of the equipotential surfaces.

(continued) ▶

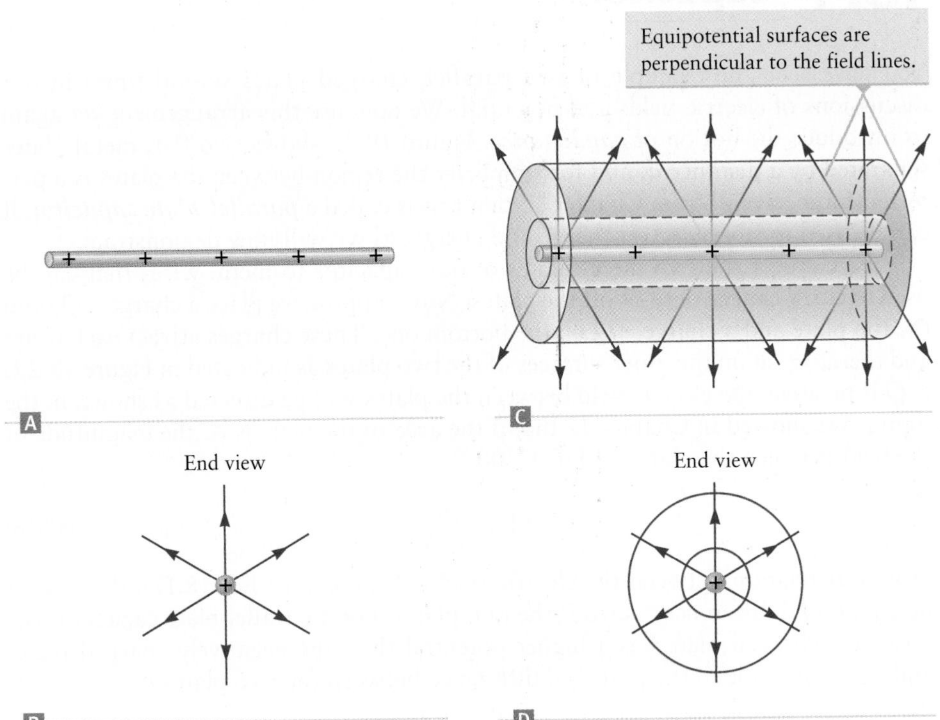

◀ **Figure 18.20** Example 18.6. **A** A long line of positive charge. **B** End view of electric field lines near the line of positive charge. **C** The equipotential surfaces are cylinders centered on the line of charge. **D** End view of the electric field lines and equipotential surfaces.

RECOGNIZE THE PRINCIPLE

Equipotential surfaces and lines are always perpendicular to the electric field. So, we first determine the direction of $\vec{E}$ and then find surfaces perpendicular to the field.

SKETCH THE PROBLEM

In Chapter 17, we showed that the electric field produced by a line of charge emanates radially outward from the line as sketched in Figure 18.20B.

IDENTIFY THE RELATIONSHIPS AND SOLVE

Equipotential surfaces must always be perpendicular to the electric field lines, so the equipotential surfaces are cylinders centered on the line of charge as shown in parts C and D of Figure 18.20.

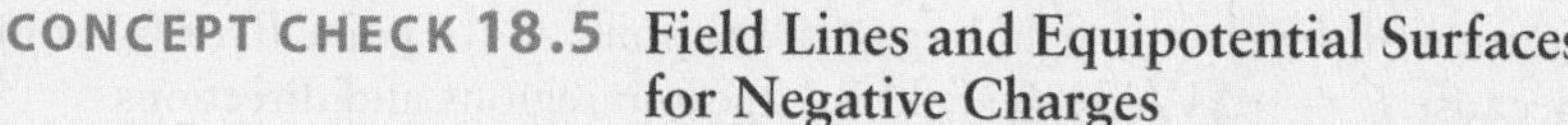

► ***What does it mean?***

Because equipotential surfaces are, by their definition, always perpendicular to $\vec{E}$, they are often the same as or very similar to the Gaussian surfaces we would choose in applying Gauss's law.

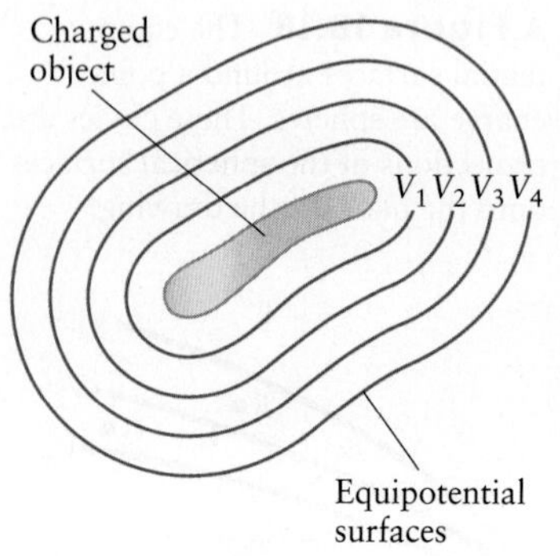

▲ **Figure 18.21** Concept Check 18.6.

CONCEPT CHECK 18.5 Field Lines and Equipotential Surfaces for Negative Charges

How would the sketches of the electric field and the equipotential surfaces in Figure 18.20 change if the charge on the line is negative instead of positive?

CONCEPT CHECK 18.6 Relating the Electric Field to the Equipotential Surfaces

Figure 18.21 shows a family of equipotential surfaces. Sketch qualitatively the corresponding electric field lines. If $V_1 > V_2 > V_3 > V_4$, is the object in Figure 18.21 positively charged or negatively charged?

18.4 Capacitors

You have seen the example of two parallel, charged plates several times in our discussions of electric fields and potential. We now use this arrangement yet again to introduce the notion of ***capacitance***. Figure 18.22 shows two flat, metal plates separated by a distance d, and for simplicity the region between the plates is a perfect vacuum. Recall from Chapter 17 that this is called a ***parallel-plate capacitor***. It can be used to store electric charge and energy, as we will now demonstrate.

Let's connect the two metal plates of our capacitor to metal wires that can be used to carry charge onto or off the plates. Now suppose we place a charge $+Q$ onto the top plate and a charge $-Q$ on the bottom one. These charges attract each other and therefore sit on the inner surfaces of the two plates as indicated in Figure 18.22. If Q is positive, the electric field between the plates will be directed as shown in the figure. We showed in Chapter 17 that if the area of the plate is A, the magnitude of the field between the plates is (Eq. 17.18)

$$E = \frac{Q}{\varepsilon_0 A} \tag{18.28}$$

From our relation between the electric field and potential (Eq. 18.17), there must be a potential difference "across" the two plates. For a parallel-plate capacitor, the positively charged plate has a higher potential than the negatively charged plate, and the magnitude of the potential difference between the two plates is

$$\Delta V = Ed$$

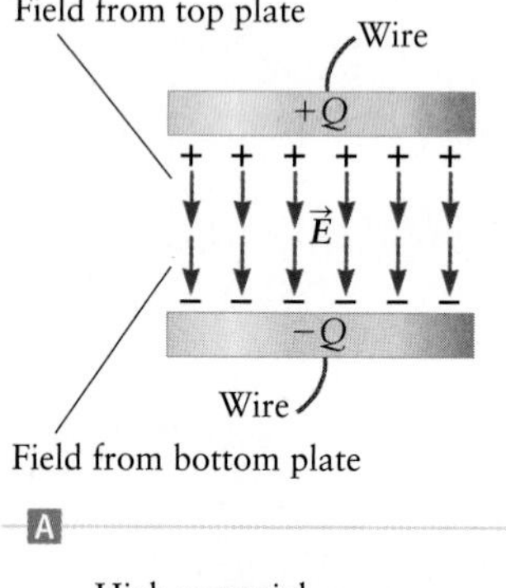

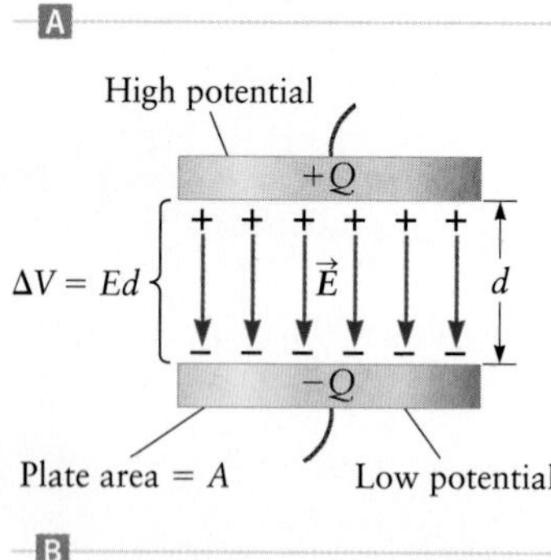

▲ **Figure 18.22** A A parallel-plate capacitor consists of two parallel metal plates. Wires connected to each plate allow charge to be placed on or taken off the plates. B The electric field between the plates is related to the potential difference between the plates.

where d is the distance between the plates. Inserting E from Equation 18.28, we find

$$\Delta V = Ed = \frac{Qd}{\varepsilon_0 A} \tag{18.29}$$

Hence, the potential difference across the capacitor plates is proportional to the charge Q. We now define a quantity C called the capacitance with

Definition of capacitance

$$C = \frac{Q}{\Delta V} \quad \text{so that} \quad \Delta V = \frac{Q}{C} \tag{18.30}$$

Comparing Equations 18.29 and 18.30, C in this case is given by

Capacitance of a parallel-plate capacitor

$$C = \frac{\varepsilon_0 A}{d} \quad \text{(parallel-plate capacitor)} \tag{18.31}$$

where A is the area of a single plate and d is the plate separation.

There are many other capacitor designs with different shapes for the plates. All capacitors, though, employ two metal plates of some sort, and in *all* cases the charge on the capacitor plates is proportional to the potential difference across the plates as in Equation 18.30. The actual value of the capacitance depends on the design; Equation 18.31 applies to the parallel-plate design, whereas other plate geometries lead to other formulas for C.

According to Equation 18.30, capacitance is the ratio of electric charge to the difference in electric potential across the plates, so the SI unit of capacitance is coulombs per volt, or C/V. This combination of units is called the farad (F) in honor of Michael Faraday, 1791–1867.

CONCEPT CHECK 18.7 Forces on a Capacitor

Does placing additional charge on the plates of the capacitor in Figure 18.22 (increasing Q) tend to push the plates apart or tend to pull them together?

Storing Energy in a Capacitor

Capacitors find many applications in electronic circuits, including radios, computers, and MP3 players. All these uses depend on the ability of a capacitor to store electric charge and on the relation between charge and potential difference (voltage), Equation 18.30.

When there is a nonzero potential difference between the two plates of a capacitor, energy is stored in the device. It is similar to the potential energy present in a system of two point charges as in Figure 18.2. To calculate the potential energy associated with the charges $+Q$ and $-Q$ on the two plates of a capacitor, we imagine that the capacitor is initially uncharged ($Q = 0$). We then transfer small amounts of charge ΔQ (perhaps by moving one electron at a time) from one plate to the other in Figure 18.23A. The capacitor plates are initially uncharged, so the

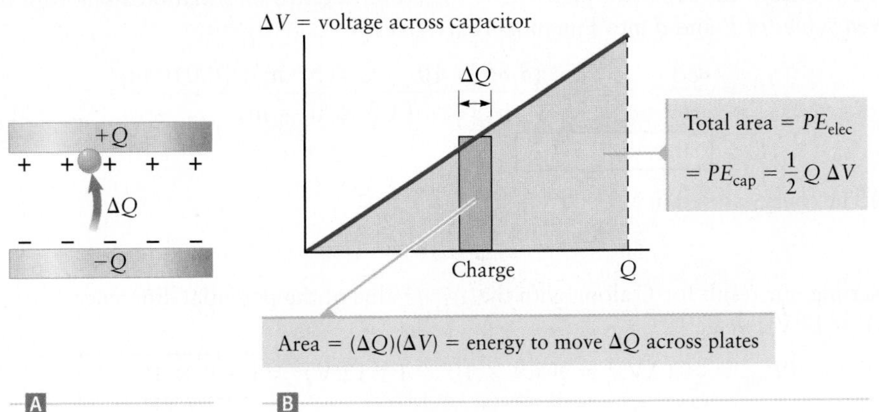

◀ **Figure 18.23** A When charge is transferred from one capacitor plate to the other, the value of Q increases. B The voltage across the capacitor plates increases with Q. The area under the graph of ΔV versus Q is equal to the potential energy stored in the capacitor PE_{elec}.

potential difference is initially $\Delta V = 0$ and the electric field between the plates is zero. It therefore takes no energy to transfer the first packet of charge. As we continue to transfer charge between the plates, however, the total charge on the plates grows and, because Q and ΔV are proportional (Eq. 18.30), the potential difference across the plates grows too. This dependence is sketched in Figure 18.23B.

To move an amount of charge ΔQ through a potential difference ΔV requires an energy $(\Delta Q)(\Delta V)$, which corresponds to the area shown by the shaded rectangle in Figure 18.23B. The total energy stored in the capacitor is equal to the energy required to move all the packets of charge ΔQ from one plate to the other, and corresponds to the total area under the ΔV–Q line in Figure 18.23B. The area of this triangle is $\frac{1}{2}Q\,\Delta V$, where ΔV is now the final potential difference and Q is the final charge. Hence, the energy stored in a parallel-plate capacitor is

Energy stored in a capacitor

$$PE_{\text{cap}} = \tfrac{1}{2}Q\,\Delta V \tag{18.32}$$

The definition of capacitance (Eq. 18.30) can be written as either $Q = C\,\Delta V$ or $\Delta V = Q/C$. Using these relations in Equation 18.32 leads to

$$PE_{\text{cap}} = \frac{1}{2}C(\Delta V)^2 = \frac{1}{2}\frac{Q^2}{C} \tag{18.33}$$

The relations for the stored energy in Equations 18.32 and 18.33 apply to all types of capacitors, not just parallel-plate capacitors. The energy stored in a capacitor is *potential energy*. It can be extracted from the capacitor and transformed into other forms of energy, or it can be used to do mechanical work.

EXAMPLE 18.7 Capacitance and Energy Storage in a Typical Capacitor

Consider a parallel-plate capacitor that is about the size of your fingernail (Fig. 18.24). The plates are squares with edges of length $L = 1.0$ cm, separated by $d = 10\ \mu\text{m} = 1.0 \times 10^{-5}$ m, which is about the diameter of a human hair. **(a)** Find the capacitance. **(b)** If the potential across the capacitor is $\Delta V = 12$ V, what is the energy stored?

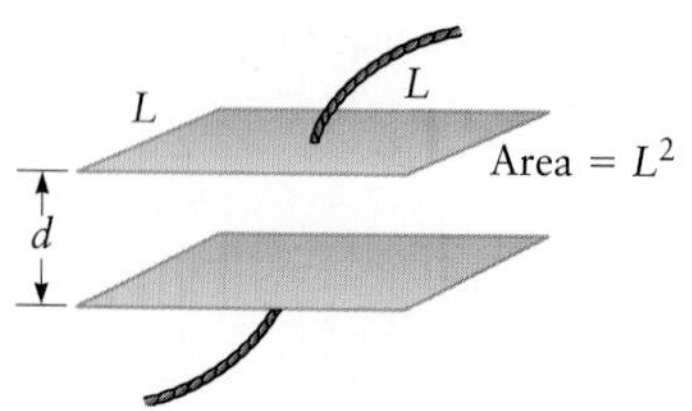

▲ Figure 18.24 Example 18.7. A parallel-plate capacitor with two square plates.

RECOGNIZE THE PRINCIPLE

These calculations involve a direct application of our results for a parallel-plate capacitor and the energy stored in a capacitor. The values we find will be interesting.

SKETCH THE PROBLEM

The capacitor and its dimensions are shown in Figure 18.24.

IDENTIFY THE RELATIONSHIPS AND SOLVE

(a) The area of the capacitor plates is $A = L^2$. Inserting this information along with the given values of L and d into Equation 18.31 gives

$$C = \frac{\varepsilon_0 A}{d} = \frac{\varepsilon_0 L^2}{d} = \frac{[8.85 \times 10^{-12}\ \text{C}^2/(\text{N}\cdot\text{m}^2)](0.010\ \text{m})^2}{(1.0 \times 10^{-5}\ \text{m})}$$

$$C = \boxed{8.9 \times 10^{-11}\ \text{F}}$$

(b) The energy stored in a capacitor is (Eq. 18.33)

$$PE_{\text{cap}} = \tfrac{1}{2}C(\Delta V)^2$$

Inserting our result for C along with the given value of the potential difference ($\Delta V = 12$ V), we get

$$PE_{\text{cap}} = \tfrac{1}{2}C(\Delta V)^2 = \tfrac{1}{2}(8.9 \times 10^{-11}\ \text{F})(12\ \text{V})^2 = \boxed{6.4 \times 10^{-9}\ \text{J}}$$

▸ What does it mean?
The value of C for this capacitor is only 89 picofarads (1 picofarad $= 10^{-12}$ F). Typical capacitors used in electronic circuits are much less than 1 F, but the value of C for our fingernail-size capacitor is still small compared with most capacitors. We'll discuss ways to increase the value of C in the next section. The energy stored in this capacitor is also very small. As a comparison with familiar mechanical energies, PE_{cap} is much less than the change in gravitational potential energy when a mosquito ($m \approx 10^{-6}$ kg) falls about 1 cm! One practical use of a capacitor is to store energy. To store useful amounts of energy, we must design capacitors with much larger values of C.

EXAMPLE 18.8 Using a Capacitor as a Computer Memory Element

Modern computer memories use parallel-plate capacitors to store information, and these capacitors are the basic elements of a random-access memory (RAM) chip. Assume one of these capacitors has plates with area $L \times L$, where $L = 0.10\ \mu$m ($= 1.0 \times 10^{-7}$ m), and a plate separation $d = 10$ nm ($= 1.0 \times 10^{-8}$ m). **(a)** Find the capacitance of such a capacitor. **(b)** Calculate the amount of charge that must be placed onto the plates to obtain a potential difference of 5.0 V across them. **(c)** How many electrons does this charge correspond to?

RECOGNIZE THE PRINCIPLE

Our RAM capacitor is just a parallel-plate capacitor, and we can find its capacitance using $C = \varepsilon_0 A/d$ (Eq. 18.31). Using the given value of ΔV with this value of C, we can then calculate the charge on the capacitor plates.

SKETCH THE PROBLEM

Our simplified model of a computer memory capacitor is described by Figure 18.24.

IDENTIFY THE RELATIONSHIPS AND SOLVE

(a) The capacitance of a parallel-plate capacitor is (Eq. 18.31)

$$C = \frac{\varepsilon_0 A}{d} = \frac{[8.85 \times 10^{-12}\ \text{C}^2/(\text{N}\cdot\text{m}^2)](1.0 \times 10^{-7}\ \text{m})^2}{1.0 \times 10^{-8}\ \text{m}} = \boxed{8.9 \times 10^{-18}\ \text{F}}$$

(b) Potential, charge, and capacitance are related by $Q = C\,\Delta V$ (Eq. 18.30), so obtaining a voltage of 5.0 V requires a charge

$$Q = C\,\Delta V = (8.9 \times 10^{-18}\ \text{F})(5.0\ \text{V}) = \boxed{4.5 \times 10^{-17}\ \text{C}}$$

(c) The charge on a single electron has magnitude e, so this value of Q corresponds to

$$N = \frac{Q}{e} = \frac{4.5 \times 10^{-17}\ \text{C}}{1.60 \times 10^{-19}\ \text{C/electron}} = \boxed{280\ \text{electrons}}$$

▸ What does it mean?
The number of electrons found in part (c) is not very large. In fact, the latest computer memory capacitors are somewhat smaller than we have assumed here, so they employ an even smaller number of electrons.

Capacitors in Series

Figure 18.25 shows a capacitor-type combination of three metal plates. This is just our usual parallel-plate capacitor (Fig. 18.22) with an added metal plate in the middle, and for simplicity we assume this middle plate is separated from the top and bottom plates by a vacuum. Let's find the capacitance of this structure; our calculation will lead to some useful results for our work with electric circuits in Chapter 19.

A calculation of the capacitance relies on the definition of C as the charge on the capacitor plates divided by voltage (Eq. 18.30). We imagine that charges $+Q$ and $-Q$ are placed on the outermost plates (Fig. 18.25A), and we want to find the total potential difference ΔV_{total} between these two plates. The ratio $Q/\Delta V_{\text{total}}$ is then the capacitance. The potential difference ΔV_{total} will depend on the charges on the outer plates, which are just $+Q$ and $-Q$, and on the charges on the inner plate, which we now need to find.

The inner plate is neutral, with a net (excess) charge of zero. The charge $+Q$ on the top plate, however, will attract the mobile negative charges (electrons) in the inner plate. This attraction will give a net negative charge on the upper surface of the inner plate, leaving a net positive charge on the lower surface of the inner plate; the different surfaces of the three plates will then be charged as shown in Figure 18.25B. Comparing this figure with our sketch of a single capacitor (two plates) in Figure 18.22, we see that the three plates in Figure 18.25 act as two "consecutive" capacitors, each with charges $\pm Q$. The potential difference ΔV_{total} across the entire structure—that is, between the top and bottom plates—equals the potential difference ΔV_{top} across the top two plates plus the potential difference ΔV_{bottom} across the bottom two plates. The top two plates form a capacitor, so we can write $\Delta V_{\text{top}} = Q/C_{\text{top}}$, where C_{top} is the capacitance of the upper capacitor, with a similar result for ΔV_{bottom}. We have

$$\Delta V_{\text{total}} = \Delta V_{\text{top}} + \Delta V_{\text{bottom}} = \frac{Q}{C_{\text{top}}} + \frac{Q}{C_{\text{bottom}}} \tag{18.34}$$

Capacitance is defined as the charge divided by the total potential difference, so

$$C_{\text{total}} = \frac{Q}{\Delta V_{\text{total}}} = \frac{Q}{\left(\dfrac{Q}{C_{\text{top}}} + \dfrac{Q}{C_{\text{bottom}}}\right)}$$

$$C_{\text{total}} = \frac{1}{\left(\dfrac{1}{C_{\text{top}}} + \dfrac{1}{C_{\text{bottom}}}\right)} \tag{18.35}$$

This result can be rewritten in an equivalent (and easier to remember) form:

$$\frac{1}{C_{\text{total}}} = \frac{1}{C_{\text{top}}} + \frac{1}{C_{\text{bottom}}} \tag{18.36}$$

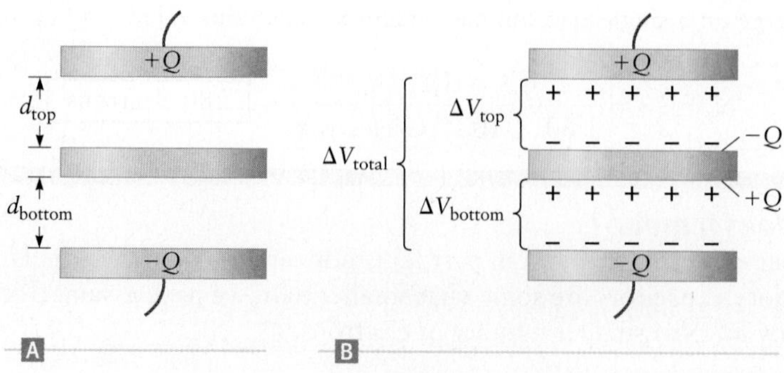

▶ **Figure 18.25** A Three parallel metal plates act as two capacitors sharing a common central plate. B When a charge $+Q$ is added to the top plate and $-Q$ to the bottom plate, these charges move to the "inner" surfaces as shown and induce charges $-Q$ and $+Q$ on the two surfaces of the inner plate.

An arrangement of two "consecutive" capacitors as in Figure 18.25 is referred to as ***capacitors in series***. The result in Equation 18.36 tells us that when any two capacitors (which we can denote by C_1 and C_2) are placed in series, they are completely *equivalent* to a *single* capacitor whose capacitance C_{equiv} is given by

$$\frac{1}{C_{\text{equiv}}} = \frac{1}{C_1} + \frac{1}{C_2} \quad \text{(capacitors in series)} \tag{18.37}$$

Equivalent capacitance of capacitors in series

We have derived this result for the case of two parallel-plate capacitors, but it applies to *any* type of capacitor. This result for capacitors in series will be very useful in our work with electric circuits in Chapter 19.

The result for capacitors in series from Equation 18.37 is shown schematically in Figure 18.26. Here we introduce the standard notation for drawing a capacitor in an electric circuit. The two parallel lines represent the two plates of a parallel-plate capacitor; this notation is used to represent all types of capacitors.

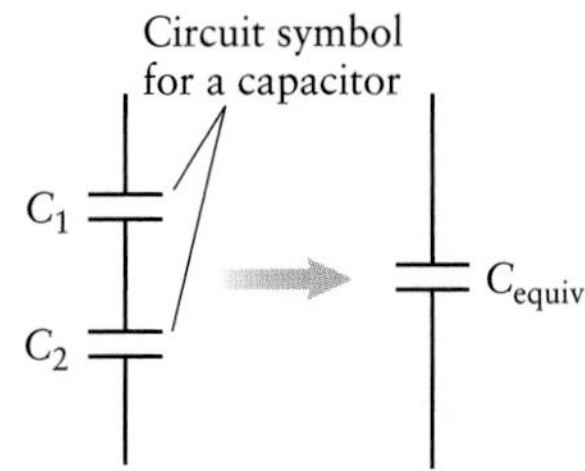

▲ **Figure 18.26** In a circuit diagram, capacitors are shown as two parallel lines, representing the plates of the capacitor. The top and middle plates in Figure 18.25 behave as a single capacitor, and the middle and bottom plates behave as a second capacitor. These two capacitors are equivalent to a single capacitance C_{equiv}.

Capacitors in Parallel

Figure 18.26 shows one particular way of connecting two capacitors. Another way is illustrated in Figure 18.27, which shows two capacitors C_1 and C_2 connected in ***parallel***. The capacitors C_1 and C_2 in Figure 18.27A are equivalent to a single capacitor C_{equiv}. To find the value of this equivalent capacitance, we imagine that charges Q_1 and Q_2 are placed on the two capacitors as shown in that figure. The voltage across each capacitor is related to its charge and capacitance, and, from Equation 18.30, we have

$$C_1 = \frac{Q_1}{\Delta V_1} \quad \text{and} \quad C_2 = \frac{Q_2}{\Delta V_2} \tag{18.38}$$

The top plates of these two capacitors are connected by a metal wire. A piece of metal in equilibrium is an equipotential object, so all the metal inside the upper dashed contour in Figure 18.27B is at the same potential. The two upper plates of C_1 and C_2 are thus at the same potential. Likewise, the lower plate of C_1 is at the same potential as the lower plate of C_2. Hence, the potential difference is the same across the two capacitors, and $\Delta V_1 = \Delta V_2$. For our equivalent capacitor (Fig. 18.27C), the total capacitance is equal to the ratio of the total charge and the potential difference across the plates with (from Eq. 18.30)

$$C_{\text{equiv}} = \frac{Q}{\Delta V}$$

The total charge on the equivalent capacitor is $Q = Q_1 + Q_2$, so we have[2]

$$C_{\text{equiv}} = \frac{Q_1 + Q_2}{\Delta V} = \frac{Q_1}{\Delta V} + \frac{Q_2}{\Delta V} \tag{18.39}$$

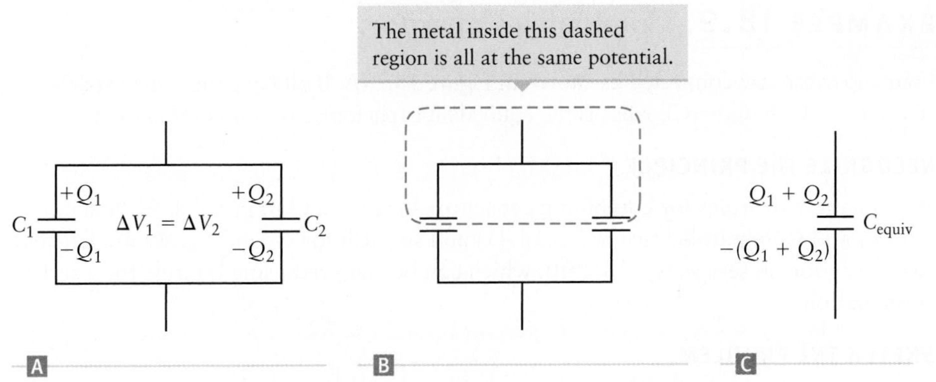

The metal inside this dashed region is all at the same potential.

◀ **Figure 18.27** Two capacitors connected in parallel behave as a single equivalent capacitor C_{equiv}.

[2]Notice that the charges Q_1 and Q_2 on the two capacitors need *not* be equal. As we saw with the lightning rod in Figure 18.15, two metal objects at the same potential need not have the same charge.

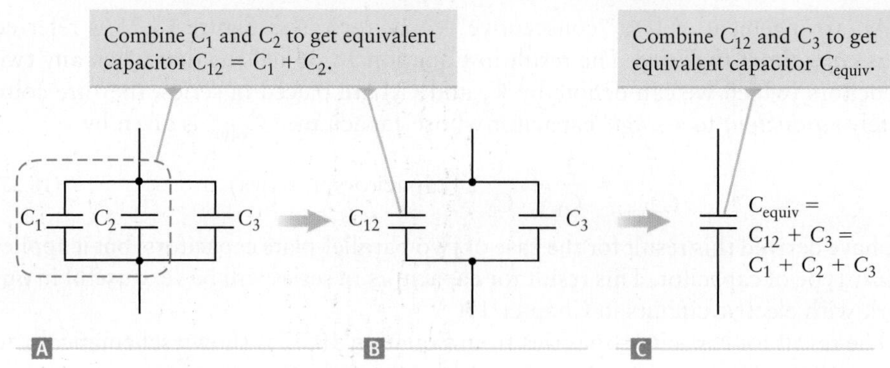

► **Figure 18.28** Three (or more) capacitors connected in parallel behave as a single equivalent capacitor.

The voltage across the equivalent capacitor is also equal to the voltage across the individual capacitors, $\Delta V = \Delta V_1 = \Delta V_2$. Using this information along with Equation 18.38 leads to

$$C_{equiv} = \frac{Q_1}{\Delta V_1} + \frac{Q_2}{\Delta V_2}$$

Equivalent capacitance of capacitors in parallel

$$C_{equiv} = C_1 + C_2 \quad \text{(capacitors in parallel)} \qquad (18.40)$$

Equation 18.40 is the general rule for how capacitors combine in parallel to act as a single equivalent capacitor.

Combinations of Three or More Capacitors

Equations 18.37 and 18.40 describe how two capacitors combine in series or in parallel. What about combinations of three (or more) capacitors? To see how to deal with such cases, consider three capacitors in parallel as shown in Figure 18.28A. We can first combine capacitors C_1 and C_2 to get an equivalent capacitor C_{12} (Fig. 18.28B); using Equation 18.40, $C_{12} = C_1 + C_2$, which reduces the problem to two capacitors (C_{12} and C_3) in parallel. We can then apply Equation 18.40 again to obtain the effective capacitance of C_{12} and C_3 to get the equivalent capacitance of all three capacitors $C_{equiv} = C_{12} + C_3 = C_1 + C_2 + C_3$. The general combination rule for many capacitors in parallel is

$$C_{equiv} = C_1 + C_2 + C_3 + \cdots \quad \text{(many capacitors in parallel)} \qquad (18.41)$$

For multiple capacitors in series, a similar approach gives

$$\frac{1}{C_{equiv}} = \frac{1}{C_1} + \frac{1}{C_2} + \frac{1}{C_3} \cdots \quad \text{(many capacitors in series)} \qquad (18.42)$$

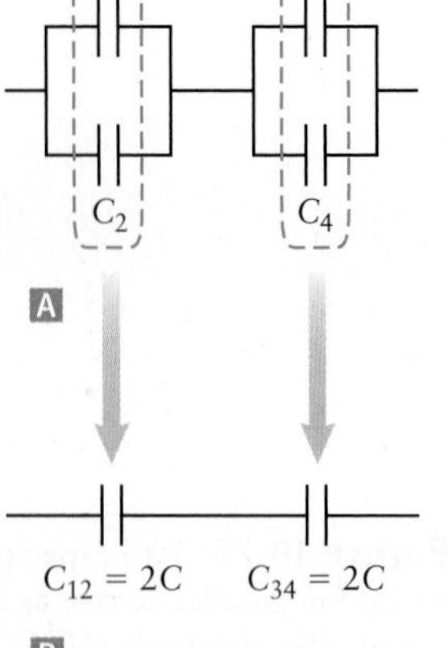

▲ **Figure 18.29** Example 18.9.

EXAMPLE 18.9 Combining Capacitors

Four capacitors are connected as shown in Figure 18.29A. If all have the same capacitance ($C_1 = C_2 = C_3 = C_4 = C$), what is the equivalent capacitance of this combination?

RECOGNIZE THE PRINCIPLE

We can apply the rules for combining capacitors in series and in parallel. We first combine C_1 and C_2 in parallel (using Eq. 18.41) and so forth for C_3 and C_4. We are left with two capacitors in series (Fig. 18.29B), which can be analyzed using the rule for a series combination.

SKETCH THE PROBLEM

Figure 18.29 shows the problem. We first combine C_1 and C_2 to get C_{12}, and C_3 and C_4 to get C_{34}. We then combine these results to get C_{equiv}.

IDENTIFY THE RELATIONSHIPS AND SOLVE

On the left in Figure 18.29B, we have combined C_1 and C_2 in parallel to get an equivalent capacitance C_{12}. According to Equation 18.41, we have

$$C_{12} = C_1 + C_2 = 2C$$

Likewise, the parallel combination of C_3 and C_4 is

$$C_{34} = C_3 + C_4 = 2C$$

The final step is to combine C_{12} and C_{34} in series (Fig. 18.29C) to get the total capacitance. Applying the rule for combining capacitors in series (Eq. 18.42) gives

$$\frac{1}{C_{\text{equiv}}} = \frac{1}{C_{12}} + \frac{1}{C_{34}} = \frac{1}{2C} + \frac{1}{2C}$$

$$C_{\text{equiv}} = \boxed{C}$$

▸ *What does it mean?*

We can deal with most combinations of capacitors by applying the parallel and series combination rules to obtain a single equivalent capacitance.

A

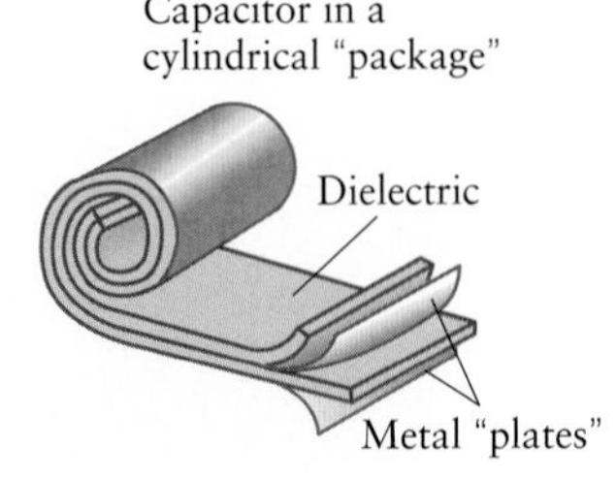

B

▲ **Figure 18.30** A Capacitors can be constructed in many different ways and come in different shapes and sizes. B Many capacitors consist of two metal foils ("plates") wrapped up in a cylinder, with a dielectric between the foils.

18.5 Dielectrics

Most real capacitors are not as simple as the parallel-plate capacitor analyzed in Examples 18.7 and 18.8. Figure 18.30A shows a number of real capacitors. They all contain two metal "plates" separated by a thin insulating region, but notice that many of the casings are cylindrical. These cylinders contain two thin, metal foils (the "plates") wrapped like a jellyroll. In the parallel-plate capacitors we have considered so far, the region between the plates is a vacuum. In most real capacitors, however, the space between the plates is filled with a solid material called a *dielectric* as illustrated in Figures 18.30B and 18.31A.

The Term *Dielectric* Is Used to Describe Any Insulating Material

Inserting a dielectric material between the plates of a capacitor changes the value of the capacitance in proportion to a factor called the ***dielectric constant*** of the material, denoted by κ. If C_{vac} is the value of the capacitance with a vacuum in the gap between the plates, the capacitance with the dielectric present is

$$C_d = \kappa C_{\text{vac}} \tag{18.43}$$

Note that κ is a dimensionless factor, so it has no units. Except for some very unusual materials, κ is greater than or equal to 1, so the presence of the dielectric *increases* the capacitance.

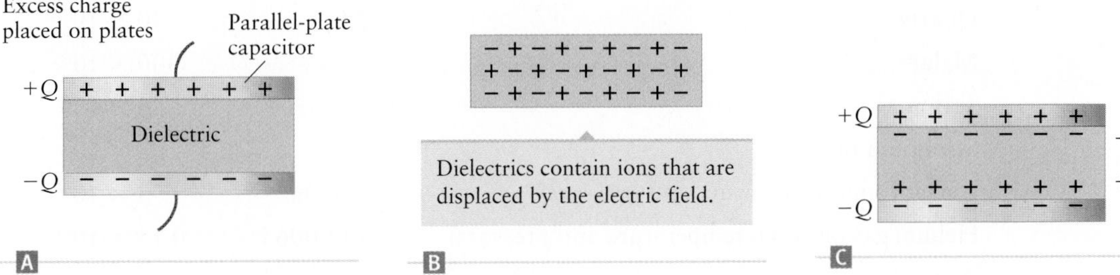

▲ **Figure 18.31** A When a dielectric is placed between the plates of a capacitor, the value of the capacitance increases. B Most dielectrics are ionic. C The electric field inside the capacitor induces charge $\mp Q_d$ on the surfaces of the dielectric.

Why Does a Dielectric Change the Capacitance?

Charge on the plates of a capacitor produces an electric field between them. This electric field extends into a dielectric placed between the plates, causing electric forces on charges in the dielectric. "Good" dielectrics—that is, materials with large values of κ—are usually highly ionic, meaning that they contain positive and negative ions bound together as sketched in Figure 18.31B. The electric force on each ion causes the ion to shift slightly according to the direction of $\vec{E}$. In the dielectric in Figure 18.31, the negative ions move slightly toward the top capacitor plate (because they are attracted to the charge $+Q$) and the positive ions shift a small amount toward the bottom plate (because they are attracted to the charge $-Q$). This slight shift in the ions' positions leads to a small amount of negative charge on the top surface of the dielectric and a corresponding small amount of positive charge on the bottom surface of the dielectric.

An equivalent way to view a capacitor with a dielectric inside is shown in Figure 18.31C. The metal plates have charges $\pm Q$ on their inner surfaces, while the surfaces of the dielectric carry charge $\mp Q_d$. If we lump these two together, we have a total charge $+Q - Q_d$ at the top of the capacitor and $-Q + Q_d$ at the bottom. The dielectric thus reduces the magnitude of the charge on the capacitor.

The field between the plates of a parallel-plate capacitor containing a dielectric is given by

$$E_d = \frac{Q}{\kappa \varepsilon_0 A} \tag{18.44}$$

where Q is the charge on just the metal plates. Since $\kappa \geq 1$, the electric field inside the capacitor is smaller with the dielectric present than when the gap is filled with a vacuum. (Compare with Eq. 18.28.) Hence, the potential difference across the plates is also smaller, again by a factor of κ. Capacitance is defined as the ratio of the external charge placed on the capacitor plates divided by the potential difference. Because the potential difference ΔV is reduced by a factor of κ, the ratio $C = Q/\Delta V$ is larger by the same factor, which takes us to Equation 18.43.

These results apply to *any* type of capacitor. Adding a dielectric (1) increases the capacitance by a factor equal to the dielectric constant κ and (2) reduces the electric field inside the capacitor by the same factor. The actual value of the dielectric constant depends on the material. Materials that are very ionic tend to have the largest

TABLE 18.1 Values of the Dielectric Constant and the Dielectric Breakdown Field for Some Common Materials

Substance	κ	E(breakdown), V/m
Mica	5	100×10^6
Glass	6	14×10^6
Paper	4	16×10^6
Plexiglas	3.4	40×10^6
Quartz	3.8	40×10^6
Mylar	3	300×10^6
Teflon	2	60×10^6
Strontium titanate	230	8×10^6
Air (at room temperature and pressure)	1.0006	3×10^6
Helium gas (at room temperature and pressure)	1.000065	0.15×10^6
Water	80.4	
Glycerine	42.5	
Benzene	2.28	

values of κ. Table 18.1 lists the values of κ for solids, liquids, and gases; all three phases of matter exhibit dielectric behavior.

Effects of Very Large Electric Fields

As more and more charge is added to a capacitor, the electric field between the plates increases in proportion to Q (Eq. 18.44). For a capacitor containing a dielectric, this field can be so large that it literally rips the ions in the dielectric apart, an effect called ***dielectric breakdown***. The resulting free ions are able to move through the material much like electrons in a metal. Because there is a large electric field present (the field that caused the dielectric breakdown), these ions move rapidly toward the oppositely charged capacitor plate and destroy the capacitor.

The value of the electric field at which dielectric breakdown occurs depends on the material (Table 18.1) and sets a practical limit on the maximum voltage that can be placed across a capacitor in applications. Dielectric breakdown can occur whenever the electric field is large (not just in capacitors), so it comes into play in other phenomena such as lightning, as we'll discuss in Section 18.6.

EXAMPLE 18.10 Dielectric Breakdown of a Capacitor

A typical capacitor in a laptop computer or MP3 player has a capacitance of around 1.0 nF ($= 1.0 \times 10^{-9}$ F). Assume it is a parallel-plate capacitor with plate area $A = 1.0\ \text{cm}^2$ using mica as the dielectric. Find (**a**) the spacing between the plates and (**b**) the voltage at which dielectric breakdown occurs (i.e., the voltage at which the capacitor will fail).

RECOGNIZE THE PRINCIPLE

For a parallel-plate capacitor containing a dielectric, the capacitance is given by

$$C_d = \kappa C_{\text{vac}} = \frac{\kappa \varepsilon_0 A}{d} \tag{1}$$

(a combination of Eqs. 18.31 and 18.43). Given the value of C along with the plate area A and the dielectric constant, we can solve Equation (1) to find the spacing d between the plates. We can then calculate the electric field between the plates and compare it with the dielectric breakdown field for mica.

SKETCH THE PROBLEM

Figure 18.31A describes our capacitor; the dielectric in this case is mica.

IDENTIFY THE RELATIONSHIPS AND SOLVE

(**a**) Rearranging Equation (1) to solve for the spacing d between the plates, we find

$$d = \frac{\kappa \varepsilon_0 A}{C_d}$$

From Table 18.1, the dielectric constant of mica is $\kappa = 5$. Inserting this value and the given values of A and C_d, we get

$$d = \frac{\kappa \varepsilon_0 A}{C_d} = \frac{(5)[8.85 \times 10^{-12}\ \text{C}^2/(\text{N}\cdot\text{m}^2)](0.010\ \text{m})^2}{1.0 \times 10^{-9}\ \text{F}}$$

$$d = 4 \times 10^{-6}\ \text{m} = \boxed{4\ \mu\text{m}}$$

(**b**) The dielectric breakdown field of mica is $E = 100 \times 10^6$ V/m (from Table 18.1). Using the relation between the electric field and potential difference between the plates (Eq. 18.17) leads to a breakdown voltage of magnitude

$$\Delta V = |-Ed| = (100 \times 10^6\ \text{V/m})(4 \times 10^{-6}\ \text{m}) = \boxed{400\ \text{V}}$$

(continued) ▶

► What does it mean?

Our result for d in part (a) is much smaller than the thickness of a human hair! Even so, pieces of mica this thin are easily obtained. The voltages used in consumer electronics (such as an MP3 player) are typically less than 10 or 20 V, so this capacitor will not be in danger of failure.

18.6 Electricity in the Atmosphere

▲ **Figure 18.32** Example of a lightning strike.

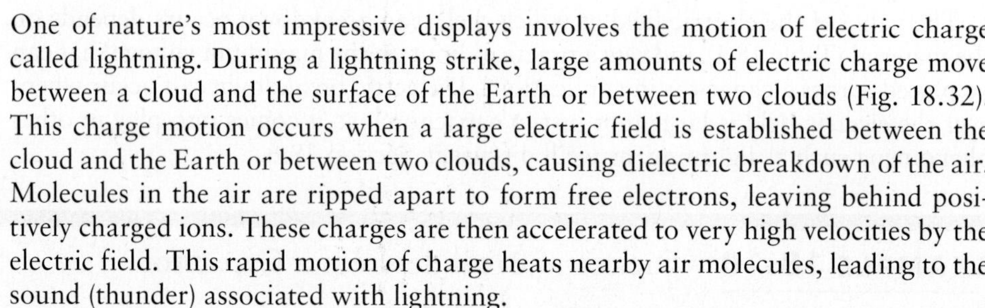

One of nature's most impressive displays involves the motion of electric charge called lightning. During a lightning strike, large amounts of electric charge move between a cloud and the surface of the Earth or between two clouds (Fig. 18.32). This charge motion occurs when a large electric field is established between the cloud and the Earth or between two clouds, causing dielectric breakdown of the air. Molecules in the air are ripped apart to form free electrons, leaving behind positively charged ions. These charges are then accelerated to very high velocities by the electric field. This rapid motion of charge heats nearby air molecules, leading to the sound (thunder) associated with lightning.

Most of the charge motion in lightning involves electrons because they are much lighter than the positive ions that are produced in dielectric breakdown of air. Do these electrons travel from a cloud to the Earth, or vice versa? Experiments show that the electric field of a lightning strike is directed from the Earth to the cloud[3] as shown in Figure 18.33. Recall that $\vec{E}$ points in the direction a positive test charge would move. Thus, once dielectric breakdown occurs in the air, electrons travel from the cloud to the Earth's surface. A typical lightning bolt contains a charge of about $|Q| = 20$ C, corresponding to

$$N = \frac{|Q|}{e} = \frac{20\text{ C}}{1.60 \times 10^{-19}\text{ C/electron}} \approx 1 \times 10^{20}\text{ electrons/bolt} \qquad (18.45)$$

You might think that these electrons would take the shortest path and travel along a straight line to the Earth, but they actually follow a jagged path, giving a lightning bolt its familiar appearance (Fig. 18.32). The jaggedness is due to the complicated nature of dielectric breakdown in air.

Perhaps the most basic question concerning lightning is, Where does the electric field in Figure 18.33 come from in the first place? Interestingly, this problem is still not very well understood, but the correct qualitative explanation is probably as follows. During a thunderstorm, there are large temperature and pressure differences between low and high altitudes. These differences cause water droplets, dust, and small ice crystals to move vertically through the atmosphere. During the later phases of a thunderstorm, the water droplets and ice crystals move downward, acquiring charge as they "rub" against molecules and ions in the air (Fig. 18.33B). This process is similar to what we found in Chapter 17 when we saw that rubbing a piece of amber with fur causes the amber to acquire a negative charge. In a thunderstorm, rubbing gives the water droplets and ice crystals a negative charge that they carry to the bottom of the cloud, leaving the top charged positively. As negative charge accumulates at the bottom of the cloud, it repels electrons from the Earth's surface because the ground is a good conductor of charge and electrons can move freely from place to place in the soil. This process causes the Earth's surface to be positively charged (Fig. 18.33A), producing an electric field qualitatively similar to the electric field between two parallel plates of a capacitor. This electric field produces the dielectric breakdown, which is the lightning bolt.

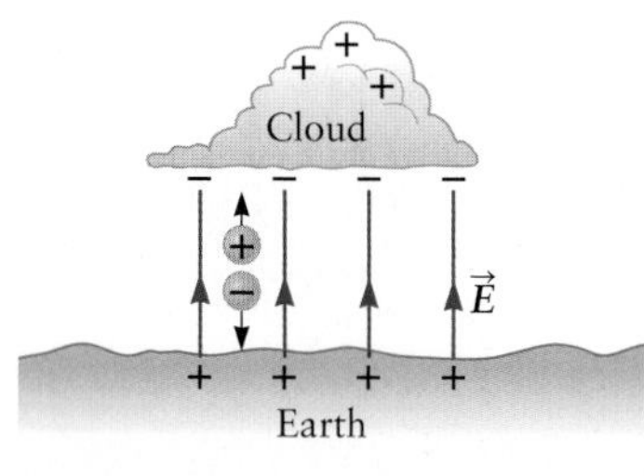

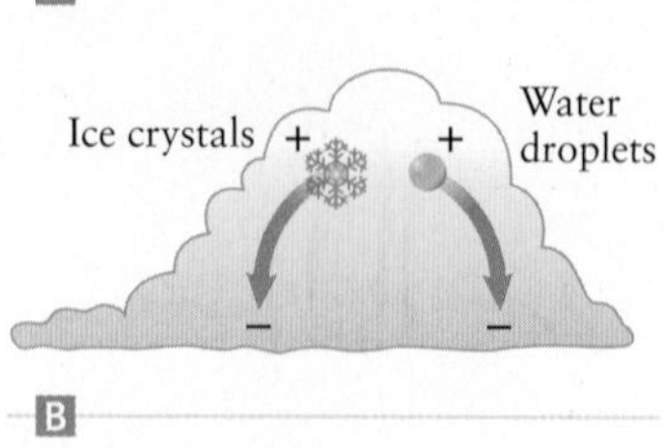

▲ **Figure 18.33** A During an electrical storm, the top of a storm cloud is positively charged, while the bottom is negatively charged. The resulting electric field induces a positive electric charge at the Earth's surface. B The cloud's charge is produced by the motion of water droplets and ice crystals as they are pushed along by the wind.

[3] In fair weather, there is a small electric field in the opposite direction (downward, toward the Earth).

EXAMPLE 18.11 Ⓡ Electric Field and Potential of a Lightning Strike

Consider a thunderstorm involving the bottom surface of a cloud that is $h = 500$ m above the surface of the Earth. If a lightning bolt travels from this cloud to the Earth, estimate the electric potential of the bottom of the cloud just prior to the lightning bolt, relative to "ground" potential $V = 0$ at the Earth's surface.

RECOGNIZE THE PRINCIPLE

A lightning bolt occurs when there is dielectric breakdown of the air. So, just before the lightning bolt occurs, the electric field between the cloud and the Earth must equal the dielectric breakdown field of air.

SKETCH THE PROBLEM

Figure 18.34 shows the problem.

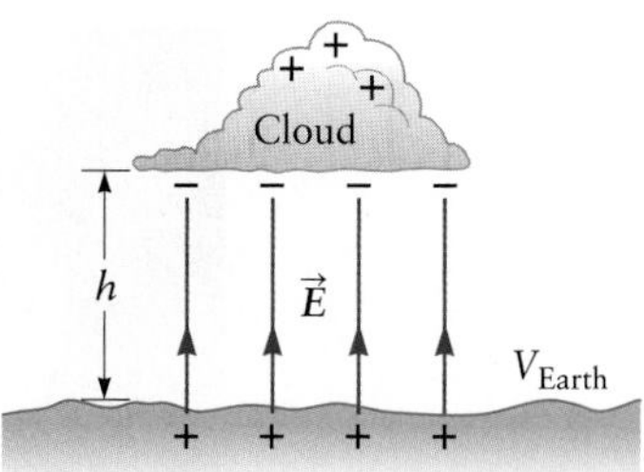

▲ **Figure 18.34** Example 18.11.

IDENTIFY THE RELATIONSHIPS

Using the relation between the electric field and the potential (Eq. 18.18) gives

$$E = -\frac{\Delta V}{\Delta x} = -\frac{V_{cloud} - V_{Earth}}{h}$$

The Earth is at "ground" potential ($V_{Earth} = 0$), so

$$V_{cloud} = -Eh$$

SOLVE

Inserting the dielectric breakdown field of air (Table 18.1) and the height of the cloud, we find

$$V_{cloud} = -E_{breakdown}h = -(3 \times 10^6 \text{ V/m})(500 \text{ m}) \approx \boxed{-2 \times 10^9 \text{ V}}$$

► ***What does it mean?***

The negative sign for V_{cloud} means that the cloud's lower surface is negatively charged. Typical household voltages are between 1 V and 120 V, so the potential associated with lightning is *as much as a billion times* larger than the voltages found around the house!

18.7 Ⓧ Biological Examples and Applications

In the 1790s, an Italian scientist, Luigi Galvani, discovered that sparks from static electricity can cause muscular contraction in a frog. Galvani's work was the first evidence that the nerves and muscles in frogs use an electrical potential and the movement of charge as part of normal physiological function. The same is true in other animals, including human beings.

The heart is perhaps the most important muscle in the body. Like the muscles in a frog's leg, the heart can be made to twitch using an externally applied potential. For example, a device called a ***defibrillator*** is used to shock the heart into a normal pattern of beating. A common diagnostic procedure is the ***electrocardiogram***, called an ECG or EKG. An EKG uses a collection of electrodes to monitor the potential at various places on a person's chest (Fig. 18.35A). The electric potentials the body sends to various parts of the heart form a rhythmic pattern that causes different chambers of the heart to contract at different times, and the different electrode potentials in an EKG can reveal if a person's heart is following a healthy

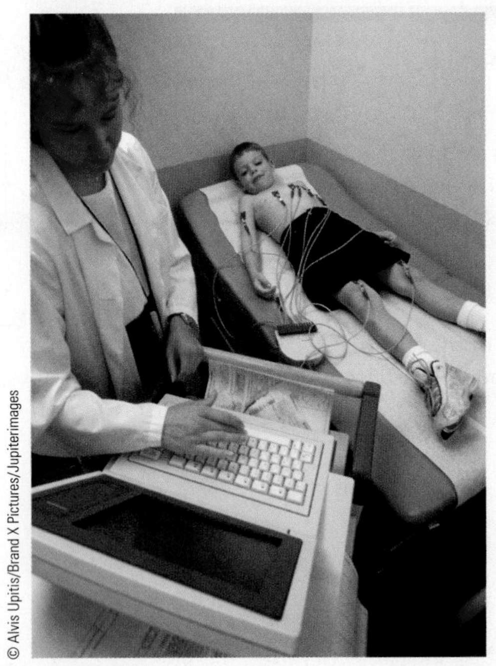

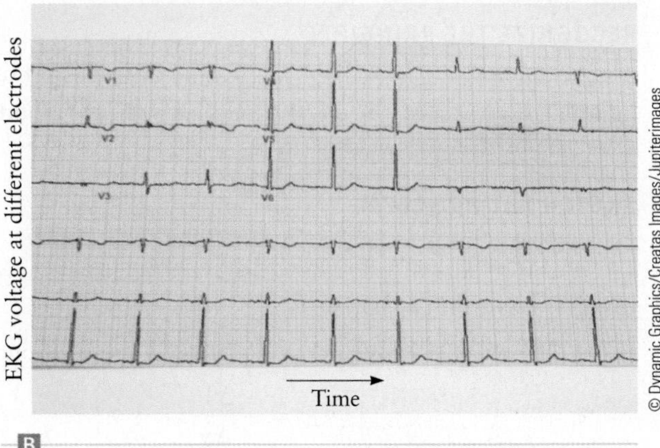

◀ **Figure 18.35** A A person having an EKG. B In an EKG, the electric potential as a function of time is recorded from electrodes at many different spots on the body. This recording shows when and how strongly different parts of the heart muscle are contracting. Typical EKG potentials are around 1 mV (10^{-3} V), about 10,000 times smaller than the potential produced by the battery of an MP3 player.

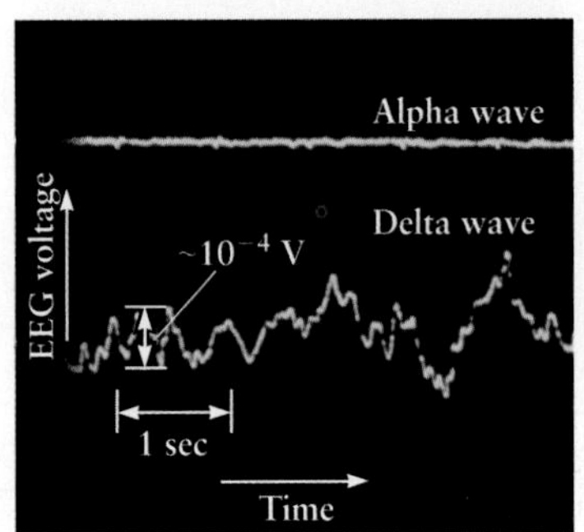

▲ **Figure 18.36** Typical EEG signals. The top signal was taken when the person was awake and alert and is called an alpha wave. The bottom EEG was recorded when the person was sleeping (a delta wave). Typical EEG potentials are about 10^{-4} V, somewhat smaller than in an EKG.

rhythm (Fig. 18.35B). There is no way to use physics to calculate what a healthy rhythm should look like, but your doctor can tell from experience.

A related procedure is the ***electroencephalogram*** (EEG), in which electrodes on the scalp are used to detect potentials generated in the brain. These potentials oscillate with time, and the frequency of an EEG signal depends on the person's activity. The brain of an awake person generates what are called alpha waves with frequencies of about 10–14 Hz, whereas during deep sleep one finds delta waves with a typical frequency of 2–4 Hz (Fig. 18.36). Physiologists have also identified beta and gamma waves, which are generated in other brain states and which have other frequencies. The absence of any EEG signal is usually taken as evidence that a person is "brain dead."

18.8 Electric Potential Energy Revisited: Where Is the Energy?

A certain amount of potential energy is associated with two point charges as a result of the electric interaction between them. We have described this potential energy as a property of the system of two charges. Another way to view and think about this electric potential energy, however, is that potential energy is stored *in the electric field* itself. That is, whenever an electric field is present in a particular region of space, potential energy is located in that region. In almost all respects, this viewpoint is equivalent to the one we have taken in previous sections, so it does not affect any of our results to this point. Let's now calculate this field energy in a familiar geometry.

Consider two flat, metal plates separated by a vacuum. If the plates have area A and are separated by a distance d, the electric field between the plates is (Eq. 18.28)

$$E = \frac{Q}{\varepsilon_0 A} \tag{18.46}$$

where the charge on the plates is $\pm Q$. The magnitude of the potential difference between the plates is $\Delta V = Ed$. We also found that the electric potential energy stored in the capacitor is (Eq. 18.32)

$$PE_{\text{elec}} = \tfrac{1}{2}Q\,\Delta V$$

Combining these last two results, we get

$$PE_{elec} = \tfrac{1}{2}QEd$$

We now eliminate Q using Equation 18.46, leading to

$$PE_{elec} = \tfrac{1}{2}(\varepsilon_0 EA)Ed = \tfrac{1}{2}\varepsilon_0 E^2(Ad)$$

The factor (Ad) is the volume between the plates, so we can also write

$$PE_{elec} = \tfrac{1}{2}\varepsilon_0 E^2 \times (\text{volume}) \tag{18.47}$$

In words, Equation 18.47 says that the electric field energy in the region between the capacitor plates is equal to $\frac{1}{2}\varepsilon_0 E^2$ multiplied by the volume of the region. Hence, the energy per unit volume, denoted by u_{elec}, is

$$u_{elec} = \tfrac{1}{2}\varepsilon_0 E^2 \tag{18.48}$$

We call u_{elec} the ***energy density*** in the electric field.

Although our derivation is for a parallel-plate capacitor, this result for u_{elec} applies for any arrangement of charges, including point charges. It tells us that potential energy is present wherever an electric field is present. Philosophically, this result is a major break from our previous view that potential energy is due to the interaction of charges. One might argue that electric fields are caused by charges, so the field E in Equation 18.48 is actually due to a charge. If this argument is correct, aren't we back with our original view that electric potential energy is always a property of a system of charges? In most cases that is true, but in one extremely important situation it is not: when dealing with an electromagnetic wave. We know from Chapter 12 that waves transport energy. Electromagnetic waves transport energy from place to place, even through regions of space where *no charges are present* or even close by. The energy carried by such a wave is carried by electric and magnetic fields, and the electric part of this energy is given by Equation 18.48. Our original view that electric potential energy always involves pairs of charges is thus *not* the entire story. We'll come back to this problem in Chapter 23 when we discuss electromagnetic waves.

Summary | CHAPTER 18

Key Concepts and Principles

Electric potential energy

The electric force is a conservative force, so there is an ***electric potential energy*** PE_{elec}. The potential energy of two point charges separated by a distance r is

$$PE_{elec} = \frac{kq_1q_2}{r} \quad \text{(18.6) (page 517)}$$

Electric potential

The ***electric potential*** V (also called just the ***potential***) is proportional to the electric potential energy. If the electric potential energy of a charge q at a particular location is PE_{elec}, the electric potential at that point is

$$V = \frac{PE_{elec}}{q} \quad \text{(18.10) (page 521)}$$

with V measured in units of ***volts*** (V).

(Continued)

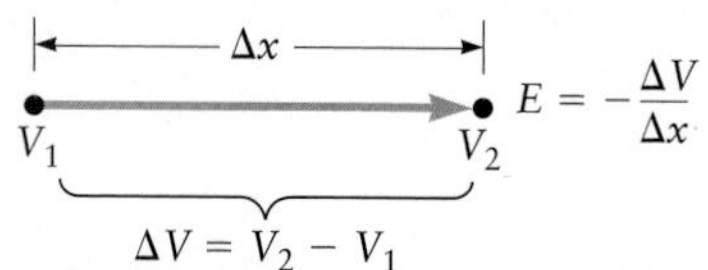

Relation between the electric field and the electric potential

Suppose the potential changes by an amount ΔV over a distance Δx. The component of the electric field along this direction is then

$$E = -\frac{\Delta V}{\Delta x} \qquad \textbf{(18.18) (page 524)}$$

The electric field thus has units of volts per meter (V/m).

Parallel-plate capacitor

Area $= A$

$+Q$

d

$-Q$

$C = \frac{\varepsilon_0 A}{d}$

Capacitors

Two parallel metal plates form a ***capacitor***. The ***capacitance*** C of this structure determines how easily charge can be stored on the plates. The charge on a capacitor is related to the magnitude of the ***potential difference*** between the plates by

$$\Delta V = \frac{Q}{C} \qquad \textbf{(18.30) (page 533)}$$

This relation holds for any type of capacitor. For a parallel-plate capacitor, the capacitance is

$$C = \frac{\varepsilon_0 A}{d} \quad \text{(parallel-plate capacitor)} \qquad \textbf{(18.31) (page 533)}$$

Energy stored in a capacitor and in an electric field

The electric potential energy stored in a capacitor is

$$PE_{\text{cap}} = \tfrac{1}{2}Q\,\Delta V = \tfrac{1}{2}C(\Delta V)^2 = \frac{1}{2}\frac{Q^2}{C} \qquad \textbf{(18.32) and (18.33) (page 534)}$$

The energy per unit volume stored in an electric field E is

$$u_{\text{elec}} = \tfrac{1}{2}\varepsilon_0 E^2 \qquad \textbf{(18.48) (page 545)}$$

Applications

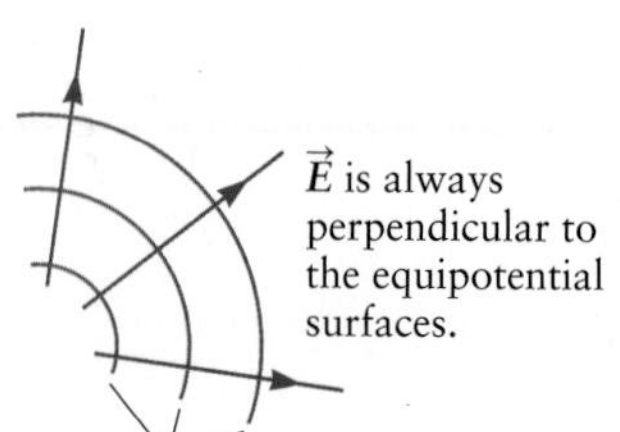

Equipotential surfaces

An ***equipotential surface*** is a surface on which the electric potential is constant. The electric field $\vec{E}$ at a particular location is always perpendicular to the equipotential surface at that spot.

Electric potential in a metal

For a piece of metal in equilibrium, the electric potential is constant throughout the metal, while the electric field is zero inside the metal. The electric field just outside a charged metal is largest near places where the radius of curvature is smallest (i.e., near sharp corners). This result explains why a lightning rod works.

Capacitors in series and parallel

When two capacitors are connected in series, they act as a single equivalent capacitance with a value

$$\frac{1}{C_{equiv}} = \frac{1}{C_1} + \frac{1}{C_2} \quad \text{(capacitors in series)} \qquad (18.37)\ \textbf{(page 537)}$$

When two capacitors are connected in parallel, the equivalent capacitance is

$$C_{equiv} = C_1 + C_2 \quad \text{(capacitors in parallel)} \qquad (18.40)\ \textbf{(page 538)}$$

CAPACITORS IN SERIES

C_1, C_2 → C_{equiv}

CAPACITORS IN PARALLEL

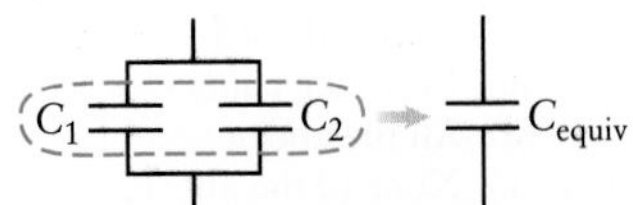

Questions

SSM = answer in *Student Companion & Problem-Solving Guide* ⊗ = life science application

1. If the electric field is zero in a particular region of space, what does that tell you about the electric potential in that region? Is the potential zero, constant, or something else? Explain.

2. Will the *electric field* always be zero at any point where the *electric potential* is zero? Why or why not?

3. Will the *electric potential* always be zero at any point where the *electric field* is zero? Why or why not?

4. SSM Make sketches of the equipotential surfaces around (a) a point charge, (b) an infinite line of charge, (c) an infinite plane of charge, (d) a finite line of charge, and (e) a charged plate of finite size.

5. Experiments show that in a particular region of space the electric field is uniform (i.e., constant). A separate experiment finds that if an electron is released from rest at the origin, it moves along the $+x$ direction. Make a sketch showing the direction of the original electric field (before the electron was inserted) and the associated equipotential surfaces.

6. A particle of positive charge is released from rest and is found to move as a result of an electric force. Does the particle move to a region of higher or lower potential energy? Does the particle move to a region of higher or lower electric potential?

7. Repeat Question 6 for an electron.

8. A charged particle is released from rest in an electric field, and the electric force is the only force acting on the particle. Will the particle's trajectory always follow (i.e., be parallel to) the electric field lines? Explain why or why not. *Hint*: Which of the following quantities are *always* parallel to the electric field lines: (a) the force on the particle, (b) the particle's acceleration, (c) the particle's velocity?

9. Two particles are at locations where the electric potential is the same. Do these particles necessarily have the same electric potential energy? Explain.

10. ⊗ Many species of fish generate electric fields and use these fields to sense their surroundings. If another entity with a dielectric constant different from that of water comes near the fish, the electric field lines generated by the fish are altered, and sensory organs on the fish can detect these slight changes. The electric field of most of these fishes is that of a dipole. Figure Q18.10 shows lines of equal potential around such a fish as seen looking down on the top of the fish. Assume the fish's head is positively charged. (a) Sketch the electric field associated with these equipotential lines. (You may want to make a photocopy of the figure to proceed.) (b) If a sphere of metal (perhaps a metal fishing weight) were placed near the side of the fish, sketch how the field lines in the vicinity of the metal would change.

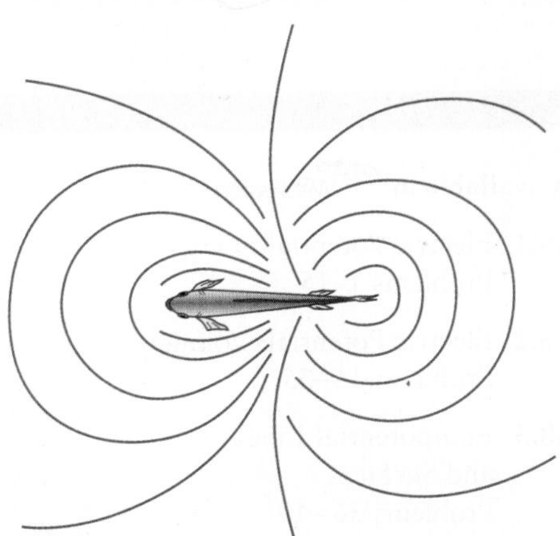

Figure Q18.10

11. Figure Q18.11 shows points on several equipotential surfaces, with $V_A > V_D$. (a) A positive test charge starts at point A and moves to point C. If the test charge starts and ends at rest, which of the following statements applies? (i) A positive amount of work must be done on the test charge by an external force. (ii) The work done on the test charge by an external force must be negative. (iii) No work is done on the test charge. Repeat this question for moving the test charge (b) from C to E, (c) from B to E, and (d) from D to F.

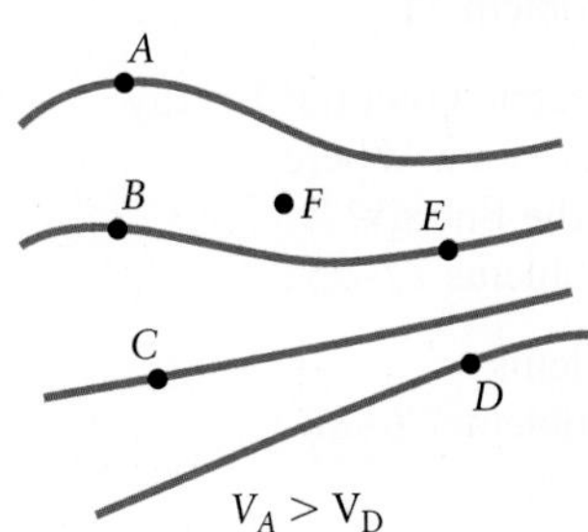

Figure Q18.11

12. SSM A charged particle reaches a speed v when accelerated through a potential difference of 5 V. How many times faster would it be going if accelerated through a potential difference of 20 V?

13. Consider the energy density in the electric field at a distance d away from a positive point charge. If the positive point charge is then replaced with a negative point charge of equal magnitude, how does the energy density change at that same point?

14. To increase the energy stored in a capacitor, what might you do? Explain your reasoning. *Hint*: More than one answer may be correct.
(a) Increase the charge on the capacitor.
(b) Insert a dielectric between the plates while holding the voltage fixed.
(c) Move the plates closer together while keeping the charge fixed.
(d) All the above.
(e) None of the above.

15. Compared with the applied electric field, is the electric field inside a dielectric material (a) smaller by a factor of κ, (b) larger by a factor of κ, or (c) larger by a factor of κ^2? Or, does it (d) depend on the shape of the dielectric? Explain your reasoning.

16. Derive the result for the equivalent capacitance of many capacitors in series in Equation 18.42.

17. Does a charged capacitor have a different net charge than an uncharged one? If not, explain the difference between a charged capacitor and an uncharged one.

18. A parallel-plate capacitor is connected to a battery such that a constant electric potential difference is produced between the plates. If the plates are moved farther apart, which of the following quantities change? Do they get larger or smaller?
(a) The magnitude of the electric field
(b) The amount of charge on a plate
(c) The capacitance
(d) The energy stored in the capacitor

19. A certain amount of charge $\pm q$ is placed on the plates of a capacitor, and the plates are then disconnected from the outside world. If the capacitor plates are pulled apart, does the amount of electric potential energy stored in the capacitor increase or does it decrease? Explain your answer in terms of the principle of conservation of energy.

Problems available in ENHANCED **WebAssign**

Section 18.1 Electric Potential Energy
Problems 1–15

Section 18.2 Electric Potential: Voltage
Problems 16–35

Section 18.3 Equipotential Lines and Surfaces
Problems 36–40

Section 18.4 Capacitors
Problems 41–57

Section 18.5 Dielectrics
Problems 58–66

Section 18.6 Electricity in the Atmosphere
Problems 67–70

Section 18.7 Biological Examples and Applications
Problem 71

Section 18.8 Electric Potential Energy Revisited: Where Is the Energy?
Problems 72–75

Additional Problems
Problems 76–88

List of Enhanced Problems

Problem number	Solution in *Student Companion & Problem-Solving Guide*	Reasoning & Relationships Problem	Reasoning Tutorial in ENHANCED WebAssign
18.10	✓		
18.25	✓		
18.32	✓		
18.37	✓	✓	✓
18.39		✓	
18.40		✓	✓
18.46		✓	✓
18.53	✓		
18.60		✓	✓
18.62		✓	
18.64		✓	✓
18.66	✓		
18.67	✓	✓	✓
18.68		✓	✓
18.69		✓	
18.74	✓		
18.81		✓	✓
18.86	✓		
18.87		✓	✓
18.88		✓	✓

CHAPTER

19

◀ *Electric circuits are at the heart of smart phones, MP3 music players, and other devices and are thus an indispensable part of everyday life.* (© RayArt Graphics/Alamy)

Electric Currents and Circuits

In Chapters 17 and 18, we considered electric forces, fields, and energy in many different situations involving static (stationary) arrangements of electric charge. We now turn our attention to the *motion* of charge and the concept of ***electric current***. When charge can flow in a closed path (a "loop"), we describe the path as an electric ***circuit***. Electric circuits are at the heart of all modern electronic devices such as televisions, cell phones, and computers. In this chapter, we'll examine some basic properties of circuit elements called resistors and capacitors. Once again, conservation principles will be very useful, and the principles of conservation of energy and charge will lead to general rules for analyzing electric circuits. In Chapter 20, we'll study how the moving charges that form electric currents give rise to magnetic phenomena.

OUTLINE

19.1 Electric Current: The Flow of Charge

When charges move from one place to another, we say that there is an ***electric current***. The term *current* probably comes from the analogy with the flow of water in a river. The strength of a river's current (Fig. 19.1A) depends on both the speed of the water molecules and the number of water molecules involved in the flow. In a similar way, we can consider the flow of electrons in a wire as sketched in Figure 19.1B. Electrons can move freely through a metal (much like the water in a river), and we will always use the term *wire* to refer to a metal wire.

The strength or magnitude of the current of water in a river is measured by the amount of water that moves down the river (Fig. 19.1A) in a particular amount of time. Electric current is denoted by the symbol I and defined as the amount of charge Δq that passes a given point on the wire per unit time Δt:

Definition of electric current

$$I = \frac{\Delta q}{\Delta t} \tag{19.1}$$

For historical reasons, current is defined in terms of net positive charge flow (Fig. 19.1C), even though electrons are responsible for the current in a metal. We'll discuss how to use Equation 19.1 to deal with a current carried by electrons shortly.

Current is measured in units of charge divided by time, or coulombs per second (C/s). This unit has been given the name ***ampere*** (A), in honor of André-Marie Ampère (1775–1836), who played a major role in discovering of the laws of magnetism.

$$\text{SI unit of electric current: } 1\text{ A} = 1\text{ C/s} \tag{19.2}$$

The ampere is a primary unit in the SI system. Other units involving electricity, such as the coulomb and the volt, are defined in terms of the ampere. For example, the coulomb is defined as the amount of charge carried by a current of 1 A in 1 s.

In Figure 19.1, we have drawn the electric current as confined to a wire, but an electric current is present whenever there are moving charges. For the case of a lightning bolt, in which charges move through the atmosphere, we can still discuss and measure the electric current carried by the bolt, even though the moving charges are not confined to a wire.

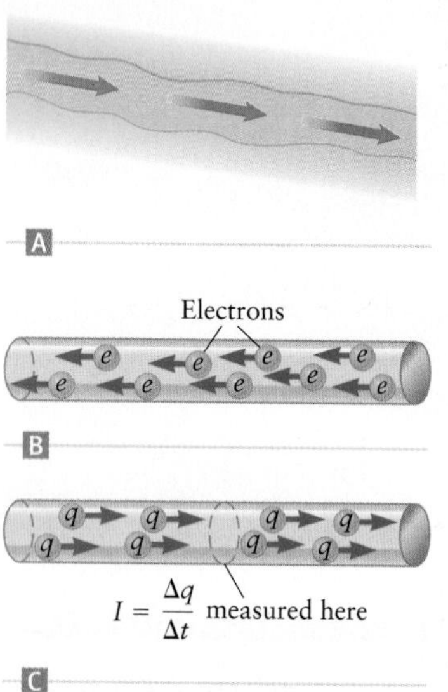

▲ **Figure 19.1** A When water flows, it forms a current as water molecules move along a river or stream. B When electrons move through a wire, they form an electric current as charge moves along the wire. C The electric current I is defined as the rate at which net positive charge passes by a point in the wire.

Which Way Does the Charge Move?

In our definition of current in Equation 19.1, Δq is the *net* electric charge that passes a particular point during a time interval Δt. This amount of charge could consist of a few particles each carrying a large electric charge or of many particles each having a small amount of charge. A current can also be produced by the motion of a combination of positive and negative charges such as a collection of positive and negative ions in solution. By convention, an electric current's direction is defined as shown in Figure 19.2. If the current is carried by positive charges moving with a velocity $\vec{v}$, the direction of the current is *parallel* to $\vec{v}$. On the other hand, if current is carried by negative charges, the direction of I is *opposite* to the charges' velocity. Thus, the direction of the current carried by electrons in a metal wire (an extremely common and important case) is *opposite* to the electron velocity (Fig. 19.2A).

A positive electric current can thus be caused by positive charges moving in one direction *or* by negative charges moving in the opposite direction. A measurement of the current alone cannot distinguish between these two alternatives.[1] Current in a metal is carried by electrons. In a liquid or gas, there will usually be both positive

[1] A phenomenon called the *Hall effect* does distinguish between these two possibilities. The Hall effect relies on magnetic forces, as described in Chapter 20.

and negative mobile ions present; if so, the current is carried by a combination of positive ions moving in one direction and negative ions moving in the opposite direction (Fig. 19.2B).

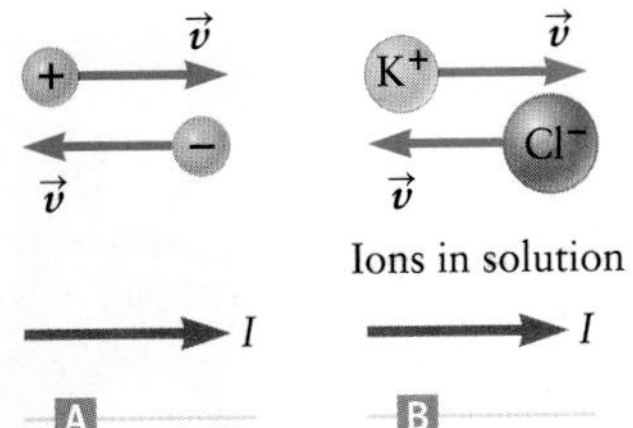

▲ **Figure 19.2** A Current is directed to the right when either a positive charge moves to the right or a negative charge moves to the left. B In a solution containing positive and negative ions, motion of these ions produces an electric current.

EXAMPLE 19.1 How Many Electrons Does It Take to Light a Lightbulb?

A typical household lightbulb requires a current of about $I = 0.50$ A. If this light is turned on for 1 hour, how many electrons pass through the lightbulb?

RECOGNIZE THE PRINCIPLE

According to the definition of electric current, $I = \Delta q/\Delta t$, so the total amount of charge that passes through the lightbulb in a time Δt is $\Delta q = I\,\Delta t$.

SKETCH THE PROBLEM

Figure 19.1B describes the problem.

IDENTIFY THE RELATIONSHIPS AND SOLVE

Our lightbulb is carrying current for 1 hour, so $\Delta t = 1 \text{ h} = 3600$ s, and

$$\Delta q = I\,\Delta t = (0.50 \text{ A})(3600 \text{ s}) = 1800 \text{ C}$$

which is the total charge in coulombs. To get the number of electrons, we divide by the magnitude of the electron's charge:

$$N = \frac{\Delta q}{e} = \frac{1800 \text{ C}}{1.6 \times 10^{-19} \text{ C/electron}} = \boxed{1.1 \times 10^{22} \text{ electrons}}$$

► ***What does it mean?***

Let's compare this number with the total number of atoms in a penny. A modern penny has a mass of about 3 g, is composed of mostly zinc, and contains about 8×10^{23} electrons. So, the number of electrons that pass through our lightbulb in 1 hour is about 1% of the number of electrons in a penny.

19.2 Batteries

Electric Current and Potential Energy

Current is the flow of charge, but how do we set the charge in motion? From mechanics, we know that the current of water in a river (Fig. 19.3A) is connected with gravitational potential energy. The kinetic energy of the moving water is derived from the change in gravitational potential energy as water moves from the high end of the river to the low end.

Electric current is produced in a similar way. For charge to move along a wire, the electric potential energy at one end of the wire must be higher than the electric potential energy at the other end. Suppose the electric current in Figure 19.3B is carried by positive charges moving to the right in the copper cylinder representing the wire. The electric potential energy is related to the potential V by $V = PE_{\text{elec}}/q$ (Eq. 18.10). In dealing with current and electric circuits, electrical engineers and physicists often refer to this potential as simply the "voltage." For a positive charge, a region of high potential energy is also a region of high voltage. The current in Figure 19.3B is thus directed from a region of higher voltage to a region of lower voltage,

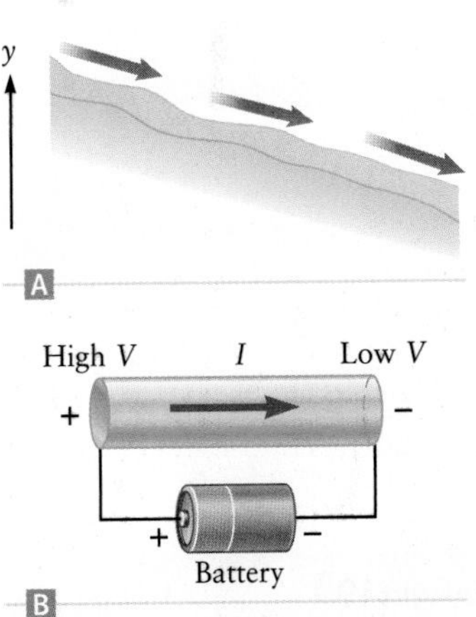

▲ **Figure 19.3** A Water flows downhill, from a region of high gravitational potential energy to a region of lower potential energy. B Electric charges in a wire move from a region of high electric potential energy to a region of lower potential energy. This electric potential difference can be produced by a battery.

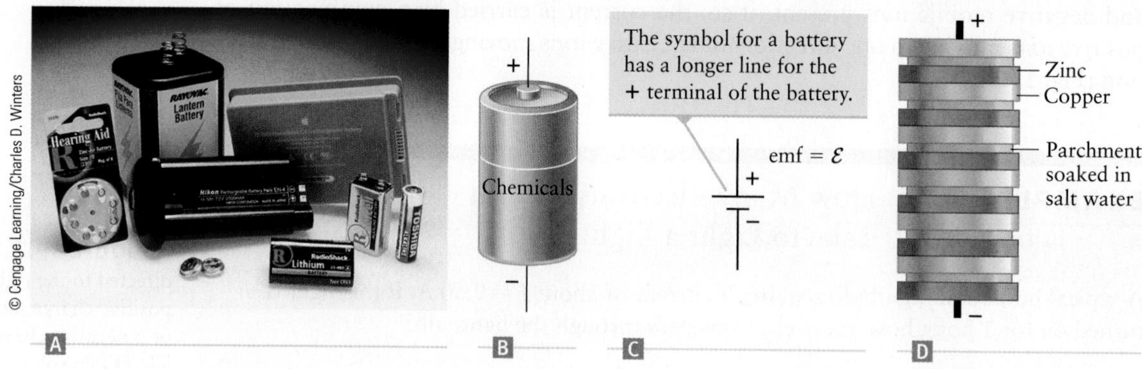

▲ **Figure 19.4** A Some commonly available batteries. B A battery uses chemical potential energy to produce an electric potential difference across its terminals. C Circuit diagrams use this symbol to denote a battery. The longer line represents the positive terminal of the battery. D Design of the battery constructed by Volta.

that is, from high to low potential. On the other hand, the current in Figure 19.3B might be carried by electrons (negative charges) moving to the left in the copper wire. Since q is negative, the potential energy PE_{elec} for electrons will be higher at the right end than at the left, and the electrons move from a region of lower voltage to a region of higher voltage in going from right to left in Figure 19.3B. Either way, *the direction of the current I is always from high to low potential*, regardless of whether this current is carried by positive or negative charges.

Figure 19.3B shows the relative voltages at the two ends of the wire using the + and − symbols. The source of this potential difference is often a device called a ***battery*** (Fig. 19.4A). A typical battery possesses two terminals, labeled "positive" (+) and "negative" (−); these terms refer to the relative electric potential at the terminals. If the battery leads are attached to the ends of a wire as sketched in Figure 19.3B, there will be a potential difference "across" the wire, producing a current.

Construction of a Battery

Alessandro Volta (1745–1827) and his contemporaries discovered and developed the first batteries (Insight 19.1). Before that time, all studies of electricity used static electricity produced by, for instance, rubbing amber with fur. Although some clever rubbing machines generating quite high electric potentials were constructed, the availability of batteries made possible many new experiments in electricity and magnetism. These experiments and the availability of batteries also led to the first practical applications of electricity such as the telegraph. Batteries illustrate conservation of energy in an interesting way as they convert chemical energy to electrical energy using ***electrochemical reactions***. The details of these reactions are beyond the scope of this text. For most of our work, we can take the "black box" view of batteries shown in Figure 19.4B. A typical battery contains two pieces of metal called electrodes, attached by metal wires to the battery terminals. The electrodes are in contact with specially chosen chemicals, and electrochemical reactions involving the electrodes and these chemicals produce an electric potential difference between the two electrodes. Different types of batteries employ different electrochemical reactions and are packaged in many different forms. In drawing an electric circuit, all types of batteries are represented by the symbol shown in Figure 19.4C, with terminals labeled + (positive potential) and − (negative potential).

Insight 19.1
VOLTA'S BATTERIES
In Chapter 18, we mentioned Luigi Galvani's experiments with frogs and electricity. He found that frog muscle lying on an iron plate twitched when he touched it with a brass scalpel. This discovery was the basis for Volta's first batteries (Fig. 19.4D), which consisted of alternating sheets of zinc and copper separated by a piece of parchment soaked with salt water. (Salt water is similar to the fluid inside a frog.)

The potential difference between a battery's terminals is called an ***electromotive force***, or ***emf*** (pronounced "ee-em-eff"). This term was adopted before it was realized that batteries actually produce an *electric potential difference* (and not a force). We follow standard practice and denote this emf by $\mathcal{E}$, and we will often refer to it as simply a voltage. The value of $\mathcal{E}$ produced by a battery depends on the particular chemical reactions it employs and how the electrodes are arranged. Typical values are a few volts, but it is possible to connect batteries in various ways to produce much higher voltages.

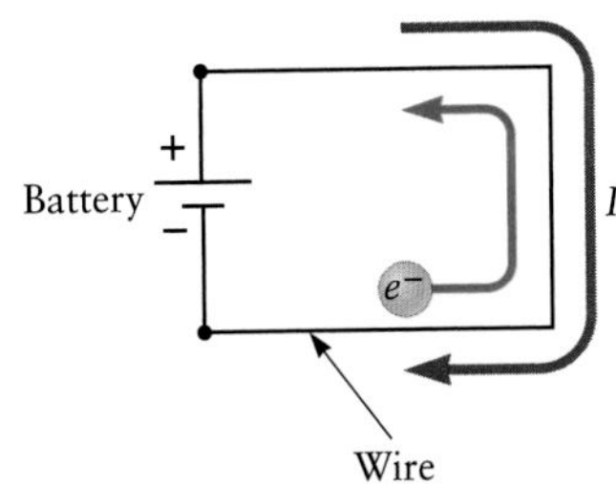

▲ **Figure 19.5** Because electrons are negatively charged, their velocity is directed opposite to the current.

Ideal Batteries and Real Batteries

When a battery is attached to a wire as sketched in Figure 19.5, electrons move out of the negative terminal of the battery through the wire and into the positive battery terminal. Hence, charge moves from one battery terminal to the other through this external ***circuit***. At the same time, ions inside the battery move between the electrodes as part of the electrochemical reaction we mentioned in connection with Figure 19.4B; this reaction moves charge internally between the electrodes so that no net charge accumulates on the battery terminals while the current is present.

An ***ideal battery*** ***always*** maintains a fixed potential difference (a constant emf) between its terminals. This emf is maintained no matter how much current flows from the battery. This picture of an ideal battery is useful, and we'll employ it in much of our work on electric circuits. *Real* batteries, however, have two practical limitations. One is that the emf decreases when the current is very high. The value of the current at which this decrease becomes significant depends on the battery size and design. The decrease in emf at high current occurs because the electrochemical reactions inside the battery do not happen instantaneously. A certain amount of time is needed for charge to move internally between the electrodes. Another limitation is that a real battery will "run down"; it will not work forever. Under normal operation, there is a certain total amount of charge that can be drawn from the battery, and when this total is reached, the battery ceases to function. This limit is reached when the chemicals used in the electrochemical reaction are exhausted. At that point, all the available chemical energy has been extracted from the battery.

EXAMPLE 19.2 Using a Capacitor as a Source of Electric Energy

Batteries and capacitors can both store energy. In a battery, the energy is stored as chemical energy, and in a capacitor, the energy is stored as electric potential energy. In both cases, this energy can be used to produce an electric current. In some applications in which very high currents are needed for a very short time (such as a camera flash), capacitors are preferred over batteries. A typical capacitor used in this way might have a capacitance of 1.0 F and an initial potential difference of 3.0 V between its plates. Compare the energy stored in this capacitor with the energy stored in an AA battery, which is about 7000 J.

RECOGNIZE THE PRINCIPLE

From Chapter 18, the energy stored in a capacitor is given by

$$PE_{\text{elec}} = \tfrac{1}{2}C(\Delta V)^2$$

(Eq. 18.33). The values of C and ΔV are given, so we can calculate PE_{elec}.

SKETCH THE PROBLEM

No sketch is necessary.

IDENTIFY THE RELATIONSHIPS AND SOLVE

Inserting the given values of capacitance and potential difference leads to

$$PE_{\text{elec}} = \tfrac{1}{2}C(\Delta V)^2 = \tfrac{1}{2}(1.0\text{ F})(3.0\text{ V})^2 = \boxed{4.5\text{ J}}$$

(continued) ▶

▶ ***What does it mean?***
The energy stored in this capacitor is *much* less—by a factor of about 1500—than the amount of energy stored in an AA battery. A very large amount of chemical energy can be stored in a small volume, if one uses the proper chemicals! Although the capacitor stores less energy, its advantage in applications is that its energy can be extracted very quickly, which is essential for a camera flash.

19.3 Current and Voltage in a Resistor Circuit

Figures 19.3B and 19.5 show simple electric ***circuits***, arrangements in which charge is able to flow around a closed path. In these examples, this path starts at one battery terminal, passes through a wire, continues to the other battery terminal, and then returns to the original terminal as it travels through the interior of the battery. In the previous section, we discussed the part of the path that is within the battery. Now let's focus on the current external to the battery and consider how this current is related to the electric potential difference provided by the battery.

Figure 19.6A shows an atomic-scale view of the charge motion within the wire. Assuming we are dealing with a metal wire, these charges are electrons and their motion is similar to the motion of gas molecules studied in our work on kinetic theory in Chapter 15 (Fig. 19.7A). The electrons frequently collide with one another and with the stationary ions within the metal, resulting in the zigzag trajectories sketched in Figure 19.6A. If no electric field is present, the net result of these zigs and zags is that the average electron displacement after many collisions is zero; there is no net movement of charge and hence no current.

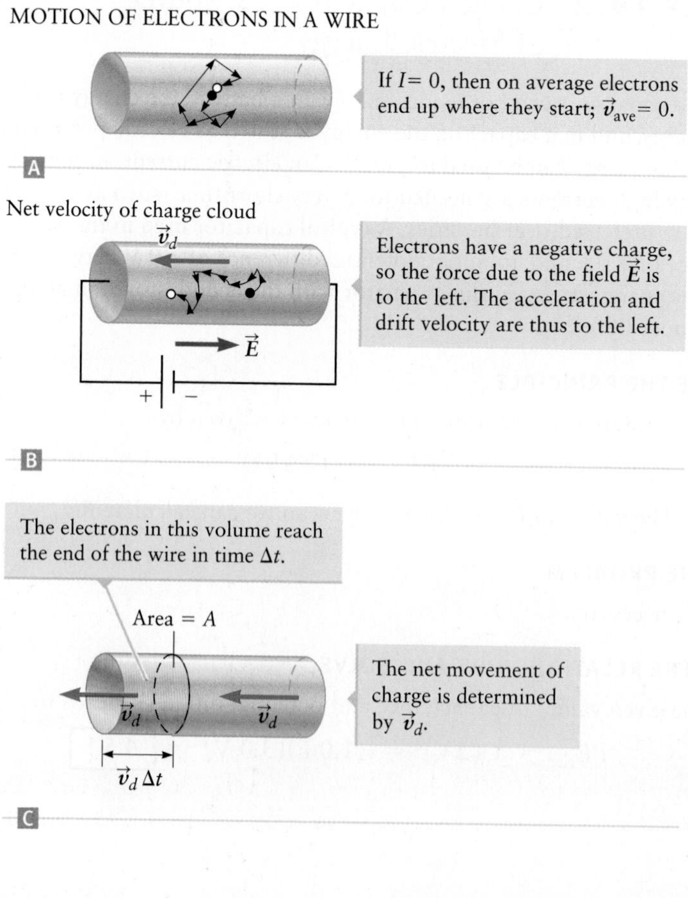

MOTION OF ELECTRONS IN A WIRE

If $I = 0$, then on average electrons end up where they start; $\vec{v}_{ave} = 0$.

A

Net velocity of charge cloud

$\vec{v}_d$

$\vec{E}$

+ −

Electrons have a negative charge, so the force due to the field $\vec{E}$ is to the left. The acceleration and drift velocity are thus to the left.

B

The electrons in this volume reach the end of the wire in time Δt.

Area = A

$\vec{v}_d$ $\vec{v}_d$

$\vec{v}_d \Delta t$

The net movement of charge is determined by $\vec{v}_d$.

C

▶ **Figure 19.6** The total current in a wire is due to the motion of electrons. A When $I = 0$, the *average* electron velocity is zero, but the *instantaneous* velocity of any particular electron is not zero. B An electric field accelerates *all* the electrons in the same direction. C This acceleration leads to a drift velocity that is the same for all electrons in the wire.

When a battery is connected to the ends of the wire, an electric potential difference is generated between those ends. To this point, we have denoted a potential difference as ΔV, using Δ (the Greek letter delta) to emphasize that we are dealing with a "difference" measured between two locations. When working with circuits, however, it is common practice to simply use V (instead of ΔV). This notation should cause no confusion as long as we choose a reference point in the circuit at which we take the potential to be zero (our reference potential). For the wire in Figure 19.6B, we can choose this reference point to be at one end of the wire. If the electric potential across the wire has a magnitude V, the electric field within the wire is

$$E = \frac{V}{L} \tag{19.3}$$

where L is the length of the wire.

Equation 19.3 might seem to contradict Chapter 17, where we argued that the electric field inside a metal is zero. However, the electric field inside a metal is zero only when the charges in the metal are in static equilibrium, which is *not* the case here. For a metal to carry a current, there must be a nonzero electric field inside, providing the force that maintains the current.

ANALOGY TO GAS MOLECULES

Quiet conditions (no wind); no net motion on average.

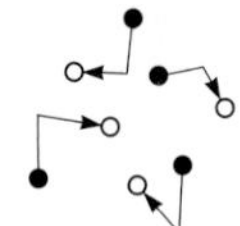

A

Windy day; net motion parallel to $\vec{v}_{wind}$.

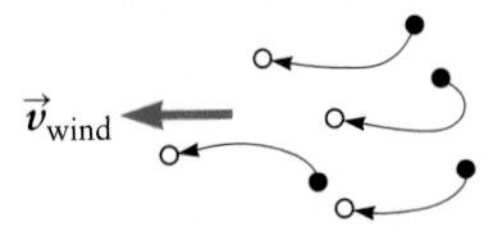

B

▲ **Figure 19.7** A Gas molecules in a room follow zigzag trajectories as they collide with other gas molecules, but the net (average) displacement is zero. B On a windy day, the average velocity of a gas molecule is equal to the wind velocity, even though the instantaneous velocity during the zigs and zags in A is much greater.

Drift Velocity and Current

The electric field in the wire in Figure 19.6B produces a force that pushes all the electrons to the left. Hence, the zigzag "swarm" of electrons now moves to the left with a ***drift velocity***, $\vec{v}_d$. *All* electrons in the wire experience this extra "drifting" motion, giving a net movement of negative charge to the left in Figure 19.6B and thus an electric current to the right. The drift motion of the electrons is similar to the motion of gas molecules in a gentle wind. In that case, the molecular motion is a combination of the zigzag trajectories in Figure 19.7A with the wind's velocity, producing a "cloud" of molecules moving as shown in Figure 19.7B. The wind velocity of the molecules is thus similar to the drift velocity of the electrons.

To calculate the electric current in a wire, it is useful to focus on just the drift motion as shown in Figure 19.6C. The current is equal to the amount of charge that passes out the end of the wire per unit time. In the time interval Δt, all electrons move (on average) a distance $v_d\,\Delta t$, so all electrons that are within this distance from the left end will (on average) leave the wire in a time Δt. The number of electrons that leave is equal to the number of electrons in the gray volume in Figure 19.6C. If the number of electrons per unit volume is n, for a wire with cross-sectional area A, the number of electrons that exit the wire is

$$N = nAv_d\,\Delta t \tag{19.4}$$

Each electron carries a charge $-e$, so the total charge that leaves the wire is

$$\Delta q = N(-e) = -nAev_d\,\Delta t \tag{19.5}$$

The current is given by $I = \Delta q/\Delta t$, so we can divide both sides of Equation 19.5 by a factor of Δt to get

$$I = -neAv_d \tag{19.6}$$

Ohm's Law

Equation 19.6 tells how the current is related to the drift velocity, but what we really want to know is how I is related to the potential difference across the wire. That will enable us to deduce some fundamental rules of circuit analysis. To address that question, we must consider the drift velocity v_d in a little more detail. Recall our discussion of the drag forces on a bacterium in a fluid (Chapter 3, Eq. 3.23). An electron moving through a metal (Fig. 19.6A) is similar to a bacterium moving through a fluid. The bacterium undergoes many collisions with fluid molecules, just as electrons experience many collisions in the course of their zigzag trajectories. The collisions produce a drag force directed opposite to the velocity:

$$\vec{F}_{\text{drag}} = -b\vec{v}$$

In Chapter 3, we showed that this drag force leads to a terminal velocity proportional to the force that pushes the *E. coli* through a fluid (Eq. 3.25). In the same way, the drag force on electrons leads to a drift velocity proportional to the force pushing the electrons through the metal. This electric force is proportional to E, so the drift velocity is proportional to the magnitude of the electric field:

$$v_d \propto E \tag{19.7}$$

According to Equation 19.3, E is itself proportional to the voltage across the wire, so the drift velocity is also proportional to V:

$$v_d \propto V \tag{19.8}$$

Inserting Equation 19.8 into our expression for the current (Eq. 19.6) gives

$$I \propto V \tag{19.9}$$

The constant of proportionality between I and V in Equation 19.9 involves a quantity called the electrical ***resistance*** R of the wire, defined by

Ohm's law

$$I = \frac{V}{R} \tag{19.10}$$

This relation between resistance, current, and voltage is called ***Ohm's law***, and it applies to a very wide range of materials. According to Equation 19.10, R can be measured in units of voltage/current = volts/amperes = V/A. This combination of units has the name "ohm" and is denoted by the symbol Ω.

The value of a wire's resistance depends on its composition, size, and shape. It is useful to define a quantity ρ called ***resistivity***, where the value of ρ depends on the material used to make the wire, but not on its size or shape. The resistance of a cylindrical wire of length L and cross-sectional area A (Fig. 19.8) is related to its resistivity by

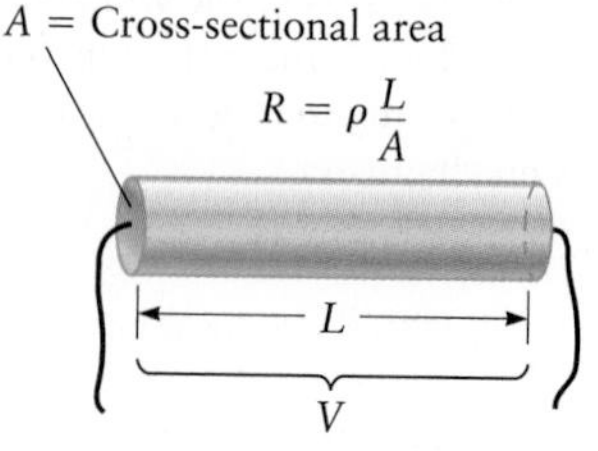

▲ **Figure 19.8** The resistance R of a cylindrical wire is related to its resistivity ρ by $R = \rho L/A$.

$$R = \rho \frac{L}{A} \tag{19.11}$$

The value of ρ is different for different metals; copper has a different resistivity than aluminum or gold, for example. Values of ρ can also depend on the metallurgical history (how the particular metal is heat treated or bent, etc.). Typical resistivities of some common materials are listed in Table 19.1.

Ohm's law (Eq. 19.10) predicts a linear relationship between current and voltage. However, even though it is called a "law," it is not a fundamental law of physics in the sense of Newton's laws of mechanics. While the relation between electric current and voltage for many materials (including metals) is accurately described by Ohm's law, there are some important exceptions. For example, devices called diodes and

TABLE 19.1 Resistivities of Some Common Materials at Room Temperature

Material	ρ ($\Omega \cdot$ m)	Material	ρ ($\Omega \cdot$ m)
Metals:		Insulators:[a]	
Copper	1.7×10^{-8}	Glass	1 to 1000×10^9
Aluminum	2.7×10^{-8}	Rubber	1 to 100×10^{12}
Gold	2.2×10^{-8}	Semiconductors:[a]	
Silver	1.6×10^{-8}	Silicon	0.1 to 100
Lead	22×10^{-8}	Germanium	0.001 to 1

[a]The resistivities of insulators and semiconductors depend strongly on their purity.

transistors do not obey Ohm's law. These devices are essential for making amplifiers, electronic switches, and computer memory elements, and are thus extremely important in many applications. Even though Ohm's law does not apply to every type of circuit element, the lessons we'll learn from studying resistors and Ohm's law in this chapter are the foundation of all circuit theory.

EXAMPLE 19.3 The Resistance of a Piece of Copper Wire

The wires commonly used in household wiring are made of copper and have a typical radius $r = 1.0$ mm. (**a**) Find the resistance of a piece of this wire of length $L = 5.0$ m. (**b**) Assume a battery with an emf of 12 V is attached to the ends of this wire. What is the current?

RECOGNIZE THE PRINCIPLE

We can use the relation between resistance and resistivity, $R = \rho L/A$, to find the resistance. Given the voltage across the wire, Ohm's law then gives the current.

SKETCH THE PROBLEM

Figure 19.8 describes the problem.

IDENTIFY THE RELATIONSHIPS AND SOLVE

(**a**) We use $R = \rho L/A$ (Eq. 19.11) to find the resistance of the wire. The wire's cross-sectional area is $A = \pi r^2 = 3.1 \times 10^{-6}\ \text{m}^2$. Using this area and the length of the wire along with the resistivity of copper from Table 19.1, we find

$$R = \rho \frac{L}{A} = (1.7 \times 10^{-8}\ \Omega \cdot \text{m}) \frac{5.0\ \text{m}}{3.1 \times 10^{-6}\ \text{m}^2} = \boxed{0.027\ \Omega}$$

(**b**) The corresponding current can be found from Ohm's law (Eq. 19.10):

$$I = \frac{V}{R} = \frac{12\ \text{V}}{0.027\ \Omega} = \boxed{440\ \text{A}}$$

► ***What does it mean?***

We'll compare these results for resistance and current to other common cases shortly. However, you should suspect that the result in part (**b**) is a very large current. We'll also see that the result in part (**a**) is a small resistance.

▲ **Figure 19.9** Resistors come in a wide variety of shapes and sizes. Larger ones are generally able to carry higher currents without overheating.

Wires and Resistors

Figure 19.9 shows a photo of some common types of electrical ***resistors***. The metal wires we considered above are types of resistors, but resistors that follow Ohm's law can be made in other shapes and from other materials. Each of these objects has a particular resistance R that determines how the current through and voltage "across" the resistor are related. If we connect the two ends of a resistor to the terminals of a battery, the voltage V across the resistor is equal to the battery's emf, hence, $V = \mathcal{E}$. Inserting into Ohm's law (Eq. 19.10), we find a current $I = \mathcal{E}/R$ through the resistor. This electric circuit is represented schematically in the circuit diagram in Figure 19.10A. The battery is represented using the symbol [battery symbol], and the resistor is represented in a circuit diagram by the symbol [resistor symbol]. Table 19.2 lists the symbols for these and several other common circuit elements.

The straight lines in Figure 19.10A represent metal wires that attach the resistor to the battery's terminals. (The wires may not be straight in the actual circuit as

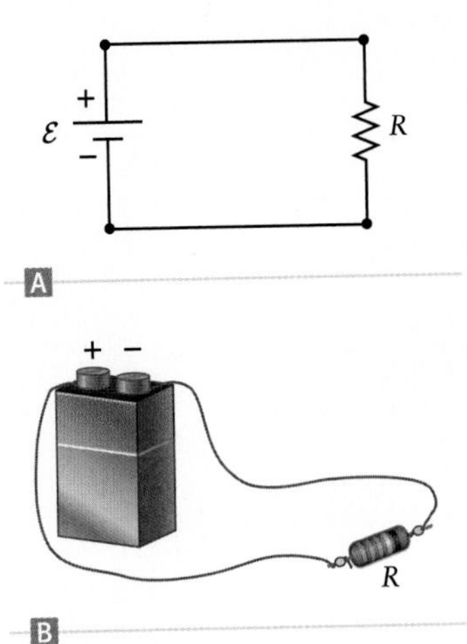

▲ **Figure 19.10** A Circuit diagram for a battery connected to a resistor. B Sketch of the actual physical circuit represented by the diagram in A.

TABLE 19.2 Symbols for Some Commonly Used Circuit Elements

Circuit Element	Symbol	Actual Appearance	Circuit Element	Symbol	Actual Appearance
Battery	+ −	© Cengage Learning/ Charles D. Winters	Lightbulb		© David Graeme-Baker/Alamy; © Anton Prado/Alamy
Resistor		©INSADCO Photography/Alamy	Wire		© Cengage Learning/ Charles D. Winters
Capacitor		© Hemera Technologies/Photo Objects/Jupiterimages	Connection between wires		© Cengage Learning/ Charles D. Winters
Switch		© Hemera Technologies/Photo Objects/Jupiterimages	Electrical ground		

shown in Fig. 19.10B, but bending a wire in this way does not significantly affect its resistance.) Metal wires are commonly referred to as simply "wires" or "leads." We saw in Example 19.3 that a metal wire has a nonzero resistance. Hence, strictly speaking, we should also refer to a wire as a resistor. However, the value of R for a typical metal wire is *much smaller* than for the resistors used in electronic devices, so when analyzing circuits it is a very good approximation to assume all wires are ideal having $R = 0$. We can account for the small but nonzero resistance of a real wire when necessary.

Dots in a circuit diagram represent connections between different circuit elements. In Figure 19.10A, they represent the points at which the wires attach to the battery and where the wires attach to the resistor. It is sometimes useful to show the dots when one wants to be clear about where the different parts of a circuit begin and end. When that is not important or is obvious, it is usually simpler to omit the dots.

Speed of an Electric Current

If we attach a battery to a metal wire at time $t = 0$, when will electrons start to "flow" out of the end of the wire? From Figure 19.6, you might expect that this time is determined by the drift velocity, but that is *not* correct. To see why, let's calculate v_d in a typical case. The relation between the current and the drift velocity is (Eq. 19.6)

$$I = \frac{\Delta q}{\Delta t} = -neAv_d$$

Solving for v_d leads to

$$v_d = -\frac{I}{neA} \tag{19.12}$$

For a household wire such as the copper wires that connect to a lightbulb, the current is typically $I = 1$ A. The cross-sectional area is about $A = 3 \times 10^{-6}$ m^2 (a cylindrical wire with a radius of 1 mm), and the density of electrons in copper is about[2] 2×10^{26} electrons/m^3. Inserting these values into Equation 19.12 gives

$$v_d = -\frac{I}{neA} = -\frac{1 \text{ A}}{(2 \times 10^{26} \text{ electrons/m}^3)(1.60 \times 10^{-19} \text{ C})(3 \times 10^{-6} \text{ m}^2)}$$

$$v_d = -0.01 \text{ m/s}$$

where the negative sign means that the drift velocity is directed opposite to the current since the electron has a negative charge. The magnitude of this drift velocity is quite small; you can easily walk at a speed greater than v_d. A typical wire can be many meters long, so electrons moving at this drift velocity could take minutes or even hours to complete their path through a circuit. You might therefore expect to notice a considerable delay from the moment you "throw a switch" to the time a lightbulb begins to emit light.

You have no doubt done this experiment, so you know that there is in fact *no* perceptible time delay from the time you push a switch and a light turns on. How do we reconcile this observation with the very small value of the drift velocity? Consider an analogy with the flow of water through a garden hose when the hose is initially full of water. When water from a faucet enters one end of the hose, water leaves the other end almost immediately. Similarly, when the first electrons enter one end of a wire, *other* electrons leave the opposite end of the wire almost instantaneously once a potential difference is established across the wire's ends. In fact, the speed of an electric current is equal to the speed of electromagnetic radiation in the wire, which is nearly the speed of light in a vacuum. (We'll return to this subject in Chapter 23.)

19.4 DC Circuits: Batteries, Resistors, and Kirchhoff's Rules

An electric circuit is a combination of connected elements such as batteries and resistors forming a complete path through which charge is able to move. Calculating the current in a circuit is a job called ***circuit analysis***. In this chapter, we consider only the simplest "DC" circuits. The letters DC stand for "direct current," in contrast to circuits with an "AC" or "alternating current." Roughly speaking, an AC circuit is one in which the current oscillates at a particular frequency, while all other cases fall into the category of DC circuits. The current in a DC circuit is constant (independent of time) or approaches a constant if we wait long enough. Our main goal for the rest of this chapter is to learn how to analyze DC circuits.

What Is a "Circuit"?

In the circuit in Figure 19.10, two wires connect the battery to the resistor. Why do we need two wires? Why can't we carry charge from the battery to the resistor with a single wire? The need for two wires can be understood from analogy with the "water

[2]This value assumes there is one current-carrying electron for each copper atom in the wire, which is a good approximation—accurate to within a factor of two or so—for copper and most metals.

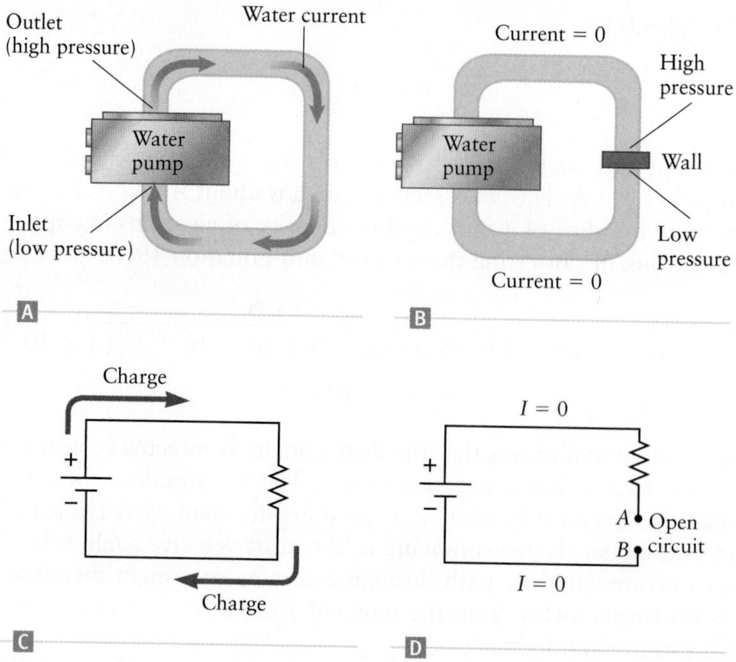

► **Figure 19.11** A The pressure difference produced by a water pump can cause a current of water around a closed loop of pipe. B If a wall is inserted into the pipe, the current of water *everywhere* stops. A pressure difference then exists across the wall. C A battery uses an emf to produce a current of charge around a complete circuit. D If a wire in the circuit is cut, the current stops *everywhere* in the circuit.

pipe circuit" in Figure 19.11, in which a water pump is used to push water through a pipe. This pipe carries water from the outlet (high-pressure) side of the pump and returns it to the inlet (low-pressure side), and the water current has the same magnitude throughout the circuit. This water circuit could be broken by inserting a wall into the pipe, which prevents water from flowing around the circuit and makes the current zero. Notice that the current is zero on *both* sides of the wall. The pressure produced by the water pump is now felt as a pressure difference between the two sides of the wall.

The same ideas apply to the electric circuit in Figure 19.11C. The electric current can be viewed as the motion of positive charge[3] out of the high-potential (positive) terminal of the battery; this charge travels around the circuit passing through the resistor and returns to the low-potential (negative) terminal of the battery. As with the water circuit, the electric current is the same at all points in the circuit: I has the *same* value when this charge is "heading into" and "returning from" the resistor. If we now cut the wire (Fig. 19.11D), charge cannot return to the battery and we have an ***open circuit***. The current is now zero *throughout the circuit*, and the potential difference generated by the battery appears across the open circuit (points A and B in Fig. 19.11D), just as the pressure difference from the water pump appears across the wall in Figure 19.11B.

The important message from Figure 19.11 is that *there must be a complete circuit for charge flow*. Simply connecting one wire from a battery to a resistor cannot produce a current. *There must always be a return path* for the current to return to the voltage source. This fact is illustrated in Figure 19.12A, which shows a bird sitting on an electric power line. A power line acts as a voltage source for your house and is at high potential. When a bird comes in contact with a power line, the bird is also at a high potential. If the bird does not simultaneously touch the ground, however, there is no return path to ground through the bird, and the current through the bird is negligible. We thus have essentially an open circuit with no current (Fig. 19.12B). The health hazards of electricity are determined by the current (Section 19.8); as long as very little current flows through the bird, its health is not endangered.

Ground
$V = 0$

A

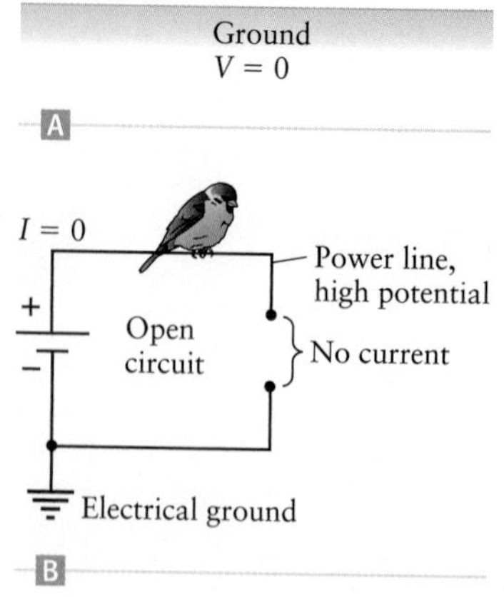

▲ **Figure 19.12** If a bird on a power line does not touch the ground, the circuit is not completed and the current is zero. The bird is spared.

[3]For a metal wire, the charge would be carried by electrons (negative charges) going in the opposite direction, but that does not change our argument or conclusions.

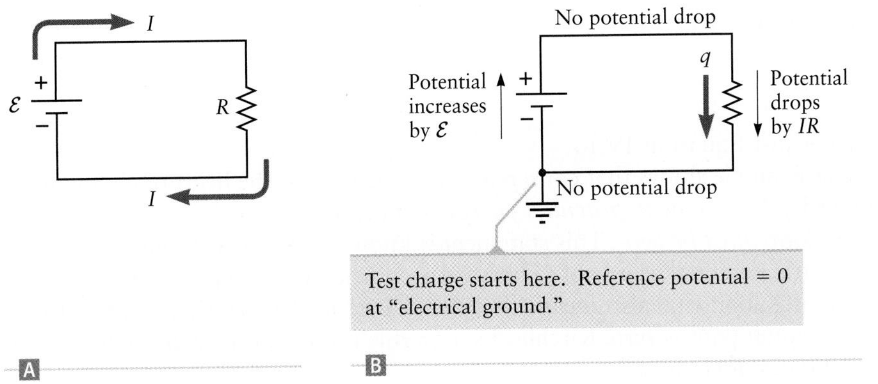

◀ **Figure 19.13** A This battery produces a clockwise current through the resistor. B The electric potential changes as we move from one part of the circuit to another. The sum of all these voltage changes must be zero when we move completely around the circuit.

One-Loop Circuits: Kirchhoff's Loop Rule

The circuit in Figure 19.13 shows a single ideal battery with an emf $\mathcal{E}$ connected to a single resistor with resistance R. From Ohm's law (Eq. 19.10), the current in this circuit is

$$I = \frac{\mathcal{E}}{R} \tag{19.13}$$

Current through a resistor

In words, we say that the battery produces an electric potential difference $\mathcal{E}$ "across" the resistor. We also refer to $\mathcal{E}$ as the "voltage across the resistor."

The electric current in Figure 19.13A consists of charges that travel along a complete loop, passing through the battery, through the attached wires and resistor, and back to the battery. Let's consider the electric potential energy of a test charge q as it moves through the circuit. Electric potential energy is related to the potential by $PE_{elec} = qV$, so we are really asking how the electric potential varies around the circuit. Suppose the test charge begins at the negative battery terminal. We can use this spot as the reference point for the potential and thus assign $V = 0$ at this point. We also recall the convention that $V = 0$ at an electrical "ground," so we call that our "ground point." In a circuit diagram, ground is indicated by the triangular symbol indicated in Figure 19.13B. Our charge q thus begins at the lower left in the figure with a potential $V = 0$ and then travels through the battery, where the electrochemical reaction increases its potential by an amount $\mathcal{E}$, the battery's emf. When the charge reaches the wire on the top edge of the circuit, it thus has a potential $\mathcal{E}$. We assume the wires in this circuit are ideal, so their resistance is zero and the potential is the same everywhere within this top wire. The potential at the "top" end of the resistor is thus still $\mathcal{E}$. The charge q then passes through the resistor, moving from a point of high potential (the top end of the resistor) to a point of lower potential (the bottom of the resistor). According to Equation 19.10, the potential *decreases* by IR, so the voltage "drops" an amount IR as the charge moves between these two ends of the resistor. The final step takes the charge through the bottom wire back to the battery, and because it is another ideal resistanceless wire, there is no potential change along this wire.

At the completion of these four steps, the charge has returned to its starting point. The net change of the potential energy and hence also the net change in the electric potential must be *zero* when the charge returns to where it began, at the bottom of the battery. We thus have

$$\text{change in the potential around the circuit} = \overbrace{+\mathcal{E}}^{\text{potential increase from the battery}} - \overbrace{IR}^{\text{potential drop across the resistor}} = 0 \tag{19.14}$$

This leads directly to

$$I = \frac{\mathcal{E}}{R}$$

which is just Equation 19.13.

Kirchhoff's loop rule

This example shows that conservation of energy is at the heart of circuit analysis. In words, *the change in potential energy of a charge as it travels around a complete circuit loop must be zero.* This statement is known as ***Kirchhoff's loop rule.*** Because the electric potential energy of a charge is proportional to the potential ($PE_{elec} = qV$), Kirchhoff's loop rule also means that the change in the electric potential around a closed-circuit path is zero. Kirchhoff's loop rule is one of our main tools for analyzing all types of circuits.

Dissipation of Energy in a Resistor

In Figure 19.13, the test charge gained energy (an increase in potential) when it passed through the battery and lost energy (a decrease in potential) as it passed through the resistor. If the potential energy of a charge decreases when it moves through a resistor, where does this energy go? The answer is that the energy of the charge is converted to heat energy inside the resistor. We say that this energy is ***dissipated*** as heat. In Figure 19.13B, the potential of the charge q decreases an amount $V = IR$ as it travels through the resistor. The energy of this charge thus decreases by an amount

$$\text{energy decrease} = qV$$

From our definition of current, $q = I\,\Delta t$, so

$$\text{energy decrease} = (I\,\Delta t)V$$

This energy shows up as heat energy in the resistor; hence,

$$\text{dissipated energy} = (I\,\Delta t)V$$

The amount of energy dissipated per unit time is the dissipated ***power*** P, so we get

Power dissipated in a circuit element

$$P = \frac{\text{dissipated energy}}{\Delta t} = VI \qquad (19.15)$$

Using our relation between voltage, current, and resistance ($I = V/R$, Eq. 19.10), Equation 19.15 can be written as

Power dissipated in a resistor

$$P = I^2R = \frac{V^2}{R} \qquad (19.16)$$

When there is current in a resistor, a certain amount of power given by Equations 19.15 and 19.16 is dissipated as heat. This energy comes from the chemical energy of the battery. So, the circuit in Figure 19.13 converts chemical energy (in the battery) to heat energy (in the resistor). We should also note that Equation 19.15 applies to *all* circuit elements, not just resistors. The power dissipated or transferred from a current I to a circuit element is always equal to the product VI, where V is the voltage across the circuit element.

Insight 19.2
ENERGY DISSIPATED IN A RESISTOR
The energy dissipated in a resistor is converted to heat, showing up as a temperature increase of the resistor and its surroundings. This is similar to the conversion of mechanical energy into heat via friction.

Resistors in Series

Figure 19.14A shows a slightly more complicated circuit with a battery connected to two resistors. This arrangement, in which the current passes first through one resistor (R_1) and then another (R_2), is called ***resistors in series.*** All the charge that flows from the battery must pass through both resistors consecutively. Let's apply

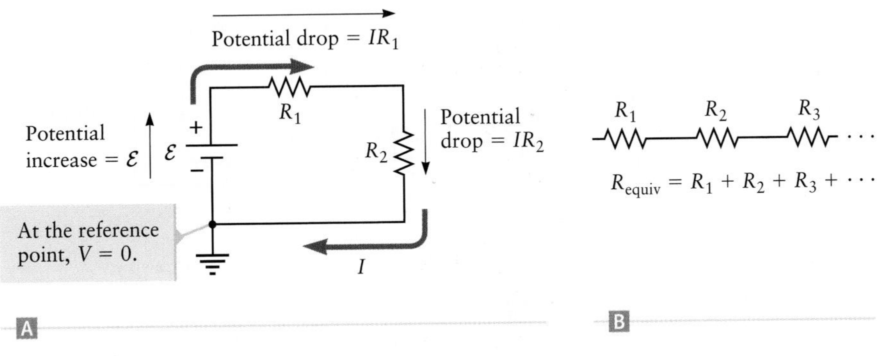

◀ **Figure 19.14** A To apply Kirchhoff's loop rule, we add up all the changes in electric potential as we move around the circuit. B When resistors are connected in series, they are equivalent to a resistor of value R_{equiv} as given in Equation 19.18.

Kirchhoff's loop rule to this circuit, starting from a reference point (ground) at the negative terminal of the battery, where $V = 0$. If a charge q begins there and travels clockwise around the circuit, its potential first increases by $\mathcal{E}$ as it passes through the battery. The potential then decreases by an amount IR_1 as it passes through the first resistor and decreases by IR_2 when it passes through the second resistor. According to Kirchhoff's loop rule, the sum of these changes around the closed circuit loop must be zero:

$$\text{change in the potential around the circuit} = \underbrace{+\mathcal{E}}_{\text{potential increase in battery}} - \underbrace{IR_1}_{\text{potential drop across } R_1} - \underbrace{IR_2}_{\text{potential drop across } R_2} = 0$$

Solving for the current, we find

$$I(R_1 + R_2) = \mathcal{E}$$

$$I = \frac{\mathcal{E}}{R_1 + R_2} \tag{19.17}$$

For the circuit with a single resistor (Fig. 19.13), we found $I = \mathcal{E}/R$ (Eq. 19.14). The combination of two resistors in Figure 19.14 is thus equivalent to a single resistor with resistance $R_{equiv} = R_1 + R_2$. In general, we can connect any number of resistors in series as in Figure 19.14B and they will be equivalent to a single resistor with

$$R_{equiv} = R_1 + R_2 + R_3 + \cdots \quad \text{(resistors in series)} \tag{19.18}$$

Insight 19.3

WHAT IS AN "EQUIVALENT" CIRCUIT?

When we say that a resistor R_{equiv} is "equivalent" to a certain arrangement of several resistors, we mean that if that arrangement of resistors is replaced by R_{equiv}, the current through the rest of the circuit is unchanged. This notion of equivalence applies to many types of circuit elements, including capacitors.

CONCEPT CHECK 19.1 Potential in a One-Loop Circuit

Consider the one-loop circuit in Figure 19.15. What is the potential at point A?

(a) $V_A = \mathcal{E}$ (b) $V_A = -\mathcal{E}$ (c) $V_A = 0$

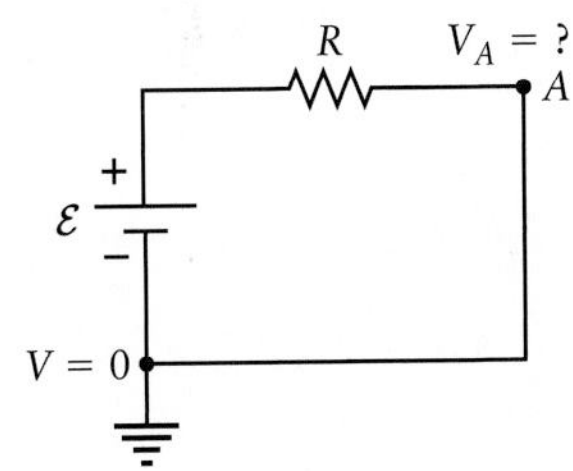

▲ **Figure 19.15** Concept Check 19.1.

CONCEPT CHECK 19.2 Potential in a One-Loop Circuit (Revisited)

One of the wires in Figure 19.15 is cut, resulting in the open circuit shown in Figure 19.16. Now what is the potential at point A?

(a) $V_A = \mathcal{E}$ (b) $V_A = -\mathcal{E}$ (c) $V_A = 0$

CONCEPT CHECK 19.3 Potential in a One-Loop Circuit (Yet Again)

What is the potential difference between points B and C in Figure 19.16?

(a) $V_B - V_C = \mathcal{E}$ (b) $V_B - V_C = -\mathcal{E}$ (c) $V_B - V_C = 0$

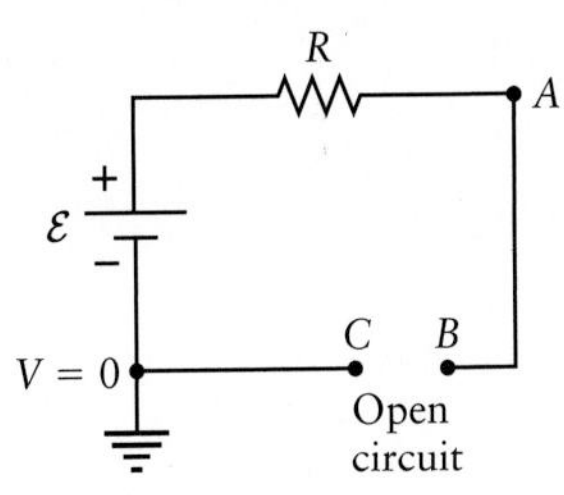

▲ **Figure 19.16** Concept Checks 19.2 and 19.3.

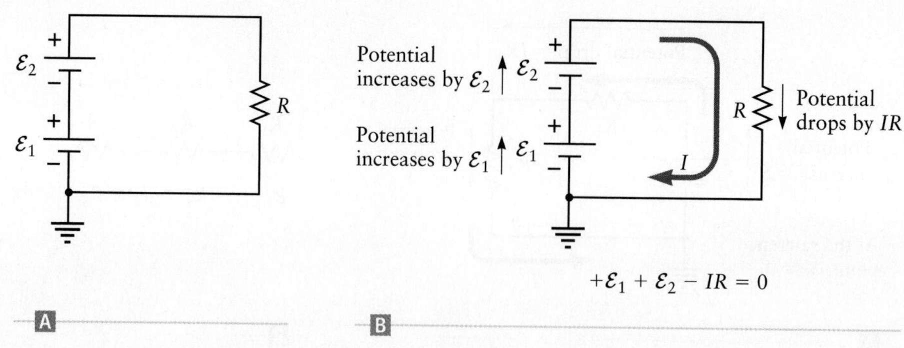

▶ **Figure 19.17** Example 19.4. Combining batteries in series.

EXAMPLE 19.4 Batteries in Series

Consider the circuit in Figure 19.17A, in which two batteries with emfs $\mathcal{E}_1$ and $\mathcal{E}_2$ are connected in series to a single resistor. What is the current in this circuit?

RECOGNIZE THE PRINCIPLE

To apply Kirchhoff's loop rule, we start at a particular point in the circuit (the reference point at ground is a convenient choice) and move around the circuit, adding up the changes in electric potential. According to Kirchhoff's loop rule, the total change in the potential around one complete loop is zero.

SKETCH THE PROBLEM

Figure 19.17B shows how the potential changes as we move around the circuit.

IDENTIFY THE RELATIONSHIPS

We start at the reference ground point (the negative terminal of battery 1) and move clockwise in Figure 19.17B. There is an increase $\mathcal{E}_1$ in the potential when we pass through battery 1 and another increase $\mathcal{E}_2$ as we pass through battery 2, followed by a potential drop of magnitude IR as we move through the resistor.

SOLVE

According to Kirchhoff's loop rule, the sum of these changes in potential must be zero:

$$\begin{matrix}\text{change in potential} \\ \text{around the circuit}\end{matrix} = \mathcal{E}_1 + \mathcal{E}_2 - IR = 0$$

Solving for the current, we find

$$\boxed{I = \frac{\mathcal{E}_1 + \mathcal{E}_2}{R}}$$

▶ ***What does it mean?***

This combination of two *batteries in series* is equivalent to a single battery with emf $\mathcal{E}_{\text{equiv}} = \mathcal{E}_1 + \mathcal{E}_2$.

Multibranch Circuits

Figure 19.18 shows a circuit in which the current from the positive battery terminal can return to the negative terminal along either of two paths: one path goes through resistor R_1, and the other path is through R_2. These different current paths are called ***branches***, and this figure shows an example of a multibranch circuit. The particular arrangement in Figure 19.18 is also referred to as ***resistors in parallel.*** The currents through these different branches need not be equal (unless $R_1 = R_2$), so now we have to solve for two unknown currents. In Figure 19.18A, we label the current through resistor 1 as I_1 and the current through resistor 2 as I_2. The current through the other parts of the

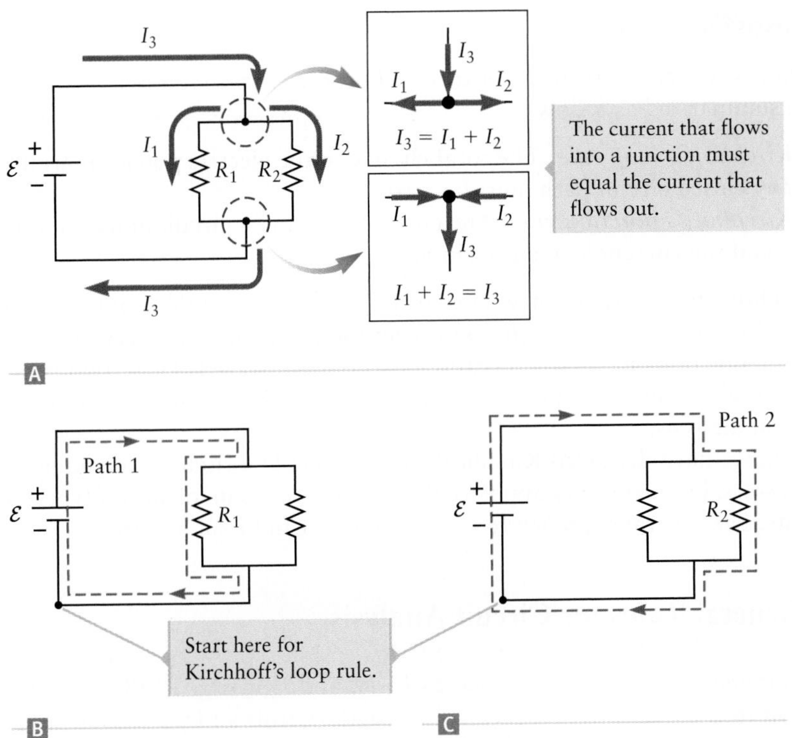

◀ **Figure 19.18** A A circuit with two resistors connected in parallel. According to Kirchhoff's junction rule, the total current entering a junction must equal the total current leaving the junction. B and C Kirchhoff's loop rule still holds, so the sum of all changes in potential along any complete path (either path 1 or path 2) around a circuit must be zero.

circuit (the wires that connect to the battery terminals) is I_3. These three currents are not independent. Figure 19.18A shows expanded views of the points where these three currents are connected. Such points are called circuit "junctions" or "nodes." Electric charge cannot be created or destroyed at a junction (or anywhere else), so *the amount of current entering a junction must be equal to the amount of current leaving it*. This fact is called ***Kirchhoff's junction rule***, and for this particular circuit it means that

Kirchhoff's junction rule

$$I_3 = I_1 + I_2 \tag{19.19}$$

To analyze this circuit and solve for all the currents, we must also apply Kirchhoff's loop rule. We start, as usual, from the negative terminal of the battery and add the changes in electric potential as we move around a closed loop. Beginning from the negative battery terminal, two closed loops take us around this circuit and back to the battery as shown in parts B and C of Figure 19.18. Following the dashed loop in part B (path 1) in which we move through resistor 1, the changes in potential are

$$\begin{matrix}\text{change in potential}\\ \text{around path 1}\end{matrix} = \underbrace{+\mathcal{E}}_{\text{battery emf}} - \underbrace{I_1R_1}_{\text{potential drop across } R_1} = 0 \tag{19.20}$$

The dashed loop in Figure 19.18C (path 2) takes us through resistor 2, and the changes in potential in this case are

$$\begin{matrix}\text{change in potential}\\ \text{around path 2}\end{matrix} = \underbrace{+\mathcal{E}}_{\text{battery emf}} - \underbrace{I_2R_2}_{\text{potential drop across } R_2} = 0 \tag{19.21}$$

Solving for the two branch currents I_1 and I_2, we get

$$I_1 = \frac{\mathcal{E}}{R_1} \quad \text{and} \quad I_2 = \frac{\mathcal{E}}{R_2} \tag{19.22}$$

The total current through the battery is then (according to Eq. 19.19)

$$I_3 = I_1 + I_2 = \frac{\mathcal{E}}{R_1} + \frac{\mathcal{E}}{R_2} \tag{19.23}$$

Insight 19.4
APPLYING KIRCHHOFF'S LOOP RULE
In all our applications of Kirchhoff's loop rule so far, we have taken the starting point at the ground point of the circuit, but that is not absolutely necessary. Because we add up changes in the potential as we make a complete loop around the circuit, we could take any point as a starting point without affecting the final result.

Kirchhoff's Rules

Kirchhoff's two rules are at the heart of *all* problems in circuit analysis. These rules can be summarized as follows.

1. ***Kirchhoff's loop rule:*** **The total change in the electric potential around any closed circuit path must be zero.**
2. ***Kirchhoff's junction rule:*** **The current entering a circuit junction must equal the current leaving the junction.**

Kirchhoff's rules are not new laws of physics, but are applications of more fundamental ideas. The loop rule follows from the principle of conservation of energy, while the junction rule is true because charge is conserved. These two general and fundamental conservation principles lead directly to Kirchhoff's very practical rules for circuit analysis.

So far, we have discussed Kirchhoff's rules using DC circuits containing batteries and resistors, but these rules apply to *all* types of circuits involving *all* types of circuit elements. The general approach to circuit analysis can be summarized as follows.

PROBLEM SOLVING General Rules for Circuit Analysis

1. **RECOGNIZE THE PRINCIPLE.** All circuit analysis is based on Kirchhoff's loop and junction rules.
2. **SKETCH THE PROBLEM.** Start with a circuit diagram and identify all the different circuit branches. The currents in each of these branches are usually the unknown variables in the problem.
3. **IDENTIFY** the different possible closed loops and junctions in the circuit.
   - Apply Kirchhoff's loop rule to each closed loop to get a set of equations involving the branch currents (such as I_1, I_2, and I_3 in Fig. 19.18) and quantities such as the battery emfs and the resistances in each loop.
   - Apply Kirchhoff's junction rule to each junction to get equations relating the different branch currents.
4. **SOLVE** the resulting equations for the current in each branch.
5. Always *consider what your answer means* and check that it makes sense.

Using Kirchhoff's Rules

Circuit elements can be connected in many different ways, so many different conceivable loops and branches are possible. While all such circuits can be attacked and solved using the above "General Rules for Circuit Analysis," keep the following points in mind.

When applying Kirchhoff's loop rule, pay attention to the sign of the voltage drop across a circuit element. For example, when navigating clockwise around the loop in Figure 19.19A, the potential *drops* as we move through the resistor. The potential

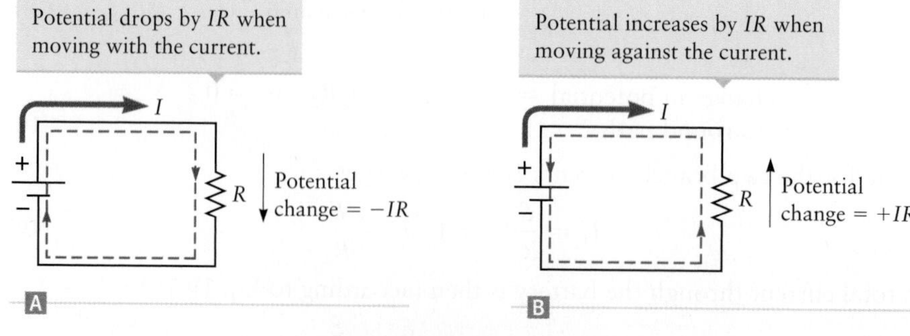

▶ **Figure 19.19** A The potential *drops* when passing through a resistor in the direction parallel to the current. B The potential *increases* when moving through a resistor in the direction opposite ("against") the current.

change across the resistor is thus $-IR$ because we are moving in a direction parallel to the direction of the current I. On the other hand, when applying Kirchhoff's loop rule, we are also free to move counterclockwise around the loop (Fig. 19.19B). The potential change in moving through the resistor is then $+IR$ because we are now moving from the low-potential end to the high-potential end. When writing Kirchhoff's loop equation, we can go around a loop in either direction; changing this direction changes the sign of every term in Kirchhoff's loop equation, so we are left mathematically with the same equation.

The signs of the potential changes in Figure 19.19 depend on the assumed direction of the current. Here and in our other examples with Kirchhoff's loop rule, we assume a direction for the current and stick with it as we write the loop rule equations. After deriving these equations, we can solve for the value of I. If I is found to be positive, our assumed direction for the current was correct, whereas if I has a negative value, the current is opposite to our assumed direction.

When applying Kirchhoff's rules to find the currents through all branches in a circuit, the number of equations must (according to the rules of algebra) equal the number of unknown currents. The repeated use of Kirchhoff's rules can sometimes lead to "repeated" equations as shown in Figure 19.20. When solving for branch currents, use Kirchhoff's rules to get a set of independent equations, with the number of equations equal to the number of unknowns.

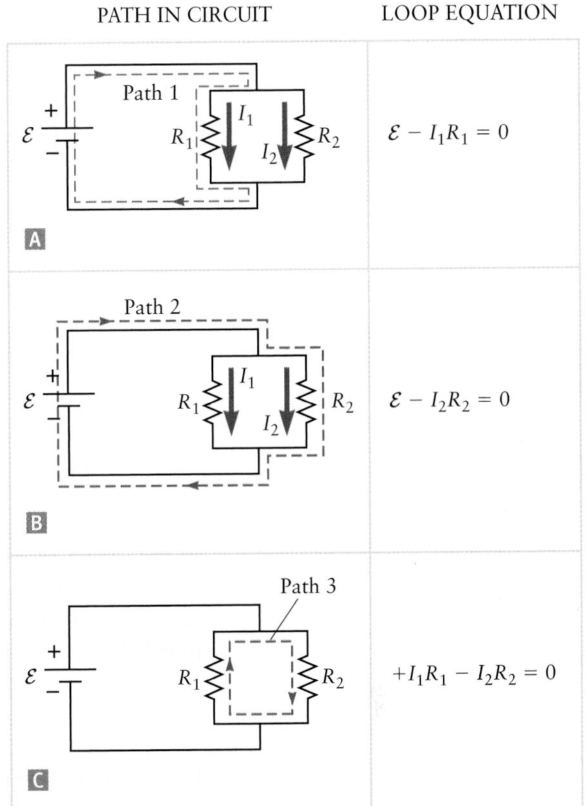

▲ **Figure 19.20** Kirchhoff's loop rule can be applied for all three of these loops around the circuit. The loop equation in C is the difference of the loop equations in A and B, however, so there are only two independent loop equations.

When we analyzed the circuit in Figure 19.18, we had three unknowns, the current in the three branches I_1, I_2, and I_3. To solve for these three unknowns, we needed three equations, which we obtained by applying the junction rule once and the loop rule twice. This circuit contains two junctions at the top and bottom of resistors R_1 and R_2, though. Could we apply the junction rule a second time to get another equation? Indeed, we could apply the junction rule to both of these junctions, obtaining the equations $I_3 = I_1 + I_2$ (Eq. 19.19) for the upper junction and $I_1 + I_2 = I_3$ for the lower junction. These two equations are identical, so the second application of the junction rule gave no new mathematical information. In general, if a circuit contains N junctions, the junction rule can be applied $N - 1$ times to obtain $N - 1$ independent equations.

Let's also consider how many times we can apply the loop rule to the circuit in Figure 19.18. Figure 19.20 shows three possible loops around the circuit, and next to each path is the resulting loop equation. We used the loop equations from paths 1 and 2 in our previous analysis of this circuit (Eqs. 19.20 and 19.21). The third loop equation—obtained using path 3 in Figure 19.20—is just the difference of the loop equations for paths 1 and 2. Hence, this third loop equation is not an independent equation. In general, a loop equation will be independent of other loop equations if it contains at least one circuit element that is not involved in the other loops.

Resistors in Parallel

When we analyzed the circuit in Figure 19.14A, we found that two resistors R_1 and R_2 in series are equivalent to a single resistor of value $R_{\text{equiv}} = R_1 + R_2$. The resistors R_1 and R_2 can thus be replaced by this equivalent resistor without affecting the rest of the circuit; in particular, the current in the circuit will be the same. The resistors

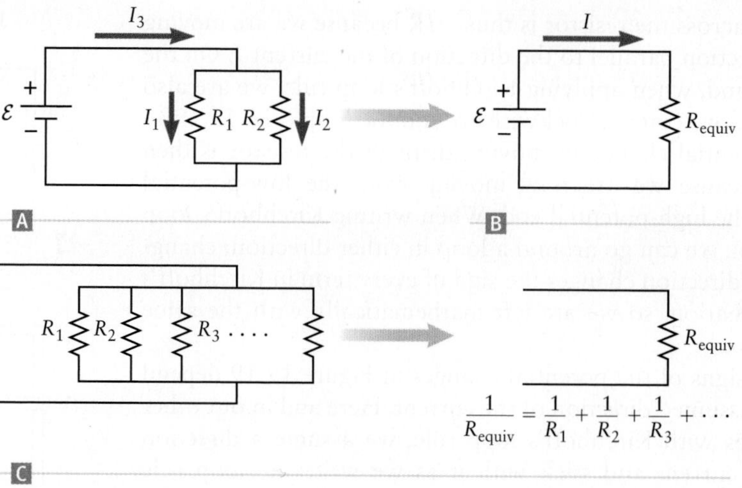

▶ **Figure 19.21** Many resistors connected in parallel are equivalent to a resistance of value R_{equiv} as given by Equation 19.25.

in Figure 19.21 are connected in parallel, and it is also possible to replace the parallel resistors R_1 and R_2 by a single equivalent resistor R_{equiv}. To find the value of this equivalent resistor, recall our result for the total current through the battery in the parallel circuit (Eq. 19.23):

$$I_3 = \frac{\mathcal{E}}{R_1} + \frac{\mathcal{E}}{R_2} = \mathcal{E}\left(\frac{1}{R_1} + \frac{1}{R_2}\right)$$

For the equivalent circuit in Figure 19.21B, we have

$$I = \frac{\mathcal{E}}{R_{\text{equiv}}}$$

where I is again the current through the battery. For the circuits in parts A and B of Figure 19.21 to be equivalent, the current through the battery must be the same, so we have

$$\mathcal{E}\left(\frac{1}{R_1} + \frac{1}{R_2}\right) = \frac{\mathcal{E}}{R_{\text{equiv}}}$$

which gives

$$\frac{1}{R_{\text{equiv}}} = \frac{1}{R_1} + \frac{1}{R_2} \tag{19.24}$$

This result can be extended to any number of resistors in parallel (Fig. 19.21C). In the general case, the value of the equivalent resistance is

Resistors in parallel

$$\frac{1}{R_{\text{equiv}}} = \frac{1}{R_1} + \frac{1}{R_2} + \frac{1}{R_3} + \cdots \quad \text{(resistors in parallel)} \tag{19.25}$$

We'll demonstrate this handy result in the next example.

EXAMPLE 19.5 Resistors in Series and in Parallel

Consider the circuit in Figure 19.22A. Find the current through resistor R_1.

RECOGNIZE THE PRINCIPLE

We could attack this problem using Kirchhoff's rules, applying the loop and junction rules to derive equations involving the currents through the different branches. In this case, however, we can instead use the rules for combining resistors in series and in parallel (Eqs. 19.18 and 19.25). These equations were derived from Kirchhoff's rules, so those rules are still the foundation of our calculation.

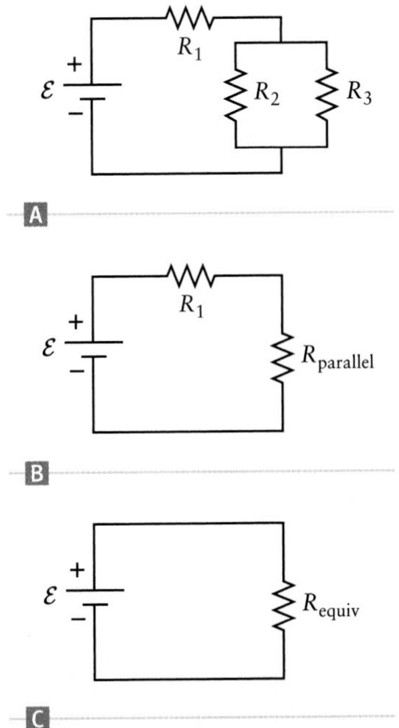

▲ **Figure 19.22** Example 19.5. A The original circuit. B R_2 and R_3 have been replaced by an equivalent resistor $R_{parallel}$. C R_1 and $R_{parallel}$ were combined in series to get a single resistance R_{equiv} that is equivalent to the three original resistors.

SKETCH THE PROBLEM

The three parts of Figure 19.22 show how the parallel and series addition rules can be applied to obtain a single equivalent resistor R_{equiv}.

IDENTIFY THE RELATIONSHIPS AND SOLVE

Resistors R_2 and R_3 in Figure 19.22A are connected in parallel, so we can replace them with a single resistor $R_{parallel}$ as shown in Figure 19.22B, with (Eq. 19.25)

$$\frac{1}{R_{parallel}} = \frac{1}{R_2} + \frac{1}{R_3}$$

This can be rearranged to get

$$R_{parallel} = \frac{R_2 R_3}{R_2 + R_3}$$

We now have an equivalent circuit (Fig. 19.22B) containing two resistors R_1 and $R_{parallel}$ in series. We can replace these two resistors by a single new resistor R_{equiv} as shown in Figure 19.22C. To find the value of R_{equiv}, we use the rule for combining resistors in series (Eq. 19.18) to get

$$R_{equiv} = R_1 + R_{parallel}$$

Using our result for $R_{parallel}$ gives

$$R_{equiv} = R_1 + R_{parallel} = R_1 + \frac{R_2 R_3}{R_2 + R_3}$$

We have now arrived at a much simpler equivalent circuit that contains a battery with a single resistor R_{equiv}. The current in this circuit is given by our familiar result

$$I = \boxed{\frac{\mathcal{E}}{R_{equiv}}}$$

▶ ***What does it mean?***

Complicated combinations of resistors can often be analyzed by applying the rules for series and parallel resistors as we have done here. For some circuits, though, one has to start directly with Kirchhoff's rules instead as we'll see in Example 19.6.

CONCEPT CHECK 19.4 Combining Resistors

Three resistors are connected as shown in Figure 19.23. What is the equivalent resistance of this combination?

(a) R (b) $3R$ (c) $R/3$

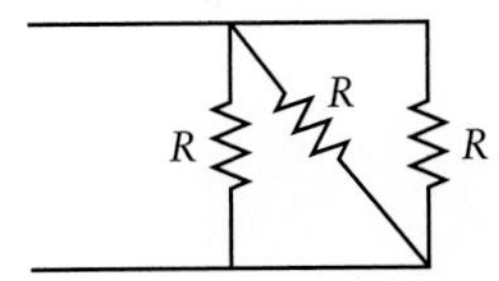

▲ **Figure 19.23** Concept Check 19.4.

EXAMPLE 19.6 A Multiloop Circuit

Consider the circuit in Figure 19.24A. Apply Kirchhoff's rules to find the current I_2 though the central branch. For simplicity, assume all three resistors have the same resistance R.

RECOGNIZE THE PRINCIPLE

This circuit cannot be simplified using the rules for series and parallel combinations of resistors because two of the branches contain both a battery and a resistor. The rules for resistors in series or in parallel in Equations 19.18 and 19.25 were derived assuming the branches contain only resistors. We must therefore start with Kirchhoff's two rules, following Step 1 of the approach outlined in the "General Rules for Circuit Analysis" given earlier.

(continued) ▶

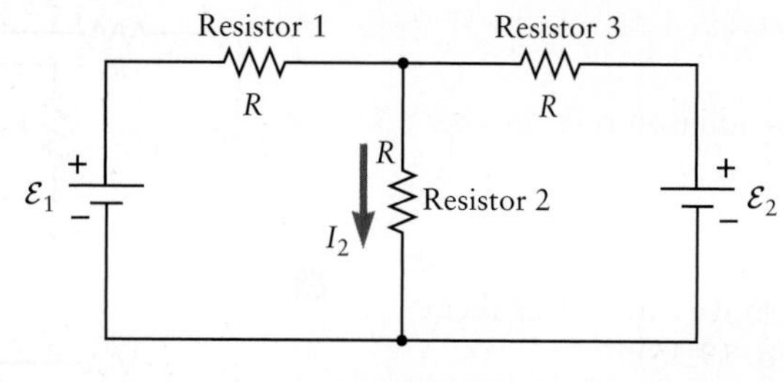

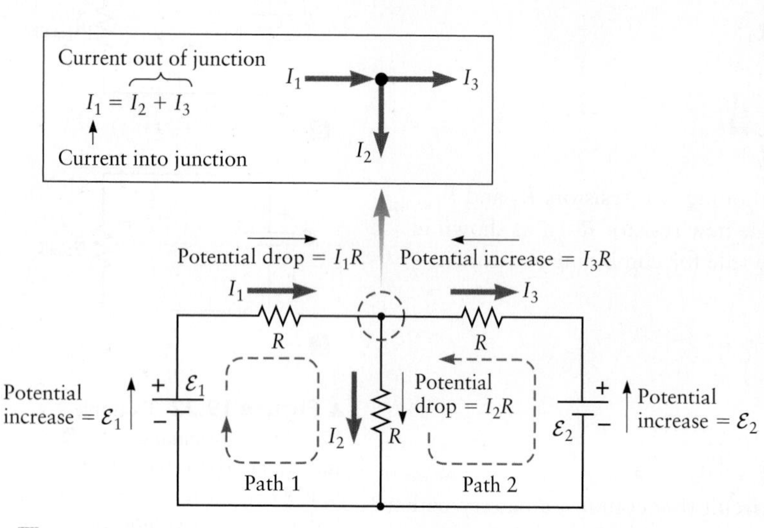

▲ Figure 19.24 Example 19.6. A A multiloop circuit with three identical resistors. B We label the currents through the three branches I_1, I_2, and I_3 and apply Kirchhoff's rules to solve for these currents.

SKETCH THE PROBLEM

Step 2: We begin by labeling the currents through all branches of the circuit in Figure 19.24B and assigning positive directions for each current as indicated by arrows in the figure. We do not know at this point if the currents will be in the directions indicated by these arrows, but we can proceed with these assumed directions and solve for I_1, I_2, and I_3. A negative value for one of the currents means that current is directed opposite to the corresponding arrow in Figure 19.24B.

IDENTIFY THE RELATIONSHIPS

Step 3 of the strategy: We express Kirchhoff's loop rule for the two loops indicated by the dashed paths in the figure. Starting at the negative terminal of battery 1, we travel around path 1 in a clockwise direction. There is a potential increase of $\mathcal{E}_1$ at the battery, a drop of I_1R as we pass through resistor 1, and another potential drop I_2R when passing through resistor 2. These potential changes must add up to zero, so

$$\mathcal{E}_1 - I_1R - I_2R = 0 \quad (1)$$

We next travel around loop 2, starting at the negative terminal of battery 2 and moving around this loop in the counterclockwise direction, which takes us first through the battery with a potential increase $\mathcal{E}_2$. We then pass through a resistor "against" the assigned direction of I_3. We are moving from the assumed low-potential end of the resistor to the high-potential end, so now we have a potential *increase* of I_3R. Continuing around loop 2, we have a potential drop I_2R on passing through the resistor in the central branch that carries current I_2. Summing all the potential changes along path 2, Kirchhoff's loop rule gives

$$\mathcal{E}_2 + I_3R - I_2R = 0 \quad (2)$$

We now have two equations, (1) and (2), and three unknowns (the currents through the three branches). We need one more equation (because the number of equations must equal the number of unknowns), and we can get it using Kirchhoff's junction rule. Applying this rule at the junction indicated at the top in Figure 19.24B, we find

$$I_1 = I_2 + I_3 \quad (3)$$

which gives us three equations that we can solve for all three unknown currents.

SOLVE

Step 4: Adding Equations (1) and (2) gives

$$\mathcal{E}_1 - I_1R - I_2R + \mathcal{E}_2 + I_3R - I_2R = 0 \quad (4)$$

Collecting the terms involving R leads to

$$\mathcal{E}_1 + \mathcal{E}_2 + R(-I_1 - I_2 + I_3 - I_2) = 0$$
$$\mathcal{E}_1 + \mathcal{E}_2 = (I_1 + 2I_2 - I_3)R$$

We can use Equation (3) to eliminate I_3 and find

$$\mathcal{E}_1 + \mathcal{E}_2 = (I_1 + 2I_2 - I_3)R = (I_1 + 2I_2 - I_1 + I_2)R$$
$$\mathcal{E}_1 + \mathcal{E}_2 = 3I_2R$$

Solving for the current I_2 gives our final result:

$$I_2 = \boxed{\frac{\mathcal{E}_1 + \mathcal{E}_2}{3R}}$$

▶ *What does it mean?*

If we want to find the currents in the other branches, we could substitute this result for I_2 into Equation (1) to obtain I_1 and so forth, which would give us the complete solution for this circuit.

EXAMPLE 19.7 Ⓡ Parallel and Series Combinations of Resistors (Revisited)

You are given an unlimited number of identical resistors with $R = 10\ \text{k}\Omega$. (**a**) How could you use several of these resistors to construct an equivalent resistance with $R_{\text{equiv}} = 60\ \text{k}\Omega$? (**b**) How could you use several of these resistors to construct an equivalent resistance with $R_{\text{equiv}} = 2.5\ \text{k}\Omega$?

RECOGNIZE THE PRINCIPLE

We begin with some observations about the rules for combining resistors in series and in parallel. The general series combination rule is (Eq. 19.18)

$$R_{\text{equiv}} = R_1 + R_2 + R_3 + \cdots \tag{1}$$

For a series combination of resistors, the equivalent resistance is thus always *larger* than *any* of the individual resistors. The parallel-resistor combination rule is (Eq. 19.25)

$$\frac{1}{R_{\text{equiv}}} = \frac{1}{R_1} + \frac{1}{R_2} + \frac{1}{R_3} + \cdots \tag{2}$$

In words, this says that the equivalent resistance is always *smaller* than *any* of the individual resistances.

SKETCH THE PROBLEM

No figure is necessary.

IDENTIFY THE RELATIONSHIPS AND SOLVE

(**a**) The problem asked for a total resistance of 60 kΩ using components with $R = 10\ \text{k}\Omega$; according to Equation (1), we must connect $\boxed{\text{six}}$ of the component $\boxed{\text{resistors in series}}$.

(**b**) If we connect n of these resistors in parallel, Equation (2) becomes

$$\frac{1}{R_{\text{equiv}}} = n\left(\frac{1}{R}\right)$$

Solving for n and inserting the desired value of the equivalent resistance, we get

$$n = \frac{R}{R_{\text{equiv}}} = \frac{10\ \text{k}\Omega}{2.5\ \text{k}\Omega} = 4$$

Hence placing $\boxed{\text{four}}$ of our $\boxed{\text{10-k}\Omega\text{ resistors in parallel}}$ gives an equivalent resistance of 2.5 kΩ.

▶ *What does it mean?*

Connecting resistors in series always gives a total (equivalent) resistance that is larger than the resistance of any of the component resistors. A parallel connection always gives an equivalent resistance smaller than any of the component resistances.

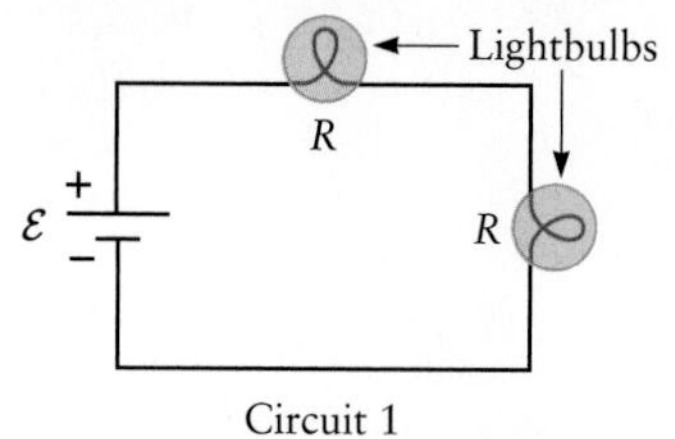

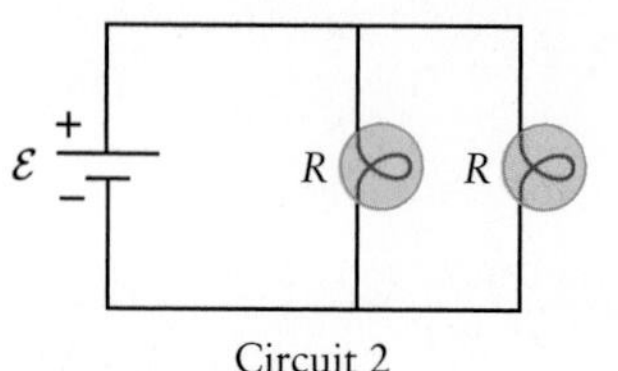

▲ **Figure 19.25** Concept Check 19.5.

CONCEPT CHECK 19.5 Comparing Series and Parallel Circuits

Incandescent lightbulbs, like the ones used in many flashlights, act as resistors, and their light intensity depends on the current: a high current produces a high light intensity, while a small current gives a small light intensity. Figure 19.25 shows two circuits with identical batteries and identical lightbulbs (each with resistance R). In which circuit do the lightbulbs shine brightest?

Real Batteries

In Section 19.2, we mentioned that an ideal battery always maintains a constant voltage across its terminals. Real batteries fall short of this ideal. The voltage provided by a real battery decreases at high currents, in part because of the time associated with the electrochemical reactions and the movement of charge inside the battery. This behavior can also be understood using the equivalent circuit picture in Figure 19.26A.

A real battery is equivalent to an ideal battery in series with a resistor R_{battery}, called the ***internal resistance*** of the battery. The current through the battery passes through this internal resistance, and the effect can be seen by analyzing the circuit in Figure 19.26B. The internal resistor R_{battery} and the external resistor R are connected in series; applying our relation for resistors in series (Eq. 19.18), the current through this circuit is

$$I = \frac{\mathcal{E}}{R + R_{\text{battery}}}$$

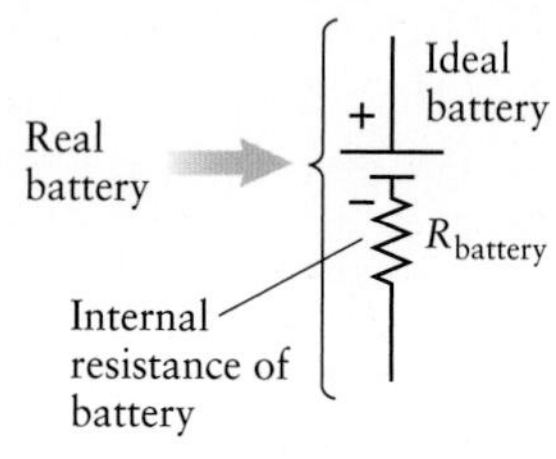

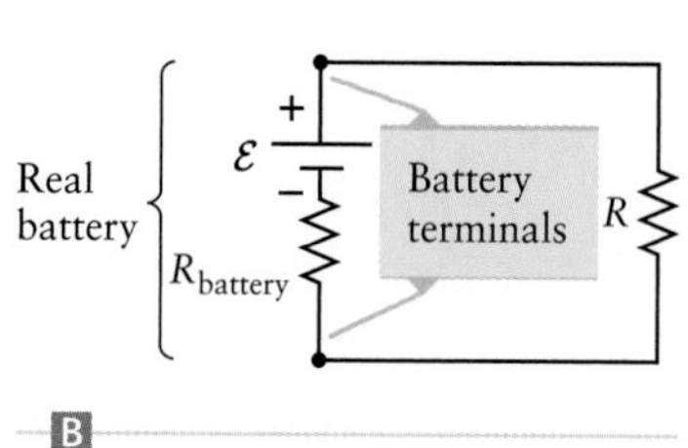

▲ **Figure 19.26** A All real batteries have a nonzero internal resistance R_{battery}. B We can think of a real battery as an ideal battery in series with a resistance R_{battery} and apply Kirchhoff's rules to this circuit model.

Usually, we are interested in the voltage (i.e., the potential difference) that appears across the battery terminals. This voltage is also equal to the potential drop across the external resistor, which is IR. We thus have

$$\text{potential of real battery} = IR = \mathcal{E}\left(\frac{R}{R + R_{\text{battery}}}\right) \tag{19.26}$$

If the internal resistance of the battery is small compared to the external resistor ($R_{\text{battery}} \ll R$), the factor in parentheses in Equation 19.26 is approximately 1 and the voltage across the terminals is nearly equal to $\mathcal{E}$, the value for an ideal battery. However, if the internal resistance is not small compared with R, the voltage across the battery terminals can be much less than $\mathcal{E}$. Hence, in practical applications, the internal resistance of the battery should be small compared with the resistance of the circuit to which the battery is attached.

CONCEPT CHECK 19.6 A Resistor Puzzle

You are given three resistors, all having the same resistance so that $R_1 = R_2 = R_3 = R$. Two of these resistors are connected in series with a battery as sketched in Figure 19.27. How could you connect resistor R_3 in this circuit so as to *increase* the current through resistor R_2?

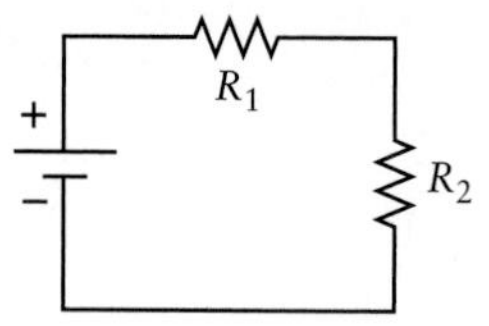

▲ **Figure 19.27** Concept Check 19.6.

19.5 DC Circuits: Adding Capacitors

In Chapter 18, we introduced capacitors and showed that when charges $\pm q$ are placed onto the plates of a capacitor, the potential difference V across the capacitor plates has a magnitude (Eq. 18.30)

$$V = \frac{q}{C} \tag{19.27}$$

Here, as in all our work on circuits in this chapter, V is the potential *difference* across the circuit element (in this case, the capacitor). In Chapter 18, we denoted this potential difference by ΔV, but now we follow the convention widely used in circuit analysis and drop the "Δ."

Applying Kirchhoff's Rules to a Circuit with a Capacitor

Kirchhoff's rules for circuit analysis apply to all kinds of circuits, including those with capacitors. Figure 19.28 shows an RC circuit containing a resistor and a capacitor in series with a battery. Recall that a capacitor is represented in a circuit diagram as two parallel plates (Fig. 18.26); this symbol is designed to remind you of a parallel-plate capacitor, even though most real capacitors do not have this simple geometry. To apply Kirchhoff's loop rule, we start at the negative terminal of the battery and proceed clockwise around the circuit. The potential increases by $\mathcal{E}$ as we pass through the battery and then decreases by an amount IR as we move across the resistor. The potential across a capacitor is given by Equation 19.27; because the top capacitor plate in Figure 19.28 is assumed to be positively charged ($+q$), the potential drops as we move from the top to the bottom capacitor plate (as we travel clockwise around the circuit). Collecting these terms and writing Kirchhoff's loop rule for this circuit, we get

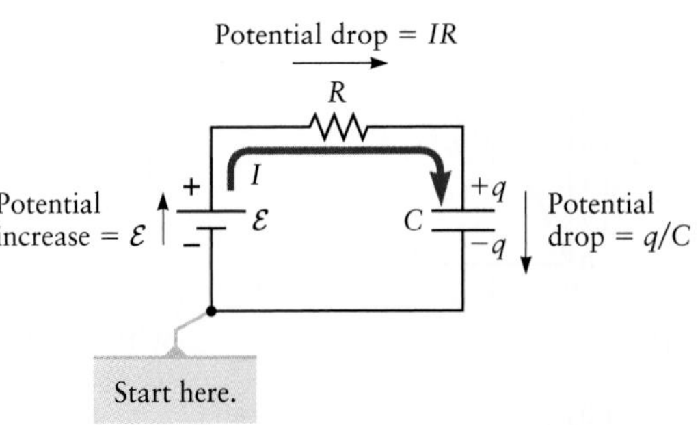

▲ **Figure 19.28** The circuit symbol for a capacitor consists of two parallel lines representing the two plates of a parallel-plate capacitor. When charges $\pm q$ are placed on the capacitor plates, an electric potential difference $V = q/C$ is established between the two plates.

$$\text{change in potential around the circuit} = \underbrace{+\mathcal{E}}_{\text{potential increase in battery}} - \underbrace{IR}_{\text{potential drop across } R} - \underbrace{\frac{q}{C}}_{\text{potential drop across } C} = 0$$

As in our previous applications of Kirchhoff's rules, we want to solve this equation to find the current in the circuit. Now, though, we have a new complication: the value of q will change with time because current carries charge onto or off the capacitor plates. As a result, both I and q will be time dependent.

RC Circuits: A Qualitative Analysis

We can get a complete solution for I and q in an RC circuit using the relation between current and charge $I = \Delta q/\Delta t$ (Eq. 19.1) along with Kirchhoff's loop rule, but the mathematics is more complicated than we wish to tackle in this book. We will therefore resort to a qualitative analysis based on the general relationship between current and the charge on a capacitor.

Figure 19.29 shows a circuit that contains a single capacitor along with a resistor, a battery, and a *switch* denoted by S. A simple switch can be open, in which case it does not allow any current to pass, or closed, allowing current to pass with no resistance, like an ideal wire. Suppose the capacitor is initially uncharged and the switch is open (Fig. 19.29A). In this case, the current is zero and there is no voltage across the capacitor because $q = 0$. We now close the switch at $t = 0$. Immediately after the switch is closed, there is a clockwise current as indicated in Figure 19.29B, carrying a positive charge to the top plate of the capacitor. The charge accumulating on the top plate repels charge from the bottom plate, causing charge to flow into the negative battery terminal and leaving the bottom plate of the capacitor with a negative charge. At any particular instant, the capacitor plates are charged $\pm q$, and according to Equation 19.27 there is a nonzero voltage across the capacitor.

As time passes, the current carries more and more charge onto the capacitor, and we say that the capacitor is "charged" by the current. As q increases, so does the

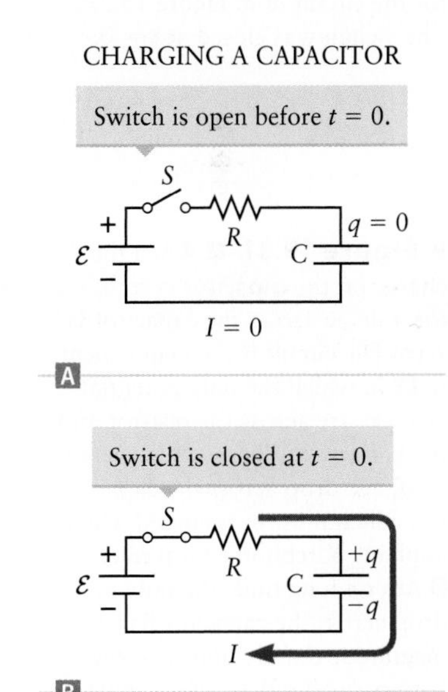

▲ **Figure 19.29** A An RC circuit containing a battery, a resistor, a capacitor, and a switch. When the switch is open, no current can pass through it or through the rest of the circuit. B When the switch is closed, a current is established, taking charge to (or from) the plates of the capacitor.

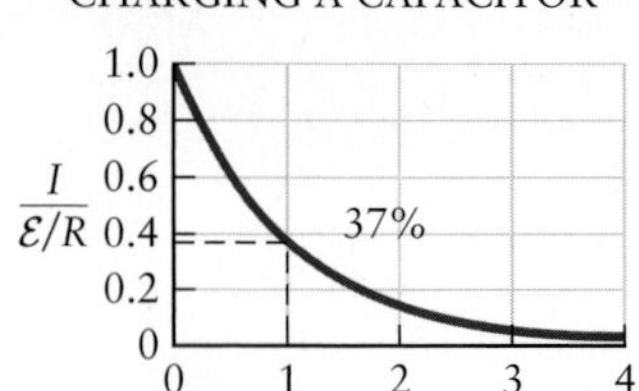

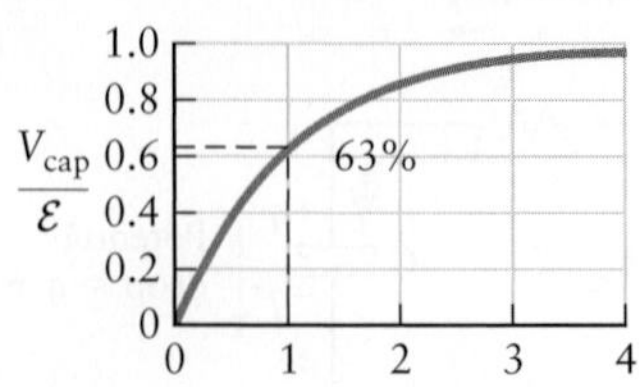

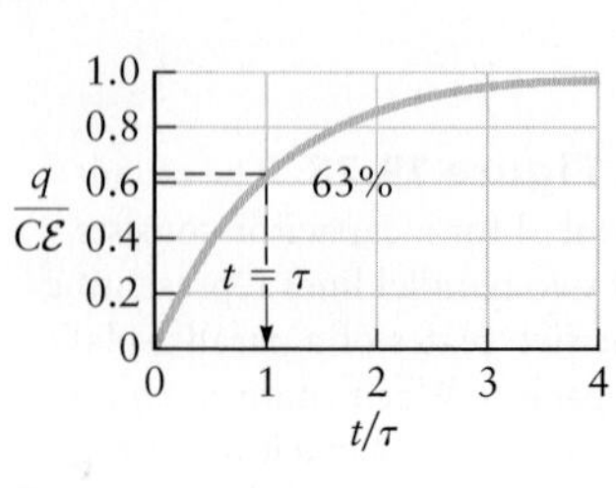

▲ **Figure 19.30** Variation of A the current, B the voltage across the capacitor, and C the charge on the capacitor as functions of time for the circuit in m Figure 19.29. The switch was closed at $t = 0$.

voltage across the capacitor. Figure 19.30 shows how the current through the circuit and the voltage across the capacitor vary with time. The voltage across the capacitor increases and asymptotically approaches $\mathcal{E}$, the battery emf. At the same time, the current through the circuit decreases to zero. Mathematically, the current in the circuit is described by

$$I = \frac{\mathcal{E}}{R} e^{-t/\tau} \tag{19.28}$$

while the voltage across the capacitor is

$$V_{cap} = \mathcal{E}(1 - e^{-t/\tau}) \tag{19.29}$$

From our relation between the voltage across and charge on a capacitor (Eq. 19.27), we get

$$q = CV_{cap} = C\mathcal{E}(1 - e^{-t/\tau}) \tag{19.30}$$

which is also plotted in Figure 19.30. The quantity τ is called the ***time constant*** and is given by

$$\tau = RC \tag{19.31}$$

The results in Equations 19.28 through 19.31 along with Figure 19.30 show how the charge on and voltage across a capacitor vary with time as the capacitor is "charged up." The amount of time required for this process depends on the time constant τ and hence depends on the product of R and C. After the switch is closed for an amount of time equal to τ—that is, for a period equal to "one time constant"—the current decreases to $e^{-t/\tau} = e^{-1} = 0.37 = 37\%$ of its value at $t = 0$. Likewise, V_{cap} and q increase to $1 - e^{-1} = 0.63 = 63\%$ of their asymptotic values after this time.

Although we will not derive the results for the current, voltage, and charge in Equations 19.28 through 19.31, we can still use Kirchhoff's loop rule to understand the circuit behavior just after the switch is closed (Fig. 19.31A). The initial charge on the capacitor is zero, and just after the switch is closed the charge will still be very small. If q is small, the voltage across the capacitor is also small, so $V_{cap} \approx 0$ just after closing the switch. In terms of circuit analysis, the capacitor at this moment is nearly equivalent to a wire because there is almost no potential drop as we pass from one side of the capacitor to the other. The equivalent circuit thus looks as shown in

► **Figure 19.31** A When the charge on the capacitor is zero, the voltage across the capacitor is zero. The circuit is then equivalent to B, in which the only potential changes are across the resistor and the battery. C When $q > 0$, there is a voltage drop across the capacitor, which must be included when applying Kirchhoff's loop rule. D After a long time, the potential drop across the capacitor has a magnitude $\mathcal{E}$ and "opposes" the battery potential, so the current is zero.

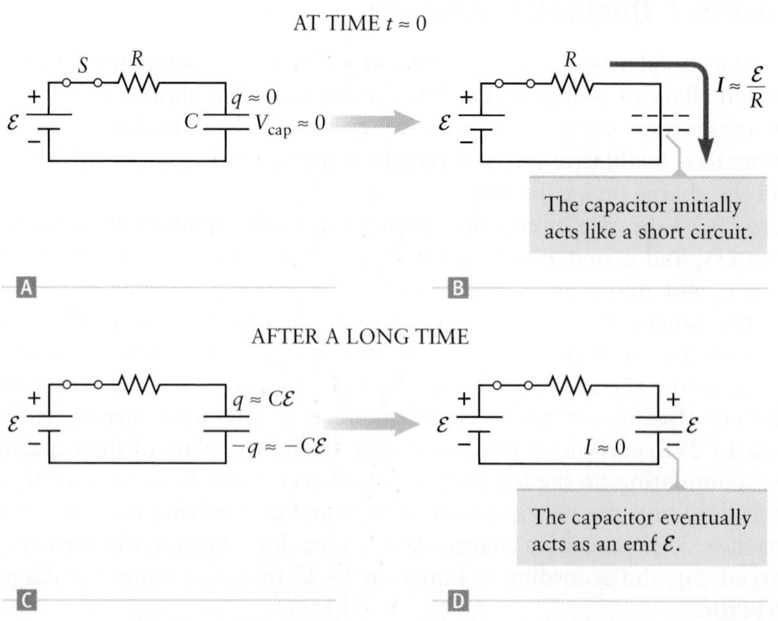

Figure 19.31B, containing a battery and a resistor, and the current is $I = \mathcal{E}/R$. This current is the initial ($t = 0$) value in Figure 19.30A (and Eq. 19.28).

We can analyze the circuit behavior when t is large—that is, a long time after the switch is closed—in a similar way. As t increases, the charge on the capacitor in Figure 19.31 becomes larger and larger. Eventually, it reaches a value for which the voltage across the capacitor is $V_{cap} \approx \mathcal{E}$, the battery emf. At this time, the equivalent circuit appears as shown in Figure 19.31D; the battery with emf $\mathcal{E}$ is now connected to the capacitor across which there is a voltage also equal to $\mathcal{E}$. However, the *polarity* of the capacitor voltage is such that it *opposes* the battery emf. We essentially have two voltage sources connected in opposition, and the net result is no current in the circuit. That is why the current approaches zero when t is large in an RC circuit (Fig. 19.30A).

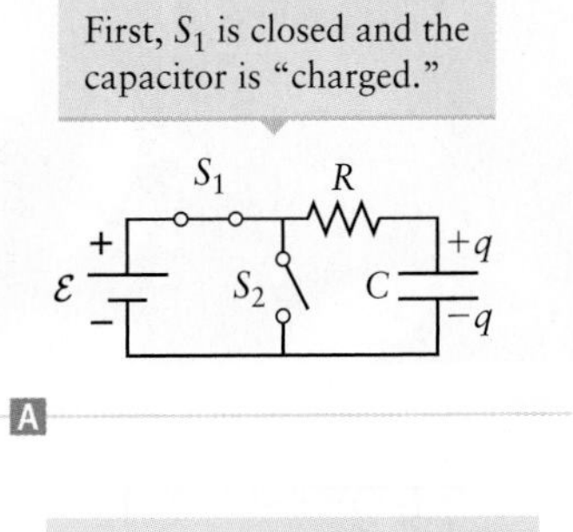

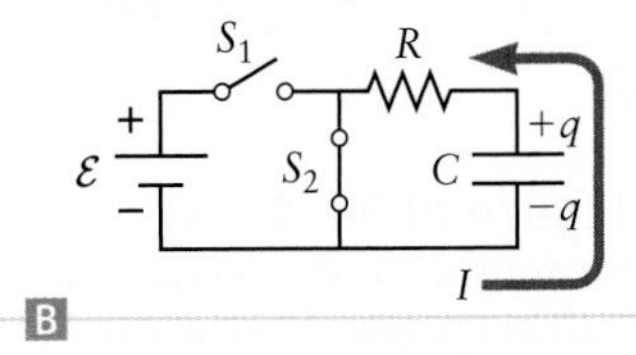

▲ **Figure 19.32** A Charging and B discharging a capacitor.

Discharging a Capacitor

Let's now consider the circuit behavior when a capacitor is discharged. An appropriate circuit is shown in Figure 19.32; it is identical to our original circuit (Fig. 19.29) except for an added second switch S_2. The capacitor can be charged by keeping switch S_1 closed for a very long time with switch S_2 open (Fig. 19.32A), and we have seen that the charge on the capacitor is then $q = C\mathcal{E}$ (Eq. 19.30). Now let's open switch S_1 and close S_2. This produces a counterclockwise current through the loop on the right in Figure 19.32B, carrying charge from the (initially) positive capacitor plate at the top to the (initially) negative plate at the bottom. After a long time, the charge on both plates is zero and the capacitor is "discharged." If switch S_2 is closed at $t = 0$ (i.e., we start our clock when the capacitor begins to discharge), the discharge process is described by

$$I = -\frac{\mathcal{E}}{R}e^{-t/\tau} \tag{19.32}$$

$$V_{cap} = \mathcal{E}e^{-t/\tau} \tag{19.33}$$

$$q = C\mathcal{E}e^{-t/\tau} \tag{19.34}$$

These results are plotted in Figure 19.33. Notice that the current is *negative* as the capacitor discharges. The current's direction is *opposite* to the direction found when the capacitor was charged (Fig. 19.29). The time constant for discharging is again given by $\tau = RC$ (Eq. 19.31). The same time constant describes both charging and discharging.

DISCHARGING A CAPACITOR

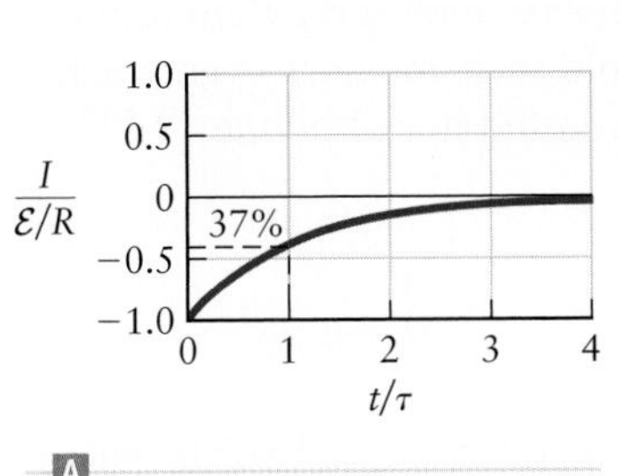

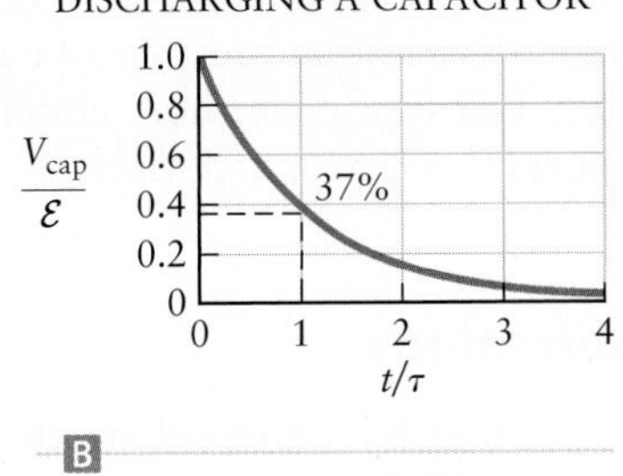

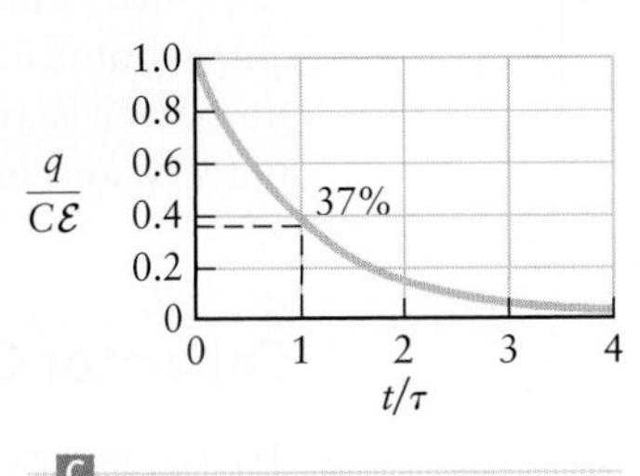

▲ **Figure 19.33** For the circuit in Figure 19.32, A the current through the capacitor, B the voltage across the capacitor, and C the charge on the capacitor as the capacitor discharges. Switch S_2 was closed at $t = 0$.

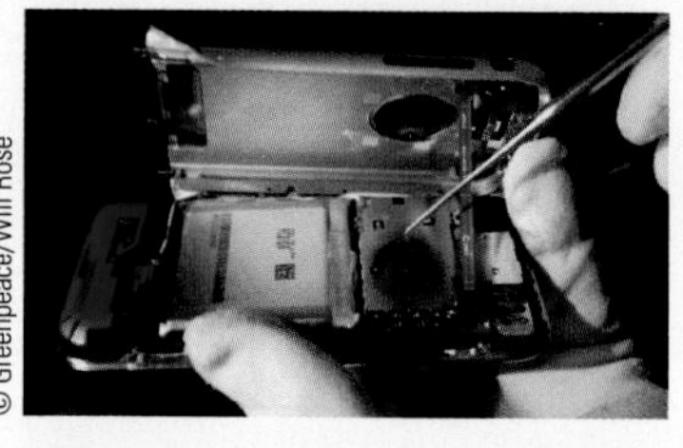

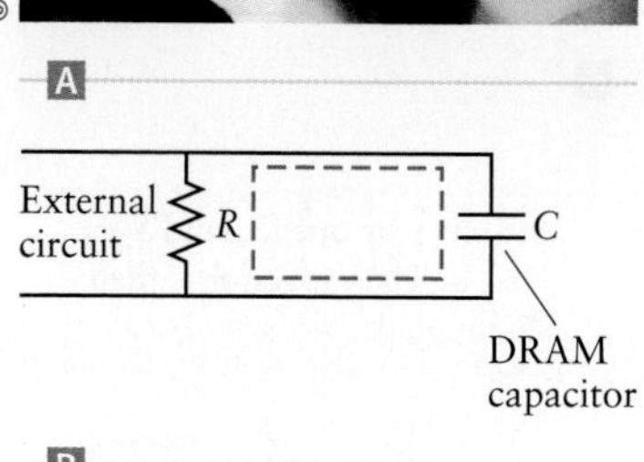

▲ **Figure 19.34** A A real capacitor such as the ones used in a DRAM integrated circuit can be modeled B as an ideal capacitor in parallel with an ideal resistor. Typical DRAM integrated circuits are a few cm on a side.

EXAMPLE 19.8 Discharge Time for a Capacitor

A key quantity when charging and discharging a capacitor is the time constant τ. The value of τ can vary widely because values of R in practical circuits can range from less than 1 Ω to more than 10^{12} Ω, and the capacitance can vary from as much as 1 F to 10^{-12} F or even less. Capacitors are central elements of the dynamic random access memory (DRAM) integrated circuit chips used in cell phones (Fig. 19.34A). In normal operation, a DRAM capacitor is connected to a resistor (Fig. 19.34B), and this RC combination is attached to other circuitry on the chip that produces a current to charge or discharge the capacitor. The charge on the capacitor is used to store information in binary form; the presence of a certain amount of charge denotes a "1" in the memory element, and the absence of charge denotes a "0." The RC circuit loop in Figure 19.34B is identical to the loop on the right in Figure 19.32B, so the resistance in parallel with the capacitor provides a discharge path. This discharging is actually a bad thing because immediately after charge is placed on the DRAM capacitor it starts to discharge, and the information stored in the memory element is eventually lost. Consider a DRAM chip for which $C = 10^{-15}$ F and $R = 10^{13}$ Ω. Calculate the discharge time for this RC circuit.

RECOGNIZE THE PRINCIPLE

The solution requires only that we evaluate the time constant τ.

SKETCH THE PROBLEM

The problem is described in Figure 19.34B.

IDENTIFY THE RELATIONSHIPS AND SOLVE

The time constant $\tau = RC$ (Eq. 19.31), so we have

$$\tau = RC = (10^{13}\ \Omega)(10^{-15}\ \text{F}) = 10^{-2}\ \text{s} = \boxed{0.01\ \text{s}}$$

▸ ***What does it mean?***

The voltage across the capacitor in an RC circuit decays in a time approximately equal to τ, so this DRAM capacitor will "remember" its contents for only about 0.01 s. For this reason, a DRAM chip contains circuitry that recharges the capacitor (the technical term is "refresh") at regular intervals. Typical refresh rates are once every 0.001 s, which is much shorter than τ, and the capacitor is recharged before the memory information is lost. The refresh circuit requires a power supply to recharge the capacitors, so when your computer is turned off the DRAM information is lost. It is possible to make memory chips that do not need to be refreshed. Called *static* RAM chips, these are more expensive to make than other chips, but are useful in devices that must be able to hold their memory even when no power is available. This type of memory chip is also used in cell phones, MP3 players, and other portable electronic devices.

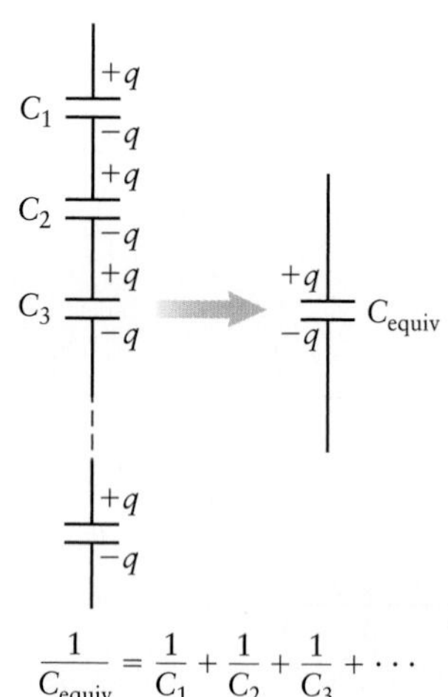

$$\frac{1}{C_{equiv}} = \frac{1}{C_1} + \frac{1}{C_2} + \frac{1}{C_3} + \cdots$$

▲ **Figure 19.35** When many capacitors are connected in series, they behave as a single equivalent capacitance given by Equation 19.35.

Capacitor Combinations

Two or more capacitors may be connected in either series or in parallel arrangements as considered in Chapter 18. Let's compare those results with the rules for combining resistors in Section 19.4.

Several capacitors connected in series are shown in Figure 19.35. Assume an external circuit has placed charge $+q$ on the upper plate of C_1 and charge $-q$ on the bottom plate of the lowest capacitor. In Chapter 18, we saw that charges $\mp q$ are then

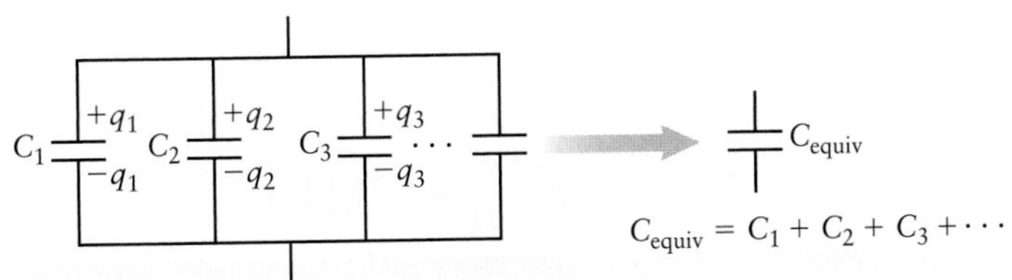

◀ **Figure 19.36** When many capacitors are connected in parallel, they behave as a single equivalent capacitance given by Equation 19.36.

induced on the corresponding inner capacitor plates. We also showed (Eq. 18.42) that this arrangement is equivalent to a single capacitor with

Equivalent capacitance of capacitors in series

$$\frac{1}{C_{equiv}} = \frac{1}{C_1} + \frac{1}{C_2} + \frac{1}{C_3} + \cdots \quad \text{(capacitors in series)} \tag{19.35}$$

which has the same form as the rule for combining resistors in *parallel* (Eq. 19.25). This form also tells us that the equivalent capacitance C_{equiv} is *smaller* than *any* of the individual capacitances $C_1, C_2, \ldots$.

A combination of two or more capacitors connected in parallel is also equivalent to a single capacitor (Fig. 19.36). In Chapter 18, we derived the rule for combining capacitors in parallel (Eq. 18.41):

Equivalent capacitance of capacitors in parallel

$$C_{equiv} = C_1 + C_2 + C_3 + \cdots \quad \text{(capacitors in parallel)} \tag{19.36}$$

Hence, the equivalent capacitance is just the sum of the individual component capacitances and is therefore *greater* than *any* of the individual values. Notice that the rule for combining capacitors in parallel has the same form as the rule for combining resistors in *series* (Eq. 19.18).

EXAMPLE 19.9 Ⓡ An *RC* Circuit in Your Car

For safety reasons, the interior dome light of most modern automobiles is designed to come on when you lift the exterior door handle; it then stays on for a short period of time before turning off again. Assuming the timing circuit for the light is controlled by an *RC* circuit, estimate the value of the capacitor used.

RECOGNIZE THE PRINCIPLE

If the light's timing is controlled by an *RC* circuit, the time constant τ will depend on the values of R and C (Eq. 19.31). Since we want to find C, we first need to estimate the values of R and τ. Here the resistance R is that of the car's interior lightbulb.

SKETCH THE PROBLEM

No figure is necessary.

IDENTIFY THE RELATIONSHIPS

The circuits in an automobile use a 12-V battery. Many new cars use interior lights that employ a light-emitting diode (LED), because LED lights use much less power than the incandescent bulbs found in older cars. From a search on the Internet (or a visit to an auto parts store), we find that these LED lights have a power rating of around $P = 1.0$ W. Using the relation between power, voltage, and resistance for a resistor (Eq. 19.16), we can write

$$R = \frac{V^2}{P} = \frac{(12 \text{ V})^2}{1.0 \text{ W}} = 140 \ \Omega \tag{1}$$

For the time constant, we estimate the light will dim in about 30 s, so we take that for the value of τ. (Check it out the next time you open a car door.)

(continued) ▶

SOLVE

Using the above values of R and τ along with the relation $\tau = RC$ leads to

$$C = \frac{\tau}{R} = \frac{30 \text{ s}}{140 \ \Omega} = \boxed{0.21 \text{ F}}$$

▶ *What does it mean?*

This value of C is rather large, but such capacitors are available. While other circuits could be designed to accomplish the same function, an RC circuit would be a very simple way to do it.

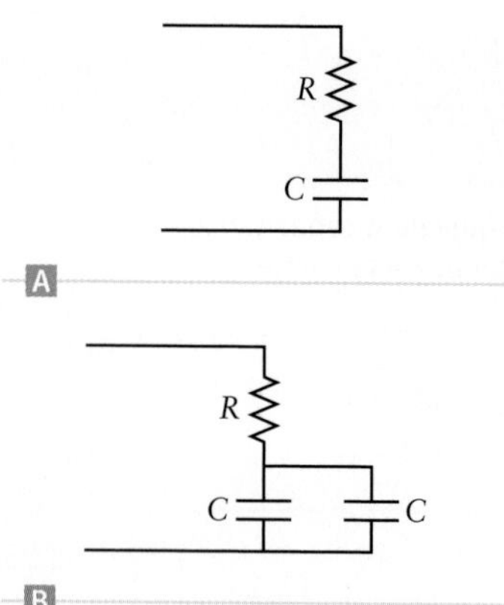

R
C
A
R
C
C
B

▲ Figure 19.37 Concept Check 19.7.

CONCEPT CHECK 19.7 What Is the RC Time Constant?

Suppose the RC time constant for the combination in Figure 19.37A is 2.4 ms. A second capacitor is then placed in parallel with the original capacitor as in Figure 19.37B. If these capacitors are identical, what is the new value of the time constant τ?

(a) 2.4 ms (b) 4.8 ms (c) 1.2 ms

19.6 Making Electrical Measurements: Ammeters and Voltmeters

We can use the circuit analysis rules described in Sections 19.4 and 19.5 to calculate the current in a particular circuit branch or the voltage across a particular circuit element, but how does one actually measure these currents and voltages? A device that measures current is called an ***ammeter***, while voltage is measured with a ***voltmeter***. Ammeters and voltmeters are both circuit elements; let's now consider how to incorporate them into a circuit analysis.

The purpose of an ammeter is to measure the current through a circuit branch, which can be accomplished by arranging for all the current to pass through the ammeter. Hence, an ammeter must be connected *in series* with the desired circuit branch as shown in Figure 19.38A. An ideal ammeter should have absolutely no effect on the circuit; it should measure the current without changing its value. From the properties of resistors in series, this means that an ammeter must have a very low resistance, ideally zero. Although an ammeter with a resistance of exactly zero is difficult to construct, in practice it is only necessary that the ammeter's resistance be much smaller than the resistance of the rest of the circuit branch. For the circuit in Figure 19.38A, the resistance of the ammeter should be much less than the resistance R_3.

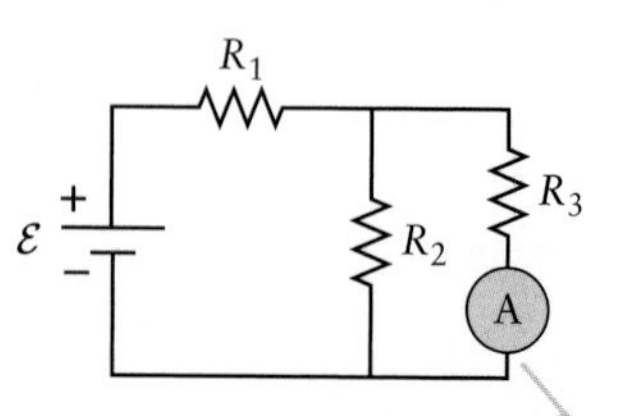

R_1
R_3
$\mathcal{E}$
+
−
R_2
A
An ammeter must be inserted in series with the branch whose current is to be measured.
A

The purpose of a voltmeter is to measure the voltage *across* a particular circuit element (or elements). The voltmeter must therefore be connected in a *parallel* arrangement as shown in Figure 19.38B. Here a voltmeter is connected in parallel with resistor R_3 so that it can measure the voltage across this resistor. The voltmeter should measure this voltage without affecting its value; a voltmeter should therefore have a resistance much larger than the circuit element to which it is attached so that very little current passes through it. An ideal voltmeter has an infinite resistance so that the current through the voltmeter is exactly zero. Real voltmeters never have an infinite resistance, but in practice it is possible to make the resistance of a voltmeter very large; values of 10^{12} Ω or even more can be achieved.

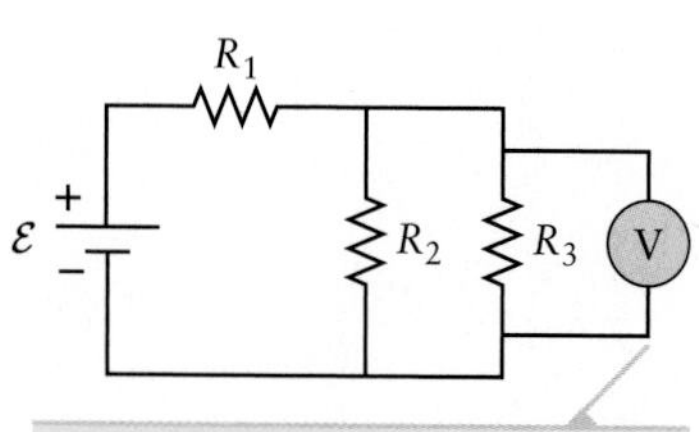

R_1
$\mathcal{E}$
+
−
R_2
R_3
V
A voltmeter is inserted in parallel with the circuit element whose voltage drop (or increase) is to be measured.
B

▲ Figure 19.38 A This ammeter is placed in series with R_3 so that it can measure the current through R_3. B This voltmeter is in parallel with R_3 so that it can measure the voltage across that resistor.

19.7 *RC* Circuits as Filters

It is often desirable to reduce or "filter out" time-dependent fluctuations in a voltage signal. Circuits that do so are called ***filters*** and can be constructed with a simple RC

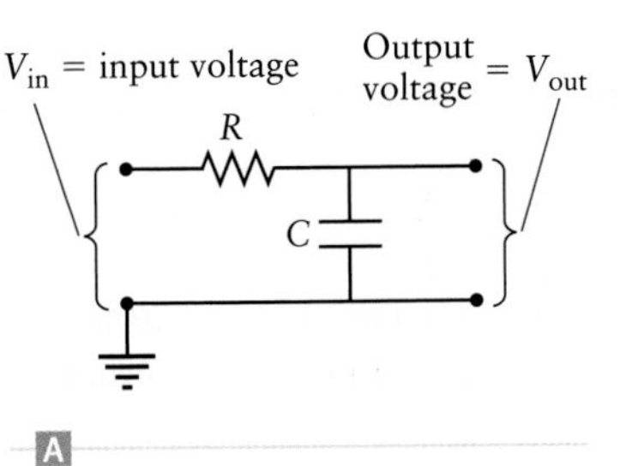

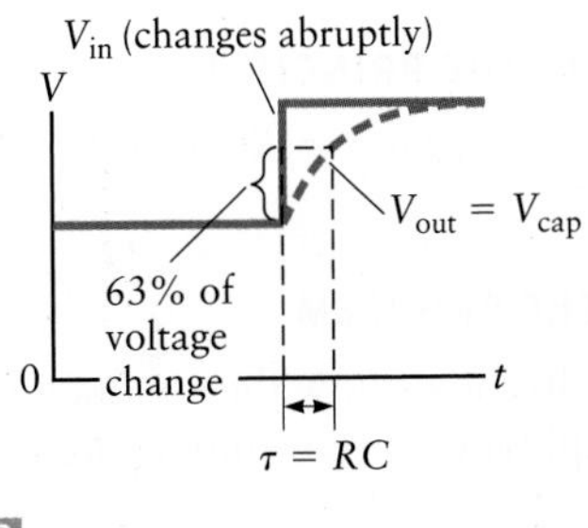

◀ **Figure 19.39** A An *RC* filter circuit takes an input voltage signal and produces an output voltage. B Comparison of the input (solid) and output (dashed) voltages. Shown is a low-pass filter that "filters out" rapid changes in the input voltage and only passes slow changes to the output.

combination. Suppose the input voltage in Figure 19.39A is initially constant. In an *RC* circuit, the capacitor will initially have a charge q such that the voltage across the capacitor is equal to the input voltage V_{in}. Now suppose at a certain time V_{in} increases very quickly as shown in Figure 19.39B. This increase causes the charge on the capacitor to change to a new value, as described by the time constant τ of the *RC* circuit. If the time constant is long, the voltage across the capacitor increases much more slowly than the abrupt change of the input voltage. In this way, the voltage across the capacitor, which serves as the output voltage for the circuit, removes rapid changes or fluctuations in the input voltage and replaces them with more gradual variations.

A filter circuit is useful in many applications. For example, the noise in a radio signal ("static") can be removed using an *RC* filter, leaving the music signal largely unaffected. The amount of filtering depends on the value of $\tau = RC$, which can be varied over a wide range by the appropriate choice of R and C.

Electric Currents in Nerves

An interesting example of an *RC* filter is found in the nervous system of the human body and in many other animals. Many nerves are long and thin, much like a wire, carrying electric currents from one part of the body to another. Nerve fibers consist of a thin layer of lipid molecules enclosing a conducting solution containing a mix of ions (K^+, Na^+, Cl^-, and others) as sketched in Figure 19.40A. This figure describes the simplest nerve fibers as found in squids and other invertebrates. The conducting solution inside the fiber acts as a resistor and the lipid layer acts as a capacitor, so this structure and biochemical composition causes a nerve fiber to behave as an *RC* filter circuit, just as in Figure 19.39.

Sensory cells provide the input signal to some nerve fibers. For example, when a squid is touched by a predator, the squid's sensory cells produce an abrupt increase in the input voltage to a nerve fiber, similar to that sketched for V_{in} in Figure 19.39B. The output signal from the *RC* nerve filter increases more slowly than the input. It is in the squid's best interest for V_{out} to increase as rapidly as possible so that the sensory signal will reach the brain quickly and the squid can react rapidly. The *RC* time constant should therefore be as small as possible. Nerves that carry signals for which a rapid response is needed have values of R and C that make the time constant very small, guaranteeing a rapid response. The value of τ is estimated for a realistic nerve fiber in Example 19.10.

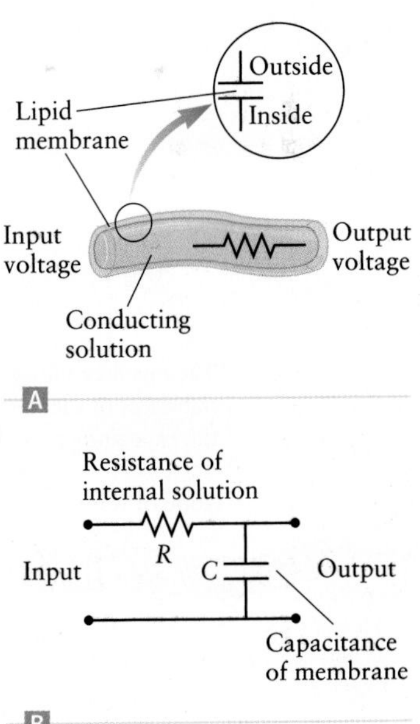

▲ **Figure 19.40** A A nerve fiber is electrically equivalent to the *RC* filter circuit in B. The membrane acts like a parallel-plate capacitor, and the resistance is provided by the ionic solution inside the nerve fiber.

EXAMPLE 19.10 Filtering by a Squid Nerve Fiber

Consider a squid nerve fiber with radius $r = 0.4$ mm and a length of 5 cm. What is the time constant τ for this *RC* filter? The resistivity of the internal solution is about $\rho = 2\ \Omega \cdot \text{m}$, the thickness of the lipid layer is $d = 10$ nm, and the dielectric constant of the lipid is $\kappa = 10$.

(continued) ▶

RECOGNIZE THE PRINCIPLE

The time constant is $\tau = RC$ (Eq. 19.31), so to find τ we must first calculate the values of R and C.

SKETCH THE PROBLEM

Figure 19.40 shows how the conducting fluid inside a nerve fiber acts as a resistor and how the lipid layer can be modeled as a capacitor, thus giving an RC circuit.

IDENTIFY THE RELATIONSHIPS

The resistance of the nerve fiber can be found using $R = \rho L/A$ (Eq. 19.11). Inserting the resistivity along with the given values of the length and cross-sectional area ($A = \pi r^2$) of the fiber into Equation 19.11, we obtain

$$R = \rho\frac{L}{A} = (2\ \Omega\cdot\text{m})\frac{0.05\text{ m}}{\pi(4\times10^{-4}\text{ m})^2} = 2\times10^5\ \Omega$$

To estimate the capacitance, we approximate the lipid layer as a parallel-plate capacitor. This capacitor's "plate" area is the cylindrical surface area of the fiber, $A_{cap} = 2\pi rL$. (This cylindrical surface area is *not* the same as the cross-sectional area A of the fiber, which we used to find the resistance.) Using the expression for the capacitance of a parallel-plate capacitor filled with dielectric (Eqs. 18.31 and 18.43), we find

$$C = \frac{\kappa\varepsilon_0 A}{d} = \frac{(10)[8.85\times10^{-12}\text{ C}^2/(\text{N}\cdot\text{m}^2)][2\pi(4\times10^{-4}\text{ m})(0.05\text{ m})]}{1.0\times10^{-8}\text{ m}} \quad (1)$$

$$C = 1\times10^{-6}\text{ F}$$

SOLVE

The time constant is then

$$\tau = RC = (2\times10^5\ \Omega)(1\times10^{-6}\text{ F}) = \boxed{0.2\text{ s}}$$

► ***What does it mean?***

According to our results for an RC filter in Figure 19.39, this time constant is the approximate response time of a squid nerve fiber.

Designing a Better Nerve Fiber

The time constant calculated in Example 19.10 is a rough measure of how fast the nerve transmits its information to the squid's brain. To keep this reaction time short, τ should be small. The squid's nerve fiber attains a fairly small time constant by reducing the resistance R as a consequence of the fiber's relatively large radius. Another way to keep the time constant short is to make the capacitance C small; that approach is used in many of the nerve fibers in vertebrates, including people. Vertebrate nerve fibers are often coated with a relatively thick layer of a protein called myelin (Fig. 19.41). This coating increases the distance between the capacitor plates—the value of d in Equation (1) of Example 19.10—making C smaller. As a result, the value of the time constant is reduced, leading to a fast response time for these nerve fibers. A faster response time is always better for the nervous system (and for survival).

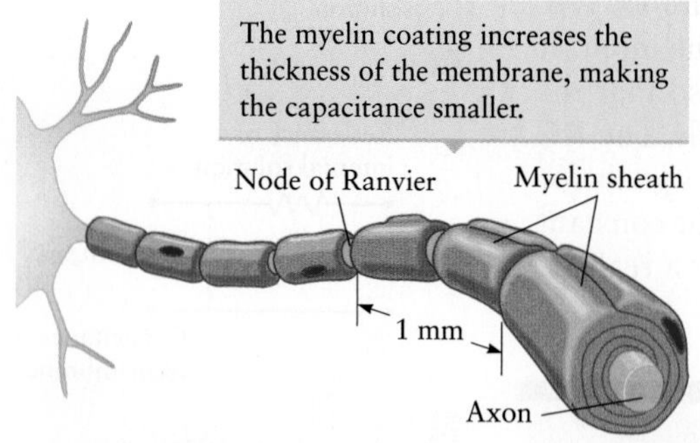
The myelin coating increases the thickness of the membrane, making the capacitance smaller.
Node of Ranvier
Myelin sheath
1 mm
Axon

▲ **Figure 19.41** Some vertebrate nerve fibers have very thick membranes due to a coating of myelin that increases the plate spacing of the membrane capacitor (Fig. 19.40) and hence reduces the capacitance. They are called myelinated fibers.

19.8 Electric Currents in the Human Body

Your body is a moderately good conductor of electricity. Taken as a whole, your body's resistance measured for conduction from the fingers of one hand to the other is about 1500 Ω when the body is dry and typically 500 Ω when wet. The electric potential at a wall socket is 120 V, so if you inadvertently touch a high-potential house wire with one hand and ground with the other, the current through your body is approximately

$$I = \frac{V}{R} \approx \frac{120 \text{ V}}{1500 \ \Omega} = 0.080 \text{ A} = 80 \text{ mA} \quad \text{(dry)}$$

$$\approx \frac{120 \text{ V}}{500 \ \Omega} = 240 \text{ mA} \quad \text{(wet)} \tag{19.37}$$

This current is carried through different parts of the body; some flows in the skin, and some can flow through your heart or other organs.

The effect of current on the human body is described in Table 19.3. A current of 1 mA is felt as a "tingling" sensation, but has no harmful effect.[4] External currents of 10 mA to 20 mA can cause your muscles to contract involuntarily, which is what happens when a person "can't let go" of a high-potential wire. Your heart is a muscle, and currents above about 100 mA can cause your heart to beat irregularly or even stop. Comparing these values with the currents calculated in Equation 19.37, we see that the current produced by contact with a wall socket can be a definite health hazard.

Large currents can be beneficial in certain situations. For example, your heart can sometimes produce a series of rapid and irregular beats, a condition called fibrillation. A large and brief external current applied by a medical device called a defibrillator can "shock" the heart back into a normal rhythm.

TABLE 19.3 Effect of Electric Current on the Human Body

Current (mA)	Effect on Body
1	Threshold for sensing (can "feel it tingle")
5	Harmless
10–20	Involuntary muscle contraction
50	Pain
100–200	Disrupts the heart
300	Burns

19.9 Household Circuits

Electric circuits are everywhere around us; for example, your residence is filled with devices (computers, lights, televisions, etc.) connected to the electric wiring in the building, which in turn is connected to the electric power lines provided by your utility company. In this chapter, we have focused on DC electric circuits involving batteries. However, the voltage provided by your utility company is not constant; rather, it is an AC voltage oscillating at a frequency of 60 Hz. This is the frequency in the United States; the frequency may be different in other countries. We'll discuss AC circuits in Chapter 22. Here we will mention some electrical safety issues that apply to both AC and DC circuits.

Safety in Home Circuits

Figure 19.42 shows some components used in household circuits, including different types of electrical connectors ("plugs") that can be used to connect to an outlet. Most modern outlets and plugs (Fig. 19.42A) contain three connections: two are flat strips, and the third is round. The round one is connected to ground potential, whereas the other two connections carry current to and from devices in your house. The connector in Figure 19.42B is called a polarized plug; it has two flat connectors of different size so that it can be inserted into the wall outlet in only one way. The larger of the two flat strips connects to a low potential that is very close to ground potential.

In normal operation, an appliance such as a hair dryer is plugged into a wall outlet like the one in Fig. 19.42C; current passes from the outlet, through the hair dryer, and back to the outlet, completing the electric circuit. The magnitude of this current can be large enough to pose a health hazard if it were to flow through your body. For example, a wire in the hair dryer might accidentally touch the handle while you are holding it, allowing a large amount of current to pass to the hair dryer

[4]Your tongue is especially sensitive to small currents. Some people are known to test for a dead battery using their tongue (not recommended!).

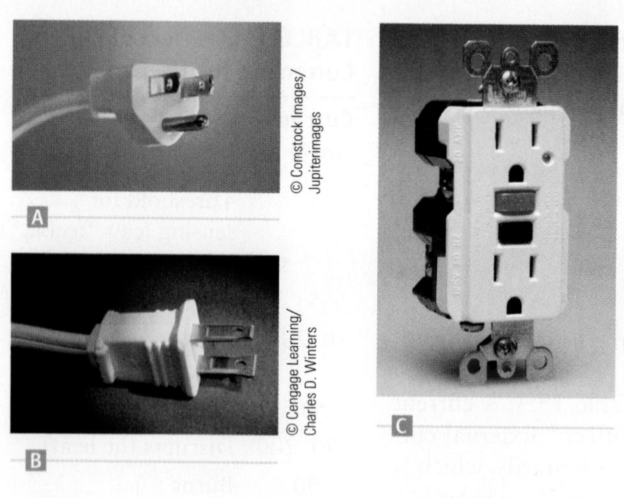

© Comstock Images/Jupiterimages

© Cengage Learning/Charles D. Winters

© Phil Degginger/Alamy

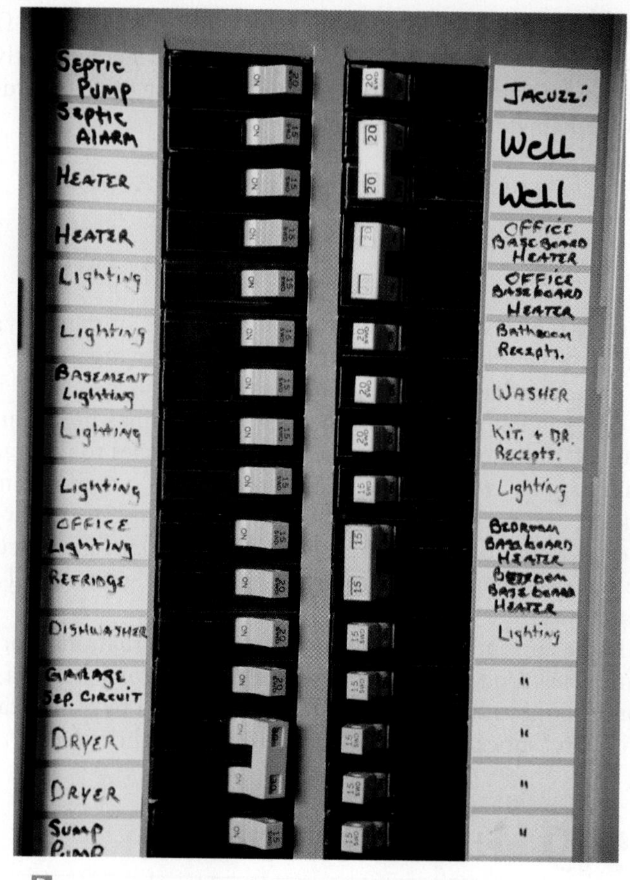

© Comstock Images/Jupiterimages

▲ **Figure 19.42** Several safety devices are incorporated in household circuits, including A three-pronged plugs, B polarized plugs, C outlets, and D circuit breakers.

handle and through you. To prevent damage to the household circuitry and to a user, safety devices are inserted into the circuit connecting the wall outlet to the utility company's power line. Older houses employ ***fuses***, in which current passes through a thin metal strip. This strip acts as a resistor with a small resistance, so under normal conditions only a small amount of power is dissipated in the fuse. If a failure causes the current to become very large, heat dissipated in the fuse is sufficient to melt and break the metal strip, stopping the current. Newer houses employ ***circuit breakers*** (Fig. 19.42D), which also stop the current when it exceeds a predetermined limit. A circuit breaker—unlike a fuse, which must be replaced after its metal strip melts—contains a switch that can be "reset" for continued use after the offending circuit has been repaired.

Electrical devices also contain safety features. The ground connection from the wall outlet is usually connected via the round piece of the plug to the outer case of an appliance, so even if there is an internal failure, a large and dangerous voltage does not appear on any surfaces that can be touched by a user. Older appliances sometimes do not use a three-pronged plug, so they cannot make use of this ground connection. In these appliances, the low-potential side of the polarized plug is often connected to the case, providing almost as much protection as a connection to ground. What do we mean by "almost"? When a device without a ground connection fails (perhaps a hair dryer), there may be a large amount of current I through the low-potential plug in the wall outlet (Fig. 19.43). The resistance R of the wires to your hair dryer is small but not zero, so this current causes the voltage at the low-potential side of the hair dryer to increase to a value IR relative to ground. If I is very large, this voltage may be large enough to pose a health hazard, especially if it is connected to the hair dryer's handle so that a large current could then flow through a person holding the hair dryer. This arrangement is thus not as safe as using a three-pronged

$V_{\text{hair dryer}} = IR$
(R = resistance of the wire)

Polarized plug

I

Ground potential
$V = 0$

If this wire touches the case of the hair dryer, the potential of the case will be large even if $V = 0$ at the wall socket.

▲ **Figure 19.43** A hair dryer with a polarized plug is not as safe as one with a three-pronged plug. If the hair dryer with a polarized plug fails, it may pose a safety hazard.

plug, which connects the handle of the hair dryer to ground potential through a separate wire. This separate "ground wire" in the hair dryer is designed so that it should never carry a large current, even if the dryer fails, and the handle should never be at a dangerous potential.

19.10 Temperature Dependence of Resistance and Superconductivity

In Section 19.3, we explained the atomic-scale origin of resistivity: when electrons move through a metal, they collide, especially with ions in the metal (Fig. 19.6). As temperature increases, these ions vibrate with larger and larger amplitudes, causing more frequent collisions and an increase in resistance. For many metals near room temperature, this temperature dependence of resistivity can be described by

$$\rho = \rho_0[1 + \alpha(T - T_0)] \tag{19.38}$$

where α is called the ***temperature coefficient of resistivity***. In Equation 19.38, ρ_0 is the resistivity at a reference temperature T_0. Usually, one picks T_0 to be room temperature, so ρ_0 is just the resistivity listed in Table 19.1. According to Equation 19.38, resistivity varies linearly with temperature; because the resistance of a wire is proportional to ρ (Eq. 19.11), the resistance of a metal also varies linearly with temperature. This linear variation usually breaks down at very low temperatures, and the resistivity of metals such as copper approaches a nonzero value at very low temperatures (Fig. 19.44A).

For some metals, the resistivity does not quite follow the low-temperature behavior found for copper. In some cases, such as mercury and lead, the resistivity drops abruptly and is zero below a certain temperature as sketched in Figure 19.44B. The value of this "critical" temperature depends on the metal. Metals for which the resistivity is zero below a critical temperature are called ***superconductors***. Such behavior is most commonly found in metals cooled to within a few degrees of absolute zero, although there are a few examples of superconductivity at temperatures above 100 K. These materials have many important applications; for example, they can be used to construct very powerful magnets as we'll discuss in Chapter 20.

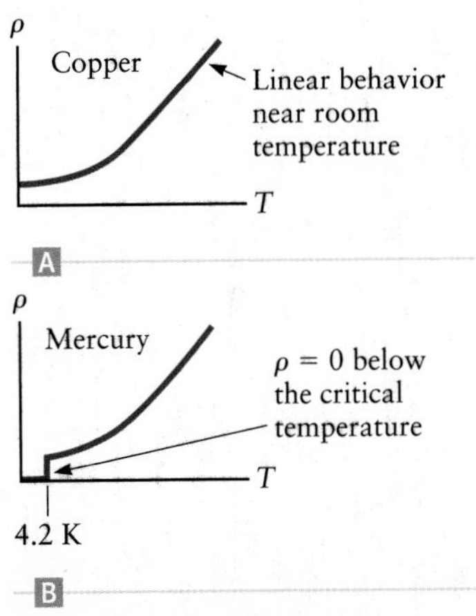

▲ **Figure 19.44** Qualitative behavior of the resistivity of A copper and B mercury as a function of temperature.

Summary | CHAPTER 19

Key Concepts and Principles

Electric current

Electric current I is the amount of charge that passes a particular point per unit time.

$$I = \frac{\Delta q}{\Delta t} \quad \text{(19.1) (page 550)}$$

where I is measured in units of ***amperes*** (A), with 1 A = 1 C/s.

Electrical resistance

The ***resistance*** R of a material of length L and cross-sectional area A is

$$R = \rho\frac{L}{A} \quad \text{(19.11) (page 556)}$$

(Continued)

where ρ is the ***resistivity*** of the material (Table 19.1). For a circuit element called a ***resistor***, the voltage across the circuit element and the current through the element are related by ***Ohm's law***:

$$V = IR \qquad (19.10)$$ (page 556)

KIRCHHOFF'S LOOP RULE

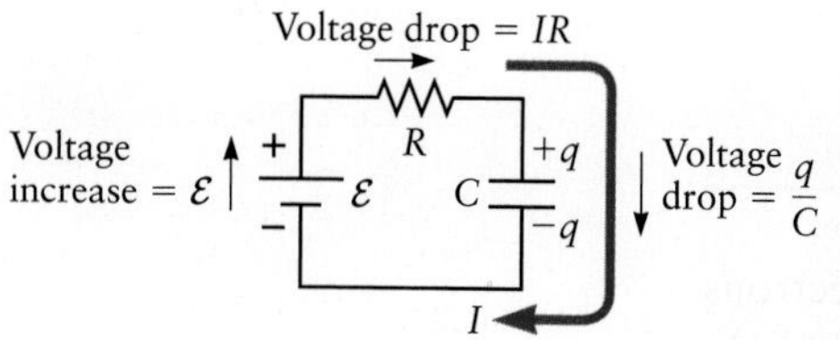

KIRCHHOFF'S JUNCTION RULE

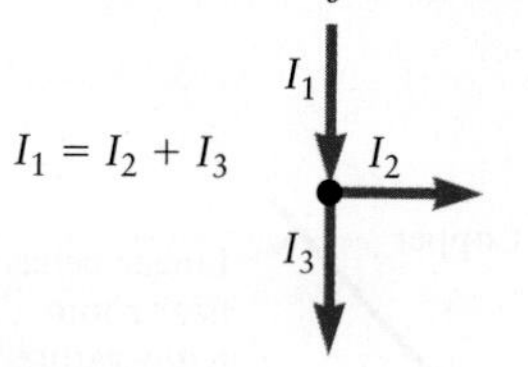

Kirchhoff's rules

Kirchhoff's rules tell how to determine the current throughout an electrical circuit. These two rules are based on the principles of conservation of energy and charge:

1. ***Kirchhoff's loop rule***: The total change in the electric potential around any closed circuit path must be zero.
2. ***Kirchhoff's junction rule***: The current entering a circuit junction must equal the current leaving the junction.

Voltage drop across a capacitor

If a capacitor carries a charge $\pm q$, the voltage across the capacitor is

$$V = \frac{q}{C} \qquad (19.27)$$ (page 572)

where C is the capacitance.

Applications

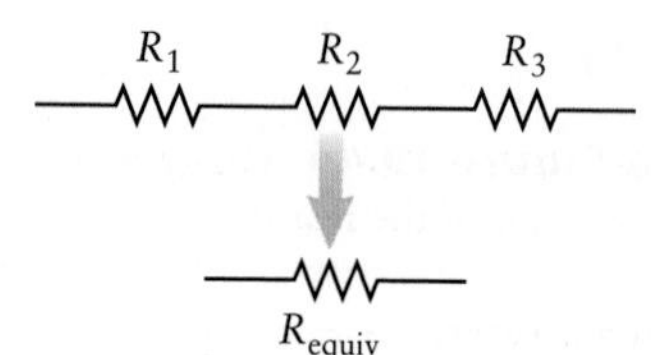

Series resistors: $R_{\text{equiv}} = R_1 + R_2 + \cdots$

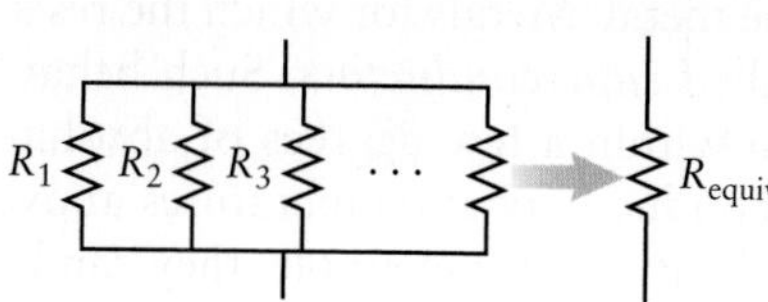

Parallel resistors: $\dfrac{1}{R_{\text{equiv}}} = \dfrac{1}{R_1} + \dfrac{1}{R_2} + \cdots$

Batteries

Batteries act as sources of energy in an electric circuit. The electric potential difference between the terminals of a battery is called the ***electromotive force*** or ***emf*** $\mathcal{E}$ produced by the battery.

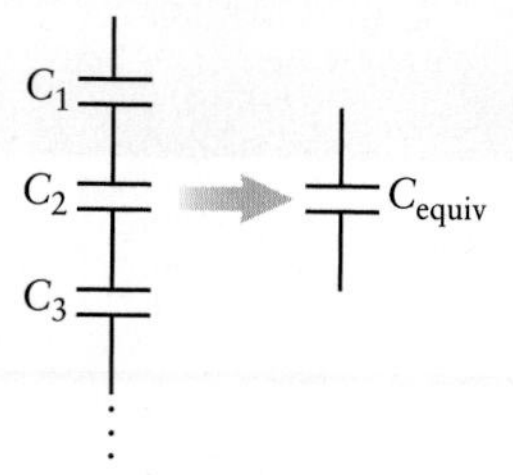

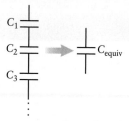

Series capacitors: $\dfrac{1}{C_{\text{equiv}}} = \dfrac{1}{C_1} + \dfrac{1}{C_2} + \cdots$

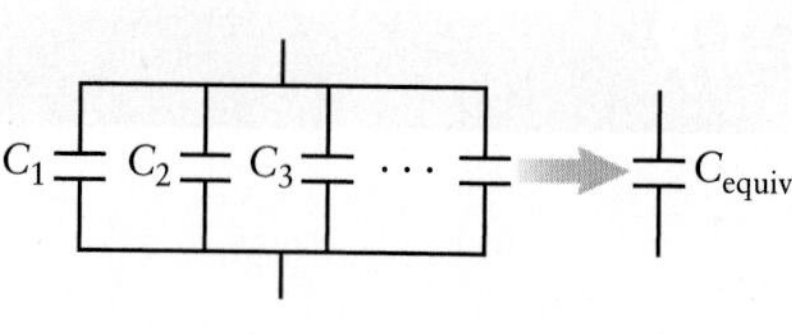

Parallel capacitors: $C_{\text{equiv}} = C_1 + C_2 + \cdots$

Charge carriers

The current in a metal wire is carried by mobile electrons that move at the ***drift velocity***.

Circuit analysis

Kirchhoff's rules tell how resistors and capacitors behave when connected in ***series*** and in ***parallel*** arrangements.

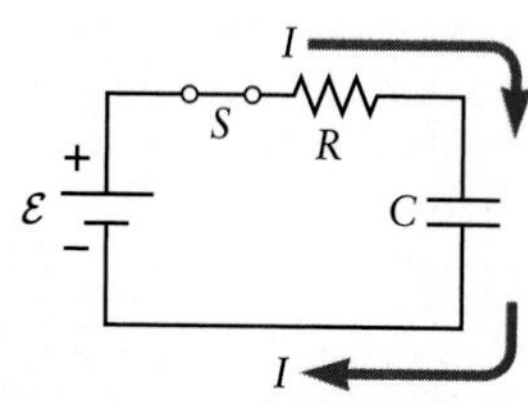

Charging a capacitor

When the switch is closed in an RC circuit, the current through the circuit and the voltage across the capacitor vary with time according to

$$I = \frac{\mathcal{E}}{R} e^{-t/\tau} \qquad (19.28)$$ (page 574)

$$V_{\text{cap}} = \mathcal{E}(1 - e^{-t/\tau}) \qquad (19.29)$$ (page 574)

where $\tau = RC$ is the ***time constant*** for the circuit.

Questions

SSM = answer in *Student Companion & Problem-Solving Guide*

= life science application

1. Discuss how Kirchhoff's rules for circuit analysis are related to conservation principles.

2. Explain why a complete circuit is necessary for a nonzero current to exist.

3. In a metal, is an electric current carried by (a) protons, (b) electrons, or (c) both protons and electrons?

4. In a solution of salt water (NaCl dissolved in water), is an electric current carried mainly by (a) electrons, (b) protons, (c) Na^+, (d) Cl^-, (e) Na^+ and Cl^-, or (f) electrons and protons?

5. Is the ratio of the voltage across a conductor to the current through the conductor called the (a) power, (b) capacitance, or (c) resistance?

6. Is the product of the voltage across a conductor and the current through the conductor called the (a) power, (b) capacitance, or (c) resistance?

7. In Chapter 17, we argued that the electric field inside a metal in equilibrium is always zero. In this chapter, we found that there is a nonzero electric field inside a current-carrying wire. How can both of these statements be correct?

8. Which of the plots in Figure Q19.8 shows how the current depends on voltage for an ordinary resistor?

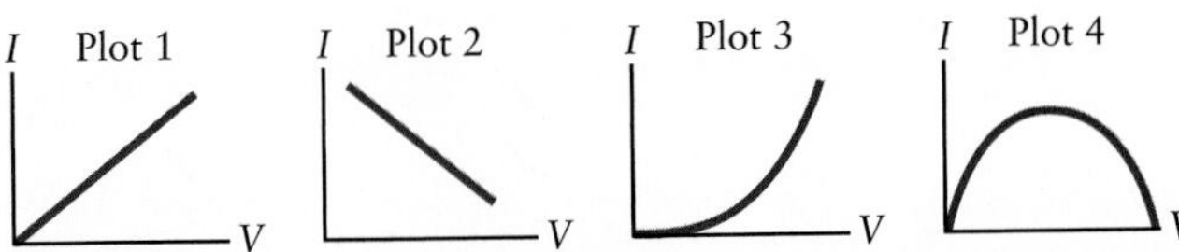

Figure Q19.8

9. SSM An incandescent lightbulb contains a filament that has a certain electrical resistance R. The brightness of the bulb depends on the current and increases as the current through the filament is increased. Consider the following situations in Figure Q19.9.
   (a) Two identical lightbulbs are connected as shown to a single ideal battery as in circuit 2. Will the brightness of one of these bulbs be greater than, less than, or the same as the brightness of the bulb in circuit 1?
   (b) Repeat part (a) for the bulbs in circuits 3, 4, and 5, comparing their brightness to that of the bulb in circuit 1. Explain your answers.

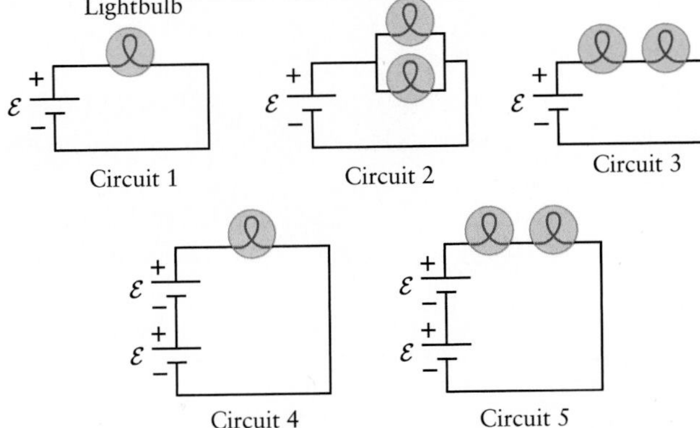

Figure Q19.9

10. If two circuit elements are connected in series, which of the following statements is true?
   (a) The voltage across the circuit elements is always the same.
   (b) The current through the circuit elements is always the same.
   (c) The current is always largest through the first circuit element.

11. If three circuit elements are connected in parallel, which of the following statements is true?
   (a) The voltage across the circuit elements is always the same.
   (b) The current through the circuit elements is always the same.
   (c) The power is always the same in all three circuit elements.

12. You are given four resistors, each with resistance R. Devise two ways to connect these resistors to get a total equivalent resistance greater than R.

13. You are given four resistors, each with resistance R. Devise two ways to connect these resistors to get a total equivalent resistance less than R.

14. You are given four resistors, all with resistance R. Devise a way to connect these resistors to get an equivalent resistance of $2.5R$.

15. A piece of wire with a constant cross-sectional area has a total resistance R_0. The wire is cut into three pieces of equal length, which are then reconnected to form an equivalent resistor with resistance $R_0/2$. Draw a circuit diagram showing how the three pieces of wire can be connected to give this equivalent resistance.

16. You are given four capacitors, each with capacitance C. Devise two ways to connect these capacitors to get a total equivalent capacitance greater than C.

17. You are given four capacitors, each with capacitance C. Devise two ways to connect these capacitors to get a total capacitance less than C.

18. SSM In Example 19.5, we used the rules for combining resistors in series and in parallel to analyze the circuit in Figure 19.22. Take a different approach and use Kirchhoff's rules to derive equations for the currents through the different circuit branches. Show that these equations lead to the same result found in Example 19.5.

19. Figure Q19.19 shows two circuits which contain the same circuit elements, but the positions of the resistors are interchanged. Will that affect the current? Justify your answer using Kirchhoff's rules.

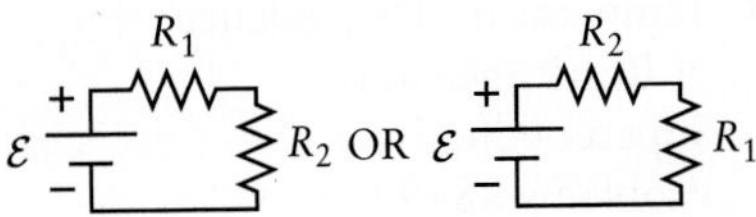

Figure Q19.19

20. In Example 19.2, we discussed how a capacitor can be used to store electrical energy. Investigate how capacitors are used in real camera flash circuits and discuss why they are better for this application than batteries.

21. In Example 19.1, we stated an estimate for the number of tungsten atoms in the filament of a lightbulb. Carry out a calculation of this number and compare your result with our estimate.

22. If one terminal of a battery is connected to an object, does any charge flow from the battery to the object? If there is charge flow, explain why it is very brief.

23. The circuit in Figure Q19.23 has a nonzero current that will carry electrons from one end of the resistor through the wire at the top to the positive terminal of the battery. We know that electrons always move from

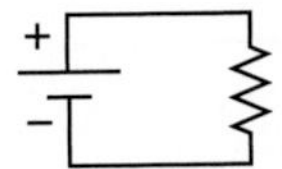

Figure Q19.23

a region of low electric potential to a region of higher potential. We also know that an ideal wire has zero resistance, so there can be no potential drop across an ideal wire. Explain how to reconcile these two facts.

24. When the author built his current house, he ran wires in the walls to be used for stereo speakers. He ran wires from several different rooms to a central location, but he forgot to mark the wires so he could tell which wires emerge in which room. Explain how he was able to solve this problem and identify which wire is which.

Problems available in ENHANCED WebAssign

Section 19.1 Electric Current: The Flow of Charge
Problems 1–5

Section 19.2 Batteries
Problems 6–8

Section 19.3 Current and Voltage in a Resistor Circuit
Problems 9–21

Section 19.4 DC Circuits: Batteries, Resistors, and Kirchhoff's Rules
Problems 22–59

Section 19.5 DC Circuits: Adding Capacitors
Problems 60–75

Section 19.6 Making Electrical Measurements: Ammeters and Voltmeters
Problems 76–83

Section 19.7 *RC* Circuits as Filters
Problem 84

Section 19.8 Electric Currents in the Human Body
Problems 85–87

Section 19.9 Temperature Dependence of Resistance and Superconductivity
Problems 88–90

Additional Problems
Problems 91–113

List of Enhanced Problems

Problem number	Solution in *Student Companion & Problem-Solving Guide*	Reasoning & Relationships Problem	Reasoning Tutorial in ENHANCED WebAssign
19.5	✓		
19.7	✓		
19.19	✓		
19.41		✓	✓
19.42		✓	✓
19.43		✓	✓
19.44		✓	✓
19.50	✓		
19.55		✓	
19.60	✓		
19.74		✓	✓
19.75		✓	✓
19.79	✓		
19.84		✓	
19.86	✓		
19.89	✓		
19.97		✓	✓
19.99	✓		
19.104		✓	
19.105		✓	✓
19.107		✓	✓
19.111		✓	✓
19.112		✓	✓
19.113		✓	

CHAPTER 20

▶ *There are many applications of magnetism, including trains that are levitated magnetically. (© Amazing Images/Alamy)*

Magnetic Fields and Forces

OUTLINE

In Chapters 17 and 18, we studied how an electric field can be produced by a charge or a collection of charges and how electric fields lead to forces on an electric charge. We now consider the same kinds of questions for magnetic fields and forces. Magnetic fields are produced by *moving* electric charges, and a magnetic field produces a force on *moving* charges. We have two main jobs in this chapter: (1) to understand the sources of magnetic fields and the fields they produce and (2) to see how to calculate the magnetic force on a charged particle. These two problems are intertwined. We begin with a qualitative discussion of how a magnetic field can be produced, and then we describe magnetic forces. We later return to a more quantitative treatment of how to calculate the magnetic field.

Magnetic forces act on charged particles, suggesting that magnetism and electricity are connected in some way. In Chapters 21 and 22, we show that electric fields can in fact serve as a source of magnetic fields and vice versa. This close connection gives rise to electromagnetic waves and light, topics we'll explore in Chapters 23 through 26.

20.1 Sources of Magnetic Fields

The first observations of magnetic fields involved ***permanent magnets***. The ancient Greeks, Romans, and Chinese all discovered that the mineral magnetite has the ability to attract or repel pieces of other magnetic materials, including iron and other pieces of magnetite. Permanent magnets are used in applications ranging from compass needles and "refrigerator magnets" to speakers, motors, and computer hard disks. Figure 20.1 shows a ***bar magnet***, a permanent magnet made in the shape of a "bar," as well as small pieces of iron (called iron "filings") sprinkled around the bar magnet. These filings are themselves small, needle-shaped, permanent magnets. The needle directions indicate the direction of the magnetic field $\vec{B}$ near the bar magnet. Figure 20.1 shows some of the magnetic field lines deduced from the pattern of iron filings.

Figure 20.2 compares the pattern of magnetic field lines near a bar magnet with the pattern of electric field lines near an electric dipole, showing that the two patterns are quite similar. We have also indicated the ***magnetic poles*** at the ends of the bar magnet in Figure 20.2A. The magnetic poles are called north (N) and south (S), and are analogous to the positive and negative electric charges in Figure 20.2B. This analogy helps explain why the iron filings line up parallel to the magnetic field lines in Figure 20.1. Each iron filing is a small bar magnet, with its own north and south poles. The north pole of an iron filing is attracted to the south pole of the large bar magnet and is repelled from the north pole. At the same time, the south pole of an iron filing is attracted to the north pole of the large bar magnet and is repelled from the south pole. These magnetic forces cause an iron filing to align parallel to $\vec{B}$ as shown in Figure 20.2A. This is just like the situation with electric charges and electric fields studied in Chapter 17 and shown in Figure 20.2B.

The unit of magnetic field is a derived unit of measure called the tesla (T) in the SI system. We'll show below how it is related to other SI units. The magnetic field close to one of the poles of a strong bar magnet has a magnitude of about 1 T. The Earth also acts approximately as a bar magnet, with a magnetic field of approximately 50 μT ($= 5 \times 10^{-5}$ T) near the Earth's surface. The Earth's magnetic field is responsible for the alignment of a compass needle (Section 20.9).

We have already noted that the pattern of magnetic field lines of a bar magnet is similar to the pattern of electric field lines near two electric charges. However, the two field patterns in Figure 20.2 *differ* in one very important way: the magnetic field lines *inside* the bar magnet point from the south pole toward the north pole, whereas between the two charges of the electric dipole the electric field lines are in the *opposite* direction. As a result, the magnetic field lines form closed "loops." While electric field lines start on positive charges and end on negative changes, ***magnetic field lines always close on themselves***, forming closed loops. This general property of magnetic fields is not limited to the magnetic field produced by a bar magnet.

Permanent magnets come in many shapes and sizes; the "horseshoe" magnet is another common example. A horseshoe magnet can be made by simply bending a bar magnet, and some of the resulting magnetic field lines are shown in Figure 20.3. There are north and south poles at the ends of the horseshoe, and the field lines again form closed loops as they circulate through the horseshoe. Permanent magnets with horseshoe shapes are used in many applications, including motors and generators.

In the introduction to this chapter, we mentioned that magnetic fields are produced by moving electric charges. This statement also applies to the bar magnets and

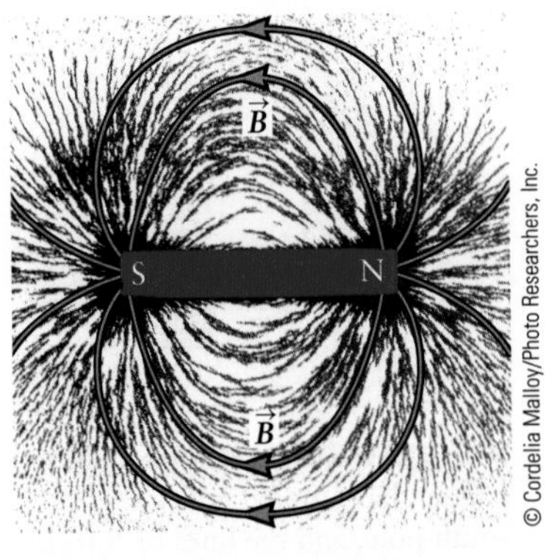

▲ **Figure 20.1** Iron filings are small slivers of iron. When placed near a bar magnet, they align along the magnetic field lines produced by the magnet. A few of these field lines are shown here. The letters "N" and "S" label the poles of the bar magnet.

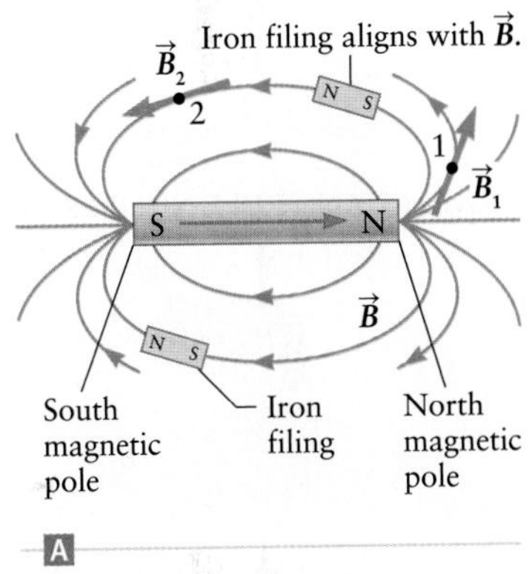

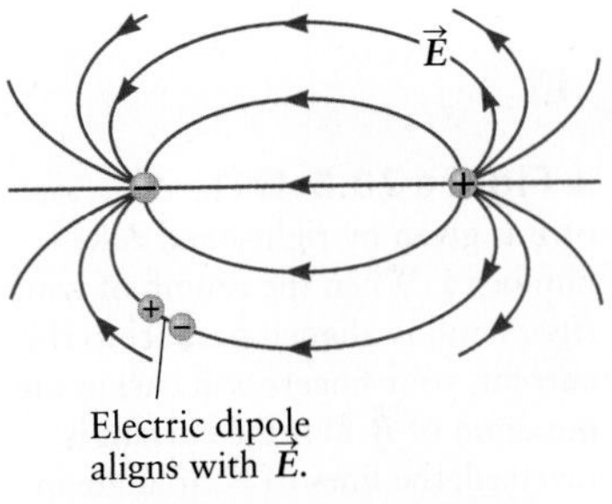

The electric field lines of an electric dipole are similar to the magnetic field lines of a bar magnet.

▲ **Figure 20.2** A Magnetic field lines near a bar magnet. B Electric field lines near an electric dipole.

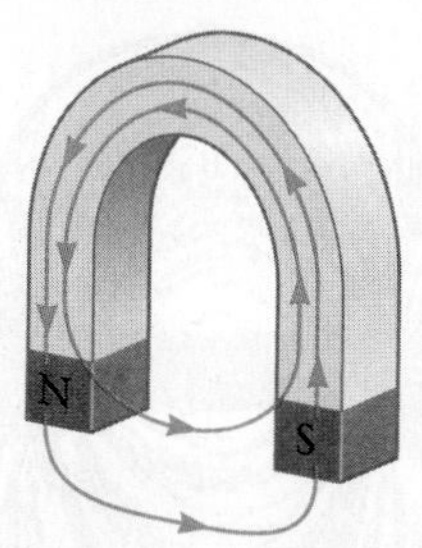

▲ **Figure 20.3** A horseshoe magnet has a north pole and a south pole, and the lines of $\vec{B}$ form closed loops.

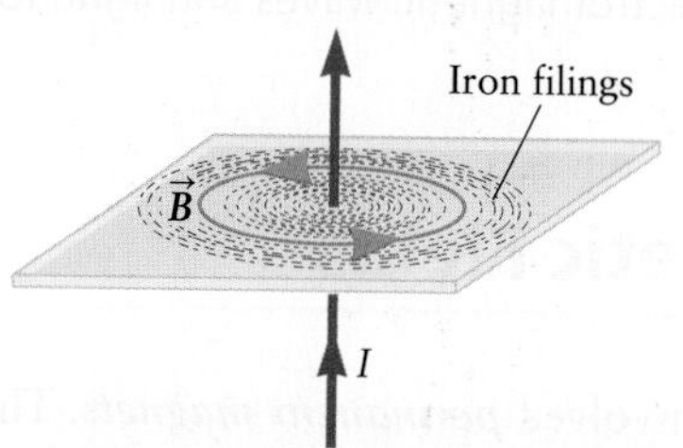

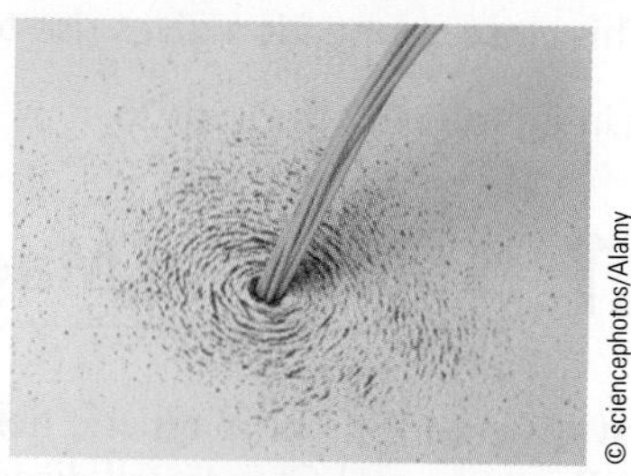

▲ **Figure 20.4** The magnetic field lines produced by a straight current-carrying wire form circles around the wire. The field lines can be visualized using iron filings. The magnitude of the field decreases as one moves away from the wire.

horseshoe magnets in Figures 20.1 through 20.3. The moving charges in these cases are electrons that move ("orbit") around the atoms in the magnetic material. We'll discuss the nature and size of the resulting currents, along with other properties of permanent magnet materials, in Section 20.8.

Magnetic Fields Produced by an Electric Current

Since an electric current consists of moving charges, an electric current will also produce a magnetic field. Figure 20.4 shows long, straight wires carrying an electric current. Each wire passes through a hole in a piece of paper on which iron filings have been sprinkled. We have seen how iron filings line up parallel to the magnetic field lines produced by a bar magnet; their alignment in Figure 20.4 indicates the pattern of magnetic field lines near each wire. For the case of a straight wire, the magnetic field lines form circles as sketched in the figure. The "strength" of the magnetic field (the magnitude of $\vec{B}$) decreases as one moves away from the wire.

So, magnetic field lines circulate around a current-carrying wire, but how do we determine in which direction they circulate? The direction of the magnetic field produced by a current is given by the ***right-hand rule*** illustrated in Figure 20.5. To apply the right-hand rule, point the thumb of your *right* hand in the direction of the current, with your thumb parallel to the wire. Curling the fingers of your *right* hand around the wire then gives the direction of $\vec{B}$. We call this method ***right-hand rule number 1***. (We'll encounter another right-hand-rule procedure below.)

If the direction of the current is reversed, the direction of $\vec{B}$ is also reversed. Figure 20.5B shows the same wire as in Figure 20.5A but with the current in the opposite direction. Applying right-hand rule number 1 shows that the lines of $\vec{B}$ now encircle the wire in the opposite sense.

Right-hand rule number 1 gives the direction of $\vec{B}$ but does not tell us its magnitude. We'll learn how to find the magnitude of $\vec{B}$ in Section 20.7.

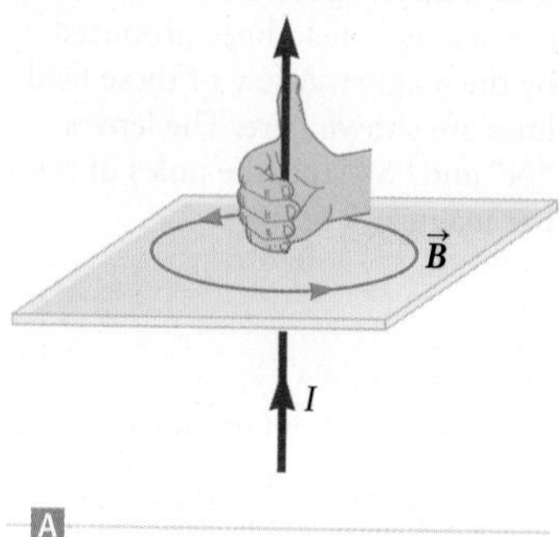

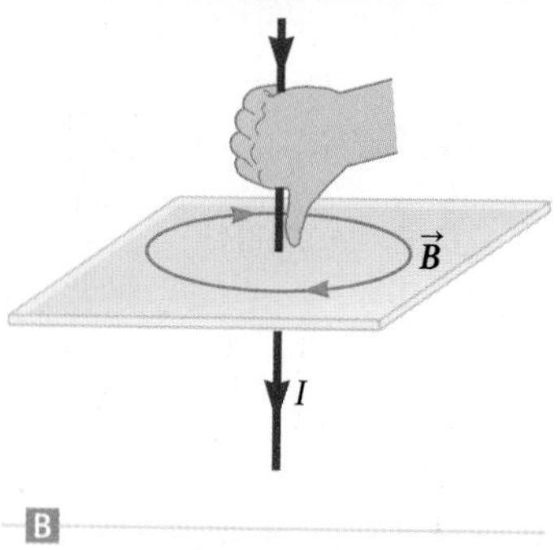

▲ **Figure 20.5** A The direction of $\vec{B}$ is given by right-hand rule number 1. When the thumb of your *right* hand is aligned parallel to the current, your fingers will curl in the direction of $\vec{B}$. B If the current is reversed, the lines of $\vec{B}$ circulate in the opposite direction.

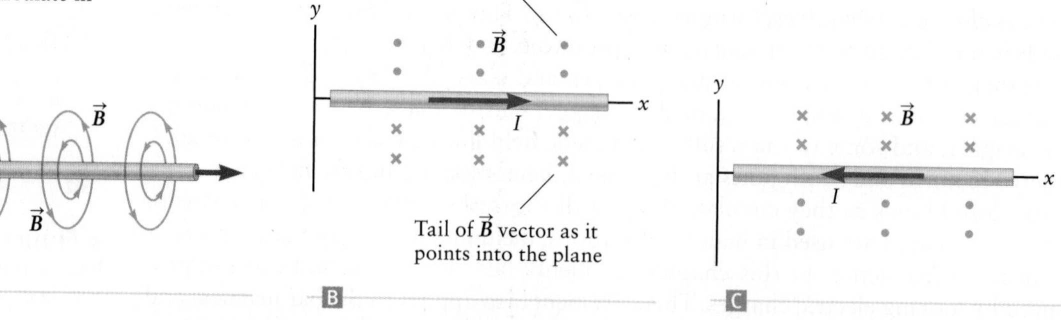

▲ **Figure 20.6** A Magnetic field near a long current-carrying wire. B and C In a two-dimensional plot, lines of $\vec{B}$ directed out of the plane are indicated by dots (the tip of the $\vec{B}$ vector), whereas lines of $\vec{B}$ pointing into the plane are denoted by crosses (the tail of the $\vec{B}$ vector).

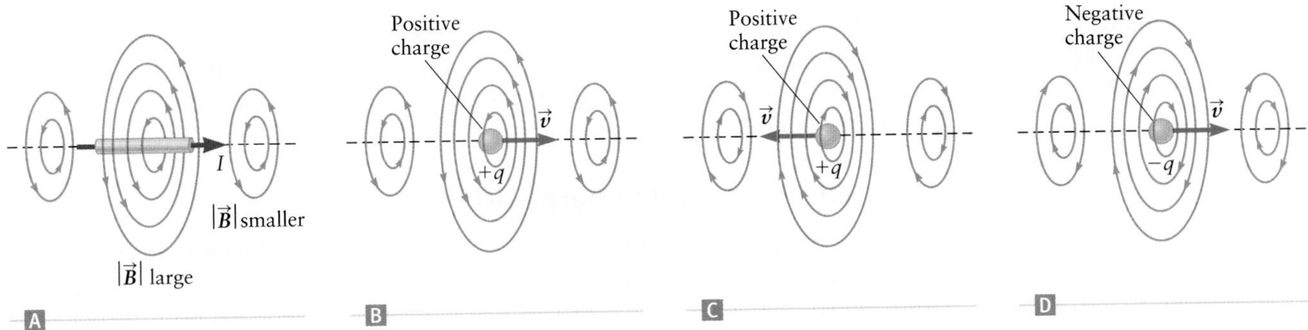

▲ **Figure 20.7** A Magnetic field lines produced by a short section of a current-carrying wire. B–D Magnetic fields produced by moving charges.

Plotting Magnetic Fields and Field Lines

Since the magnetic field lines produced by a current-carrying wire encircle the wire, the relationship between I and $\vec{B}$ is inherently three dimensional, and three-dimensional perspective plots likes those in Figures 20.4 and 20.5 will be extremely useful in our studies of magnetic fields. In many cases, however, we'll consider two-dimensional plots in which $\vec{B}$ points either into or out of the plane of the drawing. For example, Figure 20.6 shows the magnetic field of a long, straight wire. The three-dimensional representation of Figure 20.6A shows how the field lines encircle the wire, whereas Figure 20.6B shows a two-dimensional plot of the same field. In the region above the wire, the magnetic field $\vec{B}$ points *out of* the x–y plane, whereas the field lines point *into* the x–y plane below the wire. In Figure 20.6, a large "dot" (•) denotes the tip of the $\vec{B}$ vector when it points out of the plane and a "cross" (×) denotes the tail of the $\vec{B}$ vector when it points into the plane.

Figure 20.6C shows a similar two-dimensional plot of the magnetic field produced by a current-carrying wire, but here the current is in the $-x$ direction. The magnetic field lines again encircle the wire, but are now directed out of the plane in the area below the wire and into the plane above the wire.

In Figures 20.5 and 20.6, we have drawn long, straight wires, but right-hand rule number 1 can also be used to find the direction of the magnetic field lines from a short piece of wire as sketched in Figure 20.7A. We again put the thumb of our right hand along the direction of I, and curling our fingers then gives the direction of $\vec{B}$. The magnetic field lines again form circles (as for a long wire), but now the magnitude of $\vec{B}$ falls off as we go farther from the short wire along the x axis in either direction.

The electric current in a short current-carrying wire can be modeled as a collection of positive electric charges moving with velocity $\vec{v}$ parallel to the current direction (Fig. 20.7B). Each individual moving charge produces a magnetic field whose direction is given by right-hand rule number 1. Place your thumb of your right hand along the direction of $\vec{v}$; your fingers will then curl in the direction of the magnetic field (Fig. 20.7B). Figure 20.7C shows what happens to the magnetic field if the direction of the current is reversed. Here we have positive charges with $\vec{v}$ directed to the left, and applying right-hand rule number 1 shows that the field lines circulate opposite to those in Figure 20.7B. A positive charge moving to the left produces the same magnetic field as a negative charge moving to the right. Figure 20.7D shows such a moving negative charge; its magnetic field lines are the same as those produced by the positive charge in Figure 20.7C.

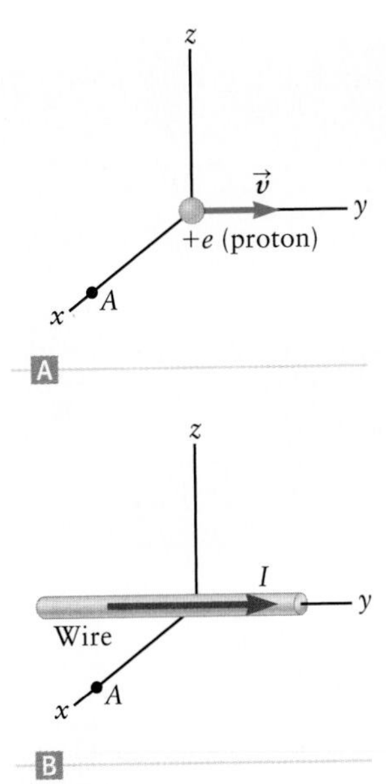

▲ **Figure 20.8** Concept Check 20.1.

CONCEPT CHECK 20.1 Direction of the Magnetic Field

Part 1: A proton travels at a constant velocity, with $\vec{v}$ directed along the $+y$ direction as shown in Figure 20.8A. At a particular moment, the proton is at the origin. Is the direction of the magnetic field at point A (a) along the $+x$ direction, (b) along the $-x$ direction, (c) along the $+z$ direction, or (d) along the $-z$ direction?

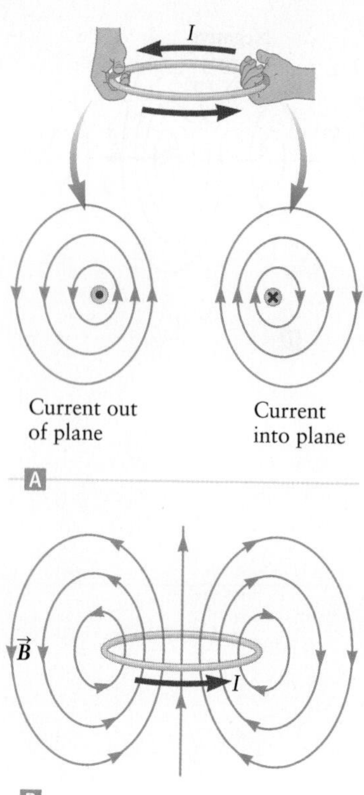

▲ **Figure 20.9** A Application of right-hand rule number 1 to find direction of the field lines near the two sides of a current loop. B Magnetic field lines produced by a full current loop.

Part 2: The proton is replaced by a long, straight wire carrying a current in the $+y$ direction as shown in Figure 20.8B. Is the direction of the magnetic field at point A (a) along the $+x$ direction, (b) along the $-x$ direction, (c) along the $+z$ direction, or (d) along the $-z$ direction?

Magnetic Fields and Superposition

So far, we have considered the magnetic field lines produced by a bar magnet, by a current in a straight wire, and by a moving electric charge. The qualitative pattern of the magnetic field lines in most other situations can be understood using these results along with the ***principle of superposition.***

Principle of superposition: **The total magnetic field produced by two or more different sources is equal to the sum of the fields produced by each source individually.**

We can apply the principle of superposition to find the direction of the magnetic field produced by a "current loop." Consider a circular loop of wire carrying a current I as shown in Figure 20.9A. This case might seem very special and idealized, but current loops are actually found in many applications. We treat the loop as many small pieces and apply right-hand rule number 1 to find the field from each piece. The magnetic field lines encircle each segment of the wire, and for the current direction shown here, $\vec{B}$ is directed upward through the center of the loop and downward outside the loop. We then use the principle of superposition and add up the fields from all these pieces, giving the overall pattern of magnetic field lines in Figure 20.9B. The lines of $\vec{B}$ circulate through the current loop, all passing inside the loop and then circling around the outside.

The overall pattern of field lines from a current loop is very similar to the magnetic field produced by a bar magnet (Fig. 20.2A). The similarity between the magnetic fields of a current loop and a bar magnet is no accident. The field of a bar magnet and other permanent magnets is due to atomic-scale current loops (Section 20.8).

EXAMPLE 20.1 Magnetic Field from Two Parallel Wires

Use the principle of superposition to sketch *qualitatively* the magnetic field lines near two long, straight wires, each parallel to the x axis and lying in the x–y plane. Assume each wire carries a current I, but with the currents in *opposite* directions.

RECOGNIZE THE PRINCIPLE

According to the principle of superposition, the total field is the sum of the fields from each wire. We use this principle along with our previous results for the magnetic field produced by a single wire (Figs. 20.5 and 20.6). Recall that for a single wire the field lines encircle the wire according to right-hand rule number 1 and the magnitude of $\vec{B}$ decreases as one moves away from the wire.

SKETCH THE PROBLEM

Figure 20.10A shows the magnetic field produced by one of the wires, and Figure 20.10B shows the field produced by the other.

IDENTIFY THE RELATIONSHIPS

The sum of the fields from the two wires is shown qualitatively in Figure 20.10C. In the region between the wires, the fields from both wires pass upward out of the plane of the drawing, so the total field in this region is also directed out of the plane. In the region

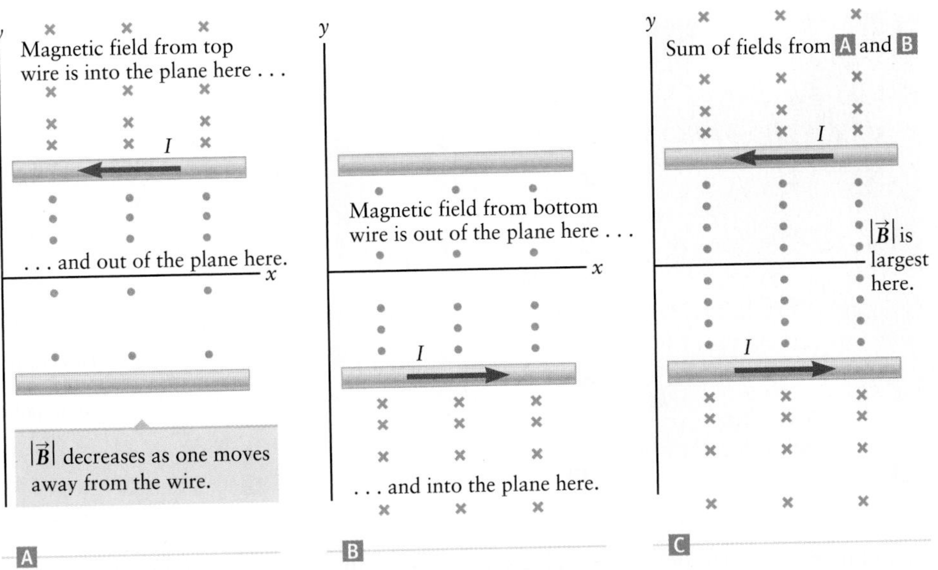

◀ **Figure 20.10** Example 20.1. Magnetic field due to A the top wire alone, B the bottom wire alone, and C both wires.

above the top wire, the field from the top wire is directed into the plane while the field from the bottom wire is out of the plane, so these fields are in opposite directions.

SOLVE

The top wire is closest to points in the upper region of Figure 20.10C, so the field of the top wire is larger (in magnitude) than the field from the bottom wire in that region. The sum of the two fields at the top of Figure 20.10C therefore points into the plane. The same reasoning shows that the field in the region below the bottom wire in Figure 20.10C is also directed into the plane.

▶ ***What does it mean?***

The principle of superposition can be used to find the pattern of magnetic field lines in virtually all situations. Most cases can be analyzed as a collection of bar magnets, straight wires, or current loops; the total magnetic field is then the sum of the fields produced by all the individual parts of the collection.

20.2 Magnetic Forces Involving Bar Magnets

In Section 20.1, we described the magnetic fields produced by various sources. Let's now consider the magnetic *forces* that these fields produce. Figure 20.11A shows a bar magnet situated in a constant magnetic field; when we say that $\vec{B}$ is constant, we mean that the magnetic field has a constant direction and a constant magnitude. To describe the total force on the bar magnet, we must include two forces, acting on the two poles of the bar. The force on the north pole is directed parallel to $\vec{B}$, while the force on the south pole is antiparallel to $\vec{B}$. For the case of a constant magnetic field, these two forces are equal in magnitude and in opposite directions. However, these two forces produce a nonzero torque on the bar magnet, which acts to align the bar magnet along the field lines. That is why iron filings (Figs. 20.1 and 20.4) align with the magnetic field. (Recall that each iron filing is a small bar magnet.)

It is interesting to compare the magnetic forces and torque on the bar magnet in Figure 20.11A with the electric forces on an electric dipole as sketched in Figure 20.11B. The figure shows an electric dipole formed by positive and negative charges $+q$ and $-q$ attached to the ends of an uncharged rod. This electric dipole is placed

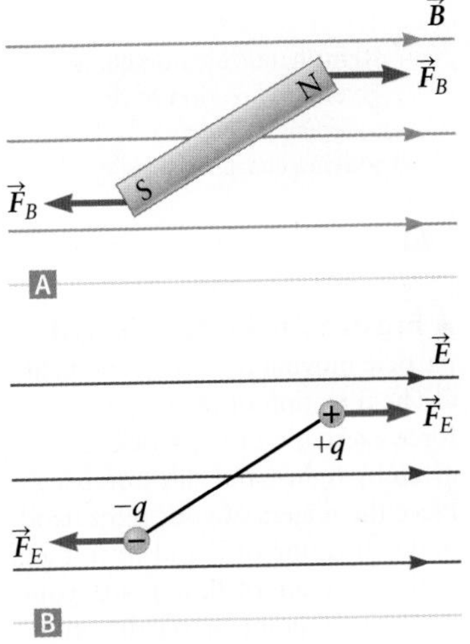

▲ **Figure 20.11** A Magnetic forces on a bar magnet placed in a uniform magnetic field. B Electric forces on an electric dipole placed in a uniform electric field.

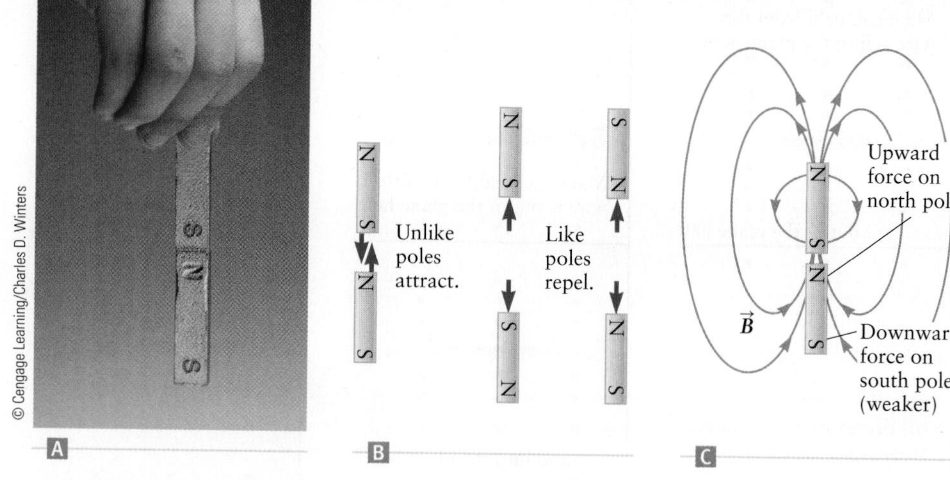

▶ **Figure 20.12** A and B Forces between two bar magnets. Unlike magnetic poles attract, and like poles repel. C The force on the north pole of the bottom magnet is upward, parallel to the field $\vec{B}$ of the upper magnet. The force on the south pole of the bottom magnet is downward. Since the field from the top magnet becomes smaller as one moves downward, the total (net) force on the bottom magnet is upward.

in a constant electric field $\vec{E}$, which exerts a force directed to the right on $+q$ and a force directed to the left on $-q$. These forces thus produce a torque on the electric dipole, just like the torque on the bar magnet in Figure 20.11A. This similarity illustrates that a bar magnet possesses a ***magnetic moment*** and also suggests that we can view the north and south poles of the bar magnet as a sort of "magnetic charge." The analogy with electric charge then implies that the north pole of one bar magnet should attract the south pole of another magnet (Fig. 20.12A), just as positive and negative electric charges attract each other. Likewise, we expect that "like" magnetic poles (two north poles or two south poles) will repel each other (Fig. 20.12B), just as found with two "like" electric charges (those of the same sign).

The forces between magnetic poles and between two bar magnets can also be analyzed and understood in terms of how the magnetic field produced by one magnet exerts a force on the other magnet. Figure 20.12C shows two bar magnets as well as the magnetic field produced by the top magnet. The magnetic field at the north pole of the lower magnet is directed upward, producing an upward force on this pole. The magnetic field at the bottom end of the lower magnet is also directed upward, which gives a downward force on the south pole. The field is weaker at the lower end, however, because it is farther from the upper magnet. As a result, the total (net) force on the lower bar magnet is upward, and the two magnets attract.

Figures 20.11 and 20.12 show some of the similarities among electric dipoles and bar magnets. However, there are some important differences between the electric and magnetic cases. In particular, north and south magnetic poles *always* occur in pairs. We'll see why it is not possible to obtain an isolated magnetic pole when we study what goes on inside a bar magnet in Section 20.8.

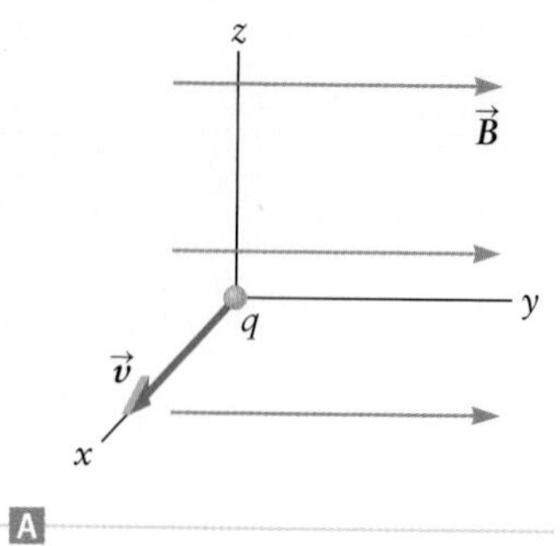

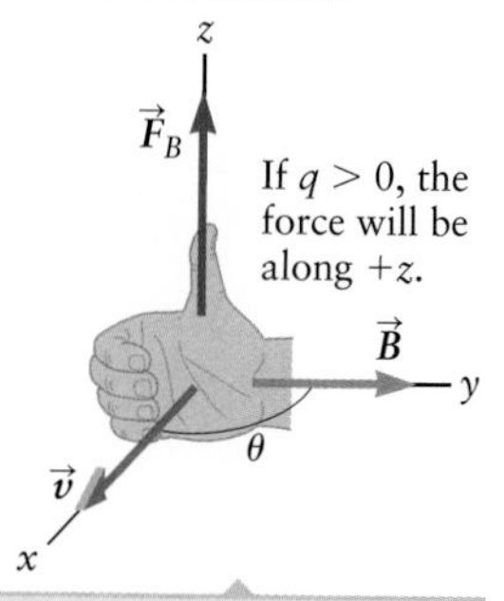

▲ **Figure 20.13** A A charged particle moving in a magnetic field. B The direction of the magnetic force exerted on the particle is given by right-hand rule number 2. Place the fingers of your right hand in the direction of $\vec{v}$ and wrap them in the direction of $\vec{B}$. If $q > 0$, your thumb then points in the direction of the magnetic force.

20.3 Magnetic Force on a Moving Charge

Magnetic forces act on electric charges, such as individual electrons, protons, or ions. The magnetic force depends on the velocity of the charge, and it is nonzero only if the charge is in motion. This is very different from gravity and the electric force, both of which are completely independent of velocity. The dependence of magnetic forces on velocity has some deep philosophical implications that we'll discuss in Section 20.11.

Figure 20.13A shows a particle with positive charge q moving with velocity $\vec{v}$. Here $\vec{v}$ is along the x direction, and the magnetic field $\vec{B}$ is parallel to y. The magnetic force on this charge is along the $+z$ direction, with a magnitude

Magnetic force on a charged particle

$$F_B = qvB \sin\theta \tag{20.1}$$

where θ is the angle between $\vec{v}$ and $\vec{B}$. In Figure 20.13, this angle is 90°, so $\sin\theta$ in Equation 20.1 is equal to 1. Notice that Equation 20.1 gives only the *magnitude* of the magnetic force. The *direction* of the force can be found using another right-hand rule illustrated in Figure 20.13B. Point the fingers of your right hand in the direction of $\vec{v}$ and curl your fingers in the direction of $\vec{B}$, taking the smallest angle between $\vec{v}$ and $\vec{B}$. Your thumb then points in the direction of the magnetic force $\vec{F}_B$. This procedure gives the direction of the force when q in Equation 20.1 is positive. For a positive charge ($q > 0$) with $\vec{v}$ and $\vec{B}$ directed as shown in Figure 20.13, the magnetic force is along the $+z$ direction. On the other hand, if q is negative, such as for an electron, the force direction is reversed and in this example is along $-z$. For a negative charge q, the value of F_B in Equation 20.1 is negative, so you can think of this as just reversing the direction of the magnetic force.

We'll call the procedure for finding the direction of the magnetic force on a moving charge in Figure 20.13B ***right-hand rule number 2***. This procedure is similar to right-hand rule number 1 (Fig. 20.5), which gives the direction of the magnetic field from a current. You should practice using both rules.

Compared with other fundamental forces such as the gravitational and electric forces, the magnetic force has several remarkable features. One is the dependence of $\vec{F}_B$ on velocity that we already noted. Also, the direction of $\vec{F}_B$ is always perpendicular to both the magnetic field and the particle's velocity. That is quite different from the gravitational and electric fields for which the force and the field are always parallel. Indeed, we originally introduced these fields as being simply proportional in magnitude and parallel in direction to the gravitational and electric forces. We now find that magnetic fields and forces are *not* related in such a simple way.

CONCEPT CHECK 20.2 Direction of the Magnetic Force

Part 1: Figure 20.14 shows a positive charge q moving with a nonzero speed v. If the magnetic field is directed to the right, is the direction of the magnetic force (a) into the plane of the drawing, (b) out of the plane of the drawing, (c) to the left, or (d) is the magnetic force zero?

Part 2: If the particle stops, is the direction of the magnetic force (a) into the plane of the drawing, (b) out of the plane of the drawing, (c) to the left, or (d) is the magnetic force zero?

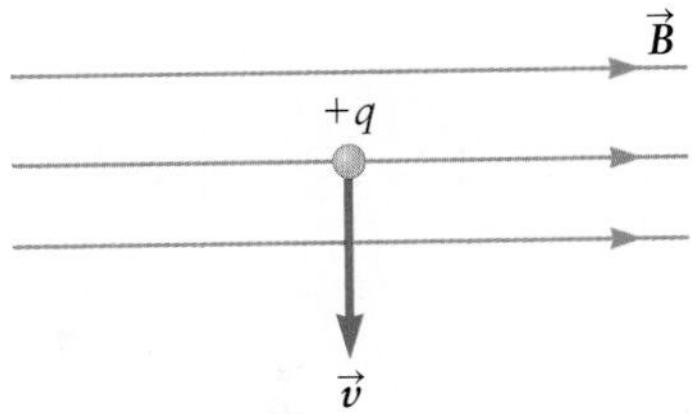

▲ **Figure 20.14** Concept Check 20.2.

Magnetic Force on an Electron: How Big Is It?

To get a feeling for the size of magnetic forces, consider an electron moving near a bar magnet. The field very close to a bar magnet can be as large as 1 T or 2 T, so let's take $B = 1$ T for this example. The electron's speed depends on the situation; let's assume a typical terrestrial speed of 50 m/s (about 100 mi/h). The charge of an electron is $q = -e = -1.60 \times 10^{-19}$ C. For simplicity, we assume the angle θ between $\vec{v}$ and $\vec{B}$ is 90° and insert these values for q, v, and B into Equation 20.1. The magnitude of the force is

$$F_B = qvB\sin\theta = (1.60 \times 10^{-19}\ \text{C})(50\ \text{m/s})(1\ \text{T})\sin(90°)$$
$$= 8 \times 10^{-18}\ \text{C} \cdot \text{m} \cdot \text{T/s} \quad (20.2)$$

Here, F_B must have units of force, so in the SI system the combination of units in Equation 20.2 must be equal to newtons (N). In terms of just the units, Equation 20.2 is thus

$$1\ \text{N} = 1\ \text{C} \cdot \text{m} \cdot \text{T/s}$$

We can now rearrange to solve for the tesla in terms of the primary SI units:

$$1\text{ T} = 1\,\frac{\text{N}\cdot\text{s}}{\text{C}\cdot\text{m}} = 1\,\frac{(\text{kg}\cdot\text{m/s}^2)\cdot\text{s}}{\text{C}\cdot\text{m}} \tag{20.3}$$

$$1\text{ T} = 1\,\frac{\text{kg}}{\text{C}\cdot\text{s}} \tag{20.4}$$

Returning to Equation 20.2, we find that the force on the electron is

$$F_B = 8 \times 10^{-18}\text{ N} \tag{20.5}$$

To judge if this force is large or small, let's compare F_B with the gravitational force on the electron. For an electron near the Earth's surface,

$$F_{\text{grav}} = mg = (9.11 \times 10^{-31}\text{ kg})(9.8\text{ m/s}^2) = 8.9 \times 10^{-30}\text{ N} \tag{20.6}$$

Comparing with F_B in Equation 20.5, we see that the gravitational force on an electron is smaller than the magnetic force in this case by a factor of

$$\frac{F_{\text{grav}}}{F_B} = \frac{8.9 \times 10^{-30}}{8 \times 10^{-18}} \approx 1 \times 10^{-12}$$

So, this magnetic force is *much* larger that the gravitational force. Of course, this comparison depends on the electron's velocity and the magnetic field, but one typically finds that magnetic forces are very much larger than gravitational forces.

EXAMPLE 20.2 Magnetic Force on an Electron

Consider an electron moving as part of the current in a metal wire. We learned in Chapter 19 that such an electron moves, on average, with the drift velocity (Eq. 19.12), which is typically $v = 0.01$ m/s. Suppose there is a magnetic field of magnitude $B = 2.0$ T perpendicular to the wire. (A field of this magnitude is found near the end of a "strong" bar magnet.) **(a)** What is the magnitude of the magnetic force on the electron? **(b)** The wire considered in Chapter 19 (Eq. 19.12) had an electric field of about 0.02 V/m. How does the electric force on the electron compare with the magnetic force?

RECOGNIZE THE PRINCIPLE

The magnetic force on the electron has magnitude $F_B = qvB \sin\theta$ (Eq. 20.1), while the electric force is $F_E = qE$.

SKETCH THE PROBLEM

Figure 20.15 shows the problem.

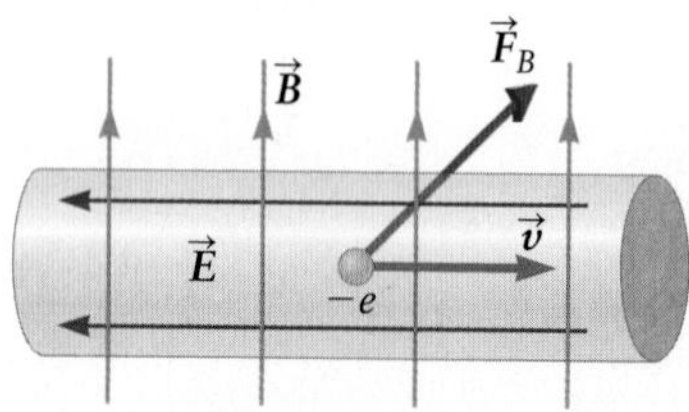

▲ **Figure 20.15** Example 20.2. An electron moving in a wire in the presence of a magnetic field and an electric field.

IDENTIFY THE RELATIONSHIPS

The electron's velocity and the magnetic field are perpendicular, so $\sin\theta = 1$. The charge on the electron is $q = -e$; hence, the *magnitude* of the magnetic force is

$$|F_B| = evB \tag{1}$$

The magnitude of the electric force is

$$|F_E| = eE \tag{2}$$

SOLVE

Evaluating Equations (1) and (2) with the given values of v, B, and E, we find

$$|F_B| = evB = (1.6 \times 10^{-19}\ \text{C})(0.01\ \text{m/s})(2.0\ \text{T}) = \boxed{3 \times 10^{-21}\ \text{N}}$$

$$|F_E| = eE = (1.6 \times 10^{-19}\ \text{C})(0.02\ \text{V/m}) = \boxed{3 \times 10^{-21}\ \text{N}}$$

► *What does it mean?*

In this case, F_B and F_E are approximately equal in magnitude. Since these forces depend on the speed of the electron and on the magnitudes of B and E, it is possible to find cases in which either F_B or F_E dominates. Notice that both of these forces are *much* larger than the weight of the electron (8.9×10^{30} N; Eq. 20.6), showing again that in most situations, electric and magnetic forces are much larger than the gravitational force.

Motion of a Charged Particle in a Magnetic Field

Let's now consider the direction of $\vec{F}_B$ and how this force affects the motion of a charged particle in a little more detail. For simplicity, we assume the magnetic field is constant, so the magnitude and direction of $\vec{B}$ are the same everywhere. Figure 20.16A shows a charged particle moving parallel to the direction of $\vec{B}$. In this case, the angle between $\vec{v}$ and $\vec{B}$ is $\theta = 0$; the factor $\sin\theta$ in Equation 20.1 is then zero, so the magnetic force in this case is also zero. *If a charged particle has a velocity parallel to $\vec{B}$, the magnetic force on the particle is zero.*

Figure 20.16B shows a charged particle moving perpendicular to $\vec{B}$. Now we have $\theta = 90°$, so $\sin\theta = 1$. The magnitude of the magnetic force is thus $F_B = qvB$, and the force is perpendicular to $\vec{v}$. In Chapter 5, we learned that when a particle experiences a force of constant magnitude perpendicular to its velocity, the result is *circular motion*. Hence, *if a charged particle is moving perpendicular to a constant magnetic field, the particle will move in a circle.* This circle lies in the plane perpendicular to the field lines. We can calculate the radius of the circle by recalling that for a particle to move in a circle of radius r, there must a force of magnitude mv^2/r directed toward the center of the circle (Eq. 5.6). Here the force producing circular motion is the magnetic force, so we have

$$F_B = \frac{mv^2}{r}$$

Inserting F_B from Equation 20.1 and using $\sin\theta = 1$ leads to

$$qvB = \frac{mv^2}{r}$$

Solving for r gives

$$r = \frac{mv^2}{qvB} = \frac{mv}{qB} \tag{20.7}$$

Let's calculate the value of r for the electron we considered in connection with Equation 20.5. In that example, we had $v = 50$ m/s and $B = 1$ T. Inserting those values into Equation 20.7, we find

$$r = \frac{mv}{qB} = \frac{(9.11 \times 10^{-31}\ \text{kg})(50\ \text{m/s})}{(1.60 \times 10^{-19}\ \text{C})(1\ \text{T})} \tag{20.8}$$

$$r = 3 \times 10^{-10}\ \text{m} \tag{20.9}$$

which is *quite* a small radius, approximately the radius of an atom. A faster-moving electron (larger v) will move in a circle with a larger radius.

In most cases, the velocity of a charged particle is not perfectly parallel or perpendicular to $\vec{B}$. Figure 20.16C shows the particle trajectory in such cases: the particle

MOTION OF A POSITIVELY CHARGED PARTICLE

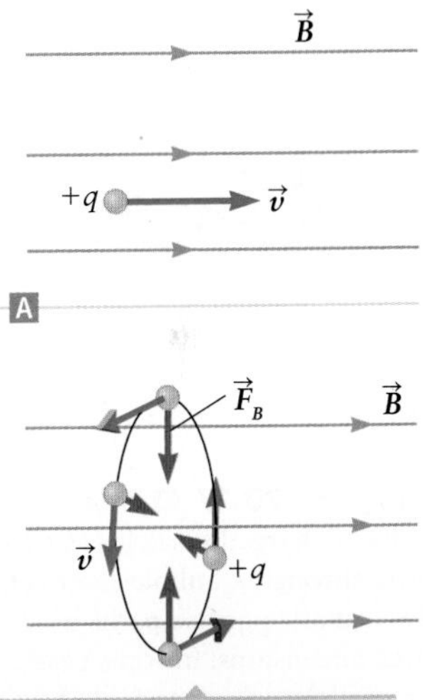

When $\theta = 90°$, the charged particle moves in a circle that lies in a plane perpendicular to the plane of this page.

B

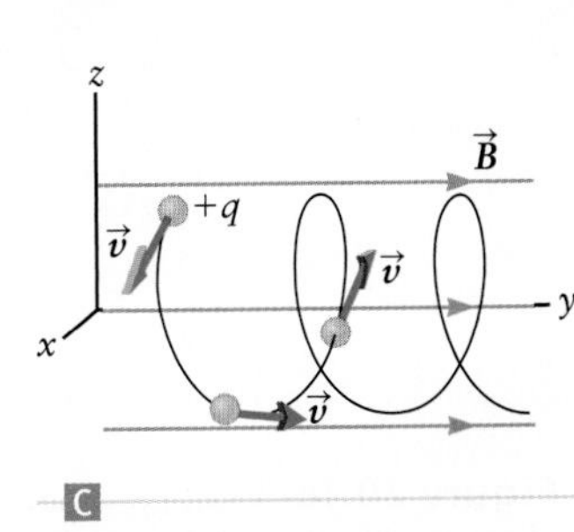

▲ **Figure 20.16** **A** When the velocity of a charged particle is parallel to the magnetic field $\vec{B}$, the magnetic force on the particle is zero. **B** When $\vec{v}$ makes a right angle with $\vec{B}$, the particle will move in a circle. **C** When the velocity has nonzero components parallel and perpendicular to $\vec{B}$, the particle will move along a helical path.

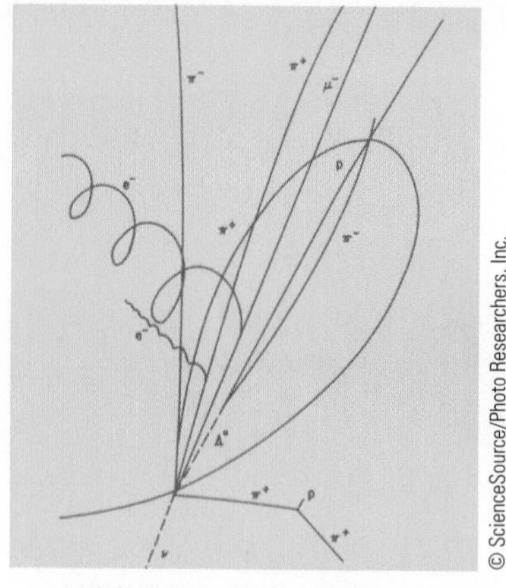

© ScienceSource/Photo Researchers, Inc.

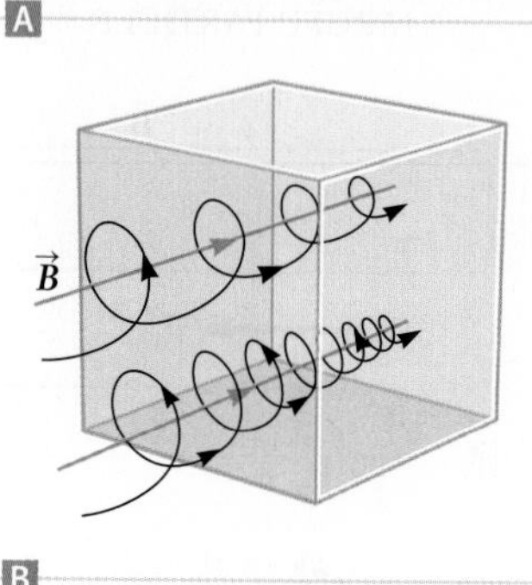

▲ **Figure 20.17** A Photo of the paths of charged particles as they move through a bubble chamber. B Schematic particle paths in three dimensions. In some cases, the particles lose energy as they move, causing the trajectories to spiral inward. As the particle speed decreases, the radius of the spiral trajectory also decreases. Several trajectories of this kind are visible in part A.

moves in a helical (spiral) path, with the axis of the helix parallel to $\vec{B}$. The helical motion can be understood as a combination of the results in parts A and B of Figure 20.16. For the component of $\vec{v}$ parallel to the magnetic field, there is no magnetic force and this component of the particle's velocity is constant. The perpendicular velocity component produces results as in Figure 20.16B: circular motion in a plane perpendicular to the field lines. The combination of circular motion in the x–z plane and a constant velocity along y gives a helical trajectory (Fig. 20.16C). Hence, *a charged particle will spiral around the magnetic field lines.*

A dramatic example of this spiral motion is shown in Figure 20.17A. This photograph is from a bubble chamber, a device used to study the trajectories of electrons, protons, and other particles. When charged particles travel through certain fluids at very high speed, they leave behind a trail of ionized fluid molecules, visible as a trail of bubbles. The entire bubble chamber is placed in a large magnetic field (Fig. 20.17B), causing the tracks to be helical. Measuring the radius of these helical tracks gives information about a particle's charge, mass, and velocity through the relation in Equation 20.7.

CONCEPT CHECK 20.3 Bubble Chamber Trajectories

Suppose an electron and a proton with the same velocity are traveling in the same direction when they enter the bubble chamber in Figure 20.17. Each moves in a helical trajectory inside the chamber. How does the radius of the electron's helix compare with the radius of the proton's helix?

(a) The helix followed by the electron has a smaller radius.
(b) The helix followed by the electron has a larger radius.
(c) The electron and proton helices have the same radius.

EXAMPLE 20.3 Direction of the Magnetic Force on a Moving Charge

Consider a positively charged particle traveling parallel to a current-carrying wire as sketched in Figure 20.18A. What is the *direction* of the magnetic force on this particle?

RECOGNIZE THE PRINCIPLE

Right-hand rule number 2 gives the direction of the magnetic force on a moving charge. To apply this rule, we must first establish the direction of the magnetic field produced by the wire, which we can find using right-hand rule number 1.

SKETCH THE PROBLEM

Figure 20.18 shows the direction of the current and the direction of $\vec{v}$, the particle's velocity.

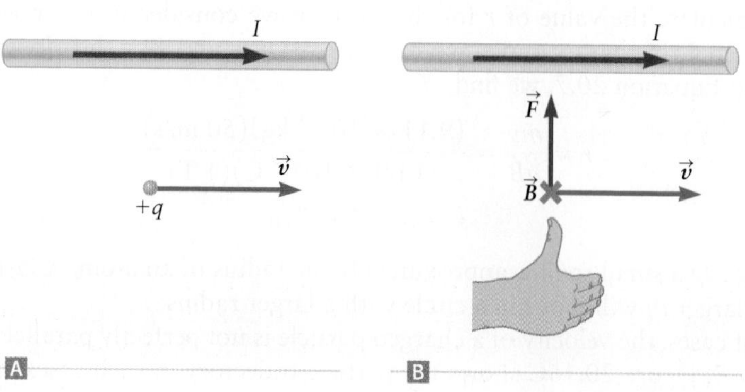

► **Figure 20.18** Example 20.3. Finding the direction of $\vec{F}_B$.

IDENTIFY THE RELATIONSHIPS AND SOLVE

From our work with current-carrying wires and right-hand rule number 1 (Figs. 20.5 and 20.6), the magnetic field lines encircle the wire and are directed *into the plane* of the drawing at the particle's location (Fig. 20.18B). To find the magnetic force on this particle, we now apply right-hand rule number 2, placing our fingers along $\vec{v}$ and curling them in the direction of $\vec{B}$ as shown in Figure 20.18B. The thumb then points in the direction of the force, *toward* the wire as shown in the figure.

▶ *What does it mean?*

To find the magnetic force on a moving charged particle, we must first find the direction of the magnetic field; here we used right-hand rule number 1 to find the direction of $\vec{B}$. We then applied right-hand rule number 2 to find the direction of the magnetic force. To deal with magnetic forces and fields, you *must* be able to apply the right-hand rules. Practice is strongly recommended.

CONCEPT CHECK 20.4 Finding the Direction of $\vec{B}$

A positively charged particle is moving with velocity $\vec{v}$ as shown in Figure 20.19. If the magnetic force on this particle is directed out of the plane of the drawing, is the possible direction of $\vec{B}$ (a) along direction 1 in the figure, (b) along direction 2, (c) along direction 3, or (d) directed into the plane of the drawing?

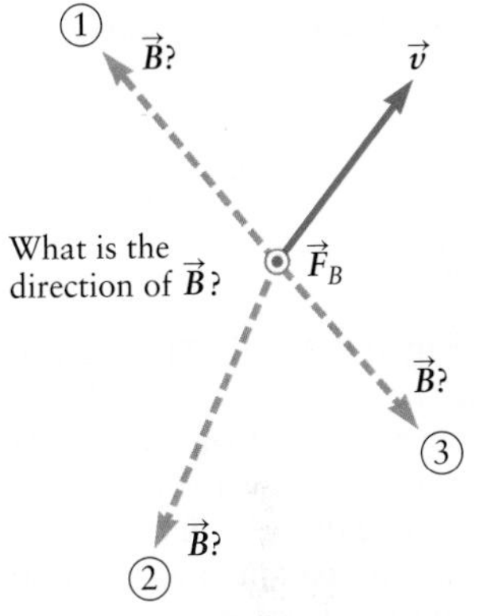

▲ **Figure 20.19** Concept Check 20.4.

Applying the Right-Hand Rules

We have introduced two right-hand rules and applied them in several example calculations. Let's state them once more for future reference.

Right-hand rule number 1: Finding the direction of $\vec{B}$ from an electric current.

1. **Place the thumb of your right hand along the direction of the current.**
2. **Curl your fingers; they will then give the direction of $\vec{B}$ as the field lines encircle the current.**

Right-hand rule number 2: Finding the direction of the magnetic force on a moving charge q.

1. **Point the fingers of your right hand along the direction of $\vec{v}$.**
2. **Curl your fingers in the direction of $\vec{B}$. Always curl through the smallest angle that connects $\vec{v}$ and $\vec{B}$.**
3. **If q is positive, the magnetic force on q is parallel to your thumb. If q is negative, the magnetic force is in the opposite direction.**

20.4 Magnetic Force on an Electric Current

An electric current consists of a collection of moving charges, so our result for the magnetic force on a single moving charge (Eq. 20.1) can be used to find the magnetic force on a current-carrying wire. This force is important in many applications, including electric motors. Consider a current-carrying wire placed in an external magnetic

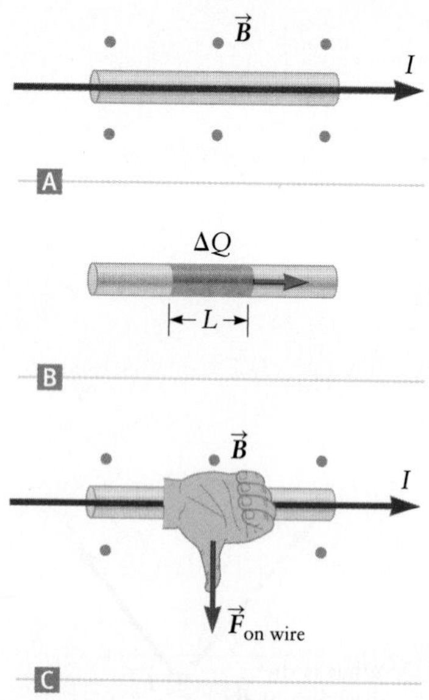

▲ **Figure 20.20** A A current-carrying wire in an external magnetic field. B The current in the wire is due to the motion of charge ΔQ along the wire. C The magnetic force on this moving charge produces a force on the wire. The direction of the force is given by right-hand rule number 2. Place the fingers of your right hand in the direction of the current and wrap them in the direction of $\vec{B}$. Your thumb then points in the direction of the magnetic force.

field as sketched in Figure 20.20A, with $\vec{B}$ constant and perpendicular to the wire. This magnetic field is produced by some external source; it is *not* produced by the current in the wire. We can calculate the magnetic force on the wire due to this external field by adding up the magnetic forces on all the moving charges in the wire.

Let's focus on the segment of the wire of length L in Figure 20.20B. The current in this segment is

$$I = \frac{\Delta Q}{\Delta t} \tag{20.10}$$

where ΔQ is the electric charge that passes by one end of the wire segment in a time Δt; this is just our usual relation between charge and current from Chapter 19. Note that ΔQ is also the charge contained in the shaded region in Figure 20.20B. Applying Equation 20.1, the magnetic force on this moving charge is

$$F_B = qvB = (\Delta Q)vB \tag{20.11}$$

The velocity of the charge is just

$$v = \frac{L}{\Delta t}$$

Inserting this into Equation 20.11 gives

$$F_B = (\Delta Q)vB = (\Delta Q)\frac{L}{\Delta t}B$$

Using the relation between current and charge in Equation 20.9, we get

$$F_B = \frac{\Delta Q}{\Delta t}LB = ILB$$

This force on the moving charge is really a force on the wire, so we arrive at

$$F_{\text{on wire}} = ILB$$

In Figure 20.20, we assumed the magnetic field is perpendicular to the wire. If $\vec{B}$ makes an angle θ with the wire, one finds

$$F_{\text{on wire}} = ILB\sin\theta \tag{20.12}$$

The direction of $F_{\text{on wire}}$ is given by (you guessed it) a right-hand rule as illustrated in Figure 20.20C. Begin with the fingers of your right hand in the direction of the current and curl them in the direction of the field. Your thumb then points in the direction of $\vec{F}_{\text{on wire}}$, downward in Figure 20.20C. The external magnetic field pulls this wire downward as long as the current is directed to the right. The magnetic force on a current is due to the force on a moving charge, so the right-hand rule used here is another example of right-hand rule number 2, with the direction of $\vec{v}$ for a moving charge replaced by the direction of the current.

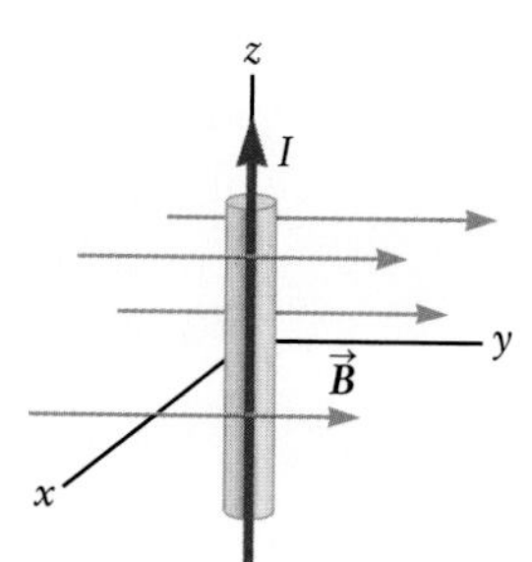

▲ **Figure 20.21** Concept Check 20.5.

CONCEPT CHECK 20.5 Finding the Magnetic Force on a Wire

The wire in Figure 20.21 carries a current along the $+z$ direction. If there is an external magnetic field in the $+y$ direction, is the direction of the magnetic force on the wire along (a) $+x$, (b) $-x$, (c) $+y$, (d) $-y$, (e) $+z$, or (f) $-z$?

EXAMPLE 20.4 Magnetic Force between Two Wires

Consider two parallel wires, each carrying current I as sketched in Figure 20.22A. Find the direction of the magnetic force that the top wire exerts on the bottom one. Then find the direction of the force that the bottom wire exerts on the top one.

RECOGNIZE THE PRINCIPLE

To find the force exerted on the bottom wire, we must first determine the direction of the magnetic field produced by the top wire using right-hand rule number 1. We then apply right-hand rule number 2 to find the force produced by this field on the bottom wire.

SKETCH THE PROBLEM

Figure 20.22B shows the direction of the magnetic field produced by the top wire at the location of the bottom wire as well as the application of right-hand rule number 2 to find the force on the bottom wire. Figure 20.22C shows the application of right-hand rule number 2 to find the force on the top wire.

IDENTIFY THE RELATIONSHIPS AND SOLVE

From right-hand rule number 1, the magnetic field produced by the top wire is directed into the plane of the drawing near the bottom wire as shown in Figure 20.22B. (See also Fig. 20.6.) Applying right-hand rule number 2 then gives a magnetic force on the bottom wire directed upward, and the bottom wire is *attracted* to the top wire.

Now consider the force on the top wire. The magnetic field produced by the bottom wire in the vicinity of the top wire is directed out of the plane of the drawing (Fig. 20.22C), and applying right-hand rule number 2 shows that the force exerted on the top wire is directed downward, toward the bottom wire. Hence, the force on the top wire is downward and the force on the bottom wire is upward. Two wires carrying parallel currents are attracted to each other.

▶ ***What does it mean?***

The magnetic forces exerted by each wire on the other are an action–reaction pair of forces. Hence, in accord with Newton's third law (the action–reaction principle), these forces must be in opposite directions, which is precisely what we have found. You should expect that these forces must have equal magnitudes; we'll show this in Section 20.7.

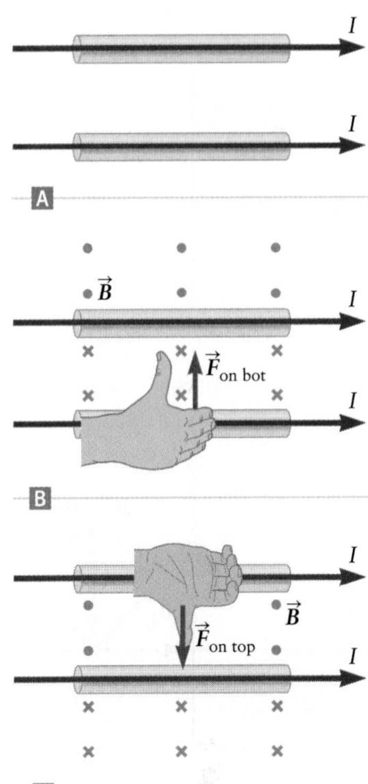

▲ **Figure 20.22** Example 20.4. A Two parallel current-carrying wires. B The magnetic force on the bottom wire. C The magnetic force on the top wire.

CONCEPT CHECK 20.6 How Does the Force Depend on the Direction of I?

Consider again the two wires in Figure 20.22 but now suppose the currents are in opposite directions, with I_{top} to the right and I_{bottom} to the left. Which of the following statements is correct?

(a) The force on the bottom wire is still upward in Figure 20.22, but the force on the top wire is now also upward.
(b) The force on the bottom wire is now downward and the force on the top wire is upward, so the overall effect is a repulsive force.
(c) There is no change.

EXAMPLE 20.5 Magnetic Force on a Wire

Figure 20.23A shows three straight, parallel wires, each carrying current I directed out of the plane of the drawing. All three wires are the same distance from the origin. What is the direction of the magnetic force on wire 3?

RECOGNIZE THE PRINCIPLE

We must first find the magnetic field at the location of wire 3 due to the currents in the other two wires. We then apply right-hand rule number 2 to find the direction of the force on wire 3.

(continued) ▶

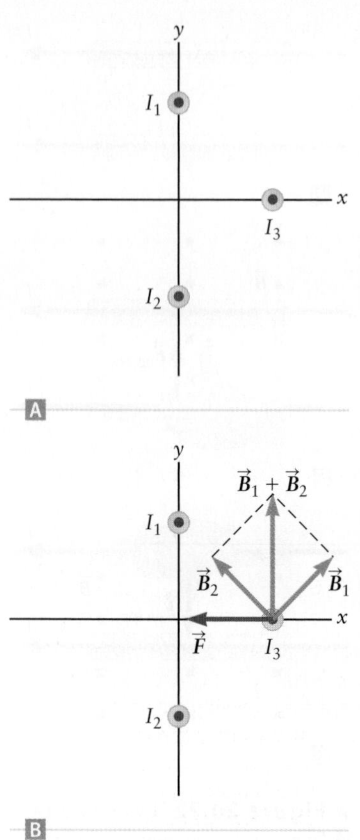

▲ **Figure 20.23** Example 20.5.

SKETCH THE PROBLEM

Figure 20.23A shows the problem.

IDENTIFY THE RELATIONSHIPS

Applying right-hand rule number 1, we find the direction of field $\vec{B}_1$ due to wire 1 and of $\vec{B}_2$ due to wire 2. They are shown in Figure 20.23B (compare with Figure 20.5A). The total field $\vec{B}_1 + \vec{B}_2$ is also shown.

SOLVE

Using the direction of $\vec{B}_1 + \vec{B}_2$ and right-hand rule number 2, we find that the force on wire 3 is to the left, toward the origin in Figure 20.23B.

► ***What does it mean?***

This problem is another example of the application of right-hand rule numbers 1 and 2 along with the principle of superposition. You should also recall that there is an attractive force between two parallel wires that carry currents in the same direction (Example 20.4). In this example with three wires, wire 3 is attracted to each of the other wires and the net force is directed toward the origin in Figure 20.23.

20.5 Torque on a Current Loop and Magnetic Moments

In Section 20.2, we saw that when a bar magnet is placed in a uniform magnetic field, there is a torque on the magnet (Fig. 20.11A). A magnetic field can also produce a torque on a current loop. Figure 20.24 shows a square current loop with sides of length L carrying a current I in a constant magnetic field. In Figure 20.24A, the field is parallel to the plane of the loop and the forces on all four sides of the loop are shown. These forces can all be found using right-hand rule number 2. In sides 2 and 4, the current is either parallel or antiparallel to $\vec{B}$, so the forces are zero. The forces on sides 1 and 3, denoted $\vec{F}_1$ and $\vec{F}_3$, respectively, are in opposite directions in Figure 20.24A and produce a torque around the loop's axis. The lever arms for these two forces are both $L/2$, and magnitude of this torque is

$$\tau = F_1\left(\frac{L}{2}\right) + F_3\left(\frac{L}{2}\right)$$

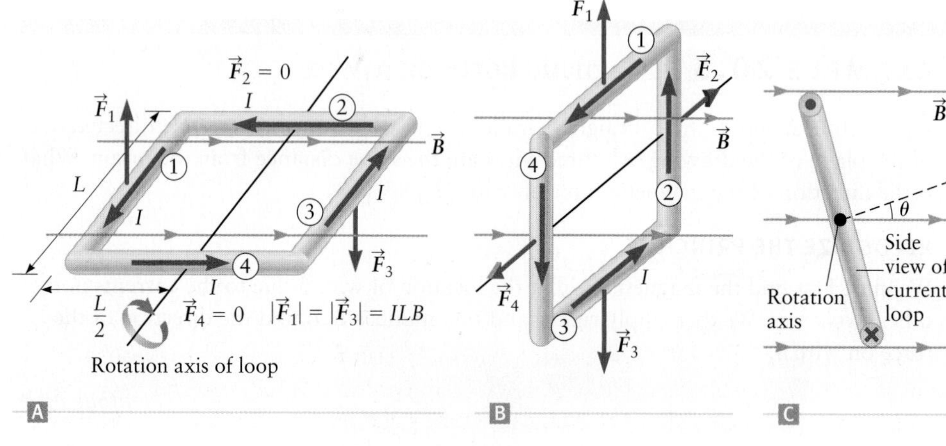

► **Figure 20.24** Torque on a current loop. A The magnetic forces on the sides of this loop tend to rotate the loop clockwise as viewed from the front, along the axis. B When the plane of the loop is perpendicular to $\vec{B}$, the torque is zero. C The torque on a current loop depends on the angle θ the field makes with the direction perpendicular to the loop.

The force on each wire segment is $F_1 = F_3 = ILB$ (from Eq. 20.12), so

$$\tau = F_1\left(\frac{L}{2}\right) + F_3\left(\frac{L}{2}\right) = 2ILB\left(\frac{L}{2}\right)$$

$$\tau = IL^2B \qquad (20.13)$$

Figure 20.24B shows the forces on the loop when its plane is perpendicular to the magnetic field. Now there is a nonzero force on all four sides of the loop; the four forces are of equal magnitude $F_1 = F_2 = F_3 = F_4 = ILB$ but are in different directions, and the total force on the loop is zero. The total torque is now also zero. Comparing parts A and B of Figure 20.24, we find that the torque on a current tends to align the plane of the loop perpendicular to $\vec{B}$.

So far, we have considered the torque when the plane of the loop is perpendicular or parallel to the field. When the angle between these two directions is θ (Fig. 20.24C), the torque is

$$\tau = IL^2B \sin \theta$$

The factor of L^2 is just the area of the square loop. For loops with different shapes, including circular, the torque on a loop of area A is

$$\tau = IAB \sin \theta \qquad (20.14)$$

The torque on a current loop is very similar to the torque on a bar magnet; both the loop and the magnet tend to "line up" with the magnetic field. Both act as a ***magnetic moment***, with the direction of the magnetic moment either along the axis of the bar magnet or perpendicular to the current loop as shown in Figure 20.25. The magnetic torque tends to align the magnetic moment parallel to $\vec{B}$, and the strength of the torque depends on the magnitude of the magnetic moment. For a current loop, the magnitude of the magnetic moment equals IA, the product of the current and loop area.

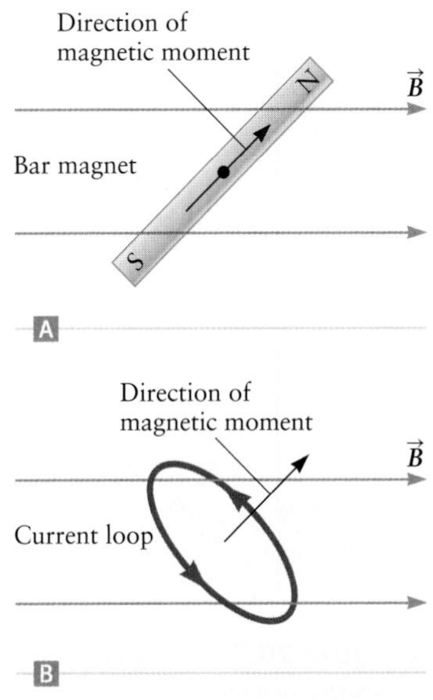

▲ **Figure 20.25** A The magnetic moment of a bar magnet is along the axis of the magnet, whereas B the magnetic moment of a current loop is perpendicular to the loop. In both cases, a magnetic field produces a torque that tends to align the magnetic moment with the field.

CONCEPT CHECK 20.7 Torque on a Current Loop

Figure 20.26 shows a current loop in a magnetic field. Is there a torque on the loop and, if so, in what direction?

(a) There is a torque that tends to rotate the loop in direction (a), clockwise.
(b) There is a torque that tends to rotate the loop in direction (b), counterclockwise.
(c) The torque is zero.

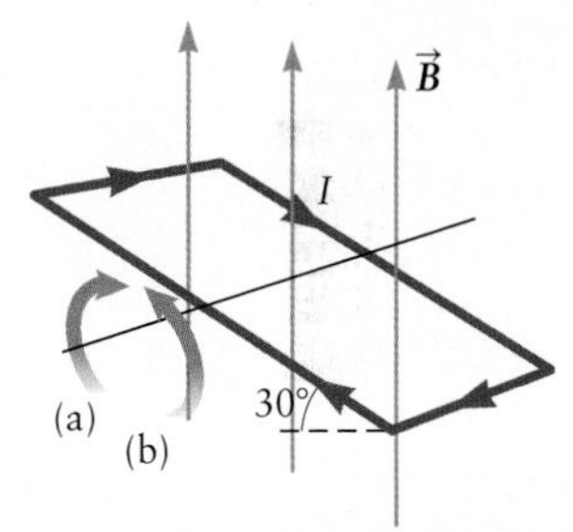

▲ **Figure 20.26** Concept Check 20.7.

20.6 Motion of Charged Particles in the Presence of Electric and Magnetic Fields

We considered the magnetic force on a charged particle in a magnetic field in Section 20.3 and found that various types of motion, including circular and helical trajectories, are possible. Also of interest is the motion of an electric charge in the presence of a *combination* of magnetic and electric fields. We'll analyze two such cases in this section.

The Mass Spectrometer

Scientists often want to separate ions according to their mass or charge. For example, archaeologists use a technique called carbon dating to determine an object's age. This technique is based on the observations that there are two isotopes of carbon atoms—denoted ^{12}C and ^{14}C (called "carbon 12" and "carbon 14," respectively)—and that the relative amounts of ^{12}C and ^{14}C in certain types of materials such as an animal bone can be used to determine the bone's age. (See Section 30.5.) One way to

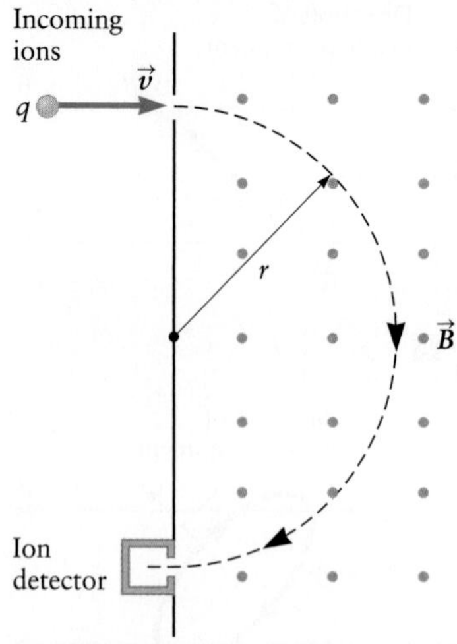

▲ **Figure 20.27** Design of a mass spectrometer.

distinguish ^{12}C from ^{14}C is through the difference in their masses, which can be done using a mass spectrometer.

Figure 20.27 shows the essential parts of a mass spectrometer. The charges are typically ions such as $^{12}C^+$ or $^{14}C^+$ that enter from the left, and for simplicity we'll assume they all have the same speed v. In practice, the ions would be accelerated to the desired speed by an electric field in the region just to the left of the slit at the top in Figure 20.27. The ions then move through a region in which the magnetic field is perpendicular to their velocity. In Figure 20.27, the magnetic field is out of the plane and perpendicular to the plane of the drawing, producing a magnetic force that causes the ions to move in a circle (as in Fig. 20.16B). The radius of this circle is (Eq. 20.7)

$$r = \frac{mv}{qB} \tag{20.15}$$

where q is the charge of the ion.

Suppose the incoming ions are of two types, $^{12}C^+$ and $^{14}C^+$, with different masses m_{12} and m_{14}. According to Equation 20.15, the different values of m lead to different values of r, so the two types of ions follow different circular arcs through the mass spectrometer. A mass spectrometer has an ion detector positioned at a certain radius r as shown in Figure 20.27. The value of either v or B is then adjusted until the detector gives a maximum ion current, indicating that the ion's circular trajectory matches the radius of the detector. For our hypothetical example with $^{12}C^+$ and $^{14}C^+$, the mass spectrometer could be used to measure the ion currents for $^{12}C^+$ and $^{14}C^+$ separately and thereby determine the relative amounts of the two isotopes in an object, which would then be used to find the object's age.

Another application of mass spectrometers is to "fingerprint" unknown molecules. Molecules of unknown composition are first ionized—by bombarding them with electrons, for instance—producing several different types of ions that are then fed to a mass spectrometer. The spectrometer then measures the values of v, B, and r corresponding to each ion. Equation 20.15 can be rearranged to give

$$\frac{q}{m} = \frac{v}{rB}$$

Because v, B, and r are all measured, the mass spectrometer gives q/m, the ratio of charge to mass for the ions. Chemists use this information to deduce the composition of each ion and hence the composition of the original molecule.

Insight 20.1

APPLICATIONS OF MASS SPECTROMETERS TO GENOMICS

Mass spectrometers are now being used in work on genomics and proteomics. For example, a protein of unknown sequence is first "cut" into fragments (e.g., peptides) using biochemical techniques. The fragments are then analyzed with a mass spectrometer that gives the ratios of q/m for the different ions (using Eq. 20.15). This information, together with the pattern of mass spectrometer intensities, is compared with mass spectrometer data on known peptides. With knowledge of the peptide "identities," the sequence of the original protein can then be determined.

EXAMPLE 20.6 Using a Mass Spectrometer

A mass spectrometer (Fig. 20.27) is used in a carbon dating experiment. The incoming ions are a mixture of $^{12}C^+$ and $^{14}C^+$ with a velocity $v = 1.0 \times 10^5$ m/s, and the magnetic field is $B = 0.10$ T. The ion detector is first positioned to find the value of r for $^{12}C^+$ and then moved to find the value of r for $^{14}C^+$. How far must the detector move?

RECOGNIZE THE PRINCIPLE

The location of the detector depends on r, the radius of the ion's circular path. The velocity v and field B are given, and the ions $^{12}C^+$ and $^{14}C^+$ each have charge $+e$. Equation 20.15 gives the value of r in terms of v, B, and m, the mass of the ion. If we can also find the mass m of each ion, we can then calculate r.

SKETCH THE PROBLEM

Figure 20.27 shows the problem. The ions follow a circular arc from where they enter the mass spectrometer to where they enter the detector.

IDENTIFY THE RELATIONSHIPS

The value of r in the mass spectrometer depends on the mass of the ion, so we need to find the masses of $^{12}C^+$ and $^{14}C^+$. The ion $^{12}C^+$ contains six protons, six neutrons, and five electrons. Consulting the table of fundamental constants on the inside front cover, we find a total mass of

$$m_{12} = 6m_p + 6m_n + 5m_e$$

$$m_{12} = 6(1.67 \times 10^{-27}\text{ kg}) + 6(1.67 \times 10^{-27}\text{ kg}) + 5(9.11 \times 10^{-31}\text{ kg})$$

$$m_{12} = 2.0 \times 10^{-26}\text{ kg}$$

A similar calculation gives the mass of $^{14}C^+$: $m_{14} = 2.3 \times 10^{-26}$ kg.

SOLVE

From Equation 20.15, the radius in the mass spectrometer for $^{12}C^+$ is

$$r_{12} = \frac{m_{12}v}{qB}$$

and the charge is $q = +1.60 \times 10^{-19}$ C. Inserting the given values, we find

$$r_{12} = \frac{m_{12}v}{qB} = \frac{(2.0 \times 10^{-26}\text{ kg})(1.0 \times 10^5\text{ m/s})}{(1.60 \times 10^{-19}\text{ C})(0.10\text{ T})} = 0.13\text{ m}$$

The same approach gives

$$r_{14} = 0.14\text{ m}$$

The circular trajectory followed by the $^{12}C^+$ ions (Fig. 20.27) will have a radius r_{12}, which is thus slightly smaller than the circular trajectory of radius r_{14} followed by the $^{14}C^+$ ions. To switch from one type of ion to the other, the ion detector in Figure 20.27 must move a distance equal to the difference in the diameters of the circular trajectories, so it must move a distance

$$\Delta r = 2(r_{14} - r_{12}) \approx \boxed{0.02\text{ m}}$$

or about 2 cm.

▸ ***What does it mean?***

The difference between r_{12} and r_{14} is not large, so to get a more accurate value we would need to employ more significant figures in our calculation. Even so, we can see that the trajectories of the two ions would differ by about 2 cm, which can easily be measured in a modern mass spectrometer.

▲ **Figure 20.28** The Hall effect gives a way to measure if a current is caused by the motion of A positive charges traveling to the right or B negative charges traveling to the left. In both cases, the current is directed to the right.

Hall Effect: How Do We Know If the Charge Carriers Are Positive or Negative?

In Chapter 19, we learned that a particular value and direction of the electric current could be produced by a positive charge moving in one direction (i.e., to the right in Fig. 20.28A) or by a negative charge moving in the opposite direction (to the left in Fig. 20.28B). Because the current in a metal is carried by electrons (negative charges), the picture in Figure 20.28B should apply, but how do we really know? Is there some measurement we can perform that distinguishes between these two possibilities?

The answer to this question was given more than a century ago by Edwin Hall.[1] He suggested placing a current-carrying wire in a magnetic field directed perpendicular to the current. Suppose the current is produced by positive charges moving to the right (Fig. 20.29A). If $\vec{B}$ is directed into the plane of the drawing, the magnetic force on the moving charges is directed upward, causing the positively charged carriers to be deflected toward the wire's upper edge. The result is an excess of positive charge on the top edge of the wire and a deficit of positive charge on the bottom, giving an electric potential difference between the top and bottom edges. In Figure

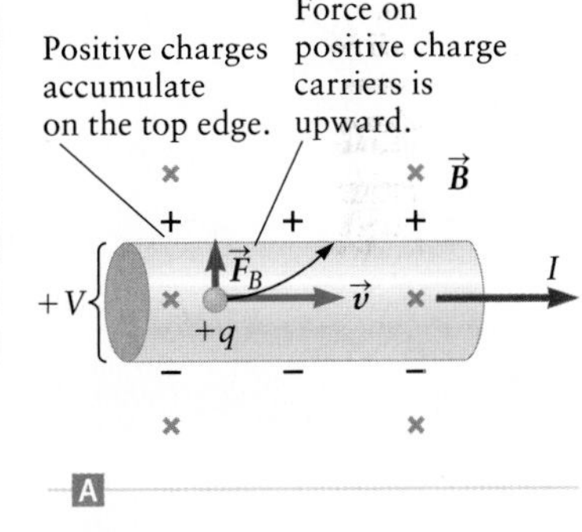

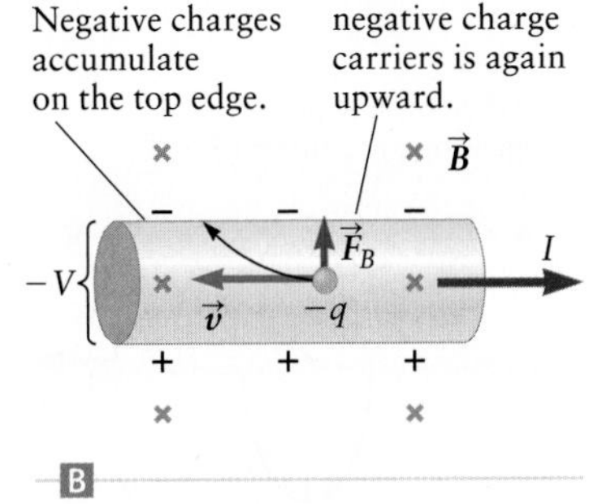

▲ **Figure 20.29** Hall's experiment. Motion of electric charges in a wire if the current is carried by A positive charges traveling to the right, or B negative charges traveling to the left.

[1]Hall had the idea for this experiment when he was a student! It is still a good homework problem.

20.29A, the upper edge has a positive electric potential relative to the bottom, which can be measured by attaching a voltmeter to the wire's top and bottom edges.

Now consider what happens if the current is carried instead by negative charges (electrons) moving to the left (Fig. 20.29B). Applying right-hand rule number 2 for negative charges, we find that the magnetic force is again directed toward the top edge of the wire. The magnetic force is thus in the *same* direction as with the positive charge carriers in Figure 20.29A, and the electrons in Figure 20.29B are deflected toward the upper edge of the wire, leading to a *negative* potential on that edge relative to the bottom.

Hall's experiment thus gives a way to determine the sign of the charge carriers: we simply measure the potential difference between the two edges of the wire. Hence, we can indeed distinguish between a positive charge moving in one direction and a negative charge moving in the opposite direction. That is how we know that the current in metals such as copper and gold is carried by electrons.

20.7 Calculating the Magnetic Field: Ampère's Law

In Chapter 17, we discussed two approaches for calculating the electric field, Coulomb's law and Gauss's law. There are also two ways to calculate the magnetic field produced by a current. One approach, which treats each small piece of a wire as a separate source of $\vec{B}$, is similar in spirit to Coulomb's law but is mathematically too complicated for us to deal with here. Another approach, which uses ***Ampère's law***, is very useful when the magnetic field lines have a simple symmetry. This approach is similar to Gauss's law for electric fields, which is most useful when the electric field is highly symmetric.

Consider a long, straight current-carrying wire encircled by magnetic field lines as shown in Figure 20.30 and imagine traveling around a closed path that also encircles the wire. Ampère's law relates the magnetic field *along* this path to the electric current *enclosed by* the path. Let's travel along the circular path in Figure 20.30 taking steps of length ΔL and let $B_\parallel$ be the component of the magnetic field parallel to these steps. Ampère's law states that

Ampère's law

$$\sum_{\substack{\text{closed}\\\text{path}}} B_\parallel \, \Delta L = \mu_0 I_{\text{enclosed}} \tag{20.16}$$

where I_{enclosed} is the total current passing through the surface surrounded by the chosen closed path. The quantity μ_0 is a constant of nature called the ***permeability of free space*** whose value is

$$\mu_0 = 4\pi \times 10^{-7}\ \text{T} \cdot \text{m/A} \tag{20.17}$$

Ampère's law thus relates the magnetic field *on the perimeter* of a region to the current that *passes through* the region. It is similar to Gauss's law for the electric field (Chapter 17), which relates the electric field *on a surface* to the electric charge *enclosed by* the surface.

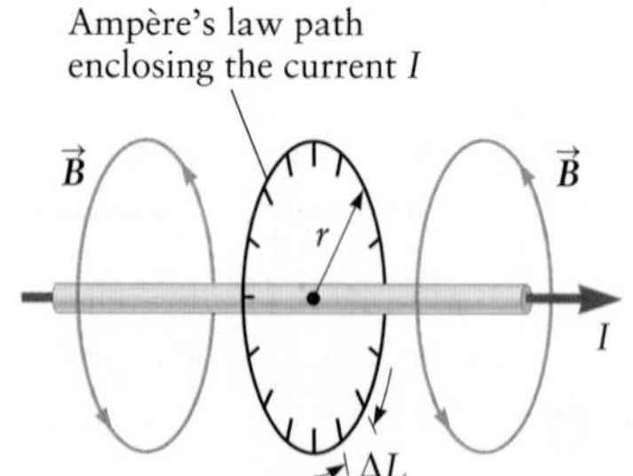

▲ **Figure 20.30** Ampère's law relates the magnetic field along a closed path to the current enclosed by that path. The path shown here is a circle of radius r.

Applying Ampère's Law: The Magnetic Field Produced by a Straight Wire

The sum $\sum B_\parallel \, \Delta L$ on the left-hand side of Ampère's law (Equation 20.16) involves the field at all points along the Ampère's law path. If the value of $B_\parallel$ varies as one moves along the path, Ampère's law can be impossible to apply in practice. On the other hand, in some cases $B_\parallel$ is constant all along the chosen path; Ampère's law is then very handy.

For a magnetic field near a long straight wire, we know that the magnetic field lines form circles. From the symmetry in Figure 20.30, $B_\parallel$ must have the same value at all points along a circular path centered on the wire. We can thus write Ampère's law as

$$\sum_{\substack{\text{closed}\\\text{path}}} B_{\parallel}\,\Delta L = B_{\parallel} \sum_{\substack{\text{closed}\\\text{path}}} \Delta L = \mu_0 I_{\text{enclosed}}$$

The key here is that $B_{\parallel}$ is the same all along the path, so we can move it in front of the summation. If the circular path has a radius r, the total length of the path is $\sum_{\text{closed path}} \Delta L = 2\pi r$, and Ampère's law gives

$$\sum_{\substack{\text{closed}\\\text{path}}} B_{\parallel}\,\Delta L = B_{\parallel} \sum_{\substack{\text{closed}\\\text{path}}} \Delta L = B_{\parallel}(2\pi r) = \mu_0 I_{\text{enclosed}} \tag{20.18}$$

The right-hand side of Ampère's law involves the total current passing through the chosen surface, which is just the current I in the wire, so we have

$$B_{\parallel}(2\pi r) = \mu_0 I$$

Although $B_{\parallel}$ is the component of the field along the circular path, in this case it is also the total field, so we can denote it simply as B and find

$$B = \frac{\mu_0 I}{2\pi r} \tag{20.19}$$

Magnetic field near a long, straight wire

With this result, we now have the full solution for the magnitude and direction of the magnetic field produced by a long, straight wire.

Let's use the result in Equation 20.19 to find the magnitude of the field in a typical case. Suppose the current is $I = 1$ A (a common value in household appliances) and we are a distance $r = 1$ cm $= 0.01$ m from the wire. Inserting these values in Equation 20.19, we get

$$B = \frac{\mu_0 I}{2\pi r} = \frac{(4\pi \times 10^{-7}\ \text{T}\cdot\text{m/A})(1\ \text{A})}{2\pi(0.01\ \text{m})} = 2 \times 10^{-5}\ \text{T}$$

which is about half as large as the Earth's magnetic field. A substantial current of 1 A thus produces only a modest magnetic field, even at this small distance from the wire. This result for B also suggests that the direction of a compass needle can be noticeably affected by the magnetic field from the current in a nearby wire.

EXAMPLE 20.7 Force between Two Current-Carrying Wires

Consider two long, straight, parallel wires each carrying a current $I = 2.0$ A, with the currents in the same direction. **(a)** Find the magnetic field at one wire produced by the other wire if the wires are separated by a distance $r = 1.0$ mm. Assume for simplicity that the wire diameters are much less than their separation. **(b)** What is the magnitude of the magnetic force exerted by one of these wires on a 1.0-m-long section of the other wire?

RECOGNIZE THE PRINCIPLE

To find the force, we must first find the magnetic field produced by one wire at the location of the other. We can find the magnitude of B using Ampère's law applied to a wire with a current I (Fig. 20.30). Once we have B, we can use our previous result for the magnetic force on a current.

SKETCH THE PROBLEM

Figure 20.22B describes the problem; it shows the current directions and the magnetic field produced by the top wire.

(continued) ▶

IDENTIFY THE RELATIONSHIPS AND SOLVE

(**a**) From our Ampère's law result (Eq. 20.19), the magnetic field due to the top wire is

$$B = \frac{\mu_0 I_{\text{top}}}{2\pi r} \tag{1}$$

where I_{top} is the current in the top wire. Inserting the given values of I_{top} and $r = 1.0$ mm, we get

$$B = \frac{\mu_0 I_{\text{top}}}{2\pi r} = \frac{(4\pi \times 10^{-7}\ \text{T}\cdot\text{m/A})(2.0\ \text{A})}{2\pi(1.0 \times 10^{-3}\ \text{m})} = \boxed{4.0 \times 10^{-4}\ \text{T}}$$

(**b**) The field from the top wire is into the plane of the drawing at the location of the bottom wire (Fig. 20.22B), so the angle between $\vec{B}$ and the current in the bottom wire is $\theta = 90°$. The force on the bottom wire is (Eq. 20.12)

$$F_{\text{on bot}} = I_{\text{bot}} L B \sin\theta$$

where I_{bot} is the current in the bottom wire and $L = 1.0$ m is the length of the section of the bottom wire on which we are asked to find the force. Substituting for B from part (a) and inserting $\sin\theta = 1$ (since $\theta = 90°$) gives

$$F_{\text{on bot}} = I_{\text{bot}} L B \sin\theta = (2.0\ \text{A})(1.0\ \text{m})(4.0 \times 10^{-4}\ \text{T})(1) = \boxed{8.0 \times 10^{-4}\ \text{N}}$$

▸ ***What does it mean?***

Although these two wires carry substantial currents and are quite close together, the magnetic force between them is still very modest. The wires in a typical household appliance or in the walls of your house carry similar currents and are sometimes spaced this closely. Our result for the force on the bottom wire shows that the magnetic force will not do any damage in such a case.

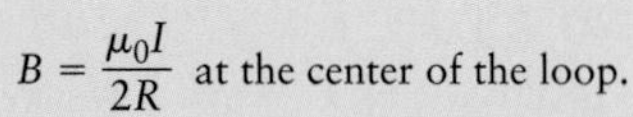

$B = \frac{\mu_0 I}{2R}$ at the center of the loop.

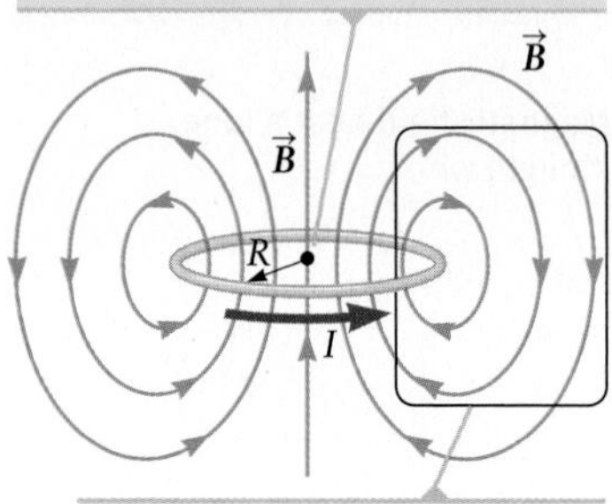

$B_{\parallel}$ is not constant on this path (or on any other closed path around a current loop).

▲ **Figure 20.31** In some cases, such as this current loop, Ampère's law does not give a (mathematically) simple way to calculate $\vec{B}$. For a circular current loop, the magnitude of the magnetic field at the center of the loop is $B = \mu_0 I/2R$.

Field from a Current Loop

The mathematics of Ampère's law is manageable when one can find a path along which the magnetic field is constant. Unfortunately, that is not possible in many cases such as the current loop sketched in Figure 20.31, which shows the pattern of field lines described in Section 20.1. There is no closed path on which $B_{\parallel}$ (the component of the magnetic field along the path) is constant, so there is no simple way to apply Ampère's law to find $B_{\parallel}$ in this situation. Other methods for calculating $\vec{B}$ can be applied, and it is found that the field at the center of a circular current loop of radius R is

Magnetic field at the center of a current loop

$$B = \frac{\mu_0 I}{2R} \tag{20.20}$$

The field at the center of the loop is perpendicular to the plane of the loop (Fig. 20.31) as given by right-hand rule number 1. The field lines encircle the loop as in Figure 20.9B.

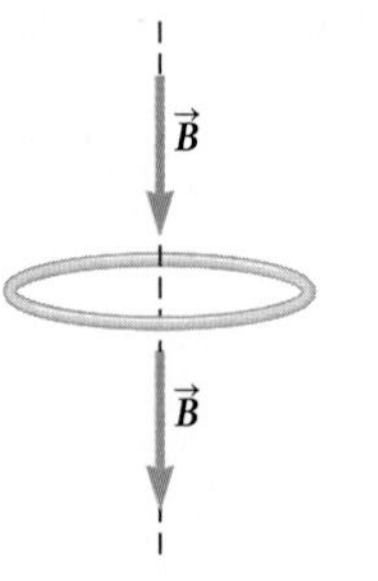

▲ **Figure 20.32** Concept Check 20.8.

CONCEPT CHECK 20.8 Direction of the Field Near a Current Loop

A current in the circular loop in Figure 20.32 produces a magnetic field on the axis that points downward as shown. As viewed from above, is the direction of the current in the loop (a) clockwise or (b) counterclockwise?

Field Inside a Solenoid

The pattern of magnetic field lines produced by a circular current loop is very similar to the field produced by a bar magnet. One handy feature of the current loop is that its field is adjustable simply by varying the current I. One disadvantage is that for reasonable values of I the field of a single current loop is much smaller than that of a

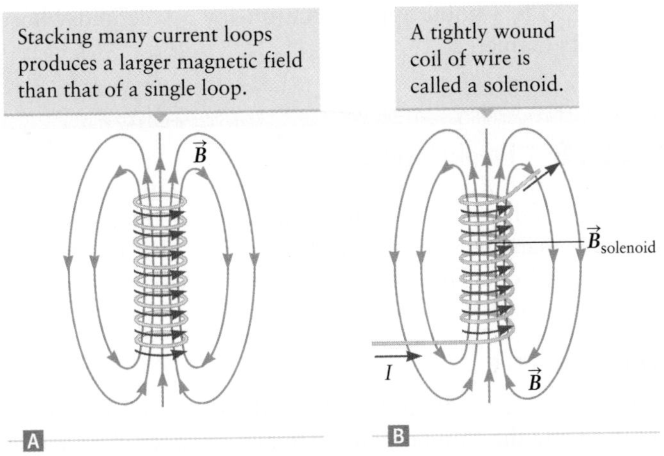

Stacking many current loops produces a larger magnetic field than that of a single loop.

A tightly wound coil of wire is called a solenoid.

◀ **Figure 20.33** A Stacking many current loops produces a magnetic field similar to that of a bar magnet. B Winding a wire in a "tight" helix produces a field similar to that in part A.

bar magnet, but this disadvantage can be overcome by stacking many current loops together. Figure 20.33A shows such a stack of current loops and the pattern of field lines they produce. This field can be understood in terms of the field of a single current loop (Fig. 20.31) along with the principle of superposition.

Usually, it is not practical to stack many separate current loops. Instead, a long piece of wire is wrapped in a very tight helix as shown in Figure 20.33B. Such a helical winding of wire is called a ***solenoid***. The field lines circulate around the outside of the solenoid, emerging from one end and reentering at the other (Fig. 20.33B). Outside the solenoid, the field lines "expand" to fill a very large volume, so $\vec{B}$ is very much smaller outside than inside the solenoid. For a very long solenoid, it is a good approximation to assume the field is constant inside and zero outside. With this approximation, let's apply Ampère's law along the path shown in Figure 20.34. This path is a rectangle with one side (part 1) inside the solenoid and parts 2, 3, and 4 outside. We therefore split the sum on the left-hand side of Ampère's law into four parts:

$$\sum_{\substack{\text{closed}\\\text{path}}} B_{\parallel}\,\Delta L = \sum_{\text{part 1}} B_{\parallel}(1)\,\Delta L + \sum_{\text{part 2}} B_{\parallel}(2)\,\Delta L + \sum_{\text{part 3}} B_{\parallel}(3)\,\Delta L + \sum_{\text{part 4}} B_{\parallel}(4)\,\Delta L \tag{20.21}$$

where $B_{\parallel}(1)$ is the field along part 1, and so on. Because the field outside the solenoid is approximately zero, $B_{\parallel}(2) \approx B_{\parallel}(3) \approx B_{\parallel}(4) \approx 0$ and we get

$$\sum_{\substack{\text{closed}\\\text{path}}} B_{\parallel}\,\Delta L = \sum_{\text{part 1}} B_{\parallel}(1)\,\Delta L \tag{20.22}$$

Here, $B_{\parallel}(1)$ is the field inside the solenoid, which is a constant along part 1 of the path; we can denote it as B_{solenoid}. For a solenoid of length L, Equation 20.22 leads to

$$\sum_{\substack{\text{closed}\\\text{path}}} B_{\parallel}\,\Delta L = \sum_{\text{part 1}} B_{\text{inside}}\,\Delta L = B_{\text{solenoid}} \sum_{\text{part 1}} \Delta L = B_{\text{solenoid}} L$$

Inserting this result into Ampère's law (Eq. 20.16) gives

$$B_{\text{solenoid}} L = \mu_0 I_{\text{enclosed}} \tag{20.23}$$

where I_{enclosed} is the total current passing through the surface enclosed by our Ampère's law path. A helical coil with N windings cuts through this surface N times, so if I is the current through the coil, then $I_{\text{enclosed}} = NI$. Inserting into Equation 20.23 leads to

$$B_{\text{solenoid}} L = \mu_0 N I$$

Solving for B_{solenoid}, we get

$$B_{\text{solenoid}} = \frac{\mu_0 N I}{L} \tag{20.24}$$

Magnetic field inside a long solenoid

Insight 20.2

AMPÈRE'S LAW IS ALWAYS TRUE (BUT MAY BE DIFFICULT TO APPLY)

In Figure 20.31, there is no path around the current loop for which the magnetic field component $B_{\parallel}$ is constant along the entire path. For this reason, Ampère's law does not lead to a simple way of finding the magnetic field in this case, but it *is still true* in such cases. Even though $B_{\parallel}$ may vary in a complicated fashion as one moves along a closed path, the left-hand side of Ampère's law (Eq. 20.16) is always equal to μ_0 multiplied by the total current through the surface enclosed by the path.

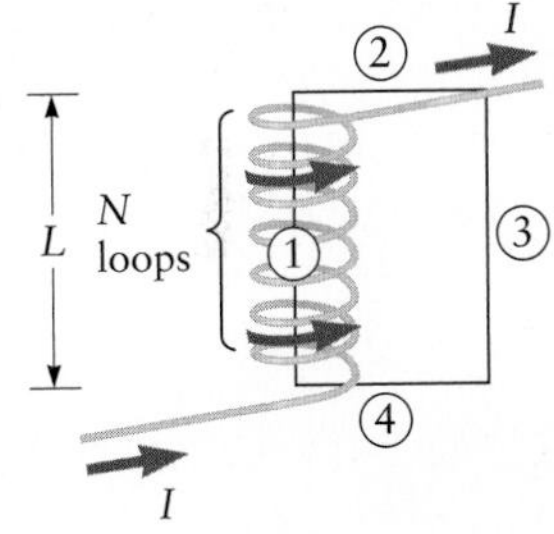

▲ **Figure 20.34** Application of Ampère's law to find the magnetic field inside a solenoid; we use the black rectangle as our Ampère's law path. The field outside a solenoid is very small (approximately zero), so only the inside field contributes to the sum of the field components in Ampère's law (Eq. 20.21).

This result for $B_{solenoid}$ is a good approximation for a solenoid whose length L is much greater than its diameter.

EXAMPLE 20.8 Designing a Solenoid

You are given the task of designing a solenoid that must produce a field in its interior that is 20 times larger than the Earth's field ($B_{Earth} \approx 5 \times 10^{-5}$ T). This solenoid is to be 10 cm long and 1.0 cm in diameter, with $N = 500$ turns. What current must the wire be able to carry without failing?

RECOGNIZE THE PRINCIPLE

We have already used Ampère's law to find the field inside a solenoid (Eq. 20.24). We can use that result to find the current required to give the desired field in our solenoid.

SKETCH THE PROBLEM

Figure 20.34 describes our solenoid.

IDENTIFY THE RELATIONSHIPS

Equation 20.24 gives $B_{solenoid}$ in terms of the number of turns N, the length L, and the current I. Setting $B_{solenoid}$ equal to $20 \times B_{Earth}$, we find

$$B_{solenoid} = \frac{\mu_0 N I}{L} = 20 B_{Earth}$$

SOLVE

Solving for I and inserting the given values of the various quantities gives

$$I = \frac{L(20B_{Earth})}{\mu_0 N} = \frac{(0.10 \text{ m})(20)(5 \times 10^{-5} \text{ T})}{(4\pi \times 10^{-7} \text{ T} \cdot \text{m/A})(500)} = \boxed{0.16 \text{ A}}$$

▶ What does it mean?

This current can be safely carried by a wire with a diameter no smaller than about 0.2 mm. A tightly wrapped helix with 500 turns each 0.2 mm thick would be about 10 cm long, so this solenoid could indeed be built. If, however, we wanted a solenoid of this size to produce a field as large as that found near a common bar magnet (for which $B = 1$ T), a *much* larger current would be required. In fact, the current required in that case would probably melt the wire of the solenoid! To get such large fields, a redesign of the solenoid is necessary, a problem we will consider further in the next section.

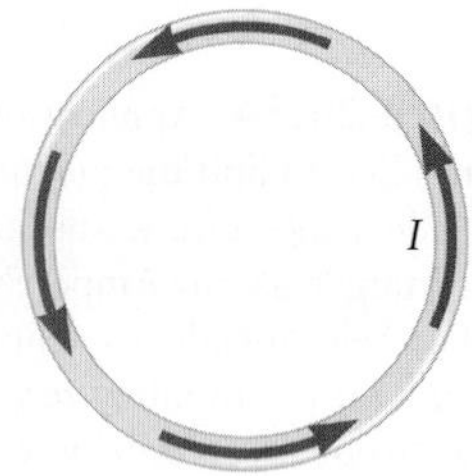

▲ **Figure 20.35** Concept Check 20.9. Will this current loop explode or collapse?

CONCEPT CHECK 20.9 Forces on a Current Loop

Consider a single current loop (Fig. 20.35) carrying a current I. It represents one loop of a solenoid. The magnetic field from one part of the loop will lead to a magnetic force on other parts of the loop. Will these magnetic forces tend to (a) make the loop collapse, (b) make the loop explode, or (c) have no effect?

20.8 Magnetic Materials: What Goes On Inside?

We have mentioned several times that magnetic poles *always come in pairs*. To understand why, we must consider the atomic origin of permanent magnetism. We have already noted that the pattern of magnetic field lines produced by a bar magnet is similar to the field produced by a current loop. The next example compares the magnitude of the field of a typical permanent magnet with the field near a current loop.

EXAMPLE 20.9 Comparing the Magnetic Fields of a Permanent Magnet and a Current Loop

Let's calculate how much current is required to make the field at the center of a current loop comparable to the magnetic field near one of the poles of a permanent magnet. The field near a typical bar magnet is about $B_{\text{bar mag}} = 1$ T. Consider a single circular wire loop of radius $R = 0.50$ cm carrying a current I. What value of I is required to produce a field equal to $B_{\text{bar mag}}$ at the center of the loop?

RECOGNIZE THE PRINCIPLE

The magnetic field at the center of a current loop depends on the current and the radius of the loop and is given by Equation 20.20. We can apply this relation to a current loop with the given radius and then find the value of the current needed to give a field equal to $B_{\text{bar mag}}$.

SKETCH THE PROBLEM

Figures 20.2A and 20.9 show the fields of a bar magnet and a current loop.

IDENTIFY THE RELATIONSHIPS

The magnetic field at the center of the loop is (Eq. 20.20)

$$B = \frac{\mu_0 I}{2R}$$

Setting this result equal to $B_{\text{bar mag}}$ and rearranging to solve for I, we get

$$I = \frac{2RB_{\text{bar mag}}}{\mu_0}$$

SOLVE

Inserting the given values for R and $B_{\text{bar mag}}$ gives

$$I = \frac{2(0.0050 \text{ m})(1 \text{ T})}{4\pi \times 10^{-7} \text{ T} \cdot \text{m/A}} = \boxed{8000 \text{ A}}$$

► ***What does it mean?***

This current is *very* large. A current loop made from a copper wire of any reasonable size would melt! A permanent magnet contains atomic-scale current loops, and it is amazing that the atomic-scale currents can be this large.

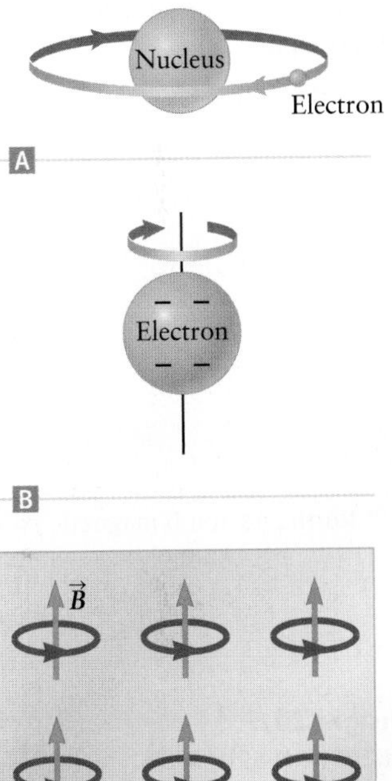
Nucleus
Electron
A
Electron
B
$\vec{B}$
In a permanent magnet (such as iron), all the atoms act as small aligned magnets.
C

▲ **Figure 20.36** A An electron in an atom acts as a small current loop, producing a magnetic field. B A spinning charged particle also acts as a current loop, with the different regions of charge circulating around the rotation axis. C The total magnetic field of a permanent magnet is the sum of the fields from parts A and B and from the alignment of the fields produced by many atoms.

Permanent Magnets: A Microscopic View

How can a permanent magnet generate such a strong magnetic field, especially when no obvious current is present? Atoms consist of a nucleus surrounded by one or more electrons, and in a classical picture the electrons are in constant motion (as in the Bohr model of the atom; Chapter 29). The motion of an electron around a nucleus can be pictured as a tiny current loop with a radius of approximately the radius of the atom (Fig. 20.36A). The direction of the resulting magnetic field for a single atom is determined by the orientation of the current loop along with right-hand rule number 1 (Fig. 20.9).

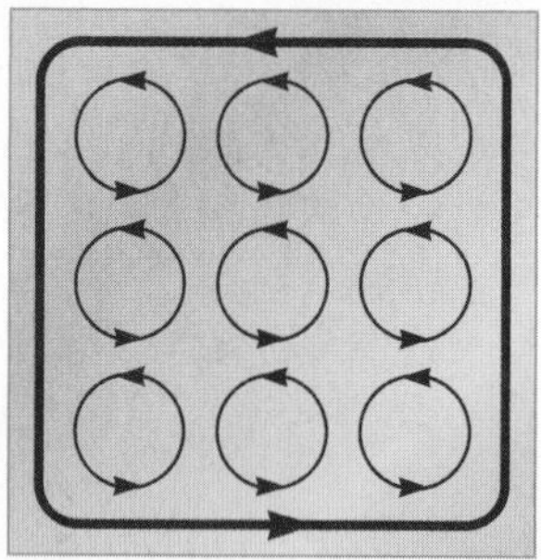

The net effect is the same as that of one large current loop around the perimeter.

▲ **Figure 20.37** A collection of aligned atomic current loops in a "slice" of magnetic material is equivalent to one large current loop around the outside and hence one large magnet.

An electron also produces a magnetic field due to an effect called ***electron spin***. It is useful to imagine the electron as a spinning ball of negative charge. A spinning charge acts as a circulating electric current, producing a magnetic field as shown in Figure 20.36B. In analogy with the properties of a current loop, we also say that the electron has a ***spin magnetic moment*** just as the current loop in Figure 20.25B has a magnetic moment. And, like a bar magnet, when an electron is placed in a magnetic field it will tend to align its spin magnetic moment with the magnetic field.

This picture of an electron is very simplified; a correct explanation of electron spin requires quantum theory, which confirms that an atom can produce a magnetic field through either the electron's orbital current loop or from the electron's spin. The total magnetic field produced by a single atom is the sum of these two fields.

Figure 20.36C shows a collection of atomic current loops, with each current loop representing a single atom. This collection of "small" current loops then acts as one "large" current loop (Fig. 20.37), producing the magnetic field of a magnetic material. Although the current in each atomic loop is very small, there are *many* atoms in a typical bar magnet, and the sum of all these atomic current loops does indeed give a net current similar to the value of I found in Example 20.9.

Lining Up the Atomic Magnets to Make a Permanent Magnet

Figure 20.36 shows how the electrons in an atom are capable of producing a magnetic field. Not all atoms will actually be magnetic, however, because the current loops of different electrons may point in different directions so that their magnetic fields cancel. The total magnetic field produced by a solid depends on how all the atomic magnetic fields are aligned. In some solids, the atomic magnets are not aligned; their individual fields point in different directions and the total field is zero. In materials that form permanent magnets, many atomic-scale magnetic fields align, pointing in the same direction to give a large net magnetic field.

North and south magnetic poles cannot be separated.

This picture of a permanent magnet as a collection of aligned atomic-scale current loops explains why it is not possible to isolate the north and south magnetic poles of a magnet. One might think that the north and south poles of a permanent magnet could be separated by simply cutting the magnet in half. Figure 20.38 shows why this approach fails. Although one can certainly cut a bar magnet in half, each of the resulting pieces still produces the magnetic field of a "complete" bar magnet and hence still possesses a north pole and a south pole.

Insight 20.3
ELECTRON SPIN AND QUANTUM THEORY

Imagining an electron as a spinning ball of charge as in Figure 20.36B provides a simple picture of the electron's intrinsic spin magnetism. In the years before quantum theory was developed, many physicists tried to construct mechanical models of an electron based on a spinning ball of charge, but these attempts failed to fully explain all the properties of an electron. Physicists now believe that an electron is a true point particle, so the failure of mechanical spinning models for a ball of radius $R = 0$ is not surprising. While electron spin and several other properties of electrons can only be explained fully using quantum theory (Chapter 28), the classical picture of a spinning ball in Figure 20.36 still provides useful qualitative insights.

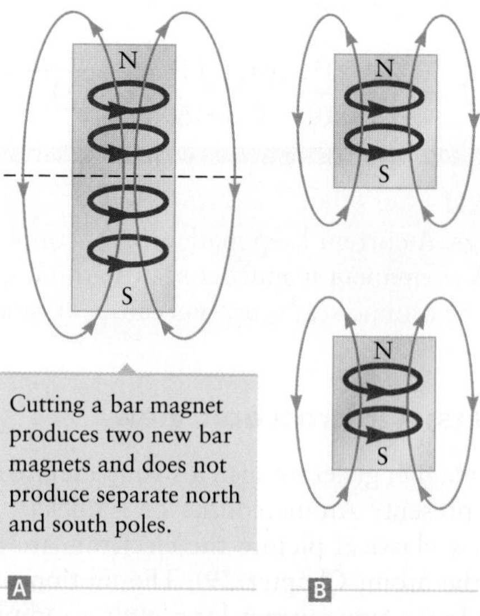

Cutting a bar magnet produces two new bar magnets and does not produce separate north and south poles.

▲ **Figure 20.38** A Cutting a bar magnet in half B produces two separate bar magnets.

Induced Magnetism and Magnetic Domains

Although permanent magnets can be formed from certain materials such as iron and magnetite, not all pieces of iron or magnetite behave as bar magnets. How can that be? Our atomic-scale view of a permanent magnet in Figure 20.36C shows the atomic bar magnets all aligned in the same direction, but it is also possible for the atomic magnets in different regions within a magnetic material to point in different directions. One possibility is shown in Figure 20.39A, which shows a material in which the atomic magnets on one side all point up and those on the other side all point down. This arrangement is equivalent to two bar magnets, side by side as in Figure 20.39B. Some magnetic field lines are also shown and are seen to be confined almost totally within the material. An observer outside the material thus sees a very small field, so this material would not appear to be a permanent magnet.

The two oppositely aligned regions in Figure 20.39 are called ***magnetic domains***. Figure 20.40A shows a material in which the two domain regions are about equal in size. Application of a magnetic field from a bar magnet can cause one of these regions to grow at the expense of the other. In Figure 20.40B, the domain aligned with the magnetic field grows at the expense of the other domain, and the material itself now acts like a bar magnet. This behavior occurs whenever you stick a "refrigerator magnet" to your refrigerator. A refrigerator panel is made of steel and usually has magnetic domains that alternate in direction as shown in Figure 20.41. This is similar to the two oppositely directed domains in Figure 20.39A, so your refrigerator appears to be nonmagnetic. When a magnet is brought near, however, the domains in the refrigerator panel align (Fig. 20.41); the panel now acts like a bar magnet and attracts the nearby magnet. That is why a refrigerator magnet sticks to a refrigerator due to a magnetic force, even though the refrigerator by itself does not appear to be magnetic.

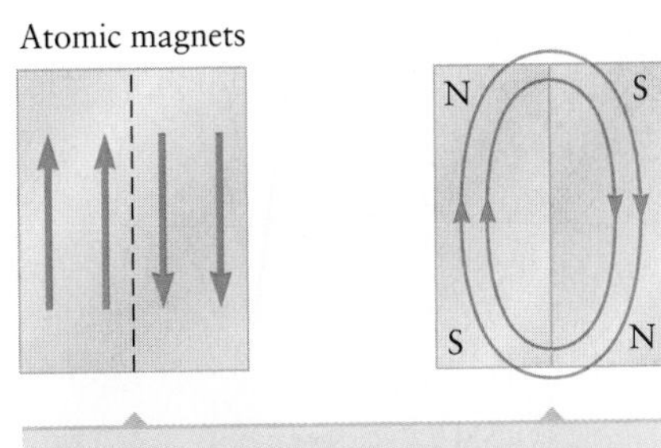
Atomic magnets
N
S
S
N
The atomic magnets in different parts of a permanent magnet can point in different directions. Here the result is two side-by-side bar magnets.
A
B

▲ **Figure 20.39** A If the atomic magnets within a permanent magnet are aligned in opposite directions, the magnetic field outside can be very small. B To an observer on the outside, this material would appear to be nonmagnetic.

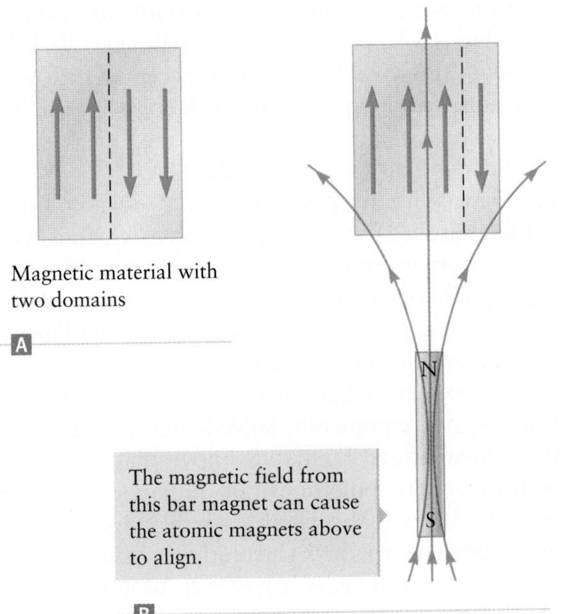
Magnetic material with two domains
A
N
S
The magnetic field from this bar magnet can cause the atomic magnets above to align.
B

▲ **Figure 20.40** A Oppositely aligned magnetic domains. B A weak magnetic field can cause the atomic magnets in a magnetic material to reorient. Some domains then grow at the expense of the others. The material at the top now behaves like a permanent magnet.

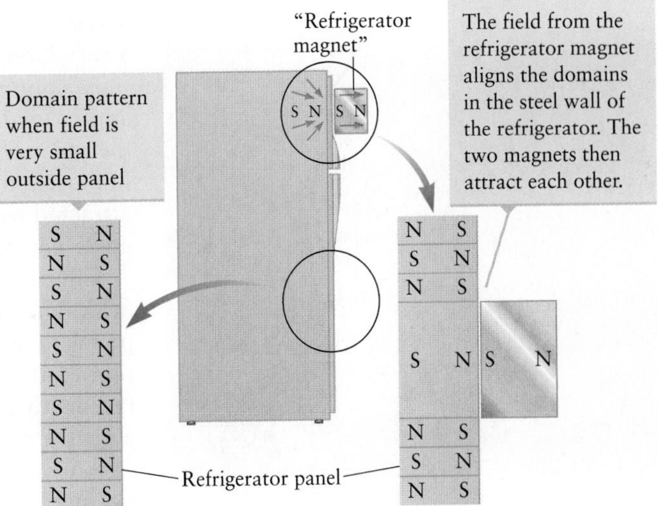
"Refrigerator magnet"
The field from the refrigerator magnet aligns the domains in the steel wall of the refrigerator. The two magnets then attract each other.
Domain pattern when field is very small outside panel
S N S N
S N
N S
S N
N S
S N
N S
S N
N S
S N
N S
N S
S N
N S
S N S N
N S
S N
N S
Refrigerator panel

▲ **Figure 20.41** A "refrigerator magnet" is a permanent magnet that "sticks" to a refrigerator panel through a magnetic force. The refrigerator panel contains many domains oriented in opposite directions. The field of the refrigerator magnet aligns the domains so that these two magnets then attract each other.

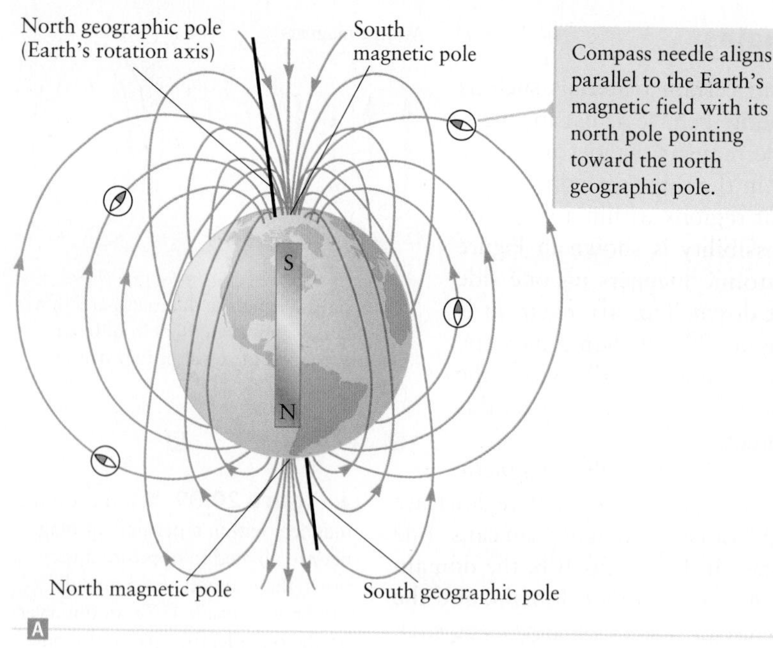

Compass needle aligns parallel to the Earth's magnetic field with its north pole pointing toward the north geographic pole.

© Cengage Learning/Charles D. Winters

▲ **Figure 20.42** A Magnetic field of the Earth. B A compass needle is a small bar magnet and thus aligns parallel to the magnetic field. If there are no other sources of magnetic field nearby, the north end of a compass needle points toward the Earth's north geographic pole.

20.9 The Earth's Magnetic Field

The largest magnet on the Earth is the Earth itself. About 400 years ago, William Gilbert proposed that a compass needle orients along a north–south direction in response to the Earth's magnetic field (Fig. 20.42A). A compass needle (Fig. 20.42B) is a small bar magnet mounted so that it can swivel freely about its center and align with the field at its location. From experiments with compass needles and some insightful reasoning, Gilbert suggested that the Earth acts as a very large magnet oriented as shown in Figure 20.42A.

A compass needle aligns with its north *magnetic* pole pointing approximately toward the Earth's north *geographic* pole, which is why the poles of a bar magnet were given the names "north" and "south." Because a north magnetic pole of one magnet (the compass needle) is attracted to the south magnetic pole of a second magnet, the Earth's *north geographic pole* is actually a *south magnetic pole*!

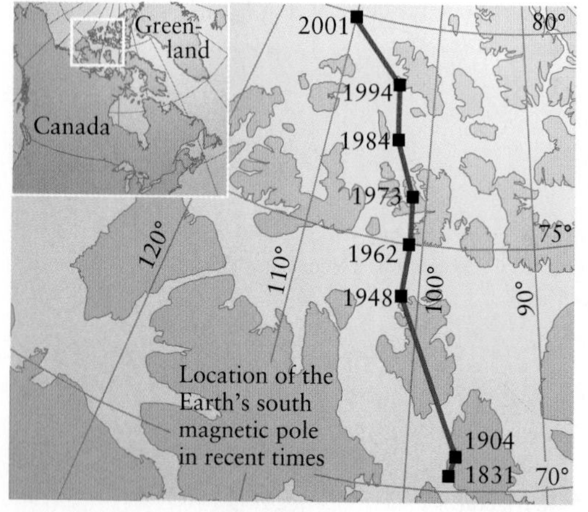

▲ **Figure 20.43** The Earth's magnetic poles move from year to year. The south magnetic pole (which is located near the north geographic pole) was near latitude 70° in northern Canada only 100 years ago.

Gilbert was not able to explain what causes the Earth's magnetic field. That problem is still not completely solved, but several clues point to a qualitative explanation. First, we know that the location of the Earth's south magnetic pole does not quite coincide with the north geographic pole. In fact, the Earth's south magnetic pole actually moves slowly from day to day. Currently, it moves about 40 km/year, and within the last 100 years it has even been located in northern Canada (Fig. 20.43). Second, geological studies show that the Earth's field has completely reversed direction many times during the planet's history. The last such geomagnetic reversal occurred about 780,000 years ago. So, in the years immediately prior to that time, a compass needle would have pointed toward the Earth's *south geographic* pole.

This behavior of the Earth's magnetic field is probably caused by electric currents in the core. Recall (Chapter 12) that the Earth's core contains a liquid. This liquid conducts electricity, and the spin of the Earth about its axis causes the liquid to circulate much like the current in a conducting loop (Fig. 20.44). Circulation within the Earth's core is thought to have a complicated flow pattern that varies with time, and these variations are believed to cause the wandering of the Earth's magnetic poles.

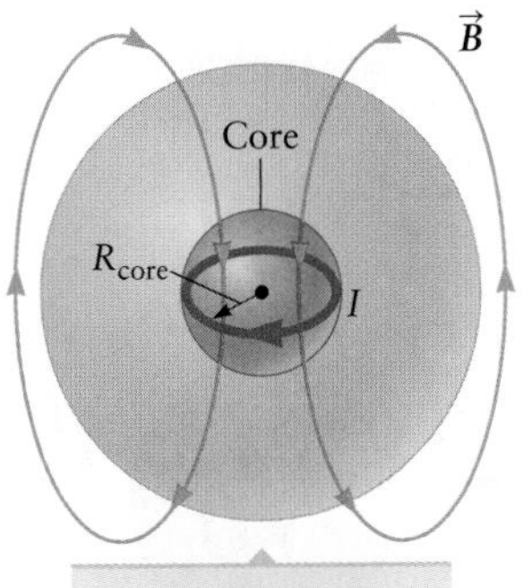
$\vec{B}$
Core
R_{core}
I
Liquid in the Earth's core circulates as a large current loop.

▲ **Figure 20.44** The Earth's magnetic field is believed to be caused by an electric current in the liquid core.

CONCEPT CHECK 20.10 Direction of a Compass Needle

A compass is usually held "horizontally" (so that the needle lies in a horizontal plane). However, the Earth's magnetic field has both horizontal and vertical components (Fig. 20.42). Your friend is an arctic explorer and pitches his tent directly over the Earth's south magnetic pole. Will his compass needle point (a) toward the Earth's north geographic pole, (b) downward, or (c) upward?

EXAMPLE 20.10 Ⓡ Estimating the Electric Current in the Earth's Core

The strength of the Earth's magnetic field has an average value on the surface of about 5×10^{-5} T. Model this magnetic field by taking the Earth's core to be a simple current loop, with a radius equal to the radius of the core. What electric current must this current loop carry to generate the Earth's observed magnetic field? The Earth's core has a radius of approximately $R_{core} = 3 \times 10^6$ m (Fig. 20.44).

RECOGNIZE THE PRINCIPLE

To get an approximate value of the current, we treat the core as a simple current loop of radius R_{core}; Equation 20.20 then gives the resulting magnetic field at the center of the loop. The field at the core's center will be larger than the field at the Earth's surface, but it will give an approximate value for the field seen by a compass at the surface.

SKETCH THE PROBLEM

Figure 20.44 shows the problem.

IDENTIFY THE RELATIONSHIPS

The field produced by our model current loop is

$$B = \frac{\mu_0 I}{2R_{core}}$$

(Eq. 20.20). We can rearrange this equation to solve for the current:

$$I = \frac{2R_{core}B}{\mu_0}$$

SOLVE

Inserting the radius of the core and setting B equal to the Earth's field (both given above), we find

$$I = \frac{2R_{core}B}{\mu_0} = \frac{2(3 \times 10^6 \text{ m})(5 \times 10^{-5} \text{ T})}{4\pi \times 10^{-7} \text{ T} \cdot \text{m/A}} = \boxed{2 \times 10^8 \text{ A}}$$

▸ ***What does it mean?***

This current is enormous. A scientific explanation of the Earth's field must explain how such a large current is maintained. Physicists still do not have a full solution for this problem.

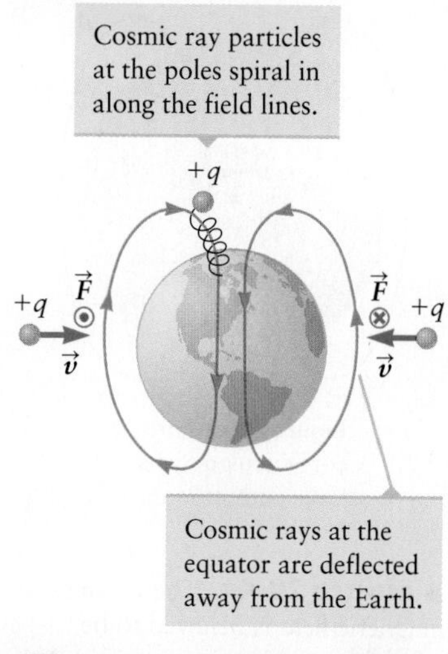
Cosmic ray particles at the poles spiral in along the field lines.
+q
+q
+q
Cosmic rays at the equator are deflected away from the Earth.
A

Cosmic Rays and the Earth's Magnetic Field

We usually think of "space"—the regions far above the Earth's atmosphere between the Earth, the Moon, and the Sun—as empty, but these regions actually contain many charged particles, including electrons and protons, that are emitted by the Sun or that reach us from outside our solar system. These particles are called ***cosmic rays***. Because they are charged, their motion is affected by the Earth's magnetic field. Figure 20.45A shows two positively charged cosmic-ray particles that approach the Earth's equator. Let's consider the magnetic force on the particle that approaches from the right. In this region, the magnetic field is directed parallel to the Earth's surface in a geographic south-to-north direction; applying right-hand rule number 2 shows that the magnetic force on this particle is directed perpendicular to the plane of this drawing and parallel to the Earth's surface in a west-to-east direction as shown in Figure 20.45A (into the plane of the drawing). Hence, the Earth's field deflects positively charged cosmic rays near the equator so that they tend to move in an easterly direction, away from the Earth's surface.

Cosmic-ray particles also bombard the Earth at the poles. In polar regions, the magnetic field is perpendicular to the surface, so these particles follow a helical path and spiral around the field lines (as in Fig. 20.16C). The Earth's magnetic field thus "steers" cosmic rays so that they tend to strike at the poles as opposed to locations at lower latitudes. As they travel through the atmosphere, these particles collide with and ionize air molecules, causing the molecules to emit light. This phenomenon is observed as the ***aurora borealis***, also called the "northern lights" (Fig. 20.45B). A similar phenomenon at the south geographic pole is called the ***aurora australis*** ("southern lights").

B

▲ **Figure 20.45** A The motion of a cosmic-ray particle is affected by the Earth's magnetic field. B The northern lights are caused by cosmic rays that ionize atoms and molecules as the cosmic rays travel through the atmosphere.

20.10 Applications of Magnetism

So far, we have mentioned only a few applications of magnetism, including mass spectrometers and refrigerator magnets. In this section, we describe how magnetism is used by doctors, engineers, archaeologists, and even bacteria.

Blood-Flow Meters

An arrangement similar to the Hall effect experiment is used by surgeons to measure the blood velocity in arteries. Blood contains many ions that move at some speed v along the artery (Fig. 20.46A). When a magnetic field is directed perpendicular to the artery, the ions are deflected toward the sides of the artery, with positive ions deflected to one side and negative ions deflected to the opposite side (Fig. 20.46B). The result is an electric potential difference across the artery that can be measured with a voltmeter. Flowmeters based on this principle are used to measure the blood velocity. These devices work best when the voltmeter leads are placed very close

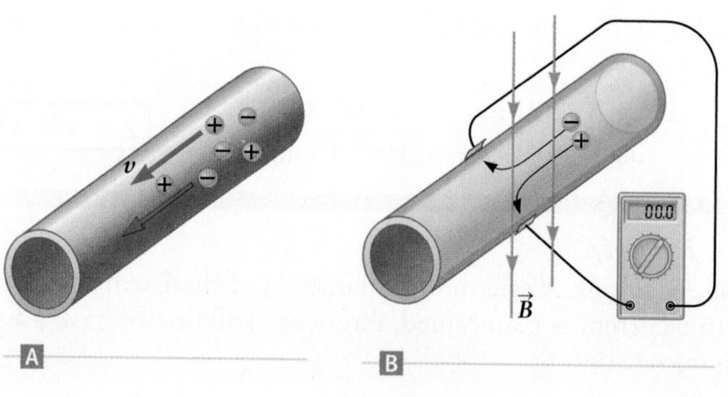
00.0
A
B

► **Figure 20.46** A Blood flowing through an artery contains many ions. B The magnetic force deflects the positive and negative ions to opposite sides of the artery. A blood-flow meter uses the electric potential difference across the artery to measure the speed of the blood. Notice the similarity to the Hall effect, Figure 20.29.

to the walls of the artery, so they are typically used in surgery when the artery is exposed. We'll consider the operation of a flowmeter and estimate the magnitude of the induced emf in Question 18.

Relays

We have already seen how a solenoid (Fig. 20.33) produces a magnetic field when an electric current passes through it. In Example 20.8, we designed such an "electromagnet" and found that even with a substantial current the field is rather modest. The field produced by a solenoid can be made much larger if the solenoid is filled with a magnetic material. The small field produced by the current can align the atomic magnets in the material (the magnetic domains described in Section 20.8), giving a large magnetic field due mainly to the magnetic material. In this way, the large field of the solenoid can be controlled with a modest current.

The notion that a small magnetic field or current can control a large magnetic field is used to make an electromechanical switch called a ***relay***, sketched in Figure 20.47. A relay is a type of "remote control" switch that uses a solenoid to exert a magnetic force on the movable part of a switch (a steel bar). A small current through the solenoid can thus control a much larger current through the switch. Although a switch is perhaps not a very "exciting" application, relays are widely used in a safety device called a ground fault interrupter (GFI), which is found in virtually all new homes; GFIs are discussed in more detail in Chapter 21.

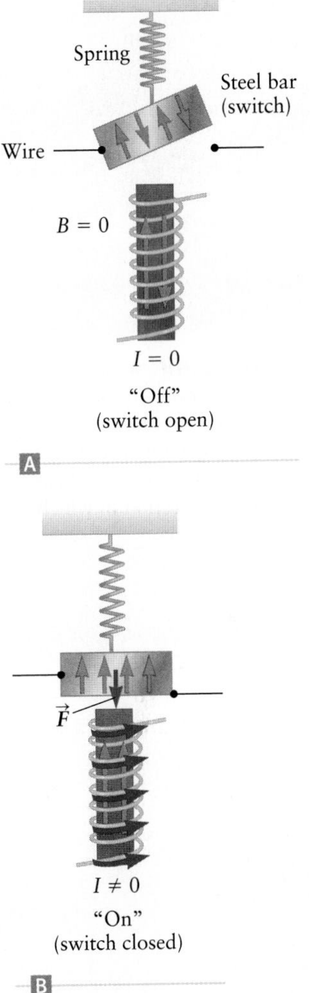

▲ **Figure 20.47** A relay uses the magnetic field from a solenoid to close a mechanical switch. This solenoid has been filled with a magnetic material whose domains are aligned by the field from the solenoid's current, producing a much larger field than is possible from the current alone.

Making an Electric Motor

In Figure 20.24, we described how a magnetic field can produce a torque on a current loop. If the loop is attached to a rotatable shaft, we have the "beginnings" of an electric motor. In a practical motor, the simple current loop is replaced by a solenoid containing a magnetic material (Fig. 20.48). When current passes through the solenoid, the material inside forms a bar magnet, with the orientation of the poles determined by the direction of the current. The north and south poles of this magnet are attracted to the poles of a second stationary magnet (shaped like a horseshoe magnet), producing a torque on the solenoidal magnet and causing the shaft to rotate (Fig. 20.48A). The torque on the solenoidal magnet is thus very similar to the torque on a current loop. The advantage of using the solenoid is that it gives a much larger torque, for the same reason that a bar magnet produces a much larger magnetic field than a current loop.

To make a working motor, we need to add one more "twist" to the design. When the solenoidal magnet rotates just past the horseshoe magnet, the direction of the current is reversed, switching the poles on the rotating magnet (Fig. 20.48B). These poles are now repelled from the poles of the horseshoe magnet, so the shaft continues to rotate. The current is reversed again every time the solenoid passes the poles of the horseshoe magnet, thus providing a continuing torque on the solenoidal magnet, and the motor's shaft continues to rotate.

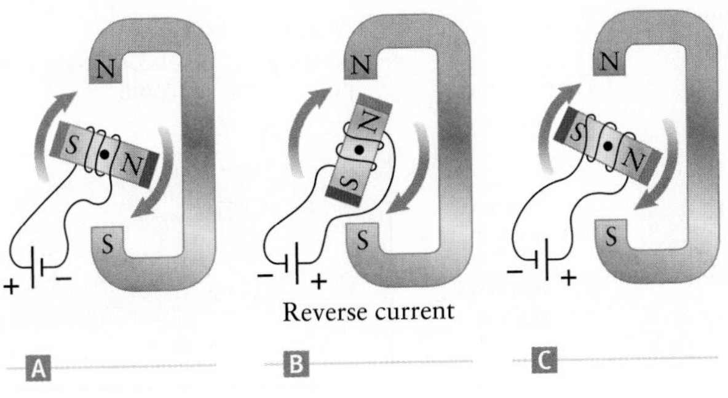

◀ **Figure 20.48** This electric motor uses the torque on a solenoid magnet to cause a shaft to rotate.

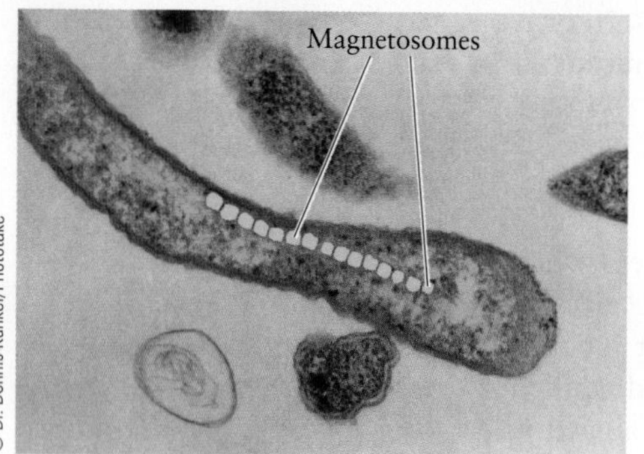

▲ **Figure 20.49** Some bacteria contain small grains of iron (shown here in yellow). It is believed that the bacteria use these iron magnets as a sort of compass to help determine the direction in which they swim.

Electric motors are a very important and widespread application of magnetism. A closely related application is the ***electric generator***. A generator produces an electric current by rotating a coil between the poles of a magnet; hence, a generator is a sort of "motor in reverse" as we'll discuss in Chapter 21. A combination of electric motor plus generator is at the heart of some of today's fuel-efficient "hybrid" automobiles such as the Toyota Prius and the Honda Insight.

Magnetic Bacteria

In addition to doctors and engineers, even bacteria seem to have an application for magnetism. ***Magnetotactic*** bacteria possess small grains of iron called magnetosomes, which are usually arranged as a "necklace" as shown in Figure 20.49. Magnetically, each grain acts like a bar magnet, and the north and south poles of the grains align just as found for bar magnets.

Bacteria appear to use their magnetosomes to orient themselves with respect to the Earth's field. In the northern hemisphere, the Earth's field has a downward component as shown in Figure 20.42A; magnetic bacteria "prefer" to swim downward—toward the bottom of a lake, for instance—where the concentrations of oxygen and other nutrients are most favorable. These bacteria thus use magnetism to determine up from down.

Magnetic Dating

In Section 20.9, we mentioned that the direction of the Earth's field varies slightly from year to year. Occasionally, the direction even reverses completely. The most recent reversal occurred 780,000 years ago, and the dates of many previous reversals are also known (Fig. 20.50A). Archaeologists and geologists use these field reversals to "date" past events. When a volcano erupts, it produces molten rock, some of which solidifies to form magnetite and other magnetic materials. When molten, these materials are not magnetic because their atomic magnets point in random directions. Magnetite and other materials only become permanent magnets when they cool to normal surface temperatures. The orientation of these permanent magnets is set by the direction of the Earth's field at the time of cooling; hence, these materials record the direction of the Earth's field when they last cooled.

To understand how magnetic reversals are used for dating, consider the hypothetical deposits of material in Figure 20.50B. The material nearest the surface (layer 1) is magnetized parallel to the current magnetic field lines, with $\vec{B}$ pointing toward the north geographic pole. As one digs deeper, one finds a layer (2) magnetized in the opposite direction and then a third layer (3) magnetized parallel to layer 1. This

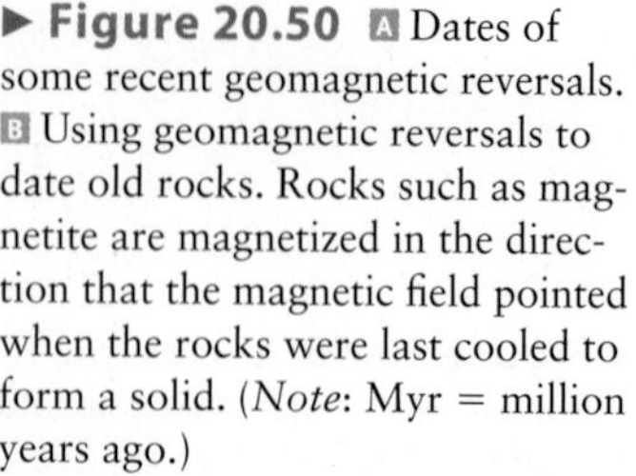

► **Figure 20.50** A Dates of some recent geomagnetic reversals. B Using geomagnetic reversals to date old rocks. Rocks such as magnetite are magnetized in the direction that the magnetic field pointed when the rocks were last cooled to form a solid. (*Note*: Myr = million years ago.)

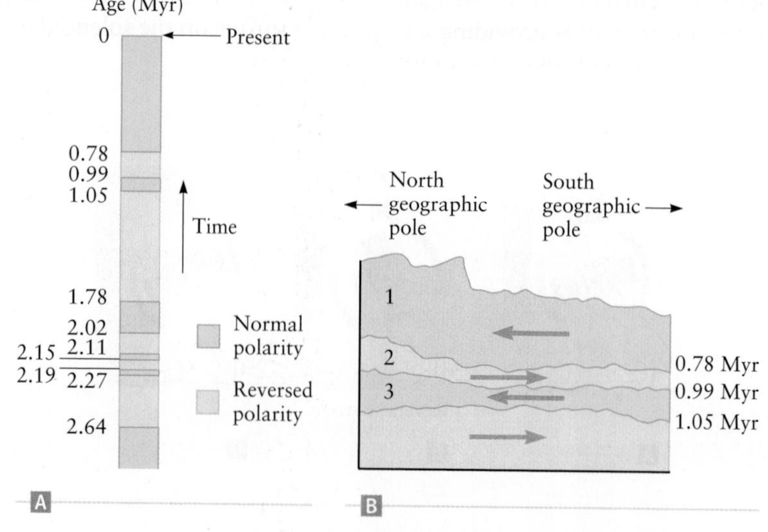

layering can be understood in terms of the geomagnetic reversals in Figure 20.50A. The transition from layer 1 to layer 2 dates from the most recent reversal 780,000 years ago, whereas the transition from layer 2 to layer 3 occurred 990,000 years ago. Hence, archaeological artifacts found in layer 2 must have an age between 780,000 and 990,000 years.

The dates of more than 25 geomagnetic reversals going back more than five million years are known, so this approach can be used to date many artifacts of interest to archaeologists, anthropologists, and geologists.

20.11 The Puzzle of a Velocity-Dependent Force

A remarkable feature of the magnetic force exerted on a moving charged particle is that it depends on the particle's velocity. Neither gravity (Newton's law of universal gravitation, Eq. 5.16) nor the electric force (Coulomb's law, Eq. 17.3 or 17.5) depend on velocity. The velocity dependence of the magnetic force has a very interesting consequence. Consider a particle of charge $+q$ moving with velocity $\vec{v}$ perpendicular to a magnetic field as in Figure 20.51A. A stationary observer watching the particle's motion would see a force on the particle as indicated in the figure. (The observer would notice this force from the particle's acceleration.) This observer (observer 1) would explain the force using $F_B = qvB \sin \theta$ and right-hand rule number 2.

Now suppose this scene is viewed by a second observer (observer 2) who is moving at velocity $\vec{v}$. Observer 2 is moving along with the charged particle, so *relative to observer 2* the particle is not moving (Fig. 20.51B). This observer would also notice a force on the particle because she would also see that the particle is accelerated. But, according to this observer, the velocity of the particle is zero, so she would predict that the magnetic force on the particle is zero!

Our two observers would both agree that the particle is accelerated, but only observer 1 would say that there is a magnetic force acting on the particle. How does observer 2 interpret this scene? Such puzzles will always arise when the force depends on velocity. Einstein's theory of relativity provides the way out of this dilemma. In Chapter 27, we show that according to observer 2 there is an *electric field* present, so she will say that the force on the particle is an *electric force*. That is, observer 1 will claim that the force in Figure 20.51 is magnetic, whereas observer 2 will measure a force of the same magnitude and direction but will claim that it is electric. There is thus a deep connection between electric and magnetic forces. This connection is a crucial part of the theory of ***electromagnetism***, which we'll explore in the next chapter.

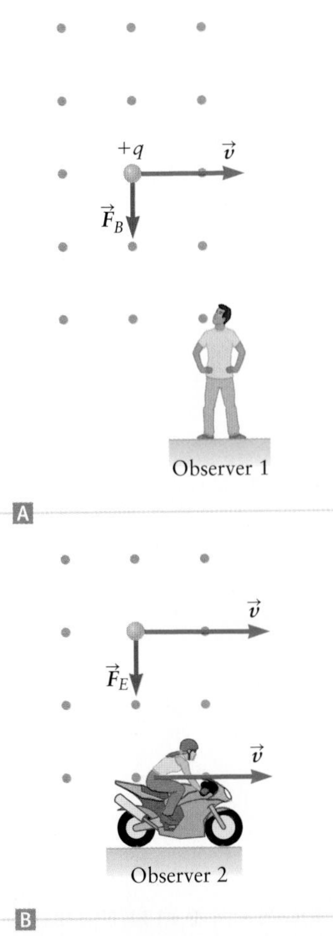

▲ **Figure 20.51** When a charged particle moves through a magnetic field, a force is exerted on the particle. The interpretation of this force is different for different observers, however. A According to observer 1, the charge $+q$ has velocity $\vec{v}$ and therefore experiences a nonzero magnetic force given by Equation 20.1. B Observer 2 moves at the same velocity as charge $+q$, so relative to this observer this charge is at rest. According to observer 2, the magnetic force on q is zero.

Summary | CHAPTER 20

Key Concepts and Principles

Sources of magnetic field

There are two ways to produce a ***magnetic field***: a permanent magnet and an electric current. A bar magnet and a current loop produce similar patterns of magnetic field lines.

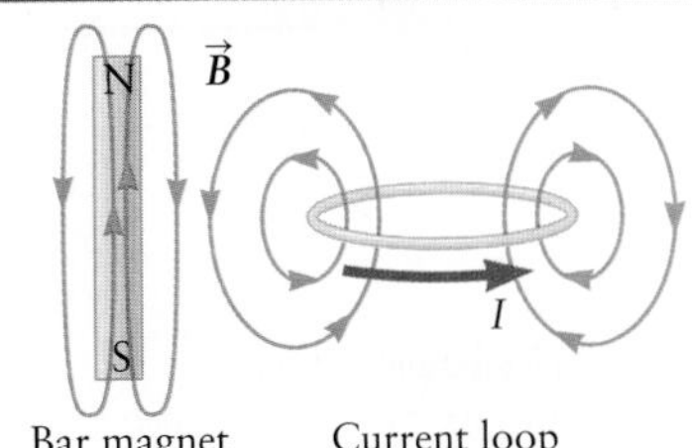

(*Continued*)

Permanent magnets and magnetic poles

The atoms in a permanent magnet material act as small bar magnets; their magnetic fields are produced by the circulating current of the atomic electrons and the spin magnetism of these electrons. A bar magnet contains a ***north magnetic pole*** and a ***south magnetic pole***.

Ampère's law

Ampère's law can be used to calculate the magnetic field produced by a current. This law relates the field directed along a closed path on the edge of a surface with the current that passes through the surface:

$$\sum_{\substack{\text{closed}\\\text{path}}} B_{\parallel}\,\Delta L = \mu_0 I_{\text{enclosed}} \qquad \textbf{(20.16) (page 606)}$$

where B is measured in teslas (T). The constant μ_0 is the ***permeability of free space*** and has the value

$$\mu_0 = 4\pi \times 10^{-7}\ \mathrm{T \cdot m/A} \qquad \textbf{(20.17) (page 606)}$$

When applied to a long, straight wire carrying a current I, Ampère's law gives the field a distance r from the wire as

$$B = \frac{\mu_0 I}{2\pi r} \qquad \textbf{(20.19) (page 607)}$$

RIGHT-HAND RULE NUMBER 1

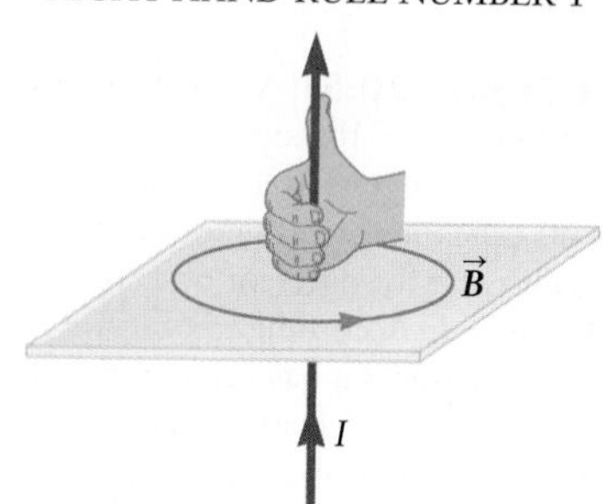

Right-hand rule number 1

To find the direction of the field produced by a current,

1. Place the thumb of your right hand along the direction of the current.
2. Curl your fingers; they will then give the direction of $\vec{B}$ as the field lines encircle the current.

Magnetic force on a moving charge

For a positive charge q moving at speed v, the magnitude of the magnetic force exerted on the charge is

$$F_B = qvB \sin\theta \qquad \textbf{(20.1) (page 594)}$$

where θ is the angle between $\vec{v}$ and $\vec{B}$.

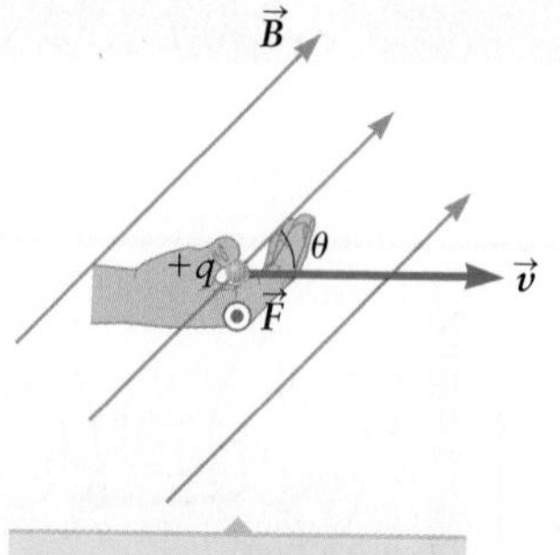

The magnetic force is directed out of the plane.

Right-hand rule number 2

To find the direction of the magnetic force on a moving charge,

1. Point the fingers of your right hand along the direction of $\vec{v}$.
2. Curl your fingers in the direction of $\vec{B}$. Always curl through the smallest angle that connects $\vec{v}$ and $\vec{B}$.
3. If q is positive, the magnetic force on q is parallel to your thumb. If q is negative, the magnetic force is in the opposite direction.

Applications

Magnetic fields due to a current

The magnetic field at the center of a current loop is

$$B = \frac{\mu_0 I}{2R} \qquad (20.20)\ (\text{page } 608)$$

A ***solenoid*** consists of N tightly wound turns of wire of finished length L. The field inside a solenoid is equal to

$$B_{\text{solenoid}} = \frac{\mu_0 N I}{L} \qquad (20.24)\ (\text{page } 609)$$

A permanent magnet is sometimes inserted inside a solenoid; in this case, the total field will be the field due to the current (Eq. 20.24) plus the field from the permanent magnet.

Hall effect

The ***Hall effect*** involves a current-carrying wire that is placed in a magnetic field perpendicular to the current. The magnetic force on the current leads to an electric potential difference between the edges of the wire, which can be used to determine the sign (positive or negative) of the charges that carry the current.

The Earth's magnetic field

The Earth is a large magnet, with a south magnetic pole located near the north geographic pole (and a north magnetic pole located near the south geographic pole). The Earth's field is produced by large, circulating electric currents in the core. The Earth's magnetic field aligns compass needles. The location of the Earth's magnetic poles varies with time, and throughout geological history (see Section 20.10) the Earth's field has often reversed its direction.

The north geographic pole is near the south magnetic pole.

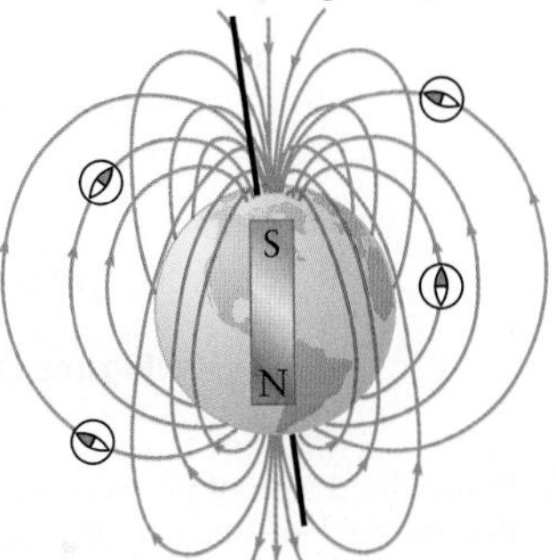

Questions

SSM = answer in *Student Companion & Problem-Solving Guide* ⓧ = life science application

1. The photo in Figure Q20.1 shows a beam of electrons moving from right to left in a glass tube in which a vacuum has been established. This beam is deflected downward by a magnetic field. What is the direction of this magnetic field at the center of the tube?

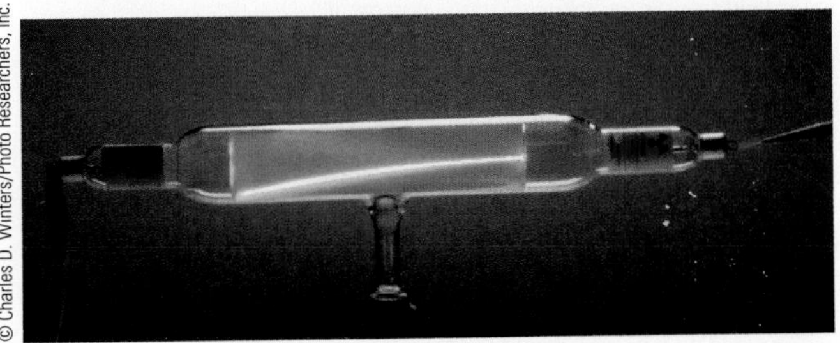

Figure Q20.1

2. A long straight current-carrying wire is placed in a region where there is a magnetic field $\vec{B}$ that has a constant direction. If there is no magnetic force on the wire, what can you say about the direction of $\vec{B}$ relative to the wire?

3. Consider the moving cosmic-ray particle on the right in Figure 20.45A and suppose it has a negative charge. What is the direction of the magnetic force on this particle?

4. A positively charged particle moves with velocity $\vec{v}$ in a region where there is a magnetic field $\vec{B}$ and a nonzero magnetic force $\vec{F}$ on the particle. Consider how these three vectors are oriented with respect to one another. (a) Which pairs of these vectors are always perpendicular? (b) Which pairs can have any angle between them?

5. Pairs of wires that allow current to flow into and out of a circuit are often twisted together. Figure Q20.5 shows a common type of cable with such twisted pairs. What is the advantage of pairing the wires that carry input currents with wires carrying output currents?

Figure Q20.5 An Ethernet cable.

6. SSM Make a qualitative sketch of the magnetic field near two bar magnets that are placed side by side as shown in Figure Q20.6.

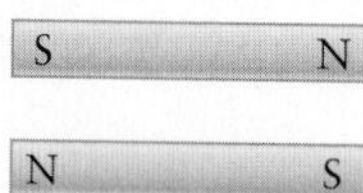

Figure Q20.6

7. Figure Q20.7 shows the direction of $\vec{B}$ in a particular region of space. The density (i.e., spacing) of the crosses and dots indicate qualitatively the magnitude of $\vec{B}$. What is the possible source of this field?

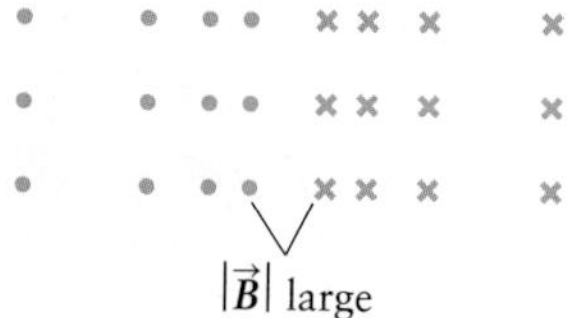

Figure Q20.7

8. In our discussion of the solenoid in Figure 20.33B, we claimed that the field outside the solenoid is much smaller than the field inside and that it is a good approximation to assume the field outside a very long solenoid is zero. Explain how this approximation can be consistent with the property that magnetic field lines always form closed loops.

9. Two parallel wires are oriented perpendicular to the page as shown in Figure Q20.9. The wires carry equal currents, and the direction of the current is into the page for each. The vertices of an equilateral triangle are marked by the two wires and a point P as shown. (a) What is the direction of the magnetic field produced by the two wires at point P? (b) A third wire, parallel to the other two and carrying the same current in the same direction, is placed at point P. Determine the direction of the force on the wire at this point.

I_1 I_2 P

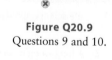

Figure Q20.9 Questions 9 and 10.

10. Repeat all parts of Question 9, but this time assume the current I_2 is directed out of the page and opposite to current I_1.

11. The medical technique called ***magnetic resonance imaging*** (MRI) uses the magnetic field produced by a large solenoid to obtain images from inside living tissue. These solenoids are large enough that a person can fit inside. Assuming the inside field is 1 T, design an MRI solenoid. Estimate the necessary length and radius of the solenoid assuming the wire can carry currents as large as 100 A. How many turns must this solenoid have to produce the desired field?

12. **The rail gun.** The simple configuration in Figure Q20.12 provides a means for launching projectiles, as long as the projectile can conduct electricity. Two copper rails are connected to a battery such that the current flows as indicated in the figure when a conducting rod is laid across the rails. What will happen to the conducting rod? If it begins to move, in what direction does it go? Consider (a) the direction of the magnetic field produced in the vicinity of the rod by the currents in the rails and rod and (b) the direction of the current in the rod. How could one make the rod move in the opposite direction?

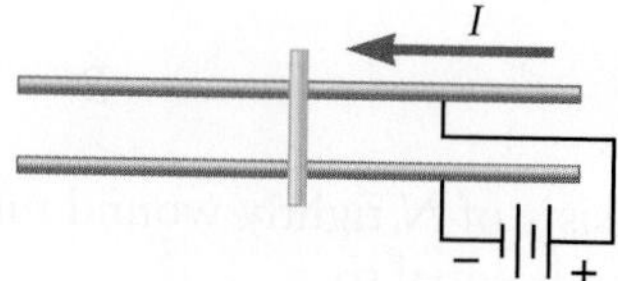

Figure Q20.12

13. A bubble chamber is a device that makes the path of a charged particle visible. When a charged particle with sufficient kinetic energy passes through a liquid, it will knock electrons off the atoms in its path. If the liquid is at a temperature above its boiling point, these ionized atoms become the nucleation point for a bubble to form, thereby marking the path of the particle. For the images in Figure Q20.13, the magnetic field was directed out of the page. (a) Is the curve of the track in image 1 (on the left) consistent with that of a high-speed electron or of a high-speed proton? (b) The particle trajectory in image 1 spirals inward. Explain why. *Hint*: Each ion-creating encounter is a collision.

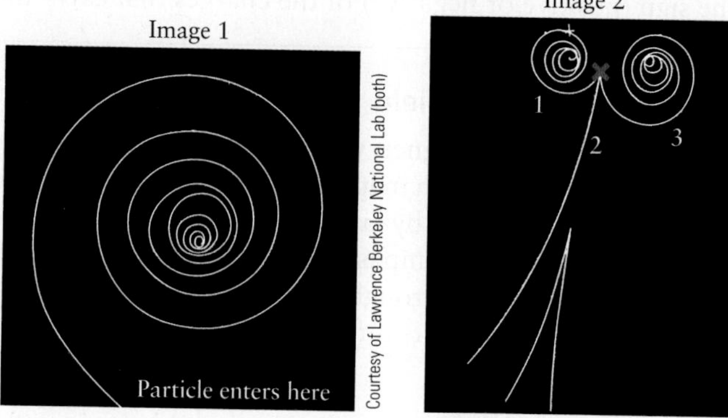

Figure Q20.13 Questions 13 and 14.

14. SSM The bubble chamber image on the right in Figure Q20.13 (image 2) shows three particles emanating from a point marked with a red × near the top of the image. The bubble chamber is in a magnetic field oriented so that the field lines point out of the page. (a) Identify the sign of the charge on each particle. (b) Rank the initial speeds of the particles from least to greatest. How do you know? Assume the particles all have the same mass and charges of the same magnitude.

15. The Earth's magnetic field is produced by the flow of electrons within its molten interior. If the flow is in the form of a ring of current in the plane of the equator, in which direction must the electrons flow to produce the observed polarity of the Earth's field?

16. Two 20-penny nails are wrapped with insulated wire in the shape of a solenoid. They are then laid down end to end with the heads almost touching as shown in Figure Q20.16. If a battery is

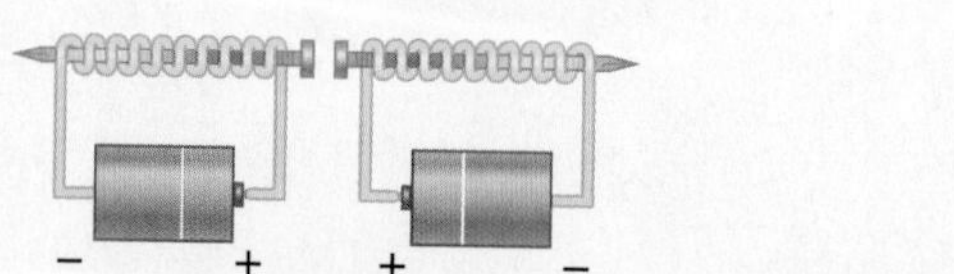

Figure Q20.16

connected to the solenoid of each nail as shown, will the nails be attracted or will they repel each other? Explain.

17. Does a current-carrying wire placed in a magnetic field always experience a magnetic force? Explain.

18. Blood plasma has a high content of Na^+ and Cl^- ions, and as blood follows through a vessel, the ions travel along at the same speed. A magnetic field is applied such that the field is perpendicular to the vessel as shown in Figure Q20.18. (a) What effect will the magnetic field have on the distribution of positive ions? (b) What effect will the magnetic field have on how the negative ions are distributed? (c) Consider the overall distribution of charge in the presence of the magnetic field. Will an electric field be present? If so, in what direction is it oriented?

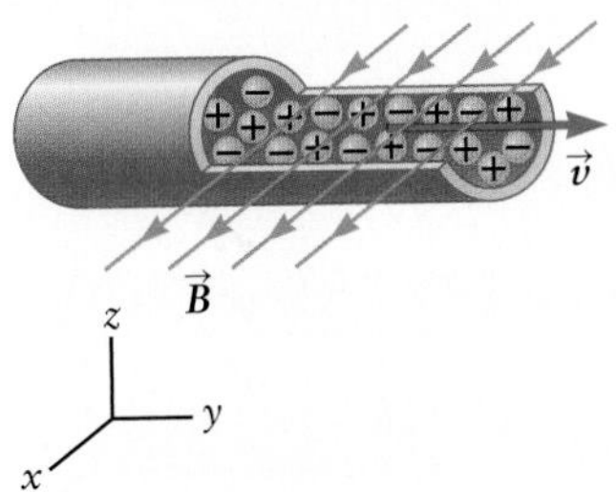

Figure Q20.18

19. Figure Q20.19 shows in part A an electric charge q moving past a stationary bar magnet and in part B a bar magnet moving past a stationary electric charge. In which case will there be a magnetic force on the charge? Will there be a force on the charge in the other case? Explain.

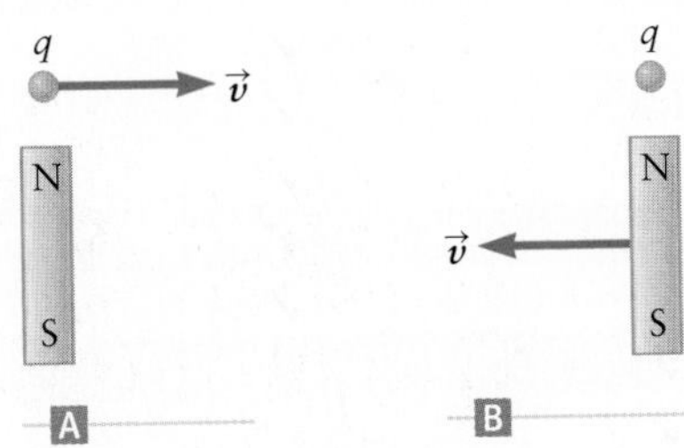

Figure Q20.19

20. A square loop with constant current I is placed in a uniform magnetic field. Explain why the total force on the loop is zero, no matter how the loop is oriented relative to the field direction.

21. Figure Q20.21 shows a straight wire of length L carrying a current I in a magnetic field of magnitude B. In all cases, the direction and magnitude of the field are the same everywhere. In which case is the magnetic force on the wire largest? In which case is it smallest? Order the situations from largest to smallest according to the magnitude of the force.

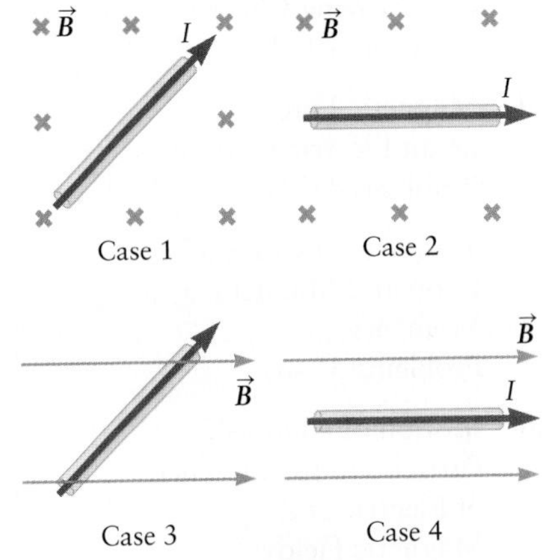

Figure Q20.21

22. A wire carries a current in the $-z$ direction. If there is a magnetic force in the $+x$ direction, what might be the direction of the magnetic field, (a) $+x$, (b) $-x$, (c) $+y$, (d) $-y$, (e) $+z$, or (f) $-z$?

Problems available in ENHANCED WebAssign

Section 20.1 Sources of Magnetic Fields
Problems 1–6

Section 20.2 Magnetic Forces Involving Bar Magnets
Problems 7–9

Section 20.3 Magnetic Force on a Moving Charge
Problems 10–44

Section 20.4 Magnetic Force on an Electric Current
Problems 45–56

Section 20.5 Torque on a Current Loop and Magnetic Moments
Problems 57–61

Section 20.6 Motion of Charged Particles in the Presence of Electric and Magnetic Fields
Problems 62–67

Section 20.7 Calculating the Magnetic Field: Ampère's Law
Problems 68–87

Section 20.8 Magnetic Materials: What Goes on Inside?
Problems 88–89

Section 20.9 The Earth's Magnetic Field
Problems 90–91

Section 20.10 Applications of Magnetism
Problems 92–93

Section 20.11 The Puzzle of a Velocity-Dependent Force
Problem 94

Additional Problems
Problems 95–103

List of Enhanced Problems

Problem number	Solution in *Student Companion & Problem-Solving Guide*	Reasoning & Relationships Problem	Reasoning Tutorial in ENHANCED WebAssign
20.5	✓		
20.7	✓		
20.14		✓	✓
20.17	✓		
20.22	✓		
20.32		✓	✓
20.33		✓	
20.35		✓	
20.49		✓	✓
20.50		✓	✓
20.55	✓		
20.58	✓		
20.65	✓		
20.72	✓		
20.74		✓	✓
20.84		✓	✓
20.88		✓	
20.89		✓	
20.90		✓	
20.91	✓	✓	
20.93	✓		
20.100	✓		
20.101		✓	✓
20.102		✓	
20.103		✓	

CHAPTER

21

▶ *The solid-body electric guitar was invented by Les Paul (1915–2009; shown here). He is the only person who has been inducted into both the Rock and Roll Hall of Fame and the Inventor's Hall of Fame. (© Lynn Goldsmith/Corbis)*

Magnetic Induction

OUTLINE

Are electricity and magnetism separate phenomena, or are they related in some way? Perhaps the most obvious hint that they are connected is that electric and magnetic forces both act only on particles carrying an electric charge. Another clue is that moving electric charges (an electric current) create a magnetic field. In this chapter, we explore another connection between electricity and magnetism: a changing magnetic field creates an electric field. This effect, called ***magnetic induction,*** links $\vec{E}$ and $\vec{B}$ in a truly fundamental way. Magnetic induction is also the key to many practical applications of magnetism, including motors, generators, cell phones, and electric guitars.

21.1 Why Is It Called *Electromagnetism*?

In 1820, Danish physicist Hans Christian Ørsted discovered that an electric current in a wire can exert a force on a compass needle. (Such magnetic forces were discussed in Chapter 20.) An electric current is produced by an electric field and a compass needle is a small bar magnet, so Ørsted's discovery also showed that an electric field can lead to a force on a magnet. He therefore concluded that *an electric field can produce a magnetic field.*

When this connection between $\vec{E}$ and $\vec{B}$ was announced in 1820, it was a complete surprise and stimulated many further experiments to probe the connection between electricity and magnetism. Scientist Michael Faraday reasoned that if an electric field can produce a magnetic field, perhaps a magnetic field can produce an electric field. Faraday attempted to observe such an "induced" electric field using an experiment like the one sketched in Figure 21.1A in which a bar magnet is positioned close to a loop of wire. This loop is connected to a lightbulb, so if there is current in the loop, the bulb will light up and thus act as a detector of the electric current. (Faraday didn't use a lightbulb—they weren't invented yet—instead, he used an ammeter to detect the current, with the same results.)

Faraday positioned a bar magnet near the loop of wire in Figure 21.1A so that the magnetic field would penetrate through and around the loop. If this magnetic field produces an electric field, the electric field would cause an electric current in the loop, lighting the bulb. Faraday was probably very disappointed that this experiment was not successful. No current was detected in the wire loop when the bar magnet was at rest near the loop. He did, however, discover that when the bar magnet is *in motion* either toward or away from the loop, the bulb does emit light, indicating an electric field within the wire (Fig. 21.1B). This electric field is produced *only when the magnet is moving.* If the magnet is stationary (as in Fig. 21.1A), the current and hence also the electric field are always zero.

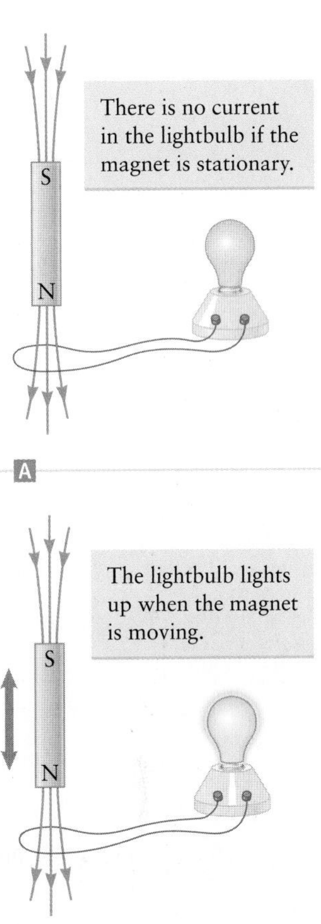
There is no current in the lightbulb if the magnet is stationary.
S
N
A
The lightbulb lights up when the magnet is moving.
S
N
B

▲ **Figure 21.1** A When a bar magnet is placed near a loop of wire, a magnetic field penetrates the loop. B When the bar magnet is in motion, an electric current is induced in the loop, lighting the lightbulb.

Faraday devised another approach to this experiment in which a solenoid is positioned near a loop of wire containing a lightbulb (Fig. 21.2). Recall from Chapter 20 that when an electric current passes through a solenoid, a magnetic field is produced with field lines similar to those produced by a bar magnet. Faraday wanted to know if this magnetic field could also produce a current in his wire loop. He passed an electric current I through the solenoid by connecting it to a battery through a switch. Faraday found that when the current through the solenoid is constant (Fig. 21.2A), there is no current (the bulb did not light), but the bulb does light up the moment the switch is opened or closed (Fig. 21.2B). An electric current is produced during those instants in which the current through the solenoid is *changing* from zero to some nonzero value (when the switch was closed) or from a nonzero value to zero (as the switch was opened).

Faraday's experiments with the bar magnet (Fig. 21.1) and the solenoid (Fig. 21.2) show that *a changing magnetic field produces an electric field.* An electric field produced in this way is called an *induced* electric field, and this phenomenon is called ***magnetic induction.***

Magnetic induction

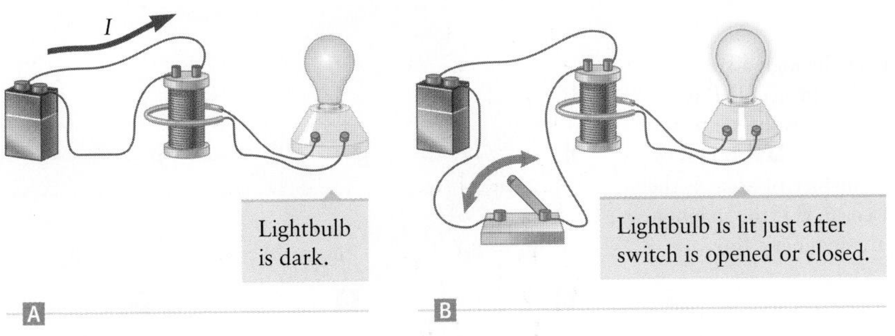
I
Lightbulb is dark.
A
Lightbulb is lit just after switch is opened or closed.
B

◀ **Figure 21.2** A When a current-carrying coil is placed near a loop of wire, the magnetic field produced by the coil penetrates the loop. When this magnetic field is constant, no current is induced in the loop. B When the magnetic field is changing with time (i.e., just as the switch is closed or opened), an electric current is induced in the loop.

21.2 Magnetic Flux and Faraday's Law

Faraday's experiments in Figures 21.1 and 21.2 demonstrated qualitatively the effect of magnetic induction. Faraday also formulated ***Faraday's law,*** which tells how to calculate the induced electric field in different situations. To use this law, we must first introduce a quantity called ***magnetic flux.***

Suppose there is a magnetic field in some region of space and let A be the area of a particular surface as sketched in Figure 21.3A. If lines of $\vec{B}$ cross this surface, there is a magnetic flux through the surface. Magnetic flux is very similar to the concept of electric flux that we encountered in Chapter 17 in connection with Gauss's law.

Let's now calculate the magnetic flux in a few specific cases. Figure 21.3A shows a case in which the magnetic field has constant magnitude B (the same everywhere in this region of space). The surface area A is flat, and $\vec{B}$ is perpendicular to the surface. The magnetic flux, denoted by Φ_B, is given by

$$\Phi_B = BA \quad \text{(when } \vec{B} \text{ is perpendicular to a surface)} \tag{21.1}$$

Qualitatively, you can think of the magnetic flux as a "count" of the number of field lines that pass through the surface.

Two other common situations are shown in parts B and C of Figure 21.3. In both cases, $\vec{B}$ has constant magnitude and constant direction. In Figure 21.3B, the magnetic field lines cut through the indicated surface, making an angle θ with the direction perpendicular (normal) to the surface. Here the magnetic flux is

Definition of magnetic flux

$$\Phi_B = BA\cos\theta \quad \text{(when } \vec{B} \text{ makes an angle } \theta \text{ relative to the normal to a surface)} \tag{21.2}$$

Since θ is not zero, the flux in Figure 21.3B is smaller than when the field lines cut perpendicularly through the surface as in Figure 21.3A (for the same values of B and A). When the magnetic field is parallel to the plane of the surface (Fig. 21.3C), the magnetic flux is zero:

$$\Phi_B = 0 \quad \text{(when } \vec{B} \text{ is parallel to a surface)} \tag{21.3}$$

This case—with $\vec{B}$ parallel to the surface—is actually a special case of Equation 21.2, with $\theta = 90°$. Likewise, when $\vec{B}$ is perpendicular to the surface (Fig. 21.3A), $\theta = 0$ and the expression for the flux in Equation 21.2 reduces to Equation 21.1.

For simplicity, in the three examples in Figure 21.3 we have assumed the magnetic field is the same at all points on a flat surface. However, Φ_B can be defined in a similar way for any surface. Any complicated surface can be broken into small regions that are each approximately flat, and Equation 21.2 can be applied to find the flux through each small region. The total flux is then the sum of the fluxes through all the individual pieces of the surface being considered.

When calculating electric flux in our applications of Gauss's law in Chapter 17, we were interested in closed surfaces, that is, surfaces that completely enclose a particular region of space. In magnetic induction, however, the surfaces of interest are *open* surfaces, like the ones in Figure 21.3. Such open surfaces always possess an edge or perimeter, which we will soon see plays a key role in magnetic induction.

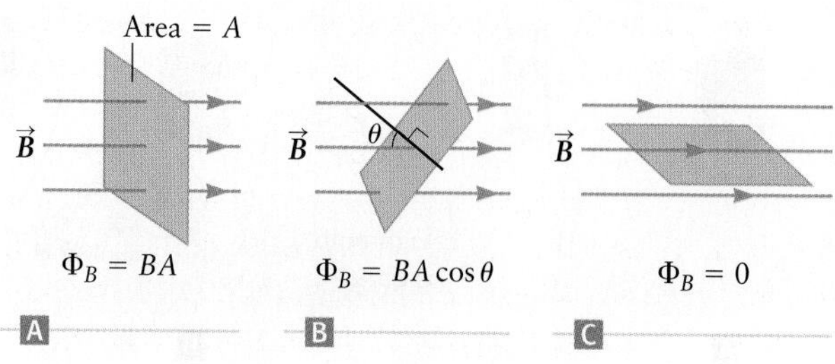

▶ **Figure 21.3** A When a magnetic field passes through a surface of area A, there is a magnetic flux through the surface. If $\vec{B}$ is perpendicular to the surface and the field's magnitude is the same at all points on the surface, the magnetic flux is $\Phi_B = BA$. B If $\vec{B}$ makes an angle θ with a line drawn perpendicular to a surface of area A, the magnetic flux through this surface is $\Phi_B = BA\cos\theta$. C If the magnetic field is parallel to the surface so that $\theta = 90°$, the magnetic flux through the surface is zero.

The units of magnetic flux can be deduced from Equation 21.2. The right-hand side of this equation has units of magnetic field multiplied by area, so magnetic flux has units of $T \cdot m^2$. This unit is named the ***weber*** (Wb):

$$\text{unit of magnetic flux: } 1 \text{ Wb} = 1 \text{ T} \cdot \text{m}^2 \quad (21.4)$$

The weber is the official SI unit of magnetic flux and is used in some texts, but many physicists just use the combination of units $T \cdot m^2$.

Faraday's Law

Faraday's induction experiments in Figures 21.1 and 21.2 demonstrate that a changing magnetic field produces a current in a wire loop. ***Faraday's law*** tells us how to calculate the potential difference that produces this induced current. This law is written in terms of the magnetic flux and the electromotive force (emf) induced in the wire loop. Recall from Chapter 19 that an emf (denoted by $\mathcal{E}$) is not a force; rather, it is an electric potential difference, like the one that appears across the terminals of a battery. Faraday's law states that a changing magnetic flux produces an electric potential difference $\mathcal{E}$ given by

$$\mathcal{E} = -\frac{\Delta\Phi_B}{\Delta t} \quad (21.5)$$

Faraday's law

This simple formula contains two important ideas. The first is that the *magnitude* of the induced emf equals the rate of change of the magnetic flux:

$$|\mathcal{E}| = \left|\frac{\Delta\Phi_B}{\Delta t}\right| \quad (21.6)$$

The second important "idea" is the negative sign in Equation 21.5, which is so important that it has its own name: ***Lenz's law***. We will devote Section 21.3 to a discussion of this negative sign and explain why it deserves to be called a law of physics. Here, we'll consider some properties and situations that depend on just the magnitude of the induced emf (Eq. 21.6). After we understand that part of Faraday's law, we will be better prepared to understand the negative sign in Equation 21.5 and the importance of Lenz's law.

The two sides of Equation 21.6 can be explained in the context of the experiment in Figure 21.4. Here the magnetic field produced by a bar magnet passes through a loop of wire that encloses an area A. This arrangement is similar to the one in Figure 21.1 except for two small changes. First, the wire loop is very small, so the magnetic field is constant across the area A and the field is assumed to be perpendicular to the plane of the loop. The magnetic flux is then simply $\Phi_B = BA$ (Eq. 21.1). However, the value of B changes as the bar magnet is moved closer or farther away from the loop, so the flux changes with time as the magnet moves. Second, we have replaced the lightbulb in Figure 21.1 with a voltmeter, and for simplicity we assume the voltmeter's input terminals are very close together and close to the wire loop so that the perimeter of the loop still defines our area A.

Let's now consider the terms in Faraday's law (Eq. 21.6) and what they mean.

1. The quantity $\mathcal{E}$ on the left-hand side is the induced emf that appears in the wire loop. The value of $\mathcal{E}$ would be indicated by the voltmeter in the wire loop circuit and has units of volts (V). This emf is an induced electric potential difference and is related to the electric field directed along and inside the wire loop. This induced potential difference produces the current in the lightbulb in Figures 21.1 and 21.2.
2. The right-hand side of Faraday's law contains the change in magnetic flux $\Delta\Phi_B$ during a short time interval Δt, divided by Δt. The ratio $\Delta\Phi_B/\Delta t$ is just the rate of change of the magnetic flux with time. In words, Faraday's law says that an emf (i.e., a voltage or electric potential difference) is produced in a circuit by *changes* in the magnetic flux through the circuit. A constant flux, no matter

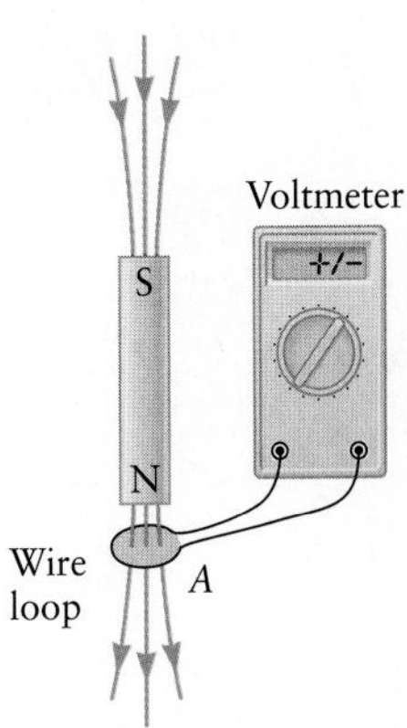

▲ **Figure 21.4** If a voltmeter is attached to the wire loop in Figure 21.1 in place of the lightbulb, it can be used to measure the emf (i.e., the voltage) induced in the loop.

how large, does not produce an induced voltage. Magnetic induction requires that the flux through the wire loop change with time. In Faraday's successful experiments (Figs. 21.1B and 21.2B), this was accomplished by arranging for the magnitude of $\vec{B}$ to change with time. According to Equation 21.2, however, the magnetic flux also depends on the area A and the angle θ. Changes in A and θ can thus also produce a changing flux and hence an induced emf.

What about the *direction* of the induced current in a wire loop? The lightbulbs in Figures 21.1 and 21.2 were either lit or not, indicating current or no current, but that does not tell the current direction. This direction can be read from the *sign* of the voltmeter reading in Figure 21.4, which is at the heart of Lenz's law. The voltmeter reading (positive or negative) specifies the direction of the induced emf and hence the direction of the induced current and electric field, all of which are important in practical applications of Faraday's law.

EXAMPLE 21.1 Magnitude of $\mathcal{E}$ Produced by Moving a Bar Magnet

Faraday's law tells us that a changing magnetic flux produces an induced emf, but how large is this emf in typical cases? Consider a wire loop of area $A = 1\text{ cm}^2$ initially far from any sources of magnetic field. A strong bar magnet with $B = 1$ T near its poles is suddenly inserted into the loop, filling the loop completely (Fig. 21.5). If the time required to insert the magnet is $\Delta t = 0.1$ s, what is the approximate value of the emf induced in the loop?

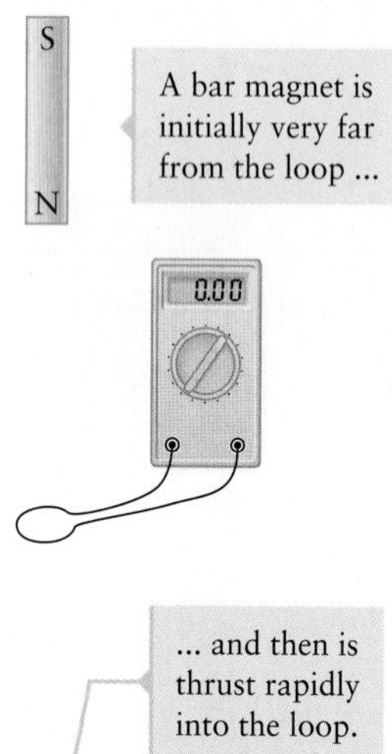
S
A bar magnet is initially very far from the loop ...
N
0.00

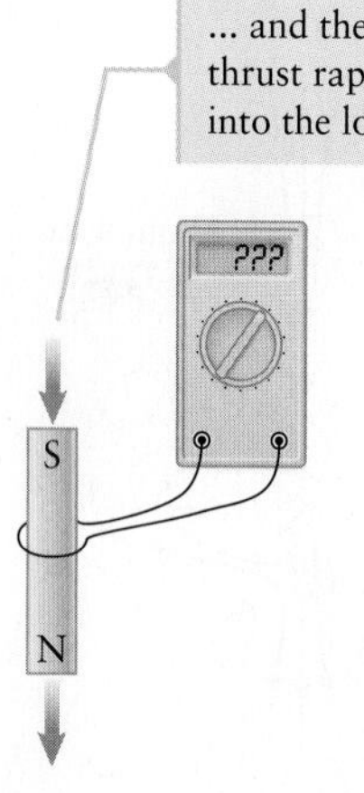
... and then is thrust rapidly into the loop.
???
S
N

▲ **Figure 21.5** Example 21.1. If a bar magnet is inserted into this wire loop, the voltmeter will measure a nonzero induced electric potential.

RECOGNIZE THE PRINCIPLE

The magnitude of an induced emf $\mathcal{E}$ is given by Faraday's law, with $\mathcal{E}$ equal to the change of the magnetic flux divided by the time over which this change of flux occurs.

SKETCH THE PROBLEM

Figure 21.5 shows the initial and final positions of the bar magnet, which correspond to the initial and final flux through the loop.

IDENTIFY THE RELATIONSHIPS

The change in magnetic flux is $\Delta\Phi_B = \Phi_f - \Phi_i$ (the final flux minus the initial flux). Because the wire loop is initially far from any magnetic fields, the initial flux through the loop is approximately $\Phi_i = 0$. After inserting the bar magnet, the field through the loop is $B \approx 1$ T, so the final flux (according to Eq. 21.1) is $\Phi_f = BA$ and the change in flux is

$$\Delta\Phi_B = \Phi_f - \Phi_i = BA \tag{1}$$

This flux change occurs over a time interval of $\Delta t = 0.1$ s (as given above).

SOLVE

Inserting the change in flux from Equation (1) and the value of Δt into Faraday's law, we find the magnitude of the emf to be

$$|\mathcal{E}| = \left|\frac{\Delta\Phi_B}{\Delta t}\right| = \frac{BA}{\Delta t} = \frac{(1\text{ T})(1 \times 10^{-4}\text{ m}^2)}{(0.1\text{ s})} = \boxed{0.001\text{ V}}$$

► *What does it mean?*

This voltage is rather modest. The emf of a typical flashlight battery is 1.5 V, for instance, and the voltage in a typical household circuit is about 120 V, so this induced emf would not be a practical replacement for the battery in a radio or MP3 player. An induced voltage of this size can be measured quite easily, however, and could be used in other applications. (See Example 21.10.)

> **CONCEPT CHECK 21.1** Faraday's Law
> Suppose the area A of the wire loop in Example 21.1 (Fig. 21.5) is increased and the bar magnet is also enlarged so that it still fills the loop. If the rest of the experiment (the values of B and Δt) stays the same, how is the magnitude of the induced emf affected?
> (a) The induced emf is larger.
> (b) The induced emf is smaller.
> (c) The induced emf does not change because the magnetic field is the same.

Key Features of Faraday's Law

According to Faraday's law, the emf induced along a closed path such as a wire loop depends on changes in the magnetic flux through the area enclosed by that path. This fact leads to a number of important consequences.

1. Only *changes* in Φ_B matter. The magnetic flux might be very large, but if it changes only a small amount, the induced emf will still be small.
2. Rapid changes in magnetic flux produce larger values of $\mathcal{E}$ than do slow changes. This dependence on the *rate* at which the flux changes means that the induced emf plays a key role in alternating-current (AC) circuits (Chapter 22).
3. The magnitude of $\mathcal{E}$ is proportional to the rate of change of the flux $\Delta\Phi_B/\Delta t$. If this rate of change is constant, then $\mathcal{E}$ is constant. However, achieving a constant rate of change $\Delta\Phi_B/\Delta t$ requires that Φ_B must eventually become very large, which is usually not possible in practice. So, in most applications, the rate of change of Φ_B is not constant and the induced emf varies with time, leading again to AC voltages (Chapter 22).
4. The induced emf is present even if there is no current in the path enclosing an area of changing magnetic flux. Hence, the voltmeter in Figure 21.4 will register an induced voltage whether or not a current is present in the loop.

Magnetic Flux through a Changing Area

Changes in the magnetic flux can be produced by changes in the magnitude of $\vec{B}$, by changes in area A of the surface, and by changes in the angle θ that the surface makes with respect to $\vec{B}$. An example in which this area is changing is sketched in Figure 21.6A, which shows a metal bar of length L that can slide on two frictionless, horizontal metal rails. The metal rails are connected through a resistor on the left, so the rails plus the bar form a closed electric circuit. A magnetic field $\vec{B}$ is directed perpendicular to the plane of the rails and bar, with constant magnitude and direction throughout. Suppose the bar is being pulled to the right by a force from an external agent (perhaps a person is pulling on it), making the bar move with a constant speed v. Let's use Faraday's law to find the induced emf in this circuit.

We apply Faraday's law using the closed path indicated by the dashed loop in Figure 21.6A. The magnetic field is constant and perpendicular to the enclosed area, so the magnetic flux is (Eq. 21.1)

$$\Phi_B = BA = BwL \tag{21.7}$$

where w is the distance from the left side of the circuit to the bar. Because the bar is moving to the right, this distance and hence the magnitude of the flux increase with time. The speed of the bar is $v = \Delta w/\Delta t$ and is constant, so the flux in Equation 21.7 changes with time according to

$$\frac{\Delta\Phi_B}{\Delta t} = BL\frac{\Delta w}{\Delta t} = BLv \tag{21.8}$$

Inserting Equation 21.8 into Faraday's law (Eq. 21.6), we find the magnitude of the induced emf,

$$|\mathcal{E}| = \frac{\Delta\Phi_B}{\Delta t} = BLv \tag{21.9}$$

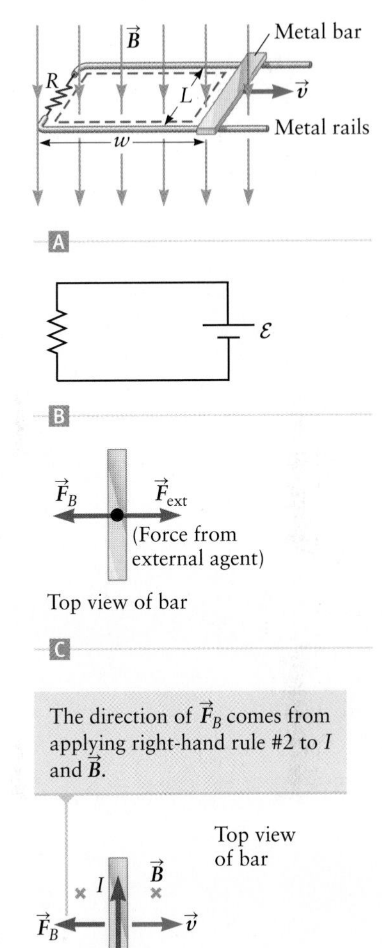

▲ **Figure 21.6** A The magnetic flux through a loop changes with time if the area of the loop changes. As the movable bar on the right slides along the rails, the enclosed area changes. B The equivalent circuit is a battery connected to a resistor. The battery voltage represents the induced emf in the sliding bar. C Forces acting on the sliding bar. D The emf induced in the bar produces a current through the circuit. The resulting magnetic force $\vec{F}_B$ on the sliding bar opposes its motion.

This induced emf acts much like a battery in the "equivalent" circuit shown in Figure 21.6B, with an emf given by Equation 21.9. The magnitude of the current I in this circuit is $I = \mathcal{E}/R$ (Eq. 19.10). Inserting our result for $\mathcal{E}$, we have

$$I = \frac{BLv}{R} \tag{21.10}$$

Recall from Chapter 19 that in circuits like the one in Figure 21.6B, power is dissipated in the resistor. The dissipated power is given by $P = I^2R$ (Eq. 19.16); hence,

$$P = I^2R = \left(\frac{BLv}{R}\right)^2 R = \frac{B^2L^2v^2}{R} \tag{21.11}$$

Conservation of Energy and the Induced emf

Let's now summarize in words what the result in Equation 21.11 means.

1. Since the metal bar in Figure 21.6A is moving, the area of the loop indicated by the dashed red line increases with time, producing a change in the magnetic flux through this loop.
2. According to Faraday's law of induction, an emf is induced in the circuit composed of the metal bar plus rails.
3. This induced emf acts just like a battery with an emf (or voltage) $\mathcal{E}$, producing an electric current in the circuit given by Equation 21.10.
4. This current leads to a power dissipation P in the circuit.

Where does this power come from? In mechanics, power equals the product of force and velocity, so to answer this question we must consider the forces acting on the metal bar. Actually, two forces act on the metal bar in Figure 21.6. One is a magnetic force due to the current in the bar; we calculated the magnitude of this current in Equation 21.10. The current direction is specified by Lenz's law as explained in Section 21.3, from the lower end of the rod toward the upper end (Fig. 21.6D). The magnetic field is perpendicular to the bar, so from Equation 20.11, the magnitude of the magnetic force on the bar is

$$F_B = ILB \tag{21.12}$$

We now apply right-hand rule number 2 from Chapter 20 to find the direction of this force. An upward current in Figure 21.6D corresponds to the motion of positive charges moving upward in the figure. If you place the fingers of your right hand upward along the bar and then curl them toward the direction of the magnetic field, your thumb must point to the left as indicated by the vector $\vec{F}_B$ in parts C and D of Figure 21.6. Hence, the magnetic force on the bar *opposes* the bar's motion.

Now we consider the second force acting on the bar. This force is from the external agent mentioned at the start of the problem, and we have assumed this agent makes the bar move to the right with constant velocity. According to Newton's laws of mechanics, if the metal bar is moving with a constant velocity, the total force on the bar must be zero. Hence, the force of the external agent $\vec{F}_{ext}$ must be equal in magnitude and opposite in direction to the magnetic force in Equation 21.12. We thus have

$$F_{ext} = ILB \tag{21.13}$$

directed to the right (Fig. 21.6C). The power exerted by the external agent is then (Chapter 6) $P_{ext} = F_{ext}v$. Inserting F_{ext} from Equation 21.13 leads to

$$P_{ext} = F_{ext}v = ILBv$$

We know the current I from Equation 21.10. Inserting that gives

$$P_{ext} = \left(\frac{BLv}{R}\right)LBv = \frac{B^2L^2v^2}{R} \tag{21.14}$$

which is identical to the electrical power dissipated in the resistor, Equation 21.11. Hence, the mechanical power "put into" the bar by the external agent is converted into electrical power "delivered to" the resistor. Energy is thus converted from one form (mechanical) to another (electrical), and the total energy is conserved.

In Chapter 6, we saw how the principle of conservation of energy can be deduced for mechanical systems from Newton's laws of motion. Now we have found that conservation of energy is also obeyed by electromagnetic phenomena. Indeed, it is believed that the principle of energy conservation is obeyed by *all* physical processes and by all laws of physics.

EXAMPLE 21.2 Using Faraday's Law to Design a Generator

A device that converts mechanical energy into electrical energy is called a ***generator.*** The system in Figure 21.6A is a very simple type of generator. A person pulling on the rod exerts a force on the sliding bar (doing mechanical work), converting mechanical energy into electrical energy as shown by the current in the circuit. Suppose we make a generator as in Figure 21.6A, with a bar of length $L = 1$ m and a magnetic field $B = 1$ T. To be useful for producing power for your house, the generator's output voltage (the induced emf) should be approximately 100 V. (The voltage in household wiring is typically 120 V.) How fast must the bar travel to produce this value for the induced emf? Is this way of making a generator practical?

RECOGNIZE THE PRINCIPLE

According to Faraday's law, the magnitude of the emf induced across the bar equals the rate of change of the magnetic flux $\Delta\Phi_B/\Delta t$, which we calculated above in our discussion of Figure 21.6. We found that $\Delta\Phi_B/\Delta t$ is proportional to the speed v of the bar. So, given the desired value of $\mathcal{E}$, we can find the required value of v.

SKETCH THE PROBLEM

Figure 21.6 describes the problem.

IDENTIFY THE RELATIONSHIPS AND SOLVE

Faraday's law led to the magnitude of the induced emf in Equation 21.9:

$$|\mathcal{E}| = \frac{\Delta\Phi_B}{\Delta t} = BLv$$

Rearranging to solve for v gives

$$v = \frac{|\mathcal{E}|}{BL}$$

Inserting the given values of $\mathcal{E}$, B, and L, we have

$$v = \frac{100\text{ V}}{(1\text{ T})(1\text{ m})} = \boxed{100\text{ m/s}}$$

▸ *What does it mean?*

Is this approach for generating electric power practical? This velocity is about 200 mi/h. If we turn the generator on for 10 s, the horizontal rails would have to be at least $vt = (100\text{ m/s})(10\text{ s}) = 1000$ m long! That would be an extremely impractical generator.

Designing a Practical Electrical Generator

The electrical generator in Example 21.2 could certainly be built, producing a constant emf as long as the bar slides at a constant speed, yet we concluded in Example 21.2 that such a DC generator is not a practical device. How can we make the rate of change of flux $\Delta\Phi_B/\Delta t$ large enough to give a useful emf? The key is to use rotational motion instead of the linear motion of the bar in Figure 21.6. Generators based on this idea are found in nearly all electric power plants, including nuclear power plants, hydroelectric plants, and those that burn coal or natural gas.

A simple generator based on rotational motion is sketched in Figure 21.7. A permanent magnet produces a constant magnetic field in the region between its two poles (north and south), and a wire loop is located in this region. This loop has a fixed area but is mounted on a rotating shaft, so the angle θ between the field and the plane of the loop changes as the loop rotates. According to the definition of magnetic flux, Φ_B is largest when the loop is perpendicular to the field (Fig. 21.7A) and zero when the plane of the loop is parallel to the field (Fig. 21.7B).

In practice, this generator would be driven mechanically by turning the shaft, which could be accomplished with a steam turbine (as in most power plants) or with a waterfall or windmill. If the shaft rotates with a constant angular velocity, the magnetic flux through the loop varies sinusoidally with time, leading to an induced emf that also varies sinusoidally with time (Fig. 21.7C).

The generator sketched in Figure 21.7 sounds reasonable, but is it really practical? Let's estimate the magnitude of the emf that could be produced by such an AC (alternating-current) generator. If the shaft rotates with period T, the loop will turn from its perpendicular orientation in Figure 21.7A to the parallel orientation in Figure 21.7B in time $T/4$ because this rotation is one-fourth of a complete turn of the rotating shaft. During this time, the magnetic flux changes from $\Phi_i = BA$ (Fig. 21.7A) to $\Phi_f = 0$ (Fig. 21.7B). The change in magnetic flux is

$$\Delta\Phi_B = \Phi_f - \Phi_i = -BA \tag{21.15}$$

during time $\Delta t = T/4$. Inserting Equation 21.15 into Faraday's law (Eq. 21.6) gives

$$|\mathcal{E}| = \left|\frac{\Delta\Phi_B}{\Delta t}\right| = \frac{BA}{T/4} = \frac{4BA}{T} \tag{21.16}$$

For a realistic generator, we might have $B = 1$ T (typical for a permanent magnet), and we take the area of the loop to be $A = 1\ \text{m}^2$ as a fair comparison with the DC generator in Example 21.2. The period T depends on the angular speed of the shaft; a rotation rate of 1000 rpm is common (a car engine runs typically in the 2000–4000 rpm range). The corresponding period is

$$T = \frac{1\ \text{min}}{1000\ \text{rev}} \times \frac{60\ \text{s}}{1\ \text{min}} = 0.060\ \text{s}$$

Inserting all these results into Equation 21.16 gives

$$|\mathcal{E}| = \frac{4BA}{T} = \frac{4(1\ \text{T})(1\ \text{m}^2)}{0.060\ \text{s}} = 70\ \text{V}$$

This voltage is quite useful, so this design would indeed be practical. The voltage could easily be increased by increasing the rotation rate of the shaft or by making the area larger.

The basic design in Figure 21.7 is used in most real electrical generators, but it is still important to recall that because $\Phi_B = BA\cos\theta$, we can produce a change in magnetic flux and hence an induced emf in four different ways (Fig. 21.8): (1) if the magnitude of the magnetic field B changes with time, (2) if the area A changes with time, (3) if the loop rotates so that the angle θ between $\vec{B}$ and the loop changes with time (as in the generator in Fig. 21.7), or (4) if the loop moves from one region to another and the magnetic field is different in the two regions.

Horseshoe magnet

N

Area = A

$\vec{B}$

S

$\Phi_B = BA$ = maximum flux

A

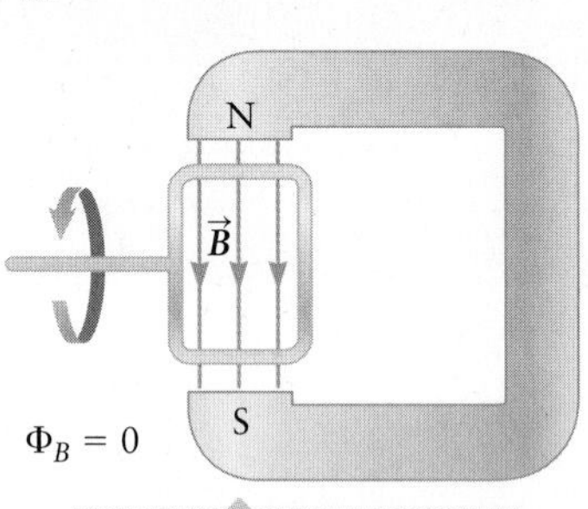

The flux is zero when the loop is in this orientation.

B

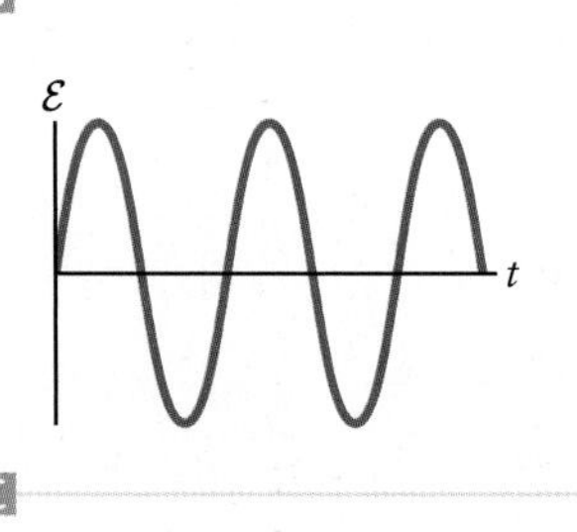

C

▲ **Figure 21.7** Design for an electrical generator based on Faraday's law of induction. A A wire loop is inserted between the poles of a magnet, creating a magnetic flux through the loop. If the loop is perpendicular to $\vec{B}$, the flux has a magnitude BA. B When the loop is parallel to $\vec{B}$, the flux is zero. C If the loop rotates about its axis, the magnetic flux will vary with time, producing a time-varying induced voltage in the loop as shown here.

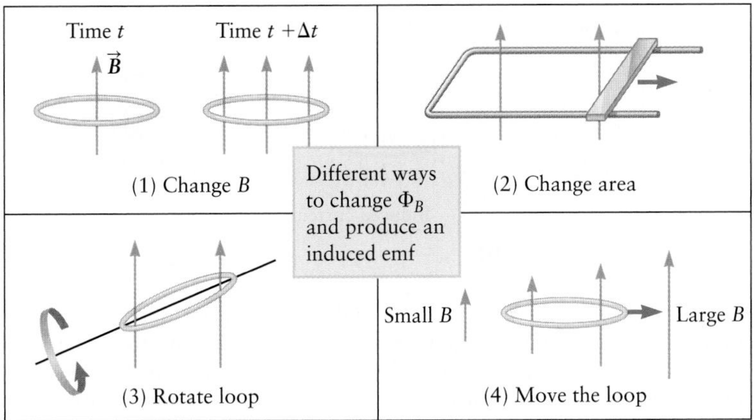

Time t
Time $t + \Delta t$
$\vec{B}$
(1) Change B
(2) Change area
Different ways to change Φ_B and produce an induced emf
Small B
Large B
(3) Rotate loop
(4) Move the loop

◀ **Figure 21.8** Different ways to change Φ_B and produce an induced emf.

CONCEPT CHECK 21.2 Designing a Better Generator

The generator in Figure 21.7 employs a loop that rotates with period T. Suppose the rotation rate is increased so that the loop is made to rotate faster and the period is reduced by a factor of two. How will that affect the magnitude of the induced emf?

(a) The magnitude of the induced emf will not change.
(b) The magnitude of the induced emf will increase by a factor of two.
(c) The magnitude of the induced emf will decrease by a factor of two.

21.3 Lenz's Law and Work–Energy Principles

In Section 21.2, we applied Faraday's law to calculate the magnitude of the induced emf in several different situations. Let's now consider the *sign* of this emf. Faraday's law reads (Eq. 21.5)

$$\mathcal{E} = -\frac{\Delta \Phi_B}{\Delta t} \tag{21.17}$$

Algebraically, the sign of $\mathcal{E}$ is determined simply by the sign of $\Delta\Phi_B$ together with the negative sign on the right-hand side of Equation 21.17. However, the magnetic flux Φ_B is produced by a field that is perpendicular to the surface of interest, whereas the induced current is along the perimeter of this surface (Fig. 21.9). The emf on the left-hand side of Faraday's law (Eq. 21.17) thus leads to a current *perpendicular* to the flux direction on the right-hand side. Hence, the two sides of Faraday's law involve quantities with different and perpendicular directions, which complicates the meaning of the negative sign. The simplest way to deal with this complication is to make use of a principle called ***Lenz's law.*** This principle gives an easy way to determine the sign of the induced emf $\mathcal{E}$, and we'll see that it is closely connected with the principle of conservation of energy.

In many situations, the induced emf of Faraday's law produces a current leading to an induced magnetic field. Lenz's law is a statement about the direction of this induced field:

Lenz's law

The magnetic field produced by an induced current always opposes any changes in the magnetic flux.

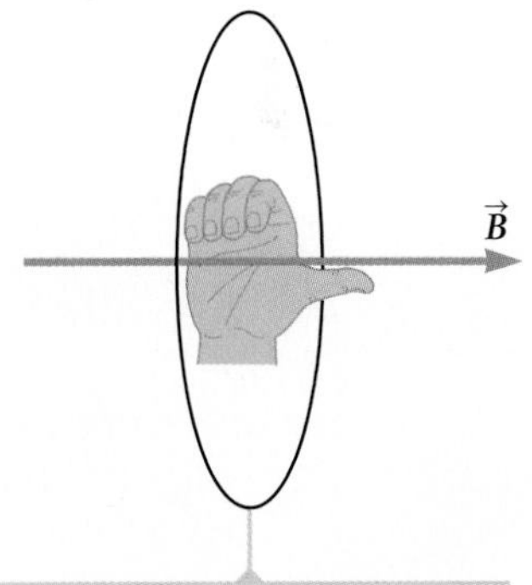

$\vec{B}$
The induced emf is directed along the perimeter of the flux surface. The induced current is thus perpendicular to $\vec{B}$.

▲ **Figure 21.9** The magnetic field that produces the magnetic flux in Faraday's law is perpendicular to the plane in which the induced emf is sensed. Lenz's law gives the direction of the induced emf, that is, either clockwise or counterclockwise around the perimeter of the surface of interest.

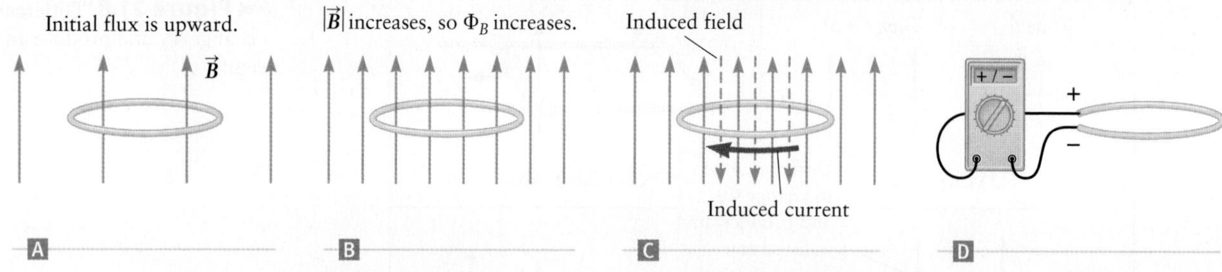
Initial flux is upward.
$\vec{B}$
$|\vec{B}|$ increases, so Φ_B increases.
Induced field
Induced current
+ / −
+
−
A
B
C
D

▲ **Figure 21.10** A This magnetic field is directed through the plane of the loop, giving a flux "upward" through the loop. B If the magnitude of B increases with time, the flux increases. C According to Lenz's law, the induced emf in the loop leads to a magnetic field that opposes this change in flux. A downward magnetic field is created by a clockwise induced current (red arrow). D Even if the loop is broken, there is still an induced emf $\mathcal{E}$, which could be measured by attaching a voltmeter to the two open ends of the loop. The voltmeter reading indicates the polarity (+ or −) of $\mathcal{E}$.

Applying Lenz's Law

Let's apply Lenz's law to an arrangement in which the magnetic field passes through a metal loop with the field initially directed upward through the loop (Fig. 21.10A). We also suppose the magnitude of the field $|\vec{B}|$ increases with time (Fig. 21.10B) and ask two questions: (1) Is there a current induced in the loop? (2) What is the direction of this current?

Since we are concerned with the current in the loop, we let it be the perimeter of our Faraday's law surface. A magnetic flux is directed upward through this surface. The magnitude $|\vec{B}|$ is increasing with time, so Φ_B is also increasing. This changing flux produces an induced emf, so there will indeed be an induced current in the loop. According to Lenz's law, the magnetic field produced by this induced current must *oppose the change* in flux, so the induced magnetic field must be *downward*. From our work on current loops (right-hand rule number 1 in Chapter 20), the induced current must be *clockwise* in Figure 21.10C when looking at the loop from above to give a field in this direction.

The crucial feature of Lenz's law is that the induced field *opposes changes* in the flux as illustrated in Figure 21.11, which again shows a circular wire loop in a perpendicular magnetic field. Now, however, we assume the magnitude of the field is decreasing with time, so the "upward" flux is getting smaller. This change in flux again produces an induced emf, causing an induced current around the loop. The induced magnetic field produced by this current opposes the *change* in flux, so now the induced field is directed upward and the induced current is counterclockwise when viewed from above the loop.

We have stated Lenz's law in terms of the magnetic field produced by the induced current, but what if there is no induced current? For example, Figure 21.10D shows a loop that contains a small break, so current cannot flow all the way around the

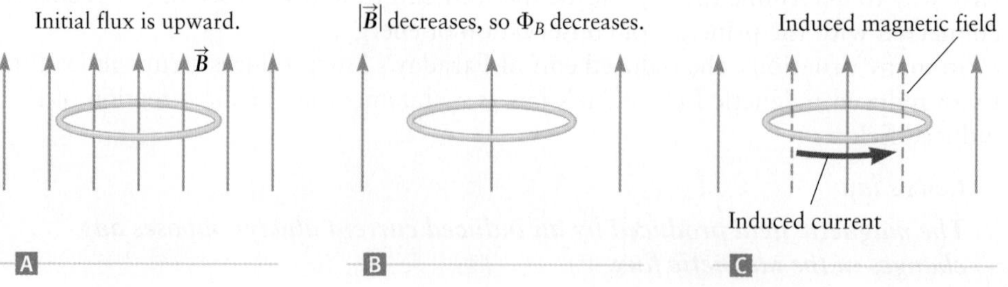
Initial flux is upward.
$\vec{B}$
$|\vec{B}|$ decreases, so Φ_B decreases.
Induced magnetic field
Induced current
A
B
C

▶ **Figure 21.11** A $\vec{B}$ is directed upward through the loop, so the magnetic flux is also in the "upward" direction. B If the magnitude of the field decreases with time, the flux through the loop decreases. C The induced current opposes this change, so I_{induced} produces a field that is upward. The required direction for the induced current is counterclockwise as viewed from above (according to right-hand rule number 1).

loop. Since there can be no current through this loop, it cannot produce a magnetic field to oppose the changing flux, yet there is still an induced emf. This emf could be measured by attaching a voltmeter across the open ends of the loop. In such cases, Lenz would say that the induced emf always attempts to oppose the change in flux.

PROBLEM SOLVING Applying Lenz's Law: Finding the Direction of the Induced emf

1. **RECOGNIZE THE PRINCIPLE.** The induced emf always opposes changes in flux through the Lenz's law loop or path.
2. **SKETCH THE PROBLEM,** showing a closed path that runs along the perimeter of a surface crossed by magnetic field lines.
3. **IDENTIFY** if the magnetic flux through the surface is increasing or decreasing with time.
4. **SOLVE.** Treat the perimeter of the surface as a wire loop; suppose there is a current in this loop and determine the direction of the resulting magnetic field. Find the current direction for which this induced magnetic field opposes the change in magnetic flux. This current direction gives the sign (i.e., the "direction") of the induced emf.
5. Always *consider what your answer means* and check that it makes sense.

EXAMPLE 21.3 Using Lenz's Law to Find the Direction of an Induced Current

Consider the sliding metal bar in Figure 21.12. The magnetic field is constant and is directed into the plane of the drawing. If the bar is sliding to the right, use Lenz's law to find the direction of the induced current.

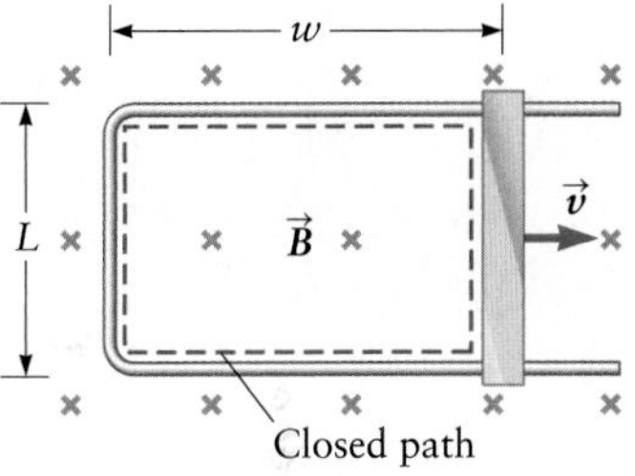

▲ **Figure 21.12** Example 21.3.

RECOGNIZE THE PRINCIPLE

According to Lenz's law, the induced current produces a field that opposes the change in flux through the circuit loop.

SKETCH THE PROBLEM

Following our "Applying Lenz's Law" problem-solving strategy (step 2), the sketch in Figure 21.12 shows a dashed, rectangular path. We are interested in the current induced in this circuit, so we must consider the flux through this rectangle.

IDENTIFY THE RELATIONSHIPS

Step 3: The magnetic field is directed into the plane of this drawing, so the flux through the rectangular surface is directed into the plane. The area of this chosen surface is wL, and the magnetic flux through the surface is $\Phi_B = BwL$. Because the bar is sliding to the right, Φ_B is increasing and is downward.

SOLVE

The induced emf produces an induced current that opposes the downward increase in flux, so the induced magnetic field must be directed *upward*. Applying right-hand rule number 1 (Chapter 20), this field direction is produced by a *counterclockwise* induced current, from the bottom to the top of the moving bar.

► ***What does it mean?***

The flux through a given area may be "upward" or "downward," and its magnitude may be increasing or decreasing with time. The induced emf always *opposes any changes* in the flux.

CONCEPT CHECK 21.3 Applying Lenz's Law

Suppose the metal bar in Figure 21.12 is moving to the left instead of to the right as assumed in Example 21.3. How would this change affect the direction of the induced current?

(a) There would be no change. The current through the loop will still be counterclockwise, directed from the bottom of the moving bar toward the top.
(b) The direction of the current will be reversed, directed from top to bottom through the bar.
(c) The current will be zero.

EXAMPLE 21.4 Moving Magnets and Induced Currents

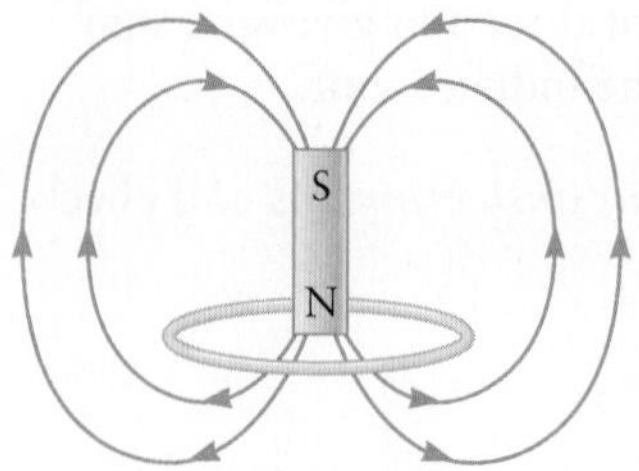

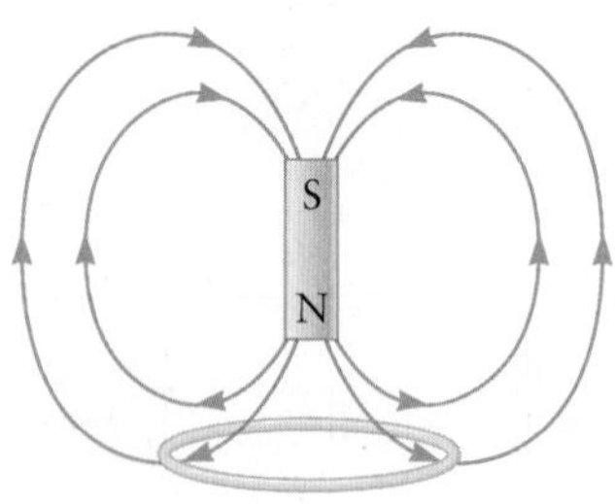

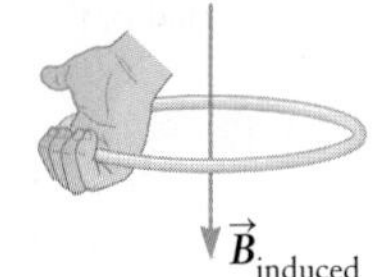

▲ **Figure 21.13** Example 21.4.

A bar magnet is positioned perpendicular to and with one end just above the surface of a metal loop as shown in Figure 21.13A. If the bar magnet is pulled rapidly upward, what is the direction of the current induced in the loop?

RECOGNIZE THE PRINCIPLE

According to Lenz's law, the induced current will oppose any change in the flux through the loop. With the bar magnet oriented as shown in Figure 21.13A, the initial flux through the loop is downward.

SKETCH THE PROBLEM

Figure 21.13A shows the problem. Following our "Applying Lenz's Law" problem-solving strategy, we will use the loop in Figure 21.13A as our Lenz's law path.

IDENTIFY THE RELATIONSHIPS

Step 3 of the problem-solving strategy: The initial flux Φ_i through the loop is downward. As the magnet is pulled upward, the flux is still downward, but the magnitude of the flux will decrease as the magnet moves upward because some of the field lines passing through the loop in Figure 21.13A will no longer pass through it as the magnet moves upward (Fig. 21.13B).

SOLVE

Step 4: To oppose this decrease in downward flux from the bar magnet, the induced current in the loop must give a downward flux, requiring a current that is clockwise when viewed from above. Figure 21.13C shows the application of right-hand rule number 1 with this clockwise current (right thumb in direction of current, fingers curl to give downward field direction inside the loop). Notice that the magnetic field here is the field produced by the induced current in the loop.

▸ *What does it mean?*

Problems involving Lenz's law usually require the application of right-hand rule number 1 to find the direction of the magnetic field produced by a current loop.

EXAMPLE 21.5 A Child's Toy Based on Magnetic Induction

Consider the simple toy shown in Figure 21.14A, consisting of two wire loops. The lower loop is connected through a switch to a battery. Both loops are mounted on a vertical wooden rod, and the upper loop is able to slide freely. The current through the lower loop is initially zero. (The switch is open.) The switch is then closed, producing a counterclockwise current through the lower loop when viewed from above. (**a**) What is the direction of the current induced in the upper loop? (**b**) Will the upper loop be attracted to or repelled from the lower loop?

RECOGNIZE THE PRINCIPLE

When the current is switched on in the lower loop, the flux through the upper loop changes. According to Lenz's law, the current induced in the upper loop will oppose this change in flux.

SKETCH THE PROBLEM

Following the "Applying Lenz's Law" problem-solving strategy (step 2), the sketch in Figure 21.14B shows the magnetic field produced by the counterclockwise current in the lower loop. This field is upward through the upper loop.

IDENTIFY THE RELATIONSHIPS

Step 3: The current in the lower loop produces a magnetic field that gives an upward flux through the upper loop. Hence, the flux increases upward when the switch is closed.

SOLVE

(**a**) Continuing with the "Applying Lenz's Law" problem-solving strategy (step 4), the field induced in the upper loop must be downward to oppose the increase in upward flux. This requires a *clockwise* induced current in the upper loop as viewed from above, as sketched in Figure 21.14B.

(**b**) We now have two current-carrying loops. We can determine the direction of the force on the upper current loop in several different ways. For example, the currents through the two loops are in opposite directions, so the loops act as two oppositely directed bar magnets and thus *repel* each other. In addition, in Concept Check 20.6 we found that the force between two parallel current-carrying wires is repulsive if the currents are in opposite directions. The two current loops in Figure 21.14 are like parallel wires, so they repel each other.

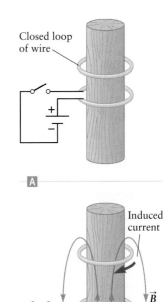

▲ **Figure 21.14** Example 21.5.

▶ ***What does it mean?***

The force on the upper loop will give it an upward acceleration and hence a nonzero velocity and a nonzero kinetic energy. Electrical energy from the lower loop is transferred to the mechanical energy of the upper loop, but how does this energy actually get to the upper loop? We'll discuss the amazing answer in Section 21.8.

CONCEPT CHECK 21.4 **Lenz's Law and a Decreasing Magnetic Field**

Consider a loop of wire in a horizontal plane with a magnetic field directed downward through the loop (Fig. 21.15). If the magnitude of the field is decreasing, is the direction of the current induced in the loop (a) counterclockwise as viewed from above or (b) clockwise as viewed from above?

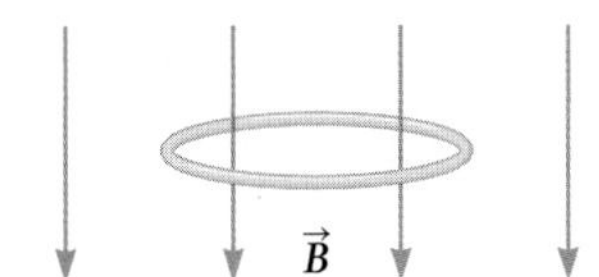

▲ **Figure 21.15** Concept Check 21.4.

Lenz's Law and Conservation of Energy

Lenz's law involves the negative sign on the right-hand side of Faraday's law (Eqs. 21.5 and 21.17). This negative sign is *very* important. To see why, we return to the

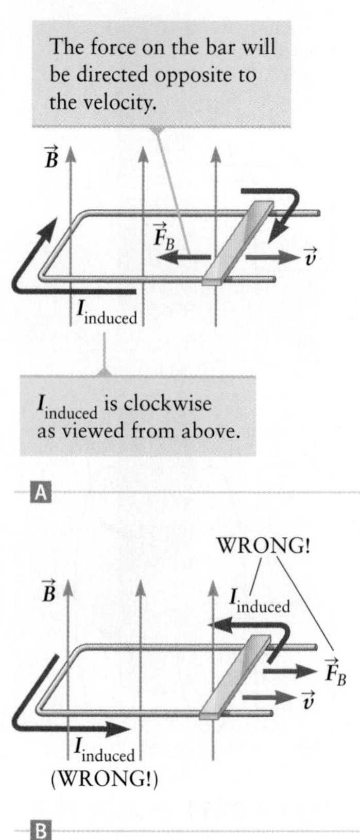
The force on the bar will be directed opposite to the velocity.
$\vec{B}$
$\vec{F}_B$
$\vec{v}$
$I_{induced}$
$I_{induced}$ is clockwise as viewed from above.
A
WRONG!
$\vec{B}$
$I_{induced}$
$\vec{F}_B$
$\vec{v}$
$I_{induced}$
(WRONG!)
B

▲ **Figure 21.16** When a conductor such as this metal bar moves through a magnetic field, the induced emf produces a current through the loop formed by the bar and rails. There is then a magnetic force on the sliding bar. A The negative sign in Faraday's law guarantees that the magnetic force on the bar opposes its motion; that is, $\vec{F}_B$ is directed opposite to the bar's velocity $\vec{v}$. B If Faraday's law did not contain a negative sign, the force would be parallel to $\vec{v}$, causing the bar to accelerate. This situation does *not* occur!

example of a metal bar sliding on two horizontal rails in Figure 21.16. A constant magnetic field is directed upward, and for simplicity there is no friction between the bar and the rails. The bar is initially at rest and is then given a push so that it starts to move with a very small velocity $\vec{v}$ to the right.

Let's use Lenz's law to analyze the resulting motion. The surface of interest is enclosed by the rails and the metal bar, and has an area A. The magnetic flux through this area is $\Phi_B = BA$. The bar is moving to the right, so this area and the flux through it are increasing. According to Lenz's law, the induced current opposes this increase in upward flux; hence, the induced current is *clockwise* as viewed from above, producing a downward induced magnetic field. The resulting current through the bar will be as shown in Figure 21.16A. Because this current-carrying bar is in a magnetic field, a magnetic force $F_B = ILB$ is exerted on the bar, in the direction given by right-hand rule number 2 (Chapter 20). Figure 21.16A shows that this force is directed to the left, so it *opposes the motion* of the bar. The force associated with magnetic induction will thus tend to bring the bar to a stop.

Now imagine what the behavior would be if the "polarity" of Lenz's law were reversed, that is, if we changed the negative sign in Faraday's law to a positive sign. This seemingly innocent change would have a very profound effect. The direction of the induced current would be as shown in Figure 21.16B. The induced current drawn here is counterclockwise (as viewed from above), and $\vec{F}_B$ would be in the same direction as the velocity of the bar. Hence, giving the bar a slight push and a small velocity would lead to a force $\vec{F}_B$ that *increases* this velocity. The bar would thus accelerate to the right in Figure 21.16B, with no limit to its velocity! This acceleration would violate the principle of conservation of energy. We thus conclude that the correct version of Lenz's law, which states that the induced magnetic field always *opposes* changes in the flux, is actually *a consequence of the principle of conservation of energy*!

Mathematically, Lenz's law is just the negative sign in Faraday's law, so the principle of conservation of energy is contained in Faraday's law. This result is amazing because nowhere in Faraday's law or in the other laws of electricity and magnetism is there any explicit mention of energy or its conservation. When these laws are analyzed, however, we find that energy is indeed conserved in all electric and magnetic phenomena, suggesting that the principle of conservation of energy is in some sense a "deeper" principle of physics than Faraday's law or even Newton's laws of mechanics. In fact, as previously mentioned, physicists believe that *all* laws of physics must satisfy the principle of conservation of energy.

21.4 Inductance

So far, when calculating the magnetic flux in applications of Faraday's law we have assumed we can ignore the contribution of the induced field to the flux. This assumption is often correct, but in some situations we can't ignore the induced flux. Consider a coil of wire (a solenoid) connected to a battery through a switch as in Figure 21.17. Most real circuits and coils of wire have some nonzero electrical resistance R, shown as a separate resistor in the circuit. If there are no magnets or other sources of magnetic field nearby, the magnetic flux through the coil from all external sources is zero. Suppose the switch is initially open, so the current in the circuit is initially zero. If we close the switch, what happens?

When the switch is closed, the current in the coil changes suddenly. Figure 21.17B shows just the coil from Figure 21.17A, the current I produced by the battery, and the resulting magnetic field. The magnetic field was zero before the switch was closed and is nonzero afterwards, and according to Faraday's law, this change in magnetic flux induces an emf in the coil. Lenz's law dictates that the induced emf must oppose the increase in flux, so the induced magnetic field is directed opposite to the magnetic field in Figure 21.17B. The induced current thus opposes the original applied current I produced by the battery (Fig. 21.17C). Figure 21.17D shows an equivalent way to understand what is happening. The induced emf can be viewed as a battery with a potential $\mathcal{E}$ and a polarity *opposite* to the polarity of the real battery in the circuit. This opposing emf $\mathcal{E}$ produces the opposing current I induced in Figure 21.17C.

The coil in Figure 21.17 is a type of circuit element called an ***inductor.*** Many inductors are constructed as small solenoids, but almost any coil or loop of wire will act as an inductor. Whenever the current through an inductor changes such as when the switch in Figure 21.17 is closed, a voltage is induced in the inductor that *opposes this change*. This phenomenon is called ***self-inductance*** because the changing current through a coil produces an induced current in the same coil. The induced current opposes the original applied current, as dictated by Lenz's law.

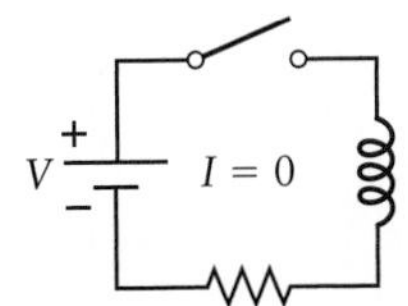

A

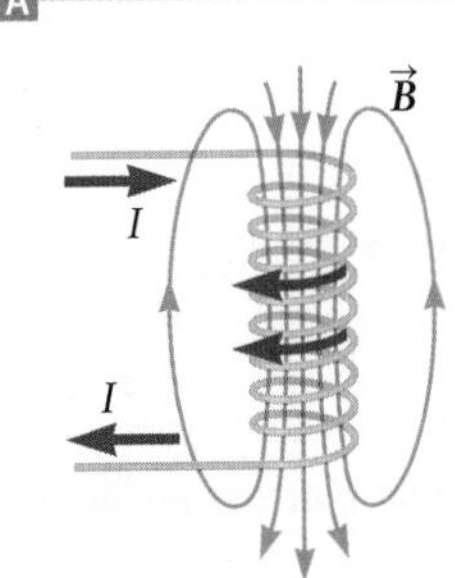

B

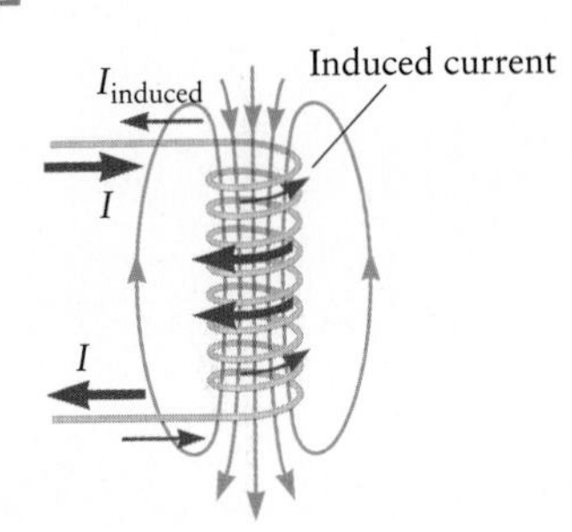

C

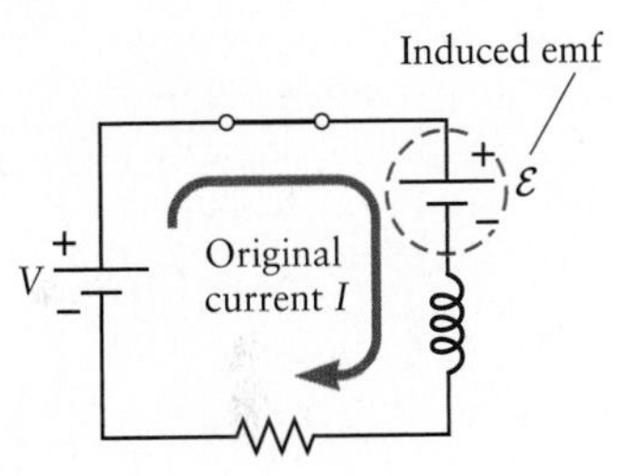

D

▲ **Figure 21.17** A A simple circuit containing a battery, switch, solenoid coil, and resistor. When current enters the coil just after the switch is closed, the rapidly increasing magnetic field lines thread through the coil as shown in B. This results in an increasing magnetic flux, inducing an opposing emf and current in the coil as shown in C. D The induced emf can be viewed as a battery ($\mathcal{E}$) with a polarity that opposes changes in current through the coil. A circuit element that acts in this way is called an inductor.

Inductance of a Solenoid

Let's now use Faraday's law to give a more quantitative description of inductance, using the solenoid in Figure 21.18. The induced emf across the solenoid is

$$\mathcal{E} = -\frac{\Delta\Phi_B}{\Delta t} \tag{21.18}$$

We assume the solenoid contains N loops (also called "turns"), and has length ℓ and cross-sectional area A. The magnetic field through the solenoid is (Eq. 20.24)

$$B = \frac{\mu_0 NI}{\ell} \tag{21.19}$$

This fields runs through the entire solenoid, whose loops are stacked as shown in Figure 21.18. The area of a single loop is A, so the magnetic flux through one loop is $\Phi_1 = BA$. The N loops are stacked on top of one another, and the flux Φ_1 passes through each loop. The total flux through the solenoid is thus

$$\Phi_B = N\Phi_1 = NBA = \frac{\mu_0 N^2 IA}{\ell} \tag{21.20}$$

Here, N, A, and ℓ are constants, so changes in flux are produced by changes in the current I. Inserting this result for the flux into Faraday's law (Eq. 21.18), we get

$$\mathcal{E} = -\left(\frac{\mu_0 N^2 A}{\ell}\right)\frac{\Delta I}{\Delta t}$$

The factor in parentheses is called the ***inductance*** L of the solenoid:

$$L = \frac{\mu_0 N^2 A}{\ell} \quad \text{(inductance of a solenoid)} \tag{21.21}$$

The voltage across the solenoid is thus

$$\mathcal{E} = -L\frac{\Delta I}{\Delta t} \tag{21.22}$$

▶ **Figure 21.18** If each loop of a solenoid has an area A, the flux through each loop is BA. With N loops stacked up together, the total flux through all the loops is $\Phi_B = NBA$. Changes in the total flux determine the induced emf in Faraday's law.

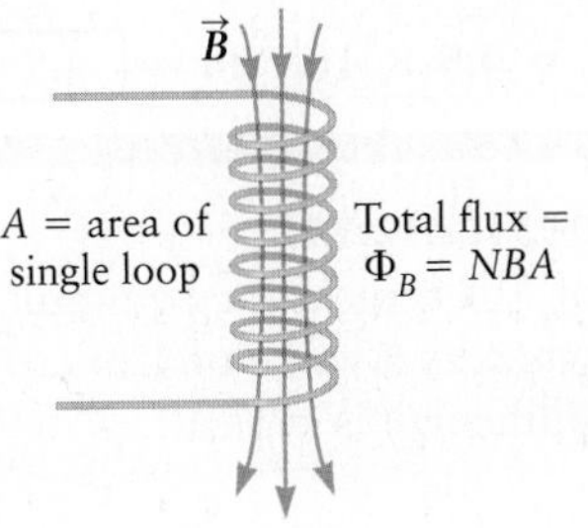

Although the result for inductance in Equation 21.21 applies only for the particular case of a solenoid, the basic notion of inductance is contained in Equation 21.22. By its definition, the inductance L relates the induced emf to changes in the current. This result applies to all coils or loops of wire. Whenever a current produces a magnetic flux through a circuit element, the element will possess an inductance L. The value of L depends on the physical size and shape of the circuit element and on factors such as the number of loops and the area of each loop, just as we found for a solenoid. When the current in an inductor changes with time, the emf across the inductor is proportional to the inductance as given by Equation 21.22. The negative sign in Equation 21.22 indicates the polarity of the voltage. It is conventional practice to speak in terms of the potential "drop" across the inductor, which is given by

Potential drop across an inductor

$$V_L = L\frac{\Delta I}{\Delta t} \tag{21.23}$$

Because V_L is defined as the voltage *drop,* there is no negative sign in Equation 21.23. This result for V_L will be needed when we construct a circuit theory that deals with inductors.

The unit of inductance is the ***henry*** (H). Equation 21.23 relates the unit of inductance to the units of voltage, current, and time. Rearranging Equation 21.23 as $L = V_L(\Delta t/\Delta I)$, the units are related by

$$1\text{ H} = 1\text{ V}\cdot\text{s/A} \tag{21.24}$$

EXAMPLE 21.6 Inductance of a Typical Coil

In a practical electric circuit, inductors are often coils of wire in the form of a solenoid. Consider a solenoid of length $\ell = 2.0$ cm consisting of 1000 turns of wire. Each turn is a circular loop with radius $r = 0.50$ cm. Calculate the value of L for this inductor.

RECOGNIZE THE PRINCIPLE

The inductance of a solenoid depends on its length, its cross-sectional area, and the number of turns. We can find the value of the inductance L by applying Equation 21.21. The point of this calculation is to get a feeling for the value of L in typical cases, which will be useful when we analyze circuits with inductors in Section 21.5.

SKETCH THE PROBLEM

Figure 21.18 describes the problem.

IDENTIFY THE RELATIONSHIPS AND SOLVE

Inserting the given values of the length and number of turns, with a loop area $A = \pi r^2$, we find

$$L = \frac{\mu_0 N^2 A}{\ell} = \frac{(4\pi \times 10^{-7}\text{ T}\cdot\text{m/A})(1000)^2\pi(5.0\times 10^{-3}\text{ m})^2}{0.020\text{ m}}$$

$$L = 4.9 \times 10^{-3}\text{ H} = \boxed{4.9\text{ mH}}$$

▶ *What does it mean?*

This value of L is typical for common inductors. An inductance with $L = 1$ H is very large, whereas a value of 1 microhenry (10^{-6} H) is rather small. An inductance of 1 mH (millihenry) is typical.

Insight 21.1

THE CONCEPT OF *SELF*-INDUCTANCE

Self-inductance poses an interesting conceptual question: How can a current "act on" or affect itself? After all, the electric field of an electron does not act on the electron that produced the field in the first place. With an inductor, changes in current are acting "back" on the current itself through the induced emf.

An electric current is produced by the motion of many charges, and the induced emf is actually due to the force exerted by one charge (from the current in one part of the inductor) on another charge (the current in a different part of the inductor). When viewed in this way, self-inductance is just the effect of one electric charge on another.

Mutual Inductance

In an inductor (i.e., a "coil"), the magnetic field produced by the coil produces an induced current in that coil. It is also possible for the magnetic field produced by one coil to produce an induced current in a second coil. Figure 21.19 shows two solenoidal coils, one wrapped around the other. The inner coil (1, shown in black) is connected to a battery and switch. When the switch is closed and the current I_1 increases, this coil produces a magnetic field that gives an increasing flux in the outer coil (2, shown in gray). According to Faraday's law, the changing flux in coil 2 produces an induced emf in that coil. Hence, the current I_1 in coil 1 produces an emf in coil 2, even though the two coils are not "directly" connected by any wires. They are, however, "indirectly" connected through the magnetic flux. This effect is called ***mutual inductance*** because it involves the "mutual" behavior of two separate coils. Mutual inductance is used in devices called transformers, which we'll discuss in Chapter 22.

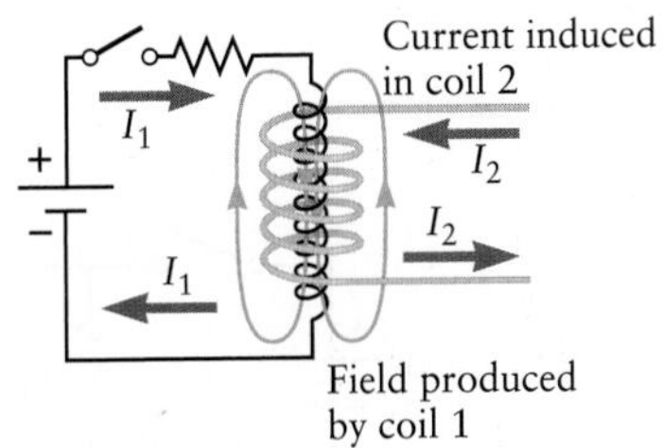

▲ **Figure 21.19** Two coils wound around each other exhibit a mutual inductance. Changes in the current in one coil (coil 1) induce an emf in the other (coil 2).

21.5 *RL* Circuits

We can apply Kirchhoff's circuit rules from Chapter 19 to circuits that contain inductors by including the voltage drop across an inductor given in Equation 21.23. The basic behavior of such circuits can be described using the following rules.

General behavior of *RL* circuits

Qualitative behavior of DC circuits with inductors

DC circuits may contain resistors, inductors, and capacitors. The voltage source in a DC circuit is a battery or some other source that provides a constant voltage across its output terminals.

1. **Immediately after any switch is closed or opened, the induced emfs keep the current through all inductors equal to the values they had the instant before the switch was thrown.**
2. **After a switch has been closed or opened for a very long time, the induced emfs are zero.**

The first rule above can be understood from the process shown in Figure 21.17. *Whenever there is an attempt to change* the current through an inductor (i.e., by closing or opening a switch), the induced emf across the inductor always opposes this change. Momentarily, this opposition prevents the current through the inductor from changing, and the current just after a switch is thrown is equal to the current just before. This result applies only for the *instant* after a switch is thrown, and is simply a statement of Lenz's law as applied to inductors.

After some time, the current through the inductor will change, eventually approaching a constant value. If I is constant, $\Delta I/\Delta t$ in Equation 21.23 is zero and all currents and magnetic fields are then constant. Hence, the voltage across the inductor is zero, leaving us with DC circuit rule 2.

These two rules apply only to DC circuits; in these circuits, the only time-dependent changes are initiated by the closing or opening of switches. Another important class of circuits, called AC circuits, contain power sources whose electric potentials vary sinusoidally with time. We'll discuss those circuits in Chapter 22, when we generalize our circuit rules to apply to both AC and DC circuits.

Let's use our two circuit analysis rules to study the behavior of the DC circuit sketched in Figure 21.20 on page 644, which contains a battery, a switch, two resistors, and an inductor. The presence of the resistors and inductor make it an "*RL* circuit." The general circuit symbol for an inductor is a solenoid-like coil.

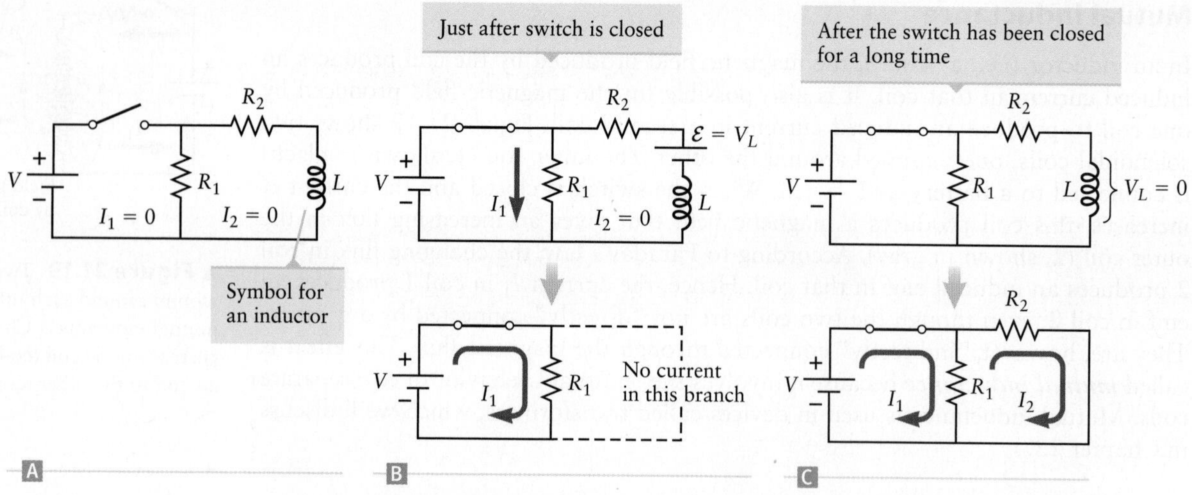

▲ **Figure 21.20** A An RL circuit. B Just after the switch is closed, the emf induced in the inductor opposes any increase in current through it. At this instant, the current I_2 is zero. The equivalent circuit *just after* the switch is closed is shown below. C After the switch has been closed for a long time, the magnetic flux through the inductor is constant (not changing with time), so the induced emf is zero. The inductor then acts as a wire (a short circuit).

In Figure 21.20A, the switch has been open for a very long time, so the current is zero through both branches of the circuit. At $t = 0$, the switch is closed, inducing a potential across the inductor that is equivalent to a battery with emf $\mathcal{E} = V = V_L$ (Fig. 21.20B). To find the values of I_1 (the current through resistor 1) and I_2 (the current through resistor 2) just after the switch is closed, we apply rule 1 from the boldfaced list above. At the instant after $t = 0$, the induced emf keeps the current through the inductor at the value it had before the switch was closed. Hence, $I_2 = 0$ just after $t = 0$; at that moment, the entire circuit is equivalent to the lower circuit in Figure 21.20B, and the circuit behaves as if L and R_2 have been removed. The result is a circuit with a battery connected to a single resistor carrying current $I_1 = V/R_1$.

Now let's find the currents after the switch in Figure 21.20 has been closed for a very long time. We apply rule 2 highlighted above and find that the voltage across the inductor is now zero, which is equivalent to replacing the inductor by a simple wire that has zero resistance (lower circuit in Fig. 21.20C). The voltage across each resistor is now the battery emf V, so the currents in the two circuit branches are

$$I_1 = \frac{V}{R_1} \quad \text{and} \quad I_2 = \frac{V}{R_2} \tag{21.25}$$

Insight 21.2

BUILDING A REAL INDUCTOR

We calculated L for a simple solenoid inductor in Equation 21.21, but most practical inductors are constructed by wrapping a wire coil around a magnetic material. From Chapter 20, we know that a small magnetic field can align the domains of a magnetic material, producing a much larger magnetic field. Filling a coil with magnetic material greatly increases the magnetic flux through the coil and therefore increases the induced emf. According to the definition of inductance in Equation 21.22, the presence of magnetic material thus increases the inductance. For this reason, most inductors contain a magnetic material inside, which produces a larger value of L in a smaller package.

EXAMPLE 21.7 Current in an RL Circuit

Consider again the circuit in Figure 21.20 and assume the switch has been closed for a very long time, so the currents through R_1 and R_2 are as given in Equation 21.25 with the directions shown in Figure 21.20C. Find the directions of the currents through R_1 and R_2 the instant after the switch is reopened.

RECOGNIZE THE PRINCIPLE

Whenever a switch is opened or closed, the induced emf across an inductor always acts to keep the current the same as it was just before the switch was thrown. This principle is circuit rule 1 of "Qualitative behavior of DC circuits with inductors," page 643.

SKETCH THE PROBLEM

The current I_2 through the inductor remains at the value it had just prior to the change of the switch (Eq. 21.25), and it will also keep the same direction. Hence, at the instant the switch is opened the current through the branch containing R_2 and the inductor is directed clockwise as sketched in Figure 21.21B.

IDENTIFY THE RELATIONSHIPS AND SOLVE

Applying the equivalent circuit in Figure 21.21B, this current must flow "upward" through R_1, opposite to the direction found when the switch was closed in Figure 21.20B.

► ***What does it mean?***

In this example, the initial current through L has a nonzero value; the induced emf maintains this nonzero current momentarily, even after the switch is opened. At that instant, the battery is disconnected leaving the circuit shown in Figure 21.21B with a nonzero current even though there is no battery! Where does the energy supplying this current come from? We'll explain that in Section 21.6 when we consider how an inductor can store energy.

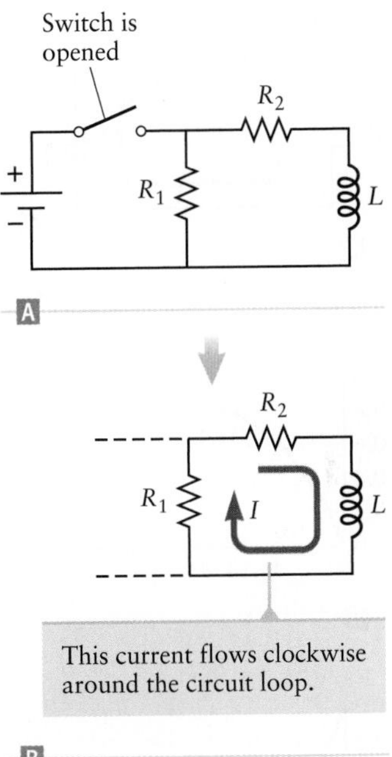

▲ **Figure 21.21** Example 21.7.

CONCEPT CHECK 21.5 Analyzing an *RL* Circuit (1)

Consider the circuit in Figure 21.22. The current is initially zero in both branches. The switch is then closed. Is the current through R_1 just after closing the switch (a) $I = V/R_1$, (b) $I = V/R_2$, (c) $I = V/(R_1 + R_2)$, or (d) $I = 0$?

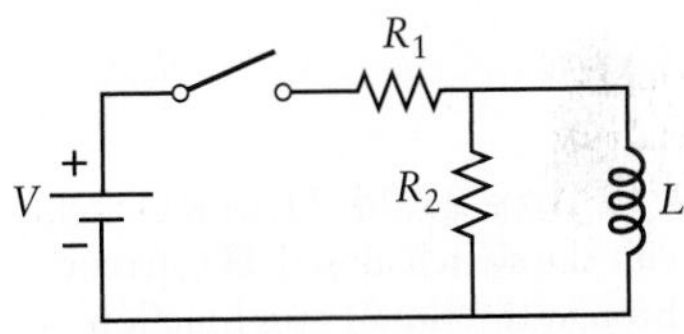

▲ **Figure 21.22** Concept Check 21.5.

CONCEPT CHECK 21.6 Analyzing an *RL* Circuit (2)

Consider the circuit in Figure 21.23. The switch is in position 1 for a very long time and then changed to position 2. What is the direction of the current through the inductor just after the switch is changed?

(a) The current is to the right.
(b) The current is to the left.
(c) The current is zero just after the switch is thrown.
(d) The direction depends on the values of R_1, R_2, and R_3.

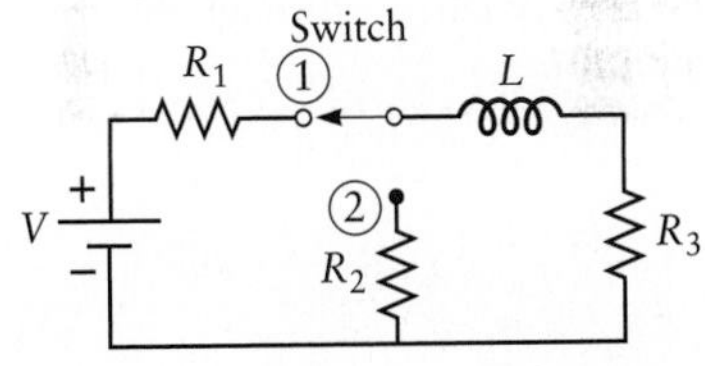

▲ **Figure 21.23** Concept Check 21.6.

Quantitative Behavior of an *RL* Circuit

Our qualitative rules for dealing with inductors in DC circuits tell how to find the current the moment just after a switch is closed or opened and how to find the current after a switch has been closed or open for a very long time. The current between these two times can also be calculated using Kirchhoff's rules for circuit analysis. Figure 21.24A shows the simplest *RL* circuit, with a battery connected through a switch to a resistor R and an inductor L. The switch is open for a long time and then is closed at $t = 0$. The quantitative behavior of the current through and voltage across the inductor is shown in parts B and C of Figure 21.24 on page 646. The current starts at $I = 0$ because the induced emf keeps the current through the inductor momentarily at the value it had before the switch was thrown (in accord with our circuit rule 1). After a very long time, the voltage across the inductor falls to zero (in accord with rule 2) and the current approaches $I = V/R$. Mathematically, the current is given by

$$I = \frac{V}{R}(1 - e^{-t/\tau}) \tag{21.26}$$

where e^x is the exponential function and τ is called the ***time constant*** of the circuit. For an *RL* circuit in which a single resistor is in series with a single inductor as in Figure 21.24, the time constant is

$$\tau = \frac{L}{R} \tag{21.27}$$

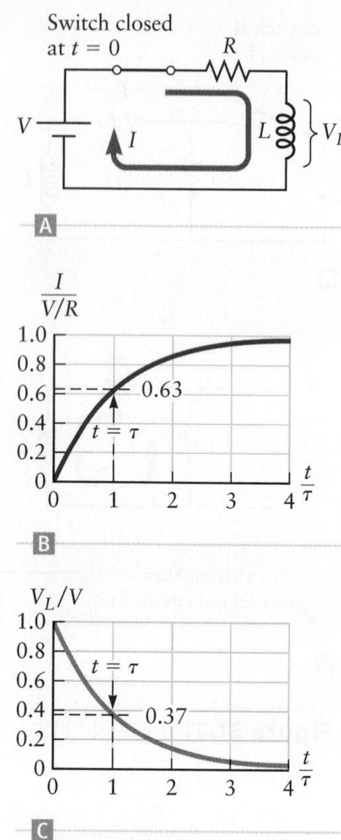

▲ **Figure 21.24** A An RL circuit with the switch closed. B Current through the circuit as a function of time. The switch was closed at $t = 0$. C Voltage across the inductor as a function of time. When this voltage is positive, it *opposes* the flow of current through the inductor.

The corresponding behavior of the voltage across the inductor is

$$V_L = Ve^{-t/\tau} \tag{21.28}$$

The time constant τ is the approximate time scale over which the current and voltage change in an RL circuit. From the properties of the exponential function (see Appendix B), the current in Equation 21.26 reaches approximately 63% of its final value when $t = \tau$ and the voltage across the inductor (Eq. 21.28) falls to approximately 37% of its initial value after this time. After three time constants ($t = 3\tau$), the current reaches 95% of its final value. The time constant τ in Equation 21.27 applies only to an RL circuit. It is *different* from the time constant we encountered for RC circuits in Chapter 19.

EXAMPLE 21.8 Value of τ for a Typical RL Circuit

Consider an RL circuit with $R = 4000\ \Omega$ and $L = 3.0$ mH; these values are typical of those used in many electronic circuits such as radios and computers. Find the time constant for a circuit that uses these components.

RECOGNIZE THE PRINCIPLE

The time constant for an RL circuit is just the ratio L/R. The purpose of this example is to determine a "typical" value of the time constant τ. Is it hours, seconds, or microseconds? The value of τ is important in applications (see below).

SKETCH THE PROBLEM

No figure is necessary.

IDENTIFY THE RELATIONSHIPS AND SOLVE

We evaluate the time constant τ from Equation 21.27 using the given values of R and L:

$$\tau = \frac{L}{R} = \frac{3.0 \times 10^{-3}\ \text{H}}{4000\ \Omega} = 7.5 \times 10^{-7}\ \text{s} = \boxed{0.75\ \mu\text{s}}$$

▸ *What does it mean?*

For this circuit, the time constant is approximately 1 microsecond (1 μs). In many applications of RL circuits (Chapter 22), this time constant is associated with the period of an oscillation, so the period of such an oscillating circuit would correspond to a frequency of about

$$f = \frac{1}{\tau} \approx \frac{1}{1 \times 10^{-6}\ \text{s}} = 1 \times 10^6\ \text{Hz}$$

(1 megahertz). This frequency lies in the middle of the "AM band" on your radio, so such an RL circuit could be useful in an AM radio.

21.6 Energy Stored in a Magnetic Field

The RL circuit in Figure 21.25A is the same circuit we analyzed in Figures 21.20 and 21.21. We know that if the switch has been closed for a long time, currents I_1 and I_2 are established through the resistors as given in Equation 21.25. If the switch is then opened at $t = 0$, there is a current in the circuit loop formed by R_1, R_2, and L (Fig. 21.25B); the time dependence of this current is shown qualitatively in Figure 21.25C. After the switch is opened, the battery is completely disconnected from this circuit. Current thus exists even with the battery removed, and electrical energy is dissipated as heat in the resistors during this time. Where does this energy come from

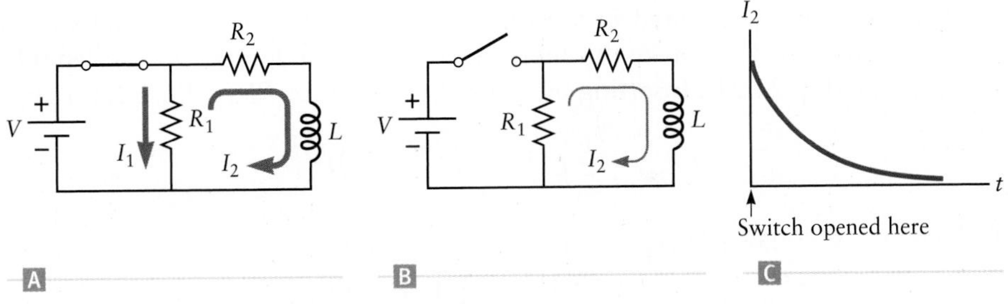

◀ **Figure 21.25** A If the switch in this RL circuit is closed for a long time, there will be current in the inductor. B When the switch is opened, the current through the inductor decays with time as plotted in C.

if the battery has been disconnected? The answer is that this energy was originally (at $t = 0$) stored in the magnetic field of the inductor.

One way to analyze the energy stored in this magnetic field is to consider the energy required to establish the current in an inductor in the first place. Suppose the voltage across the inductor in Figure 21.26A has a constant value V. Using Equation 21.23, the voltage drop across the inductor is $V = L\,\Delta I/\Delta t$. Solving for the change in current, we have

$$\Delta I = \frac{V}{L}\Delta t \tag{21.29}$$

Hence, if V is constant and the current starts from zero at $t = 0$, Equation 21.29 means that the current through the inductor I must increase linearly with t (Fig. 21.26B), and

$$I = \frac{V}{L}t \tag{21.30}$$

The power P dissipated in a circuit element is equal to the product of the voltage across that element and the current (Eq. 19.15). Hence, the instantaneous power in the inductor—the rate at which energy is delivered to the inductor at time t—is

$$P = VI = V\left(\frac{V}{L}t\right) = \left(\frac{V^2}{L}\right)t \tag{21.31}$$

This linear relation between P and t is plotted in Figure 21.26C. We can use the result of Equation 21.31 to find the total energy that has been delivered to the inductor when the current reaches a particular value I_0 at time t_0. We denote this energy as PE_{ind} since we will soon see that it is a kind of potential energy. This total potential energy PE_{ind} is equal to the average of the power in Figure 21.26C multiplied by the total time t_0. The instantaneous power P varies linearly with time, so the average power is half the maximum power. Using Equation 21.31, we get

$$P_{\text{ave}} = \frac{1}{2}\frac{V^2 t_0}{L}$$

$$PE_{\text{ind}} = P_{\text{ave}}t_0 = \frac{V^2 t_0^2}{2L} \tag{21.32}$$

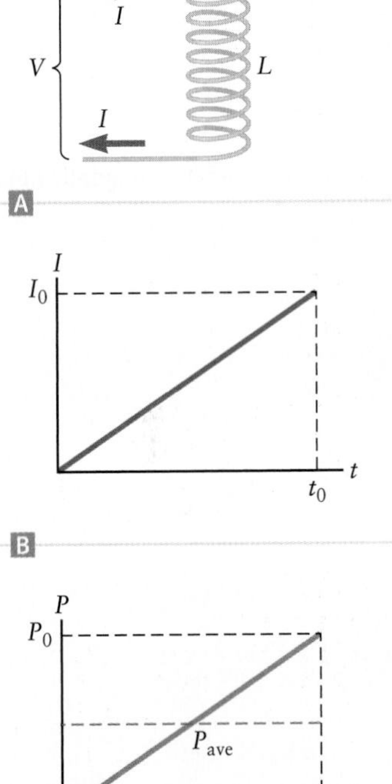

▲ **Figure 21.26** A Voltage across and current through an inductor. B If the voltage across the inductor is constant, the current increases linearly with time (Eq. 21.30). C The power delivered to the inductor also increases linearly with time.

From Equation 21.30, we find $I_0 = Vt_0/L$, which we rearrange to give $t_0 = I_0L/V$. Inserting into our expression for PE_{ind} leads to

$$PE_{\text{ind}} = \frac{V^2}{2L}\left(\frac{I_0L}{V}\right)^2 = \frac{1}{2}LI_0^2$$

where I_0 is the final current through the inductor; we can just call it I and then write our final result as

Energy stored in an inductor

$$PE_{\text{ind}} = \tfrac{1}{2}LI^2 \tag{21.33}$$

Equation 21.33 is the general result for the magnetic energy PE_{ind} stored in an inductor. It has a form that is very similar to the electric energy stored in a capacitor, $PE_{\text{cap}} = \frac{1}{2}CV^2$ (Eq. 18.33). In the case of a capacitor, we saw that this energy is stored in the electric field that exists between the capacitor plates. In a similar way, the energy PE_{ind} is stored *in the magnetic field* of the inductor, and we can rewrite Equation 21.33 to make this more apparent. The inductance of a solenoid is $L = \mu_0 N^2 A/\ell$ (Eq. 21.21). Inserting into our expression for PE_{ind} (Eq. 21.33) gives

$$PE_{\text{ind}} = \frac{1}{2}LI^2 = \frac{1}{2}\left(\frac{\mu_0 N^2 A}{\ell}\right)I^2$$

which can be rearranged as

$$PE_{\text{ind}} = \frac{1}{2}\left(\frac{\mu_0 N^2 A}{\ell}\right)I^2 = \frac{1}{2\mu_0}\left(\frac{\mu_0 NI}{\ell}\right)^2(A\ell)$$

We have rewritten the potential energy in this way because the factor involving I in parentheses is the solenoid's magnetic field ($B = \mu_0 NI/\ell$, Eq. 21.19), whereas the last factor $A\ell$ is the volume of the solenoid (the length times the cross-sectional area). So we get our final result,

Energy stored in a magnetic field

$$PE_{\text{mag}} = \frac{1}{2\mu_0}B^2 \times \text{volume} \tag{21.34}$$

In Equation 21.34, we have now denoted the energy as PE_{mag}. It is the energy contained in the magnetic field of the inductor and applies not only to an inductor but to *any* region of space that contains a magnetic field. This energy is stored *in the magnetic field.* Because a magnetic field can exist in a vacuum, this potential energy can exist even in "empty" regions containing no matter! The analogous result also applies to electric fields and electric potential energy and will be important in understanding electromagnetic waves (Chapter 23).

Another way to express the potential energy in Equation 21.34 is to say that there is a certain *energy density* (energy per unit volume) in the magnetic field. From Equation 21.34, we find

$$\text{energy density} = u_{\text{mag}} = \frac{PE_{\text{mag}}}{\text{volume}} = \frac{1}{2\mu_0}B^2 \tag{21.35}$$

This expression for the magnetic energy density is very similar in form to the energy density contained in an electric field. In Chapter 18, we found that $u_{\text{elec}} = \frac{1}{2}\varepsilon_0 E^2$ (Equation 18.48), so there is an interesting parallel between our results for magnetic and electric energies. In both cases, the energy density is proportional to the square of the field.

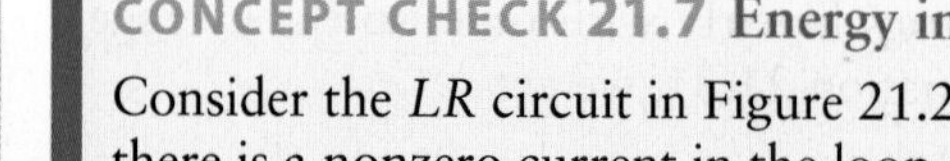

CONCEPT CHECK 21.7 Energy in an *LR* Circuit

Consider the *LR* circuit in Figure 21.25. After the switch is closed for a long time, there is a nonzero current in the loop containing the inductor and energy is stored in the inductor's magnetic field. After the switch is opened, the current eventually decays to zero and there is no magnetic field in the inductor. Where does the energy that was originally stored in the inductor go?

(a) It goes back to the battery.
(b) It is dissipated as heat in the resistors.
(c) It causes a spark in the switch.

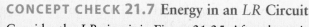

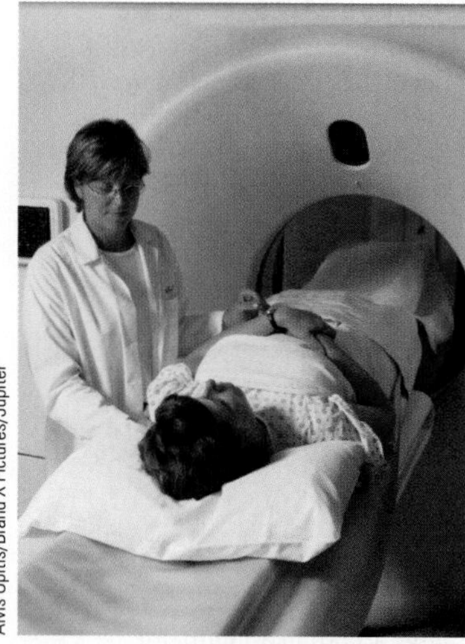

Alvis Upitis/Brand X Pictures/Jupiter

▲ **Figure 21.27** Example 21.9. The magnets used for magnetic resonance imaging (MRI) are very large, with a large amount of stored magnetic energy.

EXAMPLE 21.9 Magnetic Field Energy: How Large Is It?

To get a feeling for the magnitude of the energy stored in a magnetic field, consider the energy stored in a magnet used for magnetic resonance imaging (MRI) (Fig. 21.27). A typical MRI magnet is a solenoid providing a field of about $B = 1.5$ T. **(a)** What is the approximate energy stored in this magnetic field? **(b)** If the current through the magnet during normal operation is $I = 100$ A, what is its inductance?

RECOGNIZE THE PRINCIPLE

The motivation for this example is to understand how the potential energy stored in a magnetic field compares with cases familiar from mechanics such as the gravitational potential energy or the kinetic energy of a moving object. To find the energy stored in an MRI magnet, we need to know its magnetic field and its inside volume. The volume is not given, but we can estimate its value because a person must be able to fit inside.

SKETCH THE PROBLEM

Figure 21.27 shows a typical MRI magnet. We can estimate the volume from this photo.

IDENTIFY THE RELATIONSHIPS

We can find the stored magnetic energy using our result for the potential energy PE_{mag} (Eq. 21.34).

SOLVE

(a) From personal experience (and the photo in Fig. 21.27), we estimate that a typical MRI magnet is about 1.5 m long with an inside diameter of about 0.5 m. Its volume $= A\ell = (\pi r^2)\ell$, and inserting our estimates for r and ℓ gives

$$\text{magnet volume} = (\pi r^2)\ell = \pi(0.25\text{ m})^2(1.5\text{ m}) = 0.3\text{ m}^3$$

Using this volume in Equation 21.34 along with the given value of B, we find

$$PE_{mag} = \frac{1}{2\mu_0}B^2 \times \text{volume} = \frac{1}{2(4\pi \times 10^{-7}\text{ T}\cdot\text{m/A})}(1.5\text{ T})^2(0.3\text{ m}^3)$$

$$PE_{mag} = \boxed{3 \times 10^5\text{ J}}$$

(b) The magnetic energy found in part (a) is stored in an inductor because the MRI magnet is just a solenoid. Therefore, $PE_{mag} = PE_{ind}$ is related to the inductance through Equation 21.33. Using our calculated value of PE_{mag} and the given operating current, we have

$$\tfrac{1}{2}LI^2 = PE_{mag}$$

$$L = \frac{2PE_{mag}}{I^2} = \frac{2(3 \times 10^5\text{ J})}{(100\text{ A})^2} = \boxed{60\text{ H}}$$

This inductance is large, much larger than we found for the small coil in Example 21.6.

► ***What does it mean?***

The amount of energy found in part (a) is quite large; it is approximately equal to the kinetic energy of a car of mass 1000 kg with a velocity of 50 mi/h (23 m/s)! An MRI magnet is carefully designed to be sure this energy is dealt with safely. In addition, patients undergoing an MRI exam must take precautions such as removing all magnetic objects from their pockets.

A

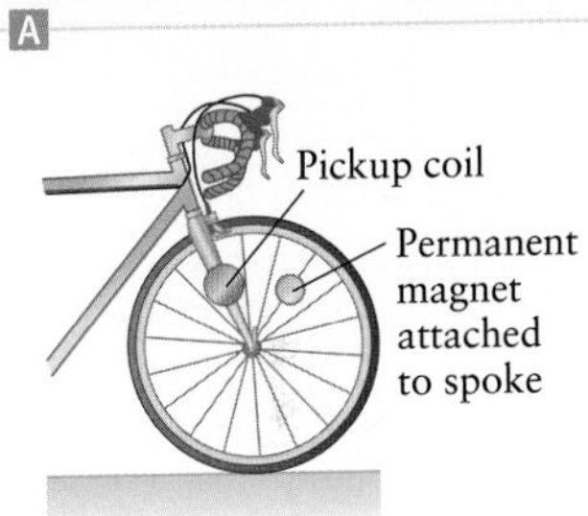

B

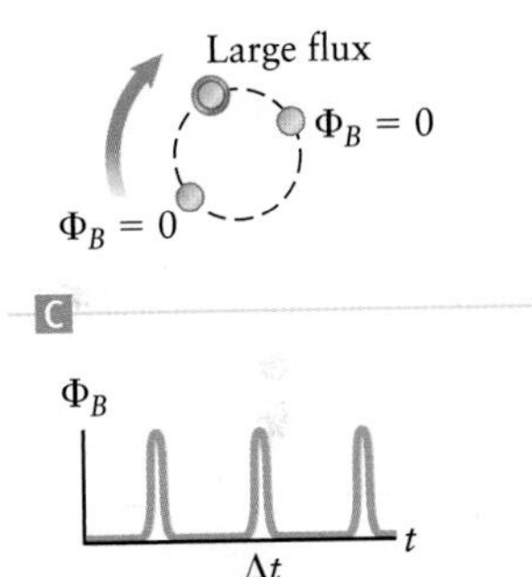

C

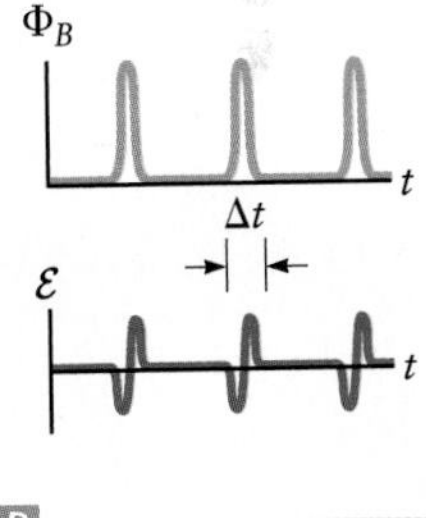

D

▲ **Figure 21.28** A This bicycle odometer uses Faraday's law to count the rotations of a bicycle wheel. B A small, permanent magnet is attached to a spoke, and a pickup coil is mounted on the bicycle frame. C When the magnet passes near the pickup coil, the time-varying magnetic flux induces an emf in the coil. D Qualitative sketch of the flux through the pickup coil as a function of time. This time-dependent flux produces an induced emf that varies with time as a series of pulses. The odometer counts these pulses.

21.7 Applications

In this section, we consider a few applications of Faraday's law.

Bicycle Odometers

Most serious bicyclists use an odometer to monitor their speed and distance traveled. Figure 21.28A shows a photo of the odometer control unit, and a sketch of the working elements is shown in Figure 21.28B. A small, permanent magnet is attached to one of the spokes of a wheel, and a "pickup" coil is mounted on the axle support. On the author's bike, this pickup unit is contained in a piece of opaque plastic with a small coil of wire inside. As the wheel turns, the permanent magnet on the spoke travels in a circular path that takes it past the pickup coil once during each revolution of the wheel. When this magnet is close to the pickup coil (Fig. 21.28C), its magnetic field penetrates the coil, giving a large flux, whereas the flux through the coil is zero when the magnet is far from the coil. A qualitative sketch of the flux through the coil as a function of time is shown in the upper graph of Figure 21.28D. Because a pulse occurs each time the magnet passes over the pickup coil, there is one pulse in this plot for each revolution of the wheel.

From Faraday's law, we know that the time-varying magnetic flux through the pickup coil induces an emf in the coil. A corresponding plot of this induced emf is shown in the lower graph of Figure 21.28D. Because $\mathcal{E} = -\Delta\Phi_B/\Delta t$, the induced emf is negative while the flux is increasing (Lenz's law) and positive as Φ_B falls back to zero. The pickup coil is connected to a computer that keeps track of the number of pulses in the emf signal and also measures the time between pulses. This information is then displayed as the bicycle's speed and distance traveled.

EXAMPLE 21.10 Ⓡ Design of a Bicycle Odometer

Let's calculate the emf (voltage) induced in the odometer pickup coil in Figure 21.28 to show that this odometer design is actually practical. For a typical permanent magnet, the field very near one of the poles is about 1 T. In practice, the permanent magnet attached to the spoke must not be too close to the frame of the bicycle (so they don't hit each other), which reduces the field at the pickup coil. To be conservative, we assume a field $B = 0.01$ T at the coil. The odometer pickup on the author's bicycle has an area of about $A = 0.1$ cm^2 (about 3 mm on a side). We also assume the coil of wire inside consists of $N = 100$ loops.

RECOGNIZE THE PRINCIPLE

According to Faraday's law, the voltage induced in the pickup coil is equal to the rate of change of the magnetic flux as the odometer magnet moves past. To find $\Delta\Phi_B/\Delta t$, we need to know B and the area of the coil, which are both given. We also need to know Δt, which is the approximate time the magnet "overlaps" with the coil. We can estimate Δt from the angular speed of the bicycle wheel.

SKETCH THE PROBLEM

Following our "Applying Lenz's Law" problem-solving strategy, Figure 21.28C shows the pickup coil. We will use this area when calculating the induced emf.

IDENTIFY THE RELATIONSHIPS

To find the change in flux, we need to calculate Φ_B when the odometer magnet is directly over the pickup coil and when it is far away. When the magnet far away, the flux is zero; when it is over the coil, the magnetic flux through one loop of the coil is BA. So, for N loops, the total flux is $\Phi_B = NBA$. Inserting our approximate values for N, B, and A, we get

$$\Phi_B = NBA = (100)(0.01\text{ T})(1 \times 10^{-5}\text{ m}^2) = 1 \times 10^{-5}\text{ T}\cdot\text{m}^2 \tag{1}$$

which is the flux at the peaks in Figure 21.28D and also the change in flux $\Delta\Phi_B$ when the magnet passes by the pickup coil.

To apply Faraday's law, we must now estimate the time Δt over which this flux change takes place. This time interval is indicated in Figure 21.28D; it depends on the speed of wheel, and it decreases as the bicycle travels faster. A slowly moving bicycle might travel at 5 mph, which is about 2 m/s. The wheels of the author's bike have a radius of about $R = 50$ cm, so the angular velocity at this speed is (from Eq. 8.43)

$$\omega = \frac{v}{R} = \frac{2 \text{ m/s}}{0.5 \text{ m}} = 4 \text{ rad/s}$$

The magnet on the spoke is approximately halfway from the edge of the wheel at $R_{\text{magnet}} = 0.25$ m, so the magnet's speed along its circular path around the axle is

$$v_{\text{magnet}} = \omega R_{\text{magnet}} = (4 \text{ rad/s})(0.25 \text{ m}) = 1 \text{ m/s}$$

The diameter of the permanent magnet is about $d = 1$ cm (a value taken from the author's bike), so the time it takes for the magnet to move past the sensor is approximately

$$\Delta t = \frac{d}{v_{\text{magnet}}} = \frac{0.01 \text{ m}}{1 \text{ m/s}} = 0.01 \text{ s}$$

SOLVE

Inserting these results for $\Delta\Phi_B$ from Equation (1) and Δt into Faraday's law gives the approximate magnitude of the induced emf:

$$|\mathcal{E}| = \frac{\Delta\Phi_B}{\Delta t} = \frac{1 \times 10^{-5} \text{ T} \cdot \text{m}^2}{0.01 \text{ s}} = \boxed{0.001 \text{ V}}$$

► ***What does it mean?***

A voltage of this size can be easily measured (as your electrical engineering friends can verify), so this odometer design is indeed practical.

CONCEPT CHECK 21.8 **Changing Speeds with the Bicycle Odometer**

If the speed of the bicycle in Example 21.10 is increased, how does the amplitude of the induced emf in the pickup coil change?

(a) The induced emf gets smaller.
(b) The induced emf gets larger.
(c) The induced emf does not change.

Ground Fault Interrupters

A ground fault interrupter (GFI) is a safety device used in many household circuits (Fig. 21.29A). A GFI is often built into an electric power outlet in bathrooms, kitchens, and other places where there is an increased risk of accidental contact with a substantial electric potential. A GFI uses Faraday's law together with an electromechanical relay (Fig. 20.47). Suppose a hair dryer is plugged into a wall socket that contains a GFI as in Figure 21.29B. This connection contains two current-carrying wires, with current directed from the wall socket to the hair dryer through one wire and back to the wall socket through the other wire. It is just like the circuits we have drawn using batteries; all have two wires connecting the battery to the rest of the circuit. During normal (safe) operation of the hair dryer, the two connecting wires carry equal currents, but in the event of an accident, some current might pass through your body and then to ground. In that case, the current in the return wire in Figure 21.29B is much smaller than the current in the outgoing wire. A GFI senses this difference and immediately turns off power to the entire circuit so that no permanent harm is done.

A GFI accomplishes its task by using a relay to control the current through the circuit. In Chapter 20, we saw that relays use the current through a coil to exert a force on a magnetic metal bar. This bar is part of a switch, so the current through the coil controls whether the switch is open or closed. The relay in a GFI contains

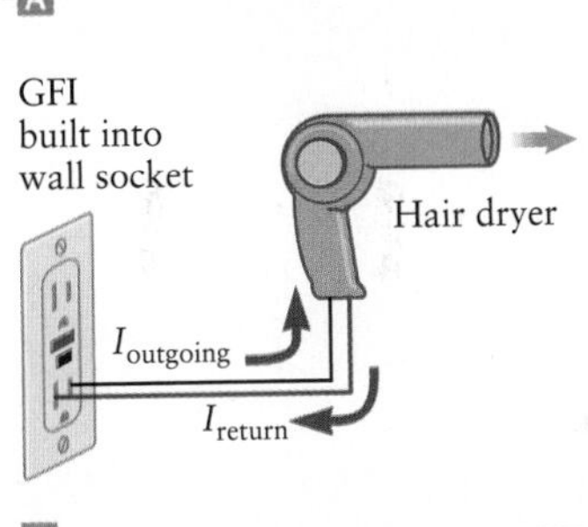

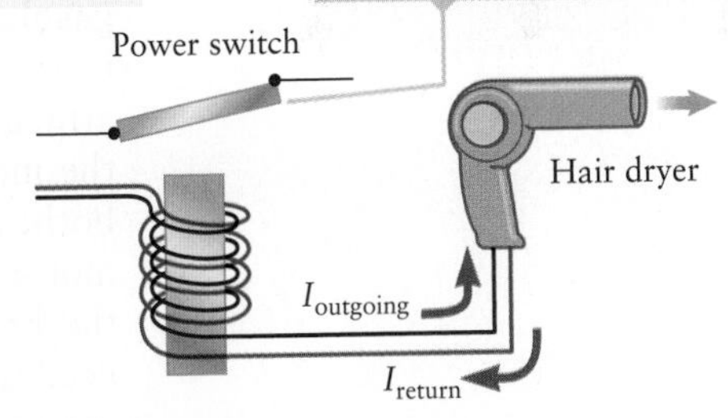

▲ **Figure 21.29** A Ground fault interrupters (GFIs) use a specially designed inductor to detect circuit failure. B Most GFI units are built into a wall socket, so you plug an appliance, such as a hair dryer, directly into the GFI. C Two coils of wire, one carrying the outgoing current and another carrying the return current, are used to power a relay in the GFI. If an appliance fails, the current through the return coil drops and the relay is activated, turning off current to the appliance.

two separate coils; the current through one coil is from the outgoing wire to the hair dryer, and the current through the other coil is from the return circuit. The two coils are wound in opposition so that their magnetic fields are in opposite directions. During normal (safe) operation, the currents in the two coils are equal and their fields cancel, giving zero magnetic field in the relay. The relay is designed so that its switch is closed in this case, allowing current in the hair dryer circuit. If there is an accident, however, the current in the return coil is smaller than the outgoing current, leading to a nonzero magnetic field from the coil. The relay switch then opens (Fig. 21.29C), turning off the current to the hair dryer.

Electric Guitars

An electric guitar (Fig. 21.30A) uses Faraday's law to sense the motion of the strings. A guitar string is made from steel, a magnetic material. The string passes near a "pickup" coil wound around a permanent magnet. The field from this magnet causes nearby regions of the string to be magnetized as shown in Figure 21.30B. This induced magnetism in the steel of the guitar string is just like the induced magnetism in the steel panel of a refrigerator or of any other magnetic material containing magnetic domains (see Fig. 20.41). The magnetized region of the string then produces its own magnetic field, which gives a magnetic flux through the pickup coil. As the string vibrates, this flux changes with time, leading to an induced emf in the coil. This emf is sent to an amplifier, and the resulting signal can then be played through speakers.

Generators, Motors, and Hybrid Cars

In Chapter 20, we showed how a magnetic field can exert a torque on a current loop that can be used to make a motor and thus produce mechanical energy (Fig. 20.48). The "opposite" process takes place in a generator, which uses an externally applied mechanical torque on a current loop to produce an electric current and thus produce electrical energy. A motor and generator therefore provide examples of the conservation of energy and the conversion of energy from one type to another, mechanical to electrical and vice versa. This conversion back and forth is used in some hybrid cars to minimize the amount of energy "lost" to friction and hence improve fuel economy.

A hybrid car contains two motors (Fig. 21.31); one motor is a conventional gasoline-powered motor (also called an "engine"; see Chapter 16), and the other is an electric motor powered by batteries. A hybrid car also contains a generator attached to the wheels, usually the front wheels, which are the ones connected to the motors. Starting from a standstill, either the gas motor or electric motor (or both) accelerates the car. At low speeds, the electric motor is the more efficient motor, so the gas motor usually shuts off automatically for those periods. When the brakes are applied, the generator is switched on and the emf it produces is used to recharge the batteries. In an ordinary car, applying the brakes causes the car's kinetic energy to be converted to heat energy through friction. A hybrid car "recaptures" this energy, storing it in the batteries so that it can be used later. A

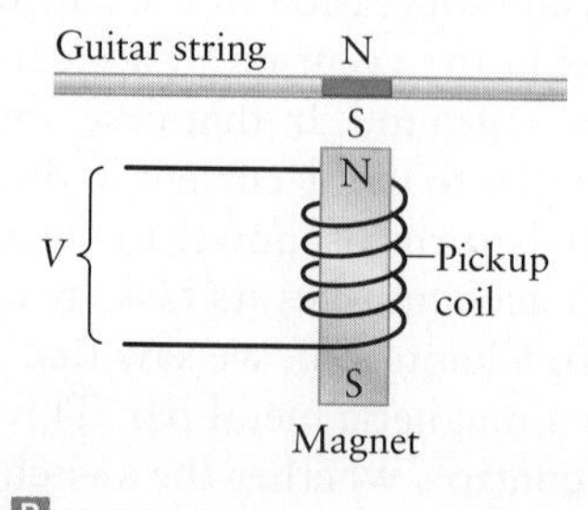

► **Figure 21.30** A The strings of an electric guitar are made of steel, a magnetic material. B When the string vibrates, it produces a time-varying magnetic flux through the pickup coil (called a guitar "pickup"), producing a voltage in the coil that is converted into sound by the guitar's amplifier. In part A, the pickups are the small circular regions, with three pickups under each string.

Courtesy of Nick Giordano

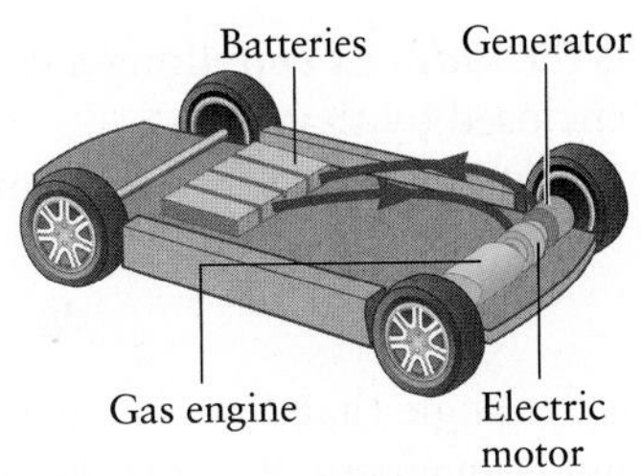
Batteries
Generator
Gas engine
Electric motor

◀ **Figure 21.31** A hybrid car such as this Toyota Prius contains a gasoline-powered motor along with an electric motor, a generator, and batteries.

hybrid car is thus a good practical example of the conversion between mechanical energy and electrical energy.

21.8 The Puzzle of Induction from a Distance

A solenoid carrying a current produces a large magnetic field in its interior and a very small field outside. By making the solenoid very long, the field outside can be made extremely small. Suppose a very long solenoid is inserted at the center of a single loop of wire as sketched in Figure 21.32. The field from this very long solenoid *at the outer loop* is essentially zero. Even so, the field inside the solenoid at the center of the loop still produces a magnetic flux through the loop's inner portion.

The solenoid field is proportional to the current I (Eq. 21.19). Let's assume I is changing so this field is changing with time. The flux through the outer loop then changes with time, inducing an emf. This emf can be used to power an electric circuit (we might attach the loop to a lightbulb), so there is electrical energy associated with this induced emf. You should suspect that this energy was somehow transferred from the solenoid to the outer loop, but how is this energy transferred across the empty space between the two conductors? What's more, there is no magnetic field from the solenoid at the outer loop, so how does the solenoid have any influence at all on that loop?

The answer to this puzzle is that energy is carried from the solenoid to the outer loop by an ***electromagnetic wave.*** Just as with mechanical waves (Chapters 11 and 12), electromagnetic waves transport energy from one place to another. We'll explore electromagnetic waves and how they propagate in Chapter 23.

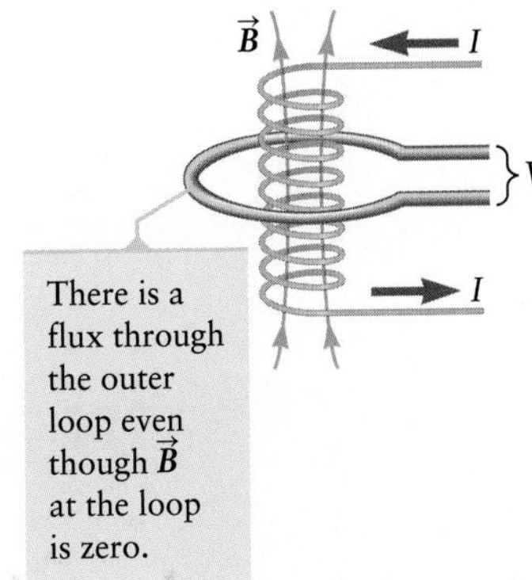
$\vec{B}$
I
V
I
There is a flux through the outer loop even though $\vec{B}$ at the loop is zero.

▲ **Figure 21.32** This long, thin solenoid produces a magnetic flux through the larger-diameter loop that encircles it. If this flux varies with time, an emf is induced in the outer loop, even though there is essentially no magnetic field there.

Summary | CHAPTER 21

Key Concepts and Principles

Magnetic flux and Faraday's law

A changing magnetic flux produces an electric field. This phenomenon is called ***induction.*** The induced electric potential is given by ***Faraday's law,***

$$\mathcal{E} = -\frac{\Delta\Phi_B}{\Delta t} \quad \text{(21.5) (page 629)}$$

(Continued)

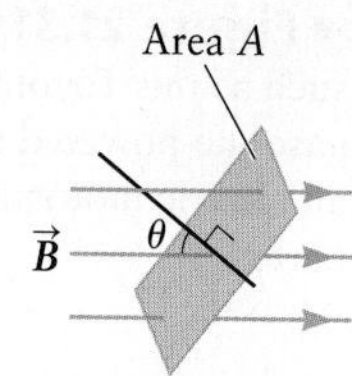

Magnetic flux
$\Phi_B = BA\cos\theta$

where $\mathcal{E}$ is the ***induced emf*** along a closed path and Φ_B is the ***magnetic flux*** through the area enclosed by that path.

The magnetic flux due to a field of magnitude B through a surface of area A is given by

$$\Phi_B = BA\cos\theta \qquad (21.2)\ \textbf{(page 628)}$$

where θ is the angle that $\vec{B}$ makes with the direction normal to the surface.

A changing magnetic flux can be produced if B is changing, the size of the area is changing, or the angle between $\vec{B}$ and the area is changing. In all three cases, the change in flux produces an induced emf according to Faraday's law.

Lenz's law

Lenz's law says that the emf induced by a changing magnetic flux always tends to produce a magnetic field that opposes the change in flux. The key to Lenz's law is that the induced emf always opposes externally imposed changes.

Energy stored in a magnetic field

The energy stored in a magnetic field B in a particular volume of space is

$$PE_{mag} = \frac{1}{2\mu_0}B^2 \times \text{volume} \qquad (21.34)\ \textbf{(page 648)}$$

Applications

Inductors

An ***inductor*** is a circuit element such as a coil or solenoid that generally consists of one or more loops of wire. When applied to an inductor with inductance L, Faraday's law gives the voltage drop across the inductor as

$$V_L = L\frac{\Delta I}{\Delta t} \qquad (21.23)\ \textbf{(page 642)}$$

There is a nonzero voltage across an inductor only when the current is changing. The voltage induced across the inductor opposes changes in the applied current.

DC circuits with inductors

The qualitative behavior of a DC circuit containing inductors can be described using two general rules:

1. Immediately after any switch is closed or opened, the induced emfs keep the current through all inductors equal to the values they had the instant before the switch was thrown.
2. After a switch has been closed or opened for a very long time, the induced emfs are zero.

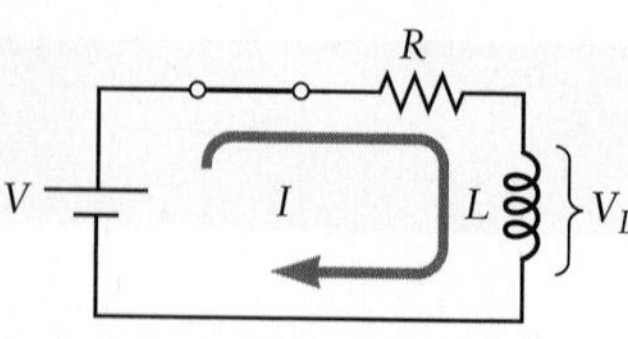

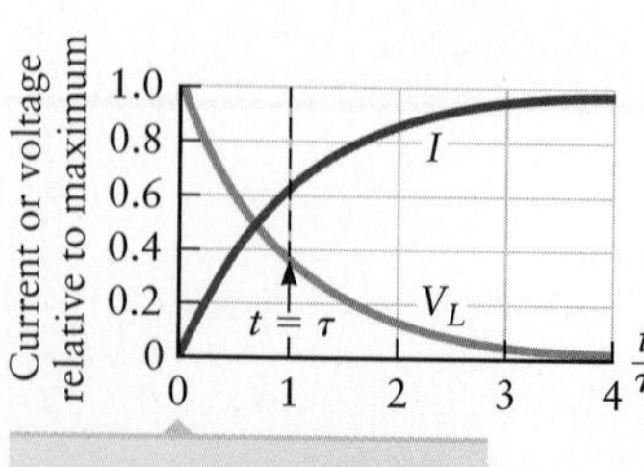

Switch is closed at $t = 0$.

Time constant for an RL circuit

The time dependence of an RL circuit is characterized by the ***time constant*** τ.

$$\tau = \frac{L}{R} \qquad (21.27)\ \textbf{(page 645)}$$

When the switch in an LR circuit is closed at $t = 0$, the current varies as

$$I = \frac{V}{R}(1 - e^{-t/\tau}) \qquad (21.26)\ \textbf{(page 645)}$$

Energy stored in an inductor

The energy stored in an inductor (in its magnetic field) is

$$PE_{\text{ind}} = \tfrac{1}{2}LI^2 \qquad \text{(21.33) (page 647)}$$

Questions

SSM = answer in *Student Companion & Problem-Solving Guide* = life science application

1. Describe an example in which the magnetic flux through a surface is zero and the magnetic field is not zero.

2. An aluminum disk is held between the poles of a powerful magnet and then pulled out. The person pulling on the disk feels an opposing force due to currents induced through Faraday's law. Explain why the magnitude of the force becomes larger as the disk is pulled faster.

3. SSM Modern coin vending machines use a configuration of magnets to help sort appropriate currency from fake or foreign coins. When a coin is dropped through the slot, it passes within a few millimeters of the magnets as it falls. Light sensitive sensors detect the passage of the coin as it breaks a beam at the top and bottom of the magnet assembly. How might the magnets help reject fake or foreign coins? How would the time of passage of a zinc coin compare with that of a coin made with copper in the presence of the magnetic field? (Modern pennies are copper-plated zinc, and modern dimes and quarters have a layer of copper sandwiched between the face and reverse.)

4. The inductance L of a solenoid is proportional to the square of the number of turns, whereas the energy stored in the solenoid PE_{ind} is proportional to L. (a) If the length of the solenoid is doubled, by what factors do L and PE_{ind} change? (b) Explain physically why the factors in part (a) are larger than 2.

5. The switch in Figure Q21.5 is initially closed for a very long time. Use Faraday's law to explain why there can be sparks when the switch is opened. *Hint*: Assume the switch contacts involve two pieces of metal that are pulled apart.

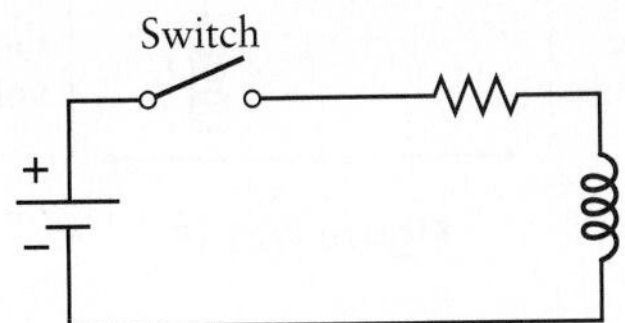

Figure Q21.5

6. Many trains use "magnetic brakes." Explain using Faraday's law how a "braking force" might be produced. *Hint*: Consider the sliding bar in Figure 21.6 and imagine that it is part of the train.

7. A moving charged particle produces an electric current and hence a magnetic field. Suppose a positively charged particle passes by a circular current loop as sketched in Figure Q21.7, with the loop and the particle's path lying in the same plane. Make a qualitative sketch of the current through the loop as a function of time. Be sure to indicate the time at which the particle is closest to the current loop.

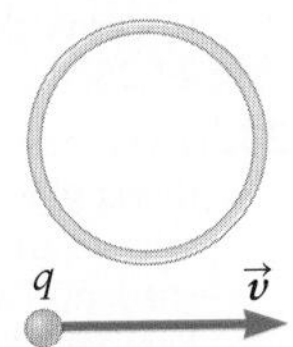

Figure Q21.7

8. A conducting loop is pulled at constant velocity through a region in which the magnetic field is constant, both inside and outside the loop (Fig. Q21.8). Explain why the induced emf is zero.

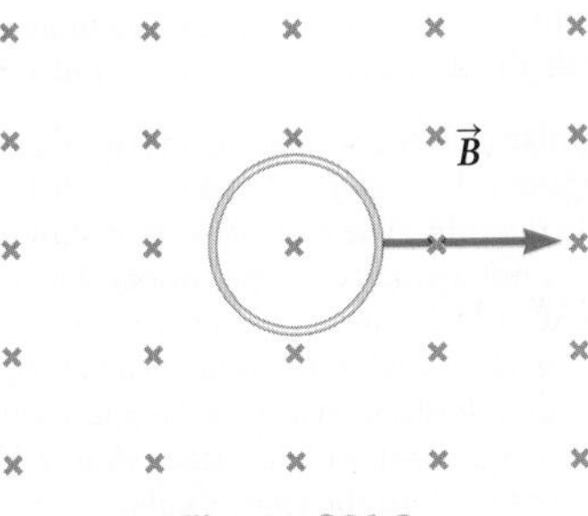

Figure Q21.8

9. Faraday and many other scientists expected that there should be a certain symmetry between electric and magnetic phenomena. Is there a magnetic analog of electric charge? That is, is there such a thing as "magnetic charge"?

10. You are given a length L of wire and told to form it into a current loop so as to produce the maximum possible emf when the loop is placed in a time-varying magnetic field. You investigate the three loop geometries in Figure Q21.10: a long, thin rectangle; a square; and a circle. Which one will give the largest emf?

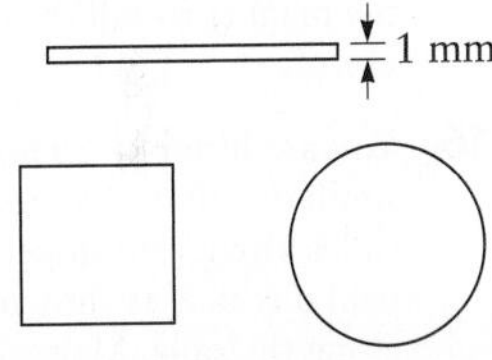

Figure Q21.10

11. Consider a circular path as sketched in black in Figure Q21.11. This path forms the boundary for a disk and for a hemisphere as well as many other possible surfaces. If the magnetic flux through the disk is $\Phi_{B,\text{ disk}}$ and the flux through the hemisphere is $\Phi_{B,\text{ hemi}}$, how are these two fluxes related? Give a physical reason to explain your answer.

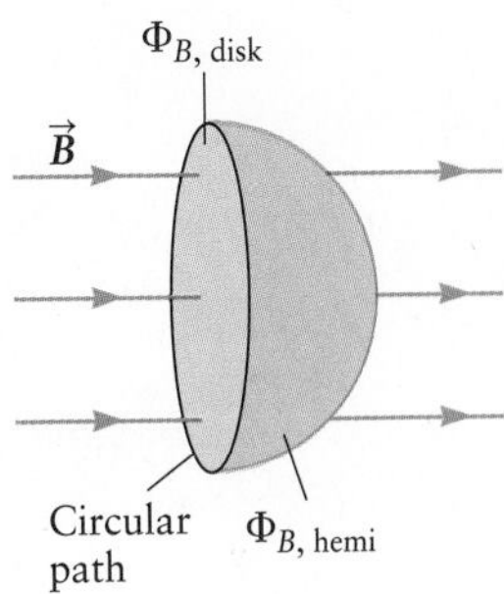

Figure Q21.11

12. A loop of copper moving with constant velocity passes through regions of magnetic field as indicated in Figure Q21.12. (a) Determine the direction of the current (if any) induced in the copper loop at each point, 1 through 5, along its path. Assume the field is uniform and of constant magnitude inside each of the two dashed rectangles and zero outside. (b) Rank the currents according to their magnitude from greatest to least for points 1 through 5. Indicate ties, if any.

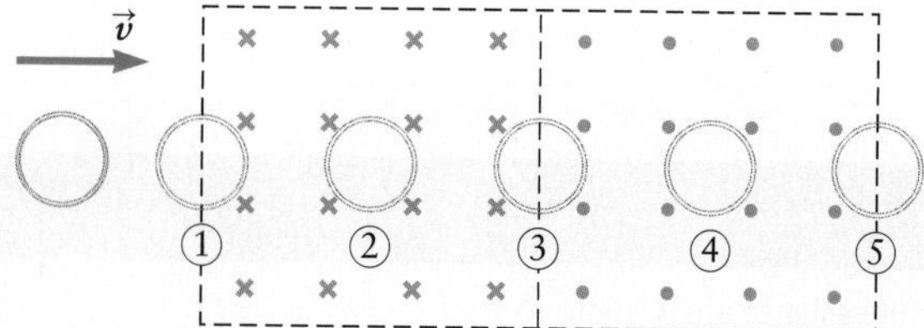

Figure Q21.12 Questions 12 and 13.

13. Consider again the copper loop in Figure Q21.12, but this time assume the loop has constant acceleration. Again determine the direction of the current (if any) induced in the copper loop at each point 1 through 5 along its path. Again assume the field is uniform inside the two dashed rectangles and zero outside.

14. SSM A popular classroom demonstration of Lenz's law involves a vertical copper tube and a permanent magnet that fits very loosely inside the tube. The instructor first shows that a permanent magnet is not attracted to the copper. The magnet is then dropped down the tube and takes a surprisingly long time to come out the bottom end of the tube. When repeating the demonstration, a student looks down the tube and sees that the magnet actually moves at a constant and rather slow velocity and does not come in contact with the tube. Explain how this demonstration illustrates Lenz's law. Why doesn't the magnet come to a complete stop?

15. The demonstration described in Question 14 often has some additional components, sometimes including tubes of aluminum and lead with the exact dimensions of the copper tube. The magnet falls at a different rate through the different tubes. Why? Rank the tubes from least to greatest according to the time interval for the magnet to fall through the same distance. Table 19.1 may be useful.

16. You are hired by an amusement park and asked to design a ride similar to that sketched in Figure Q21.16. A horizontal metal rod slides along two sloped, frictionless rails with a constant magnetic field directed as shown. A rider sits on the bar and slides down along the rails. Magnetic induction leads to a force on the bar. Use Lenz's law to show that this force is directed upward along the rails and hence keeps the rider from reaching excessive speeds.

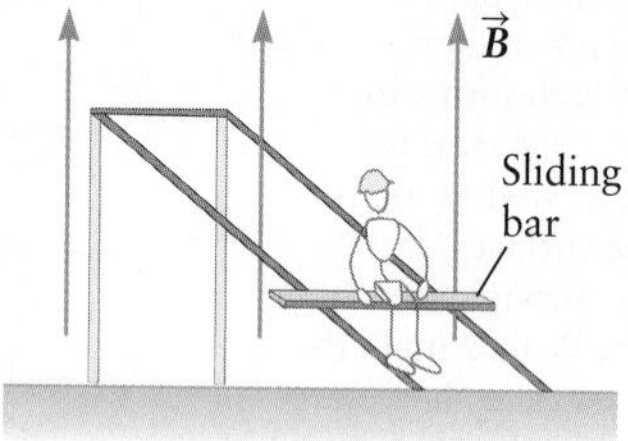

Figure Q21.16

17. When a large, time-varying current flows through a coil, a piece of steel placed within the coil will quickly become red hot (Fig. Q21.17). Why? If a piece of copper were placed in the coil, would the effect be the same? Explain.

Courtesy of Paul Peng, www.penguinslab.com

Figure Q21.17

18. **Induction range.** An alternative to gas or electric stove burners, which have an energy transfer efficiency of less than 50%, are magnetic induction cooktops (Fig. Q21.18), which can be 90% efficient. The "burner" surface does not get hot at all; only the pan gets hot. The main component of an induction range is a coil just under the surface. Describe how this device might work, taking into account that steel and iron cookware must be used. Although the surface never gets hot, is this method of heating safer than gas or conventional electric heating elements? (Consider the effect on a fork or spoon left misplaced while the burner is on near the coil.)

Courtesy of Thermador Home Appliances

Figure Q21.18 Chocolate in the metal cooking pan is melted while the chocolate in direct contact with the induction burner remains cool and solid.

19. In Chapter 19, we explained how an RC circuit could be used as a filter. The RL circuit in Figure Q21.19 can also be used as a filter. Give a qualitative argument to explain why this circuit will filter out rapid fluctuations in the input voltage.

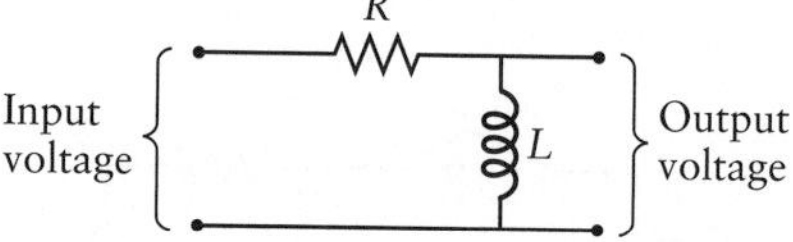

Figure Q21.19

20. In an electric circuit containing inductors, the largest induced voltage usually occurs just after a switch is opened or closed. Explain why.

21. The induced emf is an electric potential difference, so it should have units of volts (V). Show that the right-hand side of Faraday's law (Eq. 21.5) does indeed have the correct units.

22. Use the result for the current in an RL circuit in Equation 21.26 to derive the voltage across the inductor as given in Equation 21.28.

23. The roads near many traffic lights contain buried sensors that detect when a car is present (waiting at a stop light). These sensors are large loops of wire just beneath the surface of the road. (a) Explain how they can use the induced emf described by Faraday's law to detect the presence of a car. *Hint*: Cars contain a lot of steel, which is magnetic. (b) Explain why these sensors sometimes fail to detect a motorcycle and always fail to detect bicycles.

Problems available in Enhanced WebAssign

List of Enhanced Problems

Problem number	Solution in *Student Companion & Problem-Solving Guide*	Reasoning & Relationships Problem	Reasoning Tutorial in Enhanced WebAssign
21.9	✓	✓	✓
21.10		✓	✓
21.11		✓	✓
21.14	✓		
21.17	✓		
21.24	✓		
21.33	✓		
21.41	✓		
21.48	✓		
21.50		✓	✓
21.51		✓	
21.52	✓	✓	✓
21.54		✓	✓
21.56		✓	✓
21.57		✓	✓
21.58	✓		
21.71		✓	✓
21.72		✓	

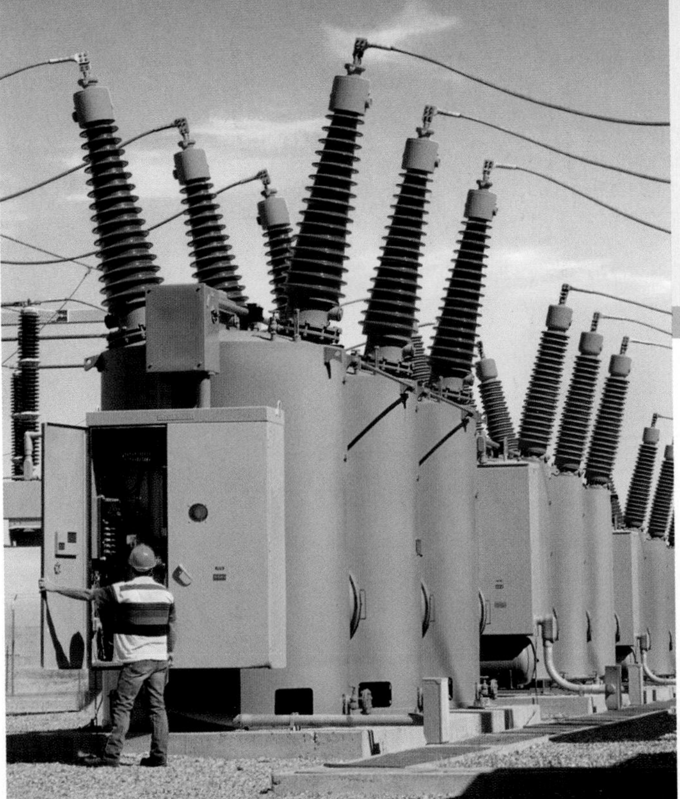

CHAPTER

22

◀ *Electrical energy is delivered to your house using AC voltages. The flow of this energy is controlled by transformers (shown here) and by other devices described in this chapter.* (© *Lester Lefkowitz/Photographer's Choice/Getty*)

Alternating-Current Circuits and Machines

We discussed circuits containing resistors and capacitors in Chapter 19 and those with inductors in Chapter 21. The source of electrical energy in those circuits was a battery, which provides a constant voltage at its output terminals (Fig. 22.1A). If a battery-powered circuit contains only resistors, the current will be independent of time. In circuits containing capacitors (*RC* circuits) or inductors (*RL* circuits), the current can vary with time but always approaches a constant value a long time after closing or opening a switch. Such circuits are called ***DC circuits***, where "DC" stands for ***direct current***. In an ***AC circuit***, the battery is replaced by a device that produces an electric potential (i.e., a voltage) that varies with time (Fig. 22.1B). Such a device is called an AC voltage source, where "AC" stands for ***alternating current***.

An AC voltage source is represented in a circuit diagram by the symbol —(~)—. The voltage at the output terminals of this device varies sinusoidally with time, with frequency f and amplitude V_{max} (also called the peak value) as shown in Figure 22.1B. The electrical energy used in your house is provided by an AC source. If you were to measure the voltage at the terminals of

OUTLINE

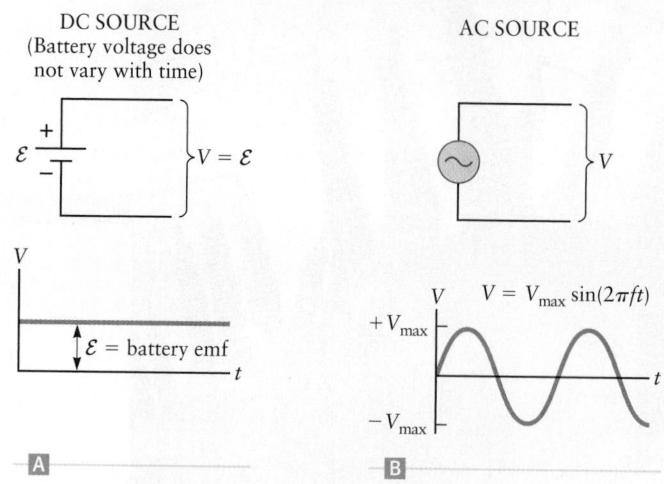

▶ **Figure 22.1** A A battery produces a constant electric potential (the emf) at its terminals. B The electric potential (voltage) at the terminals of an AC voltage source varies sinusoidally with time.

an electrical socket in a house in the United States, you would find a time dependence like that shown in Figure 22.1B with a frequency of 60 Hz. The famous inventor Thomas Edison originally proposed the use of DC electrical systems for houses and cities, but around 1880 other engineers, including one named George Westinghouse, pushed for an AC system. In the end, Westinghouse's proposal won out, and a 60-Hz system was adopted in the United States in the 1890s.

In this chapter, we'll explore the advantages of AC power. We begin with a study of AC circuits containing the basic circuit elements: resistors, capacitors, and inductors. After describing some of the properties of AC circuits, we can then explain why we use alternating current instead of direct current in our homes.

22.1 Generation of AC Voltages

One reason an AC system was adopted for the large-scale distribution of electrical energy is that generating an AC voltage is simpler than constructing and maintaining a large DC source. Most sources of AC electrical energy employ a generator based on magnetic induction. Figure 21.7 showed the general design of an AC generator. At a hydroelectric facility, water is used to turn a mechanical shaft; a steam turbine serves a similar function at a nuclear or coal-burning power plant (Fig. 22.2). This shaft is part of the generator and holds a coil with many loops of wire. The coil is positioned between the poles of a permanent magnet, so the magnetic flux through the coil varies with time as the shaft turns. According to Faraday's law of induction (Chapter 21), this changing flux induces a voltage in the coil, which is the generator's output. Faraday discovered his law of induction in 1831, and the first AC generators based on this principle were constructed in 1832! It did not take long to make practical use of Faraday's discovery.

Bureau of Reclamation

▲ **Figure 22.2** Large AC generators are used to produce AC voltages at hydroelectric dams and at other types of electric power plants.

The approach in Figure 21.7 produces an AC voltage with a frequency equal to the rotation frequency of the shaft, and electrical energy is produced as long as the shaft is rotating. In contrast, for a DC source such as a battery, one would have to replace the battery often (or replenish the chemicals in the battery), and for many practical applications this battery would need to be quite large.

Devices that are commonly referred to as "generators" of AC electrical energy actually just convert the mechanical energy of the rotating shaft into electrical energy (an electric potential difference that can produce a current in a circuit).

The principle of conservation of energy still applies, so for an ideal generator the amount of electrical energy dissipated in an attached circuit is equal to the mechanical energy used to rotate the generator shaft. In the same way, when we refer to the circuit element —(~)— as a "source" of electrical energy, you should realize that this device simply enables the transfer of electrical energy from a generator to an attached circuit.

EXAMPLE 22.1 Generating an AC Voltage

Let's analyze the large AC generators used in facilities such as the Hoover Dam power station near Las Vegas. Estimate the amplitude of the voltage (V_{max} in Fig. 22.1B) produced by rotating a coil containing one loop of wire in the generator. Assume the generator's magnetic field has magnitude $B = 1.0$ T (typical of a large magnet) and the coil has area $A = 5.0$ m^2. (Generators like the one in Fig. 22.2 are large.) Because this voltage is destined for household use, assume a frequency $f = 60$ Hz.

RECOGNIZE THE PRINCIPLE

Generators are based on Faraday's law. As the coil rotates, the magnetic flux through it changes with time. This time-varying flux induces a voltage in the coil, which is the AC voltage produced by the generator.

SKETCH THE PROBLEM

Figure 22.3 shows how the magnetic flux changes as the loop in our coil rotates.

IDENTIFY THE RELATIONSHIPS AND SOLVE

The magnetic flux through a generator coil is largest when the face of the coil is perpendicular to the magnetic field (Fig. 22.3A). This maximum flux is $\Phi_B = BA$, where B is the magnitude of the magnetic field and A is the area of the coil (Eq. 21.1). As the coil rotates one half turn, the flux alternates in sign from $+BA$ to $-BA$ (Fig. 22.3B) because the angle θ between the field and the coil's surface varies between 0° and 180°. The total flux change is

$$\Delta\Phi_B = -2BA$$

If the rotation period of the shaft is T, this flux change occurs in a time $\Delta t = T/2$. The rotation period is related to the frequency by $T = 1/f = 1/60$ s.

From Faraday's law (Eq. 21.6), the magnitude of the emf induced in the coil is

$$|\mathcal{E}| = \left|\frac{\Delta\Phi_B}{\Delta t}\right|$$

Strictly speaking, this equation applies only for small flux changes $\Delta\Phi_B$ over small time intervals Δt, but we can still use it to get an approximate value for the generator emf. Inserting our estimates for $\Delta\Phi_B$ and Δt, we get

$$|\mathcal{E}| = \left|\frac{\Delta\Phi_B}{\Delta t}\right| = \frac{2BA}{T/2} = \frac{2(1.0 \text{ T})(5.0 \text{ m}^2)}{(1/120) \text{ s}}$$

$$|\mathcal{E}| = \boxed{1200 \text{ V}}$$

► ***What does it mean?***

A coil with just a single loop of wire thus produces quite a large voltage for these values of the frequency and field. A real generator will use a coil with many loops of wire, producing an even larger voltage, typically 25,000 V to 100,000 V or even more, much larger than used in household circuits. This large voltage is then converted to household levels using a transformer as explained in Section 22.9.

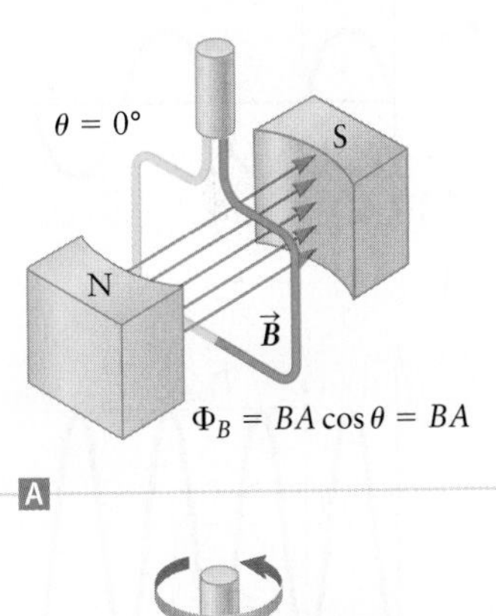

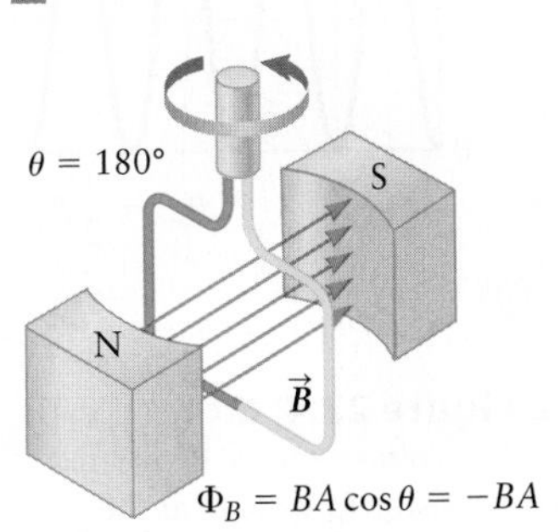

▲ **Figure 22.3** Example 22.1. As the wire loop makes a half turn around the rotation axis, the flux changes from $\Phi_B = +BA$ to $\Phi_B = -BA$.

CONCEPT CHECK 22.1 Generating a Larger Voltage

Consider the Hoover Dam generator discussed in Example 22.1. Suppose you wish to produce an AC voltage with an amplitude of 50,000 V. Which of the following approaches would work? Explain your answer.

(a) Connect many generators in series.
(b) Connect many generators in parallel.
(c) Add more turns to the coil.

AC Circuits and Simple Harmonic Motion

An old-fashioned but very descriptive unit of frequency is the "cycle per second." In terms of the time dependence of an AC voltage, a frequency of 1 Hz means that the voltage completes one full cycle each second. The voltage variation in Figure 22.1B should remind you of the simple harmonic oscillators in Chapter 11. Because of this similarity, the voltage and current in an AC circuit are said to "oscillate." We'll also find (Sections 22.5 and 22.6) a close connection between circuits with capacitors and inductors (called *LC* circuits) and simple harmonic oscillators.

CONCEPT CHECK 22.2 AC Power in the United States

For household circuits in the United States, the AC frequency is 60 Hz. How long does it take the voltage to complete one cycle?

(a) 60 s (b) 0.060 s (c) 0.033 s (d) 0.017 s

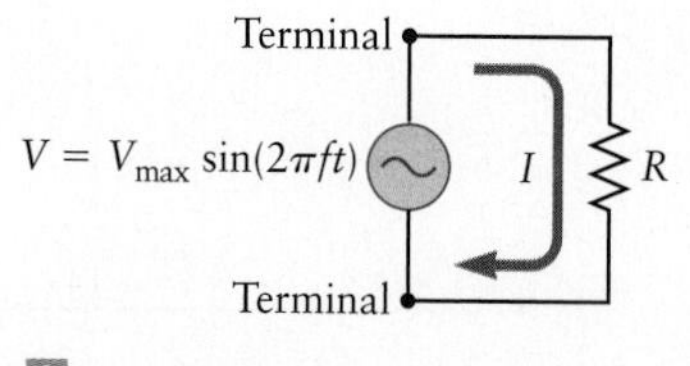

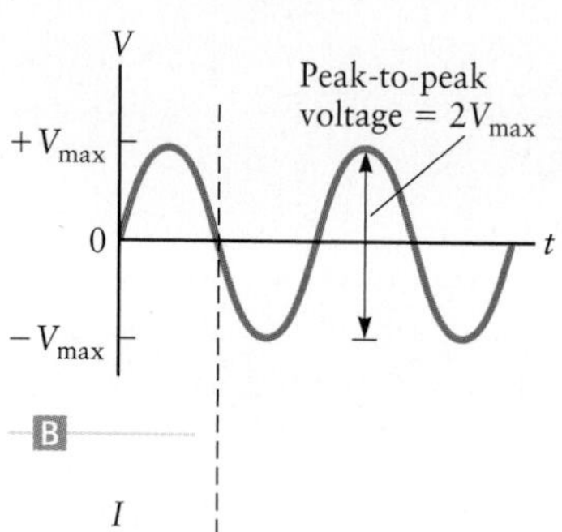

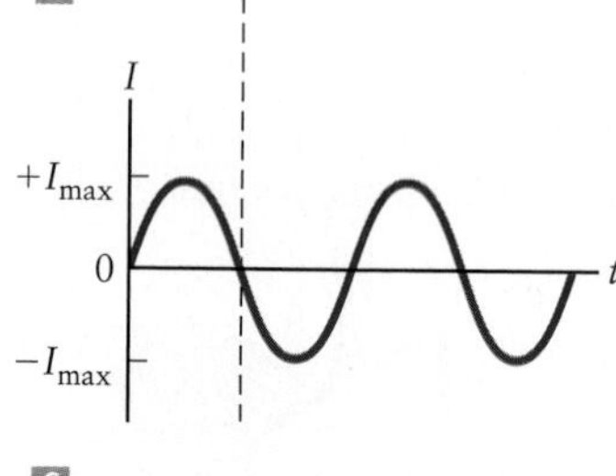

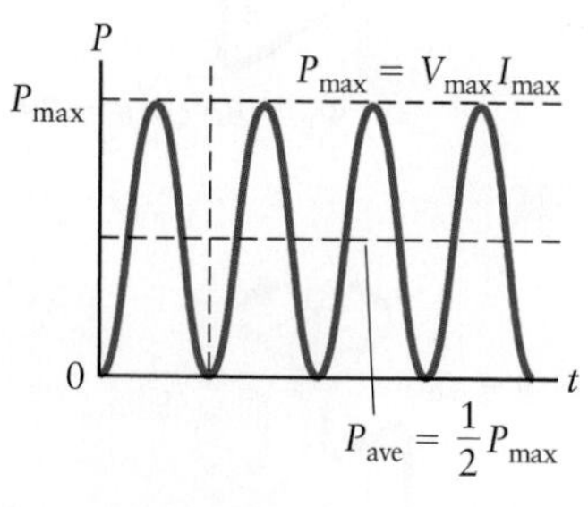

▲ **Figure 22.4** A Circuit with an AC voltage source connected to a resistor. B Voltage at the output terminals of the source as a function of time. C Current in the circuit as a function of time. D Instantaneous power dissipated in the resistor as a function of time. The average power (i.e., averaged over many cycles of the oscillation) is one-half the maximum power.

22.2 Analysis of AC Resistor Circuits

Now that we've seen how AC voltages can be generated, let's consider how to calculate the current in an AC circuit. We begin with the simplest possible circuit consisting of an AC generator and a resistor R as sketched in Figure 22.4A.

If f is the frequency and V_{max} is the amplitude of the AC voltage, the output voltage has the form

$$V = V_{max}\sin(2\pi ft) \tag{22.1}$$

This is the *instantaneous* potential difference between the terminals of the AC source and is plotted as a function of time in Figure 22.4B. In this circuit, the voltage across the output terminals is also equal to the voltage across the resistor. To find the resulting current I through the resistor, we can use ***Ohm's law*** (Eq. 19.10):

$$I = \frac{V}{R} \tag{22.2}$$

In Chapter 19, we applied Ohm's law for resistors in DC circuits, but it also holds for resistors in an AC circuit, where V and I are the instantaneous voltage and current. You can think of an AC voltage source as equivalent to a battery whose emf varies in time according to Equation 22.1. At each instant in time, the current through this AC circuit is given by $I = V/R$, just as for a DC circuit containing a battery and a resistor. As time passes, the output voltage V of the AC source varies sinusoidally, so the current through the resistor is similarly time dependent. Inserting V from Equation 22.1 into Ohm's law (Eq. 22.2), we find

$$I = \frac{V_{max}}{R}\sin(2\pi ft) \tag{22.3}$$

which is shown in Figure 22.4C. The current thus oscillates with time, with the same frequency as the voltage. The voltage amplitude V_{max} is constant, so we can denote the current amplitude as I_{max} and write

$$I = I_{max} \sin(2\pi ft) \quad \text{with } I_{max} = \frac{V_{max}}{R} \tag{22.4}$$

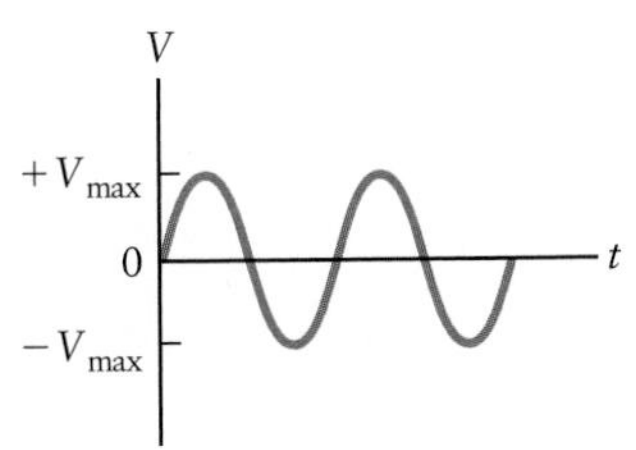

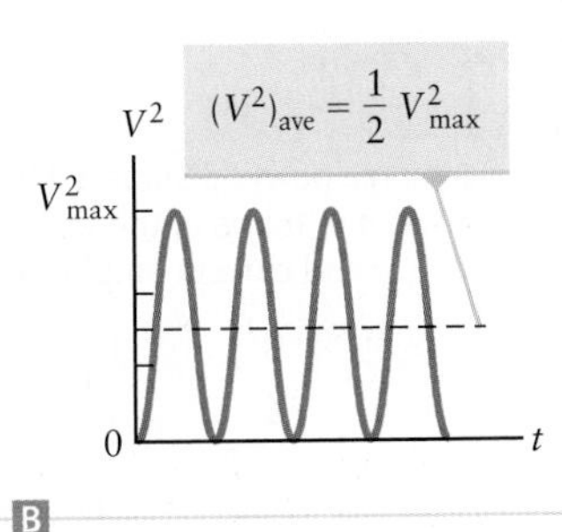

▲ **Figure 22.5** Variation of an AC voltage V and V^2 as functions of time.

Root-Mean-Square Values

The result in Equation 22.4 might be the current in your television or some other appliance. The appliance manufacturer usually specifies how much current and voltage the device is designed for. However, the voltage and current vary with time, so what numbers should the manufacturer provide? One possibility is to provide some kind of average value, but from Figure 22.4B and C we see that the average values of V and I taken over many oscillation cycles are both zero, so these values are not useful as specifications. That is one reason the notion of "root-mean-square" was adopted. Figure 22.5 shows a plot of the AC voltage $V = V_{max} \sin(2\pi ft)$ as well as how the square of the voltage varies with time, and we see that the average of V^2 is not zero. Using Figure 22.5, the average of V^2 is related to the voltage amplitude V_{max} by

$$(V^2)_{ave} = \tfrac{1}{2}V^2_{max} \tag{22.5}$$

We now define a quantity called the root-mean-square (rms) voltage, denoted as V_{rms}, through the relation $V_{rms} = \sqrt{(V^2)_{ave}}$. Here the word *mean* indicates an average, and "root" refers to a square root. We thus have

$$V_{rms} = \sqrt{(V^2)_{ave}} \tag{22.6}$$

square *root* — *square* — *mean* (average)

In words, V_{rms} is the square *root* of the *mean* (average) of the voltage *squared.* Inserting Equation 22.5 into the definition of the root-mean-square gives

$$V_{rms} = \sqrt{\frac{1}{2}V^2_{max}} = \frac{V_{max}}{\sqrt{2}} \tag{22.7}$$

Definition of rms voltage

So, the rms value of the voltage is equal to the voltage amplitude V_{max} divided by $\sqrt{2}$; hence, $V_{rms} \approx 0.71 \times V_{max}$. (See Fig. 22.6.)

The root-mean-square value can be defined for any quantity that varies with time, including the current in an AC circuit. So, for the oscillating current in Equation 22.4 we have

$$I_{rms} = \sqrt{\frac{1}{2}I^2_{max}} = \frac{I_{max}}{\sqrt{2}} \approx 0.71 \times I_{max} \tag{22.8}$$

Definition of rms current

The root-mean-square values of the voltage and current are widely used to specify the properties of an AC circuit. For example, in the United States the voltage in a household circuit has the value "120 V AC." This voltage is actually the rms value of the voltage across the terminals of a wall socket.

Power in an AC Circuit

The instantaneous power dissipated in a resistor is equal to the product of the instantaneous voltage across the resistor and the instantaneous current through the resistor, which is just an extension of Equation 19.15 to AC circuits. Hence, the instantaneous power dissipated in the resistor in Figure 22.4A is

$$P = VI \tag{22.9}$$

where V and I are the results in Equations 22.1 and 22.4. Because V and I in this circuit both vary with time, the power P is also a function of time. Inserting our results for V and I into Equation 22.9 gives

$$P = [V_{max} \sin(2\pi ft)][I_{max} \sin(2\pi ft)] = V_{max}I_{max} \sin^2(2\pi ft) \tag{22.10}$$

which is plotted in Figure 22.4D. The instantaneous power oscillates between a maximum value of $V_{max}I_{max}$ and a minimum value of zero. These minima occur at moments when the instantaneous voltage and current are zero.

Since the AC voltage and hence the instantaneous power oscillate rapidly (in the United States, the voltage has a frequency of 60 Hz), it makes sense to average the power over many cycles of its oscillation. This average power will tell how much electrical energy (and how much money) is needed to operate a circuit or appliance, such as a television, for a particular amount of time. Figure 22.4D shows that the average power P_{ave} equals half the maximum power. According to Equation 22.10, the maximum power is $P_{max} = V_{max}I_{max}$, so the average power is

Average power dissipated in a resistor in terms of the voltage and current amplitudes

$$P_{ave} = \frac{V_{max}I_{max}}{2} \tag{22.11}$$

Using our results for the rms voltage V_{rms} and current I_{rms} (Eqs. 22.7 and 22.8), we can write the average power in Equation 22.11 as

$$P_{ave} = V_{rms}I_{rms} \tag{22.12}$$

One advantage of writing the average power in this way is that Equation 22.12 has precisely the same mathematical form as the power in a DC circuit (Eq. 19.15).

When dealing with the power dissipated in a resistor, the relation for the power (Eq. 22.12) can be simplified using Ohm's law, $I = V/R$. Since the Ohm's law relation between current, voltage, and resistance (Eq. 22.2) holds for instantaneous values, it also holds for rms values. We can thus write

$$I_{rms} = \frac{V_{rms}}{R}$$

Combining this relation with Equation 22.12 leads to

Average power dissipated in a resistor in terms of the rms voltage or current

$$P_{ave} = \frac{V_{rms}^2}{R} = I_{rms}^2 R \tag{22.13}$$

EXAMPLE 22.2 Power and Voltage in a Household Circuit

The standard AC voltage found in homes in the United States has a frequency of 60 Hz and is often referred to as "120-V power," which is the value of the rms voltage; that is, $V_{rms} = 120$ V. What is the corresponding value of V_{max} for the voltage at an electrical outlet?

RECOGNIZE THE PRINCIPLE

The amplitude V_{max} of an AC voltage is also the "peak" value of the voltage (Fig. 22.6), and V_{max} and V_{rms} are just two ways of describing the same AC voltage. They are related through the basic time dependence of an AC voltage $V = V_{max} \sin(2\pi ft)$ and Equation 22.7. The point of this example is to get an intuitive understanding of the relation between rms and peak values.

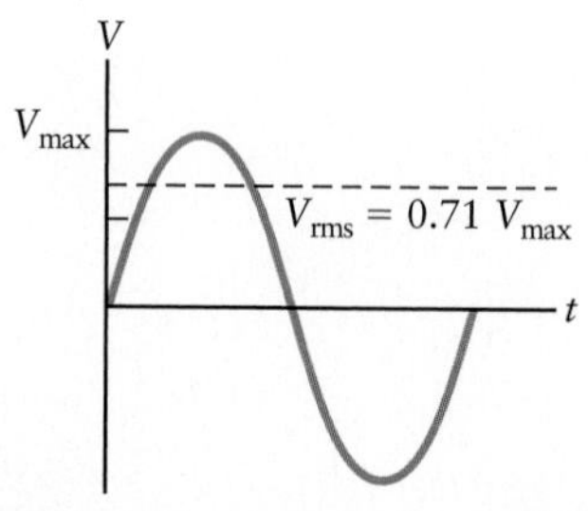

▲ **Figure 22.6** Example 22.2.

SKETCH THE PROBLEM

Figure 22.6 shows an AC voltage and indicates V_{max}.

IDENTIFY THE RELATIONSHIPS AND SOLVE

Applying Equation 22.7, we have

$$V_{rms} = \frac{V_{max}}{\sqrt{2}}$$

Solving for V_{max} and inserting the given value of V_{rms} gives

$$V_{max} = \sqrt{2}V_{rms} = \sqrt{2}(120\text{ V}) = \boxed{170\text{ V}}$$

▸ *What does it mean?*

Figure 22.6 could represent a plot of an AC voltage with $V_{max} = 170$ V (the value found in this example). A DC voltage equal to the rms value of 120 V would give the same average dissipated power in the resistor.

EXAMPLE 22.3 Voltage, Current, and Power in a Lightbulb

As a circuit element, we can assume an incandescent lightbulb functions just like a resistor (Fig. 22.7), so the relations between the voltage, current, and power in a resistor apply also to a lightbulb. Household lightbulbs are specified by their power rating. For example, one might be rated as a "60-watt" bulb. This power rating refers to the average dissipated power P_{ave}, so $P_{ave} = 60$ W when a 60-W bulb is used in a normal household "120-V" circuit. Find the resistance of a 60-W lightbulb.

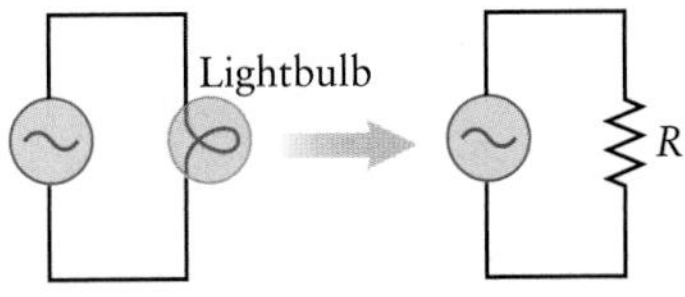

▲ Figure 22.7 Example 22.3. In circuit analysis, an incandescent lightbulb can be modeled as a resistor.

RECOGNIZE THE PRINCIPLE

The AC voltage in a "120-V" household circuit has the rms value $V_{rms} = 120$ V. When connected to a lightbulb, it gives the same average power as if we had instead connected a battery with a constant emf of $V_{DC} = 120$ V. For a given value of the voltage, the average power dissipated in a resistor depends on the value of the resistance R. For a DC voltage, the relation is $P_{ave} = V_{DC}^2/R$, whereas for an AC voltage, it is $P_{ave} = V_{rms}^2/R$. The point of this example is to get a feeling for the value of the resistance of a common household "appliance."

SKETCH THE PROBLEM

Figure 22.7 describes the problem.

IDENTIFY THE RELATIONSHIPS

For an AC circuit, the average power dissipated in a resistor is (Eq. 22.13) $P_{ave} = V_{rms}^2/R$.

SOLVE

Solving for R, we get

$$R = \frac{V_{rms}^2}{P_{ave}}$$

For the AC voltage in a household circuit, $V_{rms} = 120$ V, which we can use along with the given value of P_{ave} for the lightbulb to get

$$R = \frac{V_{rms}^2}{P_{ave}} = \frac{(120\text{ V})^2}{60\text{ W}} = \boxed{240\ \Omega}$$

▸ *What does it mean?*

The average power is important because it determines the amount of energy used by the lightbulb over a long time, and that is how the power company computes your bill. The instantaneous intensity of a lightbulb depends on the instantaneous power dissipated in the filament. According to Figure 22.4D, the power oscillates, so the intensity must also oscillate. Hence, an incandescent lightbulb must "flicker." Why don't we notice this flickering? (*Hint*: See Concept Check 22.3.)

TABLE 22.1 Useful Quantities and Notation for Describing AC Circuits

Voltage	Current	Power
V = instantaneous voltage $= V_{max}\sin(2\pi ft)$	I = instantaneous current	P = instantaneous power $= VI$
V_{max} = voltage amplitude	I_{max} = current amplitude	P_{max} = power amplitude
V_{rms} = rms voltage $= \dfrac{V_{max}}{\sqrt{2}}$	I_{rms} = rms current $= \dfrac{I_{max}}{\sqrt{2}}$	—
$V_{ave} = 0$	$I_{ave} = 0$	P_{ave} = average power $= V_{rms}I_{rms}$

When dealing with AC circuits, it is important to distinguish between instantaneous and average values of the voltage, current, and power. The amplitudes of the AC variations and the rms values are also important. Table 22.1 collects these terms and summarizes the relations between them.

CONCEPT CHECK 22.3 Flickering Lightbulbs

In Example 22.3, we noted that the intensity of an ordinary incandescent lightbulb flickers with time in the same way that the instantaneous power in a resistor varies with time (Fig. 22.4D). In the United States, the AC voltage across the lightbulb has a frequency of $f = 60$ Hz. Is the frequency of the flickering (a) 30 Hz, (b) 60 Hz, (c) 120 Hz, or (d) 240 Hz? How does the frequency of the flickering compare with the response time of your eyes?

CONCEPT CHECK 22.4 Traveling in Europe

In most European countries, the AC voltage available at a wall outlet has an rms value of about 220 V and a frequency of 50 Hz. Is the peak value (the amplitude) of this AC voltage (a) 220 V, (b) 170 V, or (c) 310 V?

Describing AC Voltages and Currents: Phasors

The sinusoidal variation of the voltage in an AC circuit,

$$V = V_{max}\sin(2\pi ft) \tag{22.14}$$

is very similar to that for circular motion (Chapter 5) and simple harmonic motion (Chapter 11). This similarity leads to a useful graphical way to analyze AC circuits. Imagine an arrow of length V_{max} as sketched in Figure 22.8. The arrow's tail is tied to the origin, while the tip moves along a circle of radius V_{max}. If this arrow makes an angle θ with the horizontal axis, the vertical component of the "voltage arrow" is

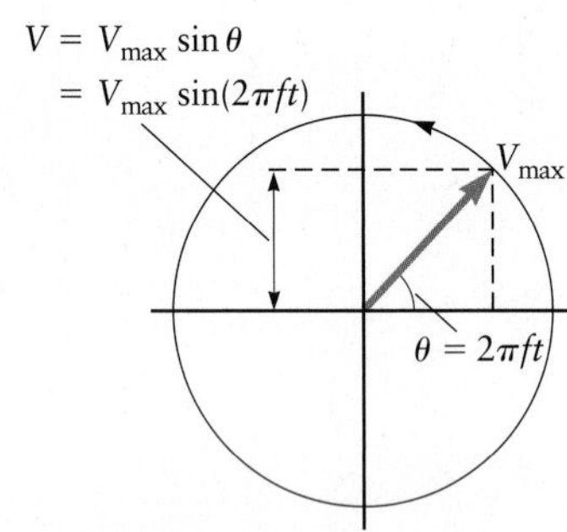

▲ **Figure 22.8** Phasor diagram of the voltage in an AC circuit. The voltage is represented by a "phasor," an arrow of length V_{max} that rotates in the coordinate plane. The instantaneous voltage in the circuit is equal to the vertical component of this arrow.

$$V = V_{max}\sin\theta \tag{22.15}$$

If the tip of the arrow undergoes uniform circular motion, the angle θ varies with time according to

$$\theta = 2\pi ft \tag{22.16}$$

where f is the rotation frequency. Inserting this result into Equation 22.15, we get $V = V_{max}\sin(2\pi ft)$, which is just the voltage in an AC circuit, Equation 22.14. The rotating arrow representing the voltage in Figure 22.8 is called a ***phasor***. A phasor is *not* a vector; the voltage in an AC circuit is *not* a vector in the way that quantities such as position and velocity are vectors. Phasor diagrams simply provide a convenient way to illustrate and think about the time dependence in an AC circuit.

The current in an AC circuit can also be represented by a phasor. Figure 22.9A shows a resistor connected to an AC voltage source. A phasor diagram for this circuit is given in Figure 22.9B, which shows the voltage arrow as it rotates around

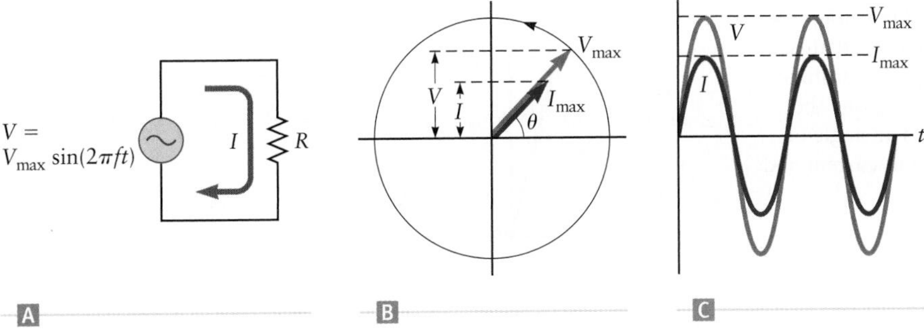

◀ **Figure 22.9** A An AC circuit containing a single resistor. B Phasors representing the voltage across the resistor and the current through the resistor. C Plots of the instantaneous voltage and current as functions of time.

the origin as well as the current as a rotating arrow of length I_{max}. In Equation 22.4, we found that $I = I_{max} \sin(2\pi ft) = I_{max} \sin\theta$, so the current arrow rotates in synchrony with the voltage arrow. These two arrows—that is, these two phasors—always make the same angle θ with the horizontal axis as time passes. Because the voltage and current phasors rotate together, we say that they are *in phase*. When the instantaneous voltage V is at one of its maxima, the instantaneous current I is also a maximum (Fig. 22.9C). Likewise, the minima of V and I occur at the same times, and the voltage and current also pass through zero at the same time.

For the resistor circuit in Figure 22.9A, the phase relation is very simple: the voltage and the current always oscillate "in synchrony" with each other. The current and voltage relationships in circuits with capacitors and inductors are usually not this simple, as we will show in the next two sections.

22.3 AC Circuits with Capacitors

Figure 22.10A shows an AC circuit containing a single capacitor. We want to calculate the current when the capacitor is attached to an AC generator with $V = V_{max} \sin(2\pi ft)$. The plates of the capacitor carry charges $+q$ and $-q$, related to the potential difference V across the capacitor by $q = CV$ (which is just Eq. 18.30 with the potential difference or "voltage" across the capacitor denoted simply by V instead of ΔV). This relation between q and V holds at all times, so q is the *instantaneous* charge on the capacitor, just as V is the instantaneous voltage across the capacitor. Since V is also equal to the voltage produced by the AC source, we have

$$q = CV = CV_{max} \sin(2\pi ft) \quad (22.17)$$

Parts B and C of Figure 22.10 show the voltage and charge as functions of time. The capacitor's voltage and charge are *in phase* with each other; that is, their maxima occur at the same times.

To understand the time dependence of the current, recall the definition of electric current from Chapter 19:

$$I = \frac{\Delta q}{\Delta t} \quad (22.18)$$

The instantaneous current I equals the rate at which charge flows onto the capacitor plates. On a graph of q versus t, the current is the *slope* of the q–t plot. The qualitative behavior of this slope is illustrated in Figure 22.10C, which shows one value of t when the slope $\Delta q/\Delta t$ is zero and another value of t when the slope is a maximum. By estimating the slope in a graphical way, we can construct the qualitative plot of the current I as a function of time shown in Figure 22.10D. The current has its largest value when the charge on the plates is zero. On the other hand, the current is zero when the charge has its largest values (in magnitude). Comparing the general plots of q and I, we see that while q is a sine function (Eq. 22.17), the current is proportional to a cosine function:

$$I = I_{max} \cos(2\pi ft) \quad (22.19)$$

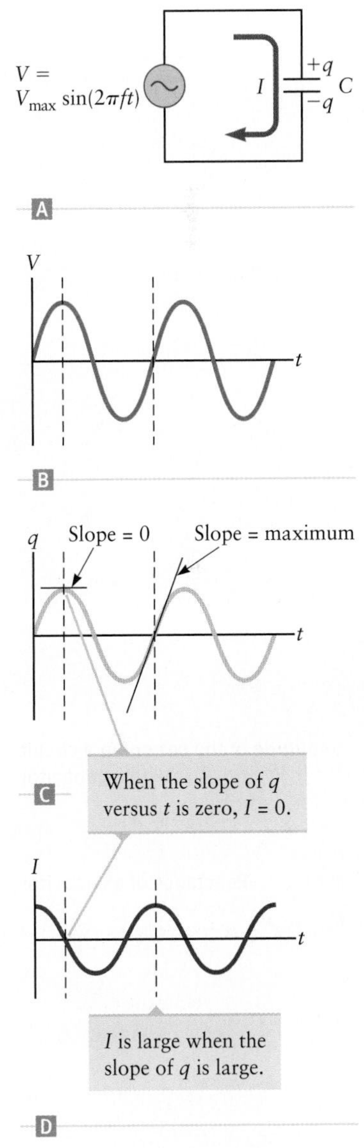

▲ **Figure 22.10** A An AC circuit with a capacitor. B Voltage, C charge on the capacitor, and D current as functions of time for the circuit in part A.

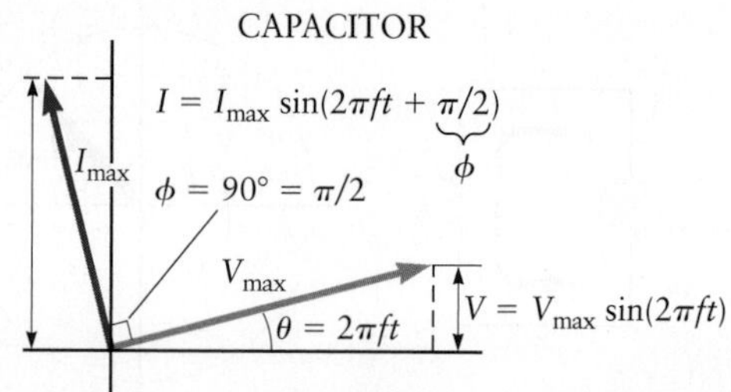

▶ **Figure 22.11** Phasor diagram showing the voltage across and the current through a capacitor. The current is represented by a rotating arrow of length I_{max}. The instantaneous current is the component of this arrow along the vertical axis. The current arrow always makes an angle of 90° ($\pi/2$) with the voltage, so the instantaneous current and voltage are always out of phase by 90°.

The sine and cosine functions have the property that for any angle α, $\cos(\alpha) = \sin(\alpha + \pi/2)$. We can therefore write the current in the equivalent forms

$$I = I_{max} \sin(2\pi ft + \pi/2)$$

$$I = I_{max} \sin(2\pi ft + \phi) \quad \text{with } \phi = \pi/2 \tag{22.20}$$

A phasor diagram for voltage and current in the capacitor circuit in Figure 22.10A is given in Figure 22.11. We show the voltage as a rotating arrow (a phasor) making an angle $\theta = 2\pi ft$ with the x axis. The tip of this voltage phasor moves in a circle, just as in the resistor circuit (Fig. 22.9B). The current in the capacitor circuit is represented by an additional rotating arrow of length I_{max} (which we have not yet calculated), and according to Equation 22.20 this phasor makes an angle $2\pi ft + \pi/2 = \theta + \pi/2$ with the horizontal axis. The current is therefore *out of phase with the voltage.* Unlike the behavior we found in the resistor circuit, the maximum values of the voltage and current *do not occur at the same time.* The angle $\pi/2$ is called the ***phase angle*** ϕ between V and I. Since the angle $\pi/2$ corresponds to 90°, for this circuit the voltage and current are $\phi = 90°$ out of phase.

The graphical analysis in Figure 22.10 shows that the current and voltage are out of phase for a capacitor. This contrasts with the case in resistors, for which I and V are always in phase (Fig. 22.9). Why is a capacitor different from a resistor? They are different because the *charge* on a capacitor is proportional to the voltage (Eq. 22.17), whereas for a resistor the *current* is proportional to V (Eq. 22.2). This difference, together with the relation between current and charge, $I = \Delta q/\Delta t$, makes the behavior of the AC capacitor circuit different from that of an AC resistor circuit.

Equation 22.20 and the phasor diagram in Figure 22.11 give the phase of the current through a capacitor. To find the peak value (the amplitude) I_{max}, we must calculate the slope of the q–t plot in Figure 22.10. This calculation requires more mathematics than we can include here. The result is

Amplitude of the current in a circuit with a capacitor

$$I_{max} = \frac{V_{max}}{X_C} \tag{22.21}$$

The factor X_C is called the ***reactance*** of the capacitor and is given by

Reactance of a capacitor

$$X_C = \frac{1}{2\pi fC} \tag{22.22}$$

The relation between the peak current and voltage for a capacitor in Equation 22.21 has the same mathematical form as the corresponding relation for a resistor $I = V/R$. Here, X_C (for a capacitor) and R (for a resistor) are both measures of how much the circuit "resists" a current. Both resistance and reactance are measured in ohms (Ω), but X_C differs from R in one very crucial way: the reactance of a capacitor X_C *depends on frequency.* As a result, the peak value of the current through a capacitor, I_{max} in Equation 22.21, depends on frequency. The reason for this frequency dependence can be traced back to the relation between current and charge $I = \Delta q/\Delta t$. Current I is the rate of change of the charge q on the capacitor. If the frequency is increased, the charge oscillates more rapidly and Δt is smaller, giving a larger current. That is why the peak current in a capacitor is larger at high frequencies and why the reactance X_C becomes smaller at high frequencies.

EXAMPLE 22.4 Properties of a Typical Capacitor

Many common electronic devices, such as the MP3 player in Figure 22.12, contain AC circuits with resistors and capacitors. Typical resistors in these applications have $R = 100\ \Omega$, and typical capacitors have $C = 1000$ pF (1.0×10^{-9} F). Find the frequency at which the reactance of one of these capacitors is equal to the resistance of one of these resistors.

RECOGNIZE THE PRINCIPLE

The reactance X_C of a capacitor depends on its capacitance and the frequency. By adjusting the frequency, we can attain a reactance of 100 Ω. We'll see in Section 22.7 how this frequency dependence is used in applications of RC circuits; for example, it is used to tune a radio to different stations.

SKETCH THE PROBLEM

No figure is needed.

IDENTIFY THE RELATIONSHIPS

The reactance of a capacitor is (Eq. 22.22) $X_C = 1/(2\pi fC)$. We want to find the value of f at which this reactance is equal to R, so we set

$$X_C = \frac{1}{2\pi fC} = R$$

SOLVE

Solving for the frequency and substituting $R = 100\ \Omega$, we get

$$f = \frac{1}{2\pi RC} = \frac{1}{2\pi(100\ \Omega)(1.0 \times 10^{-9}\text{ F})} = 1.6 \times 10^{6}\text{ Hz} = \boxed{1.6\text{ MHz}}$$

► ***What does it mean?***

The reactance of a capacitor, and hence the current in an AC circuit containing a capacitor, depend on frequency. For this combination of R and C, the frequency at which $X_C = R$ lies at the upper end of the AM radio band.

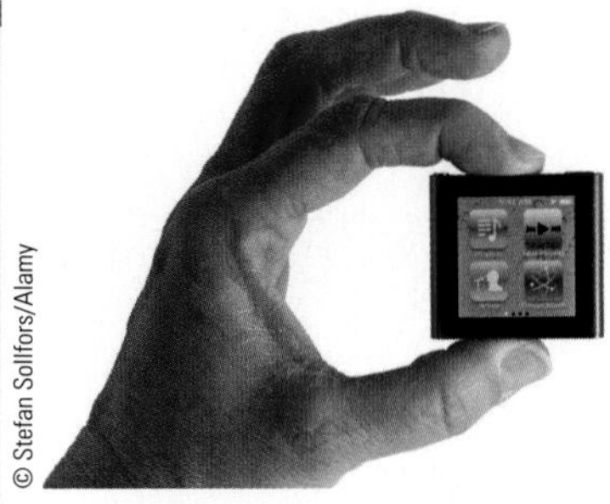
© Stefan Sollfors/Alamy

▲ **Figure 22.12** Example 22.4. Devices like this MP3 player make use of RC circuits.

CONCEPT CHECK 22.5 The Effect of Frequency on a Capacitor's Reactance

Consider an AC circuit containing a single capacitor C. If the frequency is increased by a factor of 4, how does the amplitude of the current change?

(a) The current increases by a factor of 4.
(b) The current increases by a factor of 16.
(c) The current decreases by a factor of 4.

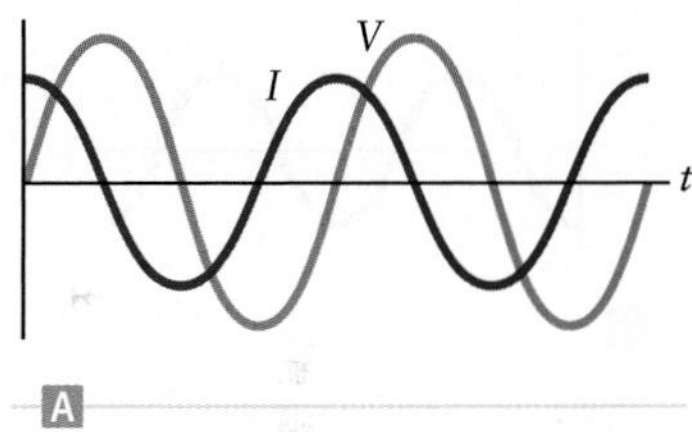
V
I
t
A

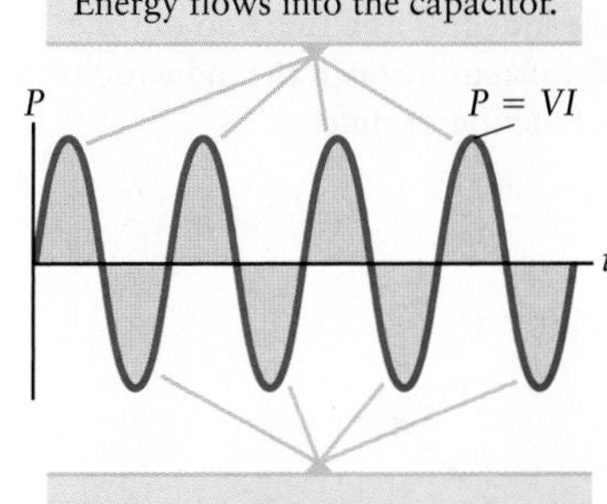
Energy flows into the capacitor.
P
P = VI
t
Energy flows from the capacitor back to the generator.
B

▲ **Figure 22.13** A For a capacitor in an AC circuit, the instantaneous power P is the product of V (a sine wave) and I (a cosine). B The result is another sine wave. At some portions of the cycle P is positive, whereas at other times it is negative. The average power over one complete cycle is zero.

Power in a Circuit with Capacitors

The instantaneous power dissipated in a resistor is $P = VI$ (Eq. 22.9). The derivation of this relation in Chapter 19 (Eq. 19.15) calculated the power from the change in the potential energy of a charge as it moved through an electric potential difference V across a resistor. In fact, the general argument applies to *any* circuit element. Let's now use that result for P to find the instantaneous power in a circuit in which an AC voltage source is connected to a capacitor. Inserting the results for V (Eq. 22.14) and I (Eq. 22.19) into our expression for P, we find

$$P = VI = V_{max}I_{max}\sin(2\pi ft)\cos(2\pi ft) \quad (22.23)$$

This result is plotted in Figure 22.13B, which shows how P oscillates between positive and negative values. These oscillations and the negative values of P can be traced to the oscillating behaviors of the sine and cosine functions in Equation 22.23 (Fig. 22.13A). The voltage and current across the capacitor both oscillate with time, so there are portions of each cycle during which current is flowing into the generator and "against" the generator voltage. When the current and voltage have opposite signs, the power $P = VI$ is negative. At these moments, energy is being transferred from the capacitor back to the generator. In fact, during half of each cycle, energy flows from the generator to the capacitor (P is positive in Fig. 22.13), whereas during the other half of the cycle energy flows from the capacitor back into the generator (P is negative). The net result—the average power—is zero.

We say that electrical energy is *dissipated* in a resistor because the electrical energy supplied by the voltage source is converted to heat energy and hence is lost from the circuit. In a purely capacitive circuit, however, the energy is not converted to heat. Instead, energy is *stored* in the capacitor as electric potential energy, and this stored energy is later returned to the AC source. Hence, in a capacitor circuit we should really speak of the power P that is *delivered to* the capacitor. When P is positive, energy is being added to the capacitor, whereas when P is negative, energy is being removed from the capacitor. That is possible because a capacitor can store energy, but a resistor cannot.

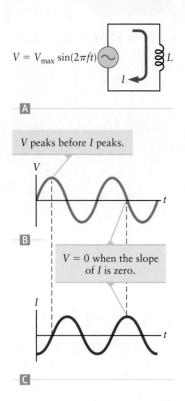

▲ **Figure 22.14** A An AC circuit with an inductor. B Voltage across the inductor as a function of time. C Current through the inductor as a function of time.

22.4 AC Circuits with Inductors

We next consider a circuit consisting of an AC generator and a single inductor as shown in Figure 22.14A. Our job again is to calculate the current in this circuit. From Chapter 21, the voltage across an inductor is proportional to the rate of change in the current. If L is the inductance and ΔI is the change in the current over a short time interval Δt, the voltage drop across the inductor is

$$V = L\frac{\Delta I}{\Delta t} \tag{22.24}$$

This is also equal to the output of the AC voltage source attached to the inductor, so $V = V_{max}\sin(2\pi ft)$. In words, Equation 22.24 states that V is proportional to the slope of the I–t relation. Because we are given V (Fig. 22.14B), we can infer how I varies with t; to do so, we must construct an I–t plot with a slope that satisfies Equation 22.24. We have sketched the answer qualitatively in Figure 22.14C. Here, $V = 0$ at times when the slope of I is zero, and V is largest and positive when I has its largest positive slope.

From Figure 22.14, we can see that V oscillates according to a sine function, whereas I oscillates with time according to a cosine function

$$I = -I_{max}\cos(2\pi ft) \tag{22.25}$$

We can use the identity $\cos\alpha = -\sin(\alpha - \pi/2)$ to write Equation 22.25 as

$$I = I_{max}\sin(2\pi ft - \pi/2)$$

$$I = I_{max}\sin(2\pi ft + \phi) \qquad \text{with } \phi = -\pi/2 \tag{22.26}$$

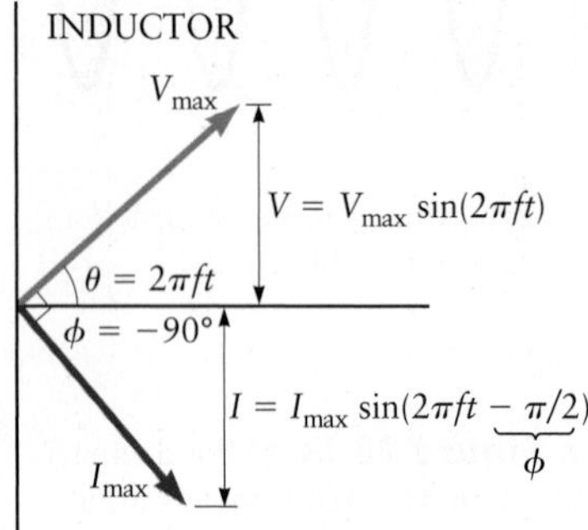

▲ **Figure 22.15** Phasor diagram of the voltage and current in an AC circuit with an inductor. The current through an inductor is 90° out of phase with the voltage.

Figure 22.15 shows the current and voltage relationship for this inductor circuit in a phasor diagram. The current phasor makes an angle $\phi = -\pi/2 = -90°$ with the voltage, with the negative sign arising from Equation 22.26 and from our standard convention of measuring positive angles in the counterclockwise direction in the coordinate plane. In words, we say that the current is 90° out of phase with the voltage. This phasor diagram is very similar to what we found for a capacitor. For both capacitors and inductors, the current is 90° out of phase with the voltage. For a capacitor the phase angle is $\phi = +90°$, whereas for an inductor it is $\phi = -90°$.

The current and voltage are out of phase in an inductor because the voltage across it is proportional to the rate of change of the current, $V = L(\Delta I/\Delta t)$ (Eq. 22.24). This is in contrast to the case with a resistor, for which the current is proportional to the voltage (Eq. 22.2) so that I and V are in phase with each other.

Equation 22.26 and the phasor diagram in Figure 22.15 give the phase of the current through an inductor. An exact calculation of the peak current I_{max} gives

$$I_{max} = \frac{V_{max}}{X_L} \tag{22.27}$$

where X_L is called the ***reactance*** of the inductor and is given by

$$X_L = 2\pi f L \tag{22.28}$$

The relation between the peak current and voltage for an inductor is very similar to the corresponding relations for a capacitor (Eq. 22.21) and a resistor (Eq. 22.2). Again, X_L is a measure of how strongly an inductor "resists" a current. Just as for a capacitor, the reactance of an inductor depends on frequency. Unlike with a capacitor, however, the reactance X_L of an inductor becomes larger as the frequency is increased. (Recall from Eq. 22.22 that X_C decreases at high frequencies.) This behavior of X_L can be traced back to the basic properties of an inductor. According to Lenz's law (Chapter 21), an inductor opposes changes in the flux through its coils. Increasing the frequency increases the rate of change of the flux and hence increases this opposing emf. For this reason, the current through an inductor decreases at high frequencies.

We have collected the results for the behavior of AC circuits with a resistor, a capacitor, and an inductor in Table 22.2.

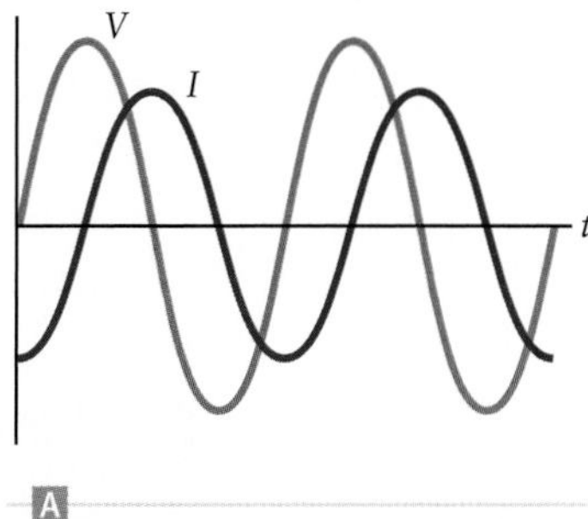

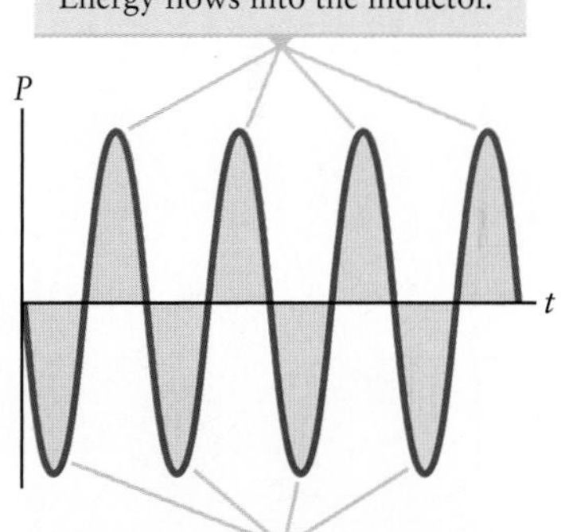

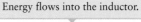

▲ **Figure 22.16** A The instantaneous power P delivered to the inductor is the product of V (a sine wave) and I (a cosine). B At some portions of the cycle P is positive, whereas at other times it is negative. The average power over one complete cycle is zero.

CONCEPT CHECK 22.6 Reactance of an Inductor

An inductor has a reactance $X_L = 2000\ \Omega$ when measured at a frequency of 1500 Hz. The frequency is then changed, and the reactance is found to be 6000 Ω. Is the new frequency (a) 500 Hz, (b) 1000 Hz, (c) 3000 Hz, or (d) 4500 Hz?

Power in a Circuit Containing an Inductor

We can calculate the instantaneous power in a circuit containing an inductor using the result for the current in Equation 22.25 together with the general result for power in a circuit element $P = VI$. We find

$$P = VI = -V_{max} I_{max} \sin(2\pi f t)\cos(2\pi f t) \tag{22.29}$$

This result is shown in Figure 22.16. The power delivered to the inductor oscillates between positive and negative values, and the average power over a complete cycle is zero. The generator delivers energy to the inductor during part of the AC cycle, and the inductor delivers an equal amount of energy back to the generator during other parts of the cycle. Energy is stored in the inductor as magnetic potential energy by virtue of the inductor's magnetic field.

TABLE 22.2 Properties of AC Circuits

Circuit Element	Resistance or Reactance to Current Flow	Amplitude of Current through Circuit Element	Phase Relation between Current and Voltage	Average Power
Resistor	R	$I = V/R$	I and V are in phase	$P_{ave} = V_{rms} I_{rms} = \frac{V_{max} I_{max}}{2}$
Capacitor	$X_C = \frac{1}{2\pi f C}$	$I = V/X_C$	I and V are out of phase by 90°	$P_{ave} = 0$
Inductor	$X_L = 2\pi f L$	$I = V/X_L$	I and V are out of phase by 90°	$P_{ave} = 0$

EXAMPLE 22.5 Analyzing an *LR* Circuit

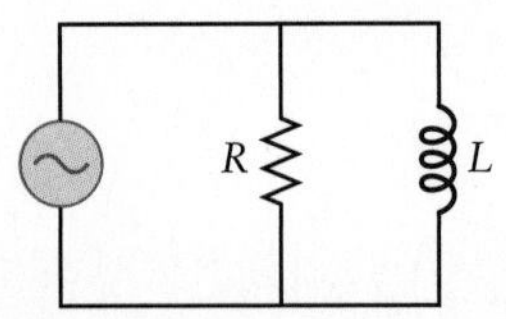

▲ **Figure 22.17** Example 22.5.

The circuit in Figure 22.17 shows an AC source of frequency $f = 3000$ Hz connected to a resistor with $R = 200\ \Omega$ in parallel with an inductor whose inductance L is unknown. If the amplitude of the current in the inductor is 6.0 times larger than the amplitude of the current in the resistor, find L.

RECOGNIZE THE PRINCIPLE

The AC source in Figure 22.17 is connected across both the resistor and the inductor. If the amplitude of the AC voltage is V_{max}, the current amplitude in the resistor is $I_R = V_{max}/R$, while for the inductor the amplitude of the current is $I_L = V_{max}/X_L$, where X_L is the reactance of the inductor. We need to find the value of X_L that will make $I_L = 6.0 \times I_R$. From there, we can find L.

SKETCH THE PROBLEM

Figure 22.17 shows the circuit.

IDENTIFY THE RELATIONSHIPS

Setting $6.0 \times I_R = I_L$ and using the relation for the reactance of an inductor ($X_L = 2\pi fL$) leads to

$$6.0 \times \frac{V_{max}}{R} = \frac{V_{max}}{X_L} = \frac{V_{max}}{2\pi fL} \tag{1}$$

SOLVE

Solving Equation (1) for L and inserting the given values of R and f, we get

$$2\pi fL = \frac{R}{6.0}$$

$$L = \frac{R}{6.0 \times 2\pi f} = \frac{200\ \Omega}{6.0 \times 2\pi(3000\ \text{Hz})} = \boxed{1.8 \times 10^{-3}\ \text{H}}$$

▸ ***What does it mean?***

The reactance of an inductor depends on frequency. The value of L found here is 1.8 mH, a typical value in many applications.

22.5 *LC* Circuits

Most useful circuits contain multiple circuit elements, so now let's explore what happens when resistors, capacitors, and inductors are connected together in different ways. In this section, we consider one particularly important example, the *LC* circuit.

Figure 22.18 shows the simplest possible *LC* circuit, consisting of a single inductor connected to a single capacitor. There is no AC generator in this circuit, but we start with a charge $+q$ on one capacitor plate and $-q$ on the other (Fig. 22.18A). We have thus "placed" an energy $PE_{cap} = q^2/(2C)$ in the circuit. The charge on the capacitor produces a voltage $V_C = q/C$ across the capacitor, and since the capacitor's leads are connected directly to the inductor, V_C is also the voltage across the inductor. In qualitative terms, the capacitor tries to "push" a current through the inductor as the charge on one capacitor plate attempts to flow through the inductor to the other plate. An inductor always resists changes in current, however; it does so through an induced emf and Lenz's law. Hence, at $t = 0$ there is an induced emf across the inductor opposing any current, and $I = 0$ at that instant.

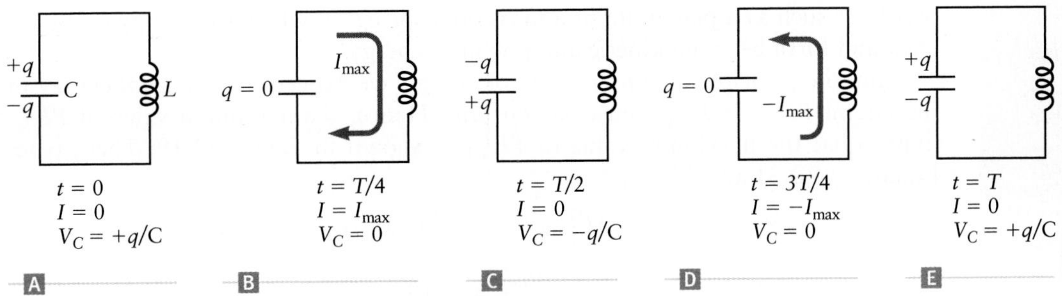

▲ **Figure 22.18** In an LC circuit with an initial charge, the current in the circuit and the voltage across the capacitor oscillate with a period T. This circuit behaves as a simple harmonic oscillator.

This state of affairs with $I = 0$ is only maintained for an instant at $t = 0$; the current is nonzero after $t = 0$, just as for RL circuits (Chapter 21). So, at times $t > 0$ charge moves from one capacitor plate to the other and there is a current in the inductor. Eventually, the charge on each capacitor plate falls to zero (Fig. 22.18B) and the corresponding voltage across the capacitor is $V_C = q/C = 0$. At this point you might expect that the current through the inductor would fall to zero, but the inductor again *opposes change* in the current, so the induced emf now acts to *maintain* the current at a nonzero value! This current continues to transport charge from one capacitor plate to the other even after reaching the point when $q = 0$, causing the capacitor's charge and voltage to reverse sign (Fig. 22.18C). The voltage across the capacitor then peaks at a negative value, leading to a current in the opposite direction through the inductor (Fig. 22.18D) as charge again moves from one plate to the other. Eventually, the charge on the capacitor plates returns to the original values found at $t = 0$ (Fig. 22.18E). The circuit has now returned to its original state (compare parts A and E of Fig. 22.18), and the process starts over again. The voltage and current in this circuit thus ***oscillate*** between positive and negative values. In fact, this circuit behaves as a ***simple harmonic oscillator*** with the current and charge given by

$$q = q_{max}\cos(2\pi ft) \quad \text{and} \quad I = I_{max}\sin(2\pi ft) \qquad (22.30)$$

These results are identical to the time dependence of the position and velocity of a simple harmonic oscillator such as a mass on a spring from Chapter 11 (Eqs. 11.9 and 11.10).

Energy Conservation and the Oscillations in an *LC* Circuit

Capacitors and inductors can both store energy; the energy in a capacitor is stored in its electric field and depends on the charge, $PE_{cap} = \frac{1}{2}(q^2/C)$, while the energy in an inductor is stored in its magnetic field and depends on the current, $PE_{ind} = \frac{1}{2}LI^2$. Hence, as the charge and current oscillate in Figure 22.18, the energies stored in the capacitor and inductor also oscillate. We can compute these energies using the results for q and I from Equation 22.30 and find

$$PE_{cap} = \frac{1}{2}\frac{q^2}{C} = \frac{1}{2}\frac{q_{max}^2}{C}\cos^2(2\pi ft) \qquad (22.31)$$

and

$$PE_{ind} = \frac{1}{2}LI^2 = \frac{1}{2}LI_{max}^2\sin^2(2\pi ft) \qquad (22.32)$$

These results are shown in Figure 22.19. The energy oscillates back and forth between the capacitor and its electric field (PE_{cap}) and the inductor and its magnetic field (PE_{ind}). This behavior is very similar to that of a mechanical simple harmonic

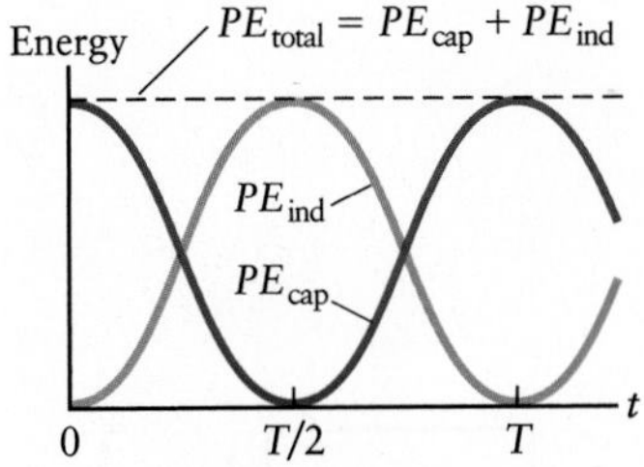

▲ **Figure 22.19** Plots of the energy in an LC circuit as a function of time, where PE_{cap} is the electric potential energy stored in the capacitor and PE_{ind} is the energy stored in the magnetic field of the inductor. The total energy $PE_{cap} + PE_{ind}$ is constant.

oscillator such as a pendulum or a mass on a spring, in which the energy oscillates back and forth between kinetic and potential energy.

From the principle of conservation of energy, we expect that the *total* energy in the circuit $PE_{cap} + PE_{ind}$ must be constant. Hence, the maximum value of PE_{cap} must equal the maximum value of PE_{ind} as shown in Figure 22.19. Then, from Equations 22.31 and 22.32,

$$(PE_{cap})_{max} = (PE_{ind})_{max}$$

$$\frac{1}{2}\frac{q_{max}^2}{C} = \frac{1}{2}LI_{max}^2$$

$$I_{max} = \frac{1}{\sqrt{LC}}q_{max} \tag{22.33}$$

This result tells us how the amplitudes of the current and charge oscillations in an LC circuit are related.

▲ **Figure 22.20** Metal detectors use changes in the resonant frequency of an LC circuit to sense when a piece of metal is nearby.

Frequency of Oscillations in an *LC* Circuit

Let's now consider the *frequency* of the oscillations in an LC circuit. In the circuit in Figure 22.18, the instantaneous voltages across the capacitor and inductor are always equal. Hence, the magnitude of the voltage across the capacitor must equal the magnitude of the voltage across the inductor. In terms of the reactances, these voltages are $V_C = IX_C$ (Eq. 22.21) and $V_L = IX_L$ (Eq. 22.27). Equating these magnitudes leads to

$$|V_C| = |IX_C| = |V_L| = |IX_L|$$

The reactances are always positive, so we can cancel the factors of I to get

$$X_C = X_L \tag{22.34}$$

This condition was derived by assuming the current in the LC circuit is oscillating and hence applies *only* at the oscillation frequency. Equation 22.34 thus says that at the oscillation frequency the reactances of the capacitor and inductor are equal.

Inserting the results for X_C and X_L from Equations 22.22 and 22.28 leads to

$$X_C = \frac{1}{2\pi fC} = X_L = 2\pi fL$$

The frequency here is the ***resonant frequency*** of the circuit. It is the oscillation frequency of the charge and current in Figure 22.18. Denoting this frequency as f_{res}, we get

Frequency of an LC oscillator

$$f_{res} = \frac{1}{2\pi\sqrt{LC}} \tag{22.35}$$

The oscillation frequency of an LC circuit thus depends on the values of L and C.

This frequency plays an important role in many applications of AC circuits. For example, a metal detector uses changes in the inductance of an LC circuit to sense the presence of a metal object. A metal detector used by beachcombers (Fig. 22.20) has a large coil of wire for its inductor. When the coil is placed near a metal object, the inductance of the coil changes, changing the resonant frequency of the circuit. The electronic unit of the metal detector is designed to register an audible signal when that happens. Metal detectors used for screening people at airports work in the same way.

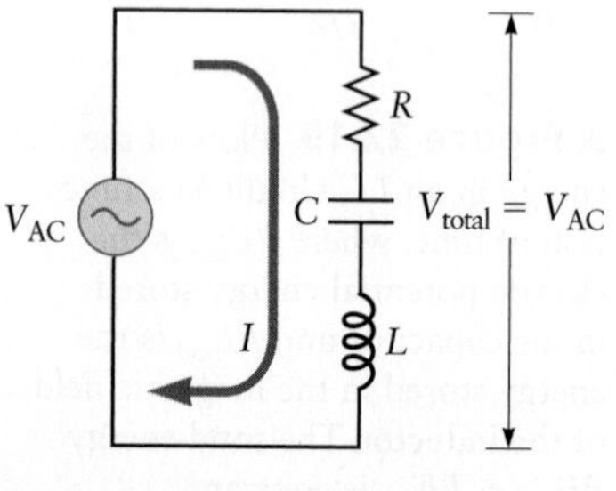

22.6 Resonance

V_{AC} R C $V_{total} = V_{AC}$ I L

▲ **Figure 22.21** An LCR circuit.

Our analysis of an LC circuit in Section 22.5 explains how this circuit can act as a simple harmonic oscillator. Let's now consider a more realistic circuit containing an inductor, capacitor, and resistor along with an AC voltage source (Fig. 22.21). Our

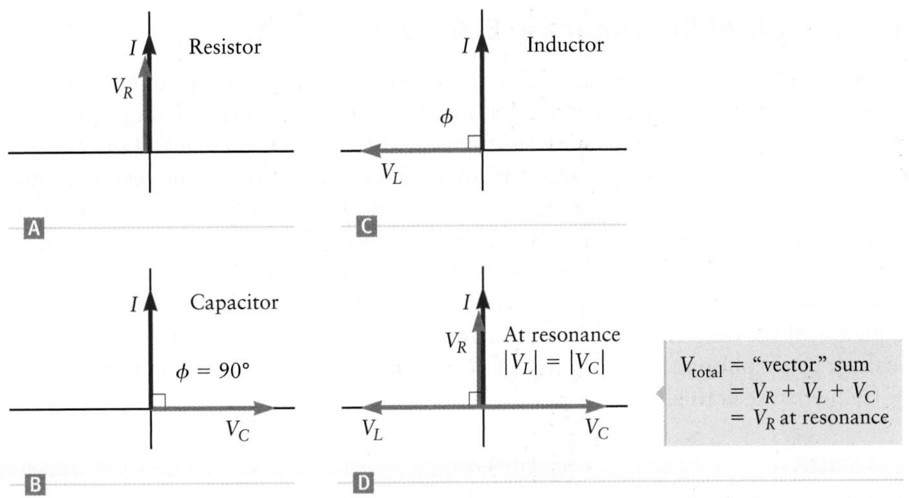

◀ **Figure 22.22** A–C Phasor diagrams of the current and voltage across different components in an *LCR* circuit. D The total voltage across the circuit is the phasor sum (i.e., the "vector" sum) of the voltages across the resistor, the capacitor, and the inductor. Notice that V_C and V_L cancel at resonance.

goal is to understand how the current in this *LCR* circuit depends on the amplitude of the AC voltage and on frequency.

To analyze the *LCR* circuit in Figure 22.21, we appeal to Kirchhoff's loop rule (Chapter 19) and move around the circuit, adding up the voltage changes across the AC voltage source (V_{AC}), the resistor (V_R), the capacitor (V_C), and the inductor (V_L). According to the loop rule, the total change in the potential must be zero, so

$$V_{AC} = V_L + V_C + V_R \tag{22.36}$$

The simple form of Equation 22.36 hides one crucial point—these voltages are not all in phase—so we must take account of both their magnitudes and phase angles when we add them.

Figure 22.22 shows a series of phasor diagrams for the current in our *LCR* circuit and the voltage across each circuit element. The inductor, capacitor, and resistor are in series, so the current is the same through each one, and all the current phasors in Figure 22.22 have the same orientation. Figure 22.22A is for the resistor; in this case, the current and voltage are in phase with each other and the phasor arrows for I and V_R are parallel. This diagram is similar to Figure 22.9B except that we have placed these arrows along the vertical axis. Parts B and C of Figure 22.22 show phasor diagrams for the voltages across the capacitor and inductor. From Sections 22.3 and 22.4, these voltages are 90° out of phase with the current. Figure 22.22D shows all three voltage phasors; the total voltage across all three circuit elements is the sum of these three phasors. The key point is that the phasors for V_L and V_C are in *opposite directions*, so they tend to cancel. In fact, we found in deriving Equation 22.34 that the two phasors V_C and V_L have the same magnitude (i.e., the same "length") when the frequency equals the resonant frequency f_{res}. Hence, at this frequency the voltages across the capacitor and inductor *cancel*. Only the resistor is left to "resist" the flow of current at this frequency. This perfect cancellation only occurs at f_{res}, so the current is highest at this frequency.

Figure 22.23 shows how the current through an *LCR* circuit varies with frequency. Here, I is largest at f_{res} and decreases at both higher and lower frequencies. This is an example of ***resonance*** and is very similar to the behavior of driven mechanical harmonic oscillators that we discussed in Chapter 11 (Fig. 11.29). By itself, an *LC* circuit acts as a simple harmonic oscillator with a resonant frequency given by Equation 22.35. Adding a resistor to the circuit (Fig. 22.21) adds damping and gives a damped harmonic oscillator. When this oscillator is "driven" by attaching it to an AC voltage source, the resulting current depends on frequency. The current is largest when the frequency of the AC source matches the resonant frequency of the circuit, hence the peak in I in Figure 22.23 when $f = f_{res}$.

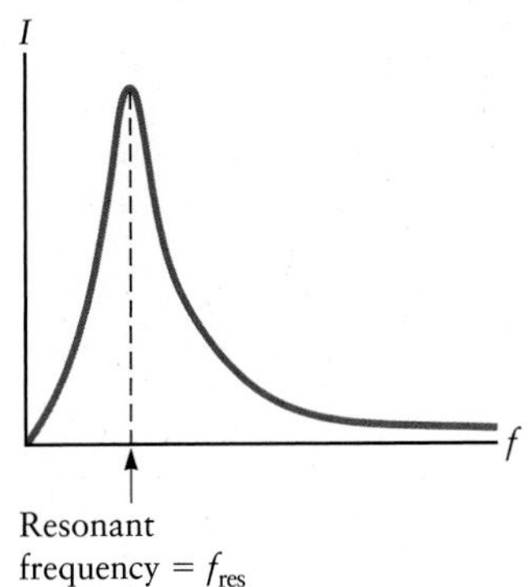

▲ **Figure 22.23** Current I in an *LCR* circuit as a function of frequency.

Applications of Resonance in Electronic Circuits

The resonant behavior of an *LCR* circuit is used in radios, cell phones, and other similar applications. The input to a radio comes from an antenna that "picks up" signals from many different stations at many different frequencies. This composite signal is the AC voltage source for an *LCR* circuit. This circuit responds most strongly—that is, it gives the largest current—at frequencies that are near its resonant frequency. When you tune your radio, you are changing the value of the capacitance in an *LCR* circuit so that the resonant frequency matches the frequency of the station you want to listen to. The resulting current is then dominated by the frequency of the desired station; this current is amplified and ultimately sent to your speakers or earphones. In this way, *LCR* circuits are used to construct devices that are frequency selective.

EXAMPLE 22.6 Resonant Frequency of an *LC* Circuit

AM radio stations broadcast at frequencies in a range centered near 1.0 MHz (1.0×10^6 Hz). If the inductance in an *LC* circuit is $L = 50\ \mu\text{H}\ (5.0 \times 10^{-5}\ \text{H})$, what is the value of the capacitance that gives a resonant frequency at the middle of the AM range?

RECOGNIZE THE PRINCIPLE

We want to choose the value of the capacitance so that the resonant frequency equals the AM frequency $f = 1.0 \times 10^6$ Hz. The purpose of this problem is to see what particular values of L and C are needed to get the desired resonant frequency. The answer will tell us if an *LC* circuit is really practical for this application.

SKETCH THE PROBLEM

No figure is needed.

IDENTIFY THE RELATIONSHIPS

The resonant frequency of an *LC* circuit is given by (Eq. 22.35)

$$f_{\text{res}} = \frac{1}{2\pi\sqrt{LC}}$$

Rearranging to solve for C, we get

$$\sqrt{LC} = \frac{1}{2\pi f_{\text{res}}}$$

$$LC = \frac{1}{4\pi^2 f_{\text{res}}^2}$$

$$C = \frac{1}{4\pi^2 f_{\text{res}}^2 L}$$

SOLVE

Inserting the given values of f_{res} and L yields

$$C = \frac{1}{4\pi^2(1.0 \times 10^6\ \text{Hz})^2(5.0 \times 10^{-5}\ \text{H})} = \boxed{5.1 \times 10^{-10}\ \text{F}}$$

► *What does it mean?*

This value for the capacitance is easily achieved with typical capacitors, so an *LC* circuit with a resonant frequency in the AM band is quite practical. Similar circuits are used in televisions, cell phones, and other applications in which one needs to distinguish between electrical signals according to their frequency.

22.7 AC Circuits and Impedance

Figure 22.24A shows an AC voltage source connected to an inductor. A typical inductor is constructed using a coil of wire, and real wires have nonzero resistance. A real inductor can therefore be modeled as an ideal inductor L in series with a resistor R as sketched in Figure 22.24B. We can calculate the current I in this circuit using phasors. The voltage across the inductor is $V_L = IX_L$, and the voltage across the resistor is $V_R = IR$. These two circuit elements are connected in series, so their currents are equal and the sum of their voltages equals the AC source voltage. It is tempting to just add our expressions for V_L and V_R, but that would *not* give the correct result. The voltages V_L and V_R must be added as phasors, so we must consider their phases when we take this sum.

Figure 22.24C is a phasor diagram showing the current I along with the voltages across L and R. For simplicity, we have placed the current phasor along the horizontal axis; this phasor describes the current through both the inductor and resistor because they are in series. The voltage across the resistor is then a phasor that also lies along the horizontal direction. (Compare with Fig. 22.9B.) The voltage across the inductor makes a phase angle $\phi = 90°$ with respect to the current in the "counterclockwise" sense (as in Figure 22.15), so the phasor representing V_L is along the vertical axis. The total voltage V_{total} across the inductor and resistor combined is the sum of these two voltage phasors (Fig. 22.24D) and has an amplitude

$$V_{\text{total}} = \sqrt{V_R^2 + V_L^2} \tag{22.37}$$

Inserting our results for V_R and V_L gives

$$V_{\text{total}} = \sqrt{V_R^2 + V_L^2} = \sqrt{(IR)^2 + (IX_L)^2} = I\sqrt{R^2 + X_L^2}$$

$$V_{\text{total}} = I\sqrt{R^2 + (2\pi fL)^2} \tag{22.38}$$

We now define a quantity called the ***impedance*** Z of the circuit through the relation

$$V_{\text{total}} = IZ \tag{22.39}$$

Impedance of an AC circuit

From Equation 22.38, we find

$$Z = \sqrt{R^2 + (2\pi fL)^2} \quad \text{(impedance for an } RL \text{ circuit)} \tag{22.40}$$

The impedance is a measure of how strongly a circuit "impedes" current. For a single resistor, the impedance is simply equal to the resistance. Likewise, for a single capacitor or inductor, the impedance is equal to the reactance of the capacitor or inductor. For combinations of these circuit elements, the impedance is given by a more complicated combination of these quantities. The result in Equation 22.40 applies only for a circuit consisting of a single resistor in series with a single

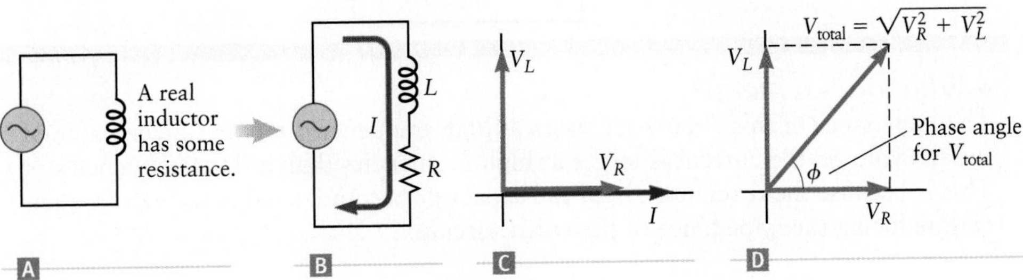

▲ **Figure 22.24** Real inductors have at least a small amount of resistance, so A an AC circuit with an inductor can be modeled as B a circuit with an ideal inductor in series with a resistor. C Phasor diagram for the current in the circuit and the voltages across the resistor and the inductor. D The total voltage across the inductor and the resistor is the phasor sum of the voltages in part C.

inductor. The impedance of other circuits can be calculated by adding phasors, as we'll see in other examples below.

Figure 22.24D contains another important lesson. In all the AC circuits we have analyzed so far, the phase angle between the current and voltage was either zero or ±90°. For an *LR* circuit, the phase angle has a value between zero and 90°, depending on the relative values of V_R and V_L, which themselves depend on *R*, *L*, and *f*. For most AC circuits, the phase angle depends on frequency and also on the specific values of the resistances, inductances, and capacitances in the circuit.

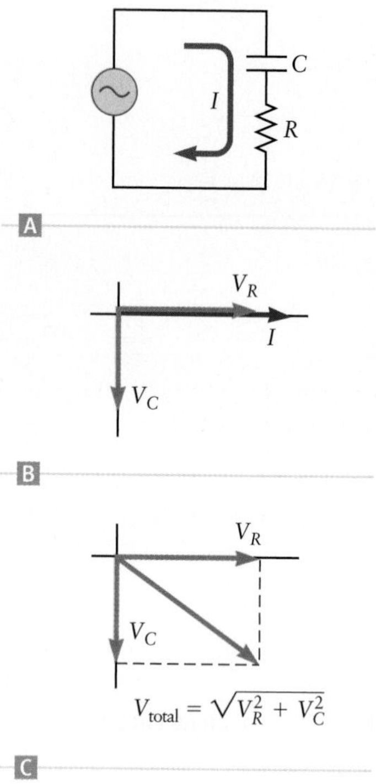

▲ **Figure 22.25** Example 22.7.

EXAMPLE 22.7 Phasor Analysis of an *RC* Circuit

Consider the circuit in Figure 22.25, with a capacitor in series with a resistor. If the frequency of the AC source is *f*, find an expression for the impedance of this circuit. Express your answer in terms of *C*, *R*, and *f*.

RECOGNIZE THE PRINCIPLE

The impedance is proportional to the total voltage (Eq. 22.39). To find the total voltage across the capacitor plus the resistor, we must add their voltages as phasors. The phasor diagrams in this case are similar to the case with an inductor (Fig. 22.24) except that the phasor representing V_C is downward along the vertical axis (Fig. 22.25B); compare this figure with the phasor diagram for a capacitor alone in Figure 22.11.

SKETCH THE PROBLEM

Figure 22.25 shows the circuit along with the corresponding phasor diagrams for the current and voltage across the capacitor and resistor.

IDENTIFY THE RELATIONSHIPS

We add the phasors representing V_R and V_C as shown in Figure 22.25C. The phasor arrows form a right triangle, so we have

$$V_{total} = \sqrt{V_R^2 + V_C^2}$$

Inserting $V_R = IR$ and $V_C = IX_C = I/(2\pi fC)$ gives

$$V_{total} = \sqrt{V_R^2 + V_C^2} = \sqrt{(IR)^2 + \left(\frac{I}{2\pi fC}\right)^2} = I\sqrt{R^2 + \left(\frac{1}{2\pi fC}\right)^2}$$

SOLVE

Comparing this expression with the definition of impedance $V = IZ$ (Eq. 22.39), we find that for this circuit the impedance is

$$\boxed{Z = \sqrt{R^2 + \left(\frac{1}{2\pi fC}\right)^2}}$$

► ***What does it mean?***

The impedance of this circuit decreases at high frequencies, so for a given value of the AC voltage, the current is larger at high frequencies than at low frequencies. This is because the reactance X_C of the capacitor becomes smaller at high frequencies, reducing the impedance of the entire circuit.

Impedance of an *LCR* Circuit

A phasor analysis for an *LCR* circuit is shown in Figure 22.22D at the resonant frequency only. We can find the impedance for the general case using a method like that in Figures 22.24 and 22.25. Beginning anew in Figure 22.26A, we again place the current phasor along the horizontal axis, so the phasor for the resistor's voltage is also in the horizontal direction. The phasor representing the voltage across the inductor is then upward along the vertical axis, and the phasor for the capacitor's voltage is downward. Collecting the horizontal and vertical components in Figure 22.26B, we add them to get the total voltage across the circuit:

$$V_{total} = \sqrt{(IR)^2 + I^2(X_L - X_C)^2} = \sqrt{(IR)^2 + I^2\left(2\pi fL - \frac{1}{2\pi fC}\right)^2}$$

$$V_{total} = I\sqrt{R^2 + \left(2\pi fL - \frac{1}{2\pi fC}\right)^2}$$

The impedance is defined through $V_{total} = IZ$, so for this circuit we get

$$Z = \sqrt{R^2 + \left(2\pi fL - \frac{1}{2\pi fC}\right)^2} \quad (22.41)$$

If the voltage provided by the AC generator is $V_{total} = V_{max}\sin(2\pi ft)$, the amplitude of the current is $I_{max} = V_{max}/Z$.

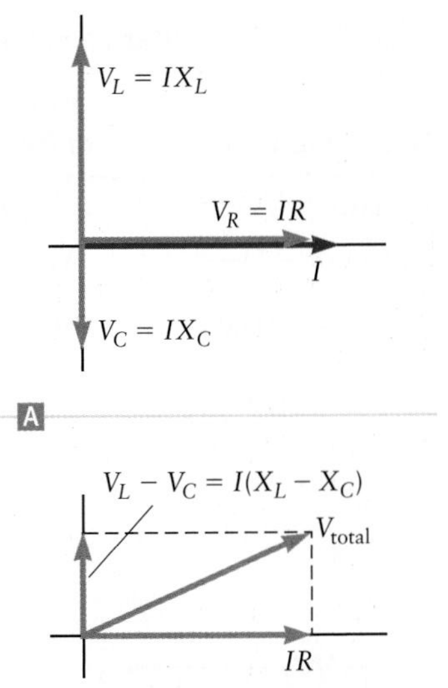

▲ **Figure 22.26** A Phasor diagram of the current in a series *LCR* circuit and the voltages across all three circuit elements for the general case in which $X_L \neq X_C$. Notice how the voltages across the capacitor and inductor are out of phase (since the phasors for V_C and V_L are in opposite directions; compare with Fig. 22.22). B Phasor sum of the voltages.

CONCEPT CHECK 22.7 Phase Angle in an *LCR* Circuit at Resonance

Is the phase angle between the current and voltage in an *LCR* circuit at resonance (a) 0, (b) 90°, or (c) 180°?

22.8 Frequency-Dependent Behavior of AC Circuits: A Conceptual Recap

We can make better sense of the circuit theory in this chapter, with its many new concepts including reactance and impedance, if we have a clear qualitative picture of why the frequency is so important in AC circuits.

Circuit Behavior at Low and High Frequencies

Resistors in an AC circuit behave very much like resistors in a DC circuit. For both types of circuits, the current in a resistor is related to the voltage by $I = V/R$ independent of the frequency, and the current is always *in phase* with the voltage (Fig. 22.9). In contrast, the current through a capacitor or an inductor is *frequency dependent*, a key aspect of AC circuits that makes them useful in applications.

The frequency-dependent behavior of AC circuits is due to the frequency dependence of the reactances X_C and X_L. For a capacitor,

$$I_{max} = \frac{V_{max}}{X_C} \quad \text{(for a capacitor)} \quad (22.42)$$

and X_C is a measure of how strongly the capacitor "resists" current. When the frequency is low, the oscillation period is long and more charge is "pushed" on and off a capacitor during each AC cycle. This larger charge produces a larger electric field in opposition; hence, the reactance X_C is largest at low frequencies. From Equation 22.42, the current through a capacitor is *smallest* at *low frequencies*. At very low frequencies, a capacitor approaches an "open circuit" and passes very little current, whereas at very high frequencies, a capacitor resembles a short circuit, passing current like a simple ideal wire with no resistance.

For an inductor, we have

$$I_{max} = \frac{V_{max}}{X_L} \quad \text{(for an inductor)} \quad (22.43)$$

TABLE 22.3 Qualitative Behavior of Capacitors and Inductors at Low and High Frequencies

Circuit Element	Resistance or Reactance	Behavior at Very Low Frequencies	Behavior at Very High Frequencies
Resistor	R	Independent of frequency	Independent of frequency
Capacitor	$X_C = \dfrac{1}{2\pi f C}$	$X_C \approx \infty$; very little current passes	$X_C \approx 0$; current passes freely
Inductor	$X_L = 2\pi f L$	$X_L \approx 0$; current passes freely	$X_L \approx \infty$; very little current passes

and X_L is a measure of how strongly an inductor "resists" current. Inductance is a result of the laws of Faraday and Lenz because an inductor opposes changes in current. The rate of change of I is largest at high frequencies, making the reactance X_L largest at high frequencies. From Equation 22.43, this fact makes the current through an inductor *smallest* at *high frequencies*. At very low frequencies, an inductor passes current freely, like a simple wire with no resistance, whereas at very high frequencies, an inductor approaches an open circuit and passes little or no current.

These qualitative observations are summarized in Table 22.3. We can use these results to get a general sense of the behavior of a circuit that contains multiple circuit elements. Consider the circuit in Figure 22.27A, an RL circuit for which we indicate the "input" and "output" voltage signals. Although we show an AC generator providing the input voltage V_{in}, this voltage might instead be provided by an antenna (as in a radio) or another AC circuit. The voltage across the inductor is the "output" voltage V_{out} and is typically connected to another portion of the circuit that is not shown here. Let's use our qualitative rules from Table 22.3 to estimate the output voltage at very low and at very high frequencies.

When the input frequency is very low, the reactance of the inductor is very small. The inductor thus acts as a simple wire, so we have redrawn the circuit with the inductor replaced by a wire in Figure 22.27B. From Chapter 19, we know that there is zero voltage drop across the ends of the wire, so V_{out} in Figure 22.27B is zero. Hence, even though the input voltage is generally not zero, the output voltage of this circuit is very small at low frequencies.

In the limit of very high frequencies, an inductor acts as an open circuit (Table 22.3) as sketched in Figure 22.27C. Because an open circuit passes no current, the current through the resistor and the voltage drop across it are approximately zero. The output voltage is then equal to the input voltage, $V_{out} = V_{in}$. The output voltage of the circuit in Figure 22.27A is therefore zero at low frequencies but equal to the input voltage at high frequencies. This circuit thus acts as a ***high-pass filter***, allowing high-frequency signals to pass through to the output while low frequencies are blocked.

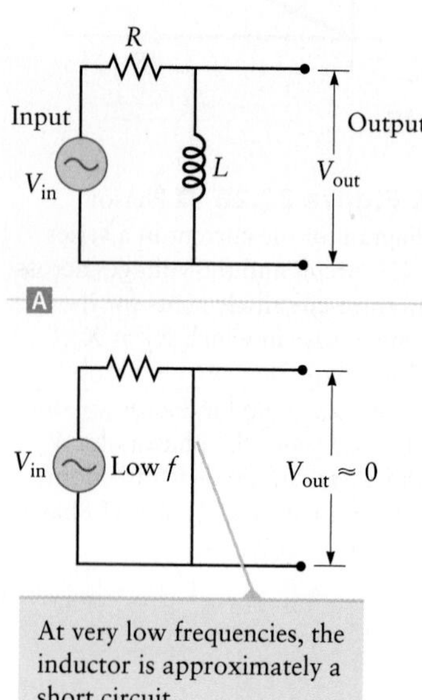

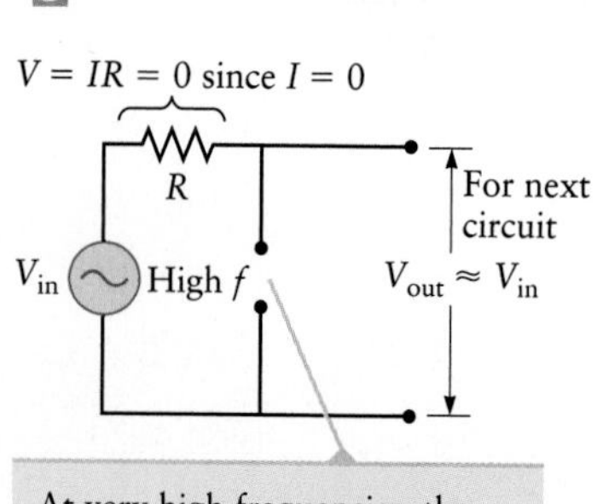

▲ **Figure 22.27** Qualitative analysis of A an LR circuit. B At very low frequencies, the reactance of the inductor X_L is very small and the inductor acts as a short circuit (i.e., a simple wire). C At very high frequencies, X_L is very large, so the inductor acts as an open circuit. The original circuit thus removes low frequencies from the output and passes high frequencies.

EXAMPLE 22.8 Ⓡ An RC Circuit at Low and High Frequencies

Figure 22.28A shows an RC circuit. Use the qualitative frequency-dependent behavior of a capacitor as given in Table 22.3 to estimate the output voltage at (**a**) very low and (**b**) very high frequencies.

RECOGNIZE THE PRINCIPLE

The qualitative behavior at low and high frequencies can be found using the frequency dependence of the reactance of a capacitor (Table 22.3), which we can use to draw "equivalent circuits" for low and high frequencies. At low frequencies, the capacitor is replaced by an open circuit (Fig. 22.28B), whereas at high frequencies, the capacitor is replaced by a short circuit (Fig. 22.28C).

SKETCH THE PROBLEM

Figure 22.28 shows the circuit, along with equivalent circuits at low and high frequencies.

IDENTIFY THE RELATIONSHIPS AND SOLVE

(a) At very low frequencies, the capacitor acts approximately as an open circuit (Fig. 22.28B) and the current is very small. The output voltage then equals the input voltage, and we have

$$V_{out} = \boxed{V_{in} \text{ at low frequencies}}$$

(b) At very high frequencies, the capacitor acts as a short circuit, so we replace it by a wire in Figure 22.28C. The output voltage is equal to the voltage across this short circuit, which is zero. Hence,

$$V_{out} = \boxed{0 \text{ at high frequencies}}$$

► ***What does it mean?***
This circuit prevents high frequencies from reaching the output (since $V_{out} = 0$ at high frequencies), so this circuit acts as a *low-pass filter*.

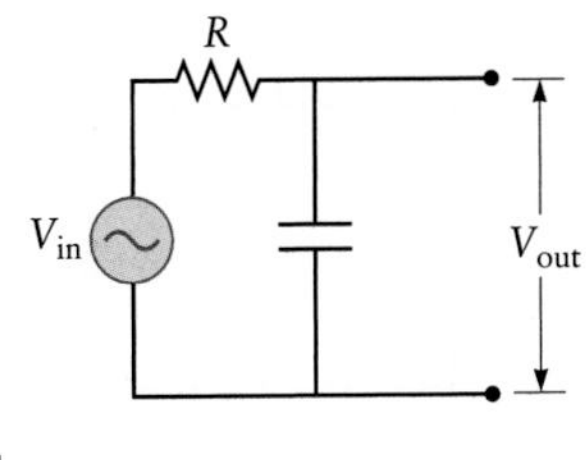

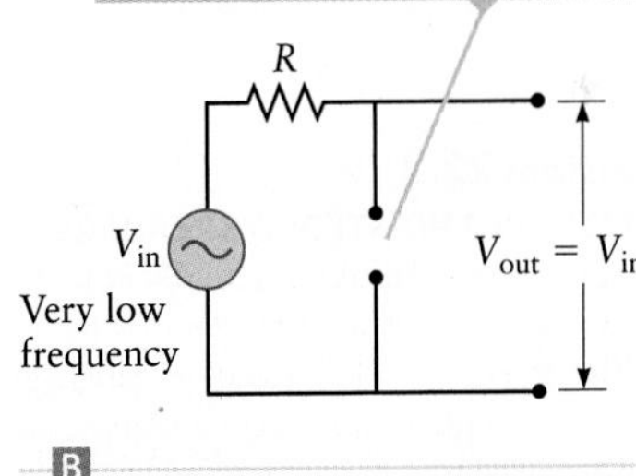

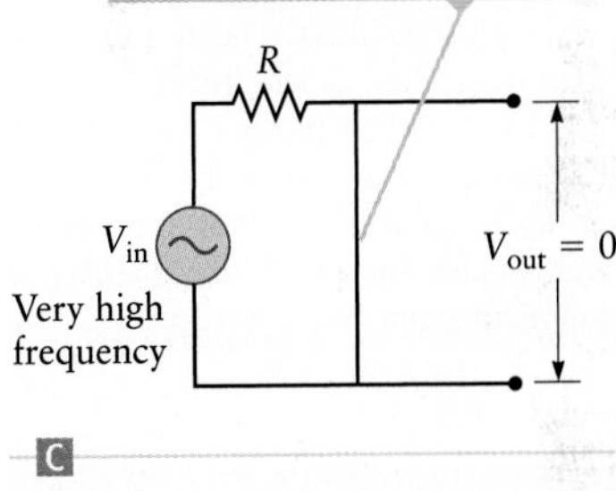

▲ **Figure 22.28** Example 22.8. Qualitative analysis of the frequency dependence of an *RC* circuit.

When Are We at Low or High Frequency?

The results from Table 22.3 are extremely useful in gaining a qualitative understanding of an AC circuit's behavior at low and high frequencies, but how do we know when a particular frequency is low enough or high enough to apply the limiting values of X_C and X_L from Table 22.3? For an *RL* circuit (Fig. 22.27), the input frequency must be compared with the corresponding *RL* time constant. The time constant of an *RL* circuit is given by $\tau_{RL} = L/R$ (Eq. 21.27). Qualitatively, you can think of this time constant as the period of an oscillation and define a corresponding frequency by

$$f_{RL} = \frac{1}{\tau_{RL}} = \frac{R}{L} \tag{22.44}$$

The high-frequency limit in Table 22.3 applies when the input or source frequency in an *RL* circuit is much greater than f_{RL}, whereas the low-frequency limit corresponds to frequencies much lower than f_{RL}. For qualitative estimates, any frequency greater than about $10 \times f_{RL}$ falls into the high-frequency limit, whereas the low-frequency limit is below about $f_{RL}/10$.

The approach is similar for an AC circuit containing a capacitor and a resistor. In this case, we have an *RC* time constant given by $\tau_{RC} = RC$ (Eq. 19.31), which leads to a corresponding frequency

$$f_{RC} = \frac{1}{\tau_{RC}} = \frac{1}{RC} \tag{22.45}$$

The high-frequency limit applies to input frequencies greater than about $10 \times f_{RC}$; the low-frequency limit is below about $f_{RC}/10$.

As a final case, we consider an AC circuit containing a capacitor and an inductor. In this case, the resonant *LC* frequency $f_{res} = 1/(2\pi\sqrt{LC})$ (Eq. 22.35) determines the boundary between the high- and low-frequency limits in Table 22.3.

Insight 22.1
APPLICATIONS OF A LOW-PASS FILTER
A low-pass filter can be constructed by replacing the inductor in Figure 22.27A with a capacitor (see Example 22.8). A low-pass filter is used in radios and MP3 players. A music (audio) signal often contains "static," which comes from unwanted high-frequency components added to the desired music. These high frequencies can be filtered out using a low-pass filter.

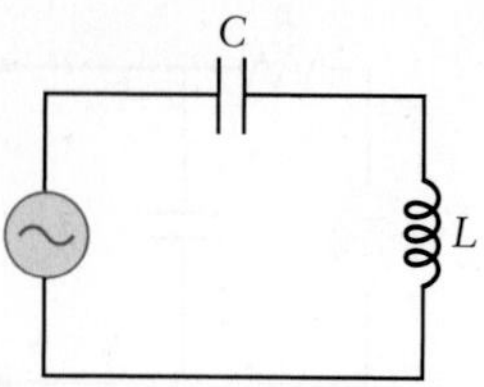

▲ **Figure 22.29** Concept Check 22.8.

CONCEPT CHECK 22.8 Frequency Dependence of an LC Circuit

Consider the LC circuit in Figure 22.29. The voltage from the AC source is $V = V_{max} \sin(2\pi ft)$, with V_{max} held fixed while the frequency f is adjustable. Which of the following statements is correct?

(a) At very high frequencies, the voltage across the inductor is greater than the voltage across the capacitor.
(b) At very high frequencies, the voltage across the inductor is less than the voltage across the capacitor.
(c) The voltage across the inductor is equal to the voltage across the capacitor at all frequencies.

Insight 22.2
FILTERS AND STEREO SPEAKERS
Many stereo speakers actually contain two separate speakers. One speaker (the tweeter) is designed to perform well at high frequencies, whereas the other (the woofer) is designed for low frequencies. The input signal to the speaker system passes through an AC circuit called a crossover network, which is a combination of low-pass and high-pass filters. The output of the high-pass filter is sent to the tweeter, and the low-pass filter is connected to the woofer. In this way, the signal sent to each speaker contains only those frequencies for which the speaker is most efficient.

22.9 Transformers

Transformers are devices that can either increase or decrease the amplitude of an applied AC voltage. The ability to "transform" an AC voltage from small to large, or from large to small, is used by electric companies to transmit electrical energy efficiently over long distances. Transformers are also found in many household devices, including the "power bricks" that connect your laptop computer to a wall socket (Fig. 22.30).

A simple transformer consists of two solenoidal coils with the loops arranged so that all or most of the magnetic field lines and flux generated by one coil pass through the other coil. This can be accomplished in several ways, two of which are illustrated in Figure 22.31. In Figure 22.31A, two coils are placed end to end, whereas Figure 22.31B shows one coil wrapped around the other. The wires are covered with a nonconducting layer ("insulation") so that current cannot flow directly from one coil to the other. Even so, an AC current in one coil will induce an AC voltage across the other one. According to Faraday's law, the magnitude of the voltage drop across a coil is related to the magnetic flux through the coil by

$$V = \frac{\Delta\Phi}{\Delta t} \tag{22.46}$$

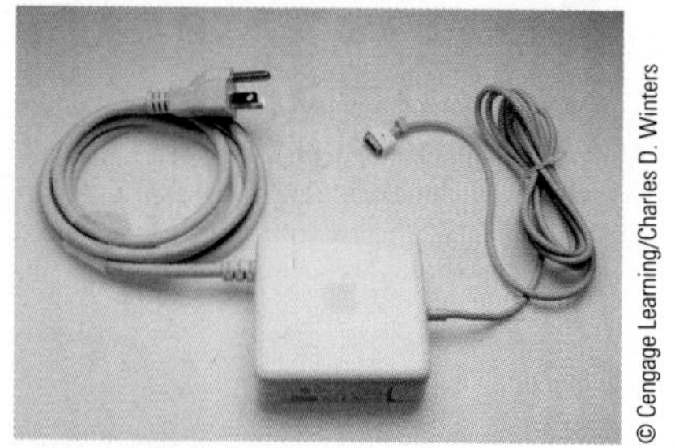

▲ **Figure 22.30** A power brick used with a laptop computer employs a transformer.

In a transformer, an AC voltage source is typically attached to one of the coils—denoted as the "input" coil—and the other is called the "output" coil. Equation 22.46 applies for both coils, so we can write

$$V_{in} = \frac{\Delta\Phi_{in}}{\Delta t} \qquad V_{out} = \frac{\Delta\Phi_{out}}{\Delta t} \tag{22.47}$$

The coils in Figure 22.31 are designed so all the field lines that pass through one coil also pass through the other. If the coils are identical (having the same diameters, lengths, and numbers of turns), the flux is the same through each coil and $\Phi_{in} = \Phi_{out}$. The fluxes in Equation 22.47 are thus the same, so the voltages are also equal:

$$V_{in} = \frac{\Delta\Phi}{\Delta t} = V_{out} \tag{22.48}$$

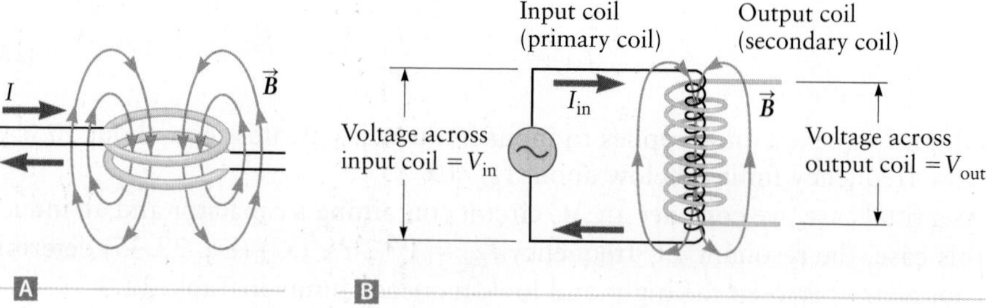

► **Figure 22.31** Two simple transformers. A End-to-end coil arrangement. B In practice, the coils are often concentric, and are wrapped so tightly that their diameters are essentially the same.

We have thus managed to "transfer" a voltage from the input coil to the output coil without any direct connection between the two. However, there is an indirect connection through the magnetic flux.

With the transformers in Figure 22.31, the output voltage is equal to the input voltage, but that could have been accomplished more simply by just connecting the input and output wires. How can we *change* the voltage? Consider again the two coils in either part of Figure 22.31, but this time let the output coil have twice as many turns as the input coil. If the flux through the input coil is Φ_{in}, the flux through the output coil will be $\Phi_{out} = 2\Phi_{in}$ because there are twice as many loops in the output coil. We can use this relation between the fluxes in Equation 22.47 to get a relation between the input and output voltages:

$$V_{out} = \frac{\Delta\Phi_{out}}{\Delta t} = 2\frac{\Delta\Phi_{in}}{\Delta t} = 2V_{in}$$

In this case, the output voltage is twice as large as the input voltage.

We can apply the same analysis to the general case of an input coil with N_{in} turns and an output coil with N_{out} turns. For a coil with N turns and a total flux Φ, the flux through a single turn is $\Phi_1 = \Phi/N$; this relation applies to both the input coil and the output coil. For the transformers in Figure 22.31 (and for most practical transformers), the flux through one turn of the output coil equals the flux through one turn of the input coil. The input and output fluxes are then related by

$$\Phi_{out} = \frac{N_{out}}{N_{in}}\Phi_{in}$$

Using this result with Equation 22.47, the output voltage is

$$V_{out} = \frac{N_{out}}{N_{in}}V_{in} \quad (22.49)$$

Output voltage of a transformer

Since the ratio N_{out}/N_{in} can be larger or smaller than unity, we can thus "transform" the input voltage to a different value, either larger or smaller than the input voltage. That is why the device is called a transformer. Transformers are based on Faraday's law, so, to apply Equation 22.47, the flux must be changing with time. Hence, transformers cannot be used to "transform" a DC voltage.

The transformer designs in Figure 22.31 are simplest to understand, but the different arrangement shown in Figure 22.32 is used in most practical applications. The central regions of both coils are filled with a magnetic material such as iron. The field from the input coil aligns the atomic magnets in the magnetic material, producing a much larger flux than would be produced by the coil's field alone. This larger flux then yields larger voltages at both the input coil and the output coil. The ratio V_{out}/V_{in} is not affected by the presence of magnetic material in the transformer, however, and is still given by Equation 22.49.

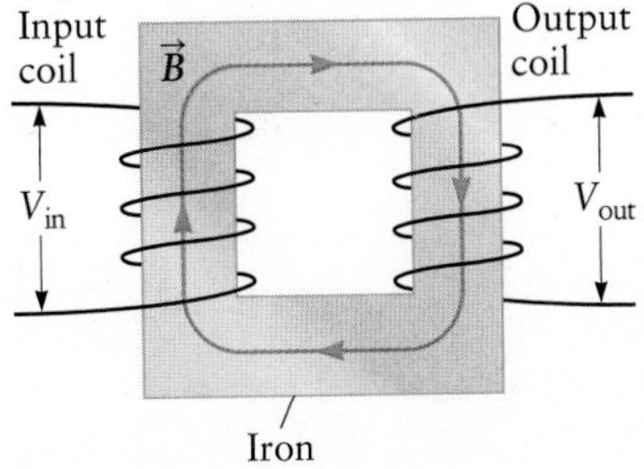

▲ **Figure 22.32** Many transformers use a magnetic material such as iron to "carry" the flux from the input coil to the output coil.

CONCEPT CHECK 22.9 How Does a Transformer Transform Frequency?

A transformer has an output coil with three times as many turns as the input coil. According to Equation 22.49, the output voltage is thus three times larger than the input voltage. If the input frequency is f, is the output frequency (a) f, (b) $3f$, or (c) $f/3$?

Applications of Transformers

Transformers are used in many household appliances to convert the AC voltage at a wall socket (120-V rms) to the smaller voltages needed in radios, computers, and other devices that commonly require DC voltages around 9 V or 12 V. These DC voltages are produced in a two-step process by power bricks like the one in Figure 22.30 or by power supplies built into an appliance. The first step uses a transformer to convert the 120 V from a wall socket to a value near 9 V or 12 V. The second step is to convert this smaller AC voltage to a DC voltage, using a circuit called a "rectifier." We don't have space here to describe how a rectifier works, but its output is a DC voltage, similar to the output of a battery. That is why a power brick can substitute for batteries for your laptop computer.

Transforming Voltage and Power

According to Equation 22.49, the output voltage of a transformer can be much larger than the input voltage by just arranging for the number of turns in the output coil to be much larger than the number of turns in the input coil. Electrical power delivered to a circuit is the product of the voltage and current (Eq. 22.9). So, if V_{out} is much larger than V_{in}, what happens to the output power? Is the output power also much larger than the input power? We can apply the principle of conservation of energy to answer these questions.

A transformer is like an inductor because it stores energy in the magnetic field produced by the two coils. This energy is *stored*; it is not "lost" or dissipated as heat as in a resistor. According to energy conservation principles, the energy delivered through the input coil must either be stored in the transformer's magnetic field or transferred to the output circuit. Over many AC cycles, the stored energy is constant, so the power delivered to the input coil must *equal* the output power. The input power is $P_{in} = V_{in}I_{in}$, whereas the output power is $P_{out} = V_{out}I_{out}$. Hence, if V_{out} is transformed to a value larger than V_{in}, the output current is simultaneously transformed to a smaller value than the input current.

This argument applies only to ideal transformers. The coils of a real transformer always have a small electrical resistance, causing some power dissipation. Hence, for a real transformer, the output power is always less than the input power, but usually by only a small amount.

22.10 Motors

In Chapter 20, we described how the magnetic torque on a current loop can be used to make a DC motor (Section 20.10). Another way to make a motor is shown in Figure 22.33; this approach also uses the torque on a current loop, but now we use an AC voltage source to "power" the motor. The AC source is connected to a coil (called the input coil) wound around a horseshoe magnet. The coil's field aligns the domains in the horseshoe magnet so that this magnet's field is parallel to the field due to the coil. The resulting field is much larger than the field from the coil alone and circulates through the horseshoe magnet as shown in the figure. As the current

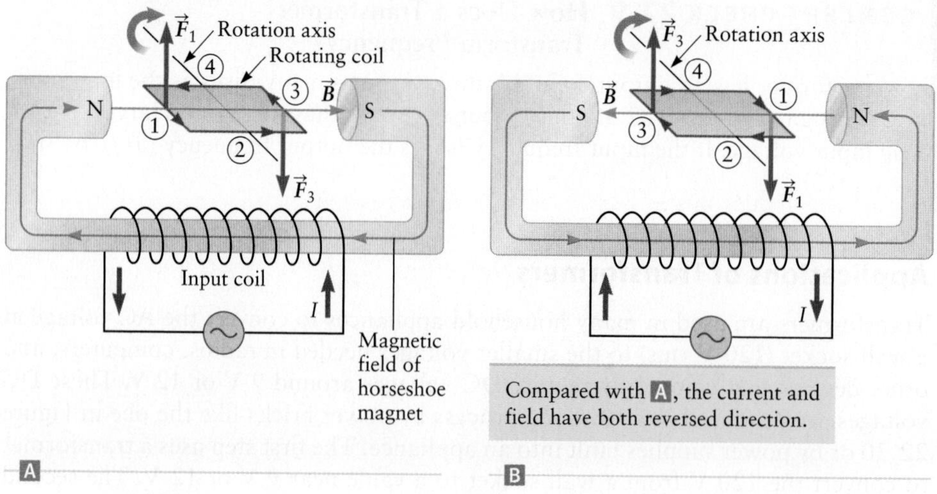

▶ **Figure 22.33** Design of an AC motor.

in the input coil oscillates, the direction of the magnetic field from the horseshoe magnet alternates back and forth as indicated in Figure 22.33.

A second coil of wire is mounted between the poles of the horseshoe magnet and attached to a rotating shaft. The rotating coil is connected to a source of DC voltage (i.e., a battery), so this coil carries a DC current. Figure 22.33A shows the rotating coil in a particular position and the north pole of the horseshoe magnet on the left. From Chapter 20, when a current-carrying wire—that is, a section of the rotating coil—is placed in a magnetic field, it experiences a force. Applying right-hand rule number 2 to calculate the direction of this force, we find the results shown in Figure 22.33A. The force $\vec{F}_1$ on section 1 of the coil is directed upward, whereas the force $\vec{F}_3$ on section 3 is down. There is no force on sections 2 and 4 because the current in those sections is parallel or antiparallel to the magnetic field. The forces $\vec{F}_1$ and $\vec{F}_3$ produce a torque on the coil, causing it and the shaft to rotate. Hence, the motor turns.

After a short time, the coil rotates to the position in Figure 22.33B and at the same time the AC current in the input coil changes direction so that the north pole of the horseshoe magnet is now on the right. Applying right-hand rule number 2 to calculate the forces on the current loop, we find that $\vec{F}_1$ is now downward and $\vec{F}_3$ is upward. Comparing the two parts of Figure 22.33, we see that the torques on the coil are in the same direction; both torques are clockwise as viewed from the front of the drawing. Hence, the motor shaft continues to turn in a clockwise manner.

As the coil continues to rotate, the field from the horseshoe magnet continues to change direction so that the torque exerted on the coil is always clockwise. In this way, the oscillations of an AC current and field make the shaft rotate. Practical AC motor designs are more complicated than shown here, with typically three separate magnets (mounted at different angles) and a slightly more complicated rotating coil, but the principle is the same as shown in Figure 22.33.

EXAMPLE 22.9 Torque Produced by an AC Motor

Consider an AC motor like the one in Figure 22.33 and let the rotating coil be square with sides of length $L = 10$ cm, carrying current $I = 1.0$ A. If the field of the horseshoe magnet is $B = 1.5$ T, what is the magnitude of the torque on the coil when it is oriented as shown in Figure 22.33A?

RECOGNIZE THE PRINCIPLE

The torque exerted on the coil is produced by the forces $\vec{F}_1$ and $\vec{F}_3$ as shown in more detail in Figure 22.34. In each case, we have a current-carrying wire in a magnetic field perpendicular to the current, so from Chapter 20 each force has magnitude $F = ILB$. The directions of these forces are given by right-hand rule number 2 and are shown in Figure 22.34; each gives a torque with a lever arm $L/2$ (the distance from the rotation axis to the point where the forces act). From Chapter 8 (Eq. 8.9), the torque associated with $\vec{F}_1$ is thus

$$\tau_1 = F_1 \times (\text{lever arm}) = \frac{F_1 L}{2}$$

In a similar way, the torque from $\vec{F}_3$ is

$$\tau_3 = F_3 \times (\text{lever arm}) = \frac{F_3 L}{2}$$

These torques combine to give the total torque on the shaft.

SKETCH THE PROBLEM

Figure 22.34 shows the problem.

(continued) ▶

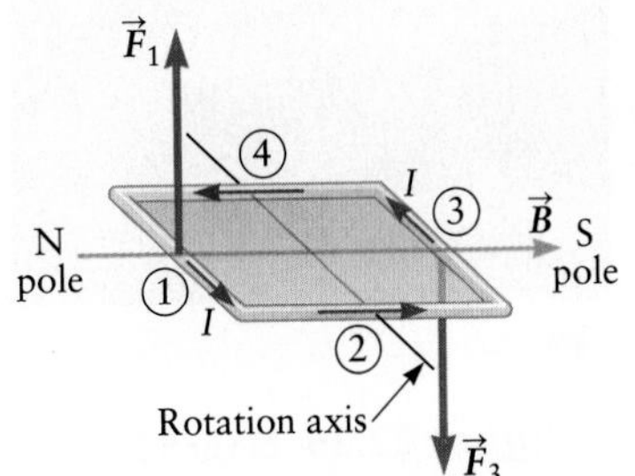

▲ **Figure 22.34** Example 22.9. Torque on the rotating coil of the AC motor in Figure 22.33A. The coil is composed of four straight sections.

IDENTIFY THE RELATIONSHIPS

The torques τ_1 and τ_3 both tend to make the shaft rotate clockwise, so the magnitude of the total torque is the sum

$$\tau_{\text{total}} = \tau_1 + \tau_3 = \frac{F_1 L}{2} + \frac{F_3 L}{2} = \frac{L}{2}(F_1 + F_3) \quad (1)$$

To compute the total torque from Equation (1), we must find the forces F_1 and F_3. The force on a wire in a magnetic field B perpendicular to the current is $F = ILB$ (from Chapter 20). Both sections of the wire have length L and carry the same current I, so we find

$$F_1 = ILB = F_3$$

Inserting this expression into Equation (1), we get

$$\tau_{\text{total}} = \frac{L}{2}(F_1 + F_3) = \frac{L}{2}(2ILB) = IL^2B$$

SOLVE

Inserting the given values of the various quantities gives

$$\tau_{\text{total}} = IL^2B = (1.0\text{ A})(0.10\text{ m})^2(1.5\text{ T}) = \boxed{0.015\text{ N}\cdot\text{m}}$$

► *What does it mean?*

This torque is not very large, so this design would make a very weak motor. In more practical motors, the coil would consist of many turns (hundreds or thousands or more), giving a much larger torque.

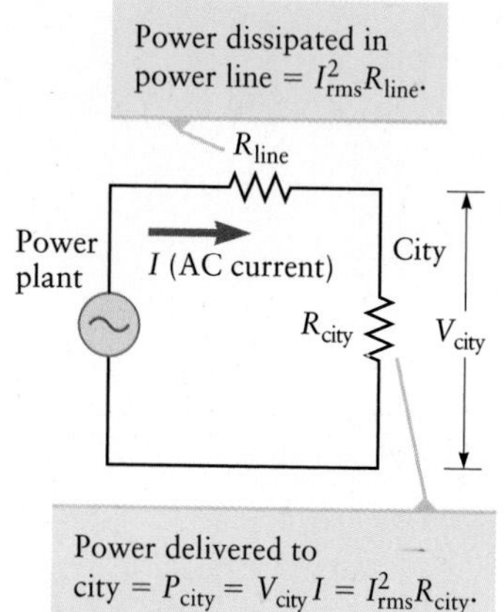

▲ **Figure 22.35** A Circuit diagram showing a power line carrying current from an electric power plant to a city. B Photo of a transformer used to transform the high voltage at a power line to a lower voltage suitable for use in a house.

22.11 What Can AC Circuits Do That DC Circuits Cannot?

At the beginning of this chapter, we mentioned that Thomas Edison proposed to wire cities and buildings with DC circuits, whereas George Westinghouse and others pushed for an AC system. Westinghouse won the argument, but why? What advantages does alternating current have over direct current for such applications? One advantage discussed in Chapter 21 is that large AC generators are easily built using a rotating shaft and permanent magnets. The biggest advantage of AC circuits over DC circuits, however, is in the systems that distribute electrical power across long distances. Most electrical power is generated at large facilities such as hydroelectric dams or coal-burning power plants. The electrical power produced at these facilities must then be distributed to distant cities as shown schematically in Figure 22.35A. The power plant acts as an AC generator, sending current through power lines (large, overhead wires) to the destination. These wires have a small but nonzero electrical resistance R_{line}, so if the power line carries an rms current I_{rms}, there is some power $P_{\text{ave}} = I_{\text{rms}}^2 R_{\text{line}}$ dissipated in the power line resistance. Because electrical energy costs money, the power company prefers to minimize the energy dissipated in the power line. The resistance of the power line usually can't be changed easily; hence, the only way to reduce the dissipated power is to make I_{rms} as small as possible. The power delivered to the destination city is $P_{\text{city}} = V_{\text{city}} I$, so if the power company decreases the current in the power line, it must increase the voltage so as to transmit the same amount of power to the city. This voltage increase is accomplished using a transformer.

The normal AC household voltage is "120-V," an rms value. In Example 22.2, we showed that this value actually corresponds to a voltage amplitude $V_{\text{max}} = 170$ V. A much larger AC voltage is used for power transmission. Typical power line voltages have $V_{max} = 500{,}000$ V or even higher. When the power line reaches your house, a transformer reduces the AC voltage amplitude to the value $V_{max} = 170$ V required by your household appliances (Fig. 22.35B). Even with a power line voltage of 500,000 V, typically 5% to 10% of the energy that leaves a power plant is dissipated in the resistance of the power line. The fraction lost would be much greater if the power line were operated at lower AC voltage amplitudes.

Transformers can readily change AC voltage amplitudes, but there is no simple and efficient way to increase and decrease DC voltage levels without dissipating significant amounts of energy. That is the main reason alternating current was adopted for the electrical power system.

EXAMPLE 22.10 Power Dissipated in an Electrical Transmission Line

An AC power line operates with voltage $V_{rms} = 500{,}000$ V and carries AC current with $I_{rms} = 1000$ A. **(a)** What is the average power carried by the power line? **(b)** If 10% of this power is dissipated in the power line itself, what is the resistance of the power line? **(c)** This power line is now operated with voltage $V_{rms} = 250{,}000$ V and current $I_{rms} = 2000$ A. The product $V_{rms}I_{rms}$ is the same as in **(a)**, so the power carried by the line is the same. What percentage of this power is now dissipated in the power line?

RECOGNIZE THE PRINCIPLE

The total power in any circuit element is $P = VI$, so the power carried by the power line is $P_{ave} = V_{rms}I_{rms}$ (Eq. 22.12). Some of this power is dissipated in the line, and the rest is delivered to the city. The power line can be modeled as a resistor R_{line}, and the power dissipated in this resistance is $P_{line} = I_{rms}^2 R_{line}$. We can use these relations for P_{ave} and P_{line} to **(a)** find the average power carried by the line, **(b)** find the line resistance, and **(c)** analyze the effect of changing the line voltage.

SKETCH THE PROBLEM

Figure 22.35A describes the problem.

IDENTIFY THE RELATIONSHIPS AND SOLVE

(a) The total power carried by the power line is

$$P_{total} = P_{ave} = V_{rms}I_{rms} = (500{,}000\text{ V})(1000\text{ A})$$

$$P_{total} = 5.0 \times 10^8\text{ W} = \boxed{500\text{ MW}}$$

(b) Most of this power is delivered (and sold) to the city, whereas 10% is dissipated in the power line. As a resistor, the line dissipates an amount of power $P_{line} = I_{rms}^2 R_{line}$, which is equal to 10% of P_{total}, or $0.10 \times P_{total}$. We thus have

$$P_{line} = I_{rms}^2 R_{line} = (0.10)P_{total}$$

Solving for R_{line},

$$R_{line} = \frac{(0.10)P_{total}}{I_{rms}^2} = \frac{(0.10)(5.0 \times 10^8\text{ W})}{(1000\text{ A})^2} = \boxed{50\ \Omega}$$

(c) If the power line voltage is reduced by a factor of two (to 250,000 V) while the current is increased by the same factor (to 2000 A), the total power delivered is unchanged and we still have $P_{total} = 5.0 \times 10^8$ W; see part (a). The power dissipated in the line is now

$$P_{line} = I_{rms}^2 R_{line} = (2000\text{ A})^2(50\ \Omega) = 2.0 \times 10^8\text{ W}$$

(continued) ▶

The percentage of the total power now dissipated in the line is

$$\text{percent dissipated} = \frac{P_{\text{line}}}{P_{\text{total}}} \times 100 = \frac{2.0 \times 10^8 \text{ W}}{5.0 \times 10^8 \text{ W}} \times 100$$

$$\text{percent dissipated} = \boxed{40\%}$$

► ***What does it mean?***

A much larger fraction of the energy is "lost" in the power line in part (c), resulting in much smaller profits for the power company. The fraction of the energy dissipated in the line decreases when the system operates at higher voltages and hence lower currents. That is why power lines are operated with very high AC voltages.

Summary | CHAPTER 22

Key Concepts and Principles

Root-mean-square (rms) voltage

The ***instantaneous voltage*** V in an AC circuit varies with time as

$$V = V_{\text{max}} \sin(2\pi f t) \qquad \textbf{(22.1) (page 662)}$$

where V_{max} is the voltage ***amplitude*** and f is the frequency. The ***root-mean-square*** (rms) voltage is

$$V_{\text{rms}} = \frac{V_{\text{max}}}{\sqrt{2}} \qquad \textbf{(22.7) (page 663)}$$

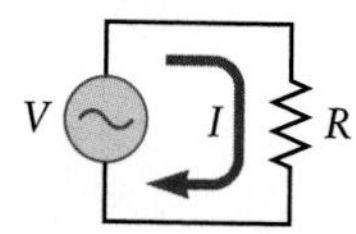

AC voltage and current in a resistor: phasors

In an AC circuit containing a resistor, the instantaneous current is

$$I = \frac{V}{R} = \frac{V_{\text{max}}}{R} \sin(2\pi f t) \qquad \textbf{(22.2) and (22.3) (page 662)}$$

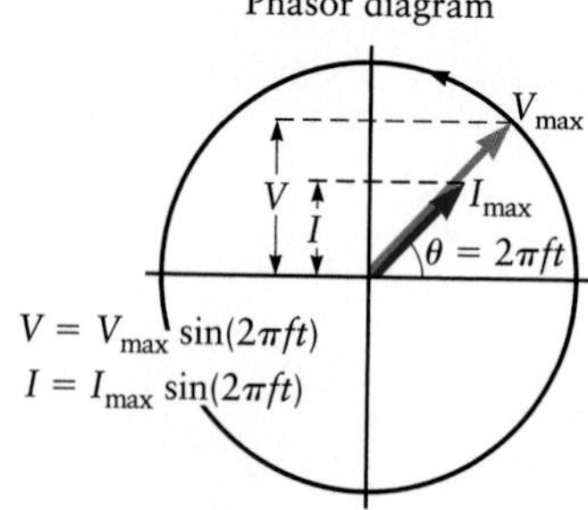

This current is ***in phase*** with the voltage. The time dependences of V and I are represented in a ***phasor diagram***.

The ***average power*** dissipated in the resistor is

$$P_{\text{ave}} = V_{\text{rms}} I_{\text{rms}} \qquad \textbf{(22.12) (page 664)}$$

where I_{rms} is the rms current.

Reactance of a capacitor in an AC circuit

In an AC circuit containing a capacitor, the current and voltage are out of phase as shown in a phasor diagram. The amplitude of the current is

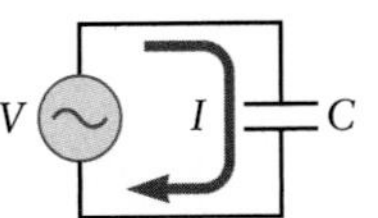

$$I_{max} = \frac{V_{max}}{X_C} \qquad (22.21) \text{ (page 668)}$$

(page 668)

where X_C is the ***reactance*** of the capacitor given by

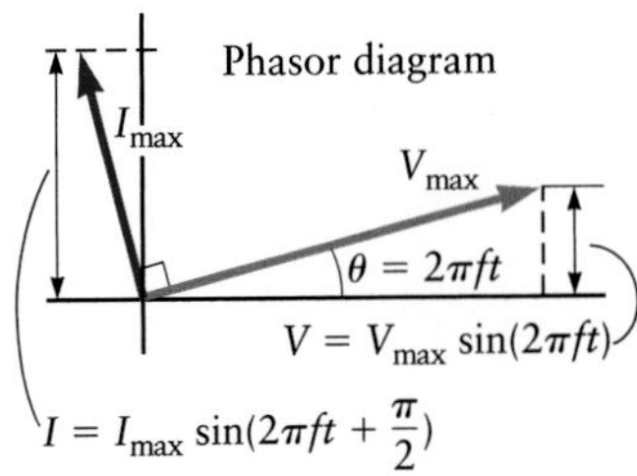

$$X_C = \frac{1}{2\pi fC} \qquad (22.22) \text{ (page 668)}$$

The average power delivered to the capacitor is $P_{ave} = 0$.

Reactance of an inductor in an AC circuit

In an AC circuit containing an inductor, the current and voltage are again out of phase as shown in a phasor diagram The amplitude of the current is

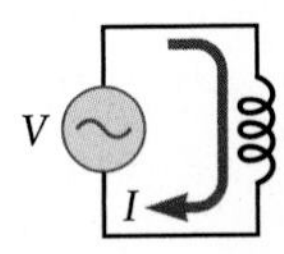

$$I_{max} = \frac{V_{max}}{X_L} \qquad (22.27) \text{ (page 671)}$$

where X_L is the ***reactance*** of the inductor and is given by

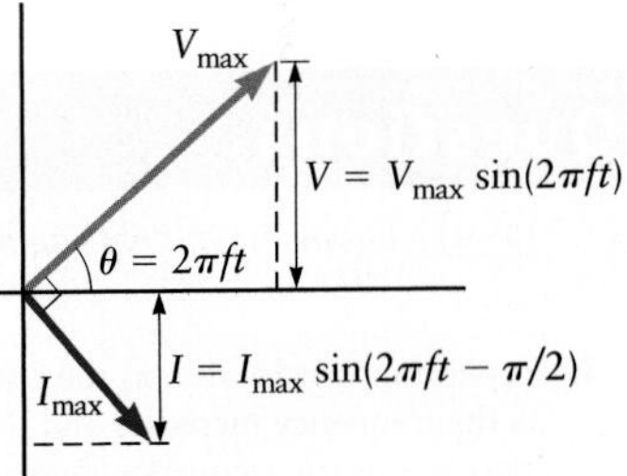

$$X_L = 2\pi fL \qquad (22.28) \text{ (page 671)}$$

The average power delivered to the inductor is $P_{ave} = 0$.

For both capacitors and inductors, the circuit behavior *depends on the frequency*.

LC circuits

An LC circuit acts as a simple harmonic oscillator; both the charge on the capacitor and the current through the circuit oscillate with time:

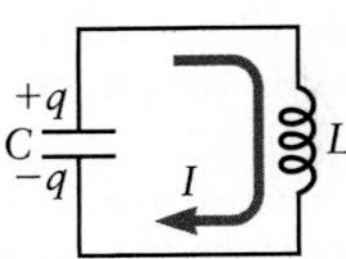

$$q = q_{max} \cos(2\pi ft) \quad \text{and} \quad I = I_{max} \sin(2\pi ft) \qquad (22.30) \text{ (page 673)}$$

The frequency of the oscillations is

$$f_{res} = \frac{1}{2\pi\sqrt{LC}} \qquad (22.35) \text{ (page 674)}$$

Impedance

The amplitude of the current in a general AC circuit is related to the voltage amplitude by

$$V = IZ \qquad (22.39) \text{ (page 677)}$$

where Z is the ***impedance*** of the circuit. The impedance Z can be found from an AC circuit analysis using phasors.

(*Continued*)

Applications

Frequency-dependent behavior of AC circuits

Reactances of capacitors and inductors are frequency dependent. At very low frequencies a capacitor acts as an open circuit and an inductor approximates a simple wire (a short circuit). In the opposite limit, at very high frequencies a capacitor acts as a short circuit and an inductor acts as an open circuit. These frequency dependences are due to the dependences of the reactances (Eqs. 22.22 and 22.28) on frequency and lead to applications such as high- and low-pass filters.

Transformers

A ***transformer*** changes the amplitude of an AC voltage. It is an essential element of power transmission circuits.

Generators and motors

Alternating-current machines can convert rotational kinetic energy into electrical energy (a ***generator***) and electrical energy into rotational motion (a ***motor***).

Questions

SSM = answer in *Student Companion & Problem-Solving Guide* = life science application

1. Explain in words why (a) the reactance of a capacitor decreases as the frequency increases and (b) the reactance of an inductor increases as the frequency increases.

2. An AC circuit contains a capacitor in series with a resistor. If the frequency of the voltage source is increased while its amplitude is held fixed, does the current amplitude increase, decrease, or stay the same?

3. A resistor is connected in series with an inductor (Fig. Q22.3). If the frequency of the AC voltage source is increased while its amplitude V_{max} is kept fixed, does the current amplitude increase, decrease, or stay the same?

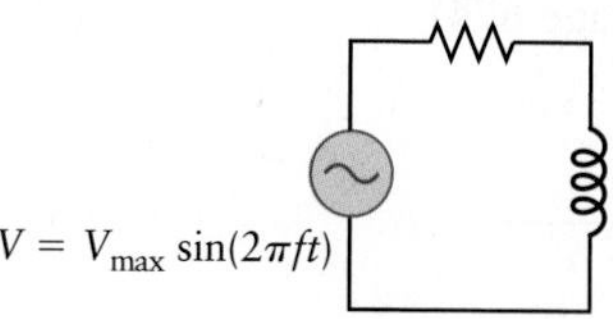

Figure Q22.3

4. What is in the box? Figure Q22.4 shows an AC voltage source connected to a resistor in series with an unknown circuit element hidden in a box. The box might contain a single resistor, a single capacitor, or a single inductor. (a) If the magnitude of the impedance as measured between points A and B increases as the frequency increases, what is in the box? (b) If the magnitude of the impedance as measured between points A and B decreases as the frequency increases, what is in the box? (c) If the magnitude of the impedance as measured between points A and B does not change as the frequency increases, what is in the box?

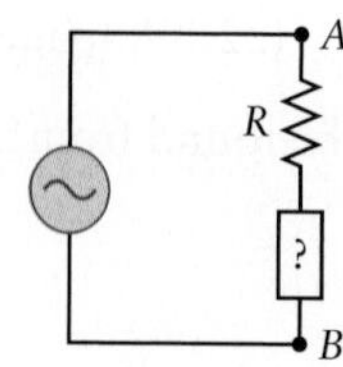

Figure Q22.4

5. An AC source with a frequency of 60 Hz is connected to a lightbulb. The instantaneous brightness of the lightbulb depends on the magnitude of the instantaneous current. Explain why the brightness of a bulb can flicker (i.e., oscillate with time) and find the frequency of the flicker. Why don't we notice the flicker?

6. The AC generator described in Figure Q22.6 (and in Section 22.1) uses a rotating coil to produce an AC voltage through electromagnetic induction as described by Faraday's law. (a) Explain how this approach can produce a voltage whose frequency is equal to the rotation frequency of the shaft. (b) Explain how you could modify the rotating coil so that the frequency of the AC voltage is twice the rotation frequency of the shaft. (c) Explain how you could modify the magnets in Figure Q22.6 so that the frequency of the AC voltage is four times the rotation frequency of the shaft.

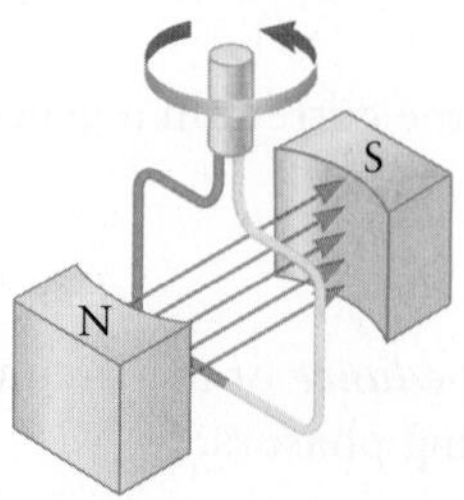

Figure Q22.6

7. Discuss why it is relatively easy to change the amplitude of an AC voltage to a higher or lower value, but why it is difficult to change its frequency.

8. Most AC voltage sources produce a voltage that varies sinusoidally with time, but some produce a "square wave" as sketched in Figure Q22.8. Use the general definition of rms voltage to find the relation between the amplitude of the square wave and the rms voltage. That is, if the square wave amplitude is V_{sq}, find V_{rms}.

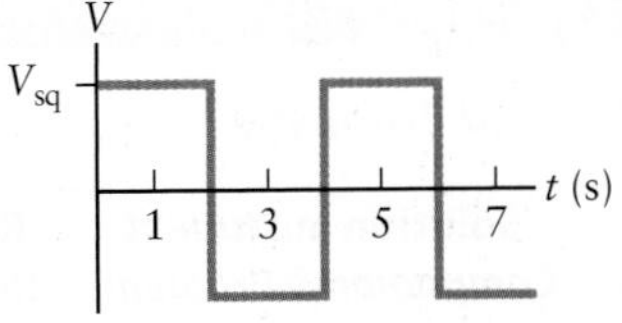

Figure Q22.8

9. Consider two AC voltage sources connected in series with voltages $V_1 = V_{1,\,max}\sin(2\pi ft)$ and $V_2 = V_{2,\,max}\sin(2\pi ft + \pi/4)$, and $V_{1,\,max} = V_{2,\,max} = 30$ V. Show that these two sources in series are equivalent to a single source with $V_{series} = V_{equiv}\sin(2\pi ft + \phi)$ and find the values of V_{equiv} and ϕ.

10. If the voltage across a capacitor is $V = V_{max}\sin(2\pi ft)$, the charge on the capacitor is also proportional to $\sin(2\pi ft)$ (see Eq. 22.17). The current through the capacitor in this case is given by $I = I_{max}\sin(2\pi ft + \pi/2)$; hence, the phase angle is $\phi = \pi/2$ radians relative to the voltage. Suppose we begin instead by assuming the capacitor voltage is proportional to a cosine function $V = V_{max}\cos(2\pi ft)$. Use a graphical argument to show that the current is now given by a function of the form $I = I_{max}\cos(2\pi ft + \pi/2)$ so that the phase of the current relative to the voltage is again $\phi = \pi/2$ rad.

11. Show that the reactance X_C of a capacitor has units of ohms. Explain why it makes sense to compare the reactance of a capacitor with the resistance of a resistor.

12. The reactance of an inductor is given by $X_L = 2\pi fL$. Show that this reactance has units of ohms. Explain why it makes sense to compare the reactance of an inductor with the resistance of a resistor.

13. SSM Consider the circuit in Figure Q22.13. (a) Draw equivalent circuits that apply at high and low frequencies. (b) Estimate the ratio V_{out}/V_{in} at high and low frequencies. (c) Explain how this circuit acts as a filter that blocks high frequencies but lets low frequencies pass. It is one type of "low-pass" filter.

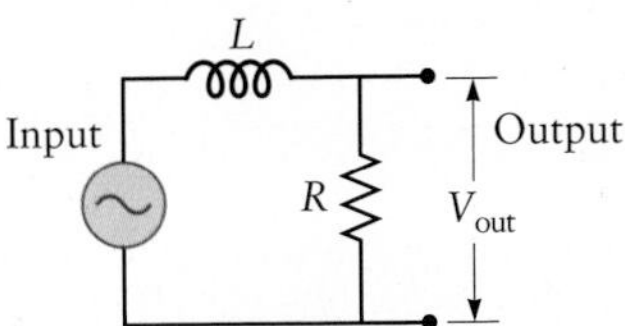

Figure Q22.13

14. A capacitor is connected in parallel with an inductor, and this circuit is connected to an AC voltage source of frequency f. It is found that the impedance of this circuit increases if the frequency is increased by a small amount. Is f above or below the resonant frequency of the circuit? Explain.

15. An LC circuit is called an oscillator. What, physically, is oscillating in such a circuit? What variables are periodic?

16. The equations describing an LC oscillator are very similar in form to those describing the mechanical harmonic oscillator of a mass on a spring. Identify the term in the equation for the mechanical oscillator that plays the role of C. What term plays the role of L? How are these corresponding pairs of variables similar?

17. An LCR circuit with a resonant frequency of 40 Hz is connected to a standard U.S. wall outlet of 120 V AC. Does the current in the circuit "lag" or "lead" the emf? That is, does the current have its maximum before or after the maximum in the voltage?

18. How is a phasor different from a vector? Is the emf a vector quantity?

19. In actual transformers (Fig. Q22.19), the iron cores are actually comprised of thin sheets of metal laminated together with insulating layers in between. Why would the cores be made this way?

Figure Q22.19 Transformer with a laminated core.

20. Although transformers are AC devices, is there some way to test a transformer to see if it works with a battery? Explain.

21. SSM **Boosting and bucking transformers.** Sometimes an electrical device may need an input voltage that is a few percent higher or lower than the supply voltage. A ***boosting transformer*** connects the output coil with the input voltage (here, a 120-V AC source) as shown on the left in Figure Q22.21. (a) Draw the sine waveforms of the input and output coils of the transformer and show how they add to produce the 150-V output. (b) If the leads from the output coil are connected as shown on the right in Figure Q22.21, a ***bucking transformer*** configuration results. (It has this name because two voltages are combined in "opposition" to make the output.) What output voltage would you expect from the circuit on the right in Figure Q22.21? Again draw the sine waveforms of the input and output coils, and the final output waveform. (Note that when the voltages from the input and output coils are combined, you must consider the phases of these voltages relative to each other.)

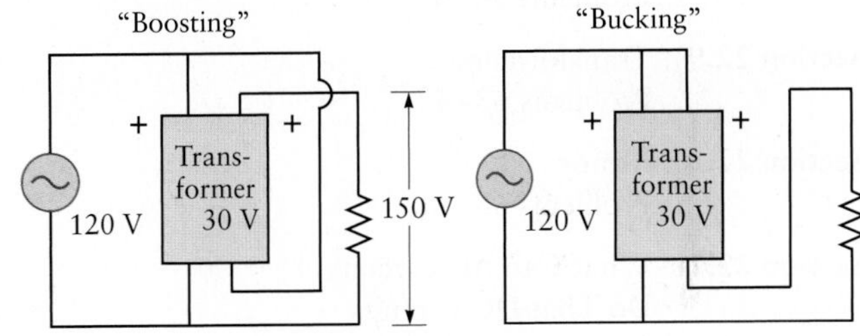

Figure Q22.21

22. Some resistors are made from a coil of wire. What is the best way to model the circuit behavior of such a resistor? Choose one answer and explain.
(a) As an ideal resistor in series with an ideal inductor?
(b) As an ideal resistor in parallel with an ideal inductor?
(c) As an ideal resistor in parallel with an ideal capacitor?

23. For an ideal generator, the electrical energy generated equals the work W done in making the shaft rotate. For a real generator, is the electrical energy greater than, equal to, or less than W? Explain.

24. **Distributed circuits.** When applying Kirchhoff's loop rule to an AC circuit, one assumes the current in the loop is the same through all circuit elements at any particular instant. Changes in current travel at the speed of light (see Chapter 23), so a certain amount of time is actually required for the AC changes to propagate from one part of a circuit to another. (a) Estimate how much time Δt is needed for these changes to travel across a circuit 5 cm in size. Assume the propagation speed in a metal wire is approximately 3×10^8 m/s. (b) Use your value of Δt to find the lowest frequency at which the current can be a maximum in one part of the circuit and a minimum at another part. Cases like this one at high frequencies are called ***distributed circuits*** and must be analyzed using a generalized version of Kirchhoff's laws, which accounts for the propagation time.

Problems available in ENHANCED WebAssign

Section 22.1 Generation of AC Voltages
Problems 1–7

Section 22.2 Analysis of AC Resistor Circuits
Problems 8–17

Section 22.3 AC Circuits with Capacitors
Problems 18–31

Section 22.4 AC Circuits with Inductors
Problems 32–40

Section 22.5 *LC* Circuits
Section 22.6 Resonance
Problems 41–49

Section 22.7 AC Circuits and Impedance
Problems 50–53

Section 22.8 Frequency-Dependent Behavior of AC Circuits: A Conceptual Recap
Problems 54–62

Section 22.9 Transformers
Problems 63–67

Section 22.10 Motors
Problems 68–70

Section 22.11 What Can AC Circuits Do That DC Circuits Cannot?
Problems 71–73

Additional Problems
Problems 74–88

List of Enhanced Problems

Problem number	Solution in *Student Companion & Problem-Solving Guide*	Reasoning & Relationships Problem	Reasoning Tutorial in ENHANCED WebAssign
22.1		✓	✓
22.3	✓		
22.9	✓		
22.25	✓		
22.37	✓		
22.47	✓		
22.50	✓		
22.58	✓		
22.69	✓		
22.73	✓		
22.78		✓	✓
22.79		✓	✓
22.80		✓	✓
22.85	✓		

CHAPTER

23

► *Your cell phone uses electromagnetic radiation to communicate with other phones.* (© *Martin Barraud/Alamy*)

Electromagnetic Waves

OUTLINE

Theoretical understanding of electricity and magnetism developed rapidly during the early 1800s and seemed to be complete by about 1850. The laws of Coulomb and Gauss describing electric forces and fields and of Ampère and Faraday describing magnetic forces and fields were verified in many experiments and provided the basis of new technologies as key parts of the industrial revolution. Despite this success, some deep questions at the heart of electromagnetic theory remained unanswered. One concerned the nature of electric and magnetic field lines and the notion of action at a distance. Consider two electric charges q_1 and q_2 located in distant space, far from all other matter. The electric field associated with q_1 is felt by (i.e., produces a force on) q_2. Now suppose q_1 moves at some instant in time to a new location. How fast will the electric field lines move along with q_1? Is the motion of q_1 felt *instantaneously* at q_2, or is there some delay, and if so, what is the delay time?

Physicist James Clerk Maxwell pondered some of these questions in the mid-1800s. His work led to what is probably the greatest physics discovery of the 19th century, the discovery of electromagnetic waves. Electromagnetic waves are used in many applications, including radios, radar, microwave ovens, cell phones, and the wireless Internet. Life today would certainly be very different without these applications of electromagnetic waves.

23.1 The Discovery of Electromagnetic Waves

An intriguing aspect of Faraday's law is that a time-varying magnetic field gives rise to an electric field. That is, a magnetic field *by itself* can produce an electric field. Many physicists then wondered, can an electric field by itself produce a magnetic field? Maxwell hypothesized that it should indeed be possible. He proposed a modification of Ampère's law in which a time-varying electric field produces a magnetic field. It turns out that Maxwell's new version of Ampère's law does not affect any of our work with Ampère's law in Chapter 20. But, in addition to giving a new way to create a magnetic field, this new version of Ampère's law gives the equations of electricity and magnetism an interesting symmetry.

Symmetry of E and B

- **The correct form of Ampère's law (due to Maxwell) says that a changing electric flux produces a magnetic field. Since a changing electric flux can be caused by a changing E, we can also say that *a changing electric field produces a magnetic field.***
- **Faraday's law says that a changing magnetic flux produces an induced emf, and an emf is always associated with an electric field. Since a changing magnetic flux can be caused by a changing B, we can also say that a *changing magnetic field produces an electric field.***

There is thus a sort of "circularity" as illustrated in Figure 23.1: an electric field can (if it is changing) produce a magnetic field, and a magnetic field can (if it is changing) produce an electric field. Can this "loop" or oscillation involving E and B continue, with the new electric field producing a new magnetic field, producing a new electric field, and so on? The answer is yes; self-sustaining oscillations involving electric and magnetic fields are indeed possible. In fact, such oscillations are an ***electromagnetic wave.*** Electromagnetic waves are also referred to as ***electromagnetic radiation.***

A key feature of the circularity in Figure 23.1 is that both the electric and magnetic fields must be changing with time. This requirement also affects how an electromagnetic wave can be created. Typically, a time-varying electric field is set up by an AC current in a wire or by moving electric charges, leading to the induced electric and magnetic fields in Figure 23.1. We'll come back to this idea in Sections 23.4 and 23.5.

Maxwell was the first to recognize that the equations of electromagnetism lead to the existence of electromagnetic waves, and he worked out the properties of these waves in great mathematical detail. Maxwell's prediction of the existence of electromagnetic waves was ahead of the experiments of the time. In fact, there was no experimental proof of the existence of electromagnetic waves until Heinrich Hertz demonstrated the generation and detection of radio waves in 1887, eight years after Maxwell's death. Hertz also showed that these waves travel at the speed of light (as predicted by Maxwell). Many other workers contributed to the work on radio and related technologies, most notably Nikola Tesla and Guglielmo Marconi. Their work forms the basis for many important modern devices, including cell phones and Wi-Fi computer networks.

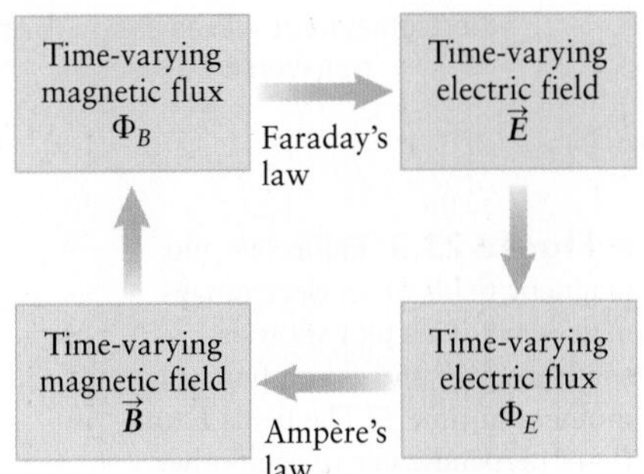

▲ **Figure 23.1** In schematic terms, a changing (time-varying) electric field produces a magnetic field, while a changing (time-varying) magnetic field produces an electric field.

A full mathematical discussion of electromagnetic waves is more than we can include in this book, but we can use our understanding of Faraday's law to understand

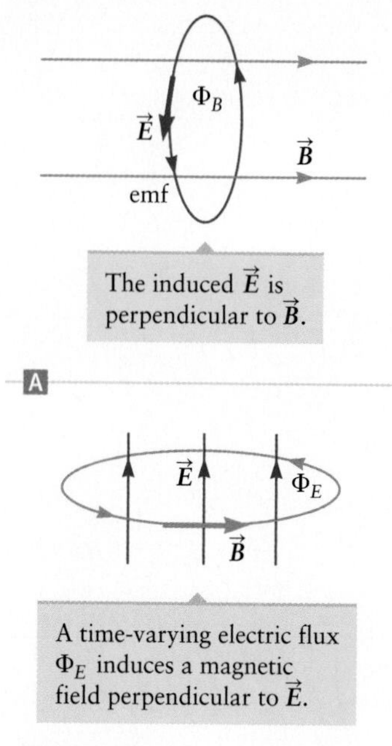

▲ **Figure 23.2** A According to Faraday's law, a changing magnetic field produces an electric field. The induced electric field is perpendicular to the magnetic field. B A changing electric field produces a magnetic field, and these two fields are again perpendicular.

one important property of these waves. According to Faraday's law, a changing magnetic flux through a given area produces an induced emf directed along the perimeter of that area (Fig. 23.2A). This induced emf is associated with an induced electric field through our usual connection between electric potential and field (Chapter 18). Hence, the direction of the induced $\vec{E}$ is *perpendicular* to the magnetic field $\vec{B}$ that produced it. In a similar way, the magnetic field $\vec{B}$ produced by a changing electric field is perpendicular to the electric field that produced it (Fig. 23.2B). As a result, the electric and magnetic fields in an electromagnetic wave are perpendicular to each other. In addition, the direction of propagation of an electromagnetic wave is perpendicular to both $\vec{E}$ and $\vec{B}$. The directions of $\vec{E}$ and $\vec{B}$ and the propagation direction all depend on how the wave is generated, but these three directions are always perpendicular to one another.

Let's now summarize three ***fundamental properties of all electromagnetic waves.***

- **An electromagnetic wave involves *both* an electric field and a magnetic field.**
- **These fields are *perpendicular* to each other.**
- **The propagation direction of the wave is *perpendicular to both $\vec{E}$ and $\vec{B}$.***

23.2 Properties of Electromagnetic Waves

In the previous section, we gave some qualitative arguments concerning the existence and properties of electromagnetic waves. The laws of electricity and magnetism can be used to derive *all* the properties of these waves. Here are some of the key results.

Electromagnetic Waves Are Transverse Waves

We mentioned above that the fields associated with an electromagnetic wave are perpendicular to each other and to the propagation direction. These relationships are illustrated in Figure 23.3A, which shows a snapshot of $\vec{E}$ and $\vec{B}$ for an electromagnetic wave at one instant in time. In this snapshot, the electric field is directed along x, the magnetic field is along y, and the wave propagates in the $+z$ direction.

To interpret Figure 23.3A, first focus on the electric field at the origin. Here, $\vec{E}$ is parallel to the x axis while the wave travels along z, and the tips of the electric field vectors form a pattern similar to a wave on a string. If we took another snapshot a short time later, we would find that the entire electric field pattern has moved a short distance along the z direction while maintaining the same wave profile. The behavior of the magnetic field in Figure 23.3A is similar, but $\vec{B}$ is along the y direction, so it is perpendicular to both $\vec{E}$ and the propagation direction. Because both fields are perpendicular to the propagation direction, this is a ***transverse wave.***

Electromagnetic waves are transverse waves.

The propagation direction can be determined using the procedure shown in Figure 23.3B. Place the fingers of your right hand along the direction of $\vec{E}$ and curl

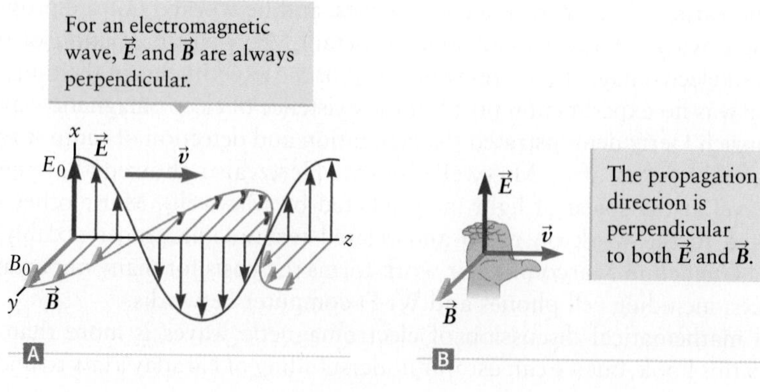

► **Figure 23.3** A Electric and magnetic fields of an electromagnetic wave. This plot shows a snapshot of $\vec{E}$ and $\vec{B}$ at a particular moment in time. B The fields $\vec{E}$ and $\vec{B}$ are perpendicular to each other. The propagation direction $\vec{v}$ of the wave is along the direction defined by a procedure similar to right-hand rule number 2 in Chapter 20.

your fingers in the direction of $\vec{B}$. Your thumb then points along the propagation direction.

CONCEPT CHECK 23.1 Finding the Propagation Direction

At a particular instant, the electric field associated with an electromagnetic wave is directed along $+y$ and the magnetic field is along $+z$. Is the propagation direction (a) $+x$, (b) $-x$, (c) $+y$, or (d) $+z$?

Light Is an Electromagnetic Wave

The speed of an electromagnetic wave can be expressed in terms of two universal constants that appear in the theories of electricity and magnetism: the permittivity of free space ε_0 and the magnetic permeability of free space μ_0. The speed of an electromagnetic wave in a vacuum (free space) is commonly denoted by the symbol c:

$$\text{speed of an electromagnetic wave} = c = \frac{1}{\sqrt{\varepsilon_0 \mu_0}} \tag{23.1}$$

Speed of electromagnetic waves in a vacuum

Inserting the known values of ε_0 and μ_0 into Equation 23.1, we find

$$\begin{aligned} c &= \frac{1}{\sqrt{\varepsilon_0 \mu_0}} = \frac{1}{\sqrt{[8.85 \times 10^{-12}\ \mathrm{C^2/(N \cdot m^2)}](4\pi \times 10^{-7}\ \mathrm{T \cdot m/A})}} \\ &= 3.00 \times 10^8\ \mathrm{m/s} \end{aligned} \tag{23.2}$$

This value for the speed c is precisely equal to the speed of light. When Maxwell derived this result in 1864, the nature of light was the subject of much debate. Most physicists suggested it was a wave while a few believed it was a particle, but no one could say what kind of wave or particle it might be. Maxwell had the answer: ***light is an electromagnetic wave.*** He thus showed that the equations of electricity and magnetism also provide the theory of light.

It is difficult to overestimate the importance of Maxwell's discovery. Besides explaining the nature of light, his work led to the discovery of other types of electromagnetic waves, including radio waves. This work also led eventually to the work of Einstein and his theory of relativity (Chapter 27). Maxwell's contributions to electromagnetic theory were so important that the basic equations of electricity and magnetism (the laws of Coulomb, Gauss, Faraday, and Ampère) are now known as ***Maxwell's equations.***

Electromagnetic Waves Can Travel through a Vacuum, All at the Same Speed

Although Maxwell worked out many of the basic properties of electromagnetic waves, including the value for their speed c, some results must have surprised even him. Recall from Chapter 12 that all mechanical waves travel in a medium of some kind. For waves on a string, the medium is the string, and sound waves travel in a material such as air. Many physicists searched for the corresponding medium for an electromagnetic wave. These waves can travel through material substances such as air or glass, but they can also travel through a vacuum, which is certainly *not* a material substance. Hence, electromagnetic waves are not like the mechanical waves we studied in Chapter 12. Roughly speaking, you can think of electric and magnetic fields as the "media" through which electromagnetic waves travel. These fields can exist in a vacuum, so electromagnetic waves can also travel through a vacuum.

Maxwell's theory does not specify the frequency or wavelength of an electromagnetic wave. Rather, frequency and wavelength are determined by the way the wave is produced. In Section 23.4, we'll describe examples of electromagnetic waves, including radio waves and X-rays. They *all* travel with speed c through a vacuum.

Electromagnetic Waves Travel Slower Than *c* in Material Substances

When an electromagnetic wave travels through a material substance such as glass or water, the speed of the wave depends on the properties of the substance. In such cases, the wave speed is always less than c, the speed in a vacuum (Eq. 23.2). This difference in wave speed explains how a lens can focus light (Chapter 24). What's more, the speed of an electromagnetic wave inside a substance depends on the wave's frequency. So, for example, blue light travels through glass at a lower speed than red light. This effect is called ***dispersion*** and gives rise to rainbows and other interesting phenomena.

CONCEPT CHECK 23.2 Time Delay in a Cell Phone Conversation

A cell phone signal that travels from coast to coast in the United States must travel approximately 3000 miles (about 5×10^6 m). This signal is carried by different types of electromagnetic radiation (typically radio and microwaves) at a speed close to c. Is the approximate time required for your cell phone signal to propagate from coast to coast (a) 0.5 s, (b) 3 s, (c) 0.02 s, or (d) 0.0005 s? Can you notice this time delay?

23.3 Electromagnetic Waves Carry Energy and Momentum

When we first discussed waves in Chapter 12, we stated that waves transport energy without transporting matter. With an electromagnetic wave, the energy is carried by the associated electric and magnetic fields. We can understand how these fields carry energy by considering what happens when an electromagnetic wave encounters a charged particle. Figure 23.4 shows the fields associated with an electromagnetic wave along with an electric charge $+q$ that is in the path of the wave. The charge will experience an electric force $\vec{F}_E = q\vec{E}$; as the wave passes by and the electric field oscillates along the x axis, the force on the charge will also oscillate. If this charge is able to move (which is true for electrons and ion cores in a solid), the electric force will do work on the charge, increasing its kinetic energy. In this way, energy is transferred from the wave to the charge; hence, the wave does indeed carry energy. You might also wonder about the magnetic force on the charge in Figure 23.4; in most cases, the magnetic force is much smaller than the electric force, but as shown below the magnetic field is crucial for understanding how an electromagnetic wave can carry momentum.

In Chapters 18 and 21, we discussed the potential energy contained in the electric and magnetic fields. For an electric field, the energy is proportional to E^2 (Eq. 18.48), and the potential energy per unit volume is

$$u_{\text{elec}} = \frac{1}{2}\varepsilon_0 E^2 \tag{23.3}$$

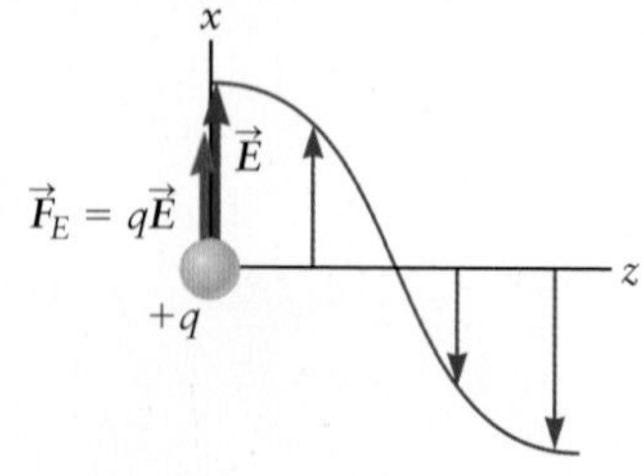

▲ **Figure 23.4** When an electromagnetic wave "strikes" an electric charge, the electric field associated with the wave produces a force on the charge. This force oscillates in direction as the wave passes by.

For a magnetic field, the potential energy is proportional to B^2 (Eq. 21.35), and the magnetic potential energy per unit volume is

$$u_{\text{mag}} = \frac{1}{2\mu_0}B^2 \tag{23.4}$$

These results apply also to electromagnetic waves, and the total energy per unit volume carried by an electromagnetic wave is the sum of its electric and magnetic energy densities, $u_{\text{elec}} + u_{\text{mag}}$. As the wave propagates, the electric and magnetic fields oscillate (Fig. 23.3), so u_{elec} and u_{mag} also oscillate. It also turns out that the electric and magnetic energy densities are equal; $u_{elec} = u_{mag}$. Applying this relation when the fields have their peak values E_0 and B_0 (Fig. 23.3A) and using Equations 23.3 and 23.4, we get

$$\frac{1}{2}\varepsilon_0 E_0^2 = \frac{1}{2\mu_0} B_0^2 \tag{23.5}$$

In words, Equation 23.5 says that the peak values of the fields E_0 and B_0 are proportional. Rearranging, we get

$$E_0 = \frac{1}{\sqrt{\varepsilon_0 \mu_0}} B_0$$

The factor $1/\sqrt{\varepsilon_0\mu_0}$ is just the speed of light in a vacuum c (Eq. 23.1); hence,

$$E_0 = cB_0 \tag{23.6}$$

Relation between E and B for an electromagnetic wave

EXAMPLE 23.1 Electric and Magnetic Fields of a TV Signal

The signal from a radio station is an electromagnetic wave and is detected through its electric field. In a typical case, the electric field needed to give a "good" radio signal is about 1×10^{-2} V/m, the amplitude (peak value) E_0 associated with the wave. What is the corresponding magnetic field amplitude associated with this electromagnetic wave?

RECOGNIZE THE PRINCIPLE

For an electromagnetic wave, the amplitude of the electric and magnetic fields are proportional. Because the electric field amplitude E_0 is given, we can find the magnetic field amplitude B_0. The aim of this problem is to understand the magnitudes of these fields for a typical application.

SKETCH THE PROBLEM

No sketch is needed.

IDENTIFY THE RELATIONSHIPS

The relation between the peak electric and magnetic fields in an electromagnetic wave is $E_0 = cB_0$ (Eq. 23.6).

SOLVE

Inserting the given value of E_0, we find

$$B_0 = \frac{E_0}{c} = \frac{1 \times 10^{-2}\ \text{V/m}}{3.00 \times 10^8\ \text{m/s}} = \boxed{3 \times 10^{-11}\ \text{T}}$$

▶ ***What does it mean?***

This value for B_0 certainly seems much smaller than the electric field E_0 (because the numerical values are very different in magnitude), but a careful comparison requires us to consider the relative magnetic and electric forces in a typical case. We'll do so after we discuss how an antenna works in Section 23.5; that analysis will show that the electric force is (in that case) much greater than the magnetic force. Thus, an electromagnetic wave is usually detected through its electric field and not by the effect of its magnetic field.

Intensity of an Electromagnetic Wave

The strength of an electromagnetic wave is usually measured in terms of its intensity, with units of watts per square meter. The intensity is the amount of energy the wave transports per unit time across a surface of unit area and equals the energy density multiplied by the wave speed (because a faster wave delivers energy at a faster rate). The wave speed is just c, the speed of light. Denoting the intensity by I, we have

$$I = \text{total energy density} \times c \quad (23.7)$$

The peak electric and magnetic energy densities are given in Equation 23.5. The total energy density carried by the wave is proportional to the sum of these two terms:

$$\text{total energy density} \propto \frac{1}{2}\varepsilon_0 E_0^2 + \frac{1}{2\mu_0}B_0^2 \quad (23.8)$$

We already said that the electric and magnetic contributions to the energy are equal, so we can rewrite Equation 23.8 as

$$\text{total energy density} \propto \frac{1}{2}\varepsilon_0 E_0^2 + \frac{1}{2\mu_0}B_0^2 = \varepsilon_0 E_0^2$$

Inserting into Equation 23.7, we get $I \propto \varepsilon_0 E_0^2 c$. In our derivation, we have dealt with E_0 and B_0, which are the peak values of the fields. To find the average intensity, we must average over a complete cycle of oscillation of the fields; this average intensity is half the peak intensity, so our final result for the intensity is

Intensity of an electromagnetic wave

$$I = \frac{1}{2}\varepsilon_0 c E_0^2 \quad (23.9)$$

The intensity of an electromagnetic wave is thus proportional to the square of the electric field amplitude. Since $E_0 = cB_0$, the intensity is also proportional to the square of the magnetic field amplitude.

EXAMPLE 23.2 The Electric Field of Sunlight

On a typical summer day, the intensity of sunlight at the Earth's surface is approximately $I = 1000\ \text{W/m}^2$. What is the amplitude of the electric field associated with this electromagnetic wave?

RECOGNIZE THE PRINCIPLE

The intensity of an electromagnetic wave is proportional to E_0^2, where E_0 is the amplitude of the electric field. Given the intensity, we can thus find E_0. This calculation will give an understanding of the size of the electric field associated with an important source of electromagnetic radiation.

SKETCH THE PROBLEM

No sketch is necessary.

IDENTIFY THE RELATIONSHIPS

The electric field amplitude and the intensity are related by $I = \frac{1}{2}\varepsilon_0 c E_0^2$ (Eq. 23.9). Rearranging to solve for E_0 gives

$$E_0 = \sqrt{\frac{2I}{\varepsilon_0 c}}$$

SOLVE

Inserting the value of the intensity of sunlight, we find

$$E_0 = \sqrt{\frac{2(1000\ \text{W/m}^2)}{[8.85 \times 10^{-12}\ \text{C}^2/(\text{N}\cdot\text{m}^2)](3.00 \times 10^8\ \text{m/s})}} = \boxed{870\ \text{V/m}}$$

Insight 23.1

APPLICATIONS OF SOLAR ENERGY: SOLAR CELLS

Example 23.2 mentions that the intensity of sunlight on a typical sunny day is about $1000\ \text{W/m}^2$. Several approaches can be used to "capture" this energy. A ***solar cell*** converts the energy from sunlight into electrical energy that can be used to run the circuits in your house or to charge a battery. A typical house uses energy at an average rate of a few thousand watts. Because sunlight has an intensity of about $1000\ \text{W/m}^2$, you might think that a solar collector that captures sunlight over an area of only a few square meters would be sufficient. Current photovoltaic cells, however, capture only about 15% of the energy that strikes them, and one must also allow for nights and cloudy days during which the Sun's intensity is much lower than its peak value. The problem of making better and more practical solar cells is an important engineering challenge.

► ***What does it mean?***
This electric field is large, especially compared with the field found in a typical electric circuit. This large value for E_0 explains how sunlight can produce large effects, such as sunburn!

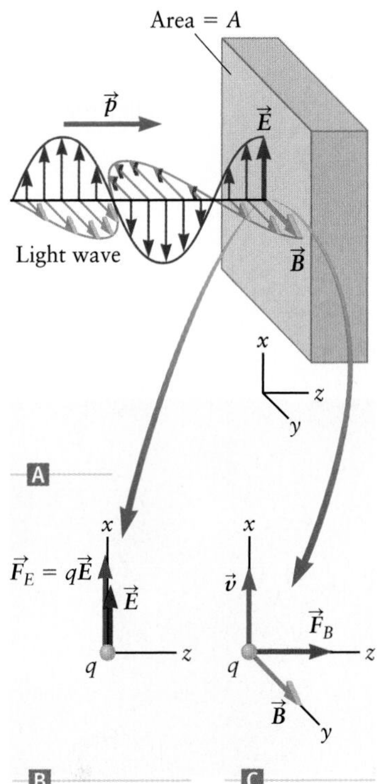

▲ **Figure 23.5** A An electromagnetic wave carries momentum. B and C This momentum can be understood in terms of the force exerted on charges in a material from the electric and magnetic fields of the wave. When electromagnetic radiation is absorbed by an object, the momentum of the object increases such that the total momentum of the system (radiation plus object) is conserved.

Electromagnetic Waves Carry Momentum

In mechanics, a particle's momentum is equal to the mass of the particle times its velocity (Eq. 7.1). An electromagnetic wave does not have a mass, however, so how can it have momentum? The momentum of a particle is related to the force that the particle exerts in a collision with another object. Let's therefore consider the force that an electromagnetic wave exerts when it is absorbed by (i.e., collides with) an object as sketched in Figure 23.5. If the object is at rest before the collision, all the initial momentum is carried by the electromagnetic wave (Fig. 23.5A). Because the wave is absorbed by the object, all the final momentum is carried by the object. The absorption of the wave occurs through the electric and magnetic forces on charges in the object. Figure 23.5B and C shows the effects on a positive charge q (which might be an ion core) at the surface of the object. When the wave is absorbed, momentum is transferred to the charge q. Since total momentum is conserved in a collision, we can write

$$p_i(\text{wave}) = p_f(\text{charge } q) \tag{23.10}$$

To analyze the effect of the electromagnetic wave on q, let's first consider the electric force. The electric field of the wave $\vec{E}$ is upward along $+x$ in Figure 23.5B, so the electric force $\vec{F}_E = q\vec{E}$ is also upward (assuming q is positive). This force accelerates the charge, giving it a velocity that is also upward. Now consider the magnetic force on this moving charge in Figure 23.5C. The magnetic field of the wave gives a magnetic force of magnitude $F_B = qvB$, and, according to right-hand rule number 2, this force is to the right along $+z$. As the electric field $\vec{E}$ oscillates up and down, the velocity in Figure 23.5B also oscillates up and down, but the magnetic force F_B in Figure 23.5C is always to the right because when $\vec{E}$ reverses direction, so does $\vec{B}$.

We have thus shown that when a charge q absorbs an electromagnetic wave, there is a force on the charge in the direction of propagation of the original wave. According to the impulse theorem (Eq. 7.5), the force F_B on the charge q is related to the charge's change in momentum Δp along the z direction during a time Δt by

$$F_B = \frac{\Delta p}{\Delta t} \tag{23.11}$$

Hence, the charge's final momentum along the z direction in Figure 23.5 is given through Equation 23.11. According to the principle of conservation of momentum, the final momentum of the charge must equal the initial momentum of the electromagnetic wave (Eq. 23.10), so the wave does indeed have momentum!

A full calculation of the momentum p of an electromagnetic wave shows that p is proportional to the total energy E_{total} carried by the wave, with the result

$$p = \frac{E_{\text{total}}}{c} \tag{23.12}$$

Momentum carried by an electromagnetic wave

Radiation Pressure

The analysis in Figure 23.5 shows that when an electromagnetic wave is absorbed by an object, it exerts a force on the object. Recalling that pressure is equal to the force per unit area, we can define a quantity called the ***radiation pressure*** $P_{\text{radiation}}$, which is equal to this electromagnetic force divided by the area. If I is the intensity of the wave, an analysis based on the ideas in Figure 23.5 gives

Radiation pressure

$$P_{\text{radiation}} = \frac{F}{A} = \frac{I}{c} \tag{23.13}$$

In words, when an electromagnetic wave is absorbed by an object, the object experiences a radiation pressure $P_{\text{radiation}}$. As an example, the radiation pressure from sunlight could be used to propel a spacecraft, as we explore in the next example.

EXAMPLE 23.3 The Momentum Carried by Sunlight

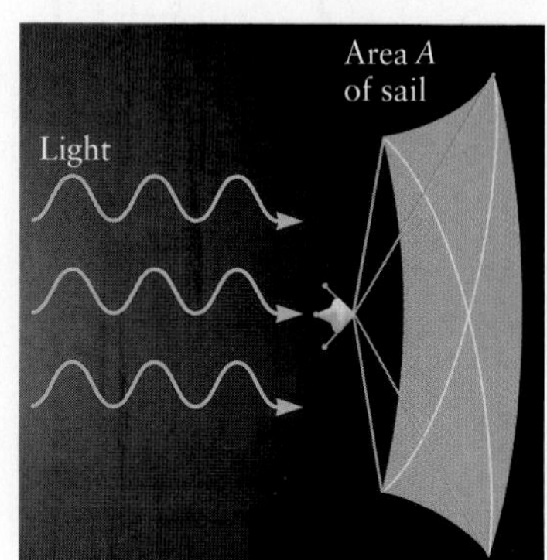

▲ **Figure 23.6** Example 23.3. A solar sail absorbs momentum from sunlight and uses the resulting force to accelerate a spacecraft.

A spacecraft designer wants to use radiation pressure to propel a spacecraft as sketched in Figure 23.6. Near the Earth, the intensity of sunlight is about 1000 W/m^2 and the exposed area of the spacecraft is 500 m^2. If the sunlight is completely absorbed, what is the force on the spacecraft?

RECOGNIZE THE PRINCIPLE

The radiation pressure produced by an electromagnetic wave is proportional to its intensity. From the definition of pressure, the resulting force is equal to the radiation pressure multiplied by the area of the spacecraft.

SKETCH THE PROBLEM

Figure 23.6 shows the problem.

IDENTIFY THE RELATIONSHIPS

The radiation pressure is related to the intensity by $P_{\text{radiation}} = I/c$, and the force on the spacecraft is $F = P_{\text{radiation}}A$. We thus have

$$F = P_{\text{radiation}}A = \frac{IA}{c}$$

SOLVE

Inserting the given values of the intensity and the area leads to

$$F = \frac{(1000 \text{ W/m}^2)(500 \text{ m}^2)}{3.00 \times 10^8 \text{ m/s}} = \boxed{1.7 \times 10^{-3} \text{ N}}$$

► ***What does it mean?***

This force is rather small; it is roughly equal to the weight of a penny at the Earth's surface. Achieving a larger force would require a larger exposed area (a larger "solar sail"). This approach does have one advantage, however: the fuel is free. Spacecraft that rely on solar sails for propulsion have not yet been built, but the effects of this force on the motion of several spacecrafts have been observed.

CONCEPT CHECK 23.3 Radiation Pressure and Reflected Light

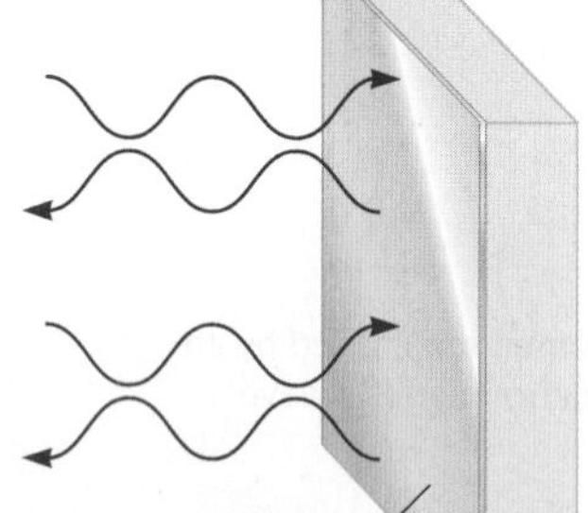

▲ **Figure 23.7** Concept Check 23.3.

When light is absorbed by an object, the radiation pressure is $P_{\text{radiation}} = I/c$ (Eq. 23.13). Suppose the exposed surface of an object has instead a reflective coating so that it reflects electromagnetic waves (Fig. 23.7). Which of the following statements is true? *Hint*: Compare absorption (Fig. 23.5) to a completely inelastic collision and reflection (Fig. 23.7) to an elastic collision.

(a) The force on the object due to radiation pressure is the same whether the light is absorbed or reflected.
(b) The force on the object due to radiation pressure is greater by a factor of two when the light is reflected.
(c) There is no force on the object due to radiation pressure when the light is reflected.

23.4 Types of Electromagnetic Radiation: The Electromagnetic Spectrum

All electromagnetic waves travel through a vacuum at the same speed c. The value of c has been measured with *great* precision and has the value

$$c = 2.99792458 \times 10^8 \text{ m/s} \tag{23.14}$$

In fact, c is actually *defined* to have the value in Equation 23.14, and the value of the SI standard meter is then derived through the value of c and the unit of time, the standard second. In most calculations, we will not need all the significant figures in Equation 23.14; the value $c = 3.00 \times 10^8$ m/s is accurate enough for the work in this book.

Electromagnetic waves are classified according to their frequency f and wavelength λ. The frequency, wavelength, and speed of an electromagnetic wave are related by

$$f\lambda = c \tag{23.15}$$

which is similar to the general relation we found for mechanical waves in Chapter 12 (Eq. 12.2); as the frequency is increased, the wavelength decreases, and vice versa. Figure 23.8 lists a number of different types of electromagnetic waves, arranged according to their frequencies and wavelengths. In this listing, the frequency increases from bottom to top; frequency and wavelength are inversely related (Eq. 23.15), so the wavelength increases from top to bottom.

This range of all possible electromagnetic waves is called the ***electromagnetic spectrum***. There is no strict lower or upper limit for electromagnetic wave frequencies, which vary over an extremely large range, from below a few hertz (Hz) for ultralow-frequency radio waves to above 10^{22} Hz for gamma rays. The range of frequencies assigned to the different types of waves is somewhat arbitrary. For example, the ultraviolet and X-ray regions in Figure 23.8 overlap as do the regions assigned

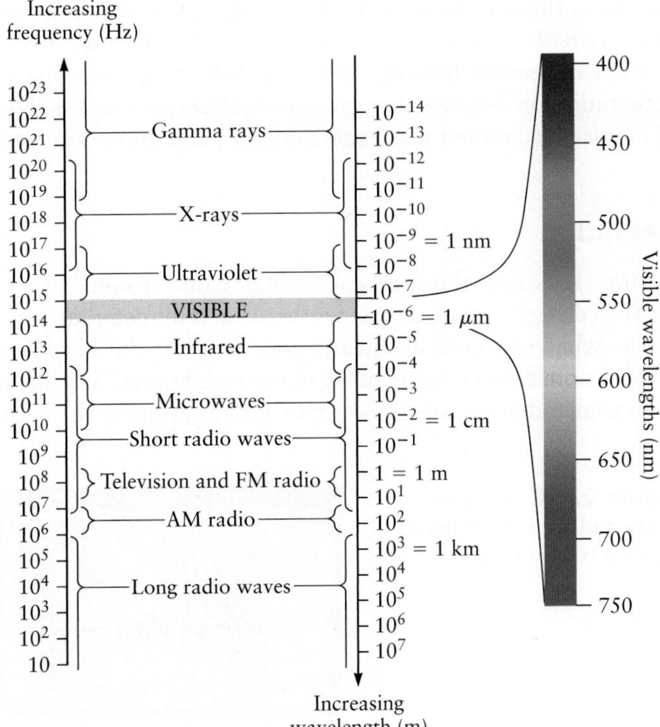

◀ **Figure 23.8** The electromagnetic spectrum. Frequency increases from bottom to top, whereas wavelength increases from top to bottom. Near the middle of the spectrum is a narrow section corresponding to visible light; this section is expanded at the right, showing also how the color of visible light varies with frequency and wavelength.

to microwaves and radio waves. The names of these different regions were chosen based on how the radiation in each frequency range interacts with matter and on how it is generated. We now describe some properties of these waves, starting at the low frequency end of the electromagnetic spectrum.

A

B

▲ **Figure 23.9** A A set of parallel conductors arranged with the proper spacing acts as an antenna that can efficiently emit or absorb electromagnetic radiation. This antenna is designed for radio waves. B Some antennas are shaped as concave dishes. These antenna dishes, used at microwave frequencies, are covered to keep out the weather, but still allow microwaves to pass.

Radio Waves

Radio waves have frequencies from as low as a few hertz to about 10^9 Hz. The corresponding wavelengths vary from 10^8 m (around 10^5 miles!) to a few centimeters. Radio waves are usually produced by an AC circuit attached to an antenna. A simple wire can function as an antenna (as discussed in Section 23.5), but antennas containing multiple conducting elements (Fig. 23.9A) or shaped as concave "dishes" (Fig. 23.9B) are usually more efficient and are much more common than those consisting of a single wire. The frequency of a radio wave generated in this way is equal to the frequency of the current in the attached AC circuit. Radio waves can be detected by an antenna similar to the one used for generation. These waves are used to carry conventional radio and television signals as well as signals for some cell phones and pagers.

In our discussion of the induced electric and magnetic fields in Figure 23.1, we emphasized that an electromagnetic wave is produced by *changing* electric and magnetic fields. For radio waves, the AC current in the antenna is produced by time-varying (oscillating) electric fields. When the AC current in an antenna oscillates with time, the moving charges in the antenna are accelerated as they oscillate. Many other sources of electromagnetic radiation involve similar time-varying electric and magnetic fields produced by *accelerated* charges. In general, *when an electric charge is accelerated, it produces electromagnetic radiation.*

Microwaves

Microwaves have frequencies between about 10^9 Hz and 10^{12} Hz. The corresponding wavelengths vary from a few centimeters to a few tenths of a millimeter. This radiation interacts strongly with molecules such as H_2O, making microwaves useful in the kitchen. Microwave ovens (Fig. 23.10) generate radiation with a frequency near 2.5×10^9 Hz. Waves of this frequency are absorbed strongly by H_2O molecules, causing the atoms in those molecules to vibrate just as masses connected by springs can vibrate. The term *absorption* means that some of the microwave energy is transferred to the water molecules, heating your food. Vibrating molecules can also emit electromagnetic radiation because they contain accelerated charges. For example, a vibrating H_2O molecule contains accelerated charges and thus emits electromagnetic radiation.

Infrared Radiation

Infrared radiation falls into the frequency range from around 10^{12} Hz to about 4×10^{14} Hz (wavelengths from a few tenths of a millimeter to a few microns). This radiation is what we sense as "heat," and much of the blackbody radiation from objects near room temperature falls in the infrared range. Satellite photographs sensitive to infrared radiation are thus useful for monitoring the Earth's weather

► **Figure 23.10** A microwave oven uses radiation in the microwave range to heat food.

(Fig. 23.11). Infrared radiation interacts strongly with molecules and is absorbed by most substances. Infrared radiation can be produced by vibrating molecules or when electrons undergo transitions within an atom (Chapter 29).

Visible Light

The radiation we are able to detect with our eyes falls into the rather narrow frequency range from about 4×10^{14} Hz to near 8×10^{14} Hz, corresponding to wavelengths from about 750 nm to 400 nm. This radiation is called ***visible light***. Light at the low end of this frequency range (long wavelengths) is sensed as the color red; increasing frequencies then correspond to yellow, green, and blue, while the high-frequency (short-wavelength) end of the visible range is violet. A mixture of all these colors such as the light from a flashlight produces "white" light.

The speed of an electromagnetic wave inside a material substance has a value c' that is less than the speed c in a vacuum. Visible light travels through air at a speed about 0.03% smaller than c, whereas in water the speed of light is about 33% less than c. This variation of the speed of light from one material to another leads to many effects in optics, including the focusing of light by a lens (Chapter 24). The speed of light inside a material also depends on the frequency of the radiation. This effect is called ***dispersion*** because it causes a prism to "disperse" the different colors in white light (Fig. 23.12). Dispersion is also responsible for rainbows.

Visible light can be produced when electrons undergo atomic transitions (Chapter 29). Many sources of visible light such as the flashlight in Figure 23.12 produce radiation with a range of frequencies, but it is also possible to produce light consisting essentially of a single frequency. Such radiation is called ***monochromatic***, meaning "single color." This term is usually used when describing visible light, but can also be applied to radiation in other parts of the spectrum. Lasers (Fig. 23.13) produce monochromatic radiation. We'll explain how lasers work in Chapter 29.

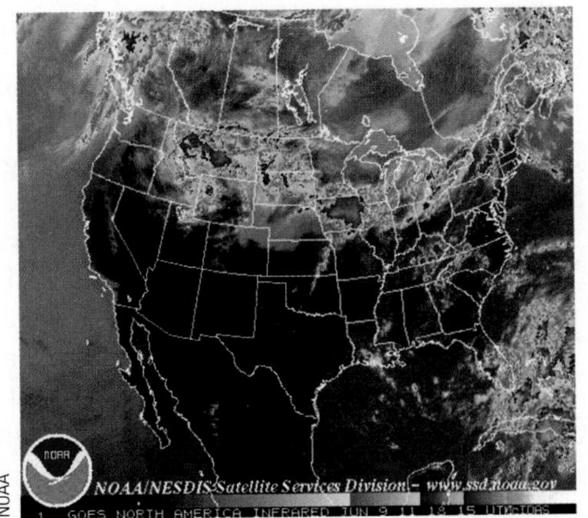

NOAA

▲ **Figure 23.11** Blackbody radiation (typically in the infrared range) indicates the temperature of an object. In this satellite photo, the temperature of the Earth's surface and atmosphere is measured by the wavelength at which the blackbody radiation has the highest intensity. (See Fig. 14.33.) In this image, different temperatures are displayed as different colors.

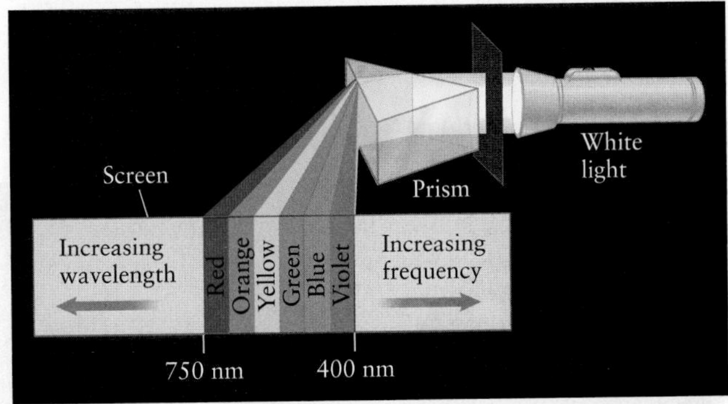

◀ **Figure 23.12** White light from a flashlight is a combination of many different frequencies (i.e., colors). After passing through a prism, the different colors propagate in different directions. Newton used a prism to show that white light contains different colors.

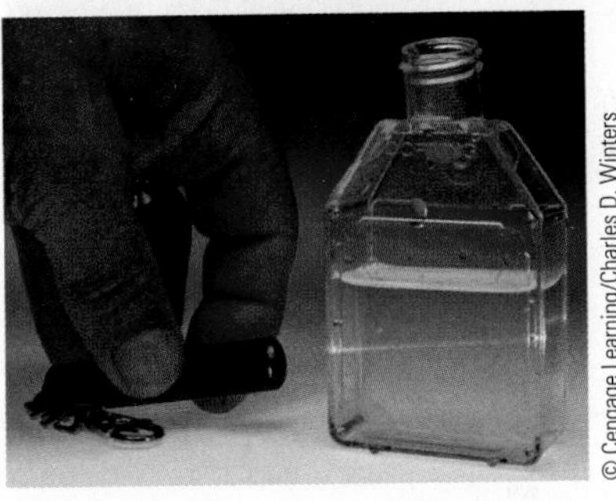

© Cengage Learning/Charles D. Winters

A

© Hank Morgan/Photo Researchers, Inc.

B

◀ **Figure 23.13** A A "key chain" laser. B A research laser. Both lasers produce monochromatic radiation (light of a single color).

© bilderlounge/Jupiterimages

▲ **Figure 23.14** Sunbathers use ultraviolet radiation from the Sun to get tan. Too much exposure to UV radiation can be hazardous to your health.

Ultraviolet Light

Proceeding to frequencies above the range of visible light takes us to ***ultraviolet (UV) light***, with frequencies from about 8×10^{14} Hz to 10^{17} Hz; the associated wavelengths are from about 3 nm to 400 nm. We cannot detect this radiation with our eyes.

Ultraviolet radiation has one major health benefit: it stimulates the production of vitamin D in the body. Nearly all the other effects of UV radiation are detrimental. Excessive exposure to UV light (Fig. 23.14) can cause sunburn, skin cancer, and cataracts. Many skin cancers (the type known as carcinoma) are treatable and often not fatal, but malignant melanoma is much more deadly. Persons who receive only two serious sunburns as children are found to have about twice the normal probability of developing melanoma as adults compared with those who did not.

X-Rays

X-rays have frequencies from about 10^{17} Hz to around 10^{20} Hz. Wilhelm Röntgen discovered X-rays in 1895 and found that they can be used to form images from inside living tissue. Only one month after he announced this discovery, X-rays were being applied in medicine and other fields, and Röntgen became a worldwide celebrity. (He received the first Nobel Prize in Physics soon after, in 1901.) An early X-ray image of a human hand is reproduced in Figure 23.15A. The soft tissue, bones, and even a ring on one of the fingers are clearly visible. This image is possible because X-rays are weakly absorbed by skin and other soft tissue and strongly absorbed by dense materials such as bone, teeth, and metal. This difference in absorption provides the contrast seen in Figure 23.15B and makes it possible for X-rays to detect broken bones.

The X-ray images in Figure 23.15 are essentially "shadows" of the tissue that the X-rays pass through. All X-ray images were produced in this way until the 1970s, when a technique called computed axial tomography (abbreviated CT or CAT) was developed. Figure 23.16A shows how photographs taken at two different angles can be combined to reveal information about the thickness of an object. For example, if two bones overlap in an X-ray image, the bone behind is obscured. This effect is illustrated in Figure 23.16A, where a basketball is positioned in front of the person; here the basketball is obscured in the image on the screen behind. For any single image,

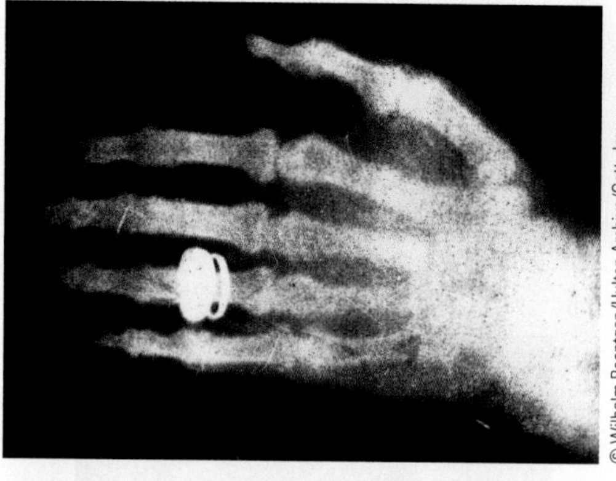

© Wilhelm Roentgen/Hulton Archive/Getty Images

A

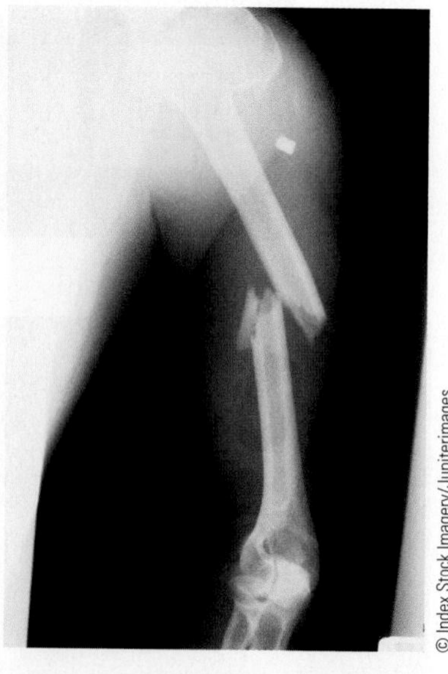

© Index Stock Imagery/Jupiterimages

B

▲ **Figure 23.15** X-rays can pass through skin, but are absorbed strongly by dense material such as bone and metal.

there will always be parts of the person's body that are obscured in some way. This problem can be overcome by comparing two images taken from different angles; in Figure 23.16A, the basketball is visible when viewed with the X-ray source at the right. A CT scan (Fig. 23.16B) takes many X-ray images at many different angles and then employs a sophisticated computer analysis to combine these images into a three-dimensional representation of the object. The results can be used to measure accurately the shape and size of bones, tumors, and so forth within the body.

Although X-ray imaging has enormous health benefits, excessive exposure to X-ray radiation can have harmful effects, including burns. At high exposures, X-rays can cause cancer and other serious diseases.

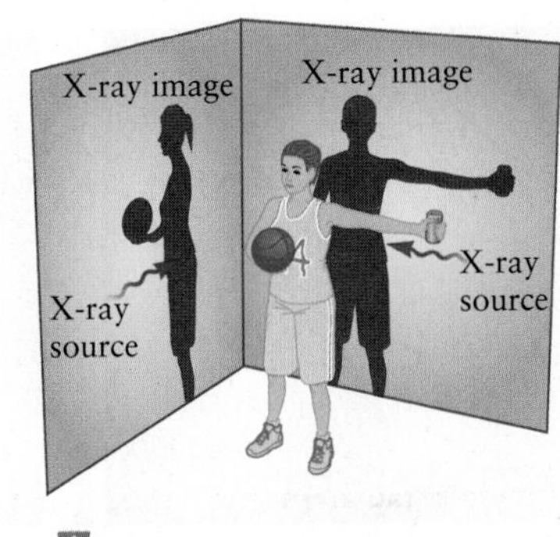

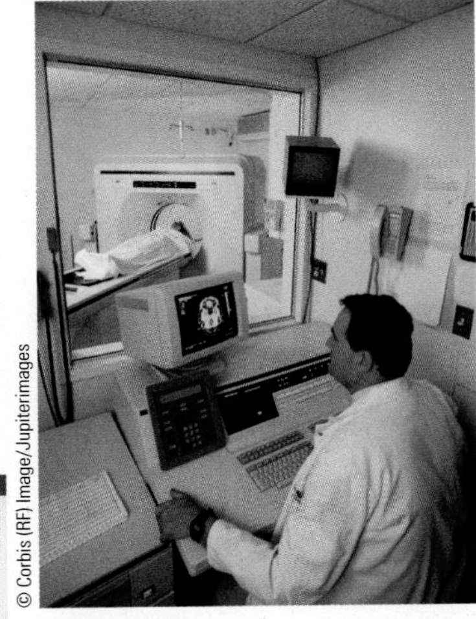

© Corbis (RF) Image/Jupiterimages

▲ **Figure 23.16** A Shadow images taken at different angles can show the thickness of an object and can also reveal when one part of an object is in the shadow of another part. B Example of a CT scan apparatus as found in most hospitals.

Gamma Rays

Gamma rays lie at the high-frequency end of the electromagnetic spectrum, with frequencies above about 10^{20} Hz and wavelengths less than about 10^{-12} m. (As with other regions of the electromagnetic spectrum, there is no precise boundary between the X-ray and gamma ray parts of the spectrum.) Gamma rays are produced by processes occurring inside the nucleus of an atom (Chapter 30). Gamma rays are produced in nuclear power plants and the Sun. Gamma rays also reach the Earth from sources outside our solar system.

EXAMPLE 23.4 Wavelength of X-rays

The X-rays used in your dentist's office or local hospital have a frequency of approximately $f = 2.0 \times 10^{18}$ Hz. What is the wavelength of these X-rays, and how does that wavelength compare to the size of an atom?

RECOGNIZE THE PRINCIPLE

The frequency and wavelength of an electromagnetic wave are related through $f\lambda = c$. Since the wave speed c is known, we can use the given value of f to find the wavelength. These X-rays are created by processes within the atoms of metals such as copper, molybdenum, and tungsten.

SKETCH THE PROBLEM

No sketch is necessary.

IDENTIFY THE RELATIONSHIPS

Using the relation between speed, frequency, and wavelength (Eq. 23.15), we get $\lambda = c/f$.

SOLVE

Inserting the given value of the frequency and the known value of c yields

$$\lambda = \frac{c}{f} = \frac{3.00 \times 10^8 \text{ m/s}}{2.0 \times 10^{18} \text{ Hz}} = \boxed{1.5 \times 10^{-10} \text{ m}}$$

► ***What does it mean?***

The radius of an atom is about 5×10^{-11} m, so the wavelength of these X-rays is approximately the diameter of an atom. As mentioned above, these X-rays are produced by atomic processes (so-called atomic transitions; see Chapter 29). Hence, a process involving electrons in atoms produces X-rays whose wavelengths are comparable to the size of an atom.

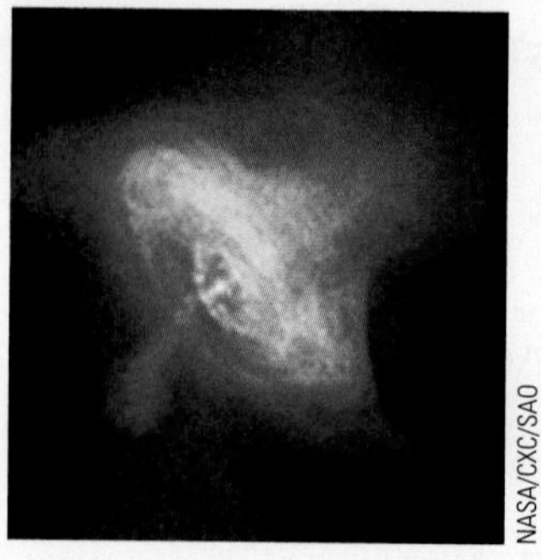
NASA/CXC/SAO

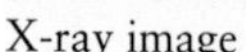
X-ray image

Palomar Observatory

Visible light

2MASS/UMass/IPAC-Caltech/NASA/NSF

Infrared

NRAO/AUI/NSF

Radio image

▲ **Figure 23.17** Photographs of electromagnetic radiation emitted by the Crab Nebula, taken using different types of radiation. The colors in these images indicate the intensity of the radiation and not its true color.

CONCEPT CHECK 23.4 Types of Electromagnetic Radiation

Place the following types of electromagnetic radiation in order according to their wavelength. Start with the radiation with the longest wavelength.

(a) Infrared (b) X-rays (c) Blue light
(d) Radio (e) Microwaves (f) Red light

Astronomy and Electromagnetic Radiation

There are many applications of electromagnetic radiation, each using the type of radiation that best matches the need. For example, most cell phones use microwaves, the optical fibers used to carry signals between computers use infrared and visible light, and tanning salons use UV radiation. One application—astronomy—uses virtually all types of electromagnetic radiation. Many objects in the universe such as stars and supernovas emit radiation over a wide range of frequencies, from radio waves to visible light to X-rays and gamma rays. For example, Figure 23.17 shows images of an object called the Crab Nebula. It is the remnants of a supernova, a star that exploded in the year 1054. The images shown here were constructed using different types of radiation. Comparisons of the different images give astronomers important information about how mass is distributed in the nebula and how it generates radiation.

23.5 Generation and Propagation of Electromagnetic Waves

Let's now consider in detail how one particular class of electromagnetic waves—radio waves—can be generated. Figure 23.18 shows an AC voltage source connected to two wires. These wires act as an ***antenna***. As the voltage of the AC source oscillates, the electric potentials of the two wires also oscillate, and we indicate these potentials with positive and negative signs. These signs also indicate electric charges present on the two wires; these charges flow from the voltage source onto and off the wires as the AC voltage alternates from positive to negative, back to positive, back to negative, and so on.

Figure 23.18A shows how the potential varies with time (just our usual AC voltage source), and the other parts of the figure show "snapshots" taken at different times as the potential difference between the two antenna wires oscillates. In part B of Figure 23.18, the potential difference is large; there are many negative charges on the upper wire and many positive charges on the bottom wire. These charges produce a large electric field directed upward as shown. As time passes (Fig. 23.18C), the potential difference becomes smaller in magnitude, so the amount of charge on the two wires decreases and the electric field is smaller. At the same time, the large electric field from part B has begun to propagate away from the wires. In Figure 23.18D, the potential difference is zero and the electric field at the wires is zero, and the fields generated at previous times (in parts B and C) continue to propagate away. Parts E through H of Figure 23.18 show the charge on the wires and the corresponding electric fields as time passes and the AC potential oscillates through one complete cycle.

The behavior of the electric field in Figure 23.18 is similar to the behavior of a wave on a string. When one end of a string is vibrated up and down, a wave is generated that propagates away along the string. Here the electric field acts in some ways like the string, and the AC voltage source generates a vibrating electric field at the antenna, similar to shaking the end of a string.

The pictures in Figure 23.18 also explain some other important features of electromagnetic waves. First, the charges move between the AC source and the antenna wires in an oscillatory way. From our work on the motion of a simple harmonic oscillator (Chapter 11), we know that these charges are accelerated, confirming the claim we made in Section 23.2 that electromagnetic waves are produced whenever charges have a nonzero acceleration. Second, the charges in Figure 23.18 undergo harmonic motion at a frequency equal to the frequency of the AC voltage source, which is also the frequency of the electromagnetic wave that is generated. Whenever an electromagnetic wave with frequency f is observed, we can say that it was generated by an electromagnetic oscillator (such as an LC circuit) that oscillates at the same frequency f.

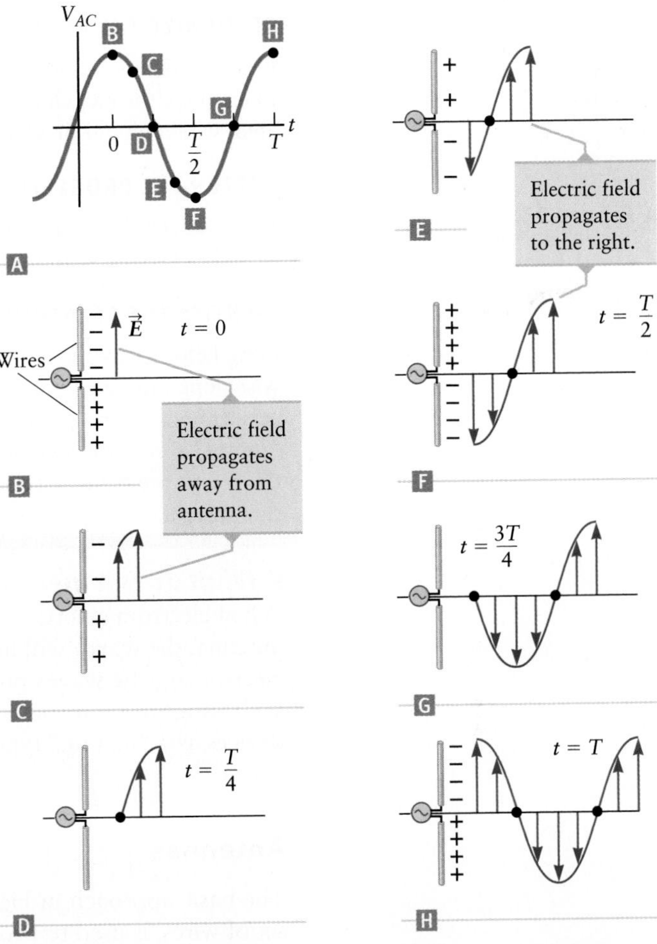

▲ **Figure 23.18** Electric field near a dipole antenna. A Variation of the AC voltage across the antenna as a function of time. B through H Snapshots of the electric field produced by the antenna as a function of time. As the electric field at the antenna oscillates, an electromagnetic wave propagates away from the antenna.

CONCEPT CHECK 23.5 What Is the Direction of $\vec{B}$?

In Figure 23.18, we considered how charge on the wires produces electric fields that then generate an electromagnetic wave. At the same time, the motion of charges on and off the wires also gives rise to the magnetic field associated with the wave. The electromagnetic wave in Figure 23.18 propagates to the right. What is the direction of the magnetic field $\vec{B}$ near the antenna in the snapshot in Figure 23.18B? (a) Directed to the right, (b) directed to the left, (c) directed out of the plane (toward the reader), or (d) directed into the plane (away from the reader)?

EXAMPLE 23.5 Noise on Your Radio

When listening to an AM radio, you can sometimes hear "static" noise when a nearby light switch is flicked on and off. Explain how the switch can produce electromagnetic waves, even if the radio is battery-powered so that no AC source is involved. Are these waves produced when the switch is open, is closed, or during the process of opening or closing?

(continued) ▶

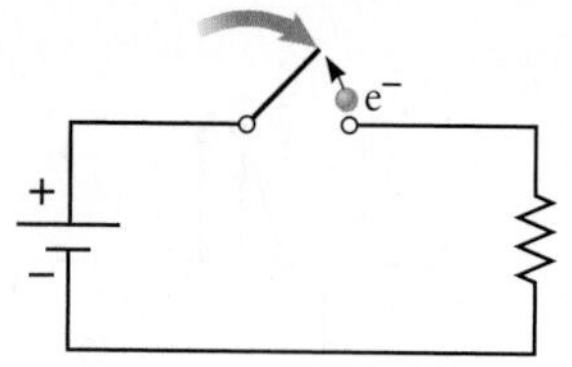

e⁻
+
−

▲ **Figure 23.19** Example 23.5.

RECOGNIZE THE PRINCIPLE

Electromagnetic waves are produced whenever electric charges are accelerated. We need to examine the switching process to see if there are any times when charges are accelerated, which would produce electromagnetic radiation that might be detected by a radio.

SKETCH THE PROBLEM

Figure 23.19 shows a switch being closed. Just before this switch is completely closed, there is a "spark"—that is, a current of electrons—between the two contacts.

IDENTIFY THE RELATIONSHIPS AND SOLVE

Long before a switch is closed, the current is zero, so no charges are accelerated. Likewise, long after a switch is closed, the current is constant, so again no charges are being accelerated (although they will be moving with a constant drift velocity). When a spark "jumps" across the switch contacts, however, the electrons that will form the current in the circuit are being accelerated. Electromagnetic waves will therefore be produced at that instant.

▸ ***What does it mean?***

When electromagnetic waves are produced by an AC source connected to an antenna, the waves will have same frequency as that of the AC source. In contrast, electromagnetic waves produced by the switch in Figure 23.19 will have a very wide range of frequencies and may be detected by your cell phone, radio, and other devices as a "hissing" type of noise.

Antennas

The basic approach in Figure 23.18, in which an AC voltage source is attached to a set of wires, is used to produce the waves that carry radio, television, and cell phone signals. The wires in Figure 23.18 are one type of antenna, but there are many other designs, some containing many wires (Fig. 23.9A); the simple case with two wires is called a ***dipole antenna***. The reason for the name "dipole" can be seen from Figure 23.20, which shows a dipole antenna with the wires parallel to the vertical (z) direction. At any particular moment, the two wires are oppositely charged, forming a sort of electric dipole. For the antenna in Figure 23.20, the electric field is along z and the electromagnetic waves propagate away in the x and y directions and in other directions in the x–y plane.

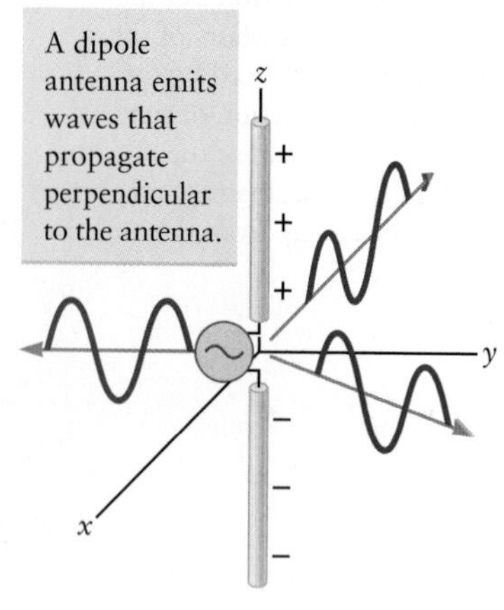

A dipole antenna emits waves that propagate perpendicular to the antenna.
z
+
+
+
y
−
−
−
x

▲ **Figure 23.20** Radiation from a dipole antenna propagates outward in directions perpendicular to the antenna's axis. There is essentially no radiation generated "off the ends" of the antenna. This antenna radiates strongly in all directions in the x–y plane, but does not emit radiation in the $+z$ or $-z$ directions.

In our discussion of antennas and the electric fields they produce (Figs. 23.18 and 23.20), we have emphasized the electric fields near the antennas. There is also an electromagnetic wave propagating inside the antenna wires, with an electric field that oscillates between positive and negative values as one moves away from the AC source. For a very long antenna, these oscillations tend to cancel, so a very long antenna does not necessarily produce a "stronger" electromagnetic wave than a short antenna. This cancellation becomes significant when one half of the antenna is longer than about $\lambda/4$ (one quarter of the wavelength). For this reason, most dipole antennas have a total length of $\lambda/2$.

Other antennas are more complicated than the dipole design, but the size or length of an antenna is usually comparable to the wavelength of the radiation. That is why the antenna for your cell phone is so small (Fig. 23.21A; see also the photo at the beginning of this chapter), whereas the antennas for an AM radio station are quite large (Fig. 23.21B).

Figures 23.18 and 23.20 describe how an antenna generates an electromagnetic wave, but the same antennas can be used to detect these waves. Figure 23.22 shows an electromagnetic wave interacting with a dipole antenna that might be in your cell phone. The electric field associated with the wave exerts a force on the electrons in the antenna, producing a current and an induced voltage across the antenna wires.

▲ **Figure 23.21** In most applications, an antenna's length is comparable to $\lambda/2$. A For a cell phone, the wavelength is short, so the antenna (which typically runs along and inside one edge of the case) is also short. B The wavelength of the radiation used by an AM radio station is much longer, so its antenna is very large.

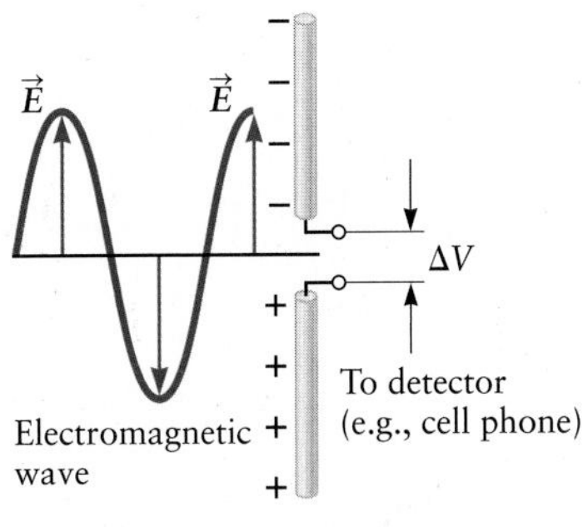

▲ **Figure 23.22** A dipole antenna can be used to detect an electromagnetic wave. The electric field of the wave induces a voltage across the wires of the antenna. This induced voltage is then amplified and processed by your cell phone or radio.

This voltage is then the input to an AC circuit inside your cell phone. That AC circuit amplifies the voltage signal and ultimately sends it to the phone's speakers.

Propagation and Intensity

The sketch of charge motion in Figure 23.18 shows how accelerated charges give rise to an electromagnetic wave. This picture is useful for understanding many radiation sources, but the charge motion is sometimes more complicated. For example, an incandescent lightbulb produces light due to the vibrations of electrons and ion cores in the hot filament of the bulb. Unlike the case in Figure 23.18, these vibrations in the filament are not confined to one particular direction; rather, charges vibrate along all different directions. As a result, the bulb produces light that propagates outward in all directions, unlike the case in Figure 23.18, in which the waves propagate only in the horizontal direction.

In many other cases, a source produces waves that propagate outward in all directions. The "ideal" case of a very small source producing spherical wave fronts is called a "point source." These wave fronts describe how energy propagates outward from the source. In Chapter 12, we saw that the intensity of a spherical wave falls with distance as (Eq. 12.11)

$$I \propto \frac{1}{r^2} \tag{23.16}$$

The intensity falls as one moves away from the source because there is a constant amount of energy and power associated with each wave front, whereas the wave front area ($A = 4\pi r^2$) increases with distance. (See Fig. 12.13, page 348, or Fig. 17.13, page 493.) Intensity is the power per unit area, so the intensity decreases as this constant amount of energy is spread over an ever-increasing area. The decrease of intensity with distance is thus a direct consequence of the principle of conservation of energy. The relation between intensity and distance in Equation 23.16 applies to many cases, including the falloff of intensity for a lightbulb, the falloff of intensity of sunlight with distance through the universe, and the "strength" (i.e., intensity) of a radio signal from a distant radio station or space satellite.

EXAMPLE 23.6 Intensity of Sunlight

The intensity of sunlight near the Earth is about 1000 W/m^2. Since the Earth is very far from the Sun, the Sun acts approximately as a point source of radiation. What is the intensity of sunlight at the surface of Mars?

RECOGNIZE THE PRINCIPLE

Because the Sun acts as a point source of spherical waves, its intensity falls with distance as $1/r^2$. We can use the known distances between the Sun and the Earth and the Sun and Mars to get both the relative intensity and the absolute intensity at Mars.

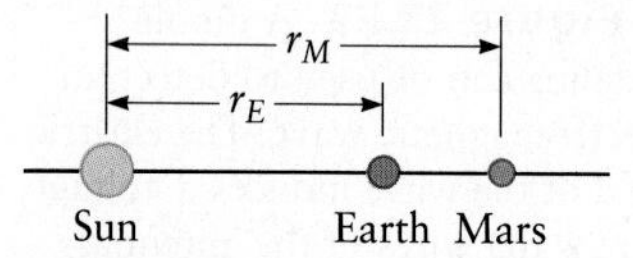

▲ **Figure 23.23** Example 23.6.

SKETCH THE PROBLEM

Figure 23.23 describes the problem.

IDENTIFY THE RELATIONSHIPS

We denote the Sun–Earth distance as r_E and the Sun–Mars distance as r_M. Using Equation 23.16, we can write the intensity at Earth as $I_E = K/r_E^2$ and the intensity at Mars as $I_M = K/r_M^2$. The constant of proportionality K is the same in these two cases, so taking the ratio, we have

$$\frac{I_M}{I_E} = \frac{K/r_M^2}{K/r_E^2} = \frac{r_E^2}{r_M^2} \tag{1}$$

From Table A.3 of solar system data in Appendix A, we find $r_E = 1.5 \times 10^8$ km and $r_M = 2.3 \times 10^8$ km.

SOLVE

Rearranging Equation (1) to solve for the intensity at Mars gives

$$I_M = I_E\left(\frac{r_E}{r_M}\right)^2$$

Inserting our values of the intensity at the Earth and the distances r_E and r_M, we get

$$I_M = (1000 \text{ W/m}^2)\left(\frac{1.5 \times 10^8 \text{ km}}{2.3 \times 10^8 \text{ km}}\right)^2 = \boxed{430 \text{ W/m}^2}$$

▶ ***What does it mean?***

The intensity of sunlight at Mars is only 43% of the value at the Earth. As a result, the surface temperature of Mars is much less than that of the Earth.

23.6 Polarization

When discussing an electromagnetic wave, it is important to know its propagation direction since that tells the direction in which energy is traveling. We know that the electric field must be perpendicular to the propagation direction, but there are still many different possible directions for $\vec{E}$ that are perpendicular to the wave direction. The direction of $\vec{E}$ is important because it determines how the wave interacts with matter (including antennas).

Figure 23.20 shows how a dipole antenna can be used to produce an electromagnetic wave with an electric field $\vec{E}$ directed along $\pm z$. To indicate that the electric field is always parallel to z, we say that this wave is ***linearly polarized*** along the z direction. Electromagnetic radiation that is linearly polarized is in many ways the simplest case to describe. Most light is not polarized in this way, however. For example, sunlight and the light from a flashlight are both ***unpolarized***. Light from these sources is a combination of many individual waves, all with different field amplitudes and polarization directions.

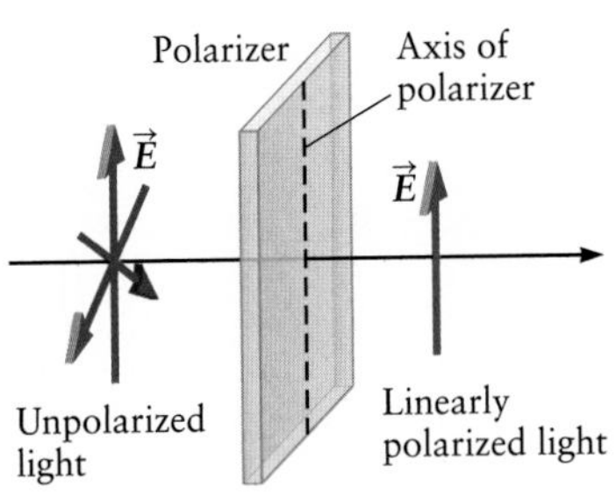

▲ **Figure 23.24** When unpolarized light is incident on a polarizer, only the component with the electric field parallel to the axis of the polarizer is transmitted.

Polarizers

Polarized light can be created using a device called a ***polarizer***, one type of which is shown in Figure 23.24. It consists of a thin, plastic film that allows an electromagnetic wave to pass through only if the electric field of the wave is parallel to a particular direction called the axis of the polarizer. The polarizer absorbs radiation with electric fields that are not along the axis. Hence, when unpolarized light (perhaps from a flashlight) strikes a polarizer, the light that comes out is linearly polarized.

To evaluate the intensity of the light transmitted by a polarizer, let's first consider what happens when linearly polarized light strikes a polarizer. In Figure 23.25A, the incident light is polarized parallel to the axis of the polarizer and the outgoing electric field is equal in amplitude to the incoming field, so $E_{out} = E_{in}$; all the incident energy is thus transmitted through. In Figure 23.25B, the incident light is polarized perpendicular to the polarizer axis and no light is transmitted, so $E_{out} = 0$. In Figure 23.25C, the incident light is polarized at an angle θ relative to the axis of the polarizer. The outgoing light is now polarized along the direction defined by the polarizer axis with an electric field amplitude

$$E_{out} = E_{in} \cos\theta$$

In words, only the component of the electric field along the polarizer axis is transmitted. Since intensity is proportional to the square of the electric field amplitude (Eq. 23.9), we can also write

$$I_{out} = I_{in} \cos^2\theta \qquad (23.17)$$

which is called ***Malus's law***. When $\theta \neq 0$, the outgoing intensity is thus smaller than the incident intensity.

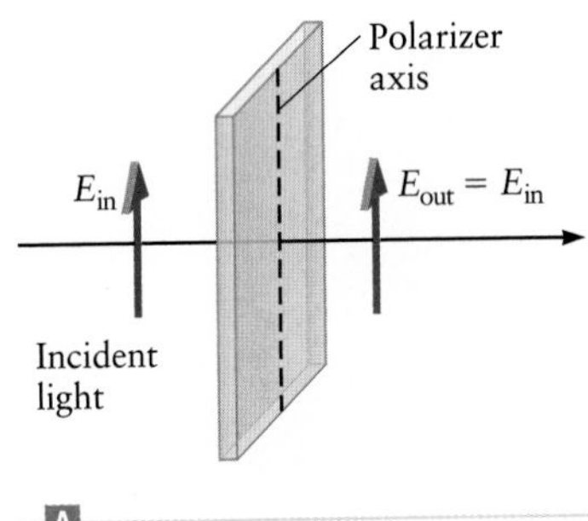

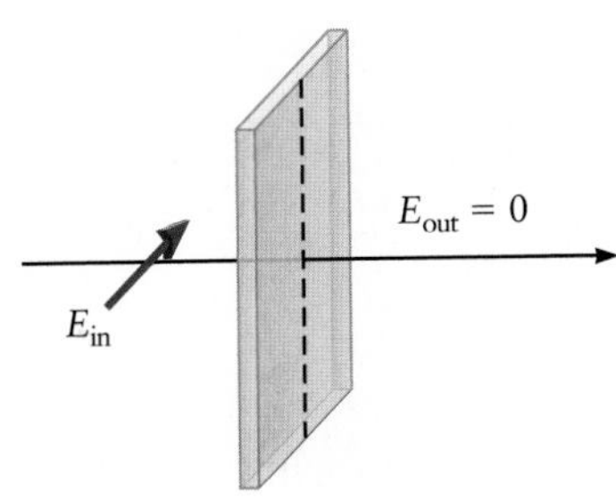

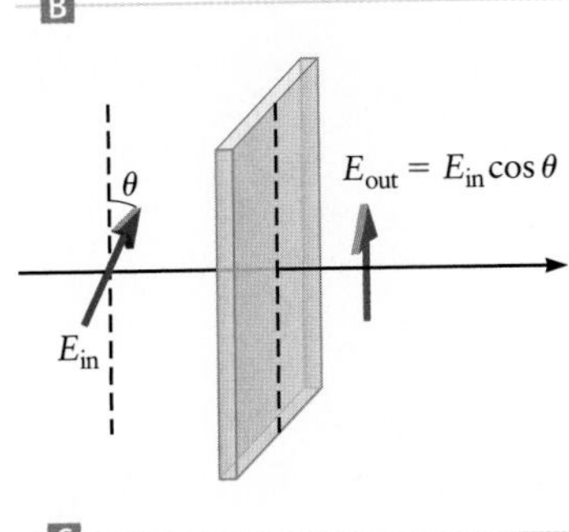

▲ **Figure 23.25** When linearly polarized light is incident on a polarizer, the transmitted intensity depends on the angle θ between the incident electric field E_{in} and the polarizer axis.

EXAMPLE 23.7 Electric Field and Intensity Transmitted through a Polarizer

Linearly polarized light is incident on a polarizer, with an angle θ between the incoming electric field and the axis of the polarizer (Fig. 23.25C). Find the value of θ for which the transmitted intensity is half the incident intensity.

RECOGNIZE THE PRINCIPLE

When the incident light is linearly polarized, only the component of the electric field along the polarizer axis is transmitted, and the transmitted intensity is given by Malus's law (Eq. 23.17). We can use this relation to find the value of θ for which I_{out} is half of I_{in}.

SKETCH THE PROBLEM

Figure 23.25C describes the problem. It shows the incident and transmitted electric fields.

IDENTIFY THE RELATIONSHIPS

From Malus's law, we have $I_{out} = I_{in} \cos^2\theta$, which we can rearrange to read

$$\frac{I_{out}}{I_{in}} = \cos^2\theta$$

(continued) ▶

SOLVE

We want the transmitted intensity to be half the incident intensity; hence, $I_{out}/I_{in} = \frac{1}{2}$, which leads to

$$\cos^2\theta = \frac{I_{out}}{I_{in}} = \frac{1}{2}$$

$$\cos\theta = \frac{1}{\sqrt{2}}$$

$$\theta = \boxed{45°}$$

▸ *What does it mean?*

When the angle θ between the polarization direction of the incident light and the axis of the polarizer is not zero, the transmitted intensity is smaller than the incident intensity. The transmitted light is still linearly polarized, but with a different polarization direction (now parallel to the polarizer axis).

Malus's law can be used to understand what happens when unpolarized light strikes a polarizer. Unpolarized light can be thought of as a collection of many separate light waves, each linearly polarized but with different and random polarization directions. Each separate wave is transmitted through the polarizer according to Malus's law. The average outgoing intensity is then the average of I_{out} for all the incident waves. We can thus write

$$I_{out} = (I_{in}\cos^2\theta)_{ave}$$

The average here is an average over all possible polarization angles for the incoming light waves. The factor $\cos^2\theta$ varies between 0 and 1, and its average is $\frac{1}{2}$, so for unpolarized light the intensity transmitted through a polarizer is

$$I_{out} = \tfrac{1}{2}I_{in} \qquad \text{(unpolarized light)} \tag{23.18}$$

An interesting application of Malus's law is sketched in Figure 23.26A, which shows light that is initially unpolarized before it passes through two polarizers in a row. After the light passes through polarizer 1, its intensity is reduced by half (Eq. 23.18), and it is now linearly polarized along z, the axis of polarizer 1. The axis of polarizer 2 is perpendicular to z, so the final transmitted intensity is zero.

Figure 23.26B shows a similar arrangement, but now we have inserted a third polarizer in the middle. The axis of polarizer 3 makes a 45° angle with the z axis, so some of the light from polarizer 1 is transmitted through polarizer 3. Moreover, after passing through polarizer 3 this light is polarized at a 45° angle with respect to polarizer 2. Hence, the final transmitted intensity after passing through all three polarizers is nonzero. Comparing this conclusion to the experiment in Figure 23.26A leads to the following "puzzle." In Figure 23.26A, the final transmitted intensity I_{final} is zero. In Figure 23.26B, we have added an extra polarizer, and now I_{final} is no longer zero! Although that may seem surprising, it can be understood by using two facts

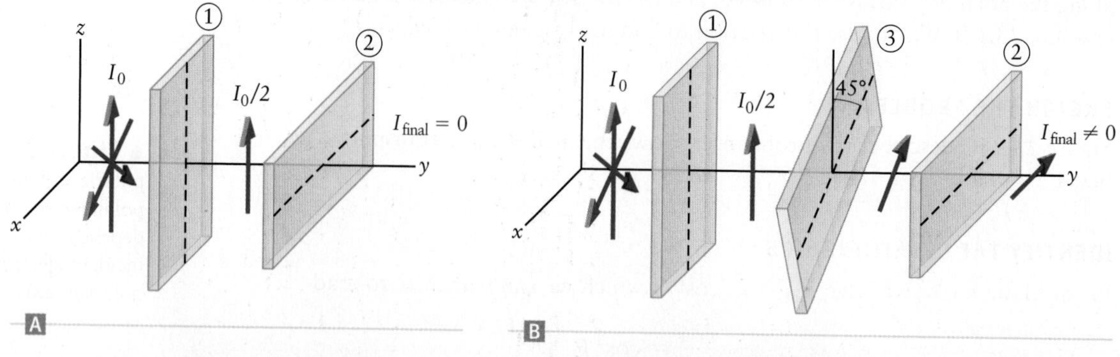

▲ **Figure 23.26** A Unpolarized light is incident on two polarizers whose axes are perpendicular. The final transmitted intensity is zero. B A third polarizer is placed between the polarizers in part A. Now the transmitted intensity is not zero!

about polarizers. First, when analyzing light as it passes through several polarizers in succession, you must always analyze the effect of one polarizer at a time. Second, the light transmitted by a polarizer is always linearly polarized, with a polarization direction determined solely by the polarizer axis. The transmitted light wave has no "memory" of its original polarization.

CONCEPT CHECK 23.6 Polarizers in "Series"

If the intensity of the initial unpolarized light beam in Figure 23.26B is I_0, is the final transmitted intensity (a) $I_0/2$, (b) $I_0/4$, (c) $I_0/8$, or (d) $I_0/16$?

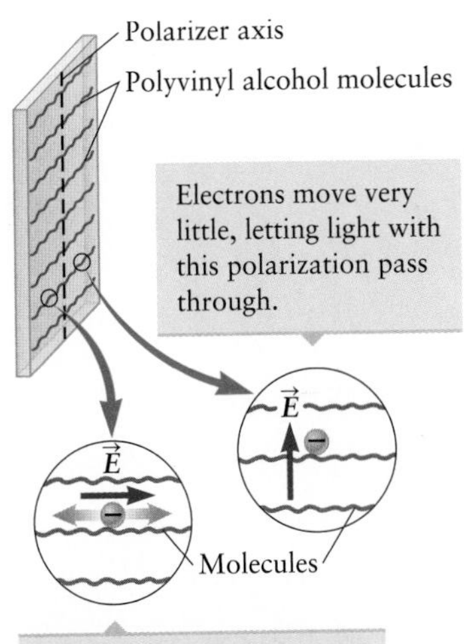

▲ **Figure 23.27** A film containing long molecules of polyvinyl alcohol aligned parallel to one another acts as a polarizer. The axis of the polarizer is perpendicular to the molecular direction.

How Does a Polarizer Work?

Polarizers have some interesting applications (including sunglasses and in the displays in watches and other devices) and can be made in several different ways. Most applications use a sandwich structure with certain types of long molecules (usually polyvinyl alcohol dyed with iodine) placed between two thin sheets of plastic. When the molecules are aligned parallel to one another, the sheet acts as a polarizer, with the axis of the polarizer *perpendicular* to the direction of the molecules (Fig. 23.27).

Electrons in the polarizer molecules respond to electric fields as shown in Figure 23.27. When the electric field of an electromagnetic wave is parallel to the molecular chains, the electrons move back and forth along the molecule as the direction of $\vec{E}$ oscillates, producing an AC electric current along the molecular chains. This current leads to the absorption of energy, just like a resistor dissipates (absorbs) energy in an AC electric circuit. Light with a polarization direction parallel to the molecular direction is therefore absorbed. On the other hand, when $\vec{E}$ is perpendicular to the molecular direction, each electron is bound to a particular molecule and moves very little because it cannot "jump" from one molecule to another. The electrons then absorb very little energy, letting light with this polarization pass through. That is why the polarization axis is perpendicular to the molecular direction.

Polarization by Scattering and Reflection of Light

Light from the Sun is unpolarized, but sunlight can become polarized after it is scattered by molecules in the atmosphere through a series of events (Fig. 23.28). The

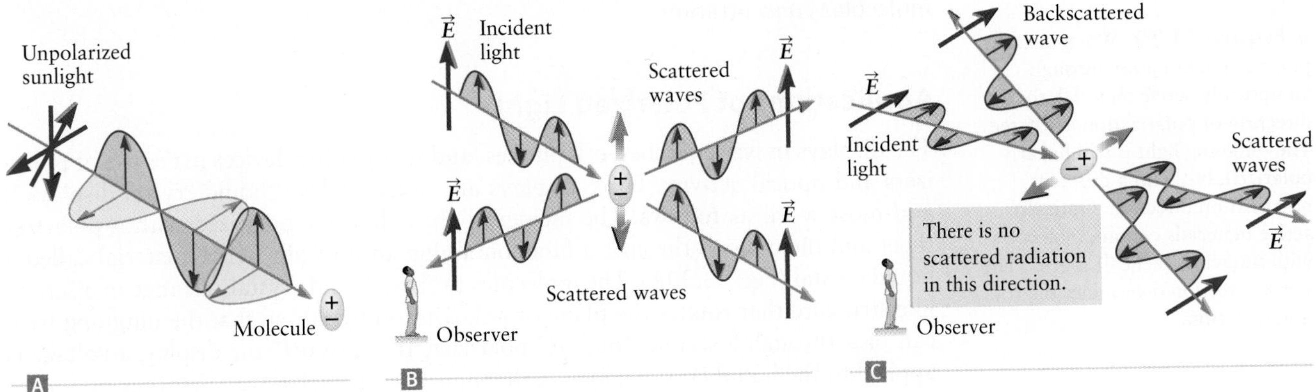

▲ **Figure 23.28** Polarization by scattering of light. **A** Sunlight is unpolarized. When light strikes molecules in the atmosphere, the electrons and ion cores oscillate in response to the light's electric field. **B** The vibrating charges act as a dipole antenna and produce scattered electromagnetic waves. The electric field of the scattered waves is parallel to the vibration direction. Here there is a scattered (and polarized) wave in the direction of the observer. **C** A dipole antenna does not emit radiation "off its ends," so radiation with this polarization does not propagate toward the observer in the position shown. Since the only radiation that reaches the observer has the electric field direction shown in part B, the scattered radiation is polarized.

molecules act as antennas; the positive ion cores and electrons respond to sunlight by oscillating back and forth in the direction of the electric field, like the charges in the antenna wires in Figure 23.18. Parts B and C of Figure 23.28 show this motion for light polarized in two different directions. In Figure 23.28B, the polarization direction and the direction of $\vec{E}$ are vertical; the ion cores and electrons thus vibrate up and down and produce new outgoing waves that are also polarized vertically. These outgoing waves are called ***scattered waves.***

Figure 23.28C shows how the ion cores and electrons move when the incoming light is polarized horizontally. The ion cores and electrons now oscillate in a horizontal direction and again produce scattered light waves. Because dipole antennas do not radiate "off their ends" (compare with Fig. 23.20), no scattered light reaches the observer looking along the vibration direction.

Now let's combine the results in parts B and C of Figure 23.28 to understand how unpolarized sunlight is scattered. The component of this light polarized vertically (Fig. 23.28B) is scattered to the observer, but none of the horizontally polarized scattered light propagates in the observer's direction. Hence, the light that reaches the observer is said to be ***polarized by reflection*** (or by scattering). You can observe this polarization by looking at the sky through a polarizer and comparing to the case without a polarizer.

Similar effects can occur whenever unpolarized light is scattered. For example, sunlight scattered by dust particles on a windshield is polarized. This scattered light, called "glare," is usually a nuisance for a driver. It can be removed using the polarizers that are built into most sunglasses.

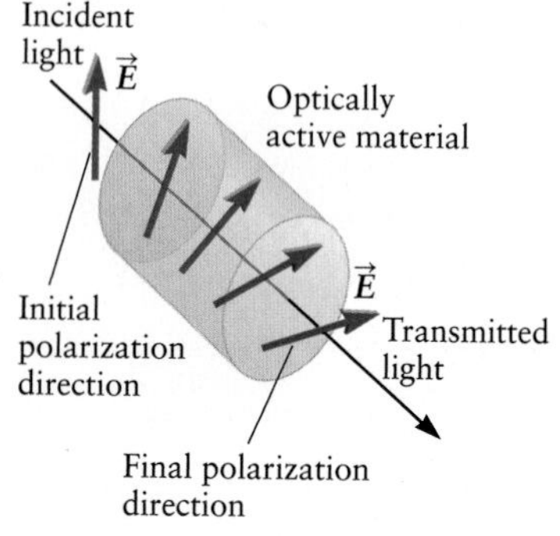

▲ **Figure 23.29** When linearly polarized light passes through an optically active material, the direction of polarization is rotated. The outgoing light is still linearly polarized, but with a different polarization direction. Optically active materials contain molecules with screwlike (helical) shapes. The screw direction determines the rotation direction.

Optical Activity

When linearly polarized light passes through certain materials, the polarization direction is rotated as shown in Figure 23.29. This effect is called ***optical activity.*** Materials that are optically active often contain molecules with a screwlike or helical structure. DNA is one of many examples; in fact, simple sugars such as fructose and dextrose are optically active molecules. In words, the electric field of linearly polarized light is "dragged along" the screw direction of the molecules. Dextrose and fructose molecules are both helical, but one forms a right-handed helix and the other is left-handed, so they rotate the polarization in different directions. This effect is used in studies of sugar molecules in solution. The rotation direction gives information on the type of sugar present, while the amount of rotation is used to deduce the molecular concentration.

Applications of Polarized Light

The displays in wristwatches, cell phones, and many other devices make use of polarizers and optical activity. These displays are called LCDs ("liquid crystal displays") and most work as follows. The incident light is linearly polarized with a polarizer sheet and then passes through a film containing an optically active material called a liquid crystal (Fig. 23.30A). The molecules in these liquid crystals arrange in a screwlike structure that rotates the plane of polarization by 90° so that the outgoing light can pass through a second "output" polarizer. To "turn off" the display, a voltage is applied to the liquid crystal, aligning the molecules in such a way that they no longer rotate the direction of polarization (Fig. 23.30B). The light that comes out of the liquid crystal is then polarized at 90° relative to the output polarizer, so no light is transmitted and the display is dark. By applying different voltages to different areas of the liquid crystal, a pattern of light and dark regions corresponding to different letters or numbers can be formed, resulting in the letters and numbers you see in the display of your wristwatch, cell phone, or MP3 player, or even the displays at gasoline station pumps. You can tell that the light from a display is polarized by placing another polarizer such as the polarizers in your sunglasses in front as shown in Figure 23.31. Try it.

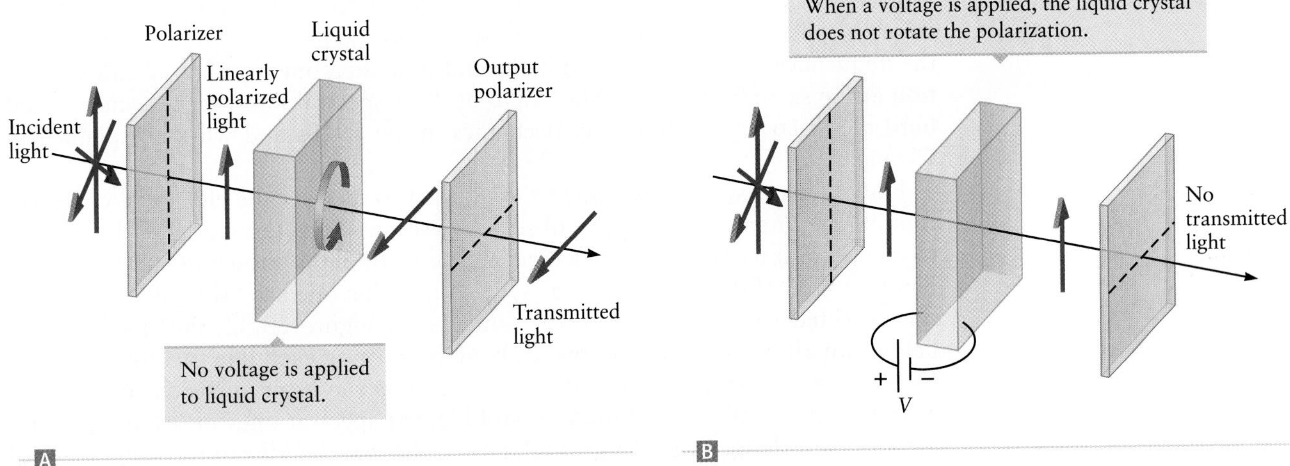

▲ **Figure 23.30** Light passing through a liquid crystal display. A When no voltage is applied to the liquid crystal layer, it rotates the plane of polarization by 90°, allowing light to pass through the output polarizer. B When a voltage is applied to the liquid crystal, it no longer rotates the plane of polarization and no light is transmitted through the output polarizer.

A

B

▲ **Figure 23.31** A liquid crystal display on a wristwatch as viewed through a polarizer. The output light is polarized (Fig. 23.30A), so A when the plane of polarization is at a right angle with the axis of a polarizer placed in front of the display, no light is transmitted and the display is dark. B When the polarizer is rotated, light from the display passes through.

CONCEPT CHECK 23.7 Polarization of a Wave

We have seen that electromagnetic waves can be unpolarized and that when an unpolarized electromagnetic wave passes through a polarizer, a linearly polarized electromagnetic wave is produced. Is it possible for a sound wave in air to be unpolarized? Explain.

23.7 Doppler Effect

In Section 13.6, we discussed the Doppler effect, a phenomenon in which the frequency of a wave is affected by the relative motion of the source and observer. The Doppler effect was used by astronomer Edwin Hubble in his studies of the motion of stars and galaxies in the universe in the early 1900s. Light is emitted and absorbed by atoms of different elements at the surface of the star. From the work of physicists in the 1800s, Hubble knew that each atom emits and absorbs light of certain characteristic colors. Figure 23.32 (top) shows some of the colors absorbed by hydrogen, magnesium, and sodium on the surface of our Sun. Each vertical dark line in this plot corresponds to a color absorbed by these atoms, and the location of each line along the horizontal axis corresponds to a certain wavelength of light.

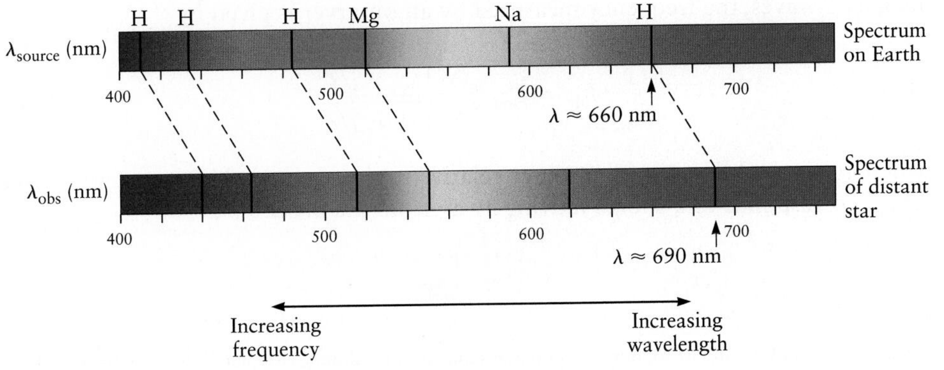

◄ **Figure 23.32** Spectral lines showing absorption by hydrogen (H), magnesium (Mg), and sodium (Na) atoms. (Emission lines occur at the same wavelengths.) Top: Spectrum found when the atoms are at rest relative to the observer. Bottom: Spectral lines of the same element as we observe from a distant galaxy. Each line is shifted toward longer wavelengths (i.e., lower frequencies).

This type of plot is called a ***spectrum*** of the starlight, and each vertical line (each color) is called a ***spectral line***. A similar set of spectral lines is found if one measures the frequencies emitted by an atom; one finds that an atom absorbs and emits radiation at the same frequencies. The origin of these spectral lines was not understood until the discovery of quantum mechanics in the 1920s and will be explained in Chapter 29.

Hubble was using measurements of spectra to identify the elemental composition of distant galaxies when he observed cases in which the spectra did not seem to correspond to any known elements. One example is shown at the bottom of Figure 23.32. Hubble showed that spectra like this one are "shifted" versions of the expected spectra from known elements. In Figure 23.32, the spectral lines below can all be derived from those above by shifting each line to a longer wavelength (lower frequency), and the shift factor is the same for each line. In this example and in all others found by Hubble, the spectral lines of distant galaxies are shifted to longer wavelengths relative to the wavelengths of the same spectral lines emitted by atoms on Earth. Because red light is at the long-wavelength end of the visible part of the electromagnetic spectrum (Fig. 23.8), this effect is called a ***red shift***.

To explain this result, Hubble proposed that the galaxy emitting the red-shifted light was moving away from the Earth. The Doppler effect then causes the frequency of the light as measured by an observer on Earth to be shifted to lower frequencies,[1] similar to the Doppler shift for the sound emitted by an ambulance siren when the ambulance is going away from you. The size of the frequency shift can be used to determine the velocity of the emitting galaxy. After many galaxies were studied in this way, it was found that most galaxies in the observable universe are moving away from the Earth.

Early universe

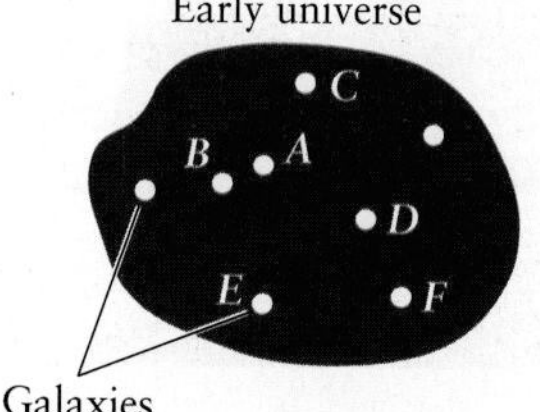

Later universe

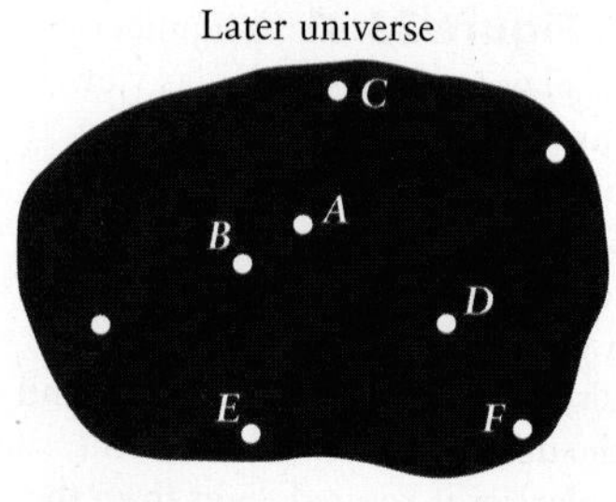

▲ **Figure 23.33** Schematic of an expanding universe. As the universe expands, the distance between any two points (i.e., between any two galaxies) increases with time. The relative velocity between two nearby points is less than the relative velocity between two widely separated points.

At first glance, you might think that if most or all of the galaxies are moving away from us, the Earth must be at the "center" of the universe, but Hubble found that this picture is *not* correct. He showed that the farther a galaxy is from the Earth, the faster it is receding from the Earth, leading to the picture in Figure 23.33 of an expanding universe at two different times. The galaxy at point *B* might, for example, contain the Earth. As time passes, moving from early in the universe's history to later, the distance from *B* to galaxy *A* and from *B* to all other galaxies increases. Hence, *as viewed from the Earth*, all other galaxies are moving away from us. At the same time, observers in other locations find that other galaxies are also moving away from them. For example, the distances from galaxy *F* to galaxies *B*, *D*, and *E* also increase as time passes, so observers at *F* would also say that all other galaxies are moving away from them.

Measurements of the red shift are now used to determine the motion of many different objects in the universe and are the basis of much of modern astronomy and astrophysics. The analysis of red shift experiments is based on how the Doppler effect causes the frequency seen by an observer f_{obs} to be shifted relative to the frequency of the source f_{source}. In Chapter 13, we derived relations for the Doppler shift for sound waves. Due to the effect of relativity (Chapter 27), however, the Doppler shift relations for light are different. For Doppler shifts of light and other electromagnetic waves, the frequency measured by an observer is given by

$$f_{obs} = f_{source}\frac{\sqrt{1 - v_{rel}/c}}{\sqrt{1 + v_{rel}/c}} \quad (23.19)$$

where v_{rel} is the velocity of the source relative to the observer. Here a positive value of v_{rel} corresponds to a source moving away from the observer.

[1]Hubble's red shift is similar to, but not quite the same as, an ordinary Doppler shift. See Chapter 31.

EXAMPLE 23.8 Ⓡ Doppler Shift of a Distant Galaxy

Figure 23.32 shows how several spectral lines from a distant galaxy are Doppler-shifted due to the galaxy's motion relative to the Earth. Use the data in this figure to estimate how fast this galaxy is moving relative to the Earth.

RECOGNIZE THE PRINCIPLE

When a light source is moving away from an observer, the wavelength of the light as measured by the observer is longer than the wavelength emitted by the source. Hence, the frequency seen by the observer f_{obs} is lower than the source frequency f_{source}.

SKETCH THE PROBLEM

Figure 23.32 shows the wavelengths at the observer λ_{obs} (bottom spectrum) and source λ_{source} (top spectrum). From the data given in the figure, we can read the values $\lambda_{obs} = 690$ nm and $\lambda_{source} = 660$ nm.

IDENTIFY THE RELATIONSHIPS AND SOLVE

If the velocity of the source galaxy relative to an observer on the Earth is v_{rel}, the frequencies at the observer and source are related by (Eq. 23.19)

$$f_{obs} = f_{source}\frac{\sqrt{1 - v_{rel}/c}}{\sqrt{1 + v_{rel}/c}}$$

Rearranging leads to

$$\frac{\sqrt{1 - v_{rel}/c}}{\sqrt{1 + v_{rel}/c}} = \frac{f_{obs}}{f_{source}}$$

$$\frac{1 - v_{rel}/c}{1 + v_{rel}/c} = \left(\frac{f_{obs}}{f_{source}}\right)^2$$

Because $f = c/\lambda$, we also have

$$\frac{1 - v_{rel}/c}{1 + v_{rel}/c} = \left(\frac{\lambda_{source}}{\lambda_{obs}}\right)^2 \quad (1)$$

Defining $y = (\lambda_{source}/\lambda_{obs})^2$, we get

$$1 - v_{rel}/c = y(1 + v_{rel}/c)$$

Next we solve for v_{rel}/c:

$$(v_{rel}/c)(1 + y) = 1 - y$$

$$v_{rel}/c = \frac{1 - y}{1 + y} = \frac{1 - 0.91}{1 + 0.91} = \boxed{0.05}$$

where we have inserted the value of $y = (\lambda_{source}/\lambda_{obs})^2 \approx 0.91$ derived from the red shift data in Figure 23.32. We thus get

$$v_{rel} \approx \boxed{2 \times 10^7 \text{ m/s}}$$

The positive value of v_{rel} indicates that the source is moving away from the observer.

► ***What does it mean?***

The value of v_{rel} is about 5% of the speed of light. Hubble discovered that the speed of a galaxy is related to its distance from the Earth by what is now known as Hubble's law. This law can be written as

$$d = \frac{v_{rel}}{H_0}$$

(continued) ►

where H_0 is called Hubble's constant, which has the value $H_0 = 2.2 \times 10^{-18}\ \text{s}^{-1}$. Inserting our value of v_{rel} into Hubble's law gives

$$d = \frac{v_{rel}}{H_0} = \frac{2 \times 10^7\ \text{m/s}}{2.2 \times 10^{-18}\ \text{s}^{-1}} = 1 \times 10^{25}\ \text{m}$$

This distance is incredibly large; it can be appreciated better if it is expressed in light-years, the distance that light travels in one year. We find

$$d = 1 \times 10^9\ \text{light-years}$$

The spectrum in Figure 23.32 thus involves light that spent about one billion years traveling to the Earth!

23.8 Deep Concepts and Puzzles Connected with Electromagnetic Waves

Maxwell's work and the subsequent experimental studies of electromagnetic waves by Hertz, Tesla, Marconi, and others led to profound changes in the way we think about electromagnetism and the laws of physics.

Fields Are Real

We first introduced the notion of a *field* in our discussion of gravitation. When analyzing the force of gravity, lines of force led naturally to the concept of gravitational field lines. The gravitational field is certainly a convenient way to visualize and think about gravity, but, in the opinion of Newton and most of his contemporaries, forces were "real" and field lines were just a helpful picture.

We took a similar approach in our work with electromagnetism. There we started with electric forces given by Coulomb's law and then noticed that these forces could be conveniently described using the notion of electric field lines. Magnetic forces and magnetic fields could be understood in a similar way, but the question remains, "Are these fields real, or are they just a useful mathematical tool?"

The existence of electromagnetic waves means that electric and magnetic fields *are indeed real*. There is no other way to explain how an electromagnetic wave can propagate through a vacuum, carrying energy and momentum. An analogy with the gravitational field also suggests the existence of ***gravitational waves***. Physicists are now working to detect the gravitational waves believed to come to us from distant galaxies.

How Can Electromagnetic Waves Travel through a Vacuum?

Although Maxwell's theoretical work (based on the equations of electricity and magnetism) showed that electromagnetic waves can travel through a vacuum, physicists at the time had difficulty accepting this result. All mechanical waves travel through a material medium, and most physicists believed that there must be a similar medium called the ***ether*** to support electromagnetic waves. It was believed that the ether permeated all space, including material substances and the vacuum, too. The ether was assumed to have some amazing properties; for example, an object moving through a vacuum would experience no frictional force due to the ether. It was claimed that electromagnetic waves were mechanical waves through the ether. There were many experimental attempts to study the ether and its properties, but eventually (around 1900) the existence of the ether was disproved. Since there is no ether, electromagnetic waves need no medium for travel. An electromagnetic wave can carry energy even through empty space (a vacuum). This finding opened the way to Einstein's work and his theory of relativity (Chapter 27), which completely changed the way physicists view the universe. Maxwell's work was thus a catalyst for much of 20th-century physics.

Electromagnetic Waves and Quantum Theory

What is light? This question was debated by scientists for many centuries. Newton and others proposed that light is made up of particles, whereas other physicists suggested that light is a wave. Maxwell's work seemed to demonstrate conclusively that light is indeed a wave, but that is not the end of the story. We mentioned earlier that Newton's laws break down when applied to very small objects such as electrons and protons and that such objects must be described using quantum theory (Chapters 28 and 29). Maxwell's equations must also be modified to describe light and other electromagnetic waves at the quantum level. Physicists now know that light has *both* wavelike *and* particle-like properties. The "particles" of light are called ***photons***.

Summary | CHAPTER 23

Key Concepts and Principles

A time-varying magnetic field produces an electric field

According to Faraday's law, a changing magnetic flux Φ_B produces an induced emf $\mathcal{E}$ with $\mathcal{E} = -\Delta\Phi_B/\Delta t$. An emf $\mathcal{E}$ is just an electric potential difference (voltage) ΔV, and an electric field is always associated with a voltage. Hence, Faraday's law means that *a changing magnetic field produces an electric field.* Maxwell showed that *a changing electric field produces a magnetic field.* This interdependency of E and B leads to the existence of ***electromagnetic waves***.

Properties of electromagnetic waves

Electromagnetic waves are ***transverse waves***: their electric and magnetic fields are perpendicular to each other, and both $\vec{E}$ and $\vec{B}$ are perpendicular to the propagation direction. These waves can travel through a vacuum and through many material substances. In a vacuum, they travel at speed c:

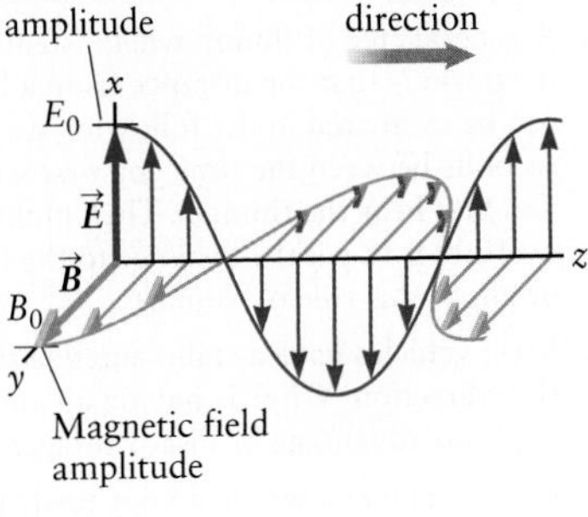

$$c = \frac{1}{\sqrt{\varepsilon_0\mu_0}} = 3.00 \times 10^8 \text{ m/s} \qquad \textbf{(23.1) and (23.2) (page 697)}$$

The ***intensity*** of an electromagnetic wave (I) is related to the amplitude of the electric field E_0 by

$$I = \tfrac{1}{2}\varepsilon_0 c E_0^2 \qquad \textbf{(23.9) (page 700)}$$

The electric and magnetic field amplitudes are related by

$$E_0 = cB_0 \qquad \textbf{(23.6) (page 699)}$$

In addition to carrying energy, electromagnetic waves carry momentum, leading to ***radiation pressure***. When light (or any other electromagnetic wave) is absorbed by an object, the radiation pressure on the object is

$$P_{\text{radiation}} = \frac{I}{c} \qquad \textbf{(23.13) (page 702)}$$

(Continued)

Maxwell's equations

The equations of electromagnetism—the laws of Gauss, Coulomb, Ampère, and Faraday—are called ***Maxwell's equations.***

Applications

The electromagnetic spectrum

Electromagnetic waves are classified according to their frequency. This frequency ranges from a few Hz for ***radio waves*** to 10^{22} Hz and above for ***gamma rays.*** In between we have ***microwaves, infrared radiation, visible light, ultraviolet (UV) light,*** and ***X-rays*** (Fig. 23.8).

Polarization

Electromagnetic waves can be ***polarized.*** If the electric field vector points along a single direction, the wave is ***linearly polarized,*** whereas light from many sources (such as the Sun) is ***unpolarized.*** Light can be polarized by transmission through a polarizer film or by scattering and reflection.

Doppler effect

The frequency of an electromagnetic wave depends on the relative motion of the source and observer, similar to the ***Doppler effect*** we studied in connection with sound in Chapter 13. This effect is used by astronomers to study the motion of distant galaxies.

Questions

SSM = answer in *Student Companion & Problem-Solving Guide* = life science application

1. A general rule of thumb when listening to and watching a thunderstorm is that the distance from a listener to the lightning strike can be estimated in the following way: First, count the number of seconds between the time you first see the lightning and the time you first hear the thunder. Then multiply this time (in seconds) by 1000 ft to get the distance to the lightning strike. Explain the origin of this rule of thumb.

2. Most vehicles have a radio antenna that is oriented along the vertical direction. Why? Is having an antenna mounted at a slanting angle an advantage or disadvantage? Explain.

3. Electromagnetic waves do not easily penetrate far into metal surfaces. Discuss why that is the case in terms of the known properties of metals, atoms, and electrons. This is one reason it is sometimes difficult to get radio or cell phone signals when you are inside a closed elevator or in a building with a lot of metal in its walls.

4. Microwave ovens heat food items by bathing them in electromagnetic waves, yet you can watch your food cook through a window in the door (Fig. Q23.4). How is it possible for the door, with only a thin layer of metal with a grid of holes, to contain the powerful electromagnetic radiation?

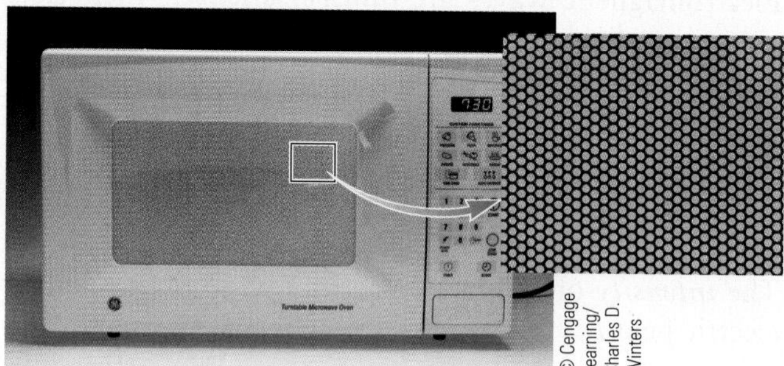

© Cengage Learning/Charles D. Winters

Figure Q23.4

5. A method for spacecraft propulsion called a "solar sail" has been proposed. (See Example 23.3 and Fig. 23.6.) It would use a sail exposed to the Sun in much the same way a sailboat's sail uses wind. The sail on such a spacecraft might reflect light, or it might absorb light. Which type of sail would yield the largest force on the sail? Explain your answer.

6. Consider the radiation pressure on an object in outer space. Explain why, all else being similar, this pressure has a larger effect on the motion of a small particle than on a large particle. *Hint*: Consider and compare the mass and the force of radiation pressure on two spherical objects composed of the same material but with different radii.

7. Explain how an AC circuit can generate electromagnetic radiation.

8. Consider the RC circuit in Figure Q23.8. The switch is closed at $t = 0$ and then kept closed for a very long time. Will this circuit emit electromagnetic radiation (a) just after the switch is closed? (b) After a very long time? Explain your answers. Can you think of ways you might have observed radiation when you opened or closed a switch?

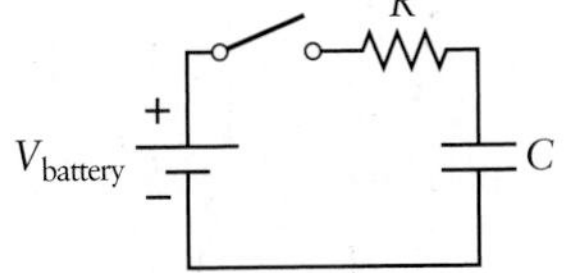

Figure Q23.8

9. In the United Kingdom (and many other European countries), residents are not allowed to use a TV without purchasing a yearly license. Fines of 1000 British pounds are levied against those who are caught using a TV without a license. The law is enforced through devices that can detect a TV in use from 20 m away or more. Figure Q23.9 shows a van that contains one of these devices. (Note the antennas on top of the van.) How might this device work? *Hint*: All TVs have LRC circuits in them.

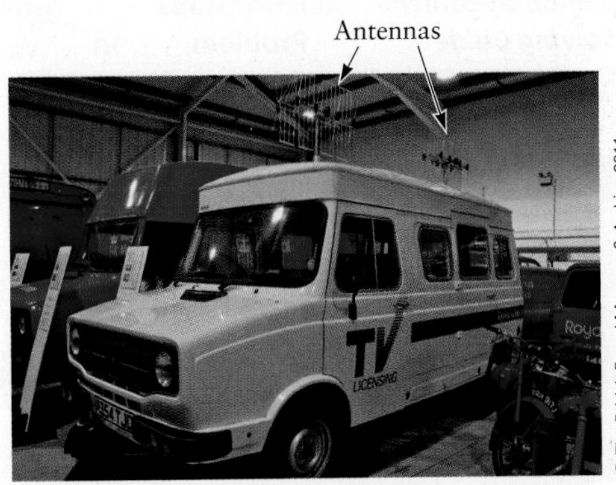

Figure Q23.9

10. SSM The glare off horizontal surfaces that you see while driving a car on a sunny day can be minimized if you wear sunglasses with polarizer films built into the lenses so that light polarized in a certain direction is not allowed to reach your eyes. Discuss how such sunglasses can reduce glare. Is the polarization axis of the glasses vertical or horizontal?

11. Suppose you are given a polarizer, but the axis of polarization is not marked. Devise an experiment you could do to determine the polarization axis.

12. Photographers often use a "haze filter" to remove (or at least minimize) the effects of haze on their photographs. Explain how these filters work.

13. It is a very dusty day, and you are standing by the side of the road looking directly across the road as a car is approaching (Fig. Q23.13). Due to the dust particles in the air, you can see light from the car's headlights as it scatters from the particles. Will this scattered light be (partially) polarized? If so, will the polarization axis be horizontal or vertical?

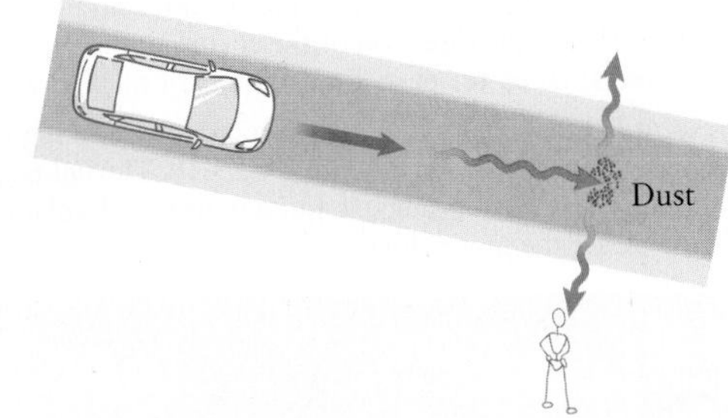

Figure Q23.13

14. Light from bulbs and other forms of indoor lighting is typically unpolarized, whereas radio waves picked up by radios or cell phones are mostly polarized. Explain why these two types of electromagnetic radiation are different in this way.

15. Can sound waves be polarized? What about ocean waves or waves propagating on a string? What fundamental characteristic allows waves to have a polarization?

16. Consider a bright, sunny day at the beach. Why is it that polarizing sunglasses work well at reducing the glare off the sand and water except when you lie on your side (or tilt your head toward a shoulder)?

17. In Figure 23.5, we analyzed the force exerted by the electric and magnetic fields of an electromagnetic wave on an electric charge at the surface of a material when the electric field has the direction shown in Figure Q23.17A. That analysis considered the force on a positive charge at a particular instant in time, with particular directions for the electric and magnetic fields. Since the mobile charges in a metal are negative (i.e., electrons) and the electric and magnetic fields oscillate in direction, it is important to consider this force for negative charges and other field directions. (a) Assume the charge is instead negative (i.e., an electron). Show that the combination of electric and magnetic fields shown in Figure Q23.17A leads to a force on the electron with a component to the right. (b) After a short time, the electric field of the wave at the location of the electron changes direction and is pointing down as shown in Figure Q23.17B. Consider again the combined electric and magnetic forces on the electron and show that it again has a component to the right. The conclusion of this analysis is that although the electric and magnetic fields oscillate in direction, there is over time always a component of the force along the direction of propagation (except at instants when the force is zero). This force is connected with the wave's momentum.

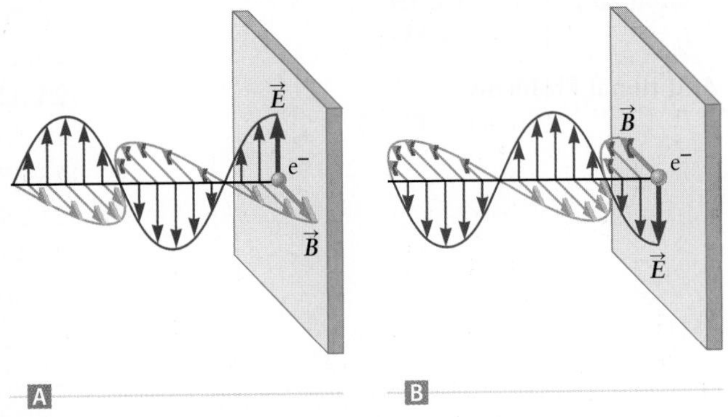

Figure Q23.17

18. Figure 23.18 shows the direction of the electric field produced by an antenna. Sketch the direction of the corresponding magnetic field near the antenna for parts B through H of the figure.

19. Fluorescent lights will glow faintly if brought under a high-tension power line as shown in Figure Q23.19. The lights glow even though they are not plugged into any fixture. Explain why they glow under the power lines.

Figure Q23.19

20. Figure Q23.20 shows linearly polarized light striking a polarizer, with different angles between the incoming electric field $\vec{E}_{in}$ and the polarizer axis. Is there a force on the polarizer due to the incoming light? If so, what is the direction of the force and in which case is the magnitude of the force largest? Explain.

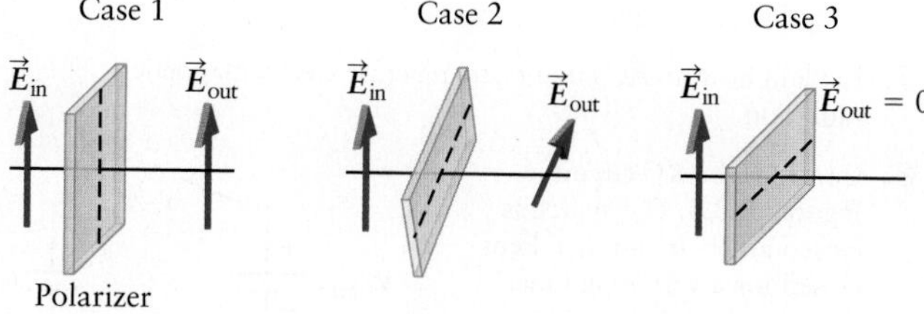

Figure Q23.20

21. Air traffic controllers at an airport use radar to monitor the location of nearby planes. Their radar units deduce this distance from the time it takes a radio wave to travel from their radar antenna to the airplane and back. The radar signal also gives information about the size and type of the plane. How do they do it?

Problems available in ENHANCED WebAssign

Section 23.2 Properties of Electromagnetic Waves
Section 23.3 Electromagnetic Waves Carry Energy and Momentum
Problems 1–22

Section 23.4 Types of Electromagnetic Radiation: The Electromagnetic Spectrum
Problems 23–30

Section 23.5 Generation and Propagation of Electromagnetic Waves
Problems 31–38

Section 23.6 Polarization
Problems 39–42

Section 23.7 Doppler Effect
Problems 43–48

Additional Problems
Problems 49–65

List of Enhanced Problems

Problem number	Solution in *Student Companion & Problem-Solving Guide*	Reasoning & Relationships Problem	Reasoning Tutorial in ENHANCED WebAssign
23.1		✓	✓
23.5	✓		
23.7		✓	✓
23.8		✓	✓
23.9		✓	
23.10		✓	✓
23.14	✓		
23.15		✓	✓
23.17		✓	✓
23.18	✓		
23.20		✓	
23.21		✓	✓
23.22		✓	
23.24		✓	
23.26		✓	✓
23.27	✓		
23.32		✓	
23.33		✓	✓
23.37	✓		
23.39	✓		
23.43	✓		
23.50		✓	✓
23.51	✓		
23.55		✓	
23.59	✓	✓	✓
23.62		✓	
23.64		✓	
23.65		✓	

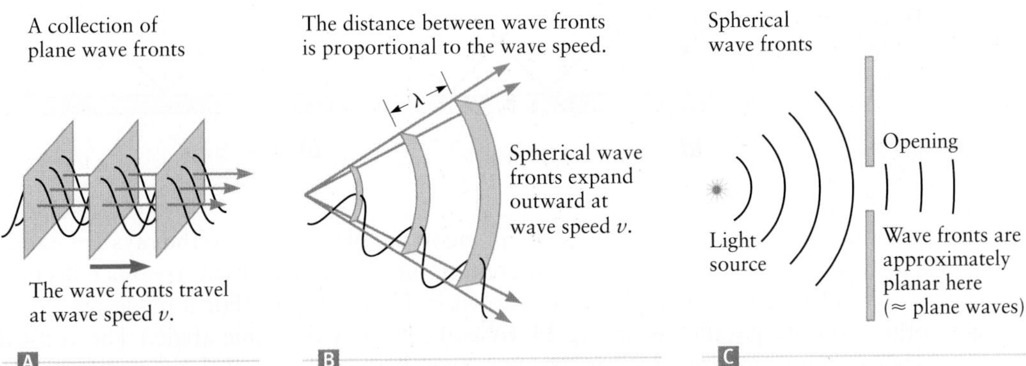

▲ **Figure 24.2** The shape of a wave front is determined by the crests and troughs of the wave. **A** A plane wave traveling from left to right. The blue arrows denote the corresponding rays; the wave fronts are planes perpendicular to the rays. **B** When a wave is emitted from a point source, the wave fronts are spherical. **C** When a point source is very far away, the wave fronts over any small region are nearly planar and can be used to make an approximate plane wave.

object reach your eye.[1] When your eye combines these rays to form an image, your brain extrapolates the rays back to their point of origin. The method of following individual rays as they travel from an object to your eye or to some other point is called ***ray tracing***. Ray tracing involves the extensive use of geometry; hence, ray optics is also called "geometrical optics."

Figure 24.3A shows rays from only two points on the object and only a few rays from each point. It would be more realistic to draw a very large ("infinite") number of rays originating from each point on an object. Figure 24.3B shows how rays from another point on the object reach your eye. The light waves associated with *all* these rays contribute to the image formed by your eye. In most of our ray diagrams, we draw only a few rays from the top or bottom of the object, but the complete image is formed by light that originates from all points on the object.

To understand how images are formed, we must consider (1) what happens to light rays when they reflect from a surface such as a mirror or a piece of glass and (2) what happens to light rays when they pass across a surface separating one material from another such as when they pass from air into a piece of glass. In each case, we must distinguish between a flat surface and a curved surface. We'll first consider flat surfaces in Sections 24.2 and 24.3, and then deal with curved surfaces in later sections.

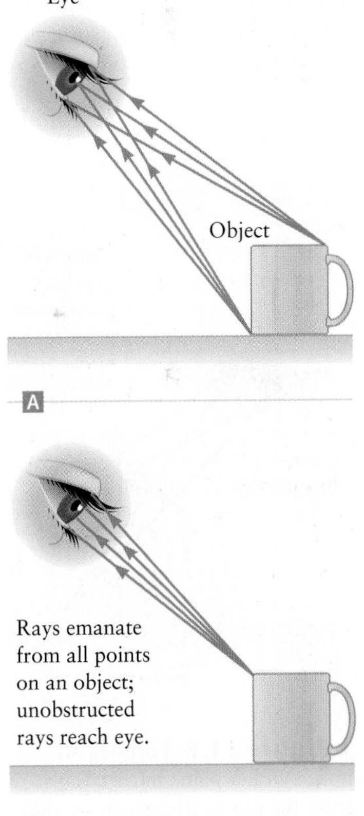

▲ **Figure 24.3** **A** An optical image is formed in your eye by rays of light that emanate from an object. **B** Many rays (an infinite number) emanate from each point on an object.

24.2 Reflection from a Plane Mirror: The Law of Reflection

Light rays travel in straight lines until they strike something. What happens next depends on what the rays strike. One possibility is that the rays are reflected as sketched in Figure 24.4A on page 728, showing the reflection of rays from a ***plane mirror***, a flat surface that reflects all or nearly all the light that strikes it.

[1]We note again that rays themselves do not "travel" but rather show the direction of propagation of the corresponding light wave.

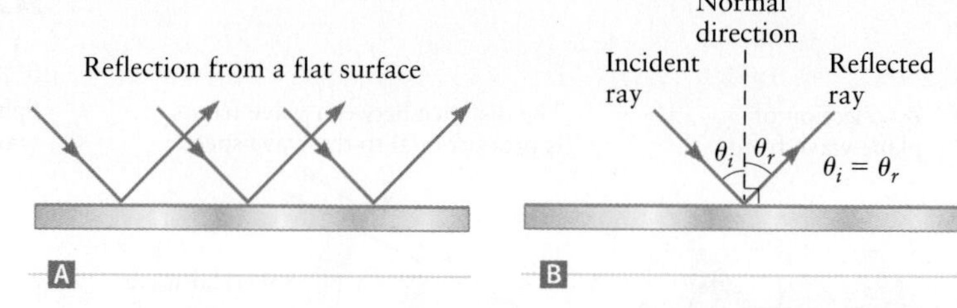

▶ **Figure 24.4** A Rays reflecting from a flat surface. A surface can be considered flat if its roughness is smaller than λ. This mirror is called a plane mirror. B The angle of reflection is equal to the angle of incidence.

Figure 24.4A describes the case of an incoming plane wave, so the rays are parallel and strike the surface at many different points. We can characterize this reflection by considering only a single ray as shown in Figure 24.4B. (For a plane wave, the other rays are parallel as in Fig. 24.4A and reflect at the same angle.) The vertical dashed line in Figure 24.4B is drawn perpendicular (also called "normal") to the mirror, and we can measure the directions of the incoming and outgoing rays relative to this perpendicular direction. The incoming ray is called the ***incident*** ray, and the angle it makes with the surface normal is the angle of incidence θ_i. The outgoing ray is called the ***reflected*** ray, and its angle with the normal is the angle of reflection θ_r. For reflection from a flat surface, the angle of incidence is equal to the angle of reflection,

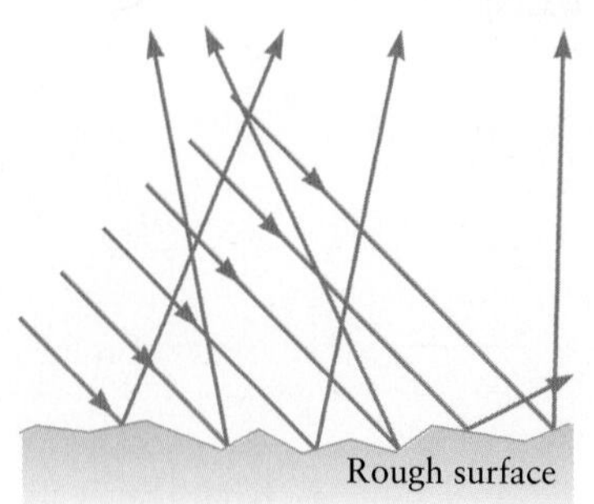

▲ **Figure 24.5** Rays are reflected in different directions by different parts of a rough surface, forming a diffuse reflection.

$$\theta_i = \theta_r \tag{24.1}$$

This relation, called the ***law of reflection***, can be derived from our general principle concerning the reversibility of the propagation of light. Suppose the angle of reflection is smaller than θ_i. If we were to reverse the outgoing ray, its angle of reflection would be even smaller and the new reflected ray would not retrace the original incoming ray. The only way the reflection process can be reversible is for the angle of reflection to equal the angle of incidence (Eq. 24.1).

Reflections from a perfectly flat mirror (Fig. 24.4) are called ***specular*** reflections. If the reflecting surface is rough, we must consider reflections from all the individual pieces of the surface. In that case, an incident plane wave will give rise to many reflected rays propagating outward in many different directions (Fig. 24.5). This reflection is called a ***diffuse*** reflection.

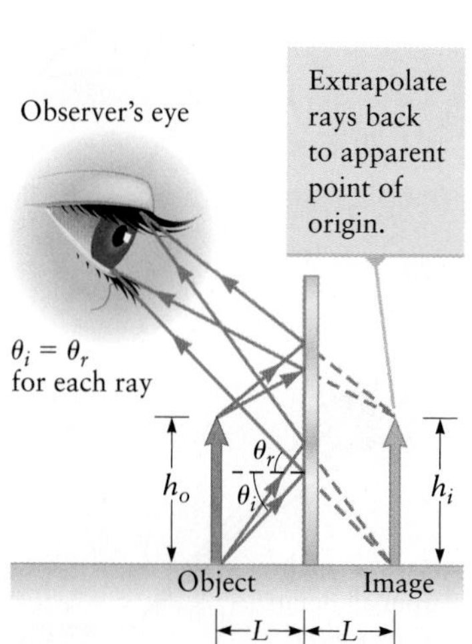

▲ **Figure 24.6** Forming an image with a plane mirror. Rays from the object (the arrow on the left) reflect from the mirror and reach your eye. These rays appear to have come from the image, on the right of the mirror. It is a virtual image because light does not actually pass through the image point. This sketch shows only rays from the top and bottom of the object, but rays from other points on the object also contribute to the image.

Image Formation by a Plane Mirror

When you view an object through its reflection in a mirror, you are viewing an image of the object. An image formed by a plane mirror is shown in Figure 24.6. Here we show a particularly simple object (an arrow), but the object could be a tree or a person. We will usually use an arrow as our example object because we'll often be interested in the orientation of the image relative to the object. As in Figure 24.3, an infinite number light rays emanate from each point on the object, although we only show two rays emanating from the top and bottom. Some of the emanated rays strike the mirror and are reflected so as to reach your eye. To your eye, the location of this image—that is, the location of the arrow—is the point from which these rays appear to emanate. This point can be found by extrapolating the rays back in a straight-line fashion to their apparent point of origin as shown by the dashed lines in Figure 24.6. This is our first example of ***ray tracing***. Since each ray obeys the law of reflection (Eq. 24.1), we can use that together with some geometry to determine the image location. A few rays are shown in Figure 24.6 for an object located a distance L in front of the mirror. When the rays that reach the observer's eye are extrapolated back to their apparent common point of origin, the image is found to be a distance L behind the mirror.

Figure 24.6 shows that the size of an image formed by a plane mirror is equal to the size of the object; that is, the height of the image h_i is equal to the height of the object h_o. Also, the image point is located behind the mirror, so light does *not* actually pass through the image. For this reason, the image is called a ***virtual image***.

EXAMPLE 24.1 Reflection from a Corner Cube

Figure 24.7A shows a combination of two plane mirrors connected to form a right angle. In three dimensions, one can form a similar structure called a ***corner cube*** by combining mirrors to make three faces of a cube. The light ray in Figure 24.7A is incident at an angle θ_i relative to the direction normal (perpendicular) to mirror 1. Find the angle of the outgoing light ray θ_{out} relative to this normal direction.

RECOGNIZE THE PRINCIPLE

To analyze reflection from a single plane mirror, we must first determine the angle of incidence, which is the angle the incident ray makes with the direction normal to the mirror. According to the law of reflection, the angle of reflection is then equal to the angle of incidence. Here we must use this procedure first for the reflection from mirror 1 and then a second time for the reflection from mirror 2.

SKETCH THE PROBLEM

Figure 24.7A shows the incident ray along with the normal direction for mirror 1 (shown as the dashed line). Reflection from mirror 1 gives an outgoing ray making an angle θ_r with the normal to mirror 1 as sketched in Figure 24.7B. This ray is then the incident ray for reflection from the bottom mirror.

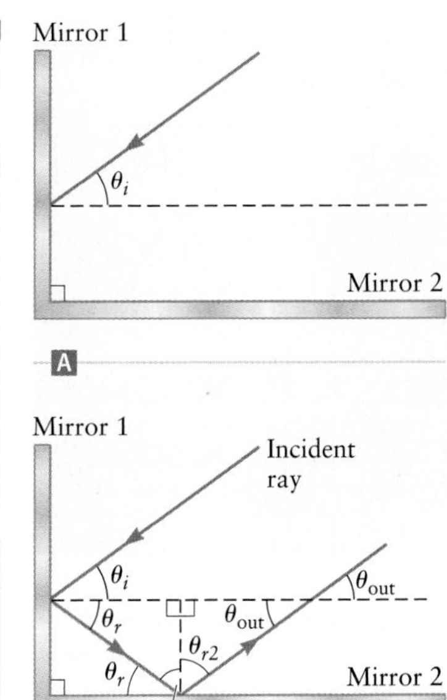

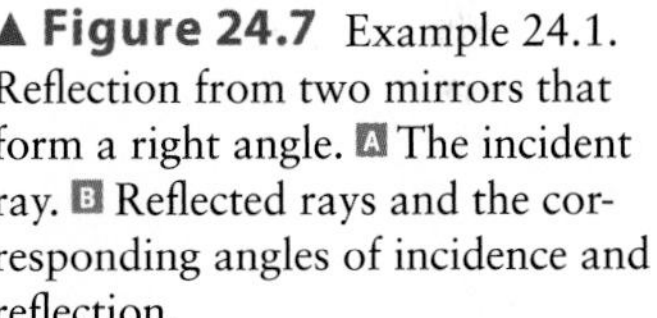

▲ **Figure 24.7** Example 24.1. Reflection from two mirrors that form a right angle. A The incident ray. B Reflected rays and the corresponding angles of incidence and reflection.

IDENTIFY THE RELATIONSHIPS

The angle θ_i in Figure 24.7A is the angle of incidence in the law of reflection, Equation 24.1. The angle of reflection from this first surface is thus $\theta_r = \theta_i$. The two reflecting surfaces (the two mirrors) are at right angles, so the dashed lines showing the directions normal to each mirror in Figure 24.7B also form a right angle. The interior angles of a triangle add up to 180°, so the angle of incidence for the second reflection is

$$\theta_{i2} = 90° - \theta_r = 90° - \theta_i$$

Applying the law of reflection to the ray as it reflects from mirror 2, we find

$$\theta_{r2} = \theta_{i2} = 90° - \theta_i \quad (1)$$

SOLVE

From the geometry in Figure 24.7B, the angle the outgoing ray makes with the horizontal is

$$\theta_{out} = 90° - \theta_{r2}$$

Combining this with Equation (1) gives

$$\theta_{out} = 90° - \theta_{r2} = 90° - (90° - \theta_i)$$

$$\theta_{out} = \boxed{\theta_i}$$

The outgoing ray is thus parallel to the incident ray.

► *What does it mean?*

An interesting feature of this result is that the outgoing ray is *always* parallel to the incident ray for *any* value of θ_i. A combination of mirrors called a corner cube, the three-dimensional version of Figure 24.7, has a similar property: its reflected rays are always parallel to the incident rays. Such devices, called ***retroreflectors***, are useful in many applications. For example, the Apollo astronauts who visited the Moon left a number of corner cube reflectors on its surface. When laser light from the

(continued) ►

Insight 24.1

WHAT DETERMINES IF A SURFACE WILL REFLECT LIGHT?

The way a surface does or does not reflect light depends on the material making up the surface. The electric and magnetic fields of the light wave exert forces on electrons and ions near the surface. A typical mirror consists of a metal surface, and the motion of electrons and ions induced by the electric and magnetic fields produces the reflected light wave.

▲ **Figure 24.8** Reflective clothing contains tiny corner cubes, each of which reflects light back parallel to the incident rays as in Figure 24.7B. This cyclist's vest seems to "glow" in the dark because it reflects light (such as from a car's headlights) back toward its source.

Earth strikes one of these corner cubes, the reflected light travels in a path parallel to the incident ray and back to the Earth. The time delay between incident and reflected light pulses can be used to measure the Earth–Moon distance with high precision. In principle, a single, flat mirror would reflect light in the same way, but the mirror would have to be aligned very precisely; otherwise, the reflected light would miss the Earth. With a retroreflector, the reflected light is always parallel to the incident ray, no matter what the alignment. A more down-to-Earth application of retroreflectors is found on many roads. Most roadside reflectors, and even some paints, contain tiny crystals that act as corner cubes to reflect the rays from your headlights back to you, thus giving a noticeable indication of the edges of the road. Reflective clothing works the same way (Fig. 24.8).

CONCEPT CHECK 24.1 Reflection from Two Mirrors

Consider two plane mirrors that make an angle of 60° with each other (Fig. 24.9). An incident ray is parallel to the bottom mirror. This ray reflects from the mirror at the left and then strikes the bottom mirror. What angle of incidence θ_{i2} does the reflected ray make with the bottom mirror, (a) 30°, (b) 45°, or (c) 60°?

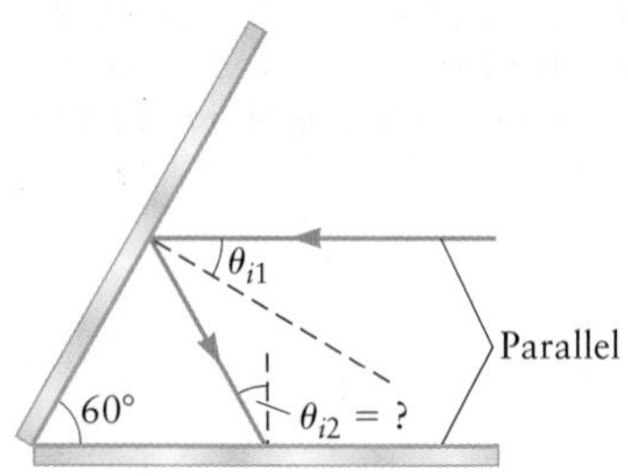

▲ **Figure 24.9** Concept Check 24.1.

24.3 Refraction

When a light ray strikes a transparent material like a windowpane, some of the light is reflected (according to the law of reflection described in Section 24.2), and the rest passes into the material. Figure 24.10 shows a light ray incident on a piece of glass with two flat, parallel sides. The ray that travels into the glass is called a ***refracted*** ray.

The Index of Refraction and Snell's Law

The direction of the refracted ray is measured using the angle θ_2 this ray makes with the surface normal. The value of θ_2 depends on the incident angle (which we now denote as θ_1) and also on the speed of light within the material. The speed of light in vacuum is $c = 3.00 \times 10^8$ m/s. However, when light travels through a material substance, the associated electric and magnetic fields interact with the atoms of the substance, and this interaction affects the speed of the wave. As a result, the speed of light inside a substance such as glass, water, or air is less than the speed of light in vacuum. Table 24.1 lists the speed of light in several substances.

Figure 24.11A shows a plane wave initially traveling in a vacuum and then striking the flat surface of a piece of glass; this sketch illustrates how the change in the speed of light in going from vacuum to glass affects the direction of the rays inside the glass. Here we show several rays along with some associated wave fronts in both the vacuum and the glass. The incoming rays make an angle θ_1 relative to the surface normal, and the incident wave fronts are perpendicular to these rays. The incident light travels at speed c, the speed of light in a vacuum. The light inside the glass, however, moves at speed v, which is less than c.

From Section 24.1 (Fig. 24.2), the perpendicular distance between wave fronts is proportional to the wave speed. The wave fronts in Figure 24.11A are drawn at successive, equally spaced moments in time, so the spacing between wave fronts is ct on the vacuum side and vt in the glass. Figure 24.11B shows an expanded view of two adjacent wave fronts, and two right triangles are indicated. On the vacuum side, the right triangle shaded in red has sides L and ct, with angle θ_1 adjacent to side L. From the geometry of this triangle, we have

$$\sin \theta_1 = \frac{ct}{L} \tag{24.2}$$

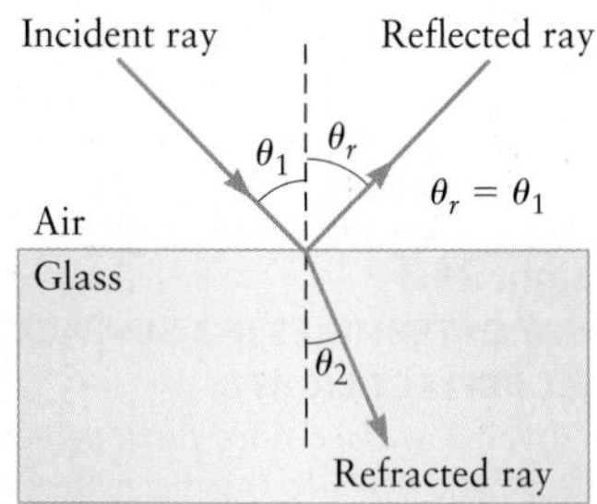

▲ **Figure 24.10** An incident light ray is both reflected and refracted when it strikes the surface of a transparent object such as a piece of glass. A portion of the light energy is reflected with $\theta_r = \theta_1$, whereas the rest is refracted as it passes into the glass.

TABLE 24.1 Speed of Light and Index of Refraction for Several Substances for Room Temperature and Yellow Light (λ = 550 nm)

Substance	Speed of Light (m/s)	Index of Refraction	Substance	Speed of Light (m/s)	Index of Refraction
Vacuum	2.998×10^8	1.0000 (exactly)	Pyrex glass	2.04×10^8	1.47
Air	2.997×10^8	1.0003	Plexiglas (plastic)	2.01×10^8	1.49
Water (liquid)	2.26×10^8	1.33	Regions in the eye:		
Water (ice)	2.29×10^8	1.31	vitreous humor	2.23×10^8	1.34
Benzene	2.00×10^8	1.50	aqueous humor	2.26×10^8	1.33
Quartz crystal	2.05×10^8	1.46	cornea	2.17×10^8	1.38
Diamond	1.24×10^8	2.42	lens	2.13×10^8	1.41

The corresponding triangle shaded in yellow on the glass side has edge lengths L and vt and an angle θ_2 adjacent to side L, which leads to

$$\sin\theta_2 = \frac{vt}{L} \tag{24.3}$$

We can rearrange each of these relations to solve for L, resulting in $L = ct/\sin\theta_1$ and $L = vt/\sin\theta_2$. Setting them equal, we find

$$\frac{ct}{\sin\theta_1} = \frac{vt}{\sin\theta_2}$$

The factor of t can be canceled, giving

$$\sin\theta_1 = \frac{c}{v}\sin\theta_2 \tag{24.4}$$

The ratio c/v is called the ***index of refraction*** of the glass and is denoted by

$$n = \frac{c}{v} \tag{24.5}$$

We can thus rewrite Equation 24.4 as

$$\sin\theta_1 = n\sin\theta_2 \tag{24.6}$$

In Figure 24.11, we assumed there is vacuum on the incident side and glass on the refracted side. A common situation involves light passing between two substances such as air and glass. If the two substances have indices of refraction n_1 and n_2 (Fig. 24.12), the incident and refracted angles are related by

$$n_1\sin\theta_1 = n_2\sin\theta_2 \tag{24.7}$$

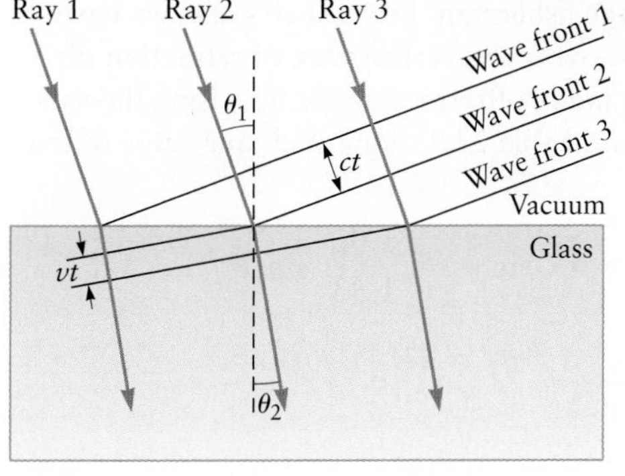

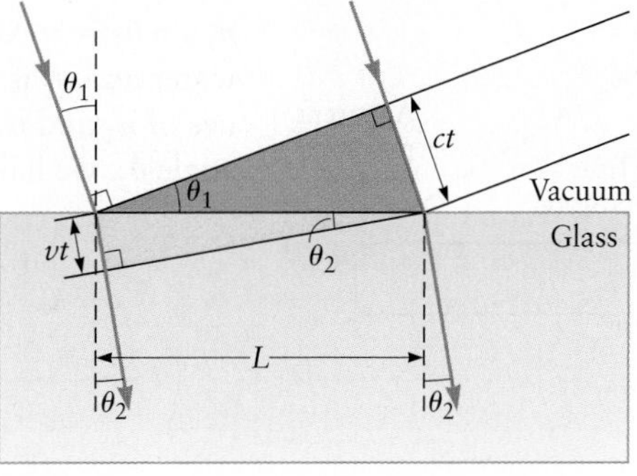

◀ **Figure 24.11** A Refraction occurs due to a difference in the wave speeds in two regions, which causes the rays to change direction at the interface. B Derivation of Snell's law. The speed of light in a vacuum is c, and the speed of light in glass is v.

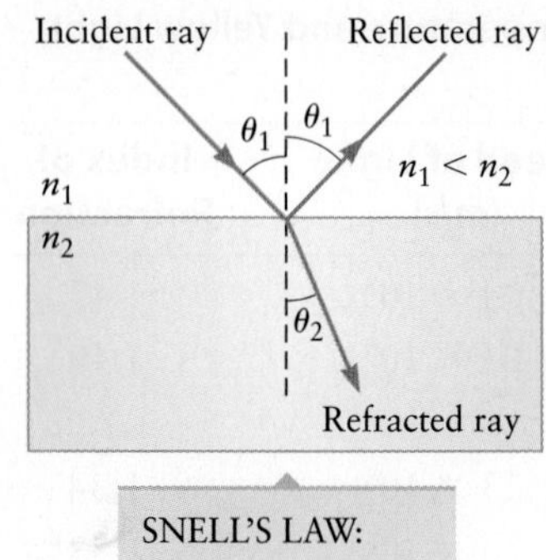

▲ **Figure 24.12** Snell's law gives the relationship between the angle of incidence and the angle of refraction.

This relation between the incident and refracted angles is called ***Snell's law***. Table 24.1 lists the indices of refraction for some common substances. Since n is the ratio of two speeds (Eq. 24.5), the index of refraction is a dimensionless number.

CONCEPT CHECK 24.2 Using Snell's Law

If the materials in Figure 24.12 are air and water, which is which? Is material 1 air or water?

Insight 24.2

REVERSIBILITY OF LIGHT RAYS AND REFRACTION

In Figure 24.12, light is incident at an angle θ_1 from side 1, leading to a refracted angle θ_2 given by Snell's law (Eq. 24.7). If the light had instead been incident from side 2 (from the bottom of the figure) at an incident angle θ_2, applying Snell's law would give a refracted angle θ_1. Hence, in accord with the general principle of reversibility of propagation direction stated in Section 24.1, the refraction of light is reversible.

EXAMPLE 24.2 Refraction on a Pier

A lobster fisherman looks just over the edge of a pier and spots a lobster resting at the bottom. This fishing spot is $d = 3.5$ m deep. Judging from the angle at which he spots the lobster, the fisherman thinks the lobster is a horizontal distance $L_{app} = 6.0$ m from the shore, but when he drops his trap at that location, he does not catch the lobster. What is the true horizontal distance of the lobster from the shore?

RECOGNIZE THE PRINCIPLE

The fisherman determines the lobster's apparent position by extrapolating the ray that meets his eye back to its apparent point of origin at the bottom of the harbor. This extrapolation is shown by the dashed line in Figure 24.13. However, because light from the lobster is refracted when it passes from the water into the air, the actual position of the lobster is much closer to the shore. We can use Snell's law to find the angle θ_1 that the true ray from the lobster makes with the surface normal, and from that we can locate the true position of the lobster.

SKETCH THE PROBLEM

Figure 24.13 shows the problem.

IDENTIFY THE RELATIONSHIPS

We denote the angle of incidence (on the water side of the water–air interface) as θ_1 and the angle of refraction (on the air side of the interface) as θ_2. The angle θ_2 is related to L_{app}, the apparent distance of the lobster from the shore, by

$$\tan\theta_2 = \frac{L_{app}}{d}$$

because L_{app} and d are two sides of the right triangle OAB in Figure 24.13. Inserting the given values $d = 3.5$ m and $L_{app} = 6.0$ m, we find

$$\theta_2 = \tan^{-1}\left(\frac{6.0 \text{ m}}{3.5 \text{ m}}\right) = 60°$$

which is also the angle of refraction for the ray that travels from the water to the fisherman. From Snell's law, we have $n_1 \sin\theta_1 = n_2 \sin\theta_2$, where n_1 is the index of refraction of water and n_2 is the index of refraction for air. Using the values of n_1 and n_2 from Table 24.1 along with our value of the angle θ_2, we have

$$\sin\theta_1 = \frac{n_2}{n_1}\sin\theta_2 = \frac{1.00}{1.33}\sin(60°)$$

$$\theta_1 = 41°$$

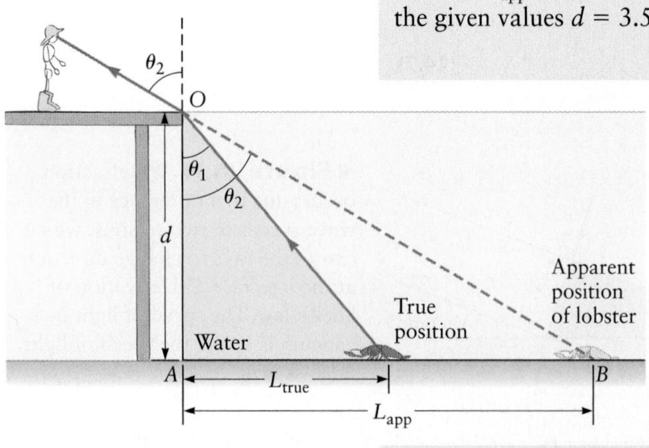

▲ **Figure 24.13** Example 24.2.

SOLVE

Now that we have the value of θ_1, we can find the lobster's true distance from the shore, L_{true}. Using the right triangle shaded in green in Figure 24.13, we have

$$\tan\theta_1 = \frac{L_{\text{true}}}{d}$$

$$L_{\text{true}} = d\tan\theta_1 = (3.5\text{ m})\tan(41°) = \boxed{3.0\text{ m}}$$

▶ What does it mean?

The lobster is thus much closer to shore than it appears to the naive fisherman. This effect is illustrated in Figure 24.14, which shows a pencil inserted into a glass of water. The pencil appears to bend where it enters the water due to refraction of light rays as they leave the water.

▲ **Figure 24.14** This pencil is partially immersed in water. It appears to "bend" at the water's surface due to refraction of rays that pass from the water into the air.

Applying Snell's Law

Figure 24.15 shows an example of reflection and refraction at a plane surface. Here light is incident from the region with the *larger* index of refraction ($n_1 > n_2$). For instance, substance 1 (the lower substance in Fig. 24.15) might be water or glass and substance 2 could be air.

Snell's law reads the same whether light begins in the substance with the larger or smaller index of refraction,

$$n_1 \sin\theta_1 = n_2 \sin\theta_2 \tag{24.8}$$

Possible angles of incidence always lie between zero and 90°, so $0° \leq \theta_1 \leq 90°$ and $0° \leq \theta_2 \leq 90°$. For these angles, the function $\sin\theta$ increases as θ is increased. Hence, according to Equation 24.8, the side with the *larger* index always has the *smaller* angle. In words, we say that light is refracted *toward* the normal direction when moving into the substance with the larger index of refraction (Fig. 24.12). Light is refracted *away* from the normal direction when moving into the substance with the smaller index of refraction (Fig. 24.15).

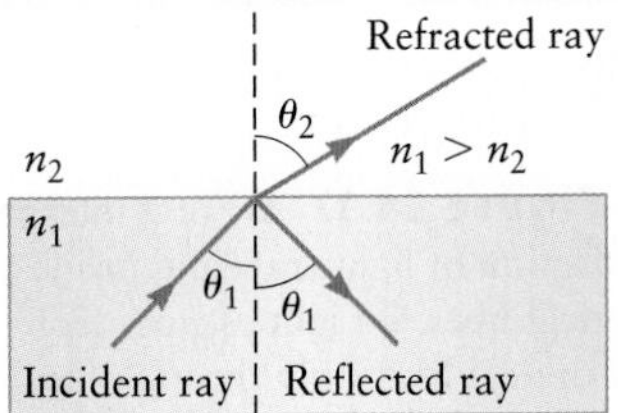

▲ **Figure 24.15** Reflection and refraction when light travels from a region of higher to a region of lower index of refraction ($n_1 > n_2$). Compare with Figure 24.12, in which $n_1 < n_2$.

CONCEPT CHECK 24.3 Snell's Law

Light travels from a vacuum ($n_1 = 1.00$) into a plate of glass with $n_2 = 1.55$, with an angle of incidence of 30°. Is the angle of refraction (a) larger than 30° or (b) smaller than 30°? Also, use Snell's law to calculate the angle of refraction.

Total Internal Reflection

Let's now consider light incident from the side with the larger index of refraction in Figure 24.15 (the bottom side) a little more carefully. Because the index n_1 is greater than n_2, the angle of refraction θ_2 is greater than θ_1 and light is refracted *away* from the normal direction. As the incident angle θ_1 is made larger and larger, the refracted angle θ_2 increases. Eventually, θ_2 will reach 90° and the refracted ray will emerge *parallel* to the surface. The value of the incident angle at which that occurs is called the ***critical angle***. If the incident angle θ_1 in Figure 24.15 is increased beyond the critical angle, Snell's law has no solution for the refracted angle θ_2. Physically, there is no refracted ray, and all the energy from the incident light ray is reflected. This behavior, called ***total internal reflection***, is possible only when light is incident from the side with the larger index of refraction.

Total internal reflection

Courtesy of The Harvard Science Center. © The President & Fellows of Harvard College.

▶ **Figure 24.16** An example of total internal reflection. Light from a laser at the lower left is reflected at the air–water interface. There is no refracted ray.

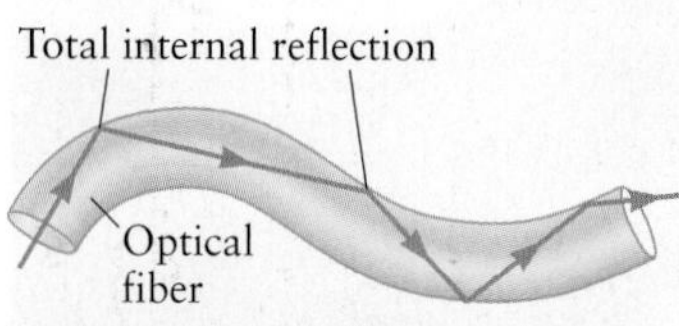

Light is confined to an optical fiber by total internal reflection.

A

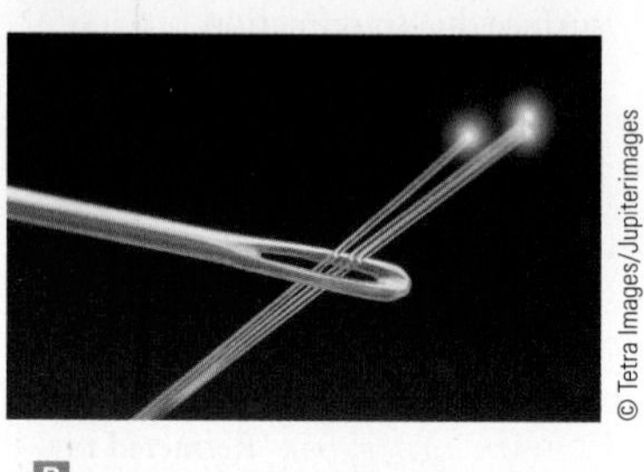

B

▲ **Figure 24.17** A Total internal reflection of light traveling inside an optical fiber. B Light escapes from an optical fiber only through the end.

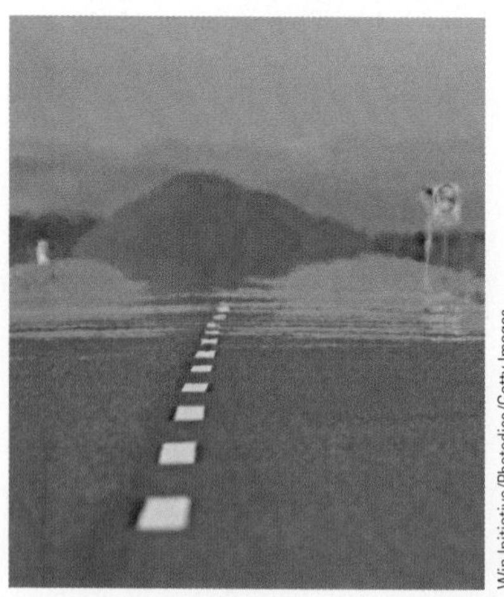

▲ **Figure 24.18** A mirage forms when there is a layer of warm air next to the ground. Light from these road signs is reflected from this layer. (Note how the reflected image is inverted.)

Figure 24.16 shows an example of total internal reflection. Here light from a laser travels into the side of a tank of water. The light is reflected at the interface between water and air, but the angle of incidence exceeds the critical angle, so there is no refracted ray. Hence, no light gets through this otherwise "transparent" interface!

We can find the value of the critical angle using Snell's law. When the angle of incidence equals the critical angle, the angle of refraction in Figure 24.15 is $\theta_2 = 90°$ (because the direction of the refracted ray would be parallel to the interface). Inserting this angle into Snell's law (Eq. 24.8) and using $\sin(90°) = 1$, we get

$$n_1 \sin \theta_1 = n_2 \sin \theta_2 = n_2 \sin(90°) = n_2$$

$$\sin \theta_1 = \frac{n_2}{n_1}$$

Denoting the critical angle as θ_{crit} $(= \theta_1)$ gives

$$\sin \theta_{crit} = \frac{n_2}{n_1}$$

$$\theta_{crit} = \sin^{-1}\left(\frac{n_2}{n_1}\right) \tag{24.9}$$

The value of θ_{crit} thus depends on the ratio of the indices of refraction of the substances on the two sides of the interface. When the angle of incidence is equal to or greater than the critical angle, light is reflected completely at the interface. Amazingly, this phenomenon can occur even when the interface between the two substances would seem to be completely transparent (such as between water and air).

Total internal reflection is used in fiber optics. Optical fibers are composed of specially made glass and are used to carry telephone and computer communications signals. These signals are sent as light waves, using total internal reflection as sketched in Figure 24.17A to keep the light from "leaking" out the sides of the fiber. Optical fibers are designed to carry signals over very long distances, so to minimize signal loss it is important that these reflections be as perfect as possible. (The model of an optical fiber shown here is highly simplified; we'll describe a more realistic picture in Chapter 26.) A mirage is another example of total internal reflection (Fig. 24.18). A mirage is formed by total reflection of light between layers of cool air and warm air.

CONCEPT CHECK 24.4 When Is Total Internal Reflection Possible?

In which of the following cases is it possible to have total internal reflection?

(a) Light traveling from water into air
(b) Light traveling from air into water
(c) Light traveling from glass into water
(d) Light traveling from water into diamond
(e) Light traveling from benzene into quartz (*Hint*: See Table 24.1.)

EXAMPLE 24.3 Total Internal Reflection between Water and Air

Find the critical angle for total internal reflection between water and air.

RECOGNIZE THE PRINCIPLE

We used Snell's law to derive the result for the critical angle in Equation 24.9. The purpose of this problem is to get a feel for typical values of the critical angle.

SKETCH THE PROBLEM

Figure 24.19A shows light incident on the interface between water and air. The incident ray is at the critical angle, so the refracted ray (if it existed) would be parallel to the interface and $\theta_2 = 90°$.

IDENTIFY THE RELATIONSHIPS

We can find the critical angle using the result for θ_{crit} in Equation 24.9:

$$\theta_{crit} = \sin^{-1}\left(\frac{n_2}{n_1}\right) = \sin^{-1}\left(\frac{n_{air}}{n_{water}}\right)$$

SOLVE

Inserting the values $n_{air} = 1.0003 \approx 1.00$ and $n_{water} = 1.33$ from Table 24.1, we get

$$\theta_{crit} = \sin^{-1}\left(\frac{n_{air}}{n_{water}}\right) = \sin^{-1}\left(\frac{1.00}{1.33}\right) = \sin^{-1}(0.75) = \boxed{49°}$$

► ***What does it mean?***

The incident ray in Figure 24.19A is drawn at $\theta_{crit} = 49°$, showing that the incident ray does not have to be "close" to the interface to achieve total internal reflection. You can test this result the next time you go swimming. If you are underwater and look in a direction for which the angle of incidence is greater than the critical angle (Fig. 24.19B), the water–air interface will appear as a perfect mirror. The swimmer in Figure 24.19B will see the fish reflected in this interface mirror. Try it.

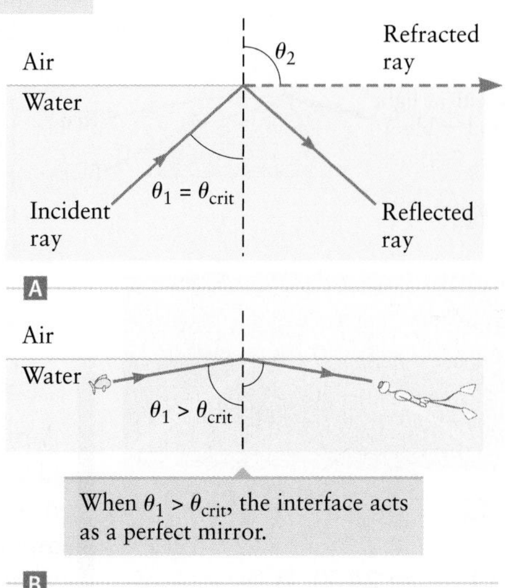

▲ **Figure 24.19** Example 24.3. **A** Total internal reflection at an air–water interface. **B** When a swimmer (on the right) looks at light coming from the air–water interface, all light with an angle of incidence greater than the critical angle is reflected. There are no rays incident from the air that give rays in the water at this angle; the interface thus acts as a perfect mirror when viewed at this angle.

Dispersion

The direction of a refracted ray depends on the indices of refraction of the materials on the two sides of an interface, and these indices of refraction depend on the speed of light in each material (Eq. 24.5). The speed of light in a vacuum has the constant value c for all wavelengths, that is, for all colors. However, when light travels in a material, the speed depends on the color of the light. This dependence of wave speed on color is called ***dispersion***, and the variation of the speed of light with wavelength for several materials is shown in Figure 24.20.

For example, the index of refraction for red light in quartz is $n_{red} = 1.46$ and that for blue light is $n_{blue} = 1.47$. This small difference between n_{red} and n_{blue} makes the angle of refraction slightly different for red and blue light in quartz (and in other materials, too). This effect is used by a ***prism*** to separate a beam of light into its component colors.

Prisms are typically composed of glass, with a triangular cross section (Fig. 24.21 on page 736). In Figure 24.21A, a beam of light is incident on one surface of a prism; we suppose this light is composed of two colors (red and blue), and we use two rays with different colors to denote the incident light. Each ray is refracted once when it enters the prism and again when it exits. Because the refractive index of glass depends on the color of the light, incident beams of different color have slightly

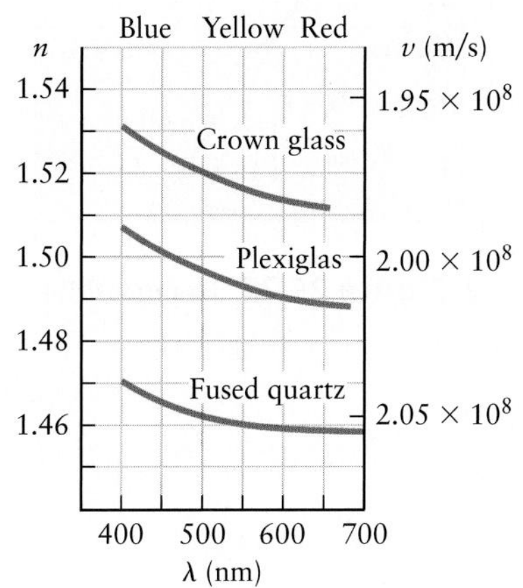

▲ **Figure 24.20** Variation of the speed of light (right axis) and the index of refraction (left axis) as a function of wavelength for several materials.

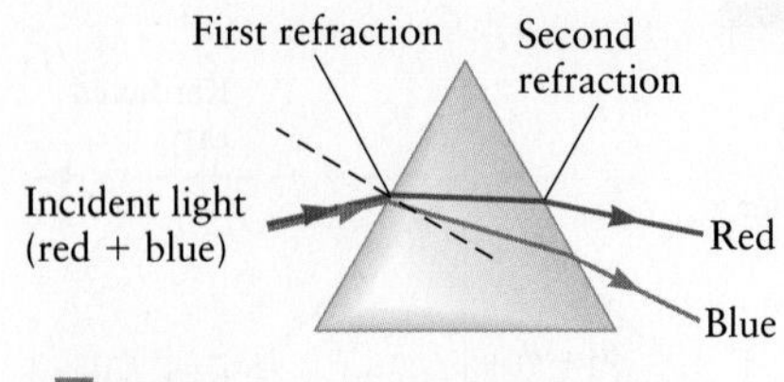

A

B

▲ **Figure 24.21** A The angle of refraction depends on the wavelength (i.e., the color) of the light. Blue light and red light incident from the left will therefore refract at different angles and emerge from the prism propagating in different directions. The amount of dispersion shown here is exaggerated for clarity. B This enables a prism to separate different colors in an incident beam of white light.

different angles of refraction. Blue light has a larger index of refraction (Fig. 24.20) than red light, so the blue ray inside the glass makes a smaller angle with the normal direction on the left than does the red ray. Hence, inside the glass the red and blue components travel in different directions. This difference in propagation direction is increased at the second refraction, when the light leaves the prism. Figure 24.21B shows a case in which white light (containing rays of all colors) is incident from the lower left. The outgoing rays at the lower right show how the prism has "dispersed" the light into different directions according to color. That is why the variation of the index of refraction with wavelength (Fig. 24.20) is called dispersion.

CONCEPT CHECK 24.5 Refraction of Rays with Different Colors

A mixture of red light and blue light is incident from vacuum onto a thick Plexiglas slab. If $n_{red} = 1.49$ and $n_{blue} = 1.51$, which color will have the larger angle of refraction, (a) the red light or (b) the blue light? Or, will they (c) have the same angle of refraction? *Hint*: You should be able to answer this question without working out values of the angles of refraction.

EXAMPLE 24.4 Dispersing Light by Refraction

Figure 24.22 shows light incident from air onto a flat quartz slab. This light is a mixture of red and blue, denoted in the figure by the two incident rays with these colors. Both rays have an angle of incidence $\theta_1 = 45.0°$. Find the angles of refraction for the two rays in the quartz plate. Assume the indices of refraction for red and blue light in quartz are $n_{\text{quartz, red}} = n_{2,\text{ red}} = 1.459$ and $n_{\text{quartz, blue}} = n_{2,\text{ blue}} = 1.467$. Give your answers to three significant figures.

RECOGNIZE THE PRINCIPLE

We can find the angle of refraction for each color using Snell's law. The point of this example is to get a feeling for how much the two rays are "dispersed." How large is the difference between the directions of the two rays?

SKETCH THE PROBLEM

Figure 24.22 shows the problem. The index of refraction in quartz is greater for blue light than for red light, so the blue ray is refracted closer to the normal.

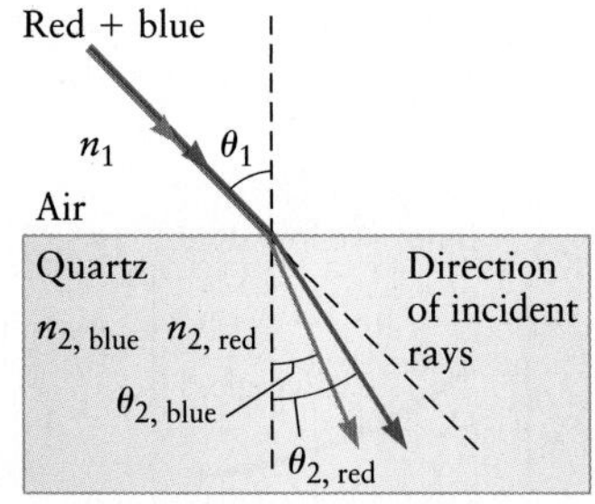

▲ **Figure 24.22** Example 24.4.

IDENTIFY THE RELATIONSHIPS

The indices of refraction for red and blue light in quartz are given. In air, the indices of refraction for both colors are the same to four significant figures (Table 24.1): $n_{\text{air, red}} = n_{\text{air, blue}} = n_1 = 1.000$.

SOLVE

Using Snell's law (Eq. 24.7), for the red light ray, we have

$$n_1 \sin\theta_1 = n_{2,\text{ red}} \sin\theta_{2,\text{ red}}$$

Solving for $\theta_{2,\text{ red}}$, we get

$$\sin\theta_{2,\text{ red}} = \frac{n_1}{n_{2,\text{ red}}}\sin\theta_1 \quad (1)$$

Inserting the values of the indices of refraction and θ_1 gives

$$\sin\theta_{2,\text{ red}} = \frac{n_1}{n_{2,\text{ red}}}\sin\theta_1 = \frac{1.000}{1.459}\sin(45.0°)$$

$$\theta_{2,\text{ red}} = \boxed{29.0°} \quad (2)$$

We have kept three significant figures because, as we'll see below, the difference between the refracted angles for the two rays will be small.

We compute the angle of refraction for the blue light ray in the same way. Equation (1) leads to

$$\sin \theta_{2,\text{ blue}} = \frac{n_1}{n_{2,\text{ blue}}} \sin \theta_1 = \frac{1.000}{1.467} \sin(45.0°)$$

$$\theta_{2,\text{ blue}} = \boxed{28.8°} \qquad (3)$$

▸ What does it mean?

The angles of refraction of the two rays differ by only a small amount, just 0.2°, but that is enough to give the dispersion of light by a prism in Figure 24.21B. (To find the angle of the outgoing ray for a prism, we must also include the refraction of the light rays at the second surface of the prism.) Also, the angle of refraction is *smaller* for blue light (see Concept Check 24.5), so the blue ray is *closer* to the normal direction. Thus, refraction causes blue light to be deflected *more* (with respect to the incoming direction) than red light.

24.4 Reflections and Images Produced by Curved Mirrors

In Section 24.2, we saw that a plane mirror can produce an image of an object (Fig. 24.6). All images formed by plane mirrors are the same size as the original object. A curved mirror, however, can achieve a ***magnified*** image, an image that appears larger or smaller than the original object. Magnified images are used in many applications, ranging from telescopes to a car's rearview mirror, so let's now consider how to produce one.

Ray Tracing for Spherical Concave Mirrors

Consider the spherical mirror in Figure 24.23A in which the surface of the mirror forms a section of a spherical shell (like a beach ball). The radius R of this sphere is called the radius of curvature of the mirror. The mirror's ***principal axis*** (parts B and C of Fig. 24.23) is the line that extends from the center of curvature C to the center of the mirror. The mirror in Figure 24.23 is ***concave***, curving toward objects placed in front of it.

Figure 24.23B shows incoming light rays directed parallel to the principal axis of the mirror. If these rays are close to the principal axis, then after reflecting at the surface of the mirror they all pass through the single point F indicated in Figure 24.23B. (We'll discuss the behavior of rays far from the axis in Section 24.8.) This point F, called the ***focal point*** of the mirror, is located a certain distance f from the mirror surface along the principal axis. The distance f is called the ***focal length*** of the mirror. Because the propagation of light is reversible, rays that originate at the focal point reflect from the mirror and propagate outward, parallel to the principal axis as in Figure 24.23C. The mirror thus "works" both ways: either focusing parallel light rays at the focal point F or generating a set of parallel rays from light that originates at point F and strikes the mirror.

When an object is placed in front of a spherical mirror, light from the object is reflected from the mirror and forms an ***image***. This is illustrated in Figure 24.24,

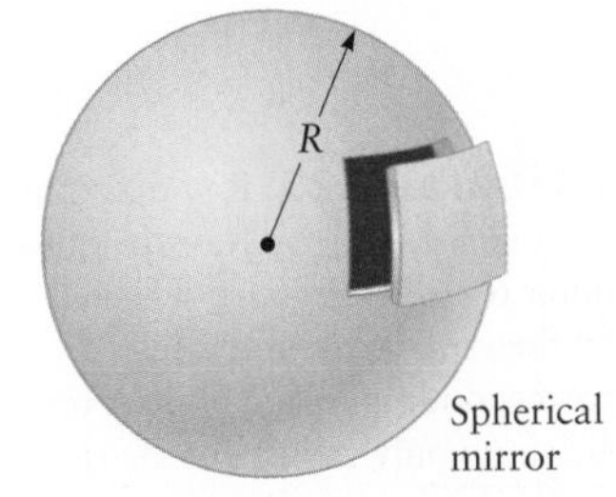

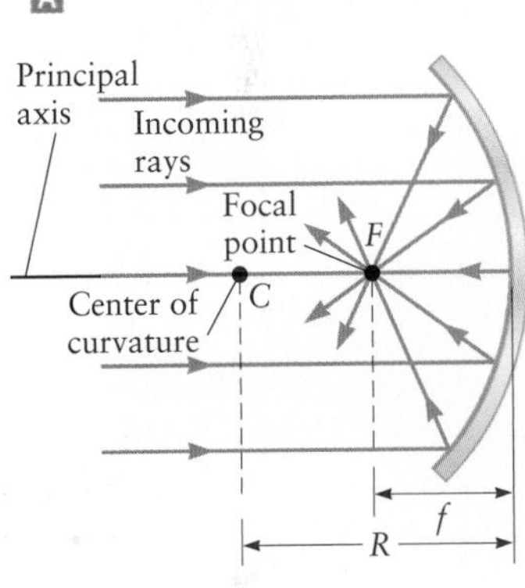

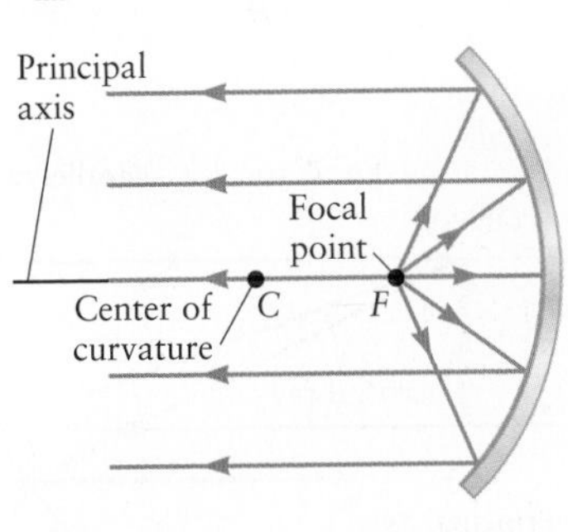

▲ Figure 24.23 A A spherical mirror matches the shape of a sphere. The radius R of this sphere is called the radius of curvature of the mirror. B If the incoming rays are parallel to the principal axis of the mirror, all the reflected rays converge at the focal point of the mirror. C If a light source is placed at the focal point, the outgoing rays reflected from the mirror are parallel to the axis.

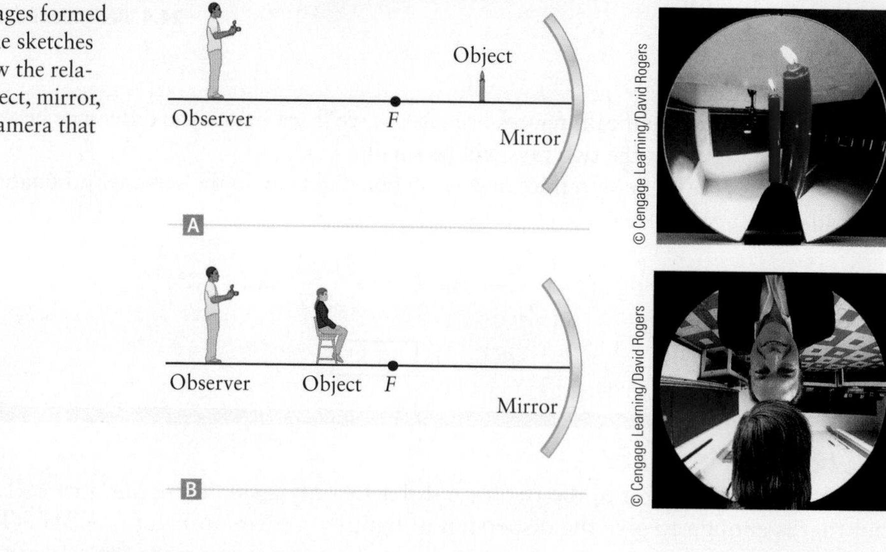

► **Figure 24.24** Images formed by a concave mirror. The sketches next to each image show the relative positions of the object, mirror, and observer (i.e., the camera that took the photo).

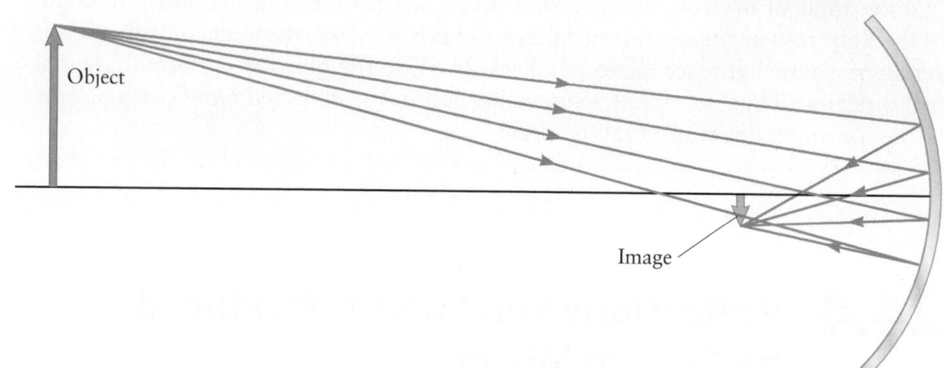

► **Figure 24.25** Ray diagram for a curved mirror. Rays from each point on the object are reflected by the mirror and intersect at a corresponding image point. Here we show only rays from the tip of the object.

which shows photos of objects placed in front of a concave mirror. Accompanying each photo is a schematic showing the object, the mirror, and the "observer" (the camera that took the photo). In Figure 24.24A, the object—a candle—is placed fairly close to the mirror. The observer can see both the original candle and its image formed by the mirror. Here the image appears larger than the true candle. In Figure 24.24B, the object—a person's head—is somewhat farther from the mirror. The observer now sees the back of the person's head along with an image of the person's face. Note that the face is upside down; we call this an ***inverted image***. Let's now apply ray tracing to explain how these images arise.

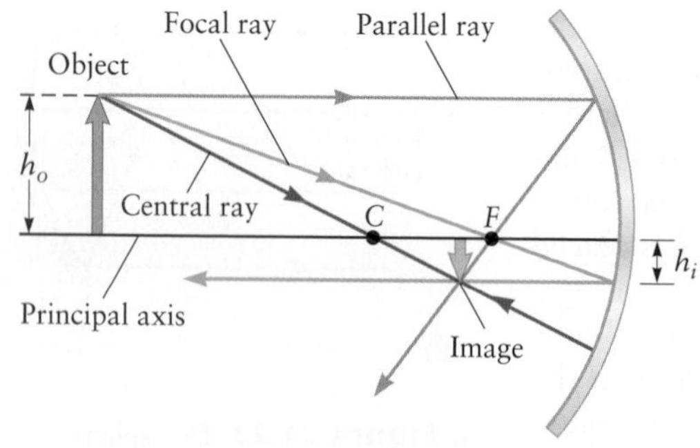

▲ **Figure 24.26** Using ray tracing to find the image formed by a concave spherical mirror. For an object outside the focal length, the image is real, inverted, and reduced. Rays emanating from the tip of the object intersect at the tip of the image. *Important note*: The colors of these rays do *not* indicate the color of the light.

Figure 24.25 is a ray diagram for an object placed in front of a curved mirror; here the object is the upward-pointing arrow. The figure shows rays that emanate from the tip of the arrow on the left. Although these rays reflect from different parts of the mirror, they all intersect at a single point, which is the tip of the arrow's image. A similar result is found for rays from other points on the object.

We can use the following ray-tracing procedure to find the location of the image formed by a spherical mirror. Figure 24.26 shows an object far from the mirror, meaning that the center of curvature and the focal point are between the object and the mirror (corresponding to the case in Fig. 24.24B.) The upward-pointing arrow on the left again represents the object, and for simplicity we place the bottom of the object arrow on the principal axis. Although many rays (an infinite number) emanate from each point on the object, three ray directions are particularly easy to draw. Figure 24.26 shows these three rays as they emanate from the tip of the object and intersect at the tip of the image.

One of these rays—called the ***focal ray***—begins at the object's tip and passes through the focal point F. After reflection from the mirror, this ray travels parallel to the principal axis (compare with Fig. 24.23C). A second ray—called the ***parallel ray***—begins at the object's tip and is initially parallel to the principal axis, so after reflection it passes through the focal point (compare with Fig. 24.23B). The third ray—called the ***central ray***—begins at the object's tip and passes through the center of curvature C of the mirror. The central ray is directed along the radius of the spherical mirror and is therefore perpendicular to the mirror's surface. By the law of reflection, this ray reflects back on itself and passes back through C. These three rays all intersect (are "focused") at the tip of the arrow's image. Other rays that emanate from the tip of the object also intersect at the corresponding point on the image (see Fig. 24.25), but the focal, parallel, and central rays drawn from the object's tip provide the easiest way to locate the tip of the image and reveal its nature. These three rays are very useful in many applications of ray tracing, so we have drawn them with different colors here and in other figures. These ray colors are *not* the color of the light, but rather are intended to help you identify these rays.

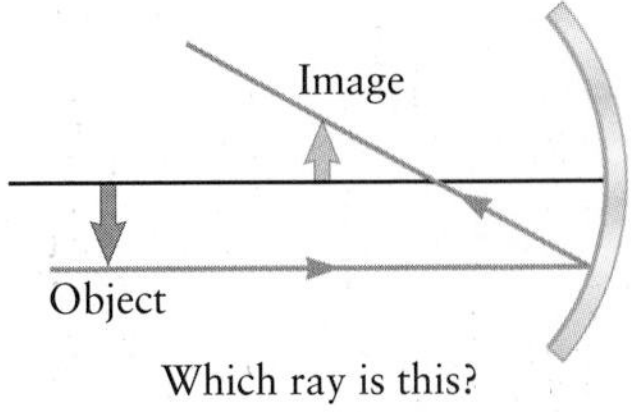

▲ **Figure 24.27** Concept Check 24.6.

CONCEPT CHECK 24.6 Which Ray Is It?

Figure 24.27 shows an object in front of a spherical mirror along with the image and one ray that travels from the tip of the object to the tip of this image. Which ray is it, (a) the focal ray, (b) the parallel ray, or (c) the central ray?

Properties of an Image

The image in Figure 24.26 is smaller than the object and is inverted (upside down). The ratio of the height of the image h_i to the height of the object h_o is called the ***magnification***, m:

Magnification

$$m = \frac{h_i}{h_o} \tag{24.10}$$

By convention, the image height h_i in Figure 24.26 is taken to be negative since the image falls below the principal axis; the magnification in this example is thus also negative. Images smaller than the corresponding objects are said to be ***reduced*** ($|m| < 1$).

Real image

The rays that form the image in Figure 24.26 all pass through a point on the image. We call the image ***real***. A real image differs from the virtual image in Figure 24.6 in two ways:

1. Light rays only *appear* to emanate from a virtual image; they do not actually pass through the image. For a real image, the rays *do* pass through the image.
2. An object and its real image are both on the same (front) side of the mirror. A virtual image is located behind the mirror, while the object is in front.

Forming a Virtual Image

We can use ray tracing to find the image of an object placed close to a spherical mirror—that is, closer to the mirror than the focal point and the center of curvature—as shown in Figure 24.28. Here we draw the focal ray, the parallel ray, and the central ray as they emanate from the object's tip (the arrow to the left of the mirror). You should remember, however, that there are actually many rays (an infinite number) that emanate from the tip of the arrow, reflect from the mirror, and then converge at the image. In Figure 24.28, the focal, parallel, and central rays do not intersect at any point on the left (the front) of the mirror, but if we extrapolate the rays back behind the mirror, we find that these extrapolations all intersect at a single "image point." If we were to observe the light reflected by the mirror as shown at the lower left in Figure 24.28, the rays would appear to emanate from this image point behind the mirror. This image is virtual because light does not actually pass through any point on the image. The object and its image are on different sides of the mirror; the image is upright (h_i is positive) and enlarged.

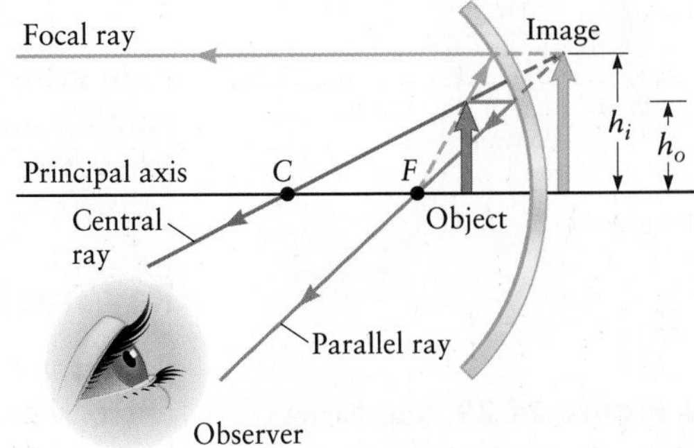

▲ **Figure 24.28** If an object is close to a concave mirror (inside the focal point), the outgoing light rays appear to emanate from behind the mirror, producing a virtual, upright image similar to the image in Figure 24.24A.

CONCEPT CHECK 24.7 Magnification of an Image

Consider the virtual image formed by the mirror in Figure 24.28. Which of the following statements is correct?

(a) The magnification is positive and less than 1.
(b) The magnification is positive and greater than 1.
(c) The magnification is negative with an absolute value less than 1.
(d) The magnification is negative and with an absolute value greater than 1.

Rules for Ray Tracing with Mirrors

Figures 24.26 and 24.28 show two examples of how to use ray tracing to construct the image produced by a spherical concave mirror. The general approach can be summarized as follows.

Ray tracing applied to spherical mirrors

1. **Construct a figure showing the mirror and its principal axis. The figure should also show the focal point and the center of curvature.**
2. **Draw the object at the appropriate point. One end of the object will often lie on the principal axis.**
3. **Draw three rays that emanate from the tip of the object:**
   (a) **The *focal ray* (or its extrapolation) passes through the focal point. After reflection, this ray will be parallel to the axis.**
   (b) **The *parallel ray* is initially parallel to the axis. After reflection, this ray (or its extrapolation) passes through the focal point.**
   (c) **The *central ray* (or its extrapolation) passes through the center of curvature of the mirror. After reflection, this ray passes back through the tip of the object.**
4. **The point where the focal, parallel, and central rays (or their extrapolations) intersect is the image point. This point may be in front of the mirror giving a *real image*, or you may need to extrapolate the rays back behind the mirror to locate a *virtual image*.**
5. **This ray-tracing procedure can be repeated for any desired point on the object, thus locating other points on the image. It is usually sufficient to consider just the tip of the image.**

In many of our ray diagrams, we use different colors to clearly denote the focal, parallel, and central rays. This use of color is to help you distinguish the different rays; these ray colors are *not* the color of the light used to form the image!

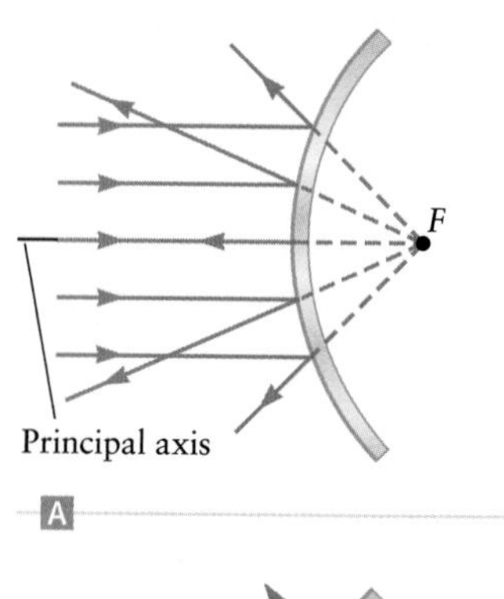

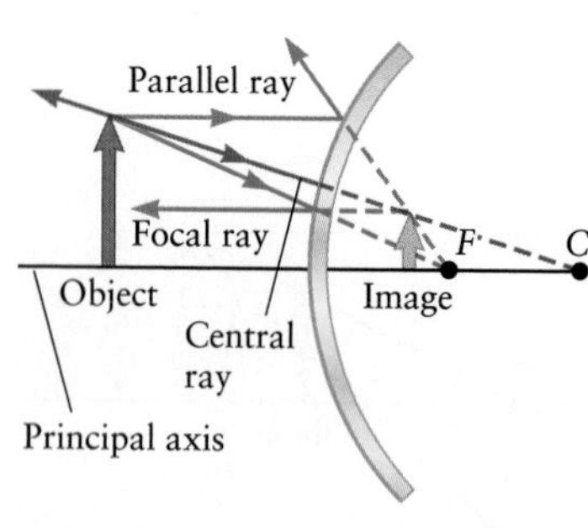

▲ **Figure 24.29** Ray diagram for a convex mirror. **A** If the incoming light is parallel to the axis of a convex mirror, the reflected rays appear to emanate from a focal point F behind the mirror. **B** Ray tracing with a convex mirror. The image is behind the mirror (a virtual image).

Ray Tracing for Spherical Convex Mirrors

A mirror that curves away from the object as in Figure 24.29 is called a ***convex*** mirror. For a convex mirror, both C and the focal point lie *behind* the mirror. The location of the focal point is illustrated in Figure 24.29A, which shows a collection of parallel rays incident on the mirror. After striking the convex surface, the reflected rays diverge from the mirror axis. These rays, however, appear to come from a virtual image point behind the mirror, the focal point F. That is, all these rays extrapolate back to the focal point F, which lies behind the mirror.

Figure 24.29B shows an application of ray tracing to a convex mirror following the suggested ray tracing rules listed above. This sketch shows the principal axis of the mirror along with the center of curvature C and the focal point F. We also show the focal ray, the parallel ray, and the central ray. Notice that the focal ray is initially directed toward the focal point, but because F is behind the mirror, this ray is reflected before reaching the focal point. These three rays extrapolate to a point of intersection that lies *behind* the mirror; hence, we have a virtual image. This image is upright, so the image height h_i and magnification m are both positive.

Finding the Location of the Image: The Mirror Equation

While ray tracing is an instructive way to analyze image formation, it is not a convenient way to find the image location quantitatively. Suppose we know the radius of curvature of a mirror and the distance from the object to the mirror. How can we calculate the location of the image? To answer this question, we must consider the geometry of the reflected rays. Figure 24.30A shows this geometry for a concave mirror, concentrating on two particular rays: the central ray that passes through C and a ray that reflects from the point where the principal axis meets the mirror. (This second ray is *not* a focal or parallel ray.) These two rays intersect at the image point, giving the height of the image h_i. The image here is inverted, so by our sign convention h_i is negative and h_o is positive. The figure also defines the distance s_o from the object to the mirror and the distance s_i between the image and the mirror.

The two triangles shaded in yellow in Figure 24.30A are redrawn in part B. They are both right triangles with the same values for the interior angle θ, so the ratios of their sides are equal, giving

$$\frac{|h_o|}{s_o} = \frac{|h_i|}{s_i} \tag{24.11}$$

We have used absolute values of the object and image heights because h_o and h_i can be negative if they lie below the principal axis, and Equation 24.11 holds for the magnitudes of the sides of the triangles shaded in yellow in Figure 24.30B. We next consider the pair of similar triangles crosshatched in red in Figure 24.30A. These triangles are redrawn in Figure 24.30C; the ratios of their sides are equal, giving

$$\frac{|h_o|}{s_o - R} = \frac{|h_i|}{R - s_i} \tag{24.12}$$

We now take the ratios of Equations 24.11 and 24.12:

$$\frac{|h_o|/s_o}{|h_o|/(s_o - R)} = \frac{|h_i|/s_i}{|h_i|/(R - s_i)}$$

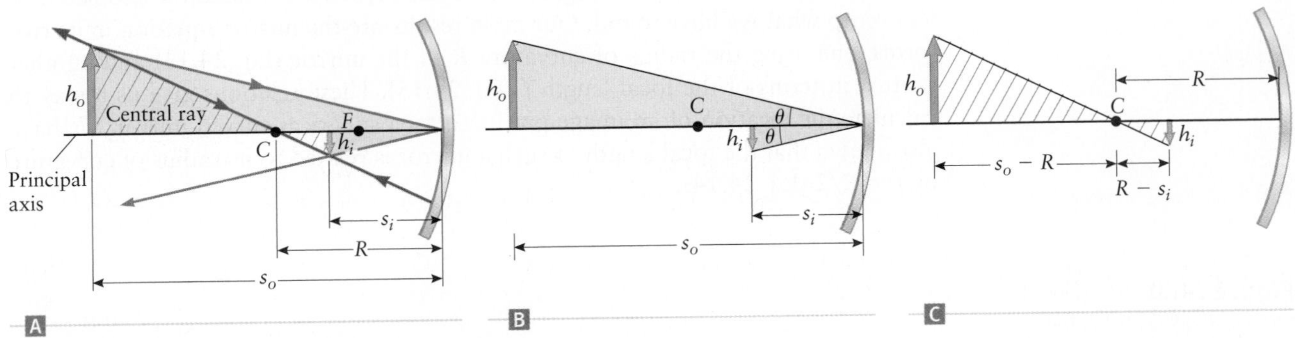

▲ **Figure 24.30** Geometry for deriving the mirror equation. **A** Finding the location of the image formed by a concave mirror. **B** Two similar triangles from part A. **C** Two other similar triangles from part A.

Canceling the common factors of $|h_o|$ and $|h_i|$ and doing some rearranging leads to

$$\frac{R - s_o}{s_o} = \frac{s_i - R}{s_i}$$

$$s_i(R - s_o) = s_o(s_i - R)$$

$$s_iR - s_is_o = s_os_i - s_oR$$

$$s_iR + s_oR = 2s_is_o$$

Dividing both sides by the factor s_is_oR, we finally get

Mirror equation in terms of radius of curvature

$$\frac{1}{s_o} + \frac{1}{s_i} = \frac{2}{R} \tag{24.13}$$

which is called the ***mirror equation***. If we know a mirror's radius of curvature R and are given the location of an object s_o, we can use the mirror equation to calculate s_i, the location of the image.

The Mirror Equation and Focal Length

Figure 24.31 shows rays for an object very far from a mirror so that the rays that strike the mirror are all approximately parallel to the axis. This object is a distance $s_o \approx \infty$ from the mirror, so we call it an object "at infinity." Inserting an infinite object distance into Equation 24.13, we get

$$\frac{1}{\infty} + \frac{1}{s_i} = \frac{2}{R}$$

Because $1/\infty$ is zero, the image distance for an object at infinity is $s_i = R/2$. A collection of parallel rays would thus reflect from the mirror and intersect at the point a distance $R/2$ from the mirror. By definition, rays that are parallel to the principal axis all reflect and pass through the focal point F (compare with Fig. 24.23B), so we have just shown that the focal point is a distance $R/2$ from the mirror. This distance is called the ***focal length***, f, of the mirror:

Focal length for a spherical mirror

$$f = \frac{R}{2} \tag{24.14}$$

Equation 24.14 enables us to rewrite the mirror equation in terms of the mirror's focal length:

Mirror equation in terms of focal length

$$\frac{1}{s_o} + \frac{1}{s_i} = \frac{1}{f} \tag{24.15}$$

Our analysis of the rays in Figure 24.30 has involved considerable geometry, so let's recap what we have found. Our main results are the mirror equation in its two forms, one using the radius of curvature R of the mirror (Eq. 24.13) and another written in terms of the focal length f (Eq. 24.15). These relations give us a way to calculate the location of an image produced by a spherical curved mirror. We have also shown that the focal length of such a mirror is related to its radius of curvature by $f = R/2$ (Eq. 24.14).

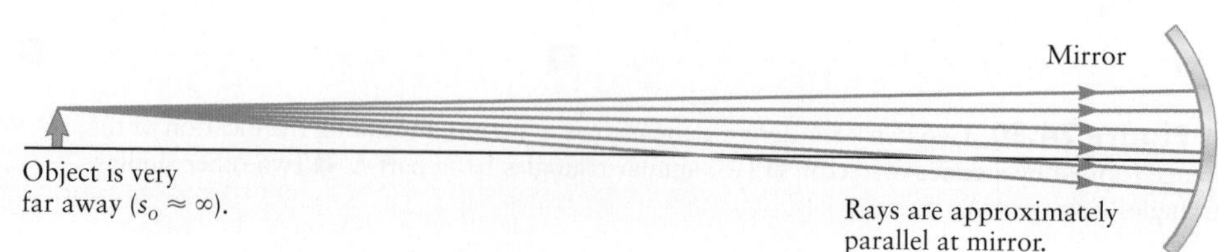
Mirror
Object is very far away ($s_o \approx \infty$).
Rays are approximately parallel at mirror.

▶ **Figure 24.31** When an object is very far away ($s_o \approx \infty$), the rays are approximately parallel to the principal axis. The object is said to be "at infinity."

Sign Conventions

We derived the mirror equation for a concave mirror, but that relation also applies to a convex mirror provided we use the following sign conventions for the object and image distances s_o and s_i, the focal length f, and the object and image heights h_o and h_i.

Sign conventions for the mirror equation

1. **All diagrams with mirrors should be drawn with the light rays incident on the mirror from the left.**
2. **The object distance s_o is positive when the object is to the left of the mirror and negative if the object is to the right (behind the mirror[2]).**
3. **The image distance s_i is positive when the image is to the left of the mirror and negative if the image is to the right (behind the mirror). Hence, $s_i > 0$ for a real image and $s_i < 0$ for a virtual image.**
4. **The focal length is positive for a concave mirror and negative for a convex mirror. Hence, for a spherical mirror we have**

$$\text{concave mirror: } f = \frac{R}{2} \qquad \text{convex mirror: } f = -\frac{R}{2} \tag{24.16}$$

5. **The object and image heights h_o and h_i are positive if the object/image is upright and negative if it is inverted.**

CONCAVE MIRROR:
f positive

h_o and h_i positive — s_o positive — s_o negative — F

Principal axis

h_o and h_i negative — s_i positive — s_i negative

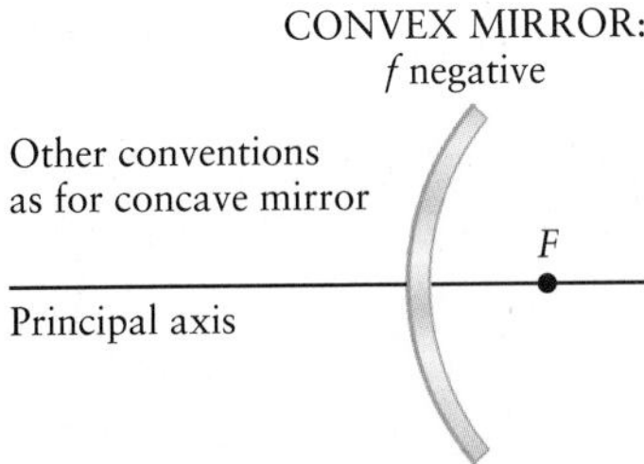

▲ **Figure 24.32** Sign conventions for application of the mirror equation.

We'll follow these sign conventions (summarized in Fig. 24.32) throughout this book. Other books may follow different sign conventions, however, so you should always check the sign conventions before applying relations involving s_o, s_i, and so forth from other books.

Insight 24.3
WHAT IS THE "BEST" SHAPE FOR A CURVED MIRROR?
In our derivation of the mirror equation (Eq. 24.13), we assumed a spherical mirror, but other mirror shapes give better focusing properties for rays that are far from the principal axis. Mirrors with a parabolic shape are used in many high-performance telescopes because this shape gives smaller aberrations than a spherical mirror. (See Section 24.8.) It is more difficult and expensive to make a surface with a parabolic shape than a spherical one, however.

CONCEPT CHECK 24.8 Object and Image Distances

Figure 24.27 shows an object and its image as formed by a curved mirror. What are the signs (positive or negative) of (a) the object distance s_o, (b) the image distance s_i, and (c) the focal length f?

Applications of the Mirror Equation

Figure 24.33 shows an object placed a distance $s_o = 2.5$ m in front (to the left) of a concave mirror with $R = 1.5$ m. Let's find the location of the image and also determine if the image is real or virtual. The distances s_o and s_i along with the three rays in our standard ray-tracing diagram are shown in Figure 24.33. (Compare with Fig. 24.26.) While this diagram gives information about the image, the mirror equation provides quantitative answers.

We first use the given value of R to find the focal length $f = R/2 = 0.75$ m and show the position of the focal point F in the ray diagram. (We could use the form of the mirror equation in Eq. 24.13 to find the image location s_i without first finding f, but it is usually a good idea to include the focal point F in all ray diagrams.) The mirror equation can then be used to find the image location. Solving Equation 24.15 for $1/s_i$, we get

$$\frac{1}{s_o} + \frac{1}{s_i} = \frac{1}{f}$$

$$\frac{1}{s_i} = \frac{1}{f} - \frac{1}{s_o}$$

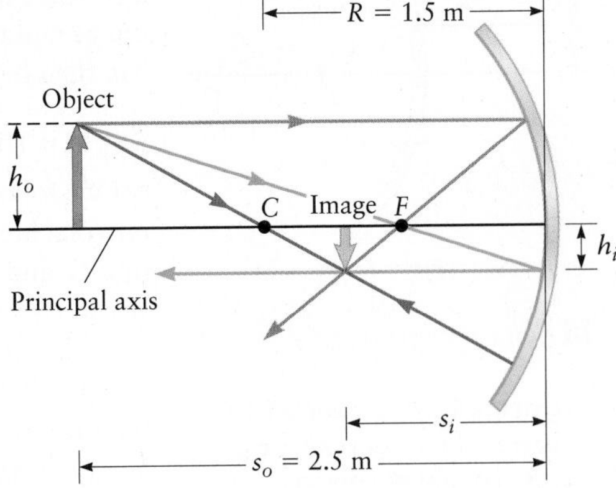

▲ **Figure 24.33** Using ray tracing together with the mirror equation to find the location of an image.

[2]So far, all our objects have been in front of the mirror, but when two or more mirrors and lenses are combined, it is possible for an object to be behind a mirror. We'll see such examples in Chapter 26.

Substituting values for the focal length and object distance gives

$$\frac{1}{s_i} = \frac{1}{0.75\text{ m}} - \frac{1}{2.5\text{ m}} = 0.933\text{ m}^{-1}$$

$$s_i = 1.07\text{ m}$$

Here we keep three significant figures to avoid rounding problems later in this example. Since s_i is positive, we expect to find a real image on the left (in front) of the mirror, which is confirmed by the ray diagram. The ray diagram also shows that the image is inverted, so according to our sign conventions the image height h_i is negative. To find the value of h_i, we recall the triangles in Figure 24.30B, which led to Equation 24.11; now we must be careful about the signs of h_o and h_i. The object is above the principal axis, so according to our sign conventions h_o is positive. Since h_i is below the principal axis, h_i is negative. We thus have

$$\frac{-h_i}{s_i} = \frac{h_o}{s_o} \tag{24.17}$$

Solving for the magnification $m = h_i/h_o$ (Eq. 24.10), we get

$$m = \frac{h_i}{h_o} = -\frac{s_i}{s_o} \tag{24.18}$$

$$m = -\frac{1.07\text{ m}}{2.5\text{ m}} = -0.43$$

The negative value of m means that we have an inverted image (as already noted). Because $|m| < 1$, the image is reduced as shown in Figure 24.33.

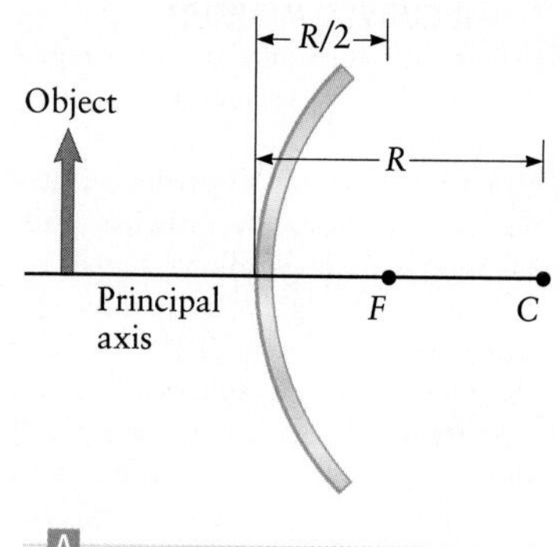

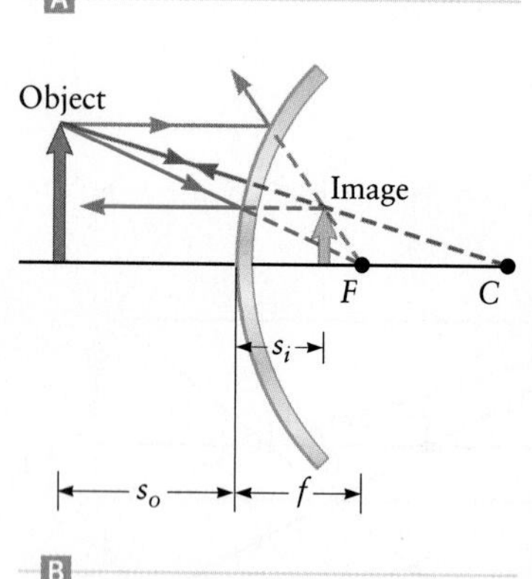

▲ **Figure 24.34** Example 24.5. Ray diagram for a convex mirror. **A** A sketch of just the object, mirror, focal point, and center of curvature, in preparation for ray tracing. **B** Same as part A, but now with the parallel, focal, and central rays included.

EXAMPLE 24.5 Applying the Mirror Equation to a Convex Mirror

Consider again the problem sketched in Figure 24.33, but now assume the mirror is convex. (**a**) Construct the ray diagram. (**b**) Is the image inverted or upright? (**c**) Find the image location s_i. (**d**) Find the size of the image. Assume the same values of R and s_o as in Figure 24.33 and assume the object height $h_o = 1.0$ cm.

RECOGNIZE THE PRINCIPLE

A ray diagram should always be your first step in problems involving image formation. We first construct a ray diagram, similar to the ray diagram we used in dealing with a concave mirror. We next use our ray diagram to apply the sign conventions for the image and object distances and heights and for the focal length. The mirror equation can then be applied to find the image's location and properties.

SKETCH THE PROBLEM

(**a**) We follow our recommended procedure for ray tracing applied to spherical mirrors. Step 1: Figure 24.34A shows the mirror with its principal axis, center of curvature C, and focal point F. According to our sign conventions, the focal length for a convex mirror is negative, so F is behind the mirror. The focal length is (from Eq. 24.16) $f = -R/2 = -0.75$ m. Step 2: We add the object to the diagram. Step 3: Add the focal ray, parallel ray, and central ray (Fig. 24.34B). The focal ray is initially directed toward the focal point; after reflecting from the mirror, it travels parallel to the axis. The parallel ray initially travels parallel to the principal axis and then reflects from the mirror so that it appears to emanate from the focal point. The central ray is initially directed toward the center of curvature of the mirror and reflects back toward the object. Step 4: Extrapolations of the rays in Figure 24.34B intersect at the image point, which is behind the mirror.

IDENTIFY THE RELATIONSHIPS AND SOLVE

(**b**) The ray diagram in Figure 24.34B shows that the image is behind the mirror (a virtual image), is $\boxed{\text{upright}}$, and is smaller than the object.

(**c**) We next use our sign conventions for image location. The object is in front of the mirror, so $s_o = +2.5$ m (positive). Using this value in the mirror equation (Eq. 24.15) along with the focal length determined above, we get

$$\frac{1}{s_i} = \frac{1}{f} - \frac{1}{s_o} = \frac{1}{-0.75 \text{ m}} - \frac{1}{2.5 \text{ m}}$$

$$s_i = \boxed{-0.58 \text{ m}}$$

The negative value of s_i indicates that the image is to the right of the mirror (behind), so it is a virtual image as we saw from the ray diagram in Figure 24.34B.

(**d**) We can find the image height h_i using Equation 24.17; we get

$$-\frac{h_i}{s_i} = \frac{h_o}{s_o}$$

Note that in this example, h_i is positive while s_i is negative. Solving for h_i and using our result for s_i gives

$$h_i = -h_o\left(\frac{s_i}{s_o}\right) = -h_o\left(\frac{-0.58 \text{ m}}{2.5 \text{ m}}\right) = +0.23h_o = \boxed{0.23 \text{ cm}} \quad (1)$$

where we have used the given value of $h_o = 1.0$ cm.

According to our sign conventions, a positive value of h_o corresponds to an upright object as confirmed in Figure 24.34B. A positive value of h_o in Equation (1) gives a positive value for h_i, so we again find that the image is upright.

► ***What does it mean?***

When an object is placed in front of a convex mirror, the solution of the mirror equation for s_i is always negative, so the image will *always* be virtual. In addition, if the object is upright, the image will always be upright. This property is handy in many applications, one of which is discussed in Example 24.6.

EXAMPLE 24.6 Ⓡ Designing a Car Rearview Side Mirror

The external rearview side mirrors on most cars carry the phrase "Objects in the mirror are closer than they appear." (**a**) What does this phrase tell us about the magnification of the image? (**b**) Is this mirror convex or concave? (**c**) Estimate the radius of curvature of a typical rearview mirror.

RECOGNIZE THE PRINCIPLE

(**a**) The "apparent size" of an object gets smaller as the object is moved farther and farther away. When viewed by its reflection from a mirror, this apparent size is also just the image size. When judging the distance from your eye to an object, your brain compares the known size of the object (as stored in your memory) with the size of the image (the apparent size). So, when you judge an object to be farther away than it actually is, the image must be *smaller* than expected for that particular object distance. Hence, the image formed by the car's rearview side mirror must have a $\boxed{\text{magnification } |m| < 1}$.

SKETCH THE PROBLEM

(**b**) The next step is to draw a ray diagram, but should this diagram be for a concave mirror or a convex mirror? A car's external rearview mirror *always* produces an upright

(continued) ►

(noninverted) image, for any object distance. (You should check for yourself.) We have seen that the image produced by a concave mirror can be upright or inverted, depending on the object distance (Figs. 24.26 and 24.28), so a rearview mirror cannot be concave. A convex mirror always produces a virtual image, as can be seen from the ray diagrams in Figure 24.29 and Example 24.5. So, a rearview mirror must be $\boxed{\text{convex}}$, and the ray diagram will look like Figure 24.34B.

IDENTIFY THE RELATIONSHIPS

(c) Because $f = -R/2$ (Eq. 24.16 for a convex mirror), we can find the radius of curvature of a rearview mirror if we can estimate the focal length. If we know something about the object and image distances, the value of f can be found from the mirror equation. We have already argued that the magnification is less than 1, but how much less? A value of $m = 0.1$ would produce a very small image, whereas (based on the author's experience) the image is only a little smaller than the true object size. We therefore estimate a magnification of about $m \approx \frac{1}{3}$ for a car that is $s_o = 30$ m from the mirror. These estimates are rough, but they should not be off by more than a factor of three. (How could you check these values for yourself?) The magnification is related to the object and image distances by Equation 24.18, giving $m = -s_i/s_o$ or $s_i = -ms_o$. Inserting this information into the mirror equation, we find

$$\frac{1}{s_o} + \frac{1}{s_i} = \frac{1}{f} = \frac{1}{s_o} - \frac{1}{ms_o}$$

Solving for f, we get

$$f = s_o\left(\frac{m}{m-1}\right)$$

SOLVE

Using our estimates for m and s_o gives

$$f = s_o\left(\frac{m}{m-1}\right) = (30\text{ m})\left(\frac{\frac{1}{3}}{\frac{1}{3}-1}\right) = -15\text{ m}$$

The radius of curvature of the mirror is thus

$$R = -2f = \boxed{30\text{ m}}$$

▸ ***What does it mean?***

This large value of R means that the mirror has just a very slight curvature, which is why it is easily mistaken for a plane mirror. Why use a curved mirror in the first place? Why not just use a flat mirror? A curved mirror gives a larger "field of view," allowing the driver to see objects over a wider angle and thus spot cars and other things that might be missed with a plane mirror (Fig. 24.35).

▲ **Figure 24.35** Example 24.6. View through a car's rearview side mirror.

24.5 Lenses

A mirror uses reflection to form an image; a ***lens*** uses refraction to accomplish the same thing. Typical lenses are composed of glass or plastic, and the refraction of light rays as they pass from air into the lens and then back into air causes the rays to be redirected as shown in Figure 24.36. Refraction occurs at the two separate surfaces as shown in Figure 24.36A, but for simplicity our ray diagrams will usually be drawn as in parts B and C of the figure, showing a single deflection of each ray as it passes through the lens.

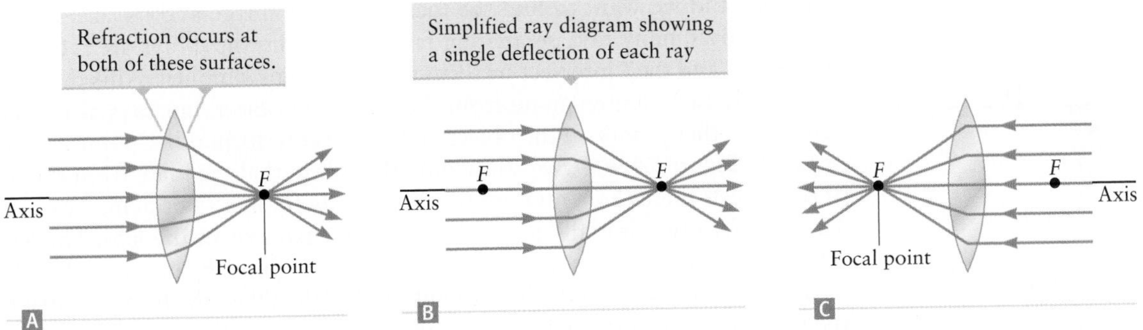

▲ **Figure 24.36** A Light rays are refracted at *both* surfaces of a lens. When parallel light rays strike a converging lens, the refracted rays converge at the focal point F of the lens. B For simplicity, most ray diagrams with lenses show only a single "deflection" of a ray as it passes through the lens. C If light strikes the lens from the opposite side, the rays converge at a second focal point F on the left side of the lens.

In the simplest case, the lens surfaces are spherical (Fig. 24.37). A careful treatment of the refraction of rays that are close to the principal axis shows that if the incident rays are parallel to the principal axis of the lens, then after passing through the lens the rays all pass through a point on the axis called the ***focal point***, F. This is true for parallel rays incident from either side of the lens (parts B and C of Fig. 24.36). Hence, there are two focal points, and they are at identical distances from the lens.

Lenses can be classified according to the curvatures of their two surfaces (Fig. 24.38). Given the radius of curvature of each surface (Fig. 24.37), it is possible to use Snell's law to calculate the focal length. (We'll see how that works in Section 24.6.) For most applications, we only need to know the value of the focal length f and whether the lens is ***converging*** or ***diverging***. The lenses in Figure 24.36 are converging lenses: all the incoming rays intersect at the focal point on the opposite side. A ray diagram for a diverging lens is shown in Figure 24.39. In that case, parallel rays from the left diverge after passing through the lens, but they all extrapolate back to the focal point on the left (the same side as the incident rays).

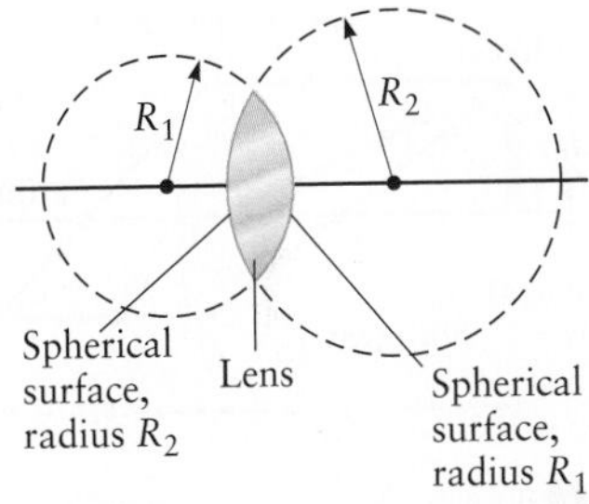

▲ **Figure 24.37** A spherical lens: both surfaces follow the shape of a sphere. The radii of curvature of the lens are the radii R_1 and R_2 of the corresponding spheres.

Forming an Image with a Lens: Ray Tracing

To analyze the images produced by a lens, we start with a ray-tracing analysis. Let's begin with the case of a converging lens. Figure 24.40A on page 748 shows the object along with the focal points on the left (F_L) and right (F_R) sides of the lens. In a typical case, we know the distance s_o from the object to the lens and the object's

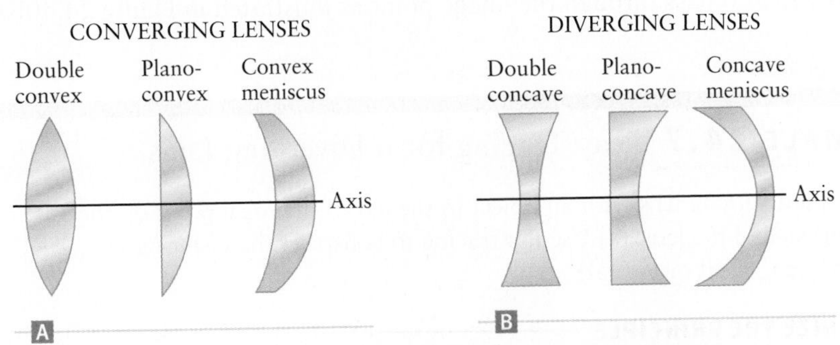

▲ **Figure 24.38** Some examples of A converging lenses and B diverging lenses.

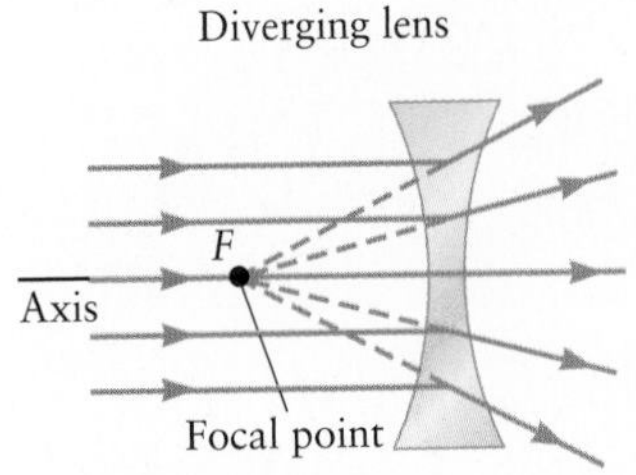

▲ **Figure 24.39** With a diverging lens, parallel incident rays from the left are refracted away from the axis. The rays on the right appear to emanate from the point F on the left side of the lens.

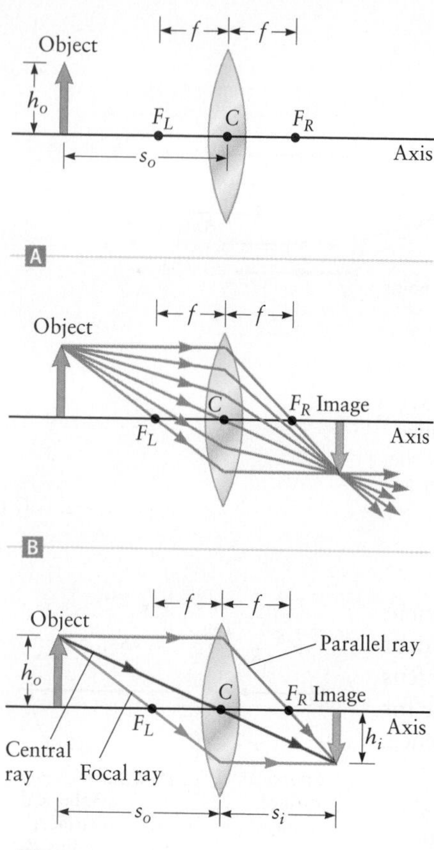

▲ **Figure 24.40** Ray tracing with a converging lens when the object is outside the focal point F_L. **A** Sketch showing the object, lens, and focal points. **B** Some typical rays from the tip of the object, intersecting at the tip of the image. **C** Ray diagram showing just the parallel ray, focal ray, and central ray, all emanating from the tip of the object. In this case, the image is real.

height h_o, and we want to find the location of the image and its size. Figure 24.40B shows many rays emanating from the object; these rays are refracted by the lens and then intersect to form the image. For simplicity, this diagram shows only rays that emanate from the tip of the object, but rays also emanate from other points on the object and intersect to form other points on the image. In Figure 24.40C, we show only three rays that emanate from the tip of the object; these three rays are particularly useful in a ray tracing analysis. The ***parallel ray*** is initially parallel to the principal axis of the lens. The lens refracts this ray so that it passes through the focal point F_R on the right. The ***focal ray*** passes through the focal point on the left. When the focal ray strikes the lens, the ray is refracted to be parallel to the axis on the right. The ***central ray*** passes straight through the center of the lens (denoted as C). If the lens is very thin (a good approximation for many lenses), this ray is not deflected by the lens. These three rays come together at the tip of the image on the right of the lens. In this case, the image is inverted, and because the rays pass through the image, the image is real.

This approach to ray tracing with lenses is described by the following rules.

Ray tracing applied to lenses

1. **Construct a figure showing the lens, principal axis, and focal points on both sides of the lens.**
2. **Draw the object at the appropriate point. One end of the object will generally lie on the axis.**
3. **Draw three rays that emanate from the tip of the object:**
   (a) **The *parallel ray* is initially parallel to the axis. After passing through the lens, this ray or its extrapolation passes through one of the focal points.**
   (b) **The *focal ray* is directed at the other focal point. After passing through the lens, this ray is parallel to the axis.**
   (c) **The *central ray* passes through the center of the lens and is not deflected.**
4. **These three rays or their extrapolations intersect at the image. If the rays actually pass through the image, the image is *real*; if they do not, the image is *virtual* (just as with images formed by a mirror). When a lens forms a real image, the object and image are on opposite sides of the lens (Fig. 24.40C). When a lens forms a virtual image, the object and image are on the same side of the lens (Example 24.7).**

Our approach to ray tracing emphasizes the parallel, focal, and central rays because these rays are the simplest ones to draw. However, all other rays that pass through the lens will also pass through the image point as illustrated in Figure 24.40B.

EXAMPLE 24.7 Ray Tracing for a Diverging Lens

Consider a diverging lens with an object to the left of the focal point on the left (incident) side of the lens. (**a**) Use ray tracing to construct the resulting image. (**b**) Is the image real or virtual?

RECOGNIZE THE PRINCIPLE

These questions can both be answered with a ray diagram, which we can construct using the rules for ray tracing applied to lenses.

SKETCH THE PROBLEM

Figure 24.41 shows the problem, including a full ray diagram.

IDENTIFY THE RELATIONSHIPS AND SOLVE

(a) Follow the suggested steps for making a ray diagram with lenses. Step 1: Figure 24.41A shows the lens, its axis, and the focal points. Step 2: Add the object to the drawing. Step 3: Figure 24.41B shows the parallel ray, focal ray, and central ray, drawn according to our ray-tracing rules. Step 4: These rays do not intersect at any point to the right of the lens. If the rays on the right are extrapolated back to the left side of the lens, however, the extrapolations intersect at the image point at the tip of the image.

(b) The image is virtual because the rays do not pass through the image point.

► ***What does it mean?***

For all problems involving image formation, you should always begin with a ray diagram. Our ray tracing rules apply to both converging and diverging lenses.

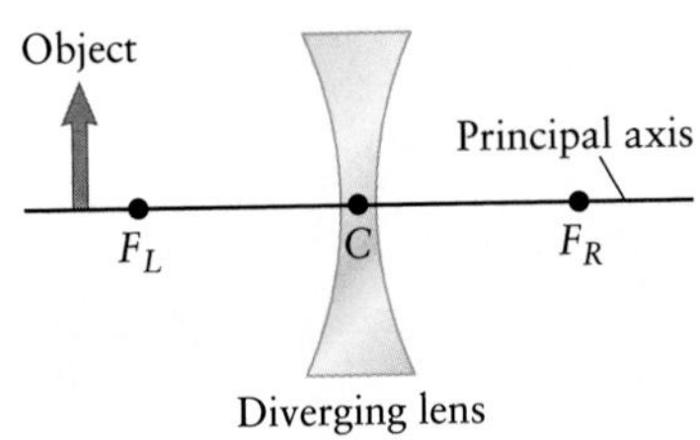

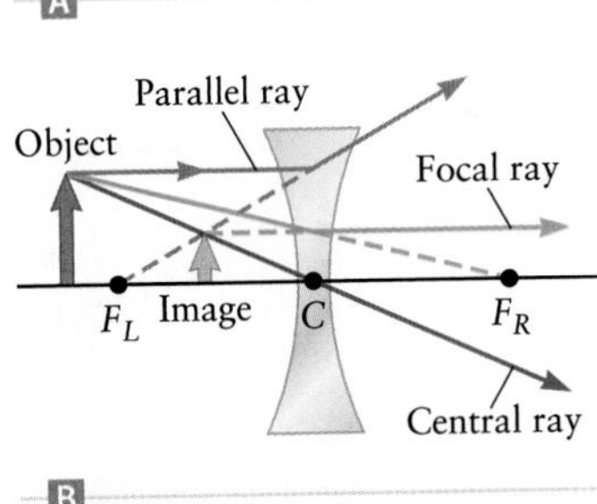

▲ **Figure 24.41** Example 24.7. Ray tracing with a diverging lens.

Finding the Image Location for a Lens

Ray tracing enables us to describe qualitatively the image formed by a lens. We now want to derive a quantitative approach for calculating the image location. As with our work with the mirror equation, we must adopt sign conventions for dealing with object and image distances, heights, and so forth. The conventions we'll follow are shown in Figure 24.42 and are listed below.

Sign conventions for dealing with lenses

1. **We always assume light travels through the lens from left to right, so the object is always located to the left of the lens.**
2. **The object distance s_o is *positive* when the object is to the *left* of the lens. With our assumption in convention 1, s_o is always positive.**[3]
3. **The image distance s_i is *positive* if the image is to the *right* of the lens and is *negative* if it is on the *left* side of the lens.**
4. **The focal length f of a converging lens is positive, whereas it is negative for a diverging lens.**
5. **The object height h_o is positive if the object extends above the axis and is negative if the object extends below.**
6. **The image height h_i is positive if the image extends above the axis and is negative if the image extends below.**

CONVERGING LENS: f positive

h_o and h_i positive | s_o positive | s_i negative | s_i positive

Principal axis

h_o and h_i negative

DIVERGING LENS: f negative

Other conventions as for converging lens

Principal axis

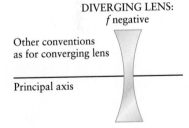

▲ **Figure 24.42** Sign conventions for application of the thin-lens equation.

Once again, there are several different sign conventions for working with lenses, so be careful when comparing our equations with those from other books.

CONCEPT CHECK 24.9 Object and Image Distances

Figure 24.41B shows an object and its image as formed by a diverging lens. What are the signs (positive or negative) of (a) the object distance s_o, (b) the image distance s_i, and (c) the focal length f?

CONCEPT CHECK 24.10 Ray Tracing with a Diverging Lens

Figure 24.43 shows a ray diagram with a diverging lens, including two rays labeled 1 and 2. (a) Identify these rays as the parallel ray, focal ray, or central ray. (b) Will the image be in region I, II, or III?

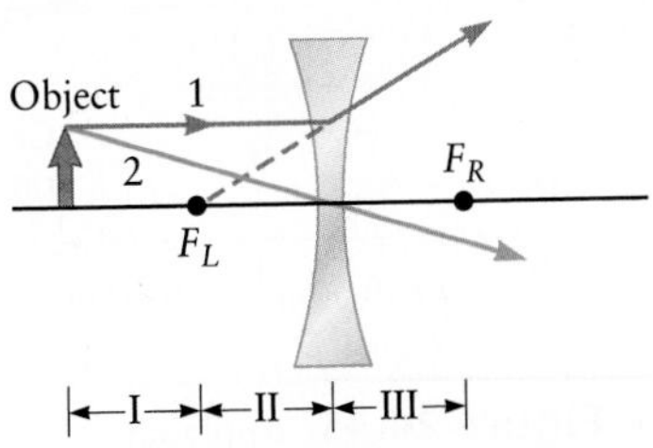

▲ **Figure 24.43** Concept Check 24.10.

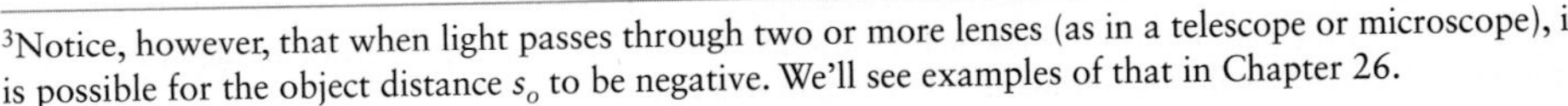

[3]Notice, however, that when light passes through two or more lenses (as in a telescope or microscope), it is possible for the object distance s_o to be negative. We'll see examples of that in Chapter 26.

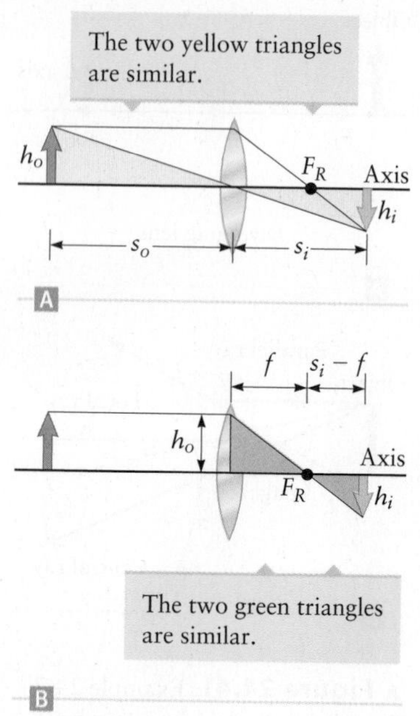

▲ **Figure 24.44** Geometry for deriving the thin-lens equation.

The Thin-Lens Equation

Figure 24.44A shows a ray diagram for a converging lens. We can use the geometry of this sketch to find a mathematical relation for locating the image. Within this figure, we identify two pairs of similar right triangles. One pair is shaded in yellow in Figure 24.44A; one yellow triangle has sides of length h_o and s_o, and the other yellow triangle has sides of length h_i and s_i. The similarity of these triangles leads to

$$\frac{h_o}{s_o} = \frac{-h_i}{s_i} \tag{24.19}$$

The negative sign here is due to our convention that h_i is negative because the image in this case points downward (extends below the axis). Another pair of similar triangles is shaded in green in Figure 24.44B and leads to

$$\frac{h_o}{f} = \frac{-h_i}{s_i - f} \tag{24.20}$$

with the same sign convention for the image height as in Equation 24.19. Taking the ratio of Equations 24.19 and 24.20 gives

$$\frac{h_o/s_o}{h_o/f} = \frac{h_i/s_i}{h_i/(s_i - f)}$$

Canceling the factors of h_o and h_i, we have

$$\frac{f}{s_o} = \frac{s_i - f}{s_i}$$

and cross multiplying, we get

$$s_i f = s_o(s_i - f)$$

We can now rearrange to obtain

$$s_i f + s_o f = s_o s_i$$

Dividing both sides by a factor of $f s_o s_i$ leads to our final result,

Thin-lens equation

$$\frac{1}{s_o} + \frac{1}{s_i} = \frac{1}{f} \tag{24.21}$$

Equation 24.21 is called the ***thin-lens equation.*** It can be used to calculate the image properties of both converging lenses and diverging lenses, provided we use the sign conventions described above.[4] Notice that the thin-lens equation is actually identical to the mirror equation (Eq. 24.15).

The thin-lens equation (Eq. 24.21) gives a relation between the object and image distances s_o and s_i, and can thus be used to find the image location. Equation 24.21 can also be used to obtain the magnification. As with mirrors, the magnification is the ratio of the image height to the object height,

$$m = \frac{h_i}{h_o} = -\frac{s_i}{s_o} \tag{24.22}$$

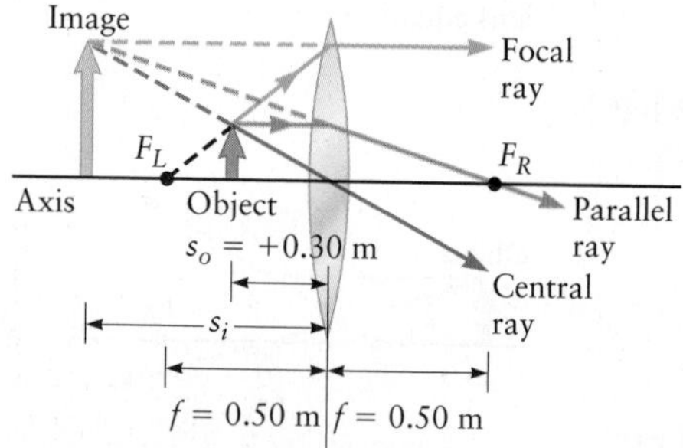

▲ **Figure 24.45** Applying the thin-lens equation to find the location of an image formed by a converging lens.

Applying the Thin-Lens Equation

Let's now see how to use the thin-lens equation together with a ray diagram to locate and describe the image formed by a lens. Consider a converging lens with $f = +0.50$ m and an object at $s_o = +0.30$ m. We have included the positive signs here to emphasize that f is positive (a converging lens) and that s_o is positive because the object is on the left side of the lens. Figure 24.45 shows our standard ray diagram and is sketched to scale. We can find the image distance from the thin-lens equation,

$$\frac{1}{s_o} + \frac{1}{s_i} = \frac{1}{f}$$

[4]As the name "thin lens" implies, "thick" lenses also exist, and their image properties are described by a different relation. We will not consider thick lenses in this book.

Solving for $1/s_i$ first, we get

$$\frac{1}{s_i} = \frac{1}{f} - \frac{1}{s_o} = \frac{s_o - f}{fs_o}$$

Rearranging to find s_i gives

$$s_i = \frac{fs_o}{s_o - f} \tag{24.23}$$

Inserting the given values of f and s_o, we obtain

$$s_i = \frac{fs_o}{s_o - f} = \frac{(0.50 \text{ m})(0.30 \text{ m})}{0.30 \text{ m} - 0.50 \text{ m}} = -0.75 \text{ m}$$

According to our sign convention, this negative value for s_i means that the image is on the left side of the lens, so it is a virtual image. This result is confirmed by the ray diagram in Figure 24.45.

CONCEPT CHECK 24.11 Image Formed by a Converging Lens

Is the image in Figure 24.45 (a) real and upright, (b) virtual and upright, (c) real and inverted, or (d) virtual and inverted?

EXAMPLE 24.8 Image Formed by a Diverging Lens

Suppose the converging lens in Figure 24.45 is replaced by a diverging lens with a focal length $f = -0.50$ m. (**a**) If the object is again at $s_o = +0.30$ m (i.e., on the left side of the lens), what is the location of the image? (**b**) Is this image real or virtual?

RECOGNIZE THE PRINCIPLE

We begin with a ray diagram, which will give us a qualitative understanding of the image. We then proceed to apply the thin-lens equation to find the actual value of s_i.

SKETCH THE PROBLEM

Figure 24.46 shows a ray diagram constructed by following our suggested procedures for ray tracing with lenses. Step 1: Figure 24.46A shows the lens and the focal points on the left and right. Step 2: Add the object to the drawing. Step 3: Figure 24.46B shows three rays. The parallel ray is initially parallel to the principal axis; after passing through the diverging lens, this ray is refracted away from the principal axis. Extrapolating this ray back to the left-hand side, we see that it passes through the focal point on the left, F_L. The focal ray is directed from the object toward the focal point to the right of the lens, F_R; this ray is refracted by the lens and then travels parallel to the principal axis. The central ray passes through the center of the lens and is not deflected. Step 4: These rays do not intersect at any point on the right side of the lens. Extrapolating them back, however, we find intersection at a common image point to the left of the lens. This ray diagram tells us much about the image and provides a check on results obtained with the thin-lens equation.

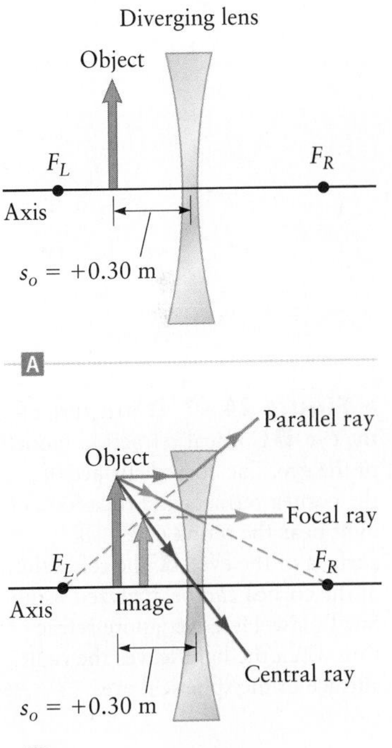

▲ **Figure 24.46** Example 24.8.

IDENTIFY THE RELATIONSHIPS AND SOLVE

(**a**) We apply the thin-lens equation (Eq. 24.21) and solve for s_i just as we did in deriving Equation 24.23:

$$s_i = \frac{fs_o}{s_o - f}$$

(*continued*) ▶

Substituting the given values of the focal length ($f = -0.50$ m) and object distance ($s_o = +0.30$ m), we get

$$s_i = \frac{fs_o}{s_o - f} = \frac{(-0.50\text{ m})(0.30\text{ m})}{0.30\text{ m} - (-0.50\text{ m})} = \boxed{-0.19\text{ m}}$$

(**b**) According to the ray diagram, light does not pass through this image point, so the image is virtual. This conclusion is confirmed by our analysis; since the image distance s_i is negative, the image is indeed on the left side of the lens and is $\boxed{\text{virtual}}$.

► *What does it mean?*

Because the focal length f for a diverging lens is negative, when s_o is positive (an object on the left side of the lens), the thin-lens equation will always give a negative value for s_i and thus a virtual image. This result is different from a converging lens, which can give either a real or a virtual image, depending on how far the object is from the lens.

Applications of the thin-lens equation are mathematically straightforward, provided you carefully follow the sign conventions in Figure 24.42. Constructing the image using ray tracing is *always* advisable, however, because it provides a check on results from the thin-lens equation and gives an intuitive understanding of the image properties.

24.6 How the Eye Works

Lenses and mirrors are used in many different kinds of optical instruments, including microscopes and telescopes (Chapter 26). In a sense, the oldest optical instrument is the eye. A cross section of a vertebrate eye is shown in Figure 24.47A; the eye is a complicated structure with several different regions, but its basic function can be understood using the principles of refraction and the properties of lenses. Light enters through the cornea, passes through a liquid region called the aqueous humor, through a lens, into a gel-like region called the vitreous humor, and then to the retina

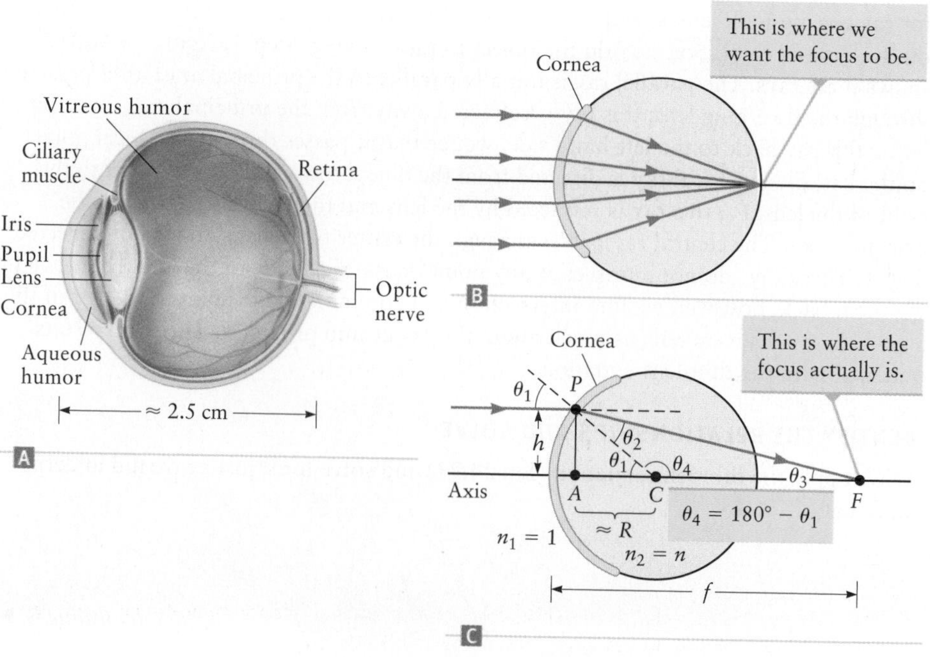

► Figure 24.47 A Structure of the eye. B Corneal refraction model of the eye. The curved surface of the cornea refracts and thus focuses light near the retina (the back surface of the eye). C The refraction at the cornea can be analyzed using Snell's law. Here we ignore refraction when the light leaves the right surface of the spherical eye.

at the back surface of the eye. The retina contains light-sensitive cells that convert light into electrical signals carried to the brain by the optic nerve.

A Simple Model of the Eye: Corneal Refraction

Different regions of the eye have different indices of refraction, so light rays are refracted and change direction as they pass from air into the cornea and then from region to region within the eye. The indices of refraction of different regions inside the eye are all in the range of approximately 1.33 to 1.40 (Table 24.1), so the biggest index change and largest refraction occur when light first enters the cornea. In the simplified model in Figure 24.47B, we approximate the eye as spherical and treat all the different parts of the eye as a single region with a single index of refraction. In this model, the only refraction occurs at the surface of the cornea, and we want to calculate how this refraction focuses light. We are especially interested in the focal length of this corneal "lens" and how close (or how far away) the focal point is to the retina. Although this illustration is certainly a simplified version of the real eye, it will still help us understand how the eye works. We'll then consider improvements to the model that make it closer to the real thing.

Figure 24.47C shows a ray diagram for refraction by the cornea. The incident angle is θ_1, the angle the incoming ray makes with the normal to the surface. Here we consider a single incident ray that initially travels parallel to the principal axis; we want to find where its refracted ray intersects the axis since that will give the location of the focal point *F*. Because we have assumed the surface of the eye is spherical, the normal direction (the dashed line through points *P* and *C*) passes through the center of the eye at *C*. For simplicity, we assume the incoming ray is near the axis, so the height *h* and the angle θ_1 are small. The refracted angle is given by Snell's law (Eq. 24.7), $n_1 \sin\theta_1 = n_2 \sin\theta_2$. The outside region is air, so $n_1 = n_{air} \approx 1.00$, and we denote the index of refraction inside the eye as simply *n*. Since we assume the angles are small, $\sin\theta_1 \approx \theta_1$ (and so forth for θ_2). We now have

$$n_1 \sin\theta_1 = n_{air} \sin\theta_1 \approx (1.00)\theta_1 = \theta_1$$

$$n_2 \sin\theta_2 = n \sin\theta_2 \approx n\theta_2$$

Inserting into Snell's law gives

$$\theta_1 = n\theta_2$$

$$\theta_2 = \frac{\theta_1}{n} \qquad (24.24)$$

To locate the focal point and find the focal length *f*, we use the geometry in Figure 24.47C to find the angle θ_3. The sum of the interior angles of triangle *PFC* is 180°, so we have

$$\theta_2 + \theta_3 + \theta_4 = \theta_2 + \theta_3 + 180° - \theta_1 = 180°$$

and thus $\theta_3 = \theta_1 - \theta_2$. Inserting the result for θ_2 from Equation 24.24 gives

$$\theta_3 = \theta_1 - \theta_2 = \theta_1 - \frac{\theta_1}{n} = \theta_1\left(\frac{n-1}{n}\right) \qquad (24.25)$$

We next consider the two right triangles *PFA* and *PCA*. From the first of these triangles, we find[5] $\tan\theta_3 = h/f$. The angle θ_3 is small, so we can also use the approximation $\tan\theta_3 \approx \theta_3$ to get

$$\theta_3 \approx \tan\theta_3 = \frac{h}{f}$$

$$h \approx f\theta_3 \qquad (24.26)$$

[5]We assume point *A* is very close to the spherical refracting surface. This approximation is good when the rays are close to the axis (as we have assumed) and *h* is small.

Applying the same approach to triangle PCA leads to

$$\theta_1 \approx \tan\theta_1 = \frac{h}{R}$$

$$h \approx R\theta_1 \tag{24.27}$$

Setting Equations 24.26 and 24.27 equal gives $f\theta_3 = R\theta_1$. The focal length of the corneal "lens" is then

$$f = R\frac{\theta_1}{\theta_3}$$

and inserting the result for θ_3 from Equation 24.25, we find

$$f = R\frac{\theta_1}{\theta_1(n-1)/n}$$

$$f = \frac{n}{n-1}R \tag{24.28}$$

Properties of the Spherical Eye

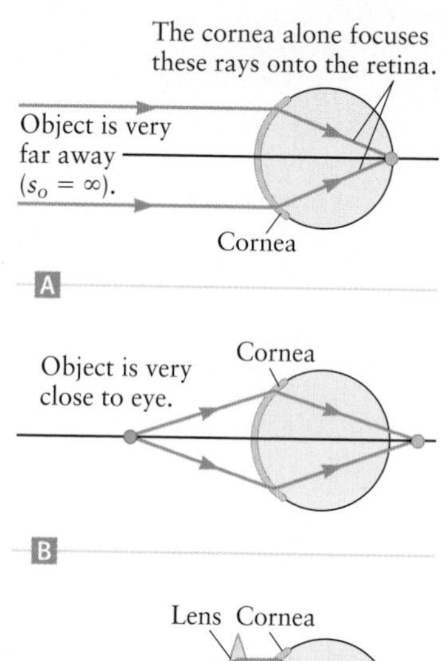

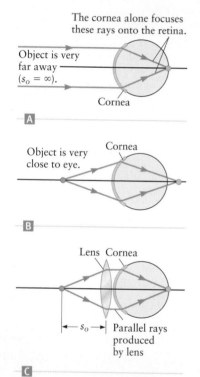

▲ **Figure 24.48** Function of the eye's lens. **A** In this model, light from a very distant object is focused by the cornea alone on the retina. **B** When an object is very close to the eye, the refracted rays do not fall on the retina, but instead intersect behind it. **C** If we place a lens in front of the eye, light from a nearby object is focused by the combination of the eye's lens and the cornea so that an image is formed at the retina.

Equation 24.28 gives the distance from the refracting surface (the cornea) to the focal point for our simplified model of the eye in Figure 24.47C. Where does this focal point lie? Using the average index of refraction of the eye, which is about $n = 1.4$, we find

$$f = \frac{n}{n-1}R = \frac{1.4}{1.4-1}R = \frac{1.4}{0.4}R = 3.5R \tag{24.29}$$

In our simple model of a spherical eye, the retina is a distance $2R$ from the front surface of the cornea, so the focal point found in Equation 24.29 is about $1.5R$ *behind* the retina.

The result above shows that this simple spherical eye would not function very well because it does not form an image at the retina; indeed, that explains why the eye actually deviates somewhat from a spherical shape. In the more realistic picture of Figure 24.47A, the cornea protrudes from the otherwise spherical profile of the eye. This protrusion makes the radius of curvature of the cornea smaller than the radius of the main spherical portion of the eye. The quantity R in Equation 24.29 should be the radius of curvature of the *cornea*, so allowing for this more realistic shape will bring the focal point closer to the retina. Our simple model has thus given some insight into why the cornea is shaped the way it is.

Refining the Model: Adding the Lens

Let's now add the eye's lens to our model. The lens is actually inside a real eye (Fig. 24.47A), and light passes through this lens after being refracted by the cornea. We will instead consider a slightly simpler model in which the lens is in front of the cornea and just outside the eye. Our model will thus not apply strictly to a real eye, but it will yield some useful insights and can be a good way to understand the combination of an eye with a contact lens. So, in our simple model, the incoming light first passes through and is partially focused by the lens and then enters the cornea, which finishes the focusing job. We now ask two questions: "What kind of lens is needed for the eye to function properly (i.e., converging or diverging)?" and "What is the shape of the lens (i.e., its radius of curvature)?"

We begin with ray diagrams for two different cases. In Figure 24.48A, the eye is focused on an object at infinity (i.e., on an object that is very far away). The incoming rays are parallel to the axis, and we assume the cornea has the correct radius of curvature so that these rays are focused on the retina. Hence, we assume the focusing for distant objects is done solely by the cornea using the refraction focusing of our first model in Figure 24.47B. The ray diagram in Figure 24.48B shows why we need the added lens. Here the object has been moved very close to the eye; the incoming rays are now diverging from the axis, and the refracted rays produced by the cornea alone intersect at a point ***behind*** the retina. We need the extra focusing of the added lens in Figure 24.48C to focus the image on the retina.

Now let's consider what properties this lens must have to make the ray diagram in Figure 24.48C work. We place the object a distance s_o in front of the lens and suppose the lens produces parallel outgoing rays (i.e., an image at infinity) as shown in the figure. We know from Figure 24.48A that the cornea by itself can focus these parallel rays on the retina. Hence, the lens should be designed such that an object at s_o focuses at infinity (so that $s_i = \infty$), producing parallel rays for the cornea. Using the thin-lens equation (Eq. 24.21) with an assumed focal length f_{lens}, we have

$$\frac{1}{f_{\text{lens}}} = \frac{1}{s_o} + \frac{1}{s_i} = \frac{1}{s_o} + \frac{1}{\infty} = \frac{1}{s_o} \tag{24.30}$$

and thus

$$f_{\text{lens}} = s_o \tag{24.31}$$

The eyes of a person with normal vision are able to focus on objects as close as $s_o = 25$ cm from each eye. The lens in Figure 24.48C must therefore have a focal length $f_{\text{lens}} = 25$ cm. Since f_{lens} is positive, the lens is a converging lens.

How must this lens be shaped to give this value for the focal length? The focal length is related to the radii of curvature of the two surfaces of a lens by a relationship[6] called the ***lens maker's formula***. For a converging lens with two convex surfaces with radii of curvature R_1 and R_2, this formula reads

$$\frac{1}{f} = (n-1)\left(\frac{1}{|R_1|} + \frac{1}{|R_2|}\right) \tag{24.32}$$

Lens maker's formula

where n is the index of refraction of the lens. Equation 24.32 applies only to the double convex lens shape in Figure 24.37 and assumes the lens is in air. More general versions of the lens maker's formula apply to other lens shapes and to situations such as a lens under water. The absolute value terms in Equation 24.32 emphasize that both these terms are positive for a double convex lens. For other lens shapes, one or both of these terms may be negative and hence describe a diverging lens with a negative value for f.

Applying the Lens Maker's Formula

Let's now apply the lens maker's formula to estimate the radius of curvature of the lens in our model of the eye in Figure 24.48C where $f_{\text{lens}} \approx 25$ cm. Although the lens maker's formula involves the radii of curvature for both sides of the lens separately, in the spirit of our approximate treatment we assume these radii are equal and $R_1 = R_2 = R$. Using this assumption in Equation 24.32 gives

$$\frac{1}{f_{\text{lens}}} = (n-1)\left(\frac{1}{|R_1|} + \frac{1}{|R_2|}\right) = \frac{2(n-1)}{R}$$

or

$$f_{\text{lens}} = \frac{R}{2(n-1)}$$

[6]The lens maker's formula can be derived by considering the refraction from the two surfaces of the lens in succession, similar to the calculation we did for our model eye in Figure 24.47C.

where we have now dropped the absolute value signs and assume R is positive. Solving for R, we find

$$R = 2(n - 1)f_{\text{lens}} \tag{24.33}$$

Inserting $f_{\text{lens}} = 25$ cm given above and using the index of refraction for the eye's lens $n = 1.41$ (Table 24.1), we have

$$R = 2(n - 1)f_{\text{lens}} = 2(1.41 - 1)(0.25\text{ m}) = 0.21\text{ m} \tag{24.34}$$

The radius of curvature of the lens is thus several times larger than the diameter of the eye (about 2.5 cm = 0.025 m; Fig. 24.47A). The front surface of the lens within the eye, which is responsible for most of the focusing, has a rather flat surface, so its radius of curvature is much larger than the radius of the eye.

In addition to allowing the eye to focus on a nearby object, the eye's lens plays another crucial role. When the object is at a particular location, the focal length of the eye's lens has a value such that the final image from the cornea is properly focused on the retina. If the object is moved closer to or farther away from the eye, the focal length of the lens must change to keep the final image in focus. This change of shape is accomplished using muscles attached to the edges of the lens (Fig. 24.47A) to pull on those edges and change the radius of curvature. The lens must be sufficiently flexible to allow its shape to change accordingly. As a person grows older, the lens becomes less flexible and the eye can no longer focus properly for objects that are both far away and very close. That is why the author wears glasses.

EXAMPLE 24.9 Designing the Lenses for a Pair of Glasses

Suppose a lens in a pair of eyeglasses has a focal length $f = 30$ cm (a typical value; see Chapter 26). If the lens is made of plastic with $n = 1.50$ and both surfaces have the same shape, what is the radius of curvature R of the lens?

RECOGNIZE THE PRINCIPLE

The focal length of a lens is related to its shape by the lens maker's formula. The purpose of this example is to find the radius of curvature for a typical lens.

SKETCH THE PROBLEM

No figure is necessary.

IDENTIFY THE RELATIONSHIPS

The lens maker's formula for a converging lens with the same radius of curvature for both surfaces is (Eq. 24.32)

$$\frac{1}{f} = (n - 1)\left(\frac{1}{|R|} + \frac{1}{|R|}\right) = \frac{2(n - 1)}{R} \tag{1}$$

where in the last step we have dropped the absolute value signs and assumed R is positive.

SOLVE

Solving Equation (1) for R and inserting the given value for f, we find

$$R = 2(n - 1)f = 2(1.50 - 1)(30\text{ cm}) = \boxed{30\text{ cm}}$$

▸ What does it mean?

This radius of curvature is so large that the surfaces of the lens will seem to be only slightly curved, which is true for most eyeglass lenses.

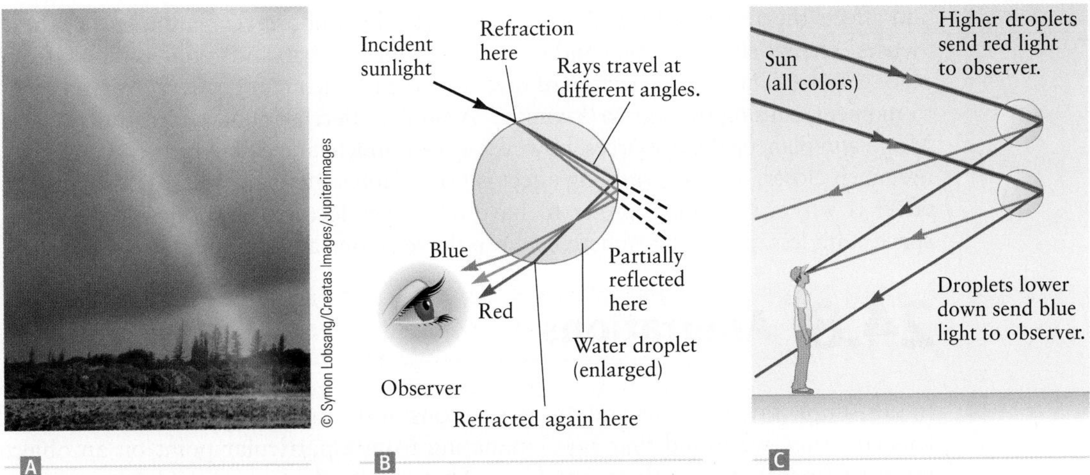

▲ **Figure 24.49** A A typical rainbow. Rainbows are caused by refraction, dispersion, and reflection of light from water droplets in the atmosphere. B When light enters a droplet, refraction and dispersion separate different colors just as in a prism. After reflecting from the back of the droplet (on the right), these rays are separated further by refraction and dispersion when they leave the droplet. They finally emerge from the droplet at different angles. C Different colors of the rainbow come from different parts of the sky. The colors of these rays *do* represent the color of the light.

24.7 Optics in the Atmosphere

Rainbows

In Section 24.3, we showed that refraction of light by a prism causes rays to change direction and that the amount light is deflected depends on the color of the light. A similar effect is responsible for rainbows (Fig. 24.49A), but in this case the refracting object is a water droplet. Water droplets in the atmosphere are illuminated by light from the Sun, and they disperse and redirect light to an observer according to the ray diagram in Figure 24.49B. An incident ray from the Sun is refracted when it enters a droplet; the refracted angles depend on the color of the light, so rays for different colors travel at different angles inside the droplet. When light reaches the back surface of the droplet, a portion is reflected, giving a set of reflected rays. (The rest is refracted and propagates out the back of the droplet.) The reflected rays are refracted when they leave the droplet; the refracted angles again depend on the color, so the final rays leave the droplet at different angles. The rays in Figure 24.49B enter droplets at one particular angle, but other rays will enter at different angles, so the outgoing rays will emerge over a range of angles. However, for each color there is one particular outgoing angle for which the outgoing rays have the greatest intensity, and this "best" outgoing angle depends on color due to dispersion. As a result, the different colors of a rainbow appear at different (angular) positions in the sky (Fig. 24.49C).

Why Stars Twinkle

The index of refraction of air depends on the pressure and temperature. Wind and other effects cause the air temperature and pressure, and hence the index of refraction, to fluctuate from place to place in the atmosphere. These fluctuations cause each small region of air to act as a lens or prism, deflecting light. As an example, Figure 24.50 shows starlight as it travels through the atmosphere and through one pocket of cool air on its way to an observer on the Earth. This pocket of air has a different index of refraction than the air around it, so rays from the star are refracted as they enter and leave this region, just as light is refracted by a lens. This "lens" of cool air thus deflects rays from the star

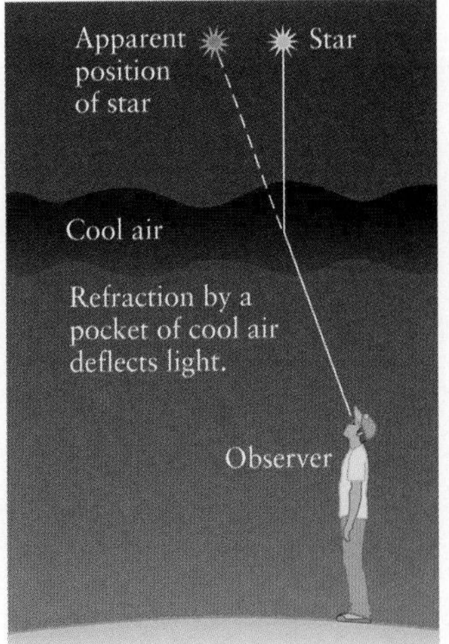

▲ **Figure 24.50** Pockets of cool air or warm air refract starlight as it travels to an observer. These pockets shift with time, so the apparent position of the star changes with time and the stars appear to "twinkle." The amount a ray is deflected is much smaller than drawn here.

and affects the apparent location of a star as seen by an observer on the Earth. These "lenses" change shape, coming and going as air—both warm and cool—is blown from place to place. Thus, the deflection of rays from a particular star changes from moment to moment, causing the star to "twinkle." A similar effect is found with light from the Moon and planets; these sources, however, appear much larger than stars (because they are much closer), so the twinkling effect is not as apparent as for stars. This twinkling effect is why astronomers prefer to have telescopes located on mountaintops or in space, with less of the atmosphere between the telescope and the star.

24.8 Aberrations

In our simple ray diagrams and our derivations of the mirror equation and the thin-lens equation, we found that rays emanating from a particular point on an object intersect (focus) at a single image point. That result relied on several assumptions, and although these assumptions are usually good approximations, they are still just approximations. The images formed by most real mirrors and lenses suffer from effects called ***aberrations***. Aberrations are deviations from the ideal images we have encountered so far; the most common is when all the light rays from a point on an object do not quite focus at a single image point.

Chromatic Aberration

Chromatic aberrations are related to the dispersion of light into different colors as in a prism. Because the index of refraction of a lens depends on the color of light, different colors are refracted by different amounts.[7] For a plane surface such as the prism in Figure 24.21, the way the rays of different colors disperse can be calculated using Snell's law (Example 24.4). For a lens, we can use the lens maker's formula to estimate the effect of chromatic aberration. According to this formula (Eq. 24.32), the focal length f depends on the index of refraction of the lens. Hence, f is different for different colors and the location of an image is different for each color as shown in Figure 24.51. Many optical instruments (Chapter 26) employ multiple lenses, and it is sometimes possible to compensate for chromatic aberration by arranging the aberrations of separate lenses to cancel each other. Even with this approach, however, most optical systems have at least a small amount of chromatic aberration.

Spherical Aberration

Since it is relatively simple to make an accurate spherical shape, spherical surfaces for mirrors and lenses are convenient. In this chapter, we have assumed the surfaces of curved mirrors and lenses are spherical and have showed that the resulting focal lengths depend on the radii of curvature. Ray diagrams such as Figures 24.26 and 24.41B rely on the assumption that the rays drawn are very close to the axis of the mirror or lens. Mathematically, this assumption allows us to make the approximations $\sin\theta \approx \theta$ and $\tan\theta \approx \theta$, and if those approximations are accurate, the rays will focus at a single image point. (Those approximations were used in derivations of the mirror equation and the thin-lens equation.) They are, however, only approximations; in any real situation, rays that are not exactly on the axis will not focus exactly at the ideal image points we found for mirrors or lenses. Such deviations are called ***spherical aberrations*** because they are due to the spherical shape of the mirror or lens.

Certain curved surfaces with nonspherical shapes give better focusing properties than those of spherical shapes. For this reason, high-precision optical components often employ parabolically shaped surfaces. Components of this shape cost more to manufacture, but they give better image properties for off-axis rays.

Blue
Red
F_{red}
F_{blue}
Red
Blue

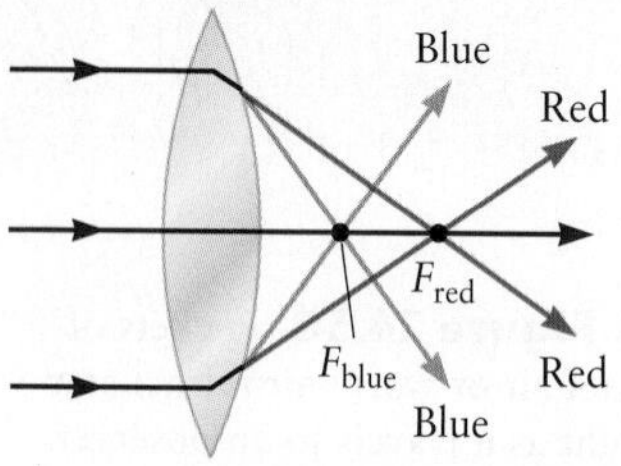

▲ **Figure 24.51** Chromatic aberration produced by a converging lens. Rays of different wavelengths (colors) focus at different points.

[7]Note that mirrors do not exhibit chromatic aberration, since dispersion is associated with refraction and not reflection.

Summary | CHAPTER 24

Key Concepts and Principles

Ray optics (also called geometrical optics)

In the *ray optics* approximation, light travels along straight lines called *rays* that change direction when light is reflected or refracted. Ray optics gives an accurate description of light propagation when light passes through openings or strikes objects that are large compared to its wavelength.

Reflection and refraction

For *reflection* and *refraction*, we measure the direction of a ray using the angle θ made by the ray with the direction normal (perpendicular) to the surface of interest.

According to the *law of reflection*, the angle of incidence is equal to the angle of reflection:

$$\theta_i = \theta_r \qquad \text{(24.1) (page 728)}$$

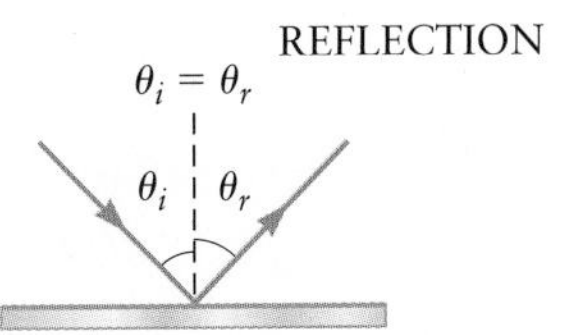

Refraction is described by *Snell's law*,

$$n_1 \sin\theta_1 = n_2 \sin\theta_2 \qquad \text{(24.7) (page 731)}$$

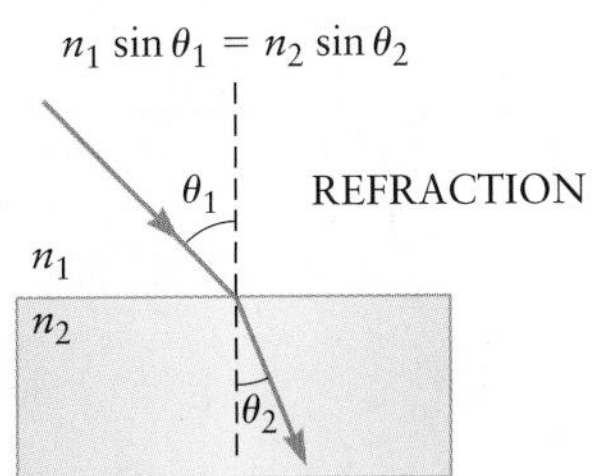

where the *index of refraction* $n = c/v$ is the ratio of the speed of light in a vacuum (c) to the speed of light in the refracting material (v). By applying the law of reflection and Snell's law, we can calculate how light interacts with plane mirrors, curved mirrors, lenses, and other refracting objects such as flat plates and water droplets.

Applications

Ray tracing

Ray tracing is the process of using light rays that emanate from an object to derive the location and other properties of an image. Important properties of an image include the following:

- *Real image*: Light from the object passes through the image point.
- *Virtual image*: Light rays do not pass through the image point. The image point is found by extrapolating rays back to a common point behind a mirror or in front of a lens.
- *Upright image*: The image has the same orientation as the object.
- *Inverted image*: The image is upside down compared with the object.
- *Enlarged image*: The image is larger than the object (magnification $|m| > 1$).
- *Reduced image*: The image is smaller than the object ($|m| < 1$).

Magnification

Magnification is the ratio of the image height h_i to the object height h_o,

$$m = \frac{h_i}{h_o} \qquad \text{(24.10) (page 739)}$$

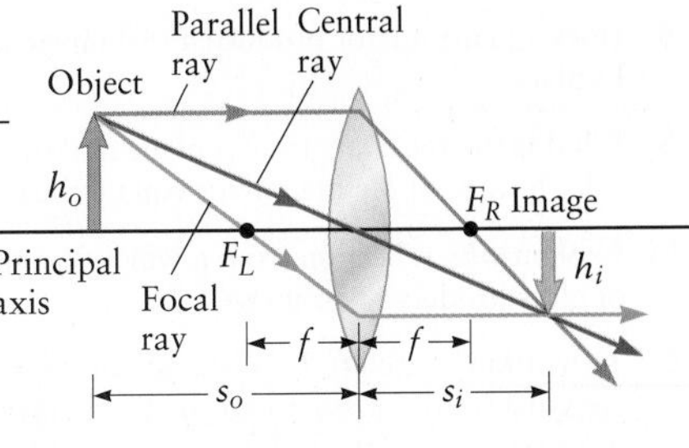

According to our *sign conventions* for h_i and h_o, the magnification can be either positive or negative. See pages 743 and 749, and Figures 24.32 and 24.42, for a description of these sign conventions.

(Continued)

Mirror equation

The *mirror equation* relates the object distance, the image distance, and the *focal length* for a curved mirror:

$$\frac{1}{s_o} + \frac{1}{s_i} = \frac{1}{f} \quad (24.15)\ (\text{page } 742)$$

For a concave spherical mirror, f is given by

$$f = \frac{R}{2} \quad (24.14)\ (\text{page } 742)$$

where R is the radius of curvature. For a convex spherical mirror, f has the same magnitude, but is negative. See Figure 24.32 for a description of the mirror sign conventions.

Thin-lens equation

The images formed by a lens are described by the *thin-lens equation*,

$$\frac{1}{s_o} + \frac{1}{s_i} = \frac{1}{f} \quad (24.21)\ (\text{page } 750)$$

where f is the focal length of the lens. For a converging lens, f is positive; for a diverging lens, it is negative. See Figure 24.42 for a description of the sign conventions for lenses.

Aberrations

Real lenses do not form perfect images. *Chromatic aberrations* are deviations due to the dependence of the refractive index on color. *Spherical aberrations* result from rays that do not propagate close to the principal axis of a lens or mirror with spherical surfaces.

Questions

SSM = answer in *Student Companion & Problem-Solving Guide* (X) = life science application

1. An angler observes a fish whose image is 10 m from the shore. Is this fish closer to or farther from the shore than its image? Explain. *Hint*: Draw a ray diagram.
2. Where is the image formed for an object in front of a plane mirror?
3. Does a plane mirror produce a real image or a virtual image? Explain.
4. What is the focal length of a plane mirror? Explain. *Hint*: Consider how to apply the mirror equation to a plane mirror.
5. Explain why reflection from a window consisting of a single pane of glass produces two images.
6. The author's satellite TV dish receives signals from a satellite in geosynchronous orbit about the Earth. (For this orbit, the period is precisely one day, so the satellite is always "at" the same point in the sky.) The signal is picked up with a receiver located 40 cm from the dish's center as shown Figure Q24.6. Explain why this "dish" is concave and estimate its radius of curvature.

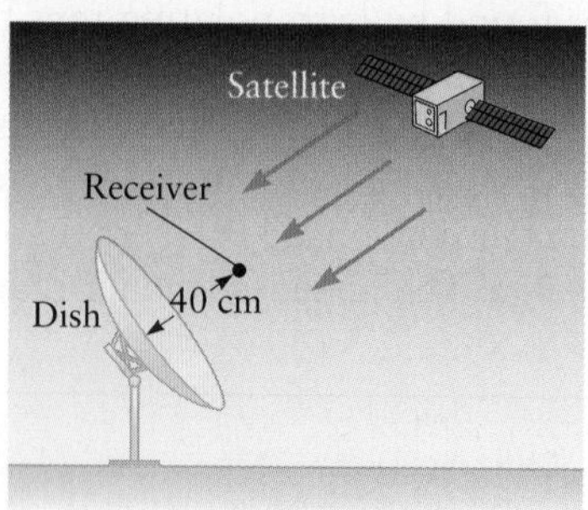

Figure Q24.6

7. When finding the image produced by a lens via ray tracing, it was suggested that you should always draw a ray that goes through the focal point (the focal ray). (In some cases, an extrapolation of the focal ray passes through the focal point.) Which of the following statements gives the most accurate description of this ray?

(a) The focal ray passes through the center of the lens.
(b) The focal ray always leaves the lens parallel to the axis of the lens.
(c) The focal ray always leaves the object parallel to the axis of the lens.
(d) The focal ray always passes through a virtual image.
(e) There are two focal rays, one for each focal point.

8. In the ray diagram in Figure Q24.8, a combination of blue ($n = 1.60$) and red ($n = 1.50$) light strikes the surface of a flat glass plate. Because of dispersion, there are two different refracted rays inside the glass. Which of the following statements correctly describes this situation? *Hint*: More than one statement applies.

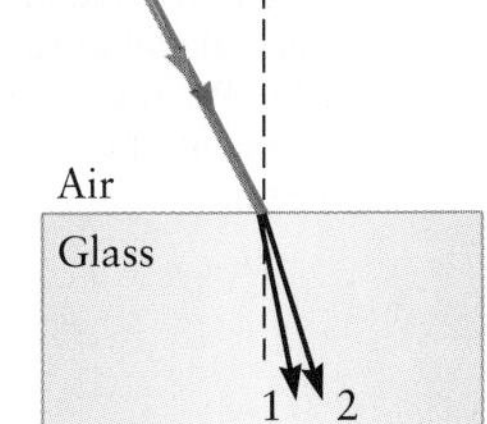

Figure Q24.8

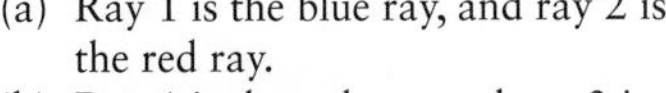

(a) Ray 1 is the blue ray, and ray 2 is the red ray.
(b) Ray 1 is the red ray, and ray 2 is the blue ray.
(c) Different rays reflect at different angles due to their color.
(d) This diagram shows an example of total internal reflection.
(e) The index of refraction in the glass is larger for red light.

9. Consider a diverging lens with an object in front of the lens. Prove that the image is always virtual.

10. Explain how a refraction experiment could be used to measure the index of refraction of a material as a function of the frequency of the light.

11. Imagine a light ray as it passes from one material to another (as in Fig. 24.11) and consider the electric and magnetic fields associated with these rays at the boundary between the materials. At (i.e., very near) the boundary, the component of the electric field directed parallel to the boundary must be the same on the two sides of the boundary. Use this fact to argue how the frequency of the light wave must change (or not change) when you cross the boundary.

12. Light travels from air into plastic. Which two of the following statements are correct? The results from Question 11 will be useful here.
(a) The frequency of the light is greater in air than in plastic.
(b) The frequency of the light is smaller in air than in plastic.
(c) The frequency of the light is the same in both air and plastic.
(d) The wavelength of the light is greater in air than in plastic.
(e) The wavelength of the light is smaller in air than in plastic.
(f) The wavelength of the light is the same in both air and plastic.

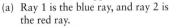

13. SSM The image formed by a plane mirror appears to be a "reverse" of the object, with left and right interchanged, as illustrated in Figure Q24.13. It is actually more accurate to think of this reversal as front to back. (a) Use a ray diagram to explain why this front-to-back reversal occurs. (b) Use your ray diagram to explain why a top-to-bottom reversal does *not* also occur.

© Image Source/Jupiterimages

Figure Q24.13

14. When you look at yourself at the center of two plane mirrors forming a 90° corner, your image is not reversed from right to left. In other words, you see yourself like others see you. Using a ray diagram, explain this observation.

15. Figure Q24.15 shows a water fountain in which many streams of water flow in curved arcs. It appears that light is also flowing along these arcs, along with the water. Explain what is actually happening.

© Robert Landau/Alamy

Figure Q24.15

16. Under water, the human eye cannot focus. Why?

17. In many convenience stores, a mirror is placed in a top corner of the store to allow the clerk to have a full view of the store. Is this mirror spherical concave or convex? Explain your answer.

18. Figure Q24.18 is a picture of a mirror one might find in a bathroom. The mirror is actually three mirrors with different curvatures so that you can view your image in different ways. The largest of the mirrors produces a normal image, whereas the smaller, two-sided mirror is used to magnify your image by two different amounts. Discuss the curvatures (plane, concave, convex) of the three mirrors required to produce these three images.

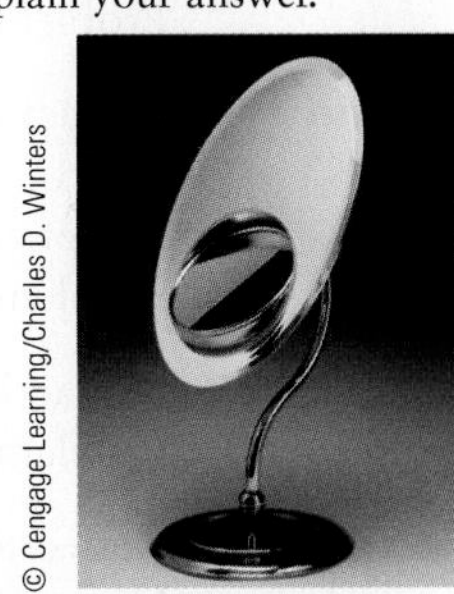
© Cengage Learning/Charles D. Winters

Figure Q24.18

19. When you observe a fish in a spherical fishbowl, the fish appears larger than it actually is. Why?

20. Funhouse mirrors distort your image so that you look shorter, taller, fatter, or thinner than you actually are. For the three images shown in Figure Q24.20, describe the curvature of the mirror.

© picture box-uk.com/Alamy

© Judity Aronson/Peter Arnold, Inc./Photolibrary

© Scott Camazine/Alamy

Figure Q24.20

21. A magnifying glass produces an enlarged image that is upright and virtual as shown in Figure Q24.21. Does it use a converging or a diverging lens? How close must you hold the magnifying glass to the object to view the image in this way?

22. Explain how a person swimming under water could observe total internal reflection and how she would know when she saw it.

23. Explain how you could measure the index of refraction of a liquid.

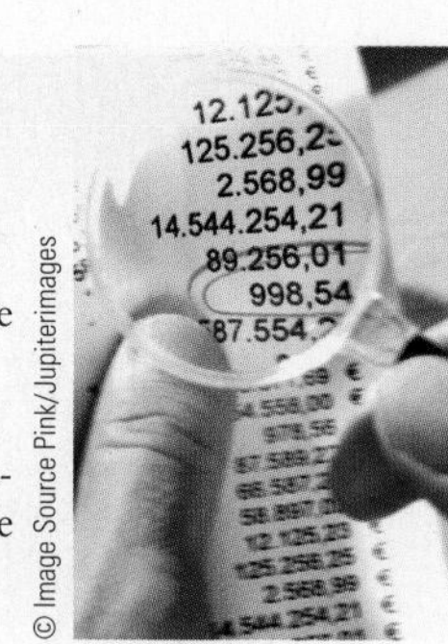

Figure Q24.21

24. Explain why dispersion is essential for producing a rainbow.

25. Explain how you could use Snell's law to measure the speed of light in a piece of plastic.

26. SSM A plastic rod becomes invisible when inserted in a container of vegetable oil as shown in Figure Q24.26. What can you conclude about the plastic rod and the vegetable oil?

Figure Q24.26

Problems available in 

Section 24.1 Ray (Geometrical) Optics
Problem 1

Section 24.2 Reflection from a Plane Mirror: The Law of Reflection
Problems 2–12

Section 24.3 Refraction
Problems 13–38

Section 24.4 Reflections and Images Produced by Curved Mirrors
Problems 39–60

Section 24.5 Lenses
Problems 61–74

Section 24.6 How the Eye Works
Problems 75–78

Section 24.7 Optics in the Atmosphere
Problems 79–82

Section 24.8 Aberrations
Problems 83–84

Additional Problems
Problems 85–110

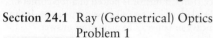

List of Enhanced Problems

Problem number	Solution in *Student Companion & Problem-Solving Guide*	Reasoning & Relationships Problem	Reasoning Tutorial in Enhanced WebAssign
24.1	✓		
24.7		✓	
24.9	✓		
24.11		✓	
24.29	✓		
24.36		✓	✓
24.37		✓	✓
24.51	✓		
24.56		✓	✓
24.74	✓		
24.78		✓	
24.79	✓		
24.82		✓	
24.88	✓	✓	
24.91	✓		
24.107		✓	✓
24.109		✓	✓
24.110		✓	✓

CHAPTER

25

► *This photograph shows two water waves undergoing interference with each other, producing regions of high and low wave intensity. Interference is the central subject of this chapter. (© 1986 Richard Megna—Fundamental Photographs)*

Wave Optics

OUTLINE

Is light a particle, or is it a wave? This question was debated for many centuries. Newton reasoned that since light travels in straight-line paths (as we found in our ray diagrams in Chapter 24), it must consist of particles moving with constant velocity. Over time, however, experiments revealed several properties of light that can only be explained with a wave theory. Maxwell's theory of electromagnetism then convinced physicists that light is an electromagnetic wave. In this chapter, we discuss properties of light that depend on its wave nature, the field known as ***wave optics***. We also show how geometrical optics—ray tracing and the focusing of light by mirrors and lenses—fits together with the wave properties of light.

The wavelength of light (λ) plays a key role in determining when we can or cannot use geometrical optics. When discussing image formation and the properties of mirrors and lenses over distances much larger than λ, geometrical optics is extremely accurate, but for all discussions of interference effects and the propagation of light through small openings (openings whose sizes are comparable to or smaller than λ), wave optics is required. Hence, in their appropriate regimes, geometrical optics and wave optics both describe light.

There is still more to this story: new properties of light were discovered only a few decades after Maxwell. These properties could not be understood with Maxwell's theory and led to the development of the quantum theory of light in which light has properties of both particles *and* waves! We'll say a little more about this quantum theory at the end of this chapter and a lot more about it in Chapter 28. The quantum theory of light builds on Maxwell's theory, so everything we learn in this chapter about the wave properties of light is true also in the quantum theory.

25.1 Coherence and Conditions for Interference

How is a wave different from a particle? How can we really prove that light is a wave phenomenon? One property is unique to waves: only waves can exhibit ***interference***. We described the interference of waves on a string and water waves in Chapter 12 and of sound in Chapter 13. In this chapter, we study the interference of light waves.

Let's first consider when light from two or more sources can and cannot exhibit interference. In particular, can any two light waves exhibit interference effects? If not, how can we tell in advance if light from two or more sources is capable of exhibiting interference? For example, can light from the Sun exhibit interference with light from a car's headlight?

We can answer these questions if we first consider the interference of sound waves emitted by two speakers as sketched in Figure 25.1A. The speakers emit sound waves of wavelength λ that come together and combine at a listener's location. If the sources emit identical waves and if the distances from the sources to the listener are equal ($L_1 = L_2$), the two waves undergo the same number of oscillations on their way to the listener. In this case, we say that the waves are ***in phase*** (Fig. 25.1B). When two waves are in phase, their maxima occur at the same time at a given point in space.

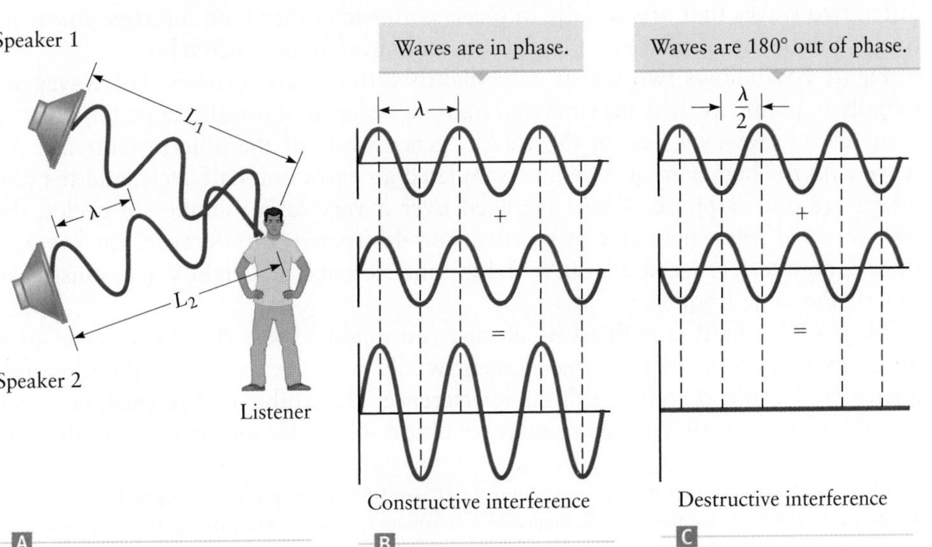

◀ **Figure 25.1** A Sound waves emitted by two separate sources (two speakers) combine when they reach a listener. These waves travel distances L_1 and L_2 on their way to the listener, where L_1 and L_2 may or may not be equal. B Here the two waves arrive in phase, and their sum is larger than either of the individual waves. C If the two waves are 180° out of phase when they reach the listener, their sum is zero, and no sound (and no energy) arrives at the listener.

The total wave displacement at the listener's location is the sum of the displacements of the two individual waves. If two waves are in phase, the sum of their displacements is large and the waves interfere constructively to produce a large amplitude and a large intensity.

Now suppose the listener moves relative to the speakers so that the distances L_1 and L_2 traveled by the two waves are no longer equal. If these distances differ by an amount $\lambda/2$, one of the waves travels an extra one-half wavelength on the way to the listener, and the maximum of one wave will coincide with the minimum of the other (Fig. 25.1C). We say that these waves are[1] ***out of phase***; in this example, the sum of the two wave displacements at the listener's location is zero and there is ***destructive*** interference.

Figure 25.1 illustrates several crucial aspects common to all interference experiments, including those involving light.

General conditions for interference. Two waves can interfere if all the following conditions are met.

1. **Two or more interfering waves travel through different regions of space over at least part of their propagation from source to destination.**
2. **The waves are brought together at a common point.**
3. **The waves must have the same frequency and must also have a *fixed phase relationship*. Thus, over a given distance or time interval the phase difference between the waves remains constant. Such waves are called *coherent*.**

Coherence

Interference conditions 1 and 2 are illustrated in Figure 25.1, but condition 3 and the notion of coherence warrants more discussion. When the ear detects a sound wave or the eye detects a light wave, it is actually measuring the wave intensity averaged over many cycles. For example, the frequency of blue light is about 7×10^{14} Hz. The eye cannot follow the variation of every cycle; instead, the eye averages the light intensity over time intervals of about 0.1 s, which corresponds to approximately 7×10^{13} cycles, and then sends this average signal to the brain. Suppose the waves in Figure 25.1 are two interfering light waves. For the waves to interfere constructively and give a large total intensity, they must stay in phase during the time the eye is averaging the intensity. Likewise, for the two waves in Figure 25.1C to interfere destructively, they must stay out of phase during this averaging time. Each scenario is possible only if the two waves have precisely the same frequency. If their frequencies differ, two waves that are initially in phase with each other (and interfere constructively) will be out of phase at a later time (and interfere destructively).

Figure 25.2 shows two waves with slightly different frequencies. The waves are initially in phase, as their maxima and minima occur at about the same times. After many cycles, however, one of the waves "gets ahead" of the other; eventually, the wave with the higher frequency has completed an extra one-half cycle, and the two waves are out of phase. When averaged over a very large number of cycles, the two waves display neither constructive nor destructive interference; on average, there is no interference at all. So, to exhibit interference, two light waves must have exactly the same frequency.

Most visible light is emitted by atoms; you might expect that light waves produced by two atoms of the same element would be "identical" with the same frequency and could therefore exhibit interference. To exhibit interference, however, two waves must satisfy an additional condition. Consider atoms in a gas that are

[1]Mathematically, we say that the waves in Figure 25.1C are 180° out of phase, corresponding to a shift of one wave relative to another by $\lambda/2$. Figure 25.1 considers waves that are either perfectly in phase or completely (180°) out of phase, but other relationships (specified by other phase angles) are also possible.

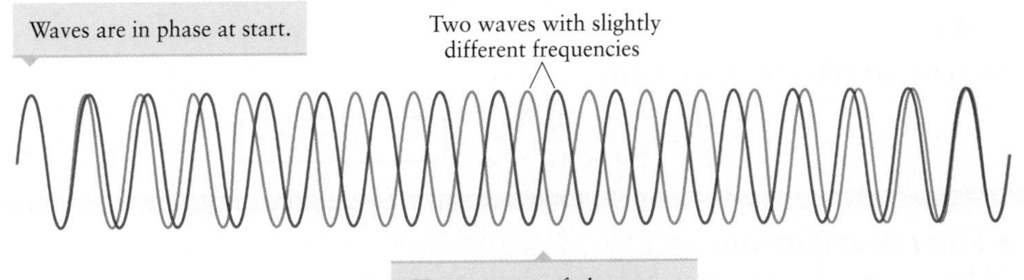
Waves are in phase at start.
Two waves with slightly different frequencies
Waves are out of phase now.

◀ **Figure 25.2** Two waves with slightly different frequencies. Initially, they are in phase, but as time passes they become out of phase. The in-phase and out-of-phase portions average out over many cycles, so on average there is no interference. *Important note*: The different colors here do *not* indicate different colors of light, but rather denote the amplitudes of the different waves.

emitting light. These atoms undergo collisions with other atoms in the gas and with the walls of the container, and each collision causes the phase of the emitted light to jump abruptly to a new value. In other words, each collision "resets" the light wave, starting its oscillation over again. These collisions are different for each atom and spoil interference effects because they cause waves that may have originally been in phase to suddenly become out of phase and vice versa. Experiments show that phase jumps in typical light waves occur about every 10^{-8} s; hence, light waves from two different atoms of a given element are not coherent. (Light from atoms of the same element within a laser is an exception; see Chapter 29.) One way to eliminate the effect of phase jumps is to derive both waves from the same source. In Section 25.2, we'll explain how to do that.

CONCEPT CHECK 25.1 Conditions for Interference

Consider an (attempted) experiment to observe interference of light from two different helium–neon lasers. The light from these lasers is emitted by neon atoms. Is it possible to observe interference with these two light sources? Explain why or why not.

EXAMPLE 25.1 Coherence Time for a Light Wave

The phase of light emitted by an atom changes (jumps) after a typical time interval $\Delta t \approx 10^{-8}$ s, called the ***coherence time***. How many cycles of a light wave occur during this coherence time? Assume the light is blue, with $f = 7.0 \times 10^{14}$ Hz.

RECOGNIZE THE PRINCIPLE

A wave's frequency is inversely related to its period T, with $f = 1/T$, and the period is the time it takes to undergo one cycle of oscillation. Given the frequency, we can find the period and from that calculate how many cycles are completed in the coherence time Δt.

SKETCH THE PROBLEM

No sketch is necessary.

IDENTIFY THE RELATIONSHIPS

Recall some basic facts about oscillations and waves from Chapters 11 and 12. The period of this light wave is (using Eq. 11.1)

$$T = \frac{1}{f} = \frac{1}{7.0 \times 10^{14}\ \text{Hz}} = 1.4 \times 10^{-15}\ \text{s}$$

which is the time required to complete one cycle of the wave. The number of cycles N contained in the coherence time Δt is thus

$$N = \frac{\Delta t}{T}$$

(continued) ▶

SOLVE

Inserting our values of Δt and T gives

$$N = \frac{\Delta t}{T} = \frac{1 \times 10^{-8}\text{ s}}{1.4 \times 10^{-15}\text{ s}} = \boxed{7 \times 10^6 \text{ cycles}}$$

▶ ***What does it mean?***

Although the coherence time Δt is very short, the frequency of visible light is very high, so there are more than one million oscillations during this brief coherence time. The precise value of Δt depends on the type of atom and its environment; the value 10^{-8} s is only a typical estimate.

25.2 The Michelson Interferometer

An optical instrument called a ***Michelson interferometer*** is based on interference of reflected waves (Fig. 25.3). This clever device, invented by Albert Michelson in the late 1800s, played an important role in the discovery of relativity (Chapter 27). The device contains two reflecting mirrors mounted at right angles. At least one of these mirrors is movable; here the mirror on the far right can be moved along the horizontal axis. A third partially reflecting mirror called a "beam splitter" is mounted at a 45° angle relative to the other two. The beam splitter reflects half the light incident on it and lets the other half pass through.

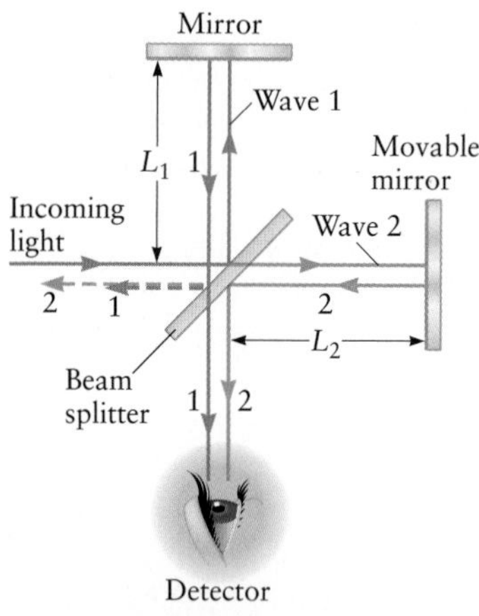

▲ **Figure 25.3** Schematic diagram of a Michelson interferometer. Light is incident from the left and split into two separate waves (beams) by partial reflection at the beam splitter in the center. These waves reflect from the mirrors at the top and right and arrive back at the beam splitter, where they recombine and travel together to the detector at the bottom. (Waves 1 and 2 also recombine and travel back to the left, but these beams are not used in the experiment.) The colors of the rays do *not* indicate the color of the light.

Light incident from the left in Figure 25.3 strikes the beam splitter and is divided into two waves. One of these waves (denoted 1) travels to the mirror at the top where it is reflected and then returns to the beam splitter. When it reaches the beam splitter, it is again split into two waves, one of which travels downward in Figure 25.3 to the detector as wave 1. The other wave from the beam splitter (wave 2) travels to the mirror on the right, where it is also reflected back to the beam splitter. When wave 2 returns to the beam splitter, a portion is reflected to the detector at the bottom of the figure. Waves 1 and 2 thus combine at the detector; they interfere after following very similar paths as they travel through the interferometer. First, both waves are reflected once by the beam splitter. Second, both waves pass through the beam splitter once. Third, both waves are reflected once by a mirror. The *only* difference between the two waves is that they travel different distances between their respective mirrors and the beam splitter. After the waves are created by the beam splitter, wave 1 travels a round-trip distance $2L_1$ and wave 2 travels a round-trip distance $2L_2$ before they recombine. The implications for interference are shown in Figure 25.4. The ***path length difference*** is

$$\Delta L = 2L_2 - 2L_1 \tag{25.1}$$

If we assume $L_2 > L_1$, wave 2 travels an extra distance ΔL and undergoes a number of extra oscillations before it reaches the detector. Each oscillation occurs over a distance equal to the wavelength λ, so the number of extra oscillations made by wave 2 is

$$N = \frac{\Delta L}{\lambda} \tag{25.2}$$

If N is an integer, then N complete oscillations fit into the extra path length and the two waves are in phase when they recombine, producing constructive interference. On the other hand, if N is half integral (i.e., if $N = \frac{1}{2}, \frac{3}{2}, \ldots$), wave 2 travels an "extra" one-half wavelength relative to wave 1 and the maxima of one wave will coincide with the minima of the other as in Figure 25.1C, giving destructive interference. The interference conditions for a Michelson interferometer are thus

$$\Delta L = m\lambda \text{ (constructive interference)}$$

$$\Delta L = (m + \tfrac{1}{2})\lambda \quad \text{(destructive interference)} \tag{25.3}$$

where m is an integer (a value 0, ±1, ±2, etc.). If the interference is constructive, the light intensity at the detector is large, whereas if the interference is destructive, the detector intensity is zero. Figure 25.5 shows how the light intensity at the detector varies as a function of ΔL. The places on the intensity curve where the interference intensity is greatest are called bright "fringes," and the places where the intensity is zero are called dark "fringes."

The Michelson interferometer uses reflection to satisfy the general requirements for interference from Section 25.1. First, reflection at the central beam splitter produces two separate waves that travel through different regions of space. Second, these waves are brought back together by the beam splitter so that they recombine at the detector. Third, because they are produced from a common incident wave, the two interfering waves are coherent.

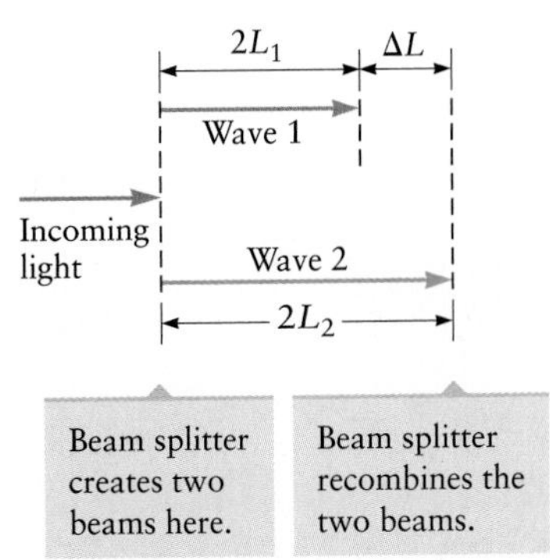

▲ **Figure 25.4** A Michelson interferometer with the waves redrawn. The two waves travel distances $2L_1$ and $2L_2$, respectively, as they travel between the beam splitter and the mirrors. The path length difference ΔL leads to interference; compare with Figure 25.1A.

Using a Michelson Interferometer to Measure Length

The wavelengths of light emitted by various sources are known. For example, a helium–neon (He–Ne) laser emits light with a wavelength of approximately $\lambda_{\text{He-Ne}} = 633$ nm. Recall that 1 nm $= 1 \times 10^{-9}$ m and that the wavelength of visible light is in the range of about 400 nm to 750 nm (about one thousand times less than the thickness of a human fingernail).

Knowing the value of $\lambda_{\text{He-Ne}}$, an experimenter can use a Michelson interferometer to measure a displacement in the following way. Using light from the laser, the mirror on the far right in Figure 25.3 is adjusted to give constructive interference and its position is noted; this corresponds to one of the bright fringes in Figure 25.5. The mirror on the far right is then moved horizontally, changing the path length difference ΔL, and the intensity at the detector moves along the curve in Figure 25.5. The intensity changes from a high value to zero and back to a high value every time the path length difference ΔL changes by one wavelength of the light. According to Equation 25.3, moving the mirror from one bright fringe to the next bright fringe corresponds to starting at the condition for constructive interference with $\Delta L = m\lambda_{\text{He-Ne}}$ and increasing the path length difference to $\Delta L = (m + 1)\lambda_{\text{He-Ne}}$. If the interferometer mirror is moved so as to pass through N cycles from bright to dark and back to bright, the change in the path length difference is

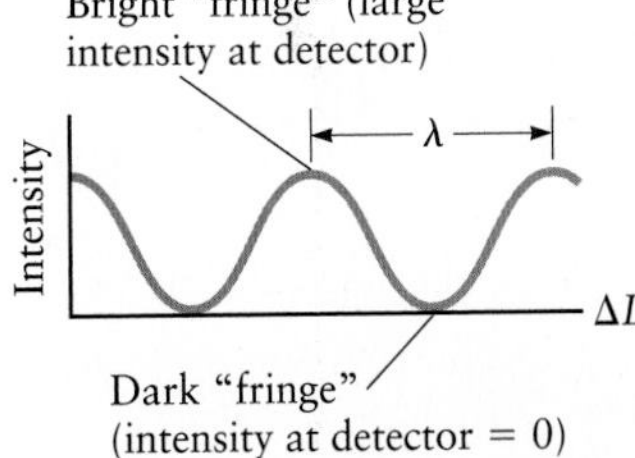

▲ **Figure 25.5** The light intensity at the detector of a Michelson interferometer oscillates as a function of the path length difference $\Delta L = 2L_2 - 2L_1$, which is varied by moving one or both mirrors.

$$(\Delta L)_{\text{change}} = N\lambda_{\text{He-Ne}} \tag{25.4}$$

When the mirror moves a distance d, the distance traveled by the light changes by $2d$ because the light travels back and forth between the beam splitter and the mirror (Fig. 25.3). Hence, if the mirror moves through N bright fringes, the distance d traveled by the mirror satisfies

$$(\Delta L)_{\text{change}} = 2d = N\lambda_{\text{He-Ne}}$$

so that

$$d = \frac{N}{2}\lambda_{\text{He-Ne}} \tag{25.5}$$

The accuracy of this measurement of d depends on the accuracy with which the wavelength $\lambda_{\text{He-Ne}}$ is known. For this reason, physicists have devoted a lot of effort to measuring the wavelength of light from certain light sources. In fact, the wavelength produced by a specially constructed helium–neon laser is known to be

$$\lambda_{\text{He-Ne}} = 632.99139822 \text{ nm} \tag{25.6}$$

Suppose this light source is used in a Michelson interferometer and one of the mirrors is moved a distance d such that exactly N = 1,000,000 bright fringes are counted.

According to Equation 25.5, we have $d = N\lambda_{\text{He-Ne}}/2$. Since N can be counted directly and the wavelength in Equation 25.6 is also known with high accuracy, this approach provides a very precise way to measure length. Moreover, people in different laboratories can use helium and neon to construct lasers that produce the wavelength given in Equation 25.6, so they can compare their separate length measurements with high accuracy. The Michelson interferometer thus makes possible a very convenient standard for the measurement of length.

CONCEPT CHECK 25.2 Analyzing the Michelson Interferometer

A Michelson interferometer using light from a helium–neon laser is adjusted to give an intensity maximum at the detector. One of the mirrors is then moved a very short distance so that the intensity changes to zero. How far was the mirror moved?

(a) $\lambda_{\text{He-Ne}}/4$ (b) $\lambda_{\text{He-Ne}}/2$ (c) $\lambda_{\text{He-Ne}}$ (d) $2\lambda_{\text{He-Ne}}$

EXAMPLE 25.2 Applications of a Michelson Interferometer: Detecting Gravitational Waves

An experiment called the LIGO (for *l*aser *i*nterferometer *g*ravitational wave *o*bservatory) is designed to detect very small vibrations associated with gravitational waves that arrive at the Earth from distant galaxies. The LIGO experiment involves several Michelson interferometers; in one of these interferometers, the mirrors are placed a distance $L_2 = 4$ km (kilometers) from the beam splitter. Suppose changes in the interference pattern corresponding to 0.1% (1/1000) of a cycle in Figure 25.5 can be detected. What is the change in the mirror–beam splitter distance that can be detected by the LIGO experiment? Assume the light used in LIGO has wavelength $\lambda = 500$ nm.

RECOGNIZE THE PRINCIPLE

For a general interference experiment, the path length difference ΔL must change by one wavelength λ to go from a bright fringe to a dark fringe and then back to a bright fringe. A 0.1% change of a fringe thus corresponds to a $\lambda/1000$ change in ΔL. When a mirror of a Michelson interferometer moves a distance d, the path length difference changes by $2d$. We can first find the value of d that gives a 0.1% fringe change and from that get the change of the mirror–beam splitter distance.

SKETCH THE PROBLEM

Figures 25.3 through 25.5 show a Michelson interferometer and the interference conditions.

IDENTIFY THE RELATIONSHIPS

The change in the path length difference equals $2d$, so each complete intensity cycle from bright to dark to bright in Figure 25.5 corresponds to moving the mirror a distance $\lambda/2$. If we can detect 1/1000 of this change, the corresponding displacement of the mirror is

$$d = \frac{\lambda/2}{1000} = \frac{\lambda}{2000} \tag{1}$$

SOLVE

Inserting the given value of the wavelength into Equation (1), we get

$$d = \frac{\lambda}{2000} = \frac{(500 \times 10^{-9}\text{ m})}{2000} = \boxed{2.5 \times 10^{-10}\text{ m}}$$

Insight 25.1

HIGH-PRECISION INTERFEROMETRY

The LIGO interferometers can actually do much better than we have assumed in Example 25.2. In fact, the real LIGO interferometers can detect mirror movements more than one million times smaller than our result for *d*!

The goal of the LIGO experiment is to detect the very tiny mirror movements physicists think will be caused by gravitational waves that originate outside our galaxy. It is believed that supernovas (exploding stars) produce rapid motion of large amounts of mass, resulting in gravitational waves. When these waves reach the Earth, they should deform the Earth, leading to movement of the LIGO mirrors that can be detected through changes in the interference pattern. By using a long distance between the beam splitter and the mirrors, the LIGO interferometers are sensitive to *very* small percentage changes in that distance.

▶ What does it mean?
The value of d found here is 0.25 nm—approximately the diameter of an atom—which is quite impressive. This result also shows that when adjusting the mirrors of a Michelson interferometer to actually observe interference, an experimenter must be able to move the mirrors with great precision. Such precision can be accomplished with carefully designed screw adjustments (a low-tech approach, but there are other ways, too).

25.3 Thin-Film Interference

Soapy water is normally transparent and colorless, but the photo of a thin soap film in Figure 25.6 shows very bright colors resulting from interference of reflected waves from the film's two surfaces. The colorful bands are called "fringes" and correspond to the locations of constructive and destructive interference for light waves with different wavelengths (colors) in the incident light.

To understand where these colors come from, consider Figure 25.7A, in which a thin soap film rests on a flat glass surface. For simplicity, assume the incident light is monochromatic, that is, that it has a single wavelength. (Later we'll consider the behavior with white light, which is light containing many different colors.) The upper surface of the soap film in Figure 25.7A is similar to the beam splitter in the center of the Michelson interferometer, reflecting part of the incoming light (ray 2) and transmitting the rest into the soap layer after refraction at the air–soap interface. This transmitted light is partially reflected at the bottom surface, producing the wave that travels back into the air as ray 3. The two outgoing rays denoted as 2 and 3 meet the conditions for interference: they travel through different regions (one travels the extra distance through the soap film), they recombine when they leave the film as parallel rays, and they are coherent because they originated from the same light source and initial wave.

To determine whether these interfering waves are in phase or out of phase with each other, we must consider what happens in the extra distance traveled by ray 3. For simplicity, let's assume the incident and reflected rays are all approximately normal (perpendicular) to the film (Fig. 25.7B). For a film of thickness d, the extra distance traveled by ray 3 is just $2d$. We must next account for the index of refraction

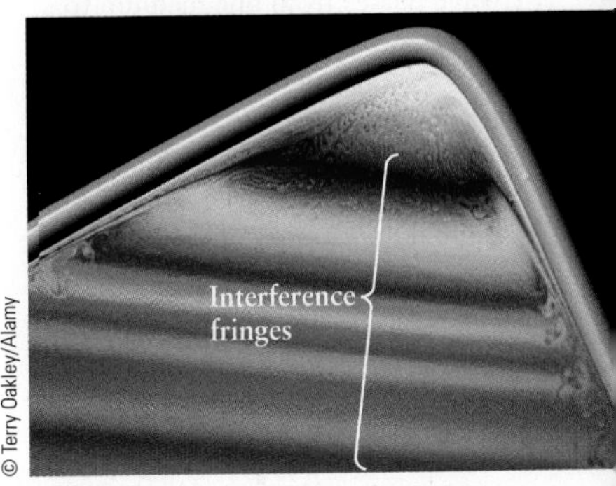

▲ **Figure 25.6** When a soap film is viewed with white light (as from the Sun), one observes colored interference fringes (colored bands). These colors are due to constructive interference; different colors (different wavelengths) give constructive interference for different values of the film thickness and for different viewing angles.

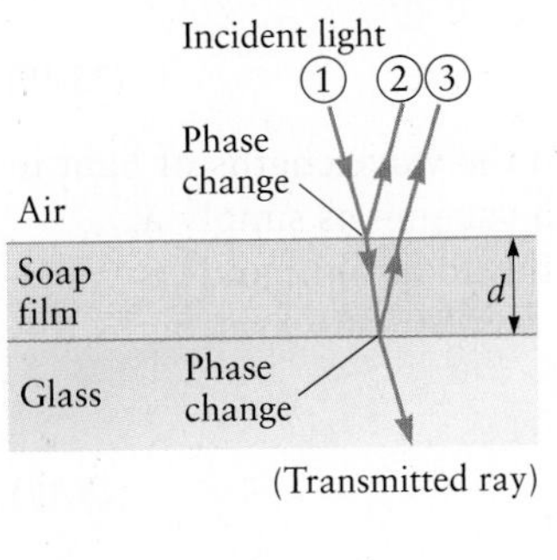

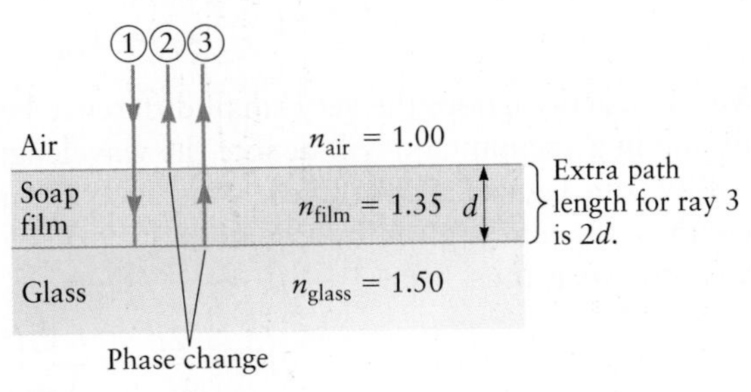

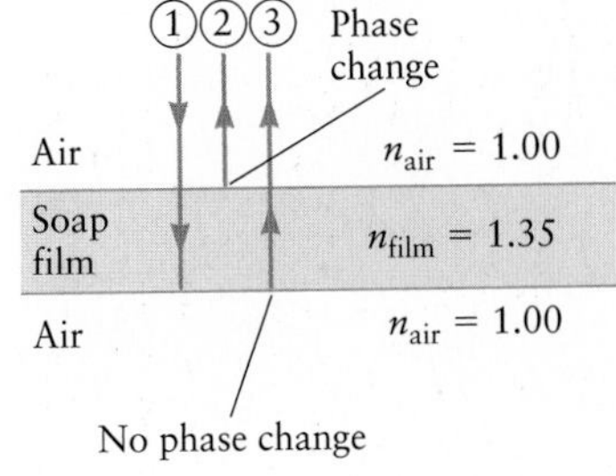

▲ **Figure 25.7** Thin-film interference. **A** Incident light (beam 1) is partially reflected at the air–soap film interface. The refracted light enters the film and is partially reflected at the bottom surface. Beams 2 and 3 have traveled different distances and can interfere either constructively or destructively, depending on the thickness d of the soap film. Beams 2 and 3 both undergo a 180° phase change when they are reflected (Fig. 25.9). Some light is also transmitted through the bottom surface of the glass. **B** Same as part A but with the incident and reflected rays normal to the film. **C** When the bottom layer is air, there is again interference of the two reflected beams, but now there is no phase change of beam 3 when it reflects at the bottom surface.

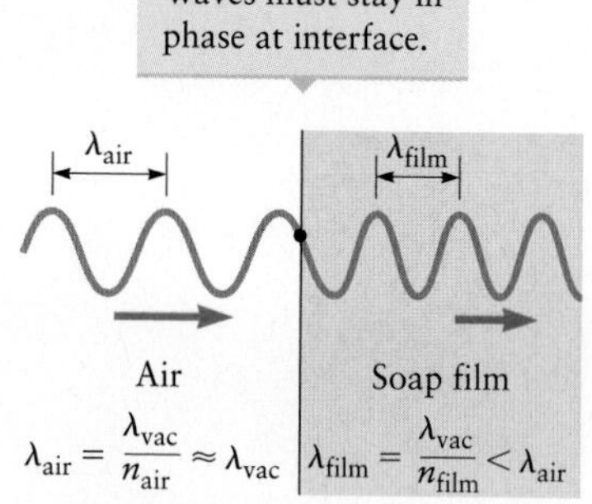

▲ **Figure 25.8** When a light wave passes from one medium to another (e.g., from air on the left into a soap film on the right), the waves on the two sides must stay in phase at the interface. That can only happen if the frequency is the same on both sides of the interface.

of the soap film. Recall that light propagates in a vacuum at speed $c = 3.00 \times 10^8$ m/s. The frequency f and wavelength $\lambda = \lambda_{vac}$ are related by

$$\lambda_{vac} f = c \tag{25.7}$$

When light travels through a substance, its speed is reduced to $v = c/n$, where n is the index of refraction of the substance (Chapter 24). A soap film is mostly water and its index of refraction is about $n_{film} = 1.35$, so light slows down inside the soap film by 35% relative to c. The product of the wavelength inside the film λ_{film} and the frequency inside the film f_{film} equals the wave speed. (Compare with the case in vacuum, Eq. 25.7.) We then have

$$\lambda_{film} f_{film} = \frac{c}{n_{film}} \tag{25.8}$$

Comparing with Equation 25.7, we see that either the wavelength or the frequency (or both) must change when light travels from a vacuum into the film. It turns out that the wavelength changes but the frequency does *not*. This can be seen from Figure 25.8, which sketches how the electric field associated with a light wave varies as the wave travels from air into a film. The fields just outside and just inside the film must stay in phase; otherwise, the wave fronts would "pile up" at the boundary. This is similar to the propagation of a wave on a string that travels between two strings of different diameters; the knot between the two strings then corresponds to the interface between air and the soap film in Figure 25.8. The strings on both sides of the knot must move with the same frequency; otherwise, the knot would come apart. Hence, the wave must have the same frequency in both strings.

Since the frequency of light does not change when it enters the soap film in Figure 25.7, its wavelength must change. Inserting $f_{film} = f$ (the frequency in vacuum) into Equation 25.8, the wavelength in the film is

$$\lambda_{film} = \frac{v}{f} = \frac{c/n_{film}}{f}$$

Using Equation 25.7, this leads to

$$\lambda_{film} = \frac{\lambda_{vac}}{n_{film}} \tag{25.9}$$

and the wavelength of light inside the film is thus shorter than its wavelength in a vacuum. The same is true for light traveling in air, with $\lambda_{air} = \lambda_{vac}/n_{air}$. Since $n_{air} = 1.0003$, the wavelength of light in air is very close to its wavelength in a vacuum:

$$\lambda_{air} \approx \lambda_{vac} \tag{25.10}$$

We can usually ignore the very small difference between the wavelengths of light in air and in a vacuum and will denote the wavelength in a vacuum as simply λ.

Ray 3 in Figure 25.7B travels a distance $2d$ inside the soap film, and its wavelength in the film is given by Equation 25.9. The extra number of wavelengths this wave travels is thus

$$N = \frac{2d}{\lambda_{film}} = \frac{2d}{\lambda/n_{film}} \tag{25.11}$$

The value of N will help determine how rays 2 and 3 interfere. Before we can say if the interference is constructive or destructive, however, we must consider how the reflections at the surfaces of the soap film affect the waves.

Effect of Reflection on the Phase of a Light Wave

The result in Equation 25.11 is analogous to what we found for the extra number of cycles in one of the interfering waves in a Michelson interferometer (Eq. 25.2).

We might therefore expect similar conditions for interference as for a Michelson interferometer. However, we also must consider one additional piece of physics. In Chapter 12, we found that under certain conditions a reflected wave is inverted relative to the incident wave; this happens for waves on a string (Fig. 12.18) when the reflecting end of the string is tied to a rigid wall. Similarly, when a light wave reflects from a surface it *may* also be inverted, corresponding to a phase change of 180° as illustrated in Figure 25.9A. This phase change on reflection is found whenever the index of refraction on the incident side (n_1) is less than the index of refraction on the opposite side (n_2). On the other hand, if the index of refraction is larger on the incident side (Fig. 25.9B), the reflected light wave is not inverted and there is no phase change.

Phase change on reflection

In a thin-film interference experiment, we must take account of the *total* phase difference between the two interfering light waves; the total phase difference includes the phase difference due to the extra distance traveled by ray 3 plus phase changes that occur due to reflection. In Figure 25.7B, we have a soap film on top of a glass layer; to three significant figures, the indices of refraction are $n_{air} = 1.00$, $n_{film} \approx 1.35$, and $n_{glass} \approx 1.50$. These values are typical and will vary slightly depending on the color of the light and the composition of the film and glass, but for a soap film on glass we'll always have $n_{air} < n_{film} < n_{glass}$. Hence, there are phase changes for both reflections at the soap film interfaces. The wave reflected at the upper surface of the film (wave 2) undergoes a 180° phase change because $n_{air} < n_{film}$, and the wave reflected at the bottom surface (wave 3) undergoes a 180° phase change because $n_{film} < n_{glass}$. Both interfering waves undergo the same phase change due to reflection, so their *relative* phase is not affected by the reflections and the nature of the interference is determined only by the extra path length $2d$ traveled by wave 3. Hence, we finally arrive at interference conditions similar to those found with the Michelson interferometer, with the only difference being the extra factor of n_{film} due to the change in wavelength inside the film.

Thin-Film Interference Case 1: When Both Waves Reflected by a Thin Film Undergo a Phase Change. In this case, the number of extra cycles traveled by the ray inside the film completely determines the nature of the interference. If the number of extra cycles N equals an integer, there is constructive interference; if N is a half integer, there is destructive interference. Using Equation 25.11, we have

$$2d = \frac{m\lambda}{n_{film}} \quad \text{(constructive interference)}$$

$$2d = \frac{(m + \frac{1}{2})\lambda}{n_{film}} \quad \text{(destructive interference)} \tag{25.12}$$

These interference conditions apply whenever the indices of refraction are related as in Figure 25.7B, with $n_{air} < n_{film} < n$ (substance below film).

Thin-Film Interference Case 2: When Only One Wave Reflected by a Thin Film Undergoes a Phase Change. A different case, with different interference conditions, is shown in Figure 25.7C. Here there is air above and below the soap film, as is the situation in a soap bubble. When ray 1 reflects from the top surface, there is a phase change because $n_{air} < n_{film}$, so the index of refraction is smaller on the incident side of the interface. For the reflection at the bottom surface of the film, however, there is no phase change because now the index is larger on the incident side. One of the interfering waves (2) thus undergoes a phase change due to reflection, and the other (3) does not. Figure 25.1C shows that a 180° phase change is equivalent to a shift of the wave by $\lambda/2$. This is equivalent to inserting an extra half cycle into the wave, which changes the conditions for achieving constructive interference and destructive interference.

The reflected wave undergoes a 180° phase change when $n_1 < n_2$.

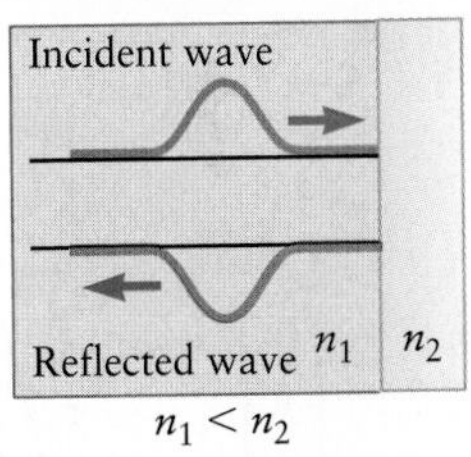

There is no phase change when $n_1 > n_2$.

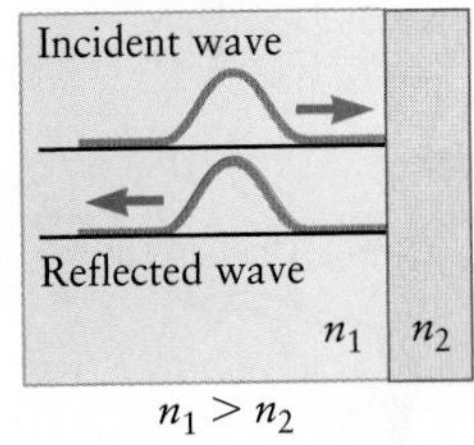

▲ **Figure 25.9** When light strikes an interface at which the index of refraction changes, part of the wave is reflected even when both substances are transparent. **A** When the index of refraction n_1 of the first substance is less than the index of refraction n_2 of the second substance, the phase of the reflected wave changes by 180° and the reflected wave is inverted. **B** On the other hand, if $n_1 > n_2$, the phase of the reflected wave does not change and the wave is not inverted.

© Science & Science/Alamy

▲ **Figure 25.10** The colors of many insect wings are due to thin-film interference. For a wing of a particular thickness, constructive interference and hence a bright interference fringe occurs for only one or a few values of the wavelength, giving the wing its color when viewed in reflected light.

Now only one of the interfering waves (wave 2 in Fig. 25.7C) undergoes a 180° phase change due to reflection, and wave 3 must travel an extra half wavelength to get back in phase with wave 2. The interference conditions are thus

$$2d = \frac{(m + \frac{1}{2})\lambda}{n_{\text{film}}} \quad \text{(constructive interference)}$$

$$2d = \frac{m\lambda}{n_{\text{film}}} \quad \text{(destructive interference)} \qquad (25.13)$$

Thin-Film Interference with White Light

Thin-film interference is responsible for the colors often seen in thin layers of water or oil on flat pavement and in free-standing films (Fig. 25.6). This effect also gives the bright colors found on the wings of many insects and birds (Fig. 25.10). Those cases usually involve sunlight (so-called white light), whereas in the situation analyzed in Figure 25.7 we assumed light had a single wavelength. We can understand what happens with white light using Figure 25.11A, which shows light of just two different colors incident on a thin film. We also assume the condition for constructive interference is satisfied for blue light at normal incidence (on the left in the figure). The wavelength of red light is longer than the wavelength of blue light, so for the same angle of incidence the condition for constructive interference will usually *not* be satisfied for red light at the same time. However, if the angle of incidence θ_i for red light is increased (on the right in Fig. 25.11A), the path length difference between the two light waves increases. It is therefore possible to have constructive interference with red light, but *for a different angle of incidence* than for blue light.

When an observer looks at a soap film, white light from the sky illuminates the film over a range of angles (Fig. 25.11B). Each different color component of the sunlight can exhibit constructive interference for a different angle of incidence, giving a series of interference fringes with different colors and producing brightly colored fringes even with a soap film of constant thickness.

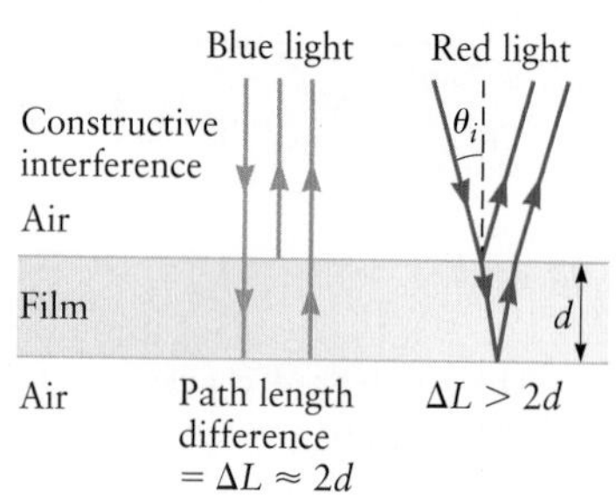

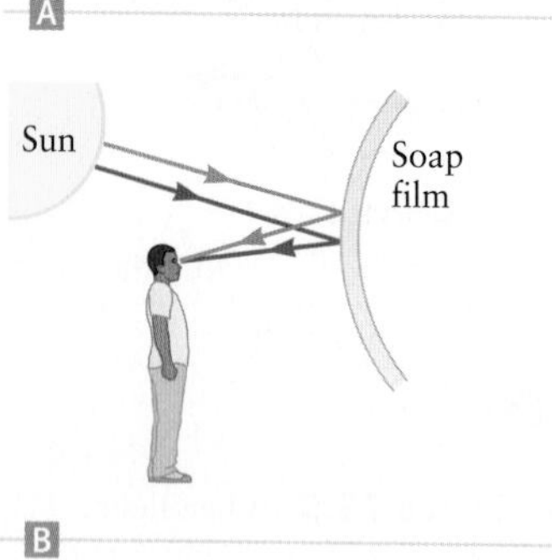

▲ **Figure 25.11** A On the right, light is incident on a thin film at an angle away from the normal direction. The path length difference between the reflected waves is greater than for light at normal incidence (on the left). B When you view a soap film (perhaps as part of a soap bubble) with reflected sunlight, different colors give constructive interference at different angles, producing the "colored" interference fringes in Figure 25.6.

EXAMPLE 25.3 Color and Thickness of a Soap Film

Consider a bubble formed from a thin, soapy film that looks blue when viewed at normal incidence. Estimate the thickness of the film. Assume its index of refraction is $n_{\text{film}} = 1.35$ and blue light has a wavelength $\lambda_{\text{blue}} = 400$ nm. Also assume the film is so thin that thinner films are *not* able to give constructive interference.

RECOGNIZE THE PRINCIPLE

The blue color of the soapy film comes from interference of waves reflected from the two film surfaces as sketched in Figure 25.7C. The wave reflected from the top surface (ray 2) undergoes a phase change on reflection because the index of refraction on the incident side (air) is less than the index of the film. The wave reflected from the bottom surface (ray 3) does not undergo a phase change on reflection because the index on the incident side (the film) is greater than the index of the opposite side (air). (See Fig. 25.9.)

SKETCH THE PROBLEM

Figure 25.7C describes thin-film interference for a soap film with air on both sides.

IDENTIFY THE RELATIONSHIPS

The interference conditions that apply with a phase change at only one of the reflections are given by Equation 25.13. Since the film has a blue color, there must be constructive interference for blue light of the given wavelength $\lambda_{\text{blue}} = 400 \text{ nm} = 4.0 \times 10^{-7}$ m.

Applying Equation 25.13 for constructive interference, we have

$$2d = \frac{(m + \frac{1}{2})\lambda_{blue}}{n_{film}} \quad (1)$$

where m is an integer that can have a value 0, 1, 2, ... Equation (1) thus has many solutions for the film thickness d, depending on the value of m. How can we determine m? The problem states that thinner films do not give constructive interference. The smallest possible value of d corresponds to $m = 0$, so that case must apply for the film.

SOLVE

Using $m = 0$ in Equation (1), we find

$$2d = \frac{(m + \frac{1}{2})\lambda_{blue}}{n_{film}} = \frac{\lambda_{blue}}{2n_{film}}$$

Solving for d and inserting values for the wavelength and index of refraction gives

$$d = \frac{\lambda_{blue}}{4n_{film}} = \frac{4.0 \times 10^{-7}\ \text{m}}{4(1.35)} = 7.4 \times 10^{-8}\ \text{m} = \boxed{74\ \text{nm}}$$

► ***What does it mean?***

This value is about five hundred times smaller than the diameter of a human hair. It is amazing that a soap film this thin can be strong enough to form a bubble.

EXAMPLE 25.4 Measuring the Diameter of a Thin Fiber

Figure 25.12 shows how thin-film interference can be used to measure the diameter of a thin fiber such as a strand of hair. Two glass plates are in contact at one end but are held apart at the other end by the fiber. Light incident from above is reflected at the top and bottom of the "film" of air between the plates, producing interference. When viewed at normal incidence and with a small wedge angle θ, the path length difference between the interfering light waves is equal to twice the plate separation. This separation increases from left to right in Figure 25.12A, so the interference alternates between constructive and destructive. The resulting interference pattern is sketched as a function of position x along the plates in Figure 25.12B. The light intensity is zero at the left end and passes through nine bright fringes (places where the intensity is a maximum), and the intensity at the right end next to the fiber is again zero. Find the diameter of the fiber. Assume the wavelength of the light is $\lambda = 600$ nm.

RECOGNIZE THE PRINCIPLE

The interference involves waves like ray 2 that reflect at the glass–air interface and waves like ray 3 that reflect at the air–glass interface. Ray 2 does not undergo a phase change at this reflection because $n_{air} < n_{glass}$, similar to the reflection from the bottom surface in Figure 25.7C. Ray 3 undergoes a 180° phase change on reflection, however, as found for the waves in Figure 25.7B. Hence, Equation 25.13 gives the correct interference conditions for this case.

SKETCH THE PROBLEM

Figure 25.12 shows the problem.

IDENTIFY THE RELATIONSHIPS

The condition for destructive interference in this case is (Eq. 25.13)

$$2d = \frac{m\lambda}{n_{film}} \quad \text{(destructive interference)} \quad (1)$$

(continued) ►

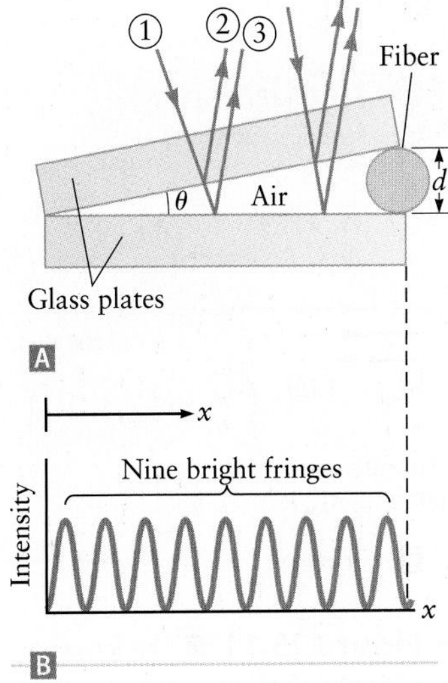

▲ **Figure 25.12** Example 25.4. **A** Two glass plates separated at one end by a thin fiber. **B** Reflected light intensity as a function of position along the plates.

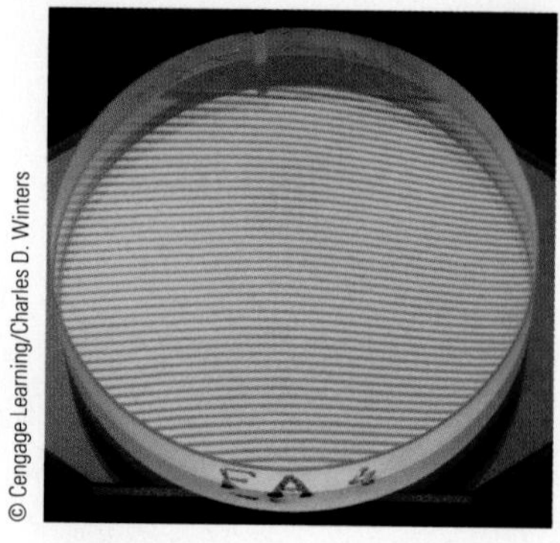

▲ **Figure 25.13** Photo of interference fringes (the horizontal bands) produced by two glass plates separated by a small amount at one end.

To pass from one dark fringe in the interference pattern to the next dark fringe, the path difference between the interfering waves must change by exactly one wavelength. In terms of the condition for destructive interference in Equation (1), this change corresponds to going from m to $m + 1$. We are told that there are nine complete fringes in Figure 25.12B, so if the separation between the plates on the right-hand side is equal to the diameter of the fiber d, the condition for destructive interference becomes

$$2d = \frac{m\lambda}{n_{\text{film}}} = \frac{9\lambda}{n_{\text{air}}}$$

In this case, the "film" between the plates is filled with air.

SOLVE

Solving for d, we get

$$d = \frac{9\lambda}{2n_{\text{air}}} = \frac{9(6.0 \times 10^{-7}\ \text{m})}{2(1.00)} = \boxed{2.7 \times 10^{-6}\ \text{m}}$$

► ***What does it mean?***

Figure 25.13 shows a photograph of the interference fringes produced by an air wedge. Each bright fringe is a "strip" on which the intensity is high, corresponding to the intensity peaks in Figure 25.12B. Similar interference methods are used to measure extremely small distances or heights in many applications, including the reading and writing of information on CDs and DVDs.

CONCEPT CHECK 25.3 Reflection from an Air Wedge

Explain why the waves that reflect from the far left edge of the plates in Figure 25.12A exhibit destructive interference (zero intensity) at $x = 0$. The puzzle: At the far left edge, the plates are in contact, so the path length difference between rays 2 and 3 is zero. Why is there still destructive interference?

A

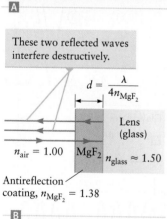

These two reflected waves interfere destructively.

$d = \frac{\lambda}{4n_{\text{MgF}_2}}$

Lens (glass)

$n_{\text{air}} = 1.00$

MgF_2

$n_{\text{glass}} \approx 1.50$

Antireflection coating, $n_{\text{MgF}_2} = 1.38$

B

▲ **Figure 25.14** A The lenses on most binoculars and cameras are coated with antireflection coatings. Although these coatings are transparent, they give the lens a dark tint. B Thin-film interference in an antireflection coating composed of MgF_2.

Antireflection Coatings

A partially reflecting mirror splits a light wave into two separate waves, which may later recombine and interfere. Nearly any flat piece of glass will act in this way. Try looking through the nearest window; even though the window glass is designed to let light pass through, a small amount (about 4%) of the light is reflected. This reflected light is not a problem for an ordinary window, but is bad for some applications. For example, light in a typical camera must pass through several lenses and filters; if 4% of the intensity is reflected at each interface, it would take only a few surfaces to reduce the light intensity by 25% or more, which would be bad for the camera's performance. To avoid this problem, most camera lenses have ***antireflection coatings*** that make a lens appear slightly dark in color when viewed in reflected light as in Figure 25.14A.

Antireflection coatings are often made from MgF_2, which has an index of refraction of about $n_{\text{MgF}_2} = 1.38$. Incident light reflects from the front surface of the MgF_2 film as in Figure 25.14B and again from the MgF_2–glass interface. There is a 180° phase change at both reflections (as with the soap film on glass in Fig. 25.7B), so the interference conditions in Equation 25.12 apply and destructive interference occurs when

$$2d = \frac{(m + \frac{1}{2})\lambda}{n_{\text{MgF}_2}} \qquad m = 0, 1, 2, \ldots$$

The smallest possible value of d and hence the thinnest MgF_2 film that gives destructive interference corresponds to $m = 0$. We thus get

$$2d = \frac{\lambda}{2n_{\text{MgF}_2}}$$

If we want to design an antireflection coating of MgF_2 to operate for red light ($\lambda = 6.0 \times 10^{-7}$ m), we have

$$d = \frac{\lambda}{4n_{MgF_2}} = \frac{6.0 \times 10^{-7}\text{ m}}{4(1.38)} = 1.1 \times 10^{-7}\text{ m} = 110\text{ nm}$$

An antireflection coating will work best at one particular wavelength in the visible range; ideally, it will give "perfect" destructive interference at only one wavelength, but it will also give "partially" destructive interference at nearby wavelengths. Antireflection coatings that function well over the entire range of visible wavelengths are made using multiple layers that give "perfect" destructive interference at different wavelengths.

CONCEPT CHECK 25.4 Color of a *Very* Thin Soap Film

Consider a *very* thin soap film, much thinner than the one discussed in Example 25.3. If the film is suspended in air, what color will it be?

25.4 Light through a Single Slit: Qualitative Behavior

The photo at the start of this chapter shows two interfering water waves. The analogous experiment with light can be performed by letting light shine onto a pair of slits in an opaque screen. To understand the resulting double-slit interference pattern, we must first consider carefully what happens when light passes through a single slit. In this section, we discuss the qualitative behavior; in the two following sections, we then give a more detailed analysis.

The basic experiment is sketched in Figure 25.15A. Light passes through a slit and then illuminates a screen. For simplicity, the incident light is assumed to be a plane wave, represented by the parallel beams on the left side of the slit. When the slit is very wide, the light that reaches the screen is a projection of the slit. If we plot the light intensity versus position along the x axis marked on the screen in Figure 25.15A, we find a "step" in intensity approximately equal to the width w of the slit as shown at the top of Figure 25.15B, but a very careful look reveals that the edges of this shadow are slightly "smeared out." As the slit is narrowed, the intensity pattern on the screen spreads out. If we keep reducing the slit width so that it is comparable to or even smaller than λ, the intensity pattern spreads more and more, and some additional intensity maxima become noticeable.

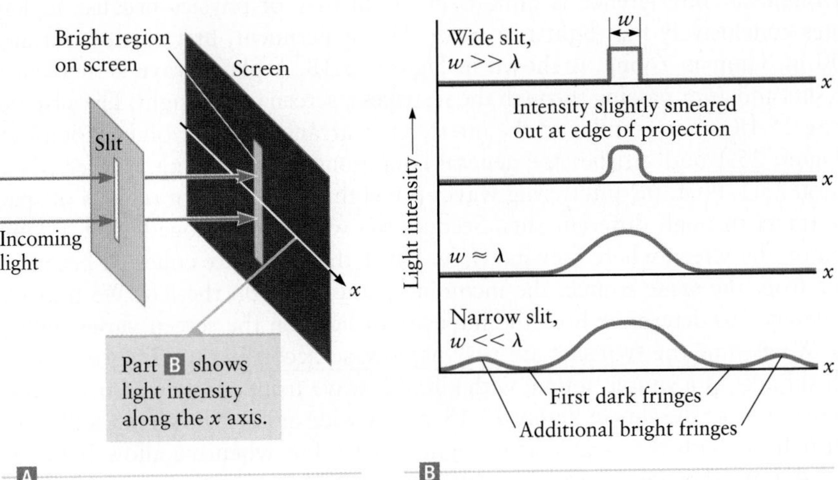

◀ **Figure 25.15** A Light passing through a slit and onto a screen. B If the slit is very wide compared with the wavelength λ, an ordinary shadow (projection of the slit) appears on the screen (top). If the slit width is comparable to or smaller than the wavelength, the wave intensity is spread over a larger region of the screen. As the slit width is made narrower, the amount of spreading increases.

This behavior occurs for all types of waves, including light and sound; it is easy to observe with a water wave because the wavelength of a water wave is large enough for individual waves to be seen easily by eye. In Figure 25.16, a water wave is incident from the left and passes through an opening whose width is approximately equal to λ. The outgoing wave on the right spreads out as a broad beam, analogous to striking a screen over a broad region as in the lowermost pattern in Figure 25.15B. Figures 25.15 and 25.16 illustrate the phenomenon of ***single-slit diffraction***. Diffraction is the bending or spreading of a wave when it passes through an opening. In single-slit diffraction, the opening is just a single slit.

Single-slit diffraction

To understand the qualitative behavior of how light passes through a very narrow slit, it is useful to draw the wave fronts and rays for the incident and diffracting waves. Figure 25.17 shows the results. In this case, the outgoing wave radiates from the slit over a very wide range of angles. This is an example of ***Huygens's principle***, which can be stated as follows.

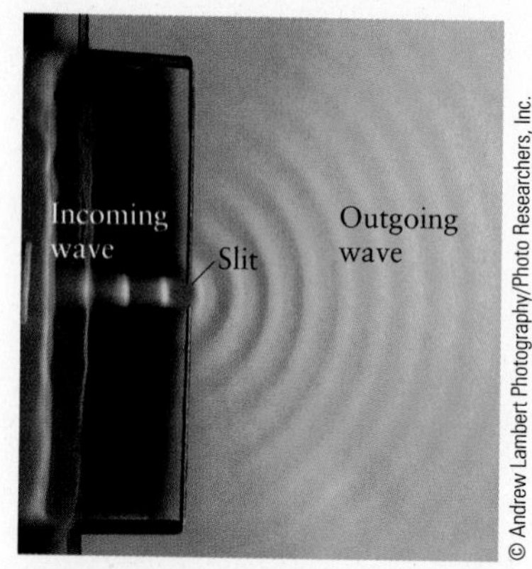

▲ **Figure 25.16** Diffraction of a water wave as it passes from left to right through a single narrow slit, producing an outgoing wave on the right. The outgoing wave front is approximately spherical.

Huygens's principle

All points on a wave front can be thought of as new sources of spherical waves.

When the incoming wave encounters the narrow openings in Figures 25.16 and 25.17, each point inside the opening acts as a new source of spherical waves. In Figure 25.17, the opening is *very much smaller* than the wavelength and acts as a single point source of waves, producing the simple outgoing spherical wave fronts shown.

We will not attempt to derive or justify Huygens's principle. It was proposed (by Huygens) in about 1680, long before there was a detailed theory of light. This principle gives a qualitative way to understand how a very narrow slit can convert a plane wave into a spherical one.

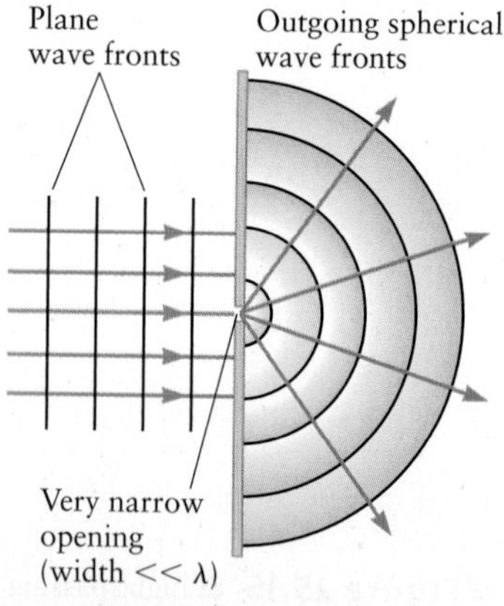

▲ **Figure 25.17** A plane wave incident on a single narrow slit. According to Huygens's principle, the slit acts as a new wave source, producing a spherical wave that propagates away from the slit.

25.5 Double-Slit Interference: Young's Experiment

We next consider the behavior of light passing through two very narrow slits, a problem called ***double-slit interference***. We jump to the double-slit case before finishing the discussion of the single slit from Section 25.4 because a full analysis of light passing through a single slit is in some ways more complicated than that for two *very narrow* slits. When both slits are very narrow, each acts as a simple point source of new waves, and the outgoing waves from each slit are like the simple spherical waves in Figure 25.17.

Double-slit interference is famous in the history of physics because it demonstrates conclusively that light is a wave. This experiment, first carried out around 1800 by Thomas Young, is shown in Figure 25.18. A plane wave is incident onto two slits and after passing through them strikes a screen on the right. The situation in Figure 25.18 is very similar to the interference arrangement involving sound waves in Figure 25.1 and satisfies the general conditions for observing interference from Section 25.1. First, the interfering waves travel through different regions of space as they travel through different slits. Second, the waves come together at a common point on the screen where they interfere. Third, the waves are coherent because they come from the same source, the incoming plane wave on the left. We thus expect interference to determine how the intensity of light on the screen varies with position. We assume the two slits are very narrow, so according to Huygens's principle each slit acts as a simple source with circular wave fronts as viewed from above. If we cover one of the slits in Figure 25.18, a very wide area of the screen is illuminated with light as with the single slit in Figure 25.17, but when we allow light to pass through both slits, we get the very different result shown in Figure 25.18. The light intensity on the screen alternates between bright and dark, signaling regions of constructive interference and destructive interference.

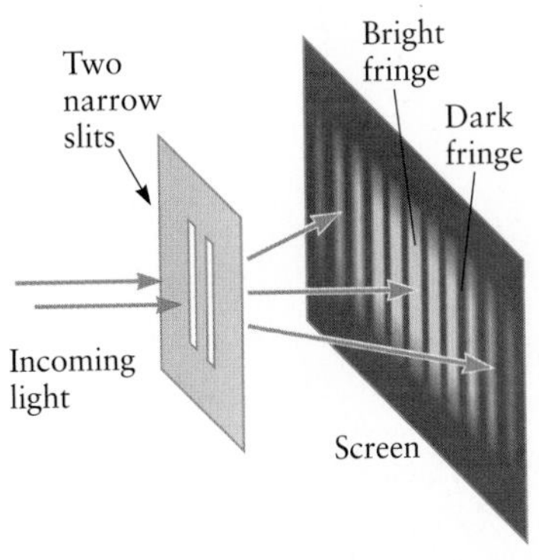

▲ **Figure 25.18** Interference of light produced by two slits. The pattern on the screen at the right is a series of bright and dark interference fringes.

Where Are the Fringes?

To analyze the interference pattern in Figure 25.18, we must determine the path length between each slit and the screen. Consider the path lengths for waves that arrive at point P on the screen in Figure 25.19. For simplicity, we assume the screen is very far away from the slits, so W is very large. Because W is large, the angles θ and θ' that specify the directions from the slits to point P are approximately equal, and we denote both by θ.

If the slits are separated by a distance d, the extra path length traveled by the wave from the bottom slit is ΔL, where

$$\Delta L = d \sin \theta \qquad (25.14)$$

as shown in Figure 25.19. If this extra path length is equal to an integral number of complete wavelengths, the two waves will be in phase when they strike the screen; interference is then constructive and the light intensity is large. On the other hand, if the extra path length is equal to $\lambda/2, 3\lambda/2, 5\lambda/2, \ldots$, interference is destructive and the intensity at the screen is zero. The condition for constructive interference and a bright fringe in the interference pattern is thus

Double-slit interference conditions

$$d \sin \theta = m\lambda \quad m = 0, \pm 1, \pm 2, \pm 3, \ldots \qquad (25.15)$$

(constructive interference)

whereas the condition for destructive interference and a dark fringe is

$$d \sin \theta = (m + \tfrac{1}{2})\lambda \quad m = 0, \pm 1, \pm 2, \pm 3, \ldots \qquad (25.16)$$

(destructive interference)

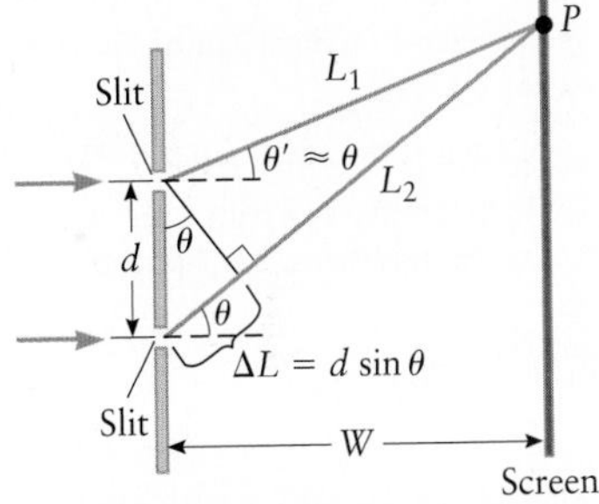

▲ **Figure 25.19** Analysis of double-slit interference. Waves that pass through the two slits travel different distances L_1 and L_2 on their way to point P on the screen. (Compare with Fig. 25.1.) Not to scale.

The double-slit intensity pattern at the screen is shown greatly enlarged in Figure 25.20. The angle θ varies as we move along the screen. At certain values of θ, the condition for constructive interference is satisfied, producing a maximum in the intensity (a bright interference fringe). Each bright fringe satisfies Equation 25.15 with a different value of the integer m. The value $m = 0$ gives $\theta = 0$ and corresponds to the center of the screen. Moving up or down from this point gives bright fringes with $m = +1, +2, +3, \ldots$ and $m = -1, -2, -3, \ldots$. Negative values of m indicate that the path to those points on the screen from the lower slit is shorter than the path to those points from the upper slit.

Now let's consider what the interference pattern will look like for particular values of the separation d between the slits and the perpendicular distance W to the screen. Suppose the slits are $d = 0.10$ mm apart and the screen is located $W = 50$ cm $= 0.50$ m from the slits. If we use light from a red laser with $\lambda = 630$ nm, what is the separation between the bright fringe at the center of the screen ($\theta = 0$) and the nearest bright fringe on either side in Figure 25.20? Two such bright fringes correspond to $m = 0$ (the central fringe) and $m = 1$ in Equation 25.15. The angle

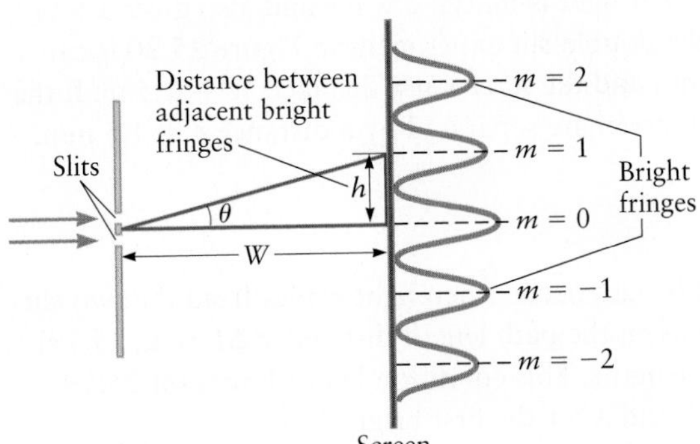

◀ **Figure 25.20** Double-slit interference. Values of θ from Equation 25.15 give the locations of bright fringes on the screen.

θ measured at the slits between the central fringe and the first fringe above it thus satisfies $d \sin \theta = \lambda$, and we have

$$\sin \theta = \frac{\lambda}{d} \tag{25.17}$$

The wavelength is much smaller than the spacing d between the slits, so the ratio on the right-hand side of Equation 25.17 is very small. Hence, $\sin \theta$ is also very small, and we can use the approximation $\sin \theta \approx \theta$ (where the angle is measured in radians) to get

$$\sin \theta \approx \theta \approx \frac{\lambda}{d} \tag{25.18}$$

To find the separation between bright fringes on the screen, we use the right triangle in Figure 25.20 with sides of length W and h. The spacing h between the two fringes corresponding to $m = 0$ and $m = 1$ is

$$h = W \tan \theta$$

Using the approximation for small angles $\tan \theta \approx \theta$, we get

$$h = W\theta$$

Our result for θ from Equation 25.18 then gives

$$h = W\theta = W\frac{\lambda}{d} \tag{25.19}$$

Inserting the values of W, d, and λ given above, we find

$$h = W\frac{\lambda}{d} = (0.50 \text{ m})\frac{6.3 \times 10^{-7} \text{ m}}{1.0 \times 10^{-4} \text{ m}} = 3.2 \times 10^{-3} \text{ m} = 3.2 \text{ mm} \tag{25.20}$$

This fringe spacing is large enough to be seen easily by the naked eye.

Our analysis of Young's double-slit experiment and the result in Equation 25.20 shows that the bright and dark fringes associated with interference are observable, provided the slits are fairly close together. The results for h in Equations 25.19 and 25.20 show that a slit separation of about 0.1 mm (or smaller) is a good choice. If the slit spacing is larger than a few millimeters or so, the fringes are quite close together and can be difficult to observe.

Insight 25.2

REMEMBER THE APPROXIMATIONS FOR SMALL ANGLES

Because the wavelength of light is small, the conditions for constructive interference and destructive interference often involve small angles. For small values of the angle θ, the sine and tangent functions can be approximated by

$$\sin \theta \approx \theta \quad \text{and} \quad \tan \theta \approx \theta$$

These relations are true only if θ is measured in radians. See Appendix B.

CONCEPT CHECK 25.5 Double-Slit Interference

Suppose the distance between the slits in Figure 25.18 is reduced. How would this change affect the interference pattern?

(a) The bright fringes would be more closely spaced on the screen.
(b) The bright fringes would be more widely spaced on the screen.
(c) The fringe spacing on the screen would not change.

EXAMPLE 25.5 Measuring the Wavelength of Light

Young's double-slit experiment shows that light is indeed a wave and also gives a way to measure the wavelength. Suppose the double-slit experiment in Figure 25.20 is carried out with a slit spacing $d = 0.40$ mm and the screen at a distance $W = 1.5$ m. If the bright fringes nearest the center of the screen are separated by a distance $h = 1.5$ mm, what is the wavelength of the light?

RECOGNIZE THE PRINCIPLE

In a double-slit experiment, the bright fringes occur when light waves from the two slits interfere constructively. That happens when the path length difference ΔL (Fig. 25.19) is equal to an integral number of wavelengths. This condition led to Equation 25.19, which gives a relation between h, W, d, and λ for the first bright fringe.

SKETCH THE PROBLEM

Figures 25.19 and 25.20 describe the problem.

IDENTIFY THE RELATIONSHIPS

Equation 25.19 relates the slit spacing d and the perpendicular distance W to the screen, along with h and λ. Rearranging to solve for the wavelength gives

$$\lambda = \frac{hd}{W}$$

SOLVE

Inserting the given values of h, d, and W, we find

$$\lambda = \frac{hd}{W} = \frac{(1.5 \times 10^{-3} \text{ m})(4.0 \times 10^{-4} \text{ m})}{1.5 \text{ m}} = 4.0 \times 10^{-7} \text{ m} = \boxed{400 \text{ nm}}$$

► ***What does it mean?***

In Young's double-slit experiment, the values of h, W, and d can all be measured with an ordinary ruler, but together they enable us to measure the wavelength λ, which is much too small to be measured with a ruler. While technology has certainly advanced since Young's studies of light more than two hundred years ago, methods involving interference are still widely used in precision measurements of the wavelength of light.

Interference with Monochromatic Light

The conditions for interference listed in Section 25.1 (page 766) state that the interfering waves must have the same frequency and hence the same value of the wavelength λ. Light with a single frequency is called ***monochromatic*** because it consists of a single color. Light from many sources, such as the Sun, contains a distribution of wavelengths. Such light sources are generally not useful in double-slit interference experiments. Suppose a light source emits mainly at two colors, that is, at two wavelengths λ_1 and λ_2. The angles that give bright fringes in the double-slit experiment (Eq. 25.15) depend on the value of λ. Different values of the wavelength thus give bright fringes at different locations on the screen, and places on the screen where wavelength λ_1 gives a bright fringe will often coincide with places where wavelength λ_2 gives a dark fringe. If a light source emits a broad distribution of wavelengths, the total intensity pattern will be a "washed-out" sum of bright and dark regions, and no bright or dark fringes will be visible. For this reason, interference experiments with light usually work best with monochromatic waves.

25.6 Single-Slit Diffraction: Interference of Light from a Single Slit

We began our treatment of how light passes through small openings by first discussing the qualitative behavior for a single slit (Section 25.4), including how the intensity pattern changes as we go from a very wide to a very narrow slit (Fig. 25.15). We next discussed the behavior with two slits (Section 25.5) that are narrow enough so that each acts as a simple point source of new spherical waves (Huygens's principle). With the insight gained from the double-slit case, we are ready to tackle the behavior of a single slit in a more general way. We'll now see how to deal with slits that are narrow enough to exhibit diffraction but not so narrow that they can be treated as a single point source of waves.

Analysis of Single-Slit Diffraction

Figure 25.21A shows light incident on a single slit of width w. The intensity pattern at the screen contains a central very bright fringe with dark fringes alongside as well as other fringes where the intensity peaks at values much smaller than the intensity of the central fringe. We want to explain the origin of these fringes and calculate where bright and dark fringes occur on the screen. The key to this calculation is Huygens's principle, which states that each point on a wave front acts as a new source of waves. All points across the slit in Figure 25.21 act as wave sources, and these different waves interfere at the screen.

Where Are the Dark Fringes?

We divide the slit into two halves and denote points in one half by A and those in the other half by B (Fig. 25.21B). Each of these points is a source of Huygens (i.e., spherical) waves, including point A_1 at one edge of the slit and the corresponding source point B_1 in the other portion of the slit. Figure 25.21C shows the distance from these points to a point P on the screen and also shows the path length difference ΔL. We again assume the screen is very far away from the slit, so all the angles denoted by θ are approximately equal. Suppose the path length difference is $\Delta L = \lambda/2$ so that the waves from A_1 and B_1 interfere destructively when they reach a particular point P on the screen. If this condition for destructive interference is met for points A_1 and B_1, the waves from A_2 and B_2 will also interfere destructively and so on for points A_3 and B_3 and for *all* similar pairs of points within the slit, producing a dark fringe on the screen. From Figure 25.21C, we see that $\Delta L = (w/2)\sin\theta$, so we have

$$\frac{w}{2}\sin\theta = \Delta L = \frac{\lambda}{2}$$

which gives the angle for the first dark fringe above the center of the screen. The first dark fringe below the center corresponds to a negative value of θ, with

$$\frac{w}{2}\sin\theta = -\frac{\lambda}{2}$$

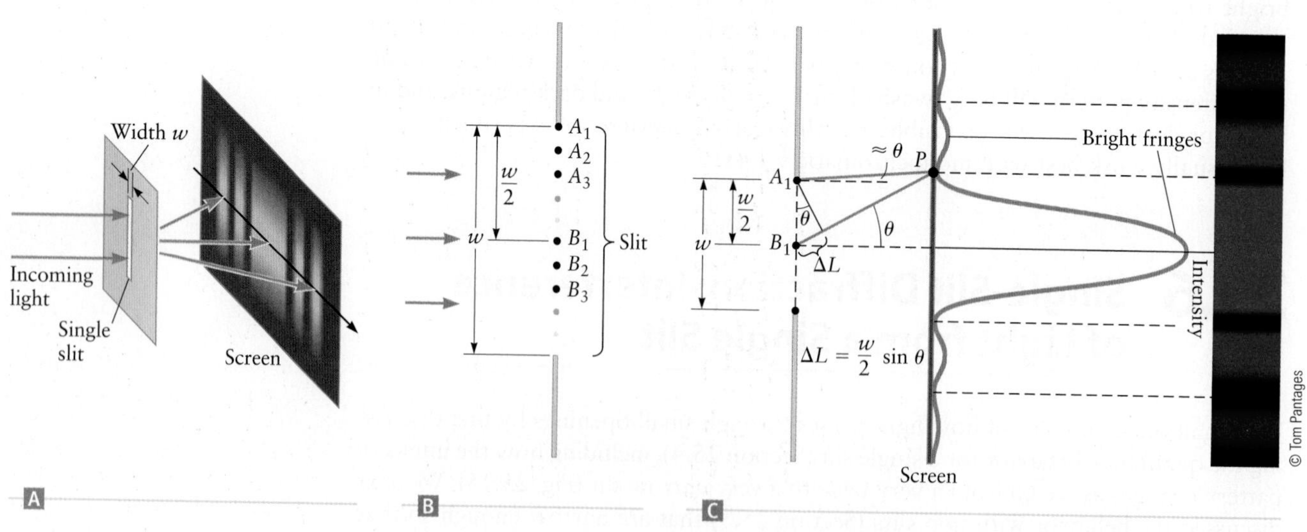

▲ **Figure 25.21** **A** Light passing through a single slit and then on to a screen. **B** Each point within the slit $A_1, A_2, \ldots, B_1, B_2, \ldots$ acts as a source of spherical waves. **C** Analysis of the single-slit diffraction pattern. The slit widths in parts B and C are greatly expanded to show the geometry of the interfering waves. In a scale drawing, the slit widths would be much *smaller* than the width of the central diffraction peak in part C.

The first dark fringes are thus at the angles

$$w \sin\theta = \pm\lambda \quad \text{single-slit diffraction: first dark fringes (destructive interference)} \tag{25.21}$$

The position of the second dark fringe can be found using Figure 25.22. In Figure 25.22A, we divide the slit into four regions and denote the points in these four regions by A, B, C, and D. Figure 25.22B shows the path lengths from the top point in each region (A_1, B_1, C_1, and D_1) to a point P on the screen. Suppose the path length difference for the Huygens waves from A_1 and B_1 is $\Delta L = \lambda/2$, so these waves interfere destructively. The path length difference for C_1 and D_1 has the same value, so these waves will also interfere destructively. (Because the figure is not to scale, the two path length differences ΔL for the pairs AB and CD appear to have very different lengths. However, for a small slit width w, they will be very close in length.) The path length difference will be the same for waves from other pairs of points A_2 and B_2, C_2 and D_2, and so on (Fig. 25.22A), so these waves will also interfere destructively, and the total intensity on the screen is zero. For the case of the second dark fringe, we thus have $\Delta L = (w/4)\sin\theta$. Setting this expression equal to $\lambda/2$, the interference condition in this case is

$$w \sin\theta = \pm 2\lambda \quad \text{single-slit diffraction: second dark fringes (destructive interference)}$$

where the negative sign again corresponds to a fringe below the center of the screen. The same argument can be used to find the third, fourth, . . . dark fringes by dividing the slit into six, eight, . . . regions. The result for the mth dark fringe is

Single-slit interference conditions: destructive interference

$$w \sin\theta = \pm m\lambda \quad \text{single-slit diffraction: } m\text{th dark fringes (destructive interference)} \tag{25.22}$$

where $m = 1, 2, 3, \ldots$. The values of θ from Equation 25.22 give the angles of all dark fringes in the diffraction pattern.

Analyzing the Complete Diffraction Pattern

The angles that give the intensity peaks in Figures 25.21 and 25.22 are not as easy to calculate as the angles of the dark fringes (Eqs. 25.21 and 25.22). However, the full intensity pattern can be calculated by adding up all the Huygens waves that come from all the source points A_1, B_1, A_2, B_2, and so forth. The resulting intensity curve

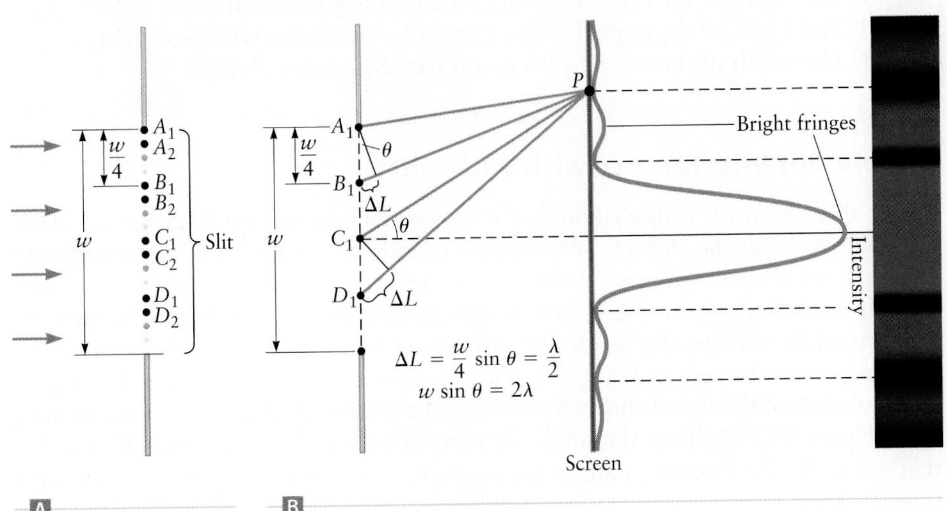

◀ **Figure 25.22** A To locate the second dark fringe in the intensity pattern, we imagine the slit as split into four regions, A, B, C, and D. Points in these regions can be grouped in pairs, A_1 and B_1, A_2 and $B_2, \ldots$, C_1 and $D_1, \ldots$, similar to the pairing in m Figure 25.21B. B Analysis of the location of the second dark fringe in a single-slit diffraction pattern. The slit widths are greatly expanded to show the geometry of the interfering waves. In a scale drawing, the slit widths would be much *smaller* than the width of the central diffraction peak in part B.

is sketched on the right in Figures 25.21C and 25.22B. There is a central bright diffraction fringe (the central maximum), with other "bright" fringes much lower in intensity. The width of the central bright fringe is given approximately by the angular separation of the first dark fringes on either side. The angles for these first dark fringes are given by Equation 25.21:

$$\sin\theta = \pm\frac{\lambda}{w}$$

If the angle θ is small (as it usually is), then $\sin\theta \approx \theta$, which gives

$$\theta = \pm\frac{\lambda}{w} \quad \text{(first dark fringes)} \tag{25.23}$$

These two dark fringes occur on either side of the central bright fringe in Figure 25.21A, so the *full* angular width of the central intensity maximum is 2θ:

$$\text{full angular width of central bright fringe} = 2\,\frac{\lambda}{w} \tag{25.24}$$

As a rough rule of thumb, the central bright fringe extends over the angular range given by Equation 25.24. If the slit is much wider than the wavelength, this angle is small. For example, if the slit width is equal to the thickness of a typical fingernail ($w \approx 0.5$ mm $= 5 \times 10^{-4}$ m) and we consider red light ($\lambda = 630$ nm), the angle is

$$\text{full angular width} = 2\,\frac{\lambda}{w} = \frac{2(6.3 \times 10^{-7}\text{ m})}{5 \times 10^{-4}\text{ m}} = 0.0025 \text{ rad} = 0.14° \tag{25.25}$$

This angle is very small, so the light beam would essentially pass straight through the slit with almost no effect from diffraction. The central diffraction fringe is much broader when the slit width approaches λ. For example, suppose the slit width is only five times larger than the wavelength; for the red light considered in Equation 25.25, we get $w = 5\lambda = 3.2 \times 10^{-6}$ m $= 3.2$ μm (about 300 times smaller than 1 mm). The central bright diffraction fringe would then correspond to an angle

$$\text{full angular width} = \frac{2\lambda}{w} = \frac{2\lambda}{5\lambda} = \frac{2}{5} = 0.40 \text{ rad} = 23°$$

so the spreading of the light beam due to diffraction is now quite substantial.

CONCEPT CHECK 25.6 Single-Slit Diffraction

A single-slit diffraction experiment is performed with red light, and the width of the central diffraction fringe is measured. The experiment is then repeated with blue light. Which of the following statements describes the result?

(a) The width of the central diffraction fringe is greater with blue light.
(b) The width of the central diffraction fringe is smaller with blue light.
(c) The width of the central diffraction fringe does not change.

Double-Slit Interference with Wide Slits

In our discussion of Young's double-slit experiment in Section 25.5, we assumed for simplicity that the slits are very narrow (widths much less than the wavelength of light); such slits produce an intensity pattern that varies approximately sinusoidally with position on the screen as shown in Figure 25.20. When the slits are not extremely narrow, the single-slit diffraction pattern produced by each slit is combined with the sinusoidal double-slit interference pattern as sketched in Figure 25.23. Part A of the figure shows the double-slit pattern with two very narrow slits, while Figure 25.23B shows the single-slit pattern for a wide slit. For two double slits that are wide, the intensity pattern is essentially the product of these two patterns (Fig. 25.23C).

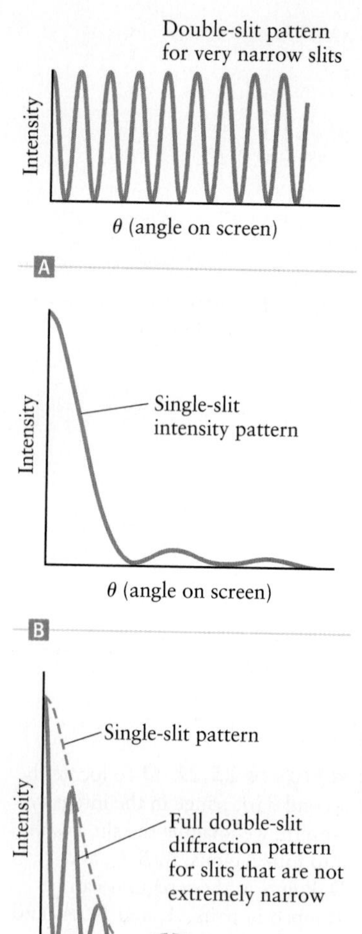

▲ **Figure 25.23** The complete interference pattern produced by two slits is a combination of A the double-slit pattern and B the single-slit pattern. The combination of these two gives the intensity pattern in C.

25.7 Diffraction Gratings

In the past few sections, we have discussed the interference patterns found when light passes through either a single slit or a double slit. As you should expect, similar interference effects also occur with three or more slits. In fact, physicists have found that systems with hundreds or even thousands of slits are quite useful; you probably own one without even knowing it as we'll explain below.

Figure 25.24 shows an arrangement of many slits called a ***diffraction grating***. When light passes through a diffraction grating, a pattern of light and dark fringes is produced on the screen. To analyze this pattern, we'll assume for simplicity that the slits are each very narrow (so that each one produces a single outgoing wave) and the screen is very far away. The rays connecting each slit with a point P on the screen are then approximately parallel, and all make an angle θ with the horizontal axis in Figure 25.24. If the slit-to-slit spacing is d, the path length difference for rays from two adjacent slits is $\Delta L = d \sin \theta$, which is the same as in the case of a double slit (Fig. 25.19 and Eq. 25.14). If ΔL is equal to $0, \pm\lambda, \pm 2\lambda, \pm 3\lambda, \ldots$, two adjacent rays will interfere constructively at the screen. Since the path length difference is the same for every pair of adjacent rays, *all* rays will interfere constructively. The condition for a bright fringe is thus

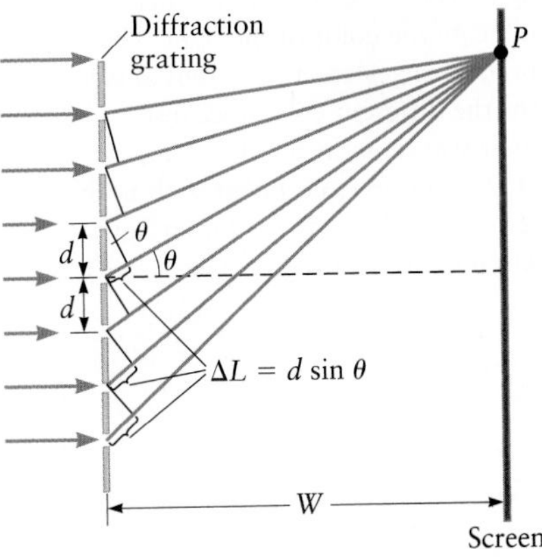

▲ **Figure 25.24** Interference experiment with a diffraction grating.

$$d \sin \theta = m\lambda \quad m = 0, \pm 1, \pm 2, \ldots \tag{25.26}$$

constructive interference for a diffraction grating

Diffraction from a grating: conditions for constructive interference

The condition for a bright fringe from a diffraction grating is identical to the condition for constructive interference from a double slit (Eq. 25.15). The angles that give constructive interference and the positions of the bright fringes on the screen are thus the same in the two cases. However, the overall intensity pattern depends on the number of slits in the grating. Figure 25.25 shows the intensity pattern found with two slits (just the double-slit pattern from Fig. 25.20), three slits, ten slits, and one hundred slits in the grating. The main intensity peaks occur at the same angles for 2, 3, or any number of slits. As the number of slits increases, however, smaller peaks occur between the main peaks. Most importantly, the main peaks become narrower as the number of slits increases, and when the number of slits is large, these peaks are *extremely* narrow. Compare the intensity pattern for a double slit (top of Fig. 25.25) with the result for a grating containing one hundred slits (bottom). The intensity peaks are *much* narrower with the grating.

Using a Grating to "Separate" Colors

The narrow intensity peaks produced by a grating make it a powerful tool for "separating" different light waves according to their wavelengths, that is, according to their colors. Figure 25.26A on page 786 shows light composed of two different colors, red and green, incident on a grating from below. The grating produces an intensity pattern on the screen for each color, with the bright fringes given by the interference conditions in Equation 25.26. The two colors have different wavelengths, so the fringes for red and green light are at different angles and thus different places on the screen as illustrated in Figure 25.26B. The diffraction fringes appear here as spots on the screen because the laser illuminates only a small spot on the diffraction grating. Gratings are widely used to analyze the colors in a beam of light.

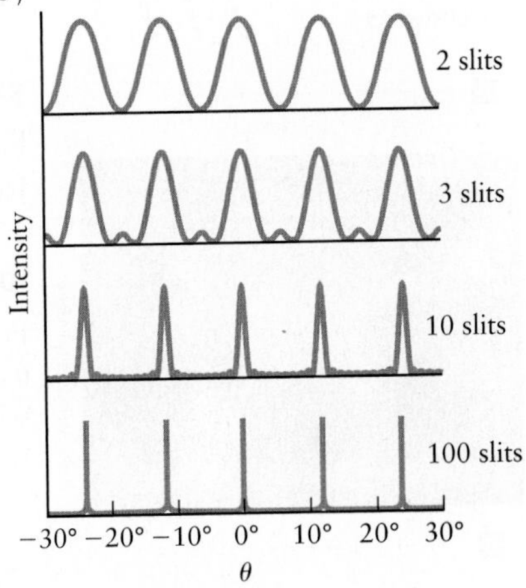

▲ **Figure 25.25** Interference patterns with 2, 3, 10, and 100 equally spaced slits. The main intensity peaks (called the principal maxima) become narrower as the number of slits is increased. With 100 slits, these peaks are very narrow.

Diffraction and the "Color" of a CD

The diffraction gratings in Figures 25.24 and 25.26 contain straight slits arranged with a constant spacing. Similar diffraction effects occur in other situations in which many light waves interfere, such as with light reflected by a compact disc (CD). A

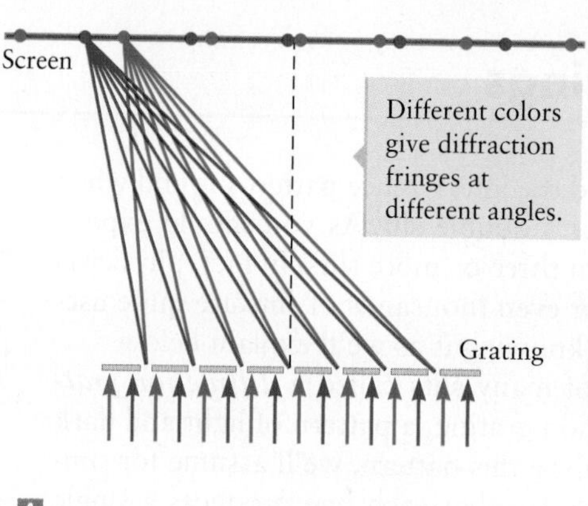
Screen
Different colors give diffraction fringes at different angles.
Grating
A

Fringes on screen
Diffraction grating
B

▶ **Figure 25.26** A Sketch of a diffraction experiment with a grating using red light and green light. The colors of the rays here *do* indicate the color of the light. The bright fringes are at different angles for the different colors because their wavelengths are different. B Diffraction experiment with both red laser light and green laser light shining on a grating.

CD contains "pits" in a reflecting layer, with the pits arranged in arcs ("tracks") as sketched in Figure 25.27A. Light reflected from these arcs acts as sources of Huygens waves, just like the slits in a grating. The reflected waves exhibit constructive interference (bright fringes) at certain angles, depending on the color (wavelength) of the light. Light reflected from a CD therefore has the colors "separated" as shown in Figure 25.27B.

Interference of reflected light
Pits
Light is reflected from pits in the CD surface.
A

B

▲ **Figure 25.27** A The pits in a CD have a regular spacing. Reflections from these pits act as separate but coherent sources of light, just like the slits of a grating, and produce diffraction fringes at different angles for different colors. B Hence, CDs appear brightly colored when viewed with reflected light.

EXAMPLE 25.6 Diffraction of Light by a Grating

A diffraction experiment is carried out with a grating. Using light from a red laser ($\lambda = 630$ nm), the diffraction fringes are separated by $h = 0.15$ m on a screen that is $W = 2.0$ m from the grating. Find the spacing d between slits in the grating.

RECOGNIZE THE PRINCIPLE

The bright fringes in a grating's diffraction pattern occur when the waves from adjacent slits interfere constructively. The condition for constructive interference is given by Equation 25.26.

SKETCH THE PROBLEM

Figure 25.28 shows the problem and indicates the fringe spacing h and the distance W from the grating to the screen.

IDENTIFY THE RELATIONSHIPS

From the geometry of the diffraction pattern in Figure 25.28, we can calculate the angle θ corresponding to the first bright diffraction fringe. We can then use Equation 25.26 with $m = 1$ to find the slit spacing. From Figure 25.28, h and W form two sides of a right triangle containing θ, which leads to

$$\tan\theta = \frac{h}{W}$$

SOLVE

Inserting the given values of h and W and solving for θ gives

$$\tan\theta = \frac{0.15\text{ m}}{2.0\text{ m}} = 0.075$$

$$\theta = 0.075\text{ rad} = 4.3°$$

From Equation 25.26, the angle for the first diffraction fringe (corresponding to $m = 1$) is

$$d \sin \theta = m\lambda = \lambda$$

Rearranging to find d and using our value of θ, we get

$$d = \frac{\lambda}{\sin \theta} = \frac{630 \times 10^{-9}\text{ m}}{0.075} = 8.4 \times 10^{-6}\text{ m} = \boxed{8.4\ \mu\text{m}}$$

► ***What does it mean?***
This value is typical for the slit spacing in a grating and is somewhat smaller than the thickness of a human hair. The fabrication of slits with this very small spacing is not a simple matter, but physicists have been making diffraction gratings since the 1800s.

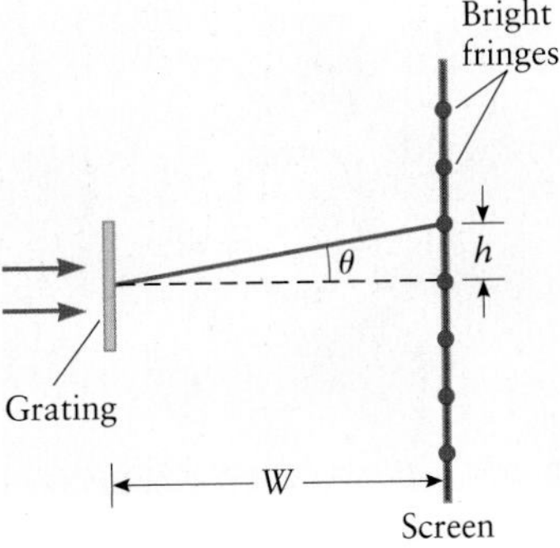

▲ **Figure 25.28** Example 25.6.

Diffraction of X-rays by a Crystal

Diffraction effects similar to those with light also occur with other types of electromagnetic waves (and with any other type of wave, such as sound). The atoms in a crystal are arranged in a periodic array, forming planes (Fig. 25.29A). These planes reflect electromagnetic radiation, leading to interference in the same way that a CD can act as a grating for reflected light (Fig. 25.27).

In a diffraction grating designed to work with light, the spacing between slits is typically 10^{-5} m to 10^{-4} m, which is 100 to 1000 times larger than the wavelength of light. For diffraction from the atomic planes in a crystal, the effective slit spacing is the distance between atomic planes, which is typically 3×10^{-10} m and hence much smaller than the spacing in a diffraction grating. To observe diffraction using atomic planes, one therefore needs to use wavelengths much shorter than visible light, and X-rays with wavelengths on the order of 10^{-10} m are a convenient choice. Experiments of this type are called ***X-ray diffraction***. By measuring the angles that give constructive interference of reflected X-rays, one can determine the spacing between atomic planes, just as the spacing between the slits in a diffraction grating can be determined from the diffraction pattern produced by a grating (Example 25.6). Because the atomic planes consist of atoms instead of simple slits, however, the intensity peaks appear as diffraction "spots" instead of fringes. An example is shown in Figure 25.29B, which shows the intensity pattern produced by diffraction of X-rays from a crystal of DNA. Because each DNA molecule has a helical structure, a crystal of DNA molecules gives a more complicated array of diffraction spots than most crystals. Through careful analysis of this pattern (along with other evidence), James Watson and Francis Crick deduced the double helix structure of DNA.

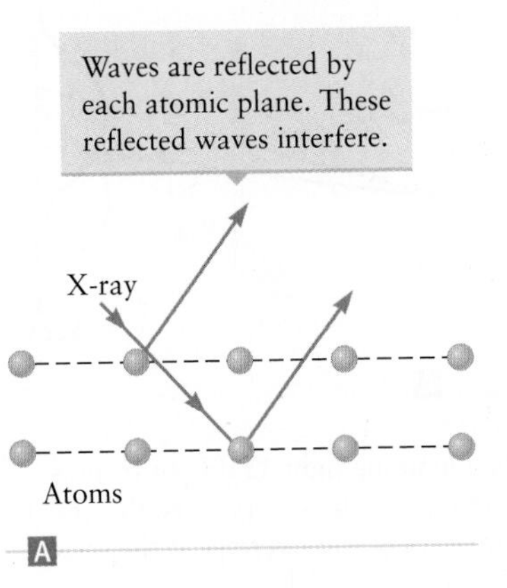

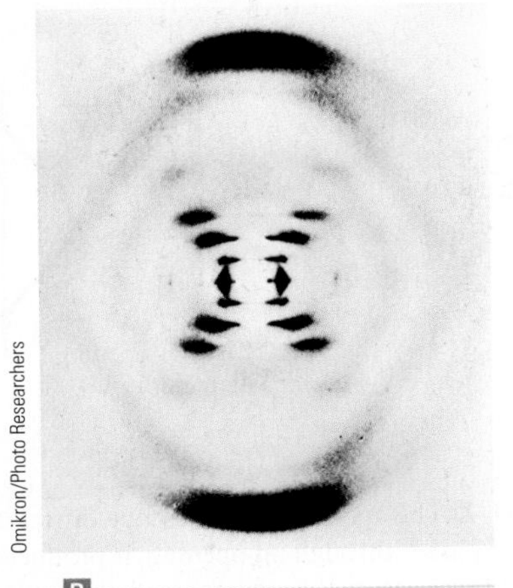

◄ **Figure 25.29** A The atomic planes in a crystal reflect electromagnetic waves, which interfere in a manner similar to that found with visible light and a grating. Diffraction experiments with a crystal are usually done with X-rays. The angles at which X-rays interfere constructively can be used to deduce the distance between atomic planes and the arrangement of atoms in the crystal. B This X-ray diffraction pattern for a DNA crystal led to the discovery of the double-helix structure of DNA. [See J. D. Watson and F. H. C. Crick, "Molecular structure of nucleic acid," *Nature* 4356:737 (1953).]

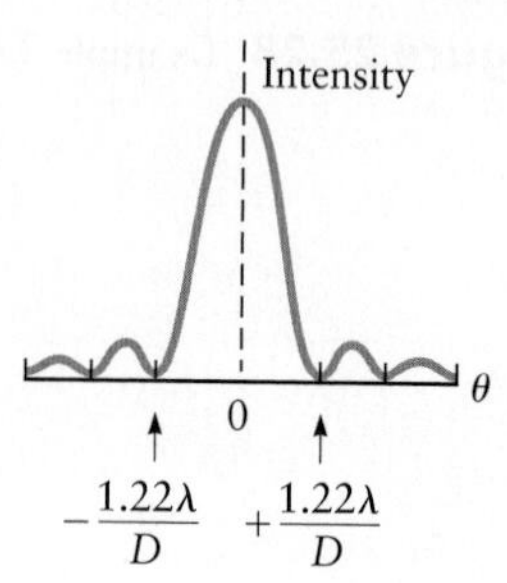

▲ **Figure 25.30** A When a diffraction opening is circular, a circular diffraction spot is created. B The intensity curve has a shape similar to that of a single slit, with the first intensity minima at an angle given by Equation 25.27.

25.8 Optical Resolution and the Rayleigh Criterion

Diffraction causes light to spread out after passing through a narrow slit. A similar diffractive spreading is found whenever a beam of light passes through an opening of any shape, including the aperture of a telescope or microscope or the pupil of your eye. The result in Equation 25.23 gives the angle between the central bright fringe and the first dark fringe for the case of a slit. For a circular opening of diameter D, the result is almost the same, but the circular geometry leads to an additional numerical factor of 1.22:

$$\theta = \frac{1.22\lambda}{D} \tag{25.27}$$

where we assume D is much larger than the wavelength. In Equation 25.27, θ is the angle between the midpoint of the central bright maximum and the first minimum in the diffraction pattern. Figure 25.30A shows the diffraction pattern produced by a circular opening, and Figure 25.30B shows the light intensity as a function of angle on the screen.

This result is important in many situations. For example, suppose you are looking through a telescope at a distant star. Light from the star must pass through a circular opening (an aperture) in the telescope before it reaches a screen or your eye, as sketched in parts A and B of Figure 25.31. Diffraction at the opening produces a circular diffraction spot like the one in Figure 25.30. Now suppose there are actually two light sources (two stars) viewed again through the telescope. If these two stars are at slightly different angles with respect to the central axis of the opening, your eye will see them at slightly different viewing angles. Because they are two separate light sources, their waves are incoherent and do not interfere. Light from each source produces its own diffraction pattern, however, resulting in two diffraction spots that may overlap at your eye.

Figure 25.32 shows how the two light sources in Figure 25.31C would actually appear through a telescope. Both sources produce a circular diffraction spot, and if the two sources are sufficiently far apart, two separate diffraction spots are observable as shown in Figure 25.32A. If the sources are too close together, however, their diffraction spots will overlap so much that they appear as a single spot as in Figure 25.32C. As a rough criterion, two sources will be resolved as two distinct sources of light if their angular separation is greater than the angular spread of a single diffraction spot. This result is called the ***Rayleigh criterion***. For a circular opening, the Rayleigh criterion for the angular resolution is (using Eq. 25.27)

$$\theta_{\min} = \frac{1.22\lambda}{D} \tag{25.28}$$

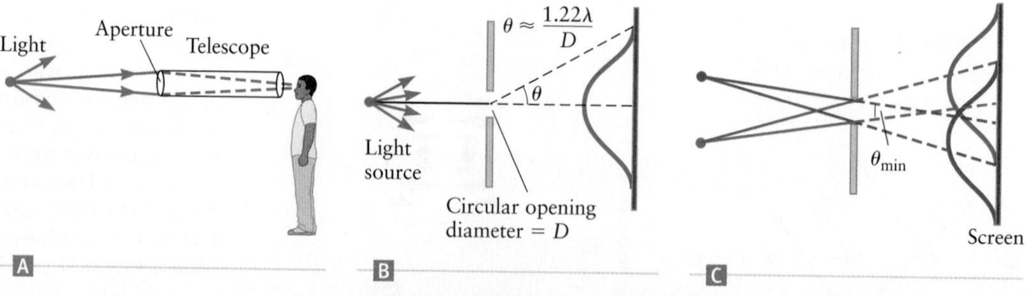

▲ **Figure 25.31** A The aperture of a telescope diffracts the incoming light. B If light from a single point source passes through an opening, it produces a single diffraction spot like the one in Figure 25.30. C If there are two nearby sources, their diffraction spots will overlap on the screen.

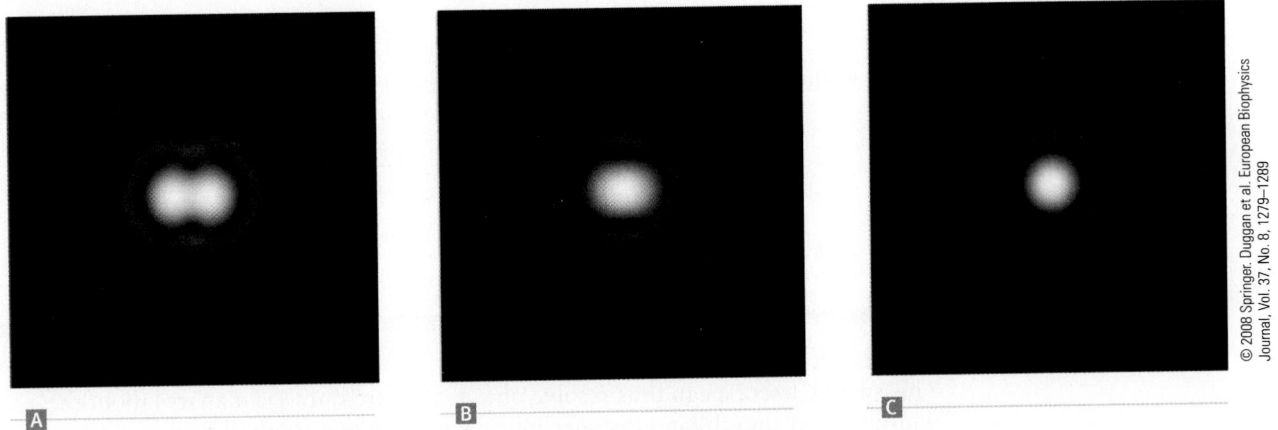

▲ **Figure 25.32** Light from two point sources demonstrates the Rayleigh criterion. A When the sources are far enough apart, they can easily be resolved as two separate light sources. B As the sources are brought close together, their images overlap. Here they are on the verge of being resolved as separate sources. This corresponds to approximately the angular separation given by the Rayleigh criterion (Eq. 25.28). C When the two sources are very close together, they cannot be resolved as separate sources.

In words, Equation 25.28 says that two objects will be resolved when viewed through an opening of diameter D (which might be a telescope or your eye) if the light rays from the two objects are separated by an angle at least as large as θ_{min}.

EXAMPLE 25.7 Rayleigh Criterion for a Marksman

A rifle used for target shooting has a small telescope so that the shooter can accurately aim at his target. Suppose the telescope has a diameter $D = 1.5$ cm and the shooter is looking at two small objects that are $L = 200$ m away. If the shooter is just barely able to tell that there are two objects (and not one), how far apart are the objects? Assume red light with $\lambda = 600$ nm.

RECOGNIZE THE PRINCIPLE

According to the Rayleigh criterion, if the two objects are just barely resolved, they must be separated by an angle that is at least as large as

$$\theta_{min} = \frac{1.22\lambda}{D} \quad (1)$$

SKETCH THE PROBLEM

Figure 25.33 shows the problem. We want to find the distance h between the two objects.

IDENTIFY THE RELATIONSHIPS

From Figure 25.33, the angle of separation between the two objects is

$$\tan\theta = \frac{h}{L} \approx \theta$$

where in the last step we have assumed the angle is small so $\tan\theta \approx \theta$.

(continued) ▶

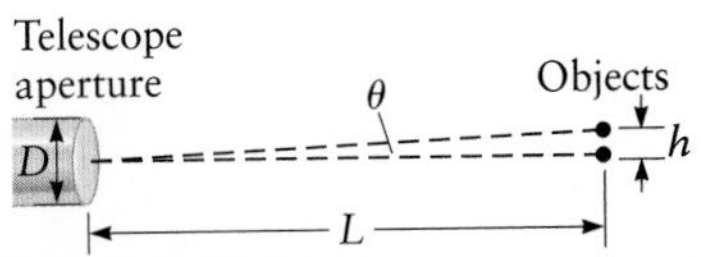

▲ **Figure 25.33** Example 25.7. Applying the Rayleigh criterion to a small telescope.

SOLVE

Setting θ equal to θ_{min} as given by the Rayleigh criterion (Eq. 1) and solving for h, we find

$$\theta \approx \frac{h}{L} = \theta_{min} = \frac{1.22\lambda}{D}$$

$$h = \frac{1.22\lambda L}{D} = \frac{1.22(6.0 \times 10^{-7}\ \text{m})(200\ \text{m})}{(1.5 \times 10^{-2}\ \text{m})} = \boxed{1.0 \times 10^{-2}\ \text{m}}$$

► ***What does it mean?***

This small telescope can thus resolve objects that are only 1 cm apart (about the width of your thumb) at a distance much greater than a football field!

EXAMPLE 25.8 Limits of Resolution of the Eye

The Rayleigh criterion sets a limit on the resolution of the eye. Light enters the eye through a circular opening called the pupil, resulting in a diffraction spot when the light is focused on the retina. (**a**) Estimate the limiting angular resolution of the eye as measured at the retina for green light with $\lambda = 500$ nm. *Note*: The wavelength λ in the Rayleigh criterion is the wavelength after the waves have passed through the opening, inside the eye in this example. We must therefore allow for a reduced wavelength due to the eye's index of refraction, similar to Equation 25.9 for a film. Assume $n_{eye} = 1.3$. (**b**) Using your result for the angular resolution in part (**a**), suppose two very tiny pieces of dust are separated by 50 μm (approximately the diameter of a human hair) and are located a distance $L = 25$ cm from your pupil. Can you resolve them as two separate dust particles?

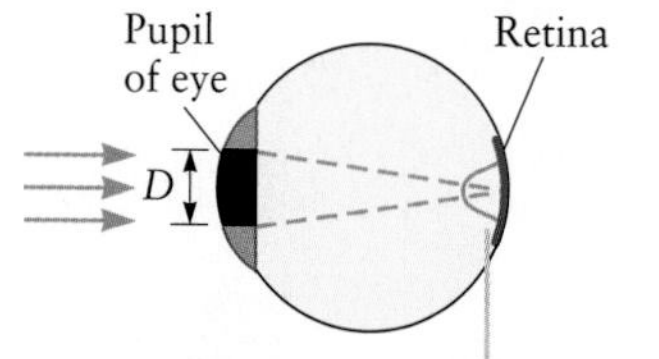

Diffraction causes the image from a point object to spread over an angle θ_{min} on the retina.

▲ **Figure 25.34** Example 25.8. The angular resolution of the eye is determined by the Rayleigh criterion.

RECOGNIZE THE PRINCIPLE

The pupil is approximately circular and diffracts light, just like the aperture in Figure 25.30A. A point source of light will be imaged as a "spot" on the retina; the size of this spot determines how well your eye can resolve two tiny objects that are close together. Using the Rayleigh criterion, we can determine how this limits the resolution of the eye.

SKETCH THE PROBLEM

(**a**) Figure 25.34 shows a sketch of the eye and the diffraction spot on the retina.

IDENTIFY THE RELATIONSHIPS

The angular resolution limit from the Rayleigh criterion is

$$\theta_{min} = \frac{1.22\lambda_{eye}}{D}$$

(Eq. 25.28), where λ_{eye} is the wavelength of the light as it travels inside the eye. By analogy with Equation 25.9, we have

$$\lambda_{eye} = \frac{\lambda}{n_{eye}} \quad (1)$$

where λ is the wavelength outside the eye (given as 500 nm). We must estimate D to apply the Rayleigh criterion. Looking at your own eye in a mirror, you can estimate the pupil diameter as about $D = 5$ mm.

SOLVE

Substituting λ_{eye} from Equation (1) and the estimated pupil diameter leads to

$$\theta_{min} = \frac{1.22(\lambda/n_{eye})}{D} = \frac{1.22(5.0 \times 10^{-7}\text{ m})/1.3}{5 \times 10^{-3}\text{ m}} = \boxed{9 \times 10^{-5}\text{ rad}} \quad (2)$$

The units here are radians because the expression for θ_{min} is a dimensionless ratio.

▸ ***What does it mean?***

The angular resolution of the human eye is quite good; two beams of light separated by only $\theta_{min} = 9 \times 10^{-5}$ rad $= 0.005°$ will be seen as separate spots on the retina. Now let's apply the result from part (**a**) to determine how well the eye can resolve two small pieces of dust.

SKETCH THE PROBLEM

(**b**) The geometry is shown in Figure 25.35. The dust particles are a distance L from the eye and are separated by a distance h.

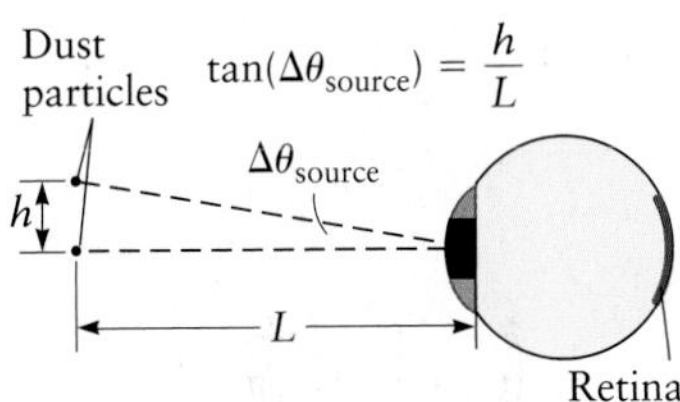

▲ **Figure 25.35** Example 25.8. Application of the Rayleigh criterion. Can your eye resolve the dust particles?

IDENTIFY THE RELATIONSHIPS

When you view the dust particles, your retina is detecting light that the particles reflect. The dust particles thus act as separate sources of light, similar to Figure 25.31C. From Figure 25.35, we see that light waves from the dust have an angular separation at the retina of $\Delta\theta_{source}$, where

$$\tan(\Delta\theta_{source}) \approx \frac{h}{L}$$

Because the dust particles are very close together, $\Delta\theta_{source}$ is very small and we can use the small-angle approximation $\tan(\Delta\theta) \approx \Delta\theta$.

SOLVE

We are left with

$$\tan(\Delta\theta_{source}) \approx \Delta\theta_{source} \approx \frac{h}{L}$$

Inserting the values of h and L, we find

$$\Delta\theta_{source} = \frac{h}{L} = \frac{50 \times 10^{-6}\text{ m}}{0.25\text{ m}} = 2.0 \times 10^{-4}\text{ rad} \quad (3)$$

The limiting value of θ_{min} in Equation (2) is smaller than the angular separation of the dust particles $\Delta\theta_{source}$ in Equation (3). Hence, the eye can resolve the pieces of dust as two separate objects.

▸ ***What does it mean?***

Our estimation of the eye's resolution using the Rayleigh criterion is only approximate. We have ignored factors such as the separation of the detector cells (called rods and cones) in the retina, but even so our estimate matches the actual resolution of the eye to within about a factor of two. You can check for yourself by carefully placing two strands of hair side by side and pushing them as close as possible, but not so close that they appear as one strand of hair. Then use a microscope or magnifying glass to measure the distance between the strands.

CONCEPT CHECK 25.7 The Rayleigh Criterion and the Diameter of a Telescope

The angular resolution of a telescope is commonly measured in terms of the smallest angular separation θ_{min} between two objects that the telescope can resolve (Fig. 25.31C). If the diameter of the telescope aperture is increased, does θ_{min} (a) get smaller, (b) get larger, or (c) stay the same?

Limits on Focusing

According to the ray optics approximation of Chapter 24, a perfect lens will focus a very narrow parallel beam of light to a precise point at the focal point of the lens. However, the ray optics approximation ignores diffraction. If the lens has a diameter D, it acts just like the openings in Figures 25.30 and 25.31 and according to the Rayleigh criterion produces a diffracted beam spread over a range of angles up to $\theta \approx 1.22\lambda/D$ (Eq. 25.27). This spreading is modeled in Figure 25.36, which shows the original beam (drawn in green) traveling along the principal axis of a lens. The diffraction spreading is modeled by the second incident beam (drawn in red), which makes an angle $\theta \approx 1.22\lambda/D$ relative to the original beam. Without diffraction, the rays would all intersect at a point on the axis a distance f from the lens, where f is the focal length. Diffraction spreads this point over a disk of radius r as shown near the intersection of the rays drawn in red. From the geometry in Figure 25.36, we find $\tan\theta = r/f$. The angle θ is usually small, so we can use the small-angle approximation to get $\tan\theta \approx \theta \approx r/f$, which can be rearranged to solve for r:

$$r = f\theta \tag{25.29}$$

Inserting θ from Equation 25.27 gives

$$r = f\theta = f\frac{1.22\lambda}{D} \tag{25.30}$$

The lens thus focuses parallel incoming light to a "smeared-out" circular region of radius r. According to Equation 25.30, the value of r depends on the diameter D of the lens and its focal length f, but these two quantities are not independent of each other. The lens maker's formula from Chapter 24 tells how f is related to the radius of curvature R of the lens. It is not physically possible to make the diameter of the lens opening larger than $2R$. (See Fig. 24.37.) For a lens with an index of refraction n and two convex surfaces both with a radius of curvature R, the focal length is

$$\frac{1}{f} = \frac{2(n-1)}{R}$$

as obtained from the lens maker's formula (Eq. 24.32) with $R_1 = R_2 = R$. We have just argued that the upper limit on the opening of a lens is $D = 2R$, so the limit on focal length is

$$\frac{1}{f} = \frac{2(n-1)}{D/2}$$

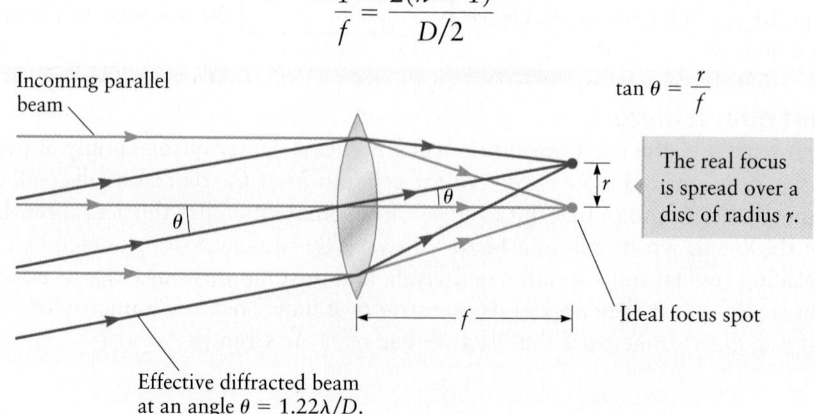
Incoming parallel beam
$\tan\theta = \frac{r}{f}$
The real focus is spread over a disc of radius r.
r
θ
θ
f
Ideal focus spot
Effective diffracted beam at an angle $\theta = 1.22\lambda/D$.

▶ **Figure 25.36** Focusing light from a distant object. The lens acts as an opening as in Figure 25.31, causing the image to be broadened ("smeared out") by diffraction. This sets limits on focusing and resolution of the lens. *Note*: The colors of these rays do *not* represent the colors of the light waves.

which leads to

$$f = \frac{D}{4(n-1)}$$

For a typical lens material such as glass, $n \approx 1.5$ (Table 25.1), and we get

$$f \approx \frac{D}{2} \tag{25.31}$$

Inserting this into Equation 25.30 gives the radius of the focal region (also called the focal disk) as

$$r = f\frac{1.22\lambda}{D} \approx \frac{D}{2}\frac{1.22\lambda}{D} = \frac{1.22\lambda}{2} \tag{25.32}$$

Our result for r is only approximate because it depends on our estimate for how the lens opening D depends on its radius of curvature R. In the spirit of an estimation problem, we can thus write

$$r \approx \lambda \tag{25.33}$$

In words, Equation 25.33 says that the wave nature of light limits the focusing qualities of even a perfect lens. *It is not possible to focus a beam of light to a spot smaller than approximately the wavelength.*

This fundamental limit on focusing determines the ultimate performance of microscopes and other optical instruments, as we'll explain in Chapter 26. Equation 25.33 defines yet again the boundary between the regimes of ray optics and wave optics. As a general rule, the ray approximation of geometrical optics can only be applied at size scales much greater than the wavelength. When a slit or a focused beam of light is made so small that its dimensions are comparable to λ, we must expect that diffraction effects will be important.

TABLE 25.1 Index of Refraction of Some Common Substances for Yellow Light

Substance	Index of Refraction
Air	1.0003
Water	1.33
Glass	1.5[a]
MgF_2	1.38

[a]Typical value. The precise value depends on composition.

25.9 Why Is the Sky Blue?

The wave nature of light leads to interesting effects when light reflects from very small objects. When light reflects from a plane surface, the angle of reflection is equal to the angle of incidence (the law of reflection, Eq. 24.1), but this result is true only when the size of the reflecting surface is much greater than the wavelength. Figure 25.37 shows the opposite situation, when λ is larger than the reflecting object. The reflected waves now travel away in all directions and are called ***scattered*** waves. This behavior is found for all types of waves, including sound (Fig. 13.15).

The amplitude of the scattered wave depends on the size of the scattering object relative to the wavelength. For particles much smaller than the wavelength of light, the amplitude of the scattered wave grows rapidly as the wavelength is decreased (Fig. 25.38). Blue light and ultraviolet light are thus scattered much more strongly than red light. This effect is called ***Rayleigh scattering.***

Figure 25.38 helps answer the very simple everyday question, "Why is the sky blue?" The Earth's atmosphere is composed mainly of oxygen and nitrogen molecules, which are both colorless, so one might expect the atmosphere to be colorless. We know that the atmosphere is blue on a normal clear day, however, and Figure 25.39 shows why. The light that we see from the sky overhead is sunlight scattered

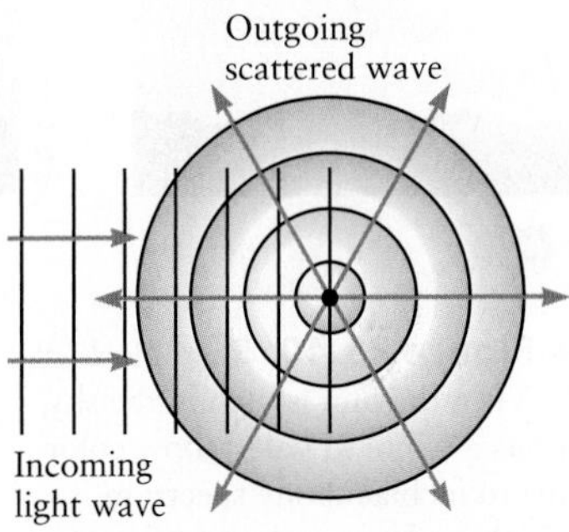

▲ **Figure 25.37** When a plane light wave strikes an object that is much smaller than the wavelength, a spherical wave propagates outward in all directions from the object. This outgoing wave is called scattered light. A similar phenomenon occurs for sound waves; see Section 13.5 and Figure 13.15.

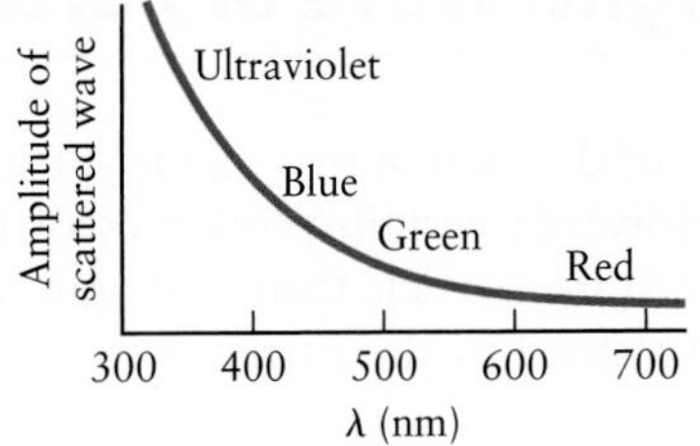

◀ **Figure 25.38** When light is scattered by very small particles in the atmosphere, the amplitude of the scattered wave in Figure 25.37 increases as the wavelength decreases.

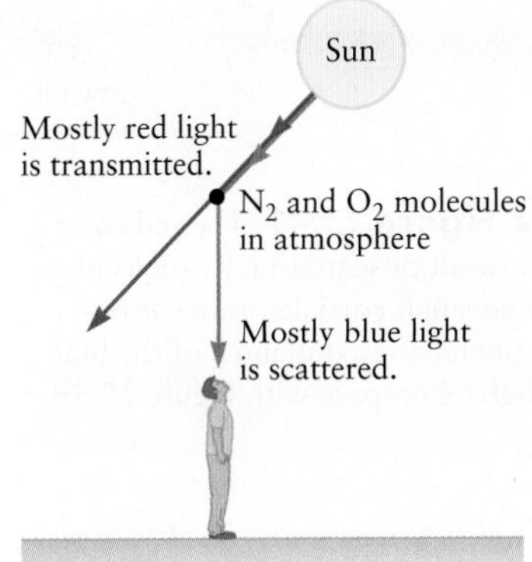

▲ **Figure 25.39** Scattering of sunlight by N_2, O_2, and other molecules in the air is responsible for the color of the atmosphere. Since these molecules are smaller than the wavelength of visible light, they scatter short-wavelength light most strongly, giving the sky its blue color.

by N_2, O_2, and other molecules in the atmosphere. These molecules are much smaller than the wavelength of visible light, so (according to Fig. 25.38) they scatter blue light more strongly than red light, giving the atmosphere its blue color.

EXAMPLE 25.9 Color of the Sun

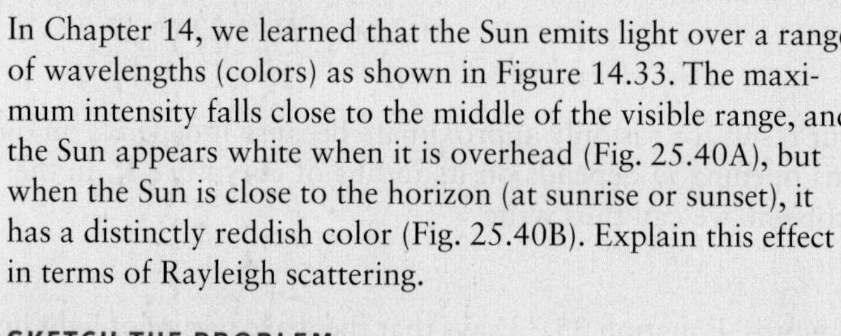

▲ **Figure 25.40** Example 25.9. A When the Sun is high in the sky, it has a white to pale yellow color due to its blackbody spectrum, which peaks near the middle of the visible range. B When the Sun is low on the horizon, it appears more reddish in color.

In Chapter 14, we learned that the Sun emits light over a range of wavelengths (colors) as shown in Figure 14.33. The maximum intensity falls close to the middle of the visible range, and the Sun appears white when it is overhead (Fig. 25.40A), but when the Sun is close to the horizon (at sunrise or sunset), it has a distinctly reddish color (Fig. 25.40B). Explain this effect in terms of Rayleigh scattering.

SKETCH THE PROBLEM

Figure 25.41 shows light traveling from the Sun to an observer on the Earth.

IDENTIFY THE RELATIONSHIPS AND SOLVE

When the Sun is near the horizon, the sunlight you observe travels a greater distance through the atmosphere than when the Sun is directly overhead. Rayleigh scattering by molecules and other small particles in the air is stronger for short wavelengths (blue light) than for long wavelengths (red light). By the time the light from a rising or setting Sun reaches an observer, a substantial fraction of the blue component has been scattered away. The remaining light (Fig. 25.41) contains a greater proportion of red light than when the Sun is high in the sky.

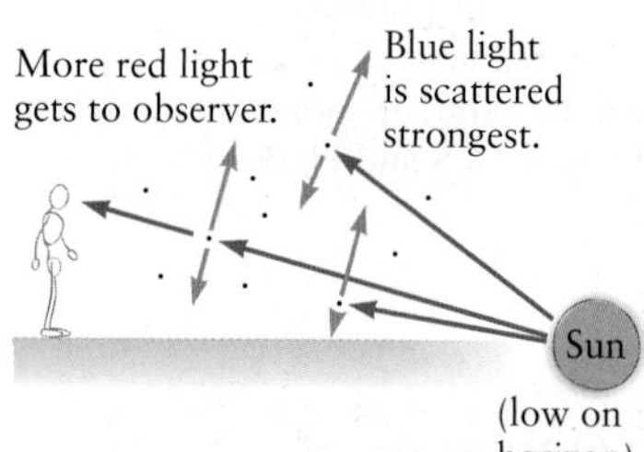

▲ **Figure 25.41** The red color is a result of scattering by molecules and small particles in the atmosphere, removing most of the blue light. Compare with Figure 25.39.

► ***What does it mean?***

The perceived color of an object depends on both the wavelengths the object emits and what happens to those wavelengths as they travel to an observer. The scattering of light can greatly affect the "color" of an object.

CONCEPT CHECK 25.8 Color of the Atmosphere

Why does a forest fire tend to give the sky a pronounced reddish color?

(a) The small particles of soot and molecules produced by a fire usually have a red color.

(b) The small particulates from a fire scatter red light from the Sun less than blue light.

(c) Heat from the fire removes water vapor from the air.

25.10 The Nature of Light: Wave or Particle?

Observations of double-slit interference and of diffraction by a single slit show convincingly that light has wave properties. However, certain properties of light and optical phenomena can *only* be explained with the particle theory of light. One very common effect that can only be correctly explained by the particle theory of light is color vision.

Origin of Color Vision

Human and many other vertebrate eyes are sensitive to color; that is, they are able to distinguish light waves according to their wavelength. This is accomplished by cells in the retina called ***cones***.[2] There are three types of cones. Some are most sensitive to red light—that is, they send a large signal to the brain when exposed to red light and a much smaller signal with either blue or green light—and other cones are sensitive to mainly blue or green light (Fig. 25.42).

How can we explain the behavior of different cone cells? The energy carried by an ordinary wave depends on its amplitude; a large amplitude corresponds to a high intensity and vice versa. So, we should always be able to arrange for two light waves with different colors (say, red and blue) to carry the same total energy just by adjusting their relative intensities; but, even when two such waves carry the same total energy into the eye, the red cone cells still respond mainly to red light and the blue cone cells respond mainly to blue light. How can a cone cell be much more sensitive to light of one color (i.e., one wavelength) than another despite such differences in intensity? How can a red cone cell be almost completely insensitive to blue and violet light, or a blue cone cell insensitive to red light? Studies of these and related questions led to the quantum theory of light (Chapter 28). According to that theory, particles of light called ***photons*** carry an amount of energy that depends on frequency, with a blue photon carrying more energy than a red photon. A complete explanation of color vision involves the properties of photons along with the quantum properties of molecules inside cone cells, and we'll come back to this problem when we describe quantum theory in Chapter 28.

So, although Newton did not realize it, the evidence for his particle theory of light was literally right in front of him. No one else realized it either until about 1905, when the beginnings of a quantum theory of light were first developed by Einstein. A central part of quantum theory is that light is both a particle *and* a wave, a concept called "particle-wave duality."

The study of light has been a central thread in the development of physics as a science, connecting Newton and mechanics to the theory of waves, to Maxwell's work on electromagnetism, and all the way to quantum theory. In Chapter 28, we'll see how quantum theory extends the concept of particle-waves to electrons, protons, and other objects that a classical physicist (Newton!) would have classified as particles. The discovery that light has properties of both classical particles and waves thus opened the way to 20th-century physics.

Insight 25.3
WHY IS THE SKY BLUE AND NOT VIOLET?
The Rayleigh scattering of light by molecules in the atmosphere gets stronger as the wavelength decreases. We have claimed that this scattered light gives the sky its blue color (Fig. 25.39), but violet light has a shorter wavelength than blue light, so why isn't the sky violet instead? Two factors combine to make the sky blue instead of violet. First, the Sun emits more strongly in the blue than in the violet range. Second, our eyes are more sensitive to blue light than to violet light. Hence, even though violet light is scattered more strongly than blue light, we still perceive the sky as blue.

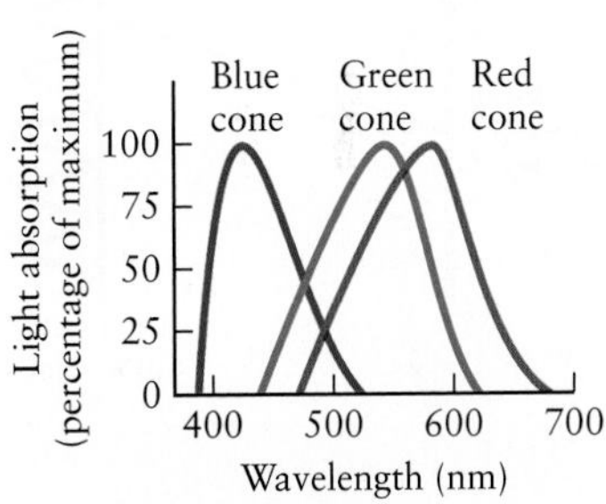

▲ **Figure 25.42** Color vision is due to light detectors in the eye called cones. The three types of cone cells are sensitive to light from different regions of the visible spectrum.

[2]Cells in the retina called ***rods*** also detect light, but rods are not sensitive to color.

Summary | CHAPTER 25

Key Concepts and Principles

Conditions for interference

Light is a wave and exhibits interference if certain conditions are met.

1. Two or more waves must travel through different regions of space.
2. The waves must be brought together at a common point.
3. The waves must be ***coherent***, which means they have the same frequency and have a fixed phase relationship.

(*Continued*)

Constructive and destructive interference

Interference of two waves is ***constructive*** if the waves are ***in phase*** and is ***destructive*** if the waves are 180° ***out of phase***.

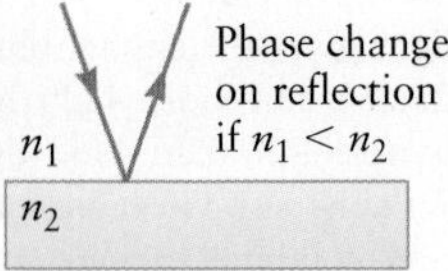

Phase change upon reflection

When light reflects from a surface, the phase of the reflected wave changes by 180° if the index of refraction on the incident side is smaller than the index on the opposite side. This effect, called a ***phase change on reflection***, is important in thin-film interference.

Huygens's principle

All points on a wave front act as sources of new spherical waves. Huygens's principle explains how light passing through a narrow opening is ***diffracted***.

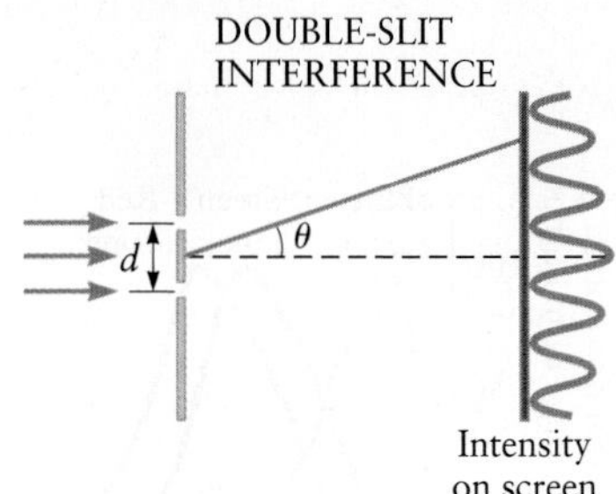

Double-slit interference

Young's double-slit experiment demonstrates that light passing through two narrow slits can interfere. The interference pattern on a distant screen shows constructive or destructive interference, depending on the outgoing angle. Constructive interference (a bright fringe in the interference pattern) is found when

$$d \sin \theta = m\lambda \quad m = 0, \pm 1, \pm 2, \pm 3, \ldots \qquad \text{(25.15) (page 779)}$$

Destructive interference resulting in a dark fringe is found when

$$d \sin \theta = (m + \tfrac{1}{2})\lambda \quad m = 0, \pm 1, \pm 2, \pm 3, \ldots \qquad \text{(25.16) (page 779)}$$

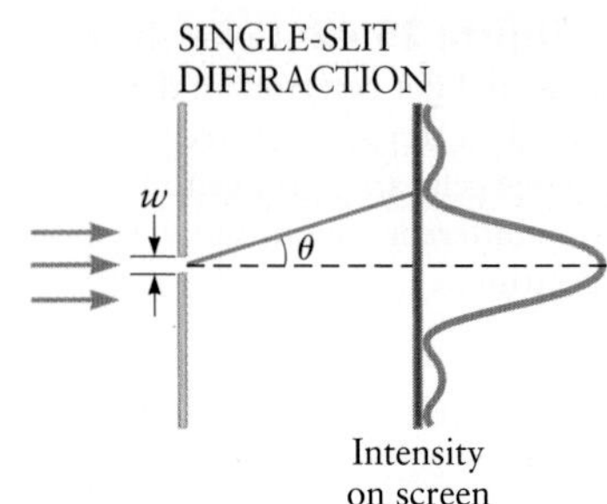

Single-slit diffraction

Single-slit diffraction occurs when light passes though a narrow opening. Dark fringes (destructive interference) occur when

$$w \sin \theta = \pm m\lambda \quad m = \pm 1, \pm 2, \ldots \qquad \text{(25.22) (page 783)}$$

Applications

Michelson interferometer

The ***Michelson interferometer*** uses the interference of light waves to measure small displacements. It was first used in fundamental studies of light and tests of relativity (Chapter 27) and is still widely used in many types of experiments.

Thin-film interference

Light waves reflected from the two surfaces of a thin film can interfere to produce brightly colored interference fringes. The relative phases of the interfering waves are affected by the different path lengths traveled by the two waves and by any phase changes on reflection.

Rayleigh criterion

When light passes through a circular opening, diffraction causes the light beam to spread. The ***Rayleigh criterion*** states that two light beams can be resolved only if their directions differ by an amount as large as or greater than

$$\theta_{min} = \frac{1.22\lambda}{D} \qquad (25.28) \text{ (page 788)}$$

This result means that it is impossible to focus a light beam to a spot much smaller than the wavelength.

Rayleigh scattering

Rayleigh scattering of light by very small particles in the sky produces the color of the sky. This scattering is strongest for shorter wavelengths and gives the sky its normal blue color.

Questions

SSM = answer in *Student Companion & Problem-Solving Guide*

(X) = life science application

1. Is it possible for light of two different colors to exhibit constructive or destructive interference? Explain.
2. Why don't you observe interference between light waves produced by two lightbulbs?
3. Figure Q25.3 shows the bands of color that are produced when a thin soap-bubble film is exposed to white light. What mechanism produces the colors seen here? Is the bubble of uniform thickness? If not, where is the bubble thinnest? Thickest? How do you know?

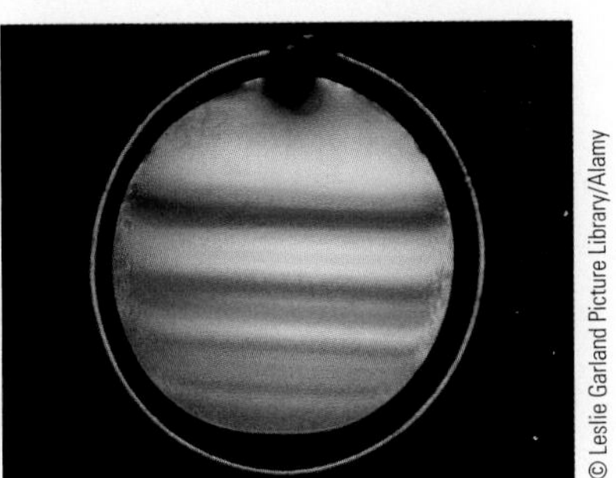
© Leslie Garland Picture Library/Alamy

Figure Q25.3

4. If a soap film suspended in air is extremely thin, what color will it be when viewed in reflected light?
5. Explain why two light waves can interfere only if they are coherent.
6. Yellow light is used to make a double-slit interference pattern with two very narrow slits. If the same slits are used with blue light, will the separation between two adjacent bright fringes be larger or smaller?
7. SSM How does a double-slit interference pattern change when the following changes are made? Do the interference fringes become closer together or farther apart? Explain your answers.
(a) The two slits are brought closer together.
(b) The two slits are moved farther apart.
(c) The wavelength is increased.
(d) The wavelength is decreased.
(e) The entire experiment is immersed in water.
8. Green light is used to make a single-slit diffraction pattern with a particular slit. If the same slit is used with red light, will the width of the central bright fringe be wider or narrower?
9. Explain how a thin coating on a lens (called an antireflection coating) can reduce the amount of reflected light by using destructive interference. Most expensive lenses have these coatings.
10. Constructive interference occurs when the crests of two coherent waves overlap (Fig. 25.1B). What happens when the troughs of two coherent waves overlap? Is the interference constructive or destructive? Explain.
11. Design an experiment in which light from a helium–neon laser ($\lambda = 630$ nm) can be used to measure (a) the width of a single slit and (b) the distance between two nearby slits.
12. In Section 25.2, we assumed the Michelson interferometer was operated with monochromatic light (light of a single wavelength). Suppose a source of white light is used instead. It is also possible to have constructive interference and hence a bright fringe with white light. What value(s) of the path length difference ΔL is required for constructive interference and thus a bright fringe with white light?
13. Some small, bright lightbulbs on a strand (Fig. Q25.13) are grouped in pairs according to color and then spaced 5 cm apart. (a) From a long distance, it becomes difficult to tell if each color is coming from one bulb or from a pair. Why? (b) There is a distance from the lights where the pairs of blue bulbs can be resolved but the pairs of red bulbs would look like one point. What feature of the red and blue light would account for this difference?

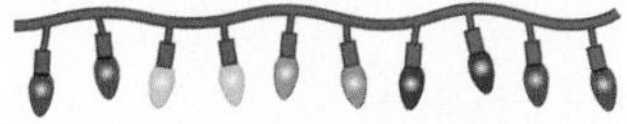

Figure Q25.13

14. SSM The star in the bottom of the Northern Cross asterism is actually a binary pair. The photograph in Figure Q25.14 came from a telescope with a mirror 10 inches in diameter. Which feature of the telescope is more important in enabling it to resolve the pair when your eye sees them as one point of light, (a) magnification or (b) the size of the mirror or primary lens? Explain your answer. You can assume fluctuations in the atmosphere as described in Section 24.7 are negligible in this case.

© Richard Yandrick (cosmicimage.com)

Figure Q25.14

15. A guitar player is at a rock concert and watches the fingers of the lead guitarist from a few rows back (Fig. Q25.15). She notes that she can see the guitarist's chord positions clearly when a yellow

or blue spotlight falls on him through the smoke, but when a bright purple light is used, the scene becomes blurry. Why can she see less detail when the purple light is used?

Figure Q25.15

16. **Eyeshine.** The eyes of cats, foxes, raccoons, and many other mammals brightly reflect light that shines into them back toward the source and makes the animal's eyes seem to glow as seen in Figure Q25.16. The reflection occurs within a thin membrane behind the retina called the *tapetum lucidum*, which allows the animal's eye more sensitivity at night. There is no metallic material within the eye, and the membrane does not look metallic under magnification. Describe a mechanism that might produce the bright reflection.

Figure Q25.16

17. You are designing a billboard that will be viewed at night. The letters used on this billboard are made of an array of bright lights like that shown in Figure Q25.17. If you want to make the sign legible from the greatest distance, which color should you choose for the lights? Why?

Figure Q25.17

18. Which single-slit diffraction experiment will give the widest central bright fringe, (a) a 900-nm slit with light of wavelength 500 nm or (b) a 600-nm slit with light of wavelength 450 nm?

19. The sky has a blue color due to the scattering of light by molecules in the air (Fig. 25.39). Explain why the sky's color can be a deeper blue when viewed through polarizing sunglasses.

20. Figure 25.20 shows the intensity pattern produced in a double-slit interference pattern with two very narrow slits. Suppose one of the slits is covered so that light can only pass through one slit. Sketch what the new intensity pattern would look like.

Problems available in ENHANCED **WebAssign**

Section 25.1 Coherence and Conditions for Interference
Problems 1–9

Section 25.2 The Michelson Interferometer
Problems 10–15

Section 25.3 Thin-Film Interference
Problems 16–23

Section 25.4 Light through a Single Slit: Qualitative Behavior
Problem 24

Section 25.5 Double-Slit Interference: Young's Experiment
Problems 25–33

Section 25.6 Single-Slit Diffraction: Interference of Light from a Single Slit
Problems 34–38

Section 25.7 Diffraction Gratings
Problems 39–47

Section 25.8 Optical Resolution and the Rayleigh Criterion
Problems 48–56

Section 25.9 Why Is the Sky Blue?
Problem 57

Additional Problems
Problems 58–67

List of Enhanced Problems

Problem number	Solution in *Student Companion & Problem-Solving Guide*	Reasoning & Relationships Problem	Reasoning Tutorial in ENHANCED WebAssign
25.6	✓		
25.14	✓		
25.22	✓		
25.24	✓		
25.28	✓		
25.35	✓	✓	
25.43	✓		
25.46		✓	✓
25.48		✓	✓
25.49		✓	✓
25.51	✓		
25.53		✓	✓
25.55		✓	✓
25.57		✓	
25.59	✓		
25.67		✓	

CHAPTER 26

▶ *The Hubble Space Telescope is one of the most advanced optical instruments ever constructed. In this chapter, you'll find out how it works. (NASA)*

Applications of Optics

OUTLINE

It is difficult to imagine a world without cameras, microscopes, CDs and DVDs, and (especially) eyeglasses and contact lenses. These and many other devices are based on the principles of optics developed in the last two chapters. In this chapter, we describe a few of these optics-based inventions and analyze how they work.

Eyeglasses were invented in the late 1200s and are perhaps the oldest optical instruments. Microscopes and telescopes were developed around 1600, and CDs and DVDs were developed in the past few decades. This continues to be an area of rapid technological change, with new optical devices constantly being invented.

26.1 Applications of a Single Lens: Contact Lenses, Eyeglasses, and the Magnifying Glass

In Section 24.6, we described how light from an object is refracted by the cornea and other regions in the eye to form an image on the retina. For simplicity, we can model the entire eye as a single lens with a focal length f_{eye}. Figure 26.1A shows a ray diagram for a "normal" eye; light emanating from a point on the object is focused to a corresponding point on the retina. The object is a distance s_o from the front of the eye, and the image is formed on the retina at an image distance $s_i = d_{eye}$, where d_{eye} is the diameter of the eye ($d_{eye} \approx 2.5$ cm in a typical case). We can find the focal length f_{eye} using the thin-lens equation (Eq. 24.21),

$$\frac{1}{s_o} + \frac{1}{s_i} = \frac{1}{f} \qquad (26.1)$$

Thin-lens equation

Substituting $s_i = d_{eye}$ and $f = f_{eye}$ gives

$$\frac{1}{s_o} + \frac{1}{d_{eye}} = \frac{1}{f_{eye}} \qquad (26.2)$$

and solving for the focal length of the eye leads to

$$f_{eye} = \frac{s_o d_{eye}}{s_o + d_{eye}} \qquad (26.3)$$

The object distance s_o in Equation 26.3 depends on the location of the object. In young adults, the eye can typically focus on objects as near as about 25 cm. This distance, called the ***near-point*** distance, is denoted as s_N. Objects closer to the eye than s_N cannot be focused on the retina; rays from such an object extrapolate to an intersection point behind the retina as shown in Figure 26.1B. This ray diagram corresponds to placing an object, such as this book, very close to your eyes; the page then appears "out of focus" on your retinas. (Try it!)

Near point of the eye

A typical young adult can also focus on objects that are very far away, that is, with $s_o \approx \infty$ (at "infinity"). Hence, the value of s_o in Equation 26.3 can vary from s_N to infinity. To satisfy Equation 26.3 at these two values of the object distance and at all values in between, the eye must adjust its focal length, which it does by using muscles that deform and change the shape of the eye's lens (Fig. 24.47).

Let's estimate the focal length of the eye when an object is at the typical near-point distance $s_N = s_o = 25$ cm. Inserting this value along with $d_{eye} \approx 2.5$ cm (as mentioned above) into Equation 26.3 gives

$$f_{eye,\,near} = \frac{s_N d_{eye}}{s_N + d_{eye}} = \frac{(25 \text{ cm})(2.5 \text{ cm})}{25 \text{ cm} + 2.5 \text{ cm}} = 2.3 \text{ cm} \qquad (26.4)$$

On the other hand, for at object at infinity we have $s_o \approx \infty$, which when inserted into Equation 26.2 leads to[1]

$$\frac{1}{f_{eye,\,distant}} = \frac{1}{s_o} + \frac{1}{d_{eye}} = \frac{1}{\infty} + \frac{1}{d_{eye}} = \frac{1}{d_{eye}}$$

$$f_{eye,\,distant} = d_{eye} = 2.5 \text{ cm} \qquad (26.5)$$

Hence, as an object moves from the near point to a point very far away, the eye must adjust its focal length from $f_{eye} = 2.3$ cm to 2.5 cm, about a 10% change.

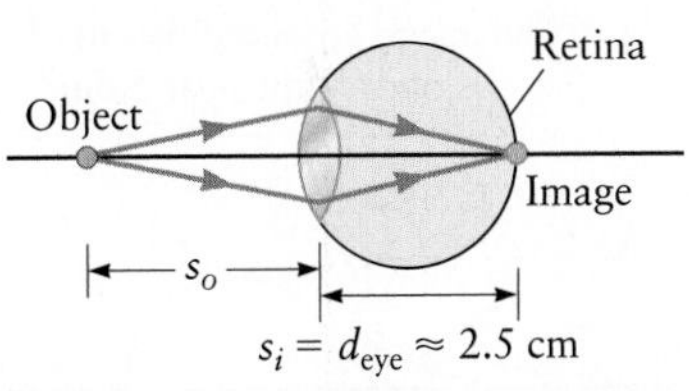
Retina
Object
Image
s_o
$s_i = d_{eye} \approx 2.5$ cm
A

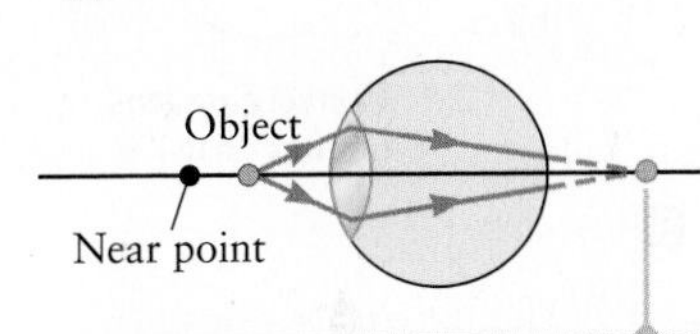
Object
Near point
When an object is placed inside the near point, the focus point is behind the retina.
B

▲ **Figure 26.1** A When properly focused, the eye forms an image on the retina. B If an object is located closer than the near point, the eye is unable to produce an image at the retina. Instead, the refracted rays extrapolate to a point behind the retina.

Eyeglasses and Contact Lenses: Adding a Lens in Front of the Eye

People who are nearsighted or farsighted often use eyeglasses or contact lenses, which are simply lenses placed in front of the eye. To understand how eyeglasses and contact lenses work, we must deal with optical "systems" that contain two or more lenses; here one lens is in the eye and the other is the eyeglass or contact lens.

In Chapter 24, we used ray tracing along with the mirror equation and the thin-lens equation to analyze the images produced by curved mirrors and lenses. In a typical case, an object is placed in front of a lens or mirror and the location of the image is then calculated. For a system containing two lenses, the general approach is as follows.

[1]We could get the same result by inserting $s_o = \infty$ into Equation 26.3 and carefully canceling terms in the numerator and denominator.

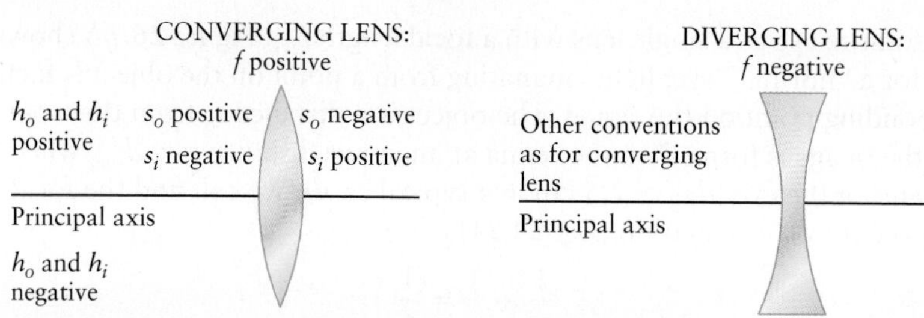

▶ **Figure 26.2** Sign conventions for working with the thin-lens equation.

RAY DIAGRAM FOR A FAR-SIGHTED PERSON

Without glasses, the image is behind the retina.

s_o

Object

Image

Near point

s_N

A

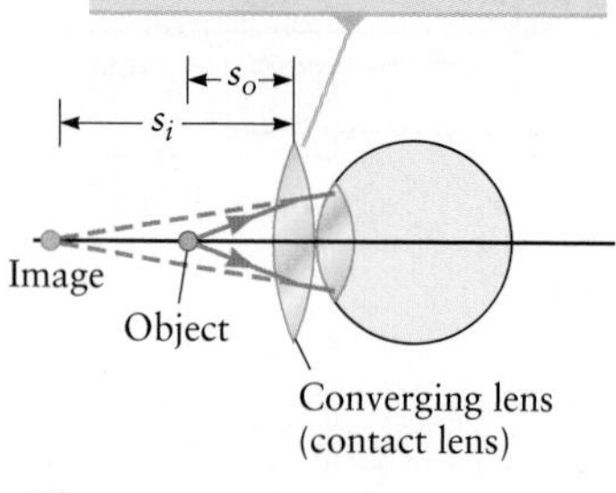

B

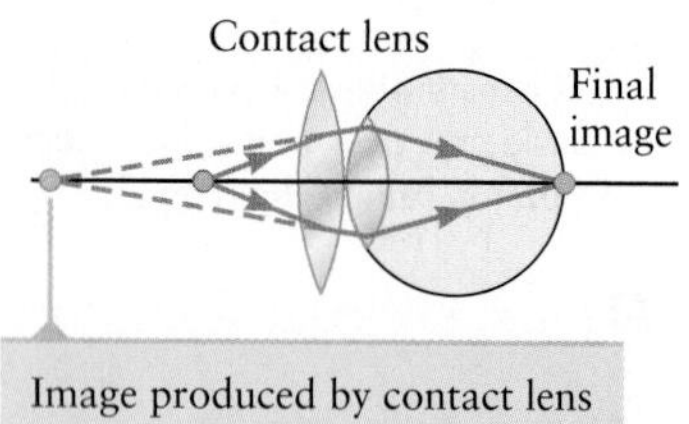

C

▲ **Figure 26.3** A For a far-sighted person, the near point is farther away than is convenient for reading or other close work. Rays from an object within the near point focus behind the retina. B When a converging lens (such as a contact lens) is placed in front of the eye, a virtual image is produced by the lens. C The image produced by the contact lens acts as the object for the eye. If this image is at or outside the near-point distance, the eye can form a final image on the retina.

Analyzing an optical system with two (or more) lenses

1. **Draw a picture showing the object of interest and the lenses in the problem.**
2. **Use the rules for ray tracing along with the thin-lens equation from Chapter 24 to find the location and magnification of the image produced by the first lens in the system.**
3. **The *image produced by the first lens* then acts *as the object for the second lens* in the system.**
4. **Use ray tracing and the thin-lens equation a second time to find the image produced by the second lens in the system.**

We'll illustrate this general approach in a number of examples throughout this chapter. Figure 26.2 shows our sign conventions (from Fig. 24.42) for applying the thin-lens equation to a single lens. A positive image distance s_i corresponds to an image on the right of the lens, whereas a negative value of s_i describes an image to the left of the lens (a "virtual image"). Notice also that a positive object distance s_o corresponds to an object to the left of the lens. In all examples in Chapter 24, the object distance was positive. Our sign conventions also allow for a negative value of s_o, which means that the object is to the right of the lens (a "virtual object"). Negative values of s_o occur in systems with two lenses, when the image produced the first lens is "beyond" (to the right of) the second lens. We'll now illustrate the key concepts by working through a series of examples.

Designing a Pair of Contact Lenses

As you age, the lenses in your eyes usually become less flexible and are not able to change focal length over the range found in Equations 26.4 and 26.5. The focal length is instead confined to a narrower range of values, reducing your ability to focus. In some cases, the near-point distance increases and is greater than the near-point distance for a normal eye; objects located closer than this near-point distance cannot be focused on the retina (Fig. 26.3A), and you must move things (such as this book) to a greater distance to see them clearly. You are then ***farsighted***. To compensate, you can place a lens (a contact lens) immediately in front of each eye. The required focal length of this lens depends on a person's near-point distance.

Suppose the near-point distance for a certain individual is $s_N = 75$ cm, which is much greater than the "normal" near-point distance of 25 cm. We wish to design a contact lens that enables this person to focus on objects at the normal near-point distance. The contact lens is placed just in front of the eye as in Figure 26.3B; it forms the first lens of our "system," and the eye acts as the second lens. We want this system of two lenses to be able to take an object at $s_o = 25$ cm from the contact lens and produce a final image on the retina. We follow our general rules for analyzing optical systems with two lenses and analyze the problem one lens at a time. The contact lens produces an image (Fig. 26.3B) that then acts as the object for the second "lens," the eye, but where should the image formed by the contact lens be located? We are given that the near-point distance for this eye is 75 cm, which means that the eye alone can focus an object that is 75 cm away to produce an image on the retina. So, if the image produced by the contact lens is at this distance (or greater, Fig. 26.3C), the eye can produce a final image at the retina. We thus want the contact lens to take an object at $s_o = 25$ cm and produce an image that is 75 cm to the left of the eye. Since the contact lens is touching the eye, the image distance is $s_i = -75$ cm. This image distance is negative because it is to the left of the contact lens and is thus a virtual image; see the sign conventions in Fig. 26.2.

We now apply the thin-lens equation (Eq. 26.1) to the contact lens; we substitute $s_o = 25$ cm and $s_i = -75$ cm to find

$$\frac{1}{f_{\text{lens}}} = \frac{1}{s_o} + \frac{1}{s_i} = \frac{1}{+25\text{ cm}} + \frac{1}{-75\text{ cm}} \tag{26.6}$$

and solve for f_{lens}:

$$f_{\text{lens}} = 38\text{ cm} \tag{26.7}$$

which is the focal length required for the contact lens. Since f_{lens} is positive, this lens is a converging lens.

If the person's near-point distance is larger than 75 cm, Equation 26.6 shows that a lens with a shorter focal length is needed. In words, we say that a person with a large near-point distance needs a "stronger" lens than a person with a short s_N. The strength of a lens is measured in terms of its ***refractive power***, where

$$\text{refractive power} = \frac{1}{f_{\text{lens}}} \tag{26.8}$$

The unit of refractive power is m^{-1}, which is called a ***diopter***, and

$$1\text{ diopter} = 1\text{ m}^{-1}$$

The contact lens with the focal length in Equation 26.7 thus has a refractive power of $1/f_{\text{lens}} = 1/(0.38\text{ m}) = 2.7$ diopters.

Let's now consider what type of contact lens is needed by a ***nearsighted*** person. Such a person is unable to focus light from distant objects on the retina as shown in the ray diagram in Figure 26.4A. The incoming rays from an object very far away are approximately parallel to the axis (an object at "infinity"); a nearsighted eye produces an image in front of the retina.

A nearsighted person is able to properly focus objects that are within a certain distance. Suppose a person can focus objects within 2.0 m on the retina. What kind of contact lens does this person need so that an object that is very far away ($s_o = \infty$) gives a final image at the retina? It is given that the eye alone can focus an object that is 2.0 m away on the retina, so the contact lens must take an object at $s_o = \infty$ and give an image at $s_i = -2.0$ m (Fig. 26.4B).

Applying the thin-lens equation to the contact lens, we have $s_o = \infty$ and $s_i = -2.0$ m and get

$$\frac{1}{f_{\text{lens}}} = \frac{1}{s_o} + \frac{1}{s_i} = \frac{1}{\infty} + \frac{1}{-2.0\text{ m}}$$

$$f_{\text{lens}} = -2.0\text{ m}$$

which is the focal length required for the contact lens. Since f_{lens} is negative, it is a diverging lens.

RAY DIAGRAM FOR A NEAR-SIGHTED PERSON

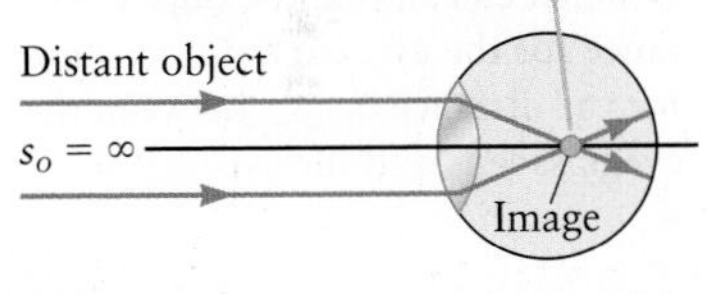
Without glasses, the image is in front of the retina.
Distant object
$s_o = \infty$
Image
A

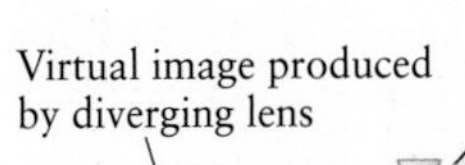
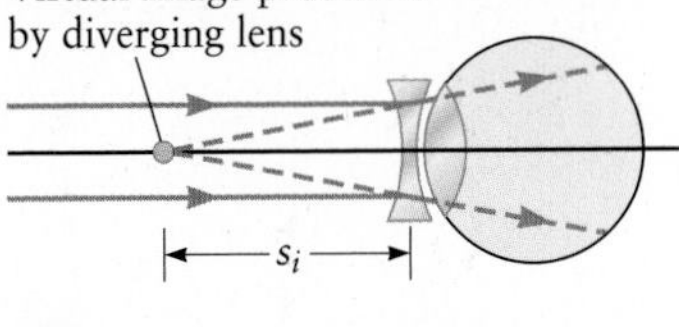
Virtual image produced by diverging lens
s_i
B

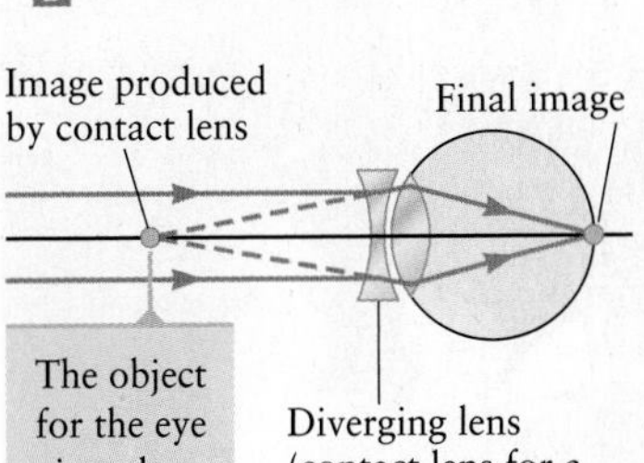
Image produced by contact lens
Final image
The object for the eye gives the image on the retina.
Diverging lens (contact lens for a near-sighted person)

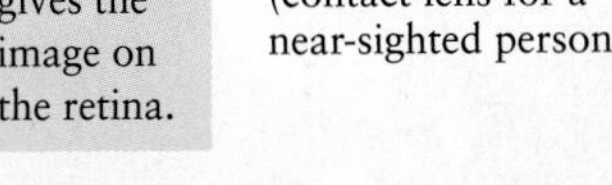
C

▲ **Figure 26.4** A When a person is nearsighted, rays from an object very far away (at infinity) focus in front of the retina. B Placing a diverging lens in front of the eye produces a virtual image to the left of the lens. C The image formed by the lens acts as the object for the eye. The nearsighted eye is able to focus this image on the retina.

CONCEPT CHECK 26.1 What Kind of Image Is It?

Figures 26.3 and 26.4 show ray diagrams for different types of contact lenses. Which of the following statements correctly describes the images formed by these contact lenses?

(a) A properly designed contact lens always forms a real image.
(b) A properly designed contact lens always forms a virtual image.
(c) A properly designed contact lens can form a real or virtual image, depending on the type of lens.

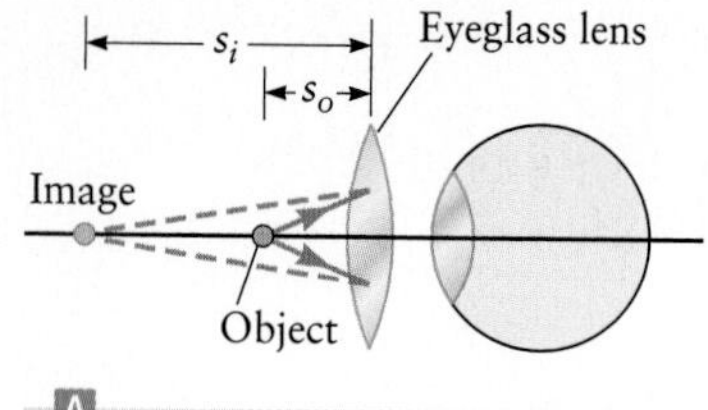

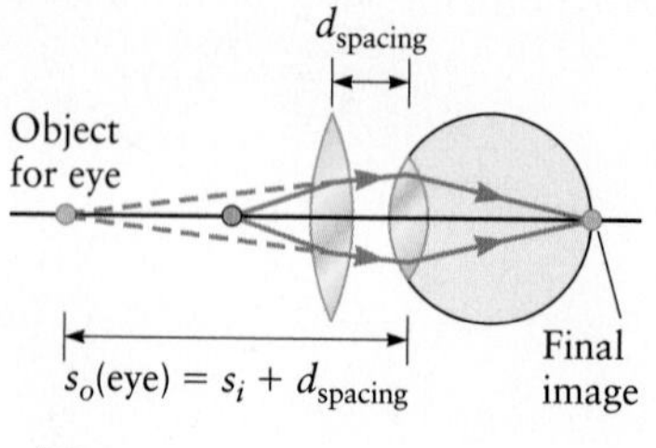

▲ **Figure 26.5** Image produced by the combination of an eyeglass lens plus the eye. **A** The eyeglass lens produces an image that then acts as the object for the eye. **B** When calculating the object distance for the eye, we must account for the distance $d_{spacing}$ between the eyeglass lens and the eye.

CONCEPT CHECK 26.2 Designing Contact Lenses

A farsighted person has an extremely long near-point distance. If the person wishes to read a newspaper at a normal distance, what focal length do you suggest for his contact lenses?

(a) 25 cm (b) 400 cm (c) 4.0 diopters

Designing a Pair of Eyeglasses

The problem of selecting the proper lens for a set of eyeglasses is very similar to the contact lens problems considered in Figures 26.3 and 26.4. The main difference is that the eyeglass lens is a short distance in front of the eye, rather than touching it (Fig. 26.5). This distance must be accounted for when relating the image distance from the eyeglass lens and the object distance for the eye. The rest of the analysis is the same as with the contact lens problem and is illustrated in Example 26.1.

EXAMPLE 26.1 What Kind of Eyeglass Lenses Do You Need?

A certain farsighted person has a near-point distance of 75 cm. In Figure 26.3, we considered a contact lens that would enable this person to read a book located at the "normal" near-point distance of 25 cm. The same person now wants a set of eyeglasses that he can use to read this book at 25 cm. What is the required focal length of the eyeglass lenses? Assume the glasses are positioned about 2 cm in front of the person's eye.

RECOGNIZE THE PRINCIPLE

With an object 25 cm in front of the eye, we want the eyeglass lens to produce an image 75 cm in front of the eye. This image then forms the object for the eye of our farsighted person, and we know that for this object distance he will be able to focus the final image on his retina.

SKETCH THE PROBLEM

The problem is described by the ray diagrams in Figure 26.5.

IDENTIFY THE RELATIONSHIPS

Since the eyeglass lens is $d_{spacing} = 2$ cm in front of the eye and the object is 25 cm in front of the eye, the object and image distances *for the eyeglass lens* are $s_o = 25 - d_{spacing} = 23$ cm and $s_i = -73$ cm (Fig. 26.5A). This image is virtual (s_i is negative) because light does not actually pass through this point.

SOLVE

Using the thin-lens equation with $s_o = 23$ cm and $s_i = -73$ cm, we get

$$\frac{1}{f_{lens}} = \frac{1}{s_o} + \frac{1}{s_i} = \frac{1}{23 \text{ cm}} + \frac{1}{-73 \text{ cm}}$$

$$f_{lens} = \boxed{34 \text{ cm}}$$

Since f_{lens} is positive, this lens is a converging lens.

► ***What does it mean?***

Comparing this result for f_{lens} with the result in Equation 26.7 for a contact lens, we see that the eyeglass lenses have a focal length about 10% shorter. This is why your prescription for contact lenses is slightly different than for your eyeglasses.

▲ **Figure 26.6** A magnifying glass produces an upright virtual image.

The Magnifying Glass

The magnifying glass is one of the most basic of all optical devices and is a component in microscopes, telescopes, and other instruments. The simplest magnifying glass is a single lens, used to view things that are too small to examine easily with the naked eye (Fig. 26.6). To analyze this optical "instrument," we must again consider a system consisting of two lenses—the magnifying glass lens and the eye behind it—so the problem is similar to our analysis of contact lenses and eyeglasses. With a magnifying glass, the goal is to produce a greatly magnified image at the retina. We thus want the image on the retina to be as large as possible.

Consider first the case without a magnifying glass. To examine a small object, the natural thing to do is to place the object as close as possible to your eye. Figure 26.7A shows how the position of an object affects the size of its image on the retina. As the object is brought closer and closer to the eye from point *A* to *B*, the image on the retina grows larger and the apparent size of the object increases, but there is a limit to how close you can bring the object and still keep it in focus. The largest clearly focused image for the unaided eye results when the object is at the near point. The object's apparent size when it is located at the near point can be measured using the angle θ defined in Figure 26.7B; we'll call it the image angle.

To make the object even larger on the retina than in Figure 26.7B, we need to move it closer to the eye and hence inside the near point. The eye by itself cannot then produce an image on the retina; it needs the help of the magnifying glass lens.

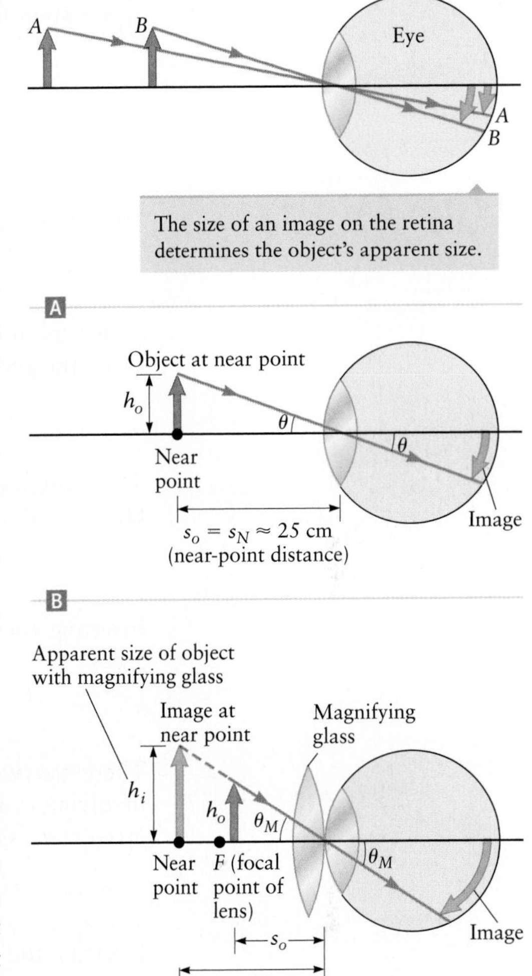

▲ **Figure 26.7** A The size of an object as perceived by the eye depends on the size of the image on the retina. If an object is brought closer to the eye (from *A* to *B*), the rays from its tip make a larger angle with the axis and the image on the retina becomes larger. B The closest object distance at which the eye can focus the image on the retina is the near point, which for a typical young person is approximately 25 cm from the eye. C A magnifying glass produces an enlarged image at the near point of the eye. The eye focuses the light from this image on the retina.

Image Properties with a Magnifying Glass

A magnifying glass is a single converging lens placed in front of the eye as shown in Figure 26.7C. The object is positioned inside the focal length of this lens, that is, so that the focal point *F* of the lens is to the left of the object in Figure 26.7C. We showed in Chapter 24 (Fig. 24.45) that this position leads to an upright virtual image at the far left in Figure 26.7C. As far as the eye is concerned, light from the object emanates from this virtual image, and if this image produced by the magnifying glass is at the near point, the eye is able to focus this light onto the retina. Comparing the ray diagrams in Figures 26.7B and C shows that the image angle with the magnifying glass θ_M (Fig. 26.7C) is greater than the image angle for the eye alone (θ in Fig. 26.7B), so the image *on the retina* is enlarged by the magnifying glass.

The enlargement of the image on the retina is given by the ***angular magnification*** m_θ, defined by

$$m_\theta = \frac{\theta_M}{\theta} \tag{26.9}$$

The angular magnification is related to (but slightly different from) the linear magnification *m* defined in Chapter 24. The linear magnification involves the height of the object and image, whereas the angular magnification involves the angle the image makes on the retina.

We can calculate the angular magnification m_θ by applying the thin-lens equation (Eq. 26.1) to the lens of the magnifying glass (Fig. 26.7C):

$$\frac{1}{s_o} + \frac{1}{s_i} = \frac{1}{f} \tag{26.10}$$

As always, we must be careful to follow the proper sign conventions. Because the magnifying glass is a converging lens, its focal length f is positive. The object is on the left side of the lens, so the object distance s_o is positive; the image is also on the left side of the lens, so the image distance s_i is negative. For simplicity, let's assume the magnifying glass is very close to the eye; the image is at the near point and hence $s_i = -s_N$ in Equation 26.10. Inserting all this information leads to

$$\frac{1}{s_o} - \frac{1}{s_N} = \frac{1}{f}$$

$$\frac{1}{s_o} = \frac{1}{f} + \frac{1}{s_N} = \frac{s_N + f}{f s_N}$$

Solving for s_o, we get

$$s_o = \frac{s_N f}{s_N + f} \tag{26.11}$$

We can now find the image angle with and without the magnifying glass in place. From the geometry in parts B and C of Figure 26.7, we have

$$\tan\theta = \frac{h_o}{s_N} \quad \text{and} \quad \tan\theta_M = \frac{h_i}{s_N}$$

These angles are usually small, so we can use the approximations $\tan\theta \approx \theta$ and $\tan\theta_M \approx \theta_M$, leading to

$$\theta = \frac{h_o}{s_N} \quad \text{and} \quad \theta_M = \frac{h_i}{s_N} \tag{26.12}$$

Inserting these angles into the definition for the angular magnification (Eq. 26.9) gives

$$m_\theta = \frac{\theta_M}{\theta} = \frac{h_i}{h_o}$$

There are two right triangles in Figure 26.7C, one involving h_o and s_o and another involving h_i and s_N. These triangles are similar (having the same interior angles), so $h_i/s_N = h_o/s_o$. The angular magnification is thus

$$m_\theta = \frac{h_i}{h_o} = \frac{s_N}{s_o}$$

Inserting the result for s_o from Equation 26.11 gives our final result,

$$m_\theta = \frac{s_N}{s_o} = \frac{s_N}{(s_N f)/(s_N + f)}$$

$$m_\theta = \frac{s_N + f}{f} = \frac{s_N}{f} + 1 \tag{26.13}$$

Practical magnifying glass lenses have focal lengths that are smaller than the near-point distance by a factor of 10 or more, so the ratio s_N/f in Equation 26.13 is much greater than 1. Hence, to a good approximation we can write

Magnification of a magnifying glass

$$m_\theta = \frac{s_N}{f} \tag{26.14}$$

For a typical magnifying glass, this result is accurate to 10% or better (and we can always use the result for m_θ in Eq. 26.13 if greater accuracy is needed). This expression for m_θ is also found when a magnifying glass is used as the eyepiece of a microscope or telescope, which we'll explore in Example 26.2.

The near-point distance s_N varies from person to person, so the magnification actually obtained with a particular magnifying glass (having a particular value of f) will depend on the person. The "typical" near-point distance for a young adult (25 cm) is used as a convention when specifying magnifying glasses:

$$m_\theta = \frac{s_N}{f} = \frac{25\text{ cm}}{f}$$

For a magnifying glass specified by the manufacturer as providing $m_\theta = 10$ (also known as a "10X" magnifying glass), the focal length is thus

$$f = \frac{25\text{ cm}}{m_\theta} = \frac{25\text{ cm}}{10} = 2.5\text{ cm}$$

The angular magnification of a magnifying glass or eyepiece is typically 10 or 20.

EXAMPLE 26.2 Angular Magnification for an Object at the Focal Point

Figure 26.8A shows an object at the focal point of a magnifying glass. This arrangement is slightly different from the one we considered in Figure 26.7C, where the object was inside the focal point and the image was at the near point. Now the image from the magnifying glass is at infinity. What is the angular magnification? Express your answer in terms of the focal length of the lens and the near-point distance.

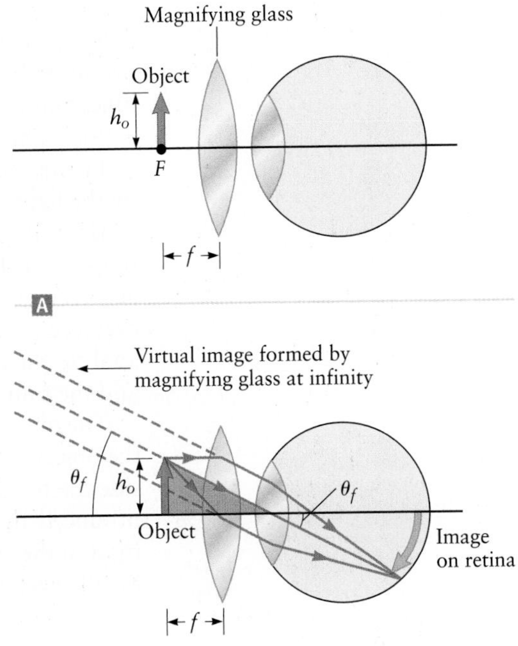

▲ **Figure 26.8** Example 26.2. A An object at the focal point of a magnifying glass. B The lens forms a virtual image an infinite distance (at "infinity") in front of the eye. The light from this virtual image forms a parallel bundle of rays that are then focused by the eye to form an image on the retina.

RECOGNIZE THE PRINCIPLE

To find the angular magnification, we must compare the angle θ that the object makes with the principal axis of the eye when the object is at the near point (without the magnifying glass; Fig. 26.7B), with the angle θ_f when the object is at the focal point of the magnifying glass (Fig. 26.8B). The angular magnification is (compare with Eq. 26.9)

$$m_\theta = \frac{\theta_f}{\theta} \qquad (1)$$

SKETCH THE PROBLEM

Figure 26.8B shows a ray diagram. The object is at the focal point of a converging lens. From Chapter 24, we know that the lens will form an image at infinity ($s_i = \infty$); hence, the rays between the magnifying glass and the eye are parallel as shown in Figure 26.8B.

IDENTIFY THE RELATIONSHIPS AND SOLVE

Consider the right triangle shaded in red in Figure 26.8B. We assume the magnifying glass is very close to the eye, so the red triangle has sides f and h_o and contains the angle θ_f, with $\tan\theta_f = h_o/f$. Using the small-angle approximation ($\tan\theta \approx \theta$) gives $\theta_f = h_o/f$. From Figure 26.7B, an object at the near point viewed with the unaided eye has an angular size θ with $\tan\theta = h_o/s_N$. Using the small-angle approximation again gives $\theta = h_o/s_N$. Inserting these results for θ_f and θ into Equation (1), we get a familiar result (Eq. 26.14):

$$m_\theta = \frac{\theta_f}{\theta} = \frac{h_o/f}{h_o/s_N} = \boxed{\frac{s_N}{f}}$$

▸ *What does it mean?*

As viewed by the eye, the object in Figure 26.8 appears to be at a location very far away (at "infinity") because the magnifying glass produces a collection of parallel rays that enter the eye. The magnifying glass is a component in many other optical instruments, including microscopes and telescopes, where it also produces a set of parallel rays that are viewed by an observer. This result for the magnification will help us analyze the role of the magnifying glass in those instruments.

CONCEPT CHECK 26.3 The Image from a Magnifying Glass
Does a magnifying glass form a real image or a virtual image? Explain.

26.2 Microscopes

A magnifying glass is very useful for viewing small objects, but there is a practical limit to the magnification that can be obtained with a single lens. Lenses with a focal length of less than a few millimeters are difficult to make, which leads to a limit on the angular magnification (Eq. 26.14) of around 100 or so, and most practical magnifying glasses have angular magnifications of only 10 or 20. A higher magnification can be achieved using two lenses arranged as a ***compound microscope***. The image produced by one lens is used as the object for the second lens, and the image produced by the second lens is then viewed by the eye of an observer. The total magnification is the product of the magnifications of the two lenses; an angular magnification as high as $m_\theta = 1000$ is readily attained in this way.

Figure 26.9A shows a compound microscope and the usual vertical path followed by the light, and Figure 26.9B shows a ray diagram with the light traveling from left to right according to our usual convention. The two lenses in a compound microscope are called the ***objective*** and the ***eyepiece***. The objective is a converging lens, so according to our sign conventions (Fig. 26.2) it has a positive focal length f_{obj}. The eyepiece is simply a magnifying glass; it also has a positive focal length f_{eyepiece}. To analyze the image produced by a compound microscope, we first apply ray tracing and the thin-lens equation to find the image produced by the objective lens.

The object to be studied is placed just outside (i.e., just to left) of the focal point of the objective lens (Fig. 26.9B). From Chapter 24 (Fig. 24.40C), an object located outside the focal length of a converging lens gives a real and inverted image. The image produced by the objective in Figure 26.9B is therefore drawn a distance s_i to the right of the objective lens. The image produced by the objective lens forms the object for the eyepiece. The eyepiece then behaves in the same way as the magnifying glass

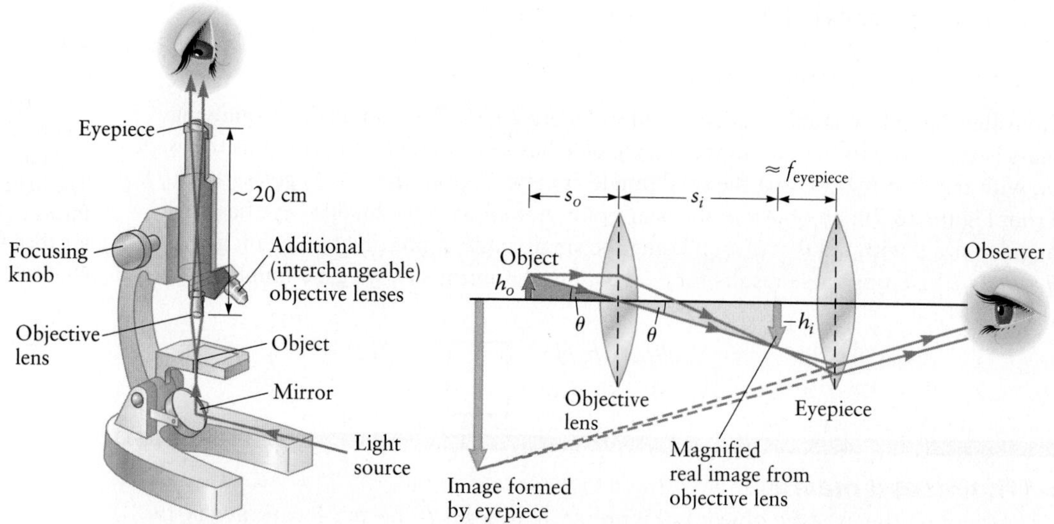

▲ **Figure 26.9** A A compound microscope contains two lenses. The object of interest is placed close to the objective lens, and the observer's eye is close to the eyepiece. B Ray diagram for a compound microscope. The object is placed just outside the focal point of the objective lens, producing a magnified real image. The image from the objective lens is then the object for the eyepiece, which produces a final virtual image much larger than the original object.

discussed in Section 26.1. The distance between the objective lens and the original object is adjusted so that the image produced by the objective falls at the focal point of the eyepiece, which gives a final virtual image for the observer. (Compare the rays produced by the eyepiece in Fig. 26.9B with the corresponding rays in Fig. 26.8B.)

To find the total magnification of the microscope, we must first calculate s_i. Figure 26.9B shows two similar triangles, one shaded in red with sides h_o and s_o and another shaded in yellow with sides $-h_i$ and s_i. (The negative sign in front of h_i is due to our sign convention for lenses, which says an inverted image has a negative value for its height.) Both of these right triangles have an interior angle θ, so

$$\frac{h_o}{s_o} = \tan\theta = \frac{-h_i}{s_i}$$

Since the object is placed very close to the focal point of the objective, $s_o \approx f_{\text{obj}}$. In Figure 26.9B, the image produced by the objective is larger than the original object. The ratio of the image and object heights is called the ***linear magnification*** (Chapter 24). For the objective lens in Figure 26.9B, we have a linear magnification $m_{\text{obj}} = h_i/h_o$. Putting it all together, we find

$$m_{\text{obj}} = \frac{h_i}{h_o} = -\frac{s_i}{f_{\text{obj}}} \quad (26.15)$$

The final virtual image produced by the eyepiece is a greatly magnified version of the original object. The total magnification of the microscope is the product of the linear magnification of the objective and the angular magnification of the eyepiece. If the eyepiece has a focal length f_{eyepiece}, its angular magnification according to Equation 26.14 is $m_{\theta,\text{ eyepiece}} = s_N/f_{\text{eyepiece}}$, where $s_N = 25$ cm is the value (by convention) of the near-point distance. Combining this expression with Equation 26.15, we have

$$m_{\text{total}} = m_{\text{obj}}m_{\theta,\text{ eyepiece}} = -\frac{s_i}{f_{\text{obj}}}\frac{s_N}{f_{\text{eyepiece}}} \quad (26.16)$$

Magnification of a compound microscope

The negative value for m_{total} means that the image is inverted.

For a typical student microscope like the one in Figure 26.9A, the focal lengths might be $f_{\text{obj}} = 4$ mm and $f_{\text{eyepiece}} = 2.5$ cm (a so-called 10× eyepiece), and the distance between the objective and the eyepiece is about 20 cm, so s_i is about 0.18 m (20 cm minus f_{eyepiece}). Inserting these values into Equation 26.16 gives

$$m_{\text{total}} = -\frac{s_i}{f_{\text{obj}}}\frac{s_N}{f_{\text{eyepiece}}} = -\frac{(0.18\text{ m})(25\text{ cm})}{(0.004\text{ m})(2.5\text{ cm})} = -450$$

which is quite a useful value.

Advances in Microscope Design

The index of refraction of the glass used to make lenses is usually slightly different for light of different colors, which, according to the lens maker's formula (Eq. 24.32), makes the focal length f slightly different for different colors and affects the focusing properties of a microscope. When light of a particular color is in focus, light of a different color will not quite be in focus. This chromatic aberration can be minimized by making the objective lens out of several separate lenses, composed of different types of glass with different indices of refraction. Clever optical design approximately cancels the aberrations from these individual lenses. Such lenses are called ***achromatic*** and are used in most modern high-quality microscopes.

Although most microscopes are thus not quite as simple as the one sketched in Figure 26.9A, the basic result for the total magnification in Equation 26.16 still applies, but now f_{obj} is the focal length of the objective lens system and f_{eyepiece} is the focal length of the complete eyepiece. Hence, to obtain a high magnification, these two focal lengths should be as short as possible. This aspect of microscope design is explored in Example 26.3.

EXAMPLE 26.3 Ⓧ Ⓡ Designing a Compound Microscope

You are given the task of designing a compound microscope like the one in Figure 26.9A. The eyepiece is to have an angular magnification $m_{\theta,\,\text{eyepiece}} = 10$, and you want the total magnification to be $m_{\text{total}} = -250$. To achieve this magnification, what should be the focal length f_{obj} of the objective lens?

RECOGNIZE THE PRINCIPLE

The total magnification of a compound microscope depends on the focal lengths of the objective and eyepiece and the distance between the two lenses. We are not given this distance, but we can estimate a typical value from Figure 26.9A. We can also find the focal length of the eyepiece from its magnification (which is given).

SKETCH THE PROBLEM

Figure 26.9 shows the geometry of a compound microscope.

IDENTIFY THE RELATIONSHIPS

The total magnification of a compound microscope is (Eq. 26.16)

$$m_{\text{total}} = -\frac{s_i}{f_{\text{obj}}}\frac{s_N}{f_{\text{eyepiece}}} \tag{1}$$

The desired magnification is given, and by convention the near-point distance is $s_N = 25$ cm. The angular magnification of the eyepiece is also given, and from Equation 26.14 we have $m_{\theta,\,\text{eyepiece}} = s_N/f_{\text{eyepiece}}$. Hence, to use Equation (1) to find f_{obj} we need to estimate s_i in Figure 26.9B. This distance is nearly equal to the distance between the objective and the eyepiece (because s_i is usually much larger than the focal length of the eyepiece); from Figure 26.9A, we see that this distance is also the approximate length of the body of the microscope. From our experience with student microscopes (and Fig. 26.9A), we estimate $s_i = 20$ cm. Solving for f_{obj} leads to

$$f_{\text{obj}} = -\frac{s_i}{m_{\text{total}}}\frac{s_N}{f_{\text{eyepiece}}}$$

Using $m_{\theta,\,\text{eyepiece}} = s_N/f_{\text{eyepiece}}$ this result can also be written as

$$f_{\text{obj}} = -\frac{s_i}{m_{\text{total}}}m_{\theta,\,\text{eyepiece}} \tag{2}$$

SOLVE

Inserting the values of the various quantities gives

$$f_{\text{obj}} = -\frac{s_i}{m_{\text{total}}}m_{\theta,\,\text{eyepiece}} = -\left(\frac{20\text{ cm}}{-250}\right)(10) = 0.80\text{ cm} = \boxed{8.0\text{ mm}}$$

► ***What does it mean?***

According to Equation (1), if we want a higher total magnification, we need an objective with a smaller focal length. Because the object must be placed a distance $s_o \approx f_{\text{obj}}$ from the approximate *center* of the objective lens (see Fig. 26.9B), however, a practical limit arises. A real lens is at least several millimeters thick, so we cannot make f_{obj} smaller than that. As a result, the practical limit on the magnification of a compound microscope is about 1000.

Insight 26.1

FOCUSING A COMPOUND MICROSCOPE

A typical compound microscope like the one in Figure 26.9A has several interchangeable objective lenses, allowing the user to obtain different magnifications. According to the result for the total magnification in Equation 26.16, if we wish to increase the magnification, we must make the focal length of the objective very short. The object is positioned just outside the focal point of the objective (Fig. 26.9B), and since f_{obj} is small, the object must be very close to the objective lens. For an objective lens with a magnification of −100, the distance between the object and the surface of the objective lens can be less than 1 mm. This distance is adjusted with the microscope's focusing knob.

Ultimate Resolution of a Microscope

Example 26.3 mentioned some practical limits on the total magnification of a compound microscope. Might a very clever physicist devise a different microscope design that can achieve much greater magnifications, or is there a fundamental limit to the magnification that any microscope can achieve? One can indeed use additional lenses, more powerful eyepieces, and so forth and make the magnitude of the total magnification greater than the value of 500 to 1000 that is typical for compound microscopes. Even with such a high magnification, however, there is still a fundamental limit to the ***resolution*** that can be achieved with *any* microscope that relies on focusing.[2] This limit is due to the diffraction of light passing through the aperture of the microscope (i.e., through the lens openings). In Section 25.8, we showed that even when a perfect lens is used to focus light, diffraction prevents the size of the focused spot from being less than a value approximately equal to the wavelength of the light (Fig. 25.36 and Eq. 25.33).

The same diffraction effect limits the resolution achievable with a microscope. Suppose a microscope is used to observe two very closely spaced features on an object. Figure 26.10 shows light from these two features as it travels through the objective lens. The aperture of the objective leads to a slight spreading of the outgoing light rays from the two features, just the "inverse" of the spreading found when focusing light to a spot. An analysis similar to the one given in Section 25.8 shows that it is possible to resolve the outgoing light from the two features only if they are separated by a distance greater than approximately the wavelength λ. If they are closer, it is not possible to tell that there are two separate features. Hence, the ultimate optical resolution of a microscope is approximately equal to the wavelength of the light that is used. Applications requiring the best possible resolution use blue or even ultraviolet light because these colors have the shortest wavelengths compared with other colors of visible light.

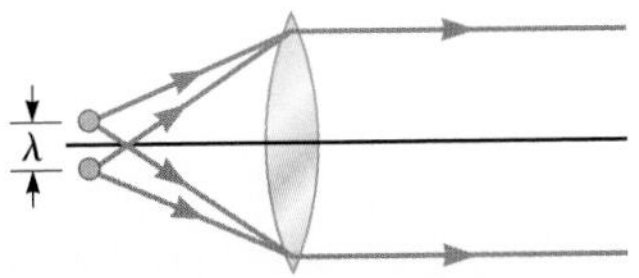
λ
Objects are indistinguishable in the outgoing beam.

▲ **Figure 26.10** When two objects are close together, their diffraction spots overlap. If they are too close together (separated by about λ or less), they appear as a single object instead of two separate objects when viewed through a lens.

The Confocal Microscope

Biologists examining samples of cells or tissue often want to resolve two nearby features according to their depth within a semitransparent object. Figure 26.11A shows a schematic diagram of a ***confocal microscope.*** This instrument is designed so that features at only one particular depth form the final image, which is accomplished by placing a very small opening, called a pinhole or aperture, in front of the observer. This pinhole allows mainly light from the selected depth to reach the observer and blocks most of the light from other depths. By combining images made at different depths, a confocal microscope yields a three-dimensional image of a sample. Confocal microscopes provide an extremely powerful way to examine features inside transparent objects such as a living cell (Fig. 26.11B).

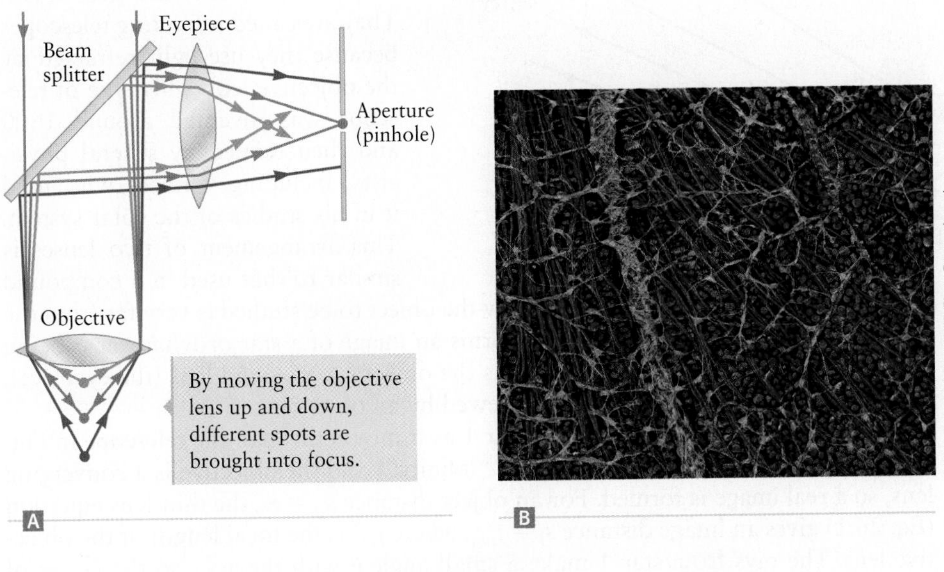
Eyepiece
Beam splitter
Aperture (pinhole)
Objective
By moving the objective lens up and down, different spots are brought into focus.
A
B
© Thomas Deerinck NCMIR/Photo Researchers, Inc.

◄ **Figure 26.11** **A** A confocal microscope is similar in design to the compound microscope in Figure 26.9, but with the addition of a small pinhole (aperture) in front of the observer. This pinhole selects light from a specific distance below the objective lens. By scanning the objective lens back and forth and also up and down, one can construct a three-dimensional image. The colors of these rays do not indicate the color of the light, but instead denote rays from points at different depths within the object being studied. **B** Confocal light micrograph of the retina. The tissue has been fluorescently stained, and the colors shown do not correlate with the colors of the rays in part A. Optic nerve fibers are shown in red and blood vessels are shown in blue. The region shown here is about 0.6 mm on a side.

[2]A microscope that does not rely on focusing is not restricted by this limit; one such microscope is described in Section 26.7.

Although the optical principles of a confocal microscope have been understood since the time of Galileo, modern computers and lasers make confocal microscopes practical. The pinhole in front of the observer blocks much of the light from the object, so a very intense light source is needed. For this reason, most confocal microscopes use lasers, which have been available for this kind of application only since the 1960s. Second, collecting many images at different depths requires that the objective lens be scanned side to side and also up and down in Figure 26.11A. Computers are essential for automating the scanning and for analyzing the large amount of data needed to assemble a three-dimensional image.

CONCEPT CHECK 26.4 Making a Better Microscope

You want to increase the magnification of a particular compound microscope. Which of the following changes would achieve that goal? (More than one answer may be correct.)

(a) Increase the focal length of the objective lens.
(b) Decrease the focal length of the objective lens.
(c) Increase the focal length of the eyepiece.
(d) Decrease the focal length of the eyepiece.
(e) Increase the spacing between the objective lens and the eyepiece.

26.3 Telescopes

When you observe two stars with a telescope, light rays extending from the stars to you are nearly parallel to one another. Figure 26.12 shows light rays emanating from two stars; here rays from star 1 make a very small angle θ with the axis of a telescope, whereas light from star 2 travels parallel to the axis. The purpose of the telescope is to increase the angular separation of rays from these two stars so that your eye can distinguish one star from the other.

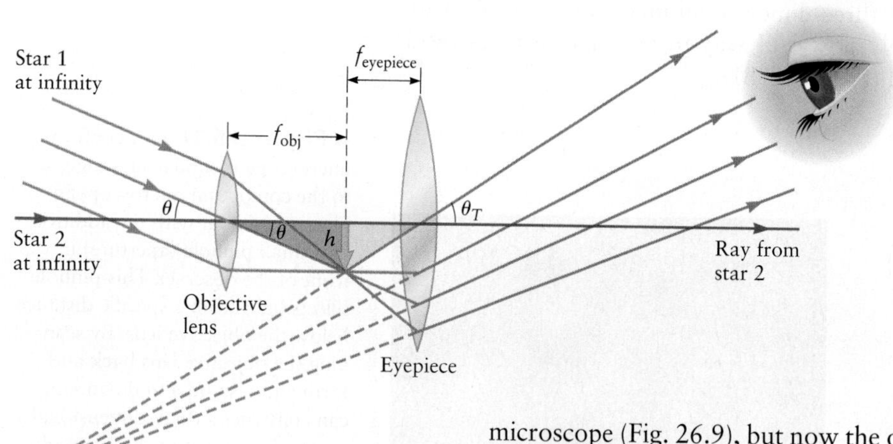

▲ **Figure 26.12** A refracting telescope uses two lenses. Light from a distant object is focused very close to the focal point of the objective lens. The eyepiece then forms a final virtual image for the observer. This ray diagram is *not* to scale.

Refracting Telescopes

The simplest telescopes employ two lenses called an objective and an eyepiece as sketched in Figure 26.12. They are called ***refracting*** telescopes because they use light refracted by the objective lens. This type of telescope was invented around 1600 and then refined by several physicists, including Galileo, who used it in his studies of the solar system. This arrangement of two lenses is similar to that used in a compound microscope (Fig. 26.9), but now the object to be studied is very far from the objective. The objective lens forms an image of a star or whatever is being studied. This image then acts as the object for a second lens (the eyepiece), which forms the final image viewed by an observer.

Let's consider light from star 1 as it moves through the telescope in Figure 26.12. This star is very far away (at "infinity") and the objective is a converging lens, so a real image is formed. For an object distance $s_o = \infty$, the thin-lens equation (Eq. 26.1) gives an image distance $s_i = f_{obj}$, where f_{obj} is the focal length of the objective lens. The rays from star 1 make a small angle θ with the axis, so the image of star 1 is a small distance h off the principal axis. The telescope's eyepiece is located such that the image formed by the objective is very close to the focal point of the eyepiece. The image from the objective is now the object for the eyepiece; since this new object is a distance f_{eyepiece} from the eyepiece lens, the rays emanating from this object are refracted by the eyepiece to form a "bundle" of nearly parallel rays that are perceived by the observer.

The telescope's magnification is determined by the angles that rays from star 1 make with the axis; before reaching the telescope (on the far left in Fig. 26.12), this angle is θ; after passing through the eyepiece, it is θ_T. The total magnification of the telescope is the ratio of these two angles,

$$m_\theta = \frac{\theta_T}{\theta} \tag{26.17}$$

Equation 26.17 actually gives an angular magnification because it involves the two angles, but it is usually called just the "magnification." A large magnification m_θ means that rays from the two stars in Figure 26.12 are separated by a large amount and hence an observer can more easily resolve that there are actually two separate stars present.

To calculate m_θ, consider the red right triangle in Figure 26.12. It has sides of length h and f_{obj} and contains the angle θ, with $\tan\theta = h/f_{\text{obj}}$. Using the small-angle approximation ($\tan\theta \approx \theta$), we get

$$\theta = \frac{h}{f_{\text{obj}}} \tag{26.18}$$

The height h is also one side of the yellow right triangle in Figure 26.12 containing the angle θ_T and leads to $\tan\theta_T = h/f_{\text{eyepiece}}$. The angle θ_T is greatly exaggerated in Figure 26.12. In practice, this angle is still small, so we can again use the small-angle approximation ($\tan\theta_T \approx \theta_T$), yielding

$$\tan\theta_T \approx \theta_T = \frac{h}{f_{\text{eyepiece}}} \tag{26.19}$$

Inserting these results for the angles θ and θ_T into Equation 26.17, we have

$$m_\theta = \frac{\theta_T}{\theta} = \frac{h/f_{\text{eyepiece}}}{h/f_{\text{obj}}}$$

$$m_\theta = \frac{f_{\text{obj}}}{f_{\text{eyepiece}}} \tag{26.20}$$

Since one usually wants the magnification to be as large as possible, Equation 26.20 suggests that the focal length of the objective lens should be as large as possible. That is why long refracting telescopes are usually preferred, and such telescopes with lengths of 10 m or more were common in the 1700s and 1800s (Fig. 26.13).

© UC Regents/Lick Observatory

▲ **Figure 26.13** The magnification of a refracting telescope is proportional to the focal length of the objective lens, which motivated people to construct some very long telescopes. This one, at Lick Observatory in California, is more than 9 m long.

EXAMPLE 26.4 Ⓡ Finding the Magnification of a Refracting Telescope

The refracting telescope in Figure 26.13 has a total length of approximately 9.0 m. Suppose this telescope has an eyepiece with a magnification of 20. Estimate the total angular magnification of this telescope. *Hint*: The eyepiece is just a magnifying glass.

RECOGNIZE THE PRINCIPLE

The total magnification of a refracting telescope equals the ratio of the focal length of the objective lens f_{obj} to the focal length of the eyepiece f_{eyepiece} (Eq. 26.20):

$$m_\theta = \frac{f_{\text{obj}}}{f_{\text{eyepiece}}} \tag{1}$$

(continued) ▶

The eyepiece is a magnifying glass, and given its magnification we can find $f_{eyepiece}$ (Eq. 26.14). From our discussion of the refracting telescope, the sum of the two focal lengths is approximately equal to the telescope's total length, which gives us a way to find f_{obj}.

SKETCH THE PROBLEM

Figures 26.12 and 26.13 describe the problem.

IDENTIFY THE RELATIONSHIPS

Equation 26.14 gives the magnification of the eyepiece as $m_\theta = s_N/f_{eyepiece}$, where $s_N = 25$ cm is the normal value of the near-point distance. Rearranging to solve for $f_{eyepiece}$ leads to $f_{eyepiece} = s_N/m_\theta$. Inserting the given value $m_\theta = 20$ gives

$$f_{eyepiece} = \frac{s_N}{m_\theta} = \frac{25 \text{ cm}}{20} = 1.25 \text{ cm}$$

where we keep an extra significant figure to avoid problems with roundoff errors below. According to the ray diagram for a refracting telescope in Figure 26.12, the sum of f_{obj} and $f_{eyepiece}$ is equal to distance between the lenses, and this distance is approximately the total length of the telescope, which is given as 9.0 m. We thus have $f_{eyepiece} + f_{obj} = 9.0$ m, which leads to

$$f_{obj} = (9.0 \text{ m}) - f_{eyepiece} = (9.0 \text{ m}) - (1.25 \text{ cm}) \approx 9.0 \text{ m}$$

SOLVE

Inserting our values of the focal lengths in Equation (1), the total magnification is

$$m_\theta = \frac{f_{obj}}{f_{eyepiece}} = \frac{9.0 \text{ m}}{0.0125 \text{ m}} = \boxed{720}$$

► ***What does it mean?***

The standard way to increase the magnification of a refracting telescope is to increase f_{obj}, which, as already noted, led to the construction of some very long telescopes in the 1700s and 1800s. When the telescope length is very long, however, the lenses tend to move or vibrate relative to each other, spoiling the performance. Most research telescopes today are reflecting telescopes, described below.

EXAMPLE 26.5 Galileo's Telescope

Figure 26.12 shows one popular design for a refracting telescope using two converging lenses. A ray-tracing analysis (Fig. 26.12) shows that it produces an inverted image. Figure 26.14A shows another way to arrange two lenses to make a telescope; here the lens on the left is a converging lens (focal length $+f_{obj}$), whereas the lens on the right is a diverging lens (focal length $-|f_{div}|$) positioned so that the rightmost focal points of the two lenses coincide. Use ray tracing to show that this telescope produces an *upright* image.

RECOGNIZE THE PRINCIPLE

A telescope is designed to look at objects far away, so we place our object far to the left in Figure 26.14B. Given this position for the object, we first use ray tracing to find the image produced by the objective lens. That image is then the object for the second (diverging) lens.

SKETCH THE PROBLEM

Figure 26.14 shows the problem. The rightmost focal points of the two lenses overlap. Since the object is very far away (at infinity), the rays incident from the left are approximately parallel.

IDENTIFY THE RELATIONSHIPS AND SOLVE

Figure 26.14B shows a ray-tracing analysis of the image formed by the objective (left) lens. Ray 1 is the parallel ray (in blue); it is initially approximately parallel to the principal axis and is refracted toward the focal point. Ray 2 is the central ray (in red); it passes straight through the center of the objective lens. These two rays meet at the far right, defining the location of the image formed by the objective lens. We have also added a third ray (3, in purple), which is *not* the focal ray for the objective lens that we used in previous ray-tracing work. Instead, this ray is drawn to pass straight through the center of the lens on the right.

The image formed by the objective lens is now the object for the diverging lens on the right. Ray 1 is directed toward the focal point of this lens, so it is refracted so as to be parallel to the principal axis on the far right as shown in Figure 26.14C. Ray 3 passes through the center of the lens on the right, so it is not deflected.

To find the final image, we must find the point where rays 1 and 3, or their extrapolations, intersect. Extrapolating ray 1 back to the left of the diverging lens, as shown by the dashed blue line in Figure 26.14C, reveals the final image just to the right of the objective lens. It is an upright image as initially claimed.

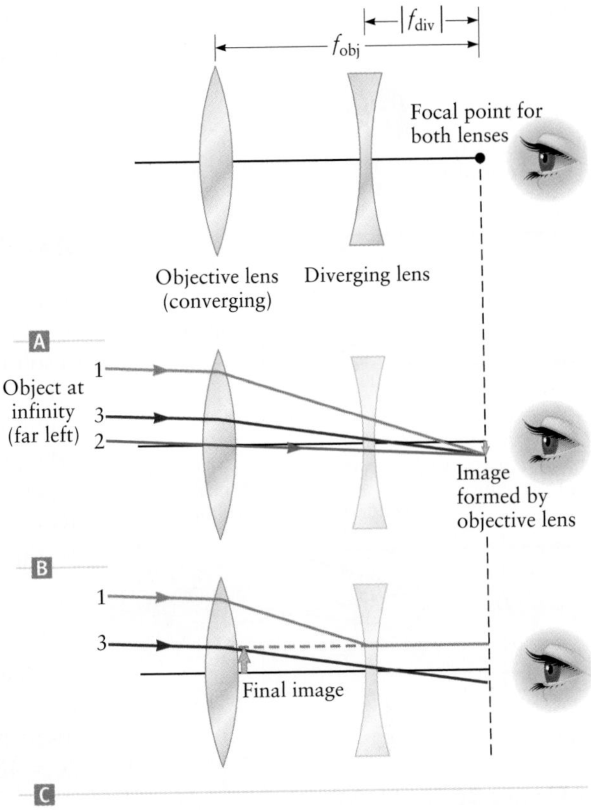

▶ ***What does it mean?***

Since this type of telescope forms an upright image, it is useful in applications such as binoculars, a sailor's spyglass, or the spotting telescope for a rifle. The inverted image formed by the telescope in Figure 26.12 would not be nearly as useful in such applications.

▲ **Figure 26.14** Example 26.5. Analyzing Galileo's telescope.

Reflecting Telescopes

The refracting telescope was an important experimental tool in the history of physics and astronomy. Galileo and others used it to observe the motion of planets and moons in the solar system, and their findings refuted the Earth-centered theory of the universe. Newton was also a serious astronomer, and during his work he noticed the chromatic aberration produced by a single objective lens (Section 24.8). This aberration led him to devise another telescope design, shown in Figure 26.15A; it is called a ***reflecting telescope*** because it replaces the objective lens in Figure 26.12 with a mirror. The particular telescope design in Figure 26.15A is called a Newtonian reflector.

Newton realized that, unlike lenses, a mirror does not have any chromatic aberration, providing a big advantage for reflecting telescopes. Another advantage is that it is easier to make a large, high-quality mirror than an equally large lens of the same quality, because the lens requires two carefully made surfaces and the glass must be absolutely transparent. Also, for a given diameter, a mirror is lighter and easier to support than a comparable lens, which is important because modern telescopes have mirrors that are 10 m in diameter or even larger. A lens of this size would sag considerably under its own weight and be useless for a refracting telescope.

Magnification of a Reflecting Telescope

Figure 26.15A (page 816) shows how light travels through a Newtonian reflecting telescope. By itself, the concave mirror at the far right would form a real image of a distant object at a point very close to the focal point of the mirror (see Fig. 24.23B). A second mirror is positioned in front of the focal point of the concave mirror, reflecting the light to an eyepiece and on to an observer. The Newtonian reflector's magnifying

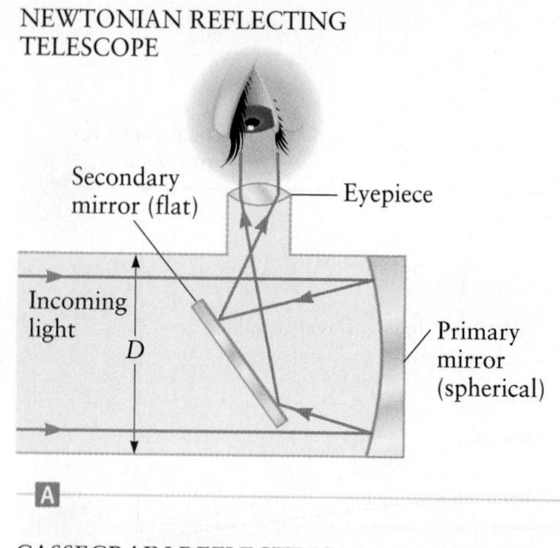

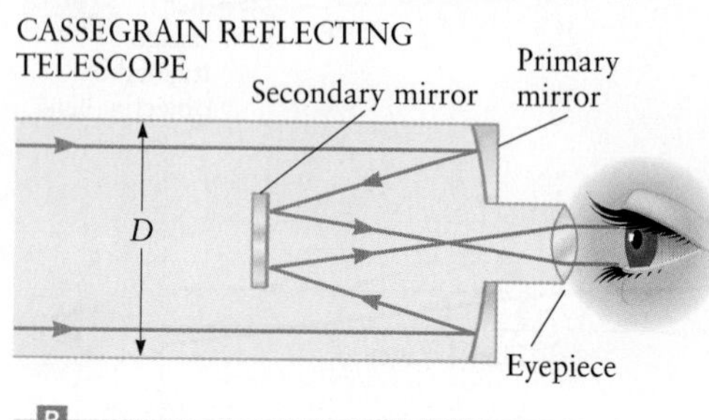

▲ **Figure 26.15** A reflecting telescope uses a concave mirror (called the primary mirror) in place of the objective lens. A The design shown here was invented by Newton and also uses a flat mirror (called the secondary mirror). B A Cassegrain reflecting telescope uses a slightly different design.

elements are similar to those in the refracting telescope, so the total angular magnification is given by a result similar to Equation 26.20. If the concave mirror has a focal length f_M, the total magnification is

$$m_\theta = \frac{f_M}{f_{\text{eyepiece}}} \quad (26.21)$$

Resolution of a Telescope

The resolution of a telescope determines how close together in angle two stars can be and yet still be seen as two separate stars (Fig. 26.12). The resolution of a telescope is limited by two main factors: diffraction at the telescope's aperture (the "opening") and atmospheric turbulence. As in a compound microscope (Section 26.2), diffraction spreads the light rays when they enter a telescope. The size of a telescope aperture is normally equal to the diameter D of the objective lens or primary mirror, and the Rayleigh criterion for the limiting angular resolution set by diffraction is (Eq. 25.28)

$$\theta_{\text{min}} = \frac{1.22\lambda}{D} \quad (26.22)$$

As an example, one of the largest reflecting telescopes in the world, with a mirror diameter $D = 10$ m, is at the Keck Observatory in Hawaii. Inserting this diameter in Equation 26.22 and assuming blue light ($\lambda = 400$ nm) gives

$$\theta_{\text{min}} = \frac{1.22\lambda}{D} = \frac{1.22(4 \times 10^{-7}\text{ m})}{10\text{ m}} = 5 \times 10^{-8}\text{ rad} \quad (26.23)$$

which is a very small angle. To get an idea of how small this angle θ_{min} is, let's convert it to degrees. For historical reasons, astronomers work in degrees, and fractions of a degree called arc minutes and arc seconds, with

$$1° = 60\text{ arc minutes} = 60'$$
$$1\text{ arc minute} = 60\text{ arc seconds} = 60''$$

Converting from radians in Equation 26.23 to arc seconds, we find

$$\theta_{\text{min}} = 5 \times 10^{-8}\text{ rad} \times \frac{360°}{2\pi\text{ rad}} \times \frac{60'}{1°} \times \frac{60''}{1'} = 0.01'' \quad (26.24)$$

Unfortunately, most telescopes do not attain the resolution limit θ_{min} set by diffraction. Starlight must pass through the atmosphere before reaching an observer on the Earth, and turbulent motion of the air causes fluctuations in the refractive index from place to place. These fluctuations act like lenses and refract the incoming light from the star (Fig. 24.50). These "lenses" are constantly changing, so the direction of the starlight changes too, making the star "twinkle." Twinkling is a random process, causing light rays from a star to fluctuate through an angle that depends on the wind speed and other meteorological factors. For a location at the Earth's surface, the angular spread caused by these atmospheric fluctuations is typically

$$\Delta\theta \approx 1\text{ arc second} = 1'' \quad (26.25)$$

This is much larger than the resolution limit due to diffraction (Eq. 26.24), so most ground-based telescopes are limited more by atmospheric fluctuations than by diffraction.

The value of $\Delta\theta$ due to the atmosphere becomes smaller at high altitudes because there is less air between the Earth's surface at high altitudes and outer space, which is why most observatories are placed on mountaintops. Even at a good observatory, however, the resolution limit due to the atmosphere is rarely smaller than $\Delta\theta = 0.5''$.

Insight 26.2
THE CASSEGRAIN REFLECTING TELESCOPE
Nowadays, many reflecting telescopes (especially those used by hobbyists) follow the Cassegrain design (Fig. 26.15B). Light from a large concave mirror (the primary mirror) strikes a second mirror and is reflected along the axis and through a small hole in the center of the primary mirror. The light then travels through an eyepiece and to the observer. The Hubble Space Telescope is a Cassegrain reflecting telescope.

This is still much larger than the typical diffraction limit. If even the 10-m Keck telescope considered in Equation 26.24 is limited more by the atmosphere than by diffraction, why did astronomers bother making such a large telescope? The answer is that the amount of light a telescope collects for an image depends on the telescope's size. Although the resolution of a ground based 10-m telescope is not any better than the resolution of a ground based 5-m telescope, the larger telescope collects light from a larger area and can thus observe fainter stars.

It is interesting to calculate the telescope diameter at which the diffraction limit for $\Delta\theta$ is equal to the atmospheric limit. Converting the atmospheric limit of about $\Delta\theta = 1$ arc second to radians gives

$$\Delta\theta = 1'' \times \frac{1^\circ}{3600''} \times \frac{2\pi \text{ rad}}{360^\circ} = 5 \times 10^{-6} \text{ rad}$$

Setting $\Delta\theta$ equal to the diffraction limit (Eq. 26.22) and solving for the diameter of the telescope, we find

$$\frac{1.22\lambda}{D} = 5 \times 10^{-6} \text{ rad}$$

$$D = \frac{1.22\lambda}{5 \times 10^{-6} \text{ rad}}$$

Assuming $\lambda = 400$ nm (blue light) gives a diameter of

$$D = \frac{1.22\lambda}{5 \times 10^{-6} \text{ rad}} = \frac{1.22(4 \times 10^{-7} \text{ m})}{5 \times 10^{-6} \text{ rad}}$$

$$D = 0.1 \text{ m} = 10 \text{ cm} \tag{26.26}$$

So, if you have a telescope with this diameter (or larger), your resolution is as good as *any* telescope located near sea level. Many amateur astronomers have telescopes that are at least this large.

This discussion shows that the resolution of a telescope located on the Earth's surface is severely limited by the atmosphere, which has prompted astronomers to move their telescopes to orbits around the Earth. The best known is the Hubble Space Telescope (see the photo at the beginning of this chapter). The Hubble telescope's resolution is limited by diffraction, as considered in Example 26.6.

▲ **Figure 26.16** This reflecting telescope on Mount Palomar has a primary mirror with a diameter of 5.0 m.

Insight 26.3

SIZE AND WEIGHT OF THE HALE REFLECTING TELESCOPE

The Hale Observatory on Mount Palomar contains a typical reflecting telescope used in astronomical research (Fig. 26.16). The mirror of the Hale telescope has a diameter of 5.0 m and a thickness of about 0.5 m, with a total mass of approximately $m = 25{,}000$ kg. The mirror's weight is thus $mg \approx 2.5 \times 10^5$ N, which is the weight of about 25 average-sized cars! The mirror's weight is so great that the supporting structure must be specially designed to prevent the mirror from sagging or bending as the telescope is rotated to view different parts of the sky.

EXAMPLE 26.6 Resolution of the Hubble Space Telescope

The Hubble Space Telescope is a Cassegrain reflector (Fig. 26.15B), with a mirror diameter $D = 2.4$ m. How does its resolution compare with that of a ground-based telescope of the same size? Assume the observer is using blue light with $\lambda = 400$ nm.

RECOGNIZE THE PRINCIPLE

Because there is no atmosphere around the Hubble telescope, its resolution is limited only by diffraction and is given by Equation 26.22. Given the diameter of the Hubble's mirror, we can find its resolution limit. A ground-based telescope of the same size would be limited to about one arc second of resolution by the atmosphere (Eq. 26.25).

SKETCH THE PROBLEM

Figure 26.15B shows the design of a Cassegrain reflecting telescope. The resolution is determined by the diameter D of the primary mirror.

(continued) ▶

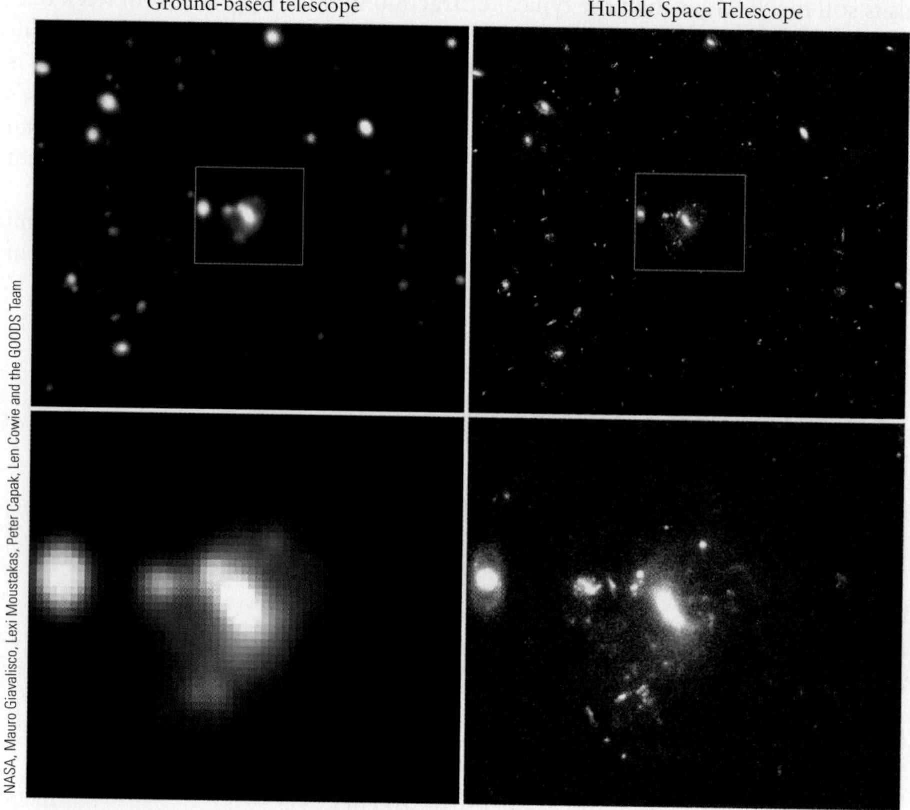

▶ **Figure 26.17** Example 26.6. Because it is above the Earth's atmosphere, the Hubble Space Telescope has much better resolution than a ground-based telescope. These images are of the same regions of space; much more fine detail is visible in the Hubble telescope's images when compared with images obtained with the 8.2-m Subaru Telescope located in Hawaii.

IDENTIFY THE RELATIONSHIPS AND SOLVE

The resolution limit of the Hubble telescope is given by $\theta_{min} = 1.22\lambda/D$ (Eq. 26.22). Inserting the given values $D = 2.4$ m and $\lambda = 400$ nm $= 4.0 \times 10^{-7}$ m, and expressing the final answer in radians, we get

$$\theta_{min} = \frac{1.22\lambda}{D} = \frac{1.22(4.0 \times 10^{-7}\text{ m})}{2.4\text{ m}}$$

$$\theta_{min} = \frac{1.22(4.0 \times 10^{-7}\text{ m})}{2.4\text{ m}} \times \frac{360°}{2\pi\text{ rad}} \times \frac{3600''}{1°} = 0.042''$$

This value is the Hubble telescope's resolution limit as set by diffraction. The atmospheric limit for a telescope on the Earth is about 1 arc second = 1″, so the Hubble telescope's resolution is better than ground-based resolution by a ratio of about

$$\frac{1''}{0.042''} = \boxed{24}$$

The Hubble's resolution is $\boxed{\text{24 times better}}$ than that of the ground-based telescope.

▶ ***What does it mean?***

The superior resolution of the Hubble Space Telescope means that it can resolve much more detail in photos of distant galaxies as illustrated in Figure 26.17, which compares photos of a particular region of space taken using the Hubble telescope and a ground-based telescope. Telescopes in space thus have a natural advantage over conventional telescopes on the Earth's surface. We'll next see how an "unconventional" ground-based telescope can overcome the resolution limit caused by atmospheric fluctuations.

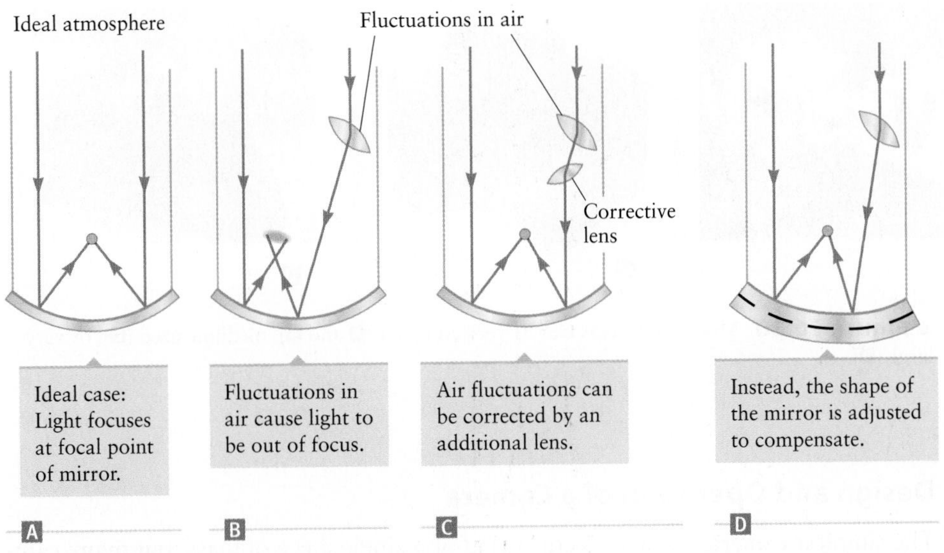
Ideal atmosphere
Fluctuations in air
Corrective lens
Ideal case: Light focuses at focal point of mirror.
Fluctuations in air cause light to be out of focus.
Air fluctuations can be corrected by an additional lens.
Instead, the shape of the mirror is adjusted to compensate.
A
B
C
D

◀ **Figure 26.18** Adaptive optics. A Under ideal conditions, a telescope focuses light from a star onto a spot that is limited only by diffraction. B Under real conditions, fluctuations in the atmosphere refract some of the incoming light, producing a larger focus "spot." C These fluctuations can in theory be corrected by using a lens to "undo" the refraction of atmosphere. Since the atmosphere fluctuates rapidly with time, the focal length of such a (hypothetical) lens must be rapidly adjusted to match the atmospheric fluctuations. D A practical approach is to adjust the shape of the mirror instead of adding an extra lens.

Adaptive Optics

The distortion of starlight by the atmosphere is illustrated very schematically in Figure 26.18. Figure 26.18A shows a hypothetical case with no atmospheric distortion, and the light is focused perfectly by the telescope's mirror. Figure 26.18B shows a fluctuating pocket of air acting as a lens, refracting some of the light rays in a different direction. Any focusing done by one lens can always be undone by a second lens, so a very clever astronomer could place a second lens in front of the telescope that would completely correct for distortion caused by the atmosphere (Fig. 26.18C). Atmospheric distortion changes as the winds and so forth change, however, so the corrective lens must change too. In practice, this corrective procedure is not done with a lens, but rather through very small changes in the shape of the mirror, made by pushing and pulling on the mirror's mounting platform (Fig. 26.18D). The mirror shape is adjusted many times each second to match fluctuations in the atmosphere by monitoring the image of a "reference star." The technology of building telescopes with adjustable mirrors to compensate for atmospheric distortion is called ***adaptive optics***. (For an example, see the photograph on page 899.)

A "reference star" is an object (usually a very distant star) known to appear as a point source at the telescope. If there were no distortion from the atmosphere, light from the reference star would focus to an image point limited only by diffraction. As the atmosphere causes the image of the reference star to be smeared out, the telescope mirror is adjusted to make this image as perfect as possible. Rapid adjustment of the mirror shape requires modern computers. The technology of adaptive optics is fairly new and is still being improved, but in some cases it is currently possible to reduce atmospheric distortion by a factor of three or more.

26.4 Cameras

The simplest cameras (Fig. 26.19) consist of a single lens positioned in front of a light-sensitive detector. In an old-fashioned camera, the detector is called ***film***, but the vast majority of cameras sold today use an electronic detector called a ***charge-coupled device*** (CCD). A CCD records information on the light intensity and color, which can then be transferred to a computer (and shared with friends). The camera's lens forms an image on the CCD, and an aperture (the shutter) is opened for a short time to allow sufficient light energy to enter. We'll describe how a CCD works below; first we consider the general optics of a camera.

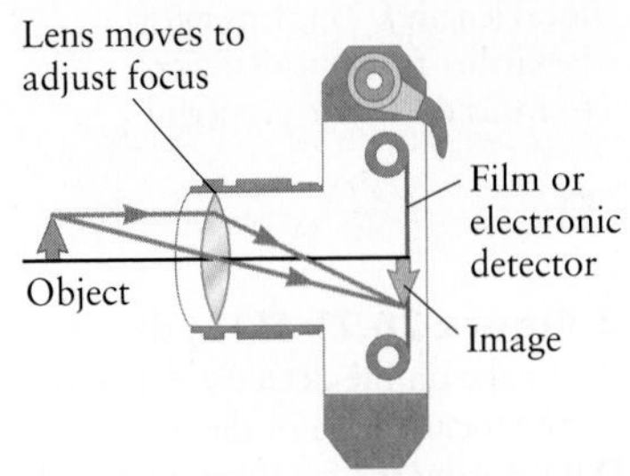
Lens moves to adjust focus
Film or electronic detector
Object
Image

▲ **Figure 26.19** A camera is focused by moving the lens relative to the film or an electronic (CCD) detector.

Comstock/Thinkstock/Getty Images

Glow Images/Getty Images

© Cre8tive Studios/Alamy

© Ryan McGinnis/Alamy

▲ **Figure 26.20** Modern cameras can be very compact (A and B), medium sized (C), or very large (D).

Design and Operation of a Camera

The simplest cameras use lenses consisting of a single piece of glass, but many camera lenses contain multiple lens elements. In either case, their overall behavior is like a single converging lens, and the focus is adjusted by moving the lens relative to the detector as indicated in Figure 26.19. Cameras come in many shapes and sizes, ranging from ultrasmall cameras built into many cell phones to cameras with very large lenses used by professional photographers (Fig. 26.20). The standard lens for a camera like the one in Figure 26.20C has a focal length of about $f = 40$ mm, but the lenses in this and the other cameras in Figure 26.20 can have both larger and smaller values of f. To understand how these different focal lengths affect a photograph, consider the ray diagram in Figure 26.21, which shows the image formed by an object that is very far from the camera. The thin-lens equation (Eq. 26.1) gives a relation between the object distance s_o, the image distance s_i, and the focal length f. We have

$$\frac{1}{s_o} + \frac{1}{s_i} = \frac{1}{f} \tag{26.27}$$

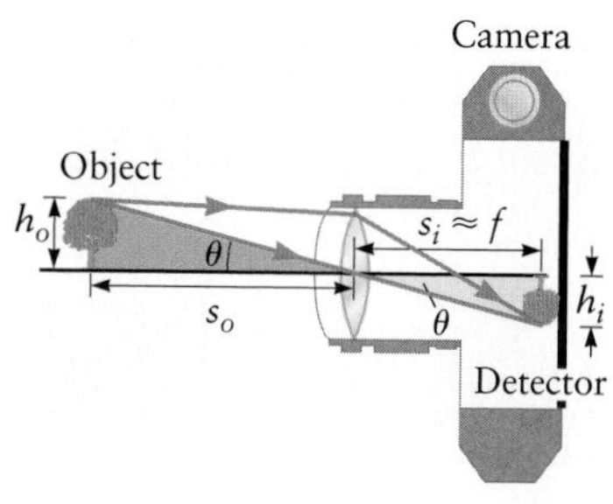
Camera
Object
h_o
θ
$s_i \approx f$
s_o
θ
h_i
Detector

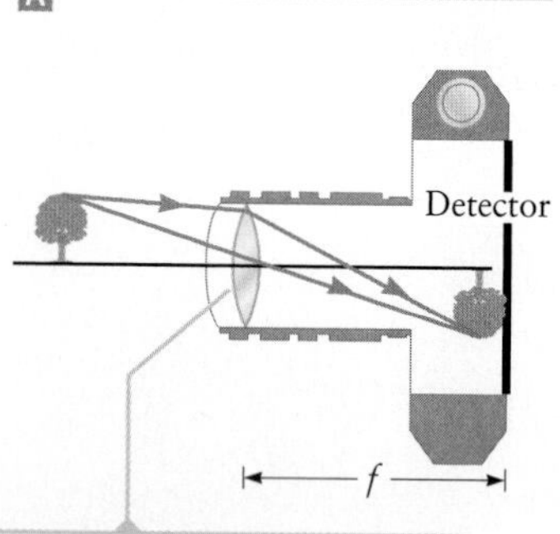
Detector
f
A telephoto lens has a large focal length f. This lens must be farther from the detector to focus the image properly.

▲ **Figure 26.21** A The size of the image on the detector depends on the focal length of the lens. B Increasing the focal length by using a telephoto lens produces a larger image. These drawing are not to scale. The object (the tree) is typically far from the camera.

The object (the tree) in Figure 26.21A is very far from the camera, so s_o is much larger than both f and s_i. The term $1/s_o$ in Equation 26.27 is thus much smaller than the other terms, and to a very good approximation we have

$$\frac{1}{s_i} \approx \frac{1}{f}$$

and

$$s_i \approx f \tag{26.28}$$

This result tells us that the lens should be positioned a distance of approximately f from the CCD detector. It also enables us to calculate the size of the image on the detector. There are two right triangles in Figure 26.21A involving the angle θ. One side of the red triangle is the height h_o of the object and yields

$$\tan\theta = \frac{h_o}{s_o} \tag{26.29}$$

A similar analysis of the right triangle containing the image distance s_i and the image height h_i (shaded in yellow in Fig. 26.21A) leads to

$$\tan\theta = \frac{-h_i}{s_i} \tag{26.30}$$

where the negative sign comes from our sign convention for geometric optics (Fig. 26.2) that a downward-pointing image has a negative height. Inserting $s_i \approx f$ from Equation 26.28 and setting the results for $\tan\theta$ in Equations 26.29 and 26.30 equal gives

$$\frac{h_o}{s_o} = -\frac{h_i}{f}$$

The linear magnification of the image is the ratio of the image height to the object height, $m = h_i/h_o$, and we find

$$m = \frac{h_i}{h_o} = -\frac{f}{s_o} \tag{26.31}$$

The magnification is negative since the image is inverted. The object distance s_o is much larger than the focal length, so the magnification is much less than 1 and the image of a tree or other large object can thus fit on the detector.

This result for the magnification tells us when a lens with a large value of the focal length (i.e., a telephoto lens) would be handy. Telephoto lenses (Fig. 26.20D) usually contain several component lenses, and the overall focal length can be adjusted by changing the spacing between these lenses. Suppose a normal lens with $f = 40$ mm is used in Figure 26.21A, while Figure 26.21B shows the corresponding diagram using a telephoto lens with $f = 200$ mm. With a longer focal length, the lens must be positioned farther from the detector (because $s_i \approx f$, Eq. 26.28), giving a larger magnification for a given value of the object distance s_o (Eq. 26.31). A telephoto lens is thus convenient for photographs of distant objects.

CONCEPT CHECK 26.5 The Image Produced by a Camera

Does a camera produce a real or a virtual image? Explain why a camera could not work if it formed a virtual image.

EXAMPLE 26.7 Designing a Camera

You are asked to design a camera using a lens with $f = 40$ mm and want to determine how much the lens will have to move to change the focus. Your camera must focus on objects very far away and also on objects that are as close as 1.00 m. Calculate how much the distance between the lens and the detector must change to adjust the focusing from one case to the other.

RECOGNIZE THE PRINCIPLE

The camera's lens must form an image on the detector for two different object distances, $s_o = \infty$ and $s_o = 1.00$ m. The focal length f is given, so we must find the image distances that correspond to these two object distances. We can do so using the thin-lens equation.

SKETCH THE PROBLEM

Figure 26.21A shows the problem.

IDENTIFY THE RELATIONSHIPS

We already analyzed this problem for the case of a very distant object and found $s_i = f$ (Eq. 26.28). We thus only need to consider the case with $s_o = 1.00$ m and find the corresponding image distance. From the thin-lens equation, we have

$$\frac{1}{s_o} + \frac{1}{s_i} = \frac{1}{f}$$

Rearranging to solve for the image distance yields

$$\frac{1}{s_i} = \frac{1}{f} - \frac{1}{s_o}$$

and doing some algebra gives

$$\frac{1}{s_i} = \frac{1}{f} - \frac{1}{s_o} = \frac{s_o}{s_of} - \frac{f}{s_of} = \frac{s_o - f}{s_of}$$

(continued) ▶

SOLVE

We can now solve for s_i:

$$s_i = \frac{s_o f}{s_o - f}$$

Inserting the given value of f and s_o = 1.00 m for the smallest object distance, we find

$$s_i = \frac{s_o f}{s_o - f} = \frac{(1.00 \text{ m})(0.040 \text{ m})}{1.00 \text{ m} - 0.040 \text{ m}} = 0.042 \text{ m} = 42 \text{ mm}$$

which is the image distance when the object is very close; it is also the distance between the lens and the detector. When the object is very far away, we had $s_i = f = 40$ mm. So, we only need to move the lens (42 − 40) mm = 2 mm farther from the detector to focus properly. To change the focus between an object at infinity and one at 1.00 m, you need only move the lens $\boxed{2 \text{ mm}}$.

▸ *What does it mean?*

Most cameras for amateurs have an autofocus mechanism that moves the lens relative to the detector, whereas some cameras allow you to move it yourself. Either way, the lens typically needs to move only a few millimeters to bring an image into focus.

Detecting Light with a CCD: What Is a Pixel?

The digital camera has developed rapidly in recent years and is now the camera of choice for most amateurs and professionals. Digital cameras are made possible by a light-sensitive detector called a charge-coupled device (CCD), which uses a type of capacitor to detect light and record its intensity. (A few digital cameras employ what is called a CMOS detector, but the operating principle is similar to that of a CCD.) A CCD is fabricated in an integrated circuit chip; the CCD chips in your cell phone camera are much smaller than your thumbnail (Fig. 26.22A), while those in a professional camera (Fig. 26.20D) are several centimeters on a side. A CCD chip contains many capacitors arranged in a grid over its surface as sketched in parts B and C of Figure 26.22. The structure of the capacitors in Figure 26.22B is somewhat simplified, showing only the two capacitor plates and a dielectric material in between. When light strikes the chip, it is absorbed in the dielectric layer and ejects some electrons from their normal chemical bonds. These electrons then move to a capacitor plate, leading to a voltage across the capacitor that is closest to where the light was absorbed. This voltage, denoted V_{CCD} in Figure 26.22B, is then detected by additional circuitry, and its value is stored in a computer memory in the camera. The

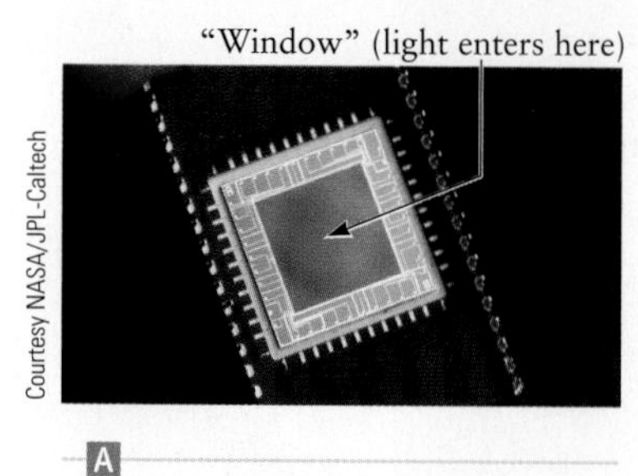

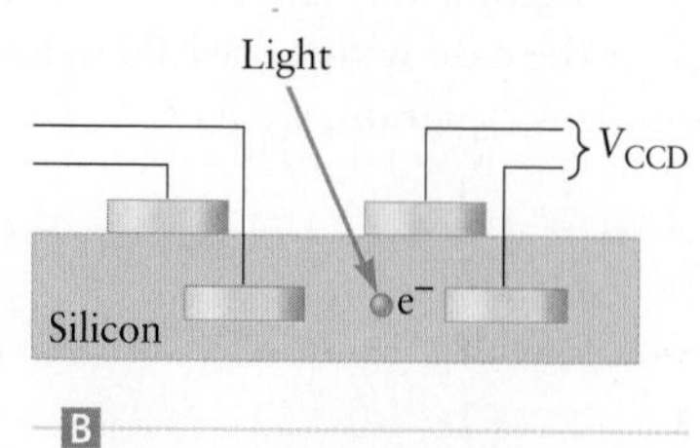

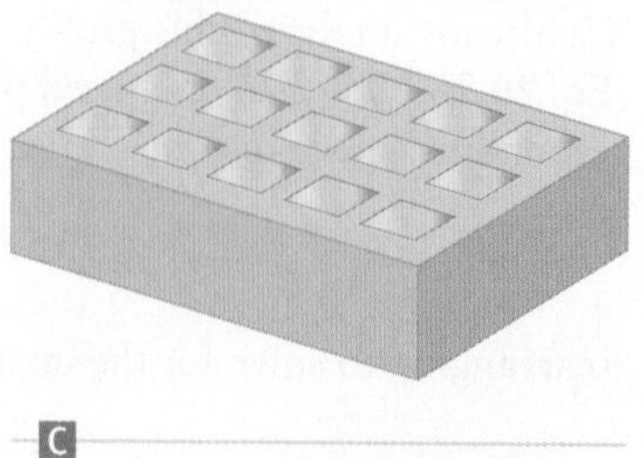

▲ **Figure 26.22** A A CCD integrated circuit chip. Light enters through the "window" on the chip. B The CCD chip contains many capacitors constructed on a piece of silicon. When light is absorbed, an electron is "released" from one of the covalent bonds in the silicon, leading to a charge (and hence a voltage) on the nearest capacitor. C A CCD chip contains an array of millions of these capacitors.

magnitude of V_{CCD} depends on the light intensity because a greater intensity ejects more electrons and gives a larger voltage. The pattern of voltages on the capacitors in the grid (Fig. 26.22C) thus gives the light intensity as a function of position on the surface of the CCD, which is just the image.

The capacitors in Figure 26.22B do not distinguish the color of the light; a charge on the capacitors indicates that light was absorbed, but does not tell its color. One way to measure the color is to combine the information from four adjacent capacitors as sketched in Figure 26.23. A filter is placed in front of each of the capacitors; one of the filters allows red light to pass through to the CCD capacitor behind it, and other filters allow blue light and green light to pass. From the voltages on the four capacitors, the camera's computer can estimate the average color over this region. The CCD contains many such regions, called ***pixels*** (an abbreviation for "picture element"). The image produced by the CCD chip is thus stored by the camera as a set of intensity and color values for each pixel.

An important specification for a digital camera is the number of pixels in each photograph. A larger number of pixels indicates a finer level of detail in the photograph, so generally speaking, better cameras produce photos with larger numbers of pixels.

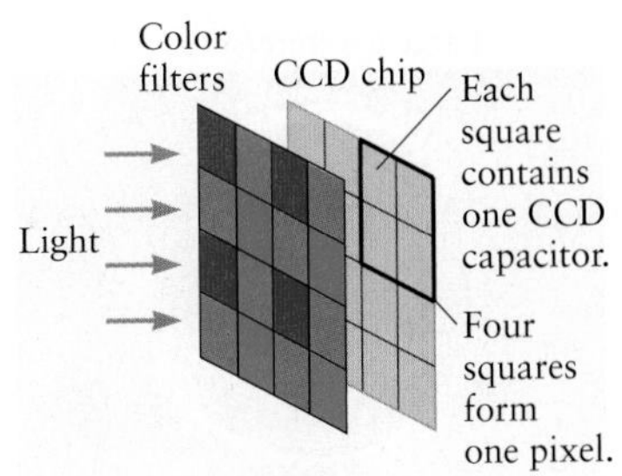

▲ **Figure 26.23** A CCD detector detects colors by using filters in front of the CCD capacitors. Each filter (red, green, or blue) lets only one color of light pass to the capacitor behind it, so each capacitor is sensitive to one particular color. The voltages from four adjacent capacitors are analyzed by the camera's computer to find the overall color of the light striking that region. These four adjacent capacitors form a pixel.

Optics of Cell Phone Cameras

The optics and ray diagrams for the cameras in Figure 26.20 are all very similar, but these cameras have very different sizes. How can the camera in your cell phone be so thin? The CCD detectors in these thin cameras are only a few mm on a side, which is much smaller than in the larger cameras in parts C and D of Figure 26.20. We can rearrange our result for the magnification in Equation 26.31 to solve for the magnitude of the focal length:

$$f = \left| s_o \frac{h_i}{h_o} \right|$$

The image must, of course, always fit on the detector, so if the CCD detector is very small, the image height h_i must also be very small. For a given value of the object distance s_o and the object distance h_o, the focal length f of the lens must therefore be much shorter for a cell phone camera than for a large camera. The CCD detector in a cell phone camera is typically 10 times smaller on each side than for the cameras in parts C and D of Figure 26.20, so the focal length is also about 10 times shorter.

In our analysis of the camera in Figure 26.21, we also showed that the distance s_i between the lens and the detector is approximately equal to the focal length f (Eq. 26.28). Since it has a short focal length, the lens in a cell phone camera must be close to the detector, which is why such a camera can be very thin. The same considerations apply to ultracompact (pocket-size) cameras (Fig. 26.20B) and the camera in an MP3 player (Fig. 26.24).

Changing the Magnification of a Camera

According to Equation 26.31, the magnification of a camera can be changed by moving the lens relative to the detector. (Compare again with Fig. 26.21.) For a digital camera, this magnification is called "optical zoom." It is also possible to use a "digital zoom" to enlarge an image. The digital zoom process constructs the entire photo using just the image data from near the center of the CCD grid. Hence, a photo taken with the digital zoom uses fewer pixels and has poorer resolution than a photo obtained with the digital zoom turned off.

AP Photo/Jeff Chiu

▲ **Figure 26.24** Digital cameras can be very compact.

f-Number

Two important camera settings are the shutter speed and the ***f-number***. The shutter "speed" is the amount of time the detector is exposed to light from the object, and the *f*-number is associated with the camera's ***aperture***, an opening that controls the

© Cengage Learning/Charles D. Winters (both)

▲ **Figure 26.25** The adjustable opening in front of a camera lens is called the aperture.

area of the lens in use (Fig. 26.25). A large aperture admits light from a large area of the lens, whereas a small aperture admits only a small amount of light (passing through the center of the lens) to the detector. There are thus two ways to change the total amount of light energy that reaches the detector: changing the shutter speed or changing the diameter of the aperture. The f-number is the ratio of the focal length to the aperture diameter D,

$$f\text{-number} = \frac{f}{D} \tag{26.32}$$

A larger aperture (a larger value of D) lets more light reach the CCD and gives a smaller f-number. A large f-number (small D) allows less light to get to the CCD. To have a properly exposed photograph (not too bright and not too dark), the total light energy must lie within a certain range. If you reduce the shutter speed and thus keep the shutter open longer, you must compensate by using a smaller aperture and hence a larger f-number.

CONCEPT CHECK 26.6 Aperture and Shutter Speed for a Camera

A photographer is taking photos using an f-number of 4 and a shutter speed setting that keeps the aperture open for $1/100 = 0.010$ s. The photographer then changes the f-number to 16. How long should the shutter now be open so as to have the same amount of light energy reach the detector, (a) the same amount of time, (b) 0.0025 s, (c) 0.04 s, or (d) 0.16 s?

Depth of Focus

Because of the trade-off between shutter speed and f-number, a photographer is often faced with the question of how to choose these two settings.[3] Using a short shutter speed reduces problems with vibrations when holding a camera and also

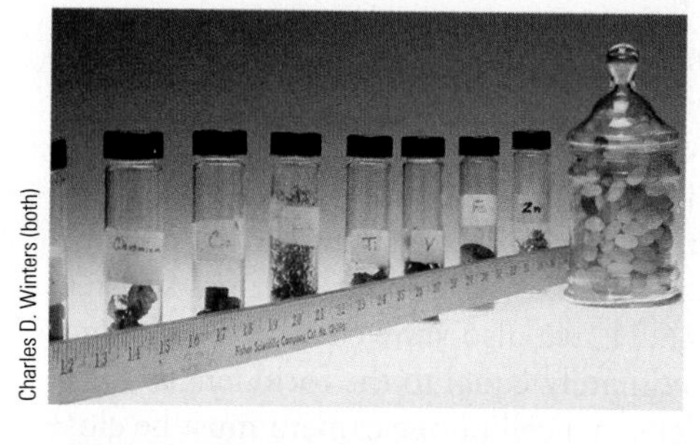

Charles D. Winters (both)

► **Figure 26.26** Changing the aperture of a camera changes the f-number. A A larger f-number gives a larger depth of focus. Here objects from a range of distances in front and behind the best focus distance are all well focused on the film. B With a small f-number, only objects a particular distance from the camera are in focus. Objects farther away or closer than this distance are now far out of focus.

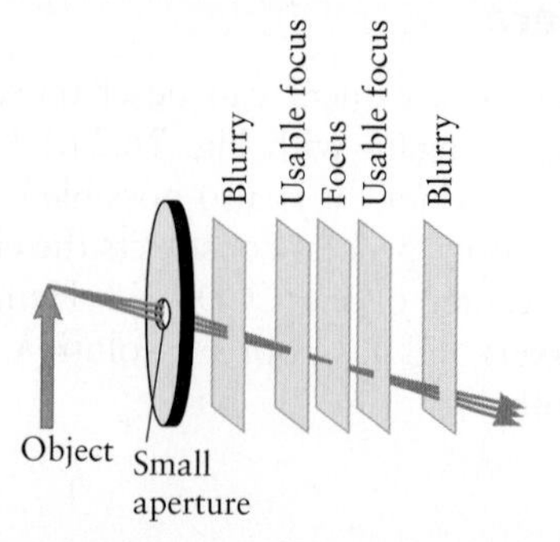

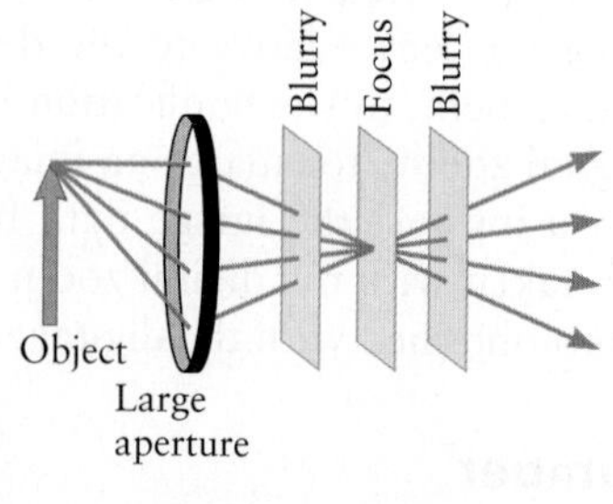

► **Figure 26.27** Ray diagrams for A a small aperture (large f-number) and B a large aperture (small f-number). With the small aperture the rays diverge more slowly from the central ray, making the depth of focus larger.

[3]Some inexpensive cameras make this choice for you; others allow you to make choices such as "indoor" or "action" and then set the shutter speed and f-number accordingly.

makes it easier to photograph moving objects. To get enough light to the detector when using a short shutter speed, you need a small *f*-number (large aperture). The *f*-number, however, also affects a property called the ***depth of focus*** illustrated in Figure 26.26, which shows two photographs of the same scene using the same camera. Figure 26.26A was taken with a large *f*-number, and all the objects are fairly well focused. Figure 26.26B was taken with a smaller *f*-number, and only objects in the center of the scene are in focus, whereas those farthest from and closest to the camera are badly out of focus. A photographer would say that the depth of focus is much larger in Figure 26.26A than in Figure 26.26B. Having a large depth of focus means that objects that are not at the best focusing point will produce acceptable images.

The *f*-number and depth of focus are related as shown by the ray diagrams in Figure 26.27. When the aperture is large (a small *f*-number; Fig. 26.27B), some rays make a large angle with the central ray, so they diverge more quickly as one moves away from the image point. With a small aperture (Fig. 26.27A), the diverging angle is smaller, and the blurring of images away from the best focus is smaller than with a large aperture.

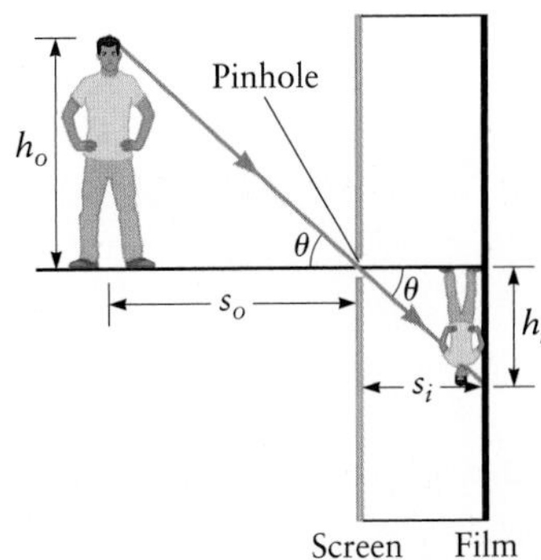

▲ **Figure 26.28** A pinhole camera forms an image without a lens. The lens of a conventional camera (Fig. 26.21) is replaced by a small pinhole. The magnification is determined by the ratio of the image distance s_i and the object distance s_o.

The Pinhole Camera

The ray diagrams in Figure 26.27 show that the "sharpness" of a photograph improves as the aperture is made smaller and smaller. In fact, if we make the aperture extremely tiny, we can make an extremely good image even without a lens! The ***pinhole camera*** uses this principle. This device, invented in the 1500s and sometimes called the ***camera obscura***, is sketched in Figure 26.28. Light from the object passes through the aperture (a tiny hole) in a screen and forms an image when it strikes a piece of film that lies behind the screen. (Most pinhole cameras use film rather than a CCD.) Since the aperture is very small, the intensity at the film is very low and the film must be exposed to the image for a long time. Pinhole cameras are easy to make and can be used to view very intense light sources safely. For example, they are a safe way to view a solar eclipse and to search for sunspots (Fig. 26.29).

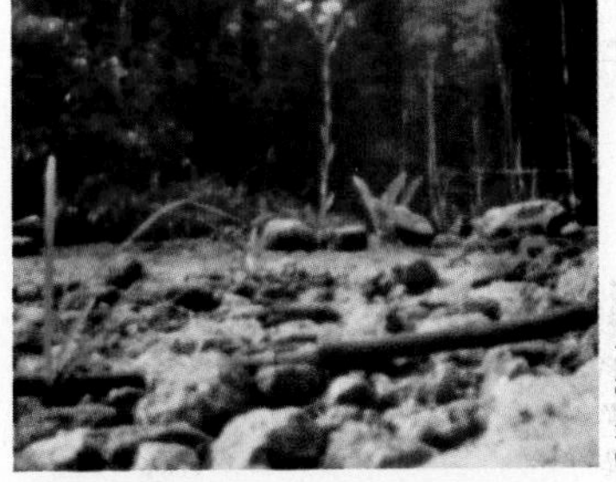

A

B

▲ **Figure 26.29** A Landscape photograph taken with a pinhole camera. B Viewing the Sun during an eclipse through a pinhole.

EXAMPLE 26.8 Magnification of a Pinhole Camera

A pinhole camera can be made from a shoebox. Assume the distance from the pinhole to the film is $s_i = 20$ cm (about the size of a shoebox). If the object is $s_o = 20$ m from the camera, what is the magnification of the camera?

RECOGNIZE THE PRINCIPLE

There is no lens in a pinhole camera, so the object and image distances do not depend on any focal lengths and the thin-lens equation does not apply. We must rely on the basic definition of magnification as the ratio of the image height to the object height.

SKETCH THE PROBLEM

Figure 26.28 shows the problem and defines the image and object distances as well as the image and object heights.

IDENTIFY THE RELATIONSHIPS

Consider the two right triangles with interior angle θ in Figure 26.28, one on the object side of the pinhole with sides s_o and h_o, and the other on the image side of the pinhole with sides s_i and h_i. These triangles have the same interior angles, so the ratios of their corresponding sides are equal and we have

$$\frac{h_o}{s_o} = \frac{-h_i}{s_i}$$

The negative sign here comes from our sign convention that an inverted image is assigned a negative height (Fig. 26.2).

(continued) ▶

SOLVE

The magnification is $m = h_i/h_o$; hence,

$$m = \frac{h_i}{h_o} = -\frac{s_i}{s_o}$$

Inserting the given values for s_o and s_i gives

$$m = -\frac{s_i}{s_o} = -\frac{0.20 \text{ m}}{20 \text{ m}} = \boxed{-0.010}$$

► *What does it mean?*

The negative value for the magnification shows that the image is inverted, as we knew from Figure 26.28. The image is also greatly reduced ($|m| \ll 1$).

26.5 CDs and DVDs

Many optical devices, including eyeglasses, microscopes, telescopes, and cameras, form images using lenses and mirrors and can therefore be analyzed using the principles of geometrical optics. We now consider several applications that make use of the principles of wave optics from Chapter 25. The CD and DVD are applications of this type. CDs and DVDs operate through similar principles, so we'll focus mainly on the CD.

Figure 26.30 shows the structure of a CD like the one used to hold your favorite video game; it is like a sandwich with the thickest layer composed of a plastic similar to Plexiglas. The bottom surface of the plastic in Figure 26.30A is smooth, but the top surface contains a pattern of pits used to encode information. This top surface is coated with a thin layer of aluminum to make it reflecting and then covered with a protective layer of lacquer (a varnishlike coating), and the label of the CD is placed on top of the lacquer.

The pits are arranged in a long spiral "track" with sequential pits as close as 0.83 μm and neighboring paths on the spiral about 1.6 μm apart (Fig. 26.30B). Information encoded in the pits is read by reflecting a laser beam from the aluminum surface. Laser light passes in and out through the bottom surface of the plastic in Figure 26.30A and reflects from the aluminum layer. The pits influence this reflection through the thin-film interference effect illustrated in Figure 26.31. When the laser beam is positioned directly over the edge of a pit (the green rays in Fig. 26.31A), some of the light reflects from the bottom of the pit, and the rest reflects from the level outside the pit. These reflected beams are very similar to reflections from a soap film (Section 25.3). Destructive interference occurs if the difference in path lengths of the two reflected beams is $\lambda/2$, $3\lambda/2$, $5\lambda/2$, and so on.[4] The pit depth is designed to produce destructive interference, so there is *no* reflected light when the laser beam is over a pit edge, whereas the intensity is large when the laser beam is over the center

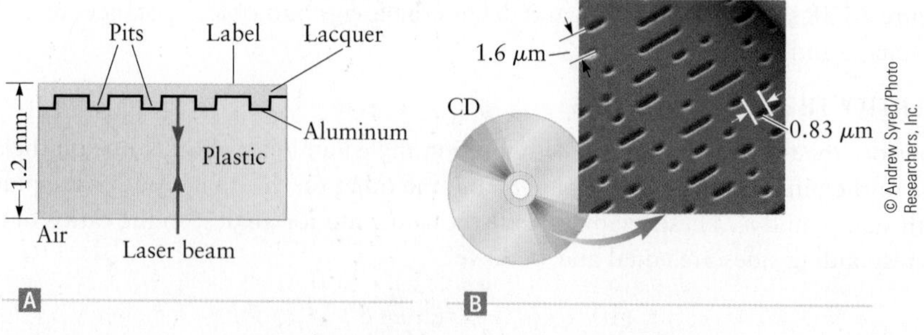

► Figure 26.30 The information on a compact disc is encoded in a series of pits buried within the CD. **A** Light from a laser is incident from below and reflects from the aluminum layer. **B** The pits are arranged in a long spiral track.

[4]There is also a phase change of the light wave on reflection from the aluminum surface. This phase change occurs for both reflected light beams in Figure 26.31A and hence does not affect their *relative* phases.

of a pit or is outside a pit. As the laser beam travels along a track, the reflected light intensity varies between zero and a large value depending on the presence or absence of pit edges as sketched in Figure 26.31B. These high and low values of the intensity correspond to ones and zeros in a binary encoding of information on the CD.

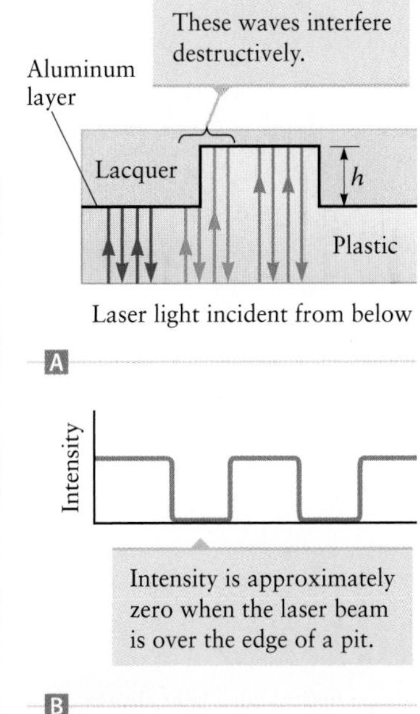
These waves interfere destructively.
Aluminum layer
Lacquer
h
Plastic
Laser light incident from below
A
Intensity
Intensity is approximately zero when the laser beam is over the edge of a pit.
B

▲ **Figure 26.31** A Light from a laser comes in from below a CD; see Figure 26.30. *Note*: The colors of these rays do *not* represent the color of the light. Light waves reflected from the top and bottom of a pit edge interfere destructively, so there is no reflected intensivty at the edge of a pit (where the interfering waves are denoted in green). Reflections from elsewhere on the CD (denoted by the red and blue rays) give a large intensity. B The reflected intensity thus encodes the presence or absence of a pit edge, representing a binary 1 or 0.

EXAMPLE 26.9 Designing a CD

The CD drive in a video game player uses laser light with $\lambda_{air} = 780$ nm in air (in the infrared part of the spectrum). What is the smallest pit depth h in Figure 26.31 that will result in destructive interference for a laser beam positioned at the edge of a pit? The index of refraction of the plastic portion of the CD is $n_{plastic} = 1.55$.

RECOGNIZE THE PRINCIPLE

To have destructive interference with the smallest possible pit depth, the path lengths of the two reflected light beams must differ by

$$\Delta L = \frac{\lambda_{plastic}}{2}$$

where $\lambda_{plastic}$ is the wavelength of the light when it travels in the plastic. Recall that the wavelength of light in a substance depends on the index of refraction of the substance (Eq. 25.9). The index of refraction of air is $n_{air} = 1.00$, so for light traveling within the CD we have

$$\lambda_{plastic} = \frac{\lambda_{air}}{n_{plastic}} \tag{1}$$

SKETCH THE PROBLEM

Figure 26.31 shows the problem. The two interfering waves are represented by the green rays near the pit edge on the left.

IDENTIFY THE RELATIONSHIPS

Using the given values of λ_{air} and $n_{plastic}$ in Equation (1) gives

$$\lambda_{plastic} = \frac{\lambda_{air}}{n_{plastic}} = \frac{780 \text{ nm}}{1.55} = 503 \text{ nm}$$

The path length difference is equal to *twice* the depth h of a pit (Fig. 26.31A), so to get destructive interference we must have

$$h = \frac{\Delta L}{2} = \frac{\lambda_{plastic}}{4}$$

SOLVE

Inserting our value for $\lambda_{plastic}$, we find

$$h = \frac{\lambda_{plastic}}{4} = \frac{503 \text{ nm}}{4} = 126 \text{ nm} = \boxed{1.26 \times 10^{-7} \text{ m}}$$

► ***What does it mean?***

The pit depth in a CD is very small, about five hundred times smaller than the thickness of a typical human hair. Covering the reflecting side of the pits with a layer of plastic protects them from scratches that would otherwise ruin the CD.

To store as much information as possible on a CD, the area of a pit must be as small as possible. Here again we encounter a limit set by wave optics. The minimum size of a focused spot of light is approximately equal to the wavelength (Eq. 25.33). CDs use the light from a laser with $\lambda_{air} = 780$ nm and $\lambda_{plastic} = 503$ nm (see Example 26.9), so the CD pits cannot be smaller than about $\lambda_{plastic}$ in width and length, as you can confirm from Figure 26.30B. DVDs can store more information than CDs because DVD players use a green laser with $\lambda_{plastic} = 420$ nm. The shorter wavelength allows the focus spot to be smaller, which in turn allows the pits in a DVD to be more closely spaced than on a CD. New DVD designs have also been developed that can store even more information by using multiple aluminum layers in different layers of pits that are probed with laser beams from both sides instead of in a single layer as in current DVDs. These new-style DVDs can store at least 30 times more information than a CD.

CONCEPT CHECK 26.7 Designing a DVD

The depth of a pit in a CD or DVD is based on the wavelength of the light used to read the information. If a CD uses light with $\lambda_{plastic} = 503$ nm and a DVD uses $\lambda_{plastic} = 420$ nm, how does the depth of a pit in a DVD compare to the depth in a CD?

(a) The pits in a DVD are deeper by a factor of $503/420 = 1.20$.
(b) The pits are the same depth.
(c) The pits in a DVD are shallower by a factor of $420/503 = 0.83$.

26.6 Optical Fibers

In an optical fiber, light always undergoes total internal reflection.

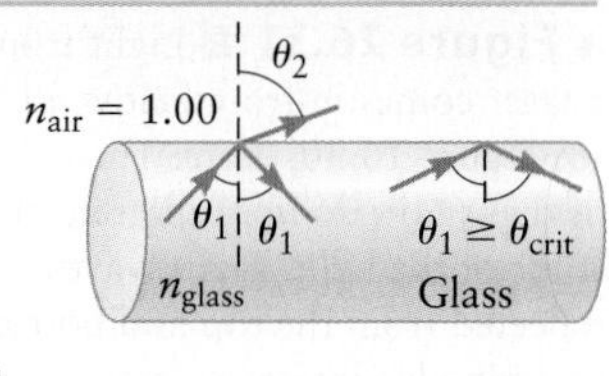

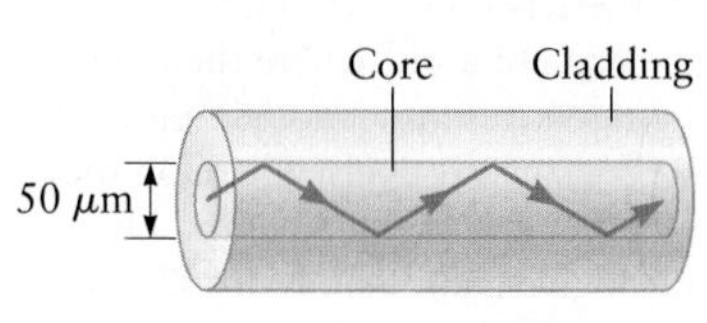

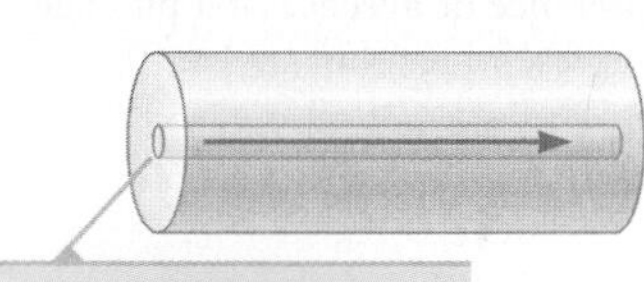

The diameter of the core of a modern optical fiber is comparable to λ.

▲ **Figure 26.32** A Light traveling inside a glass fiber is reflected at the fiber's surface. The angle of incidence is always greater than the critical angle for total internal reflection, so the light does not leave the fiber. B Real optical fibers consist of two (or more) types of glass with different indices of refraction. The light is confined to the core of the fiber. C When the diameter of the core is comparable to the wavelength (as in current fibers), the light wave travels directly along the axis of the fiber.

Optical fibers are flexible strands of glass that conduct and transmit light using total internal reflection. We described them briefly in Section 24.3, but provide more details here. Recall that light undergoes partial reflection and refraction when it encounters an interface between materials with different indices of refraction. Figure 26.32A shows a simplified model of an optical fiber. Although the fiber is cylindrical, we assume its diameter is large compared to the wavelength of light so that reflection of light at the glass–air interface is just like reflection from a flat interface. If a light ray traveling inside the fiber meets the sidewall of the fiber with an angle of incidence θ_1, there is a reflected ray with an angle of reflection equal to θ_1. If it exists, the refracted ray in the air travels at an angle θ_2. This angle of refraction is given by Snell's law (Eq. 24.7),

$$n_{glass} \sin \theta_1 = n_{air} \sin \theta_2 \tag{26.33}$$

which can be arranged to get

$$\sin \theta_2 = \frac{n_{glass}}{n_{air}} \sin \theta_1 \tag{26.34}$$

The value of n_{glass} is typically around $n_{glass} \approx 1.5$, while $n_{air} \approx 1.00$. Hence, the ratio n_{glass}/n_{air} is greater than 1, so if $\sin \theta_1$ is large enough, the right-hand side of Equation 26.34 is greater than 1. Since the sine of an angle cannot be larger than 1, Snell's law does not have any solution for the angle of refraction θ_2 in this case. This is the phenomenon of total internal reflection described in Section 24.3, where we showed that there is no refracted ray when the angle of incidence is greater than a certain critical angle θ_{crit}. Whenever the angle of incidence is greater than the critical angle, all the light is reflected at the surface, which allows optical fibers to carry light over long distances without any leakage of light energy out of the fiber.

The optical fiber in Figure 26.32A is only a simplified design. All practical optical fibers consist of at least two different types of glass as sketched in Figure 26.32B. The central core is surrounded by an outer layer called the cladding. The core and the cladding are both made of glass, but with different compositions and different indices of refraction. The index of refraction of the cladding is smaller than the index of the core, $n_{cladding} < n_{core}$, enabling total internal reflection of light within the core. In an application, light from a laser enters the fiber at one end, and the entry angle is arranged so that the light always undergoes total internal reflection at the boundary between the core and the cladding. Signals (such as a telephone signal) are carried from one end of the fiber to the other by pulses of laser light within the core. These fibers can be very long (many kilometers), and a single fiber can carry many more simultaneous signals than is possible with a conventional metal wire.

Some early optical fibers were designed as in Figure 26.32B, and although this approach works, it has the following flaw. The laser light carrying the signal consists of pulses that encode information in a binary fashion (ones and zeros, corresponding to "on" and "off" values of the light intensity). There is a certain delay between the time a pulse enters the fiber and the time it exits at the other end. For the optical fiber in Figure 26.32B, a pulse is carried by a collection of light waves that reflect at different angles as they move along the fiber. Since they propagate at different angles, some waves travel a greater total path length than others; all travel at the same speed, so some reach the end of the fiber before others, causing the pulses to be smeared out in time, thus limiting the amount of information the fiber can carry.[5]

This problem is overcome using the design sketched in Figure 26.32C. This fiber is similar to the one in Figure 26.32B except that the diameter of the core is much smaller. In fact, the core diameter is so small that ray optics can no longer be used to describe how light travels along the fiber. An analysis using wave optics shows that a light wave travels directly along the axis of the core, with no side-to-side reflections as for the larger core in Figure 26.32B. Fibers with very small diameter cores are called "single-mode" fibers because there is only one way for light to propagate in the fiber. These fibers greatly reduce the smearing of light pulses, allowing the fibers to carry more information than the "multimode" fiber in Figure 26.32B.

CONCEPT CHECK 26.8 What Happens When You Bend an Optical Fiber?

The simplest optical fibers consist of a single strand of glass. If such a fiber is straight, the angle of incidence θ_i for a ray inside the glass is large enough that there is total internal reflection and no light leaves the fiber (Fig. 26.33A). However, if the fiber is bent through a large enough angle, the angle of incidence can become small enough that some light does "escape" from the fiber (Fig. 26.33B). If, starting from a straight fiber, the bending angle of the fiber is gradually increased, which color of light escapes first?

(a) The longest wavelengths (red light) escape first.
(b) The shortest wavelengths (blue and violet light) escape first.
(c) All wavelengths escape at the same time.

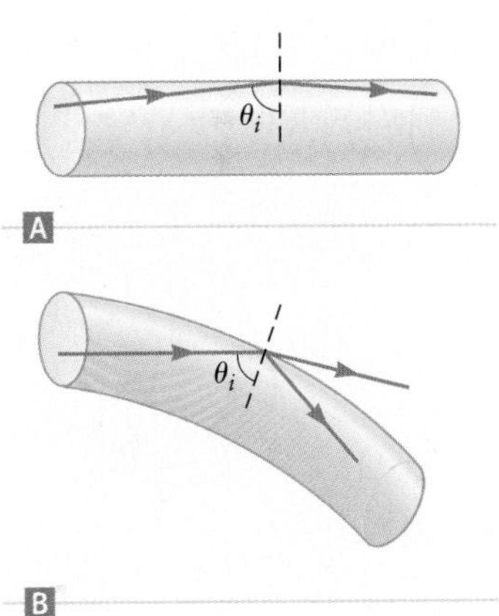

θ_i
A
θ_i
B

▲ **Figure 26.33** Concept Check 26.8.

EXAMPLE 26.10 Analysis of an Optical Fiber

Consider an optical fiber like the one in Figure 26.32B, with length $L = 10$ km and core diameter $d = 40\ \mu$m. Suppose the index of refraction of core and cladding are $n_{core} = 1.80$ and $n_{cladding} = 1.40$. Assuming light travels along the fiber as rays as sketched in Figure 26.34, estimate (**a**) the minimum and (**b**) the maximum travel times of light rays along the fiber.

RECOGNIZE THE PRINCIPLE

The minimum travel time will be for light rays parallel to the axis of the fiber, which can be calculated from the length of the fiber and the speed of light. The maximum travel time occurs for rays at the critical angle for total internal reflection (Fig. 26.34), and we can find that angle using Snell's law. Using geometry, we can compare the minimum and maximum travel distances and therefore the minimum and maximum travel times.

(continued) ▶

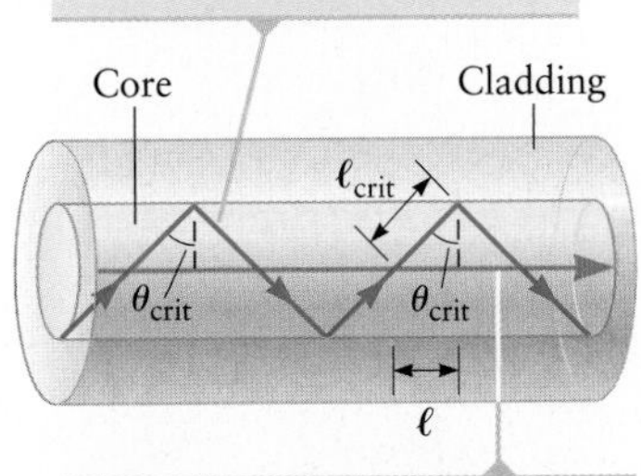

Light on this path travels the maximum distance and time.
Core
Cladding
ℓ_{crit}
θ_{crit}
θ_{crit}
ℓ
Light traveling straight through the fiber travels the minimum distance and time.

▲ **Figure 26.34** Example 26.10.

[5]Shorter pulses transmit information at a higher rate because more of them can be sent in a given amount of time, but the time between pulses must be longer than the smearing in time to avoid losing information; hence, pulse smearing limits the transmission rate.

SKETCH THE PROBLEM

Figure 26.34 shows the problem. One ray (shown in blue) lies along the axis of the fiber, and the other ray (shown in green) makes an angle θ_{crit} (the critical angle) with the direction normal to the wall of the fiber.

IDENTIFY THE RELATIONSHIPS AND SOLVE

(a) We first find the minimum travel time. The speed of light in a vacuum is $c = 3.00 \times 10^8$ m/s, but in the core this speed is reduced to $v_{core} = c/n_{core}$. The time for a light ray to travel the length of the fiber L is

$$t_{min} = \frac{L}{v_{core}} = \frac{L}{c/n_{core}}$$

Inserting the given values, we find

$$t_{min} = \frac{L}{c/n_{core}} = \frac{10 \times 10^3 \text{ m}}{(3.00 \times 10^8 \text{ m/s})/1.80} = 6.0 \times 10^{-5} \text{ s} = \boxed{60\ \mu\text{s}} \quad (1)$$

(b) The maximum travel distance depends on the critical angle θ_{crit} (Fig. 26.34). The critical angle for reflection at the interface between the core and the cladding can be found by applying Snell's law (Eq. 26.33) to the green ray in Figure 26.34. At the critical angle, we have an angle of refraction $\theta_2 = 90°$ in the cladding. We get

$$n_{cladding} \sin \theta_2 = n_{core} \sin \theta_{crit}$$

Inserting our value of θ_2 and the indices of refraction, we can solve for θ_{crit}:

$$n_{cladding} \sin \theta_2 = n_{cladding}(1) = n_{core} \sin \theta_{crit}$$

$$\sin \theta_{crit} = \frac{n_{cladding}}{n_{core}} = \frac{1.40}{1.80} = 0.78$$

$$\theta_{crit} = 0.89 \text{ rad} = 51°$$

The path length for rays that make an angle θ_{crit} with the normal to the interface is longer than the path parallel to the axis of the fiber. The critical-angle path in Figure 26.34 has a length ℓ_{crit}, whereas the path along the axis has a length ℓ. From the geometry in Figure 26.34,

$$\ell_{crit} = \frac{\ell}{\sin \theta_{crit}}$$

The corresponding travel time for light along this path is thus longer than the minimum (direct) travel time by a factor

$$\frac{\ell_{crit}}{\ell} = \frac{1}{\sin \theta_{crit}}$$

Light traveling along the axis takes a time t_{min} as found in Equation (1), so light at the critical angle takes a longer time:

$$t_{max} = t_{min}\left(\frac{1}{\sin \theta_{crit}}\right)$$

Using our results above for t_{min} and $\sin \theta_{crit}$, we find

$$t_{max} = \frac{t_{min}}{\sin \theta_{crit}} = \frac{60\ \mu\text{s}}{0.78} = 7.7 \times 10^{-5} \text{ s} = \boxed{77\ \mu\text{s}}$$

► ***What does it mean?***

The difference in travel times is thus

$$\Delta t = t_{max} - t_{min} = 77\ \mu\text{s} - 60\ \mu\text{s} = \boxed{17\ \mu\text{s}}$$

The value of Δt is nearly one-third of the minimum travel time, so the laser pulses will be quite smeared out. That is why single-mode fibers were invented.

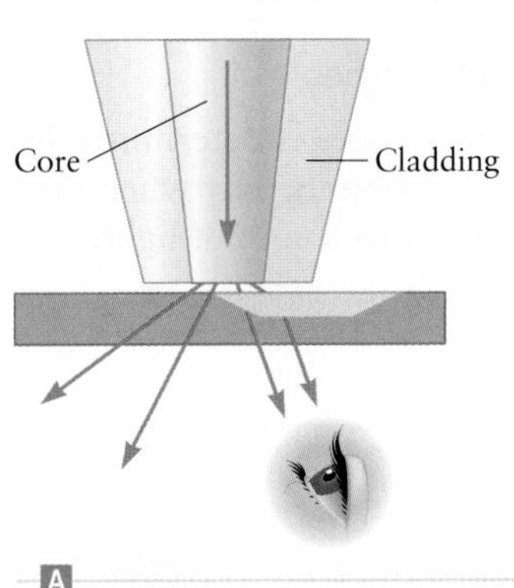

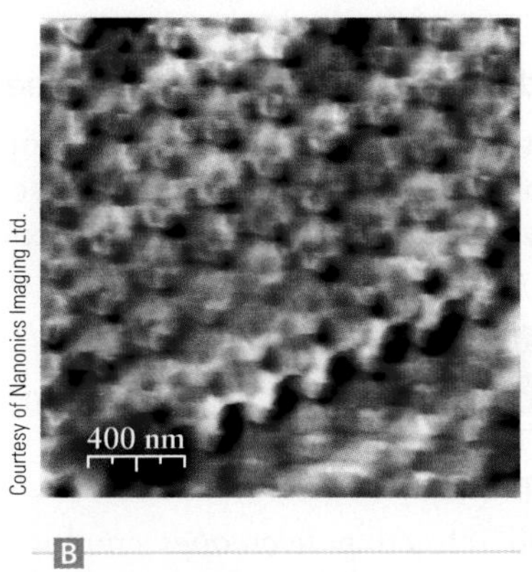

◀ **Figure 26.35** A The diameter of an optical fiber tip can be much smaller than the wavelength of light, producing a very tiny region just outside the tip where the light intensity is high. The tiny region at the fiber's tip is used in a near-field optical microscope by scanning it across the surface of the object of interest. B Image of a specially patterned layer of gold on the surface of silicon taken with a near-field scanning optical microscope. The scale bar at the lower left is 400 nm, the approximate wavelength of blue light. The small features resolved in this image are thus much smaller than the wavelength of visible light.

26.7 Microscopy with Optical Fibers

The wavelength of light sets a limit on the resolution of a compound microscope because light cannot be focused to a spot size smaller than approximately λ. However, light can still be used to perceive details smaller than λ using an approach called ***near-field scanning optical microscopy*** (NSOM). The NSOM technique uses an optical fiber to illuminate a very tiny region at the end of the fiber. The light is most intense at the opening of the tip, which can be much smaller than the wavelength. The tip is positioned very close to the object to be studied and is then is scanned over one of its surfaces (Fig. 26.35A). Because the tip is so close to the object, only a very small area near the tip is strongly illuminated; the resolution is then determined by the spacing from the surface and the tip diameter, both of which can be much smaller than the wavelength. By measuring the scattered or reflected light, one can construct an image of the surface based on the scattered intensity as a function of the tip's position.

Current NSOM fibers have openings of 25 nm or even smaller, more than a factor of 10 smaller than the wavelength of visible light. The resolution of the NSOM microscope is constantly improving, and the method is now being used to create images of objects as small as individual molecules! Conventional microscopes are limited by the ability of a lens to focus,[6] so their resolution is set by the Rayleigh criterion (Fig. 26.10) and is approximately equal to the wavelength of the light that is used. An NSOM microscope can thus resolve much smaller objects and features than is possible with a conventional microscope.

Summary | CHAPTER 26

Key Concepts and Principles

Optics and the eye

The eye is an essential part of many optical instruments, including magnifying glasses, microscopes, and telescopes. The purpose of these devices is to produce an image on the retina.

A ***magnifying glass***, also called an ***eyepiece***, is a converging lens that produces an upright virtual image when an object is placed near its focal point. This image appears at the ***near point*** of the eye, which is approximately $s_N = 25$ cm from the front of the eye in a normal young adult.

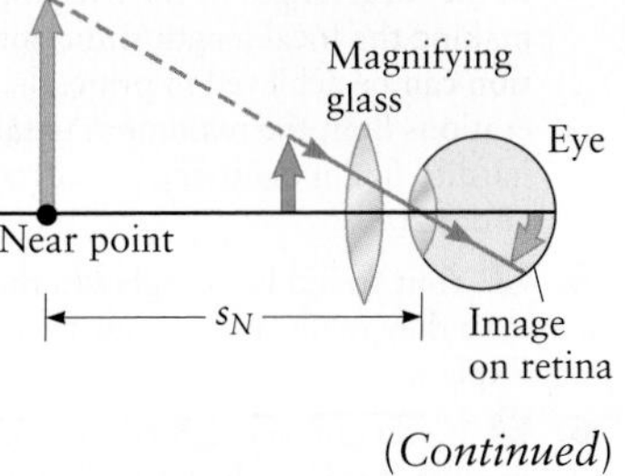

(Continued)

[6]Although very recent research suggests that radically new types of lenses (from radically new materials) may be able to beat the limit set by the Rayleigh criteria.

Diffraction and the resolution of microscopes and telescopes

The ultimate resolution of compound microscopes and telescopes is determined by diffraction. A compound microscope can resolve two nearby objects only when they are separated by a distance greater than approximately the wavelength of light.

Applications

Optical instruments

Microscopes and refracting *telescopes* employ (at a minimum) an *objective* lens together with an *eyepiece*. In a reflecting telescope, the objective lens is replaced by a mirror; this design has some practical advantages when the telescope is very large.

A *camera* produces an image on a light-sensitive detector, such as a CCD. Focusing of a camera is done by moving the lens relative to the detector.

Although microscopes, telescopes, and cameras were all invented long ago, optical technology is still improving. CDs, optical fibers, and near-field scanning optical microscopes have all been developed recently (during the author's lifetime!), and all can be understood using the principles of geometrical and wave optics.

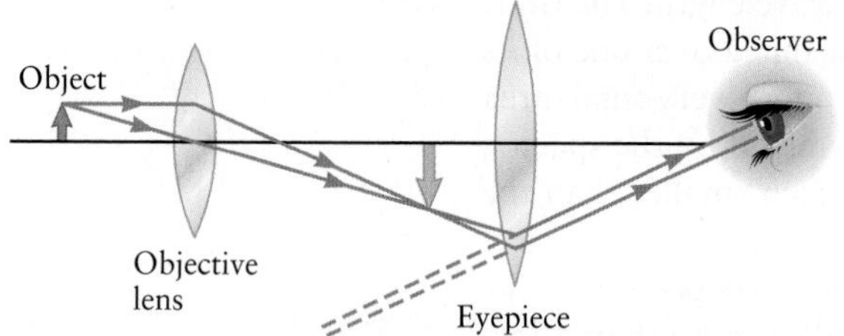

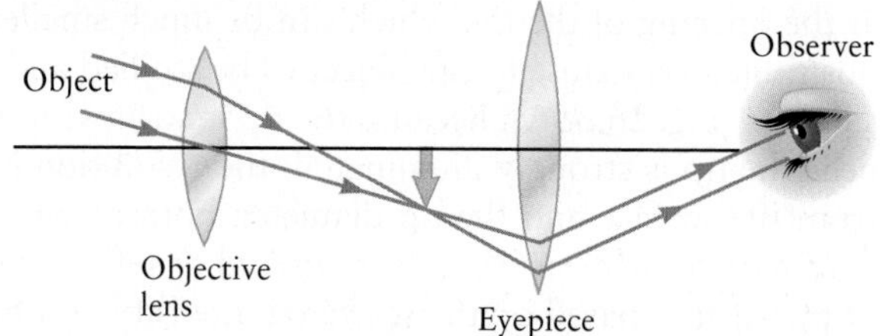

Questions

SSM = answer in *Student Companion & Problem-Solving Guide* = life science application

1. Explain why a magnifying glass uses a convex (converging) lens.
2. Consider two telescopes, one a reflector and the other a refractor. Assume both telescopes are of equal diameter, are of equal magnification, and have the same focal length. In what ways is the reflector superior to the refractor and vice versa?
3. Will a particular magnifying glass produce the same magnification for any user? Why or why not? If a magnifying glass is marked 5×, what does that imply?
4. In Section 26.1, we showed that the angular magnification of a magnifying glass is equal to the ratio of the near-point distance to the focal length of the lens ($m_\theta = s_N/f$, Eq. 26.14). Hence, by making the focal length sufficiently short, a very large magnification can be achieved in principle. However, other practical considerations limit the maximum useable value of m_θ. Explain why it is hard to find magnifying glasses with magnifications greater than about 20.
5. Your friend has eyeglasses that make his eyes appear larger than they really are. Is your friend nearsighted or farsighted? Explain.
6. If you are nearsighted, should your eyeglasses contain converging lenses or diverging lenses? What if you are farsighted?
7. Bifocals are eyeglasses that contain two different lenses for each eye (Fig. Q26.7). Each of these lenses is made from one piece of glass. One lens (lens 1) is used when looking at things far away, and the other (lens 2) is used for reading. The author uses bifocals because, without glasses, his eyes can only focus correctly on objects that are about 1 m away. (a) Is lens 1 a converging lens or a diverging lens? (b) Is lens 2 a converging lens or a diverging lens?

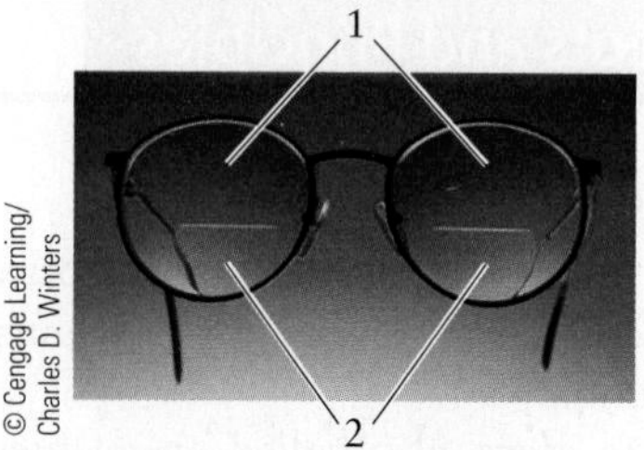

Figure Q26.7

8. Surgery can be used to correct the vision of a person who is nearsighted. Should the surgeon increase or decrease the curva-

ture of the cornea? What should the surgeon do if the person is farsighted?

9. SSM For typical cameras, the f-number can be selected from a set of fixed values such as 2, 2.8, 4, 5.6, 8, 11, and 16. Explain why this combination of f-numbers is used. *Hint*: The diameter of the aperture is inversely proportional to the f-number (Eq. 26.32), while the total intensity of the light that reaches the film is proportional to the area of the aperture.

10. Is the final image produced by a refracting telescope real or virtual? Explain.

11. Many of the refracting telescopes designed and built in the 18th and 19th centuries were extremely long (Fig. 26.13). Explain why they were made in this way; that is, what desirable property of the telescope becomes enhanced when it is made very long? What are the disadvantages of this design approach?

12. What kind of telescope did Galileo build and use in his studies? Do some investigating and try to find answers to the following questions.
(a) How long was his telescope?
(b) What was the total magnification?
(c) What was the focal length of the objective lens?
(d) What was the magnification of the eyepiece?

13. Will the image produced by the contact lenses designed in Section 26.1 (and Eq. 26.7) be upright or inverted? What about the image produced by glasses for a nearsighted person? Explain.

14. When an optometrist determines your eyeglass prescription, she always places a set of "test" lenses in front of your eyes as if they were your real glasses. The optometrist is interested in learning what lenses work best for you, but she is also measuring several other properties of your eyes. What are they?

15. Why does squinting often allow you to read at a distance? For example, even if you forget to wear corrective lenses, forming small slits by squinting your eyes often allows you to read a clock across a room or a sign across a street. Why does squinting help?

16. Some recent water fountain landscape installations include arcs of water exhibiting laminar flow like those shown in Figure Q26.16. The arc of water is illuminated at the base, and the light appears to follow along with the flow of the stream into the pond. Why does the light stay mostly within the arcing stream of water?

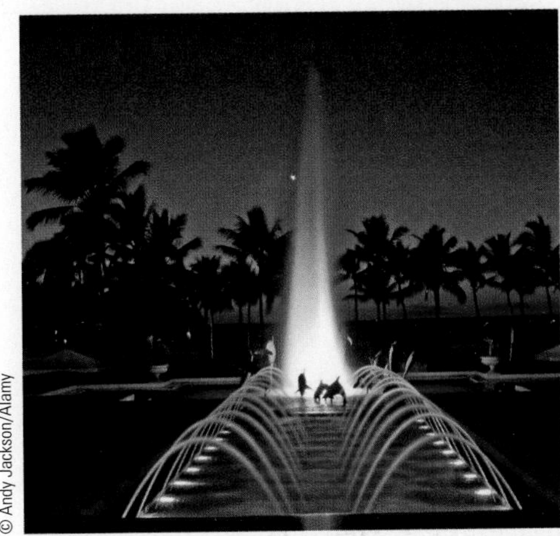

Figure Q26.16

17. The largest telescopes are all the reflector variety. Describe and explain two main design considerations in the construction of large-diameter telescopes that favor a reflector over a refractor.

18. A magnifying glass uses a lens with a focal length of magnitude $|f| = 300$ cm. Is f positive or negative?

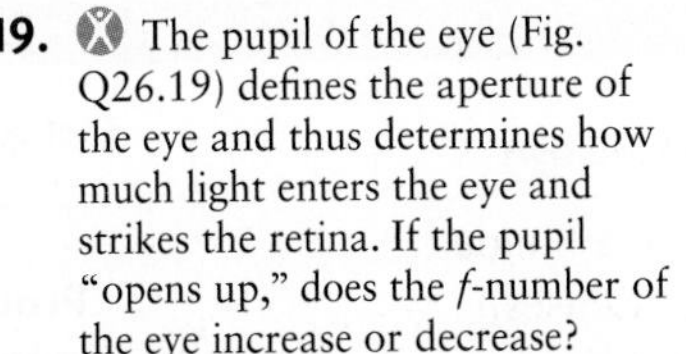

19. The pupil of the eye (Fig. Q26.19) defines the aperture of the eye and thus determines how much light enters the eye and strikes the retina. If the pupil "opens up," does the f-number of the eye increase or decrease?

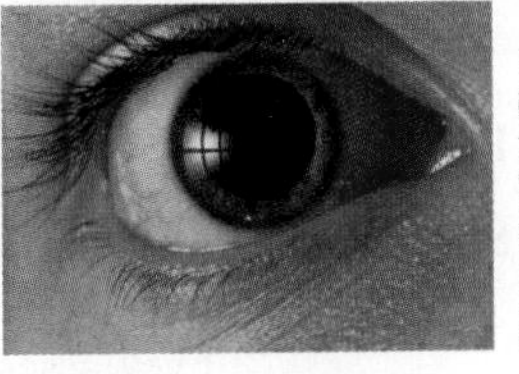

Figure Q26.19

20. Does a camera use a diverging lens or a converging lens? Does a camera produce a real image or a virtual image?

21. SSM A photographer intends to take a photograph of a country landscape with a large depth of field such that the flowers in the foreground of the picture and a barn in the distance are both in sharp focus. What conditions are more suitable to such a photo, a bright and sunny day or a dim and overcast day? Explain your choice.

22. DVD players use a laser with a higher frequency than that of CD players. Why can DVD players read CDs, but CD players cannot read DVDs?

23. You are asked to design an optical fiber. Should the refractive index of the core material be larger or smaller than that of the surrounding cladding? For the fastest transmission times, do you want a high or a low refractive index for the core material?

24. You have two unlabeled laser discs, one a CD and the other a DVD. Describe how you might use a laser pointer to tell which is which.

25. Small binoculars or opera glasses (Fig. Q26.25) are useful in seeing the stage in detail. No prisms or mirrors are used, and yet the magnified image in these glasses is not inverted. Are the eyepieces converging lenses or diverging lenses?

Figure Q26.25

26. The index of refraction of air increases with density. (Recall that the index of refraction of air at standard atmospheric pressure and temperature is $n = 1.0003$ and the index of refraction of the vacuum is $n = 1$, an exact value.) Consider light from a distant star that grazes the Earth's atmosphere as sketched in Figure Q26.26. Will the variation of the index of refraction with density cause that light to follow rays that "bend" away from the Earth (ray 1 in the figure) or bend toward the Earth (ray 3), or will they not be deflected at all? Explain.

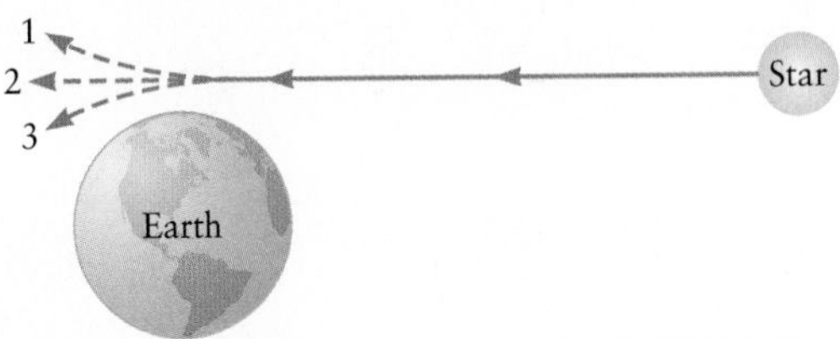

Figure Q26.26

Problems available in ENHANCED WebAssign

Section 26.1 Applications of a Single Lens: Contact Lenses, Eyeglasses, and the Magnifying Glass
Problems 1–21

Section 26.2 Microscopes
Problems 22–27

Section 26.3 Telescopes
Problems 28–34

Section 26.4 Cameras
Problems 35–44

Section 26.5 CDs and DVDs
Problems 45–47

Section 26.6 Optical Fibers
Problems 48–50

Additional Problems
Problems 51–62

List of Enhanced Problems

Problem number	Solution in *Student Companion & Problem-Solving Guide*	Reasoning & Relationships Problem	Reasoning Tutorial in ENHANCED WebAssign
26.3	✓		
26.7		✓	✓
26.8		✓	✓
26.13		✓	
26.17	✓		
26.20		✓	
26.21		✓	
26.25	✓		
26.31		✓	✓
26.32	✓	✓	
26.39		✓	
26.42	✓		
26.45	✓		
26.46		✓	
26.47		✓	✓
26.48	✓		
26.51		✓	✓
26.52		✓	
26.56		✓	
26.57		✓	
26.59	✓		
26.60		✓	✓
26.61		✓	✓
26.62		✓	

CHAPTER 27

▶ *The motion of light and matter through the universe has some surprising features that are explained by Einstein's theories of relativity. This photo shows the galaxy Centaurus A, which has a black hole at its center. Jets of high-energy particles (the opposing bright, light-colored streaks) are being ejected from the region near the black hole. (NASA/CXC/CfA/R. Kraft et al)*

Relativity

OUTLINE

Newton discovered his laws of mechanics around 1700. Over the next two centuries, these laws were applied to a wide variety of problems and seemed to work perfectly. With the development of Maxwell's equations in the late 1800s, the physics of electromagnetism and light fell into place. By 1900, many physicists thought that physics was completely understood with nothing more to be done, but how wrong they were! Two completely new and unanticipated subfields of physics—relativity and quantum theory—were discovered in the first decades of the 1900s. Relativity was developed mainly by Albert Einstein (although other physicists had important roles), and his theory of relativistic mechanics has profoundly changed the way we think about space and time.

There are actually two types of relativity theory. ***Special relativity*** is concerned with objects and observers moving with constant velocity, the case we will consider through most of this chapter. ***General relativity*** applies to situations in which an object or observer is accelerated, and has some surprising implications for the understanding of gravitation.

27.1 Newton's Mechanics and Relativity

In physics, the term ***relativity*** arises when we describe a situation from two different points of view. Figure 27.1 shows a person (Ted) standing on an open railroad car. In Figure 27.1A, the car is at rest while Ted throws a ball straight up into the air and then catches it when it falls back down to him. Figure 27.1B shows a similar event, but now the railroad car is moving with constant speed v to the right and a person on the ground next to the car (Alice) is watching. According to Ted—that is, *relative to Ted*—the ball travels straight up and then back down, just as in Figure 27.1A. Relative to Alice, the ball has a horizontal component of velocity equal to that of the railroad car and follows a parabolic trajectory.

Inertial reference frame: a set of coordinate axes that move at constant velocity relative to other inertial reference frames

These two observers are in different ***reference frames***. You can think of a reference frame as a set of coordinate axes. In this example, Alice is at rest relative to the x–y axes, whereas Ted would describe the motion of the ball using the coordinate axes x'–y' that travel along with his railroad car.

Reference frames that move with constant velocity, like the reference frames in Figure 27.1, are called ***inertial reference frames***. The idea that the laws of motion should be the same in all inertial frames is called the ***principle of Galilean relativity***. Newton's laws of mechanics obey Galilean relativity; that is, if Newton's laws of mechanics are obeyed in one inertial reference frame, they are obeyed in all inertial frames. For example, Newton's second law reads

$$\sum \vec{F} = m\vec{a} \tag{27.1}$$

where $\vec{a}$ is the acceleration of a particle of mass m and $\sum \vec{F}$ is the total force on the particle. Adding or subtracting a constant velocity does not change the acceleration of a particle. In Figure 27.1, Ted would say that the horizontal component of the ball's velocity (the component along x') is zero, but Alice would disagree; in her reference frame, the horizontal (x) component of the ball's velocity is v. Both Ted and Alice would agree, however, that the ball's acceleration is directed downward (along either y' or y) with magnitude g, and both would agree that the ball's horizontal acceleration is zero. They would also agree that the only force on the ball is the force of gravity, so both observers (each in a different inertial reference frame) would say that Newton's second law (Eq. 27.1) is obeyed. This result is similar to what we found in our analysis of the airplane accelerometer in Chapter 4; there again, observers in different inertial reference frames both found that Newton's second law is obeyed.

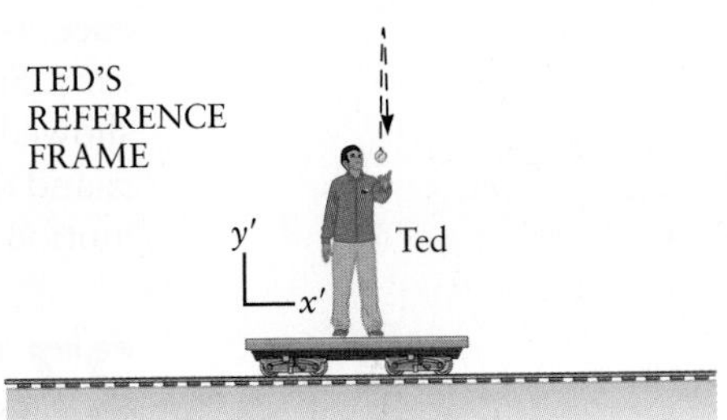

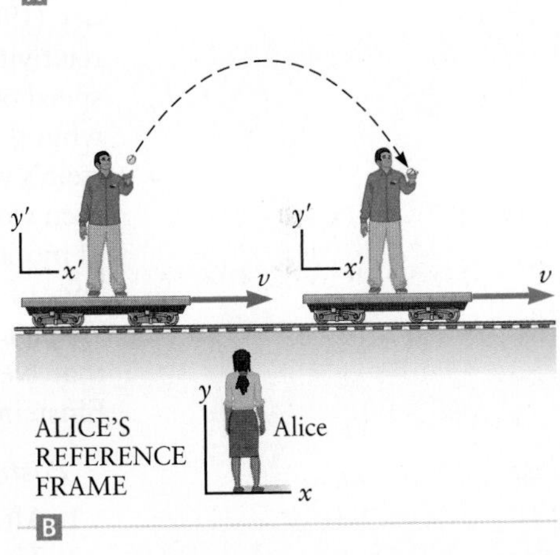

▲ **Figure 27.1** A When Ted throws a ball upward or drops a ball downward, he observes that the ball's motion is purely along the vertical (y') direction. B When Ted's railroad car has speed v relative to a second observer (Alice), the ball undergoes projectile motion with a nonzero displacement along both x and y in Alice's reference frame. However, as viewed by Ted using his reference frame and coordinates x' and y', the ball's motion is still purely vertical as in part A.

The principle of Galilean relativity agrees with our everyday intuition and the experiment in Figure 27.1, and it was believed to be an exact law of physics for more than two hundred years. The first hint of a problem came from Maxwell's work in electromagnetism. According to Maxwell's equations, the speed of light in a vacuum is (Eq. 23.1)

$$c = \frac{1}{\sqrt{\varepsilon_0 \mu_0}} \tag{27.2}$$

Maxwell also showed that this result is independent of the motion of both the source of the light and the observer. To see why this finding is surprising, suppose Ted is again moving with speed v relative to Alice when he turns on a flashlight

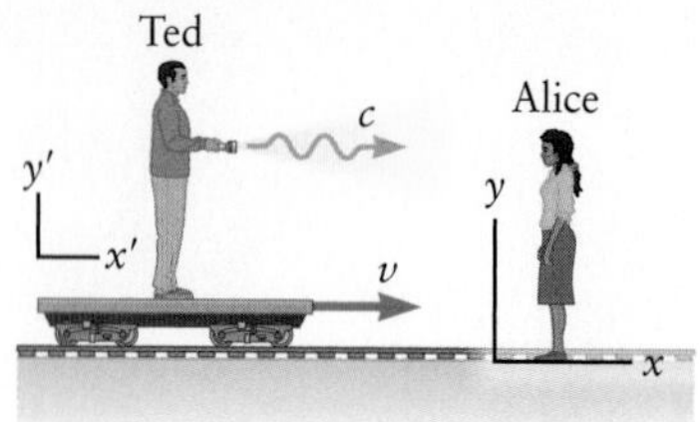

▲ **Figure 27.2** Ted moves at speed v relative to Alice when his flashlight emits a light pulse that moves at speed c relative to him. What is the speed of the light pulse relative to Alice?

that generates a light pulse traveling to the right in Figure 27.2. This is a one-dimensional situation, so we need consider only the component of velocity along the horizontal (x) direction. The light wave in Figure 27.2 has speed c relative to Ted and his flashlight. Since Ted's speed relative to Alice is v, Newton's mechanics predicts that the speed of the light wave relative to Alice should be $c + v$. According to Maxwell's theory, however, Ted and Alice should *both* observe the light wave to move with speed c. Maxwell's theory of electromagnetism is thus not consistent with the predictions of Galilean relativity for observers in different inertial frames.

Which prediction is correct? Initially, Maxwell and other physicists believed there must be a problem with Maxwell's theory, but an experimental test is difficult. For example, if Ted in Figure 27.2 is moving at 100 m/s (about 200 mi/h), Galilean relativity predicts that Alice will observe a speed of

$$v(\text{Alice}) = c + v(\text{Ted}) = [(3.00 \times 10^8) + 100]\ \text{m/s} \tag{27.3}$$

This differs from c by less than one part in one million, and such a small difference would be very hard to measure using the technology available in Maxwell's time. Successful experiments were carried out only well after Maxwell's death and showed that Maxwell's theory has it right: the speed of light in a vacuum is *always* c, and the prediction of Galilean relativity for how the speed of light depends on the motion of the source is *wrong*.

27.2 The Postulates of Special Relativity

The theory Einstein developed to analyze the predicament of Ted and Alice in Figure 27.2 is called ***special relativity***. According to Einstein, his work on this theory was not motivated by any particular experiment; indeed, at the time he developed his theory (1905), there were not yet any clear experimental results to show that Galilean relativity was wrong. Instead, Einstein suspected that Maxwell's result—that the speed of light is the same in all reference frames—is correct, and he then worked out what that implies for the other laws of physics such as Newton's laws. In a sense, Einstein's work was similar to that of Newton: both proposed basic laws of physics and then worked out the consequences. For Newton, these basic laws were his three laws of motion. For Einstein, the basic laws are now known as two "postulates" about the laws of physics. We'll first state the postulates and then spend most of the rest of this chapter working out the consequences. Experiments performed after 1905 showed that Newton's theory breaks down for fast-moving objects (such as light) and that Einstein's theory gives a correct description of motion in such a regime.

Postulates of Special Relativity

1. **All the laws of physics are the same in all inertial reference frames.**
2. **The speed of light in a vacuum is a constant, independent of the motion of the light source and all observers.**

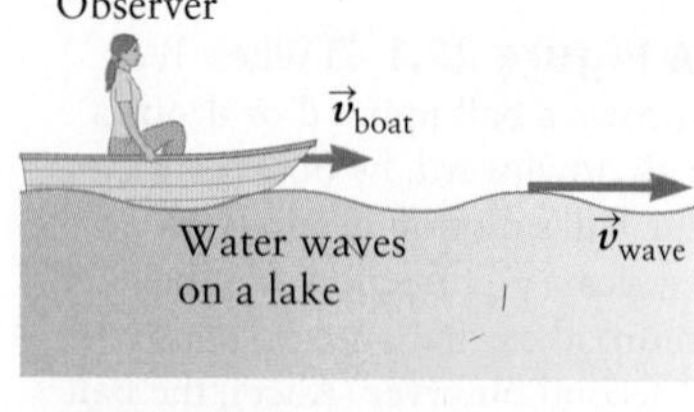

▲ **Figure 27.3** According to our intuition, water waves on a lake obey Galilean relativity. The velocity of these waves relative to the water (the wave medium) is $\vec{v}_{\text{wave}}$, while the velocity of a boat relative to the water is $\vec{v}_{\text{boat}}$. In this example, the velocity of the waves relative to the boat is $\vec{v}_{\text{wave}} - \vec{v}_{\text{boat}}$.

The first postulate can be traced to the ideas of Galileo and Newton on relativity, but this postulate goes further than Galileo because it applies to *all* physical laws, not just mechanics. The second postulate of special relativity is motivated by Maxwell's theory of light, which we have seen is *not* consistent with Newton's mechanics. Newton would have predicted that the speed of the light pulse relative to Alice in Figure 27.2 is $c + v$ (Eq. 27.3), whereas Ted (who is in a different inertial reference frame) would measure a speed c. According to the second postulate of special relativity, both observers will measure the *same* speed c for light.

Postulate 2 also tells us something very special about light. Light is a wave, and all the other waves we have studied travel in a mechanical medium. For example, Figure 27.3 shows an observer traveling in a boat with velocity $\vec{v}_{\text{boat}}$ relative to still water. If there are waves traveling at velocity $\vec{v}_{\text{wave}}$ relative to still water and if the boat is traveling along the wave propagation direction, Newton's mechanics (and

Galilean relativity) predict that the velocity of the waves *relative to this observer* is $\vec{v}_{\text{wave}} - \vec{v}_{\text{boat}}$. An observer who is stationary relative to the water would thus measure a different wave speed than the observer in the boat. This is *not* what the two observers in Figure 27.2 find; they measure the *same* speed for the light wave.

The conclusion from Figures 27.2 and 27.3 is that our everyday experience with conventional waves *cannot* be applied to light. According to postulate 2 of special relativity, the speed of a light wave is independent of the velocity of the observer. What role does the medium have in this case? Light does not depend on having a conventional material medium in which to travel. For a light wave, the role of the medium is played by the electric and magnetic fields, so a light wave essentially carries its medium with it as it propagates (which is why light can travel through a vacuum). The lack of a conventional medium is difficult to reconcile with one's intuition, and Maxwell's results were therefore slow to be accepted. Experiments, however, show that nature does work this way; the speed of light *is* independent of the motion of the observer. This example is just one of many for which conventional intuition fails in the study of special relativity.

Finding an Inertial Reference Frame

Inertial reference frames play a crucial role in special relativity. We have stated that an inertial frame is one that moves with constant velocity, but relative to what? Newton believed that the heavens and stars formed a "fixed" and absolute reference frame to which all other reference frames could be compared. We now know that stars are also in motion, so we need a better definition of what it means to be "inertial." Nowadays, we define an inertial reference frame as one in which Newton's first law holds. Recall that Newton's first law states that if the total force on a particle is zero, the particle will move with constant velocity, in a straight line with constant speed. So, we can test for an inertial reference frame by observing the motion of a particle for which the total force is zero. If the particle moves with constant velocity, the reference frame used to make the observation is an inertial frame.

This definition of how to find an inertial reference frame is thus tied to Newton's first law and the concept of inertia, so the definition may seem a bit circular. However, the notion of Galilean relativity also asserts that Newton's other laws of mechanics are valid in all inertial frames. Hence, Newton's second law ($\sum \vec{F} = m\vec{a}$) and third law (the action–reaction principle) should apply in all inertial frames, too.

The Earth as a Reference Frame

Inertial reference frames move with constant velocity; hence, their acceleration is zero. Since the Earth spins about its axis as it orbits the Sun, all points on the Earth's surface have a nonzero acceleration. Strictly speaking, then, a person standing on the surface of the Earth is not in an inertial reference frame. However, the Earth's acceleration is small enough that it can be ignored in most cases, so in most situations we can consider the Earth to be an inertial reference frame.

27.3 Time Dilation

Einstein's two postulates seem quite "innocent." The first postulate—that the laws of physics must be the same in all inertial reference frames—is in accord with Newton's laws, so it does not seem that this postulate can lead to anything new for mechanics. The second postulate concerns the speed of light, and it is not obvious what it will mean for objects other than light. Einstein, however, showed that these two postulates together lead to a surprising result concerning the very nature of time. He did so by considering in a very careful way how time can be measured.

Let's use the postulates of special relativity to analyze the operation of the simple clock in Figure 27.4. This clock keeps time using a pulse of light that travels back

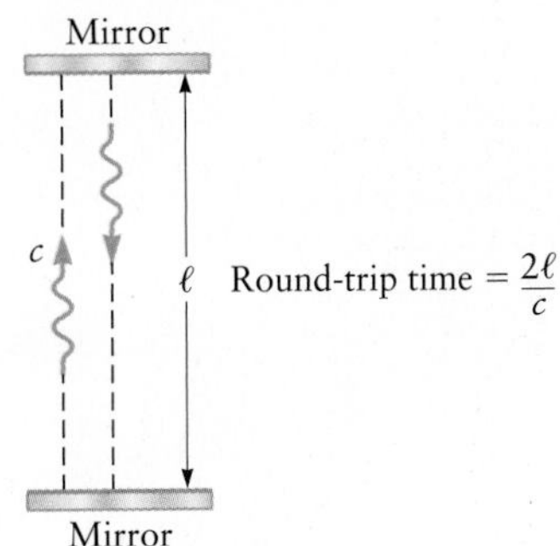

▲ **Figure 27.4** A light clock. Each round-trip motion of the light pulse between the two mirrors corresponds to one tick of the clock.

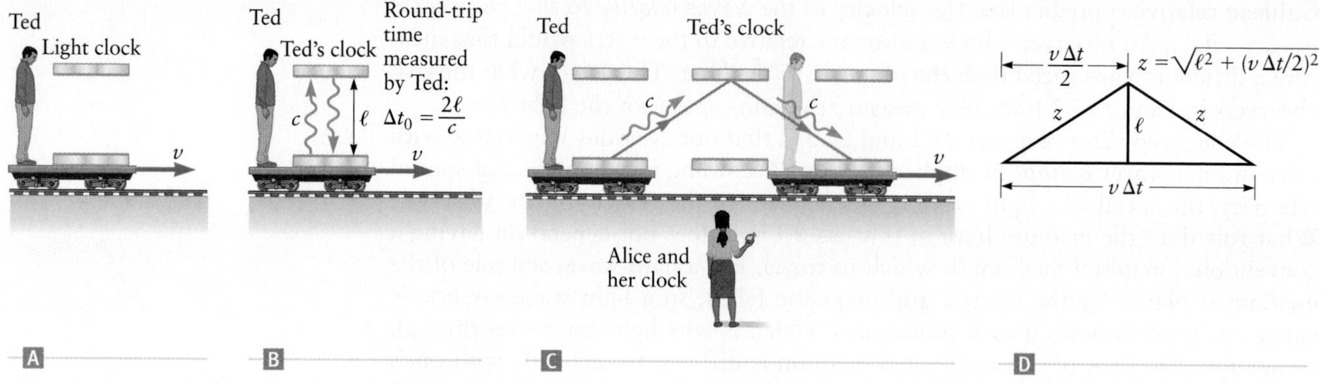

▲ **Figure 27.5** A A light clock traveling with Ted on his railroad car. B According to Ted, light pulses travel back and forth in the clock just as in Figure 27.4. Each tick of the clock takes a time $\Delta t_0 = 2\ell/c$. According to Ted, the operation of the clock is the same whether or not his railroad car is moving. C Motion of the light pulses in Ted's clock as viewed by Alice, who is at rest on the ground. D According to Alice, the round-trip travel distance for a light pulse is $2z$, where $z = \sqrt{\ell^2 + (v\,\Delta t/2)^2}$, which is longer than the round-trip distance 2ℓ seen by Ted in part B.

and forth between two mirrors. The mirrors are separated by a distance ℓ, and light travels between them at speed c. A light pulse thus makes one round trip through the clock in a time $2\,\ell/c$. That is the time required for the clock to "tick" once.

Analysis of a Moving Light Clock

We now place our light clock on Ted's railroad car in Figure 27.5A, so the clock is moving at constant velocity with speed v relative to the ground. Let's analyze the clock from Ted's viewpoint—that is, in Ted's reference frame—as we ride along on the railroad car in Figure 27.5B. In this reference frame, the operation of the clock is identical to that shown in Figure 27.4; the light pulse simply travels up and down between the two mirrors. The separation of the mirrors is still ℓ, so the round-trip time is still $2\ell/c$. If Δt_0 is the time required for the clock to make one "tick" as measured by Ted, we have

$$\Delta t_0 = \frac{2\ell}{c} \tag{27.4}$$

A second observer, Alice, is standing on the ground watching Ted's railroad car travel by and sees things differently. In her reference frame, Ted's clock is moving horizontally, so from her point of view the light pulse does not simply travel up and down between the mirrors, but must travel a longer distance as shown in Figure 27.5C. According to postulate 2 of special relativity, the speed of light is the same for Alice as it is for Ted. Because light travels a longer distance in Figure 27.5C than in Figure 27.5B, according to Alice the light will take longer to travel between the mirrors. Let's use Δt to denote the round-trip time as observed by Alice; that is the time it takes the clock to complete one "tick" in Alice's reference frame. We can find Δt by using a little geometry. In a time Δt (one "tick"), Alice sees the clock move a horizontal distance $v\,\Delta t$. The path of the light pulse forms each hypotenuse z of the back-to-back right triangles in Figure 27.5D. Using the Pythagorean theorem,

$$z^2 = \ell^2 + \left(\frac{v\,\Delta t}{2}\right)^2 \tag{27.5}$$

Since z is half the total round-trip distance, Alice finds $z = c\,\Delta t/2$, or

$$z^2 = \frac{c^2(\Delta t)^2}{4} \tag{27.6}$$

Combining Equations 27.5 and 27.6 gives

$$\frac{c^2(\Delta t)^2}{4} = \ell^2 + \frac{v^2(\Delta t)^2}{4}$$

We next solve for Δt:

$$(\Delta t)^2 = \frac{4\ell^2}{c^2} + \frac{v^2}{c^2}(\Delta t)^2$$

$$(\Delta t)^2\left(1 - \frac{v^2}{c^2}\right) = \frac{4\ell^2}{c^2}$$

$$(\Delta t)^2 = \frac{4\ell^2/c^2}{1 - v^2/c^2}$$

Taking the square root of both sides and using Equation 27.4 finally leads to

$$\Delta t = \frac{\Delta t_0}{\sqrt{1 - v^2/c^2}} \tag{27.7}$$

Recall that Δt and Δt_0 are the times required for one tick of the light clock as observed by Alice and Ted, respectively. In words, Equation 27.7 thus says that these times are *different*! The operation of this clock *depends on the motion of the observer.* Let's now consider the implications of Equation 27.7 in more detail.

Moving Clocks Run Slow

According to Equation 27.7, the ratio of Δt (the time measured by Alice) to the time Δt_0 (measured by Ted) is

$$\frac{\Delta t}{\Delta t_0} = \frac{1}{\sqrt{1 - v^2/c^2}} \tag{27.8}$$

Time dilation: moving clocks run slow.

Assuming v is less than the speed of light c (discussed further below), the factor on the right-hand side is always greater than 1. Hence, the ratio $\Delta t/\Delta t_0$ is larger than 1, and Alice measures a longer time than Ted does. In words, a moving clock will, according to Alice, take longer for each tick. Hence, special relativity predicts that *moving clocks run slow.* This effect is called ***time dilation.*** Postulate 1 of special relativity states that *all* the laws of physics must be the same in *all* inertial frames. We could use a light clock to monitor time, or we could use any process in any reference frame. Since Equation 27.8 holds for light clocks, it must therefore apply to *any* type of clock or process, including biological ones.

This result seems very strange; our everyday experience does not suggest that a clock (such as your wristwatch) traveling in a car gives different results than an identical clock at rest. If Equation 27.8 is true (and experiments definitely show that it *is* correct), why haven't you noticed it before now? Figure 27.6 shows a plot of the ratio $\Delta t/\Delta t_0 = 1/\sqrt{1 - v^2/c^2}$ as a function of the speed v of the clock. At ordinary terrestrial speeds, v is much smaller than c and the ratio $\Delta t/\Delta t_0$ is very close to 1. For example, when v = 50 m/s (about 100 mi/h), the ratio is

$$\frac{\Delta t}{\Delta t_0} = \frac{1}{\sqrt{1 - v^2/c^2}} = \frac{1}{\sqrt{1 - (50 \text{ m/s})^2/(3.00 \times 10^8 \text{ m/s})^2}}$$

$$\frac{\Delta t}{\Delta t_0} = 1.000000000000014 \tag{27.9}$$

The result in Equation 27.9 is extremely close to 1, and for all practical purposes the times measured by Ted and by Alice are the same. For typical terrestrial speeds, the difference between Δt and Δt_0 is completely negligible.

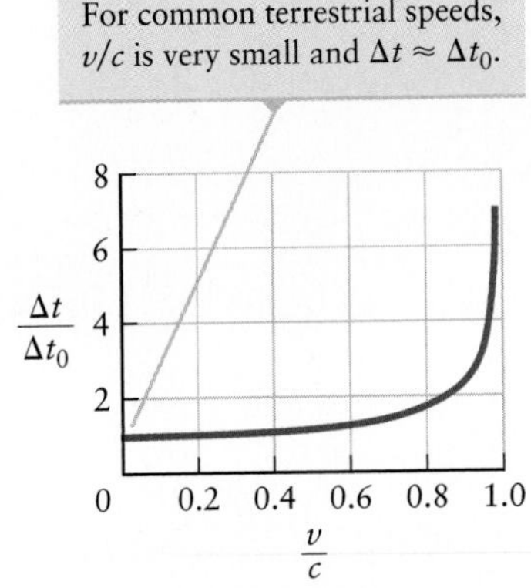

▲ **Figure 27.6** Time dilation factor $\Delta t/\Delta t_0$ as a function of v/c.

Insight 27.1

RELATIVISTIC CALCULATIONS WHEN V IS SMALL

The factor $\sqrt{1 - v^2/c^2}$ arises often in special relativity. When v is small compared with c, this factor is very close to 1. In fact, the difference between it and 1 can be so small that your calculator may have trouble dealing with it. In such cases, the approximations

$$\sqrt{1 - v^2/c^2} \approx 1 - \frac{v^2}{2c^2}$$

and

$$\frac{1}{\sqrt{1 - v^2/c^2}} \approx 1 + \frac{v^2}{2c^2}$$

are very handy and are quite accurate at terrestrial speeds. In practice, they can be used whenever v is less than about $0.1c$ (10% of the speed of light). (See Figs. 27.6 and 27.15.) We sometimes also have expressions like $1/(1 + A)$, where A is very small. In such cases, we can use the approximation

$$\frac{1}{1 + A} \approx 1 - A$$

Nature's Speed Limit

A curious feature of the time-dilation factor in Equation 27.8 is that its value is imaginary when v is greater than the speed of light. For example, if we insert $v = 2c$ into Equation 27.8, we get

$$\frac{\Delta t}{\Delta t_0} = \frac{1}{\sqrt{1 - (2c)^2/c^2}} = \frac{1}{\sqrt{-3}}$$

This result is an imaginary number! Does it mean that special relativity predicts that some time intervals Δt can be imaginary numbers? No, it does not. Speeds greater than c have never been observed in nature. In Section 27.9 we will discuss work and energy in special relativity and explain why it is not possible for an object to travel faster than the speed of light.

EXAMPLE 27.1 Time Dilation for an Astronaut

The astronauts who traveled to the Moon in the Apollo missions hold the record for the highest speed traveled by people, with $v = 11{,}000$ m/s. What is the ratio $\Delta t/\Delta t_0$ for the Apollo astronauts' clock?

RECOGNIZE THE PRINCIPLE

The Apollo astronauts have the role of Ted in Figure 27.5 because we are interested in a clock that travels with them, while an observer on the Earth has the role of Alice. The time measured by the astronaut's clock thus reads the time interval Δt_0, and an observer on the Earth measures a longer time interval Δt.

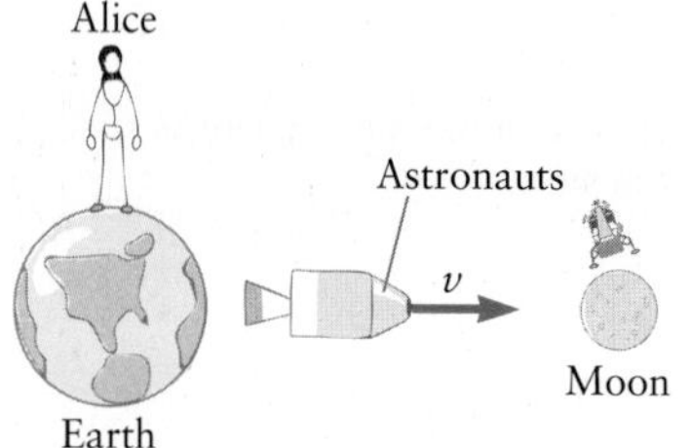

▲ **Figure 27.7** Example 27.1.

SKETCH THE PROBLEM

This problem is described by Figure 27.7. We assume the astronauts are carrying a light clock with them to the Moon, just as Ted traveled with a clock in Figure 27.5.

IDENTIFY THE RELATIONSHIPS

We can find $\Delta t/\Delta t_0$ using our analysis of Figure 27.5 (and Eq. 27.8), substituting $v = 1.1 \times 10^4$ m/s and the known speed of light.

SOLVE

Inserting the given values, we have

$$\frac{\Delta t}{\Delta t_0} = \frac{1}{\sqrt{1 - v^2/c^2}}$$

$$= \frac{1}{\sqrt{1 - (1.1 \times 10^4 \text{ m/s})^2/(3.00 \times 10^8 \text{ m/s})^2}} = \boxed{1.00000000067}$$

When v is such a small fraction of the speed of light, you may be limited by the number of significant figures your calculator can display. In such cases, you can use one of the approximations given in Insight 27.1. The second approximation gives

$$\frac{\Delta t}{\Delta t_0} \approx 1 + \frac{v^2}{2c^2} = 1 + \frac{(1.1 \times 10^4 \text{ m/s})^2}{2(3.00 \times 10^8 \text{ m/s})^2} \approx \boxed{1 + 6.7 \times 10^{-10}} \quad (1)$$

► ***What does it mean?***

Time dilation is a very small effect, even at this (relatively) high speed, yet it is possible to make clocks that are accurate enough to observe the small amount of slowing down in Equation (1). This result for Δt applies to all clocks, including the biological clocks of the Apollo astronauts. Hence, these astronauts aged slightly less than other people who stayed behind!

Proper Time

We derived Equation 27.8 from an analysis of a light clock, but the result applies to all time intervals measured with any type of clock. The time interval Δt_0 for a particular clock or process is measured by an observer *at rest relative to the clock* (Ted in Fig. 27.5) and is called the ***proper time***. Ted is moving on his railroad car in Figure 27.5, so the clock is moving along with Ted. Hence, Ted is at rest relative to this clock and he measures the proper time. On the other hand, Alice in Figure 27.5 is moving *relative to the clock*, so she does not measure the proper time. The time interval Δt measured by a moving observer is always longer than the proper time.

When an observer is at rest relative to a clock or process, the start and end of the process occur at the same location for this observer. For the light clock in Figure 27.5B, Ted might be standing next to the bottom mirror, so from his viewpoint the light pulse starts and ends at the same location. By comparison, for Alice in Figure 27.5C the light pulse begins at the bottom mirror when the clock is at the left and the pulse returns to this mirror when the clock is in a different location. The proper time is always the shortest possible time that can be measured for a process, by any observer.

CONCEPT CHECK 27.1 Measuring Proper Time

Ted is traveling in his railroad car with speed v relative to Alice, who is standing on the ground nearby (Fig. 27.8). Ted is playing with his yo-yo and uses a clock on the railroad car to measure the time it takes for the yo-yo to complete one up-and-down oscillation. The yo-yo is also observed by Alice, by a bird flying nearby (also with speed v), and by an astronaut who is cruising by at a very high speed v_{astro}. Which observer measures the proper time for the yo-yo's period, (a) Ted, (b) Alice, (c) the bird, or (d) the astronaut? (More than one answer may be correct.)

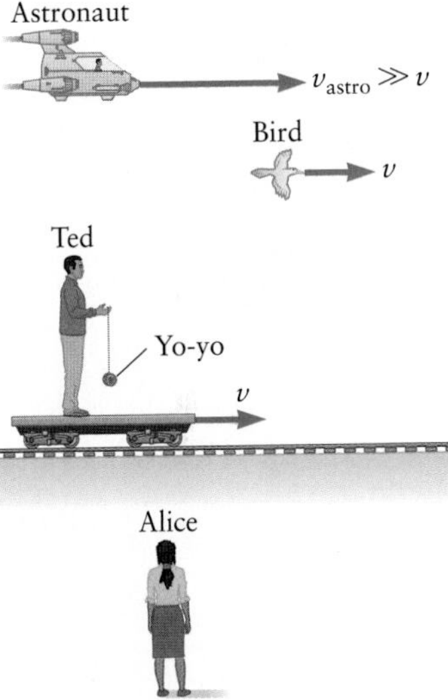

▲ **Figure 27.8** Concept Check 27.1.

EXAMPLE 27.2 Time Dilation for a Muon

Subatomic particles called muons are created when cosmic rays collide with atoms in the Earth's atmosphere. Muons created in this way have a typical speed $v = 0.99c$, very close to that of light. Muons are unstable, with an average lifetime of about $\tau = 2.2 \times 10^{-6}$ s before they decay into other particles. That is, physicist 1 at rest relative to the muon measures this lifetime τ. Another (physicist 2) studies the decay of muons that are moving through the atmosphere with a speed of $0.99c$ relative to her laboratory (Fig. 27.9). What lifetime would physicist 2 measure?

RECOGNIZE THE PRINCIPLE

The muon acts as a sort of "clock" in which its lifetime corresponds to one "tick." Our results for a light clock, including Equation 27.8, apply to this muon "clock" since the results of special relativity apply to *all* physical processes. A clock moving along with the muon measures the proper time Δt_0, just as Ted in Figure 27.5B measures the proper time of a clock that travels along with him in his railroad car. The muon is moving relative to physicist 2, so that physicist is just like Alice in Figure 27.5C and measures a longer time Δt for the muon's lifetime.

SKETCH THE PROBLEM

Figure 27.9 shows the problem schematically.

IDENTIFY THE RELATIONSHIPS

Applying the time dilation result from Equation 27.7, we have

$$\Delta t = \frac{\Delta t_0}{\sqrt{1 - v^2/c^2}}$$

The lifetime measured by a clock at rest relative to the muon is $\Delta t_0 = \tau$.

(continued) ▶

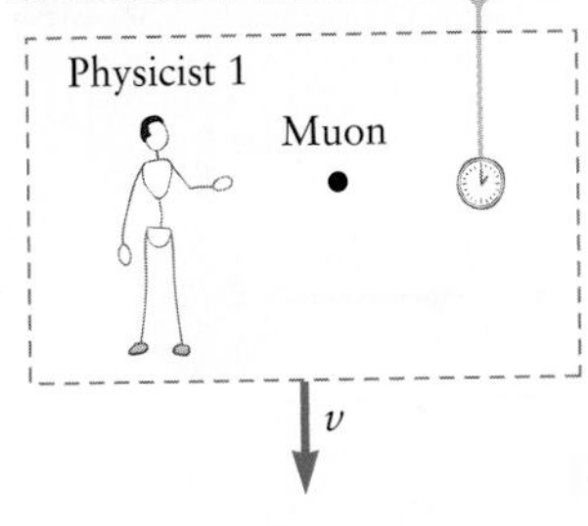

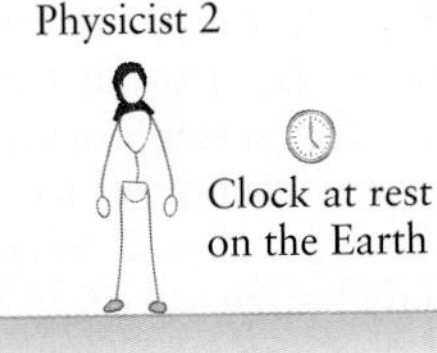

▲ **Figure 27.9** Example 27.2.

SOLVE

Using the given value $v = 0.99c$, the lifetime of the moving muon is

$$\Delta t = \frac{\tau}{\sqrt{1 - v^2/c^2}} = \frac{\tau}{\sqrt{1 - (0.99c)^2/c^2}}$$

$$\Delta t \approx \boxed{7.1 \times \tau}$$

▸ What does it mean?

According to physicist 2, the moving muon exists for a much longer time than a muon at rest. Experiments show that moving muons do indeed "live" longer before decaying than muons at rest. This is another surprising result of special relativity.

The Twin Paradox

Example 27.2 describes the effect of time dilation on the lifetime of a muon, but a similar result applies in other cases, including the lifetime of a person. Consider an astronaut (Ted) who is on a mission to travel to the nearby star named Sirius[1] and then return to the Earth (Fig. 27.10A). Sirius is 8.6 light-years (ly) from the Earth, which means that light takes 8.6 years to travel from the Earth to Sirius. Ted must therefore travel in a very fast spaceship to complete his mission before he is too old to be an astronaut, so NASA has given him a ship that travels at a speed of $0.90c$. Ted's mission is being tracked by his twin, Alice, who stays behind on the Earth. One of her jobs is to study how Ted ages during the trip. According to Alice, the round-trip distance is 2×8.6 ly $= 17.2$ ly, so the trip takes a time Δt with

$$\Delta t = \frac{17.2 \text{ ly}}{0.90c} = 19 \text{ years} \tag{27.10}$$

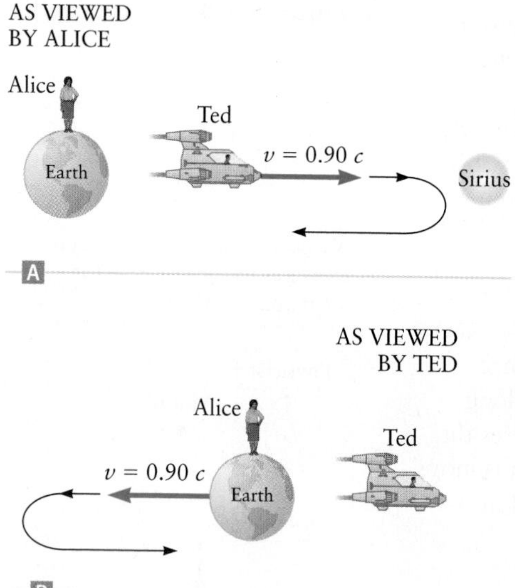

▲ **Figure 27.10** The twin paradox. **A** Astronaut Ted travels to the distant star Sirius and then returns while his trip is monitored by his twin, Alice. **B** As viewed by Ted (in his reference frame), Alice and the Earth take a journey in the direction opposite to that in part A. Not to scale!

Alice also knows about time dilation and realizes that Ted's body acts as a clock (just like the muon in Example 27.2). Ted's body clock measures the proper time Δt_0 since it travels along with him. From our work on time dilation, we know that the proper time is shorter than Alice's time in Equation 27.10, with (from Eq. 27.7)

$$\Delta t_0 = (\Delta t)\sqrt{1 - v^2/c^2} \tag{27.11}$$

Inserting $v = 0.90c$ and our value of Δt gives

$$\Delta t_0 = (\Delta t)\sqrt{1 - v^2/c^2} = (19 \text{ years})\sqrt{1 - (0.90c)^2/c^2} = 8.3 \text{ years}$$

which is the time the trip takes according to Ted's body clock. Ted thus ages only 8.3 years, whereas his twin, Alice, ages 19 years during the trip. Ted will therefore be younger than his twin at the end of the trip! This result can be understood in simple terms from the basic statement about time dilation: moving clocks (in this case, Ted himself) run slow.

It is interesting to ask now how Ted views the trip. According to Ted, Alice and the Earth both travel at a speed $0.90c$ *relative to his spaceship*, returning to him at the same speed at the end of his journey as sketched in Figure 27.10B. Ted then reasons that he can apply the results for time dilation described above to calculate that, according to his clock, Alice's trip will take 19 years while her body clock will age by the proper time $\Delta t_0 = 8.3$ years. Ted thus argues that when they get back together, Alice will be younger than he is!

This problem is called the twin "paradox," as it appears that time dilation has led to contradictory results. Alice and Ted cannot both be right; only one can be the younger twin at the end of the journey. Alice's analysis is the correct one: Ted

[1]Sirius is actually a double star, but that does not affect the mission; Ted gets to visit both stars.

ages less than she does during the trip. The mistake in Ted's analysis is that special relativity applies only to inertial reference frames. Alice stays on the Earth, so she is always in an inertial reference frame and she can apply the results of special relativity. On the other hand, Ted spends some of his time in an accelerating reference frame when his spaceship turns around at Sirius to return to the Earth, and during this time he cannot use special relativity to analyze how Alice ages. That is why Ted's conclusion is wrong and is the resolution of this apparent paradox.

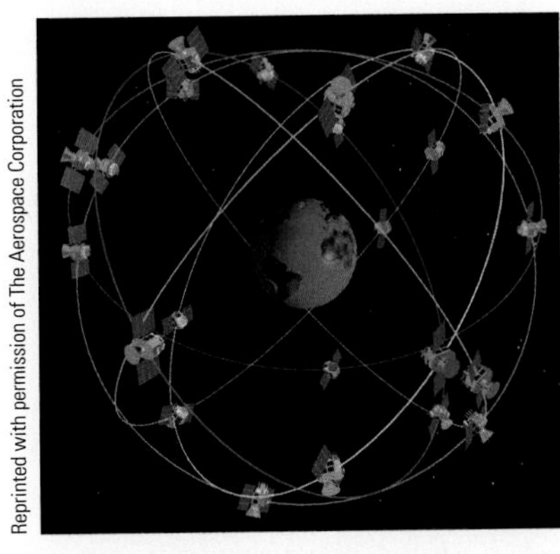

▲ **Figure 27.11** The Global Positioning System (GPS). Approximately 30 GPS satellites send down signals from orbit to the Earth. These signals are used by GPS receivers to determine their location with an accuracy of better than 10 m.

How Do We Know That Time Dilation Really Happens?

Our applications of time dilation to analyze the decay of a muon in Example 27.2 and the aging of two twins in the twin paradox (Fig. 27.10) are good examples of special relativity, but they don't seem very relevant to everyday life. A similar application of time dilation, however, does have important practical applications. The Global Positioning System (GPS) consists of about 30 satellites that orbit the Earth twice each day (Fig. 27.11). These satellites send signals to receivers on the Earth, and the receivers use the signals to "triangulate" their position with an accuracy of about 10 m. Each satellite contains a very accurate clock, and the GPS receivers compare their clocks with the time signal from each satellite to do this triangulation. The GPS satellite clocks are moving in orbit, so they run slow by the factor $\sqrt{1 - v^2/c^2}$ according to Equation 27.11. The GPS satellites move at a speed of about 4000 m/s, much less than the speed of light; even so, the GPS clocks run slow by about 7×10^{-6} s per day. The satellite signals travel at the speed of light, so the corresponding distance (which is just this time multiplied by c) is more than 2000 m, which is much larger than the theoretical accuracy of the GPS system. Only by accounting for the effect of time dilation on the satellite clocks can the GPS system successfully determine a position with an uncertainty of only 10 m.

27.4 Simultaneity Is Not Absolute

Two events are simultaneous if they occur at the same time. Our everyday experiences and intuition suggest that the notion of simultaneity is "absolute"; that is, two events are either simultaneous or they are not, for all observers. However, determining whether two events are simultaneous involves the measurement of time, and our studies of time dilation show that different observers do not always agree on measurements involving clocks and time intervals. Let's consider what special relativity implies for the notion of simultaneity by analyzing the situation in Figure 27.12; Ted is standing at the middle of his railroad car, moving with speed v relative to Alice,

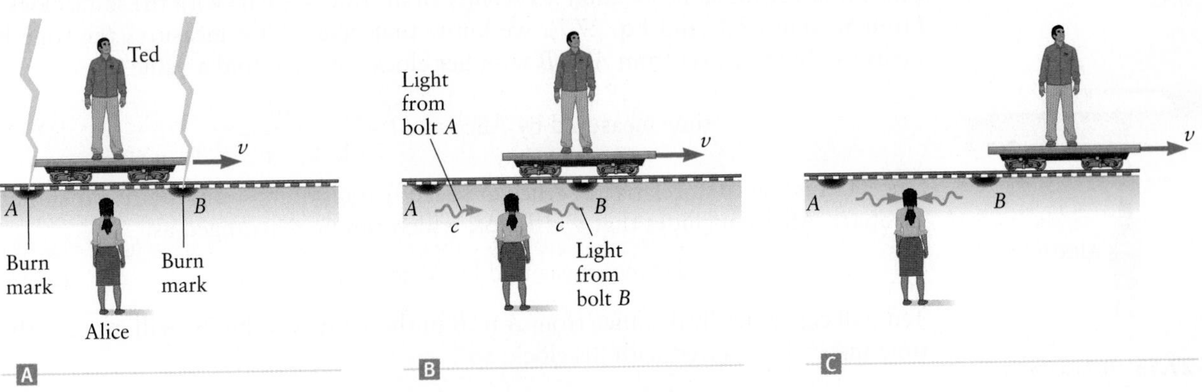

▲ **Figure 27.12** An experiment to study simultaneity. **A** Two lightning bolts strike at the ends of Ted's moving railroad car, leaving burn marks on the ground. **B** According to Alice, the lightning bolts are simultaneous. She comes to this conclusion because she is midway between the two burn marks and **C** the light pulses from the two bolts reach her at the same time.

Insight 27.2

SOME THINGS ARE NOT "RELATIVE" IN SPECIAL RELATIVITY

Observers in different reference frames may disagree about some things, such as the length of a meterstick or if two events are simultaneous, but there are certain things on which they always agree. For example, two observers will always agree on the *order* of two events that occur at the same place. Consider two light beams that are traveling toward Ted (like the two light pulses in Fig. 29.12B) and suppose these two beams have different colors, red and blue. Also suppose Ted has two clocks that are sensitive to these colors and that each clock stops when hit by the appropriately colored light beam. So, one clock stops when the red light beam arrives at Ted, and the other clock stops when the blue beam arrives. Ted can then tell which beam arrived first by simply comparing the readings of the two stopped clocks. Moreover, Alice or any other observer could also use the two stopped clocks to tell which beam arrived first and would see the same clock readings as Ted. Hence, *all* observers will agree on which light beam arrived first, that is, on the order of these events.

when two lightning bolts strike the ends of the car. The lightning bolts leave burn marks on the ground (points A and B), which indicate the locations of the two ends of the car when the bolts struck. We now ask, "Did the two lightning bolts strike simultaneously?"

We first ask this question of Alice, who notices that she is midway between the burn marks at A and B. Alice also observes that the light pulses from the lightning bolts reach her at the same time (Fig. 27.12C). Alice therefore concludes that the lightning bolts struck Ted's railroad car at the same time. *As viewed by Alice*, the bolts are simultaneous.

What does Ted have to say? Ted stands at the middle of his railroad car, so (like Alice) he is also midway between the places where the bolts strike. Hence, if the two events are simultaneous *as viewed by Ted*, the light pulses should reach him at the same time. Do they? Alice can also answer this question! Ted's railroad car is moving to the right, and Alice realizes that since Ted is moving, the flash emitted from B will reach him *before* the flash from point A in Figure 27.12B reaches him. Two observers must always agree on the order of two events that occur at the same point in space. (See Insight 27.2.) Hence, Ted will agree that the light pulse from B reaches him first. Ted therefore concludes that the lightning bolt at B struck *before* the bolt at A. The two lightning bolts in Figure 27.12 are therefore simultaneous for one observer (Alice) but not for another observer (Ted). The question of simultaneity is thus "relative" and can be different in different reference frames.

Time dilation and the relative nature of simultaneity mean that measurements of time intervals and judgments about simultaneity depend on the motion of the observer. That is very different from Newton's picture, in which "time" is an absolute, objective quantity, the same for all observers.

27.5 Length Contraction

In the past few sections, we have seen that special relativity forces us to give up the notion of absolute time. Time is just one aspect of a reference frame; reference frames also involve measurements of position and length. What does relativity have to say about these quantities?

Let's consider how Ted and Alice might work together to measure a length or distance. Suppose Alice marks two spots A and B on the ground and measures these spots to be a distance L_0 apart on the x axis (Fig. 27.13). Ted travels along the x direction at constant speed v, and as he passes point A he reads his clock. Ted reads his clock again when he passes point B and calls the difference between the two readings Δt_0. This is a proper time interval because Ted measures the start and finish times at the same location (the center of his railroad car) with the same clock. From Section 27.3 (and Eq. 27.7), we know that when Alice measures the time it takes for Ted to travel from A to B with her clock, she will find a value

$$\text{time measured by Alice} = \Delta t = \frac{\Delta t_0}{\sqrt{1 - v^2/c^2}} \tag{27.12}$$

Since Ted is traveling relative to Alice at speed v and points A and B are a distance L_0 apart, Alice concludes that $v = L_0/\Delta t$, which can be rearranged as

$$\text{distance measured by Alice} = L_0 = v\,\Delta t \tag{27.13}$$

Ted will calculate the distance from A to B in the same way, but he will use Δt_0, the time interval measured with his clock, so

$$\text{distance measured by Ted} = L = v\,\Delta t_0 \tag{27.14}$$

Comparing Equations 27.13 and 27.14, we see that since Δt is different from Δt_0 due to time dilation, the lengths measured by Alice and Ted will also be different. Using Equation 27.11, we get

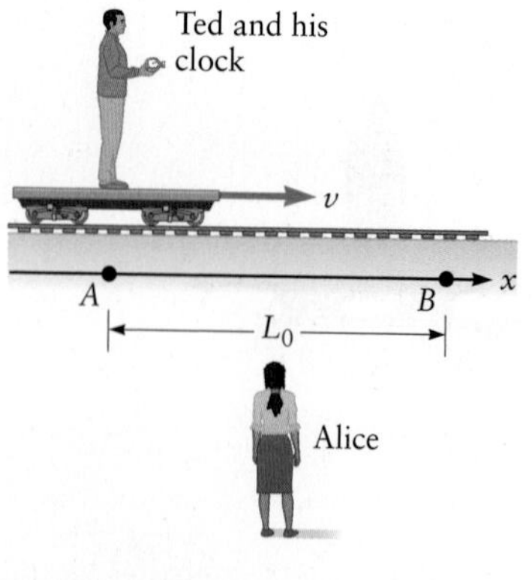

Ted and his clock

v

A B x

L_0

Alice

▲ **Figure 27.13** Ted can measure the distance between points A and B by using a clock on his railroad car to measure the time Δt_0 it takes for him to travel from A to B, together with his known speed v.

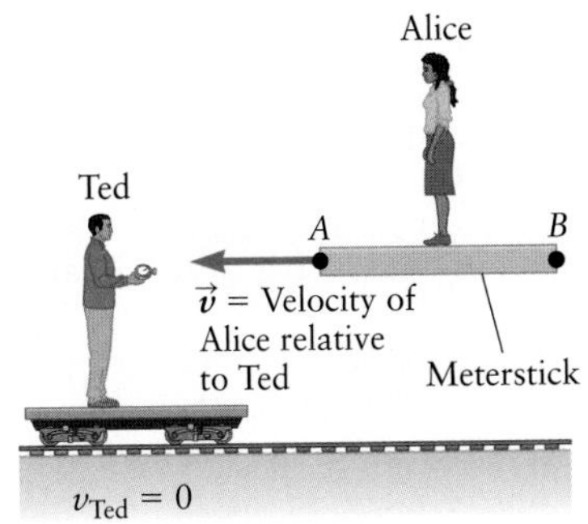

◀ Figure 27.14 A hypothetical experiment in which Ted is at rest and Alice travels on a meterstick with speed v relative to Ted. Ted finds that the moving meterstick is shorter than the length measured by Alice. This situation is very similar to Figure 27.13 because the meterstick (whose ends represent the points A and B) is at rest relative to Alice, while it has speed v relative to Ted.

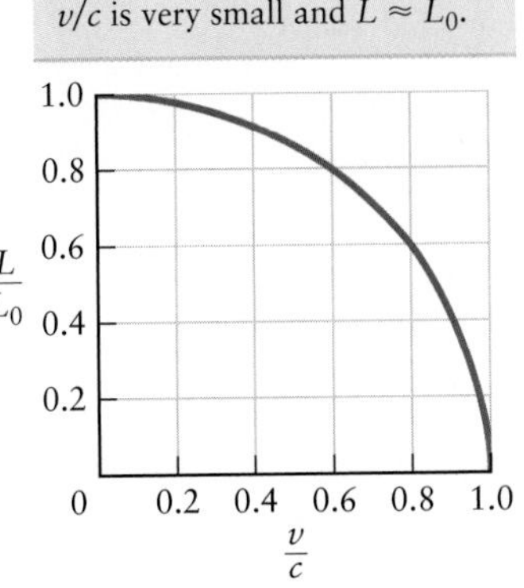

▲ Figure 27.15 Length contraction factor $L/L_0 = \sqrt{1 - v^2/c^2}$ as a function of v/c.

$$L = v\,\Delta t_0 = v(\Delta t\sqrt{1 - v^2/c^2})$$

$$L = L_0\sqrt{1 - v^2/c^2} \tag{27.15}$$

Hence, the length L measured by Ted is *shorter* than Alice's length L_0. This effect is called ***length contraction***.

Proper Length

Suppose points A and B are the two ends of a meterstick and Ted is measuring the length of the meterstick as it moves past him (Fig. 27.14). The meterstick is at rest relative to Alice, who might be standing on it, and she measures the meterstick in Figure 27.14 to have a length of exactly $L_0 = 1$ m. Ted measures a length L (Eq. 27.15), which is shorter than L_0. We can thus say that *moving metersticks are shortened*. The quantity L_0 is called the ***proper length*** because it is measured by an observer (Alice) *who is at rest relative to the meterstick*.

Length contraction is described by the factor

$$\frac{L}{L_0} = \sqrt{1 - v^2/c^2} \tag{27.16}$$

Length contraction: moving metersticks are contracted.

This factor is plotted in Figure 27.15; it is very close to 1 when the speed is small and approaches zero as v approaches c.

Notice that we denote the proper time by Δt_0 and the proper length by L_0, in both cases using a subscript zero. In both cases, they are measurements made by an observer who is at rest relative to the "thing" being measured. Proper time is measured by an observer (Ted in Fig. 27.5B) who is at rest relative to the clock used for the measurement. Proper length is measured by an observer (Alice in Fig. 27.14) who is at rest relative to the object whose length is being measured.

Proper time and proper length

EXAMPLE 27.3 Length Contraction of a Moving Car

Consider a race car measured by Ted to be 4.0 m long when moving past him at a speed of 100 m/s (about 200 mi/h). How long is the car when the race is finished and the car is stopped?

RECOGNIZE THE PRINCIPLE

This problem is an example of length contraction, with the race car playing the role of the meterstick in Figure 27.14. The length measured during the race (when $v = 100$ m/s) by an observer (Ted) watching from trackside is the contracted length $L = 4.0$ m (Fig. 27.16A). The length of the race car at the end of the race (when it is not moving) is the proper length L_0; that is the length that would be measured by an observer at rest relative to the car, such as Ted in Figure 27.16B.

(continued) ▶

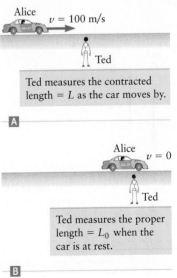

Alice $v = 100$ m/s
Ted
Ted measures the contracted length $= L$ as the car moves by.
A

Alice $v = 0$
Ted
Ted measures the proper length $= L_0$ when the car is at rest.
B

▲ Figure 27.16 Example 27.3.

SKETCH THE PROBLEM

Figure 27.16 describes the problem.

IDENTIFY THE RELATIONSHIPS

Length contraction is described by Equation 27.15; we have $L = L_0\sqrt{1 - v^2/c^2}$. Rearranging to solve for the proper length gives

$$L_0 = \frac{L}{\sqrt{1 - v^2/c^2}}$$

SOLVE

Inserting the given values of L and v, we find

$$L_0 = \frac{L}{\sqrt{1 - v^2/c^2}} = \frac{4.0 \text{ m}}{\sqrt{1 - (100 \text{ m/s})^2/(3.00 \times 10^8 \text{ m/s})^2}}$$

$$L_0 = \boxed{4.00000000000022 \text{ m}} \quad (1)$$

Your calculator may not be able to show this many zeros to the right of the decimal point. If that is the case, we can again use the results in Insight 27.1 and write

$$L_0 = \frac{L}{\sqrt{1 - v^2/c^2}} \approx L\left(1 + \frac{v^2}{2c^2}\right)$$

Inserting the values of v and c gives

$$L_0 \approx L\left(1 + \frac{v^2}{2c^2}\right) = (4.0 \text{ m})\left[1 + \frac{(100 \text{ m/s})^2}{2(3.00 \times 10^8 \text{ m/s})^2}\right] \approx \boxed{4.0 + 2.2 \times 10^{-13} \text{ m}}$$

► ***What does it mean?***

We had to include many places to the right of the decimal point in Equation (1) to show the effect of length contraction in this example. Length contraction is quite negligible in this case and is always extremely small when v is a typical terrestrial speed. Length contraction is usually only important and measurable when v approaches the speed of light c.

EXAMPLE 27.4 Length Contraction and a Moving Muon

Length contraction can be important for subatomic particles such as muons because these particles often move at very high speeds. Example 27.2 discussed muons created by cosmic rays in the Earth's atmosphere. Consider a muon created at point A in Figure 27.17 moving downward toward the Earth at speed $v = 0.99c$, decaying after a time $\tau = 2.2 \times 10^{-6}$ s. How far will the muon travel relative to the Earth before it decays? Give the answer from (**a**) the point of view of the muon and (**b**) from the point of view of an observer, Alice, standing on the ground.

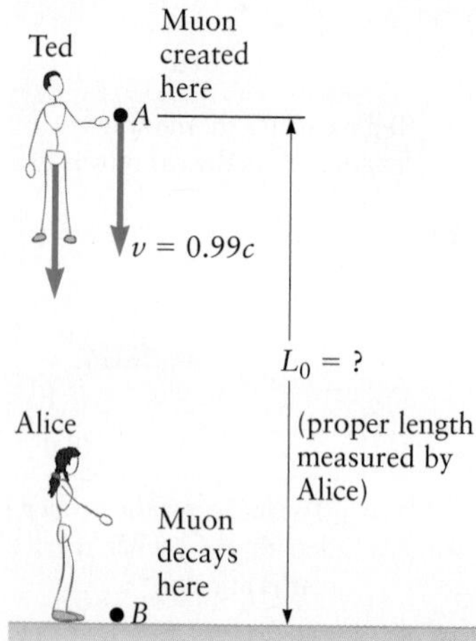

▲ **Figure 27.17** Example 27.4.

RECOGNIZE THE PRINCIPLE

The muon travels a distance = $v \times$ its lifetime before it decays. Different observers measure different values for this lifetime and hence find different values for the distance traveled. For part (**a**), consider an observer, Ted, riding along with the muon so that the muon is at rest relative to Ted. The muon exists for a time = $\Delta t_0 = \tau$, which is a proper time because Ted records both the muon's creation and its decay on his clock at the same location (Ted's location). According to Ted, the Earth moves toward him and the muon through a distance $v\,\Delta t_0 = v\tau$ during the muon's lifetime. For part (**b**), let Alice be an observer on the ground. She measures a longer muon lifetime than τ due to time dilation (Example 27.2), so she will measure a longer travel distance than Ted. Alice measures the proper length between the muon's starting point at A and ground level at B because she is at rest relative to those points.

SKETCH THE PROBLEM

Figure 27.17 shows the problem.

IDENTIFY THE RELATIONSHIPS AND SOLVE

(**a**) Riding along with the muon, Ted measures the travel distance as $L = v\,\Delta t_0$. Inserting the given values of v and $\Delta t_0 = \tau$, we get

$$L = v\,\Delta t_0 = 0.99c\,\Delta t_0 = 0.99(3.00 \times 10^8 \text{ m/s})(2.2 \times 10^{-6} \text{ s})$$

$$L = \boxed{650 \text{ m}} \quad (1)$$

(**b**) From Example 27.2, Alice observes the lifetime of the moving muon as

$$\Delta t = \frac{\Delta t_0}{\sqrt{1 - v^2/c^2}} = \frac{\tau}{\sqrt{1 - (0.99c)^2/c^2}} = 7.1\tau \quad (2)$$

During this time, the muon travels a distance $L_0 = v\,\Delta t$. Inserting the values of v and Δt leads to

$$L_0 = v\,\Delta t = (0.99c)(7.1\tau) = 0.99(3.00 \times 10^8 \text{ m/s})(7.1)(2.2 \times 10^{-6} \text{ s})$$

$$L_0 = \boxed{4600 \text{ m}} \quad (3)$$

► ***What does it mean?***

Suppose Alice finds that the muon just reaches the ground when it decays (point B in Fig. 27.17). Ted will agree with her finding because he is traveling with the muon and will find himself on the ground when it decays. Comparing the answers for parts (**a**) and (**b**), we see that the distance traveled by the muon is different for Ted and Alice. How can they agree that the muon reaches the ground? Ted and the muon observe the *contracted* length between A and B, which is (Eq. 27.15)

$$L = L_0\sqrt{1 - v^2/c^2} \quad (4)$$

Inserting the values of L from Equation (1) and L_0 from Equation (3) along with the speed of the muon gives

$$\text{left side of Equation (4)} = 650 \text{ m}$$

$$\text{right side of Equation (4)} = (4600 \text{ m})\sqrt{1 - (0.99c)^2/c^2} = 650 \text{ m}$$

Equation (4) is thus satisfied by the results in parts (**a**) and (**b**) due to the effect of length contraction. Hence, Ted and Alice agree that the muon reaches the ground when it decays. Ted explains it in terms of length contraction and claims the muon started only 650 m above the Earth's surface at point A. Alice explains the result in terms of time dilation and Equation (2).

CONCEPT CHECK 27.2 Length Contraction and Muons

Suppose the speed of the muon in Example 27.4 is increased to $0.995c$. As viewed by Ted (and the muon), how far does the muon travel before it decays, (a) 6600 m, (b) 4600 m, or (c) 660 m?

What Is "Relative" and What Is Not?

The postulates of special relativity seem quite "innocent," but they have forced us to give up the notions of absolute time and space. A large number of experiments have shown that time dilation and length contraction actually do occur. However, we have seen in Examples 27.1 and 27.3 that at ordinary terrestrial speeds these effects are negligibly small. The reason can be traced to the factor $\sqrt{1 - v^2/c^2}$

that appears in calculations of both time dilation and length contraction. At the typical terrestrial speed of 100 m/s (about 200 mph), this factor is approximately

$$\sqrt{1 - v^2/c^2} = \sqrt{1 - (100 \text{ m/s})^2/(3.00 \times 10^8 \text{ m/s})^2} \approx 1 - 6 \times 10^{-14}$$

The effects of relativity are proportional to how much this factor differs from 1. We can thus continue to use our normal intuition at everyday "low" speeds, but must realize that these intuitive notions of time and length break down for objects moving at speeds approaching the speed of light.

27.6 Addition of Velocities

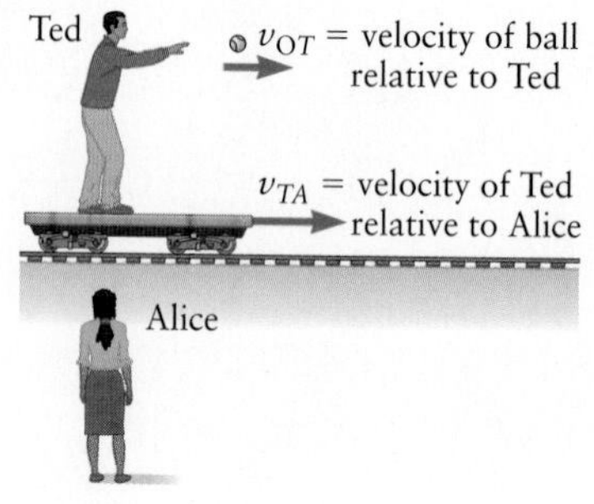

▲ **Figure 27.18** While he is moving with velocity v_{TA} relative to Alice, Ted throws a baseball with velocity v_{OT} relative to his railroad car. According to Galilean relativity and Newton's mechanics, the velocity of the ball relative to Alice is $v_{OT} + v_{TA}$. Special relativity gives a different result. Compare this situation with the light from Ted's flashlight in Figure 27.2.

The second postulate of special relativity—that the speed of light is the same for all observers—involves speed, so let's consider how the speed and velocity of an object appear to observers in different reference frames. The "object" could be a particle, a conventional wave (such as a water wave), or a light pulse. In Figure 27.18, Ted is traveling on his railroad car at constant speed v_{TA} relative to Alice when he throws an object, which might be a baseball. We now use the subscript TA to indicate that it is the velocity of Ted as measured by Alice, that is, the velocity of Ted relative to Alice. In this section, we will be concerned mainly with the component of the velocity along the x axis. Because this component can be positive or negative, we'll refer to it as the *velocity* rather than the speed. For simplicity, we ignore gravity and assume the object (the baseball in Fig. 27.18) travels horizontally, parallel to the velocity of the railroad car. If the velocity of the baseball *relative to Ted* is v_{OT}, what is the velocity v_{OA} of the ball relative to Alice who is at rest on the ground? (Notice how we again use subscripts to indicate which velocity we are referring to and who is measuring it.)

Newton's answer to this question was discussed in Section 27.2 and also in Chapter 4. For the baseball in Figure 27.18, Newton would predict

$$v_{OA} = v_{OT} + v_{TA} \tag{27.17}$$

or, in words,

velocity of the object (the ball) relative to Alice (v_{OA}) =
velocity of the object relative to Ted (v_{OT}) + velocity of Ted relative to Alice (v_{TA})

This result agrees with our everyday intuition, and it would be quite an accurate description for a person throwing a real baseball from a real railroad car. Equation 27.17, however, is inconsistent with the postulates of special relativity when the speeds are very high. To see the problem, suppose an object's speed relative to Ted is $v_{OT} = 0.9c$ and that Ted's railroad car is traveling very fast with $v_{TA} = 0.9c$. Inserting these values into Equation 27.17 gives a velocity $v_{OA} = v_{OT} + v_{TA} = 1.8c$, which is greater than the speed of light. Indeed, nothing in Newton's mechanics prevents Ted or a baseball or any other object from traveling at a speed greater than c. For an object or reference frame traveling at speed v, the factor that appears in time dilation (Eq. 27.7) and length contraction (Eq. 27.15) is $\sqrt{1 - v^2/c^2}$. If $v > c$, the result is an imaginary number, suggesting that Newton's formula for the addition of velocities (Eq. 27.17) cannot be correct for speeds close to the speed of light.

Relativistic Addition of Velocities

It is possible to analyze the situation in Figure 27.18 using the postulates of special relativity. We will not give the derivation, but simply state the relativistic result for the addition of velocities:

Relativistic addition of velocities

$$v_{OA} = \frac{v_{OT} + v_{TA}}{1 + \dfrac{v_{OT}v_{TA}}{c^2}} \tag{27.18}$$

This relation involves the component along a particular direction (x) of three different velocities and should be interpreted as follows:

1. v_{OT} is the velocity of an object relative to an observer (e.g., Ted).
2. v_{TA} is the velocity of one observer (Ted) relative to a second observer (Alice).
3. v_{OA} is the velocity of the object relative to the second observer (Alice).

When the velocities v_{OT} and v_{TA} are much less than the speed of light, the factor $v_{OT}v_{TA}/c^2$ in the denominator in Equation 27.18 is much less than 1. In such cases, the relativistic addition of velocities formula (Eq. 27.18) gives nearly the same result as Newton's mechanics (Eq. 27.17). For example, suppose Ted is able to throw the baseball in Figure 27.18 with velocity $v_{OT} = 0.10c$, and his railroad car is moving at velocity $v_{TA} = 0.10c$ relative to Alice. Inserting these values into Equation 27.18 gives

$$v_{OA} = \frac{v_{OT} + v_{TA}}{1 + \dfrac{v_{OT}v_{TA}}{c^2}} = \frac{0.10c + 0.10c}{1 + \dfrac{(0.10c)(0.10c)}{c^2}} = 0.198c \tag{27.19}$$

We have kept an extra significant figure here to emphasize that the result is very close to the prediction of Newton's formula (Eq. 27.17), which gives

$$v_{OA} = v_{OT} + v_{TA} = 0.10c + 0.10c = 0.200c$$

This answer differs from the relativistic result (Eq. 27.19) by only a small amount (approximately 1%). Hence, for speeds less than about 10% of the speed of light, the Newtonian velocity addition formula works very well (although it is not exact!).

The results are quite different for objects traveling near the speed of light. Suppose $v_{OT} = 0.90c$ and $v_{TA} = 0.90c$. Inserting these values into Equation 27.18, we find

$$v_{OA} = \frac{v_{OT} + v_{TA}}{1 + \dfrac{v_{OT}v_{TA}}{c^2}} = \frac{0.90c + 0.90c}{1 + \dfrac{(0.90c)(0.90c)}{c^2}} = 0.994c \tag{27.20}$$

We have given the result to three significant figures to emphasize that v_{OA} is just slightly smaller than c, which is quite different from Newton's prediction for this case, which is $v_{OA} = v_{OT} + v_{TA} = 1.8c$. Experiments with particles moving at very high speeds show that the relativistic result is correct.

Relativistic Velocities and the Speed of Light as a "Speed Limit"

The relativistic formula for the addition of velocities (Eq. 27.18) applies when the velocity of the object is parallel (or antiparallel) to the relative velocity of the two observers $\vec{v}_{TA}$ as is the case for Ted, Alice, and the baseball in Figure 27.18. A slightly different result applies when the object is moving perpendicular to $\vec{v}_{TA}$. It is again found that if v_{OT} and v_{TA} are both less than c, then v_{OA} is also less than c. In general, *if an object has a speed less than c for one observer, its speed is less than c for all other observers.* Since no experiment has ever observed an object with a speed greater than the speed of light, c is again a sort of universal "speed limit."

As a final application of Equation 27.18, suppose the object leaving Ted's hand in Figure 27.18 is not a baseball but a pulse of light. Its velocity is then $v_{OT} = c$. What speed will Alice find? Inserting $v_{OT} = c$ into Equation 27.18 gives

$$v_{OA} = \frac{v_{OT} + v_{TA}}{1 + \dfrac{v_{OT}v_{TA}}{c^2}} = \frac{c + v}{1 + \dfrac{cv}{c^2}} = \frac{c + v}{1 + \dfrac{v}{c}}$$

Multiplying the numerator and denominator by c leads to

$$v_{OA} = \frac{c(c + v)}{c\left(1 + \dfrac{v}{c}\right)} = \frac{c(c + v)}{c + v} = c \tag{27.21}$$

Alice thus finds a speed $v_{OA} = c$ regardless of Ted's speed v. In words, if an object moves at the speed of light for one observer, it moves at the speed of light for all observers. The only known "object" that can move at this speed is light; hence, this result is just a restatement of postulate 2 of special relativity.

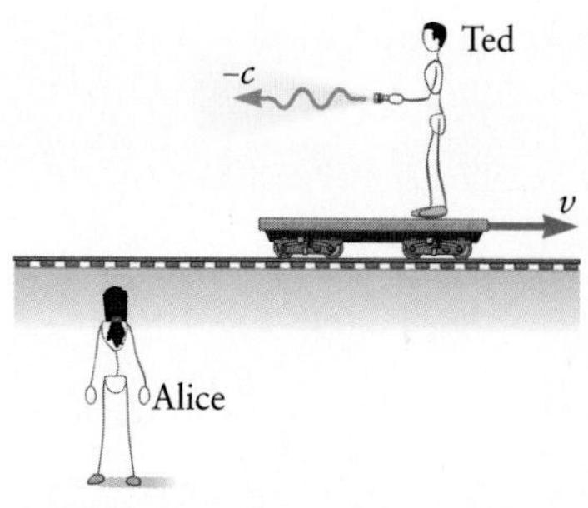

▲ **Figure 27.19** Example 27.5.

EXAMPLE 27.5 Addition of Velocities

Ted is moving at velocity $+v$ in his railroad car when he aims his flashlight backward as shown in Figure 27.19. What velocity will Alice find for the light from Ted's flashlight?

RECOGNIZE THE PRINCIPLE

According to the second postulate of special relativity, the speed of light must be the same for all observers, so Alice must measure a speed c for the light from Ted's flashlight. This result can also be obtained by applying the addition of velocities relation of special relativity.

SKETCH THE PROBLEM

The problem is described by Figure 27.19.

IDENTIFY THE RELATIONSHIPS

Light from Ted's flashlight moves at velocity $v_{OT} = -c$ relative to Ted, whereas Ted moves at a velocity $v_{TA} = v$ relative to Alice. We wish to find the velocity of v_{OA} of the light relative to Alice.

SOLVE

Inserting these velocities into Equation 27.18 gives

$$v_{OA} = \frac{v_{OT} + v_{TA}}{1 + \dfrac{v_{OT}v_{TA}}{c^2}} = \frac{-c + v}{1 - \dfrac{cv}{c^2}} = \frac{-c + v}{1 - \dfrac{v}{c}}$$

Multiplying the numerator and denominator by c leads to

$$v_{OA} = \frac{c(-c + v)}{c\left(1 - \dfrac{v}{c}\right)} = \frac{-c(c - v)}{c - v} = \boxed{-c}$$

► ***What does it mean?***

The speed measured by Alice is again equal to the speed of light. The negative sign means that the light pulse is moving in the $-x$ direction in Alice's reference frame.

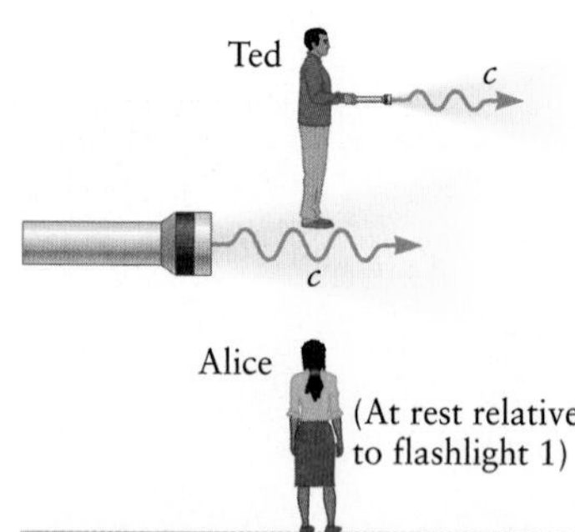

▲ **Figure 27.20** Concept Check 27.3. What is the speed of the light beam from Ted's flashlight relative to Alice?

CONCEPT CHECK 27.3 Riding on a Light Beam

Einstein said that some of the ideas for relativity theory occurred to him when he tried to imagine how the universe would look to a person "riding on a beam of light." Consider an experiment in which Ted rides on a light pulse with both Ted and the light traveling at speed c (Fig. 27.20). Ted is carrying his own flashlight and uses it to send a second light pulse to the right. According to postulate 2 of special relativity, the second light pulse travels at speed c relative to Ted. Is the speed of the second pulse as measured by Alice (a) $2c$, (b) c, or (c) 0?

27.7 Relativistic Momentum

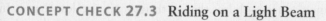

According to Newton's mechanics, a particle of mass m_0 moving with speed v has a momentum

$$\text{Newton's mechanics: } p = m_0 v \tag{27.22}$$

In Chapter 7, we showed that when there are no external forces on a system of particles, the total momentum of the system is conserved. The principle of conservation of momentum is believed to be satisfied by all the laws of physics, including the theory of special relativity. In Equation 27.22, however, there is a problem with the expression for momentum.

We can write Newton's result for the momentum of a single particle as

$$p = m_0 \frac{\Delta x}{\Delta t} \tag{27.23}$$

From our analyses of time dilation and length contraction, we know that measurements of both time Δt and length Δx can be different for observers in different inertial reference frames. Should we use the proper time Δt_0 or the proper length Δx_0 to calculate the momentum? Einstein showed that we should use the proper time Δt_0 to calculate the momentum, which amounts to using a clock that travels along with the particle. At the same time, we should use the measurement of length Δx taken by an observer who watches the particle move by with speed v. Because $\Delta t_0 = \Delta t\sqrt{1 - v^2/c^2}$ (from Eq. 27.7), the result is

$$\text{special relativity: } p = m_0 \frac{\Delta x}{\Delta t_0} = m_0 \frac{\Delta x}{\Delta t \sqrt{1 - v^2/c^2}}$$

and since $v = \Delta x/\Delta t$, we get

$$\text{special relativity: } p = \frac{m_0 v}{\sqrt{1 - v^2/c^2}} \tag{27.24}$$

Relativistic momentum

Einstein showed that when the momentum is calculated using Equation 27.24, the principle of conservation of momentum is obeyed exactly, even for particles moving at high speeds (i.e., close to the speed of light).

When a particle's speed is small compared with the speed of light, the relativistic momentum (Eq. 27.24) becomes

$$(\text{relativistic momentum when } v \ll c) = \frac{m_0 v}{\sqrt{1 - v^2/c^2}} \approx \frac{m_0 v}{\sqrt{1 - 0}} = m_0 v$$

The relativistic momentum is thus equal to Newton's result for the momentum for "slowly" moving objects. As v approaches the speed of light, however, the relativistic result for momentum is very different from Newton's result as shown in Figure 27.21, which compares the two expressions for momentum (Eqs. 27.22 and 27.24).

Newton's result gives a linear relation between p and v. In contrast, the result from special relativity shows that p becomes extremely large as the particle's speed approaches c. A particle's momentum can be increased by the application of forces, and there is no limit to how large the momentum can be. However, even when the momentum is very large, the particle's speed never quite reaches the speed of light. Hence, when the speed is large, a particle responds to forces and impulses *as if its mass had increased* relative to m_0.

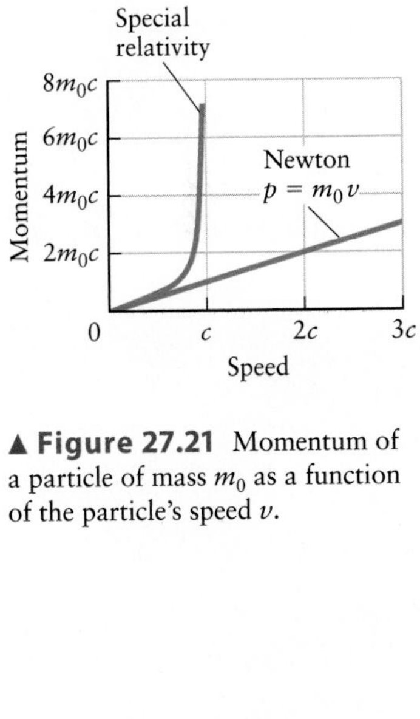

▲ **Figure 27.21** Momentum of a particle of mass m_0 as a function of the particle's speed v.

EXAMPLE 27.6 Relativistic Momentum of an Electron

In experiments to study the properties of subatomic particles, physicists routinely accelerate electrons to speeds quite close to the speed of light. For an electron moving with speed $v = 0.99c$, compare its actual momentum to the momentum predicted by Newton's mechanics. Express your answer as the ratio $p(\text{special relativity})/p(\text{Newton})$.

RECOGNIZE THE PRINCIPLE

Since the electron's speed is very close to the speed of light, special relativity will have a very large effect. According to Figure 27.21, the true momentum (as predicted by special relativity) is greater than the momentum predicted by Newton's mechanics.

(continued) ▶

SKETCH THE PROBLEM

Figure 27.21 shows the difference between the classical (Newton's) momentum and the true momentum (from special relativity).

IDENTIFY THE RELATIONSHIPS

This example is a direct application of the relativistic formula for momentum for an object (an electron) with speed $v = 0.99c$. We take the ratio of the relativistic momentum in Equation 27.24 to p in Equation 27.22. The mass of the electron cancels, leaving only terms that depend on v and c.

SOLVE

Taking the ratio of Equations 27.24 and 27.22 gives

$$\frac{p(\text{special relativity})}{p(\text{Newton})} = \frac{m_0 v/\sqrt{1 - v^2/c^2}}{m_0 v} = \frac{1}{\sqrt{1 - v^2/c^2}}$$

Inserting the given value of v, we find

$$\frac{p(\text{special relativity})}{p(\text{Newton})} = \frac{1}{\sqrt{1 - v^2/c^2}} = \frac{1}{\sqrt{1 - (0.99c)^2/c^2}} = \boxed{7.1}$$

▸ ***What does it mean?***

When an electron moves at 99% of the speed of light, the true momentum is more than seven times larger than that predicted by Newton. When a particle's speed approaches the speed of light, the effects of special relativity become very significant.

CONCEPT CHECK 27.4 Relativistic Momentum of a Proton

Consider again Example 27.6, but now the particle is a proton instead of an electron. The mass of a proton is about 1800 times greater than the mass of an electron. If the proton is moving at a speed $v = 0.99c$, is the ratio of p(special relativity)/p(Newton) for the proton

(a) the same as found for an electron with the same speed,
(b) greater than that found for an the electron with the same speed, or
(c) less than that found for an electron with the same speed?

27.8 What Is "Mass"?

The concept of "mass" is basic to all physics. When we first discussed mass in Chapter 1, we appealed to your intuition and suggested in very rough terms that mass is the "amount of matter" carried by a particle. This definition is admittedly imprecise, but it is surprisingly difficult to give a better one. According to Newton's second law, the acceleration of a particle of mass m_0 is given by

$$a = \frac{\sum F}{m_0} \tag{27.25}$$

so m_0 is just the constant of proportionality that relates acceleration and force. Newton's laws implicitly assume this constant of proportionality is really a constant,[2] but how do we know that this is true?

In the early part of the 1900s, experiments began to probe the regime in which a particle's speed approaches c, and these experiments showed that Newton's second law breaks down at such high speeds. When the postulates of special relativity are applied to this problem, we find that acceleration and force are related instead by

[2]Assuming, of course, the particle does not break apart, for example.

$$a = \frac{\sum F}{m_0/(1 - v^2/c^2)^{3/2}} \quad (27.26)$$

Comparing this relation with Newton's second law (Eq. 27.25), we have replaced m_0 by the factor $m_0/(1 - v^2/c^2)^{3/2}$. That is,

$$m_0 \rightarrow \frac{m_0}{(1 - v^2/c^2)^{3/2}} \quad (27.27)$$

At low speeds ($v \ll c$), the quantity on the right approaches m_0, so the two acceleration equations (Eqs. 27.25 and 27.26) are indistinguishable. However, as v approaches the speed of light, the denominator in Equation 27.27 becomes very small. When $v \approx c$, the acceleration in Equation 27.26 is thus very small, even when the force is very large. Hence, the particle responds to a force *as if it had a mass larger than* m_0.

We encountered a similar effect when we considered relativistic momentum. The relativistic result for p in Equation 27.24 can be obtained by replacing the mass m_0 in Newton's expression (Eq. 27.22) by the term $m_0/\sqrt{1 - v^2/c^2}$. Hence, at high speeds, the particle responds to impulses and forces "as if" its mass were larger than m_0.

So, when dealing with both acceleration and momentum, a particle behaves "as if" it had a mass larger than expected from its behavior at low speeds. For this reason, physicists have introduced the term ***rest mass***, denoted by m_0, for the mass measured by an observer who is moving very slowly relative to the particle (so that $v \approx 0$ in Eqs. 27.24 and 27.26). Some authors call it the "proper mass" in analogy with the proper length of a meterstick at rest, but the term *rest mass* is much more widely used in physics, so we'll use it in this book.

When dealing with momentum and acceleration for particles moving at very high speeds, we can solve any problem we might encounter by simply using the results from special relativity in Equations 27.24 and 27.26. These results mean that Newton's simple notions about mass break down for particles moving at high speeds. The best way to describe the "mass" of a particle is through its rest mass m_0, which describes the motion of a particle for an observer who is moving very slowly relative to the particle.

27.9 Mass and Energy

As we saw in the last two sections, the concepts of mass and momentum are not as simple as Newton had assumed. You should therefore not be surprised that we must also reexamine our concepts of work and energy. That will lead us to what is probably the most famous equation in physics.

In Chapter 6, we derived a relation between mechanical work and kinetic energy and found that the kinetic energy of a particle of mass m_0 moving at speed v is

$$\text{Newton's kinetic energy: } KE = \tfrac{1}{2}m_0v^2 \quad (27.28)$$

We can derive the corresponding relativistic expression for kinetic energy by applying work–energy ideas. According to the work–energy theorem (Chapter 6), the work done on a particle is equal to the change in the particle's kinetic energy. The work done by an applied force F is $W = F\,\Delta x$, where Δx is the particle's displacement. Using the relation between F and acceleration according to special relativity (Eq. 27.26), we get

$$W = F\,\Delta x = \frac{m_0 a}{(1 - v^2/c^2)^{3/2}}(\Delta x) \quad (27.29)$$

which is similar to our derivation of Newton's kinetic energy in Chapter 6 (Eqs. 6.5 through 6.8). Simplifying the right-hand side of Equation 27.29 to find W involves more mathematics than we can include here, so we'll simply state the answer for a particle that starts at rest and comes to a final speed v. In this case, the work done equals the particle's relativistic kinetic energy KE, and the result is

Relativistic kinetic energy

$$KE = \frac{m_0c^2}{\sqrt{1 - v^2/c^2}} - m_0c^2 \quad (27.30)$$

Equation 27.30 looks quite different from Newton's result ($KE = \frac{1}{2}m_0v^2$), but the two are actually closely related. To see the connection, we first rewrite Equation 27.30 as

$$KE = \frac{m_0c^2}{\sqrt{1 - v^2/c^2}} - m_0c^2 = m_0c^2\left(\frac{1}{\sqrt{1 - v^2/c^2}} - 1\right)$$

According to Insight 27.1, for small v ($v \ll c$) we have

$$\frac{1}{\sqrt{1 - v^2/c^2}} \approx 1 + \frac{v^2}{2c^2}$$

Inserting this result into our expression for the kinetic energy gives

$$KE \approx m_0c^2\left(1 + \frac{v^2}{2c^2} - 1\right)$$

$$KE \approx \tfrac{1}{2}m_0v^2 \quad (27.31)$$

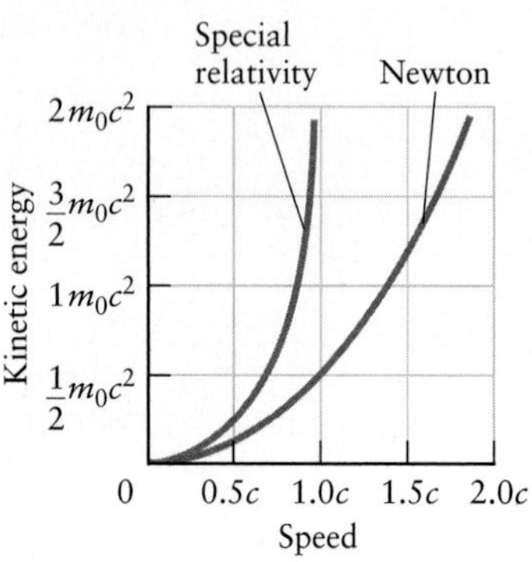

▲ **Figure 27.22** Kinetic energy of a particle as a function of its speed v. Even if the particle has a very large kinetic energy, its speed never quite reaches the speed of light.

which is identical to Newton's kinetic energy. Hence, for particles moving at speeds much less than the speed of light, we get our familiar result for KE as we should have expected. As v approaches c, however, the relativistic result for the kinetic energy has a different behavior than does Newton's expression. Figure 27.22 shows the kinetic energy as a function of the particle speed v (Eq. 27.30). At high speeds, v runs into the "speed limit" c. Although the kinetic energy can be made very large, the particle's speed never quite reaches the speed of light. As the speed increases, the particle responds as if it had a larger mass than at low speeds.

The expression for the kinetic energy in Equation 27.30 is the change in energy of a particle due to work done on the particle. We can also think of it as the difference between the final energy and the initial energy,

$$KE = \overbrace{\frac{m_0c^2}{\sqrt{1 - v^2/c^2}}}^{\text{final energy}} - \overbrace{m_0c^2}^{\text{initial energy}}$$

The last term, m_0c^2, is a constant, however. What is the meaning of this initial energy term? Einstein proposed that this term is the energy of the particle, even when it is at rest. We now call it the ***rest energy*** of the particle:

Rest energy

$$\text{rest energy} = m_0c^2 \quad (27.32)$$

In words, Equation 27.32 asserts that a particle has an amount of energy m_0c^2 even when its speed v is zero. When a particle is in motion, it also has kinetic energy; to get the total energy of the particle, we must then add the rest energy to the kinetic energy to get the total relativistic energy TE:

Total relativistic energy

$$TE = \frac{m_0c^2}{\sqrt{1 - v^2/c^2}} \quad (27.33)$$

The result for the rest energy in Equation 27.32 forces us to rethink our notion of mass. The rest energy relation of special relativity implies that ***mass is a form of energy***. This suggests that it is possible to convert an amount of energy m_0c^2 into a particle of rest mass m_0 or to convert a particle of rest mass m_0 into an amount of energy m_0c^2. So, the principle of conservation of energy must be extended to include this type of energy.

Since the speed of light is a very large number, the magnitude of the rest energy can be very large even when the rest mass m_0 is small. A rest mass of only 1 kg corresponds to a rest energy of

$$m_0c^2 = (1 \text{ kg})(3.00 \times 10^8 \text{ m/s})^2 = 9 \times 10^{16} \text{ J} \tag{27.34}$$

which is a tremendous amount of energy. Approximately 3×10^{17} J of energy is consumed each day in the United States. According to Equation 27.34, this energy could be obtained by converting a rest mass of just 3 kg into energy. Special relativity tells us that this much energy is available in principle, but it does not tell us how to actually convert mass into energy. Nuclear reactions (Chapter 30) are one way to do so.

CONCEPT CHECK 27.5 Relativistic Kinetic Energy

A particle has a kinetic energy that is twice its rest energy. What is the speed of the particle, (a) $c/2$, (b) $c/4$, or (c) $(2\sqrt{2}\, c)/3$?

EXAMPLE 27.7 A Remarkable Proton

Cosmic rays are particles that arrive at the Earth from outer space. Some of them have extremely high energies. In 1991, a special cosmic-ray detector in Utah observed a cosmic-ray proton with a kinetic energy of 3.2×10^{20} eV, which is about 50 J. (Recall the definition of electron-volt from Eq. 18.19; 1 eV = 1.60×10^{-19} J.) The rest energy of a proton is 938 MeV. What is the length contraction factor $\sqrt{1 - v^2/c^2}$ for this proton?

RECOGNIZE THE PRINCIPLE

The kinetic energy KE of a particle depends on the length contraction factor (Eq. 27.30). We can therefore use the given value of KE to find this factor.

SKETCH THE PROBLEM

No figure is necessary.

IDENTIFY THE RELATIONSHIPS

The kinetic energy of this proton is *much* greater than its rest energy, so to a good approximation the kinetic energy will be equal to the total energy (Eq. 27.33)

$$TE = \frac{m_0c^2}{\sqrt{1 - v^2/c^2}} \tag{1}$$

The rest energy is just m_0c^2.

SOLVE

The length contraction factor is just the ratio of the rest energy m_0c^2 to the total energy in Equation (1). We have

$$\frac{\text{rest energy}}{TE} = \frac{m_0c^2}{m_0c^2/\sqrt{1 - v^2/c^2}} = \sqrt{1 - v^2/c^2}$$

Inserting the given values of TE and the rest energy, we find

$$\sqrt{1 - v^2/c^2} = \frac{938 \times 10^6 \text{ eV}}{3.2 \times 10^{20} \text{ eV}} = \boxed{2.9 \times 10^{-12}}$$

▶ *What does it mean?*

The length contraction factor tells how much a moving meterstick would appear shortened to an observer moving at the speed of this proton (Eq. 27.16). As a "practical" example, the average distance from the Earth to the Moon is 380,000 m. When viewed by an observer moving with this proton, the Earth–Moon distance would be only $380{,}000 \times 2.9 \times 10^{-12}$ m = 0.0011 mm, which is less than the diameter of a human hair!

Insight 27.3

THE SPEED OF LIGHT IS A UNIVERSAL SPEED LIMIT

Several results of special relativity suggest that speeds greater than c are not possible.

1. The factor $\sqrt{1 - v^2/c^2}$ that appears in time dilation and length contraction is an imaginary number if $v > c$.
2. The relativistic momentum of a particle (Eq. 27.24) becomes infinite as $v \rightarrow c$, suggesting that an infinite force or impulse is required for a particle to reach the speed of light.
3. The total energy of a particle (Eq. 27.33) becomes infinite as $v \rightarrow c$, suggesting again that an infinite amount of mechanical work is required to accelerate a particle to the speed of light.

This notion that c is a "speed limit" is *not* one of the postulates of special relativity. Postulate 2 of special relativity states only that the speed of light is the *same* for all inertial reference frames. It is amazing that combining postulate 2 with postulate 1 (that the laws of physics are the same in all inertial frames) leads to the conclusion that it is not possible for a particle to travel faster than the speed of light.

CONCEPT CHECK 27.6 The Fastest Proton Ever Known

The proton described in Example 27.7 is the fastest particle ever observed. How does its kinetic energy compare to that of a brick ($m = 5$ kg) that falls from a distance of 1 m?

(a) The brick has more energy.
(b) The proton has more energy.
(c) They have about the same energy.

EXAMPLE 27.8 Mass–Energy and Nuclear Weapons

According to a Los Alamos National Laboratory report in 1985, the nuclear bomb dropped on Hiroshima, Japan, in 1945 released an energy of about 54 TJ = 54×10^{12} J. This estimate was derived from measurements of the blast pressure and the intensity of gamma ray radiation emitted by the blast. (This value is only approximate, due to uncertainties in these measurements. Values ranging from 54 TJ to 75 TJ are sometimes quoted.) This energy came from the conversion of rest mass to energy. Approximately how much rest mass was converted to energy at Hiroshima?

RECOGNIZE THE PRINCIPLE

This problem involves the relativistic relation between mass and energy. We can use the relation (rest energy) = m_0c^2 from Equation 27.32 to find the amount of rest mass that was converted to energy in the explosion.

SKETCH THE PROBLEM

No sketch is necessary.

IDENTIFY THE RELATIONSHIPS AND SOLVE

Setting the rest energy for an object with rest mass m_0 equal to the energy released by the nuclear bomb, we have

$$m_0c^2 = 54 \times 10^{12} \text{ J}$$

Solving for the rest mass we find

$$m_0 = \frac{54 \times 10^{12} \text{ J}}{c^2} = \frac{54 \times 10^{12} \text{ J}}{(3.00 \times 10^8 \text{ m/s})^2} = \boxed{6.0 \times 10^{-4} \text{ kg}}$$

► *What does it mean?*

Less than 1 g of matter was converted to energy in this explosion! Even a small amount of rest mass contains a very large rest energy. The energy released in a nuclear explosion is often expressed in terms of the equivalent mass of trinitrotoluene (TNT) required to release the same amount of energy. Because of uncertainties in the precise amount of energy released by TNT, the convention adopted at the end of World War II for the specification of the yield from an atomic bomb is that 1 kiloton of TNT releases exactly 10^{12} calories = 4.184 TJ. Here mass is measured in metric tons, with 1 metric ton = 1000 kg. When measured in these terms, the energy released by the nuclear explosion at Hiroshima was the same as would be produced by the explosion of about 13 kilotons of TNT. This convention is behind the commonly heard expression "a 13-kiloton bomb."

It is now known that the energy released by 1 kiloton of TNT is actually about 4.7 TJ (not 4.2 TJ as quoted above, but the convention described above is still used when discussing the yield of a nuclear explosion). This means that 1 kg of TNT actually releases about 4.7 MJ, which is quite a substantial amount of energy for an object about the size of a grapefruit. The energy released at Hiroshima was thus about the same as would be released by the simultaneous explosion of more than a million of such pieces of TNT.

Mass–Energy Conversion and Chemical Reactions

The conversion of mass into energy is an important effect in nuclear reactions (Example 27.8 and Chapter 30), but it also occurs in other cases. Consider a chemical reaction in which a hydrogen atom is dissociated. The ionization energy of a hydrogen atom is 13.6 eV, meaning that when a proton and electron are bound together in a hydrogen atom their total energy is 13.6 eV lower than when they are separated. The principle of conservation of energy then implies that

total energy of hydrogen atom =

total energy of electron plus total energy of proton − 13.6 eV

If all these particles are observed at rest, the total energies of the hydrogen atom and of the electron and proton are just the rest energies. We thus have

$$m_0(\text{hydrogen atom})c^2 = m_0(\text{electron})c^2 + m_0(\text{proton})c^2 - 13.6 \text{ eV} \quad (27.35)$$

The mass of a hydrogen atom must therefore be less than the sum of the masses of an electron and proton. So, mass is *not conserved* when a hydrogen atom dissociates. For hydrogen dissociation (Eq. 27.35), we get

$$\Delta m_0 c^2 = 13.6 \text{ eV}$$

Expressing Δm_0 in SI units, we find

$$\Delta m_0 = \frac{13.6 \text{ eV}}{c^2} = \frac{(13.6 \text{ eV})(1.60 \times 10^{-19} \text{ J/eV})}{(3.00 \times 10^8 \text{ m/s})^2} = 2.4 \times 10^{-35} \text{ kg}$$

The mass of a proton is 1.7×10^{-27} kg, so Δm_0 is only about 0.000001% of the mass of the proton. We can therefore ignore such changes in this and all other chemical reactions.

EXAMPLE 27.9 Energy Released in Electron–Positron Annihilation

A positron is a subatomic particle with precisely the same mass as an electron, but with a positive charge $+e$. A positron is an example of ***antimatter*** (Chapter 31), and it is the antimatter "twin" of an electron. When an electron comes into close contact with a positron, they can annihilate each other, converting all their rest mass into energy. Typically, this annihilation reaction results in electromagnetic radiation. How much energy is released when an electron and a positron annihilate each other?

RECOGNIZE THE PRINCIPLE

In electron and positron annihilation, all the mass of both the electron and the positron is converted into energy. Both of these particles have rest mass $m_0 = m_e$ (the mass of the electron), so we can use this known mass to find the amount of energy released from the relation (rest energy) $= m_0c^2$.

SKETCH THE PROBLEM

Figure 27.23 describes the problem.

IDENTIFY THE RELATIONSHIPS

The (rest) mass of an electron is $m_e = 9.11 \times 10^{-31}$ kg (Appendix A, Table A.1), the same as that of a positron. Equation 27.32 gives the total rest energy of the two particles:

$$E_{\text{total}} = \text{rest energy} = 2(m_e c^2)$$

(continued) ▶

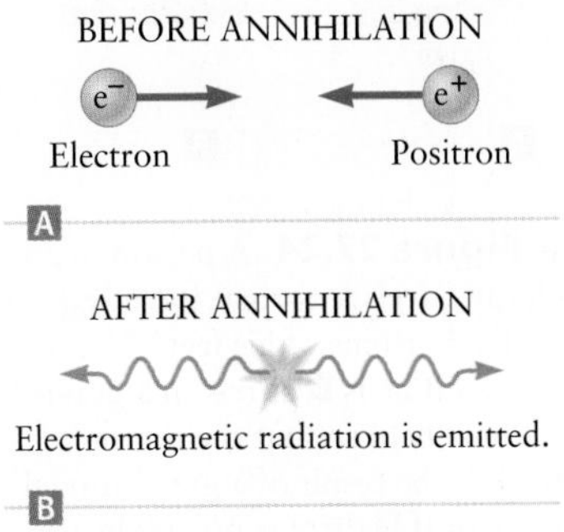

▲ **Figure 27.23** Example 27.9. In electron–positron annihilation, all the rest mass of the two particles is converted to energy.

Insight 27.4

CONSERVATION PRINCIPLES

In Newton's mechanics, mass is a conserved quantity, so the total mass of a closed system cannot change. Special relativity tells us that mass is in fact not conserved. The principle of conservation of energy must instead be extended to include the rest energy (Eq. 27.32). What about our other conservation rules? Momentum is still conserved in collisions, but we must use the relativistic expression for momentum in Equation 27.24. Electric charge is also conserved, but it is possible to create or annihilate charges (as in Example 27.8) as long as the total charge does not change.

SOLVE

All this rest energy is released in the annihilation process. Inserting the values of m_e and c gives

$$E_{\text{total}} = 2(m_e c^2) = 2(9.11 \times 10^{-31}\text{ kg})(3.00 \times 10^8\text{ m/s})^2 = \boxed{1.6 \times 10^{-13}\text{ J}}$$

► *What does it mean?*

When an electron and positron annihilate, their combined electric charge also disappears. Charge is conserved because the total charge of the two particles is $+e - e = 0$ before the annihilation. Although rest mass is not conserved, all reactions must conserve both electric charge and total energy.

CONCEPT CHECK 27.7 Mass and Energy

Ted takes a walk on a night so dark that he needs his flashlight to find his way home. Is the total mass of the flashlight (including its batteries) when Ted gets home (a) less than, (b) greater than, or (c) exactly the same as its mass just before he left?

27.10 The Equivalence Principle and General Relativity

Special relativity is concerned with the laws of physics in inertial reference frames; these are reference frames that move at constant velocity. A ***noninertial*** reference frame is one that has a nonzero acceleration. Physics in noninertial frames is described by ***general relativity***. Most of the results of special relativity, such as time dilation and length contraction, can be worked out using algebra, but general relativity involves much more complicated mathematics. We will therefore only describe the main ideas and results of the general theory.

General relativity is based on a postulate known as the ***equivalence principle:***

Equivalence principle

The effects of a uniform gravitational field are identical to motion with constant acceleration.

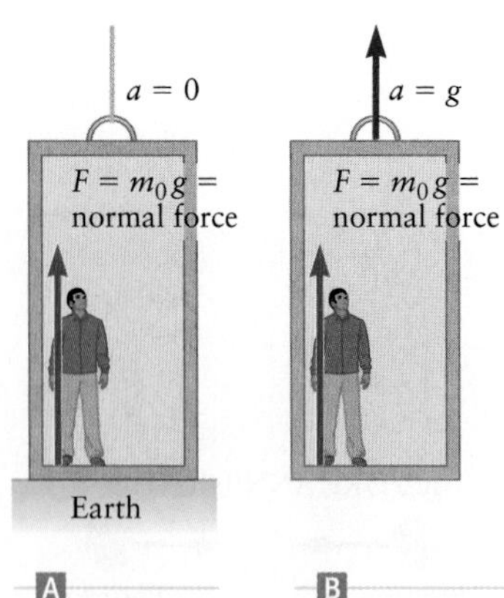

▲ **Figure 27.24** A person in an elevator feels a normal force $m_0 g$ on the bottoms of his feet. He cannot tell if he is A at rest in a gravitational field so that the normal force is the result of a gravitational force or if B there is no gravitational field and the elevator has an acceleration $a = g$. According to the equivalence principle, there is no way to tell the difference between these two situations.

An example of this equivalence is shown in Figure 27.24. In Figure 27.24A, Ted stands in an elevator compartment near the Earth's surface, with the elevator at rest. Ted feels the normal force exerted by the floor on his feet; he has mass m_0, so this force has magnitude $m_0 g$. From his study of physics, Ted deduces that g is a result of the Earth's gravity and hence that he is situated in a gravitational field.

Figure 27.24B shows the same elevator, but now it is located in distant space, far from any planets or stars, and the elevator compartment has an acceleration $a = g$. Since Ted's acceleration is a, Newton's second law tells us that a force $F = m_0 a = m_0 g$ is exerted by the elevator floor on Ted's feet.[3] According to the equivalence principle, there is no way for Ted to tell the difference between the effects of the gravitational field in Figure 27.24A and the accelerated motion in Figure 27.24B; they are completely equivalent.

This equivalence argument has the following profound consequences.

1. *Equivalence of inertial and gravitational mass.* The concept of mass enters Newton's mechanics in two places. The mass in Newton's second law is called the ***inertial mass*** m_{inertial}, while the masses in the law of gravitation (Eq. 5.16) are called ***gravitational mass*** m_{grav}. Newton assumed these two types of mass

[3]We assume the elevator's speed is much smaller than the speed of light, so the difference between Newton's second law (Eq. 27.25) and the relativistic result in Equation 27.26 is negligible.

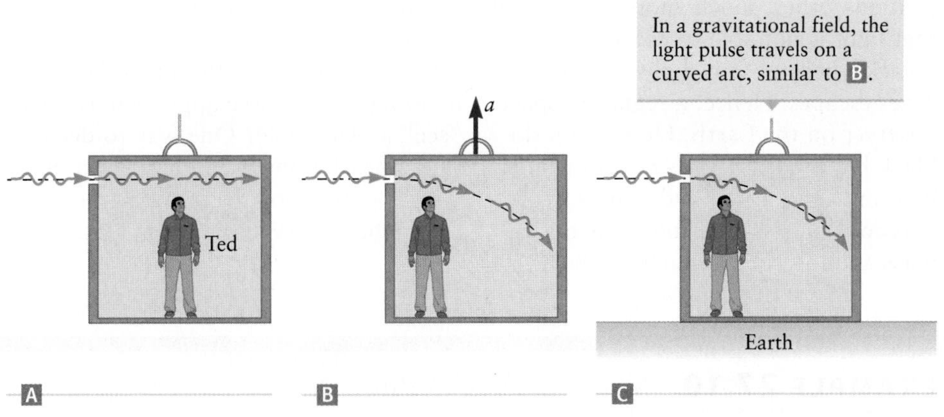
In a gravitational field, the light pulse travels on a curved arc, similar to B.
a
Ted
Earth
A
B
C

◀ **Figure 27.25** A light ray moving through the elevator compartment in Figure 27.24. A When the acceleration is $a = 0$, the light travels in a straight line through the elevator. In B, the elevator is accelerating upward. According to an observer in the elevator, the light beam travels in a downward arc. C According to the equivalence principle, the situation in part B is identical to the effect of a gravitational field. Hence, a gravitational field must also deflect a light beam.

are the same, but why should that be? In the elevator in a gravitational field (Fig. 27.24A), the force exerted on Ted is due to gravity, so $F = m_{grav}g$. In the accelerated elevator in Figure 27.24B, we have $F = m_{inertial}a = m_{inertial}g$. According to the equivalence principle, these two forces are precisely the same, so *the inertial mass must equal the gravitational mass.* We can also turn this argument around: the observation (by Newton and many others) that the inertial and gravitational masses are the same leads us to the equivalence principle.

2. *Deflection of light by gravity.* Figure 27.25 shows another experiment with Ted's elevator. A light pulse travels into the elevator car through a window on the left. In Figure 27.25A, the elevator is in distant space and has zero acceleration ($a = 0$). An observer (Ted) in the elevator then sees the light pulse travel in a straight-line path across the car. In Figure 27.25B, the elevator is accelerating upward; Ted now finds that *relative to him* the light beam travels on a downward arc. A nonaccelerated observer outside the elevator would say that the light pulse travels in a straight line, but *relative to the elevator* the light travels along a curved path. In Figure 27.25C, the experiment is repeated in a gravitational field and with no acceleration. According to the equivalence principle, the results must be the same as in the accelerated case in Figure 27.25B. So, Ted must again observe that the light follows a curved path, but now this curvature is produced by the gravitational field. In words, we say that light is "bent" by gravity. This conclusion is a straightforward result of the equivalence principle, but when Einstein proposed that light is bent by a gravitational field his prediction was met with considerable skepticism. Einstein's ideas were confirmed in experiments in 1919, in which light passing near the Sun during an eclipse (Fig. 27.26) was found to be deflected by the predicted amount. This result made Einstein an international celebrity.

Black Holes

The most massive object in our solar system is the Sun, but even its large gravitational field only deflects light a small amount (Fig. 27.26). However, the universe

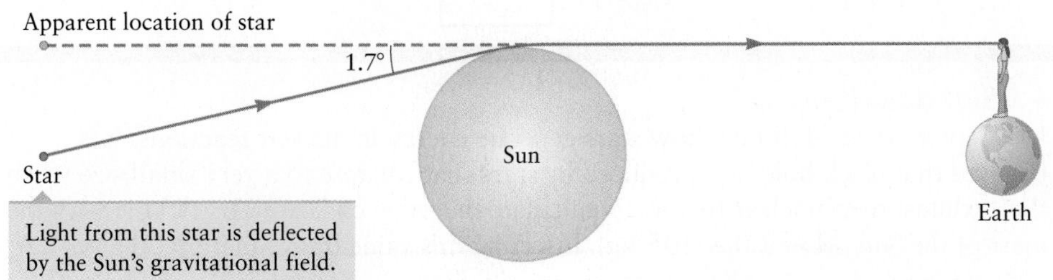
Apparent location of star
1.7°
Star
Sun
Earth
Light from this star is deflected by the Sun's gravitational field.

◀ **Figure 27.26** Deflection of light by the Sun's gravitational field. The deflection is largest for light beams that travel nearest the Sun. Light that just grazes the surface of the Sun is deflected through an angle of about 1.7°. This effect is easiest to measure when the Sun's light is blocked by the Moon during a solar eclipse, and was first measured by a team of astronomers in 1919.

contains other, much more massive objects. ***Black holes*** contain so much mass that light is not able to escape from their gravitational attraction. They are called "black" because any light that passes nearby is attracted to the black hole and never escapes. Hence, a region of space containing a black hole appears dark to an observer on the Earth. How, then, do we "see" a black hole? One way to detect a black hole is through its effect on the motion of nearby objects. Stars near a black hole move along curved trajectories due to the gravitational attraction between massive objects. Astronomical observations of these curved trajectories give the mass and location of the black hole.

EXAMPLE 27.10 Size of a Black Hole

One way to measure the size of a black hole is to study the motion of light emitted a distance R from its center. If R is small, the gravitational force of the black hole is strong enough that all light is attracted to the black hole and cannot escape. On the other hand, if R is large, the light can escape. This problem is similar to finding the escape speed for a rocket launched from the surface of a planet into outer space (Chapter 6), with light now playing the role of the rocket. We calculated the speed required for a rocket to escape from a planet of radius R in Equation 6.22. Use that result to find the value of R at which light can just barely escape from a black hole of mass M.

RECOGNIZE THE PRINCIPLE

In Chapter 6, we considered the motion of a particle with speed v that is "fired" into space. We found that such a particle could escape to an infinite distance if its speed exceeds the escape speed

$$v_{\text{escape}} = \sqrt{\frac{2GM}{R}} \tag{1}$$

where M is the mass of the Earth and R is the initial distance from the Earth's center. The same result for the escape speed also applies to light near a black hole. Since light moves at speed c, it can escape from a black hole if the escape speed is less than c.

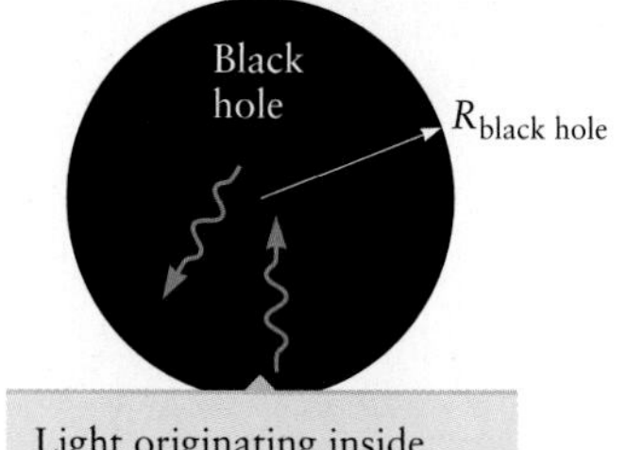
Light emitted outside the black hole will escape.
Black hole
$R_{\text{black hole}}$
Light originating inside $R_{\text{black hole}}$ is not able to escape from the black hole.

▲ **Figure 27.27** Example 27.10.

SKETCH THE PROBLEM

Figure 27.27 describes the problem.

IDENTIFY THE RELATIONSHIPS AND SOLVE

Setting $v_{\text{escape}} = c$ in Equation (1) and solving for R leads to

$$v_{\text{escape}} = c = \sqrt{\frac{2GM}{R}}$$

$$c^2 = \frac{2GM}{R}$$

$$R = \boxed{\frac{2GM}{c^2}} \tag{2}$$

▸ *What does it mean?*

In Chapter 30, we'll discuss how stars generate energy in nuclear reactions. It is believed that black holes are produced by stars that collapse to a very small size when they exhaust their nuclear fuel. Let's calculate the value of R for a black hole with the mass of the Sun ($M \approx 2.0 \times 10^{30}$ kg). Inserting this value into Equation (2) gives

$$R = \frac{2GM}{c^2} = \frac{2(6.67 \times 10^{-11}\ \text{N}\cdot\text{m}^2/\text{kg}^2)(2.0 \times 10^{30}\ \text{kg})}{(3.00 \times 10^8\ \text{m/s})^2} \approx 3000\ \text{m}$$

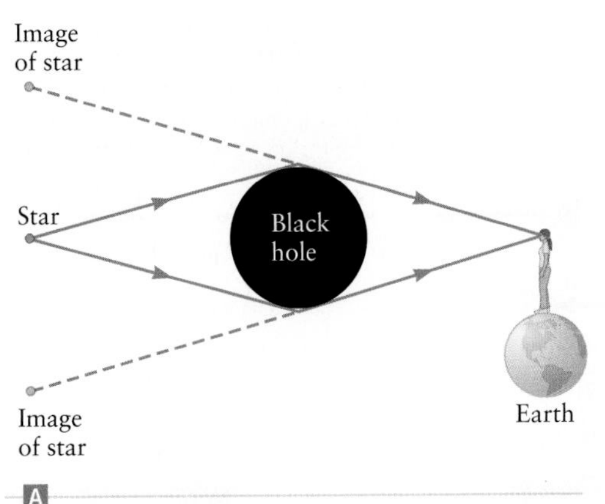

◀ **Figure 27.28** A A very massive object such as a black hole acts as a gravitational "lens." Here light rays from a star pass on both sides of a black hole and arrive at an observer, forming two separate images of the star. There will also be rays that come out of the plane of the drawing, producing additional images or a ring of images. B Example of gravitational lensing observed with the Hubble Space Telescope. Each circular arc of light (several are indicated by the arrows) comes from a single star. These arcs are due to light rays that travel out of the plane in part A.

Considering the mass and (current) radius of the Sun, this size is incredibly small. For comparison, a typical airport runway is also about 3000 m in length. A black hole is thus an extremely dense object.

Detecting a Black Hole: Gravitational Lensing

A black hole can be detected through an effect called ***gravitational lensing***. Figure 27.28A shows light from a star as it passes by a black hole on its way to an observer on the Earth. If the black hole is between the star and the Earth, light from the star can pass by either side of the black hole and still be bent by gravity so as to reach the Earth. This figure shows a view in only one plane; light from the star can also pass above and below the plane of the drawing. Hence, light from a single star can produce multiple images and even a set of images along a ring or arc. Astronomers have identified several such gravitational lenses; some examples are shown in Figure 27.28B. The positions and shapes of the images can be used to deduce the mass of the black hole. From this and other types of astronomical observations, a number of black holes have been found. In fact, a black hole resides at the center of the Milky Way, our galaxy.

CONCEPT CHECK 27.8 Size of the Black Hole at the Center of the Milky Way

Astronomers have found that the black hole at the center of the Milky Way has a mass about three million times larger than our Sun. Hence, this black hole has mass $M = (3 \times 10^6)\,M_{\text{Sun}} = 6 \times 10^{36}$ kg. Is the radius of this black hole (a) 3000 m, (b) 9×10^9 m, (c) 3×10^9 m, or (d) 9.5×10^6 m? Is it larger or smaller than our Sun? *Hint*: Use the result in Example 27.10.

27.11 Relativity and Electromagnetism

Special relativity is all about how the physical world looks to inertial observers moving at different speeds. So far, our two observers, Ted and Alice, have focused on several problems in mechanics. Let's now consider a problem in electromagnetism and give them the job of calculating the force between an infinite line of charge with charge per unit length λ and a point charge $+q$ located a distance r away.

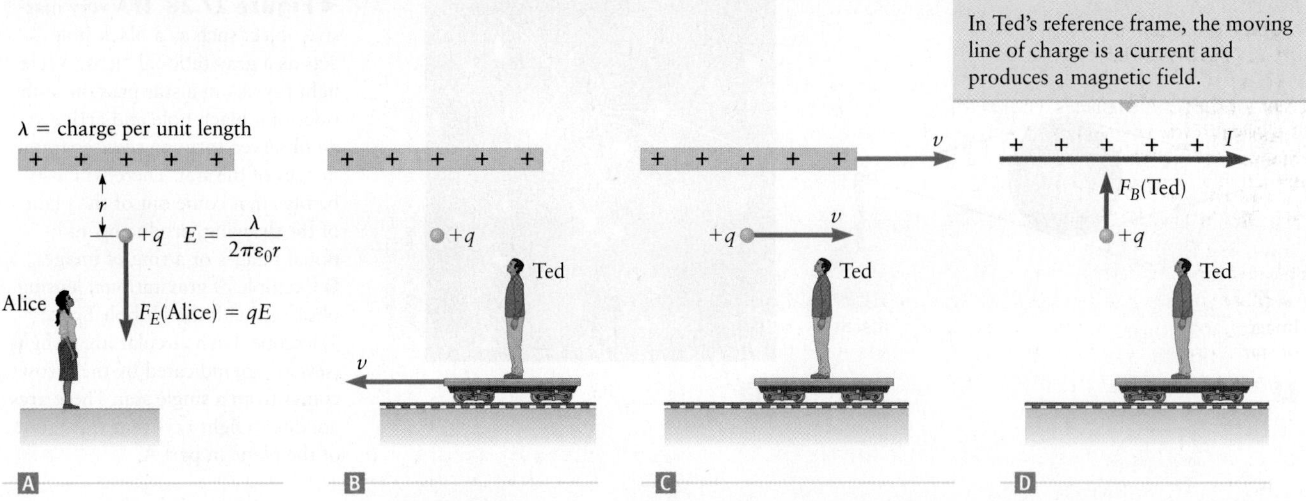

▲ **Figure 27.29** A An infinite line of charge at rest relative to Alice. This line produces an electric field $E = \lambda/(2\pi\varepsilon_0 r)$ and hence an electric force on the test charge $+q$. B The same situation viewed by Ted as he moves past on his railroad car. C In Ted's frame of reference, both the charged line and the test charge move with speed v. D According to Ted, the moving charged line is equivalent to an electric current. This current produces a magnetic field and thus also a magnetic force on the test charge.

Figure 27.29A shows Alice's analysis; she is at rest relative to both the charged line and the point charge. Alice applies results from Chapter 17 and finds that the line of charge produces an electric field $E = \lambda/(2\pi\varepsilon_0 r)$ directed away from the line, so the force on the point charge is

$$F_E(\text{Alice}) = \frac{q\lambda}{2\pi\varepsilon_0 r} \tag{27.36}$$

Ted analyzes this situation from his moving railroad car traveling to the left at speed v in Figure 27.29B, so from his point of view the charged line and the point charge are both moving to the right with speed v as shown in Figure 27.29C. In Ted's reference frame, the moving charged line acts as a current, so according to him this line produces both an electric field *and* a magnetic field. Since the test charge $+q$ is moving through this magnetic field, Ted concludes that there is also a magnetic force on this test charge. The current is to the right in Figure 27.29D, so the magnetic field at the test charge is directed into the page; applying right-hand rule number 2 (Chapter 20) gives a magnetic force $F_B(\text{Ted})$ directed toward the line.

Ted will thus say that there is an electric force *and* a magnetic force on the particle. On the other hand, according to Alice there is *only* an electric force. Who is correct? The answer is that they are *both* correct!

How can Ted possibly agree with Alice? According to Ted, the moving charged line is contracted (length contraction) as discussed in Section 27.5. Squeezing this charge into a smaller length *increases* the charge per unit length and thus *increases* the electric force calculated by Ted when he uses Equation 27.36. Thus, Ted observes a *larger* electric field and a *larger* electric force directed away from the line of charge than that found by Alice. However, the magnetic force observed by Ted is directed *toward* the line. A careful analysis shows that this magnetic force precisely cancels the extra electric force, so Ted and Alice agree on the total force on the particle. Alice will attribute it completely to an electric force, whereas Ted will attribute it to a combination of electric and magnetic forces.

This example shows that the values of the electric and magnetic fields in a particular situation depend on the motion of the observer; here Ted finds a nonzero magnetic field, while for Alice the magnetic field is zero. It is interesting that Maxwell's equations already contain this effect; his equations of electromagnetism were already consistent with the theory of special relativity, but Maxwell (and others) didn't fully appreciate it until Einstein discovered special relativity.

27.12 Why Relativity Is Important

Generally speaking, relativistic effects become important for objects or observers whose speed approaches the speed of light. Although there are a few everyday situations in which the effect of special relativity is important (such as the Global Positioning System, or GPS, discussed in Section 27.3), relativity plays no significant role in most terrestrial situations. Even so, relativity is an extremely important part of physics for the following reasons.

1. The relation between mass and energy and the possibility that mass can be converted to energy (and vice versa) mean that mass is not conserved. Instead, we have a more general view of energy and its conservation. The three hallmark conservation principles of physics are thus (a) conservation of energy, (b) conservation of momentum, and (c) conservation of charge. It is believed that *all* the laws of physics must obey these three principles.
2. The rest energy of a particle is huge, which has important consequences for the amount of energy available in processes such as nuclear reactions.
3. Relativity changes our notions of space and time. At the start of this book, we mentioned that time and position are two primary quantities in physics and that it is not possible to give precise definitions of such quantities. The best we can do is use our intuition to understand the meaning of space and time. Now we have found that this intuition breaks down when applied to special relativity. While this may seem surprising, there is no reason why an intuition developed from terrestrial experience should be reliable when applied to stars and black holes.
4. Relativity plays a key role in understanding how the universe was formed and how it is evolving. Black holes can't be understood without relativity.
5. Relativity shows that Newton's mechanics is not an exact description of the physical world. Instead, Newton's laws are only an approximation that works very well in some cases, but not in others. So, we shouldn't discard Newton's mechanics, but we should understand its limits.

Summary | CHAPTER 27

Key Concepts and Principles

The postulates of special relativity

Special relativity is based on two postulates:

1. The laws of physics are the same in all inertial reference frames.
2. The speed of light in a vacuum is constant, independent of the motions of the source and observer.

Time dilation and length contraction are consequences of these two postulates.

(*Continued*)

General relativity

General relativity is based on the ***equivalence principle***: the effects of a uniform gravitational field are identical to motion with constant acceleration.

One consequence of the equivalence principle is that light rays are deflected by a gravitational field.

Applications

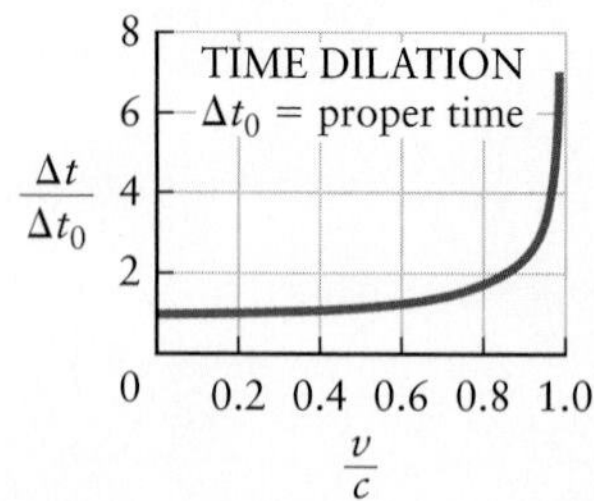

Time dilation

The ***proper time*** Δt_0 is the time interval between two events as measured by an observer for which the two events occur at the same location (i.e., an observer at rest relative to the events). A second observer moving at a relative speed v measures a longer time,

$$\Delta t = \frac{\Delta t_0}{\sqrt{1 - v^2/c^2}} \qquad \text{(27.7) (page 841)}$$

In words, we say that *moving clocks run slow.*

Length contraction

The ***proper length*** of an object is the length L_0 measured by a person at rest relative to the object. If a second observer is moving with a speed v relative to the object, that observer will measure a shorter length,

$$L = L_0\sqrt{1 - v^2/c^2} \qquad \text{(27.15) (page 847)}$$

An equivalent situation occurs when the object is moving with speed v relative to an observer. In words, we say that *moving metersticks are contracted.*

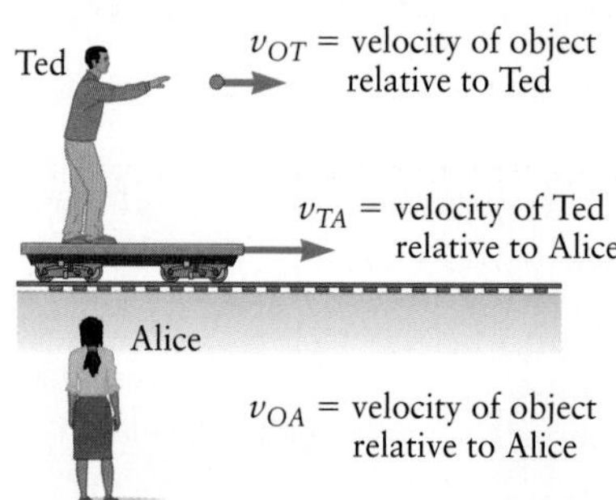

Relativistic addition of velocities

Observer Alice is at rest while Ted moves with velocity v_{TA} relative to Alice. Ted observes an object moving with velocity v_{OT} relative to him. When $v_{TA} \ll c$ and $v_{OT} \ll c$, the velocity of the object relative to Alice is given by the Galilean addition of velocities relation

$$v_{OA} = v_{OT} + v_{TA} \qquad \text{(27.17) (page 850)}$$

For speeds that are not small compared with c, we must use the relativistic addition of velocities relation,

$$v_{OA} = \frac{v_{OT} + v_{TA}}{1 + \dfrac{v_{OT}v_{TA}}{c^2}} \qquad \text{(27.18) (page 850)}$$

If both $v_{TA} < c$ and $v_{OT} < c$, then $v_{OA} < c$. No object can move faster than c.

Relativistic momentum

In Newton's mechanics, the momentum of a particle of mass m_0 moving with velocity v is

$$\text{Newton's mechanics: } p = m_0 v \qquad \text{(27.22) (page 852)}$$

Special relativity gives a different result:

$$\text{special relativity: } p = \frac{m_0 v}{\sqrt{1 - v^2/c^2}} \qquad \text{(27.24) (page 853)}$$

where m_0 is called the ***rest mass*** of the particle.

Relativistic energy

In Newton's mechanics, the kinetic energy of a particle of mass m_0 is

$$\text{Newton's kinetic energy: } KE = \tfrac{1}{2}m_0v^2 \qquad (27.28)\ \text{(page 855)}$$

In special relativity, one considers instead the total energy of a particle:

$$\text{total relativistic energy: } TE = \frac{m_0c^2}{\sqrt{1 - v^2/c^2}} \qquad (27.33)\ \text{(page 856)}$$

When $v = 0$, the total energy is $TE = m_0c^2$, which is called the ***rest energy*** of the particle. This result implies that rest mass is a form of energy. When $v > 0$, the total energy is larger than the rest energy; the energy in excess of the rest energy is the ***relativistic kinetic energy***,

$$KE = \frac{m_0c^2}{\sqrt{1 - v^2/c^2}} - m_0c^2 \qquad (27.30)\ \text{(page 856)}$$

Questions

SSM = answer in *Student Companion & Problem-Solving Guide* ⓧ = life science application

1. It is impossible for a particle of rest mass m_0 to travel at a speed greater than c. Is there an upper limit to the momentum or kinetic energy of the particle? Explain why or why not.
2. You are traveling in a windowless spacecraft, far from any planets or stars. Describe an experiment you could do to tell whether you are in an inertial or a noninertial frame.
3. SSM A constant force F is applied to a spacecraft of mass m. The spacecraft is initially at rest. Sketch the speed of the spacecraft as a function of time according to Newton's laws (a) without allowing for relativity and (b) allowing for relativity. Compare the two results and explain why they are different.
4. Give an example of an inertial reference frame.
5. Give four examples of noninertial reference frames.
6. The speed of light inside a substance depends on the index of refraction n. For water, $n \approx 1.33$. (a) What is the speed of light in water? (b) Is it possible for a particle to travel faster than your result in part (a)? Explain why or why not.
7. Length contraction applies to the length of an object measured in the direction of motion. Consider the length of an object measured in a direction perpendicular to its velocity. Give an argument that explains why there is no contraction along the perpendicular direction.
8. A particle is "extremely" relativistic if its speed is very close to c. What is the ratio of the total energy to the momentum for such a particle?
9. Two astronauts travel on spaceships that are both moving with constant velocities, but at different speeds. On which of the following quantities will the astronauts agree? In each case, give a reason for your answer.
   (a) The length of an object (e.g., a third spaceship)
   (b) The time interval between two events
   (c) The speed of light in a vacuum
   (d) The speed of a moving object (e.g., a third spaceship)
   (e) The relative speed of the observers
10. At a typical everyday speed such as 65 mi/h, does length contract, time dilate, and mass increase by amounts that could be easily measured? Explain.
11. A wind-up toy has a coil spring that stores spring potential energy. As the toy is wound, does its mass increase, decrease, or stay the same? Explain.
12. SSM Travelers onboard a spaceship are moving at $0.9c$ from one star to another. The stars are at rest relative to each other, and this is the speed of the spaceship relative to the stars. Which of the following statements are true? Assume observers on the spaceship or on the stars are in inertial reference frames.
    (a) Observers at rest with respect to the stars measure a shorter distance between the stars than the travelers do.
    (b) The observers see the travelers moving in slow motion relative to themselves.
    (c) The observers see the travelers age more slowly than they themselves do.
    (d) The travelers measure a shorter distance between the stars than the observers do.
    (e) The travelers believe they are aging more slowly than the observers are.
13. Two spaceships, each traveling at $0.5c$ relative to an observer on the Earth, approach each other head on. One ship fires its laser beam weapon at the other ship. What does the other ship measure for the speed of the laser light?
14. Suppose an object moves away from a plane mirror at a speed of $0.9c$. Does the image recede from the source at a speed of $1.8c$? Explain.
15. You are standing on the Earth and observe a spherical spacecraft fly past at very high speed. Sketch the shape of the ship as observed by you.
16. Two identical clocks are synchronized. One stays on the Earth while the other goes in orbit around the Earth for 1 year as measured by the clock on the Earth. Which of the following statements are true after both clocks are again on the Earth? Explain. *Hint*: For answers (d) and (e), you should explain what it means for a clock's reading to be "wrong."
    (a) The clocks are still synchronized.
    (b) The clock that made the trip runs slower after it returns.
    (c) The clock that made the trip does not have the same time as the clock that stayed on the Earth.
    (d) The clock that stayed on the Earth has the "wrong" time.
    (e) The clock that orbited the Earth has the "wrong" time.
17. If you are traveling very close to the speed of light and hold a mirror in front of your face, what will you see?
18. If the Earth were compressed to the density of a black hole, would it be about the size of (a) an atom, (b) a grape, (c) an orange, or (d) a basketball?

Problems available in ENHANCED WebAssign

Section 27.1	Newton's Mechanics and Relativity Problem 1
Section 27.2	The Postulates of Special Relativity Problems 2–4
Section 27.3	Time Dilation Problems 5–12
Section 27.4	Simultaneity Is Not Absolute Problems 13–14
Section 27.5	Length Contraction Problems 15–23
Section 27.6	Addition of Velocities Problems 24–32
Section 27.7	Relativistic Momentum Problems 33–39
Section 27.8 **Section 27.9**	What Is "Mass"? Mass and Energy Problems 40–50
Section 27.10	The Equivalence Principle and General Relativity Problems 51–52
Additional Problems	Problems 53–72

List of Enhanced Problems

Problem number	Solution in *Student Companion & Problem-Solving Guide*	Reasoning & Relationships Problem	Reasoning Tutorial in ENHANCED WebAssign
27.1	✓		
27.4	✓		
27.11	✓		
27.13	✓		
27.22	✓		
27.27	✓		
27.35	✓		
27.47	✓		
27.52		✓	✓
27.53		✓	✓
27.60		✓	✓
27.61	✓	✓	✓
27.63		✓	
27.65	✓		
27.70		✓	
27.72		✓	

CHAPTER

28

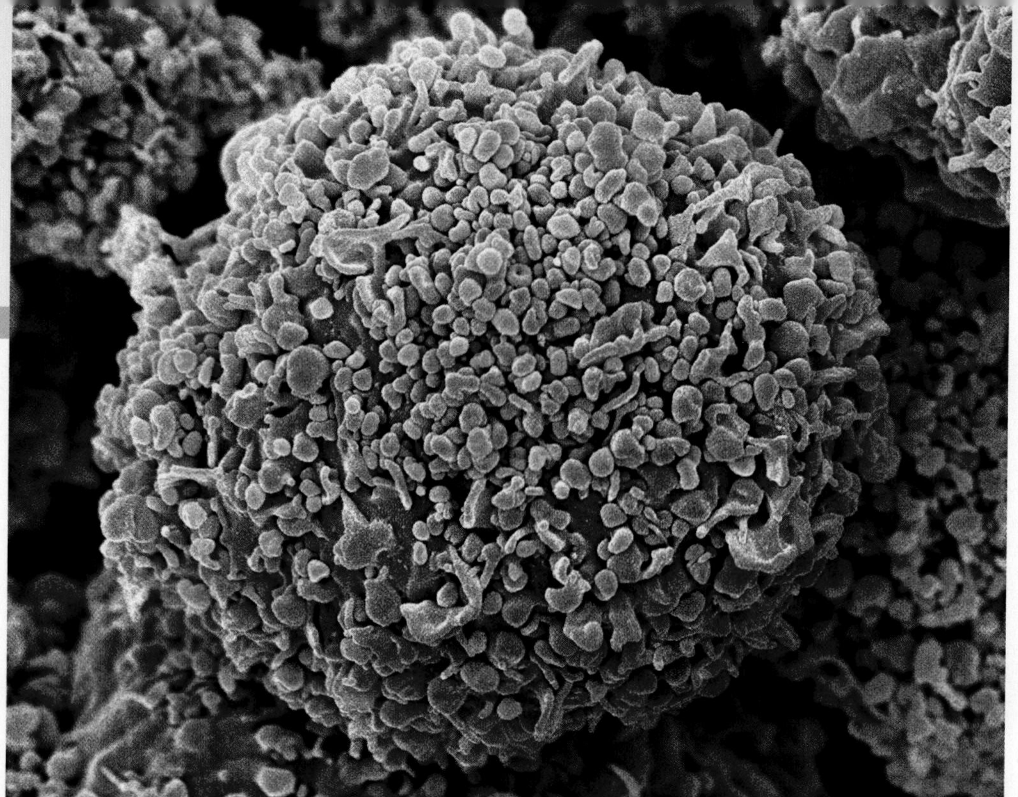

► *This image showing a breast cancer cell was obtained using a scanning electron microscope (SEM). An SEM uses the wave properties of electrons, predicted by quantum theory, to form images, similar to the way images are formed using an optical microscope. (Steve Gschmeissner/Photo Researchers, Inc.)*

Quantum Theory

OUTLINE

In this text, our work on mechanics, electricity, and magnetism has been based on two pillars of ***classical*** physics: Newton's laws and Maxwell's equations. Classical physics deals with the macroscopic world, including baseballs, automobiles, and planets. Newton's laws fail when applied to the atomic-scale world of electrons and atoms. Similarly, Maxwell's equations correctly describe electromagnetic phenomena in the macroworld, but fail when applied at the atomic scale. This atomic-scale world is called the ***quantum*** regime. In this chapter, we describe how quantum behavior is different from anything we have seen before. "Quantum" refers to a very small increment or parcel of energy. These parcels, called ***quanta***, are a central aspect of the quantum world.

28.1 Particles, Waves, and "Particle-Waves"

In the macroworld of Newton and Maxwell, energy can be carried from one point to another by only two types of "objects": particles (such as baseballs or bullets) and waves (such as sound or light). In the macroworld, it is easy to tell what is a particle

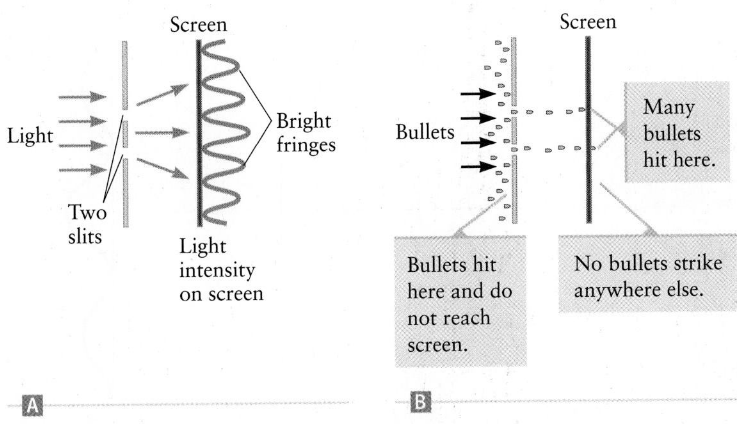

◀ **Figure 28.1** A When light passes through a double slit, constructive and destructive interference produce bright and dark fringes on the screen, evidence for the wave nature of light. B Classical particles such as bullets do not undergo interference when they pass through a double slit.

and what is a wave. In the quantum regime, however, it is not possible to make a simple distinction between particles and waves.

How Do Waves and Particles Differ?

Consider the double-slit experiment in Figure 28.1A, in which light is incident on an opaque barrier containing two very narrow openings. In Chapter 25, we learned that a double-slit interference pattern, consisting of a series of bright and dark fringes, is formed on the screen on the far right. The bright fringes are produced by constructive interference between waves that pass through the two slits, and the dark fringes are at locations where there is destructive interference. A similar result would be found with other types of waves, including sound and water waves.

Figure 28.1B shows another double-slit experiment, this time using bullets instead of light. Our macroworld intuition correctly predicts what will happen in this case: only bullets that pass through one or the other of the two slits will reach the screen on the far right; the other bullets (the ones that strike the barrier) are stopped by the barrier. The pattern of bullets on the screen is quite different from the pattern found with light waves; the bullet pattern corresponds only to the "shadows" of the slits. There is no constructive or destructive interference with bullets or other classical particles such as baseballs or rocks.

The behaviors sketched in Figure 28.1 illustrate some important ***properties of waves and particles in the classical regime***:

- **Waves exhibit interference; particles do not.**
- **Particles often deliver their energy in discrete amounts. For example, when a bullet in Figure 28.1B strikes the screen, all its energy is deposited on the screen (assuming the bullet does not bounce off). Energy arrives at the screen in discrete parcels, with each parcel corresponding to the kinetic energy carried by a single bullet.**
- **The energy carried or delivered by a wave is not discrete, but varies in a continuous manner. From Chapter 23, the energy carried by a wave is described by its intensity, which equals the amount of energy the wave transports per unit time across a surface of unit area. Hence, for the light wave in Figure 28.1A, the amount of energy absorbed by the screen depends on the intensity of the wave and the absorption time. The amount of absorbed energy can thus take on any nonnegative value. Classically, wave energy is not delivered in discrete parcels.**

An Interference Experiment with Electrons

In the world of classical physics, the experiment in Figure 28.1 can be used to distinguish between particles and waves. According to classical physics, only these two types of behavior are possible: waves exhibit interference; particles do not.

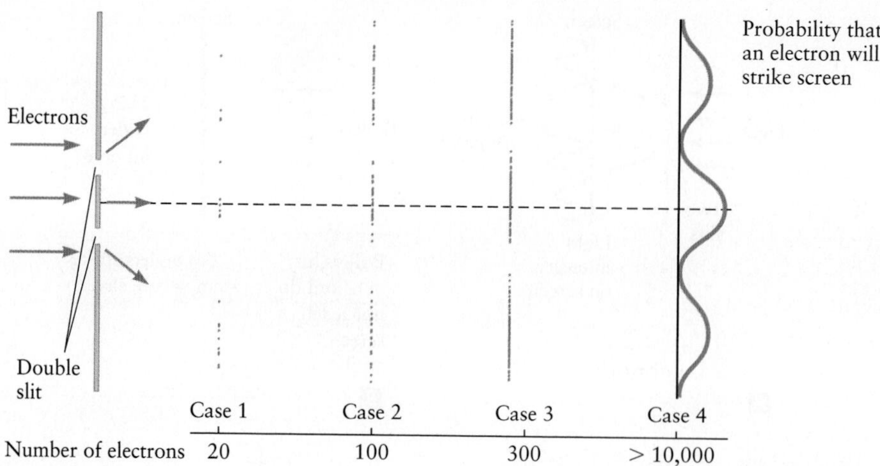

▶ **Figure 28.2** An interference experiment with electrons. Each dot in cases 1 through 3 indicates where an electron strikes the screen. Case 1: pattern formed by the first 20 electrons. Case 2: pattern formed by the first 100 electrons. Case 3: After 300 electrons, it is clear that electrons tend to arrive at certain places on the screen and not at others. Case 4: When the experiment is carried out with a very large number of electrons, the probability for an electron to reach the screen forms a double-slit interference pattern, similar to the pattern for light in Figure 28.1A.

However, this tidy separation of particles versus waves is *not* found in the quantum world. Figure 28.2 shows a hypothetical double-slit experiment performed with electrons. A beam of electrons, all with the same speed, are incident from the left and pass through a pair of slits. The electrons then travel on to a screen on the right, which records where each electron strikes. Case 1 in Figure 28.2 shows the result after 20 electrons have passed through the slits. Each dot shows where a particular electron arrived at the screen, and the arrival points seem to be distributed randomly. If this experiment is repeated with another 20 electrons, the precise arrival points would be different, but the general appearance would be the same. If we proceed with the experiment and wait until 100 electrons have arrived, we find the result shown in case 2. The arrival points are still spread out, but we can now see that the electrons are more likely to strike at certain points than at others. By the time 300 electrons have reached the screen in case 3, it is clear that electrons are much more likely to hit certain points on the screen. Case 4 on the far right shows the results after a very large number of electrons have passed through the slits. This sketch shows the probability that electrons will arrive at different points. This probability curve has precisely the same form as the variation of light intensity in the double-slit interference experiment in Figure 28.1A. The experiment shows that electrons undergo constructive interference at certain locations on the screen, giving a large probability for electrons to arrive at those locations. At other places, the electrons undergo destructive interference, and the probability for an electron to reach those locations is very small or zero.

Classical theory says that the interference exhibited by electrons in Figure 28.2 is possible only for waves. This experiment also shows aspects of particle-like behavior since the electrons arrive one at a time at the screen, with each dot in cases 1 through 3 of the figure corresponding to the arrival of a single electron as it deposits its parcel of energy at the screen.

The behavior in Figure 28.2 is characteristic of the quantum regime and shows that electrons behave in some ways as *both a classical particle and a classical wave*. In fact, *all* objects in the quantum world behave in this way. Electrons, protons, atoms, and molecules have all been found to give the results shown in Figure 28.2. Moreover, even light waves exhibit particle-like behavior. The clear-cut distinction between particles and waves thus breaks down in the quantum regime. It may be more helpful for your classical intuition to call these things "particle-waves," all of which exhibit the following ***properties of waves and particles in the quantum regime***:

- ***All* objects, including light and electrons, can exhibit interference.**
- ***All* objects, including light and electrons, carry energy in discrete amounts. These discrete "parcels" are called quanta.**

Let's now explore the implications of these ideas for light and for electrons.

28.2 Photons

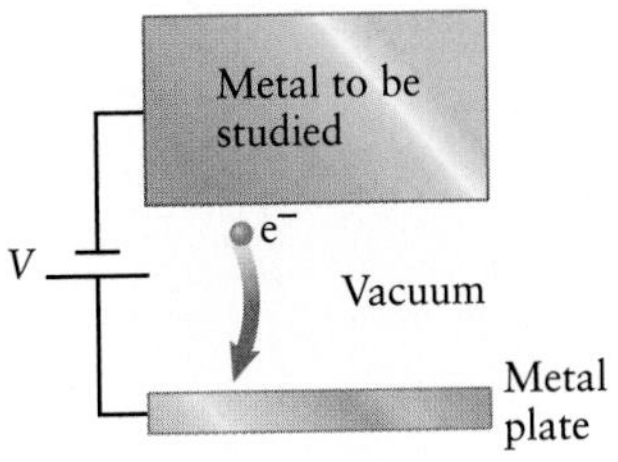

▲ **Figure 28.3** Measuring the work function W_c of a metal by applying an electric potential. When $eV > W_c$, electrons are ejected from the metal on top and jump to the metal plate below. Measuring the smallest V able to eject electrons[1] gives the value of W_c.

Around 1800, Thomas Young performed his double-slit interference experiment, which provided the first clear evidence that light is a wave. Maxwell's theory of electromagnetic waves was worked out about 60 years later, so physicists then had a detailed theory of light as an electromagnetic wave and believed that the nature of light was well understood. In the 1880s, however, studies of what happens when light is shined onto a metal gave some very puzzling results that could not be understood with the wave theory of light. To appreciate these experiments, we must first recall that a metal contains electrons that are able to move around within the metal. However, these electrons are still bound as a whole to the metal because of their attraction to the positive charges of the metal's atomic nuclei; that is why these electrons do not just "fall out of" the metal. The minimum energy required to remove a single electron from a piece of metal is called the ***work function***, W_c. The value of the work function is different for different metals; it can be measured in principle[1] by applying an electric potential difference between the metal of interest and a second metal plate (all in vacuum) and observing the value of the potential at which electrons first leave the metal (Fig. 28.3). If V is the electric potential at which electrons begin to jump across the vacuum gap in Figure 28.3, the work function is $W_c = eV$, where e is the magnitude of the electron's charge ($e = 1.60 \times 10^{-19}$ C).

The work function has units of energy, so it can be measured in joules. Since the experiment in Figure 28.3 involves electrons, it is sometimes convenient to measure W_c in units of electron-volts (eV). (Recall that 1 eV $= 1.6 \times 10^{-19}$ J; Eq. 18.19.) Values of the work function in both units are listed in Table 28.1.

The Photoelectric Effect

Another way to extract electrons from a metal is by shining light onto it. Light striking a metal is absorbed by the electrons, and if an electron absorbs an amount of light energy greater than W_c, it is ejected from the metal as sketched in Figure 28.4A. This phenomenon is called the ***photoelectric effect.***

Experimental studies of the photoelectric effect show that no electrons are emitted unless the light's frequency is greater than a critical value f_c. When the frequency is above f_c, the kinetic energy of the emitted electrons varies linearly with frequency f as shown in Figure 28.4B. Physicists initially tried to explain these results using the classical wave theory of light, but there were two difficulties with the classical

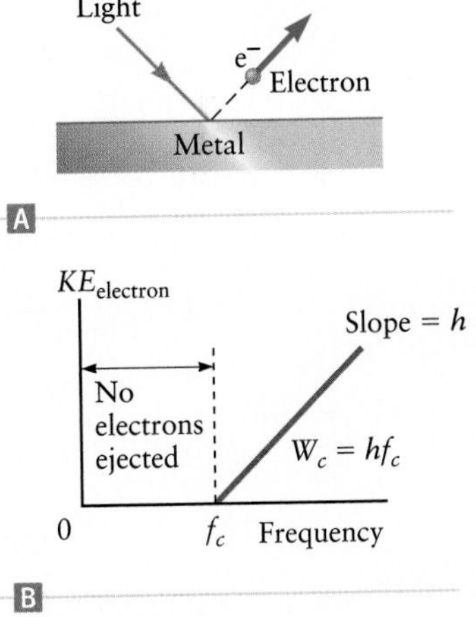

▲ **Figure 28.4** A The photoelectric effect: electrons are ejected when light strikes a metal. B Experiments show that the kinetic energy of the ejected electrons ($KE_{electron}$) depends on the frequency of the light. When the frequency is below a certain critical value f_c, no electrons are ejected. The work function is related to the critical frequency by $W_c = hf_c$, where h is Planck's constant defined in Equation 28.2.

TABLE 28.1 Work Function of Several Metals

Metal	Work Function (J)	Work Function (eV)
Na (sodium)	3.8×10^{-19}	2.4
Al (aluminum)	6.9×10^{-19}	4.3
Ca (calcium)	4.5×10^{-19}	2.8
Fe (iron)	6.9×10^{-19}	4.3
Cu (copper)	7.0×10^{-19}	4.4
Mo (molybdenum)	6.7×10^{-19}	4.2
Ag (silver)	6.9×10^{-19}	4.3
Pt (platinum)	1.0×10^{-18}	6.4
Pb (lead)	6.4×10^{-19}	4.0

[1]This approach illustrates the principle of the work function, but in practice the measurement of the work function is not quite this simple, due to the way an electron interacts with the ion cores and other electrons on its way out of the metal, especially at the surface.

explanations. First, experiments show that the critical frequency f_c is *independent* of the intensity of the light. According to classical wave theory, the energy carried by a light wave is proportional to the intensity, so it should always be possible to eject electrons by increasing the intensity to a sufficiently high value. Experiments found that when the frequency is below f_c, however, there are no ejected electrons no matter how great the light intensity. Second, the kinetic energy of an ejected electron is independent of the light intensity. Classical theory predicts that increasing the intensity will cause the ejected electrons to have a higher kinetic energy, but the experiments show no such result. In fact, the experiments show that the electron kinetic energy depends on the light's frequency (Fig. 28.4B) instead of the intensity.

The classical wave theory of light is thus not able to explain the photoelectric effect experiments. Albert Einstein surprised the physics world in 1905 when he offered the following explanation. He proposed that light carries energy in discrete quanta, now called ***photons***, and that each photon carries a parcel of energy

$$E_{\text{photon}} = hf \tag{28.1}$$

where h is a constant of nature called ***Planck's constant***, which has the value

$$h = 6.626 \times 10^{-34}\ \text{J} \cdot \text{s} \tag{28.2}$$

and f is the frequency of the light.

Planck's constant had been introduced a few years earlier by Max Planck to explain another unexpected property of electromagnetic radiation (the blackbody radiation spectrum, page 878). Let's now see how Einstein's photon theory explains the photoelectric effect.

Einstein suggested that a beam of light should be thought of as a collection of particles (photons), each of which has an energy that depends on frequency according to Equation 28.1. If the intensity of a monochromatic (single-frequency) light beam is increased, the number of photons is increased but the energy carried by each photon does not change. This theory explains the two puzzles associated with photoelectric experiments. The absorption of light by an electron is then just like a collision between two particles, a photon and an electron. The photon (according to Eq. 28.1) carries an energy hf that is absorbed by the electron. If this energy is less than the work function, the electron is not able to escape from the metal. For monochromatic light, increasing the light intensity increases the number of photons that arrive each second, but if the photon energy hf is less than the work function, even a high intensity will not eject electrons. The energy of a single photon—and hence the energy gained by any particular electron—depends on frequency f but not on the light intensity. Einstein's theory also explains why the kinetic energy of ejected electrons depends on light frequency but not intensity. The critical frequency in the photoelectric effect (Fig. 28.4B) corresponds to photons whose energy is equal to W_c:

$$hf_c = W_c$$

Such a photon has barely enough energy to eject an electron from the metal, but the ejected electron then has no kinetic energy. If a photon has a higher frequency and thus a greater energy, the extra energy above the work function goes into the kinetic energy of the electron. We have

$$KE_{\text{electron}} = hf - hf_c = hf - W_c \tag{28.3}$$

which is the equation of a straight line in a plot of KE_{electron} versus f. This linear behavior is precisely what is found in experiments as shown in Figure 28.4B. The slope of this line is the factor multiplying f in Equation 28.3, which is just Planck's constant h. Thus, photoelectric experiments give a way to measure h, and the values found agreed with the value known prior to Einstein's theory.

Photons Carry Energy and Momentum

Einstein's photon theory asserts that light energy can only be absorbed or emitted in discrete parcels, that is, as single photons. From Chapter 23, a light wave with an energy E also carries a certain amount of momentum $p = E/c$ (Eq. 23.12). Identifying the energy E with E_{photon}, the momentum carried by a single photon is

$$p_{photon} = \frac{hf}{c} \tag{28.4}$$

Momentum of a photon

Wavelength is related to frequency by $f\lambda = c$, so this can also be written as

$$p_{photon} = \frac{h}{\lambda} \tag{28.5}$$

The quantum theory of light thus predicts that "particles" of light called photons carry a discrete amount (a quantum) of both energy and momentum. Notice that we put the term *particles* in quotation marks. We did so because photons have two important properties that are quite different from classical particles: photons do not have any mass (!), and they exhibit interference effects (as in the double-slit interference experiment in Fig. 28.1A). Photons are thus unlike anything we have encountered in classical physics.

Let's use Equation 28.1 to calculate the energy carried by a single photon. The result depends on the frequency of the light, so consider the light from a green laser pointer. This light has a wavelength of about $\lambda = 530$ nm, so its frequency is

$$f = \frac{c}{\lambda} = \frac{3.00 \times 10^8 \text{ m/s}}{530 \times 10^{-9} \text{ m}} = 5.7 \times 10^{14} \text{ Hz}$$

Inserting this frequency in Equation 28.1 gives

$$E_{photon} = hf = (6.63 \times 10^{-34} \text{ J} \cdot \text{s})(5.7 \times 10^{14} \text{ Hz}) = 3.8 \times 10^{-19} \text{ J} \tag{28.6}$$

which is a *very* small amount of energy, much smaller than is normally encountered in the macroworld (as shown in Example 28.1).

EXAMPLE 28.1 Number of Photons Emitted by a Lightbulb

The light emitted by an ordinary lightbulb consists of photons, each of which has an energy given by Equation 28.1. Consider a 15-W lightbulb and for simplicity assume it emits all its energy as green light with $\lambda = 530$ nm as considered in Equation 28.6. How many photons does the lightbulb emit in 1.0 s?

RECOGNIZE THE PRINCIPLE

This example is an application of the relation between photon energy and frequency. The power rating P of the lightbulb gives the rate at which it emits energy. Power is the energy emitted per unit time, so from the value of P we can get the energy emitted in 1.0 s. We know the energy of a single photon from Equation 28.6, so we can then find the total number of photons emitted in this time.

SKETCH THE PROBLEM

Figure 28.5 shows a lightbulb emitting photons. The energy emitted each second equals the energy of a single photon times the number of photons emitted in 1 second.

Each photon
carries an energy
$E_{photon} = hf$

▲ **Figure 28.5** Example 28.1.

IDENTIFY THE RELATIONSHIPS

The lightbulb has a power rating of $P = 15$ W $= 15$ J/s, so the total energy emitted in $t = 1.0$ s is

$$E_{total} = Pt = (15 \text{ J/s})(1.0 \text{ s}) = 15 \text{ J}$$

This energy is carried by N photons, each of which has the energy E_{photon} found in Equation 28.6. We thus have $E_{total} = NE_{photon}$.

(continued) ▶

SOLVE

Solving for N, we get

$$N = \frac{E_{\text{total}}}{E_{\text{photon}}} = \frac{15\text{ J}}{3.8 \times 10^{-19}\text{ J}} = \boxed{3.9 \times 10^{19}\text{ photons}}$$

▸ *What does it mean?*

In this example as in most ordinary situations, the number of photons involved is extremely large, so the typical observer would not notice that the energy in a light beam actually does arrive as discrete quanta.

CONCEPT CHECK 28.1 Energy of a Radio Frequency Photon

Your favorite AM radio station has a frequency near $f_{\text{AM}} \approx 1$ MHz, whereas the author's favorite FM station has $f_{\text{FM}} \approx 100$ MHz. Radio waves are electromagnetic radiation, just like visible light, so both AM and FM signals are carried by photons. Which of the following statements is true?

(a) The AM photons have a higher energy than the FM photons.
(b) The FM photons have a higher energy than the AM photons.
(c) They are both radio waves, so they have the same photon energy.

EXAMPLE 28.2 The Photoelectric Effect with Aluminum

A photoelectric experiment is planned using aluminum (Al). **(a)** What is the minimum photon frequency that will eject electrons? **(b)** If the experiment is performed using blue light with $\lambda = 450$ nm, will there be any ejected electrons?

RECOGNIZE THE PRINCIPLE

(a) The minimum photon frequency f_c required to eject an electron is proportional to the work function. **(b)** We can determine if a photon will eject an electron by comparing its frequency with f_c.

SKETCH THE PROBLEM

No sketch is necessary.

IDENTIFY THE RELATIONSHIPS

The energy of a photon of frequency f is $E_{\text{photon}} = hf$ (Eq. 28.1), so we can use the value of the work function for Al from Table 28.1 to find the minimum photon frequency. We can then compare this frequency with the energy of a photon of blue light to decide if such a photon will eject an electron. (The variation of the frequency and wavelength across the electromagnetic spectrum is also shown in Fig. 23.8.) From Table 28.1, the work function for Al is $W_c = 4.3$ eV (electron-volts), or, expressed in units of joules, $W_c = (4.3\text{ eV})(1.60 \times 10^{-19}\text{ J/eV}) = 6.9 \times 10^{-19}$ J. The work function W_c equals the energy of a photon that will just barely eject an electron, so

$$hf_c = W_c$$

SOLVE

(a) Solving for f_c gives

$$f_c = \frac{W_c}{h} = \frac{6.9 \times 10^{-19}\text{ J}}{6.63 \times 10^{-34}\text{ J}\cdot\text{s}} = \boxed{1.0 \times 10^{15}\text{ Hz}}$$

(**b**) The frequency of a photon with $\lambda = 450$ nm $= 4.5 \times 10^{-7}$ m is

$$f = \frac{c}{\lambda} = \frac{3.00 \times 10^8 \text{ m/s}}{4.5 \times 10^{-7} \text{ m}} = 6.7 \times 10^{14} \text{ Hz}$$

This frequency is smaller than f_c from part (**a**), so this photon will not have enough energy to eject an electron.

► ***What does it mean?***

To eject an electron requires a photon with a higher frequency (and hence shorter wavelength) than the blue photon considered here. So, only ultraviolet light can eject electrons from aluminum.

CONCEPT CHECK 28.2 The Photoelectric Effect

A photoelectric effect experiment finds that the minimum photon frequency required to eject an electron from a metal is half that found for copper. What is the work function of the metal, (a) 4.4 eV, (b) 3.5 eV, (c) 2.2 eV, or (d) 8.8 eV?

EXAMPLE 28.3 Photons and Chemical Reactions

Light plays a role in ***photochemical reactions***. For example, the absorption of light can cause a hydrogen atom to dissociate, a process in which the atom absorbs a single photon and the electron is ejected. If the energy required for dissociation is 13.6 eV, what is the smallest photon frequency that can cause this reaction?

RECOGNIZE THE PRINCIPLE

This example is another application of the relation between photon frequency and energy. The minimum energy required for dissociation is called the ionization energy, which for hydrogen is $E_{\text{ionization}} = 13.6$ eV. To dissociate a hydrogen atom, the energy of an incoming photon must be equal to or greater than this value.

SKETCH THE PROBLEM

Figure 28.6 shows the problem.

IDENTIFY THE RELATIONSHIP

The photon frequency must satisfy the condition

$$E_{\text{photon}} = hf \geq E_{\text{ionization}} \tag{1}$$

We know $E_{\text{ionization}}$, so we can solve for the minimum photon frequency.

SOLVE

The lowest photon frequency that satisfies Equation (1) is

$$f = \frac{E_{\text{ionization}}}{h}$$

We are given the value $E_{\text{ionization}} = 13.6$ eV. Converting to units of joules, we use the conversion factor listed on the page facing the inside front cover and get $E_{\text{ionization}} = (13.6 \text{ eV})(1.60 \times 10^{-19} \text{ J/eV}) = 2.2 \times 10^{-18}$ J, which leads to

$$f = \frac{E_{\text{ionization}}}{h} = \frac{2.2 \times 10^{-18} \text{ J}}{6.63 \times 10^{-34} \text{ J} \cdot \text{s}} = \boxed{3.3 \times 10^{15} \text{ Hz}}$$

This frequency lies in the ultraviolet range and is not visible to the eye.

(continued) ►

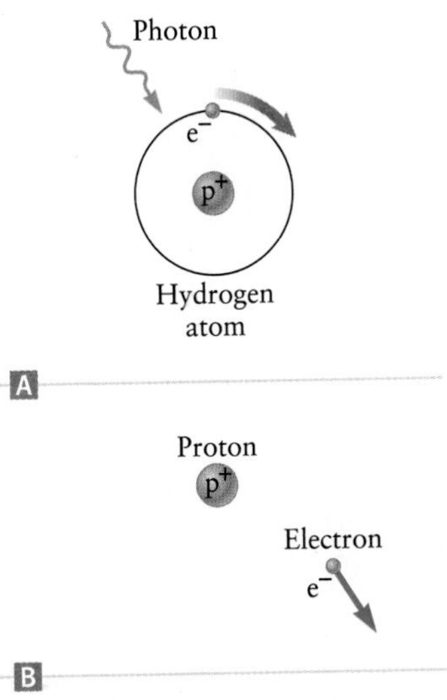

▲ **Figure 28.6** Example 28.3. Ionization of a (highly schematic) hydrogen atom. See Chapter 29 for a more accurate description of atomic structure. **A** Incoming photon. **B** The atom's proton and electron have dissociated.

▸ What does it mean?

Photochemical reactions are quite common in nature. In photosynthesis, for example, the absorption of a photon initiates a chain of reactions that leads to the storage of a portion of the photon's energy within a cell.

EXAMPLE 28.4 Photons from Your Cell Phone

Different cellular telephones use different frequencies in their communication with their networks. Many phones use a frequency near $f = 1.9$ GHz. What is the energy of a photon generated by such a cell phone?

RECOGNIZE THE PRINCIPLE

This problem is an application of the relation between frequency and photon energy. The answer has some interesting implications for the health of cell phone users.

SKETCH THE PROBLEM

No figure is necessary.

IDENTIFY THE RELATIONSHIPS AND SOLVE

The energy of a photon with frequency f is $E_{\text{photon}} = hf$ (Eq. 28.1). Inserting the given value of the frequency, we find

$$E_{\text{photon}} = hf = (6.63 \times 10^{-34}\ \text{J} \cdot \text{s})(1.9 \times 10^{9}\text{Hz}) = \boxed{1.3 \times 10^{-24}\ \text{J}}$$

▸ What does it mean?

It is interesting to express this photon energy in units of eV. We find (see the conversion in Example 28.3) $E_{\text{photon}} \approx 8 \times 10^{-6}$ eV, which is *much* smaller than the energy required to break atomic bonds or ionize atoms (Example 28.3). This answer suggests that the radiation from cell phones should not be a health hazard. On the other hand, some medical studies suggest that this radiation may be dangerous. This important question is still being studied.

Blackbody Radiation

Planck's constant has an important role in Equation 28.1 and the photon theory of light. This constant was introduced by Max Planck in 1901 when he was studying the problem of ***blackbody radiation***. Section 14.8 introduced the concept of a blackbody and described how the absorption and emission of light by a blackbody depends on its temperature. As an example, the oven in Figure 28.7A emits blackbody radiation over a range of wavelengths (λ) and frequencies as sketched in Figure 28.7B. To the eye, the color of the cavity is determined by the wavelength at which the radiation intensity is largest.[2] This wavelength is denoted as $\lambda_{\max}$ in Figure 28.7B and is determined by the temperature T of the blackbody through Wien's law (Eq. 14.14),

$$\lambda_{\max} = \frac{2.90 \times 10^{-3}\ \text{m} \cdot \text{K}}{T} \tag{28.7}$$

Insight 28.1

DETECTING SINGLE PHOTONS

Although the energy carried by a photon of visible light is small (see Eq. 28.6), it is nevertheless possible to detect single photons. Since the energy carried by a photon increases as the frequency increases, it is easiest to detect single photons of high-frequency radiation such as visible light, X-rays, and gamma rays. It is much more challenging to detect individual infrared, microwave, and radio frequency photons because they carry much less energy.

[2] In Figure 28.7B, $\lambda_{\max}$ falls in the visible range so it determines the perceived color. In some cases $\lambda_{\max}$ falls outside the visible range. For the oven in Figure 28.7A, $\lambda_{\max}$ is in the infrared; in that case the oven appears red because most of the blackbody intensity within the visible range is in the red portion of the spectrum.

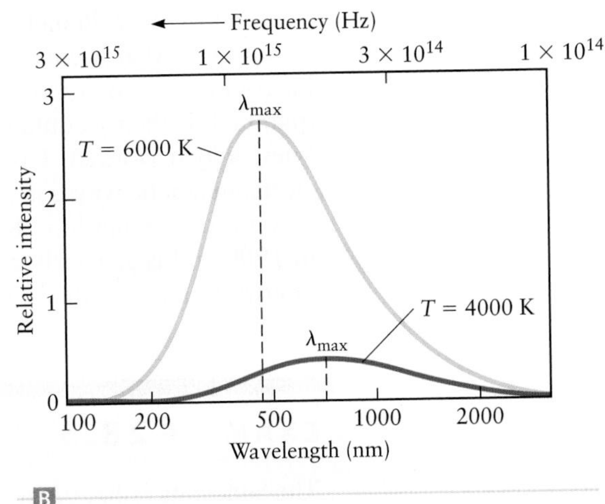

▲ **Figure 28.7** A This oven is an approximate blackbody. Light emitted from an ideal blackbody follows the blackbody spectrum in B. The wavelength λ_{max} at which the blackbody radiation intensity is largest depends on the temperature of the blackbody.

Here T is measured in kelvins and the numerical constant in Equation 28.7 is a combination of several fundamental constants. According to Wien's law, the value of λ_{max} and hence the color of the blackbody depend on the temperature. A hotter object (higher T) has a smaller value of λ_{max}, and the entire blackbody curve in Figure 28.7B shifts to shorter wavelengths when the temperature is increased. A flame that appears blue is therefore hotter than one that is red.

Experiments prior to Planck's work showed that the blackbody intensity falls to zero at both long and short wavelengths, corresponding to low and high frequencies, with a peak in the middle at the location given by Wien's law (Eq. 28.7). Planck tried to explain this behavior. At that time, physicists knew that electromagnetic waves form standing waves as they reflect back and forth inside an oven's cavity. Such standing waves are just like the standing waves we discussed in Chapter 12, where we found that standing waves on a string have frequencies following the pattern $f_n = nf_1$, where f_1 is the fundamental frequency and $n = 1, 2, 3, \ldots$. Standing electromagnetic waves in a cavity follow the same mathematical pattern. There is no limit to the value of n (as long as it is an integer), so the frequency f_n of a standing electromagnetic wave in a blackbody cavity can be infinitely large. According to classical physics, each of these standing waves carries energy, and as their frequency increases, so does their energy. As a result, the classical theory predicts that the blackbody intensity should become *infinite* as the frequency approaches infinite values!

Planck's Hypothesis and Quanta of Electromagnetic Radiation

It was obvious to Planck and other physicists of the time that any theory predicting that the intensity of blackbody radiation is infinite could not possibly be correct. Such a theory would be in conflict with the experimental intensity curves in Figure 28.7B because the true blackbody intensity falls to zero at high frequencies and would also violate common sense. It is very hard to imagine how all objects could emit an infinite amount of energy and still be consistent with our ideas about conservation of energy. Despite much effort, however, physicists were not able reconcile classical theory with the observed blackbody behavior. This disagreement was called the "ultraviolet catastrophe" since the predicted infinite intensity is found at high frequencies and high-frequency visible light falls at the ultraviolet end of the electromagnetic spectrum.

Planck proposed to resolve this catastrophe by assuming the energy in a blackbody cavity must come in discrete parcels (quanta), with each parcel having an energy $E = hf_n$, where f_n is one of the standing wave frequencies and h is the constant given in Equation 28.2. Planck showed that this assumption fixes the theory of blackbody radiation so that it correctly produces the blackbody spectrum in Figure 28.7B, but he could give no reason or justification for his assumption about standing wave quanta. His theory could fit the experiments perfectly, but no one (including Planck) knew *why* it worked. That question was answered in part by Einstein: the standing electromagnetic waves in a blackbody cavity consist of photons with quantized energies given by Equation 28.1. Planck's theory of blackbody radiation was developed in 1900 and is generally cited as the beginning of quantum theory, giving Planck an important place in the history of physics.

EXAMPLE 28.5 Temperature of the Sun

The Sun emits radiation over a wide range of wavelengths, following an approximate blackbody curve. To the eye, the Sun appears yellow, and yellow light has a wavelength near $\lambda_{max} = 5.5 \times 10^{-7}$ m (550 nm). What is the approximate temperature of the Sun?

RECOGNIZE THE PRINCIPLE

The color of a blackbody, such as the Sun, depends on its temperature. The color is determined by the wavelength at which the intensity has its maximum, which is the wavelength λ_{max} given by Wien's law.

SKETCH THE PROBLEM

Figure 28.7B gives the blackbody spectrum at two temperatures, and shows how this spectrum shifts to shorter wavelengths as the temperature of the blackbody increases.

IDENTIFY THE RELATIONSHIPS

According to Wien's law (Eq. 28.7), the location of the intensity peak is related to the temperature of the blackbody by

$$\lambda_{max} = \frac{2.90 \times 10^{-3}\ \text{m} \cdot \text{K}}{T}$$

We know λ_{max} for the Sun, so we can find its temperature T.

SOLVE

Rearranging Wien's law to solve for the temperature gives

$$T = \frac{2.90 \times 10^{-3}\ \text{m} \cdot \text{K}}{\lambda_{max}}$$

Inserting the observed value for λ_{max}, we find

$$T = \frac{2.90 \times 10^{-3}\ \text{m} \cdot \text{K}}{5.5 \times 10^{-7}\ \text{m}} = 5.3 \times 10^3\ \text{K} = \boxed{5300\ \text{K}}$$

▶ *What does it mean?*

The temperature of any star can be determined from its blackbody spectrum (and value of λ_{max}). Stars with a reddish color are thus cooler than the Sun, whereas stars that have a blue color are hotter.

CONCEPT CHECK 28.3 Ⓧ Blackbody Radiation from the Body

The human ear emits blackbody radiation. At what wavelength is the intensity of blackbody radiation from the ear largest, (a) 0.93 m, (b) 3.2×10^{-5} m, (c) 9.1×10^{-6} m, or (d) zero? *Hint*: The temperature of the human body is approximately 320 K.

Particle-Wave Nature of Light

We have discussed two phenomena, the photoelectric effect and blackbody radiation, that can only be understood in terms of the particle nature of light. While light thus has some properties like those of a classical particle, it also has wave properties *at the same time.* Electromagnetic radiation still exhibits some of the characteristic properties of classical waves, including interference. Hence, light has both wave-like and particle-like properties. In Figure 28.2, we described a double-slit experiment with electrons and described how the electrons arrive one at a time at the screen. Precisely the same behavior is found with light. Consider a double-slit experiment performed with light at very low intensity and suppose the screen responds to the arrival of individual photons by emitting light from the spot where the photon strikes. (The properties of this ***fluorescent*** screen are explained in Chapter 29.) The results would then appear just as in Figure 28.2, with the full interference pattern at the far right becoming visible only after many photons have reached the screen, thus demonstrating both the particle nature of light (the arrival of individual photons) and the wave nature (interference) in the same experiment.

28.3 Wavelike Properties of Classical Particles

In Section 28.1, we described in a general way the wavelike properties of electrons and introduced the concept of "particle-waves." The notion that the properties of both classical waves and classical particles are present at the same time is also called ***wave–particle duality***. Let's now discuss what wave–particle duality means in more quantitative terms. By the early 1920s, the photon theory of light was well established, but physicists were still struggling with how to describe particles such as electrons in the quantum regime. In particular, physicists did not yet realize that electrons are capable of the interference behavior sketched in Figure 28.2. This possibility was first proposed by Louis de Broglie in 1924, who suggested that *all* classical particles have wavelike properties. At the time, this idea must have seemed crazy because the experimental evidence for interference with electrons (similar to Fig. 28.2) had not yet been discovered. That did not stop de Broglie, however, who developed his theory in analogy with the behavior of photons. We have seen that a photon has a momentum given by (Eq. 28.5)

$$p = \frac{h}{\lambda}$$

De Broglie turned this result around and suggested that if a particle has a momentum p, its wavelength is

$$\lambda = \frac{h}{p} \tag{28.8}$$

If a particle has a wavelength, it should be able to exhibit interference just as waves do, so the test of de Broglie's hypothesis was to look for interference involving classical particles. Such an experiment is easiest if the wavelength is long, and according to Equation 28.8, a long wavelength implies a small value of momentum p. For a classical particle, $p = mv$, so a particle with a very small mass is required, and the lightest known particle (at that time) was an electron. It is thus not surprising that the first observation of wavelike behavior with a classical particle came from an experiment done with electrons by Clinton Davisson and Lester Germer (Fig. 28.8). A beam of electrons was aimed at a crystal target whose atoms formed a regular array (as is typical of many solids), acting as a series of slits for the electrons. Hence, it is really a "multislit" interference experiment similar to the diffraction gratings for light discussed in Chapter 25. Just as with a diffraction grating, the Davisson–Germer experiment exhibits interference when the wavelength of the

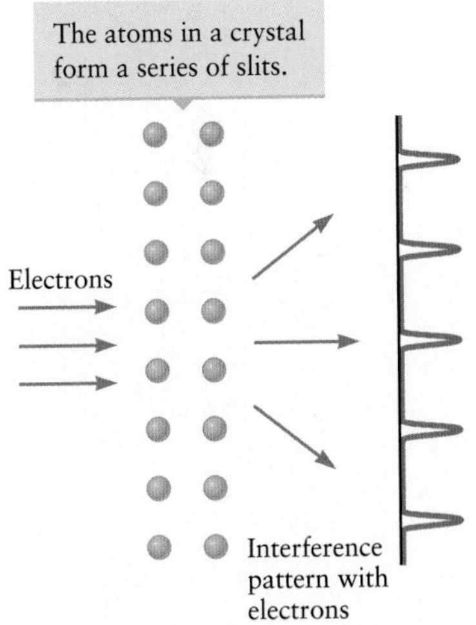

▲ **Figure 28.8** Schematic of an interference experiment with electrons, similar to the one first done by Davisson and Germer in 1927. When electrons pass through a thin crystal, the atoms in the crystal act as series of slits, producing interference and demonstrating the wave nature of electrons. Davisson and Germer's studies actually employed reflected electrons, which also form an interference pattern.

electrons is comparable to the spacing between the slits (i.e., the spacing between atoms). In Figure 28.8, we sketch the interference of electrons that pass through a crystal. Davisson and Germer actually looked at the electrons reflected by a crystal, but the basic idea is the same. There is interference in both cases, but the reflection experiment is simpler to carry out.

Davisson and Germer's experiment showed conclusively that electrons have wave-like properties with a wavelength given by de Broglie's theory (Eq. 28.8).

EXAMPLE 28.6 Wavelength of an Electron

In their studies of interference with electrons, Davisson and Germer used electrons with a kinetic energy of about $KE = 50 \text{ eV} = 8.0 \times 10^{-18}$ J. (**a**) What was the wavelength of these electrons? Express your answer in nanometers (1 nm = 10^{-9} m). (**b**) The spacing between atoms in a typical crystal is about 0.3 nm. How does this spacing compare with the wavelength of the electrons used by Davisson and Germer?

RECOGNIZE THE PRINCIPLE

Given the kinetic energy, we can find the speed of the electrons, and from that, we can get their momentum ($p = mv$). We then use the de Broglie relation (Eq. 28.8) for the wavelength of a particle-wave to calculate the electron wavelength.

SKETCH THE PROBLEM

No sketch is necessary.

IDENTIFY THE RELATIONSHIPS

The kinetic energy of an electron of mass m and speed v is $KE = \frac{1}{2}mv^2$. Rearranging to solve for v gives

$$v = \sqrt{\frac{2(KE)}{m}}$$

SOLVE

(**a**) Inserting the mass of an electron and the given value for the kinetic energy, we find

$$v = \sqrt{\frac{2(KE)}{m}} = \sqrt{\frac{2(8.0 \times 10^{-18} \text{ J})}{9.11 \times 10^{-31} \text{ kg}}} = 4.2 \times 10^6 \text{ m/s} \quad (1)$$

The wavelength of an electron with this speed is

$$\lambda = \frac{h}{p} = \frac{h}{mv}$$

Inserting our result for v gives

$$\lambda = \frac{h}{mv} = \frac{6.63 \times 10^{-34} \text{ J}\cdot\text{s}}{(9.11 \times 10^{-31} \text{ kg})(4.2 \times 10^6 \text{ m/s})} = 1.7 \times 10^{-10} \text{ m} = \boxed{0.17 \text{ nm}}$$

(**b**) This wavelength is similar to the spacing between atoms in a solid (which is typically 0.3 nm). Hence, the conditions for observing interference are indeed satisfied, which is why the Davisson–Germer experiment was successful.

▶ *What does it mean?*

The natural "slits" formed by atoms in a crystal (Fig. 28.8) are very useful for studying interference effects with particles such as electrons, neutrons, and even atoms. Notice also that for electrons the speed found in Equation (1) is low enough that the classical (Newton's law) relations for the kinetic energy and momentum are accurate, and the relativistic results from Chapter 27 are not required.

Why Don't You See Wave Interference with Baseballs?

De Broglie's ideas were first confirmed with electrons, but he proposed that they apply to all classical particles. Equation 28.8 does indeed hold for all particles, including electrons, protons, and even baseballs, all of which exhibit wavelike properties under suitable conditions. Davisson and Germer demonstrated this with electrons, but the experiment is more difficult with other particles for the following reason. Consider a particle (such as an electron or proton) with mass m, speed v, and kinetic energy KE. Using the classical expression for the kinetic energy, we have $KE = \frac{1}{2}mv^2$. Rearranging, we can solve for v:

$$v = \sqrt{\frac{2(KE)}{m}}$$

The momentum of the particle is then

$$p = mv = m\sqrt{\frac{2(KE)}{m}} = \sqrt{2m(KE)}$$

We can use this momentum in de Broglie's relation (Eq. 28.8) to get

$$\lambda = \frac{h}{p} = \frac{h}{\sqrt{2m(KE)}} \tag{28.9}$$

This result shows that for a fixed value of the kinetic energy, the wavelength becomes smaller as the mass of the particle is increased. The mass of a proton is about 1830 times greater than the mass of an electron, so assuming the same kinetic energy, the proton's wavelength will be shorter than that of an electron by a factor of approximately $\sqrt{1830} \approx 43$.

Will an interference experiment work with the much shorter wavelength of such protons? For a diffraction grating (Chapter 25), the condition for constructive interference is

$$d \sin \theta = n\lambda$$

where d is the slit spacing, θ is the diffraction angle, and $n = 0, 1, 2, \ldots$ is an integer. Had Davisson and Germer used protons, λ would have been much smaller (by a factor of 43); this small size could be compensated for by increasing n by the same factor, which corresponds to the 43rd maximum in an interference pattern such as Figure 28.2. In other words, with protons there would be 43 diffraction maxima squeezed in the same angle as one of the diffraction peaks found with electrons. For this reason, it is more difficult, though not impossible, to observe interference with protons.

In general, it becomes more and more difficult to observe interference as the mass of the particle is increased because if all else is kept fixed, a larger mass leads to a shorter wavelength (Eq. 28.9). In principle, one could observe interference with baseballs, but this experiment has not yet been done. (See Example 28.7.) This should remind you of the difference between geometrical optics and wave optics. Geometrical optics (Chapter 24) applies when the wavelength is very short, and in this regime light moves in straight lines just like classical particles. On the other hand, wave optics (Chapter 25) applies when the wavelength is relatively long and comparable to the size of any slits or openings, and in that case interference effects are important.

CONCEPT CHECK 28.4 The de Broglie Wavelength and the Mass of a Particle

An electron and a proton have the same de Broglie wavelength. Which of the following statements is true?

(a) The electron and proton have the same kinetic energy.
(b) The electron's kinetic energy must be larger than the proton's kinetic energy.
(c) The electron's kinetic energy must be smaller than the proton's kinetic energy.

CONCEPT CHECK 28.5 Wavelengths of Electrons and Protons

An electron and a proton have the same speed. Which has the longer de Broglie wavelength?

(a) The electron (b) The proton (c) They have the same wavelength.

EXAMPLE 28.7 De Broglie's Theory and the Wavelength of a Baseball

What is the wavelength for a baseball traveling at a speed of 100 mi/h (approximately 45 m/s)? This speed is attained by some major league pitchers. The mass of a baseball is approximately 0.14 kg.

RECOGNIZE THE PRINCIPLE

De Broglie's theory connects the momentum and wavelength of a particle-wave. The momentum of a baseball is $p = mv$, so given the mass and speed we can find its momentum. The wavelength of the baseball is then given by the de Broglie relation.

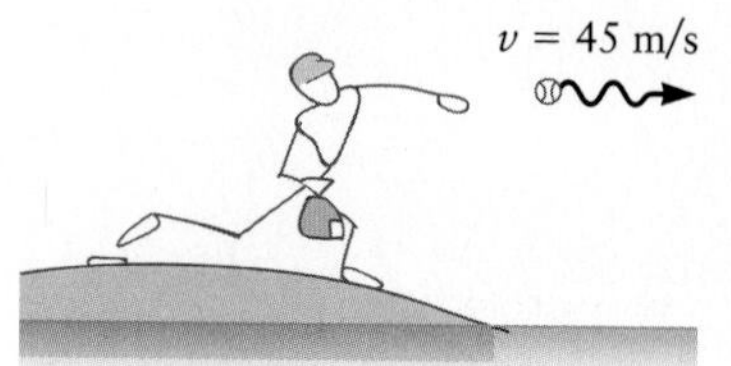

▲ **Figure 28.9** Example 28.7. A baseball is a particle-wave, but its wavelength is *very* short as given by the de Broglie relation.

SKETCH THE PROBLEM

Figure 28.9 describes the problem. While a baseball is a particle-wave, its wavelength is *very* short.

IDENTIFY THE RELATIONSHIPS AND SOLVE

Combining de Broglie's relation for the wavelength of a particle-wave (Eq. 28.8) with $p = mv$ leads to

$$\lambda = \frac{h}{p} = \frac{h}{mv}$$

Using the given values of m and v for the baseball, we find

$$\lambda = \frac{h}{mv} = \frac{6.63 \times 10^{-34}\ \text{J} \cdot \text{s}}{(0.14\ \text{kg})(45\ \text{m/s})} = \boxed{1.0 \times 10^{-34}\ \text{m}}$$

▶ ***What does it mean?***

This wavelength is *extremely* short. For comparison, the diameter of a proton is about 10^{-15} m, so the wavelength of a baseball is smaller than the size of a proton by the enormous factor of 10^{19}. Extremely small values of the wavelength are found for other macroscopic objects such as hockey pucks, planets, and people. For this reason, the wave properties of macroscopic objects are almost always completely negligible and the particle description of Newton's mechanics works with very high accuracy. The wavelike behavior of a baseball and of most other macroscopic objects will probably never be observed.

28.4 Electron Spin

We have seen that electrons behave in some ways like classical particles and in other ways like classical waves. Are there any "new" or "extra" properties that show up in the quantum regime? That is, do electrons have any properties not found in either classical particles or waves? The answer is yes: electrons have another quantum property that involves their magnetic behavior. In Section 20.8 (page 610), we mentioned that an electron has a magnetic moment, a property associated with the phenomenon of ***electron spin.*** Roughly speaking, the electron's magnetic moment can be understood by picturing the electron as a spinning ball of charge. This spinning ball of charge acts as a collection of current loops, producing a magnetic field and thus acting like a small bar magnet as shown in Figure 28.10B.

Since an electron behaves as a small bar magnet, it is attracted to or repelled from the poles of other magnets. The magnetic properties of an electron were revealed in an experiment first performed by Otto Stern and Walther Gerlach. A simplified version of their experiment is sketched in Figure 28.11, showing a beam of electrons passing near one pole of a large (macrosize) magnet. The electrons are attracted to or repelled from the large magnet according to the direction of the electron's north and south "poles." Two examples are shown in Figure 28.11; one electron is attracted to the large magnet's north pole and is thus deflected upward, and the other electron is repelled and deflected downward. The surprise in this experiment is that the outgoing electrons form just two beams. If an electron behaves like a bar magnet, a line connecting the north and south poles defines the axis of the magnet. These axes are indicated by arrows in Figure 28.11, with the arrowheads at the north ends of the magnets. These arrows also denote the direction of the spin magnetic moment of the electron, which is similar to the magnetic moment of a bar magnet or current loop as discussed in Chapter 20.

Classical theory predicts that the magnetic moment of an electron may point in any direction (up, down, sideways, or at any angle), and if that is true, the electrons in a Stern–Gerlach experiment should be deflected not just as two beams, but over a range of angles. However, experiments always give just two outgoing beams of electrons (Fig. 28.11). Hence, two and only two orientations for the electron magnetic moment with respect to the direction of the applied magnetic field are possible. In words, the electron magnetic moment is ***quantized***, with only two possible values. The Stern–Gerlach experiment in Figure 28.11 shows that the quantization of the electron's magnetic moment applies to both the ***direction*** and ***magnitude*** of its magnetic moment.

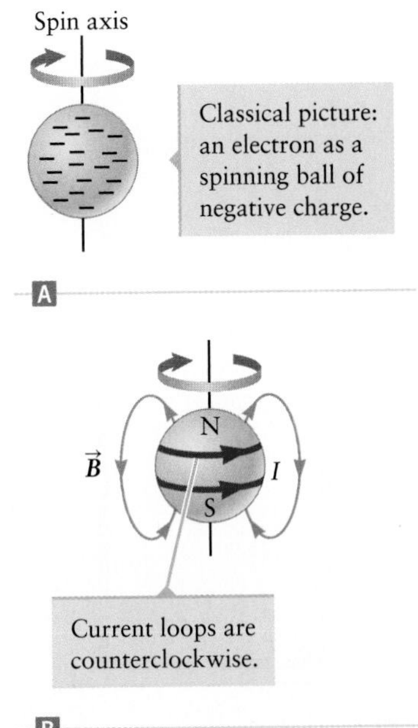

▲ **Figure 28.10** A Classical picture of the electron's spin. B A spinning ball of charge (the electron) acts like a collection of current loops, producing a magnetic field.

Quantization of Electron Spin

To understand the implications of the Stern–Gerlach experiment, we consider again the picture in Figure 28.10 showing how an electron can act as a small bar magnet. Classically, we can picture an electron as a spinning ball of electric charge and acting as a collection of current loops. The electron in Figure 28.10A spins clockwise as viewed from the top of the figure, so the circulating charge acts as a counterclockwise current loop as viewed from above (because the electron is negatively charged). These current loops then produce a magnetic field in the same way that the current loops in Section 20.1 produce a magnetic field, and the result is called the ***spin magnetic moment*** of the electron. We also say that the electron has ***spin angular momentum***. This angular momentum is similar to the angular momentum of a macroscopic spinning ball.

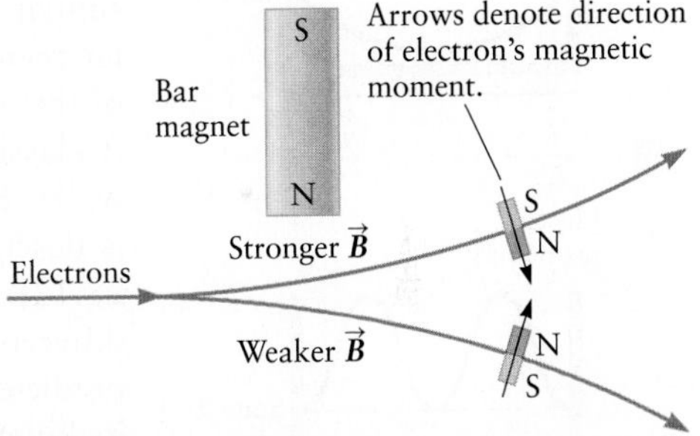

▲ **Figure 28.11** When a beam of electrons passes near one end of a bar magnet, it is found that the spin magnetic moment is either "up" or "down," so the electrons are either attracted to or repelled from the bar magnet. According to quantum theory, the electron's magnetic moment can only point in these two directions, up or down, but *not* "sideways."

This classical picture explains how the electron can have a magnetic moment, but it does *not* explain the Stern–Gerlach experiment. The rotation axis of a classical particle such as a spinning baseball can point in any direction, so the classical picture in Figure 28.10 suggests that an electron's magnetic moment can be oriented in any direction. This classical picture also predicts that the electron's magnetic moment can be large, small, or even zero, depending on the angular speed of the electron since a classical ball can spin fast, slow, or not at all. However, the Stern–Gerlach result indicates that there can be only two values for the electron magnetic moment! Physicists now understand that this property is a result of quantum theory: the electron's spin axis and the magnitude of its magnetic moment are both quantized, with only two possible values. These two states are called spin "up" and spin "down." In the Stern–Gerlach experiment, one of the outgoing beams of electrons in Figure 28.11 corresponds to electrons with spin up, whereas electrons in the other beam have spin down.

Other quantum particles, such as protons and neutrons, also have a spin angular momentum and a resulting magnetic moment. In all cases, they are quantized, although the details can be more complicated than for an electron.

The quantization of spin angular momentum and the spin magnetic moment is a purely quantum effect, and the simple classical model in Figure 28.10 gives only an intuitive picture. There is no classical explanation for why spin is quantized (i.e., why there are only two outgoing beams in the Stern–Gerlach experiment). All we can say is that is the way electrons work. Recognizing the existence of electron spin is essential for understanding the structure of atoms and the organization of the periodic table of the elements (Chapter 29).

28.5 The Meaning of the Wave Function

Newton's laws of mechanics are based on the idea that the motion of a particle can be described completely in terms of the particle's position, velocity, and acceleration. After all our work with Newton's mechanics, it is easy to take this notion for granted, but it is only an assumption about how the laws of physics are structured. This assumption works well in the classical world as proven by the outstanding success of Newton's laws in describing the mechanics of macroscale objects, but it fails in the quantum world of particle-waves. In a quantum description, the motion of a particle-wave is described by its ***wave function***. The full mathematics of the wave function is beyond the scope of this book, but we can describe a few of its properties.

The wave function can be calculated using an equation developed by Erwin Schrödinger. He was one of the inventors of quantum theory and in 1933 received the Nobel Prize in Physics for this work. Schrödinger's equation plays a role similar to Newton's laws of motion. In quantum theory, one uses the Schrödinger equation to find the wave function and how it varies with time, in much the same way that one uses Newton's laws of mechanics to find the position, velocity, and acceleration of a particle.

In many situations, the solutions of the Schrödinger equation are mathematically similar to standing waves. As an example, consider an electron confined to a particular region of space as sketched in Figure 28.12A. In classical terms, you can think of this region as a very deep canyon or box from which the electron cannot escape. A classical particle would travel back and forth inside the box, bouncing from the walls. The wave function for a particle-wave (such as an electron) in such a box is described by standing waves, similar to the standing waves on a string. Figure 28.12B shows two possible wave function solutions corresponding to electrons with different kinetic energies. The wavelengths of these standing waves are different, as predicted by the de Broglie picture, since the wavelength of an electron depends on its kinetic energy (Eq. 28.9). In a classical wave-on-a-string picture, these two solutions correspond to two standing waves with different wavelengths. (See Fig. 12.24.)

After finding the wave function for a particular situation such as for the electron in Figure 28.12B, one can calculate the position and velocity of the electron. However, the results do not give a simple single value for x. Instead, the wave function can be used to calculate the *probability* of finding the electron at different locations in space. This probability for each of the wave functions in Figure 28.12B is shown in Figure 28.12C for our electron in a box. The probability of finding the electron at certain values of x is large in some regions and small in others, corresponding to the antinodes and nodes of the standing wave. The probability distribution (how the probability varies with position in the box) is different for each wave function.

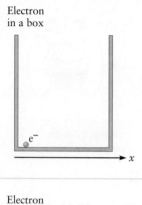

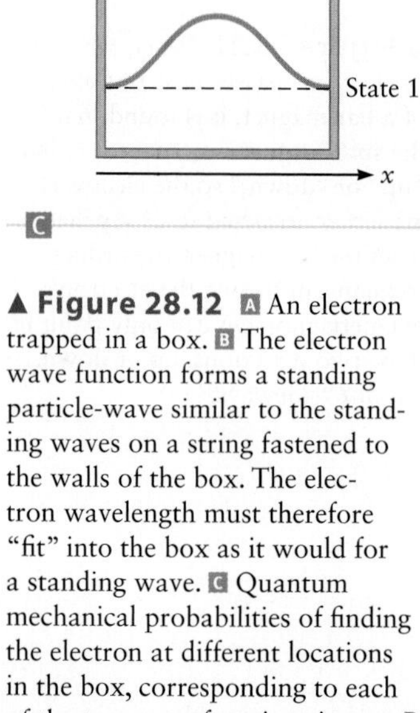

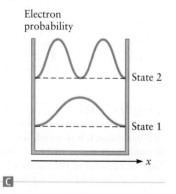

▲ **Figure 28.12** A An electron trapped in a box. B The electron wave function forms a standing particle-wave similar to the standing waves on a string fastened to the walls of the box. The electron wavelength must therefore "fit" into the box as it would for a standing wave. C Quantum mechanical probabilities of finding the electron at different locations in the box, corresponding to each of the two wave functions in part B.

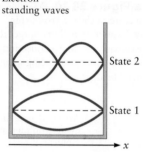

> CONCEPT CHECK 28.6 Energy of an Electron in a Box
> Figure 28.12B shows the wave functions for two different electron standing waves. Which of these electrons has the higher kinetic energy?

The Heisenberg Uncertainty Principle

In classical physics, we are used to calculating or measuring the position and velocity of a particle with great precision, but for a particle-wave, quantum effects place fundamental limits on this precision. For example, the electron in Figure 28.12 is described by a standing wave. The associated probability of finding the electron at different places in the box can be calculated from its wave function, yet what is the position of this electron? In a quantum sense, the standing waves in Figure 28.12B *are* the electron, so there is an inherent uncertainty in its position. There is some probability for finding the electron at virtually any spot in the box, and the uncertainty Δx in the electron's position is approximately the size of the box.

We can gain insight into the uncertainty Δx from the experiment described in Figure 28.13, in which electrons are incident from the left on a narrow slit. This is very similar to the case of single-slit diffraction we studied with light waves in Section 25.6, and the electron wave is diffracted as it passes through the slit. In quantum terms, the diffraction pattern gives a measure of how the wave function of the electron is distributed throughout space after it passes through the slit. The electrons coming in from the left in Figure 28.13 are all traveling parallel to the y axis, so their initial momentum is purely along the y direction. However, after being diffracted by the slit, an electron is likely to strike the screen some distance away from the y axis. Hence, a diffracted electron acquires a nonzero momentum along x.

Figure 28.13 also shows how the width Δx of the slit affects the interference pattern. The wide slit in Figure 28.13A produces a narrow electron distribution along the screen (the x direction in the figure). The narrow slit in Figure 28.13B produces a broader distribution as the wave function is more spread out along x.

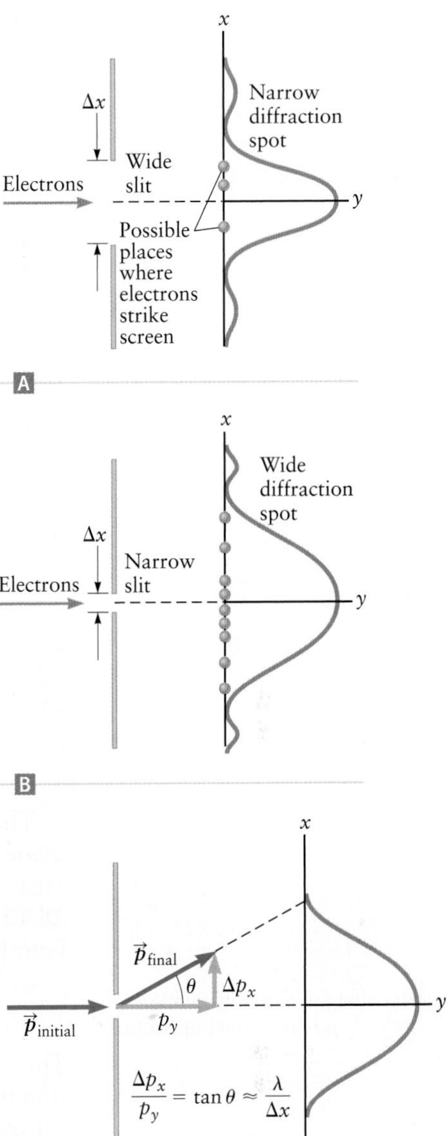

▲ **Figure 28.13** Single-slit diffraction with electrons. **A** If an electron passes through a wide slit, the diffraction spot on the screen is narrow. **B** If the slit is made narrower, the diffraction spot becomes wider. **C** The spreading of an electron wave as it passes through a slit gives a way to relate the uncertainty Δx in the electron's position to the uncertainty Δp in its momentum.

Relating the Uncertainties in Position (Δx) and Momentum (Δp)

The electrons pass through the slit, so the position of the electron at the moment it passes through the slit is known with an uncertainty Δx equal to the width of the slit. Since an electron may strike the screen over a range of angles defined approximately by the central diffraction maximum, the outgoing electrons have a spread in their momentum along x and we can say that there is some uncertainty Δp_x in the x component of the momentum. Most of the electrons arrive somewhere within the width of the central diffraction maximum, so the final momentum along the x direction is uncertain by an amount approximately equal to the x component of the momentum of an electron at the edge of the diffraction spot. The uncertainty in momentum Δp_x is related to the uncertainty in position Δx. A wide slit (Fig. 28.13A) has a large value of Δx and a smaller value of Δp_x (because the diffraction "spot" is relatively narrow). On the other hand, a narrow slit (Fig. 28.13B) has a small uncertainty in position Δx but a larger uncertainty in momentum Δp_x.

We can derive a quantitative relationship between Δx and Δp_x from Figure 28.13C. The diffraction angle θ is contained in a right triangle with sides p_y and Δp_x, where p_y is the momentum along y. We have

$$\frac{\Delta p_x}{p_y} = \tan\theta$$

If the angle is small, we can use the approximation $\tan\theta \approx \theta$, resulting in

$$\frac{\Delta p_x}{p_y} = \theta \tag{28.10}$$

The analysis of single-slit diffraction in Chapter 25 (Eq. 25.23) showed that for light of wavelength λ passing through a slit of width w, the first minimum in the diffracted intensity occurs at an angle

$$\theta = \frac{\lambda}{w}$$

This angle θ also describes the spread of the electron diffraction spot in Figure 28.13C, where the slit width w is equal to the uncertainty in the electron's position Δx. In the electron case, we have

$$\theta = \frac{\lambda}{\Delta x} \tag{28.11}$$

Setting the results in Equations 28.10 and 28.11 equal leads to

$$\frac{\Delta p_x}{p_y} = \frac{\lambda}{\Delta x}$$

or

$$\Delta x \, \Delta p_x = \lambda p_y \tag{28.12}$$

If we assume the uncertainty in the momentum Δp_x is small, then p_y is approximately equal to the total momentum, which we can denote by simply p. We will also follow standard notation and drop the subscript on Δp_x, referring to it as just the spread (i.e., uncertainty) in the electron's momentum. According to the de Broglie relation (Eq. 28.8), $p = h/\lambda$. Inserting that into Equation 28.12 gives

$$\Delta x \, \Delta p = \lambda p = \lambda \frac{h}{\lambda} = h$$

Thus, the uncertainty Δx in the electron's position is connected with the uncertainty Δp in its momentum in the x direction. Our analysis was only approximate; a more careful treatment gives nearly the same result,

$$\Delta x \, \Delta p = \frac{h}{4\pi} \tag{28.13}$$

The uncertainties Δx and Δp in Equation 28.13 are the absolute limits set by quantum theory. In many situations, there are additional contributions to these uncertainties resulting from experimental or measurement error. Hence, Equation 28.13 gives a lower limit on the product of Δx and Δp. We can allow for this lower limit by writing

Heisenberg uncertainty relation for position and momentum

$$\Delta x \, \Delta p \geq \frac{h}{4\pi} \tag{28.14}$$

This is called the ***Heisenberg uncertainty principle***, in this case expressed as a relation between position and momentum. We derived this relation between Δx and Δp for the special case in Figure 28.13, but it holds for *any* quantum situation and for any particle-wave (not just electrons).

Position, Momentum, and Energy of a Particle: How Accurately Can We Know Them?

In classical physics, we can in principle calculate the position and momentum of a particle to any desired accuracy using Newton's laws. Assuming there are no mistakes in our mathematics, we can find x and p to any desired *classical* accuracy. The Heisenberg uncertainty principle, however, dictates that in the quantum regime the uncertainties in x and p are connected. Under the very best of circumstances, the product of Δx and Δp is a constant, proportional to Planck's constant h. So, if you measure a particle-wave's position with great accuracy (small Δx), you must accept a large uncertainty Δp in its momentum. On the other hand, if you know the momentum very accurately (small Δp), you must accept a large position uncertainty Δx. You cannot make both uncertainties small at the same time.

The same ideas that lead to Equation 28.14 can also be used to derive a relation between the uncertainties in the energy ΔE of a particle and the time interval Δt over which this energy is measured or generated. The Heisenberg energy–time uncertainty principle is

Heisenberg uncertainty relation for energy and time

$$\Delta E \, \Delta t \geq \frac{h}{4\pi} \tag{28.15}$$

If the energy of a particle or system is measured over a time period Δt, Equation 28.15 leads to a minimum uncertainty in the measured energy. This minimum uncertainty is negligibly small for a macroscale object, but it can be important in atomic and nuclear reactions.

Philosophical and Practical Implications of the Uncertainty Principle

The Heisenberg uncertainty principle has profound implications for our physical description of the universe. Quantum theory and Heisenberg's uncertainty relation in Equation 28.14 mean that there is always a trade-off between the uncertainties (i.e., precision) in x and p. It is not possible, *even in principle*, to have perfect knowledge of both x and p, which suggests that there is always some inherent "uncertainty" in our knowledge of the physical universe. The best we can do is compute the probability for finding a particle (e.g., an electron) at a particular position or the probability that a particle will have a certain momentum. Quantum theory thus says that the world is inherently *unpredictable*, in sharp contrast to the perfect predictability of classical physics.

We should also ask what effect the uncertainty principle has on the use of Newton's mechanics. The value of Planck's constant is very small ($h = 6.63 \times 10^{-34}$ J · s). For any macroscale object such as a baseball, a planet, or even an amoeba, the uncertainties in the real measurement will always be much larger than the inherent uncertainties due to the Heisenberg uncertainty relation. (For a numerical example, see Example 28.8.) As a general rule, the Heisenberg uncertainty relations only become important on the scale of electrons, atoms, and other microscale objects.

Insight 28.2

THE HEISENBERG UNCERTAINTY PRINCIPLE AND THE THIRD LAW OF THERMODYNAMICS

According to the third law of thermodynamics (Chapter 16), it is not possible to reach the absolute zero of temperature. In a classical kinetic theory picture, the speed of all particles would be zero at absolute zero, and there is nothing in classical physics to prevent that. In quantum theory, however, the Heisenberg uncertainty principle tells us that the uncertainty in the speed of a particle cannot be zero because $\Delta x\, \Delta p \geq h/(4\pi)$. Hence, the uncertainty principle provides a justification for the third law of thermodynamics, revealing a surprising connection between thermodynamics and quantum theory.

EXAMPLE 28.8 Applying the Heisenberg Uncertainty Principle to a Brick

Consider a typical macroscale object such as a brick with $m = 1.0$ kg sitting on a table. A very careful measurement of the brick's location has an uncertainty of $\Delta x = 0.1$ nm (approximately the diameter of an atom). What is the corresponding uncertainty in the speed of the brick?

RECOGNIZE THE PRINCIPLE

This example involves a direct application of the Heisenberg uncertainty relation between the uncertainty in position and momentum. A given uncertainty Δx in the brick's position leads to a minimum uncertainty Δp in the brick's momentum. Since $p = mv$, there is a corresponding uncertainty in the brick's speed.

SKETCH THE PROBLEM

Figure 28.14 shows the problem.

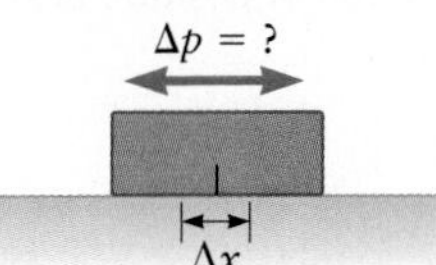

▲ **Figure 28.14** Example 28.8. Applying the Heisenberg uncertainty principle to a brick.

IDENTIFY THE RELATIONSHIPS

According to the Heisenberg uncertainty principle (Eq. 28.14),

$$\Delta x\, \Delta p \geq \frac{h}{4\pi} \tag{1}$$

Because we know Δx, we can solve for Δp. The momentum is $p = mv$ and we are given the brick's mass, so we can then find the uncertainty Δv in its speed.

SOLVE

Rearranging Equation (1) to solve for Δp gives

$$\Delta p \geq \frac{h}{4\pi(\Delta x)}$$

(continued) ▶

An uncertainty in the momentum corresponds to an uncertainty in the speed according to

$$\Delta p = m\,\Delta v \geq \frac{h}{4\pi(\Delta x)}$$

Solving for Δv and inserting the given values of m and Δx gives

$$\Delta v \geq \frac{h}{4\pi m\,\Delta x} = \frac{6.63 \times 10^{-34}\ \text{J}\cdot\text{s}}{4\pi(1.0\ \text{kg})(0.1 \times 10^{-9}\ \text{m})}$$

$$\Delta v \geq \boxed{5 \times 10^{-25}\ \text{m/s}}$$

► *What does it mean?*

This uncertainty is *extremely* small. For all practical purposes, this quantum limit on the uncertainty is so small that it could never be measured. For bricks and other macroscale objects, the limits set by the Heisenberg uncertainty principle are unobservable, but they can be important in atomic and nuclear physics experiments, as we explore in Example 28.9.

EXAMPLE 28.9 Applying the Uncertainty Principle to a Hydrogen Atom

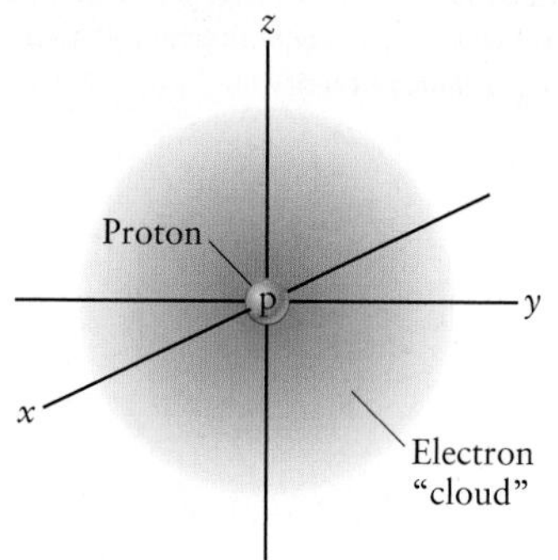

▲ **Figure 28.15** Example 28.9. Simple model of a hydrogen atom. The "cloud" represents the probability for finding the electron as a function of position near the proton.

A simple picture of the hydrogen atom has an electron moving around a proton as a sort of "electron cloud," with the density of the cloud proportional to the electron probability (Fig. 28.15). We'll see how this cloud picture is related to the electron's wave function in Chapter 29. Even without detailed knowledge of the wave function, however, we can still apply the Heisenberg uncertainty principle (Eq. 28.14), taking the uncertainty Δx in the electron's position equal to the diameter of the cloud. **(a)** If the diameter of a hydrogen atom is 1.0×10^{-10} m (= 0.10 nm), what is the minimum uncertainty Δp in the electron's momentum? **(b)** The momentum of the electron must be at least comparable to the uncertainty Δp, so we can use Δp as an estimate for the total momentum ($p \approx \Delta p$). With this approximation, use your result for Δp to find the electron's kinetic energy and compare it with the known ionization energy of a hydrogen atom, 13.6 eV.

RECOGNIZE THE PRINCIPLE

Part **(a)** is a direct application of the Heisenberg uncertainty relation between the uncertainty in position and momentum, with Δx equal to the given diameter of a hydrogen atom. For part **(b)**, we take p to be equal to the uncertainty Δp in the momentum, and from that we can get the kinetic energy of the electron.

SKETCH THE PROBLEM

Figure 28.15 describes the problem.

IDENTIFY THE RELATIONSHIPS AND SOLVE

(a) Heisenberg's uncertainty relation (Eq. 28.14) gives

$$\Delta x\,\Delta p \geq \frac{h}{4\pi}$$

Solving for Δp, we find

$$\Delta p \geq \frac{h}{4\pi\,\Delta x} = \frac{6.63 \times 10^{-34}\ \text{J}\cdot\text{s}}{4\pi(1.0 \times 10^{-10}\ \text{m})} = \boxed{5.3 \times 10^{-25}\ \text{kg}\cdot\text{m/s}}$$

(**b**) The momentum of an electron is $p = mv$, while the kinetic energy is $KE = \frac{1}{2}mv^2$. Combining these two relations gives

$$KE = \frac{p^2}{2m}$$

Using $p \approx \Delta p$ from part (**a**) and the known mass of an electron, we find

$$KE = \frac{p^2}{2m} = \frac{(5.3 \times 10^{-25}\ \text{kg} \cdot \text{m/s})^2}{2(9.11 \times 10^{-31}\ \text{kg})} = \boxed{1.5 \times 10^{-19}\ \text{J}}$$

Converting this energy to units of electron-volts (eV) gives

$$KE = (1.5 \times 10^{-19}\ \text{J})\left(\frac{1\ \text{eV}}{1.60 \times 10^{-19}\ \text{J}}\right) = \boxed{0.94\ \text{eV}}$$

▸ ***What does it mean?***

This kinetic energy is a little more than ten times smaller than the ionization energy of the hydrogen atom (13.6 eV). That these energies are roughly the same order of magnitude suggests that an accurate theory of the hydrogen atom must be based on quantum theory (the main topic of Chapter 29).

28.6 Tunneling

According to classical physics, the electron in Figure 28.12 is trapped in the box and cannot escape. However, a quantum effect called ***tunneling*** allows such an electron to escape under certain circumstances. Figure 28.16A shows the wave function for this electron, with the region near the wall of the box shown on an expanded scale. Quantum theory allows the electron's wave function to penetrate a short distance *into* the wall. We say that the wave function extends a short distance into the ***classically forbidden*** region because according to Newton's mechanics the electron must stay completely inside the box and cannot go into the wall. In most cases, the quantum penetration distance is short and of little consequence, but if two boxes are very close together so that the wall between them is very thin, the wave function can extend from the inside of one box and through the wall to the box on the other side (Fig. 28.16B). Since the wave function extends through the wall, the electron has some probability for passing through the wall. If bullets were quantum objects, it would be like a bullet passing through a wall without leaving a hole!

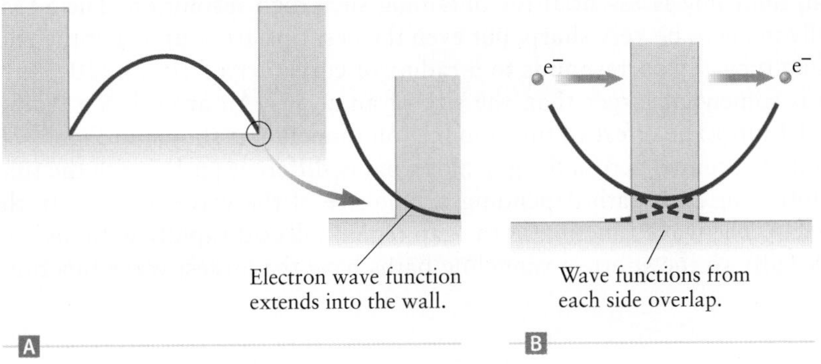
e⁻
e⁻
Electron wave function extends into the wall.
Wave functions from each side overlap.
A
B

◂ **Figure 28.16** A An electron wave function (standing wave) such as for the electron in a box in Figure 28.12B. The electron wave function actually penetrates a short distance into the walls of the box, although the wave function is very small in this region. B If the walls of two boxes are brought close together, the wave functions from the two boxes can overlap, allowing an electron to tunnel from one box to the other.

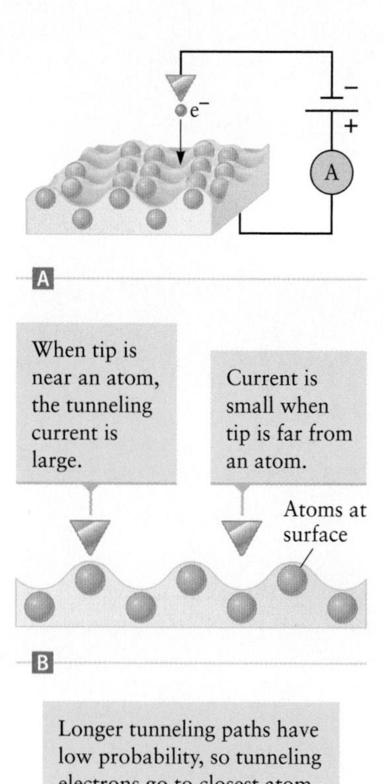
e−
−
+
A
A
When tip is near an atom, the tunneling current is large.
Current is small when tip is far from an atom.
Atoms at surface
B
Longer tunneling paths have low probability, so tunneling electrons go to closest atom.

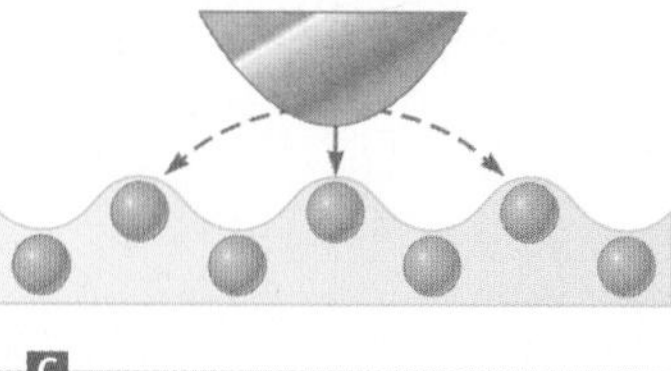
C

▲ **Figure 28.17** A In a scanning tunneling microscope (STM), a very sharp tip is brought very close to a surface, detecting the probability for an electron to tunnel between the tip and the surface by measuring the tunneling current. B The tunneling probability is largest for the shortest tunneling paths, so there is a large tunneling current when the tip is directly over an atom. C The STM is most sensitive to atoms that are directly below the tip.

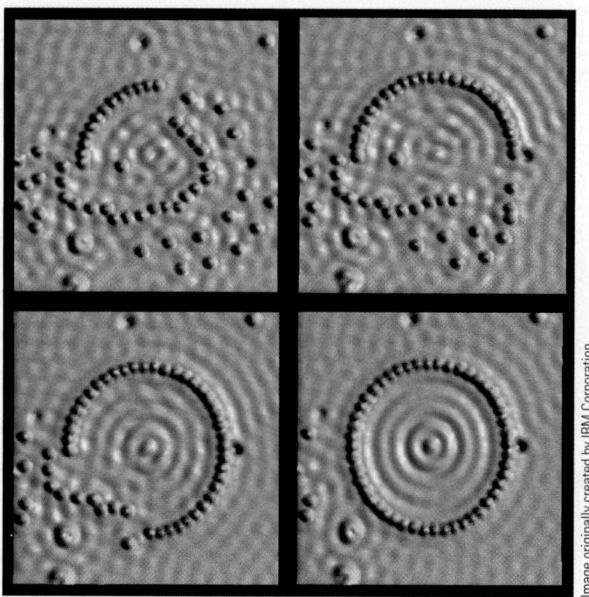

► **Figure 28.18** STM images of iron atoms on the surface of copper. The electric field from an STM tip can be used to push atoms around on the surface of a metal. In the lower right photo, the iron atoms are arranged to form a circle.

Image originally created by IBM Corporation

Using Quantum Mechanics to Make a New Kind of Microscope

Tunneling is important in certain nuclear decay processes (Chapter 30) and in the motion of certain molecules. It is also used in the operation of a ***scanning tunneling microscope*** (STM). The basic design of an STM is shown in Figure 28.17A. A very sharp metal tip is positioned near a conducting surface, and an electric potential difference is applied between the tip and the surface. If the separation is large, the space (usually a vacuum) between the tip and the surface acts as a barrier for electron flow. This barrier is similar to the wall in Figure 28.16 since it prevents electrons from leaving the metal (i.e., the box). However, if the tip is brought very close to the surface, electrons may tunnel between them, producing a tunneling current. As the tip is scanned over the surface, it moves from regions where it is directly over an atom to other regions where the nearest atoms are much farther away as in Figure 28.17B. The tunneling current is largest when the tip is closest to an atom because the wave function overlap needed for tunneling is largest in this case. An STM image is based on variations in the size of the tunneling current as the tip is scanned across the surface, with high image intensity when the current is large and vice versa. A typical STM image is shown in Figure 28.18, with each "bump" corresponding to an individual iron atom on the surface of a piece of copper. The electric field between the tip and the surface atoms can also be used to push atoms around on the surface. In this experiment the STM tip was used to arrange iron atoms that were initially scattered around the surface (upper left) into a circle (lower right).

Tunneling plays a dual role in the operation of an STM. First, the detector current is produced by tunneling, so without tunneling there would be no image. Second, tunneling is essential for obtaining such high resolution. The STM tip is specially made to be very sharp, but even the best tips are rounded at the end. This rounding typically corresponds to a radius of curvature of 10 nm (10^{-8} m) or so, which is somewhat larger than the size of an atom. (An atom has a diameter of about 0.1 nm.) The effect of tip rounding on tunneling is shown in Figure 28.17C. Electrons can tunnel across the gap along many different paths, with the tunneling probability for each path depending on the size of the wave function in the tunneling gap. The wave function in the gap region falls off rapidly with distance (see Fig. 28.16B), so the shortest tunneling paths have the largest wave functions near the tip. The vast majority of tunneling electrons follow the shortest path, reaching a very small area on the surface, an area much smaller than the tip radius. The STM can then form images of individual atoms, even though the tip is larger than they are.

28.7 Detection of Photons by the Eye

Human vision depends on the refractive properties of the eye as discussed in Chapters 24 and 26, treating light with Maxwell's wave theory, but a complete understanding of human vision also depends on the particle theory of light. Light is detected at the back surface of the eye, in a region called the retina. (Review Fig. 24.47.) The retina contains two kinds of light-sensitive cells, called ***rods*** and ***cones***. When these cells absorb light, they generate an electrical signal that travels through the optic nerve to the brain. The rod cells are more sensitive to low light intensities and are used predominantly at night, while the cone cells are responsible for color vision.

Can the Eye Detect Single Photons?

Detection of low-intensity light is the job of the rod cells. When light enters your eye, only about 10% of it actually reaches the retina. The other 90% is reflected or absorbed by the cornea and other parts of the eye. Hence, for every ten photons that enter the eye, only one (on average) strikes the retina. Experiments show that the absorption of only a single photon by a rod cell causes the cell to generate a small electrical signal. The signal of an individual rod cell is not sent directly to the brain, however. Instead, the eye combines the signals from many rod cells before passing that combination signal along the optic nerve. Measurements show that for the combination signal to register at the brain, approximately five photons must be absorbed by rod cells within a period of about 0.1 s. So, although an individual rod cell is sensitive to single photons of visible light, your eye must receive approximately 50 photons within about 0.1 s for the brain to know that light has actually arrived at the eye.

Color Vision

The retina contains three types of cone cells, which respond to light of different colors (Fig. 28.19). The brain deduces the color of light by combining the signals from all three types of cones. For example, blue light triggers a large response from the blue cone cells and much smaller responses from the green and the red cone cells. A large signal from the blue cones together with small signals from the green and red cones tell the brain that the incident light lies in the blue part of the spectrum.

The key property of the cones is that each type of cone cell is most sensitive to light with a particular color—that is, frequency—independent of the light intensity. For example, the green cone cells are most sensitive to green light and less sensitive to both red light (a lower frequency) and blue light (a higher frequency), even if the red or the blue light is very intense.

The correct explanation of color vision relies on two aspects of quantum theory. First, light arrives at the eye as photons whose energy depends on the frequency of the light. According to Equation 28.1, a photon of blue light carries more energy than a photon of green light, which in turn carries more energy than a photon of red light. When an individual photon is absorbed by a cone cell, the energy of the photon is taken up by a pigment molecule within the cell. Second, the energy of a pigment

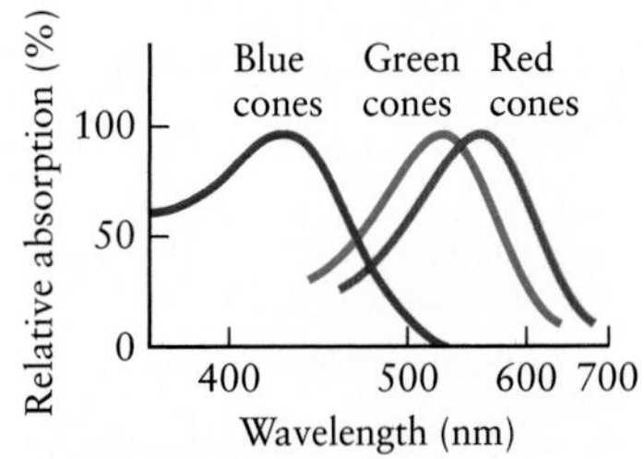

▲ **Figure 28.19** Pigment molecules inside the three types of cone cells absorb light preferentially at certain wavelengths (i.e., at certain frequencies).

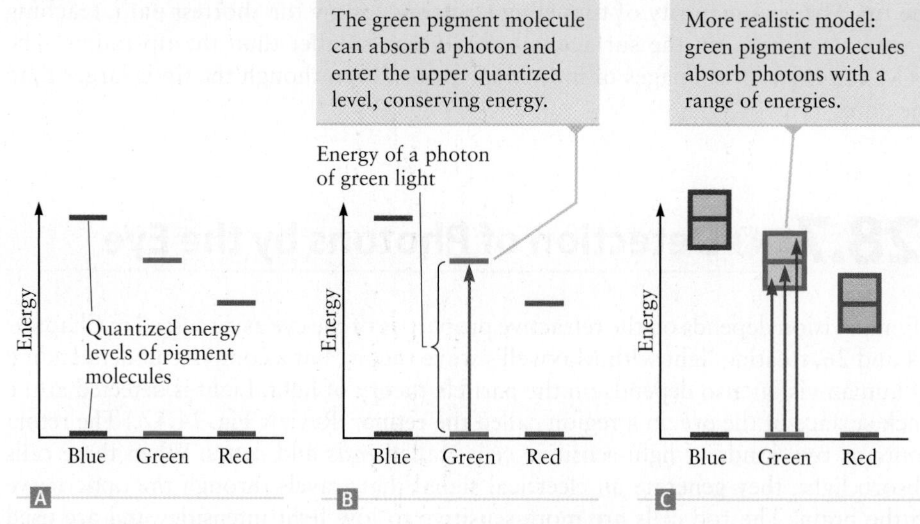

▶ **Figure 28.20** A Pigment molecules have quantized energies. B The molecules preferentially absorb photons whose energy matches the difference between two energy levels. For this reason, a green pigment molecule readily absorbs photons of green light, but not photons of red light or blue light. C More realistic energy-level diagram. Each pigment molecule responds to a range of photon energies. These ranges overlap somewhat and give the broad absorption curves in Figure 28.19.

molecule is quantized, just as the energy of a photon is quantized. Figure 28.20A shows a simplified sketch of the quantized energies of blue, green, and red pigments. When a photon of the correct frequency is absorbed by a green pigment molecule, the molecule ends up in the upper energy level in Figure 28.20B, initiating a series of chemical reactions that eventually send an electrical signal to the brain. This photon absorption is possible because the difference in energy levels in the green pigment matches the energy of the photon. The quantized energy levels of a pigment molecule thus allow only photons with a certain energy to be absorbed.

According to the simplified energy-level diagram in Figure 28.20A, a pigment molecule can absorb a photon only if the photon energy precisely matches the pigment energy level. A more realistic energy diagram is shown in Figure 28.20C. The pigment molecule sits among other molecules in the retina, and interactions between the pigment and these molecules enable the absorption of photons over a range of energies, not just at the single value of the energy indicated in Figure 28.20A. The pigment molecule still responds most strongly to frequencies near the center of its preferred range, but it also gives a nonzero (although weaker) response to photons with both higher and lower energies, corresponding to nearby colors. The other pigments are sensitive to different energy ranges, and by combining the response of all three types of pigments, the brain is still able to determine the color of the incident light. Hence, quantum theory and the existence of quantized energies for both photons and pigment molecules lead to color vision.

28.8 The Nature of Quanta: A Few Puzzles

Quantum theory forces us to rethink our classical view of the world. This rethinking leads to many new concepts, but a few of the central foundations of classical physics remain. The principles of conservation of energy, momentum, and charge are believed to hold true under *all* circumstances, but we must now allow for the existence of quanta. According to our understanding of the photon (Eq. 28.1), its energy comes in discrete quantized units. Electric charge also comes in quantized units, a result we have accepted throughout our study of electricity and magnetism. Indeed, the notion that charge is quantized in units of $\pm e$ is so familiar that it is easy to forget that the electron's existence was not known to Ampère, Faraday, Maxwell, and the other physicists who developed the theory of electricity and magnetism.

Even with the discovery of quantum theory and the true nature of electrons and photons as particle-waves, many puzzles remain, including (but not limited to) the following.

1. The relation between gravity and quantum theory is a major unsolved problem. No one really knows how Planck's constant enters the theory of gravitation or what a quantum theory of gravity looks like.
2. We know that electric charge is quantized in units of $\pm e$, but why are there two kinds of charge, positive and negative? In addition, why do the positive and negative charges come in the same quantized units? Physicists currently have no answer.
3. What new things happen in the regime where the micro- and macroworlds meet? According to quantum theory, all objects—even baseballs and bricks—behave as particle-waves. The quantum behavior of these macroscale objects is for all practical purposes unobservable, however, as we have seen in Examples 28.7 and 28.8. What about objects that are much smaller than baseballs and bricks? The quantum behavior of objects that are intermediate in size between the micro- and macroscales is now being intensively studied. There are also some interesting and unanswered questions about how quantum theory and the uncertainty principle apply to living things. For example, does the unpredictability associated with the Heisenberg uncertainty principle have any connection with consciousness and human free will? Some physicists think so, but the answer to this question is far from settled.

Summary | CHAPTER 28

Key Concepts and Principles

Photons

Light energy is carried in quantized parcels called ***photons***. For light with frequency f, the energy of a photon is

$$E_{\text{photon}} = hf \qquad (28.1)\ (\text{page } 874)$$

The quantity h is ***Planck's constant***, with the value

$$h = 6.626 \times 10^{-34}\ \text{J}\cdot\text{s} \qquad (28.2)\ (\text{page } 874)$$

Wave–particle duality

Electrons and other particles have a dual nature. They behave in some ways like classical particles and in other ways like classical waves. We call them ***particle-waves***. The de Broglie wavelength of a particle-wave with momentum p is

$$\lambda = \frac{h}{p} \qquad (28.8)\ (\text{page } 881)$$

This relation applies to electrons and all other particle-waves.

Electron spin

Electrons have ***spin angular momentum*** and behave as small bar magnets, with small magnetic moments. The direction of an electron magnetic moment is ***quantized***; it can only point "up" or "down" (i.e., parallel or antiparallel to a particular direction). The magnitude of this magnetic moment is also quantized.

(Continued)

Wave functions

Schrödinger's equation is used to calculate the behavior of objects such as electrons in the quantum regime. The solution of Schrödinger's equation gives the *wave function* of an object. The probability of finding the object at different points in space can be calculated from its wave function.

Heisenberg uncertainty principle

The *Heisenberg uncertainty relations* give fundamental limits on the accuracy with which the position, momentum, and energy of an object can be determined, with

$$\Delta x\ \Delta p \geq \frac{h}{4\pi} \qquad \text{(28.14) (page 888)}$$

and

$$\Delta E\ \Delta t \geq \frac{h}{4\pi} \qquad \text{(28.15) (page 888)}$$

Applications

Photoelectric effect

The *photoelectric effect* is a process in which light is absorbed by a metal. An electron is ejected from the metal when the photon energy is equal to or greater than the work function of the metal. This experiment can only be explained by the photon theory of light.

Tunneling

Tunneling is a process in which an electron (or any other type of particle-wave) passes through a barrier such as a wall that is classically impenetrable. Tunneling cannot be explained by Newton's mechanics, but is accounted for by quantum theory.

Questions

SSM = answer in *Student Companion & Problem-Solving Guide*   (life science icon) = life science application

1. Estimate the de Broglie wavelength of a car that (a) has a speed of 100 mi/h, (b) has a speed of 10,000 mi/h, and (c) is at rest.
2. You have two lightbulbs: one gives off green light with a very low intensity, and the other emits red light with a very high intensity. (a) Which one emits photons with a higher energy? (b) If these two lightbulbs emit the same intensity, which one emits more photons each second?
3. Consider a microscope that uses electrons instead of photons. Under what conditions will this microscope have better resolution than a microscope that uses visible light? *Hint*: Consider the electron's energy and wavelength.
4. What has greater energy, an ultraviolet photon or an X-ray photon?
5. Physicists often conduct experiments in which photons are counted one by one as they arrive at a detector. Explain why these experiments are relatively easy for photons of visible light, even easier for X-rays, and hardest for radio waves.
6. No electrons are ejected when a dim source of red light is directed onto a metallic surface. The intensity of the red light source is increased by a factor of 100. Is it now possible for electrons to be ejected via the photoelectric effect? Why or why not?
7. The (old-fashioned) film used for black-and-white photography is not affected by infrared light, is somewhat sensitive to red light, and exposes rapidly to blue light to the extent that blue filters are sometimes used to obtain an appropriate contrast. Give a possible explanation for why the film has this color dependency.
8. **Sunscreen.** Blocking ultraviolet radiation is effective in preventing sunburn. Using the concept of photons, describe why human skin is sensitive to ultraviolet frequencies, but much less so to those of visible light.
9. An electron and a proton have the same kinetic energy. Which one has the larger momentum?
10. Apply the Heisenberg uncertainty principle to a car. If you are asked to measure the position of the car, estimate the best accu-

racy you could expect to achieve with a ruler. Then calculate the minimum possible uncertainty in the car's momentum. Do you think it is feasible to measure the momentum with this accuracy?

11. SSM Consider the following types of radiation: visible light, infrared radiation, gamma rays, X-rays, radio waves, and ultraviolet radiation. From smallest to largest, order them according to (a) their frequency, (b) their wavelength, and (c) their photon energy.

12. The peak (maximum-intensity) wavelength emitted by a glowing piece of metal is found to decrease as time passes. Is the metal being heated, or is it being cooled?

13. A blacksmith heats a piece of metal in a furnace. The metal initially glows red and later yellow, and then it gets white hot while glowing brighter as more heat is applied. Is it possible for the metal to get "violet hot" and glow purple if it is hot enough? Why or why not?

14. Astronomers classify the color of stars as red, yellow, white, and blue. Why are green stars never seen? *Hint*: Consider how the eye perceives a mixture of different colors of light.

15. Which of the following experiments or phenomena are evidence (a) for the wave nature of light, and (b) which are evidence for the existence of photons?
(i) Single-slit diffraction
(ii) Photoelectric effect
(iii) Double-slit interference
(iv) Color vision

16. Explain why the wave nature of baseballs and cars is not apparent in everyday life. Provide a calculation to justify your answer.

17. Explain why the particle nature of light is not apparent in everyday life.

18. In Section 28.4, we gave a classical explanation of the electron's spin magnetic moment using the picture of an electron as a spinning ball of electric charge. The neutron has zero net charge, yet it also has a magnetic moment. Explain how a spinning ball with zero net charge can still have a magnetic moment due to its spin.

19. Color vision can only be understood using a particle (photon) model of light. Compare color vision to the photoelectric effect, explaining how they are similar and how they differ.

20. SSM Human color vision is based on the absorption of light by three different types of cone cells, which are most sensitive to red light, green light, and blue light. (a) Explain how the output of these cells can be used to determine the color of the incident light. (b) Could an eye with just two types of cone cells be used to give color vision? Explain why or why not. If your answer is yes, are there any examples in nature?

21. Suppose the human eye were so sensitive that it could detect a single photon. With input of just a single photon, could our system of rods and cones detect the color of this light?

22. Explain why the existence of a cutoff frequency in the photoelectric effect cannot be explained with the wave theory of light.

23. Explain why a classical picture cannot account for the Stern–Gerlach experiment with electrons (Fig. 28.11). The Stern–Gerlach experiment can also be done with nuclei, and it is found that certain nuclei give rise to three outgoing beams. What does that imply for the number of possible directions for the magnetic moment of that particular nucleus?

24. Planck's constant was first used to explain what phenomenon?

Problems available in ENHANCED WebAssign

Section 28.1 Particles, Waves, and "Particle Waves"
Section 28.2 Photons
Problems 1–29
Section 28.3 Wavelike Properties of Classical Particles
Problems 30–48
Section 28.4 Electron Spin
Problem 49
Section 28.5 The Meaning of the Wave Function
Problems 50–56
Section 28.6 Tunneling
Problems 57–58
Section 28.7 Detection of Photons by the Eye
Problems 59–60
Additional Problems
Problems 61–74

List of Enhanced Problems

Problem number	Solution in *Student Companion & Problem-Solving Guide*	Reasoning & Relationships Problem	Reasoning Tutorial in ENHANCED WebAssign
28.1		✓	✓
28.6		✓	✓
28.11		✓	✓
28.18	✓		
28.19		✓	
28.23		✓	
28.24		✓	✓
28.27	✓		
28.34		✓	
28.40		✓	
28.41	✓		
28.45		✓	✓
28.47	✓		
28.49	✓		
28.56	✓		
28.58	✓		
28.60	✓		
28.62	✓	✓	
28.64		✓	
28.66		✓	
28.68		✓	✓
28.73		✓	✓
28.74		✓	

CHAPTER

29

◀ *Lasers are an important application of quantum theory. This laser at the European Southern Observatory's Very Large Telescope is used to make atoms in the atmosphere emit light through fluorescence, a process described in this chapter. This light then acts as a sort of "reference star" for telescopes, allowing the use of adaptive optics (Chapter 26) to improve telescope resolution. (European Southern Observatory/Photo Researchers, Inc.)*

Atomic Theory

Matter is composed of atoms, and atoms are themselves assembled from electrons, protons, and neutrons. These facts are such a fundamental part of elementary science courses that it is hard to imagine a world without atoms. You should realize, however, that atoms were discovered *after* Galileo, Newton, Maxwell, and most of the other physicists mentioned so far had passed from the scene. Indeed, electrons were discovered barely one hundred years ago, and the discovery of protons and neutrons came even later. In this chapter, we explain how key aspects of quantum theory explain the way atoms are put together. With this understanding, we can understand why different elements have different properties and explain the organization of the periodic table.

OUTLINE

Insight 29.1
CHEMISTRY, ATOMS, AND THE LAW OF DEFINITE PROPORTIONS
In the early 1800s, chemists discovered that when a compound is completely broken down into its constituent elements, the masses of the constituents always have the same *proportions*, no matter what the quantity of the original substance. This similarity can be explained by assuming the constituents are atoms that combine in definite proportions to make molecules. For example, when water is broken down into hydrogen and oxygen, the number of hydrogen atoms released is always twice the number of oxygen atoms, the mass of hydrogen released is always one-ninth of the mass of the original water, and the mass of oxygen gives the remaining eight-ninths of the mass. This is an example of the ***law of definite proportions***.

29.1 Structure of the Atom: What's Inside?

By about 1890 or so, it was apparent to most physicists and chemists that matter is composed of atoms, and it was also widely believed that atoms are indivisible. At that time, about 90 elements were known, suggesting there were the same number of fundamental building blocks of matter. We now know that picture is not quite correct. At last count, there are more than 110 elements, but each is composed of just three different types of particles: ***electrons***, ***protons***, and ***neutrons***.

We encountered electrons and protons in our work on electricity (Chapter 17); they are subatomic particles that carry electric charges of $-e$ and $+e$, respectively. Neutrons are a third type of subatomic particle; they have no net electric charge and have a mass approximately the same as that of a proton. A key goal of atomic theory is to explain how just these three building blocks combine to make so many different kinds of atoms.

An understanding of how electrons, protons, and neutrons combine to form atoms required the invention of a new type of mechanics, called ***quantum mechanics***. Some key aspects of quantum mechanics were described in Chapter 28, where the notion of a particle-wave was introduced. In this chapter, we apply the ideas of quantum theory to understand the structure of the atom.

Plum-Pudding Model of the Atom

Electrons were the first of the building-block particles to be discovered. In about 1897, experiments involving electricity showed that electrons carry a negative charge and that they are contained within atoms. These experiments also showed that electrons have a very small mass (compared with an atom). Since atoms are electrically neutral—that is, they have zero net charge—there must be some source of positive charge inside them. It was initially suggested that the positive charge is distributed as a sort of "pudding," with electrons suspended within positively charged pudding throughout the atom as sketched in Figure 29.1A. This analogy is known as the "plum-pudding" model of the atom, a name chosen by English physicists who proposed the model and named it after a popular dessert of the time.

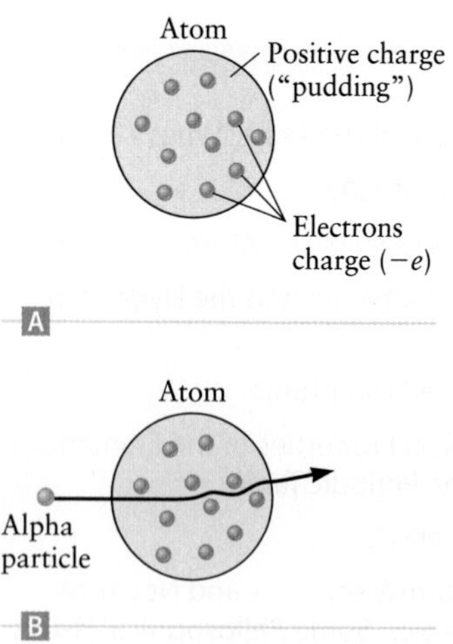

▲ **Figure 29.1** A According to the plum-pudding model, the positive charge in an atom is distributed continuously, and the electrons sit in this "pudding." B The pudding was thought to have a very low density, so the plum-pudding model predicts that an alpha particle should be deflected very little when it collides with (passes through) an atom.

The plum-pudding model of the atom raises an interesting conceptual question. A neutral atom has a total electric charge of zero, so the negative charge of the electrons must be exactly canceled by the positive charge of the pudding. Hence, an atom must contain a precisely measured amount of positive pudding, but how that could be accomplished was not clear. To learn more about the positive charge in the atom, physicists studied how atoms collide with other atomic-scale particles. The most famous of these experiments was carried out by Ernest Rutherford and his students Hans Geiger[1] and Ernst Marsden. They arranged for alpha particles to collide with a thin sheet of gold atoms. At that time, it was known that alpha particles are emitted spontaneously from certain radioactive elements, but the details of radioactivity were not understood. We now know that an alpha particle is actually the nucleus of a helium atom (Chapter 30).

Even though the true nature of an alpha particle was not understood at the time of Rutherford's work, prior experiments had shown that alpha particles behave as simple atomic-scale "bullets" when they collide with an atom. Rutherford's idea was to fire these "bullets" at an atom and then infer how the atom is put together by observing what comes out. Electrons have a very small mass, and Rutherford expected that the positively charged pudding would have a low density since (according to the plum-pudding model) this charge is spread throughout the entire atom. Using his knowledge of collisions and conservation of momentum (Chapter 7), Rutherford expected that the relatively massive alpha particles would pass freely through the plum-pudding

[1]Inventor of the Geiger counter. (See Insight 30.3.)

atom as sketched in Figure 29.1B. For example, in a collision between a car (the alpha particle) and a Ping-Pong ball (an electron), the car is deflected only a very small amount because the mass of the car is much greater than that of the Ping-Pong ball. When he carried out his experiment, Rutherford found that most of the alpha particles did indeed pass freely through the atom, but he also found that a small number of alpha particles were deflected a very *large* amount, some even bouncing directly "backward" toward the source of alpha particles. This surprising result could not be explained by the plum-pudding model. After much study, Rutherford realized that all the positive charge in an atom must be concentrated in a very small volume, with a mass and density about the same as an alpha particle. Most alpha particles in a collision experiment miss this dense region and pass right through the atom. Occasionally, though, an alpha particle collides with the dense region, giving a large deflection. Hence, Rutherford concluded that atoms contain a ***nucleus*** that is positively charged and has a large mass, much greater that the mass of an electron.

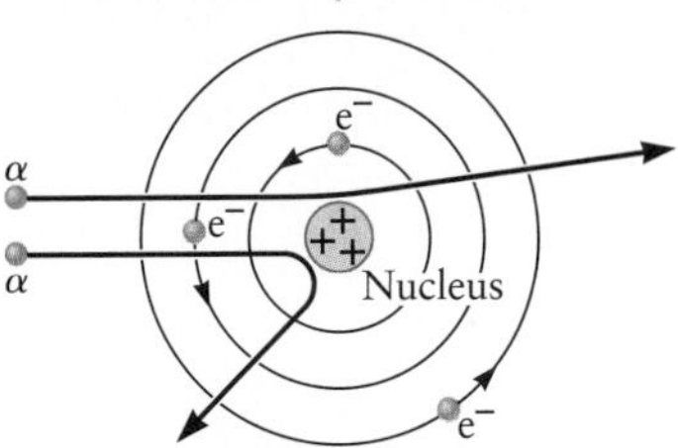
Nucleus *not* drawn to scale. The nucleus is actually about 100,000 times smaller than the region occupied by electrons.
α
α
e⁻
e⁻
e⁻
Nucleus

▲ **Figure 29.2** Planetary model of the atom, with all the positive charge located in the nucleus. When alpha particles are fired at an atom, they occasionally collide with the nucleus and are deflected through very large angles, but since the alpha particle and the nucleus are both very small, most alpha particles pass right through the atom.

The Atomic Nucleus

After his discovery of the nucleus, Rutherford suggested that the atom is a sort of miniature solar system, with electrons orbiting the nucleus just as planets orbit the Sun (Fig. 29.2). This is called the ***planetary model of the atom.*** According to this model, the electrons must move in orbits to avoid "falling" into the nucleus as a result of the electric force.

We now know that the atomic nucleus contains protons and that the charge on a proton is precisely equal to $+e$. The charge carried by an electron is $-e$, so the total charge carried by a hydrogen atom (symbol H), which contains one electron along with a nucleus consisting of a single proton, is precisely zero. The total charge of neutral atoms of all other elements is also zero, so the number of protons in the nucleus is equal to the number of electrons in the neutral atom. This number, called the ***atomic number*** of the element, is denoted by Z.

Atomic number Z = number of protons in the nucleus

Except for the H nucleus, which is a single proton, all other nuclei contain neutrons. As its name implies, the neutron is a neutral particle, carrying zero net electric charge. The neutron was not discovered until the 1930s, and we'll discuss some of its properties in Chapter 30. Even without a detailed understanding of the neutron, however, it is easy to understand why all nuclei (except H) contain both protons and neutrons. Protons are positively charged particles, repelling one another according to Coulomb's law, so a hypothetical nucleus containing just two or more protons would fly apart spontaneously. Protons are attracted to neutrons by an additional force that overcomes this Coulomb repulsion and thereby holds the nucleus together (described in Chapter 30).

Rutherford's discovery of the nucleus led to the planetary model of the atom, according to which the Coulomb force exerted by the nucleus on the electrons keeps the electrons in orbit. Physicists then tried to apply Newton's mechanics to the motion of electrons as they orbit within the atom. Their attempts were not successful as these "planetary" properties of electrons do not correctly predict atomic behavior. Even so, this approach provides some useful insights into the correct quantum theory that was developed later. With that in mind, let's estimate some atomic properties using this planetary model.

The Planetary Model: Energy of an Orbiting Electron

Figure 29.3 shows the planetary model as applied to hydrogen. It is the simplest atom, containing just[2] one proton and one electron. For simplicity, let's assume the electron's orbit is circular with a radius r. To estimate the properties of this orbit, we note that r is the approximate size of the hydrogen atom. Chemists have found

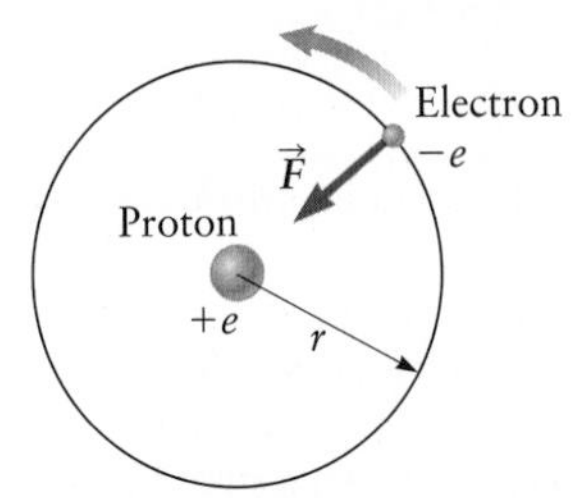
Electron
$-e$
$\vec{F}$
Proton
$+e$
r

▲ **Figure 29.3** Planetary model of the hydrogen atom. A single electron is in orbit around a proton.

[2]It is possible to form atoms called deuterium and tritium, which each contain one proton and one electron. A deuterium nucleus also contains one neutron, and a tritium nucleus contains two neutrons.

that the typical distance between hydrogen atoms in various molecules suggests a radius of about $r \approx 1.0 \times 10^{-10}$ m. Since the electron in Figure 29.3 is moving in a circle, there must be a force directed toward the center of the circle. For an electron of mass m moving with speed v, this attractive force has a magnitude (Eq. 5.6)

$$F = \frac{mv^2}{r} \tag{29.1}$$

The force in Figure 29.3 is the electric force (from Coulomb's law) exerted by the proton on the electron, which is given by (Eq. 17.3)

$$F = \frac{kq_{\text{proton}}q_{\text{electron}}}{r^2} \tag{29.2}$$

The charge on an electron is $q_{\text{electron}} = -e$, while the charge on a proton is $+e$. Inserting into Equation 29.2 gives $F = k(+e)(-e)/r^2 = -ke^2/r^2$. The negative sign indicates an attractive force, and we can drop it when considering just the magnitude of F. Inserting this result for F in Equation 29.1 leads to

$$\frac{ke^2}{r^2} = \frac{mv^2}{r}$$

The charge and mass of the electron are known and we have an estimate for the orbital radius r, so we can now solve for the electron's speed v. We get

$$v^2 = \frac{ke^2}{mr} \tag{29.3}$$

$$v = \sqrt{\frac{ke^2}{mr}} \tag{29.4}$$

Inserting values for the various quantities, including our estimate for r, we find

$$v = \sqrt{\frac{ke^2}{mr}} = \sqrt{\frac{(8.99 \times 10^9 \text{ N} \cdot \text{m}^2/\text{C}^2)(1.60 \times 10^{-19} \text{ C})^2}{(9.11 \times 10^{-31} \text{ kg})(1.0 \times 10^{-10} \text{ m})}}$$

$$v = 1.6 \times 10^6 \text{ m/s} \tag{29.5}$$

This orbiting electron has quite a high speed (approximately 4×10^6 mi/h!). Given this value of the speed, the associated kinetic energy is

$$KE = \tfrac{1}{2}mv^2 = \tfrac{1}{2}(9.11 \times 10^{-31} \text{ kg})(1.6 \times 10^6 \text{ m/s})^2 = 1.2 \times 10^{-18} \text{ J}$$

We'd like to compare this roughly estimated value of the electron's orbital kinetic energy to some energy value associated with a hydrogen atom in a chemical reaction. Chemists measure the ***ionization energy***, the energy required to remove an electron from an atom in the gas phase. In mechanical terms, the ionization energy is the energy needed to completely separate the two charges; that is, to take the electron in hydrogen from its original orbit to an infinite distance from the proton. For ease of comparison, let's convert the kinetic energy found above from joules to units of electron-volts (eV). From Equation 18.19, 1 eV $= 1.60 \times 10^{-19}$ J, which leads to

$$KE = (1.2 \times 10^{-18} \text{ J})\left(\frac{1 \text{ eV}}{1.60 \times 10^{-19} \text{ J}}\right) = 7.5 \text{ eV} \tag{29.6}$$

In fact, the measured ionization energy of a hydrogen atom is 13.6 eV, so the electron kinetic energy found with the planetary model is indeed of the same order of magnitude as the ionization energy. For this reason, Rutherford's planetary model received strong initial support. In Example 29.1, we'll consider another aspect of the planetary model.

EXAMPLE 29.1 Ionization Energy of a Hydrogen Atom

In our analysis of the planetary model of the hydrogen atom leading to Equation 29.6, we considered only the kinetic energy of the electron. To calculate the total energy required to remove the electron from the proton, we must also consider the potential energy. Assuming again the electron's orbit has a radius $r = 1.0 \times 10^{-10}$ m, find the electric potential energy in this model of the hydrogen atom. What is the change in this potential energy when the atom is ionized?

RECOGNIZE THE PRINCIPLE

The electric potential energy of two charged particles depends on the value of each charge and their separation. In the planetary model, the electron's orbit is circular, so the separation between the charges for the atom is just the radius of the orbit. When the atom is ionized, the electron is very (infinitely) far from the proton.

SKETCH THE PROBLEM

Figure 29.4 shows the problem. The electron is initially in orbit a distance r_i from the proton (Fig. 29.4A), with a speed found in Equation 29.4. When the atom is ionized, the electron is at a distance $r_f = \infty$ from the proton (Fig. 29.4B).

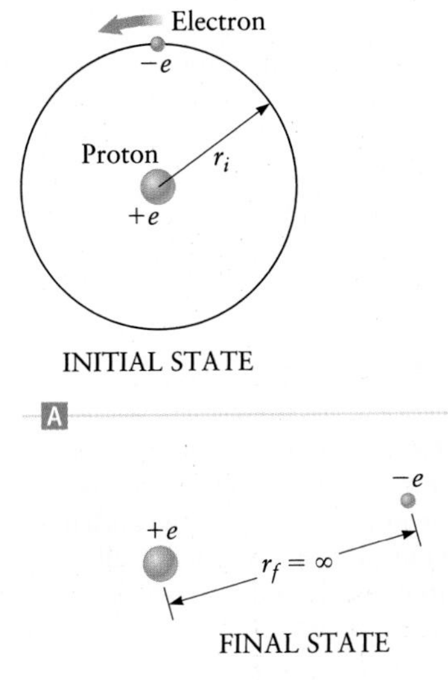

▲ **Figure 29.4** Example 29.1. Calculating the potential energy of an electron in the planetary model of the hydrogen atom.

IDENTIFY THE RELATIONSHIPS

The electron is initially in orbit at $r_i = 1.0 \times 10^{-10}$ m. The electric potential energy is given by (Eq. 18.6)

$$PE_{\text{elec}, i} = \frac{kq_{\text{proton}}q_{\text{electron}}}{r_i}$$

The charge on the proton is $q_{\text{proton}} = +e$ and the charge on the electron is $-e$, so

$$PE_{\text{elec}, i} = \frac{kq_{\text{proton}}q_{\text{electron}}}{r_i} = \frac{-ke^2}{r_i} \tag{1}$$

SOLVE

Using our estimate for r_i, we find

$$PE_{\text{elec}, i} = \frac{-ke^2}{r_i} = \frac{-(8.99 \times 10^9 \text{ N} \cdot \text{m}^2/\text{C}^2)(1.60 \times 10^{-19} \text{ C})^2}{1.0 \times 10^{-10} \text{ m}}$$

$$PE_{\text{elec}, i} = -2.3 \times 10^{-18} \text{ J}$$

Converting to units of electron-volts as in Eq. 29.6 gives

$$PE_{\text{elec}, i} = (-2.3 \times 10^{-18} \text{ J})\left(\frac{1 \text{ eV}}{1.60 \times 10^{-19} \text{ J}}\right) = -14 \text{ eV}$$

When the atom is ionized, the separation between the electron and proton is infinite. Using Equation (1), the corresponding potential energy is $PE_{\text{elec}, f} = 0$. Hence, the change in potential energy when the atom is ionized is

$$PE_{\text{elec}, f} - PE_{\text{elec}, i} = \boxed{+14 \text{ eV}}$$

► ***What does it mean?***

Because the value of r_i was not known precisely, this result is only a rough estimate, but it does confirm that the kinetic and potential energies predicted by the planetary model are both close to the measured ionization energy of hydrogen, which is 13.6 eV. We'll return to this problem in Section 29.3.

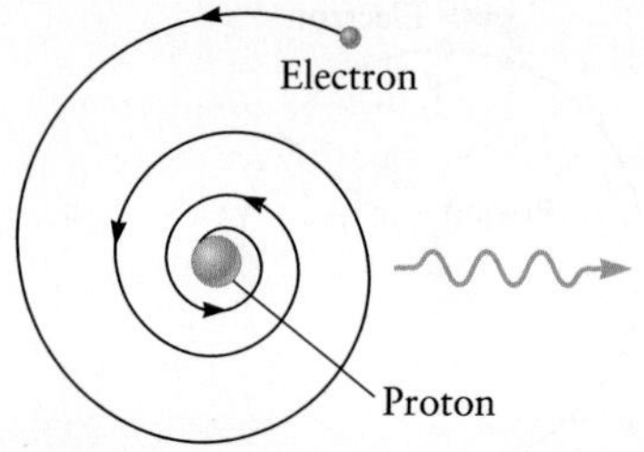

▲ **Figure 29.5** Maxwell's theory of electromagnetic waves predicts that an orbiting electron emits radiation, causing it to lose energy and spiral into the nucleus. If this picture were correct, all atoms would collapse, which is clearly *not* the case.

Problems with the Planetary Model of the Atom

Calculations with Rutherford's planetary model (like the one in Example 29.1) were initially encouraging, but some fundamental problems with this model soon became apparent. The biggest problem concerns the stability of an electron orbit. Atoms such as hydrogen can be extremely stable. If chemical reactions are avoided, an individual atom can "last" indefinitely. The electrons in Rutherford's planetary model, however, are undergoing accelerated motion, and in Chapter 23 we saw that accelerated charges emit electromagnetic radiation that carries away energy. If the electron in a hydrogen atom loses energy in this way, it will spiral inward to the nucleus (Fig. 29.5). Hence, according to classical physics (i.e., according to Newton and Maxwell), Rutherford's atom is inherently unstable!

A careful analysis in terms of Newton's laws shows that an electron in the planetary model must spiral into the nucleus in a very small fraction of a second. Hence, according to classical physics, such planetary model atoms cannot exist. Despite much trying, physicists found no way to fix the planetary model to make the atoms in this model stable.

Quantum theory avoids the problem of unstable electron orbits by replacing them with discrete energy levels, just like the discrete energy levels of an electron in a box (Fig. 28.12) and the discrete energies of the pigment molecules involved in color vision (Section 28.7). Quantum theory rejects the notion of the electron as a simple particle that obeys Newton's laws. Instead, an electron is a particle-wave described by a wave function. Atomic electrons lose or gain energy only when they undergo a transition between energy levels.

29.2 Atomic Spectra

We have just claimed that an electron in an atom can exist only in discrete energy levels. The best evidence for this claim comes from the radiation an atom emits or absorbs when an electron undergoes a transition from one energy level to another. This radiation is key to another question that was studied intensely in the late 1800s: what gives an object its color? Physicists of that time knew about blackbody radiation (Chapters 14 and 28), including the relationship between the temperature of an object and its color. This relationship is described by Wien's law (Eqs. 14.14 and 28.7). The approximate blackbody spectrum of the Sun shown in Figure 29.6A is a continuous curve with a smooth distribution of intensity over a wide range of wavelengths and frequencies.

The spectrum in Figure 29.6A describes the radiation emitted by the Sun in a general way. Careful observations, however, show that the Sun's spectrum also con-

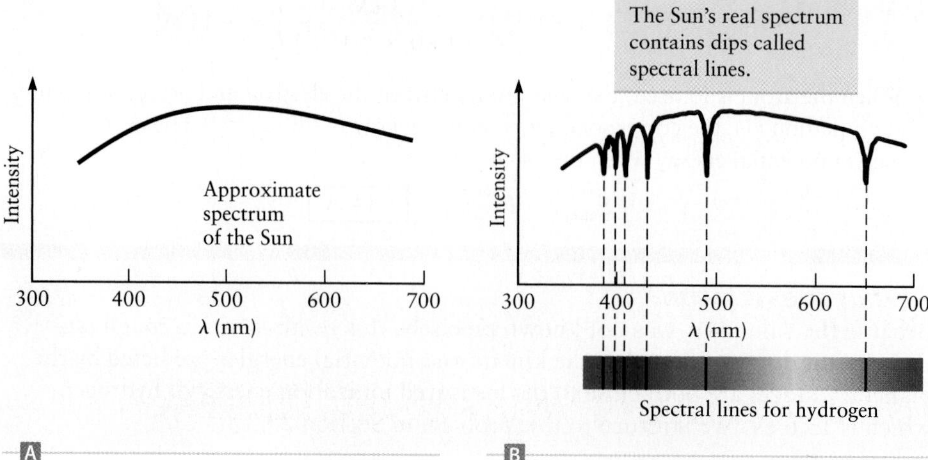

► **Figure 29.6** A The ideal blackbody spectrum of the Sun is a smooth function of the wavelength. B The actual spectrum of the Sun contains sharp "dips" on top of the blackbody spectrum in part A. Another way of displaying these dips is to use a prism to disperse sunlight (Chapter 24); the dips then show up as dark "lines," having the same wavelengths as the dips in the spectrum.

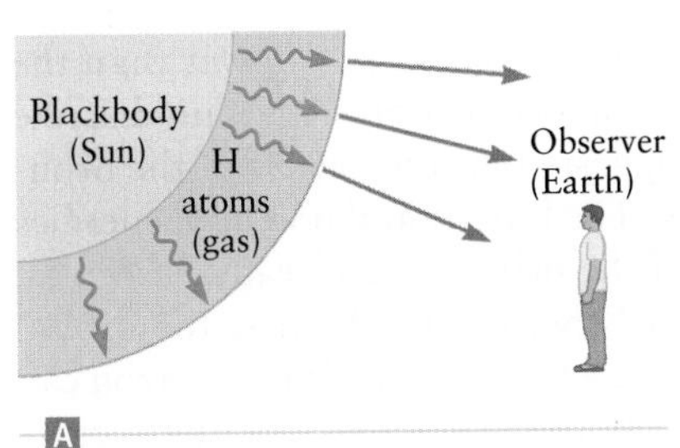

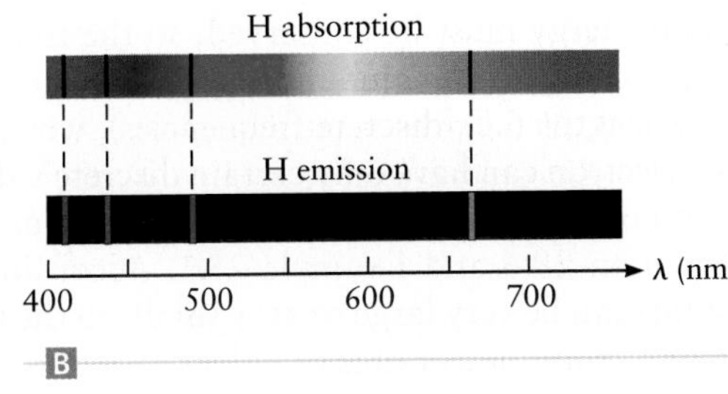

Figure 29.7 A When the Sun's blackbody radiation passes through its atmosphere, atoms in the atmosphere absorb at certain discrete frequencies, producing dark spectral absorption lines and causing the dips in the spectrum in Figure 29.6B. B Absorption and emission spectra for hydrogen. The absorption and emission lines occur at the same wavelengths. This plot shows just the spectral lines for hydrogen that lie between 400 nm and 700 nm. Figure 29.6B shows these and some other lines with wavelengths just below 400 nm.

tains sharp dips superimposed on the otherwise smooth blackbody curve as shown schematically in Figure 29.6B. These dips are called spectral "lines" because of their appearance when the spectrum is dispersed by a prism or diffraction grating (Chapters 24 and 25). Examination of the Sun's spectrum with a prism gives a band of colors extending from red to blue as shown below the graph in Figure 29.6B. The dips in the graph of the figure show up as dark lines in the band of colors, indicating wavelengths at which the light intensity is much lower than the expected blackbody value.

The origin of the dark spectral lines is illustrated in Figure 29.7A; when light from a pure blackbody source passes through a gas, atoms in the gas absorb light at certain wavelengths, producing dips in the spectrum at those wavelengths. Some of the dark lines in the Sun's spectrum are produced when hydrogen atoms in the Sun's atmosphere absorb light. This can be confirmed by an experiment using a blackbody source and hydrogen atoms on the Earth, which gives the spectral lines shown in the top band in Figure 29.7B; these lines have precisely the same wavelengths as those found in the Sun's spectrum.

The dark spectral lines are called ***absorption lines*** because they result from the absorption of light, in this case by hydrogen atoms. These same atoms can also be made to *emit* light. When the spectrum of light emitted by atoms is analyzed, it is found that the emission occurs only at certain wavelengths. The ***emission lines*** for hydrogen are shown in the lower band in Figure 29.7B, demonstrating that these emission and absorption lines occur at the same wavelengths.

Another example of spectral lines is given in Figure 29.8, which shows the emission and absorption spectra for sodium (Na) atoms. Here again the emission and absorption lines occur at the same wavelengths. The pattern of spectral lines is different for each element. In fact, by analyzing the wavelengths at which these lines occur, physicists in the 1800s determined the composition of the Sun's atmosphere.

These observations of atomic spectra lead to several questions. Why do absorption and emission occur only at certain special wavelengths? Why do the absorption and emission lines for a particular element occur at the same wavelengths? What determines the pattern of the absorption and emission wavelengths, and why are they different for different elements?

Many physicists attempted to answer these questions by applying Newton's mechanics to Rutherford's planetary model of the atom, but all ran into the following problem. When a photon is emitted by an atom, the photon carries away a certain amount of energy; for a photon with frequency f, we have (Eq. 28.1)

$$E_{\text{photon}} = hf \tag{29.7}$$

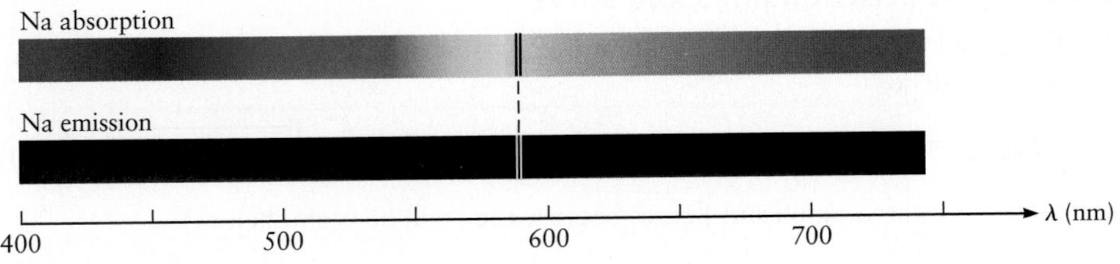

Figure 29.8 Sodium (Na) has two closely spaced emission lines in the yellow part of the visible spectrum. Sodium atoms also absorb strongly at the same wavelengths as the emission lines.

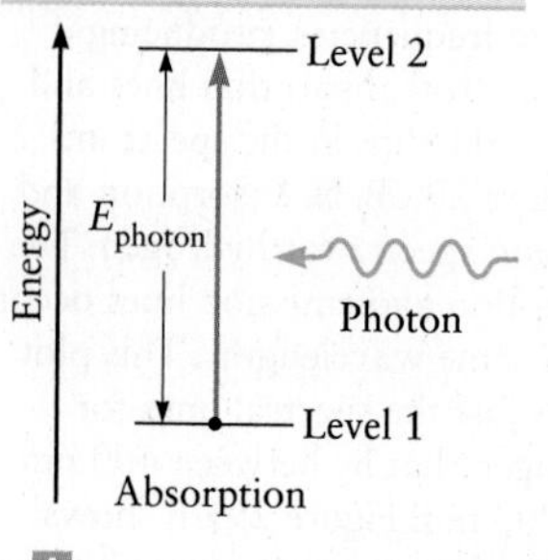

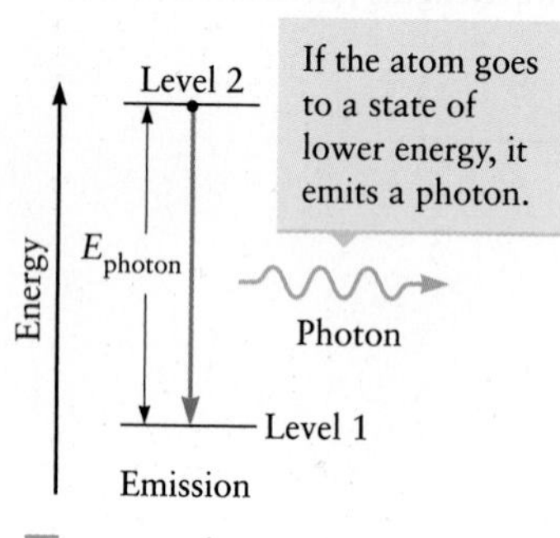

▲ **Figure 29.9** The discrete absorption and emission lines of an atom are due to discrete atomic energy levels. An atom A absorbs or B emits a photon when it makes a transition between energy levels.

Total energy must be conserved, so the final energy of the atom is lower than the initial energy by an amount E_{photon}. Atomic emission occurs only at certain discrete wavelengths (i.e., discrete frequencies), which suggests that the energy of the orbiting electron can have only certain discrete values. The kinetic and potential energies of an orbiting electron depend on the radius r of the orbit as we already derived in Equation 29.6 and Example 29.1. According to Newton's mechanics, the orbital radius can be very large or very small, so the total energy of an orbiting electron can have a continuous range of values. There is no way for this planetary orbit picture to give discrete electron energies and no way to explain the existence of discrete spectral lines. This problem is resolved in quantum theory through the description of the electron's state in terms of a wave function instead of an orbit.

Atoms Have Quantized Energy Levels

Spectra like those in Figures 29.7 and 29.8 show that under certain conditions, the photons emitted or absorbed by an atom have only certain discrete wavelengths or frequencies. The implication is that the energy of the atom itself can have only certain discrete values. That is, the energy of an atom is ***quantized***. In the language of quantum theory, we say that the energy of an absorbed or emitted photon is equal to the difference in energy between two discrete atomic energy levels (Fig. 29.9). Through Equation 29.7, the frequencies of the emission and absorption lines give the spacing between atomic energy levels. Hence, the study of atomic spectra gives a direct window into the atom's structure.

EXAMPLE 29.2 Energy of Photons Emitted by the Hydrogen Atom

Studies of the spectral lines of hydrogen show that the highest frequency of electromagnetic radiation emitted by a hydrogen atom is about 3.28×10^{15} Hz. (**a**) What is the energy of one of the corresponding photons? (**b**) Does this light fall in the infrared, visible, or ultraviolet part of the electromagnetic spectrum? *Hint*: The approximate short wavelength end of the visible range is at $\lambda \approx 400$ nm. (**c**) How does this photon energy compare with the ionization energy of the hydrogen atom, 13.6 eV?

RECOGNIZE THE PRINCIPLE

For part (**a**), this problem asks for a calculation of photon energy given the frequency, which can be done with Equation 29.7. For part (**b**), if we are given either the frequency or wavelength, we can use results from Chapter 23 and Figure 23.8 to find where this radiation falls within the electromagnetic spectrum. For part (**c**), the comparison will be apparent once we know the photon energy and convert it to electron-volts.

SKETCH THE PROBLEM

A spectral line is connected with photon emission or absorption between two discrete energy levels. Figure 29.10 shows this schematically for a hydrogen atom.

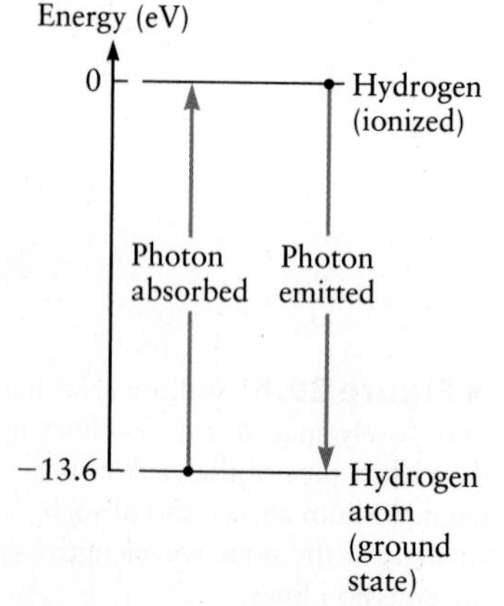

▲ **Figure 29.10** Example 29.2.

IDENTIFY THE RELATIONSHIPS AND SOLVE

(**a**) This problem begins with the relation $E_{photon} = hf$ (Eq. 29.7). Inserting the given value of f leads to

$$E_{photon} = hf = (6.63 \times 10^{-34}\ \text{J} \cdot \text{s})(3.28 \times 10^{15}\ \text{Hz}) = \boxed{2.17 \times 10^{-18}\ \text{J}} \quad (1)$$

(Here we keep three significant figures because they will be useful below.)

(**b**) To determine where this light falls in the electromagnetic spectrum, let's calculate its wavelength. The speed of light is $c = 3.00 \times 10^8$ m/s, so we have

$$\lambda = \frac{c}{f} = \frac{3.00 \times 10^8 \text{ m/s}}{3.28 \times 10^{15} \text{ Hz}} = 91.5 \text{ nm}$$

According to the hint (and also Fig. 23.8), the visible range ends at about $\lambda \approx 400$ nm (corresponding to violet light), so this photon falls well into the $\boxed{\text{ultraviolet}}$.

(**c**) To compare the energy of the photon in Equation (1) with a typical atomic energy, it is useful to convert this energy to units of electron-volts. We get

$$E_{\text{photon}} = (2.17 \times 10^{-18} \text{ J})\left(\frac{1 \text{ eV}}{1.60 \times 10^{-19} \text{ J}}\right) = 13.6 \text{ eV}$$

Hence, this photon energy $\boxed{\text{is equal to}}$ the ionization energy.

▸ ***What does it mean?***
We have quoted the value of E_{photon} to three significant figures because it is a very special energy value in chemistry and physics. This result for E_{photon} is precisely equal to the ionization energy of hydrogen, the energy required to remove the electron from a hydrogen atom and thus "break" the atom apart. Our result for E_{photon} confirms the close connection between the way an electron is bound in an atom and the way the atom emits light. These processes involve the same two energy levels as sketched in Figure 29.10. The levels in hydrogen differ in energy by 13.6 eV, and by convention the upper level, corresponding to an ionized atom, has an energy $E = 0$.

29.3 Bohr's Model of the Atom

The experimental facts described in the previous two sections showed that Rutherford's planetary model of the atom—and indeed *any* model based on Newton's mechanics—is a failure. The correct quantum theory of the atom did not "appear" instantaneously; there were first some intermediate proposals that eventually led to a complete quantum theory. One of these intermediate proposals is known as the ***Bohr model***, invented by Danish physicist Niels Bohr (1885–1962). (Bohr later engaged Albert Einstein in famous debates over the philosophical implications of quantum theory.) Although Bohr's model has some flaws, it also contains some of the most important and most revolutionary ideas of the correct quantum theory, and is an important conceptual step along the way.

Quantum States and Energy-Level Diagrams

Bohr based his thinking on the planetary model. For simplicity, he first assumed the electron orbits are circular, which allowed him to use the results for the electron kinetic and potential energies that we derived in Section 29.1 for a hypothetical hydrogen atom. However, to explain the discrete spectral lines, Bohr needed several changes to the standard results based on Newton's laws. First, Bohr postulated that only certain specific (and special) values of the orbital radius r are permitted. If only certain discrete values of r are allowed, the electron's kinetic and potential energies can have only certain values, so the total energy of the electron—the total energy of the atom—can have only certain discrete values. In the modern language of quantum theory, the energy of Bohr's atom is ***quantized***, and we refer to each of these allowed orbits as a ***quantum state*** of the atom. One way to view these energies and

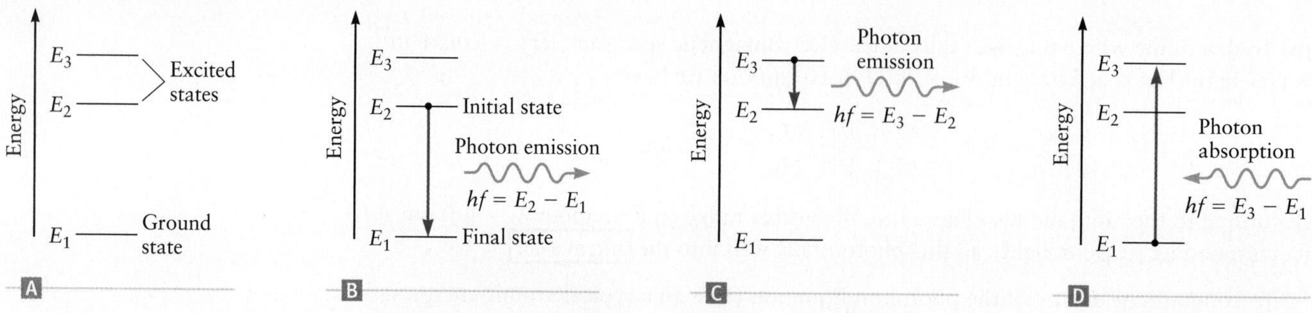

▲ **Figure 29.11** A Hypothetical energy-level diagram for an atom with three quantum states. B An atom that is initially in state 2 can undergo a transition to a state with lower energy (state 1) and emit a photon. C Photon emission due to a transition between excited states. D An atom can absorb a photon and undergo a transition to a state of higher energy.

Quantized atomic states are shown on an energy-level diagram.

states is in terms of an ***energy-level diagram.*** Figure 29.11A shows a hypothetical diagram for an atom that can be in one of three different states. In this energy-level diagram, energy is plotted on the vertical axis and each state is represented by a horizontal line.

With this postulate of discrete energy levels, Bohr could explain the discrete spectral emission and absorption lines described in Section 29.2. Let's assume an atom is initially in the quantum state with energy E_2 in Figure 29.11B, corresponding to a particular orbital radius r_2. This electron may "jump" to a different orbit with energy E_1 as shown on the energy-level diagram by a vertical arrow that connects the initial and final states. After doing so, the atom has decreased in energy by an amount $\Delta E = E_2 - E_1$. According to the principle of conservation of energy, this energy cannot just vanish; total energy is conserved as the atom emits a photon with precisely this energy ΔE. The discrete set of atomic energy levels produces the discrete frequencies found in the emission spectrum of an atom (Figs. 29.7 and 29.8). We say that the atom described in Figure 29.11B was initially in an ***excited state*** and made a transition to the ***ground state***, the state of lowest possible energy for this atom. It is also possible for an atom to emit a photon and undergo a transition between two excited states such as the states with energies E_3 and E_2 in Figure 29.11C.

Transitions between discrete energy levels also explain the absorption spectrum of an atom. The energy-level diagram in Figure 29.11D shows an atom that starts in the ground state with energy E_1 and then absorbs a photon as it undergoes a transition to an excited state with energy E_3. To conserve energy, the absorbed photon's energy must be $\Delta E = hf = E_3 - E_1$, the energy gained by the atom. This picture explains why the frequencies of the emission lines are precisely the same as the frequencies of the absorption lines. Both are determined by the difference in energy between two atomic states.

CONCEPT CHECK 29.1 Spectral Lines and Energy Levels

The spectra for Na in Figure 29.8 show two closely spaced lines in both absorption and emission. Consider now a different (and hypothetical) atom. Is the *minimum* number of atomic energy levels required to give spectral lines at two or more different frequencies (a) two, (b) three, or (c) four?

Quantization of Angular Momentum Leads to Quantized States with the Correct Energies

Central to the Bohr model is the postulate that electrons can orbit at only certain allowed values of the radius r, but what determines these values? Bohr proposed that the orbital angular momentum L of the electron could only have certain values given by the relation

$$L = n\frac{h}{2\pi} \quad (29.8)$$

where $n = 1, 2, 3, \ldots$ is an integer and h is Planck's constant. According to Equation 29.8, the allowed values of angular momentum are quantized in units of $h/(2\pi)$. Let's now see how Bohr's proposal for the angular momentum leads to quantized energy levels.

From Chapter 7, the angular momentum of a particle of mass m traveling with speed v in a circular orbit of radius r is

$$L = mvr \quad (29.9)$$

Inserting this into Bohr's angular momentum relation, Equation 29.8, gives

$$L = mvr = n\frac{h}{2\pi}$$

Rearranging to solve for the electron's orbital speed v leads to

$$v = \frac{nh}{2\pi mr} \quad (29.10)$$

Bohr kept enough of Rutherford's planetary model to use the relationship between speed and radius for an electron moving in a circular orbit that we found in Section 29.1 (Eq. 29.3),

$$v^2 = \frac{ke^2}{mr}$$

Inserting the result for v from Equation 29.10, we obtain

$$v^2 = \left(\frac{nh}{2\pi mr}\right)^2 = \frac{ke^2}{mr}$$

We can now solve for the orbital radius r:

$$\frac{n^2h^2}{4\pi^2m^2r^2} = \frac{ke^2}{mr}$$

$$r = n^2\left(\frac{h^2}{4\pi^2mke^2}\right) \quad (29.11)$$

The quantity in parentheses in Equation 29.11 is a combination of fundamental constants—Planck's constant, the mass and charge of an electron, and the constant k from Coulomb's law—so this term is also a constant. The variable n is an integer, so the factor n^2 on the right side of Equation 29.11 can have the values 1, $2^2 = 4$, $3^2 = 9$, and so on. In words, Equation 29.11 says that the orbital radius of an electron in a hydrogen atom can have only these particular quantized values. The smallest value of this radius is found when $n = 1$ and is called the ***Bohr radius*** of the hydrogen atom. We'll consider its value in Example 29.3.

Bohr radius of the hydrogen atom

Quantized Energies of the Bohr Atom

We can now continue with Bohr's approach and calculate the corresponding atomic energies using results from Section 29.1. The kinetic energy of an orbiting electron is $KE = \frac{1}{2}mv^2$. Inserting the result for v^2 from Equation 29.3, we get

$$KE = \frac{1}{2}mv^2 = \frac{1}{2}m\frac{ke^2}{mr} = \frac{1}{2}\frac{ke^2}{r} \quad (29.12)$$

The potential energy of this electron–proton pair is due to the electric force and is

$$PE_{\text{elec}} = -\frac{ke^2}{r} \quad (29.13)$$

The total energy is thus

$$E_{\text{tot}} = KE + PE_{\text{elec}} = \frac{1}{2}\frac{ke^2}{r} - \frac{ke^2}{r} = -\frac{1}{2}\frac{ke^2}{r} \quad (29.14)$$

Using the orbital radius r in Equation 29.11, we find

$$E_{\text{tot}} = -\frac{1}{2}ke^2\left(\frac{1}{r}\right) = -\frac{1}{2}ke^2\left(\frac{4\pi^2 mke^2}{n^2h^2}\right)$$

After some rearranging, we get

$$E_{\text{tot}} = -\left(\frac{2\pi^2k^2e^4m}{h^2}\right)\frac{1}{n^2} \tag{29.15}$$

The factor in parentheses is a constant (because it is a combination of fundamental constants). The only variable is the integer n, which can have values $n = 1, 2, 3, \ldots$. Hence, the energy of the hydrogen atom in the Bohr model can have only certain *quantized* values corresponding to $n = 1, 2, 3, \ldots$ in Equation 29.15. In this way, Bohr's postulate about quantized angular momentum in Equation 29.8 leads to quantized energy levels for the atom.

The values of the energy levels predicted from Equation 29.15 can be used to derive the frequencies of the absorption and emission lines of hydrogen. To see how that works, we first find the values of E_{tot} predicted by Equation 29.15. Inserting the values of the various fundamental constants leads to

$$E_{\text{tot}} = -\left(\frac{2\pi^2k^2e^4m}{h^2}\right)\frac{1}{n^2}$$

$$E_{\text{tot}} = -\left[\frac{2\pi^2(8.99 \times 10^9\ \text{N}\cdot\text{m}^2/\text{C}^2)^2(1.60 \times 10^{-19}\ \text{C})^4(9.11 \times 10^{-31}\ \text{kg})}{(6.63 \times 10^{-34}\ \text{J}\cdot\text{s})^2}\right]\frac{1}{n^2}$$

$$E_{\text{tot}} = -\frac{2.17 \times 10^{-18}\ \text{J}}{n^2}$$

Expressing this result in units of electron-volts, we find

Allowed energies of hydrogen in the Bohr model

$$E_{\text{tot}} = -\frac{13.6\ \text{eV}}{n^2} \tag{29.16}$$

We have given the value of E_{tot} to three significant figures so that we can more accurately compare it with the measured ionization energy of hydrogen.

The first few values in Equation 29.16 are plotted on the vertical axis in the energy-level diagram in Figure 29.12. The horizontal lines show levels corresponding to $n = 1, 2, 3, \ldots$. The level with $n = 1$ has the lowest energy and is the ground state of Bohr's hydrogen atom. Inserting $n = 1$ into Equation 29.16 gives $E_1 = -13.6$ eV. As n increases to 2, 3, . . . , the energies increase, with $E_2 = -3.4$ eV, $E_3 = -1.5$ eV, and so on, and at $n = \infty$, we have $E_\infty = 0$. All the levels except the highest one thus have negative energies. How can an orbiting electron have a negative energy? The energy in Equation 29.16 is the total mechanical energy of Bohr's atom, the sum of the kinetic and potential energies. The kinetic energy is positive, but the potential energy is negative since the Coulomb force exerted between the proton and electron is attractive, and our convention for electric potential energy has $PE_{\text{elec}} = 0$ when the electron is infinitely far from the proton. The level with $E_{\text{tot}} = 0$ in Figure 29.12 and Equation 29.16 corresponds to completely removing the electron from the proton, or ionization of the atom. The energy required to take an electron from the ground state E_1 and remove it from the atom is the ionization energy. Bohr's theory thus predicts an ionization energy of 13.6 eV, in excellent agreement with the measured value.

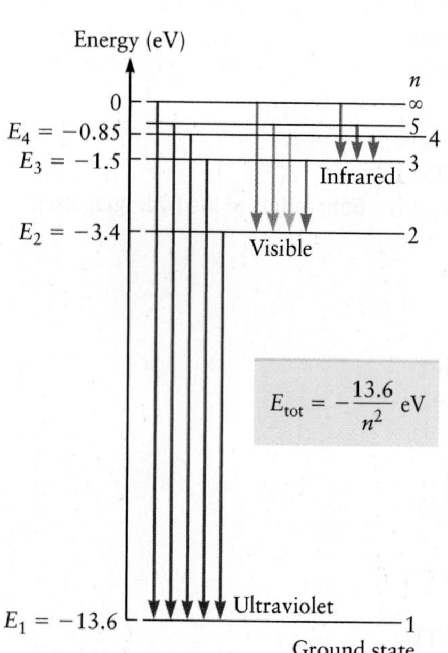

▲ **Figure 29.12** Energy-level diagram for the Bohr model of hydrogen.

The arrows in Figure 29.12 show some possible atomic transitions leading to emission of a photon. The energies of these transitions and the energies of the associated photons are equal to the difference in the energies of the levels at the start and end of each transition. Equation 29.16 predicts that there are an infinite number of energy levels (because $n = 1, 2, \ldots, \infty$), giving an infinite number of spectral emission lines. All these emission frequencies agree with the experimentally observed values, adding convincing evidence that Bohr was indeed on the right track. More work was needed, but Bohr's ideas took physics much of the way toward the correct quantum theory of the atom.

Insight 29.2

QUANTUM THEORY AND KINETIC THEORY OF A GAS

In our work on kinetic theory (Chapter 15), we claimed that the collisions between atoms in a gas are elastic (i.e., that kinetic energy is conserved). Quantum theory explains why. At room temperature, the kinetic energy of the colliding atoms is smaller than the spacing between the ground state and the excited states (Fig. 29.12), so a collision does not involve enough energy to cause a transition to a higher level. The atoms thus stay in their ground states and none of their kinetic energy is converted to potential energy of the atomic electrons.

EXAMPLE 29.3 The Bohr Model and the Size of a Hydrogen Atom

In Section 29.1, we used an approximate orbital radius $r = 1.0 \times 10^{-10}$ m for the hydrogen atom as estimated from measurements of the atomic spacing in molecules and solids. Calculate r from the Bohr model for the lowest three energy levels, that is, the levels corresponding to $n = 1$, 2, and 3.

RECOGNIZE THE PRINCIPLE

The orbital radius r in the Bohr model depends on the value of n (Eq. 29.11). The radius is smallest for the quantum state with $n = 1$ and increases as n increases.

SKETCH THE PROBLEM

No sketch is needed to start the problem, but a plot of the results for the orbits is given in Figure 29.13.

IDENTIFY THE RELATIONSHIPS AND SOLVE

The result of Bohr's theory for the orbital radius is (Eq. 29.11)

$$r = n^2\left(\frac{h^2}{4\pi^2 mke^2}\right)$$

Inserting the values of the constants in parentheses, we find

$$r = n^2\left[\frac{(6.63 \times 10^{-34}\ \text{J}\cdot\text{s})^2}{4\pi^2(9.11 \times 10^{-31}\ \text{kg})(8.99 \times 10^9\ \text{N}\cdot\text{m}^2/\text{C}^2)(1.60 \times 10^{-19}\ \text{C})^2}\right]$$

$$r = n^2 \times 5.3 \times 10^{-11}\ \text{m}$$

Evaluating for the given values of n leads to

$$n = 1\ \text{(ground state)}: r = 5.3 \times 10^{-11}\ \text{m} = \boxed{0.053\ \text{nm}} \quad (1)$$

$$n = 2: r = 2.1 \times 10^{-10}\ \text{m} = \boxed{0.21\ \text{nm}}$$

$$n = 3: r = 4.8 \times 10^{-10}\ \text{m} = \boxed{0.48\ \text{nm}}$$

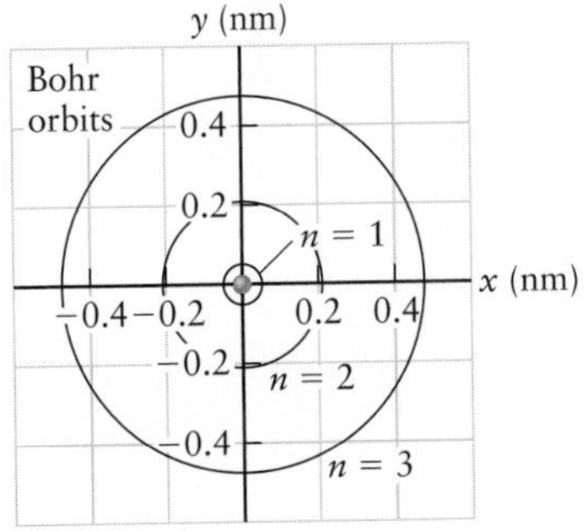

▲ **Figure 29.13** Example 29.3. Bohr orbits for the $n = 1$, 2, and 3 states of hydrogen. Note that the nucleus is not drawn to scale.

► ***What does it mean?***

The value of r for the ground state given in Equation (1) differs by only a factor of two from the value we used in our earlier calculations with the planetary model (Example 29.1), so our earlier estimates were fairly accurate. The value of r in Equation (1) is the value of the Bohr radius of the hydrogen atom and is used extensively in many estimates in atomic theory.

EXAMPLE 29.4 Bohr Theory and the Frequency of an Emission Line

Consider the transition in the Bohr model of a hydrogen atom from the $n = 2$ level to the $n = 1$ level. What is the wavelength of the photon emitted when the atom makes this transition?

(continued) ►

RECOGNIZE THE PRINCIPLE

The wavelength of an emission line is related to its frequency, which is proportional to the difference in energy of the two levels. We thus need to find the energy separation of the $n = 2$ and $n = 1$ levels, which we can get from the result for the Bohr model energy in Figure 29.12 or by using Equation 29.16.

SKETCH THE PROBLEM

Figure 29.12 shows the Bohr model energy levels.

IDENTIFY THE RELATIONSHIPS

The magnitude of the transition energy is $\Delta E = E_2 - E_1$, and that equals the photon energy hf. We thus have

$$hf = \Delta E = E_2 - E_1$$

$$f = \frac{\Delta E}{h}$$

The wavelength is then

$$\lambda = \frac{c}{f} = \frac{hc}{\Delta E}$$

SOLVE

We can read the values of E_1 and E_2 from Figure 29.12 and find

$$\Delta E = E_2 - E_1 = (-3.4 \text{ eV}) - (-13.6 \text{ eV}) = 10.2 \text{ eV}$$

Converting from electron-volts to joules gives

$$\Delta E = 10.2 \text{ eV} \times \frac{1.60 \times 10^{-19} \text{ J}}{1 \text{ eV}} = 1.63 \times 10^{-18} \text{ J}$$

The wavelength is thus

$$\lambda = \frac{(6.63 \times 10^{-34} \text{ J} \cdot \text{s})(3.00 \times 10^{8} \text{ m/s})}{1.63 \times 10^{-18} \text{ J}} = 1.22 \times 10^{-7} \text{ m} = \boxed{122 \text{ nm}}$$

▶ ***What does it mean?***

This wavelength lies in the ultraviolet; recall that the visible range ends at a wavelength of about 400 nm. This radiation would not be visible to the human eye.

Generation of X-rays by Atoms

The ionization energy of the hydrogen atom is 13.6 eV. We have seen that this is the energy of the hydrogen emission line with the highest photon energy, corresponding to a transition from the state with $E = 0$ to E_1 in Figure 29.12. The wavelength of this photon is $\lambda = 91.5$ nm (Example 29.2), which lies in the ultraviolet part of the electromagnetic spectrum. Although it is the highest photon energy that can be emitted by a hydrogen atom, other atoms can emit much more energetic photons. As the charge on the nucleus is increased (by going from hydrogen to helium and to heavier atoms), the magnitude of the electric potential energy of an electron also increases. As a result, for helium and heavier atoms the energy required to remove an electron from the $n = 1$ Bohr orbit (i.e., a tightly bound, "inner-shell" electron) is much larger than for hydrogen. The corresponding photon energies then increase into the X-ray region. Many applications employ X-ray photons generated when an electron undergoes a transition from the E_2 state to the E_1 state (Fig. 29.14A). These photons are denoted as "K_α X-rays," and their energies for a few elements are listed in Table 29.1. The X-ray photons used by your dentist or doctor are produced by these atomic transitions, usually involving atoms of tungsten or molybdenum (Fig. 29.14B).

TABLE 29.1 Energy of K_α X-rays Produced by a Few Elements

Element	Atomic Number (Z)	K_α X-ray Energy (keV)
Ne	10	0.85
Al	13	1.6
Cu	29	8.0
Mo	42	17
W	74	58

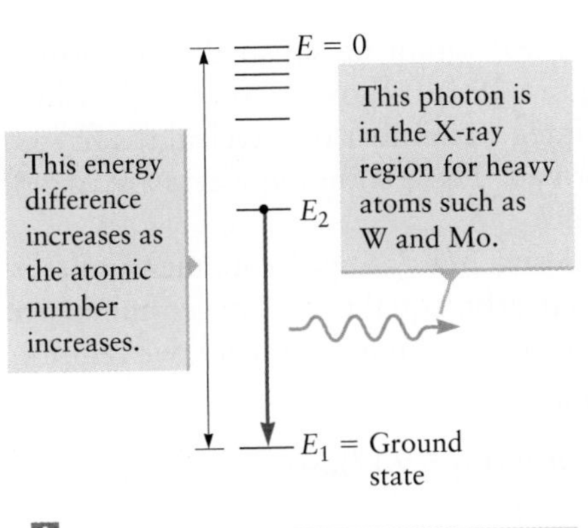

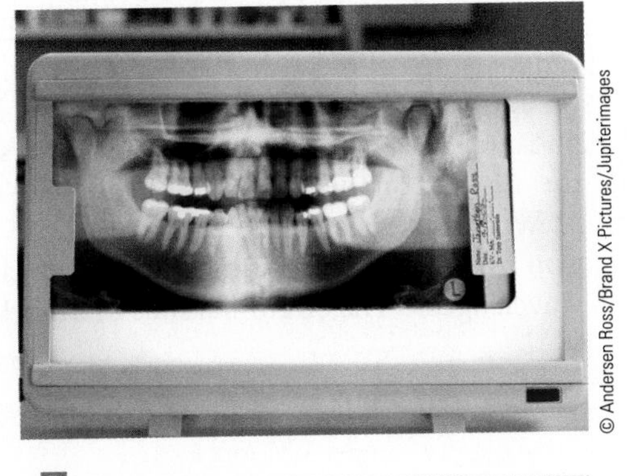

© Andersen Ross/Brand X Pictures/Jupiterimages

◀ **Figure 29.14** A For heavy atoms, the emission lines can lie in the X-ray part of the electromagnetic spectrum. B The X-rays used by a dentist are photons generated by atomic transitions of atoms such as (typically) molybdenum and tungsten.

CONCEPT CHECK 29.2 **K_α X-rays and the Energy Levels of an Atom**

The energy of a K_α X-ray of copper (Cu) is 8.0 keV (Table 29.1). According to the Bohr model, is the energy required to remove an electron from the $n = 1$ Bohr orbit of a copper atom (and ionize the atom) (a) equal to, (b) less than, or (c) greater than 8.0 keV?

Continuous Spectra

The existence of discrete spectral lines is an essential part of quantum theory. In certain situations, however, the absorption may not involve discrete photon frequencies. Figure 29.15 shows a hydrogen atom that is initially in its ground state and then absorbs a photon. This photon has an energy greater than 13.6 eV, so it has more than enough energy to remove (eject) the electron from the atom. The extra energy goes into the kinetic energy of the ejected electron. Because the ejected electron's final kinetic energy can have a continuous range of values, the absorbed photon can also have a continuous range of energies. Hence, when an atom is ionized, the absorbed energy can have a range of values, producing what is called a continuous spectrum. Not everything is quantized in the Bohr model or in quantum mechanics.

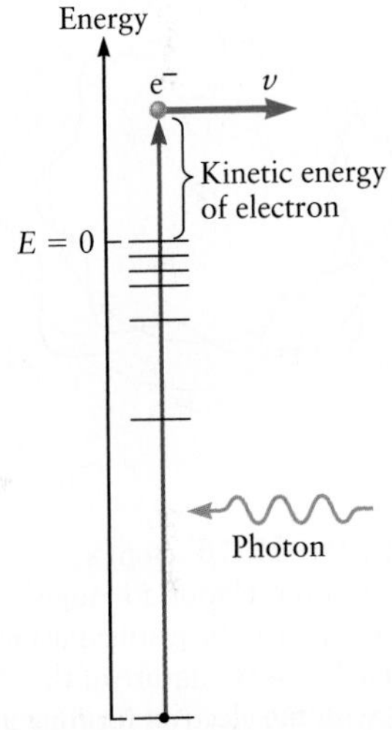

▲ **Figure 29.15** When an atom absorbs a photon and is ionized, the final kinetic energy of the electron is not quantized. Hence, the photon can have a range of energies, and the absorption energy in this case is *not* quantized.

Why Is Angular Momentum Quantized?

Bohr's theory provided the first successful calculation of the frequencies of the spectral lines of any atom, but at first no one knew *why* it worked. What is the Bohr model telling us about the quantum world? Bohr's key assumption, that the angular momentum of the electron is quantized (Eq. 29.8), was completely new. This assumption was needed to give quantized atomic levels, without which the Bohr model would have many of the same problems as Newton's mechanics and the planetary model. Bohr's assumption about angular momentum can be understood using the de Broglie theory of particle-waves (but note that the de Broglie theory came about 10 years after Bohr's work). De Broglie proposed (Chapter 28) that electrons and all other particles have a wave character, with a wavelength λ given by (Eq. 28.8)

$$\lambda = \frac{h}{p} \tag{29.17}$$

where p is the momentum of the particle. Figure 29.16 shows such a particle-wave orbiting a nucleus, corresponding to an electron moving in a circular orbit. The allowed electron orbits in the Bohr model correspond to standing waves that fit precisely into the orbital circumference. To form a standing wave, the circumference of the orbit must be an integer multiple of the electron wavelength. Figure 29.16A shows a hypothetical case in which the electron wavelength does not match the orbit; for this value of r and λ, it is not possible to form a standing wave, and Bohr's

This wave does not join with itself for this particular wavelength, so this orbital radius is not allowed in the Bohr model.

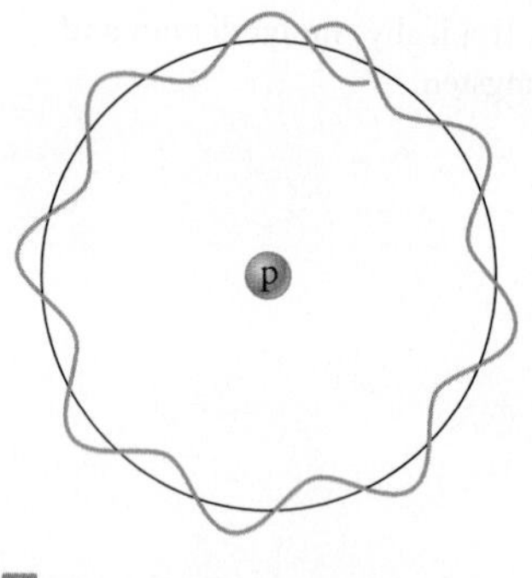

Allowed standing wave orbits in the Bohr model.

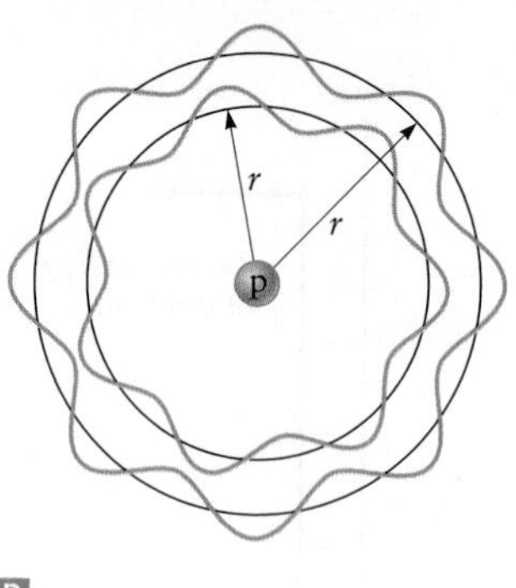

▲ **Figure 29.16** Bohr's quantization relation for angular momentum can be justified using the particle-wave nature of the electron, with the electron forming a standing wave in its orbit about the nucleus. The Bohr theory requires that an integer multiple of electron wavelengths equal the orbital circumference. In A, this condition is not met, so this orbital radius is not allowed in the Bohr theory for the wavelength shown. B Two allowed orbits. Note that the protons are not drawn to scale.

condition for the angular momentum (Eq. 29.8) cannot be satisfied. This orbital radius is not allowed in the Bohr model. Figure 29.16B shows two cases in which the radius and wavelength match so that an integral number of wavelengths fit into one circumference and a standing wave is formed. These standing waves are two of the allowed quantized states of the Bohr model.

We can use the picture in Figure 29.16 to derive Bohr's angular momentum relation by setting the circumference of the circular orbit equal to an integer number of wavelengths. The circumference of an orbit is $2\pi r$, so if n is an integer, we require

$$2\pi r = n\lambda$$

The wavelength is given by de Broglie as $\lambda = h/p$ (Eq. 29.17), so

$$2\pi r = n\lambda = n\frac{h}{p} \tag{29.18}$$

The momentum of a particle is $p = mv$; inserting this into Equation 29.18 gives

$$2\pi r = n\frac{h}{p} = n\frac{h}{mv}$$

Rearranging and noting that the angular momentum is $L = mvr$, we arrive at

$$L = mvr = n\frac{h}{2\pi} \tag{29.19}$$

which is precisely Bohr's condition for the angular momentum, Equation 29.8.

Problems with the Bohr Model: Where Do We Go Next?

The success of the Bohr model of the hydrogen atom inspired Bohr and others to apply the same ideas to other atoms, and the Bohr theory was found to work well when only a single electron is present, as in the ions He^+ and Li^{2+}. This model does not, however, correctly explain the properties of atoms or ions that contain two or more electrons. Physicists eventually concluded that the Bohr model is not the correct quantum theory, but rather a "transition theory" that helped pave the way from Newton's mechanics to modern quantum mechanics. The Bohr model gives very useful insights into why an atom has quantized states, but the correct quantum theory contains even more radical ideas.

29.4 Wave Mechanics and the Hydrogen Atom

Bohr based his work on Newton's mechanics, to which he added a few new assumptions about the nature of electron motion, but his approach still relied on classical ideas based on electrons moving in mechanical orbits. The quantum theory developed by Schrödinger, Heisenberg, and others in the 1920s provided a much more radical break with Newton. Modern quantum mechanics, also called ***wave mechanics***, is based not on mechanical variables such as position and velocity, but on wave functions and probability densities as explained in Section 28.5. Although we will not go into the details here, one solves a problem in quantum mechanics using ***Schrödinger's equation***. The solution of this equation gives the wave function, including its dependence on position and time. In this section, we describe the wave function and some features of the quantized states of the hydrogen atom. In Section 29.5, we'll apply these ideas to multielectron atoms to explain the structure of the periodic table of the elements.

In Bohr's theory, the energy levels are described by a single integer with allowed values $n = 1, 2, 3, \ldots$ corresponding to different quantized electron states. The integer n is called a ***quantum number***. In Schrödinger's quantum theory, a full description of the quantum states of electrons in an atom requires *four* quantum

TABLE 29.2 Quantum Numbers for Allowed Electron States in an Atom

Quantum Number	Name	Possible Values
n	Principal quantum number	$n = 1, 2, 3, \ldots$
ℓ	Orbital quantum number	$\ell = 0, 1, 2, \ldots, n - 1$
m	Orbital magnetic quantum number	$m = -\ell, -\ell + 1, \ldots, 0, \ldots, \ell - 1, \ell$
s	Spin quantum number	$s = -\frac{1}{2}$ or $+\frac{1}{2}$

numbers. These quantum numbers are collected and described in Table 29.2. Each allowed electron energy level is specified by a set of values for all four quantum numbers and corresponds to one of the quantum states.

The Four Quantum Numbers for Electron States in Atoms

Quantum numbers for electron states in an atom

The four quantum numbers for electron states in atoms are n, ℓ, m, and s and are defined as follows.

***n* is the *principal quantum number*.** It can have the values $n = 1, 2, 3, \ldots$. This quantum number is roughly similar to Bohr's quantum number. As n increases, the average distance from the electron to the nucleus increases. (See also Example 29.3.) The states with a particular value of n are referred to as a "shell." For example, all the states with $n = 2$ make up the "$n = 2$ shell."

***ℓ* is the *orbital quantum number*,** with allowed values $\ell = 0, 1, \ldots, n - 1$. The angular momentum of the electron is proportional to ℓ. An electron state with $\ell = 0$ is called an "*s* state," while states with $\ell = 1$ are called "*p* states." The shorthand letters for other states are (in order) "*d*" ($\ell = 2$) and "*f*" ($\ell = 3$) as listed in Table 29.3. States with $\ell = 0$ have zero angular momentum.

***m* is the *orbital magnetic quantum number*,** with allowed values $m = -\ell$, $-\ell + 1, \ldots, -1, 0, +1, \ldots, +\ell$. Intuitively, you can think of m as giving the direction of the angular momentum of the electron in a particular state.

***s* is the *spin quantum number*.** While the other quantum numbers all have integer values, the spin quantum number of an electron is $s = +\frac{1}{2}$ or $-\frac{1}{2}$, often referred to as "spin up" and "spin down." This quantum number gives the direction of the electron's spin angular momentum as discussed in Section 28.4.

Warning! Do not confuse the spin quantum number s with the notion of an "*s* state" ($\ell = 0$); they refer to completely different quantum numbers. This notation came about for historical reasons, and we are now (unfortunately) stuck with it.

TABLE 29.3 Decoding the Configuration Shorthand for States with Various Orbital Quantum Numbers

Orbital Quantum Number	Configuration Letter
$\ell = 0$	*s*
$\ell = 1$	*p*
$\ell = 2$	*d*
$\ell = 3$	*f*
$\ell = 4$	*g*
$\ell = 5$	*h*

Electron Shells and Probability Distributions

A particular quantized electron state is specified by all four of the quantum numbers n, ℓ, m, and s. Solution of the Schrödinger equation also gives the wave function of each quantum state, and from the wave function we can calculate the probability for finding the electron at different locations around the nucleus. The electron probability distributions for a few states of the hydrogen atom are shown in Figure 29.17; these probability plots are often called "electron clouds."

The ground state of hydrogen is specified by $n = 1$. According to the rules in Table 29.2, the only allowed values of the orbital and orbital magnetic quantum numbers for $n = 1$ are $\ell = 0$ (an "*s* state") and $m = 0$, but the electron can be in either the spin-up ($s = +\frac{1}{2}$) or spin-down ($s = -\frac{1}{2}$) state. The probability of finding the electron at a particular location does not depend on the value of s, so the spin-up and spin-down probabilities are the same. A plot of the electron probability

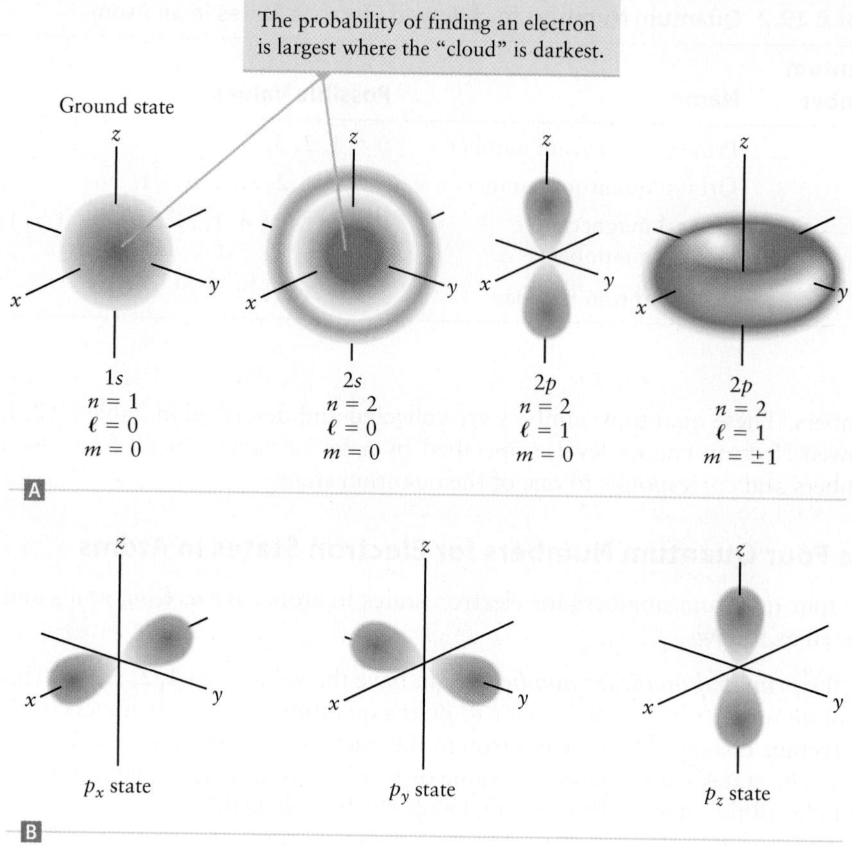

► **Figure 29.17** A Electron probability distributions for the 1*s*, 2*s*, and 2*p* states of hydrogen, calculated using the Schrödinger equation. The 2*p* states can be pictured in two different ways. In part A, the 2*p* electron probability clouds for $m = \pm 1$ are shown with "doughnut" shapes. In B, the same 2*p* states are redrawn in an equivalent form popular in chemistry, called p_x, p_y, and p_z states.

for the ground state of hydrogen is shown at the far left of Figure 29.17A. The electron probability distribution forms a spherical "cloud" around the nucleus. For this state, the probability is largest at the nucleus (where the cloud is "darkest" in Fig. 29.17A), so the electron has the highest probability of being found at or near the nucleus. In this ground state, the electron occupies the level with the lowest possible energy. There are two such states, with $n = 1$, $\ell = 0$, $m = 0$, and $s = +\frac{1}{2}$ or $-\frac{1}{2}$. Together they make up the $n = 1$ "shell" and are called the 1*s* states of the atom.

Figure 29.17 also shows the electron probability for the states with principal quantum number $n = 2$, called the $n = 2$ shell. The probability for the state with $n = 2$, $\ell = 0$, and $m = 0$ has a spherical component (a spherical "shell" where the probability is large) near the nucleus and a second "layer" (another spherical shell where the probability is high) farther away. The spin quantum number for electrons in this state can again have the values $s = +\frac{1}{2}$ and $-\frac{1}{2}$, forming two 2*s* states of the atom. The $n = 2$ shell also contains states with $\ell = 1$, which are called the 2*p* states. The value of the orbital quantum number ℓ specifies the "subshell," so the 2*s* and 2*p* states are said to be in the same shell (because they have the same value of *n*) but in different subshells. According to the quantum number rules in Table 29.2, there are 2*p* states with $m = +1$, 0, and -1. There are two ways to display the probability clouds for the 2*p* states with $m = \pm 1$. Many physicists prefer to show them as a doughnut-like electron cloud (on the far right in Fig. 29.17A). However, the two states with $m = \pm 1$ can also be combined to make what chemists call p_x and p_y states (Fig. 29.17B). While it is not obvious, both plots show the *same* quantum states. The chemists' way of displaying these electron probabilities is useful for understanding the directionality of chemical bonding.

For the hydrogen atom, the electron energy depends only on the value of *n* and is independent of the values of ℓ, *m*, and *s* (except for some tiny effects we will not discuss in this book). That is not the case for atoms containing more than one electron, as we'll see in the next section.

29.5 Multielectron Atoms

The quantum states we have discussed for hydrogen can be used to describe the states of atoms containing more than one electron; they are called ***multielectron atoms***. The electron energy levels of these atoms follow the same pattern as found for hydrogen, with the same quantum numbers listed in Tables 29.2 and 29.3. The electron probability distributions are also similar. There are two main quantitative differences between the electron states in a multielectron atom and in hydrogen. First, the values of the electron energies are different for different atoms. For example, the energy of the 1*s* state in helium is different from the energy of the 1*s* state in hydrogen. Second, the spatial extent of the electron probability clouds varies from element to element. For example, the 1*s* electron probability "cloud" for helium is closer on average to the nucleus than it is for hydrogen.

While the electron levels of all atoms are thus similar to those of hydrogen, one crucial feature is important for atoms with more than one electron: *each quantum state can be occupied by only one electron*. That is, each electron in an atom must occupy its own quantum state, different from the states of all other electrons. This is called the ***Pauli exclusion principle***. Since each quantum state is characterized by a unique set of quantum numbers, each electron is described by a unique set of quantum numbers.

Pauli exclusion principle

Figure 29.18 shows how electrons are distributed among the possible energy levels in several different cases. The diagram in Figure 29.18A shows the levels of an "empty" atom, that is, an atom before we have added any electrons. For an "empty" atom and for hydrogen, all quantum states for a given value of Z and with a particular value of the principal quantum number n have almost exactly the same energy, so we only show a single horizontal line for each value of n.

In Figure 29.18B, we apply the energy-level diagram to hydrogen, in which there is only one electron to consider. We show this electron in the ground state (occupying the lowest energy level) as an arrow, with the direction of the arrow denoting the value of the electron's spin. (Here we have arbitrarily shown the electron as spin up, but it could just as well be spin down.) In words, the diagram in Figure 29.18B indicates that for a hydrogen atom in the ground state, the electron occupies the $n = 1$, $\ell = 0$, $m = 0$, $s = +\frac{1}{2}$ energy level. This is also called a "$1s^1$" ***electron configuration***. This notation is a useful shorthand and is explained in Equation 29.20.

$$\text{electron configuration for H: } 1s^1 \quad (n = 1;\ \text{one electron};\ \ell = 0) \tag{29.20}$$

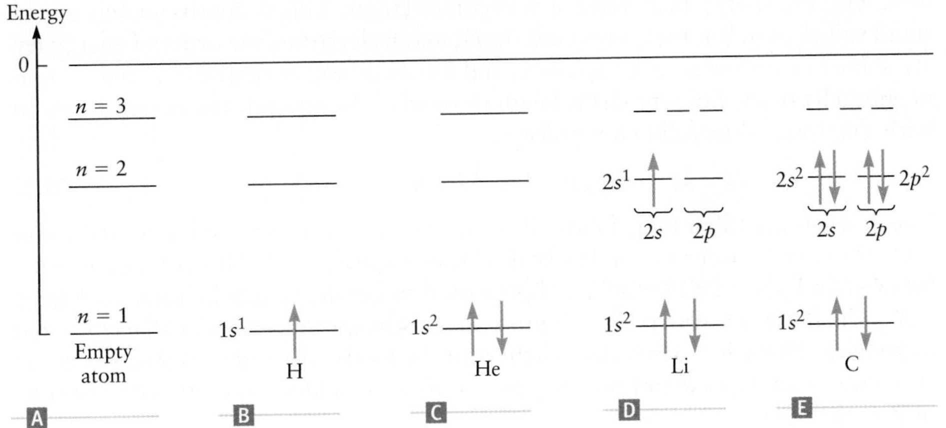

◀ **Figure 29.18** Filling the energy levels of an atom. **A** Empty levels. **B** Electron configuration of a hydrogen atom in its ground state. Configurations for **C** helium (He), **D** lithium (Li), and **E** carbon (C). These diagrams show the *order* of energy levels for each type of atom (each value of Z). The energies of these levels (e.g., the ground state and the excited states) are different for different atoms.

The first number in the configuration shorthand indicates the value of the principal quantum number n, while the letter indicates the value of ℓ using the code in Table 29.3. In Equation 29.20, the letter "s" indicates that $\ell = 0$. The superscript indicates that one electron is in the energy level with these values of n and ℓ. The value of the electron spin is not explicitly given in this notation.

Parts C through E of Figure 29.18 give energy-level diagrams for several other atoms. Helium (He) has two electrons, occupying levels with the quantum numbers $n = 1$, $\ell = 0$, $m = 0$, and $s = \pm\frac{1}{2}$, which satisfies the Pauli exclusion principle because the electrons have different values of their spin quantum number. The electrons are indicated by the two arrows in Figure 29.18C, one pointing up (spin up, denoting $s = +\frac{1}{2}$) and the other pointing down (spin down, denoting $s = -\frac{1}{2}$). In the configuration shorthand, we have

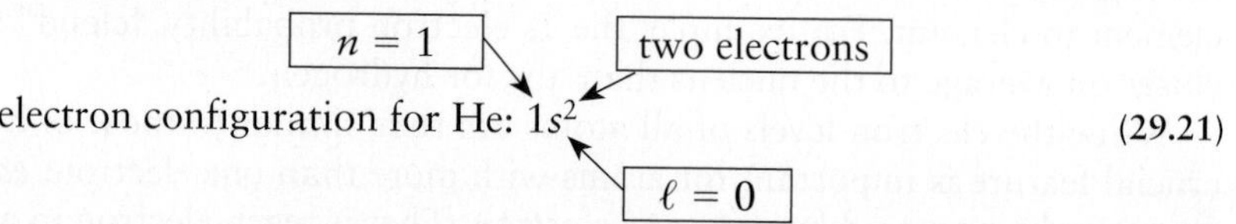

$$\text{electron configuration for He: } 1s^2 \tag{29.21}$$

This configuration differs from that of hydrogen (Eq. 29.20) only in the superscript; here the superscript 2 indicates that in helium there are two electrons in the energy levels with $n = 1$ and $\ell = 0$.

The energy-level diagram for the ground state of lithium (Li) is shown in Figure 29.18D; a lithium atom has three electrons, indicated by the three arrows. As with helium, two of these electrons occupy the levels with $n = 1$, $\ell = 0$, $m = 0$, and $s = \pm\frac{1}{2}$. According to the rules in Table 29.2, there are no other possible energy levels with $n = 1$. Since each electron must have its own unique set of quantum numbers, the third electron in lithium must occupy a higher energy level. This third electron is indicated by the arrow at the $n = 2$ level in Figure 29.18D. The lowest energy level in a shell has $\ell = 0$, so Figure 29.18D shows this third electron in the $2s$ state. The full set of quantum numbers for this electron is thus $n = 2$, $\ell = 0$, $m = 0$, and $s = +\frac{1}{2}$. The corresponding configuration for lithium is then

$$\text{electron configuration for Li: } 1s^2 2s^1$$

Notice that the configuration shorthand does not indicate the value of m for any of the electrons.

Carbon (C) contains six electrons. The third, fourth, fifth, and sixth electrons in a carbon atom occupy states with $n = 2$ as indicated in Figure 29.18E. The corresponding configuration is

$$\text{electron configuration for C: } 1s^2 2s^2 2p^2$$

There are two electrons (spin up and spin down) in levels with $n = 1$ and $\ell = 0$; two electrons (spin up and spin down) in levels with $n = 2$, $\ell = 0$ (the $2s$ levels); and two electrons (spin up and spin down) in levels with $n = 2$ and $\ell = 1$ (the $2p$ levels).

The energies of levels in the diagrams in Figure 29.18 depend mainly on the value of n, with the energy increasing as n becomes larger. This is approximately true at small values of n, but for atoms containing many electrons, the order of energy levels is more complicated as n increases, and for shells higher than $n = 2$, the energies of subshells from different shells begin to overlap. In general, the energy levels fill with electrons in the following order:

$$1s \quad 2s \quad 2p \quad 3s \quad 3p \quad 4s \quad 3d \quad 4p \quad 5s \quad 4d \quad 5p \quad 6s \quad 4f \tag{29.22}$$

The $1s$ levels are filled first, followed by the $2s$ levels, then the $2p$ levels, and so on, with the overall order of the levels shown in Figure 29.19. This filling pattern is followed in Figure 29.18 and can also be used to get the results in Table 29.4 (page 921), which lists the electron configurations for the ground states of all elements up to sodium. However, when one reaches the $3d$ levels, the order of levels becomes even more complicated and can vary as electrons are added in going from one value of Z to the next.

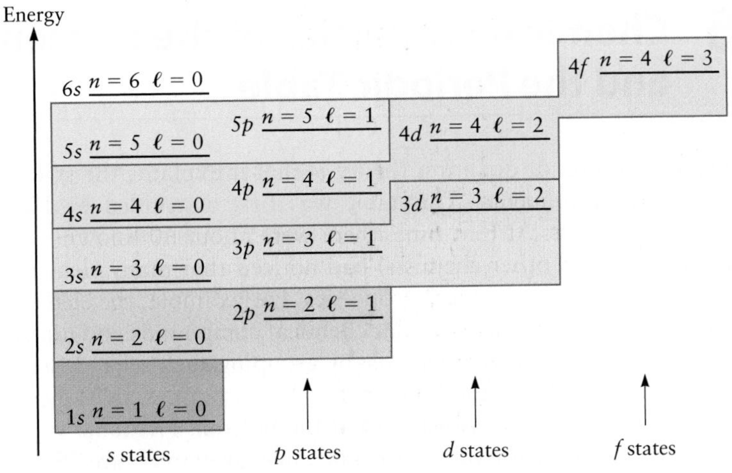

◀ **Figure 29.19** Order of energy levels in an atom.

CONCEPT CHECK 29.3 What Element Is It?

The ground state of an atom has the configuration $1s^2 2s^2 2p^6 3s^2 3p^5$. What element is it?

EXAMPLE 29.5 Electron Configuration and Energy-Level Diagram for Silicon

A silicon atom contains 14 electrons. Determine its electron configuration and construct the corresponding diagram of occupied levels.

RECOGNIZE THE PRINCIPLE

According to the Pauli exclusion principle, no two electrons can occupy the same quantum state, that is, have the same set of quantum numbers. To find the configuration of silicon, we assign the 14 electrons to the 14 quantum states that have the lowest energy.

SKETCH THE PROBLEM

We start with Figure 29.19 and add electrons to the lowest energy levels. Adding 14 electrons gives the diagram of filled energy levels in Figure 29.20.

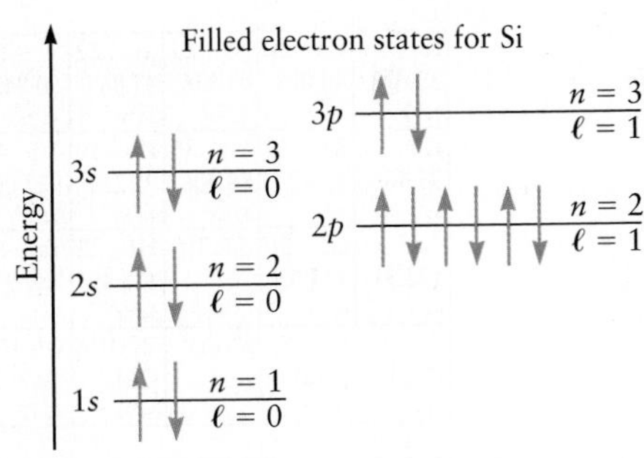

▲ **Figure 29.20** Example 29.5.

IDENTIFY THE RELATIONSHIPS

We fill the lowest 14 energy levels in Figure 29.19 according to the rules in Table 29.2. The 1*s*, 2*s*, and 3*s* levels can each hold two electrons (one electron with spin up and another with spin down), and the 2*p* and 3*p* states can each hold six electrons.

SOLVE

The 14 electrons in silicon will fill completely the 1*s* (two electrons), 2*s* (two electrons), 2*p* (six electrons), and 3*s* (two electrons) energy levels, leaving two electrons for the 3*p* state. The resulting configuration is $\boxed{1s^2 2s^2 2p^6 3s^2 3p^2}$.

▶ ***What does it mean?***

The energy-level diagram in Figure 29.19 is the key to understanding the electron configurations of all the elements. The filled electron states of a particular atom are found by simply adding electrons to the lowest available energy levels.

29.6 Chemical Properties of the Elements and the Periodic Table

One important triumph of quantum theory is that it explains the structure of the periodic table of the elements. This table was first assembled by Dmitry Mendeleyev in the late 1860s. At that time, there were about 60 known elements, and Mendeleyev (along with other chemists) had noticed that many elements could be grouped according to their chemical properties. For example, the elements lithium, sodium, and potassium form very similar chemical compounds and undergo similar reactions. Mendeleyev organized his table by grouping such related elements in the same column. He also placed the lightest (i.e., low atomic weight) elements at the top of each column and the heaviest ones at the bottom. His table was essentially identical to the modern version of the periodic table given in Figure 29.21. The main difference was that Mendeleyev's table had a number of "holes" (i.e., openings) because many elements had not been discovered yet.

While Mendeleyev discovered an intriguing regularity in the elements and their properties, he could not explain why the periodic table has the form it does. That is, why does it contain just two elements in the first row, eight in the second and third rows, and even more elements in the lower rows? The answer is contained in the organization of electron energy levels given in Table 29.4 and Figure 29.19. To appreciate this answer, we must first understand how the electron energy levels and the electron configuration of an atom are responsible for its chemical properties.

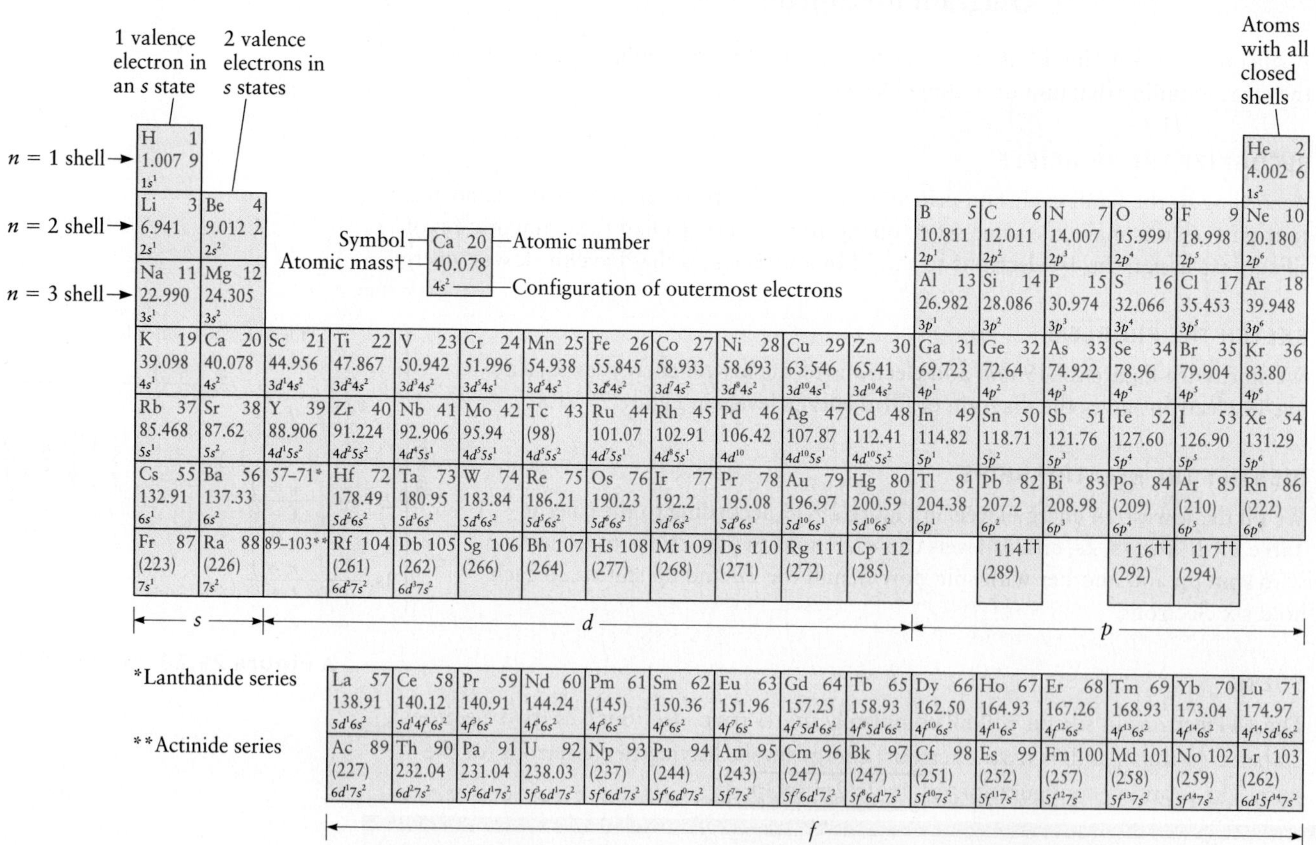

Note: Atomic mass values given are averaged over isotopes in the percentages in which they exist in nature. For a description of the atomic data, visit physics.nist.gov/PhysRefData/Elements/per_text.html.

† For an unstable element, mass number of the most stable known isotope is given in parentheses.

†† Elements 114, 116, and 117 have not yet been named.

▲ **Figure 29.21** Periodic table of the elements.

TABLE 29.4 Electron Configurations for Several Elements

Element	Number of Electrons	Electron Configuration	Quantum Numbers of Occupied Electron Levels
H (hydrogen)	1	$1s^1$	$n = 1, \ell = 0, m = 0, s = +\frac{1}{2}$
He (helium)	2	$1s^2$	$n = 1, \ell = 0, m = 0, s = +\frac{1}{2}$ $n = 1, \ell = 0, m = 0, s = -\frac{1}{2}$
Li (lithium)	3	$1s^2 2s^1$	Same as He plus $n = 2, \ell = 0, m = 0, s = +\frac{1}{2}$
Be (beryllium)	4	$1s^2 2s^2$	Same as He plus $n = 2, \ell = 0, m = 0, s = +\frac{1}{2}$ $n = 2, \ell = 0, m = 0, s = -\frac{1}{2}$
B (boron)	5	$1s^2 2s^2 2p^1$	Same as Be plus $n = 2, \ell = 1, m = 0, s = +\frac{1}{2}$
C (carbon)	6	$1s^2 2s^2 2p^2$	Same as Be plus $n = 2, \ell = 1, m = 0, s = +\frac{1}{2}$ $n = 2, \ell = 1, m = 0, s = -\frac{1}{2}$
N (nitrogen)	7	$1s^2 2s^2 2p^3$	Same as C plus $n = 2, \ell = 1, m = 1, s = +\frac{1}{2}$
O (oxygen)	8	$1s^2 2s^2 2p^4$	Same as C plus $n = 2, \ell = 1, m = 1, s = +\frac{1}{2}$ $n = 2, \ell = 1, m = 1, s = -\frac{1}{2}$
F (fluorine)	9	$1s^2 2s^2 2p^5$	Same as O plus $n = 2, \ell = 1, m = -1, s = +\frac{1}{2}$
Ne (neon)	10	$1s^2 2s^2 2p^6$	Same as O plus $n = 2, \ell = 1, m = -1, s = +\frac{1}{2}$ $n = 2, \ell = 1, m = -1, s = -\frac{1}{2}$
Na (sodium)	11	$1s^2 2s^2 2p^6 3s^1$	Same as Ne plus $n = 3, \ell = 0, m = 0, s = +\frac{1}{2}$

When an atom participates in a chemical reaction, some of its electrons combine with electrons from other atoms to form chemical bonds. The bonding electrons are those that are most easily removed from an atom; hence, they are the electrons occupying the highest energy levels. For example, the ground state of lithium has the configuration $1s^2 2s^1$ (Fig. 29.18D and Table 29.4). The $2s$ electron occupies the state with the highest energy and is more weakly bound than the two electrons in the $1s$ state. This $2s$ electron is called a ***valence*** electron. In contrast, the two electrons in the $1s$ state form a ***closed shell***, a shell with all possible states filled. These $1s$ electrons have a much lower energy than the $2s$ electron. The $1s$ electrons are much more difficult to remove from the atom and do not participate in bonding (in chemical reactions).

The same ideas apply to other atoms. For example, the bonding of a sodium atom involves the $3s$ electron (Table 29.4). The remaining electrons form a closed $1s$ shell, a closed $2s$ subshell, and a closed $2p$ subshell. The configuration of the closed shells is thus $1s^2 2s^2 2p^6$. Likewise, potassium also has a single bonding electron, occupying the $4s$ state outside a $1s^2 2s^2 2p^6 3s^2 3p^6$ configuration. The elements lithium, sodium, and potassium thus have the same number of valence electrons (one), all in s states. As a result, these elements have similar chemical bonding properties, which is why Mendeleyev placed them in the same column in the periodic table (Fig. 29.21).

CONCEPT CHECK 29.4 Electron Configuration of an Ion

What is the electron configuration of the ion O^{-2}?

(a) $1s^2 2s^2 2p^4$ (b) $1s^2 2s^2 2p^6$ (c) $1s^2 2s^2$ (d) $1s^2 2s^2 2p^2$

Closed Shells and the Periodic Table

The last column in the periodic table includes helium, neon, and argon (He, Ne, and Ar). Atoms of these elements all contain completely filled shells. Since they have no valence electrons, it is very difficult for them to form chemical bonds. Such elements are largely inert, almost never participating in chemical reactions. Atoms of these elements also interact very weakly with each other, resulting in low condensation and freezing temperatures. For this reason, they are called ***inert gases*** or ***noble gases.*** In Mendeleyev's time, elements were discovered through their chemical reactions and compounds, so very little was known about these elements when Mendeleyev did his work in the 1860s. In fact, most inert gases were discovered after Mendeleyev's time, so his version of the periodic table did not even contain that column.

Structure of the Periodic Table

In his periodic table, Mendeleyev grouped elements into *columns* according to their common bonding properties and chemical reactions. These properties rely on the valence electrons and can thus be traced to the electron configurations in Table 29.4. What, though, determines the number of elements in each *row* of the table, and why do different rows have different numbers of elements? Each row in the periodic table corresponds to a particular value of the principal quantum number n. The row containing hydrogen and helium corresponds to $n = 1$, the row beginning with lithium has $n = 2$, and so forth. The number of elements in a given row is equal to the number of electrons needed to fill completely a particular shell or subshell. Because the $n = 1$ shell can hold only two electrons, this row contains just two elements. The $n = 2$ shell can hold eight electrons (two in the $2s$ states and six more in the $2p$ states), so this row contains eight elements. The number of elements in all the other rows[3] can be found using the rules for allowed quantum numbers in Table 29.2.

EXAMPLE 29.6 Filling the 4*f* Subshell

What is the maximum number of electrons that can occupy the $4f$ subshell of an atom?

RECOGNIZE THE PRINCIPLE

Each electron state in the $4f$ subshell must have a unique set of quantum numbers, and the allowed quantum numbers are determined by the rules in Table 29.2.

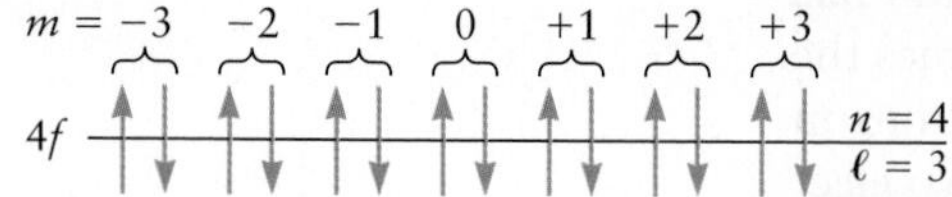

▲ **Figure 29.22** Example 29.6.

SKETCH THE PROBLEM

No figure is necessary, but the result is illustrated in Figure 29.22.

IDENTIFY THE RELATIONSHIPS AND SOLVE

An f subshell corresponds to $\ell = 3$ (Table 29.3). According to Table 29.2, the allowed values of m, the orbital magnetic quantum number, are $m = -\ell, -\ell + 1, \ldots, \ell - 1, \ell$. If $\ell = 3$, we then have $m = -3, -2, -1, 0, 1, 2, 3$, for a total of seven different possible values of m. For each of these different values of m, the spin quantum number can be

[3]The first three rows of the periodic table correspond to filling levels with the same principal quantum number, but starting with the fourth row, subshells with different values of n are filled as one moves across a row. This is because of the more complicated order of energy levels when $n > 3$ as shown in Figure 29.19.

either $s = +\frac{1}{2}$ (spin up) or $s = -\frac{1}{2}$ (spin down), so the total number of allowed states in the $4f$ subshell is $2 \times 7 = \boxed{14}$.

► ***What does it mean?***

Figure 29.22 shows the electron states for a filled $4f$ subshell. From the periodic table (Fig. 29.21), the elements constructed by adding electrons to this subshell are called the lanthanide series. These elements have many important applications, including in lasers, catalysts, and permanent magnets.

CONCEPT CHECK 29.5 Valence Electrons for Barium

How many valence electrons does barium have, and what energy levels do they occupy?

(a) One valence electron in the $6s$ state
(b) Two valence electrons in $6s$ states
(c) No valence electrons

CONCEPT CHECK 29.6 Valence Electrons of an Atom

Which atoms from this group—(a) Sr, (b) F, (c) Mg, and (d) K—have the same number of valence electrons?

29.7 Applications

The existence of discrete spectral lines played a major role in the development of quantum theory and also makes possible a number of important applications.

Atomic Clocks and the Definition of the Second

Atomic clocks are used as global time standards (Fig. 29.23). These clocks are based on the accurate measurement of certain spectral line frequencies. Since all atoms of a given element are identical, they all have precisely the same set of spectral lines at precisely the same frequencies.

Atomic clocks have been constructed using several different types of atoms, with cesium (Cs) being a popular choice. Two of the cesium levels differ by an energy of approximately 3.8×10^{-5} eV ($\approx 6.1 \times 10^{-24}$ J). These two levels are much closer in energy than the typical spacing of a few electron-volts seen for the levels in hydrogen. The spectral emission line produced when a cesium atom undergoes a transition between these two levels is very strong, and the frequency of this spectral line can be measured with *extremely* high accuracy.

A

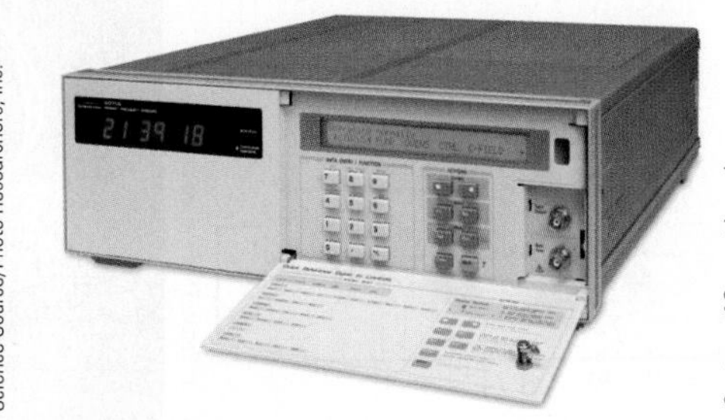

B

◄ **Figure 29.23** A Early atomic clock. This clock dates from the 1950s. B Modern atomic clocks are much more compact and fit on the corner of a desk.

Insight 29.3

DEFINING THE "SECOND"

The second is now defined in terms of the frequency of a particular atomic transition in Cs. Before 1960, the second was defined in terms of the length of a solar day, and 1 second was 1/86,400 of the length of what astronomers call a "mean solar day." A mean solar day is the time it takes (on average) for the Earth to complete one rotation about its axis. A major difficulty with this definition is that the Earth's rotation rate is gradually slowing. For this reason, the modern definition is much more useful in scientific work.

To make an atomic clock, a collection of cesium atoms in the gas phase is placed in a container. The atoms are excited by heating or other means so that many of them are in the higher of the two energy levels mentioned above. These atoms then undergo transitions to the lower state, all emitting photons with the same frequency. Separate calibration experiments have shown that this frequency is

$$f_{\text{Cs clock}} = 9{,}192{,}631{,}770 \text{ Hz} \qquad (29.23)$$

This frequency is used as the "tick rate" of the atomic clock. A cesium clock thus ticks 9,192,631,770 times per second, forming a very accurate clock because the rate of ticking is extremely fast and is known very accurately. What's more, all cesium clocks have the same ticking rate, making them ideal for performing and comparing scientific measurements made in different laboratories at different times. In fact, in 1960 Equation 29.23 was adopted as the SI definition of the second. One second is now *defined* to be the time it takes a cesium clock to complete precisely 9,192,631,770 ticks.

Atomic clocks are widely used in situations where very accurate time measurements are required. Applications such as GPS (Global Positioning System), which involves precisely timed radio signals from a set of satellites orbiting the Earth, are made possible by such clocks.

Fluorescent Lights

There are two main types of lightbulbs in use today. The ***incandescent bulb***, developed by Thomas Edison, contains a very thin wire filament that carries a large electric current. The electrical energy dissipated in the filament heats it to a high temperature, and the filament then acts as a blackbody and emits radiation. The other type of lightbulb uses a gas of atoms in a glass container. In common terms, bulbs filled with a gas are often called either "neon" bulbs or "fluorescent" bulbs, but the spectra emitted by neon and fluorescent bulbs are somewhat different. A neon bulb contains a gas of Ne atoms. An electric current passes through the gas, producing ions and high-energy electrons as shown schematically in Figure 29.24A. The electrons, ions, and neutral atoms in the gas undergo many collisions, causing many of the Ne atoms to be in excited states. These atoms then decay back to their ground state, emitting light in the process as discussed in Section 29.2. Neon lamps thus emit light only at certain discrete wavelengths, that is, only with certain colors. By using different types of atoms, including neon and sodium, these lightbulbs can be used to make brightly colored signs (Fig. 29.24B).

Fluorescent lightbulbs (Fig. 29.24C) also contain a gas of atoms, often mercury (Hg), inside a glass bulb, but the inside of the bulb is coated with a fluorescent material. The Hg atoms are excited and emit photons in the same way as the Ne atoms in a neon bulb emit light. Mercury, however, emits strongly in the ultraviolet, and those photons are not detected by the eye, so Hg atoms alone would not make a useful lightbulb. This problem is overcome by the fluorescent material; the photons

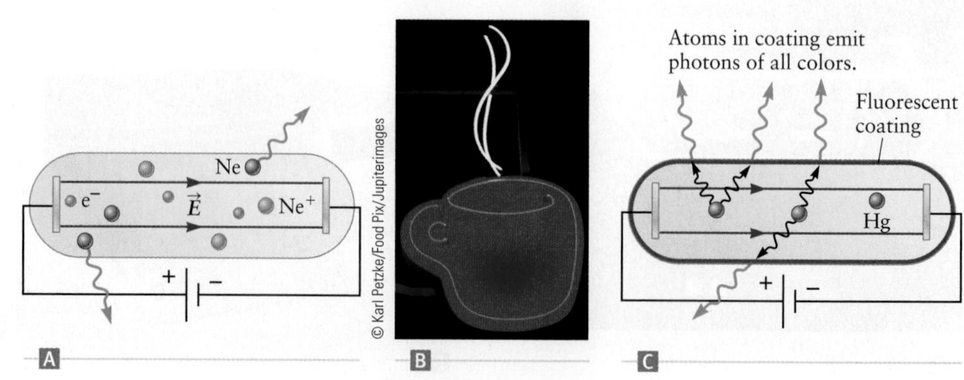

▶ **Figure 29.24** A Neon bulbs contain a gas of Ne atoms in a glass tube. The atoms are excited by an electric current, and light is emitted when the Ne atoms undergo transitions from an excited state back to the ground state. B Neon bulbs emit red light and are used in many advertisements. The other colors seen in a "neon" sign come from atoms of other elements. C In a fluorescent lamp, the ultraviolet photons from an atom (here Hg) are absorbed by a coating on the inside of the tube and then reemitted as photons with a wide range of wavelengths in the visible range.

emitted by Hg are absorbed by the fluorescent coating, exciting those atoms and molecules to higher energy levels. When these atoms and molecules undergo transitions to lower energy states, they emit new photons. The fluorescent material is designed to emit photons throughout the visible spectrum, so a fluorescent lamp produces "white" light, that is, light of virtually all colors. Fluorescent bulbs were invented in the late nineteenth century, but their use has expanded greatly since about 2000 due to advances in the design of compact fluorescent bulbs.

EXAMPLE 29.7 Photons Emitted by a Sodium Lamp

Sodium (Na) lamps emit light when the atoms undergo a transition from an excited $3p$ level to a $3s$ level. There are actually two levels in the $3p$ group with different values of the quantum numbers m and s and with slightly different energies, explaining why sodium emits two spectral lines that are very close in frequency (Fig. 29.8). The spectral line with a wavelength of $\lambda = 589.0$ nm is the stronger (more intense) of the two lines. What is the energy of these photons? Express your answer in units of electron-volts and compare it with the ionization energy of a hydrogen atom.

RECOGNIZE THE PRINCIPLE

This problem involves the relation between photon wavelength and energy. (See also Example 29.2.) We use the given photon wavelength λ to find the frequency f. The energy of the photon can then be found using $E_{\text{photon}} = hf$ (Eq. 29.7).

SKETCH THE PROBLEM

Figure 29.25 shows the transitions that produce the two yellow Na spectral lines.

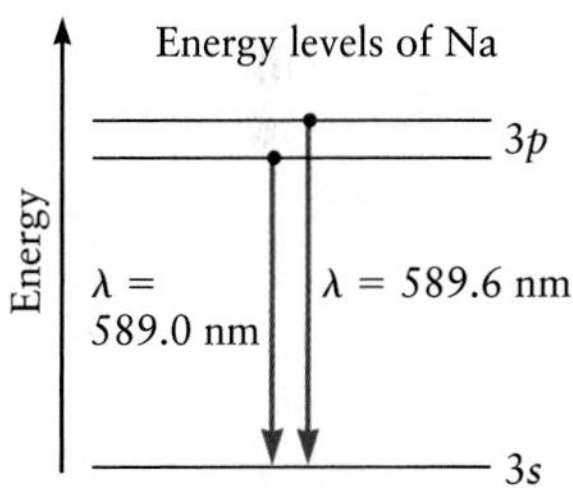

▲ **Figure 29.25** Example 29.7. These three energy levels are involved in the yellow spectral lines emitted by a sodium atom.

IDENTIFY THE RELATIONSHIPS

The frequency of a photon produced by a Na lamp is $f = c/\lambda$, where the wavelength is given above. Combining this equation with Equation 29.7, we find that the energy, frequency, and wavelength of a photon are related through

$$E = hf = \frac{hc}{\lambda}$$

SOLVE

Inserting the given value of the wavelength yields

$$E = \frac{hc}{\lambda} = \frac{(6.63 \times 10^{-34}\ \text{J}\cdot\text{s})(3.00 \times 10^{8}\ \text{m/s})}{5.890 \times 10^{-7}\ \text{m}} = 3.4 \times 10^{-19}\ \text{J}$$

which expressed in units of electron-volts is

$$E = (3.4 \times 10^{-19}\ \text{J})\left(\frac{1\ \text{eV}}{1.60 \times 10^{-19}\ \text{J}}\right) = \boxed{2.1\ \text{eV}} \quad (1)$$

The ionization of energy of H is 13.6 eV, so the energy of this spectral line in Na is much smaller.

► *What does it mean?*

The energy of this photon emitted by Na lies in the yellow part of the visible spectrum, making it useful for lightbulbs. So-called high-pressure sodium bulbs are used in applications such as street lights, where very high intensities are required. The high pressures used in these bulbs cause them to emit light over a range of wavelengths (not just yellow).

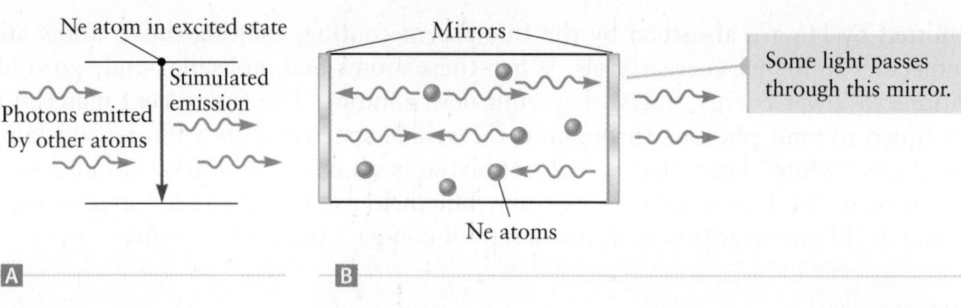

▶ **Figure 29.26** A The emission of photons by an atom can be stimulated by the presence of other photons with the same frequency. B A laser uses mirrors to reflect light back and forth within the laser tube. A small amount of light is allowed to escape through one of the mirrors; that is the light you see coming from a laser.

Lasers

Lasers are another application of quantum and atomic theory and are commonplace in early-twenty-first-century life. Lasers depend on the coherent emission of light by many atoms, all at the same frequency.

In an ordinary fluorescent lightbulb, the transitions from the excited atomic states to lower energy levels occur in a random fashion. That is, once an atom is put into an excited state, it is impossible to predict when it will emit a photon, and the emitted photons are radiated randomly in all directions. In this ***spontaneous emission*** process, each atom emits photons independently of the other atoms. In a laser, an atom undergoes a transition and emits a photon in the presence of many other photons that have energies equal to the atom's transition energy (Fig. 29.26A). A process known as ***stimulated emission*** causes the light emitted by this atom to propagate in the same direction and with the same phase as surrounding light waves. Such a light source is called a laser, an acronym for *l*ight *a*mplification by *s*timulated *e*mission of *r*adiation. The light from a laser is thus a coherent source (Chapter 25) and is very useful for experiments involving the interference of light.

The design of many lasers is similar to that of a neon lightbulb, but there are mirrors at the ends of the bulb (called the laser "tube"). Light emitted by the gas atoms is reflected by the mirrors and travels back and forth along the tube (Fig. 29.26B). One of these mirrors is designed to let a small amount (typically a few percent) of the light pass through; that is the light you see from the laser. Just as in a neon bulb, atoms in a laser tube are in excited energy levels (produced typically by passing current through the gas). The excited atoms emit photons that stimulate the emission of more photons with the same energy, and all these photons are reflected repeatedly inside the tube. Photons resulting from stimulated emission are precisely in phase with the stimulating radiation. In this way, a laser can produce a very intense beam of light that is highly directional.

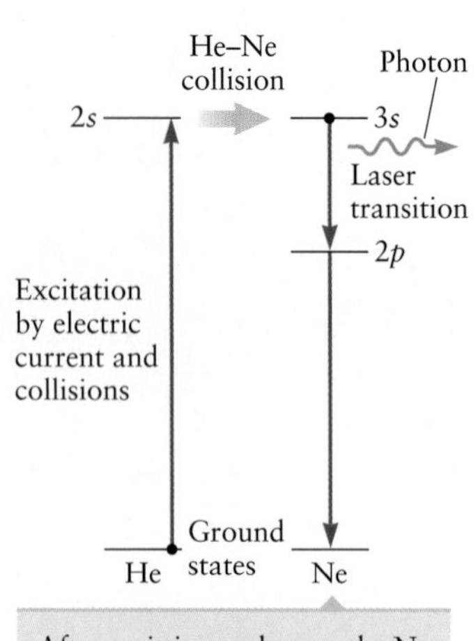

▲ **Figure 29.7** Energy-level diagram for a helium–neon laser.

Lasers can be made with a variety of different atoms. A popular design uses a mixture of He and Ne gas and is called a helium–neon (He–Ne) laser. It was one of the first laser types invented (in 1960), making use of the energy levels shown in Figure 29.27. Helium atoms are excited by an electric current to an energy level about 23 eV above the ground state. There is an energy level of Ne at nearly the same energy, and when an excited He atom collides with a Ne atom, there is a high probability that the He atom's energy will be transferred to the Ne atom, leaving the Ne atom in an excited state, with an electron in its 3*s* level. The laser action makes use of photons emitted when electrons in these excited Ne atoms decay to a 2*p* level as sketched in Figure 29.27. After emitting a photon, the Ne atoms decay quickly to their ground state and are ready to begin the process again.

The photons emitted by a helium–neon laser have a wavelength of about 633 nm, which is in the red part of the visible spectrum. These lasers are used in many physics laboratories, often in lecture demonstrations. Another common type of laser is based on light produced by light-emitting diodes (called LEDs). These lasers typically have a wavelength near 650 nm (which is also in the red part of the spectrum) and are used in optical barcode scanners at many stores. Figure 29.28 shows some contemporary devices that use lasers.

CONCEPT CHECK 29.7 Operation of a CO_2 Laser

The helium–neon laser described in Figure 29.27 emits visible light, but some lasers emit photons in the infrared. A popular type of infrared laser uses photons emitted by CO_2, which have a wavelength of $10.6\ \mu m = 1.06 \times 10^{-5}$ m. This wavelength is longer than that of the Na photon in Example 29.7 by a factor of 18. Which of the following statements is correct?

(a) The energy of a photon from a CO_2 laser is greater than that of a photon from a Na lamp by a factor of 18.
(b) The energy of a photon from a CO_2 laser is less than that of a photon from a Na lamp by a factor of 18.

© Toshifumi Kitamura/AFP/Gettin Images

A

Sam Ogden/Photo Researchers, Inc.

B

▲ **Figure 29.28** Some applications of lasers. A Some flatscreen televisions use laser light sources. B A bar-code reader.

® The Force between Two Atoms

How much force does it take to pull two atoms apart? This question is important because physicists have recently developed ways to manipulate individual atoms, and one application of this work is to construct (or deconstruct) molecules one atom at a time. This question is also relevant for understanding the atomic force microscope (AFM) in which atoms from a sharp tip are scanned near atoms on the surface of another object. The AFM uses the force between atoms to produce an image of the surface (Chapter 11). An accurate quantum mechanical calculation of the force between two atoms would be very complex, but we can estimate it using what we know about the relation between potential energy and force, and using typical values of the ionization energy of an atom.

Consider the two hypothetical atoms in Figure 29.29A and assume they are bound together to form a molecule. The binding energy of a molecule is the energy required to break the chemical bond between the two atoms. This energy comes from the "sharing" of valence electrons, and we can estimate this energy from the spacing between energy levels in one of the atoms. Qualitatively, the sharing of valence electrons lowers the energy of one or more of the electrons by an amount that is approximately equal to the spacing between energy levels in an atom. For a hydrogen atom, this spacing is about 1 eV to 10 eV, and it will be greater for heavier atoms, so a typical bond energy is about 10 eV. That is the change in potential energy when comparing the molecule to two unbound atoms, so we have $\Delta PE \approx 10$ eV.

If this atom is now pulled apart by separating the atoms a distance Δx, there is an associated force F between the atoms. From the relation between potential energy and force (Sections 6.3 and 6.8), the magnitude of the force F is

$$F = \left|\frac{\Delta PE}{\Delta x}\right|$$

The radius of a hydrogen atom is about 0.05 nm (Example 29.3), so separating the atoms in Figure 29.29B a distance $\Delta x \approx 1$ nm should be large enough to break the chemical bond. Our estimate for the force between two atoms is thus

$$F = \left|\frac{\Delta PE}{\Delta x}\right| \approx \frac{10\ \text{eV}}{1\ \text{nm}}$$

An energy of 10 eV is equal to $(10\ \text{eV})(1.60 \times 10^{-19}\ \text{J/eV}) = 1.6 \times 10^{-18}$ J. The force is thus

$$F \approx \frac{1.6 \times 10^{-18}\ \text{J}}{1\ \text{nm}} = 1.6 \times 10^{-9}\ \text{N} \tag{29.24}$$

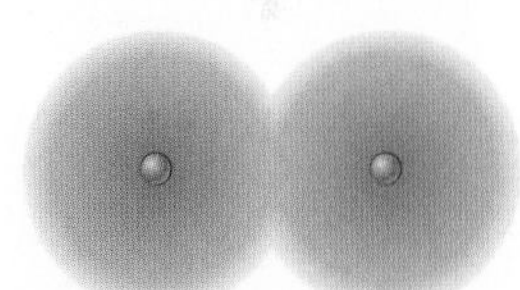

Energy of a typical molecular bond is $|PE| \approx 10$ eV.

A

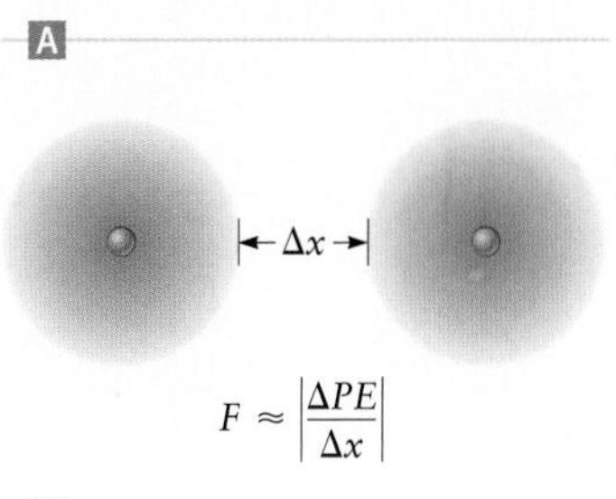

B

▲ **Figure 29.29** Estimating the force between two atoms.

This value is only an estimate for the force, but atomic force microscope experiments show that the force between two atoms is indeed close to that found in Equation 29.24. The keys to our calculation were the connection between force and potential energy, and our knowledge of the potential energy associated with a chemical bond.

29.8 Quantum Mechanics and Newton's Mechanics: Some Philosophical Issues

Newton's mechanics fails when applied to objects such as electrons and atoms. In that regime, quantum theory is required. Does that mean we should discard Newton's laws and redo all the physics in this book using quantum theory? To answer this question, we first note that Newton's laws work extremely well in the classical regime, that is, when applied to macroscopic objects such as baseballs and planets. Quantum theory can also be applied to such macroscopic objects, giving results that are virtually identical to Newton's mechanics. How can quantum theory, which is based on a particle-wave description of all objects, be consistent with Newton's particle theory for objects such as baseballs?

The answer depends on the wavelength of the particle-waves. The wavelength λ and momentum p of a particle-wave are connected by de Broglie's relation,

$$\lambda = \frac{h}{p} \tag{29.25}$$

An electron is a very light object (i.e., has a small mass), so its momentum is small. Equation 29.25 therefore tells us that λ for an electron is large. It is more accurate to say that the electron's wavelength is *relatively* large because we have seen that the electron wavelength in an atom is typically 0.1 nm (1×10^{-10} m). For an object such as a baseball, p is much larger than for an electron, so the wavelength is *much* smaller. For example, a baseball has a mass $m = 0.14$ kg and a speed of typically 50 m/s (about 100 mi/h). Inserting these values in Equation 29.25, the de Broglie wavelength of a baseball is

$$\lambda = \frac{h}{p} = \frac{h}{mv} = \frac{6.63 \times 10^{-34}\ \text{J} \cdot \text{s}}{(0.14\ \text{kg})(50\ \text{m/s})} = 9.5 \times 10^{-35}\ \text{m} \tag{29.26}$$

which is an *extremely* short wavelength. When a wave such as light has a very short wavelength, it moves in a straight line described by a simple ray (Chapter 24). The same is true of a baseball. It will move as a simple ray; that is, it will move as a simple particle. The wave properties of the baseball are only apparent on length scales comparable to the wavelength in Equation 29.26. Hence, for all practical purposes a baseball behaves as a classical particle, and its wave properties will never be observable. That is why Newton's mechanics works so well for classical objects. Such objects always have extremely short wavelengths, making the quantum theory description in terms of particle-waves unnecessary.

Where Quantum Theory Meets Newton

If Newton's mechanics works well for macroscopic objects while quantum theory is needed to describe behavior at the atomic scale, does anything interesting happen where these two regimes meet? Physicists are now actively studying this problem. One interesting question concerns the quantum behavior of a living entity. Objects such as viruses are both very small and "alive." Do the quantized energy levels of a living thing lead to any new behavior? What if the living thing is conscious? Can we understand the wave function of biological objects such as a brain? These intriguing questions are but a few of the ones physicists and biologists are now studying.

Summary | CHAPTER 29

Key Concepts and Principles

The nuclear atom

Atoms contain ***electrons***, ***protons***, and ***neutrons***. The positive charge in an atom is located in the ***nucleus***, which contains protons and neutrons. The ***Rutherford planetary model*** pictures the atom as a miniature solar system in which electrons orbit the nucleus.

Atomic spectra

Atoms emit and absorb light (and other electromagnetic radiation) at certain discrete frequencies. These ***spectral lines*** result from transitions between discrete energy levels of an atom. When an electron undergoes a transition from a quantum state with energy E_2 to another with energy E_1, a photon of energy

$$hf = E_2 - E_1$$

is emitted. Here f is the frequency of the photon.

Bohr model

The ***Bohr model*** of the atom was an early attempt at incorporating quantum ideas into an atomic theory. A key assumption of the Bohr model is that the orbital angular momentum of an electron is quantized with

$$L = n\frac{h}{2\pi} \qquad \text{(29.8) (page 909)}$$

where n is an integer ($n = 1, 2, 3, \ldots$) and h is Planck's constant. This assumption implies that electrons form standing waves as they orbit the nucleus, with wavelengths given by the de Broglie theory of particle-waves ($\lambda = h/p$, where p is the momentum; Eq. 28.8). The Bohr model leads to discrete atomic energy levels with energies

Allowed standing wave orbits in the Bohr model

$$E_{\text{tot}} = -\left(\frac{2\pi^2k^2e^4m}{h^2}\right)\frac{1}{n^2} \qquad \text{(29.15) (page 910)}$$

The Bohr model predicts the correct values of the energy levels of hydrogen, but does not correctly describe atoms with two or more electrons.

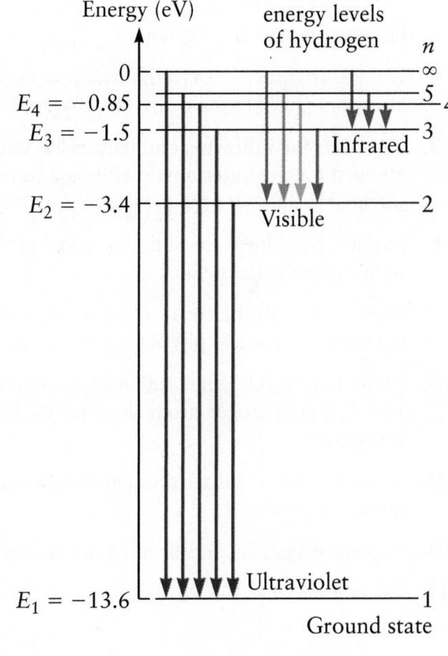

The four quantum numbers of the atom

The correct quantum theory is based on the Schrödinger equation. Each energy level in an atom is described by four ***quantum numbers.***

1. n is the ***principal quantum number.*** Possible values are $n = 1, 2, 3, \ldots$.
2. ℓ is the ***orbital quantum number.*** Allowed values are $\ell = 0, 1, \ldots, n - 1$.
3. m is the ***orbital magnetic quantum number.*** Allowed values are $m = -\ell, -\ell + 1, \ldots, -1, 0, +1, \ldots, \ell$.
4. s is the ***spin quantum number.*** Allowed values are $s = +\frac{1}{2}$ ("spin up") and $s = -\frac{1}{2}$ ("spin down").

(*Continued*)

The Pauli exclusion principle

Each electron in an atom occupies a separate energy level and thus has a unique set of quantum numbers.

Applications

Atomic clocks and lasers use transitions between discrete energy levels to produce photons with precisely defined frequencies. They are used in many applications, including the Global Positioning System, bar-code readers, and CD and DVD players. The force between two atoms is also large enough to be measured in an atomic force microscope (AFM), which uses this force to form an image of atoms on the surface of an object.

Questions

SSM = answer in *Student Companion & Problem-Solving Guide*   X = life science application

1. Explain why He, Ne, and Ar have similar chemical properties.
2. Explain why some periodic tables list H twice, once above Li and again above the column containing F.
3. Consider the He^+ ion. Like a hydrogen atom, He^+ has one electron, but it has two protons, so the charge of the nucleus is $+2e$. Do you expect the ionization energy of He^+ to be greater than or less than that of H? Explain.
4. Is there an upper limit to the wavelength of light that a hydrogen atom can emit? Explain why or why not.
5. SSM Of the following configurations, which ones are *not* allowed for the outermost shell of an atom by quantum theory, (a) $4s^4$, (b) $3d^7$, (c) $4f^9$, or (d) $2d^3$? Explain why not.
6. Explain why the plum-pudding model of the atom is inconsistent with experimental results.
7. Explain why Rutherford's planetary model of the atom is inconsistent with classical physics and what we know about atoms.
8. In the Bohr model, the total energy of an electron in the ground state in the hydrogen atom is -13.6 eV. Explain why this energy is negative.
9. Show that $4\ell + 2$ electrons can occupy a subshell with orbital quantum number ℓ.
10. Explain why F, Cl, and Br have similar chemical properties.
11. Explain why He, Ne, Ar, and Kr rarely form stable molecules.
12. Explain why the Pauli exclusion principle is not needed to understand the allowed quantum states of the electron in a hydrogen atom.
13. Derive a general expression for the de Broglie wavelength of an electron for the Bohr model in state n.
14. Neon is one of many gases that emit light when exposed to electrodes across which there is a large electric potential difference. Neon bulbs emit red light, but other gases emit other colors. Use the Internet to look up what kinds of atoms are used in bulbs that emit red, green, blue, and yellow light and explain why different gases can emit light of different colors.
15. Explain how the elements Cr through Cu in the periodic table can have fairly similar chemical properties, even though they are not arranged in the same column of the periodic table.
16. Explain why it is difficult to chemically separate the elements Ce (cerium) through Lu (lutitium) from each other. *Hint*: Consider the electrons that determine the bonding properties of these atoms. What states do these electrons occupy?
17. SSM What would the spectrum of hydrogen look like if the electrons could move between orbits of any radius in Bohr's model?
18. What experiment would you do to determine if there is oxygen in a distant star?
19. What experiment would you do to determine if there is oxygen in the atmosphere of a planet going around a distant star?
20. When illuminated with ultraviolet light, many minerals glow in the visible part of the electromagnetic spectrum. Why?

Problems available in ENHANCED WebAssign

Section 29.1 Structure of the Atom: What's Inside?
Problems 1–3

Section 29.2 Atomic Spectra
Problems 4–7

Section 29.3 Bohr's Model of the Atom
Problems 8–21

Section 29.4 Wave Mechanics and the Hydrogen Atom
Problems 22–25

Section 29.5 Multielectron Atoms
Problems 26–31

Section 29.6 Chemical Properties of the Elements and the Periodic Table
Problems 32–33

Section 29.7 Applications
Problems 34–35

Additional Problems
Problems 36–62

List of Enhanced Problems

Problem number	Solution in *Student Companion & Problem-Solving Guide*	Reasoning & Relationships Problem	Reasoning Tutorial in ENHANCED WebAssign
29.2	✓		
29.6	✓		
29.20	✓		
29.25	✓		
29.29	✓		
29.32	✓		
29.35	✓		
29.54		✓	
29.55	✓		
29.57		✓	
29.60		✓	
29.61		✓	✓
29.62		✓	✓

CHAPTER

30

▶ *Nuclear reactors are an important source of electricity in many countries. This photograph shows the core of a reactor. The blue light emanating from the core is due to Cherenkov radiation, produced by charged particles moving at very high speeds through the core. These particles are a product of nuclear fission, one of this chapter's topics. (Science Source/Photo Researchers, Inc.)*

Nuclear Physics

OUTLINE

In Chapters 28 and 29, we described the development of quantum theory and discussed the quantum properties of electrons and photons. Rutherford's discovery of the atomic nucleus raised the obvious question, "How is the nucleus itself put together?" Every nucleus is composed of just two different building blocks, particles called ***protons*** and ***neutrons***. Different elements have different numbers of these two particles in their nuclei.

The protons and neutrons in most nuclei are bound together very tightly. Because a great deal of energy is needed to break apart a nucleus, the nucleus can usually be treated as a simple, unbreakable "particle" when dealing with phenomena such as chemical reactions and atomic spectra. In ***nuclear reactions***, however, a nucleus breaks apart or is assembled when smaller nuclei and particles combine. Nuclear reactions can release very large amounts of energy that can be applied for great societal benefit, although they can also be used destructively. In this chapter, we describe the structure and properties of the nucleus and explain some applications of nuclear physics.

30.1 Structure of the Nucleus: What's Inside?

The discovery of the nucleus is usually credited to Rutherford as a result of his scattering experiments described in Chapter 29. The experiments of Rutherford and others showed that all nuclei are composed of two particles, protons and neutrons. These two types of particles, called ***nucleons***, are both much more massive than an electron. Their masses are

The nucleus is composed of nucleons (protons and neutrons).

$$m_p = 1.673 \times 10^{-27}\text{ kg} \quad \text{(mass of a proton)} \tag{30.1}$$

$$m_n = 1.675 \times 10^{-27}\text{ kg} \quad \text{(mass of a neutron)} \tag{30.2}$$

The values of m_p and m_n are thus very similar, and both particles are about 1800 times more massive than an electron. The neutron is electrically neutral, whereas the proton carries a positive charge $+e$. The charge on an electron is $-e$, so the total charge carried by one electron plus one proton is precisely zero.

Every nucleus can be specified by the numbers of protons and neutrons it contains. The number of protons in a nucleus is called the ***atomic number***, Z. Because each element has a particular and unique number of protons in its nucleus, each has a characteristic value of Z. For example, the atomic number of He (helium) is $Z = 2$, while the atomic number of O (oxygen) is $Z = 8$. Since atoms are electrically neutral, Z is also equal to the number of electrons in an atom.

Definition of atomic number

The number of neutrons in a nucleus is usually denoted by the symbol N. The value of N for a particular element can vary. For example, He nuclei can have $N = 1$, $N = 2$, or $N = 4$, and other values are also possible. The number of neutrons in atomic nuclei generally increases as the atomic number increases.

The ***mass number***, A, of a nucleus is the sum of the number of protons and neutrons, so

Definition of mass number

$$A = Z + N \tag{30.3}$$

All this information is described in the following shorthand notation:[1]

mass number → $^{A}_{Z}\text{X}$ ← symbol for element
atomic number →

$$^{A}_{Z}\text{X} \tag{30.4}$$

The letter "X" in Equation 30.4 denotes an element whose chemical symbol should be substituted from the periodic table. As an example, a He nucleus containing $N = 2$ neutrons is denoted

A = 4 (2 protons + 2 neutrons) → $^{4}_{2}\text{He}$ ← symbol for element (helium)
Z = 2 (2 protons) →

The various helium nuclei having $N = 1$, 2, or 4 are written as

$$^{3}_{2}\text{He} \qquad ^{4}_{2}\text{He} \qquad ^{6}_{2}\text{He}$$

and are called different ***isotopes*** of He. All isotopes of a particular element have the same numbers of protons and electrons. Chemical properties are determined mainly by the bonding of electrons, so the chemical properties of different isotopes of a given element are essentially identical.

Isotopes have the same Z but different values of N.

Isotopes and Atomic Mass

The ***atomic mass*** of an atom is the mass of the nucleus plus Z electrons. Since different isotopes of an element have different numbers of neutrons, their atomic masses will differ. The periodic table lists an average value of the atomic mass for

[1]Sometimes the atomic number is omitted; the value of Z can always be determined from the identity of the element.

each element based on the ***natural abundance*** of each isotope. For example, several different isotopes of carbon are found in nature, with $A = 11, 12, 13$, and 14, and the isotope with $A = 12$ has the greatest abundance. The average atomic mass for a natural sample of pure carbon (based on the abundance of the different isotopes) is listed in the periodic table as the atomic mass of the element. The listed value is the mass in grams of 1 mole of carbon atoms.

CONCEPT CHECK 30.1 Nucleus of Carbon

Use the notation described in Equation 30.4 to write the symbol for the nucleus of carbon that contains 8 neutrons.

EXAMPLE 30.1 Identifying a Nucleus

A particular nucleus contains $N = 13$ neutrons and has mass number $A = 25$. What element is it? Write the chemical symbol for this nucleus.

RECOGNIZE THE PRINCIPLE

The chemical identity of an element is determined by its atomic number Z, the number of protons in its nucleus.

SKETCH THE PROBLEM

We follow the shorthand description of a nucleus in Equation 30.4.

IDENTIFY THE RELATIONSHIPS

The mass number A is the sum of Z and N (Eq. 30.3), so $A = Z + N$, and we have been given the values of A and N.

SOLVE

The number of protons in the nucleus is thus

$$Z = A - N = 25 - 13$$

$$Z = 12$$

Using the periodic table (inside the back cover of this book), the element with $Z = 12$ is $\boxed{\text{magnesium (Mg)}}$. The symbol for this nucleus is $\boxed{{}^{25}_{12}\text{Mg}}$.

▶ *What does it mean?*

This element is not the only isotope of magnesium found in nature. The isotopes ${}^{24}_{12}\text{Mg}$ and ${}^{26}_{12}\text{Mg}$ are also stable. In fact, approximately 79% of the magnesium nuclei found in nature are ${}^{24}_{12}\text{Mg}$. (See Table A.4.)

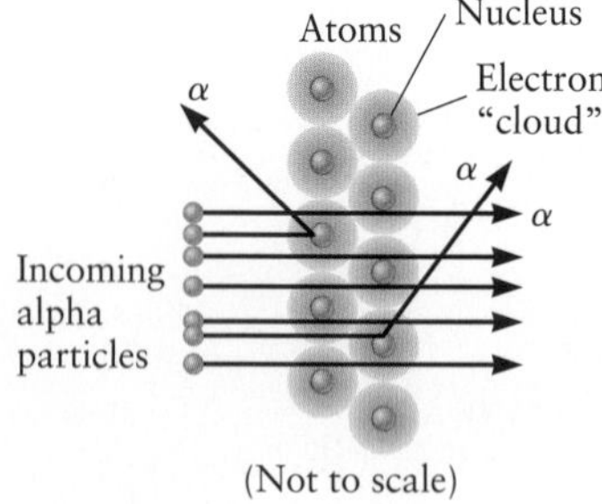

▲ Figure 30.1 The nucleus of an atom is much smaller than the radius of the atom (i.e., the volume occupied by the electrons). When alpha particles are fired at an atom, there is only a small chance that they will collide with the nucleus; most alpha particles pass straight through. Because the mass of a nucleus is comparable to or larger than the mass of an alpha particle, the alpha particle can be scattered over a wide range of directions.

Size of the Nucleus

Rutherford discovered the nucleus by firing alpha particles at atoms and observing how the alpha particles were deflected (Fig. 30.1). At the time, Rutherford did not know exactly what an alpha particle was; we now know that it is the nucleus ${}^{4}_{2}\text{He}$. (We'll explain how Rutherford obtained alpha particles in Section 30.2.) As far as Rutherford was concerned, alpha particles were simply very small particles with a mass approximately equal to the mass of a helium atom. His experiments used them as tiny "billiard balls" that could bounce off atoms just as real billiard balls are used in collisions in a game of pool. Rutherford found that most alpha particles in his experiment passed right through the atoms in his target. Only a very tiny fraction of them were deflected. The implications of this result can be appreciated from Figure 30.1. The probability for an alpha particle to collide with the nucleus depends on the diameters of the alpha particle and nucleus. Since the probability of a collision is very small (as established by Rutherford's experiment), both the alpha particles and the nucleus must be much smaller than the overall size of the atom. By measuring the collision probability and assuming alpha particles are about as small as the nucleus, Rutherford could deduce the size of the nucleus.

Rutherford's collision experiments showed that the diameter of a typical nucleus is a few femtometers (1 femtometer = 1 fm = 1×10^{-15} m). Recall that the diameter of an atom is about 1×10^{-10} m (Chapter 29). The nucleus is thus smaller than the atom by a factor of about $(1 \times 10^{-15})/(1 \times 10^{-10}) = 1 \times 10^{-5} = 1/100{,}000$. In other words, the nucleus is about 100,000 times smaller than an atom!

Scattering experiments like those of Rutherford along with other studies have shown that most nuclei have an approximately spherical shape, with a radius r that depends on the number of nucleons A it contains. It is found that

$$r = r_0 A^{1/3} \tag{30.5}$$

The constant $r_0 \approx 1.2$ fm $= 1.2 \times 10^{-15}$ m. Rearranging Equation 30.5 gives

$$A = \left(\frac{r}{r_0}\right)^3$$

Since the volume of a sphere is $\frac{4}{3}\pi r^3$, the number of nucleons A is proportional to the volume of the nucleus, in agreement with the picture of nucleons packed together like tiny hard spheres in Figure 30.2. As A increases, the radius of this nuclear "sphere" increases according to Equation 30.5.

Insight 30.1

RUTHERFORD'S SCATTERING EXPERIMENT AND CONSERVATION OF MOMENTUM

Experiments like those of Rutherford, in which a projectile particle is fired at a target particle, are called ***scattering experiments***. They are a very powerful way to probe a nucleus or other subatomic particle and have been used to study the internal structure of protons and neutrons (Chapter 31). Rutherford's observation that some alpha particles are "backscattered"—that is, scattered back in the direction from which they came—implies a very massive nucleus, and can be understood from the conservation of momentum and energy in an elastic collision. When a heavy object (such as a bowling ball) collides with a very light one (a marble), the bowling ball is not deflected significantly from its original path. When a marble scatters from a bowling ball, however, backscattering is possible; the marble may reverse its direction. Rutherford's results thus mean that the nucleus of an element such as gold (which he used as a target) is much more massive than an alpha particle.

Electric Potential Energy of a Nucleus

The diameter of a typical nucleus is a few femtometers, so the protons in a nucleus are very close together. Such closely spaced positive charges have a large electric potential energy. For two point charges q_1 and q_2 separated by a distance r, the electric potential energy is (Eq. 18.6)

$$PE_{\text{elec}} = \frac{kq_1q_2}{r}$$

For two protons in a He nucleus, we have $q_1 = q_2 = +e$, and estimating $r = 1$ fm, we get

$$PE_{\text{elec}} = \frac{k(+e)(+e)}{r} = \frac{(8.99 \times 10^9\ \text{N}\cdot\text{m}^2/\text{C}^2)(1.60 \times 10^{-19}\ \text{C})^2}{1 \times 10^{-15}\ \text{m}}$$

$$PE_{\text{elec}} = 2 \times 10^{-13}\ \text{J} \tag{30.6}$$

To see if this energy is small or large on an atomic scale, let's express PE_{elec} in units of electron-volts (eV). (Recall from Chapter 18 that 1 eV = 1.60×10^{-19} J.) We find

$$PE_{\text{elec}} = (2 \times 10^{-13}\ \text{J})\left(\frac{1\ \text{eV}}{1.60 \times 10^{-19}\ \text{J}}\right) = 1 \times 10^6\ \text{eV} = 1\ \text{MeV} \tag{30.7}$$

In Chapter 29, we found that a typical atomic-scale energy is around 10 eV. (The ionization energy of the hydrogen atom is 13.6 eV.) Equation 30.7 indicates that a typical nuclear-scale energy is larger than that by a factor of approximately 100,000! The potential energy associated with a nucleus is thus *much* larger than the energies associated with electrons, molecules, and chemical reactions.

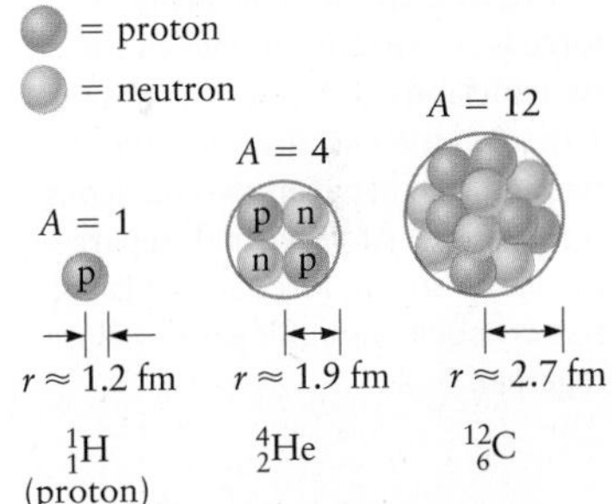

▲ **Figure 30.2** The nuclear radius r increases as the number of protons and neutrons increases, implying that protons and neutrons are closely packed within the nucleus.

Forces in the Nucleus

The electric potential energy in Equation 30.7 is positive since all protons are positively charged. The protons in a nucleus thus repel one another with a very large force. To make a nucleus stable, there must be another even larger force—what physicists call the ***strong force***—at work in the nucleus that is capable of producing an enormous attractive force between nucleons, strong enough to overcome the electrical repulsion. The strong force attracts any two nucleons to each other. Attraction due to the strong force occurs between a pair of protons, a pair of neutrons, and

The strong force is an attractive force between nucleons.

THE STRONG FORCE ACTS BETWEEN ANY TWO NUCLEONS

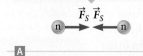

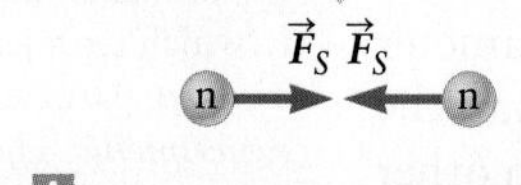
Two neutrons are attracted to each other by the strong force.
$\vec{F}_S$ $\vec{F}_S$
n
n
A

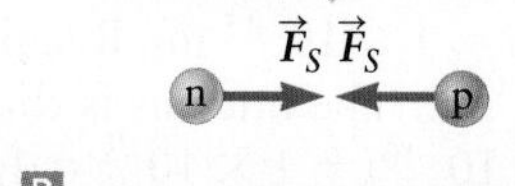
A neutron and a proton are attracted by the strong force.
$\vec{F}_S$ $\vec{F}_S$
n
p
B

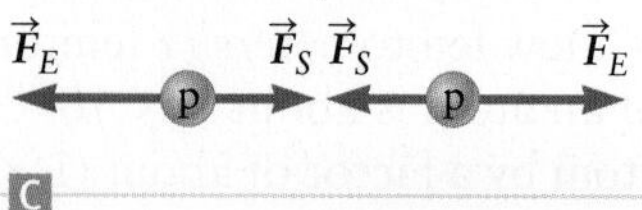
The total force between two protons is the sum of the electric and strong forces.
$\vec{F}_E$
$\vec{F}_S$ $\vec{F}_S$
$\vec{F}_E$
p
p
C

▲ **Figure 30.3** The strong nuclear force leads to an attraction between any pair of nucleons. The neutron has a net electric charge of zero, so the dominant force between A a pair of neutrons or B a neutron and a proton is the strong force. C Since protons are charged, there is a repulsive electric force between two protons along with the attractive strong force. If only two protons are involved, the electric force is larger in magnitude than the strong force, so a nucleus containing just two protons is not stable.

a proton and a neutron (Fig. 30.3). The strong force has a very large and approximately constant magnitude when the two nucleons are about 1 fm apart, but it is negligible when they are separated any farther (Fig. 30.4).

The strong force gives a large attractive force between nucleons and holds the nucleus together, but it does not act on electrons. So, when an electron approaches a nucleus, it experiences essentially just an electric force (Coulomb's law) due to the protons in the nucleus.[2] The strong force does not play any role in the binding of electrons in an atom.

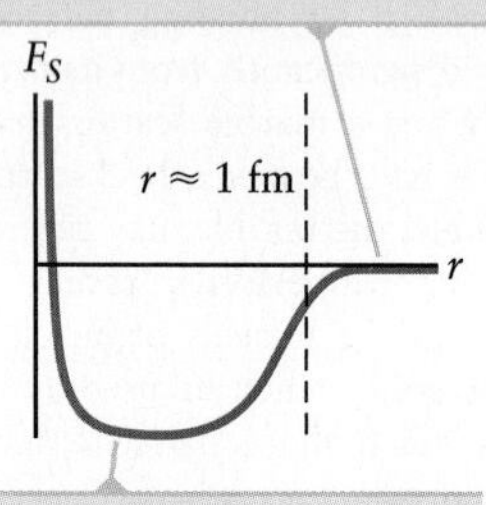
The strong force between nucleons is negligible when the separation is larger than about 1 fm.
F_S
$r \approx 1$ fm
r
The strong force is attractive for separations in this range.

▲ **Figure 30.4** The strong force is important only when the two nucleons are closer than about 1 fm. At larger separations, the strong force between two nucleons is negligible. At very small separations, the strong force is repulsive, so two nucleons can't get too close together.

Stability of the Nucleus

The simplest nucleus is $^1_1$H (hydrogen), consisting of just a single proton. All other nuclei contain both protons and neutrons. For example, there are several isotopes of helium, including $^3_2$He and $^4_2$He, all containing neutrons. Why is there not a nucleus of helium that contains just two protons? The reason is that the strong force between two protons is not large enough to overcome the repulsive electric force in that case, so a nucleus containing just two protons is not stable (Fig. 30.3C). The addition of one or more neutrons provides the extra attractive force needed to overcome the Coulomb repulsion of the protons. If a neutron is positioned between two protons (Fig. 30.5), the strong force attracts the neutron to each of the protons. The presence of this neutron also increases the separation between the protons, reducing the repulsive Coulomb force. Neutrons are thus essential for the stability of the nucleus. The number of neutrons generally increases as the atomic number increases; as more and more protons are added to the nucleus, more and more neutrons are needed to keep all the protons sufficiently separated (at a "safe distance") so that the Coulomb force does not cause the nucleus to fly apart.

The picture in Figure 30.5 indicates that neutrons are an essential "glue" that holds the nucleus together. Based on this picture, you might also think that adding more glue (neutrons) should always result in a more stable nucleus, but adding more and more neutrons will eventually lead to an unstable nucleus. This result can be explained by the quantum theory of the nucleus called the ***nuclear shell model***, which gives the energy levels of nucleons inside a nucleus, similar to the energy levels of electrons within an atom (Chapter 29).

Mass and Energy Scales in Nuclear Physics

We have seen that the electric potential energy of protons in a nucleus is extremely large, on the order of MeV (Eq. 30.7). The strong force in the nucleus must overcome this positive potential energy, so its associated potential energy is negative and is also on the order of MeV.

[2]An electron also experiences another, much weaker force, called the "weak force," which we'll discuss in Chapter 31.

Another important energy scale in nuclear physics is the rest energy associated with the proton and neutron. According to the theory of relativity (Chapter 27), the rest energy of a particle of mass m is

$$E = mc^2 \tag{30.8}$$

where c is the speed of light. For a proton, we find

$$E_p = m_p c^2 = (1.67 \times 10^{-27}\ \text{kg})(3.00 \times 10^8\ \text{m/s})^2 = 1.50 \times 10^{-10}\ \text{J}$$

$$E_p = 938\ \text{MeV} \tag{30.9}$$

The mass and rest energy are related though Equation 30.8, which can be rearranged to read $m = E/c^2$. Nuclear physicists often specify the mass of a particle in units of (energy)/c^2 = MeV/c^2. In these units, the mass of a proton is (using Eq. 30.9)

$$m_p = 938\ \text{MeV}/c^2 \tag{30.10}$$

The masses of the proton, neutron, and electron are collected in Table 30.1 in units of kilograms and MeV/c^2 and also in terms of the ***atomic mass unit***, denoted by "u." This unit is defined as 1/12th of the mass of a carbon atom with the nucleus $^{12}_{6}\text{C}$; this nucleus contains 6 neutrons and 6 protons. An atomic mass unit is related to other units of mass by

$$1\ \text{u} = 1.66 \times 10^{-27}\ \text{kg} = 931.5\ \text{MeV}/c^2 \tag{30.11}$$

Atomic mass units are handy in nuclear physics because 1 u is approximately equal to the mass of a nucleon (either a proton or a neutron).

The values in Table 30.1 include more significant figures than we have commonly used in this book because small changes in mass in a nuclear reaction correspond to large amounts of released energy, which can be calculated using the relation between mass and energy in Equation 30.8. When working problems in nuclear physics, we need to keep track of small differences and changes in mass.

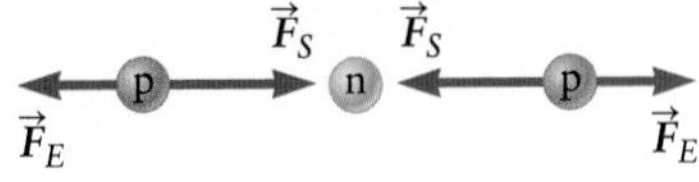
The strong force attracts these protons to a neutron located between them.
$\vec{F}_S$
$\vec{F}_S$
p
n
p
$\vec{F}_E$
$\vec{F}_E$

▲ **Figure 30.5** A neutron located between two protons acts as a sort of "glue" to bind them all together. Each proton is attracted to the neutron, and since the protons are now farther apart than they would be if the neutron were not present, the repulsive electric force is smaller than in Figure 30.3C. As a result, a nucleus with two protons and one neutron is stable. The nucleus shown here is $^{3}_{2}\text{He}$.

30.2 Nuclear Reactions: Spontaneous Decay of a Nucleus

In a nuclear reaction, one or more nuclei and nucleons "react" to form new nuclei, perhaps emitting nucleons and releasing energy. For example, a collision with an alpha particle can sometimes cause a nucleus to break into fragments. This reaction is called an ***induced*** nuclear reaction because the alpha particle caused the decay. We'll discuss several important examples of induced reactions when we describe the processes of nuclear fission and fusion in Section 30.3, but let's first discuss reactions that occur ***spontaneously***.

Radioactivity

Certain nuclei decay spontaneously into two or more particles. This process is called ***radioactive decay***, and the term ***radioactivity*** refers to the process in which a nucleus spontaneously emits either particles or radiation.

TABLE 30.1 Properties of the Proton, Neutron, and Electron

Particle	Charge	Mass (kg)	Mass (atomic mass units[a] = u)	Mass (MeV/c^2)
Proton	$+e$	1.6726×10^{-27}	1.0073	938.27
Neutron	0	1.6749×10^{-27}	1.0087	939.57
Electron	$-e$	9.109×10^{-31}	5.486×10^{-4}	0.5110

[a] 1 u corresponds to 931.5 MeV/c^2.

When a nucleus decays, it can emit particles with mass such as electrons or neutrons, or it can emit photons (electromagnetic radiation). At first, physicists did not know the identities of the particles emitted in radioactive decay, so these unknown decay products were initially given the names *alpha*, *beta*, and *gamma*. Eventually, the nature of these particles was deduced from experiments, but these somewhat mysterious names are still widely used. Even though their identity was initially unknown, these three types of decay products could be distinguished according to their properties, including their electric charge and their ability to penetrate matter.

Loosely speaking, we sometimes refer to alpha, beta, and gamma radiation emitted in radioactive decay. The term *radiation* is reminiscent of electromagnetic radiation (Chapter 23). We now know that only gamma radiation, also called "gamma rays," is actually a form of electromagnetic radiation, while alpha and beta "radiation" consist of particles with mass.

Alpha Particles. An alpha particle is composed of two protons and two neutrons. It is thus the nucleus of a He atom and is also denoted ${}^{4}_{2}\text{He}$. An alpha particle produced by nuclear decay usually does not have any bound electrons, so alpha particles almost always have a charge of $+2e$ (the charge of two protons). A typical radioactive decay involving an alpha particle is that of the radium nucleus ${}^{226}_{88}\text{Ra}$. This process is written in a form very similar to a chemical reaction:

Alpha decay: emission of an alpha particle from a nucleus

$$\begin{array}{ccccc} \text{parent nucleus} & & \text{daughter nucleus} & & \text{alpha particle} \\ {}^{226}_{88}\text{Ra} & \rightarrow & {}^{222}_{86}\text{Rn} & + & {}^{4}_{2}\text{He} \end{array} \tag{30.12}$$

This decay reaction (Fig. 30.6A) begins with a ***parent nucleus*** of ${}^{226}_{88}\text{Ra}$ containing 88 protons and $N = A - Z = 226 - 88 = 138$ neutrons. This nucleus spontaneously decays into two particles called ***decay products.*** One decay product is an isotope of radon, ${}^{222}_{86}\text{Rn}$, which contains $N = A - Z = 222 - 86 = 136$ neutrons. In this reaction, ${}^{222}_{86}\text{Rn}$ is called the ***daughter nucleus.*** The other decay product in Equation 30.12 is an alpha particle, ${}^{4}_{2}\text{He}$. The total number of nucleons is *conserved* in the reaction. Since the reaction products contain 222 nucleons (${}^{222}_{86}\text{Rn}$) plus 4 nucleons (${}^{4}_{2}\text{He}$), there are a total of 226 nucleons (protons plus neutrons) at the start of the reaction and the same number at the end. This particular reaction also conserves the numbers of protons and neutrons separately. That is, there are 88 protons at the start (in ${}^{226}_{88}\text{Ra}$) and 88 at the end (86 in the radon daughter nucleus and 2 in the alpha particle). You can view this reaction as simply ejecting two protons and two neutrons (the alpha particle) from the original radium nucleus. In his scattering experiments, Rutherford obtained his alpha particles from the alpha decay of nuclei such as ${}^{214}_{84}\text{Po}$.

The alpha decay reaction in Equation 30.12 has important medical consequences. It contributes to the (ill) health of miners and can also lead to a health hazard in homes. For more details, see Section 30.4.

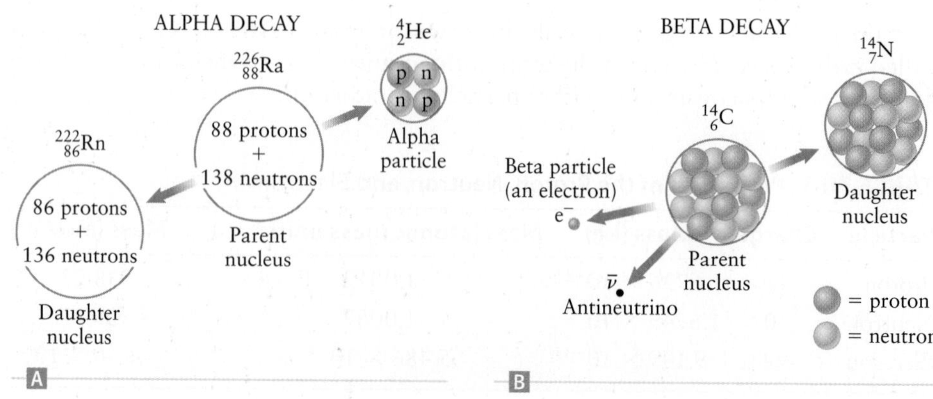

▶ Figure 30.6 **A** Example of alpha decay: the original parent nucleus (${}^{226}_{88}\text{Ra}$) splits to form ${}^{222}_{86}\text{Rn}$ and an alpha particle (${}^{4}_{2}\text{He}$). **B** Example of beta decay: the original ${}^{14}_{6}\text{C}$ nucleus produces ${}^{14}_{7}\text{N}$ plus an electron (a beta particle) and an antineutrino ($\bar{\nu}$).

Beta Particles. There are two varieties of beta particles. One variety is negatively charged and is simply an electron. The other type is identical to an electron except it has a positive charge of precisely $+e$. This positive beta particle is called a ***positron*** (denoted e^+, with the superscript indicating its positive charge) and is an example of ***antimatter*** (Chapter 31).[3] Electrons and positrons have the same mass and are both point charges.

A typical radioactive decay producing a beta particle is

Beta decay: emission of a beta particle (an electron or positron) from a nucleus, along with an antineutrino or a neutrino

$$\underset{\text{nucleus}}{\overset{\text{parent}}{{}^{14}_{6}\text{C}}} \rightarrow \underset{\text{nucleus}}{\overset{\text{daughter}}{{}^{14}_{7}\text{N}}} + \underset{\text{(electron)}}{\overset{\text{beta particle}}{\text{e}^-}} + \overset{\text{antineutrino}}{\bar{\nu}} \tag{30.13}$$

In this reaction (shown schematically in Figure 30.6B), the emitted beta particle is an electron denoted by e^-; the superscript negative sign indicates that the particle is negatively charged. This reaction also produces a particle called an ***antineutrino***, denoted by $\bar{\nu}$. The antineutrino is another example of antimatter. We'll discuss neutrinos and antimatter further in Chapter 31.

As with alpha decay, beta decay also conserves the total number of nucleons. It is useful to rewrite Equation 30.13 to indicate explicitly both the number of protons and neutrons and the electric charge at the start and end of the reaction:

$$\text{number of protons: } 6 \rightarrow 7$$
$$\text{number of neutrons: } 8 \rightarrow 7$$
$${}^{14}_{6}\text{C} \rightarrow {}^{14}_{7}\text{N} + \text{e}^- + \bar{\nu}$$
$$\text{charge: } +6e \rightarrow +7e - e + 0 \tag{30.14}$$

Hence, this reaction converts a neutron in the parent ${}^{14}_{6}\text{C}$ nucleus into a proton, conserving the total number of nucleons (neutrons plus protons). Electric charge is also conserved since the total charge is $+6e$ at the beginning and at the end of the reaction. The positive charge of the new proton is balanced by the creation of an electron.

Gamma Decay. Gamma decay produces gamma "rays," which are photons, quanta of electromagnetic radiation denoted by the Greek letter γ. One example of a nuclear decay that produces gamma rays is

Gamma decay: emission of a gamma ray photon from an excited nucleus

$$\underset{\text{(in excited state)}}{\overset{\text{parent nucleus}}{{}^{14}_{7}\text{N}^*}} \rightarrow \underset{\text{nucleus}}{\overset{\text{daughter}}{{}^{14}_{7}\text{N}}} + \underset{\text{ray}}{\overset{\text{gamma}}{\gamma}} \tag{30.15}$$

The asterisk on the left in this equation denotes that the nucleus is in an excited state. In Chapter 29, we saw that an atom in excited state can undergo a transition to a state of lower energy and emit a photon in the process. In the same way, a nucleus that is initially in an excited state can undergo a transition to a state of lower energy and emit a gamma ray photon. A particular nucleus is often placed in an excited state as a result of alpha or beta decay. For example, a ${}^{14}_{7}\text{N}^*$ nucleus in an excited state (as on the left in Eq. 30.15) can be produced by beta decay (although we did not indicate it in Eq. 30.13). A highly simplified energy-level diagram for the reaction in Equation 30.15 is shown in Figure 30.7. This diagram shows only the initial and final states in Equation 30.15. The decay from the excited to the ground state in Figure 30.7 produces a gamma ray with an energy equal to the difference in the energies of the two levels.

A more realistic energy-level diagram, this time for the excited states of ${}^{60}_{28}\text{Ni}$, is shown in Figure 30.8. A ${}^{60}_{28}\text{Ni}^*$ nucleus can emit gamma rays with many different energies, depending on which excited and final states are involved. For the states shown here, the emitted gamma rays have energies of a few MeV.

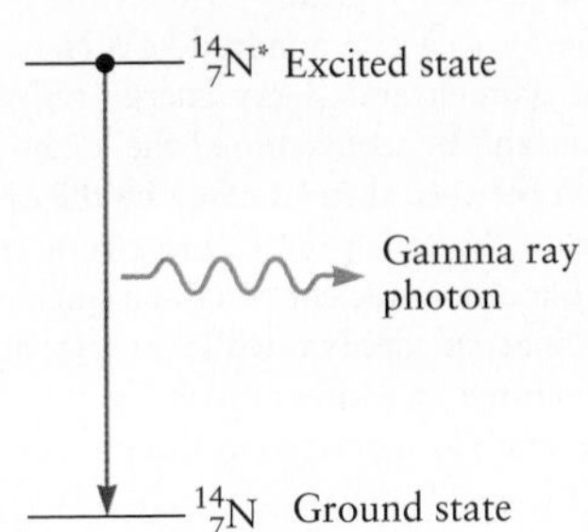

▲ **Figure 30.7** A nucleus in an excited state (${}^{14}_{7}\text{N}^*$) can decay to its ground state (${}^{14}_{7}\text{N}$) by emitting a gamma ray photon.

[3]Here and throughout this chapter we follow the standard notation of nuclear physics and denote particles such as an electron by a "nonitalic" letter.

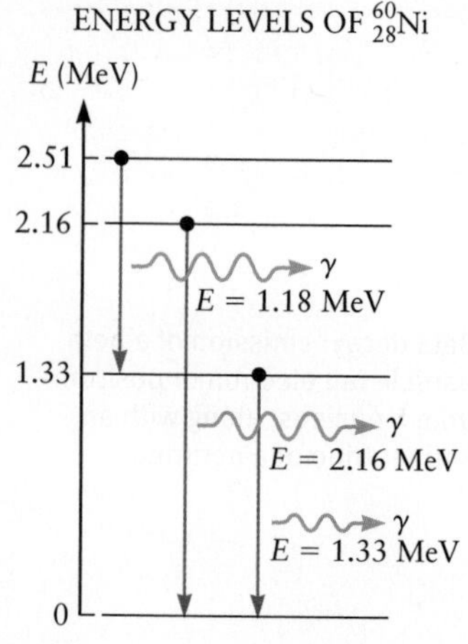

▲ **Figure 30.8** Energy-level diagram showing a few of the energy levels of the nucleus $^{60}_{28}\text{Ni}$. This nucleus can emit gamma ray photons with several different energies, depending on the initial and final states.

The energy of a gamma ray photon depends on the particular nuclear decay that produces it, but typical gamma rays have energies from 10 keV (10^4 eV) to 100 MeV (10^8 eV) or even higher. Recall that the spectral emission of atoms (Chapter 29) involved photons with energies of typically 10 eV. This comparison shows again that the overall energy scale of nuclear decay—and of nuclear processes in general—is much greater than a typical atomic scale energy.

CONCEPT CHECK 30.2 Wavelength of a Gamma Ray Photon

According to the energy-level diagram for $^{60}_{28}\text{Ni}$ in Figure 30.8, this nucleus can emit a gamma ray photon with an energy of 1.33 MeV. What is the wavelength of this photon?

(a) 9.3×10^{-13} m $= 9.3 \times 10^{-4}$ nm
(b) 2.7×10^{-31} m $= 2.7 \times 10^{-23}$ nm
(c) 2.3×10^{2} m

How does the wavelength of the gamma ray compare with the wavelength of visible light?

Conservation Rules in Nuclear Reactions

The processes of alpha, beta, and gamma decay can be understood in terms of the following conservation principles.

Conservation of Mass–Energy. The principle of conservation of energy applies to *all* physical processes, including nuclear reactions. The total energy at the start of a reaction such as the gamma decay in Equation 30.15 must equal the total energy at the end. Indeed, this principle allows us to use an energy-level diagram (as in Figs. 30.7 and 30.8) to calculate the energy of the gamma ray produced in such a reaction. In addition, we must account for the results of special relativity and allow for the conversion of mass to energy and vice versa. We'll discuss some specific examples later in this section.

Conservation of Momentum. All nuclear reactions must conserve momentum. The application of momentum conservation in nuclear physics is similar to our work on collisions in Chapter 7.

Insight 30.2

GAMMA RAYS AND X-RAYS

The terms *gamma ray* and *X-ray* both refer to portions of the electromagnetic spectrum. Typical gamma ray energies can be as low as 10 keV and as large as 100 MeV or even higher. X-ray energies are typically in the range of 100 eV to a few hundred keV. Hence, the gamma and X-ray energy ranges overlap. By convention, the distinction between them is made by the origin of the radiation. Gamma rays are produced in nuclear reactions (such as radioactive decay), while X-rays are generated by process involving atomic electrons as described in Chapter 29.

Conservation of Electric Charge. All nuclear reactions conserve electric charge, but the total number of charged particles may change in a reaction. For example, the beta decay in Equation 30.14 produces a proton and an electron. The total charge does not change, but the number of charged particles increases.

Conservation of Nucleon Number. The number of nucleons—that is, the number of protons plus neutrons—does not change. For example, in the beta decay reaction in Equation 30.14, a neutron is converted into a proton, but the total number of neutrons plus protons is unchanged by the reaction.

Radioactive Decay Series

When a nucleus undergoes radioactive decay through the emission of an alpha particle or a beta particle, it is converted into another type of nucleus. An important example is the alpha decay of $^{238}_{92}\text{U}$:

$$^{238}_{92}\text{U} \rightarrow {}^{234}_{90}\text{Th} + {}^{4}_{2}\text{He} \tag{30.16}$$

This decay converts uranium (U) into thorium (Th). The ability to convert one element into another was a goal of many chemists and alchemists throughout the Middle Ages. This conversion is precisely what happens in alpha and beta decay, but it is not an economical approach for producing any element on a commercial scale (and certainly not for producing the gold so eagerly sought by medieval alchemists).

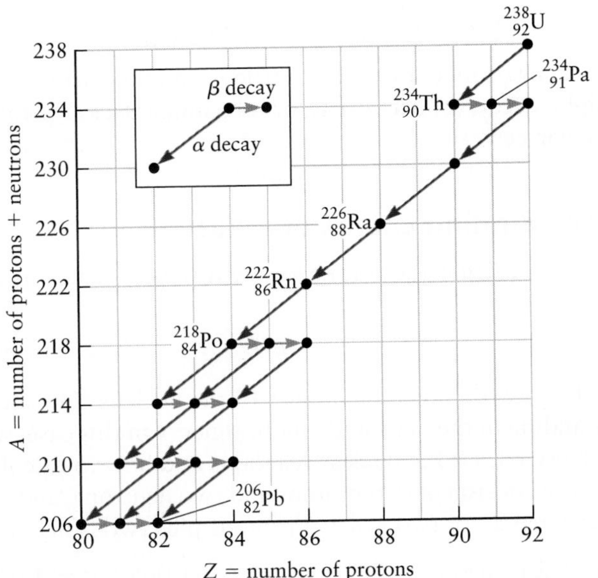

◀ Figure 30.9 Sequence of nuclear decays starting with the parent nucleus $^{238}_{92}\text{U}$. The horizontal arrows denote beta decay, and the diagonal arrows show alpha decay.

Although Equation 30.16 describes a complete nuclear reaction, the decay reaction often continues. That is, after $^{238}_{92}\text{U}$ decays into $^{234}_{90}\text{Th}$, this thorium nucleus decays further. Figure 30.9 shows a sequence of nuclear decays starting from $^{238}_{92}\text{U}$, with each decay indicated by an arrow. A horizontal arrow pointing to the right denotes a beta decay similar to the one in Equation 30.13, in which the number of protons, Z, increases by one. For example, the second decay in Figure 30.9 is the reaction

$$^{234}_{90}\text{Th} \rightarrow {}^{234}_{91}\text{Pa} + e^- + \bar{\nu} \tag{30.17}$$

in which thorium (Th) is converted into protactinium (Pa) plus an electron and an antineutrino. The diagonal arrows in the figure denote alpha decay; for example, the diagonal arrow at the upper right links $^{238}_{92}\text{U}$ to $^{234}_{90}\text{Th}$ in the reaction in Equation 30.16. Each alpha decay reduces the proton number by two and the neutron number by two. Some nuclei, such as $^{218}_{84}\text{Po}$, can undergo either beta decay or alpha decay, so both a horizontal arrow and a diagonal arrow originate at these nuclei.

In the radioactive decay series in Figure 30.9, $^{238}_{92}\text{U}$ is the original parent nucleus while many unstable daughter nuclei are produced, and the decay proceeds until reaching the nucleus $^{206}_{82}\text{Pb}$. This nucleus is a stable final product of the decay of $^{238}_{92}\text{U}$. This particular decay series is especially important; among the daughter nuclei are $^{226}_{88}\text{Ra}$ (radium), which is one of the radioactive nuclei discovered by Pierre and Marie Curie (Fig. 30.10), and $^{222}_{86}\text{Rn}$ (radon), which is a significant health hazard.

▲ Figure 30.10 Marie Curie was a pioneer in the study of radioactivity. She received two Nobel Prizes for her work: one in physics (along with her husband, Pierre, and French physicist Henri Becquerel) in 1903 and a second prize in chemistry in 1911.

CONCEPT CHECK 30.3 Alpha Decay of Radium

Which of the following reaction equations correctly describes the alpha decay of $^{223}_{88}\text{Ra}$?

(a) $^{223}_{88}\text{Ra} \rightarrow {}^{227}_{90}\text{Th} + {}^{4}_{2}\text{He}$
(b) $^{227}_{90}\text{Th} \rightarrow {}^{223}_{88}\text{Ra} + {}^{4}_{2}\text{He}$
(c) $^{223}_{88}\text{Ra} \rightarrow {}^{219}_{86}\text{Rn} + {}^{4}_{2}\text{He}$

CONCEPT CHECK 30.4 Beta Decay

Consider a nucleus that undergoes beta decay and emits an electron. Which of the following statements is true?

(a) The atomic number Z decreases by one.
(b) The atomic number Z decreases by two.
(c) The atomic number Z increases by one.
(d) The atomic number Z increases by two.
(e) Z and A are unchanged.

CONCEPT CHECK 30.5 Making a New Element

Which type of decay changes the parent nucleus into a nucleus of a different element: (a) alpha decay, (b) beta decay, or (c) gamma decay? (More than one answer may be correct.)

Calculating the Binding Energy of a Nucleus

Recall from Chapter 27 that the rest energy of a particle (such as a nucleus) with rest mass m is

$$E = mc^2 \tag{30.18}$$

Let's now consider how this relation applies to the nucleus ${}^4_2\text{He}$, an alpha particle. With an understanding of the mass and binding energy in this case, we will be ready to discuss the energy released in nuclear reactions and how to calculate it.

An alpha particle consists of two protons and two neutrons, and from Table 30.1 we can add up the masses of two isolated protons plus two isolated neutrons to get

$$m(2 \text{ protons} + 2 \text{ neutrons}) = 2(1.0073 \text{ u}) + 2(1.0087 \text{ u}) = 4.0320 \text{ u} \tag{30.19}$$

Here we have included many more significant figures than usual because small mass differences will be extremely important in our final result. The result in Equation 30.19 is larger than the mass of an alpha particle, which is 4.0015 u. (This is the mass of just a helium nucleus, a helium atom without the two electrons; see Tables A.4 and 30.1.) The difference is due to the binding energy of the alpha particle, E_{binding}. The mass difference is

$$\Delta m = 4.0015 - 4.0320 \text{ u} = -0.0305 \text{ u} \tag{30.20}$$

Applying the relation between mass and energy from special relativity (Eq. 30.18), the mass difference in Equation 30.20 corresponds to an energy

$$E_{\text{binding}} = (\Delta m)c^2$$

To evaluate E_{binding}, we note that 1 atomic mass unit (1 u) corresponds to 931.5 MeV/c^2 (Eq. 30.11). Using this conversion factor with the value of Δm in Equation 30.20 gives

$$E_{\text{binding}} = (-0.0305 \text{ u})\left(\frac{931.5 \text{ MeV}/c^2}{\text{u}}\right)c^2 = -28.4 \text{ MeV} \tag{30.21}$$

This binding energy is negative, indicating that an alpha particle has a *lower* energy than a collection of separated protons and neutrons. An alpha particle is therefore more stable than a collection of separate protons and neutrons and so will not decay spontaneously into protons and neutrons.

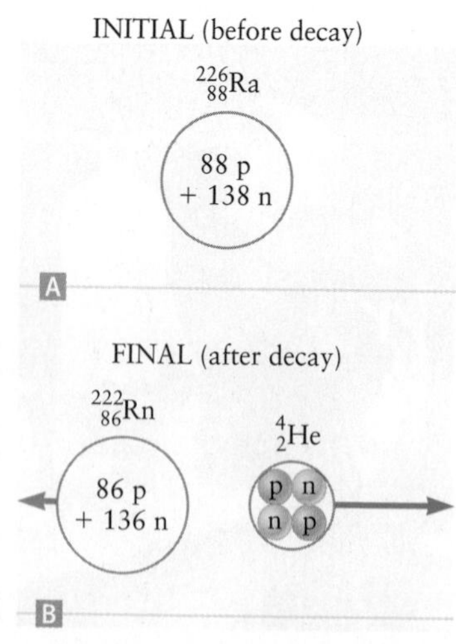

▲ **Figure 30.11** When a ${}^{226}_{88}\text{Ra}$ nucleus undergoes alpha decay, it emits an alpha particle (${}^4_2\text{He}$) and a ${}^{222}_{86}\text{Rn}$ nucleus. This decay leaves these two particles with some kinetic energy. To conserve momentum, the reaction products are emitted in opposite directions.

Finding the Energy Released in a Nuclear Reaction

Let's now apply our understanding of nuclear binding energy as illustrated for an alpha particle in Equations 30.19 through 30.21 to calculate the energy released in a nuclear decay process. The alpha decay reaction of radium plays an important role in many decay processes like the one diagramed in Figure 30.6A:

$${}^{226}_{88}\text{Ra} \rightarrow {}^{222}_{86}\text{Rn} + {}^4_2\text{He} \tag{30.22}$$

$$m = 226.0254 \text{ u} \rightarrow 222.0176 \text{ u} + 4.0026 \text{ u}$$

$$\text{total mass} = 226.0254 \text{ u} \rightarrow 226.0202 \text{ u}$$

(Note that the masses here and in other examples below include the electrons for each atom, unlike Eq. 30.20 where the mass value for the alpha particle was for the nucleus only.) For simplicity, Figure 30.11A shows the parent nucleus ${}^{226}_{88}\text{Ra}$ as being at rest. Will the reaction products ${}^{222}_{86}\text{Rn}$ and ${}^4_2\text{He}$ also be at rest after the reaction? The answer is no; a nucleus such as ${}^{226}_{88}\text{Ra}$ will decay spontaneously only if the total rest energy of the final products is less than the rest energy of the original parent nucleus. Total energy must be conserved, so the reaction in Equation 30.22 will release energy, usually in the form of kinetic energy of the decay products. (A portion of the energy is sometimes also used to put a daughter nucleus into an excited state, similar to the excited $^{14}_{7}N^*$ nucleus produced by beta decay; Eqs. 30.13 and 30.15.) That is why Figure 30.11B shows the $^{222}_{86}Rn$ and $^{4}_{2}He$ nuclei traveling away after the reaction.

To calculate the energy released in the decay of $^{226}_{88}Ra$, we must consider the masses of the nuclei in Equation 30.22. These masses are listed in Table A.4 and are given in Equation 30.22. The mass at the start of the reaction is $m_{start} = m(^{226}_{88}Ra) =$ 226.0254 u, whereas the total mass at the end is $m_{end} = m(^{222}_{86}Rn) + m(^{4}_{2}He) =$ 226.0202 u. The change in mass in the reaction is thus

$$\Delta m = m_{end} - m_{start} = 226.0202 - 226.0254 \text{ u} = -0.0052 \text{ u} \qquad (30.23)$$

which corresponds (from Eq. 30.18) to an energy $E_{reaction} = \Delta mc^2$. Using the definition of the mass unit u, we get

$$E_{reaction} = \Delta mc^2 = -(0.0052 \text{ u})\left(\frac{931.5 \text{ MeV}/c^2}{1 \text{ u}}\right)c^2 = -4.8 \text{ MeV} \qquad (30.24)$$

This energy is negative since the nuclei of the reaction products have a lower rest energy than the parent nucleus. This reaction releases 4.8 MeV, typically as the kinetic energy of the products. We'll explore this issue in the next example.

EXAMPLE 30.2 Speed of an Alpha Particle

The alpha decay of $^{226}_{88}Ra$ releases 4.8 MeV (Eq. 30.24). If one-tenth of this energy goes into kinetic energy of the emitted alpha particle, what is the speed of the alpha particle? Use the classical relation for the kinetic energy ($KE = \frac{1}{2}mv^2$).

RECOGNIZE THE PRINCIPLE

The kinetic energy of the alpha particle is related to its speed by $KE = \frac{1}{2}mv^2$, so given the kinetic energy we can find v.

SKETCH THE PROBLEM

Figure 30.11B describes the problem.

IDENTIFY THE RELATIONSHIPS

We are given that the alpha particle has a kinetic energy of

$$KE = \tfrac{1}{2}m_\alpha v^2 = \tfrac{1}{10}(4.8 \text{ MeV}) = 4.8 \times 10^5 \text{ eV} \qquad (1)$$

We must convert this energy to SI units (joules) so that we can obtain v in meters per second. The mass of the alpha particle is given in atomic mass units, and we must also convert this to SI units (kilograms) using Equation 30.11.

SOLVE

Rearranging Equation (1) to solve for v gives

$$v = \sqrt{\frac{2KE}{m_\alpha}} \qquad (2)$$

Converting KE to joules gives

$$KE = (4.8 \times 10^5 \text{ eV})\left(\frac{1.60 \times 10^{-19} \text{ J}}{1 \text{ eV}}\right) = 7.7 \times 10^{-14} \text{ J}$$

(continued) ▶

Converting m_α to kilograms, we have

$$m_\alpha = (4.00\text{ u})\left(\frac{1.66 \times 10^{-27}\text{ kg}}{1\text{ u}}\right) = 6.6 \times 10^{-27}\text{ kg}$$

Substituting these values in Equation (2), we find

$$v = \sqrt{\frac{2KE}{m_\alpha}} = \sqrt{\frac{2(7.7 \times 10^{-14}\text{ J})}{6.6 \times 10^{-27}\text{ kg}}} = \boxed{4.8 \times 10^6\text{ m/s}}$$

► *What does it mean?*

While this speed is only about 2% of the speed of light, the alpha particle still carries a lot of energy compared to the binding energy of electrons in an atom or molecule. For this reason, nuclear decay products can do significant damage when they travel through human tissue (Section 30.4).

Half-life

An important aspect of spontaneous nuclear decay is the decay time; that is, if you are given a $^{226}_{88}\text{Ra}$ nucleus, how long will you have to wait for it to decay by the process in Equation 30.22? Experiments show that individual nuclei decay one at a time, at *random* times. This randomness is a feature of quantum theory; there is no classical analogy.

Although one cannot predict when a particular nucleus will decay, one can give the *probability* for decay. This probability is specified in terms of the ***half-life*** of the nucleus, $T_{1/2}$. Suppose a total of N_0 nuclei are present at time $t = 0$. Half of these nuclei will decay during a time equal to one half-life $t = T_{1/2}$. Hence, half of the original nuclei will remain at $t = T_{1/2}$ as shown graphically in Figure 30.12. If we then wait until another half-life has passed—that is, until $t = 2T_{1/2}$—the number of original, undecayed, nuclei will fall to $N_0/4$. Half of the nuclei decay during each time interval of length $\Delta t = T_{1/2}$.

The decay curve in Figure 30.12 is described by the exponential function. We define the ***decay constant*** λ in such a way that

$$N = N_0 e^{-\lambda t} \tag{30.25}$$

The value of the decay constant for a particular isotope can be determined through experimental measurements of N as a function of time and using Equation 30.25. The decay constant λ is closely related to the half-life. From the definition of $T_{1/2}$, the number of nuclei is $N = N_0/2$ at $t = T_{1/2}$. Inserting these values into Equation 30.25 leads to

$$\frac{N_0}{2} = N_0 e^{-\lambda T_{1/2}}$$

$$e^{-\lambda T_{1/2}} = \tfrac{1}{2} \tag{30.26}$$

Taking the natural logarithm of both sides and doing some rearranging leads to

$$\ln(e^{-\lambda T_{1/2}}) = -\lambda T_{1/2} = \ln(1/2) = -\ln 2$$

$$\lambda T_{1/2} = \ln 2$$

$$T_{1/2} = \frac{\ln 2}{\lambda} \approx \frac{0.693}{\lambda} \tag{30.27}$$

using $\ln 2 \approx 0.693$. Equation 30.27 thus relates the value of the decay constant to the half-life. This relation is an inverse one: a large decay constant means a short half-life and vice versa.

Values of $T_{1/2}$ vary widely; a half-life can be as short as a tiny fraction of a second or longer than the age of the Earth. Values for a few important nuclei are given in Table 30.2, and a more complete listing is found in Table A.4. The value of $T_{1/2}$ is important in many effects and applications of radioactivity (Sections 30.4 and 30.5).

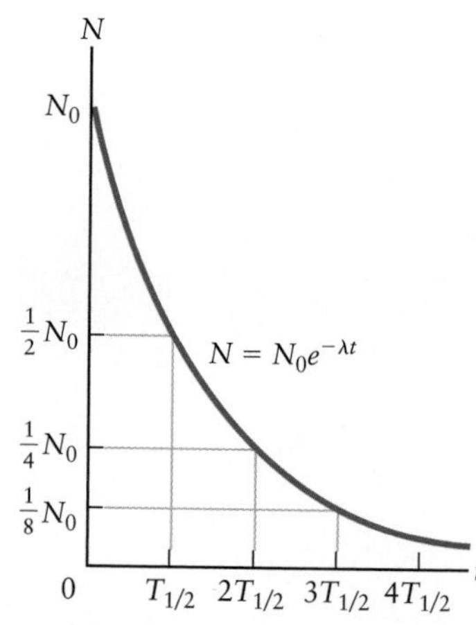

▲ **Figure 30.12** Radioactive decay is described by an exponential decay curve. The curve shows the number N of nuclei remaining after a time t. If there are N_0 nuclei at $t = 0$, only half are left at time $t = T_{1/2}$.

CONCEPT CHECK 30.6 PET Scans and the Half-life of ${}^{20}_{9}F$

The isotope ${}^{20}_{9}F$ is used in a medical procedure called positron emission tomography (PET). The half-life of ${}^{20}_{9}F$ is approximately 110 minutes. If your doctor has a sample with 16 g of pure ${}^{20}_{9}F$ at $t = 0$, how much ${}^{20}_{9}F$ will he have 330 minutes later, (a) 12 g, (b) 8.0 g, (c) 4.0 g, (d) 2.0 g, or (e) 1.0 g?

TABLE 30.2 Half-lives of Nucleons and Some Important Nuclei

Nucleus	Half-life
n (a free neutron)	10.4 min
${}^{1}_{1}H$ (a proton)	Stable
${}^{2}_{1}H$ (deuterium)	Stable
${}^{3}_{1}H$ (tritium)	12.33 yr
${}^{14}_{6}C$	5730 yr
${}^{60}_{27}Co$	5.27 yr
${}^{123}_{53}I$	13.2 h
${}^{222}_{86}Rn$	3.82 days
${}^{226}_{88}Ra$	1600 yr
${}^{235}_{92}U$	7.04×10^8 yr
${}^{238}_{92}U$	4.47×10^9 yr

EXAMPLE 30.3 Half-life and the Radioactive Decay of Tritium

Tritium is an isotope of hydrogen containing two neutrons and is denoted by ${}^{3}_{1}H$. It is a component in some fusion bombs (i.e., "hydrogen bombs") as we'll discuss in the next section. Tritium is radioactive and decays into an isotope of helium (${}^{3}_{2}He$) with a half-life of $T_{1/2} = 12.33$ yr. Suppose a fusion bomb contains 10 kg of tritium when it is first assembled. How much tritium will it contain after 30 years on the shelf?

RECOGNIZE THE PRINCIPLE

If there are N_0 tritium nuclei at $t = 0$, the number remaining after a time t is

$$N = N_0 e^{-\lambda t} \quad (1)$$

We weren't given the value of λ for tritium, but we can find it from the given half-life using the relation between λ and $T_{1/2}$ in Equation 30.27.

SKETCH THE PROBLEM

Figure 30.12 describes the problem. After a time $T_{1/2}$ passes, half of the radioactive nuclei have decayed.

IDENTIFY THE RELATIONSHIPS

We first rearrange Equation 30.27 to find the decay constant for tritium in terms of the half-life:

$$\lambda = \frac{\ln 2}{T_{1/2}} = \frac{0.693}{12.33 \text{ yr}} = 0.056 \text{ yr}^{-1}$$

SOLVE

Using this value of λ in Equation (1), we find that the fraction N/N_0 of tritium nuclei that remain after $t = 30$ yr is

$$\frac{N}{N_0} = e^{-\lambda t} = e^{-(0.056 \text{ yr}^{-1})(30 \text{ yr})} = 0.19$$

Hence, only 19% of the original tritium nuclei will remain. The initial mass was 10 kg, so the final mass of tritium is $0.19 \times 10 \text{ kg} = \boxed{1.9 \text{ kg}}$.

▸ ***What does it mean?***

Because ${}^{3}_{2}He$ is not useful for making bombs, the decay of tritium will cause a fusion bomb to stop working after a certain period of time. As a result, fusion bombs containing tritium must be periodically "reloaded."

EXAMPLE 30.4 Isotope Abundance and ${}^{235}_{92}U$

The isotope ${}^{235}_{92}U$ has a half-life for spontaneous fission of 7.0×10^8 yr. In deposits of naturally occurring uranium, this isotope is only about 0.72% of the total fraction of

(continued) ▸

uranium nuclei. Hence, if you had a sample containing 100 g of pure uranium, only 0.72 g would be $^{235}_{92}$U. The Earth is believed to be about 4.5 billion = 4.5×10^9 years old, or about $(4.5 \times 10^9)/(7.0 \times 10^8) \approx 6$ half-lives of $^{235}_{92}$U. How many grams of $^{235}_{92}$U were in your sample when the Earth was formed 4.5×10^9 yr ago?

RECOGNIZE THE PRINCIPLE

As we go back in time, the amount of $^{235}_{92}$U increases by a factor of two for every half-life. Hence, if we go back in time by one half-life, the amount of $^{235}_{92}$U is doubled; if we go back in time another half-life, the amount of $^{235}_{92}$U doubles again, and so on.

SKETCH THE PROBLEM

Figure 30.13 shows how the amount of $^{235}_{92}$U in our sample grows as we go back in time by six half-lives.

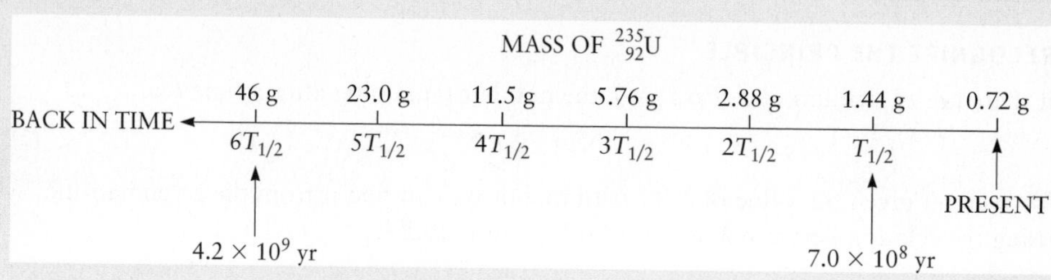

▶ Figure 30.13 Example 30.4. As we go back in time, the amount of $^{235}_{92}$U doubles every $T_{1/2} = 7.0 \times 10^8$ yr.

IDENTIFY THE RELATIONSHIPS AND SOLVE

According to Figure 30.13, the total mass of $^{235}_{92}$U after going back in time six half-lives is approximately $\boxed{46 \text{ g}}$.

We can be a bit more mathematical by noting that after one half-life, the mass increases by a factor of two, after two half-lives it increases by $2^2 = 4$, and after six half-lives it increases by $2^6 = 64$, which again gives

$$m_{\text{start}} = 2^6 m_{\text{end}} = 2^6(0.72 \text{ g}) = \boxed{46 \text{ g}}$$

▶ *What does it mean?*

When the Earth was first formed, uranium was more equally apportioned between $^{238}_{92}$U and $^{235}_{92}$U. At present, most of the original $^{235}_{92}$U has decayed. The half-life of $^{238}_{92}$U is about 4.5×10^9 yr, so only about half of the original $^{238}_{92}$U has decayed since the formation of the Earth.

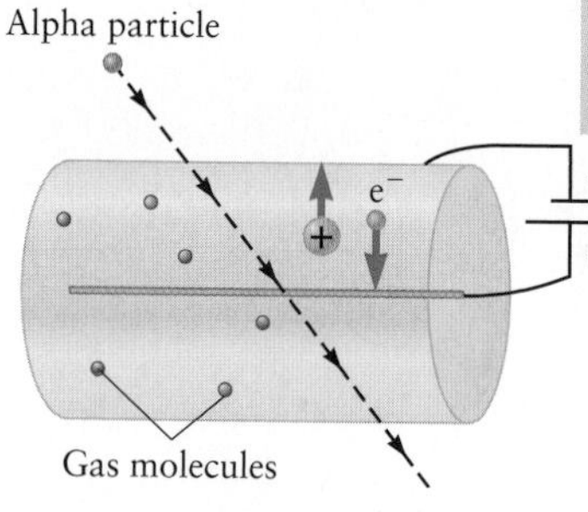

B

▲ Figure 30.14 A Activity can be measured with a Geiger counter. B Charged particles passing through the detector chamber ionize gas molecules, causing a short pulse of current to flow. Some Geiger counters give an audible "click" to indicate each current pulse.

How Do We "Measure" Radioactivity?

Each time a nucleus undergoes radioactive decay, it can produce potentially harmful products such as the alpha particle produced in the decay of $^{226}_{88}$Ra (Eq. 30.22). A nucleus with a short half-life is more likely to decay, so it is potentially more dangerous than a nucleus with a long half-life. Since a sample with more radioactive nuclei will be more dangerous than one with a small number of such nuclei, to fully assess such dangers we must also know how many nuclei are present in a sample. For this reason, the "strength" of a radioactive sample is measured using a property called the ***activity***, which can be measured with a Geiger counter (Fig. 30.14). The activity of a sample is proportional to the number of nuclei that decay in 1 second. If all other factors are similar, a sample with a high activity is more dangerous than a sample with a low activity. We often say that a sample with a high activity level is "hot."

One common unit of activity is the curie (Ci), which is defined as

$$1 \text{ Ci} = 3.7 \times 10^{10} \text{ decays/s} \tag{30.28}$$

Hence, a sample of radioactive material has an activity of 1 Ci if it exhibits 3.7×10^{10} decays each second; 1 g of naturally occurring radium has an activity of approximately 1 Ci. In practice, a sample with an activity of 1 Ci is extremely dangerous. Most studies or medical procedures with radioactive substances involve samples with activities of millicuries (mCi = 10^{-3} Ci) or microcuries (μCi = 10^{-6} Ci).

The official SI unit of activity is the becquerel (Bq), named in honor of Henri Becquerel (1852–1908), who shared the Nobel Prize in Physics for his discovery of spontaneous radioactivity. The becquerel is defined by

$$1 \text{ Bq} = 1 \text{ decay/s} \tag{30.29}$$

As time passes and nuclei in a sample decay, the number of remaining unstable nuclei drops. Hence, the activity of a sample decreases with time. We thus say (loosely speaking) that a "hot" radioactive sample (such as the nuclear reactor at Chernobyl in Ukraine) becomes "cooler" with time and therefore safer to deal with or clean up.

While a sample with a high activity level is generally more dangerous than one with low activity, there are complicating factors due to the variable amount of energy carried by decay products as we'll discuss in Section 30.4.

Insight 30.3

GEIGER COUNTERS

A Geiger counter (Fig. 30.14) is a useful tool to study and monitor radioactive materials. Invented by Hans Geiger (a student of Rutherford), the Geiger counter detects the passage of a fast-moving particle (typically produced by radioactive decay) through a gas. Such a particle ionizes gas atoms and molecules, allowing a current between the central wire and the wall of the container. Typically, this current comes as short pulses, giving the familiar "clicks" of a Geiger counter. Since each click corresponds to a single nuclear decay, a Geiger counter measures activity.

CONCEPT CHECK 30.7 How Does the Activity Level of a Radioactive Sample Change with Time?

Consider a radioactive material that decays with a half-life of $T_{1/2}$. How long does it take for the activity of the sample to decrease by a factor of four, (a) $T_{1/2}$, (b) $2T_{1/2}$, (c) $3T_{1/2}$, or (d) $4T_{1/2}$?

EXAMPLE 30.5 Activity of Tritium

You are given a sample containing exactly 1 mole of tritium ($^3_1$H). The tritium decays with a half-life of 12.33 yr. What is the activity of the sample? Express your answer in units of decays per second. *Hint*: Inserting $t = 1$ s in the exponential decay law $N = N_0 e^{-\lambda t}$ gives the number of nuclei remaining after 1 s. The activity is the number of nuclei that *decay* in the same 1-s time period.

RECOGNIZE THE PRINCIPLE

If a sample initially contains N_0 nuclei, the number remaining after $t = 1$ s is

$$N_{\text{remaining}} = N_0 e^{-\lambda t}$$

Hence, the number that decay during this time is

$$N_{\text{decays}} = N_0 - N_0 e^{-\lambda t} = N_0(1 - e^{-\lambda t}) \tag{1}$$

The activity equals the number of nuclei that decay per second.

SKETCH THE PROBLEM

Figure 30.15 shows how the number of nuclei remaining decreases with time as well as the number of nuclei that decay in $t = 1$ s.

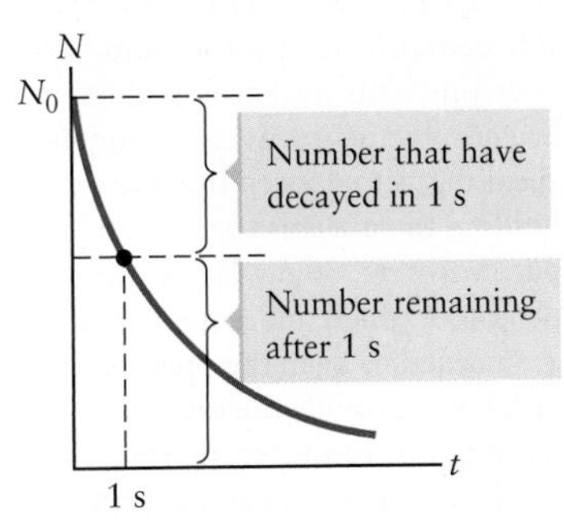

▲ **Figure 30.15** Example 30.5. Not to scale.

IDENTIFY THE RELATIONSHIPS

To evaluate Equation (1) for our given tritium sample, we must find both N_0 and the decay constant λ. The sample originally contains exactly 1 mole of tritium, so $N_0 = 6.02 \times 10^{23}$ (Avogadro's number). From Equation 30.27, the decay constant is related to the half-life through $\lambda = 0.693/T_{1/2}$.

(continued) ▶

Inserting the known half-life of tritium (12.33 yr) gives

$$\lambda = \frac{0.693}{T_{1/2}} = \frac{0.693}{12.33 \text{ yr}} = \frac{0.693}{3.9 \times 10^8 \text{ s}} = 1.8 \times 10^{-9} \text{ s}^{-1}$$

SOLVE

Inserting our values of N_0 and λ as well as $t = 1$ s into Equation (1) leads to

$$N_{\text{decays}} = N_0(1 - e^{-\lambda t}) = (6.02 \times 10^{23})[1 - e^{-(1.8\times10^{-9}\text{ s}^{-1})(1\text{ s})}]$$

The exponential factor is very close to one, so we have to keep extra significant figures in performing this calculation. We find

$$\text{activity} = N_{\text{decays}} = 1.1 \times 10^{15} \text{ per second} = \boxed{1.1 \times 10^{15} \text{ Bq}}$$

► ***What does it mean?***

Nuclei with a shorter half-life decay more rapidly than those with a long half-life. Hence, when comparing two samples, the one with a short half-life will have a higher activity (and hence be more dangerous) than the one with a long half-life, assuming the number of nuclei in the two samples is the same.[4]

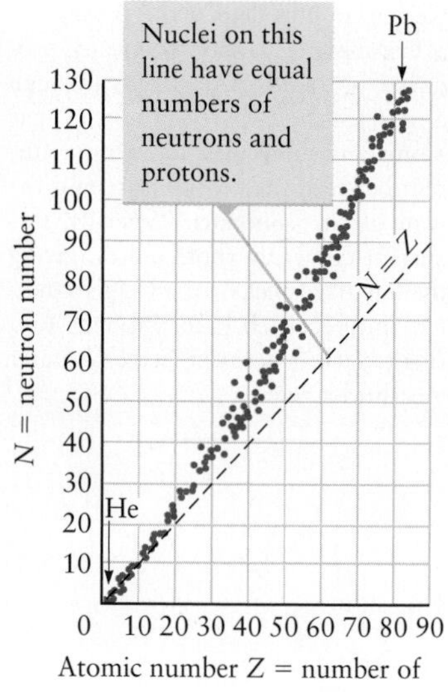

▲ **Figure 30.16** Stable nuclei usually have a greater number of neutrons than protons, so $N > Z$ for all but the lightest nuclei. This trend is especially noticeable at high Z.

30.3 Stability of the Nucleus: Fission and Fusion

To be stable, a nucleus containing two or more protons must contain neutrons. Neutrons are thus necessary for the stability of essentially all matter. Figure 30.16 shows a plot of the neutron number N as a function of the number of protons Z (i.e., the atomic number) for all known stable nuclei. The dashed line shows the function $N = Z$; nuclei that lie on this line have equal numbers of neutrons and protons, while nuclei above this line contain an excess of neutrons. Low-mass nuclei such as He and C tend to have equal numbers of protons and neutrons and are thus on or near the dashed line corresponding to $N = Z$ in Figure 30.16. Heavy nuclei, however, have many more neutrons than protons; for example, Pb (lead) nuclei contain around 40 more neutrons than protons.

The reason these additional neutrons are needed can be seen by considering the forces between protons and neutrons. Figure 30.17A shows a highly schematic nucleus with two protons and two neutrons, representing $^4_2$He. This nucleus is extremely stable, so this arrangement of protons and neutrons is particularly favored. We now add another proton to make lithium (Li). To bind this additional proton and achieve a stable arrangement of protons and neutrons similar to $^4_2$He, we need two more neutrons as shown in Figure 30.17B. This highly simplified picture leaves out the possibility that the protons in Li might rearrange to better "share" the attraction of the neutrons, just as two atoms share electrons when they form a chemical bond. In reality, $^6_3$Li and $^7_3$Li are both stable isotopes. However, this simple picture does explain why nuclei with large numbers of protons require greater numbers of neutrons to be stable, accounting for the trend that $N > Z$ for heavy nuclei in Figure 30.16. This trend is essential for understanding nuclear fission.

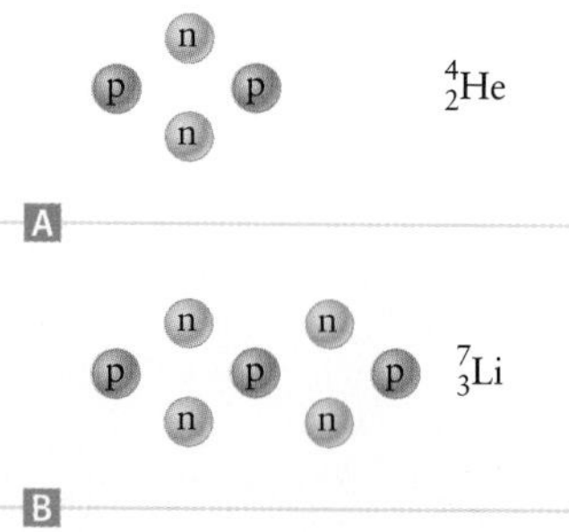

▲ **Figure 30.17** A The isotope $^4_2$He contains two protons and two neutrons. This nucleus is stable because the neutrons can (roughly speaking) sit between the protons, holding the nucleus together. (Compare with Fig. 30.5.) B If a third proton is added, more neutrons are needed to keep the protons apart and provide the needed binding due to the strong force. As the number of protons grows, extra neutrons are needed. For this reason, $^7_3$Li (pictured here) is more stable than $^6_3$Li.

Nuclear Binding Energy

We have discussed how the mass of a nucleus is related to its binding energy, and as an example we calculated the binding energy of an alpha particle (Eq. 30.21). We

[4]Notice also that our calculation gives the *average* activity during a 1-s interval, but because the half-life here is much greater than 1 s, the *instantaneous* activity will have almost exactly the same value.

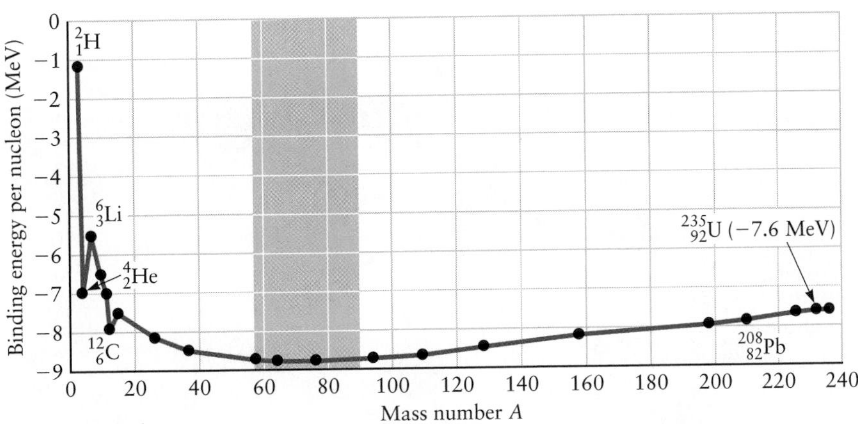

◀ **Figure 30.18** Binding energy per nucleon for nuclei with different mass numbers. A binding energy of high magnitude corresponds to a tightly bound nucleon, so nuclei falling in the shaded region of the graph are the most stable. The binding energy is negative because a nucleus is more stable and has a lower energy than a collection of separated protons plus neutrons.

found that the mass of the nucleus ^4_2He, which contains two protons and two neutrons, is less than the total mass of two isolated protons plus two isolated neutrons. The same is true for other stable nuclei; the mass of a nucleus containing N neutrons and Z protons is *less* than the total mass of N isolated neutrons plus Z isolated protons. This mass difference is connected to the binding energy of the nucleus through the relativistic relation between mass and energy ($E = mc^2$). By measuring the mass of a nucleus and then subtracting the mass of the same numbers of isolated protons and neutrons, we can find the binding energy of the nucleus. Some results for the binding energy per nucleon are shown in Figure 30.18. In the figure, the horizontal axis is mass number A. Different isotopes (nuclei having the same Z but different A) can have different binding energies; here we plot an approximate average value for each element, that is, for each value of A. Values of the binding energy are plotted as negative because each of these nuclei has a lower energy than a collection of separated protons and neutrons. The magnitude of the binding energy is largest for nuclei close to $A = 60$, which is near Fe and Ni in the periodic table, meaning that these nuclei are the *most stable* ones. This fact and the relationship between binding energy and Z are the basis for two important nuclear phenomena: fission and fusion.

In our discussions of binding energies in previous chapters, including the binding energy of an atom, the energy of a stable bound system has also been negative. For example, the gravitational potential energy of two masses is negative (Eq. 6.20), and the energy of a bound atom such as hydrogen is also negative (Fig. 29.12). When an attractive force exists between two or more particles, you must add energy to separate them. Hence, binding lowers the potential energy relative to the state in which the particles are at infinite separation.

Nuclear Fission

Figure 30.9 shows how a parent nucleus (in that case, $^{238}_{92}U$) can undergo spontaneous radioactive decay to form a series of daughter nuclei. Radioactive decay can also be *induced* by a collision between two nuclei or by the collision of a neutron with a nucleus. A particularly important example occurs when $^{235}_{92}U$ collides with a neutron. A typical reaction of this kind is

one neutron induces fission of $^{235}_{92}U$ → ; three neutrons are produced →

$$^1_0n + ^{235}_{92}U \rightarrow ^{139}_{56}Ba + ^{94}_{36}Kr + 3^1_0n \tag{30.30}$$

Here a neutron is denoted by ^1_0n since it "contains" no protons ($Z = 0$) and has a total of one nucleon ($A = 1$). The process in Equation 30.30 is called ***nuclear fission***, a reaction whose main products are two nuclei both roughly half the size of the original parent nucleus. (By contrast, in alpha and beta decay one of the decay products is very close in mass to the original parent nucleus.)

Nuclear fission

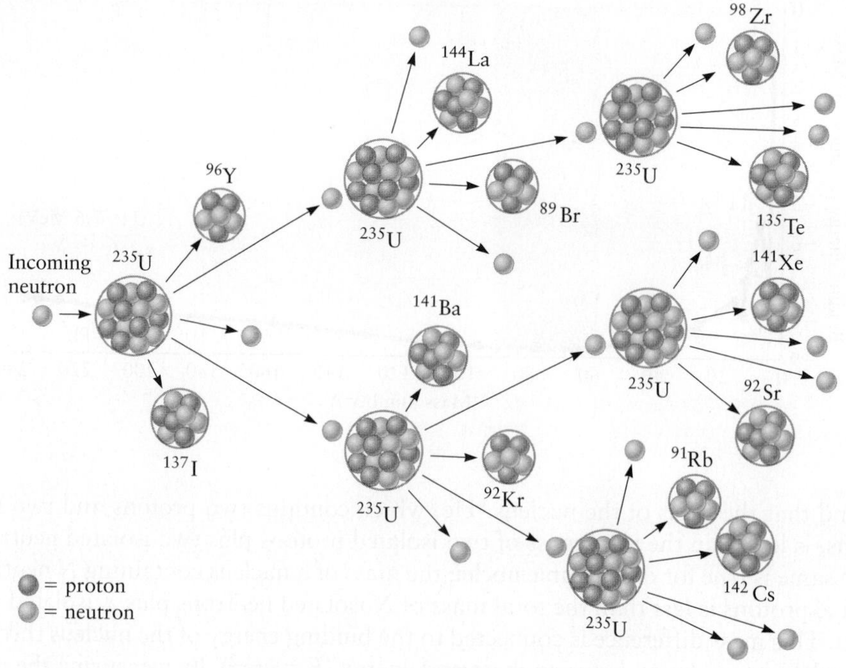

▶ **Figure 30.19** The fission of $^{235}_{92}U$ can be initiated by a collision with a neutron. Here the original $^{235}_{92}U$ nucleus splits into two fission fragments, $^{96}_{39}Y$ and $^{137}_{53}I$, along with three neutrons. These neutrons then collide with other $^{235}_{92}U$ nuclei and cause them to fission. This is an example of a chain reaction.

Fission reactions have several important features. First, there can be many different possible reaction products. For example, when $^{235}_{92}U$ collides with a neutron, it may produce $^{139}_{56}Ba$ and $^{94}_{36}Kr$ as in Equation 30.30, but it may also produce a different pair of nuclei. Another possible fission reaction involving $^{235}_{92}U$ is

$$^{1}_{0}n + ^{235}_{92}U \rightarrow ^{95}_{38}Sr + ^{138}_{54}Xe + 3^{1}_{0}n$$

As in the reaction in Equation 30.30, the collision with a neutron causes $^{235}_{92}U$ to split into two smaller nuclei and also emit several neutrons. There are approximately 90 different fission reactions involving $^{235}_{92}U$.

A second key feature of fission is that it produces more free neutrons than it uses. For example, the reaction in Equation 30.30 begins with one neutron, but there are three neutrons at the end. Some fission reactions with $^{235}_{92}U$ emit three neutrons, whereas others emit two or four. On average, the fission of a single $^{235}_{92}U$ nucleus caused by the collision with a neutron results in 2.5 neutrons at the end of the reaction. These neutrons can then induce the fission of other nearby $^{235}_{92}U$ nuclei, leading to a ***chain reaction*** as sketched in Figure 30.19. In this way, the fission of a single $^{235}_{92}U$ nucleus can induce the fission of an enormous number of additional $^{235}_{92}U$ nuclei. Each fission event produces additional free neutrons because the ratio of neutrons to protons in a heavy nucleus such as $^{235}_{92}U$ is greater than in either of the two nuclei that are produced (e.g., $^{139}_{56}Ba$ and $^{94}_{36}Kr$ in Eq. 30.30). This fact can be traced to the shape of the stability curve in Figure 30.16.

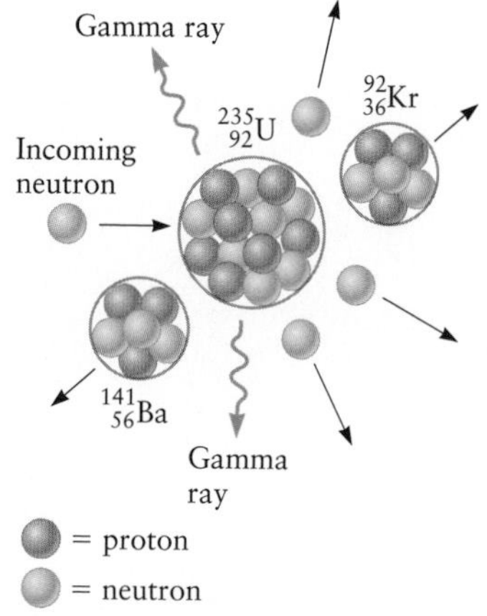

▲ **Figure 30.20** When a nucleus undergoes fission, some gamma ray photons are usually emitted along with the two fission fragments and neutrons. Typically, these particles all have large kinetic energies.

A third important feature of fission reactions is that they release energy as can be seen from the binding energy curve in Figure 30.18. The magnitude of the binding energy for $^{235}_{92}U$ is about 7.6 MeV/nucleon, while the magnitudes of the binding energies of $^{139}_{56}Ba$ and $^{94}_{36}Kr$ (the fission products in Eq. 30.30) are above 8.0 MeV/nucleon. This fission reaction releases the extra binding energy in the form of gamma rays and through the kinetic energies of the particles produced (Fig. 30.20). The shape of the binding energy curve guarantees that the magnitude of the binding energy per nucleon of $^{235}_{92}U$ is always smaller in magnitude than the binding energies of its fission products. The same is true for other nuclei that undergo fission such as $^{239}_{94}Pu$ (plutonium). That is why fission releases large amounts of energy.

CONCEPT CHECK 30.8 Fission Reaction of $^{235}_{92}\text{U}$

Consider the fission reaction

$$^{235}_{92}\text{U} + {}^{1}_{0}\text{n} \rightarrow {}^{144}_{56}\text{Ba} + {}^{90}_{36}\text{Kr} + ?{}^{1}_{0}\text{n}$$

How many neutrons does it produce?

Energy from Fission

Nuclear fission is the basis for two modern technologies: fission weapons (bombs) and nuclear reactors for producing electrical power. Both are based on nuclear chain reactions like the one in Figure 30.19. In a nuclear fission bomb, the chain of fission reactions is allowed to run out of control, rapidly releasing extremely large amounts of energy.

Figure 30.19 implies that it's easy to set off an uncontrolled chain reaction: just fire one neutron at a sample of $^{235}_{92}\text{U}$ and the neutrons produced in the initial fission event go on to split other $^{235}_{92}\text{U}$ nuclei, releasing enough energy to cause a huge explosion. Fortunately, it is not that easy. First of all, natural uranium is more than 99% $^{238}_{92}\text{U}$, an isotope that does not readily fission. The uranium used in weapons is "enriched" to increase the concentration of scarce $^{235}_{92}\text{U}$, and enrichment is a very difficult engineering process. Second, if the piece of uranium is small, many of the nuclei are close enough to the surface that many of the released neutrons escape from the sample before inducing another fission event. In such a case, the reaction may stop before a significant number of $^{235}_{92}\text{U}$ nuclei undergo fission. Increasing the size (and hence also the mass) of a uranium sample lowers the surface-to-volume ratio and lowers the probability that a free neutron will reach the surface; it is then more likely that the neutron will encounter a $^{235}_{92}\text{U}$ nucleus and induce fission. The minimum amount of nuclear fuel material needed to sustain a chain reaction is called the ***critical mass***. For uranium, the critical mass is typically about 50 kg (corresponding to a uranium sphere with a radius of about 9 cm), but depending on how the sample is shaped and how the surface is prepared it can be as small as 15 kg.

When $^{235}_{92}\text{U}$ undergoes fission, it produces two nuclei of roughly equal mass and several neutrons as described in Equation 30.30. We can estimate the amount of energy released using the binding energy curve in Figure 30.18. The binding energy per nucleon of $^{235}_{92}\text{U}$ is about −7.6 MeV, whereas for many of the nuclei produced by fission it is about −8.5 MeV per nucleon. The difference is $(-8.5 + 7.6)$ MeV = −0.9 MeV per nucleon. Since there are 235 nucleons in one $^{235}_{92}\text{U}$ nucleus, the total amount of energy released by the fission of one such nucleus is $(0.9 \text{ MeV}) \times 235 \approx 210$ MeV. To get a feeling for the scale of this value, let's calculate how much energy we can expect from the fission of 1 g of pure $^{235}_{92}\text{U}$ (about the mass of a paper clip).

The atomic mass of pure $^{235}_{92}\text{U}$ is 235 g (from the definition of the atomic mass unit), so 235 g of $^{235}_{92}\text{U}$ contains 1 mole of nuclei. Our 1-g sample thus contains

$$N_{\text{U-235}} = (1 \text{ g}) \times \frac{6.02 \times 10^{23}}{235 \text{ g}} = 2.6 \times 10^{21} \text{ nuclei}$$

Each nucleus releases about 210 MeV when fission takes place, so the total energy released by the fission of 1 g of $^{235}_{92}\text{U}$ is

$$E_{\text{fission}} = (2.6 \times 10^{21})(210 \text{ MeV})\left(\frac{10^6 \text{ eV}}{1 \text{ MeV}}\right)\left(\frac{1.60 \times 10^{-19} \text{ J}}{1 \text{ eV}}\right)$$

$$E_{\text{fission}} \approx 9 \times 10^{10} \text{ J per gram of } ^{235}_{92}\text{U} \tag{30.31}$$

This is an *enormous* amount of energy. By comparison, 1 g of the chemical explosive TNT releases about 4700 J. The energy released by 1 g of $^{235}_{92}\text{U}$ is more than 10 million times larger! The amount of energy released in a fission bomb is usually measured in terms of the mass of TNT that would release the same amount of energy. The fission bombs that were exploded in World War II each released an amount of energy equivalent to more than 10,000 *tons* of TNT.

EXAMPLE 30.6 Energy Released in a Fission Bomb

The fission bomb that was dropped on the Japanese city of Hiroshima in 1945 contained approximately 64 kg of $^{235}_{92}$U and released an amount of energy equivalent to the explosion of approximately 13,000 metric tons (1.3×10^{10} g) of TNT (called a "13-kiloton bomb"; see Example 27.8 for a description of how the explosive yield of a nuclear weapon is measured and specified). What mass of $^{235}_{92}$U underwent fission?

RECOGNIZE THE PRINCIPLE

In our analysis of fission, we found that the energy released by the fission of 1 g of $^{235}_{92}$U is about 9×10^{10} J (Eq. 30.31). We also noted that 1 g of TNT releases about 4700 J. We can use these values to compute what mass of $^{235}_{92}$U is needed to give the energy released by 13,000 metric tons of TNT.

SKETCH THE PROBLEM

Fission releases energy in the form of the kinetic energy of the reaction products plus the energy of gamma rays that are emitted (Fig. 30.20).

IDENTIFY THE RELATIONSHIPS

Let's first find the energy released by 13,000 metric tons = 1.3×10^{10} g of TNT. From the given information, we have

$$E = (1.3 \times 10^{10}\ \text{g})\left(\frac{4700\ \text{J}}{\text{g}}\right) = 6.1 \times 10^{13}\ \text{J}$$

SOLVE

Using the value of $E_{\text{fission}} \approx 9 \times 10^{10}$ J/g of $^{235}_{92}$U from Equation 30.31, this energy corresponds to

$$\text{mass of } ^{235}_{92}\text{U} = (6.1 \times 10^{13}\ \text{J})\left(\frac{1\ \text{g of } ^{235}_{92}\text{U}}{9 \times 10^{10}\ \text{J}}\right) \approx \boxed{700\ \text{g}}$$

► ***What does it mean?***

Only about 1% of the $^{235}_{92}$U in the Hiroshima weapon actually underwent fission. Even so, an enormous amount of energy was released. So-called modern nuclear weapons yield much greater amounts of energy.

Insight 30.4

SAFETY AT NUCLEAR FISSION REACTORS

Safety is always a major concern at a nuclear fission reactor, particularly those used as power reactors to generate electricity. All reactors are designed with extensive safety features to prevent a "meltdown" of the reactor core or any similar hazard. The latest reactor designs contain many "passive" safety features that do not require operator action or even electronic feedback to deal with emergency situations. For example, these safety features work even in the event of a complete electrical failure, making these reactors extremely safe. Unfortunately, many reactors now in use were designed several decades ago, and their safety is a source of ongoing concern, particularly in view of the reactor failures at the Fukushima Daiichi power plant in Japan in 2011.

Nuclear Power Plants

Nuclear fission is used in a productive and controlled way in nuclear power plants. In a power reactor, fission takes place in a region called the reactor core, shown schematically in Figure 30.21A. The important components of the core are fuel rods, control rods, and moderators.

Fuel rods contain $^{235}_{92}$U that can be added or removed from the core while the reactor is operating. The fuel rods are removed when the $^{235}_{92}$U is exhausted or when operators need to reduce the fission rate. *Control rods* contain materials (e.g., $^{113}_{48}$Cd) that absorb neutrons. They can be inserted or removed from the reactor core, allowing the reactor operators to control the number of free neutrons and thus adjust the fission rate. The *moderator* (often water) circulates through the reactor core. One of the moderator's functions is to slow the neutrons down, which increases their probability of inducing a fission event. (Roughly speaking, a slow neutron passing near a uranium nucleus has more time to induce fission.) The water also carries away heat

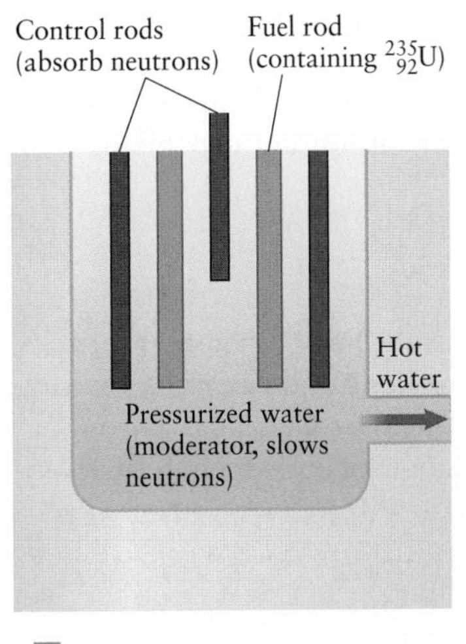

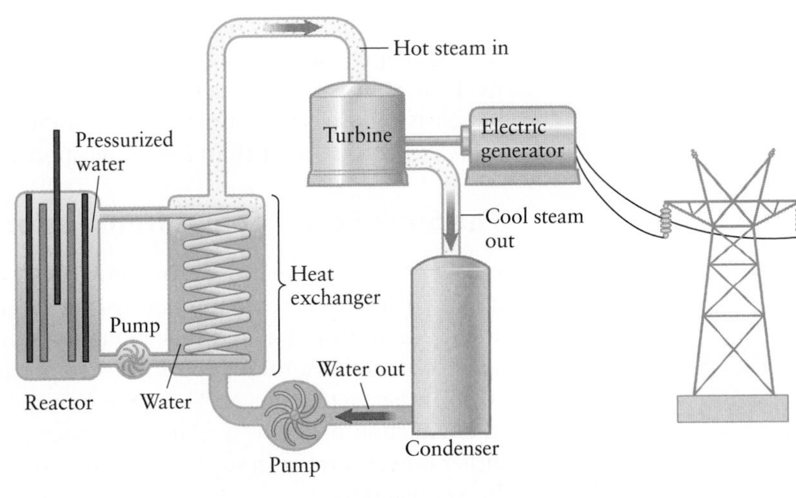

▲ **Figure 30.21** A Schematic of the core of a nuclear reactor. The fuel rods contain $^{235}_{92}U$ (or some other fissionable nuclei). Neutrons emitted during fission collide with nuclei in the moderator (here, water), slowing the neutrons. Control rods containing nuclei that readily absorb neutrons are used to control the rate of fission. B Heat generated in the moderator (water) is used to power a steam turbine and thereby generate electric power.

to a separate steam engine (Fig. 30.21B), where the energy generated by the reactor is converted to electrical energy using a generator (Figs. 21.7 and 22.3). This heat engine operates according to the principles of thermodynamics, so it requires a cold reservoir. Nuclear power plants are often located near bodies of water such as lakes that provide this cold reservoir (Fig. 30.22).

As mentioned previously, naturally occurring uranium contains a mixture of the isotopes $^{235}_{92}U$ (0.7%) and $^{238}_{92}U$ (99.3%). The fission reactions in a reactor involve only $^{235}_{92}U$, so the uranium in fuel rods is usually enriched to increase the concentration of $^{235}_{92}U$. However, this concentration is always kept low enough that the fuel rod does not exceed the critical mass, so it is not possible for the core of a nuclear reactor to explode like a nuclear fission bomb. It is possible, though, to have an accident called a "meltdown" in which the fuel rods become so hot that they and the walls of the reactor melt and the water in the core rapidly turns to steam. The result can be a conventional explosion (due to the high-pressure steam), as occurred at Chernobyl in Ukraine in 1986. This accident allowed the escape of significant amounts of radioactive material into the environment. A similar meltdown-induced explosion occurred at the Fukushima Daiichi reactor complex in Japan in 2011, which also led to the release of considerable radioactive material into nearby water and soil.

▲ **Figure 30.22** The steam engines in a nuclear power reactor must cool the steam as part of the thermodynamic cycle (Chapter 16). This cooling is accomplished with large cooling towers or with a source of cool water from an ocean or river. This photo shows the Diablo Canyon Nuclear power plant in California. The two large domes each house a reactor, and the cooling water is taken from the ocean.

Nuclear Fusion

In ***nuclear fusion***, two nuclei join together to produce one new particle. Consider two $^{6}_{3}Li$ nuclei; they have $Z = 3$ and are on the left side of the binding energy curve in Figure 30.18. If they fuse, the new nucleus has $Z = 6$ and contains a total of $A = 12$ nucleons, which is $^{12}_{6}C$. According to Figure 30.18, $^{12}_{6}C$ has a greater binding energy per nucleon (in magnitude) than $^{6}_{3}Li$ and is thus more stable. Hence, this fusion reaction releases energy:

Nuclear fusion

$$^{6}_{3}\text{Li} + {}^{6}_{3}\text{Li} \rightarrow {}^{12}_{6}\text{C} + \text{energy} \tag{30.32}$$

This energy is in the form of gamma rays plus the kinetic energy of the $^{12}_{6}C$ nucleus. We can use Figure 30.18 to estimate how much energy this reaction will release. The binding energy per nucleon is about -5.4 MeV for $^{6}_{3}Li$ and about -7.9 MeV for

$^{12}_{6}C$. Hence, the energy *released* in this fusion reaction is approximately $7.9 - 5.4 = 2.5$ MeV per nucleon. The fission of $^{235}_{92}U$ releases approximately 0.9 MeV per nucleon, so this fusion reaction releases even more energy (per nucleon) that fission, which is one reason (unfortunately) bombs employing fusion are capable of even more explosive power than fission weapons.

The Sun's Power Comes from Fusion

Nuclear fusion is the process that powers stars such as our Sun. When first formed, a typical star consists mainly of hydrogen—that is, protons and electrons—with a small amount of $^{4}_{2}He$ and much smaller amounts of heavier elements. Through a series of reactions, these protons fuse to produce alpha particles, which in turn fuse to form heavier elements. Our own Sun is hot enough to produce elements up to about carbon in this way. All the heavier elements in the universe were created in either hotter stars or in supernova explosions, which can trigger even more energetic fusion reactions.

Fusion reactions can take place in a star due to the very high temperatures and pressures in its core. These conditions cause nuclei to come into very close contact when they collide, which is necessary for fusion to occur. To create a fusion reaction on Earth, we must approach or duplicate stellar conditions, giving the initial nuclei very high speeds and keeping them very close together. In a fusion bomb, this is accomplished using a fission bomb as a trigger. That is, a bomb that uses the fission of $^{235}_{92}U$ or some other nucleus is exploded in such a way that the fusion ingredients are compressed very rapidly, initiating a fusion reaction.

Fusion has at least one potential useful application on the Earth: as the basis for a power plant. Hypothetical designs for a fusion power reactor use $^{2}_{1}H$ (deuterium) or $^{3}_{1}H$ (tritium) as the starting nuclei in a fusion reaction that releases very large amounts of energy to produce electrical power. Naturally occurring water has a small but significant concentration of deuterium, so the fuel for a fusion reactor would be very inexpensive and virtually limitless. Moreover, current designs for these reactors would produce virtually no harmful radiation or by-products. Although a practical fusion power plant has not yet been built, several exploratory engineering projects have been. The latest is the ITER (International Tokomak Engineering Reactor), due to be operational in 2016, which uses what is called a "tokomak" geometry for confining the nuclei while they undergo fusion. The ITER will still be an experimental project, but the long-term hope is that it will lead to the construction of practical fusion-based power plants that would provide abundant amounts of cheap and clean electrical energy.

EXAMPLE 30.7 Energy of a Fusion Reaction

A fusion reaction that might be used in a power reactor involves deuterium, $^{2}_{1}H$:

$$^{2}_{1}H + ^{2}_{1}H \rightarrow ^{3}_{1}H + ^{1}_{1}H \quad (1)$$

In this reaction, two deuterium nuclei fuse to form $^{3}_{1}H$ plus a proton, $^{1}_{1}H$. Use the masses of these nuclei from Table A.4 to calculate the energy released in this reaction. Express the answer in MeV.

RECOGNIZE THE PRINCIPLE

This reaction converts some of the mass of the original nuclei into energy. The amount of energy can be found from the mass–energy relation $E = \Delta mc^2$, where Δm is the difference in the mass of the masses on the left- and right-hand sides of Equation (1).

SKETCH THE PROBLEM

No sketch is necessary, but it is useful to rewrite Equation (1) to also show the rest masses of the nuclei; see Equation (2) below.

IDENTIFY THE RELATIONSHIPS

According to Table A.4, the masses in Equation (1) are

$$m({}_1^1\text{H}) = 1.007825 \text{ u}$$
$$m({}_1^2\text{H}) = 2.014102 \text{ u}$$
$$m({}_1^3\text{H}) = 3.016049 \text{ u}$$

We include extra significant figures because the change in mass in Equation (1) will be small, but will still lead to a substantial energy.

SOLVE

Let's rewrite Equation (1) to show the masses before and after the reaction:

$$\begin{aligned} {}_1^2\text{H} \quad + \quad {}_1^2\text{H} \quad &\rightarrow \quad {}_1^3\text{H} \quad + \quad {}_1^1\text{H} \\ 2.014102 \text{ u} + 2.014102 \text{ u} &\rightarrow 3.016049 \text{ u} + 1.007825 \text{ u} \\ 4.028204 \text{ u} &\rightarrow 4.023874 \text{ u} \end{aligned} \quad (2)$$

The mass difference is thus

$$\Delta m = m(\text{right side}) - m(\text{left side}) = -0.004330 \text{ u} \quad (3)$$

which corresponds to an energy of $E = \Delta mc^2$. From Equation 30.11, atomic mass units can be converted to units of MeV/c^2 using the relation 1 u = 931.5 MeV/c^2. From Equations (2) and (3), we get

$$E = (-0.004330 \text{ u})\left(\frac{931.5 \text{ MeV}/c^2}{\text{u}}\right)c^2 = \boxed{-4.033 \text{ MeV}}$$

The negative value means that this is the energy *released* by the reaction.

▸ ***What does it mean?***

Since deuterium occurs naturally in ordinary water, this source of energy is potentially cheap.

30.4 Biological Effects of Radioactivity

The biological effects of radioactivity result from the way decay products or reaction products interact with atoms and molecules. From Chapter 29, the typical binding energy of an electron in an atom is on the order of 10 eV. The energy released in a nuclear reaction is much larger, often appearing as the kinetic energy of an emitted alpha or beta particle. If one of these particles collides with an atomic electron or with a molecule, there is ample energy to break a chemical bond in molecules such as DNA, making these molecules biologically inactive or even harmful. That is how radiation damages living tissue.

The amount of damage that a particular particle is capable of doing is a complicated issue because alpha, beta, and gamma radiation have different masses and charges and therefore interact with tissue in different ways. We defined the curie (Ci) and becquerel (Bq) as units of activity that measure the number of radioactive decays per unit time in Equations 30.28 and 30.29. Some particles produced in radioactive decay carry a great deal of kinetic energy, while others carry little. To account for this difference, we introduce a unit called the ***rad***. By definition, 1 rad is the amount of radiation that deposits 10^{-2} J of energy into 1 kg of absorbing material. Hence, this unit accounts for both the amount of energy carried by the particle and the efficiency with which this energy is absorbed.

The name "rad" denotes *r*adiation *a*bsorbed *d*ose. The definition of the rad and the term *dose* both indicate that the total amount of radiation received is important when assessing biological effects, but even the rad unit does not capture the full complexity of radiation damage. Different types of particles can do different amounts of biological damage even if they deposit the same amount of energy and hence have the same strength as measured in rads. For this reason, another unit called the *relative biological effectiveness* factor is also very useful. This factor, denoted ***RBE***, measures how efficiently a particular type of particle damages tissue. Some typical *RBE* factors for various kinds of radiation from nuclear reactions are given in Table 30.3. Notice that the *RBE* value tends to increase as the mass of the particle increases. For example, the *RBE* of an alpha particle (a helium nucleus) is greater than the *RBE* of a beta particle (an electron or positron).

Another widely used measure of radiation experienced by biological organisms is the unit called the ***rem***, which is defined as

Definition of the rem

$$\text{dose in rem} = (\text{dose in rads}) \times (RBE) \tag{30.33}$$

Since the rem includes the total amount of energy delivered by the radiation (measured in rads) and the efficiency with which this radiation damages real tissue, it combines the usefulness of both the rad and the *RBE* factor. The rem is the most widely used unit in discussions of radiation exposure and damage.

TABLE 30.3 Relative Biological Effectiveness (*RBE*) for Different Types of Radiation

Radiation	*RBE*
Alpha particles	10–20
Beta particles	1.0–1.7
Gamma rays and X-rays	1.0
Slow neutrons	4–5
Protons	5
Fast neutrons	10
Heavy ions	20

EXAMPLE 30.8 Comparing the Dose from Different Types of Radiation

A person is exposed to a radiation dose of 0.25 mrad (millirad) in a chest X-ray. What dose of fast neutrons would produce the same amount of tissue damage?

RECOGNIZE THE PRINCIPLE

The amount of damage done by radiation is determined by the product of the dose (which can be measured in rads) and the *RBE* (relative biological effectiveness):

$$\text{radiation damage} = \text{dose} \times RBE \tag{1}$$

Hence, to compare the damage done by X-rays and fast neutrons, we must take their *RBE* values into account.

SKETCH THE PROBLEM

No sketch is necessary.

IDENTIFY THE RELATIONSHIPS

From Table 30.3, the *RBE* for X-rays is 1.0, while for fast neutrons we find $RBE = 10$. To achieve the same damage with these two types of radiation, the fast neutron dose must therefore be 10 times smaller than the X-ray dose.

SOLVE

If a dose of fast neutrons produces the same amount of damage as a 0.25-mrad chest X-ray, we can use Equation (1) to write

$$\text{dose(X-rays)} \times RBE(\text{X-rays}) = \text{dose(fast neutrons)} \times RBE(\text{fast neutrons})$$

which leads to

$$\text{dose(fast neutrons)} = \frac{RBE(\text{X-rays})}{RBE(\text{fast neutrons})} \times \text{dose(X-rays)}$$

Inserting *RBE* values along with the given value of the X-ray dose, we find

$$\text{dose(fast neutrons)} = \frac{1.0}{10} \times (0.25 \text{ mrad}) = \boxed{0.025 \text{ mrad}}$$

> ▶ ***What does it mean?***
> When comparing the radiation damage from different sources, one must always take the *RBE* of the radiation into account.

Some Facts about Radiation Damage

The physiology and cell biology of radiation damage is a complex and interesting subject. Here we can only mention some of the important facts.

1. When the radiation dose is low, cells are sometimes able to repair the damage, especially if the dose is absorbed over a long period of time. Hence, small amounts of radiation generally do not cause significant harm to living cells.
2. If the radiation dose is very large, cells can be completely destroyed.
3. At intermediate radiation doses, cells survive but often malfunction as a result of the damage or in their attempt to repair the damage. A typical result is that the affected cells reproduce in an uncontrolled fashion, leading to cancer.
4. Radiation damage is usually most severe for quickly dividing cells. Many types of blood and bone marrow cells fall into this category, along with cells in an embryo or fetus. Cancerous cells are also quickly dividing, so radiation can be used as a tool to selectively destroy cancer cells (Section 30.5).
5. The amount of damage can depend strongly on where the radiation source is located. For example, alpha particles have a very short range and can be stopped by a sheet of paper. So, if a source of alpha particles is outside the body, the particles will be stopped in the outer layer of skin and do relatively little damage. On the other hand, if a person somehow ingests an alpha source, the source may become lodged in his or her lungs or some other place, and we say that the person is contaminated. In this case, the alpha particles can do a great deal of damage to nearby cells, damaging their DNA and so forth.

Sadly, many of the early scientists who studied radioactivity suffered from their exposure to hazardous amounts of radiation. Röntgen (the discoverer of X-rays) died from bone cancer, and Marie Curie died from leukemia that was probably caused by her exposure to many radioactive substances. Her husband, Pierre, also suffered from radiation-induced illnesses (although he died after being run over by a horse-drawn carriage; go figure). In fact, the Curies' scientific papers are reportedly still quite radioactive.

People who work with radiation are normally required to use radiation monitors to record their exposure (Fig. 30.23). In the event of an accidental overdose, the worker can then quickly get the appropriate medical treatment.

Sources of Radioactivity in Everyday Life

Table 30.4 on page 958 lists the radiation doses of several common procedures and shows that we are all exposed to radiation during the course of everyday life. Even so, the benefits from, for instance, a chest X-ray, dental X-ray, or mammogram usually far outweigh possible risks.

There are several natural sources of radiation. Cosmic rays—a collection of many different types of particles that come to the Earth from outer space—are a significant source, especially at high altitudes. They are responsible for the increased radiation dose absorbed by astronauts and by persons at altitude (e.g., during an airplane flight or in a city such as Denver).

One of the largest natural sources of radiation exposure involves radon gas produced by the radioactive decay of $^{238}_{92}U$. There are very small concentrations of $^{238}_{92}U$ in the soil at virtually all places on this planet, and these nuclei initiate the series of decays in Figure 30.9. One of these daughter nuclei is $^{222}_{86}Rn$ (radon), and being a gas, it diffuses through pores in the surrounding soil or rock. These $^{222}_{86}Rn$ nuclei

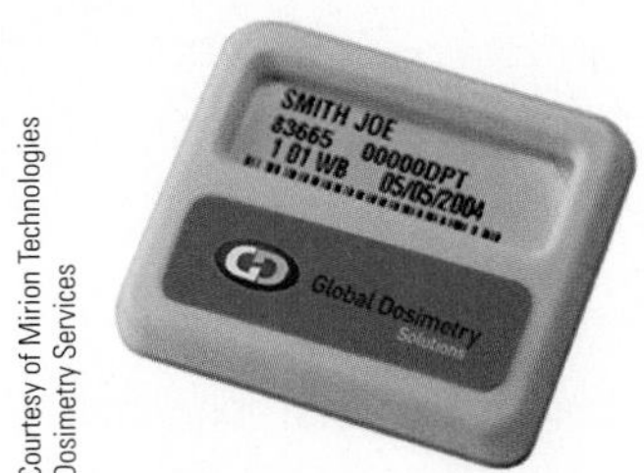

Courtesy of Mirion Technologies Dosimetry Services

▲ **Figure 30.23** The radiation and particles emitted by radioactive substances cause photographic film to be exposed. To monitor their exposure to harmful radiation, people who work with radioactive materials wear small "film badges."

▲ **Figure 30.24** Simple radon detector for home use. This detector contains a small canister of activated charcoal, which absorbs radon gas (along with other gases). After being exposed to the atmosphere for a specified period of time, the canister is sent to a laboratory for analysis.

TABLE 30.4 Radiation Doses in Perspective (Typical Values)

Activity	Dose (mrem)
Airline flight from New York City to Los Angeles	3
Dental X-ray	10
Chest X-ray	5–10
Mammogram	50–100
Approximate annual dose from natural background sources	300
Recommended maximum annual dose in addition to background dose	500
Apollo XVI astronauts	500
1-in-30 risk of cancer	10,000
Radiation sickness possible	60,000

then decay further, producing nuclei that are absorbed in the lining of the lungs of anyone who breathes the gas, leading to alpha and beta radiation that can ultimately cause lung cancer. Radon gas is a major health hazard for miners, and in some homes. It can seep into basements and crawl spaces, where it can collect to unsafe levels. Fortunately, there are simple ways to remediate this hazard, typically by improving the ventilation in the mine or home (Fig. 30.24).

EXAMPLE 30.9 Radiation Exposure from a CT Scan

In Chapter 23, we described how a CT (or CAT) scan uses X-rays to form an image of bones and soft tissue within the body. The X-ray photons in a CT scan produce radiation damage similar to that produced by gamma rays. The radiation dose for a CT scan depends on many factors, but is typically about 150 mrem for a scan of the head. How many CT scans can your doctor prescribe before the radiation damage of the scan itself becomes a potential health hazard?

RECOGNIZE THE PRINCIPLE

The "normal" dose from radiation sources in the environment (cosmic rays, the small amount of radioactive material in the soil, etc.) is about 300 mrem per year (Table 30.4). To be safe, it is recommended that the dose from other sources such as a CT scan should be less than about 500 mrem.

SKETCH THE PROBLEM

No sketch is necessary, but the data in Table 30.4 are very useful.

IDENTIFY THE RELATIONSHIPS AND SOLVE

We want the total dose from all the CT scans in a year to be no larger than the recommended maximum dose of 500 mrem. The safe number of CT scans is therefore

$$\text{number of scans} = \frac{500 \text{ mrem}}{\text{CT dose}} = \frac{500 \text{ mrem}}{150 \text{ mrem}} = \boxed{3}$$

▶ *What does it mean?*

You might now conclude that it is unsafe to have more than three CT scans per year. Our analysis is a bit simplistic, however, since a doctor must carefully weigh the potential benefits (e.g., from locating a suspected tumor) in assessing whether or not to prescribe a CT scan.

30.5 Applications of Nuclear Physics in Medicine and Other Fields

Radiation has many positive uses in medicine and other fields. We'll now describe a few of these applications.

Radioactive Tracers

A ***radioactive tracer*** is a chemical that contains radioactive nuclei. The movement of tracer nuclei through the body can be monitored by observing the radiation they emit through radioactive decay. Tracers are widely used in medicine and many other fields. In most applications, the dose of radioactivity is kept low so that it does no significant harm.

For example, in a method known as nuclear stress testing, a small amount of a radioactive liquid containing (typically) $^{201}_{81}\text{Th}$ or $^{99}_{43}\text{Tc}$ is injected into a vein near the heart. These isotopes emit gamma rays, which are detected by special cameras placed over the heart. Gamma rays from blood containing the tracer molecules cause the blood to "light up" in the images. These images thus show how blood flows in and around the heart and are used to diagnose heart disease.

The method of positron emission tomography (called PET scanning) works in a similar way. A glucose solution containing typically $^{18}_{9}\text{F}$ (the isotopes $^{11}_{6}\text{C}$, $^{15}_{8}\text{O}$, and $^{13}_{7}\text{N}$ are also used) is injected into a patient's blood. These nuclei decay by the emission of a positron, and these positrons in turn produce gamma rays when they encounter an electron. (We'll discuss how and why in Chapter 31.) These gamma rays are observed by a collection of detectors positioned around the patient (Fig. 30.25). By combining the images from many different detectors, medical professionals construct a three-dimensional image of the positions of the radioactive nuclei. These images are used to monitor blood flow near and inside the heart and brain. Some types of cancer cells absorb large amounts of glucose, so PET scans can also be used to find these cancers.

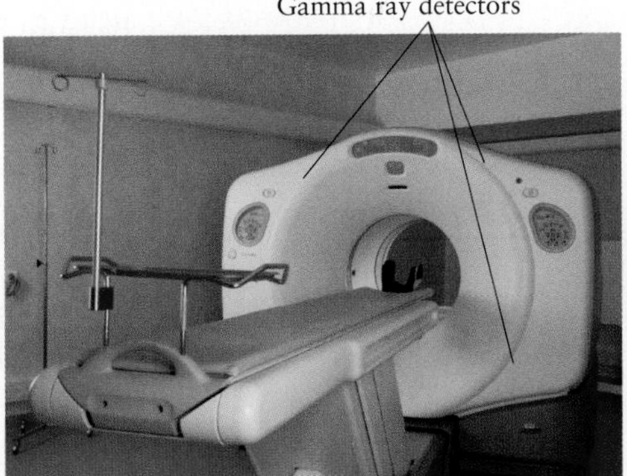

▲ **Figure 30.25** Positron emission tomography (PET scanning) is a widely used medical procedure. After ingesting a radioactive tracer, the patient enters an apparatus like the one shown here. This device measures the gamma rays emitted when positrons annihilate with electrons and determines the sites from which they are emitted.

Another use of tracers is to fight thyroid cancer. Iodine tends to accumulate in the thyroid gland, so by using a tracer containing the radioactive isotopes $^{123}_{53}\text{I}$ or $^{131}_{53}\text{I}$ and detecting the radiation they emit, one can measure how effectively the thyroid absorbs iodine. Cancerous thyroid cells can be killed using the electrons emitted in beta decay as well as by the gamma radiation emitted by these iodine nuclei.

The best choice of tracer nucleus depends on the application, but all tracers should have a short half-life so that their activity decreases to negligible levels soon after use. This means that tracer nuclei cannot be stored for future use, but must be made just before they are used. Tracer nuclei are usually made by nuclear reactions induced in a nuclear reactor core.

EXAMPLE 30.10 Formation of a Radioactive Tracer

The tracer $^{123}_{53}\text{I}$ is used to treat thyroid problems. This nucleus is produced by bombarding xenon with protons in the reaction

$$^{1}_{1}\text{H} + ^{124}_{54}\text{Xe} \rightarrow ^{123}_{53}\text{I} + ? \qquad (1)$$

Assuming the question mark represents a single type of nucleus, what is that nucleus?

(continued) ▶

RECOGNIZE THE PRINCIPLE

We can use conservation of nucleon number and conservation of charge to deduce the missing nucleus.

SKETCH THE PROBLEM

No figure is necessary.

IDENTIFY THE RELATIONSHIPS AND SOLVE

There are 125 nucleons on the left in Equation (1) (one in the hydrogen nucleus and 124 in the xenon nucleus), so there must be that number of nucleons on the right. Since the iodine nucleus has 123 nucleons, this leaves 2 nucleons for the missing part of Equation (1). Likewise, there are 55 protons on the left and 53 on the right, so these missing 2 nucleons must both be protons. The full reaction equation is thus

$$^{1}_{1}\text{H} + {}^{124}_{54}\text{Xe} \rightarrow \boxed{^{123}_{53}\text{I} + 2\,^{1}_{1}\text{H}}$$

► ***What does it mean?***

All nuclear reactions must conserve nucleon number and charge. This particular reaction simply removes one proton from $^{124}_{54}\text{Xe}$.

Using Radiation to Fight Cancer

Because quickly dividing cells are generally the ones most damaged by radiation, radiation treatments are effective in treating many types of cancers. For example, radiation is widely used to kill breast cancer cells. This therapy uses X-rays, so strictly speaking it does not involve radioactivity. Breast cancer is sometimes also treated by implanting radioactive nuclei, a treatment called brachytherapy. These nuclei (usually $^{90}_{38}\text{Sr}$) emit high-energy electrons (via beta decay), gamma rays, or both, killing nearby cancer cells. A similar treatment is used in dealing with prostate cancer in men. Radioactive "seeds" containing typically $^{125}_{53}\text{I}$ or $^{103}_{46}\text{Pd}$ are implanted in the prostate gland, and the gamma rays they emit kill nearby cancerous cells. These isotopes have relatively short half-lives, so their activity is very small after a few months.

Carbon Dating

One of the most important uses of radioactivity outside of medicine is the technique of ***carbon dating*** developed in 1947 by Willard Libby, who received the Nobel Prize in Chemistry for this work. Of the carbon nuclei occurring naturally on the Earth, approximately 99% are the isotope $^{12}_{6}\text{C}$ and about 1% are $^{13}_{6}\text{C}$. In addition to these two stable isotopes, the radioactive isotope $^{14}_{6}\text{C}$ is produced in the Earth's atmosphere when cosmic rays collide with $^{14}_{7}\text{N}$. The half-life of $^{14}_{6}\text{C}$ is $T_{1/2} = 5730$ yr as it undergoes beta decay, producing $^{14}_{7}\text{N}$ (Eq. 30.13). This decay is balanced by the cosmic-ray–induced production of $^{14}_{6}\text{C}$, resulting in a carbon isotopic ratio in the atmosphere[5] equal to $^{14}_{6}\text{C}/^{12}_{6}\text{C} \approx 1.3 \times 10^{-12}$.

Carbon-containing molecules including proteins and DNA are the building blocks of life. When an animal is living, it accumulates carbon from the environment as it eats and breathes, while plants take in atmospheric CO_2 during photosynthesis and absorb carbon-containing nutrients from the water and soil. Since there is a small amount of $^{14}_{6}\text{C}$ in the atmosphere and in their nutrients, organisms absorb $^{14}_{6}\text{C}$ until the percentage of $^{14}_{6}\text{C}$ in the organism equals the $^{14}_{6}\text{C}$ percentage in the atmosphere. When the organism dies, it stops accumulating new $^{14}_{6}\text{C}$, and the $^{14}_{6}\text{C}$ that it contained at death undergoes radioactive decay. Hence, after one half-life (5730 yr), the amount of $^{14}_{6}\text{C}$ in the organism is equal to half of what it was at the

[5]Nuclear weapons testing in the 1950s and 1960s nearly doubled the $^{14}_{6}\text{C}/^{12}_{6}\text{C}$ ratio in the atmosphere. This ratio is now returning to its pre-1950s level.

time the organism died. Likewise, after a time equal to $2T_{1/2}$, the amount of $^{14}_{6}C$ is a fourth of the original amount. The behavior of the $^{14}_{6}C$ concentration as a function of time is given by the exponential relation in Equation 30.25 and is plotted in Figure 30.26. Since the ratio $^{14}_{6}C/^{12}_{6}C$ at the time of death is known, the ratio measured today in a sample can be used to determine how much time has passed since death. In this way, the age of an animal or plant specimen can be accurately determined. This technique is now an essential part of experimental archaeology.

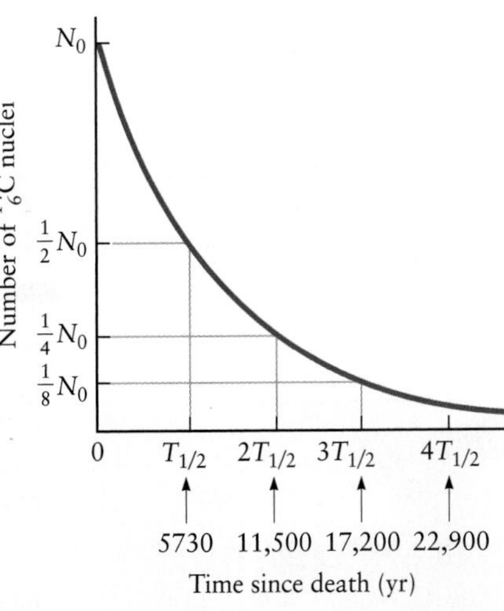

▲ **Figure 30.26** Carbon dating uses the relative concentration of $^{14}_{6}C$ (the ratio $^{14}_{6}C/^{12}_{6}C$) in a sample from a formerly living organism to deduce the time at which the organism died. After death, the number of $^{14}_{6}C$ nuclei falls with time due to its radioactive decay, with a half-life of 5730 yr.

EXAMPLE 30.11 Carbon Dating

Some ancient plant material is found to have a fraction of $^{14}_{6}C$ given by $^{14}_{6}C/^{12}_{6}C = 6.0 \times 10^{-14}$. How old is it? The $^{14}_{6}C/^{12}_{6}C$ ratio at the time of death was equal to 1.3×10^{-12} (as explained above).

RECOGNIZE THE PRINCIPLE

The radioactive decay of $^{14}_{6}C$ is described by

$$N = N_0 e^{-\lambda t} \tag{1}$$

with a half-life of 5730 yr, whereas $^{12}_{6}C$ is stable and does not decay. The $^{14}_{6}C/^{12}_{6}C$ thus decreases with time according to

$$\frac{^{14}_{6}C}{^{12}_{6}C} \text{ at time } t = \left(\frac{^{14}_{6}C}{^{12}_{6}C} \text{ at time of death}\right) \times e^{-\lambda t} \tag{2}$$

SKETCH THE PROBLEM

The decay behavior in Equations (1) and (2) is described graphically in Figure 30.26.

IDENTIFY THE RELATIONSHIPS

The given $^{14}_{6}C/^{12}_{6}C$ ratio is smaller than the ratio at the time of death by

$$\frac{^{14}_{6}C/^{12}_{6}C \text{ (at present)}}{^{14}_{6}C/^{12}_{6}C \text{ (at death)}} = \frac{6.0 \times 10^{-14}}{1.3 \times 10^{-12}} = 0.046 = 4.6\% \tag{3}$$

We can now approach this problem in two ways. One would be to use the graph in Figure 30.26 and read off the time at which the $^{14}_{6}C/^{12}_{6}C$ ratio has dropped to 4.6% of its original value, but this small value is hard to locate accurately on the vertical scale in Figure 30.26. Instead, let's use the exponential relation in Equation (1). At an unknown time t, only 4.6% of the original $^{14}_{6}C$ remains, so if N is the number of $^{14}_{6}C$ we have

$$N(\text{at present}) = N(\text{at death})e^{-\lambda t} \tag{4}$$

where λ is the decay constant for $^{14}_{6}C$. The decay constant is related to the half-life by (Eq. 30.27)

$$T_{1/2} = \frac{0.693}{\lambda}$$

SOLVE

Solving for λ, we get

$$\lambda = \frac{0.693}{T_{1/2}} = \frac{0.693}{5730 \text{ yr}} = 1.2 \times 10^{-4} \text{ yr}^{-1}$$

Rearranging Equation (4) and using the $^{14}_{6}C/^{12}_{6}C$ ratio in Equation (3) leads to

$$\frac{N(\text{at present})}{N(\text{at death})} = 0.046 = e^{-\lambda t}$$

(continued) ▶

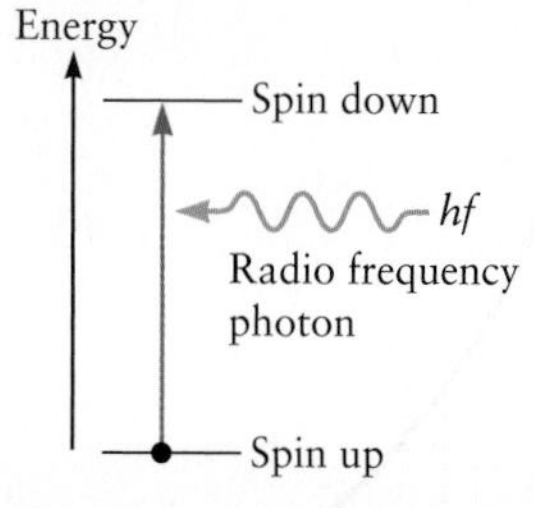

A

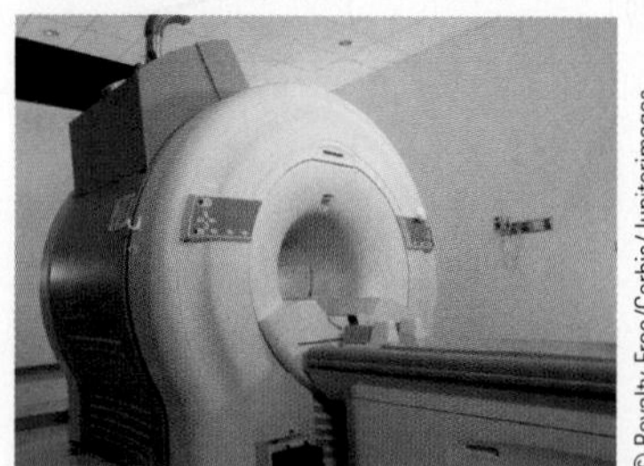
© Royalty-Free/Corbis/Jupiterimages

B

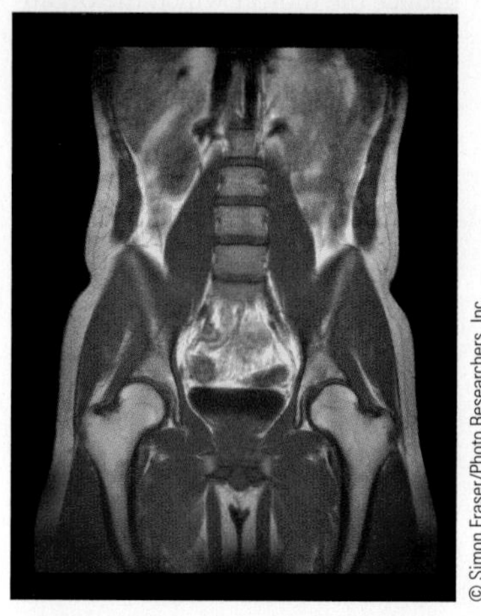
© Simon Fraser/Photo Researchers, Inc.

C

▲ **Figure 30.27** A When a proton (a hydrogen nucleus) is placed in a magnetic field, the spin-up and spin-down states have different energies. A proton can absorb a radio frequency photon and undergo a transition from one level to the other. B Magnetic resonance imaging (MRI) uses the spin magnetism of hydrogen nuclei to measure the density and other properties of tissue within the body. This procedure requires that the patient be placed inside a large magnet. C MRI image of a person's abdomen.

We now take the logarithm of both sides and use the definition of logarithms $\ln(e^x) = x$. We find

$$\ln(0.046) = -\lambda t$$

$$t = -\frac{\ln(0.046)}{\lambda} = -\frac{\ln(0.046)}{1.2 \times 10^{-4}\ \text{yr}^{-1}} = \boxed{26{,}000\ \text{yr}}$$

▸ ***What does it mean?***

Carbon dating is a precise way to date many kinds of material. However, one limitation on this technique is set by the half-life of ${}^{14}_{6}\text{C}$. If the material is very much older than the half-life, there will be very little ${}^{14}_{6}\text{C}$ left to measure. In practice, carbon dating is useful for material that is no older than about 50,000 years.

Magnetic Resonance Imaging

Magnetic resonance imaging (MRI) makes use of the magnetic properties of nuclei. Hydrogen has the simplest nucleus—a single proton—and a proton has an intrinsic spin angular momentum similar to the spin angular momentum of the electron discussed in Chapter 28. As with electrons, a proton has just two spin states, "spin up" and "spin down," and in a magnetic field these two states have different energies. The presence of a proton can be detected by observing the absorption of a photon that induces a transition from the lower- to the higher-energy spin state (Fig. 30.27A). This is just like the absorption of a photon by an atom as it undergoes a transition between energy levels (Chapter 29).

The absorption of a photon only occurs if the separation between energy levels matches the photon energy, and this energy separation depends on the magnetic field. An MRI magnet is designed so that these energies match only at a particular spot within the body, so the MRI signal gives the density of protons (and other properties) at just that spot. By then scanning this spot around the body, a three-dimensional image can be constructed, giving a unique picture of the body's internal structure.

Most applications of MRI employ a magnet with coils in a unit large enough for a person to fit inside (Fig. 30.27B). After only a few minutes, an image with remarkable resolution can be obtained (Fig. 30.27C). MRI images are now widely used to diagnose many different types of medical conditions (including the author's knee injuries). This method uses the magnetic properties of naturally occurring nuclei (usually hydrogen nuclei) in the patient and radio frequency photons (which have very low energies). As a result, there do not appear to be any health risks associated with MRI, making this technique a very safe and effective medical tool.

30.6 Questions about the Nucleus

In this chapter, we have described nuclear properties and nuclear reactions. We now ask *why* the nucleus has these properties. As usual, this "why" question is much more difficult than discovering the actual properties, but it also leads us to fundamental issues.

Conservation rules (Section 30.2) are fundamental principles of physics that apply to a wide range of situations. Nuclear physics reveals a new conservation rule, the conservation of nucleon number. There are just two types of nucleons—protons and neutrons—and *all* known nuclear reactions conserve the number of nucleons. Physicists have spent much time and effort studying this problem, and as far as we can tell, nucleon number is always conserved. Why? We don't know. All we can say is that nucleon number conservation is a property of all known processes.

Such number conservation rules do not apply to all particles. For example, the number of electrons is *not* conserved in all reactions. Beta decay (e.g., Eq. 30.13) creates an electron where one did not previously exist, so electrons are different from nucleons in some fundamental way. We'll explore this topic further in Chapter 31.

We can also ask about the structure of a nucleon. Is there anything inside a proton or a neutron? Physicists have spent years working on this question. Scattering experiments similar (in principle) to those of Rutherford have shown that nucleons are composed of particles called ***quarks***. There are several different types of quarks, all which carry an electric charge of $\pm e/3$ or $\pm 2e/3$, where $-e$ is the charge carried by an electron. Protons and neutrons are each composed of three quarks. In a proton, the quark charges add up to $+e$; in a neutron, they add up to zero. So, although a neutron is electrically neutral, it contains charged particles! In Chapter 31, we'll discuss the behavior of quarks and how they combine to make protons and neutrons, but we still don't know *why* nature is put together this way.

Summary | CHAPTER 30

Key Concepts and Principles

Nuclear structure

Nuclei contain ***protons*** and ***neutrons***, also called ***nucleons***. The number of protons in a nucleus is the ***atomic number*** Z, and the total number of protons plus neutrons is the ***mass number*** A. The number of neutrons N is thus related to Z and A by

$$A = Z + N \qquad (30.3)\ (\text{page } 933)$$

Different ***isotopes*** of a particular element contain the same number of protons, but different numbers of neutrons.

All this information is described in the following shorthand notation:

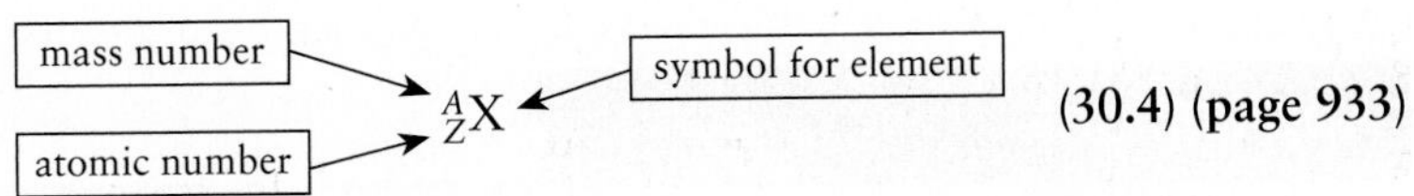

(30.4) (page 933)

Nuclear size

Nuclei have a radius r of a few femtometers (1 fm $= 10^{-15}$ m). The radius depends on the number of nucleons A in the nucleus through the relation

$$r = r_0 A^{1/3} \qquad (30.5)\ (\text{page } 935)$$

with $r_0 \approx 1.2$ fm $= 1.2 \times 10^{-15}$ m.

The strong force

Nuclei are held together by the ***strong force***, an attractive force that acts between pairs of nucleons. The strong force does not act on electrons, and it is responsible for the stability of the nucleus.

(*Continued*)

Radioactive decay

Some nuclei are unstable and decay spontaneously, a process called ***radioactive decay***. In ***alpha decay***, an alpha particle (a helium nucleus, ${}^{4}_{2}\text{He}$) is emitted. In ***beta decay***, an electron or a positron is emitted. In ***gamma decay***, a gamma ray (a high-energy photon) is produced. The ***half-life*** is the time it takes for half of the nuclei in a sample to decay. Half-lives can be less than 1 second or longer than the age of the Earth.

Nuclear fission

In nuclear ***fission***, a nucleus splits into two nuclei that are each about half the size of the original nucleus and several neutrons. This reaction releases energy; it is the basis for nuclear weapons and nuclear power plants.

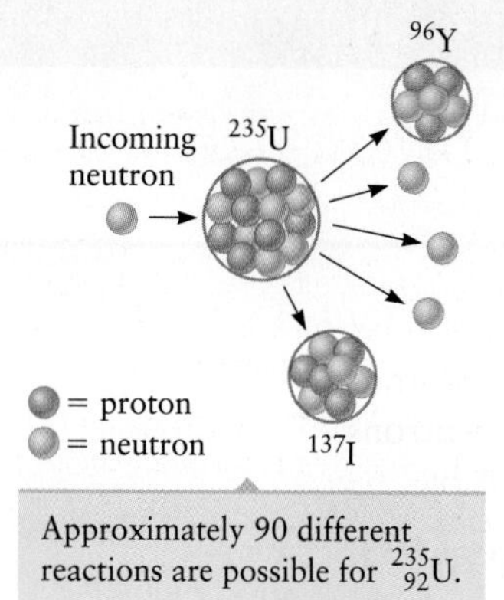

Nuclear fusion

In nuclear ***fusion***, two low-mass nuclei combine to form one heavier nucleus. Fusion reactions also release large amounts of energy and are responsible for the energy output of the Sun.

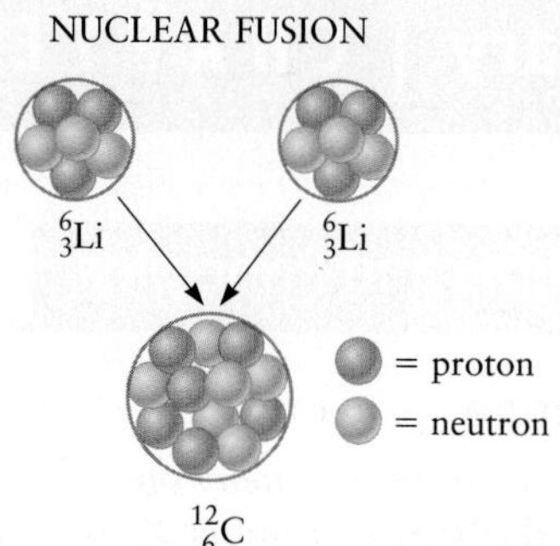

Applications

Energy in nuclear reactions

The energy released in a typical nuclear reaction is typically on the order of MeV. It is much larger than a typical atomic or molecular energy (about 10 eV), and the particles emitted by a nuclear reaction can damage living tissue.

Units of radioactivity

Several units are used to measure the amount of radioactivity in a sample.

The ***activity*** of a sample is proportional to the number of nuclei in the sample that undergo radioactive decay in 1 second. One unit of activity, the curie (Ci), is defined by

$$1\ \text{Ci} = 3.7 \times 10^{10}\ \text{decays/s} \quad (30.28)$$ (page 947)

Activity is also measured with the SI unit becquerel (Bq), with

$$1\ \text{Bq} = 1\ \text{decay/s} \quad (30.29)$$ (page 947)

Biological effects

The amount of damage caused by radiation is determined in part by the energy deposited by the radiation in tissue. A radiation dose of 1 *rad* deposits 10^{-2} J in 1 kg of material. Different types of radiation can do more or less damage. The ***RBE*** (relative biological effectiveness) factor measures this difference. The unit called the ***rem*** takes all these factors into account:

$$\text{dose in rem} = (\text{dose in rads}) \times (RBE) \qquad \textbf{(30.33) (page 956)}$$

Benefits of nuclear radiation

Radiation presents many health hazards, but it also has useful applications, including carbon dating and the treatment of many types of cancer.

Questions

SSM = answer in *Student Companion & Problem-Solving Guide* ⓧ = life science application

1. Why do different isotopes of a given element have, for most practical purposes, the same chemical properties? That is, why do different isotopes form the same molecules and compounds even though they have different numbers of neutrons and different masses?

2. Different isotopes of an element essentially share chemical properties. How, then, is it possible to separate different isotopes? *Hint*: Consider how one might use diffusion (Chapter 15) or a mass spectrometer (Chapter 20) (but not at the same time).

3. What is heavy water? Would an equal number of moles of heavy water actually weigh more than plain water?

4. Was evidence for the nucleus first obtained through (a) carbon dating, (b) the spectral lines of hydrogen, (c) alpha particle scattering, (d) cosmic rays, or (e) electromagnetic radiation?

5. What is (a) an alpha particle? (b) a beta particle? (c) a gamma ray? (d) an X-ray? (e) a daughter nucleus?

6. Is the mass of a nucleus larger or smaller than the sum of the masses of the nucleons of which it is made? Explain.

7. SSM In the mid 1800s, the elements of the periodic table were being identified and discovered, yet for many decades element 43, between molybdenum and ruthenium (Fig. Q30.7), remained unfound. Technetium (Tc) was finally discovered almost 100 years later, in 1936. (a) Use the Internet to discover what is special about this element, and why it was so hard to find. (b) Where can technetium be found today?

Cr 24	Mn 25	Fe 26
51.996	54.938	55.845
$3d^54s^1$	$3d^54s^2$	$3d^64s^2$
Mo 42	43	Ru 44
95.94	?	101.07
$4d^55s^1$		$4d^75s^1$
W 74	Re 75	Os 76
183.84	186.21	190.23
$5d^46s^2$	$5d^56s^2$	$5d^66s^2$

Figure Q30.7 Element 43.

8. Why is the element plutonium not found in significant amounts in nature (on the Earth)?

9. Compare how a fission bomb works with how a fusion bomb works. Be sure to discuss what each would use for fuel and the waste products produced by each.

10. There is a critical mass for the creation of a spontaneous chain reaction involving $^{235}_{92}$U. Why would it be difficult to make a bomb that is much larger than the critical mass? Why does this limit not exist for a fusion bomb?

11. ⓧ Explain why carbon dating can in general only be used to measure the age of material that was once alive.

12. Do you expect that $^{56}_{26}$Fe would undergo fusion? Explain why or why not. *Hint*: Use the binding energy curve in Figure 30.18.

13. Consider the radioactive decay of $^{235}_{92}$U. Construct a series of alpha and beta decays (similar to Fig. 30.9) that include the daughter nuclei $^{223}_{88}$Ra, $^{219}_{86}$Rn, and $^{211}_{83}$Bi, and terminates with $^{207}_{82}$Pb. Why can this decay series not terminate with $^{206}_{82}$Pb or $^{208}_{82}$Pb?

14. ⓧ In the treatment of cancer using radioactive "seeds" (brachytherapy), the radioactive material must be placed very close to the cancer cells. (a) Explain why cells farther away from the radioactive seeds are not significantly harmed by the emitted radiation. (b) Explain why placing the radioactive seeds outside the body is not as effective as placing them inside or next to a cancerous tumor.

15. A nucleus of $^{245}_{96}$Cm that is initially at rest undergoes alpha decay. What is the daughter nucleus? Which of the reaction products has the greater kinetic energy, the alpha particle or the daughter nucleus? *Hint*: Is momentum conserved in this process?

16. Explain why the neutrons emitted in a nuclear reaction can penetrate very thick sheets of lead, but the alpha and beta particles cannot.

17. ⓧ Which of the following objects could be dated with carbon dating?
(a) The logs used to make George Washington's house
(b) The bricks used to make George Washington's house
(c) Animal bones found in George Washington's barn
(d) The blade of George Washington's pocket knife

18. ⓧ An archaeologist friend proposes to determine the age of a dinosaur bone using carbon dating. Why does your friend need to take a physics course? That is, what is wrong with this proposal?

19. SSM Scientists in areas such as geology, paleontology, and archaeology use "radiodating" techniques (such as carbon dating) in the study of the history of the Earth and its inhabitants. There are several radiodating techniques, each using different nuclei: (i) carbon 14, (ii) potassium–argon 40, (iii) rubidium–strontium, and (iv) uranium–lead. (a) Approximately how old an object can be dated with each of these techniques? (b) Conversely, what are the youngest objects that can be reliably dated with each technique? (c) Which techniques are best for archaeology? For paleontology? *Note*: You may have to look up the half-lives of the relevant parent isotopes for the elements used in each method.

20. In a typical volume of soil that measures 1 mi^2 and 1 ft deep, there are roughly 2 g of radium, predominantly ^{236}Ra. But ^{236}Ra has a half-life of just 1600 yr, while the Earth is billions of years old. Where did this radium come from?

21. The nuclei $^{235}_{92}U$, $^{253}_{100}Fm$, and $^{14}_{6}C$ have very different half-lives. Explain how samples of these materials can still have the same activity.

22. How might the $^{239}_{94}Pu$ used in nuclear weapons be created? *Hint*: Consider reactions in which a nucleus captures a neutron, followed by beta decay.

23. The sketch in Figure 30.11 shows a nuclear decay in which two product nuclei are produced. Why do they travel away in opposite directions?

Problems available in ENHANCED WebAssign

Section 30.1 Structure of the Nucleus: What's Inside? Problems 1–20

Section 30.2 Nuclear Reactions: Spontaneous Decay of a Nucleus Problems 21–43

Section 30.3 Stability of the Nucleus: Fission and Fusion Problems 44–50

Section 30.4 Biological Effects of Radioactivity Problems 51–62

Section 30.5 Applications of Nuclear Physics in Medicine and Other Fields Problems 63–71

Additional Problems Problems 72–91

List of Enhanced Problems

Problem number	Solution in *Student Companion & Problem-Solving Guide*	Reasoning & Relationships Problem	Reasoning Tutorial in ENHANCED WebAssign
30.8	✓	✓	✓
30.20	✓		
30.25	✓		
30.34	✓		
30.48	✓		
30.50		✓	
30.56	✓		
30.57		✓	✓
30.60	✓		
30.62		✓	✓
30.67	✓		
30.71		✓	✓
30.74		✓	
30.75	✓		
30.76		✓	
30.80		✓	✓
30.82		✓	✓
30.83		✓	
30.84		✓	
30.87		✓	
30.91		✓	

CHAPTER

31

◀ *Many studies of elementary particles use enormous detectors like this one at the European Organization for Nuclear Research (called CERN) in Geneva, Switzerland. (Notice the people in the photograph, one at the left and one standing next to the detector at the bottom.) This detector is designed to study the particles created when opposing beams of protons with very high energies collide. (CMS Experiment © 2009 CERN)*

Physics in the 21st Century

Toward the end of the 19th century, some physicists believed that all the laws of physics had been discovered and that the main job left for physicists was to measure quantities like the fundamental constants and the orbital parameters of the Moon more accurately. These physicists were proven wrong with the discoveries of relativity and quantum theory, leading to a century of fantastic progress in physics. Advances in physics during the 20th century led to many important applications, including computers, space travel, cell phones, nuclear power, lasers, DVD players, and magnetic resonance imaging. What's more, physics continues to be an active research area, with no end in sight. In this chapter, we describe several topics of current research in physics. We start with elementary particle physics and consider how protons, neutrons, and other subatomic particles are put together. We then discuss some topics in the field of astrophysics, which is concerned with the origin and fate of the universe. In the last section of this chapter, we briefly describe how physicists are teaming up with biologists, chemists, and engineers to address interdisciplinary questions, including the new field of nanoscience.

OUTLINE

NASA

▲ **Figure 31.1** Some experiments use balloons to carry particle detectors to high altitudes and detect cosmic rays as they enter the atmosphere from outer space. Here NASA launches a balloon that will carry a package of instruments to the upper atmosphere.

31.1 Cosmic Rays

The Earth is constantly bombarded with particles that arrive from outer space. These particles are called ***cosmic rays,*** a name chosen when physicists mistakenly believed that cosmic rays were a type of electromagnetic radiation. Cosmic rays are actually a mix of different particles, including protons (about 89%) and alpha particles (10%); the rest are the nuclei of elements more massive than He and other types of particles. Cosmic rays are often studied in experiments in which balloons carry particle detectors to high altitudes (Fig. 31.1).

A key question for physicists is, "How are cosmic rays generated?" It is believed that some cosmic rays are created when stars collide. Cosmic rays are also created when a massive star uses up its fuel. The star then collapses (due to gravity) and triggers what is called a supernova explosion (Fig. 31.2). These explosions are also thought to be the main source of elements heavier than oxygen in the universe.

Typical cosmic-ray energies are 100 MeV (1×10^8 eV), but some particles with energies above 10^{20} eV have been detected. This is an incredibly high energy—approximately the kinetic energy of a baseball moving at 100 mi/h—and is carried by a particle with about the mass of a proton! It is still not completely understood how particles are accelerated to such high energies.

New, more sensitive detectors are continually being built, with the goal of better understanding cosmic rays and other particles that arrive from outer space. Figure

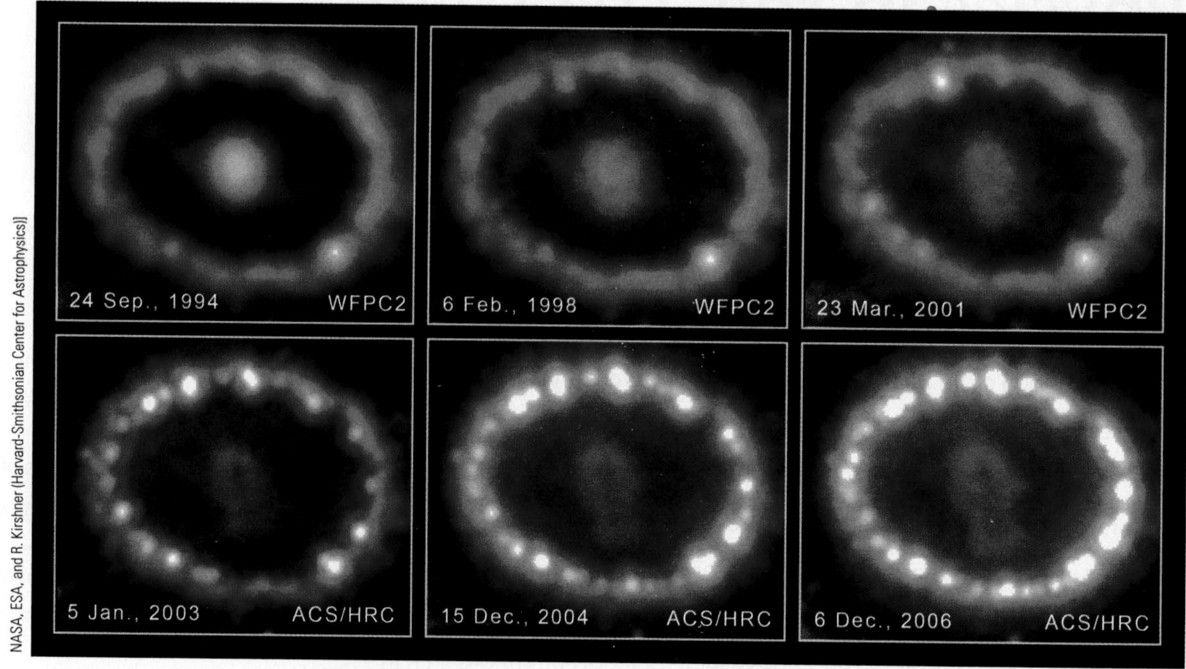

NASA, ESA, and R. Kirshner (Harvard-Smithsonian Center for Astrophysics)]

▲ **Figure 31.2** Physicists believe that cosmic-ray particles are generated in supernova explosions. These photographs of Supernova 1987A were obtained with the Hubble Space Telescope shortly after the supernova exploded in 1987. The bright ring around the star is due to a ring of gas that formed long before the explosion. When the supernova exploded, particles were ejected from the star; light from these particles is visible as violet in these photos. In the first photo (upper left), these particles were near the star (at the center of the ring). As time passed, the particles moved outward, and when they reached the gas ring, they produced the bright white spots. The bright spots at the upper-left and lower-right parts of the ring (seen most clearly in the image at the upper right) are due to stars between 1987A and the Earth.

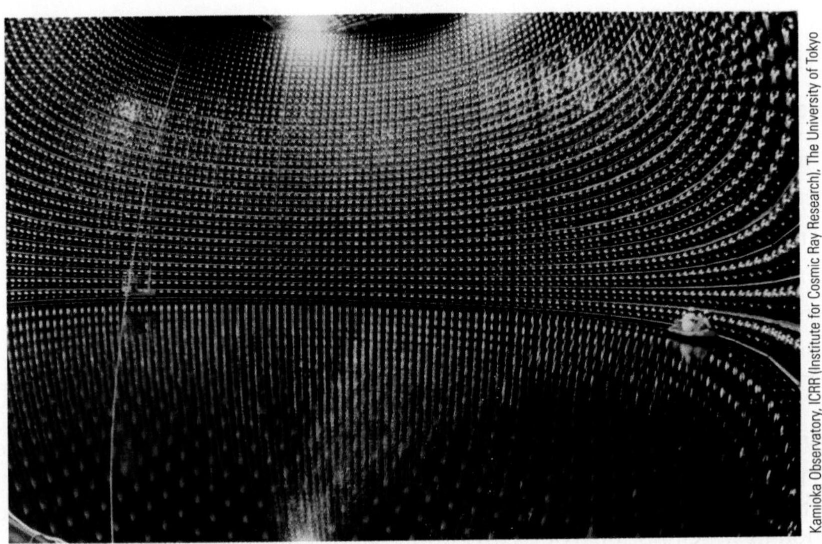

Kamioka Observatory, ICRR (Institute for Cosmic Ray Research), The University of Tokyo

◀ **Figure 31.3** This detector is called Super-Kamiokande. Located in Japan, it is designed to detect neutrinos that arrive at the Earth from space. This detector is a large container of water with walls covered with photodetectors that detect light generated when incoming particles interact with the water. This photo of the detector was taken during maintenance, when some of the water was drained out. Notice the people in a small boat (doing repair work).

31.3 shows the Super-Kamiokande detector located in Japan, built as part of a joint experiment with physicists from the United States and other countries. Super-Kamiokande consists of a large volume of water (approximately 64,000 m^3) in a container whose walls are lined with photodetectors. The photodetectors detect light produced when cosmic rays and other particles from space collide with water molecules. The Super-Kamiokande experiment is designed to study neutrinos, which are generated as a result of fusion reactions in our Sun (Chapter 30) and in supernova explosions. These neutrinos give physicists a "glimpse" of the kinds of particle reactions that take place in these systems.

31.2 Matter and Antimatter

One of the developers of quantum physics was P. A. M. Dirac; around 1930, he formulated a theory of quantum mechanics that combined the quantum theory of Schrödinger and Heisenberg with the postulates of special relativity. Dirac's theory gave two important results: (1) it predicted the phenomenon of electron spin, and in Chapter 29 we saw how electron spin is essential for explaining the periodic table of the elements, and (2) it predicted the existence of a completely new particle having the same mass as the electron and an electric charge of the same magnitude as an electron but with opposite sign. Several years later, this particle, named the ***positron***, was discovered in cosmic rays. The positron is denoted by the symbol e^+ to distinguish it from the ordinary electron, which is denoted by e^-. The superscripts indicate that the electron carries a negative charge, while the charge on the positron is positive. The positron is an example of ***antimatter***, a particle of matter having the same mass and opposite charge as the corresponding particle of ordinary matter. Other examples of antimatter particles (***antiparticles***) are the antiproton and the antineutron.

Particles such as electrons, positrons, and protons undergo reactions, much like the reactions involving nuclei. When describing these reactions, we denote each of these particles by a symbol, similar to the symbols used in Chapter 30 to denote different nuclei. Protons are denoted by p and neutrons by n; the corresponding antiparticles are denoted with an overbar, with $\bar{p}$ for the antiproton (pronounced "p bar") and $\bar{n}$ for the antineutron. Most other particles and antiparticles are denoted in a similar way, except for the positron, which is denoted by e^+ without a

TABLE 31.1 Some Properties of Electrons, Protons, Neutrons, and Their Antiparticles

Particle	Symbol	Mass (kg)	Mass (MeV/c^2)	Charge
Electron	e^-	9.109×10^{-31}	0.511	$-e$
Positron	e^+	9.109×10^{-31}	0.511	$+e$
Proton	p	1.673×10^{-27}	938	$+e$
Antiproton	$\overline{p}$	1.673×10^{-27}	938	$-e$
Neutron	n	1.675×10^{-27}	940	0
Antineutron	$\overline{n}$	1.675×10^{-27}	940	0

bar on top. For convenience, Table 31.1 lists the mass and electric charge of the electron, proton, and neutron along with the values for the corresponding antiparticles.

When an electron encounters its antiparticle (a positron), the two can undergo a reaction in which both particles are annihilated. This reaction is written as

$$e^- + e^+ \rightarrow \gamma + \gamma \tag{31.1}$$

In words, an electron plus a positron react to form two gamma rays, that is, two photons. This reaction must satisfy the conservation of energy, so the total energy before the reaction (the energies of the original electron and positron) is equal to the final energy (of the two gamma ray photons). The total initial energy includes the kinetic energies of the electron and positron plus their rest energies. According to special relativity (Chapter 27), the rest energy of an electron (rest mass m_e) is m_ec^2. The positron also has a rest mass m_e, so it has the same rest energy. By measuring the energies of the two gamma rays emitted in this reaction, one can determine the rest energies of the electron and positron and thus check the predictions of special relativity. The analysis is carried out in Example 31.1.

EXAMPLE 31.1 Electron–Positron Annihilation

An electron and positron initially at rest annihilate each other, producing two gamma rays. Assuming the gamma rays have the same energies, find the energy of one of the gamma ray photons.

RECOGNIZE THE PRINCIPLE

The annihilation reaction must conserve energy, so

$$\text{initial energy } (e^- + e^+) = \text{final energy } (\gamma \text{ rays})$$

The energy of the electron–positron pair consists of their kinetic energies (which are zero) and their rest energies.

SKETCH THE PROBLEM

See Figure 31.4. The electron and positron are at rest, so the initial momentum is zero. Since the final momentum must also be zero, the outgoing gamma ray photons are traveling in opposite directions. The two photons must also have the same frequency. This is the only way two photons can have a total momentum of zero.

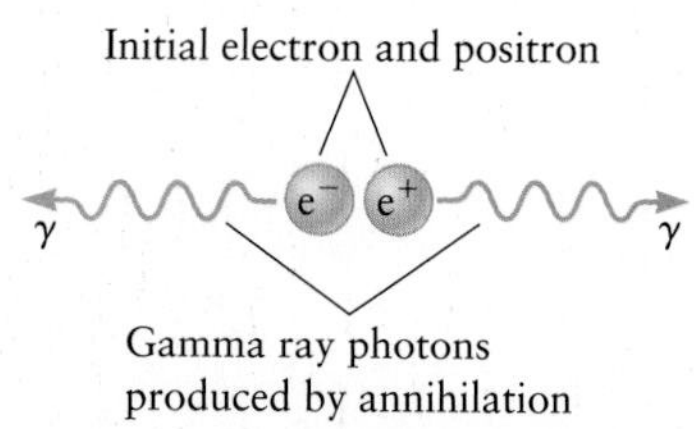

▲ **Figure 31.4** Example 31.1.

FIND THE RELATIONSHIPS

We have

$$\text{initial energy } (e^- + e^+) = KE + \text{rest energy}$$

and the initial kinetic energy is zero because the electron and positron are initially at rest. The initial rest energy is $m_ec^2 + m_ec^2 = 2m_ec^2$ since the electron and positron have the same mass m_e. We are left with

$$\text{initial energy } (e^- + e^+) = 2m_ec^2$$

SOLVE

The energy of each of the gamma ray photons must therefore be

$$E_{\text{photon}} = m_ec^2$$

From Table 31.1, we have

$$E_{\text{photon}} = \boxed{0.511 \text{ MeV}}$$

► ***What does it mean?***

This energy is *much* greater than that produced by atomic transitions. For example, in Chapter 29 we saw that the highest energy photon a hydrogen atom can emit is 13.6 eV.

Conservation Rules for Particle Reactions

The electron–positron reaction in Equation 31.1 obeys the principles of conservation of energy and conservation of momentum (Example 31.1). It also conserves electric charge, as the total electric charge is the same (zero) before and after the reaction. All other particle reactions also conserve energy, momentum, and charge. These conservation laws are nothing new—we learned about them in our work on classical physics—but we'll soon encounter several other important conservation laws that go far beyond our classical laws of physics.

CONCEPT CHECK 31.1 Electron–Positron Creation?

Your friend says that he has observed an electron–positron reaction in which a single photon forms an electron and positron as shown in Figure 31.5. He also claims that the two particles are at rest after the reaction and that the energy of the photon is equal to the sum of the rest energies of the two particles. Is this reaction (a) possible or (b) not possible? Explain why.

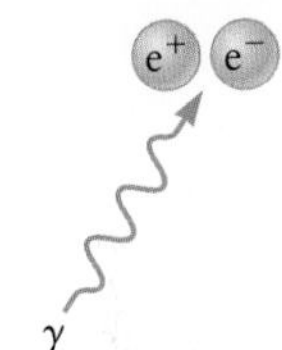

▲ **Figure 31.5** Concept Check 31.1.

The Stability of Antimatter

A positron is completely stable if it is kept away from electrons. Similarly, an antiproton is stable if it is kept away from protons. It is even possible to bring together a positron and an antiproton to form an atom of antihydrogen. Such atoms of antimatter have been studied in physics experiments, but it is very difficult to create large amounts of antimatter because it is difficult to contain it sufficiently far from regular matter to prevent annihilation.

The amount of energy released by the annihilation of an electron is very large (Example 31.1), and since a proton has a much larger mass than an electron, the energy released in the annihilation of a hydrogen atom with antihydrogen is much greater than calculated in Example 31.1. For this reason, antimatter is a popular fuel in science fiction stories. Unfortunately, it is not a practical energy source. All known methods for creating antiparticles require an amount of energy much greater than the rest energies of the particles that are created.

The Earth and our galaxy are composed almost entirely of matter. One might expect to find an equal amount of antimatter in the universe, perhaps in the form

▲ **Figure 31.6** Richard Feynman (1918–1988) was one of the developers of quantum electrodynamics (QED). He also made important contributions to the Manhattan project in World War II, to our understanding of liquid helium, and to many other areas of physics. Feynman was a very entertaining and enthusiastic teacher of science to the general public.

of antimatter galaxies, but as far as astrophysicists can tell, the amount of matter in the universe is much greater than the amount of antimatter. Physicists are now trying to understand this puzzle.

31.3 Quantum Electrodynamics

The theory of the photon as a quantum of electromagnetic energy (i.e., a quantum of electromagnetic radiation) was first discussed by Einstein in 1905 and was an important step in the development of quantum theory (Chapter 28). Einstein's discovery of the photon was only part of a complete quantum theory of light, however. His ideas were developed further by other physicists, including Richard Feynman (Fig. 31.6), Julian Schwinger, and Shinichiro Tomonaga, who developed a theory called ***quantum electrodynamics*** (QED), which combines electromagnetism and Maxwell's equations with quantum mechanics. According to QED, photons are the elementary quanta of electric and magnetic fields, and they play a role not just in electromagnetic radiation, but in all electric and magnetic forces.

Consider the interaction of two charged particles through Coulomb's law. A classical picture of this interaction is sketched in Figure 31.7A, showing the electric field produced by the charge on the left (q_1). The electric force on the right-hand charge (q_2) is due to this electric field. The QED picture of this force involves photons as sketched in Figure 31.7B; here the interaction between two charged particles is due to the exchange of photons, the elementary quanta of the electromagnetic field. We say that the electric force is "mediated" by photons.

Figure 31.7C gives a classical picture of how the exchange of particles can lead to a force between two objects. Here the "objects" are two children playing catch with a ball. The person throwing the ball experiences a recoil force during the throw while the person catching the ball experiences a recoil force when she catches it. This ball plays the role of a photon in the electromagnetic force as both the ball and the photon mediate (or "carry") a force between two objects.[1] It is believed that all the fundamental forces in nature are mediated by elementary particles.

The photon is an example of an elementary particle, but it differs in two ways from the other particles we have discussed so far. First, the photon has zero rest mass. Second, the photon is its own antiparticle, so there is no "antiphoton."

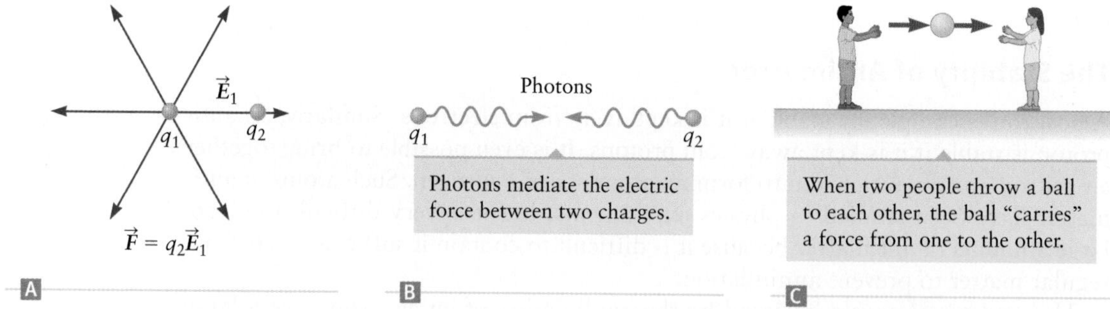

▲ **Figure 31.7** Force between two charged particles. A In the classical picture in Chapter 17, the force on one charge q_2 is due to the electric field produced by the charge q_1. B In the quantum mechanical picture of QED, the electric force is mediated (or "carried") from one charge to another by photons. C The force carried by photons is similar to the case of two people playing catch with a ball. The ball "carries" a force between the two people.

[1]This analogy is a very qualitative way to think about how photon exchange leads to forces and has its limitations. In particular, it is hard to give a classical explanation of how the exchange of particles can lead to an *attractive* force. The full theory of QED correctly describes both repulsive and attractive electromagnetic forces.

31.4 Elementary Particle Physics: The Standard Model

Studies of nuclear and particle physics began in the early 1900s, when Rutherford and other physicists conducted many experiments in which various particles were fired at atoms and nuclei (Chapters 29 and 30), revealing that nuclei are composed of protons and neutrons. Many other particles were then discovered in cosmic-ray studies, in collision experiments, and in nuclear decay processes. It is now clear that protons and neutrons are themselves composed of particles called ***quarks***, truly fundamental point particles whose charges are multiples of $\pm e/3$, one-third of the electron charge. Particles that are composed of quarks—such as protons and neutrons and their antiparticles—form a family of particles called ***hadrons***. Not all particles are members of the hadron family, however. There is a second class of particles called ***leptons***, which include electrons and positrons. The behavior of hadrons and leptons is described by what is called the ***standard model*** of elementary particles.

The standard model describes the properties of fundamental particles and the interactions between them.

Quarks Bind Together to Form Hadrons

There are six different kinds of quarks, named "up" (denoted by the symbol u), "down" (d), "charm" (c), "strange" (s), "top" (t), and "bottom" (b). These names do not refer to any physical properties of the quarks; rather, they are just whimsical names coined by physicists. Each of these quarks has a corresponding antiquark; denoted $\bar{u}$, $\bar{d}$, and so on. Some properties of quarks are listed in Table 31.2.

Quarks were first discovered in collision experiments involving protons. When a high-energy electron collides with a proton, the way the electron scatters (i.e., its outgoing direction and energy) gives information about how mass and charge are distributed inside the proton. This is similar to Rutherford's experiment (Chapter 29), in which the way alpha particles are scattered from an atom indicate that a massive nucleus is located at the atom's center. Collision experiments with protons show that there are three pointlike particles inside. Table 31.3 gives the quark composition of the proton: two up quarks and one down quark. The total charge on the proton is the sum of the charges of the constituent quarks, so the proton's charge is

$$\begin{aligned}\text{proton charge} &= 2 \times (\text{charge of u quark}) + (\text{charge of d quark})\\ &= 2\left(\frac{+2e}{3}\right) + \left(\frac{-e}{3}\right)\\ &= +e\end{aligned}$$

as expected.

TABLE 31.2 Quarks and Their Properties

Quark	Symbol	Charge	Mass (MeV/c^2)	Antiparticle
Up	u	$+2e/3$	4	$\bar{u}$
Down	d	$-e/3$	8	$\bar{d}$
Charm	c	$+2e/3$	1,500	$\bar{c}$
Strange	s	$-e/3$	150	$\bar{s}$
Top	t	$+2e/3$	176,000	$\bar{t}$
Bottom	b	$-e/3$	4,700	$\bar{b}$

TABLE 31.3 Properties of Some Baryons

Particle	Symbol	Constituent Quarks	Lifetime (s)	Mass (MeV/c^2)
Proton	p	uud	Stable	938
Neutron	n	udd	890	940
Lambda zero	Λ^0	sud	2.6×10^{-10}	1116
Sigma plus	Σ^+	uus	0.8×10^{-10}	1189
Sigma zero	Σ^0	sud	6.0×10^{-20}	1193
Sigma minus	Σ^-	dds	1.5×10^{-10}	1197
Hyperon	Ξ^-	ssd	1.6×10^{-10}	1321

Note: There are many other baryons, which are composed of other combinations of three quarks and antiquarks.

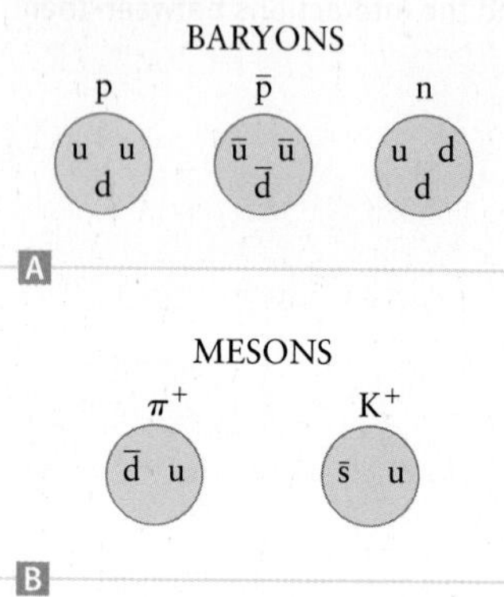

▲ **Figure 31.8** A Baryons and antibaryons are composed of three quarks. The three particles represented here are the proton (p), the antiproton ($\bar{p}$), and the neutron (n). B Mesons are composed of one quark and one antiquark. These mesons are called pions and kaons.

Quarks can combine to form a hadron in two ways. Hadrons composed of three quarks are called ***baryons*** (Fig. 31.8A). Protons and neutrons are baryons; Table 31.3 lists a few other baryons, and dozens of other baryons have been observed and their component quarks identified. It is also possible for a quark and an antiquark to combine to form a particle. Hadronic particles composed of just two quarks are called ***mesons*** (Fig. 31.8B), and a few are listed in Table 31.4.

All hadrons are composed of quarks, so the interactions between quarks determine the properties of hadrons and how they interact with one another. The two most important hadrons are the proton and neutron, so the behavior of quarks also determines the properties of nuclei. Quarks are charged, so they interact via the electric (Coulomb) force. They also interact via the strong force mentioned in Chapter 30. The strong force holds quarks together to form protons and neutrons (nucleons) and is also responsible for holding protons and neutrons together to make nuclei.

Rules for "Making" Baryons and Mesons, and Their Reactions

The standard model places restrictions on the ways quarks can combine to form baryons and mesons and on the kinds of particle reactions that are possible. One of these restrictions has to do with the allowed electric charges of baryons and mesons. The charge of any particular baryon or meson can be found by simply adding up the charges of its constituent quarks. The quark charges are listed in Table 31.2, and the corresponding antiquarks have the opposite charges (so the $\bar{d}$ has a charge of $+e/3$, which is opposite to that of the d quark, etc.). Working out the charges of the baryons and mesons in Tables 31.3 and 31.4, you'll find that they all have electric charges of $+e$, $-e$, or zero; some other baryons (not listed here) have charges $+2e$ and $-2e$. The standard model guarantees that an isolated particle cannot have a

Insight 31.1
PARTICLES AND ANTIPARTICLES

Mesons all consist of a quark and an antiquark (Table 31.4), leading to some interesting consequences. For example, the phi meson (φ) is a combination of a strange quark s and the antiquark $\bar{s}$. That is,

$$\varphi = s\bar{s}$$

The antiphi meson is then

$$\bar{\varphi} = \bar{s}\bar{\bar{s}}$$

According to the rules of antiparticles, the antiparticle of the $\bar{s}$ is just an s quark, so $\bar{\bar{s}} = s$, and because the order in which the quarks are listed does not matter, we have

$$\bar{\varphi} = \bar{s}\bar{\bar{s}} = \bar{s}s = \varphi$$

The antiparticle of the phi meson is thus itself!

TABLE 31.4 Properties of Some Mesons

Particle	Symbol	Constituent Quarks	Lifetime (s)	Mass (MeV/c^2)
Pion (pi plus)	π^+	u$\bar{d}$	2.6×10^{-8}	140
Pi zero*	π^0	d$\bar{d}$/u$\bar{u}$	8.4×10^{-17}	135
Kaon (K plus)	K$^+$	u$\bar{s}$	1.2×10^{-8}	494
Kaon (K minus)	K$^-$	$\bar{u}$s	1.2×10^{-8}	494
Phi	φ	s$\bar{s}$	1.6×10^{-22}	1020

Note: There are many other mesons, which are composed of other combinations of a quark and an antiquark.

*The π^0 is a quantum mechanical combination of the d$\bar{d}$ and u$\bar{u}$ quark states.

fractional charge, that is, a charge of $\pm e/3$ or $\pm 2e/3$. Since quarks themselves have fractional charge, the implication is that individual quarks cannot be isolated. We'll explain why when we consider the strong force in more detail below.

The standard model also predicts that ***baryon number*** must be conserved in all particle reactions. Each quark has a baryon number of $+\frac{1}{3}$, so the baryon number of a proton (which is composed of three quarks) is +1. Antiquarks have a baryon number of $-\frac{1}{3}$, so all mesons have a baryon number of zero. We'll discuss some examples of baryon number conservation below, after we describe leptons.

CONCEPT CHECK 31.2 Which of These Hadrons Are Allowed?

Which quark combination is *not* an allowed meson or baryon, (a) $c\bar{c}$, (b) bbs, (c) uud, (d) $uu\bar{d}$, or (e) $\overline{uud}$?

EXAMPLE 31.2 Rest Mass of the Proton

The proton is composed of two up quarks and a down quark (p = uud). Compare the total rest mass of these three quarks with the rest mass of the proton. Are they equal?

RECOGNIZE THE PRINCIPLE

The rest masses of the proton and the u and d quarks are listed in Tables 31.2 and 31.3. We can simply add the appropriate masses and compare the results.

SKETCH THE PROBLEM

No sketch is needed.

FIND THE RELATIONSHIPS

The u quark has a mass of about 4 MeV/c^2 and the d quark mass is around 8 MeV/c^2, so the total mass of the quarks that make up a proton is

$$\text{mass of u} + \text{u} + \text{d} = (4 + 4 + 8)\text{ MeV}/c^2 = 16\text{ MeV}/c^2 \quad (1)$$

From Table 31.3, the rest mass of the proton is

$$\text{rest mass of proton} = 938\text{ MeV}/c^2 \quad (2)$$

SOLVE

Comparing Equations (1) and (2), the rest mass of the proton is *much* greater than the combined masses of its constituent quarks, so the answer is **no**.

► ***What does it mean?***

The difference between the masses in Equations (1) and (2) is due to the quark kinetic and potential energies within the proton, together with the relativistic relation between energy and mass.

Insight 31.2

MEASURING THE MASS OF A QUARK

One way to measure the mass of a quark is through Newton's second law, $\sum F = ma$ (or the corresponding result of quantum mechanics). However, quarks are always confined in baryons or mesons, so it is impossible to study how a free quark would respond to an applied force. The best we can do is probe the motion or properties of quarks inside a baryon or meson and then deduce the quark mass using the theory of the strong force. Such an analysis must allow for the contributions of quark kinetic energy and the potential energy of the strong force, and these contributions are difficult to calculate with high accuracy. For this reason, the masses of the quarks are not known with the same level of precision as the masses of particles such as the proton, neutron, and electron.

Leptons

We have just discussed the hadron family of particles, which are all assembled from quarks. Let's now consider a second family of particles called leptons. There are six fundamental leptons (Table 31.5, page 976) plus their corresponding antiparticles. These particles group naturally into three pairs: the electron (e^-) and the electron neutrino (ν_e), the muon (μ^-) and the muon neutrino (ν_μ), and the tau (τ^-) and the tau neutrino (ν_τ). The muon (lifetime $= 2.2 \times 10^{-6}$ s) and the tau (lifetime 2.9×10^{-13} s) are not stable. Electrons are stable, while the behavior of neutrinos is more complicated. Recent experiments indicate that as they travel through space,

TABLE 31.5 Leptons and Their Properties

Particle	Symbol	Charge	Mass/c^2	Antiparticle
Electron	e^-	$-e$	0.511 MeV	e^+
Electron neutrino	ν_e	0	$0.05 \text{ eV} < m < 2 \text{ eV}$	$\bar{\nu}_e$
Muon	μ^-	$-e$	106 MeV	μ^+
Muon neutrino	ν_μ	0	< 0.19 MeV	$\bar{\nu}_\mu$
Tau	τ^-	$-e$	1780 MeV	τ^+
Tau neutrino	ν_τ	0	< 18 MeV	$\bar{\nu}_\tau$

neutrinos change from one to another of the three types of neutrinos listed in Table 31.5, an effect called "neutrino oscillations." Another interesting property of neutrinos is that they have very small masses; the best experiments to date give only an approximate mass for the electron neutrino between 0.05 and 2 eV/c^2. That is more than 100,000 times less than the mass of an electron!

Leptons play important roles in certain reactions involving hadrons, such as the decay of an "isolated" neutron (a neutron outside a nucleus):

$$n \rightarrow p + e^- + \bar{\nu}_e \tag{31.2}$$

Neutrinos were first observed in studies of the particles emitted by a nuclear power reactor. A large source of the neutrinos observed on the Earth are the nuclear fusion processes that power the Sun. Studying these solar neutrinos provides one of the best ways to understand the nuclear reactions that take place inside the Sun.

Reactions involving leptons always conserve ***lepton number.*** Each lepton has a lepton number of +1, whereas each antilepton has a lepton number of −1. Since neutron decay involves both baryons and leptons, this process must conserve both baryon number and lepton number. Notice that all baryons have a lepton number of zero and all leptons have a baryon number of zero. For Equation 31.2, we have

$$\begin{array}{lccccccc} & n & \rightarrow & p & + & e^- & + & \bar{\nu}_e \\ \text{baryon number:} & +1 & \rightarrow & +1 & + & 0 & + & 0 = +1 \\ \text{lepton number:} & 0 & \rightarrow & 0 & + & 1 & - & 1 = 0 \end{array} \tag{31.3}$$

Hence, although this reaction produces a lepton and an antilepton, it still conserves both baryon number and lepton number.

CONCEPT CHECK 31.3 Conserving Lepton Number

Muons are unstable and decay to form an electron and two neutrinos. Which of the following reactions describes this decay?

(a) $\mu^- \rightarrow e^- + \nu_\mu + \bar{\nu}_e$

(b) $\mu^- \rightarrow e^- + \nu_\mu + \nu_e$

EXAMPLE 31.3 Review Problem: Studying the Decay of a Tau Lepton

Elementary particles are produced and studied in collision experiments at laboratories like the one shown in Figure 31.9. Among the particles studied in this way is the tau lepton. Tau leptons have a lifetime of approximately $\Delta t_0 = 2.9 \times 10^{-13}$ s measured in the tau's reference frame (i.e., measured by an observer at rest relative to the tau). Suppose a certain collision experiment creates a tau particle with a speed $v = 0.99c$ relative to a physics laboratory. How far will this particle travel relative to the laboratory before it decays? Give the distance as measured in (**a**) the reference frame of the tau and (**b**) the reference frame of a physicist in the laboratory.

RECOGNIZE THE PRINCIPLE

The tau's speed is sufficiently high that we must consider the effects of special relativity. This example is a time-dilation problem very similar to Examples 27.2 and 27.4, in which a high-speed muon decayed in the Earth's atmosphere. According to an observer traveling with the tau (i.e., in the tau's reference frame), the lab moves at a speed of $0.99c$ for a time Δt_0. The physicist sees the tau moving by at $0.99c$, but observes a dilated lifetime. According to Equation 27.7,

$$\Delta t = \frac{\Delta t_0}{\sqrt{1 - v^2/c^2}} \tag{1}$$

SKETCH THE PROBLEM

Figure 31.10 shows the tau lepton along with two observers. One observer travels with the tau and measures a lifetime Δt_0. The other observer (the physicist at rest in the lab) measures a dilated lifetime Δt. (Compare with Fig. 27.9.)

FIND THE RELATIONSHIPS

(**a**) The value Δt_0 of the tau lifetime given above is the proper time because it is measured in the tau's reference frame. The distance d_0 traveled by the tau as measured in its own reference frame is

$$d_0 = v\,\Delta t_0 \tag{2}$$

(**b**) A physicist studying the moving tau measures a dilated lifetime Δt given by Equation (1). The distance traveled by the tau in the physicist's frame of reference is thus

$$d = v\,\Delta t = \frac{v\,\Delta t_0}{\sqrt{1 - v^2/c^2}} \tag{3}$$

SOLVE

(**a**) Inserting the given values of v and Δt_0 from Equation (2) leads to

$$d_0 = (0.99c)\Delta t_0 = (0.99)(3.00 \times 10^8 \text{ m/s})(2.9 \times 10^{-13} \text{ s})$$

$$d_0 = 8.6 \times 10^{-5} \text{ m} = \boxed{0.086 \text{ mm}}$$

(**b**) Inserting the values for v and Δt_0 again from Equation (3), the distance traveled in the laboratory reference frame is

$$d = \frac{(0.99c)\Delta t_0}{\sqrt{1 - (0.99c)^2/c^2}} = \frac{(0.99)(3.00 \times 10^8 \text{ m/s})(2.9 \times 10^{-13} \text{ s})}{\sqrt{1 - (0.99)^2}}$$

$$d = 6.1 \times 10^{-4} \text{ m} = \boxed{0.61 \text{ mm}}$$

▸ ***What does it mean?***

A physicist studying this particle must complete her measurements before the tau has traveled less than 1 mm! Actually, most studies of elementary particles such as the tau use measurements of the decay products (which may travel much farther) to infer the energy, momentum, and other properties of the original particle.

▲ **Figure 31.9** The large "ring" in this aerial photo shows the Large Hadron Collider, a particle accelerator at the European Organization for Nuclear Research in Geneva, Switzerland. The actual accelerator is in a tunnel located beneath this ring. This machine is designed to collide protons with energies of 7 TeV.

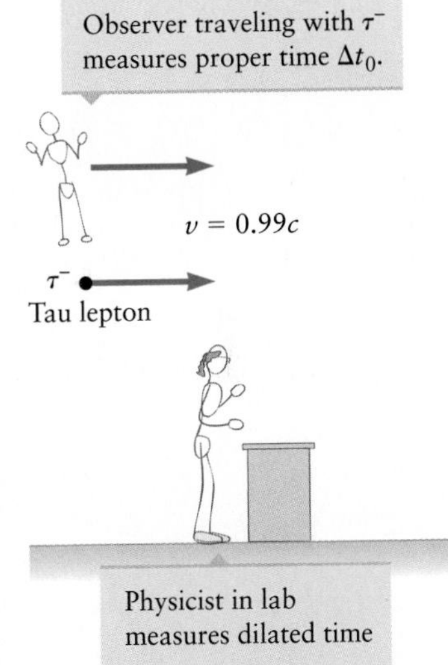

Observer traveling with τ^- measures proper time Δt_0.
$v = 0.99c$
τ^-
Tau lepton
Physicist in lab measures dilated time
$\Delta t = \frac{\Delta t_0}{\sqrt{(1-v^2/c^2)}}$.

▲ **Figure 31.10** Example 31.3.

31.5 The Fundamental Forces of Nature

In Section 31.4, we described a few of the properties of hadrons and leptons such as their charge and mass and some of the reactions they can undergo. To get a complete understanding, we must also consider the forces that act between them. There are four fundamental forces, as listed and compared in Table 31.6.

The Strong Nuclear Force Only Acts on Quarks

Quarks carry electric charge, so they experience electric and magnetic forces, but the largest force between two quarks is the strong force. The strong force binds quarks together to form particles such as protons and neutrons. After the quark model was proposed, physicists searched unsuccessfully for free particles with charge $\pm e/3$ or $\pm 2e/3$. Despite much effort, a single isolated quark has never been observed, which can be explained by the theory of the strong force and the standard model. One striking prediction of these theories is that the energy required to separate two quarks grows larger as the separation increases; eventually, at large separations, this energy is large enough to produce new quark–antiquark pairs that then bind to form baryons and mesons. This phenomenon is called ***confinement*** and is why an individual quark is never observed in isolation.

A proton is composed of two up quarks and a down quark, a combination denoted by uud. Quarks are quantum particles obeying the Pauli exclusion principle (Chapter 29), so two quarks cannot occupy the same quantum state. Quarks have been found to possess a new quantum property called ***color*** that helps satisfy the Pauli exclusion principle in the case of a proton. All quarks carry a ***color quantum number*** with the possible values red, green, and blue, like the spin quantum number of an electron. From Section 29.5, two electrons can occupy the 1*s* state of a helium atom provided that they have different spin values. In the same way, the two up quarks in a proton must have different color values.

The Strong Force Is Mediated by Gluons

In the language of quantum electrodynamics (Section 31.3), photons are said to "mediate" electromagnetic interactions because they "carry" the electromagnetic force acting on charged particles. In the same way, the strong force is mediated or carried by particles called ***gluons***, and there are eight different types of gluons. The force exerted between two quarks depends on the type of each quark.

The theory of the strong force is called ***quantum chromodynamics***, or QCD. One prediction of QCD is that baryon number is always conserved in a reaction, as

TABLE 31.6 The Fundamental Forces in Nature

Force	Force-Carrying Particle(s)	Acts on	Responsible for	Comments
Electromagnetism (QED)	Photon	All particles that carry electric charge	Electric and magnetic forces	Unified with the weak force as the electroweak force
Weak force	W^+, W^-, Z	Leptons and quarks	Beta decay and other nuclear decay processes	
Strong force (QCD)	Gluons	Quarks	Binding of quarks to form baryons and mesons	Also responsible for binding of protons and neutrons in the nucleus
Gravitation	Graviton (not yet observed)	All particles	Gravity	

we mentioned in connection with the decay of a neutron (Eq. 31.3). All known reactions obey the conservation of baryon number, now believed to be a fundamental conservation principle.

Forces Acting on Leptons: Electromagnetism and the Weak Force

Three of the fundamental leptons—the electron, muon, and tau—are charged, so they experience electric and magnetic forces. These forces are described by the theory of QED. The strong force that is so important for quarks and hadrons does not act on leptons. All leptons, however, do experience what is known as the ***weak force***, which is carried by three different particles called the W^+, the W^-, and the Z. The weak force acts on both leptons and quarks (and hence on all hadrons). This force has an extremely short range; two particles must be within about 10^{-18} m $= 0.001$ fm to experience this force. This distance is about one one-thousandth the diameter of a proton! Two particles rarely come this close together and hence the name "weak" for this force. Even so, the weak force is responsible for some important processes such as the decay of the neutron (Eq. 31.2):

$$\text{n} \rightarrow \text{p} + \text{e}^- + \bar{\nu}_\text{e}$$

The neutron and proton are composed of quarks, so we can also write this reaction as

$$\text{udd} \rightarrow \text{uud} + \text{e}^- + \bar{\nu}_\text{e} \tag{31.4}$$

Hence, the decay of the neutron converts one of the down quarks (d) in the neutron into an up quark (u). Processes in which one type of quark is converted into another type always involve the weak force.

EXAMPLE 31.4 Particle Decay and Conservation Rules

The negatively charged pion π^- is a meson composed of two quarks $\bar{\text{u}}\text{d}$. The π^- meson can decay into a muon and an antimuon neutrino:

$$\pi^- \rightarrow \mu^- + \bar{\nu}_\mu \tag{1}$$

Show that this decay conserves both charge and lepton number.

RECOGNIZE THE PRINCIPLE

To determine if this decay conserves charge, we compare the total charge before and after the decay. We analyze the conservation of lepton number in the same way.

SKETCH THE PROBLEM

No sketch is necessary.

IDENTIFY THE RELATIONSHIPS

The π^- meson has a charge of $-e$. Mesons are hadrons, so the lepton number of the π^- is zero. The μ^- muon has a charge of $-e$ and a lepton number of $+1$. The antimuon neutrino is neutral, and since it is an antiparticle, its lepton number is -1.

SOLVE

We can rewrite the decay in Equation (1) as

		π^-	$\rightarrow \mu^-$	$+ \bar{\nu}_\mu$
charge	=	$-e$	$-e$	0
lepton count	=	0	$+1$	-1

Hence, both charge and lepton number are indeed $\boxed{\text{conserved}}$.

(continued) ▶

► ***What does it mean?***

We have also seen that particle decays must conserve baryon number. In Equation (1), the final baryon number is zero because the muon and antimuon neutrino are both leptons. The π^- meson contains a $\bar{u}$ and a d quark; since the $\bar{u}$ is an antiparticle, the total baryon number of the π^- meson is zero, so baryon number is also conserved in this decay.

EXAMPLE 31.5 Annihilation of a Positron

When an electron and positron annihilate each other, the result is two photons:

$$e^- + e^+ \rightarrow \gamma + \gamma \tag{1}$$

(Recall that a photon is denoted by the symbol γ.) The apparently similar reaction

$$e^+ \rightarrow \gamma + \gamma \tag{2}$$

does *not* occur. Explain why the decay in Equation (2) is not allowed.

RECOGNIZE THE PRINCIPLE

A particle decay or reaction is allowed only if it conserves baryon number, lepton number, and electric charge.

SKETCH THE PROBLEM

No figure is necessary.

IDENTIFY THE RELATIONSHIPS AND SOLVE

There are no baryons in Equations (1) and (2), so both conserve baryon number. Equation (1) conserves lepton number since the lepton number of an electron is +1, whereas the lepton number of a positron is −1 since it is the antiparticle of the electron. The lepton number of a photon is zero (since it is not a lepton), so Equation (2) does *not* conserve lepton number. Equation (1) conserves electric charge since the total charge is zero before and after the decay. Equation (2) does *not* conserve electric charge since the initial charge is $+e$ while the final charge is zero.

► ***What does it mean?***

It is believed that baryon number, lepton number, and charge are conserved by *all* processes in nature.

Insight 31.3

WHAT DOES IT MEAN WHEN TWO FORCES ARE "UNIFIED"?

Two forces are "unified" when it is shown that they are two aspects of the same "underlying" force. For example, in Section 27.11 we showed that the electromagnetic force on a moving charge can be viewed as either an electric force or a magnetic force, depending on the relative motion of the charge and an observer. Hence, we say that these are two aspects of the *electromagnetic* force. Maxwell's theory gives a unified theory of electromagnetism.

Unification of the Weak Force and Electromagnetism

Prior to Maxwell, physicists thought that electricity and magnetism were two completely different types of forces, but Maxwell showed that they are really two aspects of the same phenomenon (hence the name "electromagnetism"). Maxwell thus provided a unified theory of electricity and magnetism. In the same way in the 1970s, Sheldon Glashow, Abdus Salam, and Steven Weinberg developed a theory that unifies electromagnetism and the weak force. This combined theory of the ***electroweak force*** explains phenomena that cannot be accounted for by either separate theory. This theory predicted the existence of new particles that carry the electroweak force; these particles—the W^+, the W^-, and the Z (Table 31.6)—were then found in subsequent experiments.

Gravitation

Newton's theory of gravitation (Chapter 5) was extended by Einstein in his work on general relativity (Chapter 27). Gravitation is another fundamental force in nature. It plays an important role in our everyday terrestrial lives and is also responsible for the motion of the planets and stars in the universe. Even so, in some ways gravity is much weaker than the other fundamental forces of nature. For example, the electric force between two electrons is much larger in magnitude than the gravitational force between them.

The particles that mediate the electromagnetic, strong, and weak forces have all been observed and their properties measured (Table 31.6). It is believed that the gravitational force is also carried by a particle; although this particle has not yet been directly observed, physicists believe it must exist and have already given it a name: the ***graviton***.

EXAMPLE 31.6 Comparing the Electromagnetic and Gravitational Forces

The two up quarks in a proton are separated by a distance $r \approx 1$ fm ($= 1 \times 10^{-15}$ m). These quarks each have an electric charge $+2e/3$, so they repel each other through an electric force obeying Coulomb's law. They also attract each other through a gravitational force. What is the ratio of the magnitudes of the electric and gravitational forces between these two quarks?

RECOGNIZE THE PRINCIPLE

The gravitational force between two point particles of mass m separated by a distance r is

$$F_{\text{grav}} = \frac{Gmm}{r^2}$$

while the electric force between two point charges q is

$$F_{\text{elec}} = \frac{kqq}{r^2}$$

We wish to calculate $F_{\text{grav}}/F_{\text{elec}}$ for two quarks in a proton.

SKETCH THE PROBLEM

Figure 31.11 shows a classical picture of two u quarks separated by a distance r. The electric force and the gravitational force between them are both proportional to $1/r^2$.

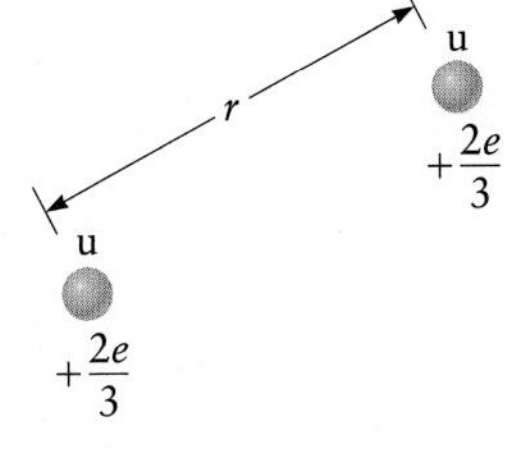

▲ **Figure 31.11** Example 31.5.

IDENTIFY THE RELATIONSHIPS

Let m_u be the mass of an up quark. The force of gravity between two up quarks is then

$$F_{\text{grav}} = \frac{Gm_u m_u}{r^2}$$

The charge on an up quark is $q = +2e/3$ (Table 31.2), so the electric force between two up quarks is

$$F_{\text{elec}} = \frac{kqq}{r^2} = \frac{k(2e/3)^2}{r^2} = \frac{4ke^2}{9r^2}$$

The ratio of these two forces is

$$\frac{F_{\text{grav}}}{F_{\text{elec}}} = \frac{Gm_u^2/r^2}{4ke^2/(9r^2)} = \frac{9Gm_u^2}{4ke^2} \tag{1}$$

(continued) ▶

To evaluate Equation (1), we need the mass of the u quark, m_u. From Table 31.2, the mass of the up quark is $m_u = 4\ \text{MeV}/c^2$. Converting this value to kilograms, we get

$$m_u = \frac{4\ \text{MeV}}{c^2} = \left[\frac{4 \times 10^6\ \text{eV}}{(3.00 \times 10^8\ \text{m/s})^2}\right]\left(\frac{1.60 \times 10^{-19}\ \text{J}}{1\ \text{eV}}\right) = 7 \times 10^{-30}\ \text{kg}$$

SOLVE

Inserting our value of m_u into Equation (1) along with values of the other factors leads to

$$\frac{F_{grav}}{F_{elec}} = \frac{9Gm_u^2}{4ke^2} = \frac{9(6.67 \times 10^{-11}\ \text{N}\cdot\text{m}^2/\text{kg}^2)(7 \times 10^{-30}\ \text{kg})^2}{4(8.99 \times 10^9\ \text{N}\cdot\text{m}^2/\text{C}^2)(1.60 \times 10^{-19}\ \text{C})^2}$$

$$\frac{F_{grav}}{F_{elec}} = \boxed{3 \times 10^{-41}}$$

▶ ***What does it mean?***

The value of r canceled when we computed the ratio F_{grav}/F_{elec}, so the ratio has this value for all values of r. The gravitational force between two up quarks is thus always *much* smaller than the electric force. That is why gravity is considered the weakest of the fundamental forces.

The Four Forces in Nature and Their Unification

We have referred to the four fundamental forces in nature, but as indicated in Table 31.6 the electromagnetic and weak forces are two aspects of the same electroweak force. So, one could also say that there are only three fundamental forces in nature. In addition, many physicists are attempting to unify the electroweak and strong forces; such a ***grand unified theory*** does not yet exist and is the subject of much current research. Going one step further, some physicists suspect that all forces in nature can be unified into one "theory of everything." A candidate for such a theory is string theory in which all particles are stringlike objects in a high-dimensional space. The details of such unified theories are still very uncertain. Einstein was working on this problem up until his death in 1955 and many others are working on it now, so progress has been slow.

31.6 Elementary Particle Physics: Is This the Final Answer?

At an intuitive level, scientists believe that nature is inherently "simple" and that the fundamental forces and particles should reflect this simplicity. However, adding up the number of quarks and leptons in Tables 31.2 and 31.5 along with the photon and other force-carrying particles in Table 31.6 brings the total number of fundamental particles to more than two dozen. Some physicists (including the author) suspect that a simple universe should not contain this many "fundamental" particles. For this reason, many physicists believe that our current understanding of fundamental particles and forces is not complete. The search for the ultimate laws of physics is thus still in progress. While debate over the correct laws of physics continues, it should not overshadow the theories we *do* have. These theories very accurately describe the quantum properties of atoms and molecules, the behavior of nuclei, the production of energy in the Sun, the behavior of the laser in your DVD player, the motion of your car when you apply the brakes, and the propagation of the electromagnetic waves used by your cell phone. Our current theories may turn out to be only approximate, just as Newton's mechanics is an approximate theory, but our present theories do work *extremely* well.

31.7 Astrophysics and the Universe

The Big Bang and the Expansion of the Universe

How did the universe begin, and how will it end? Did it even have a beginning? Will it ever end? Questions about the origin and fate of the universe have deep implications for philosophy and many other fields. Such questions came within reach of physics when astronomers observed that the spectrum of light from distant galaxies can be used to determine the motion of those galaxies relative to the Earth. In Chapter 13, we discussed the Doppler effect, which causes the frequency of a wave to shift to either higher or lower values when the source and observer are in relative motion. As a result, the frequency of light or other electromagnetic radiation is shifted upward when the source and observer are moving toward each other and is shifted to a lower frequency when they are moving apart. (See also Chapter 23 and Example 23.8.)

In the early 1900s, astronomers discovered that light from all galaxies visible from the Earth is shifted to lower frequencies compared with what would be observed if their sources were at rest. This effect is called a ***red shift*** because red light is at the low frequency end of the visible spectrum. Using this measured frequency shift together with the Doppler effect formula, one can calculate the *apparent* velocity of a particular galaxy relative to the Earth. (We'll explain why we use the term *apparent* shortly.) Astronomer Edwin Hubble collected such data for many galaxies and plotted their apparent velocities as a function of their distances from the Earth, obtaining the result in Figure 31.12. His data showed that galaxies farthest away have the largest apparent velocities, and all galaxies are moving away from the Earth.

Hubble's results in Figure 31.12 (and much data gathered more recently) follow a linear behavior, with the apparent velocity v of a galaxy proportional to its distance d from the Earth:

$$v = H_0 d \tag{31.5}$$

Hubble's law

This is called ***Hubble's law***, and H_0 is called the Hubble constant. The best current estimate for the Hubble constant is $H_0 = 21\ (\text{km/s})/(10^6\ \text{ly})$, where $1\ \text{ly} = 1$ light-year is the distance light travels in 1 year. So, a galaxy that is 10^6 ly (one million light-years) from the Earth is moving away at an apparent velocity of 21 km/s.

Hubble's law is evidence for the ***Big Bang*** model of the universe, which states that the universe originated with an explosion. At the instant of this explosion all the universe was an infinitesimal point, and since this "bang," the universe has

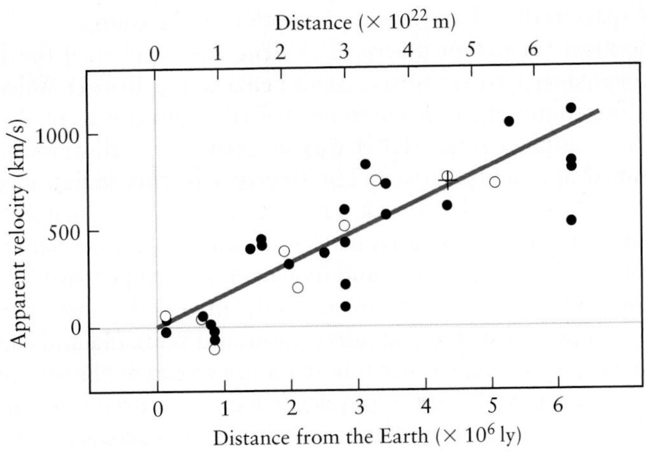

◀ **Figure 31.12** Edwin Hubble's original data showing how the apparent velocities of different galaxies increase with distance from the Earth. Galaxies farthest from the Earth have the largest velocities relative to the Earth. The bottom scale shows distance in units of light-years (ly); 1 ly is the distance light travels in 1 year.

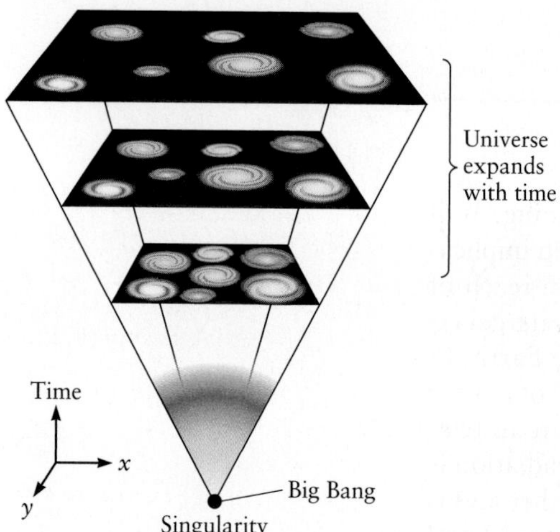

▲ **Figure 31.13** Schematic of the Big Bang. As time passes, the universe is expanding, carrying galaxies farther apart.

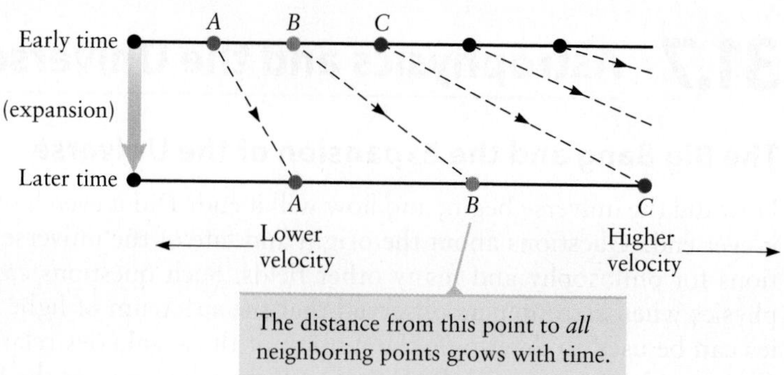

▲ **Figure 31.14** Schematic of how the relative distance between any two points in an expanding universe increases with time. As time increases (going from the top line to the bottom), the distance between any pair of points increases. For an observer at any point, all other points are moving farther away, and each observer perceives that he or she is at the "center" of the expansion.

expanded as sketched in Figure 31.13. All galaxies appear to be moving away from us, which would seem to suggest that the Earth is at the center of it all. Figure 31.13, however, shows that an observer at any location in an expanding universe will find that all galaxies are moving away from her. Galaxies with the largest apparent velocities relative to the Earth are farthest away from the Earth. Since light takes a certain amount of time to travel from these galaxies to the Earth, this light was emitted when the universe was younger than at present. In this way, studying the most distant galaxies gives a picture of the universe shortly after it was formed.

Now let's explain why we refer to the quantity v in Hubble's law as an *apparent* velocity. In our normal work in mechanics, we define and measure the velocity of an object in terms of its position relative to a fixed coordinate system, that is, relative to "markers" in space such as a coordinate axis. With an expanding universe (Fig. 31.14), these markers are themselves expanding along with the universe. That is, the "fabric" of space itself expands with time, carrying us and all other galaxies along with it. Although in some ways we can think of the expanding universe as involving "motion," it is a very different type of motion than in normal mechanics, which is why astronomers usually refer to the astronomical red shift as giving an "apparent" velocity. This picture also gives a more accurate way to understand the astronomical red shift. Light emitted by a distant galaxy has a certain wavelength. As that light travels to the Earth, the universe expands, "stretching" space and thereby increasing the wavelength of the light. This increase is similar to a Doppler red shift, but now the shift to a longer wavelength (lower frequency) is due to the expansion of space rather than the simple motion of the source.

▲ **Figure 31.15** Arno Penzias and Robert Wilson were using this antenna to study how radio and microwaves propagate through the atmosphere. They discovered blackbody radiation arriving at the Earth from outer space, with an apparent blackbody temperature of approximately 3 K.

One of the most important pieces of evidence in support of the Big Bang theory was discovered by two scientists, Arno Penzias and Robert Wilson, who were studying satellite communications systems at Bell Laboratories in the 1960s. They found that their antenna (Fig. 31.15) was detecting a small amount of radiation even when aimed at "empty" space. The intensity of this radiation was the same as that produced by a blackbody at a temperature of approximately 3 K, and this radiation from space is now known as the cosmic microwave background radiation. The existence of this radiation and its blackbody temperature are key features supporting the Big Bang model. According to the model, this radiation was emitted when the universe was only a few hundred thousand years old and then red-shifted due to subsequent expansion so that it now has an effective blackbody temperature of 3 K. In more recent experiments, physicists have measured tiny variations in the intensity of the cosmic microwave background radiation across the sky (Fig. 31.16).

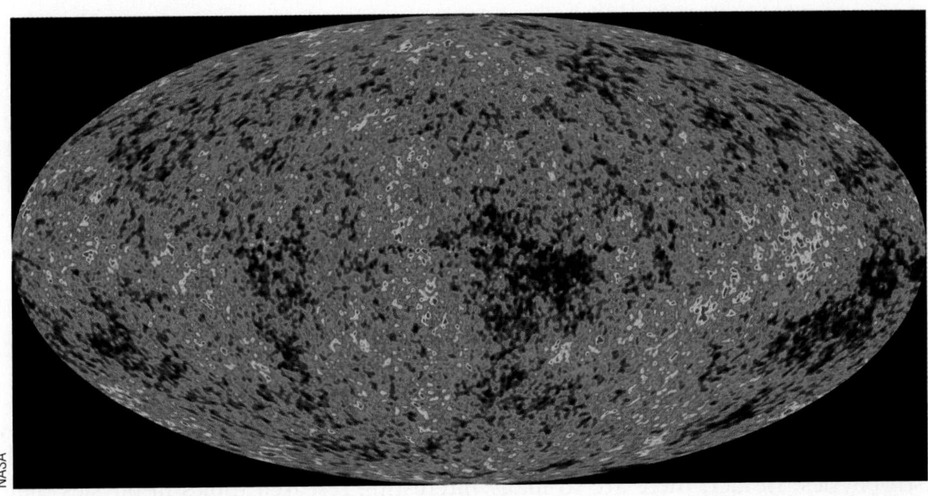

◀ **Figure 31.16** Recent experiments have studied the cosmic background radiation in great detail. This plot shows how the radiation intensity varies slightly from point to point in the sky, with the red regions being hottest and the blue regions coolest. The temperature variations are quite small, with the hottest regions only about 10^{-5} K higher in temperature than the average. These variations were produced by slight variations of the density of matter throughout space shortly after the Big Bang. These density variations eventually became stars and galaxies.

These variations give information on the state of the early universe when stars and galaxies were first forming.

Fate of the Universe

Hubble's law together with the cosmic microwave background radiation and other observations are strong evidence that the Big Bang occurred approximately 14 billion years ago. Physicists are also interested in the ultimate fate of the universe. For many years, it was believed that there were just two possibilities, labeled curves 1 and 2 in Figure 31.17. Curve 1 shows a universe that continues to expand forever, but with the rate of expansion slowing due to the gravitation attraction between stars. On the other hand, the gravitational attraction may be large enough so that the entire universe collapses back to an infinitesimal point; this model, called the "big crunch," is illustrated by curve 2 in Figure 31.17.

These "options" for the ultimate fate of the universe depend on how much matter it contains; if the amount of mass is large enough, the mutual gravitational attraction will lead to the big crunch. This issue has prompted astronomers to do careful measurements of the total mass contained in the universe, and the results show that the universe contains a considerable amount of ***dark matter***, matter that does not emit or absorb enough radiation to be observed directly. While dark matter cannot be "seen" with an ordinary telescope, its presence can still be measured by studying how it exerts a gravitational force on nearby galaxies. In fact, the best current measurements suggest that as much as 80% to 90% of the mass in the universe is dark matter! The nature of dark matter is not yet known. Speculations are that it could be composed of hitherto undiscovered fundamental particles.

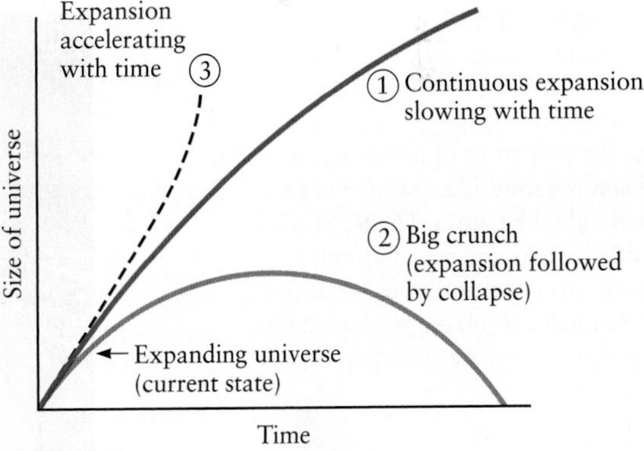

▲ **Figure 31.17** Different possible scenarios for the fate of the universe. Curve 1 shows a universe that expands forever, while curve 2 shows a universe that collapses back to a point (the big crunch). Our universe is now on the far left side of this plot and is expanding. Curve 3 shows a universe that is expanding at an ever-increasing rate ("accelerating"). Recent observations suggest that our universe is following a curve like curve 3.

Another big surprise in recent years has been the discovery that the universe is actually described by curve 3 in Figure 31.17, which shows that the expansion of the universe is *accelerating*. According to the standard Big Bang model, all matter was given an initial velocity by the Big Bang. If there were no forces on a particular galaxy, it would continue forever with a constant velocity and the universe would expand forever at a constant rate. The effects of gravity should eventually cause the relative speed of a galaxy to decrease with time (as in curves 1 and 2 in Fig. 31.17), but the most recent observations indicate that galaxies are instead accelerating.

The explanation of this puzzle is not yet known. Some physicists have suggested that the acceleration is due to a phenomenon called ***dark energy***. Currently, there is no agreed-upon theory of dark energy or even a precise definition of what it is.

The rough idea is that an entirely new form of energy is present in "empty" space and that this energy is imperceptible to most observations ("dark"). The density of this energy is very low and is normally not noticed. In the very large volume of the universe, however, dark energy produces significant effects and is responsible for repelling matter and for the accelerating expansion of the universe. The existence of dark energy and its properties are now being debated, so the picture of fundamental forces and elementary particles given in this chapter may soon need to be revised.

31.8 Physics and Interdisciplinary Science

Physicists are actively studying many areas besides elementary particle physics and astrophysics. In fact, there are so many interesting research topics in physics that there is no way we can even list them all. We'll mention just a few of these topics and give only very brief descriptions.

Nanoscience is the study of objects with sizes in the range of roughly 1 nm to 1000 nm. It spans the regime of atoms and molecules up to objects such as cells, which can be seen with an optical microscope. There are several things that make this "nanoworld" interesting. Physicists have developed ways to manipulate matter atom by atom (Fig. 31.18), thus making things not found in nature, including materials that are both stronger and lighter than steel. The new properties of nanoscale objects are beginning to lead to entirely new kinds of applications. For example, computers that store information using the quantum properties of individual electrons are now being developed; such computers, if successful, could be much faster and more powerful than those we have today.

▶ **Figure 31.18** Image of Fe atoms on the surface of Cu, obtained using a scanning tunneling microscope (STM), one of the new tools of nanoscience. Each pyramid-like "bump" is an individual Fe atom. The STM is able to form images of atomic-scale structure and to move atoms from place to place, producing the atomic arrangement shown here.

Another promising area is ***biophysics***, in which physics principles and methods are applied to problems in biology. This work is giving important new insights into a wide range of problems, including photosynthesis, molecular motors, how the nose detects odors, and how the brain works. Some of these advances involve new techniques from nanoscience. For example, nanoscientists may soon be able to probe and assemble DNA and other biomolecules atom by atom (Fig. 31.19), which could lead to the design of custom drugs, breakthrough medical therapies, new species of plants with higher yields than previously possible, and much more.

Physics in the 21st century thus has much to offer humanity and promises to be a very exciting journey.

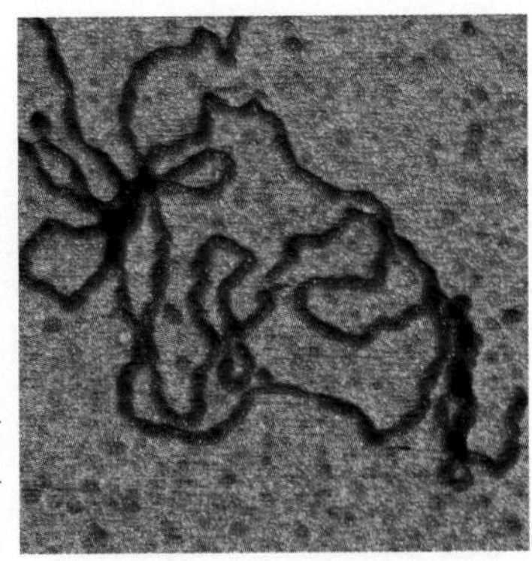

Courtesy of Asylum Research, Santa Barbara, CA.

▲ **Figure 31.19** Image of strand of DNA obtained using an atomic force microscope. The "bumps" on the DNA strand are individual base pairs.

Summary | CHAPTER 31

Key Concepts and Principles

Fundamental particles

There are two families of ***fundamental particles***:

There are six types of ***leptons***:
electron (e^-)
electron neutrino (ν_e)
muon (μ^-)
muon neutrino (ν_μ)
tau (τ^-)
tau neutrino (ν_τ)

All leptons are point particles.

All ***hadrons*** are composed of ***quarks***.
There are six types of quarks:

up (u)	strange (s)
down (d)	top (t)
charm (c)	bottom (b)

Quarks are point particles.
Hadrons composed of three quarks are called ***baryons***. They include protons (denoted by p and composed of two up quarks and one down quark, uud), and neutrons (n, quark composition udd).
Hadrons composed of two quarks are called ***mesons***.

Matter and antimatter

For each fundamental particle, there is a corresponding antiparticle that has the same mass but opposite electric charge. For example, the antiparticle of the electron is the positron. The electron has charge $-e$, whereas the positron has charge $+e$.

(*Continued*)

Fundamental forces

There are four fundamental forces in nature, and it is believed that each is carried or "mediated" by one or more particles.

1. *Electromagnetism* is described by Maxwell's equations. The quantum theory of this force is called ***quantum electrodynamics*** (QED). The electromagnetic force is carried by the photon.
2. The ***strong force*** is an attractive force between quarks and holds quarks together to form baryons (including protons and neutrons) and mesons. The strong force also attracts protons and neutrons and holds them together in the nucleus. The strong force is carried by particles called ***gluons*** and is described by ***quantum chromodynamics*** (QCD).
3. The ***weak force*** acts on all leptons and hadrons. The weak force is much smaller in magnitude than the strong force, but is responsible for processes such as the beta decay of a nucleus. The weak force is carried by the W^+, W^-, and Z particles.
4. ***Gravity*** is the weakest of the fundamental forces. There is no quantum theory of gravity yet, but it is believed that the gravitational force is transmitted by a particle called the ***graviton***. Gravitons have not yet been observed.

The ***standard model*** is a quantum mechanical theory for the behavior of the electromagnetic, strong, and weak forces and the interactions between leptons and hadrons.

Unification of forces

Electromagnetism is an example of a "unified" theory because it reveals electricity and magnetism as two aspects of a single type of force. The quantum theory of electromagnetism (QED) has also been unified with the weak force. These combined forces are called the ***electroweak*** force. Many physicists suspect that there is a single unified theory that includes all the fundamental forces, but this theory has not yet been discovered.

Applications

Measurements of the ***red shift*** of light from distant galaxies show that the universe is now expanding and provides evidence for the ***Big Bang theory*** of the universe.

Questions

SSM = answer in *Student Companion & Problem-Solving Guide*   Ⓧ = life science application

1. In Figure 31.4, an electron and a positron annihilate and produce electromagnetic radiation in the form of two gamma ray photons, each of energy 0.511 MeV. Explain why it is not possible for this reaction to give a single gamma ray photon with energy $2 \times (0.511\text{ MeV}) = 1.022\text{ MeV}$.
2. Explain why there are no doubly charged mesons, that is, mesons with an electric charge of $+2e$ or $-2e$.
3. How many different hadrons are stable outside the nucleus? *Hint*: They are all listed in Table 31.3.
4. Consider the hypothetical reaction in Figure Q31.4 in which an electron (e^-) annihilates a positron (e^+), with the positron initially at rest. Can this reaction result in a single emitted photon? Explain why or why not. *Hint*: Does this process conserve momentum?

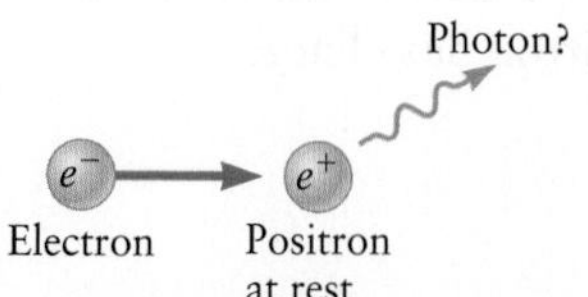

Figure Q31.4

5. The Σ^+ (mass 1189 MeV/c^2) is composed of two up quarks (u) and one strange quark (s). The mass of an up quark is about 4 MeV/c^2, and the mass of a strange quark is about 150 MeV/c^2. Why don't these mass values "add up" to equal the mass of the Σ^+ as one might expect? *Hint*: Does the mass of the Σ^+ come mainly from the masses of these quarks or from the potential energy associated with binding? Explain.
6. Design an experiment that could distinguish a neutron from an antineutron.
7. Are mesons composed of (a) a total of three quarks and antiquarks, (b) three leptons, or (c) a quark and an antiquark?

8. Are baryons composed of (a) three quarks, (b) three leptons, or (c) two quarks?

9. How many quarks and antiquarks are present in an antimeson?

10. Conservation rules rule out certain decays. (a) Which conservation rule explains why a single electron cannot decay into a pair of photons? (b) Which conservation rule explains why a single electron cannot decay into a pair of neutrinos?

11. SSM A particle with nonzero velocity enters a region of space that has a uniform magnetic field directed into the page as depicted in Figure Q31.11. The resulting path (trajectory) followed by the particle will depend on the particle's charge and mass. Sketch qualitatively the trajectory of an electron, a positron, a proton, and a neutron. Assume all have the same initial velocity.

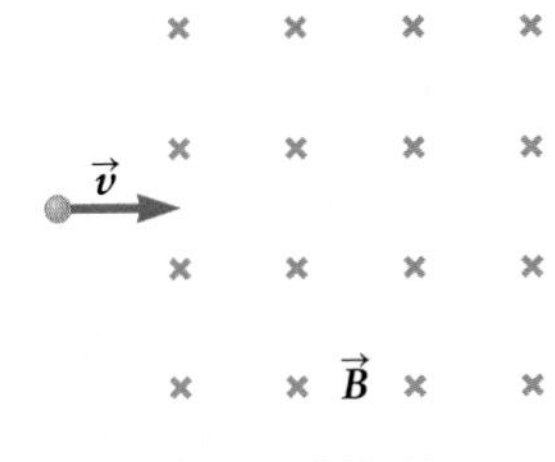

Figure Q31.11

12. At sea level, a significant number of cosmic-ray particles are muons with relativistic velocities. If the muons have a mean lifetime of only 2.2 microseconds, where do they come from and how are they produced?

13. Describe some processes that produce antimatter. *Hint*: Consider radioactive decay processes and interactions involving cosmic rays. How common are anti-electrons (positrons) on the Earth? How common are antimuons on the Earth?

14. It has been suggested that Newton's law of gravity brought about the first unification of physics principles. In what way does his law of gravity unify the motion of heavenly bodies with the motion of earthly objects?

15. Unification of physical laws often leads to new concepts and novel insights. Describe some of the consequences from the unification of the physics of electricity with that of magnetism. How was the concept of light altered?

16. (a) Where did the Big Bang occur? (b) Does the universe have a center? Does the universe have edges? If not, explain how that could be possible.

17. Consider the science of some hypothetical intelligent life-forms inhabiting a planet in a galaxy many millions of light-years away from the Earth. Would they measure the same Hubble constant as we do?

18. Given the current age of the universe and a telescope powerful enough, how far away is the most distant object we could observe? What kind of object is it?

19. SSM Due to the cosmological red shift, the photons that make up the cosmic microwave background are, as the name implies, in the microwave region of the electromagnetic spectrum. In what region of the electromagnetic spectrum were these photons when they were first released?

Problems available in ENHANCED WebAssign

Section 31.1 Cosmic Rays
Problems 1–4

Section 31.2 Matter and Antimatter
Problems 5–9

Section 31.4 Elementary Particle Physics: The Standard Model
Problems 10–16

Section 31.5 The Fundamental Forces of Nature
Problems 17–18

Section 31.7 Astrophysics and the Universe
Problems 19–22

Additional Problems
Problems 23–29

List of Enhanced Problems

Problem number	Solution in *Student Companion & Problem-Solving Guide*	Reasoning & Relationships Problem	Reasoning Tutorial in ENHANCED WebAssign
31.4	✓		
31.9	✓		
31.12	✓		
31.17		✓	✓
31.18	✓		
31.21	✓		
31.23	✓	✓	✓

APPENDIX

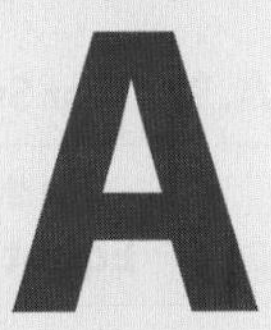

Reference Tables

TABLE A.1 Some Useful Physical Constants and Data

Quantity	Symbol	Value
Universal gravitational constant	G	$6.674 \times 10^{-11}\ \text{N} \cdot \text{m}^2/\text{kg}^2$
Acceleration due to gravity near the Earth's surface	g	$9.8\ \text{m/s}^2$
Speed of light in vacuum	c	$2.997\,924\,58 \times 10^{8}\ \text{m/s}$
Mass of the electron	m_e	$9.109\,382 \times 10^{-31}\ \text{kg}$
Mass of the proton	m_p	$1.672\,621\,6 \times 10^{-27}\ \text{kg}$
Mass of the neutron	m_n	$1.674\,927\,2 \times 10^{-27}\ \text{kg}$
Magnitude of the electron charge	e	$1.602\,176\,5 \times 10^{-19}\ \text{C}$
Permittivity of free space	ε_0	$8.854\,187\,817 \times 10^{-12}\ \text{C}^2/(\text{N} \cdot \text{m}^2)$
Coulomb constant	$k = \dfrac{1}{4\pi\varepsilon_0}$	$8.987\,551\,788 \times 10^{9}\ \text{N} \cdot \text{m}^2/\text{C}^2$
Permeability of free space	μ_0	$4\pi \times 10^{-7}\ \text{T} \cdot \text{m/A}$
Boltzmann's constant	k_B	$1.380\,650 \times 10^{-23}\ \text{J/K}$
Planck's constant	h	$6.626\,069 \times 10^{-34}\ \text{J} \cdot \text{s}$
Avogadro's number	N_A	$6.022\,142 \times 10^{23}\ \text{mol}^{-1}$
Universal gas constant	R	$8.314\,472\ \text{J/(mol} \cdot \text{K)}$
Stefan–Boltzmann constant	σ	$5.670\,40 \times 10^{-8}\ \text{W/(m}^2 \cdot \text{K}^4)$
Atomic mass unit	u	$1.660\,538\,8 \times 10^{-27}\ \text{kg}$

Note: 2006 CODATA recommended values.

TABLE A.2 SI Base Units

Quantity	Unit	Symbol	Quantity	Unit	Symbol
Length	meter	m	Electric current	ampere	A
Mass	kilogram	kg	Temperature	kelvin	K
Time	second	s	Luminous intensity	candela	cd
			Amount of substance	mole	mol

TABLE A.3 Solar System Data

Body	Radius at Equator (km)	Mass (kg)	Average Density (kg/m³)	Orbital Semimajor Axis (km)	Orbital Period (years)	Orbital Eccentricity
Mercury	2,440	3.30×10^{23}	5430	5.79×10^{7}	0.241	0.205
Venus	6,050	4.87×10^{24}	5240	1.08×10^{8}	0.615	0.007
Earth	6,370	5.97×10^{24}	5510	1.50×10^{8}	1.000	0.017
Mars	3,390	6.42×10^{23}	3930	2.28×10^{8}	1.881	0.094
Jupiter	71,500	1.90×10^{27}	1330	7.78×10^{8}	11.86	0.049
Saturn	60,300	5.68×10^{26}	690	1.43×10^{9}	29.46	0.057
Uranus	25,600	8.68×10^{25}	1270	2.87×10^{9}	84.01	0.046
Neptune	24,800	1.02×10^{26}	1640	4.50×10^{9}	164.8	0.011
Pluto	1,200	1.3×10^{22}	1750	5.91×10^{9}	247.7	0.244
Sun	6.95×10^{5}	1.99×10^{30}	1410			
Moon	1,740	7.35×10^{22}	3340	3.84×10^{5}	27.3 days	0.055

Note: From NASA Planetary Fact Sheet (2006).

TABLE A.4 Properties of Selected Isotopes

Atomic Number Z	Element	Chemical Symbol	Chemical Atomic Mass (u)*	Mass Number A	Atomic Mass (u)	Percent Abundance	Half-Life (if radioactive) $T_{1/2}$	Primary Decay Modes†
—	(Proton)	p		1	1.007276			
0	(Neutron)	n		1	1.008 665		10.4 min	β^-
1	Hydrogen	H	1.007 94	1	1.007 825	99.988 5		
	Deuterium	D		2	2.014 102	0.011 5		
	Tritium	T		3	3.016 049		12.33 years	β^-
2	Helium	He	4.002 602	3	3.016 029	0.000 137		
				4	4.002 603	99.999 863		
3	Lithium	Li	6.941	6	6.015 122	7.5		
				7	7.016 004	92.5		
4	Beryllium	Be	9.012 182	7	7.016 929		53.1 days	EC
				9	9.012 182	100		
5	Boron	B	10.811	10	10.012 937	19.9		
				11	11.009 306	80.1		
6	Carbon	C	12.010 7	10	10.016 853		19.3 s	EC, β^+
				11	11.011 434		20.4 min	EC, β^+
				12	12.000 000	98.93		
				13	13.003 355	1.07		
				14	14.003 242		5 730 years	β^-
7	Nitrogen	N	14.006 7	13	13.005 739		9.96 min	EC, β^+, β^-
				14	14.003 074	99.632		
				15	15.000 109	0.368		
8	Oxygen	O	15.999 4	15	15.003 065		122 s	EC, β^+
				16	15.994 915	99.757		
				18	17.999 160	0.205		

*Atomic mass values given are averaged over isotopes in the percentages in which they exist in nature.

†EC denotes decay by electron capture, SF denotes spontaneous fission.

Atomic Number Z	Element	Chemical Symbol	Chemical Atomic Mass (u)*	Mass Number A	Atomic Mass (u)	Percent Abundance	Half-Life (if radioactive) $T_{1/2}$	Primary Decay Modes†
9	Fluorine	F	18.998 403 2	19	18.998 403	100		
10	Neon	Ne	20.179 7	20	19.992 440	90.48		
				22	21.991 385	9.25		
11	Sodium	Na	22.989 77	22	21.994 437		2.60 years	EC, β^+
				23	22.989 770	100		
				24	23.990 963		14.96 h	β^-
12	Magnesium	Mg	24.305 0	24	23.985 042	78.99		
				25	24.985 837	10.00		
				26	25.982 593	11.01		
13	Aluminum	Al	26.981 538	27	26.981 539	100		
14	Silicon	Si	28.085 5	28	27.976 926	92.229 7		
15	Phosphorus	P	30.973 761	31	30.973 762	100		
				32	31.973 907		14.26 days	β^-
16	Sulfur	S	32.066	32	31.972 071	94.93		
				35	34.969 032		87.3 days	β^-
17	Chlorine	Cl	35.452 7	35	34.968 853	75.78		
				37	36.965 903	24.22		
18	Argon	Ar	39.948	40	39.962 383	99.600 3		
19	Potassium	K	39.098 3	39	38.963 707	93.258 1		
				40	39.963 999	0.011 7	1.28×10^9 years	EC, β^+, β^-
20	Calcium	Ca	40.078	40	39.962 591	96.941		
21	Scandium	Sc	44.955 910	45	44.955 910	100		
22	Titanium	Ti	47.867	48	47.947 947	73.72		
23	Vanadium	V	50.941 5	51	50.943 964	99.750		
24	Chromium	Cr	51.996 1	52	51.940 512	83.789		
25	Manganese	Mn	54.938 049	55	54.938 050	100		
26	Iron	Fe	55.845	56	55.934 942	91.754		
27	Cobalt	Co	58.933 200	59	58.933 200	100		
				60	59.933 822		5.27 years	β^-
28	Nickel	Ni	58.693 4	58	57.935 348	68.076 9		
				60	59.930 790	26.223 1		
29	Copper	Cu	63.546	63	62.929 601	69.17		
				65	64.927 794	30.83		
30	Zinc	Zn	65.39	64	63.929 147	48.63		
				66	65.926 037	27.90		
				68	67.924 848	18.75		
31	Gallium	Ga	69.723	69	68.925 581	60.108		
				71	70.924 705	39.892		
32	Germanium	Ge	72.61	70	69.924 250	20.84		
				72	71.922 076	27.54		
				74	73.921 178	36.28		
33	Arsenic	As	74.921 60	75	74.921 596	100		

*Atomic mass values given are averaged over isotopes in the percentages in which they exist in nature.

†EC denotes decay by electron capture, SF denotes spontaneous fission.

Atomic Number Z	Element	Chemical Symbol	Chemical Atomic Mass (u)*	Mass Number A	Atomic Mass (u)	Percent Abundance	Half-Life (if radio-active) $T_{1/2}$	Primary Decay Modes†
34	Selenium	Se	78.96	78	77.917 310	23.77		
				80	79.916 522	49.61		
35	Bromine	Br	79.904	79	78.918 338	50.69		
				81	80.916 291	49.31		
36	Krypton	Kr	83.80	82	81.913 485	11.58		
				83	82.914 136	11.49		
				84	83.911 507	57.00		
				86	85.910 610	17.30		
37	Rubidium	Rb	85.467 8	85	84.911 789	72.17		
				87	86.909 184	27.83	4.75×10^{10} years	β^-
38	Strontium	Sr	87.62	86	85.909 262	9.86		
				88	87.905 614	82.58		
				90	89.907 738		158 s	β^-
39	Yttrium	Y	88.905 85	89	88.905 848	100		
40	Zirconium	Zr	91.224	90	89.904 704	51.45		
				91	90.905 645	11.22		
				92	91.905 040	17.15		
				94	93.906 316	17.38		
41	Niobium	Nb	92.906 38	93	92.906 378	100		
42	Molybdenum	Mo	95.94	92	91.906 810	14.84		
				95	94.905 842	15.92		
				96	95.904 679	16.68		
				98	97.905 408	24.13		
43	Technetium	Tc		98	97.907 216		4.2×10^{6} years	β^-
				99	98.906 255		2.1×10^{5} years	β^-
44	Ruthenium	Ru	101.07	99	98.905 939	12.76		
				100	99.904 220	12.60		
				101	100.905 582	17.06		
				102	101.904 350	31.55		
				104	103.905 430	18.62		
45	Rhodium	Rh	102.905 50	103	102.905 504	100		
46	Palladium	Pd	106.42	104	103.904 035	11.14		
				105	104.905 084	22.33		
				106	105.903 483	27.33		
				108	107.903 894	26.46		
				110	109.905 152	11.72		
47	Silver	Ag	107.868 2	107	106.905 093	51.839		
				109	108.904 756	48.161		
48	Cadmium	Cd	112.411	110	109.903 006	12.49		
				111	110.904 182	12.80		
				112	111.902 757	24.13		
				113	112.904 401	12.22	7.7×10^{15} years	β^-
				114	113.903 358	28.73		

*Atomic mass values given are averaged over isotopes in the percentages in which they exist in nature.

†EC denotes decay by electron capture, SF denotes spontaneous fission.

Atomic Number Z	Element	Chemical Symbol	Chemical Atomic Mass (u)*	Mass Number A	Atomic Mass (u)	Percent Abundance	Half-Life (if radio-active) $T_{1/2}$	Primary Decay Modes†
49	Indium	In	114.818	115	114.903 878	95.71	4.4×10^{14} years	β^-
50	Tin	Sn	118.710	116	115.901 744	14.54		
				118	117.901 606	24.22		
				120	119.902 197	32.58		
51	Antimony	Sb	121.760	121	120.903 818	57.21		
				123	122.904 216	42.79		
52	Tellurium	Te	127.60	126	125.903 306	18.84		
				128	127.904 461	31.74	2.2×10^{24} years	β^-
				130	129.906 223	34.08	7.9×10^{20} years	β^-
53	Iodine	I	126.904 47	127	126.904 468	100		
				129	128.904 988		1.6×10^{7} years	β^-
54	Xenon	Xe	131.29	129	128.904 780	26.44		
				131	130.905 082	21.18		
				132	131.904 145	26.89		
				134	133.905 394	10.44		
				136	135.907 220	8.87	$\geq 2.36 \times 10^{21}$ years	β^-
55	Cesium	Cs	132.905 45	133	132.905 447	100		
56	Barium	Ba	137.327	137	136.905 821	11.232		
				138	137.905 241	71.698		
57	Lanthanum	La	138.905 5	139	138.906 349	99.910		
58	Cerium	Ce	140.116	140	139.905 434	88.450		
				142	141.909 240	11.114	$> 2.6 \times 10^{17}$ years	β^-
59	Praseodymium	Pr	140.907 65	141	140.907 648	100		
60	Neodymium	Nd	144.24	142	141.907 719	27.2		
				144	143.910 083	23.8	2.3×10^{15} years	α
				146	145.913 112	17.2		
61	Promethium	Pm		145	144.912 744		17.7 years	EC, α
62	Samarium	Sm	150.36	147	146.914 893	14.99	1.06×10^{11} years	α
				149	148.917 180	13.82		
				152	151.919 728	26.75		
				154	153.922 205	22.75		
63	Europium	Eu	151.964	151	150.919 846	47.81		
				153	152.921 226	52.19		
64	Gadolinium	Gd	157.25	156	155.922 120	20.47		
				158	157.924 100	24.84		
				160	159.927 051	21.86		
65	Terbium	Tb	158.925 34	159	158.925 343	100		
66	Dysprosium	Dy	162.50	162	161.926 796	25.51		
				163	162.928 728	24.90		
				164	163.929 171	28.18		
67	Holmium	Ho	164.930 32	165	164.930 320	100		

*Atomic mass values given are averaged over isotopes in the percentages in which they exist in nature.

†EC denotes decay by electron capture, SF denotes spontaneous fission.

Atomic Number Z	Element	Chemical Symbol	Chemical Atomic Mass (u)*	Mass Number A	Atomic Mass (u)	Percent Abundance	Half-Life (if radio-active) $T_{1/2}$	Primary Decay Modes†
68	Erbium	Er	167.6	166	165.930 290	33.61		
				167	166.932 045	22.93		
				168	167.932 368	26.78		
69	Thulium	Tm	168.934 21	169	168.934 211	100		
70	Ytterbium	Yb	173.04	172	171.936 378	21.83		
				173	172.938 207	16.13		
				174	173.938 858	31.83		
71	Lutecium	Lu	174.967	175	174.940 768	97.41		
72	Hafnium	Hf	178.49	177	176.943 220	18.60		
				178	177.943 698	27.28		
				179	178.945 815	13.62		
				180	179.946 549	35.08		
73	Tantalum	Ta	180.947 9	181	180.947 996	99.988		
74	Tungsten	W	183.84	182	181.948 206	26.50		
				183	182.950 224	14.31		
				184	183.950 933	30.64	$> 3 \times 10^{19}$ years	α
				186	185.954 362	28.43		
75	Rhenium	Re	186.207	185	184.952 956	37.40		
				187	186.955 751	62.60	4.1×10^{10} years	β^-
76	Osmium	Os	190.23	188	187.955 836	13.24		
				189	188.958 145	16.15		
				190	189.958 445	26.26		
				192	191.961 479	40.78		
77	Iridium	Ir	192.217	191	190.960 591	37.3		
				193	192.962 924	62.7		
78	Platinum	Pt	195.078	194	193.962 664	32.967		
				195	194.964 774	33.832		
				196	195.964 935	25.242		
79	Gold	Au	196.966 55	197	196.966 552	100		
80	Mercury	Hg	200.59	199	198.968 262	16.87		
				200	199.968 309	23.10		
				201	200.970 285	13.18		
				202	201.970 626	29.86		
81	Thallium	Tl	204.383 3	203	202.972 329	29.524		
				205	204.974 412	70.476		
				208	207.982 005		3.053 min	β^-
				210	209.990 066		1.30 min	β^-

*Atomic mass values given are averaged over isotopes in the percentages in which they exist in nature.

†EC denotes decay by electron capture, SF denotes spontaneous fission.

Atomic Number Z	Element	Chemical Symbol	Chemical Atomic Mass (u)*	Mass Number A	Atomic Mass (u)	Percent Abundance	Half-Life (if radio-active) $T_{1/2}$	Primary Decay Modes†
82	Lead	Pb	207.2	204	203.973 029	1.4	$\geq 1.4 \times 10^{17}$ years	α
				206	205.974 449	24.1		
				207	206.975 881	22.1		
				208	207.976 636	52.4		
				210	209.984 173		22.2 years	α, β^-
				211	210.988 732		36.1 min	β^-
				212	211.991 888		10.64 h	β^-
				214	213.999 798		26.8 min	β^-
83	Bismuth	Bi	208.980 38	209	208.980 383	100		
				211	210.987 258		2.14 min	α, β^-
84	Polonium	Po		210	209.982 857		138.38 days	α
				214	213.995 186		164 μs	α
85	Astatine	At		218	218.008 682		1.5 s	α, β^-
86	Radon	Rn		222	222.017 570		3.823 days	α, β^-
87	Francium	Fr		223	223.019 731		22 min	α, β^-
88	Radium	Ra		226	226.025 403		1 600 years	α
				228	228.031 064		5.75 years	β^-
89	Actinium	Ac		227	227.027 747		21.77 years	α, β^-
90	Thorium	Th	232.038 1					
				228	228.028 731		1.912 years	α
				232	232.038 050	100	1.40×10^{10} years	α
91	Protactinium	Pa	231.035 88	231	231.035 879		3.28×10^{4} years	α
92	Uranium	U	238.028 9	232	232.037 146		69 years	α
				233	233.039 628		1.59×10^{5} years	α
				235	235.043 923	0.720 0	7.04×10^{8} years	α, SF
				236	236.045 562		2.34×10^{7} years	α, SF
				238	238.050 783	99.274 5	4.47×10^{9} years	α
93	Neptunium	Np		237	237.048 167		2.14×10^{6} years	α
94	Plutonium	Pu		239	239.052 156		2.412×10^{4} years	α
				242	242.058 737		3.75×10^{5} years	α
				244	244.064 198		8.1×10^{7} years	α, SF

*Atomic mass values given are averaged over isotopes in the percentages in which they exist in nature.

†EC denotes decay by electron capture, SF denotes spontaneous fission.

Sources: Chemical atomic masses are from T. B. Coplen, "Atomic weights of the elements 1999," a technical report to the International Union of Pure and Applied Chemistry and published in *Pure and Applied Chemistry* 73(4):667–683 (2001). Atomic masses of the isotopes are from G. Audi and A. H. Wapstra, "The 1995 update to the atomic mass evaluation," *Nuclear Physics* A595(4):409–480 (December 25, 1995). Percent abundance values are from K. J. R. Rosman and P. D. P. Taylor, "Isotopic compositions of the elements 1999," a technical report to the International Union of Pure and Applied Chemistry and published in *Pure and Applied Chemistry* 70(1):217–236 (1998). Data on decay modes and half-lives are from Lawrence Berkeley Laboratory Isotopes Project Listing and the National Nuclear Data Center, Brookhaven National Laboratory.

APPENDIX

B Mathematical Review

B.1 Mathematical Symbols

The following symbols are used to express mathematical relationships.

Symbol	Meaning	Example	Explanation
$+$	Addition	$a + b = c$	a plus b equals c
$-$	Subtraction	$a - b = c$	a minus b equals c
$\times$ or $\cdot$	Multiplication	$a \times b = c$ $a \cdot b = c$	a times b equals c
$/$ or $\div$	Division	$a/b = c$ $a \div b = c$	a divided by b equals c
$>$	Greater than	$a > b$	a is greater than b
$<$	Less than	$a < b$	a is less than b
$\geq$	Greater than or equal to	$a \geq b$	a is greater than or equal to b
$\leq$	Less than or equal to	$a \leq b$	a is less than or equal to b
$\approx$	Approximately equal to	$a \approx b$	a is approximately equal to b
$\lim_{x \to 0}$	Limit as x approaches 0	$y = \lim_{x \to 0} f$	y equals the value of f as the variable x approaches zero

B.2 Algebra and Working with Equations

The basic rules of algebra are reviewed in Section 1.6. When solving an equation to find the value of a single variable, you should rearrange the equation to put the *unknown* variable (the variable you wish to solve for) on one side of the equation and the *known* quantities on the other side. For example, suppose the variable a in the equation

$$4a + 7 = T - 9$$

is the unknown. The goal is to isolate a on one side of the equal sign (usually the left). In this example, we can subtract 7 from both sides to get

$$4a = T - 16$$

We next divide both sides by the constant factor 4 to get the solution

$$a = \frac{T}{4} - 4$$

When dealing with two equations containing two unknowns, the usual approach is to use one equation to eliminate one of the unknowns from the other equation. For example, suppose a and T are unknown in the equations

$$3a = T - 9 \qquad \text{(B2.1)}$$

$$2a = -T + 24 \qquad \text{(B2.2)}$$

We can rearrange Equation B2.2 to get

$$T = -2a + 24$$

We then substitute this expression for T in Equation B2.1 to find

$$3a = T - 9 = (-2a + 24) - 9 = -2a + 15 \quad \text{(B2.3)}$$

We can now solve for a:

$$5a = 15$$
$$a = 3$$

This value can then be used in Equation B2.1 to find the value of T; we get $T = 18$.

Another approach when working with two equations and two unknowns is to simply add the two equations. For example, if we add Equations B2.1 and B2.2, we will eliminate T:

$$\begin{aligned} 3a &= T - 9 \\ 2a &= -T + 24 \\ \hline 5a &= -9 + 24 \end{aligned}$$

which is another way to get Equation B2.3.

The equations considered in Equations B2.1 through B2.3 all involve the unknowns (e.g., a and T) raised to the first power. In a ***quadratic equation***, the unknown quantity is raised to the second power. If x is an unknown, an equation of this type is

$$ax^2 + bx + c = 0$$

The solution is given by the ***quadratic formula***,

$$x = \frac{-b \pm \sqrt{b^2 - 4ac}}{2a} \quad \text{(B2.4)}$$

There are two solutions as indicated by the appearance of the $\pm$ sign.

B.3 Scientific Notation

Scientific notation is a convenient way to express extremely large or extremely small numbers. For example, the number 640,000 is written as 6.4×10^5 in scientific notation. To express a number in scientific notation, move the decimal point in the original number to obtain a new number between 1 and 10. Count the number of places the decimal point has been moved; this number will become the exponent of 10 in scientific notation. If you started with a number greater than 10 (such as 150,000,000,000), the exponent of 10 is positive (1.5×10^{11}). If you started with a number less than 1 (such as 0.000055), the exponent is negative (5.5×10^{-5}).

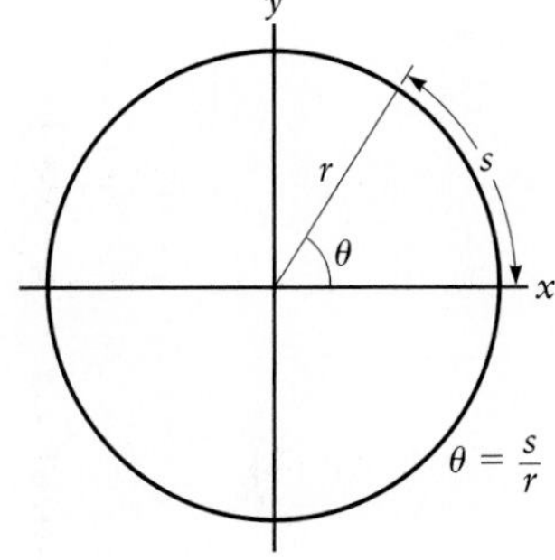

▲ **Figure B.1** The angle θ measured in radians equals the ratio of the length s measured along the circular arc divided by the radius r of the circle.

B.4 Geometry and Trigonometry

One way to define and measure angles is shown in Figure B.1. The angle θ equals the ratio of the arc length s measured along the circle divided by the radius r of the circle:

$$\theta = \frac{s}{r} \quad \text{(B4.1)}$$

This relation gives the value of θ in ***radians***. To convert to degrees, we use the fact that one complete "trip" around the circle in Figure B.1 corresponds to both 2π radians and to 360°. We thus have

$$\theta \text{ (in radians)} \times \frac{360°}{2\pi \text{ radians}} = \theta \text{ (in degrees)} \quad \text{(B4.2)}$$

Figure B.2 shows a right triangle with sides x, y, and r, where r is the hypotenuse. According to the ***Pythagorean theorem***,

$$x^2 + y^2 = r^2 \quad \text{(B4.3)}$$

The trigonometric functions ***sine***, ***cosine***, and ***tangent*** are defined as

$$\sin\theta = y/r \quad \text{(B4.4)}$$
$$\cos\theta = x/r \quad \text{(B4.5)}$$
$$\tan\theta = y/x \quad \text{(B4.6)}$$

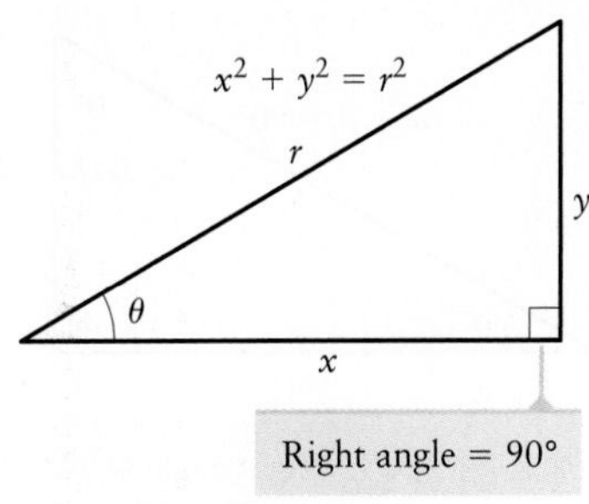

▲ **Figure B.2** The Pythagorean theorem is a relation between the lengths of the sides of a right triangle.

The Pythagorean theorem (Eq. B4.3) implies that

$$\sin^2\theta + \cos^2\theta = 1 \tag{B4.7}$$

for any value of the angle θ. Equations B4.3 through B4.6 also lead to a number of other trigonometric relations (called ***identities***), some of which are the following:

$$\sin(-\theta) = -\sin\theta \qquad \cos(-\theta) = \cos\theta \qquad \tan(-\theta) = -\tan\theta$$
$$\sin(\theta + 90°) = \sin(\theta + \pi/2) = \cos\theta$$
$$\cos(\theta + 90°) = \cos(\theta + \pi/2) = -\sin\theta$$
$$\sin(\alpha \pm \beta) = \sin\alpha\cos\beta \pm \cos\alpha\sin\beta$$
$$\cos(\alpha \pm \beta) = \cos\alpha\cos\beta \mp \sin\alpha\sin\beta \tag{B4.8}$$

The relations in Equation B4.8 hold for any values of the angles θ, α, and β. If $\alpha = \beta$, the relation for $\sin(\alpha + \beta)$ becomes

$$\sin(\alpha + \alpha) = \sin(2\alpha) = \sin\alpha\cos\alpha + \cos\alpha\sin\alpha$$

We can also replace the angle α with θ; rearranging then leads to

$$\sin\theta\cos\theta = \tfrac{1}{2}\sin(2\theta) \tag{B4.9}$$

Equations B4.4 through B4.6 tell how to calculate the sine, cosine, and tangent of an angle from the lengths of the sides of the associated right triangle. We can also use this approach to find the value of the angle. For example, if we know the values of y and r, we can solve for the angle as

$$\sin\theta = y/r$$
$$\theta = \sin^{-1}(y/r) \tag{B4.10}$$

The function $\sin^{-1}$ is the ***inverse*** sine function, also known as the ***arcsine*** ($=$ arcsin $= \sin^{-1}$). In words, Equation B4.10 says that θ equals the angle whose sine is y/r. Scientific calculators all contain this function (as a single "button"). If the values of y and r are known, you must only take the ratio y/r and then compute the arcsine of this value to find θ in Equation B4.10. Likewise, there are also functions for the inverse cosine ($\cos^{-1}$ = arccos) and inverse tangent ($\tan^{-1}$ = arctan):

$$\theta = \cos^{-1}(x/r) \quad \text{and} \quad \theta = \tan^{-1}(y/x) \tag{B4.11}$$

We often need to deal with triangles like the one in Figure B.3, which shows a right triangle with interior angles α and β. The sum of the three interior angles of a triangle is always 180°. For the triangle in Figure B.3, one of the angles is 90° (it is a right triangle); hence,

$$\alpha + \beta = 90° \tag{B4.12}$$

Two angles whose sum is 90° are called complementary angles. From the definitions of the sine and cosine functions, for any pair of complementary angles α and β we have

$$\sin\alpha = \cos\beta$$
$$\cos\alpha = \sin\beta$$

Two approximate relations involving trigonometric functions are also useful. When the angle θ is small,

$$\sin\theta \approx \theta \quad \text{and} \quad \tan\theta \approx \theta \tag{B4.13}$$

which are good approximations (accurate to a few percent or better) when θ is smaller than about 30°. Notice that the relations in Equation B4.13 apply only when the angle is measured in radians.

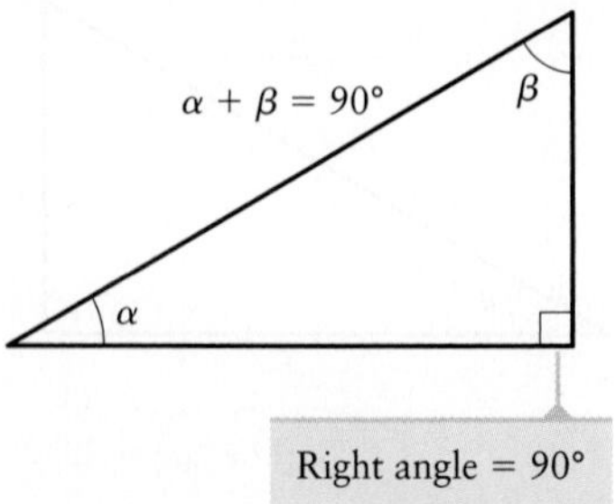

▲ **Figure B.3** The sum of the three interior angles of a right triangle is 180°. Hence, $\alpha + \beta = 90°$.

B.5 Vectors

A vector quantity has both a magnitude and a direction. There are two ways to do calculations with vectors; one is a graphical approach, which is useful for approximate or qualitative calculations, and the other involves the components of the vectors.

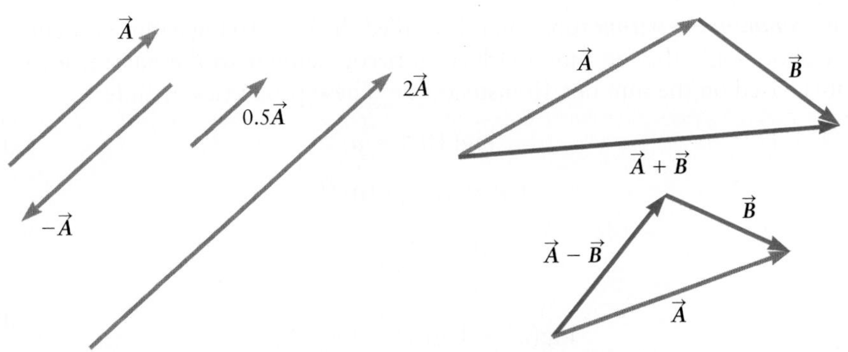

◀ **Figure B.4** Examples showing the effect of multiplying a vector by a scalar, adding two vectors, and subtracting one vector from another.

The graphical approach is illustrated in Figure B.4, which shows two vectors $\vec{A}$ and $\vec{B}$ along with their sum $(\vec{A} + \vec{B})$ and difference $(\vec{A} - \vec{B})$. Multiplying a vector by a scalar (e.g., 0.5 or 2) changes the length but does not change the direction. Multiplying by a negative number (e.g., -1) reverses the direction. The vectors $-\vec{A}$, $0.5\vec{A}$, and $2\vec{A}$ are also shown in Figure B.4.

The components of a vector can be obtained using trigonometry. The length of a vector $\vec{A}$, also called its ***magnitude***, is denoted by either A or $|\vec{A}|$. If this vector makes an angle θ with the x axis, the ***components*** are

$$A_x = A \cos\theta = |\vec{A}| \cos\theta \text{ and } A_y = A \sin\theta = |\vec{A}| \sin\theta \quad \text{(B5.1)}$$

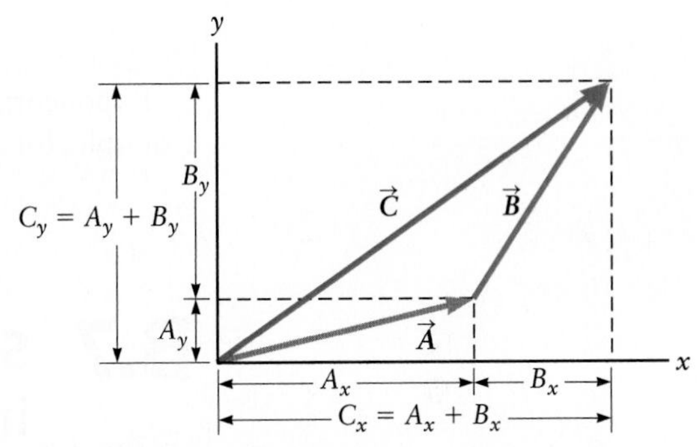

▲ **Figure B.5** To compute the sum of two vectors, we add their components. Here we add the x components of $\vec{A}$ and $\vec{B}$ to get the x component of $\vec{C}$. We follow the same procedure to find the y component of $\vec{C}$.

The sum of two vectors can be found by adding components, as illustrated in Figure B.5. If

$$\vec{C} = \vec{A} + \vec{B} \quad \text{(B5.2)}$$

in terms of the components of these vectors we then have

$$C_x = A_x + B_x \quad \text{and} \quad C_y = A_y + B_y \quad \text{(B5.3)}$$

B.6 Exponential Functions and Logarithms

Exponential functions and logarithms play a role in many areas of physics, including the behavior of fluids (Chapter 10), sound (Chapter 13), electric circuits (Chapters 19 and 22), and radioactive decay (Chapter 30).

The number $e = 2.71828\ldots$ plays an important role. Useful numerical values involving the number e are $e^{-1} \approx 0.37$ and $1 - e^{-1} \approx 0.73$.

There are two versions of the logarithm function. The ***natural logarithm*** function is denoted by the symbol "ln" and is related to the exponential function by

$$\ln(e^a) = a \quad \text{(B6.1)}$$

where a is a number. Other useful properties of these functions are

$$\begin{aligned} e^a \times e^b &= e^{a+b} \\ \frac{e^a}{e^b} &= e^{a-b} \\ \ln(a) + \ln(b) &= \ln(ab) \\ \ln(a) - \ln(b) &= \ln(a/b) \\ \ln(a^b) &= b \ln a \end{aligned} \quad \text{(B6.2)}$$

The *common logarithm* function (also called the base 10 logarithm) is denoted by the symbol "log"; the log function has properties similar to the natural logarithm, but it is based on the number 10 instead of e. These properties include

$$\log(10^a) = a \tag{B6.3}$$

$$10^a \times 10^b = 10^{a+b}$$

$$\frac{10^a}{10^b} = 10^{a-b}$$

$$\log(a) + \log(b) = \log(ab) \tag{B6.4}$$

$$\log(a) - \log(b) = \log(a/b)$$

$$\log(a^b) = b \log a$$

Exponential functions may also involve base numbers besides e and 10. For example, for any number y, one has

$$y^a \times y^b = y^{a+b} \quad \text{and} \quad \frac{y^a}{y^b} = y^{a-b} \tag{B6.5}$$

B.7 Some Relations Useful in Special Relativity

In special relativity, factors such as v/c and v^2/c^2, where c is the speed of light and v is the speed of an object, appear often. The speed of light is very large, so the ratio v/c is usually very small. When that is the case, certain algebraic expressions can be simplified. For example, when A is a very small number, the approximation

$$\frac{1}{1+A} \approx 1 - A \tag{B7.1}$$

is very accurate. If we replace A by v/c, we find

$$\frac{1}{1+v/c} \approx 1 - v/c \tag{B7.2}$$

and in a similar way, we get

$$\frac{1}{1+(v/c)^2} \approx 1 - (v/c)^2 \tag{B7.3}$$

Another expression that arises in special relativity, especially in problems involving time dilation and length contraction, is $\sqrt{1-(v/c)^2}$. When v/c is very small, to a good approximation we have

$$\sqrt{1 - v^2/c^2} \approx 1 - \frac{v^2}{2c^2} \tag{B7.4}$$

and

$$\frac{1}{\sqrt{1 - v^2/c^2}} \approx 1 + \frac{v^2}{2c^2} \tag{B7.5}$$

The accuracy of these approximations depends on the value of v/c. If $v/c = 0.1$, the approximation in Equation B7.2 is good to about 1%, whereas Equations B7.4 and B7.5 are accurate to about 0.005%. The accuracy improves for smaller values of v/c.

Answers to Concept Checks

Chapter 1

1. Significant Figures

Quantity	Length or Distance in Meters (m)	Number of Significant Figures
Diameter of a proton	1×10^{-15}	1
Diameter of a red blood cell	8×10^{-6}	1
Diameter of a human hair	5.5×10^{-5}	2
Thickness of a piece of paper	6.4×10^{-5}	2
Diameter of a compact disc	0.12	2
Height of the author	1.80	3
Height of the Empire State Building	443.2	4
Distance from New York City to Chicago	1,268,000	4
Circumference of the Earth	4.00×10^{7}	3
Distance from Earth to the Sun	1.5×10^{11}	2

2. Using Prefixes and Powers of 10

Quantity	Length or Distance in Meters (m)	Same Length or Distance Using Prefix
Diameter of a proton	1×10^{-15}	1 femtometer = 1 fm
Diameter of a red blood cell	8×10^{-6}	8 micrometers = 8 μm
Diameter of a human hair	5.5×10^{-5}	55 micrometers = 55 μm
Thickness of a sheet of paper	6.4×10^{-5}	64 micrometers = 64 μm
Diameter of a compact disc	0.12	1.2 decimeters = 1.2 dm = 12 centimeters = 12 cm
Height of the author	1.80	1.80 meters = 1.80 m
Height of the Empire State Building	443.2	443.2 meters = 443.2 m
Distance from New York City to Chicago	1,268,000	1.268 megameters = 1.268 Mm
Circumference of the Earth	4.00×10^{7}	40.0 megameters = 40.0 Mm
Distance from Earth to the Sun	1.5×10^{11}	150 gigameters = 150 Gm

3. (a), (c), and (d)

Chapter 2

1. (a) is inconsistent, (b) is inconsistent, (c) may be consistent.
2. Yes, at $t \approx 1.6$ s in Figure 2.9.
3. (a) graph 3; (b) graph 2; (c) graph 1.
4. (b) and (d)
5. (b)
6. (a) and (c) are not action–reaction pairs.

Chapter 3

1. In Figure 3.7, $\vec{F}_{grav}$ and $\vec{N}$ are *not* an action–reaction pair. The reaction force to $\vec{N}$ is the contact force of the person on the floor.
2. The driver can drive with an acceleration smaller than $a_{crate, max}$ for a very long time and thereby reach a very large speed, and the crate will not slip.
3. (a) Opposite; the velocity is to the right, while the acceleration is directed to the left.
   (b) Parallel; $\vec{v}$ and $\vec{a}$ are both directed to the right.
   (c) Parallel; both $\vec{v}$ and $\vec{a}$ are directed downward.
4. Only statement (a) is true.
5. (b)
6. The bumper experiences a force equal to three times the weight of the car!
7. (d)

Chapter 4

1. (b)
2. (b)
3. (a) The acceleration is *never* zero. (b) $v_y = 0$ at the instant the ball is at the highest point on its trajectory. (c) The force is *never* zero. (d) The speed is *never* zero.
4. The acceleration of m_1 is *not* the same in the two cases.
5. (b)
6. Suppose a spacecraft is sitting at rest in deep space. Now consider the spacecraft as viewed from a noninertial reference frame; that is, it is viewed by an observer who has an acceleration $+a$ along a particular direction. As viewed by this observer, the spacecraft will appear to have an acceleration $-a$. Since there are no forces acting on the spacecraft, Newton's first law is violated in this noninertial reference frame.

Chapter 5

1. Statements (a), (b), and (c) are all correct.
2. (c)
3. (c)
4. The observers in (a) and (b) are moving at constant velocities, so their accelerations are zero. Hence, they are both inertial frames. In (c) and (d), the frames are accelerating, so they are noninertial frames.
5. (d)
6. The answer is shown in the accompanying figure.

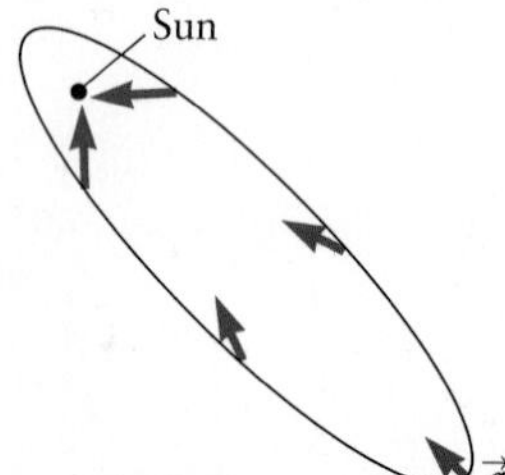

7. (b)

Chapter 6

1. (1) corresponds to case (a); (2) corresponds to case (c); (3) corresponds to case (b).
2. (c)
3. As the roller coaster in Figure 6.13 moves higher on the track, approaching point B, its velocity will decrease. If it is just barely able to make it to point B, it will do so with an extremely small velocity. In such a case, the kinetic energy at point B will be zero. Applying the conservation of energy condition then gives

$$PE_A + KE_A = PE_B + KE_B = PE_B$$
$$KE_A = \tfrac{1}{2}mv_A^2 = PE_B - PE_A = mgh_B - mgh_A$$

   which can then be solved for h_B.
4. (1) and (2)
5. No reasonable bicycling hill would require 1000 Calories to climb it.
6. (3)

Chapter 7

1. (a)
2. (b)
3. (b)
4. (b)
5. (f)
6. (a)
7. (c)
8. (c)
9. (a)

Chapter 8

1. (h)
2. (c)
3. (b)
4. The torques are (a) negative, (b) positive, (c) positive, and (d) positive.
5. (a)
6. (b)
7. (d)
8. (a)

Chapter 9

1. (b)
2. (d)
3. (c)
4. (c)
5. The larger ball has a greater angular momentum.
6. In both cases, the system is able to rotate even though the angular momentum is zero and there are no external torques.

Chapter 10

1. (a)
2. (c)
3. (a)
4. (b)
5. (b)
6. (a)
7. (a)

Chapter 11

1. The acceleration of an object undergoing simple harmonic motion is nonzero unless the object is at the equilibrium position. There are forces on a bungee jumper due to the bungee cord and gravity. In general these forces do not cancel, and the acceleration of the jumper is also nonzero.
2. (d)
3. (a) Point B (b) Point B
4. Kinetic energy
5. (c)
6. Air drag and friction between the mass and the table
7. (c)

Chapter 12

1. (c)
2. (a) Wave 4 (b) Wave 2 (c) Wave 3
3. (b)
4. The frequency and amplitude depend on the way the wave is generated. The wavelength depends on both the medium and the way the wave is generated because it depends on both the frequency and the speed.
5. Five nodes and four antinodes.

Chapter 13

1. (b)
2. (c)
3. (d)
4. (b)
5. (a)
6. (a)
7. (b)

Chapter 14

1. The body is composed mainly of water, so its specific heat is close to that of water.
2. (c)
3. The coefficient of thermal expansion α determines how the length of an object changes when it is heated. It applies to the length, width, and height, and if all three change a large amount, the corresponding changes in the volume and hence the value of β are also large.
4. The frequency will go down.
5. Figure 14.25A shows how heat conduction is due to the transfer of energy between vibrating atoms. Solids have a higher density of atoms than gases, so (if all else is similar) solids can conduct a greater amount of heat.
6. (d)
7. (c)

Chapter 15

1. (d)
2. (a)
3. (a)
4. (b)
5. (c)
6. (c)
7. (c)

Chapter 16

1. (c)
2. (c)
3. (a)
4. Both. Because bear fur is very thick, very little heat flows between the bear and its environment, so this situation is approximately adiabatic. During hibernation, the bear produces very little heat because its metabolic rate is very low. A balance between the heat it creates and the heat flow out through its fur keeps the bear approximately isothermal.
5. (a) W is positive. (b) W is negative. (c) $W = 0$. (d) W is positive. (e) W is negative.
6. (a) Process 3 could be an isothermal expansion. (b) Process 4 could be an adiabatic compression. (c) Process 1 is an isobaric expansion. (d) Process 2 is an isobaric compression.
7. W is positive during steps $1 \rightarrow 2$ and $2 \rightarrow 3$. W is negative during steps $3 \rightarrow 4$ and $4 \rightarrow 1$.
8. Process (a) is a heat engine; process (b) is a refrigerator.

Chapter 17

1. (b)
2. (a)
3. (a)
4. Yes, two electric dipoles
5. (d)
6. (b)
7. (c)
8. (b)

Chapter 18

1. (d)
2. (c)

3. (a) and (c)
4. (a)
5. If the line carries a negative charge, the electric field lines are directed radially and point toward the line. The equipotential surfaces are unchanged (cylinders centered on the line of charge).
6. A qualitative sketch of the field lines is as shown. Electric field lines are always perpendicular to equipotential surfaces. Because the electric field is directed from regions of high potential to regions of low potential, the object at the center must have a positive charge.

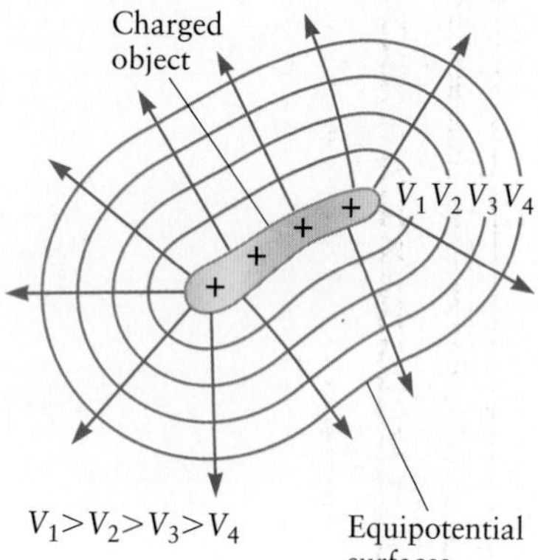

7. Pull them together

Chapter 19

1. (c)
2. (a)
3. (a)
4. (c)
5. Circuit 2
6. Connect R_3 in parallel with R_1.
7. (b)

Chapter 20

1. Part 1: (d), Part 2: (d)
2. Part 1: (b), Part 2: (d)
3. (a)
4. (a)
5. (b)
6. (b)
7. (a)
8. (a)
9. (b)
10. (b)

Chapter 21

1. (a)
2. (b)
3. (b)
4. (b)
5. (c)
6. (a)
7. (b)
8. (b)

Chapter 22

1. (a) and (c)
2. (d)
3. (c)
4. (c)
5. (a)
6. (d)
7. (a)
8. (a)
9. (a)

Chapter 23

1. (a)
2. (c) This is much smaller than the maximum recommended delay for phone service, which is about 0.2 s, and is why you can have a pleasant conversation with a friend on the opposite coast.
3. (b)
4. Using Figure 23.8, we get the following order: (d) radio, (e) microwaves, (a) infrared, (f) red light, (c) blue light, (b) X-rays.
5. (c)
6. (c)
7. Sound is a longitudinal wave, so it cannot be polarized.

Chapter 24

1. (a)
2. Material 1 is air
3. (b); 19°
4. Total internal reflection is possible in cases (a), (c), and (e) because in these cases light is traveling into the material with the smaller index of refraction.
5. (a)
6. (b)
7. (b)
8. (a) Positive; (b) positive; (c) positive
9. (a) Positive; (b) negative; (c) negative
10. (a) Ray 1 is the parallel ray; ray 2 is the central ray (b) Region II
11. (b)

Chapter 25

1. It is *not* possible to observe interference with light from two *different* helium–neon lasers. Light from the two lasers is emitted by different atoms, and light emitted by independent sources is not coherent.
2. (a)

3. Wave 2 in Figure 25.12A is reflected when it reaches the air interface at the bottom of the uppermost glass plate. The index of refraction is larger for glass than for air, similar to the reflection in Figure 25.9B, so there is no phase change at this reflection. On the other hand, wave 3 reflects back into air from the air–glass interface at the top of the lower plate, with a phase change of 180°. At the far left edge of the wedge in Figure 25.12 the path length difference between the two reflected waves is zero, so the only phase difference comes from the 180° phase change of wave 3. The interference is thus destructive, and the leftmost fringe is dark.
4. black
5. (b)
6. (b)
7. (a)
8. (b)

Chapter 26

1. (b)
2. (a)
3. As long as the object is closer than the focal point, the image produced will be on the same side of the lens as the object and will be virtual.
4. (b) and (d)
5. Light must focus on the CCD detector (or film) to be recorded clearly, so all cameras must produce a real image.
6. (d)
7. (c)
8. (a)

Chapter 27

1. (a) and (c)
2. (c)
3. (b)
4. (a)
5. (c)
6. (c)
7. (a)
8. (b). *R* is about 10 times the radius of the Sun.

Chapter 28

1. (b)
2. (c)
3. (c)
4. (b)
5. (a)
6. The electron in state 2 has the higher kinetic energy.

Chapter 29

1. (b)
2. (c)
3. Chlorine (Cl)
4. (b)
5. (b)
6. (a) and (c)
7. (b)

Chapter 30

1. ${}^{14}_{6}C$
2. (a) The wavelength of blue light is approximately 400 nm, so gamma rays have a *much* shorter wavelength.
3. (c)
4. (c)
5. (a) and (b)
6. (d)
7. (b)
8. 2

Chapter 31

1. (b)
2. (d)
3. (a)

Index

Note: Information presented in tables and figures is denoted by t and f respectively.

C

D

E

N

Q

R

S